POLYMERIC MATERIALS SCIENCE AND ENGINEERING

VOLUME 79

FALL MEETING
AUGUST 23-27, 1998
BOSTON, MASSACHUSETTS

PROCEEDINGS OF THE AMERICAN CHEMICAL SOCIETY
DIVISION OF POLYMERIC MATERIALS: SCIENCE AND ENGINEERING

© 1998 American Chemical Society

Please visit PMSE's website at: http://www.umd.umich.edu/~rpotts/acs/

or

link to the PMSE website from: http://www.acs.org

PMSE Volume 79
ISBN 8412-3580-5
ISSN 074-0515
Printed in U.S.A.

Polymer Source, Inc.
Offering Deuterated Amphiphilic Block Copolymers

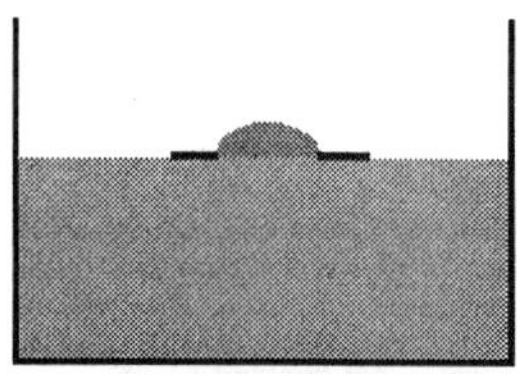

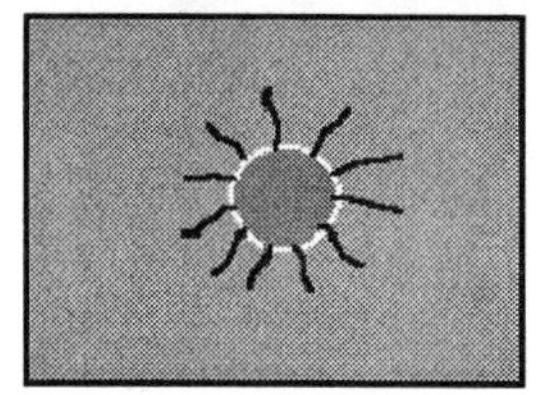

Poly(deuterated styrene-b-acrylic acid)

Poly(deuterated styrene-b-sodium acrylate)

Deuterated Poly(styrene-b-ethylene oxide)

Deuterated Poly(methyl methacrylate-b-ethylene oxide)

Deuterated Poly(ethylene-b-ethylene oxide)

Contents

Preface

It is truly an honor to present the proceedings of the Fall 1998 meeting of the ACS Division of Polymeric Materials: Science and Engineering, held in Boston, Massachusetts. These proceedings contain papers on the presentations being given, as well as much other useful information, including contact information for PMSE officers and a schedule of future meetings, symposia and workshops. This information can also be obtained on line by pointing your browser to:
http://www-personal.umd.umich.edu/~rpotts/acs/.

PMSE is sponsoring several symposia on polymer science and engineering in Boston on a range of topics. As a special added bonus we have three workshops on topics for which we also have symposia: biomaterials, scattering and toughening. A brief description of each symposium follows.

- **Biostable Polymers for Medical Use:** This symposium will address the biostability of several polymers used in medical applications such as drug delivery, cardiovascular devices and orthopaedic implants. Various aspects of the structure, properties and performance of polyurethanes, silicones, polyethylene and other polymeric biomaterials will be discussed.

- **Biomedical Applications of Water-Soluble Polymers and Hydrogels:** Design, synthesis, characterization, testing and applications as implants and drug release systems of water-soluble polymers and hydrogels will be presented by international authors.

- **Water-Soluble/Water-Swellable Polymers:** Synthesis and characterization of acrylamide, PVP, acrylate, selected carbohydrates, dendrimers and oxide containing block copolymers will be highlighted. The swellable and gel polymer sessions will characterize similar compositions via a number of different techniques.

- **Scattering from Polymers:** All aspects will be covered. This meeting will emphasize scattering from block copolymers, crystalline polymers, complex fluids, multicomponent systems, polymeric surfaces and applications to processing. Experimental and theoretical aspects will be addressed, as well as applications to polymer technology.

- **Toughening of Plastics:** these symposia will addresses recent theoretical and experimental developments in the understanding of toughening mechanisms in plastics. Other papers will describe novel toughened materials and approaches, including mechanical behavior.

- **Dynamically Vulcanized Alloys:** This symposium will be dedicated to a special class of elastomer-thermoplastic blends prepared by dynamic vulcanization. This is the process wherein an elastomer is vulcanized in situ during melt blending with a thermoplastic. The dynamically vulcanized alloys discussed in this symposium will consist of at least one olefinic component. The synthesis, kinetics, thermo-dynamics, structure-properties, deformation and fracture behavior as well as rheology and applications will be discussed.

- **General Papers/New Concepts in Polymeric Materials:** There will be four sessions on new concepts. The three oral sessions will cover "Synthesis with Controlled Architecture", "Biosystems, Crosslinked Polymers and Gas Transport" and "Surface Engineering and Bulk Properties".

Finally, PMSE is pleased to offer two award symposia

- **Unilever Award Symposium:** The Unilever Award for Outstanding Graduate Research in Polymer Science and Engineering will be awarded to James J. Watkins for his Ph.D. research on preparation of polymer composites by polymerization in supercritical carbon dioxide-swollen polymers. The remainder of the half-day symposium will focus on polymer processing in supercritical fluids.

- **Tess Award Symposium: Tess Award Symposium:** Loren W. Hill of Solutia, Inc., Springfield, Massachusetts, is the winner of the Tess Award for 1998. Dr. Hill is being recognized for his outstanding work in the area of correlating mechanical properties of polymers and their coating characteristics.

PMSE preprints would not be possible were it not for the efforts of many dedicated individuals, most of whom are volunteers. Special thanks go to symposia chairs, who coordinate contributions from each author. Colossal thanks go to the PMSE Administrative Assistant, Eileen Ernst. To say that she deserves a lot of the credit for organizing these preprints and the PMSE meeting activities is an understatement. And, of course, this would all not be possible were it not for the individual contributors themselves. I hope you all will appreciate and enjoy reading this volume, as I will.

Peter G. Edelman
PMSE Vice-Chair and Preprint Editor

In my message in the Spring Proceedings, I asked you to give me your feedback about how PMSE is working. Several of you sent me useful ideas. In this Fall message I want to point out some of the opportunities you have to make a difference in PMSE.

- <u>Vote in the PMSE Election</u>: Shortly, you should receive the Fall Edition of the PMSE Newsletter, which contains your ballot for electing officers. Read the biographies of the candidates, choose those you would like to serve in the various positions and send in your vote. Only 200 to 300 members have voted in recent elections, so your vote can make a difference as to who holds an office.

- <u>Nominate Candidates for PMSE Awards</u>: The division presents several awards each year. If you have someone in mind that deserves one of the awards, contact the appropriate award chair (see the PMSE Executive Committee List, page V.) The award chairs are Ken Edwards (Distinguished Service), Bob Brady

(Tess) and Richard Turner (Cooperative Research). If you know of any graduate students who are doing great work, encourage them to contact Elsa Reichmanis about the ICI Award. This is one of the main ways PMSE gets recognized by the outside world, so we need to continue honoring the best of the best.

- <u>Support Polymer Education</u>: PMSE is always looking for volunteers to help in our various efforts in polymer education. We are supporting activities to improve the way polymers are taught and thought about from grade school through graduate school and beyond. This is primarily done through the POLYED and IPEC Committees, and we welcome more volunteers in these organizations. Those in industry can help by encouraging your companies to provide financial donations for these efforts. Most importantly, you can help by suggesting new ways to promote education on polymers.

- <u>Contribute to Technical Programming</u>: The technical programming of PMSE at the national ACS meetings is continually cited as one of the best features of PMSE. Once again, this is strong at the Boston meeting. One of the strengths of our programming is the number of joint symposia with other ACS divisions, such as POLY, and societies, such as the Society of Plastics Engineers and the Optical Society of America. Every member is encouraged to make suggestions on possible symposia topics and to volunteer to organize a symposium. We especially want to hear about fields of polymer science and engineering that you feel are being neglected by our programming. Please direct your comments to the Program Chairs: Rick Register, Jean Fréchet or Jay Dias.

- <u>Celebrate Our 75th Anniversary</u>: At the upcoming Spring meeting in Anaheim, you will have the chance to help recognize the past 75 years of excellence in PMSE. There will be two special symposia on these topics, one looking back at our history and one looking at the future of polymer science. We will publish a new volume, *Applied Polymer Science – 21st Century*, covering the past and present of polymer science. I hope that you can attend the banquet in Anaheim, where we will honor past awardees and officers.

Finally, let me say that it has been an honor and a great pleasure to serve you as Chairman in 1998. I can't imagine a better group to work with. Thank you for your efforts.

David J. Lohse, Chair

Missions, Values and Principles

MISSION

PMSE is a division of the American Chemical Society providing a forum for the exchange of technical information on the chemistry of polymeric materials including plastics, paints, adhesives, composites and biomaterials. Our mission is to meet the needs of our members and the community at large through technical programming, educational outreach and publication.

VALUES

Fundamental to the success of the Division are these basic values:

People. Our members are the source of our strength. Member involvement and volunteer effort are essential to our success.

Products. Our principal product is the sharing of technical information; our technical program and our publications are the tangible results of our efforts. We should strive to improve them continuously. As our products are viewed, so is the Division viewed.

Finances. Financial stability is essential to the long-term viability of the Division. We have a responsibility to each other as members to ensure the long-term financial health of the Division.

GUIDING PRINCIPLES

Quality. The quality of our products and services must be our number one priority.

Customer Focus. Our customers are our members, the technical community, the educational community and the public at large. Our focus must be on meeting the needs of our customers.

Value. Our products and services must offer good value to our customers and must be provided at affordable and sustainable cost.

Continuous Improvement. We must strive for excellence and work toward meeting our customers' needs at affordable cost.

Membership Involvement. The success of the Division requires the involvement of members in all of our activities: volunteer contribution of time and resources is a key.

Uncompromised Integrity. The conduct of the Division must command respect for its integrity and for its positive contributions to the technical community and the public good.

OFFICERS, 1998

POSITION: Chair

David J. Lohse
Exxon Res. & Engineering Co.
Corporate Research Laboratories
Route 22 East
Annandale, NJ 08801

WORK: (908) 730-2541
FAX: (908) 730-2536
HOME: (908) 526-6196
e-mail: djlohse@erenj.com

POSITION: Chair-Elect
CHAIR: Program Development

David A. Cocuzzi
Akzo Nobel Coatings Inc.
1313 Windsor Avenue
P.O. Box 489
Columbus, OH 43216-0489

WORK: (614) 294-3361
FAX: (614) 421-4360

POSITION: Vice Chair
CHAIR: Meeting Arrangements

Peter G. Edelman
Chiron Diagnostics
63 North Street
Medfield, MA 02052

WORK: (508) 359-3551
FAX: (508) 359-3243
HOME: (508) 520-6557
e-mail: peter.edelman@chirondiag.com

POSITION: Secretary, 1997-1998
CHAIR: Member Services; Workshops

Christopher Ober
Department of Materials Science
 & Engineering
Cornell University
327 Bard Hall
Ithaca, NY 14853-1501

WORK: (607) 255-8417
FAX: (607) 255-6575
e-mail: cober@msc.cornell.edu

POSITION: Treasurer, 1998-1999
CHAIR: Finance and Investment

Larry F. Charbonneau
Hoechst Research & Technology,
 U.S., LLC
86 Morris Avenue
Summit, NJ 07901

WORK: (908) 522-7813
FAX: (908) 522-7835
HOME: (973) 543-2406
e-mail: lfc@sumhcc1.hcc.com
          Lcharbonneau@compuserve.com

POSITION: Past Chair
CHAIR: Nominations

Frank N. Jones
Coatings Research Institute
Eastern Michigan University
430 West Forest Street
Ypsilanti, MI 48197

WORK: (734) 487-2203
FAX: (734) 483-0085
HOME: (734) 769-5296
e-mail: frank.jones@emich.edu

POSITION: Administrative Assistant

Eileen Ernst
2088 Berwyn Street
Union, NJ 07083

WORK: (908) 851-9465
FAX: (908) 810-0723
HOME: (908) 851-9465
e-mail: eileen.ernst@worldnet.att.net

COUNCILORS

POSITION: Councilor, 1997-1999

Clara D. Craver
Consultant
Goose Creek Lake
Highway Y
P.O. Box 265
French Village, MO 63036-0265

WORK: (573) 358-2589
FAX: (573) 358-2589

POSITION: Councilor, 1997-1999
CHAIR: Investment

John H. Lupinski
Consultant
9318 Old Courthouse Road
Vienna, VA 22182-2014

WORK: (703) 281-9450

POSITION: Councilor, 1998-2000
CHAIR: Ad Hoc Committee on Program
Development; Finance and Investment

Sandy S. Labana
111 North Brady
Dearborn, MI 48124

HOME: (313) 562-6246
FAX: (313) 562-6247
e-mail: slabana@aol.com

POSITION: Councilor, 1996-1998
CHAIR: Polymer Education

Theodore Provder
ICI Paints
16651 Sprague Road
Strongsville, OH 44136-1739

WORK: (440) 826-5289
FAX: (440) 826-5233
HOME: (440) 235-3680
e-mail: ted_provder@ici.com

ALTERNATE COUNCILORS

POSITION: Alt. Councilor, 1997-1999
CHAIR: Financial Planning

John L. Gardon
Akzo Nobel Coatings Inc.
P.O. Box 7062
Troy, MI 48007-7062

WORK: (810) 637-8502
FAX: (810) 649-7367
HOME: (810) 642-5130

POSITION: Alt. Councilor, 1998-2000
CO-CHAIR: Long-Range Planning

Michael Jaffe
N.J. Center for Biomaterials and
   Medical Devices
Department of Chemistry
Rutgers University
Piscataway, NJ 08854-8087

WORK: (732) 445-0788
FAX: (732) 445-0790
HOME: (973) 761-1184
e-mail: aerobatkat@worldnet.att.net

POSITION: Alt. Councilor, 1998-2000
CO-CHAIR: Audit; Long Range Planning

David R. Bauer
Ford Motor Company
MD-3182, SRL
P.O. Box 2053
Dearborn, MI 48121-2053

WORK: (313) 594-1756
FAX: (313) 323-1129
e-mail: dbauer3@ford.com

POSITION: Alt. Councilor, 1997-1999
CHAIR: Educational Outreach;
Educational Funding

George R. Pilcher
Akzo Nobel Coatings Inc.
P.O. Box 489
Columbus, OH 43216-0489

WORK: (614) 294-3361
FAX: (614) 421-4360
HOME: (614) 895-3191

MEMBERS-AT-LARGE, 1997-1998

CHAIR: Membership

Jeffrey H. Helms
Ford Motor Company
MD-3182, SRL
P.O. Box 2053
Dearborn, MI 48121-2053

WORK: (313) 337-1098
FAX: (313) 323-1129
e-mail: jhelms@ford.com

CHAIR: Roy W. Tess Award

Robert F. Brady
Naval Research Laboratory
Chemistry Division, Code 6123
Washington, DC 20375-5342

WORK: (202) 767-2268
FAX: (202) 767-0594
email: brady2@ccf.nrl.navy.mil

CO-CHAIR: Professional Development
Short Courses

Alec B. Scranton
Dept. of Chemical Engineering
A202 Engineering Building
Michigan State University
East Lansing, MI 48824

WORK: (517) 353-3516
FAX: (517) 432-1105

CO-CHAIR: Professional Development
Short Courses

Thomas Neenan
GelTex Pharmaceuticals, Inc.
303 Bear Hill Road
Waltham, MA 02154

WORK: (617) 290-5888, Ext. 52
FAX: (617) 290-5890
e-mail: tneenan@geltex.com

Russ Gaudiana
Polaroid Corporation
730 Main Street - 1A
Cambridge, MA 02139-1563

WORK: (617) 386-5779
FAX: (617) 386-8172
e-mail: gaudiar@polaroid.com

Loren Hill
Monsanto Company
730 Worcester Street
Springfield, MA 01151

WORK: (413) 730-2308
FAX: (413) 730-2506
HOME: (413) 596-4463
e-mail: lwhill@ccmail.monsanto.com

Peter Green
Dept. of Chemical Engineering
Campus Mail Code: C0400
University of Texas
Austin, TX 78712

WORK: (512) 471-3188
FAX: (512) 471-7681
e-mail: green@che.utexas.edu

MEMBERS-AT-LARGE, 1998-1999

CHAIR: Ford Travel Grant

Peggy Cebe
Dept. of Physics & Astronomy
Tufts University
Science and Technology Center
Room 208
4 Colby Street
Medford, MA 02155

WORK: (617) 627-3365
FAX: (617) 627-3744
e-mail: peggy@cebe.phy.tufts.edu

CO-CHAIR: Technical Program

Richard A. Register
Dept. of Chemical Engineering
Engineering Quadrangle
Princeton University
Princeton, NJ 08544

WORK: (609) 258-4691
FAX: (609) 258-0211
HOME: (609) 275-8257
e-mail: register@pucc.princeton.edu

Leslie H. Sperling
Lehigh University
5 East Packer Avenue
Bethlehem, PA 18015-3194

WORK: (610) 758-3845
FAX: (610) 758-4244
e-mail: lhs0@lehigh.edu

CHAIR: Newsletter/Publicity

John W. Gilmer
Research Laboratories
Eastman Chemical Company
Kingsport, TN 37662

WORK: (423) 229-8637
FAX: (423) 229-4558
e-mail: jwgilmer@eastman.com

CHAIR: By-Laws

Kirk J. Abbey
Lord Corporation
Thomas Lord Research Center
110 Lord Drive
P.O. Box 8012
Cary, NC 27512-8012

WORK: (919) 469-2500, ext. 2367
FAX: (919) 469-9688
HOME: (919) 821-5329
e-mail: kirk_abbey@lord.com

Mary E. Galvin
University of Delaware
Materials Science Program
Spencer Laboratories
Room 201D
Newark, DE 19716

WORK: (302) 831-0873
FAX: (302) 831-4545
e-mail: megalvin@udel.edu

Christine Landry
Eastman Kodak Company
Building 82
Floor 6/KP/02116
Rochester, NY 14650-2116

WORK: (716) 722-3683
FAX: (716) 477-6498
HOME: (716) 223-7265
e-mail: clandry@kodak.com

Douglas A. Wicks
Bayer Corporation
100 Bayer Road
Pittsburgh, PA 15205

WORK: (412) 777-7851
FAX: (412) 777-2940
HOME: (412) 344-4142
e-mail: doug.wicks.b@bayer.com
        doug.wicks@worldnet.att.net

COMMITTEE CHAIRS

CO-CHAIR: Technical Program

A. Jay Dias
Exxon Chemical Company
Baytown Polymers Center
5200 Bayway Drive
Baytown, TX 77520-2101

WORK: (281) 834-5255
FAX: (281) 834-1665
HOME: (281) 486-0125
e-mail: ajdias@erenj.com

CO-CHAIR: Technical Program

Jean M. J. Fréchet
Department of Chemistry
University of California
Berkeley, CA 94720-1460

WORK: (510) 643-3077
FAX: (510) 643-3079
email: frechet@cchem.berkeley.edu

CHAIR: Macromolecular Secretariat;
Cooperative Research Award

S. Richard Turner
Research Laboratories
Eastman Chemical Company
Kingsport, TN 37662

WORK: (423) 224-0489
FAX: (423) 229-4558
email: srturner@eastman.com

CHAIR: ICI Student Award; Industrial
Relations

Elsa Reichmanis
Bell Laboratories
Lucent Technologies
600 Mountain Avenue
Room 1D-260
Murray Hill, NJ 07974

WORK: (908) 582-2504
FAX: (908) 582-3609
HOME: (908) 232-3078
e-mail: er@bell-labs.com

CHAIR: Doolittle Award

Yu-Chin Lai
Contact Lens Research &
  Development
Bausch & Lomb, Inc.
1400 North Goodman Street
P.O. Box 450
Rochester, NY 14692-0450

WORK: (716) 338-8711
FAX: (716) 338-5304
e-mail: ylai@bausch.com

COMMITTEE: Ford Travel Grant

Benny D. Freeman
North Carolina State University
Dept. of Chemical Engineering
Campus Box 7905
Raleigh, NC 27695-7905

WORK: (919) 515-2460
FAX: (919) 515-3465
e-mail: benny_freeman@ncsu.edu

CHAIR: Regional Meetings Coordination

Stephen McCarthy
Plastics Engineering Department
University of Massachusetts at
  Lowell
Lowell, MA 01854

WORK: (508) 934-3417
FAX: (508) 458-4141
HOME: (508) 957-1765
e-mail: mccarthy@cae.uml.edu

CHAIR: Biotechnology Secretariat

Graham Swift
Rohm & Haas Company
772 Norristown Road
Spring House, PA 19477

WORK: (215) 641-7756
FAX: (215) 619-1642
e-mail: rsiygs@rohmhaas.com

CHAIR: ChemTech Advisory Board

James F. Kinstle
550 Mt. Zion Road
Apartment 307
Florence, KY 41042

HOME: (317) 692-3020

CHAIR: Distinguished Service Award

Kenneth N. Edwards
Dunn-Edwards Corporation
4885 East 52nd Place
Los Angeles, CA 90040

WORK: (213) 771-3330
FAX: (213) 771-9029
HOME: (818) 241-8514
e-mail: KNEatDE@aol.com

CO-CHAIR: Joint Polymer Education

Charles E. Carraher
Florida Atlantic University
500 N.W. 20th Street
Boca Raton, FL 33431

WORK: (561) 297-2107
FAX: (561) 297-2759 or 2985
HOME: (954) 755-2768
email: carraher@fau.edu

CO-CHAIR: Joint Polymer Education;
Director, POLYED National Information
Center

John P. Droske
Department of Chemistry
University of Wisconsin at
  Stevens Point
Stevens Point, WI 54481

WORK: (715) 346-3703
FAX: (715) 346-2640

CO-CHAIR: Long Range Planning, 1997

Ray A. Dickie
Ford Motor Company
MD-3083, SRL
P.O. Box 2053
Dearborn, MI 48121-2053

WORK: (313) 337-4059
FAX: (313) 621-0646
HOME: (248) 349-1251
email: rdickie1@ford.com

CHAIR: Advertising

Virginia A. Dais
368 E. Saginaw Road
Sanford, MI 48657

WORK: (517) 687-2452
HOME: (517) 687-2452
email: BVLDais@aol.com

CHAIR: Books

Paul L. Valint, Jr.
Bausch & Lomb, Inc.
1400 North Goodman Street
P.O. Box 450
Rochester, NY 14603-0450

WORK: (716) 338-6414
FAX: (716) 338-6316
e-mail: pvalint@bausch.com

COMMITTEE: Newsletter/Publicity

James W. Taylor
S.C. Johnson Polymer
P.O. Box 902, MS 712
Sturtevant, WI 43177-0902

WORK: (414) 631-4044
FAX: (414) 631-4039
HOME: (847) 856-0434
e-mail: JWTaylor@scj.com

CO-CHAIR: Workshops

Vladimir Tsukruk
Chairman, CMD Department
Western Michigan University
College of Engineering & Applied
  Sciences
Kalamazoo, MI 49008

WORK: (616) 387-6510
FAX: (616) 387-6517
e-mail: vladimir@wmich.edu

CHAIR: Speaker's Bureau

Donald N. Schulz
Exxon Res. & Engineering Co.
Corporate Research Laboratories
Route 22 East
Annandale, NJ 08801

WORK: (908) 730-2526
FAX: (908) 730-2536
e-mail: dnschul@erenj.com

EX-OFFICIO MEMBERS

CHAIR: Materials Secretariat

Theodore Davidson
109 Poe Road
Princeton, NJ 08540-4121

WORK: (609) 924-6146
FAX: (609) 683-8828
e-mail: TedDav@aol.com

Roy W. Tess
1615 Chandelle Lane
Fallbrook, CA 92028-1707

WORK: (760) 728-3695
HOME: (760) 728-3695

E. E. McSweeney
17 Nuevo Leon
Port St. Lucie, FL 34952

HOME: (561) 335-9073

PMSE Committee Organization

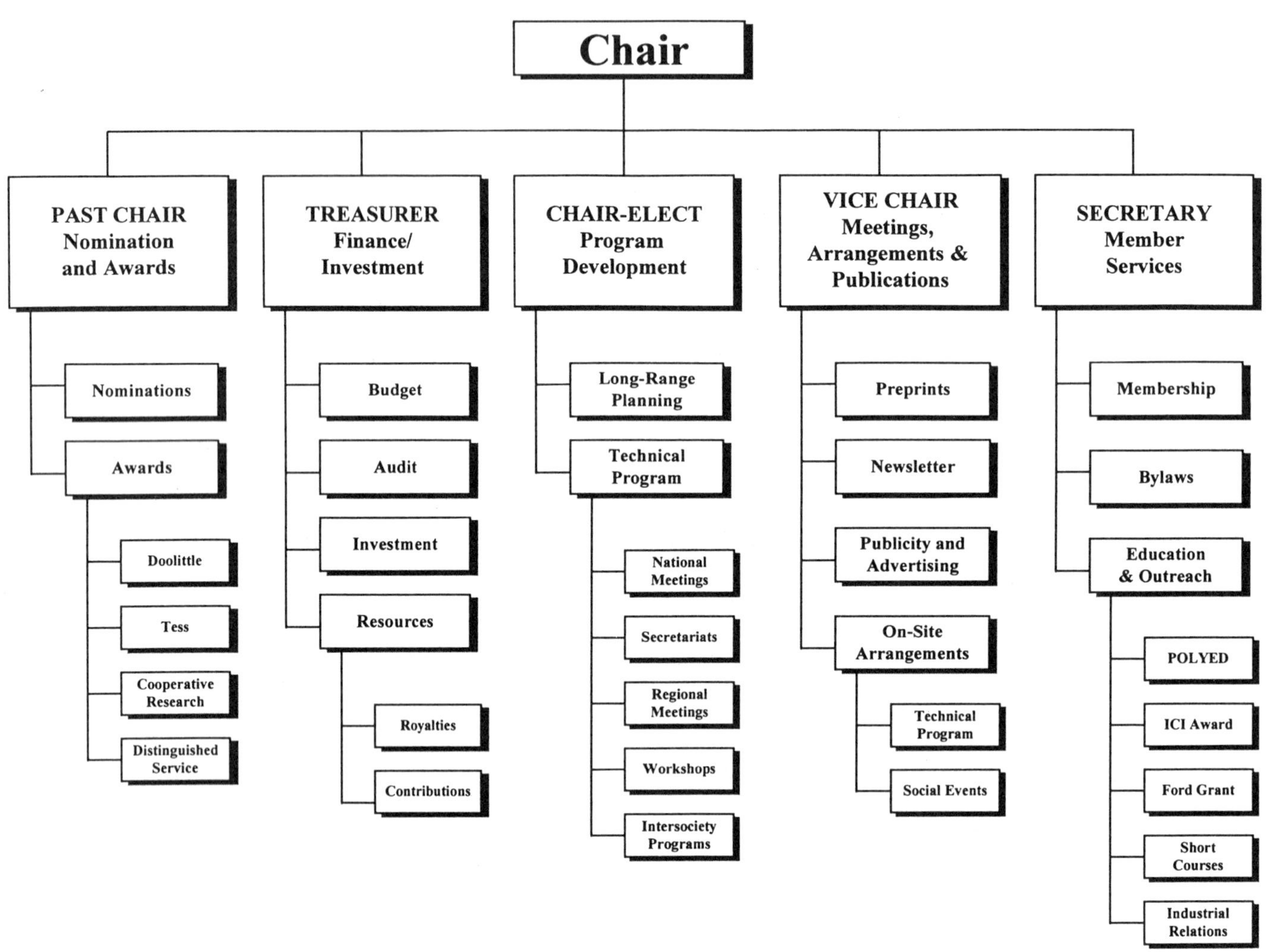

AN INVITATION TO JOIN THE DIVISION OF

POLYMERIC MATERIALS: SCIENCE & ENGINEERING

PMSE was founded in 1924 as the Paint and Varnish Division, and until recently was known as the Division of Organic Coatings and Plastics Chemistry. Over the years, the interests of the division's members have come to include adhesives, biomedical polymers, composites, polymers for electronic applications, plastics and many other areas of applied polymer science and technology in addition to traditional coatings-related topics.

The principal objective of the division is to promote interest in and understanding of applied polymer science. This is achieved through regular meetings, publications, professional contacts and discussions. Since its founding, the division has regularly participated in ACS National Meetings. The division frequently joins with other ACS divisions in joint and co-sponsorship of symposia and is an active participant in the Macromolecular, Biotechnology and Materials Science Secretariats. PMSE also sponsors topical workshops in conjunction with its technical meetings to provide tutorial reviews of various technologically significant fields. Program planning and the operation of the division are carried out by member volunteers; your participation is encouraged and welcomed.

Proceedings of the division's meetings are sent to members shortly before the spring and fall National ACS Meetings. Combined, these two preprint volumes bring over 1000 pages of current technical information to members.

You are cordially invited to join the division as a full or affiliate member. ACS members are entitled to full membership. Non-ACS members may join the division as an affiliate. Affiliates share all PMSE membership benefits and receive the proceedings of the division, but may not vote or hold office in the division.

ACTIVITIES

Meetings

- International and National Symposia
- Specialty Subject Symposia
- Workshops

Publications

- Proceedings of Division Meetings distributed prior to national meetings.
- Newsletter published semiannually
- Sponsorship for ACS publications

Awards

- **DISTINGUISHED SERVICE AWARD** for outstanding contributions to the long-term welfare of the Division.

- **FORD TRAVEL GRANT** sponsored by the Ford Motor Company to provide partial travel support to graduate student women or under represented minority men to attend and present their research at the National ACS Meeting.

- **ARTHUR K. DOOLITTLE AWARD** for the best paper presented at division meetings.

- **ROY W. TESS AWARD IN COATINGS** to recognize outstanding contributions in coatings science, technology and engineering.

- **COOPERATIVE RESEARCH AWARD IN POLYMER SCIENCE AND ENGINEERING** sponsored by Eastman Kodak Company, in recognition of sustained research between polymer scientists working on behalf of industry and those in either the academic community or at national laboratories.

- **ICI STUDENT AWARD IN APPLIED POLYMER SCIENCE** established by ICI PLC to promote graduate research in polymer science.

APPLICATION FOR MEMBERSHIP

❏ DR.   ❏ MR.   ❏ MRS.   ❏ MS.

NAME: _______________________________________

ADDRESS: ____________________________________

____

____

____

Area Code   Telephone Number

(Please use address where you receive C&EN)

I wish to join the PMSE Division in the following classification:

❏ ACS Member ($20.00/year)
❏ ACS Student Member ($10.00/year)
❏ ACS Emeritus Members ($10.00 year)

ACS Membership Number ___________

❏ Division Affiliate ($25.00/year)
❏ National Affiliate ($25.00/year)

For Students: Full-time enrollment at ________________________ qualifies me for student status.

Return this form with your remittance (payable to the **Division of Polymeric Materials: Science and Engineering**, drawn in U.S. funds on a U.S. bank) to:

Dr. Jeffrey H. Helms
Ford Motor Company
P.O. Box 2053, MD-3182
Dearborn, MI 48121-2053
(313) 337-1098 (Work)
(313) 323-1129 (FAX)
email: jhelms@ford.com

POLYED Sponsors

Supporters of POLYED, *The Polymer Education Committee*

The Division of Polymeric Materials: Science and Engineering acknowledges the contributions from our industrial supporters on behalf of the Polymer Education Committee, POLYED. POLYED is a consortium of groups interested in science education, in general, and polymer education in particular.

The major purpose of POLYED is the nurturing of education at all levels from kindergarten through post graduate, including professionals seeking continuing education. Some of the numerous POLYED programs include education symposia, teacher training, the industrial teachers program, media and short course cataloguing, polymer demonstration kits, summer employment programs and awards for a variety of functions, including polymer curriculum development, undergraduate travel, organic chemistry, excellence in polymer education by a high school teacher and the ICI Student Award. The Polymer Education Newsletter (PEN) is mailed to all departments of chemistry and chemical engineering in the United States and Canada to keep them abreast of these activities.

To obtain more information regarding POLYED programs, please contact:

> Professor John Droske
> The Undergraduate Education Center
> University of Wisconsin-Stevens Point
> Stevens Point, WI 54481
>
> **Phone**: (715) 346-3703
> **Fax**: (715) 346-2640

To become a POLYED supporter, please send a check for $1,000.00 to:

> Dr. Larry Charbonneau
> Hoechst Research & Technology, U.S., LLC
> 86 Morris Avenue
> Summit, New Jersey 07901
> **Phone**: (908) 522-7813
> **Fax**: (908) 522-7835
> **e-mail**: lfc@sumhcc1.hcc.com

We appreciate your support!

The technical program of the Division at the national meetings is a central element to achieving our goal of providing the membership with easy access to the latest developments in polymer science. Our ability to do this depends not only on the efforts of the many volunteers who organize and run symposia at the meetings, but also on financial support from several sources. We are grateful to those companies and organizations who have provided funds for the symposia that will be presented at the Boston meeting:

Dow Chemical Company

Exxon Research & Engineering Company

Howmedica

ICI

Petroleum Research Fund

Rohm and Haas

Unilever

Roy W. Tess Award In Coatings

The Officers and the Award Committee of the Division of Polymeric Materials: Science and Engineering (PMSE) of the American Chemical Society are pleased to announce that Dr. Loren W. Hill of Solutia, Inc., Springfield, Massachusetts, will receive the Roy W. Tess Award in Coatings for 1998.

Dr. Hill has made seminal contributions to our understanding of curing mechanisms in thermoset coatings. He was the first to develop practical techniques for performing dynamic mechanical analysis on free coatings films, and he used this method to measure crosslink density of the films. He also developed straightforward methods for calculating theoretical crosslink density, a difficult task in view of the complex mixture of polymers used in coatings. Using these values for measured and theoretical crosslink density with spectroscopic methods, he related polymer chemistry to empirical coatings properties and used these insights to develop optimal materials and formulations. Subsequently, Dr. Hill used similar methods to develop a new understanding of processes which occur during weathering of thermoset resins.

In other work on the rheology and morphology of water-reducible coatings, Dr. Hill explained the processes which occur during loss of water, and this paradigm remains the standard for water evaporation. Additionally, his studies of crosslinking have shown how kinetic considerations set limits on the maximum possible pot life for a coating designed to cure at a given time and temperature. This fundamental work led to new chemistry which has found commercial application in environmentally-benign high solids coatings for automobiles.

Dr. Hill holds a Bachelor of Science degree from North Dakota State University and a doctorate degree in chemistry from Pennsylvania State University. He was a member of the faculty at North Dakota State University from 1965 to 1980, rising from Assistant Professor to Professor in the Polymers and Coatings Department. Dr. Hill joined the Specialty Resins Division of Solutia (formerly Monsanto) as Research Fellow in 1980, and was named in 1987 to his present position of Senior Fellow.

Dr. Hill is the author of more than forty technical papers. In 1991 he presented the Mattiello Lecture to the Federation of Societies for Coatings Technology on "Structure-Property Relationships of Thermoset Coatings." Dr. Hill has also won an Elias Singer Award (1995) and a Roon Award (First Prize, 1981). He serves on the Editorial Review Boards of the *Journal of Coatings Technology* and *Progress in Organic Coatings,* and was chairman of the Gordon Research Conference on Coatings and Films in 1985. He is a sought-after lecturer on the chemistry of thermosetting resins in Europe, Asia and the United States, where he is valued for his skill in getting to the essence of a complex subject and explaining it understandably.

Dr. Hill will receive the Tess Award from Dr. David J. Lohse, Chairman of the PMSE Division, on Tuesday, August 25,

1998 during the 216th meeting of the American Chemical Society in Boston, Massachusetts. Dr. Hill will present an Award Address at that time. An evening reception in honor of Dr. Hill, co-sponsored by Solutia, Inc. and the Shell Chemical Company, will follow the Award Symposium.

The Tess Award is presented annually by the Division of Polymeric Materials: Science and Technology in recognition of outstanding contributions to coatings science and technology. It is funded by a grant to the Division from Dr. and Mrs. Roy W. Tess. The purpose of the Award is to encourage interest and progress in coatings, and to recognize significant contributors to the field. The Award consists of a plaque and a cash prize.

Distinguished past recipients of the Tess Award include:

William D. Emmons	(1986)	Ray A. Dickie	(1992)
Marco Wismer	(1987)	Larry F. Thompson	(1993)
Zeno W. Wicks, Jr.	(1988)	Werner Funke	(1994)
Theodore Provder	(1989)	John L. Gardon	(1995)
Walter K. Asbeck	(1990)	John K. Gillham	(1996)
Kenneth L. Hoy	(1991)	Werner J. Blank	(1997)

UNILEVER AWARD FOR OUTSTANDING GRADUATE RESEARCH

The recipient of the 1998 Unilever Award for Outstanding Graduate Research is James J. Watkins, who received his doctorate under the guidance of Professor Thomas J. McCarthy at the University of Massachusetts, Amherst.

His doctoral dissertation involved high pressure chemical synthesis, materials characterization and chemical engineering polymer processing aspects of a new approach to make advanced polymer materials. The approach takes advantage of the observation that virtually all polymers are swollen to some extent in supercritical CO_2, in order to use this medium for reactions to modify the swollen polymer. For example, a polymer composite may be formed by in situ polymerization of monomers dissolved in the swollen polymer, providing the possibility for the synthesis of a wide range of multicomponent compositions. Control of the morphology of the composite is provided through reaction temperatures, times, concentrations, etc.

The Unilever Award, which will be presented at the Boston meeting of the American Chemical Society (August 22-26, 1998) consists of a $2,000 prize, a plaque and travel expenses. This award,

administered by the Polymer Education Committee of the Polymer Chemistry and Polymeric Materials Science and Engineering Divisions, was established in 1991 and is sponsored by Unilever, a global manufacturer of consumer products, foods and specialty chemicals. The award recognizes and encourages outstanding graduate research in the design, synthesis and physical chemistry of polymers.

American Chemical Society
Division of Polymeric Materials: Science and Engineering

Announces a New Award for Women or Under-Represented Minorities

FORD TRAVEL GRANT

PMSE is pleased to announce a new award sponsored by the Ford Motor Company, which will provide $500.00 of partial travel support to graduate student women or under-represented minority men (Black, Hispanic or Native American) to attend and present their research at the National ACS Meetings. Four awards will be made yearly, two at each of the National ACS Meetings. Awards will be made competitively among those students who have applied for the award and who have concurrently submitted papers to the PMSE program.

Eligibility:

- Women or under-represented minority men (Black, Hispanic, Native American) graduate students performing research in fields related to Polymer Science and Engineering.

- U.S. citizens or permanent residents.

- Full-time graduate students at the time of application.

- Concurrent preprint submission to PMSE. A full length preprint must be submitted for the Ford Travel Grant, even for poster presentations.

Selection Criteria:

- Awardees will be selected by a committee appointed by PMSE.

- Selection will be based on the quality of technical work, written preprint and recommendation of student's research advisor.

Application:

Eligible students must submit <u>three copies</u> of the documents listed below:

1. Copy of the preprint which was submitted to PMSE describing their research. Preprint should be written in the style conforming to ACS submission procedures. Guidelines are published in PMSE Proceedings. Poster presenters must submit a full length PMSE preprint to be considered for this award. Any work submitted for this award must have the graduate student as first author and presenter.

2. The Application Forms certifying eligibility, with advisor's signature.

3. Letter of recommendation from research advisor.

4. Deadline for receipt of application materials will coincide with the deadline for submission of PMSE preprints.

Send Application Materials To:

Professor Peggy Cebe
Chairwoman, Ford Travel Grant
Physics Department, Tufts University
Science and Technology Center, Room 208
4 Colby Street
Medford, MA 02155

PHONE: (617) 627-3365
FAX: (617) 627-3744
e-mail: peggy@cebe.phy.tufts.edu

American Chemical Society
Division of Polymeric Materials: Science and Engineering

FORD TRAVEL GRANT

NAME: ______________________________   **SCHOOL:** ______________________________

ADDRESS: ______________________________   **MAJOR:** ______________________________

______________________________   **DEGREE:** __________ M.S. __________ Ph.D.

______________________________   **DEGREE DATE** (*expected*): ______________________________

PHONE: ______________________________   **TODAY'S DATE:** ______________________________

E-MAIL: ______________________________

Certification of Eligibility

The student named above meets the eligibility criteria for the Ford Travel Grant:

1. ☐ Female ☐ Minority Female   2. ☐ U.S. Citizen

   ☐ Minority Male (Black, Hispanic, Native American)   ☐ Permanent Resident

3. ☐ PMSE preprint submitted concurrently   4. ☐ Full-Time Graduate Student

______________________________   ______________________________
Research Advisor's Name Printed   Research Advisor's Signature

Recommendation of Research Advisor

(Please write a brief statement about the qualification of the applicant. You may continue on the reverse side, or attach a separate statement.)

____

____

____

____

COOPERATIVE RESEARCH AWARD IN POLYMER SCIENCE AND ENGINEERING SPONSORED BY THE EASTMAN KODAK COMPANY

Nominations are invited for the Division of Polymeric Materials: Science and Engineering's **Cooperative Research Award in Polymer Science and Engineering Sponsored by the Eastman Kodak Company.** The Division believes that this award is unique, relevant and timely to today's research environment. It is given to recognize and encourage sustained cooperative research between industrial and academic or industrial and national laboratory scientists. The cooperative research must be of significant importance to polymer science and technology.

The award was established in the fall of 1992 and is supported by a generous gift from the Eastman Kodak Company. It consists of $2,000, a plaque and a travel allowance to attend the meeting at which the award is presented. The awardee(s) is expected to give a lecture at the meeting.

The nominee(s) for this award must have a documented (patents, publications, etc.) record of sustained, intensive **cooperative and collaborative** research across the university/industry or national laboratory/industry interface. This research must be of significant importance to polymer science and technology. This award can be an individual award or can be shared between an academic and industrial scientist or between a national laboratory and an industrial scientist.

The 2000 award will be presented at the Spring 2000 ACS meeting at the PMSE Awards Luncheon. Nominations for the 2000 award will be accepted at any time before June 1, 1999 and should be sent to S. Richard Turner, Research Laboratories, Eastman Chemical Company, Kingsport, TN 37662. The nomination should include a brief curriculum vitae of the candidate(s) and significant evidence of the collaborative research including supporting letter(s) explaining the significance of the research.

Previous winners of the award include:

1994	Jean M. J. Fréchet, Hiroshi Ito, C. Grant Willson
1995	Leo Mandelkern, C. Stanley Speed, Ferdinand C. Stehling
1996	Ray H. Baughman
1997	Henry K. Hall, Jr.
1998	Lynda K. Johnson, Maurice Brookhart

1998 ICI STUDENT AWARD IN

APPLIED POLYMER SCIENCE

Applications are invited for the 1999 **ICI Student Award in Applied Polymer Science** sponsored by the Polymer Education Committee of the ACS Divisions of Polymeric Materials: Science and Engineering and Polymer Chemistry.

Graduate students, either currently in graduate school or not more than one year beyond graduation, are invited to submit a research paper for presentation at the 1999 Fall ACS Meeting in New Orleans, Louisiana. The paper should conform to the preprint format of the PMSE Division and be commensurate with the regulations and customs of papers presented in PMSE/ACS programs.

Up to **six** finalists are normally selected based on scientific merit of the submitted papers. Out-of-pocket expenses up to **$600** will be available to each finalist to attend the ACS meeting and present his/her paper in the ICI Student Award Symposium within the PMSE program. The paper must be given by the student, not by the co-author/thesis advisor. The awardee will be chosen by an anonymous evaluation committee which will audit the presentations.

The Award, consisting of $750 and a plaque, will be presented at the PMSE Division luncheon at the following Spring meeting. All finalists will receive a one-year complimentary membership in the PMSE Division.

Instructions regarding application procedures are available from :

> Elsa Reichmanis
> Chairperson, ICI Award Committee
> Bell Laboratories, Room 1D-260
> Lucent Technologies
> 600 Mountain Avenue
> Murray Hill, New Jersey 07974

NOTE: Applications and preprints for the 1998 Award must be submitted by **March 1, 1999.**

ARTHUR K. DOOLITTLE AWARD

The Arthur K. Doolittle Award, established by the Union Carbide Corporation, is given to the authors of an outstanding paper presented before the Division at either the Spring or Fall Meeting of the ACS. A prize in the amount of $1,000 is financed with the gift of royalties from A. K. Doolittle's book, *Technology of Solvents and Plasticizers*. All papers appearing in the Preprint Book are evaluated on the basis of content, with emphasis on originality and development of new concepts, and on the quality of presentation. Recipients are selected by an anonymous panel of judges appointed by the Chairman of the Doolittle Award Committee.

1981 — W. H. Ray and F. J. Schork (University of Wisconsin), "On-Line Monitoring of Emulsion Polymerization Reactor Dynamics."

1982 — A. G. MacDiarmid, P. J. Nigrey, D. F. Macinnes, Jr., D. P. Nairns and A. J. Heeger (University of Pennsylvania), "Electrochemistry of Polyacetylene (CH): Lightweight Rechargeable Batteries Using (CH) as the Cathode-Active Material."

1983 — W. J. Hall, R. L. Kruse, R. A. Mendelson and Q. A. Trementozzi (Monsanto Company), "New Styrene-Maleic Anhydride Terpolymer Blends."

1984 — T. Inabe, J. F. Lomas, J. W. Lyding, C. R. Kannewurf and T. J. Marks (Northwestern University), "Metal-Like Metallomacrocyclic Polymers: Chemical, Physical and Processing Studies."

1985 — J. T. Koberstein (Princeton University) and T. P. Russell (IBM Research Laboratories), "Simultaneous Synchrotron X-ray Scattering and Differential Scanning Calorimeter Characterization of Polymers."

1986 — J. M. J. Fréchet, F. M. Houlihan (University of Ottawa) and C. G. Willson (IBM Research Laboratories), "Polycarbonates Derived from o-Nitro-benzyl Glicidyl Ether: Synthesis and Radiation Sensitivity."

1987 — I. V. Yannas, E. Lee, A. Ferdman, D. P. Orgill, E. M. Skrabut (Massachusetts Institute of Technology) and G. F. Murphy (Brigham Young Women's Hospital), "De Novo Synthesis of Skin."

1988 — M. G. Allen and S. D. Senturia (Massachusetts Institute of Technology), "Measurement of Polyimide Interlayer Adhesion Using Macrofabricated Structures."

1989 — H. Ito (IBM Research Division), M. Ueda and M. Ebina (Yamagata University), "A Copolymer Approach to the Design of Sensitive Deep UV Resist Systems with High Thermal Stability and Dry Etch Resistance."

1989 — J. L. West, J. W. Doane, Z. Domingo (Kent State University) and P. Ukleja (Southeastern Massachusetts University), "Electro-Optics of Polymer Dispersed Liquid Crystals: Infrared Applications."

1990 — G. D. Friends, J. F. Kunzler (Bausch and Lomb, Inc.), "Hydrogels Based Upon N-Vinyl Pyrrolidone and 4-t-Butyl-2-Hydroxycyclohexylmethacrylate."

1990 — J.-S. Kim, V. C. Yang (The University of Michigan), "Biopolymer Couples with Heparin-Antidote for In Vivo Heparin Removal."

1991 — J. L. Koenig (Case Western Reserve University), "Recent Advances in FT-IR of Polymers."

1991 — M. R. Schure (Rohm & Haas Company), "Optimization of Channel Dimensions Used in Sedimentation Field-Flow Fractionation through Computer Simulation."

1992 — V. Volksen, M. I. Sanchez, J. W. Labadie (IBM Almaden Research Center) and T. Pascal (CNRS, France), "Base-Catalyzed Cyclization of ortho-Aromatic Amide Alkyl Esters: A Novel Approach to Chemical Imidization."

1992 — Y. C. Chung and N. Leventis (Molecular Displays, Inc.), "Preparation and Characterization of a New Viologen Polymer/Redox Conductive Oxide Layered Material."

1993 — M. K. George, R. P. N. Veregin, P. M. Kazimaier, G. R. Hamer (Xerox Corporation), "Narrow Molecular Weight Resin by Free Radical Process."

1994 — J. E. G. Lipson (Dartmouth College), "Born-Green-Yvon Integral Equation Treatment of Fluids and their Mixtures."

1995 — J. F. G. A. Jansen, E. W. Meijer (Eindhoven University of Technology), E. M. M. de Brabander-van den Berg (DSM Research, The Netherlands), "The Dendritic Box and Bengal Rose."

1996 — J. J. Watkins, T. J. McCarthy (University of Massachusetts-Amherst), "Chemistry in Supercritical Carbon Dioxide-Swollen Polymers."

1997 — C. J. Hawker, E. E. Malmstrom (IBM), C. W. Frank, J. P. Kampf (Stanford University), C. Mio, J. Prausnitz (University of California - Berkeley), "Dendritic Macromolecules: Hype or Unique Materials."

1998 — J. G. Curro, J. D. Weinhold (Sandia National Laboratories), "Application of Integral Equation Theory to Polyolefin Liquids and Blends."

POLYMERIC MATERIALS: SCIENCE & ENGINEERING
1998 WORSHOPS - BOSTON

RESERVE YOUR PLACE IN THESE IMPORTANT WORKSHOPS TODAY!

WHY ATTEND?

- Continuing education for chemists, technologists, engineers and other professionals in research, development, production, application and characterization.

- Workshops designed to inform both persons with experience in a field as well as newcomers. Workshops run by renowned specialists in their fields.

- Learn the fundamentals at the workshops and hear about the latest advances at the companion PMSE symposia.

WORKSHOPS AT THE BOSTON ACS MEETING (August 23, 1998)*

1. **Biomedical Polymers.** *(Prof. Sam Huang, University of Connecticut, (860) 486-4627, FAX: (860) 486-2981, shuang@ims.uconn.edu).* Workshop will consist of fundamentals of design and synthesis, properties and lifetimes, surface characterization, biocompatibility, implants, tissue engineering and biosensitive applications.

2. **Small Angle Scattering in Polymer Processing and Characterization.** *(Dr. John Barnes, NIST, (301) 975-6786, FAX: (301) 977-2018, john.barnes@nist.gov).* The American Crystallographic Association and the International Union of Crystallography will conduct a tutorial session for students and workers in analytical laboratories who require an understanding of how SAS methods can expand their ability to characterization polymers.

3. **Toughening of Plastics.** *(Prof. Albert Yee, The University of Michigan, (734) 763-2445, FAX: (734) 763-4788, afyee@engin.umich.edu).* This workshop will provide some of the fundamental basis required to understand the fracture and failure processes of plastics and rational toughening strategies. Fundamentals to rheological and interfacial control of blend morphology will be covered.

*Registration: $475.00 prior to 7/15/98; $525 after 7/15/98. Discount registration for students and post-docs: $100.00 prior to 7/15/98.

PROSPECTIVE WORKSHOPS AT THE ANAHEIM ACS MEETING (MARCH 21, 1999)‡

1. Polymer Materials Reliability
2. Conductive Polymers

‡ Registration fees higher after February 15, 1999, check web site for details.

For updates, new topics and scheduling, browse our website at:
http://www.msc.cornell.edu/~cober/workshops/

REGISTRATION INFORMATION:

Christopher K. Ober
Department of Materials Science & Engineering
Cornell University
Ithaca, New York 14853-1501
Fax: (607) 255-6575
Phone: (607) 255-8417
e-mail: cober@msc.cornell.edu

Vladimir Tsukruk
College of Engineering & Applied Science
Western Michigan University
Kalamazoo, Michigan 49008
Fax: (616) 387-6517
Phone: (616) 387-6510
e-mail: vladimir@wmich.edu

Future Programs

SYMPOSIA SCHEDULE

Technical Program Chairs:

Richard A. Register
Department of Chemical Engineering
Engineering Quadrangle
Princeton University
Princeton, New Jersey 08544-5263
PHONE: (609) 258-4691
FAX: (609) 258-0211
email: register@pucc.princeton.edu

Jean M. J. Fréchet
Department of Chemistry
University of California - Berkeley
Berkeley, California 94720-1460
PHONE: (510) 643-3077
FAX: (510) 643-3079
email: frechet@cchem.berkeley.edu

A. Jay Dias
Exxon Chemical Company
Baytown Polymers Center
5200 Bayway Drive
Baytown, Texas 77520-2101
PHONE: (281) 834-5255
FAX: (281) 834-1665
email: ajdias@erenj.com

ANAHEIM **- (March 21-25, 1999)**
Original and two (2) copies of 150-word abstract (original on ACS abstract form) and original and three (3) copies of PMSE preprint (original on PMSE preprint paper) due to the symposium organizer by November 15, 1998. ACS abstract forms may be obtained in hard copy from symposium organizer or downloaded from: http://www.acs.org/meetings/abstract/abinfo.html; PMSE preprint forms are available only in hard copy and may be obtained from the symposium organizer.

75 Years of Progress in Applied Polymer Science. J. Edward Glass, North Dakota State Univ., Dept. of Polymers & Coatings, Dunbar Hall, Rm. 136, Fargo, ND 58105, (701) 231-7128, FAX (701) 231-8439, email: glass@plains.nodak.edu.

Future of Applied Polymer Science. Michael Jaffe, NJ Ctr. For Biomaterials & Medical Devices, Dept. of Chem., Rutgers Univ., Piscataway, NJ 08854-8087, (732) 445-0788, FAX (732) 445-0790, email: aerobatkat@worldnet.att.net.

Synthesis of Novel Polymeric Materials. Virgil Percec, Dept. of Macromolecular Sci., Case Western Res. Univ., Cleveland, OH 44106-7202; (216) 368-4242, FAX (216) 368-4202, email: vxp5@po.cwru.edu, Jean M. J. Fréchet, Dept. of Chem., Univ. of California-Berkeley, Berkeley, CA 94720-1460, (510) 643-3077, FAX (510) 643-3079; email: frechet@cchem.berkeley.edu.

Advanced Catalysis: New Polymer Syntheses and Modifications. *(Cosponsored POLY, PMSE is primary sponsor)*. Bruce Novak, Dept. of Poly. Sci. & Engg., Univ. of Mass., Amherst, MA 01003, (413) 545-2160, FAX (413) 545-0764, email: novak@polysci.umass.edu; Lisa S. Boffa, Exxon Res. & Engg. Co., Rm. LC124, Rt. 22 East, Annandale, NJ 08801, (908) 730-2240, FAX (908) 730-2536, email: lsboffa@erenj.com.

Polymers for Batteries and Fuel Cells. Thomas A. Zawodzinski, Los Alamos Nat. Lab., P.O. Box 1663, MS D429, MST-11, Los Alamos, NM 87545, (505) 667-0925, FAX (505) 665-4292, email: zawod@lanl.gov; John P. Ferraris, Dept. of Chem., Univ. of Texas-Dallas, P.O. Box 830688, Mailstop BE26, Richardson, TX 75083-0688, (972) 883-2905, FAX (972) 883-2925, email: ferraris@utdallas.edu; Gary E. Wnek, Dept. of Chem. Engg., Virginia Commonwealth Univ., Richmond, VA 23284-3028, (804) 828-7789, FAX (804) 828-4269, email: gewnek@saturn.vcu.edu.

Emulsion Polymers. Dennis M. Wilson, S.C. Johnson Polymer 8310 16th St., P.O. Box 902, Sturtevant, WI 53177-0902, (414) 631-4190, FAX (414) 631-4039, email: dmwilson@scj.com; Mohamed S. El-Aasser, Lehigh Univ., Emulsion Polymers Inst., 111 Research Dr., Iacocca Hall, Bethlehem, PA 18015-4791, (610) 758-4470, FAX (610) 758-6245, email: mse0@lehigh.edu; Carrington Smith, Air Products & Chemicals, Inc., 7201 Hamilton Blvd., Allentown, PA 18195-1501, (610) 481-6496, FAX (610) 481-6460, email: smithcd2@apci.com.

Bioinspired Materials: Properties, Processing and Self-Assembly. *(Cosponsored BTEC)*. Graham Swift, Rohm & Haas Co., 727 Norristown Rd., P.O. Box 904, Spring House, PA 19477-0904, (215) 641-7756, FAX (215) 619-1642, email: rsiygs@rohmhaas.com; Buddy D. Ratner, Dept. of Bioengineering, Univ. of Washington, Box 351720, Seattle, WA 98195, (206) 685-1005, FAX (206) 616-9763, email: ratner@uweb.engr.washington.edu.

Chiral Polymers. *(Cosponsored POLY; POLY is primary sponsor)*. Samuel J. Huang, Univ. of Connecticut, Inst. of Mats. Sci., 97 N. Eagleville Rd., Storrs, CT 06269-3136, (860) 486-4627, FAX (860) 486-4745, email: shuang@mail.ims.uconn.edu; Dominic V. McGrath, Univ. of Connecticut, Dept. of Chem., U-60, 215 Glenbrook Rd., Storrs, CT 06269-4060, (860) 486-3673, FAX (860) 486-2820, email: dmcgrath@nucleus.chem.uconn.edu.

General Papers/New Concepts in Polymeric Materials. Peter G. Edelman, Chiron Diagnostics, 63 North St., Medfield, MA 02052, (508) 359-3551, FAX (508) 359-3243, email: peter.edelman@chirondiag.com.

Cooperative Research Award Symposium

NEW ORLEANS **- August 22-26, 1999**
Original and two (2) copies of 150-word abstract (original on ACS abstract form) and original and three (3) copies of PMSE preprint (original on PMSE preprint paper) due to the symposium organizer by April 1, 1999. ACS abstract forms may be obtained in hard copy from symposium organizer or downloaded from: http://www.acs.org/meetings/abstract/abinfo.html; PMSE preprint forms are available only in hard copy and may be obtained from the symposium organizer.

Adhesion. Ray A. Dickie, Ford Motor Co., MD-3083 SRL, P.O. Box 2053, Dearborn, MI 48121-2053, (313) 337-4059, FAX (313) 621-0646, email: rdickie1@ford.com; Anthony J. Kinloch, Dept. of Mech. Engg., Imperial College of Sci., Technology & Medicine, Exhibition Rd., London, SW7 2BX, UK 0171-594-7082, FAX 0171-823-8845, email: a.kinloch@ic.ac.uk.

Polymeric Surfactants. *(Cosponsored POLY; PMSE is primary sponsor)*. Robert K. Prud'homme, Dept. of Chem. Engg., Princeton Univ., Princeton, NJ 08544-5263, (609) 258-4577, FAX (609) 258-0211, email: prudhomm@princeton.edu; Saad A. Khan, Dept. of Chem. Engg., North Carolina State Univ., Campus Box 7905, Raleigh, NC 27695, (919) 515-4519, FAX (919) 515-3465, email: khan@ds1.che.ncsu.edu.

Film Formation. Theodore Provder, ICI Paints, 16651 Sprague Rd., Strongsville, OH 44136-1739, (440) 826-5289, FAX (440) 826-5233, e-mail: ted_provder@ici.com.

Polymeric Materials Science and Technology of Micro- and Nano-Patterning Devices and Structures. Omkaram Nalamasu, Bell Laboratories, Lucent Technologies, 600 Mountain Ave., Murray Hill, NJ 07974, (908) 582-6458, FAX (908) 582-6255, email: nlms@lucent.com; Don Hofer, IBM Almaden Research Ctr., 650 Harry Rd., San Jose, CA 95120; Elsa Reichmanis, Rm. 1D-260, Bell Laboratories, Lucent Technologies, Murray Hill, NJ 07974, (908) 582-2504, FAX (908) 582-3609, email: er@bell-labs.com; Seiichi Tagawa, Dept. of Beam Mats. Sci. & Tech., Inst. of Scientific & Industrial Research, Osaka University, Osaka, Japan, 81-6-879-8500, FAX 81-6-876-3287, email: tagawa@sanken.osaka-u.ac.jp

Smart Surfaces. Jeffrey T. Koberstein, Univ. of Conn., Inst. of Mats. Sci., 97 N. Eagleville Rd., Storrs, CT 06269-3136, (860) 486-4716, FAX (860) 486-4745, email: jeff@mail.ims.uconn.edu.

Semicrystalline Polymers. Hervé Marand, Dept. of Chem., Virginia Polytechnic Inst. & State Univ., Hahn Hall, Rm. 2103, Blacksburg, VA 24061-0212, (540) 231-8227, FAX (540) 231-8517, hmarand@chemserver.chem.vt.edu; Srivatsan Srinivas, Exxon Chem. Co., Baytown Polymer Ctr., 5200 Bayway Dr., Baytown, TX 77520-5200, (281) 834-2932, FAX (281) 834-2480.

Polymeric Materials in Separations. *(Cosponsored MACR).* Benny D. Freeman, Dept. of Chem. Engg., North Carolina State Univ., 315 Riddick Hall, Campus Box 7905, Raleigh, NC 27695-7905, (919) 515-2460, FAX (919) 515-3465, email: benny_freeman@ncsu.edu; Stephen S. Kelley, Ctr. for Renewable & Chem. Technologies & Mats., National Renewable Energy Lab., 1617 Cole Blvd., Golden, CO 80401, (303) 384-6123, FAX (303) 384-6103, email: kelleys@tcplink.nrel.gov.

Polymers in Display Applications. *(Cosponsored POLY; POLY is primary sponsor).* T. A. Tervoort, Inst. for Polymers, ETH Zentrum, UNO C15, Universitatstrasse 41, CH-8092, Zurich, Switzerland, 41 1 632 6188, FAX 41 1 632 1178, email: tervoort@ifp.mat.ethz.ch; C. W. M. Bastiaansen, Eindhoven Polymer Laboratories, Eindhoven University of Technology, P.O. Box 513, 5600 MB Eindhoven, The Netherlands, 31 40 247 4915, FAX 31 40 243 6999, email: tgtkkb@chem.tue.n1.

General Papers/New Concepts in Polymer Materials. Christopher K. Ober, Dept. Mats. Sci. & Engg., Cornell Univ., 327 Bard Hall, Ithaca, NY 14853-1501, (607) 255-8417, FAX (607) 255-6575, email: cober@msc.cornell.edu.

ICI Student Award in Applied Polymer Science. Elsa Reichmanis, Bell Laboratories, Lucent Technologies, Rm. 1D-260, 600 Mountain Ave., Murray Hill, NJ 07974, (908) 582-2504, FAX (908) 582-3609, email: er@bell-labs.com.

Tess Award Symposium

SAN FRANCISCO - March 26-31, 2000

Practical Polymer Synthesis. Donald N. Schulz, Exxon Res. & Engg. Co., Rt. 22 East, Annandale, NJ 08801, (908) 730-2526, FAX (908) 730-2536, email: dnschul@erenj.com; Paul L. Valint, Jr., Bausch & Lomb, Inc., 1400 N. Goodman St., P.O. Box 450, Rochester, NY 14603-0450, (716) 338-6414, FAX (716) 338-6316, e-mail: pvalint@bausch.com; Abhi O. Patil, Exxon Res. & Engg. Co., Rt. 22 East, Annandale, NJ 08801, (908) 730-2639, FAX (908) 730-2536, email: aopatil@erenj.com.

Polymer Nanocomposites. Emmanuel P. Giannelis, Dept. of Materials. Sci. & Engg., Cornell Univ., Bard Hall, Ithaca, NY 14853-1501, (607) 255-9680, FAX (607) 255-2365, email: epg2@cornell.edu; Ramanan Krishnamoorti, Dept. of Chem. Engg., Univ. of Houston, Houston, TX 77204-4792, (713) 743-4312, FAX (713) 743-4323, email: ramanan@bayou.uh.edu.

Structure, Dynamics and Organization of Polymers in the Solid State by Magnetic Resonance. *(Cosponsored POLY; PMSE is primary sponsor).* Jeffery L. White, Exxon Chem. Co., Baytown Polymer Ctr., 5200 Bayway Dr., Baytown, TX 77520-5200, (281) 834-5022, FAX (281) 834-2395, email: jeffery.l.white@chemical.exxon.sprint.com; Peter A. Mirau, Bell Laboratories, Lucent Technologies, Rm. 1T-206, Murray Hill, NJ 07974, (908) 582-5841, FAX (908) 582-3609, email: mirau@bell-labs.com.

Vibrational Spectroscopy of Polymer Surfaces and Interfaces. Marek W. Urban, North Dakota State Univ., Dept. of Polymers & Coatings, Fargo, ND 58105, (701) 231-7859, FAX (701) 231-8439, email: urban@plains.nodak.edu.

Molecularly Ordered Networks. Elliot P. Douglas, Univ. of Florida, Dept. of Mats. Sci. & Engg., Gainesville, FL 32511, (352) 846-2836, FAX (352) 392-3771, email: edoug@mse.ufl.edu; Steven H. McKnight, Army Res. Laboratory, Attn: AMSRL-WM-MA, Aberdeen Proving Ground, MD 21005, (410) 306-0699, FAX (410) 306-0676, email: shm@arl.mil.

Rheology and Mechanical Properties of Metallocene Polyolefins. *(Cosponsored Society of Plastics Engineers; PMSE is primary sponsor).* Cesar A. Garcia-Franco, Exxon Chem. Co., Baytown Polymers Ctr., 5200 Bayway Dr., Baytown, TX 77520-5200, (281) 834-2447, FAX (281) 834-1516, email: cesar.a.garcia-franco@chemical.exxon.sprint.com; Scott H. Wasserman, Union Carbide Corp., P.O. Box 670, Bound Brook, NJ 08805-0670, (732) 563-5068, FAX (732) 563-5603, email: wassersh@ucarb.com; Savvas Hatzikiriakos, Dept. of Chem. Engg., Univ. of British Columbia, 2216 Main Mall, Vancouver, BC V6T 1Z4, (604) 822-3107, FAX (604) 822-6003, email: hatzikir@unix2.ubc.ca

Frontiers for Polymer Science the 21st Century. *(Cosponsored MACR).* S. Richard Turner, Research Laboratories, Eastman Chem. Co., Kingsport, TN 37762, (423) 224-0489, FAX (423) 229-4558, email: srturner@eastman.com.

Organotransition Metal-Catalyzed Polymerizations. *(Cosponsored POLY; POLY is primary sponsor).* Christopher Gorman, Box 8204, Dept. of Chem., North Carolina State Univ., Raleigh, NC 27695-8204, (919) 515-4252, FAX (919) 515-8920, email: chris_gorman@ncsu.edu.

General Papers/New Concepts in Polymeric Materials. Christopher K. Ober, Dept. of Mats. Sci. & Engg. Cornell Univ., 327 Bard Hall, Ithaca, NY 14853-1501, (607) 255-8417, FAX (607) 255-6575, email: cober@msc.cornell.edu.

Cooperative Research Award Symposium

WASHINGTON - August 20-25, 2000

Materials for Transportation. Ray A. Dickie, Ford Motor Co., MD-3083 SRL, P.O. Box 2053, Dearborn, MI 48121-2053, (313) 337-4059, FAX (313) 621-0646, email: rdickie1@ford.com.

Coatings for Transportation. Frank N. Jones, Coatings Res. Inst., Eastern Mich. Univ., 430 W. Forest St., Ypsilanti, MI 48197, (734) 487-2203, FAX (734) 483-0085, email: frank.jones@emich.edu.

UV Degradation of Polymers: Accelerated Tests. David R. Bauer, Ford Motor Co., MD-3182 SRL, P.O. Box 2053, Dearborn, MI 48121-2053, (313) 594-1756, FAX (313) 323-1129, email: dbauer3@ford.com.

Fire and Polymers. Gordon L. Nelson, Florida Inst. of Tech., College of Sci. & Liberal Arts, 150 W. University Blvd., Melbourne, FL 32901-6975, (407) 674-7260, FAX (407) 984-8864, email: nelson@fit.edu.

Transducer-Active Polymers: Responsive Materials in Chemical and Biological Sensors, Actuators and Controlled Release. Anthony Guiseppi-Elie, ABTECH Scientific, Inc., 1273 Quarry Commons Dr., Yardley, PA 19067, (215) 321-3256, FAX (215) 321-7099, email:guiseppi@abtechsci.com; Norman F. Sheppard, Jr., Gamera Bioscience, 200 Boston Ave., Medford, MA 02155, (781) 306-0827, Ext. 16, FAX (781) 306-0837, email: nsheppard@gamerabioscience.com; Gary E. Wnek, Virginia Commonwealth Univ., Dept. of Chem. Engg., Richmond, VA 23284-3028, (804) 828-7789, FAX (804) 828-4269, email: gewnek@saturn.vcu.edu. Visit this symposium's website at: *http://www.abtechsci.com/TAP2000.html*

Organic Thin Films for Photonic Applications. *(Cosponsored POLY and Optical Society of America; PMSE is primary sponsor).* Robert J. Twieg, Dept. of Chem., Kent State Univ., Kent, OH 44242 (330) 672-2791, FAX (330) 672-3816, email: rtwieg@lci.kent.edu.

Sustainable Chemistry for the Polymer Industry. *(Cosponsored POLY; POLY is primary sponsor).* Graham Swift, Rohm & Haas Co., 727 Norristown Rd., P.O. Box 904, Spring House, PA 19477-0904, (215) 641-7756, FAX (215) 619-1642, email: rsiygs@rohmhaas.com; Richard A. Gross, Polytechnic Univ., 6 Metrotech Ctr., Brooklyn, NY 11201, (718) 260-3180, FAX (718) 260-3125, email: rgross@poly.edu.

General Papers/New Concepts in Polymeric Materials

ICI Student Award in Applied Polymer Science

Tess Award Symposium

SAN DIEGO - April 1-5, 2001

Macromolecular Self-Assembly at Surfaces and Interfaces. Paula T. Hammond, Dept. of Chem. Engg., MIT, 77 Massachusetts Ave., Rm. 66-550, Cambridge, MA 02139, (617) 258-7577, FAX (617) 253-9695, email: hammond@mit.edu.

Functional Condensation Polymers. Graham Swift, Rohm & Haas Co., 727 Norristown Rd., P.O. Box 904, Spring House, PA 19477-0904, (215) 641-7756, FAX (215) 619-1642, email: rsiygs@rohmhaas.com; Charles E. Carraher, Dept. of Chem., Florida Atlantic Univ., 500 NW 20th St., Boca Raton, FL 33431, (561) 367-2107, FAX (561) 367-2759, email: carraher@fau.edu.

General Papers/New Concepts in Polymeric Materials

Cooperative Research Award Symposium

CHICAGO - August 26-30, 2001

Water-Borne Coatings. J. Edward Glass, North Dakota State Univ., Dept. of Polymers & Coatings, Fargo, ND 58105, (701) 231-7128, FAX (701) 231-8439, email: glass@plains.nodak.edu; John Padget, ICI.

Polymers at Additives. Donald N. Schulz, Exxon Res. & Engg. Co., Rt. 22 East, Annandale, NJ 08801, (908) 730-2526, FAX (908) 730-2536, email: dnschul@erenj.com; Abhi O. Patil, Exxon Res. & Engg. Co., Rt. 22 East, Annandale, NJ 08801, (908) 730-2639, FAX (908) 730-2536, email: aopatil@erenj.com.

Particle Size Assessment and Characterization. Theodore Provder, ICI Paints, 16651 Sprague Rd., Strongsville, OH 44136-1739, (440) 826-5289, FAX (440) 826-5233, email: ted_provder@ici.com.

Dendrimers and Hyperbranched Polymers. Jean M. J. Fréchet, Dept. of Chem., Univ. of California - Berkeley, Berkeley, CA 94720-1460, (510) 643-3077, FAX (510) 643-3079, email: frechet@cchem.berkeley.edu.

General Papers/New Concepts in Polymeric Materials

ICI Student Award Symposium

Tess Award Symposium

TENTATIVE LOCATIONS/DATES OF FUTURE NATIONAL ACS MEETINGS

Orlando	April 7-12, 2002
Boston	September 8-13, 2002
New Orleans	March 23-28, 2003
New York	September 7-12, 2003
Anaheim	March 28 - April 2, 2004
Philadelphia	August 22-27, 2004
San Diego	March 13-18, 2005
Washington	August 28 - September 2, 2005

Instructions for Authors

As a precondition for presenting a paper in the Division of Polymeric Materials: Science and Engineering, a standard ACS abstract and (with one exception noted below) a manuscript for the Proceedings volume must be submitted to the indicated Symposium Chairman <u>in strict accordance with published deadlines</u>.

CONTACTS

All communication should be addressed to the Symposium Chairman, as listed in Future Programs. For the New Concepts and General Papers sessions, the Vice Chairman of the Division acts as Symposium Chairman.

ACS ABSTRACT PREPARATION

Standard ACS abstract forms MUST be used. They are available from the Symposium Chairmen or can be downloaded directly from the ACS website by pointing your browser to: *http://www.acs.org/meetings/abstract/abinfo.html*. The ACS abstract form gives detailed instructions. Please make sure that all sections of the form are completed and that the title and author listing (Sections C and D of the form) correspond exactly to that listed in the abstract proper. Please submit the *original and two copies of the form*. The postcards on the abstract forms will be returned to authors only if U.S. postage is provided. An exception can be made for authors from other countries upon request.

PROCEEDINGS MANUSCRIPT PREPARATION

The Division of Polymeric Materials: Science and Engineering Proceedings include preprints of papers presented in Division-sponsored symposia. We are proud of this service and intend to keep the quality of the Proceedings high. Time does not permit editing of the submitted manuscripts, thus the content and quality of each paper are the responsibility of the authors. Publication in the Proceedings <u>does not</u> exclude publication elsewhere.

Manuscripts are generally required for all presentations before the PMSE Division. There is one exception: Manuscripts are not required for presentations in award symposia; they are strongly encouraged. Please follow the detailed instructions below for preparation of Proceedings manuscripts.

<u>Content and Style</u>: The editorial format and style should be that of a technical paper and should contain as much experimental detail and discussion as possible within the allowed length. References may be given as specified by the "ACS Style Guide" or in the style used in the *Journal of Polymer Science*. Figures should be captioned and numbered. Tables should be clear and large enough to allow for legibility after reduction to 68% of original size. The manuscript should <u>not</u> contain an abstract. Papers are expected to be science and technology based. Commercial references and the use of trade names must be limited to those essential for initial characterization of materials and/or equipment; subsequent references and discussion must employ generic terms.

<u>Paper</u>: Special paper is provided and MUST be used. The text area of the special paper for each page is divided in two columns; each column is 128 × 342 mm (approximately 5-1/16 × 13-1/2 inches). The text for each column can conveniently be printed on legal sized paper and affixed to the form with rubber cement. The manuscript may be folded, with minimum creasing, for mailing.

<u>Manuscript Length and Page Charge</u>. One-page (2-column) manuscripts are encouraged. Manuscripts of up to 2 pages (4 columns) will be published without charge. Manuscripts longer than 4 pages (8 columns) are not acceptable. A charge of $100 per page will be billed for the third and fourth pages. Submission of a 3 or 4 page manuscript is regarded as a commitment to pay page charges.

<u>Format</u>: Typing should be single spaced; 12 cpi prestige elite or 12 pt times roman fonts reproduce well and are preferred. The manuscript must be ready for photo-offset reproduction, allowing for reduction to approximately 68% of original size.

- At the very top of the first column, the title and authors names, addresses and zip codes should be given. These MUST agree with what is given on the ACS abstract form.

- Figures should be pasted directly onto the manuscript pages. Figures smaller than 7 × 12 cm in the original are likely to be unreadable. Figure captions and tables should not be reduced before mounting on preprint forms. Remember, all submitted material will be reduced to approximately 68% of original size during production of the book.

- Photographs required for effective presentation of results should be pasted directly onto manuscript pages. Photos are expensive to reproduce; they should be used sparingly.

- All pages should be numbered in the upper right hand corner of the margin with a non-reproducing blue pencil.

- The *original preprint manuscript and 3 copies* should be sent to the Symposium Chairman (or to the Vice Chairman of the Division in the case of New Concepts and General Papers sessions). It is strongly preferred that the 3 copies be *reduced* to 21.6 × 27.9 cm (8-1/2 × 11 inch) or 29.6 × 21 cm size.

WITHDRAWAL OF PAPERS

Submission of an abstract and a Proceedings manuscript constitutes a commitment on the part of the author(s) to present the paper at its designated time. Requests to withdraw from publication cannot be honored if received after the Division's deadline for submission of ACS abstracts and preprint manuscripts. It is prudent to obtain proper publication clearance by the sponsoring organization <u>prior to submission</u>. The Division cannot assume responsibility for confidentiality of submitted material.

SCHEDULING OF PRESENTATIONS

The Division schedules Symposia in accord with ACS regulations, which require "level programming" throughout the meeting. A Final Program, with the time assigned to each presentation, is published in *Chemical and Engineering News* five to six weeks before each meeting.

COPYRIGHT

Each volume of the Proceedings is copyright, as a whole, by the American Chemical Society. Rights to individual papers remain with the author(s). Thus, permission of the original author is required for reproduction of material appearing in these volumes. The Proceedings should be cited as the origin of the material.

Program Listing

SUNDAY MORNING		**SECTION A**

GENERAL PAPERS/NEW CONCEPTS IN POLYMERIC MATERIALS
Organizer: Peter G. Edelman

Session 1
Synthesis with Controlled Architecture
Presiding: Martin Baumert

Pg.	Time	No.	
	8:15		Introductory Remarks.
1	8:20	1.	Anionic Photopolymerization Using Acylferrocene Derivatives. Y. Yamaguchi, B. J. Palmer, T. Wakamatsu, D. B. Yang, C. R. Kutal
3	8:50	2.	Poly(isobutylene-g-styrene) Graft Copolymers by Quasiliving Atom Transfer Radical Grafting of Styrene. T. Fónagy, B. Iván, M. Szesztay
5	9:20	3.	Synthesis and Characterization of Oxazoline-Terminated Polymers Via Controlled Radical Polymerization as Intermediates for Reactive Processing. M. Baumert, J. Zimmermann, R. Mülhaupt
7	9:50	4.	Control of Polymer Architecture by a Special Chain Transfer Catalyst. Z. Guan, C. Jackson
9	10:20	5.	Synthesis of Long Chain Branched Syndiotactic Polystyrene. Y-B. Huang, T. H. Newman, M. S. Chahl
11	10:50	6.	6,6'-Bisfunctionalized 2,2'-Bipyridine Metal Complexes as Supramolecular Initiators for Functional Poly(oxazolines). U. S. Schubert, O. Nuyken, G. Hochwimmer
13	11:20	7.	Curing of Partially Brominated Poly(isobutylene-co-4-methylstyrene) Elastomers Using Phenolic Resins: A Mechanistic Investigation. M. Jayaraman, J. M. J. Fréchet, A. J. Dias, H. C. Wang
15	11:50	8.	Hydroquinone, Quinone and Amine Functionalized Polyolefins. A. O. Patil, S. Zushma

SUNDAY MORNING		**SECTION B**

Session 2
Biostable, Crosslinked and Gas Transport
Presiding: Thomas J. Kennedy

Pg.	Time	No.	
	8:15		Introductory Remarks.
17	8:20	9.	Hyperbranched Grafting on a Polyethylene Surface. J. G. Franchina, D. E. Bergbreiter
19	8:50	10.	Nanoporous Ceramic Coated Structures from Block Copolymers. V. Z-H. Chan, E. L. Thomas, V. Lee, R. D. Miller
21	9:20	11.	Siloxane Functionalized Isocyanurate Coatings. H. Ni, A. D. Skaja, R. A. Sailer, M. D. Soucek
23	9:50	12.	Surface Coverage, Rheology and Morphology of Copolymer Modified PP/PS Blends. H. Asthana, K. Jayaraman
25	10:20	13.	Polymerization and Dynamic Testing of Difunctional Acrylic Metal Salts in a Polybutadiene Golf Ball Center. T. J. Kennedy
27	10:50	14.	Structure Development of Spherulitic Domain in Polymer Thin Films. T. Huang, T. Tsuji, A. D. Rey, M. R. Kamal
29	11:20	15.	Equilibrium Order-To-Disorder Transition Temperature in Block Copolymers. G. A. Floudas, T. Pakula, G. Velis, S. Sioula, N. Hadjichristidis
31	11:50	16.	Ultrasonic Methods for Characterization of Polymeric Materials. I. Alig, D. Lellinger, S. Tadjbakhsch

SUNDAY AFTERNOON		**SECTION C**

Session 3
Surface Engineering and Bulk Properties
Presiding: Kelly A. Mowery

Pg.	Time	No.	
	1:15		Introductory Remarks.
33	1:20	34.	Polymeric Diazeniumdiolates for Fabricating Thromboresistant Electrochemical Sensors Via Nitric Oxide Release. K. A. Mowery, M. H. Schoenfisch, M. E. Meyerhoff, J. E. Saavedra, L. K. Keefer
35	1:50	35.	Synthesis and Characterization of Aliphatic Polyesters With Hydrophilic Pendant Groups Using Enzymes. B. J. Kline, E. J. Beckman, A. J. Russell

TUESDAY EVENING   **SECTION A**

JOINT PMSE/POLY POSTER SESSION

Poster Session
Presiding: Peter G. Edelman

Pg.	Time	No.	
84	6:00	163.	Lifetime and Property Changes of Kevlar Fiber Yarn Under Cyclic Loading. Y. Q. Rao, R. J. Farris

---

SUNDAY MORNING   SECTION C

DYNAMICALLY VULCANIZED ALLOYS
Organizers: Sudhin Datta, Maria D. Ellul

Session 1
Plenary Lecture and New Thermoplastic Elastomers
Presiding: Sudhin Datta

Pg.	Time	No.	
86	8:00	17.	Dynamic Vulcanization – A Major Innovation in TPEs. S. Abdou-Sabet
88	9:00	18.	An Oil Resistant Thermoplastic Vulcanizate from Epoxidized Natural Rubber and Polypropylene. J. Patel, A. J. Tinker
90	9:40	19.	New Highly Crosslinked TPEs Based on VNB-EPDM. M. D. Ellul, P. S. Ravishankar
92	10:20	20.	Innovative Thermoplastic Vulcanizates: Formation, Morphology, Processing and Properties. H. G. Fritz
94	11:00	21.	Novel Thermoplastic Vulcanizates Based on Compatibilized Plastic Phase. K. Venkataswamy
96	11:40	22.	Analysis of TPV Extrusion Chemical Foaming Process. Y. Wang, H. Cai, L. Freitas, B. Dion, R. Brzoskowski

---

SUNDAY AFTERNOON   SECTION B

Session 2
Kinetics
Presiding: D. Hazelton

Pg.	Time	No.	
98	1:00	29.	Kinetics of Vulcanization-Induced Phase Separation in Blends of Polypropylene/Ethylene Propylene Diene Terpolymer. T. Kyu, A. Ramanujam, K. J. Kim
100	1:40	30.	Chemistry of Phenol-Formaldehyde Resin Crosslinking of EPDM. M. van Duin, H. A. E. de Keijzer
102	2:20	31.	Kinetic Model of Dynamic Vulcanization. E. V. Prut, N. A. Yerina

---

Pg.	Time	No.	
104	3:00	32.	Reactive Compatibilization of SAN/EPR Blends: Morphology Development and Interfacial Grafting Kinetics. C. Pagnoulle, C. E. Koning, L. Leemans, R. Jérôme
106	3:40	33.	Aging of Thermoplastic Elastomers. E. V. Prut, N. A. Yerina, L. V. Kompaniets

---

MONDAY MORNING   SECTION C

Session 3
Deformation and Fracture
Presiding: R. Brzoskowski

Pg.	Time	No.	
108	8:30	53.	Deformation Mechanism of Thermoplastic Vulcanizates Investigated by Combined FTIR- and Stress-Strain Measurements. M. Soliman, M. van Dijk, M. van Es
109	9:10	54.	Strength Properties of Dynamically Vulcanized Elastomer-Polypropylene Blends. M. D. Ellul, G. J. Lake, A. G. Thomas
111	9:50	55.	Fracture Behavior of Dynamically Vulcanized Thermoplastic Elastomers. A. J. Lesser, N. A. Jones
113	10:30	56.	Effect of Composition and Processing Conditions on the Development of Structural Hierarchy in PP/EPDM Blends. M. Cakmak, S. Cronin
115	11:10	57.	Linear Viscoelasticity of Dynamically Vulcanized Alloys: Kinetic Emulsion Network Model Studies. Y. F. Wang, M. D. Ellul

---

MONDAY AFTERNOON   SECTION C

Session 4
Rheology and Applications
Presiding: R. Patel

Pg.	Time	No.	
118	1:00	77.	Controlling Morphology, Impact Strength and Rheological Properties in Dynamically Crosslinked Polypropylene/EPDM Elastomer Blends. I. Fortelný, Z. Kruliš
120	1:40	78.	Viscoelastic Properties of Thermoplastic Elastomers. E. V. Prut, L. V. Kompaniets, N. A. Yerina
122	2:20	79.	Dynamic Mechanical and Rheological Behavior of Thermoplastic Elastomers. E. V. Prut, N. A. Yerina, L. M. Chepel, A. N. Zelenetskii

SUNDAY AFTERNOON SECTION A

ICI STUDENT AWARD IN APPLIED POLYMER SCIENCE
Organizer: Elsa Reichmanis
Presiding: Elsa Reichmanis

MONDAY MORNING SECTION A

TOUGHENING OF PLASTICS
Organizers: Albert F. Yee, Hung-Jue Sue, Raymond A. Pearson

Session 1
Rubber Toughening
Presiding: Albert F. Yee

MONDAY AFTERNOON SECTION B

Session 2
Rigid-Rigid Toughening
Presiding: Ali S. Argon, Presiding

Toughening of Plastics -- Poster Session
Presiding: Raymond A. Pearson

Pg.	Time	No.	
160	7:00	82.	Design of Mechanical Behavior in PMMA-Modified Epoxy Resins by Control of Phase Separation. P. M. Remiro, C. C. Riccardi, <u>I. Mondragon</u>
162	7:00	83.	Role of Rubber Particle Size Distribution of High Impact Polystyrene (HIPS). <u>M. Demirors</u>
164	7:00	84.	Designing Super High Impact PS/PB Bulk Blends: Optimal Particle Size and Diblock Length. <u>D. Mathur</u>, E. B. Nauman
166	7:00	85.	Strengthening and Toughening of Polymeric Materials Via Controlled Molecular Orientation. <u>C. K-Y. Li</u>, H-J. Sue
168	7:00	86.	New Semi-IPN System of Bismaleimide. D. J. T. Hill, P. Pomery, A. K. Whittaker, <u>Z. Xia</u>
170	7:00	87.	Semi-Crystalline Blends: Enhancing Low Temperature Mechanical Properties. <u>K. A. Chaffin,</u> F. S. Bates, P. Brant
171	7:00	88.	Dynamic Dielectric and Rheological Detection of Phase Separation and Reaction Advancement in PPE/DGEBA-MCDEA Blends. <u>D. Kranbuehl</u>, J. Rogozinski, A. Gilchriest, C. Ambler, S. Poncet, J. P. Pascault, H. Sautereau, G. Boiteux, J. F. Gerard, G. Seytre
173	7:00	89.	Monte Carlo Simulations to Predict the Behavior of Polymer Blends Containing the Block Copolymers. <u>S. Kandasamy</u>, E. B. Nauman
175	7:00	90.	Diffusion Induced Coalescence in the Late Stage Ripening of Polymer Blends. <u>A. P. Russo</u>, E. B. Nauman
177	7:00	91.	Toughening of Polycarbonate/Polyamide-6 Blends. <u>P. Arjunan</u>, R. M. Mininni
178	7:00	92.	Micromechanical Deformation Processes in PA12-Layered Silicate Nanocomposite. <u>G-M. Kim</u>, G. M. Michler, P. Reichert, J. Kresseler, R. Mülhaupt
180	7:00	93.	The Effect of Particle Size in PP/EPR Blends on the Brittle/Ductile Transition Temperature at Low and High Test Speed. <u>P. T. S. Dijkstra</u>, A. van der Wal, R. J. Gaymans, J. Huétink

Pg.	Time	No.	
181	7:00	94.	The Brittle/Ductile Transition of Polycarbonate as a Function of Test Speed. <u>J. P. F. Inberg</u>, M. J. J. Hamberg, R. J. Gaymans
182	7:00	95.	The Effects of the Cavitation Resistance of Rubber on the Toughening of Modified Epoxies. <u>J. Kim</u>, S. Lim, M. Park, M. Ko, C. R. Choe
184	7:00	96.	Phosphine Oxide Containing Copolymers Toughened Epoxy. <u>J. Wang</u>, S. Wang, H. Kang, Q. Ji, O. Kwon, J. E. McGrath, T. C. Ward
186	7:00	97.	Thermally Stable Elastomeric Tougheners for Polycyanurate Thermosetting Systems. <u>D. S. Porter</u>, S. Bhattacharjee, T. C. Ward
188	7:00	98.	Effect of Interfacial Chain Structure on Toughening Behavior in Rubber Modified Poly(methyl methacrylate). <u>J-H. Kim</u>, J-H. An, Y-J. Park, H-J. Ha
190	7:00	99.	Impact Fracture Toughness of Propylene/1-Pentene Random Copolymers. <u>I. Tincul</u>, D. J. Joubert, A. H. Potgieter
192	7:00	100.	Toughening of Rigid Silicone Resins. <u>B. Zhu</u>, D. E. Katsoulis, J. R. Keryk, F. J. McGarry
194	7:00	101.	The Influence of Morphology and Concentration on Toughness in Dispersions Containing Preformed Acrylic Elastomer Particles in an Epoxy Matrix. H. Raghavan, D. Hunston, <u>J. He</u>, D. K. Hoffman
196	7:00	102.	Toughness of PBT/BT Co-Polymer Blends. <u>J. A. Donovan</u>, P. Chang

TUESDAY MORNING   **SECTION A**

Session 3
Novel Toughening Concepts/Systems
Presiding: Hung-Jue Sue

Pg.	Time	No.	
198	8:15	103.	Rate Dependence of the Fracture Behavior in a Rubber-Modified Epoxy. J. Du, P. C. Niven, M. D. Thouless, <u>A. F. Yee</u>
200	9:00	104.	Inorganic Particle Toughening: Micro-Deformation in Epoxies Filled with Glass Beads. <u>J. Lee</u>, A. F. Yee
201	9:30	105.	On Plastic Zone Growth in BN-Filled, Rubber-Modified Epoxy Polymers. <u>R. A. Pearson</u>, M. F. DiBerardino

WEDNESDAY MORNING SECTION D

Session 5
Presiding: Peter G. Edelman

MONDAY MORNING SECTION D

UNILEVER AWARD FOR OUTSTANDING GRADUATE RESEARCH IN POLYMER SCIENCE SYMPOSIUM IN HONOR OF JAMES J. WATKINS
Organizer: Warren T. Ford

Session 1
Polymer Processing in Supercritical Fluids
Presiding: Mark A. McHugh

TUESDAY MORNING **SECTION C**

SCATTERING FROM POLYMERS
Organizers: Peggy Cebe, David J. Lohse, Benjamin S. Hsiao

Session 1
Blends and Multicomponent Systems
Presiding: Peggy Cebe

TUESDAY AFTERNOON **SECTION B**

Session 2
Solutions
Presiding: David J. Lohse

JOINT PMSE/POLY POSTER SESSION

Poster Session
Presiding: Peggy Cebe

<table>
<tr><td>Pg.</td><td>Time</td><td>No.</td></tr>
</table>

314 6:00 164. SAXS and Rheological Studies on the Order-Disorder Transition in Mixtures of Polystyrene-b-Polyisoprene-b-Polystyrene and Low Molecular Weight Polystyrene. <u>S. H. Lee</u>, K. Char

316 6:00 165. Successive Crystallizations from the Melt Followed by Selective Dissolution (SCM-SD): A Fractionation Method for Polyolefines. C. Vandermiers, P. Damman, <u>M. D. Dosière</u>

317 6:00 166. <<X-ray>>, A Software for the Analysis of Two-Dimensional WAXD and SAXS Patterns of Polymers Recorded with Image Plates. <u>M. D. Dosière</u>, D. Villers, F. Delava, O. Hernaut

318 6:00 167. Crystalline Structure and Morphology of Microphases in Compatible Mixtures of Tetrahydrofuran (PTHF)-Methyl Methacrylate (PMMA) Diblock Copolymer and Polytetrahydrofuran. <u>L-Z. Liu</u>, B. Chu

320 6:00 168. The Nature of Secondary Crystallization in Poly(ethylene terephthalate). <u>Z-G. Wang</u>, B. S. Hsiao, B. B. Sauer, W. G. Kampert

322 6:00 169. Real-Time SAXS Study of Crystallization of Metallocene Poly(propylene). <u>P. S. Dai</u>, P. Cebe, R. G. Alamo, L. Mandelkern, M. Capel

324 6:00 170. Morphology Development in Polypropylene Homopolymer Tacticity Mixtures: Isotactic/Syndiotactic Blends. <u>S. Kim</u>, Z-G. Wang, R. A. Phillips, F. Yeh, B. S. Hsiao

326 6:00 171. Time-Resolved Light Scattering Studies on the Morphology Development in PP/EPR Blends. <u>J. K. Lee</u>, S. W. Lim, K. H. Lee, C. H. Lee B. S. Jin

328 6:00 172. Neutron Spin Echo Spectroscopy at the NIST Center for Neutron Research. <u>N. Rosov</u>, S. Rathgeber, M. Monkenbusch

330 6:00 173. Kinetics of Order to Order Transitions of SI and SIS Block Copolymers by Synchrotron SAXS and Rheology. <u>J. K. Kim</u>, H. H. Lee, K. B. Lee

Pg. **Time** **No.**

332 6:00 174. Deformation-Induced Structural Changes in Poly(urethaneurea) Film. F. Yeh, <u>B. S. Hsiao</u>, B. B. Sauer

334 6:00 175. SAXS and WAXD Study of Mesostructure of Molybdenum Oxide Formed by a Block Copolymer Matrix. <u>Y. Xie</u>, T. Liu, L-Z. Liu, B. Chu

336 6:00 176. Real-Time SAXS and Thermal Analysis Study of Multiple Melting Behavior in PEEK. <u>G. Georgiev</u>, P. S. Dai, E. Oyebode, P. Cebe, M. Capel

338 6:00 177. Characterization of Charged PAMAM Dendrimer Interactions in Solution by Small Angle Neutron Scattering. G. Nisato, <u>R. Ivkov</u>, B. J. Bauer, E. J. Amis

340 6:00 178. Dynamic Structure Development During Isothermal Crystallization and Melting of Linear Polyethylenes. <u>J. L. Lopez</u>, Z-G. Wang, B. S. Hsiao, P. J. Armistead

342 6:00 179. Phase Behavior of Styrene-Isoprene Diblock Copolymers in Selective Solvents. <u>C. Lai</u>, W. B. Russel, R.A. Register

342 6:00 180. Block Copolymer Micellar Networks and Mesophases. Shear and Temperature Dependence. <u>K. Mortensen</u>, K. Almdal, R. Kleppinger, N. Mischenko, H. Reynaers

342 6:00 181. Lamellar Versus Supramolecular Scattering in Semi-Crystalline Polymers. <u>J. D. Barnes</u>

342 6:00 182. Morphology of Microfibrillar Reinforced Composite Materials Based on Polymer Blends -- Structural Changes as Revealed from Time-Resolved SAXS/WAXD. <u>M. I. Sarkissova</u>, G. Groeninckx, M. Koch, H. Reynaers

342 6:00 183. Probe Diffusion in High Molecular Weight Hydroxypropylcellulose. Solutionlike and Melt-like Regimes <u>K. A. Streletzky</u>, G. D. J. Phillies

342 6:00 184. Dynamics of Myosin Gel Formation Analyzed by Light Scattering. <u>J-I. Kawahara</u>, Y. Kobayashi, K. Kubota, H. Takaya

343 6:00 185. Effect of Solvent Quality on the Molecular Dimensions of PAMAN Dendrimers. <u>B. J. Bauer</u>, A. Topp, D. A. Tomalia, E. J. Amis

343 6:00 186. Small Angle Neutron Scattering Investigations and Molecular Dynamics Simulations of Hybrid Linear Dendritic Star Copolymers in Polar and Nonpolar Solvents. <u>M. S. Diallo</u>, D. Yu, J. M. J. Fréchet, P. Miklis, W. A. Goddard

WEDNESDAY MORNING   **SECTION C**

Session 3
Crystalline Polymers
Presiding: Benjamin S. Hsiao

WEDNESDAY AFTERNOON   **SECTION C**

Session 4
Crystalline Polymers and Effects of Stress
Presiding: Marcel D. Dosière

THURSDAY MORNING   **SECTION C**

Session 5
Block Copolymers
Presiding: Peter Kofinas

THURSDAY AFTERNOON   SECTION C

Session 6
Block Copolymers and Effects of Stress
Presiding: David J. Lohse

TUESDAY MORNING   SECTION D

ROY W. TESS AWARD SYMPOSIUM IN HONOR OF LOREN W. HILL
Organizers: Frank N. Jones, Charles R. Hegedus
Presiding: Frank N. Jones, Charles R. Hegedus

WEDNESDAY MORNING   SECTION A

BIOSTABLE POLYMERS FOR MEDICAL USE
Organizers: Ed DiDomenico, Anuj Bellare

Session 1
Presiding: Ed DiDomenico

Pg.	Time	No.	
	8:00		Introductory Remarks.
485	8:05	222.	Implantable Polymeric Biomaterials: Degradation Issues and Approaches to Assuring Their Long-Term Performance. A. J. Coury
486	9:00	223.	Synthesis and Use of Silicones. A. L. Himstedt
488	10:00	224.	Biostable Polymers? Measuring and Predicting Stability? L. W. Schroeder
490	10:30	225.	Perspectives on Polyurethane Biostability and Biodegradation. J. M. Anderson, A. Hiltner, T. Collier, J. Tan, M. Shive, S. Hasan, M. Wiggins, M. Schubert, A. Mathur
491	11:00	226.	Stable Polymer Matrix for Controlled Release Dosage Form. K. Ainaoui, J. M. Vergnaud
493	11:30	227.	Controlled Release with Dosage Forms Made of a Core and Shell with Different Drug Concentration. E. M. Ouriemchi, J. M. Vergnaud

WEDNESDAY AFTERNOON   SECTION A

Session 2
Presiding: Orhun K. Muratoglu

Pg.	Time	No.	
495	1:00	249.	Ultrahigh Molecular Weight Polyethylene in Joint Replacement Prostheses: Intrinsic and Extrinsic Factors Affecting Wear Performance. M. Spector, W. M. Jones, A. Bellare
497	2:00	250.	Prevention of Oxidation in the Manufacturing Process of UHMWPE Implants. D. C. Sun, S. S. Yau, G. Schmidig, P. Pereira, C. Stark, J. H. Dumbleton
498	2:30	251.	The Effect of Molecular Weight Between Cross-Links (M_c) on the Wear Behavior of Cross-Linked UHMWPE. O. K. Muratoglu, C. R. Bragdon, D. O. O'Connor, M. Jasty, R. Gul, W. H. Harris
499	3:00	252.	Characterization of Gamma Irradiated UHMWPE After an Oxidative Challenge. J. V. Hamilton, K. Greer, M. B. Schmidt, C. Urian, P. Joshi, A. Crugnola

Pg.	Time	No.	
501	3:30	253.	Insight Into the Molecular Network in Irradiated UHMWPE. D. C. Sun, A. Chopra, C. Stark, J. H. Dumbleton
502	4:00	254.	Wear Behavior of Carbon Fiber Reinforced PEEK Composites in Total Joint Replacements. A. Wang, R. Lin, C. Stark, J. H. Dumbleton
504	4:30	255.	Design Considerations for the In Vivo Stability of Hollow Fiber Based Cell Encapsulation Devices. D. Rein, T. Gunrud, M. Shoichet

THURSDAY MORNING   SECTION A

Session 3
Presiding: Anuj Bellare

Pg.	Time	No.	
506	8:00	272.	Infrared Spectroscopic Study of Oxidation in Polyurethane Hard Segments. S. R. Skorich, M. Benz
508	8:30	273.	Using NMR for Structure Determination of Polyurethanes. J. T. Ippoliti
509	9:00	274.	Mechanically Assisted Corrosion of MP35N Alloy. N. Istephanous, J. Gilbert, G. Martinez, J. Wiser, U. Ko, T. Irwin, D. Untereker
512	9:30	275.	In Vitro Macrophage/Iron/Stress System for the Study of PEU Biodegradation. J. Casas, Q. Zhao, M. Donovan, D. Untereker
514	10:00	276.	The Chemistry of Silicone Based Biomaterials and Why They Continue to be Materials of Choice for Healthcare Applications. D. J. Petraitis
515	10:30	277.	Poly(ethylene oxide) – Polysulfone Block Copolymer Membranes. N. K. Barrera, S. M. Fagen, K. A. Gagnon, L. F. Hancock
516	11:00	278.	Synthesis and Characterization of Novel Dendrimer-Based Gadolinium Complexes as MRI Contrast Agents for the Vascular System. B. Radüchel, H. Schmitt-Willich, J. Platzek, W. Ebert, T. Frenzel, B. Misselwitz, H. J. Weinmann
518	11:30	279.	Ring Opening Polymerization Via Microwave Irradiation. X. Fang, S. J. Huang, D. A. Scola

<u>Session 4</u>
Presiding: Arthur J. Coury

<u>Pg.</u>	<u>Time</u>	<u>No.</u>	
520	1:00	294.	The Biostability of an 80A Aliphatic Polycarbonate Polyurethane. <u>K. B. Stokes</u>
522	1:30	295.	Elastomers for Implantable Applications-Silicone or Polyurethane? <u>M. Skalsky</u>
523	2:00	296.	Degradation Resistant Polyurethane Compositions. <u>G. F. Meijs</u>, P. A. Gunatillake, S. J. McCarthy, D. J. Martin, L. Poole-Warren, K. Schindhelm
524	2:30	297.	Biodurable Polyurethanes in Vascular Grafts. <u>M. Szycher</u>, A. E. Edwards, R. J. Dempsey
525	3:00	298.	In Vivo Stability of Poly(ether urethaneurea) Blood Sacs. L. Wu, J. T. Garrett, D. M. Weisberg, <u>J. Runt</u>, G. Felder, G. Rosenberg
526	3:30	299.	Improved Polymer Biostability Via Oligomeric End Group Incorporated During Synthesis. <u>R. S. Ward</u>, Y. Tian, K. A. White
528	4:00	300.	New Ion-Selective Polymer Membranes for Chemical Sensing Based on Swellable Microspheres Suspended in a Hydrogel. R. Seitz, <u>H. Wang</u>

ANIONIC PHOTOPOLYMERIZATION USING ACYLFERROCENE DERIVATIVES

Yoshikazu Yamaguchi[a, b], Bentley J. Palmer[c], Tomoyo Wakamatsu[d],
D. Billy Yang[e] and Charles Kutal*[a]

[a] Department of Chemistry, University of Georgia, Athens, GA 30602
[b] JSR Corporation, Tsukuba Research Laboratory
   25 Miyukigaoka, Tsukuba, Ibaraki 305-0841, Japan
[c] Schlumberger-Dowell, 150 Gillingham Lane, Sugar Land, TX 77478
[d] Loctite Corporation, 1001 Trout Brook Crossing, Rocky Hill, CT 06067
[e] Avery Research Center, 2900 Bradley Street, Pasadena, CA 91107-1599
*Corresponding author

INTRODUCTION

Most UV curable resin formulations contain a photoinitiator system that generates a radical and/or an acid for the initiation of radical and/or cationic polymerization.[1] Recently we reported that trans-$Cr(NH_3)_2(NCS)_4^-$ (abbreviated R^- to denote the anion of Reinecke's salt) and $Pt(acac)_2$ ($acac^-$ is acetylacetonate) photoinitiate the polymerization of neat ethyl α-cyanoacrylate (CA) via an anionic mechanism.[2,3,4] Upon irradiation, each complex generates an anionic species which rapidly attacks the carbon-carbon double bond of the electrophilic monomer to yield a resonance-stabilized carbanion. Polymerization then proceeds via the repetitive addition of CA to the propagating anionic chain.

Continuing our exploration of anionic photoinitiation, we have found that several members of the ferrocene family, acetylferrocene (ACF), benzoylferrocene (BF), and 1,1'-dibenzoylferrocene (DBF) can photoinitiate the polymerization of CA. These organometallic sandwich complexes dissolve in a broad range of nonaqueous solvents and absorb strongly in the ultraviolet and, in some cases, in the visible wavelength regions (Figure 1). Several ferrocene complexes have a long history of use in photoinitiator systems that generate radical or cationic initiating species.[5,6] In this paper we report the first evidence that acylferrocenes can function as effective anionic photoinitiators.

EXPERIMENTAL

Materials

Commercially available samples of ACF, BF, and DBF were further purified by sublimation or recrystallization from n-hexane. Neat CA (99.9% purity) was obtained from Loctite Corp. as a colorless liquid that contained hydroquinone and methanesulfonic acid as scavengers for radical and basic impurities, respectively. Ethyl α-cyanopropionate (CP) was purchased from TCI America and used as received in the photochemical studies of the acylferrocenes.

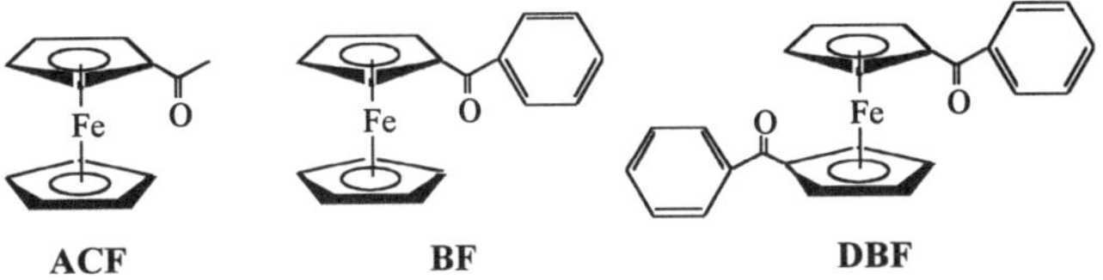

Photopolymerization study of CA containing ferrocenes

A 2-ml sample of CA containing ACF, BF, DBF, or K^+R^- was irradiated with stirring in a 1-cm methacrylate cell. The light source was a 200-W high-pressure mercury-arc lamp whose output was passed through Pyrex glass (>290 nm) or a narrow bandpass interference filter (436 nm or 546 nm). The kinetics of photoinitiated polymerization were studied by real-time Fourier transform infrared (RT-FTIR) spectroscopy.[7,8] Thin films of CA containing DBF were coated onto acid-treated silicon wafers and irradiated with a 100-W high-pressure mercury lamp having a power of 27 mW/cm^2 at 366 nm.

Confirmation of long-lived anion

A 3-ml solution of DBF in CP was irradiated at 546 nm for 480 s. Immediately thereafter, 0.5g of photolyte and 2.0g of CA were mixed in a test tube. The time required for the mixture to become so viscous that it does not flow was noted. A second aliquot of the photolyte was tested similarly after being kept in the dark for 24 hours at room temperature.

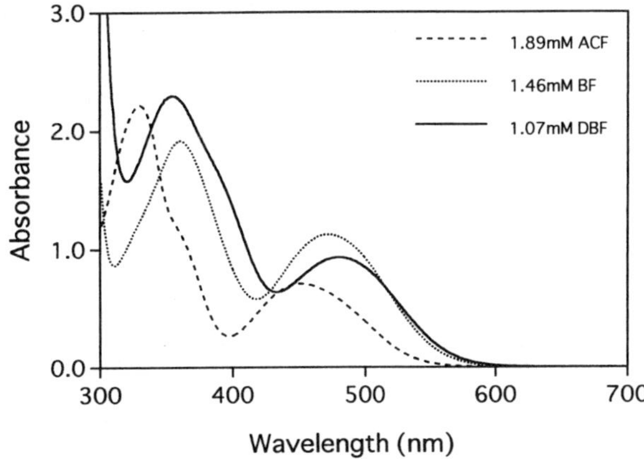

Figure 1. Electronic absorption spectra of ACF (---), BF (...), and DBF (—) in ethyl α-cyanopropionate. Concentration = 400 ppm.

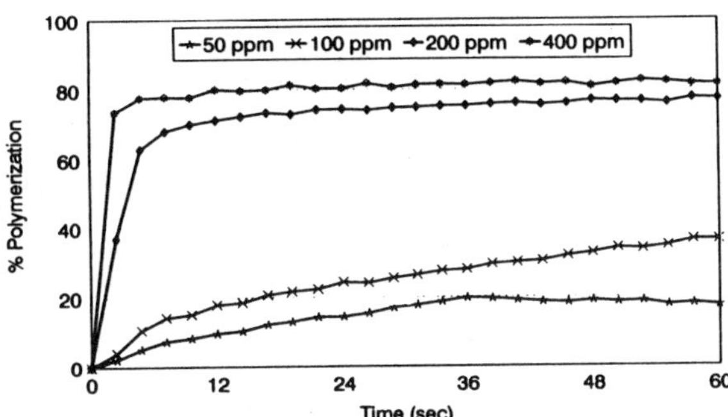

Figure 2. Plots of percentage polymerization vs. time of irradiation for samples of CA containing indicated amounts of DBF

RESULTS AND DISCUSSION

While neat CA monomer is unaffected by prolonged (>6500 s) exposure to Pyrex-filtered light, irradiated solutions containing an acylferrocene or K^+R^- polymerized to a viscous liquid and then solidified. Results in Table 1 show that this photoinitiated process occurs most rapidly for samples containing DBF, which is effective at wavelengths well into the visible region. As shown in Figure 2, polymerization commences within the first few seconds following exposure to light and reaches a plateau within 10-15 s at the highest photoinitiator concentration. The trend toward faster rates at higher photoinitiator concentrations reflects the greater fraction of light absorbed by the sample and the corresponding increase in the concentration of the photogenerated species responsible for initiating polymerization. Disappearance quantum yields (ϕ_{dis}) were determined for BF and DBF in CP solutions irradiated at 546 nm. The % decomposition was calculated from the decrease in intensity of the long-wavelength absorption band of each complex (at 480 and 472 nm for DBF and BF, respectively). The results showed that DBF (ϕ_{dis} = 0.28) is considerably more photosensitive than BF(ϕ_{dis} = 0.035), suggesting that the greater effectiveness of the former complex as an anionic photoinitiator arises from its more efficient photochemical generation of the active initiating species.

In order to investigate the nature of this reactive species, studies of photoinitiated polymerization were conducted in the presence of potential inhibitiors. Methanesulfonic acid (MSA) and hydroquinone (HQ) were added to the CA samples as scavengers for basic and radical impurities, respectively. Argon gas was bubbled into the solution for 30 min. in order to exclude O_2 from the samples. As listed in Table 2, neither added HQ nor O_2 affects the photopolymerization of CA solutions containing acylferrocenes. From these observations, it seems unlikely that photogenerated radicals initiate the polymerization of CA in these systems. In contrast, MSA strongly inhibits polymerization. We attribute this

Table 1. Photoinitiated Polymerization Studies of CA containing Ferrocenes

Photoinitiator (PI)	Concentration mmol/L (ppm)	λ_{ex} nm	Photopolym. Time[a], sec.
ACF	1.89 (404)	>290	552
BF	1.46 (404)	>290	176
BF	1.90 (523)	436[b]	>900
DBF	1.07 (403)	>290	3.1
DBF	1.61 (605)	436[b]	38
DBF	2.40 (902)	546[c]	31
K^+R^{-d}	1.19 (404)	>290	64

[a]Irradiation time required for the sample to become so viscous that the 8-mm stirring bar ceased to spin. [b]Sample absorbance = 1.0; light intensity = 7.8 x 10^{-8} einstein/s. [c]Sample absorbance = 0.7; light intensity = 1.2 x 10^{-7} einstein/s. [d] R^- is <u>trans</u>-$Cr(NH_3)_2(NCS)_4^-$.

Table 2. Inhibition of CA Photopolymerization[a]

Photoinitiator (PI)	Concentration PI mM (ppm)	MSA[b] mM	of HQ[c] mM	Photopolym. Time, sec.
ACF	2.19(500)	0.05	9.5	421
ACF	2.19(500)	5.40	9.5	>1890
ACF	2.19(500)	0.05	29	460
BF	2.36(685)	0.05	9.5	196
BF	2.36(685)	5.40	9.5	>600
BF	2.36(685)	0.05	29	219
BF	1.45(400)	0.05	9.5	165
BF	1.45(400)	0.05	9.5	174 [d]
DBF	0.584(230)	0.05	9.5	4.6
DBF	0.584(230)	5.40	9.5	>610
DBF	0.584(230)	0.05	41	5.3
DBF[e]	2.40(902)	0.05	9.5	31
DBF[e]	2.40(902)	0.05	9.5	31 [d]

[a]Unless indicated otherwise, the sample was irradiated with the Pyrex-filtered output (>290 nm) of a 200-W high-pressure mercury lamp. [b]Methanesulfonic acid (base scavenger) [c]Hydroquinone (radical scavenger) [d]Irradiated after 30-min. Ar purge. [e]Sample irradiated at 546 nm.

inhibition by strong acid to the ability of protons to scavenge photogenerated anions and/or anionic sites on the propagating polymer chain.

Figure 3 shows (a) the spectral changes that accompany the 546-nm photolysis of DBF in CP, and (b) the polymerization time of mixtures of CA and the photolyte as a function of irradiation time and concentration of MSA. Polymerization time is seen to decrease with increasing irradiation time up to ~480 s. At that point, the resulting plateau reflects the almost total photodecomposition of the photoinitiator. Repeating this experiment for a second sample that had been irradiated 480 s, but kept in the dark for 24 hours prior to addition to CA, yielded the identical polymerization time (17 min.). This finding demonstrates that the photolysis of DBF generates a long-lived anionic initiator. Efforts to identify this photogenerated species are currently underway. One likely possibility is a carboxylate anion, since Kemp et al. reported the generation of benzoate from BF via the photoaquation of the carbonyl group accompanied by ring-metal and ring-carbonyl cleavage (eq. 1; S is solvent).[9] Benzoate effectively initiates CA polymerization as evidenced by the immediate solidification that occurs when a suspension of sodium benzoate in acetonitrile comes in contact with the monomer.

$$BF \xrightarrow[H_2O]{h\nu,\, S} (\eta 5\text{-}C_5H_5)Fe(S)_3^+ + C_6H_5COO^- + C_5H_6 \qquad (1)$$

CONCLUSION

Our results provide the first evidence that acylferrocenes can initiate anionic photopolymerization. Among the complexes investigated, DBF exhibited excellent performance even at wavelengths well into the visible wavelength region. Further studies of these and related members of the ferrocene family are underway with the goal of elucidating the key photochemical and thermal steps responsible for anion formation.

ACKNOWLEDGEMENT

We thank the National Science Foundation (Grant DMR-9122653 to C. K.) and JSR Corp. for financial support and Loctite Corp. for a gift of ethyl α-cyanoacrylate.

References

1. Davidson, R. S. *J. Photochem. Photobiol. A: Chem.* **1993**, *73*, 81.
2. Kutal, C.; Grutsch, P. A.; Yang, D. B. *Macromolecules* **1991**, *24*, 6872.
3. Palmer, B. J.; Kutal, C.; Billing, R.; Hennig, H. *Macromolecules* **1995**, *28*, 1328.
4. Lavellee, R. J.; Palmer, B. J.; Billing, R.; Hennig, H.; Ferraudi, G.; Kutal, C. *Inorg. Chem.* **1997**, *36*, 5552.
5. Allen, D. M. *J. Photograph. Sci.* **1976**, *24*, 61.
6. Yang, D. B.; Kutal, C. in *Radiation Curing: Science and Technology*, Pappas, S. P., Ed.; Plenum Press: New York, 1992, Ch. 2.
7. Allen, N. S.; Hardy, S. J.; Jacobine, A. F.; Glaser, D. M.; Yang, D. B.; Wolf, D.; Catalina, F.; Navaratnam, S.; Parsons, B. J. *J. Appl. Polym. Sci.* **1991**, *42*, 1169.
8. Yang, D.B. *J. Polym. Sci., Part A*, **1993**, *31*, 199
9. Ali, L. H.; Cox, A.; Kemp, T. J. *J. Chem. Soc. Dalton* **1973**, 1468.

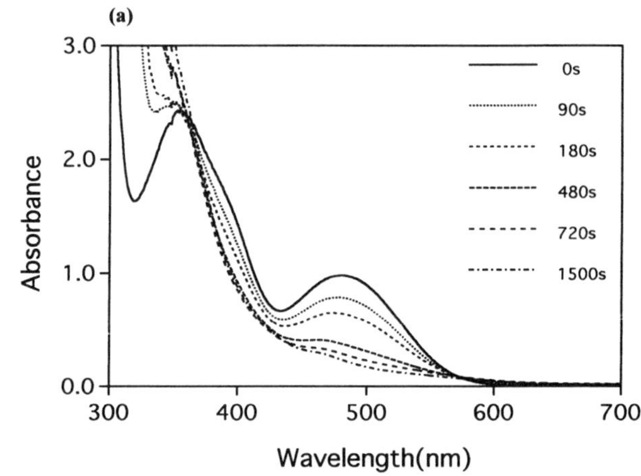

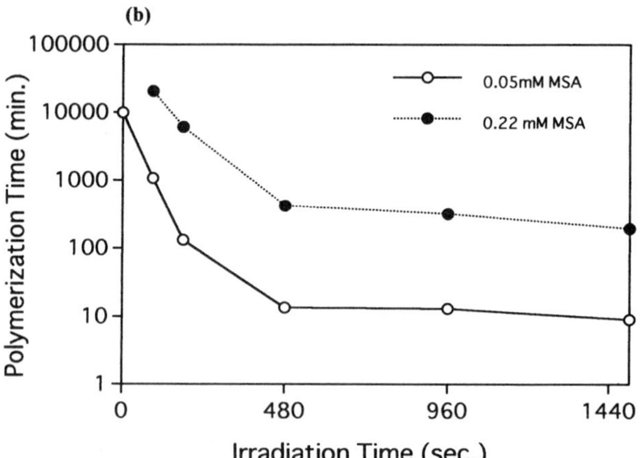

Figure 3. (a) Spectral changes accompanying the 546 nm photolysis of DBF in CP upon irradiation at 546 nm. (b) Polymerization times of mixtures of CA monomer and the photolyte as a function of irradiation time and concentration of MSA. Concentration of DBF=1.07 mM (400 ppm), Light intensity; 1.3×10^{-7} einstein/s

POLY(ISOBUTYLENE-*g*-STYRENE) GRAFT COPOLYMERS
BY QUASILIVING ATOM TRANSFER RADICAL GRAFTING
OF STYRENE

Tamás Fónagy[1,2], Béla Iván[1,2] and Márta Szesztay[1]

[1]Material Science of Polymers Group, Institute of Chemistry, Chemical
Research Center of the Hungarian Academy of Sciences, H-1525
Budapest, Pusztaszeri u. 59-67, P.O.Box 17, Hungary, and [2]Department of
Chemical Technology and Environmental Chemistry, Eötvös Lóránd
Science University, H-1518 Budapest, P.O.Box 32, Hungary

Introduction

Quasiliving polymerizations [1,2] have undergone significant advances
during the last decade. As it has been recently analyzed in a
comprehensive way, practically all the existing polymerizations proceed
with involvement of quasiliving equilibria, i. e. equilibria between *active*
(propagating, living) and *inactive* (nonpropagating, nonliving) polymer
chains by anionic, cationic, free radical, group transfer, ring-opening
metathesis and coordinative mechanisms (Refs. 1,2 and references
therein). One of the most promising recent quasiliving polymerization
systems is quasiliving free radical polymerization (QLFRP). These are
mediated by stable radicals [3-5], cobalt-porphynato complex [6], or
transition metal catalyzed atom transfer processes [7-10]. In QLFRPs
termination, which is unavoidable in conventional free radical
polymerizations, can be suppressed by creating an equilibrium between
active radicals and inactive (terminated, nonliving) species, and by
keeping simultaneously concentration of radicals low by shifting the
equilibrium toward the formation of inactive chains.

It has been found that not only conventional low molecular weight
initiators but macromolecules containing initiating functional groups can
also be utilized by QLFRPs leading to block [11-14] and graft [15,16]
copolymers. Commercial polymers with suitable pendant functional
groups are of great interest as macroinitiators for the synthesis of graft
copolymers with controlled architectures and properties. Polyisobutylene
(butyl rubber) is one of the most attractive choice as backbone for such
applications due to the advantageous chemical, thermal and oxidative
stability and physical properties of this polymer. Earlier, chlorobutyl
rubber was tested by Kennedy and Charles [17] for carbocationic grafting
of styrene with limited success because of gelation. Recently, Nuyken and
coworkers [18] has prepared amphiphilic graft copolymer by grafting 2-
methyl-2-oxazoline from poly(isobutylene-*co*-p,m-chloromethylstyrene).
One of the recent industrial developments is the production of
poly(isobutylene-*co*-p-methylstyrene-*co*-p-bromomethylstyrene) (PIB) copolymers (EXXPRO®) by carbocationic copolymerization followed by
bromination. This new functional polymer has created wide interest for a
variety of new processes and applications [19].

This paper reports on the synthesis of poly(isobutylene-*g*-styrene)
(PIB-*g*-PSt) by quasiliving ATRP of styrene using the commercially
available PIB as macroinitiator.

Experimental part

Materials. The poly(isobutylene-*co*-p-methylstyrene-*co*-p-bromomethyl-
styrene) (EXXPRO 96-3; free sample by Exxon Chemical Co.) copolymer
(PIB) was dissolved in toluene and the solution was passed through a
column filled with neutral Al_2O_3. The solvent was removed by rotavap and
the polymer was dried to constant weight at room temperature in vacuo.
Styrene (Aldrich) was distilled under reduced pressure at room
temperature. Xylene was passed through a chromatography column filled
with neutral Al_2O_3. CuBr (98%, Aldrich) and 2,2'-bipyridine (bpy) (99%,
Aldrich) was used as received.
Polymerizations. Polymerizations were carried out both in bulk and in
solution. In bulk polymerization PIB was dissolved in styrene before CuBr
and bpy were added to the reaction mixture. In solution polymerization
styrene was added together with CuBr and bpy after PIB was dissolved in
xylene. Oxygen free condition was obtained by three freeze-thaw cycles,
then the polymerization reactor was placed in an oil bath at 100 °C. After a
predetermined polymerization time, the polymerization system was
removed from the oil bath and cooled to room temperature. The solution

was diluted with chloroform followed by precipitation into methanol. The
resulting polymer was dissolved in chloroform and the solution was
passed through a column filled with neutral Al_2O_3. The solvent was
removed by rotavap and the polymer was dried until constant weight at
room temperature in vacuo.
Characterization. Change in the molecular weight distribution was
followed by gel permeation chromatography (GPC) with a
Waters/Millipore liquid chromatograph equipped with a Waters 510
pump, Ultrastyragel columns of pore sizes 1×10^5, 1×10^4, 1×10^3 and 500 Å,
and a Viscotec parallel differential refractometer/viscometer detector.
Tetrahydrofuran was used as eluent with a flow rate of 1.5 ml/min. [1]H
NMR analyses were performed by Varian 200 and 400 MHz equipments.
Differential scanning calorimetry analyses were performed on a Mettler
4000 TA thermal analytical system.

Results and Discussion

New complex macromolecular architectures prepared by combining
different polymerization methods are of great interest. Block and graft
copolymers with elastic and glassy segments are expected to exhibit
thermoplastic elastomer behavior. On the basis of these thoughts,
synthesis of poly(isobutylene-*g*-styrene) (PIB-*g*-PSt) graft copolymer by
quasiliving ATRP of styrene using a new commercial polymer,
poly(isobutylene-*co*-p-methylstyrene-*co*-p-bromomethylstyrene) (PIB) as
macroinitiator was attempted by us. The randomly distributed
bromobenzyl groups along the polymer chain were expected to initiate
grafting of styrene from the polyisobutylene backbone by quasiliving
ATRP of styrene as shown in Scheme 1.

Scheme 1. Quasiliving ATRP of styrene initiated by PIB.

The ratio of bromobenzyl initiator groups and catalyst (CuBr/3bpy) was
1:1. The amount of catalyst was calculated from the composition of PIB.
The composition was determined from the ^{1}H NMR spectrum of PIB
(Figure 1) by integration of aliphatic methyl (**a**), aromatic methyl (**b**) and
bromomethylene protons (**c**).

Polymerization of styrene under ATRP conditions with the
PIB/CuBr/bpy initiating system yielded 11% and 16% monomer
conversion in bulk in 4 hours and in xylene solution in 23 hours
polymerization time, respectively. The formation of polystyrene chains
was confirmed by ^{1}H NMR spectroscopy and GPC measurements. Figures
1 shows the ^{1}H NMR spectra of the starting random copolymer
macroinitiator and the resulting graft copolymer obtained in bulk,
respectively. The relative number of aliphatic (**a**) and aromatic methyl (**b**)
protons does not change in the course of styrene polymerization, and
therefore these NMR signals can be used as reference. It can be seen by
comparing the two spectra that the relative intensity of aromatic protons
significantly increased as a consequence of the polymerization process.
The function of bromobenzyl groups as initiator moieties was also
confirmed by ^{1}H NMR spectra. The relative intensity of the
bromomethylene signal (**c**) decreased after polymerization of styrene. This
indicates that initiation of styrene occurred by the bromobenzyl groups of
the PIB macroinitiator yielding PIB-*g*-PSt graft copolymer. However, the

efficiency of initiation was not quantitative in bulk for 4 hours, but improvement was found by increasing either the polymerization time to 23 hours or the catalyst/initiator ratio to 5:1. Table 1 summarizes the results of several grafting experiments.

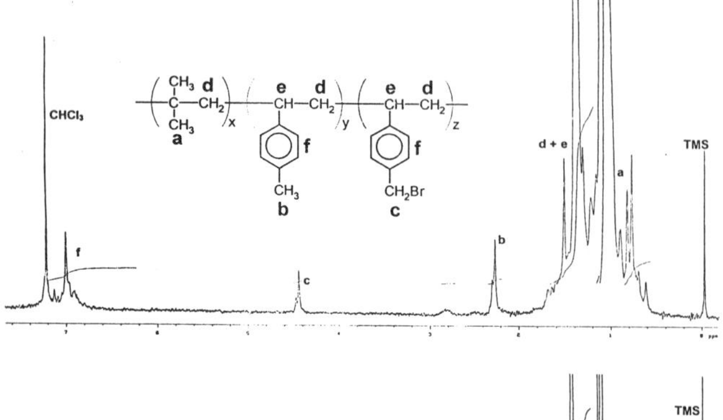

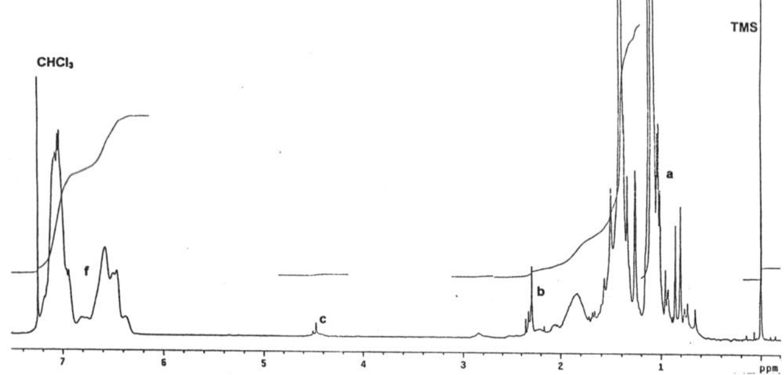

Figure 1. ^{1}H NMR spectra of PIB (upper) and PIB-*g*-PSt obtained by quasiliving ATRP of styrene in bulk for 4 hours using 1:1 initiator/catalyst ratio (lower) (solvent: CDCl$_3$).

Table 1. Initiator/catalyst ratio (n$_i$/n$_c$) of solution ATRP of styrene initiated by PIB at 100 °C, polymerization time (t), styrene conversions, the compositions of PIB-*g*-PSt (wt%(PSt)), initiation efficiency (f) by bromobenzyl groups, styrene conversions and the average length of pendant polystyrene chains (n; see Scheme 1).

n$_i$/ n$_c$	t (h)	conv. (%)	wt % (PSt)	f (%)	n
1/1	11	3	6	12	10
1/1	23	16	26	81	9
1/5	11	22	32	80	13

Results obtained by ^{1}H NMR spectroscopy do not exclude the formation of polystyrene as homopolymer. Figure 2 shows the GPC traces of PIB and the resulting graft copolymer. The peak of the PIB-*g*-PSt copolymer compared with the starting PIB shifted toward lower elution volumes, i. e. higher molecular weights. This suggests that every PIB chain was transformed to PIB-*g*-PSt graft copolymer, on the one hand.

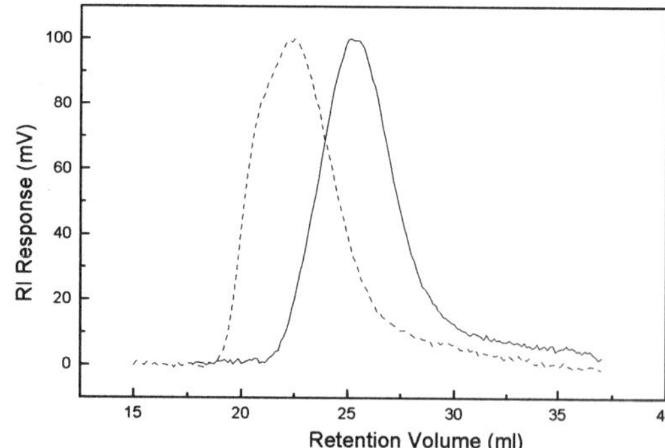

Figure 2. GPC traces of the starting PIB (——) and the poly(isobutylene-*g*-styrene) graft copolymer (- - -).

Since there is only one peak, which excludes the formation of polystyrene as homopolymer, the increase of molecular weight can be explained by grafting of styrene from the polyisobutylene backbone, that is by the formation of poly(isobutylene-*g*-styrene) graft copolymer in the absence of detectable homopolymerization, on the other hand.

Preliminary investigations indicate that the mechanical properties of PIB-*g*-PSt graft copolymers depend significantly on the polystyrene/polyisobutylene ratios. PIB-*g*-PSt with 28 wt% polystyrene does not possess special mechanical properties, but a graft copolymer with 6 wt% polystyrene can be stretched to about 500% of its original dimension without any irreversible deformation. This indicates thermoplastic elastomer (TPE) behavior, which can be explained by phase separated structure of the graft copolymer, resulting in physical crosslinks. The phase separated structure was confirmed by differential scanning calorimetry (DSC) analysis of PIB-*g*-PSt. Distinct glass transition temperatures of polystyrene at 77 °C and polyisobutylene at −59.7 °C were observed indicating advantageous phase separation in these new graft copolymers. Further investigations are in progress in our laboratory.

Conclusions

Quasiliving ATRP of styrene using commercial poly(isobutylene-*co*-p-methylstyrene-*co*-p-bromomethylstyrene) as macroinitiator yielded poly(isobutylene-*g*-styrene). Mild reaction conditions in the presence of CuBr/bpy catalyst system resulted in controlled grafting process both in bulk and in xylene solution. The PIB-*g*-PSt has phase separated structure, which provides unique mechanical properties in case of appropriate composition. The graft copolymer can be stretched to 500 per cent of its original dimension without any irreversible deformation, indicating that the investigated grafting process by quasiliving ATRP of styrene leads to well-defined PIB-*g*-PSt thermoplastic elastomers.

Acknowledgements

Partial support by the National Scientific Research Fund (OTKA T14910) and technical assistance by Mrs. E. Tyroler are gratefully acknowledged.

References

1. B. Iván, *Makromol. Chem., Makromol. Symp.* **67**, 311 (1993)
2. B. Iván, *Macromol. Symp.* **88**, 201 (1994)
3. M. K. Georges, R. P. N. Veregin, P. M. Kazmaier, G. K. Hamer, *Macromolecules* **26**, 2987 (1993)
4. T. Fukuda, T. Terauchi, A. Goto, K. Ohno, Y. Tsujii, T. Miyamoto, S. Kobatake, B. Yamada, *Macromolecules* **29**, 6393 (1996)
5. D. Colombani, M. Steenbock, M. Klapper, K. Müllen, *Macromol. Rapid Commun.* **18**, 243 (1997)
6. B. B. Wayland, G. Poszmik, S. L. Mukerjee, M. Fryd, *J. Am. Chem. Soc.* **116**, 7943 (1994)
7. J.-S. Wang, K. Matyjaszewski, *J. Am. Chem. Soc.* **117**, 5614 (1995)
8. M. Kato, M. Kamigaito, M. Sawamoto, T. Higashimura, *Macromolecules* **28**, 1721 (1995)
9. V. Percec, B. Barboiu, A. Neumann, J. C. Ronda, M. Zhao, *Macromolecules* **29**, 3665 (1996)
10. D. M. Haddleton, C. Waterson, P. J. Derrick, C. B. Jasieczek, A. J. Shooter, *Macromolecules* **30**, 2190 (1997)
11. B. Gao, X. Chen, B. Iván, J. Kops, W. Batsberg, *Polym. Bull.* **39**, 559 (1997)
12. X. Chen, B. Iván, J. Kops, W Batsberg, *Polym. Prepr.* **38(1)**, 715 (1997)
13. S. Coca, K. Matyjaszewski, *Macromolecules* **30**, 2808 (1997)
14. S. G. Gaynor, K. Matyjaszewski, *Macromolecules* **30**, 4241 (1997)
15. C. J. Hawker, D. Mecerreyes, E. Elce, J. Dao, J. L. Hedrick, I. Barakat, P. Dubois, R. Jérôme, I. Volksen, *Macromol. Chem. Phys.* **198**, 155 (1997)
16. R. B. Grubbs, C.J. Hawker, J. Dao, J. M. Fréchet, *J. Angew. Chem., Int. Ed. Engl.* **36**, 270 (1997)
17. J. P. Kennedy, J. J. Charles *J. Appl. Polym. Sci.* **30**, 119 (1977)
18. O. Nuyken, J. R. Sanchez, B. Voit, *Macromol. Rapid Commun.* **18**, 125 (1997)
19. H. C. Wang, K. W. Powers *Elastomerics* February, 22 (1992)

SYNTHESIS AND CHARACTERIZATION OF OXAZOLINE-TERMINATED POLYMERS VIA CONTROLLED RADICAL POLYMERIZATION AS INTERMEDIATES FOR REACTIVE PROCESSING

*Martin Baumert, Jörg Zimmermann, Rolf Mülhaupt**

Freiburger Materialforschungszentrum und Institut für Makromolekulare Chemie der Albert-Ludwigs-Universität Freiburg, Stefan-Meier-Str. 21, D-79104 Freiburg i. Br., Germany

Introduction

The synthesis of well-defined polymer architectures and polymers with narrow molar mass distribution in conventional polymerization processes, typical for free radical polymerization, represents a key challenge in industrial and academic research. Recent advances have led to the development of controlled radical polymerization as a very versatile new synthetic tool to tailor new polymers which are not available by means of conventional living ionic polymerization. Examples for such polymers are new block copolymers, e.g., polystyrene-*block*-poly(styrene-*co*-acrylonitrile)[1,4], new hyperbranched polymers and dendrimers as well as various polymers with new functional end groups[2,3,4]. In order to introduce end groups functional initiators are applied in TEMPO-based controlled radical polymerization processes. Hawker prepared adducts of styrene, TEMPO and benzoylperoxide or other functionalized initiators, which he used as initiators[3]. Recently, it was demonstrated that well-known functional azo initiators can be used successfully in conjunction with TEMPO to produce polystyrene and poly(styrene)-*block*-poly(styrene-*co*-acrylonitrile) with one functional end group[4]. The use of functional azo initiators circumvents the need for preparing special TEMPO adduct-based initiators and is very versatile because a wide range of functional azo initiators are commercially available. Here we report the synthesis of novel bis(1,3-oxazolin-2-yl)-functionalized azo initiators which were applied in controlled radical polymerization to introduce oxazoline end groups. Oxazoline-functional polymers represent attractive intermediates in reactive processing application[5]. Oxazolines can react with carboxylic acids within a few minutes at elevated temperature to afford esteramide coupling in very high yield[6]. This clean reaction can be monitored by means of FTIR spectroscopy[7]. Oxazolines were introduced as end groups in various liquid rubbers using end group conversion reactions[8]. Sivaram reported the synthesis of monooxazoline-terminated polymethacrylate using ionic oxazoline initiators[9]. However, there are no reports on oxazoline-functional azo initiators which are of special interest for applications in controlled radical polymerization processes and production of new oxazolines as intermediates in reactive processing.

Experimental
Materials

Styrene was refluxed and distilled over LiAlH$_4$, Acrylonitrile was distilled over CAH$_2$. 2,2´-azobis[2-methyl-N-(2-hydroxyethyl)-propionamid (Wako Chemicals), 4,4'-azo-bis(4-cyanopentane carboxylic acid), TEMPO, KOH, SOCl$_2$ were obtained from FLUKA and used without further purification.

Synthesis of bis-oxazoline azo initiators

2,2'-Azobis[2-(1,3-oxazolin)propane] (**OxaI**): 2,2'-azobis[2-methyl-N-(2-hydroxyethyl)-propionamide] was reacted with SOCl$_2$ at low temperature. Then the resulting 2,2'-azobis[2-methyl-N-(2-chloroethyl)-propionamide] was treated with KOH to form **OxaI**.

4,4'-Azobis-4-cyanopentanecarboxylic acid-[5-pentyl-2-(1,3-oxazoline)]ester (**OxaII**) was synthesized by an esterfication reaction of 2-(hydroxypentyl)-1,3-oxazolin[10] with 4,4'-azobis(4-cyanopentanecarboxylic acid).

Polymerizations

2,2'-Azobis[2-(1,3-oxazolin)propane] (0.188g) and TEMPO (0.116g) were dissolved in 20mL styrene and degassed by two freeze-pump-thaw-cycles using argon as inert gas. The flask was immersed into an oil bath at 140°C for 12 hours, taking samples in intervals. The samples were dissolved in THF and precipitated from methanol and dried in vacuum at 40°C for 24 hours.

Characterization

Gel permeation chromatography (GPC) (Knauer Mikrogelset A14) was used to determine molar mass and molar mass distributions, M_w/M_n, conversion was determined by ^{1}H-NMR-spectroscopy (300MHz, CDCl$_3$, Bruker ARX 300). FTIR-spectroscopy was performed on a Bruker IFS 88.

Results and Discussion

Synthesis of oxazoline-terminated polymers

Two novel bis(1,3-oxazolin-2-yl)functional azo-initiators were prepared: 2,2'-Azobis[2-(1,3-oxazolin)propane] (**OxaI**) and 4,4'-Azobis-4-cyanopentanecarboxylic acid-[5-pentyl-2-(1,3-oxazoline)]ester (**OxaII**). As displayed in Figures 1 and 2 controlled radical polymerization of styrene was performed in bulk under argon atmosphere with **OxaI** and **OxaII** and TEMPO at oil bath temperatures of 140°C to produce monooxazoline-terminated polystyrene.

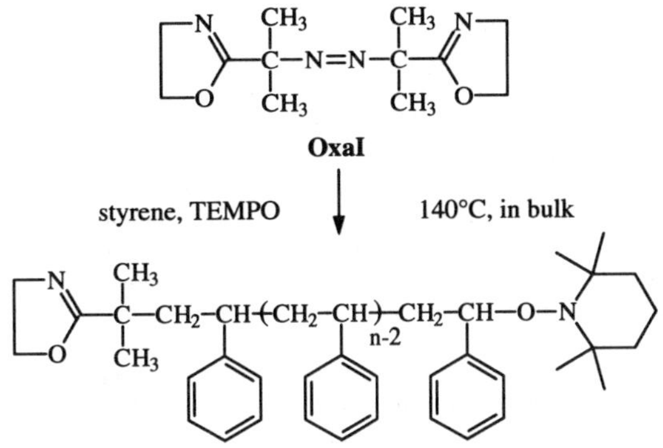

Figure 1. *Synthesis of monooxazoline-terminated polystyrene*

A typical plot of molar mass versus conversion is shown for **OxaII** in Figure 3, where a linear dependence of molar mass versus conversion can be seen. Typical polydispersities M_w/M_n were found to be around 1.2 similar to those observed for benzoyl peroxide initiated TEMPO processes.

Also monooxazoline-terminated poly(styrene-*co*-acrylonitrile) was synthesized. In free radical polymerization in the absence of TEMPO recombination affords polystyrene with two oxazoline end groups. Oxazoline azo initiators can be used to initiate free radical polymerization of various acrylics to produce a wide range of mono- and difunctional oxazoline-terminated polymers.

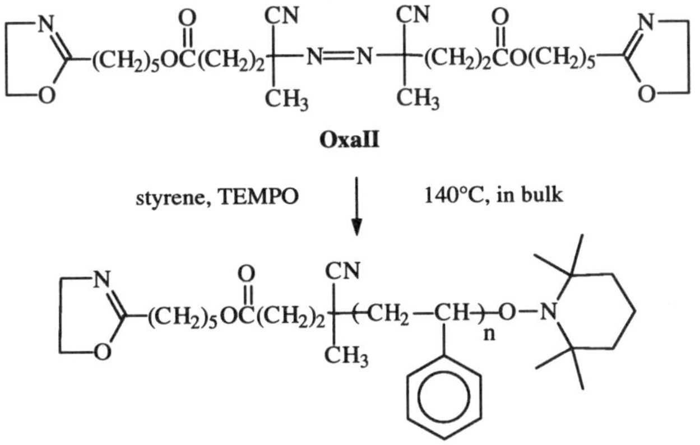

Figure 2. *Synthesis of monooxazoline-terminated polystyrene*

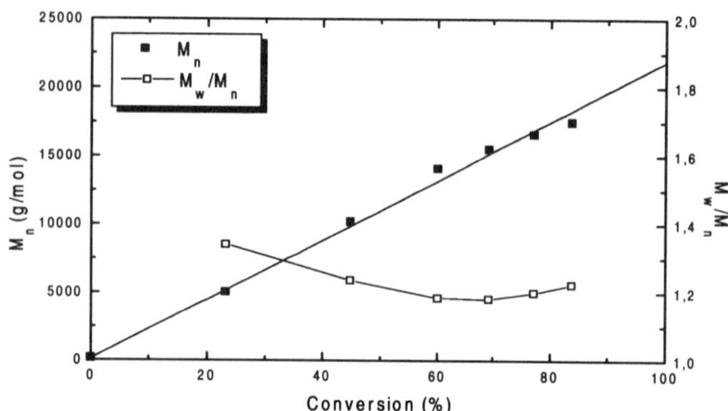

Figure 3. *Oxazoline-terminated polystyrene: correlation of number-average molar mass M_n and polydispersities M_w/M_n with conversion; $M_n(calc)=25\,000g/mol$, [TEMPO]/[**OxaII**]=1.4; 140°C*

Reaction of monooxazoline-terminated polymers with monocarboxy-terminated polymers

Novel mono (1,3-oxazolin-2-yl)-terminated polystyrene PSOxaI was applied as a component of melt-phase coupling reaction with carboxylate-terminated polystyrene (CTPS) which was also produced by controlled radical polymerization using bis-carboxylic-functional azo initiators and TEMPO. The coupling reaction was performed in bulk at 200°C for 60 minutes (Figure 4). FTIR spectroscopy was applied to monitor this reaction, reflected by conversion of oxazoline ($1654cm^{-1}$) and carboxylic acid groups ($1761cm^{-1}$) and formation of carboxylic ester and amide ($1740cm^{-1}$ and $1674cm^{-1}$). A typical FTIR spectrum measured in 1 minutes intervals is displayed in Figure 5.

This coupling reaction of oxazoline-terminated polymers, which are readily available by means of controlled radical polymerization in the presence of novel oxazoline-functional azo-initiators and TEMPO, demonstrates the attractive potential of such intermediates in reactive processing and diversification of polymeric materials.

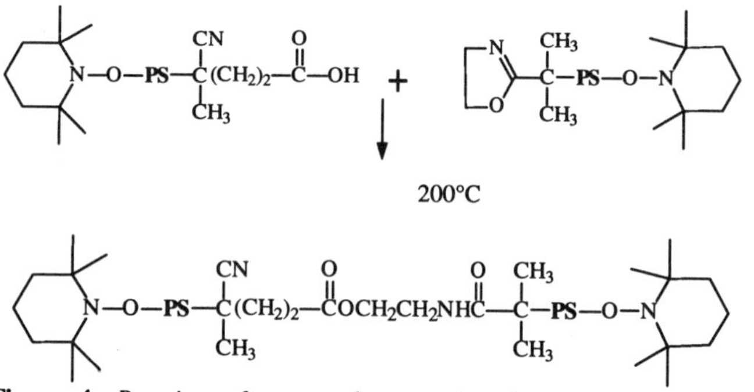

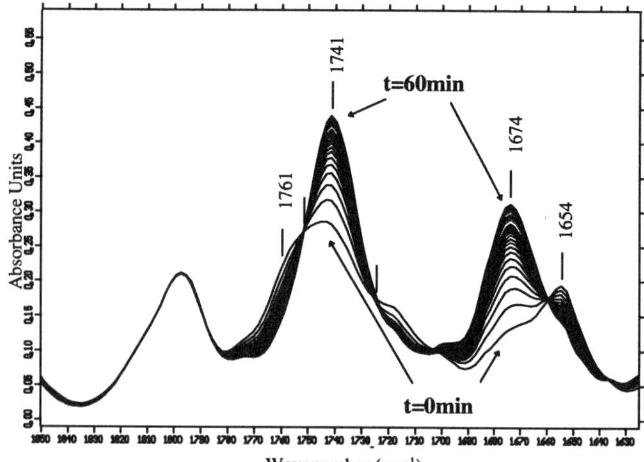

Figure 4. *Reaction of monocarboxy-terminated with monooxazoline-terminated polystyrene at 200°C*

Figure 5. *FTIR-spectra of the reaction of CTPS (Mn=6000g/mol) and PSOxaI ($M_n=3600g/mol$) at 200°C, 1 minutes intervals*

Acknowledgments

The authors thank the Bundesminister für Bildung und Forschung for financial support as part of the project no.03N10280. Also we thank BASF AG, Ludwigshafen, especially Dr. Geprägs, for their support.

References

(1) Fukuda, T.; Terauchi, T.; Goto, A.; Tsuji, Y.; Miyamoto, T.; Shimizu, Y.*Macromolecules* **1996**, *29*, 3050

(2) Hawker, C. J.; Fréchet, J. M. J.; Grubbs, R. B.; Dao, J. *J. Am. Chem. Soc.* **1995**, *117*, 10763; Grubbs, R. B.; Hawker, C. J.; Dao, J.; Fréchet, J. M. J. *Angew. Chem., Int. Ed. Engl.* **1997**, *36*, 270; Leduc, M. R.; Hawker, C. J.; Dao, J.; Fréchet, J. M. J. *J. Am. Chem. Soc.* **1996**, *118*, 11111; Ohno, K; Tsuji, Y.; Miyamoto, T.; Fukuda, T.; Goto, M.; Kobayashi, K.; Akaike, T. *Macromolecules* **1998**, *31*, 1064; Matyjaszewski, K.; Shigemoto, T.; Fréchet, J. M. J.; Leduc, M. R. *Macromolecules* **1996**, *29*, 4167; Keoshkerian, B.; Georges, M. K.; Boils-Boissier, D. *Macromolecules* **1995**, *28*, 6381

(3) Hawker, C.J.; Barclay, G.G.; Orellana, A.; Dao, J,; Devonport, W. *Macromolecules* **1996**, *29*, 5245

(4) Baumert, M; Mülhaupt, R. *Macromol. Rapid Commun.* **1997**, *18*, 787

(5) Liu, N.C.; Baker, W.E. in S. Al-Malaika (ed.) „Reactive Modifiers for Polymers", Blackie Academic & Professional, London**1997**, p. 163

(6) Müller, P; Wörner, C.; Mülhaupt, R. *Macromol. Chem. Phys.* **1995**, *196*, 1918; Müller, P.; Wörner, C.; Mülhaupt, R. *Macromol. Chem. Phys.* **1995**, *196*, 1929

(7) Schäfer, R.; Kressler, J.; Mülhaupt, R.*Acta Polymerica* **1996**, *47*, 170

(8) Wörner, C.; Müller, P.; Mülhaupt, R. *Polymer* **1998**, *39*, 611

(9) Sivaram, S.; Khisti, R. S. *Macromol. Rapid. Commun.* **1991**, *12*, 435

(10) Synthesis according to Hölderle, M.; Bar, G.; Mülhaupt, R. *J. Polym. Sci. Part A: Polym. Chem.* **1997**, *35*, 2539

CONTROL OF POLYMER ARCHITECTURE BY A SPECIAL CHAIN TRANSFER CATALYST

Zhibin Guan and Christian Jackson*

DuPont Central Research & Development, Experimental Station,
P. O. Box 80328, Wilmington, DE 19880-0328

Introduction

Dendritic and hyperbranched polymers have attracted much attention recently because of their unusual shapes and physical properties.[1] Although hyperbranched polymers are not as perfect as dendrimers, they resemble the properties of dendrimers and usually can be synthesized by one pot polymerization. So far, most of the hyperbranched polymers were synthesized by condensation type of polymerization of AB_2 type of monomers.[2] It is demonstrated only recently that hyperbranched polymers can also be synthesized by chain growth polymerization of vinyl monomers. The first example is Frechet's "self-condensing" polymerization of a vinyl monomer by cationic polymerization to afford hyperbranched polymers.[3] The second one is the preparation of a hyperbranched polystyrene by "self-condensing" free radical polymerization of a tetramethylpiperidinyloxy (TEMPO) derivatized styrene monomer.[4] These represents the first few examples of the synthesis of hyperbranched polymers by addition polymerization. These are elegant approaches, however, special monomers have to be designed and synthesized first for the hyperbranched polymer formation.

Free radical polymerization still remains one of the most widely used polymer production method. The major advantages of free radical polymerization are the generality of monomer choice and the robustness of reaction conditions. It should be very interesting if some more general method can be developed for the preparation of hyperbranched or controlled branched polymers via direct free radical polymerization of commercially readily available monomers.

Here, we describe a novel approach for the synthesis of hyperbranched polymers by direct free radical polymerization of divinyl monomers which are used commercially as cross-linkers for free radical polymerizations. The key for the success of this approach is the control of propagation by a specialty chain transfer (SCT) catalyst. The synthesis and characterization of hyperbranched polymethacrylate controlled by a SCT catalyst is discussed in this paper.

Experimental

Polymerization Procedure. A typical polymerization was done as the following: 10 mg of COBF and 50 mg of 2,2'-Azobis(2,4-dimethylvaleronitrile) (VAZO-52) were added to a solution of 5.0 mL of ethylene glycol dimethacrylate (EGDMA) (de-inhibitored by passing through Al_2O_3 column) and 5.0 mL of 1,2-dichloroethane. The solution was degassed by freeze-thawing three times. The polymerization was conducted at 52 °C for 24 hrs. After finishing polymerization, the solution was precipitated into a large excess of petroleum ether. Oily polymer was recovered and dried under vacuum over night (63% yield). The polymer was analyzed by ^1H and ^{13}C NMR and GPC.

NMR Analysis. NMR spectra were obtained on a Varian Unity 400 Mhz spectrometer in a 5 mm probe in $CDCl_3$.

GPC Analysis. GPC analysis was performed on Waters GPC with light scattering and viscometer detectors.

Results and Discussion

Concept. The structure of the cobalt SCT catalyst used in this study is shown in Figure 1. From previous studies, a series of cobalt complexes cause very efficient catalytic chain transfer reaction to free radical polymerization of vinyl monomers through β-H abstraction.[5,6] Gridnev, Ittel, Fryd and Wayland *et al* have conducted systematic studies on the free radical polymerizations of vinyl monomers in the presence of a cobalt SCT catalyst.[7] In these studies, the SCT catalysts were used to control the molecular weight. This catalyst was shown to have very high catalytic chain transfer efficiency for free radical polymerization of methyl methacrylate (MMA) with chain transfer constant $C_s \sim 10^3$.[8] Therefore, with relatively small amount of the catalyst, MMA polymerization can be controlled to form very short oligomers such as dimer, trimer, and tetramer, etc.

We proposed that if a dimethacrylate monomer such as ethylene glycol dimethacrylate (EGDMA) is used as the monomer and if the SCT catalyst concentration is at the concentration specifically for the trimer formation of methacrylate, the controlled propagation of EGDMA should result in a hyperbranched polymer (Scheme 1):

Scheme 1. Concept of hyperbranched free radical polymer synthesis controlled by a SCT catalyst.

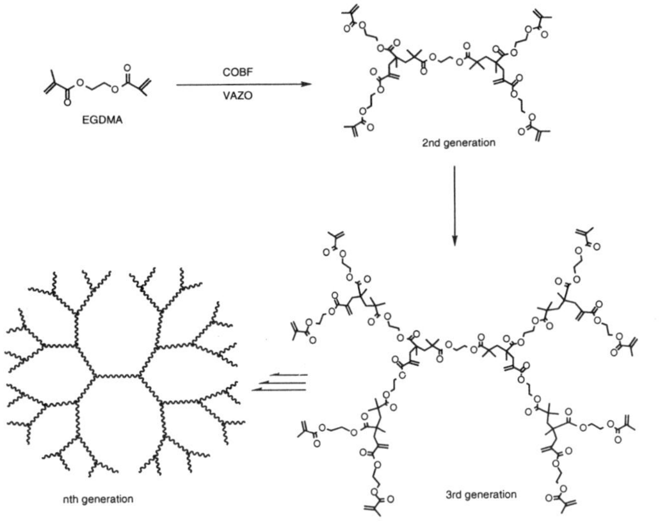

Polymerization. Based on previous studies and estimated by Mayo equation, a series of EGDMA polymerizations were carried out with [EGDMA]/[COBF] around 1000 at which MMA polymerization gives mainly trimer. From this series of experiments, it was observed that both the conversion and the molecular weight of the polymers obtained increase with the decrease of COBF concentration. Bulk polymerization generally results in higher conversion and gives polymer with higher molecular weight. Soluble polymers were obtained from direct free radical polymerization of

EGDMA as long as the concentration of the special chain transfer catalyst, COBF, is sufficiently high. For solution polymerization, the concentration of COBF can be lower than that in bulk polymerization in order to avoid gellation. The chemical structure of the polymer was confirmed by NMR and the molecular weight and intrinsic viscosity were measured by GPC with light scattering and viscometer detectors.

Characterizations. Both ^{1}H and ^{13}C NMR spectra agree well with the hyperbranched structure of the polymer. Figure 2 compares the ^{1}H NMR spectra of both the EGDMA monomer and a Poly(EGDMA). The assignments of the peaks are: 1.1 ppm (backbone -CH_3), 1.93 ppm (-CH_3 of the peripheral methacrylate group), 2.0-2.61 ppm (backbone -CH_2 -), 4.15-4.40 ppm (-OCH_2CH_2O-), 5.50-5.60 ppm (overlap of vinyl H_a of peripheral methacrylate and internal vinylidene groups), 6.10 ppm (vinyl H_b of methacrylate), 6.22 ppm (vinyl H_b of internal vinylidene). The relative ratios of the peaks agree well with those calculated from the structure shown in Scheme 1. ^{13}C NMR also agrees with the expected hyperbranched structure.

NMR spectra provide strong evidence to support the hyperbranched structure of the polymers formed. Two sets of vinyl protons were observed on ^{1}H NMR spectrum: one set belonging to the peripheral unreacted methacrylate groups and the other set due to the internal vinylidene groups resulted from the chain transfer reaction. There are two sets of carbonyl peaks on ^{13}C NMR spetrum, too. The one at 167.5 ppm is due to the peripheral methcarylate and the other one at 175-178 ppm is due to backbone ester groups.

The intrinsic viscosities of the hyperbranched poly(EGDMA) are much lower compared to linear PMMA with the same molecular weight. For example, the [η] of a hyperbranched poly(EGDMA) with M_n of 29000 is only 0.027 dL/g, while for a linear PMMA with the same M_n, the [η] is 0.122 dL/g. Another interesting feature is that there is no depedence of intrisic viscosity on the molecular weight for the hyperbranched poly(EGDMA). This strongly suggests that the hyperbranched polymers assume spherical structures in solution.

Acknowledgement. ZG would like to express grateful appreciation to Dr. C. Roe and L. A. Howe for the DEPT and the proton-carbon correlation NMR experiments; to J. Wang for technical assistance; to Drs. A. Gridnev and C. Berge for providing the COBF catalyst; and to Drs. O. Webster, M. Spinu, H. E. Simmons III, S. D. Ittel, A. Gridnev, B. E Smart, Y. Kim, L. Wilczek, and M. Fryd, C. Berge, etc., for many helpful discussions.

Figure 1. Structure of the COBF catalyst used in this study.

Table 1. Polymerization for EGDMA controlled by COBF[a]

ID#	COBF (mg)	Conv. (%)	$\overline{M_n}$	$\overline{M_w}/\overline{M_n}$	[η] (dL/g)
1[b]	20	19	7000	1.69	0.021
2[b]	10	34	10,000	2.27	0.022
3[b]	5	67	9000	3.98	0.023
4[c]	20	36	9000	1.81	0.020
5[c]	10	65	21500	4.37	0.026

[a] All polymerization were conducted under argon at 55 °C.

[b] The polymerization solutions for 1-3 were composed of 5.0 mL of EGDMA, 5 mL of 1,2-dichloroethane, 50 mg of VAZO-52, and different amount of COBF.

[c] Run 4 and 5 were the same as 1-3 except that no solvent was used.

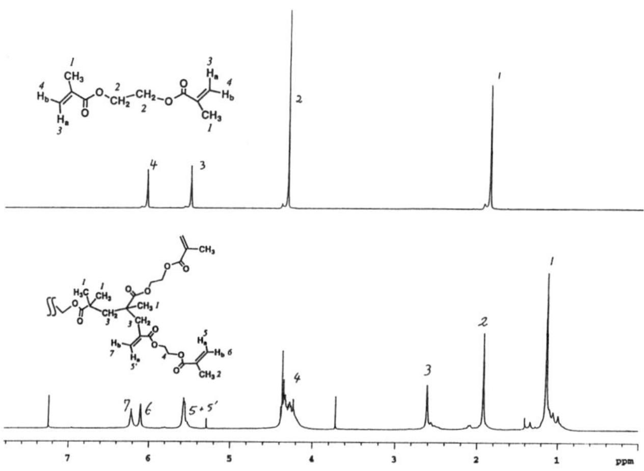

Figure 2. ^{1}H NMR spectrum of poly(EGDMA).

References:

(1) Tomalia, D. A.; and Durst, H. D. *Top. Curr. Chem.* **1993**, *165*, 193..

(2) a) Kim, Y. H. and Webster, O. W. *J. Am. Chem. Soc.* **1990**, *112*, 4592; Kim, Y. H. and Webster, O. W. *Macromolecules* **1992**, *25*, 5561. b) Hawker, C. J.; Lee, R.; and Frechet, J. M. J. *J. Am. Chem. Soc.* **1991**, *113*, 4583.

(3) Frechet, J. M. J. et al. *Science* **1995**, *269*, 1080.

(4) Hawker, C. J.; Frechet, J. M. J.; Grubbs, R. B.; and Dao, J. *J. Am. Chem. Soc.* **1995**, *117*, 10763.

(5) Enikolopyan, N.S. et al, *J. Poly. Sci., Polym. Chem. Ed.* **1981**, *19*, 879.

(6) Davis, T.P. et al *Trends in Polymer Science* **1995**, *3(11)*, 365. **1981**, *19*, 879.

(7) Gridnev, A.A.; Ittel, S.D.; Fryd, M. and Wayland, B.B. *Organometallics*, **1996**, *15*, 222. *(1)*, 772.

(8) Sanayei, R. A. and O'Driscoll, K.F. *J. Makromol. Sci. Chem.* **1989**, *A26(8)*, 1137.

SYNTHESIS OF LONG CHAIN BRANCHED SYNDIOTACTIC POLYSTYRENE

Yi-Bin Huang, Thomas H. Newman and Mark S. Chahl
Engineering Plastics R&D
The Dow Chemical Company
Midland, Michigan 48667

INTRODUCTION

Syndiotactic polystyrene (sPS) was first synthesized in 1985[1]. It is completely different from conventional, atactic polystyrene in both physical properties and manufacturing method. Synthesized from styrene monomer using metallocene catalysts, SPS has the highest melting point (270°C) of any single-monomer polymerization product. Based on its crystalline nature, SPS offers much higher heat resistance and chemical resistance than the conventional styrenic polymers. And, because SPS features very low moisture absorption, it provides excellent dimensional stability. Other benefits include low specific gravity, exceptional electrical properties and high modulus. SPS is expected to compete with liquid crystal polymers, polyphenylsulfide, polyamides and polyesters in applications such as injection molded electrical and electronic parts, automotive parts, optical films, capacitor films, fibrous filter elements and food packaging containers.

To date, all commercial sPS resins have been composed of linear sPS molecules. Similar to other linear polymers, linear sPS has low melt strength which poses limits on some end-use fabrication technologies. Long chain branched polymer, on the other hand, has significantly higher melt strength than its linear counterpart[2]. In this study, we attempted to synthesize long chain branched sPS (LCB-sPS) by copolymerizing a di-functional monomer (such as distyryl ethane or divinyl benzene) with styrene in the presence of sPS polymerization catalysts. Then, the branching in the obtained LCB-sPS was characterized by using a high temperature gel permeation chromatograph (HT-GPC) equipped with refractive index, intrinsic viscosity, and light scattering detectors. We also compared the melt strength, shear viscosity, and film tear strength of the LCB-sPS with that of linear sPS.

EXPERIMENTAL

LCB-sPS synthesis was carried out in a Teledyne kneader-mixer followed by a vertical tank reactor. Styrene monomer was mixed with a small amount of distyryl ethane in toluene and fed into the Teledyne kneader-mixer. The distyryl ethane was prepared following a literature procedure[3]. A catalyst solution of methylaluminoxane and a titanium metallocene complex was also fed into the reactor system. After polymerization, the polymer was devolatilized in a twin-screw extruder and then pelletized. In a separate trial, commercially available divinyl benzene was used instead of the laboratory-synthesized distyryl ethane. The divinyl benzene used is DVB80 (produced at Dow Chemical) which is composed of 80 weight% divinyl benzene and 20 weight% ethyl benzene.

For branching analysis, the polymer samples were dissolved in trichlorobenzene at 160°C for 2 hours and then injected into a HT-GPC operated at 145°C. The HT-GPC instrument was a Polymer Labs PL-GPC210 system equipped with three detectors: a differential refractive index detector, a two-angle light scattering detector, and a differential viscometer detector.

Melt strength was measured according to a literature technique[2] with the test conditions of 1 in/min plunger speed, 50 ft/min winder rate, and two temperatures (280 and 290°C). Shear viscosity was measured at 300°C using a Kayeness Galaxy V rheometer according to ASTM method D3835. For the evaluation of film tear strength, the polymer pellets were first converted to 250-micron-thick webs using a Killion single-screw extruder, and then the webs were biaxially stretched at 110°C to become 25-micron-thick films using an Iwamoto film stretcher. The films were then annealed at 220°C for 1 minute. Tear test was conducted at both room temperature and 200°C according to ASTM method D1938.

RESULTS AND DISCUSSION

The three LCB-sPS samples synthesized in this study are described in Table 1 together with a linear sPS sample as control. The linear sPS was produced similarly to the LCB-sPS except that no difunctional monomer was used when the linear sPS was produced. We aimed at the same weight-average molecular weight (Mw) target for all these samples. As can be seen in Table 1, their Mw's were very close. Their molecular weight distributions (MWD), determined by the HT-GPC, are shown in Fig. 1. The peak molecular weights of the three LCB-sPS samples were lower than that of the linear sPS sample. However, as seen in the high MW tails in Fig. 1, the LCB-sPS samples contain significantly larger molecules (speculated to be highly branched molecules) than the largest linear sPS molecule. These larger LCB-sPS molecules essentially offset the discrepancies in the peak molecular weight and bring the LCB-sPS Mw's close to that of the linear sPS.

Fig. 2 plots intrinsic viscosity (measured by the viscometer detector) against absolute molecular weight (determined by the light scattering detector) in a log-log scale. In this plot, called the Mark Houwink plot, the linear sPS follows the straight line very closely while all the three LCB-sPS samples show curvatures. At the high molecular weight portion, the LCB-sPS molecules contain many branches and as such have smaller hydrodynamic volume and consequently lower viscosity than the linear sPS of the same molecular weight. Therefore, the curvatures in Fig. 2 clearly evidenced the existence of branching in these LCB-sPS samples. Based on the degree of deviation from the straight line, one can rank the three LCB-sPS samples in the order of descending branching density: LCB1, LCB3 (very closely behind LCB1), and LCB2.

Melt strength data, summarized in Table 2, shows that the long chain branching indeed significantly improved the melt strength of sPS. The ranking of the LCB-sPS samples in melt strength is identical to the above based on branching density, indicating a strong correlation between the melt strength and the branching density in sPS. Shear viscosity data, plotted in Fig. 3, shows lower shear viscosity of the LCB-sPS sample than the linear sPS. In addition, the LCB-sPS samples are more shear thinning, which is more obvious at the high shear rate. Both the lower shear viscosity and the more shear thinning are considered advantageous for melt processing of thermoplastics.

Tear strength data measured from biaxially stretched films are summarized in Table 3. At room temperature (well below the 100°C Tg of sPS), the LCB-sPS film showed 40% higher tear strength than the linear sPS film. At 200°C (well above the Tg of sPS), the LCB-sPS film showed more than 200% higher tear strength than the linear sPS film. This significant improvement in tear strength could be very important for both the fabrication/processing and end use of sPS films.

CONCLUSION

For the first time, LCB-sPS has successfully been synthesized and demonstrated to offer higher melt strength, lower shear viscosity, and higher film tear strength than linear sPS.

ACKNOWLEDGEMENTS

The authors would like to thank A. J. Ribes for performing branching analysis, W. Liang for measuring melt strength, and S. Wu for conducting film tear tests.

REFERENCES

1. N. Ishihara, T. Seimiya, M. Kurmoto, M. Uoi, *Macromolecules* **19**, 2464, 1986.
2. S. K. Goyal, *Plastics Engineering* **51**, 2, 25, 1995.
3. W.-H. Li, K. Li, H. D. H. Stover, and A. E. Hamielec, *Journal of Polymer Science A* **32**, 2023, 1994.

Table 1 Sample description

Sample ID	Molecular Structure	Branching Agent	Mw (Kg/mole)
LCB1	LCB	250ppm DSE	313
LCB2	LCB	100ppm DVB	260
LCB3	LCB	600ppm DVB	298
Linear	Linear	None	300

Table 2 Melt strength data

Sample ID	Melt Strength (dyne) at 280°C	Melt Strength (dyne) at 290°C
LCB1	5,400	3,700
LCB2	3,100	2,200
LCB3	4,500	3,200
Linear	1,900	1,300

Table 3 Tear strength data

Room Temperature Tear Test

Sample	Tear Strength (dyne/μm)
LCB3	104
Linear	73

200°C Tear Test

Sample	Tear Strength (dyne/μm)
LCB3	310
Linear	96

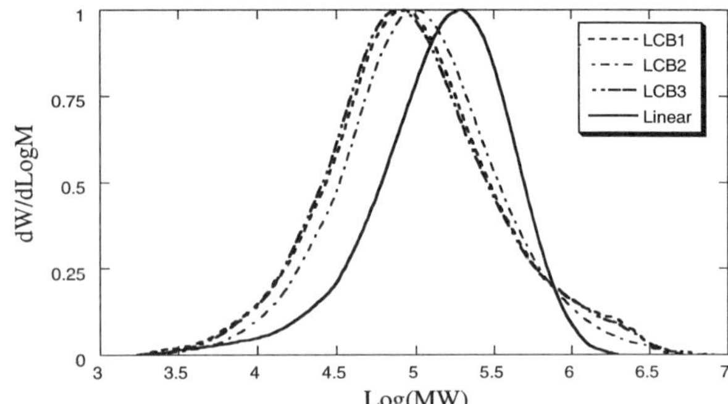

Figure 1 Molecular Weight Distribution

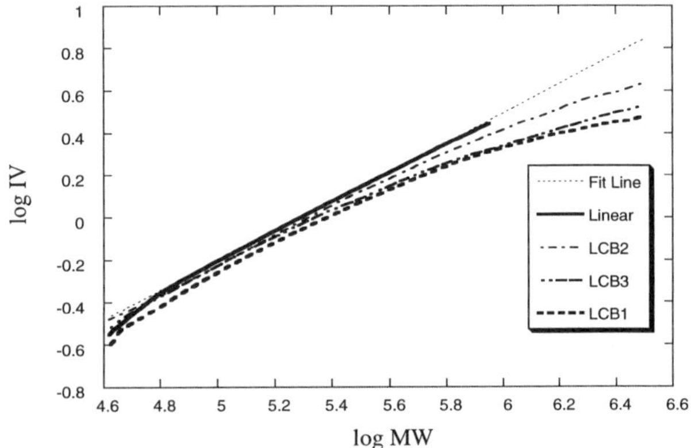

Figure 2 Mark Houwink Plot

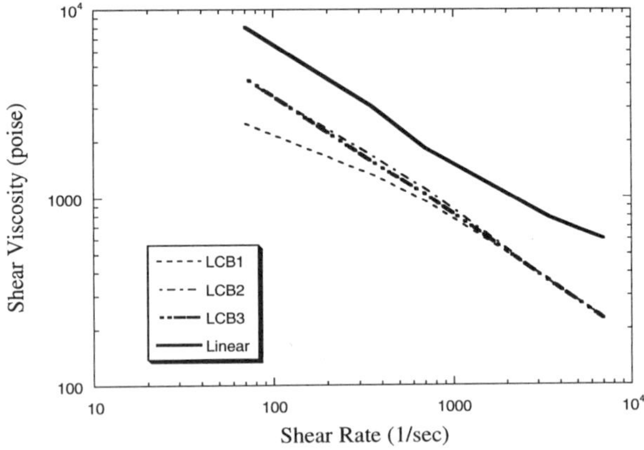

Figure 3 Shear Viscosity at 300°C

6,6'-*Bis*functionalized 2,2'-Bipyridine Metal Complexes as Supramolecular Initiators for functional Poly(oxazolines)

Ulrich S. Schubert*, Oskar Nuyken and Georg Hochwimmer

Lehrstuhl für Makromolekulare Stoffe, Technische Universität München Lichtenbergstr. 4, 85747 Garching, Germany

Introduction

Heterocyclic ligands with metal complexing abilities are of special interest in the fields of inorganic, analytical and polymer chemistry for decades.[1-3] The research in supramolecular chemistry focused in the last years especially on bipyridine, *oligo*(bipyridine), and terpyridine molecules and the corresponding metal complexes (Figure 1).[4-7]

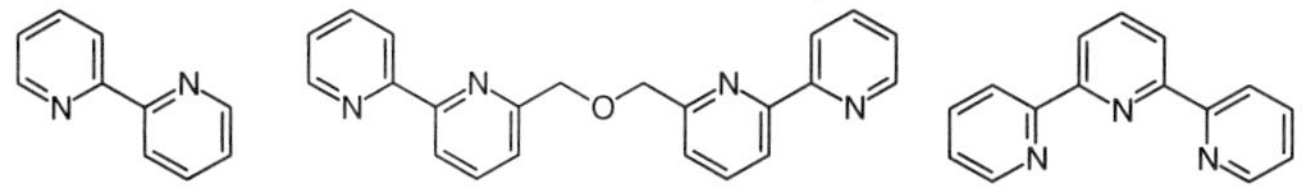

Figure 1.

However, only a few soluble and functionalized derivatives are known which allow the incorporation of these units into synthetic polymers. Only a couple of polymers with bipyridines and *bis*(bipyridines) in the main chain and in side chains were described.[8-14] This reflects the mayor problem in this area: difficulties in the synthesis of the heterocyclic monomers, low solubility of certain ligands and difficulties in the polymerization itself.

In this paper, we describe an extension of our work using 6,6'-*bis*-functionalized 2,2'-bipyridines as supramolecular initiators in the preparation of polyoxazoline oligomers (Figure 2).[15]

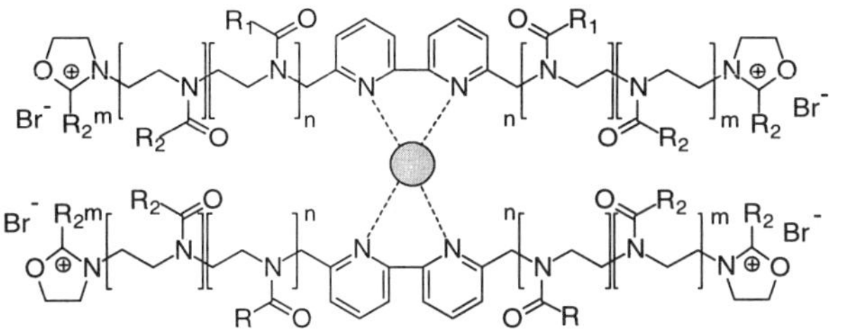

Figure 2. Star shaped polyoxazoline without termination (• = Cu(I)).

In contrast to the work by *Fraser et al.* using 4,4'-functionalized bipyridine complexes as initiators[16,17] we choose the 6,6'-*bis*substituted bipyridine building block due to their special interest in supramolecular chemistry.[4-7] These ligands may be useful in polymer chemistry to construct synthetic polymer materials with ordered architectures.[10] The selective complexation for silver(I) ions may also offer possibilities for the construction of selective metal complexation ion exchange columns and membranes.

Experimental part

Preparation of 6,6' *Bis*-(trimethylsilyl)-2,2'-bipyridine: To diisopropyl-amine (8.6 mmol) in dry THF at -78 °C was added *n*-BuLi (7.5 mmol) and the solution was stirred at -78°C for 8 min. The mixture was then warmed up to 0 °C and was cooled back to -78 °C. 6,6'-Dimethyl-2,2'-bipyridine (3.4 mmol) was dissolved in dry THF and was transferred to the LDA (lithium diisopropylamide) solution. After stirring 1 h at -78 °C trimethylsilyl chloride (9.1 mmol) was added and the reaction was quenched with absolute methanol after 45 s. Saturated aqueous NaHCO₃ was added and the product was extracted into EtOAc, washed with brine and dried over Na₂SO₄.

Preparation of 6,6' *Bis*-(bromomethyl)-2,2'-bipyridine: To the dissolved TMSbyp compound (2.43 mmol) and C₂Br₂F₄ (9.8 mmol) in dry DMF was added CsF (9.8 mmol). The mixture was stirred for 4 h at room temperature. The reaction was poured into H₂O/EtOAc and the layers were seperated. The aqueous layer were extracted with EtOAc. The combined layers where washed with brine and dried over Na₂SO₄.

Preparation of Polymers: The metallo-supramoleuclar initiators were synthesized according to previous published procedures.[15] A typical procedure for the polymerization of 2-oxazolines was as follow: The dry monomer was added under nitrogen to the dissolved initiator in dry CH₃CN. The mixture was stirred and heated at 80 °C for 24 h, then piperidine was added and the mixture was stirred for another hour at 80 °C. The mixture was allowed to cool to room temperature and the solvent was evaporated, the polymer was dissolved in CH₂Cl₂ and precipitated in ether. The synthesis of the blockcopolymers were analogue to the synthesis of the homopolymers, only the reaction of the phenyloxazoline monomer was carried out at 110 °C. Terpyridine terminated polymers where obtained by addition of a solution of the terpyridine compound in CH₃CN/CH₃OH (1:1) to the reaction mixture and stirring for 2.5 d at 80 °C. Afterwards the solvent was removed by vacuo and the remaining polymer was dissolved in CH₂Cl₂ and precipitated in diethylether. This procedure was repeated 3 times in order to remove unreacted terpyridine compound.

Results and discussion

Motivated by the recently published experiments performed by *Fraser et al.*[16,17] with 4,4'-*bis*functionalized 2,2'-bipyridine metal complexes as novel initiators for the living cationic polymerization of 2-oxazolines we investigated the behavior of the corresponding 6,6'-*bis*-functionalized copper(I) complexes as metallo-supramolecular initiators.[15] Such bipyridines are of great interest in supramolecular chemistry due to their ability to form well-defined architectures, like double helicates, as well as to their selective complexation behavior. Most suitable monomers for the applications in mind are 2-oxazolines, due to their variability in structure and due to the living nature of the polymerization of many 2-oxazoline derivatives.[18-20] First, we synthesized the basic 6,6'-(*bis*bromo-methyl)-2,2'-bipyridine unit according to published procedures by the reaction of 6,6'-dimethyl-2,2'-bipyridine with NBS.[21,13] However, the drawbacks of the radical bromination of heterocyclic ligands motivated us to explore a new strategy: Transferring the recently published LDA/TMS procedure[22] to 6,6'-(dimethyl)-2,2'-bipyridines we were able to obtain the desired product in high yield without critical purification procedures (Scheme 1).

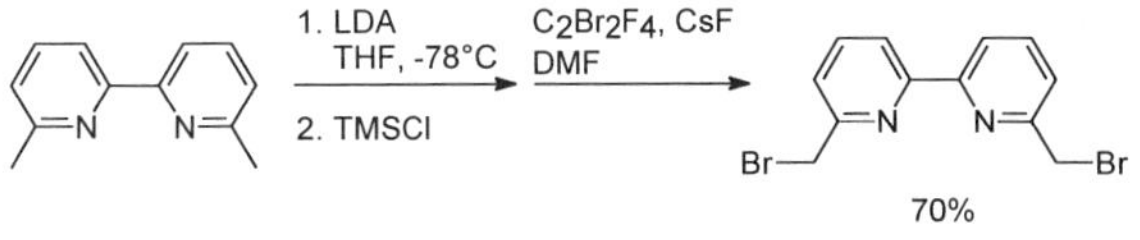

Scheme 1. Syntheses of the 6,6'-(*bis*bromomethyl)-2,2'-bipyridine ligand.

The [Cu(I)(6,6'-(*bis*bromomethyl)-2,2'-bipyridine)₂](PF₆) complex was synthesized in a simple complexation reaction in 95% yield.[15] The metallo-supramolecular initiator was then reacted with 2-ethyl-2-oxazoline in dry acetonitrile at 80 °C for one day. The resulting polymer was isolated after or without termination with piperidine (Scheme 2).

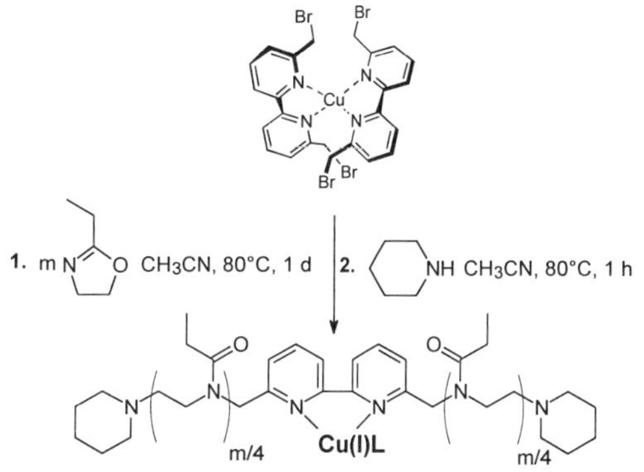

Scheme 2. Schematic representation of the polymerization.

The living character of the polymerization was proved by experiments with various monomer to initiator concentrations. These experiments illustrated the linear relationship between the average molecular weight M_n, and the [monomer]/[Cu(I)(6,6'-(*bis*bromomethyl)-2,2'-bipyridine)$_2$](PF$_6$) [M]/[I], complex ratio for the fragmented polymers (Figure 3). [I] refers to initiating CH$_2$Br group.

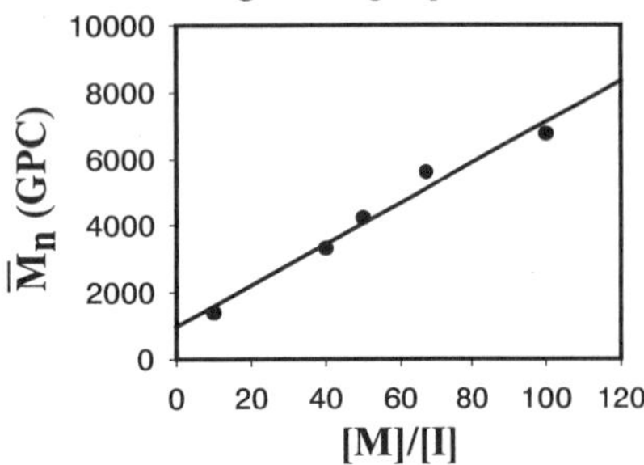

Figure 3. Plot of the average molecular weight M_w versus [M]/[I].

The central metal ion can be removed out of the polymer using aqueous basic or acidic conditions, resulting in uncomplexed polymers with a free central metal binding unit. Also the copper containing star shaped polymer fragments fully or to a large extent on the GPC columns liberating the uncomplexed bipyridine centered polymer with half of the molar mass of the corresponding complex star shaped polymer.

Test reactions with the non brominated complex[23] as well as the copper salt itself using identical conditions did not yield any polymer.[15] However, reaction of the 6,6'-(*bis*bromomethyl)-2,2'-bipyridine without copper ions with 2-oxazolines also initiated the polymerization. In all cases a shoulder could be observed in the GPC.[15] Obviously, the nitrogen atoms of the bipyridines can act as termination reagents. The complexing metal ion is therefore not responsible for the initiating behavior itself, but it acts as a protecting group for the nitrogen atoms of the bipyridines. These findings were confirmed by preliminary experiments using bipyridines as termination reagents (added at the beginning of the polymerization) and methyl *p*-toluenesulfonate as initiators (Figure 4).

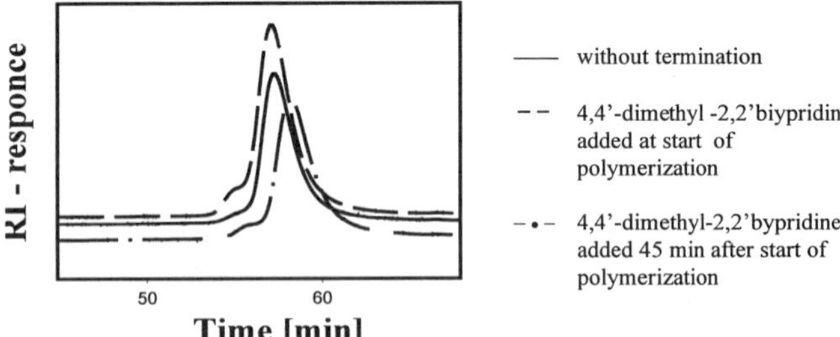

Figure 4. GPC curves for the polyoxazolines initiated methyl *p*-toluenesulfonate and terminated with bipyridines.

In addition, we synthesized block copolymers using ethyl and phenyl monomers, confirmed by NMR and GPC.[15] Extended studies using different monomers to obtain well-defined amphiphilic block copolymers are currently in progress.

The termination of the living end groups during the polymerzation opens possibilities for incorporation of other interesting units. We choose a different metal binding ligand functionalized with a amino group as termination agent to construct a novel polymer material with specific metal binding sides in the center and at the end blocks. As first example we synthesized a polymer with a 6,6'-functionalized bipyridine center block and terpyridine end blocks offering selective complexing units for Ag(I) and Fe(II) metal ions in the same polymer chain (Figure 5).

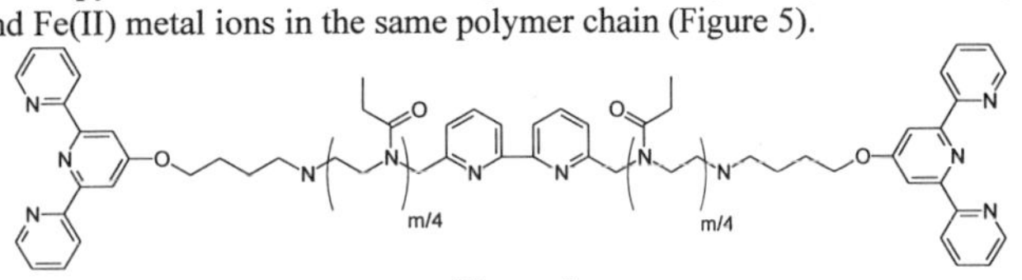

Figure 5

Conclusions

We describe our attempts to use 6,6'-*bis*functionalized 2,2'-bipyridines as metallo-supramolecular initiators for the living cationic polymerization of 2-oxazolines. In agreement with the data published by *Fraser et al.* for the 4,4'-system, we observed a linear relationship between the average molecular weight M_n and the [monomer]/[Cu(I)(6,6'-(*bis*bromomethyl)-2,2'-bipyridine)$_2$](PF$_6$) complex ratio for the fragmented polymers. In addition to the known synthetic strategy towards 6,6'-brominated methyl bipyridines we could successfully apply the LDA-TMS procedure for this position. The central metal ions can be removed easily using basic or acidic conditions as well as shear forces. However, the metal ions play a crucial role during polymerization: The bipyridine units can act also as termination agents and therefore the metal ion serve as protecting group. The termination agents offers possibilities for the incorporation of different metal binding units. As a first example we synthesized a polymer with a 6,6'-functionalized bipyridine center block and terpyridine end blocks, offering selective complexing units for different metal ions like Ag(I) and Fe(II). This strategy allows the construction of novel materials with special metal binding abilities with a wide range of possible applications.

Acknowledgments

The research was supported in parts by the *Bayerisches Staatsministerium für Unterricht, Kultus, Wissenschaft und Kunst* (Bayerischer Habilitations-Förderpreis for USS) and the *Fonds der Chemischen Industrie*.

References

1. N. F. Curtius, *Coord. Chem. Rev.* **3**, 3 (1968)
2. G. R. Newkome, J. D. Sauer, J. M. Roper, D. C. Hager, *Chem. Rev.* **77**, 513 (1977)
3. E. Buhleier, W. Wehner, F. Vögtle, *Chem. Ber.* **111**, 200 (1978)
4. J.-M. Lehn, "*Supramolecular Chemistry - Concepts and Perspectives*", VCH, Weinheim 1995
5. D. S. Lawrence, T. Jiang, M. Levett, *Chem. Rev.* 2229 (1995)
6. G. R. Newkome, W. E. Puckett, G. E. Kiefer, V. K. Gupta, Y.-J. Xia, M. Coreil, M. A. Hackney, *J. Org. Chem.* **47**, 4116 (1982)
7. J.-M. Lehn, A. Rigault, *Angew. Chem.* **100**, 1121 (1988); *Angew. Chem., Int. Ed. Engl.* **27**, 1095 (1988)
8. J.-M. Lehn, *Makromol. Chem., Makromol. Symp.* **69**, 1 (1993)
9. Y. Chujo, K. Sada, T. Saegusa, *Macromolecules* **26**, 6315 (1993)
10. Y. Chujo, N. Naka, M. Krämer, K. Sada, T. Saegusa, *J. Macromol. Sci., Pure Appl. Chem.* **32**, 1213 (1995)
11. C. D. Eisenbach, U. S. Schubert, *Macromolecules* **26**, 7372 (1993)
12. C. D. Eisenbach, U. S. Schubert, G. R. Baker, G. R. Newkome, *J. Chem. Soc., Chem. Commun.* 69 (1995)
13. C. D. Eisenbach, A. Göldel, M. Terskan-Reinold, U. S. Schubert, in "*Polymeric Materials Encyclopedia*", vol. 10, J. C. Salamone, Ed., CRC Press, Boca Raton 1996, p. 8162 ff.
14. P. K. Ng, X. Gong, W. T. Wong, W. K. Chan, *Macromol. Rapid Commun.* **18**, 1009 (1997)
15. G. Hochwimmer, O. Nuyken, U. S. Schubert, *Macromol. Rapid Commun.*, in press
16. J. J. S. Lamba, C. L. Fraser, *J. Am. Chem. Soc.* **119**, 1801 (1997)
17. J. J. S. Lamba, J. E. McAlvin, B. P. Peters, C. L. Fraser, *Div. Polym. Chem., Polymer Preprints* **38(1)**, 193 (1997)
18. Y. Chujo, T. Saegusa, in "*Ring Opening Polymerization*", D. I. Brunelle, Ed., Hanser, Munich 1993, chap. 8 and references therein
19. S. Kobayashi, *Progr. Polym. Sci.* **15**, 751 (1990)
20. O. Nuyken, G. Maier, A. Groß, H. Fischer, *Macromol. Chem. Phys.* **197**, 83 (1996)
21. M. M. Harding, U. Koert, J.-M. Lehn, A. Marquis-Rigault, C. Piquet, J. Siegel, *Helv. Chim. Acta* **74**, 594 (1991)
22. C. L. Fraser, N. R. Anastasi, J. J. S. Lamba, *J. Org. Chem.* **62**, 9314 (1997)
23. P. J. Burke. D. R. McMillin, W. R. Robinson, *Inorg. Chem.* **19**, 1211 (1980)

Curing of Partially Brominated Poly(isobutylene-co-4-methylstyrene) Elastomers using Phenolic Resins: A Mechanistic Investigation.

M. Jayaraman, J. M. J. Fréchet
Department of Chemistry
University of California, Berkeley, CA 94720-1460
A. J. Dias, H. C. Wang
Exxon Chemical Company, Baytown Polymer Center
5200 Bayway Drive, Baytown, TX 77520

Introduction

The crosslinking chemistry of elastomers has been widely investigated and new and improved curing conditions have been developed based on the results of those investigations. The objective of this work is to model the reaction between partially brominated Poly(Isobutylene-co-4-Methylstyrene) 1 and a phenolic resin crosslinker 2 in the presence of Zn salts using small molecule analogs. Such model studies have proven to be successful in elucidating the mechanism of crosslinking in various polymeric

1

2

systems including similar elastomers and a variety of negative-tone photoresists.[1]

p-Isopropylbenzylbromide (PIBB) 3 and *o*-hydroxybenzylalcohols 4 & 5 were chosen as model compounds for the elastomer and the resin respectively. Isooctane was used as the solvent to simulate the non-polar conditions found within the largely aliphatic brominated poly(isobutylene-co-4-methylstyrene) copolymers. The resin analog was deuterated at the benzylic position in order to differentiate it from the elastomer analog.

3

4

5

Experimental

Most experiments were carried out by dissolving 0.25 mmol of the model compounds in ~1 ml of isooctane followed by addition of the desired powdered catalyst: coated ZnO (Kadox). This dispersion was then transferred to an NMR tube and the reaction was then monitored by recording ^{2}H and ^{1}H NMR spectra at 80° and 100°C.

Results and Discussion

PIBB is known to react with itself in a non polar medium in the presence of ZnO via electrophilic aromatic substitution.[2] The resulting oligomeric products show a new benzylic resonance at $\delta = 3.9$ as well as the original benzylic bromide resonance at $\delta = 4.4$ in the ^{1}H NMR.

When the same reaction was carried out in the presence of equimolar amount of **4**, three new peaks were observed in the ^{2}H NMR : $\delta = 5.2$, $\delta = 4.55$ and $\delta = 3.9$ as well as a peak at $\delta = 4.65$ corresponding to starting material **4**. As the reaction proceeds the peak at $\delta = 3.9$ grows while the other peaks diminish. Monitoring by ^{1}H NMR shows a new peak at $\delta = 3.9$, while the peak $\delta = 4.4$ corresponding to the benzylic bromide remains as the major peak until the end of the reaction.

The above observations are rationalized as follows. Peaks at $\delta = 5.2$ and $\delta = 4.55$ in the ^{2}H NMR indicate the formation of benzyl-phenyl ether and dibenzyl ether linkages respectively. The peak at $\delta = 3.9$ corresponds to the desired "cure" product formed by electrophilic aromatic substitution of the resin analog on PIBB. The data from ^{1}H NMR further indicates that the PIBB is reacting with itself in a competing "cure" process.

^{2}H NMR data

^{1}H NMR data

">

Replacing resin analog **4** by **5** affords similar results though two significant differences are observed. The reaction rate is higher with compound **5** and the reaction of PIBB with itself is minimized. The above observations can be attributed to the higher solubility of **5** in the reaction medium when compared to **4**. Another significant point is that the "cure" reaction of the resin analog with PIBB is faster than the "cure" reaction of PIBB with itself. The nature of the crosslinks is probably dictated by the miscibility of the components and may well be diffusion controlled. Decreasing the reaction temperature to 80°C not only led to a lowering of the rate of the curing reaction but also suppressed completely the reaction of PIBB with itself.

In conclusion, the model studies carried out with the deuterated analogs of phenolic cure agents have shed light on the nature of the crosslinks between brominated poly(isobutylene-co-4-methylstyrene) and these phenolic resins. The results have also shown possible ways to control the curing using reaction temperature and homogeneity of the formulation.

References

1) a) Stover, H. D. H.; Matuszczak, S.; Wilson, C. G.; Fréchet, J. M. J. *Macromolecules*, **1991**, *24*, 1741. b) Fréchet, J. M. J.; Matuszczak, S.; Reck, B.; Stover, H. D. H. *Macromolecules*, **1991**, *24*, 1746. c) Lee, S. M.; Fréchet, J. M. J.; Wilson, C. G. *Macromolecules*, **1994**, *27*, 5154. d) Lee, S. M.; Fréchet, J. M. J. *Macromolecules*, **1994**, *27*, 5160.

2) Bielski, R.; Fréchet, J. M. J.; Fusco, J. V.; Powers, K. W.; Wang, H. C. *J. Polym. Sci., Polym. Chem. Ed.* **1993**, *31*, 755.

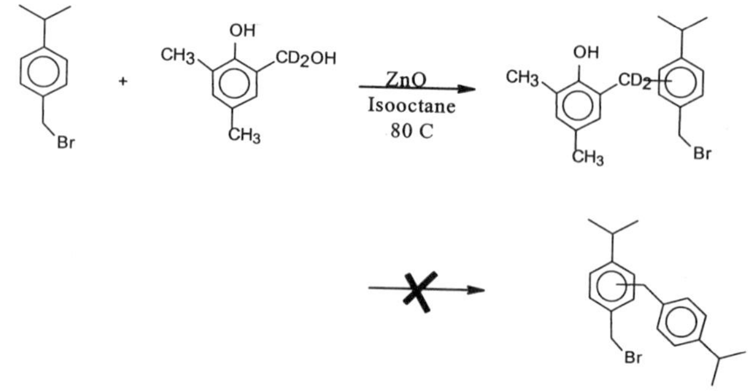

In order to test the validity of the proposed crosslinking pathway in the polymeric system the experiments were repeated with low molecular weight brominated poly(isobutylene-co-4-methylstyrene) instead of PIBB. The results observed were analogous to those with PIBB, but the reaction was slower in the polymeric system. This can be explained by the increase in viscosity and the resulting decrease in mobility upon using the polymer.

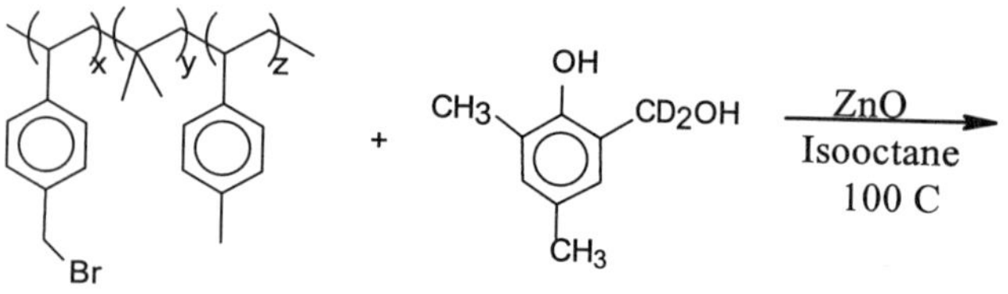

EXXPRO™

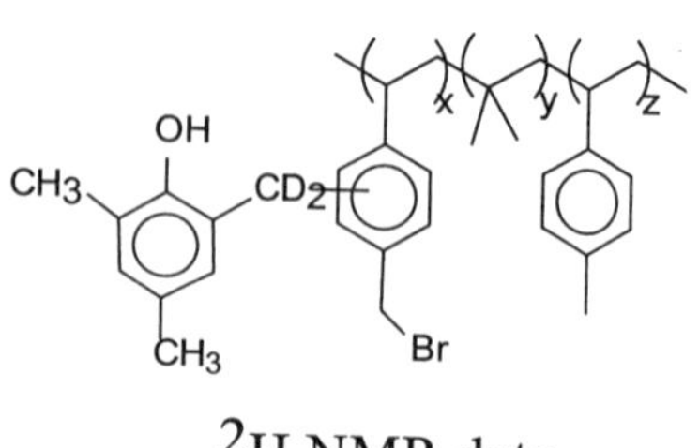

^{2}H NMR data

Hydroquinone, Quinone and Amine Functionalized Polyolefins

Abhimanyu O. Patil and Steve Zushma*

Corporate Research Laboratory,
Exxon Research & Engineering Company,
Route 22 East, Clinton Township, Annandale, NJ 08801

The chemical modification of polymers is of great scientific and technological interest.[1] For example, the modification of polyolefins to impart polar functionality has been pursued for many years. Increased polarity is important for the use of polyolefins in solution (dispersancy) or in solid phase (adhesion, printability). For example, alkylation of phenols by unsaturated hydrocarbons has been studied extensively[2] and are widely used in the pharmaceutical industry and in petroleum products.[3] Low molecular weight polyisobutylene polymers with one unsaturated double bond per chain, have been alkylated with phenol. The phenol functionalized polymer, can be subsequently reacted with amines and formaldehyde to synthesize Mannich base dispersant molecules for lube applications. In these dispersants, the polyisobutylene chain is responsible for the molecule's solubility, and the polar amine head group keeps the oxidation products and sludge particles suspended in the oil.

In turn, metallocene technology, with its single-site catalysts has opened up new ways to make polyolefins.[4] Ethylene/α-olefin copolymers produced with metallocene catalysts have narrow compositional and molecular weight distributions. Typically, there is one olefin (especially vinylidene), and one saturated end group per chain. The terminal olefin is produced by chain-transfer reactions involving either β-hydride elimination, β-hydride transfer to ethylene, or metalation followed by rearrangement.[5] Such double bonds could allow a higher level of functionalization in these polymers.

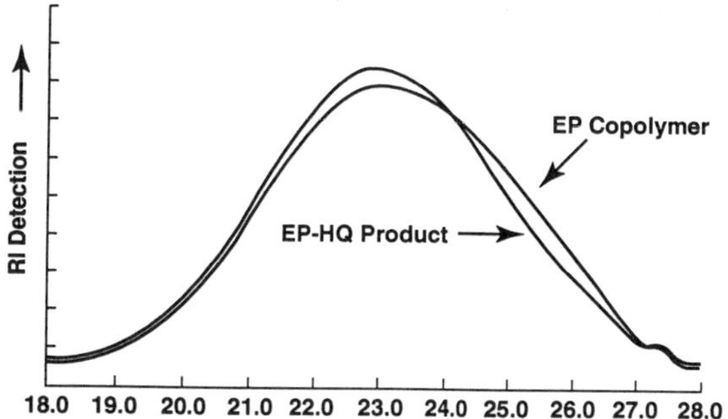

1-Octadecene has been alkylated with hydroquinone using Amberlyst-15 catalyst.[6] Using a similar alkylation procedure, a low molecular weight ethylene/propylene copolymer (I) (Mn 870) with 95% vinylidene terminal double bonds was alkylated with hydroquinone.[7,8] The FTIR spectra of the starting EP copolymer showed double bond absorption peaks at 3074, 1650 and 887 cm⁻¹. These peaks disappeared on alkylation and a new absorption peak appeared at 3620 cm⁻¹, due to hydroquinone grafted EP copolymer (II). ^{13}C NMR spectra of the starting ethylene/propylene copolymer showed predominantly, two types of vinylidene double bonds with peaks at 109.5, 145.7, 111.2, 144.2 ppm.[5c] ^{13}C NMR spectra of the product showed disappearance of all four terminal olefin peaks, and new peaks appeared in the aromatic region (115 to 150 ppm).

Gel permeation chromatography (GPC) of the starting EP copolymer and hydroquinone alkylated product using RI and UV detector are shown in Figure 1. The GPC trace with RI detector is similar both in the starting polymer, and the product. The GPC trace using UV detector showed no absorption for starting EP copolymer. The alkylated product, however, shows UV absorption suggesting hydroquinone attached to the polymer chain. Hydroquinone alkylated with ethylene/propylene copolymer oxidizes to the corresponding quinone form (III) even at room temperature. The IR spectrum of the sample stored at room temperature for 1 month showed a decrease in the absorption peak at 3620 cm⁻¹ due to the hydroxyl group and a new carbonyl absorption peak appeared at 1655 cm⁻¹, suggesting substantial conversion of hydroquinone bound EP copolymer to the corresponding quinone bound EP copolymer (III).

Quinones are known to react with amines under mild conditions to obtain aminoquinones.[9] Quinone attached ethylene/propylene copolymer was reacted with amine (2:1 mole quinone/amine ratio), such as diethylenetriamine (DETA) to obtain amino-functionalized polymers (IV).[11] IR spectrum of the product showed a broad peak at 3250 cm⁻¹ due to amine along with carbonyl peak at 1655 cm⁻¹. The product was found to be an effective lube dispersant as measured by the dispersancy bench test.[12]

In Quinone-amine reaction, the technologically important science question is, whether one can carry the reaction to completion using just 1,4-benzoquinone and 1-octylamine in 1:2 mole ratio, in a such way that the resultant hydroquinone by-product reoxidizes back to 1,4-benzoquinone with air. Only then, one could use this reaction effectively.

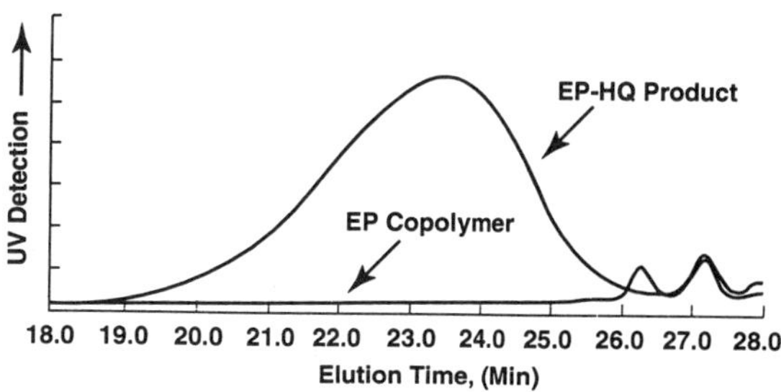

Figure 1. GPC of the starting EP copolymer and hydroquinone reacted product (top scan: RI detector) (bottom scan: UV detector)

Quinone-Amine Reaction: Model study

As a model, we have studied the reaction of 1,4-benzoquinone with 1-octylamine at room temperature. In this reaction, 1 mole of 1,4-benzoquinone reacts with 2 moles of 1-octylamine to obtain 1 mole of 2,5-dioctylamino-1,4-benzoquinone product. However, in this process, 2 moles of 1,4-benzoquinone is reduced to form 2 moles of hydroquinone by-product. To use this reaction effectively, one needs to oxidize the hydroquinone by-product back to benzoquinone. In earlier studies, calcium hypochloride was chosen as the oxidizing agent despite some issues associated with this reagent.[10]

1,4-benzoquinone has one carbonyl carbon signal at 186.3 ppm for two equivalent carbonyl carbons, and one signal at 135.5 ppm for the four equivalent double bond carbons. The product spectrum of 2,5-dioctylamino-1,4-benzoquinone shows one carbonyl carbon signal at 178.5 ppm for two equivalent carbonyl carbons. The double bond peaks, however, are split into two separate resonances. The signal at 151.8 ppm is due to the carbon with alkylamine group attached, while the response at 93.1 ppm is due to the carbon with one hydrogen attached (from couple spectra). The ^{13}C NMR resonance at the 42.2 ppm is due to the C1 carbon of the 1-octylamine in the starting amine, moves to 43.1 ppm in the product. Thus, ^{13}C NMR can be conveniently used to follow this reaction with respect to all reagent disappearance and product formation.

To follow the course of the reaction, we ran the reaction in a NMR tube using 1:2 mole ratio of 1,4-benzoquinone and 1-octylamine in benzene solvent. Figure 2 (#1) shows the ^{13}C NMR spectrum of the reaction mixture after 15 minutes. As shown in the figure, all benzoquinone carbonyl carbon peaks at 186.3 ppm and the signal at 135.5 ppm (representing the four double bond carbons) have completely disappeared and a new carbonyl peak due to the disubstuted product at 178.5 ppm and the peak at 151.8 ppm and 93.1 ppm due to the double bond carbon appeared. There is no sign of monoadduct formation, since that product would have two separate carbonyl carbon signals. There are, however, peaks due to unreacted 1-octylamine that are present, along with the product peaks, suggesting the presence of substantial unreacted octylamine in the mixture. The ^{13}C NMR also shows two resonances at 151.4 ppm due to aromatic carbons (with two equivalent hydroxyl groups) and at 117.2 ppm (due to four equivalent aromatic carbons). These results indicate formation of 1,4-hydroquinone as a by-product in the reaction. The reaction mixture also shows several small signals, especially in the double bond, and carbonyl regions.

Figure 2 (#2) shows the ^{13}C NMR spectrum of the reaction mixture after 3 hours. The spectrum is very similar to the one shown in Figure 2 (#1), except relative hydroquinone peaks and carbon next to amine of 1-octylamine are diminishing in intensity, while the product resonances are increasing in intensity. Figure 2 (#3) shows the ^{13}C NMR spectrum of the reaction mixture after 30 hours. Again, the spectrum is very similar to the one before, except the relative hydroquinone resonances and C1 carbon next to the amine of 1-octylamine have disappeared further more and the product peaks are increasing. Figure 2 (#4) show ^{13}C NMR spectra of the reaction mixture after 72 hours. The ^{13}C NMR spectrum of the reaction mixture after 72

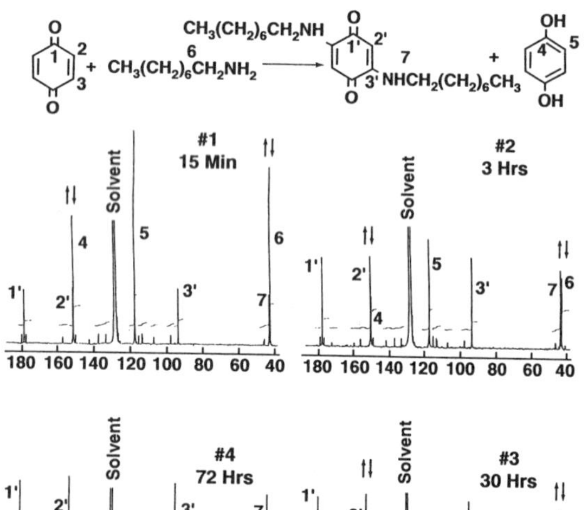

Figure 2. ^{13}C NMR spectrum of the reaction mixture of 1,4-benzoquinone, 1-octylamine after 15 minutes (#1), 3 hours (#2), 30 hours (#3) and 72 hours (#4).

hours is very similar to that the pure product. Thus, one can obtain the desired product 2,5-dioctylamino-1,4-benzoquinone in high yields using just 1:2 mole ratio of the reactants, 1,4-benzoquinone with octylamine, using air as an oxidizing agent.

Conclusion

Mannich base dispersants are important class of lubricant additives. These additives typically are prepared by reaction of low molecular weight polyisobutylene alkylated phenols with amines and formaldehyde. In contrast, we have alkylated low molecular weight ethylene/propylene copolymer having terminal double bonds with hydroquinone. These hydroquinone alkylated polymers were oxidized to their corresponding quinone form. We have shown that, 1,4-benzoquinone can react with 1-octylamine at room temperature using 1:2 mole ratio, in a such way that the resultant hydroquinone by-product reoxidizes back to 1,4-benzoquinone with air. Unlike Mannich base lube dispersants, the quinone bound ethylene/propylene copolymers were directly reacted with amines, *without formaldehyde*, to obtain novel functionalized polymers for lube applications.

Acknowledgment

I would like to thank Greg Springstun for experimental help and Debbie Sysyn and B. Liang for NMR spectra.

References and Notes

(1) Frechet, J. M. J. *Science* (Washington, D. C.) **1994**, 263(5154), 1710-15. Marechal, E. In *Comprehensive Polymer Science: The Synthesis, Characterization, Reactions, and Applications of Polymers*; Allen, G., Bovington, J. C., Eds.; Pergamon Press: Oxford, 1989; Vol. 6, pp 1-47.
(2) For a review on alkylation of phenols, see Shuikin, N. I.; Viktorova, E. A. *Russ. Chem. Rev.* **1960**, 29, 560.
(3) "Mannich Bases Chemistry and Uses", Tramontini, M.; Angiolini, L. CRC Press, Boca Raton, Florida, 1994.
(4) For general overview on polyolefins by metallocene catalyst, see Brintzinger, H. H.; Fischer, D.; Mulhaupt, R.; Rieger, B.; Waymouth, R. W. *Angew Chem., Int. Ed. Engl.* **1995**, 34, 1143. Kaminsky, W. *Macromol. Chem. Phys.* **1996**, 197(12), 3907-3945. Hamielec, A. E.; Soares, J. B. P. *Prog. Polym. Sci.* **1996**, 21(4), 651-706. Soga, K. *Macromol. Symp.* **1996**, 10, 281-8. Montagna, A. A. *Chemtech* 44, October 1995. Galimberty, M.; et. al *Macromol. Symp.* **1995**, 89, 259.
(5) a) Kaminsky, W.; Ahlers, A.; Moller-Lindenof, N. *Angew. Chem., Int. Ed. Engl.* **1989**, 28, 1216. b) Resconi, L.; Piemontesi, F.; Franciscono, G.; Abis, L.; Fiorani, T. J. *Am. Chem. Soc.* **1992**, 114, 1025. c) Randall, J. C.; Rucker, S. P. *Macromolecules* **1994**, 27, 2120.
(6) a) Nobutaka, O. *JP 06157383 A2 940603 CAN 121:230480*; b) *JP 06157384 A2 940603.. CAN 121:230479*; c) Malloy, T. P.; Engel, D. J. *US 4323714 A 820406. CAN 97:23455*
(7) a) Rossi, A.; Odian, G; Zhang, J. *Macromolecules* **1995**, 28(6), 1739; b) Rossi, A.; Zhang; Odian, G, J. *Macromolecules* **1996**, 29(7), 2331.
(8) A typical example of the hydroquinone alkylation of ethylene/propylene copolymer is shown as follows: To a 250 mL round-bottom flask equipped with water condenser, a dropping funnel was charged 6.32 g (0.057 moles) of hydroquinone, 15 g of Amberlyst-15 (0.0705 mole/equi) and 50 ml heptane. The mixture was heated to 98 °C with stirring under nitrogen. A 50 g quantity of ethylene/propylene copolymer (Mn 870, 0.057 moles) with 95% vinylidene terminal double bonds was then added showly to the above solution and the mixture was refluxed for 6 hours under nitrogen. The reaction mixture was then filtered, the filtrate evaporated on rotary evaporator to obtain the product.
(9) There are several reports on quinone-amine chemistry, see Erhan, S. J. "The Polymeric Materials Encyclopedia: Synthesis, Properties and Applications., CRC Press, Boca Raton, **1996**, 7311. Nikles, D. E.; Cain, J. L.; Chacko, A. P.; Liang, J.; Webb, R. I. "The Polymeric Materials Encyclopedia: Synthesis, Properties and Applications., CRC Press, Boca Raton, **1996**, 7303.
(10) Nithianandam, V. S.; Erhan, S. *Polymer* **1991**, 32, 1145.
(11) Alkylated quinone adduct (2g, 2x10^{-3} moles) was dissolved in 50 mL heptane and 0.103 g (1x10^{-3} moles) of diethylenetriamine (DETA) was added. The solution was stirred at room temperature for 72 hours. The product was diluted with 50 mL heptane and filtered. The solvent was removed initially by nitrogen stripping and then by high vacuum.
(12) Patil, A. O. *U. S. Patent 5,576,274* (November 19, 1996).

Hyperbranched Grafting on a Polyethylene Surface

Justine G. Franchina, David E. Bergbreiter

Texas A&M University, Department of Chemistry, PO Box 300012,
College Station, TX 77842-3012

Introduction:

Organic thin film chemistry is important in many areas.[1-5] Our prior work has shown that robust thin films can be synthesized on a gold surface using hyperbranched grafting with a α,ω-diaminopoly(*tert*-butyl acrylate).[2] This hyperbranched chemistry is based upon the repetitive covalent attachment of α,ω-diaminopoly(*tert*-butyl acrylate) to a gold surface originally modified with mercaptoundecanoic acid. The multiple cycles of attachment of the acrylate polymer lead to increasingly rapid growth and eventually linear growth of the thin polymer film on the gold surface as measured using ellipsometry.

In prior work,[6,7] our group has also examined graft modification of polyethylene surfaces by anionic and radical chemistry and has studied the effect of derivatizing the oxidized surfaces with various spectroscopic probes at these surfaces. Here we describe the synthesis of hyperbranched film grafts on polyethylene surfaces by covalent attachment of α,ω-diaminopoly(*tert*-butyl acrylate) polymer. These derivatized films have been investigated using X-ray photoelectron spectroscopy (XPS), contact angle analysis, and attenuated total reflectance-IR (ATR-IR) spectroscopy.

Experimental:

The α,ω-diaminopoly(*tert*-butyl acrylate) (NH$_2$-PTBA-NH$_2$) polymer was synthesized by radical polymerization of 15 mL of distilled *tert*-butyl acrylate using 0.380 g of the initiator 4,4'-azobis(4-cyanovaleric acid) in 60 mL of 1,4-dioxane under a nitrogen atmosphere for 15 h at 120 $^\circ$C. The polymer was recovered by precipitation into a methanol:water/1:1 mixture. The diacid terminated polymer was purified by two precipitations in methanol: water/1:1. The diacid polymer was characterized using ^{1}H-NMR spectroscopy, FT-IR spectroscopy and gel permeation chromatography (GPC). The diacid polymer was then converted to a bisamine amide by allowing 5 g of the diacid polymer to first react with 0.324 g of 1,1'-carbonyldiimidazole (CDI) for 5 h in 30 mL of CH$_2$Cl$_2$ and then with 0.2 mL of ethylene diamine for 15 h. The reaction mixture was washed with water and saturated brine (3x50 mL), the organic layer was then dried over MgSO$_4$ and the CH$_2$Cl$_2$ solvent was removed at reduced pressure using a rotary evaporator. The diamine polymer was characterized using ^{1}H-NMR and FT-IR spectroscopy.

The polyethylene used was a linear medium density polymer (Fortiflex J36-25-142) provided by Soltex in the form of a blown film (density of 0.936g/mL). The polyethylene films were cut into square pieces, ½" by ½" which were extracted using CH$_2$Cl$_2$ in a Soxhlet for 15 h. The films were dried under vacuum, at 0.1 atm, for 15 min. The films were then oxidized in a mixture of CrO$_3$/H$_2$O/H$_2$SO$_4$ for 5 min at 90 $^\circ$C to form an oxidized film.[7] After the films were removed from the oxidizing solution they were washed with water and then acetone. The films were allowed to air dry. The oxidized films were extracted using CH$_2$Cl$_2$in a Soxhlet for 15 h. The films were dried under vacuum for 15 min at 0.1 atm. ATR-IR spectroscopy was used to confirm oxidation of the surface, by the appearence of a carbonyl signal at 1710 cm^{-1}. Grafting of the diamine terminated polymer on to this oxidized polyethylene film substrate then takes place through a sequence of activation, substitution and hydrolysis. The starting acid endgroup of the oxidized film surface was activated using ethyl chloroformate (ECF, 0.1 mL) and N-methyl morpholine (NMM, 0.1 mL) in 7 mL of DMF for 15 min. The surface was then washed with ethyl acetate and dried with a stream of nitrogen. The activated film was then placed in a solution of 0.25 g of NH$_2$-PTBA-NH$_2$ in 6 mL of DMF for 1 h. After 1 h the film was washed with ethanol and dried with a stream of nitrogen and the film was then extracted for 15 h using CH$_2$Cl$_2$ in a Soxhlet. After extraction, the film was dried under vacuum for 15 min at 0.1 atm. The film was then hydrolyzed in a saturated benzene solution of p-toluene sulfonic acid monohydrate (p-TsOH) at 55 $^\circ$C for 1 h and 30 min. The 1PAA film was washed with ethanol and dried with a stream of nitrogen, then dried further under vacuum for 1h at 0.1 atm. The activation, grafting and hydrolysis steps were repeated to construct the 2PAA and 3PAA films.

Derivatization of the oxidized films as well as the PAA grafts(1PAA, 2PAA, 3PAA) was studied. During this sequence, the carbonyl containing films were activated as indicated previously with ECF and NMM and then placed in one of the following solutions of DMF and 1H,1H-pentadecafluorooctylamine (NH$_2$CH$_2$(CF$_2$)$_6$CF$_3$) (0.1mM), dansyl amine (15 mg/6 mL), N-6-(Aminohexyl)-1-pyrenebutanamide (15 mg/6 mL) or modified, primary amine terminated, p-methyl red (15 mg/6mL). After the film was removed from each derivitization solution it was washed with ethanol and dried with a stream of nitrogen and then extracted for 15 h using CH$_2$Cl$_2$ in a Soxhlet. The films were dried under vacuum for 15 min at 0.1 atm.

Derivatization of the 1PAA or 3PAA films with the fluorous polyacrylate copolymer consisting of a fluorous acrylate and N-acryloxy succinamide (TAN-c-NASI) was accomplished by first reacting the hydrolyzed surface with a diamine such as hexane diamine or diamine terminated polyethylene glycol(NH$_2$-PEG$_{2,000}$-NH$_2$). The aminated surfaces were allowed to react with the copolymer in a solution of α,α,α-trifluorotoluene at 70 $^\circ$ for 5 h. The films were sonicated for 15 min in 1,1,2-trichlorotrifluoroethane to remove physisorbed TAN-c-NASI.

Deprotonation of the oxidized or hydrolyzed films was accomplished by stirring the film in NaOH/EtOH for 1h. The film was then washed with ethanol and dried with a stream of nitrogen, then dried under vacuum for 1 h at 0.1 atm. The films were successfully reprotonated using HCl/EtOH.

Solutions of poly-D-lysine, a β-cyclodextrin with primary amines or 4-poly(amidoamine) (PAMAM) dendrimer in pH 7.0-7.2 buffer were made. When the 3PAA films were allowed to stand in the solutions for 30 min. and then removed, washed with ethanol and dried with a stream of nitrogen, the entrapment of polycationic salts in the films was accomplished.

To form a fluorous monolayer at the base of a hyperbranched film, the first graft on the polyethylene surface was accomplished by putting the film in a DMF solution of NH$_2$CH$_2$(CF$_2$)$_6$CF$_3$ (0.1 M) for 15 min. Then the film was allowed to react with a DMF solution of NH$_2$-PTBA-NH$_2$ (250 mg/6 mL) for 1 h. After this treatment, the film was washed with ethanol and dried with nitrogen, then extracted using CH$_2$Cl$_2$ in a Soxhlet for 15 h. The film was dried under vacuum at 0.1 atm. The normal hyperbranched grafting was then continued with hydrolysis, activation and grafting to form a 3 PTBA layer. After each grafting step a piece of the film was removed and set aside so there was a piece of film to be investigated by XPS for the 1PTBA, 2PTBA and 3PTBA layers.

Results and Discussion:

Once the α,ω-diaminopoly (*tert*-butyl acrylate) was synthesized using the chemistry described (equation 1), it was characterized using ^{1}H-NMR spectroscopy, GPC, FT-IR spectroscopy and titration. The precipitation of the polymer was designed to remove any low molecular weight material so that the growth of layers on the polyethylene surface would be consistent. The polydispersity of the polymer was 1.8 as determined by GPC and the M_n was 15,000 as determined by titration with 0.01M NaOH. The ^{1}H-NMR spectrum indicated that no monomer was present. The FT-IR spectrum indicated amidation of the diacid polymer had occurred based on the introduction of an amide I peak at 1650 cm^{-1}. The dominant peak was the ester peak at 1730 cm^{-1}. In the equation below, the polymer is drawn as a diacid (or diamine) based on termination through coupling. Some monoacid or monoamine polymer is also probably present.

$$\underset{(CH_3)_3CO}{CH_2=CH-CO_2C(CH_3)_3} \xrightarrow[\substack{1,4\ dioxane \\ 120\ ^\circ C\ 20h}]{\substack{4'4\text{-Azobis-} \\ (4\text{-cyanovaleric acid})}} HOOC(CH_2)_2C(CN)(CH_3)\text{-}(CH_2CH)_n\text{-}C(CH_3)(CN)(CH_2)_2COOH$$

$$\downarrow \substack{1\ \ 1,1'\text{-carbonyldiimidazole} \\ 5h\ \ CH_2Cl_2 \\ 2\ \ NH_2(CH_2)_2NH_2 \\ 15h}$$

$$NH_2(CH_2)_2NHCO(CH_2)_2C(CN)(CH_3)\text{-}(CH_2CH)_n\text{-}C(CH_3)(CN)(CH_2)_2CONH(CH_2)_2NH_2$$

Equation 1.

Once the α,ω-diaminopoly(*tert*-butyl acrylate) was fully characterized it was covalently bound to the oxidized polyethylene film using the chemistry previously used for hyperbranched grafting on gold (scheme 1).[2] Spectroscopic studies indicated that the chemistry shown in this scheme occurs.

In the ATR-IR spectrum of the oxidized film, the carbonyl peak is present at 1710 cm^{-1}; in the grafted film the carbonyl peak is at 1728 cm^{-1}. The unoxidized film is hydrophobic (θ_a=105± 3 °) as is the NH$_2$-PTBA-NH$_2$ grafted film (θ_a=100-125± 3 °), but the hydrophilic oxidized film has a contact angle of θ_a=65± 3 °. The error of the contact angle measurements is assumed to be ± 3 degrees. The growth of the ester and amide carbonyls was observed by ATR-IR spectroscopy from layer to layer during the synthesis. A corresponding change in concentration of oxygen and carbon was seen in XPS. The ATR-IR spectrum of the activated film also contained anhydride peaks at 1805 and 1780 cm^{-1}, a further indication that the synthesis was proceding as expected.

Several experiments help establish that hyperbranched covalent grafting occurs from the surface. First, deprotonation/protonation of the oxidized/ hydrolyzed film leads to the disappearance and reappearance of the acid carbonyl signal at 1710 cm^{-1}. The contact angle also changes in this experiment from θ_a=65 ° (protonated) to 24 ° (deprotonated). The protonated and deprotonated 3PAA layer have contact angles, θ_a=48 ° and 15 ° respectively and similar reversible changes in their IR spectra are seen. A second experiment was conducted to investigate whether the coverage of the film was complete. In this experiment, the original oxidized film was first fluorinated and then sequential grafts up to a 3PTBA layer then covered these fluorocarbon groups. The concentration of fluorine on the surface diminished by 20 percent from the first layer to the second and then fluorine was barely detectable in the third layer using XPS.

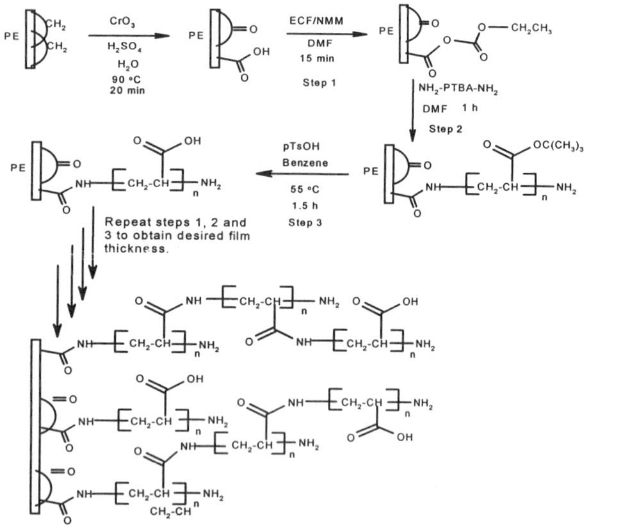

Scheme 1. Synthesis of 3PAA layer using activation, grafting and hydrolysis

As with the gold hyperbranched films, the polyethylene films can be derivatized by various amine terminated molecules including fluorous amines and copolymers as well as colored and fluorescent dyes. The resulting fluorous derivatized films are very hydrophobic, with advancing water contact angles of 135 °. The colored and fluorescent films are evenly colored and evenly fluorescent. The *p*-methyl red modified films show an increase in color intensity. The 1PAA layer has the lowest dye concentration and therefore has the least intense color, whereas the 3PAA film has the most intense color.

The 1PAA and 3PAA layers have also been derivatized using the fluorous copolymer TAN-c-NASI. The appearance of C-F peaks in the ATR-IR spectrum at 1240 and 1200 cm-1 confirmed the derivatization of the film. The contact angle of this film increased to θ_a=145 ° and the fluorine content of the film increased while the oxygen content decreased as measured by XPS. As a control experiment, 6-aminohexanol was attached to the 1PAA and 3PAA films and the resulting films were then allowed to react with the TAN-c-NASI polymer under the same conditions as were used for the diamine modified hyperbranched grafted films. There was no change in the ATR-IR spectrum after reaction and the contact angle did not increase. This control experiment indicates that the amine functionalized films are covalently bound to the fluorous copolymer and the fluorous copolymer is not merely entrapped in the film.

The ATR-IR spectrum of the hydrolyzed film was distinctly different from the non-hydrolyzed film with broadening and shift of the carbonyl and loss of the C-O-C stretch at 1152 cm^{-1} (figure 1). This is another illustration that grafting is taking place on the film surface and the polymer is not physisorbed to the polyethylene.

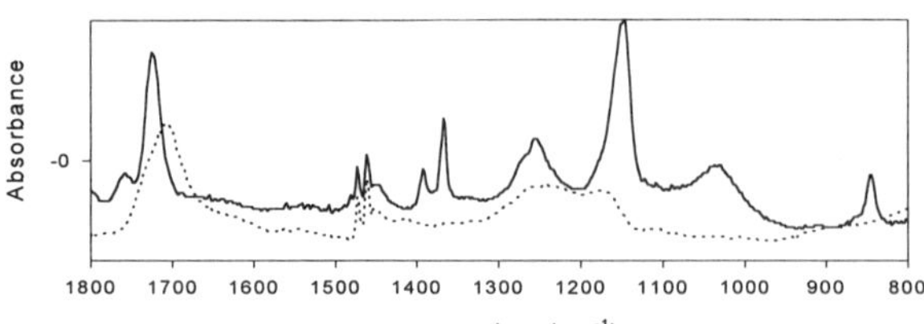

Figure 1. ATR-IR spectrum of 3PTBA film (top) and 3PAA film (bottom), comparing the carbonyl region (1650-1750 cm^{-1}) and C-O-C (1152 cm^{-1}) stretch

Entrapment of polycationic molecules has been reported for hyperbranched films on a gold surface.[8] This phenomenon has been investigated by liquid ellipsometry. In the case of hyperbranched poly(acrylic acid) grafts on gold, it is known that the film thickness of the hyperbranched film increases substantially upon deprotonation and when polycationic molecules are introduced, the increase in thickness can be maintained in the dry state. While, polyethylene can not be studied using ellipsometry, similar intercalation still seems to occur based on ATR-IR spectroscopy. The reversible changes in the carbonyl region in the 1710 cm^{-1} to 1560 cm^{-1} range in the ATR-IR spectrum of 3PAA/PE suggest the 3PAA grafts on polyethylene behave like similar grafts on gold. When a 3PAA/PE film is put in to a pH 7 solution with the PAMAM dendrimer, beta-cyclodextrin or poly-D-lysine it appears that the polycationic molecules are incorporated, based on ATR-IR spectroscopy. The resulting films' ATR-IR spectra show amide and amine bands and in the case of the cyclodextrin, the C-O-C stretching bands increase. These results are expected due to the incorporation of macromolecules that have a large number of amides, amines or ether bonds. The incorporation of the salts in the film is also supported by XPS.

Conclusion:

Hyperbranched grafting using α,ω-diaminopoly(*tert*-butyl acrylate) can be utilized to build robust thin films on an oxidized polyethylene surface. The hyperbranched film can be derivatized in a number of ways. Once derivatization occurs with amine terminated molecules, predictable changes were seen in the contact angle, the ATR-IR spectrum and in the film's XPS spectrum. The completeness of the graft coverage has been investigated and it appears that that grafting procedure covers the polyethylene surface completely.

References:
(1) Swalen, J.D.; Allara, D.L.; Andrade, J.D.; Chandross, E.A.; Garoff; S.; Israelachvili, J.; McCarthy, T.J; Murray, R.; Pease, R.F; Rabolt, J.F; Wynne, K.J.; Yu, H. *Langmuir* **1987**, *3*, 932-950.
(2) Bruening, M.L.; Zhou, Y.; Aguilar, G.; Agee, R.; Bergbreiter, D.E; Crooks, R.M. *Langmuir* **1997**, *13*, 770-778. Zhou, Y.; Bruening, M.L.; Bergbreiter, D.E; Crooks, R.M; Wells, M. *J. Am. Chem. Soc* **1996**, *118*, 3773-3774.
(3) Decher, G. *Science* **1997**, *277*, 1232-1237.
(4) Blodgett, K. *J. Am. Chem. Soc.* **1935**, *57*, 1007-1022.
(5) Lui, Y.; Bruening, M.; Bergbreiter, D.E.; Crooks, R.M *Angew. Chem. Int. Ed. Engl.* **1997**, *36*, 2114-2120. Lui, Y.; Zhao, M.; Bergbreiter, D.E.; Crooks, R.M. *J. Am. Chem. Soc* **1997**, *119*, 8720-8721.
(6) Bergbreiter, D.E.; Gray, H.N.; Srinivas, B. *Macromolecules* **1994**, *27*, 7294-7301. Bergbreiter D.E; Xu, G-F; Zapata Jr., C. *Macromolecules* **1994**, *27*, 1597-1602.
(7) Holmes-Farley, S.R.; Reamy, R.H.; Nuzzo, R.; McCarthy, T.J.; Whitesides, G.M. *Langmuir* **1987**, *3*, 799-815.
(8) Peez, R.F.; Dermody, D.L.; Franchina, J.G.; Jones, S.J; Bruening, M.L.; Bergbreiter, D.E; Crooks, R.M submitted to *Langmuir* **1998**, *14*.

Nanporous Ceramic Coated Structures from Block Copolymers

Vanessa Z-H Chan and Edwin L. Thomas
Department of Materials Science and Engineering, Massachusetts Institute
of Technology, Cambridge, MA
Victor Lee and Robert D. Miller,
IBM Almaden Research Center, San Jose, CA
Apostolos Avgeropoulos and Nikos Hadjichristidis
Department of Chemistry, University of Athens, Athens, Greece

Introduction

The formation of nanoporous structures from block copolymers by ozonolysis and reactive ion etching, has been investigated by many groups recently. In these studies, the polymers under investigation have been hydrocarbon based, typically polystyrene-polydiene, where the polyisoprene is selectively removed by ozone [1,2] or the double bond in the polyisoprene block is reacted with a heavy metal to provide etch selectivity with the polystyrene block to an oxygen plasma. [3] Applications as nanolithographic templates and as templates for nickel catalysts have been envisioned by these researchers.

Researchers have also investigated the use of silicon-containing polymers as bilevel resists for lithography. Gabor et al. have investigated the hydrosilylation of a number of poly(styrene)-poly(butadiene) or poly(styrene)-poly(isoprene) block copolymers. [4] The O_2-RIE etch rate selectivity of the silylated block relative to polyimide was impressive (42 : 1). However, the stability of the functionalized butadiene copolymers was limited and the hydrosilylation of the isoprene phase of the block copolymers could not be driven to completion.

In this paper, we report the preparation of ceramic nanostructures from a silicon-containing block copolymer precursor. The advantages of this material over those used by previous researchers is their potential to be converted to a ceramic due to the silicon present in the pentamethyldisilylstyrene, P(PMDSS) block. Since the silicon is intrinsically present in the monomer, the etch selectivity is also intrinsic in the block copolymer, so no post-polymerization chemistry is necessary.

Experimental

The block copolymer used in this study is a triblock copolymer, PI-P(PMDSS)-PI, that forms the double gyroid morphology (cubic space group Ia3d) at equilibrium. The block copolymer was polymerized by living anionic polymerization and its synthesis is reported elsewhere. [5] The PI-P(PMDSS)-PI triblock has a weight fraction of PI block chains of 33.8%, a molecular weight of 24K-98K-26K and a polydispersity of 1.12. The P(PMDSS) homopolymer used in this study was prepared by similar means and had a molecular weight of 38K and a polydispersity of 1.06.

The triblock was bulk cast by making 5% by weight solutions of the triblock in toluene and allowing the solvent to slowly evaporate over a one week period. The sample was then annealed at 120°C in a vacuum oven for two weeks and cryo-microtomed.

Unstained sections of the 24/98/26 PI/P(PMDSS)/PI as prepared by microtoming, were exposed to a flowing 2% ozone atmosphere for one hour at room temperature in order to preferentially oxidize the polyisoprene block. The PI-oxidized fragments were then removed by soaking the samples in deionized water overnight and the samples examined under TEM at 100 keV.

Oxygen reactive ion etching was performed on the P(PMDSS) homopolymer. 5% by weight solutions of the homopolymer was made in toluene and spun onto 1" silicon wafers still containing their native oxide. The etching experiments were carried out in a low pressure, magnetically enhanced inductively coupled plasma etcher. Etching was performed with a base pressure of 10 mTorr and an oxygen flow rate of 40 sccm. The top r.f. generator was set at 250 W and the bottom r.f. generator at 50 W. Etching was carried out at 0° C for 60 seconds.

The atomic concentrations of silicon, carbon and oxygen were estimated by X-ray photoelectron spectroscopy. Analysis was done on an ESCA spectrometer with a monochromatic Al K-alpha source. 1000 eV survey spectra were recorded at 150 eV pass energy with a 1000 μm spot size for each sample. High resolution spectra (50 eV pass energy, 20-30 eV window) were recorded for carbon, oxygen and silicon.

Variable angle XPS studies of the samples were conducted in order to estimate possible surface contamination. Data were taken at two detector-to-sample angles : 10°and 60°. The 10° data provide information on approximately the top 15 Å of the film, while the 60° data sample the top 75Å.

Depth profiles of the C, Si and O concentrations were obtained by Auger electron spectroscopy (AES) using a commercially available scanning Auger microprobe. The electron energy was 10 keV and the electron current was 1 A. Sample sputtering for depth profiling was accomplished with a 3 keV argon ion beam over a 2 x 2 mm² raster.

Results and Discussion

When thin films of a bulk sample containing the tricontinuous morphology were exposed to ozone, the polyisoprene block was preferentially removed. The contrast in Figure 1 arises from the selective removal of the PI regions. The structure seen in the bulk morphology is preserved after exposure to ozone.

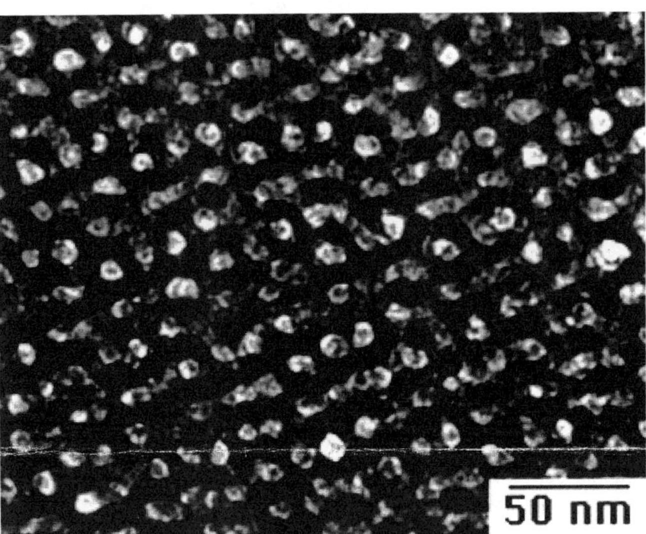

Figure 1 : Bright-field TEM of the unstained double gyroid sample after exposing to an ozone atmosphere for one hour. Contrast is due to the removal of the PI channels.

The chemical composition of both P(PMDSS) before and after etching was examined by X-ray photoelectron spectroscopy, (XPS). The binding energy of the Si2p peak was monitored closely as it reflects the chemical environment of the silicon. From literature, it is known that the binding energy of Si in a hydrocarbon environment lies between 100.5-101.8 eV and in an environment of SiOx the binding energy is at 103.5 eV.

By XPS it was shown that in the unetched samples, the silicon exists mainly in a carbon-rich environment, as seen by the atomic concentration of silicon species with a binding energy peak centered at 100 eV, in Figure 2. In addition, there is evidence of some SiO_x on the surface of the as-cast polymer. The oxygenated species appears to be surface-localized as the relative amount of Si in this environment measured at lower detection angles (10°) is larger than that at higher detection angles (60°). Once the films are exposed to an oxygen plasma, the amount of carbon decreases significantly as expected since carbon in the polymer is removed by the oxygen plasma. This is reflected by the dramatic increase of the silicon/carbon ratio with etching and by depth profiling in AES. The most obvious change in the silicon XPS spectra upon etching is the shift in the Si2p binding energy from 100.5/101.8 to 103.5 eV, indicating that the silicon near the surface of the film is converted to SiO_x. When the data for the etched sample is modeled as a thin surface layer of SiOx on top of an underlying carbon-rich film (i.e.

all detected oxygen is silicon-bound, not carbon-bound), a ratio of O/Si of 1.8 is calculated at the outer surface for the etched samples.

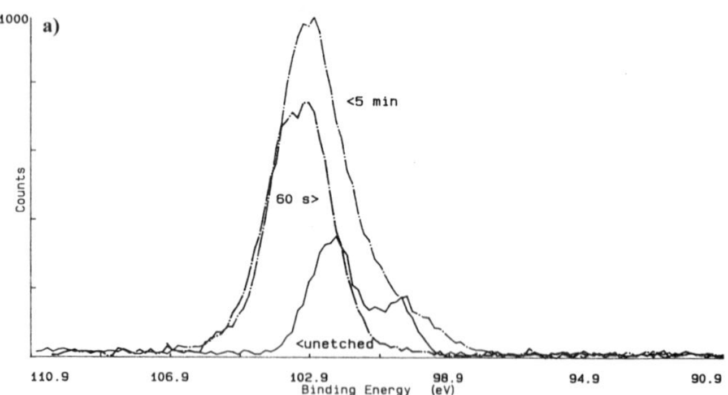

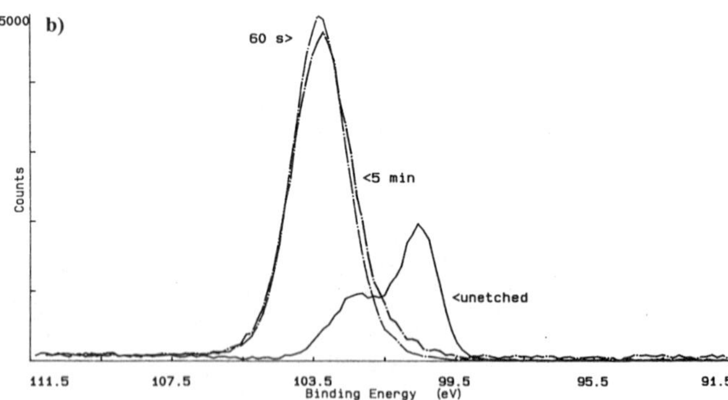

Figure 2: X-ray photoelectron spectroscopy data of P(PMDSS) before and after oxygen reactive ion etching at 0°C. Data taken at sample-to-detector angle of a) 10° b) 60°

The AES data obtained for P(PMDSS) etched for 60 seconds at 0°C are consistent with the XPS results. Depth profiling (argon ion sputtering), reveals that the ratio of silicon to oxygen decreases from that of SiO_2 to the elemental distribution of the unetched polymer.

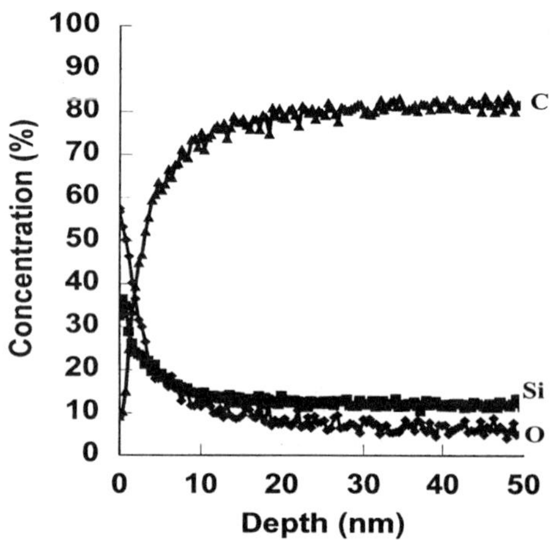

Figure 3 : AES Spectra of P(PMDSS) etched for 60 seconds at 0°C

The surface free energies as calculated from contact angle measurements on P(PMDSS) before and after reactive ion etching for 60 seconds at 0°C were 42.0 ergs/cm^2 and 70.8 ergs/cm^2 respectively. These energies corroborate the XPS and AES data: the higher surface free energy of the RIE treated sample is consistent with an increase in surface polarity upon conversion of the silicon in the polymer to silicon oxide. For comparison, literature values for the surface free energy of polystyrene and silica are 34 ergs/cm^2 and 78 ergs/cm^2 respectively. [6]

Thus, by exposing this block copolymer to an oxygen plasma after ozonolysis, it should be possible to convert the silicon in the P(PMDSS) matrix to silica thereby forming a 3-D nanoporous ceramic from a polymer precursor. More extensive studies are being carried out on the conversion of the silicon in the P(PMDSS) matrix to SiOx by either O_2-RIE and/or ozone.

Conclusion

We have shown that it is possible to produce porous 3-D nanostructures from block copolymers which can ultimately be converted to a ceramic (SiOx) by a combination of ozone and O_2-RIE. The use of these materials as low dielectric constant materials, membranes and photonic band gap materials is currently under investigation.

Acknowledgments

The authors would like to thank Dolores Miller for her help with XPS, Vaughn Deline for his help with AES and George Tyndall with his help on contact angle measurements. The authors would also like to acknowledge CPIMA (Center for Polymer Interface and Macromolecular Assemblies) and AFOSR-AASER F49620-97-1-0385 for partial funding of this research. VZC would also like to thank the National Science Foundation and IBM for graduate research fellowships.

References

[1] Mansky P., Harrison, C.K., Chaikin P.M., Register R.A., Yao, M., *Appl. Phys. Lett.,* 68 (18), 2586, 1996.
[2] Hashimoto T., Tsutsumi K., Funaki Y., *Langmuir,* 13, 6869-6872, 1997.
[3] Park M., Harrison C., Chaikin P.M., Register R.A., Adamson D.H., *Science,* 276, 1401, 1997.
[4] Gabor A.H., Ober C.K., "Microelectronics Technology: Polymers for Advanced Imaging and Packaging", ACS Symposium Series, 614, Chapter 19, 1995.
[5] Avgeropolous A., Chan V.Z-H., Lee V.Y., Ngo D., Miller R.D., Hadjichristidis N., Thomas E.L., accepted to Chemistry of Materials.
[6] Arklas B., "Tailoring Surfaces with Silanes", ChemTech, Critical Surface Tension, Vol. 7, 1977.

Siloxane Functionalized Isocyanurate Coatings

Hai Ni, Allen D. Skaja, Robert A. Sailer and Mark D. Soucek

Department of Polymers and Coatings
North Dakota State University, Fargo, ND 58105

Introduction

Polyurethane and Polyurea coatings are widely used in the furniture, automobile, and aerospace industries. Polyurethane or polyurea coatings are derived from reaction of an isocyanate with a hydroxyl functionalized oligomer or a amino functionalized oligomer respectively. Isocyanate groups can also react with water to afford "moisture curing polyurea coatings". The chemistry of the moisture curing process is depicted in equations 1-3. After reaction with water, an unstable carbamic acid intermediate is formed which spontaneously decarboxylates into an amine. This amine reacts with   another isocyanate group to form the urea crosslink.[1] The reaction rate is controlled by temperature, humidity and catalysts.

$$R\text{-}NCO + H_2O \longrightarrow R\text{-}\underset{\underset{H}{|}}{N}\text{-}\overset{\overset{O}{\|}}{C}\text{-}OH \qquad (1)$$

$$R\text{-}\underset{\underset{H}{|}}{N}\text{-}\overset{\overset{O}{\|}}{C}\text{-}OH \longrightarrow R\text{-}NH_2 + CO_2 \qquad (2)$$

$$R\text{-}NH_2 + R\text{-}NCO \longrightarrow R\text{-}\underset{\underset{H}{|}}{N}\text{-}\overset{\overset{O}{\|}}{C}\text{-}\underset{\underset{H}{|}}{N}\text{-}R \qquad (3)$$

Alkoxy silane groups also have a moisture curing mechanism as depicted in equations 4-6. The alkoxy group can react with water to form a silanol and an alcohol. The silanol group can then condense to form siloxane groups with elimination of water or alcohol.

$$\equiv Si\text{-}OR + H_2O \longrightarrow \equiv Si\text{-}OH + ROH \qquad (4)$$

$$\equiv Si\text{-}OH + \equiv Si\text{-}OH \longrightarrow \equiv Si\text{-}O\text{-}Si \equiv + H_2O \qquad (5)$$

$$\equiv Si\text{-}OR + \equiv Si\text{-}OH \longrightarrow \equiv Si\text{-}O\text{-}Si \equiv + ROH \qquad (6)$$

Further hydrolyzing and condensation reaction forms silicon oxide network producing silicon ceramer. The processes are determined by many factors including pH, temperature and solvent. Brinker and Scherer[2] summarized the effects of these factors on the hydrolysis and condensation processes, and concluded that the dominant factor was pH. Alkoxy silane hydrolysis is very slow in neutral condition, however it can be accelerated either with acid or base catalyst.

Organofunctional silanes, R'-Si-(OR)$_3$ possess a siloxane group –Si-(OR)$_3$ and a organic group R', which have been used as coupling agent for composites, and adhesion promoter in coatings.[3-5] Alkoxy silane primers have been applied to metal substrates including aluminum to improve the corrosion resistance. Walker[4] used organofunctional silanes such as n-beta aminoethyl aminopropyl-trimethoxysilane and γ-glycidoxypropyltrimethoxy-silane were as pretreatment primers for epoxide and polyurethane coatings. Initial, wet and recovered adhesion on aluminum was dramatically increased. Witucki[5] summarized the utilizing methods of organofunctional silane as a primer: (1) surface treatment, (2) additive into coating, (3) reactive intermediate for silicone resin synthesis and organic resin modification. Surface treatment and additive of are effective in adhesion enhancement for many coating system.

Alkyl silane functionalized urethane prepolymer was synthesized through aminosilane and isocyanate terminated urethane oligomer by Surivet.[6] The hydrolysis and condensation mechanism was investigated using ^{29}Si NMR and GPC. Similar to sol-gel precursor, the siloxane attached oligomer can hydrolyze and condense to form polymer. New alkoxy silane functionalized isocyanate trimer was prepared in our group using 1,6-hexamethylene diisocyanate (HDI) isocyanurate and an aminosilane. Both isocyanate and alkoxy silane groups of the functionalized HDI isocyanurate crosslinked via a moisture curing mechanism (eq.1-6).

Experimental

The HDI isocyanurate was obtained from Rhone Poulenc Inc. and Bayer Corporation. The 3-aminopropyltriethoxysilanne (3-APTES) was obtained from Aldrich, and was used without further purification. All solvents were dried over molecular sieves. A bare aluminum, alloy 3003 H14 was obtained from Q-Panel Lab Products.

The mono-functionalization reaction of HDI isocyanurate with 3-APTES is shown as follows (eq.7). The molal ratio is 1:1 and the reaction carried on under nitrogen at ambient temperature. Mono-functionalized isocyanurate (1) further reacted with methanol (molal ratio, 1:2) producing model compound (2) as shown in eq.8.

The resultant product (2) was characterized by ^{1}H NMR, δ 2.86 ppm (-CONHC$\underline{H}_2$-) and δ 3.35 ppm (-COOC$\underline{H}_3$) as shown in Figure 1, ^{13}C NMR, δ 148.8 ppm (=N$\underline{C}$ON=), δ 157.1 ppm (-NH$\underline{C}$OOCH$_3$) and δ 158.7 ppm (-NH$\underline{C}$ONH-) as shown in Figure 2, ^{29}Si NMR, δ -45.0 ppm (CH$_2$$\underline{Si}$(OC$_2H_5$)$_3$) as shown in A of Figure 3.

Model compound (2), 1.00g was dissolved in ethanol, 3.00g. Water (mole ratio, H$_2$O/Si = 0.7) and catalyst, p-Toluenesulfonic acid (mole ratio, H$^+$/Si = 0.05) were added into the solution, and the solution was kept at 25°C for 18hours and then concentrated with vacuum and characterized with ^{29}Si NMR as shown in B of Figure 3.

The coating was prepared through mixing HDI isocyanurate and mono-functionalized isocyanurate. The films were cased on aluminum panels.

Results and Discussion

Model Compound Study: The monofunctionalized reaction was characterized previously using ^{13}C NMR.[7] The ^{1}H NMR in Figure 1 showed that the new group –COOCH$_3$ (δ 3.35 ppm) was attached to 1 and ^{13}C NMR in Figure 2 indicated that the bonding formation of –NHCOOCH$_3$ (δ 157.1 ppm).

The peak at –52.8 ppm was assigned to the dimer[3] with the structure as below, which meant the siloxane group of 2 can hydrolyze and condense to form inorganic portion in coating film. The isocyanate group of 2 provides linkage with polyurea through reactions (1) and (2).

$$-H_2C\text{-}\underset{\underset{OR}{|}}{\overset{\overset{OR}{|}}{Si}}\text{-}O\text{-}\underset{\underset{OR}{|}}{\overset{\overset{OR}{|}}{Si}}\text{-}CH_2\text{-} \qquad R = C_2H_5$$

Mechanical Properties: The sol-gel coating films were prepared using functionalized isocyanurate and HDI isocyanurate. The adhesion was well improved through the siloxane group as indicated in Table 1. The other properties do not change much. Functionalized isocyanurate modified the interface between film and substrate significantly; but siloxane content is very small and the other properties were controlled by polyurea.

Rheology: It can be from Table 2 that crosslink density increased for functionalized isocyanurate over the parent isocyanurate. This was caused by the additional two functionality of functionalized isocyanurate. The Tg and storage modulus minimum, E'$_{minimum}$ are a function of crosslink density. The polyurea provide a high crosslink density, and thus excellent barrier property for corrosion resistance. The transition of the elastic modulus (E') after Tg in

Figure 4 may be caused by crystallinity or other factors, which will be investigate in future work.

Corrosion Resistance: The corrosion resistance tests are being investigated using Electrochemical Impedance Spectroscopy (EIS).

Table 1. Mechanical Properties

Functionalized Isocyanurate /Isocyanurate (w/w)	Crosshatch Adhesion	Pull-off Adhesion, lb_f/in^2	Impact resistance, in./lb	Tukon hardness KHN
0/100	0B	47.5±17.1	24.3±1.5	10.3±0.9
10/90	5B	173.3±65.6	20.8±4.6	12.0±0.2
25/75	5B	213.3±89.4	21.8±3.1	11.3±0.3
35/65	5B	227.5±76.8	22.5±4.8	10.8±0.6

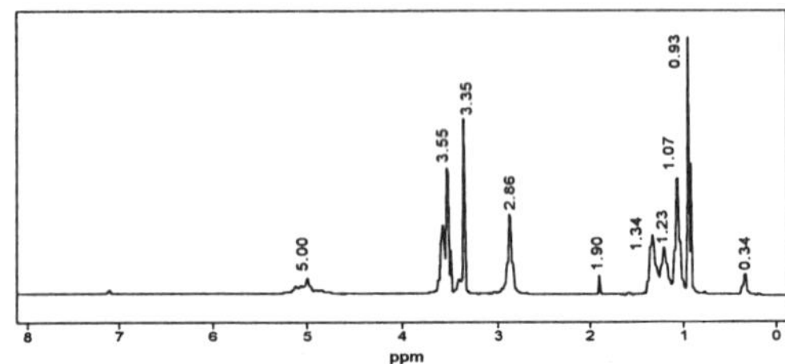

Figure 1. ^{1}H NMR Spectrum of **2**.

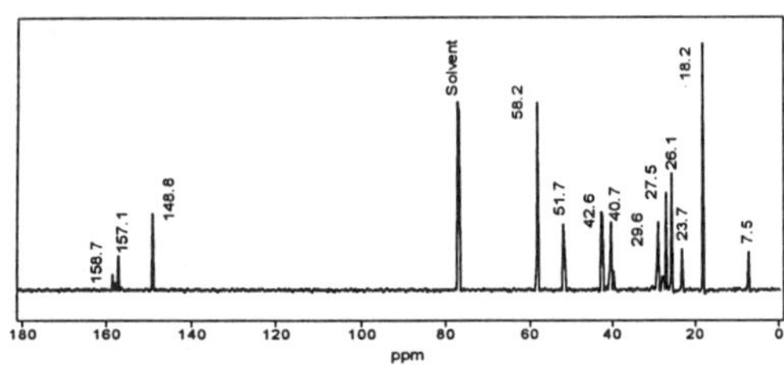

Figure 2. ^{13}C NMR Spectrum of **2**

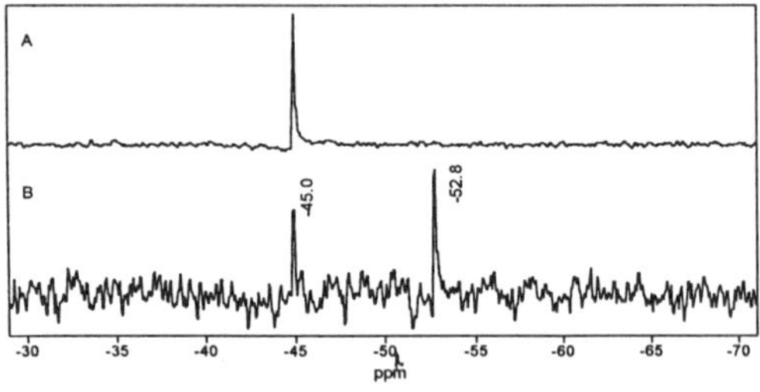

Figure 3. ^{29}Si NMR spectra; (A): 3-aminopropyltriethoxysilane, (B): hydrolysis and condensation compositions 18h.

Conclusion

(1) Model compound study indicated that siloxane group of functionalized isocyanurate can hydrolyze and condense producing inorganic portion. (2) Siloxane functionalized isocyanurate enhanced adhesion significantly. (3) Crosslink density was increased by functionalizing isocyanurate with an alkoxy silane

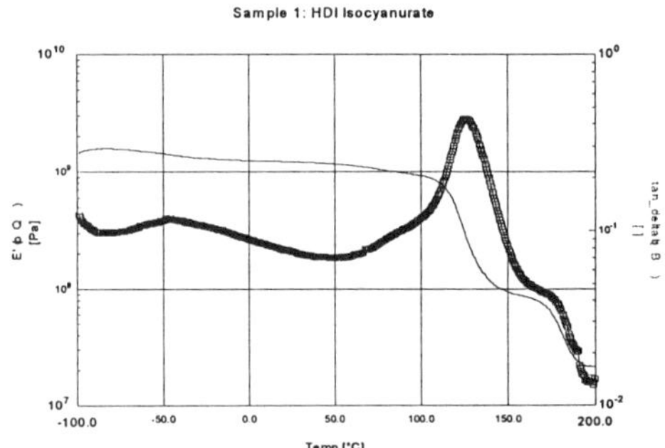

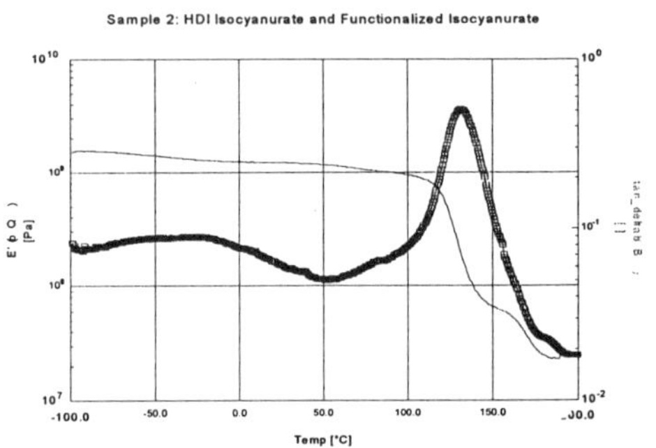

Figure 4. DMTA Data of Coating Film. Sample 1, Pure HDI Isocyanurate; Sample 2, 25% Functionalized Isocyanurate (**1**) and 75% HDI Isocyanurate.

Table 2: Viscoelastic Properties

Sample*	Tg (°C)	E'x 10^{-8} (Pa) (Minimum)	Crosslink Density x10^{-3}(mole/m^3)
1	127	0.21	1.74
2	130	0.22	2.29

*Samples 1 and 2 are the same as Figure 4.

Acknowledgements

Financial support for this research from Air Force Office of Scientific Research is gratefully acknowledged.

References

(1). Wicks, Z. W.; Jones, F. N.; Pappas, S. P. *Organic Coatings Science and Technology, Volume I: Film Formation, Components, and Appearance;* John Wiley and Sons, Inc.: New York, **1992**; PP189-191.

(2). Brinker, C. J.; Scherer, G. W. *Sol-Gel Science;* Academic Press: New York, **1990**; Chapter 3.

(3). Wicks, Z. W.; Jones, F. N.; Pappas, S. P. *Organic Coatings Science and Technology, Volume II: Application, Properties, and Performance;* John Wiley and Sons, Inc.: New York, **1992**; PP163-164.

(4). Walker, P. *J. Coat. Technol.* **1980**, 52(670), 49.

(5). Witucki, G. L. *J. Coat. Technol.* **1993**, 65(822), 57.

(6). Surivet, F. S.; Lam, T. M.; Pascault, J. P.; Pham, Q. T. Macromolecules **1992**, 25, 4309.

(7). Ni, H.; Skaja, A. D.; Sailer, R. A.; Soucek, M. D. *Polymer Preprint* 1998, 39(1), 367 (Papers presented at Dallas, Texas).

SURFACE COVERAGE, RHEOLOGY AND MORPHOLOGY OF COPOLYMER MODIFIED PP/ PS BLENDS

Himanshu Asthana and K. Jayaraman
Department of Chemical Engineering
Michigan State University
E. Lansing MI 48824

INTRODUCTION

Polymer blends are frequently compatibilized by the addition of a block copolymer or by an in-situ reaction between functionalized polymers. Compatibilization refers to producing a dispersion with smaller particles; this phenomenon has been associated with two mechanisms: a reduction in interfacial tension as well as stabilization against coalescence[1,2]. Compatibilization by external agents also has a significant impact on the rheology of the blends[3,4]. Brahimi et al.[3] report that the rheology of the copolymer-modified blend responds in a non-monotonic fashion to amount of the block copolymer. But the system used by Brahimi et al. had HIPS as a component, which by itself has a complex morphology. Germain et al.[4] did not vary the amount of copolymer added but observed that depending on the shear rate, the block copolymer may serve as a lubricant or make the disperse phase more rigid.

It is possible to infer interfacial tension from the dynamic moduli of the molten polymer blend, using measured particle sizes, with the help of a theory developed by Palierne[5] for linear viscoelasticity of emulsions with nearly spherical particles. The task is easier if the viscosity ratio and relaxation times of the components are chosen carefully. This approach has been used to study the connection between morphology and rheology for a variety of blends by Graebling et. al.[6], Graebling et al.[7], and by Asthana and Jayaraman[8]. Brahimi et al.[3] have concluded that for their compatibilized blends consisting of HIPS as the matrix, HDPE as the disperse phase and a compatibilizer, the Paliene theory was unsatisfactory.

The objectives of the present study are to compare the rheology and morphology of copolymer modified PP/PS blends with increasing amounts of S-EP block copolymer and attempt to determine trends in interface mobility in these systems. The trends in rheology for this system will be compared with trends observed for another reactively compatibilized system[8].

MATERIALS AND METHODS

Blends were prepared with polypropylene (Profax 6523, Montell Polyolefins, M_n 54000) as the matrix, polystyrene (Polycom Huntsman, M_n 220000) as the dispersed phase and -S-EP- di block copolymer (Kraton G1702, Shell Chemicals) as the compatibilizing agent. Two sets of blends contained 10 wt % PS and 20 wt % PS with the copolymer amount in each set ranging over 0 wt %, 2 wt %, 4 wt % and 6 wt %. The blending was carried out in a ZSK 30 twin screw extruder at a temperature of 180 °C for all the zones.

Rheological characterization was carried out on the RMS-800 rheometer using a 50 mm parallel plate arrangement. The instrument was purged with dry nitrogen during measurements to minimize the effects of degradation. All the measurements were carried out at strains of 10-15% and a temperature of 180 °C.

The samples for morphology were prepared by placing the pre-notched disks in the oven of the RMS 800, heating them to 180°C and placing them directly in liquid nitrogen. Those were fractured under liquid nitrogen and examined in a Phillips Electroscan 2020 environmental scanning microscope. Water vapor at a pressure of 2-3 torr was used as the imaging gas. The size measurements were made directly from the micrographs and the volume average radii were calculated.

RESULTS AND DISCUSSION

Blend Morphology

Figs. 1a through 1d show the micrographs for the PP/ PS (90/ 10) blend compatibilized with varying amounts of the compatibilizer. The volume average particle sizes as measured from these micrographs are tabulated below.

Table 1: Dispersed phase sizes in the various blends.

	R, μm
PP/ PS (90/ 10)	1.12
PP/ PS/ 1702 (90/ 10/ 2)	0.9
PP/ PS/ 1702 (90/ 10/ 4)	0.8
PP/ PS/ 1702 (90/ 10/ 6)	0.6
PP/ PS (80/ 20)	1.86
PP/ PS/ 1702 (80/ 20/ 2)	1.0
PP/ PS/ 1702 (80/ 20/ 4)	1.0
PP/ PS/ 1702 (80/ 20/ 6)	0.8

It is seen that the particle size increases in the case of the non-compatibilized blends as the polystyrene content is raised from 10 wt. % to 20 wt. %. This trend can be attributed to greater coalescence at higher concentrations of the disperse phase, in the non-compatibilized blends. The data shows that after compatibilization this effect is reduced. The probability of coalescence is reduced because the mobility of the interface is reduced[2]. The rheology data below allow us to examine the degree of reduction in interface mobility at various levels of block copolymer.

Rheology

Fig. 2 presents a comparison of the storage moduli of the compatibilized and non-compatibilized 90/10 blends. It is observed that the storage modulus of the blend with 2 wt. % SEP (90/10/2) falls below the 4uncompatibilized blend (90/10/0). By increasing the amount of the compatibilizer, it is seen that storage moduli curves for the blend with 4 wt. % SEP (90/10/4) and 6 wt. % SEP (90/ 10/ 6) rise above that of the uncompatibilized blend (90/10/0). In the cases where the storage moduli shift up, the modulus at high frequencies crosses over either component as shown in Fig. 3. This particular feature signifies an immobilized interface. So with the block copolymer as compatibilizer, the interface becomes immobilized only above a significant level of copolymer. In a related study on directly compatibilized blends by in-situ reaction of nylon 6 with maleated polypropylene Asthana and Jayaraman[8] observed that the morphology was stabilized at a very low extent of reaction. The high frequency response also indicated that the interface was immobilized at a very low extent of reaction.

An important effect of adding the compatibilizer is altering the nature of the interface. The compatibilizer diffuses to the 'bare' interface and the blocks of the compatibilizer 'hook-up' with the chains of the matrix and the dispersed phase. This diffusion continues until the interface is saturated with such linkages. Brahimi et. al.[3] observed that the interface with the tapered block copolymer compatibilizer did not get saturated. This led to a continuous fall in the storage modulus curves. For the straight di-block copolymer the interface was saturated at 1% compatibilizer level which led to an increase after the initial decrease. In this study, the storage modulus curve for the 90/ 10/ 2 blend falls compared to the 90/ 10 blend but then further addition of the compatibilizer causes the curves to move up. The reason is that the interface is possibly saturated with the -[S-EP]-copolymer and the interface is fully immobilized.

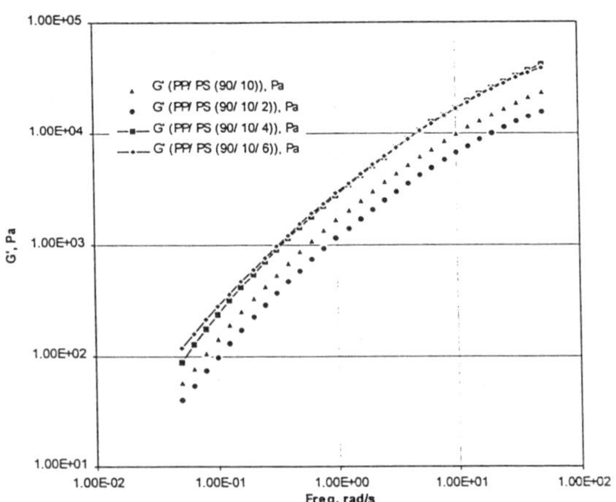

Figure 2: Comparison of the storage moduli of the 90/ 10 PP/ PS blends with 0, 2, 4, 6 wt.% SEP di-block copolymer.

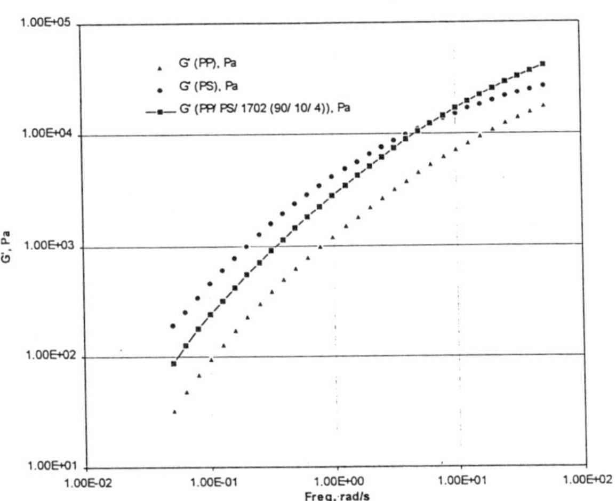

Figure 3: Comparison of the storage moduli of the PP/ PS/ 1702 (90/ 10/ 4) blend with its components.

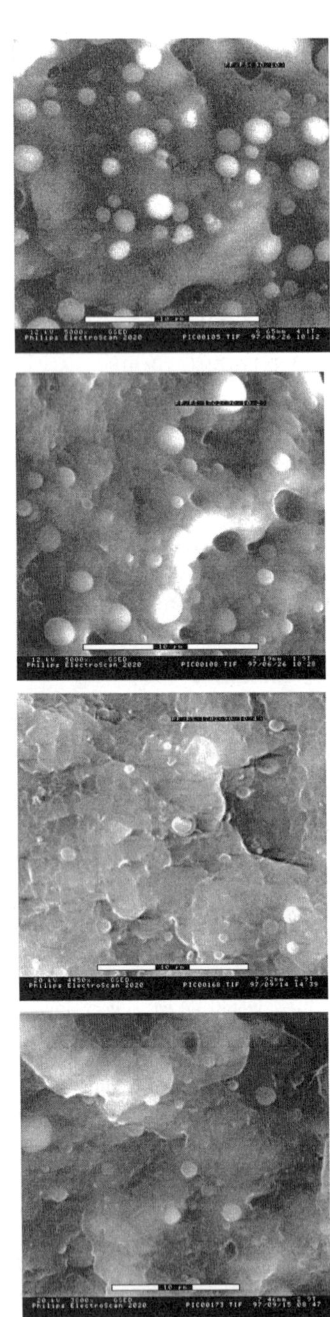

Figure 1a-d (top to bottom): Scanning electron micrographs of 90/ 10 PP/ PS blends with 0, 2, 4, 6 wt.% SEP di-block copolymer.

BIBLIOGRAPHY

1. Macosko C.W.; Guegan P.; Khandpur A.k.; Nakayama A.; Marechal P.; Inoue T. *Macromolecules* **1996**, *29*, 5590-5598.
2. Sundraraj U.; Macosko C.W. *Macromolecules* **1995**, *28*, 2647.
3. Brahimi B.; Ait-Khadi A.; Ajji A.; Jerome R.; Fayt R. *J. Rheology* **1991**, 35(6).
4. Germain Y.B.; Ernst B.; Genelot O.; Dhamani L. *J. Rheology* **1994**, 38(3).
5. Palierne J.F. *Rheologica Acta* **1990**, *29*:204-214.
6. Graebling D.; Muller R. *Colloids and Surfaces* **1991**, 89-103.
7. Graebling D.; Muller R.; Palierne J.F. *Macromolecules* **1993**, *26*, 320-329.
8. Asthana H.; Jayaraman K. *submitted to Macromolecules.*

Polymerization and Dynamic Testing of Difunctional Acrylic Metal Salts in a Polybutadiene Golf Ball Center

Thomas J. Kennedy III

Spalding Sports Worldwide, Research and Development Department, 425 Meadow Street, Chicopee, MA 01021

Introduction

The construction of the golf ball has changed many times from when the first featherie was played on the links of the Old Course at St. Andrews, Scotland. The cover has changed from leather to gutta percha rubber to balata rubber to ionomer blends. The golf ball core has changed from goose feathers boiled in oil to solid gutta percha to a wound center to a solid, rubber center. The golf ball center is the focus of this paper, in particular, the solid golf ball center and the acrylates that form the center's crosslinked structure.

It is well known in the present golf ball art that the modern solid center of a golf ball is a highly crosslinked, high cis polybutadiene rubber that exhibits exceptional rebound at high speed. The co-agent that is utilized in the free radical crosslinking of the polybutadiene is typically zinc diacrylate. The reaction is initiated with heat and a peroxide, such as dicumyl peroxide, with reaction exotherm exceeding 400°F. This paper will discuss experimental data utilizing co-agents other than zinc diacrylate.

The combination of high cis polybutadiene and zinc diacrylate (ZDA) was first utilized in the late 1960's. Zinc dimethacrylate (ZDMA) was also in use at this time, and, in fact, preceded the use of ZDA. ZDMA is much less sensitive to moisture in a rubber formulation than is ZDA. However, in a low moisture system (<0.5%), ZDA was preferable due to the greater resilience that is obtained by cross-linking the rubber with the smaller molecule.

The ZDA reaction process has undergone many refinements from its first use in the late 1960's and is still the current co-agent of choice for solid golf ball centers. ZDA is a condensation reaction product of zinc oxide and acrylic acid. This reaction is usually carried out in an alkane, such as hexane. Co-agents with different metal centers are examined in this paper. Metal oxides that are in the same family as zinc, mercury oxide and cadmium oxide, were not chosen because of their toxicity and cost. Instead, difunctional materials with metal centers from the 2A column in the periodic chart were utilized, magnesium oxide and calcium oxide specifically. These materials were chosen because of their low toxicity and low cost.

A surfactant / mixing aid is incorporated in the reaction to facilitate dispersion of the reactants. The material utilized in this set of experiments is zinc stearate. Zinc stearate is the material of choice due to its compatibility with the polybutadiene rubber formulation.

Experimental

Three separate formulations were produced as co-agents for cross-linking high cis polybutadiene rubber (Fig. 1)

Fig. 1

Co-Agent Formulations

	1	2	3
HEXANE	400	400	400
Zinc Oxide	50		
Magnesium Oxide		50	
Calcium Oxide			50
Zinc Stearate	2.5	2.5	2.5
Acrylic Acid	89	180	129

The metal oxides and acrylic acid were balanced stoichiometrically 1:1. The hexane was charged into a jacketed reactor and the agitator started at a rate of about 350 rpm. The reactor was held at 95°F with a heated water bath. The metal oxide and zinc stearate were added under agitation. Once the powders were dispersed, the acrylic acid was added drop-wise at the rate of two drops per second. This rate was utilized to avoid the build up of reaction product at the site where the acrylic acid was entering the solution.

This reaction was first attempted with out the zinc stearate, or any other surfactant / mixing aid. Large agglomerates, or "rocks" formed on the bottom of the reactor and caused the mixing blade to stop. The addition of the zinc stearate solved this problem. Zinc stearate was chosen due to its compatibility with the rubber formulation.

The reaction proceeded smoothly for all three batches. A slight exothermic rise in temperature was seen (about 4°F). A fluffy, white product was produced. After the addition of the acrylic acid was complete, the agitator was allowed to mix for 30 minutes more. This allowed the reaction to go to completion.

The reaction product was filtered and dried in a vacuum oven overnight.

Results and Discussion

The reaction products all had the appearance of a white, fluffy powder with an acrid smell.

The three materials were examined under a microscope at 1000x to determine the morphology of the crystals that were formed during the reaction.

All of the materials appeared to be round-ish in shape (Fig. 2). This compares to the needle like crystals that are indicative of the commercial ZDA product (Fig. 3).

Fig. 2 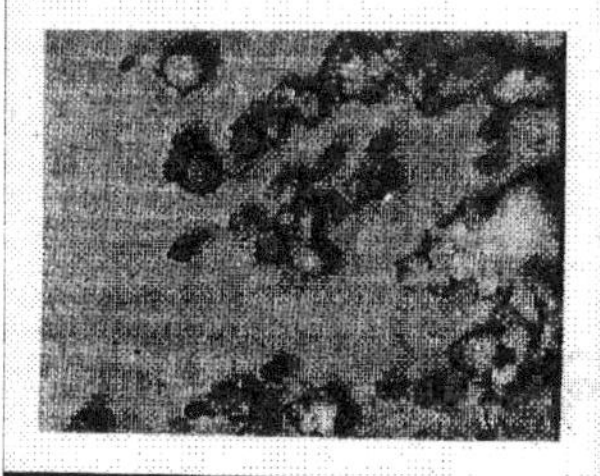

Fig. 3 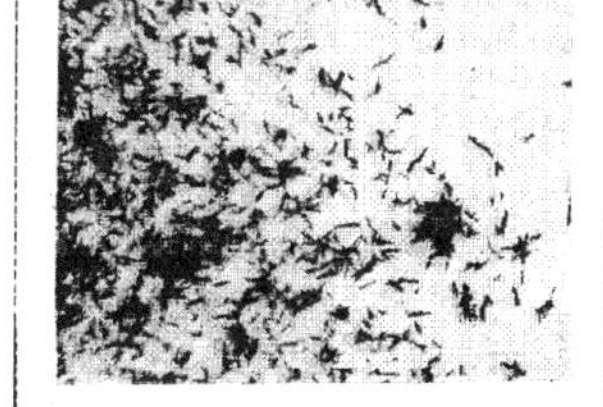

FTIR analysis showed the experimental ZDA material to be chemically identical to the commercial ZDA material.

Golf ball centers were molded utilizing the experimental ZDA, magnesium diacrylate (MDA) and calcium diacrylate (CDA) at 24 parts by weight in a polybutadiene (Mooney viscosity of 30-50) center formula and compared to the commercial ZDA control at 24 parts by weight. These centers were tested for hardness (PGA compression) and resilience at 135 feet per second (coefficient of restitution - COR) (Fig. 4).

Fig. 4

GOLF BALL CENTER PHYSICALS

	Compression	C.O.R.
Commercial ZDA	74	0.789
Experimental ZDA	63.2	0.775
Experimental MDA	59.7	0.762
Experimental CDA	57.3	0.754

The commercial material had the highest COR and the hardest compression, both desirable attributes for a high performance golf ball center. The experimental materials were lower in COR and hardness. The experimental ZDA was the fastest and hardest material, followed by MDA and then CDA.

Small samples of the materials were molded and tested on a dynamic mechanical thermal analyzer (DMTA). The parts were run at 86°F and from 0.3 to 30 Hertz (Fig. 5). The commercial ZDA had the highest modulus, with the experimental ZDA next highest. The MDA and CDA followed, in that order. These results mirrored the physical results obtained from the molded centers.

Fig. 5

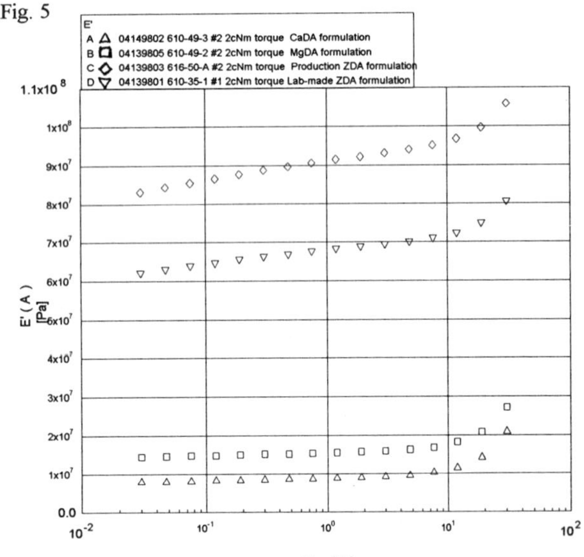

Conclusions

The morphology of the experimental materials is quite different versus the commercial materials. This is most likely due to several factors during the reaction; the rate of stirring, the surfactant that is utilized and the rate of addition of the acid. Future experiments will address this difference in crystal shape.

The resilience properties of the various polybutadiene / acrylic acid metal salt polymers is surprising. It was thought that smaller ionic centers would give a tighter crosslink structure and thus more resilience. This was obviously not the case. Part of the reason may be due to the outer electron structure in zinc. Further investigation will include the hydroxides of magnesium and calcium as well as trivalent (aluminum complexes) and tetravalent (zirconium complexes).

The technique of DMTA seems well suited to predict the properties of the cured polybutadiene / acrylic acid metal salt polymer. The testing methodology will be refined along with a time/temperature study to assess the effects on the loss versus storage modulus (tan delta).

Finally, a reaction profile of the uncured polybutadiene / acrylic acid metal salt polymer will be tested in a heated cell on an FTIR to determine any differences in the reaction profiles.

Acknowledgments

The author would like to thank Jack Neill and Sam Sico for their invaluable help with this paper.

References

1. Ferry, John D. , *Viscoelastic Properties of Polymers*, John Wiley & Sons, 1980.
2. Hayes, Robert A. and Conrad, Wendell R., *U.S. Patent 4,500,466*, 1985.
3. Morrison, Robert T. and Boyd, Robert N., *Organic Chemistry*, Allyn and Bacon Inc, 1977.
4. Cochran, Alastair and Stobbs, John, *Search for the Perfect Swing*, Triumph Books, 1996.
5. Pollitt, Duncan H. and Reich, Murray H., *U.S. Patent 4,056,269* , 1972.
6. Terry, Mark , Interface Corp., personal communication , June, 1994.

Structure Development of Spherulitic Domains in Polymer Films

Tao Huang, Tomohiro Tsuji, A.D. Rey, and M.R. Kamal

Department of Chemical Engineering, McGill University
Montreal, Quebec, H3A 2B2, Canada

Introduction

The nature of the microstructure of a polymer strongly affects its functionality in optical and electronic applications. There is great need for reliable predictive techniques to estimate the properties and understand the behavior of materials, especially in relation to the role of crystalline morphology. The properties of the solid polymer depend on both the supermolecular and grain structures formed in nucleation-growth process, in which the details are not well understand so far, comparing to the studies of phase separation.[1]

In this work, the extensive investigation on post-nucleation and spherulitic domain growth have been carried out experimentally and computationally in order to understand and predict the structure development in polymer thin films. Cross-scale computational model has been developed to predict and analyze nucleation-growth processes, grain patterns of polymer spherulitic structures, and the internal lamellae organization. We take spatial correlation, geometry and topology of the structural evolution into account to elucidate nucleation-growth dynamics. Novel domain-spatial correlation function simultaneously probe the time-dependent domain-size distribution and the spatial correlation of nuclei throughout the entire process. Scaling relationships have been found during early free growth stage. These results do not only significantly advance our understanding of polymer structure formation, but also provide the reliable predictive techniques to optimize material functionality and processing controllability.

Experimental Studies

Experiments were carried out with a common polymer, isotatic polypropylene (iPP, molecular weight M_η=250,000). A polymer thin film was formed between two glass slides and by pressing the top slide to form a 10 μm thick polymer film. A Leitz polarized microscope, equipped with a hot stage for polymer film solidification was used in the direct observation experiments. In this isothermal solidification study, the temperature is controlled within ±0.1°C. JAVA-Jandel Scientific video measurement and image processing system was directly connected to the microscope via a CCD camera, with proposed digital image analysis programs. We focus on the post-nucleation stage, during which the size of the nucleus is greater than 1 μm and visible under the optical microscope for real-time in-situ observation and accurate real-space measurement. A nearly linear nucleation law is found after an induction period and before impingement from experimental measurements. With increasing transformation, the rate decreases to zero as impingement becomes significant. The experimental data of the nucleation rate can be explained by the Kashchiev's formula before impingement becomes significant.[2] The growth velocities measured from different spherulites in the same experimental run confirmed the linear growth before impingement under isothermal crystallization conditions.[2]

The spatial features of post-nucleation stage are the new nuclei packing and the impingement, which can be captured by the nuclei pair distances distribution. Experimental results show that the mean distance to the next nuclei decreases with increasing time. The size of the depletion zone surrounding each nucleus also decreases with time. The mean distance to the next nuclei is more distinct at higher area fractions with nuclei packing more densely. The partial pair correlation function for new nuclei to the old nuclei shows that the new nuclei tend to appear randomly near old nuclei. At any time during the late stage, where the statistical self-similarity of the spatial order achieved, the density function of nearest neighbor distance distribution can be easily derived by considering the impingement at all nuclei sites. Figure 1 shows the corresponding experimental results of the normalized nearest neighbors distances Λ_{NN} normalized with the scale of the average nearest neighbor distance δ_{NN}. The most important feature of the normalized nearest neighbors distance Λ_{NN} is the self-similar fashion. It indicates that the normalized nearest neighbors distances Λ_{NN} are selected with self-similarity for $\Lambda_{NN} < 1$; however, they are not for $\Lambda_{NN} > 1$. The deviation decreases by more nuclei production with increasing time.

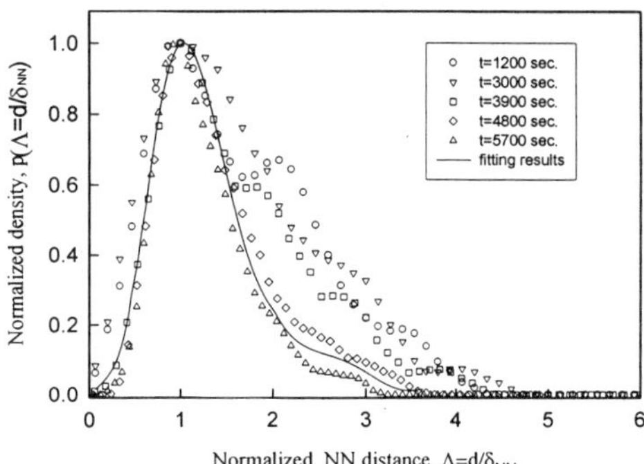

Figure 1. The normalized distribution function of the density of nearest neighbors distances of iPP spherulite nucleation-growth at isothermal condition Tc=406K.

Computational Modeling

A novel physical model is presented for prediction and characterization of the development of semi-crystalline polymer microstructure across mesoscopic length and macroscopic grain length scales. We assume that there are no macroscopic heat/mass transfer in the system. Also, we assume that the fluctuations and a long range interaction between growth domains are negligible. Considering the nucleation and growth in space, the core origin position is represented by X_i(i=1,...N_c), where N_c is a total number of cores. The relation between a certain core and its neighbors can be represented by the distance function of the neighbor cores, which is given by $|X_i-X_j|$, where $i \neq j$. Thus, an envelop profile for the multiple spherulitic growth domains is given by:

$$f_i(\theta,t) = \int_0^t \nu(t')\prod_{i \neq j}\{1 - \delta(A)\}dt' \tag{1}$$

where $\nu(t')$ is growth velocity vector, the term, $\prod\{1 - \delta(A)\}$, describes the impingement dynamics of near nearest neighbor domains; for the unit vector along one domain center coordinate on an envelop, $\vec{u} = (\cos\theta, \sin\theta, 0)^T$, the parameter A is compensation factor, $A=\{[x_{ix}+\cos\theta f_i(\theta,t)-x_{ix}+\cos\theta f_j(\theta,t)]^2+[x_{iy}+\sin\theta f_i(\theta,t)-x_{iy}+\sin\theta f_j(\theta,t)]^2\}^{1/2}$.

In order to improve our understanding of the internal structure of polymer spherulites, it is important to analyze quantitatively the orientation order of lamellae within spherulite. On the mesoscopic lamellar scale, we assume that the lamellae configuration represents the spherulitic internal texture. Based on the experimental TEM image of the lamellae configuration of polyethylene spherulite, a unit vector $\vec{l}$ is introduced as local averaged lamellae director in order to describe the spherulitic internal structure,. We assume that the lamellae director is an unit vector of the major axis direction and changes along the length of a mono-lamellae. There are three possible configurations: (1) splay energy, F_s, which represents the tilting a lamellae toward another; (2) bend energy, F_b, which represents the angle between two adjacent lamellae; and (3) twist energy, F_t, which arises by counterotation at the lamellae ends. In this model, their distortions are else defined. The curvature energy of the lamellae director field is similar to Frank energy equation in liquid crystal:[2]

$$F_s = C_1(\nabla \bullet \vec{l})^2, \; F_t = C_2(\vec{l} \bullet \nabla \times \vec{l})^2, \; F_b = C_3\left|\vec{l} \times \nabla \times \vec{l}\right|^2 \tag{2}$$

where C_1, C_2 and C_3 are elastic coefficients. In 2D polymer films, we consider both splay and bend for normal spherulite; and twist for banded spherulite.

The orientation angle of the lamellae director is

$$\theta = \arcsin(I / I_0) / 2 \tag{3}$$

where I is the intensity of polarized light passing through the spherulites.

Computational modeling has been performed based on the coupling of above equations to predict and analyze the lamellae organization based on the mesoscopic lamellae director and grain patterns of semi-crystalline polymer spherulitic structures. The simulation results and its scientific visualization present the spherulitic textures of semi-crystalline polymer as Maltese cross gray images. It provides information not only the functional data on envelop profiles of the spherulites but also the spherulite internal structure.

Figure 2 shows that typical results. Within the spherulite, the computational results indicate that the characteristic of lamellae is the orientation deviations caused strongly by the large splay distortion, and by the bend distortion in some regions. It is clearly shown that both energies at the spherulite grain boundaries are higher than within the spherulite grains. The energy profiles agree exactly with the corresponded vector plot of the mesoscopic lamellae director. This is due to the mis-matching of the orientation of the mesoscopic lamellar directors at the grain boundary between two neighboring spherulites.

Dynamic comparison between large scale computational and experimental results have been also accomplished by quantitatively matching the simulated growth patterns to the experimental images.[2] Furthermore, this model is being extended to modeling the structure development in real polymer processing.

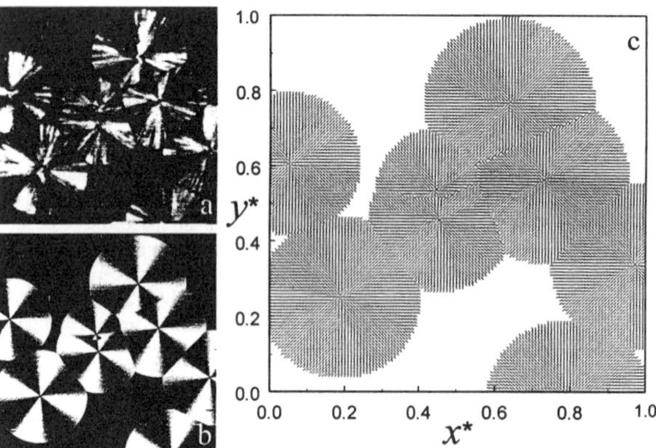

Figure 2. Typical results of the experimental image in (a), the computational pattern in (b), and the vector plot of the mesoscopic lamellae director in (c).

Structural Evolution Dynamics

In this section, we take spatial correlation, geometry and topology of the structural evolution into account to elucidate nucleation-growth dynamics. It is an important feature of the structural evolution has not been identified or taken into account so far. Based on this argument, we proposed the following new concepts and approaches on the studies of the dynamic domain-spatial correlation function and the topological anticorrelation.

(1). Domain-Spatial Correlation Function: The new domain-spatial correlation function is defined as: $G_i(r,t)$, for an arbitrarily chosen domain i with position vector $\vec{x}$, as the origin of the domain core center, is defined by counting the domains whose position vectors lie within a distance Δr from a circle of radius r with center at the origin at time t. Considering the whole system with the total number of domains, N_c the domain-spatial correlation function, $G(r,t)$, is the ensemble averages of $G_i(r,t)$ for all domain center positions of over all domains placed at the origin, which yields:

$$G(r,t) = \left\langle \frac{1}{\int \psi(\vec{x},t)d\vec{x}} \int \delta(r - |\vec{x}_1 - \vec{x}_j|) \psi(\vec{x}_j,t)d\vec{x}_j \right\rangle \tag{4}$$

where $\delta(r-|x_i-x_j|)$ is number density function and $\psi(x,t)$ is the order parameter:

$$\delta(r - |\vec{x}_i - \vec{x}_j|) = \begin{cases} 1, & r - |\vec{x}_i - \vec{x}_j| = 0 \\ 0, & r - |\vec{x}_i - \vec{x}_j| \neq 0 \end{cases} ; \quad \psi(\vec{x},t) = \begin{cases} 1, & \vec{x} \in domain\ at\ t \\ 0, & \vec{x} \notin domain\ at\ t \end{cases}$$

Figure 3 shows the domain-spatial correlation function for experiment of continuous nucleation under isothermal crystallization condition at T_c = 406K.

The left part of Figure 3 is the smooth curves of the domain size distribution when $r < R_g(t)$. It represents the growth domains with different sizes and

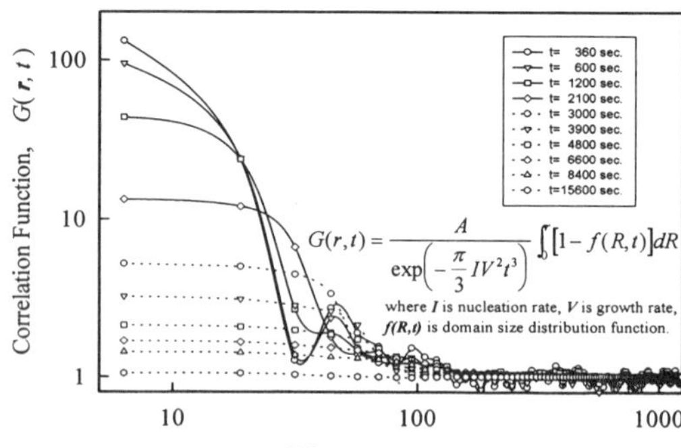

$$G(r,t) = \frac{A}{\exp\left(-\frac{\pi}{3}IV^2 t^3\right)} \int_0^r [1 - f(R,t)]dR$$

Figure 3. Dynamic domain-spatial correlation functions for experiment of continuous nucleation under isothermal crystallization condition, Tc=406K.

some local impingement structures. The right part of Fig.3 is the spatial pair-correlation function of domain core centers when $r > R_g(t)$ The first peaks shift slightly to the left with increasing time, which means that the average inter-nuclei distances decrease with increasing time due to the nucleation of new domains. $G(r,t)$ approaches one in an oscillatory manner at very large r, which means that there is no long range order. The scaling relation, $G(r,t)/G(r=0,t)$, as a function of $r/R_g(t)$, where $R_g(t)$ is the location of the first minimum of $G(r,t)$ has been found during early free growth stage.[2]

(2). Topological Anticorrelation: The topological nature of structure evolution in the nucleation and growth process is determined by the nearest neighbors configuration

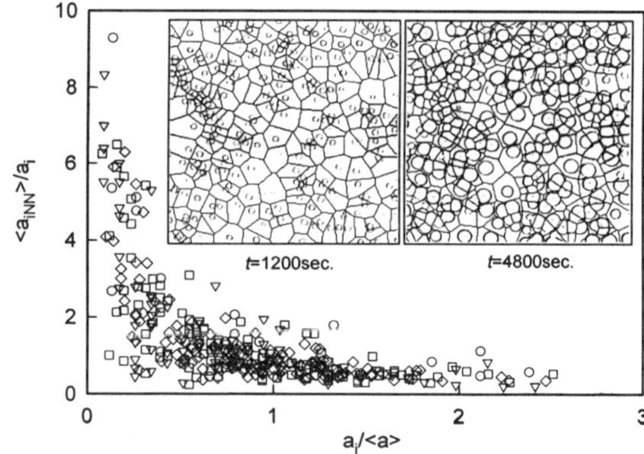

Figure 4. Topological anticoorelation function of the nearest neighbors area; the insert images show the evolution of the nearest neighbors configuration .

by Voronoi characterization of nucleation and growth patterns. Two nuclei are defined as nearest neighbors if they belong to adjacent cells of the Voronoi diagram (fine lines), see the insert images in Figure 4. Figure 4 shows that for each domain of area a_i in a given pattern with average area $<a>$, the average area of its nearest neighbors $<a_{NN}>$ relative to a_i shows the ascertain correlation in the areas assumed by adjacent domains. The important result is: for domains smaller than average area $<a>$, $<a_{NN}>$ exceeds a_i and vice versa. This is referred as "pronounced anticorrelation", which can be explained by the maximum entropy formalism.[2] This topological nature play an important role in the understanding of the structural evolution in nucleation and growth.

Reference

(1) Langer,J.S. in *Solids Far From Equilibrium*, Ed by C. Godreche, 1989.
(2) Tao Huang, Ph.D thesis, McGill University, 1997.

Equilibrium Order-to-Disorder Transition Temperature in Block Copolymers

G. Floudas[1], T. Pakula[2], G. Velis[3,1], S.Sioula[3,1], N. Hadjichristidis[3,1]

[1] Foundation for Research and Technology-Hellas (FO.R.T.H.), Institute of Electronic Structure and Laser, P.O. Box 1527, 711 10 Heraklion Crete, Greece

[2] Max-Planck Institut für Polymerforschung, Postfach 3148, D-55021 Mainz, Germany

[3] Department of Chemistry, University of Athens, Zografou 15771, Athens, Greece

The order/disorder process in amorphous block copolymers bears many similarities with the crystallization/melting process in semicrystalline homopolymers. From the thermodynamic point of view, the latter is a first order transition whereas the former is a fluctuation induced first order transition [1]. Experiments have been performed to explore the small thermodynamic effects associated with the fluctuation induced transition in amorphous block copolymers. These experiments have shown a small discontinuity in the density [2] and an endothermal peak in the specific heat [3-5] at the T_{ODT}.

In previous publications [6-8] it has been shown that amorphous block copolymers can be supercooled in just the same way that semicrystalline homopolymers can and that the ordering process (nucleation and growth) is reminiscent to the crystallization process (of the Avrami type). The rate of nucleation and growth was found to depend on the supercooling and for a given supercooling to the architecture of the block copolymer. For star-shaped block copolymers the process was shown to slow-down with respect to linear block copolymers as a result of the hindered diffusion.

In the present work we report on another sound similarity between semicrystalline homopolymers and amorphous block copolymers, namely, that in the latter systems there exist many apparent order-to-disorder transition temperatures but only a single equilibrium order-to-disorder transition temperature (T_{ODT}^{o}) just as in the former there is a single equilibrium melting temperature (T_{m}^{o}).

For this purpose we employed a linear AB diblock copolymer (f_{PS}=0.5, MW=13600) and a non-linear miktoarm star block copolymer (f_{PS}=0.3, MW=25200) of the AB$_3$ type [9], where A is polystyrene and B is polyisoprene. The phase state of the two copolymers was studied by SAXS and TEM. The SI diblock copolymer is lamellar-forming whereas in the asymmetric SI$_3$, cylinders of the minority phase (S) are formed in the polyisoprene matrix. The relevant thermodynamic parameters were extracted by bringing the systems to their disordered phase and by applying SAXS. For the SI the mean-field structure factor [10] was used to fit the profiles above T_{ODT}^{o} and the χ parameter was extracted (χ=54/T-0.079) [11]. For the SI$_3$ again the generalized (for A$_n$B$_m$ star block copolymers) mean-field structure factor was employed [12]. Rheology was used for the remaining studies.

Fig. 1 shows the result of the isochronal measurements of the storage (G') and loss (G") moduli at ω=1rad/s obtained on heating (symbols) and subsequent cooling (lines) of the SI diblock copolymer. The abrupt change of the moduli signify an apparent ODT. There is a pronounced hysteresis on cooling which extends to some 10°C below the transition and which is the signature of metastability. Subsequent isothermal frequency sweeps have been performed in order to identify frequencies below some critical frequency ω$_c$, separating the high from the low-frequency regimes. A frequency below ω$_c$ was chosen for the kinetic experiments. The sample was then heated to 90°C and the time evolution of G' and G" was monitored immediately after the temperature jumps to the different final temperatures, and the results for G' are shown in the inset to Fig.1.

Both moduli show an "S" shape with short and long-time plateaus reflecting the quenched disordered and ordered states, respectively. The volume fraction of the ordered phase, as it develops with time (φ(t)), was extracted using simple mechanical models [7][8] which provide limits for the mechanical response of a composite material as a function of the properties of the constituent phases (ordered and disordered); φ(t) was then fitted to the Avrami equation φ(t)=1-exp(-zt^n) [13], with z being the rate constant and n the Avrami exponent. The results revealed that nucleation and growth constitute the underlying mechanism, as reported earlier in detail. As seen in the inset to Fig.1, the short- and the long-time plateaus as well as the step in between depend on the quench depth, the latter being smaller for shallow quenches.

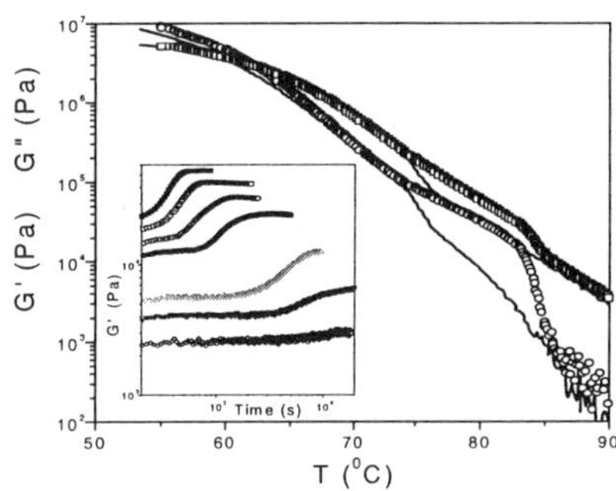

FIG. 1 Isochronal measurement of the storage (○) and loss (□) moduli for the diblock copolymer SI, taken with ω=1rad/s and a strain amplitude of 2% on heating (symbols) and subsequent cooling (lines)(heating rate=0.2K/min). In the inset the kinetic experiments are shown for the different quenches: (◇):83, (▽):82,(△):81,(●):79, (○):78, (□):77 and (■):76° C

After the long-time plateaus were reached we performed isochronal temperature scans (at ω=1rad/s) aiming to disorder the system. Should there be a single T_{ODT} all curves would overlap. However, as shown in the composite plot of Fig.2, for every final state there is a different T_{ODT}; the lower the ordering temperature the lower the apparent ODT. Furthermore, the step in G' -which in Fig. 2 is shown by the vertical data points- during the isothermal/isochronal experiments is well within the

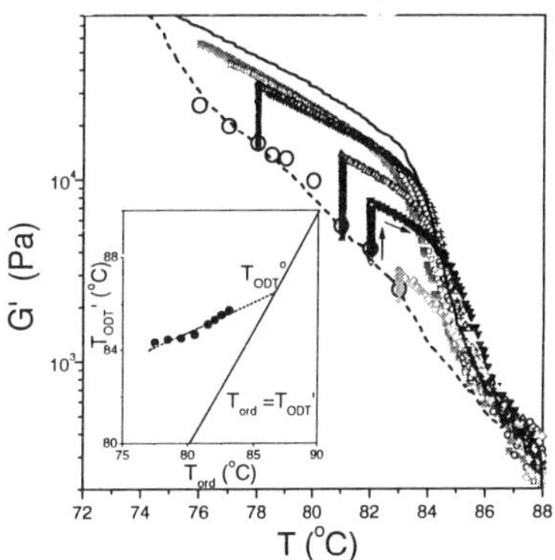

FIG. 2 Composite plot showing the result of the isochronal T-scans (heating rate=0.2K/min) following the ordering kinetics of Fig.1. The symbols are as in Fig.1. The solid and broken lines, respectively, are the isochronal heating and cooling curves. The moduli of the quenched disordered states are also shown (○). In the inset, the modified Hofmann-Weeks plot is shown which is used to extract the equilibrium order-to-disorder transition temperature (T_{ODT}^{o}).

hysteresis envelope which is also shown with lines. The short-time plateau values from the kinetic experiments in Fig.1 are also included as circles in Fig.2 and show that the quenched disordered states correspond fairly well to the isochronal cooling curve (dashed line in Fig.2) and to an extrapolation from the high T disordered state. The pronounced dependence of the ODT on the ordering temperature is shown in the inset to Fig.2, where we plot the apparent transition temperature (T_{ODT}') as a function of the corresponding ordering temperature (T_{ord}). The corresponding plot for semicrystalline polymers is known as the Hofmann-Weeks plot where the apparent melting temperature (T_{m}') is plotted as a function of the crystallization temperature (T_{c}). The solid line in the figure signifies the T_{ord}=T_{ODT}' and the equilibrium ODT can be obtained by extrapolation (dashed line). The extrapolation used in Fig.2 is based on the underlying Gibbs-Thomson equation [14] which, for

$$T_{m}' = T_{m}^{o}\left(1 - \frac{2\sigma_e}{\Delta H_f}\frac{1}{d}\right)$$ (1)

semicrystalline copolymers, gives the melting point depression due to the finite thickness of the crystal (d). In eq. 1, σ_e is the surface free energy and ΔH_f is the heat of fusion. Some typical values of these parameters for the semicrystaline poly(ethylene oxide) (PEO) are 20erg/cm^2 and 200J/g, respectively.

The results presented above are not restricted to diblock copolymers. To show the universality of these findings we employed an asymmetric and non-linear miktoarm star block copolymer. The choice of the system is based on the wider range of metastability of star as compared to linear block copolymers [8].The isochronal temperature scans for G' are shown in Fig. 3. There is a clear dependence of the T_{ODT}' on the ordering temperature which is more pronounced when compared to the SI. In the inset, the apparent T_{ODT} is again plotted as a function of the ordering temperature and reveals that the equilibrium ODT is at about 99°C, that is, some 4°C higher than the apparent transition one would obtain by ordering the system at 85°C ($\Delta T=T_{ODT}^o-T_{ODT}'=14°C$).

Motivated by the similarities between the order/disorder process and the crystallization/melting process we first, simply re-write eq.1 for the former process, as

$$L=\left(\frac{2\sigma}{\Delta H}\right)\left(\frac{T_{ODT}^o}{T_{ODT}^o-T_{ODT}'}\right) \tag{2}$$

where σ is now the interfacial tension and ΔH is the change of enthalpy upon mixing. Eq 2 implies a strong effect of supercooling on sizes (L) of ordered regions. We can obtain a rough estimate of this effect

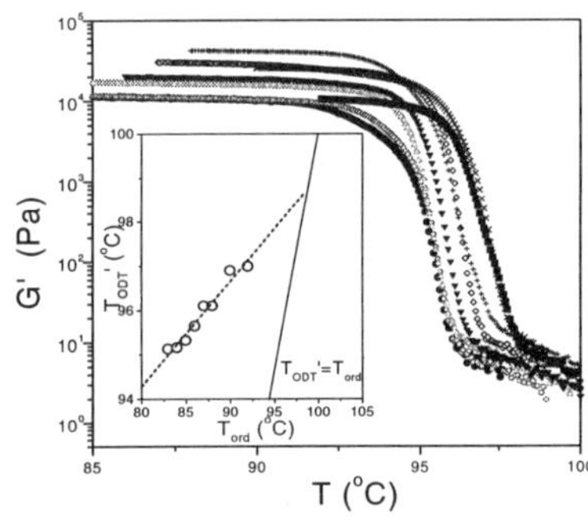

FIG. 3. Isochronal (ω=1rad/s) temperatute scans (heating rate=0.2 K/min) for SI$_3$ obtained by heating the ordered phase formed at the kinetic experiments: ($\bullet$):83, ($\bigcirc$):84, ($\triangle$):85, (∇):86, ($\Diamond$):87, (+):88, (X):90, ($\blacksquare$):92°C. In the inset the corresponding Hofmann-Weeks plot is shown and the T_{ODT}^o is obtained by extrapolation.

for the diblock copolymer by using the values of T_{ODT}^o and ΔH. The interfacial tension for a weakly phase separated symmetric diblock copolymer has been calculated both in the mean field theory approximation and in the fluctuation picture. The theory [15] is based on a classical droplet theory for the nucleation of lamellar phases from a metastable disordered melt and within the mean-field picture, the interfacial tension is

$$\sigma=\frac{8}{12}\frac{\tau^{3/2}}{R_g^2 u}\kappa_B T \tag{3}$$

where $\tau=(2/c^2)(\chi_s N-\chi N)$, $u=\lambda N^{-1/2}$ and $N=(a^6/v^2)N$ is the Ginzburg parameter, with a and v being the statistical segment length and volume, respectively, and the parameters c and λ are defined in [1]. Using f=0.5, c=1.1, λ=106.2, a=6.35Å, v=144Å^3, N=150, χ=54/T-0.079, T_{ODT}^o=86.7°C and ΔH~2J/g, we obtain 1erg/cm^2 for a supercooling of 10°C. This value is a factor of ten lower than the corresponding value in PEO but the ΔH is reduced by a factor of 100. It is because of the large decrease in the latter that the size effects which are observed correspond to L>d (typically L~10d) which are of the order of the grain size.

In applying eq. 1 we should keep in mind that it has been derived for lamellar crystallites where only the interactions at the lamellar surfaces play a role. In the case of grains, however, the whole intergrain surface should be considered. We can now generalize eq. 1 by reformulating the problem for the case of grains in ordered copolymers. The transition temperature is defined as $T_{ODT}'=\Delta H/\Delta S$, where ΔH (ΔS) is the enthalpy (entropy) difference between the ordered and disordered states. Under the assumption that the finite grain size causes the enthalpy reduction related to the disorder at the intergrain surfaces (constant entropy approximation), $\Delta H_o/T_{ODT}^o=\Delta H/T_{ODT}'$, where ΔH_o refers to the enthalpy of an infinitely large grain. The enthalpy reduction is related to a kind of disorder (defects) at the grain boundaries. This effect should be proportional to the surface to volume ratio of individual grains and the remaining enthalpy limiting the stability of grains can be expressed as $\Delta H_o(1-(s/v^*)d)$, where s and v* are, respectively, the grain surface and volume and d is the intergrain layer thickness. Based on these considerations the grain dimensions, L, are:

$$L=\beta\frac{T_{ODT}^o}{T_{ODT}^o-T_{ODT}'} \tag{4}$$

where β is a function of the geometrical characteristics of the grain and of the intergrain layer thickness. Eq. 4 can be considered as a generalized Gibbs-Thomson equation for grain sizes and knowledge of β would allow a determination of the grain size dependence on supercooling. This pronounced grain size dependence has been observed in experiments which probe the development of birefringence [16] and optical anisotropy [17] following quenches from the disordered phase. From the former study and the inset to Fig.2 (T_{ODT}'=0.25T_{ord}+64) we deduce β~1.3 μm. which gives reasonable grain sizes on supercooling.

Finally, we comment on the consequences of these findings for block copolymers. For experiments made far away from the T_{ODT}^o there should be no considerable effects, however, for experiments made in the vicinity of the transition (weak segregation) strong coupling is expected. The equilibrium transition temperature, for example, is needed if kinetic experiments are performed since knowledge of the quench depth, ΔT, is essential. Furthermore, shear-induced transitions might also be affected in a narrow temperature interval below T_{ODT}^o.[18]

References

[1] G. H. Fredrickson and E. Helfand, J. Chem. Phys. **87**, 697 (1987).

[2] H. Kasten and B. Stühn, Macromolecules **28**,4777 (1995).

[3] D. A. Hajduk, S. M. Gruner, S. Erramili, R. A. Register and L. J. Fetters, Macromolecules **29**, 1473 (1996).

[4] B. Stühn, J. Polym. Sci. Polym. Phys. Ed. **30**, 1013 (1992).

[5] G. Floudas, N. Hadjichristidis, M. Stamm, A.E. Likhtman and A.N. Semenov, J. Chem. Phys. **106**, 3318 (1997).

[6] J. H. Rosedale and F. S. Bates, Macromolecules **23**, 2329 (1990).

[7] G. Floudas, et al. Macromolecules **27**, 7735 (1994).

[8] G. Floudas, S. Pispas, N. Hadjichristidis, T. Pakula and I. Erukhimovich, Macromolecules **29**, 4142 (1996).

[9] Y. Tselikas et al. J. Chem. Phys. 105, 2456 (1996).

[10] L. Leibler, Macromolecules, **13**, 1602 (1980).

[11] G. Floudas, et al. J. Chem. Phys. **104**, 2083 (1996).

[12] G. Floudas, N. Hadjichristidis, Y. Tselikas, and I. Erukhimovich, Macromolecules **30**, 3090 (1997).

[13] M. J. Avrami, J. Chem. Phys. **7**, 1103 (1939); **8**, 212 (1940); **9**, 177 (1941).

[14] L. Mandelkern and R.G. Alamo, *Physical Properties of Polymers Handbook*, edited by J. E. Mark (AI Press, Woodbury, N.Y., 1996).

[15] G. H. Fredrickson and K. Binder, J. Chem. Phys. **91**, 7265 (1989).

[16] N.P. Balsara et al. Macromolecules **25**, 6072 (1992).

[17] G. Floudas et al. Macromolecules 28, 2359 (1995)

[18] G. Floudas, T. Pakula, G. Velis, S. Sioula, N. Hadjicristidis J. Chem. Phys. in press

ULTRASONIC METHODS FOR CHARACTERIZATION OF POLYMERIC MATERIALS

Ingo Alig, Dirk Lellinger and Sascha Tadjbakhsch
Deutsches Kunststoff-Institut, Schlossgartenstrasse 6
D-64289 Darmstadt, Germany

INTRODUCTION

Ultrasonic techniques have been shown to be an excellent tool for non-destructive testing or imaging since decades. Although the relations between material properties and acoustical parameters have also been studied for a long time[1], ultrasonic devices are not very frequently used for material characterization. With the development of high frequency digital and computer technique it was possible to overcome some of the reasons for the limited applications of ultrasonic methods for material characterization or monitoring of processing.

When propagated in polymeric materials, acoustic waves are influenced by the structure as well as by molecular relaxations. From velocity and attenuation of longitudinal and/or shear waves it is possible to estimate the viscoelastic properties[2,3] of polymeric materials. Besides from the viscoelastic properties of polymer melts[4] one can characterize semi-crystalline polymers[5].

Ultrasonic methods have been successfully applied to monitoring of polymer processing[6], chemical reactions (polymerization or curing of thermosets[7,8]), film formation[9], glue processes or crystallization in polymers[10,11].

Although there is a large number of investigations using longitudinal waves experiments applying shear waves are rare. We present some experimental results using a shear wave reflection technique[12] during isothermal film formation from aqueous dispersions, isothermal crystallization in semi-crystalline polymers and temperature dependent investigations near the glass transition in amorphous polymer films. Furthermore we present an example for the application of longitudinal ultrasonic waves in the extrusion process[13].

EXPERIMENTAL

Instrumentation. The shear wave reflection technique for time and temperature dependent measurements of the complex dynamic shear modulus (G*=G'+iG") was recently described in[12]. The ultrasonic spectroscopic system for measurements of the longitudinal ultrasonic velocity and attenuation in the extrusion process is similar to those described before[14].

Materials. The polychloroprene films investigated were formed from aqueous latex dispersions (see Table 1) supplied by the BAYER AG (Leverkusen). The content of 1,4-*cis* sequences of CR1 was higher (5.2%) compared to CR2 and CR3 (3.8%), leading to a wholly amorphous film as found by X-ray scattering.

Table 1. Characterization of the Polychloroprene Samples

sample	1,4-trans		1,4-cis	1,2	3,4	Gel. %	T_g [°C]	solids content
	HT[a]	HH+TT[a]						
CR1	80.5%	12.0%	5.2%	1.2%	1.1%	68	-41.5	58.0%
CR2	83.0%	11.5%	3.8%	1.0%	0.8%	0	-43.0	46.6%
CR3	83.0%	11.5%	3.8%	1.0%	0.8%	89	-41.5	54.4%

a: HT: head to tail, HH: head to head, TT: tail to tail

The two n-butyl acrylate/styrene copolymers (BS1 and BS2) with different monomer ratio and the ethylhexyl acrylate/methyl methacrylate copolymer (AC) were prepared from commercial aqueous dispersions (BASF, Ludwigshafen). The glass transition temperatures T_g for these materials obtained by DSC were -4°C and 23°C for BS1 and BS2 respectively and -46°C for AC.

The materials used for the extrusion experiments were commercial samples supplied by BASF AG (Ludwigshafen).

RESULTS AND DISCUSSION

Film formation and crystallization. In Fig 1 the application of an ultrasonic shear wave reflection technique[12] for investigation of isothermal film formation and crystallization kinetics of an amorphous (CR1) and two semi-crystalline polychloroprene samples (CR2 and CR3) with different gel content is demonstrated. The measurements of the dynamic shear modulus have been performed at a frequency of 5.32 MHz. The process of film formation during the evaporation of water is expressed by a step-wise increase of the shear modulus. For the semi-crystalline samples a further increase, which is due to crystallization, can be observed. The nucleation and crystallization is triggered by the film formation while it is suppressed in the latex particles. Film formation and crystallization are delayed for the

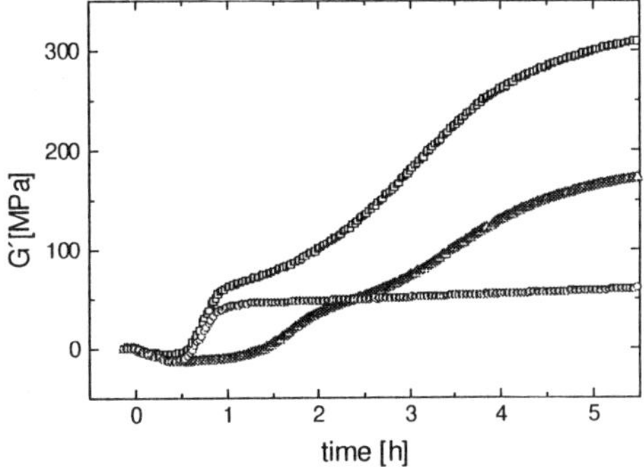

Fig 1: Time dependence of G' for aqueous dispersions of an amorphous (CR1: O) and two semicrystalline polychloroprenes (CR2: □, CR3: Δ) during film formation and crystallization at 25°C (frequency: 5.32 MHz; final film thickness: about 120 μm.

sample with high gel content (CR3) and its minor final modulus is explained by a lower degree of crystallinity.

The time dependent increase of the shear modulus due to the growth of spherulites has been analyzed by the Avrami equation[15] combined with the Kerner model[16] for the modulus of a two-phase composit (spherulites in an amorphous matrix). Assuming that the "composition of the spherulites" does not change with crystallization time, the volume fraction of the crystalls $\phi_c(t)$ in the Avrami equation can be replaced by the volume fraction of the spherulites $\phi_s(t)$. The time dependence of $\phi_s(t)$ for isothermal crystallization is then given by:

$$\phi_s = 1 - \exp\left(-kt^n\right). \tag{1}$$

From the polarization microscopy it is evident that polychlorprene crystallizes in spherulites. We suggest a two phase model consisting of an amorphous phase and spherulites, where the spherulites can be treated as

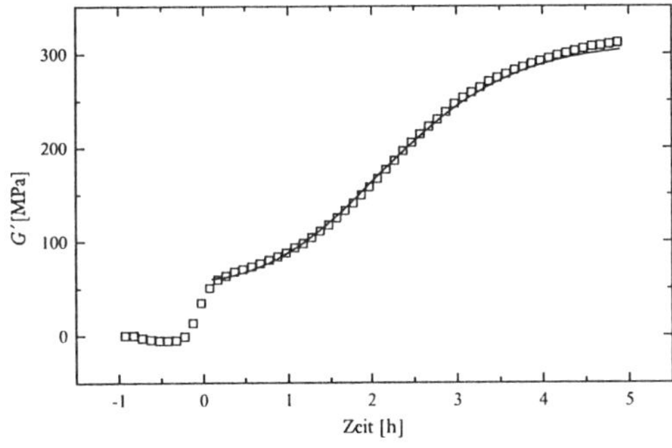

Fig 2: G' during isothermal crystallization at 25°C in a polychloroprene film (CR2). The data are fitted (solid line) by a combination of the Avrami equation and the Kerner model.

homogeneous spheres because the inhomogenities in the spherulites are small compared to the ultrasonic wavelength. Therefore we have chosen the Kerner model which predicts the elastic properties of a composite existing of spherical particles included in a matrix:

$$G = G_m \frac{\phi_m G_m + \left(\gamma + \phi_s\right)G_s}{\left(1 + \gamma\phi_s\right)G_m + \gamma\phi_m G_s} \quad \text{with} \quad \gamma = \frac{8 - 10\nu_m}{7 - 5\nu_m}. \tag{2}$$

ϕ_m is the volume fraction of the amorphous matrix outside of the spherulite and v_m is its Poisson's ratio. The indices m and s refer to the matrix and the spherulite phase, respectively. For data analysis the Kerner model was extended to a complex shear modulus[10]. In Fig 2 the fit of a combination of equations (1) and (2) to the G' data of CR2 is shown. To exclude the effect of film formation and secondary crystallization only the time interval from one hour to three hours was considered for the fit (the time scale in Fig 2 is shifted by 56 minutes compared to Fig 2). For the parameters G_m'=60 MPa, G_s'=300 MPa and $v_m = 0.499$ were used (for details see[10]). The best fit was obtained with the Avrami parameters $n = 1.93$ and $k = 2 \cdot 10^{-8}$. The value of the Avrami exponent n is consistent with heterogeneous nucleation and three-dimensional growth which is controlled by diffusion processes[17].

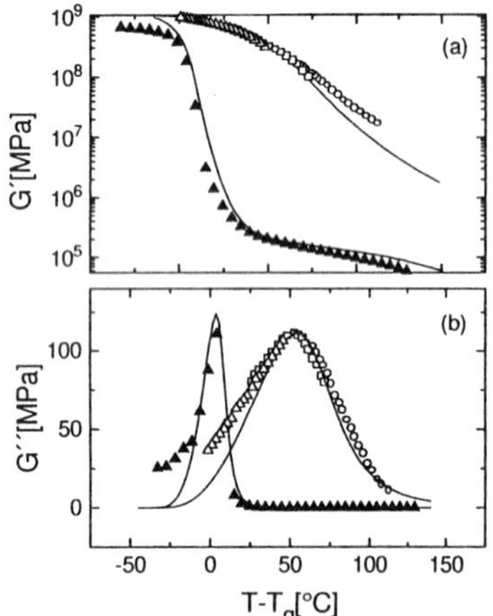

Fig 3: Temperature dependence of G' (a) and G" (b) for acrylic acid type copolymers (O: AC; □: BS1; Δ: BS2) measured with ultrasonic (open symbols) and conventional rheology (filled symbols). The solid lines are calculated using a combination of HN and VF equation.

Comparison of ultrasonic and conventional rheology. In Fig 3 the temperature dependence of G' and G" measured with "ultrasonic" (3.5 MHz) and conventional rheology (1Hz, only BS2) are compared for the acrylic acid type copolymers. To extend the temperature range a "sort of a master plot" was constructed for the ultrasonic data of BS1, BS2 and AC using $T-T_g$ instead of T. The G' and G" data from both methods can be fitted by the same set of parameters using the Havriliak-Negami function (HN) incorporating the Vogel-Fulcher (VF) equation for the temperature dependence of relaxation times[9]. The complex shear modulus can be expressed by the empirical HN function[18]:

$$G^* = G_\infty + (G_0 - G_\infty)\frac{1}{\left[1 + \left(i\omega\tau\right)^\alpha\right]^\gamma}, \qquad (3)$$

where $\omega=2\omega f$ is the angular frequency, G_∞ is the modulus for $\omega \to \infty$ and G_0 is the relaxed modulus ($\omega \to 0$). The exponents α and γ ranging between 0 and 1, are parameters describing the shape of the loss curve. The temperature dependence of the relaxation time for the main transition can be expressed by the VF[19] equation:

$$\tau = \tau_0 \exp(B / T - T_0), \qquad (4)$$

where T_0 is the Vogel temperature and B and τ_0 are parameters. The agreement between both methods is good. This suggests an almost thermo-rheological simplicity of the samples and demonstrates the capacity of the ultrasonic rheometer. For details of the fits and the parameters see[9].

Monitoring of the extrusion process. In Fig 4 an example for the application of a longitudinal ultrasonic technique for monitoring of the viscosity in the extrusion process is shown. The excess attenuation relative to a polyethylene sample (PE) was measured by an in-line pulse transmission technique were the two transducers were positioned behind the end of the

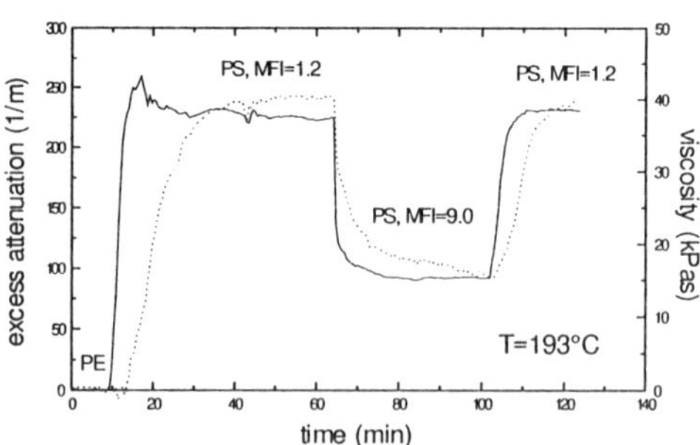

Fig 4: Comparison of the excess attenuation (solid line) and the shear viscosity (dashed line), estimated by an in-line rheometer, during the extrusion of two different polystyrenes (PS). Polyethylene (PE) was used for calibration. The melt viscosities of the PS samples are expressed by the mold flow index (MFI).

extruder screw[13]. At the same position the viscosity was measured by an in-line slot die rheometer. Fig 4 shows changes of both quantities for serveral sample changes. A clear correlation between the melt viscosity and the ultrasonic attenuation of PS is obvious and can be related to the different glass transition temperatures. With this technique it was also possible to find an relation between ultrasonic attenuation and/or velocity and the elastomer or filler content in polymer melts during extrusion[13]

CONCLUSION

We have given some examples for application of ultrasonic shear and longitudinal waves to demonstrate the feasibility of ultrasonic methods in polymer research including monitoring and control.

Acknowledgement. The work has been supported by the Bundesminister für Wirtschaft through the Arbeitsgemeinschaft industrieller Forschungsvereinigungen e.V., AiF-Grant no. 10835 and no. 10943.

BIBLIOGRAPHY

1. W.P. Mason (Ed.) in: *Physical Acoustics*, Vol. I, pt. A, (Academic Press, New York, 1964).
2. I. Alig in: *Handbuch der Kunststoffprüfung*, edited by H. Schmiedel (Hanser Verlag, München, 1992), chapter 10.2, 393
3. J.D. Ferry: *Viscoelastic Properties of Polymers*, (Wiley & Sons, New York 1961).
4. I. Alig, F. Stieber, A.D. Bakhramov, Yu.S. Manucarov and V.A. Solovyev, *Polymer* **30**, 842 and 877, 1989
5. Perepechko, I.I. in: *Akusticheskie metody issledovaniya polimerov* (Khimiya, Moskwa, 1973).
6. R. Gendron, M.M. Dumoulin and L. Piche, Polym. Mater. Sci. Eng. **72**, 23, 1995
7. I. Alig, D. Lellinger and G.P. Johari, *J. Polym. Sci B* **30**, 791, 1992
8. M.G. Parthun and G.P. Johari, *J. Chem. Phys.* **102**, 6301, 1995
9. I. Alig, S. Tadjbakhsch and A. Zosel, *J. Polym. Sci B* **3**, 1998 (in press).
10. I. Alig and S. Tadjbakhsch, S., *J. Polym. Sci B* (submitted).
11. I. Alig, S. Tadjbakhsch and G. Floudas, *Macromolecules* (submitted).
12. I. Alig, D. Lellinger, J. Sulimma and S. Tadjbakhsch, S. *Rev. Sci. Instrum.* **68**, 1536, 1997
13. I. Alig, D. Lellinger, R. Lamour and J. Ramthun, *Kunststoffe* (submitted).
14. I. Alig and D. Lellinger, *J. App. Phys.* **72**, 5565, 1992
15. M.J.J. Avrami, *Chem. Phys.* **7**, 1103, 1939; **8**, 212, 1940; **9**, 177, 1941
16. E.H. Kerner, *Proc. Phys. Soc.*, **69B**, 808, 1956
17. B. Wunderlich, in *Macromolecular Physics, Vol II* (Academic Press, New York, 1976)
18. S. Havriliak and S. Negami, *Polymer* **8**, 161, 1967
19. H. Vogel, *Phys. Z.* **22**, 645, 1921, Fulcher, G.S. *J. Am. Chem. Soc.*, **8**, 789, 1925

POLYMERIC DIAZENIUMDIOLATES FOR FABRICATING THROMBORESISTANT ELECTROCHEMICAL SENSORS VIA NITRIC OXIDE RELEASE

Kelly A. Mowery, Mark H. Schoenfisch and Mark E. Meyerhoff
Department of Chemistry, The University of Michigan,
Ann Arbor, MI 48109-1055

Joseph E. Saavedra[†] and Larry K. Keefer[‡]
[†]SAIC-Frederick and [‡]Laboratory of Comparative Carcinogenesis,
National Cancer Institute, Frederick Cancer Research and Development
Center, Frederick, MD 21702

INTRODUCTION

Although intra-arterial sensors for various electrolytes (Na^+, K^+ and Ca^{2+}), blood gases (pH, O_2 and CO_2) and metabolites (glucose and lactate) have been developed, their use for continuous patient monitoring is often complicated by vasoconstriction of the blood vessel around the implanted sensor and a lack of biocompatibility that leads to platelet deposition on the sensor surface[1]. As a result, decreased blood flow and cellular metabolism of adhered platelets create a micro-environment that differs from that of bulk blood and, ultimately, leads to sensor output values that do not correlate well with those obtained from discrete blood samples analyzed *in vitro*. However, due to the potent platelet anti-aggregation[2] and vasodilating[3] properties of nitric oxide (NO), there is evidence that the localized release of NO at the sensor surface may overcome both of these problems and enable the development of more thromboresistant sensors[4].

Initial investigations of this concept were based on the use of (Z)-1-{N-methyl-N-[6-(N-methylammoniohexyl)amino]}diazen-1-ium-1,2-diolate (MAHMA/N_2O_2)[5]. This diazeniumdiolate NO-donor molecule can be incorporated as a uniform dispersion of solid particles into a plasticized matrix of poly(vinyl chloride) (PVC) or Tecoflex polyurethane (PU) to create sensor membranes capable of generating NO for several days when exposed to aqueous solution[4]. Films prepared in this manner exhibit a favorable NO release, excellent biocompatibility and utility as sensing membranes for pH and K^+ ion-selective electrodes (ISEs) and amperometric O_2 sensors without affecting sensor performance. However, MAHMA/N_2O_2 and its corresponding decomposition products (N,N'-dimethylhexanediamine and the corresponding nitrosamine) were found to leach from the polymer matrix into an aqueous soaking solution[6]. Therefore, while these previous studies demonstrated initial feasibility of the concept of improved sensor thromboresistivity via NO release, alternative methods of preparing NO release polymer films are necessary to overcome potential toxicity concerns.

In an effort to generate NO exclusively from within a polymer matrix and eliminate amine leaching, several polymeric NO donors have been developed, including methoxymethyl piperazine-PVC/N_2O_2, (mompipPVC/N_2O_2, equation 1) and linear polyethylenimine/N_2O_2 (LPEI/N_2O_2, equation 2). The NO release profiles of films prepared with these polymeric NO-donors indicate that the films are capable of generating NO for several days. It will be shown that potentiometric pH sensors

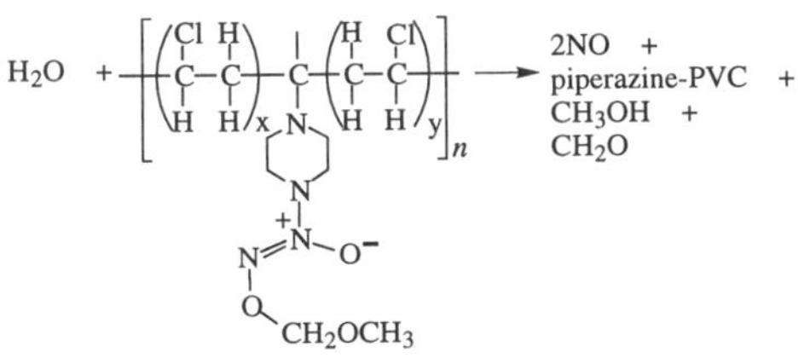

$$H_2O + \left[\begin{array}{c} Cl\ H \\ |\ | \\ C-C \\ |\ | \\ H\ H \end{array}_x N \begin{array}{c} H\ Cl \\ |\ | \\ C-C-C \\ |\ | \\ H\ H \end{array}_y\right]_n \longrightarrow \begin{array}{l} 2NO\ + \\ \text{piperazine-PVC}\ + \\ CH_3OH\ + \\ CH_2O \end{array} \quad \textbf{(1)}$$

$$H_2O + \longrightarrow 2NO + LPEI \quad \textbf{(2)}$$

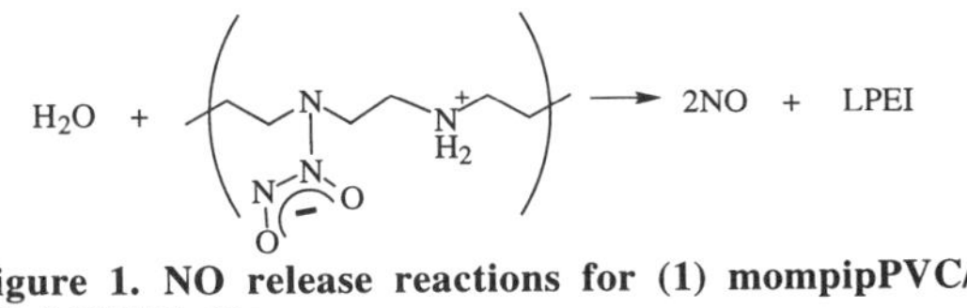

Figure 1. NO release reactions for (1) mompipPVC/N_2O_2 and (2) LPEI/N_2O_2.

prepared with these donors demonstrate the same analytical response properties as conventional polymeric membrane ISEs. In addition, preliminary *in vitro* platelet adhesion studies with mompipPVC/N_2O_2 reveal a significant decrease in platelet adhesion and activation compared to blank PVC controls.

EXPERIMENTAL

Polymeric diazeniumdiolate NO-donors were synthesized in a manner similar to that previously described for other diazeniumdiolate molecules[5,7,8,9]. MompipPVC/N_2O_2 polymer films for NO release measurements were then prepared by dissolving the polymer and the plasticizer, dioctyl sebacate (DOS), in 3.0 ml THF, and casting the resulting solution in a 2.5 cm i.d. glass ring placed on a glass plate. Likewise, single layer films containing LPEI/N_2O_2 were prepared by dissolving LPEI/N_2O_2, Tecoflex polyurethane and DOS in THF/MeOH (2:1), and casting the cocktail in a 2.5 cm i.d. glass ring. To prepare membranes for electrochemical pH sensors, the ionophore tridodecylamine (TDDA) and the lipophilic anionic site additive, potassium tetrakis(4-chlorophenyl)borate (KTpClPB) were added to the membrane cocktails prior to casting.

NO release profiles were studied by measuring the amount of NO generated indirectly as nitrite using a flow-injection analysis system with biamperometric detection[10]. Due to the rapid reaction of NO with water and oxygen to form almost exclusively nitrite[11], the accumulated NO_2^- concentration in a PBS soaking solution, pH 7.4, is directly proportional to the amount of NO released from the polymer film. One cm i.d. disks of the polymer films were used except in the case of mompipPVC/N_2O_2 doped films, for which 2.5 cm i.d. disks were employed.

To demonstrate that the NO releasing films also remain permselective for protons, potentiometric response studies were carried out according to standard procedures[12] using Philips IS-560 electrode bodies.

Potential thrombogenicity was assessed *in vitro* according to a procedure previously reported[12]. Briefly, membranes were incubated with platelet rich sheep plasma for 2 h at 37 °C with gentle rotation, fixed with a 2.0 % (w/v) glutaraldehyde solution, and examined using a scanning electron microscope.

RESULTS AND DISCUSSION

In order to assess the feasibility of using polymeric diazeniumdiolate NO donors for preparing thromboresistant polymer

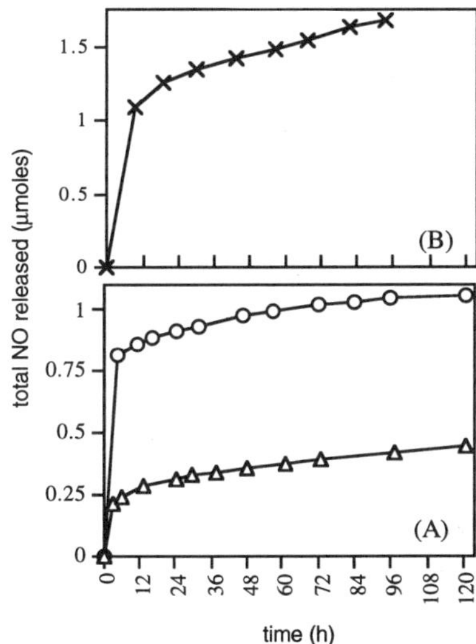

Figure 2. NO release profiles for polymer films soaked in PBS, pH 7.4, at 37 °C. Compositions by wt.%: (x) 50% mompipPVC-N_2O_2/50% DOS; (Δ) 62.2% PU/32.3% DOS/5.5% LPEI/N_2O_2; (o) 55.8% PU/31.7% DOS/12.5% LPEI/N_2O_2.

membrane ISEs, initial experiments were conducted to measure the NO release rate from doped films. The results shown in Figure 2 illustrate that all of the polymer films release NO for several days, although the total amount of NO generated is dependent on the amount and type of donor used. MompipPVC/N_2O_2 films release less NO than LPEI/N_2O_2 doped films, probably due to the low loading of methoxymethylpiperazine/N_2O_2 on the polymer backbone. Furthermore, as expected, increasing the amount of LPEI/N_2O_2 in the polymer results in a corresponding increase in the amount of NO generated.

Experiments using polymeric diazeniumdiolates to prepare H^+-selective ISEs reveal that the presence of these new NO donors does not compromise the sensitivity or selectivity of the resulting electrochemical sensors. Indeed, typical calibration curves for ISEs prepared with each of the polymeric NO donors are shown in Figure 3. For all cases, Nernstian response is obtained and selectivity over interferents (i.e., Na^+) is unaffected by the presence of the diazeniumdiolate group or concomitant NO generation. Additional experiments need to be conducted to determine whether or not these donors can also be used for other types of ISEs (i.e., K^+ and Ca^{2+}) and amperometric oxygen sensors.

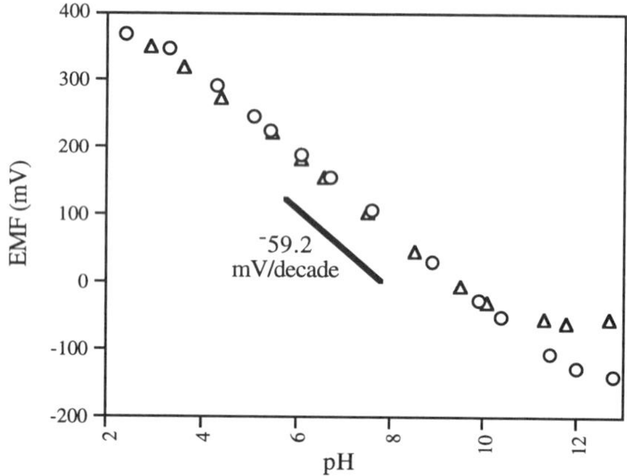

Figure 3. Potentiometric response of pH-selective ISEs. Membrane compositions by wt.%: (Δ) 46% mompipPVC/N_2O_2/50% DOS/2.4% TDDA/1.6% KTpClPB; (o)59.5% PU/31.5% DOS/2.0% TDDA/1.3% KTpClPB/5.7% LPEI/N_2O_2.

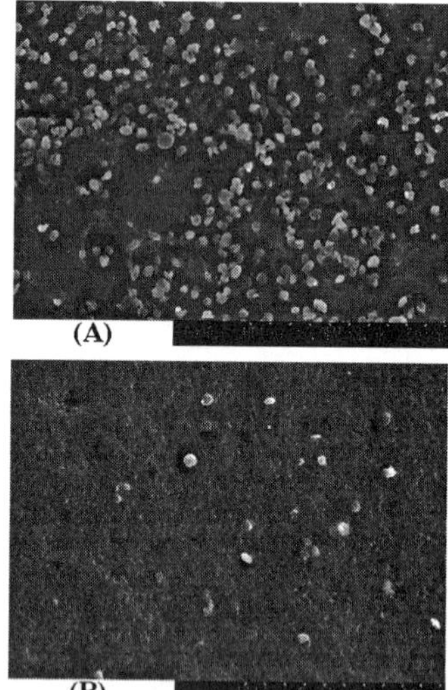

Figure 4. Scanning electron micrographs of: (A) 60% PVC/40% DOS and (B) 60% mompipPVC/N_2O_2/40% DOS (w/w) polymer films after exposure to sheep platelet rich plasma at 37 °C.

While polymer films doped with the NO releasing adduct display properties similar to blank films with regard to ion-sensing, the films have significantly different thrombogenic properties based on the *in vitro* results shown in Figure 4. Blank PVC controls have many platelets adhered to the membrane surface following exposure to platelet rich sheep plasma, and the platelets are highly activated, as indicated by the extent of pseudopod formation (Figure 4A). In contrast, there is a decrease in number of platelets adhered to mompipPVC/N_2O_2 samples (Figure 4B). When adhered platelets are present on these samples, a significant decrease in platelet activation is indicated by the spherical shape of the platelets.

CONCLUSIONS

The approach described herein should be applicable for improving the biocompatibility of previously reported potentiometric pH, CO_2 and K^+ sensing catheters, and is likely to be applicable to amperometric devices and optical sensors as well. Unlike previously reported polymers containing a dispersion of diazeniumdiolate NO-donor molecules[6], covalent attachment of the N_2O_2-functional group to the polymer matrix eliminates possible toxicity concerns associated with leaching of the donor amine molecule. The use of sensors which release NO may eliminate the need for systemic anticoagulation while effectively dilating the local arteries for precise blood gas and electrolyte measurements. These same polymeric materials may also prove valuable in the development of more thromboresistant extracorporeal tubing (for cardiopulmonary bypass surgery, ECMO, etc.) that could eliminate the need for systemic heparin anticoagulation for patients undergoing such procedures. Studies in this direction are currently in progress in collaboration with researchers at the University of Michigan medical school.

REFERENCES

1. Meyerhoff, M.E. *Trends in Anal. Chem.* **1993**, *12*, 257-266.
2. Mellion, B.T., Ignarro, L.J., Ohlstein, E.H., et al. *Blood* **1981**, *57*, 946.
3. Radomski, M.W., Palmer, M.J. and Moncada, S. *Br. J. Pharmac.* **1987**, *92*, 639.
4. Espadas-Torre, C.; Oklejas, V.; Mowery, K.; Meyerhoff, M.E. *JACS*, **1997**, *119*, 2321-2322.
5. Keefer, L.; Nims, R.; Davies, K.; and Wink, D. *Meth Enzym.* **1996**, *268*, 281-293.
6. Mowery, K. and Meyerhoff, M. "Thromboresistant Ion-Selective Electrodes via Nitric Oxide Release Polymeric Membranes." Paper 480. Presented at The Pittsburgh Conference, New Orleans, LA, March **1998**.
7. Smith, D. J.; Chakravarthy, D.; Pulfer, S.; Simmons, M. L.; Hrabie, J. A.; Citro, M. L.; Saavedra, J. E.; Davies, K. M.; Hutsell, T. C.; Mooradian, D. L.; Hanson, S. R.; Keefer, L. K. *J. Med. Chem.* **1996**, *39*, 1148-1156.
8. Saavedra, J.; Keefer, L.; Roller, P. and Atamatsu, M. "Biopolymer-Bound Nitric Oxide-Releasing Compositions, Pharmaceutical Compositions Using the Same and Methods of Treating Biological Disorders Using the Same." U.S. Patent 5 632 981, **1997**.
9. Keefer, L. and Hrabie, J. "Implants, Protheses and Stents Comprising Polymer-Bound Nitric Oxide/Nucleophile Adducts Capable of Releasing Nitric Oxide." U.S. Patent 5 676 963, **1997**.
10. Hulanicki, A., Matuszewski, W. and Trojanowicz, M. *Anal. Chim. Acta* **1987**, *261*, 119.
11. Ignarro, L. J.; Fukuto, J. M.; Griscavage, J. M.; Rogers, N. E.; Byrns, R. E. *Proc. Natl. Acad. Sci. USA* **1993**, *90*, 8103-8107.
12. Espadas-Torre, C. and Meyerhoff, M.E. *Anal. Chem.* **1995**, *67*, 3108.

SYNTHESIS AND CHARACTERIZATION OF ALIPHATIC POLYESTERS WITH HYDROPHILIC PENDANT GROUPS USING ENZYMES

Billie J. Kline, Eric J. Beckman, and Alan J. Russell

Department of Chemical and Petroleum Engineering
University of Pittsburgh
Pittsburgh, PA 15261

INTRODUCTION

Research in the area of aliphatic polyesters has been motivated by the need for efficient synthesis of functional polyesters. By generating a linear polyester with a reactive pendant group, the number of potential applications increases greatly. Traditional, unsubstituted aliphatic polyesters are biodegradable and hydrophobic, however the sensitivity of the polyester toward water can be manipulated by adding a pendant hydroxyl or carboxyl group to each repeat unit of the polyester. Additionally, the pendant groups can be reacted to form novel comb polyesters.

Functional polyesters are difficult to produce chemically and the synthesis involves the formation of a monomer possessing a protected functional group. To attain the substituted polyester, tedious multi-step reaction pathways comprising polymerization and deprotection of the functional group are necessary.[1-3] Several difficulties can arise as a result of the complex synthesis including low yield of the monomer, polymer degradation during the deprotection step, and unreliable amounts of deprotected pendant groups.

The necessity for functional group protection during polymerization lies in the nonspecific nature of chemical synthesis. Polyester networks have been produced when the reaction of glycerol and various dicarboxylic acids has been performed.[4,5] However, the straightforward synthesis of a multi-hydroxyl compound with a diester or a diacid can take place through the use of a biocatalyst. One of the major advantages of enzymes in addition to their activity at mild conditions is its intrinsic specificity which provides them with the ability to introduce interesting characteristics such as chirality to the reaction product.[6,7] The goal of this work is to produce linear polyesters with pendant hydroxyl groups through an enzymatic one-step synthesis.

EXPERIMENTAL

Materials

Divinyl adipate (DVA) was acquired as a kind gift from Union Carbide Corporation. Novozym 435® was purchased from Novo Nordisk Bioindustries (Denmark). All other chemicals necessary for the reaction, polymer recovery and characterization were obtained from Aldrich (St. Louis, MO).

Procedure

DVA and an equimolar amount of the multi-hydroxyl compound are added to a vial which is incubated at 50°C until DVA is melted. Candida antarctica lipase B (commercially Novozym 435®) is then added to the reaction mixture and the vial is placed in an incubator/shaker at a temperature of 50°C and an agitation speed of 250 rpm. After 24 hours, the reaction contents are dissolved using tetrahydrofuran (THF) and the enzyme is separated from the reaction solution by filtration. The product polyester is recovered by evaporating the THF.

Characterization

Molecular weight averages were obtained using gel permeation chromatography (GPC) performed on a Waters 150CV gel permeation chromatograph with THF as the mobile phase. Polystyrene standards were used for calibration purposes. Matrix-assisted laser desorption/ionization – time of flight (MALDI-TOF) mass spectrometry was performed using a PerSeptive Biosystems Voyager Elite DE mass spectrometer in linear mode. The samples were prepared by first, spotting a solution of sodium iodide (0.1 mg/ml in distilled water) onto the target plate. Then, 2 µl of the polyester solution (20 mg/ml in THF) was added to 18 µl of a dithranol solution (5 mg/ml in THF) and this mixture was spotted onto the target plate on top of the dried NaI solution. Titrimetric analysis for hydroxyl content was performed according to a method outlined by the Bayer Corporation (Method Number 414-C, Miles' Quality Laboratories).

RESULTS AND DISCUSSION

Figure 1 shows the reaction of glycerol with DVA to produce linear polyester with hydroxyl pendant groups. The byproduct of the reaction is vinyl alcohol which tautomerizes to acetaldehyde. While this side reaction aids polymerization by shifting the reaction in the forward direction, acetaldehyde has previously been reported to deactivate some types of lipases. Novozym 435® was chosen as the lipase for these reactions due to its stability in the presence of acetaldehyde.[8] Previous work has also shown that this enzyme was successful in catalyzing the reaction of DVA with 1,4-butanediol under solvent-free conditions.[9]

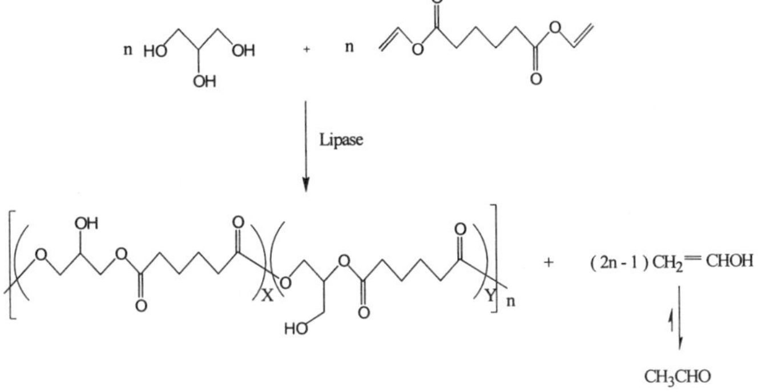

Figure 1. Lipase-catalyzed polycondensation between glycerol and divinyl adipate.

Reactions of various triols with DVA resulted polyesters of varying molecular weights shown in Table 1. The MALDI-TOF spectra of these polyesters confirm the presence of linear hydroxyl-substituted polyester indicated by peaks that are separated by the appropriate molecular weight for the polyester repeat unit. For example, the molecular weight of the repeat unit for the glycerol/DVA polyester is 202 Daltons. No evidence of network formation was present. Furthermore, hydroxyl number analysis of the polyester has confirmed the presence of approximately one hydroxyl group per repeat unit. Attempts to control the hydroxyl content of the polyester by adding 1,4-butanediol to the reaction mixture were successful.

Table 1. Molecular Weight Distribution Data for Triol/DVA Reactions

Triol Type	M_w	PDI
Glycerol	11569	2.44
1,2,4-Butanetriol	8060	2.23
1,2,6-Trihydroxyhexane	14300	3.35

Reaction Conditions: ~1wt% enzyme, 24 hours

Figure 1 shows that two different repeat units exist when DVA and glycerol react and three repeat units exist when the triol is asymmetric. The actual structure of the polyester depends on the regioselectivity of the enzyme as well as the relative reactivity of the primary and secondary hydroxyl groups. Determining whether the enzyme exhibits regioselectivity can be accomplished be identifying which two hydroxyl groups react on the triol. DVA was reacted with a series of diols containing primary and secondary hydroxyl groups. The degrees of polymerization of the resulting polyesters were compared with that of the polyester formed from the reaction of the triols with DVA. The results, shown in Table 2, clearly indicate that 1,2-propanediol reacts at a much slower rate than 1,3-propanediol. If the secondary hydroxyl groups on glycerol possesses the same reactivity as that on the diol it can be concluded that glycerol and the polyester consists of approximately 95% of the repeat unit denoted by X in Figure 1.

Table 2. Primary versus Secondary Hydroxyl Reactivity

Diol/Triol	M_w	Degree of Polymerization
1,2-Propanediol	823	6.0
1,3-Propanediol	12480	91
Glycerol	10160	70

Degree of polymerization was calculated by dividing the M_w by the average molecular weight of the two monomers.

CONCLUSIONS

The inherent specificity of the enzyme has been exploited to achieve the straightforward synthesis of aliphatic polyester with pendant hydroxyl groups. Since the synthesis only requires one step to achieve polymer, this work constitutes an improvement over chemical synthesis of functional polyesters. The reaction of the ester group takes place at the primary hydroxyl groups such that the pendant groups are essentially of the same structure. Hydroxyl number analysis confirms that the polyesters contain approximately one hydroxyl group per repeat unit. The degree of polyester functionality can predictably controlled by adding specified amounts of 1,4-butanediol to the reaction mixture. Current work involves the further characterization of the hydroxyl-substituted polyester as well as the enzymatic synthesis of aliphatic polyesters with other types of pendant groups.

ACKNOWLEDGMENT

This work was funded by the National Renewable Energy Laboratories / Department of Energy.

REFERENCES

(1) Tian, D.; Dubois, P.; Granfils, C.; Jérôme, R. *Macromolecules* **1997**, 30, 406-409.

(2) Vert, M.; Lenz, R. W. *Polym. Prepr. (Am. Chem. Soc., Div. Polym. Chem.)* **1979**, 20, 608-611.

(3) Barrera, D. A.; Zylstra, E.; Lansbury, P. T.; Langer, R. *J. Am. Chem. Soc.* **1993**, 115, 11010-11011.

(4) Kiyotsukuri, T.; Kanaboshi, M.; Tsutsumi, N. *Polym. Int.* **1994**, 33, 1-8.

(5) Nagata, M.; Kiyotsukuri, T.; Ibuki, H.; Tsutsumi, N.; Sakai, W. *React. Funct. Polym.* **1996**, 30, 165-171.

(6) Margolin, A. L.; Crenne, J.-Y.; Klibanov, A. M. *Tetrahedron Lett.* **1987**, 1607-1610.

(7) Wallace, J. S.; Morrow, C. J. *J. Polym. Sci., Part A: Polym Chem.* **1989**, 27, 2553-2567.

(8) Weber, H. K.; Faber, K. *Methods Enzymol.* **1997**, 286, 509-518.

(9) Chaudhary, A. K.; Lopez, J.; Beckman, E. J.; Russell, A. J. *Biotechnol Prog.* **1997**, 13, 318-325.

FORMATION OF A POLYMERIZABLE PHOSPHOLIPID HYDROGEL FROM A TRANSIENT NANOTUBULE PHASE.

S. Svenson, P. B. Messersmith, Division of Biological Materials, Northwestern University, 311 E. Chicago Ave., Chicago, Illinois 60611

A polymerizable hydrogel based on 1,2-bis(tricosa-10,12-diynoyl)-sn-glycero-3-phosphocholine and 1,2-dinonanoyl-sn-glycero-3-phosphocholine was studied with regard to its use as scaffold for the formation of artificial bone tissue replacements. Equimolar mixtures of the phospholipids were suspended in hot water and allowed to cool to room temperature. Nanotubules are formed within about 2 h from roughly spherical lipid particles. They posses an uniform diameter of 50-60 nm, lengths between 10 and >100 microns, and are stable for several hours. The tubules eventually transform into a network of entangled helical threads, which appears to occur via planar intermediates and typically results in physical gelation of the suspension. The nanotubule phase and the network gel can be polymerized by exposure to UV radiation at 254 nm. UV-irradiated samples exhibit an intense red color, indicating polymerization of the diacetylenic phosphocholine.

Cycloaliphatic Epoxide Crosslinked Acrylic Latex Coatings and the Crosslinking Conditions

Shaobing Wu, Mark D. Soucek*
Department of Polymers & Coatings
North Dakota State University
Fargo, ND 58105

Introduction

Conventional latex coatings primarily rely on the coalescence of thermoplastic polymer particles to provide mechanical and resistance properties. Coalescence is a result of physical entanglement of polymer molecules, and consequently it is not an adequate substitute for the chemical crosslinking. In order to provide the similar performance of solventborne thermosetting coatings, crosslinking technology has been adopted into the latex coatings.

A number of crosslinkers and crosslinking technologies have been developed for the latex coatings in recent years [3-6]. Polycarbodiimides [4], aziridines [5], and oxazlines [6] are the typical crosslinkers used for thermosetting waterborne coatings. Due to increasing concerns on the environment and health, needs for new crosslinkers always arise for lower toxic hazards and higher performance. In this study, hydroxyl or carboxyl functional acrylic latexes were synthesized, and a cycloaliphatic diepoxide was developed to be a alternative crosslinker for the acrylic latex coatings. The effects of crosslinking conditions on the coatings properties are reported herein.

Experimental

Materials were used without further purification. Methyl methacrylate (MMA), butyl acrylate (BA), methacrylic acid (MAA), 2-hydroxyethyl methacrylate (HEMA), ammonium persulfate, and p-toluene sulfonic acid (p-TsOH) were purchased from Aldrich. Cycloaliphatic diepoxide (UVR-6105) and surfactants (Triton-200 and Tergital-XJ) were supplied by Union Carbide Corporation.

Latex synthesis was carried out in a four-neck fluted round bottom flask (500 mL). The glass transition temperature of the latex polymers was designed to be 0 °C. The recipes for the synthesis of hydroxyl or carboxyl functional latexes are listed in Table 1.

Table 1. Recipes for the Synthesis of the Latexes *.

Latex	MMA	BA	HEMA	MAA	Ammonium Persulfate	Triton-200	NaHCO₃
HO **	72.74	90.94	16.33	0.0	0.81	8.04	0.4
COOH	71.56	97.64	0.0	10.80	0.81	8.04	0.4

* All the ingredient quantities are listed in grams.
** The content of the hydroxyl and the carboxyl functional monomers of the two latexes are equivalent in moles (0.1256 moles).

The monomers were mixed and then slowly added into a stirred water solution (0.2 g NaHCO₃, 7.96 g Triton-200, and 96 g water) to obtain a monomer pre-emulsion. The reactor was first charged with 77 g water, 14.0 g the pre-emulsion, 0.203 the initiator, 0.2 g NaHCO₃, and 0.08 g Triton-200 at 80 °C. The content were then heated to 80 °C for 30 min. The pre-emulsion was fed to the reactor over a period of 3.0 h using a addition pump (ISMATEC REGLO-100). The initiator was concurrently added to the reactor at a constant rate using a universal syringe pump (Valley Scientific Model 575). The polymerization temperature was maintained at 80 °C under nitrogen for 3 h.

After the addition of all the ingredients, the reaction mixture was heated for an additional 2 h at 85 °C. Then, the reactor contents were cooled to 30-35 °C, and neutralized with ammonia (pH of 8.5). The latex was obtained after filtration of the contents through a 300 mesh screen to remove any residual coagulum.

The cycloaliphatic diepoxide (UVR-6105) was emulsified using a mixture of anionic and non-ionic surfactants (Triton-200 and Tergital-XJ). A stoichiometric amount of the epoxide emulsion was then added into the latexes. The hydroxyl functional latex coatings contained 0.6 wt % ammonia neutralized toluene sulfonic acid (p-TsO⁻ NH₄⁺) as the catalyst. All the latex coatings were cast on aluminum or glass substrates using a drawdown bar (5 mils). After coalescence, both hydroxyl and carboxyl functional latex coatings were cured at three temperatures (130, 150, or 170 °C) for three different durations (60, 105, or 165 min). To study the crosslinker effect on the coatings properties, 87.5 and 112.5 % of the stoichiometric amount of the epoxide were also added to formulate the coatings. The gel content, water absorption, and the mechanical properties (hardness, adhesion, etc.) of the crosslinked coatings were measured 24 h after baking.

Results & Discussion

The crosslinking reactions of carboxyl or hydroxyl functional latex coatings with cycloaliphatic epoxide were monitored using Differential Scanning Calorimetery (DSC). Figure 1 depicts the crosslinking reaction of the carboxylic group with the cycloaliphatic diepoxide. The first scan indicates the reaction of the diepoxide with phthalic acid. The resultant exotherm identifies the reaction temperature for the ester formation. The second scan is the carboxyl functional latex reaction with the diepoxide as shown in Scheme 1, and the third scan proves that the reaction is complete. From this data, it is clear that the crosslinking reaction (ester formation) starts at 110 °C and is complete after 175 °C. Inferred (IR) spectra were consistent with the result.

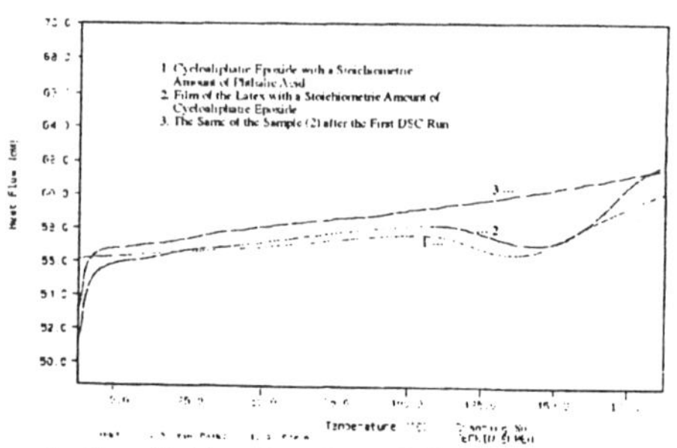

Figure 1. Crosslinking Reactions of Carboxyl Group with Cycloaliphatic Diepoxide Group and Carboxyl Functional Latex Coatings (MAA 15 wt %) with Cycloaliphatic Diepoxide (DSC).

The DSC scans of the hydroxyl functional latex coating, the Tone diol with cycloaliphatic epoxide, and cycloaliphatic epoxide, are illustrated in Figure 2, respectively. An effective temperature for the crosslinking reaction of the hydroxyl latex coating with the crosslinker appeared to be approximately at 180-200 °C. The homopolymerization of cycloaliphatic epoxide occurred at a higher temperature (240-260 °C). IR spectra were consistent with the result. Similarly, the exothermic peak and the onset temperature of the Tone diol with the crosslinker further confirmed the crosslinking reaction of the hydroxyl groups with cycloaliphatic epoxide.

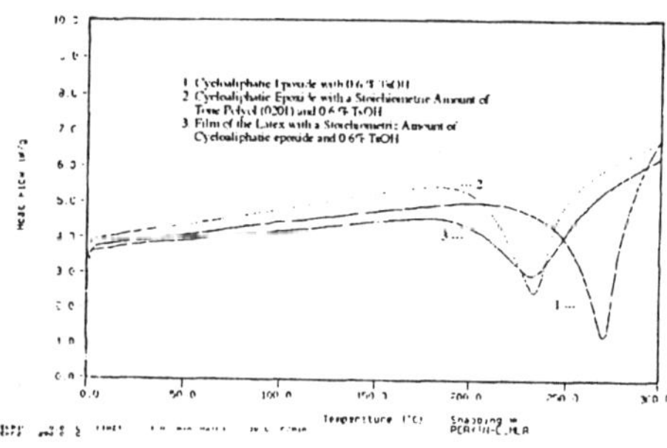

Figure 2. Crosslinking Reactions of Hydroxyl Group with Cycloaliphatic Diepoxide Group and Hydroxyl Functional Latex Coatings (HEMA 22.7 wt %) with Cycloaliphatic Diepoxide (DSC).

The gel content and the water absorption of the crosslinked carboxyl functional latex coatings as a function of crosslinking temperatures are shown in Figures 3 and 4. The gel contents of the coatings with the crosslinker were substantially higher than the coatings without the crosslinker. The water absorption of the coatings with or without the crosslinker also reflected the crosslinking effect. As the crosslinking temperature was increased, the gel content did not appear to significantly increase. The water absorption showed a small decrease as the temperature was increased. It may be assumed that a crosslinking temperature of 130 °C was high enough to force the carboxyl-epoxide reaction to completion, which is consistent with the DSC result.

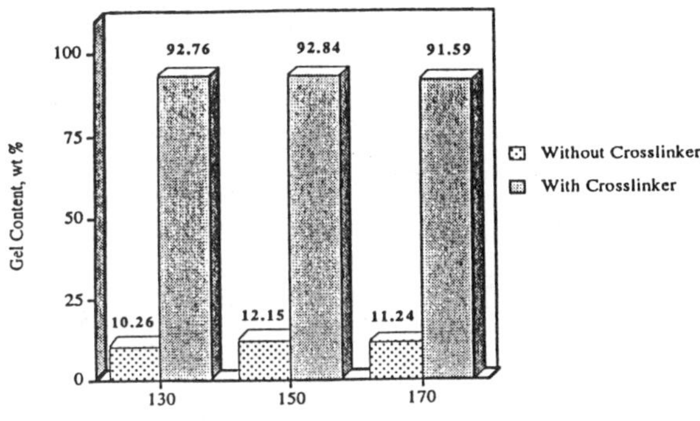

Figure 3. Gel Content of the Carboxyl Latex Coatings as a Function of Crosslinking Temperature (MAA 6.0%).

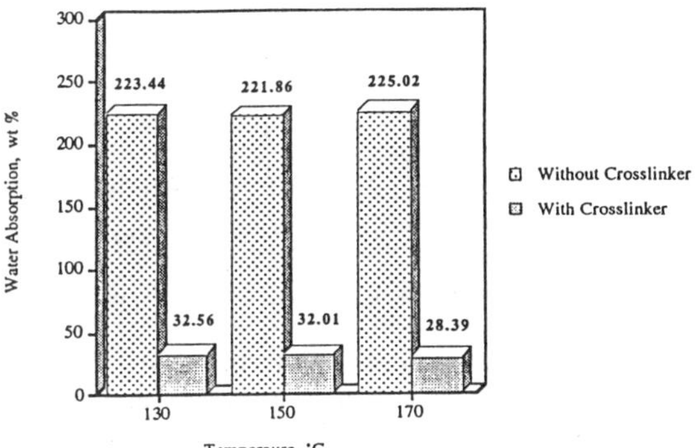

Figure 4. Water Absorption of the Carboxyl Latex Coatings as a Function of Crosslinking Temperature (MAA 6.0%).

Similarly, for the hydroxyl functional latex coatings, the epoxide crosslinking increased the gel content and decreased the water absorption of the coatings. However, the gel content and the water absorption of the crosslinked hydroxyl functional latex coating were significantly affected by the crosslinking temperatures. As the temperature was increased from 130 to 170 °C, the gel content greatly increased from 65 to 92 wt %, and the water absorption substantially decreased from 181 to 39 wt %. It may be assumed that a higher crosslinking temperature of 170 °C be required force the hydroxyl-epoxide reaction to completion, which is consistent with the DSC result.

The gel content and the water absorption of both crosslinked carboxyl and hydroxyl functional latex coatings were also affected by the baking time. As the baking time was increased, a small increase in the gel content was observed. A minimum cure time of 105 min was derived for both carboxyl and hydroxyl latex coatings crosslinked at 130 and 170 °C, respectively.

Figure 5 showed the gel content of the epoxide crosslinked carboxyl latex coatings as a function of cycloaliphatic epoxide. As the percentage of the crosslinker was increased, the gel content increased. The water absorption also was observed to significantly decrease. In addition, the adhesion of the coating to the substrate increased as the crosslinker was increased, as shown in Figure 6. Similar trends were showed for the hydroxyl functional latex coatings.

Conclusion

Cycloaliphatic diepoxide was an effective crosslinker for both carboxyl or hydroxyl functional acrylic latexes. The crosslinking reactions of the carboxyl or hydroxyl latex coatings with the epoxide were significantly affected by the crosslinking temperature, time, and the amount of the crosslinker used. The crosslinking reaction of the carboxyl functional latex coatings required a lower temperature (130 °C) than the hydroxyl functional latex coatings (170 °C). At the effective crosslinking temperatures, a minimum crosslinking time was required for both of the latex coatings to complete the crosslinking reactions. Addition of more than a stoichiometric amount of the crosslinker to the

coatings was favorable for the enhancement of overall coatings properties.

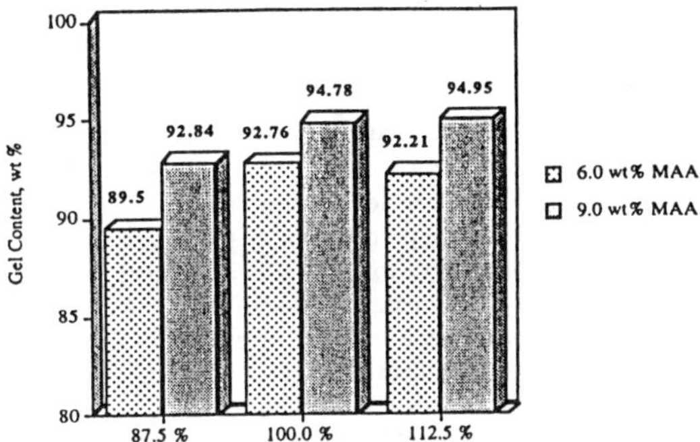

Figure 5. Gel Content of the Carboxyl Latex Coatings as a Function of the Crosslinker.

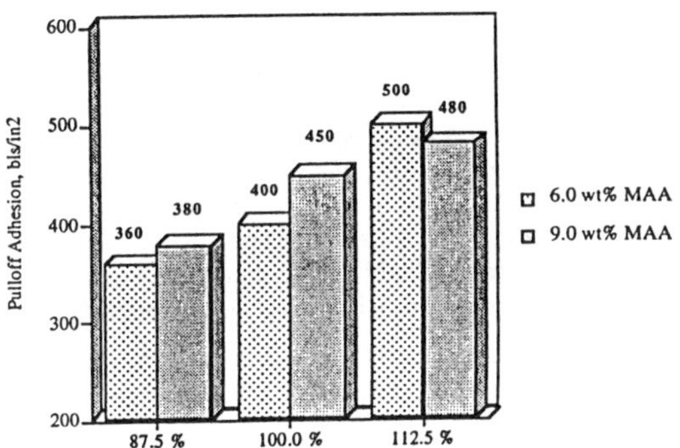

Figure 6. Pull-off Adhesion of the Carboxyl Latex Coatings as a Function of the Crosslinker.

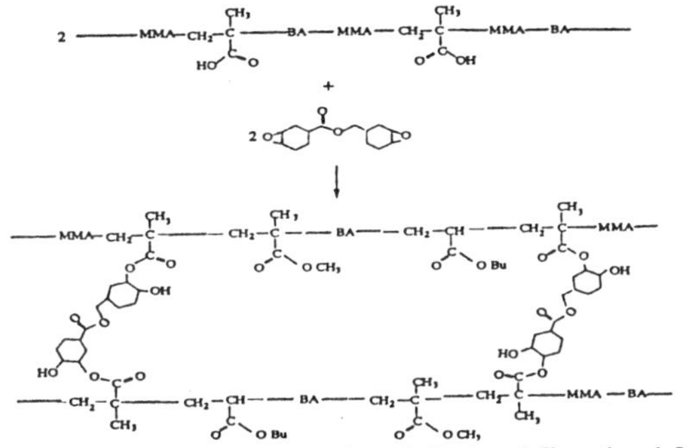

Scheme 1. Crosslinking Reaction of Carboxyl Functional Latex with Cycloaliphatic Diepoxide

References

1. Richey, B.; Wood, T., *Surf. Coat. Int.*, **1994**, 1, 26.
2. Bouvy, A.; John, G. R. *Polym. Paint Colour J.*, **1991**, V181, 285.
3. Eaton, R. F.; Lamb, K. T. *J. Coat. Tech.*, **1996**, 68(860), 49.
4. Brown, W.; *Surf. Coat. Int.*, **1995**, 6, 238.
5. Moles, P., *Polymers & Paints Color J.*, **1991**, 181(267), 4279.
6. Moor, G.; Zhu, D-W., *Surf. Coat. Int.*, **1995**, 9, 376.
7. Eslinger, R. D., *J. Coat. Tech.*, **1995**, 67, 850

Unsaturated Polyetherester (PEER™) Resin Modified with Bisphenol A Epichlorohydrin Diepoxy

Gangfeng Cai*, Lau S. Yang and Jeffrey A. Klang
ARCO Chemical Company
http://www.arcochem.com
3801 West Chester Pike, Newtown Square, PA 19073
*E-mail: cnsgxcc@arcochem.com

INTRODUCTION

In the unsaturated polyester industry, there are two major classes of high-performance resins: isophthalic corrosion-resistant resins and vinyl ester resins[1]. The isophthalic resins, which incorporate isophthalic acid, give thermosets with excellent physical properties and good chemical resistance. However, they are quite susceptible to hydrolysis in a caustic environment. Vinyl ester resins made from methacrylic acid and epoxy resins give thermosets with an excellent overall balance of properties, including high tensile, flexural strengths and excellent corrosion resistance. Unfortunately, vinyl ester resins are much more costly. In addition, bonding of vinyl esters to another commercial resin is often difficult because of compatibility problems.

A new type of unsaturated polyetherester resin (PEER™ resins, products of ARCO Chemical Company) showed low viscosity, good physical properties and good water resistance for many composites applications[2-9]. To further improve the chemical corrosion resistance toward harsh environments such as aqueous acid or caustic solutions, we modified the PEER resin with a bisphenol A based epoxy resin[10]. The epoxy modified PEER resin maintains low viscosity, high tensile elongation, high heat distortion temperature, it also gains much improved physical properties and corrosion resistance. This resin offers the composites industry a new low-VOC material that cures similarly to an ordinary unsaturated polyester with better corrosion resistance than an isophthalic resin, especially in caustic environment.

EXPERIMENTAL

I. Synthesis of Epoxy Modified PEER Resin (Scheme I)

To a 12-L reactor equipped with a stirrer, a nitrogen sparger, and a condenser were added poly(propylene oxide) polyol (Mn=2000, 4725 g), propylene glycol (1475 g), maleic anhydride (3800 g) and p-toluene sulfonic acid (10.0 g) . The mixture was heated to 160-250 °C and maintained at the temperature until an acid number of 95-105 mgKOH/g was reached. Propylene glycol (200 g) was added and the reaction mixture was heated to an acid number of 50-90 mgKOH/g. Hydroquinone (1.0 g) and triethylamine (2.0 g) were added after the mixture was cooled to less than 160 °C. Bisphnol-A epoxy resin (EPON 828, product of Shell Chemical Company, 1500 g) was added and the mixture was heated at 150 °C for 2 to 5 hrs to an acid number of 10-45 mgKOH/g. The resin product was cooled to 120 °C and mixed with styrene (35 wt.%).

II. Characterization of Epoxy Modified PEER Resin

Average molecular weight and molecular weight distribution was measured with gel permeation chromatograph using poly(propylene oxide) diols as standards. Acid number is determined by dissolving the polymer in THF and titrating it with a 0.1 N KOH/water solution. Tensile and flexural properties of cured polymers were done according to ASTM methods D-638 and D-790. Heat distortion temperature was measured according to ASTM D-648.

III. Preparation of Clear Cast from Epoxy Modified PEER Resin

The PEER resin (often contain 35 wt.% styrene) was used as is or diluted to 40 to 50 wt.% styrene. A 0.12-0.15 wt.% of cobalt naphthenate solution (6% cobalt naphthenate in mineral sprits) was diluted with a small amount of styrene before adding to the PEER resin. A 1.2-1.5 wt.% of methyl ethyl ketone peroxide, such as LUPERSOL DDM-9 initiator (product of Atochem) was finally added. The well-mixed resin solution (degassed if necessary) was then poured into a mold. The resin was cured at room temperature overnight, and post-cured at 100 °C for 5 hrs.

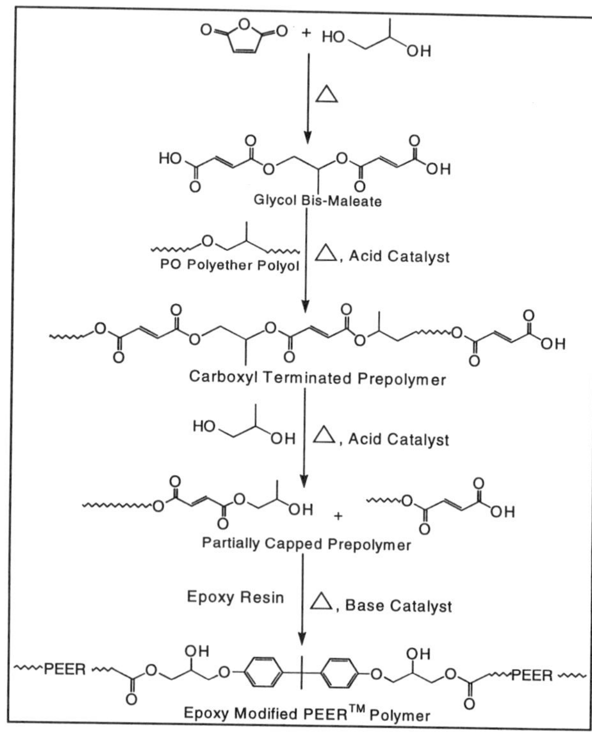

Scheme I. Synthesis of Epoxy Modified PEER™ Resin

IV. Corrosion Resistance of PEER Thermosets

Five standard flexural strength test specimens (4"×½"×1/8") were immersed in distilled water (or 5 wt.% KOH , or 5 wt.% HCl aqueous solution) in a sealed glass tube and were heated at 100 °C in an oil bath for 6 days. The specimens were cooled, removed from water, and wiped dry. The samples were weighed and tested for Barcol hardness within 2 hrs of removal from water. Flexural strength was tested according to ASTM D-790.

RESULTS AND DISCUSSIONS

I. Synthesis of Epoxy Modified PEER Resin

The reaction was followed by measuring the acid number and GPC. As shown in **Figure 1**, the acid number dropped quickly in the first hour showing that epoxy resin reacts quickly with the carboxylic terminated PEER prepolymer. As indicated in GPC traces (**Figure 2**), most of the epoxy resin reacted with the PEER prepolymer. The final resin (**Table 1**) has an average molecular weight about 1000-1800 and a viscosity of 500-1000 cps at 35 wt.% styrene. The average molecular weight can be controlled by adjusting the amount of propylene glycol and epoxy resin.

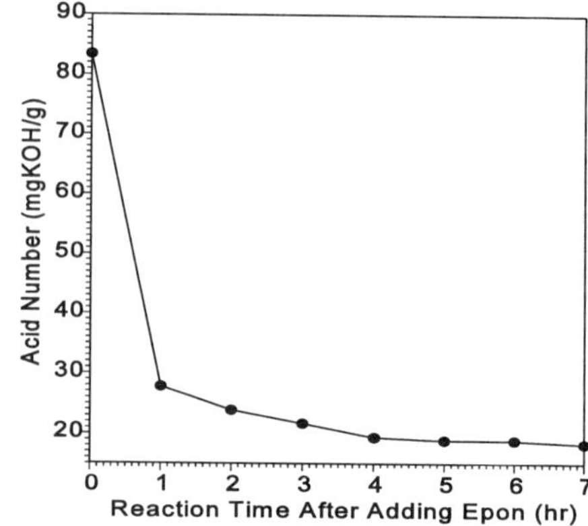

Figure 1. Variation of acid number with reaction time in the synthesis of epoxy modified PEER™ resin after adding 20 wt.% Epon-828 at 150 °C.

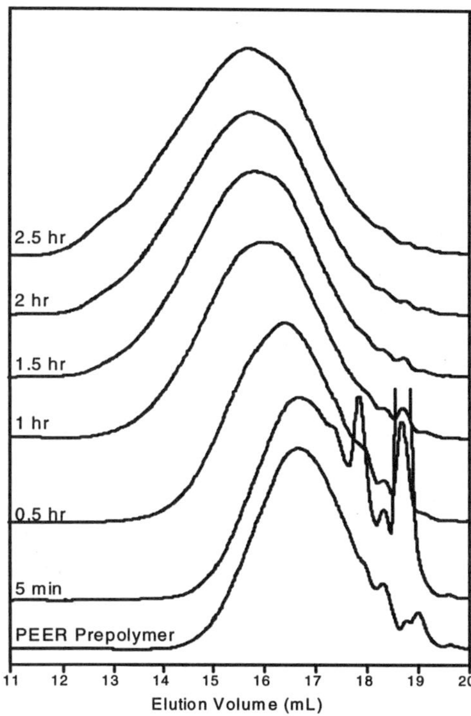

Figure 2. GPC traces of epoxy modified PEER™ resins at various stages after adding 20 wt.% Epon-828 resin at 150°C.

Table 1. Properties of Liquid Epoxy Modified PEER™ Resin

Pure Resin Acid Number, mgKOH/g	15-35
Viscosity at 77 °F, 35% styrene, cps	500-1000
Specific Gravity at 77 °F, 50% styrene	1.043
Mn	1000-1800
Mw	3000-9000
Mp	1800-4000
Mw/Mn	3-5

II. Properties of Clear Casting from Epoxy Modified PEER Resin

As illustrated in **Scheme II**, the epoxy modified PEER resin combines the advantages of vinyl ester resin and polyetherester. These include excellent flexibility, toughness, and chemical/water resistance. The PEER resin inherently has low viscosity due to its random length polyether blocks. In addition, the PEER resin has a trans/cis double bond ratio of 95/5 which contributes to high heat distortion temperature. Indeed, this is comparable with vinyl ester or isophthalic resins.

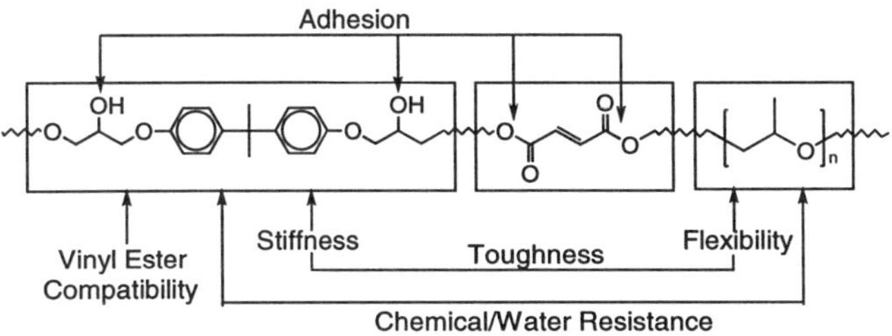

Scheme II. Functional Structure of Epoxy Modified PEER™ Resin

Table 2. Physical Properties of Clear Castings Made with Epoxy Modified PEER™, Vinyl Ester and Isophthalic Resins

Properties		Epoxy Modified PEER Resin	Vinyl Ester Resin	Isophthalic Corrosion-Resistant Resin
Viscosity, cps at 50% Styrene, 77 °F		120	350	475
Tensile Strength, psi		10400	12000	10100
Tensile Modulus, kpsi		468	492	566
Tensile Elongation, %		4.5	6.1	2.1
Heat Distortion Temperature, °F		220	226	225
Barcol Hardness		38	38	45
Flexural Strength (psi)	Before 6-Day Boil	18280	20800	22610
	Water Boil	17500	20000	17500
	5% KOH Boil	17600	20500	11200
	5% HCl Boil	16200	20500	17700
Flexural Modulus (kpsi)	Before 6-Day Boil	469	493	583
	Water Boil	458	493	578
	5% KOH Boil	435	488	568
	5% HCl Boil	452	497	575

In addition to low viscosity, as shown in **Table 2**, the epoxy modified PEER resin has physical properties comparable with a vinyl ester resin. It is tougher and more resistant to caustic environment than an isophthalic acid based polyester resin. The PEER resin maintained more than 95% of the flexural strength after 6-day boil test in both water and 5% KOH aqueous solution.

CONCLUSIONS

A novel unsaturated polyetherester resin (PEER™) resin modified with bisphenol A di-epoxide has been developed. The PEER resin shows low viscosity, good physical properties and excellent corrosion resistance, especially in a caustic environment. This new class of resin offers low VOC and high performance to the composites industry.

REFERENCES

[1] J. Selley, *Encyclopedia of Polymer Science and Engineering*, H.F. Mark, et.al. Ed., Vol. 12, p256, Wiley, New York, 1985.
[2] J. Klang and L. Yang, *US Patent 5,436,313* (1995) (to ARCO Chemical Co.).
[3] L. Yang and J. Klang, *US Patent 5,436,314* (1995) (to ARCO Chemical Co.).
[4] L. Yang and J. Klang, *US Patent 5,569,737*; (1996) (to ARCO Chemical Co.).
[5] G. Cai, S. Guo and L. Yang, *US Patent 5,580,909*; (1996) (to ARCO Chemical Co.).
[6] L. Yang and J. Klang, *US Patent 5,610,205* (1995) (to ARCO Chemical Co.).
[7] G. Cai, L. Yang and J. Klang, *US Patent 5,612,444*; (1997) (to ARCO Chemical Co.).
[8] G. Cai, S. Guo and L. Yang, *US Patent 5,633,315*; (1997) (to ARCO Chemical Co.).
[9] J. Klang, *US Patent 5,677,396*; (1997) (to ARCO Chemical Co.).
[10] L. Yang and K. Johnson, *US Patent 5,684,086* (1997) (to ARCO Chemical Co.).
[11] G. Cai, J. Klang and L. Yang, *US Patent 5,696,225*) (1997) (to ARCO Chemical Co.).

EFFECTS OF POLYURETHANE-BASED THERMOPLASTIC ADDITIVES ON THE MISCIBILITY AND CURING BEHAVIOR OF UNSATURATED POLYESTER RESINS

Yan-Jyi Huang[*], Cheng-Jou Chu, and Shwu-Chyn Lee
Department of Chemical Engineering
National Taiwan University of Science and Technology
Taipei, Taiwan 107, Republic of China

INTRODUCTION

Adding specific thermoplastic polymers as low profile additives (LPA) in the unsaturated polyester resins (UP) could lead to a reduction or even elimination of the polymerization shrinkage during the cure process[1-2]. Depending on the chemical composition and structure of UP resins and LPA employed, differing degrees of drift in styrene/UP/LPA composition as a result of phase separation during the cure would occur for the styrene/UP/LPA system[3-5]. This could greatly affect the physical and mechanical properties of cured samples.

The objective of this work is to investigate the effects of chemical structure and molecular weight of three series of thermoplastic polyurethane-based LPA on the miscibility and the curing behavior of ST/UP/LPA systems.

EXPERIMENTAL

2,4-tolylene diisocyanate (2,4-TDI) and varied polyhydroxy materials were first reacted to make the hydroxyl-terminated PU prepolymers. Subsequently, excess maleic anhydride (MA) was added to make the carboxyl-terminated thermoplastic PU-based LPAs. The five LPAs used in this study are summarized in Table1.

The UP resin[5] was made from MA and 1,2-propylene glycol (PG) with a molar ratio of 1:1.09. The acid number and hydroxyl number were found to be 28.9 and 32.0 by titration, which gives an M_n of 1820 g/mole.

For the sample solution, 10% by weight of LPA was added, while the molar ratio of styrene to polyester C=C bonds was fixed at MR = 2/1. The reaction was initiated by 1% by weight of tert-butyl perbenzoate.

To study the compatibility of ST/UP/LPA systems prior to reaction, 20g of sample solutions were prepared in 100ml separatory glass cylinders, which were placed in a constant-temperature water bath at 30°C. The phase separation time was recorded.

For the cure kinetic study, 6-10 mg sample solution was placed in a hermetic aluminum sample pan. The isothermal reaction rate profile at 110°C was measured by a DuPont 9000 differential scanning calorimeter.

In the morphological study, the fractured surface of the sample, which was cured at 110℃ for 1 hr, followed by a post-cure at 150℃ for another 1 hr in a stainless steel mold, was observed by SEM at 1000 X.

For the measurements of transition temperatures in each phase region, the sample specimen with a thickness of 1mm was polarized at 230°C under an electric field of 500V/mm over a period of 10-30 min by using a Solomat 91000 TSC/RMA apparatus. Thermally stimulated currents were recorded from -150°C to 250°C at a heating rate of 7°C/min.

RESULTS AND DISCUSSION

The upper critical solution temperatures (UCST or T_c) prior to reactions for the ST/UP/LPA ternary polymer blends were calculated by using the Flory-Huggins theory and group contribution methods. Table 2 shows that the calculated T_c was lower for the sample solution containing the polyester-based PU (i.e. PDEA-PU) than that containing the polycaprolactone-based PU (i.e. PCL-PU), which reveals that the former system would be theoretically more compatible than the latter one. For both systems, adding a higher molecular weight of PU would result in the less compatibility of the system due to a higher calculated T_c. On the other hand, the calculated T_c was the lowest for the sample containing the polyether-based PU (i.e. PPG-PU), yet it is the only system with the phase separation occurring at room temperature prior to reaction (see Table 3). This would be attributed to the negligence of exothermic effect caused by the polar interaction between PU and UP resins in the calculation of ΔG_M.

During the cure at 110°C, the sample solution containing the polycaprolactone-based PU (i.e. PCL-PU) was also more incompatible than that containing the polyester-based PU (i.e. PDEA-PU). This could be evidenced by the emergence of a shoulder in the DSC rate profile at the later stage of reaction (see Fig. 1), and the characteristics of larger microgel particles for the fractured surface (compare Figs. 2(a)-(b) and (c)-(d)) for the former system. In contrast, the sample solution containing the polyether-based PU (i.e. PPG-PU) was the most incompatible during the cure, as evidenced by the DSC rate profile and the SEM micrograph. This could not be predicted by the calculated upper critical solution temperature for the uncured ST/UP/LPA systems, but would agree with the experimental results of compatibility study at room temperature (Table 3).

For the cured LPA-containing UP resin systems with morphologies as shown in Figs. 2(a)-(e), their mechanical behavior can be approximately represented by the Takayanagi models[6], where arrays of weak LPA (R) and stiff styrene-crosslinked polyester (P) phases are indicated (see Figure 3). The subscripts 1, 2, and 3 for P phases are employed due to the distinction of styrene and UP compositions as a result of phase separation during cure, and the quantities λ, φ, ξ, and ν or their indicated multiplications indicate volume fractions of each phase.

For the systems shown in Figs. 2(a)-(d), the microgel particles (phase P_1) would be surrounded by a layer of LPA (phase R). Between the LPA-covered microgel particles, there would be some other microgel particles (phase P_2), with different compositions of ST and UP from those in phase P_1, dispersed in the LPA phase (phase R). Hence, the characteristic globule microstructure may be represented by the P-P-S model as shown in Fig. 3(a), which is a parallel combination of the three elements, i.e. P_1, R, and P_2-R in series. In contrast, for the system shown in Fig. 2(e), the microstructure consists of a stiff globule continuous phase of styrene-crosslinked polyester (phase P_1) and a weak globule LPA-dispersed phase (i.e. the hole left behind due to the solvent extraction, the globule morphology of which prior to the solvent extraction can also be represented by a P-P-S model). Hence, the upper bound of mechanical behavior for the overall morphology can be represented by a P-(P-P-S) model as shown in Fig. 3(b), which is simply a parallel combination of the continuous phase P_1 and the dispersed phase denoted by a (P-P-S) model.

Table 4 shows that for the cured sample containing the semicrystalline PCL-PU, the melting point T_m of the PU could be identified by the method of TSC. Two glass transition temperatures of PU phases, one in parallel adjacent to the P phase (i.e. $T_g(R_2)$) and the other in series adjacent to the P phase (i.e. $T_g(R_1)$), could also be identified. The interaction between R_1 and the adjacent P phases would be less than that between R_2 and the adjacent P phases, leading to a lower value of $T_g(R_1)$ than $T_g(R_2)$ (i.e. from -44°C to -53°C vs. -22°C to 6°C).

Adding PPG1-PU would lead to the most incompatible ST/UP/LPA cured system, and the interaction of either P_1 or P_2 phase with the R phase would be the most insignificant. This would result in the highest T_g for either P_1 or P_2 phase among the five cured systems. Similarly, adding PCL-PU would cause a more incompatible cured system when compared with that adding PDEA-PU, hence T_g in either P_1 or P_2 phase for the former system would be higher.

With a fixed PU type, the higher molecular weight of PU would cause a higher T_g in either P_1 or P_2 phase, which could be due to the slower phase separation rate during the cure leading to less deviation from MR = 2/1 in either the continuous or the LPA-rich phases after the cure.

CONCLUSIONS

The phase separation characteristics of ST/UP/LPA systems during the cure, as revealed by the cured sample morphology, and the DSC reaction rate profile, could be generally predicted by the calculated upper critical solution temperature for the uncured ST/UP/LPA systems based on the Flory-Huggins theory and group contribution methods. For the proposed Takayanagi mechanical models, the transition temperatures in each phase region of cured samples in the model could be identified by the method of thermally stimulated currents (TSC).

REFERENCES

1. E.J. Bartkus and C.H. Kroekel, *Appl. Polym. Symp.* **15**, 113 (1970).
2. K.E. Atkins, in "*Sheet Molding Compounds: Science and Technology*," Ch. 4, H.G. Kia ed., Hanser Publishers, New York, 1993.
3. L. Suspene, D. Fourquier, and Y.S. Yang, *Polymer* **32**, 1593 (1991).
4. Y.J. Huang and C.C. Su, *J. Appl. Polym. Sci.* **55**, 323 (1995).
5. Y.J. Huang and W.C. Jiang, *Polymer, in press*.
6. M. Takayanagi, K. Imada, and T. Kajiyama, *J. Polym. Sci. Part C* **15**, 263 (1966).

Table 1 PU-based[a] LPAs used in this study

LPA codes	diol used	M_n	NCO/OH[b]	MA added[c]	M_n of PU[d]
PCL1-PU	polycaprolactone	1250	0.875	0.0625	18770
PCL2-PU	polycaprolactone	2000	0.933	0.010	27170
PDEA1-PU	poly(diethylene adipate)	1890	0.86	0.035	2870
PDEA2-PU	poly(diethylene adipate)	4890	0.85	0.015	5040
PPG1-PU	poly(propylene glycol)	2000	0.85	0.015	9750

a:containing 2,4-TDI;b:equivalent ratio;c:by equivalent;d:byGPC (g/mole).

Table 2 Calculated T_c (°K) for ST/UP/LPA uncured systems

PU Added	PCL1-PU	PCL2-PU	PDEA1-PU	PDEA2-PU	PPG1-PU
T_c	236.9	237.4	213.4	216.5	200.4

Table 3 Phase separation time for ST/UP/LPA uncured systems at 30°C

PU Added	PCL1-PU	PCL2-PU	PDEA1-PU	PDEA2-PU	PPG1-PU
Time(min)	∞[a]	∞	∞	∞	800

a:one phase.

Table 4 Transition temperatures (°C) in each phase region of Figures 3(a) and 3(b) as determined by the method of TSC

PU	$T_g(P_1)$	$T_g(P_2)$	$T_g(P_3)$	T_m	$T_g(R_1)$	$T_g(R_2)$
PCL1-PU	140	90	-	30	-44	5
PCL2-PU	191	97	-	40	-46	-22
PDEA1-PU	127	22	-	-	-49	-28
PDEA2-PU	137	23	-	-	-50	-28
PPG1-PU	225	163	100	-	-53	6

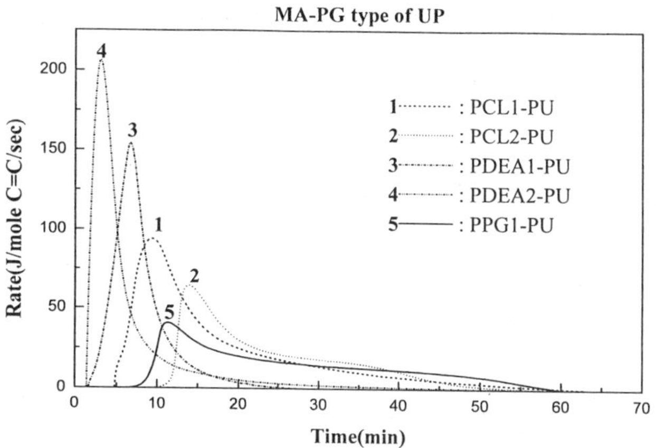

Fig. 1 Effects of LPA types on DSC reaction rate profile at 110°C

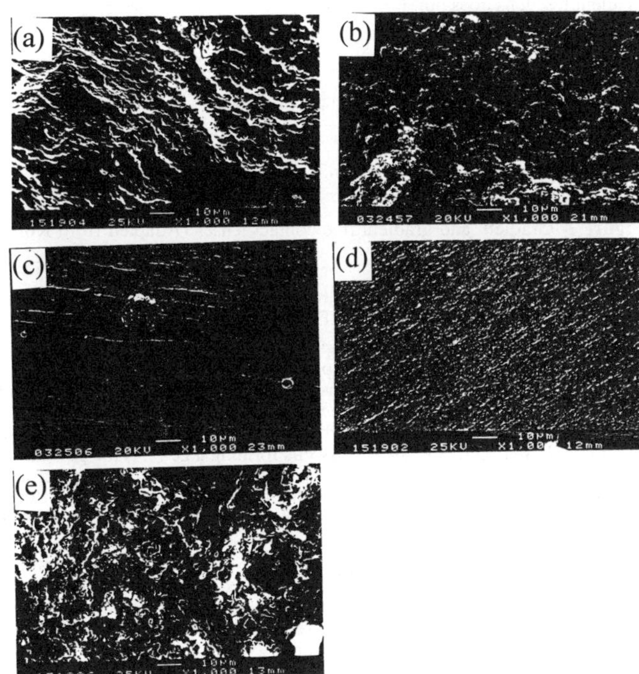

Fig. 2 Effects of LPA types on the cured sample morphology under SEM. (a)PCL1-PU,(b)PCL2-PU,(c)PDEA1-PU,(d)PDEA2-PU,and (e)PPG1-PU.

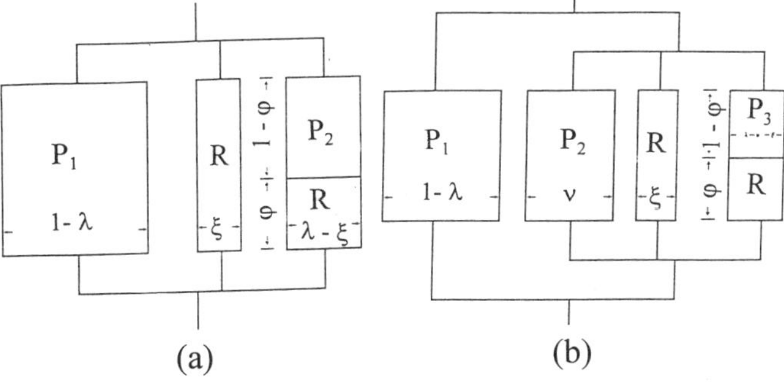

Fig. 3 The Takayanagi models for mechanical behavior of cured LPA-containing UP resin systems. (a) parallel-parallel-series (P-P-S) model, and (b) parallel-parallel-parallel-series (P-(P-P-S)) model. The area of each diagram is proportional to a volume fraction of the phase.

AN EXPERIMENTAL INVESTIGATION OF THE MECHANISM BEHIND ISOTHERMAL FRONTAL POLYMERIZATION

Lydia Lee Lewis, John A. Pojman,* Jim Owens, and Jennifer Coleman
University of Southern Mississippi
Hattiesburg, MS 39406

Introduction

Over the past two decades, gradient materials have become important in industry where they are used as fiber optic cables,[1] optical limiters,[2, 3] as well as in other applications.[4-6] These materials can be classified into two categories of gradient materials: optical limiters and Gradient Refractive INdex Materials (GRINs). Optical limiters contain a dye that limits high intensity light by absorbing wavelength(s) characteristic to the dye. They can be found in high-reflection coatings,[7] recording devices,[8] and optical switches.[4] GRINs contain a gradient in the refractive index of their composition medium, and this gradient gives the material special properties such as magnification or focalization. An example of the advantage of a gradient contact lens over a gradient free lens is shown in Figure 1 and shows the gradient materials property of focalization. The gradient lens focuses all incoming rays to the same point, whereas the gradient free lens does not.

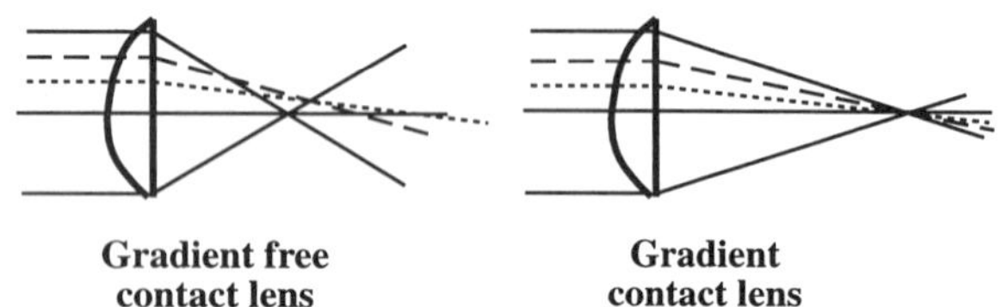

Figure 1 Gradient and gradient free contact lenses (Adapted from Koike et. al.[3])

The method of producing these gradient materials is known as isothermal frontal polymerization (IFP), or interfacial gel polymerization, and was developed by Koike for producing industrial gradient materials.[9, 10] It is a method that uses the Trommsdorff, or gel, effect[11] though a polymeric seed to convert monomer into polymer through a localized propagating reaction zone. The mechanism is shown in Figure 2 and is described as follows: Monomer and initiator from a bulk solution diffuse into the polymer seed, and the seed swells forming a gel. As the gel forms, polymerization reactions take place in it from the diffused monomer and initiator. The rate of polymerization of the particles in the gel is faster than the particles in the bulk solution due to the higher viscosity of the gel. This higher rate of polymerization is known as the gel effect.[11] As the particles polymerize in the gel, more particles are able to diffuse into the gel and polymerize creating a propagating front. The extent of propagation by the front is limited by the time before the bulk solution homogeneously polymerizes. To extend the time before homogeneous polymerization, an inhibitor may be used in the bulk solution.

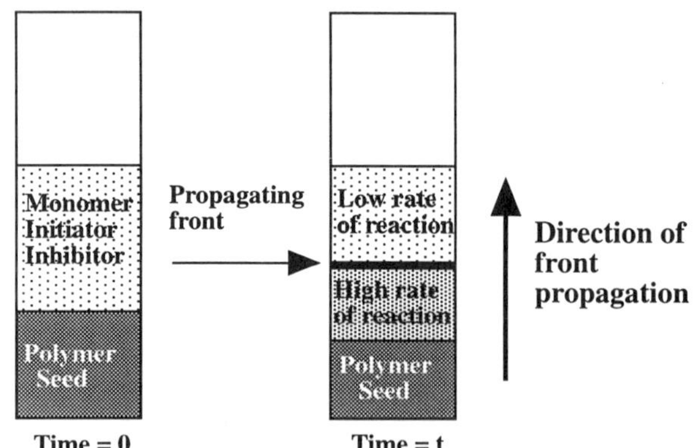

Figure 2 Propagation of IFP

The primary method of producing gradient materials via IFP is a diffusion method. The primary material, or monomer, diffuses into the seed faster than the secondary material, a dopant material such as a dye or even a different polymer. Since the primary material diffuses into the seed faster, there is more of the secondary material remaining to diffuse into the gel later, which results in a gradient. This dopant material can be placed in the bulk solution to produce GRIN materials, or can be placed in the seed to produce optical limiters.[2]

While there have been many publications on gradient materials, to our knowledge there has not been a conclusive study published on the factors affecting IFP. Our goal is to determine these factors that affect IFP, such as type of monomer and initiator and to determine the limit of propagation that can be obtained by a system (combination of monomer, initiator, and inhibitor). We intend to determine these factors of propagation for a homopolymer system by examining monomer, initiator and its concentration, inhibitor and its concentration, molecular weight and size of polymer seed, temperature, and presence and concentration of chain transfer agent. Thus, far, we have done preliminary studies on how the inhibitor and initiator and their concentrations affect the system. We will discuss how 2,2,6,6-tetramethyl-1-piperidinyloxy (TEMPO), poly-4-vinyl-phenol (P4VP), Quercetin, and C-undecylcalix[4]resorcinarene monohydrate affect the system of methyl methacrylate and initiator. Several initiators were used including 2,2'-azo-bis-isobutyronitrile and tert-butyl peroxide. The time and limits of propagation of the systems will be discussed as well as the inhibiting abilities of the various inhibitors.

Experimental Procedure

Methyl methacrylate (MMA) (99%) was purchased through Aldrich Chemical Co. It was prepared for IFP by storing over molecular sieves for several days and by removing the methylhydroquinone through use of an inhibitor removal column to prevent extraneous parameters. To ensure optical purity, which is necessary for gradient materials, it was dried using calcium hydride and then filtered through a 0.45 μm filter. Our seeds were manufactured by placing 2.0 mL of MMA in a 15X110 mm test tube and heated in a thermostated bath for roughly thirty-six hours. The bulk solution consists of MMA, initiator, and inhibitor. Initiators employed are 2,2'-azo-bis-isobutyronitrile (AIBN) (98%, Aldrich), lauroyl peroxide (LPO) (97%, Aldrich), 2,3-dimethyl-2,3-diphenyl butane (Perkadox) (AKZO Chemicals), 1,1-di-tert-butylperoxy-3,3,5-trimethyl cyclohexane (Lupersol 231) (ATO Chemicals), 2-butanone peroxide (methyl ethyl ketone peroxide, MEK) (50 wt. %, Aldrich), and tert-butyl peroxide (TBPO) (98%, Aldrich). Inhibitors employed are 2,2,6,6-tetramethyl-1-piperidinyloxy (TEMPO) (98%, Aldrich), poly-4-vinyl-phenol (P4VP) (98%, Aldrich), 3,3',4',5,7-pentahydroxyflavone (Quercetin) (Sigma), and C-undecylcalix[4]resorcinarene monohydrate (99%, Aldrich).

Results

We first tested MMA using AIBN and/or LPO with the previously mentioned inhibitors. These systems, along with their concentrations and limits of propagation, appear in Table I. The TEMPO system achieved the highest limit of propagation for the systems of similar concentrations and temperatures. A problem with TEMPO, as well as Quercetin and the resorcinarene inhibitor , is that they all diffuse into the polymer matrix during the reaction. This diffusion introduces an "impurity" to the study, as well as causing the propagation to slow and eventually stop. To prevent this problem, we proceeded to test other systems using P4VP as our inhibitor.

To determine the effect of initiator half-life on IFP, MMA with P4VP was tested with the initiators of AIBN, Perkadox, Lupersol 231, MEK, and TBPO. The reported half-lives of these initiators are 4.8 hours at 70°C (AIBN),[11] ten hours at 205°C (Perkadox),[12] ten hours at 96°C (Lupersol 231),[13] insignificant for MEKP,[14] and 218 hours at 100°C (TBPO).[11] Figure 3 shows the propagation obtained from these systems. Also from Table 1, notice that TEMPO allowed the front to propagate further than P4VP did. The systems with longer half-lives propagated further, but it is undetermined if the results of propagation are linear with respect to increasing half-lives. Further work must be done to determine this.

Table I Limits of Propagation for the inhibitors of TEMPO, P4VP, Quercetin, and C-undecylcalix[4]resorcinarene monohydrate

Inhibitor (mM)		Initiator (mM)	Temp (°C)	Length (cm)
TEMPO	(2.0)	AIBN (2.0)	60	0.8
TEMPO	(7.5)	LPO (7.5)	70	1.0
P4VP	(4.0)	AIBN (2.0)	60	0.3
P4VP	(40.0)	AIBN (2.0)	60	0.7
P4VP	(5.0)	AIBN (2.0)	60	0.9
P4VP	(15.0)	LPO (7.5)	60	1.0
P4VP	(4.0)	AIBN (2.0)	70	1.2
Quercetin	(4.0)	AIBN (2.0)	60	0.5
Resorcinarene (0.5)		AIBN (2.0)	70	0.9

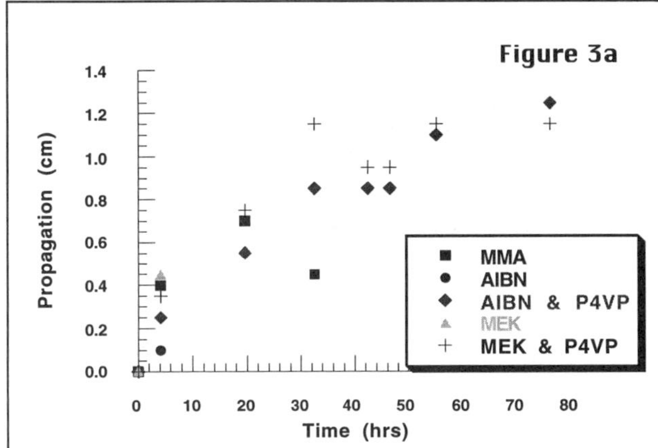

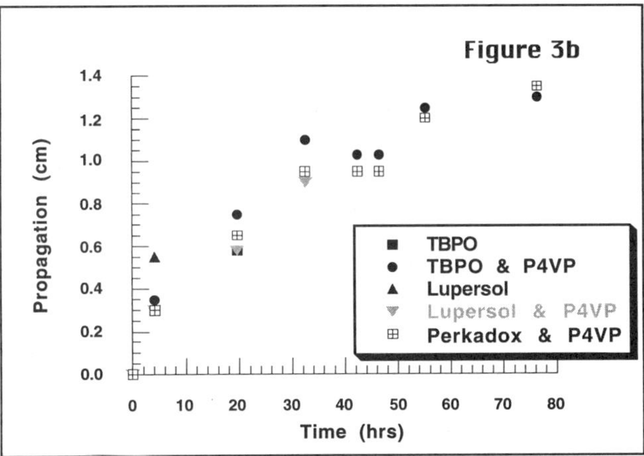

Figure 3 Propagation with respect to time of AIBN, MEK, TBPO, Lupersol, and Perkadox with and without the inhibitor P4VP

Conclusions

Industrial use of IFP to produce gradient materials necessitates the study of the factors that affect IFP. While not much is known about these factors, we have determined that the inhibitor and initiator affect the length and time of propagation. Previously in PMSE (Spring 1998), we reported that the average length of propagation with use of an inhibitor was 1.0 cm +/- 0.3 cm with an average propagation time of twenty-four hours. We can now expand the average length to 1.5 cm +/- 0.3 cm and the average propagation time to seventy-two hours, because of the use of initiators with longer half-lives and because of increased temperature. Systems that are known to exhibit IFP encompass MMA with various combinations of AIBN, LPO, Perkadox, Lupersol 231, MEK, and TBPO as initiators and P4VP, TEMPO, Quercetin, and C-undecylcalix[4]resorcinarene monohydrate as inhibitors.

Current work is underway to synthesize and test an amino-TEMPO derivative of methacrylic acid as well as a hydroxy-TEMPO derivative of methacrylic acid. Ivanov et. al. have reported propagation of up to 4 cm in systems that used an amino-TEMPO derivative.[15, 16] We also plan to study the effect of percent conversion of the seed and the effect of temperature on these systems.

References

(1) MacChesney, J. *Scientific American* **1997**, *277*, 96.
(2) Koike, Y. in *Polymers for Lightwave and Integrated Optics*; Hornak, L. A.,Ed., Marcel Dekker: New York, 1992; pp 71-104.
(3) Koike, Y.; Asakawa, A.; Wu, S. P.; Nihei, E. *Appl. Opt.* **1995**, *34*, 4669-4673.
(4) McCahon, S. W.; Tutt, L. W. *U. S. Patent* **1992**,
(5) Harter, D. J.; Sh a nd, M. L.; Band, Y. B.; Samelson, H. *U. S. Patent* **1988**,
(6) Sharp, E. J.; Miller, M. J.; William W. Clark, I.; Wood, G. L.; Salamo, G. J. *U. S. Patent* **1991**,
(7) Nakao, T.; Arimoto, A.; Ogawa, K.; Sukeda, H.; Sugiyama, H. *U. S. Patent* **1994**,
(8) Miller, M.; Salamo, G.; Clark, W. W.; Wood, G. L.; Sharp, E. J.; Monson, B. D. *U. S. Patent* **1995**,
(9) Koike, Y.; Nihei, E. *U. S. Patent* **1993**,
(10) Sasaki, K.; Koike, Y. *U. S. Patent* **1995**,
(11) Odian, G. *Principles of Polymerization*; Wiley: New York, 1991.
(12) AKZO
(13) ATO **1992**,
(14) Sanchez, J.; Myers, T. N. *Organic Peroxides*; Wiley: 1996.
(15) Smirnov, B. R. ;. M., S. S.; Lusinov, I. A.; Sidorenko, A. A.;Stegno, E. V.; Ivanov, V. V. *Vysokomol. Soedin., Ser. B*, **1993**, *35*, 161-162.
(16) Ivanov, V. V.; Stegno, . V.; Pushchaeva, L. M. *Chem. Phys. Reports* **1997**, *16*, 947-951.

Gas Permselectivity and Volumetric Properties in Polysulfone Membranes Modified by Additives

F. A. Ruiz Treviño[1] and D. R.Paul[2].

1. Departamento de Ingenierías (Ingeniería Química), Universidad Iberoamericana. Prol. Paseo de la Reforma No. 880, México, D. F. 01210.
2. Chemical Engineering Department and Center for Polymer Research. The University of Texas at Austin, Austin, TX 78712.

Introduction

The incorporation of additives into available polymers like polysulfone, PSF, is an alternative to improve their gas permeation selectivity at the expense of permeability via antiplasticization. Typically, this effect is explained in terms of a decrease in polymer fractional free volume, FFV, driven by a markedly negative deviation from volume additivity.[1,2] In this work, a mathematical model to describe the change in the specific volume of polymer-additive glassy materials upon additive incorporation is described and used to explain the mentioned effects which are characteristics of antiplasticized polymers.[3]

Experimental

Additives based on naphthalene, bisphenol A or F, and fluorene structures were incorporated into PSF to form films containing different concentrations of additive. The chemical structure, acronym and important physical properties of each additive may be found in Table 1 of reference 3. The glass transition temperature of pure additives, T_{gd}, the glass transition temperature for glassy mixtures, T_{gm}, the gas permeation at 35 °C and the volumetric properties of glassy compatible films at 30 °C, V_{mg} (30 °C), were measured and compared to the corresponding properties measured for the pure PSF.

Mathematical model for $V_{mg}(T)$

A mathematical model to estimate the specific volume of polymer-additive glassy systems, $V_{mg}(T)$, at any concentration of additive and temperature below T_{gm} is given by

$$V_{mg}(T) = w_d V_{dl}(T) + w_p V_{pl}(T) + \left(\frac{dV_{ml}}{dT} - \frac{dV_{mg}}{dT}\right)(T_{gm} - T) \qquad (1)$$

where w_i is the weight fraction of additive (d) or polymer (p), V_{il} is the specific volume for pure liquid additive or pure liquid polymer, dV_{ml}/dT and dV_{mg}/dT are the linear thermal expansion coefficients for the liquid and glassy mixture. The assumptions and parameters in (1) can easily be understood and determined experimentally. Comparisons of predicted and experimental values for different polymer-additive glassy mixtures may be found elsewhere.[4]

Results

Fig. 1 shows the effect of the amount and type of additive on the glass transition temperature of PSF; the acronym for each additive is located adjacent to the maximum concentration studied. All additives depress the T_g of PSF and the extent of the depression is determined by the T_g of the pure additive (see Table 1 of reference 3). Roughly speaking, the additives studied can be divided into three categories as seen by the influence of the additive on T_g.

Fig. 2 illustrates how incorporating the additives mentioned above change the FFV of PSF. All additives decrease the FFV of PSF. As suggested by the T_g behavior shown in Fig. 1, the additives can also be roughly classified into three categories according to how they affect FFV. The additives in category C have very little effect on FFV while those in categories B and A show progressively greater effect on FFV. Generally, this categorization is in agreement with that presented in Fig. 1.

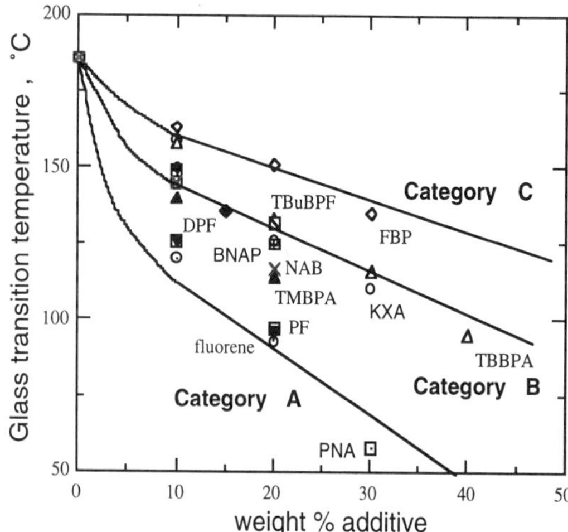

Fig. 1. Glass transition temperatures of PSF mixtures with various additives. See Table 1 of reference 3 for definition of acronyms.

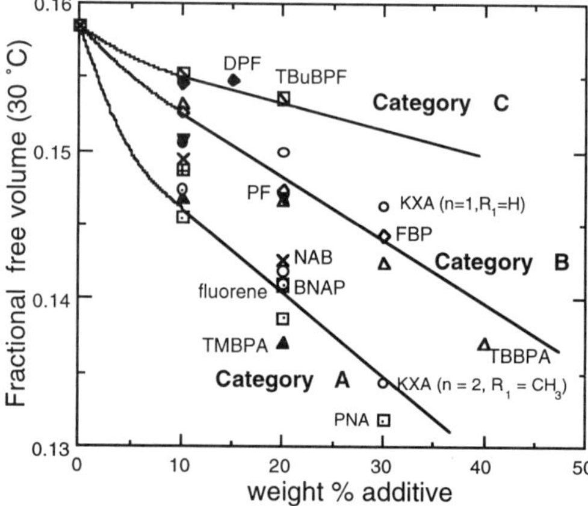

Fig. 2. Fractional free volume of PSF mixtures with the various additives.

The effectiveness of the various additives for decreasing the permeability of PSF to He and for increasing the selectivity of PSF to the gas pair He/CH4 is shown graphically in Fig. 3 where the ordinates have been normalized by the permeability and selectivity coefficients of a neat PSF membrane. Similar trends are observed for other gas pairs.[5] Following the scheme suggested earlier, the additives can be roughly classified into three categories according to their effect on gas permeability and selectivity. In both figures the categorization is in good agreement with those mentioned in Fig. 1 and 2 respectively.

Fig. 4 shows the experimental specific volume at 30 °C for mixtures (solid points) of PSF with the glassy additive KXA, an amorphous solid with a T_{gd} of 32 °C made from an alkylated naphthalene/formaldehyde condensate (Kenrich Petrochemicals Inc.). The V_{pg} and V_{dg} values shown for the pure polymer and for KXA in the glassy state are experimental values. The V_{dg} value for pure KXA is larger than the V_{pg} value of pure PSF. For PSF, the V_{mg} value increases upon addition of KXA and there is a negative departure from ideal volume additivity as defined by a straight line connecting the glassy state specific volumes of the pure components. In this Fig., the straight line represents the expected volumetric behavior of the equilibrium liquid mixture while the solid line represents the volumetric behavior for the glassy mixtures predicted by (1). The model describes the data very well. For these mixtures, the V_{mg} value approaches the equilibrium line as the concentration of KXA increases since the glass transition temperature of the mixture is closely approaching the measurement temperature. That is, the unrelaxed volume of these mixtures decreases as KXA concentration increases, but since the T_{gd} of KXA ($T_{gd} = 32$ °C) is slightly above T, the fully relaxed or equilibrium state is never reached. The magnitude of the observed negative departure from volume additivity is related to the degree of relaxation. Fig. 5 shows model calculations for the specific volume of PSF mixtures with hypothetical additives that have identical values of $V_{dl}(30$ °C), but with different T_{gd}. As may be seen, the specific volume of the mixture is predicted to contract more relative to pure PSF the lower the glass transition temperature of the diluent. This prediction is consistent with the larger reduction in the T_g, the FFV and the gas permeability of PSF caused by those additives that have the lowest T_{gd}.

Conclusions

Additives with the lowest T_{gd} caused the largest reductions in the T_{gm}, the FFV and the gas permeability of PSF. These effects are in part explained by the model proposed here which shows that the specific volume of a mixture contracts more, relative to an additive relationship between the specific volume of the pure glassy polymer and the amorphous diluent, the lower the T_{gd} of the diluent.

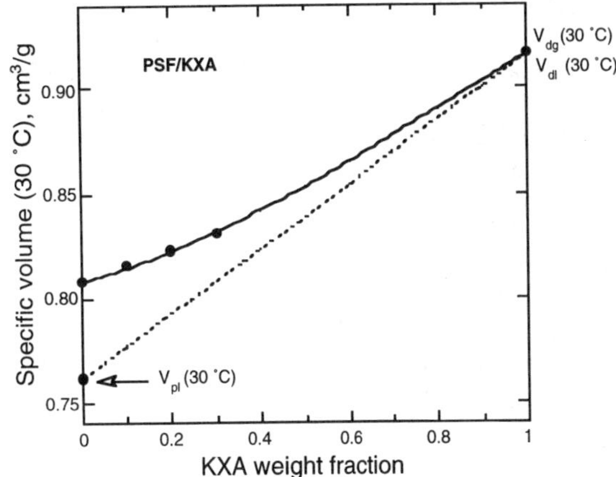

Fig. 4. Isothermal volumetric behavior for glassy mixtures of PSF, with Kenflex A or KXA, which is glassy at room temperature.

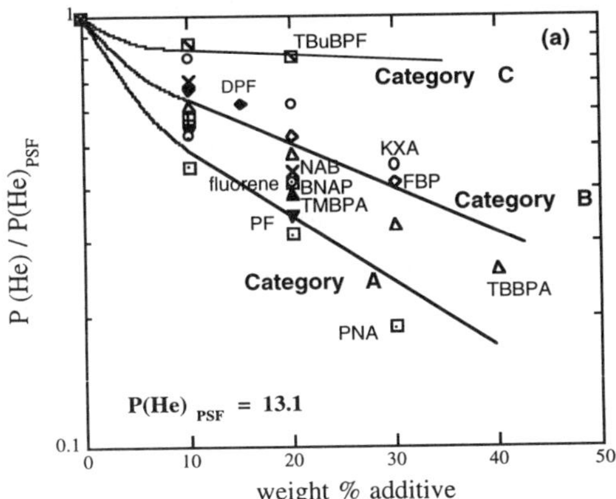

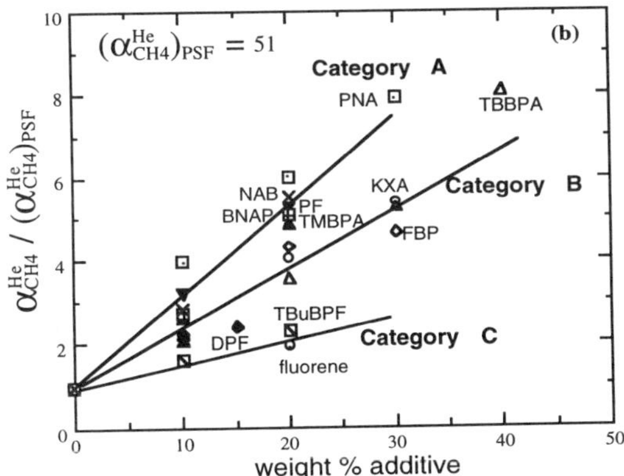

Fig. 3. Helium permeability (part a) and He/CH$_4$ selectivity (part b) of PSF-additive mixtures at 35 °C, normalized by the permeability and selectivity of pure PSF.

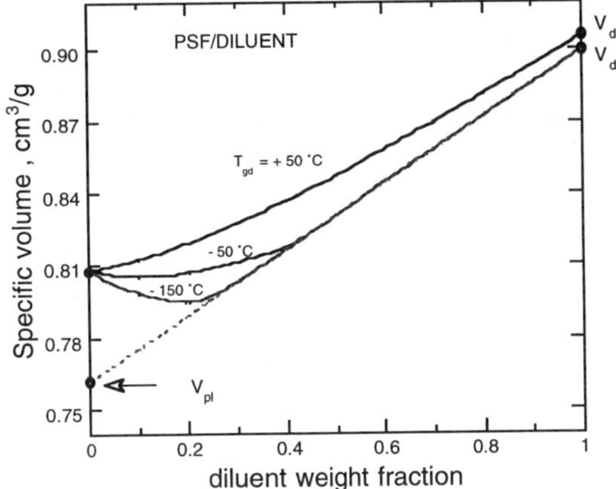

Fig. 5. Predicted volumetric behavior at 30 °C for mixtures of PSF with hypothetical diluents with different T_{gd} but with all other properties the same. The solid line is the glassy predicted behavior while the dashed line represents the equilibrium liquid mixture of PSF with the hypothetical diluents.

References

1. L. M. Robeson and J. A. Faucher, *Polymer Letters*, **7**, 35 (1969).
2. Y. Maeda and D. R. Paul, *J. Polym. Sci.: Part B: Polym. Phys.*, **25**, 957 (1987).
3. F. A. Ruiz-Treviño and D. R. Paul, *J. Appl. Polym. Sci.*, **66**, 1925 (1997).
4. F. A. Ruiz-Treviño and D. R. Paul, *Submitted to J. of Polym. Sci.; Part B: Polym. Phys.*, (1997).
5. F. A. Ruiz-Treviño, Ph. D. dissertation, The University of Texas at Austin, (1997).

STRUCTURAL CHARACTERIZATION OF 123-HIGH TEMPERATURE SUPERCONDUCTOR MATERIALS: SELECTIVE REMOVAL OF COPPER

Charles E. Carraher, Jr., Huifang Zhuang, Fernando Medina and Donald Baird
Department of Chemistry and Biochemistry and
Department of Physics
Florida Atlantic University
Boca Raton, FL 33431
Florida Center for Environmental Studies
NorthCorp Center
Palm Beach Gardens, FL 33410 and
Research Frontiers, Inc.
Woodbury, NY 11917

INTRODUCTION

Superconductivity was first observed in mercury by Heike Onnes in 1911 (1). He found that the resistivity of mercury to an electric current sharply fell to zero at about 4 K. In early 1986 George Bednorz and K. Alex Muller reported what is now know as high temperature, high Tc, superconductors (2). The material was a ceramic material containing La, Ba, Cu and O. Rapid progress continued. By February 1987 Wu and co-workers (3) had raised the threshold for superconductivity to 93 K through substitution of the La sites. For the first time superconductivity was achieved above liquid nitrogen temperatures. The superconducting phase was identified as $YBa_2Cu_3O_7$, now referred to as 123-compounds.

If the magnetic field exceeds a critical value, superconductivity is lost. The 123-compounds are called a Type II material where the applied magnetic field is exactly canceled by the induced field up to a lower critical field, Between this lower critical field and a higher critical field, the applied field is not completely balanced by the induced field, and the material is said to be in a vortex state. For the 123-materials, when the field strength reaches a certain threshold, the lines of magnetic flux actually penetrate the superconductor forming a lattice of flux lines resembling the parallel bristles of a brush.

123-Compounds are meta stable. They are thermodynamically stable in air over a temperature range of about 800 to 950 C but with an oxygen content too low to give a superconducting material. In fact, the 123-phase is first prepared with a low oxygen content, then annealed under oxygen at lower temperatures where it is oxidized. Therefore, the oxidation of $YBa_2Cu_3O_6$ to $YBa_2Cu_3O_7$ is an intercalation reaction. Products made by intercalation reactions are nearly always metastable.

123-Compounds decompose in water to CuO, barium hydroxide and Y_2BaCuO_5 evolving oxygen (4). Exposure to high-intensity UV and X-ray beams produces surface changes leading to less or inactive materials (5). The 123-material is also unstable under applied pressures greater than 100 MPa at temperatures from 700 to 950 C decomposing to other metal oxides (6). Thus, there is a continued need to produce materials with greater stability.

The 123-compounds consist of two types of polymeric copper oxide layers held together by ionic bonding metals such as barium and yttrium (7). We were part of a large effort aimed at understanding and modifying such 123 compounds. Our efforts were aimed at selectively removing particular cations and eventually replacing them with other ions in an effort to increase the Tc, to increase the amount of current that can be carried, and finally to increase the stability of the product.

In 1990, we reported the selective removal of copper using oxalic acid (8). Here we report further results of these studies.

EXPERIMENTAL

Sample preparation is described elsewhere (8).

For Auger electron spectroscopy, AES, and X-Ray Photoelectron Spectroscopy, XPS measurements all samples were powdered. The 123-sample was treated by mixing 0.10 gram of the 123-material powder with 0.30 ml of oxalic acid solution (1 gram of oxalic acid dihydrate in 15 ml deionized water). The 123-material was exposed to the oxalic acid solution for one hour employing nitrogen gas to assist in mixing. The reacted powder was washed with acetone (twice) and dried in air.

The powder was pressed into pellets with a standard KBr press. Survey scans (30-1000 eV auger KE and 1200-0 eV XPS BE) were followed by scans of the specific elements of interest. A Perkin-Elmer PHI 610 Scanning Auger run with 0.1uA, 3kV incident electron beam over a 1 um spot at a resolution of 0.6% was used to obtain the Auger spectra. A Kratos XSAM800 XPS operating with a 20 mA, 15 kV Al X-ray source was used to collect the XPS data. This instrument operates with an approximate resolution of 1 eV.

RESULTS AND DISCUSSION

STRUCTURAL ANALYSIS-Polydentate ligands, containing at least two donor atoms, are typically used to chelate with metal ions. They are less used to actually "dissolve" or remove metal ions from the solid state.

Copper and yttrium are transition metals. They form metal-ligand bonds with a number of ligands. For transition metals, the (n-1)d, ns, and np orbitals that are regarded as regular valence orbitals that can become involved with both electron donor and electron acceptor materials.

When the ligand forms a stable, water soluble metal chelate, the ligand is said to be a sequestering agent and the metal is said to be "sequestered".

Our initial efforts were aimed at preferentially removing copper since it is considered a critical component in the superconductivity of the 123- materials. We placed small amounts of 123-material into beakers and added solutions containing various ligands, including those reported to chelate copper. The ability to remove copper forming soluble products was indicated by the formation of a blue solution. Our intention was to not only remove surface copper ions, but also to remove those further in without collapsing the surrounding structure.

There are some chelating agents that produce a specific product with only one kind of metal atom. Thus, dimethyglyoxime combined with a number of metal ions, forms an insoluble product with only nickel. Some chelating agents change their selectivity according to the particular reaction conditions. In particular, pH is a critical factor in the chelation of many metals. Thus, pH changes were also studied in an effort to selectively remove the metal atoms.

We were looking at a chelating agent that will (preferentially) remove only copper to encourage the retention of the surrounding structure. Here we report results using oxalic acid.

Oxalic acid is the simplest organic dibasic acid. The planar oxalate ion behaves as a bidentate ligand forming five-membered rings with the chelated metal ion.

Surface analysis using XPS and AES focus on the initial two to five layers of the surface. XPS results of 123-materials treated with various solutions containing oxalic acid were obtained. These results were compared with XPS results for inorganic salts composed of the same elements that are present in the 123-material itself and with untreated 123-material. The C 1s spectrum has two major components with peaks at 284.6 and 288.9 eV.

The 284.6 eV peak is due to neutral carbon. The higher binding energy peak is attributed to the carbonate species and is present in both the spectrum of the "like"-test material barium carbonate and the untreated 123-material.

The barium $3d_{5/2}$ spectrum of the 123-material and barium carbonate has two peaks, one at 778.3 and the second at 780.1 eV. The 778.3 eV peak is attributed to the barium species contained in the superconducting phase while the 780.1 eV peak corresponds to the barium in surface impurities which are partly due to the presence of barium carbonate.

The O 1s spectrum has three peaks-528.5, 531.3, and 533.3 eV. The 528.5 eV peak corresponds to the oxygen in the superconductor. The 531 eV peak is due to impurities such as carbonate, hydroxide, and/or strongly adsorbed oxygen species. (The "like"-compound is copper II oxide.) The relatively small peak at 533 eV is due to the presence of C-O impurities.

The Cu $2p_{3/2}$ spectrum has one peak at 933.3 eV with a prominent shake-up satellite structure that is typical of the Cu II state observed in 123 materials. (The "like"-compound is copper II oxide.)

The Y 3d peaks are attributed to two types of yttrium compounds-yttrium III oxide from the superconductor phase, and a yttrium compound formed as an impurity phase. (Yttrium III oxide was used as the "like"-compound.)

Following is a discussion of the results of a 123 material treated with a 0.53 M oxalic acid solution for one hour to effect a 5% decrease in overall superconductivity. The XPS with respect to the barium 3d photoelectron peaks showed a slight shifting to higher energies from 795 and 779 eV to 796 and 780 eV consistent with the presence of chelated oxalate.

The superconductive oxygen absorption region (O 1s) at 528 eV is greatly decreased. Also, the type of oxygen grouping is changed. The 529 eV band in yttrium oxide and in copper II oxide is missing. The oxalate O 1s band appears within the same region as the band (about 531 eV) present in most of the samples due to the presence of impurities, so it is difficult to determine changes associated with it. The band at 933 eV with a shake-up satellite between 940 and 945 eV is typical of the copper II ($2p_{3/2}$) state observed in the 123-material. These bands are smaller in the treated materials in comparison with the untreated materials.

The Y 3d band at 157-158 eV in the untreated sample is decreased and shifted to a higher binding energy (159) for the treated sample consistent with binding by the oxalate. Finally, the 289 eV band characteristic of carbonate and oxalate carbon (1s) is increased for the treated sample.

AES showed that the surface concentration of copper had greatly decreased; that the concentration of barium had increased; and that the concentration of yttrium remained about the same for the treated 123 sample in comparison with the untreated 123 material.

XPS and AES spectral results are consistent with the presence of bound oxalate; a decreased surface concentration of copper; and a decrease in the intensity of peaks associated with superconductivity for the sample treated with oxalic acid.

EXPOSURE RESULTS-It was found that aqueous oxalic acid solutions with a suitable concentration (about 0.8 to 1.0 M) easily and controllably reduces the magnetic susceptibility of the 123-compound. Figure 1 contains test results as a function of time for one of these concentrations.

Another series of experiments was carried out using aqueous oxalic acid solutions (0.53 M, pH about 1) in which the amount of chelating agent was changed. Figure 2 contains representative results.

In general, decrease in magnetic susceptibility increased as the concentration of oxalic acid increased. Further, the presence of oxalic acid allows easier control of magnetic susceptibility in comparison with water itself. The mechanism for the decline in magnetic susceptibility is probably different. In water, collapse of the infra-structure probably occurs while in the case of oxalic acid, preferential removal of copper occurs.

REFERENCES

1. H. K. Onnes, Leiden Comm., 120b, 122b, and 124c (1911).
2. J. G. Bednorz and K. A. Muller, Z. Phys, B64, 189 (1986).
3. M. K. Wu, J. R. Aushburn, C. J. Torng, P. H. Hor, R. L. Meng, L. Gao, Z. Huang, Y. Q. Wang, and C. W. Chu, Phys. Rev. Lett., 58(9), 908 (1987).
4. M. F. Yan, R. L. Barns, H. M. O'Bryan, P. K. Gallagher, R. C. Sherwood, and S. Jin, Appl. Phys. Lett., 51(7), 532 (1987).
5. Y. Chang, M. Onellion, D. W. Niles, and G. Margaritondo, Phys. Rev. B, 36, 3986) (1987).
6. B. C. Hendrix, T. Abe, J. C. Borofka, and J. K. Tien, Appl. Phys. Lett., 55(3), 313 (1989).
7. C. Carraher, **Polymer Chemistry: An Introduction**, Dekker, NY, 1996, 4th Ed.
8. C. Carraher, H. Zhuang, F. Medina, d. Baird, R. Pennisi, B. D. Landreth, and F. Nounou, Polymeric Materials: Science and Eng., 633 (1990).

We are pleased to acknowledge the assistance of Robert Pennisi, Bobby Dean Landreth and Faye Nounou, Motorola, Inc., Advanced Manufacturing Technologies, Ft. Lauderdale, FL 33322 and support by DARPA Grant No. MDA972-88-J-1006.

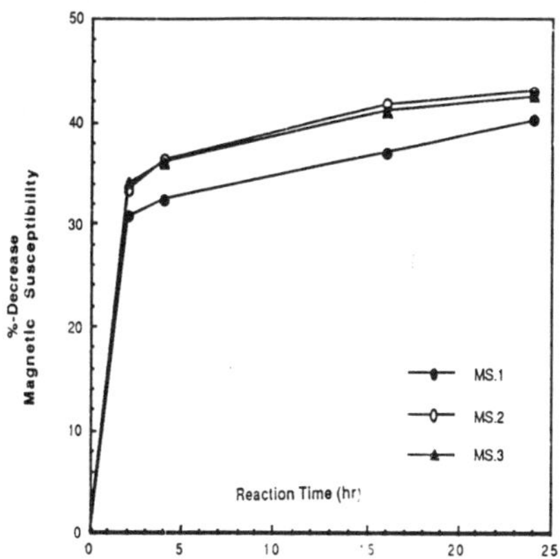

Figure 1. Reproducibility using a 0.79 M aqueous solution of oxalic acid.

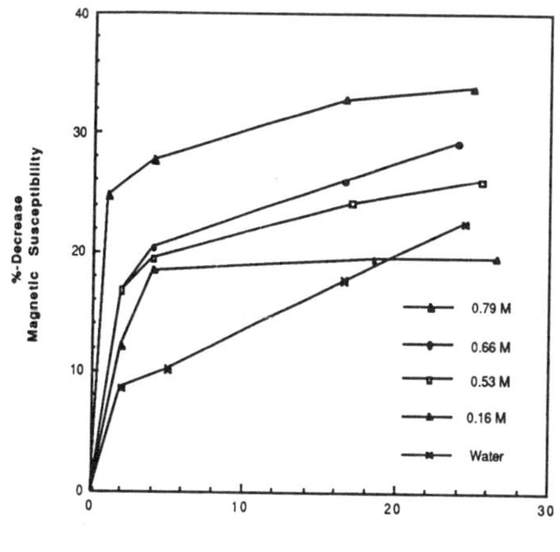

Figure 2. The effect of oxalic acid solutions of varied concentrations on the magnetic susceptibility of the 123-compound.

SYNTHESIS AND STRUCTURAL CHARACTERIZATION OF VANADOCENE-CONTAINING POLYESTERS

Charles E. Carraher, Jr., and Ernest B. Randolph

Florida Atlantic University

Department of Chemistry and Biochemistry

Boca Raton, FL 33431

and

Florida Center for Environmental Studies

NorthCorp Center

Palm Beach Gardens, FL 33410

and

DeForest Enterprises, Inc.

Amtec Center

6421 Congress Avenue

Boca Raton, FL 33487

INTRODUCTION

Vanadocene dichloride, bis(cyclopentadienyl) vanadium IV dichloride, is a member of the metallocene family with a structure similar to that of the Group IVB metalocenes. While the synthesis and characterization of a number of transition metal polymers has been reported (for instance 1 and 2), little has been reported on the synthesis of vanadocene-containing polymers.

Vanadium and vanadocene-containing products are employed for a variety of applications (for instance 3-9). Vanadium is employed as a corrosion inhibitor. Stainless steel owes its anticorrosive properties to the presence of small amounts of vanadium. the vanadium metal resists oxidation/corrosion at temperatures near 700 C, at which point rapid oxidation occurs (4).

Organic vanadium compounds are widely used as catalysts in the production of a variety of monomers, polymers, and copolymers (for instance 5-9). The new so-called soluble stereoregulating catalysts employ metallocenes as the basic catalytic active-site including the use of vanadium-containing compounds.

Recently we reported the synthesis of vanadocene-containing polyethers based on the reaction between vanadocene dichloride and diols (10).

$$Cp_2VCl_2 + HO-R-OH ---> -(V(Cp)_2-O-R-O)-$$

Here we report on the initial synthesis of the analogous vanadocene-containing polymers based on the reaction between vanadocene dichloride and salt of diacids.

$$Cp_2VCl_2 + HO-\underset{O}{C}-R-\underset{O}{C}-OH ---> -(V(Cp)_2-O-\underset{O}{C}-R-\underset{O}{C}-O-)-$$

EXPERIMENTAL

Reagents were used as received (vanadocene dichloride, 95%, Aldrich Chemical Co., Milwaukee; terephthalic acid (97, Aldrich); isopthalic acid (99%, Aldrich), sebacic acid (98%), and adipic acid (98%).

Physical characterization was carried out as follows. Infrared spectra were recorded using a Mattson Instruments Galaxy 4020 FT-IR Spectrophotometer employing potassium bromide pellets. UV-Vis spectra were obtained using a Perkin-Elmer Lambda 2 Spectrophotometer. High resolution electron impact positive ion mass spectral analyses were carried out at the Midwest Center for Mass Spectrometry in Lincoln, NB using a Dupont 21-496-B double focusing mass spectrometer. The samples were inserted as solids in a glass ampoule using a direct insertion probe. The samples were rapidly heated to about 450 C.

Thermogravimetric thermograms were obtained using a Model 951 Dupont Thermogravimetric Analyzer using a heating rate of 20 C/min from room temperature to 800 C using both air and nitrogen. Differential Scanning Calorimetry was accomplished using a Dupont DSC Cell Base II connected to a Dupont Model 990 Thermal Analyzer console using a heating rate of 20 C/min from room temperature to 450 C in both air and nitrogen. Molecular weight measurements were obtained using a Brice-Phoenix BP3000 Universal Photometer. The index of refraction was obtained using a Bauch & Lomb Abbe Model 3-L refractometer. Elemental analysis was carried out by Gailbraith Laboratories, Knoxville, TN.

Polymerizations were carried out employing a one quart Kimax emulsifying jar supported by a Waring Model 1120 blender with a "no load" speed of about 18,000 RPM. The following general procedure was utilized. The organic layer contains 1.00 millimole vanadocene dichloride dissolved in 100 ml chloroform (or 50 ml benzene and 50 ml chloroform for the reaction with isopthalic acid). The aqueous layer consists of 2.00 millimoles of NaOH and 1.00 millimole diacid are contained in 100 ml DI water. The blender as run for about 30 secs.

RESULTS AND DISCUSSION

Products derived from salts of aliphatic diacids were produced in small (typically <5%) yield whereas reactions employing aromatic salts (terephthalic acid, 25%; isopthalic acid, 25%) gave better product yields. For the reaction with terephthalic, phase transfer agents (12-crown-4 and 18-crown-6) were also used but the product yield remained about the same. The low yields for reactions employing aliphatic Lewis bases is similar to that observed for the analogous reaction except with diols (10).

Vanadocene dichloride itself exhibits bands at 3095 (C-H aromatic stretch; all bands given in cm^{-1}), 1616 and 1440 (C=C stretch) and 830 (C-H out-of-plane rock). A weak band at 489 corresponds to the ring to metal stretch characteristic of metallocenes. The V-Cl stretch appears at about 385-300, below the capability of the instrument (11). The product from terephthalic acid shows bands at 3095 (characteristic of aromatic C-H stretching), 1028 (derived from an aromatic C-H in-plane stretch) and bands at 1616 and 1560 corresponding to aromatic ring stretching. The presence of a band at 490 is assigned to the vanadium-Cp stretching. The presence of a new band at 931 is assigned to the V-O stretch.

Bonding to the metal by the acid can be of two general types- bridging and non-bridging as shown below.

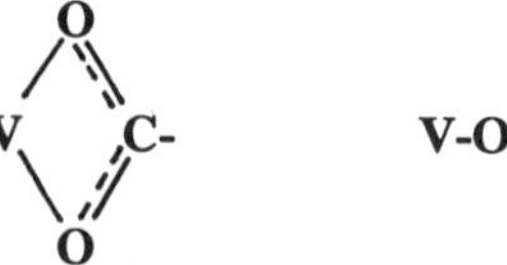

Bridging   Non-bridging

Bands associated with bridging appear at about 1570 (asymmetric) and 1430 (symmetric bridging) while bands associated with non-bridging appear at about 1650-1610 (asymmetric) and 1250 (symmetric). The products contain bands at 1645, 1560, 1400 and 1250 consistent with the presence of both types of bonding.

The mass spectral analysis of Group IV B metallocene dichlorides has been studied by Dillard and Wailes (12,13) and Carraher and co-workers (for instance 14,15). For the product from reaction with the disodium salt of terephthalic acid, the most intense ion fragments (all mass fragments are given in m/e=1) are found at 70 (CH-CH-COO), 83 (O-V-O), 109 (O-V-O-CH-CH), 121 (terephthalic acid minus one COOH), 236/237 (V(Cp)$_2$COOC), 266 (V-OOC-C$_6$H$_4$-COO-V), 314 (OOC-C$_6$H$_4$-COO-VCp-OC), 561 (VCp-OOC-C$_6$H$_4$-COOVCpOOC-C$_6$H$_4$-COO). Additional lower mass ion fragments are present that are consistent with the formation of one and two COO linkages with V.

Both products show an absence of ion fragments from 35-38 consistent with the absence of unreacted and entrapped Cl. The mass spectral analyses are consistent with the proposed structure.

The products are oligomeric to low polymers. For the product from isophthalic acid, the weight average molecular weight is 11,000 (dn/dc = 1.49) corresponding to an average degree of polymerization of 30. For the product from terephthalic acid, the weight average molecular weight is 3,500 (dn/dc = 1.479) corresponding to an average degree of polymerization of about 10.

The TGA thermogram for the product from terephthalic acid is similar in air and nitrogen with initial weight loss occurring at about 100 C, probably occurring through the loss of small amounts of the Cp groups. Total weight loss to 800 C is about 10%. While the TGA thermograms are similar, the DSC thermograms are not. In air and nitrogen there are endothermic regions to about 200 C. In air, there is an exothermic region from about 200 C to the end (about 450 C) while in nitrogen the thermogram is less exothermic, beginning at about 350 C and continuing to the end.

For the product from isophthalic acid, weight loss again begins in both air and nitrogen around 50 C and continues to abut 600 C after which the weight loss continues but more slowly. Total weight loss in air is about 10 % and in nitrogen it is about 15% at 800 C. The DSC thermograms are similar to those reported above for terephthalic acid.

REFERENCES

1. C. Carraher, A. Rivalta and J. Haky, PMSE <u>74</u>, 149 (1996).
2. C. Carraher, A. Li, J. Kloss and A. Lombardo, PMSE <u>70</u>, 38 (1994).
3. G. Curran, J. Biol. Chem. <u>210</u>, 765 (1954).
4. D. H. Blankenhorn, GP <u>29</u>, 134 (1964).
5. W. Rostoker, "Metallurgy of Vanadium", Wiley, NY 1958.
6. W. Eberhard, "Macromolecular Chemistry", M. Dekker, NY 1970.
7. R. B. King, Org. Metal Poly. Synthesis <u>1</u>, 185 (1965).
8. National Academy of Science, "Trends in the Use of Vanadium", Publication NMA B-267, Washington, DC, page 46, 1970.
9. N. G. Gaylord, "Linear and Stereoregular Addition Polymers", Interscience, NY, page 571, 1980.
10. C. Carraher and E. Randolph, PMSE, <u>77</u>, 497 (1997).
11. D. L. Kepert, "The Early Transition Metals, Academic Press, NY, page 210, 1972.
12. J. Dillard and R. J. Kiser, J. Organometallic Chemistry, <u>16</u>, 259(1969).
13. P. Wailes, "Organometallic Chemistry of Ti, Zr, and Hf, Academic Press, NY, page 53, 1974.
14. C. Carraher and L. Reckleben, PMSE <u>69</u>, 314 (1993).
15. C. Carraher, J. Haky, A. Rivalta and D. Sterling, PMSE <u>70</u>, 329 (1994).

Table 1. Selected infrared spectral assignments for the product from vanadocene dichloride and terephthalic acid. All assignments are given in 1/cm.

Wavenumber	Intensity	Assignment
1645	Broad, strong	C=O
1616	Weak	Aromatic ring stretch
1560	Sharp, strong	Aromatic ring stretch
1028	Sharp, weak	In-plane C-H stretch
931	Weak	V-O stretch
745, 829	Sharp, strong	C-H Aromatic rock
490	Weak	Metal-aromatic ring stretch

Table 2. Selected infrared spectral assignments for the product from vanadocene dichloride and terephthalic acid.

Wavenumber	Intensity	Assignment
1645	Wide, strong	C=O
1560	Small	V-OCO Bridging
1429	Strong, wide	Aromatic C-H bend
1280-1285	Sharp, weak	C=O
900-950	Broad, strong	V-OCO Non-bridging
930	Weak, sharp	V-O stretch
873	Wide, strong	C-H In-plane rock
461	Weak, sharp	Metal-aromatic ring stretch

INFLUENCE OF KINETIN AND KINETIN-CONTAINING POLYMERS ON THE GERMINATION AND SEEDLING DEVELOPMENT OF SAWGRASS AND CATTAIL FOR EVERGLADES RESTORATION

Charles E. Carraher, Jr., Miyuki Nagata, Herbert Stewart, Shi Li Miao, Shawn M. Carraher, Anupan Gaonkar, Chance Highland, and Fengmai Li

Florida Atlantic University
Department of Chemistry and Biochemistry
Department of Biological Sciences
Boca Raton, FL 33431

Florida Center for Environmental Sciences
NorthCorp Center
Palm Beach Gardens, FL 33410

Everglades System Research Division
South Florida Water Management District
West Palm Beach, FL 35406

Indiana State University
Department of Management and Finance
Terre Haute, IN 47809

INTRODUCTION

The Florida Everglades is called the "River of Grass". The "grass" is actually a sedge called sawgrass (<u>Cladium jamaicense</u> Crantz). The Everglades is a subtropical freshwater wetland ecosystem that has undergone much shrinkage since the arrival of Europeans. It is now about 180 miles long and 90 miles wide. It is dominated by monospecific sawgrass marshes that historically covered about 70% of the Everglades (1,2).

The Everglades is undergoing rapid change including continued encroachment by urban development and by agricultural practices and displacement of the sawgrass by cattail (mainly <u>Typha domingensis</u> Pers.).

Various legislative mandates have recognized the need to halt the recent changes and to re-establish the Everglades as an ongoing, self sustaining, and healthy ecosystem (for instance Douglas Act-Chapter 373.4592 and Everglades Forever Act-Chapter 94-115). A major part of this restoration involves the "River of Grass".

Sawgrass helps in the trapping of sediments, retaining rainwater, assists in recharging groundwater, decreases the tendency for flooding, and filters out toxins and other water pollutants assisting in providing an ongoing, sufficient, and safe water supply for the some 5 million people in south Florida.

The natural re-establishment of sawgrass after fires and similar disasters is known to be difficult. One of the major difficulties of re-establishing sawgrass is low seed germination (3,4). Thus, a major effort has focused on increasing the germination rate of sawgrass.

Much of the restoration will involve plants grown from seeds-either under controlled conditions, or through broadcasting the seeds as from an airboat or airplane.

Plants contain a number of so-called plant growth regulators including cytokinins. Cytokinins promote cytokinesis or cell division, protein synthesis, etc. Structurally, cytokinins are substituted adenine compounds. Cytokinins are a group of chemical agents including synthetic (such as kinetin) and natural (such as zeatin and zeatin riboside).

EXPERIMENTAL

Synthesis of the polymer was achieved using the interfacial polycondensation process. Briefly, the organostannane dichloride was dissolved in 50 ml of chloroform or carbon tetrachloride and added to a solution of kinetin (Aldrich) contained in 50 ml of water containing NaOH. The blender was run for 30 seconds and the precipitate recovered, washed with organic and water washes to remove unreacted materials. The product was dried in the open. Infrared spectral analysis was achieved using a Mattson Model 4020 Fourier Transform Infrared Spectrometer utilizing KBr pellets. Mass spectra were provided by the Midwest Center for Mass Spectroscopy, Lincoln, NE, with partial support by the National Science Foundation, Biology Division, Grant No. DIR9017262, using high resolution electron impact mass spectrometry. UV-VIS-NIR were recorded using a Cary 14 Spectrophotometer. Molecular weight analyses were carried out using a Brice-Phoenix Model OM-4000 Light Scattering Photometer. Refractive indexes were determined utilizing a Busch and Lomb Abbe refractometer.

The germination experiments were carried out in the FAU greenhouse that was modified to moderate the excessive temperatures found during the summer months in south Florida. Regular "tap" water was used. The water level was maintained to the level of the seeds and represented what is called a "saturated" situation. Watering was done so that water would come through the bottom of the trays minimizing the movement of the treatments.

The seeds underwent a preparatory procedure previously developed by us including washing with Chlorox(TM) (5). For cattails, viable seeds were identified as those that sank to the bottom of the blender that was used to separate the seed from the fruit parts (6). For sawgrass, mature seeds were identified by visual observation. Immature seeds were discarded along with bracts and other fruit parts using a mortar and pestle. Mature seeds were tested for viability using a dye staining technique (7). Viability was about 46% for the sawgrass seeds.

Rectangular plastic trays offering a soil surface area of about 137 square inches were used. Holes were drilled in the bottom to allow exchange of water. Four inches of potting soil (Jungle Growth Potting Soil) was added to allow for seedling development to occur without need for immediate replanting.

A combined seed arrangement was used where 100 seeds of each cattail and sawgrass were used with sawgrass seeds planted on one half of the tray and cattail seeds planted on the other half of the tray. Both cattail and sawgrass seeds were obtained from the area know as Water Conservation Area 2A, impacted area (F1). The seeds were treated with talc-mixtures containing varying amounts of test material corresponding to levels where responses have been previously reported for other seeds (8).

The trays were placed in "table-top" trenches. The water level in the trenches was maintained at about four inches, presenting what is referred to as "saturated" conditions. The water-filled trenches act to moderate temperature changes more closely approximating field conditions.

RESULTS AND DISCUSSION

In general, controlled release polymers can offer longer "shelf life", greater retention of the active agent (particularly in nature where the non-polymer form is water soluble) due to the insolubility or lowered solubility of the polymer, sustained release of the active agent, and the co-reactant could impart some special additional property to the system (such as algae control for polymers containing organotin moieties).

The evidence for controlled release of the active form of kinetin from the polymer is circumstantial and is based on the ability of the polymer to influence the germination and growth of the sawgrass and cattail. Further, small structural differences typically render plant growth regulators inactive so that the degradation of the polymer most likely gives kinetin itself.

Plant regulators can express their effects in varying ways dependent upon concentration, presence/absence of other plant growth regulators, and the particular plant species (7,9). For instance, different cytokinin concentrations give different effects. Thus, very high concentrations applied to Begonia and Bryophyllum leaf cuttings increases adventitious bud formation but reduces the quality of new shoots. Lower concentrations, but within the range generally considered to be a high concentration, promotes bud formation and inhibits root formation.

Cytokinins have the greatest effect on initiating buds and shoots from leaf cuttings and in tissue growth culture systems (7,9). It is believed that cytokinins are synthesized in the root tips, being transported throughout the plant by the xylem.

Interestingly, the use of kinetin on seed germination has not been widely reported. Our interest deals with both seed germination and on seedling development including, eventually, flowering and seed formation. Here our report focuses mainly on germination effects. Later reports will focus on seedling development and on flowering and subsequent seed formation.

The experiment was begun July 12, 1997, late summer corresponding to the approximate time when cattail and sawgrass seeds mature in the Everglades.

Table 1 contains both germination percentages for day 48 and the number of days for initial germination to occur. Typically germination for untreated sawgrass begins at day 30 to 60 consistent with that observed in the present study. For untreated cattail, germination generally begins at days 3 to 14, again consistent with that observed in the present study.

Table 2 contains the maximum germination data for the experiment run to 220 days. Table 3 contains T and P-values for the data presented in Table 2 comparing the results with the control. Differences are present for both the polymers containing kinetin and for kinetin itself. The increase in germination percentage for some of the polymers is consistent with the polymers acting as control release agents since only "free" kinetin is believed to act as a plant growth regulator. Further, this is consistent with the kinetin, and selected polymers, influencing the percentage germination of both sawgrass and cattail. In both cases, this influence is positive with respect to increasing germination fraction.

Of particular interest to the present research is the observation that cytokinins can increase the tolerance to stress (8). Eventually, cytokinin treated plants may be able to withstand the many stresses related to temperature, water availability and water level present in the Everglades.

REFERENCES

1. J. H. Davis, Geological Bulletin 25, Florida Geological survey (1943).
2. C. M. Loveless, Ecology, <u>40</u>, 1, 1959.
3. T. R. Alexander, Soil and Crop Sci. Soc. Florida Proceedings, <u>31</u>, 72 (1971).
4. C. Forsberg, Physiologia Plantarum, <u>19</u>, 1105 (1966).
5. H. Stewart, S. L. Miao, M. E. Colbert and C. E. Carraher, Wetlands, <u>17</u>, 116 (1997).
6. S. J. McNaughton, Ecology, <u>49</u>, 367 (1968).
7. H. Hartman, D. Kester, F. T. Davies, and R. Geneve, <u>Plant Propagation: Principles and Practices</u>, 6th Ed., Prentice Hall, Upper Saddle River, NJ, 1997.
8. W. T. Thomson, <u>Agricultural Chemicals, Book III-- Fumigants, Growth Regulators, Repellents, and Rodenticides</u>, Thomson Publications, Fresno, CA, 1988-89).
9. F. B. Salisbury and C. Ross, <u>Plant Physiology</u>, 4th Ed., Wadsworth Pub. Co., Belmont, CA, 1992.

Table 1. Germination percentages at day 48 and date for appearance of first seedling.

Treatment Organotin/ppm	% Germination Sawgrass/Cattail	Days Inception Sawgrass/Cattail
Me$_2$Sn/0.1	3/14	30/<7
Me$_2$Sn/10	1/16	30/<7
Me$_2$Sn/1000	4/31	22/<7
Et$_2$Sn/0.1	4/32	27/14
Et$_2$Sn/10	2/19	34/<7
Et$_2$Sn/1000	6/26	30/<7
Bu$_2$Sn/0.1	6/28	24/<7
Bu$_2$Sn/10	1/10	30/14
Bu$_2$Sn/1000	1/17	44/20
Kinetin/0.1	3/24	34/<7
Kinetin/10	6/24	30/<7
Kinetin/1000	4/10	24/<7
Control	1/13	27/20

Table 2. Maximum germination percentages.

Treatment Organotin/ppm	%-Germination Sawgrass	%-Germination Cattail
Me$_2$Sn/0.1	14	16
Me$_2$Sn/10	9	23
Me$_2$Sn/1000	17	38
Et$_2$Sn/0.1	12	34
Et$_2$Sn/10	21	29
Et$_2$Sn/1000	16	37
Bu$_2$Sn/0.1	16	28
Bu$_2$Sn/10	10	20
Bu$_2$Sn/1000	12	17
Kinetin/0.1	15	37
Kinetin/10	19	36
Kinetin/1000	20	36
Control	12	23

Table 3. T and P-Values for data given in Table 2.

Treatment Organotin/ppm	Sawgrass T / P	Cattail T / P
Me$_2$Sn/0.1	.38/.37	.38/.37
Me$_2$Sn/10	1.0/.21	1.0/.21
Me$_2$Sn/1000	6.9/.01	4.8/.02
Et$_2$Sn/0.1	.14/.45	1.5/.13
Et$_2$Sn/10	1.7/.12	4.3/.02
Et$_2$Sn/1000	1.5/.13	2.6/.06
Bu$_2$Sn/0.1	1.2/.17	3.0/.05
Bu$_2$Sn/10	.46/.34	1.7/.11
Bu$_2$Sn/1000	.14/.45	1.6/.13
Kinetin/0.1	3.5/.03	1.9/.10
Kinetin/10	3.2/.04	15/.004
Kinetin/1000	3.9/.03	2.6/.06

We are pleased to acknowledge assistance of the South Florida Water Management District.

Hyperbranched Polymerization From Silicon Surfaces

Peggy-Jean Prest and Jeffrey S. Moore

Contribution from the Departments of Chemistry and Materials Science and Engineering and The Beckman Institute for Advanced Science and Technology, University of Illinois, Urbana, IL 61801.

Introduction:

The modification of silicon surfaces with hyperbranched polymers is of interest in the areas of chromatography (GC, CE, and HPLC), lithography, and sensor technology, due to the ability to control the high density of chemical functionality at interfaces.[1] In chromatography, the functionalization of the stationary phase with covalently attached organic molecules provides a means to alter the flow of particles.[2] Organic molecules attached to surfaces can also be patterned to act as resists in lithography. For example, self-assembled monolayers (SAMs) of long-chain alkanethiolates and alkylsiloxanes with hydrophobic terminal groups can protect the underlying substrates from dissolution in certain types of aqueous etchants.[3] The ability to functionalize surfaces with organic moieties through covalent bonds is of fundamental importance for optimizing these technologies.

Alkylsilanes containing hydrolyzable bonds have been shown to react with hydrated surfaces.[4] This reaction is believed to proceed through the formation of silanols as intermediates, which react laterally or with surface OH groups to form a network polymer which is covalently bound (to some degree) to the surface.[4]

The goal of this project is to modify silicon surfaces in air with covalently attached phenylacetylene hyperbranched polymers using methods of organic chemistry. Highly branched polymers bound to a surface may provide novel interfacial properties due to the high number of terminal functional groups.[5] The chemistry of phenylacetylene hyperbranched polymers has been extensively studied both in solution and on solid support,[6] but has yet to be studied on silicon substrates. This provides an opportunity to determine the effect that silicon surfaces have on hyperbranched polymerizations. As the silicon surfaces are modified, rigorous characterization, including the techniques of XPS (ESCA), FTIR, and ellipsometry will be performed in order to understand the chemistry at the surface, and to begin to control it.

Experimental:

The substrates for the experiments were silicon (100) wafers (test grade, *n*-doped, with a resistivity between 1 and 10 Ω/cm) purchased from Silicon Sense, Nashua, NH. The substrates were cut into 1 x 1 cm^2 pieces for further modifications. Bromophenyltrimethoxysilane (mixed isomers, **1**) was purchased from Geleste, Inc. Other starting materials were purchased from Aldrich and used without further purification.

The substrates were cleaned by the RCA method[7] using 10 minute washes followed by rinses in deionized water. Washes consisted of first a 1:2 H_2O_2 : H_2SO_4 wash, then a 5:1:1 H_2O : NH_4OH : H_2O_2 wash, and a 6:1:1 H_2O : HCl : H_2O_2 wash. The washes removed residual hydrocarbon impurities to yield hydrophilic surfaces. Functionalized surfaces (**2**) were formed by immersing the cleaned substrates into freshly prepared solutions of 5% of **1** dissolved in dry 2% triethylamine/toluene, and heating at reflux for 4 hours.

NMR spectra were recorded on a Varian Unity 400 or 500 MHz spectrometer using $CDCl_3$ as the solvent. XPS measurements were performed by a Perkin-Elmer/PHI XPS 5400 spectrometer using Al K_α radiation (15 kV, 500 W). Photoelectrons were energy analyzed using a hemispherical analyzer at a constant pass energy of either 35.7 or 17.9 eV. Ellipsometry measurements were performed with a Gaertner Scientific (Chicago, IL) Model L116C ellipsometer employing a He-Ne laser (632.8 nm), set at an angle of incidence of 70°. Measurements were made at 5-10 points randomly picked across the sample and averaged. A two-layer transparent film model was used for the thickness calculations. The refractive index of the samples was fixed at 1.45.

Acetylenic compounds were synthesized by the general procedure presented by Sonogashira.[8] 1,3,5-triiodobenzene (**3**) was prepared from 1, 3, 5-tribromobenzene to yield 13.7 g of a white powder (30.0 mmol, 94%). ^{1}H NMR (500 MHz, $CDCl_3$) δ 8.0 (s, 1H); ^{13}C NMR (125 MHz, $CDCl_3$) δ 144.4, 95.3 ppm. 3, 5-Diiodophenyl-acetylene (**4**) was synthesized from **3** to yield 1.5 g of a white powder[6] (4.5 mmol, 53%). ^{1}H NMR (400 MHz, $CDCl_3$) δ 8.03 (t, *J* = 1.5 Hz, 1H), 7.78 (d, *J* = 1.6 Hz, 2H), 3.15 (s, 1H); ^{13}C NMR (100 MHz, $CDCl_3$) δ 144.4, 95.3 ppm.

Results and Discussion:

In order to modify surfaces in a controlled manner, it is critical to attach molecules to the surface covalently, and to quantitatively characterize subsequent reactions from the surface. Surface molecules define the location and density of initiation sites for polymerization. Compound **1** was attached to silicon (100) wafers in dry toluene in the presence of triethylamine to yield the SiO_2 derivatives, **2**. Wafers were characterized using XPS, and the bromine 3d peak was observed at 70 eV, as shown in figure 1. Wafers prior to functionalization had a thickness of 26 Å, and after functionalization with **1** had a thickness of 32 Å.

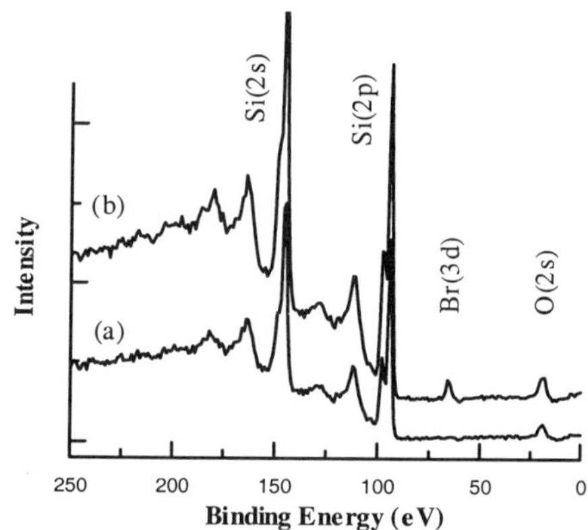

Figure 1. X-ray photoelectron spectroscopy survey scans of a) an unfunctionalized wafer and b) a wafer functionalized with **2**. The bromine 3d peak is a tag to determine the functionalized wafer.

To verify substrate stability under the Sonogashira[6,8] palladium coupling conditions, triethylamine and piperidine were exposed to **2** in control experiments as shown in figure 2. After 5.5 hours, the atomic concentration of the bromine present on the wafer had decreased in the triethylamine solution to 50% of its original value. After 13 hours in triethylamine, there was no characteristic evidence for the presence of bromine on the silicon wafer. In contrast, after 16 hours in piperidine, 20% of the bromine present on the wafer still remained. Therefore, subsequent Sonogashira palladium coupling reactions were performed in toluene with a small amount of piperidine.

When **2** was allowed to react under Sonogashira conditions[6,8] with **4**, the end group of the predicted hyperbranched polymer, iodine, was detected on the surface by XPS. In the control experiment (**4** reacted with an unfunctionalized wafer), iodine peaks were also observed. The hyperbranched polymer[6] in solution was insoluble to organic solvents and remained on the surface. Characterization and determination of the binding relationship between the surface and the polymer remains a challenge. Other schemes are currently being studied in order to obtain a high degree of control over polymerization and more complete characterization.

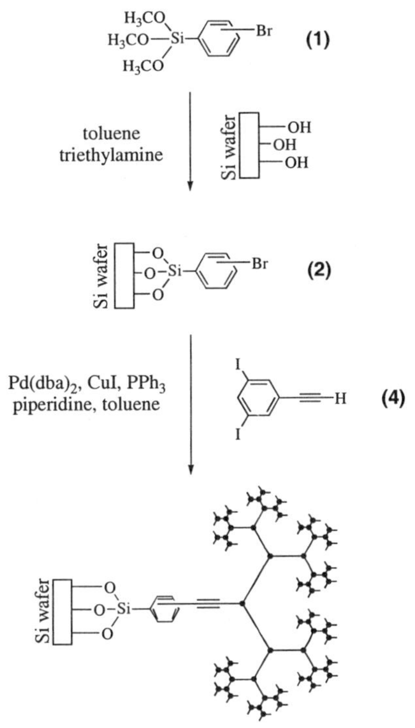

Figure 2. Hyperbranched polymerization from hydroxylated silicon wafers.

Conclusions:

Bromophenyltrimethoxysilane (**1**) can be covalently attached to silicon wafers and used as initiators for polymerization. A detailed study of polymeric growth initiated from functional surfaces (**2**) is currently underway along with rigorous surface characterization.

Acknowledgements:

This work was funded by the U. S. Department of Energy through the Materials Research Laboratory at the University of Illinois, the U. S. Army Research Office under grant DAAG55-97-0126, and Chromatography Research Supplies, Inc. A portion of this research was carried out in the Center for Microanalysis of Materials, University of Illinois, which is supported by the U. S. Department of Energy under grant DEFG02-96ER45439. NMR data was collected on spectrometers in the University of Illinois School of Chemical Sciences Varian/Oxford Instrument Center for Excellence in NMR Laboratory. We acknowledge NIH grant 1-S10-RR-10444-01, the Keck Foundation and the Beckman Institure for their contributions to this facility.

References:

1. Ullmann, A. *An Introduction to Ultrathin Organic Films*; Academic Press: New York, 1991.
2. Wirth, M. J.; Fatunmbi, H. O. *Anal. Chem.* **1993**, *65*, 822-826. Colon, L. A.; Guo, Y.; Fermier, A. *Analytical Chemistry News & Features* **1997**, 461A-467A. Pesek, J. J.; Matyska, M. T.; Sandoval, J. E.; Williamsen, E. J. *J. Liq. Chrom. & Rel. Technol.* **1996**, *19 (17&18)*, 2843-2865.
3. Zhao, X.-M.; Xia, Y.; Whitesides, G. M. *J. Mater. Chem.* **1997**, *7*, 1069-1074.
4. Prucker, O.; Ruhe, J. *Macromolecules* **1998**, *31*, 592-601. Haller, I. *J. Am. Chem. Soc.* **1978**, *100*, 8050-8055. Yang, Z.; Yu, H. *Adv. Mater.* **1997**, *9*, 426-429. Mao, G.; Castner, D. G.; Grainger, D. W. *Chem. Mater.* **1997**, *9*, 1741-1750. Wieringa, R. H.; Schouten, A. J. *Macromolecules* **1996**, *29*, 3032-3034. Carlier, E.; Guyot, A.; Revillon, A. *Reactive Polymers*, **1992**, *16*, 115-124. Firestone, M. A.; Shank, M. L.; Sligar, S. G.; Bohn, P. W. *J. Am. Chem. Soc.* **1996**, *118*, 9033-9041.
5. Wells, M.; Crooks, R. M. *J. Am. Chem. Soc.* **1996**, *118*, 3988-3989. Zhou, Y.; Bruening, M. L.; Bergbreiter, D. E.; Crooks, R. M.; Wells, M. *J. Am. Chem. Soc.* **1996**, *118*, 3773-3774. Zhou, Y.; Bruening, M. L.; Liu, Y.; Crooks, R. M.; Bergbreiter, D. E. *Langmuir* **1996**, *12*, 5519-5521.
6. Bharathi, P.; Patel, U.; Kawaguchi, T.; Pesak, D. J.; Moore, J. S. *Macromolecules*, **1995**, *28*, 5955-5963. Bharathi, P.; Moore, J. S. *J. Am. Chem. Soc.* **1997**, *119*, 3391-3392.
7. Kern, W. *Semiconductor International*, **1984**, 94-99.
8. Sonogashira, K.; Tohda, Y.; Hagihara, N. *Tetrahedron Lett.* **1975**, 4467-4470. Takahashi, S.; Kuroyama, Y.; Sonogashira, K.; Hagihara, N. *Synthesis*, **1980**, 627-630.

Separation of Random Copolymer by Phase Fluctuation Chromatography

Yunmei Xu and Iwao Teraoka

Polytechnic University, 333 Jay Street, Brooklyn, NY 11201

Introduction

Statistical copolymers, produced by copolymerizing monomer mixtures, have a molecular weight distribution and a chemical composition distribution (CCD). Physical properties of a copolymer are determined by the averages of molecular weight and chemical composition as well as by their distributions. Earlier, high osmotic pressure chromatography (HOPC) was developed in our lab and has been used to fractionate various polymers according to the molecular weight in large quantities.[1-3] Now we show a novel chromatographic method to fractionate the copolymer by the chemical composition.

There are several methods to fractionate a random copolymer into fractions of a narrow CCD, for instance, solvent-nonsolvent method[4] and gradient-elution HPLC.[5] These methods, however, work on dilute solutions of the copolymer, and hence their processing capacity is limited. Our chromatographic method rather takes advantage of properties specific to a concentrated solution of the copolymer, and therefore has a large processing capacity. We term the method phase fluctuation chromatography (PFC). Separation results under various conditions will be compared for styrene-acrylonitrile (SAN) copolymer as a model copolymer.

Principle of Separation

A concentrated solution of the copolymer tends to separate into many microdomains, each of which is enriched with components of a narrow CCD. The solution, however, usually does not phase-separate macroscopically. PFC involves partitioning of the solution between a confined space of pore and the surrounding solution. When the multi-phase solution is injected into a column packed with porous materials that have specific surface moieties, domains rich in a component with greater affinity to (or less repulsion from) the surface will be retained longer by the pore than the domains rich in the other components. The separation at high concentrations is repeated as the mobile phase is transferred along the column. The front end of the transported solution repeatedly sheds surface-preferred components as it approaches the column outlet and thus becomes purer in the other components. The early fraction will be enriched with surface-repelled components. The following portion of the mobile phase has to drive the components into the stationary phase that has already some of them, and it becomes less easy to remove the components from the mobile phase. As a result, later fractions will decrease the purity. Toward the end of the processing, components initially retained by the stationary phase will be eluted. As in HOPC, the advantage of PFC will be its ability to produce a sizable amount of purified copolymer in early fractions.

Experimental

Materials. Table 1 shows characteristics of the five SAN samples used for the present study. The molar fraction of the acrylonitrile (AN) content, x_{AN}, was calculated from its weight percentage. The weight-average molecular weight, M_w, with respect to polystyrene standards was measured in size exclusion chromatography.

Table 1. Characteristics of SAN copolymers

polymer	Source	AN wt%	x_{AN}	$M_w/1000$	h_{Ph}/h_{CN}
SAN20	Asahi	20	0.33	198	1.288
SAN25	Aldrich	25	0.39	184	0.948
SAN29	Asahi	29	0.44	195	0.772
SAN30	Aldrich	30	0.46	211	0.741
SAN40	Asahi	40	0.57	137	0.484

The AN contents of fractions collected and the original SAN samples were determined from FT-IR spectra. The peak at 2239 cm^{-1} was selected to estimate the AN content, and the peak at 1605 cm^{-1} was selected to represent the styrene content. The ratio h_{Ph}/h_{CN}, where h_{Ph} and h_{CN} are the peak heights at 1605 cm^{-1} and 2239 cm^{-1}, was found to be proportional to the molar ratio: $h_{Ph}/h_{CN} = 0.6253 \times (1 - x_{AN})/x_{AN}$. Controlled pore glasses (CPG) with different pore sizes and surface moieties were used as separating media to pack the columns.

Surface treatment of silica. Acid-washed CPG is coded as CPGxxxx-OH, where xxxx denotes the mean pore size in Å. We also prepared diphenylmethylsilane-substituted CPG (CPGxxxx-Ph) and 3-Cyanopropyldimethyl-silane-substituted CPG (CPGxxxx-CN).

Instrumentation. A standard system of HOPC was used for separation. The solution and solvent injection was at 0.3 mL/min. 15 drops were collected in each test tube in fractions 1 to 5. Fractions 6 to 8 collected 30 drops each, fractions 9 to 14 collected 50 drops each, and fraction 15 collected 150 drops.

Results and Discussion

Pore size and surface area. PFC needs porous materials with surface moieties that have a sufficiently strong influence on the solution in the entire pore volume. The larger the surface area, the more the surface moieties, and the stronger the interaction between the pore wall and the polymer. It is also necessary to avoid size exclusion effect by the pores. These two requirements are inconsistent, because porous materials with a smaller pore diameter have a greater surface area to volume ratio. To find the optimal pore size, we compared four columns packed with CPG350-OH, CPG500-OH, CPG1000-OH, and CPG3000-OH. A 16 wt % (total) solution of a mixture of SAN20 and SAN40 (1:1 w/w) in dioxane was injected into each column. When the two larger pore sizes were used, the separation was not effective. With CPG500-OH, x_{AN} decreased with fraction number. With CPG350-OH, the separation was influenced by the chain dimension, reversing the trend observed for CPG500-OH. The optimal pore size is 500 Å for the SAN samples used here. The following studies were therefore conducted by using 500 Å pore.

Surface chemistry. The resolution of PFC depends on the interactions between polymers of different chemical compositions and between polymer and the surface moieties. Here we compare separations by CPG500-Ph, CPG500-CN, and CPG500-OH. A 16 wt % solution of a mixture of SAN20 and SAN40 (1:1 w/w) in dioxane was injected into each of the three columns. Figure 1 shows h_{Ph}/h_{CN} and x_{AN} as a function of the fraction number. With CPG500-Ph and CPG500-OH, AN-rich components were eluted earlier, whereas styrene-rich components came out earlier with CPG500-CN. Apparently, diphenyl and silanol surfaces repel AN more strongly than they repel styrene.

The separation works for the pure SAN sample as well. A 16 wt % solution of SAN20 in dioxane was injected into the three columns. The results are shown in Figure 2. The tendency is similar with the one observed for the SAN20/SAN40 mixture although the difference in x_{AN} between separated fractions is much smaller. The silanol surface was not effective to separate a SAN fraction with an already narrow CCD.

Solvent. It is expected that selective solvent of the two monomers is better in PFC, because the phase separation occurs at lower concentrations. We compared dioxane and methylethylketone in the separation of a mixture of SAN29 and SAN40. The latter solvent is a nonselective solvent. It was found that methylethylketone resulted in almost no separation with CPG500-CN and only a weak separation with CPG500-Ph.

Concentration. Figure 3 demonstrates the concentration effect on the separation in methylethyketone. A mixture of SAN29 and SAN40 (1:1 w/w) was dissolved at total concentrations of 25 wt %, 21 wt %, and 15 wt %. When left for overnight, the first two solutions separated into two macroscopic phases, but the 15 wt % remained in a single phase. The 25 wt % concentration resulted in the best separation with the largest span in h_{Ph}/h_{CN}, followed by the 21 wt % solution. Separation at 15 wt % was poor. Injection of the two-phase solution resulted in a much better separation compared with the injection of the single-phase solution.

The result was similar in the separation in dioxane. These results show that the higher the concentration, the better the separation, as long as the pump and column can withstand the high back-pressure that increases with increasing concentration.

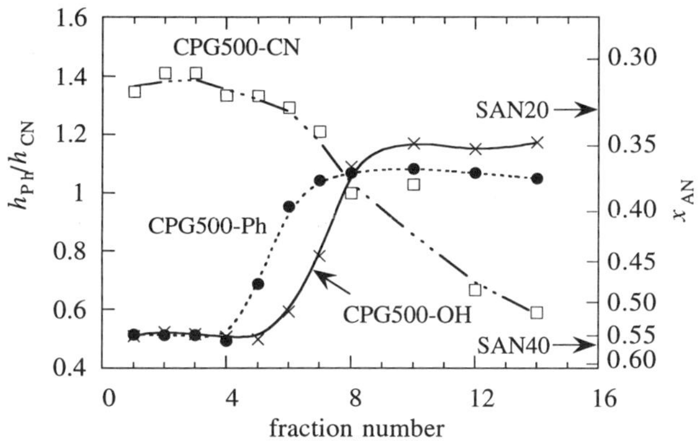

Figure 1. Surface chemistry dependence of PFC. A mixture of SAN20 and SAN40 in dioxane was separated.

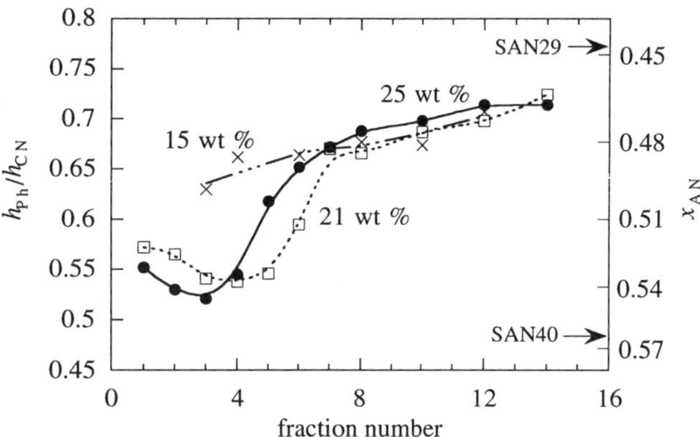

Figure 3. Concentration dependence of PFC. A mixture of SAN29 and SAN40 in dioxane at different concentrations was separated with CPG500-Ph.

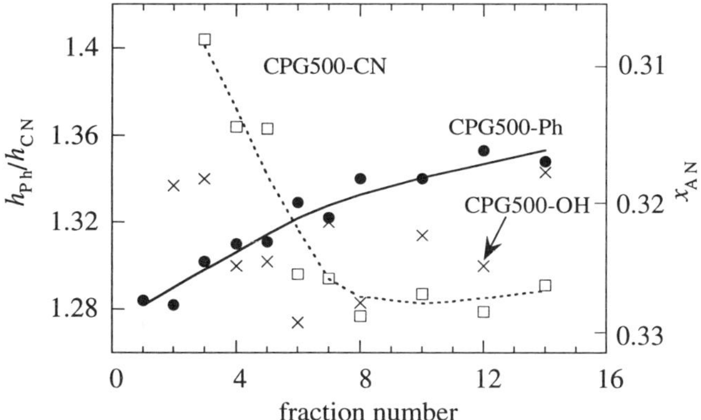

Figure 2. Surface chemistry dependence of PFC. SAN20 was separated.

References

1. Luo, M.; Teraoka, I. *Macromolecules* **1996**, *29*, 4226.
2. Teraoka, I.; Luo, M. *Trends Polym. Sci.* **1997**, *5*, 258.
3. Luo, M.; Teraoka, I. *Polymer* **1998**, *39*, 891.
4. Teramachi, S.; Tomioka, H.; Sotokawa, M. *J. Macromol. Sci. Chem.* **1972**, *A6*, 97.
5. Ogawa, T.; Sakai, M. *J. Polym. Sci., Polym. Phys. Ed.* **1981**, *19*, 1377.

Methacrylate-Functionalized Silane Coupling Agents Covalently Linked to Iron Particles. Effect of Surface Treatment on the Rheological Behavior of Magnetic Dispersions in Acrylates

Jin Young Huh, Kwang S. Jeon,
James P. Parakka, and David E. Nikles
Center for Materials for Information Technology
The University of Alabama, Tuscaloosa, Alabama 35487-0209

Magnetic tape and floppy disks are manufactured by a continuous web coating process where a magnetic dispersion is coated onto polyester base film. The coating fluids use organic solvents, including MEK, MIBK and toluene. These solvents are on the EPA's list of hazardous air pollutants and are regulated under the Clean Air Act Amendments.[1] The industry meets EPA air emissions regulations by using pollution control equipment to capture and recycle the organic solvents. We have estimated that a conventional solvent-based coating line uses more than 600 kg of organic solvents per hour.[2-3] Industry sources tell us that 93 to 95% of the solvents from the coating process are controlled, which means that 5 to 7% (or 30 to 40 kg/hr) are lost to the environment. Assuming the coating line runs 4000 hr/yr (two 8 hr shifts/day x 5 days/week x 50 weeks), then the potential emissions per year would be 120 to 170 metric tons. The object of this research project is to take that number to zero. Our approach is to replace the organic solvents with liquid acrylate monomers. The acrylates would serve as the solvent for the coating fluid and after coating and electron beam irradiation, the acrylates would polymerize the form the binder. The result would be a coating process with no VOC's.

Earlier we reported the identification of mixtures of acrylate monomers and oligomers that give electron beam cured films with adequate tensile properties and good adhesion to the polyester base films.[4,5] Recently we have been working on a serious rheology problem. We are limited in the amount of liquid monomer in the coating formulation. The information recorded on magnetic tape is held as magnetized particles. The signal to noise ratio depends on the number of particles in the data address.[6] To obtain a good SNR, the volume fraction of magnetic particles in the coating must be at least 0.30. Since the solvent in the magnetic coating fluid will remain as the binder, after curing, the volume fraction of magnetic particles in the coating fluid must be greater than 0.30. In conventional solvent-based coating processes, the coating fluids are let down to 3 to 5 volume percent, giving fluids with rheological properties useful for coating. If we are to achieve our goal of a solventless coating process, we do not have the luxury of diluting the coating fluid. In this paper we describe a promising approach to solving the rheology problem.

Experimental

All starting materials and solvents were reagent grade and used as received. The methacrylate-functionalized silane coupling agent, Z-6030, was a gift from Dow Corning. The liquid acrylates in the solventless coating formulation were CN 965 A80 (aliphatic urethane diacrylate blended with tripropylene glycol diacrylate), SR 9003 (propoxylated neopentyl glycol diacrylate), SR 9035 (highly ethoxylated trimethylolpropane triacrylate), and SR 506A (isobornyl acrylate), all obtained as gifts from Sartomer. The magnetic particles were a commercial iron particle for metal particle tape. The particles were acicular with a length of 100 nm and an aspect ratio of 6. They had an alumino-silicate coating that protected them against corrosion. The coercivity (H_c) was 1500 Oe, the saturation magnetization (σ_s) was 132 emu/g and the specific surface area was 54 m^2/g. The magnetic properties were obtained from hysteresis curves measured on a DMS model 1660 vibrating sample magnetometer. TGA curves were measured on a TA Instruments model 2970 thermogravimetric analyzer. Rheological measurements were made using a Haake Rheostress RS100 oscillating shear rheometer.

<u>Synthesis of the Branched Coupling Agent, **2**</u>. To a round bottomed flask, equipped with magnetic stirring, was added 10 mL toluene, 0.95 g cyanuric chloride, 0.673 g 2-hydroxyethyl methacrylate, and 0.4 g sodium hydroxide. The mixture was allowed to stir for 4 hr at room temperature to give intermediate **1**. To a round bottom flask, equipped with magnetic stirring was added 10 mL toluene, 0.25 g 3-aminopropyltrimethoxysilane and 0.33 g sodium hydride. The mixture was stirred for 24 hr, then the mixture from the first flask was added. The combined mixture was stirred for 1 hr at room temperature, then the product, **2**, was purified by column chromatography.

<u>Surface Treatment of the Iron Particles with the Branched Coupling Agent, **2**</u>. A mixture of 1.00 g iron particles, 0.06 g 2.0 N ammonium hydroxide, 0.20 g **2** and 10 g toluene were ball milled for three days. After three days the particles were washed four times with ethanol and four times with acetone. They were then dried at 60°C for 24 hr.

<u>Surface Treatment of the Iron Particles with the Branched Coupling Agent, **3**</u>. A mixture of 1.00 g iron particles, 0.06 g 2.0 N ammonium hydroxide, 0.20 g 3-(N-2-aminoethyl)-aminopropyltrimethoxysilane, and 10 g toluene were ball milled for 3 days. After 3 days the intermediate, **2**, was added and the ball milling continued. The particles were rinsed repeatedly with ethanol and with acetone. The treated particles had a saturation magnetization of 99 emu/g.

<u>Surface Treatment of the Iron Particles with Z-6030</u>. To a glass vial was added 1.00 g iron particles, 10 g toluene and 0.05 g Disperbyk-111. This was agitated in an ultrasonic bath for 30 s, then 0.05 g 2.0 N ammonium hydroxide and 0.20 g Z-6030 were added. The mixture was agitated on a wrist-action shaker for 3 days. After three days the particles were allowed to settle and the supernatent decanted away. The particles were washed four times with ethanol and four times with acetone. They were then dried at 60°C for 24 hr. The treated particles had a saturation magnetization of 110 emu/g.

<u>Magnetic Dispersions</u>. A mixture of 50 g iron particles, 15.5 g CN 965A80, 31.0 g SR 9003, 1.75 g SR 9035 and 1.75 g SR 506A were ball milled overnight.

Results and Discussion

Our first attempts to prepare pigmented, solventless formulations demonstrated the severity of the rheology problem. In magnetic tape the signal comes from the magnetic particles and in order to maximize the signal, the particle fraction must be as large as possible. A typical magnetic tape has a 30 volume percent of particles, which means the weight fraction of magnetic particles in the coating fluid must be 80%. The iron particles, with a specific surface area of 54 m^2/g, adsorbed all the liquid in our first formulation after predispersing for 1.5 hr in a double planetary mixer. The dispersion had the consistency of brick dust, not at all satisfactory for coating.

The commercial acrylate functionalized silane coupling agent, Z-6030, was covalently bonded to the surface of the iron particles. The original purpose was to use the magnetic particles to mechanically reinforce the binder by linking the binder to the particles. The commercial iron particles had a ceramic coating containing silica. We assumed that the surface contains silanol groups that can react with silanol groups from the coupling agent. This was standard silane coupling agent chemistry, with an important difference. The coupling agent chemistry worked out for fiberglass surfaces uses an acid catalyst, acetic acid with water. When the iron particles are exposed to these acid conditions they rapidly corroded. The key step in the reaction between the coupling agent and the particle surface was the hydrolysis of the methoxy groups. After hydrolysis the silanol groups condensed to form a silicone polymer. The polymer adsorbed onto the particle surface and condensed with surface silanols, thereby anchoring the coupling agent to the particle surface. The hydrolysis reaction can be catalyzed by acid or by base. Without particles, the reaction was followed by ^{1}H NMR spectroscopy and required five to six hours, regardless of the choice of solvent (cyclohexanone, 95% ethanol, tetrahydrofuran or toluene) or choice of catalyst (acetic acid, boric acid, or ammonium hydroxide). Since the particles are sensitive to acid, we pursued reactions catalyzed by ammonium hydroxide. After treatment the particles had a saturation magnetization of 110 emu/g, a loss of 17%. A TGA curve showed a 12% weight loss upon heating. We attribute the loss in magnetization to the increase in mass due to the surface coating, not a degradation in the particles.

When dispersions were prepared using the particles treated with Z-6030 at 80 weight percent, a thick paste was obtained. The paste had a viscosity of in excess of 10^8 cps, too high for coating, but it did show shear thinning.

Z-6030, Methacrylate-functionalized silane coupling agent.

The viscosity of a typical solvent-based coating fluid is in the range of 10^3 cps, with significant shear thinning. When the surface-treated magnetic particles were used at 50 weight percent, we obtained a coating fluid with rheological properties, Fig. 1, comparable to a conventional, solvent-based coating fluid. The solventless coating fluid also showed a shear thinning behavior, similar to that observed for conventional coating fluids.

The rheological properties of magnetic coating fluids are dominated by the magnetic attraction forces between the particles, which increase the elasticity (G') of the coating fluid. The surface treatment occupied surface sites on the particles that would otherwise adsorb acrylate monomers. The surface coating also provided a steric barrier against particles approaching each other, which decreased the strength of the magnetic attraction between particle, thus decreasing the elasticity of the fluid. To exploit this discovery, we focused attention on the synthesis of a new branched methacrylate-terminated silane coupling agent.

The branched silane coupling agent was synthesized by first reacting cyanuric chloride was reacted with two equivalents of hydroxyethyl methacrylate to give the intermediate, **1**. The intermediate was reacted with 3-aminopropyltrimethoxy-silane to give the branched coupling agent, **2**. The coupling agent was purified by column chromatography. The details of the synthesis will be reported elsewhere. Iron particles were surface treated with this coupling agent and used in a solventless formulation at a 50 weight percent loading. The resultant magnetic dispersion was extremely viscous and could not be coated. A TGA analysis of these treated particles showed only a 6% weight loss, which indicated that the

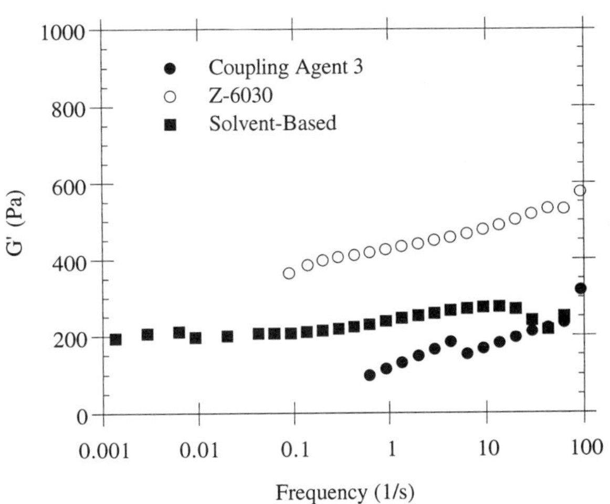

Intermediate (1)

Branched Coupling Agent (2)

amount of **2** on the particle surface was less than the particles treated with Z-6030. Perhaps the branched coupling agent was too bulky and could not completely cover the particle surface.

A different branched coupling agent was assembled on the iron surface in two stages. First 3-N-(2-aminoethyl)-propyltrimethoxysilane was condensed on the particle surface, giving a teaming with amine groups. Next the intermediate, **1**, was reacted with the amines. This gave particles with a saturation magnetization of 99 emu/g (a 25% decrease) and a TGA weight loss of 28%. Again we conclude that the decrease in magnetization was due to the increase in mass provided by the coupling agent. The magnetic dispersion containing a 50 weight percent loading of these particles has an elastic modulus (G') lower than that for the conventional solvent-based dispersion.

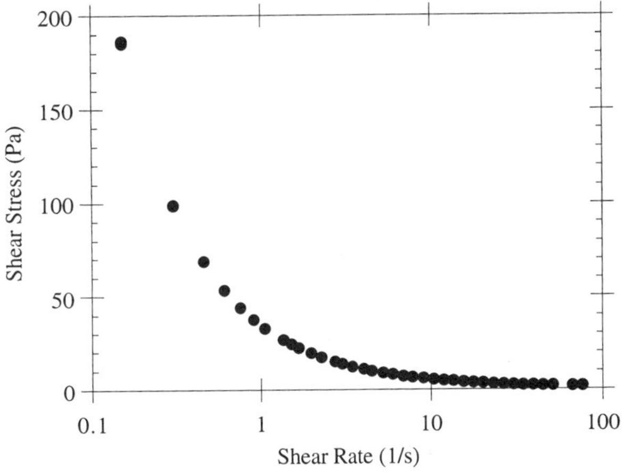

Branched Coupling Agent (3)

Conclusions

The results presented here suggest a way of solving the rheology problem for our solventless formulations. By surface treating the magnetic particles we greatly enhanced our ability to disperse the particles in the solventless acrylate formulations. The dispersions had an elasticity comparable to that for a conventional, solvent-based dispersion. We still must fully load the formulations to a level of 80 weight percent. This is the subject of further work.

Acknowledgments

This project was funded in part by The University of Alabama; Federal funds under the cooperative agreement CR822961-01-0 with the Risk Reduction Engineering Laboratory, U. S. Environmental Protection Agency; and Federal funds as part of the program of the Gulf Coast Hazardous Substances Research Center, supported under cooperative agreement R815197 with the EPA. The contents do not necessarily reflect the views and policies of the EPA. The mention of trade names or commercial products does not constitute endorsement or recommendation for use. The project used shared instrumentation purchased through the NSF Materials Research Science and Engineering Center award DMR-9400399.

References

1. *Federal Register* **1994**, 59(240): 64180-64612.
2. Cheng, S.; Fan, H.; Gogineni, N.; Jacobs, B.; Jefcoat, I. A.; Lane, A. M.; Nikles, D. E. *CHEMTECH* 1995, 25 (October), 35-41.
3. Cheng, S.; Fan, H.; Gogineni, N.; Jacobs, B.; Jefcoat, I. A.; Lane, A. M.; Nikles, D. E. *Waste Management* **1995**, 15(4), 257-264.
4. Parakka, J. P.; Nikles, D. E *Proceedings of the Division of Polymeric Materials: Science and Engineering* **1996**, 75, 297-298.
5. Ellison, M. M.; Huh, J. Y.; Power, A.; Purse J. B.; Nikles, D. E. *Proceedings of the Division of Polymeric Materials: Science and Engineering* **1997**, 76, 115-116.

Fig. 1. Plot of elastic modulus, G' (Pa) as a function of frequency for solventless dispersions containing surface treated iron particles (coupling agent 3 or Z-6030) or a conventional, solvent-based dispersion.

Fig. 2. Plot of shear stress as a function of shear rate for the solventless magnetic dispersion containing particles surface treated with Z-6030

PRELIMINARY STRUCTURAL DATA ON NEW BIODEGRADABLE POLY(ESTER AMIDE)S DERIVED FROM L- AND L,D-ALANINE

J. Puiggalí, J. E. Aceituno, N. Paredes, A. Rodríguez-Galán, M. Pelfort and J. A. Subirana

Departament d'Enginyeria Química, ETS d'Enginyers Industrials. Universitat Politècnica de Catalunya, Diagonal 647, Barcelona 08028, SPAIN

INTRODUCTION

Aliphatic Poly(ester amide)s have been suggested and recently investigated[1] as a family of polymers with optimum potential properties: mechanical and thermal due to the amide groups and degradability due to the ester groups. However, structural studies on this family of polymers are scarce. On the other hand, it is known that polymers containing α-amino acids[2,3] are more susceptible to degradation, although they have not yet found the expected applications, mainly because of preparation difficulties[4,5]. In this work we present results on Poly(ester amide)s derived from an α-amino acid as alanine in both the quiral L-configuration (PADAS100) and the racemic L,D-mixture (PADAS50).

SYNTHESIS AND CHARACTERIZATION

The poly(ester amide)s were synthesized following the procedure outlined in the Scheme and previously reported for related polymers[6,7]:

$$HO-(CH_2)_{12}-OH \qquad 2\ HOOC-\underset{CH_3}{CH}-NH_2$$

$$PTS^-\ ^+NH_3-\underset{CH_3}{CH}-COO-(CH_2)_{12}-OOC-\underset{CH_3}{CH}-NH_3^+\ PTS^- \qquad ClOC-(CH_2)_8-COCl$$

$$\left[NH-\underset{CH_3}{CH}-COO-(CH_2)_{12}-OOC-\underset{CH_3}{CH}-NH-OC-(CH_2)_8-CO\right]$$

Interfacial polymerization was optimized taking into account the proton acceptor and the concentration of the sebacoyl dichloride in the organic solvent. CCl_4 gave better results than $CHCl_3$ or toluene. It was demonstrated that hydrolysis was produced by using NaOH as proton acceptor, whereas the best results were attained with Na_2CO_3. An optimum intrinsic viscosity (measured in dichloroacetic acid at 25 °C) was respectively determined for the L-alanine (0.83 dL/g) and the L,D-alanine (1.00 dL/g) derivatives, as shown in Table 1.

STRUCTURAL DATA

A modified Statton camera (W. H. Warhus, Wilmigton, DE, USA) with nickel-filtered copper radiation ($\lambda = 1.542$ Å) was used to obtain X-ray diffraction patterns. Both polymers gave a highly oriented pattern (Figure 1) with similar spacings and intensities. Thus, all reflections can be indexed (Table 2) with a unique orthorhombic unit cell of parameters: $a = 4.86$ Å, $b = 24$ Å and $c = 30$ Å. A high resolution is obtained with the racemic sample, probably due to its high molecular weight.

Crystallizations were carried out isothermally from dilute butanediol (PADAS100) or hexanediol (PADAS50) solutions at 50 °C and 57 °C, respectively. For electron microscopy, the crystals were deposited on carbon coated grids, which were then shadowed with Pt-carbon pellets at an angle of 15°. A Philips EM-301 electron microscope operating at either 80 or 100 kV for bright field and electron diffraction modes, respectively, was used. Both polymers crystallized as lath-shaped crystals (Figure 2), which frequently appear folded over themselves due to their ribbon-like morphology. Note that their length can be more than 10 μm, whereas their width is generally less than 1 μm. Surprisingly, the best crystals were obtained with the racemic sample, as it was also demonstrated by the electron diffraction patterns. The preferred crystal growth direction can be well explained, in both polymers, assuming that hydrogen bonds are established according to only one direction which run parallel to the long crystal axis. This feature is also supported by the electron diffraction data. Thus, in all cases, the patterns show an *mm* symmetry with one parameter of the projected unit cell close to 4.8 Å (characteristic value for hydrogen bonded molecular chains) and oriented along the long axis of the crystals (Figure 3). Note also the presence of the reflection indexed as 110 that is indicative of a structure with six layers of hydrogen bonded chains in the unit cell. From the electron diffraction patterns, it may be also concluded that molecular chains are folded within the crystal, as a consequence of their molecular weight and the reduced lamellar thickness (60 Å can be estimated from shadows in the electron micrographs).

ENZYMATIC HYDROLYSIS

Enzymatic degradation studies were carried out with papain, using discs of the polymers ($\phi = 12$ mm, thickness 200-300 μm, weight 30-45 mg) prepared by melt pressing at 110 °C. The proteolytic enzyme was activated, in a 0.05 M phosphate buffer, with 34 mM L-cysteine and 30 mM ethylenediaminetetraacetic disodium salt, in a similar procedure to that proposed by Arnon[8]. A clear and steady degradation was observed for PADAS100, whereas the racemic sample was degraded at a slower rate as expected. Figure 4 shows the weight loss of discs of PADAS100 after exposure to papain. When the disc-exposure was continued beyond 35 % weight loss, disintegration of the discs occurred and finally, complete disc disappearance was observed. A parallel series of experiments performed in the same buffered solution and without papain, showed that the recovered samples remain essentially unchanged during the equivalent exposure times.

Acknowledgements. This work has been supported by research grants from CICYT (MAT 97-1013) and DGCYT (PB93-1067).

REFERENCES

1. R. J. Gaymans, J. L. de Haan, *Polymer*, **34**, 4360, 1993.
2. W. J. Bailey, B. Gapud, *Ann. N. Y. Acad Sci.*, **446**, 42, 1985.
3. K. E. Gonsalves, X. Chen, T. K. Wong, *J. Mater. Chem.*, **1**, 643, 1991.
4. G. S. Kumar, Biodegradanle Polymers: Prospects and Progress, Marcel Dekker, New York, 1987.
5. R. Duncan, J. Kopecek, *Adv. in Polym. Sci.*, **57**, 51 1984.
6. L. H. Ho, S. Huang, *Polym. Prepr. (Am. Chem. Soc. Div. Polym. Chem.)*, **33(2)**, 94, 1992.
7. N. Paredes, A. Rodríguez-Galán, J. Puiggalí, *J. Polym. Sci., Polym. Chem. Ed.*, in press.
8. R. Arnon, Methods in Enzymology (Eds. G. E. Perlmann and L. Lorant), Vol. XIX, Academic, New York, 1970, p. 226.

TABLE 1

Polymer	Concentration in the CCl$_4$ solvent	Yield (%)	$[\eta]$ (dL/g)
PADAS100	0.01	67	0.74
PADAS100	0.05	73	0.83
PADAS100	0.15	68	0.80
PADAS100	0.30	78	0.55
PADAS100	0.50	65	0.60
PADAS100	0.80	70	0.56
PADAS100	1.00	85	0.60
PADAS50	0.03	50	0.80
PADAS50	0.25	65	0.77
PADAS50	0.20	68	1.00
PADAS50	0.10	60	0.92
PADAS50	0.50	55	0.88

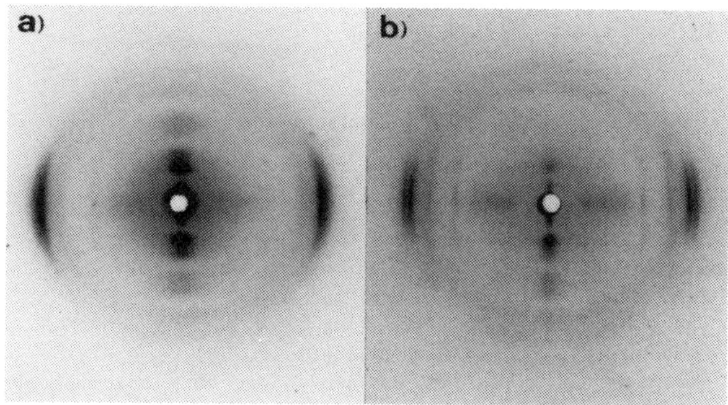

FIGURE 1. X-ray fiber diffraction patterns of (a) PADAS100 and (b) PADAS50.

TABLE 2

Index	Calcd. Spacings (Å)	Measd. Spacings (Å) PADAS100	Measd. Spacings (Å) PADAS50
ELECTRON DIFFRACTION DATA			
060	4.00	4.01 vs	4.00 vs
110	4.77	4.78 m	4.80 m
130	4.18	4.15 vs	4.15 vs
190	2.34	2.32 vw	2.34 m
200	2.43	2.43 vw	2.43 w
260	2.07	-	2.07 w
X-RAY FIBER DIFFRACTION DATA			
Index[a]	Calcd. Spacings (Å)	PADAS100	PADAS50
002	15.0	15.0 vs	15.0 vs
003	10.0	10.8 s	10.6 m
004	7.5	7.5 m	7.5 m
005	6.0	6.2 m	6.3 m
006	5.00	4.98 w	4.95 w
007	4.28	-	4.29 vw
008	3.73	-	3.71 vw
040	6.00	6.01 vw	6.00 w
042	5.61	5.59 vw	5.58 w
130	4.15	4.15 vs	4.17 vs
060	4.00	4.00 s	4.01 s
110	4.77	4.75 vw	4.78 w
112	4.55	4.56 vw	4.59 w

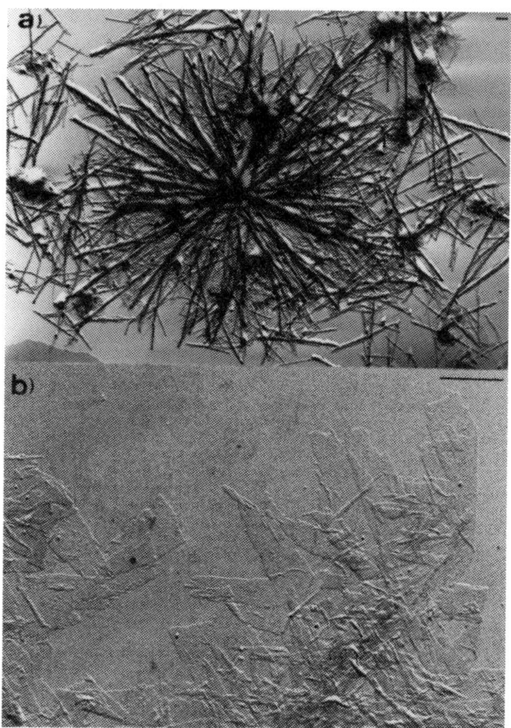

FIGURE 2. Transmission electron micrographs of crystals prepared in a diol solution of (a) PADAS100 and (b) PADAS50. Crystals show a preferred growing direction, which corresponds to the *a* axis. Scale bars: 1 μm.

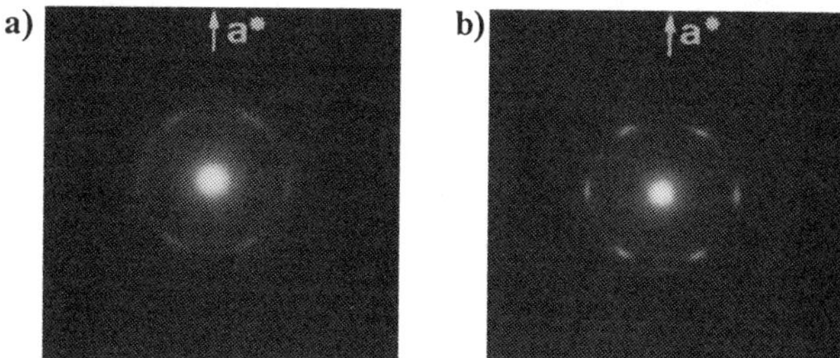

FIGURE 3. Single-crystal electron diffraction diagrams of (a) PADAS100 and (b) PADAS50. Note the 110 reflections, which clearly appear out of the *a** axis in the PADAS50 sample.

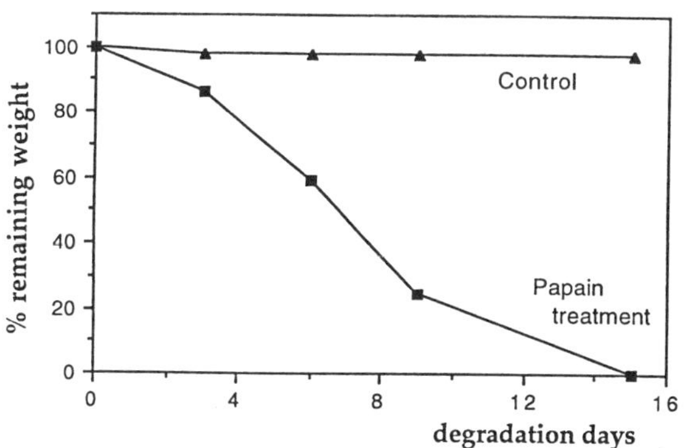

FIGURE 4. Plot of the weight changes of PADAS100 during *in vitro* enzymatic degradation with papain. Results using control buffer without papain are also reported.

Synthesis, Characterization of Acrylonitrile Copolymers, and Their Ultrafiltration Membranes

Wenli Han, Harry P. Gregor, Eli M. Pearce

Department of Chemical Engineering, Chemistry & Materials Science
Polytechnic University
Six Metrotech Center, Brooklyn, NY 11201

INTRODUCTION

Synthetic ultrafiltration(UF) membranes based upon cellulose acetate and other derivatives were well developed by the mid 1930's. Today, synthetic membranes ranging from PVDF (polyvinylidine fluoride), PAN (polyacrylonitrile), the nylons and the polysulfones. In spite of the many kinds of UF materials available today, the large scale, commercial utilization of UF has been hampered by the costs involved, largely due to the occurrence of membrane fouling, namely the adhesion of dissolved and suspended materials onto the hydrophobic polymer surfaces and the resulting flux decline. This is difficult to reverse except by extensive and often severe cleansing methods which add substantial costs to the process. Interactions between solute-membrane and solute-solute species can be broadly classified as (1) polar interactions (e.g. H-bonding), (2) interactions due to dispersion forces and (3) electrostatic interactions. Hydrophilic membrane materials are considered by some to have high surface free energies, largely due to strong polar interactions; with some hydrophobic materials such as the fluorinated hydrocarbons considered by some to have low surface free energies [1].

We have investigated the use of polymers containing hydrophilic moieties in various forms. Only by the use of the monomers and the polymers containing the strongly polar amide group of an extremely high hydrophilicity can UF materials demonstrate a high degree of non-fouling character.

EXPERIMENTAL

Synthesis of polymer

A variety of copolymers based on the different feed compositions of acrylonitrile and acrylamide were prepared at low conversion by free radical polymerization.

The copolymerization of acrylonitrile and acrylamide was carried out in a five liter three-neck round bottom flask fitted with a water bath, thermometer, overhead stirrer, gas inlet and gas outlet. The solution of monomers in water was purged with nitrogen gas for at least forty-five minutes. The initiators were added to the solution at the polymerization temperature (25 ^{0}C) under a nitrogen atmosphere. Hydroquinone (0.1 gram) was added to terminate the polymerization after about 48 hours.

Magnesium chloride (about 2 weight percentage of all monomers) was added after the solution was heated to 60 ^{0}C. The product was filtered and washed with water until the test for chloride with 1 wt. % silver nitrate aqueous solution was negative. The product was dried under vacuum to constant weight.

Characterization of polymer

Viscosity measurement were made in a thermostatted water bath at 25± 0.1 ^{0}C by means of a Cannon-Ubbelohde viscometer. A copolymer was dissolved in dimethylformamide(DMF) which had been exhaustively dried. For each polymer, the viscosity of five concentrations was measured; multiple readings were made at each concentration. Intrinsic viscosity was obtained by extrapolation of a plot of specific viscosity/concentration vs. concentration to infinite dilution using linear least squares; such analysis yield regression coefficients $\geq$ 0.999. Estimates of the copolymer molecular weight were obtained from the relationship for PAN in DMF at 25^{0}C [2]:

$$[\eta] = 0.0392 M_V^{0.75}$$

where $[\eta]$ is the intrinsic viscosity and M_V is the viscosity-average molecular weight.

Glass transition temperatures (T_g) of the polymers were determined by using a Du Pont 910 differential scanning calorimeter. The heating rate was 15 ^{0}C/min for all samples.

Membrane Casting and Characterization

The membranes described below were prepared using standard casting techniques. Ultrafiltration membranes were prepared from 3 wt% polymer solutions in DMF by casting the solution onto Hollytex, with a 5 mils (127 μm) gate opening on a casting knife, followed by immediate coagulation in 2-5 ^{0}C water to minimize skin formation. The membranes were rinsed several times in deionized water and kept at 5^{0}C until used.

The surface characterization of each polymer could be done using dried clean membrane. The intrinsic hydrophilicity of the membrane materials is determined from the advancing contact angles measured by a goniometer (Model 100-00 Rame-Hart, USA).

Prior to the contact angle measurement, the membrane sample was rinsed three times for 10 min in double-distilled water and the sample was then cleaned with double-distilled water in an ultrasound bath for two periods of 15 min. The angles were evaluated from photographs using video-enhanced image processing. The values of the contact angles are the average of the ten air bubbles (two angles per bubble), giving a total of 20 angles for each membrane sample.

The polar and dispersion force components of the surface energy were determined from the following equation;

$$1 + \cos\theta = 2\left(\gamma_s^d\right)^{0.5}\left[\left(\gamma_l^d\right)^{0.5}/\gamma_{lv}\right] + 2\left(\gamma_s^p\right)^{0.5}\left[\left(\gamma_l^p\right)^{0.5}/\gamma_{lv}\right]$$

using the reported parameters [3] and measured contact angles for water and formamide.

RESULTS AND DISCUSSION

Intrinsic viscosity measurement

Copolymers of AN and AM can be made in any range of polymer compositions. Since PAM tends to compact in the poor solvents such as DMF, the viscosities are higher with low percentages of AM and substantially lower as AM substitution increases.

The copolymers high in AN are insoluble in water but swell increasingly as AM content increases. Copolymers containing 40% of AM become partially water soluble and entirely water soluble as the percentage of AM increases further.

Glass transition temperature measurement

The glass transition temperatures also varies with copolymer composition as shown in Table II. Each of these polymers was measured in the dry state by DSC. PAN itself has a relatively low glass transition temperature but this increases sharply as AM substitution in the copolymer increases. It is presumed that strong internal hydrogen bonding in PAM is responsible for its relatively high glass transition temperature.

Surface Characterization of Membrane

As would be expected from the chemical modification, Table III shows that the polymer/water contact angles decrease with an increasing acrylamide content. The polar component of the surface energy increases with increasing acrylamide content; while the dispersion force component of all acrylamide containing membranes is less than that of the PAN membrane.

CONCLUSIONS

A variety of poly(acrylonitrile-*co*-acrylamide) polymers of different composition were synthesized by free radical copolymerization. The intrinsic viscosities in DMF were higher with low percentages of AM and substantially lower as AM substitution increased. The glass transition temperature increased sharply as AM substitution in the copolymer increased. In comparison with PAN, the AM containing copolymers yield membranes with higher polarity/hydrophilicity and smaller dispersive surface energy.

ACKNOWLEDGEMENTS

The research was funded by the Office of Naval Research (Grant N00014-93-1-1371). General assistance from Drs. Dexin Luo and Frank Mikes is greatly appreciated.

REFERENCES
1. D. K. Owens and R. C. Wendt, J. Appl. Polym. Sci., **13**, 1741 (1969)
2. J. Brandrup and E. H. Immergut, Polymer Handbook, 2nd Ed., Wiley, New York, 1975, Sec. IV.
3. F. W. Fowkes, Ind. Eng. Chem., **56**, 40 (1964)

Table I. Results of intrinsic viscosity

Copolymer	AM in Feed (mol%)	AM in Copolymer (mol%)	Conversion (%)	Intrinsic Viscosity (dl/g)
A-1	10	N/A	5	2.41
A-2	20	N/A	16	2.73
A-3	30	N/A	18	1.58
A-4	40	N/A	8	1.40
A-5	50	N/A	19	1.91
A-6	60	N/A	6	0.78

Table II. DSC Results

Polymer	AM in Feed (mol%)	Glass transition Temperature (^{o}C)
PAN	0	96
A-1	10	107
A-2	20	113
A-3	30	116
A-4	40	117
A-5	50	120
A-6	60	132
PAM	100	154

Table III. Surface Characterization

Memb.	AM in Feed (mol%)	Contact Angle (^{o})[a]		Surface Energy (erg/cm^2)[b]		
		Water	Formamide	γ_s^p	γ_s^d	γ_s
PAN	0	72	47	8	34	42
A-1	10	68	44	11	33	44
A-2	20	61	37	15	32	47
A-3	30	54	35	22	27	49

[a] Measured on membrane surface by the sessile drop method using a goniometer

[b] γ_s^p polar (H-bonded) component of surface free energy of solid (erg cm^{-2})

γ_s^d dispersion force component of surface free energy of solid (erg cm^{-2})

γ_s surface free energy of solid (erg cm^{-2})

Organic Analogs to Molecular Sieves: Heterogeneous Catalysis with the Cross-linked Inverted Hexagonal Phase

Esther Kim and Douglas L. Gin*

Dept. of Chemistry, University of California, Berkeley, CA 94720-1460

Introduction

Zeolites are crystalline, open-frame structures with uniform pore systems and channel diameters (3-10 Å) that are used as industrial catalysts.[1] Zeolites facilitate reactions by localizing reactants in their pores and providing a high local concentration of active sites (typically acidic or basic).[2] As heterogeneous catalysts, zeolites exhibit high selectivity and reactivity. However, they are not amenable to processing, and their pore size and geometry are not easily adjustable to optimize reactivity.

Researchers at Mobil recently reported the first family of tunable synthetic mesoporous sieves, named M41S.[3,4] Unlike zeolites, mesoporous sieves are amorphous, silicate-based networks formed by sol gel condensation around an organic surfactant liquid crystal template. MCM-41, one member of this family, exhibits a hexagonal array of uniform pores with diameters ranging from 15 to 100 Å. Although mesoporous sieves allow better control over pore architecture and size, they cannot match the reactivity of zeolites and are not processible.

A lyotropic liquid crystal (LLC) **1** that can be cross-linked in the inverted hexagonal phase has been developed in our group (Figure 1).[5] The resulting LLC network possesses a well-defined pore architecture, consisting of hydrophilic channels with a pore diameter of approximately 20 Å and an interchannel spacing of 40 Å. Because of the structural similarities of this system with inorganic zeolites and mesoporous sieves, we reasoned that the cross-linked inverted hexagonal phase might act as an "organic molecular sieve" and catalyze organic reactions. It could potentially combine the tunability of mesoporous sieves with the reactivity and selectivity of zeolites.

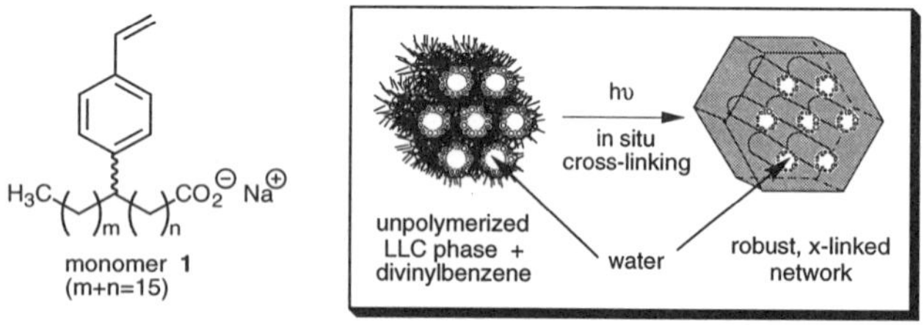

Figure 1. Polymerization of the inverted hexagonal phase of **1**.

Since the ionic functionality within the channels of our LLC network consists of slightly basic sodium carboxylates, this system was tested as a heterogeneous base catalyst for organic reactions.to test its utility as a base catalyst. The LLC network of **1** proved to be an effective base catalyst for the Knoevenagel reaction (Scheme 1). Systematic studies to determine the source of the catalytic activity were performed. The details of these studies will be presented along with comparisons to the activity of basic MCM-41 and zeolite systems.

Scheme 1. Knoevenagel condensation of benzaldehyde and ethyl cyanoacetate.

Experimental Section

Materials. Materials were obtained from commercial suppliers and used without further purification unless otherwise noted. Na-zeolite Y was obtained from Zeolyst International, Inc. Na-MCM-41 was obtained from Dr. Jacek Klinowski (Cambridge University). Gas chromatography was executed on a Hewlett-Packard 5890 GC with a 25 m (5%)-Diphenyl-(95%)-dimethyl siloxane capillary column (0.20 mm id, 0.33 μm film thickness). Monomer **1** was prepared and cross-linked in the inverted hexagonal phase according to literature procedures.[5]

Typical catalysis study using GC for experiments in THF. Ethyl cyanoacetate (0.254 mL, 2.50 mmol), benzaldehyde (0.266 mL, 2.50 mmol), and dodecane (0.142 mL, 0.625 mmol) were added to THF (5 mL). The reaction was placed in an oil bath at 67 °C. To the stirring reaction mixture, 5 mol % of the LLC network (based on the number of sodium carboxylates) was added. An atmosphere of N_2 was applied for the duration of the reaction. Aliquots were taken every half hour during the first 3 h, and then again after 4 h and 7 h. Aliquots were placed in the centrifuge for 1.5 min at 2500 rpm. The supernatant was then filtered through a glass fiber plug, and 8 μL of this solution was dissolved in 200 μL of THF. Samples were then stored in the refrigerator before ejecting into the GC.

Typical catalysis study using ^{1}H NMR for experiments in H_2O and the sodium acetate experiment. Ethyl cyanoacetate (0.133 mL, 1.25 mmol), benzaldehyde (0.127 mL, 1.25 mmol), and p-xylene (0.0766 mL, 0.625 mmol) were added to THF-d_8. The reaction mixture was placed in an oil bath at 67 °C. To the stirring reaction mixture, 5 mol % of sodium acetate was added. An atmosphere of N_2 was applied for the duration of the reaction. Aliquots were taken every half hour for the first 3 h, and then again after 4 h and 7 h. Aliquots was filtered through a glass fiber plug and diluted with additional THF-d_8. Samples were stored in the refrigerator prior to ^{1}H NMR analysis.

Results and Discussion

Previous work with the basic forms of zeolite Y and MCM-41 used the Knoevenagel condensation of ethyl cyanoacetate and benzaldehyde (Scheme 1) as a platform to gauge catalytic activity. Corma found that sodium-exchanged zeolite Y (Na-zeolite Y) afforded 19% conversion to the condensation product after 2 h when an equimolar solution of the two reagents was heated at 140 °C.[6] Kloetstra found that sodium-exchanged MCM-41 (Na-MCM-41) yielded ~90% conversion after 3 h in H_2O at 100 °C[7] (see Table 1).

Table 1. Previous results with base catalysis of the Knoevenagel condensation.

Base catalyst	Amount Used	Solvent	T (°C)	t (min)	% Conv.
Na-zeolite Y	20 wt %	none	140	120	19
Na-MCM-41	5 mol %	water	100	180	~90

We chose to extend this work to our system, using the conditions detailed for the Na-MCM-41 system. At 100 °C in H_2O, the LLC network shown in Figure 1 afforded 94% conversion in 15 min for the same reaction. Control experiments in the absence of any solid base in H_2O as well as in the absence of both solid base and H_2O also yielded varying amounts of conversion to product at 100 °C. In addition, the initial rate was too fast to measure using the analytical techniques available. Therefore, we changed the experimental conditions and used THF at 67 °C. Na-exchanged zeolite Y and Na-MCM-41 were subjected to the same experimental conditions and exhibited negligible activity under these conditions. The results of these studies are shown in Figure 2.

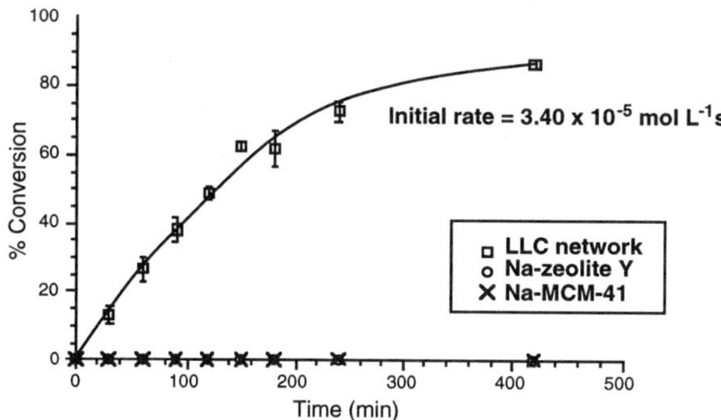

Figure 2. Comparison of heterogeneous base catalysts in THF at 67 °C.

Due to the behavior of this system in H_2O, control experiments were performed to confirm that H_2O was not playing a role in the observed activity of our catalyst. A sample extracted with supercritical CO_2 showed similar activity to the "wet" LLC network. The addition of more than a catalytic amount of H_2O to a solution of the two substrates in THF at 67 °C did not result in any conversion to the desired product. These experiments demonstrate that H_2O in the channels or the reaction mixture does not contribute to the catalytic activity.

Experiments were also performed to confirm that the catalytic activity was not simply due to the presence of carboxylates but rather to the local concentration of carboxylate groups. A control reaction with 5 mol % sodium acetate gave some conversion, as expected, but not as much as that observed with our system (Figure 3).

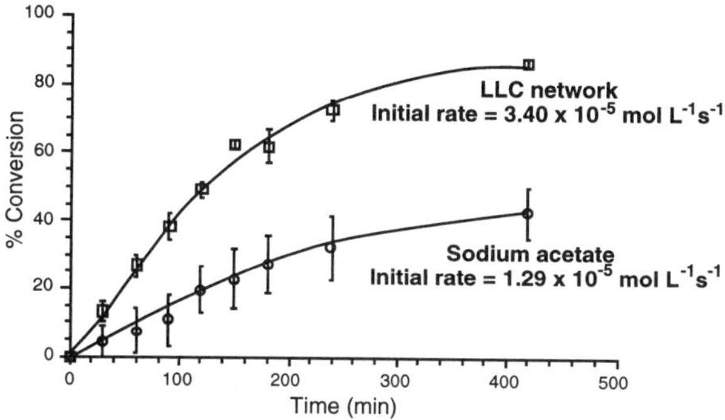

Figure 3. Comparison to heterogeneous sodium acetate.

With any heterogeneous catalyst, recyclability is an important consideration. Experiments with recycled and washed LLC network showed a 40% decrease in initial rate (Figure 4). Additional experiments showed that leaching of soluble unpolymerized monomer may account for the higher activity of the original sample. Experiments to determine the effect of diffusion showed that particle size and stirring rate do not affect the observed activity of this system.

In conclusion in H_2O at 100 °C, the cross-linked inverted hexagonal phase of **1** was shown to possess higher activity than Na-zeolite Y and Na-MCM-41. In THF at 67 °C, the LLC network exhibits activity while Na-zeolite Y and Na-MCM-41 show negligible activity under the same conditions. Control experiments show that the catalytic activity may be due to the high local concentration of carboxylates in the nanochannels of of the LLC network. Future work includes probing the nature of these active sites, as well as exploring the size selectivity of these systems.

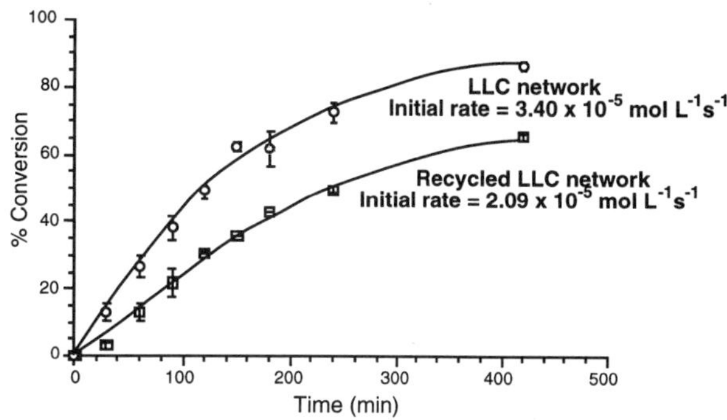

Figure 4. Activity of recycled and washed LLC network.

Acknowledgements

We would like to thank the Exxon Education Fund for funding, Zeolyst International, Inc. for the donation of Na-zeolite Y, Dr. Jacek Klinowski for the donation of Na-MCM-41, and the Phasex Corporation for CO_2-extraction.

References

(1) Chen, N. Y. In *Catalysis*; J. W. Ward, Ed.; Elsevier Science Publishers B. V.: Amsterdam, 1987; pp 153-163.

(2) For a comprehensive review of zeolites, see: "Zeolites and Related Molecular Sieves," In *Handbook of Heterogeneous Catalysis*; Ertl, G.; Knozinger, H.; Weitkamp, J., Eds.; Vol. 1; Wiley–VCH: Weinheim, 1997; pp 286–387.

(3) Kresge, C. T.; Leonowicz, M. E.; Roth, W. J.; Vartuli, J. C.; Beck, J.S. *Nature* **1992**, *359*, 710-712.

(4) Beck, J. S.; Vartuli, J. C.; Roth, W. J.; Leonowicz, M. E.; Kresge, C. T.; Schmitt, K. D.; Chu, C. T-W.; Olson, D. H.; Sheppard, E. W.; McCullen, S. B.; Higgins, J. B.; Schlenker, J. L. *J. Am. Chem. Soc.* **1992**, *114*, 10834-10843.

(5) Gray, D. H.; Hu, S.; Juang, E.; Gin, D. L. *Adv. Mater.* **1997**, *9*, 731-736.

(6) Corma, A.; Fornes, V.; Martin-Aranda, R. M.; Garcia, H.; Primo, J. *Appl. Catal.* **1990**, *59*, 237-248.

(7) Kloetstra, K. R.; van Bekkum, H. *J. Chem. Soc., Chem. Commun.* **1995**, 1005-1006.

PREPARATION AND CHARACTERIZATION OF THE ELASTICITY OF POLY(TETRAMETHYL-*m*-SILPHENYLENE-SILOXANE) NETWORKS OF EXCELLENT THERMAL STABILITY

*Ruzhi Zhang, Allan R. Pinhas, and James E. Mark**

Department of Chemistry and the Polymer Research Center, The University of Cincinnati, Cincinnati, Ohio 45221-0172

Introduction

With new advances in technology, there have been numerous demands for high-temperature elastomers for applications in areas such as advanced aerospace, defense, and computer industries.[1] Much attention in the search for high-temperature elastomers has focused on the modification of polysiloxanes, due to their excellent thermal stability and elasticity to extremely low temperatures.[2-6] Their good properties could presumably be greatly improved by incorporation of more thermally stable aromatic groups, for example, into the polydimethylsiloxane (PDMS) backbone. In the past two decades, *p*-silphenylene has been the most studied insertion group.[1,7-9] However, poly(tetramethyl-*p*-silphenylenesiloxane) (PTMPS) (Figure 1) is a crystalline polymer (T_m=127°C) and thus would exhibit rubberlike elasticity only at elevated temperatures.[7,11]

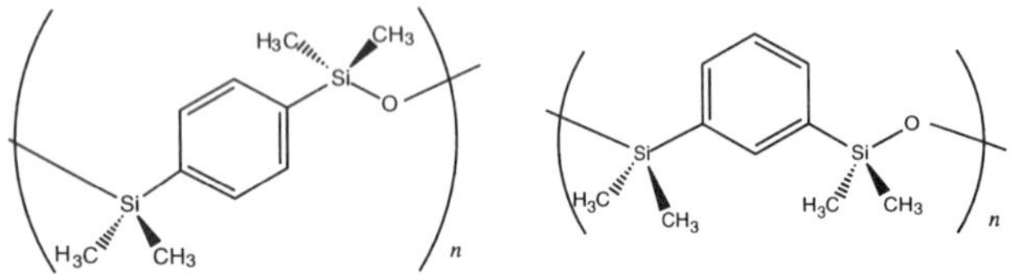

Figure 1. Structures of PTMPS and PTMMS homopolymers.

In our research, we proposed the insertion of *m*-silphenylene groups into the PDMS backbone to obtain a high-temperature resistant elastomer, as a result of the less stiffening effect and comparable thermal stability of *m*-phenylene, relative to *p*-phenylene. Recently, we reported the synthesis of poly(tetramethyl-*m*-silphenylenesiloxane) (PTMMS) through step-growth polymerization.[10-12] A thermal study showed that this homopolymer possessed excellent thermal and thermooxidative stabilities. We now extend this work to a cross-linking study of PTMMS, and the characterization and the elastomeric properties of PTMMS networks.

Experimental

Reagents. 1,3-Dichlorobenzene, 1,3-dibromobenzene, *n*-hexylamine, 2-ethylhexanoic acid, and benzophenone were purchased from the Aldrich Chemical Company. Dimethylchlorosilane was purchased from United Chemical Technologies, Inc., and the remainder of the chemicals were purchased from Fisher Scientific. Tetrahydrofuran (THF), benzene, and toluene, were distilled from potassium prior to use, and petroleum ether was distilled from molecular sieves.

Characterizations. All infrared spectra were recorded on a Perkin-Elmer Model 1600 series FTIR spectrophotometer. ^{1}H NMR (250 MHz) and ^{13}C NMR (62.9 MHz) spectra were recorded on a Bruker 250 MHz spectrometer with ^{1}H NMR referenced to tetramethylsilane at 0.00 ppm, and ^{13}C NMR referenced to the center chloroform-d peak at 77.0 ppm. Solid-state ^{13}C (100.6 MHz) NMR and ^{29}Si (79.5 MHz) NMR spectra were recorded on a Bruker 400 MHz spectrometer. The gel permeation chromatography (GPC) was performed on a Waters gel permeation chromatography unit with a differential refractometer and four in-line columns having pore sizes of 100, 500, 10^3, and 10^4 Å in toluene at 40 °C. The eluent flow rate was set at 1 mL/min, and the molecular weights were determined using polystyrene standards. Thermal data were obtained with a Rheometric Scientific PL-DSC (differential scanning calorimeter) and a Rheometric Scientific PL-STA (simultaneous thermal analysis, i.e., DSC and thermogravimetric analysis, TGA). Stress-strain isotherms in uniaxial extension were obtained in the usual manner using a

* To whom correspondence should be addressed.

home-made apparatus.[13] All measurements were conducted at room temperature, and the elongation was gradually increased to the rupture points of the samples.

Monomer Synthesis and Polymerization. *m*-Bis(dimethylsilyl)benzene was prepared via an *in situ* Grignard reaction using 1,3-dibromobenzene and dimethylchlorosilane as starting materials, and was subsequently converted to *m*-bis(dimethylhydroxysilyl)-benzene.[10] PTMMS was prepared using the method previously reported (Scheme 1).[10-11]

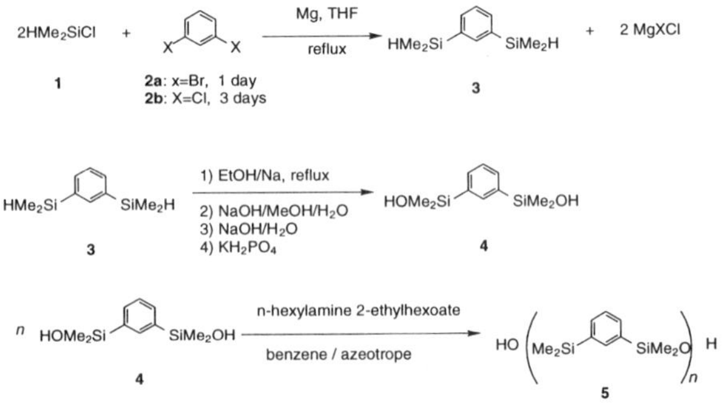

Scheme 1

Preparation of PTMMS Networks. (1)UV irradiation: In a typical procedure, 0.3 g of PTMMS and 5% benzophenone were dissolved in toluene. A strip of PTMMS film was formed by casting the mixture into an aluminum mold with the dimensions 4.2 X 1.5 X 30 mm³. The sample was irradiated by UV light under an air or argon atmosphere. Upon completion of irradiation, the sample was weighed and swollen in toluene. After swelling to equilibrium, the irradiated sample was deswollen through a series of solutions of toluene and methanol. Then the sample was dried in a vacuum oven for 1 day, and the soluble fraction of the sample was determined. (2) γ irradiation: A sample in an aluminum mold was placed in a glass vial filled with argon and sealed. The sample was irradiated with γ (^{60}Co) photons at a dose of 40 Mrad. at room temperature.

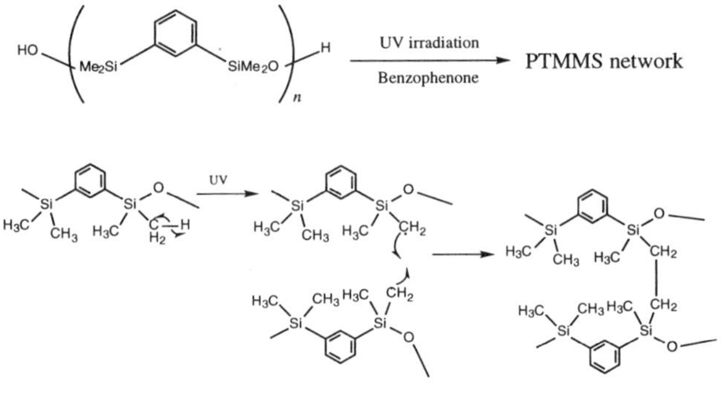

Scheme 2

Results and Discussion

The cross-linking reaction is shown in Scheme 2, and a possible radical mechanism was proposed to account for the cross-linking. The resulting elastomers were slightly yellow due to the incorporation of benzophenone into the network. The effect of cross-linking conditions on the final products was studied. Cross-linking was also carried out on samples floating on mercury. The soluble fractions of the networks were high compared to that of PDMS.

The PTMMS networks were characterized by solid-state ^{13}C and ^{29}Si NMR spectroscopy using MAS technology. The ^{13}C spectrum showed chemical shifts at 142.3, 141.4, 137.7, 130.9, and 4.83 ppm, respectively. The ^{29}Si spectrum had a peak of 1.37 ppm.

The polymer-solvent interaction parameter was estimated by means of an equilibrium swelling study in toluene. It is about 0.245 for PTMMS and 0.465 for PDMS. Due to the structure similarity of PTMMS and toluene, PTMMS has a lower value of the interaction parameter than PDMS.

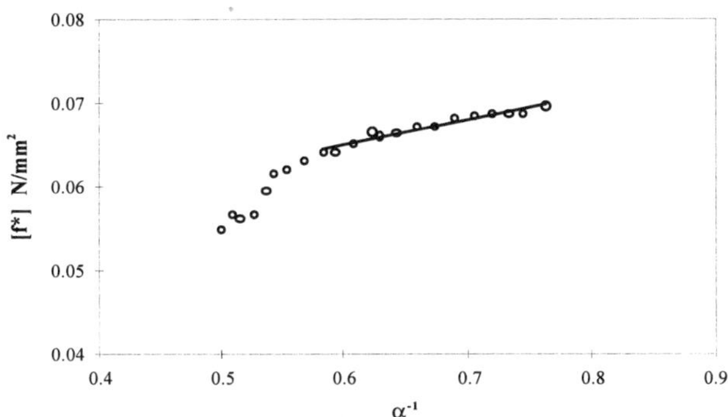

Figure 2. Mooney-Rivlin plot for extension of a PTMMS network.

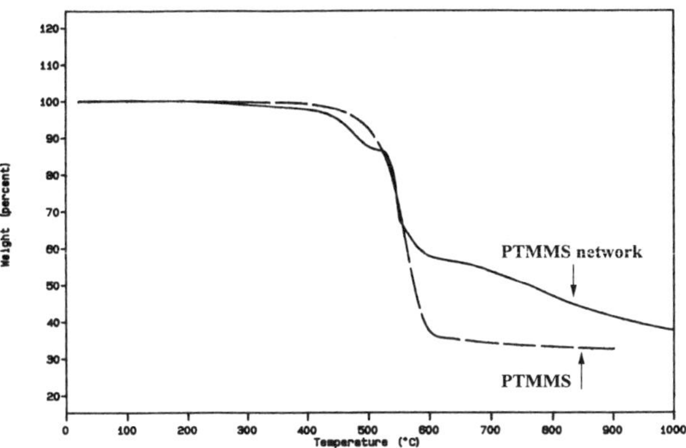

Figure 3. TGA curves for PTMMS homopolymer and network under a nitrogen atmosphere.

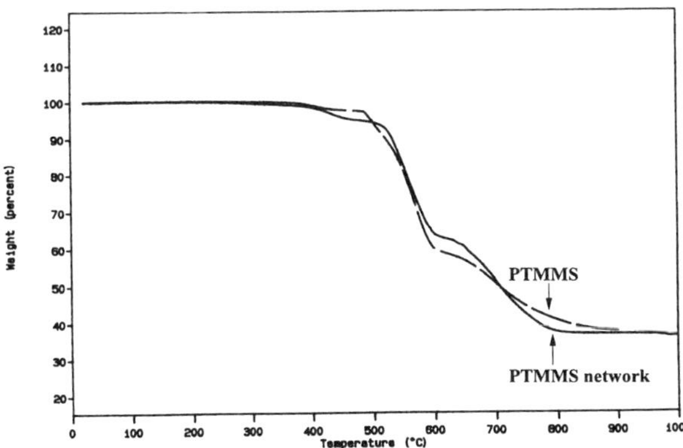

Figure 4. TGA curves for PTMMS homopolymer and network under an air atmosphere.

The thermal and thermooxidative stabilities were studied by TGA under an inert or an air atmosphere (Figures 3 and 4). The onset temperature for degradation in nitrogen is 410 °C and the residue is 37% when heated to 1000 °C at a rate of 10 K/min. In air, the onset temperature is 401 °C and the residue is 36%. The DSC study revealed that the glass transition temperature of PTMMS network is about -52 °C.

The mechanical properties of the PTMMS networks were studied by equilibrium stress-strain measurements. The data were analyzed in plots of modulus against reciprocal elongation, as suggested by the Mooney-Rivlin equation,[14]

$$[f^*] = 2\,C_1 + 2\,C_2\,\alpha^{-1} \tag{1}$$

where $2C_1 = C\upsilon kT/V$. From the Mooney-Rivlin plot in Figure 2, the chain density υ/V was found to be 38.4 mol/m^3, the cross-link density was 19.2 mol/m^3, and the molecular weight of the chain between two cross-link sites was 26300 g/mol, under the assumption that C has a value of 1/2. The elongation at rupture was in the range of 200-400%, and the ultimate strength was about 0.05 MPa.

Conclusions

Poly(tetramethyl-*m*-silphenylene-siloxane) has been successfully cross-linked by UV irradiation under an air or argon atmosphere in the presence of an organic sensitizer, i.e., benzophenone. It showed great resistance to γ irradiation in comparison to PDMS. Some mechanical properties of PTMMS networks were studied by equilibrium stress-strain measurements, and the cross-link density was estimated by means of the Mooney-Rivlin equation. The elongation at rupture was in the range of 200-400%, and the ultimate strength was about 0.05 MPa. The ultimate properties of PTMMS networks are comparable to those of PDMS networks. The PTMMS and toluene interaction parameter was estimated to be 0.245 by the equilibrium swelling study. $\chi_{1(PTMMS)}$ is slightly greater than $\chi_{1(PDMS)}$ as a result of the incorporation of phenylene groups into the PDMS backbone. The solid-state NMR spectra of PTMMS networks are almost identical to those of PTMMS itself, as expected. The TGA study revealed that the PTMMS elastomers have excellent thermal and thermooxidative stabilities.

Acknowledgments

It is a pleasure to acknowledge the financial support provided JEM by the National Science Foundation through Grant DMR 94-22223 (Polymers Program, Division of Materials Research). We also would like to thank Professor G. R. Kreishman for assistance in obtaining the solid-state NMR spectra.

References

(1)   Dvornic, P. R.; Lenz, R. W. *High Temperature Siloxane Elastomers*; Hüthig & Wepf, Basel, 1990.
(2)   Flory, P. J. *Statistical Mechanics of Chain Molecules*; Interscience: New York, 1969.
(3)   Mark, J. E.; Flory, P. J. *J. Am. Chem. Soc.* **1964**, *86*, 138.
(4)   Flory, P. J.; Crescenzi, V.; Mark, J. E. *J. Am. Chem. Soc.* **1964**, *86*, 146.
(5)   Mark, J. E.; Allcock, H. R.; West, R. *Inorganic Polymers*; Prentice Hall: New Jersey, 1992.
(6)   Clarson, S. J.; Semlyen, J. A. (eds.) *Siloxane Polymers*; PTR Prentice Hall: Englewood Cliffs, 1993.
(7)   Merker, R. L.; Scott, M. J. *J. Polym. Sci.: Part A* **1964**, *2*, 15.
(8)   Koid, N.; Lenz, R. W. *J. Polym. Sci., Polym. Symp.* **1983**, *70*, 91.
(9)   Dvornic, P. R.; Lenz, R. W. *J. Polym. Sci., Polym. Chem. Ed.* **1982**, *20*, 951.
(10)   Zhang, R. ; Pinhas, A. R. ; Mark, J. E. *Macromolecules* **1997**, *30*, 2513.
(11)   Zhang, R. ; Pinhas, A. R. ; Mark, J. E. *Polym. Prepr.* **1998**, *39*(1), 575.
(12)   Zhang, R. ; Pinhas, A. R.; Mark, J. E. *Polym. Prepr.* **1998**, *39*(1), 607.
(13)   Mark, J. E. *J. Polym. Sci., Macromol. Rev.* **1976**, *11*, 135.
(14)   Mark, J. E.; Erman, B.; *Rubberlike Elasticity: A Molecular Primer*; John Wiley & Sons, New York, 1988.

A Topological Analysis of Long-Chain Branching in Polyolefins

Danail Bonchev[1], Armen Dekmezian[2], and Eric Markel[2]

[1] Department of Marine Sciences, Texas A&M University, P. O. Box 1675, Galveston, Texas 77553, [2] Exxon Chemical, Baytown Polymer Research Center, 5200, Baytown, Texas 77522

Introduction

Polymer rheology depends strongly on the large-scale polymer architecture, i.e., on patterns of long-chain branching (LCB). Typical shapes range from 3-arm to comb-like to hyperbranched structures, and some rheological work on model, ideal systems has been published [1]. However, there is yet no mathematical framework for a unified quantitative treatment of LCB polymers. The problem of molecular branching has been the subject of many studies in theoretical chemistry over the last 25 years [2] proceeding from the concept of *molecular topology* and the mathematical formalism of graph theory [3]. Chemical structures manifest various topological patterns which are not only reflected in molecular properties but control them to a very high extent [4]. *Molecular branching* includes some structural motifs in which branches of molecular skeleton vary in their number, length, spacing, clustering, etc. Branching contributes to *molecular complexity* the importance of which for chemistry was recently recognized [5]. The degree of branching can be quantified using a variety of *topological indices* [6] which found a wide application as descriptors of molecular structure in quantitative structure-property and structure-activity relationships (QSPR/QSAR). Recently, branching descriptors started being rediscovered in polymer science [7].

The current study presents formulas derived for select branching descriptors for linear, singly-branched (star) and comb polyolefins and shows the results of a theoretical analysis of branching patterns in polyolefins. The findings of our study will be used in a subsequent part of the project related to QSPR study of polyethylenes. It is anticipated that such a general approach could provide the opportunity for finding a family of different polymer topologies that would produce the same desirable product properties.

The Topological Descriptors of Polymer Branching

Topological indices are numbers uniquely derived from a molecular graph which is a structure composed of vertices and edges representing atoms and covalent bonds. Two basic indices were selected as measures of polymer branching, the Wiener index, W [2,6,8] and the complexity index K [9,10]. The Wiener number is the sum of the distances between all vertices in molecular graph. The K index is the sum of all connected subgraphs of the molecular graph. Thus, the K index reflects the intuitive idea that the more complex the molecule, the larger the number of components to which the molecule can be decomposed.

Formulas for the Wiener Number W and the Complexity Index K of Polyolefins

$$W_{linear} = \frac{M^3}{16{,}464} \quad ; \quad K_{linear} = \frac{M^2}{392}$$

$$W_{3\text{-star}} = \frac{M^3}{21{,}168} \quad ; \quad K_{3\text{-star}} = \frac{M^2(M + 189)}{74{,}088} \quad ; \quad K_{singly\text{-branched}} = \frac{7M^2 + M_1 M_2 M_3}{2744}$$

$$W_{singly\text{-branched}} = \frac{1}{16{,}464}\left[(M_1 + M_2)^3 + (M_1 + M_3)^3 + (M_2 + M_3)^3 - M_1^3 - M_2^3 - M_3^3 \right]$$

$$K_{n\text{-comb}} = \frac{1}{392}\left(M_c^2 + \sum_{i=1}^{n} M_i^2 + 2 \sum_{i=1}^{n} M_i M_{ei1} + 2\sum_{ijk\ldots z}^{n} M_{ei1} M_{ejr} \prod_{i=1}^{z} M_i / 14^i \right)$$

$$W_{n\text{-comb}} = \frac{1}{6.14^3}\left\{ M_c^3 + \sum_{i=1}^{n} M_i^3 + 3 \sum_{i=1}^{n} M_i [Mp_i^2 + (M_c - Mp_i)^2 + M_i \cdot M_c] \right. $$

$$\left. + 3 \sum_{i=1}^{n-1} \sum_{j=i+1}^{n} M_i M_j [M_i + M_j + 2(Mp_j - Mp_i)] \right\}$$

The parameters used are the number of branches n, the molecular weight of each branch M_1, M_2, etc., that of the main chain M_c, the molecular weight Mp_i of the (left) end fragment determined by the position p_i of branch i, and the total molecular weight M. The above formulas comprise all cases of branched polyolefins except the hyperbranched ones having branched branches.

Branching Pattern Analysis

The analysis of the above formulas determined the basic patterns of increasing branching and complexity showing major differences in the topological indices of various LCB motifs relative to linear polymer:

(i)   The degree of branching (complexity) K increases with the number of branches:

$$K_{linear} < K_{3\text{-star}} < K_{2\text{-comb}} < K_{3\text{-comb}} < \ldots < K_{n\text{-comb}}$$

(ii) At the same number of branches the degree of branching K increases with the branch length, their more central position, and their clustering.

(iii) Trends (i) and (ii) are valid for both for the same total molecular weight and for a total molecular weight increasing with the number of arms.

(iv) The level of structural complexity which linear polymers achieve at high molecular weights can be reached by 3-arm stars and, particularly, by n-arm combs at lower MWs. Branch length, centrality and clustering additionally favor this trend.

Some of these patterns are illustrated in Figs. 1-3. The first figure shows the complexity index K plotted against the branch mole fraction for 3-arm stars and combs with up to ten branches at a constant total molecular weight M = 200 kD. Branch mole fraction is used on the X axis instead of branch molecular weight in order to remove composition from the calculation, such that the computed complexities apply for other polyolefins. Fig. 1 clearly indicates that the major change in polymer complexity occurs for an increase in the branch mole fraction within the 0 to 0.04 range which corresponds to arm MWs ranging from 0 to 1000 - 2000 (for PE). The further increase in the arm lengths proceeds in a very gradual manner and for small number of arms (1 to 5) a nearly plateau area is formed for those polyolefins whose branches account for more than half the total molecular weight. Another conclusion is that the number of arms affects the degree of branching much more strongly than the arm molecular weight, however, the relative importance of each additional arm is lower.

Fig. 2 compares three series of 5-arm comb polyethylenes having a total molecular weight of 150, 200, and 250 kD. It is shown that branch length in combs results in more dramatic effect than the total molecular weight. The three curves come very close to each other and although they do not result in a unique master curve, they would allow us to model molecular weight distributions in commercial polyethylenes. Analogous conclusion for the number of arms being a stronger complexity factor than the total molecular weight is drawn from other log K/branch molecular weight plots not shown here.

A topological equivalence based on equal log K values was established between 2-arm combs with branch molecular weights M_{br} = 0 to 8,000 (MW = 100,000 to 116,000) and symmetrical 3-arm stars with a molecular weight ranging from 50,000 to 600,000 (Fig. 3). This finding justifies the usage of symmetrical 3-arm stars in the modeling of 2-arm combs. The narrow "window" of M_{br} = 200 to 2500 for 3-arm combs, showing their equivalence to 3-arm stars with molecular weights up to 10^6 , evidences against the usage of the 3-arm star modeling beyond 3-arm combs. The topological equivalence found between pairs of 2-arm combs (H-polymers) and symmetrical 3-arm stars can, after calibration, be reformulated for their rheological properties.

Conclusions

This study revealed qualitatively and quantitatively the topological patterns that increase the degree of branching in stars and combs. The strongest effect comes from the number of branches, the next strongest factor is the branch length. Much weaker effects are caused by the length of the main chain, the branch clustering, and branch more central position. The use of branch mole fraction in our analysis has made our conclusions valid not only to polyethylenes, for which the calculations were initially made, but for all linear, star, and comb polyolefins except the hyperbranched ones (trees) which were not covered by our formulas. The results obtained can be applied to specific polymers provided a set of model structures are available for validation and calibration. A successful step in this direction would prompt further efforts toward a major goal: to use identified branching patterns to control polymer properties.

References

1. McLeish, T., in *Mathematical Chemistry Series*, Vol. 5, Bonchev D. and D. H. Rouvray, Eds., Gordon and Breach, Reading, U. K., in press.
2. Bonchev, D., *Theochem* **1995**, *336*,137.
3. Trinajstic, N., *Chemical Graph Theory*, 2nd edn., CRC Press, Boca Raton, FL, 1992.
4. Bonchev, D., Kamenska, V., and Mekenyan, O., *J. Math. Chem.* **1990**, *5*, 43.
5. Bonchev, D. and Seitz, W. A., in *Concepts in Chemistry: A Contemporary Challenge*, Rouvray, D. H., Ed., Research Studies Press, Taunton, U. K., 1996, p. 348-376.
6. Rouvray, D. H., *Theochem*, **1989**, *185*, 187.
7. Yan, D., Mueller, A. H. E., and Matyjaszewski, K., *Macromolecules*, **1997**, *30*, 7024.
8. Wiener, H., *J. Am. Chem. Soc.* **1947**, *69*, 17.
9. Gordon, M. and Kennedy, J. W., *J.C.S. Faraday Trans. II* **1973**, *69*, 484.
10. Bonchev, D., *SAR/QSAR Envir. Res.*, **1997**, *7*, 23.

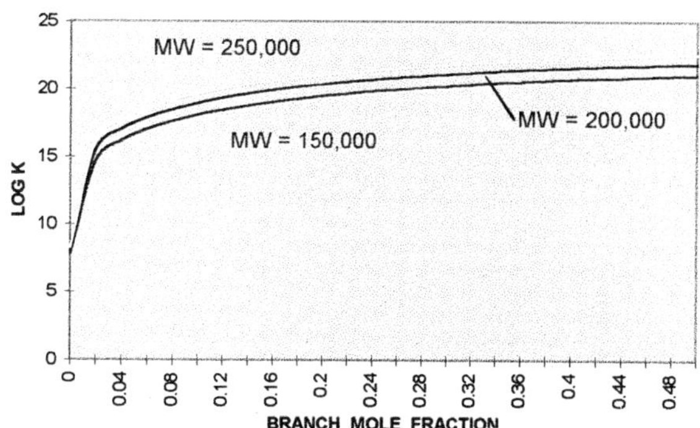

Figure 2. The topological complexity index versus branch mole fraction for three series of 5-arm comb polyethylenes with evenly distributed arms of equal length.

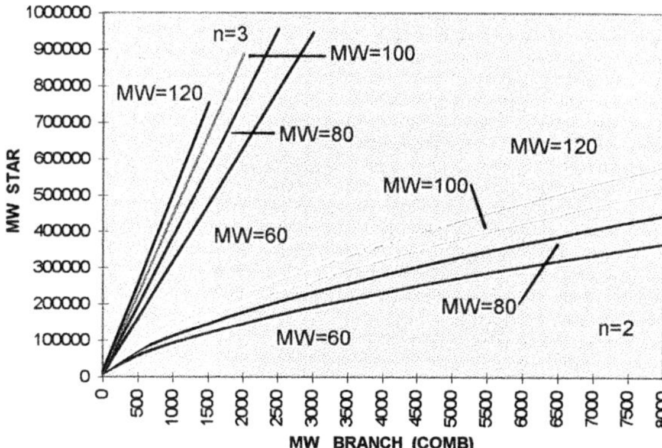

Figure 3. Topological equivalence between 2- and 3-arm comb and symmetric 3-arm star polyethylenes shown for four series with a constant backbone molecular weight.

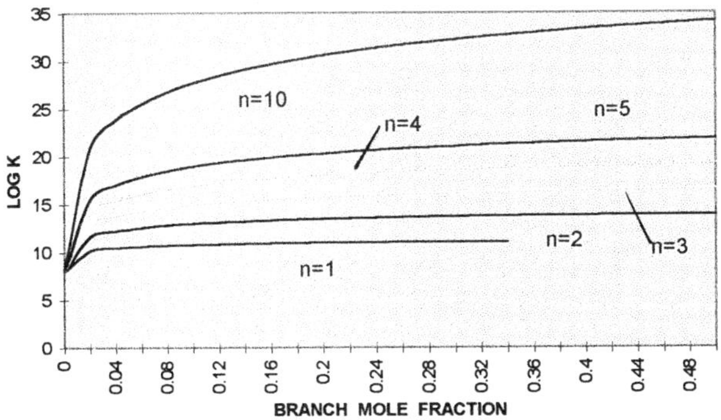

Figure 1. The topological complexity index versus branch mole fraction for six series of n-arm comb polyethylenes with evenly distributed arms of equal length (MW = 200,000).

Novel Approach for the Compatibilization of Polymer Blends and Polymeric Waste

Naomi Furgiuele[*], Klementina Khait[*], and John M. Torkelson[*,#]

[*]Department of Chemical Engineering
[#]Department of Materials Science and Engineering
Northwestern University
Evanston, IL 60208

Introduction

Polymer blend compatibilization and phase separation remains an area of significant interest for the development of blended products from virgin polymers and the utilization of recycled polymer feedstocks. Polymer recycling is a pressing area of research, considering that the United States alone produces more than 70 billion pounds of polymers annually. Separation of recycle streams into single component materials has proven to be both a labor and capital intensive process, with the recycled material often exhibiting substandard properties in comparison to products made from virgin feedstocks. The mixed waste processing results in products with poor mechanical properties because of weak adhesion between phases and/or sub-optimal microphase morphology, making it difficult or uneconomical to reuse polymeric material in conjunction with virgin material. Some systems of particular interest include polyethylene, polypropylene, and polystyrene, because they constitute 60% of thermoplastic waste and their blends typically exhibit poor adhesion between phases and thereby poor mechanical properties in the absence of interfacial modifications.

Some approaches investigated thus far that could enable the blending of mixed recycle streams into useful materials have concentrated on the addition of block or graft copolymers as interfacially active species.[1-8] These copolymers promote interfacial adhesion and may slow down coalescence during subsequent melt processing. However, the economic cost of these materials, coupled with thermodynamic limitations (block copolymers phase separate to form micelles) and slow diffusion of copolymers to the interface, make them impractical for recycling efforts. In this work, the novel processing approach to compatibilization called solid-state shear pulverization (S^3P) is applied to blends involving unmodified polystyrene (PS), polypropylene (PP), polyethylene (PE), poly(ethylene terephthalate) (PET), and poly(methyl methacrylate) (PMMA). S^3P is a continuous environmentally benign process that can be used to produce powder from both virgin and commingled polymeric waste feedstocks in an industrially-applicable manner. High shear and extensional forces applied to solid-state polymer results in the reduction of plastic pellets or flakes to a fine, well-mixed powder, as evidenced by color uniformity of products made from multi-colored feedstock. Additionally, it has been shown[9,10] that, immediately after pulverization, powders often have a significant free radical population, indicative of chain scission taking place during pulverization. Here we focus on quantifying the degree of chain scission that can occur in this process, as well as the changes in macroscopic properties (thermal and mechanical) of polymer blends, providing an indication of the compatibilizing effects of the intimate mixing resulting from the S^3P process.

Experimental

At temperatures below the polymers' melting points (for semicrystalline polymers) or glass transition temperatures (for amorphous polymers), the pulverizer compresses the polymer with a pressure equal to 0.2 to 0.7 MPa, and then pushes the material through the barrel with shear stresses from 0.03 to 5 N/mm^2 and pressure up to 50 MPa to rupture chemical bonds and create powders with free radicals.

PS ($M_N = 110,000$ and $M_W = 205,000$), PMMA ($M_N = 49,000$ and $M_W = 84,000$), and PP, PET, HDPE, LDPE, and LLDPE of various melt flow rates, were used as feedstocks. The blends were injection molded pre- and post-pulverization and then tested to assess the tensile, flexural, and impact properties. The blends were also analyzed via gel permeation chromatography (GPC) and melt flow rate (230°C with a 2.16 kg load) pre- and post-pulverization to analyze changes in the molecular weight distribution of the material. Differential scanning calorimetry (DSC) at 10°C/min was used to detect changes in the glass transitions (T_g) of the polystyrene-rich domains of some polystyrene blends upon pulverization.

Results and Discussion

Figure 1 illustrates the effect of S^3P processing on the molecular weight distribution of virgin PS. Figure 1a shows the shift in the molecular weight distribution of PS, with an original $M_N = 110,000$, as a result of S^3P processing with moderately high shearing conditions. After, the molecular weight distribution was shifted to lower molecular weights with M_N reduced to 58,000 and a significant reduction in polydispersity. Figure 1b illustrates similar processing of virgin, low molecular weight PS ($M_N = 10,500$ and $M_W = 19,000$) in which little chain scission was observed.

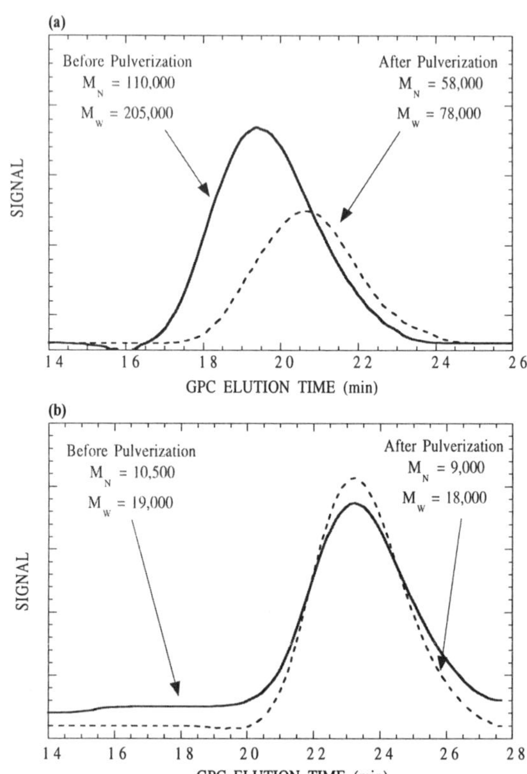

Figure 1: GPC chromatograms of PS before and after S^3P processing at moderately high shear.

Table 1 also shows similar effects as determined by melt flow rate measurements. Under the same conditions used to pulverize the PS samples in Figure 1, both high and low molecular weight PP undergo significant reduction in molecular weight as evidenced by an increase in melt flow rate. Under similar pulverization conditions, high molecular weight PMMA shows essentially no change in molecular weight distribution.

Material	Melt Flow Rate (g/10 min)
low molecular weight PP	
pre- S^3P	19.0
post- S^3P	29.0 ± 1.4
high molecular weight PP	
pre- S^3P	0.53 ± 0.03
post- S^3P	1.02 ± 0.08
PS	
pre- S^3P	2.03 ± 0.05
post- S^3P	14.2 ± 0.4

Table 1: Melt flow rate before and after S^3P processing for virgin PS and two PP samples.

These results indicate that S^3P not only can provide a means of achieving well mixed powders from solid-state polymers via a continuous process, but that some modification of the molecular-scale details of the mix is also possible. These molecular-scale modifications are significant functions of polymer species, molecular weight, and pulverization conditions, and are the subject of further study.

Figure 2 and Table 2 provide evidence that, along with the molecular-scale modifications achievable via S^3P, significant improvements in mixing and interfacial properties may be achieved in polymer blends with the S^3P process. Figure 2 provides DSC thermograms of pre- and post-S^3P-processed PS as well as melt-mixed/no-S^3P and S^3P/no-melt-mixing PP/PS (25/75) blends. Virgin and pulverized PS have T_g's of 98-100°C, as expected. Upon melt-mixing, the PP/PS (25/75) exhibits a PS-rich phase T_g of ~ 95°C indicating a small extent of PP mixing with PS. However, a much greater reduction in the PS-rich phase T_g, down to ~ 88-90°C, is achieved by the S^3P process without melt-mixing, indicating that a much more intimate mixing of the PP with the PS is achievable through S^3P.

Feedstock	Tensile Strength (psi)	% Elongation
PMMA/PS (75/25) virgin		
pre-S^3P	11,500	1.4
post-S^3P	10,700	5.7
PS/PP (75/25) virgin		
pre-S^3P	6070	1.8
post-S^3P	6580	2.6
HDPE/LDPE/PP/PET/PS (40/30/5/20/5) post-consumer		
pre-S^3P	2020	4
post-S^3P	2610	8
HDPE/LLDPE/PET (40/30/30) post-consumer		
pre-S^3P	1880	4
post-S^3P	2450	5

Table 2: Examples of mechanical property improvements for blends of virgin and post-consumer waste before and after S^3P processing.

Conclusions

Solid-state shear pulverization has demonstrated potential as an approach for the processing of polymer blends of both virgin materials and post-consumer waste. Changes in the molecular weight distributions of some processed materials show that chain scission may occur as a result of S^3P processing, confirming previous studies indicating the development of a free radical population as a result of chain scission[9-10]. DSC has shown that pulverization achieves a greater degree of intimate mixing of the PS and PP domains (as indicated by a decrease in T_g) than melt mixing the blend. Potentially, starting with this higher degree of mixing before melt processing can facilitate significant molecular-scale and macroscopically observable compatibilization of blends. Further investigation of molecular-scale modifications, microstructural characteristics, and macroscopic property improvements in S^3P processed blends is underway.

Acknowledgments

We thank John Rasmussen for assistance with S^3P processing. This work was supported by the MRSEC program of the National Science Foundation (Grant DMR-9632472) at the Materials Research Center of Northwestern University. We are also grateful to the Material Sciences Corporation, Elk Grove Village, IL for their continuing support.

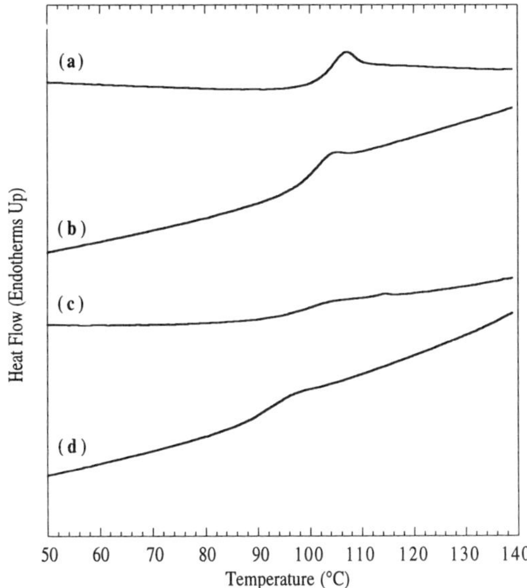

Figure 2: Glass transitions for a) PS standard, b) pulverized virgin PS, c) polystyrene-rich phases in melt mixed/no pulverization PP/PS (25/75) and d) polystyrene-rich phases in pulverized PP/PS (25/75).

Such intimate mixing may allow for improvements in mechanical properties of blends processed from S^3P-produced powders. Table 2 illustrates the change in tensile properties observed for blends injection molded pre- and post-S^3P processing. Some virgin and post-consumer blends show up to a factor of four increase in elongation and/or up to a 30% increase in tensile strength as a result of the pulverization. In addition, a PP/PS (75/25) blend showed a 15% increase in flexural strength upon pulverization.

References

1. Macosko, C.W.; Guégan, P.; Khandpur, A.K.; Nakayama, A.; Marechal, P.; Inoue, T. *Macromolecules* **1996**, *29*, 5590.
2. Paul, D.R. <u>Polymer Blends: Vol. 2</u>; Academic Press, Inc.: New York, **1978**, Ch. 12.
3. Lyatskaya, Y.; Gersappe, D.; Gross, N.A.; Balazs, A. *J. Phys. Chem.* **1996**, *100*, 1449.
4. Park, D.-W.; Roe, R.-J. *Macromolecules* **1991**, *24*, 5324.
5. Lee, M.S.; Lodge, T.P.; Macosko, C.W. *J. Polym. Sci.: Part B: Polym. Phys.* **1997**, *35*, 2835.
6. Vilgis, T.A.; Noolandi, J. *Macromolecules* **1990**, *23*, 2941.
7. Brown, H.R.; Char, K.; Deline, V.R.; Green, P.F. *Macromolecules* **1993**, *26*, 4155.
8. Kramer, E.J.; Norton, L.J.; Dai, C.-A.; Sha, Y.; Hui, C.-Y. *Faraday Discuss.* **1994**, *98*, 31.
9. Ahn, D.; Khait, K.; Petrich, M.A.; *J. Appl. Polym. Sci.* **1995**, *55*, 1431.
10. Nesarikar, A.R.; Carr, S.H.; Khait, K.; Mirabella, F.M. *J. Appl. Polym. Sci.* **1997**, *63*, 1179.

Ring-Opening Metathesis Polymerization from Surfaces

Marcus Weck[*], *Jennifer J. Jackiw*[**], *Paul S. Weiss*[**] *and Robert H. Grubbs*[*]

[*]California Institute of Technology
Arnold and Mabel Beckman Laboratories of Chemical Synthesis
Division of Chemistry and Chemical Engineering
Pasadena, CA 91125
[**]The Pennsylvania State University
Department of Chemistry
University Park, PA 16802

Introduction

The organization of well-defined molecular assemblies on solid surfaces provides a rational approach for the fabrication of interfaces with well-defined structures and compositions, especially regarding organized mono- and multilayer assemblies.[1] In particular, ultrathin ordered polymeric layers on inorganic surfaces exhibit a large potential for a variety of advanced technological applications such as microlithography,[2] biocompatibility,[3] biosensors,[4] multilayer light emitting diodes (LEDs), modification of electrode surfaces, catalysis, corrosion, lubrication, and adhesion.[1,5,6] Furthermore, theoretical models which describe the absorption of polymeric structures on surfaces and their experimental proof are of fundamental interest.[7,8] Currently, the most widely used methods to obtain these polymeric ordered thin films are vapor-deposition and Langmuir-Blodgett films.[1] However, several applications require the covalent attachment of the thin layer to the surface.

In recent years, the most widely studied method to self-assemble polymers on surfaces was the use of polymers with functionalized groups which can react with the surface.[1,9-12] Unfortunately the self-assembly reactions in these cases are ill-defined and sometimes the polymer is incompatible with the surface. Therefore, controlled polymer growth off the surface is of great interest, as it would overcome these limitations. Furthermore, a living polymerization off the surface would result not only in a well-defined system on the surface, but would also allow the construction of multilayer structures and the introduction of a large variety of functionalities at the surface. Most attempts to realize this goal have failed due to side reactions and impurities on the surface.[13]

In the last decade, ring-opening metathesis polymerization (ROMP) catalyzed by well-defined metal-alkylidines has been established as an efficient method to control a polymer's molecular structure, size and bulk properties through variations at the molecular level.[14] Recently, our group reported the synthesis of a new class of well-defined ruthenium complexes which are highly active in ROMP.[15-16] In particular, initiator **1** has been shown to polymerize a large variety of monomers in a living fashion in a variety of solvents ranging from water to benzene.[14,15,17] With these recent advances in catalyst design, ROMP is capable of overcoming the above mentioned obstacle for polymerizations on surfaces.

$$Cl_{\text{---}} \overset{PCy_3}{\underset{PCy_3}{\overset{|}{\underset{|}{Ru}}}}{=}\!\!\!\backslash^{H}_{Ph} \quad \mathbf{1}$$

Results and Discussion

In the last ten years, the use of alkylthiols self-assembled on gold surfaces has been the most studied system for the covalent attachment of thin films on surfaces.[1,18] Therefore, we decided to synthesize thiol-based-lipids which bear a strained cyclic olefin as the initiating group. For characterization purposes, and to eliminate the possibility that the cyclic olefin will 'fall' into the polymer brush and therefore not be able to initiate, the stiff lipid **2** based on phenyl-ethynylenes was synthesized in analogy to published procedures.[19,20]

Scheme 1 illustrates the general outline for the surface polymerization. Initially, the surface is functionalized using dodecanethiol. In a self-assembly step the monofunctionalized surface is exposed to a solution containing lipid **2**. Figure 1 shows an STM image of a 2000Å by 2000Å area in which the functionalized lipid **2** (apparent as protrusions) has been inserted into structural domain boundaries of the dodecanethiolate film.

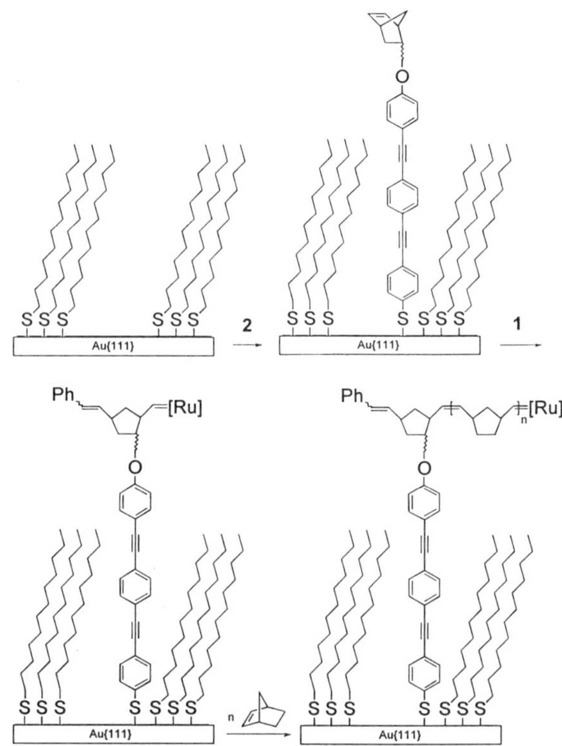

Scheme 1. General Polymerization Scheme

The regular alkylthiol matrix supports and dilutes the functionalized lipid,[19,20] preventing polymerization along the surface which would occur if the surface were fully covered with lipid **2**. The surface is then exposed to a methylene chloride solution of the catalyst **1**. The then initiated surface wafers were rinsed several times with clean solvent to remove any excess of the catalyst. The final step involves the exposure of the initiated wafers to a monomer solution which results in polymerization.

Figure 1. STM image of 2000Å by 2000Å area of a mixed monolayer containing lipid **2** in a dodecanethiolate matrix.

Characterization of these systems by scanning tunneling microscopy, X-ray photoelectron spectroscopy (XPS), ellipsometry, and infrared reflection spectroscopy will be discussed.

Acknowledgments

The authors would like to thank the NSF and ONR for financial support.

References

(1) Ulman, A. *An Introduction to Ultrathin Organic Films: From Langmuir Blodgett to Self-Assembly*; Academic Press: San Diego, **1991** and references therein.

(2) Berggren, K.K.; Bard, A.; Wilbur, J.L.; Gillaspy, J.D.; Helg, A.G.; McClelland, J.J.; Rolston, S.L.; Phillips, W.D.; Prentiss, M.; Whitesides, G.M. *Science* **1995**, *269*, 1255.

(3) Müller, W.; Ringsdorf, H.; Rump, E.; Wildburg, G.; Zhang, X.; Angermaier, L.; Knoll, W.; Liley, M.; Spinke, J. *Science* **1993**, *262*, 1706.

(4) Stelze, M.; Weismüller, G.; Sackman, E. *J. Phys. Chem.* **1993**, *97*, 2974.

(5) Wegner, G. *Adv. Mater.* **1991**, *3*, 8.

(6) Fuchs, H.; Ohst, H.; Prass, W. *Adv. Mater.* **1991**, *3*, 10.

(7) deGennes, P.-G. *Macromolecules* **1982**, *15*, 492.

(8) Konstandinis, K.; Prager, S.; Tirell, M. *J. Chem. Phys.* **1992**, *97*, 7777.

(9) Stouffer, J.M.; McCarthey, T.J. *Macromolecules* **1988**, *21*, 1204.

(10) Rozsnyai, L.F.; Wrighton, M.S. *Chem. Mater.* **1996**, *8*, 309.

(11) Erdelen, C.; Häussling, L.; Naumann, R.; Ringsdorf, H.; Wolf, H.; Yang, J.; Liley, M.; Spinke, J.; Knoll, W. *Langmuir* **1994**, *10*, 1246.

(12) Sun, F.; Castner, D.G.; Grainger, D.W. *Langmuir* **1993**, *9*, 3200.

(13) Oosterling, M.L.C.M.; Sein, A.; Schouten, J. *Polym. Commun.* **1992**, *33*, 4394.

(14) Ivin, K.J. *Olefin Metathesis*; Academic Press: London, **1996**.

(15) Schwab, P.; France, M.B.; Ziller, J.W.; Grubbs, R.H. *Angew. Chem. Int. Ed. Engl.* **1995**, *34*, 2039.

(16) Lynn, D.M.; Mohr, B.; Grubbs, R.H. *J. Am. Chem. Soc.* **1998**, *120*, 1627.

(17) Lynn, D.M.; Kanaoka, S.; Grubbs, R.H. *J. Am. Chem. Soc.* **1996**, *118*, 784.

(18) Bain, C.D.; Troughton, E.B.; Tao, Y.-T.; Evall, J.; Whitesides, G.M.; Nuzzo, R.G. *J. Am. Chem. Soc.* **1989**, *111*, 321.

(19) Bumm, L.A.; Arnold, J.J.; Cygan, M.T.; Dunbar, T.D.; Burgin, T.P.; Jones II, L.; Allara, D.L.; Tour, J.M.; Weiss, P.S. *Science* **1996**, *271*, 1705.

(20) Cygan, M.T.; Bumm, L.A.; Arnold, J.J.; Shedlock, N.F.; Dunbar, T.D.; Burgin, T.P.; Jones II, L.; Allara, D.L.; Tour, J.M.; Weiss, P.S. *J. Am. Chem. Soc.* **1998**, *120*, 2721.

Layer Network Formation of
Poly(styrene-*alt*-n-octadecyl-maleimide) in Dodecane

XIAORONG WANG, Bridgestone/Firestone Research Center,
1200 Firestone Parkway, Akron, OH 44317

Polymer blends and block copolymers exhibit ordering transitions from the molten homogeneous state to various microphase-separated states [1-4]. The simplest of these ordered phases are the lamellar layer structures [2,3]. The development of lamellar layer structures using comb-shaped polymers has received a great deal of attention over the past 20 years [4-12]. An important reason is that these polymers can form stable and well ordered nanostructures in the bulk [7-10,12] and in semi-dilute solutions [5,11]. An additional advantage is that the linkages between the polymer backbones and the side chains can be either chemical covalent bonds [4-9] or physical associations [10-12]. In principle, these layers if linked together can form polymer super-structures and layer networks. The behavior of layer networks is important to a variety of relevant situations, including the geological layers, liquid crystalline and bio-tissue structures. Some years ago de Gennes [13,14] theoretically proposed a network of the similar type discussed, but used a different approach.

In this research, the network formation of an alternating copolymer, poly(styrene-*alt*-n-octadecyl-maleimide), in dodecane is discussed. The polymer is characterized by a semi-rigid main chain with many short side chains. The mechanical moduli (G) of the network are monitored as a function of the network concentration (φ), and also as a function of the long period (L) of the network. Both relations are found to follow simple cubic power laws, i.e., $G \sim \varphi^3 \sim L^{-3}$. These behaviors are considered to be associated with the unique nature of a layer network.

MATERIALS   The copolymer, poly[styrene-*alt*-(n-octadecylmaleimide)] or SMI-C_{18}, was synthesized by the condensation reactions of n-octadecylamine and poly(styrene-*alt*-maleic anhydride). Sample gels were prepared by dissolving the SMI-C_{18} polymer in dodecane at 120°C in sealed flasks.

RESULTS AND DISCUSSION   Small angle x-ray scattering experiments of the SMI-C_{18} polymer strongly indicate that the alternating copolymer synthesized possesses ordered structures in the bulk state. Figure 1 shows the scattering intensities as a function of the scattering angle for the polymer. There are two diffraction maximums: a strong one at 2.75° corresponding to the primary diffraction and a very weak one at 5.55° corresponding to the second-order diffraction. The important feature is that the gap between these two peaks [i.e., $(2\theta_2-2\theta_1)/2\theta_1$] equals unit. This fact reflects that the ordered structures are parallel layers characterized by an one-dimensional order normal to the layers. The ordered structures are lamellar layers of alternated alkane and main chain layers. The long period (L) is the sum of the thickness of the main chain layer (L_1) and the thickness of the side chain layer (L_2).

The SMI-C_{18} polymer is a solid and brittle material and can be dissolved in a hot dodecane solvent. The solution is clear and viscous at temperatures above 80°C. However, upon cooling to about 15°C or below, the solution turns into a rubber-like gel. The small angle x-ray scattering results of the SMI-C_{18}/dodecane gels of various polymer concentrations are presented in Figure 2. In comparison with that of the dry SMI-C_{18}, two important features are found in the x-ray scattering patterns of the gels. First, the maxima in the diffraction patterns gels shift toward low angles as the polymer concentration in the gels decreases. This feature reflects that the mean distance between layers increases in the direction normal to the layers as the polymer concentration decreases. Second, the diffraction patterns of gels of various concentrations do not show the second-order diffraction. This feature indicates that the layer structures of the gels are less ordered than that of the dry polymer.

Figure 3 sketches a close look of the layer structures of the SMI-C_{18} polymer in dodecane. Now, the long period L, as measured by the x-ray scattering, is the sum of two components: the thickness of the polymer-rich layer ($L_{p.r}$) and the thickness of solvent-rich layer ($L_{s.r}$). It is noted that in general the relationship between $L_{p.r}$ of gels and (L_1+L_2) of the dry polymer would be very complicated because the main chain layer here is plasticized and the side chain layer is swelled by the dodecane solvent. However, according to the layer structures of Figure 4, the thickness of a layer is proportional to its volume fraction, therefore

$$\frac{(1-\alpha)L_{p.r}}{L} = V_p - \beta \qquad (1)$$

where α is the fraction of the dodecane solvent in the polymer-rich layer; V_p is the total volume concentration of the polymer in the system; β is the saturation concentration of the polymer in the dodecane solvent. When the system is diluted enough that the alkane side chains of two layers are fully separated, the quantity $(1-\alpha)L_{p.r}$ can be regarded as a constant. Therefore, a liner relation between 1/L and V_p would be expected based on eq. (1).

Figure 4 presents such a plot between 1/L and V_p for the SMI-C_{18}/dodecane gels. The straight line in the plot is expected and supports the layer structures proposed. The non-zero value of β indicates a small fraction of polymer is dissolved in the dodecane solvent. Based on our observations, solutions if its concentration is smaller or closes to β can not form gels upon cooling. It is considered that the volume fraction of the polymer V_p is not a good parameter to describe the network studied. It is therefore desirable to define a new quantity $\varphi = V_p-\beta$, namely the concentration of layers or the concentration of the layer network, for effectively describing the system.

The variation of the modulus (G') as a function of the network concentration (φ) is plotted in Figure 5. The data are linear in the double logarithmic scale and yield the following power relation

$$G' \sim \varphi^{2.8 \pm 0.2} \qquad (2)$$

The power relation has been tested for 20 different frequencies ranging from 0.01 to 10 Hz and over a temperature range from 0 to 15°C. Below 0°C, however, the power relation could not be tested due to the crystallization of the dodecane solvent in the gels.

Since the network concentration φ is inversely related to the long period L, one then expects that there will be a similar power relation between G' and 1/L. Figure 6 presents such a plot between G' and 1/L on a logarithmic scale. Data in this double-logarithmic scale are linear also, yielding the following power relation

$$G' \sim L^{-(2.9 \pm 0.2)} \qquad (3)$$

Again, this power relation has been tested over frequency range from 0.01 to 10 Hz and over a temperature range from 0 to 15°C.

It is well known that in dealing with gels of flexible chain networks an exponent of 2 to 2.25 is usually observed for the modulus-concentration power relation [14]. Here, an exponent of about 3 is therefore astonishing and is considered to be associated with the unique nature of the network. To address this behavior, the SMI-C_{18}/dodecane gels are modeled as layer networks and the layers are consider to be rigid plates. Deformation of the plates with relative large aspect ratios is considered to be achievable easily only through bending the objects rather than stretching or compression. Based on such a model, dimensional analysis shows [16] that the elasticity of a layer network follows simple cubic power laws

$$G \sim \varphi^3 \sim L^{-3} \qquad (4)$$

The data in Figures 5 and 6 thus give the experimental support to the prediction. This feature is entirely due to the nature of a layer network.

Acknowledgment   The author is indebted to S. Simon, Y. Ozawa and W. Hergenrother for many delightful discussions and for carefully reading the manuscript.

REFERENCES
1. M. Matsuo and S. Sagaye, in *Colloidal and Morphological Behavior of Block and Graft Copolymers*, G. E. Molau, ed., Plenum Press, New York, 1971.
2. M. W. Matsen, F. S. Bates, *Macromolecules*, **29**, 1091 (1996).
3. F. S. Bates, *Science*, **251**, 898 (1991).
4. N. A. Plate and V. P. Shibaev, *Comb-Shaped Polymers and Liquid Crystals*, Plenum Press, New York and London, 1987.
5. N. A. Plate and V. P. Shibaev, *Macromol. Rev., J. Polym. Sci.*, **8**, 117 (1974).
6. M. Ballauff, *Makromol. Chem., Rapid Commun.*, **7**, 407 (1986).
7. M. Ballauff, *Angew. Chem., Int. Ed. Engl.*, **28**, 253 (1989).

8. W. Brostow, *Polymer*, **31**(6), 979(1990).

9. S. B. Damman, F. P. M. Mercx, and C. M. Kootwijk-Damman, *Polymer*, **34**(9), 1891 (1993).

10. G. T. Brinke, J. Buokolainen, and O. Ikkala, *Europhys. Lett.*, **32**(2), 91 (1996).

11. O. Ikkala, G. T. Brinke, and J. Buokolainen, *Macromolecules*, **28**, 7088 (1995).

12. M. Antonietti, J. Conrad, and A. Thunemann, *Macromolecules*, **27**, 6007 (1994).

13. P. G. de Gennes, *Phys. Lett.* **28A**, 725 (1969).

14. P. G. de Gennes, *Scaling Concepts in Polymer Physics*, Cornell Univ. Press, Ithaca and London, 1977, Chap. V.

15. S. T. Milner, *Science*, **251**, 905 (1991).

16. X. Wang, to be published.

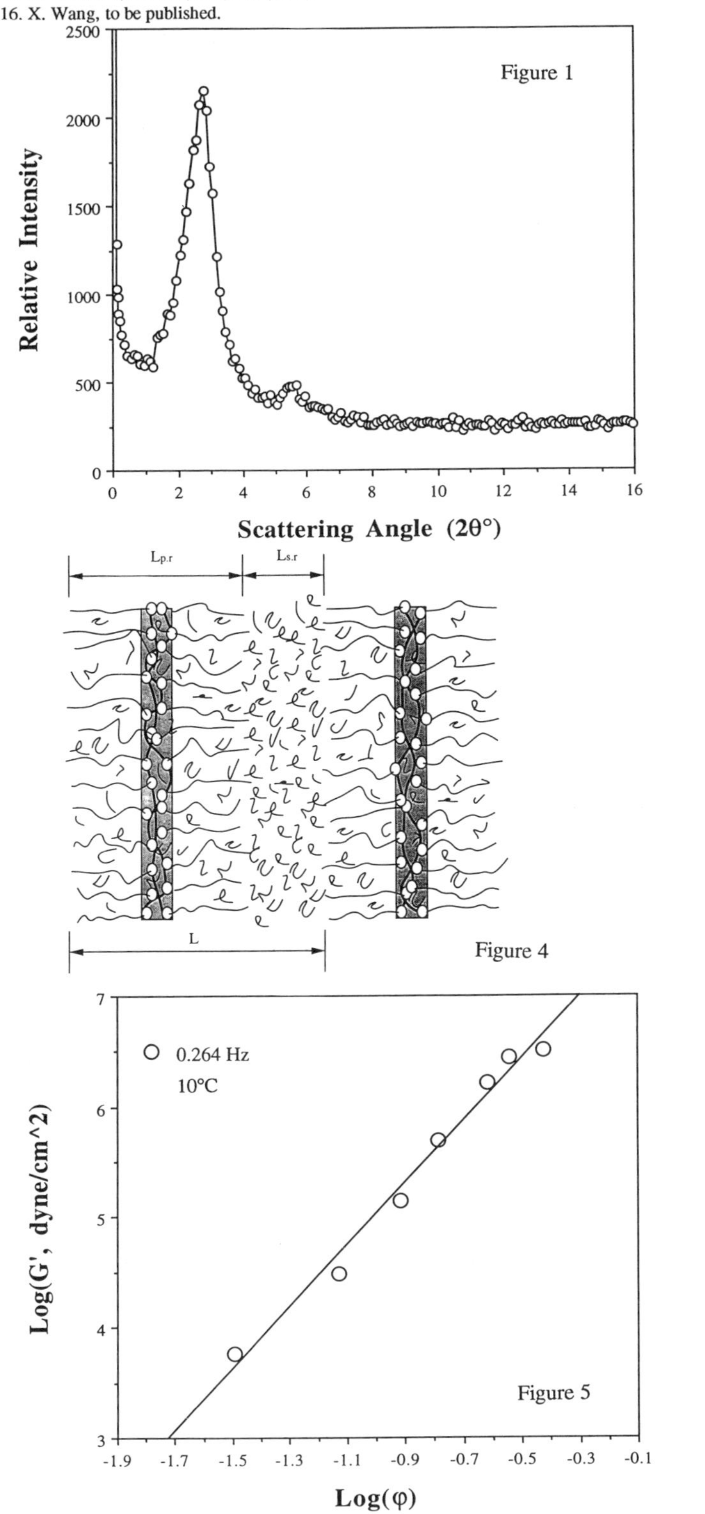

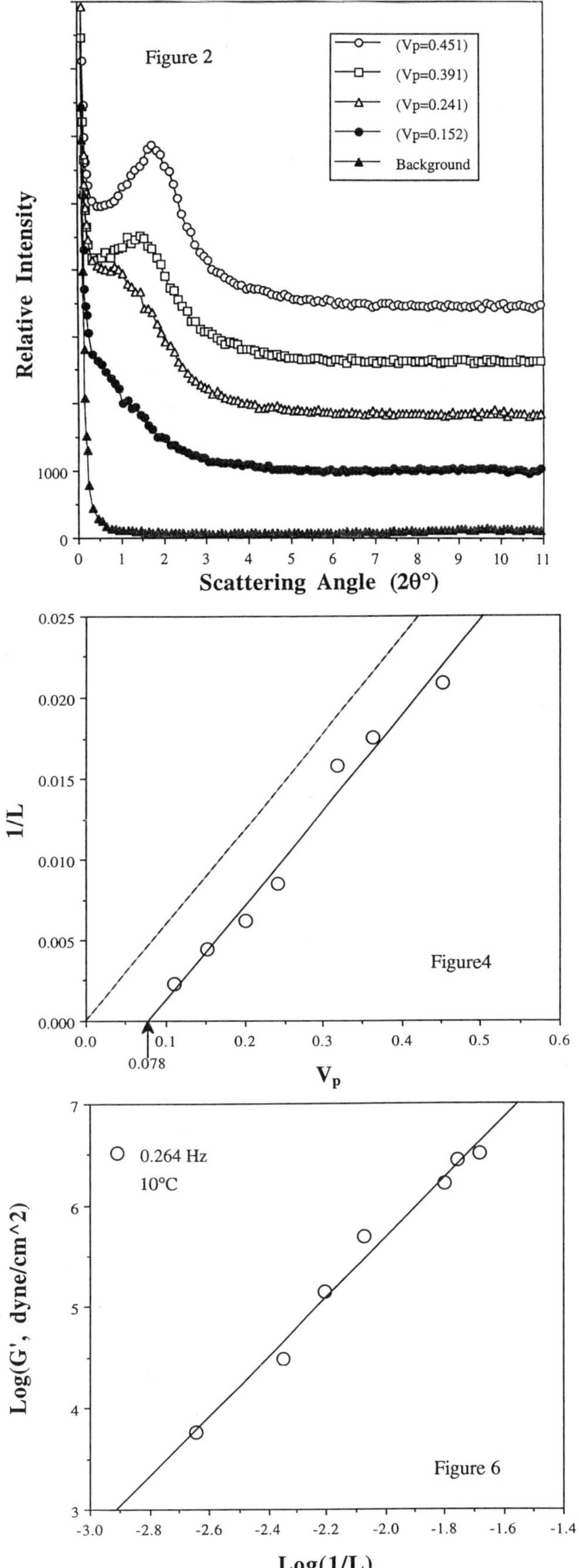

Choosing an Elementary Model for Non-Newtonian Flow

F. Rodriguez, School of Chemical Engineering, Olin Hall, Cornell University, Ithaca, NY 14853-5201

INTRODUCTION

In many instances, flow calculations for polymer solutions and melts reduce to estimations of stresses (pressure drops, torques) as a function of flow (uniaxial, rotational). The most common shear flow behavior of both melts and solutions can be summed up in two regimes: Newtonian behavior at low shear rates and shear-thinning at high shear rates. The "zero-shear viscosity," η_o, is defined in terms of the shear stress τ and the shear rate, $\dot{\gamma}$:

$$\eta_o = (\tau/\dot{\gamma})_o \qquad (1)$$

An elementary form of shear-thinning is described by the "power-law" variably represented as combination as a combination of stress and rate of shear or viscosity and either stress or rate of shear in the following equivalent formalisms:

$$\tau = (k\dot{\gamma})^n \qquad \eta = (k_1\dot{\gamma})^z \qquad \eta = (k_2\tau)^y \quad \{\text{where } \eta = (\tau/\dot{\gamma})\} \quad (2)$$

Real polymer melts and solutions can exhibit other kinds of behavior. Melts containing gel particles or fillers may show a yield stress at low shear rates rather than the well-ordered transition to Newtonian viscosity. At very high shear rates, a shear-thinning material may have an "Upper-Newtonian" viscosity, η_∞. There are some systems (especially suspensions) which exhibit shear-thickening (dilatant) behavior.

In the cases dealt with here, systems are restricted to those which have a Newtonian regime at low shear rates , and a "power-law" regime at some higher shear rates. Another restriction is that the flow models contain no time dependence.

ELEMENTARY MODELS

In order to connect the two flow regimes, a variety of models have been suggested. However, the need to describe both regimes in one equation makes almost all which have only two adjustable parameters quite limited in their applicability. Two three-parameter equations which provide a smooth transition from one regime to the other are the Cross [1,2] and Ellis [3] models:

$$\text{Cross:} \qquad \eta_o/\eta - 1 = (k_c\dot{\gamma})^c \qquad (3)$$
$$\text{Ellis:} \qquad \eta_o/\eta - 1 = (k_e\tau)^e \qquad (4)$$

In both of these models, viscosity approaches η_o when shear rate or stress becomes very small. Similarly, when shear rate or stress is very large, η_o/η is much greater than one, so that both models approach power-law behavior.

One other model which is useful for polymer solutions appears to correlate flow data with a minimum of fitted parameters, but it is not explicit in either rate of shear or stress [4-6]. It is

$$\log \eta_r /\log \eta_{ro} = 0.68 - 0.32 \, \text{erf}\{\log(\tau\dot{\gamma}/B)/2.27\sqrt{2}\} \qquad (5)$$

where η_r is the relative viscosity (solution viscosity/solvent viscosity), erf x is the error function of x, and B is an adjustable parameter. The equation also predicts an upper Newtonian flow regime. A similar equation is

$$\log \eta_r /\log \eta_{ro} = 0.68 - 0.32 \tanh \{0.4 \log(\tau\dot{\gamma}/B)\} \qquad (6)$$

where tanh x is the hyperbolic tangent of x. This equation gives a curve much like equation (5) for values of $(\tau\dot{\gamma}) < B$.

CRITERIA

Probably the most obvious reason for picking a model is to represent a set of experimental data using the minimum number of fitted constants. Reducing the data set to an equation simplifies record keeping. A well-ordered dependence of the fitted constants on some other measurable parameter allows for further condensation of the results. Polymer concentration is an obvious parameter for solutions. Temperature, molecular weight, and molecular weight distribution all may be involved.

Mathematical manipulations are not equally facile with models 3 through 6. The calculation of viscosity, stress, or shear rate from any

equation from a knowledge of another variable is simple when the equation is explicit for the variable which is sought. The Ellis model can be rearranged easily to be explicit in rate of shear as a function of stress. The Cross model can be rearranged easily to be explicit in stress as a function of rate of shear. Equations 5 and 6 do not lend themselves to such rearrangement.

In order to calculate a velocity profile for simple pipe flow, the Ellis model is more convenient than the Cross model. To derive the velocity, u, at any radius r, in a pipe of maximum radius R, having the rate of shear in explicit form simplifies the needed integrations since

$$\text{Rate of shear} = -du/dr \quad \text{and Stress} = \Delta Pr/2L \qquad (7)$$

where the pressure drop ΔP per unit of pipe length L is specified. Even so, the result is rather messy. Unlike the case of the power law where the ratio of velocity to maximum velocity at any radius is a function only of the exponent, the Ellis model velocity ratio is a function of the three constants together with an additional piece of information such as the stress at the wall. However, one does expect the velocity profile to change from the conventional quadratic Newtonian form to the power law form as one increases stress or rate of shear.

An attractive feature of the Cross model is that the constant k_c has the units of time and, as such, represents a kind of relaxation time. It is, in fact, the reciprocal of the shear rate at which the viscosity drops to half of η_o. An interesting application of the Cross model is its use in specifying commercial polyolefins. The Dow Chemical Company [7] defines a DRI, "Dow Rheology Index," as

$$\text{DRI} = (3.65 \times 10^6 k_c / \eta_o - 1)/10 \qquad (8)$$

where k_c (sec) and η_o (poise) are calculated from a nonlinear regression of experimental data fitted to the Cross model. Polyolefins produced by Dow's "Insite™" technology using metallocene catalysts show a different dependence of η_o on k_c (and DRI >0) than competitive polyolefins (with DRI = 0).

EXAMPLES

Polymer solution. Brodnyan and Kelley published an extensive set of flow curves for aqueous solutions of poly(acrylic acid) neutralized to various degrees [8]. A fully neutralized sample with a molecular weight of 1.8×10^5 (for the polyacid) and a concentration of 0.22 g/mL had a η_o of 7.4 Pa·s (74 poise). Stress and rate of shear values (Table 1) have been used to calculate the parameters needed for making plots according to equations (3) and (4). When the appropriate plots are made for equations 3 and 4, it can be seen that either model gives a fair representation of the data when subjected to objective regression analysis. The Cross model (Figure 1) is correlated with $r^2 = 0.997$. The correlation coefficient r represents [9] the "proportion of the variance in y attributable to the variance in x." The value of $r^2 = 0.990$ for the Ellis model (Figure 2) seems nearly as good, but the plot does not appear to be as linear as the Cross model.

The simplest way to fit equation (6) to the data is to choose one reasonably central point since only one parameter needs to be fitted. Both η_o and the solvent viscosity, η_s, are known already. In Figure 3, the flow data are represented by points and equations (3), (4), and (6). It is apparent that both (3) and (4) follow the data within the expected bounds of experimental reproducibility. In defense of equation (6), it should be pointed out that it has one less fitted parameter. Since non-Newtonian flow data for most polymers of commercial interest are well ordered with no inflection points, it is obvious that fitting at two points has an advantage over fitting at one. Moreover, it has been observed [4-6] that a single value of B in equation (6) will often serve over a wide range of concentrations for a particular polymer sample in a single solvent.

Polymer melt. When it comes to selecting a model for a polymer melt, equation (6) is not an option. A flow curve for a poly(ethyl acrylate) sample with a M_v of 310×10^3 was reported by Rosen [9] who used both biconical and capillary rheometers at room temperature (Figure 4). Except for the points at the lowest rate of shear, the Cross model appears to be quite satisfactory (Figure 5). One feature of both Cross and Ellis models is that small changes in viscosity can cause large deviations from the correlation

line when $\eta_o/\eta < 2$. Thus an increase in $(\eta_o/\eta - 1)$ from 0.1 to 0.2 represents only about 10% increase in viscosity whereas an increase in $(\eta_o/\eta - 1)$ from 10 to 20 represents a 100% increase in viscosity. However, both changes appear to be equal on the vertical logarithmic scale. For Rosen's flow data, the Ellis model is not at all satisfactory (Figure 6). A rather similar situation was found for the flow behavior of a poly(dimethylsiloxane) [10].

Comparing Cross and Ellis.

If we assume that the Cross model represents a set of data, we can calculate the corresponding stresses and plot $\log (\eta_o/\eta - 1)$ versus $\log (\tau)$ in order to test the linearity of the Ellis model. First, let

$$\phi = \eta_o/\eta - 1 = (k_c\dot{\gamma})^c \qquad (9)$$

At the point where $\eta_o/\eta - 1 = 1$,

$$1 = (k_c\dot{\gamma}) = (k_e\tau) \text{ independent of the exponents} \qquad (10)$$

and

$$k_c/k_e = \tau/\dot{\gamma} = \eta = \eta_o/2 \qquad (11)$$
$$k_e = 2k_c/\eta_o \qquad (12)$$

And

$$k_e\tau = k_e\dot{\gamma}\eta = (2k_c/\eta_o)\gamma\eta = (2k_c\dot{\gamma})/(1 + \phi) \qquad (13)$$

Since

$$(k_c\dot{\gamma}) = (\phi)^{1/c} \qquad (14)$$

we can write a dimensionless stress, $k_e\tau/2$, in terms of ϕ and c:

$$k_e\tau/2 = (\phi)^{1/c}/(1 + \phi) \qquad (15)$$

It can be seen that whether or not the Ellis plot derived from a linear Cross model looks linear (Figure 7) depends mainly on the range of η_o/η covered and hardly at all on the slope c. Of course, both become linear when η_o/η exceeds about 10 since the power-law behavior holds for either model in the limit of high shear rates and strresses. In the power-law range,

$$e = c/(1-c) \qquad (16)$$

CONCLUSIONS

The convenience and versatility of the Cross model for polymer solutions and melts appears to justify its present popularity. When shear thinning behavior is observed over a narrow range of shear rates, the Ellis model often may be just as good. However, neither model can represent "abnormal" flow curves in which there is an upper Newtonian regime or when there is a lower region of shear thinning. The latter case can occur in polymer melts which contain gel or filler particles, or certain liquid crystal polymers.

When polymer solutions are dealt with, there appears to be a place for the mathematically complex equations based on the error function or the hyperbolic tangent of a product of stress and rate of shear (equations 5 and 6). The need for fitting one less parameter is attractive, especially since the one parameter may correlate data for a variety of concentrations.

REFERENCES

1. M. M. Cross, J. Coll. Sci., 20, 417 (1965).
2. M. M. Cross, Rheol. Acta, 18, 609 (1979).
3. R. E. Gee and J. B. Lyon, Ind. Eng. Chem., 49, 956 (1957).
4. F. Rodriguez and L. A. Goettler, Trans. Soc. Rheol., 8, 3 (1964).
5. F. Rodriguez, Trans. Soc. Rheol., 10, 169 (1966).
6. F. Rodriguez, Chem. Eng. Commun., 33, 287 (1985).
7. "INSITE™ Technology," Form No. 305-01970-993 SMG, Dow Chemical Company, 1993.
8. J. G. Brodnyan and E. L. Kelley, Trans. Soc. Rheol., 5, 205 (1961).
9. S. L. Rosen, Ph.D. Thesis, Cornell University, Ithaca, NY, 1964 (Also reported in J. Appl. Polym. Sci., 9, 1601 (1965)).
10. F. Rodriguez, "Principles of Polymer Systems," 4th ed., Taylor & Francis, Bristol, PA, 1996, p. 267.

Table 1. Flow Data by Brodnyan and Kelley [8]. η_o = 7.4 Pa·s, Water viscosity at 20°C = 0.0010 Pa·s.

Shear rate, 1/s	Shear stress, Pa
0.076×10^3	0.411×10^3
0.133	0.631
0.206	0.858
0.306	1.17
0.579	1.85
1.47	3.02
3.85	4.64
27.4	10.0
44.7	14.9
58.3	14.9
79.2	20.3
163	26.7
292	35.2

Fig. 1 Data of Brodnyan and Kelley [8] plotted according to equation 3 (Cross). The regression line corresponds to k_c = 0.0027 s and c = 0.62 with the square of the correlation coefficient, r^2, = 0.997.

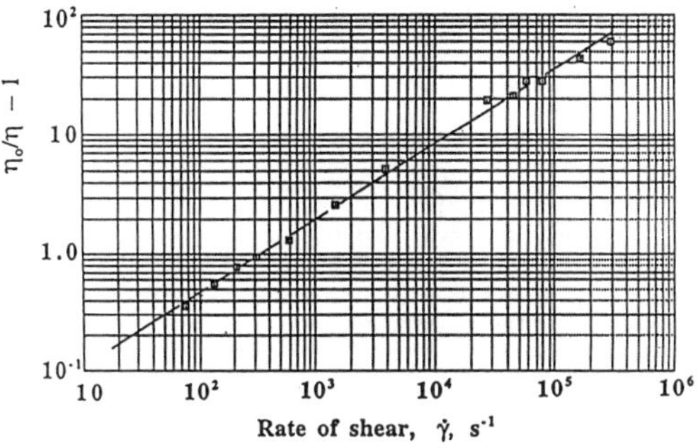

Fig. 2 Same data as in Fig. 1 plotted according to equation 4 (Ellis). The regression line corresponds to k_e = 8.95 x 10^{-4} Pa^{-1} and e = 1.196 with the square of the correlation coefficient, r^2, = 0.990.

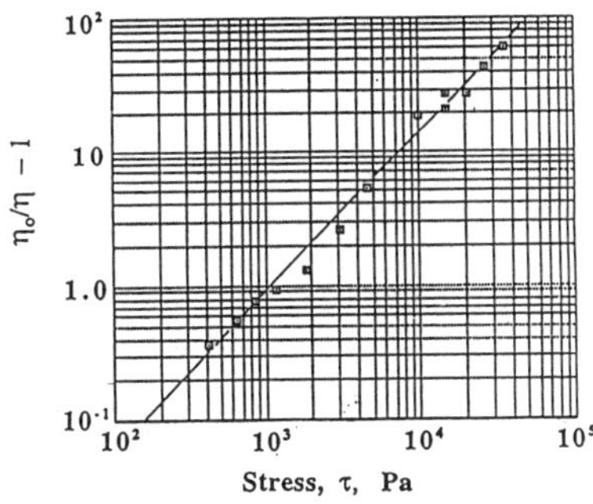

Fig. 3 Data of Brodnyan and Kelley compared with Cross (solid line), Ellis (long dashes), and equation 6 (short dashes). Although all three models can represent the data in a qualitative manner, the Cross model seems best.

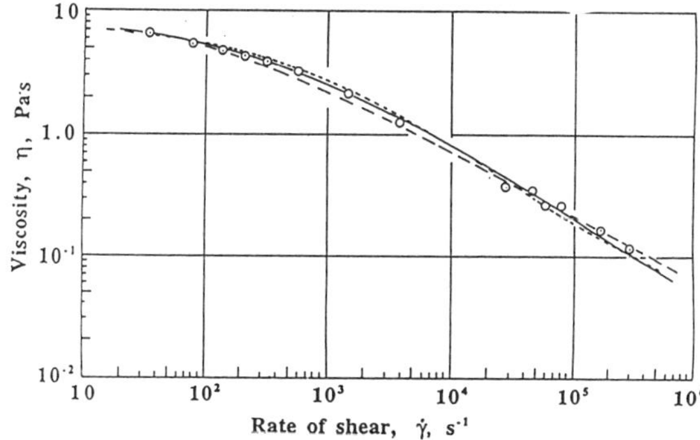

Fig. 4 Flow data obtained by Rosen [9] using biconical and capillary viscometers.

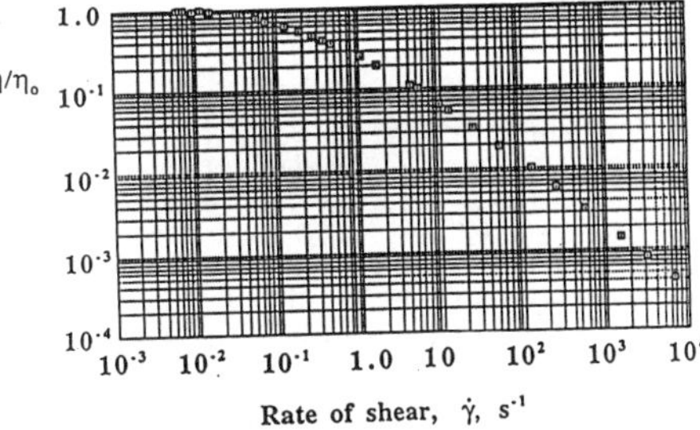

Fig. 5 Cross model plot of Rosen data.

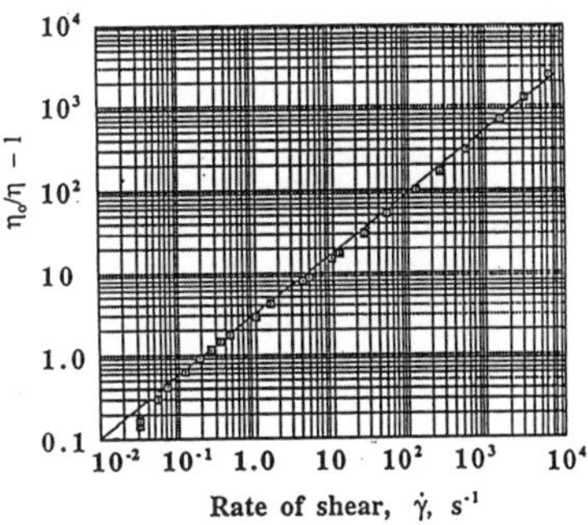

Fig. 6 Ellis model plot of Rosen data.

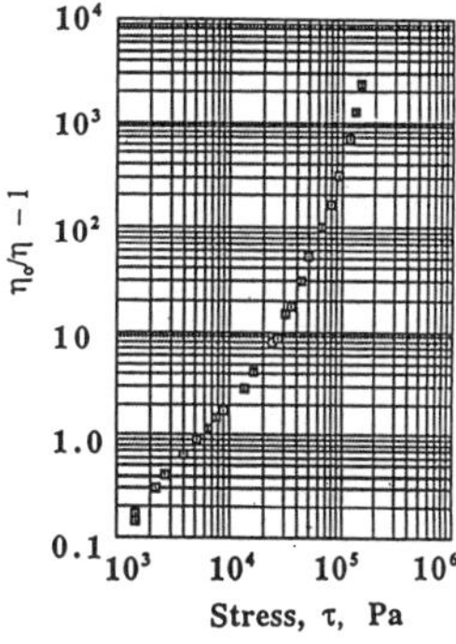

Fig. 7 Plot of a linear Cross model fluid on Ellis model coordinates according to equation 15 for four values of c. The stresses have been multiplied in order to make the shapes at low stresses more clear. The stresses are multipied by 1 for c = 0.5, 10 for c = 0.6, 100 for c = 0.7, and 1000 for c = 0.8.

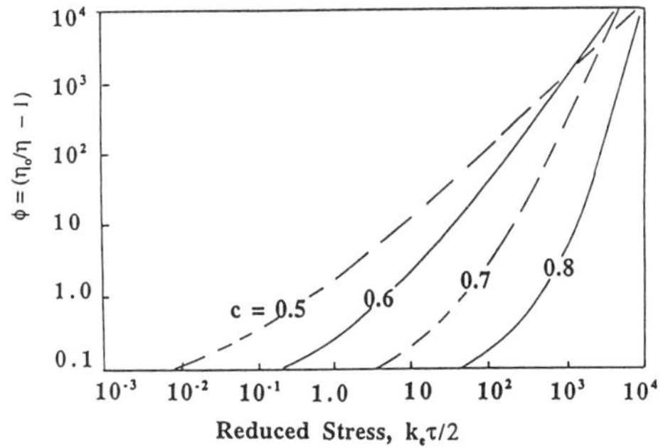

Amine-Quinone Polyurethanes and the Corrosion Protection of Iron

David E. Nikles*, J. Carlos Arroyo**, Antony P. Chacko*, Yongqi. Hu*, Jacqueline A. Nikles, Rahul Sharma*, Shane C. Street* and Garry W. Warren**

*Department of Chemistry, **Department of Metallurgical and Materials Engineering and the Center for Materials for Information Technology, The University of Alabama, Tuscaloosa, Alabama 35487-0209.

Amine-quinone polymers, having the 2,5-diamino-1,4-benzoquinone functional group, have been shown to strongly adhere to iron surfaces and inhibit corrosion of iron. This new class of polymers was first reported by Erhan and coworkers,[1] who have numerous publications on the synthesis[2] and the use of the polymers to prevent iron corrosion.[3] Muralidharan et al demonstrated the corrosion inhibition of iron and steel by an amine-quinone oligomer made by electropolymerization.[4] Mathias and coworkers have reported an alternative synthesis of amine-quinone polymers by the condensation of 2,5-di-n-butoxy-1,4-benzoquinone with 1,6-hexanediamine in amide solvents.[5] Recently Scola has used fluorinated polyaminoquinones as coupling agents to improve the adhesion of epoxy to iron and steel.[6]

We have been interested in the use of amine-quinone polymers as binders to prevent the corrosion of the iron particles used in metal particle (MP) tape. Iron particles have emerged as the only choice for future generation high capacity magnetic tapes. These particles have an oxide coating that somewhat protects them against corrosion. However this non-magnetic layer decreases the saturation magnetization for the particles. We have demonstrated that amine-quinone inhibit the corrosion on commercial iron particles.[7] Furthermore, we have demonstrated that we could replace the oxide protective layer with an amine-quinone polymer coating, providing corrosion protection and particles with a high saturation magnetization.[8]

Our recent work has focused on achieving a fundamental understanding of the mechanism of corrosion protection. In this paper we report the use of electrochemical impedance spectroscopy to compare the ability of two different amine-quinone polymers to prevent the corrosion of iron. We also report x-ray photoelectron spectroscopy evidence for chemisorption of a model compound onto an iron surface.

Amine-Quinone Polyurethanes

Amine-quinone polyurethanes were prepared by the condensation of the amine-quinone monomer, AQM-1, and an oligomeric diol with a diisocyanate.[9] Two amine-quinone polyurethanes, AQPU-15 and AQPU-100, were prepared for this study. AQPU-15 contained mole percent AQM-1, 10 mole percent polytetrahydrofuran diol (M_n = 650), and 50 mole percent TDI (80% 2,4-isomer and 20% 2,6-isomer). AQPU-15 had a soft segment T_g of -39°C. We reported earlier that AQPU-15 was effective in preventing the corrosion of commercial iron particles.[10] AQPU-100 was the result of the desire to decrease the AQM-1 content, without sacrificing corrosion inhibition. We also wanted to replace toxic TDI with less toxic methane diphenyl-4,4'-diisocyanate (MDI). Attempts to completely replace TDI with MDI gave intractable, insoluble polymers. Hard segments containing AQM-1 and MDI had a high degree of phase separation and crystallinity, which lowered the solubility. In order to obtain soluble polymers we had to decrease the MDI content. AQPU-100 contained 7.5 mole percent AQM-1, 42.5 mole percent polycaprolactone diol (M_n = 1250), 40 mole percent TDI and 10 mole percent MDI. It had a soft segment T_g of -32°C and a soft segment melting point of 42°C.

AQM-1

Electrochemical Impedance Spectroscopy

Electrochemical impedance spectroscopy (EIS) was used to further demonstrate the ability of amine-quinone polyurethanes to prevent corrosion of iron.[11] Samples used in this study were 25 mm x 25 mm square iron substrates (99.9+ purity, 0.5 mm thick), coated with a 15 mm thick polymer binder by spin coating. One coating was a standard metal particle tape binder, consisting of a commercial thermoplastic polyurethane , a polyvinylchloride wetting resin, and a polyisocyanate cross-linker. In the other samples the polyvinylchloride resin was replaced with either AQPU-15 or AQPU-100. The details of the apparatus and experimental procedure were described elsewhere.[13] Electrochemical impedance spectra were collected by applying an ac voltage with an amplitude varying from 10 to 30 mV about the measured

corrosion potential. Data were collected at frequencies ranging from 10 mHz to 100 KHz. Electrochemical impedance spectra were collected at different times during the continuous exposure to the electrolyte solution. The current response was interpreted using the equivalent circuit models, containing resistive and capacitive elements. The choice of equivalent circuit depended on the stage of coating degradation. Initially, the impedance spectrum was modeled by the resistance of the electrolyte R_s, in series with the capacitance of the coating, C_{coat}. After a period of exposure to the electrolyte, the electrolyte diffused into the coating, lowering the coating resistance, and the spectra were modeled adding a coating resistance, R_{coat} in parallel with the coating capacitance. When the electrolyte arrived at the polymer-metal interface, the adhesive bond between polymer and metal can be broken creating a metal-electrolyte interface which is the first step along the path to corrosion. The creation of the metal-electrolyte interface required a modification to the equivalent circuit to include the double layer capacitance, C_{dl}, and a charge transfer resistance, R_{ct}, representing the corrosion process. In Fig. 1 are Bode plots of data obtained for an iron sample coated with the commercial tape binder compared with the theoretical prediction obtained from the appropriate equivalent circuits. Two curves are plotted for the sample in its initial condition (i.e. 5 min exposure), one for impedance and one for phase angle as a function of ac frequency. At low frequencies the phase angle was low, indicating the film was behaving like a resistor with a resistance of about 10^8 Ω. At higher frequencies the phase lag approached 90° and the film was behaving like a capacitor. After 80 days exposure to the electrolyte, significant changes were observed in the electrochemical impedance spectrum. A new peak in the phase angle curve appeared at low frequency. This peak was the double-layer capacitance and was indicative of disbonding between the polymer and the metal surface. The appearance of this peak represents the onset of corrosion. Clearly, the electrolyte had diffused through the coating, broke the adhesive bond been the polymer and the iron surface, allowing electrolyte adsorption, which facilitated corrosion. A different behavior was observed for the iron samples with the coating containing AQPU-15, Fig. 2, or AQPU-100, Fig. 3. For AQPU-15 the initial electrochemical impedance spectrum had a higher resistance and the film was capacitive over a broader range of frequencies. After 81 days exposure to the electrolyte, the resistance of the film decreased, indicating that the electrolyte had diffused into the coating. However, there was no low frequency peak in the phase angle curve. . The results for AQPU-100, were even more dramatic. After 120 days, there still was no indication of a double-layer capacitance. The resistance was still high, indicating little or no penetration of the electrolyte into the polymer coating. A major difference between AQPU-15 and APQU-100 was that the soft segment crystallized, preventing penetration of the electrolyte into the coating. The amine-quinone polymer prevented the electrolyte from diffusing to the iron surface and forming a double-layer. The interaction between the amine-quinone polymer and the iron surface was strong enough to prevent water from breaking the adhesive bond. Includng a semicrystalline soft segment further improved the ability of the amine-quinone polymers to inhibit iron corrosion.

XPS Studies

To gain some insight into the nature of the interaction between the amine-quinone polymer and the iron surface, x-ray photoelectron spectra (XPS) were measured for the model compound, AQM-14A, adsorbed onto iron squares. AQM-14A had the 2,5-diamino-1,4-benzoquinone functional group. But it did not have the urethane, ether, or ester functional groups that would make interpretation of the XPS difficult.

AQM-14A

The N (1s) peak for bulk AQM-14A appeared at 398.6 eV, Fig. 4. The peak for AQM-14A adsorbed onto the iron surface was at 399.5 eV, a shift of 0.9 eV to higher binding energy. It was difficult to interpret any shifts in the oxygen peaks because of the surface oxide. The shift in the N (1s) binding energy provides evidence for chemisorption of AQM-14A onto the iron surface.

Acknowledgments

The project was supported in part by the corporate sponsors of the Center for Materials for Information Technology and by the Department of Defense Advanced Research Projects Agency through a grant administered by the National Industry Consortim. The project used shared instrumentation purchased through the NSF Materials Research Science and Engineering Center award DMR-9400399. The x-ray photoelectron spectrometer was purchased by NSF Academic Research Infrastructure Grant DMR-9512264.

References

1. Kaleem, K.; Chertok, F.; Erhan, S. *Prog. Organ. Coatings* **1987**, *15*, 63-71.
2. a. Kaleem, K.; Chertok, F.; Erhan, S. *J. Polym. Sci., Pt. A: Polym. Chem.* **1989**, 27, 865-871, b. Nithianandam, V. S.; Kaleem, K.; Chertok, F.; Erhan, S. *J. Appl. Polym. Sci.* **1991**, 42, 2893-2897, c. Nithianandam. V. S.; Chertok, F.; Erhan, S. *J. Appl. Polym. Sci.* **1991**, 42, 2899-2901. Vithiananddam, V. S.; Erhan, S. *J. Appl. Polym. Sci.* **1991**, 42, 2385-2389, d. Reddy, T. A.; Erhan, S. *J. Polym. Sci, Pt. A: Polym. Chem.* **1994**, 32, 557-565.
3. Nithianandam, V. S.; Reddy, T. A.; Erhan, S. *Int. J. Polymeric Mater.* **1994**, 23, 167-175.
4. a. Phani, K. L. N.; Pitchumnani, S.; Muralidharan, S.; Ravichandran, S. Iyer, S. V. K. *J. Electroanal. Chem.* **1993**, *353*, 315, b. Muralidharan, S.; Phani, K. L. N.; Pitchumnani, S.; Ravichandran, S. Iyer, S. V. K. *J. Electrochem. Soc.* **1995**, *142*, 1478.
5. Colletti, R. F.; Stewart, M. J.; Taylor, A. E.; MacNiell, N. J.; Mathias, L. J. *J. Polym. Sci., Polym. Chem.* **1991**, *29*, 1633.
6. Vaccaro, E.; Scola, D. A. *Proc. ACS Div. Polym. Mater. Sci Eng.* 1998, 78(
7. Liang, J.-L.; Nikles, D. E. *IEEE Transactions on Magnetics* **1993**, 29(6), 3649-3651.
8. Nikles, D. E.; Cain, J. L.; Chacko, A. P.; Webb, R. I. *IEEE Transactions on Magnetics* **1994**, 30(6), 4068-4070.
9. Nikles, D. E.; Liang, J.-L.; Cain, J. L.; Chacko, A. P.; Webb, R. I; Belmore, K. *J. Polym. Sci., Pt. A: Polym. Chem.* **1995**, 33, 2881-2886.
10. Chacko, A. P.; Webb, R. I.; Nikles, D. E. In "Film Formation in Waterborne Coatings: Provder, T.; Winnik, M. A.; Urban, M. K., Eds. ACS Symposium Series Volume 648, American Chemical Society, Washington, DC, 1996, 525-534.
11. Arroyo, J. C.; Warren, G. W.; Nikles, D. E. Proceeding of the 6th International Workshop on Moisture in Microelectronics, Oct. 15-17, 1996, NIST-IR 5960, Gaithersburg, MD.

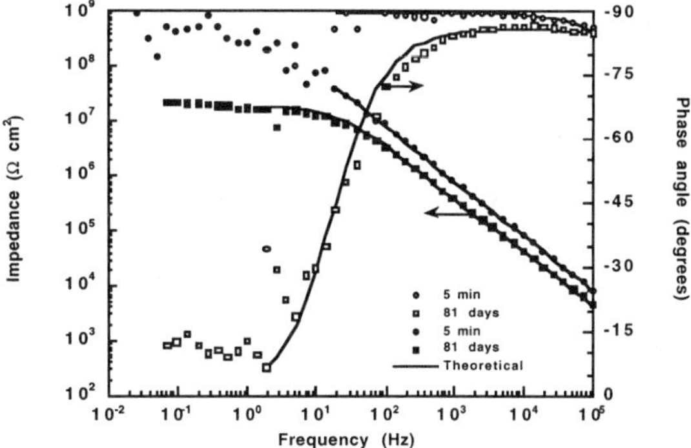

Fig. 2. Effect of exposure to 1 M NaCl electrolyte on the electrochemical impedance spectra for an iron square with a coating containing AQPU-15.

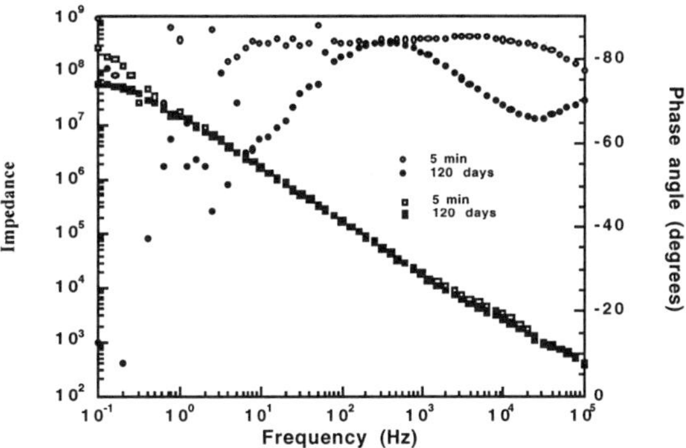

Fig. 3. Effect of exposure to 1 M NaCl electrolyte on the electrochemical impedance spectra for an iron square with a coating containing AQPU-100.

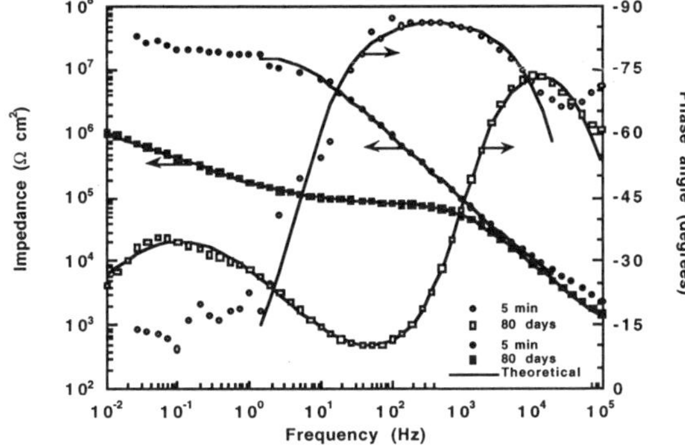

Fig. 1. Effect of exposure to 1 M NaCl electrolyte on the electrochemical impedance sectra for an iron square with a coating containing a commercial magnetic tape binder.

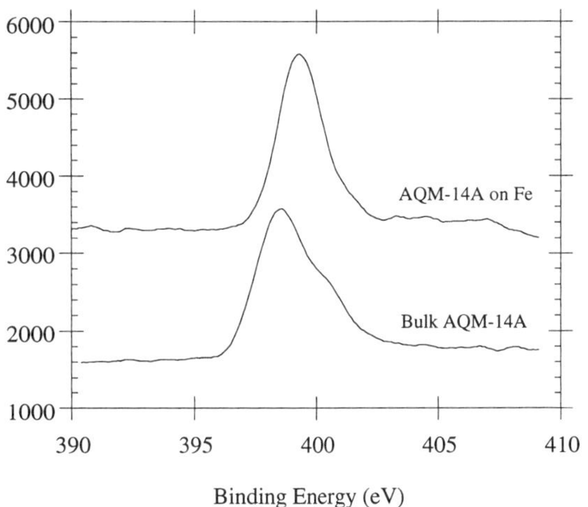

Fig. 4. N(1s) x-ray photoelectron spectra for AQM-14A, either bulk sample or adsorbed onto an iron surface.

Period-doubling behavior in propagating polymerization fronts of multifunctional acrylates

Jonathan Masere and John A. Pojman,[*]
Department of Chemistry and Biochemistry,
University of Southern Mississippi,
Hattiesburg, MS 39406
U.S.A.
http://www-chem.st.usm.edu/japgroup/japgroup.html

Introduction

Frontal polymerization, a process in which the heat released by the reaction is utilized to sustain the polymerization process has proven to be simple and cost effective in terms of energy consumption and time.[1-4] Reactants are placed in a cylinder and when the reaction is initiated at one end, a self-sustaining reaction zone propagates through the sample.

Although most of frontal polymerization occurs via a planar front, changes due to heat loss culminates in spinning or oscillatory fronts. Studies have shown that dilution of the system results in spin modes. Dilution, the bifurcation parameter, dictates the frontal behavior.

We performed experimental investigations of periodic and aperiodic fronts during frontal polymerization of 1,6-hexanediol diacrylate (HDDA) and trimethylolpropane ethoxylate triacrylate (TEMPTA). A better understanding of these modes has important ramifications if frontal polymerization is to be used for the production of materials at an industrial level. The quality of frontal polymerization products depends on the local microstructure which, in turn, is related to the temperature history. Nonplanar frontal propagation induces inhomogeneieties in the product. Ways to prevent such instabilities can only be formulated if factors responsible are known and well understood. In this paper, results of the investigation of the frontal polymerization of HDDA and TEMPTA are presented.

Experimental procedure

Experiments were conducted using 1,6-hexanediol diacrylate monomer (Aldrich) diluted with diethyl phthalate (Aldrich) as a passive diluent and a reactive diluent, benzyl acrylate (Fisher) and, trimethylolpropane ethoxylate triacrylate (Aldrich) diluted with dimethyl sulphoxide (Fisher). Lupersol 231 (ATO) was used as the initiator. The chemicals were used as received.

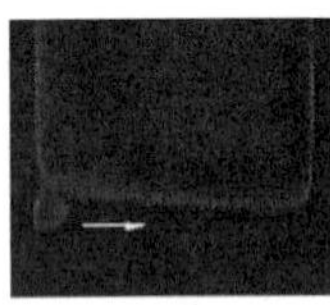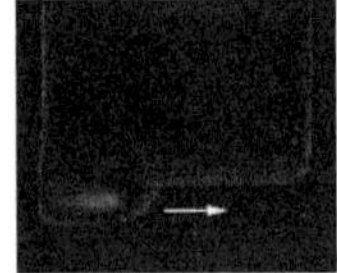
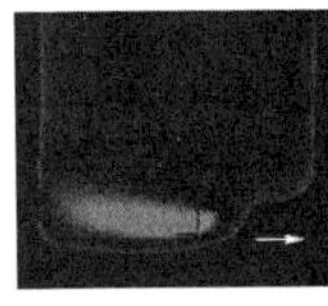

Figure 1. IR montages of a single-head HDDA spin mode front taken at 20 second intervals.

The inhibitor in the HDDA monomer was found not to interfere with the reaction. All the reactions were performed in 16 mm x 125 mm capped culture test tubes (KIMAX). A free radical indicator, Bromophenol Blue (Eastman Kodak) dissolved in dimethyl sulphoxide (DMSO), was used to visually observe reaction fronts.[5]

A solution of the reactant mixture was placed in a test tube and the polymerization reaction initiated with a soldering iron. Visual video images were recorded using a camcorder (Handyman video Hi8) and an Amber camera was used to capture Infra Red images. The images were digitized on a MacIntosh PowerPC 8500.

Results and Discussion

A single-headed spin mode is observed for the HDDA percentage range of 38-42%, Fig. 1. Using the Bromophenol Blue indicator, the spin can be observed as a pale yellow line moving into the green reactant solution in a spiral fashion. The single-head mode polymerization regime becomes double-headed when the percentage volume of HDDA is increased into the 42-46% range, Fig. 2.

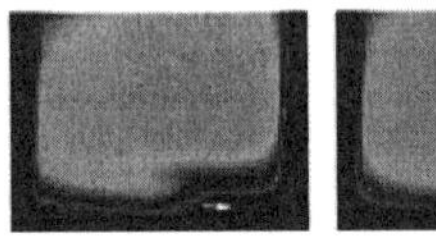
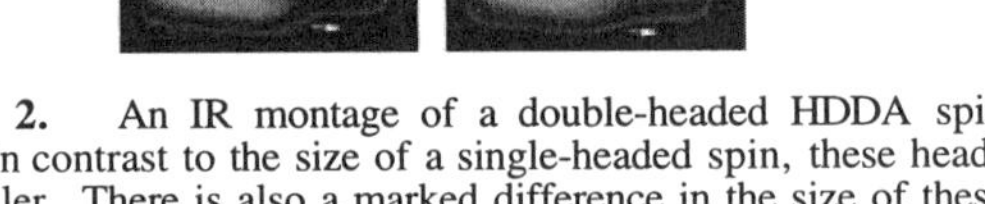

Figure 2. An IR montage of a double-headed HDDA spin mode. In contrast to the size of a single-headed spin, these heads are smaller. There is also a marked difference in the size of these heads.

Like the one-headed spin, the two-headed spin mode maintained the spiral direction from the top of the test tube all the way to completion. Four spin heads are observed upon a further increase in the HDDA percentage (46-49% range), with only two of these spots can be observed at a time, Fig. 3.

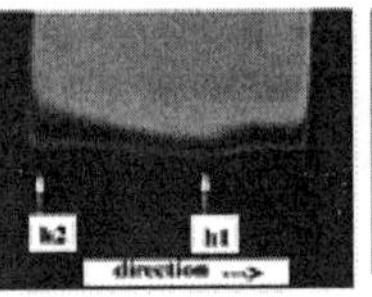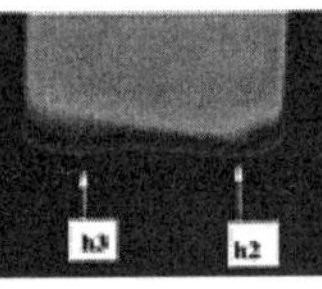

Figure 3. An IR image of two of the four hot spots of a four-headed spin mode. Only two of these spots, h1 and h2, can be observed at a time because the other two, h3 and h4, are hidden behind (a). After 20 seconds, h1 disappears and, as h2 is about to disappear, h3 appears, (b).

These spots are evenly spaced and maintain the same direction unless there is interference from bubbles, which is rare. The hot spots get even smaller in size with increasing monomer.

When TEMPTA was used as the acrylate monomer, with DMSO as the passive diluent, a doubling of spin heads was observed, a similar behaviour exhibited by the HDDA system. However, unlike the HDDA in which an increase in the percentage of the monomer composition resulted in flat front, multiple spin heads were observed with TEMPTA. Beyond four spin heads, subsequent hot spots that did not move in a unidirectional fashion were observed. The heads would collide, Fig. 4, and a reversal in

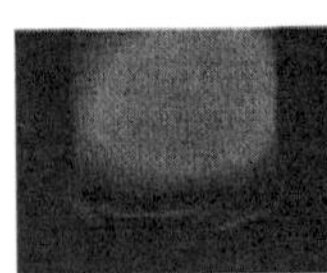

Figure 4. Montages of two TEMPTA heads approaching each other, (a). These heads get closer to each other after 20 seconds, (b). spin direction observed.

As a result of this change of direction, it was difficult to determine the exact number of spin heads. The number of these spin heads increases with an increase in the percentage composition of TEMPTA.

A further increase in TEMPTA composition resulted in the disappearance of the colliding spin heads. Instead, numerous spots were observed on the entire reactant-polymer interface which appeared rippled when the BPB indicator was used. The IR images showed bright flare-ups ahead of the progressing polymerization front, Fig. 5. The observed lack of directional consistence of the hot spots can be characterized as *chaotic*; however, we have not performed an analysis of this apparent spatiotemporal chaos.

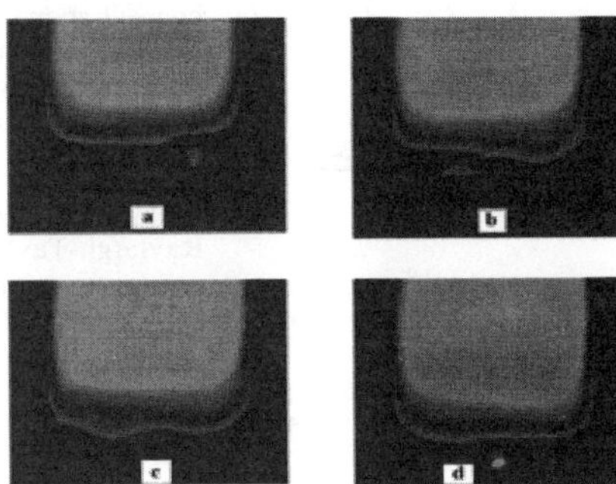

Figure 5. A multi-headed TEMPTA front also exhibiting flare-ups ahead of the jagged front.

Conclusions

In summary, we have presented results of a period-doubling sequence route to a chaotic front propagation. This study shows that the frontal polymerization exhibits complex behavior, the character of which depends on the percentage composition and the nature of both the monomer and diluent and, the heat loss to the environment around the test tube. Of all these factors, the variation of percentage composition with respect to inert diluent was the most convenient parameter for the study of the bifurcation sequence. A schematic summary is given in Fig. 6.

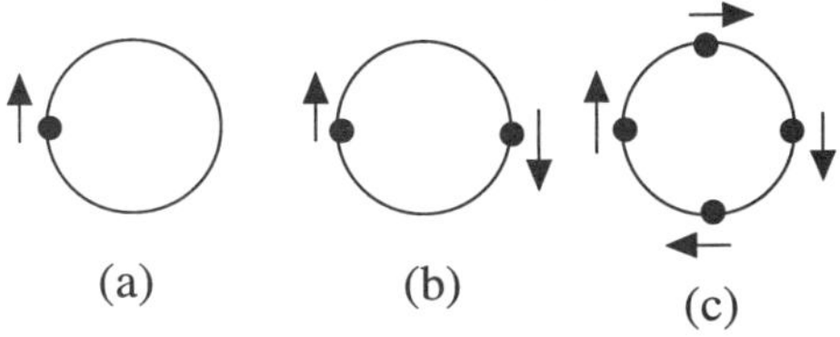

Figure 6. A schematic summary of the effect of dilution on the behavior of the HDDA fronts.

A 1-head spin is illustrated in Fig. 6 a. However, the behavior becomes more complex as the dilution is decreased. As the percentage of the HDDA increases, there comes a critical percentage at which the number of spin heads doubles, as shown if Fig. 6 b. With a further increase in HDDA percentage, another doubling of spin heads is observed which results in four spin heads, Fig. 6 c. For TEMPTA fronts, a doubling of heads is observed, Fig. 7 a-c but above the TEMPTA range that gives four hears, subsequent heads do not move in a unidirectional fashion, Fig. 7 d.

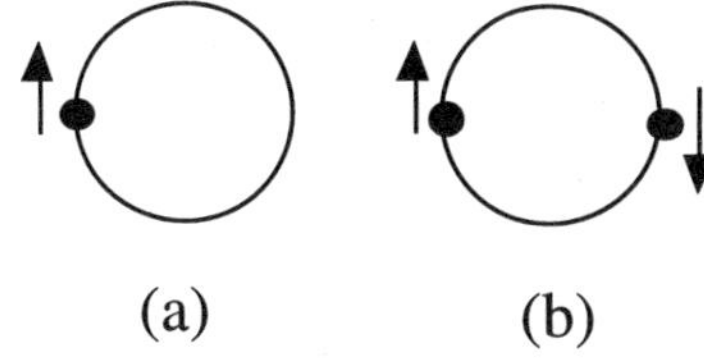

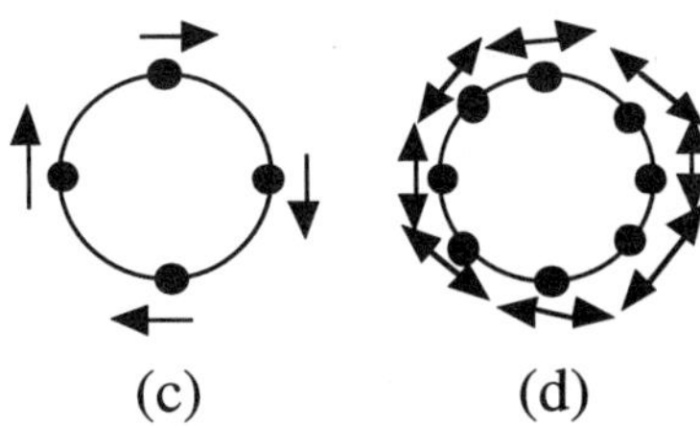

Figure 7. A schematic summary of the effect of dilution on the behavior of the TEMPTA fronts.

This trend is typical of a *period-doubling*[6,7] sequence. This study shows that the range of the percentage monomer composition, the bifurcation parameter, over which the a particular spin-head mode is sustained decreases as the number of spin heads increases. With a further increase in TEMPTA composition, spin modes become aperiodic, shown in Fig. 12d. It is interesting that Coffman *et al.*[6] observed similar behavior in the Belousov-Zhabotinsky reaction as did Shkiro *et al.*[7] when they studied the propagation of tantalum-carbon combustion fronts.

References

(1) Chechilo, N. M.; Enikolopyan, N. S. *Dokl. Phys. Chem.* **1976**, *230*, 840-843.
(2) Pojman, J. A.; Craven, R.; Khan, A.; West, W. *J. Phys. Chem.* **1992**, *96*, 7466-7472.
(3) Pojman, J. A.; Ilyashenko, V. M.; Khan, A. M. *J. Chem. Soc. Faraday Trans.* **1996**, *92*, 2825-2837.
(4) Fortenberry, D.; Pojman, J. A. *Polym. Prepr. Am Chem. Soc. Div. Polym. Chem.* **1997**, *38(2)*, 472-473.
(5) Masere, J.; Pojman, J. A. *J. Chem. Soc. Faraday Trans. in press*.
(6) Coffman, K. G.; McCormick, W. D.; Noszticzius, Z.; Simoyi, R. H.; Swinney, H. L. *J. Chem. Phys.* **1987**, *86*, 119-129.
(7) Shkiro, V. M.; Nersisyan, G. A. *Comb. Expl. Shock Waves* **1978**, *14*, 121-122.

Functionally-Graded Polymeric Materials Prepared Via Frontal Polymerization

John Pojman*,Yuri A. Chekanov, Chad Case and Tim McCardle

Department of Chemistry and Biochemistry
University of Southern Mississippi
Hattiesburg, Mississippi 39406-5043

john.pojman@usm.edu
www-chem.st.usm.edu/japgroup/JAPGROUP.html

Introduction

Functionally Gradient or Graded Materials (FGMs) are materials whose composition varies spatially in a controlled manner. In polymers, a centrifugal force approach was used to prepare a composite with spatially varying conductivity.[1] Lee et al. used centrifugal force to prepare gradients in carbon fiber reinforced epoxy composites.[2] Several works have been done on preparing gradient Interpenetrating Polymer Networks (IPNs).[3-7] Most of these methods involve producing the gradient by diffusing one component into the other pregelled component and then curing, or producing the gradient with a gradient of illumination. The diffusion method can require as much as 280 hours to produce a gradient over 10 μ.[8] Using the absorbance of light to produce a gradient is limited to less than 1 mm.[3] None of these techniques can be used to produce gradients over several centimeters in thick samples.

Graded polymeric materials have found wide use in optical applications, especially Graded Refractive INdex (GRIN) materials.[9-13] These materials are prepared via isothermal frontal polymerization (interfacial gel polymerization), which is a slow process limited to producing gradients over about 1 cm.

Another type of gradient material with definite utility is an optical limiter based on a gradient of nonlinear optical dye dissolved in a polymer matrix. An optical limiter is a device that strongly attenuates intense optical beams but allows high transmittance of low level light. Such a device would be very useful for protecting human eyes from intense laser pulses.[14] Perry et al. discussed the types of organic materials that exhibit such nonlinear absorption.[15] They have found that metallophthalocyanine (M-Pc) complexes containing heavy central atoms work well. These dyes are compatible with poly(methyl methacrylate) and dissolve in the monomer.[15] This affords the great advantage of inexpensive materials.

Miles calculated that the maximum attenuation of a light pulse can be achieved if the absorbing species is distributed as a hyperbolic function of position.[16] Perry et al. demonstrated the value of a gradient by approximating the hyperbolic distribution with slabs containing different dye concentrations at set distances.[15]

A New Approach

We have developed a new method to produce gradients in polymeric materials based on **thermal frontal polymerization**. With our approach we can produce gradients of mechanical and optical properties in acrylates, amine and cationically cured epoxy and acrylamides and acrylates in water and other high boiling point solvents.

Frontal Polymerization

Frontal polymerization is a mode of converting monomer into polymer via a localized reaction zone that propagates through the coupling of the heat released by the reaction and thermal diffusion. Frontal polymerization was first discovered at the Institute of Chemical Physics in Chernogolovka, Russia by Chechilo and Enikolopyan in 1972.[17] Such fronts can exist with free-radical polymerization of mono- and multifunctional acrylates[18-22] or epoxy curing.[23] Frontal polymerization can be achieved in solution polymerization with reactive monomers such as acrylamide, methacrylic acid and acrylic acid in solvents such as water and DMSO.[24]

Frontal polymerization reactions are relatively easy to perform. In the simplest case, a test tube is filled with the reactants. The front is ignited by applying heat to one end of the tube with an electric heater. The position of the front is obvious because of the difference in the optical properties of polymer and monomer.

Under most cases, a plot of the front position versus time produces a straight line whose slope is the front velocity. The velocity can be affected by the initiator type and concentration but is on the order of a cm/min.

The defining feature of thermal frontal polymerization is the sharp temperature gradient present in the front. Figure 1 shows three different temperature profiles for methacrylic acid for different initiators. Notice that the temperature jumps about 200 °C over as little as a few millimeters, which corresponds to polymerization in a few seconds at that point.

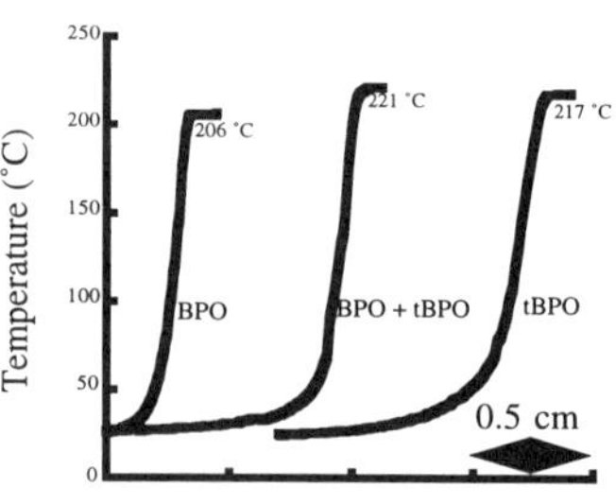

Figure 1. Temperature profiles for three different methacrylic acid polymerization fronts.

Convection

Convection caused by the large thermal and concentration gradients in the front significantly interfere with ascending fronts.[25, 26] Descending fronts with monomers producing monoacrylates can not be sustained because of the Rayleigh-Taylor instability unless a filler is added[19] or the experiment is performed in weightlessness.[27]

How does it work?

The stability of an ascending front with an infinite layer of unreacted monomer above it is different from that of a narrow layer. The system will be more stable but no theory exists for this case. We can make some rough estimates by analogy with the non-reactive case of a fluid layer of thickness L heated from below. The Rayleigh number,

$$Ra = \frac{g\Delta T\beta\, L^3}{\nu\kappa},$$

can be affected for a given frontal system by changing the thickness, L. Because of the cubic dependence, the stability is very sensitive to L.

In a thin layer, with L = 1 cm, ascending fronts will propagate with diacrylates and even monoacrylates. The "secret" is to continuously add the reactant solution on top of the front as it propagates, maintaining a nearly constant value of L.

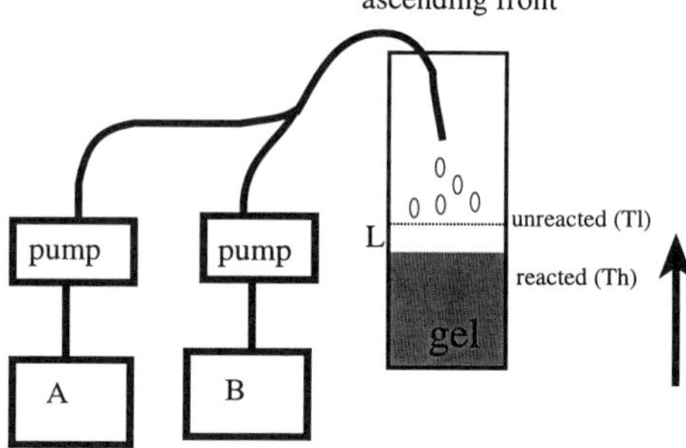

Figure 2. Process for producing FGMs with smooth and controlled gradients.

Figure 2 shows a schematic of the process in which two pumps provide monomers or resins in a ratio that can be controlled.

Experimental

Tri(ethylene glycol) dimethacrylate (TGDMA) and benzyl acrylate were used as monomers. They were stored over molecular sieves, dried over CaH_2 and then filtered before use. Lupersol 231 was used as initiator in experiments where crosslink gradient was produced. Tricaprylmethylammonium persulfate was used as an initiator in experiments producing dye gradients.[28] The main advantage of this initiator is its gasless nature under decomposition. This allowed us to avoid the use of pressure to produce bubble-free optical materials.

Aluminum phthalocyanine chloride was used as a dye. The dye was dissolved in TGDMA to its saturation point, and this solution was used as a coloring component. All chemicals are from Aldrich.

A peristaltic pump was used to supply the reactive media into the test tube in which the ascending front propagated. The inner diameter of the test tube was 12 mm. Test tube, where process of front propagation was accomplished, exposed to ambient pressure and temperature.

Results and Discussion

We have prepared materials with gradients of crosslink density (amount of benzyl acrylate in TGDMA) however, the resulting product contained many bubbles, because of formation of volatile by-products in course of Lupersol 231 decomposition. One possibility to make bubble-free product in this case is to apply pressure, but it significantly complicates the process. Another possibility is to use a gasless initiator. Tricaprylmethylammonium persulfate is such a one, and its usage allowed us to make optically clear materials.

We focused on preparation of gradients of nonlinear optical dyes. The current method for preparation of optically nonuniform materials is isothermal frontal polymerization (interfacial gel polymerization.[29, 30] While this method is rather good it still has many drawbacks in comparison with the novel method proposed. One of them is the difficulty to produce gradients over long distances.[30] Our novel method does not suffer from this limitation. Figures 3 and 4 show a sample of TGDMA with dye distributed along the longitudinal axis over several centimeters. (We can generate any desired nonuniform disitribtion, and the one shown is only an example.)

Figure 3. Dye gradient in a TGDMA. Dark regions correspond to high dye concentration.

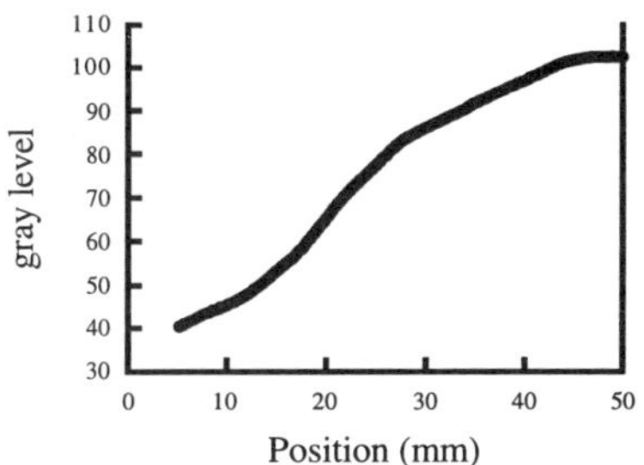

Figure 4. Gray level of along the longitudinal axis of the sample in Figure 3.

Achieving the full potential of this method requires additional experimental and theoretical investigation; however, we can point out its advantages for production of polymeric and composite materials: no need of a pressurized reactor, high throughput, and custom design of the spatial structure of the product.

Potential Applications

Our approach can be applied to preparing gradients or other types of distribution of almost any material in any polymer system that can support frontal polymerization, viz., reactive acrylates, acrylamides in solution and cationically and amine cured epoxies. For example, graded IPNs using binary frontal polymerization[31] could be prepared as well as graded rubber toughened epoxies using frontal curing.[23]

Conclusions

The novel method was developed which allows to design the spatial distribution of different materials in polymers. It was successfully applied for production of optically transparent dye gradient materials. Many advantages of the method were named and the only limitation of its application seems to be the ability of polymer to support the front.

Acknowledgment
This work was supported by a NASA grant (NAG8-1466).

References

(1) Funabashi, M.; Kitano, T. *Sen'i Gakkaishi* **1994**, *50*, 573-580.

(2) Lee, N. J.; Jang, J.; Park, M.; Choe, C. R. *J. Mater. Sci.* **1997**, *32*, 2013-2020.

(3) Murayama, S.; Kuroda, S.; Osawa, Z. *Polymer* **1993**, *34*, 3893-3898.

(4) Johnston, N. J.; Srinivasan, K.; Pater, R. H. *Int. SAMPE Symp. Exhib. (Mater. Work. You 21st Century)* **1992**, *37*, 690-704.

(5) Lipatov, Y. S.; Karabanova, L. V.; Gorbach, L. A.; Lutsyk, E. D.; Sergeeva, L. M. *Polym. Int.* **1992**, *28*, 99-103.

(6) Dror, M.; Elsabee, M. Z.; Berry, G. C. *J. Appl. Polym. Sci.* **1981**, *26*, 1741-1757.

(7) Jasso, C. F.; Hong, S. D.; Shen, M. *Adv. Chem. Ser.* **1979**, *176(Multiphase Polym.)*, 443-453.

(8) Lipatov, Y. S.; Karabanova, L. V. *J. Mater. Sci.* **1995**, *30*, 2475-2484.

(9) Zhang, Q.; Wang, P.; Zhai, Y. *Macromolecules* **1998**, *30*, 7874-7879.

(10) Koike, Y. *Polymer* **1991**, *32*, 1737-1745.

(11) Koike, Y.; Nihei, E.; Tanio, N.; Ohtsuka, Y. *Appl. Opt.* **1990**, *29*, 2686-2691.

(12) Nihei, E.; Ishigure, T.; Koike, Y. *Appl. Opt.* **1996**, *35*, 7085-7090.

(13) Q. Zhang, P., Zhang, Yan Zhai *Macromolecules* **1997**, *30*, 7874-7879.

(14) Crane, R.; Lewis, K.; Stryland, E. V.; Koshnevisan, M. Eds..*Materials for Optical Limiting*; Materials Research Society: Pittsburgh, PA, 1995.

(15) Perry, J. W.; Mansour, K.; Lee, I.-Y. S.; Wu, X.; Bedworth, P. V.; Chen, C.; Ng., D.; Marder, S. R.; Miles, P.; Wada, T.; Tian, M.; Sasabe, H. *Science* **1996**, *273*, 1533-1536.

(16) Miles, P. *Appl. Opt.* **1994**, *33*, 6965.

(17) Chechilo, N. M.; Khvilivitskii, R. J.; Enikolopyan, N. S. *Dokl. Akad. Nauk SSSR* **1972**, *204*, 1180-1181.

(18) Pojman, J. A. *J. Am. Chem. Soc.* **1991**, *113*, 6284-6286.

(19) Pojman, J. A.; Ilyashenko, V. M.; Khan, A. M. *J. Chem. Soc. Faraday Trans.* **1996**, *92*, 2825-2837.

(20) Khan, A. M.; Pojman, J. A. *Trends Polym. Sci. (Cambridge, U.K.)* **1996**, *4*, 253-257.

(21) Pojman, J. A.; Willis, J.; Fortenberry, D.; Ilyashenko, V.; Khan, A. *J. Polym. Sci. Part A: Polym Chem.* **1995**, *33*, 643-652.

(22) Pojman, J. A.; Simmons, C.; Ilyashenko, V. *Polym. Prepr. Am Chem. Soc. Div. Polym. Chem.* **1997**, *38(2)*, 420-421.

(23) Chekanov, Y.; Arrington, D.; Brust, G.; Pojman, J. A. *J. Appl. Polym. Sci.* **1997**, *66*, 1209-1216.

(24) Pojman, J. A.; Curtis, G.; Ilyashenko, V. M. *J. Am. Chem. Soc.* **1996**, *115*, 3783-3784.

(25) Bowden, G.; Garbey, M.; Ilyashenko, V. M.; Pojman, J. A.; Solovyov, S.; Taik, A.; Volpert, V. *J. Phys. Chem. B* **1997**, *101*, 678-686.

(26) McCaughey, B.; Pojman, J. A.; Simmons, C.; Volpert, V. A. *Chaos in press*

(27) Pojman, J. A.; Khan, A. M.; Mathias, L. J. *Microg. sci. technol.* **1997**, *X*, 36-40.

(28) Chekanov, Y. A.; Pojman, J. A. *Polymer Preprints (Boston)* **1998**,

(29) Ivanov, V. V.; Stegno, . V.; Pushchaeva, L. M. *Chem. Phys. Reports* **1997**, *16*, 947-951.

(30) Lewis, L. L.; Pojman, J. A. *Polym. Mater. Sci. Eng. Prep. Div. Polym. Mater. Sci. Eng.* **1998**, *78 (1)*, 38-39.

(31) Pojman, J. A.; Elcan, W.; Khan, A. M.; Mathias, L. *J. Polym. Sci. Part A: Polym Chem.* **1997**, *35*, 227-230.

Lifetime and Property Changes of Kevlar Fiber Yarn under Cyclic Loading

YuanQiao Rao, R. J. Farris
Polymer Science and Engineering Department, University of Massachusetts at Amherst, Amherst, MA 01003

High performance fibers such as Kevlar are often subjected to cyclic fatigue loading. In reality, the material's performance depends upon the properties that evolve during use rather than the initial properties. The understanding of property evolution with time is crucial to the usage of the material and can often be a probe to help understand the material's behavior. Although some efforts have been made to study single filaments of Kevlar under stress fatigue or changing humidity [1-4], the fatigue behavior of Kevlar yarn has not been addressed and no systematic study has been published to elucidate the time dependent property changes during fatigue.

In this paper, commercial Kevlar yarn was chosen as the target material. The material was subjected to a sinusoidal stress of different amplitudes using a servo-hydraulic Instron dynamic tester. The lifetime of the material under different loading condition was recorded and analyzed to give a universal law. The cyclic loading experiment was also stopped at different stages and the sample was then characterized for its thermal and mechanical properties as well as morphology using X-ray diffraction, optical and electron microscopes. The result shows that both structure and property were altered to some extent by cyclic loading.

Experimental

Kevlar 29® fiber yarn with a linear density of 1500 denier was examined in this study. Cyclic load testing of the yarn was performed using an Instron model 8511 hydraulic dynamic tester. The effective gauge length was set at roughly 150 mm. The tabs to which the samples were bonded were held between two sets of grips. During the experiment, the load was programmed to be a sinusoidal waveform. A fixed cyclic minimum load of 2 N was maintained to assure that the specimen was always in tension during the test and a frequency of 5 Hz was used. The cyclic maximum load was varied. During the cyclic loading, temperature, time, load, displacement and total number of cycles were tracked and transferred to a computer. The data were analyzed after the experiment was completed.

General characterizations were performed on single filaments and included linear density measurement using vibroscope, volumetric density by a gradient column and diameter measurement using an optical microscope. Mechanical tensile testing of yarns and single filaments was performed using an Instron model 5564 testing machine. All of the tests were performed at standard conditions of 21°C (+/- 1°C) and 65% (+/- 2%) relative humidity, with a strain rate of 10 %/min. X-ray fiber diagrams were recorded with a Simens 2D area detector using Cu Kα radiation with a beam monochromator. The pinhole size was 200 µm. Thermal mechanic measurements were done on a TA instruments TMA using nitrogen as the purging gas. Fracture topography was studied using an optical microscope (Olympus BH-2) and scanning electron microscope (Jeol JSM- 35CF). For the SEM studies, the fracture surfaces were coated with gold in vacuum.

Results and Discussion

Fatigue lifetime of Kevlar fiber yarn

By varying the stress amplitude from 65 percent to 94 percent of the ultimate strength, the fatigue lifetime of the studied Kevlar fiber yarn was recorded. A linear regression between the stress amplitude and logarithm of the number of loading cycles to failure results in a good fit. The linear curve intersects with stress amplitude axis at a value that is the same as the ultimate strength as shown by figure 1. This is consistence with the proposed lifetime law where the ultimate strength in a tensile test can be treated as the stress amplitude in a fatigue experiment in which the material only has a half

cycle of lifetime. The followed equation is used to separate the material strength from the latent fatigue characteristic,

$$\sigma_a = \sigma_f (1 - m * \ln N)$$

where σ_a is the applied cyclic stress amplitude ($\sigma_{max}.\sigma_{min}$) and N is the number of cycles to failure. Here **m** is called the fatigue strength index and σ_f is the strength in a tensile test. For better fatigue performance, high fatigue strength and a small fatigue strength index is desired.

It is worthwhile to point out that the fatigue strength indices for these fiber yarns are comparable to those obtained by Landgraf [4] who studied the fatigue of hardened steel for small strains. Clearly no so called elastic plastic fatigue transition [5] is observed in the fatigue of Kevlar yarn. The governing law is expected to hold for the whole range of stress amplitude. A model has been proposed based on the observations in this paper to support the governing equation and will be introduced separately.

Characterization of Kevlar fiber upon fatiguing

To elucidate the mechanism of fiber fatigue, retrieved dynamic data at various stages of fatigue were analyzed and different mechanical and structural characterizations have been performed.

Figure 2 shows the modulus of the fiber yarn changes with the number of loading cycles. An interesting observation is that the modulus of the fiber increased by 37 percent* during the initial phase of the cyclic loading and then remained constant. A preliminary study of the property enhancement by cyclic mechanical conditioning was conducted. After subjecting the yarn to a high amplitude fast oscillating stress (1.99 GPa) for 10 cycles, the sample was removed from the tester and allowed to recover at ambient temperature for a day before further measurement. Mechanical testing shows an initial modulus increase of 23 percent and a small increase of 5 percent in strength. Also X-ray diffraction patterns show improved orientation after this mechanical conditioning treatment. This leads to a possible new route of post-treatment of fiber by oscillating stress at elevated temperature. The modulus change upon deformation has been discussed by many researchers [6,7]. Erickson [6] has compared the modulus of the fiber at the earlier stage and later stage of creep and the difference is within 6 percent. This indicates that cyclic loading facilitate the internal structure alternation and supports a model of boundary slip induced chain axis rotation instead of a lattice deformation model. An azimuthal scan of x-ray fiber diffraction confirmed the trend as shown in Figure 3. The full width at half maximum of the peak decreased from 17.4 ° (+/- 1.2) to 13.6 ° (+/-0.6) and then no detectable change upon further cycling.

Based on the above discussion, it is important to distinguish the perfection of the structure from the damage when the material has been subjected to cyclic loading. An energy dissipation calculation has been carried. Instead of trying to integrate the stress-strain curve, an easy way was adopted by fitting the stress-time data and strain-time data into a sinusoidal function and using the phase lag to characterize the energy dissipation. Figure 4 shows that tan δ phase lag initially decreases rapidly with the number of loading cycles and then reach a constant of 0.033 around 10^3 to 10^4 cycles for a cyclic loading with a stress amplitude of 1.64 GPa. This number of loading cycles corresponds to 0.1 to 1 percent of the lifetime of the material and the same trend was found for higher stress amplitude experiments. Also for higher stress amplitude, the plateau tanδ is higher which indicates higher energy dissipation.

The development of the strain at cyclic maximum load for different stress amplitudes is shown in Figure 5. It was found that the fatigue failure occurs at the same ultimate strain as a tensile test. This illustrates that the failure is strain determined. There is a critical strain at which the fiber will fail independent of the stress history. An imbedded strain gauge in a fiber yarn would be a good indicator of the residual material lifetime.

A very intriguing finding is that the thermal shrinkage strain is strongly influenced by the fatigue history. A thermal shrinkage strain measurement was carried on a TMA (TA Instruments) on single filaments removed from the yarn at various stages of fatigue. These measurements are load sensitive and the deformation shrinkage upon heating was determined at essentially at zero (< 0.1 g/den). Figure 6 shows the shrinkage increases with the number of loading cycles at the same stress amplitude. Figure 7 correlates the maximum shrinkage strain, which is the shrinkage strain observed in the plateau region (~ 220 °C), with the number of loading cycles. The origin of the shrinkage strain is still under investigation. It appears to be a very sensitive measure of the stress history of the material.

Optical microscopy and SEM were used to check the topography of the fiber. Figure 8 shows that the characteristic feature of the fatigue-failed yarn is axial splitting which is similar to that observed in a tensile failure.

Conclusion

Unlike most bulk isotropic material, the characteristic behavior of Kevlar fiber yarn under stress controlled fatigue is a competition between structure perfection and fatigue damage. An increase in modulus and strength is observed during the initial phases of fatigue, while further fatiguing shortens the lifetime and strength of the material. The damage inside the material can be characterized by thermal shrinkage strain. A fatigue strength index **m** along with the ultimate strength defines the fatigue behavior of the material and lifetime of the material is determined by an exponential relationship with the stress amplitude.

Acknowledgments

We would like to thank the Toyobo Co. (Japan) and the Dupont Co. (U.S.A.) for financial support, MRSEC at UMass for the use of the facility. Helpful discussions with Dr. S.R. Allen (Dupont Co.) are greatly appreciated.

References

1. L. Kpnopasek and J.W.S. Hearle, J. Appl. Polym. Sci., 21, 2791(1977)
2. M. H. Lafitte and A. R. Bunsell, J. Mater. Sci., 17, 2391(1982)
3. A.R. Bundell, Journal of Materials Science,10,1300 (1975)
4. J.Z. Wang, D.A. Dillard and T.C. Ward, Journal of Polymer Science: Part B: Polymer Physics, 30, 1391 (1992)
5. R.W. Landgraf, Achievement of High Fatigue Resistance in Metals and Alloys, ASTM STP-467,1970, p3
6. W. Hertzberg, Deformation and Fracture Mechanics of Engineering Materials, John Wiley and Sons, p457
7. R.H. Ericksen, Polymer, 26, 733 (1985)
8. S.R. Allen and E.J. Roche, Polymer, 30, 996 (1989)

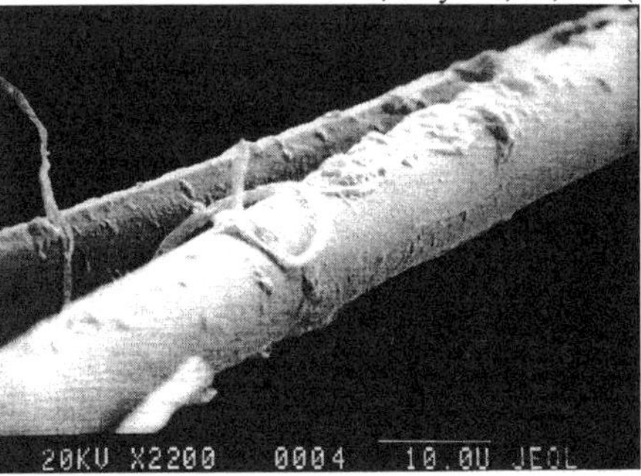

Figure 8. The end of a fatigue-fractured Kevlar fiber

---

* The modulus here was calculated based on the stress range of 0.78 GPa to 1.64 GPa

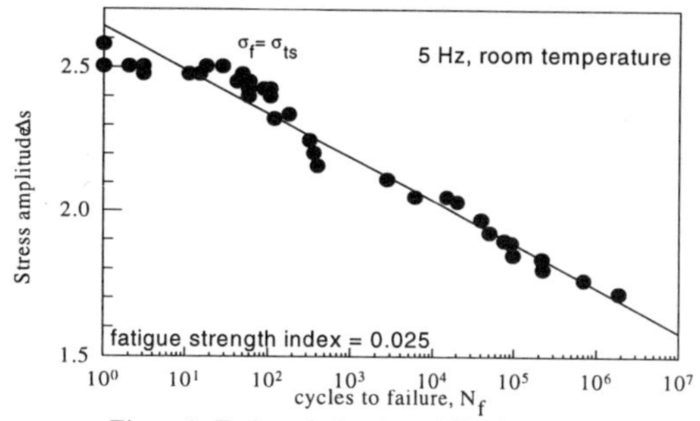

Figure1. Fatigue behavior of Kevlar 29 yarn

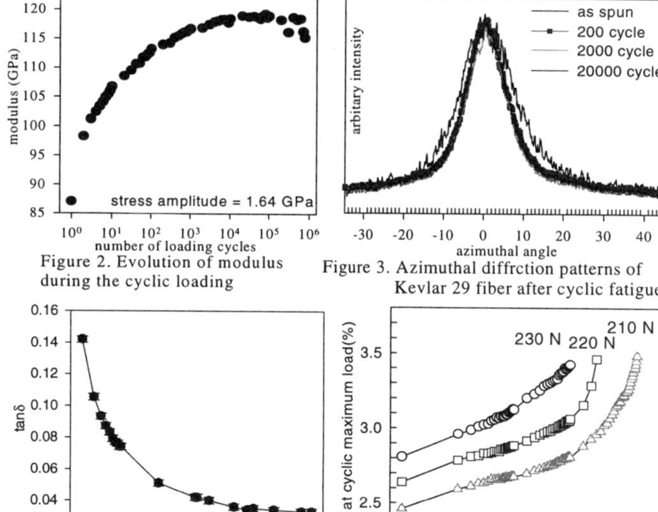

Figure 2. Evolution of modulus during the cyclic loading

Figure 3. Azimuthal diffrction patterns of Kevlar 29 fiber after cyclic fatigue

Figure 4. Phase angle evolution during fatigue

Figure 5. Changes of cyclic maximum strain with loading cycles at different δS

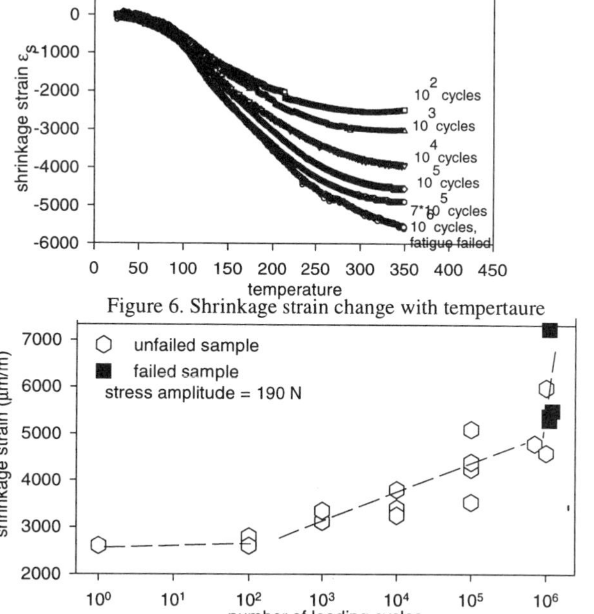

Figure 6. Shrinkage strain change with tempertaure

Figure 7. Maximum shrinkage strain changing with loading cycles

DYNAMIC VULCANIZATION–A MAJOR INNOVATION IN TPEs

Sabet Abdou-Sabet
Advanced Elastomer Systems, L.P., 388 S. Main Street,
Akron, OH 44311-1059

INTRODUCTION

Three long decades have dramatically impacted the development of thermoplastic elastomers[1]. From the late 1950s to the early 1980s, three major materials have been discovered, developed and found significant commercial success. These are thermoplastic polyurethane elastomers (TPU), block copolymers TPEs (copolyester, copolyamide, SEBS, etc.) and more recently dynamically vulcanized rubber/plastic blends (EPDM/PP TPV).

The dynamic vulcanization is a process that embraces the curing of the rubber component of a rubber/plastic blend under mixing conditions generating a fine dispension of small rubber particles in a plastic matrix. Gessler and Haslett[2] discovered dynamic vulcanization in their attempt to improve the impact strength of polypropylene through partial crosslinking of a butyl rubber component. Fisher[3] later claimed thermoplastic elastomer compositions made of ethylene-propylene diene containing terpolymer (EPDM) and isotactic polypropylene through the partial vulcanization of the EPDM phase. Significant improvement in the properties of these blends was achieved in 1978 by Coran, Das and Patel[4] by fully vulcanizing the rubber phase under dynamic shear without affecting the thermoplasticity of the blend. This concept was further improved by Abdou-Sabet and Fath[5] to achieve true rubber like properties and provided further improvement in processability that allowed the successful commercialization of these types of materials.

PREPARATION AND MORPHOLOGY

Dynamic vulcanization involves the melt mixing under shear of a semicrystalline thermoplastic with a soft elastomer plus the vulcanization of the latter. The temperature must be above the melting point of the thermoplastic and sufficiently high to activate and complete the vulcanization in a short time. Any crosslinking system can be used as long as it does not alter the characteristics of the plastic phase. Other ingredients, e.g., plasticizers, processing aids, fillers, etc. naturally used in thermoset rubber stocks are suitable additives which are incorporated for specific property modifications.

Melt mixing of the two polymers prior to vulcanization is the first step in TPV preparation. A uniform blend (on the micron scale) of two immiscible polymers is prepared where the interfacial tension is close to achieve phase separation. The larger the interfacial tension, the two polymers will become grossly incompatible and will not mix. These blends could be made into useful compositions through the incorporation of compatibilizer.[6]

The rubber-like properties and thermoplastic processability of the TPV depend directly on their morphology. The simplest model of this morphology, Figure 1, has the vulcanized rubber particles fully encased in a continuous phase of thermoplastic. As the elastomer/thermoplastic ratio is increased, the thermoplastic between the rubber particles becomes thinner and thinner. To achieve optimum properties, the crosslinked rubber particles should have an average particle size of 1-2 μ. Figure 2 illustrates the effect of particle size on the ultimate stress strain properties.

EFFECT OF CROSSLINKING ON PROPERTIES

The unique properties of the TPV also depend on the degree of crosslinking of the elastomer phase as well as the type of crosslink used. The variation of ultimate tensile strength and oil resistance can be seen in Figure 3 as a function of crosslink density or degree of cure. The type of curative also plays a key role. For EPDM/PP, the use of peroxide requires precautionary measures to prevent the degradation of the PP by the free radical generated. This can be accomplished through reactivity enhancement of the EPDM[7] or the addition of additives[8] to protect the PP. From Table 1, it is quite apparent that the phenolic curatives provide compositions with enhanced rubber like properties superior to those of the sulfur or peroxide. This superiority is clearly shown in oil resistance, compression set and processability.

EFFECT OF PLASTICIZERS

Recently, Ellul[9] demonstrated significant improvement to the low temperature properties through the use of ester plasticizers instead of the conventional process oil normally used. The use of paraffinic oil has a moderate effect on the Tg of both EPDM and PP. The use of isooctyl tallate gave a significant improvement by lowering the Tg much further, Table 2.

CONCLUSION

The combination of improved properties and ease of fabrications gives TPV a distinct niche in the spectrum of available materials for rubber applications. These materials have found wide applications in all market segments of the industrial rubber market except pneumatic tires. The breadth of uses of TPV have been covered in detail by Muhs[10]. Additionally these materials are favored over conventional thermoset due to their ease to recyclibility.

BIBLIOGRAPHY

1. N.R. Legge, Rubber Chem. Technol. <u>62</u>, 529 (1989)
2. A.M. Gessler, W.H. Haslett, U.S. Patent No. 3,037,954 (1962)
3. W.K. Fisher, U.S. Patent No. 3,806,558 (1974)
4. A.Y. Coran, B. Das, R.P. Patel, U.S. Patent No. 3,130,535 (1978)
5. S. Abdou-Sabet, M.A. Fath, U.S. Patent No. 4,311,628 (1982)
6. S. Abdou-Sabet, Y.L. Wang and E.F. Chu, Rubber and Plastic News, p. 20, (November 4, 1985)
7. M.D. Ellul, P.S. Ravishankar, PMSE Symposium, ACS Meeting, Boston (Aug. 23-27, 1998)
8. A. Matsuda, S. Shimizu and S. Abe, U.S. Patent No. 4,247,652 (1981)
9. M.D. Ellul, Plastic, Rubber & Composites Processing and Applications, Vol. 26, No. 3, p. 137, 1997
10. J.H. Muhs, "Handbook of Thermoplastic Elastomers," second edition, B.M. Walker and C.P. Rader, Eds. Chapter 12, Van Nostrand Reinhold Co., New York, 1988.

Figure 1
EPDM/PP TPV Morphology

Table 1
Effects of type of curatives

Property	Sulfur	Dimethylol Alkyl Phenol	Peroxide
Hardness, Shore D	43	44	39
Tensile Strength, MPa	24.3	25.6	15.9
100% Modulus, Mpa	8	9.72	8.07
Ultimate Elongation, %	530	350	450
ASTM #3 Oil Swell, %	194	109	225
Compression Set, 22 hrs/100°C, %	43	24	32
Processing	*	****	****
Product Range	*	****	****

*60/40 EPDM/PP Weight Ratio

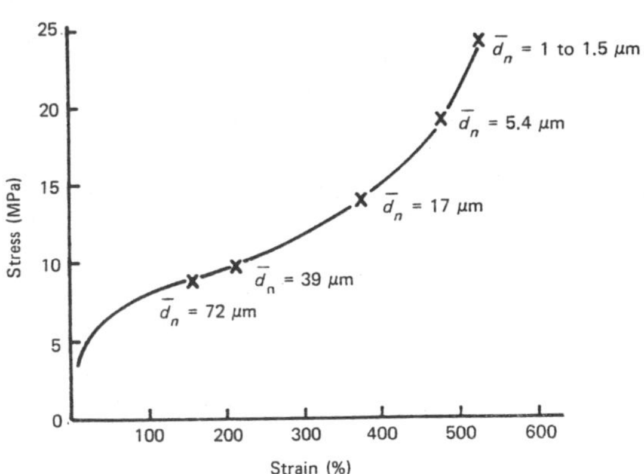

Figure 2
Effect of rubber particle on S/S properties

Table 2
Effect of Plasticizers on Tg, °C

Oil/Plasticizer	Rubber Tg, °C	Plastic Tg, °C
None[a]	−41	10
Sunpar® 150 paraffinic oil	−46	−5
Parapol® 750 polybutene	−47	−5
Cyclolube® 410 naphthenic oil	−51	−5
Amoco 9012 polypropene	−45	−10
Alkylalkylether diester glutamate (Plasthall® 7041)	−55	−11
Dioctyl sebacate (Platshall DOS)	−60	−18
Dioctyl azelate (Plasthall DOZ)	−60	−22
n-Octyl oleate (Plasthall 914)	−71	−24
Isooctyl tallate (Plasthall 100)	−75	−26

[a] EPDM, 100; PP, 219.1; carbon black, 19.28; clay, 40; phenolic curative, 10.5; plasticizer oil, 130.

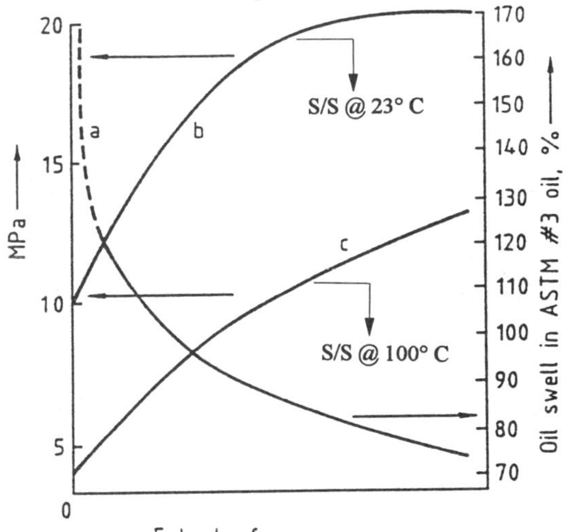

Figure 3
Effects of the degree of cure on properties of TPVs

AN OIL-RESISTANT THERMOPLASTIC VULCANIZATE FROM EPOXIDIZED NATURAL RUBBER AND POLYPROPYLENE

J Patel and A J Tinker
Tun Abdul Razak Research Centre
MRPRA, Brickendonbury, Hertford, SG13 8NL, UK

INTRODUCTION

Thermoplastic vulcanizate (TPV) is one of a number of names given to thermoplastic elastomers comprising a blend of one or more dynamically vulcanized rubber phase(s) and a thermoplastic phase. Although a large number of combinations of elastomers and thermoplastics have been investigated[1], the majority of the materials which have found utility, and hence commercial success, contain a polyolefin, particularly polypropylene (PP), as the thermoplastic component.

Epoxidized natural rubber, ENR, is a chemically-modified form of natural rubber (NR) formed by reaction with peroxycarboxylic acid to give a copolymer comprising cis-1,4-polyisoprene with cis-epoxide groups randomly distributed along the polymer backbone. Whilst a wide range of compositions can be made, ENR is available commercially at modification levels of 25 and 50mole%.

Epoxidation causes several property changes; glass transition temperature and polarity increase for instance, although the ability to undergo strain-induced crystallization is retained largely unimpaired to modification levels of 50mole%. The increase in solubility parameter is modest, from $16.7\text{MPa}^{1/2}$ to $18.8\text{MPa}^{1/2}$ for 50mole% modification, but is sufficient to impart a measure of resistance to swelling by hydrocarbon oils[2].

TPVs based on natural rubber have been known for some time[1,3]. Although these materials have their merits, these do not extend to oil resistance. TPVs based on poly(acrylonitrile-co-butadiene) rubber (NBR) are oil resistant[1] because of the high solubility parameter of NBR when the acrylonitrile content exceeds about 30% ($\delta > {\sim}19.6\ \text{MPa}^{1/2}$). However, the consequent large differences in solubility parameter between the elastomer and thermoplastic components of the blend necessitate the use of compatibilizing agents in order to reduce phase sizes and hence obtain satisfactory properties[1]. Being based on NBR, these TPVs tend to stiffen markedly on ageing at high temperatures.

A TPV has been developed from ENR and PP. Aspects of this material, thermoplastic epoxidized natural rubber (TPENR), revealed by the use of new characterization procedures, and some of its virtues are considered here.

EXPERIMENTAL

Plasticized, dynamically vulcanized blends of ENR (Epoxyprene, Kumpulan Guthrie Bhd) and PP (Propathene, ICI) were prepared in a OOC Banbury internal mixer. Dynamic vulcanization was achieved with a peroxide/coagent cure system.

Rheological properties of the blends were investigated in a Rheograph 1000 extrusion rheometer fitted with a 100x10x1mm slit die equipped with three pressure transducers in the die and one above. True viscosity and shear rate and velocity profiles were calculated by established procedures.

Sheets 2mm and 6mm thick were injection moulded in cavities described previously[3] at melt temperatures of 190-200°C.

Sections about 1μm thick were cut from 2mm thick sheet by using cryo-ultramicrotomy. The sections were stained with osmium tetroxide vapour and viewed by STEM performed on an unmodified, conventional Hitachi S-2700 SEM as described previously[4].

The PP component of TPENR was removed by careful extraction with hot xylene in a reflux apparatus[5]. Samples were prepared for 'network visualization' microscopy[6] and for swollen-state NMR spectroscopy as described previously[6].

Tensile properties and tear strength were measured in accordance with ASTM D412 and D624 on test pieces cut from 2mm thick sheet. Compression set was measured on pellets 12.7mm in diameter and 6.3mm thick in accordance with ASTM D395.

RESULTS AND DISCUSSION

The range of polymer ratios of interest in TPVs is limited at one end by the practicalities of preparing the material with an acceptable phase morphology and processability, and at the other by the modulus or hardness of the material. The former constraint generally imposes an upper limit on elastomer:thermoplastic ratio of about 80:20. The lower limit on the ratio is less well-defined because of the necessary use in practical TPVs of plasticizer, but 65:35 is a reasonable value. The phase morphology achieved in TPVs over this range of blend compositions has been the subject of considerable study and some debate. Whilst TPVs prepared from NR have been said to have a continuous rubber phase throughout this range of compositions, early investigation of TPENR demonstrated a change from a continuous rubber phase to a discrete rubber phase[5]. This conclusion, which is supported here for more recent materials, is drawn not simply from microscopy studies, which have limited utility when in settling the issue of type of phase morphology in these systems, but also from dissolution of the PP phase. At an elastomer:thermoplastic ratio close to 80:20, a sample of TPENR completely retains its integrity on removal of the PP phase, whereas at a ratio of 65:35 the dynamically vulcanized ENR phase is released as swollen, fine particles.

Determining crosslink density in a TPV is not straightforward, but a technique developed originally for the estimation of crosslink density in the individual phases of vulcanized rubber blends[7] has been used successfully to measure crosslink density in TPENR. The technique, swollen-state NMR spectroscopy, requires the correlation between a measure of peak width for a well-defined signal in the spectrum of a swollen vulcanizate and physical crosslink density as measured by an established procedure; estimation of the Mooney-Rivlin constant C1 is used here. The calibration curve for ENR is given in Figure 1. Crosslink densities for two TPENR blends of the same composition, but dynamically vulcanized with different peroxide/coagent combinations, are given in Table 1 together with some pertinent physical properties. It is evident that the lower crosslink density achieved in Blend 1 causes lower modulus and strength and poorer recovery properties. The crosslink density attained in Blend 2 gives a more useful combination of properties.

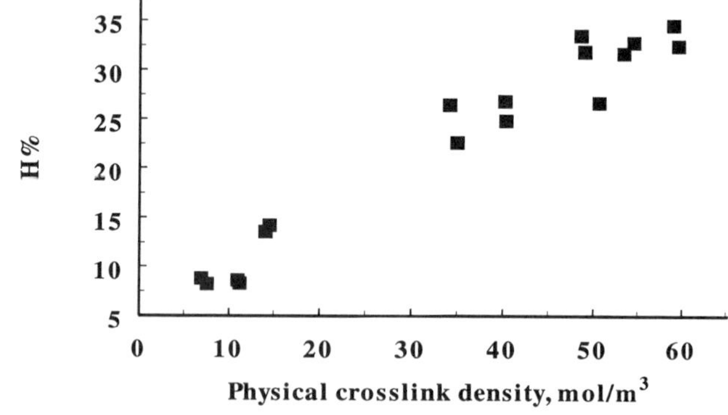

Figure 1   Dependence of olefin signal peak width, H%, in swollen-state ^{1}H NMR spectra of ENR vulcanizates on physical crosslink density.

The crosslinking and phase morphology described above receive confirmation from 'network visualization' microscopy. Following removal of the PP with hot xylene, the xylene swelling the rubber is exchanged for styrene containing inhibitor, peroxide and a small amount of plasticizer. The styrene is polymerized and the resulting composite, which may be described as a semi-IPN, is sectioned (100-150nm) and the sections stained with osmium tetroxide. When sections of similar composites prepared from conventional rubber vulcanizates are viewed in a TEM, a mesh structure is revealed and the average mesh size correlates with M_c - the average molecular weight between crosslinks[8]. For TPVs, the original phase morphology is preserved to a degree, although the space occupied by the PP is now filled with polystyrene and 'phase sizes' are increased. If the rubber phase was particulate, then the spacing of the particles within the composite bears no relation to that in the original TPV. A mesh structure is evident within the rubber region of the composite and an average mesh size may be measured. For Blend 2, the average mesh size is 22.7nm.

Table 1: Effect of ENR crosslink density on physical properties of TPENR

Blend	n_{phys} (mol/m^3)	M100 (MPa)	T.S. (MPa)	E.B. (%)	Comp. set (%)
1	15	3.0	6.0	355	37
2	33	3.7	6.5	240	24

The assertion that TPENR with a high rubber content has a rubber phase which is both continuous and crosslinked to a reasonable degree raises the question of processability. In practice, TPENRs with compositions over the entire range of interest have satisfactory processing characteristics. In common with other TPVs, they approximate to power law fluids with a strong dependence of viscosity on shear rate (Figure 2). The power law indices increase with increasing PP content and are similar to those of PP at high shear rates (Figure 2). The low power indices mean that flow behaviour approaches plug flow - much of the material moves little relative to neighbouring material and flow is concentrated close to the wall of the duct. This form of analysis assumes that the material is homogeneous. In practice, there is a noticeable enrichment of PP within a layer at the interface with the duct wall. This will undoubtedly aid flow and may well explain the concurrence of power law indices noted above.

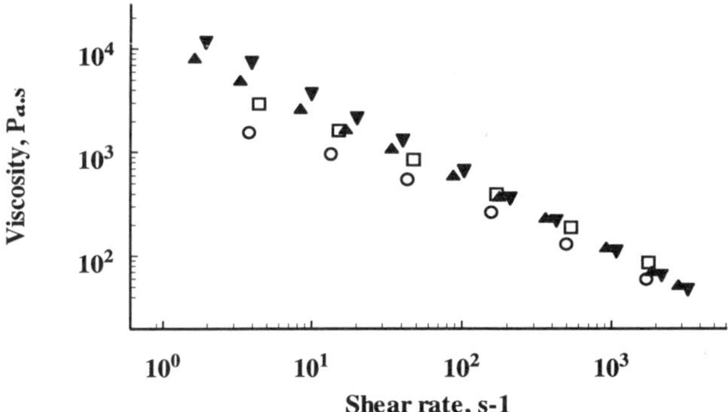

Figure 2   True viscosity as a function of true shear rate for PP, MFI 4 (□) and 13 (○), and for TPENR, ENR:PP 78:22 (▼) and 65:35 (▲).

The physical properties of TPENR are typical of those to be expected for a TPV, although compression set at elevated temperatures is more in line with expectations for a TPV based on EPDM than on a the more highly unsaturated ENR (Table 2). Despite the modest oil resistance of ENR, TPENR has good oil resistance (Table 3) - commensurate with that of a well-compounded vulcanizate based on NBR with a medium acrylonitrile content (33%), but much better heat resistance than the latter. Indeed, resistance to heat ageing is outstanding, both in the long term at 100°C (<30% change in tensile properties after 48 weeks) and in accelerated testing at temperatures up to 150°C (Table 4). Retention of properties after ageing in hot oil is also very good (Table 3). The environmental resistance is completed by resistance to ozone cracking, which is conferred by the PP phase of the blend in common with other TPVs containing highly unsaturated elastomers. The oil and heat resistance of TPENR compare very favourably with published data for a TPV based on NBR[9].

Table 2: Physical properties of TPVs and NBR vulcanizate

Material	M100 (MPa)	T.S. (MPa)	E.B. (%)	Die C tear (N/mm)	Comp. set 1d/100°C (%)
TPENR	3.7	7.1	290	26	36
TPNR[†]	3.7	8.2	365	26	46
EPDM TPV[‡]	4.6	7.0	310	22	32
NBR	3.0	17.2	420	42	17

[†] NR based TPV.   [‡] Santoprene 210-64, Advanced Elastomers L.P.

Table 3: Property retention (%) for TPENR and NBR after 3 days at 125°C

Medium	Air		ASTM #3 oil	
Material	TPENR	NBR	TPENR	NBR
Volume swell (%)	-	-	11	11
M100	101	202	90	117
T.S.	95	122	83	109
E.B.	99	65	78	88

Table 4: Percent retention of properties of TPENR on heat ageing

Condition	15d/125°C	15d/135°C	3d/150°C
M100	111	126	106
T.S.	101	100	88
E.B.	101	80	93

CONCLUSION

New characterization procedures have contributed to the understanding and development of TPENR, a TPV which shows a transition in phase morphology from co-continuous ENR and PP phases to a discrete ENR phase dispersed in a PP matrix over the range of rubber/plastic compositions of interest as TPEs. Flow is believed to be assisted by enrichment with PP at the melt/wall interface. TPENR has physical properties typical of a TPV, but offers a combination of good oil resistance and excellent heat resistance. The oil resistance is much better than that of ENR, being comparable with that available from medium acrylonitrile NBR. Heat resistance is also much better than may be expected for a material based on an elastomer which, despite chemical modification, is still highly unsaturated.

REFERENCES

1 A Y Coran, in "Thermoplastic Elastomers" ed. N R Legge, G Holden and H E Schroeder, Hansar, 1987.
2 I R Gelling, A J Tinker and Haidzir b Ab. Rahman, J Nat Rubb Res, **6**, 20 (1991).
3 A J Tinker, R D Icenogle and I Whittle, Rubber World, 199, 25 (1989).
4 P E F Cudby and B Gilbey, Rubb Chem Technol, **68**, 342 (1995).
5 C L Riddiford and A J Tinker, Paper No. 33, presented at a Meeting of the Rubber Division, ACS, Washington, D.C., October 9-12 (1990).
6 S Cook, P E F Cudby, J Patel and A J Tinker, Paper No 67, presented at a Meeting of the Rubber Division, ACS, Nashville, Tennessee, November (1992).
7 P S Brown, M J R Loadman and A J Tinker, Rubb Chem Technol, **65**, 744 (1992).
8 S Cook, P E F Cudby and A J Tinker, Paper No 3, presented at a Meeting of the Rubber Division, ACS, Nashville, Tennessee, November (1992).
9 S Abdou-Sabet, Y L Wang and E F Chu, Rubber & Plastics News, November 4 (1985).

New Highly-Crosslinked TPEs Based on VNB-EPDM

M.D. Ellul, Advanced Elastomer Systems, L.P., 388 S.Main Akron, OH 44311-1059
and P.S. Ravishankar, Exxon Chemical Co. , 5200 Bayway Drive, Baytown, TX 77520-2101

I. Introduction

VNB-EPDM had previously been suggested by Baldwin and VerStrate[1] for obtaining highly crosslinked EPDM by peroxides. However it was not possible to obtain such a polymer due to incipient gelation as a result of the long chain branching (LCB) by Ziegler polymerization of the pendent double bond, if a diene such as vinyl norbornene (VNB) is used in place of the more common ethylidene norbornene (ENB)[2]. Recently a gel-free process for incorporating VNB in the EPDM backbone was developed, resulting in a polymer with a high degree of **controlled** LCB by Ziegler polymerization of the VNB pendent double bond[3, 4, 5]. This type of LCB is not achievable in ENB-EPDM, whose branching is obtained through cationic coupling of the pendent double bond in ENB. For peroxide crosslinking, it is advantageous to use VNB, since it is a substantially more efficient termonomer, producing high states of cure at low concentrations of both the VNB diene and peroxide. Reduction in the diene content through the use of the reactive VNB termonomer improves the EPDM heat aging because of less unsaturation, while lower peroxide requirements reduces overall compound cost for the end user. Furthermore for TPEs made by dynamic vulcanization of EPDM-polypropylene, the lower the peroxide concentration necessary to produce highly crosslinked TPEs, the less the degradation of polypropylene by random chain scission.

Former attempts in the late 70's by Coran, Das and Patel to produce highly crosslinked EPDM-Polypropylene TPEs via peroxide cure system were only marginally successful with an ENB-EPDM, and a termonomer content of 10 wt% was necessary[6]. Through the use of these new VNB-EPDMs only a low level of VNB termonomer is necessary (<3 wt %). Moreover, it has become possible to make TPEs using lower peroxide levels but with a very high degree of crosslinking. The resulting materials have improved elasticity resulting in up to 50% lower compression and tension set values than those measured in resin-cured TPEs of high crosslink density[7]. Relative to resin-cured TPEs, these novel peroxide-cured VNB-EPDM TPEs have other useful attributes that include, lighter color, low moisture uptake and better chemical resistance.

II. Experimental

Materials

The VNB-EPDMs were specially synthesized by Exxon Chemical Co. Thermoplastic elastomers were prepared by blending the EPDM rubber, a polypropylene, a hydrocarbon oil, a peroxide and a coagent in an electrically heated Brabender mixer of 70 cm^3 capacity, at a mixing speed of 100 rpm and a temperature of 120°C to 180°C. Curing took place *in situ* during mixing. In this process, known as dynamic vulcanization, the rubber is vulcanized during its melt blending with polypropylene in the presence of a diluent, which becomes part of the composition. Morphology typical of these TPEs prepared by dynamic vulcanization has been described previously,[8,9,] i.e. a discontinuous array of rounded and irregular shaped rubber particles ranging in size from 0.2-5 μm in a continuous matrix of polypropylene.

III. Results and Discussion

Effect of EPDM Termonomer/Cure System Type on Set Properties of TPEs

The results of various ENB/VNB, EPDM - Polypropylene TPEs cured by peroxide/coagent (TPEs 1-5), versus a control (resin-cured) TPE analogue of high crosslink density (TPE 6), are shown in Table I. Figure 1 illustrates the much higher peroxide cure efficiency of the VNB-EPDM relative to an ENB-EPDM at similar mole % of termonomer. The ethylene concentration was comparable in both polymers, ~50 wt %. For peroxide crosslinking the preferred free radical attack is on the double bond of the pendent termonomer group in the EPDM. This leads to crosslinking. Alternatively attack could occur on the allylic hydrogen which would result in stable non-reactive free radicals. Since ENB-EPDM has **five** allylic hydrogens relative to **one** in VNB-EPDM it is not as efficient in crosslinking. Also steric considerations favor the VNB EPDM as regards crosslinking.

In a TPE, one of the practical measures of cure state is % set or residual deformation after a fixed deflection for a fixed time at a fixed temperature. It is noteworthy that about 9 weight % of ENB termonomer is necessary in a peroxide-cured TPE (TPE 2) to obtain a compression set comparable to that of a typical phenolic resin-cured commercial TPE of high crosslink density (~85 mol/m^3)[8]. In contrast, a peroxide-cured TPE based on an EPDM with 3 weight % VNB termonomer (TPEs 4 and 4A) results in up to a 50% decrease in compression set (TPE 4A). Higher VNB concentrations in the polymer do not affect further improvement in compression set (TPE 5). It is important to note that all the peroxide-cured TPEs have a lower tension set than the control (resin-cured TPE), indicating improved elasticity. The superior set properties are attributed to an overall higher cure state in the peroxide-cured TPEs, which is reflected in a lower sol fraction (Table II).

Estimates of crosslink densities were also determined using the swollen-state ^{13}C NMR technique described earlier for resin-cured TPEs[7]. A new calibration curve based on thermoset peroxide-cured VNB-EPDMs (Figure 2) was first constructed. By this method the crosslink densities of peroxide-cured TPEs described in Table I were in the range of 80 - 140 mol/m^3 , depending on the termonomer type and concentration, as well as peroxide/coagent level. The engineering properties of all the TPEs are comparable. Oil swell for peroxide-cured TPEs may increase depending on extent of polypropylene scission by peroxide.

Effect of Cure System Type on Chemical Resistance

A conventional resin-cured TPE and a novel VNB-EPDM TPE crosslinked by a peroxide and TAC coagent system were immersed in a strongly oxidizing medium of aqueous chlorine and other chemicals. The rate and amount of mass uptake of the resin-cured TPE were much higher than that of the peroxide-cured TPE (Figure 3). Visual inspection showed progressive deterioration of the former, whereas the latter did not exhibit any cracks or loss of properties in the time frame of this experiment. This is attributed to the much greater resistance to oxidation media of the direct carbon-carbon crosslink achieved in the peroxide-cured TPE.

Effect of Cure System Type on Moisture Uptake

Since peroxide curatives are non-hygroscopic relative to Lewis acid catalysts in resin-cured systems, TPEs crosslinked by the former are essentially non-hygroscopic as shown in Figure 4. This should have an impact in lowering processing costs since the drying step may be eliminated, and it would minimize scrap from moisture issues.

Effect of Cure System Type on Color

Peroxide cure systems are ideal for lighter color in dynamically vulcanized TPEs.3 As shown in Table III, the new TPEs cured by peroxide are substantially lighter in color with no yellow tint arising from the curative, as is common in conventional resin-cured TPEs.

IV. Bibliography

1. F.P. Baldwin and G. Ver Strate, Rubber Chem. Technol., **45**, 709, 1972
2. E.N. Kresge, C. Cozewith and G. Ver Strate, "Long Chain Branching and Gel in EPDM, "Rubber Division, ACS Symposium, IN, 1984
3. N.R. Dharmarajan and P.S. Ravishankar, U.S. Patent No. 5,674,613, Oct.7, 1997.
4. P.S. Ravishankar and N.R. Dharmarajan, "Advanced EPDM Elastomer for Wire and Cable Application", Paper No. 41, Rubber Divion, ACS, Cleveland, Ohio, October 21-24, 1997
5. E.P. Jourdain and P.S. Ravishankar, U.S. Patent No. 5,571, 883, Nov 5, 1996
6. A. Y. Coran, B. Das and R. P. Patel, U.S. Patent No. 4,130,535, Dec, 19, 1978
7. M.D. Ellul, D.R. Hazelton and P.S. Ravishankar, U.S. Patent No. 5,656,693, Aug. 12, 1997
8. M.D. Ellul, J. Patel and A. J. Tinker: Rubber Chem. Technol., **68**, 573, 1995
9. S. Abdou Sabet and R.P. Patel, Rubber Chem. Technol., **64**, 769, 1991

Table I
EFFECT OF TERMONOMER TYPE ON PROPERTIES OF PEROXIDE-CURED TPEs

TPE					1	2	3	4	4A	5	6
Formulation EPDM	ML(1+4) @ 125°C	C2, %	ENB, %	VNB, %	--	--	--	--	--	--	Resin-Cured Control TPE Based on ENB-EPDM
EPDM A	67	48	3.3	--	100	--	--	--	--	--	
EPDM B	90	54	9.0	--	--	100	--	--	--	--	
EPDM C	64*	54	--	1.6	--	--	100	--	--	--	
EPDM D	80*	50	--	3.0	--	--	--	100	100	--	
EPDM E	69*	50	--	4.9	--	--	--	--	--	100	
EPDM F	95	56	4	--	--	--	--	--	--	--	100
Phenolic Resin / SnCl₂	--	--	--	--	--	--	--	--	--	--	5.8
ZnO	--	--	--	--	--	--	--	--	--	--	2
Isotactic PP, MFR = 0.7, MW = 593,000 , MWD = 3.9					32	32	32	32	41	32	41
Paraffinic Hydrocarbon Oil					49	49	49	49	75	49	100
Vulcup40KE (40%) bis(tert-butylperoxy)diisopropyl benzene					1.27	1.27	1.27	1.27	2.36	1.27	--
TAC (50%), Triallyl Isocyanurate, Perkalink®, Akzo					3.3	3.3	3.3	3.3	3.3	3.3	--
Tetrakis[methylene(3,5-di-tert-butyl-4-hydroxyhydrocinnamate)]methane					0.95	0.95	0.95	0.95	0.95	0.95	--
Property	Test Method										
Hardness, Shore A	TPE - 1069	ASTM D2240			60	65	67	68	70	70	65
Stress @ 100%, MPa	TPE - 0153	ASTM D 412			1.90	2.34	2.7	2.86	4.12	2.64	2.64
Tensile Strength, MPa	TPE - 0153	ASTM D 412			4.18	5.77	5.52	5.88	6.45	4.13	5.73
Elongation @ Break, %	TPE - 0053	ASTM D 412			327	338	219	232	197	155	315
Tension Set, %	TPE - 0053	ASTM D 395			8	9	8	8	7	7	11
Compression Set, % 22 hrs @ 100°C	TPE - 0016 (Method B)	ASTM D 395			40	30	27	21	15	21	32

* ML(1+8)

Table II
Effect of VNB and ENB on Cure State of Peroxide-Crosslinked TPEs

TPE	2	3	4	5	7	6
EPDM	B	C	D	E	E	F
EPDM C2, %	52	50	50	50	50	
VNB, %	--	1.6	3.0	4.9	4.9	Resin-Cured Control TPE
ENB, %	9.2	--	--	--	--	
Vulkup 40KE, phr	1.27	1.27	1.27	1.27	2.54	
% Cured Rubber Insoluble in Cyclohexane at 23°C	92	94	96	96	98	95
Hardness, Shore A	65	67	68	68	72	65
Compression Set, % 22 hrs @ 100°C	30	27	21	21	20	32

Table III
Color of Neutral-colored TPEs in Table I

TPE-Type	Color L (black/white)	Color a (green/red)	Color b (blue/yellow)
ENB-EPDM - PP ; resin-cured	71	-2.03	13.83
VNB-EPDM -PP; peroxide-cured	82	-0.95	8.17

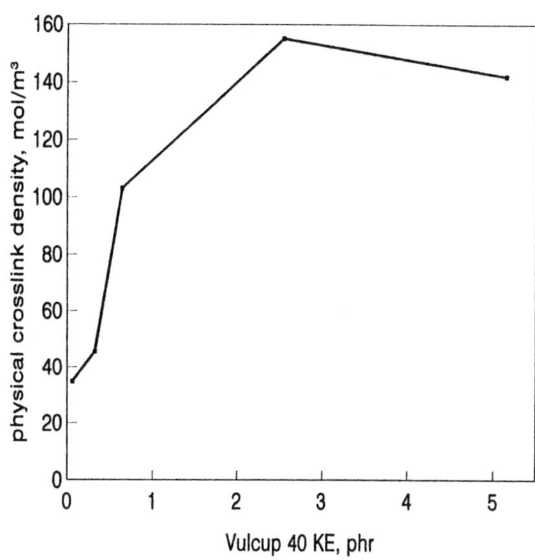
Fig.2: Physical Crosslink Density, determined from stress strain data for a 3% VNB-EPDM vulcanizate, 100 phr, with 3.3phr of TAC(50%) as a function of peroxide (Vulcup 40KE) concentration.

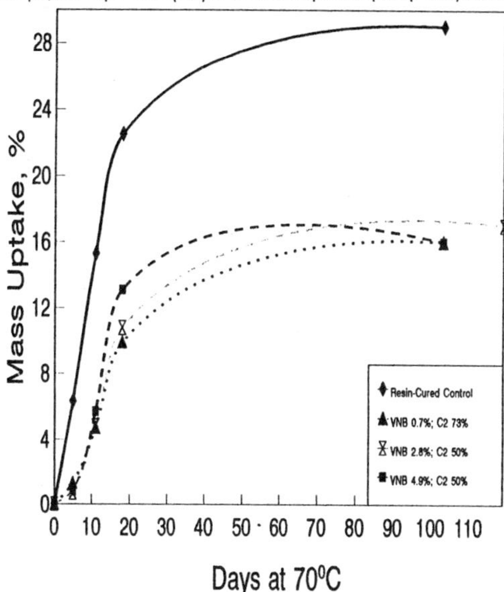

Fig. 3: Chemical resistance of peroxide cured VNB-EPDM Polypropylene TPEs in halogenated methyl hydantoins - Fluid: Clorox Automatic Bowl Cleaner 1.2g/l at 70 C, bent loop samples.

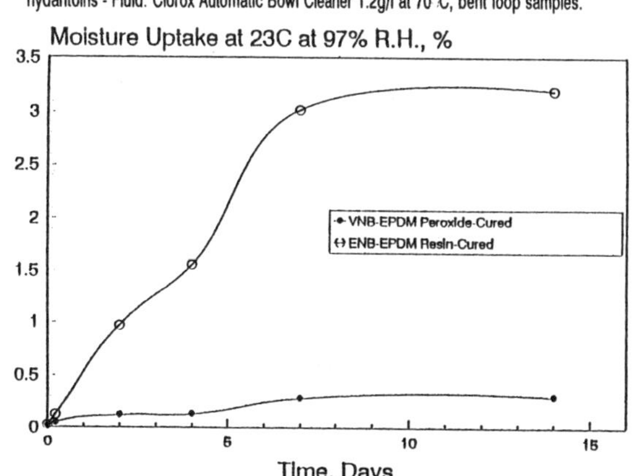

Figure 4: Effect of Cure System Type on Moisture uptake of EPDM-Polypropylene Dynamically Vulcanized TPEs

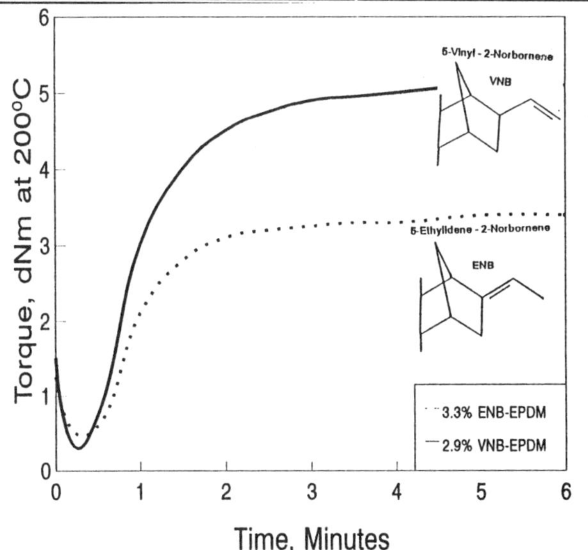

Fig. 1: Cure Characteristics of VNB/ENB EPDM by the Monsanto Moving Die Rheometer at 2000C, Compositions: 100 EPDM, 50 Paraffinic oil, 1.27 Vulcup40KE, TAC (50%), 3.3 .

INNOVATIVE THERMOPLASTIC VULCANIZATES:
FORMULATION, MORPHOLOGY,
PROCESSING AND PROPERTIES

H.G. Fritz[*]
[*]Institut fuer Kunststofftechnologie (IKT)
Universitaet Stuttgart, Boeblinger Str. 70, 70199 Stuttgart

INTRODUCTION

With two-digit annual growth rates the thermoplastic elastomers belong to the most rapidly developing polymeric systems. From this large family a new type of polyolefin blends, so-called thermoplastic vulcanizates (TPV), will be discussed. TPV is a two-phase polymer system that consists of a continuous polypropylene (PP) matrix and a synthetic rubber phase. The latter is in a form of microdisperse dispersion and highly crosslinked [1-6]. At room-temperature these blends show a rubber-like behavior, still maintaining thermoplastic processing characteristics - a significant advantage with respect to recycling.

The innovative steps that lead to materials with enhanced performance comprise [7,8]:

◇ Replacement of EPM or EPDM copolymer by saturated metallocene-catalyzed ethylene/octene-copolymer,

◇ Modification of the elastomeric phase crosslinking by the use of organosilanes, and

◇ Generation of two-phase polymers in an optimized single-stage process using a corotating intermeshing twin-screw extruder.

BLEND COMPONENTS

Synthetic Rubber Phases

The conventional EPM- and EPDM synthetic rubber phases were substituted by the newly developed ethylene/octene copolymers. The metallocene catalysts used for its polymerization generated an outstanding chain topology resulting in advanced material properties. In the experimental studies four types of EO-copolymers produced by the DOW Chem. Co. were used. These are identified in this paper as ENGAGE pelletized grades EG 8200, 8100, 8150 and 8180. They differ in molecular weight and rheological behavior.

Polypropylene Matrix

Composition and structure of the PP matrices to a high degree influence the property profiles of TPVs. PP-copolymers should be selected considering their miscibility with the elastomeric phase. When oil-swelling is a decisive quality criterion, PP-copolymers with a high degree of crystallinity α_K should be selected. The approved formulation window is defined by PP to EO ratios that range from 60:40 to 35:65. These volume concentration ratios $\lambda_3 = \Phi_{EO}/\Phi_{PP}$ have to be adjusted considering the viscosity ratio at constant shear stress $\lambda_1 = \eta_{EO}(\tau_{12})/\eta_{PP}(\tau_{12})$ to avoid the generation of undesired co-continuous phase structures.

Extender Systems

Oil is used to reduce the TPV-hardness. If such extenders are present during the peroxide-initiated grafting of organosilanes, the oil can act as radical scavenger, in consequence, the grafting rates and crosslinking density of the TPV would be reduced, leading to poor property profiles. To eliminate such a possibility, a refined paraffinic oil was selected. It is characterized by reduced volatility and migration rates that make sure that it stays within the TPV.

CROSSLINKING OF EO-COPOLYMERS BY MEANS OF ORGANOSILANES

Chemical Reactions

The conventional crosslinking of the synthetic rubber component by means of peroxides, sulphur or phenolic resins was replaced by silane grafting and crosslinking reactions. Peroxide radicals resulting from thermal multi-stage peroxide decomposition initiate hydrogen abstraction from EO-macromolecules. These free radicals enter into reaction with the vinyl double bonds of the vinyl-tri-methoxy organosilane (VTMOS)

molecules. As a result, these are grafted onto the EO chain, causing a remarkable increase of shear viscosity especially in the low shear rate range. At the processing temperature, in the presence of water and a catalyst (dibutyltin dilaurate or DBTL) the hydrolysis of the VTMOS grafts takes place. The hydrolysis and subsequent condensation lead to formation of stable Si-O-Si bridges, that form a polyfunctional network within the dispersed elastomeric phase.

Formulation Parameters

Depending on the type and concentration of organosilane, peroxide content as well as the molecular weight and molecular structure of the elastomer, high gel content and excellent crosslinking density can be obtained. However, the process must be optimized. VTMOS-concentrations of 2.5 to 3.0 per hundred of resin (phr) and VTMOS/ peroxide-ratio between 15:1 and 50:1 was found to guarantee gel contents up to 96% and high crosslinking density of $35 \cdot 10^5$ to $40 \cdot 10^5$ mol/ml. The high crosslinking density is essential for the reduction of the oil swelling (ASTM-Oil Nr. 3; 70°C/22h) of these two-phase polymeric systems. When the VTMOS peroxide ratio is too low, ≤ 15, undesirable peroxide-initiated pre-crosslinking of the rubber phase can take place, what may cause problems during morphology formation.

THE GENERATION OF TPV-TWO-PHASE POLYMERS

For the generation of the two-phase blends described above, a co-rotating close-intermeshing twin-screw extruder, (ZSK 40 from Werner & Pfleiderer) with a L/D-ratio of 56 was used. The following unit operations take place at different kneader locations:

◇ Prior to feeding, the EO copolymer is coated with the organosilane/peroxide mixture. The mixture is absorbed by the EO pellets, providing optimum pre-distribution of the grafting chemicals.

◇ In the feed zone of the ZSK 40 the EO copolymer melts, the peroxide decomposes, hence free radicals are formed on the EO chains and the grafting reaction takes place. The overall reaction rate is determined by the slowest reaction - the peroxide decomposition.

◇ The PP-copolymer coated with a heat stabilizer (that is also a radical scavenger) and a metal-ion deactivator is fed to the twin-screw extruder by means of a side-stream feeder. After PP-melting a physical blend of the TPV-components is generated. This step is critical for the morphology formation of the two-phase polymer blends.

◇ In the following a crosslinking mixture consisting of water, ethanediol and dibutyltin dilaurate (DBTL) is injected into the blended melt by means of a diaphragm pump. The crosslinking reaction has to take place under the high stresses and deformation rates that continue dispersing the elastomeric phase.

If a set of material and process parameters is fulfilled, a two-phase polymer is generated, where PP is a continuous matrix and the highly crosslinked EO-copolymer is a microdispersed phase of discrete rubber particles with diameters in the range of $0.1 \leq d_p (\mu m) \leq 2$. The formation of a co-continuous phase structure must be avoided, for such blends are no longer remeltable.

PROPERTY PROFILES

TPV-properties are determined by the type and content of the matrix resin (PP), the elastomeric phase (EO-copolymer) and of the extender (paraffinic oil), as well as by the size, size distribution and the crosslinking density of the elastomeric phase. The blend morphology is fixed during the crosslinking step and no variation should take place during the repeated moldings. Thus, to avoid subsequent changes of the blend properties during the molding, the crosslinking reaction has to be finished within the twin-screw extruder.

Mechanical Properties of TPV

For many ranges of TPV-applications (automotive industry, sealing technology) the TPV-hardness represents an important material criterion. Shore-A and Shore-D hardness are determined by the species (copolymer type, degree of cristallinity) and the content of the selected PP, elastomer and the extender system. For the preparation of soft and flexible TPV-types EO-copolymers with high octene content are recommended.

By a systematic variation of the above mentioned parameters (for PP/EO-ratios of 60:40 up to 35:65), a broad spectrum of TPV-types was generated, covering Shore-A values between 60 and 95 SHE. The formulation-dependent ultimate tensile strength values σ_B are closely correlated with TPV-hardness. With increasing hardness the tensile strength progressively increased. As can be seen from Fig. 1, the ultimate tensile strength values of the newly developed TPV-types are higher or identical with those of marketable products (MP1, MP2) based on PP/EPDM-formulations. The corresponding values of elongation-at-break, ε_B, are in the range of 300 to 700%, thus also remarkably high. Using a pre-set test temperature of 70°C and a loading time of 22 h, the compression set values DVR = 30 to 55% were measured. The compression set data increase with increasing TPV-hardness. A drop of gel content and cross-link density causes an immediate deterioration of the elastic material properties and an increase of the compression set. For constant composition of TPV, the DVR-value represents a reliable indicator of the curing process of the elastomeric phase.

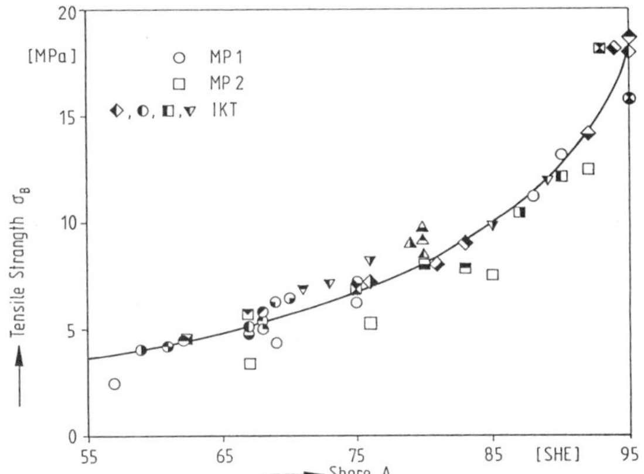

Fig. 1: Ultimate tensile strength of TPV vs Shore-A-hardness

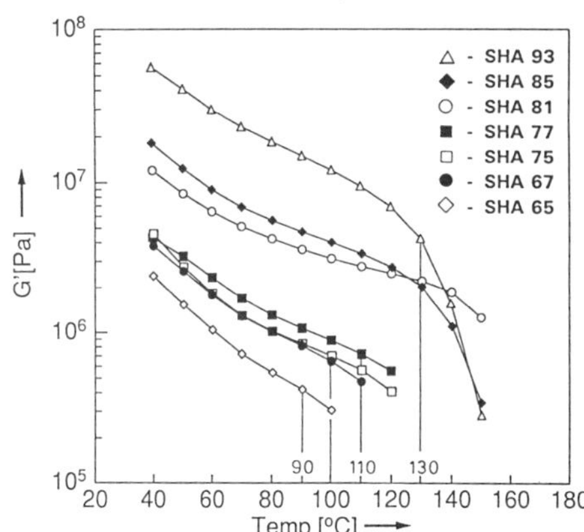

Fig. 2: Storage modulus as a function of temperature for TPV's characterized by different Shore-A values

Thermomechanical TPV-properties

For optimized TPV samples the dynamic storage (G'$_s$) and loss (G''$_s$) shear moduli were measured as a function of temperature in a mechanical spectrometer RMS 800 from Rheometrics Inc. In Fig. 2 G'$_s$ vs T is shown for seven TPV grades, characterized by different Shore-A values. The vertical lines indicate the beginning of material softening. Depending on the formulation (type of polypropylene, concentration ratios, degree of dispersion) the hardness and the softening temperature increase in parallel, e.g., the hardness from SHA 65 to 93, and the softening temperature from 90 to 130°C.

Rheological properties

The rheological properties of a physical PP/EO-blend (containing uncrosslinked EO), and of a TPV having the same composition, are fundamentally different (Fig. 3). The physical blend shows typical non-linear viscoelastic properties. By contrast, in TPV the cross-over point, where G' = G'', disappears. Independently of the frequency ω, the storage modulus is always larger than the loss modulus, G' > G''. The phase difference $\delta(\omega)$ = arctan (G''(ω)/ G'(ω)) is highly reduced indicating a significant increase of the elastic properties. While at low frequencies the complex viscosity $\eta^*(\omega)$ of the physical blends almost reaches the Newtonian plateau, for TPV this function continuously increases, owing to an apparent yield point. In spite of the complex TPV structure the Boltzmann time/temperature superposition principle is approximatively valid. Its applicability is particularly good for the shear viscosity and complex viscosities, respectively $\eta(\dot\gamma)$ and $| \eta^*(\omega) |$.

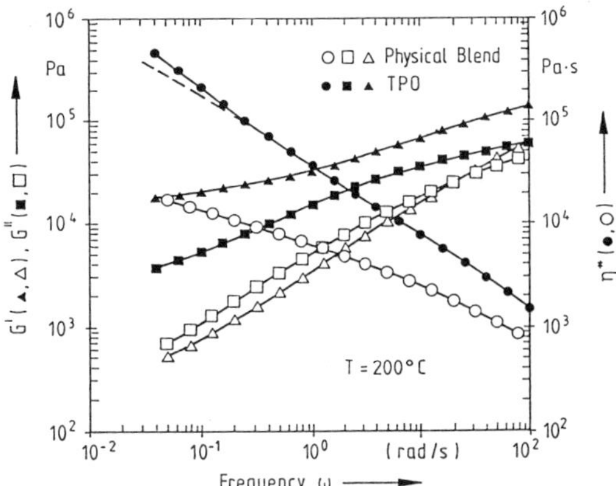

Fig. 3: Rheological behavior of a PP/EP blend and a TPV having the same formulation

SUMMARY

This paper describes the generation of innovative TPV two-phase polymeric systems composed of polypropylene matrix and ethylene/octene copolymer as the dispersed, elastomeric phase. The EO-copolymers are dynamically crosslinked by means of organosilanes. Grafting, hydrolysis and condensation crosslinking reactions are carried out in a single stage process within a corotating intermeshing twin-screw extruder. The resulting TPV-grades are characterized by attractive property profiles due to blend composition and crosslinking concept: processability without material pre-drying, an excellent surface quality, the absence of smell (as a result of the innovative rubber phase and the modified crosslinking concept), an uncomplicated coloring, and reduced cost of ingredients are the main advantages of the newly developed thermoplastic vulcanizates. High ultimate strength (σ_B) and elongation (ε_B) values as well as a broad spectrum of different Shore A grades open up interesting fields of applications. However, the main focus is on automotive industry, not least due to extremely low fogging rates.

LITERATURE

[1] Coran, A.Y., Patel, R.: *Rubber Chem. Technology* 53 (1989), p.141
[2] Hoffmann, W.: *Kautschuk + Gummi, Kunstst.* 40 (1987), p.308
[3] Coran, A.Y., Patel, R., Head, D.W.: *Rubber Chem. Technology* 60 (1987), p.1014
[4] Fritz, H.G., Anderlik, R.: *Kautschuk + Gummi, Kunstst.* 46 (1993), p. 374
[5] Anderlik, R., Fritz, H.G.: *Plaste und Kautschuk* 40 (1993), p. 1
[6] Fritz, H.G., Anderlik, R., Röthemeyer, R. and Lüpfert, S., *DE-OS 14 13*,063 A1, Application date: 22 Apr. 1991
[7] Fritz, H.G., Anderlik, R., Thumm, C., *DE-PA P, 4, 402, 943.8-43*, Application date: 31. Jan. 1994
[8] Fritz, H.G., Cai, Q.: Paper B1 in abstract book of TECHNOMER '95; Chemnitz, 11/16,17/1995

NOVEL THERMOPLASTIC VULCANIZATES BASED ON COMPATIBILIZED PLASTIC PHASE

Krishna Venkataswamy

Advanced Elastomer Systems, L.P.

388 South Main Street

Akron, Ohio 44311

INTRODUCTION

The science and technology of thermoplastic vulcanizates (TPVs) is based on melt blending an elastomer in a thermoplastic and subsequently dynamically vulcanizing the rubber phase[1,2]. The desired morphology of such a multiphase polymer system is finely dispersed, micron sized rubber particles in a continuous plastic phase[3,4]. There are several key variables governing the morphology and hence the resultant elastomer properties, i.e., the type of plastic, the type of rubber, their relative surface energies, viscosity ratios and the extent of cross-linking of the rubber phase. There are other compounding material ingredients such as process oils or plasticizers[5], fillers and additives which can modify either the plastic or rubber phases, or both, resulting in properties meeting specific performance needs.

The majority of the commercially known thermoplastic vulcanizates, as a class of thermoplastic elastomers, are based on polypropylene (PP) and ethylene-propylene diene terpolymer (EPDM)[1,2] rubber. Other elastomers researched for TPVs include polyisobutylene-co-isoprene (IIR), natural rubber (NR), acrylonitrile-co-butadiene (NBR)[2] and brominated paramethyl styrene-isobutylene copolymer[6] (BIMS). In these materials, the polypropylene plastic phase dictates the upper service temperature. The functionality of the polypropylene phase, i.e., heat distortion temperature (HDT) as well as the onset of melting which marks the loss of modulus and the melting point (T_M) of polypropylene determine the TPV temperature functionality. In order to increase the upper service temperature, polypropylene must be modified, or a different plastic phase needs to be considered. In this work, the polypropylene phase of the TPV was compatibilized with polyamide using maleic modified polypropylene to enhance the high temperature functional performance. The resultant materials are novel triblend thermoplastic vulcanizates of polyamide 6, polypropylene, a compatibilizer and a dynamically vulcanized olefinic rubber component[7]. These TPVs bridge the gap between olefinic based TPVs and engineering plastic and polar rubber based TPVs[8-10]. In the current study, the effect of compatibilizer level, compatibilizer type, and olefinic rubber type were investigated. Physical properties and high temperature performance were evaluated.

EXPERIMENTAL

Materials

The materials used for the present study to make thermoplastic vulcanizates included commercially available EPDMs, butyl rubber, BIMS, isotactic polypropylene, and curatives. Commercially available maleic modified polypropylene (mPP), polyamide 6 (PA6) and copolyamide were used to modify the polypropylene phase. The polypropylene and olefinic rubber were melt mixed along with additives and diluents in a Brabender® mixer at 180°C. The rubber phase was dynamically vulcanized by choosing a suitable curative. The completion of dynamic vulcanization was marked by a leveling of the torque. The resultant TPV was compatibilized with polyamide 6 using a maleic modified polypropylene-polyamide copolymer (PA6-mPP) compatibilizer in a separate step or in the same step at a temperature above the melting point of polyamide 6 (225°C). The resultant triblend thermoplastic vulcanizates were compression molded at 230°C.

Methods - Physical Properties and Characterization

The initial tensile properties were evaluated according to the ASTM D-412 method to characterize the mechanical properties of the triblend thermoplastic vulcanizates. Other physical properties included tension set, % compression set (CS) at 125°C (ASTM D-495), and ASTM #3 oil weight gain (OS by ASTM D-471). Dynamic modulus data was obtained using a Rheometrics RDA II dynamic mechanical analyzer to provide information about structural integrity and functional performance at elevated temperatures. Measurements were done at 1 Hz and 0.5% strain in rectangular torsion mode.

RESULTS AND DISCUSSION

PA/PP/ Olefinic Rubber Triblend Vulcanizates

Table 1 illustrates triblends of polyamide 6 with a thermoplastic vulcanizate of PP/BIMS (ratio of 33/100). PP/BIMS TPV when blended with polyamide 6 resulted in a material with poor mechanical and elastomeric properties. Similar Compositions when compatibilized with polyamide 6 - maleated PP (PA6-mPP) show improved elastomeric properties. Addition of polar plasticizers such as Ketjenflex 8 (toluene sulfanomide) to the polyamide phase improves the overall elongation significantly. Effect of PA6 and maleated PP ratio from 60/40 to 80/20 is shown in the Table 1. A PA rich compatibilizer with 80/20 ratio of PA 6 and maleated PP (sample #4) was as effective as a 60/40 compatibilizer (sample #2). Use of copolymer PP and copolyamide (sample #5) results in softer TPE composition.

Table 1: Examples of triblend TPVs

Material	1	2	3	4	5
PA 6	50	25	25	25	25(a)
PP/BIMS TPV	100	100	100	100	100(b)
PA6/mPP (60/40)	-	25	25	-	-
PA6/mPP (80/20)	-	-	-	25	25(a)
Plasticize			8		-
Property					
UTS, psi	1,550	2,580	2,250	2,680	900
%Elong	120	230	360	230	380
M100%	1,550	1,980	1,440	2,060	440
%TS	F	46	38	47	14
OS(22hr, 125C)	—	83	80	69	200
CS(22hr, 125C)	-	61	61	61	45
Hardness	43D	44D	39D	44D	70 A

a) PA is a copolyamide

b) TPV has copolymer PP plastic phase.

Effect of the Olefinic Rubber Type

Table 2 illustrates the effect of olefinic rubber type on the triblend properties. Thermoplastic vulcanizates made from olefinic elastomers such as EPDM (PP/EPDM TPV) and BIMS rubber (PP/BIMS TPV) can be compatibilized to result in triblend dynamic vulcanizates with useful properties. The triblend dynamic vulcanizates made from bromo butyl rubber (PP/Butyl TPV) do not have good elastomeric properties. The reason for this is not clear at the present time.

Table 2: Effect of olefinic rubber type

Material	1	2	3
PA6	10	10	10
PA6-mPP	10	10	10
PP/BIMS	40		
PP/EPDM		40	
PP/Butyl			40
Property			
UTS, psi	2620	2940	1920
%Elong	230	240	78
M100%	2060	2180	1790
% TS	48	40	-
CS, %	59	54	77
OS, %	61	70	72
Shore D	43	46	46

Effect of the Compatibilizer Level

The compatibilizer for these triblend vulcanizates was the reactant product of polyamide 6 and maleated polypropylene. A polyamide rich compatibilizer in the ratio of 80/20 was chosen. The properties of the triblend thermoplastic vulcanizates based on polypropylene/BIMS were studied as a function of compatibilizer level. The tensile properties are plotted as a function of compatibilizer concentration expressed in parts per hundred parts of rubber as in Figure 1. Maximum in tensile strength and percent elongation occur at a compatibilizer level around 30 phr. In these compositions, an equal amount of free polyamide 6 was mixed.

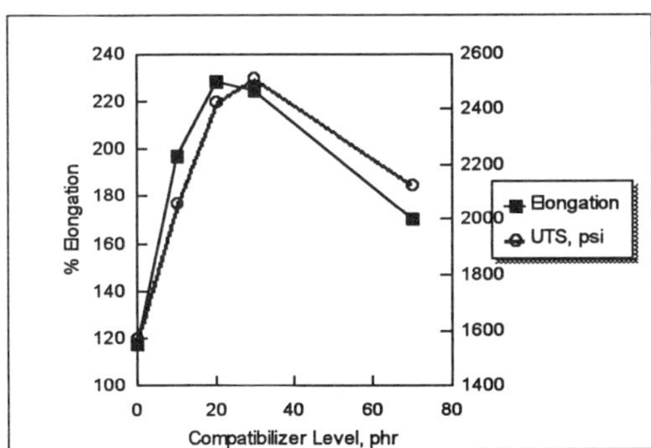

Figure 1. Effect of compatibilizer level

High Temperature Properties

The polyamide 6/polypropylene/olefinic rubber triblend thermoplastic vulcanizates show two melting points as evidenced by differential scanning calorimetry, i.e., one corresponding to polypropylene at 161°C and another corresponding to polyamide 6

at 217°C. The dynamic modulus at 1 Hz as a function of temperature for a PP/EPDM TPV of 40 Shore D hardness was compared against polyamide 6 based triblend TPVs with two hardnesses. PP/EPDM TPV starts to lose its modulus substantially after 135°C. At 150 °C this material looses functional performance. The two samples with compatibilized polyamide phase have good dynamic elastic modulus even after 150°C which is indicative of good functional performance at elevated temperatures.

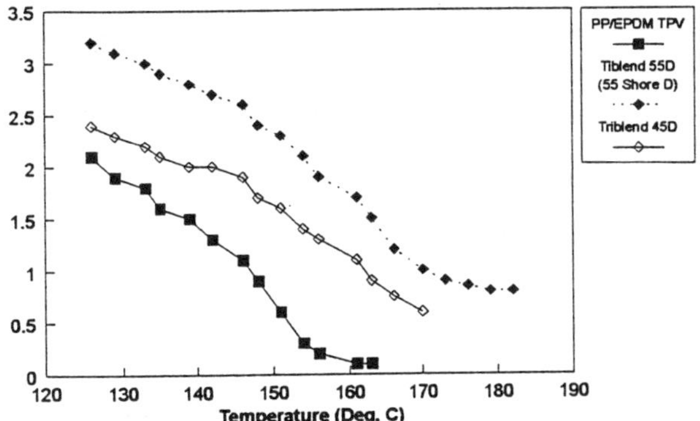

Figure 2: Dynamic elastic modulus as function of temperature

CONCLUSIONS

In thermoplastic vulcanizates based on polypropylene and olefinic rubber, the plastic phase can be compatibilized with polyamide 6 using maleated polypropylene. The thermoplastic vulcanizates based on EPDM and BIMS rubbers result in triblend TPVs which have good elastomeric properties. In this study, an 80/20 reactant product of polyamide 6 and maleated polypropylene serves well as a compatibilizer at an optimum level in the vicinity of 30 phr. The triblend thermoplastic vulcanizates have improved upper service functionality in comparison to the polypropylene based TPVs.

BIBLIOGRAPHY

1. A.Y. Coran and R. Patel, US Patent 4,104,210, 1978.
2. C.P. Rader " *Handbook of Thermoplastic Elastomers, Second Edition"*, B.M Walker and C.P.Rader Editors, Van Nostrand Reinhold, Chapt.4, 1988.
3. A.Y. Coran and R. Patel, *Rubber Chemistry and Technology*, 83, 141, 1980.
4. L.L Ban and K.S.Campo, *Rubber World*, 207, 20, 1993.
5. M.D. Ellul, Plastics, *Rubber and Composites Processing and Applications* 26(3), 137, 1997.
6. H.C.Wang, K.W.Powers, R.C.Puydak and N.R. Dharamarajan US Patent 5, 013,793, 1991.
7. Krishna Venkataswamy, US Patent 5,574,105, 1996.
8. Krishna Venkataswamy and D.S.T Wang, US Patent 5,523,350, 1996.
9. Raman Patel US Patent 5,300,573, 1994
10. A. Jha and A.K. Bhowmick , *Polymer* , 38(17), 4337, 1997

ACKNOWLEDGMENTS

The author would like to thank Advanced Elastomer Systems, L.P for permission to publish this work and Colleen McMahan and Marc Payne for their support.

ANALYSIS OF TPV EXTRUSION CHEMICAL FOAMING PROCESS

Yundong Wang, Hua Cai, Louis Freitas, Bob Dion, and Ryszard Brzoskowski, DSM Thermoplastic Elastomers, Inc., 29 Fuller Street, Leominster, MA 01453

INTRODUCTION

TPV extrusion foaming technology has been explored extensively in recent years[1-6]. Conventional chemical foaming systems were found to be capable of reducing the specific gravity of TPVs to 0.7. The use of physical foaming agents or a combination of physical and chemical foaming agents could further reduce the specific gravity to a level below 0.7. The use of water as a physical foaming agent was found to yield TPV foams with a specific gravity as low as 0.2[7-12]. Recently, DSM Thermoplastic Elastomers introduced a new chemical foaming technology which uses a water releasing compound (WRC) as a blowing agent for producing low-density foamed TPVs[13-15]. Extrusion foam profiles can be produced on conventional single screw extruders without additives or complex injection systems. The water releasing compound which is pre-compounded into Sarlink® foamable TPV composition decomposes and releases water in the extruder at temperatures above its decomposition temperature. The released water expands and foams the TPV melt into lower densities at the die exit. In this paper, we present some analysis on extrusion foaming process using Sarlink foamable compound containing WRC.

SARLINK® EXTRUSION CHEMICAL FOAMING PROCESS

The composition of Sarlink foaming grade consists of two main ingredients: (1) Sarlink TPV with specially designed rheological properties and (2) water releasing compound (WRC). The WRC is in the form of a filler which has a proper particle size and particle size distribution. The WRC is uniformly distributed within Sarlink TPV.

Sarlink low density foaming compound is fed to the extruder in a steady-state fashion through the hopper. The pellets are conveyed and melted in a way typical for standard extrusion process. The barrel temperature settings are different from standard extrusion operation and the Sarlink foaming compound is exposed to a "bell shape" temperature profile (low, high, high, low). The temperature setting in the middle part of the barrel is high enough to initiate water generation from WRC. The amount of the released water depends on the actual melt temperature and the residence time at that temperature. Conceptually, polymer melt can be viewed in this section as a multi-phase system consisting of many small isolated micro-bubbles suspended in the TPV melt. Each micro-bubble consists of a solid particle of WRC surrounded by a thin layer of water. The amount of water surrounding each WRC particle depends on thermal history of that particle. Some of these water layers can be broken during the mixing process by shear and elongational stresses into separate micro-bubbles and moved away from WRC. The above described water generation mechanism combined with the mixing mechanism in single screw extruder produces a polymer matrix which contains a high concentration of uniformly distributed micro-bubbles filled with water.

Water/TPV melt mixture has unique characteristics. It is different from conventional chemical foaming or physical foaming systems. In the conventional systems, the gas generated by the foaming agents usually has some degree of solubility in the liquid or polymer melt. With the Sarlink extrusion foaming system, the water or steam is expected to have fairly low solubility in the TPV melt. Therefore, the foaming agent remains mainly in a separate phase as micro-bubbles.

Since the micro-bubbles filled with water already exist in the system the water/TPV melt mixture will surpass the bubble initiation process and start to foam when the pressure is below a critical point. The extrudate swells rapidly right after exiting the die. The rate of swelling depends on the bubble growth rate and the initial nucleation density (number of micro-bubbles per unit volume of melt). Higher nucleation density results in higher rate of swelling, smaller cell size, and higher cell density. The driving force for bubble growth is the pressure difference between the internal vapor pressure and the ambient pressure. The bubble growth process is mainly controlled by the initial bubble size, the amount of water in each micro-bubble, the rheological properties of the TPV melt (viscosity, elasticity and interfacial tension), and the rate of pressure drop.

To prevent premature foaming of the water/TPV mixture inside the die, the pressure level in the die should be always kept high enough. The pressure of steam (vapor) in equilibrium with water (liquid) increases exponentially with the increases of temperature[16]. Therefore, the pressure required to keep the water in liquid form in the extruder and in the die increases with the increase of the melt temperature.

Premature foaming inside the die will destroy or damage the cell structure resulting in higher foam density and bad foam surface. In principle, the bubble growth process should be postponed to a point as close to the die exit as possible. This can be achieved by designing the die with a sharp converging angle and short land length. Lowering the water/TPV melt temperature will increase the viscosity of the melt, therefore increase the overall pressure level in the die, and thus postpone the bubble growth process. After enough water is released from WRC in the middle of the extruder at high temperature, the melt mixture needs to be cooled down sufficiently by a low barrel temperature at the last zone of the extruder and by a low die temperature. However, the die temperature cannot be set too low as it may result in rough foam surface. The die temperature needs to be set at a level low enough to generate enough pressure in the die and high enough to ensure a smooth foam surface.

In general, higher throughput rate gives higher shear rate and higher overall pressure level in the die. Another benefit with higher shear rate is the possibility of having increased nucleation density as explained by Lee[17]. Higher nucleation density will result in smaller cell size and higher cell density. In practice, higher flow rate or throughput rate can be achieved by increasing the screw speed. However, higher screw speed may increase the degree of shear causing the melt temperature to increase which will in turn reduce the viscosity of the melt and lower the overall level of pressure in the die. In addition, higher screw speed may reduce the residence time of the compound in the extruder thus reducing the amount of water released from WRC. The use of screws such as a low work barrier screw increases the throughput rate while maintaining a low melt temperature. These kinds of screws can be used in Sarlink extrusion chemical foaming process. For profiles with large cross sectional areas, larger size extruders need to be used to generate high enough throughput rate at a relatively low screw speed and maintain long enough residence time of the compound in the extruder to generate enough water from WRC. A detailed description in regard to the size of the extruder needed for the production of profiles of certain size can be found in references[13].

CONTROL OF FOAM DENSITY

Temperature and Residence Time

With a proper die design and machine set-up, TPV foam profiles can be produced on a conventional single screw extruder using a TPV composition containing WRC. The amount of water released from WRC is mainly controlled by the temperature settings of the extruder and the residence time of the compound in the extruder (throughput rate). The physical properties of the foams are dependent upon several factors including the durometer of TPV, cell structure, and foam density. The following examples show the basic concept of Sarlink low density chemical foaming technology.

A Sarlink chemical foaming material with a durometer of 75A was fed to a 63.5 mm single screw extruder of 24:1 L/D at a rate of 22.3 kg/hr. A low work barrier screw and a circular die with a diameter of 4.4 mm were used in the following examples.

As shown in table 1, a series of different foamed samples was obtained by simply changing the temperature settings at the second and third zones of the extruder. The higher the temperatures at the second and third zones were set, the lower the density of the foam sample became. The foamed samples were found to have fine and uniform cell structure with a smooth surface. In general, foams with finer cells are expected to be softer than foams with larger cells at the same density level[18]. Foams with finer cell structure and thin cell walls buckle more easily and therefore have lower initial compression modulus[19].

Take-up Speed

To determine the effect of take-up speed on foam density, a series of foaming experiments was performed on a 38.1 mm three zone single screw extruder with a general purpose screw having a L/D ratio of 24:1. A 2.4 mm capillary die with a 45° entrance angle and a L/D ratio of 1:1 was used in the experiments. The same foaming compound as presented above was also used in the following foaming trials.

As shown in table 2, the density of the foams was found to increase with the increase of take-up speed or draw down ratio. In this experiment, all the conditions were kept the same except for the take-up speed. The increase in density could be due to the following factors: (1) Increased cooling efficiency of the foamed strand at higher draw down ratios, (2) Increased hoop stress, therefore causing bubbles to grow to a lesser extent and (3) Increased percentage of bubbles that collapse during the take-up operation.

To study the effect of draw down ratio on the physical properties of the foams, another series of experiments was performed on the same extruder using the same die with the aim of producing foams of same density but different cell aspect ratios. Three foamed samples with similar densities but different draw down ratios were produced by adjusting the temperature setting at barrel #2 and the take-up speed. The cell aspect ratio was determined using an optical microscope and an image analysis software. Here the cell aspect ratio is defined as the ratio of the length of the long axis to the length of the short axis of the cell. As shown in table 3, the cell aspect ratio increases as the take-up speed increases. The cells are elongated in the direction of stretching. The tensile strength of the foam increases and the elongation at break decreases in the direction of drawing with the increase of draw down ratio or cell aspect ratio. As expected, anisotropic cell structure may result in higher compressive strength in the direction of cell orientation and lower compression load in the direction perpendicular to the direction of cell orientation. This could be a benefit for some sealing applications where lower compression load is needed.

CONCLUSION

With DSM's new chemical foaming technology, low density foamed TPV profiles can be produced on conventional single screw extruders without additives or complex injection systems. The density of the profile can be easily adjusted by changing the barrel temperatures and screw speed. The resulting foamed TPV profiles have uniform, fine closed cell structure with a smooth surface. The cell structure, density and physical properties of the foams can also be affected by other foaming conditions such as take-up speed. Higher take-up speed produces foams with higher density and elongated cell structure. Foams with elongated cell structure have higher tensile strength and lower elongation at break in the direction of elongation compared to the foams of the same density with spherical cell structure.

REFERENCES

1. Dumbauld, Garry L. U.S. Patent 5 070 111, 1991.
2. Shen, K. S.; Lawrence, G. K.; Peterson, D. E. *SPE ANTEC*, **1991**, 2420-2423.
3. Peterson, D. E.; Arnold, R. L.; Dumbauld, G. L. *SPE ANTEC*, **1992**, 305-309.
4. "Extrusion Water Foaming of Thermoplastic Vulcanizates", Presented by Kerkimis, Anthony N. and Kenens, Leander at the Regional Polymer Processing Society Meeting, Akron, OH, November 14-16, 1995.
5. Deseke, Otto; Meyke, Joachim; Pfeiffer, Armin, EP-A-664 197, 1995.
6. Petrakis, J.; Hoge, W.; Leenaerts, T.; Alfenaar, M.; Nieukamp, J. G. M. WO-A-95/04775, 1995.
7. DeMello, Alan J.; Hartford, Douglas W.; Mertinooke, Peter E.; Muessel, Dan C.; Halberstadt, Louis, WO-A- 92/18326, 1992.
8. Mertinooke, Peter E.; Perry, Joseph V.; Halberstadt, Louis; Muessel, Dan C. EP-B-0 369 748, 1994.
9. DeMello, Alan J.; Hartford, Douglas W.; Mertinooke, Peter E.; Muessel, Dan C.; Halberstadt, Louis, Canadian Patent 2 102 227, 1996.
10. Halberstadt, Louis; Mertinooke, Peter E.; Perry, Joseph V.; Muessel, Dan C.; US Patent 5 512 601, 1996.
11. DeMello, Alan J.; Hartford, Douglas W.; Mertinooke, Peter E.; Muessel, Dan C.; Halberstadt, Louis, US Patent 5 607 629, 1997.
12. Kenens, L. In *Proceedings of Foamplas'97*, Mainz, Germany, November 4-5, 1997; pp 209-218.
13. Brzoskowski, R.; Wang, Y. D.; La Tulippe, C.; Dion, B.; Cai, H.; Sadeghi, R. *SPE ANTEC*, **1998** (in press).
14. Niemark, R. In *Proceedings of Foamplas'97*, Mainz, Germany, November 4-5, 1997; pp 117-131.
15. Wang, Y. D. In *Proceedings of Foamplas98*, Teaneck, New Jersey, USA, May 19-20, 1998.
16. *CRC Handbook of Tables for Applied Engineering Science*, 2nd ed.; Bolz, R. E.; Tuve, G. L., Ed.; CRC Press, Inc., Boca Raton, Florida 1980; pp 24-25.
17. Lee, S. T. *SPE ANTEC*, **1991**, 1304-1307.
18. Bailey Jr., F. E. In *Handbook of Polymeric Foams and Foam Technology*; Klempner, Daniel and Frisch, Kurt C., Ed.; Hanser Publishers, Munich, Vienna, New York, Barcelona 1991; Chapter 4.
19. Skochdopole, R. E.; Rubens, L. C. *J. Cell. Plast. 1*, **1965**, 91.

Table 1. Effect of extruder temperature settings on foam density

Examples	I	II	III
Screw speed, RPM	14	14	14
Temp. at the 1st zone, °C	191	191	191
Temp. at the 2nd zone, °C	238	254	260
Temp. at the 3rd zone, °C	260	274	285
Temp. at the 4th zone, °C	166	166	166
Die temperature, °C	180	180	180
Foam density, kg/m^3	450	300	200
Foam diameter, mm	8.3	9.3	9.1
Tensile strength of the foam, MPa	1.43	0.96	0.68
Elongation at break, %	386	310	167
Compression set, % 22 hr@70°C (50 % deflection) 30 min recovery	32.2	35.3	39.0

Table 2. Effect of take-up speed on foam density

Take-up speed, m/min	Foam density, kg/m^3
6.1	270
7.6	290
9.1	340

Table 3. Effect of cell aspect ratio on physical properties of the foams

Take-up speed, M/min	Foam density, kg/m^3	Cell aspect ratio	Tensile strength at break, MPa	Elongation at break, %
4.6	318	1.5-2.5	1.08	234.1
7.6	300	2.5-4.7	1.13	105.5
9.1	305	3.5-5.5	1.25	83.4

Kinetics of Vulcanization-Induced Phase Separation in Blends of Syndiotactic Polypropylene/Ethylene Propylene Diene Terpolymer

T. Kyu, A. Ramanujam and K.J. Kim, Institute of Polymer Engineering, University of Akron, Akron, OH 44325.

Introduction

In recent years, thermoplastic elastomers (TPE) have gained considerable interest in industry and academia[1]. One of the most commercially successful thermoplastic polyolefins (TPO) includes blends of thermoplastics such as polypropylene (PP) and olefinic elastomers such as ethylene propylene diene terpolymer (EPDM) because of in-situ crosslinkability afforded by the elastomeric component and melt processability offered by the thermoplastic component[1]. These PP/EPDM melt blends are perceived to be immiscible[1]. This perception has changed when a lower critical solution temperature (LCST) type phase diagram was found in the isotactic-PP/EPDM blends[2]. A small miscibility gap existed below the LCST, but above the crystallization temperatures of i-PP in the blends. However, the mutual interference of the crystal melting transition and the LCST makes the study on kinetics of reaction induced phase separation difficult[2]. It is necessary to decouple the melting and phase separation in order to prove the existence of the LCST and to make the kinetic study easier. Syndiotactic PP has been sought for the above purposes because the melting temperature of s-PP is about 40 °C lower than that of i-PP, while their miscibility with EPDM may not change appreciably.

In this paper, phase diagrams of blends of s-PP/EPDM have been established by means of differential scanning calorimetry, light scattering, and optical microscopy. The blends exhibit a lower critical solution temperature (LCST) about $25 \sim 35$ °C above the crystal melting temperature (T_m) of s-PP. Temporal evolution of structure factors and/or phase separated domains in these blends have been investigated by time-resolved light scattering and optical microscopy following several temperature jumps from a single phase into the LCST immiscibility gap.

Vulcanization was also performed using curing agents such as phenolic resins and dicumyl peroxide at several temperatures in the single phase between the LCST and T_m of s-PP in the blends. The cross-linking reaction of EPDM by the curing agent induces instability to the blends resulting in phase separation. The effect of curing agent amount, curing temperature, and composition on kinetics of vulcanization induced phase separation have been investigated. These temporal evolutions of structure factors have been analyzed in the context of nonlinear reaction-diffusion and dynamical scaling laws. Tensile properties of these blends are evaluated in relation to the domain structures.

Experimental

s-PP was kindly supplied by Fina Chemical and Oil Co. ENB-type EPDM was obtained through the courtesy of Exxon Chemical Co. s-PP and EPDM were dissolved separately in hot xylene (at 150 °C) at a polymer concentration of 2 wt%. These polymer solutions were mixed in desired proportions. The solutions were spread onto glass slides and thin films were formed by evaporation (~ 10 μm thick) for optical and light scattering studies. As for DSC studies, the solutions were poured into Petri dishes and evaporated in a fume hood. The as-cast films and precipitates were washed in non-solvent (ethanol) for several times to remove residual xylene. The samples were vacuum-dried for 2 days.

Results and Discussion
Phase Diagram

Figure 1 shows the phase diagram of s-PP/EPDM blends. In the descending order of temperature, the phase diagram displays an LCST in the melt, depression of melting transitions (T_m) of s-PP in the blend, an upper critical solution temperature (UCST) intersected by the crystallization temperature of s-PP and dual glass transition temperatures of the constituents. The observed LCST is thermally reversible. A single phase region was found to locate between the LCST and the T_m of s-PP. The melting point of s-PP is lowered upon adding EPDM, indicative of miscible character of the pair. With continued cooling, a UCST seemingly appears when one of the components crystallizes. The unambiguous confirmation of the UCST was by no means easy because of the influence of s-PP crystallization. The evidence comes from the thermal quench experiments at high EPDM compositions (e.g. 10/90 s-PP/EPDM) where phase separation occurs without any significant influence of the s-PP crystallization. The spinodal like structure develops in all T quenches as well as during annealing near the UCST. Another evidence comes from the revelation of two distinct Tg's corresponding to the constituents.

Thermally Induced Phase separation

Several temperature jumps were undertaken from a single-phase (140 °C) to the two-phase immiscible gap above the LCST. The evolution of structure factor was monitored by time-resolved light scattering using one dimensional silicone diode array detector (Reticon Camera, EG & G). Figure 2 shows the temporal change of scattering wavenumber maximum for the 50/50 blend at several T jumps. The scattering peak moves to a low scattering angle suggesting the lack of the early stage of spinodal decomposition. The power law (growth) exponents were found to vary from 0.2 to 0.3, characteristics of the intermediate to late stages of growth. These values are in agreement with those of the classical coarsening dynamics.

Vulcanization Induced Phase Separation

The 50/50 s-PP/EPDM were cured using phenolic resins at several reaction temperatures. The amount of phenolic resins used was 10 phr of the EPDM. During curing, the molecular weight of EPDM increases, which in turn induces instability to the system, resulting in phase separation. The curing entails crosslinking of the EPDM chains into three-dimensional networks. The free energy arising from the network formation can drive the system to become unstable. The phase separation due to curing may be called hereafter vulcanization induced phase separation (VIPS). Figure 3 exhibits the log-log plots of the scattering maximum versus curing time.

Several features are noteworthy. (i) At low reaction temperatures within the single phase, the onset of growth is apparently delayed due to the slow reaction time. The growth rate appears to be subtle. It is tempting to speculate that the phase separation at low reaction temperatures may be driven by the UCST and vice versa the LCST because of the proximity of the reaction temperatures to these coexistence curves. (ii) At high curing temperatures within the single phase, the reaction rate is expedited relative to the phase separation dynamics. Hence the two-phase structure is fixed very rapidly, thereby preventing the domain growth. It is reasonable to conclude that increasing reaction temperature reduces the average domain size. We are astounded by the fact that the curing at two-phase temperatures can even give smaller domain sizes.

Mechanical Properties

Figure 4 shows the variation of tensile properties as a function of s-PP concentration in the s-PP/EPDM blends. In general, modulus and tensile strength of TPE increase with increasing thermoplastic content, but elongation at break is reduced. In the present case, the elongation at break of $50 \sim 90$ % s-PP show higher extensibility than s-PP. It is interesting to notice that the tensile strength of the high s-PP contents ($70 \sim 90$ %) turn out to be higher than that of the neat s-PP. The increase of modulus is more gradual with increasing s-PP content, but the 90 % specimen shows a higher modulus than the pure s-PP due to greater molecular orientation afforded by the higher extensibility. The toughness of the materials, as estimated from the area under a stress-strain curve corresponding to the energy required breaking the sample, is higher in the s-PP rich blends relative to the neat s-PP, characteristic of toughened plastics.

References

1. A.Y. Coran in "*Handbook of Elastomers: New Developments and Technology,*" A.K. Bowmick and H.L. Stephens Eds., Marcel Dekker, New York, 1988, p 249.
2. C.Y. Chen, M.Md.Z.W. Yunus, H.-W. Chiu, and T. Kyu, Polymer, 38, 4433 (1997).

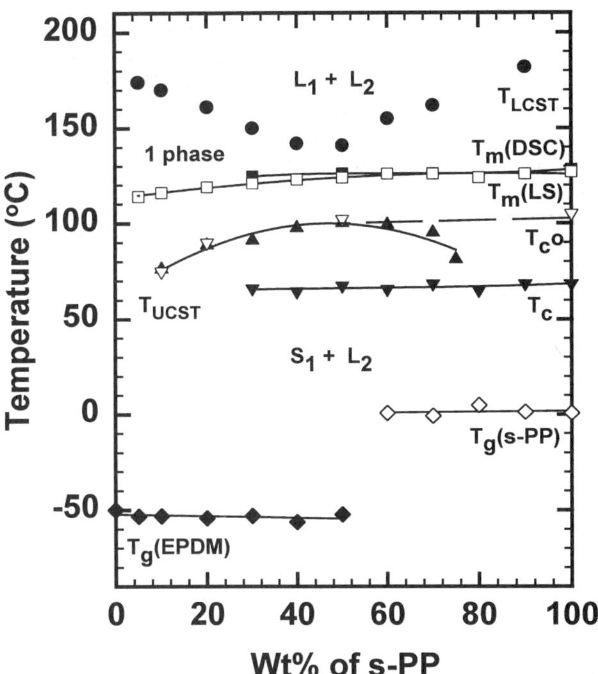

Figure 1. Phase diagram of s-PP/EPDM blends displaying lower critical solution temperature, melting transition, crystallization overlapping with upper critical solution temperature and glass transitions.

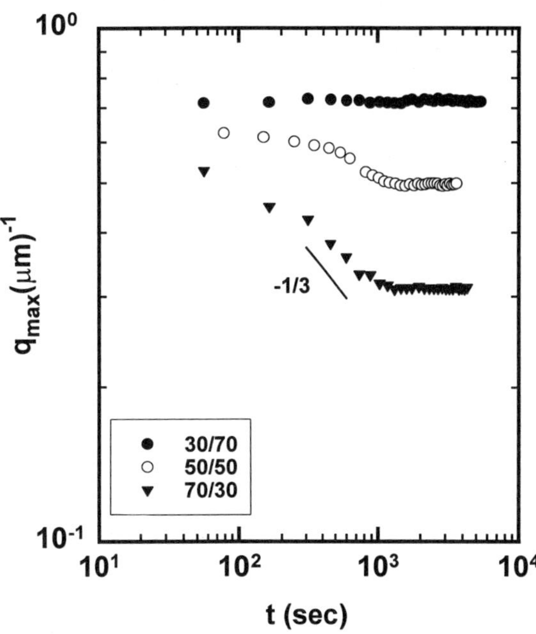

Figure 3. Temporal evolution of scattering wavenumber maxima for the 70/30, 50/50, and 30/70 blends of s-PP/EPDM upon curing with phenolic resin at 140 °C. The amount of curing agent was 10 phr to with respect to the EPDM content.

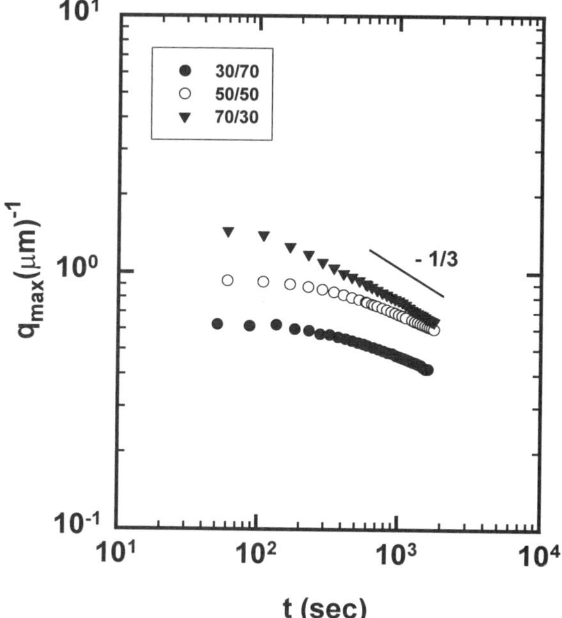

Figure 2. Temporal evolution of scattering wavenumber maxima for the 70/30, 50/50, and 30/70 blends of s-PP/EPDM, following T jumps from a single phase (128 °C) into an LCST gap at 175 °C.

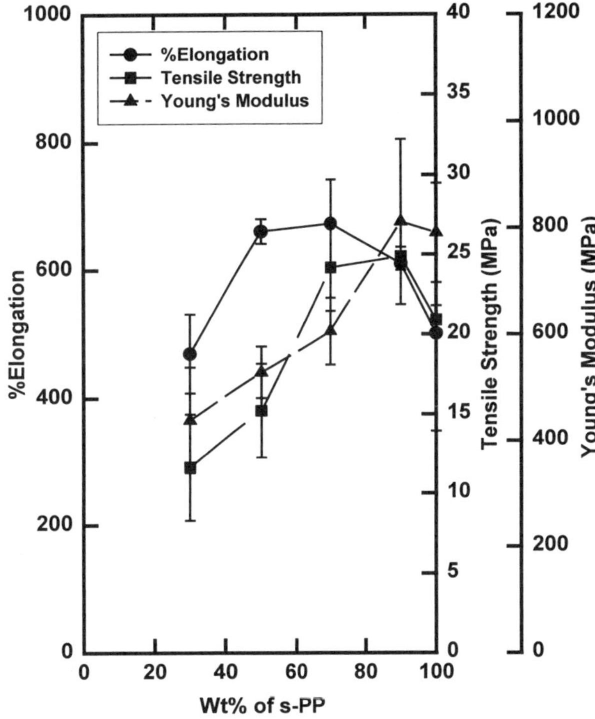

Figure 4. Dependence of tensile properties on composition of the s-PP/EPDM blends before curing with a phenolic resin.

CHEMISTRY OF PHENOL-FORMALDEHYDE RESIN CROSSLINKING OF EPDM

Martin van Duin and Henk A. E. de Keijzer
DSM Research, P.O. Box 18,
6160 MD Geleen, The Netherlands

1. Introduction

Thermoplastic vulcanizates (TPVs) comprise the fastest growing part of the elastomer market. They are prepared via dynamic vulcanization, i.e. the simultaneous blending and crosslinking in the melt, of a heterogeneous rubber/thermoplastic mixture.[1,2] The resulting product usually consists of a thermoplastic matrix and a crosslinked rubber dispersion. It allows processing like a thermoplastic, but is elastic like a crosslinked rubber. The most common commercial TPV is that based on PP/EPDM blends crosslinked with phenol-formaldehyde resins (resols). This particular crosslinking system is applied, since it results in thermally stable crosslinks and is active in the desired temperature regime.

With respect to the chemistry of resol crosslinking of unsaturated elastomers some studies have been performed in the 40's - 60's,[3] but studies using modern analytical tools are scarce. Lattimer et al.[4] showed using MS that only chroman structures are formed. Recently, we have published a study[5] involving a representative, low molecular weight model for EPDM with ENB as diene monomer, viz. hydrogenated ENB (ENBH) (Figure 1). It was shown that in an acidic environment the dimethylene ether bridges of the resol are completely degraded and that both methylene bridged and chroman crosslink structures are formed, the former finding being in conflict with the study of Lattimer et al.[4] In addition, we have shown that 2-methylolphenol (Figure 1; MP) can be used as a representative model for the resol.

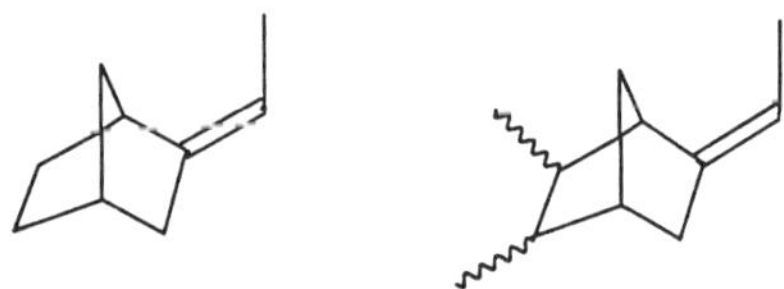

Figure 1: Low molecular weight models used for resol: MP and DMC (top) and for EPDM with ENB as diene monomer: ENBH (bottom).

Here, first of all a more detailed study of mixtures of models both for the rubber and the resol, i.e. ENBH and MP or 2,6-dimethylol-p-cresol (Figure 1; DMC), resp., is presented. DMC is a more realistic model for the resol than MP, since it contains two methylol groups and a p-alkyl substituent. The use of a specific derivatization agent for unreacted hydroxyl groups enables a sound discrimination between methylene bridged and chroman structures. In the second part a study on the chemical reactions in the ENBH/resol system at various temperatures with and without SnCl$_2$.2H$_2$O catalyst using FT-IR, GPC, GC-MS and probe-MS is put forward. Finally, a mechanism is presented combining all structures identified and explaining the effect of temperature and catalyst.

2. Experimental

All chemicals used were commercially available, with the exceptions of ENBH and DMC, which were synthesized according to references 6 and 7. Equimolar mixtures of ENBH and MP or DMC were heated in closed Wheaton vials for 1 hr at 140 °C. The products were analysed with GC-MS (HP 5790 mass spectrometer; 25 m. CPsil 5 CB column; 70 - 250 °C) after dissolving the heterogeneous products in chloroform or THF and decanting the liquid to remove the solid catalyst residue. Reaction products were also analysed with GC-MS after derivatization with N,O-bis(trimethylsilyl)acetamide (BSA). Column chromatography (Merck Kieselgel 60; 230 - 400 mesh ASTM; toluene) using an ISCO Retreiver II automated fraction collector was used for preparative separation. ^{1}H and (DEPT) ^{13}C NMR spectra were recorded on a Bruker AM400 NMR spectrometer.

Mixtures of ENBH, resol and zinc oxide (40/10/2; w/w/w) were heated both in the presence and the absence of SnCl$_2$2.2H$_2$O (2 w/w) in closed Wheaton vials for 30 min. at 150 or 225 °C. The reaction products were analysed as such with FT-IR (Perkin Elmer 1760X spectrometer; DRIFT accessory), GPC (Waters 100, 500 and 1000 Å Ultrastyragel columns in series; RI and UV detection at 254 nm; THF as eluent; 50 °C), GC-MS (Finniganmat 4600 mass spectrometer; 25 m CPsil 5 CB column; 100 - 260 °C) and Direct Insertion Probe (DIP) MS (Finnaganmat 4600; 40 - 400 °C).

3. Results and discussion

3.1. Reaction of ENBH with MP or DMC

GC-MS of the reaction products of ENBH with MP shows the formation of about 10 condensation products with an ENBH:MP molar ratio of 1:1 (m/e = 228 g/mol). Derivatization with BSA indicates the presence of condensation products with 1 and 0 hydroxyl groups, i.e. methylene bridged and chroman structures, resp. The large number of reaction products is, amongst others, due to diastereoisomerism. Preparative column chromatography yielded the two main reaction products, which were subsequently characterized with ^{1}H and (DEPT) ^{13}C NMR. The presence of methylene bridged species (phenolic OH at 5.6 ppm, no J coupling of ENBH CH$_3$ etc.) in addition to chroman species (no phenolic OH, J coupling of ENBH CH$_3$ to CH etc.) (Figure 2) was firmly established. Finally, dimers and trimers of MP were shown in the ENBH/MP reaction product, in some cases also condensated with one ENBH unit.

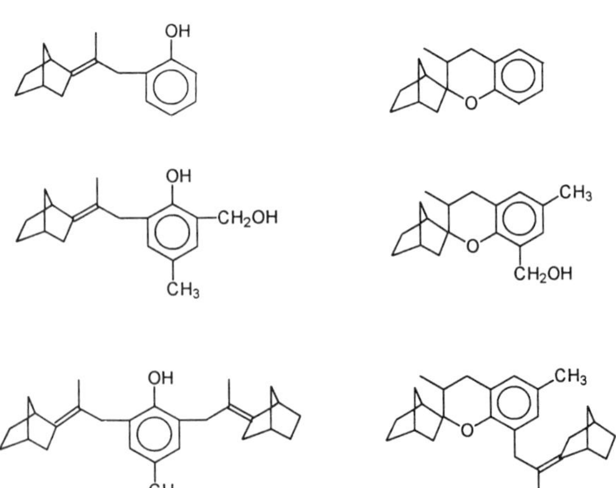

Figure 2: Reaction products of ENBH and MP (top) and of ENBH and DMC (middle and bottom) with methylene bridged as well as chroman structures.

GC-MS of the ENBH/DMC reaction products showed about 10 ENBH/DMC condensation products with a molar ratio of 1:1 as well as 2:1 (m/e = 272 g/mol: crosslink precursor; m/e = 376 g/mol: crosslink). Again, derivatization allowed the determination of the number of hydroxyl groups and, thus, discrimination between methylene bridged and chroman structures (Figure 2). DMC oligomers were also demonstrated, in some cases coupled to an ENBH moiety.

3.2. Reaction of ENBH with resol

FT-IR witnesses the degradation of the dimethylene ether bridge (1070 cm^{-1}) upon reaction of the resol with ENBH. GPC confirms the degradation of the polymeric resol (M_n = 1800 g/mol) to lower molecular structures. GC-MS and DIP-MS indicate the formation of a large number of ENBH/resol condensation products with the general formula R-isooctylphenol-[-CH$_2$-isooctylphenol-]$_{n-1}$-R'. The end groups R and R' are H, CH$_3$, CHO, CH$_2$OH and CH$_2$-C$_9$H$_{13}$ (ENBH = C$_9$H$_{14}$); the former three end groups are inert side products, whereas the latter agrees with crosslink precursors or the actual crosslinks. n = 1, 2 or 3, indicating that two ENBH molecules can be linked by mono-, di- and triphenolic units. FT-IR, GPC and MS show that the conversion of the ENBH/resol reaction increases in the series 150 °C < 150 °C/SnCl$_2$ ~ 225 °C < 225 °C/SnCl$_2$, which agrees with the expected effects of temperature and catalysts on chemical reactions.

4. Reaction mechanism

GC-MS in combination with BSA derivatization has not only allowed the identification of condensation products of ENBH with DMC with a molar ratio of 1:1 and 2:1, but also the unambiguous discrimination between methylene bridged and chroman structures. NMR provided final proof for the formation of methylene bridged in addition to chroman structures. This is in agreement with our former study[5], but remains in conflict with the study of Lattimer et al.[4] In a follow up we will explore whether differences in experimental conditions or in choice of model olefin are the cause for this discrepancy. It should be noted, however, that when only chroman structures would be formed a crosslink must consist of at least two phenolic moieties linking two ENBH molecules. Above it was shown that monophenolic crosslinks are formed, again evidencing the formation of methylene bridged structures.

The various species identified in the ENBH/MP, ENBH/DMC and ENBH/resol reaction products are presented in Figure 3. The formation of CH$_3$ and CHO end groups has previously been reported[8] and is due to a disproportionation reaction. In the presence of the acidic catalyst system SnCl$_2$.2H$_2$O the dimethylene ether bridge of the resol will degrade and formaldehyde is split off in the absence of unsaturated compounds, resulting in a novolak resin and/or a H end group. In the presence of unsaturation the benzylic cation adds to the least sterically hindered C-atom of the double bond, viz. C8 of ENBH, yielding a cation with the positive charge on C2. Then, either a methylene bridged structure with an unsaturation at C2-C8 or a chroman ring is formed. In Figure 3 only the formation of the crosslink precursor is shown. The formation of the final crosslink proceeds by degradation of the second dimethylene ether bridge attached to the phenolic moiety, resulting in coupling to a second unsaturation. The final crosslink structures may consist of two methylene bridged structures or one methylene bridged structure in combination with one chroman ring (cf. Figure 2, bottom).

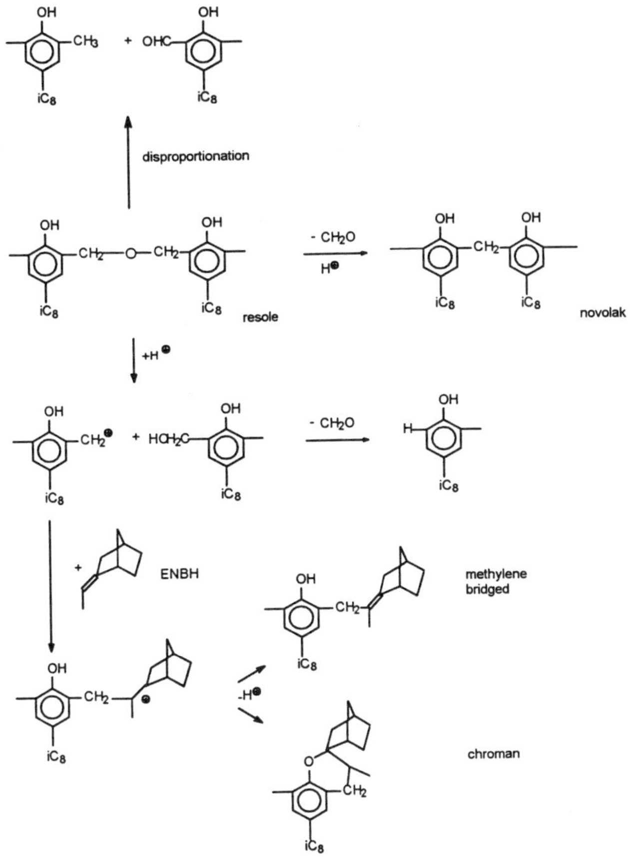

Figure 3: Reaction mechanism for formation of crosslink precursors, crosslinks and side products upon resol crosslinking.

5. References

1. A.Y. Coran, "Thermoplastic Rubber Plastic Blends", in "Handbook of Elastomers", A.K. Bhowmick and H.L. Stephens (Eds.), Marcel Dekker Inc., New York, ch. 8.
2. A.Y. Coran, Thermoplastic Elastomers based on Elastomer-Thermoplastic Blends Dynamiccally Vulcanized", in "Thermoplastic Elastomers", N.R. Legge, G. Holden and H.E. Schroeder (Eds.), Hanser Publishers, Munich, 2nd Ed., 1996, ch. 7.
3. A. Giller, Kautsch. Gummi Kunstst. 17, 3 (1964).
4. R.P. Lattimer, R.A. Kinsey, R.W. Layer and C.K. Rhee, Rubber Chem. Technol. 62, 107 (1989).
5. M. van Duin and A. Souphanthong, Rubber Chem. Technol. 68, 716 (1995).
6. J.H.M. van den Berg, J.W. Beulen, E.F.J. Duynstee and H. Nelissen, Rubber Chem. Technol. 57, 265 (1984).
7. Ullman and K. Brittner, Ber. 42, 2540 (1909).
8. K. Hultzsch, "Chemie der Phenolharze", Springer Verlag, Berlin, 1950.

Kinetic Model of Dynamic Vulcanization.

E.V. Prut, N.A. Yerina

Institute of Chemical Physics of RAS
st. Kosygin 4, Moscow, 117977, Russia

Introduction

Polymer alloys and blends are one of the most dynamic sector of the polymer industry. The creation of such polymer systems entails the potential possibility of combining the attractive qualities of each component material [1]. Much attention has been devoted in the preparation of polymeric materials by mixing components in which chemical reaction takes place[2]. This technology not only opens up new possibilities for old products but also allows preparation of blends which could not be economically made before.

Considerable attention in reactive blending has attracted the using of extruders as continuous flow reactors. In reactive blending process the chemical reaction takes place simultaneously with the mixing.process. Reactive blending combines two traditionally separate operations: the chemical reactions for formation or modification of macromolecules and the processing of the polymer for the purpose of structuring into shaped plastic products (Fig.1). Reactive blending in addition to melting involve introducing reactive agents at optimum points in the reaction sequence, homogenizing the ingredients, and allowing sufficient time for the completion of the reaction. The reactants are fed into the extruder through a feed hopper.usually. However, various liquid or gaseus reactants can be introduced at specific points in the reaction sequence by using igection ports along the extruder barrel. The reactive mixture is conveyed through the extruder and the reaction is driven to the desired degree of completion. At this point the molten polymeric productis pumped through a die and subsequently quenched, solidified, and pelletized. Thus, production and processing can be integrated in one step.

Among all the processes of the reactive blending, dynamic vulcanization is the best may to produce thermoplastic elastomeric composition. It is the process of vulcanizing the elastomers during its melt-mixing with molten plastic[3]. This process leads to the formation of a heterogeneous structure in which the crosslinked elastomer particles with dimensions of the order of 1-10 μm are dispersed in a continious thermoplastic matrix. These blends have elastomeric properties, but they are processable as thermoplastics. The rubber particles should be crosslinked to promote elasticity. Because of this requirement the usual methods for preparing elastomer-plastic blends by melt mixing are not sufficient.

Results and Discussion

The first stage of dynamic vulcanization is the preparation of a homogeneous mixture of PP and an elastomer.

The crosslinked elastomer particle dimentions should be several microns. There are two ways to add vulcanizing agents: either after the mixing of PP and the elastomer, or to the pure elastomer before mixing. In the first case, vulcanizing agents are dispersed in the blend components according to their thermodynamic affinity. Redistribution of vulcanization agents proceeds during the rection. The rate of vulcanization is determined considerably by their diffusion within the PP matrix. Different solubility and rate of diffusion of vulcanizing agents lead to their redistribution between the phases. As a result, vulcanizing agents can be retained in the compozition in free form. During subsequent processing they react and deteriorate the properties of thermoplastic elastomer (TPE).

In the second case, when vulcanizing agents are added to the pure elastomer before crosslink formation, the interaction of vulcanizing agents and elastomers is observed at the very onset of heating. However, network of chemical crosslinks should not be formed, as the dispersion of the crosslinked elastomer is hindered. Thus, if the vulcanization kinetic curve is S-shaped, the induction period τ_i should be of the same order of magnitude as the mixing time of the initial components τ_m (Fig.2) [4]. Such a condition determines the selection of vulcanizing agents and their ratio [5]. In this case the blending of vulcanizing agents to the elastomer hinders the redistribution of the vulcanizing agents between the phases due to their low diffusion rate.

The mixing is usually carried out at 180-220°C for TPE based on PP. At the same time for elastomers the vulcanization temperature does not exceed 130-160°C and so at the elevated temperatures a degradation leads to the deterioration of the mechanical properties of TPE[6]. Molecular mass of the sol-fraction durung the vulcanization of the polyisoprene elastomer at 190°C decreases (Fig.3).

Thus, the residence time of TPE in the extruder $\tau_f = L/V$ (L is extruder length, V is a longitudinal flow rate) should be less than the degradation time τ_d.

The time of limit degree of vulcanizated τ_∞ can be defined as follows (Fig.2): $\tau_\infty = \tau_i + q_\infty/W_{max}$, where q_∞ is a limit degree of vulcanization, and W_{max} is a maximum rate of vulcanization.

Now the following condition can be defined $\tau_m \leq \tau_i < \tau_\infty \leq \tau_f$. If me suppose that $\tau_m = R^2/6D_m$ (R is an extruder radius and D_m is an effective diffusion coefficient) we can estimate $R \leq (6D_m\tau_i)^{1/2}$ and $L \geq V(\tau_i + q_\infty/W_{max})$.

References

1. Polymer Blends. D.Paul, S.Newman, (eds.). Academic Press, New York (1978).
2. A.O.Baranov, A.V.Kotova, A.N.Zelenetskii, E.V.Prut. Russian Chemical Reviews <u>66</u>, 877(1997).
3. A.J.Coran in «Thermoplastic Elastomers: A Comprehensive Review» N.R.Legge, G.Holden, H.E.Schroeder, (eds). Hanser Publishers, Munich-Vienna-New York, 1987, Ch.7.
4. E.V.Prut. Chem.Phys. 1998 (in press)
5. Russian Patent N2069217
6. O.P.Kuznetsova, L.M.Chepel, G.M.Trofimova, D.D.Novikov, A.N.Zelenetskii, E.V.Prut. Polymer Sci. <u>39B</u>, 374(1997).

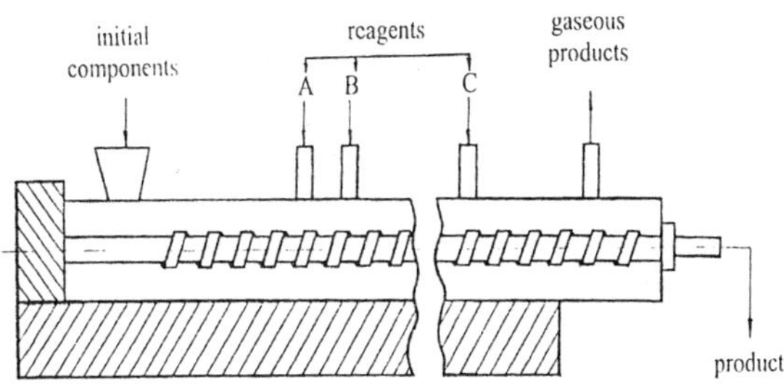

Fig.1. Schematic represenation of a reactive blending

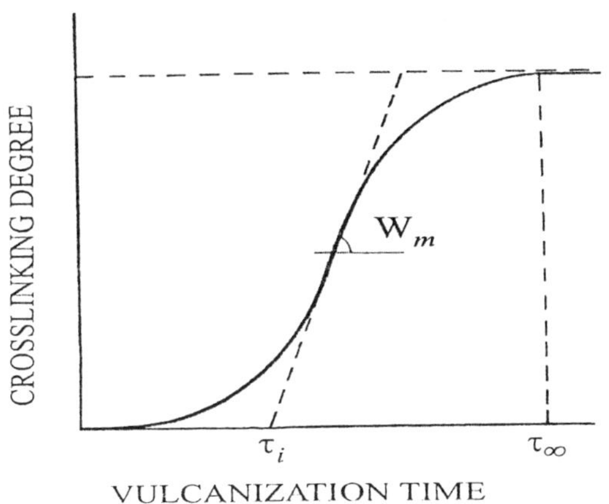

Fig.2. A typical kinetic curve of elastomer vulcanization.

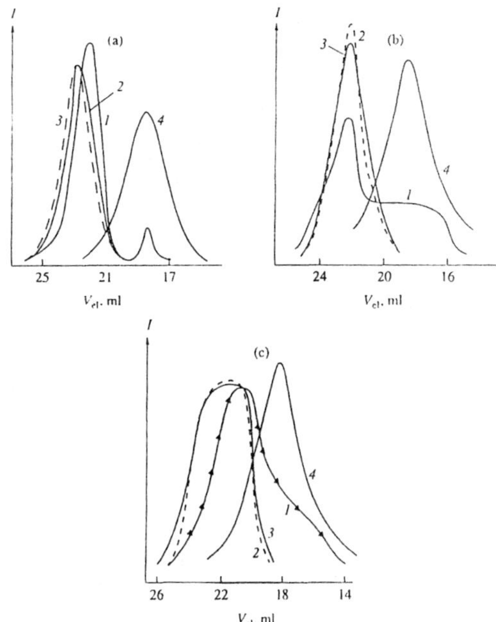

Fig.3. Gel-chromatograms of the sol-fraction of the polyisoprene elastomer at 190°C for the vulcanization time 3.5(1), 7.0(2) and 14 min at different ratio of tetramethylthiuram disulfide/sulfenamide 2.0/0.0 (a), 1.0/1.0 (b) and 0.0/2.0 (c). The curve (4) corresponds to gel chromatogram of the initial elastomer.

REACTIVE COMPATIBILIZATION OF SAN/EPR BLENDS: MORPHOLOGY DEVELOPMENT AND INTERFACIAL GRAFTING KINETICS

C. Pagnoulle*, C.E. Koning**, L. Leemans** and R. Jérôme*,
*Center for Education and Research on Macromolecules (CERM),
University of Liège, Sart-Tilman, B6, 4000 Liège, Belgium
** DSM Research, P.O. BOX 18, 6160 MD, Geleen, The Netherlands

INTRODUCTION

Since most polymers are immiscible, polyblends have usually to be compatibilized in order to improve the poor mechanical performances associated with a gross phase morphology and a low interfacial adhesion. Efficiency of block and graft copolymers in reducing the interfacial tension and improving the interfacial adhesion has been extensively discussed [1-2]. These additives can be premade or generated in situ in a reactive blending process. Reactive blending has the advantage to be straightforward and to form the compatibilizer where it has to be localized, thus at the interface.

Morphology development is the change of the blend morphology during mixing from pellet or powder sized particles to submicron droplets which exist in the final blend. In reactive blending, it may be anticipated that the compatibilization, i.e. the phase morphology and mechanical properties, will strongly depend on the kinetics and completeness of the interfacial reaction with respect to the blending time. Indeed, this kinetic control will decide on the amount and molecular architecture of the copolymer formed at the interface. Although the control of the interfacial reaction is of a prime importance, the problem of the intrinsic reactivity of the functional groups attached to the chains that have to react at the interface has not been clearly addressed. In this respect, Scott et al. have compared the reaction of maleic anhydride (MA) with amine and oxazoline (OX), respectively[3]. Unfortunately, the comparison is not straightforward since different polymer pairs have been studied, i.e. EP-MA/PS-OX and PS-MA/PA-NH_2, respectively. Changes in the mixing torque and the average particle size were delayed in time and less pronounced in case of the MA/OX reactive pair compared to the MA/NH_2 one, consistently with the lower reactivity of the former pair[3].

In this study, maleic anhydride attached to EP chains will be reacted in the melt with either carbamate or amine groups grafted onto SAN chains. In case of SAN bearing carbamate groups, the interfacial reaction is thought to be controlled by the (slow) carbamate thermolysis into primary amine. Then, compared to SAN-NH_2, the grafting of SAN-carbamate is expected to proceed slower[4]. Since carbamates are nothing but primary amine precursors at the processing temperature, this system is however ideal to address the question of the mutual reactivity of the functional groups while keeping constant polymer (EP and SAN) molecular weight and, content and distribution of each type of reactive groups in the reactive chains. This system will be studied with the purpose of compatibilizing SAN/EPDM blends.

In this work, reactive SAN-X, where X is either a carbamate or a primary amine (0.028 mol X /wt %), will be diluted by neat SAN (20 wt % SAN-X), whereas EP-g-MA will be diluted by neat EPDM (50 wt % EP-g-MA). The general SAN/Rubber weight composition was systematically 75/25. The effect of the intrinsic reactivity of the SAN reactive groups, i.e. primary amine or carbamate, will be studied in close relation to the phase morphology development. For this purpose, the morphology will be sampled from 1 to 17 minutes of mixing.

EXPERIMENTAL

Materials

SAN used in this work was the RONFALIN 2770 from DSM containing 26.5 wt % acrylonitrile. The maleic anhydride grafted rubber (EP-g-MA) EXXELOR VA 1801 from EXXON contained 0.6 wt % or 6.10^{-3} mol/wt % succinic/maleic anhydride groups. The non reactive rubber was the EPDM, KELTAN 4778 from DSM. Terpolymerization of styrene (55 mol%), acrylonitrile (43 mol%) and a carbamate containing comonomer, {1-methyl-1-[3-(1-methylethenyl)-phenyl]ethyl}carbamic acid 1,1-dimethylethyl ester (2 mol%), was initiated by AIBN in toluene at 60°C for 24 h. It was recovered by precipitation in methanol and characterized by NMR and FTIR.

The carbamate pendant groups of functional SAN were transformed into amines either in solution prior to blending[5] or in situ during the blending process[6]. The actual carbamate content was 0.028 mol/wt % or 2 mol %, and the molecular weight was 10^5 with a polydispersity index of 1.5.

Mixing conditions

Samples (20 gr) were melt blended with a laboratory two-roll mill at 200°C. A masterbatch of 50 wt % EPDM and EP-g-MA was previously melt blended, for 7 min. under moderate shearing (rolls speed= 15 and 25 rpm) and stabilized by 0.5 wt % IRGANOX 1010 from CIBA GEIGY. After mixing, the rubbery blend was cooled down to room temperature and cut into small pieces (2X5 mm). 20 wt % reactive SAN was then melt blended with neat SAN at 200°C, followed by the addition of the required amount of the (EPDM/EP-g-MA) premixture. 1 min. later, the shear rate was regularly increased (rolls speed of 0.5 and 25 rpm). The mixing conditions were then maintained. After the specified time of mixing, small piece of blend was cut from the large gap region of the roller blades and dropped directly into a bath of liquid nitrogen to freeze the morphology. Mixing times investigated here were 1 ,2 ,3 , 5, 8, 12 and 17 min. Time zero corresponds to the start of feeding of the rubber pellets.

Transmission electron microscopy

Observations were conducted with the transmission electron microscope PHILIPS M100 at an accelerating voltage of 100 kV. Thin sections (90 nm) were prepared by ultramicrotomy (ULTRACUT E from REICHERT-JUNG) at -130°C and stained by exposure to RuO_4 vapors for ca. 2 hrs.

RESULTS AND DISCUSSION

Summaries of the change in number average diameter, D_n, as a function of mixing time are represented in fig. 1. The experimental curve is clearly shifted towards longer mixing times, when SAN-carbamate is used rather than SAN-NH_2. However, whatever the type of reactive SAN, the major reduction in the dispersed phase size is found to occur at short processing times, in conjunction with the softening and melting of the rubber component (see hatched region in fig.1). For example, the number average particle diameter is reduced from ca. 5 mm (rubber pellet size) to 1 μm within 150 and 300 s of mixing in case of SAN-NH_2 and SAN-carbamate, respectively. Subsequent mixing affects the phase morphology in a way that however depends on the type of SAN reactive groups. In case of primary amine, the rubber dispersion is rapidly improved and most of the rubber particles have attained their equilibrium size after 5 min. The major effect of subsequent mixing is a reduction of the size of the largest particles, which are transformed into smaller ones (see particle diameter distribution plots at 5 and 17 min. of mixing, fig. 2), the diameter of the smallest rubber domains being unchanged. In contrast, the particle size observed with SAN-carbamate does not level off, but rather regularly decreases with mixing time (fig.1).

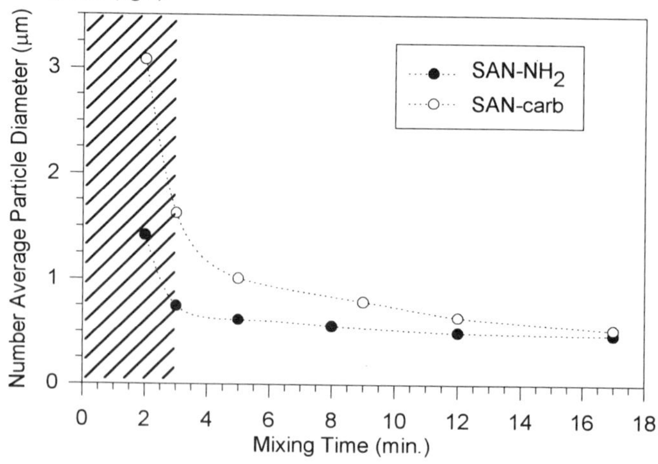

Fig. 1. Number average rubber particle diameter vs. mixing time for blends modified by either SAN-NH_2 or SAN-carbamate.

At 5 min. of mixing, the substitution of SAN-NH$_2$ with SAN-carbamate results in an enhancement in the average size of the dispersed phase particles and a broadening of the size distribution (see fig.2).

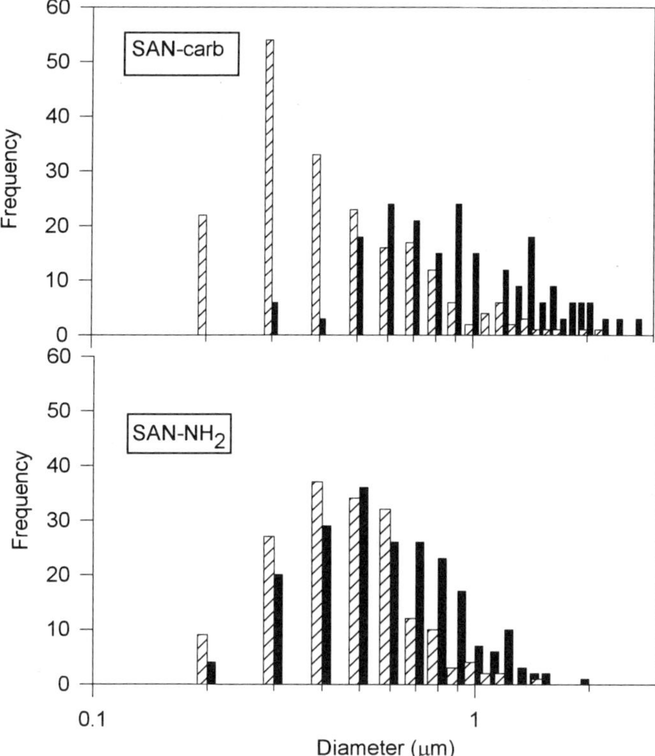

Fig. 2. Particle diameter distributions at 5 (■) and 17 min.(◨) of mixing for blends modified by either SAN-NH$_2$ or SAN-carbamate.

Upon further mixing, the rubber particle size distribution is clearly shifted towards smaller values, leading to a significant decrease of D_n (see fig. 1). For instance, 17 min. of blending are required to reach comparable particle size as the final size observed with SAN-NH$_2$.

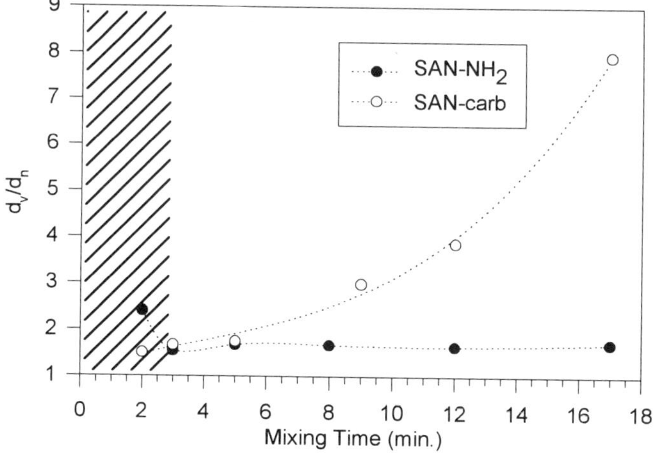

Fig.3. d_v/d_n vs. mixing time for blends modified by either SAN-NH$_2$ or SAN-carbamate.

Thus, the rubber dispersion is clearly delayed in time when SAN-carbamate is used rather than SAN-NH$_2$, consistently with the higher reactivity (towards maleic anhydride) of primary amine compared to the carbamate counterpart.

The breadth of the particle size distribution may be roughly gauged by d_v/d_n, the ratio of volume average to number average diameters. In fig. 3 is plotted d_v/d_n as a function of the mixing time. After the softening/melting process, which is associated to the main reduction in particle size, i.e. after 3 min. of mixing, d_v/d_n is basically constant in case of SAN-NH$_2$, while longer mixing times results in a broadening of the particle size distribution in case of SAN-carbamate.

CONCLUSIONS

In agreement with observations by several authors[7], the most changes in phase morphology have been found to occur during the first few minutes of mixing, thus when melting and softening of the blend components are also occurring.

Would it be so, the fast primary amine-maleic anhydride interfacial reaction is expected to mainly proceed during the phase dispersion step and to inhibit accordingly coalescence. This explains why most of the rubber domains have attained their equilibrium size after 5 min. of mixing in case of SAN bearing NH$_2$. Only little changes in rubber particle sizes are observed during subsequent mixing.

In case of SAN bearing carbamate groups, the interfacial reaction is thought to be controlled by the (slow) carbamate thermolysis into primary amine. Then, compared to SAN-NH$_2$, the grafting of SAN-carbamate occurs later, particularly after the softening/melting of the blend components, and becomes challenged by coalescence process during the so-called dispersive mixing. Interfacial grafting and coalescence are known to have quite opposite influences on the phase morphology development, the former favoring the rubber dispersion and the latter resulting in a coarsening of the morphology. As a rule, the occurrence of these two mechanisms could explain why the number average rubber particle size decreases with mixing time, while the size distribution broadens regularly. Nevertheless, further investigations are needed to more deeply understand the behavior of the slow grafting SAN-carbamate.

Thus, SAN-NH$_2$ is more efficient than SAN-carbamate in stabilizing fine and narrow distributed particle size even at long mixing times. This feature underlines, however, the decisive role of the interfacial grafting reaction during the melting/softening of the blend components, thus before coalescence takes place. So the final morphology strongly depends on the relative kinetics of the phase morphology development and the graft copolymer formation.

ACKNOWLEDGEMENTS
The authors are grateful to the "Services Fédéraux des Affaires Scientifiques, Techniques et Culturelles" for general support in the frame of the "PAI-4: Chimie et Catalyse Supramoléculaire". They also thank Dr. Ph. Maréchal (FINA research) for stimulating and helpful discussions, Mrs Dejeneffe for assistance with the cryoultramicrotome. Support for this work was provided by DSM.

REFERENCES

1. M. Van Duin., C.E. Koning, C. Pagnoulle and R. Jérôme *Prog. Pol. Sc., to be published*
2. M. Xanthos *Polym. Eng. Sci.* **28**, 1392 (1988)
3. C.E. Scott and C.W. Macosko *Polymer* **35(25)**, 5422 (1994)
4. C.Pagnoulle and R. Jérôme*Polyblends' 97 SPE RETEC, October 9-10, 1997, IMI Boucherville, Qc, Canada Oral Communication*
5. C.Pagnoulle, C.E. Koning, L. Leemans and R. Jérôme *Dutch Application No.* **1007074** (1997)
6. W.P. England, G.J. Stoddard and J.J. Scobbo *US Patent No.* **5310795** (1994)
7. C.E. Scott and C.W. Macosko *Polymer*, **35(25)**, 5422 (1994)

Aging of Thermoplastics Elastomers

E.V. Prut, N.A. Yerina, L.V. Kompaniets

Institute of Chemical Physics of RAS
St. Kosygin 4, Moscow, 117977, Russia

Introduction

Thermoplastic elastomers (TPE) based on mixtures of polyolefin elastomers with polyolefin thermoplastics have many of the properties of elastomers, but they are proccessable as thermoplastics[1].TPE compises finely divided elastomers particles dispersed in a relatively small amount of plastic. The elastomers particles should be crosslinked to promote elasticity.

Polypropylene is a semi-crystalline widely used in many engineering applications. The PP/EPDM blends is one of the systems widely studied. These blends are incompatible with poor physical and chemical interactions across the phase interface. The PP recrystallization takes place within a domain of pure resin and phase interface during the aging.

In this work we have investigated the effect of the temperature and time of aging of TPE on its mechanical properties.

Experimental

Isotactic PP had $M_w = 3,7 \times 10^5$, degree of crystallinity 50-60%, $T_m = 164\text{-}165^\circ C$. EPDM having 30% propylene content was supplied by Enicem Elastomer. Sulfur used as the vulcanizing agent.

The blends were prepared in a Berstorff twin screw extruder[2].

The samples of aging were in the forms of the pellets and films. The aging temperatures were the room temperature, $100^\circ C$ and $130^\circ C$ .The aging carried out in air. After aging the films were prepared from the pellets.

The stress strain behavior was determined by means of a Instron mechanical tester (Instron 1122).

A Du Pont DSC (Model 990) was used for the calorimetric studies.

Results and Discussions.

Fig.1,2 show the dependences of tensible strength σ_b,modulus E_o, elongation of break ε_b, melting heat ΔH_m and melting temperature T_m as the variation of the time of aging at the temperatures $100^\circ C$ and $130^\circ C$. These curves is found to change as a function of the form of samples: the films or pellets.

This may be caused by the different size of surface of samples and skin-core morphology. The skin of films consists of a thin pure PP layer.

It is interesting to note that the melting temperature T_m of the film samples aging at $130^\circ C$ increases and exceeds of $T_m$of pure PP (Fig.2e, curve 2). The melting temperature T_m of the samples prepared from the pellets aging at $130^\circ C$ is not changed (Fig.2e, curve1). The mtlting temperatures T_m of the samples aging at $100^\circ C$ reduce slightly as the time of aging increases. The crystallization temperature of PP is about $130^\circ C$. And so the increasing of T_m after the aging of films at $130^\circ C$ may be caused by recrystallization of PP and improvement of the crystalline size. In this case the melting heat ΔH_m is not changed (Fig.5d, curve 2). In other cases the melting heats ΔH_m are decreased.

Such comolicated behavior of TPE during the aging influences on the its mechanical properties (Fig.1,2). We observe that there are maximum or minimum on the curves. This may be caused by a number of factors acting independently and simultaneously.

The results of aging of films of TPE at the room temperature are shown in Tabl.1.

Tabl. 1

Aging of films of TPE at the room temperature.

aging time, days	$\sigma_b/\sigma_b^{\,o}$	$E_o/E_o^{\,o}$	$E_{100}/E_{100}^{\,o}$	$\varepsilon_b/\varepsilon_b^{\,o}$
1 hour	1.0	1.0	1.0	1.0
1	0.84	1.18	1.03	0.63
3	0.96	0.84	1.0	0.89
4	0.92	1.18	1.03	0.75
7	0.92	1.16	1.03	0.78
14	0.90	1.32	1.05	0.71
36	0.82	1.33	1.05	0.52
64	0.88	1.10	1.03	0.71

This study reveals the influences of the time and temperature of aging on the mechanical properties of TPE.

Further studies will be made in more detail and reported elsew here.

References

1. A.J.Coran in «Thermoplastic Elastomers: A Comprehensive Review» N.R.Legge, G.Holden, H.E.Schroeder,(eds). Hanser Publishers, Munich-Vienna-New York, 1987, Ch.7.
2. Russian Patent № 2069217.

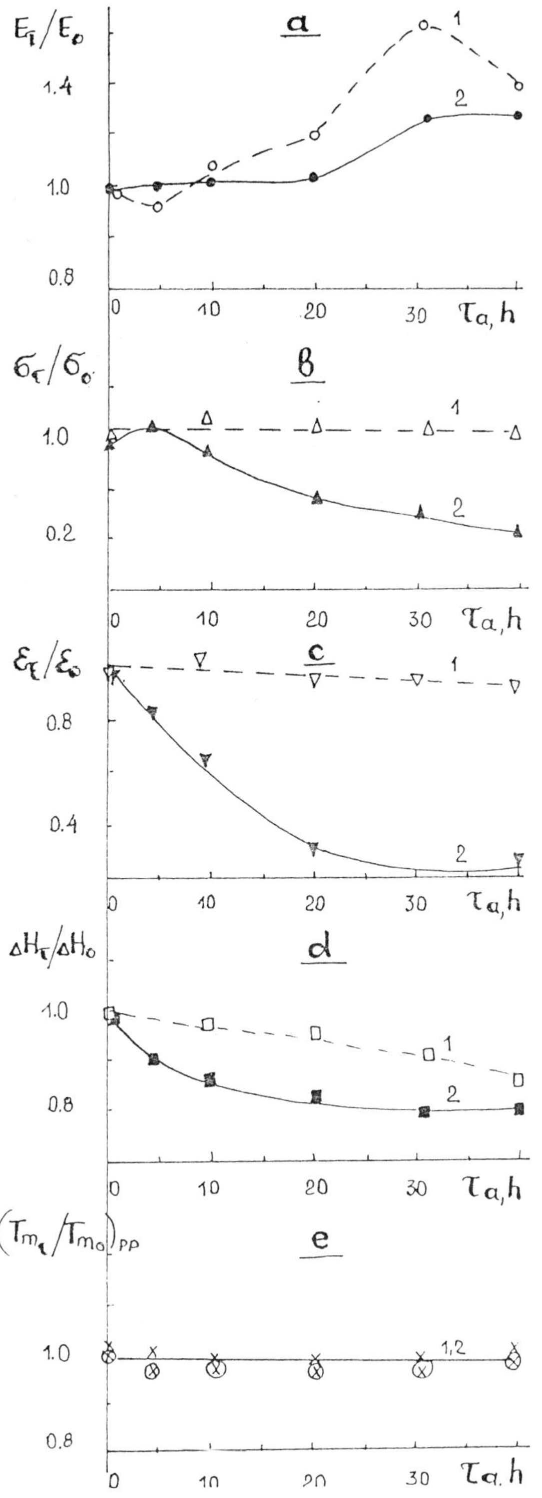

Fig.1 TPE'S properties in relative scales after aging at 100°C (curve 1- the film after aging ; curve2- the film prepared from the pellets after aging):
(a)- Young modulus E, (b)- tensile strength σ, (c)- elongation at break ε, (d)- melting heat ΔH_m, (e)- PP melting temperature T_m

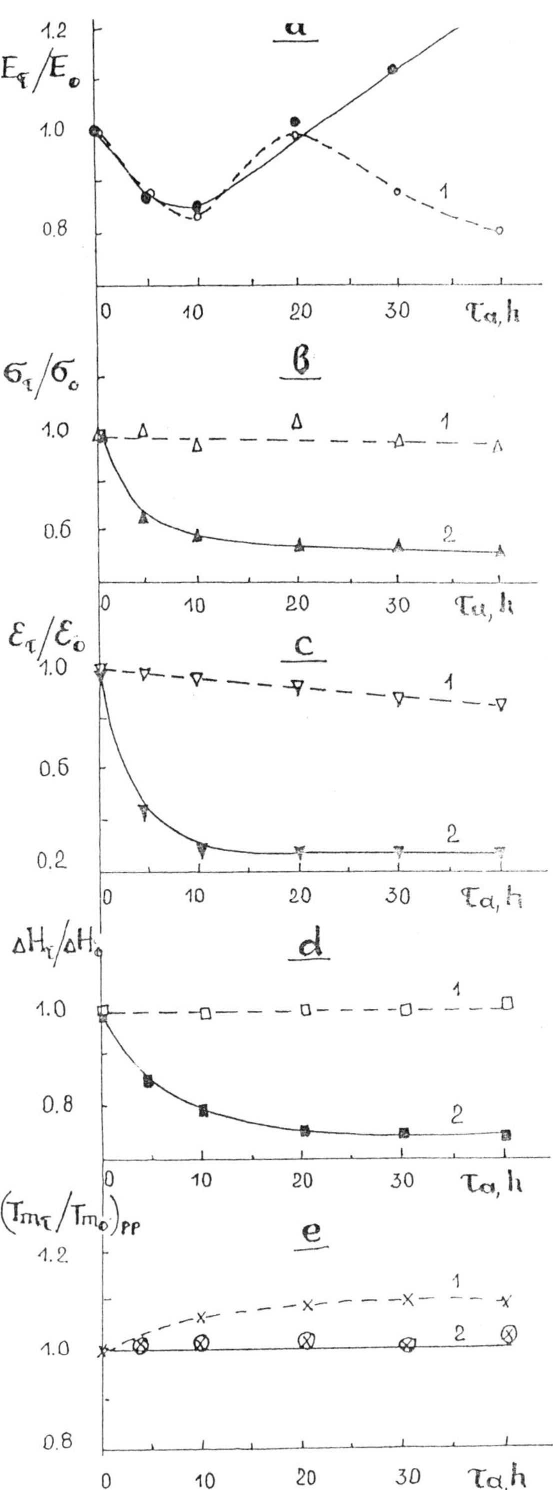

Fig.2 TPE'S properties in relative scale after aging at 130°C (curve 1- the film after aging; curve 2- the film prepared from the pellets after aging)
(a)- Young modulus E, (b)- tensile strength σ, (c)- elongation at break ε, (d)- melting heat ΔH_m, (e)- PP melting temperature T_m

DEFORMATION MECHANISM OF THERMOPLASTIC VULCANISATES INVESTIGATED BY COMBINED FTIR - AND STRESS-STRAIN MEASUREMENTS

Maria Soliman, Menno van Dijk, Martin van Es
DSM Research, P.O. Box 18 6160 MD Geleen, The Netherlands

Thermoplastic vulcanisates are blends of a thermoplastic matrix with a crosslinked rubber. The rubber is dynamically crosslinked during blending. Due to this procedure it is possible to get phase inversion and disperse a large amount (up to 80%) of rubber into a thermoplastic matrix. These materials, which are in our case based on polypropylene (PP) and EPDM, show a high elasticity upon tension and pressure. This, despite the fact, that the matrix consists of polypropylene, which should be deformed plastically. We prepared a large variety of model compounds, where we varied the amount of polypropylene, the crosslink density and the amount of aliphatic oil. The technique we used to investigate the deformation behaviour is a combination between infrared spectroscopy and stress-strain measurements. We measure in-situ the orientation of different chemical groups during stretching. To prevent overabsoprtion (the measurements are done in transmission), the measured films have a thickness of 50 μm. The material was stretched up to 200% strain and than recovered until the stress was zero again. This gave a measure for the remaining strain. One of the main advantages of this technique is the fact, that the different blend components and also crystalline and amorphous phases of the two polymers can be investigated separately. One example of a sample with 30% PP and 70% EPDM + crosslinking agent is shown in figure 1.

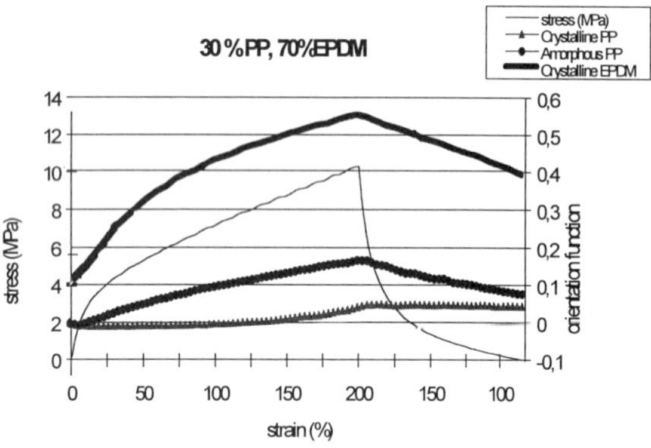

Figure1: Stress-strain curve, recovery and orientation function of TPV based on 30% PP

The stress-strain curve shows the typical behaviour of a thermoplastic elastomer: a) an initial E-modulus, which is lower than expected from a simple composite model with dispersed rubber; b) a broad yield "region" and c) a constant modulus up to 200% strain. The recovery experiment shows a remaining strain of 75%, which is also low considering the fact, that there is a thermoplastic matrix material, which should deform plastically.

It could be shown that the amorphous and crystalline EPDM regions show orientation directly proportional to the applied strain, also a relatively high orientation function with a maximum value of 0.5 can be found. This shows that the EPDM is stretched as if it is the continous phase and is also able to recover to a certain extent. This is even possible although there is a large difference in modulus between the rubber and the matrix. The amorphous PP shows a small orientation and also some recovery already from the beginning on, whereas the crystalline PP shows a very small orientation up to 300% macroscopic strain. This fact can only be explained by a model which assumes a very inhomogenous deformation of the PP matrix and a good adhesion between matrix and dispersed phase. Figure 2 shows an illustration of the idea of an inhomogenous deformation, leading to plastic deformation only in a small fraction of the available PP.

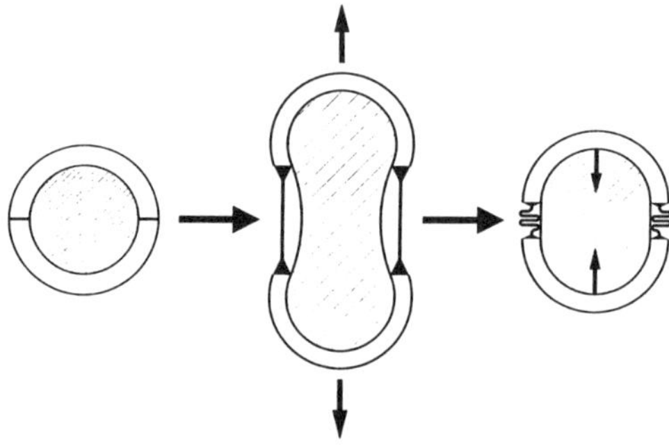

Figure 2: Deformation model for TPVs

This model shows a high plastic deformation which is localized at the boundary between different EPDM spheres. The PP in the areas not directly connected to the EPDM shows no orientation. These ideas where confirmed by TEM micrographs taken form a sample which was stretched to 200% and than relaxed. The orientation of PP lamellae was only visible at a few spots on the interphase between two EPDM spheres, whereas the major part of the lamellae was unoriented. The remaining strain could be seen due to the highly stretched EPDM particles. Further investigation should give more insight in the deformation mechanism of TPVs.

STRENGTH PROPERTIES OF DYNAMICALLY VULCANIZED ELASTOMER-POLYPROPYLENE BLENDS

M.D. Ellul, Advanced Elastomer Systems, L.P., 388 S Main St, Akron, OH 44311,
G.J. Lake, University of E London, Longbridge Road, Dagenham, Essex RM8 2AS,
A.G. Thomas, Queen Mary & Westfield College, University of London, Mile End Road, London E1 4NS

I. INTRODUCTION

Dynamically vulcanized thermoplastic elastomers (TPEs), are heterophase blends of an elastomer crosslinked *in situ* during melt mixing with a compatible thermoplastic at elevated temperatures. The resulting blends consist of finely dispersed micron-sized elastomer particles of a high crosslink density (ca. 50-120 mol/m^3) in a thermoplastic matrix. For the most part, these specialty TPEs have many of the elasticity characteristics of vulcanized rubber at ordinary temperatures but they are also sufficiently thermoplastic in nature at melt temperatures of the thermoplastic phase. They have therefore been steadily replacing traditional thermoset rubber in many applications because of their fabrication ease as well as recyclability. Several publications have appeared covering compositions (1), various physical and performance attributes of dynamically vulcanized alloys (2,3), as well as morphology (4,5) and rheology (6,7). Surprisingly, however, only recently have some detailed studies begun on fracture characteristics (8) and none has yet been reported on the rubbery fracture behavior of dynamically vulcanized TPEs consisting of blends of polypropylene with relatively large amounts of crosslinked elastomer and oil. In this paper we compare the fracture characteristics of a soft dynamically vulcanized EPDM-polypropylene TPE with its analogue using natural rubber, NR. On the basis of the behavior in normal vulcanizates, it would be expected that the strength properties of the NR-based material would be better.

II. MATERIALS

The materials used for making TPE vulcanizates for the present study included EPDM or NR, curatives for the rubber, isotactic polypropylene and paraffinic hydrocarbon oil. Ternary blends consisting of elastomer, polypropylene and oil were melt-blended in the presence of rubber curatives, in a high shear device at 170 - 185°C. The ratio of rubber; plastic; oil was approximately 40:20:40. Test sheets were prepared by compression molding for 3 minutes at 185°C. Morphology was determined by Transmission Electron Microscopy (TEM) using methods described earlier (5). In general the EPDM-PP TPE had discrete, smaller and smoother-edged particles ranging from sub micron to about 3 microns relative to the NR-PP TPE where the particles were less discrete, larger and with irregular contours (Figure 1).

III. RESULTS AND DISCUSSION
Standard Test Results

Results of various standard tests are given in Table I. A check on the isotropy and homogeneity of the 2mm thick, compression molded sheets, made by cutting test pieces from various locations and directions, showed variations in stress-strain behavior to be small for both thermoplastic elastomer vulcanizates (but see later discussion also). The ultimate strength properties are modest, although the tear strengths are rather better than the tensile failure properties and overall the EPDM-based material has some advantage. The better properties obtained with the EPDM thermoplastic elastomer blend are rather surprising in view of the behavior of conventional thermoset vulcanizates, where natural rubber has an advantage deriving from its ability to crystallize on extension.

Time-Dependent Crack Growth Tests

For these tests both trousers and tensile strip test pieces (with edge cracks) were used (10) and the tests were mainly carried out under constant load although some 'trousers' tests were carried out at imposed constant rates of growth. The two methods generally gave similar results, as did 'trousers' tests with 'unbacked' and 'backed' test pieces (a flexible but inextensible backing was used to prevent extension of the legs in some cases). The tensile strip tests, provided a check on the validity of an energy approach (9). To enable the energy release rates to be estimated for the latter stress -strain tests were carried out under the same conditions as used for the crack growth tests. The retraction curve (at the end of the test period) was used to estimate the strain energy density.

The crack growth results show the different test pieces to give similar results for each material, with the behavior of the two thermoplastic elastomer materials being very similar (Figure 2). In 'trousers' tests at relatively low energy release rates, delamination (in a plane approximately parallel to the major surfaces) was observed with both thermoplastic elastomer vulcanizates. This presumably reflected some through-thickness anisotropy arising from mold flow and caused the 'trousers' test results to be displaced towards higher energies. Results where delamination occurred are not shown in Figure 2. Overall the time-dependent crack growth characteristics of the thermoplastic elastomer materials are similar to those observed for unfilled or lightly-filled conventional thermoset vulcanizates of non-crystallizing elastomers of moderately low glass transition temperature. Therefore the unique strength behavior of NR thermosets, where time-dependent crack growth is often absent and tearing occurs in a stick-slip manner, is not translated into these dynamically vulcanized NR-PP TPEs (10).

X-ray Diffraction of Thermoplastic Elastomers based on Dynamically Vulcanized Elastomer -Thermoplastic Blends

As it was suspected that the relatively poor performance of the natural rubber blend, especially its time dependent failure behavior, was due to its failure to strain crystallize, the strain crystallization behavior was studied by wide angle X-ray diffraction.

X-ray photographs were taken of the NR-PP and EPDM-PP blends strained to about 300%. This strain is quite adequate to produce readily detectable patterns in conventional NR thermoset vulcanizates, and an example of such a pattern is shown for comparison. As expected, the EPDM material shows no pattern ascribable to oriented NR but some powder-type rings no doubt due to the PP, and possibly some inorganic filler, are visible (Figure 3). Of prime interest is that the dynamically vulcanized NR-PP blend shows no sign of the characteristic pattern associated with strain-crystallized NR. The reason for this may be the presence of a substantial quantity of oil. The strength of NR is known to be reduced by swelling (11) and Figure 4 shows an abrupt drop in breaking elongation at a swelling of about 30%, suggesting that this amount may inhibit or substantially reduce crystallization. There is some indication that degradation of NR may occur under the conditions used for mixing the TPEs which may also affect the crystallizability.

Cyclic Fatigue Tests

Fatigue tests were carried out using dumbbell test pieces cycled at a frequency of about 4 Hz. The intention was to cycle to constant maximum force (minimum zero) with initial maximum strains of 60, 100 or 140%. In fact, the lives obtained with the NR-based material were so short (Figure 5) that it was not possible to keep the force constant; this condition was achieved with the EPDM-PP blend. The difference between the two materials in Figure 5 is much greater than would be expected from the (constant force) crack growth characteristics discussed above since if the cyclic crack growth behavior is also similar for the two materials it will imply that the very low fatigue lives for the NR-PP blend are associated with the presence of much larger naturally-occurring flaws or inhomogeneities in this material. Some evidence supporting this is provided by the electron micrographs (Figure 1); the morphology of the NR TPE vulcanizate illustrates NR domain sizes up to 20 microns in size, whereas in the EPDM-based TPE, rubber domains did not exceed a few microns. The presence of larger flaws in the NR-PP blend would also be expected to affect its tensile strength. This material does show lower strength than the EPDM- PP blend (see Table I), but the difference is not as great as might be expected on the basis of the fatigue results. Test pieces for the latter tests were cut from separate sheets and maximum stresses for individual test pieces (at the same maximum strain) showed much wider variation than was observed in the isotropy check. Thus it appears that there may have been appreciable sheet-to-sheet variations, possibly associated with differences in details of the preparation.

CONCLUSIONS

Compared with thermoset NR, the thermoplastic elastomer-polypropylene blends show moderate strength properties, which are, perhaps, much as would be expected from their structure and method of preparation. Surprisingly at first sight, the blend based on NR is generally weaker than that based on EPDM, markedly so in the case of tensile fatigue resistance. The failure of NR to offer any advantage in the NR-PP TPE appears to be associated with its inability to strain-crystallize in the blend, and its inferiority to result from larger intrinsic flaws.

REFERENCES

1 A.Y. Coran, B. Das and R.P. Patel, U.S. Patent 4,130,535, 1978
2 M.T. Payne and C.P. Rader, "Thermoplastic Elastomers: A Rising Star" in "Elastomer Technology Handbook, N.P. Cheremisinoff, editor, pp 557-595; 1993, CRC Press, Boca Raton, FL
3 A.Y. Coran and R.P. Patel, "Thermoplastic Elastomers Based on Dynamically Vulcanized Elastomer-Thermoplastic Blends", in "Thermoplastic Elastomers, 2nd Edition, G. Holden, N.R. Legge, H.E. Schroeder and R.P. Quirk editors, pp 133-161, 1996, Hanser Publications USA
4 S. Abdou-Sabet and R.P. Patel, Rubber Chem. Technol. 64, 769, 1991
5 M.D. Ellul, J. Patel and A.J. Tinker, Rubber Chem. Technol. 68, 573, 1995
6 L. Goettler, J.R. Richwine and F.J. Wille, Rubber Chem. Technol. 55, 1449, 1982
7 P.K. Han and J.L. White, Rubber Chem. Technol., 68, 728 1995
8 M.D. Ellul, Plastics, Rubber and Composites Processing and Applications, 26(3), 137, 1997
9 G.J. Lake and A.G. Thomas, "Strength" in "Engineering with Rubber", A.N Gent, editor, pp. 95-127, Hanser Publications, USA, 1992
10 G.J. Lake and A.G. Thomas, "Strength Properties of Rubber", in Natural Rubber Science and Technology, A.D. Roberts editor, Oxford University Press, Oxford, Chapter 15, 1988
11 Replotted from data, due to Mullins, given in Figure 10.7, p. 263 of "The Chemistry and Physics of Rubber-like Substances", L. Bateman editor, Maclaren, London, 1962

Table I
Standard Test Results

Material	EPDM-PP TPE	NR-PP TPE	Typical NR*	Typical EPDM**
Relaxed modulus***, MPa	1.11 ± 0.03	0.86 ± 0.02	ca. 0.5	ca. 0.6
Tensile Strength, MPa	5.2 ± 0.04	4.2 ± 0.4	25-30	5
Elongation at Break, %	340 ± 25	290 ± 16	ca. 600	355
Work to Break, MJ/m³	10.3 ± 1.5	7.5 ± 0.96	ca.35	8
Tension Set, %	13.5 ± 0.4	19.3 ± 0.01	<5	3
Catastrophic tearing energy, kJm⁻²	10 ± 0.8	6.8 ±1.5	10 -15	8

*   Unfilled conventional sulfur vulcanizate
** Unfilled EPDM vulcanizate, phenolic resin cured
***Stress after one minute at 50% extension

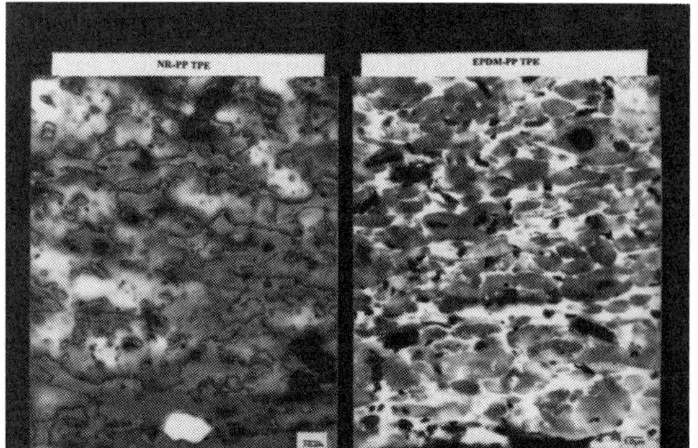

Fig.1.Transmission Electron Micrographs of NR-PP and EPDM-PP, TPEs

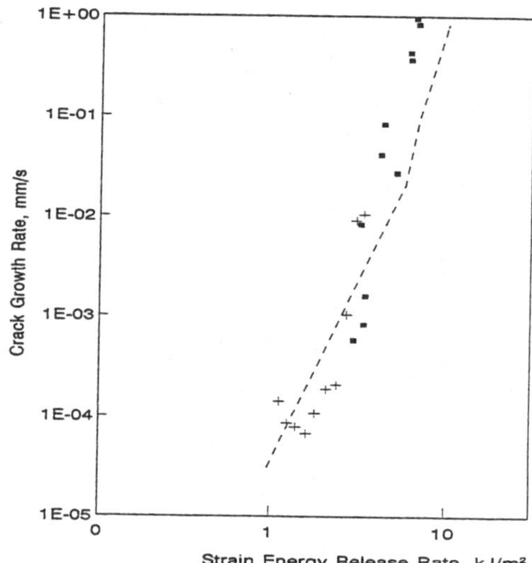

Fig.2. Time dependent crack growth results for the NR-PP TPE. 'Trousers' test results (constant force or constant rate of extension)■; results for tensile strips with edge cracks (constant force) +; the line shows the trend of the corresponding results for the EPDM-PP TPE

Fig. 3. WAXD of NR-PP and EPDM-PP, TPEs, unstrained and strained to λ = 4 and NR thermoset strained to λ = 4.

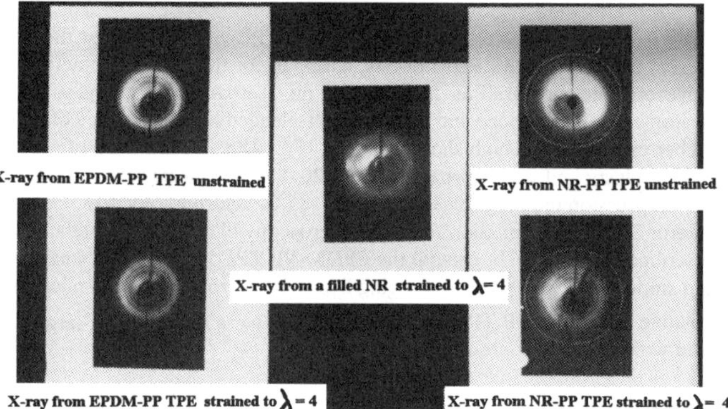

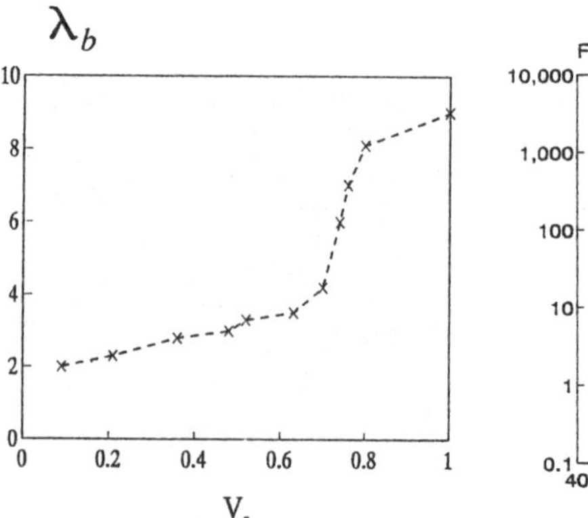

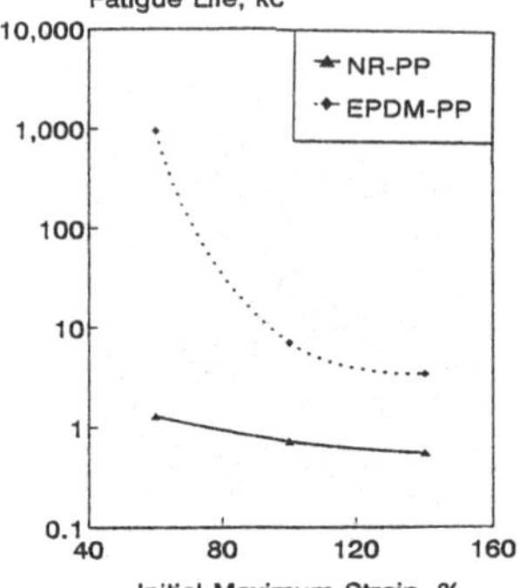

Fig.4: Effect of solvent swelling on breaking elongation of an NR thermoset vulcanizate

Fig.5: Tensile fatigue results

FRACTURE BEHAVIOR OF DYNAMICALLY VULCANIZED THERMOPLASTIC ELASTOMERS

A. J. Lesser and N. A. Jones
Polymer Science and Engineering Department
Silvio O. Conte Research Center
University of Massachusetts
Amherst, MA 01003

INTRODUCTION

Over the recent years, dynamically vulcanized Ethylene Propylene Diene Monomer rubber – Polypropylene thermoplastic elastomers (EPTPE's) have gained significant interest by the polymer community due to their complex morphologies and unique properties. Excellent discussions showing the range of morphologies and basic properties that can be obtained with EPTPE alloys have been reported by others [1, 2].

The purpose of this communication is to describe the energetics and micro-mechanisms of fracture in a model EPTPE subjected to a Mode I loading condition on a relatively thin specimen. The difficulties inherent in conducting these fracture studies are two-fold. First, the large reversible deformation that occurs in the vicinity of the crack due to the elastomeric nature of the material, significantly alters the stress singularity during loading. Second, the large irreversible deformation that occurs as a consequence of both the material behavior and specimen thickness negates the assumption of small scale yielding and requires that nonlinear fracture studies be conducted on these materials.

For ductile materials, two approaches have been used in order to characterize the fracture behavior. The most widely used parameter for characterizing the fracture in ductile materials is the J-integral approach proposed by Rice [3]. Traditionally, fracture characterized by this method required that the specimen must meet a certain size constraint in order to generate a plane-strain condition [4]. A second approach used to characterize the fracture of ductile materials is called the Essential Work method and was originally proposed by Broberg [5]. In this method, the total work of fracture is considered to be made of two components; one associated with the initiation of the instability (essential part) and the other associated with the plastic deformation in the plane-stress condition (non essential part). More recent studies by Paton and Hashemi studied the equivalence of both of these approaches to characterize the plane-stress fracture behavior [6].

This paper investigates the use of J-integral methods to characterize the fracture behavior of these materials under Mode I loading conditions. We also investigate the damage that occurs in the process zone of the crack tip and discuss its scale with relation to the morphology of the EPTPE.

EXPERIMENTAL

Material and Sample Preparation

The material discussed in this communication is a model EPTPE alloy comprised of 34/66 wt % of EPDM/isotactic polypropylene (iPP). The molecular weights of the iPP used in this alloy are $Mn=49,524$ and $Mw = 158,785$.

Samples were compression molded into plaques 3 mm thick, 200 mm long and 105 mm wide in a hot press. The polymer was melted at 200°C, pressed to 3.5MPa and cooled to room temperature by flowing cold water through pipes in the hot press while maintaining the pressure at 3.5MPa.

Center-notched fracture specimens were fabricated from the compression molded plaques by cutting a 38 mm (1.5") crack parallel to the plaques length (200 mm dimension) using a new razor blade. The specimen was then clamped along the entire specimen length using specially designed clamping fixtures. After clamping the specimen, the gauge width reduced to 64 mm.

The fracture tests were conducted on an Instron Model 1123. All tests were conducted at a crosshead speed of 51 mm/min (2"/min) and the crack lengths were measured optically with a Ziess stereo-microscope. All crack lengths were measured at the maximum load of each cycle. A typical load-displacement curve containing 9 cycles is presented in Figure 1.

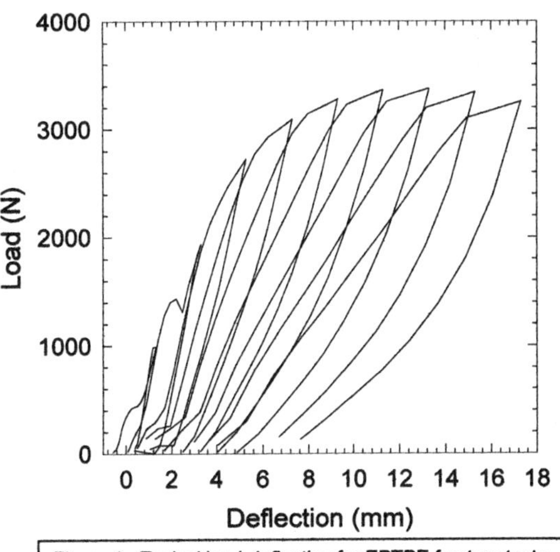

Figure 1: Typical load-deflection for EPTPE fracture tests.

The fracture energies were calculated using methods close to that described in the ASTM E813 [4]. The nonlinear energy release rate, J, was calculated by:

$$J_i = \frac{\eta U_i}{BL}$$

where U_i is the energy consumed in the propagation of the crack (area in the load-deflection curve), B is the specimen thickness, and L is the ligament length (W-a, where a is the crack length). In general, η is a dimensionless constant reflecting the geometry of the specimen. For our case, η equals 1.

RESULTS AND DISCUSSION

Fracture Behavior

The energy release rates were calculated for the model composition EPTPE as described in the experimental section. The measured energy release rates were plotted against crack extension for each cycle. These are shown typically in Figure 2. These results indicate a linear response between crack extension and energy release rate. The critical energy release rate as defined by zero crack extension was determined to be $J_{1c} = 16$ kJ/M^2 for this formulation. It should be noted that no offset was used in this estimate since the crack extensions were measured optically and no *blunting line* is necessary. The slope of the line in Figure 2 is indicative of the relative stability of the crack growth; a material with a steep slope is less likely to experience unstable crack propagation. This slope is quite often quantified by a dimensionless tearing modulus. The slope corresponding to the line in Figure 2 is dJ/da

= 4.1 kJ/M^3 and the behavior is linear up to crack extensions beyond 20 mm indicating that the crack propagation in these materials is quite stable.

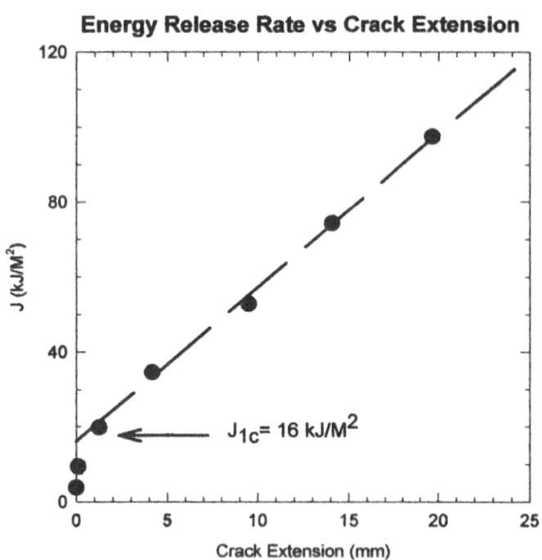

Energy Release Rate vs Crack Extension

Figure 2: Plot of Energy Release Rate (J) versus crack extension for EPTPE.

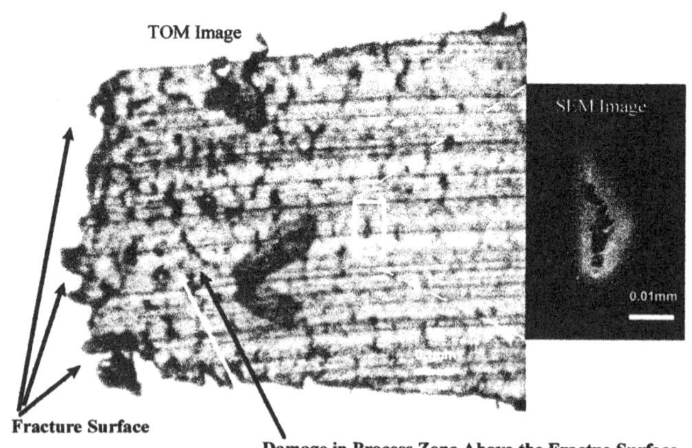

Figure 3: Composite TOM/SEM image of damage in the process zone above the fracture surface. Note that the damage appears as an array of regularly spaced crazes which increase in density as the fracture surface is approached.

Micro-mechanisms of Fracture:

It is anticipated that the fracture toughness and stability are highly dependent on the composition and morphology. Different compositions damage to different extents as the composition is altered. It is well established that the damage that proceeds and surrounds the crack consumes a significant amount of the fracture energy and often accounts for a major part of the materials resistance to fracture. Consequently, it is useful to identify the micro-mechanisms associated with the fracture process to help guide materials development.

During the fracture tests, a process zone of damage was observed in the optical stereo-microscope by the scattering of white light transmitted through the sample. In order to isolate the micro-mechanisms associated with this damage, optical sections were cut perpendicular to the fracture surface through the damaged material. These sections were cryo-cut using liquid nitrogen cooling and a glass knife, they could then be examined by Transmission Optical Microscopy (TOM). Similarly oriented block faces were microtomed smooth in a liquid nitrogen cooled ultratome, and then coated in gold and examined using Scanning Electron Microscopy (SEM).

Figure 3 shows a composite of TOM/SEM micrographs from a section cut through the damage area (note that the fracture surface is on the far left). The whitening macroscopically visible is evident here as a series of vertical black streaks perpendicular to the stress direction. The horizontal streaks are microtome knife marks, the v-shape in the lower half of the image is a second smaller section, both should be ignored. Closer inspection of the damage features by SEM showed that they are in fact crazes (see insert photo on right). The size of these crazes varies from 0.01mm up to 0.1mm. Note that these crazes are larger than the discrete elastomer domain sizes (1-3μm) within the continuous polypropylene matrix. The crazes appeared to be, in general, perpendicular to the stress direction, although the shape of the crazes are often tortuous. From the existing data it is unclear whether or not these crazes go through both the elastomer and the polymer domains, or just the polymer domains. It is evident from the image that the density of crazes is greatest near the fracture surface and gradually decreases with distance from the fracture surface. This is in accord with the diffuse nature of the damage zone in Mode I fracture tests.

CONCLUSIONS

The fracture behavior of a model EPTPE illustrates that these materials are tough materials with a critical energy release rate of 16 kJ/M^2. Moreover, they fracture in a stable fashion and show no evidence of instability up to crack extensions as high as 20 mm. The micro-mechanisms of damage occur as an array of regularly spaced crazes, which increase in density as the fracture surface is approached. Further, these crazes are of a scale that is approximately 2 orders of magnitude larger than the heterogeneous texture of the alloy (i.e., domain size of the iPP). Hence, it is expected that the crazes pass through both phases of the material, however, the interrelationships between this damage and material morphology are still under investigation.

ACKNOWLEGEMENTS

The authors would like to gratefully acknowledge the Advanced Elastomer Systems for their financial support and materials provided for this study. Also, the authors would like to thank Dr. M. D. Ellul for her many thoughtful comments and suggestions during this study.

REFERENCES

1. Ellul, M. D., Hazelton, D. R. (1994) *Rubber Chem. Tech*, **67**, 4, 582.

2. Ellul, M. D., Patel, J., Tinker, A. J. (1995) *Rubber Chem. Tech*, **68**, 4, 573.

3. Rice, J. R. (1968) *J. Appl. Mech. ASME*, **35**, 379.

4. *1988 Annual Book of ASTM Standards* (1988) **03.01**, 698.

5. Broberg., K. B. (1968) *Int. J. Fract.* **4**, 11.

6. Paton. C. A., Hashemi, S. (1992) *J. Matls. Sci.*, **27**, 2279.

EFFECT OF COMPOSITION AND PROCESSING CONDITIONS ON THE DEVELOPMENT OF STRUCTURAL HIERARCHY IN PP/EPDM BLENDS

M. Cakmak* and S. Cronin,
Polymer Engineering Institute,
University of Akron, Akron, OH 44325

INTRODUCTION

Injection molding imparts a complex thermal deformation history to polymer melts. This dependence results in a three dimensionally varying structure and corresponding properties[1,23]. This complexity further increases with a composition as a variable in a multiphase blends systems as in the case of dynamically vulcanized PP/EPDM blends. In this research, the effect of blend composition and processing conditions on the development of structural hierarchy in injection molded parts were studied through optical microscopy and matrixing microbeam X-ray diffraction technique and thermal analysis techniques.

EXPERIMENTAL

Three commercial grades with varying PP/EPDM composition were kindly provided by by Advanced Elastomer Systems. The commercial designations and their estimated PP contents are presented in the table1.

Table 1.

Commercial Name	Estimated content from DSC	PP
201-64	15%	
201-87	40%	
203-40	70%	

A Van Dorn 55-ton injection molding machine was used to produce end gated dumbbell samples. These samples were molded at three mold temperatures 30,70, 110°C and three injection speeds 0.75, 2.2 and 6.6 in./sec with 1 min. holding time. The temperature profile along the injection unit was set at 116°C, 190°C, 220°C, 220°C from first zone to the nozzle. A holding pressure of 3.45 MPa was used for all the parts.

CHARACTERIZATION

For optical observations the samples were cut normal to the flow direction at #3 location indicated as cut a in figure 1.

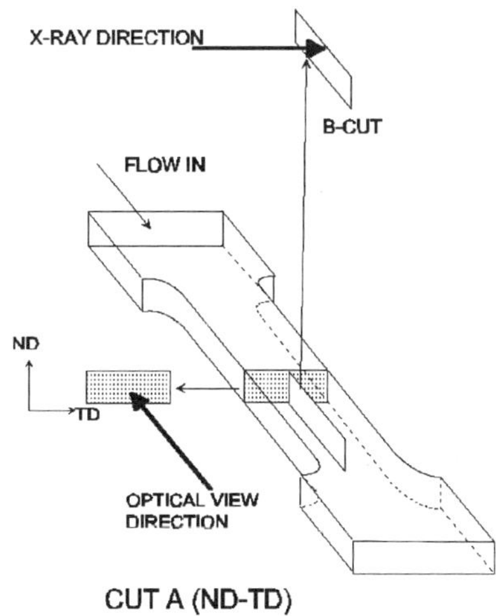

Figure 1.

The matrixing micro beam X-ray diffraction camera developed in our laboratories allows us to mount the sample cut with procedure b shown in figure1 with the cutting plane along the flow direction (FD-ND) and take WAXS patterns from skin to core regions at 100micrometer intervals without dismounting the sample. This enables us to maintain spatial registry between the consecutive patterns obtained. The WAXS patterns were digitized and orientation factors from these digitized images were obtained assuming transverse isotropy at these highly localized regions. The orientation factors for a,b,c crystallographic axes were determined from the azimuthal intensity profiles of (110) and (040) peaks using the following equation obtained through Wilchinsky's rule[4].

$$\langle \cos^2 \chi_{c,z} \rangle = 1 - 1.099 \langle \cos^2 \chi_{110,z} \rangle - 0.901 \langle \cos^2 \chi_{040,z} \rangle$$

And the Herman's orientation factors:

$$f_{b,z} = 1/2(3(\cos^2 \chi_{b,z}) - 1)$$
$$f_{c,z} = 1/2(3(\cos^2 \chi_{c,z}) - 1)$$
$$f_{a,z} = 1/2(3(\cos^2 \chi_{a,z}) - 1)$$

DSC

In order to obtain the crystallinity variation across the thickness of the samples a series of slices were obtained in FD-TD plane at the location where optical images and X-ray patterns were obtained and they were microtomed at the mid location shown in Figure 1 along the FD-TD plane and the samples were scanned with Perkin Elmer DSC7 at a heating rate of 20°C/min.

RESULTS

The DSC scans at a series of locations from skin to core of a typical 70/30 PP/EPDM sample molded at mold temperature of $T_m = 110°C$ and .75in/sec is shown in Figure 2

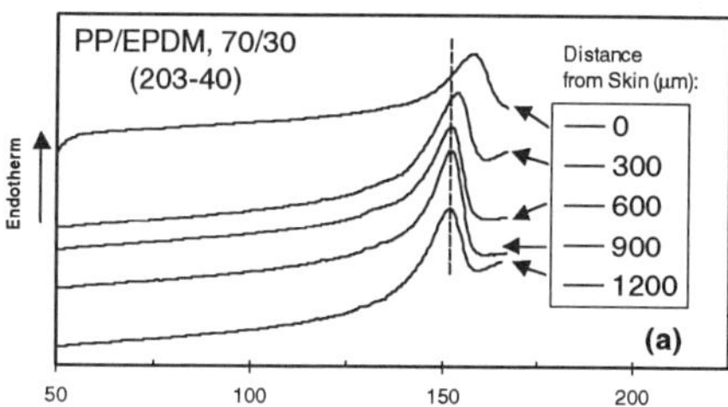

Figure 2.

The melting peak position of these curves can be observed in Figure 3 All three compositions show a decrease in the melting point from skin to core. This observation of high

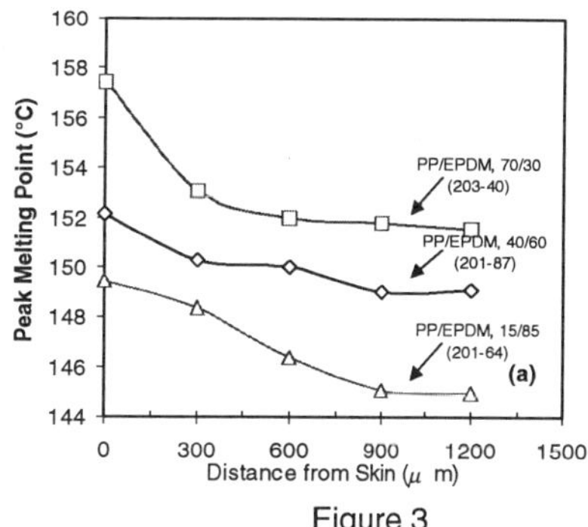

Figure 3

melting   temperature at the skin is attributed to the of highly oriented crystallites near the skin

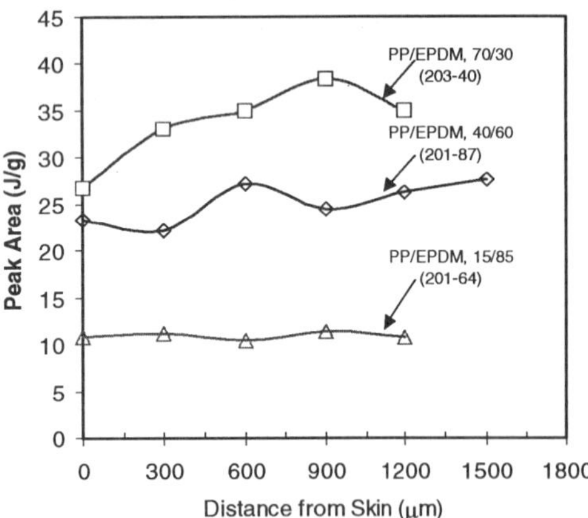

Figure 4

What is most striking in Figure 4 is that ΔH_{exp} (the area under the endothermic peak) is the lowest at the skin and increases with distance from the surface. This trend may mean either the overall crystallinity in the skin regions are low as a result of rapid cooling that occurred in the regions during injection molding, or the PP concentration varies from skin to core regions or possibly both.   In injection molding of fast crystallizing polymers like PP[5] and HDPE skin regions exhibit low crystallinity while the core regions exhibit high crystallinity due to the differences in cooling rates in these regions.

A   typical   WAXS   patterns   of   of   203-40, (PP/EPDM,70/30) molded at 70°C, with .75in/sec are shown in Figure 5.

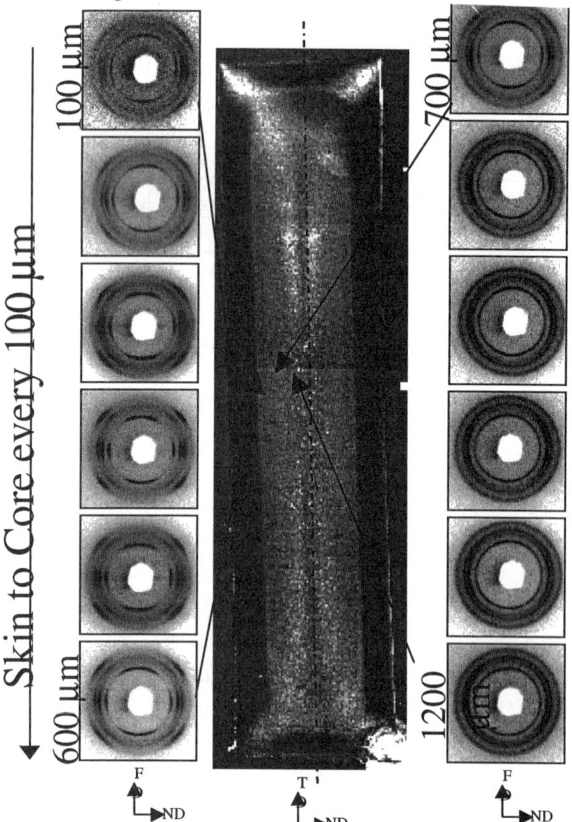

The skin regions possess slightly lower orientation level as judged from the azimuthal spread of intensities of (110) and (040) planes.   The (110) peaks also exhibit bimodal orientation behavior in this grade.   The equatorial peak (in ND) designated as c-axis oriented crystals, and the a* oriented crystallites in the double meridional peaks (in FD). It is quite evident that a transition from bimodal (c+a*) orientation regions to monomodal occurs by the disappearance of c–axis oriented crystals at the inner boundary of the dark shear regions. The spatial position of this boundary is determined when the flow front reaches the end of the cavity.

The quantitative orientation factors obtained from these patterns are shown in Figure 6.

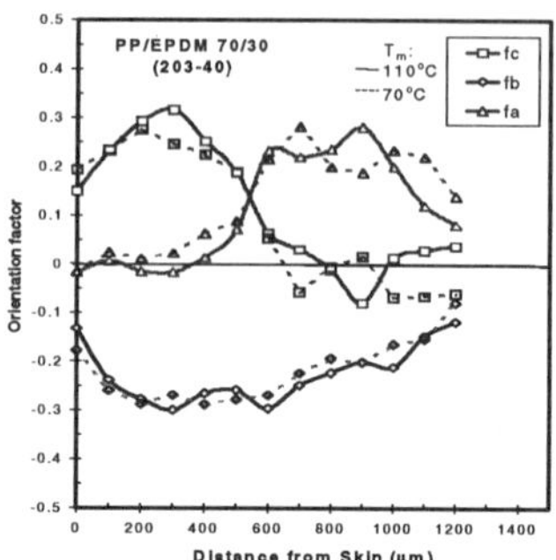

Figure 6.

The C-axis orientation factor in the flow direction first shows an increase to a maximum value that corresponds to the middle of the dark shear region, and then a decrease to a slightly negative value. On the other hand, the a-axis orientation steadily increases and reaches its maximum value at around 800µm before beginning to decrease. The crossover position between Fc and Fb with depth roughly corresponds to the inner boundary of the dark shear layers. There is very little difference in orientation behavior between samples molded at, 70°C and 110°C. The b-axes are generally concentrated in the direction roughly transverse to the flow direction.In addition lowering the mold temperature to 30° C resulted in formation of smectic structure near the mold surface (the data not shown here) near the skin region of this grade of material presumably as a result of rapid cooling. However , this structure changes to usual monoclinic structure at the interior of these parts.

CONCLUSIONS

Injection molded PP /EPDM are composed of optically dark skin layer followed by the core regions of lower nucleation density. The crystalline regions at the skin is highly oriented but disordered due to rapid quenching of the material against the mold wall. In certain blends, the decrease of mold temperature causes the skin to form disordered smectic structure. In general, the level of crystalline orientation increases towards the middle of the shear crystallized dark regions followed by rapid decay of orientation at the inner boundary of the shear crystallized region that is established at the moment flow ceases. It was also shown that the decrease of rubber fraction results in higher crystalline orientation in PP regions.

REFERENCES

[1] Y. Ulcer and M. Cakmak , Polymer 2907,**38** (1997)
[2] Y. Ulcer, M. Cakmak, C. Miao, C.M. Hsiung
     J. Appl. Poly. Sci. **60**, 669 (1996)
[3] C. M. Hsiung, M. Cakmak and Y. Ulcer,
     Polymer **37**,4555(1996)
[4] L. Alexander, "X-ray Diff. Meth. in Polymer Science", Wiley NY (1971)
[5] M. Kamal, F. May, J. Appl. Polym Sci. <u>28</u>, 1787 (1983)

Linear Viscoelasticity Of Dynamically Vulcanized Alloys: Kinetic Emulsion Network Model Studies

Yu F. Wang, Exxon Chemical Company, 5200 Bayway Dr. Baytown, Texas 77522-5200;

Maria D. Ellul, Advanced Elastomer Systems, L.P., 388 S. Main, Akron, Ohio 44311

ABSTRACT

In simple or dynamically vulcanized blends of polypropylene (PP) with EPDM, the EPDM cure state, PP MWD and Mw, PP/EPDM composition, Mw and MWD of EPDM, and process oil content can be varied to affect processability in injection molding, a major process step in shaping such materials. We have developed a predictive rheological model which incorporates molecular variables, such as PP and EPDM molecular weight distribution, process oil concentration, EPDM cure states, and blend composition into the model. This model, the Kinetic Emulsion Network (KEN) model, can *address the effects of these molecular variables and compositions on the linear viscoelastic behavior of such blends.*

From the model compound laboratory studies of 66 EPDM/34 PP blends, we have found:
• The KEN model predicts the linear viscoelastic (LVE) behavior of PP with different oil concentrations, specifically storage (G') and loss modulus (G") from the GPC trace.

• The KEN model predicts LVE behavior of PP and EPDM (cured and uncured) blends of different oil concentrations from known GPC traces of the components and oil/polymer concentration.

INTRODUCTION

Shen et al [1] modified the Rouse [2], Bueche [3], and Zimm [4] (RBZ) theory to elucidate polymer relaxation behavior of entangled systems. This is the Kinetic Network (KN) model. The bead's friction coefficient is not homogeneous but varies as the fourth power of the bead position on the chain. Thus, polymers with greater molecular weight are more "viscous" and have longer relaxation time. The friction increases from the chain end towards the center. Thus, for a linear chain with N beads (N entanglements) connecting N-1 springs, the equations of motion describing the chain dynamics are :

$$\frac{dx_i}{dt} = V_{si} - \frac{kT}{f_i}\frac{\partial \ln \psi}{\partial x_i} - \frac{3kT}{f_i<b_0^2>}(-x_{j+1} + 2x_j - x_{j-1}) - \frac{3kT}{f_i<b_0^2>}\sum_{i=1}^{N}\varepsilon_{ij}(x_j-x_i) \qquad (1)$$

Where :

$\varepsilon_{ij} = \alpha/|j-i|$ and $j\neq i$ with i being the index numbering starting from the chain end,

$<b_0^2>$ is the mean square end to end distance of a statistical segment,

f_b is the chain friction coefficient without entanglement,

f_i is the enhanced friction coefficient due to entanglement at the ith bead, $f_i = f_b (i/2)^4$,

α is the chain friction slip coefficient counting the chain slip at the entanglement point,

ψ is the chain end configuration distribution function, which describes the distribution of the distance between the end-to-end vector,

V_{si} is the surrounding medium (solvent) velocity at bead i.

Equation 1 poses an eigenvalue problem. The eigenvalues correspond to the relaxation times τ_i given by Zimm [4]:

$$\text{as } \tau_i = \frac{1}{2\sigma\lambda_i} \quad i = 1, Ne \; ; \; \lambda_i \text{ is the ith eigenvalue and } \sigma = \frac{3kT}{f_i<b_0^2>}. \qquad (2)$$

The relaxation modulus is :

$$G(t) = \frac{1}{Ne}\sum_{i=1}^{Ne}\text{Exp}(\frac{-t}{\tau_i}) \qquad (3)$$

The storage (G') and loss (G") moduli are $\dfrac{1}{Ne}G_N^0 \displaystyle\sum_{i=1}^{Ne}\dfrac{\lambda_i^2\omega^2}{1 + \omega^2\lambda_i^2}$ and $\dfrac{1}{Ne}G_N^0$

$\displaystyle\sum_{i=1}^{Ne}\dfrac{\lambda_i\omega}{1 + \omega^2\lambda_i^2}$, respectively. Ne equals to 2 N + 1.

The original approach was applicable to a binary blend system [5]. This modification can be generalized to a polydisperse system by considering such a material is composed of multiple fractions. The GPC trace is dissected into segments of molecular weights and each segment is modeled by using Equations 1 to 3 with the modifications outlined above. We have adopted the mixing rule of Schausberger [6] for the polydisperse systems :

$$G'_{blend} = \sum_{i=1}^{n}(w_i + 2\sum_{j=1}^{n}w_j)w_i\,G'_i \qquad (4)$$

where w_i and G'_i are the mass fraction and storage modulus of the component i.

The LVE behavior of polymer blends has been studied extensively in the past. Palierne [7, 8] considered linear viscoelastic behavior of emulsions of two incompatible polymers by incorporating the interfacial tension. The model has been tested on various systems: PMMA/PS/ PS-b-PMMA [8], PDMS/POE [9, 10], PS/PE [11, 12], PP/Polyamide 6 [13], and PS/LLDPE [14] blends. The emulsion model states that the complex modulus as a function of frequency, $G^*(\omega) = G'(\omega) + I\,G''(\omega)$, can be expressed as :

$$G^*(\omega) = G_m^*(\omega)\,\frac{1 + 3\sum_i \phi_i H_i(\omega)}{1 - 2\sum_i \phi_i H_i(\omega)} \qquad (5)$$

where:

$$Hi(\omega) = \frac{4\alpha/R[2G_m^* + 5G_i^*] + [G_i^* - G_m^*][16G_m^* + 19G_i^*]}{4\alpha/R[G_m^* + G_i^*] + [2G_i^* + 3G_m^*][16G_m^* + 19G_i^*]} \qquad (6)$$

α is the interfacial tension between blend components,
ϕ is the volume fraction of disperse phase,
m is the subscript refers to the matrix, the continuous phase,
i is the subscript refers to the inclusions, the disperse phase,
R is the particle radius of the disperse phase,
I is the square root of -1.

Equation 6 suggests that a blend's G' and G" can be determined by knowing the interfacial tension, particle radius or its distribution [8], and components' relaxation behavior in their neat form. *Equation 5 clearly states that the matrix phase (i.e., the continuous phase) dominates a blend's G' and G",* i.e., the LVE behavior. The term, Hi(ω), depicts the interactions between the matrix and inclusions.

It is well known that the plateau modulus of polymers decreases with the square of polymer concentration, ϕ_p, i.e. [15, 16]:
$$G_N^0 \text{ (diluted polymers)} = G_N^0 \text{(neat polymers)} \phi_p^2 \qquad (7)$$
The entanglement molecular weight, Me, is affected by the addition of diluent;
$$Me \text{ (diluted polymers)} = Me \text{ (neat polymers)} /\phi_p. \qquad (8)$$

A Combination of KN and Emulsion Model (KEN)

The PP GPC trace is used as the input to the KN model. The G' and G" are computed by the KN model. The emulsion model provides a method that combines individual phase's G' and G" with additional contributions from mechanical couplings between phases and from interfacial tension. The oil effect in PP and EPDM phase can be determined by the KN model by considering Equations 7 and 8. There are additional variables such as the cure state of EPDM, this cannot be incorporated into the KN model but will considered as a measured input variable into the KEN model. The blend LVE properties can thus be determined by the KEN model.

MATERIALS AND EXPERIMENTAL

Table 1A. Ziegler-Natta PP used in this study

	Oil* (wt %)	MWD	Mw
6088A-1	0	3.85	593 K
6088A	18	**	**

*The process oil is a paraffinic oil.
**The samples were made by adding oil to 6088A-1.

Table 1B. Uncured TPO (PP: EPDM+ = 1:2 and all blends uncured)

	Oil* (wt %)
60850B0	41.6

+ EPDM has Mw = 317 K and MWD = 3.7.

Table 1C. Cured TPE (PP: EPDM = 1:2 and all blends cured)

	Oil* (wt %)	Cure
6085C	0	high

The samples were molded at 230°C and cooled without cooling medium to room temperature to relieve thermal stress. The G' and G" were measured by the small amplitude oscillatory method at 190 °C in the frequency range of 10^{-2} to 10^{2} rad/sec range using the Rheometrics RMS800.

RESULTS AND DISCUSSIONS

The G' and G" of 6088A-1were calculated from the GPC trace by the KN model. The entanglement molecular weight, Me, was taken as 4500 and the plateau modulus is 4×10^{5} Pa from the KN model. The PP Me of 4,500 agrees with Plazek's 4,700 [17]. The validity of the KN model application to PP is verified through oil containing samples. The plateau modulus and entanglement molecular weight should vary with polymer concentration as Equations 7 and 8 prescribed. These changes were incorporated into the KN model for PP. Figure 1 shows the comparison of calculated G' and G" with the experimental data of PP containing 18 % oil.

Figure 2 shows the comparison of the measured G' and G" with the KEN model prediction of an uncured EPDM and PP blend with oil, sample 60850B0. Figure 3 shows the morphology 60850B0, obtained by TEM. From the comparison of the G' and G" at the low frequency region ($\omega < 1$ rad/sec), with G' and G" for EPDM, we can infer that the EPDM phase controls the blend (TPO [18]) relaxation behavior at the low frequency range. The LVE behavior at the low frequency range is related to polymer relaxations at a large time scale. Thus, the long time relaxation behavior is controlled by EPDM in such blends.

Figure 4 shows the KEN predicted and measured G' and G" of "fully" cured TPE without oil. The G' and G" resemble those of the cured neat EPDM. The G' and G" measurements and KEN model prediction all indicate that the EPDM is the dominant phase in the particular samples of this study. Thus, it is clear that the EPDM phase in these 66 EPDM/34 PP compositions determines the model blends LVE relaxation not PP.

SUMMARY

We have :

- Used the KN model to predict PP LVE behavior from its GPC trace. The KN model provides the linkage between the molecular composition of PP (MWD and oil concentration) with LVE properties.

- Developed a KEN model by combining the KN model with the Emulsion model to predict the LVE behavior of TPO and TPE. The KEN model can also account for cure states, oil concentrations, and phase compositions effects on TPO and TPE LVE behavior.

- In these model blends, the EPDM phase determines the blends LVE relaxation behavior.

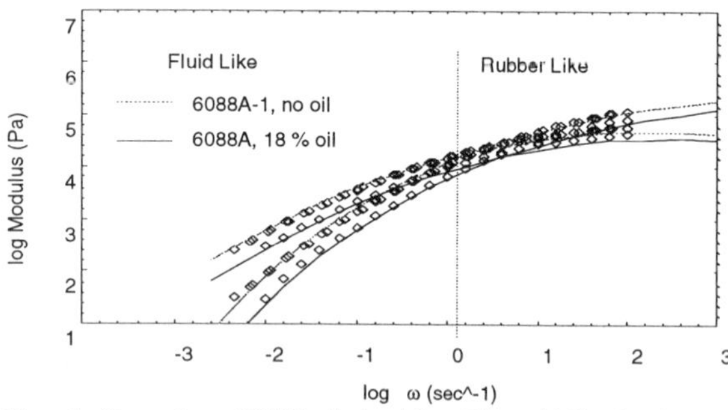

Figure 1. Comparison of G'/G" calculated from KN model (lines) and experimental data (diamonds). The dashed line is for PP with 18 % of process oil.

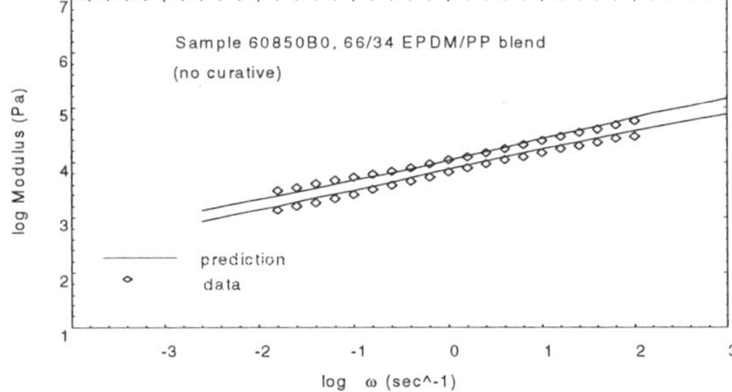

Figure 2. Comparison of G'/G" calculated from KEN model (lines) and experimental data (diamonds) for TPO, sample 60850B0, assuming the average particle size is 1 μm.

Figure 3. TEM micrograph of 60850B0, 66/34 EPDM/PP blend with 41.6 % oil.

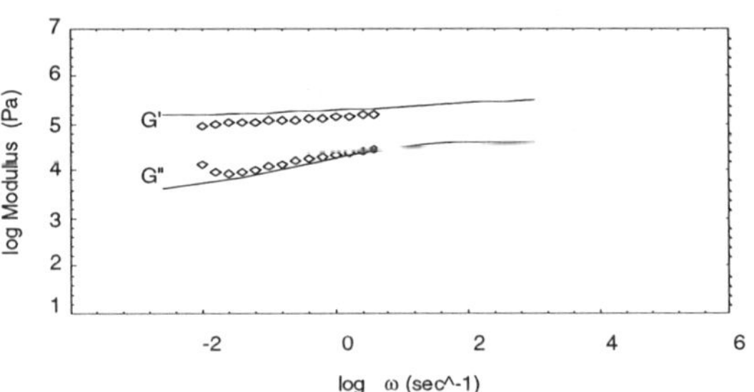

Figure 4. The comparison of G' and G" calculated from KN model (lines) and experimental data (diamonds) of fully cured TPE without oil, 6085C, 66/34 EPDM/PP, assuming average particle size of 1 μm.

REFERENCES

1   Hansen, D.R.; Williams, M.C.; Shen, M. Macromolecules, 345-354, 9(2), 1976.
2   Rouse, P.E. J. Chem. Phys. 1272,12 1953.
3   Bueche, F. J. Chem. Phys. 603, 22, 1954.
4   Zimm, B. J. Chem. Phys. 269,24,1956.
5   Soong, D.; Shyu, S.S.; Shen, M.; Hong, S.D.; Moacanin, J. J. Appl. Phys. 50(10), 6077-6082, 1979.
6   Schausberger, A.; Rheol. Acta. 25, 596-605, 1986.
7   Palierne, J.F. Rheol Acta, 29, 204-214, 1990.
8   Palierne, J.F. Macromoleculs, 26, 320-329, 1993.
9   Graebling, D.; Muller, R. J. Rheol. 34, 193-205, 1990.
10   Graebling, D.; Muller, R. Colloid Surf. 34, 89-103, 1991.
11   Brahimi, B.Ait-kadi, A. Ajji, A.; Jerome, R.; Fayt, R. J. Rheol. 35, 1069 - 1090, 1991.
12   Bousmina, M.; Bataille, P.; Sapieha, S. Schreiber, H.P. J. Rheol. 39(3), 499-517, 1995.
13   Scholz, P.; Froelich, D.; Muller, R. J. Rheol. 38(3), 481-499, 1989.
14   Lee, H.M.; Park, O.O. J. Rheol. 38(5), 1405-1425.
15   Ferry, J.D.; "Viscoelastic Properties of Polymers", John Wiley, 1980.
16   Rendell, R.W.; Ngai, K.L.; McKenna, G.B. Macromlecules, 20, 2250-2256, 1987.
17   Plazek, D.P. Macromolecules, 16,1469, 1993.
18   We use the term to describe PP/uncured EPDM blends

Controlling Morphology, Impact Strength and Rheological Properties in Dynamically Crosslinked Polypropylene/EPDM Elastomer Blends

Ivan Fortelný, Zdeněk Kruliš
Institute of Macromolecular Chemistry, Academy of Sciences of the Czech Republic, Heyrovsky Sq. 2, 162 06 Praha 6, Czech Republic

INTRODUCTION

Dynamically crosslinked polypropylene/EPDM elastomer blends, (PP/EPDM)$_C$, posses properties of tough plastics at high contents of PP and they behave as thermoplastic elastomers at low contents of PP. End use properties of tough polypropylenes and thermoplastic elastomers are strongly dependent on their morphology formed in their preparation. It is well known that PP/EP(D)M blends achieve high impact strength only when small particles of the elastomer are evenly distributed throughout the blend[1,2]. In some cases, the impact strength achieved for crosslinked PP/EPDM blends was higher than that for related uncrosslinked blends[1,3,4]. It was proposed that the higher impact strength of dynamically crosslinked blends is caused by the change in mechanical properties of the inclusions and their improved adhesion to the matrix[4]. Our experience with the study of the impact strength of PP/EP(D)M blends leads to the conclusion that the difference between the phase structure of dynamically crosslinked and uncrosslinked blends have decisive effect on the value of impact strength. Viscosity of the elastomer, which has substantial effect on the blend morphology, strongly increases during crosslinking. The course of curing affects the viscosity growth. Therefore, the phase structure and impact strength of the dynamically crosslinked blends can be controlled by a proper combination of the curing system and mixing conditions.

The main advantage of (PP/EP(D)M)$_C$ blends with higher contents of elastomer in comparison with PP/EP(D)M ones is their high yield stress in the solid state[5]. The mechanism improving yield stress apparently causes extraordinary rheological properties of molten (PP/EPDM)$_C$ blends as well. Steady viscosity at low shear rates or dynamic viscosity at low frequencies steeply grows with decreasing shear rate or frequency. The plateau occurrs in the dependence of storage modulus on frequency in the low-frequency range. At high shear rates or frequencies, rheological properties of dynamically crosslinked blends are practically the same as those of the uncrosslinked of the same composition. These effects are described for a number of blends with dynamically cured elastomers. It is accepted that a physical network is formed in dynamically crosslinked blends. So far, no idea explaining the physical ground of this network has been generally accepted. Recently, it was suggested that the physical network is formed by entanglements between different inclusions[6].

The aim of the contribution is to show that the (PP/EPDM)$_C$ blends with fine and uniform phase structure and high impact strength can be prepared by a proper control of dynamic crosslinking. The method is based on the knowledge of the phase structure development in uncrosslinked PP/EPDM blends and the course of curing. The second aim is to show that the differences between rheological properties of dynamically crosslinked and uncrosslinked blends can be explained as a consequence of entanglements between different domains of the elastomer.

EXPERIMENTAL

Commercial ethylene-propylene-diene terpolymer (EPDM) Buna AP 331 (Hulls, Germany) and isotactic polypropylene Mosten 52 422 (Chemopetrol, Czech Republic) were used.

Both uncured and dynamically cured blends were prepared by the melt mixing in the W50EH chamber of a Brabender Plasti-Corder for 10 min at 190°C and 60 rpm. In some cases, the curing system or its part was premixed in the W50EH chamber with EPDM (for 10 min at 60°C and 20 rpm) or PP (for 5 min at 190°C and 30 rpm). Dicumyl peroxide (DCP), tetramethylthiuram disulfide (TMTD) with 2-mercaptobenzothiazole (MBT) and sulfur systems were used for dynamic crosslinking.

Impact resistance of the blends was determined by measuring the notch impact strength at 0°C by Charpy method (ISO R 179/2C) and tensile impact strength at 23°C (DIN 53448). A Zwick tester having pendulum with maximum energy 4 J was used.

The morphology of the blends was observed using a JSM 35 scanning electron microscope (Jeol). The samples were cut in liquid nitrogen. The EPDM was etched off from the sample surface with heptane. This method was appropriate for uncrosslinked blends with lower contents of EPDM. Pictures of blends dynamically crosslinked and/or with higher contents of elastomer were less "readable".

Dynamic shear moduli, the absolute value of complex viscosity and steady viscosity were determined at 190°C with a Rheometrics System Four rheometer.

Polypropylene and the uncrosslinked elastomer part of dynamically crosslinked blends were extracted with trichlorobenzene at 140°C for 96 h.

RESULTS AND DISCUSSION
Morphology and impact strength

During mixing at 190°C of PP with EPDM premixed with dicumyl peroxide, strong reaction between PP and DCP appeared. The reaction caused degradation of PP accompanied by the production of gases which led to the foaming of the molten blend and prohibited preparation of test specimens. Therefore, we tried to prepare blends of PP with dynamically crosslinked EPDM. EPDM was dynamically crosslinked in the unheated chamber of a Plasti-Corder up to the steady temperature in the melt (about 8 min). The impact strength of the blends of PP with dynamically crosslinked EPDM was always lower than that of the related PP/EPDM blends and decreased with increasing amount of DCP. The decrease in impact strength was accompanied by an increase in the number and size of large elastomer particles in the blend.

We found that the phase structure development in the chamber of a Plasti-Corder for uncrosslinked PP/EPDM blends is nonuniform[2]. Some regions with fine phase structure were already found in the blends after a very short time of mixing. Simultaneously, large pieces of the elastomer surrounded by neat polypropylene appeared. The number and size of large elastomer particles decreased with increasing time and/or rate of mixing. At high rates and long times of mixing, the uniform phase structure was achieved. Impact strength of the blends increased with decreasing the number and size of large particles up to a limit related to the uniform phase structure. The time and rate of mixing necessary for the achievement of the uniform phase structure strongly increased with increasing viscosity of the dispersed phase[2,7]. It is apparent that EPDM dynamically crosslinked with DCP is highly viscous. Therefore, the preparation of its fine and uniform dispersion in the PP matrix by melt mixing in the chamber is practically impossible.

Sulfur and TMTD curing systems do not lead to degradation of PP. Therefore, they can be used for dynamic crosslinking of EPDM during its blending with PP. In this case, the phase structure of (PP/EPDM)$_C$ blends can be controlled if the phase structure development and the course of the curing are simultaneously considered. The viscosity of the elastomer phase gradually increases during curing. The phase structure of a blend at mixing is determined by the competition between the break-up and coalescence of droplets. Dynamic crosslinking leads to an increase in the droplet resistance against break-up and coalescence. At a certain degree of crosslinking, a further break-up of elastomer droplets at mixing is impossible[8]. Therefore, large elastomer particles should be broken before attaining this degree of crosslinking. On the other hand, fast crosslinking of small droplets is advantageous because it prohibits their coalescence. These requirements can be fulfilled if a curing system is premixed with PP or if PP, EPDM and the curing system are charged into the mixing chamber at once (one-step mixing). The blends prepared by one-step mixing and premixing the curing system with PP had comparable impact strengths. The blends prepared by premixing the curing system with EPDM had substantially lower impact strength.

In addition to the proper mixing protocol, an optimum composition of the curing system should be found. For the sulfur system, an admixture of a cure inhibitor improved the morphology and impact strength of (PP/EPDM)$_C$ blends[3]. For the TMTD system, it was found[3] that the optimum concentration of TMTD is 1 phr in the dispersed phase and the optimum ratio TMTD:MBT is 1:1. For (PP/EPDM)$_C$ blends prepared by one-step mixing with the optimum composition of the curing systems, a substantially higher impact strength than that for the related PP/EPDM blends was achieved (see Figure). Dependences of the notch impact strength and tensile impact strength on the mixing protocol and curing system were qualitatively the same. The tensile impact strength was somewhat more sensitive to the presence of large elastomer particles.

Dependence of the Charpy notched impact strength at 0 °C on EPDM concentration for dynamically cured and uncured PP/EPDM blends

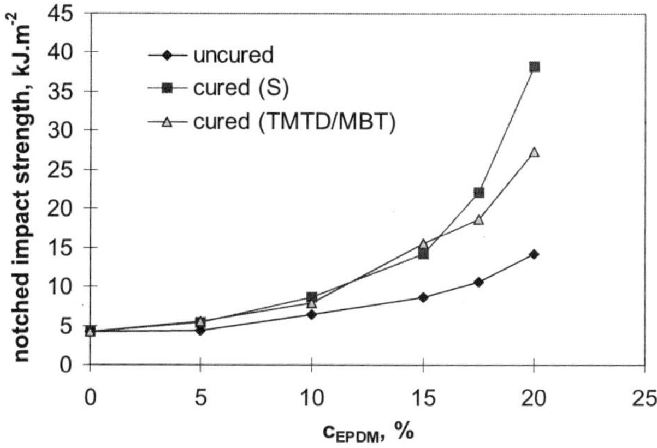

Rheological properties of the melts

Dependences of the absolute value of complex viscosity, $|\eta^*|$, and storage modulus, G', on angular frequency, ω, are the same for PP/EPDM and (PP/EPDM)$_C$ blends with the EPDM content smaller than 20%. Substantial differences appear for the blends with higher contents of EPDM. $|\eta^*|$ and G' of PP/EPDM blends lie between the values of PP and EPDM in the whole range of ω. The changes in continuity of the blend components are reflected[6] in concentration dependences of $|\eta^*|$ and G' at a constant ω. The behavior of (PP/EPDM)$_C$ blends in the range of small ω is typical of dynamically crosslinked blends: a steep increase in $|\eta^*|$ with decreasing ω and a plateau in the dependence of G' on ω[6]. These effects are stronger for high contents of the elastomer in the blend. Also the steady viscosity of (PP/EPDM)$_C$ blends steeply grows with decreasing shear rate in the range of low shear rates. Dependences of $|\eta^*|$ and G' on ω were similar for blends crosslinked by various curing systems. The blends cured with the sulfur system show somewhat higher values of $|\eta^*|$ and G' for small ω than the related blends cured with TMTD.

PP/EPDM blends were dissolved with trichlorobenzene on the molecular level (clear solution resulted). Rare gels appeared in solutions of dynamically crosslinked blends containing 30% and more EPDM. The gels broke very slowly (in days) into pieces and could be disintegrated by intensive shaking. The toughness of the gels increased with increasing EPDM content in the blends.

It follows from the preceding results that a physical network is formed in dynamically crosslinked blends with the elastomer content higher than 20%. The physical network can be disintegrated at a high stress. We believe that sulfur and TMTD curing systems cured EPDM only and did not induce any reaction with PP. Molecular forces between elastomer inclusions in dynamically crosslinked and uncrosslinked blends should be the same and no grafting of PP onto inclusions is assumed. The network is formed at a volume fraction of the elastomer where many inclusions touch their neighbors (percolation threshold is 0.156 for monodisperse spheres and somewhat higher for polydisperse ones). Entanglements are apparently formed between elastomer chains belonging to different inclusions. The relaxation time of the entanglements generally steeply increases with molecular weight of the entangled chains[9]. Therefore, relaxation times of the entanglements between touching inclusions in uncrosslinked blends and between chains in a neat uncrosslinked EPDM should be the same and comparable with relaxation times of the entanglements between PP and EPDM chains. However, entanglements between different dynamically crosslinked inclusions link chains with extremely high molecular weights. Therefore, they should also have extremely long relaxation times. We believe that in dynamically crosslinked blends with co-continuous structure, crosslinked domains exist which are joined by long-living entanglements.

We believe that this concept can explain all the observed effects. The number and relaxation time of long-lived entanglements should increase with the elastomer content in the blend. It is well known that the contribution of the entanglements (i. e., the effect of molecular weight) to the rheological properties decreases with increasing shear rate or frequency. Therefore, the entanglement concept can satisfactorily explain dependences of $|\eta^*|$ and G' on the elastomer content and ω for (PP/EPDM)$_C$ and PP/EPDM blends.

The authors wish to thank to the Academy of Sciences of the Czech Republic for support by grant No. 12/96/K.

REFERENCES
1. K. C. Dao, *Polymer*, **25**, 1527 (1984)
2. Fortelný, D. Michálková, J. Koplíková, E. Navrátilová, and J. Kovář, *Angew. Makromol. Chem.*, **179**, 185 (1990)
3. Z. Kruliš, I. Fortelný, and J. Kovář, *Collect. Czech. Chem. Commun.*, **58**, 2642 (1993)
4. M. Ishikawa, M. Sugimoto, and T. Inoue, *J. Appl. Polym. Sci.*, **62**, 1495 (1996)
5. S. Abdou-Sabet, R. C. Puydak, and C. P. Rader, *Rubber Chem. Technol.*, **69**, 476 (1996)
6. Z. Kruliš and I. Fortelný, *Eur. Polym. J.*, **33**, 513 (19977)
7. Fortelný , R. Rosenberg, and J. Kovář, *Polym. Networks Blends*, **3**, 35 (1993)
8. J. Karger-Kocsis, A. Kalló, and V. N. Kuleznev, *Polymer*, **25**, 279 (1984)
9. G. V. Vinogradov and A. Ya. Malkin,.*Rheology of Polymers*, Mir, Moscow, 1980

Viscoelastic Properties of Thermoplastic Elastomers

E.V. Prut, L.V. Kompaniets, N.A. Yerina.

Institute of Chemical Physics of RAS.
St. Kosygin 4, Moscow. 117977, Russia.

Introduction.

A class of polymer blends known as thermoplastic elastomers (TPE) is of both fundamental and applied importance[1]. The best may to produce TPE comprising vulcanized elastomer in melt - processable plastic matrixes is by the method called dynamic vulcanization. It is the process of vulcanizing elastomers during its melt - mixing with molten plastic. The combined mixing - vulcanization process leads to the formation of a heterogeneous structure in which the vulcanized elastomer particals with dimensions of the order of 1-10 μm are dispered in a continuos thermoplastic polymer matrix. The content of elastomer exceeds that of the thermoplast. The dimensions of the particles and the crosslinking density of the elastomers are factors significantly affecting the mechanical properties of TPE. They have many of the properties of elastomers, but they are processable as thermoplastics.

The mechanism of TPE deformation is insufficiently studied, and it is still unclear why the room- temperature mechanical properties of TPE are determined by the disperse phase rather than by the matrix[2].

In this work we have studied effects of the deformation rate ε and temperature T on the viscoelastic properties of TPE.

Experimental

Isotactic PP, oil-filled EPDM («Enicem Elastomers») and sulfur as the crosslinking agent were used in this study. The components were mixed at 453-473K using either a double-rotor static laboratory Brabender type mixer or a Berstorff twin screw extruder[3].

The tensile tests were carried out using an Instron tensile tester (model 1122). The moduli were determined for various deformation levels: E_0 at $\varepsilon = 1\%$, $E_1 = \sigma_1$ at $\varepsilon = 100\%$ and $E_2 = \sigma_2/3$ at $\varepsilon = 300\%$, where σ_1, σ_2 the stresses at 100% and 300% elongations respectively. The stress σ was calculated by dividing the measured load on the specimen by the original cross-sectional area of the specimen.

Results and Discussion

Fig.1a shows the stress (σ)-strain (ε) diagrams of TPE measured at various temperature T. A similar shape of the stress-strain diagrams was observed for the different deformation rates ε. All the TPE samples exhibit no «neck» formation on strecing.

The σ-ε diagrams are well linearized in the σ - $(\lambda-\lambda^{-2})$ coordinates for $\varepsilon > 3\%$ (Fig.1b). The stretch ratio was calculated

as $\lambda = \varepsilon + 1$. At small elongation ($\varepsilon < 3\%$) the curves exhibit a sharp increase passing to a less sloped straight line (see the inset in Fig.1b). These results can be described by the following equation:

$$\sigma = \sigma_0 + G (\lambda - \lambda^{-2})$$

where σ_0 is the initial stress jump and G is the modulus determined from the slope of the straight line. According to this equation the stress σ is a sum of two components: the first component rapidly increases with elongation to attain the σ_0 level and the second compontnt linearly increases with $(\lambda - \lambda^{-2})$.

This character of the σ-λ plot is related to individual contributions of PP and EPDM in the deformation behavior of TPE.

It was found that the initial stress jump σ_0 linearly decreases with increasing temperature and is independent of ε. For $T \rightarrow 418K$ $\sigma_0 \rightarrow 0$ this temperature corresponds to the onset of PP melting. We assume that the initial jump σ_0 depends on the yield limit of PP and determines by the deformation of PP.

G linearly increases to $\log\varepsilon$ and decreases with increasing temperature T. The behavior of G as a function of T and ε is inconsistent with the theory of rubber elasticity.

Fig.2 shows the dependence of E_0 and E_2 on ε and T. The values of moduli are linearly increasing functions of $\log \varepsilon$ and decreasing functions of T. The slope of the E - $\log\varepsilon$ plots corresponds to the same temperature decreases with the increasing of the deformation: $\Delta E_0/\Delta\log\varepsilon > \Delta E_1/\Delta\log\varepsilon > \Delta E_2/\Delta\log\varepsilon$. Therefore, an increase of the deformation is accompanied by decreasing dependence of the moduli on the deformation rate. These features are apparently related to the mechanisms of deformation of the PP and EPDM: the modulus of PP decreases and that of the crosslinked EPDM slightly increases as the temperature increases.

The WLF - equation is modified to following relation:
$$-1//\log a_T = a + b (1/T-T_0)$$
where a_T is the temperature shear factor, $T_0 = 293K$ is a reference temperature, a and b are the parameters.

Fig.3 shows the plots of (-1//log a_T) vs. 1/(T - 293) calculated from the experimental data for different values of modulus. The a_T values for E_0 (Fig.3) exhibit a considerable scatter and it is difficult to make definite conclusions concerning the character of the variation. The experimental points for E_1 fit the same straight line irrespective of the value of the modulus. The plots for E_2 are parallel to one another and have the same slope ($b = -12,8$). This value agrees satisfactory with ratio $c_2/c_1 = -11,5$ in WLF-equation. The value of parameters a drops with decreasing of E_2.

The WLF-equation is applicable to description of the viscoelastic properties of TPE, but the parameters of the equation depend on the stress. The lower the modulus E, the more adequately the viscoelastic properties of TPE are discribed by the WLF-equation with commonly accepted parameters. For TPE at large stresses in the nonlinear field the parameters of the WLF-equation depend on the modulus. As the modulus decreases the parameters tend to the published values[4].

We can propose the following mechanism of TPE stretching. The initial stage involves predominantly deformation of the PP matrix. In the subsequent stage the PP matrix deformes together with the dispersed EPDM phase. Domains of the crosslinked EPDM are bound together with a definite number of PP chains forming a network structure. The space between the domains and PP chains is filled with remaining PP molecules and the oil which is distributed between the PP and EPDM components according to the thermodynamic relationships. Deformation of the PP chains and PP macromolecules between the domains in the initial stage of stretching determines the initial modulus E_0 and the initial stress jump σ_0. Because of a weak bonding of domains and PP macromolecules and the presence of oil separating different regions, the subsequant deformation develops virtually independently in the PP components and the crosslinked EPDM.

References

1. A.J.Coran in «Thermoplastic Elastomers: A Comprehensive Review» N.R.Legge, G.Holden, H.E.Schroeder, (eds) Hanser Publishers, Munich-Vienna-New York, 1987, Ch.7.
2. L.V.Kompaniets, N.A.Yerina, L.M.Chepel,' A.N.Zelenetskii, E.V.Prut, Polymer Sci. <u>39A</u>, 827 (1997).
3. Russian Patent N2069217.
4. J.D.Ferry «Viscoelastic Properties of Polymers» New York-London, 1961.

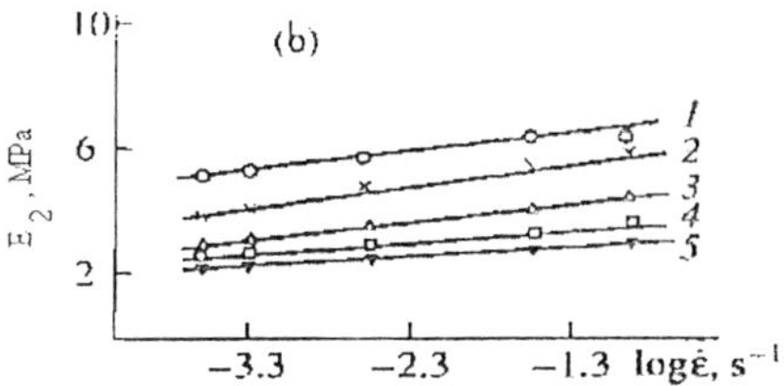

Fig.2. The plots E_o (a), E_2 (b) vs. log $\dot{\varepsilon}$: T = 293 (1). 323 (2), 348 (3), 373 (4) and 398 (5).

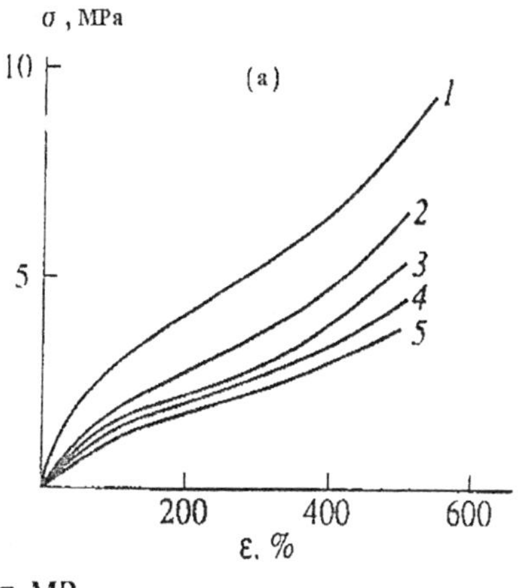

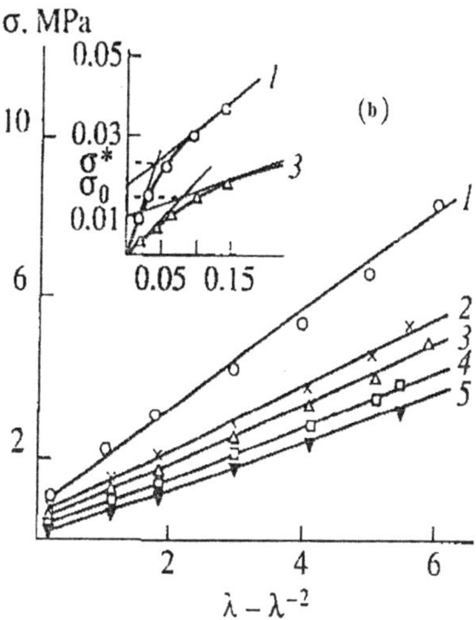

Fig. 1. Stress-strain diagrams plotted in the σ vs. ε (a) and σ vs. $(\lambda - \lambda^{-2})$ (b) coordinates: T= 293 (1). 323 (2), 348 (3), 373 (4), and 398 K (5). $\varepsilon = 2,4\times10^{-4}$ s^{-1}. The inset in (b) shows the stress-strain diagram for the small elongation (<3%).

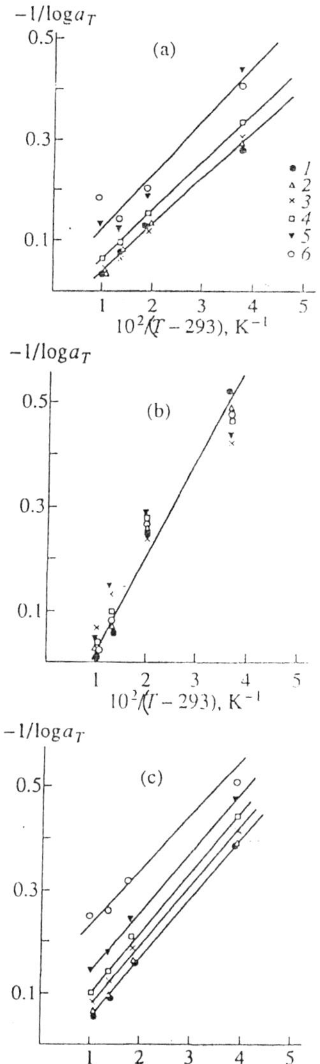

Fig.3. The plots of 1/log a_T vs. 1/(T-293) for various values of moduli: (a) E_o = 10.5(1); 10.0(2); 8.0(3); 6.0(4); 4.0(5) and 3.5 MPa (6); (b) E_1 = 3.5(1); 3.0(2); 2.6(3); 2.2(4); 1.8(5) and 1.2 MPa (6); (c) E_2 = 6.4(1); 6.0(2); 5.0(3); 4.0(4); 3.0(5) and 2.0 MPa (6).

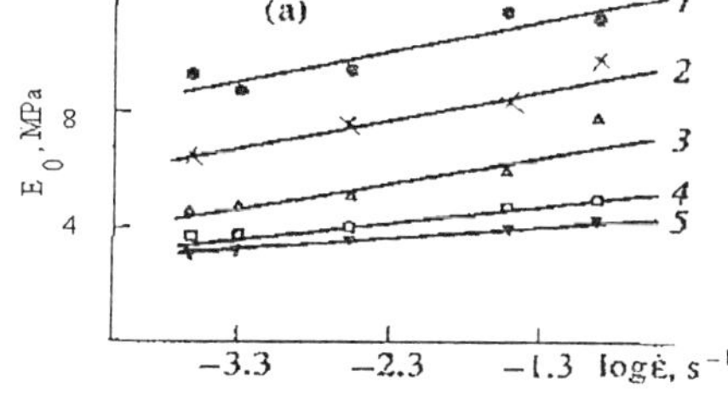

Dynamic Mechanical and Rheological Behavior Thermoplastic Elastomers.

E.V. Prut, N.A. Yerina, L.M. Chepel', A.N. Zelenetskii

Institute of Chemical Physics of RAS
st. Kosygin 4, Moscow, 117977, Russia

Introduction

Dynamic mechanical investigations are extensively used for elucidation of the phase structure of polymer blends. This is due to the fact that this method makes clear whether or not there exists a compatibility between the mixture's components[1].

Blends of PP with rubbers have been gained serios attention because they possess the processing and mechanical properties of PP and the flexibility of rubber. One of the widely studied system is PP/EPDM blend[2].

The blends of PP and EPDM where EPDM is cured have important science and technical advantages. The best way to produce PP/EPDM blends is the process of vulcanizing the elastomer during its melt - mixing with molten plastic called dynamic vulcanization[3]. Dynamic vulcanization is a route to new thermoplastic elastomers (TPE) which have elastomeric properties but can be processed as thermoplastics. TPE comprises finely divided elastomer particles (1-10 μm) dispersed in a relatively small amaunt of plastic. The elastomer particles should be crosslinked to promote elasticity.

In the present work we discuss the influence of crosslink density and oil content on dynamic mechanical and mechanical properties of TPE.

Experimental

It were used the following polymers: PP with $M_w = 3,7 \times 10^5$, crystallinity degree 50-60%, $T_m = 165^{\circ}C$; EPDM by «Enichem Elastomeri» was cured by the dynamic vulcanization method with different amounts of sulfur (S).

The blends were prepared in a «Brabender» plasticorder or in «Berstorff» twin screw extruder[4].

The mechanical tests were carried out on a «Instron 1122». Storage modulus (G'), loss modulus (G'') and damping (tanδ) of the pure components and the blends were determined by a «Du Pont 983» Dynamic Mechanical Analyzer from -150 to 140°C.

Results and Discussion

A DMA spectra of the pure PP and EPDM components are found to be in agreement with published data. The glass transition temperatures (T_g) of PP and EPDM are clearly detectable in the curver of G'', tanδ vs T. DMA spectra of blend PP/EPDM crosslinked with the different sulfur content shows the temperature peaks (Tabl.1).

Position of temperature of pure PP, EPDM and PP/EPDM, crosslinked with the different sulfur content [S].

	G', T$^{\circ}$C	G'', T$^{\circ}$C	tanδ, T$^{\circ}$C
PP	-22, 131	-84, -9, 146	-128, -81, -25, -9, 19, 76, 144
EPDM	-88, -38, 70	-75, -33	-81, -64, -54, -14, 13, 27, 70, 87
PP/EPDM [S]=0	-83, -38	-76, -32	-68, -33, 26, 76
PP/EPDM [S]=0.2phr	-83, -35	-76, -32	-69, -33, 26
PP/EPDM [S]=0.5phr	-82, -25	-73, -32	-69, -36, -35
PP/EPDM [S]=1.0phr	-80	-75, -32	-69, -36, -35
PP/EPDM [S]=1.5phr	-85	-75, -32	-71, -36, 26
PP/EPDM [S]=2.0phr	-83, -33	-75, -30	-68, -34, 26

However, the position of temperature peaks of PP/EPDM blends doesn't depend on sulfur content. The two-phase structure of PP/EPDM blends is clearly demonstrated by DMA spectra.

It is shown from the G' vs T curves of blends with the different sulfur content that the glass region of EPDM extends withe the increasing of sulfur content.

Fig 1 shows the effect of sulfur content at different temperatures. The storage modulus G' does not change at the glass transition region of EPDM (-62.5°C). In the glassy region G' at first drops and then increases. When sulfur content is equal 1.5phr there is the maximum of G' at $T<T_g$ of EPDM. At $T>T_g$ of EPDM the storage modulus G' enlarges with the increasing of sulfur content.

Various composite models such as the Halpin-Tsai's equation, Coran's equation, and Takayanagi's model have been applied to predict the storage modulus of blends. The upper bound of G' is given by

$$G' = G'_1\varphi_1 + G'_2\varphi_2 \qquad (1)$$

when G' is property of the blend G'_1 and G'_2 are the corresponding properties of PP and EPDM, respectively and φ_1 and φ_2 are the volume fractions of PP and EPDM respectively. In the case of the lower bound of G' the equation is

$$\frac{1}{G'} = \frac{\varphi_1}{G'_1} + \frac{\varphi_2}{G'_2} \qquad (2)$$

Fig.2 shows the experimental and theoretical curves of the storage modulus of the PP/EPDM blends as a function of temperature. As expected the storage modulus values lie in between those of G' obtained from eq.1 and eq.2 at $T>T_g$ of EPDM. For at $T<T_g$ of EPDM the experimental values of G' are higher than those obtained from eq.1. This may be due to the fact that the structure of EPDM in PP matrix is distinguished from of pure EPDM.

On curve of tanδ vs T there is clearly detectable peak at 68°C. The intensity of this peak depends on sulfur content (Fig.3). This intensity is found to decrease as sulfur content increases.

Oils are used in rubber to improve the mechanical properties and processability. The effect of oil in PP/EPDM blends is shown in Tabl.2.

Influence of oil content on the mechanical properties and melt flow index.

TPE*			s,	MFI,	E_0,	E_{100},	σ_b,	ε_b,
PP	EPDM	oil	phr	g/10min	MPa	MPa	MPa	%
25	75.0	0	1	-	56.9	4.4	7.1	330
25	60.0	15.0	1	0.05	47.8	4.5	12.8	533
25	45.0	30.0	1	1.60	41.6	4.2	11.0	593
25	37.5	37.5	1	5.75	39.9	3.6	11.6	641

The increasing of oil content is found to improve the melt flow index (MFI) and elongation at break (ε_b) and decrease the initial modulus (E_0).

References

1. Polymer Blends. D.Paul, S.Newman, (eds). Academic Press, New York (1978).
2. J.Karger-Kocsis, L.Kiss. Polym. Eng. Sci. 27, 254 (1987).
3. A.J.Coran in «Thermoplastic Elastomers: A Comprehensive Review» N.R.Legge, G.Holden, H.E.Schroeder, (eds). Hanser Publishers, Munich-Vienna-New York, 1987, Ch.7.
4. Russian Patent № 2069217.

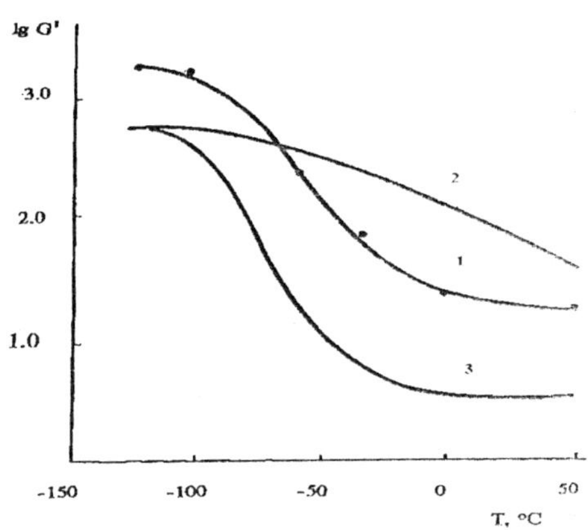

Fig.2. Experimental (1) and theoretical (2 - eq.1; 3 - eq.2) curves of storage modulus of PP/EPDM blends as a function of temperature

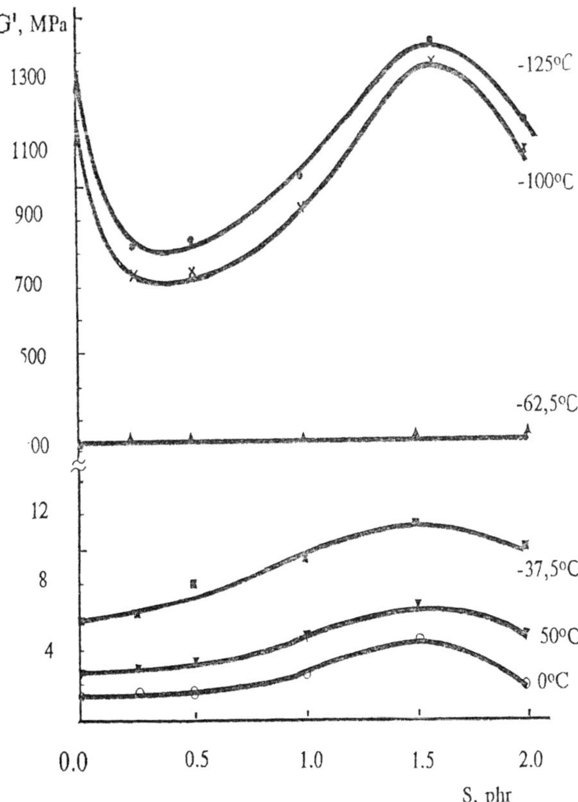

Fig.1 Effect of sulfur content on the storage modulus G' for different temperatures

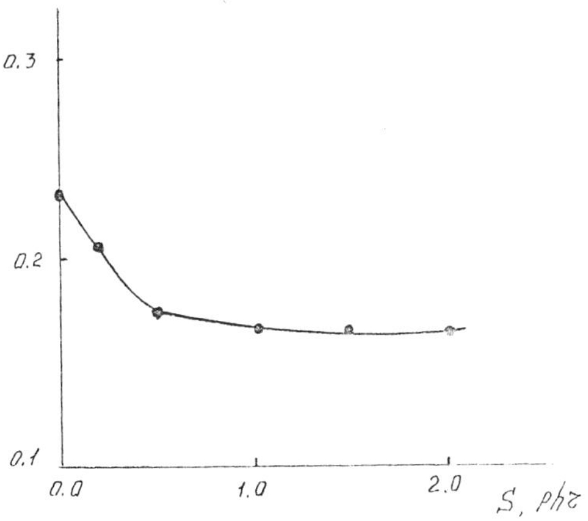

Fig.3. The peak intensity of tanδ at -68°C as a function of sulfur content

THE USE OF TPVS IN HIGH PRESSURE THERMOPLASTIC HOSE CONSTRUCTIONS

M. C. Hill*, and T. Ouhadi**

*Advanced Elastomer Systems, L.P., 388 South Main Street, Akron, OH 44311
**Advanced Elastomer Systems NV/SA, AVENUE dE BALE, 1,
BAZELLAAN,
B - 1140 BRUSSELS, BELGIUM

INTRODUCTION:

Engineered thermoset and thermoplastic elastomers have been used successfully in hose constructions for the transfer of gas, liquid, and solid materials as well as for the transmission of energy. Elastomers have been used to provide flexibility to the hose construction while improving strength for burst resistance in both static and dynamic applications. This balance of performance in elastomeric properties must also accomodate hose motion, misalignment, vibration, and portability. Flexibility of the hose is very important to enable easier routing and installation. Fittings and couplings are atttached to the hose ends to facilitate connection of the hose to a pressure source.

The vast majority of hose constructions consist of the three elements shown in Figure 1. The three components, tube, reinforcement, and cover are joined by some adhesive method to provide optimal performance. These composite constructions derive synergistic benefits from combining each material's inherent attributes. The materials used for cover and tube constructions are chosen based on their abrasion, fluid, and temperature resistance.

The requirements for the elastomeric components vary considerably depending on the end use applications. Historically, the most demanding applications have been in the transfer of hydraulic fluids. The requirements for these hose constructions are typified by high working pressures, a wide use temperature range and fluid resistance. Factors which affect the life of a hose are flexing below the minimum bend radius, twisting, kinking, crushing and abrading the construction.

The present work covers the use of flexible TPVs (thermoplastic vulcanizates) as replacement materials for other engineered thermoset and thermoplastic elastomers in hydraulic hose constructions and specialty fluid transfer applications. The standard designations for thermoset hydraulic hose constructions are SAE 100R1 and SAE 100R2 while the corresponding designations for the thermoplastic constructions are SAE 100R7 and SAE 100R8. Both the thermoset and thermoplastic hoses are rated for use from 2,500 psi to 5,000 psi working pressures at temperatures ranging from -40°F to 200°F.

Figure 1. Typical Hose Construction

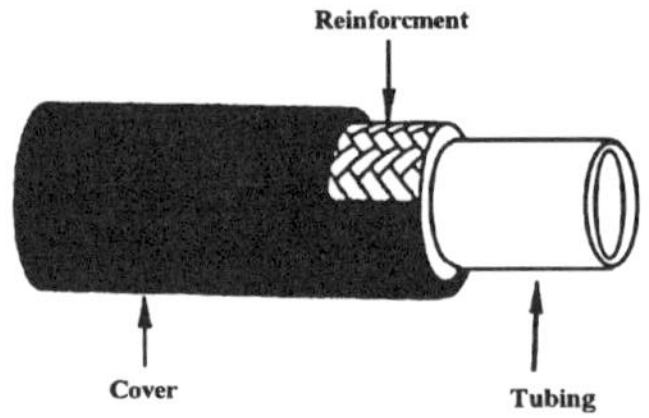

The function of the tube is to provide a pathway for different types of materials to be transferred. The tubing material must demonstrate satisfactory fluid resistance over the desired application temperature range of the hose. Failure to provide adequate fluid resistance results in weepage or permeation of the transfer medium through the tube assembly. The migration of transfer fluid through the tube often occurs during the life of the hose and represents the time at which degradation takes place and the assembly has to be replaced. In high pressure hose applications this can be a very hazardous situation and can result in an unsafe work environment and equipment failure.

Coupling retention is an important property of any given hose construction. Material properties which are considered to influence the coupling retention of a hose assembly are compression set, stress relaxation, and the modulus versus

temperature performance of the material. Good elastomeric candidates for hose tube and cover materials demonstrate low compression set, low stress relaxation and stable properties over the end use temperature range.

The primary function of the cover is to provide a barrier layer for the braided tube assembly from potential corrosion, abrasion, and other physical damage. Since the reinforcing braid in a closed braid construction is the major contributing component to the overall pressure rating of the hose it is imperative to protect it from damaging elements.

It is through the use of reinforcing braid that the hose is capable of performance at moderate to high pressures. The key to high pressure hose constructions is the linkage and adhesion of the tubing material with that of the reinforcing braid. This reinforced composite construction enables the transfer of material at high pressures over a wide temperature range.

The reinforcment braid angle is very important to the overall hose performance. The purpose for the selection and designation of the braid angle is to balance the hoop stresses with that of the end stresses for a given hose construction.

EXPERIMENTAL

The scope of this study focused on constructing a thermoplastic hydraulic hose utilizing TPV's as both tube and cover materials and evaluated various methods to promote adhesion.

The greatest barrier for widespread acceptance of TPVs in closed braid hose constructions has been due to the lack of adhesion to the reinforcing braid. TPVs, produced by the process of dynamic vulcanization, consist of crosslinked EPDM rubber particles in a continuous phase of polypropylene. The olefinic or non-polar nature of the TPV makes it difficult to bond to different surface chemistries using conventional adhesive systems. Polypropylene, for example exhibits low surface energy levels of 30-33 dyne/cm. The low surface energy of olefins and the lack of polar, functional groups make it difficult for satisfactory wetting using epoxy, urethane, and acrylic structural adhesives[1]. These adhesive types exhibit higher surface energy than polypropylene or polyethylene and, as a result, tend to bead up on the surface of the TPV rather than form a continuous adhesive film.

The low surface energies of these TPV's can be improved by using surface treatments and primers[2,3]. Mechanical pretreatments such as plasma discharge and flame treatment, and chemical / solvent treatments are used throughout the marketplace to improve adhesion to thermoplastic olefins. The overall impact of these pretreatments on the adhesive characteristics of TPVs varies, depending on many process and compositional factors. The level of rubber and its composition within the TPV formulation have a significant impact on the adhesive performance .

The effect of mechanical pretreatments is related to the degree of oxidation on the adhering surface. Oxidation of the polyolefin results in increases of the surface energy thereby making it easier to adhere and coat. The ability to oxidize a surface is also strongly dependent on the composition of the TPV. Natural rubber based TPV's have been shown to be receptive to halogenation treatments. Oils and fillers which are prevalent in the crosslinked rubber phase of a TPV will deter oxidation and surface modification.

Primers such as chlorinated polyolefins (CPO) have been used successfully as "tie-coats" to improve the paintability and adhesion of thermoplastic olefin surfaces[2,4]. These primers, which often are carried in solvent, increase the surface energy of the TPV through the polar functionality. Equally importantly, these materials impart decreased crystallinity on the surface of the TPV, resulting in improved adhesion performance using conventional coatings and structural adhesive systems[4].

RESULTS

Table 1 shows the effect of pretreatments on the surface energy of TPV materials. Untreated TPVs all exhibit very low surface energy. The corona pre-treatment was conducted on flat material stock at a line speed of 17 feet per minute. There was a marked improvement in the surface characteristics of the TPV after corona treatment. The response of the TPV to the corona treatment improved as the hardness of the material increased. This improvement can be correlated to the level of thermoplastic in the TPV. The chlorinated polypropylene primer evaluated in this study did not show the same improvement in surface energy. Based on these results corona treatment would be the choice surface pretreatment used in combination with conventional adhesive systems.

Table 1. The Effect of Surface Pre-Treatments on TPV Surfaces.

TPV	Untreated	Corona Treatment	CPO Treatment
55 Shore A	30-32 dyne/cm	40-41 dyne/cm	34-36 dyne/cm
75 Shore A	30-32 dyne/cm	41-42 dyne/cm	36-38 dyne/cm
50 Shore D	30-32 dyne/cm	64 dyne/cm	36-38 dyne/cm

The prevalent adhesive systems used within the high pressure thermoplastic hose industry are single component urethane adhesives. These moisture cure urethanes have been used for over 20 years to bond copolyetherester and TPU materials to textile reinforcment. These liquid systems are available in solvents as well as in 100% solids form. Generally it takes 72 hours for the adhesive system to fully cure with moisture to develop final adhesive properties. Once crosslinked, these urethane systems exhibit excellent fatigue and temperature resistance.

The current study evaluated an alternative approach to using liquid adhesive systems for a bonded TPV assembly for use in specialty fluid transfer applications. This work was conducted to replace solvent-based adhesives and other necessary pretreatments for the fabrication of TPV hose assemblies. Due to the low hydraulic fluid resistance exhibited by tubing made with olefinic TPV systems, a fluid resistant barrier layer was coextruded with the TPV to replace nitrile thermoset elastomers and copolyetherester materials as tubing materials. Nylon 6 was chosen as the inner portion or sleeve of the tube construction to provide the required fluid resistance. A 0.2 mm wall of the polyamide was coextruded with 0.6 mm of an 85 shore A nylon bondable grade of TPV to provide flexibility and fatigue resistance to the tube construction. The nylon bondable TPV grade had been developed to adhere directly to polyamide 6 materials during insert molding and co-extrusion. The functionality present in this TPV allows for adhesion to metal surfaces.

The coextruded tube was reinforced with brass plated steel braid commonly used in thermoset hose and belting applications. An elastomeric cover made of the same nylon bondable TPV was then crosshead extruded over the reinforced hose carcass. An alternative design utilized a coextruded tube consisting of a maleic anhydride modified polyolefin tie layer in combination with a standard olefinic TPV crosshead extruded over the reinforced tube. Anhydride and/or acid functionality provides adhesion to the metal surface while the olefinic backbone provides adhesion to the continuous thermoplastic phase of the TPV system.

Table 2 shows the adhesion performance of several combinations of TPVs to wire reinforcment using various tie layer adhesive systems. The data shows a marked improvement in adhesion using both tie layer adhesives in combination with a conventional olefinic TPV and the Nylon Bondable TPV. The "H" test data show excellent adhesion to the wire reinforcment using the modified polyolefins and the nylon bondable grade.

Table 2. TPV Adhesion Results to Wire Reinforcement

TPV	Tie Layer Adhesive	Wire Type and Size (mm)	Adhesion (N / 4 cm)
80 Shore A	none	Brass plated (0.36)	50
80 Shore A	Adhesive A	Brass plated (0.36)	520
80 Shore A	Adhesive B	Brass plated (0.36)	450
80 Shore A	Adhesive C	Brass plated (0.36)	410
85 Shore A Nylon Bondable	None	Brass plated (0.36)	390

This level of adhesion enabled these hose designs to pass stringent impulse and burst testing. A primary discovery was the excellent fluid resistance demonstrated

by the hose assembly. The multilayer hose was pressurized with hydraulic fluid at 100°C for one month without weepage or leakage of the fluid through the hose layers or at the fittings.

The use of a tie layer adhesive in combination with a newly developed series of TPV's (TPV B) makes for a promising hose cover construction. Figure 2 shows the comparison of abrasion resistance for several thermoplastic materials using a Taber Abrader. The 90 shore A hardness material, TPV B, compares very favorably in its abrasion resistance with a competitive polyether based TPU of identical hardness. TPV B also shows an improvement in abrasion resistance over more conventional TPV (TPV A) grades. This new series provides a potential alternative to thermoplastic urethane materials.

Figure 2. Comparison of Taber Abrasion Results.

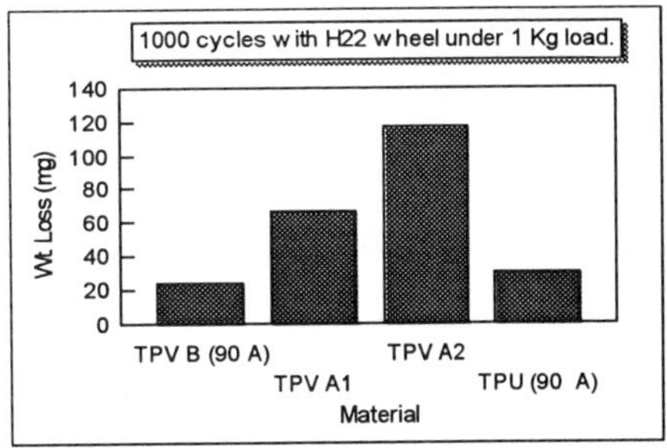

CONCLUSION

A thermoplastic hydraulic hose construction has been successfully fabricated utilizing TPV materials for the cover and as intermediate layers of the tubing assembly.

The preferred tube assembly consists of a thin layer of an impact modified polyamide 6 or a pure polyamide 6 resin to provide resistance to hydraulic oil migration and weepage. This fluid resistant layer was coextruded with a nylon bondable TPV which demonstrated excellent adhesion to polyamide materials in the melt phase. The coextrusion of a 0.2 mm thick inner layer of polyamide with the Nylon Bondable TPV as the outer tube layer provided a flexible, fluid resistant, tube construction.

The reinforcing layer between the tube and cover was comprised of metallic wires which were braided to provide satisfactory resistance to burst during exposure in moderate to high pressure environments. Adhesion between the elastomeric TPV and the wire reinforcement was achieved through use of specific tie layer adhesives and modified TPV materials. Rubber tearing bonds were achieved between the TPV and the wire through the use of select adhesive tie layers and understanding the surface characteristics of the reinforcment.

The hose assembly fabricated utilizing the polyamide and TPV tube demonstrated satisfactory performance during extended testing. The hose assembly did not exhibit any weepage of the hydraulic oil after one month testing at 100°C.

BIBLIOGRAPHY

1. I. Skeist
   Handbook of Adhesives, New York, Chapman and Hall, 1990.

2. R. A. Ryntz
   "Coating Adhesion of Low Surface Free Energy Substrates in the Automotive Industry", *Polymer News*, Vol. 18, pp.101-106, 1993.

3. D. Briggs
   "Surface Treatments for Polyolefins" in Surface Analysis and Pretreatment of Plastics and Metals, New York, Macmillian Publishing Co., 1982.

4. R. A. Ryntz,
   "Recent Advances in the Understanding of the Paint Adhesion Mechanism to Automotive Plastics",

ADVANCED THERMOPLASTIC VULCANIZATES BONDABLE TO NYLON

Reza Sadeghi, Mark O'Connell, and Dave Clare
DSM Thermoplastic Elastomers, Inc.
29 Fuller St., Leominster, MA 01453

Introduction

Power tools, electronics, appliances, and sporting goods demand soft-touch surfaces for ergonomics, personal comfort and styling. In these applications, low hardness adhesive thermoplastic vulcanizates (TPVs) can be utilized. TPVs based on polypropylene (PP) and terpolymer of ethylene, propylene and non-conjugated diene (EPDM) are non-polar in nature, which makes adhering them to polar materials such as nylon, a considerable challenge. In the past, mechanical interlocking was used to produce overmolded TPVs onto the polar substrates used in these applications. Today, a new generation of TPVs can bond to nylon 6, 66, nylon blends with ABS and/or PP. These materials can be processed via insert injection molding, two-shot injection molding, and coextrusion. These nylon bondable grades are based on Sarlink® 3000 series technology and offer characteristics similar to the 3000 series products. These new products offer excellent temperature and fluid resistance properties similar to the standard TPVs. The concept of a new TPV bondable to nylon offers design engineers vast new options in designing combinations of soft skin to hard segment applications.

Physical Properties

Table 1 shows physical properties of the two grades of the nylon bondable TPVs. These two grades are available in black, with a hardness of 65 and 75 shore A durometer. As can be seen from tensile strength, elongation at break, 100% modulus, and tear strength data, these two TPV grades have substantial cohesive strength, which is needed in the coextrusion and demolding during injection molding processes.

Sample Preparation and Peel Test Methodology

Determination of bond strength can be effectively measured by the utilization of a peel test, using insert injection overmolded samples to various nylon substrates. The nylon substrate can be prepared by injection molding a plaque (10.2 cm x 10.2 cm x 0.32 cm) containing a triangular metal insert. This triangular nylon plaque is then inserted inside a mold adjacent to a triangular metal insert (10.2 cm x 10.2 cm x 0.64 cm) of the same thickness. The TPV is molded over both the nylon substrate and metal insert. The melt temperature, mold temperature, and preheat temperature are selected to be appropriate for both compounds. A tensile bar 2.5 cm in width is cut diagonally through the overmolded section but not through the nylon substrate, as shown in Figure 1. This overmolded tensile bar is tested for peel strength using an Instron® Tensile tester equipped with a sliding grip. The rate of peel test was 50 cm/min.

This peel strength test is a tool that can be utilized for measuring relative adhesion of the material to various substrates. Overmolded samples were tested after 24 hours for peel strength. Peel strength results provided throughout this paper are average values of at least 5 peel tests.

Bond Strength Results

Table 2 shows peel strength of 65A and 75A nylon bondable TPVs insert injection overmolded at the same condition to various nylon substrates. These nylon substrates were nylon 6, nylon 6 impact-modified glass filled (J7/33), and nylon 6,6 glass filled alloy with PP. Adhesion peel strength of the 65A grades based on the above nylon substrates varies from 44 N/cm to 65 N/cm. The failure mode was cohesive for all these substrates. As can be seen from the table, the adhesion peel strength for both grades is a function of the nylon substrate types.

Heat Aging of Nylon Bondable TPVs

Heat aging exposure of thermoset elastomer compounds could result in significant changes in physical and mechanical properties. Table 3 shows heat aged physical properties of nylon bondable TPVs exposed to a temperature of 150°C for a week. Both 65A and 75A nylon bondable grades show excellent retention in tensile strength, 100% modulus and tear strength.

Figure 2 shows peel strength retention of overmolded 65A TPV with nylon 6 (J7/33) exposed to various temperatures ranging from 70°C up to 150°C for a week. This nylon bondable grade exhibits excellent retention in bond strength with exposure to high temperatures.

Fluid Resistance of Nylon Bondable TPVs

Fluid resistance of nylon bondable grades is an important parameter. Table 4 shows the effect of two types of oils at room temperature, IRM-903 oil and ASTM #1 oil, on the bond strength of 65A TPV. This grade was overmolded onto nylon 6 (J7/33) and bonded samples were submersed in the above oils for 24 and 168 hours. As can be seen from this table, 65A nylon bondable grade retains about 85% of its original bond strength. This result reconfirms this new nylon bondable grade can withstand exposure to these types of oils without significant deterioration of its cohesive and bond strength.

Processing of Nylon Bondable TPVs

Due to more emphasis on soft-touch applications and ergonomics issues, new opportunities have been created for TPVs bondable to nylon resins. With the increasing demands and utilization of these grades, we need to provide a detailed recommendation on the processing of adhesive TPVs. These materials are similar to the standard TPVs in regard to viscosities and flow properties. These materials can be processed via extrusion and injection molding processes.

A general summary of the recommended processing conditions for both insert injection overmolding and two shot injection molding is provided in this section. In the case of the insert injection-molding process, pre-heating the nylon substrate to 93 – 110°C is required to attain the adhesion needed at the interface. Moreover, in order to obtain excellent bond strength for these TPVs bondable to nylon, it is recommended to keep melt temperature in the range of 260 – 280°C for both insert injection molding and two shot injection molding processes. The processing conditions of these grades are similar to our general-purpose conditions with the exception of higher melt temperature. In the case of coextrusion, depending on the profile, the nylon bondable TPV and the nylon are at their melting points, which is the ideal case for a good adhesion.

Summary

TPV bondable to nylon has many good features and an appropriate balance of properties needed for many applications. These materials offer flexibility and ease of processing of TPV and yet offer the desired adhesion to nylon materials.

Table 1 Physical properties of the nylon bondable TPVs.

Properties & Units	Test Method	65A Hardness	75A Hardness
Color		Black	Black
Hardness, Shore A	ASTM D 2240	65A	75A
Specific gravity	ASTM D 792	0.97	0.97
Tensile Strength, Mpa. (Strong Dir.)	ASTM D 412	3.9	4.5
Elongation at Break	ASTM D 412	420	440
Modulus @ 100%, Mpa.	ASTM D 412	2.7	3.4
Tear Strength, KN/m	ASTM D 624	30.3	36.8
Compression Set, % @ 70 C/22 Hrs	ISO-815 (A)	62	67

Table 2 Peel strength of 65A & 75B TPVs to various nylon substrates (preheat temperature of 149°C).

Nylon Substrate Type	65A TPV Peel Strength, N/cm	75A TPV Peel Strength, N/cm
Nylon 6	44	32
Nylon 6 Impact Modified/ 33% glass filled (J7/33)	43	30
Nylon 6,6/ PP alloy with 33% glass filled	65	60
Failure Mode	Cohesive	Cohesive

Table 3 Heat aged physical properties retention of the nylon bondable TPVs for a week @ 150°C.

Properties & Units	Test Method	65A Hardness	75A Hardness
Color		Black	Black
Hardness changes	ASTM D 2240	+1	+3.5
Tensile Strength, Mpa. (Strong Dir.), % retained	ASTM D 412	96	92
Elongation at Break, % retained	ASTM D 412	120	80
Modulus @ 100%, Mpa., % retained	ASTM D 412	99	96
Tear Strength, KN/m, % retained	ASTM D 624	86	80
Compression Set, % @ 70 C/22 Hrs, retained	ISO-815 (A)	No change	104

Table 4 Effect of exposure to oils on the peel strength

Exposure Time to Oils at room temperature	Peel Strength, N/cm
TPV 65A Black (Control)	50
24 Hours Exposure to IRM 903 Oil	43
24 Hours Exposure to ASTM #1 Oil	42.6
168 Hours Exposure to ASTM #1 Oil	46.3

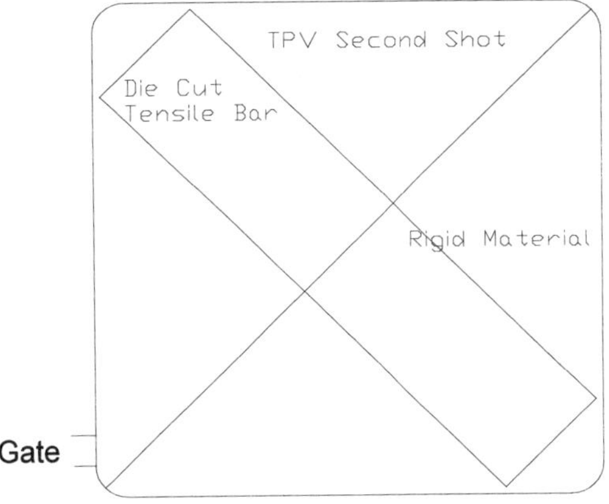

Figure 1 Nylon over molding mold

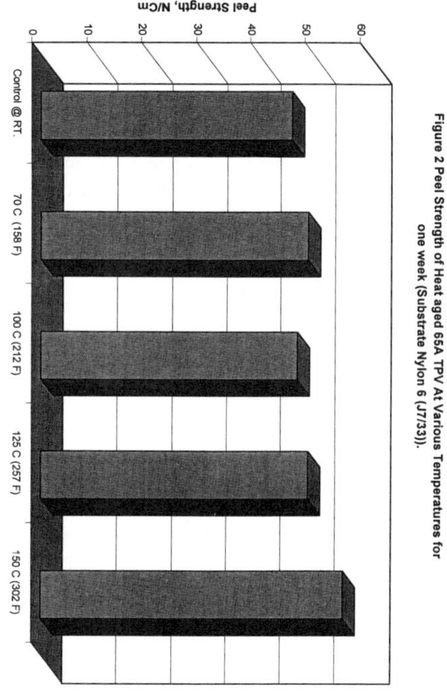

Figure 2 Peel Strength of Heat aged 65A TPV At Various Temperatures for one week (Substrate Nylon 6 (J7/33)).

Preparation of Polymeric Materials with Novel Topologies by "Living"/Controlled Radical Polymerization

*Scott G. Gaynor and Krzysztof Matyjaszewski**

Department of Chemistry, Carnegie Mellon University,
4400 Fifth Avenue, Pittsburgh, PA 15213

Introduction

Since Staudinger's[1] proposal that large macromolecules could be (and are) prepared by covalently linking many small molecules (monomers) together, and subsequently applied by Carothers in the preparation of polyamides, polymer chemists have strived to control the composition, size and/or the topology (architecture) of synthetic polymeric materials. To this end, one of the major developments in obtaining such control was the discovery of the living anionic polymerization of styrene by Szwarc in 1956.[2]

Unfortunately, the utility of living anionic polymerizations has been quite limited due to many factors: the stringent reactions conditions (no water), the relatively low number of monomers that can be polymerized, and the inability to form random copolymers of two or more monomers. To overcome these limitations, much effort has been devoted to developing "living"/controlled free radical polymerizations.[3-5] Of these systems, atom transfer radical polymerization (ATRP) has been demonstrated to be the most versatile, polymerizing styrenes,[6-8] acrylates,[9] methacrylates,[9-11] and acrylonitrile.[12] The polymers prepared by ATRP are very well-defined with degrees of polymerization dependent on $DP_n = \Delta[M]/[I]_o$, and narrow molecular weight distributions, $M_w/M_n < 1.4$, some as low as 1.05.[6]

ATRP employs the reversible activation/deactivation of an alkyl halide (R-X), by redox reaction with a transition metal halide (M_t^n) complexed with a suitable ligand. Abstraction of the halogen by the transition metal results in the formation of a radical and the transition metal halide, $X-M_t^{n+1}$. The radical initiates the polymerization of monomer, with the metal halide deactivating the propagating chain after addition of a few monomer units. The dormant polymer chain end, P-X, can be reactivated by reaction with the lower oxidation state transition metal. As long as deactivation of the propagating radical is fast, bimolecular termination between the radicals can be suppressed. This process, Scheme 1, repeats to form high polymer. The number and type of metals that can be employed is quite broad. Initially, copper[5,6,9] and ruthenium[10] were employed, but this has been expanded to include iron,[13,14] nickel,[15] and palladium.[16]

<u>**Scheme 1**</u>

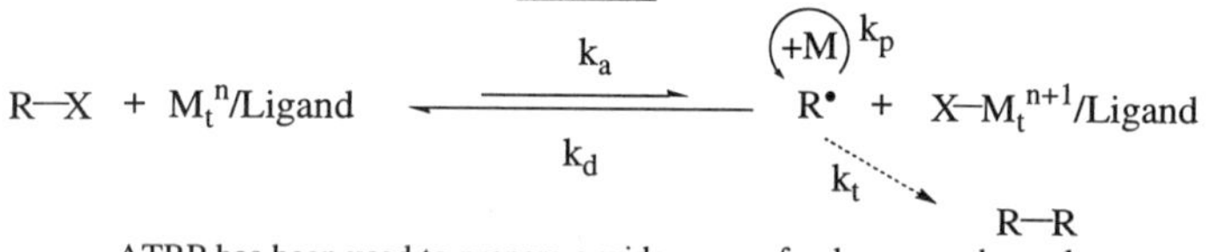

$$R\!-\!X + M_t^n/Ligand \underset{k_d}{\overset{k_a}{\rightleftharpoons}} R^{\bullet} + X\!-\!M_t^{n+1}/Ligand$$

$$R^{\bullet} \xrightarrow{(+M)\,k_p} \qquad \xrightarrow{k_t} R\!-\!R$$

ATRP has been used to prepare a wide range of polymers and copolymers such as homopolymers, random/gradient copolymers, and block copolymers.[17] This report will focus on the use of ATRP to prepare polymers of novel topology, i.e., branched, hyperbranched, graft and polymer brushes.

(Hyper)branched Polymers

In 1995, the preparation of hyperbranched polymers by cationic polymerization using AB* monomers[18] was reported.[19] Subsequently, the preparation of hyperbranched polymers by radical polymerization using TEMPO based "living"/controlled radical polymerization was reported.[20] At nearly the same time, Gaynor et al., used ATRP to polymerize the simple, commercially available, AB* monomer, p-chloromethylstyrene (CMS).[21]

Briefly, the B* site, the benzyl chloride, is activated as in a normal ATRP polymerization and can initiate the polymerization of the double bonds. After deactivation, a new activation site is formed, A*. This oligomer can now grow in one of two directions, at either A* or B*, thus introducing branching. Subsequent addition of monomer further introduces possible branching points, Scheme 2.

<u>**Scheme 2**</u>

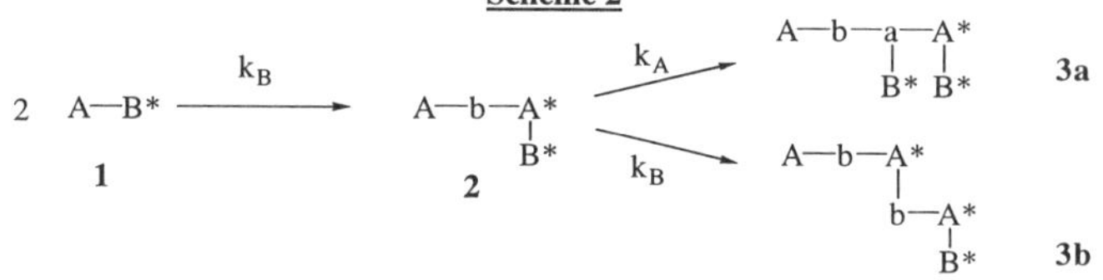

The degree of branching is governed by the relative rate of reaction at either the A* or B* site, $r = k_A/k_B$.[22] If monomer is preferentially added at one or the other, a mostly linear polymer will result. The density of the branch points can be attenuated by copolymerization of the AB* monomer with a conventional, radically (co)polymerizable monomer, as was demonstrated for the copolymerization of CMS with either styrene or methyl methacrylate.[21]

Due to the brittleness of the resulting poly(CMS), an acrylic polymer with a lower glass transition temperature was prepared. The synthesis of the acrylic AB* monomer was done by esterification of 2-hydroxyethyl acrylate with 2-bromopropionyl bromide.[23] The resulting monomer, 2-(2-bromopropionyloxy)ethyl acrylate (BPEA) was then homopolymerized using 1 mol% of copper(I)/2dTbpy relative to monomer(dTbpy = 4, 4'-di(t-butyl)-2, 2'-bipyridine), Scheme 3. The resulting polymer was determined to be hyperbranched by evaluation of the SEC traces of the polymer during the polymerization and by analysis of the ^{1}H NMR spectra, and it was determined to have a degree of branching (DB) of DB = 0.48.[24]

<u>**Scheme 3**</u>

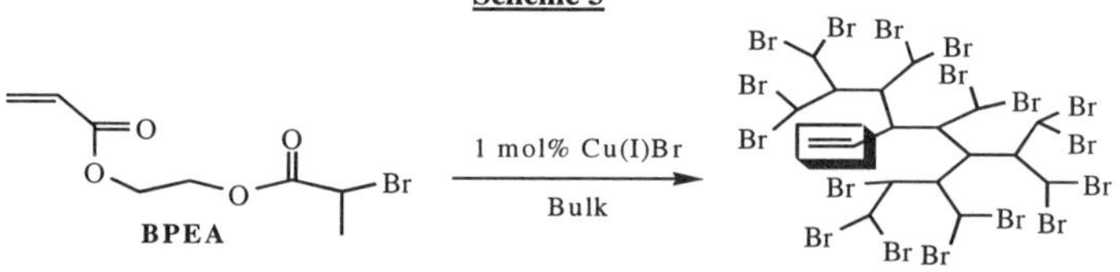

BPEA

Further study revealed that the degree of branching in these systems can be affected by changing the reaction conditions.[25] For example, it was determined that when a less soluble catalyst system was used, i.e., 2, 2'-bipyridine (bpy) as ligand, mostly linear polymers were obtained. This was due to the lowered solubility of the deactivating species in the reaction mixture allowing for relatively long, linear segments to be formed after activation of A* or B*; the addition of multiple monomer units to an active center gives an apparently high r value (k_A appears to be favored).

In addition, it was found that when a completely soluble catalyst system was employed, i.e., 4, 4'-di(5-nonyl)-2, 2'-bipyridine (dNbpy), the polymeriz-ation was extremely slow and only proceeded at high temperatures (> 100 °C). The reason for this was attributed to the initial build up of a high concentration of deactivator resulting from bimolecular termination due to the high concentration of initiating sites (B* = R-X, Scheme 1). As a result of the higher than "normal" levels of deactivator, copper(II) bromide, the equilibrium was strongly shifted towards the dormant species, resulting in a slow polymerization. The use of the dTbpy ligand limited the solubility of the deactivator and thus allowed for the polymerization to proceed at a useful rate.

<u>Graft/Brush Copolymers</u>

When an AB* monomer is copolymerized with a radically polymerizable comonomer by ATRP, a randomly branched polymer is obtained. However, if a polymerization method is used that does not react with the B* group, then linear polymers can be obtained with the AB* monomer dispersed along the polymer chain. This was achieved by copolymerization of BPEA (0.5 mol%) with butyl acrylate using AIBN at 60 °C.

This copolymer ($M_n = 215,000$; $M_w/M_n = 1.6$) was then dissolved in either styrene or methyl methacrylate and used as a macroinitiator for ATRP. The result was the formation of the graft copolymers poly(butyl acrylate-g-styrene) ($M_n = 473,000$; $M_w/M_n = 1.6$, 31 mol% styrene) and poly(butyl acrylate-g-methyl methacrylate) ($M_n = 337,000$; $M_w/M_n = 2.2$, 11 mol% MMA), both of which behaved as thermoplastic elastomers.

In a similar fashion, BPEA was homopolymerized using AIBN to obtain a linear homopolymer ($M_n = 27,300$; $M_w/M_n = 2.3$); carbon tetrabromide was added as a chain transfer agent. This was then used as a macroinitiator to prepare polymeric brushes of polystyrene or poly(butyl acrylate). Using triple detector SEC (on-line light scattering, viscometry, and differential refractometer), the molecular weights were determined to be: poly(BPEA-g-styrene) (M_n = 90,800; M_w/M_n = 2.2) and poly (BPEA-g-butyl acrylate) (M_n = 1,700,000; M_w/M_n = 1.5). When compared to the results obtained by SEC using linear polystyrene standards, (M_n = 34,200; M_w/M_n = 3.2) and (M_n = 182,000; M_w/M_n = 2.3) respectively, it indicated the successful synthesis of the polymeric brushes.

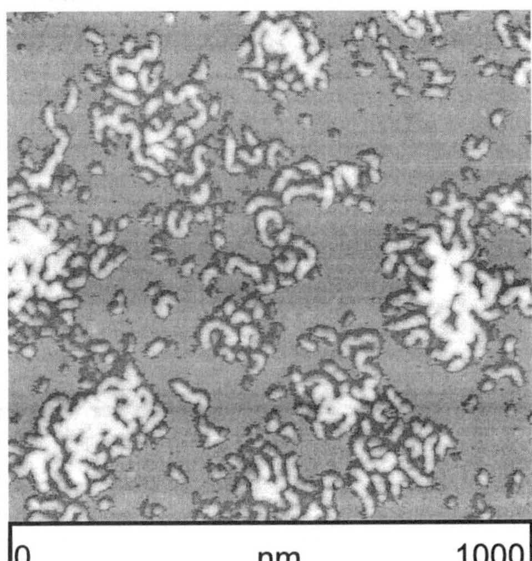

Figure 1. AFM of poly(BPEA-g-butyl acrylate).

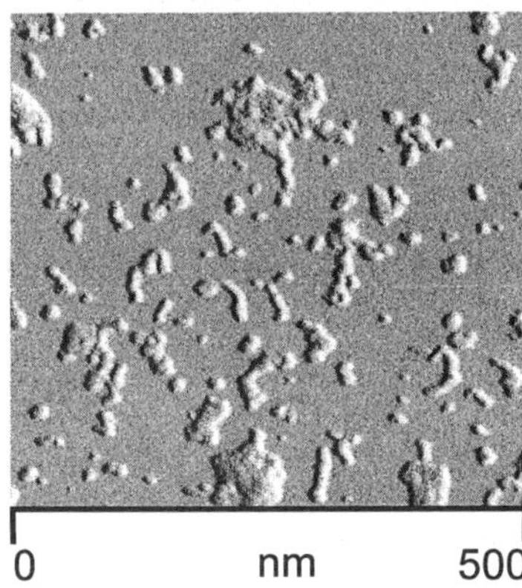

Figure 2. AFM of poly(BPEA-g-styrene).

These materials were further characterized by atomic force microscopy (AFM), in collaboration with Prof. Martin Möller at the University of Ulm, Germany. The results are displayed in Figures 1 and 2. As can be observed, the individual polymer brushes are observed as well as aggregates of the macromolecules. The individual polymer chains are the worm-like structures; aggregates of the chains can also be seen. Both the synthesis and AFM studies have not yet been optimized as this was a exploratory study of the feasibility of preparing such materials by a solely radical mechanism. The butyl acrylate copolymers had the following dimensions: height = 2.0-2.3 nm, width = 9-12 nm, length = 30-140 nm; styrene: height = 1.9-2.3 nm, width = 8-11 nm, length = 10-100 nm. It was observed that the lengths of the individual chains were of varying lengths, as would be expected, due to the high polydispersity of the backbones in the macromolecules. This is a consequence of preparing the backbone with an uncontrolled radical polymerization process, e.g., initiated using AIBN. Also, some of the chains appear to have branches, which may be a result of transfer to polymer during the radical polymerization of either the backbone or the grafts.

Although this is not the first time that such polymers have been made,[26] or observed,[27,28] it is the first time that such macromolecules have been prepared in a completely radical polymerization process. Because of the use of an uncontrolled radical polymerization to prepare the backbones, the resulting polydispersity of the macromolecules is directly observable in the AFM micrographs; note the

difference in the lengths. Due to the large size of these molecules and their inability to chain entangle, it is expected that these materials will have novel physical and mechanical properties.

<u>**Conclusion**</u>

It has been demonstrated that ATRP is a robust polymerization method for the preparation of well-defined polymers. By using ATRP to polymerize or copolymerize AB* monomers, a wide variety of well-defined topologies has been obtained. The architectural control afforded by this technique opens the way for the preparation of a wide variety of novel materials. This work also demonstrated that for the first time radical polymerizations can be highly controlled to yield polymers of novel composition, functionality and architectures.

<u>**Acknowledgment**</u>

This research was partially supported by the industrial members of the ATRP Consortium at Carnegie Mellon University, as well as the Army Research Office. SGG is grateful to the Plastics Institute of America for a fellowship. Prof. M. Möller and Sergei S. Sheiko are gratefully acknowledged for performing the AFM experiments .

<u>**References**</u>

(1)   Staudinger, H. *Chem. Ber.* 1920, *53*, 1073.

(2)   Szwarc, M. *Nature* 1956, *178*, 1168.

(3)   Solomon, D. H.; Rizzardo, E.; Cacioli, P., *US 4,581,429*, 1985

(4)   Georges, M. K.; Veregin, R. P. N.; Kazmaier, P. M.; Hamer, G. K. *Macromolecules* 1993, *26*, 2987.

(5)   Wang, J.-S.; Matyjaszewski, K. *J. Amer. Chem. Soc.* 1995, *117*, 5614.

(6)   Matyjaszewski, K.; Patten, T.; Xia, J.; Abernathy, T. *Science* 1996, *272*, 866.

(7)   Matyjaszewski, K.; Patten, T. E.; Xia, J. *J. Amer. Chem. Soc* 1997, *119*, 674.

(8)   Qiu, J.; Matyjaszewski, K. *Macromolecules* 1997, *30*, 5643.

(9)   Matyjaszewski, K.; Wang, J.-S. *Macromolecules* 1995, *28*, 7901.

(10)   Kato, M.; Kamigaito, M.; Sawamoto, M.; Higashimura, T. *Macromolecules* 1995, *28*, 1721.

(11)   Grimaud, T.; Matyjaszewski, K. *Macromolecules* 1997, *30*, 2216.

(12)   Matyjaszewski, K.; Jo, S. M.; Paik, H.-j.; Gaynor, S. G. *Macromolecules* 1997, *30*, 6398.

(13)   Ando, T.; Kamigaito, M.; Sawamoto, M. *Macromolecules* 1997, *30*, 4507.

(14)   Matyjaszewski, K.; Wei, M.; Xia, J.; McDermott, N. E. *Macromolecules* 1997, *30*, 8161.

(15)   Granel, C.; Dubois, P.; Jerome, R.; Teyssie, P. *Macromolecules* 1996, *29*, 8576.

(16)   Lecomte, P.; Drapier, I.; Dubois, P.; Teyssie, P.; Jerome, R. *Macromolecules* 1997, *30*, 7631.

(17)   Gaynor, S. G.; Matyjaszewski, K. *Polym. Prepr. (Am. Chem. Soc., Div. Polym. Chem.)* 1997, *37(1)*, 758.

(18)   A = double bond; B* = unreacted functional group; A* = functional group formed after addition of double bond followed by deactivation; a, b = consumed double bond and functional group, respectively.

(19)   Frechet, J. M. J.; Henmi, M.; Gitsov, I.; Aoshima, S.; Leduc, M.; Grubbs, R. B. *Science* 1995, *269*, 1080.

(20)   Hawker, C. J.; Frechet, J. M. J.; Grubbs, R. B.; Dao, J. *J. Amer. Chem. Soc.* 1995, *117*, 10763.

(21)   Gaynor, S. G.; Edelman, S. Z.; Matyjaszewski, K. *Macromolecules* 1996, *29*, 1079.

(22)   Yan, D.; Muller, A. H. E.; Matyjaszewski, K. *Macromolecules* 1997, *30*, 7024.

(23)   Gaynor, S. G.; Kulfan, A.; Podwika, M.; Matyjaszewski, K. *Macromolecules* 1997, *30*, 5192.

(24)   A perfectly linear polymer is defined as having a degree of branching, DB = 0. A perfectly branched polymer, i.e., a dendrimer, DB = 1. According to theoretical calculations, the maximum DB for an AB* monomer is 0.5.

(25)   Gaynor, S. G.; Matyjaszewski, K. *Macromolecule* 1997, *30*, 7042.

(26)   Tsukahara, Y.; Tsutsumi, K.; Yamashita, Y.; Shimada, S. *Macromolecules* 1990, *23*, 5201.

(27)   Sheiko, S. S.; Gerle, M.; Fischer, K.; Schmidt, M.; Moller, M. *Langmuir* 1997, *13*, 5368.

(28)   Dziezok, P.; Sheiko, S. S.; Fischer, K.; Schmidt, M.; Moller, M. *Angew. Chem. Int. Ed. Engl.* 1997, *36*, 2812.

Block copolymers as supercritical CO$_2$ developable photoresists

Narayan Sundararajan, Suresh Valiyaveettil, Kenji Ogino, Xinyi Zhou, Jianguo Wang, Shu Yang and Christopher K. Ober
Department of Materials Science and Engineering
Cornell University
Ithaca, NY 14853

Introduction

Remarkable progress in the recent miniaturization of electronic devices has been made possible by the continuous and ongoing improvements in both the chemistry of the resists as well as in every resist processing step such as exposure and development. The number of components in a chip has increased at a staggering rate of 10^2 to 10^3 per decade though the chip area by itself has not changed much[1].

The decrease in the resolvable feature size was achieved by decreasing the wavelength of the exposing radiation as well as by using certain optical techniques such as phase-shift masks. With the advent of chemically amplified resists for 193 nm ArF excimer laser lithography, optical lithography still seems the most promising technology well into the next decade[2]. However, as the target feature size keeps diminishing, the selectivity of the solvent used as the developer becomes increasingly important. Suitable solvents with very high selectivity and tunable solvating power are required. These properties are not adequately found in the aqueous liquid developers currently used. Also, the use of liquid developers in the mainstream production environment generates enormous amounts of waste streams causing great environmental concern. Hence, processes which minimize environmental contamination are highly desirable in the microelectronics industry.

Supercritical fluids (SCFs) possess many unique properties such as very good selectivity, diffusivities and viscosities comparable to gases, very low surface tension and pressure adjustable solvating capability[3]. Particular attention is given to SCF CO$_2$ because of its nontoxic, nonflammable nature, very low cost and the opportunity for recycling it using simple pressure manipulation cycles. Supercritical CO$_2$ has been commercially used for various applications such as decaffeination of tea and coffee and low volatile organ carbon (VOC) emission spray painting.[4]

Prior to reaching the stage of actually implementing these environmentally friendly processes, much research needs to be conducted on the suitable design of the resists and processing involved. These new photoresists should not only be compatible with the environmentally friendly processes but also be able to satisfy the existing goals of the semiconductor industry such as minimum image size.

Block copolymers[5] with their intrinsic property of separating resist function, present unique opportunities for satisfying some of the key requirements of these new resists such as very high sensitivity and good adhesion and also decrease the inherent compromise existing between sensitivity and development characteristics. They also provide new avenues for designing resists for improved processes because of their unique characteristics such as surface and interfacial segregation and microphase separation.

The objective of this study was to utilize the concept of block copolymers and their unique properties to provide an environmentally friendly process for the fabrication of sub-0.3 μm features using supercritical CO$_2$ development.

Experimental

Synthesis

Narrow polydispersity block copolymers of tetrahydropyranyl methacrylate (THPMA) and supercritical solubilizers such as heptafluorobutyl methacrylate (F3MA) and pentadecafluorooctyl methacrylate (F7MA) with different molar ratios were synthesized using group transfer polymerization with 1-methoxy-1-(trimethylsiloxy)-2-methyl-1-propene (MTMS) as the initiator and tetra (n-butyl) am monium biacetate as a catalyst[6].

Polymer	Vol. fraction of F3MA (%)	M$_n$	M$_w$	M$_w$/M$_n$
GTPK23	11	8520	8880	1.04
GTPK23-2	22	7260	8180	1.13
GTPK24-2	32	7430	7930	1.09
GTPK22	51	7890	8960	1.14
GTPS1	Vol.fraction of F7MA (%) 66	8950	9540	1.07

Characterization

The molecular weight of the polymers were determined by GPC using polymethyl methacrylate (PMMA) as standard. The copolymer compositions were verified by ^{1}H NMR studies. Glass transition temperatures of the copolymers were measured using DSC and DMA. The thermal deprotection characteristics were analyzed using TGA.

Lithographic processing

Polymer solutions (15wt % in PGMEA) were filtered using 0.45 μm pore size PTFE filters. The photoacid generator (PAG) concentration was 1 wt % of the polymer. The PAG was also supercritical/liquid CO$_2$ soluble. An organic anti-reflective coating was coated initially in the case of THPMA-b-F3MA polymers. The polymer solutions were spun at speeds of around 2000-3000 rpm to obtain appropriate film thicknesses and were post-apply baked at 120°C for a min. Exposures were made with a ArF excimer laser operating at 193 nm wavelength. Post-exposure bake was performed at either 120°C or 90°C for a minute. The exposed wafers were placed in an extraction vessel and developed in supercritical CO$_2$ at about 4000 psi and 40°C for the unexposed film to dissolve completely in less than 5 minutes.

Results and Discussion

Block copolymers posess the advantage of having two chemically different segments linked covalently, where each component prefers to segregate to a specific interface. The very nonpolar fluoromethacrylates, which can be soluble in supercritical CO$_2$, were chosen to be copolymerized with the acid cleavable tetrahydropyranyl methacrylate (THPMA) so that the classical chemically amplified polarity change in the presence of a proton can be utilized. In these polymers, the low surface energy fluorinated block should strongly self organize on the surface of polymer film, while the THPMA block would segregate down to the interface between the silicon wafer and the polymer film. After exposure to radiation, the THPMA block converts to a more polar methacrylic acid, which improves the adhesion between the polymer film and substrate, thus improving the ultimate resolution. By adjusting the composition of the blocks and varying the length of the fluorinated side chain, it is possible to tune the domain size and shape of each block, thus optimizing the solubility of the block copolymer in supercritical CO$_2$, its phase segregation and its surface properties.

Block copolymers such as THPMA-F3MA and THPMA-F7MA with different volume and molar ratio (Fig. 1) were synthesized by group transfer polymerization. THPMA was introduced first, initiated by MTMS with TBAB as a catalyst in THF. The temperature increased immediately indicating a fast reaction. F3MA or F7MA was then added as second block and polymerized with yields greater than 80%. GPC traces indicated the block copolymers were monodisperse and the composition was measured by ^{1}HNMR.

The optimum conditions for the dissolution of the virgin polymer before exposure were determined by evaluating the dissolution characteristics of the polymer at different pressures, temperatures, flow rates of CO_2 and times of development. After exposure, the proton generated from the photoacid generator cleaves off the acid-labile group in the THPMA component of the block copolymer and converts it to methacrylic acid. This gives rise to a polarity change which then makes the polymer insoluble in supercritical CO_2 after exposure (Fig. 2). It was also noted that the random copolymer of the THPMA-F7MA corresponding to the same chemical composition of the block copolymer (34/66 vol %) did not dissolve even at 6000 psi and 60 °C.

A plot of film thickness after development versus exposure dose gives an understanding of the sensitivity of the photoresist. The sensitivity curve for a 49/51 vol % of a block copolymer of THPMA-F3MA is shown in Figure 3. The sensitivity curve was obtained by exposing the polymeric film to an array of increasing dosage and measuring the thickness of the remaining exposed insoluble film after post-exposure bake (PEB) and development in supercritical CO_2 at the optimum conditions of temperature and pressure previously determined. The thicknesses of the films after development in supercritical CO_2 were measured using *Alphastep®*. The sensitivity of a 49/51 vol % block copolymer of THPMA-F3MAwas determined to be approximately 10 mJ/cm^2 at a post-apply bake of 120 °C for a min and a post-exposure bake of 120 °C for a min. The sensitivity of 34/66 vol % of THPMA-b-F7MA was approximately 4 mJ/cm^2. Figure 4 shows features fabricated using block copolymers of THPMA-F7MA (34/66 vol%) and THPMA-F3MA (49/51 vol%) respectively and supercritical CO_2 development.

References
1) Reichmanis, E. and Thompson, L. F., *ACS Symp. Ser.*, **1989**, *412*, 1
2) Ogawa T., *J. Photopolym. Sci. Technol.*, **1996**, *9*, 379
3) McHugh, M. A. and Krukonis, V. J., *Supercritical Fluid Extraction: Principles and Practice,* 2nd ed., Butterworth-Heinemann, Stoneham, MA **1994**
4) Dixon, D. J. and Johnston, K. P., Supercritical fluids, 452, Kirk-Othmer Encycopaedia of Chemical Technology, 4th ed., *Vol. 23*, Edrs. Kroschwi'z, J. I. and Howe-Grant, M., John Wiley & Sons, New York, **1992**.
5) Bates, F. S. and Fredrickson, G. H., *Annu. Rev. Phys. Chem.*, **1990**, *41*, 525
6) Sogah, D. Y., Hertler, W. R., Webster, O. W. and Cohen, G. M., *Macromolecules*, **1987**, *20*, 1473

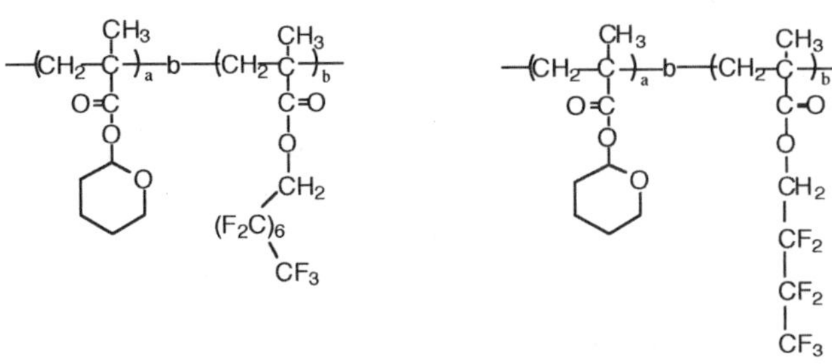

THPMA-b-F7MA　　　**THPMA-b-F3MA**

Figure 1. Block copolymers of acid cleavable tetrahydropyranyl methacrylate (THPMA) and fluoromethacrylate supercritical CO_2 solubilizers

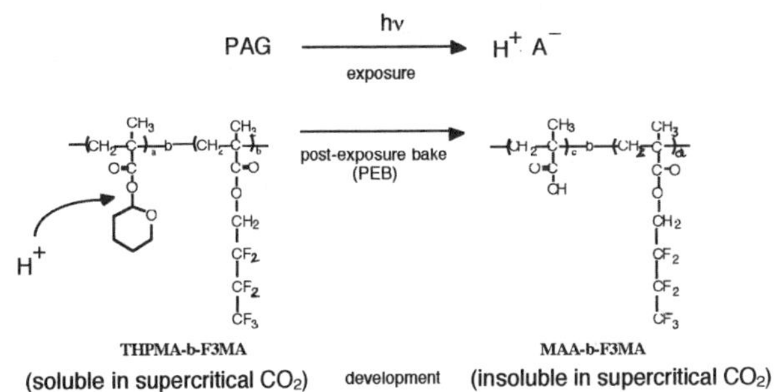

Figure 2. Imaging process of a supercritical CO_2 developable photoresist

Sensitivity curves

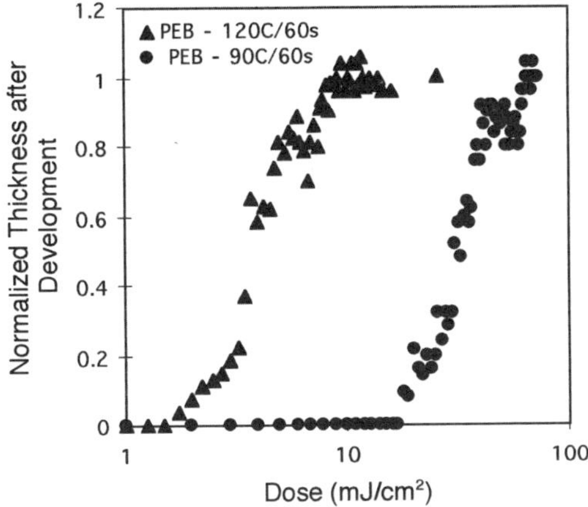

Figure 3. Sensitivity curves for a block copolymer of tetrahydropyranyl methacrylate (THPMA) and heptafluorobutyl methacrylate (F3MA)

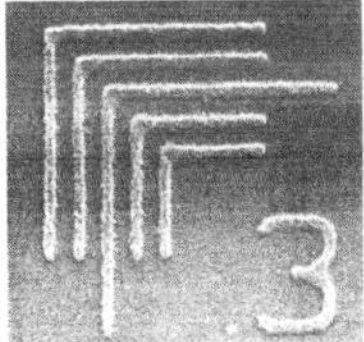

Figure 4. 0.21 μm and 0.3 μm features fabricated in block copolymers of THPMA-F7MA and THPMA-F3MA respectively using 193 nm ArF excimer laser source and developed in supercritical CO_2

Structure Development of Spherulitic Domain in Polymer Films

Tao Huang, Tomohiro Tsuji, A.D. Rey, and M.R. Kamal

Department of Chemical Engineering, McGill University
Montreal, Quebec, H3A 2B2, Canada

Introduction

The nature of the microstructure of a polymer strongly affects its functionality in optical and electronic applications. There is great need for reliable predictive techniques to estimate the properties and understand the behavior of materials, especially in relation to the role of crystalline morphology. The properties of the solid polymer depend on both the supermolecular and grain structures formed in nucleation-growth process, in which the details are not well understand so far, comparing to the studies of phase separation.[1]

In this work, the extensive investigation on post-nucleation and spherulitic domain growth have been carried out experimentally and computationally in order to understand and predict the structure development in polymer thin films. Cross-scale computational model has been developed to predict and analyze nucleation-growth processes, grain patterns of polymer spherulitic structures, and the internal lamellae organization. We take spatial correlation, geometry and topology of the structural evolution into account to elucidate nucleation-growth dynamics. Novel domain-spatial correlation function simultaneously probe the time-dependent domain-size distribution and the spatial correlation of nuclei throughout the entire process. Scaling relationships have been found during early free growth stage. These results do not only significantly advance our understanding of polymer structure formation, but also provide the reliable predictive techniques to optimize material functionality and processing controllability.

Experimental Studies

Experiments were carried out with a common polymer, isotatic polypropylene (iPP, molecular weight M_η=250,000). A polymer thin film was formed between two glass slides and by pressing the top slide to form a 10 μm thick polymer film. A Leitz polarized microscope, equipped with a hot stage for polymer film solidification was used in the direct observation experiments. In this isothermal solidification study, the temperature is controlled within ±0.1°C. JAVA-Jandel Scientific video measurement and image processing system was directly connected to the microscope via a CCD camera, with proposed digital image analysis programs. We focus on the post-nucleation stage, during which the size of the nucleus is greater than 1 μm and visible under the optical microscope for real-time in-situ observation and accurate real-space measurement. A nearly linear nucleation law is found after an induction period and before impingement from experimental measurements. With increasing transformation, the rate decreases to zero as impingement becomes significant. The experimental data of the nucleation rate can be explained by the Kashchiev's formula before impingement becomes significant.[2] The growth velocities measured from different spherulites in the same experimental run confirmed the linear growth before impingement under isothermal crystallization conditions.[2]

The spatial features of post-nucleation stage are the new nuclei packing and the impingement, which can be captured by the nuclei pair distances distribution. Experimental results show that the mean distance to the next nuclei decreases with increasing time. The size of the depletion zone surrounding each nucleus also decreases with time. The mean distance to the next nuclei is more distinct at higher area fractions with nuclei packing more densely. The partial pair correlation function for new nuclei to the old nuclei shows that the new nuclei tend to appear randomly near old nuclei. At any time during the late stage, where the statistical self-similarity of the spatial order achieved, the density function of nearest neighbor distance distribution can be easily derived by considering the impingement at all nuclei sites. Figure 1 shows the corresponding experimental results of the normalized nearest neighbors distances Λ_{NN} normalized with the scale of the average nearest neighbor distance δ_{NN}. The most important feature of the normalized nearest neighbors distance Λ_{NN} is the self-similar fashion. It indicates that the normalized nearest neighbors distances Λ_{NN} are selected with self-similarity for $\Lambda_{NN} < 1$; however, they are not for $\Lambda_{NN} > 1$. The deviation decreases by more nuclei production with increasing time.

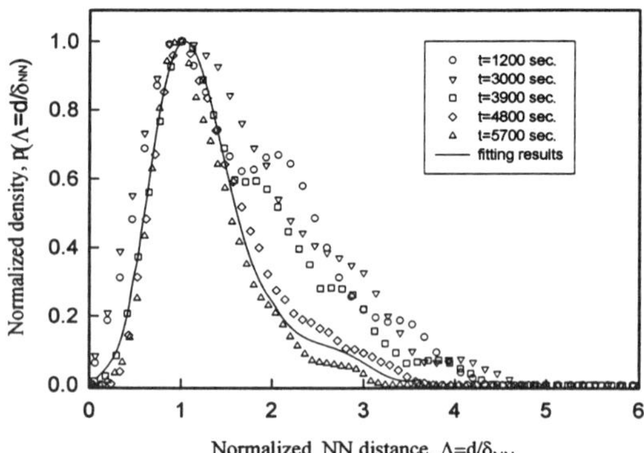

Figure 1. The normalized distribution function of the density of nearest neighbors distances of iPP spherulite nucleation-growth at isothermal condition Tc=406K.

Computational Modeling

A novel physical model is presented for prediction and characterization of the development of semi-crystalline polymer microstructure across mesoscopic length and macroscopic grain length scales. We assume that there are no macroscopic heat/mass transfer in the system. Also, we assume that the fluctuations and a long range interaction between growth domains are negligible. Considering the nucleation and growth in space, the core origin position is represented by X_i(i=1,...N_c), where N_c is a total number of cores. The relation between a certain core and its neighbors can be represented by the distance function of the neighbor cores, which is given by $|X_i-X_j|$, where i ≠j. Thus, an envelop profile for the multiple spherulitic growth domains is given by:

$$f_i(\theta,t) = \int_0^t v(t')\prod_{i \neq j}\{1 - \delta(A)\}dt' \tag{1}$$

where $v(t')$ is growth velocity vector, the term , $\prod\{1 - \delta(A)\}$, describes the impingement dynamics of near nearest neighbor domains; for the unit vector along one domain center coordinate on an envelop, $\vec{u} = (\cos\theta, \sin\theta, 0)^T$, the parameter A is compensation factor, $A=\{[x_{ix}+\cos\theta f_i(\theta,t)-x_{ix}+\cos\theta f_j(\theta,t)]^2+[x_{iy}+\sin\theta f_i(\theta,t)-x_{iy}+\sin\theta f_j(\theta,t)]^2\}^{1/2}$.

In order to improve our understanding of the internal structure of polymer spherulites, it is important to analyze quantitatively the orientation order of lamellae within spherulite. On the mesoscopic lamellar scale, we assume that the lamellae configuration represents the spherulitic internal texture. Based on the experimental TEM image of the lamellae configuration of polyethylene spherulite, a unit vector $\vec{l}$ is introduced as local averaged lamellae director in order to describe the spherulitic internal structure,. We assume that the lamellae director is an unit vector of the major axis direction and changes along the length of a mono-lamellae. There are three possible configurations: (1) splay energy, F_s, which represents the tilting a lamellae toward another; (2) bend energy, F_b, which represents the angle between two adjacent lamellae; and (3) twist energy, F_t, which arises by counterotation at the lamellae ends. In this model, their distortions are else defined. The curvature energy of the lamellae director field is similar to Frank energy equation in liquid crystal:[2]

$$F_s = C_1(\nabla \bullet \vec{l})^2, \quad F_t = C_2(\vec{l} \bullet \nabla \times \vec{l})^2, \quad F_b = C_3\left|\vec{l} \times \nabla \times \vec{l}\right|^2 \tag{2}$$

where C_1, C_2 and C_3 are elastic coefficients. In 2D polymer films, we consider both splay and bend for normal spherulite; and twist for banded spherulite.

The orientation angle of the lamellae director is

$$\theta = \arcsin(I / I_0) / 2 \tag{3}$$

where I is the intensity of polarized light passing through the spherulites.

Computational modeling has been performed based on the coupling of above equations to predict and analyze the lamellae organization based on the mesoscopic lamellae director and grain patterns of semi-crystalline polymer spherulitic structures. The simulation results and its scientific visualization present the spherulitic textures of semi-crystalline polymer as Maltese cross gray images. It provides information not only the functional data on envelop profiles of the spherulites but also the spherulite internal structure.

Figure 2 shows that typical results. Within the spherulite, the computational results indicate that the characteristic of lamellae is the orientation deviations caused strongly by the large splay distortion, and by the bend distortion in some regions. It is clearly shown that both energies at the spherulite grain boundaries are higher than within the spherulite grains. The energy profiles agree exactly with the corresponded vector plot of the mesoscopic lamellae director. This is due to the mis-matching of the orientation of the mesoscopic lamellar directors at the grain boundary between two neighboring spherulites.

Dynamic comparison between large scale computational and experimental results have been also accomplished by quantitatively matching the simulated growth patterns to the experimental images.[2] Furthermore, this model is being extended to modeling the structure development in real polymer processing.

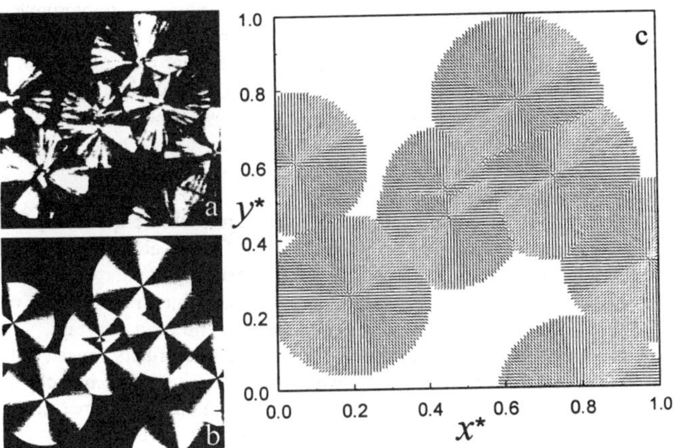

Figure 2. Typical results of the experimental image in (a), the computational pattern in (b), and the vector plot of the mesoscopic lamellae director in (c).

Structural Evolution Dynamics

In this section, we take spatial correlation, geometry and topology of the structural evolution into account to elucidate nucleation-growth dynamics. It is an important feature of the structural evolution has not been identified or taken into account so far. Based on this argument, we proposed the following new concepts and approaches on the studies of the dynamic domain-spatial correlation function and the topological anticoorelation.

(1). Domain-Spatial Correlation Function: The new domain-spatial correlation function is defined as: $G_i(r,t)$, for an arbitrarily chosen domain i with position vector $\vec{x}$, as the origin of the domain core center, is defined by counting the domains whose position vectors lie within a distance Δr from a circle of radius r with center at the origin at time t. Considering the whole system with the total number of domains, N_c the domain-spatial correlation function, $G(r,t)$, is the ensemble averages of $G_i(r,t)$ for all domain center positions of over all domains placed at the origin, which yields:

$$G(r,t) = \left\langle \frac{1}{\int \psi(\vec{x},t)d\vec{x}} \int \delta(r - |\vec{x}_i - \vec{x}_j|)\psi(\vec{x}_j,t)d\vec{x}_j \right\rangle \quad (4)$$

where $\delta(r-|x_i-x_j|)$ is number density function and $\psi(x,t)$ is the order parameter:

$$\delta(r - |\vec{x}_i - \vec{x}_j|) = \begin{cases} 1, & r - |\vec{x}_i - \vec{x}_j| = 0 \\ 0, & r - |\vec{x}_i - \vec{x}_j| \neq 0 \end{cases} ; \qquad \psi(\vec{x},t) = \begin{cases} 1, & \vec{x} \in domain\ at\ t \\ 0, & \vec{x} \notin domain\ at\ t \end{cases}$$

Figure 3 shows the domain-spatial correlation function for experiment of continuous nucleation under isothermal crystallization condition at $T_c = 406K$.

The left part of Figure 3 is the smooth curves of the domain size distribution when $r < Rg(t)$. It represents the growth domains with different sizes and

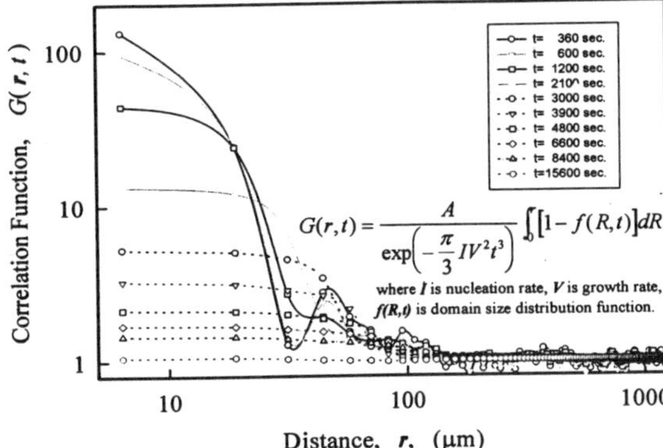

$$G(r,t) = \frac{A}{\exp\left(-\frac{\pi}{3}IV^2t^3\right)}\int[1 - f(R,t)]dR$$

where I is nucleation rate, V is growth rate, $f(R,t)$ is domain size distribution function.

Figure 3. Dynamic domain-spatial correlation functions for experiment of continuous nucleation under isothermal crystallization condition, Tc=406K.

some local impingement structures. The right part of Fig.3 is the spatial pair-correlation function of domain core centers when $r > R_g(t)$ The first peaks shift slightly to the left with increasing time, which means that the average inter-nuclei distances decrease with increasing time due to the nucleation of new domains. $G(r,t)$ approaches one in an oscillatory manner at very large r, which means that there is no long range order. The scaling relation, $G(r,t)/G(r=0,t)$, as a function of $r/R_g(t)$, where $R_g(t)$ is the location of the first minimum of $G(r,t)$ has been found during early free growth stage.[2]

(2). Topological Anticorrelation: The topological nature of structure evolution in the nucleation and growth process is determined by the nearest neighbors configuration

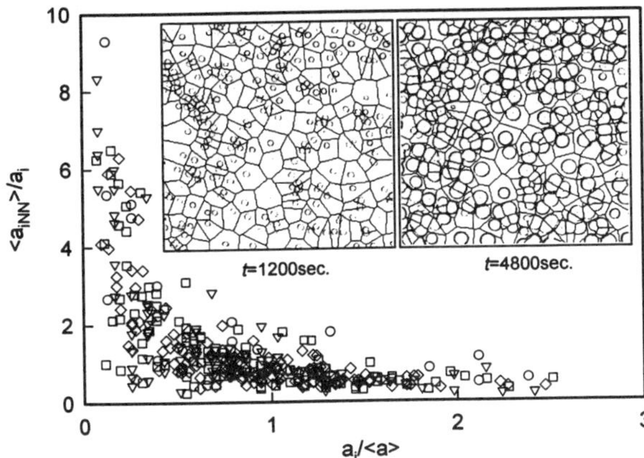

Figure 4. Topological anticoorelation function of the nearest neighbors area; the insert images show the evolution of the nearest neighbors configuration .

by Voronoi characterization of nucleation and growth patterns. Two nuclei are defined as nearest neighbors if they belong to adjacent cells of the Voronoi diagram (fine lines), see the insert images in Figure 4. Figure 4 shows that for each domain of area a_i in a given pattern with average area $\langle a\rangle$, the average area of its nearest neighbors $\langle a_{NN}\rangle$ relative to a_i shows the ascertain correlation in the areas assumed by adjacent domains.The important result is: for domains smaller than average area $\langle a\rangle$, $\langle a_{NN}\rangle$ exceeds a_i and vice versa. This is referred as "pronounced anticorrelation", which can be explained by the maximum entropy formalism.[2] This topological nature play an important role in the understanding of the structural evolution in nucleation and growth.

Reference
(1) Langer,J.S. in *Solids Far From Equilibrium*, Ed by C. Godreche, 1989.
(2) Tao Huang, Ph.D thesis, McGill University, 1997.

Phase Behavior and Morphology of Polyolefin Blends and Thermoplastic Vulcanizates in Supercritical Propane

S. J. Han[1, 4], D. J. Lohse[1], M. Radosz[2], L. H. Sperling[3, 4]
1. Exxon Research and Engineering Company, Annandale, NJ 08801
2. Department of Chemical Engineering and Macromolecular Studies Group, Louisiana State University, Baton Rouge, LA 70803-7300. 3. Department of Chemical Engineering, Lehigh University 4. Center for Polymer Science and Engineering, Lehigh University, Bethlehem, PA 18015-3194

Introduction

Blending is widely applied to polyolefin products to obtain a high level of enhanced material properties, e.g., impact modification of polypropylene with rubbery elastomers[1].   In rubber toughening applications, optimal material properties are achieved by controlling specific blend morphology.[2, 3, 4]   In common practice, melt blending in extruders is a popular choice in mixing the polymer components. However, the potential problems of melt mixing are chain degradation due to high intensity of mixing, and poor dispersion of polyolefin components each other due to melt-viscosity mismatching.

The main purpose of this work involves controlling the morphologies of polymer blends made from polypropylene and ethylene copolymers in supercritical propane via crosslinking or pressure-temperature processing paths manipulation.   The approach utilizes the phase behavior of polypropylene and ethylene copolymers in supercritical propane to determine processing conditions, and to crosslink the system with peroxide in the supercritical solution.   This results in thermoplastic vulcanizates (TPVs). The polymer blends were also prepared by rapidly expanding the supercritical solution (RESS) or isobarically cooling the supercritical solution (ICSS).   The morphology of the final polypropylene blends was investigated by SEM, optical phase contrast microscopy, and DSC.   The advantages of polymer blending in supercritical propane over conventional melting blending process are easy solvent recovery from final product, recycling of solvent providing improved environmental controls, high interspersion of components, and fine control of various processing paths in terms of temperature and pressure.

Experimental

Materials: Polypropylene, ethylene copolymers, and EPDM are synthesized from various synthetic methods using metallocene, vanadium, and anionic catalysts. The polymer microstructures were analyzed by ^{1}H NMR for the extent of branching, gel permeation chromatography for molecular weight of polymers, and differential scanning calorimetry for melting transitions. Solvent propane (99.9 % purity) was obtained from Matheson Gas Co. and *t*-butyl peroxide (99.4 % purity) was obtained from Aldrich Company. Both were used without further purification.

Apparatus: The cloud-point pressures of the polymer solution were determined with the high pressure variable volume optical batch cell[5]. The unit was modified for the RESS process by connecting a spraying unit into a sampling port of the cell.[6]   The spraying unit consists of a GC valve, 75 micron ID fused silica capillary tubing, and adapters.   The processing paths for the RESS, ICSS and crosslinking reaction condition are shown in *Figure 1*.

Results and Discussions

Phase behavior

The cloud-point pressures of the phase diagram determine the minimum pressure required to maintain a homogeneous polymer solution. *Figure 2* illustrates binary (B) and ternary (T) solution phase behavior of ethylene-butene (EB) copolymers, and polypropylene in propane. As the comonomer content (i.e. short chain branches) in EB copolymers increases, the cloud-point pressures decrease.   The same type of phase behavior was also observed in ethylene-propylene (EP) copolymers and ethylene-hexene (EH) copolymers in propane.   This solution phase behavior can be explained by shifts in the solubility parameters.   The solubility parameters of the polymers were calculated from the P-V-T data of polymers to characterize the intermolecular interaction energy.[7] As shown in *Figure 3*, the solubility parameters of ethylene copolymers decrease as the comonomer content in the ethylene copolymer increase. *Figure 2* also shows that the cloud-point pressures of ternary solution were higher than corresponding binary solutions, which is due to repulsive interactions of polymer components in the solutions. For the ternary solutions of polypropylene and EPDM in propane, the miscible region in the ternary phase diagram was very narrow, and slightly increased upon increasing the pressure from 600 bar to 700 bar as shown in *Figure 4*. This type of phase behavior occurs when the two polymer components are immiscible each other.

TPV morphology

TPVs from polypropylene and EPDM were synthesized by crosslinking with *t*-butyl peroxide in an homogeneous polymer solution at 175 °C and 650 bar.   The domains of EPDM dispersed in polypropylene were finer than corresponding polymer blends from melt mixing at 195 °C in a Brabender as shown in *Figure 5*.   The improved dispersion of EPDM is primarily due to solution mixing in supercritical propane, in comparison to melt mixing, especially affected by frequent mismatches in melt viscosities.

RESS-blends morphology

Polypropylene and ethylene copolymer were pressure-quenched from an homogeneous polymer solution at 175 °C and above 600 bar to ambient pressure through the RESS nozzle, resulting in microfibers of ID 10 ~ 30 μm.   During the RESS process, the solution in the cell was maintained one-phase to keep the polymer concentration constant.   The morphology of the thin films after melting the microfibers revealed that finer dispersion in polypropylene were obtained from more branchy EB copolymer as shown in *Figure 6*.   However, the EB copolymer domain coarsened as the films aged in the elevated temperature, suggesting thermodynamic immiscibility of two polymers dominating the polymer blends.

ICSS-blends morphology

The homogenous ternary polymer solutions of polypropylene and ethylene copolymer in propane at 175 °C were slowly cooled isobarically at 600 bar to below the solid-liquid transition of polypropylene, then recovered the polymer blends by venting the solvent to ambient condition. *Figure 7* shows phase separated surface morphology of the blends, where EB copolymers were coated onto polypropylene microspheres.   This is because polypropylene in the polymer solutions was initially crystallized while cooling below solid-liquid transition, then followed by precipitation of the EB copolymers during lowering the pressure to ambient.

Conclusions

Supercritical propane was utilized as a reaction and processing medium to control morphology of polymer blends from polypropylene and EPDM terpolymer.   The cloud-point pressures of binary solution of ethylene copolymers in propane decreased with increasing the number of short chain branches in the ethylene copolymers.   The cloud-point pressures of ternary solutions from polypropylene, ethylene copolymer, and propane were higher than those from binary solutions. The surface morphology of TPVs were microporous (i.e., microcellular foams) due to crystallization in the presence of solvent, followed by evaporation. The dispersions of EPDM in the TPVs, synthesized in supercritical propane, were finer than those from melt blending due to high interspersion in supercritical solutions.   The comonomer contents in the ethylene copolymers significantly influence the dispersion of ethylene copolymers in polypropylene processed by the RESS.   The thermodynamic immiscibility dominated the coarsening process of the polyolefin blends. The polyolefin blends from the ICSS showed a coated microspheres morphology due to thermally induced phase separation from the ternary polymer solutions.

References

1. Kresge, E. N. Chapter 20 in *Polymer Blends*, Ed. By Paul, D. R.; Newman, S. Academic Pres., (1979).
2. Jang, B. Z.; Uhlmann, D. R.; Vander Sande, J. B. *Polym. Sci. Eng.* 25, 10, 643, (1985).
3. Bucknall, C. B. *Makromol. Chem., Macromol. Symp.* 38, 1, (1990).
4. Wu, S. *Polymer* 26, 1855, (1985).
5. Chen, S.-j.; Randelman, R. L.; Seldomridge, R. L.; Radosz, M. *J. Chem. Eng. Data* 38, 211, (1993).
6. Han, S. J.; Lohse, D. J.; Radosz, M.; Sperling, L. H. *ACS, PMSE*, 75, 279, (1996).
7. Graessley, W. W.; Krishnamoorti, R.; Balsara, N. P.; Butera, R. J.; Fetters, L. J.; Lohse, D. J.; Schulz, D. N.; Sissano, J. A. *Macromolecules* 27, 3896, (1994).

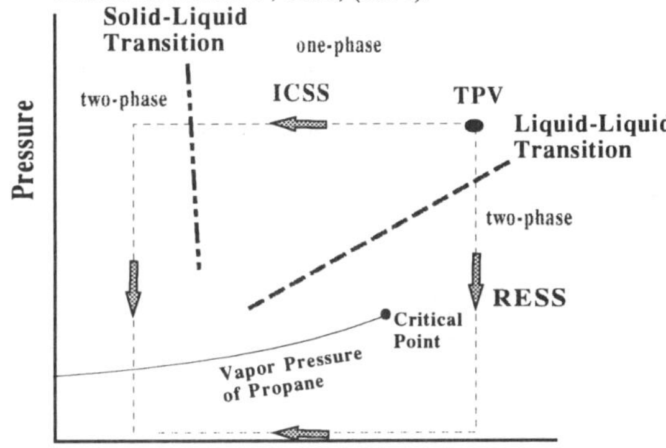

Figure 1 Development of polyolefin blends in supercritical propane

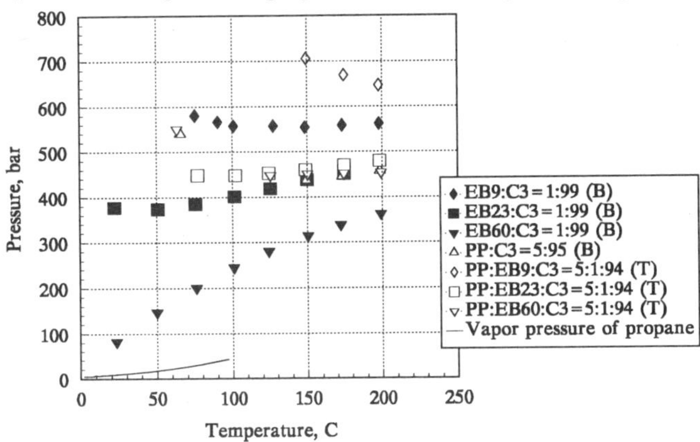

Figure 2 Cloud-point pressures of binary (B) and ternary (T) polymer solutions in supercritical propane. The copolymers are coded in terms of comonomer mole percent

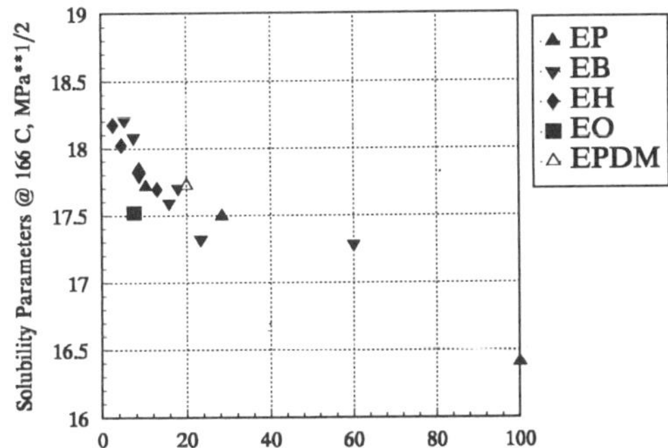

Figure 3 Solubility parameters of ethylene copolymers @ 166 C

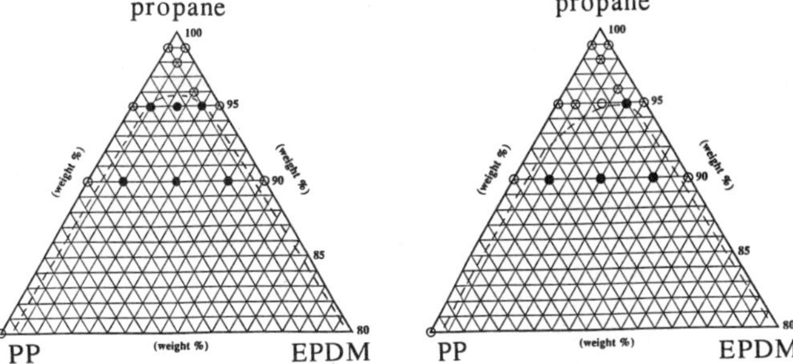

Figure 4 Ternary phase diagrams of polypropylene, EPDM, and propane @ 175 C: (a) 600 bar (b) 700 bar
open circle: one-phase, closed circle: two-phase

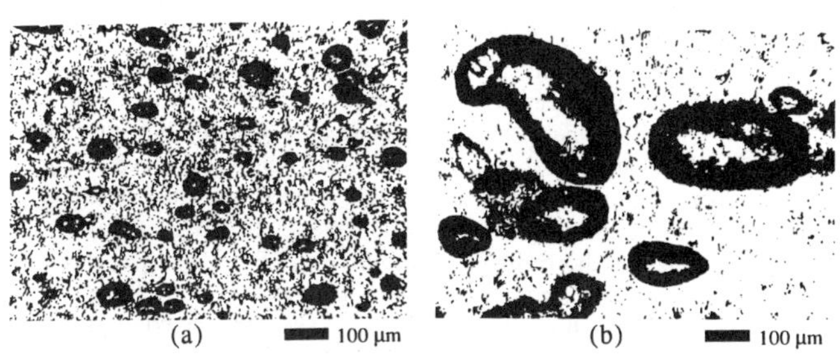

Figure 5 Phase morphology of EPDM dispersed in polypropylene (a) TPV via reaction in supercritical propane (b) melt-blends - PP:EPDM=4:1

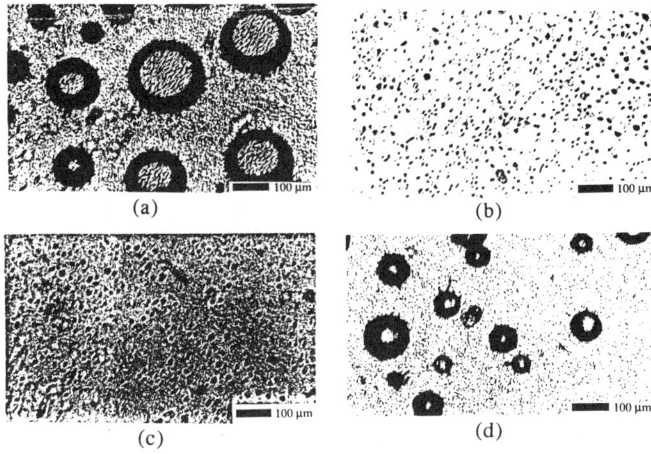

Figure 6 Phase morphology of EBs dispersed in polypropylene by RESS
(a) PP:EB9=5:1 (b)PP:EB60=5:1
(c) PP:EB23=5:1 @ 2.5 min, 195 C
(d) PP:EB23=5:1 @ 5 min, 195 C

Figure 7 Surface morphology of polymer blends by ICSS
(a) PP:EB9=5:1 (b) PP:EB23=5:1

ERFs and PDMS Networks: A Fast and Elastic Electromechanical Response

Katherine Bohon and Sonja Krause
Departments of Chemistry and Polymer Science and Engineering
Rensselaer Polytechnic Institute
Troy, NY 12180

1. Introduction:

A new electromechanical actuator has been reported recently, using the speed of electrorheological fluids (ERFs) and the elastic response and strength of polymeric gels.[1] Electrorheological fluids are colloidal suspensions that align under an electric field. The alignment causes a dramatic increase in viscosity, turning the liquid into a gel. Response time is on the order of a millisecond.[2]

ERFs have a rapid electrical response, but the randomization of the particles after the field is removed is diffusion controlled and slow. ERFs in the unactivated state are also lacking in mechanical stiffness. In our previous work it was shown that by combining an ERF with a gel, it is possible to prepare a system that can exert a force sufficient to move a light object[1]. The force exerted is associated with the ERF, and after the field is removed another force is generated as the gel elastically returns to its original state. This creates a new, strong, and fast type of electromechanical actuation.

The previous work was done with proprietary materials. In order to better understand the mechanism of the electromechanical response observed, it was important to synthesize our own ERFs and gels to better control the material properties of the system. In this work, polyaniline (pani) was synthesized and combined with trimethyl-terminated poly(dimethylsiloxane) fluid (3MPDMS) for use as an ERF. The conductivity and dielectric constants of the pani particles were controlled and varied, as well as the viscosity of the 3MPDMS phase. Matrices with both an ordered and randomly crosslinked PDMS (XPDMS) network were used. The effect of particle content, molecular weight between crosslinks (M_c), and electric field strength were also investigated in order to determine the maximum force and displacement of the electromechanical actuator.

2. Experimental:

Poly(dimethylsiloxane) networks (XPDMS) were made by both condensation (C-XPDMS) and vinyl addition (V-XPDMS) cures. This generated two distinct networks, one with a "model" polymer network of known structure and one with random crosslinks.[3] The crosslink density of the networks was varied.

ERFs were made from pani in 3MPDMS. The viscosity of the 3MPDMS was varied from 2-5000 cst. The pani particles were doped to different conductivities, which ranged from 10^{-3}-10^{-10} S/cm. The weight percent of pani particles in the ERF was varied from 5-20%.

Conductivities and dielectric constants were measured on a Hewlett-Packard Model 4192A Impedance Bridge. The pani samples were pressed into pellets under 12 tons of pressure and the top and bottom surfaces sputtercoated with 1000 A of gold for electrical contact.

XPDMS/ERF composite gels were made by mixing the XPDMS before cure with the ERF. Volume ratios of 60/40, 50/50 and 40/60 XPDMS/ERF were investigated.

Flexible electrodes were made by sputter-coating gold onto Xerox transparencies using a Dentoncoat II desktop sputter-coater.

The XPDMS/ERF gels were placed between flexible electrodes for displacement measurements, using a Trek 610C high voltage power supply as a DC voltage source. Displacement was measured using an optical microscope with a CCD camera attachment and Optimas 6.1 image analysis software. The displacement measured was the distance a single electrode was moved in response to the applied electric field. The DC fields investigated were from 2-20 kV/cm.

Compression modulus experiments were conducted on an Instron 4704 Universal Tester set up for compression testing. The instrument was modified for use with an electric field by using ceramic plates as insulation between the Instron machine and the copper electrode plates.

3. Results and Discussion:
The Electomechanical Response:

When an electric field was applied to the XPDMS/ERF composite gel between the flexible electrodes, a displacement of the flexible electrodes and an increase in the compression modulus of the composite gel was observed.

ERF Effects:

As shown in Figure 1, the viscosity of the 3MPDMS fluid in the ERF phase of the composite gel was very important. As the viscosity of the 3MPDMS in the ERF was decreased from 5000 to 2 cst, the displacement of the electrodes of the composite gel increased by an order of magnitude. Also, as the particle content of the pani in the composite gel was increased, the compression modulus increased.

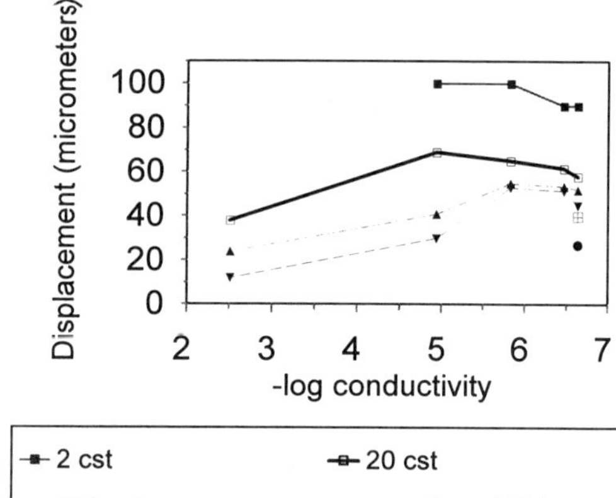

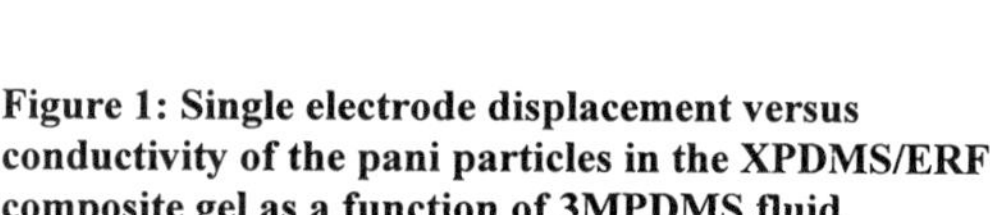

Figure 1: Single electrode displacement versus conductivity of the pani particles in the XPDMS/ERF composite gel as a function of 3MPDMS fluid viscosity. A 10 wt% pani in the ERF was used.

As seen in Figure 2, there was a dramatic increase in the compressive modulus when the pani particles were pre-aligned during the cure of the XPDMS/ERF gel composite. From Figure 1, it can also be seen that there was no increase in the electrode displacement of the pre-aligned samples.

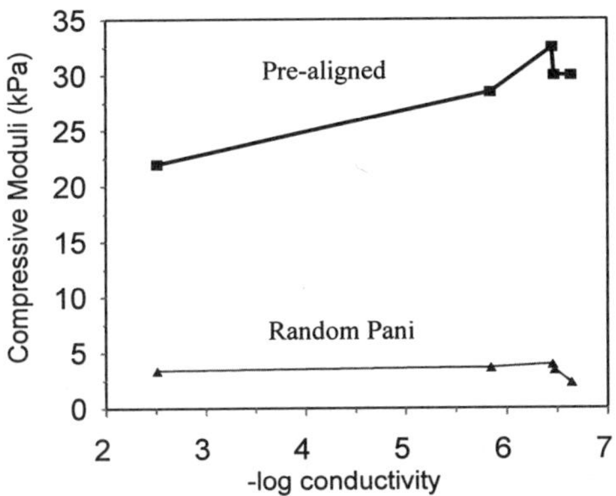

Figure 2: **Compressive moduli of the XPDMS/ERF composite gels versus the conductivity of the pani particles for random and pre-aligned pani. A 10 wt% pani in the ERF was used.**

As shown in Figure 3, when the volume percent of ERF was increased in the gel, the displacement of the electrodes also increased.

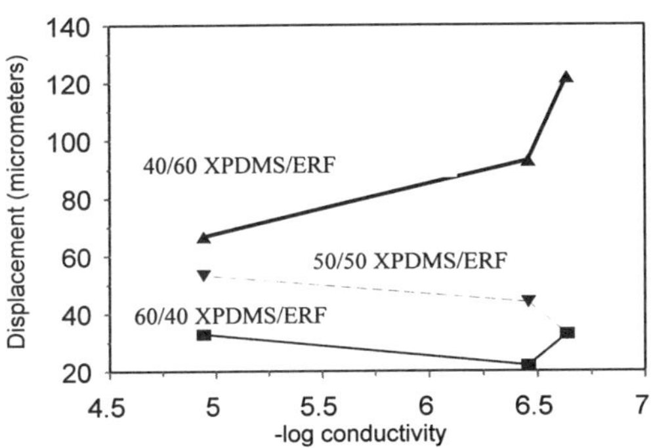

Figure 3: **Single electrode displacement versus conductivity of the pani particles as a function of ERF content in the gel. A 200 cst 3MPDMS fluid was used.**

Figures 1, 2 and 3 also show the effect of the pani conductivity on the XPDMS/ERF composite gels. As the conductivity of the pani particles in the ERF was decreased, the electrode displacement (Figure 1) and compression modulus (Figure 2) initially increased and then leveled off around 10^{-5} S/cm.

XPDMS Effects:

By using a "model" network system, where the crosslinks are found at ordered, consistent intervals (C-XPDMS), the threshold voltage of the XPDMS/ERF composite gel could be reduced by half. The same electrode displacement was generated in an ordered system at half the applied voltage as in a randomly crosslinked network (V-XPDMS). Also, in either a random or ordered network XPDMS system, an increase in the crosslink density of the XPDMS resulted in an increase in displacement.

Electric Field Effects:

The compression modulus of the composite gels increased with the square of the electric field. The force exerted by the moving electrodes was proportional to the compression modulus.

4. Conclusions:

The displacement of the flexible electrodes and the increase in compression modulus of the gel in the electric field can be attributed mostly to interaction between the particles in the ERF. The particles become polarized in response to the electric field, are attracted to one another, and attempt to minimized their interaction energy by aligning parallel to the field direction. The gel network prevents the alignment, but allows for some movement of the particles. These small movements of the particles appear to be responsible for the electrode displacement of 50 μm. In this way, the composite gel is compressed upon application of the electric field. The increase in compression modulus can also be attributed to the increased attraction of the ERF particles to one another as they become polarized in response to the applied electric field.

Experimental results have also been examined in light of the electroviscoelastic effect observed by Shiga et al.[4], where an increase in the shear and compression moduli was observed when gels containing polyelectrolyte particles were placed in an applied electric field. No particle movement or electrode displacement was seen in their systems. Also, recent work by Davis[5] and Khusid and Acrivos[6] suggests that the conductivity model of ERFs could be used for modeling the electromechanical response of the XPDMS/ERF composite gels.

Acknowledgements

The Office of Naval Research is gratefully acknowledged for financial support through grant number N00014-95-0136.

References:

1. Bohon, K., Krause, S., *Journal of Polymer Science Part B:Polymer Physics*, in press.
2. Hill, J. C., Van Steenkiste, T.H., *Journal of Applied Physics*, 1991, 70 (3), 1207.
3. Mark, J., Eisenberg, A., Graessley, W., Mandelkern, L., Samulski, E., Koenig. J., Wignall, G., <u>Physical Properties of Polymers</u>, 2nd Ed., 1993, ACS Professional Reference Book, Washington DC.
4. Shiga, T., Ohta, T., Hirose, Y., Okada, A., Kurauchi, T., *Journal of Materials Science*, 1993, 28, 1293.
5. Davis, L.C., *Journal of Applied Physics*, 1997, 81 (4), 1985.
6. Khusid, B., Acrivos, A., *Physical Review E*, 1995, 52 (2), 1669.

Small Molecule Probe Diffusion in Ultrathin Polymer Films

David B. Hall[*] and John M. Torkelson[*],[#]

[*]Department of Chemical Engineering
[#]Department of Materials Science and Engineering
Northwestern University
Evanston, IL 60208

Introduction

"Ultrathin" polymer films (films with thicknesses < 100 nm) are becoming increasingly important in applications such as coatings, membranes, and microelectronics. However, the properties of these ultrathin polymer films have remained relatively unexplored until quite recently. Ellipsometry[1-3], x-ray reflectivity[4,5], and dewetting[6] studies of substrate-supported ultrathin polymer films have suggested that the glass transition temperature, T_g, may be a function of film thickness. With decreasing film thickness, decreases in T_g (relative to "bulk" values) are found for systems in which the polymer and substrate have weak affinity for one another while increases in T_g are found for systems in which the polymer and substrate have a strong affinity for one another. These results suggest that polymer-substrate affinity can have a strong influence on polymer segmental mobility in ultrathin films, but that the range of these interactions is fairly short since significant deviations from bulk T_g are only found for films thinner than 50 nm. Polymer self-diffusion measurements[7,8], in contrast, have suggested that the range of influence of the substrate may be much larger, possibly extending over 100 nm. A lack of supporting theoretical work and also apparent contradictions regarding results from similar systems necessitate more experimental studies in this area.

In this work, fluorescence nonradiative energy transfer (NRET) is used to measure the translational diffusion of small molecule probes in thin and ultrathin polymer films with thicknesses as low as 47 nm. The small molecule probes, decacyclene and lophine (see figure 1), are large enough so that their motion is partially coupled to the cooperative segmental mobility of the polymer and thus are sensitive to changes in polymer film T_g. The role of polymer-substrate affinity is explored by measuring probe diffusion in two polymers, polystyrene (PS) and poly(isobutyl methacrylate) (PiBMA) on acid-cleaned fused quartz substrates.

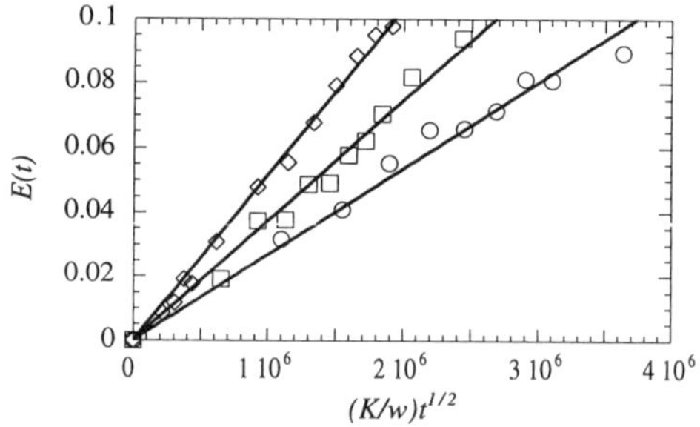

(a)   (b)

Figure 1: Small molecule probes: (a) decacyclene and (b) lophine.

Experimental

A fluorescence "donor" molecule may transfer its excited state energy to an appropriate "acceptor" molecule without emitting light in a process called nonradiative energy transfer (NRET). For NRET to occur, the donor and acceptor must be in close proximity, usually within 2 to 4 nm of one another. It is this sensitivity to small length scales that allows NRET to be used to measure very slow diffusion and diffusion in ultrathin polymer films. A complete description of the use of NRET to measure small molecule probe diffusion in thin and ultrathin polymer films is given elsewhere[9,10]. The technique involves using a "sandwich" of two polymer films, one of which contains freely doped acceptor molecules and the other contains polymer to which a fluorescence donor molecule has been covalently attached. Upon annealing above T_g, acceptor molecules will diffuse into the film containing donor-labeled polymer and the steady-state fluorescence intensity of the donor film will decrease due to NRET. Assuming Fickian diffusion of the acceptor molecules and negligible diffusion of the donor-labeled polymer during the time scale of the experiment, the efficiency of energy transfer, $E(t)$, may be related to the translational diffusion coefficient of the acceptor, D, in the following manner:

$$E(t) = \frac{I(0) - I(t)}{I(0)} = K\left(\frac{\sqrt{Dt}}{w}\right) \qquad \text{for } t < \frac{w^2}{16D} \qquad (1)$$

where t is annealing time, $I(0)$ is donor fluorescence intensity at $t = 0$, $I(t)$ is donor fluorescence intensity at t, w is donor film thickness, and K is a constant which depends on acceptor concentration.

Pyrene-labeled polymer (pyrene label content < 0.4 mol%) was employed as the NRET donor species and the small molecule probes, decacyclene and lophine, used as NRET acceptors. Acceptor concentration was less than 0.5 mol% in all cases. PS (M_n = 72,000 and M_w = 110,000) and PiBMA (M_n = 230,000 and M_w = 430,000) were synthesized via free radical polymerization. Bulk T_g's were determined using DSC at 10°C/min: PS T_g = 100°C, PiBMA T_g = 64°C. Steady-state fluorescence measurements as a function of time were performed using a Spex Fluorolog spectrophotometer equipped with a temperature controlled sample cell. Pyrene was excited using 331 nm light and emission intensity was monitored at 397 nm. Films were layered using a water transfer method and allowed to dry for at least 24 hours under vacuum. Film thickness was characterized using a Tencor P10 profilometer.

Results and Discussion

Figure 2 plots $E(t)$ as a function of $(K/w)t^{1/2}$ for decacyclene diffusion in three different PS film thicknesses at 103°C. The slope of the curve is equal to $D^{1/2}$. It is clear that D decreases with decreasing film thickness for the samples shown here. Decacyclene diffusion coefficients were measured in a variety of PS films ranging in thickness from 66 nm to 931 nm and are plotted in Figure 3(a). Each data point represents an average of at least 3 films with error bars equal to one standard deviation. The plot shows that D begins to drop from a bulk value of approximately 3×10^{-15} cm²/s when film thickness is decreased below about 150 nm to a minimum value of approximately 1×10^{-15} cm²/s for the thinnest film studied, 66 nm. Similar results are found for lophine diffusion in PS films at 103°C as shown in Figure 3(b).

Figure 2: $E(t)$ vs. $(K/w)t^{1/2}$ for decacyclene diffusion in PS films of thickness (◊) 367 nm, (□) 124 nm, and (○) 66 nm at 103°C. Straight lines are linear fits with slope equal to $D^{1/2}$.

A reduction in polymer matrix mobility can slow small molecule diffusion. It has been proposed that an increase in T_g for PS ultrathin films as compared to bulk may explain why PS self-diffusion seems to slow as film thickness decreases below 150 nm[11]. Decacyclene diffusion in PS follows a modified Williams-Landel-Ferry (WLF) temperature dependence[12]:

$$\log\left[\frac{D(T)}{D(T_g)}\right] = \frac{\xi C_1(T - T_g)}{(C_2 + T - T_g)} \qquad (2)$$

where C_1 and C_2 are the WLF parameters of the polymer and ξ is the coupling parameter ($\xi \leq 1$). The effect of a T_g shift on D may be calculated using eq. (2). Using C_1 = 15 and C_2 = 40K for PS and ξ = 0.63 for decacyclene diffusion in PS[12], a T_g increase of 2°C would cause D to drop from a bulk value of 3×10^{-15} cm²/s to a value of 1×10^{-15} cm²/s as found in 66 nm thick films. This result is qualitatively consistent with the PS self-diffusion measurements[7] in ultrathin films; however, it is inconsistent with dewetting[6] and x-ray reflectivity[4] studies of PS ultrathin films on SiO$_x$ substrates which suggest a substantial depression in T_g for this system.

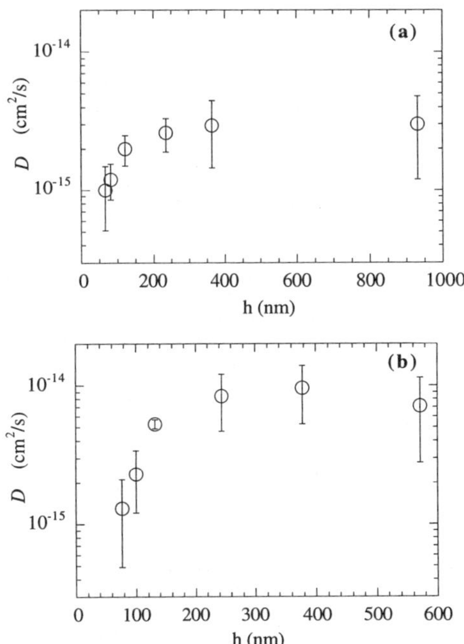

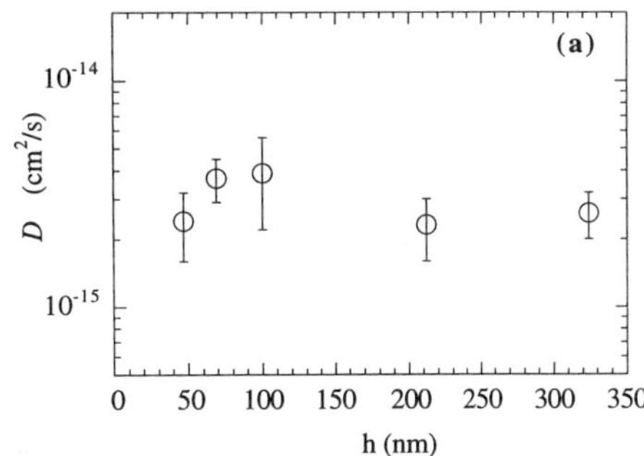

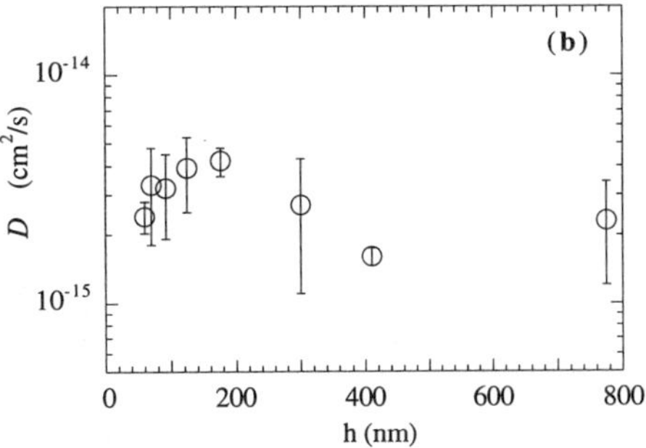

Figure 3: D as a function of film thickness, h, for (a) decacyclene and (b) lophine diffusion in PS at 103°C.

Figure 4: D as a function of film thickness, h, for (a) decacyclene and (b) lophine diffusion in PiBMA at 72°C.

PS is expected to have a weak affinity for the fused quartz substrate. If polymer-substrate interactions play an important role in polymer mobility in ultrathin films, one would expect to see an even greater drop in small molecule probe diffusion coefficient with film thickness for a system in which the polymer has a stronger affinity for the substrate. PiBMA would be expected to have a higher polymer-substrate affinity than PS since it may undergo hydrogen bonding with the polar quartz substrate. Preliminary dewetting studies of ultrathin PS and PiBMA films on fused quartz support this notion. Decacyclene and lophine diffusion coefficients in thin and ultrathin PiBMA films at 72°C are given in Figures 4(a) and 4(b) respectively. For both probes, within experimental error, there is little or no change in D with film thickness for films ranging from 47 nm to 775 nm thick. This implies that the T_g of these films is not changing measurably over the film thickness range examined here. This is consistent with second harmonic generation measurements of ultrathin films of an isobutyl methacrylate containing copolymer on quartz[13] and also ellipsometry measurements of poly(methyl methacrylate) ultrathin films on SiO_x[2].

A comparison of small molecule probe diffusion results for PS and PiBMA ultrathin films based on polymer-substrate affinity is problematic since we find a decrease in D for PS ultrathin films which have a weaker affinity for the quartz substrate than PiBMA. This suggests that for the film thicknesses that we have examined, h > 47 nm, polymer-substrate interactions are not of primary importance in determining the overall properties of the film. Ellipsometry[1-3] and x-ray reflectivity[4,5] studies support the notion that these interactions are short range and only affect polymer film properties in the thinnest of ultrathin films, less than 50 nm thick. In this case, it is unlikely that the underlying cause for the decrease in D found for ultrathin PS films is an increase in T_g. One possible explanation for the observed behavior is a difference in chain packing or density in these ultrathin PS films. Baschnagel and Binder[14] have proposed that polymer chains adopt an extended conformation parallel to solid interfaces. This would increase density in the direction perpendicular to the substrate and thus hinder polymer or small molecule mobility in this perpendicular direction without any change in T_g. PS may be more susceptible to this phenomenon than PiBMA. Further work involving small molecule probe diffusion measurements in ultrathin poly(2-vinyl pyridine) films on fused quartz is underway. Poly(2-vinyl pyridine) has an extremely strong affinity for fused quartz and measurements should shed more light on the range and importance of polymer-substrate interactions in determining the overall properties of ultrathin polymer films.

Conclusions

Small molecule probe diffusion coefficients were measured with fluorescence NRET in PS and PiBMA films as thin as 47 nm. A slowing of probe translational diffusion was found as PS film thickness is decreased below 150 nm while no corresponding change is found for PiBMA films. This result is not expected based on polymer-substrate affinity considerations. The results may be best explained in terms of differing polymer chain conformation in ultrathin films rather than any changes in T_g of the films at these thicknesses.

Acknowledgments

We thank R.D. Miller for providing the lophine probe. This work was supported by the MRSEC program of the National Science Foundation (Grant DMR-9632472) at the Materials Research Center of Northwestern University.

References

1. Keddie, J.L.; Jones, R.A.L.; Cory, R.A. *Europhys. Lett.* **1994**, *27*, 59.
2. Keddie, J.L.; Jones, R.A.L.; Cory, R.A. *Faraday Discuss.* **1994**, *98*, 219.
3. Forrest, J.A.; Dalnoki-Veress, K.; Dutcher, J.R. *Phys. Rev. E* **1997**, *56*, 5705.
4. Wallace, W.E.; van Zanten, J.H.; Wu, W.L. *Phys. Rev. E* **1995**, *52*, 3329.
5. van Zanten, J.H.; Wallace, W.E.; Wu, W.L. *Phys. Rev. E* **1996**, *53*, 2053.
6. Reiter, G. *Europhys. Lett.* **1993**, *23*, 579.
7. Frank, B.; Gast, A.P.; Russell, T.P.; Brown, H.R.; Hawker, C. *Macromolecules* **1996**, *29*, 6531.
8. X. Zheng et al., *Phys. Rev. Lett.* **1997**, *79*, 241.
9. Deppe, D.D.; Dhinojwala, A.; Torkelson, J.M. *Macromolecules* **1996**, *29*, 3898.
10. Hall, D.B.; Miller, R.D.; Torkelson, J.M. *J. Polym. Sci.: Part B: Polym. Phys.* **1997**, *35*, 2795.
11. Russell, T.P.; Kumar, S.K. *Nature* **1997**, *386*, 771.
12. Deppe, D.D. Ph. D. Dissertation, Northwestern University, Evanston, IL, 1996.
13. Hall, D.B.; Hooker, J.C.; Torkelson, J.M. *Macromolecules* **1997**, *30*, 667.
14. Baschnagel, J.; Binder, K. *J. Phys. I France* **1996**, *6*, 1271.

THE ROLE OF RUBBER PARTICLE CAVITATION IN DEFORMATION AND
FRACTURE OF TOUGHENED PLASTICS

C.B.Bucknall,[*] D.S.Ayre[*], A. Lazzeri# and D.J.Dijkstra[+]

[*]Advanced Materials Department, Cranfield University, Bedford MK43 0AL, UK

[#]Centre for Materials Engineering, University of Pisa, 56126 Italy

[+]Bayer AG, ZF-FPT, D-51368 Leverkusen, Germany

Rubber particle cavitation is now recognised as an important part of the toughening mechanism in rubber-modified plastics. An energy balance model which relates the critical conditions for cavitation to the structure and morphology of the polymer was proposed in a previous paper.[1] This model analysed the energy changes involved in forming a void in a homogeneous spherical rubber particle held at constant volume strain: *ie* without additional energy inputs from the surrounding matrix. The analysis showed that cavitation is controlled by the particle size, rubber shear modulus and applied volume strain.[2]

The conditions imposed on the system in this earlier treatment are quite restrictive. In practice, the cavitation of rubber particles embedded in a glassy matrix is always more complex. The glassy matrix not only interacts with the rubber particle to provide an additional source of energy during voiding, but also subjects the rubber to dilatational strains at an earlier stage, as a consequence of differential thermal expansion between the two phases. The particle's morphology also affects its behaviour. Therefore, in the present programme the model has been developed to account for matrix/particle interactions.[3] The energy released from the material surrounding a cavitating particle is incorporated in the analysis, and the energy changes within the complex rubber particle are re-evaluated, taking account of internal morphology, differential thermal contraction and mechanical loading.

The critical conditions for rubber particle cavitation have been determined experimentally[3] for a toughened PMMA. The toughening particles have a solid core / rubber shell morphology. A specimen in flexure has a compressive region and a tensile region, separated by a neutral surface. At yield the position of the neutral surface is dependent upon the relative compressive (C) and tensile (T) yield values. Tensile yield can occur with or without particle cavitation, depending on whether critical conditions for rubber voiding are achieved. Such a change in mechanism is reflected by a change in neutral surface position (or C/T ratio). Fig 1 demonstrates the change in yield mechanism in the RTPMMA studied, which occurs at a yield stress of 24 MPa and a test temperature of 55°C. Predictions of the energy-balance model (figs 2 & 3), using experimental yield stress data at given temperatures, suggest that the thermal strain contribution is smaller than expected, but nevertheless a necessary one. Without thermal strain energy, tensile yield at 50°C is expected to occur without cavitation.

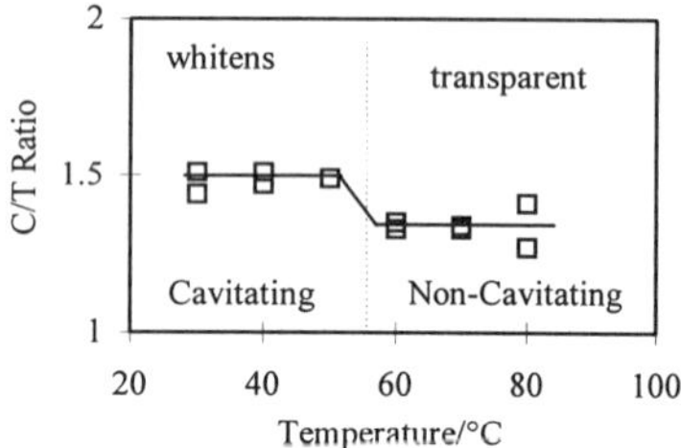

Fig. 1. Experimental results for RT-PMMA in flexure

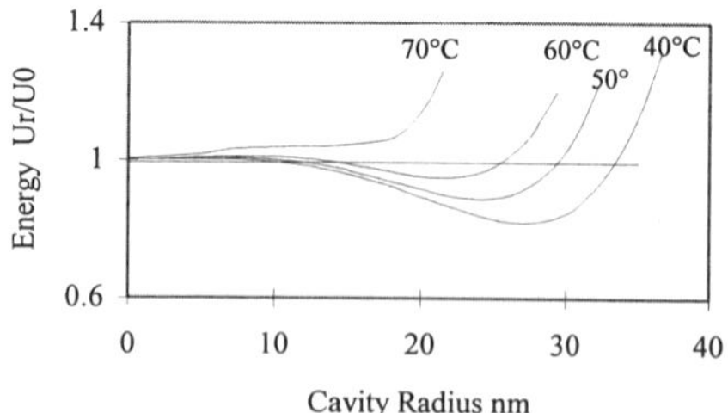

Fig. 2. Energy balance model for RT-PMMA including
thermal strain energy.

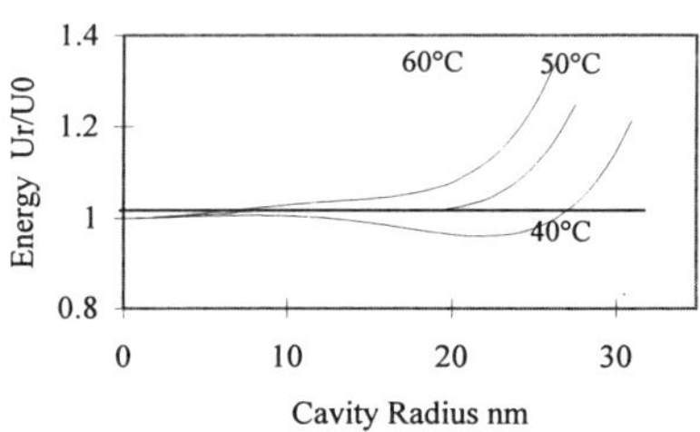

Fig. 3. Energy balance model for RT-PMMA
excluding thermal strain energy

Particle cavitation is usually accompanied by crazing and shear yielding of the glassy matrix, which make it difficult to identify the specific role of cavitation in the toughening process. It is not clear whether rubber cavitation is a precursor to matrix yielding, or simply a consequence of that yielding. This issue has been resolved in the present work using a novel experimental approach.

The most effective method is to apply dilatational stresses to the particle using differential thermal contraction, which has been shown previously to cause rubber particle cavitation in selected materials. When a critical point is reached the particles cavitate and the thermal contraction behaviour of the polymer blend changes, as shown in Fig 4a for ABS. These conditions impose no dilatational forces on, and hence no damage to, the matrix. Figure 4b demonstrates the effects of rubber particle cavitation: two of he ABS specimens contain rubber particles that have been damaged by cooling, and therefore craze much more rapidly than standard specimens. The cavitation of the rubber particles is verified by DMTA measurements of the rubber T_g (Fig. 4c). Compressive loading causes an increase in density of the rubber particles, which is reflected in an increasing T_g value. Tensile loading expands any cavities present, and thus cannot reduce the rubber density, so that tensile loading does not affect the T_g of the rubber. Alternative techniques for generating voids in the rubber and identifying their role in the toughening process are under active investigation.

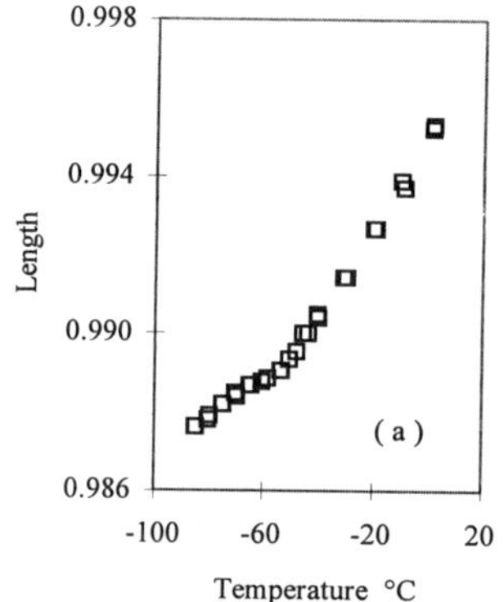

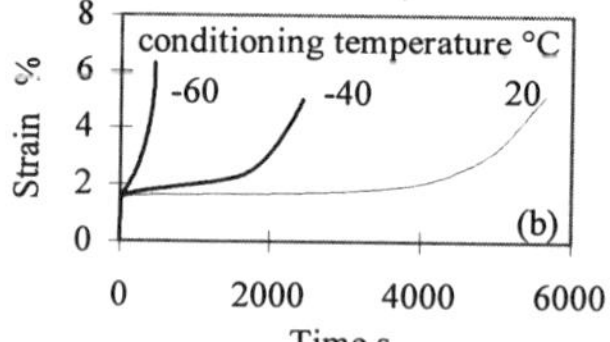

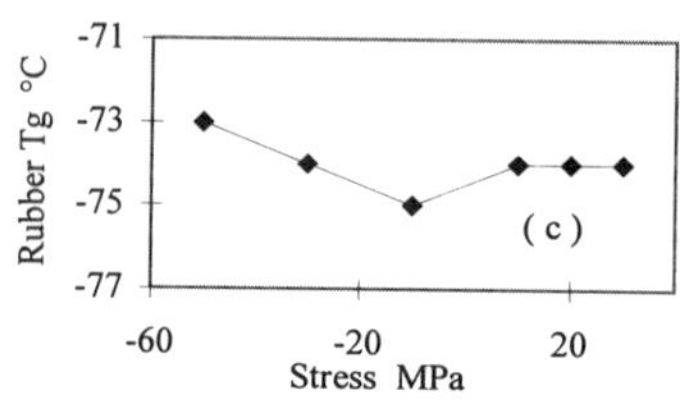

Fig. 4 Effects of cavitation in ABS observed by (a) cooling, (b) creep at 20°C,
(c) DMTA.

References
1 A.Lazzeri and C.B.Bucknall, *J. Mater. Sci.* 28 (1993) 6799.
2 C.B.Bucknall, A.Karpodinis and X.C.Zhang, *J. Mater. Sci.* 29 (1994) 3377.
3 D.S.Ayre and C.B.Bucknall, *Polymer* in press.

RECENT DEVELOPMENTS IN THE MODELLING OF DILATATIONAL YIELDING IN TOUGHENED PLASTICS

A.Lazzeri*, R. Dreassi* and C.B. Bucknall**

*CIIM, University of Pisa, 56126 Pisa, Italy
**SIMS, Cranfield University, Bedford, MK43 OAL, UK

INTRODUCTION

A model for cavitation of rubber particles and subsequent yielding in the matrix was proposed in earlier papers[1-3]. The energy-balance criterion for cavitation predicts that particle cavitate at a critical volume strain which depends upon particle size and rubber shear modulus. The model shows that cavitation accelerates both shear yielding and multiple crazing in the matrix. In this paper new developments in the modelling of dilatational yielding in rubber toughened polymers will be presented. In particular it will be shown that a variation in the rate of yielding, following cavitation, can be predicted by means of a modified version of the Gurson equation, which is obtained on the basis of a FE analysis of the stress and strain fields around a cavitated rubber particle.

EXPERIMENTAL RESULTS

Creep data and standard stress-strain curves were obtained over a range of temperatures and strain rates for moisture-equilibrated PA6.6, and PA6.6 blends containing 5, 10, 15 and 20 wt% of maleinated EPR.

The creep results demonstrated the onset of cavitation in the larger particles at a tensile stress of about 27 MPa, corresponding to a mean stress (negative pressure) of 9 MPa. Increasing the mean stress beyond this point promotes further cavitation, as smaller particles reach their critical volume strains.

No dilatation occurs in plain PA6.6, which contains no rubber. Eyring plots of σ_y/T for this material give a series of parallel straight lines obeying the Eyring equation:

$$\frac{\sigma_y}{T} = \frac{1}{\gamma V^*}\left[\frac{\Delta H}{T} + R\ln\left(\frac{\dot{\varepsilon}_y}{\dot{\varepsilon}_0}\right)\right]$$

where σ_y = yield stress at temperature T, R = gas constant, ΔH = activation energy, V^* = activation volume, $\dot{\varepsilon}_y$ = applied strain rate, $\dot{\varepsilon}_o$ = reference strain rate and γ = stress concentration factor. In plain PA6.6, the stress concentration factor is taken as 1.0, and the activation volume $V^* = 1.1$ nm^3.

Applying the same activation volume to the rubber modified blends gives a series of stress concentration factors γ which increase steadily with EP rubber content, to 2.0 nm^3 in blends containing 20wt% rubber. Plots of σ_y/T against $ln\dot{\varepsilon}$ decrease in slope as the rubber content increases. The yield behaviour also shows a transition in behaviour: plain PA6.6 necks at a strain of $\approx$10% and fractures at ε = 25%: whereas PA66 containing 20% EPR yields and draws without necking, finally breaking at $\varepsilon >$ 100%. In the blends, internal cavitation gradually replaces macroscopic necking as the mode of extension. Raising the temperature reduces the degree of cavitation, because the yield stress of the blends fall below the level at which cavitation can occur.

DISCUSSION

The rubber particles not only reduce the yield stress but also affect the deformation of nylon 6.6 in a number of other ways. When the applied stresses are high enough, the rubber particles undergo cavitation, thereby promoting dilatational shear yielding in the matrix. Both rubber particle cavitation and matrix shear deformation play an important part in the extension and failure behaviour of PA6.6-EPR blends under uniaxial tension. In standard tensile tests, dilatational yielding enable the parallel gauge portion to draw uniformly, without the necking that is characteristics of PA6.6 itself. Eyring plots of tensile creep data show that rubber particle cavitation allows the PA6.6/15 blend to deform at much higher rates than would otherwise be observed over the same range of applied stresses.

The cavitation diagrams introduced in previous papers[1,3] show graphically the complex response of rubber particles with the applied stress and their interaction with the yield and fracture behaviour of toughened plastics.

For a pure polymer, or for a rubber toughened polymer where the rubber particles do not cavitate, the yield condition is represented in this plane by a straight line. The strong dependence of the yield stress upon strain rates means that there is no unique yield line, even at a fixed temperature. Instead, yield is represented by a series of parallel lines, each corresponding to a different strain rate, each corresponding to a different strain rate.

The yield behaviour changes when the rubber cavitate. The yield condition is now represented by a curve given by a modified version of the Gurson's equation[1,3]:

$$\sigma_e = \sigma_o \sqrt{\left(1 - \frac{\mu\sigma_m}{\sigma_o}\right)^2 - 2fq_1\cosh\left(\frac{3q_2\sigma_m}{2\sigma_o}\right) + (q_1 f)^2}$$

where σ_o is the effective stress at yield when the normal stress σ_m and the void content f are both zero, while μ is the pressure coefficient of yielding. The factors q_1 and q_2 were introduced originally by Tvergaard[4] to improve the fit between Gurson's predictions and finite analysis (FE) studies on metals. A recent FE analysis[5] suggests the values q_1 = 1.375 and q_2 = 0.927 for some polymers. When $\sigma_m = 0$, rubber particles have the same effect upon yield stress as voids. However, under other combinations of effective and mean stress, the two yield curves diverge.

Figure 1 illustrates a very important principle relating to polymers capable of cavitating, namely that the strain rate resulting from the application of a given stress does not have a unique value, but varies substantially with the extent of cavitation. The three curves shown in Fig. 1 were all calculated for a polymer containing 20 vol% of rubber particles, and are based on experimental data from the present programme. If none of the particles has cavitated (f = 0) the strain rate $\dot{\varepsilon}$ = 0.1. If 50% of the particles have cavitated (f = 0.1) $\dot{\varepsilon}$ = 0.4, and if all of the particles have cavitated (f = 0.2) the strain rate $\dot{\varepsilon}$ reaches its maximum attainable value of 2.1. The three curves intersect where they cross the construction line for uniaxial tension.

For materials containing a distribution of particle sizes, there is a corresponding range of critical mean stresses, which can be represented by a cavitation zone, as in Fig. 2. Cavitation begins in the larger particles, on the low-stress side of this zone, and proceeds to affect smaller particles as the applied stress increases. Fig. 2 shows yielding under uniaxial tension, at a stress of 40 MPa, where virtually all of the particles have cavitated, and yielding follows eqn 2 with f = 0.2. In the absence of cavitation, an applied stress of 40 MPa would give a substantially reduced strain rate, amounting to less than 10% of that for the fully-cavitated nylon blend.

Yield stress data obtained over a range of strain rates at 20°C are plotted in Fig. 3 for the PA6.6/15 blend, which contains 20 vol.% rubber. They show that yielding occurs at stresses between the upper bound line, corresponding to f = 0, and the lower bound at f = 0.2. The points appear to move towards the lower bound curve as the yield stress increases with strain rate, a result that supports the theoretical prediction that the fraction of particles cavitated increases with strain rate.

The foregoing discussion based on cavitation diagrams provides a clear quantitative explanation for the increase in creep rate (over and above that obtained by extrapolating Eyring rate data) that is observed at stresses above the critical mean stress for rubber particle cavitation ($\sigma_m \approx$ 9 MPa). Tensile tests on PA6.6 blends often reach stresses much higher than 27 MPa, and it is clear from the resulting data that most of the specimens tested cavitated well below the yield point defined by the 1% offset strain. When $\sigma_y >$ 27 MPa, as in the case of blend PA6.6/10 over the range -20° to 20°C, a single mechanism controls yielding. Only at T$\,>\,$40°C do the yield stresses become low enough to permit significant plastic deformation without cavitation. Consequently, yield stress data for temperatures below 40°C form a coherent set of parallel straight lines. A similar set of straight lines is obtained for the PA6.6/20 blend between -40°C and

10°C. In this material, yield data obtained at 20°C are within the 27-30 MPa band, where cavitation effects are minimal, and the slope $[\partial(\sigma_y/T) \, / \, \partial\ln\dot{\varepsilon}]_T$ decreases.

REFERENCES

1. A. Lazzeri and C.B. Bucknall, *J. Mater. Sci.* **28** (1993) 6799.
2. C.B. Bucknall, A. Karpodinis and X.C. Zhang, *J. Mater. Sci.* **29** (1994) 3377.
3. A. Lazzeri and C.B. Bucknall, *Polymer* **36** (1995) 2255.
4. V. Tvergaard, *Int. J. Fracture* **17** (1981) 389.
5. A. Lazzeri, A. Livi, L. Bertini, unpublished results.

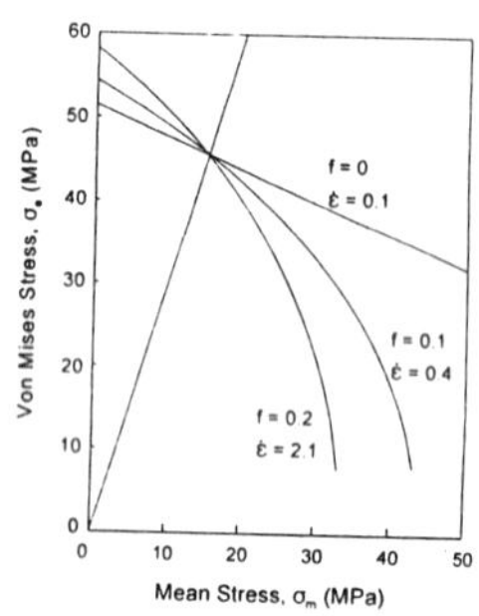

Fig. 1 - Diagram showing acceleration in strain rate following rubber cavitation.

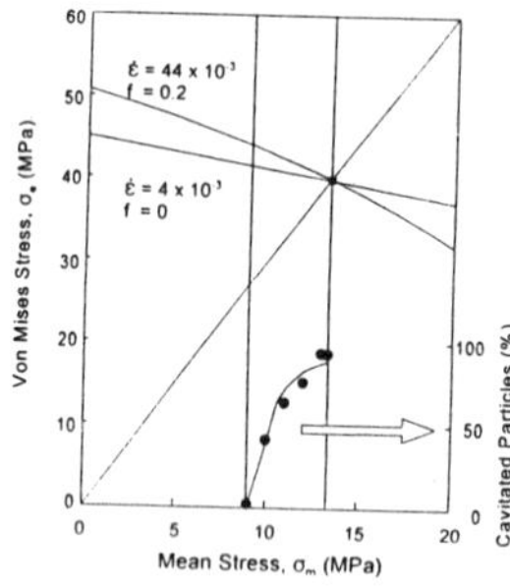

Fig. 2 - Effect of cavitation on tensile creep strain rate at 40 MPa.

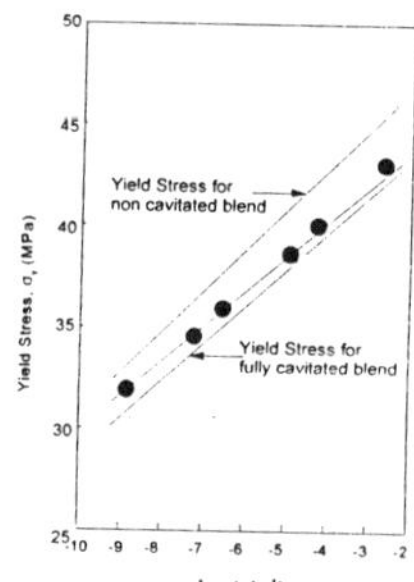

Fig. 3 - Upper bound (no cavitation) and lower bound (full cavitation) Eyring plots for PA6.6/20 vol% EPR at 20°C.

YIELD AND FRACTURE BEHAVIOR OF POROUS EPOXY

A. J. Lesser, E. D. Crawford, R. S. Kody

Polymer Science and Engineering Department
Silvio O. Conte Research Center
University of Massachusetts
Amherst, MA 01003

INTRODUCTION

Toughening glassy polymers by the introduction of an elastomeric second phase has become commonplace in many engineering formulations. The role of inelastic matrix deformation subsequent to particle cavitation in the form of localized shear bands, dilational bands, and plastic void or hole growth have been extensively discussed in the literature [1-5]. As a consequence, a variety of micro-mechanical models have been proposed to describe the effective material properties of a modified polymer including its yield [5-7] and fracture [1-3, 8, 9] behavior.

One issue limiting a thorough evaluation of these models has included results from controlled deformation experiments on fully cavitated materials (i.e., materials containing known concentrations of micro-voids). These experiments are necessary to isolate the behavior of the cavitated matrix without convoluting the result with properties of the rubbery phase, which cannot be directly measured. Some noteworthy studies have been conducted on epoxies containing pre-cavitated particles[10], non-adhering particles[11], and hollow latex particles[12,13].

The purpose of this communication is to present results on the yield and fracture behavior of a micro-voided epoxy formulation. The yield behavior is reported over stress states ranging between uniaxial compression and biaxial tension. The results from fracture studies are also presented. The experimental results are then discussed in context to current yield and fracture models.

EXPERIMENTAL

Material and Sample Preparation

The method employed follows closely to that recently reported by Hilborn[14] in which they chemically induce phase separation between a low molecular weight solvent (hexane) and the epoxy before curing. Epon 825, a diglycidyl ether of bisphenol A epoxy with an epoxide equivalent of 175 g/mol was provided by Shell Chemical Co. N-Aminoethyl piperazine (AEP) and hexane were purchased from Aldrich Chemical Co. and used as received.

Epon 825 was dried under vacuum at 80 °C for 8 hrs. After drying, 50 grams of the resin, 12.4g of AEP (stoichiometric ratio), and 15.6 g of hexane were mixed with a stir plate for 5 minutes at room temperature. After mixing, an emulsion is produced with the higher density phase consisting of epoxy, AEP, and soluble hexane, and a lower density phase consisting primarily of hexane. This mixture was then poured into either a spinning metal tube (hollow cylinders), between teflon separated glass plates (tensile bars, fracture specimens), or into test tubes (compression samples) for 8 hrs. at room temperature. This allowed gelation to occur without much of an exotherm. After gelation, the materials were cured at 60 °C for 3 hrs followed by an additional cure at 150 °C for 3 hrs. Unmodified materials were made with the same procedure without the addition of hexane.

Mechanical Characterization Methods

Room temperature tensile tests were performed on ASTM D638 Type I tensile bars at a crosshead speed of 5 mm/min equipped with an axial strain gauge. Compression tests were performed on cylinders (dia.=11.2 mm, l=22.4) at a crosshead speed of 2.2 mm/min.

Fracture toughness tests were performed on 3-point bend fracture toughness bars (100 mm x 19 mm x 3.2 mm) at crosshead speed of 5 mm/min and a span to width ratio was set to 4. The fracture behavior was analyzed by linear elastic fracture mechanics.

The biaxial tests were conducted pressurized hollow cylinders. Details of the fabrication and testing technique can be found elsewhere [15]. All samples were tested in stress states ranging from uniaxial compression to biaxial tension, at 60°C (to promote inelastic void growth), and a constant octahedral shear strain rate.

RESULTS AND DISCUSSION

Scanning electron microscopy was performed on cryogenically fast fractured samples to investigate the morphology of the micro-voided epoxy samples (see Fig. 1). Stereological information was extracted from these micrographs and an average void size of 6 μm was estimated with a corresponding volume concentration $f= 0.07$. This value is lower than that which was targeted from initial concentration levels of the mixture ($f \sim 0.20$) and supports the contention that some of the hexane is retained in the epoxy and/or evaporates during the gelation.

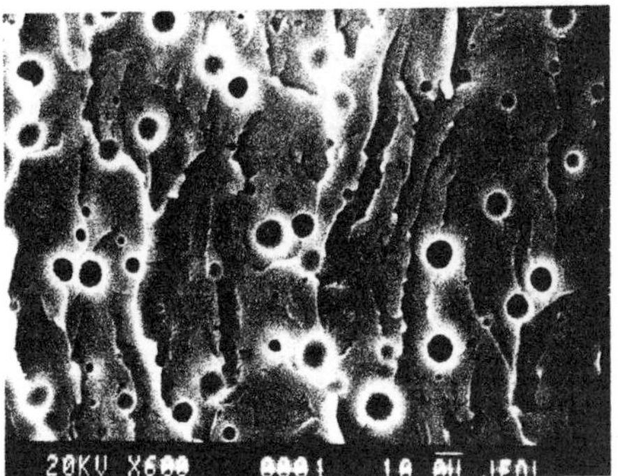

Figure 1: SEM Micrograph of microporous epoxy prepared EPON 825, AEP, and hexane.

The glass transition temperatures (T_g) of the materials were measured with a DuPont Thermal Analyzer 2000 at a heating rate of 10 °C/min. These results are presented in Table 1 along with the results from the uniaxial tension, compression and fracture toughness tests. The 28% drop in the T_g is consistent with the stereological results and further supports the contention that some hexane remains in the epoxy even after the postcure. Thus the 15% drop in the compressive yield stress and 13% increase in fracture toughness may be attributed to both the presence of microvoids and residual hexane.

Table I: Material Characteristics

Resin	T_g (°C)	E (GPa)	σ_{yc} (MPa)	K_{Ic} (MPa m$^{0.5}$)
Unmodified	126	2.8	90	1.5
Hexane Modified	93	2.3	76	1.7

Recently, Lazzeri and Bucknell have proposed a yield function to describe the yield behavior for cavitated polymers [5, 6]:

$$\Phi = \frac{\sigma_e^2}{\sigma_o^2} + \frac{\mu\sigma_m}{\sigma_o}\left(2 - \frac{\mu\sigma_m}{\sigma_o}\right) + 2f\cosh\left(\frac{3\sigma_m}{2\sigma_o}\right) - f^2 - 1 = 0 \qquad \boxed{1}$$

where f is the volume fraction of voids, σ_e and σ_m are the equivalent and mean stress, σ_o is the yield stress in the absence of hydrostatic stress, and μ is the coefficient of internal friction.

In the absence of hydrostatic stress (i.e., $\sigma_m = 0$) Eq. 1 reduces to:

$$\sigma_o' = (1 - f)\sigma_o \qquad \boxed{2}$$

where σ_o' denotes the effective yield stress of the modified polymer and should be viewed in the macroscopic sense. Eq. 2 states that, in the absence of hydrostatic stress, the apparent yield stress of the polymer should reduce directly by the void fraction amount. Additionally, Eq. 1 reduces to a linear form similar to the modified Von Mises yield model proposed by Sternstein [16] in when f=0.

$$\sigma_e = \sigma_o - \mu\sigma_m \qquad \boxed{3}$$

At increasing values of f, the yield envelope as described by Eq. 1 becomes increasingly nonlinear.

The biaxial tests on the hollow cylinders was conducted at an elevated temperature (60 C) to further drop the yield stress and promote inelastic void growth in the formulation. Fig. 1 shows the results from the biaxial tests conducted on the microvoided epoxy at 60 C.

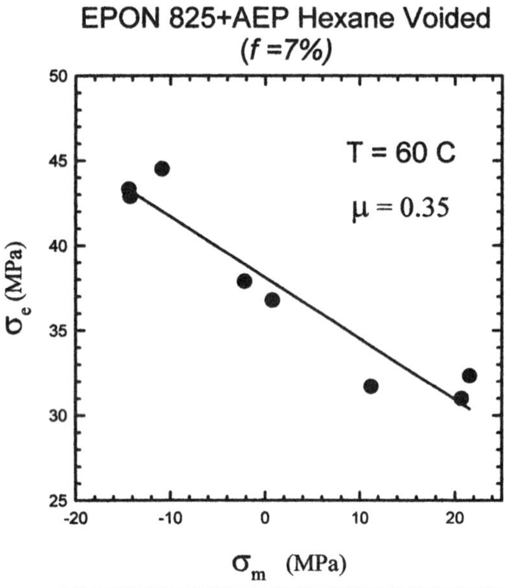

Figure 2: Plot of biaxial test results on microporous epoxy formulation conducted at 60 C. The void volume fraction is f = 7%.

Fig. 2 shows that the yield behavior of the microvoided epoxy is linear and can be modeled simply by Eq. 3 with a coefficient of internal friction determined as $\mu = 0.35$. This result is somewhat surprising and further suggests that, even though the epoxy matrix is plasticized with hexane, and the tests were conducted at elevated temperature, the void fraction is to low to promote a significant nonlinear reduction in the apparent yield stress (see Eq. 1). Consequently, a higher void volume fraction is necessary to significantly alter the effective yield response of this resin even though it is a reasonably tough formulation. This result is further supported by the fracture toughness test results, which only show a modest increase in the modified formulation. Further estimates can show that the increase in the fracture toughness and corresponding drop in yield can be attributed primarily to the plasticizing effect of the hexane remaining in the epoxy matrix and not to the void volume fraction. Hence, effective toughening would require a concentration of voids (or cavitated particles) higher than 7%.

CONCLUSIONS

It was shown that the introduction of hexane into an EPON 825-AEP formulation can produce a material with 7% microvoids with an average size of 6 μm. A consequence of using this method also introduces hexane into the matrix of the polymer, which depresses the glass transition and yield stress, and modestly increases the fracture toughness. Further, biaxial testing on these formulations showed that the voids had no significant effect on the yield behavior in stress states ranging between uniaxial compression and biaxial tension.

ACKNOWLEGEMENTS

The authors would like to gratefully acknowledge Shell Chemical Co. for the financial support provided for this study. Also, the authors would like to thank Dr.'s M. A. Masse and M. J. Modic for their many thoughtful comments and suggestions during this study.

REFERENCES

1. C. B. Bucknell, *Toughened Plastics*, Applied Science, London (1977).

2. A. J. Kinloch in *Rubber Toughened Plastics*, (Ed. C. K. Riew) Advances in Chemistry Series 222, American Chemical Society, Washington D. C., (1989), 67.

3. A. J. Kinloch, S. J. Shaw, D. L. Hunston, *Polymer*, (1983), **24**, 1355.

3. R. A. Pearson, A. F. Yee, *J. Mater. Sci.*, (1986), **21**, 2475.

4. H. J. Sue, *J. Mater. Sci.*, (1992), **27**, 3098.

5. A. Lazzeri, C. B. Bucknell, *Polymer*, (1995), **36**, 2895.

6. A. Lazzeri, C. B. Bucknell, *J. Mater. Sci.*, (1993),**28**, 6799.

7. A. C. Steenbrink, E. Van Der Giessen, P. D. Wu, *J. Mech. Phys. Solids*, (1997) **45**, 405.

8. F. J. Guild, R. J. Young, *J. Mater. Sci.*, (1989), **24**, 2454.

9. Y. Huang, A. J. Kinloch, *J. Mater. Sci.*, (1992), **27**, 2763.

10. Y. Huang, A. J. Kinloch, *Polymer*, (1992), **33**, 1330.

11. H. J. Sue, A. F. Yee, *Polymer*, (1992), **33**, 4868.

12. R. Bagheri, R. A. Pearson, *Polymer*, (1995), **36**, 4883.

13. R. Bagheri, R. A. Pearson, *Polymer*, (1996), **37**, 4529.

14. J. Kiefer, J. G. Hilborn, J. L. Hedrick, *Polymer*, (1996) **37**, 5715.

15. R. S. Kody, A. J. Lesser, *J. Mater. Sci.*, (1997) **32**, 5637

16. S. S. Sternstien, L. Ongchin, *American Chemical Society, Polym. Prep.*, (1969) **10**, 1117.

NOVEL MECHANISMS OF TOUGHENING NOTCH BRITTLE HDPE WITH RUBBERS OR CaCO$_3$ PARTICLES

A.S. Argon*, Z. Bartczak**, R.E. Cohen*

* Massachusetts Institute of Technology, Cambridge, MA 02139

** Centre of Molecular and Macromolecular Studies, Polish Academy of Sciences, 90-363, Lodz. Poland.

INTRODUCTION

Many otherwise attractive polymers that show impressive plasticity at usual slow laboratory tests at room temperature exhibit severe notch brittleness. While such behavior is readily accepted for most glassy polymers, it is often of surprise that it also afflicts semi-crystalline polymers such as Nylon, HDPE, PP, and the like. As is frequently the case, incorporation of rubber in semi-crystalline polymers has been found to alleviate the problem, - often dramatically. Early experiments, however, have indicated that neither the volume fraction of incorporated rubber nor the particle size by themselves could provide a rational mechanism for the toughening. A precise correlation for the toughening was discovered by Wu[1] who noted that toughness jumps occurred in Nylon 6.6 when the inter-particle ligament thickness Λ became less than 0.3μm. Since the correlation was based on a specific material length the explanation could not rely on field theory notions of stress concentrations or stress state which, by definition, are scale independent. In a series of detailed studies of morphology based on TEM, SEM, WAXS and unusual free-standing stretching experiments of thin films of Nylon 6 and 6.6 in the sub-micron range, Muratoglu et al[2-4] have demonstrated that the explanation lies in the special morphology of material in these ranges of thickness. In Nylon 6 (or 6.6) films of thickness less than 0.3μm, crystallized between planar rubber interfaces, a highly regular preferential crystalline form appears in which the lowest energy, lowest plastic resistance (001) crystallographic planes lie parallel to the interface. Free-standing stretching experiments of films of sub-micron thickness have demonstrated that the plastic resistance of such anisotropic films differs by a factor of 2 from that of bulk material with random crystal orientations. In blends in which the average interparticle ligament thicknesses falls below the critical dimension, oriented matrix material of reduced plastic resistance percolates throughout the blend and the overall plastic resistance decreases markedly, resulting in remarkably tough behavior. In these blends the indication was that the incorporated rubber particles produce their effect through preferential crystallization of the polymer matrix material by initiating directed crystal growth from the interface and that particle properties are not relevant. Thus, the toughening mechanism might be more general and achievable with other particles of different chemistry - provided they produce the same anisotropic crystallization. On the realization of this we have carried out an extensive study on HDPE which is known to also exhibit severe notch brittleness. In this study we have incorporated into HDPE either rubber particles or particles of CaCO$_3$, both at a volume fraction of 0.22 but in the particle size range to result in interparticle ligament thicknesses less than a critical amount. For HDPE this critical thickness was found to be 0.6μm.

EXPERIMENTAL RESULTS

Material and morphology.

The HDPE used in the present study was Dowlex 1P-10, supplied by Dow Chemicals. In the rubber modified blends several types of rubber were used consisting of: a) amorphous ethylene-propylene-diene for polymers (EPDMs); semi-crystalline EPDMs; and semi-crystalline ethylene-octane rubbers (EORs) of several descriptions. The rubbers were arranged to be well grafted but were mechanically dispersed in twin screw extruders to achieve the required quasi-uniform interparticle ligament dimensions. In blends modified by CaCO$_3$ particles three different particle sizes were used with dimensions of 0.44, 0.70 and 1.20μm. The particles were treated with calcium stearate to achieve proper dispersion without clustering. In all cases the volume fraction was 0.22. In special crystallization studies of HDPE against CaCO$_3$, calcite single crystals with (104) crystallographic surfaces were used.

Since the toughening mechanism relies on preferential crystallization, producing layers of anisotropic plastic resistance, a separate study was carried out in which the crystallization habit of PE was studied

in thin films ranging in thickness from 0.1μm to 1.2μm. The morphology was monitored in these films by means of WAXS pole figures and AFM. These studies indicated, that, similar to the findings in Nylon 6, PE lamellar crystallites formed overwhelmingly with their low energy, low plastic resistance (100) planes parallel to the rubber or calcite interfaces in discoid morphology, provided the films were less than 0.3μm thick. In films of larger thickness the lamellar crystallites deviated systematically from the planar discoid form toward the banded spherulitic forms, with twisted lamellae.[5] While not specifically probed, as was done for Nylon 6 films, the preferential discoid morphology in the thin films of HDPE was taken to signify similar anisotropic plastic resistance as was established in Nylon 6.

Izod impact experiments

Figure 1a summarizes the dependence on average interparticle ligament thickness of the Izod impact energies of all of the rubber modified HDPE blends. The figure shows a toughness jump when the ligament thickness falls under 0.6μm, or twice the distance of preferential discoid shaped crystallization of lamellae attached to the HDPE/rubber interfaces. There is no dependence on the type of rubbers used. The results of the corresponding blends modified by CaCO$_3$ particles are given in Fig. 1b. Clearly, since the CaCO$_3$ particles could be obtained only in three different sizes, the data is not as plentiful as for the rubber modified blends. Nevertheless, the toughness jump occurs exactly at the same interparticle ligament thickness of 0.6μm.[6,7]

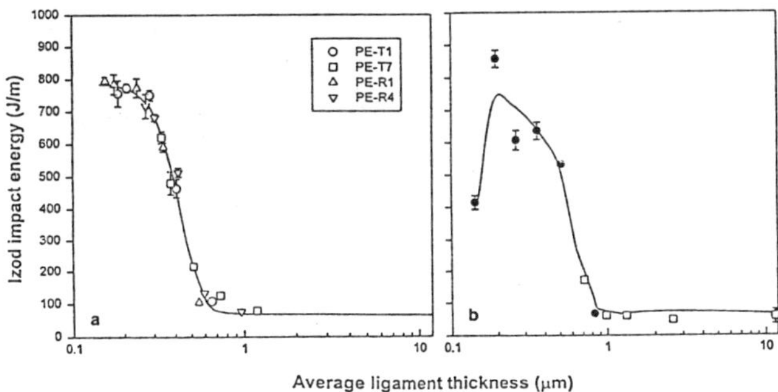

Fig. 1a) Dependence of impact energy on interparticle ligament thickness in rubber blends (from Ref. 6); b) dependence of impact energy on interparticle ligament thickness in CaCO$_3$ blends (from Ref. 7).

Associated SEM and light microscopy observations on sections of fractured samples indicated that in all cases tough behavior was accompanied by whitening resulting from cavitation in the rubber particles or decohesion of the CaCO$_3$ particles from the matrix. Such separation of the particles from the interparticle matrix material of pedigreed semi-crystalline anisotropic morphology is essential for the development of large strain plastic response in the released interparticle ligaments. If the adhesion of the particle to the matrix were too good and such separation were stifled the toughness mechanism was suppressed and the blends exhibit quite uninteresting levels of toughness, regardless of the interparticle ligament thicknesses. This is demonstrated dramatically in Fig. 2 showing the temperature dependence of Izod impact energy in blends in which the interparticle ligament dimensions are all below the critical 0.6μm thickness. In the rubber modified blends the high toughness at room temperature becomes compromised at temperatures below the respective T$_g$S of the rubbers, when the glass transition in the rubber the large elastic mismatch between the particles and the HDPE matrix is reduced to levels too low to result in cavitation of the rubber particles, which remain well adhered to the matrix. As a result the interparticle ligaments can not undergo their large strain stretching behavior and the blend remains brittle. In the CaCO$_3$-particle-modified blends, however, the particles always decohere from the deforming matrix resulting in the release of the matrix from constraint and permit it to undergo large plastic extensions at all temperatures, down to -80C with associated tough behavior. Figures 3a and 3b show the sectioned blends of rubber and CaCO$_3$-modified-blends both with 0.22 volume fraction of particles. In Fig. 3a the cavities have formed by cavitation of the rubber particles while in Fig. 3b they result from decohesion of the CaCO$_3$ particles. Otherwise the cellular morphology of the tough HDPE appear identical.[6,7]

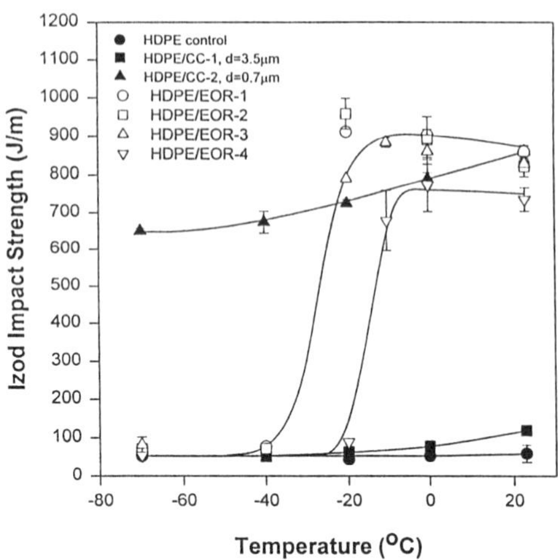

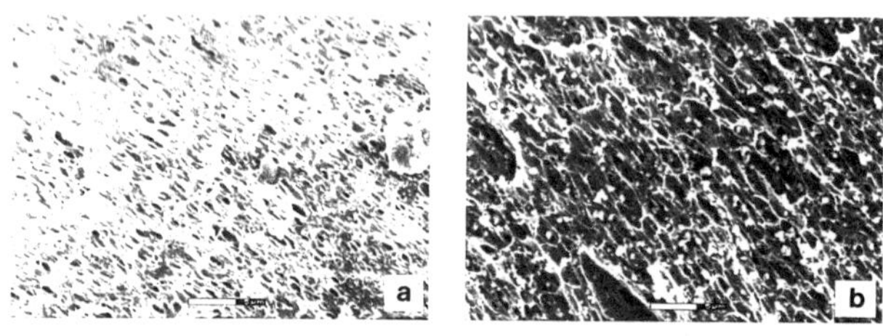

Fig. 2. Temperature dependence of impact energy in both rubber and $CaCO_3$ blends of 0.22 volume fraction of particles in the appropriate ligament thickness range (from Ref. 7).

Fig. 3. Transverse, cryo-fracture surfaces near the impact fracture surface showing cavity elongation in: a) a rubber-modified blend; b) a $CaCO_3$ - modified blend (from Ref. 7).

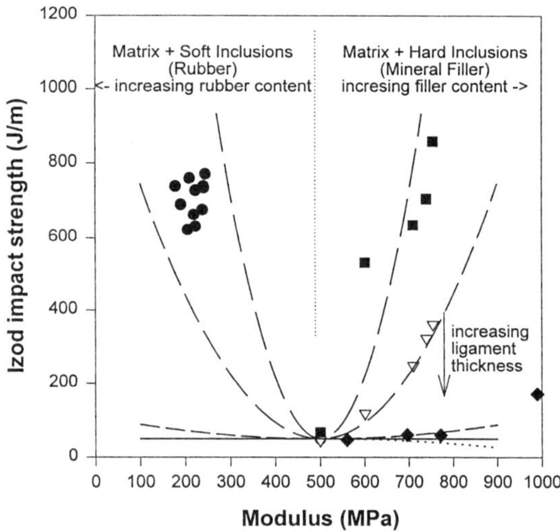

Fig. 4. Diametrically different association of toughness increase in rubber blends with 50% decrease in modulus, and in $CaCO_3$ blends with 50% increase in modulus.

DISCUSSION

The toughening experiments of HDPE where both rubber and rigid $CaCO_3$ heterogeneities were used as modifiers, the effect of these on the toughening was found to be very similar and independent of the nature and mechanical properties of the particles - provided that the interparticle ligament thicknesses were less than $0.6\mu m$. In ligaments of such small thicknesses a rather special crystallization morphology results which has strongly anisotropic plastic resistance. When interparticle ligament material with such pedigreed properties percolates through the blend the overall plastic resistance of the blend is markedly reduced and the blend can undergo large plastic flow without "precipitating" fracture from the usual adventitious inclusions that result in brittle behavior.

Clearly, this entire scenario of tough behavior relies on the plasticity of the crystalline component of the semi-crystalline polymer. This fact still appears to come as a surprise to many polymer scientists who have learned that these materials have a sizable fraction of an amorphous component and that there are tie molecules between the two with much bewildering response all dating back to the early experiments carried out on the stages of electron microscopes showing many artifacts. Recent research, much of which was carried out by us in both Nylon and HDPE,[8-12] has established that in both of these semi-crystalline polymers, with high levels of crystallinity, while plastic response starts indeed with the amorphous component this deformation ceases at overall strain levels of the order of 0.1-0.15 while the remainder of the deformation to strains of the order of 2.00, or larger, comes exclusively from the crystalline component through crystallographic slip processes.

A very pleasing finding of our study was the realization that when rigid $CaCO_3$ particles replace rubber as the modifiers, not only the same level of toughening is achieved but the elastic properties of the blend, instead of being compromised by 50%, can be enhanced by about 50%. This is shown in Fig. 4 which shows the combined effect of rubber and $CaCO_3$ particles. The same volume fraction of reinforcement achieves the same notch toughness, but the rigid particles produce an attractive improvement in elastic properties while the rubber particles result in an unattractive reduction - provided that the interparticle ligament dimension criterion is adhered to.[7]

ACKNOWLEDGMENT

This research was supported by the MRSEC Program of the NSF under award DMR-94-00334, through the Center for Materials Science and Engineering at M.I.T.

BIBLIOGRAPHY

1. S.Wu, *Polymer*, **26**, 1855, 1985; ibid., *J. Appl. Polym. Sci.*, **35**, 549, 1988.
2. O.K. Muratoglu, A.S. Argon, R.E. Cohen and M. Weinberg, *Polymer*, **36**, 921, 1995.
3. O.K. Muratoglu, A.S. Argon, R.E. Cohen and M. Weinberg, *Polymer*, **36**, 2143, 1995.
4. O.K. Muratoglu, A.S. Argon, R.E. Cohen and M. Weinberg, *Polymer*, **36**, 4771, 1995.
5. Z. Bartczak, A.S. Argon, R.E. Cohen and M. Weinberg, Polymer, submitted for publication.
6. Z. Bartczak, A.S. Argon, R.E. Cohen and M. Weinberg, *Polymer* (HDPE/rubber blends), submitted for publication.
7. Z. Bartczak, A.S. Argon, R.E. Cohen and M. Weinberg, *Polymer* (HDPE/$CaCO_3$ blends), submitted for publication.
8. L. Lin and A.S. Argon, *Macromolecules*, **25**, 4011, 1992.
9. A. Galeski, Z. Bartczak, A.S. Argon and R.E. Cohen, *Macromolecules*, **25**, 5705, 1992.
10. Z. Bartczak, A.S. Argon, and R.E. Cohen, *Macromolecules*, **25**, 5036, 1992.
11. Z. Bartczak, R.E. Cohen and A.S. Argon, *Macromolecules*, **25**, 4692, 1992.
12. B.J. Lee, A.S. Argon, D.M. Parks, S. Ahzi and Z. Bartczak, *Polymer*, **34**, 3555, 1993.

MESOSCOPIC LOCALIZED DEFORMATIONS IN TOUGHENED RUBBER BLENDS

E. van der Giessen, K.G.W. Pijnenburg and
A.C. Steenbrink*

Koiter Institute Delft, Delft University of Technology,

Delft, The Netherlands

*TNO Road-Vehicles Research Institute, Delft, The Netherlands

INTRODUCTION

It is a well established fact that the fracture toughness of polymers can be greatly enhanced by adding a dispersion of rubber particles. The toughening involves a number of competing mechanisms. The first is cavitation of the rubber particles. This relieves the stress triaxiality in the matrix polymer, so as to suppress the likelihood of matrix crazing and to promote plastic deformation in the matrix by shear yielding. Toughening is generally enhanced when a region of large plastic deformation spreads out over a large volume in the material. Our interest has been how this plastic deformation spreads out in the blend in between the (cavitating) rubber particles (mesoscopic scale).

Evidence of massive plastic deformation is given in the TEM micrographs in Figure 1 taken near the fracture surface in notched samples of ABS (with polybutadiene rubber particles). After cavitation, the

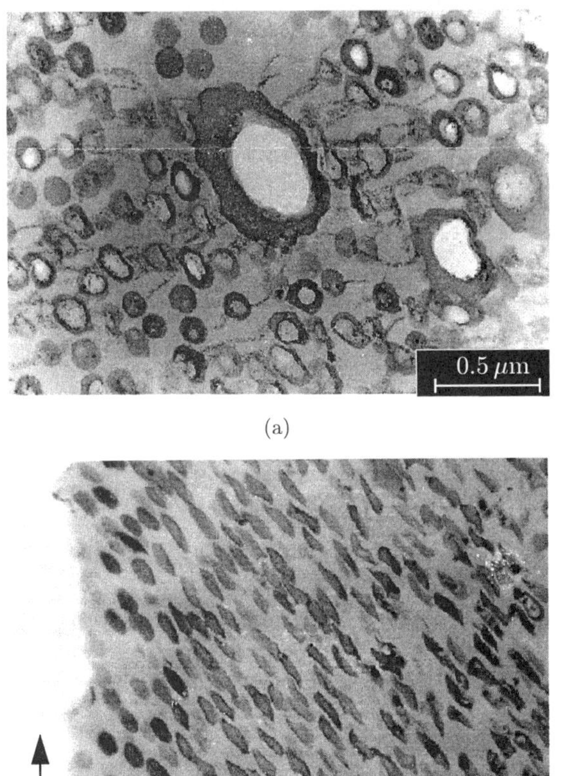

(a)

(b)

Figure 1: Deformation zones in ABS (a) at a distance of $100 \mu m$ from, and (b) near the fracture surface. From [1].

initially spherical particles in Fig. 1.a are seen to develop rather bulgy shapes. Similar shapes have been observed by other researchers and are expected to be relevant also at some distance ahead of a crack tip. The shape of the particles near the fracture surface (Fig. 1.b) is quite different, indicating that the deformation history has been significantly different.

The purpose of this paper is to investigate how these mesoscopic deformation patterns can be understood by numerical simulations of the plastic flow process.

PREDOMINANT TENSION

We first consider a model material of initially spherical rubber particles in an amorphous glassy matrix subjected to macroscopic stress state corresponding to uniaxial tension with a superimposed hydrostatic stress of 67 %. Assuming that the particles are ordered in a regular packing, we can confine attention to a unit cell with a single rubber particle. As we are interested in the development of plastic deformation after particle cavitation, we assume that the rubber particle already contains a small initial cavity of 0.2 times the particle size.

Finite element calculations have been carried out, using up-to-date large strain constitutive models for the rubber phase as well as for the glassy matrix. The model for the latter accounts for strain-rate and pressure dependent yield, followed by strain softening and, for continued straining, progressive orientational hardening (see [2, 3]).

Figure 2 shows the computed evolution of plastic deformation around one such cavitated particle in a SAN matrix. This particular case corresponds to a material with a rubber (modulus of 0.4 MPa) content of almost 10 %. Due to the intrinsic softening of the matrix, plasticity starts in the form a shear band emanating from the equator of the particle-matrix interface, which is oriented at roughly 45° from the maximal principal stress direction (Fig. 2a). With continued deformation, orientational hardening causes the shear band to propagate, but at some stage changes into another type of shear band that is oriented at approx. −30° (Fig. 2b). Note that quite some growth of the hole inside

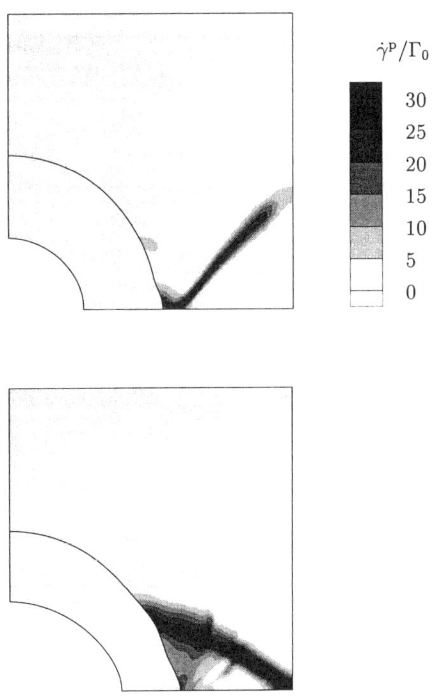

Figure 2: Distribution of instantaneous plastic shear rate $\dot{\gamma}^{\mathrm{P}}$ under predominant tension at two stages of the macroscopic strain level. Only a quarter of the rubber particle and the surrounding matrix is shown. From [4].

the rubber particle has taken place.

A rather elaborated parametric study of this problem [4] has revealed that whether or not void growth takes places depends on the rubber modulus compared to the matrix yield stress and on the macroscopic stress triaxiality. If void growth takes place, the plastic deformation always occurs (i) in the form of propagative shear bands, which (ii) belong to one of the two families of shear bands seen in Fig. 2. The details are dependent on the volume fraction of rubber and on the packing of the particles, but only mildly [5]. Qualitatively similar deformation modes were observed in planar model studies [6].

PREDOMINANT SHEARING

Next, it is assumed that the macroscopic deformation occurs by shear. This is modeled in two dimensions using a similar regular array of rubber particles as above but now subjected to macroscopic simple shear using periodic boundary conditions. Assuming that the dilatation of the particle can be neglected in this deformation mode (which has indeed been confirmed), and neglecting the shear modulus of the rubber compared to the yield stress of the matrix, we can replace the particle by a void.

Figure 3 shows the evolution with increasing shear strain of void shape and rate of deformation in the matrix material. At small strains, there are some traces of plastic flow at angles of 45° to the shear direction (Fig. 3a), but these soon disappear due to the constraint of surrounding material. Upon further shearing another type of deformation is activated in the form of shear bands parallel to the shear direction in between adjacent particles (Fig. 3b). Once the entire ligament separating two particles has yielded, the material is no longer restricted and is free to deform further. Strain hardening with continued deformation causes the material in the band to lock-up and to initiate yield in the neighboring material. Then the process repeats itself and the shear band propagates outward from the center, as is seen in the last two figures.

DISCUSSION

The rubber particles in a blend serve to some extent as markers of the deformation history at the mesoscopic level. Under predominantly tensile stressing, the initially spherical particles develop into a cigar-like shape when the stress triaxiality is low [4] or develop distinct bulges at higher triaxiality, as shown in Fig. 2. In shear, the particles evolve into the typical S-shapes seen in Fig. 3. All these differences are inherent to the different mesoscopic plastic deformation modes.

Though tentative, the predicted particle shapes in Figs. 2 and 3 are akin to those observed in Fig. 1a and Fig. 1b, respectively. This would imply that just below the fracture surface, the blend is intensely sheared in planes that are oriented at some tens of degrees away from the fracture surface. This is not consistent with classical viewpoints of plastic deformation around crack tips, but would require macroscopic localization of deformation in the blend. This localization is probably triggered by the macroscopic strain softening that is caused by the rubber cavitation and subsequent plastic void growth inside the rubber particles.

References

[1] A.C. Steenbrink, H. Janik, R.J. Gaymans, *J. Mat. Sci.* **32**, 5505–5511, 1997.

[2] M.C. Boyce, D.M. Parks and A.S. Argon, *Mech. Mater.* **7**, 15–33, 1988.

[3] A.C. Steenbrink, E. Van der Giessen and P.D. Wu, *J. Mech. Phys. Solids* **45**, 405–437, 1997.

[4] A.C. Steenbrink and E. Van der Giessen, *J. Mech. Phys. Solids*, accepted for publication.

[5] A.C. Steenbrink and E. Van der Giessen, In: *Material Instabilities in Solids* (R. de Borst & E. van der Giessen, eds.), J. Wiley & Sons, Chicester (in print).

[6] A.C. Steenbrink and E. Van der Giessen, *J. Mat. Sci.*, submitted.

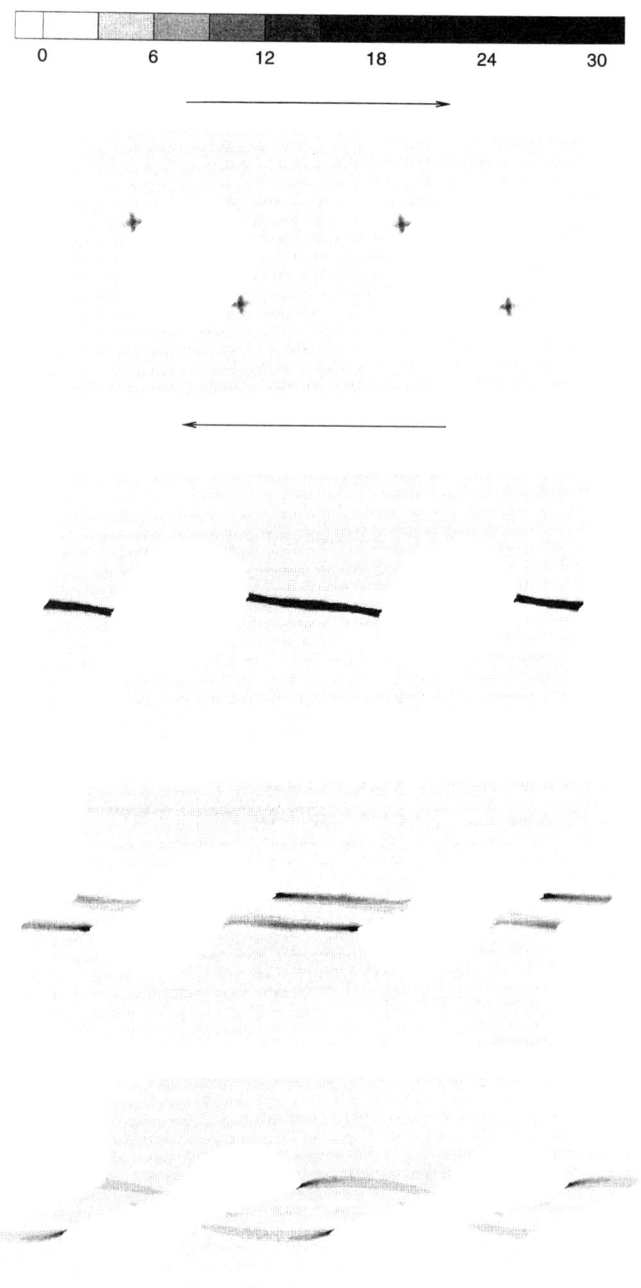

Figure 3: Distribution of instantaneous plastic shear rate $\dot{\gamma}^{\mathrm{p}}$ between particles under macroscopic shear at four stages of deformation, normalized by the applied shear rate.

THREE-DIMENSIONAL ELASTOPLASTIC FINITE ELEMENT MODELLING OF TOUGHENING PROCESSES IN RIGID-RIGID POLYMER BLENDS

Yiu-Wing Mai and Xiao-hong Chen
Centre for Advanced Materials Technology (CAMT)
Department of Mechanical & Mechatronic Engineering J07
The University of Sydney
Sydney, NSW 2006, Australia

INTRODUCTION

It is well known that rubber toughening is an effective way to enhance the toughness of many brittle polymers; but it also leads to reductions in modulus and strength simultaneously[1-3]. Chen & Mai[4,5] provided a face-centred cuboidal cell model to study the toughening mechanisms in rubber-modified epoxies by 3-D elastoplastic FEA because it is easy to simulate both rubber cavitation and matrix shear banding processes using such a staggered periodic layout.

A new generation of polymer blends with beneficial combination of toughness, strength and modulus is required to provide a wider range of engineering applications for polymer materials. ABS/PC, PA/PPO, PBT/PC and PEI/PC are typical examples of such blends. The recent introduction of the rigid-rigid polymer toughening concept was originated from the idea of rubber toughening[6]. It is realised that the rigid polymer matrix can be toughened by the existence of a second-phase rigid polymer as long as the latter can act as stress concentrators and help relieve the triaxial constraint. The substantial increase in toughness of PBT/PC blends comes mainly from the debonding-cavitation process at the PBT/PC interface which in turn promotes extensive shear deformation in both the PBT and PC; as well as from the craze-stabilising and crack-bridging mechanisms by the PC domains as observed by Wu, Mai & Yee[7].

Unlike ordinary rubber-toughened polymers, both phases of rigid-rigid polymer blends can undergo anelastic and plastic deformation. Sue, Pearson & Yee[6] have modelled the initiation of localised yielding in rigid-rigid polymer alloys based on 2-D FEA of a cylindrical inclusion embedded in a thin plate under plane-stress condition omitting the inter-particle interaction. However, these analyses do not seem to be directly applicable to the case of spherical particles embedded in a thick specimen under triaxial stress or indeed to the crack-particle interaction problem.

In this paper, three-dimensional elastoplastic finite element analysis is carried out to study the toughening processes in rigid-rigid polymer blends with initially spherical secondary-phase plastic particles in a periodic face-centred cubic layout accounting for the particle-particle interaction by a face-centred cuboidal cell model.

MICROMECHANICAL MODEL

The representative unit of the periodic FCC microstructure is shown in Fig.1. Under symmetric loading condition, only a one-eighth face-centred cuboidal cell is used to make the numerical analysis less time-consuming.

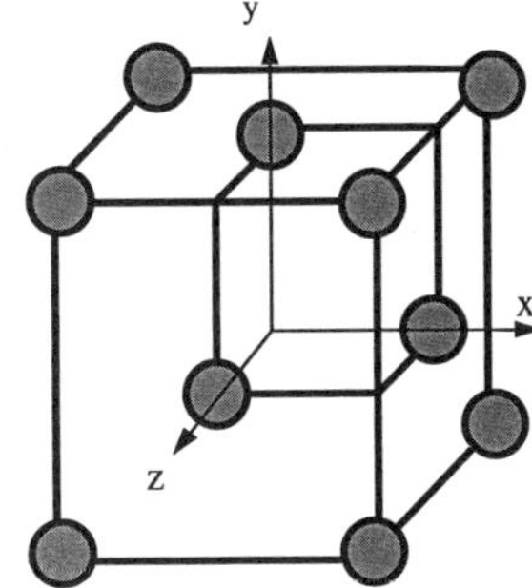

Fig.1 Face-centred cuboidal cell model.

The periodic symmetry requirements are satisfied by imposing the following constraint equations on the corresponding surfaces of the cell:

$$u_z = U_{z0}, \qquad t_x = t_y = 0 \qquad \text{at } z = a_0$$
$$u_x = U_{x0}, \qquad t_y = t_z = 0 \qquad \text{at } x = a_0 \qquad (1)$$
$$u_y = U_{y0}, \qquad t_x = t_z = 0 \qquad \text{at } y = a_0$$

and

$$u_z = 0, \qquad t_x = t_y = 0 \qquad \text{at } z = 0$$
$$u_x = 0, \qquad t_y = t_z = 0 \qquad \text{at } x = 0 \qquad (2)$$
$$u_y = 0, \qquad t_x = t_z = 0 \qquad \text{at } y = 0$$

where a_0 is the cell length, u_x, u_y and u_z are the components of displacement vector $\mathbf{u}$ along x-, y- and z- directions, t_x, t_y and t_z are the components of the surface traction vector $\mathbf{t}$ along x-, y- and z-directions, U_{x0}, U_{y0} and U_{z0} are displacement constants. The surfaces of $x = 0$, $y = 0$ and $z = 0$ are symmetric planes, while the surfaces of $x = a_0$, $y = a_0$ and $z = a_0$ are kept parallel with respect to their original shapes during deformation to satisfy the periodic symmetry requirements.

The effective stress σ^e and strain ε^e are obtained by averaging the local stress σ and strain ε in the face-centred cuboidal cell, that is,

$$\sigma^e = \frac{1}{V_\Omega} \int_{V_\Omega} \sigma dV = \frac{1}{V_\Omega} \int_{S_\Omega} \mathbf{r} \otimes \mathbf{t} dS \qquad (3)$$

$$\varepsilon^e = \frac{1}{V_\Omega} \int_{V_\Omega} \varepsilon dV = \varepsilon^0 \qquad (4)$$

where $\mathbf{r}$ is the position vector and $\mathbf{t}$ is the traction vector on the cell surface, ε^0 is the constant strain tensor dependent on the constant normal displacement on the cell surface, V_Ω and S_Ω represent the cell volume and surface, respectively. We use the equilibrium condition without a volume force, i. e. $\nabla \cdot \sigma = 0$, to obtain the last equality of equation (3).

FEA RESULTS

3-D elastoplastic finite element analysis was carried out using the ABAQUS program on an ALPHA STATION 500. We study the PEI/PC system with PC as second-phase particles and PEI as matrix, respectively. Both materials are governed by the Von Mises criterion with piece-wise linear stress-strain relations given by Sue, Pearson & Yee[6]. PC has a Young modulus of 2400 MPa and a Poisson ratio of 0.42. PEI has a Young modulus of 3400 MPa and a Poisson ratio of 0.40. The ratio of particle diameter to cell length is chosen as 0.5, 0.4 and 0.2, corresponding to spacing ratio of 0.5, 0.6 and 0.8. The volume fractions of PC particles are hence 26.18%, 13.40% and 1.68%. The PATRAN program was used to generate the mesh automatically in the one-eight face-centred cuboidal cell model. For comparison, we use the same mesh for the PEI/void system with PC particles replaced by voids. The applied stress system $T_x : T_y : T_z = 0:0:1$ corresponds to macroscopic uniaxial tension. Whereas the applied stress system $T_x : T_y : T_z = 1.5:2.0:2.5$ corresponds to a high tensile triaxial stress state ahead of the crack tip elastic-plastic boundary[8].

From the contour plots of Von Mises stress, hydrostatic pressure and direct stress, we can see that maximum Von Mises stress, dilatational stress and direct stress in the PEI matrix are all reached in the equatorial regime of the PC particles. Local dilatational stress increases while local Von Mises stress decreases with increasing stress triaxiality, which shows that a high tensile triaxial stress state is beneficial for debonding, cavitation and crazing whereas it defers the development of shear deformation. Hence, the plane-strain triaxial stress condition in front of the crack tip has to be released in order to activate the shear yielding mechanism for a thick specimen with a macro-crack.

Maximum shear stress concentration in the matrix for the PEI/PC system at three different particle volume fractions of 1.68%, 13.40% and 26.18% under uniaxial tension is shown in Fig.2. Maximum shear stress, dilatational stress and direct stress concentrations for both PEI/PC and PEI/void systems at a particle volume fraction of 1.68% under uniaxial tension are shown in Fig.3. Local stress concentration increases with increasing PC particle volume fraction. Two distinct peaks occurs in the curve of maximum shear stress concentration versus applied stress as PC particles begin to undergo anelastic and plastic deformations with the spacing between the two peaks decreasing with increasing PC particle volume fraction. The maximum shear stress concentration is constant at 1.15

at the elastic stage and rises to 1.18 as anelastic deformation occurs in the PC phase for the PEI/PC system at a low particle volume fraction of 1.68%. In contrast, the maximum shear stress concentration is as high as 1.91 for the PEI/void system at the same particle volume fraction. Sue, Pearson and Yee reported a shear stress concentration factor of ~1.2 in the elastic regime and about 1.3 as the PC inclusion begins to undergo anelastic deformation by 2-D FEM simulation of a cylindrical inclusion embedded in a thin plate under uniaxial tension[6]. The 2D results by Sue, Yee and Pearson agree qualitatively with our 3D results in the prediction of the general trend of the effects of anelastic and plastic deformation of PC particles on initiating the shear yielding mechanism. Localised plane strain shear bands can be formed under uniaxial tension even without the cavitation process because the shear stress concentration factor has reached 1.15.

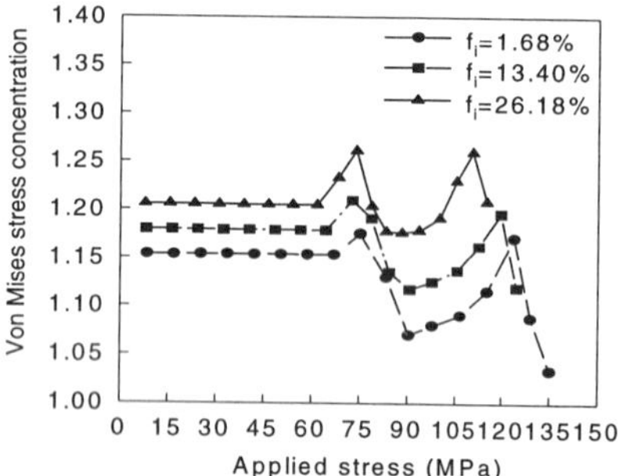

Fig.2 Maximum shear stress concentration versus applied stress for PEI/PC system at three different particle volume fractions under uniaxial tension.

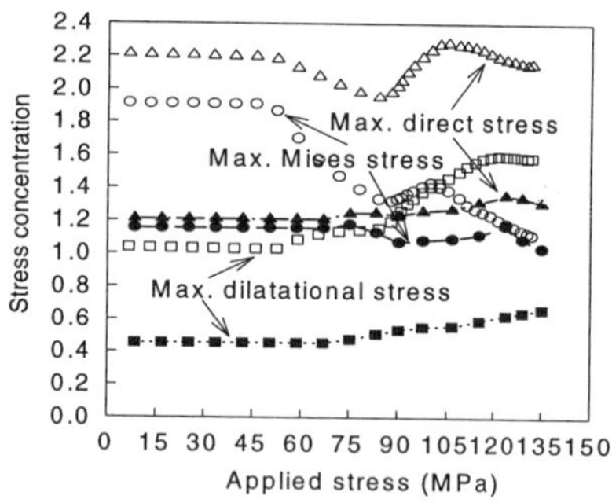

Fig.3 Maximum shear stress, dilatational stress and direct stress concentration versus applied stress for both PEI/PC and PEI/void systems at a particle volume fraction of 1.68% under uniaxial tension. (Solid symbols: PEI/PC system. Open symbols: PEI/void system).

The effective elastoplastic stress-strain relations for the PEI/PC system at three particle volume fractions of 1.68%, 13.40% and 26.18% under uniaxial tension are shown in Fig.4. Likewise the effective elastoplastic stress-strain relations for both PEI/PC and PEI/void systems at a particle volume fraction of 1.68% under uniaxial tension and the given triaxial stress system ahead of the crack tip elastic-plastic boundary are given in Fig.5.

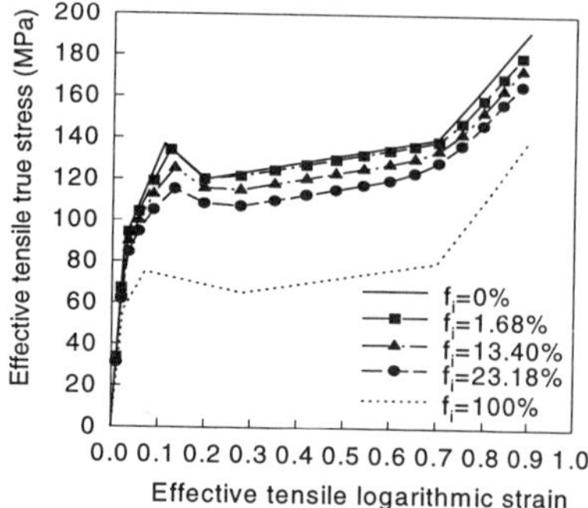

Fig.4 Effective elastoplastic stress-strain relations for PEI/PC system at three different particle volume fractions under uniaxial tension.

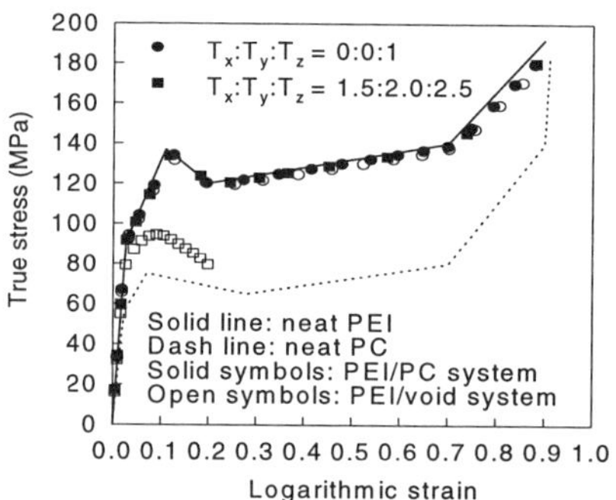

Fig.5 Effective elastoplastic stress-strain relations for both PEI/PC and PEI/void systems at a particle volume fraction of 1.68% under uniaxial tension and imposed triaxial stress in front of a crack tip.

We can see that the effective stress-strain relations for the PEI/PC system are similar to the neat PEI and PC with a strain-hardening effect. The effective stress decreases with increasing particle volume fraction. The yield behaviour of the PEI/void system is quite different to that of the PEI/PC system with a much lower effective shear yield stress under the high tensile triaxial stress associated with a crack tip although they are similar under uniaxial tension. Unstable void growth may occur in the PEI/void system at high stress triaxiality as the macroscopic strain reaches around 20%.

CONCLUSIONS

Maximum shear stress concentration is enhanced abruptly as the PC particles begin to undergo anelastic and plastic deformations. Localised plane-strain shear bands can be formed under uniaxial tension without resorting to the debonding-cavitation process; whereas the triaxial stress plane-strain condition associated with a macro-crack in a thick specimen has to be relieved so that extensive plastic deformation can be developed to produce toughening. Partial debonding-cavitation at the PEI/PC interface can promote the shear yielding mechanism while still keeping the craze-stabilising and crack-bridging functions of the PC domains.

ACKNOWLEDGMENTS

The authors wish to thank the Australian Research Council (ARC) for the continuing support of the polymer blends project. The financial support by the ARC postdoctoral fellowship and the National Natural Science Foundation of China is appreciated by Xiaohong Chen.

BIBLIOGRAPHY

1. C. B. Bucknall
Toughened Plastics, Applied Science Publishers LTD, London, 1977.
2. C. K. Riew and A. J. Kinloch
Toughened Plastics II, American Chemical Society, Washington, DC, 1996.
3. A. F. Yee and R. A. Pearson
J. Mater. Sci. **21**, 2462, 1986.
4. H. J. Sue, R. A. Pearson and A. F. Yee
Polymer Engineering and Science **31**, 793, Mid-June 1991.
5. J. S. Wu, Y. W. Mai and A. F. Yee
J. Mater. Sci. **29**, 4510, 1994.
6. X. H. Chen and Y. -W. Mai
Key Engineering Materials **137**, 115, 1998.
7. X. H. Chen, Y.-W. Mai
Key Engineering Materials **145-149**, 233, 1998.
8. J. F. Knott
Fundamental of Fracture Mechanics, John Wiley & Sons, New York, 1976.

Epoxy Toughening with in situ Polymerized Poly(Meth)Acrylate

J. M. LIEGEOIS, N. TAHIR
Laboratoire Matériaux Polymères et Composites
Université de Liège, Bât. B6
4000 Liège   Belgium

Introduction

High performance thermoplastic (TP) modified epoxy resin networks have been shown to provide a distinct advantage over elastomeric modified systems, because the fracture toughness can be increased without substantially sacrificing the heat distorsion temperature or the modulus of the resulting blend. Since the pioneering work of Bucknall and Partridge[1] on epoxies modified with poly(ethersulfone), there has been much worthwhile activity reported in the scientific literature[1-8].

However the process of blending a high molecular weight high Tg TP with the precursors of an epoxy network, has limitation in that blends containing more than ca 15 percent TP can be difficult to prepare and can result in wetting problems when applied onto reinforcing fibers for composite application.

Therefore in the present study, an attempt was made to modify the epoxy resin with a high Tg thermoplastic synthesized in situ so that mixing of monomeric species is no longer a problem nor would be the wetting of cloth or roving.

Highest possible Tg TP were selected among current acrylate polymers, namely based on isobornyl acrylate (IBA) (Tg : 94°C) and isobornyl methacrylate (IBMA) (Tg : conflicting 140-170°C). Variation of morphology was induced adding various amounts of glycidyl methacrylate (GMA) in either case.

Preparation procedures

Semi-simultaneous interpenetrating networks (semi-SINs) were made from mixes that consisted of DGEBA resin (Shell Epikote 828), p,p'-Diaminodiphenyl sulphone (Ciba HT976) both used without further purification, either IBA or IBMA and optional GMA (all Atochem) free from the storage inhibitor. A mixture of the epoxy resin and curing agent was heated at 120°C for about 40 min to homogenize with stirring. When curing agent was dissolved, the monomer mix with 1 percent based on acrylate components, tert-butyl benzoate peroxide (Akzo Trigonox C) initiator, was added. Then the blend was poured into oil heated teflon coated aluminium moulds, and purged with dry nitrogen for 5 to 10 min. The curing cycle was 120°C/1 hour, 180°C/5 hours. In all instances, the curing agent was used in stoechiometry with the total epoxy content.

Results and discussion

Compositon of materials

Are presented here, three series refering to IBA modification (10, 20, 30 w per cent) and one series refering to IBMA modification (30 w per cent), each series showing the effect of increasing amount of the grafting component GMA. The compositions shown in table 1 and 2 are total blend weight percent. Tg_E is the highest observed Tg assigned to the epoxy rich phase. Wherever visible, the lower Tg_A is assigned to the poly(meth)acrylate rich phase.

Electron microscopy

The morphology of the semi-SINs was observed by scaning electron microscopy (SEM) of the fractured samples. A clear two phase segregated morphology was seen for epoxy networks modified with P-IBA at grafting levels of 0 and 1.3% (GMA) for series S10A, at 0, 2.88 and 5.94% for series S20A and 0, 4.4 and 7.5% for series S30A. The poly(isobornyl acrylate) rich phase forms well defined spherical particles nicely dispersed throughout the matrix. On the other hand, semi-SINs with higher content of grafting agent seemed to bear only one phase as otherwise confirmed by DMTA (see below). Fig.1 shows extremes for 30 per cent total acrylate.

Interestingly, the morphology of blends with IBMA is much different from that of blends with IBA and this holds for any extent of grafting up to 7,5% GMA. The contents of spherical particles is rather limited and a more cocontinuous phases situation is observed instead (figure 2).

Dynamic viscoelastic analysis

Dynamic viscoelastic analysis can give more information on microstructures of cured resins, and provides details about molecular mixing and phase continuity. Figures 3 and 4 show the storage modulus and tanδ for series S10A and S30A pertaining to the acrylate modification.

While the unmodified epoxy resin has a glass transition Tg_E of 204°C the SINs with GMA show different Tg_E as function of composition (table 1 including data on S20A).
The reduction of Tg_E can be qualitatively explained either by dissolution of some of the P-IBA within the epoxy component, or by selective extraction of one resin partner towards the P-IBA phase.
It is indeed interesting to note that Tg_E increases again as grafting level goes up also keeping better molecular mixing [9]. At the same time, Tg_A stays under its value for the homopolyacrylate. Either plasticization by one resin partner or GMA copolymer effect can count for this.
More grafting agent is likely to maintain miscibility up to larger degree of conversion, further stabilized by a larger number of chemical links between the two networks. Moreover the unexpectedly high Tg of this one phase blend has to be related to a supposedly higher crosslink density (see G' at 30°C above Tg as compared with the control) [9].

The dynamic viscoelastic analysis of the networks modified with P-IBMA shows two phases in all instances up to 7.5 per cent. (Fig.5, table 2). Both Tg_E and Tg_A are depressed and to a larger relative extent than in the case of the acrylate modifier. It is believed that a similar behaviour is present as in the other case when segregation occured. Nevertheless a distinct difference takes place inducing more interconnected morphology being perhaps the result of a slower polymerization kinetics of the methacrylate.

Mechanical properties

Tables 1 and 2 give the fracture toughness results for the control and the modified networks. The latter appear to be very sensitive to the amount of the thermoplastic phase and to the morphology as it can be altered by the level of GMA. These results show that rather large toughness improvements can be achieved by this modification technique and this does not require any large degree of grafting between the two phases.

Conclusion

The very brittle isobornyl-acrylate and -methacrylate polymers have proven to be efficient modifiers for high Tg epoxies when they are synthesized during the resin cure process in a manner of semi-simultaneous interpenetrating polymer networks.
Toughening occurs only if phase separation is present. This effect depends on the amount of the acrylic phase and of the level of a grafting agent. The SINs with P-IBA are reinforced with a particles dispersed phase whereas a cocontinuous phases morphology plays an interesting role in the toughening of the SINs with P-IBMA.
Modulus is kept at a reasonably good level in the moderate temperature range. As far as the heat distorsion temperature is concerned together with the toughness, this technique offers a nice range of options.

References

1. Bucknall, C.B. and Partridge, I. K. *Polymer* 1983, 24, 639
2. Raghava, R. S. *J. Polym. Sci, Part B, Polym. Phys.* 1988, 26, 65
3. Bucknall, C.B. and Partridge, I. K. *Polym. Eng. Sci.* 1986, 26, 54
4. Bucknall, C.B. and Gilbert, A. H. *Polymer* 1989, 30, 213
5. Raghava, R. S. *J. Polym. Sci, Part B, Polym. Phys.* 1987, 25, 1017
6. Hedrick, J. L., Yilor, I., Wilkes, G. L. and McGrath, J. E. *Polym. Bull.* 1985, 13, 201
7. Kim, S. C. and Brown, H. R. *J. Mater. Sci.* 1987, 22, 2589
8. Cecere, J. A., Hedrick, J. L. and McGrath, J. E. *SAMPE* 1986, 31, 580
9. Tahir, N. Ph.D Thesis, University of Liège, in preparation

Table 1: Data on isobornylacrylate modification.

IBA+ GMA	GMA %	Tg_A (°C)	Tg_E (°C)	K_{IC} (MPa.m$^{1/2}$)
0	0	-	204	0.658
10 w%	0	81	191	0.974
	1.3	81	195	0.922
	4.4	none	200	0.686
20 w%	0	89	193	1.094
	2.88	75	187	1.164
	5.94	84	194	1.324
	12	none	199	0.721
30 w%	0	91	196	1.213
	4.4	83	200	1.260
	7.5	83	193	1.276
	17	none	200	0.739

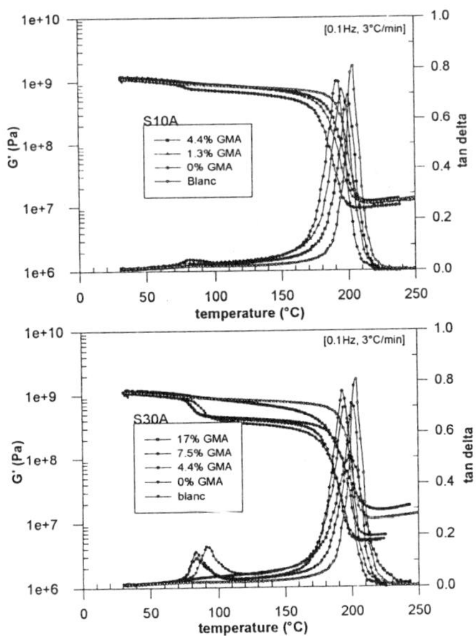

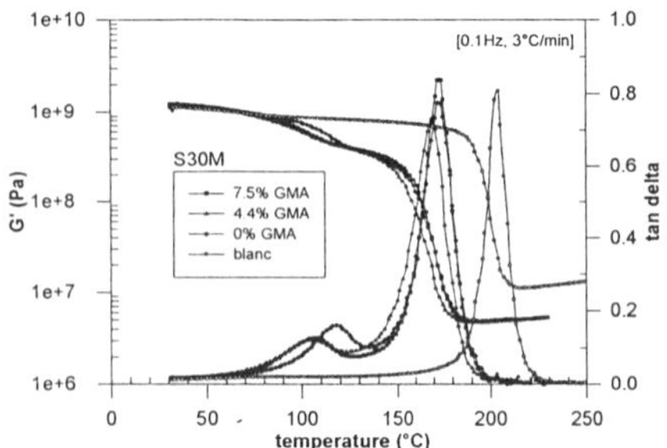

Figure 3, 4 and 5: Dynamic viscoelastic analysis for S10A, S30A, and S30M.

Table 2: Data on isobornylmethacrylate modification.

IBMA+ GMA	GMA %	Tg_A (°C)	Tg_E (°C)	K_{IC} (MPa.m$^{1/2}$)
0	0	–	204	0.658
	0	118	172	1.391
30 w%	4.4	105	170	1.250
	7.5	106	172	1.193

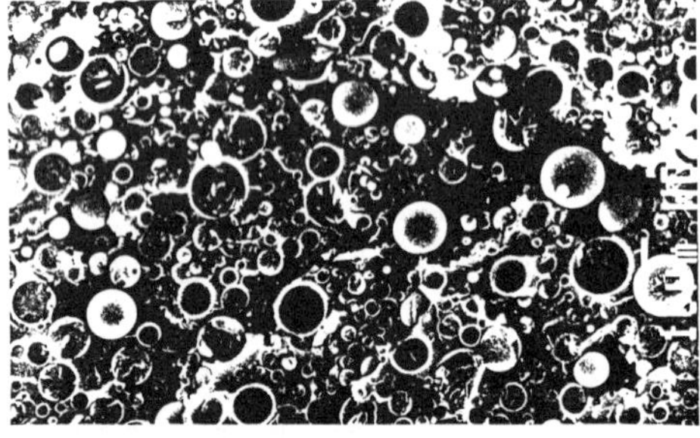

Figure 1a: 30% acrylate (IBA), low GMA.

Figure 1b: 30% acrylate (IBA), high GMA

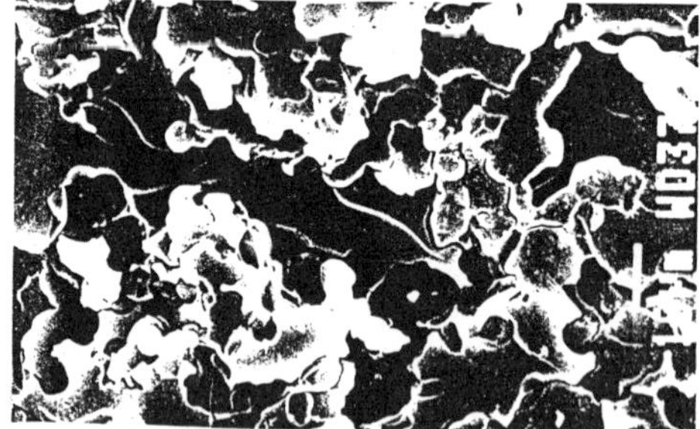

Figure 2: 30% methacrylate (IBMA), no GMA

In situ control of the secondary phase in polymethacrylate toughened epoxies

Alexandre A. BAIDAK and J. M. LIEGEOIS

Laboratoire Matériaux Polymères et Composites,
Université de Liège, Bât. B6, B-4000 Liège, Belgium

Introduction

Epoxy toughening based on rigid thermoplastic particles has been rather extensively studied in the past few years and is known to be effective[1]. A novel technique leading to poly(meth)acrylate toughened epoxies has been recently developed by Liégeois and Tahir[2]. Using simultaneous polymerisation of both the (meth)acrylate and the epoxy network, epoxies whose toughness increased by a factor of two could be obtained. This synthesis method has led to various morphologies, depending mainly on the chemistry of the system but also on the curing conditions. Among those, a two phase system with a nodular secondary phase seems to bring some nice improvements in terms of toughness.

The control of the morphology appears to be one of the key factors affecting the mechanical properties of the material. Many studies have focused on the phase separation and morphology development when an already polymerised secondary phase is added before the curing of the epoxy[3,4]. The approach followed in this study brings a new additional parameter: the control of the modifier polymerisation and its optional grafting to the epoxy matrix.

Experimental

Table 1 gives the recipe of a typical composition. The 4,4'-diaminodiphenyl sulfone (DDS) was first dissolved into the diglycidyl ether of bisphenol A (DGEBA) for 40 minutes at 120 °C with vigorous stirring. The methacrylic monomers were first passed through neutral aluminium oxide columns to remove the inhibitors. The precursors for the thermoplastic phase were then mixed and purged with dry nitrogen to remove the dissolved oxygen and then added to the epoxy. The thermoplastic content for all materials presented in this study is 20 % (wt). The temperature drop due to the addition of the cold methacrylic monomers was sufficient to prevent the radical polymerisation from starting. The mix was then poured into Teflon coated moulds and cured for one hour at 120 °C and five hours at 180 °C.

Toughness was measured on razor blade notched samples using a three point bending test. Glass transition temperatures and G^* were evaluated by using a Rheometric 700. Morphology was examined with a scanning electron microscope (SEM). Molecular weights were determined by GPC, and are expressed in PS equivalent molecular weights.

Results and Discussion

Influence of the modifier molecular weight

The first group of experiments was investigating the influence of the thermoplastic molecular weight as controlled by a transfer agent (TA series in Table 2), on the mechanical properties and the morphology development of the resulting material. The use of various amounts of the 1-dodecanethiol (DDT) as transfer agent for the radical polymerisation produced a thermoplastic phase whose Mw varied from 450 000 g/mol (TA1) to 12 000 g/mol (TA20). No significant variation on the toughness was observed (see Table 2) provided that the modifier molecular weight remained higher than a value of 90 000 g/mol. For lower molecular weights, a strong reduction was found. The visual observation of TA5 revealed a macroscopically heterogeneous sample showing regions of nearly pure epoxy. This can explain the unexpected low value for its toughness as well as its particularly high standard deviation. Though sample preparation and polymerisation were strictly identical to the other compositions, nothing could possibly explain this apparently erratic result yet.

The analysis of tanδ from DMTA measurements reflected an increasing miscibility of the two systems When thermoplastic Mw goes down (see Figure 1). At the lower end (Mw of 27 000 and 12 000 g/mol), a single broad peak for tanδ was observed indicating an enhanced miscibility of the two phases. This apparent single transition occurs at an intermediate temperature between those of the epoxy and of the polymethacrylate. Still, the SEM observations indicate the presence of a distinct nodular secondary phase. A partial miscibility of the two polymers could yield to this inward shift of their respective Tg, while keeping a two phase system. A gross estimation of the decrease in Tg may be calculated from Fox's equation. Assuming a complete miscibility, we find 163 °C compared to the 154 °C found experimentally. However, it must be kept in mind that the use of 1-dodecanethiol is also reducing the kinetic of the radical polymerisation. After 30 min at 120 °C, the conversions of TA1, TA2, TA5, TA10 and TA20 were respectively 46, 42, 32, 12 and 15%. One hour at 120°C, which is the optimal temperature for the decomposition of the thermal radical initiator, might not be sufficient to fully polymerise the methacrylate, leaving unreacted monomers dissolved in the matrix. The latter would then act as a plasticizer of the matrix with the reduction of the elastic modulus and a decrease in the Tg of the sample.

The effect of adhesion between the two phases

The purpose of second part of the study was to examine the influence of the adhesion of the secondary phase to the epoxy matrix on the toughness of the material. A previous study had already focused on the role of the glycidyl methacrylate (GMA) for grafting the thermoplastic onto the epoxy[2,5]. However, the GMA favours the mutual dissolution and chemical grafting of the thermoplastic in the matrix and consequently the loss of the nodular morphology. In this study, a different approach was tempted. The grafting between the two phases should only occur after the morphology is already defined and stabilised (i.e. in the later stage of the curing), hence after the complete *in situ* synthesis of the thermoplastic and its phase separation from the reacting epoxy. The 2-hydroxyethylmethacrylate (HEMA) was chosen as comonomer having a functional group able to react at high temperatures only (180 °C) with the growing epoxy network. Hydroxyl groups are known to react with the oxirane group of the DGEBA[6]. This reaction is slow compared to the reaction of the oxirane and the primary amine, but can be considered to occur with a non negligible rate at high temperatures. We thus assume that at 120 °C the so modified polymethacrylate remains unreactive towards the epoxy network. The study of the radical copolymerisation was first performed in solution (20 % wt in the methylethylketone). Using the terminal model of Mayo and Lewis, values of r_1 and r_2 (where 1: IMA and 2: HEMA) were found to be respectively 0.31 and 3.57, which can be described as a nearly ideal copolymerisation ($r_1.r_2 = 1.1$).

In addition to the grafting, two supplemental effects must be considered. The first is the increase of the solubility parameter of the copolymer. The poly(isobornylmethacrylate) is indeed a non polar polymer with a low solubility parameter, 16.6 $(J/cm^3)^{0.5}$ compared to ~ 21 $(J/cm^3)^{0.5}$ for the epoxy resin. The introduction of 2-hydroxyethylmethacrylate as a comonomer will decrease the difference between the two parameters, hence favouring the miscibility. Since the tan δ peak of the epoxy resin remains unchanged for all the samples (see Figure 2), molecular mixing of the two polymers must have been limited. The second effect resulting from the copolymerisation will be the decrease in the Tg of the copolymer. Conflicting data are reported for the Tg of P-HEMA homopolymer (55 – 86 °C), far below that of the P-IBMA. This last effect can be noted from the DMTA of the various samples. The introduction of 30 % wt of HEMA in the copolymer reduces the Tg_l by as much as 23 °C.

Though the effect is not extremely strong, we believe that the introduction of the hydroxyl functional group has a positive effect on the toughness of the samples, see Table 2. Figure 3 shows a SEM picture of the sample HE30. Crack is propagating from bottom to top. Spherical particles of a diameter of ca 7 µm are nicely dispersed throughout the matrix.

Conclusion

In situ control of the thermoplastic modifying phase polymerisation for the toughening of high Tg epoxies has shown to be an interesting way to control the mechanical properties and the morphology of the blend.

Low molecular weights (below 90 000 g/mol) should be avoided in order to prevent an excessive mixing of the two polymers. Miscibility increase due to Mw reduction lowers the matrix Tg and leads to a loss of the toughening effect.

Enhanced adhesion between the two phases appears to have a beneficial effect on the toughness. The replacement of 30 % of isobornylmethacrylate by the 2-hydroxyethylmethacrylate further increases the toughness by 20% with only a slight reduction of the modulus at high temperatures.

References

1. K. Riew, A. J. Kinloch, Eds, *Toughened Plastics II*, Advances in Chemistry Series, vol **252**, Washington, 1996, p 405.

2. J. M. Liégeois and N. Tahir, submitted to *Polym. Mater. Sci. Eng. Prepr.*, Boston ACS National Meeting, 1998.

3. R. J. J. Williams, B. A. Rozenberg, J. P. Pascault, Polymer Analysis Polymer Physics, Advances in Polymer Science, vol **128**, Berlin, 1997, p95.

4. B. S. Kim, T. Chiba and T. Inoue, *Polymer*, **34(13)**, 2809, 1993.

5. N. Tahir, Thesis Dissertation, Presses Universitaires, Liège, 1998, in preparation.

6. C. May, and Y. Tanaka, Eds, *Epoxy Resins Chemistry and Technology*, Marcel Dekker Inc., New York, 1973.

Table 1: Components description and typical recipe

Compound	Amount	Supplier
Epoxy network	**80 parts**	
Diglycidyl ether of Bisphenol A (epoxy group content 5360 mmol/kg)	75 % (wt)	Shell (Epikote 828 EL)
4,4' diamino diphenyl sulfone	25 % (wt)	Ciba (HT 976)
Methacrylate network	**20 parts**	
t-butyl peroxybenzoate	1 % (wt)	AKZO Chemical (trigonox C)
1-dodecanethiol	0 - 2 % (wt)	Aldrich
2-hydroxyethyl methacrylate	0 - 30 % (wt)	Aldrich
Isobornyl methacrylate	Up to 100 %	Elf Atochem

Table 2: Samples description and their properties

	DDT %	HEMA (%)	K_{IC} (+/- L *) (MPa.m$^{0.5}$)	Tg$_h$ (°C)	Tg$_l$ (°C)	Mw (g/mol)
Control	0	0	0.644 (0.058)	198	/	
TA1	0.11	0	1.249 (0.044)	175	126	451000
TA2	0.22	0	1.283 (0.072)	172	120	332000
TA5	0.51	0	0.687 (0.199)	168	141	90000
TA10	1.00	0	1.169 (0.079)	152	150	27000
TA20	2.06	0	0.768 (0.05)	154	151	12000
HE0	0.2	0	1.157 (0.102)	173	131	
HE15	0.2	15	1.305 (0.092)	170	119	
HE22	0.2	22.5	1.297 (0.089)	171	112	
HE30	0.2	30	1.393 (0.118)	171	109	

* L is the confidence limit for a probability level of 95%

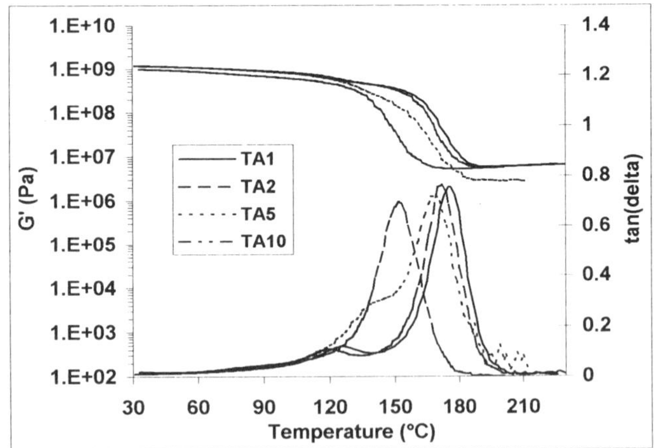

Figure 1: Influence of the modifier Mw on the dynamical mechanical properties (Mw varies from 450 000 (TA1) to 12 000 g/mol (TA20)).

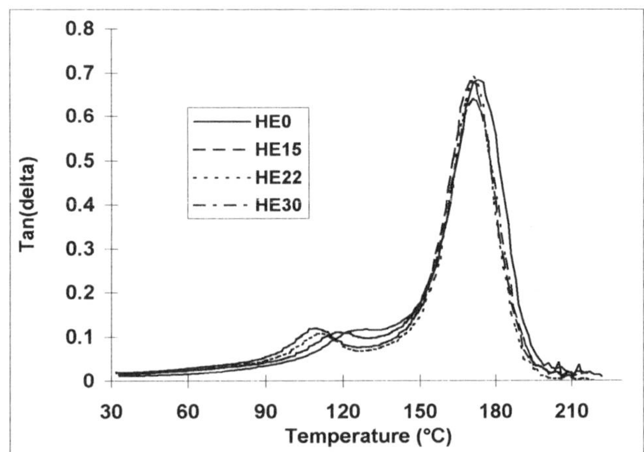

Figure 2: Influence of the HEMA content on the dynamical mechanical properties.

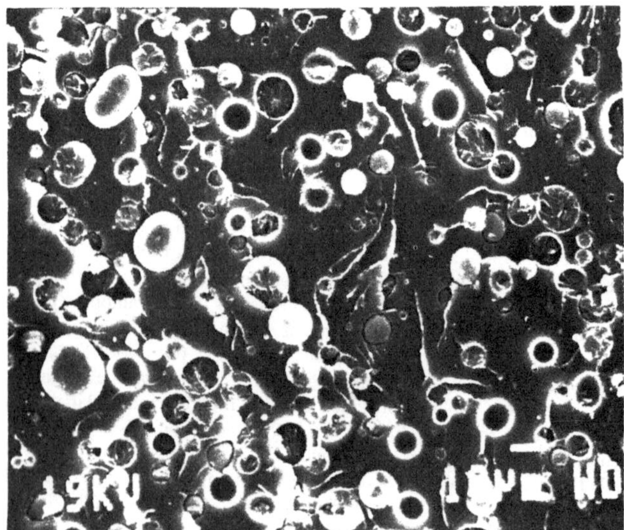

Figure 3: SEM picture of a fractured surface of sample HE30. Typical particle size is ca 7μm.

Epoxy-diamine Thermoset/Thermoplastic blends : thermoset reactions and rheological evolutions during curing

A. Bonnet[1,2], Y. Camberlin[2], J.P. Pascault[1], H. Sautereau[1]

1) Laboratoire des Matériaux Macromoléculaires, UMR-CNRS 5627,
Institut National des Sciences Appliquées, Bâtiment 403, 20 avenue Albert Einstein,
69621 Villeurbanne Cedex, France.
2) Institut Français du Pétrole, Industrial Research and Development Center,
Autoroute A7, Solaize BP3, 69390 Vernaison, France.

INTRODUCTION

Thermoset (TS)/Thermoplastic (TP) blends are materials resulting from the mixing of the TP polymer with the TS precursor (one or two monomers, a catalyst...) and the subsequent reaction of this precursor. Usually the initial mixture is homogeneous and due to the molar mass increase of the TS precursor a liquid-liquid phase separation occurs at a given conversion[1]. The beginning of the phase separation process can be determined by light transmission (LT, in this case, the Cloud Point, Cp is characterized), light scattering (LS) or small angle X-ray scattering (SAXS). Depending on the technique used, the observation scales can be different.

One important factor controlling the generated morphologies is the location of the composition of the initial blend, ϕ_{TP}°, with respect to the critical composition $\phi_{TP,crit}$. For TP with average molar masses between 10000-30000g/mol, $\phi_{TP,crit}$ in liquid epoxy and diamine monomers (as thermoset precursor) may be obtained from the Flory-Huggins model and is in the range of 10-15 wt% TP[1]. In a previous study, E. Girard-Reydet et al.[2] have demonstrated that for quantities of TP, if ϕ_{TP}° is in the range of $\phi_{TP,crit}$, 10 wt% for example, phase separation during epoxy-amine reactions proceeds by spinodal dimixing (SD) and LT or LS give correct estimation of the beginning of the phase separation process. But for off-critical composition, typically TP $\geq$ 30wt% the second phase appears through nucleation and growth (NG) mechanism, and in this case SAXS is able to observe the beginning of the phenomena before LT or LS.

It also has been shown that for TP $\leq$ 10 wt% the continuous phase is the epoxy-amine network, but for TP $\geq$ 30 wt% the continuous phase is the TP, and between 10-30 wt% more or less well defined bi-continuous structures are formed. One consequence is that TS-TP blends have been mainly studied for two reasons : i) modification of TS networks with high performance ductile TP[3], or ii) new processing routes for intractable high-temperature resistant TP polymers[4-5]. For the first application the TP concentrations are generally lower than 15 wt% and for the second one higher than 30 wt%.

Most of the studies on TS-TP blends are mainly focused on final morphologies and / or mechanical properties[1,3-6]. But the control of morphologies and properties requires a good knowledge of the different events occuring during the reaction of the TS precursor. Few studies try to investigate the effects of the TP on TS chemistry and rheology. The aim of this work is to propose a detailed analysis on the effect of a non functionalized thermoplastic, like commercial polyetherimide, PEI, on the epoxy-amine reaction on the basis of a previous complete description of the neat TS system[7], but also the characterization of the different events occurring during the phase separation process. Taking into account all these parameters, kinetics and viscosity modeling during cure will be proposed.

EXPERIMENTAL

The TS precursor is an epoxy system based on classical liquid diglycidyl ether of bisphenol A, DGEBA $\bar{n} = 0.15$ and a liquid diamine, sterically hindered 4,4'- methylenebis [3-chloro 2,6-dianiline], MCDEA with a stoechiometric ratio equal to 1. The low reactivity of this hardener was chosen in order to follow carefully the phase separation process. Two TPs have been used, mainly PEI from GE (Ultem 1000) but also PS from Atochem (Lacqurene 1070N) for a comparison.

The initial blends with low TP concentration $\leq$ 20 wt% were prepared in a two-stage process : TP was first dissolved at 140°C in the epoxy prepolymer, the diamine was then added at 90°C. For the initial blends with higher TP quantities, a co-rotating twin-screw extruder was used in a one-stage process, the TP was directly dissolved in the epoxy-amine mixture.

Kinetic studies were conducted as described previously[7]. HPLC was used to measure the extent of epoxy-amine reaction. Gelation times were also obtained during HPLC measurements, it is considered to be the time at which the presence of an insoluble fraction is first observed.

The morphology of blends was studied by transmission (TEM) or scanning (SEM) electron microscopies.

Differential scanning calorimetry was used to measure the glass transition temperature, T_g and the change in heat capacity through the glass transition temperature, ΔC_p, a heating rate of 10°C/min was used. The viscoelastic properties were measured by shear rheology with a RDA II.

RESULTS AND DISCUSSION

1) The experimental Cloud Point Curves (CPC) of PEI/DGEBA-MCDEA and PS/DGEBA-MCDEA blends before any reaction exhibit an UCST behaviour. PS is observed to be less soluble in epoxy TS precursor than PEI but at the selected curing temperature, $T_i = 135$°C both are soluble, and exhibit one T_g which can be measured or calculated by the help of Couchman equation[8].

During curing the $T_g(x)$ of the homogeneous mixture increases due to the molar mass increase of the TS precursor, but after the CP two T_gS are observed, one higher and one lower than the $T_g(x)$ of the homogeneous phase. The upper one, $T_{g\beta}$ is attributed to the PEI rich phase, the lower one, $T_{g\alpha}$ to the epoxy-amine rich phase. Both $T_{g\beta}, \Delta C p_\beta$ and $T_{g\alpha}, \Delta C p_\alpha$ have different evolutions with curing time.

2) The experimental evolutions of epoxy conversion, x, with curing time for DGEBA-MCDEA reactions with different TP concentrations have been measured. By taking into account the kinetic rate constants of the neat system previously[7] measured and the dilution effect induced by the presence of the TP, it was also possible to predict the behaviour of the blends, but in the one phase area only. We observe that for all TP concentrations before phase separation, the addition of TP leads to a dilution of reactive groups. Even if x vs time decreases with the increase of TP content, the real kinetic rate constants are not modified in the one phase area.

For low TP concentrations $\leq$ 10 wt% the fitting is rather good during all the reaction. But when the TP concentrations $\geq$ 30 wt% a sudden increase of reaction rate is observed where the phase separation begins. This effect is stronger in the case of PS compared to PEI. When the phase separation occurs, the dilution ratio changes rapidly and differently for α and β phases. These new two dilution ratios also change with reaction times and x measured is $\bar{x}$, the average epoxy conversion.

To model the behaviour after phase separation it is necessary i) to have the composition of α and β phases at each curing time, ii) to know the concentration of the dispersed phase. Microscopy and thermal measurements were performed on a blend with 48 wt% PEI to make an estimation of the composition of each phase, leading to kinetic prediction in both α and β phases. This prediction shows that gelation in the α phase occurs at a conversion close to 0.6.

3) When the TP concentration is lower than $\phi_{TP,crit}$ (10 wt%) phase separation at the CP is followed by a rapid decrease in viscosity. Gelation of the continuous α phase leads to an increase of G'. For longer times, vitrification of the α phase can be detected by a peak on δ and G' curves.

When the TP concentration is slightly higher than $\phi_{TP,crit}$ the behaviour is more complex. In the beginning of the phase separation process a gradual increase in viscosity occurs due to the bicontinuous structure and the behaviour of the β phase. Then a decrease is observed due to the fact that the bicontinuous structure is destroyed and the α phase dominates. The amplitude of this phenomenon depends on the frequency[9].

When TP composition is high enough to ensure to have a continuous TP rich phase up to the end of the phase separation process (typically 30 wt%), the onset of phase separation is accompanied by a gradual increase in viscosity.

For all TP compositions, before phase separation the viscosity of the homogeneous mixture can be described with the help of the composition and the viscosities of the two components. It increases due to the reaction and the increase of $\overline{M}_w$ of the TS precursor. After phase separation, Einstein relation can be used to describe the viscosity of a suspension taking in mind the same parameters than for kinetic modeling : phases composition and volume fraction of the dispersed phase. In the later stages of the reaction for TP concentrations $\geq$ 30 wt%, β phase behaviour dominates, and in the vicinity of its vitrification, WLF relation has to be used. Taking into account all these considerations, viscosity evolutions during cure of a blend with 33 wt% of PEI have been modeled.

CONCLUSION

Phase separation in TS / TP blends was monitored using kinetic and rheological measurements. The main parameters for modeling are phase compositions and phase concentrations.

References

1   R.J.J. Williams, B.A. Rozenberg, J.P. Pascault "Reaction induced phase separation in modified thermosetting polymers" Adv. Polym. Sci., 128, 95 (1996)

2   E. Girard-Reydet, H. Sautereau, J.P. Pascault, P. Keates, P. Navard, G. Thollet, G. Vigier, Polymer, in press (1998)

3   R.A. Pearson, in "Rubber toughened plastics I", C.K. Riew and A.J. Kinloch eds, Adv. Chem. Ser., ACS, 233, 407 (1993)

4   R.W. Venderbosch, H.E.H. Meijer, P.J. Lemstra, Polymer, 36, 2903 (1995)

5   P.J. Lemstra, J. Kurja, H.E.H. Meijer in "Materials Science and Technology" R.W. Cahn, P. Haasen and E.J. Kramer, eds, Vol.18 "Processing of polymers" H.E.H. Meijer, ed, Wiley VCH-Weinheim, 513 (1997)

6   E. Girard-Reydet, V. Vicard, J.P. Pascault, H. Sautereau, J. Appl. Polym. Sci., 65, 2433 (1995)

7   E. Girard-Reydet, C.C. Riccardi, H. Sautereau, J.P. Pascault, Macromolecules, 28, 7599 (1995)

8   P.R. Couchman, Macromolecules, 11, 1156 (1978)

9   C. Vinh-Tung, G. Lachenal, B. Chabert, J.P. Pascault in "Toughened Plastics II – Novel approaches in Science and Engineering", C.K. Riew and A.J. Kinloch, eds, Adv. Chem. Ser., ACS, 252, 59 (1994)

RIGID ROD EPOXY RESINS

Marta Giamberini, Eugenio Amendola, and Cosimo Carfagna,

Department of Materials and Production Engineering,

University of Naples "Federico II", Piazzale Tecchio 80,

Napoli Italy, 80125

Curing of a bifuntional epoxy molecule with a stiff and elongated structure with polyfunctional amines can lead to thermosets having a liquid crystalline structure (LCER). One of the more crucial factors in determining such an anisotropic structure is the curing temperature of the thermosetting systems. Thermosets with liquid crystalline structure can be also obtained by curing a blend of a rigid rod epoxy resin and a commercial epoxy resin, in a proper composition range, with a difunctional amine. Liquid crystalline epoxy thermosets exhibit higher fracture toughness with respect to the cured systems having isotropic structure. This feature can be ascribed to the microstructure of LCER, consisting of ordered domains acting as micronodules around which crack can deviate.

Anisotropic Deformation Behavior of the Cubic Double Gyroid Phase in ABA Elastomeric Triblock Copolymers.

Benita J. Dair, Edwin L. Thomas
Department of Materials Science and Engineering
Massachusetts Institute of Technology, Cambridge, MA 02139

Apostolos Avgeropoulos, Nikos Hadjichristidis
Department of Chemistry
University of Athens, Panepistimiopolis, 157 71 Zografou, Athens, Greece

Malcolm Capel,
National Synchrotron Light Source,
Brookhaven National Laboratory, Upton, NY 11973.

Introduction

Styrene-diene-styrene (SDS) triblock copolymers, in which the styrene phase is the minority component, are industrially important materials because of their outstanding thermoelastic properties. The three classic morphologies are spheres, cylinders, and lamellae, with the glassy phase discrete and continuous in 1- and 2-dimensions, respectively. Novel cubic phases with interconnected structures have recently been discovered in copolymer systems (1-4), including the ordered tricontinuous double gyroid (DG) morphology (1,2). In polyisoprene (PI) -rich SDS materials, the cubic DG morphology conforms to Ia$\overline{3}$d symmetry, and consists of two tricontinuous, interpenetrating PS networks reinforcing a rubbery PI matrix.

We investigate the large-strain mechanical properties of PI-rich (~35 volume percent styrene) triblock copolymers with DG morphology (5) (of 74,000 total molar mass and block molecular weights 14K-46K-14K) via in-situ SAXS and TEM The samples are additionally oriented and biaxially textured to permit the study of the directional dependence of the mechanical behavior of the DG.

Experimental Procedures and Discussion

Samples are oriented by a process called roll-casting (6) whereby self-assembly of microdomains occurs in the presence of a flow field. The polymer was roll-cast from a concentrated cumene solution (1 wt/2 vol), dried overnight in ambient air, and then under vacuum for an additional 2 days. The application of the flow field, which has symmetries compatible with those of either cylinders or lamellae, but not the cubic DG, induces a cylindrical PS phase for the roll-cast film. The cylinders are highly oriented, with the cylinder axis aligned along the flow direction. Since the DG structure is the equilibrium phase for this triblock composition, upon annealing (120°C for 10 days) DG nucleates and grows with the [111] oriented along the flow direction. Figure 1 shows the DG axes relative to the axes of the film. The [$\overline{1}$10] is in the plane of the film but transverse to the flow direction (i.e., along the neutral axis). Mechanical testing was performed both parallel to- and transverse to the direction of flow. Tensile samples 2mm wide were cut from the resulting 0.67 mm - thick films.

Figure 1: Axes of DG relative to the process directions of the roll-cast film

Small Angle X-ray Synchrotron experiments were conducted at Brookhaven National Laboratory on beamline X12B using 1.54 Å x-rays and sample-to-detector distances ranging from 247 cm to 260 cm. A tensile stage with an attached 50 lb. load cell was used to deform tensile specimens at strain rates of 0.57 - 1.05 min^{-1}. In-situ SAXS patterns, along with simultaneous load-deformation curves, were taken as tensile specimens were loaded up to ~ 400 - 700% and unloaded to zero load. Scattering patterns were collected using a 2D detector (7) with exposure times ranging from 5 s. to 10 min.

TEM provides a real-space correspondence to the reciprocal-space SAXS data taken during the deformation. However, in-situ TEM of a continuously-stretched bulk elastomer is impossible. To retain high strains for TEM, the elastomer is irradiated with high energy electrons while stretched. Irradiation heavily crosslinks the rubber and raises the glass transition temperature to above room temperature. Stretched samples for TEM examination were made by stretching and then irradiating with 200 Mrad of high energy electrons at the MIT High Voltage Lab. Sections 500-1000 Å thick were prepared by cryo-microtomy at -90° C using a Reichert Jung FC 4E Cryo-ultramicrotome equipped with a diamond knife. TEM samples were picked up on 600 mesh copper grids, and placed in the vapors of 4% osmium tetraoxide-water mixture for 2 hours to selectively stain the PI phase. Sections were examined in bright field using a JEOL 2000 FX TEM operated at 200 kV.

Both the X-ray beam (for SAXS) and electron beam (for TEM images) are oriented parallel to the [11$\overline{2}$].

Results and Discussion

Figures 2a and b show TEM and SAXS patterns of roll-cast-oriented double gyroid, respectively, viewing through the plane of the film (along the [11$\overline{2}$]). The flow ([111]) direction is vertical and the neutral axis ([$\overline{1}$10]) is horizontal. The high order peaks in the SAXS pattern both on- and off- equator indicates good microdomain ordering.

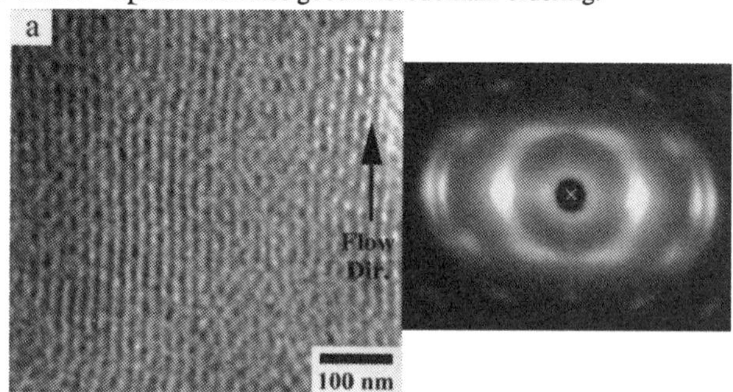

Figure 2: a) TEM of roll cast DG; b) SAXS pattern.

Figure 3 shows the stretch directions relative to the DG structure. The [111] stretch (parallel to flow direction) is along the body diagonal, which is has the highest polystyrene connectivity; this observation is consistent with the [111] having nucleated from the cylinder axes of the as-rolled material. The [$\overline{1}$10] (transverse direction/neutral axis) stretch is along the face diagonal, which has no polystyrene connectivity.

Figure 3: Stretch directions relative to the DG structure

Figure 4a shows a stress-strain curve of DG loaded in tension in the flow direction with accompanying SAXS patterns. Yield occurs at approximately 15% strain (ε), followed by necking. This is accompanied by fading of the Bragg peaks into lobe patterns, as seen in figures 4c-e. As yielding is dependent upon PS connectivity, yielding occurs from the break-up and plastic flow of the continuous PS domains. The hysteresis in the stress-strain curve is 55-60%, as energy is dissipated by

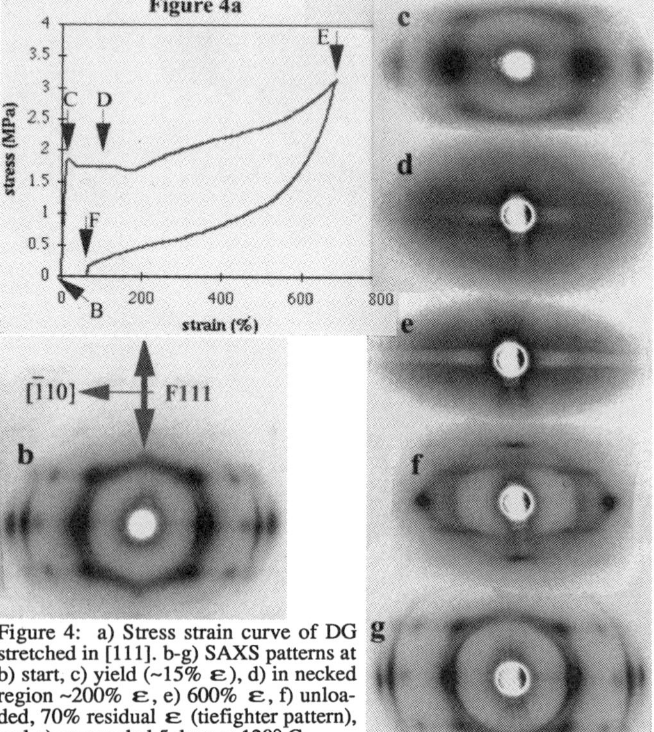

Figure 4: a) Stress strain curve of DG stretched in [111]. b-g) SAXS patterns at b) start, c) yield (~15% ε), d) in necked region ~200% ε, e) 600% ε, f) unloaded, 70% residual ε (tiefighter pattern), and g) reannealed 5 days at 120° C.

break-up and flow of the PS domains. Upon unloading, Bragg peaks return, giving rise to the characteristic "tiefighter" pattern, shown in figure 4f. Figure 4g shows the SAXS pattern of the unloaded sample, annealed at 120°C for 5 days. This pattern is identical to the original (figure 4b), indicating that the structure fully recovers. This suggests that the memory of the connected PS network was not permanently destroyed, even with loading to 650%, and the microdomain structure returns to the original state at long times or at high temperature annealing.

Figures 5a-b show TEM pictures of deformed DG stretched along the flow direction. Note from figure 2a that the undeformed material is quite homogeneous and that the PS connectivity is high, especially along the flow direction. At yield (figure 5a), the material has become heterogeneous and the polystyrene network has undergone plastic deformation. Upon unloading (figure 5b) the material has regained some of its original DG character, but the PS network is not completely restored until long times or high temperature annealing.

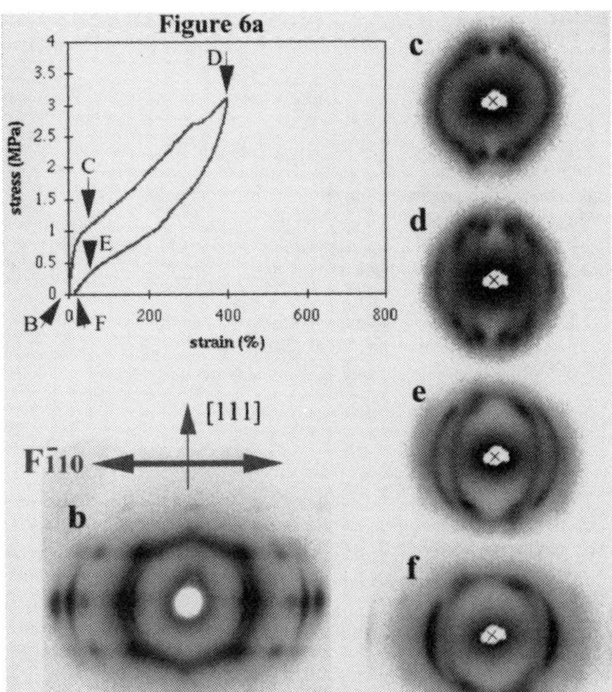

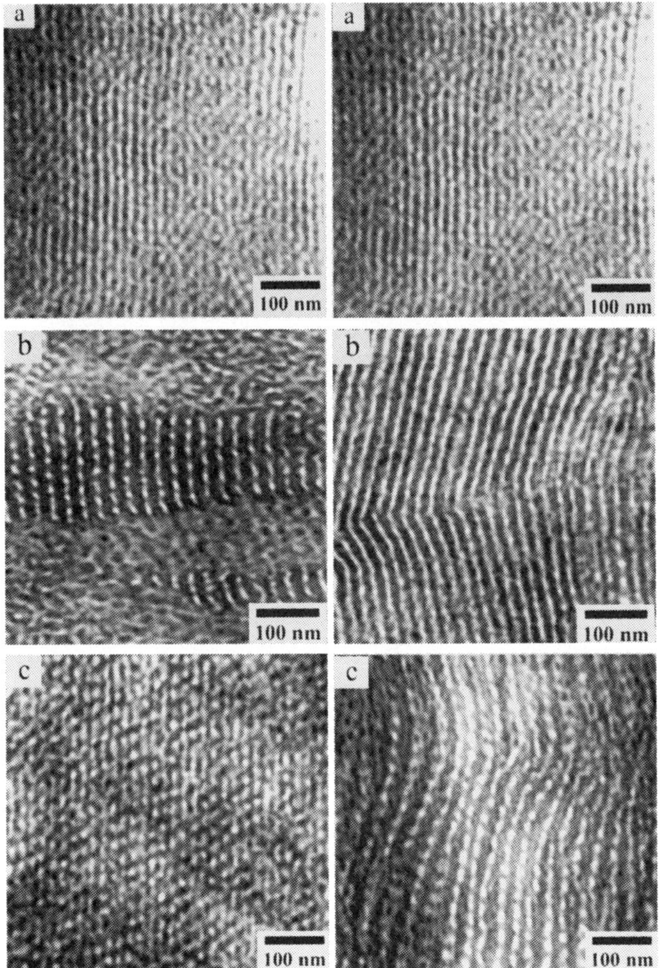

Figure 5: TEM pictures of a) roll cast DG, undeformed, b) deformed along [111] until yield (15% ε), and c) unloaded ("tiefighter") to 0 load (from 600% ε), residual ε ~70%.

Figure 7: TEM pictures of a) roll cast DG undeformed, b) deformed along [1̄10] to 65% ε), and c) deformed along [1̄10] to 130% ε),

Figure 6 shows a stress-strain curve of the DG deformed in the [1̄10] (transverse) direction with accompanying SAXS patterns. In this direction there is no PS connectivity. This characteristic exhibits itself in the following ways: There is no yielding during deformation, Bragg peaks and streaks persist even to 400% strain, the modulus is 1/5th that in the flow direction, and the hysteresis of the stress-strain curve is ~40%, (1/3 less than that in the flow direction), and the residual strain upon unloading is ~10-40%, markedly less than in the flow direction.

Figure 6: a) Stress strain curve of DG stretched in [1̄10]. b-h) SAXS patterns b) before loading, c) at 50% ε, d) at 400% ε, f) unloaded to 50% ε, h) unloaded to 0 load (~10% residual ε).

Figures 7a-b show TEM of DG deformed in the [1̄10] transverse direction. Note that the PS domains stay continuous at all strains and the material deforms more homogeneously with buckling of the PS networks.

Summary

Rollcasting of an SIS triblock produced oriented DG films, of which the mechanical properties were investigated via simultaneous in-situ load-deformation measurements and SAXS patterns, as well as TEM of stretched, deformed DG fixed by high energy electron crosslinking. We have found that the deformation behavior exhibits very different characteristics: in the flow direction, the modulus is 5 times larger than in the transverse direction, there is yielding with a distinct neck (in the transverse direction, no yielding is observed), and there is a 50% higher mechanical hysteresis than in the transverse direction. These characteristics are due to a high PS connectivity in the flow direction, whereas the PS domains in the transverse direction are discontinuous, which leads to dramatic and significantly different deformation modes, SAXS characteristics, and TEM images during large-strain deformation.

Acknowledgments

We would like to thank C. Honeker for the construction of the stretching apparatus; R. Albalak for the development of the roll-caster; C. Zimba for his cooperation at the BNL; and K. Wright for help at the MIT High Voltage Lab. This research was supported by the Air Force (AFOSR F49620-94-1-0224) and the National Science Foundation (NSF DMR 92-14853).

References

(1) Hajduk, D. A.; Harper, P. E.; Gruner, S. M.; Honeker, C. C.; Kim, G; Thomas, E. L.; Fetters, L. J., <u>Macromol.</u> **27** (1994) 4063.
(2) Schultz, M. F.; Bates, F. S.; Almdal, K.; Mortensen, K., <u>Phys. Rev. Lett.</u> **73** (1994) 86.
(3) Thomas, E. L.; Alward, D. B.; Kinning, D. J.; Martin, D. C.; Handlin, D. L.; Fetters, L. J., <u>Macromol.</u> **19** (1986) 2197.
(4) Hasegawa, H.; Tanaka, H.; Yamasaki, K.; Hashimoto, T., <u>Macromol.</u> **20** (1987) 1651.
(5) Avgeropoulos, A; Dair, B. J., Hadjichristidis, N.; Thomas, E. L., <u>Macromol.</u> **30** (1997) 5634.
(6) Albalak, R. J.; Thomas, E. L., <u>J. Polym. Sci.: Part B: Polym. Phys.</u> **31** (1993) 37.
(7) Capel, M. C.; Smith, G. C.; Yu, B., <u>Rev. Sci. Instrum.</u> **66** (1995) 2295.
(8) Odell, J. A.; Keller, A., <u>Polym. Eng. and Sci.</u> **17** (1977) 544.
(9) Honeker, C. C.; Thomas, E. L., <u>Chem. Mat.</u> **8** (1996) 1702.

DESIGN OF MECHANICAL BEHAVIOR IN PMMA-MODIFIED EPOXY RESINS BY CONTROL OF PHASE SEPARATION

P. M. Remiro[*], C. C. Riccardi[**] and I. Mondragon[***]

* Dpto. de Ciencia y Tecnología de Polímeros. Facultad Ciencias Químicas. Universidad País Vasco. Pº Manuel de Lardizábal, nº3. 20009-San Sebastián/Donostia (Spain). E_mail: popremop@sq.ehu.es
** Institute of Materials Science and Technology (INTEMA), University of Mar del Plata and National Research Council (CONICET). J. B. Justo 4302 - (7600) Mar del Plata (Argentina). E_mail: criccardi@patora.fi.mdp.edu.ar
*** Dpto. de Ingeniería Química y Medio Ambiente. Escuela Ingeniería Técnica Industrial. Universidad País Vasco. Avda. Felipe IV, 1-B, 20011-San Sebastián/Donostia (Spain). E_mail: iapmoegi@sp.ehu.es

Introduction

In the last years, many attempts have been made to improve the toughness of the inherent brittle thermosetting matrices by blending them with rubbers or thermoplastics (1-4). These modifiers are initially miscible with the resin and curing agent. The final morphology is dependent on the variation of thermodynamic and kinetic factors through curing. The size increase of the oligomeric species and the corresponding decrease in the entropic contribution to the free energy of mixing during curing causes a modifier-rich phase to segregate from the matrix at a particular conversion level (5,6). When the matrix gelifies, the primary phase separation is practically finished, but a secondary phase separation may continue inside the dispersed phase particles (7). So that, the study of chemorheology of polymerization and phase separation behavior becomes completely necessary in order to understand the final morphologies generated, and hence, their influence in the ultimate properties of the modified mixtures.

In this work, a diglycidyl ether of bisphenol-A (DGEBA) epoxy resin was modified with poly(methyl methacrylate) (PMMA) and cured with a stoichiometric amount of diaminodiphenylmethane (DDM) at 80°C for various periods of time, from 2 to 7 h. After this primary curing, samples were submitted to secondary curing at 140°C for 1.5 h and, finally, to post-curing at 200°C for 2 h. The obtained plaques ranged from completely opaque to fully transparent, thus revealing different morphologies. Mechanical properties, including fracture toughness, were measured in bending.

Experimental

The DGEBA used in this study was DER 332, kindly supplied by Dow Chemicals. It has an epoxy equivalent weight of 174.3 g/eq and and hydroxy/epoxy ratio of 0.03. The aminic curing agent, DDM HT-972, was obtained from Ciba-Geigy. The PMMA used was Altuglas GR 9E from Elf-Altochem. Its average number molecular weight was 58,000, as measured by gel permeation chromatography.

In order to prepare a stoichiometric DGEBA:DDM 1:1 mixture containing 15 wt % of PMMA, DGEBA and PMMA were dissolved together in methylene chloride. The solvent was then vaporized in an oil bath, followed by residual solvent removal at 80°C in vacuo for 24 h. A weighted amount of DDM was added, and subsequently the mixture was mechanically stirred until complete dissolution of the amine. The reacting mixture was molded into 6 mm thick plaques, and cured in an oven in the above stated conditions.

A Perkin Elmer differential scanning calorimeter, model DSC-7, was used to perform thermal measurements. It was calibrated with indium standards before use. A heating rate of 10°C/min was applied in every test. Dynamic mechanical thermal analysis (DMTA) and scanning electron microscopy (SEM) observations were also carried out to investigate the morphologies generated.

Flexural properties were determined in a three-point bending device using an Instron universal testing machine, model 4206. Tests were carried out at a cross-head speed of 1.7 mm/min. $80 \times 10 \times 6$ mm^3 specimens were machined from the molded plaques. For toughness measurements, specimens measuring $60 \times 12 \times 6$ mm^3, were notched with a saw and a precrack was initiated at the root of the notch with a razor blade. Results are the average of at least 5 measurements.

Results and Discussion

All samples were visually transparent after primary curing at 80°C. Nevertheless, in the same way than previously shown for this family of epoxy matrices modified with different percentages of PMMA (8), after the secondary curing at 140°C, samples cured for 2 or 3 h at 80°C became opaque. On the other hand, samples cured for longer times at 80°C remained transparent after post-curing.

In Table I, the glass transition temperatures (Tg) of the cured plaques are shown. The Tg for the unmodified DGEBA:DDM 1:1 and that for pure PMMA are also collected for comparison. As can be seen, plaques cured for 2 or 3 h at 80°C showed two different Tg's which changed with the curing time. On the contrary, plaques cured for 5, 6 or 7 h at 80°C only showed a single Tg, independent of the curing time. Thus, multiphase or homogeneous materials can be obtained with the same overall mixture composition only by changing the precuring time. The Tg's of the heterogeneous fully cured mixtures were comprised between those for both pure PMMA and unmodified DGEBA:DDM mixture, i. e., the obtained phases are not pure.

Table I. Glass transition temperatures for DGEBA:DDM/PMMA 1:1/15 wt % plaques submitted to different precuring times.

System	Curing time at 80°C (h)	Tg_1 (°C)	Tg_2 (°C)	Appearance
DGEBA:DDM/PMMA 1:1/15%	2	111.9	151.8	opaque
	3	111.2	158.3	opaque
	5	----	141.2	transparent
	6	----	141.2	transparent
	7	----	142.2	transparent
DGEBA:DDM 1:1	3	----	174.0	
PMMA		105.6	----	

Figure 1 shows a plot of the Tg, measured by DSC, of samples after primary curing at 80°C for different times. Initially, the Tg continuously increased with curing time until it reached a "plateau" for times longer than about 300 min. The highest Tg measured was of the order of the curing temperature. This indicates that vitrification had taken place. For that reason, the transparent plaques cured for 5, 6 or 7 h at 80°C, that vitrified, all presented almost identical Tg.

The thermal properties and transparency observed for mixtures precured for different times led to suppose miscibility, but this behavior could also correspond to the existence of very small particles in the epoxy matrix. In this way, the final morphology of these mixtures, determined by SEM, will be related to the DMTA behavior and will be discussed in terms of the conversion profile through precuring.

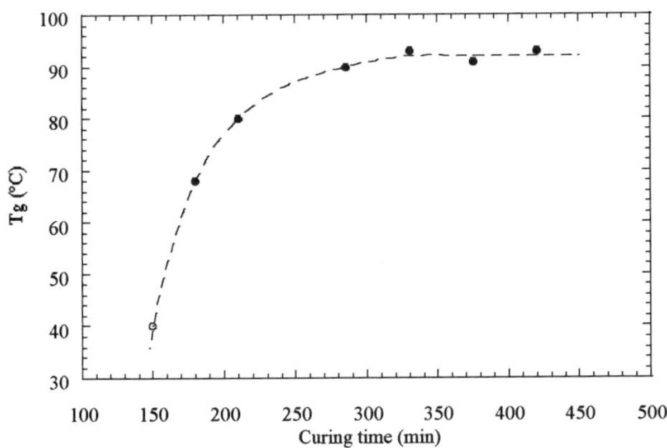

Figure 1. Tg vs. curing time for DGEBA:DDM/PMMA 1:1/15 wt % cured at 80°C.

The flexural properties measured in 3-point-bending tests, and the fracture toughness determined by the critical-stress-intensity factor, K_{Ic}, and the critical strain energy release rate, G_{Ic}, measured in single edge notch bending (SENB) tests, for the fully cured materials are shown in Tables II and III, respectively.

The opaque plaques presented similar rigidity and strength values than those corresponding for the unmodified matrix, as these systems were clearly phase separated. The immiscibility had an evident influence in the fracture toughness as G_{Ic} was higher for the completely separated mixtures.

As a conclusion of technological interest, we have demonstrated that, without making formulation changes, the final morphology, and consequently the fracture toughness of an epoxy matrix modified with PMMA can be designed by only modifying the precuring time.

References

1.- H.-J. Sue, E. Y. García Meitin, and N. A. Orchard, *J. Polym. Sci., Part B: Polym. Phys.*, **31**, 595 (1993).

2.- D. Verchere, J. P. Pascault, H. Sautereau, S. M. Moschiar, C. C. Riccardi, and R. J. J. Williams, *J. Appl. Polym. Sci.*, **42**, 701 (1991).

3.- C. B. Bucknall and Y. K. Partridge, *Polymer*, **24**, 639 (1983).

4.- A. J. MacKinnon, S. D. Jenkins, P. T. McGrail, and R. A. Pethrick, *Polymer*, **34**, 3252 (1993).

5.- C. M. Gómez and C. B. Bucknall, *Polymer*, **34**, 2111, (1993).

6.- T. Inoue, *Prog. Polym. Sci.*, **20**, 119 (1995).

7.- C. C. Riccardi, J. Borrajo, and R. J. J. Williams, *Polymer*, **25**, 5541 (1994).

8.- I. Mondragon, P. M. Remiro, M. D. Martin, A. Valea, M. Franco, and V. Bellenguer, *Polym. Int.*, (in press).

Table II. Mechanical properties in bending for DGEBA:DDM/PMMA 1:1/15 wt % plaques submitted to different precuring times.

System	Curing time at 80°C (h)	Flexural modulus (MPa)	Flexural strength (MPa)	Strain at break (%)
DGEBA:DDM/PMMA 1:1/15%	2	2410	122 [1]	15.0
	3	2670	127 [1]	14.0
	5	2750	136 [1]	11.6
	6	2760	136 [1]	12.6
	7	2750	138 [1]	10.9
DGEBA:DDM 1:1	3	2450	128 [2]	9.9

(1): At yield.

(2): At break.

Table III. Fracture toughness for DGEBA:DDM/PMMA 1:1/15 wt % plaques submitted to different precuring times.

System	Curing time at 80°C (h)	K_{Ic} (MPa m$^{1/2}$)	G_{Ic} (N/m)
DGEBA:DDM/PMMA 1:1/15%	2	1.11	470
	3	1.15	440
	5	0.95	290
	6	0.99	310
	7	1.07	360
DGEBA:DDM 1:1	3	0.88	280

Role of Rubber Particle Size Distribution In Toughness of High Impact Polystyrene (HIPS)

Mehmet Demirors
Polystyrene R&D
The Dow Chemical Co.
Midland, MI 48667
USA

1. Introduction

High Impact Polystyrene (HIPS) is probably the largest volume of rubber toughened thermoplastic. It has a unique position as a molding resin due to its balance of properties and moldability coupled with relatively low cost. As the filed of applications for this plastic increases, it becomes utilized more and more. As a consequence the demands on the product has also increased requiring a better balance of properties. In an attempt to increase the rigidity, appearance and toughness of HIPS resins, major manufacturers have been developing products that have bimodal rubber particle size (RPS) distributions. In comparison to their standard counter types, these products offer improved rigidity and appearance, while maintaining the toughness of the standard products. While impact toughening of standard rubber particle size distribution resins are extensively studied ([1-15]), there are relatively few studies that deal with bimodal RPS distribution in HIPS ([16-18]).

The role of the rubber particles in impact modification is fairly well understood ([6-8, 11, 14, 15, 19-30]). When a stress is applied to a resin with rubber particles, there is a non-uniform distribution of the stresses around the particles. At the equatorial plane the stress is concentrated upto two times the original stress while in polar regions the stress is reduced. Concentration of the stress at the equatorial plane ([11]) ensures the onset of crazing in the areas where stresses are concentrated causing local deformation. This leads to energy absorption prior to any substantial damage occurs to the bulk of the material. It has been found that optimum rubber particle size for HIPS is around 2 to 3 micron for impact toughening ([10, 13, 14, 31, 32]). Particles with sizes less than 1 micron do not generally nucleate crazes ([33, 4, 12, 34]).

While larger particles are optimum for toughness, they are usually associated with lowering of rigidity and also of appearance. On the other hand, the bimodal RPS distribution HIPS resins provide a better balance as they combine all the desirable properties of small particle HIPS, rigidity and appearance, with those of larger particle HIPS, toughness.

There has been very few articles on such bimodal rubber particle size distribution HIPS resins ([32, 17, 27, 16, 18]). Most of those articles employ a mixture of two resins of varying particle size at different weight ratios.

In the current study a complete experimentation was carried out in which both the volume fraction of larger particle size HIPS was varied as well as its rubber particle size by in situ polymerization of two feed streams and in situ solution blending of those products.

2. Experimental

This work was done on a miniplant scale, whereby a three reactor set up was employed. Both the large and small particle HIPS resins were made in situ and solution blended in the second reactor for optimum dispersion.

Small particles were made by utilizing a styrene butadiene block rubber of 40% styrene and 60% butadiene. This typically leads to a core-shell morphology([12, 16, 35]) where a thin shell of rubber encapsulates a core of polystyrene. The base product was of a very small particle size of 0.2 micron as measured by transmission electron microscopy . Larger particles are made by utilizing standard rubber. Rubber was first dissolved in styrene monomer and then polymerized in a stirred tube reactor until the polymer syrup was well beyond the phase inversion point. To this mixture was added a second polymerization mixture containing larger particles made from a standard low cis polybutadiene rubber commonly referred to as "Diene 55". After the two streams were mixed in the second reactor, the polymerization was completed and the polymer was devolatilized by flashing in a devolatilzing extruder. The particle size of the larger particles were obtained by manupilation of reactor conditions, agitation and chain transfer agent addition. The materials were granulated and all the testing were done on the granulated resin.

All the physical properties, such as tensile, notched izod impact and charpy impact were measured on injection molded specimens according to ASTM methods. An injection molding protocol was employed which yielded physical properties that were close to compression molded specimens.

Rubber particle size was measured using a Coulter Multisizer and a 50 micron tube.

Gloss measurements were done on a spiral flow plaque with sharp corners. The dimensions were 3 mm thick and 20 mm wideand the total flow path was 60 cm. The gloss measurements were made (60 degree Gardner) near the gate and at the end of the flow path reflecting the high gloss and low gloss parts of the molded specimen. Mold temperature, melt temperature, injection time were kept constant while injection pressure was used to obtain a just filled condition whereby the mold was just filled without over packing.

3. Results and Discussions

In general tensile properties followed the rule of mixtures. The tensile yield stress and tensile rupture stress of a resin was close to the volumetric average of what would have been expected from the component materials.

The notched izod impact of the base resin with 0.2 micron is fairly low, at 30 J/M. As soon as a blend with larger particles are made then there is a dramatic increase on the izod impact as shown in figure 1, for 10%, 18%, 25% and 33% large particles. As the particle size increases the izod impact reaches a maximum at around 3 to 5 microns. The highest notched izod impact of monomodal large particles are typically achieved at around 3 micron with an izod impact of around 120 J/M. Clearly most of the bimodal materials at 3 micron large particles and above have significantly higher izod impact values then the best of the large particle based monomodal HIPS resins.

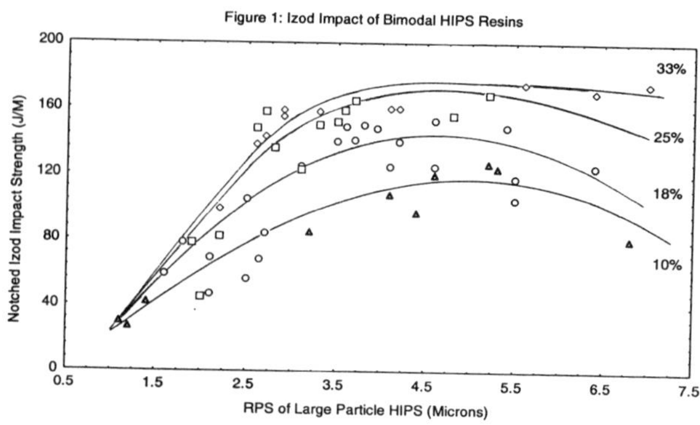

In the case of 10 and 18% large particle based bimodal resins at larger particles there is a slight drop in izod impact at 6 microns or more. As the ratio of large particles are increased the improvement in notched izod impact continues. At 33% large particles the izod impact for this type of resins are extremely high indication the synergistic behavior of such bimodal blends.

Improvement in impact strength of bimodal HIPS resins have been studied [16, 27, 32]. Systems studied basically consisted of small 0.2 micron HIPS resins blended with larger rubber particle HIPS products. The observations from these studies are in line with the current study. Bucknall [27] showed that at the same rubber phase volume fraction, these material showed similar crazing behavior and the real differences were in post yielding area. Improvements in impact strength is most likely due to the craze growth stage as well as the termination stage of the crazing process. Small particles in this study of 0.2 micron are extremely small to initiate crazing. They are also too small to stop crazes developing into cracks. Those two elements are reflected by the fact that the base products have very low izod impact strengths. In bimodal systems the primary contribution of small particles come from the fact that on the plane of the propagating craze front, the small particles reduce the yielding stress facilitating growth of crazes much faster than it would have been possible in their absence. Okamota [16] has shown through finite element analysis the stress field distribution on the plane of a craze in a bimodal system where small particles are in the stress field of larger particles. Interestingly the current study shows that the optimum improvement in impact strength is at around 3 to 5 micron range, somewhat larger than monomodal HIPS. This shows the importance of stress field overlap as the dominating parameter in toughening the bimodal systems. Larger particles extend the stress fields leading to a longer craze growth for bimodal materials. In line with previous studies, a maximum is observed at around 25 to 30% large particles in the impact behavior.
In general gloss behavior is similar to the tensile properties in that it follows a volumetric average of the components.

4. Conclusions

A series of HIPS resins with a bimodal particle size distribution has been made with a 0.2 micron base particles. As the base particles are not efficient impact improvers, addition of larger particle HIPS improves the impact strength of those resins dramatically, well above the impact strength of both resins. Studies have been conducted at four levels of large particles covering 10% to 33% by volume. The optimum particle size at which the impact toughening occurs on bimodal resins shifts to successively smaller particles as the ratio of larger particles are increased. It is believed that the improvement in impact is due to the role of stress field overlap between large and small particles. Small particles in the vicinity of larger particles and craze path improve the craze propagation process leading to improved izod impact strength. Large particles govern the craze initiation and termination stages.

5. References

1.Schmitt, J.A. *J. Polym. Sci., Part C (1970), No. 30, 437-45*
2.Bucknall, C.B. *British Plastics, 40, 84-6 (1967)* (1967).
3.Bucknall, C.B. *Adv. Poly. Sci. 27, 121-48 (1978)* (1978).
4.Bucknall, C.B. *Journal of Material Science 22*, 1341 (1987).
5.Keskkula, H. *Appl. Polym. Symp. (1970), No. 15, 51-78*
6.Keskkula, H. *Adv. Chem. Ser. (1989), 222(Rubber-Toughened Plast.), 289-99*
7.Matsuo, M., Wang, T.T. & Kwei, T.K. *J. Polym. Sci., Part A-2 (1972), 10(6), 1085-95*
8.Michler, G.H. *Acta Polymer. 44 (1993), 113-124* (1993).
9.Sauer, J.A. & Chen, C.C. *Adv. Polym. Sci. (1983), 52-53, 169-224*
10.Donald, A.M. & Kramer, E.J. *J. Mater. Sci. (1982), 17(8), 2351-8*
11.Ricco, T.P., A. and Danusso, F. *Poly. Eng. & Sci. 18 (1978), 10, 774-780* .
12.Maestrini, C., Merlotti, M., Vighi, M. & Malaguti, E. *J. Mater. Sci. (1992), 27(22), 5994-6016*
13.Boyce, M.E., Argon, A.S. & Parks, D.M. *Polymer (1987), 28(10), 1680-94*
14.Cigna, G., Matarrese, S. & Biglione, G.F. *J. Appl. Polym. Sci. (1976), 20(8), 2285-95*
15.Bubeck, R.A., Buckley, D.J., Jr., Kramer, E.J. & Brown, H.R. *J. Mater. Sci. (1991), 26(23), 6249-59*
16.Okamoto, Y., Miyagi, H., Kakugo, M. & Takahashi, K. *Macromolecules (1991), 24(20), 5639-44*
17.Mark, J.E. *Chemtracts: Macromol. Chem. (1992), 3(2), 128-31*
18.Okamoto, Y., Miyagi, H. & Mitsui, S. *Macromolecules (1993), 26(24), 6547-51*
19.Michler, G.H. *Kunststoffe 81 (6), 38-40* (1991).
20.Michler, G.H. *Acta Polym. (1985), 36(6), 325-30*
21.Keskkula, H., Schwarz, M. & Paul, D.R. *Polymer (1986), 27(2), 211-16*
22.Keskkula, H., Turley, S.G. & Boyer, R.F. *J. Appl. Polym. Sci. (1971), 15(2), 351-67*
23.Hall, R.A. & Burnstein, I. *J. Mater. Sci. (1994), 29(24), 6523-8*
24.Hall, R.A. *J. Mater. Sci. (1992), 27(22), 6029-35.*
25.Cigna, G., Lomellini, P. & Merlotti, M. *J. Appl. Polym. Sci. (1989), 37(6), 1527-40*
26.Bucknall, C.B. *Rubber Chem. Technol. (1987), 60(1), 35-4*
27.Bucknall, C.B., Cote, F.F.P. & Partridge, I.K. *J. Mater. Sci. (1986), 21(1), 301-6*
28.Bucknall, C.B., Clayton, D. & Keast, W. *J. Mater. Sci. (1973), 8(4), 514-24*
29.Bubeck, R.A., Blazy, J.A., Kramer, E.J., Buckley, D.J., Jr. & Brown, H.R. *Mater. Res. Soc. Symp. Proc. (1987), 79(Scattering, Deform., Fract. Polym.), 293-8*
30.Bubeck, R.A., Buckley, D.J., Jr., Kramer, E.J. & Brown, H.R. *Polym. Prepr. (Am. Chem. Soc., Div. Polym. Chem.) (1990), 31(2), 116-17*
31.Gilbert, D.G. & Donald, A.M. *J. Mater. Sci. (1986), 21(5), 1819-23*
32.Hobbs, S.Y. *Polym. Eng. Sci. (1986), 26(1), 74-81*
33.Donald, A.M. & Kramer, E.J. *J. Appl. Polym. Sci. (1982), 27(10), 3729-41*
34.Piorkowska, E.A., A. S.; Cohen, R.E. *Macromolecules 1990, 23, 3838-3848*
35.Echte, A.G., H. and Lutje, H. *Die Angew. Makromol. Chem. 90 (1980) 95-108* .

DESIGNING SUPER HIGH IMPACT PS/PB BULK BLENDS: OPTIMAL PARTICLE SIZE AND DIBLOCK LENGTH

D. Mathur, E. B. Nauman

The Howard P. Isermann Department of Chemical Engineering,
Rensselaer Polytechnic Institute, Troy, New York 12180

INTRODUCTION

Mixing two or more polymers to produce blends or alloys is a well established strategy for achieving desired physical properties. The most common blending techniques are mechanical, solution, polymerization, and reactive. Blends in this paper have been produced by compositional quenching (CQ), which is a form of solution blending.[1]

The amount of block required to saturate the interface and block length for maximum interfacial adhesion have been the subject of various studies.[2,3] Recent studies show that the reinforcement at the interface depends on the copolymer areal chain density at the interface, Σ, the molecular weight, M, of the block and its relation to the entanglement molecular weight M_e.[4,5] Three fracture mechanisms were proposed:

1) $M_A < M_{eA}$ or $M_B < M_{eB}$. The fracture toughness increased linearly with Σ. Copolymer chains are pulled out.
2) $M_A > M_{eA}$ and $M_B > M_{eB}$ and low Σ. The fracture toughness increases linearly with Σ. Copolymer chains fail by scission.
3) $M_A > M_{eA}$ and $M_B > M_{eB}$ and high Σ. The fracture toughness scales as Σ^2 and is greatly enhanced. There is large scale plastic deformation of the interface followed by chain scission, disentanglement or both.

The impact strength of rubber toughened styrene polymers increases with rubber phase volume. The effects of particle size are not well known except that an optimal exists between 0.8-4μm.[6,7,8] There is conflicting literature on the effects of particle morphology (occluded or solid) and adhesion level.

Bulk blends of polystyrene (PS), polybutadiene (PB), and their respective diblock copolymer (PS-PB), have been produced. The effects of diblock molecular weight, and minor phase domain size on impact strength have been determined. The emphasis is on independent control of particle size and compatibilization using CQ as a unique method to produce blends.

EXPERIMENTAL

Table 1 gives properties and sources for the starting polymers. Xylene purchased from Ashland Chemical Company was the solvent. An antioxidant, 0.1% Irganox® 1010, was added.

A 5% polymer solution was prepared in xylene. This single phase solution was heated to 220-240°C, and quenched at an absolute pressure of 5 Torr to remove the solvent. The depth of quench, dictated by the pre-flash temperature and the flash pressure, determined the particle size. Spherical particles, having a size of ~0.3μ, were obtained for the polybutadiene with the diblock at the PS & PB homopolymer interface.[9,10,11] The resulting blend was dried by further devolatilization and ground to a powder. The particle size and concentration of the diblock at the interface was then varied by ripening at 200°C for various times in a Carver press.

Izod bars obtained from the Carver press were notched and then tested with a 3 lb. hammer.[12] The broken surface was scanned in an SEM after a 100Å layer of gold was sputtered on the surface. The molded blends were cut, shaped to form a sharp pyramid, and stained in 4% aqueous osmium tetraoxide.[13] Thin sections (approx. 800 Å) were cut using a Reichert-Jung UltracutE Ultramicrotome with a diamond knife. The morphology was observed under a 100KV electron beam using a Philips CM12 Transmission Electron Microscope. The observed particle size was corrected using histogram transformation.[14] Bars were also tested for modulus using an Instron 4204 tensile testing machine with a 0.5in. extensometer attachment.

RESULTS AND DISCUSSION

The impact strength results obtained were compared against three different standards processed under the same conditions, PS, PS/PB, and HIPS. The rubber phase for the HIPS sample was 23% by volume. Young's modulus for the PS and HIPS was 3.73 and 2.18 GPa respectively. The blends reported in this paper had an overall composition of 23% which included the PB contribution of the diblock. The initial particle size of ~0.3μ was sub-optimal with respect to impact strength. Thus, blends were ripened to study the optimal particle size for impact strength.

The impact strengths for various diblock blends are shown in Figure 1 as a function of ripening time. The initial increase in impact strength is due to increasing Σ with the increasing particle diameter. This leads to increasing interfacial adhesion and higher impact. The impact strength eventually goes down because the particle size becomes larger than the optimal. The best performing system was PS/PB/730A (74.5/16.4/9.1) having an impact strength of 6.2 ft-lbf./in compared to 1.5 ft.-lbf./in for a typical HIPS. The modulus was 2.21 GPa, the same as HIPS. This combination of impact strength and modulus exceeds that of commercial, super high impact polystyrene, (e.g. Huntsman 850 has an Izod of 4.2 ft-lbf./in. with a modulus of 1.96 GPa).

Since both block segment lengths for 730A are greater than M_e, 30,000 for PS and 5,900 for PB,[15] adequate adhesion was expected. However, a lower maximum strength of 5.4 ft.-lbf./in. was obtained for PS/PB/E142 (71.6/20.3/8.1) even though the PB block length remained substantially higher than M_e. This suggests that the classic viscosity-based measurement of M_e does not provide an adequate measure of entanglements in bulk. However, the value for M_{ePS} was found to adequate. With a low molecular weight diblock, S183, poor impact strengths were obtained for the PS/PB/S183 (76.2/15.0/8.8) blend as expected.[15] The maximum impact strength for the PS/PB/F411 (74.5/16.4/9.1) blend was 4.2 ft.-lbf./in. This is attributed to the very high molecular weight of the diblock. Stearic effects limit Σ to values lower than those in the 730A blends.

The blends containing 730A as the diblock were analyzed for particle size. Figure 2 shows an optimal particle size of ~0.95μ. Micelle formation was observed at a particle size of 0.6μ.[16] Thus, the optimum in Figure 2 is a particle size effect since the surface is saturated with respect to the diblock. At sizes less than about 0.6μ, the surface may not be saturated and thus our results may be confounded between particle size and surface coverage effects.

Figures 3(a) and 3(b) show the fracture surface for the blend ripened for 3 and 180 minutes with an impact strength of 3.0 and 6.2 ft.-lbf./in respectively. Since the 730A block has high molecular weights, the mechanism of surface reinforcement shifts from chain scission to plastic deformation due to increasing Σ.[4] Figure 3(a) shows a fairly brittle fracture with small amount of plastic deformation with an areal chain density of 0.1 chains/nm^2. The holes correspond to the rubber particles. In contrast, Figure 3(b) shows a large amount of plastic deformation due to a higher surface coverage of 0.16 chains/nm^2, which is the saturation concentration.

CONCLUSIONS

This study clearly brings out the effect of diblock, and particle size on impact strength in bulk blends.

1. Diblock copolymers provide good interfacial adhesion. Reasonably high impact strengths are obtained when the block lengths are longer than the entanglement length. In fact, all the strengths were higher than

commercial HIPS (>1.5 ft.-lbf./in.), except S183. Super high impact strengths are obtained when $M_{PB} \gg M_e$.

2. The optimal particle size with 730A diblock copolymer is ~0.95μ for maximum impact strength. This is much smaller than the optimal size of ~2μ commonly cited for HIPS. The difference is likely due to the lower modulus of the blended rubber particles compared to the highly occluded particles typical of HIPS.

3. There is an optimal molecular weight for the block in a diblock copolymer. The blocks must be long enough to entangle but not so long as to lower the areal chain density due to stearic effects. For PB at least, the optimal block length is much longer than M_e as determined by solution viscosity measurements.

Table 1: Properties of Materials Used

Material	Manufacturer	Mw	PS Block (M_n)	PB Block (M_n)	M_w/M_n
1.Polystyrene GP Polystyrene	Novacor Chemical	202,00	-	-	2.2
2.Polybutadiene Diene® 55 NF	Firestone	320,000	-	-	2.4
3.HIPS Styron® 484	Dow Chemical	Close to PS	-	-	Close to PS
4.PS-PB Diblocks A.Europrene® Sol S183	Enichem	109,000	5,000	90,000	1.1
B.Stereon®730A	Firestone	154,000	35,000	113,000	1.1
C.Europrene® Sol E142	Enichem	90,000	35,000	42,000	1.1
D.Fina® 411	Fina Oil & Chemical	246,000	74,000	172,000	1.1

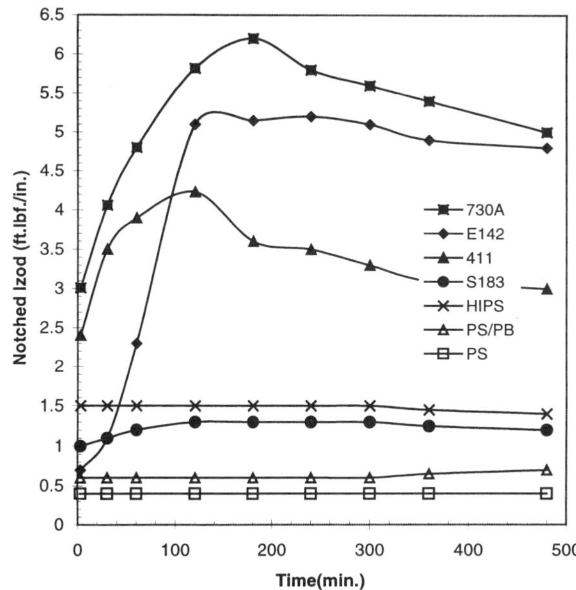

Figure 1: Impact Strength as a function of ripening time for various diblock copolymers at a temperature of 200°C

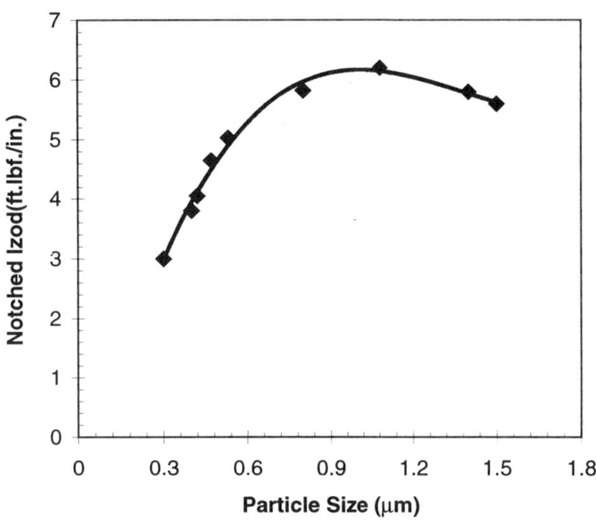

Figure 2: Effect of Particle Size on Impact Strength for PS/PB/730A system

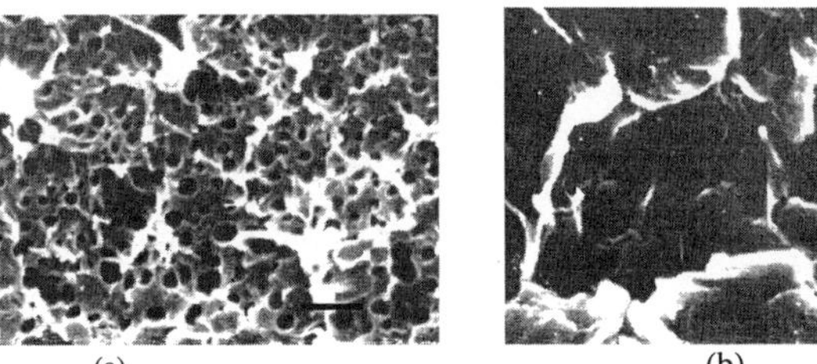

(a)   (b)

Figure 3: SEM of a fracture surface at 25°C for PS/PB/730A. Sample molded at 200°C, (a) t=3 min.; (b) t=180 min. Scale bar is 1μm.

REFERENCES

1. Nauman, E. B., et al., *Chem. Eng. Comm.*, **66**, 29 (1988)
2. Manson, J.A., and Sperling, L.H., *"Polymer Blends and Composites"*, Plenum Press, New York, (1976)
3. Riess, G, Jolivet, Y., in " *Copolymers, Polyblends, and Composites"* (N.A.J. Platzer, ed.), Advances in Chem. Ser., Vol 142, American Chemical Society, Washington D.C., (1975)
4. Crenton, C., Kramer, E. J., Hui, C.-Y., Brown, H. R., *Macromol.*, **25**, 3075 (1992)
5. Crenton, C., Kramer, E. J., Hadziioannou, G., *Macromol*, **24**, 1846 (1991)
6. Bucknall, C. B., *Toughened Plastics*, Applied Science, London, (1977)
7. Donald, A. M., Kramer, E. J., *J. Mat. Sci*, **17**, 2351 (1982) ; *J. Appl. Poly. Sci.*, **27**, 3729 (1982)
8. Hobbs, S. Y., *Poly. Eng. Sci.*, **26**, 74 (1986)
9. Nauman, E. B; He, D. Q., *Polymer*, **35**, 2243 (1994)
10. Brown, H. R., *Macromolecules*, **22**, 2859 (1989)
11. Shull, H. R., Kramer, E. J., *Macromolecules*, **23**, 4769 (1990)
12. Annual Book of ASTM Standards, 8.01, **D-256**, (1996)
13. Kato, K. *J. Electron. Microscopy*, **14(3)**, 219 & 220 (1965)
14. Goldsmith, P.L., *J. Appl. Phys.*, **18**, 813 (1967)
15. Porter, R. S., Johnson, J. F., *Chem . Rev.*, **66**, 1 (1966)
16. Mathur, D.; Hariharan, R.; Nauman, E. B., Submitted (1998)

Strengthening and Toughening of Polymeric Materials *via* Controlled Molecular Orientation

Chris K.-Y. Li and H.-J. Sue

Polymer Technology Center
Department of Mechanical Engineering
Texas A&M University
College Station, TX

INTRODUCTION

It is intuitively clear that if one can control molecular orientation in polymeric systems during the fabrication process, the physical and mechanical properties of the polymers can be greatly improved in the orientation(s) of interest. Unfortunately, the conventional polymer processing techniques such as extrusion and injection molding cannot produce a controlled, uniform molecular orientation across the thickness of the fabricated part. Only when the fabricated part is in fiber or in thin film forms, which are not useful for structural applications, will the molecular orientation be controlled in the extrusion/drawing direction. Therefore, no significant progress is reported in strengthening and toughening of polymers *via* the control of molecular orientation.

The present paper intends to utilize a new solid state extrusion process, termed the equal channel angular extrusion (ECAE) process [1-4], to control molecular orientation in polymers. The ECAE process is a simple, yet effective extrusion process to impose a uniform through-thickness simple shear deformation on polymers without altering the cross-section dimensions of the extrudate (Figure 1) [5-8]. To demonstrate the effectiveness of ECAE in altering microstructure, crystallinity, molecular orientation, various model polymers, which include linear low-density polyethylene (LLDPE) and polycarbonate (PC), were chosen for the present work. Mechanical properties, microstructural characteristics, and failure mechanisms in the model polymers, before and after ECAE, were investigated.

EXPERIMENTAL

The ECAE Process

A simple extrusion die was designed and built to extrude prefabricated polymer plaques with dimensions of 152.4mm × 152.4mm × 9.5mm (Fig. 1). The extrusion was conducted at various temperatures and rates depending on the glass transition temperature (T_g) and/or melting temperature (T_m) of the polymer extrudate. The processing conditions and the basic characteristics of the model polymers used in this research are listed in Table I.

A servo hydraulic driven mechanical test machine (MTS 810 Model 312.31) was utilized to push the sample through the ECAE die. The extrusion load and stroke data were recorded and converted into true stress-true strain data needed for polymer extrudate to pass through the ECAE die.

Microscopy and Spectroscopy

Optical microscopy was conducted using an Olympus BX60F optical microscope under transmission mode (TOM). Transmission electron microscopy (TEM) was performed using a Zeiss-10C transmission electron microscope operated at 80 kV. The LLDPE sample preparation for TEM observation was performed by exposing samples to ruthenium tetraoxide (RuO_4) vapor for approximate 2 hours at ambient temperature. Thin sections were then obtained using a Reichert Ultracut-E ultramicrotome at ambient temperature.

Wide-angle X-ray diffraction (WAXS) spectra were recorded on a Rigaku 200 X-ray diffractometer using CuK_α radiation (λ = 1.5406Å) between 5° and 55° with 0.04° (2θ) scan increments.

Fracture Toughness Measurement

The single-edge-notch three-point-bend (SEN-3PB) methodology (ASTM Standard D5045-96) was adopted to measure the plane-strain fracture toughness of PC sample. A screw-driven mechanical testing machine (Instron, Model 4411) was used to perform the SEN-3PB tests at a crosshead speed of 10 mm/min.

RESULTS AND DISCUSSION

LLDPE System

Figure 2 shows the effectiveness of ECAE to transform the banded spherulites of LLDPE into highly aligned lamellar bundles. The spherulitic structure at 0-pass (Fig. 2a) is disintegrated into an intermediate structure after 2-pass, followed by fibrillar structure formation after 4-pass (Fig. 2b). To confirm that the aligned string-like birefringent lamellar bundles are indeed lamellar bundles, TEM investigation was conducted. As shown in Figure 3, nearly parallel, well-stretched lamellae were found in the 4-pass LLDPE sample. Improvements in physical and mechanical properties of LLDPE are expected.

WAXS experiment was performed to detect whether or not the ECAE simple shear process causes any structural change in LLDPE. As shown in Figure 4, a monoclinic peak (010) emerges at the expenses of decreasing intensities of orthorhombic characteristic peaks i.e., (110) and (200). This suggests that partial crystalline structure transformation has taken place [9,10].

PC System

A significant improvement in fracture toughness is found in PC system after ECAE. The fracture toughness is improved by as much as 80% in the longitudinal direction (i.e., the crack propagates perpendicular to the extrusion direction) after a single ECAE pass (Fig. 5). However, when the crack propagates along the extrusion direction (i.e., the transverse direction), improvements in fracture toughness can only be found at low extrusion rates.

Fracture mechanism investigation was conducted to learn how the fracture toughness is enhanced [11]. As shown in Figure 6, the crack appears to propagate along the molecular orientation direction, which is about 26° away from the extrusion direction. As a result, hackling is formed.

SUMMARY

The effectiveness of the novel ECAE process in controlling molecular and microstructural orientation was demonstrated using LLDPE and PC. The results clearly indicate that the ECAE process is effective in altering the morphologies, both at the micrometer and at molecular levels, yielding significant improvement in physical and mechanical properties. It is thus conceivable that optimized morphology and mechanical property in polymeric systems can be achieved if the ECAE process is implemented.

ACKNOWLEDEMENTS

The authors would like to thank Drs. K.T. Hartwig and R.E. Goforth for their insightful discussion and experimental assistance with this work. The research grants from Dow Chemical and the State of Texas (Grant# ARP-98-32191-72380) are greatly appreciated.

REFERENCES

1. S. Middleman, *"Fundamentals of Polymer Processing"*, McGraw-Hill, New York, 1977.
2. R.G. Griskey, *"Polymer Process Engineering"*, Thomson Publishing, New York, 1995.
3. L.C. Sawyer and D.T. Grubb, *"Polymer Microscopy"*, Chapman and Hall, London, 1987.
4. A. Birley, B. Haworth, and J. Batchelor, *"Physics of Plastics"*, Hanser Pub., New York, 1992.
5. V. M. Segal, Invention Certification of the USSR, No. 575892, 1977.
6. V.M. Segal, K.T. Hartwig and R.E. Goforth, *Mater. Sci. Eng.*, **A224** (1997) 107.
7. S. Ferrasse, V.M. Segal, K.T. Hartwig and R.E. Goforth, *Metall. Mater. Trans. A* **28** (1997) 1047.
8. S. Ferrasse, V.M. Segal, K.T. Hartwig and R.E. Goforth, *J. Mater. Res.*, **12** (1997) 1253.
9. H. Kiho, A. Peterlin and P. H. Geil, *J. Appl. Phys.* **35** (1965) 1599.
10. K. Tanaka, T. Seto and T. Hara, *J. Phys. Soc. Jpn.* **17** (1962) 873.
11. W. Brostow and R.D. Corneliussen, *"Failure of Plastics"*, Hanser Pub., New York, 1986.
12. J. Spruiell, D. McCord, and R. Beuerlein, *Trans. Soc. Rheol.*, **16** (1972) 535.
13. S. A. Jabarin, *Polym. Eng. Sci.*, **31** (1991) 11.
14. A. Misra and R.S. Stein, *J. Polym. Sci., Polym. Phys. Ed.*, **17** (1979) 235.

Table I. Materials used for ECAE process evaluation.

Material	Semicrystalline or Amorphous	Crystallinity (weight %)	Density (g/cc)	T_g (°C)	Processing Temperature (°C)
LLDPE, EXXON ESCORENE® 6047	Semi-crystalline	40	0.92	-130 – -125	Room Temperature
PC, GE LEXAN® 103	Amorphous	-	1.20	140-150	100

*The PC plates were annealed to remove the pre-existing residual stresses.

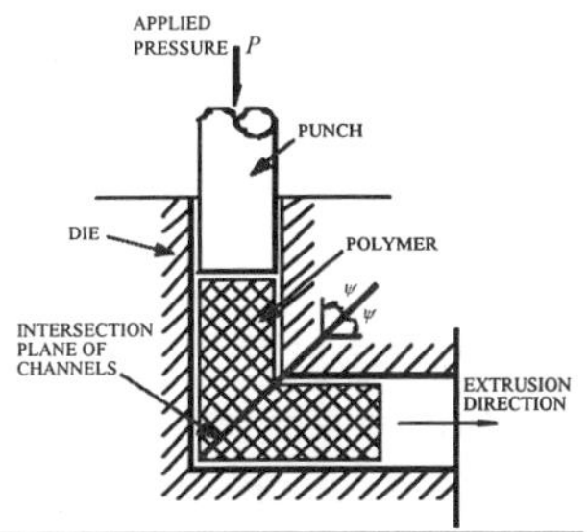

Figure 1. Schematics of the ECAE process.

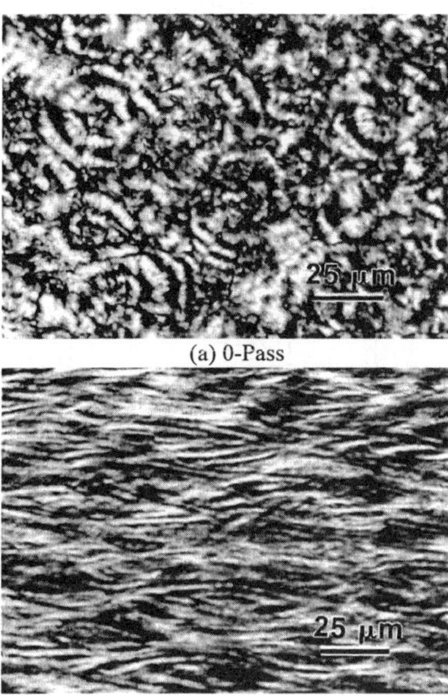

(a) 0-Pass

(b) 4-Pass

Figure 2. TOM of the ECAE-processed LLDPE under crossed-polars. (a) 0-pass, and (b) 4-pass. The arrow indicates the extrusion direction.

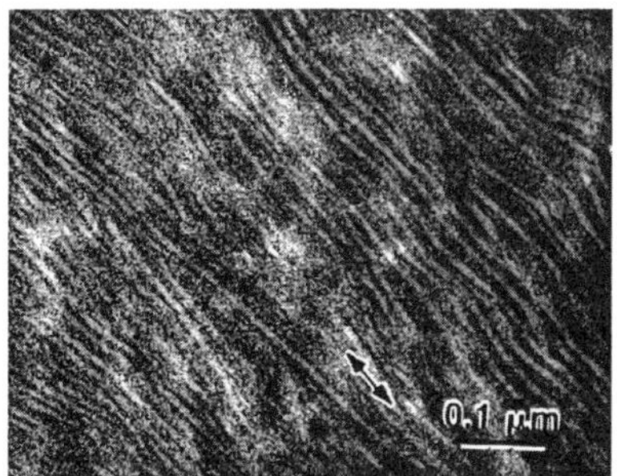

Figure 3. TEM of LLDPE which has been process for four ECAE passes. The arrow indicates the extrusion direction.

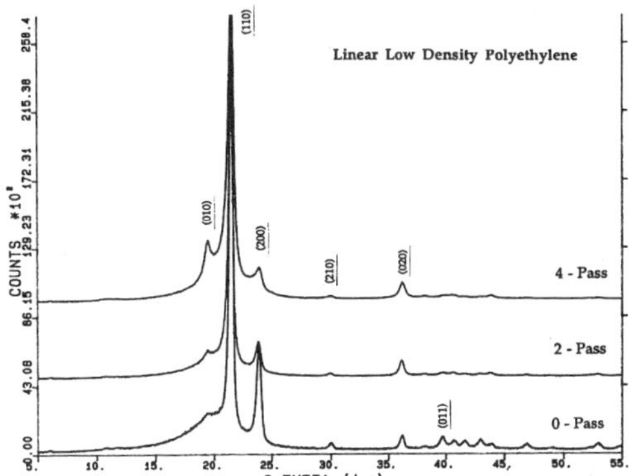

Figure 4. WAXS spectra of the ECAE-processed LLDPE.

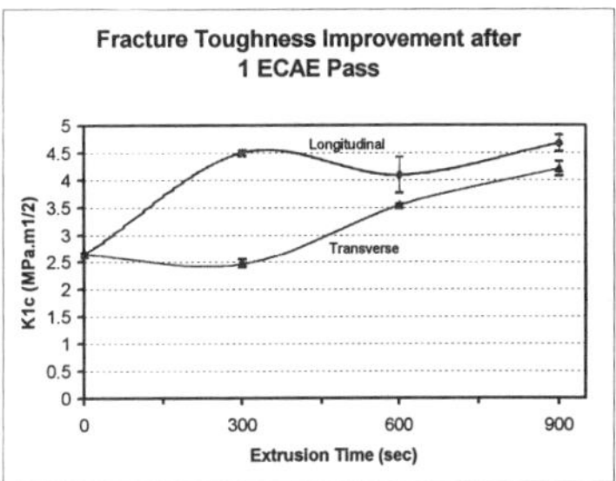

Figure 5. Fracture toughness data of the ECAE-processed PC at varied extrusion rates.

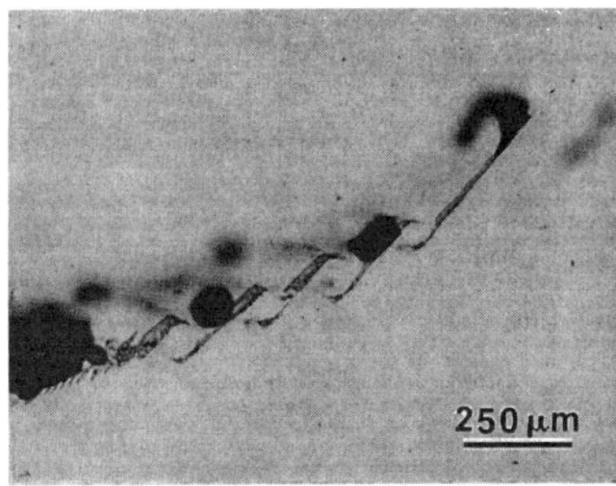

Figure 6. OM observation shows hackles propagate along the shear plane.

NEW SEMI-IPN BLEND SYSTEM OF BISMALEIMIDE

David J.T. Hill, Peter Pomery,
Andrew K. Whittaker[*], Zhiyu Xia

Department of Chemistry
*Centre for Magnetic Resonance
The University of Queensland,
St. Lucia, Qld4072, AUSTRALIA

INTRODUCTION

Bismaleimide (BMI) is a leading class of thermosetting polyimides and it has been of particular interest to the research and development in advanced composites because of its outstanding stability at elevated temperatures and retention of properties in hot/wet environments. Nevertheless, the practical application of BMI has been greatly limited by its inherent brittleness as a result of its high crosslink density. In order to overcome this disadvantage, some recognized methods such as chain extension copolymerization with diamine or diallyl compounds have been explored with considerable success in toughening BMI, but unfortunately at the price of its thermo-oxidative stability in most cases. A relatively new and effective technique for toughening is through blending with high performance thermoplastics which will bring their outstanding thermal and mechanical features into the BMI network if the blending process is appropriately controlled. In this project, PES-C, a modified poly(phenylene ether sulfone), is used to make the blends with BMI, with the purpose of taking advantage of PES-C's excellent thermomechanical properties as well as its good solubility. Various instrumental analysis methods have been utilized in studying the curing process and the thermal mechanical behaviors of the blends. The effect of UV radiation on the curing process has also been investigated.

EXPERIMENTAL

4,4'-bismaleimidodiphenylmethane was purchased from Aldrich Chemical Co. and purified by silica gel column chromatography followed by precipitation. Solution blending of BMI and PES-C was carried out in chloroform in a freeze drying process. A Bruker MSL-200 solid state NMR and optical microscopy (NIKON OPTIPHOT) were used to study the level of miscibility in the blends. A Perkin-Elmer thermoanalyser series (Mode 7.0) including DSC, TGA and DMA was used for studying the resin thermal behaviour. Specimens for three-point-bending test were prepared through compression molding and tested on an Instron 4505 in accordance with ASTM D5045. Morphological analysis of the fracture surface was carried out by SEM (JEOL JSM 890). UV aided curing was performed isothermally under nitrogen in a special heating block with UV cut-off filter (>310nm). An ORIEL Hg-Xe UV lamp (1,000w) was used as the UV source.

RESULTS AND DISCUSSION

Due to the modifying side groups in the backbone structure (Figure 1), PES-C can be easily dissolved in common chlorinated solvent which aids the process of blending. DSC and TGA results have shown an outstanding thermostability of PES-C with its Tg as high as 251°C. PES-C does not participate directly in the crosslinking reaction of the BMI monomer, and the structure of semi-Interpenetrating network (Semi-IPN) is formed upon curing.

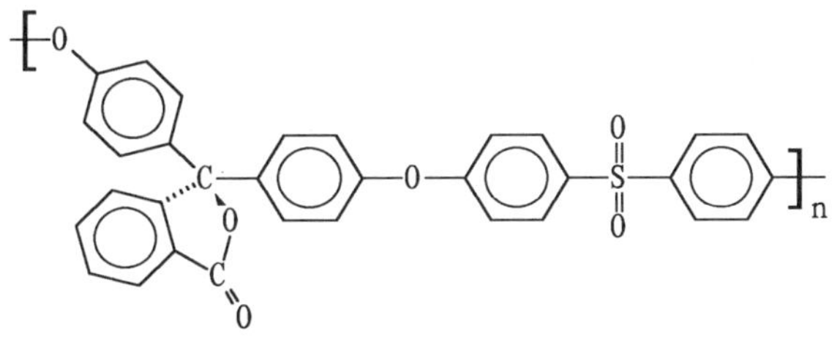

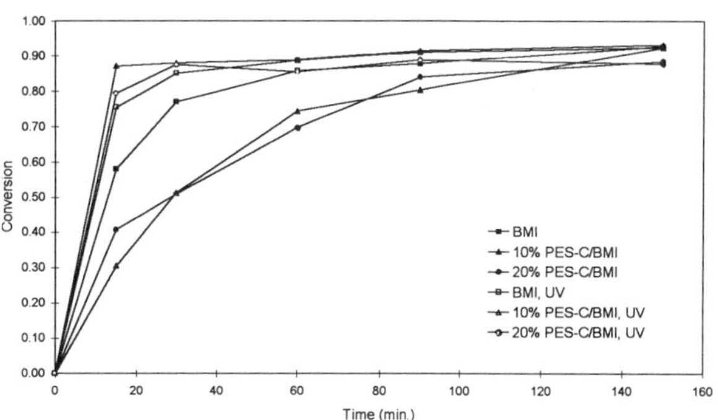

Figure 1. Molecular structure of PES-C

In order that a tailored Semi-IPN structure can be formed in the final material and thus an effective toughening of BMI can be acquired, two important steps in the preparation of the blends are worth special attention, i.e., the mixing process and the curing profile. In this connection, a freeze drying process has been chosen in an attempt to promote a good distribution of PES-C in the BMI and an intimate mixing, along with a high temperature curing profile for the fast vitrification of the matrix. DSC, optical microscopy and NMR results have confirmed a fairly intimate mixing in the blends. Investigation of the curing process using dynamic and isothermal DSC has shown that PES-C hinders the crosslinking reaction of the BMI and consequently decreases the crosslink density of the materials (Figure 2). The heat resistance properties of the material, however, have not been impaired as revealed by the thermoanalysis tests. Cure kinetics have also been examined. SEM has shown a micro-phase separation which is one of the toughening mechanisms (Figure 3). Preliminary results from mechanical test indicate a promising improvement in toughness.

UV-radiation has been tried on this new blend system so as to accelerate the free radical reaction in crosslinking. An enhancement in conversion has been observed by DSC analysis on samples cured under UV radiation. Of particular interest is the greater effect observed in the blends (Figure 2). It is believed that the PES-C acts as a photosensitizer and effectively transfers the absorbed energy to initiate the BMI through triplet state formation.

Figure 2. Cure conversion for BMI, 10% & 20% PES-C/BMI (T=220°C)

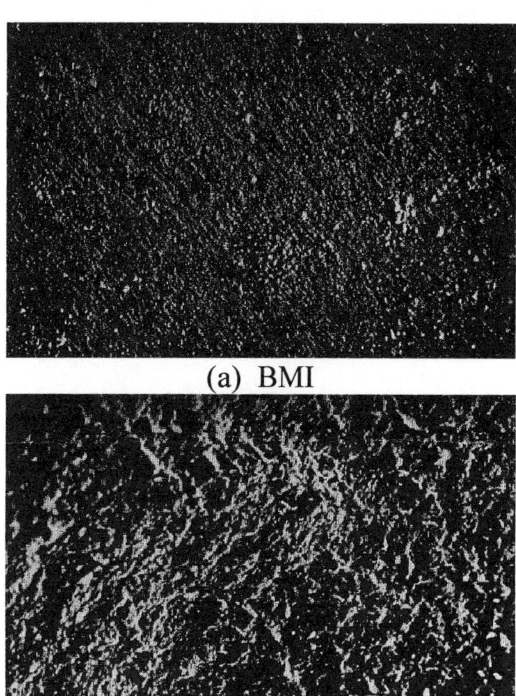

(a) BMI

(b) 10%PES-C/BMI

Figure 3. Scanning electron micrographs of fracture surface

ACKNOWLEDGMENT

Grateful acknowledgment is made to the Overseas Postgraduate Research Scholarship (OPRS) from Australian government and The University of Queensland Postgraduate Research Scholarship (UQPRS). The authors also wish to express their gratitude to Dr. Andy Goodwin of Monash University for the supply of material and helpful discussions.

SEMI-CRYSTALLINE BLENDS: ENHANCING LOW TEMPERATURE MECHANICAL PROPERTIES. K. A. Chaffin, F. S. Bates, University of Minnesota, Minneapolis, Minnesota 55455, P. Brant, Exxon Chemical Company, Baytown Polymers Center, Baytown, Texas 77520

The mechanical properties of semi-crystalline polyolefin blends are enhanced. Varying the blend constituents, in addition to introducing a polyolefin block copolymer, allows the predominate failure mechanism to be deduced. Using a combination of experimental techniques, including TEM, SAXS, and DSC, we demonstrate the importance of the initial crystalline structure to the overall mechanical properties. Upon optimizing this initial structure, injection molded specimens exhibit ductile failure below the glass transition temperature of the matrix phase.

**Dynamic Dielectric and Rheological Detection of Phase Separation
and Reaction Advancement in PPE/DGEBA-MCDEA Blends**

D. Kranbuehl, J. Rogozinski, A. Gilchriest and C. Ambler
Departments of Chemistry and Applied Science
College of William and Mary
Williamsburg, VA 23185
S. Poncet, J. P. Pascault, H. Sautereau, G. Boiteux,
J. F. Gerard, G. Seytre
INSA-CNRS
University "Claude Bernard" Lyon I
Villeurbanne, FRANCE

INTRODUCTION

The development of high temperature, high performance polymeric materials presents several major challenges. Besides possessing a high glass transition temperature the polymer must have the toughness of a thermoplastic and the strength of a thermoset. In addition the polymeric material should have an initial processable viscosity at an acceptable temperature. One of the most promising approaches to achieving these often opposing material properties is through thermoplastic-thermoset blends. In order to monitor the fabrication process of thermoplastic-thermoset blends, it is necessary to detect and characterize both the phase separation process and the reaction advancement.

This paper discusses the use of dynamic mechanical and frequency dependent dielectric measurements to monitor the time of phase separation, and buildup in Tg. The dielectric measurements can be made both in the laboratory as well as insitu in the fabrication tool in the production environment. The systems studied are 30%, 45%, and 60% PPE/DGEBA-MCDEA blends. It is observed in the dielectric measurements that the MWS charge polarization is sensitive to the onset of phase separation before the mechanical measurements. The dielectric conductivity is related to the viscosity buildup of the PPE continuous phase. The dielectric dipolar relaxation peak occurs at long times, monitoring vitrification of the $\propto$-relaxation peak in the epoxy phase. On the other hand, the mechanical relaxation peak monitors the Tg of the PPE continuous phase as epoxy diffuses out.

EXPERIMENTAL

Mixtures of 30%, 45%, and 60% poly (2,6-dimethyl-1, 4-phenylene ether) PPE distributed by General Electric as Ultem PPE800 and PPE820 with Mn=12000, MW = 25,000 g/mol were blended with the epoxy diglycidylether of bisphenol A, DGEBA, from DOW, DER 332, 348.5g/mol and a stoichiometric amount of 4,4-methylene bis (3-chloro-2, 6 diethy/aniline), MCDEA, 380 g/mol, sold by LONZA.

The DGEBA and the MCDEA were mixed at 80°C using an IKA dual blade mixer. The epoxy-amine mixture was introduced into a bi-screw corotative extruder from Clextral and extruded at 60°C. After extrusion it was determined the PPE/DGEBA-amine blend advancement to between .03 and .07.

Dynamic mechanical measurements were made using a Rheometrics RDA II in both laboratories. Frequency dependent dielectric measurements were made using either a Hewlett Packard 4192A Impedance Analyzer or a Schlumberger 1620 Impedance Analyzer at frequencies between 5 hZ and 1 MHZ. Measurements were made by imbedding planar interdigitated DekDyne sensors between layers of the PPE/DGEBA-MCDEA.

RESULTS AND DISCUSSION

Figures 1-2 show the dynamic mechanical results during at 10 rad/sec, strain < 3%, using 25 mm plate, 1.5 mm gap at 175°C for the 45% PPE blend. The values of G', G", δ and η all show distinct changes with time. First at 35 minutes, there is a sharp rise which is attributed to the onset of phase separation. It is known for PPE-epoxy blends that in mixtures of 25% or greater PPE, phase separation-inversion occurs where the PPE is the continuous phase (1-3). This results in the continuous PPE

phase becoming richer in PPE, less plasticized and the viscosity rises rapidly. An important practical consequence of this effect is that by blending a reactive epoxy with PPE the initial processing temperature for flow of this otherwise high viscosity high performance thermoplastic is significantly reduced below its Tg of 210°C. A cross over in G' and G" occurs near 3500 seconds, 58 minutes. The onset of a PPE vitrification peak for 10 rad/sec is observed in δ at 70 minutes and its peak at 86 minutes.

Mechanical measurements are difficult to make in a processing tool environment. Furthermore they reflect macroscopic force-displacement changes. Thus it is of interest to examine the ability of dielectric, electric field molecular force-displacement measurements of ions and dipoles to detect the phase separation and advancement of cure in this reactive epoxy, thermoplastic system.

Figures 3 and 4 display the dielectric insitu sensor measurements of $\log[\omega\epsilon''(\omega)]$ and ϵ'. Both ϵ'' and ϵ' show a sharp rise due to a Maxwell-Wagner-Sillars peak at 33 minutes. This peak is due to the onset of phase separation of occluded regions of a more conductive epoxy-amine surrounded by the much less conductive PPE continuous phase. As the epoxy-amine reaction progresses and its viscosity increases the occluded phase decreases in conductivity.

Corresponding the MWS relaxation time becomes longer, moving to lower frequencies. After 100 minutes the MWS effect is no longer observed at the higher frequencies; after 200 minutes at the lower frequencies. In the 100 to 300 minute region, a second series of peaks are observed due to the $\propto$-relaxation, onset of the glass transition in the epoxy phase. These dielectric $\propto$ relaxation peaks occur at longer times, 130 to 240 minutes than the mechanical 85 minute, 100 rad/sec mechanical $\propto$ peak. The dielectric peak is close to the cure kinetics of the $\propto$-relaxation vitrification peak of the pure DGEBA-MCDEA. The mechanical peak reflects both the advancement of the epoxy reaction and primarily the increasing viscosity of the PPE as epoxy-amine diffuses out of the PPE continuous phase. Additional studies were done at 150°, 200° and 225° and on 30% and 60% PPE blends. The results are similar.

CONCLUSIONS

Both mechanical and dielectric data are good experimental techniques to monitor the onset of phase separation and the reaction advancement in thermoset-thermoplastic 2 phase systems. The dielectric data detects the onset of phase separation at a time sooner than mechanical measurements and tends to monitor the onset of the glass transition of the occluded epoxy-amine phase while mechanical measurements are sensitive to the onset of the glass transition of the continuous thermoplastic phase.

REFERENCES

1. S. Ponet, Theses "Transformation of PPE with the Aid of a Reactive Solvent and an Extruder" INSA, Univ. Lyon I, 1996.
2. R, Venderbosch, H. Meijer and P. Lemstra, Polymer 35(4) 4349 (1994).
3. H. Eklind, F. Maurer, P. Stecman, Polymer 38(5) 1997.
4. B. Lestriez, A. Maasous, J. Gerard, S. Sautereau, G. Boiteux, G. Saytre, D. Kranbuehl, Polymer (in print 1998).

ACKNOWLEDGMENT: Support from CNRS in France, the NSF Foundation INT 9726207 and NSF Center of Excellence at VPI MR 912004.

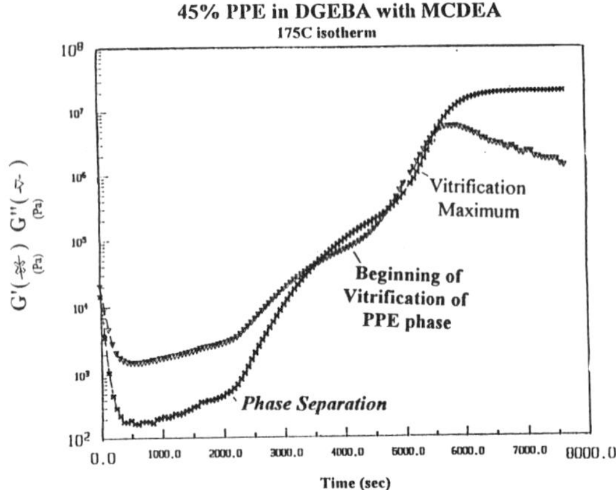

Figure 1

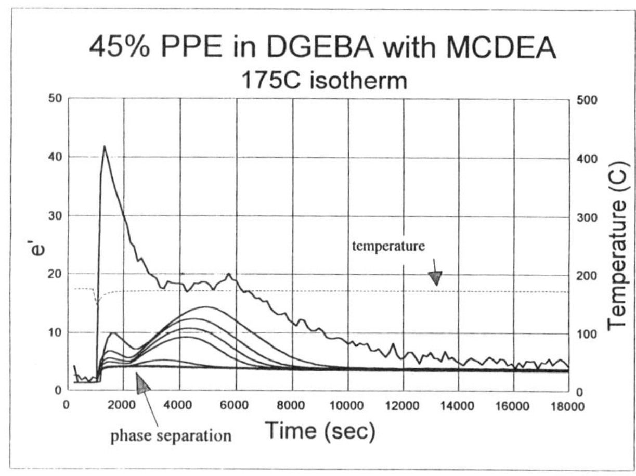

Figure 4

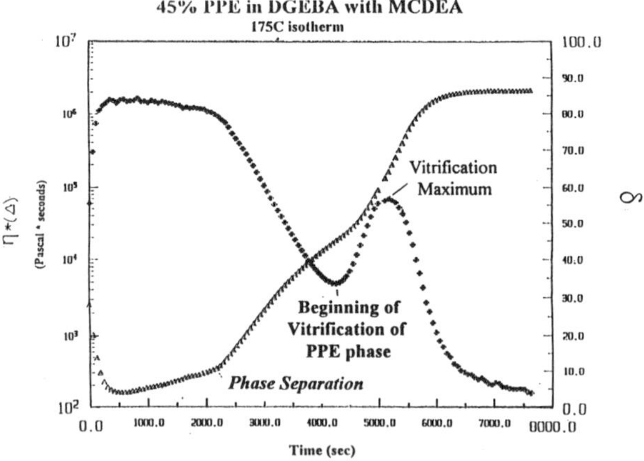

Figure 2

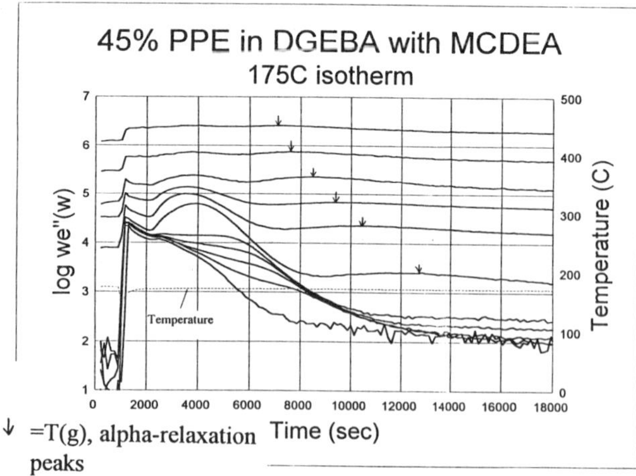

↓ =T(g), alpha-relaxation peaks

Figure 3

MONTE CARLO SIMULATIONS TO PREDICT THE BEHAVIOR OF POLYMER BLENDS CONTAINING THE BLOCK COPOLYMERS

Senthil Kandasamy and E. Bruce Nauman

The Howard P. Isermann Department of Chemical Engineering,
Rensselaer Polytechnic Institute, Troy, NY, 12180-3590

INTRODUCTION

Recently, a fast Ising- type model was used to predict the rate of domain growth in an amphiphilic system[1]. In that work, 2-D Monte Carlo simulations were performed on a ternary, non-critical polymer blend with varying concentrations of block copolymer. The model consisted of A and B homopolymers occupying single sites of the lattice, while the block copolymers occupied two adjacent lattice sites. Molecules of type A were assigned a spin of +1 and molecules of type B had a spin of -1. Spins were exchanged following a simple algorithm. The sites were chosen at random and the probability of spin exchange was given by

$$P_{exchange} = \frac{e^{-(\Delta H/RT_q)}}{1 + e^{-(\Delta H/RT_q)}}$$

where T_q was the final temperature to which the system was quenched and ΔH was the change in free energy of the chosen sites. The model, though simpler than the commonly used coarse-grained models[2,3] to describe polymer blends, predicted scaling exponents for domain growth that were in reasonable agreement with experiments. Though not able to capture the intricacies of polymer chain motion, the model provided a computationally cheap alternative for predicting certain properties of binary blends with a 50/50 block copolymer.

Reactions occuring during phase separation have theoretical and commercial interest. Recent theoretical contributions have shown that chemical reactions dramatically affect domain growth in phase separating mixtures[4-8]. The relative rates of reaction and diffusion control the domain sizes.

Effective ways to reduce interfacial tension in binary polymer blends have been to introduce a block copolymer as a third component . The block copolymers promote mixing by reducing the interfacial tension and the particle coarsening rate. Typically, the block copolymer is added to the blend as a third component . But some polymers with specific end functional groups are capable of forming the block copolymers in-situ. There have been recent theoretical and computational attempts to model the kinetics of reactions at polymeric interfaces[7,8].

SIMULATIONS WITHOUT REACTION

In this study, 2-D simulations were run at high concentrations of block copolymer to observe the resulting morphologies on a 128 × 128 square lattice. The initial state was a random configuration, corresponding to infinite temperature. Then, the system was quenched to a temperature T_q. The temperature was normalized so that $kT_q = 1$. Simulations were run to a maximum of 10^5 Monte Carlo seconds, where one Monte Carlo second (MCS) is defined as 128 × 128 attempted spin exchanges. Figures 1-4 show the morphologies for mixtures with an overall A to B ratio of 77/23 but with varying concentrations of block copolymer.

In Figures 1 and 2, both at $10^4 MCS$, white and gray represent the homopolymers while black represents the copolymer. Figure 1 shows a quench in which the homopolymer-A/ homopolymer-B/ copolymer ratio is 67/13/20. It can be seen that the block copolymer covers the surface of the minor phase. This situation is close to optimal surface coverage, and there is little diblock in the matrix phase. Figure 2 shows a 62/8/30 case in which there is an excess of block copolymer which forms micelle-like structures in the matrix. At longer simulation times, these micelles migrate towards the interface to form lamellas around the particulate

phase as expected for a symmetric diblock copolymer. This is shown in Figure 3, at $10^5 MCS$.

Figure 4 shows the case of 54/0/46, at $10^4 MCS$ where only one homopolymer is present. This homopolymer is represented by gray while white and black represent the two different blocks of the copolymer. Note that the block copolymer occupies two lattice sites. As expected, lamellas are formed, and the rate of domain growth is low.

Domain sizes at different times were calculated and averaged over 20 independent runs. The average domain size is expected to increase as t^n. A log-log fit of the domain sizes with the time of simulation gave the asymptotic scaling exponent, n. The simple binary case gave $n \simeq 1/3$, as expected. As the concentration of block copolymer was increased, the scaling exponent decreased. For a trivial case of 100% block copolymer, n reduced to zero. Results are shown in Figure 5.

SIMULATIONS WITH REACTION

Simulations were run with an initial concentration of 77/23/0 (no diblock copolymers). For each attempted spin exchange, the possibility of reaction was also considered. Adjacent units of unlike homopolymers had a probability p of coupling:

$$A + B \longrightarrow AB$$

The probability ranged from 2.5×10^{-5} to 0.25. Note that a central molecule has four nearest neighbors so that the total reaction probability could be as much as $4p$.

Morphology comparisons of reacting and non-reacting cases were made for various values of p. With fast reactions (high p), the overall composition rapidly approached 54/0/46 and the morphology at a given simulation time was effectively identical to the non-reactive case starting with 54/0/46. For a slow reaction with $p = 2.5 \times 10^{-5}$, the composition had only reached 67/13/20 after 8000 MCS. However, at this time it was essentially identical to that of the non-reactive mixture that started at 67/13/20. For the entire range of reaction rates, and times, there was never a discernable difference between the morphology of the reactive system and a non-reactive system with the same composition as the instantaneous composition of the reactive mixture

CONCLUSIONS

Monte Carlo simulations were performed using a simple fast Ising-type model to predict the phase separation behavior of polymer blends. Simulations at relatively high concentrations of block copolymer showed that as the concentration of block copolymer increased, the excess block not covering the surface, formed microphases resembling micelles, but given sufficient time , reached the interface to give lamella like structures. Extremely high concentrations of block copolymer yielded lamella like structures. The model can not predict stable micellar microphases. We would need different block ratios to be able to predict that effect. It does predict reasonable values for the scaling exponent.

Simulations with reactions in the bulk, where A and B homopolymers react to form a diblock copolymer showed that the reaction did not directly affect the morphology. It indirectly affected it by changing the composition. At all times, the system behaved as if the block copolymer formed by reaction had been present from the onset.

REFERENCES

1. Kandasamy, S. and Nauman, E.B., Accepted, *Computational and Theoretical Polymer Science*,(1998)

2. Larson. R.G., *J. Chem. Phys.* **96**, 7904 (1992)

3. Carmesin, I. and Kremer, K., *Macromolecules* **21**, 2819 (1988)

4. Glotzer, S.C., Stauffer, D. and Jan N., *Phys. Rev. Let.* **72**, 4109 (1994)

5. Glotzer, S.C., Di Marzio, E.A. and Muthukumar, M., *Phys. Rev. Let.* **74**, 2034 (1995)

6. Toxevard, S. *Phys. Rev. E.* **53**, 3710 (1996)

7. O'Shaughnessy, B. and Sawhney, U., *Phys. Rev. Let.* **76**, 3444 (1996)

8. Muller, M.,*Macromolecules* **30**, 6353 (1997)

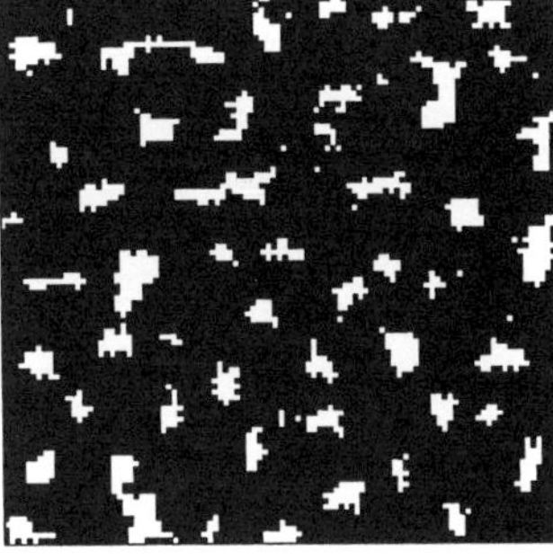

Figure 1: Morphology at a concentration of 67/13/20 and $10^4 MCS$

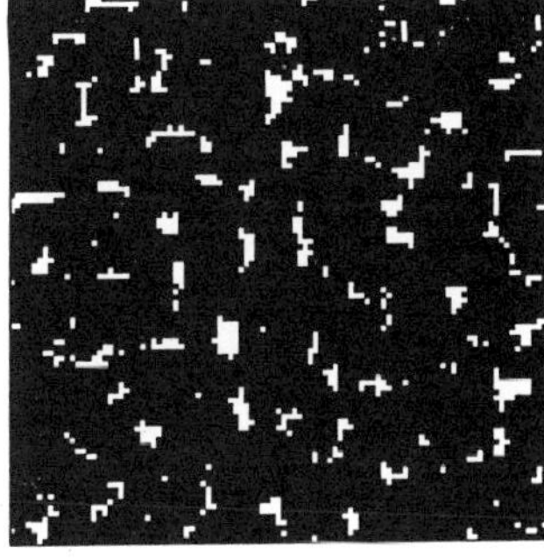

Figure 2: Morphology at a concentration of 62/8/30 and $10^4 MCS$

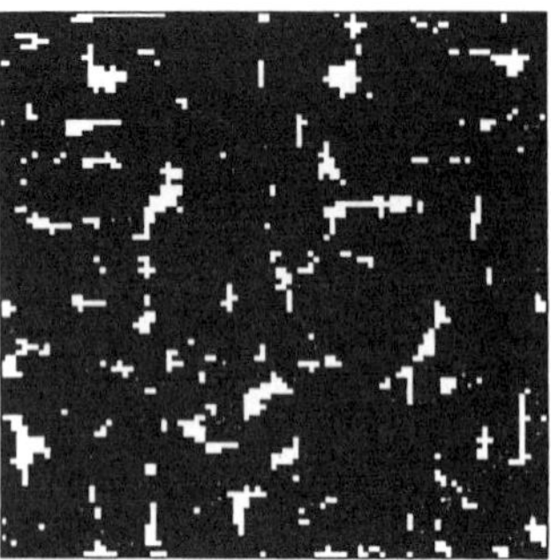

Figure 3: Morphology at a concentration of 62/8/30 and $10^5 MCS$

Figure 4: Morphology at a concentration of 54/0/46 and $10^4 MCS$

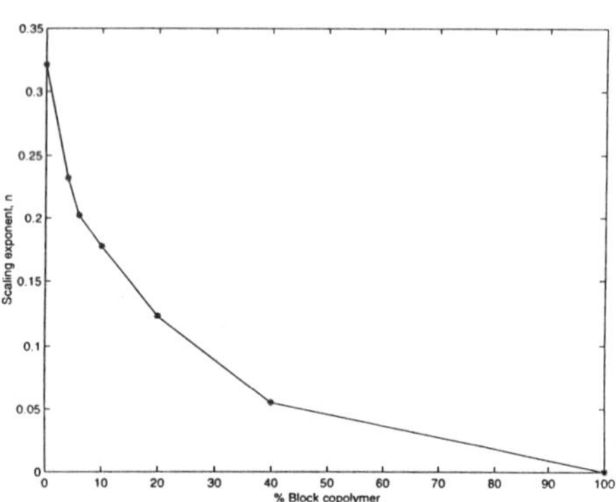

Figure 5: Plot of scaling exponent vs. block copolymer concentration

DIFFUSION INDUCED COALESCENCE IN THE LATE STAGE RIPENING OF POLYMER BLENDS

A. Peter Russo and E. Bruce Nauman

The Howard P. Isermann Department of Chemical Engineering,
Rensselaer Polytechnic Institute, Troy, New York 12180-3590

INTRODUCTION

Multiphase polymer blends are being used to satisfy the ever increasing demand for new polymers. The wide use of blends is due to their ability to be tailored to fit specific applications, usually at a lesser cost than could be achieved by creating a new polymer. Polymer blends can be made by a variety of techniques, including melt blending, reactive extrusion, and compositional quenching.[1]

One significant difference in the various blending methods is the resultant particle size and particle size distribution of the blends. These characteristics have a profound effect on blend properties such as impact strength. Optimal particle sizes exist for increasing impact strength and vary with the failure mechanism of the material.[2] However, the average particle size and particle size distribution are often not the same as those observed in the final product. This is due to phase ripening which occurs in during molding. By modelling the ripening process, it may be possible to tailor the production variables to produce a blend with optimal properties with minimal experimentation.

Ripening occurs by two mechanisms, Ostwald ripening and coalescence. Ostwald ripening is governed by diffusion while coalescence is primarily controlled by fluid mechanics. However, conventional fluid mechanics are unable to predict the entire coalescence process. This process may be separated into four steps:[3]
   (1) approach of the droplets through the continuous phase,
   (2) deformation of the droplets and drainage of the interstitial fluid,
   (3) rupture of the interstitial fluid, and
   (4) evolution of the coalesced droplet.
Fluid mechanics plays an essential role in the approach of two droplets through a fluid, as well as in the drainage of the fluid as the interparticle distance decreases. The evolution of a coalesced droplet has also been modelled using the Navier-Stokes equations.[4]

The rupture of the interstitial fluid cannot be accomplished by fluid mechanics alone. An attractive force is required which is not present in the Navier-Stokes equations. This stage of coalescence has been treated in a number of ways. The most primitive is to arbitrarily assume the continuous phase ruptures once the particles are some prescribed distance apart. This distance is known as the critical rupture thickness and is often estimated by van der Waals interactions between the particles.[5] Typical values of the critical rupture thickness range from a few tens to hundreds of Angstroms.[6]

Arbitrary assumptions of rupture provide a means of continuing the calculation, but offer no insight into the rupture process, or the time scale over which it occurs. Molecular dynamics have been used to study the interactions between individual droplet molecules and can successfully model the rupture process.[7] This approach encounters difficulty in large systems with many molecules and therefore has limited application.

THEORY

This study utilizes a continuum mechanical approach which is applicable to large systems of physical interest. A binary non-linear diffusion model of the Cahn-Hilliard type[8] was used:

$$\frac{\partial c}{\partial t} = \nabla Dc(1-c)\nabla\left(\frac{dg}{dc} - \kappa\nabla^2 c\right)$$

where c denotes the volume fraction of one component, D is the diffusivity, κ is the gradient energy parameter and g is the homogeneous free energy of mixing. Since this model uses component volume fractions in place of individual molecules, it is capable of modelling large systems. The use of a diffusion based model is also supported by the observation of a "diffusion-coupling mechanism" which leads to coalescence.[9]

RESULTS AND DISCUSSION

Numerical simulations were performed on a system consisting of two equal-sized droplets at equilibrium with a surrounding fluid. The initial size and separation of the droplets were varied, as well as the gradient energy parameter. The physical dimension of the system was 1.28 x 1.28 μm and the diffusivity was assumed to be 10^{-11} m^2/s. A degree of polymerization of 1000 and a radius of gyration of 200 Å were assumed for both components. These values were chosen to approximate the coalescence of two minor phase droplets in a polymer blend.

The results of these simulations demonstrate the existence of diffusion induced coalescence in systems where hydrodynamics are absent. Figure 1 shows the results of a typical simulation with $\kappa = 8\times10^{-19}$ m^2, corresponding to a value of the binary interaction parameter, $\chi=0.003$. A particle size of ~0.1 μm and an initial separation of 944 Å was used. As time progresses, the interstitial phase is observed to darken until it ruptures. This is followed by the relaxation of the coalesced droplet into a spherical shape which minimizes the free energy. The observed darkening and rupture is actually the interstitial phase dissolving into the droplet phase.

Coalescence times were observed to increase with increased droplet separation. However, above a critical separation, coalescence no longer occurs. For the conditions used in obtaining Figure 1, this critical separation was 1256 Å and corresponds closely to the minimum size for growth predicted by spinodal decomposition theory. By extrapolating this result to completely immiscible polymers, a critical rupture thickness of 724 Å is found for diffusion induced coalescence. In comparison, van der Waals interactions predict a critical rupture thickness of this magnitude for particles 100 μm in size or larger. Therefore, nonlinear diffusion causes coalescence at distances greater than that predicted by van der Waals interactions for the particle sizes of interest in polymer blends.

These results demonstrate the importance of diffusion in coalescence and in phase ripening. However, many polymer blends involve the addition of one or more compatibilizers. These compatibilizers may be a third polymer, or a random or block copolymer of the two blend components. The use of such compatibilizers is thought to retard the ripening process and reduce coalescence. In order to examine this phenomena, a ternary version of the Cahn-Hilliard model[8] was used. The physical parameters were the same as those used in the binary case.

The initial particles were of a core-shell morphology where the shell was a compatibilizing third polymer. The particle size was 0.113 μm with a shell thickness of ~450 Å. Coalescence occurs in a similar manner to that observed in the binary case. The interstitial fluid dissolves into the shell phase and ruptures, causing coalescence. Next, the shell evolves toward a spherical shape and pulls the cores closer. Finally, the two cores are close enough to cause the rupture of the intervening shell phase and the coalesced cores evolve as well.

This process has been observed experimentally in bulk multicomponent polymer blends using a 1.2MeV high voltage electron microscope. The diffusion-induced rupture of the interstitial phase is observed in Figure 2 for a ternary blend of polystyrene (matrix), polybutadiene (core) and a diblock copolymer (shell, not visible). The final coalescence of the cores is shown in Figure 3 for a quaternary blend of polystyrene (matrix), polybutadiene (shell), diblock (outer shell, not visible) and polypropylene (core).

Figure 4 shows the comparison of coalescence times for binary and ternary simulations. The presence of a compatibilizing shell shifts the curve toward smaller separations. This demonstrates the stabilizing effect of the third component. Core-shell particles will require more time to coalesce than single-component particles at the same separation. The critical rupture thickness has also decreased for the core-shell system, and the exponential increase in coalescence times is sharper. These findings illustrate the reduced ripening rate observed in polymer blends with core-shell morphologies.

CONCLUSIONS

The properties of polymer blends depends strongly on the average particle size and particle size distributions of the dispersed phase. Nonlinear diffusion has been shown to play a vital role in coalescence and hence, in the ripening process. By the addition of a third, compatibilizing component, coalescence times increase and the system is stabilized. These findings demonstrate the significance of diffusion in polymer blending. In order to control the ripening process and resultant particle size distributions, one must control diffusion.

ACKNOWLEDGMENT

Figures 2 and 3 were obtained by Dr. Timothy J. Cavanaugh using the equipment that is supported in part by Biotechnological Resource Grant RR-01219. This grant supports the Wadsworth Center's Biological Microscopy and Image Reconstruction Facility as a National Resource.

REFERENCES

1. Nauman, E. B. et al., *Chem. Eng. Comm.*, **66**, 29 (1988).
2. Folkes, M. J. and Hope, P. S. (eds), *Polymer Blends and Alloys*, Blackie Academic and Professional, New York (1993).
3. Fortelny, I. and Zivny, A., *Polymer*, **36**, 4113 (1995).
4. Rosenzweig, N. and Narkis, M., *Poly. Eng. and Sci.*, **23**, 32 (1983).
5. Vrij, A., *Disc. Farad. Soc.*, **42**, 23 (1966).
6. Chesters, A. K., *Trans. I. Chem. Eng.*, **69A**, 259 (1991).
7. Koplik, J. and Banavar, J. R., *Science*, **257**, 1664 (1992).
8. Nauman, E. B. and Balsara, N. P., *Fluid Phase Equil.*, **45**, 229 (1988).
9. Tanaka, H., *J. Chem. Phys.*, **103**, 2361 (1995).

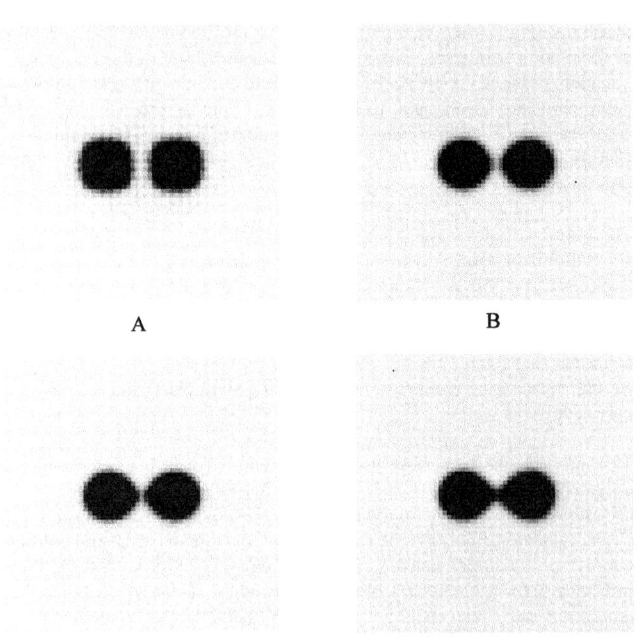

A

B

C

D

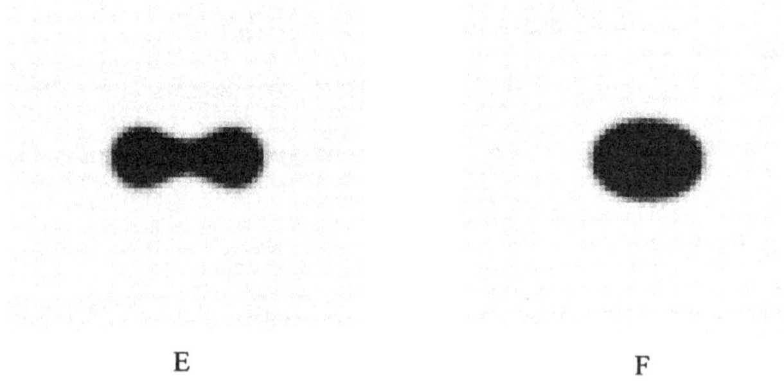

E   F

Figure 1: Diffusion induced droplet coalescence at (A) 1s, (B) 45s, (C) 48s, (D) 50s, (E) 54s, and (F) 100s.

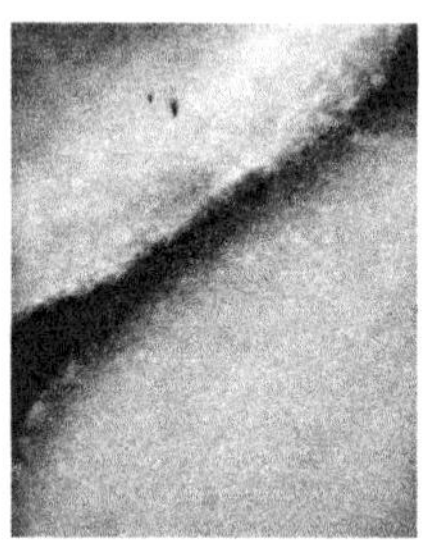

Figure 2: HVEM micrograph of shell coalescence in a polystyrene-polybutadiene-diblock copolymer blend.

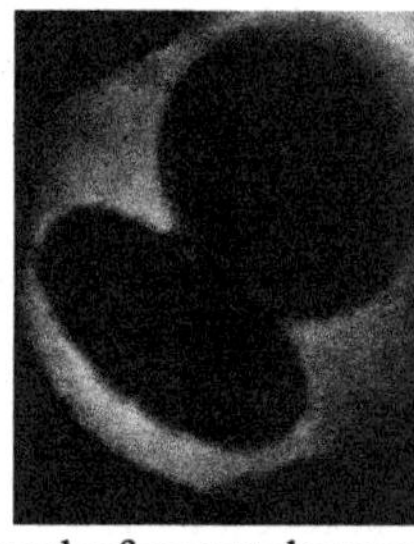

Figure 3: HVEM micrograph of core coalescence in a polystyrene-polybutadiene-diblock copolymer-polypropylene blend.

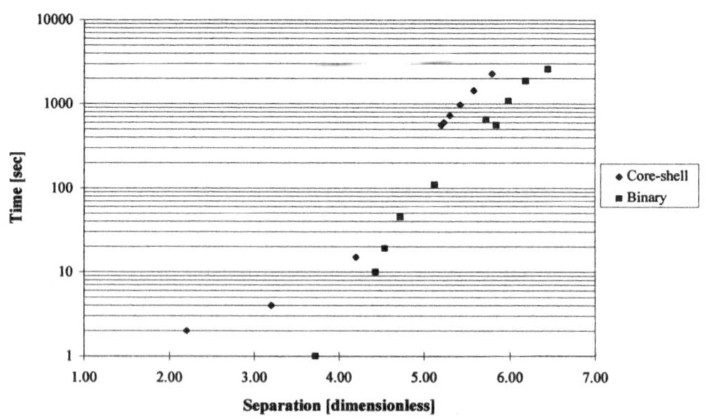

Figure 4: Comparison of simulated coalescence time with separation for binary particulate and ternary core-shell morphologies.

TOUGHENING OF POLYCARBONATE/POLYAMIDE-6 BLENDS.

P. Arjunan, Exxon Chemical Company, 5200 Bayway Drive, Baytown, TX 77522, R.M. Mininni, ISP Corporation, 1361 Alps Road, Wayne, NJ 07470

Introduction

Both Polycarbonates (PC) and polyamides (PA) are good engineering thermoplastics in their own right. However, both the above materials have some inherent deficiencies which preclude them in many end use applications. It is expected that compatibilization of PC/PA blends will improve their inherent deficiencies and result in new materials possessing improved solvent resistance, ESCR, dimensional stability, and other desired mechanical properties. Also, PC/PA blends can further be modified with suitable impact modifiers to achieve better thick section impact strength, and low temperature impact strength. The main objective of this work was to formulate new PC/PA-6 blends with improved compatibilization and impact modification that would be useful in new end-use applications.

Compatibilization of PC/PA-6 Blends

A base line study of PC/PA-6 blends indicated their incompatibility in terms of poor mechanical properties. We proposed new compatibilization approaches using unique compatibilizers that could bond chemically with PA-6 and interact physically with PC. We demonstrated the compatibilization of PC/PA-6 blends using either the methyl methacrylate/styrene/methyl maleate terpolymer (TP) or the imidized acrylic copolymers, Paraloid Exl 4151 and -4171. The above terpolymer, TP, was synthesized in our laboratory while the acrylic copolymers were commercially available from Rohm & Hass Co. The blend compatibility was confirmed by analyzing the thermal (DSC), mechanical properties, and spectroscopic (SEM) data. A synergistic effect on the tensile strength (specifically the % elongation) of the PC/PA-6/Compatibilizer alloys evidenced their compatibility. Analysis of both SEM and TEM data indicated better mixing and dispersion of PA-6 phase in PC phase of the above alloys. These compatibilized alloys also exhibited improved solvent resistance. The results are listed in Tables 1-3.

Impact Modification of PC/PA-6/Compatibilizer Alloys

The impact strength of the above alloys was improved by adding various impact modifiers. As shown in Tables 2-3, both the notched (Izod) and unnotched (Falling Dart) impact strengths were improved significantly by using a compatibilizer (Exl 4151 or TP) and an impact modifier (Paraloid Exl 3607, core-shell type, Rohm-Haas Co.). The effect of different types of impact modifiers was also investigated by using modifiers that were selective for PC phase (Paraloid Exl 3607), selective for PA-6 phase (Royaltuf 465A, Uniroyal), and a combination thereof. We also observed that the processing conditions and the material selection (types of PC and PA-6) had impact on the overall properties of the above alloys. Neither the impact strength nor the solvent resistance of PC/PA-6 blends were improved by using the above impact modifiers alone.

Solvent Resistance of the PC/PA-6/Imidized Acrylic Copolymers Alloys

The solvent resistance of the PC/PA-6 blends was poor by itself. However, the presence of a compatibilizer such as the Paraloid Exl 4151 improved the solvent resistance of the above blends. Also the impact modifiers alone did not improve the solvent resistance of the above blends. As expected the PC/PA-6 blends containing both the compatibilizers and the impact modifiers improved their solvent resistance significantly. These result are shown in Figures 1-2.

Summary and Conclusions

The PC/PA blends were incompatible and exhibited poor mechanical properties. Both the terpolymer, TP, and the imidized acrylic copolymers by their unique structures compatibilized the above blends. The PC/PA-6/Exl 4151or TP alloys improved their mechanical and solvent resistance properties due to their compatibility that was also confirmed in terms their phase morphology using SEM and TEM techniques. The use of impact modifiers alone did not improve either the impact strength or the solvent resistance of the PC/PA-6 blends. However, the above properties were improved significantly by using both a compatibilizer (Exl 4151) and an impact modifier (Exl 3607). Other factors such as the processing conditions, types of PC and PA-6 etc., also impacted the overall properties of the above alloys. The concept of compatibilization, impact modification, and the characterization methods that we used in this study are applicable to other similar polymer blend systems.

TABLE I. PC/PA-6/COMPATIBILIZER BLENDS

Blend	A	B	C
Parts Polycarbonate	60	54	54
Parts Nylon-6	40	36	36
Parts Paraloid EXL 4151	--	10	--
Parts Paraloid EXL 4171	--	--	10
Tensile Properties [1]			
% Elongation	8.4	188.2	125.2
Modulus (ksi)	128.2	134.1	132.2
Tensile Strength (ksi)	8.6	10.6	10.6
Flexural Properties [1]			
Yield Stress (ksi)	12.0	14.2	14.2
Modulus (ksi)	288.3	300.3	308.1
Toughness (in-lb/in3)	43.7	86.7	87.9
Impact Properties [1]			
Notched Izod (ft-lb/in notch)	0.42	0.96	0.42
Falling Dart (ft-lb)	1.25	1.13	1.01
Thermal Properties [2]			
Tg of Polycarbonate (C)	142.0	144.1	143.1
Tm of Nylon-6 (C)	214.1	214.0	214.2
Heat of Fusion of Nylon-6 (W/g min)	16.6	16.8	15.5

TABLE II. PC/PA-6/COMPATIBILIZER/IMPACT MODIFIER BLENDS

Blend	A	B	C	D
Parts Polycarbonate	60	54	54	48
Parts Nylon-6	40	36	36	32
Parts Compatibilizer	--	10	--	10
Parts Impact Modifier	--	--	10	10
Tensile Properties [1]				
% Elongation	7.7	144.1	11.3	133.8
Modulus (ksi)	130.5	136.3	106.4	121.0
Tensile Strength (ksi)	8.2	10.5	8.0	8.8
Flexural Properties [1]				
Yield Stress (ksi)	12.7	14.8	11.9	12.5
Modulus (ksi)	290.9	305.1	242.1	265.8
Toughness (in-lb/in3)	43.6	89.9	73.7	77.2
Impact Properties [1]				
Notched Izod (ft-lb/in notch)	0.25	0.99	0.32	10.1
Falling Dart (ft-lb)	1.30	1.69	1.38	44.1

[1] All samples were tested "dry-as-molded".

Materials:
- Polycarbonate used was Lexan 141-112 from General Electric.
- Nylon-6 used was Nivionplast 273MR from Enichem.
- Compatibilizer used was Paraloid EXL 4151 from Rohm and Haas Company.
- Impact Modifier used was Paraloid EXL 3607 from Rohm and Haas Company.

Table III: Properties of PC/PA/MMA-ST-MME Terpolymer Alloys

Blends ==>	PC/PA6[1] 60/40	PC/PA6/TP 54/36/10	PC/PA6/TP/EXL3607 48/32/10/10
Properties			
% Elongation	10.5	156.9	213.4
Flex Modulus (ksi)	403.2	401.7	346.1
Notched Izod Impact (1/8") (ft-lb/in)	0.4	0.66	0.8
Falling Dart Impact (ft-lb)	1.68	1.95	6.85
T_g, °C (PC)[2]	140.6	140.2	——
T_m, °C (PA6)[3]	213.7	217.1	——

Figure 1: PC/PA-6 Blends -
Impact Properties

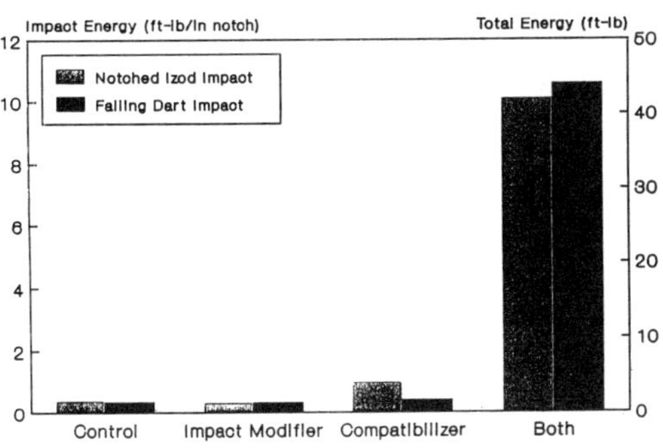

Impact modifier was Paraloid EXL3607.
Compatibilizer was Paraloid EXL4151.

Figure 2: PC/PA-6 Blends -
Tensile Properties
Pre- and Post-Chemical Aging

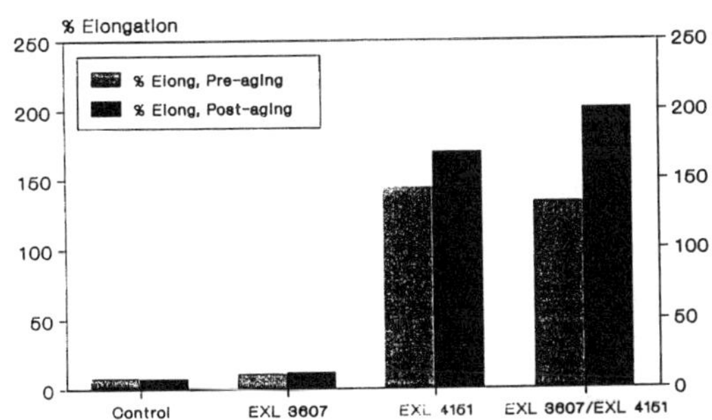

MICROMECHANICAL DEFORMATION PROCESSES IN PA12-LAYERED SILICATE NANOCOMPOSITE

G.-M. Kim, G.H. Michler*, P. Reichert**, J. Kresseler**, R. Mülhaupt***

* Institut für Werkstoffwissenschaft
Martin-Luther-Universät Halle-Wittenberg,
Geusaer Str., 06217 Merseburg, Germany
** Institut für Makromolekulare Chemie und Freiburger
Materialforschungszentrum,
Albert-Ludwigs-Univeristät Freiburg
Stefan-Meier-Str. 21, 79104 Freiburg, Germany

Introduction

As the use of polymers in structural applications increases, a new class of polymeric materials with a unique combination of high strength, modulus and toughness is being demanded. Hybrid organic-inorganic composites have been considered as a attractive representative[1]. It has been found that by incorporating a very small amount of inorganic components (only a few wt %) in the polymer matrix on a nanometer scale unexpected mechanical properties can be synergistically achieved. Frequently used inorganic filler particles are platy clay mineral particles consisting of layered silicates in a range of 1-100 nm in length and several nm in thickness (very large aspect ratio).

Although a considerable development of synthetic approaches in polymer nanocomposites has been performed in the last years, the appropriate mechanisms responsible for the improvement of mechanical properties in this subject are still not presented. In this work, the influence of the morphology of dispersed silicate layers in PA12 on the micromechanical deformation processes has been investigated by high voltage electron microscopy using *in situ* tensile technique.

Experimental

Nanocomposites. As matrix the PA12 was a commercially used polyamide (GRILAMID) from EMS-CHEMIE AG (Domat/Ems, Switzerland), and as inorganic component the synthetic three-layer silicates (SOMASIF ME100) were used for this studies that is produced by CO-OP CHEMICAL Co., Japan. In order to achieve swelling of the synthetic layer silicate 12-aminolauric acid (ALA) were used. The nanocomposites were prepared by polymerization of ALA at 280°C under pressure in the presence of swollen ME100.

Study of morphology. Morphological studies were performed using a Jeol 200 kV transmission electron microscopy (TEM). Ultra-thin sections suitable for TEM measurements were microtomed at -80 °C using a Leica Ultracut E cryo-ultramicrotome in about 80-100 nm thickness.

Investigation of micromechanical deformation processes. To study micromechanical deformation processes, samples were uniaxially deformed and investigated *in situ* by high voltage electron microscopy (HVEM, 1000 kV). All deformation tests under the microscope were performed at room temperature. The specimens for HVEM investigations were microtomed at -80°C containing thickness of about 0.5 µm.

Results and Discussion

Morphology. The studies of morphology of nanocomposite show that the silicate layers were not homogeneously dispersed but locally aggregated in the PA12 matrix, exhibiting about 1µm in dimension. It has been also observed some needle-like structured agglomerates with a length of about 100 nm and a thickness of a few nanometers. Due to mixing of ALA the defoliation of silicate layers has been occurred. Single stacks of silicate layers have an average distance of about 10 nm. Emphasizing that we have found the evidence of intercalation of the matrix material in the silicate interlayer region. Detailed morphological results have been already discussed in our previous work[2].

Micromechanical deformation processes. The results of the *in situ* electron microscopic investigation of micromechanical deformation processes will be schematically discussed in this preprint. It has been found that the main initiation site of plastic deformation is the microvoid formation at the polar region of the stacked silicate layers (**Figure 1**). Once the uniaxial stress is applied on the specimen, the stress concentrates around the stacked layers. Due to the weak interaction between the silicate layers and the polymer matrix, the debonding process easily takes place on the polar region of the stacked layers, as a consequence the microvoids nucleates on there. Hereby the orientation of the stacked layer plays a great role. Although the stacked silicate layers consist of numerous rigid platy mineral particles, they act on the whole similar to soft modifier particles during the deformation processes, since the interaction between the silicate layers and the polymer matrix is weak. In other words, during deformation processes the stacked silicate layers act as agglomerates commonly appeared in conventional composites[3].

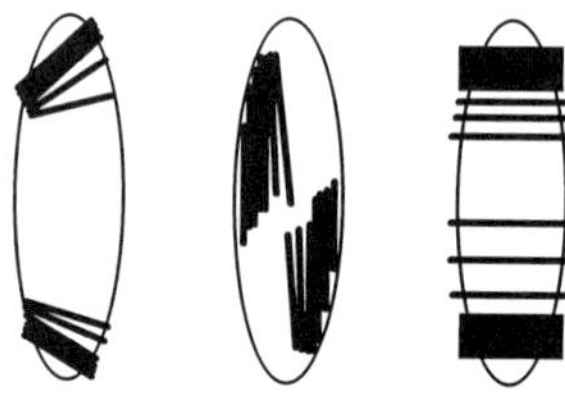

Figure 1. Schematic illustration of microvoid formation of the stacked silicate layers

According to the orientation of the stacked silicate layers dispersed in the matrix, three deformation modes can be characterized: a) splitting, b) sliding and c) opening of the layered structure (**Figure 1**).

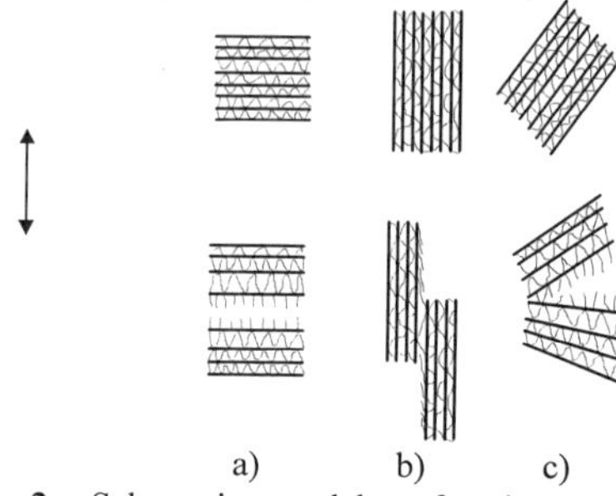

Figure 2. Schematic models of micromechanical deformation processes of the stacked silicate layer depending on the orientation (arrow indicates the load direction): a) splitting mode, b) sliding mode and c) opening mode

Once the uniaxial load is applied at a stacked layer, which is orientated perpendicular to the applied load, deformation initiates at the middle region of the stacked layer. Splitting of the layer takes place in the middle region. Interlayer distances in the stacked layer increase as far away from the middle region (**Figure 2a**). A stacked layer is oriented parallel to the applied load, sliding of bundles of the stacked layer takes place during deformation, by which the interlayer distances hold in constant (**Figure 2b**). The stacked layers are placed with a certain angle to the direction of applied load in the matrix, opening of bundles of the stacked layer occurs during deformation processes (**Figure 2c**).

Conclusions

It has been found that the main micromechanical deformation process in PA12-layered silicate nanocomposite is the microvoid formation caused by debonding at the boundary regions of the stacked silicate layers. During the deformation processes the separated bundles in the stacked layers tilt perpendicular to the direction of applied load. Depending on the orientation of the stacked layers some amount of applied energy is dissipated by splitting, sliding or opening of the separated bundles in the stacked layers. As a result the toughness of nanocomposite will be enhanced. The layered silicates bounded at the banks of microvoids are load-bearing and prevent from further growing of the microvoids, as a consequence the stiffness of the nanocomposite should be improved.

References

(1) Y. Fukushima, A. Okada, M. Kawasumi. T. Kurauchi and O. Kamigatto, Clay Miner. **1988**, 23, 27

(2) P. Reichert, J. Kressler, R. Thomann, R. Mülhaupt and G. Stöppelmann, Acta Polymer., **1998**, 49, 116.

(3) G.-M. Kim and G.H. Michler, Polymer in press.

The effect of particle size in PP/EPR blends on the brittle/ductile transition temperature at low and high test speed

P.T.S. Dijkstra, A. van der Wal, R.J. Gaymans, J. Huétink
University of Twente, Department of Chemical Technology
P.O. Box 217, 7500 AE, Enschede, The Netherlands

Introduction

Particle size has a strong influence on the toughness of rubber modified polymer systems. Several studies have shown that the toughness increases with decreasing particle size under impact conditions often leading to an optimum particle size below which rubber cavitation or craze initiation is difficult[1,2]. At lower test speeds, however, the effect of particle size is small. Gaymans et al.[3] conducted low notched tensile impact test at two speeds (10^{-3} and 1 m/s) on rubber toughened nylon. They found the brittle to ductile transition temperature T_{bd} to be indepent of particle size at low speeds, in contrast with the increase of T_{bd} with increasing particle size at high test speed. The effect of test speed on the particle size effect is the subject of this work.

Experimental

A commercial polypropylene (GE 7100, Montell; MFI 0.8 (230 °C, 21.6 N)) and a series of experimental EPRs supplied by Shell (KSLA, Amsterdam, the Netherlands) were used. The material properties of the EPR rubbers are listed in *Table 1*.

Table 1 EPR properties

Rubber code	E40L	E40M	E40H	E60L	E60M	E60H
Ethylene content [mol%]	40	40	40	60	60	60
intrinsic viscosity[a] [dl/g]	3.4	6.9	14.9	4.3	8.0	14.1

a) in decaline at 135 °C

An 80/20 vol% PP-EPR blend was prepared on a Berstorff co-rotating twin screw extruder (L/D = 33, D = 25 mm) at a barrel temperature of 230°C and a screw speed of 100 rpm. Rectangular bars and dumbbell-shaped specimens were injection moulded on a 221-55-250 Arburg Allrounder injection moulding machine. The barrel temperature and the screw speed were 230°C and 300 rpm, and the mould temperature was 40°C. The geometry of the rectangular bars (74x10x4 mm) was according to ISO 180/1A and that of the dumbbell-shaped specimens (10x3x115) according to ISO R527-1 specifications. A single-edge 45° V-shaped notch (tip radius 0.25 mm, depth 2 mm) was milled into the bars. The blends will further be denoted by the rubber they contain.

Results and discussion

The average particle sizes of the blends[4] are given in *Table 2*. The blends are listed in order of ascending weighted average particle size D_w.

Table 2 Particle sizes for the blends

Blend type	E40L	E60L	E40M	E40H	E60M	E60H
D_w [μm]	0.37	0.67	1.34	1.55	2.08	3.83

The yield stress of the blends was measured at 1 mm/s and is shown in *Figure 1*. The yield stress is independent of particle size.

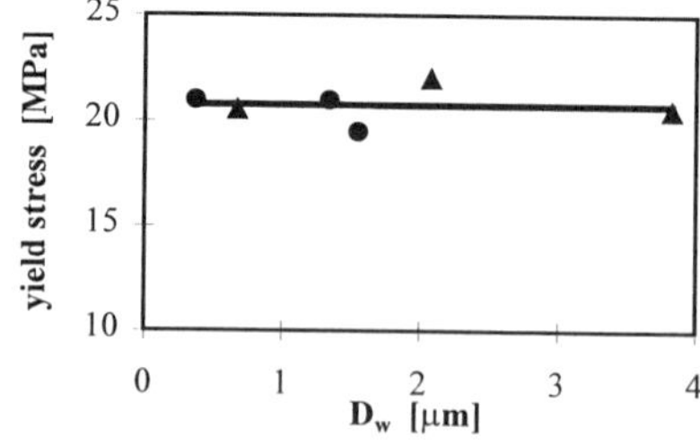

Figure 1 Yield stress versus average weighted particle size of 20 vol% PP-EPR blends (1 mm/s): ● 40 mol% ethylene, ▲ 60 mol% ethylene.

The toughness of the blend has been determined at both high and low speeds with notched Izod impact tests and SEN tensile tests[4] respectively. From these the brittle to ductile transition temperature has been determined. The effect of test speed on the particle size effect is shown in *Figure 2*. At high test speeds the T_{bd} increases considerably with increasing particle size, whereas it decreases slightly at low test speed.

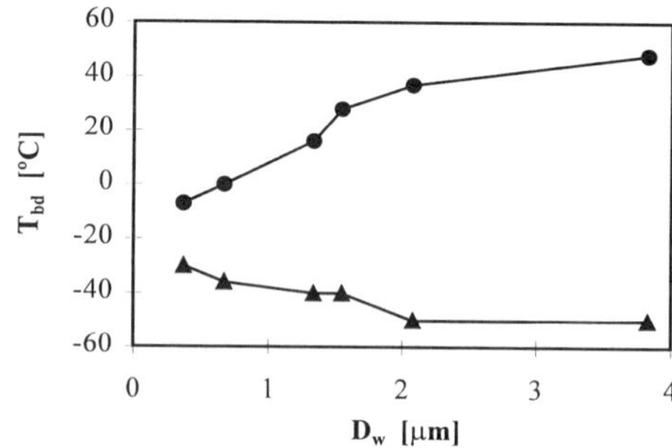

Figure 2 Brittle to ductile transition temperature (T_{bd}) as a function of weighted average particle size for 20 vol% PP-EPR blends: ● high test speed (Izod impact test), ▲ low test speed (SEN tensile test 1 mm/s).

A possible explanation is that the sensitivity for crack initiation from a cavity increases with speed. Smaller particles are then necessary to ensure ductile fracture. Coran[5] has shown with low speed tensile tests on PP/EPDM blends that bigger particles reduce ductility. Although others[1,6] have found matrix crazing to be the dominating deformation mechanism at high speeds or low temperatures, no evidence was found for crazing[7]. Extensive post-mortem SEM fractography at both high speeds (12 m/s) and low temperatures (-50 °C) indicate that matrix yielding is the dominating deformation mechanism. Continued research on this subject focusses on the preparation of blends with a wider range in particle size. Bigger particles are expected to promote embrittlement at low speed comparable to the behaviour at high speed as illustrated by the sketched graph of *Figure 3*. The wide range in particle size is prepared using

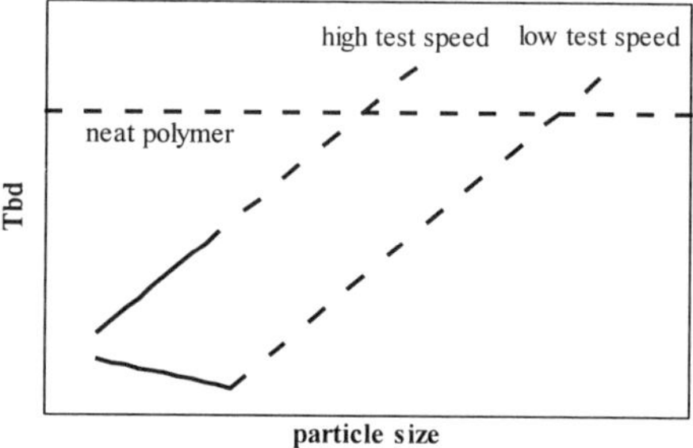

Figure 3 Expected particle size effect at different speeds

dynamic vulcanization of PP-EPDM blends[5]. Pre-vulcanization of the EPDM rubber to different crosslink densities results in different molecular weight of the rubber prior to melt blending. During melt blending the vulcanization continues and the crosslink density reaches a constant saturation level, thus minimizing the difference in rubber properties between the blends.

References

1) Jang, B.Z., Uhlmann, D.R. and Vander Sande, J.B., *J. Appl. Polym. Sci.*, **30**, 2485 (1985).
2) Oshinski, A.J., Keskkula, H. and Paul, D.R., *Polymer*, **33**, 284 (1992).
3) Gaymans, R.J., Dijkstra, K., Dam, M.H. ten, in: Toughened Plastics II, edited by Riew, C. Keith and Kinloch, Anthony J., Advances in chemistry series 252, American Chemical Society, 1996.
4) Wal, A.van der and Gaymans, R.J., *Polymer*, in press.
5) Coran, A.Y., and Patel, R., *Rubber Chem. Techn.*, **53**, 141 (1980).
6) Ramsteiner, F., *Acta Polymerica*, **42**, 584 (1991).
7) Wal, A. van der, Verheul, A.J.J. and Gaymans, R.J., *Polymer*, in press.

THE BRITTLE/DUCTILE TRANSITION OF POLYCARBONATE AS
A FUNCTION OF TEST SPEED

J.P.F. Inberg, M.J.J. Hamberg and R.J.Gaymans
University of Twente, Department of Chemical Engineering
P.O.Box 217, 7500 AE Enschede
The Netherlands

INTRODUCTION

PC fails ductile in a wide temperature range, including room temperature. However, PC can also fracture in a brittle or semi-ductile manner with a sharp notch at high test speeds. The brittle/ductile transition is very sensitive to parameters like specimen thickness, rate of loading, temperature and notch tip. For highly ductile materials like PA-blend and PP-blend, considerable warming up of the deformation zone was observed at high test speeds [1,2]. For these materials a thermal blunting process was found to take place. We studied the influence of test speed on the brittle-ductile transition of PC and look at the fracture behavior.

EXPERIMENTAL

PC Lexan 101-111 Bisphenol A from General Electric Co. was used (M_w 29,500). Prior to blending and injection molding the materials were dried. UMIST Manchester supplied the rubber particles used as deformation markers. The particles have a polybutylacrylate core and a PMMA shell (shell thickness 10 nm, particle diameter 0.35 μm). The extrusion step was carried out on a twin screw extruder. Rectangular bars (ISO 180/1, 74x10x4 mm) were injection molded. A single edge notch with a tip radius of 0.25 mm and a depth of 2 mm was milled in the bars. The fracture behavior was studied by tests referred to as single edge notch tensile (SENT) test. Test speeds range from 10^{-5} to 12 m/s, temperatures from -80 to 100°C. Fracture surfaces and deformation zones under the fracture surface were studied with SEM.

RESULTS AND DISCUSSION

The specimens tested at different speeds and temperatures showed three fracture surface types, termed surface I, II and III.

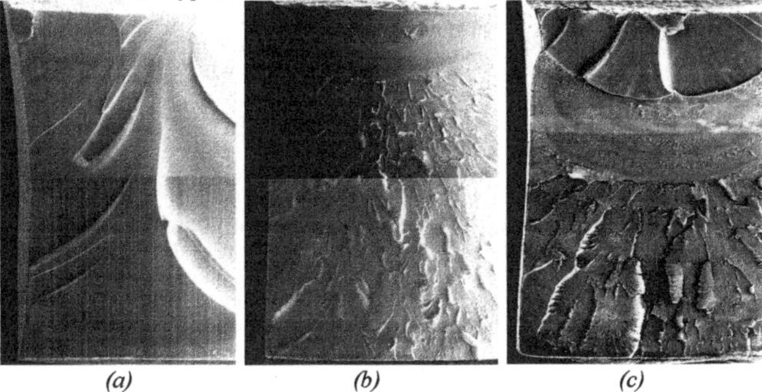

(a)   (b)   (c)

Fig. 1: *SEM micrographs of the fracture surfaces of notched polycarbonate specimens. Notch at the top.*
a) Type I, test conditions: 10^{-5} m/s, -50°C
b) Type II, test conditions: 1 m/s, -20 °C
c) Type III, test conditions: 10^{-5} m/s, -60 °C

Type I occurs for every ductile mode failure, regardless of test speed. Type II is the surface of a specimen fractured at high test speed, and below T_{bd}. Type III is the surface of a specimen fractured at low speed and below T_{bd}. This surface is a combination of surfaces of types I and II. Failure starts in a ductile mode. At some point during crack propagation failure mode changes and a surface type II is created. The size of the ductile mode area decreases with temperature. The change in fracture mode was explained by the ductile tearing instability [3]. For temperatures above T_{bd} failure occurs by massive yielding and the fracture surface is characteristic for ductile failure of PC. Below T_{bd} failure occurs by a crazing mechanism.

The deformation zone below the fracture surface was studied. Matrix deformation could be observed by measuring the deformation of the rubber particles. Draw ratio of the PC was 2-3, which is low compared to PA66-EPDM, where a draw ratio of 10-12 was observed [4]. The orientation of the particles was the same throughout the deformation zone. No melt layer was observed in the PC. Microscopy studies of the

orientation of the particles was the same throughout the deformation zone. No melt layer was observed in the PC. Microscopy studies of the deformation zones under the fracture surfaces of the PC/rubber blend support the results of the fracture surface.

For all SENT test speeds the effect of temperature on the fracture behavior was studied. The fracture process can be devided into a crack initiation and a crack propagation stage. The average crack speed is calculated from the crack propagation displacement and the clamp speed (fig. 2). In the ductile region the crack speed depends strongly on test speed; crack speed increases with test speed. The brittle crack speed appears to reach a constant level at temperatures below T_{bd}, indicated with a vertical line, but for the low speed tests not enough data is available to determine this value. The transition from brittle to ductile becomes sharper with increasing test speed.

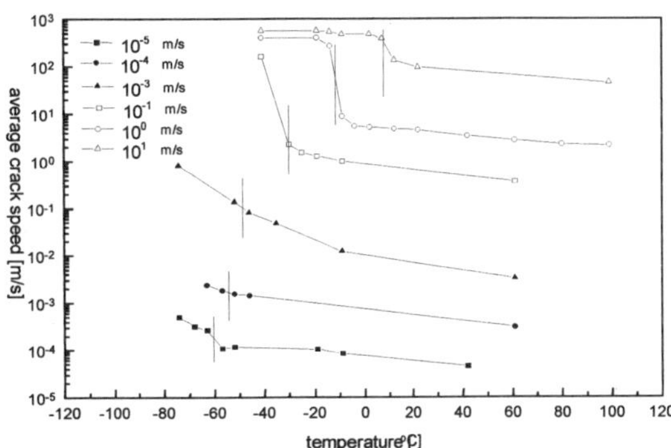

Fig. 2 *Average crack speed in notched PC specimens as a function of temperature for various clamp speeds.*

T_{bd} was determined from the test results. At high test speeds T_{bd} is clearly indicated by a sharp transition to ductile fracture and a change in fracture surface. By measuring the depth of the ductile zone as a function of temperature for various test speeds a correction for the ductile tearing instability effect could be made. The T_{bd} versus clamp speed curve shows a linear increase of T_{bd} with test speed.

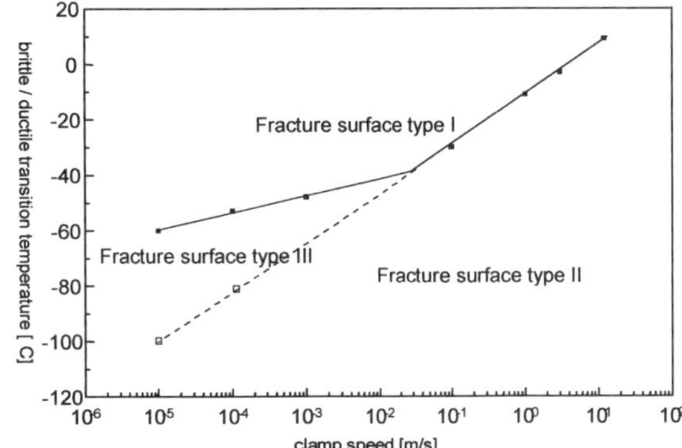

Fig. 3 *Adjusted T_{bd} versus clamp speed curve for notched polycarbonate specimens.*

Measurements were conducted with PC/rubber blend to ensure that the 1% rubber did not influence the fracture behavior. The SENT test results of the blend are almost identical to the results of the hompolymer. So the rubber particles merely act as deformation markers.

REFERENCES

1. K. Dijkstra, J. ter Laak, R.J. Gaymans, *Polymer*, **35**, 315, 1994
2. A. van der Wal, R.J. Gaymans, *Polymer*, Submitted
3. M.E.J. Dekkers, S.Y. Hobbs, *Polym.Eng.Sci.*, **27**, 1164, 1987
4. O.K.Muratoglu, A.S.Argon, R.E.Cohen, M. Weinberg, *Polymer*, **36**, 4771, 1995

The effects of the cavitation resistance of rubber on the toughening of modified epoxies

Junkyung Kim, Soonho Lim, Min Park, Moonbae Ko, and Chul Rim Choe

Korea Institute of Science and Technology.
Polymer Hybrid Center
P.O.Box 131, Cheongryang, Seoul, 130-650, Korea

Introduction

Toughness of epoxies has often been remarkably enhanced by the incorporation of rubber phase. For this reason, the toughening mechanisms in rubber-modified systems have long been the research interest of many material scientists[1-8]. For rubber toughened epoxies, it has been suggested the most important mechanisms are rubber cavitation and induced plastic deformation of surrounding matrices[3-6]. However, the roles of cavitation is still remained in argumentation. According to the hypothesis of Yee et al, the cavitation of rubber particles promotes shear banding and, therefore, is a prerequisite for massive shear yielding of matrix[8-12]. Also, they speculate that higher cavitation resistance of rubber particles may further improve the fracture toughness[11]. However, Kinloch et al hypothesized that the role of cavitation is just to initiate the plastic dilatation within the matrix and that this process is independent of shear banding[13-17]. Moreover, they insisted the cavitation resistance of rubber particles should not play a major role in toughening based on the results of efficiency of microvoids on toughening[17]. Based on the similar experimental work, Pearson[18] also conclude that the cavitation resistance of the rubber phase does not directly contribute to the toughness.

Thus, the purpose of this work is to find the direct experimental evidences for the exact role of cavitation in rubber-modified epoxies. To investigate the role of cavitation, the cavitation resistance is modified by changing the crosslinking density and the size of rubber phases. The amount of crosslinking density is changed by varying the content of crosslinking agents, such as trially isocyanate, tetramethylene glycole diacrylate, and penzoyl peroxide. The size of rubber is controlled by varying the end group of rubbers, the AN contents of rubbers, the composition of mixed rubbers, and the curing agents.

Experimental

The epoxy used in this study is diglycidyl ether of bisphenol A (DGEBA) epoxy and the curing agent are piperidine and diaminodiphenylmethan(DDM). The optimum concentration of piperidine and DDM is 5 and 20 wt%, respectively. The piperidine system is cured at 120 °C for 16 hrs, and DDM system is cured at 80 °C for 2 hrs/postcured at 150 °C for another 2 hrs. The rubbers modifiers used in this study are liquid type carboxyl- & amine-terminated butadiene-acrylonitrile rubber (CTBN & ATBN) and poly(butylacrylate-g-styrene & -g-methylmethacryalte) core-shell rubber with different crosslinking density. For crosslinking of rubber phase, trially isocyanate (TAIC), tetramethylene glycole diacrylate (TGD), and penzoyl peroxide (BPO) are used as crosslinking agent.

The fracture energy is measured by using a single-edge notched type specimen in a three-point bending geometry. For yield stress and modulus, specimens are tested in uniaxial compression. Scanning electron microscope (SEM) and transmission electron microscope (TEM) are used to investigate the fracture and phase separation behaviors. To examine a process zone at a crack tip, polarized transmission optical microscopy is used to investigate cavitation zone size and plastic deformation of the matrix near the crack tip. The specimens used for optical microscopy study are broken by double-notched four-point bending method. The thin section including this stationary crack is made by using the petrographic polishing technique. The section plain is parallel to the crack propagation direction and normal to the fracture surface.

Results and discussion

(1) core/shell rubber system with different crosslinking density

The fracture toughness of modified epoxies is highly enhanced with an inclusion of core/shell rubber. The amount of toughness enhancement has been found to be highly sensitive to content of crosslinking agent of core, ie the cavitation resistance, as seen in Figs. 1 and 2. For TAIC system, the fracture energy increases almost linearly up to 2 wt% of TAIC, and, at higher content, the toughness decreases rather slowly. However, for TGD system, the maximum toughness can be obtained when content of crosslinking agent is 1 wt%. When comparing these two system, the maximum toughness of TAIC system is higher than TGD system.

From SEM micrographs of fracture surface, when the rubber is crosslinked by TAIC, many internal cavities can be seen inside the rubber domain. However, when using TGD as a crosslinking agent, it seems debonding occurs between shell and rubber interface. The plastic dilatation of matrix around rubber phase after fracture has been found to be highly sensitive to the mount of crosslinking agents. As the content of crosslinking agent is increased, the amount of plastic dilatation of the matrix is increased. As seen on the optical micrographs of thin section perpendicular to the fracture surface, the process zone is increased with the content of crosslinking agent. The size of the process zone well corresponds to the fracture toughness of the systems.

(2) CTBN rubber system with different crosslinking density

The crosslinking density of rubber phase also can be changed by adding the radical initiator which can react with the double bond in liquid type butadiene rubber, even though, in this system, the quantitative control is quite difficult. However, when adding the BPO to rubber phase before or after mixing with epoxy, the fracture toughness is successfully enhanced with an increase of content of BPO, as seen in Fig.3.

(3) CTBN/ATBN rubber system with different domain size

The cavitation resistance has been reported to be also affected by the domain size of rubber phase[12,19]. As mentioned, the size of rubber is controlled by changing the end group of rubbers, the AN contents of rubbers, the composition of mixed rubbers, and the curing agents, because the phase separation behaviors depends on both critical degree of cure for phase separation and the curing rate. Thus, under slow curing rate, the rubber with higher AN contents corresponds to smaller domain size. However, under high curing rate, the reactive end group also suppresses the phase separation. When mixing the rubbers with different compatibility to epoxies, the size of rubber phases has been found to be strongly depend on the composition of rubbers. For fast cured system, the domain size is decreased with an increase of ATBN composition, but, in slow cure system, the domain size is increased with an increase of ATBN content. The size of rubber phase studied is shown in table 1.

The toughness of modified epoxies has been found to be sensitive to the size of rubber. As seen in Fig.4, there exists an optimum rubber particle size for maximum fracture toughness. The opimum particle size depends on the intrinsic critical stress for crack propagation, which is affected by the crosslinking density of epoxy matrices. For rubber toughened epoxies, the cavitation resistance is increased with a decrease of size of rubber phases.[12,17]

(4) the role of rubber cavitation on toughening

Above three systems all clearly exhibits the dependency of fracture toughness on the cavitation resistance. These results reveals that the cavitation of rubber phases highly contributes to the shear deformation/plastic dilatation of matrix, and that the relief of triaxial tensile stress by rubber cavitation is one of the most important steps in toughening. However, still, it is not clear that how the enhanced cavitation resistance affects the shear deformation or plastic dilatation of matrix. But, when the cavitation occurs, possibly, the octahedral shear stress is increased and the yield stress is lowered as a result of relief of triaxial tensile stress, which affects both the shear deformation and plastic dilatation of epoxy matrix. Based on the experimental results, it may be concluded that, for the maximum fracture toughness, the cavitation resistance should as high as the intrinsic critical stress for crack propagation, but should be lower. Thus, for higher crosslinked epoxy, the higher cavitation resistance is required for the maximum fracture toughness. Also, the relief of triaxial tensile stress by rubber cavitation should be occurred "on time".

References

1. Riew, C.K., Rowe, E.H., and Siebert, A.R. ACS *Adv. Chem. Ser.* 1976, **154**, 326
2. Kunz-Douglass, S., Beaumont, P.W.R., and Ashby, M.F. *J. Mater. Sci.* 1980, **49**, 442
3. Kinloch, A.J., Shaw., S.J., Tod, D.A., and Hunston, D.L. *Polymer* 1983, **24**, 1355
4. Yee, A.F. and Pearson, R.A. *J. Mater. Sci.* 1986, **21**, 2462
5. Kinloch, A.J., Finch, C.A., and Hashemi, S. *Polym. Commun.* 1987, **28**, 232
6. Pearson, R.A. and Yee, A.F. *J. Mater. Sci.* 1987, **24**, 2571
7. Levita, G., DePetris, S., Marchetti, A., and Lazzeri, A. *J. Mater. Sci.* 1991, **26**, 2348
8. Pearson, R.A. and Yee, A.F. *J. Mater. Sci.* 1991, **26**, 3828
9. Bucknall, C.B., Heather, P.S., and Lazzeri, *J. Mater. Sci.* 1989, **24**, 2255

10. Sue, H.J. *Polym. Eng. Sci.* 1991, **31**, 275
11. Yee, A.F., Li, D., and Li, X. *J. Mater. Sci.* 1993, **28**, 6392
12. Lazzeri, A. and Bucknall, C.B. *J. Mater. Sci.* 1993, **28**, 6799
13. Huang, Y. and Kinloch, A.J. *J. Mater. Sci.* 1992, **27**, 2763
14. Broutman, L.J. and Panniza, G. *Int. J. Polym. Mater.* 1991, **1**, 95
15. Huang, Y and Kinloch. A.J. *Polymer* 1992, **33**, 5338
16. Guild, F.J. and Young, R.J. *J. Mater. Sci.* 1989, **24**, 2424
17. Huang, Y. and Kinloch. A.J. *Polymer*, 1992, **33**, 1330
18. Bagheri, R. and Pearson, R.A. *Polymer* 1996, **37**, 4529
19. Bucknall, C.B., Karpodinis, A., and Zhang, X.C. *J. Mater. Sci.* 1994, **29**, 3377

Table 1. Characterization of rubber-modified epoxies

Rubbers[a]	DGEBA/DDM		DGEBA/piperidine	
	Particle size[b] (μm)	G_{Ic}[c] (J/m^2)	Particle size[b] (μm)	G_{Ic}[c] (J/m^2)
CTB162	76.5	393	71.8	510
CTBN8	2.54	430	1.63	2376
CTBN8/ATBN16 (7/3)	2.19	507	1.69	2550
CTBN8/ATBN16 (5/5)	1.60	526	1.71	2339
CTBN8/ATBN16 (3/7)	1.17	633	1.62	2638
CTBN13	0.339	573	~0.02	1083
CTBN13/ATBN16 (7/3)	0.194	673	0.420	1696
CTBN13/ATBN16 (5/5)	~0.02	761	1.03	2186
CTBN13/ATBN16 (3/7)	~0.02	753	1.43	2395
ATBN16	~0.02	675	1.67	2846

[a] The total rubber content for modified epoxies is 10wt%.

[b] The average particle size of rubbers were measured from the fast crack propagation region of 3PB specimens by using SEM micrographs.

[c] The fracture energy were determined by using single-edge notched specimens tested in three point bending.

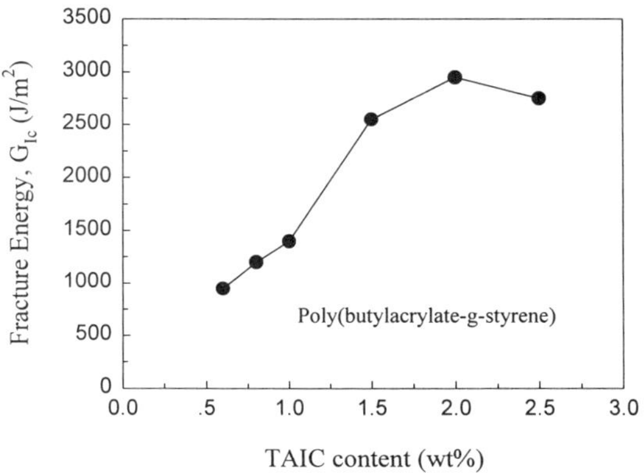

Figure 1. Effects of TAIC content on fracture toughness of modified epoxies (10 wt%).

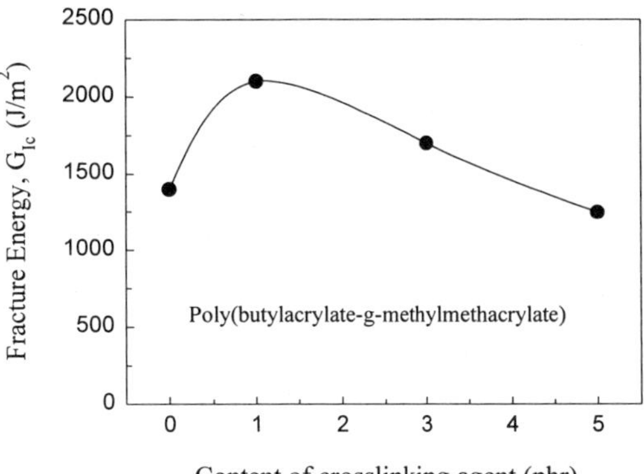

Figure 2. Effects of TGD content on fracture toughness of modified epoxies (10 wt%).

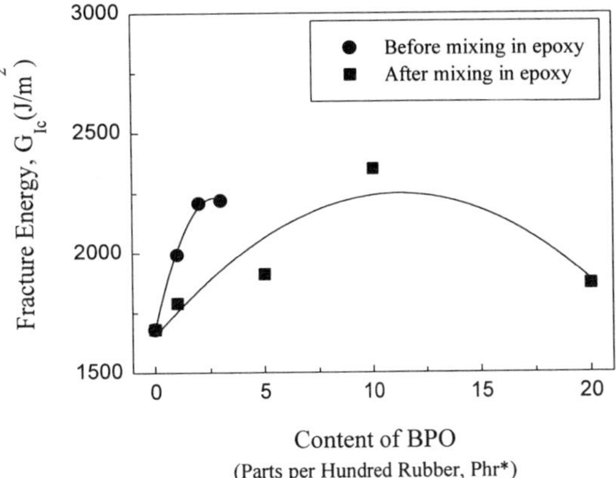

Figure 3. Effects of content of BPO on fracture toughness of modified epoxies (10 wt%).

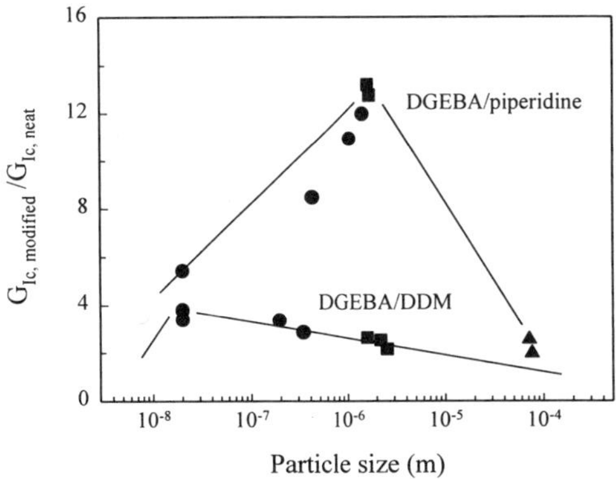

Figure 4. Effects of particle size on fracture toughness of modified epoxies (10 wt%).

PHOSPHINE OXIDE CONTAINING COPOLYMERS TOUGHENED EPOXY

J. Wang, S. Wang, H. Kang, Q. Ji, O. Kwon, J. E. McGrath, T. C. Ward

Virginia Polytechnic Institute and State University
Department of Chemistry and NSF Science and Technology Center:
High Performance Polymeric Adhesives and Composites
Blacksburg, VA 24061-0344

INTRODUCTION

A comprehensive research on toughened epoxy modified by new copolymers is reported. The change of microstructure and fracture toughness of the thermosets is determined by the phosphine oxide content in copolymers.

The toughening of epoxy resins have received much attention in high performance composites and structural adhesive applications because of some exceptional properties[1,2]. In the current research, studies were focused on a series of copolymers: poly(ether phosphine oxide)/poly(ether sulfone) (PEPO/PES), which have been synthesized with well controlled micro-structure. By adjusting the molar ratio of PEPO and PES, the copolymers can add toughness to the epoxy resins. It was also possible to gain a better understanding of the relationship between the structure and properties in the blends with copolymers and epoxies. The high Tg of these copolymer systems was maintained in the blends and fire retardant properties are expected to be improved with PEPO content.

EXPERIMENTAL

Materials

The diglycidyl ether of bisphenol-A (DGEBA) epoxy resin (Shell EPON 828, MW=380 g/mol) and the curing agent, 4,4`-diaminodiphenylsulfone (DDS, Aldrich) were used as received. Bisphenol-A terminated random copolymers of poly (ether phosphine oxide) / poly (ether sulfone) (PEPO/PEP) were prepared in our laboratory (Fig. 1). The structures of copolymers were found to be well controlled with a series of PEPO/PEP molar ratio and constant molecular weight, as shown in Table 1. Their Tgs vary from 185°C to 195°C.

PEPO   PES

Figure 1. Chemical structure of copolymer PEPO/PES

Table 1. Molar ratios and molecular weights of copolymers

ID #	PEPO (mol %)	PES (mol %)	Mn (g/mol)	Mw (g/mol)	Pd
0	0	100	26,000	48,000	1.85
10	10	90	25,000	45,000	1.88
20	20	80	26,000	49,000	1.88
30	30	70	24,000	46,000	1.88
50	50	50	24,000	44,000	1.87
75	75	25	23,000	43,000	1.87
100	100	0	25,000	46,000	1.84

The toughened epoxy resins were prepared with the method shown in the reference[3]. Two loadings of copolymers were provided, 10%wt and 20%wt. The resins were named Epx-(75:25) 10%wt, in which Epx means the epoxy part, (75:25) is the copolymer with PEPO/PEP molar ratio of 75:25, and 10%wt is the loading of copolymer.

Characterization and Property Measurements

DMTA testing was performed all toughened resins to observe the phase separation point from the information of tan δ curves at a frequency of 1 Hz. Morphology studies were also carried using SEM experiments.

Fracture toughness (K_{1C}) was measured following ASTM Standard D-5045-91.

Results and Discussion

Morphology studies of toughened epoxy resins

For each copolymer, two loadings of epoxy resins were prepared, 10%wt and 20%wt. In DMTA studies (Figure 2), tan δ curves at 1 Hz were employed. The phase separation for both loadings appeared when the content of PEPO in copolymers was as low as 20%. For those phase separated materials (PEPO:PES = 20:80, 10:90 and 0:100), the shapes of tan δ curves suggested that a higher degree of phase separation with lower PEPO contents.

For the two series of toughened resins(10% wt and 20% wt), SEM pictures further supported the DMA results. Both phase separation points appeared at the PEPO content level of 20% in copolymers despite of the copolymer loadings (10% wt and 20% wt). Specifically, for the series of 20% wt copolymer loading, the phases are inverted, in which the copolymer is the continuous phase and epoxy is the dispersed phase (Fig. 3). For phase separated resins with both loadings, the clarity of the boundary between epoxy phase and copolymer phase is poorer, with higher content of PEPO in copolymers. The same happened for particle sizes, the more PEPO, the smaller particle.

The change of morphology as shown by the clarity of boundary and particle size, is important for the performance of materials. This change mainly comes from the specific interaction by the way of hydrogen bonding between the phosphine oxide group in the copolymers and the hydroxy group in the epoxy, which essentially depends on the chemical structure of copolymers (PEPO:PES ratio) as shown in our previous works[4].

Fracture toughness of epoxy resins

The influence of these resins' difference in microstructures on their mechanical properties was reflected by the fracture toughness (K_{1C}) test as shown in Fig. 4. Epx-(20:80) (20% wt) showed the highest K_{1C} value of 1.34 MPa*m$^{1/2}$ and that for Epx-(10:90) (20% wt) is also high. In both loading series, all toughened materials had increased toughness compared to pure epoxy. All homogeneous materials performed similarly in value, while the phase separated materials behaved very different. Among two-phase materials, pure polysulfone toughened systems (Epx-(0:100)) showed low values compared to those copolymer toughened systems, and there is a maximum point along each loading curve.

CONCLUSION

By adjusting the PEPO content in copolymers, the strength of specific interaction (hydrogen bonding) between copolymers and epoxy could be controlled. As the result, the change of macrostructure, phase separation, occurred at a certain content of PEPO (20% of copolymers). Furthermore, the microstructure could also be adjusted by different PEPO content in the copolymers. Accordingly, the mechanical properties were essentially affected by the chemical structures of materials. This knowledge is one of key factors in the design of materials with expected properties.

REFERENCES

1. J. N. Sultan and F. J. McGarry, *Polym. Eng. & Sci.,* 13(1), 29 (1973)
2. H. G. Recker, *SAMPE Proceeding,* 34, 747 (1989)
3. T. H. Yoon, D. B. Priddy Jr., G. D. Lyle and J. E. McGrath, *Macromol. Symp.* 98, 673 (1995)
4. J. Wang, S. Wang, Q. Ji, O. Kwon, J. McGrath and T. Ward, *21th Adhesion Science Meeting,* Savannah, Feb., 1998

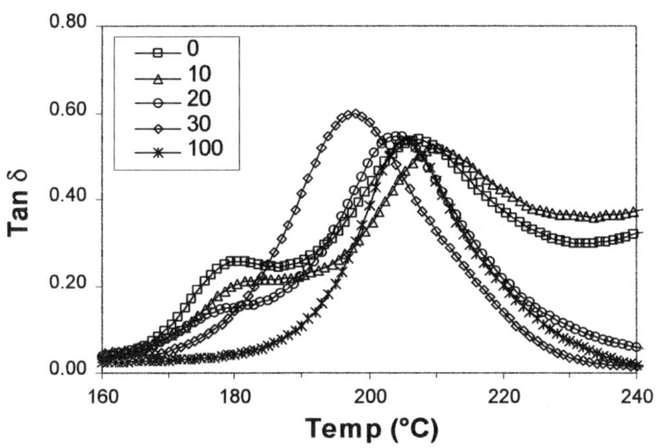

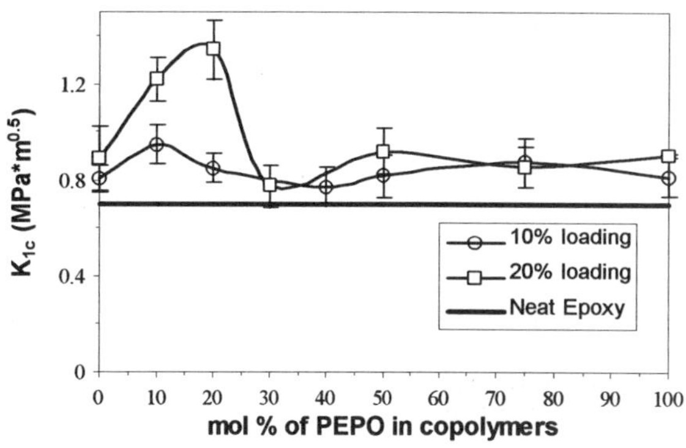

Figure 2. DMTA results of toughened epoxy resins with different compositions of copolymers (0,10,20,30,100 represent for the percentages of PEPO component in the copolymers, respectively) @ 20%wt loading

Figure 4. Fracture toughness results for copolymer toughened epoxy resins (10% wt and 20% wt loading) with various PEPO content in the copolymers

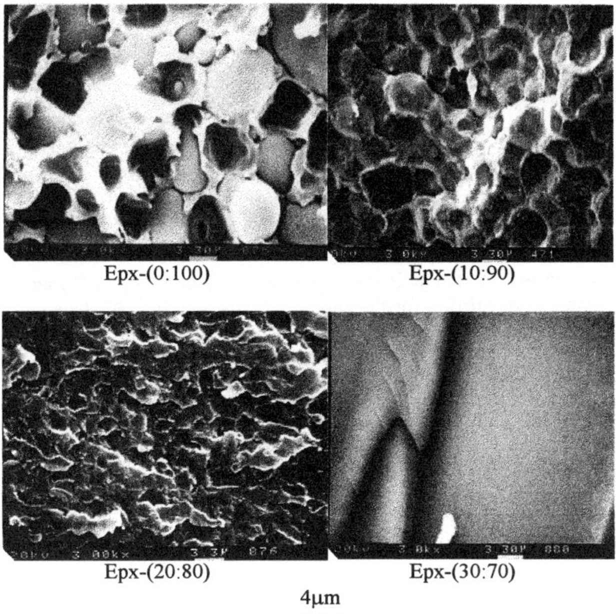

Figure 3 Morphology evolution of epoxy resins with different composition of copolymers @ 20%wt loading

THERMALLY STABLE ELASTOMERIC TOUGHENERS FOR POLYCYANURATE THERMOSETTING SYSTEMS

D. S. Porter, S. Bhattacharjee and T. C. Ward

Center for Adhesive and Sealant Science
National Science Foundation Science and Technology Center:
High Performance Polymeric Adhesives and Composites
Department of Chemistry, Virginia Tech
Blacksburg, Virginia 24061

Introduction

The usefulness of thermosetting polymers depends upon their unique properties: processibility, high temperature capabilities, good adhesive properties and low creep. However, high-performance thermosetting materials often suffer from high brittleness arising from high crosslink density. To offset this tendency, toughening agents have been employed to enhance the fracture properties. The polycyanurates are a new class of high performance thermosetting resins increasingly finding use in adhesive and aerospace applications. The advantages of these thermosetting systems are many: high operating temperatures, high stiffness, and excellent transparency to radio frequency energy due to a low dielectric constant. These resins possess some of the same drawbacks as other thermosetting systems, however, and may be unacceptably brittle if unmodified.

Fortunately, the polycyanurates are easily toughened through incorporation of a suitable thermoplastic or rubber toughener, yielding a tough, high temperature material. [1,2] Toughening through inclusion of separate, ductile particles is often accomplished via the inclusion of the toughener in the low molecular weight thermosetting resin before cure, followed by phase separation induced by the decrease in configurational entropy due to the increase in molecular weight as the polymerization progresses. This process places severe restrictions on the toughener: the pre-formed polymer must be initially miscible with the low molecular weight resin, must phase separate on curing to produce well-controlled microstructures, and must be thermally and oxidatively stable at the cure and use temperatures of the matrix resin.

The aim of this work is to investigate the efficacy of practical tougheners for polycyanurates considering the multiple requirements for optimal performance of the system. Because elastomeric modifiers are more efficient than glassy inclusions, the materials selected have low glass transition temperatures.

Materials

The cyanate resin employed in this investigation is a bisphenol-A-based dicyanate (BADCy), AroCy B-10 (Ciba-Geigy) (figure 1). Thermal activation of the polymerization is utilized, with the addition of a co-catalyst solution of aluminum acetylacetonate in nonylphenol, and the reaction continues to a high degree of conversion. The resultant polycyanurate has a high glass transition temperature, 290 °C (DMA, 1 Hz). [3] The tougheners employed include a commercial hydroxyl-terminated butadiene-acrylonitrile (HTBN) copolymer produced for the cyanate chemistry, Echo Resins ATx013, together with a series of low-T_g amorphous copolyesters. These materials were supplied by Bostik as Vitel 3200, 3300, 3500, and 3550 resins, and used as received.

Figure 1
Cyanate chemistry and
polymerization

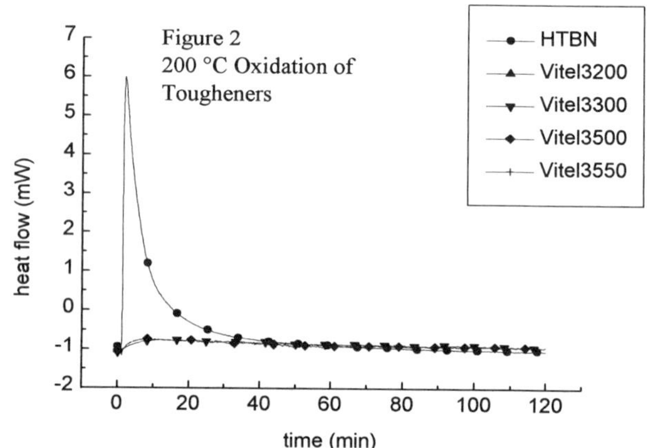

Bisphenol-A Dicyanate (BADCy)
Ciba-Geigy AroCy B-10
m.p. 78 °C

Results and Discussion

Thermal stability measurements were undertaken using thermogravimetric analysis (TGA). The onset temperature of the main degradation detected by TGA as weight loss at 10 °C/min was recorded. The degradation temperatures were 397 °C for HTBN, 368 °C for Vitel 3200, 365 °C for Vitel 3300, 385 °C for Vitel 3500 and 380 °C for Vitel 3550. Small differences are apparent; however, one of the major degradation mechanisms of the BN toughener is oxidative, and no weight loss would be apparent using this technique, though the material's properties are changing as a result of oxidation. Additional thermal stability experiments were undertaken using differential scanning calorimetry (DSC), where samples were placed in the DSC under nitrogen purge, held isothermally, and the purge gas switched to air.

The exothermic peak area appears substantially larger for HTBN than HHTBN; however, the polyester materials produced much smaller exotherms in the presence of the oxidizing environment, indicating better resistance to oxidation at temperatures from 150 °C to 250 °C. (figure 2)

The HTBN and Vitel series of tougheners were initially incorporated into the molten cyanate monomer in 5, 10, 15, 20 and 25% by weight via stirring for the case of HTBN, and in methylene chloride solution followed by subsequent solvent elimination in the case of the polyesters. The resultant solutions were catalyzed and cured in an aluminum mold at 150 °C or 210 °C to greater than 60% cure (the gel point), and postcured at 250 °C for two hours to equalize the degree of cure in all samples.

The resultant samples were examined using dynamic mechanical analysis (DMA) at 1 Hz. Phase separation was investigated through consideration of the depression of the glass transition temperature of the polycyanurate mixed homogeneously with low T_g toughener. For the case of phase separation via a nucleation and growth mechanism, incomplete phase separation may result from gelation of the thermoset during the time consuming phase separation process, where the phases would be locked in a non-equilibrium condition. A faster approach to the gel point provides less time for phase separation to take place, leaving toughener mixed homogeneously with the polycyanurate, depressing its T_g. For spinodal decomposition, the phase compositions should be near equilibrium regardless of cure temperature. Toughening with the polyester materials produces different results depending on the nature of toughener. In three cases (3300-type, 3500-type and 3550-type polyesters), the modifier depressed the glass transition temperature according to the Fox equation considering homogenous mixing of all added toughener, indicating an insignificant degree of phase separation. One type of polyester, the 3200-type, did demonstrate a much smaller degree of T_g depression as well as a separate relaxation near the T_g of the toughener due to the formation of a separate toughener-rich phase.

Increasing the amount of toughener added resulted in the magnitude of the toughener-rich phase relaxation increasing due to the formation of an increased volume fraction of toughener-rich phase. Cure temperature increases resulted in a small increase in the magnitude of the relaxation due to this phase. (figure 3) For the HTBN-toughened material, an increase in the amount of toughener increases the magnitude of the relaxation below 15% addition; beyond this point, only a single T_g is observed. Below 15% addition, an increase in the cure temperature results in a smaller magnitude for the relaxation peak resulting from a decrease in the amount of toughener that phase separates. This data suggests that the mechanism of phase separation is spinodal in the case of the polyester toughener and nucleation and growth in the metastable state for the HTBN modifier.

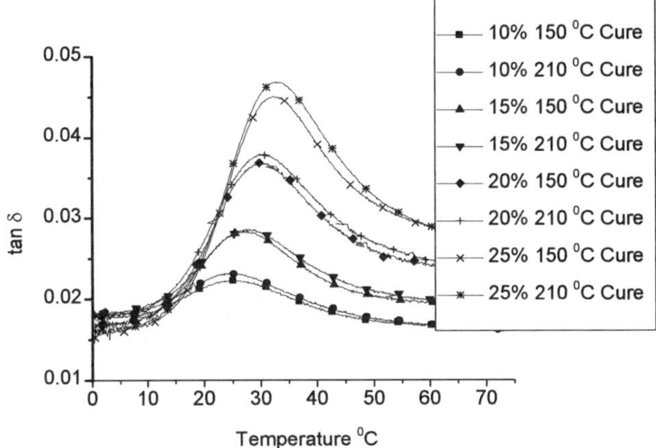

Figure 3
Loss tangent peak for toughener-rich phase with cure temperature and increasing concentration

Toughness measurements were performed in three point bending mode using the single-edge-notched-beam method. [4] Toughness of the HTBN-modified samples increases with amount of added toughener below 15%; above this the toughness is relatively unchanged from the untoughened polycyanurate. Increases in the cure temperature result in small reductions in the fracture toughness considering a constant amount of added toughener. Results for the polyester modified resin indicate that increasing the amount of toughener added produces increases in toughness for each case. (figure 4) Increasing cure temperature has little effect on the measured toughness. Improvements in toughness of the polyester toughened cyanate are comparable to the improvements resulting from the same weight fraction addition of the HTBN toughener, indicating similar efficacy.

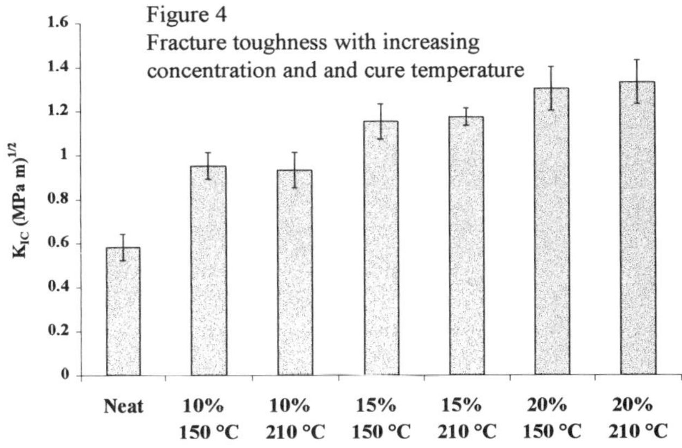

Figure 4
Fracture toughness with increasing concentration and and cure temperature

The microstructure of each sample was investigated using Atomic Force Microscopy (AFM) in tapping mode. The microstructure of each sample was examined for the presence, size and shape of any second phase. Samples which did not display a distinct mechanical relaxation did not exhibit a distinct second phase. Materials which displayed the presence of a second phase were the HTBN and Vitel 3200 polyester toughened materials. For the 3200-type polyester samples, irregularly-shaped particles were evident, which increased in size with increasing cure temperature. Particle size also increased with increasing weight fraction of toughener added. For the HTBN-toughened samples, an increase in the amount of toughener added resulted in larger and more numerous spherical particles up to 15% addition. Beyond this amount, a distinct second phase was not detectable. As cure temperature was increased, the size of the particles decreases. These observations support the idea that phase separation takes place by a spinodal mechanism in the case of the polyester toughened systems and by nucleation and growth in the metastable state for the HTBN toughened materials.

Conclusions

The inclusion of various low-T_g modifiers produced significant improvements in the fracture toughness of a bisphenol-A based polycyanurate thermosetting system. The thermal and oxidative stability of a commercial butadiene-acrylonitrile was compared to low-T_g amorphous polyester resins, which were shown to resist oxidation much better than the HTBN toughener. The 3200-type polyester phase separates while the 3300, 3500, and 3550 types do not; the HTBN toughener also produces two distinct mechanical relaxations corresponding to the glass transitions of the two phases. The magnitude of these relaxations increases with increasing toughener addition in both cases; increasing cure temperature produced a larger magnitude relaxation for the polyester-modified samples and decreasing relaxation magnitude for the HTBN-modified material. AFM revealed the presence of two distinct phases for all samples demonstrating two T_gs. These second phases increase in size with increasing addition of modifier in both cases. With an increase in cure temperature, the particle size increases in the case of the polyester-modified material and decreases for the HTBN-toughened samples. Fracture toughness increases with increasing toughener concentration in all cases, though changes in cure temperature have a relatively small effect on fracture toughness.

Acknowledgments

The authors would like to acknowledge the Center for Adhesive and Sealant Science and the Adhesive and Sealant Council Education Foundation for their generous support, as well as Steve McCartney for microscopy, the members of the PolyPkem research group for invaluable discussions, and Bostik and Echo Resins for materials.

References

[1]   D. A. Shimp and J. R. Christenson, 11th International SAMPE Conference, Basel, Switzerland, 1990, pg. 5.
[2]   S. A. Srinivasan, Ph. D. Dissertation, VPI&SU, 1994.
[3]   D. S. Porter, L. T. James and T. C. Ward, Proceedings of the 20th Annual Meeting of the Adhesion Society, February 18-21, 1997, pg. 40.
[4]   D5045-96, Annual Book of ASTM Procedures, 1996.

Effect of interfacial chain structure on toughening behavior in rubber modified poly(methyl methacrylate)

Jung-Hyun Kim, Jeong-Ho An*, Young-Jun Park, and Hyun-Jae Ha

Dept. of Chem. Eng., Yonsei University,
134 Shinchon-Dong, Seodaemoon-Ku, Seoul 120-749, KOREA
*Dept. of Polym. Sci. and Eng., Sung Kyun Kwan University,
Jangan-Ku, Suwon, Kyungido 440-749, KOREA

INTRODUCTION

Rubber modification of brittle plastics has a significant importance in commercial impact plastics. The toughening mechanisms of PMMA have been investigated by many other authors.[1,2] However, a quantitative understanding of the mechanism of how rubber modification results in improved impact properties has met with only limited success. It has been shown that the morphology of the rubber toughened composite particles provide strong influence on the mechanical behavior of the base polymer matrix.[3,4] Also, the properties along the interface, for example, interface thickness, grafting efficiency, adhesion between phases, etc., is essential for improvement in the impact strength.[5-7]

In our previous reports, we could modify the interfacial properties of the composite particles.[8] The molecular interdiffusion and microstructure developments of heterocoagulated composite particle were investigated.[9] This made it possible to vary interfacial property without changing the other physical or chemical properties. In this paper, the interfacial chain structure was discussed in relation to the effectiveness of the impact reinforcement of PMMA matrix.

EXPERIMENTAL

Materials

Monomers - styrene (ST), methyl methacrylate (MMA), butyl acrylate (BA), acrylic acid (AA), and glycidyl methacrylate (GMA) - were purified by vacuum distillation and stored in the refrigerator. Initiators, potassium persulfate (KPS), 2,2'-azobis (2-amidino propane) hydrochloride (AIBH), have been used as received. Sodium dodecyl sulfate (SDS) and reagent grade chemicals were also used without further purification. DDI water was used throughout.

Polymerization

Colloidal large particles (LP) with surface epoxy groups were prepared through semi-continuous polymerization of GMA as shells around existing PMMA seed particles. The recipe for PMMA and LP latex is given in Table 1. The SP latexes having surface carboxyl groups were prepared using the two-stage shot-growth process because particle size and surface charge density can be controlled independently. The recipe was listed in Table 2. PMMA particles for matrix were prepared through batch process (Table 3).

Particle Size and Particle Size Distribution

Particle size and distribution were determined using capillary hydrodynamic fractionation (CHDF-1100, Matec Applied Science).

Surface Charge Density

Surface charge densities of latex particles were determined using the conductometric titration after the latexes had been rigorously cleaned.

Heterocoagulation

Heterocoagulation was carried out as follows: [1] SP latex was diluted with water to 5wt% solid. [2] Then SP latex was added to LP latex slowly by tubing pump at room temperature at a rate of 1.0ml/min. [3] Then the temperature of heterocoagulated particle latex was maintained at room temperature for 6 hours for the film formation of SP on the surface of LP.

Solid Specimen Preparation

The solid specimen preparation scheme and notation for the final samples are summarized in Fig. 1. Heterocoagulated particles and matrix particles were mixed in the emulsion state. These emulsion blends were dried in a freeze dryer at -50℃. From the blends prepared, sheets of 2.0mm thickness were obtained by compression molding in a heated press. The samples were heated to 140℃ and kept 5min between the plates without any applied pressure. After this period, a pressure of 3,000 psi was applied at the same temperature for 195min.

Mechanical Property

The mechanical properties, the stress intensity factor (K_{IC}) and modulus of the final solid specimens were determined in an universal testing machine (Instron Corporation Series IX Automated Materials Testing System, USA) under static loading at a crosshead travel speed of 1.0 ml/min.

Fracture toughness, K_{IC}, was measured by the notched three-point bending test with a crosshead speed of 1mm/min according to ASTM E399.

RESULTS AND DISCUSSION

Particle Size, Particle Size Distribution, and Surface Charge Density

Measured particle size and particle size distribution for LP, SP, matrix latexes were shown in Table 4. The surface charge densities were from 1.68 to $20.52\ \mu C/cm^2$. Sample SP-A is the lowest charged case and SP-E is the highest case. The increase in surface charge density derived from AA meant that the number of carboxyl groups on SP's surface was increased.

Mechanical Properties Measurement

In this work, it could be possible by hetrocoagulation and annealing process using five kinds of SP latexes with different carboxyl charge density and a kind of epoxy functional LP latexes. These functional groups are capable of associating with one another through chemical reaction by thermal treatment.

As mentioned above, the increase in surface charge density derived from AA meant that the number of carboxyl groups on SP's surface was increased. Consequently, the heterocoagulated particles composed LP and SP-E had the highest degree of interfacial reaction between hard core and soft shell phase. On the other hand, the size was similar with heterocoagulated composed LP and the other SP's. Fig. 2 showed the variation in K_{IC} as a function of the surface charge density derived from AA of SP used in heterocoagulation process. A small increase in K_{IC} was observed up to surface charge density of the order of $12.7\ \mu C/cm^2$, followed a significant increase in K_{IC} toward the high surface charge densities.

In general, the loss of cohesion must be initiated at the weakest points and hence at the interfaces. These interfaces are hard core/soft shell interface and soft shell/matrix interface. Therefore, this interface between core and shell phase appears to play an important part in the mechanical behaviors of these materials. The degree of interfacial reactions was related to the number of SP's chains attached to LP. It is thought the greater this number, the better would be the resistance of interface between core and shell. The increase in the resistance must lead to improved toughness. Fig. 3 shows the relation of degree of chemical linkages between two phases with surface charge density. However, the moduli of the final solid specimens composed of matrix and heterocoagulated particles showed similar values due to the constant soft phase contents in the specimens (Table 5).

CONCLUSIONS

A series of HCP were prepared using two kinds of particles of different sizes: soft small anionic and hard large cationic particles. The surface of large particles and small particles was covered with epoxy groups and carboxyl groups, respectively. These functional groups are capable of associating with one another through chemical reaction by thermal treatment. To vary the degree of interfacial reaction between functional groups of the small and the large particles, five kinds of small particles with different surface carboxyl charge density were used. The critical stress intensity factor (K_{IC}) and moduli of emulsion blends were measured by universal testing machine. The moduli of solid specimens composed of matrix and heterocoagulated particles showed similar values due to constant soft phase contents in the emulsion blends. However, the higher degree of interfacial chemical linkage between two phases in HP resulted in improved toughness.

REFERENCES

1. Borgreeve, R. J., Gaymans, R. J., & Schuijer, J., *Polymer*, **28** (1987) 1489.
2. Bucknall, C. B., Partridge, I. K., & Ward, M. V., *J. Mater. Sci.*, **19** (1984) 2064.
3. Cavaille, J.Y., Jourdan, C., Kong, X.Z. & Perez, J., *Polymer*, **27** (1986) 693.
4. O'Conner, K.M., & Tsaur, S., *J. Appl. Polym. Sci.*, **33** (1987) 2007.
5. Canche-Escamillar, G., Mendizabal, E., Hernandex-Patino, M.J., Arce-Romero, S.M., & Gonazlez-Romero, V.M., *J. Appl. Polym. Sci.*, **56** (1995) 793.
6. Kim, K.R., An, J.H., & Park, C.E., *J. Appl. Polym. Sci.*, **47** (1993) 305.
7. Lambla, M., Schlund, B., Lazarus, E., & Pith, T., *Macromol. Chem. Suppl.*, **10/11** (1985) 463.
8. Park, Y.J., Ha, H.J., & Kim, J.H., *Polym. Gels & Networks*, **5** (1997) 153.
9. Park, Y.J. & Kim, J.H., *Polym. Eng. Sci.*, in press.

Table 1. The recipe for the preparation of PMMA seed and LP latexes

Components	Amount used in seed (g)	Amount used in second stage polymerization (g)
Seed	-	400.0
MMA	150.0	-
AIBH	0.9	0.1
GMA	-	10.8
Triton X-405	-	0.1
D.D.I. water	1515.0	700.0

Table 2. Shot growth polymerization recipe for preparations of SP latexes

Component	Initial charge (gram)	Shot injection (gram)
ST	40.0	Variable[1]
BA	40.0	Variable[1]
AA	0.4	Variable[2]
KPS	0.8	0.2
SDS	2.4	-
D.D.I. water	452.4	150.0

[1]: 7.9-5.6 g, [2]: 0.2-4.8 g

Table 3. The recipe for preparation of matrix particles

Component	Amount (g)
MMA	500.0
KPS	3.4
SDS	7.2
DDI water	1272

Table 4. Characteristics for small particles (SP), large latexes (LP), and heterocoagulated particles (H-SP)

Sample I. D.	D_n(nm)	D_w/D_n	$\sigma\,(\mu C/cm^2)$
LP	274.8	1.04	-
SP-A	67.9	1.06	1.68
H-SP-A	436.0	1.12	-
SP-B	70.7	1.06	6.09
H-SP-B	446.4	1.35	-
SP-C	69.7	1.04	12.70
H-SP-C	434.2	1.27	-
SP-D	68.2	1.06	14.84
H-SP-D	431.0	1.28	-
SP-E	69.7	1.06	20.52
H-SP-E	450.4	1.15	-
MP	103.0	1.07	-

Table 5. Tensile modulus values for the molded specimens

The molded specimens	Tensile modulus (Log E'. Pa)
H-SP-A	8.83
H-SP-B	8.84
H-SP-C	8.82
H-SP-D	8.83
H-SP-E	8.89

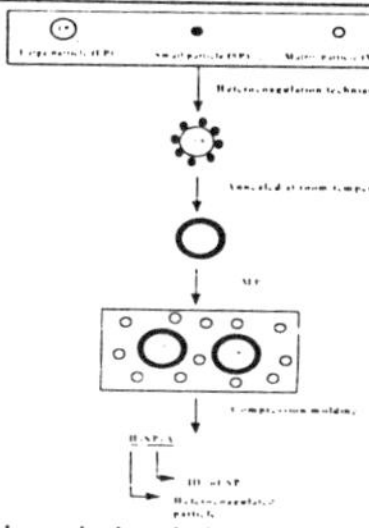

Fig. 1. Schematic description of sample preparation.

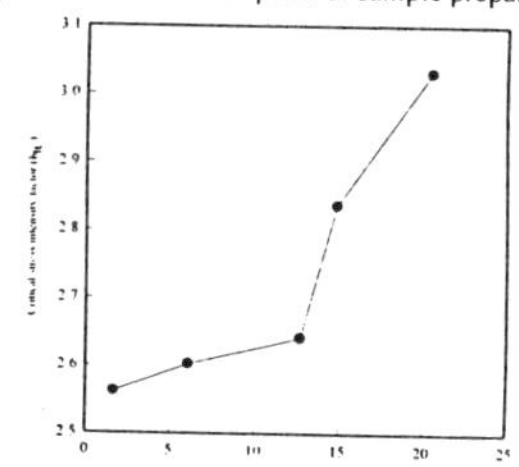

Fig. 2. K_{IC} values of the specimens containing HCP.

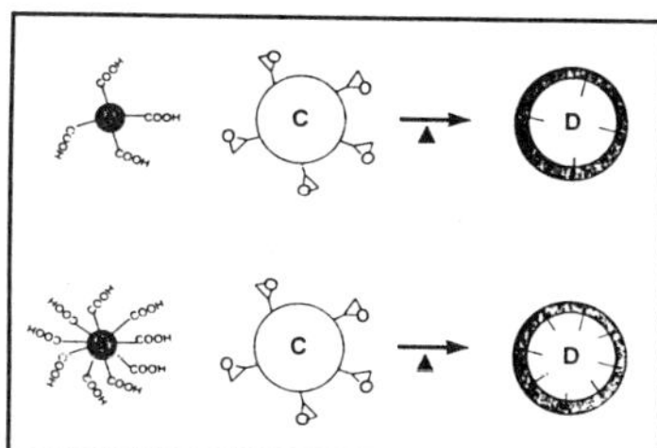

Fig. 3. Schematic diagram of formation of linkage between SP and LP.

IMPACT FRACTURE TOUGHNESS OF PROPYLENE / 1-PENTENE RANDOM COPOLYMERS

I. Tincul*, D. Joubert*, A.H. Potgieter**
*Sasol Technology, 9570 Sasolburg, Republic of South Africa
**Polifin, 2194 Randburg, Republic of South Africa

INTRODUCTION

A lot of research is being done to tailor polypropylene properties to meet specific requirements. To improve the low impact performances of the polypropylene homopolymer the trend is to use reactor-produced block copolymers and blends with high rubber content or extruded compounds of polypropylene polymers with elastomers. As the elastomer content increases the polymer behaviour under impact changes from brittle to increasingly more ductile. However, impact resistance increases at the expense of other properties. Some random copolymers of propylene with ethylene or 1-butene have moderate impact strength and stiffness and good optical properties.

We prepared propylene / 1-pentene random copolymers which have high impact strength, good tensile properties, excellent optical properties, good rheological properties and a large processing window. In this paper the impact properties of propylene / 1-pentene random copolymers related to their structure and fundamental properties are discussed.

EXPERIMENTAL

To a 10ℓ automated autoclave at 70°C containing 3000g of purified n-heptane was added triethyl aluminium cocatalyst, diphenyl dimethoxysilane external electron donor and a magnesium chloride-supported titanium trichloride catalyst containing 3.5% Ti. After 15 minutes ageing of the catalyst, a continuous flow of propylene together with 1-pentene was started and the reaction continued for 2 hours at 15 bar.

The slurry was deactivated with iso-propanol, filtered and the copolymer washed with acetone and dried in a vacuum oven at 80°C. Molecular weight was regulated with hydrogen and the 1-pentene content controlled by the propylene/1-pentene ratio.

The composition of the copolymer was determined on a Perkin Elmer FT-IR 1720X instrument on compression moulded film samples of 0.3 mm thickness moulded on a Graseby Specac press at 180°C by making use of a calibration curve obtained by plotting absorbance over film thickness against concentration of poly (1-pentene) melt-blended with polypropylene. The moderately strong peak at 969 cm^{-1} in the spectrum of PP arises from coupling vibrations, while the rocking of the CH$_3$ group at 734 cm^{-1} was used to determine the polymethylenic branch derived from 1-pentene.

Molecular masses were determined on a Waters 150 CV GP chromatograph equipped with data module and computer acquisition system. Determinations have been done on the copolymer samples dissolved at 150°C in 1,2,4-trichlorobenzene. Each tray of samples included a polystyrene standard and the NBS 1475a standard in order to check the validity of data against the calibration curve data. Polystyrene standards, spanning the MW range 3 100 000 - 1000 g/mol were used for calibration. Differential Refractive Index was used for detection.

Fusion enthalpy was measured on a Perkin-Elmer DSC-4 controlled by a TADS data station. Determinations were done under a nitrogen atmosphere at a pressure of two bar. Samples were initially heated from 50 to 200°C at 20°C/ min., held at 200°C for 1 min., cooled to 50°C at a rate of 20°C/ min. and held for 1 min. The analysis was than recorded between 50 and 200°C at a heating rate of 10°C/min.

The copolymer powder samples were stabilized against thermo-oxidative degradation by a particular technique. The monomeric antioxidant 3,5-di-t-butyl-4-hydroxy styrene at a concentration of 0.1% was melt grafted on the copolymer backbone by mixing in the chamber of a Brabender Plasticorder PL 2000/6 at 60 rpm for 8 min. at 170°C in the presence of 0.1% bis (2,4-di-t-butyl phenyl) pentaerythritol diphosphite and 0.2% calcium stearate. The efficiency of the stabilization was assessed by multiple extrusion on an Extrusiograph 19/25 unit run at 60 rpm. The melt flow index of a copolymer with a 1-pentene content of 1% does not change more than 50% at 5 extrusion passes.

Tensile properties were determined according to ASTM D 638 M, melt flow index (MFI) according to ASTM D 1238, haze according to ASTM D 1003 and Izod impact strength according to ASTM 256. Machined notches were prepared in accordance with ISO 2818. Conditioning for 24 hours and determination was done at 23°C.

Optical microscopy was performed with a Leitz Laborlux-S with 400 times enlargement fitted with a thermostatically controlled hot-stage, polarizing filter and a 35 mm camera.

RESULTS AND DISCUSSION

To investigate the relationship between the impact properties and structure of the propylene/1-pentene copolymers, samples 1 to 6 in table 1 having similar molecular weights and different 1-pentene contents were considered.

Table 1. Fundamental properties of propylene/1-pentene copolymers.

Expt. Nr.	1-Pentene (wt %)	Mw (g/mol)	Mn (g/mol)	MWD (Mw/Mn)	MFI (dg/min)
1	0	440668	85277	5.2	7.4
2	1	446923	85023	5.3	7.0
3	2	415298	83278	5.1	7.0
4	3.1	415608	76973	5.4	8.5
5	4	406325	78432	5.3	8.0
6	4.9	405880	76581	5.3	7.5
7	3.7	268557	50210	5.3	40
8	3.8	361761	67220	5.4	22
9	3.5	665082	110997	6.0	6
10	4	851702	137001	6.3	2.4

In Figure 1 the notched Izod impact strength of these copolymers are represented against 1-pentene content. It can be seen that increased impact strength is obtained when 1-pentene content is increased.

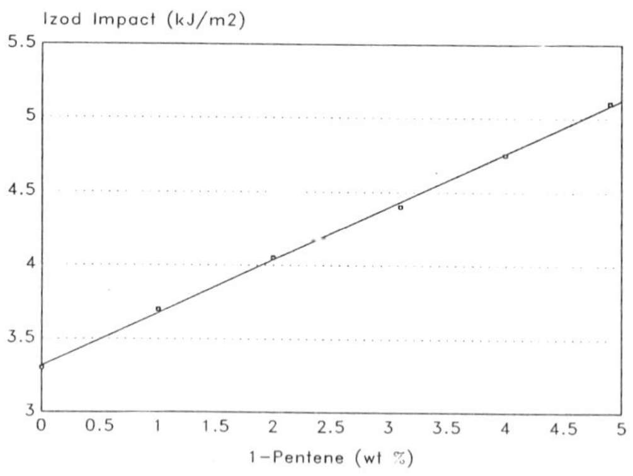

Fig 1: Increase in impact strength with 1-pentene content.

It is known from previous work on the mechanical properties of the propylene homopolymer[1] that impact strength increases with a decrease in crystallinity. Later, Alamo[2] observed that side groups larger than methyl are excluded from the polypropylene crystal, thus decreasing overall crystallinity. In Figure 2 the effect of 1-pentene content on DSC fusion enthalpy is shown. The increase in 1-pentene content (i.e. propyl branches) leads to a decrease in crystallinity of the copolymer which correlates with the decreased fusion enthalpy measured.

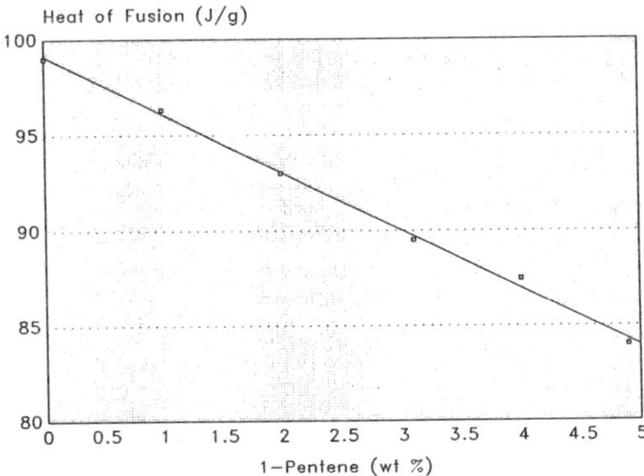

Fig 2: Effect of 1-pentene on heat of fusion.

It was expected that short range deformation properties like modulus would drop with increased 1-pentene content. Figure 3 shows this decrease with an increase in 1-pentene content. The decrease in modulus for 1-pentene copolymers is less significant than that mentioned in the literature for other random copolymers[3].

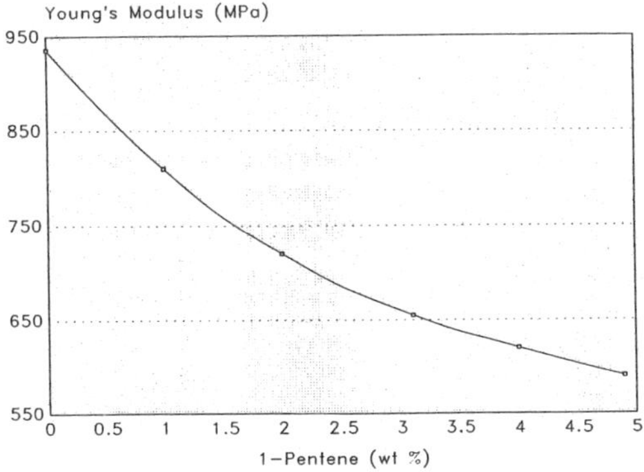

Fig 3: Decrease in Young's modulus with 1-pentene content.

It is also known[4] that the molecular weight has a significant effect on the impact strength of the propylene homopolymer. In order to investigate the impact strength relationship of the propylene/1-pentene copolymers with the molecular weight, a series of copolymers with similar 1-pentene content and different molecular weights were considered (Table 1, Expt. 7 to 10). In Figure 4 the notched Izod impact strength of propylene/1-pentene copolymers with similar 1-pentene content are plotted against Mw The effect molecular weight has on the impact strength of propylene/1-pentene copolymers can be observed.

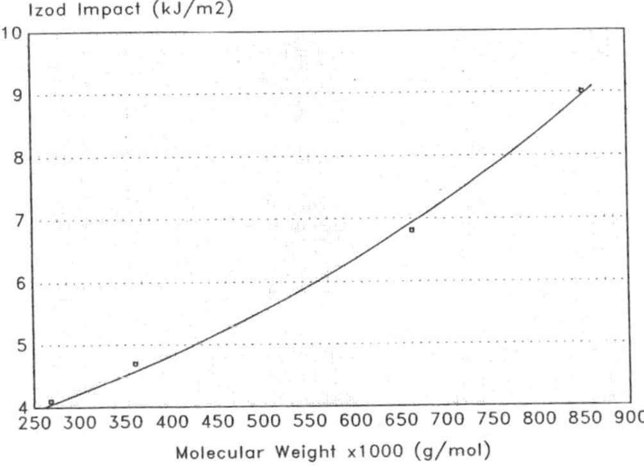

Fig 4: Increased impact strength with molecular weight

While increasing impact strength, the inclusion of 1-pentene in the polypropylene backbone does not decrease other important application properties significantly and actually improve some. Transparency increases from 54% transmittance for sample 1 to 66% for sample 6.The activation energy of flow calculated on the basis of zero shear viscosity drops from 37.6 kJ/mol for sample 1 to 34.0 kJ/mol for sample 2 and then gradually increases with increased 1-pentene content.

Figure 5 shows the difference in spherulite morphology between a propylene homopolymer and a 1-pentene copolymer. The increased amount of branches hinder spherulite growth and results in a larger degree of supercooling. The effect of this is twofold: increased spherulitic boundary area yields more tie-molecules and thus improved mechanical properties and secondly, smaller spherulites scatter less light, resulting in better optical properties.

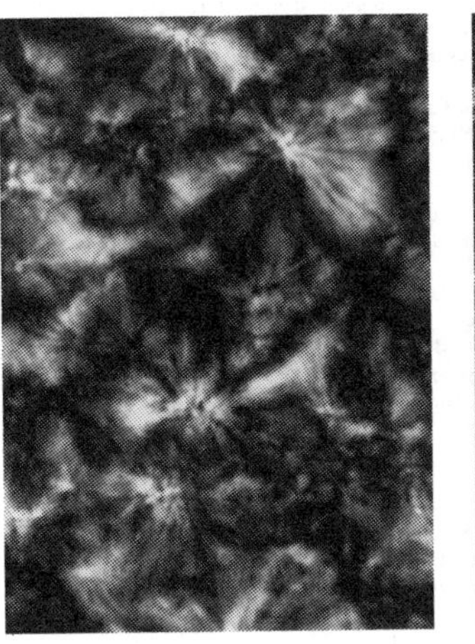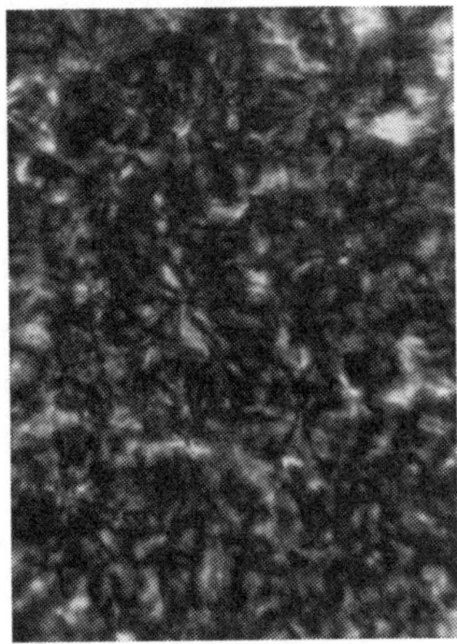

5a. PP Homopolymer   5b. 3% 1-Pentene

Fig 5: Spherulite morphology of a propylene homopolymer (5a) and a propylene/1-pentene copolymer (5b).

CONCLUSIONS

The preparation of propylene/1-pentene random copolymers opened the possibility to easily prepare propylene polymers with high impact strength while maintaining good stiffness, improved optical properties and good rheological properties. An attempt to relate the impact strength of propylene/1-pentene random copolymers to their structure, molecular weight and morphology was made. The impact strength of propylene/1-pentene copolymers increases proportionally with increased 1-pentene content and molecular weight. Overall crystallinity measurements are in accordance with mechanical results and only a moderate decrease in Young's modulus was observed with increased 1-pentene content. It is thus possible to tailor properties of propylene copolymers to obtain copolymers suitable for diverse applications. However, for a more comprehensive understanding of the failure of these copolymers under impact, some other techniques like fracture mechanics and elastic-plastic finite element analysis will give more insight into the toughening mechanism of these copolymers.

BIBLIOGRAPHY

1. J. VAN SCHOOTEN, H. VAN HOORN and J. BOERMA
    Polymer **2**(2), 161, 1961
2. R. ALAMO
    J. Phys. Chem., **88**, 6587, 1984
3. H.K.FICKER and D. WALKER
    Plastics and Rubber Proc. and Appl., **14**, 103-108, 1990
4. J.R.MARTIN, J.F.JOHNSON and A.R.COOPER
    J. Macromol. Sci.- Revs. Macromol. Chem., **C8**(1), 57-199, 1972

Toughening of Rigid Silicone Resins

Bizhong Zhu, Dimitris E. Katsoulis, John R. Keryk
Central R&D, Dow Corning Corporation,
Midland, MI 48686

Frederick J. McGarry
Department of Materials Science and Engineering,
Massachusetts Institute of Technology,
Cambridge, MA 02139

Introduction

Silicone resins are attractive for their heat, thermal oxidation and fire resistance. However rigid silicone resins are brittle and mechanically weak, which limit their uses to non structural applications. While most organic thermoset resins can be toughened by the addition of rubbers, to our knowledge no work to toughen silicone resins has been published.

Rigid silicone resins can be classified according to their cure reactions: the ones which use the condensation of silanol groups to cure, and the others which cure via hydrosilylation. The incorporation of rubbery particles directly into these resins usually does not increase their toughness, due to their highly crosslinked nature[1]. It has been shown[2,3,4] for some organic thermoset resins that the intrinsic flow capability of the matrix itself is a critical factor in determining how effectively a polymer can be toughened by rubbery particles. The toughening method[6] discussed below, i.e., reacting short PDMS segments with resin pre-polymers, seeks to selectively increase the network mobility and hence the fracture toughness of a typical rigid silicone resin.

Experimental

Tougheners. For condensation cured resins, short PDMS chains with triethoxy functional groups were used as toughners.

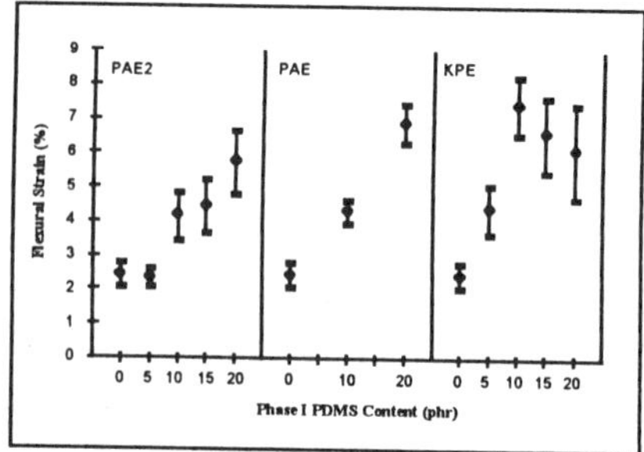

$$X=6 \text{ (PAE2), } 14 \text{ (PAE), } 55 \text{ (KPE)}$$

They were prepared by end capping silanol terminated PDMS with TEOS. For hydrosilylation cured resins the end functionality is either vinyl or hydride. The synthesized tougheners were characterized by ^{29}Si, ^{1}H NMR, and FTIR[5].

Casting Techniques and mechanical testing. To evaluate the fracture toughness of condensation cured silicone resins, thick cast plates had to be obtained to ensure plane strain test conditions. The release of water vapor during cure made it difficult to obtain such plates without voids. This problem was resolved by matching the water release rates with the diffusion rate to prevent bubble formation. Flexural properties and fracture toughness K_{Ic} and G_{Ic} were measured per ASTM D790 and D5045.

Results and Discussion

Co-reaction and Pre-reaction The condensation resin used in this study was comprised mostly of phenyl and methyl silsesquioxane units and it was commercially available as a pre-polymer bearing silanol ends. We found that there were two ways the PDMS segments could be incorporated into the resin matrix: mix the resin pre-polymer with the PDMS and cure them together (co-reaction), or react the tougheners with the resin pre-polymer first and then cure the modified resin (pre-reaction). Depending upon the nature and amounts of the R groups, silicone resins can be incompatible with PDMS segments. In these instances a co-reaction could lead to a phase separated structure with rubbery particles of 2 to 50 μm in diameter. The pre-reaction in the presence of condensation catalysts such as titanium tetrabutoxide established bonding of the PDMS ends with the resin pre-polymer and distributed the PDMS segments into the resin without phase separation. Fracture toughness of the cured resins was increased by the pre-reaction procedure, but not the co-reaction.

Chain length effect. When the PDMS segments were pre-reacted into the resin matrix, the strain at break was increased, see Figure 1. The longer chain tougheners were more effective than the shorter ones, as long as a homogeneous distribution was achieved. The toughener with an average degree of polymerization (DP) of 55, at its optimal concentration level 10 parts per hundred parts of resin, increased the strain from 2.4% to 7.5%. More than 10 parts of it caused serious residual stress built up in the cured cast plates and lowered the strain.

Figure 1. Strain at break of the cured silicone resins.

The fracture toughness of the cured resins was also increased by the PDMS segments. The toughness values in Figure 3, the energy input per unit volume applied until a rectangular specimen breaks in the three point bending mode, increase with the amounts of PDMS. Again the longer chains were better. K_{Ic} and G_{Ic}, in Figures 3 and 4, show the same trend. The best toughener, 10 parts of PDMS of average DP 55, increased the K_{Ic} from 0.25 to 0.46 MPam$^{1/2}$, and G_{Ic} from 38 to 260 J/m^2, or by 580%. The Young's modulus decreased from 2.1 to 1.4 GPa, while the flexural strength increased from 37.8 to 48.5 MPa.

Thermal stability. When a polymer is toughened by the addition of a rubber, its thermal resistance is often compromised. The toughened silicone resins, however, mostly retain the thermal stability of the untoughened resin. This was shown by TGA measurements in air and argon. In Figure 5, the TGA curves of the unmodified resin and the resin modified by 10 parts of KPE are compared. The latter loses slightly more weight at 500 °C. When the PDMS chains are shorter, e.g. with a DP of 6 or 14, no obvious difference from the unmodified resin can be detected. It is proposed that the anchoring of the chain ends of the PDMS hinders the back-biting process and helps preserve the chain at higher temperature.

Toughening mechanisms. DMA revealed that the unmodified resin had a broad glass transition from ~50 to ~160 °C. and three other damping mechanisms at ~190, -25, and -108 °C. The tan δ value of the highest temperature peak, at 190 °C, was low. A non uniform morphology of the cured resin is proposed, and this tan δ peak has been assigned to the higher phenyl T containing regions in the cured resins. The pre-reacted modifiers successfully introduced another transition at -125 °C into the cured resin, and the less restricted PDMS chain mobility increased the damping of the high phenyl T regions, as shown by peak 1 in Figure 6, the DMA spectrum of a resin toughened by 10 parts of PDMS of DP 55. Longer chains appeared to be less restricted, evidenced by a slightly lower transition temperature; and they were more effective in increasing the tan δ magnitude of the highest Tg region. On the fracture surface of the effectively toughened resin, a deformation zone approximately 10 μm in width was clearly identified, whereas no such zone was found on the surface of the unmodified resin.

Conclusions

A rigid condensation cured silicone resin has been substantially toughened by short chain PDMS segments without loss of thermal stability and with a small drop of stiffness. The less restricted PDMS chain mobility and the more effectively increased damping of the highest Tg regions contribute to the higher toughness of the toughened resins. In the toughened resins, there is evidence that plastic deformation is enhanced.

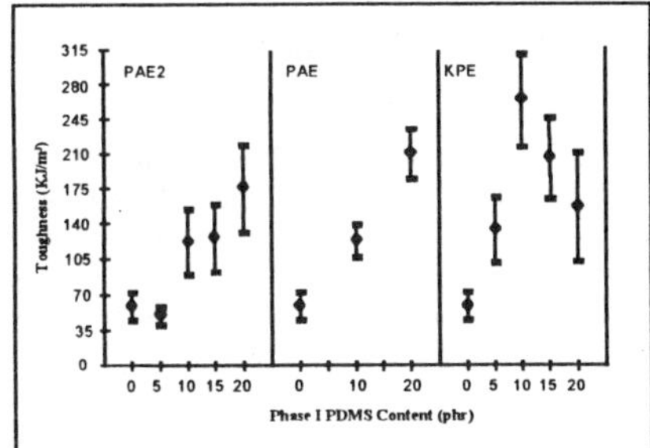

Figure 2. Toughness of the cured silicone resins.

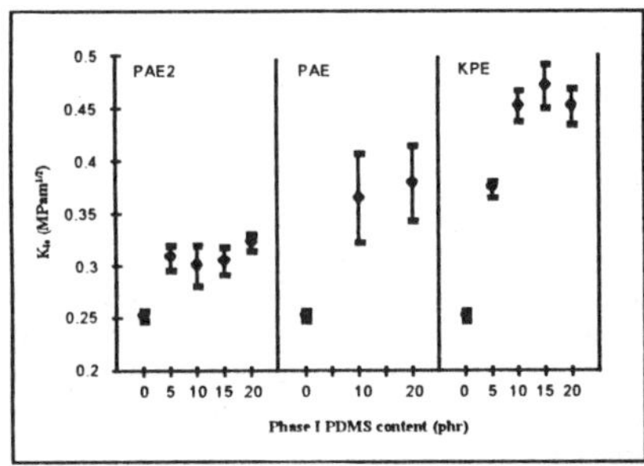

Figure 3. K_{Ic}, the critical stress intensity factor, of cured resins.

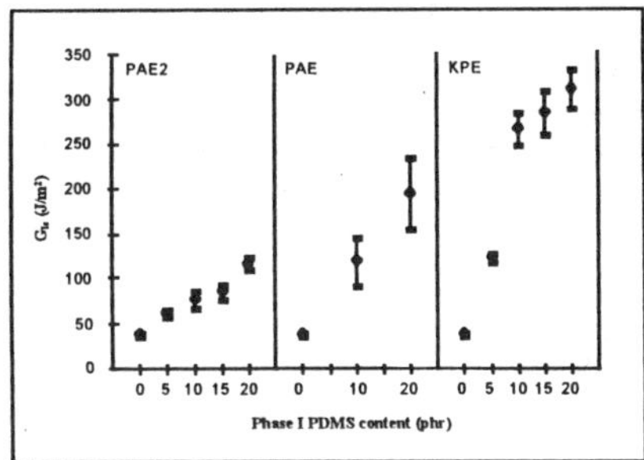

Figure 4. G_{Ic}, the critical strain energy release rate, of the cured resins.

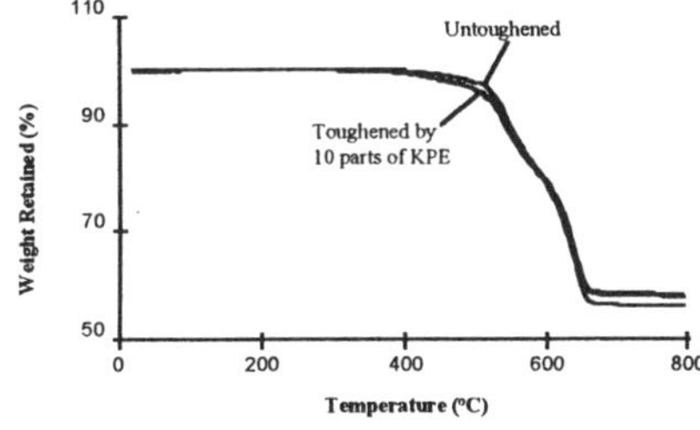

Figure 5. TG curves of toughened and untoughened resins in air.

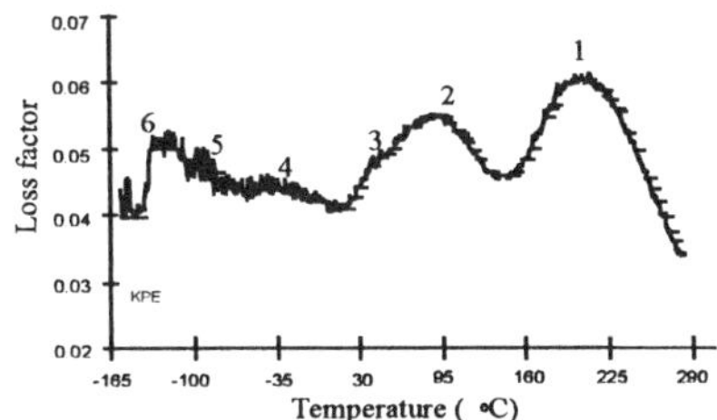

Figure 6. DMA of the resin toughened by 10 parts of PDMS of DP 55.

References

[1]. Internal Research Reports, Dow Corning Corporation.
[2]. Bos, H. L., and Nusselder, J. H., *Polymer*, 1994, **35**, 2793.
[3]. Kim, H., Keskkula, H., and Paul, D. R., *Reinforcement, Impact Modification and Nucleation of Polymers*, Society of Plastic Engineers, Regional Technical Conference, Houston, 1990, Page 33.
[4]. Chen, T. K., and Jan, Y. H., *Polym. Eng. Sci.*, 1995, **35**, 778.
[5]. Bizhong Zhu, *Ph.D. Thesis*, M.I.T., September 1997.
[6]. U.S. patent applications pending.

The Influence of Morphology & Concentration on Toughness In Dispersions Containing Preformed Acrylic Elastomer Particles In An Epoxy Matrix

D. Raghavan[1], D. Hunston[2], J. He[1], and D. Hoffman[3]
[1] Polymer Science Division, Department of Chemistry, Howard University, Washington DC 20059, USA.
[2] Polymer Division, National Institute of Standards and Technology, Gaithersburg, MD 20899, USA.
[3] The Dow Chemical Company Midland MI 48674.

ABSTRACT

Any improvement in toughness of epoxy depends on the volume fraction and morphology of the phase separated rubber particles. However, it is very difficult to control the morphology because most systems phase separates during processing. The current study addresses the issue of particle size control, particle stability and control of particle morphology by preforming acrylic rubber particles in an uncured epoxy matrix using grafted copolymer as stabilizers in an epoxy matrix. Over the wide range of rubber concentration, the morphology of the system was retained as indicated by no change in the glass transition temperature and TEM observation of microtomed RuO_4 stained samples. The results showed that the toughness of the system goes through a maximum at about 15phr by weight of rubber concentration in epoxy for rubber particles of 0.5µm. Efforts are underway to synthesize small (0.1 µm) and large (2.0 µm) preform rubber particles in epoxy so as to investigate the toughening mechanism of the composite.

INTRODUCTION

The technology to toughen a cross-linked epoxy resin without undue sacrifices in the mechanical and thermal properties by the addition of soluble reactive liquid elastomer is well known (1,2). The toughened material is usually a two phase system with small rubber particles in a glassy epoxy matrix. In many systems, including the carboxyl terminated acrylonitrile butadiene (CTBN) elastomer-epoxy composite, the rubber is initially compatible with the epoxy resin, but as curing progresses the compatibility between rubber and epoxy decreases. This eventually causes phase separation (3). The amount and type of phase separation largely depends on the formulation, curing conditions, and processing parameters. Poor phase separation can lead to a depression in the glass transition temperature of the matrix. Furthermore, the absence of micrometer size rubber particles can result in a system with minimal improvement in toughness. Although, there are a number of factors that can influence the toughening in a two phase rubber modified epoxy, the current study focuses on the effect of particle concentration at a fixed morphology (particle size and size distribution) on the toughness of the rubber modified epoxy.

To achieve this goal, the materials used here are made from a dispersion containing preformed rubber particles. A key aspect of the work is the ability to maintain a stability of the system until it cures into a solid. Grafted copolymers dispersants are used to stabilize the system. We expect that by using preformed rubber particles the particle size and size distribution can be held constant. Another advantage of the system used in this work is that the concentration of second phase material can be varied over a very wide range so the material is ideally suited to the study of volume fraction effects. The chemistry of stabilization and grafting is discussed in references 2 and 4. The dispersion used here has an average particle size of 0.5 µm in diameter. The stabilization is excellent and there is no visual evidence of non-uniformity in the dispersion. The work here uses a freshly prepared lot of the dispersion.

EXPERIMENTAL PROCEDURE

Materials

The dispersion, supplied by Dow Chemical Company, is a commercial system designated XU-71790.04L and contains 0.5955 mass fraction of epoxy monomer and 0.4045 mass fraction of preformed acrylic rubber particles. The epoxy is Tactix 123 LED, a diglycidyl ether of bisphenol A (DGEBA) type resin, while the acrylic rubber is poly(2-ethylhexylacrylate-co-glycidylmethacrylate). Although the particles size has a wide distribution, the average size is 0.5 µm. To vary the particle concentration, the dispersion was diluted with additional epoxy, Tactix 123 LED before curing with piperidine. For comparison, an unmodified epoxy was generated by curing Tactix 123 LED alone with piperidine.

Preparation of cured sample

To fabricate test specimens, the acrylic rubber dispersion was mixed with the appropriate amount of epoxy for 5 minutes to10 minutes. The mixture was degassed under vacuum at 50 °C to 60 °C until frothing stopped. The mixture was allowed to cool to room temperature, and 5 g of piperidine were added for each 100 g of epoxy in the sample. The liquid was mixed gently to minimize any air entrapment. The mixture was immediately poured into the preheated mold, cured at 120 °C for 16 hours, & slowly cooled to room temperature.

Measurements

The flexural moduli were determined in 3 point bending (ASTM D-790) using rectangular specimens. The toughness measurements were done in general accordance with ASTM E-399 using a mode-I fracture tests with compact tension specimens. Differential Scanning Calorimetric (DSC) experiments were conducted by first cooling from room temperature to -150 °C at 10 °C/min and then heating from -150 °C to 150 °C at a heating rate of 20 °C/min. For morphological studies, the samples were cryogenically sectioned to thickness of 100 nm. The sectioned samples were then exposed to RuO_4 vapors for 2 minutes and then examined in a Transmission Electron Microscope (TEM).

RESULTS AND DISCUSSION

Concentration Effect on Fracture Toughness

The critical stress intensity factors are plotted as a function of acrylic rubber concentration in Figure 1. The results are averages from measurements on at least 10 samples. The concentration of rubber is generally given as grams of rubber for each 100 grams of epoxy in the sample (designated as phr). This is equivalent to the mass ratio of rubber to epoxy (designated as mr) times 100. The unmodified epoxy has a comparatively low value of K_{IC}, a reflection of brittle characteristics of an amorphous, highly-cross-linked, epoxy system. With the addition of acrylic rubber, there is a steep increase in K_{IC} for the composite. The toughness of the system goes through a maximum at about 15 phr rubber (0.15 mr). The concentration dependence is quite similar to other systems like the CTBN-epoxy composite, even though the acrylic system is able to maintain its morphology over the entire concentration range. The observation is consistent with the toughening mechanism based on the particles initiate and assist in the yielding and plastic flow in the matrix (3,5-8). More rubber particles mean more initiation but there is a trade off because there is a decrease in the amount of matrix available to yield.

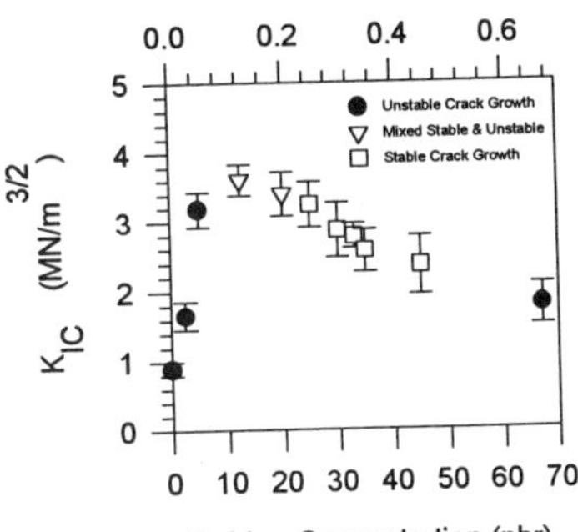

Figure 1. Critical stress intensity factor versus rubber concentration for acrylic modified epoxy.
Error bars represent standard uncertainty in data.

There is a clear transition in the fracture behavior with increasing rubber concentration from brittle-unstable failure to ductile-stable crack growth. The stress whitening seem to correlate with the fracture behavior of the composite.

Differential Scanning Calorimetry of Modified and Unmodified Epoxy

The DSC experiments indicated two deflections for the unmodified epoxy: a broad low temperature transition at ~-60 °C and a high temperature transition at 89 °C. These correspond to the known positions of the beta relaxation and the glass transition temperature for the epoxy when cured with piperidine for 16 h at 120 °C. The rubber modified epoxies gave similar results. A separate deflection for the glass transition of the acrylic elastomer was not observed, but may be hidden by the epoxy's beta relaxation. The glass transition of the epoxy matrix as a function of acrylic rubber concentration remains relatively unchanged which indicates complete phase separation and supports the idea that the morphology is constant over the whole range of rubber concentrations.

Microstructural Studies

Figure 2 shows the TEM micrograph for modified epoxy samples with 5 phr (0.05 mr) and 67 phr (0.67 mr) rubber. The micrographs clearly show two phase morphologies. There is good dispersion of rubber particles in an epoxy matrix, and unlike the CTBN-modified epoxy system, there is no phase inversion at high rubber concentrations. As best one can tell from a qualitative examination of the micrographs, there is no significant change in the morphology (particle size and size distribution) over the concentration range studied. Work is now underway to quantify the average particle size and size distribution in these samples.

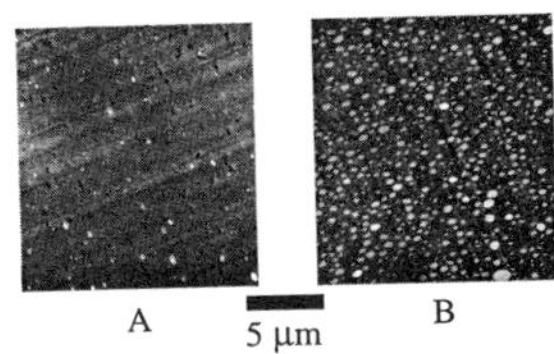

Figure 2 TEM micrograph of stained micrtomed section for samples (a) 5 phr & (b) 67 phr

CONCLUSIONS

This work studied the influence of toughener concentration on the fracture behavior of a rubber toughened epoxy over the range from 0 phr to 67 phr (0 mr to 0.67 mr). By using samples generated from dispersions of preformed rubber (acrylic) particles in liquid epoxy monomer, the average particle size and the size distribution were held constant. Measurements show that the glass transition temperature for the epoxy matrix remains unchanged over the entire concentration range which indicates complete phase separation and supports the idea of a constant morphology. TEM micrographs also support the idea of a constant morphology in these systems. As the rubber concentration increases from zero, the toughness increases to a maximum at about 15 phr (0.15 mr) rubber and then decreases gradually. This is consistent with a mechanism based on particles initiating yielding and plastic flow of the matrix.

ACKNOWLEDGMENTS

The authors thank the Polymer Division of NIST and the Department of Chemistry at Howard University for providing financial support to Mr. J. He. The authors would like to thank Dow Chemical Company for providing epoxy resin and the stabilizer and are particularly grateful to Bob (R. C.) Cieslinski and Bill (W. A.) Heeschen of Dow Chemical Company for assisting with the transmission electron microscopy study of the composite.

REFERENCES

1. D. K. Hoffman and G. C. Kolb, "Toughened Epoxy Resin via Dispersions of Acrylic Elastomer," *Polymeric Materials Science and Engineering*, 1994, **63**, 593-594.
2. D. K. Hoffman, C. Oritz, D. L. Hunston and W. G. McDonough, "The Chemistry of Toughening Epoxy Resins via Preformed Dispersions of Acrylic Elastomers." *Polymeric Materials Science and Engineering*, 1994, **70**, 7-8.
3. A. J. Kinloch, S. J. Shaw, D. A. Tod, and D. L. Hunston, "Deformation and Fracture Behavior of a Rubber-Toughened Epoxy: 1. Microstructure and Fracture Studies," *Polymer*, 1983, **24**, 1341-1354.
4. J. He, D. Raghavan, D. K. Hoffman and D. Hunston, "The Influence of Elastomer Concentration on Toughness in Dispersions Containing Preformed Acrylic Elastomer Particles in an Epoxy Matrix," submitted for publication 1998.
5. A. J. Kinloch, "Relationship Between the Microstructure and Fracture Behavior of Rubber-Toughened Thermosetting Polymers," *Advances in Chemistry Series 222*, ACS 194th. Annual Meeting, LA, 1989, 67-91.
6. A. J. Kinloch, S. J. Shaw, D. A. Tod, and D. L. Hunston, "Deformation and Fracture Behavior of a Rubber-Toughened Epoxy: 2. Failure Criteria," *Polymer*, 1983, **24**, 1355- 1363.
7. A. F. Yee, and R. A. Pearson, "Toughening Mechanisms in Elastomer-Modified Epoxies, Part 1: Mechanical Studies," *J. Mater. Sci.* 1986, 21(7), 2462-2474.
8. R. A. Pearson, and A. F. Yee, "Toughening Mechanisms in Elastomer-Modified Epoxies, Part 2: Microscopy Studies," *J. Mater. Sci.* 1986, 21(7), 2475.

TOUGHNESS OF PBT/BT CO-POLYMER BLENDS

James A. Donovan* and P. Chang**
*University of Massachusetts, Amherst, MA 01003-2210
**Sanyang Industry Co., Ltd., Taipei, Taiwan, R.O.C.

INTRODUCTION

Blending poly(butylene terephthalate) (PBT) with BT polyether copolymer significantly improved the fracture properties, if the blends were immiscible. However, miscible blends had inferior fracture properties. The BT copolymers are thermoplastic elastomers with BT hard blocks of varying length, thus, the increase in toughness was expected to be due to typical rubber toughening mechanisms: energy absorption by the rubber particles (crack bridging)[2,3], wake effects[4], cavitation around rubber particles[5,6], crack or craze termination at rubber particles[7,8], and matrix crazing or shear yielding, or both[7,9,10].

The objective was to understand the effect of microstructure, miscibility and crystallinity on the fracture properties. A toughness model, based on the formation of a plastic zone ahead of the crack tip, is developed that is compatible with the data and correlates with miscibility and crystallinity.

EXPERIMENTAL

The mechanical properties of PBT (Valox 315) and copolymer-BT/polyether (Lomod) blends were studied as a function of the hard block (BT) content of the copolymer and the homopolymer/copolymer ratio. The copolymers contained hard block (BT) contents of 36, 50, and 75 percent. The characteristics of the blend components are summarized in Table 1. All materials were injection molded and supplied by the General Electric Company.

Table 1
Characteristics of the Component Materials

Materials	wt% PBT	Hard Block Length	T_g	T_m
J6(BT36)	36	6	-50	185
J10(BT50)	50	10	-34	200
J50(BT75)	75	28	0	218
PBT	100	-	46	224

Tensile tests were performed at a strain rate of 1×10^{-3} s^{-1} and at 23°C. The tensile yield energy was the area under the stress-strain curve up to yield.

The fracture toughness, J_c, was measured with three point bend specimens by extrapolating the J at maximum load as a function of ligament length. This protocol was shown to give similar results as the ASTM procedure to determine fracture toughness for plastics[1].

RESULTS

The mechanical properties of the PBT/BT copolymer blends are shown in Figures 1 and 2 for the immiscible and miscible blends, respectively. The tensile yield strength (σ_y) and modulus (E) decreased by blending the PBT and copolymers. The fracture toughness, J_c as a function of co-polymer content is also shown in Figures 1 and 2. The toughness of the immiscible blends increased with co-polymer content, but decreased for miscible blends.

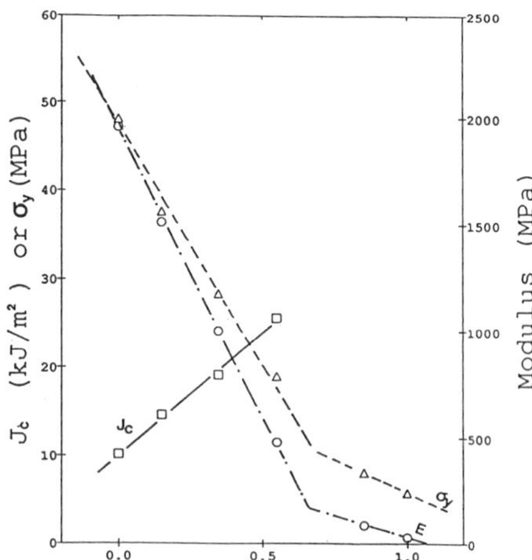

Wt. Fraction of Copolymer J6

Figure 1. Mechanical properties of immiscible blends as function of composition.

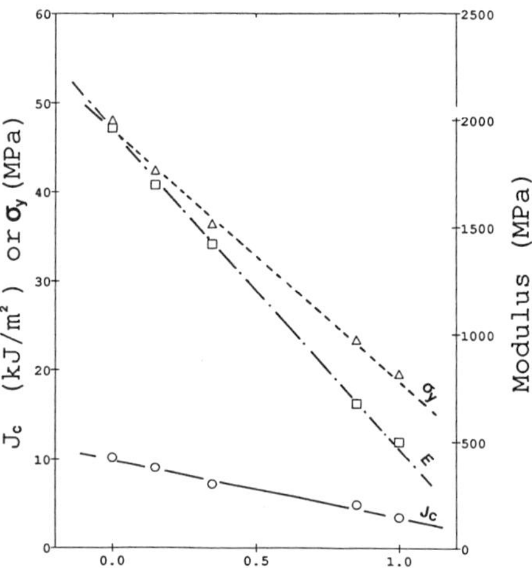

Wt. Fraction of Copolymer J50

Figure 2. Mechanical properties of miscible blends as function of composition.

DISCUSSION

The yield stress and the modulus for all blends decreased as the co-polymer content increased. But the fracture toughness increased for the blends with the short BT blocks (J6 and J10), but decreased for blends of PBT and J50, which had the longest BT block. The primary difference between these two classes of blends is that blends of the J6 and J10 co-polymer with PBT were immiscible and the J50 blends were miscible. DSC and DMTA showed that J6- and J10 blends had two T_g's and two T_m's, but J50 blends had only one T_g and T_m. Therefore, the J50 blends were not only miscible but also probably co-crystallized[1]. Transmission electron microscopy showed that the co-polymer was randomly distributed as spherical particles of various sizes in the immiscible blends.

The fracture of toughness dependence on miscibility can be better understood by considering Evans *et al.* [16] model for the increase in toughness, ΔJ_c due to the contribution of the plastic zone at the crack tip.

Their model predicts

$$J_c = J_0 + \frac{\theta J_c}{4}\,\frac{E}{\sigma_y^2}$$

where J_0 is the inherent toughness without a plastic zone, θ depends on the deformation mechanism in the plastic zone. Therefore, a graph of J_c as a function of $J_c E/\sigma_y^2$, the plastic zone radius, should be linear with a slope of $\theta/4$ and intercept of J_0. For the immiscible blends a linear relation was obtained (Fig. 3) with θ equal to 2.4 MJ/m^3 and J_0 equal to 5 kJ/m^2. This suggests that the increase in toughness of immiscible blends is primarily due

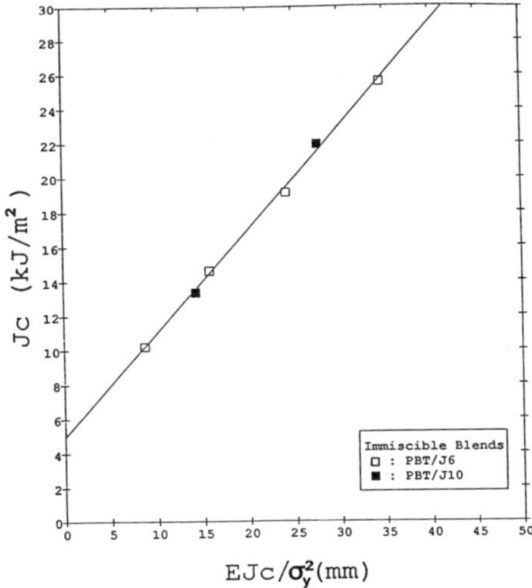

Figure 3.   Toughness of immiscible blends as a function of the plastic zone radius.

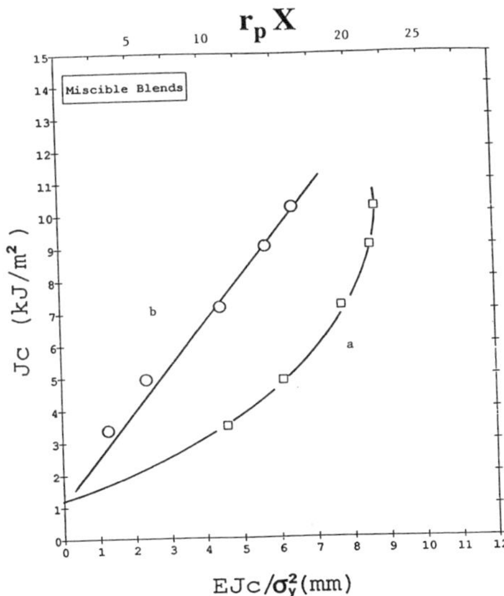

Figure 4.   Toughness of miscible blends as a function of the a) plastic zone radius and b) the plastic zone corrected for crystallinity.

to the increase in the plastic zone size. For the miscible blends J_c as a function of plastic zone radius was not linear (Fig. 4). This suggests that θ depends on composition.

Shear yielding is a possible toughening mechanism for rubber toughened polymers if the matrix is ductile. The toughening mechanism of the rubber toughened PBT, studied by Hobbs *et al.*[19], was due to the dispersed rubber particles that enhanced the local stresses and strains in the PBT matrix that caused yielding. Thus, shear yielding is the suspected deformation mechanism for the immiscible PBT/BT copolymer blends.

In the immiscible blends the co-polymer exists as a separate rubbery phase of randomly dispersed particles of various sizes. These rubbery particles act as stress amplifiers that cause the PBT matrix to yield at lower macroscopic stresses than neat PBT. Thus blending increases the plastic zone and toughness.

The value of θ obtained from Figure 3 is 2.4 MJ/m^3 is close to the energy required to cause yielding in PBT that is about 30% crystalline, 2.2 MG/m^3. This similarity suggests that the source of toughness is more material near the crack tip yielding.

For the miscible blends the toughness increased with crystallinity[1] and J_c was not linear with the plastic zone size. Fig. 4. This suggests that θ depends on composition or crystallinity. The yield energy and yield stress were proportional to the crystallinity. Therefore, if as in the case of the immiscible blends, θ is proportional to yield energy then the product of yield energy or crystallinity and the plastic zone would be the appropriate parameter to correlate J_c. Figure 4 shows this graph is linear and J_0 is the same as the estimate made from the graph of J_c as a function of plastic zone radius. The difference between the miscible and immiscible blends is that the energy density to cause yielding depends on the composition in the miscible blends, but not in the immiscible blends.

The inherent toughness, J_0 for the immiscible blends was about four times that for the miscible blends. The reasons for this are not clear, but could be related to a difference in the effective number of tie molecules and the difference in the modulii. However, a satisfactory explanation is not yet available.

CONCLUSIONS

For miscible blends the toughness <u>increased</u> with increasing PBT concentration, and crystallinity because the energy density (θ) required to form the plastic zone increased. For immiscible blends the toughness <u>decreased</u> with increasing PBT concentration because the yield stress increased. The inherent toughness, J_0 of the immiscible blends was four times that of the miscible blends. The reasons are not known.

ACKNOWLEDGMENTS
We appreciate the support of this work and supply of materials by the General Electric Company.

REFERENCES

1.   P. Chang, Ph.D. Dissertation, University of Massachusetts, (1991).
2.   Merz, E. H., Claver, G. C. and Baer, M. J., J. Polym. Sci., 22 (1956) 325.
3.   Kunz, S, Beaumont, D.W.R. and Ashby, M. F., J. Mater. Sci., 15 (1980) 1109.
4.   Nair, S. V., Subramaniam, A., Goettler, L. A., J. Mat. Sci., 32 (1997) 5347.
5.   Yee, A.F., J. Mater. Sci., 12 (1977) 757.
6.   Yee, A.F., Proc. Conf. "Toughening of Plastics", London, (1985).
7.   Bucknall, C.B., "Toughened Plastics", Applied Sci. Pub., London, 1977.
8.   Wrotecki, C., de Charentenay, F.X., "7th Deformation, Yield, and Fracture of Polymer", PRI, Cambridge, U.K. (1988).
9.   Bucknall, C.B., in "Polymer Blends", Paul, D. R. And Newman, S. Eds., Academic Press, New York, (1978).
10.   Hobbs, S.Y., Bopp, R. C. And Watkins, V.H., Polym. Engin. Sci., 23 (1983) 380.
11.   Takeways, R., Ward, I.M., and Wilding, M.A., J. Polym. Sci., Polym. Phys. Ed., 13 (1975) 799.
12.   Hall, I.H. and Pass, M.G., Polymer, 17 (1976) 807.
13.   Desborough, I.J. and Hall, I.H., Polymer, 18 (1977) 825.
14.   Haidemenopoulos, G.N., Olson, G.B., Cohen, M., and Tsuzaki, K., Scripta Metal., 23 (1989) 207.
15.   McMeeking, R.M., and Evans, A.G., J. Am. Ceramic Soc., 65 (1982).
16.   Evans, A.G., Ahmad, Z.B., Gilbert, D.G., Beaumont, P.W.R., Acta Metall., 34 (1986) 79.
17.   Rice, J.R. and Rosegren, C.F., J. Mech. Phys. Solids, 16 (1968) I.
18.   Hutchinson, J.W., J. Mech. Phys. Solids, 16 (1968) 13.
19.   S. Y. Hobbs, M.E.J. Dekkers, and V.H. Watkins, J. Mater. Sci., 23 (1988) 1219.

RATE DEPENDENCE OF THE FRACTURE BEHAVIOR IN A RUBBER-MODIFIED EPOXY

J. Du*, P. C. Niven**, M. D. Thouless** and A. F. Yee*
*Department of Materials Science and Engineering,
The University of Michigan, Ann Arbor, Michigan 48109, USA
** Department of Mechanical Engineering and Applied Mechanics, The
University of Michigan, Ann Arbor, Michigan 48109, USA

INTRODUCTION

Epoxy resins can be significantly toughened by the incorporation of second-phase rubber particles. These particles usually cavitate under the triaxial stress state around the crack tip. This cavitational process locally relieves the triaxial constraint in the surrounding matrix, and, hence, induces extensive plastic deformation in terms of void growth and shear band formation [1, 2]. This non-linear, irreversible deformation generally occurs within a process zone associated with the crack extension. Typically, a continual development of the process zone gives rise to a crack-growth resistance-curve (R-curve) where the toughness depends on the extent of crack advance [3]. The realization of a substantial portion of the R-curve ensures that a successful toughening can be obtained for the rubber-modified epoxies. However, instabilities in crack growth caused by the loading rate may result in a catastrophic failure, and, hence, prevent the optimal fracture strength from being achieved, even though a stable testing geometry, *e.g.* a double-cantilever-beam (DCB) test, is employed. Therefore, a full understanding of the influence of crack instability upon the fracture behavior is of immense importance in establishing the general principles of toughening of engineering polymers.

Previous studies on rate effects in rubber-modified epoxies have been limited to the initiation toughness [4]. No efforts have been made on the rate dependence of the R-curve behavior in these materials. Therefore, in this work, the R-curve behavior under different crosshead-displacement rates was investigated for a rubber-modified epoxy using DCB tests. The relationship between applied crack-growth driving force and crack velocity was established. Transmission optical microscopy (TOM) was used to study the evolution of the process zone. Scanning electron microscopy (SEM) was used to examine the steady-state fracture surfaces. Based on these observations, some nature of the crack instability was elucidated.

EXPERIMENTAL

The rubber-modified epoxy used in this work was formed by combining (by weight) 100 parts of a liquid diglycidyl ether of bisphenol-A epoxy (DER® 331 produced by the Dow Chemical Co.), 10 parts of a liquid carboxyl-terminated acrylonitrile-butadiene copolymer rubber (HYCAR® CTBN 1300X8 produced by the B. F. Goodrich Co.), and 5 parts of a curing agent (piperidine produced by the Aldrich Co.). The material was cured at 120 ˚C for 16 hours in a Teflon®-coated mold. After curing, DCB specimens were machined. The width of these sample was 6.35 mm, and the thickness was 40 mm. A sharp crack of length about 40 mm was formed prior to testing by tapping a razor blade into the tip of a pre-machined notch. Side grooves of depth 1.27 mm were used to keep the crack in the mid-plane of the samples.

Mechanical testing was performed using a screw-driven Instron machine with crosshead speeds of 1, 1.5 and 2 mm/min. The crack propagation was optically observed using a video-recording system. From the video recordings of the crack length and the corresponding applied load, the applied energy-release rate was evaluated. The crack velocity was also calculated from the video images.

After testing, thin sections at the mid-plane of the samples and perpendicular to the crack surfaces were obtained by petrographic thinning techniques [5]. The thinned section was then examined using an optical microscope to study the evolution of the process zone. The fracture surfaces of the samples were coated with a thin film of gold by sputtering, and then examined using a scanning electron microscope to study the steady-state fracture surfaces.

RESULTS AND DISCUSSION

For each testing rate, a R-curve behavior was observed for the 10 phr rubber-modified epoxy (Fig. 1). The applied crack-growth driving force increased with crack extension until it reached an approximately steady-state value. Eventually, a crack instability occurred. The R-curves strongly depended on the crosshead-displacement rate. The steady-state applied energy-release rate decreased with increasing the testing rate.

The features of the experimental R-curves can be directly correlated to the evolution of the process zone (Fig. 2). As the crack grew, the process zone fanned out until it reached a steady-state thickness. In addition, a linear correlation appeared to exist between the thickness of the steady-state process and the steady-state applied crack-growth driving force (Fig. 3). This is consistent with the micromechanics models of toughening in the literature [6, 7]. Furthermore, it should be noted that the crack instability was always characterized by an abrupt disappearance of the process zone. Despite the fact that the nature of the process zone (*e.g.* the thickness of the steady-state process zone) strongly depended on the testing rate, it is interesting to note that the features of the steady-state fracture surfaces (*e.g.* the volume fraction of the voids) appeared to be independent of the testing rate (Fig. 4).

Generally, the crack instability is closely related to the intrinsic connection between the applied energy-release rate and the crack velocity. For each testing rate, the applied crack-growth driving force increased with increasing crack velocity until the crack instability occurred (Fig. 5). For all crosshead speeds, the steady-state applied energy-release rate decreased with increasing crack velocity. Furthermore, there is no unique relationship between crack velocity and applied crack-growth driving force. Rather, it appears that the prior history of the development of the process zone (as dictated by the crosshead-displacement rate) determined the relationship between the crack velocity and the applied energy-release rate. Both rubber cavitation and plastic flow in the rubber-modified epoxies are time-dependent phenomena, because they are visco-elastic and/or visco-plastic in nature. This means that these deformation processes can be substantially suppressed at certain critical crack velocities in which relatively high deformation rates prevail. This appears to be the origin for the observed crack instabilities, as these instabilities are invariably associated with an abrupt cessation of the development of the process zone. The detailed mechanism of the interaction between the crack velocity and the development of the process zone is currently under further investigation.

CONCLUDING REMARKS

The R-curve behavior of the 10 phr rubber-modified epoxy strongly depends on the testing rate. The features of the experimental R-curves can be directly correlated to the evolution of the process zone. For each crosshead displacement, the applied energy-release rate increased with increasing crack velocity until the crack instability occurred. For all crosshead speeds, the steady-state applied crack-growth driving force decreased with increasing crack velocity. The crack instability was inevitably associated with an abrupt cessation of the development of the process zone, and, hence, a low toughness. This is probably caused by the suppression of both rubber cavitation and plastic flow at relatively high crack velocities.

ACKNOWLEDGMENT
This work is supported by the National Science Foundation under Grant CMS-9523078.

REFERENCES

1. A. J. Kinloch, S. J. Shaw, D. A. Tod and D. L. Hunston, *Polymer* **24** (1983) 1341.
2. A. F. Yee and R. A. Pearson, *J. Mater. Sci.* **21** (1986) 2462.
3. J. Du, M. D. Thouless and A. F. Yee, submitted to *Int. J. Fract.*.
4. B. J. Cardwell and A. F. Yee, *Polymer* **34** (1993) 1695.
5. A. S. Holik, R. P. Kambour, S. Y. Hobbs and D. G. Fink, *Microstruct. Sci.* **7** (1979) 357.
6. R. M. McMeeking and A. G. Evans, *J. Am. Ceram. Soc.* **65** (1982) 242.
7. A. G. Evans and K. T. Faber, *J. Am. Ceram. Soc.* **67** (1984) 255.

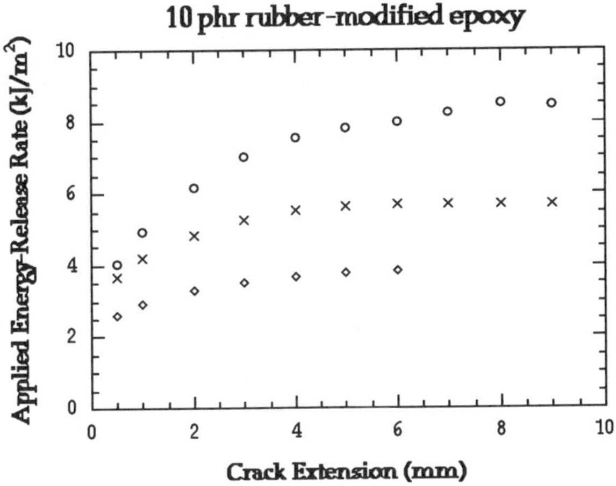

Fig. 1   Plot of applied energy-release rate against crack extension for the 10 phr rubber-modified epoxy tested under different crosshead speeds (1 mm/min - ○, 1.5 mm/min - ×, 2 mm/min - ◇).

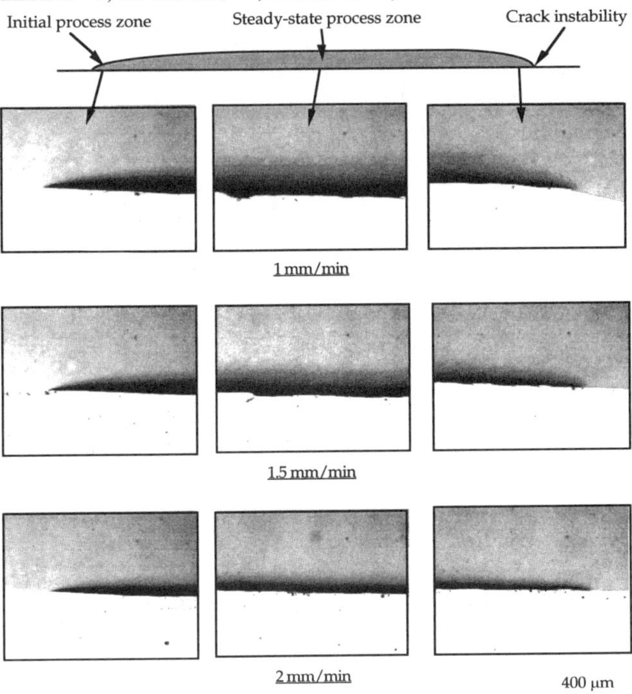

Fig. 2   TOM micrographs of evolution of process zone for the 10 phr rubber-modified epoxy tested under different crosshead speeds.

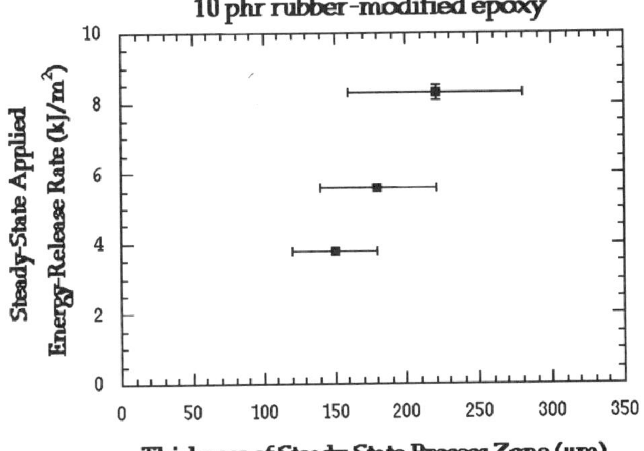

Fig. 3   Plot of steady-state applied energy-release rate against thickness of steady-sate process zone for the 10 phr rubber-modified epoxy.

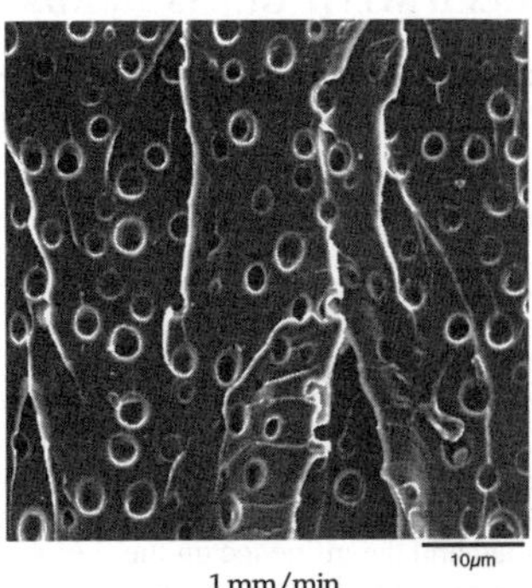

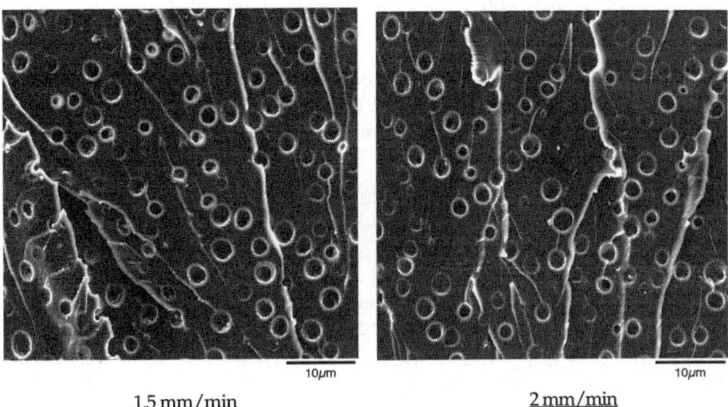

Fig. 4   SEM micrographs of steady-state fracture surfaces for the 10 phr rubber-modified epoxy tested under different crosshead speeds.

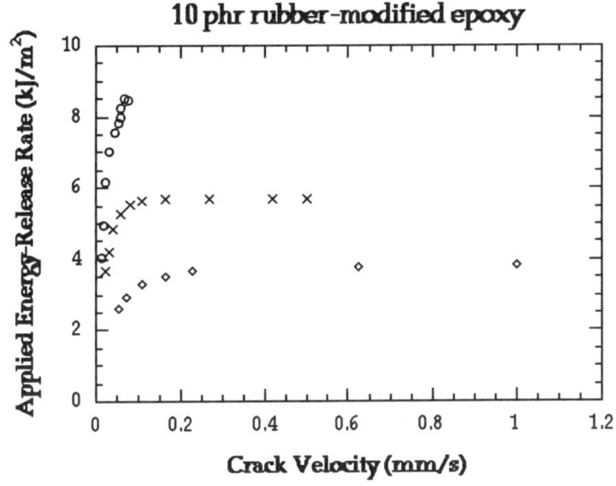

Fig. 5   Plot of applied energy-release rate against crack velocity for the 10 phr rubber-modified epoxy tested under different crosshead speeds (1 mm/min - ○, 1.5 mm/min - ×, 2 mm/min - ◇).

INORGANIC PARTICLE TOUGHENING: MICRO-DEFORMATION IN EPOXIES FILLED WITH GLASS BEADS

J. Lee and A. F. Yee
Macromolecular Science and Engineering Program
The University of Michigan, Ann Arbor, MI 48109

INTRODUCTION The action of inorganic particles in improving the mechanical properties of particulate polymer composites is a subject worthy of investigation, not only because inorganic particles have been widely used in the polymer industry but also because knowledge in this area is essential for the future development of inorganic/organic particulate composites. However, the research of past decades on even the simplest system, that of epoxies filled with glass beads, has produced few reliable generalizations, with many questions left unanswered.[1,2] One major problem is to determine the key sources of fracture toughness in the system. To address this here, the contributions of micro-deformation to toughening are examined and the influence on these of matrix ductility and interfacial strength are systematically investigated.

EXPERIMENTAL Five different kinds of glass beads, uncleaned, cleaned glass beads, and glass beads encapsulated by rubber layers of differing thickness, were prepared as described in ref 3. Also, four different kinds of diglycidyl ether of bisphenol-A epoxies, DER® 332, 661, 664, and 667 provided by DOW chemicals, were used. Their molecular weight was varied from 340 to 3600 g/mol to explore the effect of matrix ductility. Glass beads so prepared were first mixed with pre-dried epoxides under vacuum for 1 hour at different temperatures according to the types of epoxides, 80 °C for DER® 332, and 160 °C for all other epoxides. Then, the equivalent amount of 4,4'-diaminodiphenylsulfone was mixed for 20 min to 3 hours depending on the types of epoxides so as to prevent the sedimentation of glass beads by obtaining an appropriate mixture viscosity. After this, the degassed mixture was poured into a preheated metal mold and cured. The total curing schedules were 16 hours at 120 °C for DER®332 and 16 hours at 160 °C for all other epoxides. The post-curing schedules were 2 hours at 250 °C for DER®332, and 2 hours 200 °C for all other epoxides. Fracture toughness measurements and microscopy examinations were performed as cited elsewhere.[3,4]

RESULTS AND DISCUSSION Several micro-deformation processes, which may dissipate elastic strain energy, were observed in the fracture of our glass bead filled epoxies. These are: step formation; debonding of glass beads; diffuse matrix shear yielding; and micro-shear banding of the matrix. These processes are discussed in turn.

Step formation leads to an increase in surface area, which increase requires extra energy, and results in the increase of fracture toughness. However, our microscopy studies revealed that the increase of surface area was insufficient for absorption of a significant amount of energy. Moreover, it was found that the number of steps per unit area was uncorrelated with the fracture toughness. Thus, these observations lead us to believe that step formation makes only a minor contribution to toughening epoxies.

Diffuse shear yielding of the matrix was always and only observed around debonded glass beads, and therefore, these two processes are considered to act as one in our analysis. At constant glass bead content, the total energy absorption due to debonding/ diffuse shear yielding should depend on the size of the deformation zone. If this combined process makes a major contribution to fracture toughness, the change in the deformation zone size will correlate with a change in fracture toughness. However, the plot in Fig.1. (a), K_{IC} (critical stress intensity factor) versus measured debonding zone size, shows no straightforward correlation. On the other hand, the debonding zone size data in Fig. 1. (a) shows their dependence on $e_{0.8E}$ which is the strain where tangential modulus drops to 80% of initial modulus during the uniaxial tensile test of glass bead filled epoxies(a measure of debonding strength): Debonding zone size increases with decreasing $e_{0.8E}$. As a result, it can be stated that the contribution of debonding/ diffuse shear yielding to toughening is not sufficient to control the fracture toughness of our composites.

Fig. 1. (b) shows a reasonable correlation between the measured size of the micro-shear band zone and fracture toughness. It is particularly noteworthy that although the data are scattered over a large region(50 to 260μm, 1.3 - 1.5MPa-m$^{1/2}$) in Fig.1.(a), they collapse onto a smaller region(10 to 40μm, 1.3 - 1.5MPa-m$^{1/2}$) in Fig. 1. (b). These results indicate that micro-shear banding has a dominant effect on the fracture toughness of the composites through the irreversible energy absorption of shear banding. Clearly, the other processes described above may play a role and all the processes may be inter-related, but the micro-shear banding must be the dominant process in this system as a correlation arises between this and K_{IC} even in the presence of the other processes. Furthermore, Fig. 1. (b) also reveals little dependence of the micro-shear band zone size on $e_{0.8E}$. This finding is consistent with the results of our microscopy study which show that, in the fracture of glass bead filled epoxies, micro-shear bands are not always accompanied by the debonding of glass beads.

It has hitherto been believed that the effect of inorganic particle toughening decreases as matrix ductility increases.[1,2] However, in our present experiments, as matrix ductility is increased via increasing molecular weight between cross-links, the toughening effect due to the incorporation of glass beads increases with a commensurate increase in micro-shear banding. Our finding that the size of the micro-shear band zone depends on the matrix ductility, and not interfacial strength, may explain the lack of success of previous efforts to improve toughness by modifying the surface of glass beads.

CONCLUSION Micro-shear banding has been found to be a major energy dissipating process in the fracture of glass bead filled epoxies. Enhancing the micro-shear banding produces a sizable toughening effect, namely a 340% increase of fracture energy over that of unmodified epoxy for 10 vol% of glass beads.

ACKNOWLEDGMENT This research was supported by NIDR grant 2-P50-DEO9296-06.

BIBLIOGRAPHY
1. *Particulate-Filled Polymer Composites*; Rothon, R. Ed.; Longman Scientific & Technical: Harlow, 1995; Chapter 9.
2. Kinloch, A. J.; Young, R. J. *Fracture Behavior of Polymers*; Elsevier Applied: New York, 1985.
3. Lee, J.; Yee, A. F.; *Polym. Prepr., Am. Chem. Soc. Div. Polym. Chem.* **1997**, *38*, 369.
4. Sue, H. J.; Yee, A. F. *J. Mater. Sci.* **1989**, *24*, 1447.

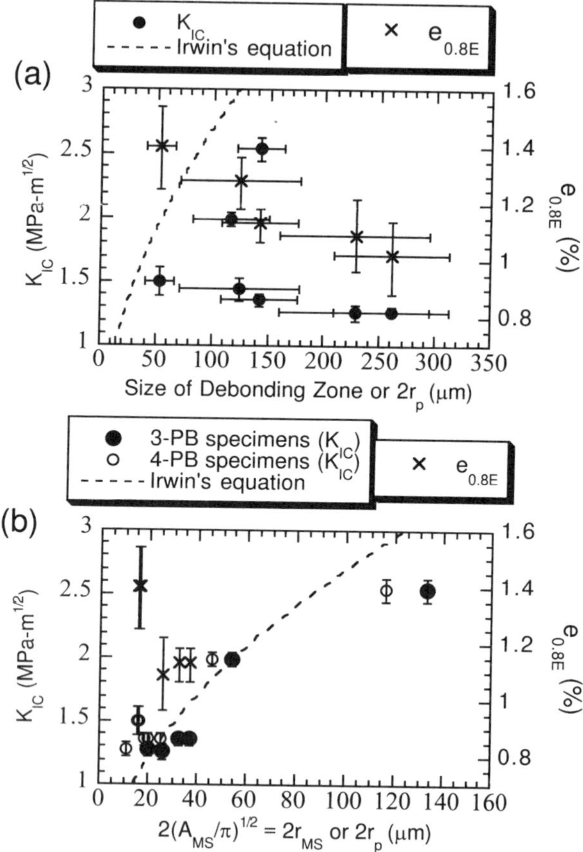

Fig. 1. Relationship between K_{IC} or $e_{0.8E}$ and debonding zone size (a) or micro-shear band zone size($2r_{MS}$) measured (b) for various 10 vol% glass bead filled epoxies. The dash line is plotted for comparison with Irwin's theoretical relation,[2] $K_{IC} = [3\pi(2r_p)]^{1/2}\sigma_y$, using σ_y (yield stress) of DER661/ DDS (88MPa), where r_p is the plastic zone size. The A_{MS} is the measured area of micro-shear band zone.

On Plastic Zone Growth in BN-Filled, Rubber-Modified Epoxy Polymers

Raymond A. Pearson and Michael F. DiBerardino
Lehigh University
Polymers Interfaces Center
Bethlehem, PA 18015 U.S.A.

INTRODUCTION

The fracture resistance of epoxy polymers can be improved by either adding compliant rubber particles or rigid inorganic fillers. Rubber particles can toughen epoxy polymers by cavitating under the high hydrostatic tensile stresses at the crack tip and subsequently triggering shear banding to occur in the epoxy matrix adjacent to the particle. Matrix shear banding and plastic void growth reduce the effective crack driving force at the crack tip, and thereby shield the crack tip. Inorganic fillers can also shield the crack tip from the applied crack driving force, however, the shielding mechanisms are very different from those triggered by compliant rubber particles. For example, crack path deflection and particle bridging are commonly observed in filled epoxy polymers.

Synergistic toughening has been seen in some epoxy systems containing both compliant rubber particles and rigid inorganic fillers [1-5]. For example, Hamid et al. [4,5] observed synergistic toughening in epoxy systems containing both compliant rubber particles and hollow glass spheres and for epoxy systems containing both compliant rubber particles and solid glass spheres. Synergistic toughening occurs when two or more toughening mechanisms interact in a positive fashion. To date, most of the papers addressing synergistic toughening have focused on spherical fillers.

The purpose of this investigation is to elucidate the toughening mechanisms in an epoxy polymer containing both compliant rubber particles and rigid disc-shaped particles. The fracture properties of the blends were determined as a function of volume fraction filler as well as the average disc size (10 microns versus 50 microns in diameter). Toughening mechanisms were revealed using scanning electron microscopy (SEM) and transmission optical microscopy (TOM).

EXPERIMENTAL

Materials. The epoxy matrix used in this investigation is a liquid diglycidyl ether of bisphenol A with a epoxide equivalent weight of 187 g/eq (DER® 331 resin). The epoxy resin is cured using 5 parts per hundred of an amine curing agent - piperidine. The epoxy polymer is modified by the addition of a carboxyl terminated copolymer of butadiene-acrylonitrile (HYCAR® CTBN 1300x8 resin) and filled with boron nitride (BN) powder. Two fine BN powders are examined: PT®-120 (10 μm in diameter and 0.5 μm thick) and PT®-110 (50 μm in diameter and 0.5 μm thick).

Plaques of the toughened epoxies were prepared following a standard procedure. First, 500 grams of DGEBA was heated to 80°C and degassed with agitation. Next, the modifiers were added with continued agitation (stirring). The resulting mixture was again degassed and 23 ml of piperidine was added at ambient pressure. The final resin mixture was again degassed and poured into a 120°C preheated, aluminum mold. The mold was then placed in a circulating air oven to cure for 16 hours at 120 C.

Yield Strength. The yield strength of the toughened epoxy polymers were measured in compression according to ASTM D695-84 guidelines. The compression specimens were 12.7 mm in length and had a 12.7 mm square cross section. Specimens were tested on a screw-driven materials testing machine at a crosshead speed of 1.27 mm/min. The compressive yield strengths reported are the average of at least five tests.

Fracture Toughness. The fracture toughness of the toughened epoxy polymers were measured using a three-point bend test described in the ASTM D5045-93 method. Fracture toughness is measured in terms of the critical stress intensity factor , K_{IC}. Tests were performed using a screw-driven materials testing machine at a crosshead speed of 10 mm/min. K_{IC} values reported are the average of a minimum of five tests.

Scanning Electron Microscopy (SEM). Fracture surfaces have been examined using a JEOL 6300 low voltage scanning electron microscope. Samples were coated with a thin layer of chromium to prevent charge build-up. The images of the fracture surfaces were obtained by capturing secondary electrons generated by a 5kV accelerating voltage.

Transmission Optical Microscopy (TOM). Sub-surface damage at crack tips were examined using an Olympus BH-2 light microscope. Samples were prepared using petrographic polishing techniques. Thin sections ranging from 40 to 100 microns were required. Both bright field and crossed polarized light were used to obtain images of the sub-surface damage.

RESULTS AND DISCUSSION

Yield Strength. The addition of BN to DGEBA epoxy results in a decrease in yield strength. The decrease in yield strength may be due the poor adhesion between BN and DGEBA. Clustering of weakly bonded particles may also act as large flaws hence reducing the overall strength. In summary, both types of BN particles results in a decrease in yield strength.

Fracture Toughness. An increase in toughness is observed up to 5 vol% BN. Figure not shown. After 5 vol% BN, the fracture toughness levels off. In general, the large, 50 microns in diameter, BN particles are better toughening agents (K_{max}= 1.9 MPa√m) than the smaller, 10 microns in diameter, BN particles (K_{max}= 1.5 MPa√m).

Figures 2-3 contain plots of K_{IC} versus volume fraction of BN filler/CTBN. Note that the volume fraction of particles for all the blends is 10 vol%. The BN/CTBN toughened epoxies containing 10 μm diameter BN particles did not display synergistic toughening whereas the BN/CTBN toughened epoxies containing 50 μm diameter BN particles did. The differences between these two systems can be explained using microscopy.

SEM. Investigation of the fracture surfaces suggests that the rubber particles are indeed cavitating in both series of BN/CTBN toughened epoxies. There is evidence of a lack of adhesion between the BN particles and the epoxy matrix (not shown). Unfortunately, both series of BN filled epoxies display poor particle-matrix adhesion so a lack of adhesion is not the reason for the lack of synergy.

TOM. Investigation of the sub-surface damage in both series of BN filled epoxies is much more revealing than SEM. TOM micrographs of the damage zones in the CTBN(7.5)/BN(2.5) toughened epoxies show that the plastic zone is significantly smaller for the toughened epoxy containing 10 μm diameter BN platelets than for the epoxy containing 50 μm platelets. A rationale for these observations is given below.

Effect of interparticle distance. The differences in the two series of CTBN/BN toughened epoxies can be explained by analyzing the interparticle spacing of the BN fillers. Our explanation is illustrated in Figure 4. The rationale is as follows. When the BN interparticle spacing is large, the BN fillers can interact with the crack tip and can actually induced plastic zone branching. When the BN interparticle distance is small, the BN fillers will also interact with the crack tip but will retard the growth of the shear bands since the particles will be in the path of the shear bands. Such a rationale can also explain why synergistic toughening is not observed at high BN filler loading.

CONCLUSIONS

Epoxy polymers modified with BN platelets improved toughness as a result of a crack path deflection mechanism. The efficiency of the toughening depends on the particle size. Composite blends of rubber particles and BN platelets exhibit synergistic toughening when 50 μm diameter platelets are used. The reason why 10 μm diameter platelets do exhibit synergistic toughening can be explained by considering the interparticle distant between BN platelets. When the spacing is small, the rigid BN particles impede the growth of shear bands in the plastic zone at the crack tip, hence, suppressing plastic zone growth. Conversely, when the interparticle distance is large, the BN can induce plastic zone branching.

ACKNOWLEDGMENTS

This work is sponsored by the National Science Foundation (Grant No. EEC-9526445) as a TIE project in conjunction with the University of New Mexico Center for Micro-Engineered Ceramics (CMEC). Continued support has been provided by the Polymer Interfaces Center at Lehigh University. The authors wish to acknowledge Mr. Richard Hill of Advanced Ceramics Corporation for supplying the BN powders used in this study.

REFERENCES

1. I. M. Low, S. Bandyopadhay, and Y. W. Mai, *Polym. Int.*, 27 (1992) 131.

2. A. J. Kinloch, D. L. Maxwell, and R. J. Young, *J. Mater. Sci.*, 20 (1985) 4169.

3. A. C. Roulin-Moloney, W. J. Cantwell, and H. H. Kausch, *Polymer Composites*, 8 (1987) 314.

4. H. R. Azimi, R. A. Pearson, and R. W. Hertzberg, *J. Appl. Polym. Sci.*, 58 (1995) 449.

5. H. R. Azimi, R. A. Pearson, and R. W. Hertzberg, *Polym. Eng. & Sci.*, 36 (1996) 2352.

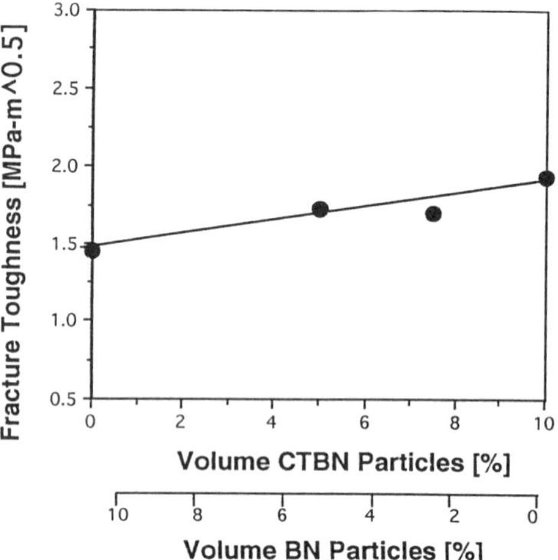

Figure 2. Synergistic toughening is NOT observed in a CTBN/BN modfied epoxy system containing containing 10 μm dia. BN platelets.

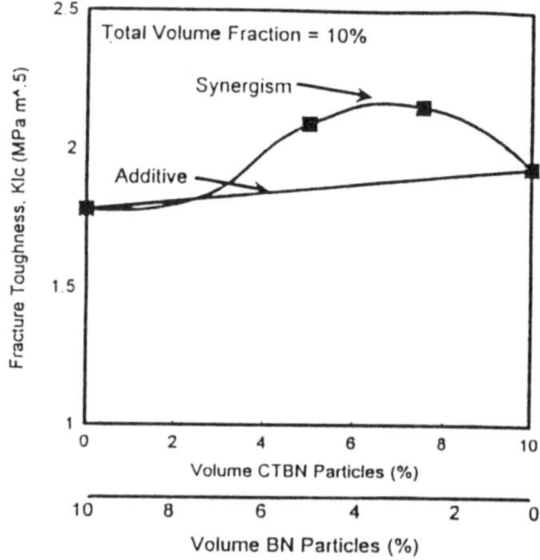

Figure 3. Synergistic toughening is observed in a CTBN/BN modfied epoxy system containing containing 50 μm dia. BN platelets.

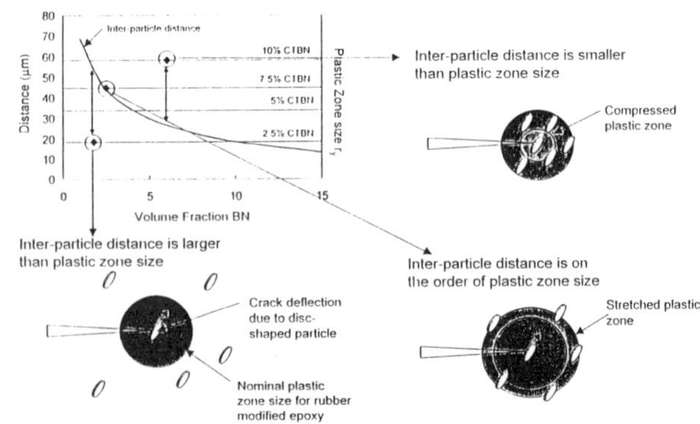

Figure 4. Process zone size/interparticle distance rationale for the occurrence of synergy.

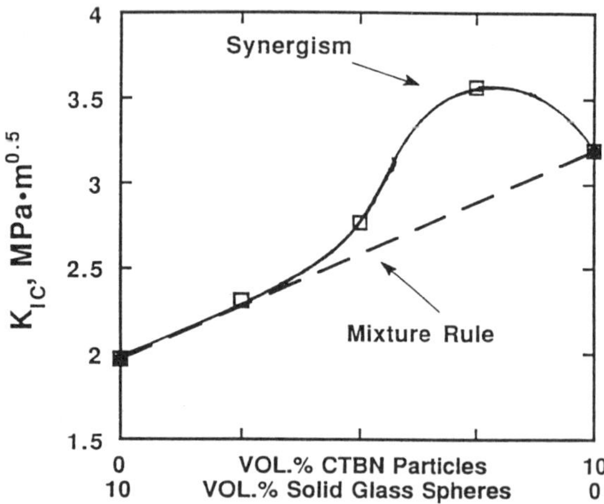

Figure 1. Synergistic toughening is observed in a hybrid epoxy composite containing rubber particles and solid glass spheres [ref. 4].

Prediction of Toughness of Notched Heterogeneous Polymer Systems

Han E.H. Meijer, Robert J.M. Smit, Marcel A.M. Brekelmans,
Leon E. Govaert

TUE, Eindhoven Univeristy of Technology;

P.O. Box 513, 5600 MB Eindhoven,

The Netherlands

It is the aim of this paper to address and elucidate the
fundamental problems concerning the deformation and fracture
of rubber-modified heterogeneous polymer systems and, as an
extreme example, we will focus on the obvious defferences in
the responses of polycarbonate (PC) and polystyrene (PS). We
will emphasize the problems of (i) the time-, temperature-, rate-
and history dependent mechanical behavior of the individual
components, which is solved by the choice of proper 3D
constitutive equations, (ii) the complexity of the microstructure,
that is taken care of by detailed finite element calculations on
RVE's Representative Volume Elements, and (iii) the relation
between micro- and macrodeformation, for which the so-called
MLFEM, Multi-Level Finite Element Method, is employed. The
paper concludes with an explanation why established and new
routes to toughen amorphous polymers are successful.

DEFORMATION AND FRACTURE OF POLYMER SILICATE NANOCOMPOSITES

J.P. Harcup and A.F. Yee
Department of Materials Science & Engineering
University of Michigan, Ann Arbor, MI48109 USA

M.K. Akkapeddi
Plastics R&D, AlliedSignal Inc., Morristown, NJ07962 USA

INTRODUCTION

A nanocomposite is a material whose distinct components are dispersed on the scale of nanometres. In this work, the deformation and fracture behaviour of polyamide 6-silicate nanocomposites has been studied. Polymer-Silicate layer nanocomposites are relatively lightweight and have excellent moduli and good barrier properties, but their fracture toughness is typically rather low. Thus the objective of this work is to investigate the sources of toughness in nanocomposites and to consider methods for improving the fracture toughness without incurring a deterioration in other properties. Two types of nanocomposite were studied, each having slightly different *microstructure* as a result of processing method. Elastomerically modified nanocomposites were also investigated given the conventional use of elastomer particles in polymer toughening.

EXPERIMENTAL

Four materials were investigated:

	Description	Silicate content (wt %)	Elastomer content (vol %)
PA	Neat Nylon 6	0	0
NCH I	Nylon Clay Hybrid	1,2,3,4	0
NCH II	Nylon Clay Hybrid	4	0
rm-NCH II	Elastomer modified NCH	4	5, 10

Table 1 - Properties of materials investigated

NCH II was subjected to a greater extent of shear mixing than NCH I - full details of the synthesis are given elsewhere [1]. The NCH II material contained microscale aggregates of partially intercalated material, whereas the NCH I contained significantly smaller particles of pure silicate. The nanostructure of the materials is shown in fig 1. The dark lines in the photograph are the monolayers of silicate. XRD revealed that there was no preferred orientation of the layers on the microscale.

Values of fracture toughness and modulus were determined using three point bend specimens and tensile bars, following the methods set out by the ASTM.

The Double Notch Four Point Bend geometry [2] was used to enable the pre-failure and post-failure analysis to be carried out on the same specimen. In the DN4PB geometry, two nearly identical pre-cracks are introduced on the same side of a four point bend specimen. Upon loading, fracture may only initiate at *one* of these pre-cracks. Consequently the failed specimen comprises the fracture surface remaining from the pre-crack which had failed, and the process zone at the tip of the pre-crack which did not fail. The process zone may be extracted and viewed using Transmission Optical Microscopy (TOM) or Transmission Electron Microscopy (TEM). The fracture surface may be viewed using Scanning Electron Microscopy (SEM) and X-ray Energy Dispersive Spectroscopy (XEDS).

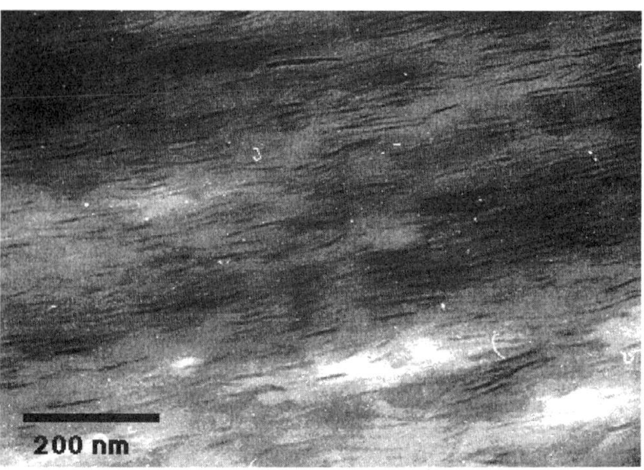

Fig 1 - Nanostructure of Polymer Silicate Nanocomposite NCH I.

The microstructures of NCH I and NCH II are shown in fig 2. In particular, these show the presence of the microscale aggregates in NCH II, which aggregates are uniformly dispersed throughout the material.

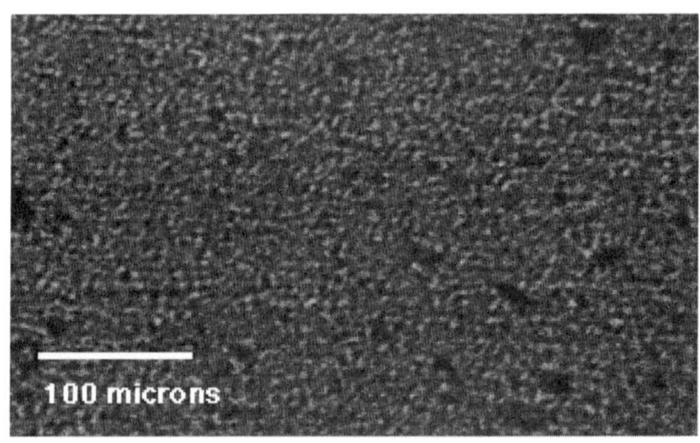

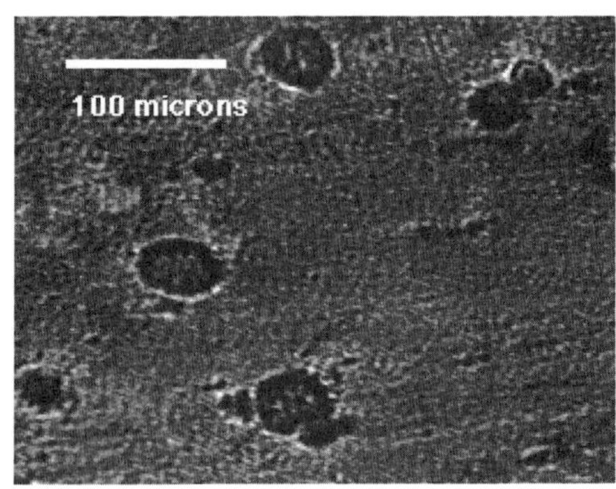

Fig 2 - TOM of microstructure a) NCH I b) NCH II

RESULTS AND DISCUSSION

The modulus of the NCH materials for various silicate contents is shown in fig 3 and the corresponding values of toughness are shown in fig 4. It is clear that the increased silicate content leads to good improvment in modulus, but also brings about a deterioration in fracture toughness. It is noteworthy that NCH II exhibits significantly better fracture toughness and modulus than NCH I.

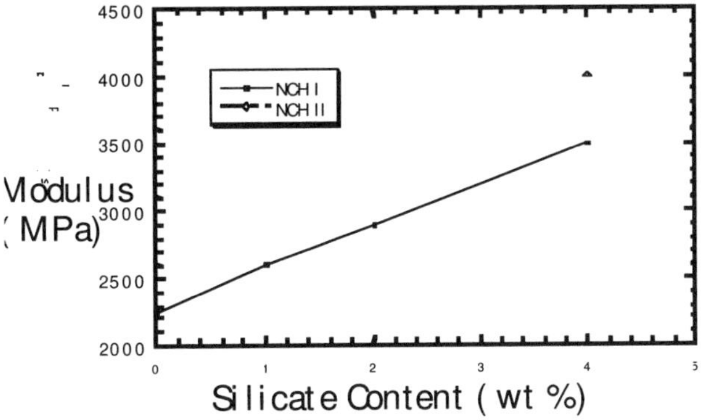

Fig 3 - Stiffness of Nanocomposite

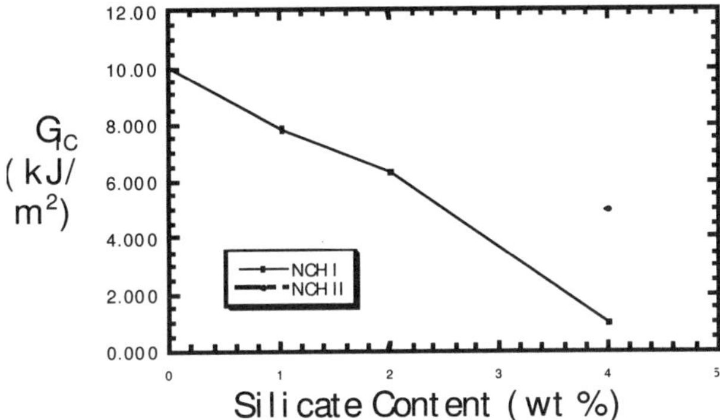

Fig 4 - Toughness of Nanocomposites

The elastomer modified material exhibited an approximately 80% increase in toughness for a 10% volume content of elastomer. However, this was accompanied by a 20% reduction in modulus.

Plastic deformation of the matrix was observed in both NCH II and the elastomerically modified material. Fig 5 shows a TOM micrograph taken from a region of a tensile bar specimen close to the failure zone - it is clear that plastic deformation occurs at the aggregates of partially intercalated material. Moreover, these aggregates appeared to have fractured under the applied load. Such features were also observed in the crack tip process zone. Consequently, whilst brittle fracture is exhibited by the NCH I, this is not the only deformation mode available to an NCH matrix - the inclusion of elastomer particles or large microscale aggregates may instead bring about plastic deformation in the matrix. The toughest materials were those which exhibited such plastic deformation on the microscale.

One interpretation of the currently available data is that loading to a level lower than the critical failure load causes the aggregates in the NCH II material to fracture, bringing about a change in the local stress state of the matrix and enabling plastic deformation to occur under the reduced constraint. This is akin to the toughening effect of elastomer particle cavitation as has been discussed for polyamide 6 previously [3].

On the basis of the work described here, it is proposed that an efficient means of toughening nanocomposite materials without incurring a reduction in modulus is to modify the *microstructure* of the nanocomposite so that the capability of the matrix to plastically deform may be realised. Further work is being carried out to investigate the sequence of events prior to fracture and the material factors which are relevant to the occurrence of the plastic deformation.

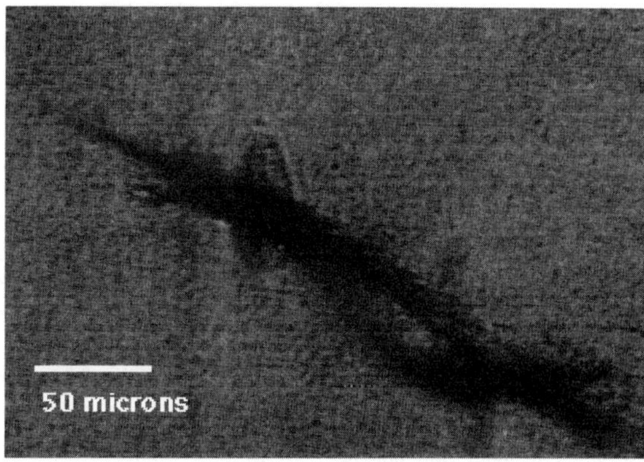

Fig 5 - TOM Micrograph showing appearance of aggregate region prior to failure

CONCLUSIONS

The following points have emerged so far:

- even a modest silicate content leads to excellent modulus improvements, but with a commensurate reduction in toughness
- the NCH matrix is capable of microscale plastic deformation (with an accompanying increase in toughness), but whether this is realised depends upon the microstructure:

    i)   elastomer particles lead to extensive plastic deformation and improvements in toughness, but the modulus is reduced

    ii)   the matrix is able to plastically deform in the material containing microscale aggregates. This is accompanied by an improvement in toughness and importantly, a reduction in modulus. Further investigation is taking place.

ACKNOWLEDGEMENTS

We would like to thank AlliedSignal Foundation for partial funding and provision of some of the resources used in this study.

REFERENCES AND BIBLIOGRAPHY

1. M.K. AKKAPEDDI Private Communication (1997)
2. H.J. SUE and A.F. YEE, *J. Mater. Sci.,* **24** (1989) 1447
3. D. LI, A.F. YEE, K.W. POWERS, H.C. WANG and T.C. YU, *Polymer,* **34** (1992) 21471

TOUGHENING OF NYLON-6 BY THE ADDITION OF REACTIVE RIGID POLYMER AND ELASTOMER

I. Kelnar*, M. Stephan**, L. Jakisch** and I. Fortelný*

*Institute of Macromolecular Chemistry, Academy of Sciences of the Czech Republic, 162 06 Prague 6, Czech Republic

**Institute of Polymer Research Dresden e.V., Hohe Strasse 6, D-01069 Dresden, Germany

INTRODUCTION

Impact resistance of ductile polymers can be enhanced by the addition of stiffer, usually brittle component like PSAN for PC[1] or poly(styrene-*co*-maleic anhydride) (SMA) for Nylon-6 (N-6) [2,3]. Though the enhancement of toughness is not as significant as for toughening by elastomers, the advantage of rigid-rigid toughening is a simultaneous increase of strength and stiffness (reduction of these parameters is a significant shortcoming of of elastomers).

If the rigid-rigid system has a sufficiently fine phase structure[4] and high interfacial adhesion[5], the dispersed brittle inclusions are able to deform plastically with the matrix and thus to absorb mechanical energy instead of brittle failure under loading. The explanation is that the global uniaxial tensile stress results in a local triaxial stress state around the inclusion, its magnitude being determined by the disparities in elastic constants between the two phases. If the inclusion has a sufficiently higher modulus and a lower Poisson's ratio than the matrix, the compressive stress evolved exceeds the brittle-to-ductile transition pressure[6] and the brittle particle becomes ductile.

Due to a significant dependence of brittle-to-ductile transition on temperature, the limited low-temperature toughness of these systems can be expected. In our previous work dealing with N-6/SMA[7], properties of SMA were modified by the reaction with aminated liquid rubber to study the range of dispersed-phase properties, where this toughening occurs. Though an increase in toughness was found, the liquid rubber (partly dispersed in N-6 matrix) caused a significant decrease in the blend modulus by influencing negatively the N-6 crystallinity and thus its stiffness.

In the present work, the use of high-molecular-weight maleated elastomers, such as ethylene-propylene rubber (EPR), dispersed in the N-6 matrix together with SMA is studied. Due to the fact that best results for the binary N-6/SMA blend have been found at 10% SMA, a low total content of dispersed phase was also chosen for ternary systems.

EXPERIMENTAL

Materials

Nylon-6 (N-6) (Ultramid B3) BASF, M_n 18 000

Poly(styrene-*co*-maleic anhydride) (Dylark 332) ARCO, maleic anhydride content 14%, M_n =180 000

Ethylene-propylene elastomer functionalized with 0.6% of maleic anhydride, (EPR) (Exxelor VA 1801) EXXON

Styrene-ethylene/butene-styrene functionalized with 2% of maleic anhydride (SEBS) (Kraton FG 1901 X), SHELL

Blend preparation

Prior to mixing, N-6 was dried at 85°C for 12 h in a vacuum oven. The blends were prepared by mixing the components in the W 50 EH chamber of a Brabender Plasti-Corder at 250°C and 50 rpm for 10 min. The material removed from the chamber was immediately molded at 250°C in a laboratory press to form 1-mm thick plates. Strips cut from these plates were used for preparation of dog-bone specimens (gauge length 40 mm) in a laboratory micro-injection molding machine (DSM). The barrel temperature was 265°C, that of mold 80°C.

Testing

Tensile tests were carried out at 22°C using an Instron 6025 apparatus at a crosshead speed of 20 mm/min. The stress at break, σ_b, and Young's modulus, E, were evaluated.

Tensile impact strength, a_t, was measured using a Zwick hammer with an energy of 4J and one-side notched specimens.

Morphological observations

were performed using scanning electron microscopy (SEM) and cryo-fractured samples. For better visualisation of the SMA phase, the samples were etched in ethyl methyl ketone for 1 h. The EPR phase was etched in n-heptane for 5 min.

RESULTS AND DISCUSSION

Binary blends. Table 1 shows the properties of the binary blends of N-6 with SMA and both reactive elastomers. With the N-6/SMA 90/10 blend, the toughness (a_t) is about 20% higher than the value for N-6. Of importance is a simultaneous increase in both σ and E. Surprisingly insignificant decrease in a_t was found at - 20°C. With the same addition of EPR or SEBS, the more than two-fold increase in a_t was accompanied by a significant decrease in other parameters.

Ternary blends. Due to the fact that the best results for N-6/SMA were found with SMA contents[3,7] below 10 % , most of the ternary blends were prepared with this or a lower SMA content. When properties of the 90/10/5 N-6/SMA/EPR system are compared with appropriate binaries, the toughness is even close to 90/10 N-6/EPR (Table 2), whereas σ_b and E are practically unchanged in comparison with N-6 or N-6/SMA. A blend with a lower amount but the same SMA/EPR ratio (90/6.7/3.3) shows then a slightly higher σ_b and lower a_t (but still significantly higher than that of N-6/SMA).

If the EPR amount was increased (with the total additive content 10%), the changes in σ_b and E were negligible, whereas the gain in a_t at ambient temperature was remarkable. Surprisingly, the a_t value at −20°C was practically unchanged in this case (Table 2).

Morphological observations of the 90/5/5 blend show that the size of both separately dispersed SMA and EPR was close to 100 nm. For SMA, it was somewhat lower, for EPR slightly over this value.

From Table 2 is further clear, that mechanical properties of analogous blends with SEBS are similar and their values slightly higher than for EPR blends. Therefore, ternary systems with SEBS seem to be even more suitable.

From the above results it can be concluded that combination of rigid SMA with a well-dispersed elastomeric component leads to certain synergism between both types of toughening. As a result, the a_t of ternary system approaches the value of binary N-6/elastomer, whereas σ_b and E are close to the values of N-6/SMA. Tentatively, due to the fact that stress fields around elastomeric inclusions must strongly overlap for toughening[8] to occur (by inducing shear yielding of the matrix), the high toughness of ternary blends is probably caused by certain synergistic interaction between (dissimilar) stress fields induced by both (relatively dilute and small) elastomeric and SMA inclusions. Further fractographic and morphological study are in progress.

Quite opposite results have been found with ternary blends containing higher amounts of both additives. For instance, systems with 10% SMA and 5, 10 and 15% EPR showed practically unchanged a_t and significantly lowered σ_b and E with increasing EPR content (Table 2). Similarly, toughness of the 70/15/15 blend was slightly and other parameters significantly lower than those of the 90/5/5 composition. (The phase structure was similar, the size of SMA inclusions in the 70/15/15 blend was ~100 nm, that of EPR ~150nm).

In this case, except for worsening of N-6/SMA properties with increasing SMA contents[3,7] , the fact that increased content of EPR does not increase a_t indicates that higher (than 15%) total content of inclusions probably reduce the effectivity of both toughening mechanisms. For example, lower stiffness of the surrounding matrix can suppress plastic deformation of SMA. This system may then behave like a "filled" N-6/ elastomer blend.

Table 1. Mechanical properties of binary blends

Composition (% wt.)	σ_b (MPa)	E (MPa)	a_t (22 °C) (kJ.m^{-2})	a_t (-20 °C) (kJ.m^{-2})
N-6/SMA 90/10	75	2465	36	32
N-6/EPR				
96.7/3.3	71	2335	44	46
95/5	65	2255	52	48
90/10	59	2100	73	61
N-6/SEBS				
95/5	64	2295	46	45
90/10	60	2145	71	55
N-6	72	2380	30	26
SMA	29	3405	-	-

Table 2. Mechanical properties of ternary blends Nylon-6/SMA/elastomer

Composition (% wt)	σ_b (MPa)	E (MPa)	a_t (22 °C) (kJ.m^{-2})	a_t (-20 °C) (kJ.m^{-2})
N-6/SMA/EPR				
90/3.3/6.7	71	2360	70	51
90/5/5	74	2415	59	49
90/6.7/3.3	75	2440	53	48
85/10/5	72	2450	71	55
80/10/10	61	2160	70	55
75/10/15	55	1980	73	57
70/15/15	53	1720	68	56
N-6/SMA/SEBS				
90/6.7/3.3	74	2445	54	53
90/5/5	76	2455	76	56
85/10/5	72	2455	69	55

CONCLUSIONS

Impact modification of Nylon-6 by a simultaneous addition of finely dispergated SMA and reactive elastomer, with the total content of both additives below 15 %, suitably combines the advantages of both rigid-rigid and elastomeric toughening. As a result, the a_t of ternary blend reaches the value of elastomer-toughened N-6, whereas σ_b and E are unchanged or even higher than those of N-6. Explanation can be mutual synergistic influencing of both toughening mechanisms.

Quite opposite results were found for blends with higher contents of both additives. In this case, increasing the content of elastomer did not increase toughness, whereas strength and stiffness were lowered significantly. To explain both effects, futher fractographic and morphological studies are in progress.

Acknowledgement
Financial support of the Academy of Sciences of the Czech Republic (Grant 12/96/K) is gratefully acknowledged.

REFERENCES

1. T. Kurauchi and T. Ohta, *J. Mater. Sci.,* **19**, 1699 (1984)
2. M. Lu, H. Keskkula and D.R. Paul, *Polymer,* **34**, 1874 (1993)
3. I. Kelnar, M. Stephan, L. Jakisch and I. Fortelný, *J. Appl. Polym. Sci.,* **66,** 555 (1997)
4. W. L. Nachlis, R. P. Kambour and W. J. MacKnight, *Polymer,* **35**, 3643 (1994)
5. R. D. Evans, L. A. MacGillivray, J. D. Boyd, W.E. Baker and T. Pith, *Polym. Networks Blends,* **6,** 181 (1996)
6. K. Matsushige, S. V. Radcliffe and E. Baer, *J. Appl. Polym. Sci.,* **20**, 1853 (1876)
7. I. Kelnar, M. Stephan, L. Jakisch and I. Fortelný, *Polym. Eng. Sci,* in press
8. S. Wu, *J. Appl. Polym. Sci.,* **35**, 549 (1988)

PP-EPDM BLENDS: CHANGING DEFORMATION WITH TEST SPEED

R.J. Gaymans and A. van der Wal
Twente University, Department of Chemical Technology
P.O. Box 217, NL-7500 AE, Enschede, The Netherlands

Introduction

PP-rubber blends are ductile to very low temperatures and considerable plastic deformation is taking place in the notch and during crack propagation. The fracture behavior of rubber toughened polypropylene strongly depends on the blend rubber content[1], particle size[2], molecular weight of the PP[3] and the test temperature[1]. The blends have as measured with a notched Izod method a sharp brittle-ductile transition. The brittle-ductile transition is characteristic for semi ductile materials.

The effect of test speed is for the strongly plastically deforming materials rather complicated since the deformation energy is to a large extend dissipated as heat[4]. With increasing test speed the deformation process changes from isothermal to adiabatic[4,5]. The yield stress is expected to increase with test speed but to decrease with temperature.

The effect of the test speed has been studied on rubber toughened nylon-6[6,7,8]. These blends were found to have at relative high test speeds an increase in fracture energy with increasing test speed. This unusual behavior was explained by a change in deformation mechanism with thermal blunting taking place at high test speeds. The strong plastic deformation at the crack tip resulted at high test speeds in the formation of a melt zone in front of the crack tip[6,8]. This melt zone just beneath the fracture surface had a thickness of 5 µm.

PP has unmodified a brittle-ductile transition temperature (T_{bd}) as measured in Izod at 70° - 120 °C. The T_{bd} decreased strongly with rubber concentration, to -40°C for a 40% blend. This is below the Tg of PP (5°C). The influence of test speed can well be studied with a single edged notched tensile test (SENT). In this way the test speed can be varied from 10^{-5} m/s - 10 m/s. With increasing test speed the temperature development in the deformation zone as measured with Infra Red Thermography was considerable[9].

Experimental

The materials used encompass a polypropylene homopolymer (Vestolen P7000, Vestolen GmbH) with an MFI of 2.4 (230 °C, 21.6 N) and an EPDM rubber (Keltan 820, DSM). A series of blends with varying rubber content was prepared by diluting a PP-EPDM (60/40 vol%) extrusion master blend. Rectangular bars and dumbbell-shaped specimen were prepared by injection molding. The blending and injection molding processes are described in detail elsewhere[1].

The SENT tests were carried out on the notched bars with a Schenck servo-hydraulic tensile machine. The injection molded bars (ISO 180/1, 74x10x4 mm) had a single-edge 45 ° V-shaped notch (depth 2 mm, tip radius 0.25 mm), milled in the specimens. The Schenck tensile apparatus is specially designed for high speed testing. The studied clamp speeds range is from 10^{-5} to 10 m/s. At high test speeds a pick-up unit is used to allow the piston to reach the desired test speed before loading the specimen. All moving parts are made of titanium in order to diminish inertia effects. Force and piston displacement are recorded using a transient recorder (sample rate 2 MHz). After completion of the test, the results are send to a computer.

The deformation mechanism was studied by post-mortem SEM analysis of the fracture zone. The sample position is shown in Figure 2. The area of interest is perpendicular to the fracture surface, and parallel to the crack growth direction. Samples were taken from the specimens with a fresh razor blade. The samples were smoothed down on a polishing machine to avoid deformation during the trimming procedure. The finishing preparation was carried out by cryotoming at -100 °C with a diamond knife. The final surface was sputter-coated with a gold layer and investigated with a Hitachi S-800 field emission SEM.

Results and Discussion

The studied PP-EPDM blends were prepared by extrusion blending. The particle sizes obtained are given in table 1.

Table 1 Particle size in PP-EPDM extrusion blends

Rubber Content	5%	10%	20%	30%	40%
Dw (µm)	0.50	0.67	0.69	0.86	0.92

The particle size increases with increasing rubber content. This particle size difference was found to have only a small effect[1]. The tensile properties and the physical properties of the blends are reported elsewhere[1].

The fracture behavior was studied as a function of test speed by a single-edge notched tensile (SENT) test. The influence of test speed is studied by varying the clamp speed from 10^{-5} to 10 m/s.

The fracture process may be divided in two stages: the 'initiation' and the 'crack propagation' stage. During 'initiation' the stress builds up at the notch tip, but is too low to enable crack propagation. Crack propagation begins at or past the stress maximum in the stress displacement curve. PP deform with this SENT test at room temperature brittle over the whole test speed range.

The stress displacement curves obtained by an SENT test for the 30 vol% blend at varying test speed are shown in Figure 1.

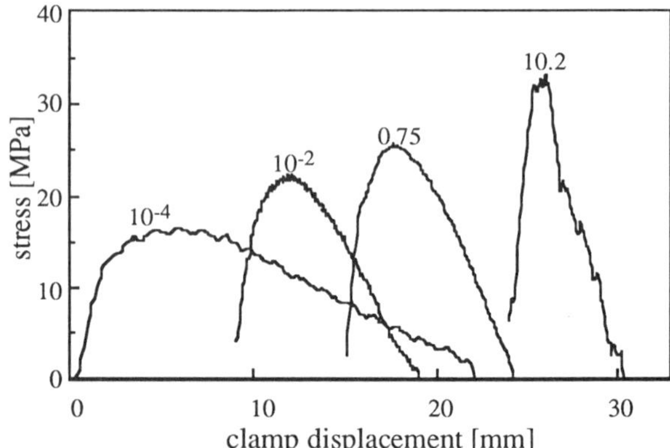

Figure 1. SENT stress-displacement PP-EPDM (70/30), test speed (m/s)

The maximum stress of this 30% blend increases with test speed and considerable plastic deformation is taking place during crack propagation. Over the whole studied test speed range this 30% blend fractures ductile. The maximum stress as function of test speed of polypropylene and the 30% blend is given in Figure 2.

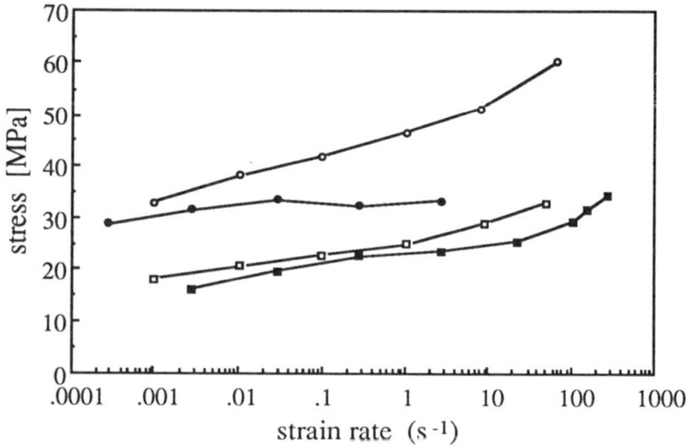

Figure 2. Maximum stress of PP (○,●) and PP-EPDM (70/30) (□,■) in tensile(○,□) and SENT (●,■) as function od strain rate

For comparison also the yield stress measured for unnotched specimens are given. PP at low strain rate (10^{-3} s^{-1}) has a maximum stress that almost equals the yield stress. This suggests that the whole cross section ahead of the notch sustains a yield stress and thus the notch is blunted. At intermediate test speed the maximum stress of PP curve drops to well below the yield stress curve and crack propagation starts before the entire cross section is bearing a yield stress. If the material fractures before yielding in the notch has taken place, then a strong decrease in fracture stress is expected. This is not observed in PP at these strain rates at 20°C and even not at -40°C[10]. The 30% blend has nearly over the whole strain rate region a maximum stress that equals the yield stress. At very high strain rates a strong increase in stress is observed. A similar increase can be seen for the unnotched tensile results. This increase seems to be typical for materials with a strong plastic deformation in the notch.

The initiation energy (IE) are high at low test speeds and decrease with increasing test speed (Figure 3a). At very high test speed the IE appears to stabilize or even to increase. The blends have over the whole studied test speed range a strong plastic deformation before the initiation of a crack.

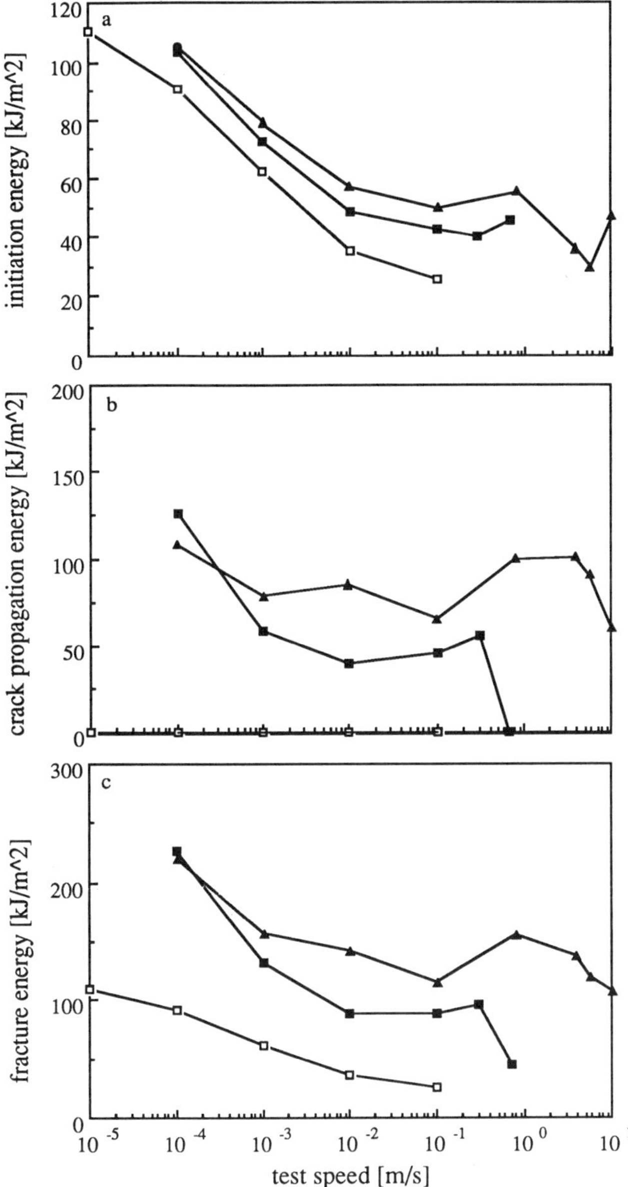

Figure 3. SENT PP-EPDM as function of test speed: □, PP; ■, 15 % EPDM; ▲, 30% EPDM

The crack propagation energy (CPE) (Figure 3b) curves are complex. At low test speeds the CPE decreases with increasing test speed as expected as with increasing test speed the fracture strain decreases. Surprisingly, at intermediate test speeds the CPE increases with increasing test speed. At very high test speeds they decrease again.

The results suggest that at low test speed with increasing test speeds, the materials turn brittle. However at intermediate test speeds for all samples at 10^{-2}-10^{-1} m/s this trend is changed to a less brittle behavior. At even higher test speeds the materials turn brittle again. The rubber content has a strong effect on where the blends start to behave brittle. The higher the rubber content the higher the speed at which the blends become brittle. The turning ductile at 10^{-2}-10^{-1} m/s suggest that at this speed the deformation process under goes a change. The change is probably as a result of an adiabatic plastic deformation. This suggests a strong warming up of the sample at high test speeds. A similar complex behavior of the CPE with test speed was found for rubber modified nylon and this was there explained as a melt blunting effect [5,8].

The total fracture energy (Figure 3c) shows the complex behavior as function of test speed too. The fracture energy is about equally dependent on the initiation and the propagation part of the deformation. Over the whole test speed range the fracture energy increases with rubber concentration and this effect is strong.

To study the asymmetric deformation process in fractured notched samples, SEM micrographs are taken of the fracture zone next to the fracture surface in the middle of the samples. The effect of test speed on the structure of the fracture zone is studied on of a 30 vol% blend of samples that all fractured ductile (Figure 4).

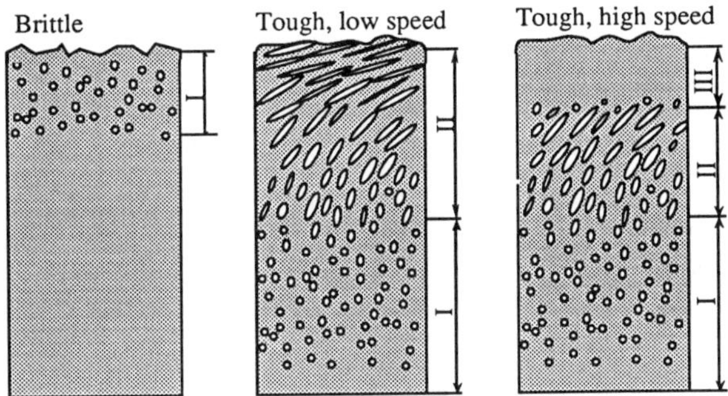

Figure 4 Cartoon of structure of the zones next to fracture plane: I, cavitation layer; II, deformation layer; III, relaxation layer.

At low test speed, a clear deformation layer leading up to the fracture surface is observed with a strong elongation of the cavities. At higher test speed ($\geq$ 0.1 m/s) the deformation of the cavities seems to be less that at low speeds and a layer without cavities appears. The lower elongation of the cavities might be due to relaxation of the blend after fracture as a result of the temperature increases at these high test speeds. A complete relaxation of semi crystalline polymers takes place above their melting temperature[5]. This layer without cavities indicates the melting of the PP before fracture and is called the melt layer. Up until about 10^{-2} m/s, the cigar-shaped voids reach up to the fracture surface. The thickness of the strongly deformed layer is about 25 µm. The appearance of the relaxation layer at high test speeds is also observed in nylon 6-EPR blends[5,6,8]. In nylon 6-EPR blends the thickness of the relaxation layer was 3-5 µm.

The appearance of the relaxation layer suggests that at the crack tip melting is taking place in polypropylene-EPDM blends at the high test speeds. The appearance of the relaxation layer coincides with the CPE increase with the test speed (Figure 4).

Conclusions

The notch deformation of PP-EPDM blends as function of test speed is complex. At low test speed the crack propagation energy (CPE) decreases rapidly with increasing test speed. However, at intermediate test speed (> 10^{-2} m/s) the CPE increase with increasing test speed. At this speed the structure of the deformation zone (of the 30% blend) next to the fracture surface changes and the samples show to have a layer that must have been molten. This suggests that during the plastic deformation at high test speeds a head of the notch a melt layer is formed and the melt layer may blunt the notch, a melt blunting process. The observed two processes might be at low test speeds a fracture process with strong plastic deformation both during initiation as during propagation process and at high speeds plastic deformation process with the formation of a melt layer ahead of a notch. This thermal blunting process considerable enhances the toughness, without this process these blends might well be brittle at high test speeds.

1　Van der Wal, A., Nijhof, R. and Gaymans, R.J., PP-EPDM blend 2, *Polymer*, in press
2　Van der Wal, A., Verheul, A.J.J and Gaymans, R.J., PP-EPDM blend 4, *Polymer*, in press
3　Van der Wal, A, Mulder, J.J., Oderekerk, J., Gaymans, R.J. PP-EPDM blend 1, *Polymer*, in press
4　Godovsky, Y.K., *'Thermophysical Properties of Polymers'*, 1992, Springer-Verlag, N.Y., Chapter 8, p. 149
5　Dijkstra, K., Laak, J. ter, and Gaymans, R.J., *Polymer*, 1994, **35**, 315
6　Dijkstra, K. and Gaymans, R.J., *J. Mater. Sci.*, 1994, **29**, 3231
7　Dijkstra, K., Wevers, H.H. and Gaymans, R.J., *Polymer*, 1994, **35**, 315
8　Janik, H., Gaymans, R.J. and Dijkstra, K., *Polymer*, 1995, **36**, 4203
9　Van der Wal, A., Gaymans, R.J., PP-EPDM blend 3, *Polymer*, in press
10 Van der Wal, A., Mulder, J.J., Thijs, H.A. and Gaymans R.J., *Polymer*, in press

Toughening of Polypropylene with Elastomers
-The Use of H-SEBS-

Junichiro Washiyama and Hiroshi Takenouchi
Montell-JPO Co., Ltd.
Kawasaki Technical Center, Automotives Team
2-3-2, Yako, Kawasaki-ku, Kawasaki, 210-8548 JAPAN

Introduction

Elastomer reinforced polypropylenes (PP) have been commonly utilized in many of motor vehicle applications, such as bumper fascia, instrument panel and door trims. Since in these applications, toughness is a principal and inevitable subject of the material design. It has been pointed out that morphologies, such as domain size and/or structure, are quite important to improve the toughness of the materials. These materials are kept at an elevated temperature in a molten state for a certain time, typically in the range of 1 to 60 minutes, upon injection molding. Hence, the size of elastomer domains increases with increasing the retention time through demixing. As a result, toughness of the materials would be inferior. In this paper, we focus on demixing characteristics in the polymer blends of PP, poly(ethylene-r-propylene), termed as EPR, and poly(styrene-b-ethylene/butene-b-styrene), termed as H-SEBS.

Experimental Section
[1] Materials

Polypropylene homopolymer was utilized as a matrix. The MFR and isotactic pentad (mmmm, measured by NMR) were 30 dg/min and 98.5 %, respectively. EPR used in this study was polymerized by means of Vanadium catalyst, the MFR and propylene content of which were 2.0 dg/min and 25 wt%, respectively. In addition, H-SEBS, which was prepared by anionic polymerization to obtain poly(styrene-b-butadiene-b-styrene) triblock copolymer followed by hydrogenation, was utilized as a compatibilizer. The MFR, styrene content and butene content in the middle (E/B) block were 10 dg/min, 20 wt% and 40 wt%, respectively. Two kinds of compounded materials, PP/EPR = 70/30 by weight and PP/EPR/H-SEBS = 70/20/10 by weight, were prepared by means of a co-rotating twin-screw type of extruder; screw diameter = 30 mm, L/D = 40. Demixing was conducted with these two materials at 220 °C for up to 4 hours in a Si-oil bath.

[2] Morphology Observation

Transmission electron microscopic (TEM) observation was conducted to determine the structure of elastomer, comprising of EPR and H-SEBS, domains with the above specimens. A small piece of the specimen was microtomed at -100 °C to obtain 100 nm slices, which was stained with RuO4 at room temperature for 2 hours. Under this condition, the EPR and H-SEBS phases were stained to give grey and dark contrast, while the PP phase was not stained to give a bright contrast, allowing us to distinguish these three phases from each other. In addition, secondary electron microscopic (SEM) observation was performed to determine the elastomer domain size as a function of demixing time. Thin sliced (~1 μm thick) specimens were treated with toluene at room temperature for 2 minutes to etch the elastomer domains, followed by gold decoration, which was then observed by SEM. Image analysis was made on the SEM photographs: at least 5000 domains for each specimen were taken into account for better statistics. The domain size, D, was then computed viz D = 2 $[S/\pi]^{1/2}$, where S denotes the area of the domain.

[3] Others

Interfacial tension between i and j phase, γ_{ij} (where i, j = PP, EPR and H-SEBS), were measured by the imbedded fiber retraction method (IFR) at 220 °C. We obtained gPP/EPR, $\gamma_{PP/H-SEBS}$ and $\gamma_{EPR/H-SEBS}$ to be 1.5, 1.0 and 0.05 dyn/cm, respectively.

Results and Discussion
[1] Morphology of PP/EPR/H-SEBS

Figure 1 shows the morphology of PP/EPR/H-SEBS before demixing. It is clear that the H-SEBS phase (dark) encapsulates the EPR one. Newmann's model for ternary blends account well for this morphology.

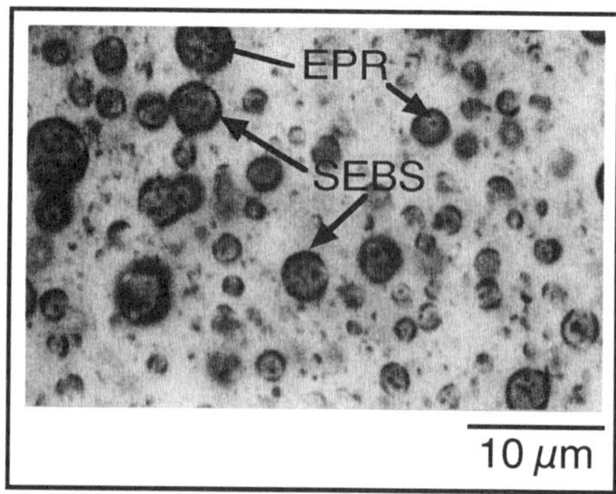

Figure 1

[2] Demixing in the PP/EPR and PP/EPR/H-SEBS Blends

Figure 2 shows the time evolution of the distribution in D in the PP/EPR blend. The peak position in D, D_{peak}, is $D_{peak} \sim 1.5$ μm at the demixing time of t=0, and the the peak shifts to a larger number, i.e. $D_{peak} \sim 2$ μm at t=10 min and 50 min. A longer t results in a broader peak width. It should be noted that large particles having D~10 μm appear when t=220 min.

As opposed to the PP/EPR blend, no particles having D~10 μm are observed by the addition of H-SEBS as shown in figure 3.

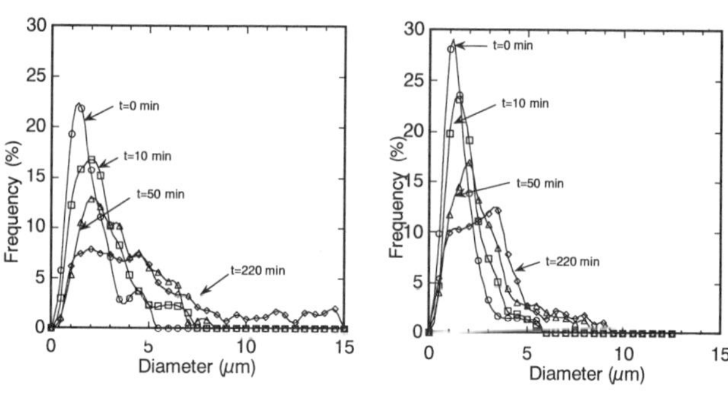

Figure 2   Figure 3

TEM micrographs reveal that H-SEBS encapsulates the EPR domains even after 220 min of demixing as shown in figure 4. The coverage of EPR domains by the H-SEBS phase gives a lower number of interfacial tension, resulting in a slower demixing rate.

[3] Mechanical Properties of the Blends

Figure 5 shows the elongation at break, eb, as a function of the retention time at 220 °C. It is clear that the addition of H-SEBS improves ε_b, in particular, ε_b shows less drop at a longer retention time.

Conclusion

The addition of H-SEBS to PP/EPR suppressed the coarsening of the elastomer domains. The mechanical properties, eb, thereby show less drop upon annealing at an elevated temperature.

Time Evolution of Morphology

PP/EPR PP/EPR/SEBS

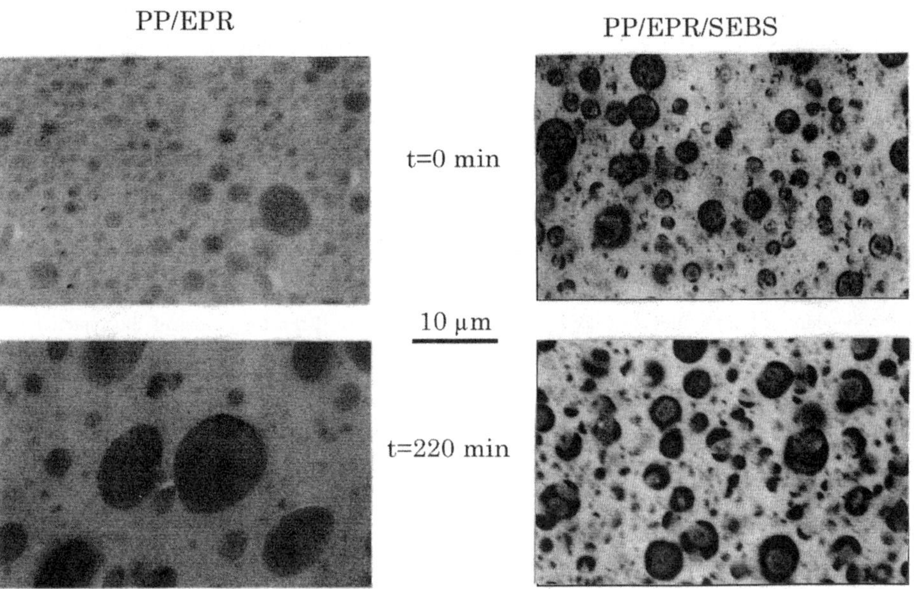

Figure 4

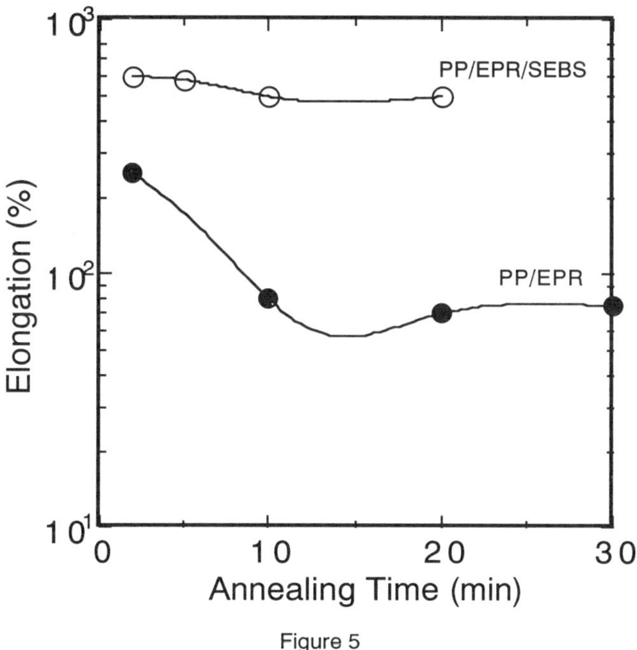

Figure 5

211

MICROFRACTURE AND MATERIAL REMOVAL IN SCRATCHING OF THERMOPLASTIC POLYOLEFINS

S. Wu, K. Sehanobish, C. Christenson, and J. Newton
Texas Polymer Center
The Dow Chemical Company
Freeport, Texas 77541

Introduction

Thermoplastic polyolefins (TPO) are rapidly used in automotives because of their advantages in specific gravity, economics and recyclability.   To extend the use of TPO in interior parts, the scratchability becomes a critical issue and scratch-resistant TPO grades have to be developed to meet different requirements from OEMs.

Scratch resistance represents a material's ability to withstand abrasive interaction with another body. However, it is not the same as abrasion resistance and wear commonly addressed in the literature. A simple case of scratch is an indentation test (horizontal speed is zero). The elastic and plastic stress analysis around the indenter have been analyzed [1,2]. General scratch test and analysis have been done for amorphous and fine grained ceramics [3], glass and silicon [4], aluminum [5], glassy polymers [6,7,8], and pigmented mineral-filled polypropylene [9].

Several factors, such as polypropylene type, impact modifier, filler type and amount, surface hardness, additives, color and pigments, and gloss level, are crucial to the increase of the scratch resistance of TPOs. However, without understanding the mechanisms and scratch process, it would be very difficult to control its effect on material performance and design material with superior scratch resistance. Despite extensive study on this subject, the fundamental understanding of the scratch process is still not well established. The purpose of this study is to observe the damage induced by scratching on samples of TPO's within a certain rigidity range, understand the material removal or microcracking mechanisms, and develop scratch-resistant TPO resins.

Test Materials and Experimental Procedures

Several thermoplastic polyolefins have been developed for improving the scratch resistance. All of these samples are blends of polypropylene, impact modifier, filler and other additives. The samples for scratch testing were injection molded with the geometry of 5x5x0.125 inches. The scratch tests were performed with the same method illustrated in detail in [9]. After the scratch test, optical and SEM microscopic analyses were used to examine the scratched surfaces and a profilometer was used to examine the scratch profile.

The physical properties of these TPOs were also examined. The tensile test conditions and sample preparation follow ASTM 638. The coefficients of friction of the specimens were measured according to ASTM D1894. The surface roughness of injection molded samples was examined with a profilometer.

Observations

Scratch resistance of TPOs depends on the residual depth and fragmentation of the scratch. This results in the scattering effect of the light from the surface. Scratching surface analysis can be used to differentiate the relative scratch resistance of materials. Two different types of scratched surfaces were observed and shown in Figure 1. Scratch tracks with partial ring marks on TPO surface are shown in Figure 1a. SEM analysis shows that these partial ring marks are induced by small cracks. The scratch tracks become more visible for Figure 1b. For samples with partial ring marks, the average width between the rings (w), the average scratch tracks width (D) changes from 0.038 mm to 0.075 mm, and 0.31mm to 0.38 mm for different TPOs under the same loading level of 7 N. When ploughing takes place, the partial ring cracks disappear and fragmentation can be observed on the scratch surfaces. In this case, only short cracks are observed on the edges of scratching (see figure 1d). This kind of fracture mechanism on the surface usually causes most damage to the material, in terms of its optical features, in comparison with plasti deformation. It makes the surface more uneven and fragmented and thi

results in scattering effect of the light from the surface.

Sliding contact fracture problem

The scratch patterns observed show that there are two different fracture mechanisms involved during scratching (see Figure 1). One is the partial ring type microcracking induced by the high tensile stress concentration just behind the traveling indenter. The initiation of cracks usually produces a stress release and therefore, lower the contact load. A contact loading oscillation can be obtained as the result of microcracking during the scratch process.

Another fracture mechanism involves material removal just ahead of the traveling indenter. This mechanism is very similar to the wear process and results in weight loss. It is speculated that if the material is not tough enough, the fracture induced by the shear stress (material removal) will take place first and the partial ring type cracks will not be produced during scratch testing. The scratch pattern shown in Figure 1d represents this scenario. If the toughness of the material is relatively high, the tensile stress induced by contact and friction just behind the traveling indenter will cause Mode I type fracture. In this case material removal will not take place and only partially ring type microcracks will appear on the surface. Scratch pattern shown in Figure 1a represents the second scenario.

The fracture process can be analyzed if we have the toughness numbers for tensile and shear induced fracture. However, shear induced fracture is a very complex problem and the test to evaluate the toughness associated with this fracture is not well defined. Thus we will focus on tensile induced fracture in this study. Besides, the scratch visibility can be reduced to a minimum if we could avoid the tensile induced fracture during scratching.   The tensile stress concentration just behind the traveling indenter has been analyzed and can be expressed as [11,12]:

$$\sigma^{tot} = \sigma^T + \sigma^N = \frac{3(4+v)\mu_f P}{16r^2} + \frac{(1-2v)P}{2\pi r^2} \qquad (1)$$

where σ^T and σ^N represent the stress contribution from friction and normal loading, P is the vertical load applied by the indenter, μ_f is the friction coefficient, v is the Poission's ratio, and r is a projected radius of the projected contact area. At the onset of microcracking, the tensile stress starts to reach the tensile strength of the material thus microcracking will take place. By using a similar approximation [11] we will have the criteria to predict the microcracking initiation:

$$\sigma^{tot} \approx E\varepsilon_{f,} \qquad (2)$$

where E is the Young's modulus and ε_f is the ultimate strain at break. It is clear from the equation (2) that we have two ways to avoid fracture taking place during scratching test. One is to reduce the local stress level and the other is to increase the modulus and ultimate strain of the material.

The equation (2) can be used to predict the ultimate strain of material. The projected radius is taken as half of the distance between the ring type microcracks measured from the scratched surfaces (see Figure 2). The results of prediction of ultimate strain to break and their reasonable correlation with the measured ultimate strain at break from tensile test is shown in Figure 2.

Material removal mechanism is not discussed in this report because shear induced fracture toughness is very difficult to measure and the elastic-plastic stress distribution just ahead of the traveling indenter is not known. Because of scratch appearance and residual indentation, a material will not have good scratch resistance if the material removal mechanism has been involved during the scratch processing. For TPO material, in general, this mechanism should be avoided in order to have good scratch resistance.

Overall Scratch Resistance

Finally, overall scratch resistance can be expressed as a function of hardness, friction coefficient and toughness. This is very similar to the suggestions in wear studies where wear resistance is considered to be directly proportional to the friction coefficient and load, and inversely proportional to hardness and practical toughness [16]. From the analysis of elastic-plastic indentation and sliding contact fracture, the measure of scratch resistance of materials can be expressed as:

$$S_R = \frac{(E\varepsilon)^{\alpha}{}_c}{(E/\varepsilon_y)^{\beta}(\mu_f)^{\gamma}} \qquad (3)$$

Where α, β, γ are parameters dependent on material, applied load, scratch speed and indenter geometric. Thus, for different materials, the controlling parameter will not be the same. It is reasonable to take a first approximation to assume that α, β, γ are all equal to one in this case. The calculation results from equation (3) for several TPOs and comparative compounds are shown in Figure 3.

It should be emphasized that scratch mechanism strongly depends on indenter geometry, applied load level and scratching speed. The correlation between the scratch resistance and a material's physical properties have to be reexamined if the test condition changes.

Summary

Scratch fracture mechanisms have been identified and hypothesized as a combination of microcracking and wearing process. Scratch resistance is described as a function of the rigidity, the coefficient of friction and the toughness of a material. Fracture analysis indicates that the microcracking process is mainly controlled by the ultimate strain at break of a material. The prediction based on local stress analysis has a good correlation with the measured ultimate strain at break from a simple tensile test.

Acknowledgments

The authors would like to thank B. Wooden for SEM analysis, S. Chum, C. Kao and C.P. Bosnyak for useful discussions.

References

1. H.Benabdallah, J.P. Chalifoux, *Polymer Testing*, **13** (1994), 377-394
2. K.L. Johnson, *J. of Mechanical Physics of Solids*, **18** (1970), 115-126
3. A.G. Evans, in 'the science of ceramic maching and surface finishing II,' Ed. By B.J. Hockey and R.W. Rice, National Bureau of Standard SP 562, 1979
4. H.J. Leu, and R.O. Scattergood, *J. Mater. Sci.*, **23** (1988) 3006-3014
5. R. Hill, The mathematical theory of plasticity, Oxford, Clarendon Press, 1950
6. L.A. Galin, Contact problems in the theory of elasticity, Raleigh, N.C., 1961
7. D.M. Marsh, Proc. Royal Soc. London, A279, 420 (1964).
8. K.L. Johnson, J. Mech. Pgys. Solids, 18, 115 (1970)
9. C.J. Studman, M.A. Moore and S.E. Jones, J. Phys. D. - Appl. Phys., 10, 617 (1977)
10. H. K. Xu, and S. Jahanmir, *J. Mater. Sci.*, **30** (1995), 2235-2247
11. A.C. Yang and T.W. Wu, *J. Material Sci.* **28**, (1995) 955
12. A.C. Yang, T.W. Wu, *J. Polymer Sci. B: Polymer Physics*, **35** (1997) 1295-1309
13. B.J. Briscoe, E. Pelillo, and S.K. Sinha, *Polymer International*, **43** (1997) 359-367
14. J. Chu, R. Rumao, and B. Coleman, Scratch and mar resistance of filled polypropylene materials, Report, 1995.
15. J. Mukcrji and P.K. Das, *J. Amer. Ceram. Soc.* **76** (1993, 2376
16. B.J. Briscoe, in 'Physical and chemical aspects of polymer surfaces,' Ed. K.L. Mittal, Plenum, New York, 1981
17. G.M. Hamilton and L.E. Goodman, *J. Appl. Mech.*, **33** (1966), 424

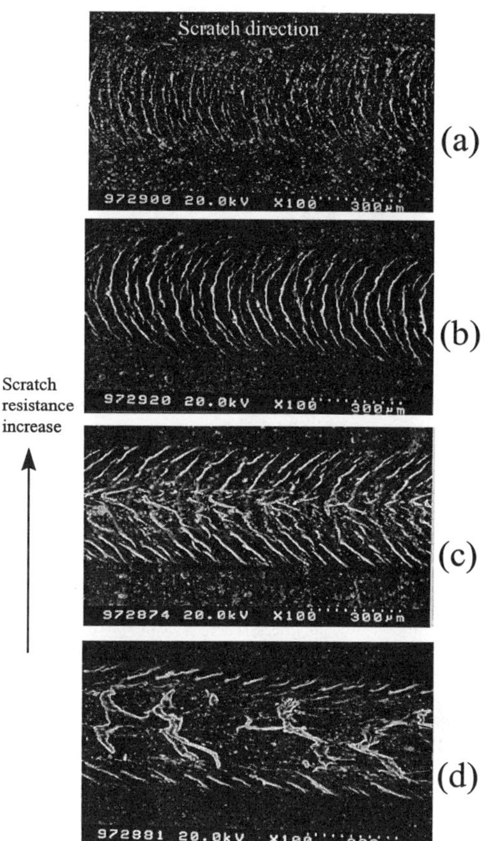

Figure 1. Scratch patterns for different TPO under the same loading level and at the ambient.

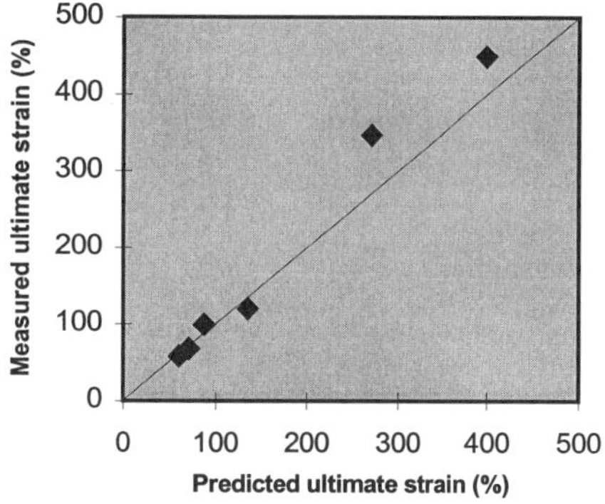

Figure 2. Correlation between measured and predicted ultimate strain at break.

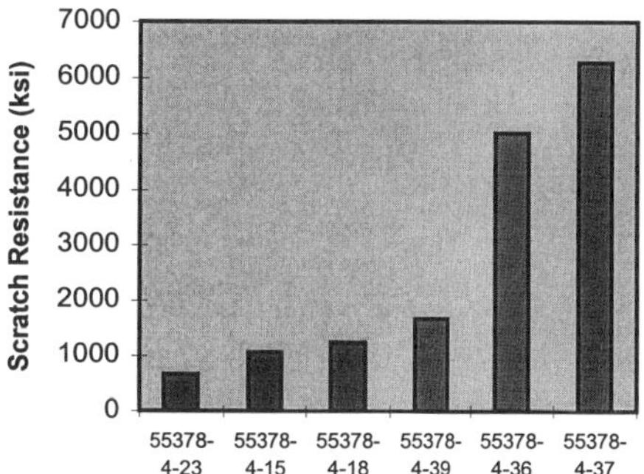

Figure 3. Scratch resistance of several TPO formulations.

Relation Between Interface Width and Toughness in a Weakly Immiscible System: Random Copolymers at Interfaces

Hugh R. Brown   R. Sellars

BHP Steel Institute, University of Wollongong, Wollongong NSW 2522, Australia

Introduction

A thin layer of copolymer applied to the interface between two immiscible polymers can significantly affect the adhesion of the interface. When the copolymer acts as a compatibilizer between the different polymers, the strength and toughness of the interface is increased. Adhesion promotion through the use of diblock copolymers has been studied extensively[1-6]: The mechanism of reinforcement with diblocks can be described as a stitching at the interface where each block of the copolymer chain blends with the more compatible phase and so each molecule provides a single stitch. The situation with random copolymers is not so clear. They are known to be effective coupling agents at an interface and it has been suggested that each copolymer molecule can form loops that penetrate into the homopolymers on both sides of the interface and so each copolymer molecule forms a number of stitches. This concept is very similar to the current understanding of the strength of interfaces between immiscible polymers, where it seems likely that the interface toughness originates from entanglement between the two materials. This entanglement happens when some loops polymer A that cross the interface are long enough to entangle with polymer B. Clearly the density of such entanglements will increase with the width of the interface. The aim of this work is to propose a model for this process.

A Simple Model of Entanglement Across an Interface

Introduction

The toughness of an interface between immiscible polymers is likely to be controlled mainly by the areal density Σ of entanglements across the interface. Σ is expected to vary with the width of the interface which itself is controlled by the interaction parameter χ between the polymers. This entanglement density has been considered by de Gennes[7] who suggested that it is controlled by the fraction of loops across the interface that are at least of length N_e, the number of repeat units between entanglements. This assumption gave an interface areal entanglement density that varied

as $\exp-(L_e/a_I)^2$ where L_e is the mean distance between entanglements along a chain, otherwise known as the tube diameter, and a_I is a measure of the interface width.

The actual nature of polymer entanglements is poorly understood. However there is good experimental evidence that the length of chain between entanglements in the bulk is that which pervades a volume that is about 10 times the volume that it actually occupies[8]. Within this model one can consider entanglements as approximately evenly distributed along a chain so it is not obvious that a loop has to be of length L_e across an interface to be entangled. An alternative approach, considered here, is to assume entanglements evenly spaced along a chain and to estimate the number of strands between entanglements that have a terminating entanglement on each side of the interface.

A second issue is connected with the effect of the interface on the value of N_e and L_e. The current author has suggested that an impenetrable interface changes the packing of adjacent chains sufficiently to increases N_e by up to a factor of nearly 4. Clearly this change in entanglement density, if it exists, may be relevant when calculating the density of entanglements across a narrow interface.

The Model

The profile of an interface between high molecular weight polymers A and B can be represented by a function of the form

$$\phi_A(z) = \frac{1}{2}\left(1 + \tanh(2z/a_I)\right)$$
$$\phi_B = 1 - \phi_A$$

where $\phi_A(z)$ is the volume fraction of polymer A at a position z relative to the center of the interface and the interface width a_I is given by

$$a_I = \frac{2b}{\sqrt{6\chi_{AB}}}$$

with b a segment dimension. This relation was first derived by Helfand and coworkers and has been extended to polymers A and B with finite molecular weight, different size and stiffness[9,10, Broseta, 1990 #247,11]. It is also consistent with a number of experimental observations[12].

Assume that a chain of polymer B has an entanglement at position z = -h on the majority B (-ve z) side of the interface within a distance L_e of the interface. The probability that the next entanglement along the chain is on the majority A side of the interface can be approximated by

$$P(h) = \int_0^{L_e-h} \frac{\phi_B(z)}{2L_e\phi_B(h)}\,dz$$

where it is assumed that the distance between entanglements is constant through the interface and that the probability of a strand crossing the interface from position h to z is controlled by the ratio of the volume fractions of the B material at those two locations. We recognize that these two assumptions are not self-consistent but assume that they give a good first order approach to the problem. If the volume density of entanglements is given by ρ_e then the areal density of B chains that cross z = 0 with adjacent entanglements on either side will be given by

$$\Sigma_B = \rho_e \int_0^{L_e} \phi_B(h)P(h)dh = \frac{\rho_e}{2L_e}\int_0^{L_e}\int_0^{L_e-h}(1 - \phi_A(z))dzdh$$

from the relation for $\phi_A(z)$ given above it can be shown that

$$\Sigma_B = L_e\rho_e\left(\frac{1}{384}\left[48\ln(2)a_r - \pi^2 a_r^2\right] - \frac{a_r^2}{32}\int_1^{1+\exp(-4/a_r)}\frac{\ln(v)}{1-v}dv\right)$$

or

$$\Sigma_B = L_e\rho_e J(a_r)$$

where the relative interface width a_r is a_I/L_e.

As expected the areal density of entanglements across the interface increases with increasing interface width and is close to saturation when the interface width is about five times the tube diameter.

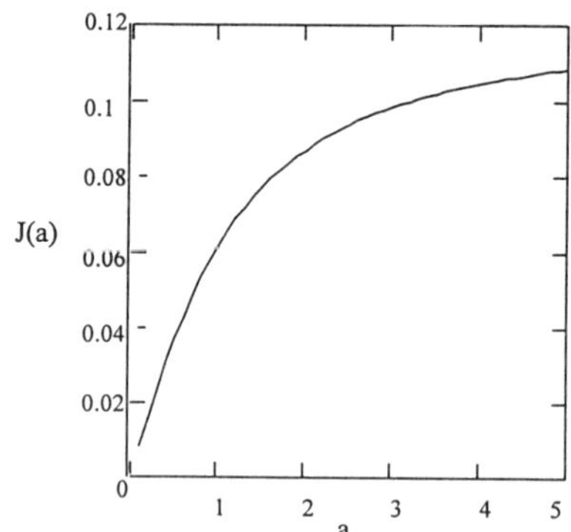

Plot of scaled areal density of entanglements against relative interface width.

">

For crazing failure in both glassy and semicrystalline polymers the interface toughness varies as Σ_B, and so will vary with interface width as J^2. It is evident from the figure that the interface toughness is predicted to be about quarter that of cohesive failure for an interface of width equal to the tube diameter and to decrease approximately linearly with interface width for narrower interfaces.

References

1)Brown, H. R. *MRS Bulletin* **1996**, *21*, 24.

2)Creton, C.; Kramer, E. J.; Hui, C.-Y.; Brown, H. R. *Macromolecules* **1992**, *25*, 3075-88.

3)Brown, H. R.; Char, K.; Deline, V. R.; Green, P. F. *Macromolecules* **1993**, *26*, 4155-63.

4)Char, K.; Brown, H. R.; Deline, V. R. *Macromolecules* **1993**, *26*, 4164-71.

5)Washiyama, J.; Kramer, E. J.; Hui, C. Y. *Macromolecules* **1993**, *26*, 2928-34.

6)Creton, C.; Brown, H. R.; Deline, V. R. *Macromolecules* **1994**, *27*, 1774-80.

7)de Gennes, P. G. *Comptes Rendues* **1989**, *308*, 1401-3.

8)Fetters, L. J.; Lohse, D. J.; Richter, D.; Witten, T. A.; Zirkel, A. *Macromolecules* **1994**, *27*, 4639-47.

9)Helfand, E.; Tagami, Y. *Polymer Letters* **1971**, *9*, 741-6.

10)Helfand, E.; Tagami, Y. *Journal of Chemical Physics* **1972**, *56*, 3592-3601.

11)Helfand, E.; Bhattacharjee, S. M.; Fredrickson, G. H. *Journal of Chemical Physics* **1989**, *91*, 7200-8.

12)Shull, K.; Mayes, A. M.; Russell, T. P. *Macromolecules* **1993**, *26*, 1993.

NETWORK STRUCTURE AND TOUGHENABILITY
RELATIONSHIP IN MODEL EPOXY SYSTEMS

H.-J. SUE

DEPARTMENT OF MECHANICAL ENGINEERING
TEXAS A&M UNIVERSITY
COLLEGE STATION, TX 77843-3123

P.M. PUCKETT AND J.L. BERTRAM

B-1603, RESINS R&D DEPARTMENT
DOW CHEMICAL U.S.A.
FREEPORT, TX 77541

INTRODUCTION

It is generally understood that low molecular weight epoxy resins, which usually exhibit good processability, when cured with tetra-functional curing agents, are brittle and cannot be effectively toughened [1-5]. Their usage for structural and electronic packaging applications is limited. On the other hand, when the epoxy monomer molecular weights are high, although they becomes much more toughenable, their processability, modulus and use temperature could be greatly compromised. As a result, they are not suitable for high performance composites, adhesives, and coatings applications.

The recent breakthrough by Bertram, et al. [6], have made it possible to produce commercially viable high T_g, high modulus, and low viscosity epoxies for high performance structural applications. The approach taken by Bertram, et al., also termed crosslinkable epoxy thermoplastics (CET) technology, is based on two fundamental concepts: (1) the utilization of epoxy monomers that contain stiff backbone(s) or bulky side group(s) to achieve high modulus and high T_g of cured epoxy [6-8] and (2) the control of *in-situ* chain extension of epoxy backbone during curing, which then leads to a low crosslink density epoxy network with great processability and toughenability. A few commercial (TACTIX[*] 695) and experimental epoxy (CET-3 and CET-4) products based on the above approaches were made available for full scale evaluation and received high acceptance by the practicing industry.

The present work, which is part of a larger effort to fundamentally understand how the T_g, modulus, toughenability, and processability, are affected by the network structure in epoxies, will focus on investigating structure-property relationships in three model epoxy systems. Approaches for producing high performance, low cost epoxies is discussed.

EXPERIMENTAL

Three sets of model epoxy monomer systems were chosen for the present study. These are: (a) diglycidyl ether of bisphenol-A (DGEBA) epoxy monomer coupled with various amount of bisphenol-A (BPA) chain extender, (b) diglycidyl ether of tetramethyl bisphenol-A (DGE-TMBA) epoxy monomer coupled with various amount of tetramethyl bisphenol-A (TMBA) as chain extender, and (c) diglycidyl ether of tetrabromo bisphenol-A (DGE-TBBA) epoxy monomer coupled with various amount of tetrabromo bisphenol-A (TBBA) as chain extender.

For each model epoxy system, the epoxy monomer and chain extender ratio is altered to vary their molecular weight between crosslinks (M_c). The tetra-functional sulfanilamide curing agent, which contains two different functional amine reactivities, was utilized to promote both chain extension and crosslinking of the epoxy network. In all cases, a curing step at 150°C for four hours and a post-cure at 200°C for two hours were chosen to cure the model epoxies. Five weight percent of preformed core-shell rubber (CSR) particles was added to all the model epoxy resins to study the toughenability of the model epoxies.

The double-notch four-point-bend (DN-4PB) [9-11] experiment as well as the single-edge-notch three-point-bend (SEN-3PB) plane strain critical stress intensity factor (K_{IC}) fracture toughness

---

measurements [12,13] were conducted to study the failure mechanisms and toughness of the model epoxies.

The dynamic mechanical behaviors of the model CET resins, having dimensions of 5.08 cm x 1.27 cm x 0.635 cm (2" x 0.5" x 0.25"), were studied using DMS (Rheometrics® RMS-805) under a torsional mode, with 5°C per step. A constant strain amplitude of 0.05% and a fixed frequency of 1 Hz were employed. The samples were analyzed at temperatures ranging from -150°C to 200°C. The temperature at which the primary Tan δ peak was located was recorded as T_g. Since it has been shown that the presence of the preformed CSR particles in epoxy does not affect the T_g and crosslink density in epoxies [11-14], the DMS behavior in CRS-modified model epoxy resins were not studied.

The detailed sample preparation procedures for reflected optical microscopy (ROM), scanning electron microscopy (SEM), and transmission electron microscopy (TEM) observations of the morphology and fracture mechanisms in model epoxies have been reported earlier [9-11]. They will not be reported here.

RESULTS AND DISCUSSION

Dynamic Mechanical Spectroscopy

The DMS plots of three model DGE-TMBA-co-TMBA epoxy resins are shown in Fig. 1. These plots are typical of the model systems investigated. The labels shown in Fig. 1, i.e., TMBA-600, TMBA-1,000, and TMBA-1,400, indicate that the theoretical distances between crosslinks of DGE-TMBA-co-TMBA epoxies are equal to those of the DGEBA systems with M_c of 600, 1,000, and 1,400. Therefore, the TMBA-600 will have an average of 60 g/mole (four additional methyl groups) higher M_c than that of the DGEBA-600 epoxy.

The DMS plots clearly show that the higher the M_c, the lower the T_g for the model epoxies. Also, the Young's modulus and the storage modulus (G') at room temperature are higher for the TMBA-1,400, followed by TMBA-1,000, and by TMBA-600 (Table I). The findings suggest that crosslink density is important for affecting T_g, but not for affecting modulus in the TMBA-based epoxies.

Fracture Toughness and Failure Mechanisms

The investigation of the fracture toughness and failure mechanisms in CSR-modified model DGEBA and TBBA epoxies show that their fracture behaviors follow those described by Yee and Pearson [1,2]. They will not be reported here. However, the fracture behavior of the model TMBA-based epoxies indicates that the highest crosslink density TMBA-600 epoxy system exhibits a higher toughenability than the lower crosslink density TMBA epoxies (Table I). This finding is quite surprising. To further confirm the above findings, fracture mechanisms investigation on CSR-modified TMBA epoxies was conducted using ROM and TEM.

The damage features in the TMBA-based modified epoxy systems, using ROM, appears to be quite different among the three TMBA epoxies (Fig. 2). The size of the light scattering cavitation zone and the size of the birefringent shear yielded zone are all bigger for the TMBA-600 system (Fig. 2a). For the TMBA-1,400 system, the croiding and/or microcracking mechanism appear to dominate the fracture process (Fig. 2c). Further investigation using TEM confirms that the TMBA-600 system exhibits the highest degree of plastic deformation, followed by TMBA-1,000. For TMBA-1,400, indeed, the croiding mechanism is found in the damage zone.

The present study reveals that a rather unusual relationship between epoxy toughenability and M_c for the TMBA-based model epoxies. The earlier study on rubber-toughened DGEBA epoxies by Yee and Pearson [15] concluded that the higher the M_c, the more toughenable the epoxy becomes. Our findings on TMBA epoxies indicated otherwise (Fig. 3). The apparent contradiction mentioned above may be due to either the differences in the crosslinkers used or the differences in the epoxy monomer backbone rigidity incorporated, or both.

SUMMARY

Model epoxy networks with variations in crosslink density and in epoxy monomer backbone rigidity were prepared to study how the nature of the epoxy network structure affects modulus, T_g, and toughness/toughenability of epoxy systems. The present findings indicate that, upon rubber toughening, the rigidity of the epoxy backbone structure, not the molecular weights between crosslinks

(M$_c$), is more relevant to the toughenability of the TMBA-based epoxies. Potential significance of the present findings for a better design of toughened thermosets for structural applications will be discussed.

REFERENCES

1. A.F. Yee and R.A. Pearson, *J. Mater. Sci.*, 21, 2462(1986).
2. R.A. Pearson and A.F. Yee, *J. Mater. Sci.*, 21, 2475(1986).
3. H. Lee and K. Neville, "Handbook of Epoxy Resins", McGraw-Hill, New York, 1967.
4. R.A. Pearson and A.F. Yee, *J. Mater. Sci.*, 24, 2571(1989).
5. W.L. Bradley, W. Schultz, C. Corleto, and S. Komatsu, "Toughened Plastics —— Science and Engineering", Ed. by C.K. Riew, Adv. Chem. Ser., 233, in press.
6. J.L. Bertram, L.L. Walker, J. R. Berman, and J.A. Clarke, US Patent 4,594,291, 1986.
7. R. Bauer, 18th International SAMPE Tech. Conf., Oct. 7-9 1986.
8. K.C. Dewhirst, US Patent 4,786,668, 1988.
9. H. J. Sue, R. A. Pearson, D. S. Parker, J. Huang, and A. F. Yee, *Polym. Preprint*, 29, 147(1988).
10. H.-J. Sue, *Polym. Eng. Sci.*, 31, 270(1991).
11. H.-J. Sue and A.F. Yee, submitted to *J. Mater. Sci.*, March 1992.
12. ASTM Standard, E399-90.
13. O.L. Towers, "Stress Intensity Factor, Compliance, and Elastic η Factors for Six Geometries", The Welding Institute, Cambridge, England, 1981.
14. ASTM Standard, E647-90.
15. R.A. Pearson and A.F. Yee, *J. Mater. Sci.*, 26, 3828(1991).

Table I

Resin	K_{1c}, MPa·m$^{0.5}$	Tg, °C	G', MPa	M$_C$, est.	E, GPa
600-TMBA	0.70±0.06	165	4.8	516	3.38
1000-TMBA	0.80±0.17	135	1.25	3628	3.45
1400-TMBA	0.52±0.15	123	0.73	4000	3.62
600-TMBA/CSR	2.02±0.20	165	4.8	516	
1000-TMBA/CSR	1.69±0.19	135	1.25	3628	
1400-TMBA/CSR	1.25±0.09	123	0.73	4000	

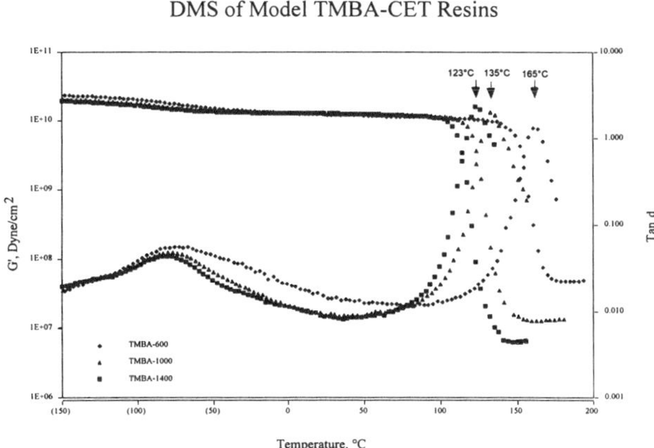

Fig. 1. DMS of three TMBA-based model epoxies.

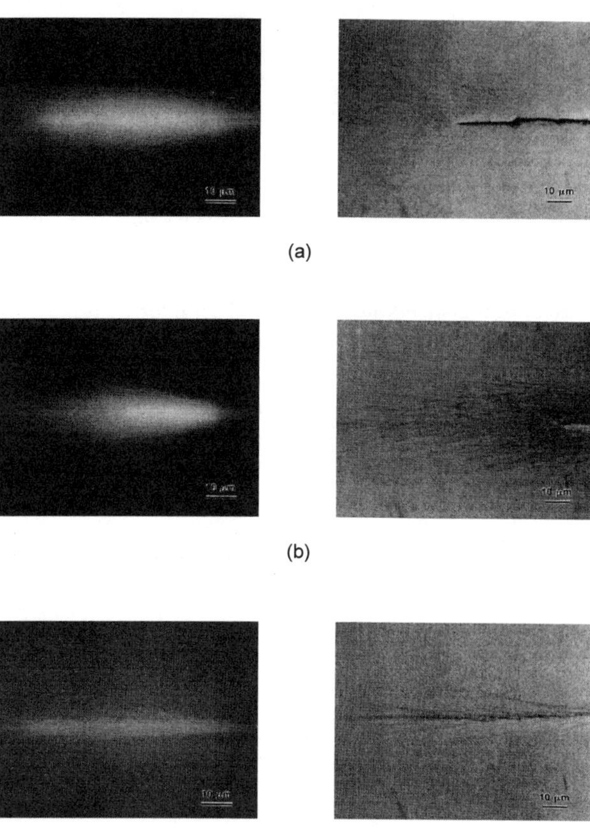

Fig. 2. ROM of CSR-modified TMBA epoxies. a) TMBA-600, b) TMBA-1,000, and c) TMBA-1,400. The dark background micrographs were taken under crossed-polars. The light background micrographs were taken under bright field imaging.

M$_c$ vs. Toughenability

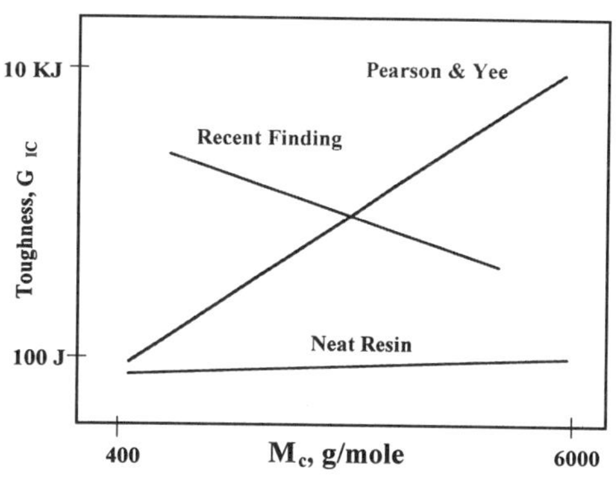

Fig. 3. Toughness is plotted against M$_c$.

INTELLIGENT TEMPERATURE-RESPONSIVE HYROGELS

Naoya Ogata
Department of Chemistry, Sophia University
Tokyo 102-0094, Japan

INTRODUCTION

Drug delivery systems are currently attracting many interests because of potential applications of pharmaceutics to human body. Stimuli-responsive hydrogels which show rapid and sharp swelling-shrinking behaviors of volumes by temperature, pH or electrical potential changes are very attracting drug carriers to release drugs from the hydrogels by environmental changes and many papers have been published for these stimuli-responsive polymeric drug carriers. The concept of the so-called intelligent drug delivery systems can be illustrated as follows:

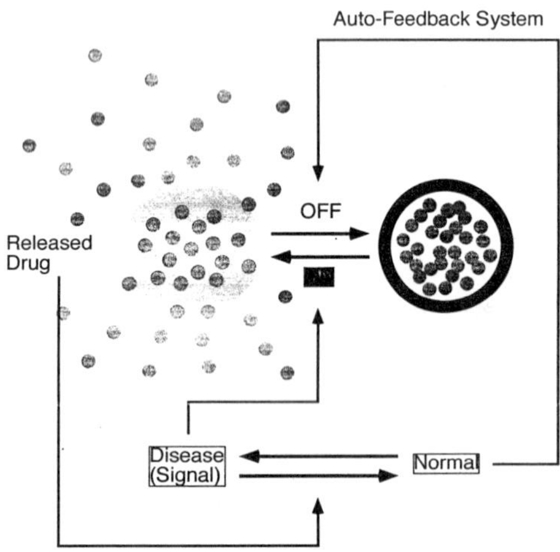

Temperature responsive hydrogels are more realistic to apply to DDS as many deseases accompany with generation of fever in our body. Much higher sensitivity to small temperature changes are required for hydrogels to release drugs in an effective way. This paper deals with structural effect of hydrogels for the rapid volume changes by using hydrogen-bonding interactions among hydrogels.

RESULTS AND DISCUSSION

Hydrogen-bonding among polymers may cleave at a certain temperature by a zipper type way as schematically shown in Fig. 1 and this hydrogen-bonding interaction may lead to a rapid volume changes of hydrogel with a high sensitivity to temperatre changes in a narrow ranges. It is quite interesting to compare the temperature sensitivity of hydrogels which have linear and comb structures with the same compositions as the structural difference may cause different entanglement effect among polymers. It is well known that nucleic acid bases in DNA form a well stacked structure along double helices through hydrogen-bonding which is temperature

sensitive. Base on thes expectations, two type polymers having comb structucre and uracil units were synthesized.

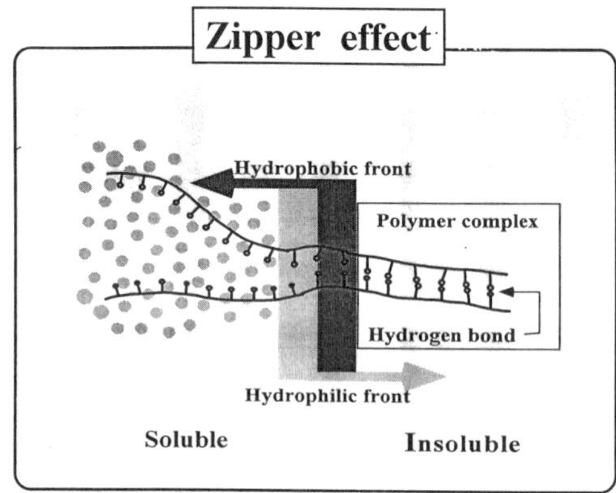

Fig. 1 Concept of zipper-type cleavage of hydrogen-bonding among polymers

Hydrogels having comb-type structures

Comb-type hydrogels based on acrylamide units were synthesized from macromonomer and methylenebisacrylamide as a crosslinking agent and the reaction route is shown as follows:

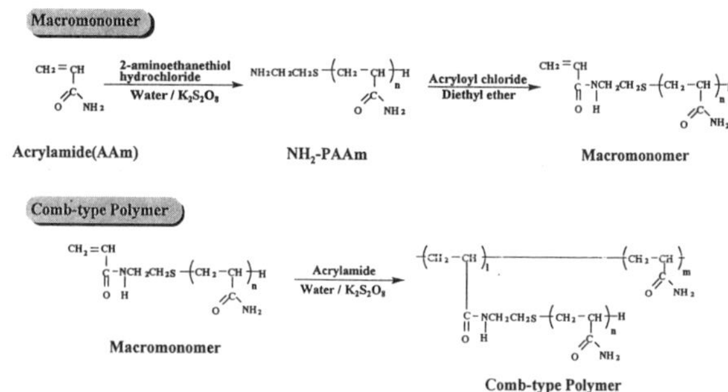

Linear poly(acrylamide) hydrogel was obtained by a radical polymerization in the presence of methylenebisacrylamide with the same feed composition as the comb-type poly(acrylamide). These hydrogels were immersed in water and acrylic acid was added to the aqueous solution to polymerize so that interpenetrating network hydrogenls (IPN) were obtained, as shown in Fig. 2. Feed compositions of the monomers and acrylic acid are shown in Table I.

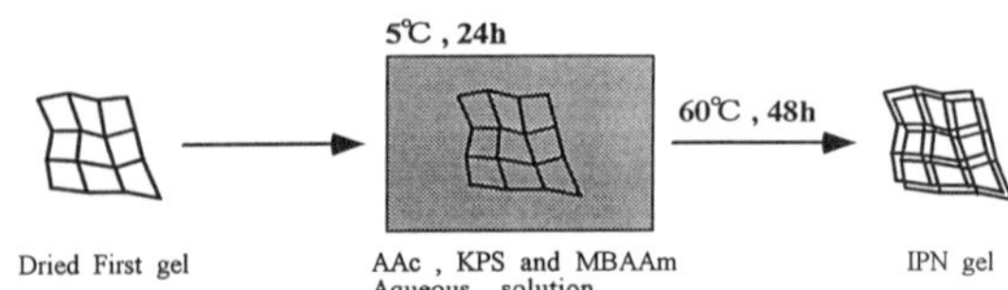

Fig. 2   Synthetic scheme of IPN hydrogels

Transmittance of aqueous solutions of linear and comb poly(acrylamide) was measured to detect solube-insouble changes of the polymers by temperature changes and swelling behaviors of these hydrogels were measured by weight changes of the hydrogels in water.

Table I Feed compositions of macromer and monomers

Feed composition for first gels

Code	Macromonomer (g)	AAm (g)	MBAAm (wt%)	KPS (g)	H$_2$O (ml)
PAAm	———	2.1296	1.0	0.06	total 15.0
comb-type PAAm	0.5513	0.5517	1.0	0.03	total 7.5

Table 2-4 Feed composition for 2nd gels in IPN

Code	AAc (g)	MBAAm (wt%)	KPS (g)	H$_2$O (ml)
PAAm / PAAc IPN	0.3138	1.0	0.016	total 4.0
comb-type PAAm / PAAc IPN	0.1719	1.0	0.024	total 6.0

Fig. 2 shows transmittance changes of aqueous solutions of mixed solution of comb-type or linear poly(acrylamide)(PAAm) and poly(acrylic acid)(PAAc) at pH=3.3 when solution temperatures were changed. It is seen in Fig. 2 that the mixture of comb-type PAAm and PAAc had a rapid change of soluble-insoluble behaviors in aqueous solution in comparison with linear PAAm/PAAc mixture.

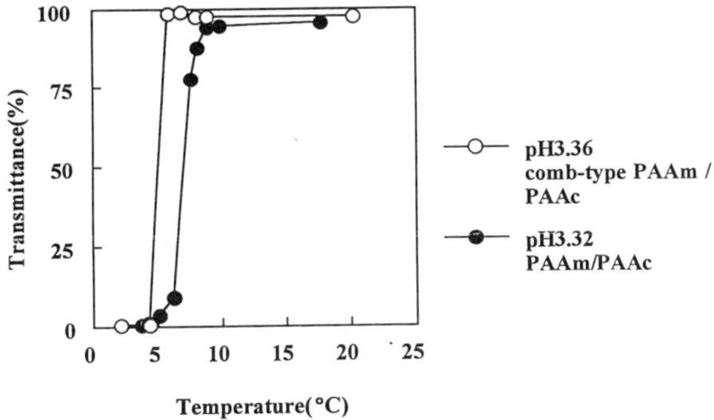

Fig. 2 Temperature dependence of transmittance changes of mixed linear-type, comb-type polymer and PAAc in HCl solutions (pH3.3).

Fig. 3 indicates volume changes of IPN hydrogels derived from comb-type and linear PAAm/PAAc. It is seen in Fig. 3 that the IPN hydrogel from comb-type PAAm has much larger volume change than that from linear PAAm.

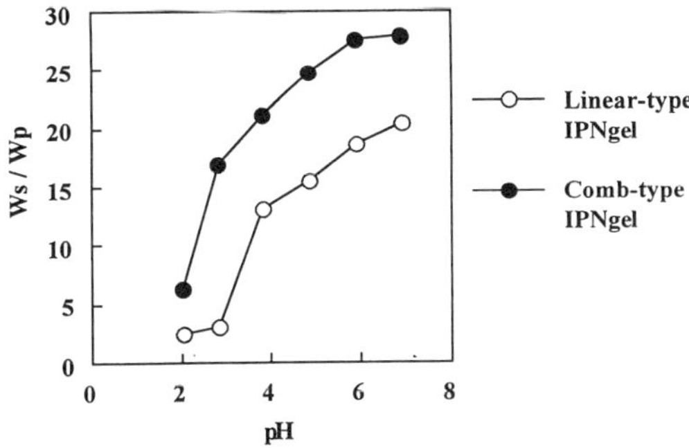

Fig. 3 pH dependence of equibrium swelling ratio of linear-type and comb-type hydrogels in buffered solurions.

Hydrogels containing uracil moiety

Acryloyl monomer having a uracil (AU) moiety was synthesized and polymerized by a radical method. The polymer (PAU) did not dissolve in water below 50°C, while it suddenly dissolve in water above 50°C and this dissolution behavior was reversible, possibly owing to the cleavage of hydrogen-bonding at 50°C. Copolymers from AU and acrylic acid (Aac) were prepared and the swelling-shrinking behaviors of the copolymers were investigated. Fig. 4 shows the swelling behavor of poly(AU-co-AAc) (1/1 molar ratio) hydrogel in responce to temperaure changes between 35° and 40°C. It is seen in Hig. 4 that the hydrogel shows reversible volume changes in responce to temeprature changes.

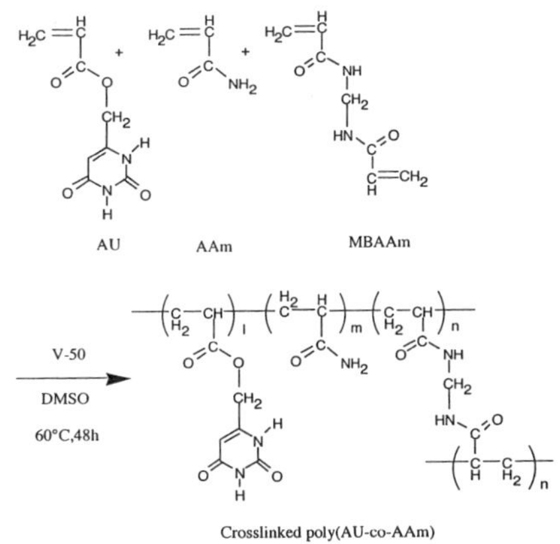

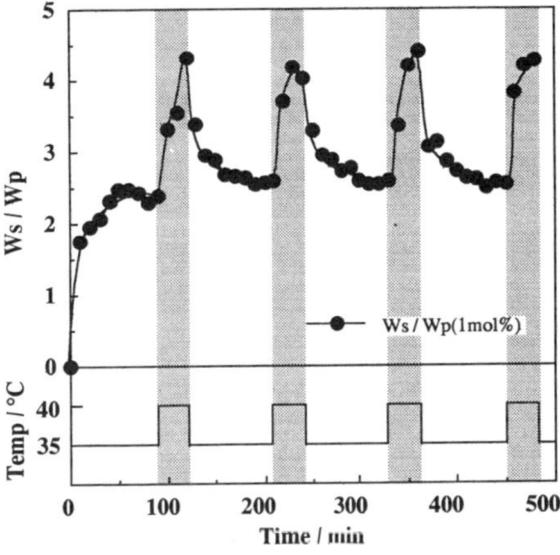

Fig. 4 Swelling-shrinking behaviors of poly(AU-co-AAc) hydrogel between 35° and 40°C

Ketoprophen was loaded to the hydrogel derived from poly(AU-co-dimethylacrylamide (DMAAm)) with a molar ratio of 1/1 which also indicated a rapid change of volume by temperature changes between 30° and 35°C. The release behaviors of Ketoprophen were exactly the same as the swelling-shrinking behaviors of the hydrogel in responce to temperature changes and it was confirmed that these hydrogels based on hydrogen-bonding interactions were useful for DDS in responce to temperature changes.

ANTIMICROBIAL PROPERTIES OF QUATERNARY AMMONIUM CELLULOSE AND CHITOSAN DERIVATIVES

William H. Daly and Melissa Manuszak Guerrini

Macromolecular Studies Group, Chemistry Department,
Louisiana State University, Baton Rouge, LA 70803-1804

Introduction

Quaternary ammonium synthetic and modified natural polymers exhibit antimicrobial activity.[1,2] The quaternary ammonium polymers are generally more active than their corresponding monomers, particularly against gram-positive bacteria.[3] The potential activity of quaternary ammonium polysaccharides is of great interest to the cosmetic industry, because these polymers have proved to be very effective adjuvants in cosmetic formulations. Chitosan derivatives have been reported to exhibit limited antimicrobial activity[4] and recently, the synthesis and antibacterial activities of quaternized DEAE- and TEAE-chitin derivatives were reported.[5] Further investigation of the potential antibacterial activity of quatermary ammonium polysaccharides is warrented. There is an unfilled need for new antimicrobial agents suitable for use as preservatives in personal care products as well as for new biocides to which bacterial strains have not yet developed resistance.

We have prepared a new diquaternary ammonium alkylcarbamoyl cellulose derivative with a defined charge density, 3-trimethylammonium-2-hydroxypropyl-N,N-dimethylammonium ethyl carbamoyl-methyl cellulose chloride (DQNNED) where the quaternary nitrogens are located at the sites of carboxymethylation of the starting polymer.[6] A comparable derivative of chitosan, 3-trimethylammonium-2-hydroxypropyl-N-chitosan (CHI-Q188) has also been prepared.[7] This paper describes the antimicrobial activities of DQNNED and CHI-Q188 against *Escherichia coli, Staphylococcus aureus* and *Pseudomonas aeruginosa* as determined using the minimum inhibitory concentration (MIC) test. The chitosan derivative exhibits antimicrobial activity an concentrations as low as 10-20 μg/mL, at least an order of magnitude lower than the concentrations at which previously reported chitosan antimicrobial agents exhibit activity.[4,5]

Experimental

Materials

The synthesis of N-(3-trimethylammonium -2-hydroxy propyl)-N,N-dimethylammonium ethyl carbamoyl methyl cellulose dichloride (DQNNED) from the corresponding carboxymethyl cellulose derivative has been described.[6] 3-Trimethylammonium-2-hydroxy propyl-N-chitosan (CHI-Q188) was prepared by treating chitosan with with 3-chloro-2-hydroxypropyl trimethylammonium chloride (Quat 188 available from Dow Chemical Co.).[7] Methylparaben, methyl-4-hydroxybenzoate, was purchased from Aldrich Chemicals and used as received.

Test Organisms

The strains of bacteria used in this study were received directly from the American Type Culture Collection (ATCC) and consisted of *Escherichia coli* (*E. coli*) 25922, *Staphylococcus aureus* (*S. aureus*) 29213 and *Pseudomonas aeruginosa* (*P. aeruginosa*) 27853. The cultures were stored frozen in a solution of nutrient broth and glycerol in order to prevent mutation. The cells were maintained by growth of a culture overnight in 5 mL of nutrient broth. The next morning, 1 mL of the culture was resuspended in 25 mL of nutrient broth at room temperature and placed in a shaker-incubator for 4 hours at 150 rpm and 37 °C to ensure a cell density of at least 10^8 cells/mL in mid-logarithmic growth phase. The optical density (OD) of the cell solution was measured at 600 nm. The suspensions were diluted with nutrient broth until the OD read 0.20 ± 0.01 for *E. coli* or *P. aeruginosa* and 0.40 ± 0.01 for *S. aureus*. These OD readings correspond to concentrations of 10^8 bacterial cells/mL of nutrient broth. Aerobic plate counts (APCs) were performed in order to determine the size of each inoculum.

Test Solutions

Aqueous solutions of 0.2% (w/v) methylparaben, 0.2% (w/v) DQNNED, or 0.2% (w/v) methylparaben and 0.2% (w/v) DQNNED were prepared with sterile deionized water. The pH of the solutions was adjusted to 7.0 ± 0.2 with 1 M sodium hydroxide. Nutrient broth was prepared by dissolving 8 g of soy trypsin digest protein per liter of deionized water. The pH of the solution was adjusted to 7.2 ± 0.2 with 1 M sodium hydroxide. The nutrient broth was autoclaved prior to use. Sterile, deionized water was used as a control for the experiments. Solutions were stored at room temperature prior to use.

Minimum Inhibitory Concentration Test

The pathogen, *E. coli, S. aureus* or *P. aeruginosa*, is added to a small well containing the polymer of interest, phosphate buffer, and nutrient broth. A range of eight concentrations may be tested at once so the minimum concentration at which the polymer will prevent growth can be determined. A concentration of 10^4 bacterial cells / mL was studied in each test. The polymer concentration tested ranged from approximately 1600 μg/mL to 2 μg/mL. A cell control and a polymer control were included in each set of tests. The well plates were incubated for 18 to 24 hours at 37°C. The cell well where the concentration of the polymer was effective at inhibiting growth will appear clear. The concentrations of the polymer at which growth occurs will appear cloudy (turbid) from the large population of cells that have grown. The experiment was repeated several times to obtain an average value for the MIC. Cell wells were judged visually for clarity of the solutions.

Sterilization Times

A modified preservative efficacy test (PET)[8] was used to determine sterilization times (STs) for the test organisms in aqueous samples. The inocula were prepared as described and added to buffered solutions of methylparaben, DQNNED, CHI-Q188 or binary mixtures of methylparaben and quaternary polysaccharide. The contents of test tubes were mixed using a Vortex Genie mixer. 100 μl of the room temperature samples were spread on agar plates at time intervals of 0 to 48 hours. Agar plates were incubated for 18 to 24 hours at 37 °C. The agar plates were examined for growth of the test organisms, and the ST was determined to be the time at which test organisms were not recovered from the solution. When no endpoint was reached in the ST experiments, because the bacteria were still alive at 48 hours, the minimum possible ST (MPST) was used. The MPST is defined as a time longer than the last time at which test organisms were recovered.[8]

Results And Discussion

Two different types of quaternary ammonium polysaccharides have been studies examined to determine their potential as antibacterial agents. The polyquat, DQNNED, derived from carboxymethyl cellulose (D.S.=0.7) was prepared with a degree of substitution of 0.55.

$$\text{CH}_2\text{OCH}_2\overset{\overset{\displaystyle O}{\|}}{\text{C}}\text{NH-(CH}_2)_3\text{-}\overset{\overset{\displaystyle R}{|}}{\underset{\underset{\displaystyle R}{|}}{\text{N}^+}}\text{—CH}_2\text{-}\underset{\underset{\displaystyle OH}{|}}{\text{CH}}\text{-CH}_2\text{-N}^+(\text{CH}_3)_3 \quad 2\,Cl^-$$

DQNNED

Treatment of chitosan with a degree of deacetylation of 88% with Quat 188 afforded practically complete substitution of the free amino substituents to product the corresponding quaternary ammonium derivative, CHI-188. Both polymeric derivatives are soluble in water over a wide pH range and are compatible with all the buffers used in these experiments.

CH₂OH ... CHI-188

The MIC results for methylparaben and the polymers tested are summarized in Table 1. The MIC for methylparaben was found to vary with the nature of the microorganizm, ranging from 640 for *E. coli* to 1120 µg / mL for *S. aureus*. *S. aureus* which is a gram positive bacteria, required twice as much methylparaben to inhibit growth as did *E. coli* or *P. aeruginosa* which are gram negative. The MIC of the cellulose quat, DQNNED, was found to be slightly lower for the gram negative strains and higher for *S. aureus*. Thus, this polymer inhibited growth about as well as methylparaben, but in the case of *S. aureus*, approximately 3.6 times more DQNNED was required to inhibit growth than for *E. coli* or *P. aeruginosa*.

The MIC of the combination of DQNNED and methylparaben was found to be 320, 160 and 800 µg / mL of each of the compounds for the inhibition of *E. coli*, *P. aeruginosa* and *S. aureus*, respectively. This effect of the combination of DQNNED and methylparaben on bacterial growth appears to be additive rather than synergistic. Thus, incorporation of a diquaternary ammonium cellulose derivative into a product will contribute to product preservation. The quaternary ammonium polymers will play dual roles in formulation as rheology modifiers and preservatives.

The MIC of the chitosan derivative,CHI-Q188, was found to be 16 µg / mL for all three bacteria tested, *E. coli*, *P. aeruginosa* and *S. aureus*; no noticable differentiation between bacterial strains was detected. This polymer is 40 times more effective against *E. coli* than methylparaben and 30 times more effective than DQNNED. CHI-Q188 is 50 times more effective against *P. aeruginosa* than methylparaben and 30 times more effective than DQNNED. Remarkably, CHI-Q188 is 70 times more effective against *S. aureus* than methylparaben and 110 times more effective than DQNNED. The level of activity observed appears to be at least an order of magnitude greater than any reported for chitin/chitosan derivatives.[5]

Neither the unquaternized aminoalkylcarbamoyl cellulose derivative (CMCamide) nor chitosan exhibited antimicrobial activity. This result indicates that the antimicrobial activity is a result of the quaternary ammonium groups with some enhancement attributed to the hydroxy group β to the quaternary ammonium functionality.

Sterilization Times

Sterilization times of *E. coli* and *S. aureus* in solutions of methylparaben, DQNNED, and a mixture of DQNNED and methylparaben were determined (Table 1). The 2000 µg / mL methylparaben solution was not rapidly bactericidal as test organisms were recovered at 48 hours. This result agrees with the results of Orth et al.[10] for ST of *E. coli* and *S aureus* by a 0.2% aqueous solution of methylparaben. Neither the solution of DQNNED nor the mixture of DQNNED and methylparaben were rapidly bactericidal. Test organisms were again recovered in the solutions at 48 hours. This result is not surprising, as quaternary polymers are known to be slow acting in killing microorganisms, and more biostatic than biocidal.[1]

Sterilization times of *E. coli* and *S. aureus* in solutions of CHI-Q188 were determined. The 20 µg / mL CHI-Q188 solution was rapidly bactericidal as test organisms were not recovered at 2 hours.

Sterilization times were determined to be 1.5 hours for *E coli* and 30 minutes for *S. aureus*.

Conclusions

The bacteristatic activity of DQNNED alone and in combination with methylparaben indicates the dual role that quaternary polymers may have in cosmetic products. They potentially may be used as thickeners or conditioners, and they may also exhibit antimicrobial activity in the preservation of the cosmetic products.

The CHI-188 chitosan derivative is uniquely active as a biocide particularly against S aureus. A wide variety of applications for this new material can be envisioned.

Acknowledgements

This work was supported by a Graduate Research Fellowship to M. Manuszak Guerrini from the Society of Cosmetic Chemists.

References

1. O. W. May, *Polymeric Antimicrobial Agents* in: *Disinfection, Sterilization and Preservation*; Fourth ed.; S. S. Block, Ed.; Lea & Febiger: Philadelphia, 1991, pp 322-333.

2. S. D. Worley and G. Sun, *Trends in Polymer Science*, **4**, 364-370 (1996).

3. J. J. Merianos, *Quaternary Ammonium Antimicrobial Compounds* in *Disinfection, Sterilization and Preservation*; Fourth ed.; S. S. Block, Ed.; Lea & Febiger: Philadelphia, 1991, pp 225-255.

4. M. Yalpani, F.Johnson and L.E. Robinson, *Antimicrobial Activity of Some Chitosan Derivatives* in Advances in Chitin and Chitosan, C. J. Brine, P. A. Sandford, J. P. Zikaki, Ed. Elsevier Applied Science, London, 1992, pp 543-548.

5. C-H. Kim, S-Y Kim and K-S. Choi, *Polym. Adv. Tech.*, **8**, 319-325 (1997).

6. D. Culberson and W. H. Daly, *Polymeric Materials, Science and Engineering*, **71**, 498 (1994).

7. .J. Macossay. *Synthesis and Characterization of Chitosan Derivatives*; Dissertation, Louisiana State University: Baton Rouge, 1995

8. D. S. Orth, M. Lutes Anderson, D. K. Smith, and S. R. Milstein, *Journal of the Society of Cosmetic Chemists*, **40**, 347-365 (1989).

Table 1. Summary of Minimum Inhibitory Concentration (MIC) and Sterilization Time Data for Quaternary Ammonium Polysaccharides.

Pathogen / Biocide	E. coli (25922)	S. aureus (29213)	P. aeruginosa (27853)
Methylparaben			
MIC, (µg/mL)	640	1120	800
Sterilization time @ 2000(µg/mL)	> 48 hr	> 48 hr	> 48 hr
DQNNED			
MIC, (µg/mL)	480	1760	480
Sterilization time @ 2000(µg/mL)	> 48 hr	> 48 hr	> 48 hr
Methylparaben and DQNNED, 1:1 ratio			
MIC, (µg/mL)	320	800	160
Sterilization time @ 2000(µg/mL)	> 48 hr	> 48 hr	> 48 hr
CHI-188			
MIC, (µg/mL)	16	16	16
Sterilization time @ 20 (µg/mL)	> 1.5 hr	> 0.5 hr	

GELATIN-BASED HYDROGELS FOR WOUND TREATMENT

E. Schacht[1], A. Van Den Bulcke[1], B. Bogdanov [1], J-P. Draye[2], and B. Delaey[2]

[1]University of Gent, Department of Organic Chemistry, Polymer Material Research Group
Institute Biomedical Technologies (IBITECH), Krijgslaan 281 S4, B-9000 Gent, Belgium
[2]Innogenetics NV, Industriepark Zwijnaarde 7, box 4, B-9052 Gent, Belgium

INTRODUCTION

Hydrogels are an interesting class of polymer materials and have been extensively studied and used for a variety of applications in the biomedical field including contact lens materials, artificial tendons, matrices for tissue engineering, drug delivery systems and others. The aim of this work is the synthesis of gelatin-based hydrogels suitable for the controlled release of drugs to enhance wound healing. When loaded with epidermal growth factors or wound repair promoting substances, the matrix can be used for the fabrication of wound dressings. Gelatin, which is the denatured form of the protein collagen, has been used in a variety of wound dressings and several methods are available for the chemical cross-linking of collagen- or gelatin-based materials. Glutaraldehyde is the most commonly used bifunctional reagent, but this reagent may result in the release of toxic products. In the present study two alternative types of gelatin-hydrogels, which uses a different, less toxic, cross-linking technology, were synthesised: In a first approach hydrogels were prepa-

$$Gel-NH_2 \; + \; Dex-\overset{O}{\underset{\|}{C}}-H \longrightarrow Gelatin-N{=}CH-Dextran$$

red by cross-linkage of gelatin with partial periodate oxidised dextran[1,2], alternatively hydrogels comprising cross-linked methacryl-amide modified gelatin were prepared.

The visco-elastic properties the release characteristics and the biosafety were evaluated and will be discussed.

$$Gel-NH_2 \; + \; CH_2{=}\overset{CH_3}{\underset{\underset{O}{\|}}{C}}-C-O-\overset{CH_3}{\underset{\underset{O}{\|}}{C}}-C{=}CH_2 \longrightarrow$$

$$Gelatin-NH-\overset{CH_3}{\underset{\underset{O}{\|}}{C}}-C{=}CH_2 \longrightarrow Crosslinking$$

MATERIALS AND METHODS

Dextran dialdehydes were prepared by reaction of sodium periodate[3]. One mm thick hydrogel films were prepared by mixing aqueous solutions of gelatin and oxidised dextran and were stored at different temperatures for various periods of time. Gelatin methacrylamide was prepared by reaction with methacrylic anhydride. UV-induced cross-linking of the vinylmodified gelatin was performed in aqueous medium in the presence of a photo-initiator.

RESULTS

Hydrogels prepared from gelatin and dextran dialdehyde (type 1) are transparent gels resulting from chemical interaction between gelatin and oxidised dextran.
It was found by rheometry that the mechanical properties, e.g. elasticity modulus, are a contribution of two factors: a) physical gelation of the gelatin component and b) chemical cross-linkage as result of interpolymer reactions[4]. The extent to which both phenomena contribute can be controlled by storage temperature and storage time[4]. Both the physical association and chemical structuring of the gelatin-dextran dialdehyde hydrogels is increased with longer storage times. In case of type 2 hydrogels prepared by radical cross-linkage of methacrylamide modified gelatin, ageing of the gels only due to the physical structuring of the gelatin chains. No chemical aging is observed (figure 1).

AFM analysis of the gelatine-dextran dialdehyde hydrogels indicated that the microstructure of the gels depends on the reaction conditions during preparation and storage. There is strong evidence that cryogenic treatment leads to enhanced chemical crosslinkage and different porosity[5].

The release of biomolecules immobilised in the gels has been studied. It has been demonstrated that peptides incorporated in the gelatin-dextran dialdehyde gels are partly immobilised in the gel matrix. On the other hand the type 2 hydrogels release the peptides quantitatively (figure 2).
The *in vitro* cytotoxicity of dextran gelatin aldehyde hydrogels was evaluated in keratinocytes, fibroblasts, and endothelial cells (3 cell types actively involved in wound healing) and was compared with the cytotoxicity induced by three commonly used dressings: OpSite (Smith and Nephew), Tegaderm, and DuoDERM. After 3-day incubations, cross-linked gelatin hydrogel and occlusive polyurethane dressings (Tegaderm and OpSite) induced a similar low toxicity in keratinocyte cultures. By contrast, DuoDERM samples already induced a strong cytotoxicity. When the incubations were prolonged up to 6 days, the gelatin hydrogel was more toxic than the polyurethane dressings, but still less toxic than DuoDERM. Cytotoxicity experiments conducted with fibroblast and endothelial cells showed that the cytotoxicity induced by either hydrogel or DuoDERM samples were less marked in fibroblast and endothelial cell cultures than in keratinocyte cultures. Dressing sample extractions with culture medium, before conducting the 6-day incubations in the presence of keratinocyte cultures, markedly decreased the toxicity of the gelatin hydrogel but did not modify the strong toxicity induced by DuoDERM samples.
Wound healing evaluations in mice, after implantation of dressing samples, showed that the gelatin hydrogel films were progressively invaded and degraded by inflammatory cells. Cellular infiltration of the hydrogel samples started at day 2 after implantation. Cuniculi were created by the inflam-matory cells (neutrophils and macrophages) which were progressively enlarged. In the proximity of the inflammatory cells, the structure of the hydrogel appeared to be spongy. Wound re-epitheliali-sation occurred on the modified hydrogels. By day 10 after implantation, wounds were largely re-epithelialised. Foreign body reaction was very mild in the presence of the hydrogel films.

CONCLUSIONS

The formation of hydrogels by cross-linkage of gelatin with oxidised dextran or by methacrylamide gelatin cross-linked by photo-initiation are attractive concepts because a proper control of reaction time and temperature allows to control the final strength of the gels. The hydrogel films were found to be an appropriate release system for delivery of biologically active EGF.
From *in vitro* cytotoxicity tests and *in vivo* implantation studies, we conclude that hydrogel films, prepared by cross-linking of gelatin with dextran dialdehydes, are biocompatible, and biodegradable when placed into the wound site.

ACKNOWLEDGEMENT

This work was supported by the Flemish Institute for Research and Development (IWT), the Fund for Scientific Research - Flanders (FWO) and the Belgian Ministry of Scientific Programming, IUAP/PAI-IV/11.

REFERENCES

1 E. Schacht, M. Nobels, S. Vansteenkiste, J. Demeester, J. Franssen & A. Lemahieu, *Polymer Gels and Networks* 1(4), 213-224 (1993)

2 E. Schacht, J.C. Vandichel, A. Lemahieu, N. De Rooze and S. Van-steenkiste; *Encapsulation and controled release*, 19-34 (1993)

3 L. Ruys, J. Vermeersch, E. Schacht, E. Goethals, P. Gyselinck, P. Braeckman and R. Van Severen; *Acta Pharm. Techn.*, <u>29</u>(2), 105-112 (1983)

4 B. Bogdanov, E. Schacht and A. Van Den Bulcke; *J. of Thermal Analysis* <u>49</u>, 847-856 (1997)

5 E. Schacht, B. Bogdanov, A. Van den Bulcke and N. De Rooze; *Reactive Polymers* <u>33</u>, 109-116 (1997)

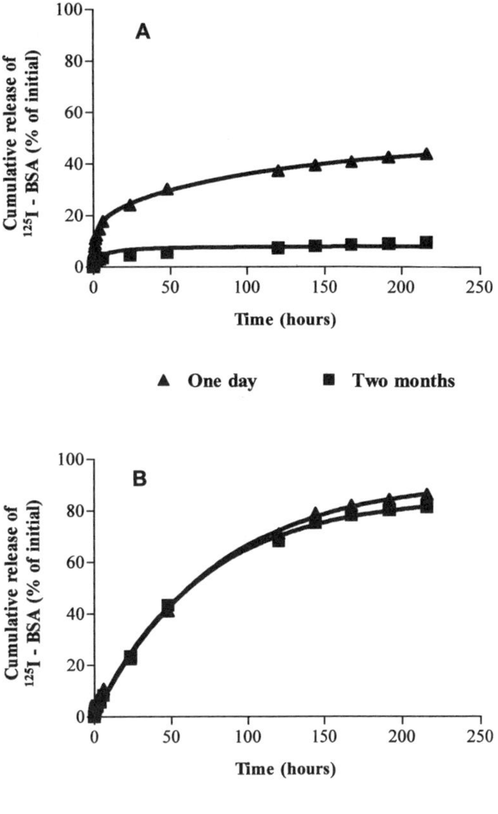

*Fig. 2: Release of ^{125}I - BSA from gelatin-dextranaldehyde (**A**), resp. gelatin-methacrylamide hydrogel (**B**) stored during one day or two months.*

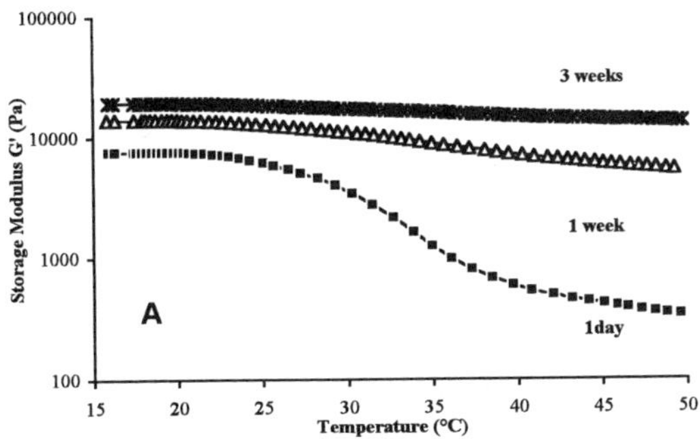

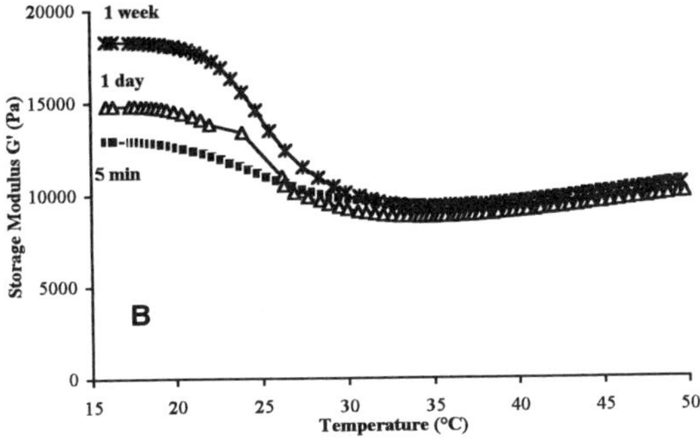

*Fig. 1: Effect of storage time on the G'-modulus of gelatin-dextran aldehyde (**A**), resp. gelatin-methacrylamide hydrogel (**B**).*

NEW BIORESORBABLE POLYANION-POLYCATION COMPLEX HYDROGELS

H. Rossignol, M. Boustta, and M. Vert

CRBA-URA CNRS 1465, University Montpellier 1, Faculty of Pharmacy, 15 ave. Charles Flahault, 34060 Montpellier, France

Introduction

The electrostatic interactions between polyanionic and polycationic macromolecules lead to polyelectrolyte complexes (PEC) which are more or less hydrophilic depending of chain structures[1]. In the field of biopolymers, such interactions are of critical significance and are involved to some extents in many phenomena related to life (immune complexes, ligand-receptor interactions, tissue structures, etc.). Artificial PEC's are compounds which are relatively close to complexes between biopolymers. For several decades, they have been studied from a basic viewpoint and with respect to applications as well. In the field of therapeutic applications, PEC's have been recognized as compounds of interest for controlled drug delivery and cell trapping, and as hydrogels as well[2]. On the other hand, complexation, also designated as condensation, of oligo and polynucleotides with polycations is presently regarded as a worthwhile approach to violate cell defenses and allow antisense and gene therapies through cell transfection of normally rejected polyanions. All these domains of applications are relevant of the concept of temporary therapy and thus of the use of bioresorbable systems which can be eliminated from the body through natural pathways at the end. The use of naturally occurring polymers to make PEC's aimed at being in contact with animal tissues and body fluids can be the source of immune response and of transfers of viral diseases in the case of seric proteins. With these respects synthetic bioresorbable polyanions and polycations to make bioresorbable PECs is of interest. For the last two decades, we have been studying a synthetic bioresorbable polyacid which is soluble in water regardless of the pH, namely poly(β-malic acid)[3]. Insofar as synthetic bioresorbable polybases are concerned, literature is rather poor. One of the most attractive compounds is the poly(β-hydroxy acid)-type polymer derived from the serine amino acid, namely poly(amino serinate) or PAS. Both polymers are chiral.

Malic acid

Poly(β-malic acid), PMLAH

Serine

Poly(amino serinate), PAS

Several groups have already attempted to synthesize the PAS polycation[4-6]. All the reported synthesis routes required protection of the amino group of the serine precursor. N-protected PAS's were prepared using substitution by benzenesulfonyl[4], trityl[5], ter-butoxycarbonyl and benzyloxycarbonyl[6-7] protecting groups. In all cases, deprotection turned to be problematic and the polyamine form was difficult to recover[5,7].

In this paper, we wish to report first a new route to PAS based on Z-protection of the amino group and mesyl activation of the carboxyl group of serine and acid-catalyze hydrolytic cleavage of the Z protecting groups. Poly(L- and DL-amino serinates) were obtained in rather large quantities. Secondly, we report preliminary characteristics of PSA-PMLA PEC's.

Experimental

Mesyl chloride was purchased from Aldrich and distillated prior to use. Acetone was commercial grade from Merck. It was distilled on potassium carbonate prior to use.

Poly(β-malic acid) was a homemade sample prepared according to a previously reported protocol[3].

N-benzyloxycarbonyl DL-serines (N-Z-DL-serine)

70 cm^3 (0.49 moles) of benzyl chloroformate were added dropwise to a gently stirred solution of 44 g (0.42 moles) of DL-serine dissolved in 200 cm^3 2M NaOH. The pH was maintained at 9 for 3 hours by occasional addition of 2M NaOH. The reaction medium was then allowed to stand at 8°C and pH 10 for 1 hour. After washing with diethyl ether, the resulting aqueous solution was acidified with HCl up to pH = 2 and the protected serine was extracted with ethyl acetate. The organic phase was washed with water up to neutrality, and dried with anhydrous MgSO$_4$. The solvent was evaporated and the solid residue was redissolved in the minimum volume of hot ethyl acetate. The mixture was allow to crystallize at room temperature. Yields were always in the range of 75%.

Z-protected poly(DL-amino serinates) DL-PASZ

Typically, 1g (4.18 x 10^{-3} moles) of N-Z-serine were dissolved in 40 cm^3 acetone in a 100 cm^3 flask.. 2 g (18.87 x 10^{-3} moles) of sodium carbonate were added to the solution and the mixture was allowed to stir for 15 min. at 20°C under nitrogen atmosphere before dropwise addition of 0.42 cm^3 (5.43 x 10^{-3} moles) mesyl chloride. The reaction medium was allowed to stand at room temperature. After 24 hours the reaction mixture was filtered. The liquid phase was evaporated under vacuum and the recovered yellow solid was washed twice with 10 cm^3 diethyl ether. The solid residue was dissolved in methane dichloride and reprecipitation was obtained by addition of diethyl ether. After drying yields in PASZ were generally in the range of 80% with respect to the monomer. The resulting PASZ DL-derivative was mass fractionated using acetone/ethanol as solvent/precipitant.

Cleavage of the Z protecting groups of DL-PASZ

Typically, 1 g (4.15 x 10^{-3} monomoles) PSAZ were dissolved in 1.5 cm^3 methylene chloride at 20°C and under nitrogen. 6.2 cm^3 of a 33% HBr solution in acetic acid were then added and the mixture was allowed to stir at room temperature, the released carbone dioxide being trapped in a sodium hydroxide solution. To reach 95% deprotection the reaction time was 10 min.. The more or less deprotected precipitate DL-PASZ$_{100-x}$Br$_x$ which was recovered upon addition of diethyl ether, was filtered and further washed twice with aliquots of pure diethyl ether.

Reprotection of DL-PASZ$_{100-x}$Br$_x$

100 mg of the copolymer to be reprotected were mixed with 2 cm^3 acetone. 300 mg sodium carbonate and 1.5 M/M excess benzyloxycarbonyl chloride were added successively. The mixture was allowed to stand 24 hours at room temperature up to complete dissolution. The filtered solution was evaporated. The recovered solid was washed with diethyl ether and purified by the solution/precipitation method using CH$_2$Cl$_2$/diethyl ether.

^{13}C nmr spectra were obtained by using various (100, 250 and 400 MHz proton) Brucker spectrometers. Size Exclusion Chromatography (SEC) analyses were performed either in THF or dioxane, depending on the compounds, using a Waters equipment. The eluant flow rate was 1 cm^3/min.. Relative molar mass data were referred to polystyrene standards. In the case of water soluble compounds, SEC was carried out using a Pharmacia setting equipped with a DEAE gel column followed by a G25 gel one. The eluant was an aqueous 0.15 M NaCl solution at a 1cm^3/min. flow rate.

Results and discussion

According to literature protected PAS polymers can be synthesized from N-protected serine after activation of the monomer precursor. The conversion of N-protected serines to corresponding β–lactones was prospected by various authors[4-7]. Heath, and several compounds such as triethylamine, betaine, various tetraalkylammonium carboxylates were tested to initiate the ring opening polymerization of the protected serine-derived β–lactones[4-6]. Some of these compounds led to N-protected poly(amino serinate)s with molar masses in the 10000 to 30000 dalton range[5-6]. In 1992 Gelbin and Kohn[7] synthesized N-protected PAS of similar molar masses via the polycondensation of hydroxybenzotriazole N-protected serinates, these intermediates being usually not isolated. In most cases the deprotection of the pendent amino groups could not be achieved. Removal of trityl protecting groups by acid-catalyzed hydrolysis and of Z protecting groups by catalytic hydrogenolysis were also reported together with some characteristics of the resulting basic compounds in solution[5-6]. However, isolation of the deprotected poly(amino serinate) was never reported. In attempts to prepare the rather large amounts of PAS necessary to investigate their potential interest in the biomedical field as drug-polymer conjugates or polyanion-polycation complexes we tried to take advantage of the methods reported in literature unsuccessfully. Therefore we looked for another method based on activation of N-protected serine derivatives under the form of a mesyl mixte anhydride according to a method proposed to prepare lactones from hydroxy acids[8-9].

Cyclisation of the N-Z-DL-serine was attempted in anhydrous pyridine but the reaction failed. Only oligomers of N-protected DL-PAS were formed. This finding suggested that direct polymerization was possible. In a second stage the reaction was conducted in acetone in the presence of sodium carbonate under heterogeneous conditions. The resulting polymeric compounds had molar masses in the range of 10000 daltons which exhibited the IR band typical of ester bonds. The mechanism of polymerization is still unknown. It is likely that polymerization proceeded through a β-lactone intermediate by intramolecular attack of the mixte anhydride by the serine alcohol group or by direct condensation after intermolecular nucleophile attack of the mixte anhydride of one serine molecule by the alcohol group of another serine molecule according to Fig. 1. The reaction appeared difficult to control and depending on various factors, namely temperature, solvent and residual water and excess of mesyl chloride, the latter being predominant.

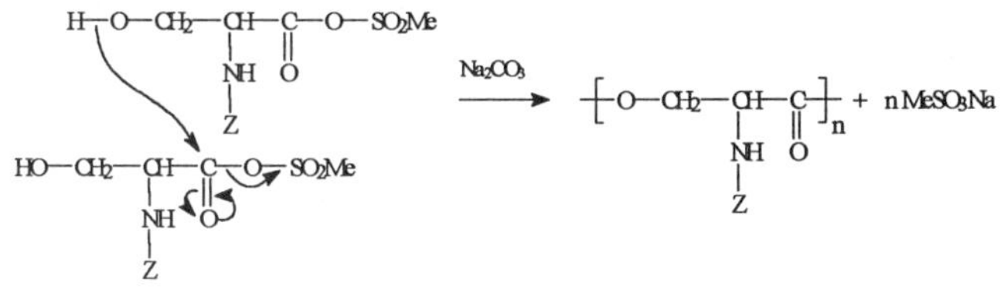

Figure 1 : Probable mechanism of polymerization of the Z-protected serine in the presence of mesyl chloride

Presently, this method leads to Z-protected PAS with molar masses in the range of 20,000 daltons with a 50% yield in one step starting from N-Z serine and in the range of 35,000 daltons in 15% yield after fractionation. Copolymerization was possible with N-boc serine but failed with N-Z threonine and N-benzene sulfonyl serine.

The deprotection of DL-PASZ by the catalytic hydrogenolysis method proposed by Zhou[6] failed. In contrast pendent urethane groups were cleaved in organic medium to yield poly(amino serinate-co-Z-protected amino serinate) copolyesters of different compositions up to 95% deprotection.

The acidolysis was conducted by HBr in acetic acid. Fig. 2 shows the modification of the nmr spectrum of DL-PASZ with time. The removal of the Z groups is shown by the appearance of a resonance at 4.5 ppm due to the methylene protons of the benzyl chloride which was formed and by the decrease of the Z proton resonances at 5.1 and 7.4 ppm. The composition of the copolymers was deduced from ^{1}H nmr spectra. [13]C nmr spectra well agreed with the copolymer structures. In order to check the preservation of the polymer chain, pendent amino groups were reprotected by reaction with benzyloxycarbonyl chloride. SEC analyses showed that the polyester chains underwent less than 10% degradation during the acid-catalyzed deprotection of N-Z-PAS.

The same protocol was applied to L-serine to make L-PASZ and L-PAS. The optically derivatives had a narraow range of solvents in agreement with isostacticity. The cytotoxicity of a 18,000 dalton PASZ5Br95 copolymer appeared c.a. one order of magnitude smaller than that of a poly-L-lysine) of similar molar mass.

The sodium salt of poly(β-malic acid), PMLANa100 (Mn = 26,000) and the bromide salt of the 95% deprotected PSA polybase derivative, PSAZ5Br95 (Mn = 15,000) were allowed to react in water in various proportions (1/1, 2/1 and 1/2) to form polyelectrolyte complexes

according to :

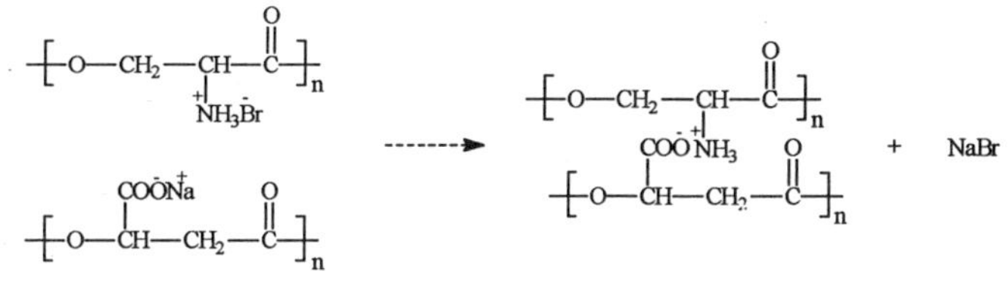

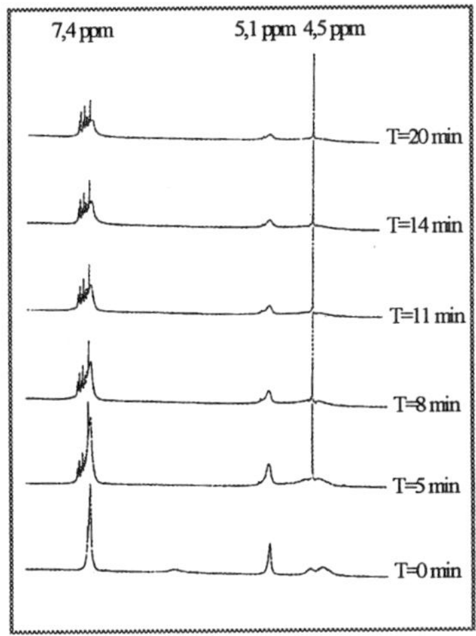

In all cases stoechiometric reaction between positive and negative charges led to a neutral complex in the presence of the polyelectrolyte in excess. After centrifugation the complex appeared as a slightly yellow paste which was soluble in DMSO, DMF and NMP only.

Figure 3 : Variations of the nmr spectra of N-Z-PAS during hydrolytic deprotection by the HBr/acetic acid method.

The 1/1 complex hydrogel was allowed to age in various aqueous media, namely pH = 7.4 phosphate buffer, water, pH=7.4 borate buffer, and saline. In all these media the complex returned to solution within a couple of days. This solubilisation is likely to be due to the hydrolytic degradation of at least one or the two polyelectrolyte components. It was faster at 37°C than at 20°C. The formation of serine was controlled by thin layer chromatography.

Conclusion

A practical route has been found to make largely deprotected PASZ100-xBrx copolymers with molar masses in the range of 20-25,000 daltons and different compositions depending on the degree of deprotection. These polybases are soluble in water provided deprotection is higher than 40 %. They do degrade in aqueous media and can be used to make polyelectrolyte hydrogel complexes which degrade quite rapidly in aqueous media including in the presence of salt and at physiological conditions.

References

1. K. Petrak, « Polyelectrolytes, Science & Technology, « Polyelectrolyte complexes », H. Masanori ed., M. Dekker Inc., 1993, p. 265

2. A.B. Scranton, B. Rangarajan and J. Klier, Adv. Polym. Sci., 122, 1 (1995)

3. C. Braud and M. Vert, , Trends in Polymer Science, 3, 57 (1993)

4. V. Jarm and D. Fles, J. Polym. Sci. Polym. Chem. Ed., 15, 1061 (1984)

5. I. Fiétier, A. Leborgne and N. Spassky, Polym. Bull. 24, 349 (1990)

6. Q.X. Zhou and J. Kohn, Macromolecules, 23, 3399 (1990)

7. M.E. Gelbin and J. Kohn, J. Am. Chem. Soc., 114, 3962 (1992)

8. W. Adam, J. Baeza and JC. Liu, J. Am. Chem. Soc., 94, 2000 (1972)

9. S. Mageswaran and M.U.S. Sultanbawa, J. Chem. Soc., Perkin Trans 1, 884 (1976)

AMPHIPHILIC BLOCK COPOLYMERS BASED ON
POLY(α-AMINO ACID) DERIVATIVES

F. Rypáček, J. Pytela[1], R. Kotva, B. Lánská and L. Machová

Institute of Macromolecular Chemistry,
Academy of Sciences of the Czech Republic
Heyrovský Sq. 2, 162 06 Prague, Czech Republic

INTRODUCTION

Amphiphilic copolymers are important components of biomaterials and pharmaceutical preparations. They provide for a hydrophilic interface at biomaterials' surfaces, thus contributing to the compatibility of biomaterials with aqueous milieu of the living body, and minimizing their interactions with the host immune system.[1] In pharmaceuticals, the polymer "drug carrier" system is required to combine the inertness in the biological systems with a high capacity for the loaded drug. For lipophilic drugs, such a system could be based on polymer micelles.[2] Polymer micelles can be formed by block copolymers of the AB or ABA type composed of hydrophilic (A) and hydrophobic (B) blocks. In aqueous media, such copolymers will associate by their hydrophobic parts, forming a hydrophobic core of the micelle, surrounded by a shell of hydrophilic blocks.[3] Aqueous solutions of copolymer micelles, based on vinylic and (meth)acrylate polymers, and their capacity to solubilize lipophilic substances have been investigated.[4] However, for a medical or pharmaceutical use, it is necessary to develop such a system on the basis of biocompatible materials. Complete biodegradability of the copolymers would be an essential requirement in such a case.

Synthetic poly(amino acids) can be rationally synthesized and their biodegradation in mammalian tissues can be well controlled by their structure.[5] We investigate the feasibility of preparing block copolymers with lipophilic and hydrophilic segments, formed by polymers with biodegradable backbone and functional side chains. To this end we studied the synthesis of block copolymers formed by hydrophilic and lipophilic derivatives of L-glutamic and L-aspartic acids. Such amphiphilic copolymers can be prepared by a stepwise ring-opening polymerization of respective N-carboxyanhydrides (NCA) with ester-protected ω-carboxylic acid groups. By using the protecting esters with different reactivity, one of the copolymer blocks can be selectively modified to expose or introduce hydrophilic groups. The properties of resulting copolymers greatly depend on the extent of termination reactions in the course of polymerization and on the selectivity of copolymer modification. The existence of reactive side chains in the polymer affects the course of polymerization. In this paper we focus on the feasibility of preparing defined block copoly(amino acid)s via the amino-propagating mechanism of NCA polymerization and we discuss the effect of reaction conditions on the final copolymer properties.

EXPERIMENTAL

Materials

N-carboxyanhydrides of γ-benzyl L-glutamate (BG-NCA) and γ-(2,2,2-trichloroethyl) L-glutamate (TCEG-NCA) were synthesized according to published methods.[6,7] 1,2-Dichloroethane (DCE) was distilled from P_2O_5 just before use. 1,6-Hexanediamine (HMDA) was purified by sublimation. 2-Amino ethanol (AE) was purified by a vacuum distillation.

Telechelic amino-terminated poly(γ-benzyl L-glutamate) (NH_2-PBG-NH_2)

The polymerization of BG-NCA in DCE (0.2 mol/l) was initiated by HMDA. A stock solution of initiator (about 1 mol/l) was prepared by dissolving HMDA in benzene prior to use. An exact concentration of

---

Present address: Research Institute of Synthetic Rubber, Kaucuk a.s. Kralupy , Czech Republic

HMDA in the stock solution was determined by titration. The polymers were isolated by precipitation to diethyl ether.

TCEG-b-BG-b-TCEG copolymer

The stock solutions of NH_2-PBG-NH_2 (43 mg/ml) and TCEG-NCA (0.3 mol/l) in DCE were equilibrated to a given polymerization temperature and mixed in amounts required to achieve the desired mole ratio of TCEG-NCA to terminal amines of NH_2-PBG-NH_2 (M/I ratio). The composition of copolymers was determined by elemental analysis, Cl values being a sensitive indicator of TCEG content.

HEG-b-BG-b-HEG copolymers: aminolysis of TCEG-b-BG-b-TCEG copolymer by 2-amino-ethanol (AE)

The kinetics of aminolysis by AE was studied in TCEG-b-BG-b-TCEG copolymer with the mole fraction of TCEG units x_T = 0.59, in dimethylacetamide (DMA). The samples withdrawn at given time intervals were quenched with 2 M HCl. The composition of the polymer in the samples was then determined by elemental analysis and NMR spectroscopy. The other part of the sample was saturated with NaCl and extracted with hexane to take up the released trichloroethanol (TCEA) and benzyl alcohol (BA). The organic phase was then analyzed by HPLC.

Methods

Conductometric titrations of amino groups were carried out with aqeous HCl in phenol/isopropyl alcohol (1:1) according to a described procedure.[8]

HPLC analysis was carried out on a reversed-phase column (Type SGX C18, 3x150 mm, Tessek, Czech Republic) in methanol/water 2:3 with UV (259 nm) and refractometric detection.

^{1}H NMR spectra of TCEG-b-BG-b-TCEG and HEG-b-BG-b-HEG copolymers were measured in deuterated trifluoroacetic acid on a Bruker WH 360 spectrometer. [BG]: δ 1.70-1.85 (β-CH$_2$), 2.15 (α-CH$_2$), 4.30 (α-CH), 4.75 (-O-CH$_2$); [TCEG]: δ 1.85-2.00 (β-CH$_2$), 2.35 (γ-CH$_2$), 4.40 (-O-CH$_2$, α-CH); [HEG]: δ 1.65-1.90 (β-CH$_2$), 2.20 (γ-CH$_2$), 3.30 (=N-CH$_2$), 4.10 (CH$_2$-O-), 4.30 (α-CH).

RESULTS

ABA block copolymers with L-glutamic acid esters of different reactivity in blocks A and B were prepared in two steps. In the first step, the central B block was synthesized by polymerization of BG-NCA using a bifunctional initiator, hexamethylenediamine (HMDA). The reaction proceeds by an aminolytic mechanism[9] and should yield NH_2-PBG-NH_2, terminated by amino groups on both chain ends. In the second step, NH_2-PBG-NH_2 was used as a macroinitiator in the polymerization of TCEG-NCA, thus yielding TCEG-b-BG-b-TCEG copolymer.

The analysis of NH_2 groups in NH_2-PBG-NH_2, obtained from the first step, revealed a significant loss of amines during the synthesis. In the samples polymerized at 25 °C, only about 60 to 65% of amine end groups were recovered with respect to the theoretical amount calculated from the M/I ratio. It was also evident, that the percentage of preserved amines was not significantly affected either by the M/I ratio or by the polydispersity of the sample. However, it decreased with reaction time and with increasing temperature.

The loss of the nucleophilic propagating species terminates the growth of the PBG chain. Based on the kinetic observation, the analysis of benzyl alcohol released during polymerization and NMR data, we concluded that the termination is caused by cyclization of terminal glutamate via the reaction of the nucleophilic end group with the γ-benzyl ester of the same glutamate unit, thus forming a substituted pyrrolidone. The amount of benzyl alcohol released in the course of polymerization was determined by HPLC analysis. The amount of benzyl alcohol liberated during the polymerization was in good correlation with the amount of lost amine end groups, found by titration.

Assuming a stronger temperature dependence of the aminolytic cyclization, relative to the reaction of amine with NCA, we tried to suppress the loss of terminal amine by lowering polymerization temperature. Table 1. shows the content of propagating amine groups in the samples withdrawn from the reaction in the course of polymerization.

The decrease in the amine content is substantially suppressed at 4°C, the amine loss being only about 10% after 24 h. Isolation of the polymer and even its storing in the dry state may cause further loss of terminal amines in NH_2-PBG-NH_2.

Nevertheless, by carrying out the polymerization of BG-NCA at a low temperature and avoiding extended reaction or storage times, telechelic NH_2-PBG-NH_2 prepolymers with high content of functional amino end groups, suitable to initiate the polymerization of other NCA, can be prepared.

Table 1: Content of amino end groups in PBG isolated in the course of HMDA-initiated polymerization of BG-NCA. (M/I = 15:1; DCE)

Time (h)	NH_2 content (%)	
	25 °C	4 °C
0	103	102
4	-	100
8	78	92
24	63	91

TCEG-NCA polymerization initiated by NH_2-PBG-NH_2

A set of NH_2-PBG-NH_2 prepolymers served as polymeric initiators in TCEG-NCA polymerization. Two characteristics of resulting TCEG-*b*-BG-*b*-TCEG copolymers were dermined: the mole fraction of TCEG units in the copolymer, x_T, and the average polymerization degree of the polyTCEG chain resulting from the initiation by one amino group, pd_T.

Although the length of polyTCEG blocks increased with increasing M/I ratio, a significant termination of the TCEG-NCA polymerization occurred at room temperature and the polymerization ceased before all the monomer could be consumed. The termination reaction appeared to follow the same mechanism as that occurring in PBG chains, i.e., the cyclization of the terminal glutamate unit, which, due to higher reactivity, is more pronounced with trichloroethyl ester than with benzyl ester. Also in these case, the extent of termination reaction can be suppressed in favor of the chain propagation by decreasing the polymerization temperature. The results from the polymerizations of TCEG-NCA carried out at 22, 12, and 4 °C are summarized in Table 2.

The copolymers obtained at lower temperatures show significantly higher x_T (and consequently pd_T) in comparison with the polymerization at room temperature. A lower solubility of the monomer at low temperatures can be overcome by a gradual addition of the monomer to the reaction.

Amphiphilic HEG-b-BG-b-HEG copolymers

The higher reactivity of TCEG ester groups, in comparison with BG, allows for a selective conversion of the polyTCEG blocks to hydrophilic blocks via the aminolysis by AE, thus yielding a block copolymer of hydrophilic N^5-(2-hydroxyethyl)-L-glutamine (HEG) and γ-benzyl L-glutamate: HEG-*b*-BG-*b*-HEG.

Table 2: Polymerization of TCEG-NCA initiated by NH_2-PBG-NH_2: the effect of temperature on the extent of termination (initiator (I): NH_2-PBG-NH_2, $\overline{M}_n$=8000; monomer (M): TCEG-NCA, DCE, 72 h)

M/I a)	x_T b)		
	22 °C	12 °C	4 °C
12.5	0.36	0.41	0.50
25	0.35	0.53	0.54
50	0.42	0.58	0.41 c)
100	0.47	0.59	0.37 c)

a) mole ratio TCEG-NCA /NH_2 groups of NH_2-PBG-NH_2
b) mole fraction of TCEG in the copolymer
c) due to low solubility of TCEG-NCA in DCE

The kinetics of aminolysis with AE was investigated with the aim to find the conditions ensuring complete removal of TCEG esters while leaving BG esters intact. The aminolysis was followed in two ways, i.e., by determining the composition of the polymer isolated from the reaction mixture, and by measuring the amounts of released alcohols, such as 2,2,2-trichloroethanol (TCEA) and benzyl alcohol (BA). The TCEG esters of the copolymer can be rapidly and completely aminolyzed under mild conditions, while the amount of liberated BA indicated almost negligible aminolysis of BG units. A comparison of the reaction rates for the aminolysis of BG and TCEG esters is shown in Fig. 1.

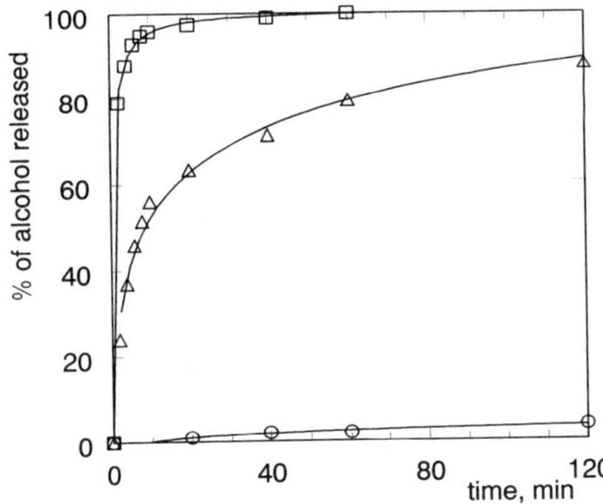

Figure 1: Aminolysis of TCEG and BG segments of TCEG-*b*-BG-*b*-TCEG copolymer by AE in DMA at 25 °C: □ - TCEA , AE/TCEG = 35; △ - TCEA, AE/TCEG = 7 ; ○ BA, AE/BG = 7.

CONCLUSIONS

The synthesis of functionalized poly(amino acid)s by polymerization of glutamate-NCA via a nucleophilic propagating end group, can be strongly affected by the reactivity of the γ-glutamate ester. However, well defined amphiphilic block copolymers can be prepared by using reactive glutamates, provided that the polymerization conditions were adjusted to minimize the termination reaction through cyclization of the terminal glutamate residue. In this way, amphiphilic block copolymers with biodegradable backbone chains can be prepared.

Acknowledgment

The financial support by grant No. A4050503 from the Grant Agency of ASCR and grant No. 203/98/00882 of the Grant Agency of Czech Republic is gratefully acknowledged.

REFERENCES

1. Freij-Larsson, C., Nylander, T., Jannasch, P. and Wesslen, B., *Biomaterials*, 17: 2199, 1996.
2. Kataoka, K., Kwon, G.S., Yokoyama, M., Okano, T. and Sakurai,Y., *J. Controlled Release*, 12: 119,1993.
3. Tuzar, Z. and Kratochvíl, P., in: Encyclopedia of Advanced Materials, Bloor, D., ed., Pergamon Press, 1994, pp.2047-2052.
4. Tian, M., Arca, E., Tuzar, Z., Webber, S.E. and Munk, P., *J. Polym. Sci. , Part B: Polym. Phys.*, 33: 1713, 1995.
5. Rypáček, F., Pytela, J., Kotva, R. and Škarda, V., *Macromol. Symp.*, 123: 9, 1997.
6. Daly, W.H. and Poché, D., *Tetrahedron Lett.*, 29: 5859, 1988.
7. Ishikawa, K. and Endo, T., *J. Polym. Sci. Part A:Polym. Chem.*: 28: 3525, 1990.
8. Lánská, B. and Šebenda, J., *Collect. Czech. Chem. Commun.*, 39: 257, 1974.
9. Kricheldorf, H.R., *Alpha-Aminoacid-N-Carboxy-Anhydrides and Related Heterocycles.* Springer Verlag, Berlin, Heidelberg, New York, 1987.

HYDROPHILIC-HYDROPHOBIC INTERPENETRATING POLYMER NETWORKS CONTAINING POLY(VINYL PYRROLIDONE)

S. Roweton and S. J. Huang
Polymer Program,
Institute of Materials Science and
Department of Chemistry,
University of Connecticut, Storrs, CT 06269

Introduction

Synthetic hydrogels are an important class of biomaterials that are designed to chemically and physically mimic natural hydrogels. There are a range of biomedical applications for these synthetic materials including wound management, controlled drug delivery, and use as materials for soft contact lenses. However, certain physical limitations have restricted the range of applications. Hydrogels are characterized by inherent mechanical weakness and solubility characteristics that are difficult to control. In addition, they cannot be processed by traditional thermal methods. Hydrophobic materials have been incorporated into hydrogels in an effort to improve the physical characteristics.

Hydrophilic-hydrophobic multicomponent polymeric systems can be created from block copolymers, graft copolymers, polymer blends, and a variety of networks. These systems possess unique properties which make them valuable as biomaterials. Hydrophilic components facilitate the retention of water necessary for many biomedical applications while hydrophobic components impart mechanical integrity, improve the control of solubility, and broaden the range of available methods for processing.

The current study involves the synthesis and characterization of novel hydrophilic-hydrophobic systems containing interpenetrating polymer networks. Hydrophilic polymeric components investigated include the nitrogen-containing polymers poly(vinyl pyrrolidone) (PVP) and gelatin. PVP has been investigated as a component in materials for soft contact lenses[1], matrices for controlled release[2], and bioadhesives.[3] Gelatin has been studied for use in a variety of biomaterials including bioadhesives[4] and constructs for tissue engineering.[5,6] Poly(ε-caprolactone) (PCL) is a polymer of biomedical significance that has been investigated for a range of biomaterials applications including constructs for cell growth.[7] For this reason, PCL is employed as the hydrophobic component in the interpenetrating network materials under investigation. Both semi-interpenetrating network (SIPN) and interpenetrating network (IPN) multicomponent materials have been synthesized. The SIPN materials contain only crosslinked PVP while the IPN materials contain both crosslinked PVP and PCL. The multicomponent systems produced are characterized by unique chemical and physical characteristics that can be tailored for a variety of applications.

Experimental

The synthesis of the PCL-PVP interpenetrating network systems is described below and represented schematically in Figure 1.

Synthesis of Semi-Interpenetrating Networks of PCL and PVP

PCL and N-vinyl-2-pyrrolidone were dissolved in THF with ethylene glycol dimethacrylate (EGDMA) and AIBN to form homogeneous solutions. These solutions were cast into Teflon molds for solvent evaporation. Dried films were heated at 70°C for 2 hours and then at 90° for 45 minutes to form semi-interpenetrating network materials containing crosslinked PVP. These materials were extracted with an excess of ethanol to remove residual N-vinyl-2-pyrrolidone monomer and non-crosslinked PVP.

Synthesis of Interpenetrating Networks of PCL and PVP

PCL and N-vinyl-2-pyrrolidone were dissolved in THF with EGDMA, AIBN, and dicumyl peroxide to form homogeneous solutions. These solutions were cast into Teflon molds for solvent evaporation. Dried films were heated at 90°C for 2 hours and then at 110°C for 45 minutes to form interpenetrating network materials containing crosslinked PVP and crosslinked PCL. These materials were extracted with an excess of ethanol to remove residual N-vinyl-2-pyrrolidone monomer and non-crosslinked PVP.

Results and Discussion

Two PCL materials, one with MW=43,000 and another with MW=80,000, were used to synthesize SIPN and IPN materials with a range of PVP content in their compositions. Initial investigations of the thermal characteristics of both the SIPN and IPN materials revealed phase separated systems with two distinct glass transition temperatures, one associated with PCL and one associated with PVP. In addition, a melting transition corresponding with the PCL melting temperature was consistently evident.

The equilibrium water content (EWC) of the PVP-PCL network systems has been determined for the range of compositions in aqueous buffered systems at pH=4.0, pH=7.4, and pH=10.4 and 37°C. EWC is calculated as $(m_1 - m_0)/m_1$ where m_0 is the weight of the dry polymer and m_1 is the weight of the swollen polymer at equilibrium. This data is summarized in Figures 2-4.

Conclusions

The synthesis of PCL-PVP interpenetrating network materials produces phase separated hydrogel systems. The characteristic equilibrium water content values for the SIPN and IPN materials are similar at pH values of 4.0, 7.4, and 10.4 and can be controlled by varying PCL content.

References

1. Perera, D. I. and Shanks, R. A. *Polymer International*, **39**, 121(1996).
2. Arcos, D., Cabanas, M. V., Ragel, C. V., Vallet-Regi, M., and San Roman, J. *Biomaterials*, **18**, 1235(1997).
3. Kao, F. Manivannan, G., and Sawan, S. P. *Journal of Biomedical Materials Research*, **38**, 191(1997).
4. Otani, Y., Tabata, Y., and Ikada, Y. *Journal of Biomedical Materials Research*, **31**, 157(1996).
5. Santin, M., Huang, S. J., Iannace, S., Ambrosio, L., Nicolais, L., and Peluso, G. *Biomaterials*, **17**, 1459(1996).
6. Koide, M., Osaki, K., Konishi, J., Oyamada, K., Katakura, T., Takahashi, A., and Yoshizato, K. *Journal of Biomedical Materials Research*, **27**, 79(1993).
7. Peluso, G., Petillo, O., Anderson, J. M., Ambrosio, L., Nicolais, L., Melone, M. A. B., Eschbach, F. O., and Huang, S. J. *Journal of Biomedical Materials Research*, **34**, 327(1997).

Acknowledgements

The authors thank Union Carbide Corporation for providing PCL sample materials.

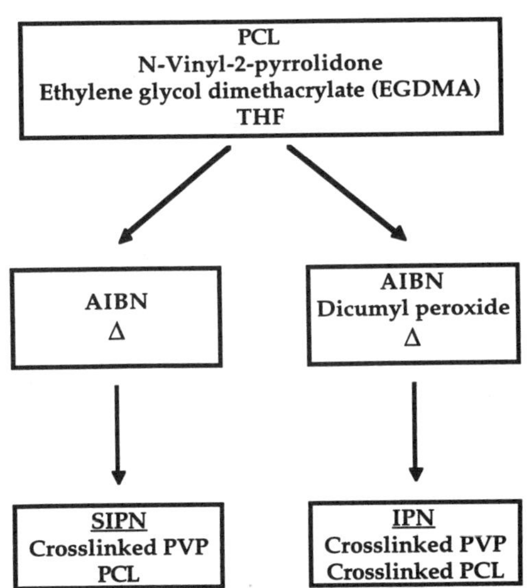

Figure 1. Synthesis scheme for hydrophilic-hydrophobic interpenetrating network materials.

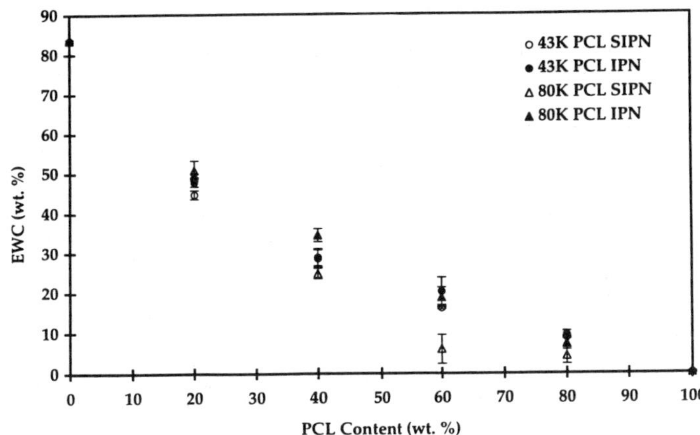

Figure 2. Equilibrium water content values for PVP-PCL systems at pH=4.0 and 37°C.

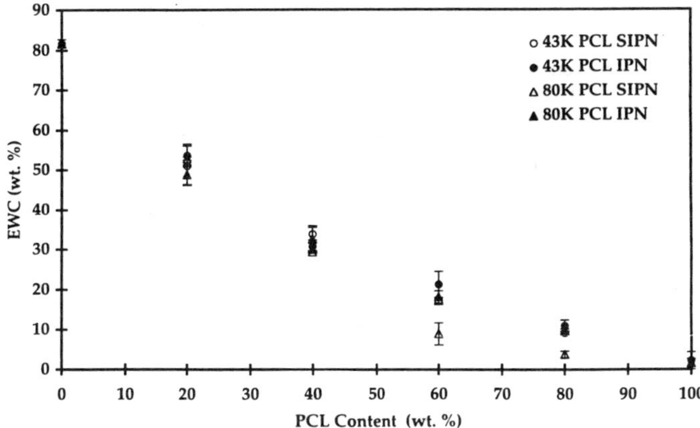

Figure 3. Equilibrium water content values for PVP-PCL systems at pH=7.4 and 37°C.

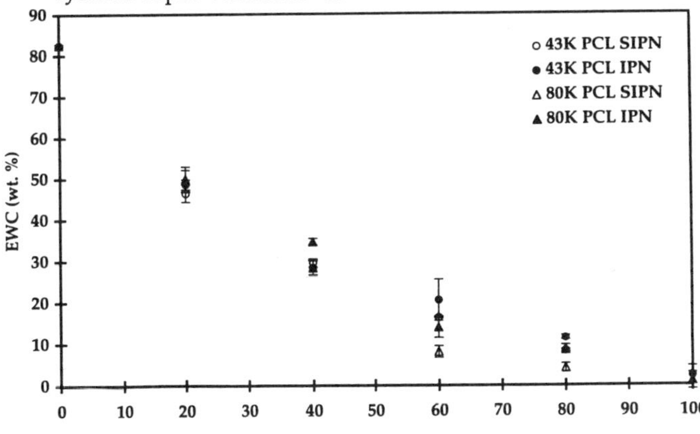

Figure 4. Equilibrium water content values for PVP-PCL systems at pH=10.4 and 37°C.

HYDROGEL NANOPARTICLES FORMED BY SELF-AGGREGATION OF HYDROPHOBIZED POLYSACCHARIDES

*J. Sunamoto**[,#], *K. Akiyoshi**, *S. Deguchi**, *Y. Kato*[#], *I. Taniguchii**,
*Y. Sasaki**, and *K. Kuroda**

Department of Synthetic Chemistry & Biological Chemistry, Graduate School
of Engineering, Kyoto University, Sakyo-ku, Yoshida Hommachi, Kyoto
6065801, Japan

Introduction

Naturally occurring polysaccharides are one of abundant and diverse families of biopolymers. The polysaccharides, which are biodegradable polyhydroxyl compounds, have unique properties such as formation of hydrogel or liquid crystal. Therefore, they have been widely utilized in pharmaceutical applications. Especially, the polysaccharide derivatives with long alkyl chains have been prepared for a variety of applications such as hydrophobic gel chromatography, enzyme immobilization, polymer drugs, and coating of cytoplasma membrane surface. Solution viscosity drastically increases in the presence of hydrophobized polysaccharide such as hydroxypropyl cellulose derivatives that are partly substituted by long hydrocarbon chain. This is due to the intramolecular and/or intermolecular association of hydrophobic groups. The property as rheological modifier is applied to the many commercial materials such as paints, inks, cosmetic and pharmaceutical.

Since 1982 we have been developing several hydrophobized polysaccharides and utilized them in biotechnology and medicine. In this article we would like to give an overview of the physicochemical characterization and several functions of such the naturally occurring polysaccharides partly hydrophobized.

Hydrophobization of Naturally Occurring Polysaccharides

In order to synthesize a nonionic hydrophobized polysaccharide, the reaction between polysaccharide and cholesteryl N-(6-isocyanatehexyl)-carbamate (1) was proposed (Figure 1). For example, an activated cholesterol derivative (1) (0.344 g, 0.62 mmol) was reacted at 80 °C for 8 h with pullulan (M_w 55000, M_w/M_n = 1.54, 4.00 g, 24.8 mmol eq. as its anhydroglucoside unit) in 100 mL of dry DMSO containing 8.0 mL of pyridine. Five hundred mL of ethanol was added to the reaction mixture and the suspension was kept overnight at 4.0 °C. The precipitates were separated, purified by dialysis against water using Seamless Cellulose Tube (VISKASE SALES Corp.), and lyophilized to give white powder; yield, 3.81 g (90 %). The degree of substitution of cholesterol group was determined by ^{1}H-NMR. When pullulan (M_w 55000) was substituted by 2.0 cholesterol groups per 100 anhydroglucoside units, it was coded as CHP-55-2.0. No degradation of the polymer during the modification was ascertained by size exclusion column chromatography (SEC) using DMF as the solvent. Various hydrophobized polysaccharides such as pullulan, dextran, mannan, and amylose were synthesized by the same procedure.

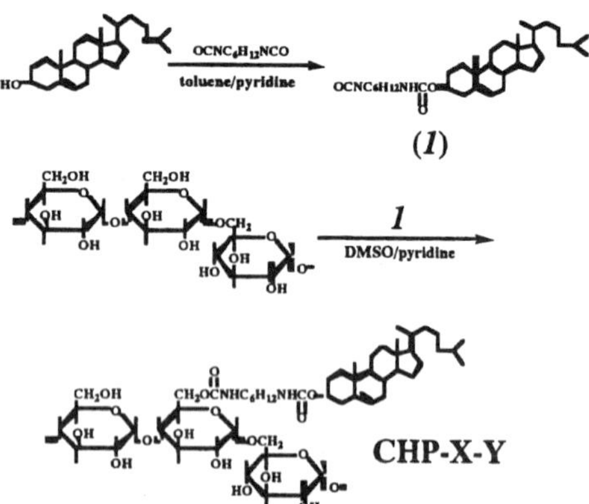

Figure 1. Synthetic rout of cholesteryl group-bearing pullulan

Self-Aggregation of Hydrophobized Polysaccharides

Parent polysaccharides such as pullulan, dextran and mannan easily dissolve in water and give an optically clear solution. Hydrophobized polysaccharides are easily soluble in DMF and DMSO, but less in water. The substitution degree of cholesterol moiety sensitively affected the water solubility of the polymer. The polysaccharides with the substitution degree of cholesterol moieties higher than 5 are not soluble in water even after ultrasonication. With these polymers, therefore, the following two methods were applied to prepare an optically clear solution. Method I (Sonication method); a hydrophobized polysaccharide (1.0–5.0 mg/mL) was suspended and swelled in pure water or an aqueous buffered solution under stirring for 24 h at 50–60 °C to give a milky suspension. This was sonicated using a probe type sonifier (TOMY, UR-200P) at 40 W and room temperature. Method II (DMSO dissolving method); a hydrophobized polysaccharide (10.0 mg) was dissolved in 10 mL of DMSO and then dialyzed with water until the DMSO concentration becomes less than 1 ppm in the final solution. The DMSO content was determined by HPLC. Both methods of the sample preparation lead to the same result.

All the hydrophobized polysaccharides studied formed monodispersive nanoparticles upon intermolecular self-aggregation. For CHP-55-2.1, the hydrodynamic radius (R_H) of the self-aggregate as determined by the dynamic light scattering (DLS) was approximately 9.5 nm. Intensity of the autocorrelation function measured fitted well to a single exponential, and its size distribution was quite uniform. The particle size decreases with an increase in the number of cholesterol groups; 11.6 nm for CHP-55-1.1; 10.2 nm for CHP-55-1.7; 9.5 nm for CHP-55-2.1; and 8.4 nm for CHP-55-3.4. Transmission electron microscopic images showed that nanoparticles formed by self-aggregation were spherical and the size distribution was rather narrow for all the cases.

The molecular weight of CHP self-aggregate was determined by static light scattering (SLS) method. For example, the weight averaged molecular weight (M_w), the root mean-square radius of gyration (R_G), and the second virial coefficient (A_2) of CHP-55-2.1 self-aggregate were found to be 5.8 x 10^5, 11.1 nm, and 2.60 x 10^{-4} (mol mL)/g², respectively. From the apparent M_w of the self-aggregate, the aggregation number was estimated to be approximately 10. The substitution degree of cholesterol moiety of CHP little affect the aggregation number.

The self-aggregates of hydrophobized polysaccharides so obtained are regarded as a nanosize hydrogel. The polymer density changes depending on the substitution degree of cholesterol moiety; 16% for CHP-55-1.1; 21% for CHP-55-1.7; 27% for CHP-55-2.1; and 50 % for CHP-55-3.4. The substitution degree of cholesterol moiety becomes higher, the more condensed and smaller particles are formed (Figure 2).

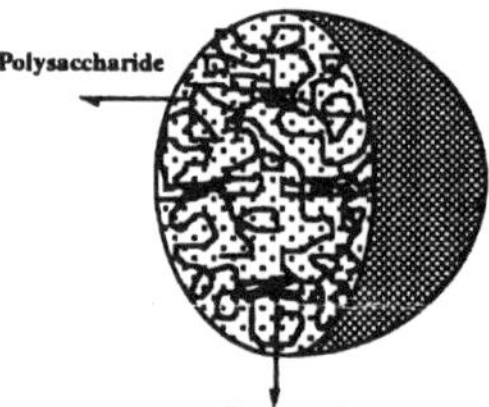

Figure 2. Schematic drawing of a hydrogel nanoparticle as formed by self-aggregation of hydrophobized polysaccharide

The self-aggregate has microdomains as provided by a rigid and hydrophobic core of cholesterol moieties and a relatively mobile corona of polysaccharide skeleton. One self-aggregate nanoparticle has several hydrophobic domains of cholesterol moieties associated. The result of SLS measurement tells us that, for example with CHP-55-1.7, one nanoparticle is formed by approximately ten CHP molecules, and approximately 60 cholesterol moieties exist in the nanoparticle. In addition, the fluorescence quenching study using pyrene-cetylpyridinium chloride system reveals that one hydrophobic domain consists of four or five cholesteryl groups. These data suggest that there exist approximately fourteen hydrophobic domains of four cholesterol groups within one nanoparticle. The domain of cholesterol association seems to be a cross-linking point for the physical hydrogel formation. The CHP self-aggregate is, in addition, a unique thermo-responsive hydrogel nanoparticle.

Macrogel Formation

Hydrophobized polysaccharide consists of a water soluble polymer backbone, on which hydrophobic groups are grafted in part. In aqueous solutions, the hydrophobes intramolecularly and/or intermolecularly associate as mentioned above. At high polymer concentrations, the hydrophobe domains that intermolecularly associated serve as transient cross-links to connect among the polymer chains. Because of this network formation, the hydrophobized polysaccharide effectively modifies the rheology of aqueous solutions or dispersions, which is one of the most important aspects in practical usage of the polymer amphiphiles. When hydrophobized polysaccharide is mixed with surfactants, the transient network is enforced by the association of the surfactants with the hydrophobe of polysaccharide.

Gelation of cholesterol-bearing pullulan (CHP) with SDS in water was then studied by rheological measurements. The apparent viscosity of the CHP (2 %(w/w))/SDS mixture increased with an increase in the SDS concentration up to approximately 1% and then decreased after a maximum.

In the presence of large amounts of SDS, the CHP self-aggregate certainly dissociated. With 3 %(w/w) CHP, a macroscopic gel was formed by the addition of SDS above 0.5%(w/w). At higher concentrations of SDS (above 4.5%(w/w)), the gel changed to a sol. The mechanism of the gelation and the transition to the sol is related to the formation of mixed aggregates between the cholesteryl groups of CHP and SDS. Due to the strong · association of the cholesteryl groups, large amounts of SDS were required to achieve the complete solubilization of cholesteryl groups. Oscillatory shear measurements were carried out for the gel of the CHP/SDS mixture. In the low frequency region (ω < 0.1 Hz), G'' (loss modulus) showed a maximum, while G' (storage modulus) reached a plateau with an intersection of the G'' curve. This is a trend typical of a Maxwellian fluid. An extremely long relaxation time (approximately 20 sec) was observed for the CHP/SDS gel at relatively low SDS concentrations. Such a long relaxation time would be ascribed to the strong association of the cholesteryl groups of CHP.

At the CHP concentration above 3.5%(w/w) , CHP itself formed a macrogel even without the surfactant. At 2 %(w/w) CHP, the addition of SDS induced the enhancement of viscosity, and the gelation was observed at 3% (w/w) CHP. The rheological measurements on the CHP/SDS gel suggested that the underlying mechanism of gelation is a cross-linking by the mixed aggregates of the cholesteryl groups of CHP and SDS. At low SDS concentrations, the mixed aggregation promotes the inter-polymer chain coupling. At rather high SDS concentrations, the mixed aggregation leads to the solubilization of cholesteryl moieties, leading to the low viscosity due to a loss of cross-links. Because of the strong association among the cholesteryl groups, large amounts of SDS were required to achieve the complete solubilization of the cholesteryl groups. A cationic surfactant, dodecyltrimethylammonium chloride, also showed a similar trend

Complexation with Small Molecules

The self-aggregate of hydrophobized polysaccharide bound various hydrophobic substances such as fluorescent probes, porphyrin, birilubin (BR) and antitumor drug such as doxorubicine and neocarzinostatin chromophore (NCS-chr). The binding constants became larger with an increase in the hydrophobicity of the guest molecule. In this case, the driving force for the complexation would be primarily a hydrophobicity of the guest. The guest molecule located in the polysaccharide corona as well as in the hydrophobic cholesterol domain of the self-aggregate.

The CHP self-aggregate is expected to be chiral binding site because the polysaccharide has many chiral centers. Induced circular dichroism (ICD) of BR has been observed by enantioselective complexation in various chiral environments. BR, in this sense, is a convenient probe to understand the chiral environment. The BR molecule was strongly incorporated into the CHP self-aggregates and exhibited negative ICD. The hydrophobic association with the cholesterol group and/or the hydrogen bonding with the saccharide moiety of CHP play an important role in the induction of the cotton effect in CD. The CHP self-aggregates certainly have chiral binding sites.

In order to simulate the function of apoprotein of NCS, NCS-chr was isolated and from the apoprotein and complexed with the CHP self-aggregate. By simply mixing the two in 0.5 M AcONa/AcOH (pH 4.7) at 2 °C under shading, the complex was easily isolated by high performance size exclusion chromatography. The particle size of the CHP/NCS-chr complex did not change during incubation for 4 days at 37 °C in 0.5 M AcONa/AcOH (pH 4.7). No leakage of NCS-chr from the complex was observed during the incubation. In vitro studies, the complex showed effective sission of DNA double strand. The biological activity of complexed NCS-chr was studied with various cancer cells such as HeLa , HepG2, and 3'-mRLh-2 cells. In all the cases, the cytotoxic activity was certainly maintained even after substitution of apoprotein with the hydrophobized polysaccharide similarly to the case of intact NCS and free NCS-chr themselves. When the galactose moiety was additionally conjugated to CHP, the obtained galactose-conjugated CHP/NCS-chr complex showed more enhanced cytotoxicity to 3'-mRLh-2 cell almost compalable to intact NCS (Figure 3). The results obtained here suggested that the apoprotein of NCS can be substituted with even a simple macromolecule like the CHP self-aggregate and that the complex between NCS-chr and the nanoparticle behaves similarly to intact NCS itself.

Complexation with Proteins and Enzymes

Various water soluble proteins were complexed by the CHP self-aggregates. The complex showed an excellent colloidal stability without any precipitation. In addition, no dissociation of the protein from the complex was observed at all even after keeping for a week at pH 7.2 and 25 °C.

The concentration of the complexed protein showed a saturation phenomenon as a function of the initial concentration of the protein added. The complexation between a guest molecule (protein) and a host molecule (the CHP self-aggregate) can be represented by the number of complexation sites (n) and the complexation constant (K). Both K and n can be estimated independently by Klotz's plot. The number of complexation site and complexation constants for various soluble proteins so obtained are summarized in Table 1.

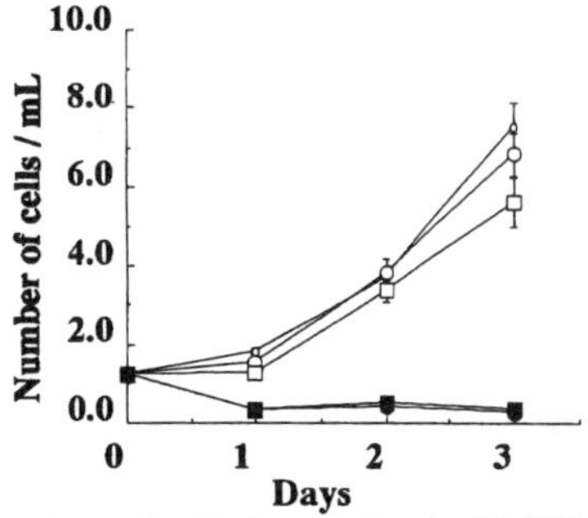

Figure 3. Effect of NCS-chr complexed with CHP or Gal(2)-7-CHP self-aggregate on the growth of 3'-mRLh-2 cells as a function of time at 37 °C in 5 % CO_2: ($\circ$) control, ($\bullet$) intact NCS, (O) free NCS-chr, ($\square$) CHP/NCS-chr complex, and ($\blacksquare$) Gal-CHP/NCS-chr complex. [NCS]=90 nM (1.0 mg/mL), [NCS-chr]=90 nM (60 ng/mL), [saccharide]=2.8 mg/mL, NCS-chr/CHP self-aggregate=26, NCS-chr/Gal(2)-7-CHP self-aggregate =29.

Tbale 1. Number of Complexation Sites (n) and Complexation Constant (K) with CHP Self-aggregates

Protein	MW	n	K/M^{-1}
Ins	5,735	9.3±0.3 (CHP-55-1.7)	(1.8 ± 0.2) × 10^6
Cyt c	12,400	4.3±1.0 (CHP-55-1.6)	(8.1±3.0)×10^3
Mb	17,800	1.9±0.3(CHP-55-1.6)	(1.9±0.8)×10^4
α-Chy	25,000	2.0±0.2(CHP-55-1.6)	(1.9±0.3) ×10^5
BSA	66,000	0.98±0.03(CHP-55-1.7)	(8.1±1.3) × 10^6

The maximum amount of the protein complexed by the CHP self-aggregate depends on the molecular weight (or size) of the protein. This suggests that the complexation seems to have a limited capacity in the nanoparticle. The complexation constant K tends to increase with an increase in the molecular weight of the protein, except Ins. The hydrogel structure of the self-aggregate also influences the complexation. The number of the complexed BSA molecule did not change with an increase in the subsutitution degree (DS) of cholesterol moiety. However, the complexation constant (K) was significantly affected by the hydrogel structure. BSA more strongly complexed with the less densely packed nanoparticle (cf. CHP-55-1.1 self-aggregate). In the case of a smaller protein, Ins, the number of complexed Ins increased with an increase in DS of cholesterol groups, though the complexation constant did not change so much. The number of crosslinks of associated cholesterol moieties in the nanoparticle increased with an increase in DS of cholesterol.

Molecular chaperone is an interesting example of controlled association between proteins in living system. Proteins are folded to their native conformations with the assistance of molecular chaperone that has a cavity for binding of protein. The molecular chaperone prevents the formation of aggregates of folding intermediate or heat denatured protein by specific binding. The bound protein is released in the aid of ATP to be refolded to an active form.

When carbonic anhydrase B (CAB) was heated up to denaturation temperature (70 °C) and then cooled down to r.t., the denatured CAB precipitated in water. On the other hand, in the presence of the CHP self-aggregate (CHP self-aggregate:CAB = 1:1), the precipitation was not observed at all even after heated up to 70 °C. CAB was quantatively complexed by the CHP self-aggregate and formed a polysaccharide-protein hybrid nanoparticle. The CHP self-aggregate certainly complexed the intermediate and refolded CAB, which is unfolded by 6M guanidine hydrochloride, during the complexation. The complexation certainly prevented the irreversible aggregation of CAB. Interestingly, CAB was almost completely released from the complex by the addition of b-cyclodextrin to the system. In addition, CAB was effectively refolded to the native form and 100% recovered its enzymatic activity upon the dissociation. This is exactly the simulation of the function of molecular chaperone with a simple artificial and nonprotein system.

Acknowledgement

This study is financially supported in part by Grant-in-Aid for Scientific Research by the Ministry of Education, Science, Culture and Sports and by the Terumo Life Science Foundation.

Poly(aspartic acid) Hydrogel

C. J. Chang and G. Swift
Rohm and Haas Company, Research Laboratory,
Spring House, Pennsylvania 19477

Introduction

Stimuli-responsive (intelligent) hydrogels have attracted much interest recently (1). Crosslinked polyelectrolytes, such as acrylic acid homopolymer and copolymers, are known to be pH- and electrolyte-sensitive hydrogels (2,3). The lightly crosslinked, partially neutralized, high molecular weight poly(acrylic acid) imparts very high swell ratio in water and body fluids and is the most commonly used superabsorbent (4). However, concerns on its poor biodegradability, toxicological profile, and biocompatibility prevent it to have a broader applications.

We now report a pH- and electrolyte- responsive superabsorbent hydrogel based on the crosslinked poly(aspartic acid) (5). It is derived from a naturally occurring non-toxic amino acid, L-aspartic acid. A hydrogel based on poly(aspartic acid) is expected to have good biodegradability and biocompatibility.

Experimental

Synthesis of Polysuccinimide (PSI). A solution like mixture was obtained by heating two parts of L-aspartic acid with one part of 85% phosphoric acid under the reflux temperature of 145° C. the mixture was then poured into a Teflon pan and placed in an oven at 180° C for one hour. The mixture was allowed to cool and the solid condensate was ground to powder. The powder was subjected to further condensation at 160° to 180° C for 2 to 16 hrs. It was identified as polysuccinimide by NMR spectroscopy with a Mw of 30,000 to 40,000 measured by aqueous GPC. Further increase in Mw was achieved by coupling washed PSI with dicyclohexyl carbodiimide in DMF (6) to give PSI of Mw 70,000 to 80,000

Synthesis of Sodium Polyaspartate. After repeated washing of PSI powder with water, the PSI was hydrolyzed with NaOH to give sodium polyaspartate following the method reported previously (7).

Synthesis of Crosslinked Sodium Polyaspartate - To 1-ounce vials containing a magnetic stirring bar were added 10.0 g of an aqueous solution of 10% sodium polyaspartate at pH 5 to 6.5. Each of the vials was then treated with 0.02 to 0.10 g of 50% aqueous solution of ethylenglycol diglycidylether (Aldrich) while stirring the mixture. The contents of each vial were transferred to a freeze-drying vial and the samples were frozen using an acetone-dry ice mixture. The samples were then freeze dried in Labcono Freeze Dry System. The samples were ground to powders, heat treated for 30 minutes at 180° C, allowed to cool under vacuum or inert (nitrogen) atmosphere.

Results And Discussion

The key to produce a poly(aspartic acid) hydrogel with a high water absorbency is to employ a proper crosslinking chemistry on a high MW poly(aspartic acid). The high MW polysuccinimide (PSI), the precursor of poly(aspartic acid), was prepared by the modifying the procedures of Neri (6) as described in the experimental section.

Unlike the previously reported poly(aspartic acid) superabsorbent preparations (8, 9), our method of crosslinking reaction did not involve the handling of a highly swollen hydrogel. This was accomplished by freeze drying an aqueous mixture of sodium polyaspartate and ethylene glycol diglycidyl ether at pH 5 - 6.5, followed by the crosslinking reaction in the solid state at 180° C for 30 minutes.

The obtained polymers gave swell ratios as high as 700 g/g in water. The swell ratio is very sensitive to the electrolyte concentration. The swelling was reduced to one fifth in 0.9% NaCl solution (Figure 1). The swelling was also reduced drastically as the pH decreased (Figure 2). The swelling was reversible. These results are comparable to those of commercial SAPs based on poly(acrylic acid).

The crosslinked, swellable poly(aspartic acid) can be irreversibly converted back to uncrosslinked, soluble poly(aspartic acid). This is carried out by treating the swollen hydrogel at pH ≥ 11. The hydrogel was destroyed in minutes when it was treated at pH ≥ 12 at ambient temperature or at pH ≥ 11 at an elevated temperature.

Conclusions

We have synthesized a novel crosslinked poly(aspartic acid), which is a pH- and electrolyte- responsive hydrogel. It can impart extremely high swelling and the swelling is reversible but can be rendered irreversible. We expect it to be useful in a variety of applications including personal care and biomedical areas.

<u>**References**</u>

1. Park, H. and Park, K. in *Hydrogels and Biodegradable Polymers for Bioapplications*; Ottenbrite, R.M.; Huang, S.J.; and Park, K., Eds.; ACS Symposium Series 627, ACS, Washington, D.C., 1996, pp. 2-10.
2. Brannon-Peppas, L. and Peppas, N.A., *Chem. Eng. Sci.* **46**, 715-722 (1991).
3. Ohmine, I. and Tanaka, T., *J. Chem. Phys.*, **77**, 5725-5729 (1982).
4. Buchholz, F.L., in *Superabsorbent Polymers*; Buchholz, F.L and Peppas, N.A., Eds.; ACS Symposium Series 573, ACS, Washington, D.C., 1994, pp. 27-38.
5. Rohm and Haas Patent Application, filed on 5/29/97.
6. P. Neri, G. Antoni, F. Benvenuti, F. Cocola, and G Gazzei, J. Med. Chem, 16, 893-897 (1973).
7. S. Wolk, G. Swift, Y. Paik, K. Yocom, R. Smith, and E. Simon, Macromolecules, 27, 7613-7620 (1994).
8. J. Donachy and C. S. Sikes, U. S. Patent 5,284, 936, 1994.
9. A. nagatomo, H. Tamatani, M. Ajioka, A, Yamaguchi, U. S. Patent 5,525,682, 1996.

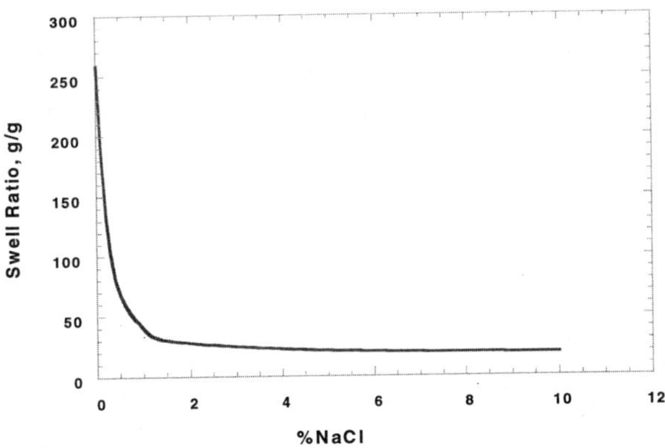

Figure 1. Salt Effect on Swell Ratio

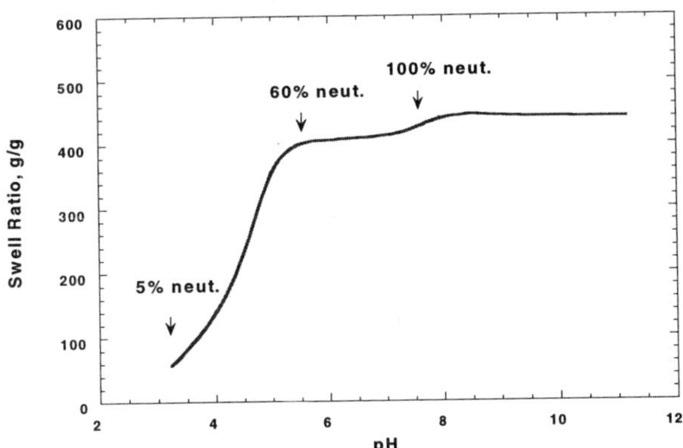

Figure 2   pH Effect on Swell Ratio in Water

Synthesis and Characterization of Poly(aspartic acid) and its Derivatives as Biodegradable Materials

Masayuki Tomida, Kyoko Oda, Masako Yoshitake and Takeshi Nakato
Tsukuba Research Center, Mitsubishi Chemical Corporation,
8-3-1, Chuo, Ami, Inashiki, Ibaraki 300-0332   JAPAN

Toyoji Kakuchi
Graduate School of Environmental Earth Science, Hokkaido University,
Sapporo 060   JAPAN

Introduction

Water-soluble polymers, such as poly(vinyl alcohol), poly(ethylene glycol) and poly(acrylic acid), are widely used as cosmetics, paper additives, dispersants and detergent builders, but they are hardly recovered or collected after use.   Of concern is the diffusion and accumulation of such nonbiodegradable water-soluble polymers in the earth's environment after its release.   Polymers with carboxylic acid groups are one of the most important water-soluble polymers, e.g., poly(acrylic acid) and poly(methacrylic acid) have been used as detergent builders, scale inhibitors and flocculants and are directly released into the earth's environment.   However, they are hardly biodegradable except for their oligomers[1-3], which will possibly produce much damage to the environment.   Therefore, the biodegradable substitutes for nonbiodegradable polymers having carboxylic acid groups are really desired in terms of the earth's environment.

Poly(amino acid) having free carboxylic-acid groups, such as Poly(aspartic acid) and poly(glutamic acid), is one of the candidates for the biodegradable water-soluble polymer.   The thermal polycondensation of aspartic acid is known to easily produce poly(aspartic acid) (PASP), so that it has a great potential to resolve environmental problems, in particular, in the field of biodegradable water-soluble polymer.[4]

There are a lot of research about the production of polysuccinimide(PSI) that is hydrolyzed to poly(aspartic acid)[5-14], but they have the disadvantages, for example, low molecular weight, difficulty of isolation and low biodegradability.

We describe the synthesis of a high-molecular-weight PSI prepared by the polycondensation of L-aspartic acid using acid catalyst in various solvents and the properties of it which are focused on hygroscopicity, calcium-ion chelating ability and water absorption of crosslinked PASP.

Experimental
Materials and Measurements

L-Aspartic acid (ASP) was obtained from the Mitsubishi Chemical Corporation (Japan).   85%-o-phosphoric acid, mesitylene, sulfolane, N,N-dimethylformamide (DMF), sodium hydroxide and deionized water were commercially obtained and used without further purification.

The molecular weight of PSI was estimated in DMF containing 20 mmol/L of LiBr by gel permeation chromatography (GPC) (column: Plgel 5 mm MIXED-C x 2, detector: refractive index, standard: polystyrene).   The residual amount of ASP after the polycondensation was measured in water containing 2.5g/L of H_3PO_4 and 31.2g/L of $NaH_2PO_4 \cdot 2H_2O$ by liquid chromatography (LC) (column: Shim-pack ISC-07/S1504, detector: UV 210nm).

Acid-Catalyzed Polycondensation of L-Aspartic Acid with Solvent

A suspension of ASP (25 g, 0.188 mol) and 85%-o-phosphoric acid (9.4 mmol) in mesitylene (56g) and sulfolane (24 g) was refluxed under a N_2 atmosphere.   Water formed in a reaction mixture was removed using a Dean-Stark trap with a reflux condenser.   After 4.5 h, the solvent was removed, the conversion of ASP of the precipitate was 98 wt%, then the precipitate was washed with water (200 mL) several times until it was neutral.   The residue was washed with MeOH (200 mL) and dried at 85 °C under reduced pressure.   The residual ASP was not detected in the residue.

The hydrolysis of PSI was carried out as follows: To a 100 mL beaker with a stirring bar PSI (3 g) and a solution of sodium hydroxide (1.4 g, equivalents per succinimide residue), and deionized water (20 mL)) under ice cooling were added.   After the mixture was stirred for 1h, the pH of the solution was adjusted by the addition of 35 % aqueous HCl.   The solution was poured into methanol (300 mL), and then the precipitate was filtered and dried at 40 °C under reduced pressure to yield PASP.

Hygroscopicity of Poly(aspartic acid)

The hygroscopicity of PASP was measured using a constant temperature and constant moistened bath (Seiwa Rikou Co. Ltd., Japan) in various temperature and humidity.

Biodegradability of Poly(aspartic acid)

The biodegradability of PASP was estimated using OECD 301C (Modified MITI Test).   A sample was treated with the standard activated sludge, which was obtained from the Chemicals Inspection & Testing Institute, Japan, at 25 ± 1 °C for 28 days.   Aniline was used as the standard to check the activity of the standard activated sludge.   Biological oxygen demand (BOD) and total organic carbon amount (TOC) are the consumption of the oxygen and total organic carbon amount during the evaluation, respectively.   Both are generally used for evaluating biodegradability.   BOD and TOC were measured using an OM3001 coulometer (Ohkura Electric Co., Ltd., Japan) and a Total Carbon Analyzer TOC-5000A (Shimadzu Corporation, Japan), respectively.   Removed TOC was calculated from the difference between the amount of the total organic carbon before and after the evaluation of biodegradability.

Calcium-ion Chelation by Poly(aspartic acid)

The calcium-ion chelating ability of the various polymers was determined using a calcium-ion electrode and an ion meter in accordance with the description in a previous paper.[15]   A sample (10 mg) was dissolved in 50 ml of an aqueous solution, which had been adjusted to give a calcium chloride concentration of 1.0×10^{-3} mol/l and a potassium chloride concentration of 0.08 mol/l.   The resulting mixture was stirred at 30 °C for 10 min and the calcium ions in the solution were determined using a calcium-ion electrode (Orion Model 93-20) and an ion meter (Orion Model 720A).

Preparation of Poly(aspartic acid) Hydrogels

PASP hydrogels were prepared by γ-irradiation (1.6 kGy/h) of PASP aqueous solutions (1-25 wt/vol%) using an irradiation system with a ^{60}Co (110 TBq) source.   The PASP solution (2 mL) was contained in a 10-mL glass vial capped under nitrogen and air.   The resultant hydrogels were swollen to equilibrium for 1 week at room temperature.   During this time, the uncrosslinked PASP was removed by changing the swelling media daily.

The ratio of gelation was calculated as

$$\text{ratio of gelation (\%)} = (W_l / W_f) \times 100$$

where W_f is the weight of PASP feed in the vial and W_l is the dry weight of the PASP hydrogel after the removal of the uncrosslinked PASP.

Results and Discussion
Synthesis of Poly(aspartic acid) and Copoly(aspartic acid)

The polycondensation of ASP with o-phosphoric acid as a catalyst was carried out under reflux using various solvents (Table 1).   The mixed solvents of mesitylene with DMF, NMP and sulfolane apparently affected the polymer yield and weight-average molecular weight (M_w).   The highest M_w, 64300, was obtained for mesitylene/sulfolane.   The difference in the yield and M_w between sulfolane and DMF should depend on the difference of the boiling points and   the difference in the M_w between sulfolane and NMP should depend on the dispersity of PSI in the solvents.

The copolycondensation of ASP and various aminocarboxylic acids with o-phosphoric acid as a catalyst was also carried out under reflux in mesitylene/sulfolane.

Hygroscopicity of Poly(aspartic acid)

The hygroscopicity of poly(aspartic acid) produced by various procedures and several references, for example L-lactic acid sodium salt and hyaluronic acid sodium salt was measured in various temperatures and humidities. All PASPs showed the very similar features and the effect of the difference of the procedures and the structures was not detected. The hygroscopicity of PASPs at 25 °C were lower than that of hyaluronic acid sodium salt and higher than that of L-lactic acid sodium salt.

Calcium-ion Chelating Ability of Poly(aspartic acid)

The chelating ability toward calcium ion of PASP is about a half of that of poly(acrylic acid). The copolymers consisting of ASP and aliphatic and linear α,ω-aminocarboxylic acids produced to improve chelating abilities toward calcium ion. The chelating abilities depended on the content of 6-aminocaproic acid in the copolymer as shown in Figures 1. The result indicates that the chelating ability remarkably increases with increasing the content of 6-aminocaproic acid in the copolymers. The highest value was 3 times higher than that of poly(acrylic acid) with a M_w of 14000.

Hydrogel of Poly(aspartic acid)

PASP hydrogels have been prepared by the γ-irradiation of PASP produced by thermal polycondensation reactions with an acid catalyst. The effects of the molecular weight of PASP, pH, concentration of PASP in the aqueous solution and dosage of γ-irradiation on the PASP hydrogel preparation were investigated. PASP hydrogels were prepared in the case when the PASP M_w was 95000, the pH of the solution was 7.5 or higher, the concentration of the solution was 5- 10 wt/vol%, the dosage of γ-irradiation was 32 kGy or more and the preparation of PASP solutions was done under N_2. On the other hand, PASP with low-molecular weight (M_w = 15000) could not formed hydrogel by γ-irradiation. The swelling of PASP hydrogels in deionized water and artificial urine was measured. The maximum swelling by deionized water was 3400 g-water/g-dry hydrogel. Biodegradation of the hydrogel was check that the hydrogel showed about 50% biodegradation for 28 days.

Conclusion

The acid-catalyzed homo- and copolycondensation of L-aspartic acid was carried out in various solvents. The high molecular-weight polysuccinimide in high yield was produced using mesitylene/sulfolane solvent system. The hygroscopicity, calcium-ion chelating ability and performance of hydrogel produced by the γ-irradiation of poly(aspartic acid) were measured. The hygroscopicity is higher than L-lactic acid sodium salt. The chelating ability remarkably increases with increasing the content of 6-aminocaproic acid in the copolymers. The highest value was 3 times higher than that of poly(acrylic acid) with a M_w of 14000. The hydrogel showed that the maximum swelling by deionized water was 3400 g-water/g-dry hydrogel and it also showed biodegradability by the activated sludge.

References

(1) Hayashi, T., Nishimura, H., Sakano, K. and Tani, Y., *Biosci. Biotech. Biochem.*, **1994**, *58*, (2), 444

(2) Hayashi, T., Mukouyama, M., Sakano, K. and Tani, Y., *Appl. Environ. Microbio.*, **1993**, May, 1555

(3) Hayashi, T., *Bioscience & Industry*, **1995**, *53*, (6), 521

(4) Kovacs, J., Koenyves, I. and Pusztai, A., *Experientia*, **1953**, *9*, 259

(5) Vegotosky, A., Harada, K. and Fox, S. W., *J. Am. Chem. Soc.*, **1958**, *80*, 3361

(6) Harada, K. *J. Org. Chem.* **1959**, *24*, 1662

(7) Kovacs, J., Kovacs, H. N., Konyves, I., Csaszar, J., Vajda, T. and Mix, H., *J. Org. Chem.*, **1961**, *26*, 1084

(8) Neri, P., Antoni, G., Benvenuti, F., Cocoda, F. and Gazze, G., *J. Med. Chem.*, **1973**, *16*, 893

(9) Koskan, L. P., *US5057597*, **1991**

(10) Koskan, L. P., *WO9214753*, **1992**

(11) Cassata, T. A. and Park, P., *US Patent 5219986*, **1993**

(12) Wood, L. L., *WO9323452*, **1993**

(13) Paik, Y. h., Swift, G. and Simon, E. S., *EP578448 A1*, **1993**

(14) Paik, Y. h., Swift, G. and Simon, E. S., *EP578449 A1*, **1993**

(15) Abe, Y., Matsumura, S. and Takahashi, J., *Oil Chemistry*, **1986**, *35*, (3), 167

Table 1. Polycondensation of L-aspartic acid (ASP) with acid catalyst in various solvent system[a]

Run	Solvent	Acid Catalyst	Yield (%)	M_w (M_w/M_n)
1	toluene	85 %-*o*-phosphoric acid	0	-
2	mesitylene	85 %-*o*-phosphoric acid	77	24800 (1.5)
3	DMF	85 %-*o*-phosphoric acid	0	-
4	sulfolane	85 %-*o*-phosphoric acid	89	19000 (1.5)
5	toluene/sulfolane	85 %-*o*-phosphoric acid	0	-
6	mesitylene/DMF	85 %-*o*-phosphoric acid	59	12900 (1.2)
7	mesitylene/NMP	85 %-*o*-phosphoric acid	96	24500 (1.6)
8	diethylbenzene/sulfolane	85 %-*o*-phosphoric acid	96	49300 (1.9)
9	mesitylene/sulfolane	85 %-*o*-phosphoric acid	96	64300 (1.9)
10	mesitylene/sulfolane	trichloroacetic acid	0	-
11	mesitylene/sulfolane	*p*-toluenesulfonic acid	89	27000 (1.5)
12	mesitylene/sulfolane	sulfuric acid	96	27900 (1.6)
13[b]	---	---	95	9000 (1.8)

[a] ASP, 25 g (0.188 mol); acid catalyst, 0.019 mol; solvent, 80 g or 56 g/24 g; time, 4.5 h; under N_2.

[b] ASP, 100 g (0.752 mol); temperature, 260 °C; time, 7.0 h; under N_2.

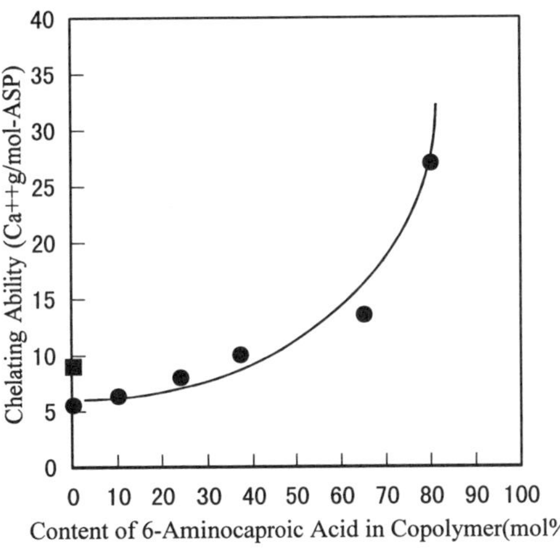

Figure 1. Chelating abilities of copolymer produced by the hydrolysis of copolysuccinimide of L-aspartic acid and 6-aminocaproic acid (●) and poly(acrylic acid) (■)

SUPERPOROUS HYDROGELS:
FAST RESPONSIVE HYDROGEL SYSTEMS

Jun Chen[1], Haesun Park[2], and Kinam Park[2]

[1]Merial Limited, WP 78-110, West Point, PA 19486
[2]Purdue University, School of Pharmacy, West Lafayette, IN 47907

Introduction

Intelligent or smart hydrogels are those which can change their volume rather abruptly upon small changes in environmental conditions, such as temperature[1], pH[2], solvent[3], electric field[4,5], light[6], or pressure[7]. While these smart hydrogels are highly useful in various applications, the typical response time usually ranges from hours to days[8,9], and this slow response time sometimes limits the usefulness of the smart hydrogels. Making smart hydrogels with fast response time would be highly beneficial in various applications. Recently, we synthesized superporous hydrogels which swell fast, in a matter of minutes, regardless of the size. This superporous hydrogels can also be used to make environment-sensitive hydrogels (e.g., temperature-sensitive hydrogels) which swell or shrink much faster than conventional hydrogels.

Experimental

Materials. Acrylic acid (AA), acrylamide (AM), N-isopropyl acrylamide (NIPAM), N,N'-methylenebisacrylamide (Bis), and ammonium persulfate (APS) were obtained from Aldrich Chemical Co. (Milwaukee, WI). Tetramethylethylenediamine (TEMED) was purchased from Bio-Rad Laboratories (Richmond, CA). All the chemicals were used as received. Pluronic® F-127 (PF127) was obtained from BASF Corporation (Parsippany, NJ). Pluronics are triblock copolymers of poly(ethylene oxide) (PEO) and poly(propylene oxide) (PPO) in the form of PEO-PPO-PEO.

Synthesis of superporous hydrogels. Superporous hydrogels were synthesized using various vinyl monomers. Poly(acrylamide-co-acrylic acid) (P(AM-co-AA)) superporous hydrogels were prepared to examine the fast swelling properties. Three types of poly(NIPAM-co-AM) superporous hydrogels were synthesized to examine the temperature-sensitive swelling and deswelling property. The molar ratios of NIPAM to AM in these superporous hydrogels were 9:1, 8:2, and 7:3, and were labeled for convenience as N90, N80, and N70, respectively. In general, to make a superporous hydrogel, monomer, crosslinker, water (if necessary), foam stabilizer, acid, polymerization initiator, initiation catalyst (if any), and foaming agent were added sequentially to a test tube (20 mm outer diameter x 150 mm in length). PF127 was used as a foam stabilizer. The test tube was shaken to mix the solution after each ingredient was added. The pH of the monomer solution was adjusted to 5 (between 4.5 and 5.5) using acrylic acid or hydrochloric acid. Addition of sodium bicarbonate, which was used as the foaming ingredient, increased the pH of the solution to accelerate the polymerization. Synthesized superporous hydrogels were removed from the tube and allowed to swell in water before drying.

Temperature sensitivity and kinetics of swelling and deswelling. Superporous hydrogels were allowed to swell in distilled water. The LCST behavior of the superporous hydrogels was characterized by recording the equilibrium swelling volume when the temperature was increased from 10°C to 75°C at a rate less than 0.2°C/min. The length and diameter of the superporous hydrogels were measured to calculate the volume at different temperatures. The normalized swelling volume was calculated from the ratio of the equilibrium swelling volume at a certain temperature to the equilibrium swelling volume at 10°C. To characterize the deswelling kinetics, the superporous hydrogels (N90) which were fully swollen at 10°C were transferred to 65°C. Their volumes were recorded at timed intervals. The swelling kinetics of the superporous hydrogels were examined by transferring the fully collapsed (i.e., no more size changes) superporous hydrogels at 65°C back to 10°C, and the volumes were also recorded at timed intervals.

SEM study. After superporous hydrogels (N90) reached the equilibrium state at 10°C or 65°C, they were quickly frozen by dropping them into liquid nitrogen and subsequently freeze-dried. The structures of the freeze-dried superporous hydrogels were examined using a scanning electron microscope (JEOL JSM-840).

Results

Fast swelling of dried superporous hydrogels. Superporous P(AM-co-AA) hydrogels were used for the study of kinetics and extent of gel swelling. Fig. 1 shows the swelling kinetics of P(AM-co-AA) hydrogels synthesized using different amount of NaHCO$_3$. Conventional hydrogels made in the absence of added NaHCO$_3$ did not swell to a measurable extent for more than an hour. When 40 mg of NaHCO$_3$ was added to the monomer mixture, the swelling of the dried hydrogels was faster than the control, but it still took about 15 min to swell. When the amount of NaHCO$_3$ was 50 mg and above, superporous hydrogels swelled to their equilibrium sizes in less than 10 min. This is most likely due to the formation of homogeneous open channels throughout the hydrogels. The equilibrium swelling occurred within 6 min as the amount of NaHCO$_3$ increased to 90 mg even when a cylindrical gel was as big as 10 mm diameter and 25 mm height. The fast swelling was accompanied with the high degree of swelling. The equilibrium swelling ratio reached 300 in several minutes without any difficulty as shown in Fig. 1.

Fast response time of superporous hydrogels. The most remarkable property of superporous hydrogels is their ability to response fast to the temperature changes. Fig. 2 shows the fast deswelling and swelling kinetics of the N90 superporous hydrogel at 10°C and 65°C. The dry weight of the superporous hydrogel used in Fig. 2 was 245 mg. When the superporous hydrogel was transferred from 10°C to 65°C, it shrank from the fully swollen state (36 cm^3) to the fully collapsed state (6.5 cm^3) in 72±14 sec. More than 90% of the volume change occurred in less than 30 sec. When transferred back to 10°C water bath, the fully collapsed superporous hydrogel reswelled to 36 cm^3 in 78±15 sec. This cycle was repeated many times without change in the thermoreversible properties of the superporous hydrogel. The time for the superporous hydrogel to shrink to the equilibrium state was faster than that for their swelling, although by only several seconds. The response times of conventional thermosensitive hydrogels were in the order of hours to days, even though the hydrogels had much smaller swollen sizes[8,9].

SEM pictures of N90 superporous hydrogels prepared at 10°C and 65°C. The diameter of most of the pores at 10°C was several hundred micrometers, while it was reduced to about 100 μm at 65°C. The pores were well connected to each other so that water can easily be removed or absorbed through such open channels. In the conventional hydrogels, however, there is no similar structures, and the swelling is governed by the diffusion process through glass polymers which is very slow.

Discussion

Of the many smart hydrogels, thermo-reversible hydrogels have been studied most extensively. Several approaches have been used to reduce the thermo-response time. Reducing the gel size has been the most commonly employed technique to achieve the fast response time. Thin hydrogels in either disc-, film-, or rod-shape usually have very fast response time[1,8-13]. Engineering the polymer structure at the molecular level is another approach. Recently, comb-type grafted hydrogels showed faster deswelling in the range of 20 min[8,10]. The swelling of these come-type hydrogels, however, was still quite slow as other hydrogels since the enhanced hydrophobic interaction resulting in the fast shrinking (or deswelling) of the swollen hydrogels did not work for swelling. Another common approach for shortening the response time is to increase porosity of the hydrogels. Monomers can be polymerized at temperatures above the lower critical solution temperature (LCST)[9,12,14]. Polymerization of monomers at temperature above the LCST can cause phase separation of the polymer, which leads to the porous structures of the synthesized hydrogels. Wu et al.[9] reported the deswelling time on very thin gel film (2 mm) prepared by this method was only a few minutes, but the reswelling time was in the order of hours. Yan et al.[14] reported that the macroporous hydrogel film (4 mm thickness) made by this method showed a large volume change in only 30 sec upon temperature changes from 20°C to 50°C. In this study, however, the gel did not reach the equilibrium state during that time period. All the macroporous hydrogels in the previous studies[1,8-15] had pore sizes less than a few micrometers. The existence of the interconnected channels formed by such pores has been speculated, but never confirmed. In our study, we were able to prepare hydrogels with pores larger than 100 μm, and for this reason, we call them superporous hydrogels. Since the pores maintain open channel structures even after drying, their swelling occurs by capillary wetting rather than by diffusion of water into the glassy matrix of dried gels. This makes the swelling so fast.

Superporous hydrogels have many potential applications. In this study, we have synthesized superporous hydrogels that can respond rapidly to the temperature change regardless of the size. Using the same technique, we can also make other fast responsive polymeric materials that can quickly respond to changes in pH, solvent, light, electric field, or pressure. The fast water-absorbing property of superporous hydrogels is ideal for the applications in baby diapers, sanitary napkins, surgical pads, and others where fast water absorption is beneficial. The fast swelling property is also essential in designing gastric retention devices for long-term oral drug delivery systems. In biotechnology, many enzymes or cells have been immobilized in hydrogels to use them as bioreactors. However, one of the major problems of such systems is the mass transfer resistance of the nonporous hydrogels. By providing much larger surface area and fast mass transport, superporous hydrogel are expected to provide higher reaction rate and faster product recovery. Superporous hydrogels can also be used in restorative surgery where the hydrogels fill the cavity and allow new tissue growth in their porous structures.

References

(1) Dong, L.-C.; Hoffman, A. S. *Jounal of Controlled Release* **1990**, 13, 21-31.

(2) Brannon-Peppas, L.; Peppas, N. A. *Chemical Engineering Science* **1991**, 46, 715-722.

(3) Hu, Y. *Macromolecules* **1993**, 26, 1761-1766.

(4) Rossi, D. D.; Suzuki, M.; Osada, Y.; Morasso, P. *J. of Intell. Mater. Syst. and Struct.* **1992**, 3, 75-95.

(5) Tanaka, T.; Nishio, I.; Sun, S.-T.; Ueno-Nishio, S. *Science* **1982**, 218, 467-469.

(6) Mamada, A.; Tanaka, T.; Kungwatchakun, D.; Irie, M. *Macromolecules* **1990**, 23, 1517-1519.

(7) Lee, K. K.; Cussler, E. L.; Marchetti, M.; McHugh, M. A. *Chemical Engineering Science* **1990**, 45, 766-767.

(8) Kaneko, Y.; Sakai, K.; Kikuchi, A.; Yoshida, R.; Sakurai, Y.; Okano, T. *Macromolecules* **1995**, 28, 7717-7723.

(9) Wu, X. S.; Hoffman, A. S.; Yager, P. *Journal of Polymer Science: Part A: Polymer Chemistry* **1992**, 30, 2121-2129.

(10) Yoshida, R.; Uchida, K.; Kaneko, Y.; Sakai, K.; Kikuchi, A.; Sakurai, Y.; Okano, T. *Nature* **1995**, 374, 240-242.

(11) Kabra, B. G.; Akhtar, M. K.; Gehrke, S. H. *Polymer* **1992**, 33, 990-995.

(12) Kabra, B. G.; Gehrke, S. H. *Polymer Communications* **1991**, 32, 322-323.

(13) Chiklis, C. K.; Grasshoff, J. M. *Journal of Polymer Science: Part A-2* **1970**, 8, 1617-1626.

(14) Yan, Q.; Hoffman, A. S. *Polymer Communications* **1995**, 36, 887-889.

(15) Huang, X.; Unno, H.; Akehata, T.; Hirasa, O. *Journal of Chemical Engineering of Japan* **1987**, 20, 123-128.

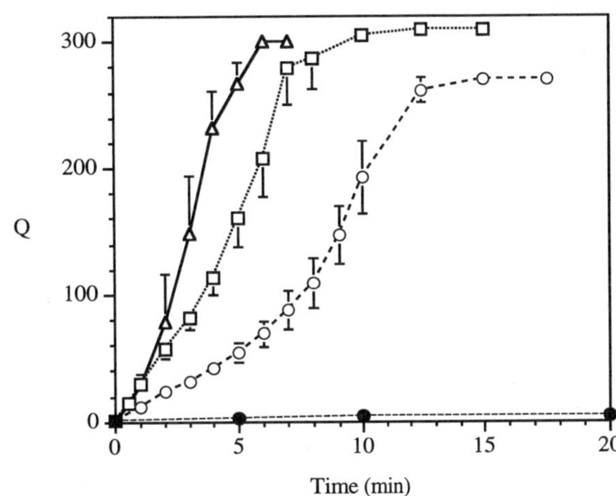

Fig. 1. Effect of the amount of NaHCO$_3$ on swelling kinetics of poly(acrylamide-co-acrylic acid) conventional and superporous hydrogels. Each hydrogel was prepared from 1.0 ml of monomer mix in a 15 mm x 85 m glass tube. The amount of added NaHCO$_3$ was 0 (●), 40 (O), 60 (□), or 90 (Δ) mg. Superporous hydrogels were dehydrated in ethanol first before drying to constant weights. n≥4.

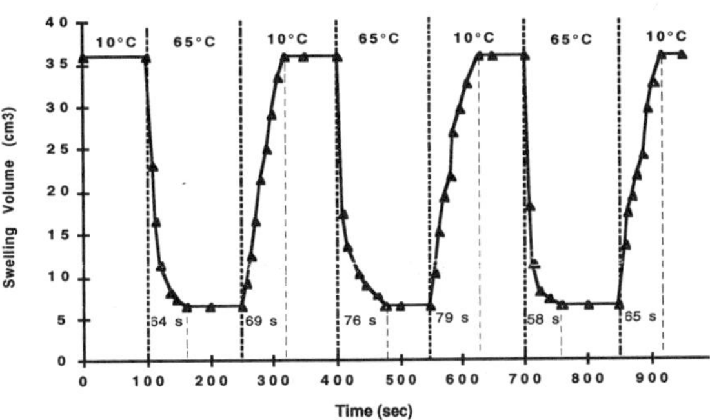

Fig. 2. Fast deswelling and swelling kinetics of the superporous hydrogel (N90) at 10°C and 65°C. The deswelling completed in 72±14 sec, and the reswelling completed in 78±15 sec. The swelling volume at fully collapsed and fully swollen state were 6.5 cm^3 to 36 cm^3, respectively.

CRYOTROPIC GELATION AS AN APPROACH TO THE PREPARATION OF SUPERMACROPOROUS HYDROGELS

Vladimir I.Lozinsky

AN Nesmeyanov Institute of Organoelement Compounds
Russian Academy of Sciences,
Vavilov st.28, 117813 Moscow, Russia

This review communication deals with the description and discussion of general peculiarities and regularities inherent in the cryotropic gelation in polymer systems, the approaches to the preparation of supermacroporous gels by means of similar cryostructuration methods are also considered.

Cryotropic gelation is the specific type of a gel-formation occurring as a result of the cryogenic influence on the initial systems potentially capable of gelling. In this case, the gelation processes proceed during the specimen freezing / frozen storage / thawing. Thus, the major characteristic feature of such gel-formation is the compulsory crystallization of a low-molecular liquid existing in an initial system. This feature differs the cryotropic gelation from the well-known chilling-induced gelation, where gel-formation takes place upon the system cooling, but without the phase transition of a solvent, likewise the instance of the gelation of solutions of gelatin or agar at the decrease in temperature below the sol-gel threshold.

In general, the spatially-structured polymer systems with the immobilized solvent, lyogels, are well-known to be obtained either through the limited swelling of the non-linked polymers, or by the swelling of xerogels, or by the gel-formation of monomer or high-polymer precursors in the medium of a solvent. In this manner one may produce the covalently-linked gels, ionotropic gels, as well as the non-covalent or physical gels. Lyogels of the more complex morphology are usually formed originating from the colloid sols. We have demonstrated that all these kinds of gel-formation can also be realised in the cryogenic variants, carrying out the respective process under appropriate conditions of freezing and thawing. The final products of the cryotropic gelation have been termed as *cryogels*, i.e., the gels formed under specific conditions of the frost-bound specimens.

When the initial solution of the gelling agents (gel precursors) freezes at moderate minus temperatures, the frozen system consists, at least, of two phases: the solvent crystals and the so-called liquid microphase, where solutes are concentrated. Just in this liquid-like inclusions due to the considerable increase in concentration of solutes the gelation phenomena can intensify, resulting in a cryotropic gelation. In fact, a cryogel is formed in a much more concentrated solution of the precursors as compared to the initial one (therefore the effect of the apparent decrease in critical concentration of gelation is usually inherent for the cryotropic gelling, if it is compared with the gel formation at positive temperatures).

After the frozen system thawing, the macropores arise instead of solvent crystals melted, so the crystals act as the porogens. Since these crystals grow until close contact with the surface of another crystal, the macropores in the cryogels are interconnected, their size depends on the solvent nature and regimes of the freezing-thawing treatment. A similar high-porous (very often, supermacroporous) structure is the characteristic feature for these gel materials, which represent, in appearance, turbid cellular gel species, sometimes of a sponge-like morphology. As a result of such a structure the swelling characteristics of the heterogeneous cryogels are distinct from the swelling behaviour of usual macrohomogeneous gels formed at positive temperatures in the medium of the liquid solvent. In addition to the solvate liquid, which is consumed by the concentrated polymer framework of a cryogel upon its swelling, a large amount of a free fluid is absorbed by the numerous microcapillaries in the body of a sample.

The cryoconcentrating significantly effects on the kinetics of the cryogelation causing, at a certain range of negative temperatures, marked acceleration of the gel-formation, in spite of the system is, as a whole physical body, in the solidified state. The temperature dependencies for the rates of similar processes are commonly being of extreme character due to the competition of accelerating and decelerating factors. Besides the influence on the kinetic parameters of the gelation, the decrease in temperature facilitates *per se* the course of exothermic reactions, also promoting the formation of various covalent and thermoreversible cryogels.

The particular examples of preparation of supermacroporous polymer hydrogels considered in this communication are as follows:

(i) The typical instance of the formation of covalently-bound cryogels by the cross-linking of soluble polymer with the low-molecular reagents is exemplified by the production of spongy chitosan cryogels through the cross-linking of high-molecular chains of the polyamino-saccharide with the dialdehydes in the medium of frozen dilute acetic acid. In this case the effect of apparent decrease in the critical concentration of gelation for the process at minus temperature (e.g., at -8°C) as compared to the same system gelation at plus temperatures (e.g., at +2 or +25°C) is clearly observed in respect to both the polymer concentration and the cross-agent added, when, due to the cryoconcentrating phenomena, the cryogels are formed on the basis of considerably less concentrated initial solutions than ordinary homogeneous covalent hydrogels of this polymer.

(ii) Another example of the formation of covalentl sponge-like polymeric hydrogels is their preparation through the free-radical copolymerization of acrylic and *bis*-acrylic monomers in the frozen aqueous medium (initiation with the persulphate/tertiary amine redox system). In this manner the cryogels, e.g., of cross-linked poly(acrylamide) are obtained, their macroporosity is mainly determined by the regimes of a cryogenic treatment: the lower the freezing temperature, the smaller the dimensions of ice crystals formed and, hence, the size of macropores in the gel material prepared. The dependencies for the rate of gelation and gel-fraction yield on the temperature within the range of -10...-30°C for this system appear to be of extreme character, with the highest rate the gelation proceeding in the vicinity of -20°C, but only 5 degrees above and below this point the rate is decreased by several times. The low-temperature quenching technique, which consists in the fast deep freezing (e.g., in a liquid nitrogen) with the subsequent heating of the sample to the pre-assigned moderate minus temperature, results in another macrostructure of the poly(acrylamide) cryogels and differences in the dynamics of gel-formation comparing the usual freezing procedure. The analysis of phase diagrams and NMR studies of the frozen reacting specimens show that the composition of the unfrozen liquid microphase differs significantly from the composition of the initial solution: because of the various solubility of the components at minus temperatures the ratio between the comonomers and the initiators are changed.

(iii) By the radical copolymerization of the frozen aqueous solutions of N,N-diethylacrylamide and N,N'-methylene-*bis*-acrylamide super-macroporous thermoresponsible cryogels are fabricated, and by the alkaline saponification of poly(acrylamide) cryogels pH-responsible cellular gels are prepared. Similar stimuli-responsible cryogels are capable of very fast "answering" of the swelling degree to the variation of ambient conditions, as the mass-transfer of a solvent in the cryogels occurs on the short distances within the very thin walls of the macropores, but the capillary-bound liquid is excreted during the fast collapse, as well.

(vi) The formation of non-covalent (physical) cryogels is illustrated by the example of preparation of poly(vinyl alcohol) cryogels. The freeze-thawing treatment of the concentrated (5-20%) water or DMSO solutions of the polymer gives rise to the viscoelastic macroporous cryogels, whose rheological characteristics and structure are determined by the concentration and properties of PVA used, and also by the regimes of a cryogenic treatment. The cryogelation of PVA proceeds virtually in the course of the thawing of the frozen system. NMR and ESR spectroscopies demonstrate the chain mobility in the frozen system only commencing from about several degrees below the melting point of the frozen solvent, and the temperature range of the most efficient gel-formation being lies 2-3 degrees below this point. Therefore, the slower the thawing rate, the more prolonged time the system resides in this "favourable" temperature range, thereby promoting the formation of more rigid cryogels.

The use of the cryotropic gelation methods as an approach to the preparation of supermacroporous hydrogels opens new opportunities for the creation of polymeric materials possessing very interesting properties. It has been suggested that these materials may be employed as promising biomedical gel materials, as the gel carriers for the immobilization of cells and enzymes, as the gel basis for some types of foodstuffs (the case of cryotropic gelation of edible biopolymers), as the high-porous supports for the heterogeneous catalysts, etc. The implementation of the freezing-thawing technology provides a rather simple way for the construction of the system of interconnected channels (macropores) in the cryogel's body without addition of the auxiliary porogens.

WATER-IN-OIL-EMULSIONS BASED ON RAPESEED OIL

M. Ruebenacker, A. Negele, K.H.Schneider, D.Moench, G. Schornick
BASF Aktiengesellschaft, D-67056 Ludwigshafen

Introduction. Water soluble high molecular weight homo- and copolymers of acrylamide are useful as retention and dewatering aids for paper manufacturing and flocculation aids for wastewater clarification and sludge dewatering. State of the art are solid products as well as liquid products mainly applied as water-in-oil-emulsions. Aliphatic hydrocarbon mixtures with a high content of iso-paraffins are normally used as oil phase. Manufacturers are increasingly confronted with questions about the fate of the mineral oil after usage. Developmental efforts were directed towards replacing mineral oil by vegetable oil with the following advantages in mind: vegetable oil is a renewable resource, biodegradable and has a flash point above 25°C. However, these water-in-vegetableoil-emulsions have to be technically comparable to current market products.

Water-in oil-emulsions contain the polymer in a highly concentrated aqueous solution finely dispersed in an inert oil phase, usually a mineral oil. The emulsion is stabilised by a w/o-emulsifier or an emulsifier mixture having a low HLB-value and being oil soluble. The polymer content of such a w/o-emulsion is in the range of 20 to 50%. The auxiliary material, i.e. oil and emulsifier, is roughly 1/3 of the total formulation. The emulsions have good flowability. A corresponding aqueous solution of such a high molecular weight polymer would be a highly viscous gel difficult to handle. Prior to usage the w/o-emulsion is inverted by a wetting agent already present in the formulation: the emulsion is simply stirred into a large amount of water. The polymer is now in the continouos outer phase.

The auxiliary material is completely introduced into the process water or wastewater which have to undergo treatment. Therefore, the research project was focussing on manufacturing w/o-emulsions with improved environmental compatibility by exchange of the mineral oil against a vegetable oil. A suitable emulsifier system turned out to be the key issue in developing a water-in-vegetable oil-emulsion.

Experimental. The following requirements apply for mineral oil based as well as vegetable oil based emulsions:
- activity/efficiency as flocculation and retention aids
- minimal costs
- low emulsion viscosity
- sedimentation stability
- invertibility & low temperature stability

Vegetable oils are triglycerides differing in composition of fatty acids and degree of unsaturation. Because of cost reasons rapeseed oil, soybean oil and sunflower oil are to be considered. Initially rapeseed oil was chosen due to the lowest content of unsaturated fatty acids.

Compared to normally used mineral oils Shellsoll® D 70 and Shelsoll® D 100 rapeseed oil has a higher flash point. Disadvantages are its higher density and viscosity. The totally different chemistry has a significant impact on the selection of the emulsifier system and on the required amount.

A low cationic copolymer based on 65% of acrylamide and 35% dimethyl- aminoethyl acrylate is used as an example to demonstrate the different steps of the developmental work.

Results and Discussion. Initially Chremophor® WOCE 5115 was used as single emulsifier yielding formulations with equally high polymer and emulsifier content (20%). Adding Span 80 as co-emulsifier the emulsifier content was reduced to 9% by increasing the polymer content to 25%. Although quite efficient the products showed disadvantges in terms of sedimentation stability and content ratio polymer/emulsifier. A new emulsifier system comprising a polymeric emulsifier and Span 80 finally led to water-in-rapeseed oil-emulsions with an emulsifier content of 5,5% and a polymer content of 40%.

These results were achieved by detailed studies of the polymerisation behavior. In a typical process for w/o-emulsions the reaction vessel is charged with the oil phase, i.e. oil and emulsifier. Then the aqueous phase is stirred into the oil phase and polymerised at 50°C using an oil soluble azo initiator. In addition to observing inner and outer temperature recording of torque as a measure for viscosity and recording of conductivity as a measure for phase behavior gave important hints.

As a result the emulsions showed a distinct difference during polymerisation depending on the emulsifier system used. A region of high viscosity is passed through in case of the combination Cremophor® WOCE 5115/Span 80 most likely causing problems during scale-up. This phenomena could be clarified by conductivity measurements. The high conductivity of the outer phase indicated a o/w-emulsion at he beginning of the polymerisation instead of the expected w/o-emulsion. Polymerisation initially starts in the outer phase resulting in a strong increase of viscosity. After conversion of about 5-10% phase inversion occurs resulting in an instant drop in viscosity and conductivity. The viscosity increase can be compensated by a higher emulsifier amount shifting the phase inversion close to polymerisation start. However, this approach will increase material costs. In contrast the polymerisation process using the system polymer emulsifier/Span 80 runs without phase inversion avoiding the previously described problems. A w/o-emulsion exists from the very beginning.

Summary. Standard water-in-oil-emulsions of high molecular weight acrylamide copolymers used as retention and dewatering aids in paper industry and as flocculants in waste water clarification and sludge dewatering exclusively contain mineral oils as oil phase. Key issue of a new environmentally compatible w/o-emulsion with mineral oil replaced by vegetable oil was a suitable and cost competitive emulsifier system. Cationic as well as anionic acrylamide copolymers were obtained under lab and pilot plant conditions by using a polymeric emulsifier. Efficiency of the products as flocculants and retention aids is similar to currently used products with mineral oil phase.

MULTIFUNCTIONAL HYDROPHILIC POLYMERS

Emo Chiellini, Ranieri Bizzarri, Paolo Bonaguidi, Paolo Talamelli and Roberto Solaro

Department of Chemistry and Industrial Chemistry
University of Pisa
via Risorgimento 35, 56126 Pisa
Italy

Introduction

In recent years, the increasing use of proteic drugs endowed with high activity, low stability and significant toxicity, has prompted the development of new dosage forms, such as polymeric release systems able to reduce unwanted side effects while improving the drug bioavailability.[1]

Following our interest in the synthesis and characterization of biodegradable polymeric systems for biomedical and pharmaceutical applications,[2-4] attention has been directed to the preparation of new segmented poly(amide)s derived from tartaric acid and α,ω-diaminooligo(oxyethylene)s. These materials appear to be well suited for the proposed applications due to the presence of natural tartaric acid residues and hydrophilic oligo(ethylene glycol) groups, which are known to elicit an antiopsonizing effect.[5] Moreover, the large number of amide linkages should guarantee for substantial mechanical properties.

However, the hydrolytic stability of amide bonds might constitute a limitation to their use in applications requiring a facile biodegradation. To overcome this possible flaw, the research was also addressed to the preparation of poly(ester-amide)s containing oxyethylene segments, able to join the features of the above poly(tartaramide)s with the ease of hydrolysis of ester bonds.

Parallel to this research line, we have undertaken the development of new functional cyclodextrin derivatives, both monomeric and polymeric, to be used in the stabilization and complexation of proteic drugs.

Experimental

Materials. All reagents were purchased from Aldrich and used without further purification. Oligo(ethylene glycol) diamines having degree of oligomerization 2-6 (**PnAM2**, n = 2-6, the hydrochlorides of monoaminooligo(ethylene glycol)s esterified with succinic acid monopentachlorophenyl ester (**PSPnNH**, n = 1-3, and monoglycidyl ether of xylitol, trimethylolethane, trimethylolpropane, and D-glucose, in which the free hydroxyl groups were reversibly protected by *iso*propylidene groups were prepared according to common synthetic procedures.

Synthesis of poly(tartaramide)s. A solution of 10.9 g (50 mmol) of **DMIPT**, 7.41 g of **P3AM2** and 1.02 g (10 mmol) of triethylamine in 100 ml of anhydrous methanol was heated at 55 °C for 48 h under magnetic stirring. After cooling at room temperature the solution was poured into 800 ml of 1:1 diethyl ether/benzene mixture. The coagulated polymer was dissolved in a small amount of dichloromethane, precipitated in 700 ml of diethyl ether and dried under vacuum to yield 11.9 g of a white solid.

Synthesis of poly(ester-amide)s in solution. A solution of 500 mg (1.0 mmol) of **PSP1NH** in 2.3 ml of anhydrous dichloromethane was placed in a 10 ml glass vial under dry nitrogen atmosphere, then 0.32 ml (3.2 mmol) of anhydrous triethylamine were added at room temperature under magnetic stirring. After 7 days the precipitate was filtered, washed with dichloromethane and dried under vacuum to yield 190 mg of a white solid.

Synthesis of poly(ester-amide)s in bulk. 400 mg of **PSP1NH** were placed in a 5 ml glass vial under dry nitrogen atmosphere, then 0.36 ml (3.6 mmol) of anhydrous triethylamine were added at room temperature under magnetic stirring. After 4 days the precipitate was filtered, washed with dichloromethane and dried under vacuum to yield 106 mg of a white solid.

Functionalization of β-cyclodextrin. A solution of 6.47 g (5.7 mmol) of β-cyclodextrin in 12 ml of 12 % aqueous NaOH was placed in a 50 ml flask, then 12.0 g (56 mmol) of *O*-glycidiy-*O*-*iso*propylidentrimethylolethane were slowly added under vigorous stirring. The stirring was continued for 4 h at 60 °C, then the solution was neutralized with dilute HCl, concentrated under vacuun and extracted with chloroform. The organic extracts were dried over anhydrous sodium sulphate, concentrated under vacuum and poured into 100 ml of diethyl ether. The precipitate was dried under vacuum to yield 10.4 g (56 %) of a white powder.

Polymerization of β-cyclodextrin. A solution of 5.0 g (4.4 mmol) of β-cyclodextrin in 8 ml of 33 % aqueous NaOH was placed in a 50 ml flask, then 3,5 ml (44 mmol) of epichlorohydrin were added under vigorous stirring (600 rpm). The stirring was continued for 3 h at 30 °C, then 100 ml of acetone were added and the solid filtered off. The acetone was removed under vacuum and after addition of 100 ml water the solution was neutralized with 1 N HCl and dialyzed over a 1 kD cut-off cellulose ester membrane. After removal of water under vacuum, 4.5 g (59 %) of a white solid were obtained.

Results and Discussion

Synthesis of oligo(ethylene glicol) diamines. Three monodispersed oligo(oxyethylene) diamines (**P4AM2**, **P5AM2**, and **P6AM2**) were prepared from corresponding glycols having degree of oligomerization 4-6. The commercially available diamines of di and tri(ethylene glycol) (**P2AM2** and **P3AM2**) were also used. The two commercially available dimethyl (R,R)-2,3-*iso*propilidentartrate (**DMIPT**) and dimethyl (R,R)-tartrate (**DMT**) were chosen as diacid monomers (**Figure 1**).

The oligo(oxyethylene) diamines were prepared either by Williamson etherification of oligo(ethylene glycol) di-*p*-toluenesulfonates with potassium 2-aminoethoxide or by a modified Gabriel reaction of oligo(ethylene glycol) dichlorides with potassium phthalimide under phase transfer conditions, followed by hydrazinolysis of the diphthalimido derivatives. The second procedure gave better yields and the work-up of the crude reaction products was much easier than the Williamson synthesis.

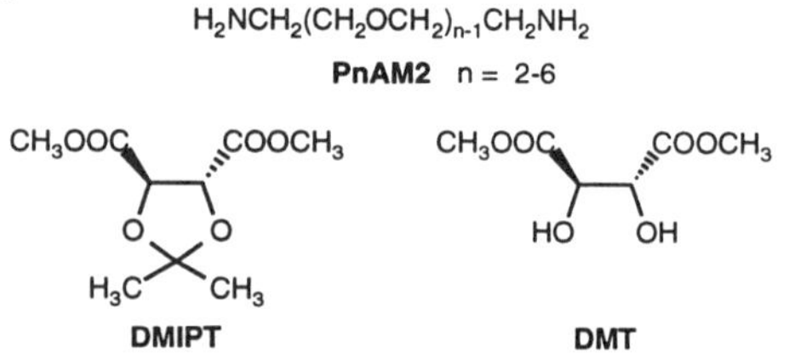

$$H_2NCH_2(CH_2OCH_2)_{n-1}CH_2NH_2$$

PnAM2   n = 2-6

Figure 1. Structures of the monomers used in the synthesis of poly(tartaramide)s.

Synthesis and Characterization of Poly(Tartaramide)s. To select the best experimental conditions, several polymerization experiments of **P3AM2** and **DMIPT** were performed in the presence of either triethylamine or 2-hydroxypyridine as catalyst and by using ethanol or 2-methoxyethanol as solvent, at temperatures included between 55 and 90 °C. The reactions were monitored within time by end group titration. Kinetic data showed that the polycondensation is autocatalytic in nature, at least for conversions lower than 70 %.

Under the adopted conditions, racemization of the chiral centers present in the tartaric acid was lower than 10 % after 72 h. Conversions and molecular weights of final polymeric products showed that methanol/triethylamine/55 °C was the best combination of experimental parameters.

The two sets of poly(tartaramides) **PnITA** and **PnTA** were then prepared starting respectively from **DMIPT** and **DMT** and adopting the best experimental conditions (**Figure 2**).

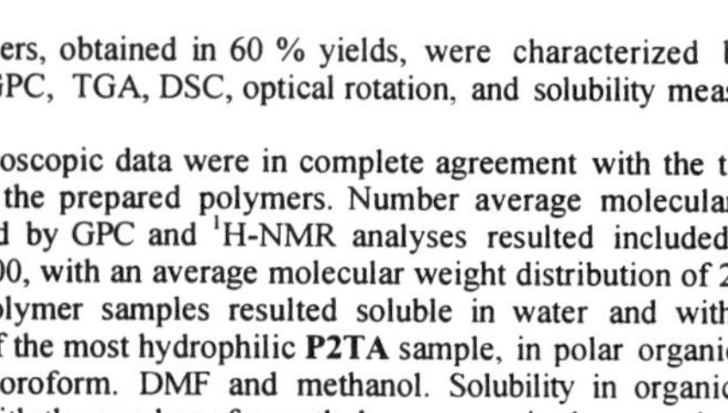

Figure 2. Structures of the prepared poly(tartaramide)s.

Polymers, obtained in 60 % yields, were characterized by FT-IR, ^{1}H-NMR, GPC, TGA, DSC, optical rotation, and solubility measurements (**Table 1**).

Spectroscopic data were in complete agreement with the theoretical structure of the prepared polymers. Number average molecular weights, as evaluated by GPC and ^{1}H-NMR analyses resulted included between 800 and 2000, with an average molecular weight distribution of 2-3.

All polymer samples resulted soluble in water and with the sole exception of the most hydrophilic **P2TA** sample, in polar organic solvents such as chloroform. DMF and methanol. Solubility in organic solvents increased with the number of oxyethylene groups in the repeating units.

In water solution all poly(tartaramide)s exhibited a strong optical rotation, thus confirming the absence of a significant racemization of the tartaric acid residues. The dependence of the molar optical rotation on the length of the oligo(oxyethylene) segment demonstrated a contribution to the molecular chirality by the hydrophilic residues. This effect can be tentatively ascribed to the presence of a partially ordered secondary structure.

Table 1. Characterization of poly(tartaramide)s.[a]

Sample	Conv. (%)	M_n	M_w/M_n	$[\Phi]$[b]	T_d[c] (°C)	T_g (°C)
P2ITA	56	1300	1.9	-54	252	48.2
P3ITA	56	1500	1.2	-81	nd	13.0
P4ITA	60	1200	1.2	-82	210	-3.1
P5ITA	57	2000	3.2	-91	nd	-10.8
P6ITA	65	1000	3.3	-88	250	-28.4
P2TA	60	800	nd	+152	183	63.9
P3TA	65	800	nd	nd	197	33.9
P5TA	60	1200	nd	+206	209	1.7

[a] nd = not determined. [b] Molar optical rotation at sodium D line, at 25 °C, in degree·M⁻¹cm⁻¹. [c] At 1 % thermal decomposition.

Thermogravimetric analysis (TGA) showed that **PnITA** and **PnTA** were stable up to about 200 °C. The presence of a dioxolane ring gave **PnITA** a larger thermal stability as compared to **PnTA**, suggesting that hydroxyl groups play a fundamental role in the onset of the polymer degradation.

DSC analysis of the prepared samples highlighted that the T_g values of **PnITA** are about 10-20 °C lower than those of **PnTA**, as expected from the absence of strongly interacting free hydroxyl groups. In addition, the T_g of **PnITA** decreased from 50 to -30 °C on increasing the number of flexible oxyethylene units from 2 to 6. **PnTA** showed a similar behavior. No crystallinity was detected in any of the polymer samples, very likely because of their low molecular weights.

Synthesis of Activated Ester-Amides. The synthesis of the hydrochlorides of monoamine oligo(ethylene glycol)s esterified with succinic anhydride (**PSPnNH**, n = 1-3), in which the poly(ether) segment is constituted by 1-3 oxyethylene units and the carboxylic group is activated by esterification with monopentachlorophenol, was performed as outlined in **Figure 3**.

Figure 3. Synthesis of ω-amminooligo(ethylene glycol) pentachlorophenyl succinates.

Polymerization experiments were performed at room temperature, both in dichloromethane solution and in bulk, in the presence of a threefold excess of triethylamine. **PSP1NH** and **PSP2NH** monomers resulted insoluble in dichloromethane giving rise to milky suspensions. However, on addition of triethylamine the suspensions turned clear and after 6 day, 92 and 73 % conversions to polycondensation products were observed, respectively. Conversions larger than 70 % were recorded when the same reactions were performed in bulk. In all cases the recovered product resulted soluble in DMSO and insoluble in water and in most common organic solvents.

The molecular weight of the polymerization products could not be determined by GPC analysis, due to their insolubility in common GPC solvents. An average degree of oligomerization of about 9 was however computed by ¹H-NMR analysis. The observed low degree of polymerization can be tentatively attributed to the insolubility of the oligomers in the reaction medium, thus preventing a further increase of the molecular weight.

Polycondensation of **PSP3NH** in dichloromethane solution for 5 days afforded only a 1:3 mixture of the intramolecular cyclization product and low molar mass oligomers. When the polymerization was carried out in bulk, a 54 % conversion to a poly(ester-amide) soluble in water and in polar organic solvents, such as chloroform, DMS, and methanol took place. An average number molecular weight of 21 kD and an average polydispersity index of 2.2 was evaluated by GPC analysis. Also in this case, an appreciable amount of the cyclization product and of low molecular weight oligomers was detected in the polymerization solution.

DSC analysis of the **PSP2NH** and **PSP3NH** polycondensation products evidenced only a well defined glass transition respectively at 7.5 and -7.0 °C, followed by a relaxation endotherm, in accordance with the amorphous nature of the samples. The **PSP1NH** polymer sample exhibited a melting peak at about 210 °C, whereas a glass transition at 17.6 °C was observed only after quenching from 200 to -50 °C. In the TGA curves, the onset of the thermal decomposition increased from 220 to about 270 °C on increasing the length of the oligo(oxyethylene) segment. The different thermal behavior of the investigated poly(ester-amide)s can be related to both the distance among polar amide groups along the polymer backbone and the flexibility of the oxyethylene segments.

Functionalization of β-cyclodextrin. A series of monoglycidyl ether of polyols, such as xylitol, trimethylolethane, trimethylolpropane, and D-glucose, in which the free hydroxyl groups were reversibly protected by *iso*propylidene groups were prepared by reaction of the protected polyols with epichlorohydrine (**EPY**). The glycidyl ethers were then reacted with β-cyclodextrin (β**CD**) to give new β**CD** derivatives soluble in both water and chloroform. Controlled removal of the *iso*propylidene groups allowed to obtain materials characterized by a very large water solubility (> 1 g/ml) and tunable hydrophylic/hydrophobic balance.

Polymerization of β-cyclodextrin derivatives. Native β**CD** and β**CD** derivatives functionalized with the glycidyl ethers of different polyols were polymerized by reaction with **EPY** in aqueous NaOH at room temperature (**Figure 4**). Oligomeric, polymeric and crosslinked products were obtained depending upon stirring speed, NaOH concentration, reaction time and β**CD/EPY** molar ratio.

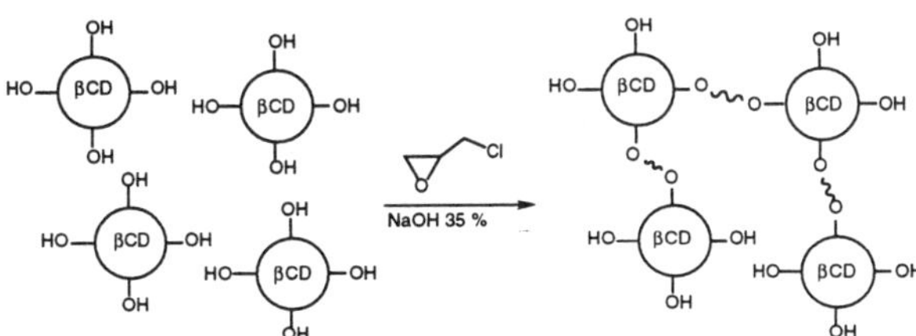

Figure 4. Schematic representation of the polymerization reaction of β-cyclodextrin with epichlorohydrine.

Careful ¹³C-NMR analysis allowed to identify the different products present in the reaction mixtures. Time evolution of the chemical composition of the reaction mixture seems to indicate that reaction of **EPY** with β**CD** primary hydroxyl groups led to the exclusive formation of oligomeric products. Polymerization and eventually crosslinking occurred only after reaction of a significant amount of secondary hydroxyl groups with **EPY**.

Acknowledgements

The partial financial support by CNR and MURST is gratefully acknowledged.

References

(1) Langer, R. *Science* **1990**, 249, 1527.
(2) Chiellini, E.; Bemporad, L.; Solaro, R. *J. Bioact. Compat. Polym.* **1994**, *9*, 152.
(3) Chiellini, E.; Solaro, R. *Adv. Mat.* **1996**, *8*, 305.
(4) Chiellini, E.; D'Antone, S.; Solaro, R. *Macromol. Symp.* **1990**, *123*, 25.
(5) Baley, F. E.; Koleske, J. V. *Poly(ethylene oxide)*, Academic Press, New York (1976).

pH-SENSITIVE HYDROGELS BASED ON CHITOSAN AND D,L-LACTIC ACID

Xin Qu, Anders Wirsén and Ann-Christine Albertsson

Department of Polymer Technology, Royal Institute of Technology
Stockholm S-100 44, Sweden

Introduction

Chitosan derived by N-deacetylation of chitin, is a biodegradable and biocompatible polymer. It has been shown recently that partially deacetylated chitins can be hydrolysed by enzymes[1]. The enzymatic digestibility depends on the degree of N-deacetylation and method of preparation. Therefore, attention has been focused recently to employ chitosan[2,3] to compose hydrogels with specific response to a biological environment. These hydrogels have a potential use in the site-specific delivery of drugs to specific regions of the gastrointestinal (GI) tract due to pH variations throughout the GI tract[4]. Moreover, chitosan itself has some pharmaceutical activities such as antacid, hypocholesterolemic activity and suppression of growth of tumour cells. These excellent properties have attracted many investigators to work in this field[5].

Graft copolymerization is one of the best methods to bring together synthetic and natural polymers in order to retain the good properties of natural polymers such as biodegradation and bioactivity, etc. In this work, lactic acid and water-soluble chitosan with a degree of deacetylation (DD) of 88% were used to synthesize the graft copolymers with hydrophobic, synthetic side chains and hydrophilic, natural main chains by direct polycondensation without using catalysts. These graft copolymers which form pH-sensitive hydrogels in aqueous solution could be used in biomedical applications, such as artificial muscles and/or switches or as a component in biochemical separation systems and controlled-release systems.

Experimental

Materials. Chitosan (CS) (Mw = 70,000) from Fluka (Switzerland) and D,L-lactic acid (LA) (99%) from Lancaster (England) were used for the preparation of CS-PLA graft copolymers. The degree of deacetylation (DD = 88%) of chitosan was determined by the IR spectroscopy method[6]. All these chemicals were used as delivered, without further purification.

Instrumentation. Wide-angle X-ray diffraction (WAXD) measurements were carried out at room temperature by using a D/MAX-YA diffractometer with a CuKa tube, 40kV, 100mA, made by Rigaku Co., Japan. The diffraction patterns were determined over a range of diffraction angle $2\theta = 3° - 40°$.

Degree of substitute(DS) of chitosan amine group was determined by formation of N-salicylidene chitosan[6]. An accurately weighed sample was slurried for 24h in 100ml 0.02M solution of salicylaldehyde in methanol/1% acetic acid aqueous solution (80/20,v/v). After 24h the mixture was filtered, a portion of the filtrate diluted 400 times and the UV absorbance at 255 nm measured to determine the residual concentration of salicylaldehyde(RSA).

The ^{1}H nuclear magnetic resonance (NMR) spectrum was recorded with a Bruker AC-400 FT-NMR spectrometer at 400 MHz. The chloroform-extracted graft copolymers were slowly dissolved in D_2O with stirring for three days. The operating temperature was 300K.

Elemantal analysis(EA). The content of C, H and N was determined for virgin and grafted chitosan samples. In this case the total nitrogen content was determined according Dumas with a Carlo Erba NA 1500 instrument. All analyses were performed by Mikro Kemi AB, Uppsala Sweden.

Preparation of CS-PLA hydrogels. Chitosan powder (1 g) dispersed in water (40 ml) was dissolved by adding predetermined amounts of lactic acid. Each solution was poured onto a Teflon dish (15.0 cm diameter) and dried first at 80°C for 4 hours to avoid evaporation of the monomer during the polymerization of LA. The chitosan lactate films formed were then under vacuum (6×10^{-2} mmHg) at 80°C for 3 hours to promote dehydration of the chitosan-poly(lactic acid copolymer (CS-PLA) salts with formation of the

corresponding amide linkages. In order to remove the homopolymers or CS-PLA salts formed during the grafting reaction, all samples were extracted with methanol in a Soxhlet apparatus for 24 hours. The thickness of the membranes used in this work was 0.17 ± 0.02 mm.

Swelling of CS-PLA hydrogels. To determine the effect of pH on hydrogel swelling, McIlvaine buffer with the same ionic strength, I = 0.5M, at various pH was used in this work [Citric acid / Na_2HPO_4 (pH 2.2-8.0)], and [0.5M NaOH / KCl (pH 12.0)]. The weight of solution absorbed by the gel was calculated from the weights of the gel before and after vacuum-drying at room temperature. The specific water or solution content was expressed by the following equation:

$$\text{Specific solution content} = (W_s - W_d)/W_d$$

where W_s and W_d are the weights of the samples in the swollen (W_s) and dry states (W_d), respectively.

Results and Discussion

Chitosan dissolved in aqueous lactic acid solution to give a homogeneous viscous polyelectrolyte solution. By heating the solution, dehydration of the chitosan amino lactate salt will occur to form amide groups between the chitosan and D,L-lactic acid, the polycondensation of lactic acid occurring at the same time[7]. Methanol was chosen as an extractant because it could remove PLA oligomers and CS-PLA salt links, as well as unreacted monomer. Chloroform-extracted samples without PLA homopolymers inside but containing salt bound side chains could partly dissolve in water, which allows the PLA side chain length to be measured by ^{1}H NMR while assuming that the salt and amide linked side chains have the same length. The degree of substitution of the available chitosan amino groups increases with the LA to CS ratio approaching a saturation level of about 16 to 17% for LA/CS≥2.(**Table 1**) Higher degrees of substitution can possibly be attained by increasing the reaction time and temperature in the amide forming dehydration step as long as undesirable degradation of the CS-PLA graft copolymers does not occur.

CHI (g),	Lactic acid (g)	LA / NH$_2$ (mol/mol)	yield (g) CCl$_3$H-extracted	DP of PLA side chains weight,	^{1}H-NMR,	Degree of substitute (%) RSA	EA
1.0	0.5	1.05	1.35	1.04	1.35	6	2.3
1.0	1.0	2.10	1.67	2.03	1.54	12	13.1
1.0	2.0	4.19	2.20	3.94	2.30	14	17.6
1.0	3.0	6.29	2.78	4.47	3.82	17	16.8
1.0	4.0	8.38	3.12	5.20	4.27	16	16.2

Table 1. Graft Copolymerization of D,L-lactic Acid onto Chitosan

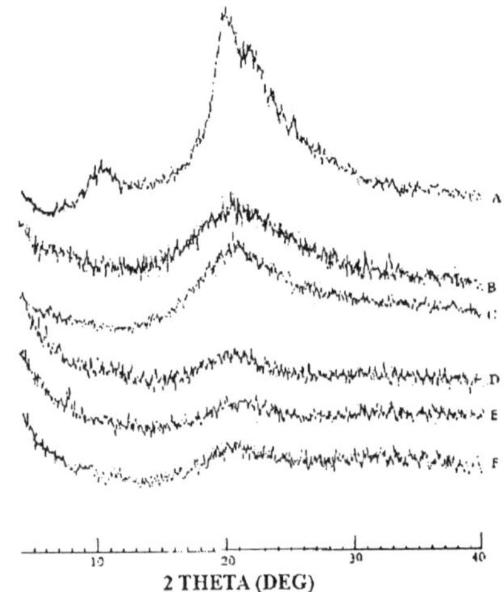

Figure 1 shows the X-ray patterns of chitosan and graft copolymers with different ratios. These WAXD patterns show that the grafting decreases the intensity at both peaks. When the weight ratio of the starting components reaches CS/LA = 1/2 (**Figure1D**), the graft polymerized samples became totally amorphous. Since D,L-lactic acid reacts with chitosan in a homogeneous solution, the grafting by PLA will take place at random along the chain, giving rise to a random copolymer. This will efficiently destroy the regularity of packing the original chitosan chains which results in the formation of totally amorphous copolymers.

These samples could form hydrogels in aqueous solutions and change their swellability according to the buffer pH. No chemical crosslinking will occur in this grafting reaction. So we can assume that swellability but insolubility of the graft copolymers is due to the physical crosslinking through hydrophobic side chains aggregation and/or intermolecular interactions through hydrogen bonds with hydrophilic main chains, which would lead to a corresponding decrease of the polymer chain mobility (see **Figure 2**). We have noticed in the literature that by combining different amounts of hydrophobic monomers with a natural substrate such as cellulose and gelatin[3], pH sensitive hydrogels with different degrees of swelling were obtained. These characteristics, together with the demonstrated biocompatibility offer excellent systems to be used as support for controlled drug delivery formulations.

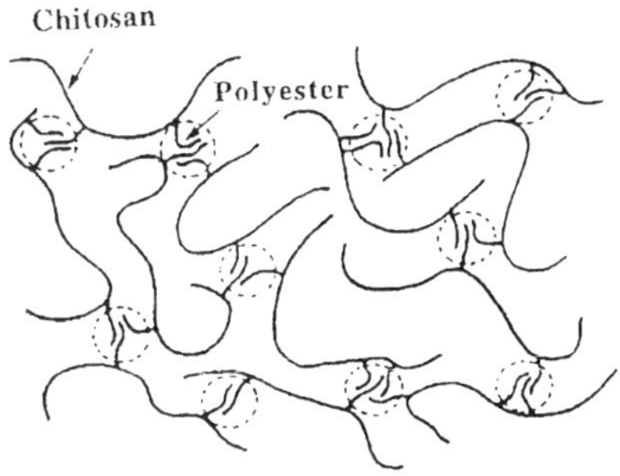
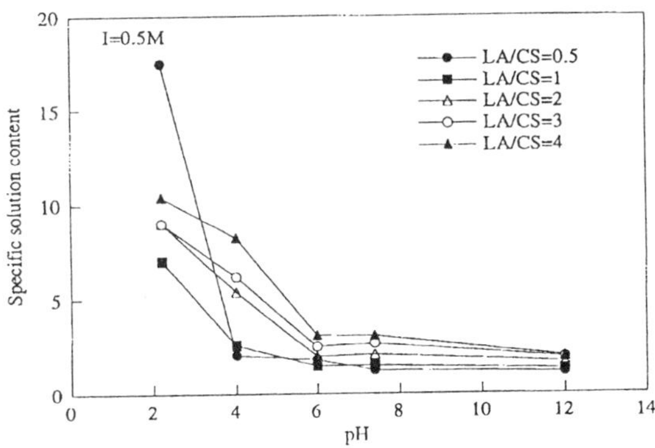

Figure 2. Structure of hydrogels in aqueous solutions.

Figure 3. Specific solution content of hydrogels as a function of buffer pH.

The effect of LA/CS weight ratio and pH value on the swellability of hydrogels are shown in **Figure 3**. The samples were allowed to imbibe buffer solutions with different pH values from 2.2 to 12.0. The ionic strength of buffers were kept constant (I=0.5M), since it will largely affect the swellability of hydrogels as mentioned. The hydrogels show lower specific solution content at basic pH than acidic pH. It is well known that a high

concentration of charged ionic groups in the gel increases swelling due to osmosis and charge repulsion. Thus, when the degree of ionisation of gel bound groups is decreased, swelling is decreased. At high pH, the degree of ionisation is reduced due to the deprotonation of the amine units of chitosan. The hydrophobic side chains aggregation and/or intermolecular interactions through hydrogen bonds with hydrophilic main chains are regenerated, leading to lower swellability of samples. When the buffer pH value is higher than 7.0, the swellabilities of all samples are similar because the amine groups have already reached their maximum protonation.

Figure 4 shows the reversibility of swelling of the hydrogel between pH2.2 and pH7.4 buffers, respectively. The film was brought into a swelling equilibrium at pH7.4 for 1 h and then transferred to a acetic acid solution at pH2.2 so that an abrupt swelling was ensured. After that, the sample was placed back into the pH7.4 buffer. This procedure was repeated several times and the specific solution content was measured as a function of time. The results demonstrated that the sample changes its ability to absorb water when the environmental pH is altered, the time for swelling is much shorter than deswelling of hydrogel. It could remain the reversibility toward the pH even after the prolonged treatment (overnight). This swelling transition is appealing since it implies that water permeability, which should increase with sample hydration, can also be converted in response to a change in environmental pH. It would be a desirable characteristic for a pH-sensitive controlled-release system.

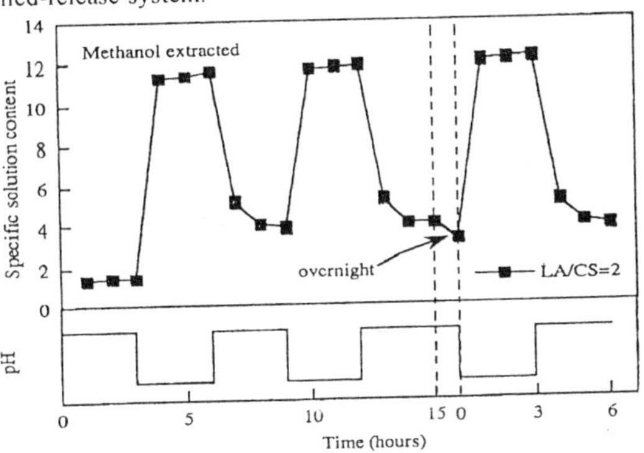

Figure 4. Specific solution content of hydrogel (LA/CS=2) as a function of time under repeated abrupt change of pH between 7.4 and 2.2.

Conclusions

The pH-sensitive and biodegradable hydrogels was synthesized by grafting D,L-lactic acid to chitosan. The formation of hydrogels is explained with hydrophobic polyester side chains interactions serving as pseudocrosslinks, which stabilize hydrogel-forming molecules against permanent deformation in buffer. The specific solution content of hydrogels decreased when the pH value of buffers were increased, and this change of swellability is reversible. These pH-sensitive hydrogels have potential use in biomedical application, such as controlled-release systems.

References
(1)   Martin G. Peter, J.M.S.–Pure Appl. Chem., **1995,** A32(4), 629.
(2)   Guan Y.L., et. al., J. Appl. Polym. Sci., **1996,** 61, 2325.
(3)   Hoffman, A.S., et.al., ACS PMSE Preprints, **1997,** 76(1).
(4)   Choi H.J., et.al., J. Appl. Polym. Sci., **1995,** 31(4) 807.
(5)   Hong X.Y., et.al., J.M.S.-Pure Appl. Chem., **1996,** A33(10) 1459.
(6)   Roberts G.A.F., Chitin Chemistry, Mcamillan, Houndmills, **1992.**
(7)   Zhang X.C. et.al., J.M.S.-Rev. Macromol. Chem. Phys. **1993,** 6(5) 411.

RHEOLOGICAL PROPERTIES OF HYALURONIC ACID BASED SOLUTIONS

L. Ambrosio, A. Borzacchiello, P.A. Netti, L. Nicolais

Institute of Composite Materials Technology-CNR and Interdisciplinary Research Center on Biomaterials (CRIB) University of Naples "Federico II", Piazzale Tecchio 80, 80125 Naples, Italy.

Introduction

Hyaluronic Acid (hyaluronan, HA) is present in all soft tissues of higher organism, and in particulary high concentrations in the synovial fluids and vitreus humor of the eye. Along with fulfilling structural roles related to its lubricanting and water retaining properties; HA plays an important role in a number of biological processes such as cell mobility and cell-cell interactions (1). Solutions of purified of high molecular weight HA demonstrate interesting rheological properties making them very actractive for viscosupplementation and viscosurgery applications. In order to improve the short residence time of HA and to explore the possibility to use this material in other biomedical field, chemical modification of HA by cross-link and coupling reactions has been obtained. The majority of researchers have used crosslinking reactions to alter the physico-chemical properties and coupling reactions to alter the biological properties. One exception is an esterification technology which permits the controlled modification of the physico-chemical properties of HA through a coupling reaction.

In the present study, solution of HA at different molecular weight, ester of HA and cross-linked HA at different level of crosslinking were investigated to determine the rheological properties.

Materials

Hyaluronic acid at different molecular weights (150.000 and 1.200.000 daltons), benzyl esters of hyaluronic acid (Hyaff11p50, 50% degree of esterification, M.W.=150.000 daltons) and ACP (Auto Crosslinked Polysaccharides), were supplied by Fidia Advanced Biopolymers S.p.A., Italy. The degree of esterification of Hyaff11p50 is determined only by the quantity of alkylating agent added.

The ACP are inter- and intra-molecular esters of HA in which part of the carboxyl groups are esterified with hydroxyl groups belonging to the same molecule and/or different molecules of the polysaccharide thus forming a mixture of lactnes and intermolecular ester bonds.

The unique feature of this crosslinking technology with respect to alternative approaches to crosslinking HA (2) is that no bridge molecules are present between the crosslinked HA chains. This ensures that only the natural components of HA are released upon degradation of ACP.

In the present study, the following derivatives, synthesized from HA of 200 kDa, are considered: ACP-5, ACP-15.The numbers reflect the relative level of crosslinking although they are not a direct measure of the percentage of carboxyl groups bound to hydroxyl residues. All the materials are white powders and give rise, upon hydration, to water adsorbing transparent gels.

Bidistilled water was added to the powders in order to obtain different weight concentrations 20 and 40 mg/ml for Hyaluronic acid and Hyaff 11p50 and 10 mg/ml for every degree of crosslinking of ACPs.

Methods

The rheological properties of Hyaluronic Acid and derivatives were evaluated on a Bohlin VOR Rheometer (Bohlin Reologi A B, Lund, Sweden) at controlled temperature of 25 °C. The measuring system was cone and plate (CP 5/30 cell).

The non-linear flow properties of the investigated materials were evaluated through steady shear measurements to determine the viscosity η as function of shear rate $\dot{\gamma}$,while the small-amplitude oscillatory shear experiments allowed the measurement of the unsteady response of the samples and hence the determination of their linear viscoelastic

properties. Moreover this technique is very useful for the determination of structure-mechanical properties relationships (3,4).

In this dynamic experiment the material is subjected to a sinusoidal shear strain :

$$\gamma = \gamma_0 \sin(\omega t)$$

where γ_0 is the shear strain amplitude, ω is the oscillation frequency and t the time. The mechanical response expressed as shear stress τ of viscoelastic materials is intermediate between an ideal pure elastic solid (obeying to the Hooke's law) and an ideal pure viscous fluid (obeying to the Newton's law) and therefore is out of phase respect to the imposed deformation as expressed by:

$$\tau = G'(\omega)\, \gamma_0 \sin(\omega t) + G''(\omega)\, \gamma_0 \cos(\omega t)$$

where $G'(\omega)$ is the shear storage modulus and $G''(\omega)$ is the shear loss modulus. G' gives information about the elasticity or the energy stored in the material during deformation, whereas G" describes the viscous character or the energy dissipated as heat .

Strain sweep tests at fixed oscillation frequency (consisting in monitoring the viscoelastic properties while logarithmically varying the strain amplitude γ_0) were previously performed on these materials to determine the strain amplitudes at which linear viscoelasticity is valid.

Results and discussion

Steady shear experiments were performed in order to evaluate the effect of the molecular weight, chemical modification and concentrations of Hyaluronic acid, Hyaff 11p50 and ACP on the shear viscosity.

Low molecular weight hyaluronic acid and Hyaff11p50 at low concentration exhibit essentially Newtonian characteristics. Pseudoplastic behaviours (shear thinning) are observed for Hyaluronic acid with high molecular weight, Hyaff11p50 at high concentration (40 mg/ml) and ACP for which the viscosity increases as function of concentration and degree of crosslinking.

The rheological analysis allows the quantitative evaluation of the viscous (G") and elastic (G') response of these materials and the correlation of these values to concentration, the molecular weigth, and chemical modification. Hyaluronic acid having low molecular weight and Hyaff11p50 at low concentration exhibit viscous behaviour G">G' in the exhamined frequency range. ACP and Hyaff11p50 at high concentration (40mg/ml) show an essentially elastic behaviuor; in fact frequency sweep tests show that the dynamic elastic modulus G' is higher than the dynamic viscous modulus G" in the examined frequency range (10^{-2}-10^1 Hz). High molecular weight hyaluronic acid shows dual behaviour (Figures 1, 2), at low shear rate the solution presents viscous behaviour G"> G' while at high shear rate an elastic behaviour G'>G" are observed.

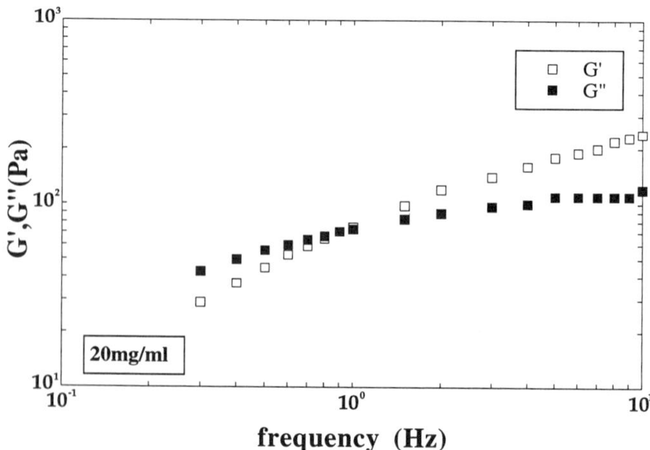

Fig.1: Mechanical spectra of Hyaluronic Acid (M.W. 1.200.000, 20mg/ml)

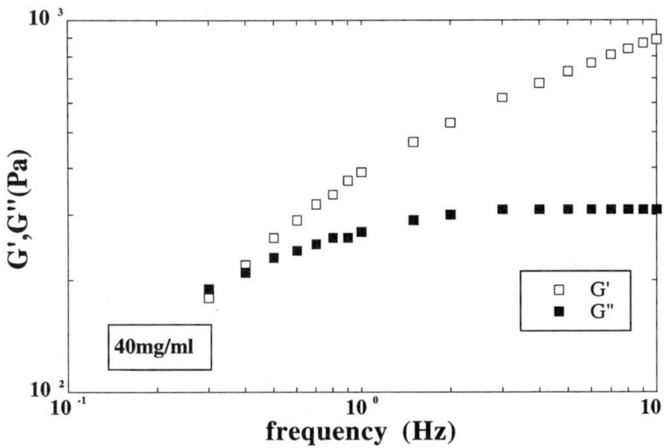

Fig.2: Mechanical spectra of Hyaluronic Acid (M.W. 1.200.000, 40mg/ml)

The "transition point" depends upon the concentration.

The solid-like behaviour of ACP is typical of "weak gel" because the elastic and viscous moduli are slightly frequency dependent; in fact , especially at higher degrees of crosslinking, G' and G" maintain a constant value and are parallel each other all over the frequency range (5,6).

This mechanical behaviour is different from that typical of HA solutions where at low frequencies the viscous character prevails, whereas at higher frequencies the elasticity overcomes the energy dissipation.

Moreover ACP's show higher elasticity than HA solutions and does not flow easily under moderate stresses, so permitting higher residence times.

In the case of ACP's the overall elastic response of the material is due to the contributions from permanent covalent crosslinks obtained through chemical reaction and from transient crosslinks, mainly due to physical interactions and entanglements formed by free polymer chains.

Conclusions

The rheological analysis permits to evaluate viscosity and the viscous and elastic response of hyaluronic acid solution and its derivative and correlate these parameters to the concentration, molecular weight, and chemical modification.

The results showed the possibility to modulate the rheological properties of HA based solution in a wide range by chemical modification without altering the biocompatibility of the HA molecules providing interesting informations to design proper substitute for a specific biomedical applications.

References

1. E.A. Balazs , P.A. Bland, J.L Denlinger, A.I. Goldman, N.E. Larsen, E.A. Leshchiner, A. Leshchiner And B. Morales, *Blood Coagulation and Fibrinolysis* **2** (1991) 173.

2. United States Patent N. 4713488 (1987)

3. M. Doi in "Materials Science and Technology, Vol. 12, Structure and Properties of Polymers", edited by E.L. Thomas, (VCH, Weinheim, 1993),p.389.

4. J. D. Ferry, in "Viscoelastic properties of polymers" (J. Wiley, New York, 1970).

5. S.B. Ross-Murphy in "Polymer Gels. Fundamentals and Biomedical Applications" edited by D. De Rossi, K. Kajiwara, Y. Osada and A. Yamauchi (Plenum Press, New York, 1991) p. 21.

6. M. Mensitieri, L. Ambrosio, L. Nicolais, D. Bellini, M.O'Regan, J. of Materials Science: Materials in Medicine, 7 (1996), 695-698.

Glycolipid Containing Polyacrylate and Polyacrylamide Copolymers

Kirpal S. Bisht, Arthur C. Watterson and †Richard A. Gross

Department of Chemistry, University of Massachusetts Lowell, One University Avenue, Lowell MA 01854; †Polytechnic University, Herman Mark Chair, Six Metrotech Center, Brooklyn, N.Y. 11202.

Introduction

Sophorolipids are microbial extracellular glycolipids produced by resting free cells of *Torulopsis bombicola*.[1] First described by Gorin *et al.*[1] in 1961, sophorolipids occur as a mixture of macrolactones and free acid structures that are acetylated to various extents at the primary hydroxyl sophorose ring positions (Figure 1).[1]

Our attention to sophorolipids arose due to their interesting structures and potential application in a wide variety of fields. Sophorolipids are routinely prepared in our laboratory by fermentation of *Torulopsis bombicola* on glucose/oleic acid mixtures.[2] Incorporation of these interesting biosurfactants in polymers provides an opportunity to develop new families of water soluble and dispersible products bearing amphiphilic side chains. Such glycolipids as polymer side chains have considerable potential for use as pharmacologically active agents.[3,4]

Considering the complex structure of sophorolipids, their modification to monovinyl compounds that will be polymerizable to linear polymers via traditional free-radical mechanisms presents considerable challenges. Chemical synthetic routes to the target monovinyl sophorolipid monomer would involve tedious protection/deprotection chemistry. To circumvent these difficulties, selective enzyme transformations were developed in this work.

This paper describes the regioselective synthesis of 6-O-acryl sophorolipid, its homopolymerization, and copolymerizations of this monomer with acrylic acid and acrylamide. The formation of an unnatural sophoromacrolactone and the monovinyl analogue was accomplished by using the lipase Novozym 435 from *Candida antarctica*

Experimental Section
Materials

General Chemicals and Procedures. All chemicals and solvents were of analytical grade and were used as received unless otherwise noted. Zeolite, gift from Rohm & Hass Co., was dried over P_2O_5 in an oven dessicator (0.1 mm Hg; 38 h; 50 °C) prior to use. Sophorolipids were synthesized by fermentation of *T. bombicola* on glucose/oleic acid mixtures following a literature procedure.[2] Prior to their use, sophorolipids were dried over P_2O_5 in a dessicator (0.1 mm Hg; 38 h; room temperature).

Porcine pancreatic lipase (PPL) Type II Crude (activity = 61 units/mg protein) and *Candida rugosa* lipase (CCL) Type VII (activity = 4570 u/mg protein) was obtained from Sigma Chemical Co. The lipases PS-30, AK and MAP-10 from *Pseudomonas cepacia*, *Pseudomonas fluorescens* and *Mucor javanicus*, respectively, were obtained from Amano enzymes (USA) Co., Ltd. (specified activities at pH 7.0 were 30000 u/g, 20000 u/g and 10000 u/g, respectively. Immobilized lipases from *C. antarctica* (Novozym-435) and *Mucor miehei* (Lipozyme IM) were gifts from Novo Nordisk Bioindustrials, Inc. All enzymes, prior to their use, were dried over P_2O_5 (0.1 mm Hg, 25 °C, 16 h).

Nuclear Magnetic Resonance (NMR). ¹H-NMR spectra were recorded using Bruker ARX-250 and DRX-500 spectrometer at 250 and 500 MHz, respectively. ¹³C-NMR spectra were recorded at 62.9 or 125.7 MHz with chemical shifts in ppm referenced relative to TMS at 0.00 ppm.

Mass Spectrometry Instrumentation. Mass analyses were performed using a Bruker Biflex™ MALDI-TOF (Bruker Instrument Inc., Bellerica, MA, USA).

Other Instrumental Measurements. Optical rotation measurements were made using a Perkin Elmer 241 digital polarimeter. Infrared Spectra were recorded using a Perkin Elmer FT-IR Spectrometer 1760 X. Samples were dissolved in THF and deposited as thin film on sodium chloride optical disks.

Synthesis of methyl 17-L-([2'-O-β-D-glucopyranosyl-β-D-glucopyranosyl]-oxy)-*cis*-9-octadecenoate (SL-Me, 1): In a typical reaction, to a 100 mL round bottomed flask equipped with a reflux condensor 10 g of dry crude sophorolipid, 100 mL 0.022 N methanolic solution of sodium methoxide were added. The reaction assembly was protected from atmospheric moisture by a $CaCl_2$ guard tube. The reaction mixture was refluxed for 3 h, cooled to room temperature (30 °C), and acidified using glacial acetic acid. The reaction mixture was concentrated by rotoevaporation and poured with stirring into 100 mL of ice cold water when the sophorolipid methylester precipitated as white solid. The precipitate was filtered, washed with ice-water and lyophilized (8.77 g, yield 95.0 %) : $[\alpha]^{25}_D$ -9.77 (c = 0.0145 g/mL) (lit.[5] $[\alpha]^{25}_D$ -17.6 (c = 4.0 g/mL in pyridine); IR (cm⁻¹) 3354 (1.7%), 2929 (2.0%), 2860 (3.4 %), 1741 (4.4 %), 1165 (4.8 %), 1074 (2.0

%), 1030 (3.2 %); ¹H NMR (500 MHz, CD₃OD) δ 1.28 (3H, d, 7.0 Hz, H-18), 1.38 (14H, brs, H- 4-7 & -12-14), 1.48 (2H, m, H-15), 1.62 (4H, p, 7.0 Hz, H-3 &-16), 2.04 (4H, dt, 7.0 Hz, H-8 &-11), 2.34 (2H, t, 7.5 Hz, H-2), 3.23 - 3.38 (4H, m, H-2",-4",-4', &-5'), 3.40 (1H, t, 8.4 Hz, H-3"), 3.48 (1H, t, 8.4 Hz, H-2'), 3.58 (1H, t, 8.4 Hz, H-3'), 3.65-3.73 (3H, m, H-6'a,-6"a, & -5"), 3.68 (3H, s, OCH₃), 3.80-3.96 (3H, m, H-6'b, -6"b & -17), 4.47 (1H, d, 7.0 Hz, H-1'), 4.66 (1H, 6.7 Hz, H1") and 5.38 (2H, m, H-9 &-10); ¹³C NMR (125.8 MHz) δ 20.89, 25.02, 25.26, 27.12, 27.17, 29.13, 29.16, 29.22, 29.39, 29.86, 29.90, 29.95, 33.85, 36.80, 50.99, 61.82, 62.13, 70.56, 70.85, 74.89, 76.76, 76.83, 77.24, 77.31, 77.76, 81.12, 101.69, 103.75, 130.45, 130.60, 175.04; MALDI-TOF *m/z* 659.84 (M + Na)⁺.

Novozym 435-catalyzed synthesis of 17-L-([2'-O-β-D-glucopyranosyl-β-D-glucopyranosyl]-oxy)-*cis*-9-octadecenoic acid 1', 6"-lactone (2). In a representative example, using a glove bag and dry argon to maintain an inert atmosphere, 1.5g of methyl 17-L-([2'-O-β-D-glucopyranosyl-β-D-glucopyranosyl]-oxy)-*cis*-9-octadecenoate (1), 2.2 g of zeolite and 2 g of Novozym 435 which had previously been dried in a vacuum dessicator (0.1 mm Hg, 25 °C, 16 h) were transferred to an oven dried 100 mL round bottomed (RB) flask. Dried and distilled THF, 50 mL was then added and the RB flask was immediately stoppered. The RB flask was then placed in a constant temperature oil bath maintained at 35 °C for 96 h and the contents were stirred magnetically. A control reaction was set up as described above except Novozym 435 was not added. Progress of the reaction was followed by TLC ($CHCl_3$: MeOH :: 7:3). The reaction was quenched by removing the enzyme and zeolite by vacuum filtration (glass fritted filter, medium-pore porosity), the enzyme was washed 3-4 times with 5 mL portions of THF, and the filtrates were combined and solvent removed *in vacuo* to give 1.45 g of the product. The crude biotransformation product (1.45 g) was purified by column chromatography over silica gel (100g, 130-270 mesh, 60 Å, Aldrich) using a gradient solvent system of chloroform-methanol (2 mL/minute) with increasing order of polarity to give 1.3 g of purified product: $[\alpha]^{25}_D$ -4.25 (c = 0.0141 g/mL); IR (cm⁻¹) 3360 (50.1 %), 2927 (44.0 %), 2855 (52.0 %), 1735 (53.1 %), 1458 (65.5 %), 1375 (64.4 %), 1174 (58.9 %), 1076 (39.3 %); ¹H NMR (500 MHz, CD₃OD) δ 1.24 (3H, d, 7.0 Hz, H-18), 1.38 (14H, brs, H- 4-7 & -12-14), 1.42 (2H, m, H-15), 1.62 (4H, p, 7.0 Hz, H-3 &-16), 2.08 (4H, dt, 7.0 Hz, H-8 &-11), 2.36 (2H, t, 7.5 Hz, H-2), 3.23 - 3.48 (4H, m, H-2",-4",-2', &-5'), 3.48 (2H, m, H-3' & -5"), 3.58 (2H, m, H-3" & -4'), 3.7 (1H, m, H-6'a), 3.85 (2H, m, H-6'b & -17), 4.20 (1H, m, H-6"a), 4.42 (1H, m, H-6'b), 4.49 (1H, d, 7.0 Hz, H-1'), 4.52 (1H, 6.7 Hz, H1") and 5.38 (2H, m, H-9 &-10); ¹³C NMR (125.8 MHz) δ 20.71, 24.70, 25.47, 26.15, 27.21, 27.88, 28.21, 28.66, 28.97, 29.57, 29.72, 29.99, 33.99, 36.49, 61.76, 63.38, 70.12, 70.40, 74.69, 75.24, 76.58, 76.67(double), 77.05, 83.87, 101.11, 105.41, 129.78, 130.29, 174.30; MALDI-TOF *m/z* 627.95 (M + Na)⁺.

17-L-([2'-O-β-D-glucopyranosyl-β-D-glucopyranosyl]-oxy)-*cis*-9-octadecenoic acid 1', 6"-lactone 6'-acrylate (3). Synthesized from **2** by Novozym 435 catalyzed acrylation with vinyl acrylate. 1.3 g of **2**, 2.0 g of zeolite, and 3 g of Novozym 435 which had previously been dried in a vacuum desiccator (0.1 mm Hg, 25 °C, 16 h) were transferred to an oven dried 100 mL round bottomed (RB) flask. Dried and distilled THF, 30 mL was then added and the RB flask was immediately stoppered. 2.5 mL of Vinyl acrylate was then added and the RB flask was then placed in a constant temperature oil bath maintained at 35 °C for 96 h and the contents were stirred magnetically. The reaction setup secluded from the light with black paper. A control reaction was set up as described above except Novozym 435 was not added. Progress of the reaction was followed by TLC ($CHCl_3$: MeOH :: 9:1). The reaction was quenched by removing the enzyme and zeolite by vacuum filtration (glass fritted filter, medium-pore porosity), the enzyme was washed 3-4 times with 5 mL portions of THF, and the filtrates were combined and solvent removed *in vacuo* to give 1.5 g of the product. The crude biotransformation product (1.5 g) was purified by column chromatography over silica gel (50g, 130-270 mesh, 60 Å, Aldrich) using a gradient solvent system of chloroform-methanol (2 mL/minute) with increasing order of polarity to give 1.2 g of purified product: $[\alpha]^{25}_D$ -2.81 (c = 0.0116 g/mL); IR (cm⁻¹) 3378 (49.6 %), 2928 (42.8 %), 2854 (50.1 %), 1735 (39.8 %), 1710 (52.0 %), 1457 (63.4 %), 1410 (61.9 %), 1354 (60.3 %), 1295 (55.3 %), 1185 (49.6 %), 1077 (29.7 %); ¹H NMR (250 MHz, CDCl₃) δ 1.22 (3H, d, 7.0 Hz, H-18), 1.38 (14H, brs, H- 4-7 & -12-14), 1.42 (2H, m, H-15), 1.66 (4H, p, 7.0 Hz, H-3 &-16), 2.08 (4H, dt, 7.0 Hz, H-8 &-11), 2.36 (2H, t, 7.5 Hz, H-2), 3.23 - 3.70 (8H, m, H-2"- 5"& -2'-5'), 3.78 (1H, q, H-17), 4.10-4.64 (6H, m, H-6', -6", -1' & -1"), 5.38 (2H, m, H-9 &-10), 5.92 (1H, dd, 10.1 & 2.0 Hz, COCH=CH₂cis), 6.25 (1H, dd, 17.0 & 10.0 Hz, COCH=CH₂), and 6.44 (1H, dd, 17.0 & 2.0 Hz, COCH=CH₂trans); ¹³C NMR (125.8 MHz) δ 20.66, 24.67, 25.48, 26.12, 27.15, 27.88, 28.18, 28.67, 28.92, 29.47, 29.61, 29.92, 33.97, 36.64, 63.41, 63.88, 70.18, 70.67, 73.99, 74.69, 75.34, 76.69, 76.92, 77.00, 83.57, 101.24, 105.28, 128.42, 129.77, 130.25, 130.58, 166.58, 174.80; MALDI-TOF *m/z* 681.90 (M + Na)⁺.

Synthesis of Poly (17-L-([2'-*O*-β-D-glucopyranosyl-β-D-glucopyranosyl]-oxy)-*cis*-9-octadecenoic acid 1', 6"-lactone 6'-acrylate). In a small flask, a solution of 250 mg of **3** in 3 mL of DMF was prepared (freeze/pump/thaw cycles), and 0.1 % (w/v) AIBN was added. The flask was then placed then kept in an oil bath maintained at 65 °C. The polymerization was continued for 16 h. The reaction was terminated by precipitating the polymer in ethyl acetate, and the white precipitate was washed with acetone to yield 200 mg (% yield) of the polymer.

Copolymerization of Acrylic acid and 3. The polymerization reaction was conducted under similar conditions as was described above for homo-polymerization of **3**. Compositional copolymers were generated by varying the monomer feed ratio, Acrylic acid : **4**, 1:1; 1:9; 1:19; 1:32.3; and 1:0.

Copolymerization of Acrylamide and 3. The polymerization reaction was conducted under similar conditions as was described above. Compositional copolymers were generated by varying the monomer feed ratio, Acrylamide : **3**, 1:1; 1:19; 1:32.3; 1:99 and 1:0.

Results and Discussion

Sophorolipids produced by the yeast *T. bombicola* grown on a mixture of oleic acid and glucose were a mixture of at least eight different compounds. The products exist mainly in the lactonic and acidic forms having acetate groups at the 6' and 6" positions (Fig.1). The strategy that was developed for the site-selective incorporation of an acryl group in the sophorolipid molecule is shown in Scheme 1. The methyl ester of the sophorolipid (SL-Me, **1**) was synthesized by refluxing the sophorolipid mixture with an alcoholic solution of sodium methoxide. The optically active methyl ester, $[\alpha]^{25}_D$ -9.77, was isolated in 95 % yield. This product was soluble in anhydrous THF and was identified on the basis of its MALDI-TOF mass spectrum [(M+Na$^+$] at m/z 659.84] as well as detailed NMR spectral data. SL-Me was described earlier[5] but its complete spectral data was not reported. The NMR spectrum of **1** was complex and chemical shifts were found from 1.2 to 5.5 ppm (Experimental Section). The assignments shown are derived from a ^{1}H-^{1}H COSY 45 NMR spectrum (not shown).

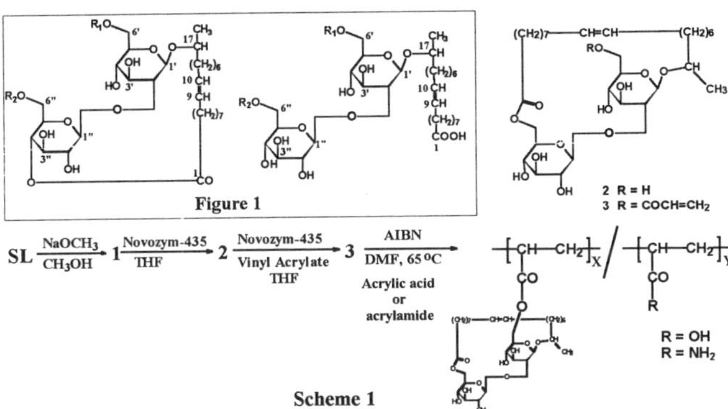

Figure 1

SL $\xrightarrow[\text{CH}_3\text{OH}]{\text{NaOCH}_3}$ **1** $\xrightarrow[\text{THF}]{\text{Novozym-435}}$ **2** $\xrightarrow[\substack{\text{Vinyl Acrylate}\\\text{THF}}]{\text{Novozym-435}}$ **3** $\xrightarrow[\substack{\text{DMF, 65 °C}\\ \\ \text{Acrylic acid}\\ \text{or}\\ \text{acrylamide}}]{\text{AIBN}}$

R = OH
R = NH$_2$

Scheme 1

The methyl ester was reacted with vinyl acrylate (excess) in dry THF. The ability of the lipases PPL, CCL, PS-30, AK, MAP-10, Novozym-435 and Lipozyme IM to catalyze this transformation was evaluated. Of all the lipases screened, Novozym-435 was found to be the biocatalyst of choice.

Surprisingly, compound **2** was isolated ($[\alpha]^{25}_D$ –4.25, m/z 627.95 (M + Na)$^+$) while attempting to synthesize the monoacryl ester (molar ratio of the SL-Me to vinylacrylate was 1:1). The compound was purified from the unreacted methyl ester by column chromatography. In the proton NMR spectrum of the compound **2** there was a significant difference in the appearance of the region 3.25 to 4.5 ppm due to the carbohydrate protons. Assignments to resonances were made from a ^{1}H-^{1}H COSY NMR spectrum. The two doublets due to the H 1' and 1" at 4.49 and 4.52, respectively, were also closer to each other than in the NMR spectrum of the methyl ester. Moreover, the ^{1}H-NMR spectrum of compound **2** did not indicate any acrylation. Also, the proton ^{1}H-NMR spectrum of **2** had no resonances corresponding to the methyl ester group. These anomalous features of the proton NMR of **2** suggested the formation of a lactone ring between the carboxylic acid end group of the fatty acid chain with one of the hydroxyl groups of the sophorose ring. The ^{1}H-NMR spectrum of **2** also showed a 0.5 ppm down field shift in the resonance position of C-6 protons suggesting participation of a C-6 hydroxyl in the formation of the lactone ring. However, conclusive determination of the position of the hydroxyl group on the sophorose ring that took part in the lactone ring formation was considerably difficult due the complexity of the ^{1}H-NMR spectrum.

The ^{13}C-NMR spectrum of **2** was also complex and not fully interpretable using only this information. Hence, the ^{13}C-NMR of **2** was edited by a DEPT-135 pulse sequence. This permitted resolution of the resonances due to the methyl, methylene, and methine carbons. The DEPT-135 spectra of **1** and **2** were compared. A downfield shift of 1.5 ppm in the resonance position of the 6" carbon that was accompanied by an upfield shift of 1.3 ppm in the resonance position of 5" carbon suggested the formation of the lactone at the 6" carbon. Interestingly, the formation of the lactone was accompanied by a downfield shift of 3.0 ppm in the resonance position of the carbon 2'. Similar observations were also made based on the theoretical calculation of chemical shift positions for **2** using ACD/C NMR lab software (version 1.0).

The Novozym-435 catalyzed acrylation of lactone **2** using vinyl acrylate in dry THF was conducted to give **3** ($[\alpha]^{25}_D$ -2.81, m/z 681.90 (M + Na)$^+$). The ^{1}H-NMR shifts at 5.92 (1H, dd, 10.1 & 2.0 Hz, COCH=CH_{2cis}), 6.25 (1H, dd, 17.0 & 10.0 Hz, COCH=CH$_2$), and 6.44 (1H, dd, 17.0 & 2.0 Hz, COCH=CH_{2trans}) confirmed monoacrylation. In addition, the ^{1}H-NMR spectrum of **3** showed a 0.7 ppm downfield shift in the resonance position of the methylene on carbon 6' when compared to the ^{1}H-NMR spectrum of **2**. The position of acrylation in **3** was conclusively identified from comparisons with the corresponding ^{13}C-NMR spectrum of **2**. A downfield shift of 1.5 ppm in the resonance position of the 6" carbon together with an upfield shift of 1.5 ppm in the resonance position of carbon 5" confirmed that the acrylation occured at the 6" hydroxyl position.

The homopolymerization of **3** and it's copolymerization with acrylic acid or acrylamide was conducted by using AIBN as the catalyst in dry DMF. The copolymer composition was adjusted by controlling the monomer feed ratio. Polyacrylamide and polyacrylic acid based copolymers were prepared having 1 to 10 mol % of **3**. The solution properties of these novel polymers are currently under evaluation in our laboratories. Comparison of the ^{1}H NMR spectra of **3** and it's corresponding homopolymer clearly reveals the disappearance of the acryl protons (see Figure 2). Analysis by ^{13}C-NMR of the diad and triad sequences for acrylic/3 and acrylamide/3 copolymers suggests that random repeat unit sequences were formed.

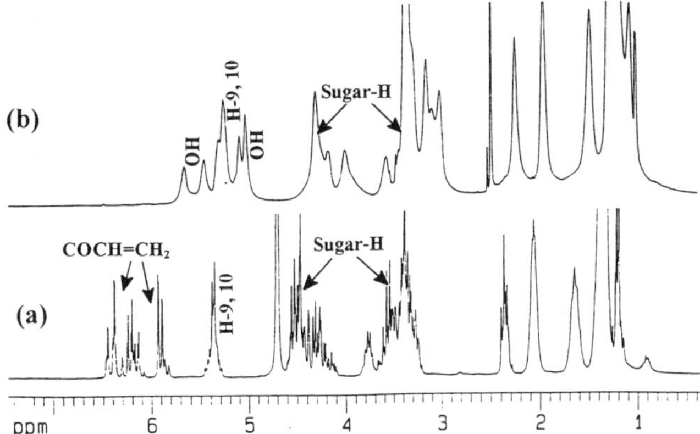

Figure 1. ^{1}H NMR spectrum of **3** (a) in CDCl$_3$, and its homopolymer (b) in DMSO d$_6$.

Conclusions. In conclusion, we have successfully carried out enzymatic transformations on sophorose lipid with remarkable regioselectivity. To the best of our knowledge, compounds **2** and **3** are new. Also, this is first report of any attempt towards incorporation of these biosurfactant analogs into polymers. The copolymer structures generated having amphiphilic side chain groups will be of great interest in solution studies that will follow. It is also noteworthy to point out that this work made use of a chemoenzymatic approach, taking the best enzymatic and chemical methodology.

Acknowledgement. We are grateful for the financial support of this work by the NSF/EPA Partnership for Environmental Research in the area of Technology for a Sustainable Environment (Grant # CTS-9525091).

References

1. Gorin, A.P.J.; Spenser, J.F.T.; Tulloch, A.P. *Can. J. Chem.* **1961**, *39*, 846.
2. Asmer, H.J.; Lang, S.; Wagner, F.; Vray, V. *J. Am. Oil Chem. Soc.* **1988**, *65*, 1640.
3. Nishio, K.; Nakaya, T.; Imoto, M. *Makromol. Chem.* **1974**, *273*, 3319.
4. Carpino, L.A., Ringsdrof, H.; Ritter, H. *Makromol. Chem.* **1976**, *177*, 1631.
5. Tulloch, A. P.; Hill, A.; Spencer, J. F. T. *Can. J. Chem.* **1968**, *46*, 3337.

CURING BEHAVIOR OF POLYPEPTIDE MOISTURE-RESISTANT ADHESIVES

*Miaoer Yu and Timothy J. Deming**

Departments of Materials and Chemistry
University of California, Santa Barbara
Santa Barbara, CA 93106

Introduction

The prolific adhesive capabilities of many marine organisms are well known. These glues are able to form permanent bonds in a few seconds to a wide variety of substrates with complex and often irregular surface coatings. In contrast, the success of synthetic adhesives in wet environments requires carefully cleaned adherends which often must also be chemically treated and/or partially dried.[1] Furthermore, preliminary immunological studies on marine adhesive proteins revealed that they are poor antigens and thus excellent candidates for *in vivo* medical applications.[2] An understanding of the materials and mechanisms used by mussels and barnacles to adhere to underwater surfaces would be valuable for the design and synthesis of superior moisture-resistant adhesives.

Natural adhesive precursor protein has been extracted from the blue mussel, *Mytilus edilus*, and, when this material was spread on culture plates and oxidized with mushroom tyrosinase, it formed an adhesive which could be used for cell attachment and growth.[3] It has also been shown to be effective as an adhesive for the bonding of wet tissue samples.[4] The adhesive protein was found to be non-toxic, however the toxicity of the enzymatic oxidizing agent was problematic. Using an alternative approach, synthetic DOPA containing polypeptides have been reported by Yamamoto and coworkers. They synthesized DOPA homopolymer as well as sequence-specific copolymers of DOPA with L-lysine and L-glutamic acid.[5] They have also reported random L-lysine/L-tyrosine copolymers[6] and the synthesis of random copolypeptides which contain as many as 18 different amino acids, including DOPA.[7] Studies on the crosslinking and adhesive properties of these polymers were limited. Initial adhesive studies were performed using iron and Al_2O_3 adherends with no oxidizing agent.[8] More detailed studies focused on L-lysine/L-tyrosine random copolymers and complex random copolymers where tyrosinase enzyme was used as the oxidizing agent.[6] The adhesive systems were studied in water and/or diluted synthetic seawater and were found to form adhesive bonds, although conversion of tyrosine residues to DOPA was inefficient. Successful oxidation was achieved only when the polymer chains were cleaved with chymotrypsin.

Scheme 1. Synthesis of Adhesive Copolypeptides.

Our efforts in this area are focused on the design and synthesis of simplified adhesive polymers which incorporate only the essential functional components of the marine proteins. We especially wanted to test the premise that functionality, and not amino acid sequence, was the only feature necessary for moisture-resistant adhesion. Using the known compositions of the natural adhesive proteins, we prepared sequentially random copolypeptides through copolymerization of a few select α-amino acid N-carboxy anhydrides (NCAs) (Scheme 1). NCAs are readily prepared from amino acids by phosgenation and can be polymerized into high molecular weight polypeptides via successive ring opening addition reactions which liberate carbon dioxide.[9] NCA polymerizations allow the preparation of multigram quantities of polymer, and the monomers readily copolymerize allowing copolymer composition to be easily varied. We also wanted to study the oxidation of these polymers in detail since this chemistry is a key feature

of the DOPA functionality. Thus we oxidized our copolymers under a variety of conditions and evaluated the effect of oxidant on both crosslinking and adhesive capabilities. The role of oxidation in these adhesives has not received much attention in the past: the complexity of the full-length proteins and ambiguities associated with oxidation of tyrosine residues to DOPA and/or quinone has hindered characterization of the oxidized polymers. Our functionally simple copolymers allowed us to make strong correlations between oxidation and physical behavior.

Experimental

General. Tandem gel permeation chromatography/light scattering (GPC/LS) was performed on a Spectra Physics Isochrom liquid chromatograph pump equipped with a Wyatt DAWN DSP light scattering detector and Wyatt Optilab DSP. Separations were effected by 10^5Å, 10^4Å, and 10^3Å Phenomenex 5μm columns using 0.10 M LiBr in DMF as eluent. NMR spectra were recorded on a Varian Gemini 200MHz spectrometer. A Mettler TA3000 thermal analysis system was used for TGA measurements on the polymers. All IR samples were prepared on NaCl plates and IR spectra were recorded on a Perkin Elmer model 1615 FTIR. Optical rotations were measured using a JASCO P1020 Digital Polarimeter equipped with a sodium lamp source. Elemental analyses were performed by the Marine Science Institute Laboratory at the University of California, Santa Barbara. Hexanes and THF were distilled from sodium/benzophenone in an inert atmosphere and stored under nitrogen.

Amino Acid N-carboxyanhydrides. N_ϵ-carbobenzyloxy-L-lysine N-carboxyanhydride, **1**, was prepared by using phosgene following literature procedures.[10] O,O'-dicarbobenzyloxy-L-DOPA N-carboxy-anhydride, **2**, was prepared from N,O,O'-tricarbobenzyloxy-L-DOPA and phosphorus pentachloride.[11]

Poly(N_ϵ-carbobenzyloxy-L-lysine$_4$-O, O'-dicarbobenzyloxy-L-DOPA$_1$). To a mixture of **1** (1.1 g, 3.7 mmol) and **2** (0.46 g, 0.93 mmol) in THF (15 mL) in a Schlenk tube, 0.1N sodium *t*-butoxide in THF (0.11 mL, 0.019 mmol) was added with stirring. The mixture was stirred for one day at room temperature, then two days at 40 °C and finally four hours at 80 °C. The product was precipitated by addition of ether and then dried *in vacuo* at room temperature overnight to give white polymer (1.22 g, 87%). GPC: M_n = 186,000, M_w/M_n = 1.35. FTIR 1772 cm^{-1} (vCO, vs), 1722 cm^{-1} (vCO, vs), 1652 cm^{-1} (Amide I, s br), 1543 cm^{-1} (Amide II, s br). $[\alpha]_D^{20}$ (DMF, c = 0.005) = +13.2.

Poly(L-lysine$_4$·HBr-L-DOPA$_1$). To a solution of **7** (1.1 g, 3.7 mmol) in TFA (~10 mL), four equivalents of 33% HBr in acetic acid (w/w) was added with stirring. The mixture was stirred for one hour. The product was precipitated by addition of ether and then dried. The crude polymer was dissolved in small amount of water-methanol (1:4) and precipitated by addition of ether and dried *in vacuo* at room temperature overnight. The polymer was dissolved in water and isolated by freeze-drying as a white fluffy solid (0.74 g, 3.6 mmol 97%). The composition of the polymer was estimated by ^{1}H NMR: DOPA content found = 19 mole %. $[\alpha]_D^{20}$ (H_2O, c = 0.005) = -39.7.

Tensile Shear Measurements. <u>Adherend preparation.</u> Aluminum adherends (5052-H32) were treated with a mixture of water, H_2SO_4 (conc.) and $Na_2Cr_2O_7$ (40:20:4) at 65-70 °C for 20 min. They were then rinsed with deionized water and air dried. Steel adherends (A366) were polished with sandpaper (220 grit), and then rinsed with hexane followed by acetone. Poly(methyl methacrylate) adherends were used without further treatment after removal of the protective paper coating from the clear plastic surface. Glass adherends were prepared by attaching glass slides to aluminum test pieces using epoxy resin, and the exposed glass surfaces were cleaned with acetone.

<u>Test sample preparation.</u> The adhesive polymer solution (40 mg/100μL solvent, unless otherwise noted) was spread on both adherend slides, which were then overlapped with two Cu wires (0.06 mm diameter) placed as spacers between the adhesive joint. The overlapped samples were clamped together for 3 hours to prevent motion and kept in a temperature controlled oven for the specified time period.

<u>Tensile shear strength measurements.</u> The tensile strength was measured at room temperature with an Instron 1123 Mechanical Testing Apparatus according to the ASTM D1002 method.[12] An Instron 5000 Lb reversible load cell was used for the measurements. *Data acquisition & control version 3.00* (© 1994 University of California Santa Barbara) was

used to monitor the data output. The adherend size was 4 in x 1 in, the loading rate was 0.05 in/min, bond line thickness was 0.06 mm, and the area of the bond was 0.39 in^2. Three samples were measured for each experiment and the average of these values was reported. All samples failed cohesively within the polymeric adhesive bond. The average standard deviation of these measurements was 9% of the average values. In general, bonds with higher shear strength had smaller standard deviations (*ca.* 4%), while weaker bonds had greater deviations (*ca.* 15%).

Results and Discussion

We have prepared simple copolypeptides of L-lysine and DOPA containing different compositions of the two monomers. We wanted to incorporate DOPA directly into the polymer to avoid complications associated with the oxidation of tyrosine residues. L-Lysine was the other major component of our copolymers since it (i) is present in large quantities in marine adhesive proteins, (ii) is thought to be involved in protein crosslinking reactions,[13] and (iii) should provide good water solubility to the copolymers when the side chain is ionized. We prepared homopolymers of L-lysine and DOPA as well as a series of copolymers of varying compositions and molecular weights. Data for the side-chain protected and deprotected polymers are given in Table 1.

Composition	Yield	M_n	M_w/M_n	H$_2$O soluble
Poly(CBZ-lysine)	94 %	147,000	1.42	-
Poly(lysine•HBr)	97 %	103,000	1.45	+
Poly(CBZ-lysine₄-DiCBZ-DOPA₁)	87 %	186,000	1.35	-
Poly(lysine•HBr₄-DOPA₁)	97 %	113,000	1.39	+

Table 1. Data for Example Polypeptides.

Di-CBZ-DOPA NCA was found to only form short chains in low yield when polymerized using sodium *tert*-butoxide initiator. As CBZ-L-lysine NCA polymerized very efficiently, this suggested that the steric crowding around the DOPA monomer resulted in poor homopolymerization. DOPA, however, was found to incorporate well in copolymerization reactions: yields were near quantitative and no short homopoly(DOPA) oligomers were formed. ^{1}H NMR and UV/vis quantification of the DOPA catechol functional groups confirmed that the bulk copolymer compositions were essentially equal to the comonomer feed compositions. Deprotection of the copolypeptides was accomplished using HBr in acetic acid. Reprotection of a deprotected lysine/DOPA copolymer with excess benzyl chloroformate followed by GPC analysis showed that there was only limited peptide chain cleavage in the deprotection step: molecular weight of the polymer after reprotection with benzyl chloroformate was *ca.* 75% of the original value. The white copolymers were all found to be soluble in aqueous buffers over a wide pH range (*ca.* 2 -12). The DOPA containing copolymers were stable in air for days under acidic and neutral conditions, but discolored rapidly in basic environments. Thermal analysis of the polymers showed negligible weight loss under N$_2$ up to *ca.* 200 °C. The polypeptides could be stored in a ⁻40 °C freezer indefinitely with no discoloration or change in properties.

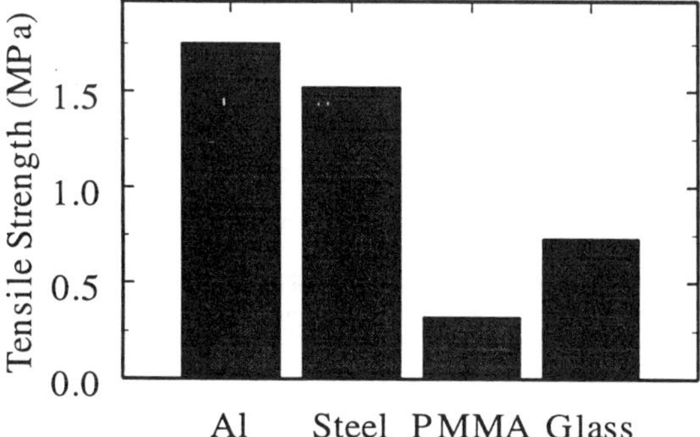

Figure 1. Adhesive Strength of Poly(lysine•HBr₄-DOPA₁) on Different Substrates.

Lap shear tensile adhesive measurements were performed using the L-lysine/DOPA binary copolymers on a variety of substrates. The survey of adherends included aluminum(5052-H32), steel (A366), plastic (PMMA), and borosilicate glass (Figure 1). The copolymers were also tested under a variety of oxidizing conditions. If allowed to cure for a sufficient period of time (1 day), all oxidizing conditions gave adhesive bonds on a given susbtrate of nearly equivalent strength, which was proportional to the amount of DOPA in the copolymer. Copolymer containing 20% DOPA formed adhesive bonds nearly ten times stronger than those formed with the control, pure poly-L-lysine. When the adhesives were allowed to cure for shorter periods of time, the choice of oxidant became significant. The use of chemical oxidizing agents (e.g. hydrogen peroxide) was found to be superior to simple aerobic oxidation.

In addition to copolymer composition and oxidant, we also studied the effects of cure temperature, copolymer molecular weight, and copolymer concentration on adhesive strength. All of these parameters had siginficant effects on adhesive bond formation and ultimate strength. The results of these experiments will be presented, as well as guidelines for preparations of effective adhesive bond formulations of these copolymers.

Conclusions

We have reported the synthesis of copolymers of DOPA and L-lysine which display moisture-resistant adhesive properties when suitably oxidized. The choice of oxidant was partially substrate specific, since in some cases, as with steel adherends, no external oxidant is required. On other substrates, different oxidants can be chosen to satisfy particular requirements for curing time. Generally, hydrogen peroxide was found to be the most convenient and versatile oxidant when fast curing is required. Our system consisting of synthetic polypeptides and chemical oxidants thus provides an alternative to natural marine adhesive proteins for formation of adhesive bonds in water. The advantages of our system include the incorporation of inexpensive oxidizing agents and the ready availability of large quantities of adhesive polymer of consistent quality. Furthermore, through adjustment of copolymer composition, molecular weight, or curing conditions, the adhesive properties of these synthetic copolypeptides can be readily tuned for specific applications.

Acknowledgment. This work was supported by an Office of Naval Research Young Investigator Award (No. N00014-96-1-0729, awarded to T.J.D.), University of California Santa Barbara Materials Department start up funds, and partially supported by the MRSEC program of the National Science Foundation under award No. DMR-9632716.

References

(1) (a) Hull, D. Introduction to Composite Materials, Cambridge Solid State Series, Cambridge University Press, Cambridge, **1981**. (b) Houwink, R.; Salomon, G. Adhesion and Adhesives, Elsevier, New York, **1967**.

(2) Green, K. "Mussel Adhesive Proteins" in Surgical Adhesives and Sealants, Siera, D. H. and Saltz, R. Eds.; Technomic, Lancaster, **1997**.

(3) Benedict, C. V.; Picciano, P. T. *ACS Symp. Ser. #385*, **1989**, 465-483.

(4) Saez, C.; Pardo, J.; Gutierrez, E.; Brito, M.; Burzio, L. O. *Comp. Biochem. Physiol.*, **1991**, *98B*, 569-572.

(5) (a) Yamamoto, H.; Hayakawa, T. *Polymer*, **1978**, *19*, 1115-1117. (b) Yamamoto, H.; Hayakawa, T. *Macromolecules*, **1983**, *16*, 1058-1063. (c) Yamamoto, H.; Hayakawa, T. *Biopolymers*, **1982**, *21*, 1137-1151. (d) Yamamoto, H.; Hayakawa, T. *Biopolymers*, **1979**, *18*, 3067-3076.

(6) (a) Yamamoto, H.; Kuno, S.; Nagai, A.; Nishida, A.; Yamauchi, S.; Ikeda, K. *Int. J. Biol. Macromol.*, **1990**, *12*, 305-310. (b) Nagai, A.; Yamamoto, H. *Bull. Chem. Soc. Japn.*, **1989**, *62*, 2410-2412.

(7) (a) Yamamoto, H.; Nagai, A. *Mar. Chem.*, **1992**, *37*, 131-143. (b) Yamamoto, H.; Nagai, A.; Okada, T.; Nishida, A. *Mar. Chem.*, **1989**, *26*, 331-338.

(8) Yamamoto, H. *J. Chem. Soc. Perkin Trans. I*, **1987**, 613-618.

(9) Kricheldorf, H. R. α-Aminoacid-N-Carboxyanhydrides and Related Heterocycles, Springer-Verlag, New York, **1987**.

(10) Bodanszky, M.;Bodanszky, A. The Practice of Peptide Synthesis, 2nd Ed. Springer: Berlin Heidelberg, **1994**.

(11) Yamamoto, H.; Hayakawa, T. *Biopolymers*, **1977**, *16*, 1593-1607.

(12) ASTM Standards on Adhesives, 5th Ed., Philadelphia, PA, **1961**.

(13) Holl, S. M.; Hansen, D.; Waite, J. H.; Schaefer, J. *Arch. Biochem. Biophys.*, **1993**, *302*, 255-258.

CELL ADHESION PROPERTIES OF AMPHIPHILIC CONETWORKS

Miklós Tolnai[1], Béla Iván[1], Attila L. Kovács[2]

[1]Institute of Chemistry, Chemical Research Center of the Hungarian Academy of Sciences, H-1525 Budapest, Pusztaszeri u. 59-67, P. O. Box 17, Hungary; [2]Department of General Zoology, Eötvös Loránd University of Sciences, H-1088 Budapest, Puskin u. 3, Hungary

SUMMARY

Four different types of amphiphilic conetworks (APCN) were synthesized by radical copolymerization of N,N-dimethyl acrylamide (DMAAm) and methacrylate-telechelic polyisobutylene (MA-PIB-MA) in different compositions with respect to their PIB content (30, 50, 60, 70 %). The cell adhesion tests were carried out by rat red blood cells in protein-free environment. Tissue culture (TC) quality polystyrene, non TC polystyrene and glass were used as control surfaces. The experiments revealed that the control surfaces adhere the red blood cells more strongly and in a larger amount than the four different APCN samples. The APCNs of different PIB content did not show any significant difference in cell adhesion. These interesting findings indicate that the hydrophilicity of nonionic surfaces of APCNs is either very similar or does not play a critical role in cell adhesion.

INTRODUCTION

Amphiphilic conetworks (APCN) are composed of hydrophilic and hydrophobic polymer segments coupled with covalent bonds. Generally applicable synthetic strategies were developed in the late 1980s [1], and a series of amphiphilic conetworks were prepared by copolymerizing methacrylate-telechelic polyisobutylene (MA-PIB-MA) with various water soluble monomers, e. g. hydroxyethyl methacrylate (HEMA) and N,N-dimethylacrylamide (DMAAm). APCNs are of great interest with respect to their biocompatibility. Recent works have focused on their potential biomedical applications, such as drug delivery systems capable of controlled release, blood compatible surfaces, catheters or replacement of certain blood vessel segments. It was found that human monocites adhere to the surface of DMAAm containing APCNs in a significantly smaller amount than to HEMA containing surfaces, glass or polyethylene [2]. These investigations also showed that the amount of human plasma proteins (fibrinogen, immunoglobulin G, albumine, Haegemann factor, factor VIII) adsorbed on the surface of the above mentioned APCNs is smaller than on glass and silicon rubber. These facts indicate that certain APCNs might be suitable for producing blood contacting devices.

It is described [3] that the adhesion of different types of cells shows relation with the hydrophilicity and the net charge of the surface. In the first phase of adhesion ionic interactions between the cell and the artificial surface are supposed to be the determining factor. According to the general view, there is a correlation between the hydrophilicity and wettability of the artificial surface and cell adhesion. However, the cell attaching qualities of hydrophilic nonionic surfaces have not been sufficiently studied in a protein-free environment yet. In order to investigate this phenomenon, we have selected the simplest possible experimental system, i. e. interactions of surfaces with red blood cells. The process of cell attachment is very complex with systems containing different types of proteins e.g. fibronectin, laminin, fetuin etc.[3].

The present work is the first phase of a series of systematic experiments on interactions between cells and APCN surfaces. In this study APCNs composed of DMAAm and PIB in different proportions were investigated. Both PIB and PDMAAm are nontoxic, therefore poly (N,N-dimethylacrylamide)-*l*-polyisobutylene (PDMAAm-*l*-PIB) networks are presumably suitable for biomaterial applications. This paper reports on the influence of hydrophilicity on the adhesion of rat red blood cells.

EXPERIMENTAL

Synthesis of PDMAAm-*l*-PIB networks

The synthesis of PDMAAm-*l*-PIB networks was carried out by radical copolymerisation of MA-PIB-MA (M_n = 11200, M_w/M_n = 1.07) and DMAAm as described [1] (Table 1.) Four samples, A-11.2-30, A-11.2-50, A-11.2-60 and A-11.2-70 were prepared (A and the numbers stand for DMAAm, the average molecular weight of PIB divided by 1000, and PIB content in weight %, respectively).

Table 1. The amounts of reactants used in the synthesis of PDMAAm-*l*-PIB amphiphilic conetworks.

Conetwork	PIB (g)	Solution of PIB in THF (ml)	MA end group (mole*10e4)	DMAAm (mole*10e3)	AIBN (mole*10e4)
A-11.2-30	0.3	1.44	0.54	7.07	7.60
A-11.2-50	0.5	2.54	0.89	5.05	3.88
A-11.2-60	0.6	2.98	1.07	4.04	2.48
A-11.2-70	0.7	3.35	1.25	3.03	1.40

Blood preparation

To 500 µl blood 50 µl heparin (Kőbányai Pharmaceutical Co., Budapest, Hungary) was added. Blood cells were washed free of plasma components by four times pelleting in suspension buffer (SB) [5] with the help of a Beckmann Spinchrone R centrifuge (1600 rpm, 4°C, 4 min). After the fourth washing the cell suspension was adjusted to 100 mg/ml wet weight in SB. The cells were used in adhesion tests in this concentration.

Preparation of the PDMAAm-*l*-PIB samples

Each type of conetwork was cut into quadrates of 1mm x 1mm x 0,5mm. Then the quadrates were placed on cover glasses. The network pieces were fixed onto the glass surface by polycyanoacrylate glue to prevent floating. The cover glasses were placed into 35 mm glass Petri-dishes. Before the cell adhesion tests the remnants of cyanoacrylate were extracted by distilled water. After removing the water, the quadrates were washed 3 times with 3x1 ml SB solution. Finally all the samples were allowed to swell in SB solution for 20 minutes. Tissue culture (TC) quality and non TC polystyrene and glass were used as control surfaces. TC and non TC quality polystyrene were obtained from the bottom and cover pieces of corning culture dishes, respectively (SIGMA, C6296).

Spreading tests

The hydrophilicity of the surfaces was estimated by adding 25 µl distilled water to the surface, and then the diameter of the drops was measured.

Cell adhesion tests

The SB solutions were removed and each Petri-dish was filled with 2 ml blood cell suspension. The red blood cells were allowed to settle down and attach to the testing surfaces for 1 hr. The non adhered cells were removed by suction with water pump while holding the dishes at an angle of 45°. Then new SB solution was carefully poured into the dishes while keeping them at the same angle. The average coverage of the sample surfaces was estimated in an invertoscope repeatedly after a series of washings until the amount of cells adhered to the surface was constant. A five grade estimation scale was used: totally covered surface being 100%, totally uncovered being 0% with 20% divisions.

RESULTS AND DISCUSSION

The drop size of water on TC quality polystyrene and on glass was 5.5 mm. On non TC quality polystyrene, it was 4.5 mm. The APCN surfaces show no definite spreading values since the water drop keeps spreading slowly. The initial value is over 6 mm even on the most hydrophobic APCN. This indicates that the hydrophilicity of the PDMAAm-l-PIB APCNs changes when contacting with water, and more time is needed to reach equilibrated conditions.

TC quality surfaces do not loose significant amount of cells during serial washing as shown in Table 2. APCNs start from a lower level since only about half of their surface area was covered initially as summarized in Table 3. Subsequent washing reduce the covering to a very low level below 20 %. However, there is no difference between the APCN samples of different hydrophilic content.

These results show that the simple system of applying red blood cells to various surfaces of polymers is capable of revealing conspicuous differences in cell adhesion properties. Attachment to hydrophilized ionic TC surface is high and strong. According to the spreading tests APCNs are the most hydrophilic amongst the test surfaces. Their cell attachment

capability, in spite of their hydrophilic nature, is very low. This indicates that the hydrophilicity does not seem to correlate exactly with cell attachment. Further work is needed to study the correlation between the surface characteristics and cell adhesion. The APCNs are good model materials for studying these interactions, because there are many possibilities to modify their surface (e. g. treating with serum, or changing their composition to ionic side group containing monomers etc.).

Table 2. The average coverage (%) of the reference surfaces as a function of washing (PSt = polystyrene).

No. of washing	PSt (electric discharge treated ionic surface)	PSt	Glass
0	100	60-80	80-100
1	100	60-80	80-100
2	100	60-80	60-80
3	100	60-80	60-80
4	100	40-60	60-80
5	100	40-60	60-80
6	100	40-60	40-60

Table 3. The average coverage (%) of APCN surfaces as a function of washing.

N° of washing	A-11.2-30	A-11.2-50	A-11.2-60	A-11.2-70
0	40-60	40-60	40-60	40-60
1	20-40	20-40	40-60	20-40
2	20-40	20-40	40-60	0-20
3	20-40	0-20	40-60	0-20
4	0-20	0-20	0-20	0-20
5	0-20	0-20	0-20	0-20

REFERENCES

1. B. Iván, J.P. Kennedy and P.W.Mackey, in: „Polymeric Drugs and Drug Delivery Systems", R. L. Dunn and R. M. Ottenbrite (Eds.), ACS Symp. Ser. 469, Washington D. C., 1990, pp. 194-202, (b) ibid. , 1990, pp. 203-212

2. B. Keszler, J. P. Kennedy, N. P. Ziats, M. R. Brunstedt, S. Stack, J. K. Yun, J. M. Anderson, Polym. Bull. 29,681 (1992)

3. B. Griffiths, in Animal Cell Culture A Practical Approach, R. I. Freshney (Ed.), IRL Press Ltd. Oxford, 1986

4. J. P. Kennedy, J. Macromol. Sci.-Pure Appl. Chem., A31(11), 1771 (1994)

5. P. O. Seglen, Experimental Cell Research, 82, 391 (1973)

PROTEIN AND LIPID INTERACTIONS WITH HYDROGEL CONTACT LENSES: IN VITRO AND IN VIVO STUDIES

Brian J Tighe, Lyndon Jones, and Valerie Franklin.

Aston Biomaterials Research Unit, Aston University, Birmingham B4 7ET , UK.

INTRODUCTION

The development of hydrogels as biomaterials owes a great deal to the pioneering work carried out in the contact lens field, where a range of chemical structures has evolved varying, principally in charge and equilibrium water content. In the broader context the eye is an ideal body site to study the interactions of hydrogels with protein and lipids in a complex biochemical system, both because of ease of access and the fact that materials can be worn for a set period of time under known conditions before removal for analysis takes place.

In this paper a comparison is made of two types of neutral hydrophilic functional group in non-ionic hydrogel contact lenses and between two levels of anionicity in ionic hydrogel contact lenses. Additionally the results are used to demonstrate differences in patient-to-patient and eye-to-eye behaviour.

The deposition of contact lenses with substances derived from the tear fluid (commonly referred to as "spoilation" in the contact lens literature) is a well known phenomenon, leading in extreme cases to clinical complications. There is, currently, inadequate understanding of the relationship between deposition behaviour of lenses and their chemical structure and water content, which for existing materials varies from 38% to 85%. Two principal strategies are available to increase the water content of hydrogels above that of poly (2-hydroxyethyl methacrylate) or polyHEMA, which has a water content of 38%. Small quantities of charged groups such as methacrylic acid or larger amounts of more hydrophilic, neutral groups such as polyvinyl alcohol (PVA) or N-vinyl pyrrolidone (NVP) are added to polyHEMA or methyl methacrylate to raise their equilibrium water contents to 60% or greater. The Food and Drug Administration (FDA) currently classifies contact lens materials into four groups, depending upon their charge and water content. Two of these groups are relevant here: ionic materials with water contents in excess of 55% being categorised as Group IV and non-ionic materials having water contents in excess of 55% as Group II. Generic hydrogel compositions are identified by the USAN "filcon" nomenclature used here.

This paper compares two neutral hydrophiles, the hydroxyl group and the pyrrolidone ring, in lenses worn for six months and two levels of ionicity together with the effect of pyrrolidone ring incorporation in lenses worn for one month periods, in two controlled crossover clinically controlled trials.

MATERIALS AND METHODS[1]

Comparison of FDA Group II Materials.
Two Group II lens materials were chosen to ensure that one of the constituents in each material remained consistent (methyl methacrylate) and that equilibrium water contents were similar. Twelve subjects (six male and six female) were fitted by one of the investigators (LJ) with the two test lenses, which were "Lunelle ES70", in which N-vinyl pyrrolidone is the neutral hydrophile, from Essilor and "Excelens" (Atlafilcon), which is based upon poly vinyl alcohol as the neutral hydrophile, from CIBA-Vision. The subjects were divided into two equal groups, each of which wore one of the lens types on a daily wear basis every day for six months in conjunction with the same conventional overnight care and storage system. At the six month visit a cross-over took place, with all subjects repeating the above visit schedule with the other lens material.

Comparison of FDA Group IV Materials
Twenty-one subjects (six male and fifteen female) were fitted with the two test lenses, which were Vistakon "Surevue" (Etafilcon, based on HEMA and methacrylic acid) and CIBA-Vision "Focus" (Vifilcon, based on HEMA, polyvinyl pyrrolidone and methacrylic acid). All subjects used Bausch & Lomb's "ReNu" disinfecting system, utilising a "rub and rinse" step prior to disinfection. The subjects were divided into two roughly equal groups, each of which wore a pair of lenses for one month. At the end of the one month period a cross-over took place.

Design of studies utilised a randomised, double-masked paradigm, in which neither practitioner nor patient were aware of the lens type being used at the time of the clinical visits. These were not clinical studies but simply clinically controlled wearing and collection procedures to ensure that subsequent comparisons could be made with on the basis of uniform wear protocols. To ensure that no transfer of skin lipids onto the lens surface took place at the end of the wearing schedule, each lens was removed with tweezers and placed in a glass vial containing sterile non-preserved saline. The vials were then capped, labelled with the subject's initials and study number and refrigerated prior to analysis.

Lipid Analysis - Fluorescence Spectrophotofluorimetry
Lipoidal deposition was evaluated using a fluorescence technique on a modified Hitachi F-4500 fluorescence spectrophotometer. This is a non destructive technique which relies on the low level fluorescence of lipoidal species following excitation by UV light. Lenses are placed in distilled water in a specially designed cylindrical quartz cell which allows reproducible orientation of the lens to the incident light beam to be achieved. An incident (excitation) beam wavelength of 360nm is used and the height of the resultant fluorescence (emission) peak monitored at approximately 440nm. The height of the emission peak is monitored and provides an estimate of the relative level of deposition.

Baseline fluorescence for the lens materials was evaluated by examining a blank, unworn lens of each material as described above. This background trace was then subtracted from the result achieved with each worn lens to accurately assess the degree of deposited material.

Surface Protein Analysis - Fluorescence Spectrophotofluorimetry
Surface protein deposition was evaluated in a similar manner to that for lipids, using non-destructive fluorescence spectrophotofluorimetry. In this case an incident (excitation) beam wavelength of 280nm was used and the height of the resultant fluorescence (emission) peak was monitored at approximately 340nm. Baseline fluorescence for each material was subtracted as described above.

Total Protein Analysis - Ultra Violet Spectroscopy
Total protein concentration was measured using UV absorbance on a Hitachi U2000 spectrophotometer. The measurement was carried out in matched quartz cells, against an unworn contact lens of the same material and power at 280nm.

RESULTS AND DISCUSSION
The surface lipid and surface protein for each of the worn lenses was determined. The results were first plotted as left eye versus right eye in which the value obtained for the right eye of each lens was plotted against that of the lens concurrently worn in the left eye. Values for lipid levels are shown for all four materials in Figure 1. Similarly, the surface protein levels for all four materials are shown in Figure 2.

It is clear that values for left and right eye for the same material worn for identical time periods show significant variations and that for protein, these are not appreciably less than the patient-to-patient values. The material differences are significant with the two FDA Group IV materials attracting the highest protein levels despite their shorter wear periods, with Etafilcon, which contains the highest level of ionicity attracting significantly more than the less anionic Vifilcon. Lipid levels are highest for the two N-vinyl pyrrolidone-containing materials Vifilcon and ES 70. The values for ES 70, worn for six months are markedly higher than for any of the other materials as can be seen from Figure 1. Left eye-right eye variations are very patient specific and much greater patient-patient variations occur with lipid. Separate experiments show that lipid penetrates into the polymer matrix of the hydrogel lenses up to a depth of around five microns and that similar, but not identical levels of deposition occur on posterior and anterior surfaces. Experimental values shown here are for anterior surfaces throughout.

Separate protein penetration experiments were also carried out and these indicate that no significant protein penetration occurs with either of the FDA Group II materials Excelens and ES 70. This is also reflected in the substantially linear plot of total protein against surface protein for these materials (Figure 3).

Protein penetration measurements on both of the FDA Group IV materials show protein penetration concentrated within ten microns of both anterior and posterior surfaces. The fact that protein penetrates into the matrix of these materials is illustrated by incorporating results of surface versus total protein for Group IV Vifilcon together with those for the two Group II materials in the form of plot presented in Figure 3. The combined results are shown in Figure 4. Here the fact that not all Group IV protein is located on the lens surface is reflected by the fact the points lie in a cluster to the right of the line associated with the Group II materials. By calibrating the surface protein technique a plot showing actual protein levels associated with both anterior and posterior surfaces against total protein associated with surface and matrix may be produced. Separate experiments established that the protein associated with Group IV lenses is substantially lysozyme and that it is readily extracted from the lens matrices and has intact biological activity.

CONCLUSIONS
• Eye-to-eye and patient-to-patient variations are both important but are more marked with lipid than with protein.
• Lipid accumulation and penetration occurs with all materials but is enhanced in materials that contain N-vinyl pyrrolidone.
• Both surface-located and matrix-located protein are enhanced by increasing ionicity in FDA Group IV materials.
• Matrix protein is located within ten microns of the lens surface and consists predominantly of mobile biologically active lysozyme.

REFERENCES
1. Jones, L; Franklin, V; Evans, K and Tighe, BJ (1996) Spoilation and Clinical Performance of Monthly vs Three Monthly Disposable Contact Lenses Opt Vis Sci 73 , 16-21.

Legends:
Figure 1. Surface lipid measured by fluorescence showing left eye versus right eye variations for Excelens, Etafilcon, Vifilcon and ES 70.

Figure 2. Surface protein measured by fluorescence showing left eye versus right eye variations for Excelens, Etafilcon and Vifilcon and ES 70.

Figure 3. Surface protein measured by fluorescence plotted against total protein (uv) for FDA Group II materials (Excelens and ES 70).

Figure 4. Surface protein measured by fluorescence plotted against total protein (uv) for FDA Group II materials (Excelens and ES 70) together with Group IV materials (Vifilcon and Etafilcon).

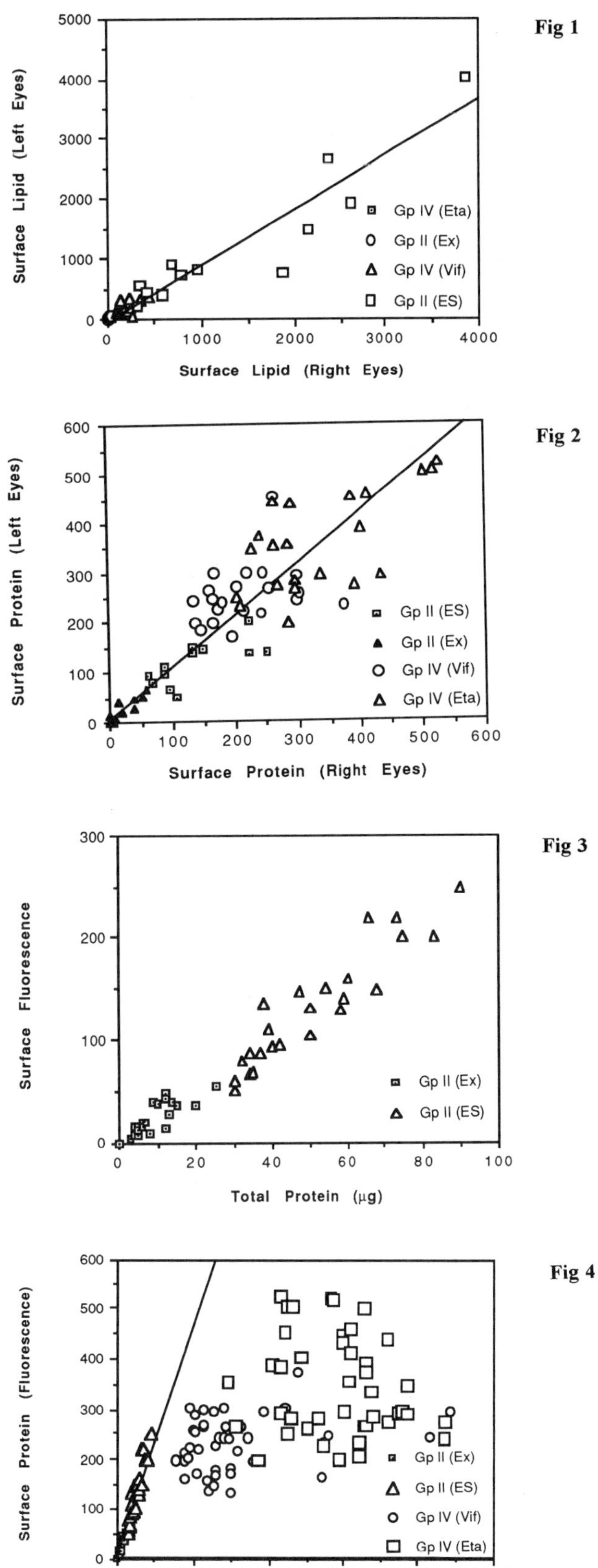

HYDROGEL-BASED MEMBRANES FOR BIOARTIFICIAL ORGANS

David Hunkeler
Laboratory of Polymers and Biomaterials,
Swiss Federal Institute of Technology, CH-1015 Lausanne, Switzerland

INTRODUCTION

The development of biological microencapsulation systems have included pioneering efforts by Chang [1], Sun [2] and Sefton [3]. The latter two have focused on the immunoisolation of pancreatic islets for the formation of a bioartificial pancreas. Thin film polymer membranes comprised of water-insoluble thermoplastics, symplexes and hydrogel copolymers have been prepared, and several recent reviews detail the technological aspects involved in cell or islet encapsulation [4-6] and bioartificial organs [7]. Unfortunately the fragile nature of islets, and the specificity of the capsule processing conditions to the properties of the often viscoelastic polymer fluid has constrained the number of polymers which have been rigorously evaluated. Indeed, most researchers have been limited to the poly-L-lysine-alginate [2] and lysine-chitosan [8] systems which are based on the ionotropic gelation of alginate with polyvalent cations, typically calcium or barium. However, although lysine-alginate produces quite stable membranes it has relatively poor mechanical properties. Ionotropic gelling alternatives for alginate, as an inner polymer, have thus far been limited to the cationic chitosan and blends of alginate with other polysaccharides such as carrageenan, carboxymethylcellulose or dextran sulfate [9]. Furthermore, it has been speculated that a family of capsule chemistries will need to be available in order to provide alternatives in the event that the primary immunoisolation material is rejected by a given patient. This problem is likely to be particularly acute for Type-I diabetics since they typical contract the disease for over 40 years. Therefore, in an attempt to identify alternatives to the classical systems listed, we have undertaken a massive screening of polyelectrolytes in an attempt to make molecular inferences as to the complexation mechanism. The evaluation has included 36 polyanions and 40 polycations in 1235 binary combinations. Naturally occurring polysaccharides, biopolymers and synthetic macromolecules were evaluated as a function of their molecular weight, composition, charge spacing, type of charge (permanent, induced), rigidity and aromaticity of the backbone, location of the functional group attachment and hydrogen bonding affinity.

RESULTS

Cytotoxicity Screen.

Cytoxocity was measured, in blind studies [10], by the DNA content of rat insulinoma cells and a visual indication of the shape and surface adhesion following incubation for various periods in polyelectrolyte solutions of differing concentrations. Synthetic polyanions exhibited a moderate level of cell cytotoxicity, with some notable exceptions such as polyacrylic acid and carboxy modified polyacrylamide. The majority of polycations tested seem to be quite non-toxic in contrast to the synthetic quaternary ammoniums which generally exhibited strong cytotoxicity. The only exception to this trend was a polydimethylaminoethylacrylate-co-acrylamide, which unfortunately had to be eliminated from further consideration due to the extreme viscosities caused by its high molecular weight ($>10^6$ daltons). It is, however, possible that this polymer could be utilized at lower molecular weights or as a component of a polycation blend. Given these results it appears that the optimal polymer for contact with cell suspensions is a natural polyanion. In the subsequent discussion six natural polyanions will show particular effectiveness in membrane formation (Alginate, Carrageenan, Cellulose Sulfate, Gellan, Hyaluronic Acid, Xanthan) and these are recommended for further development as the inner constituents of a capsule coacervation system.

Categorization of Polymer Effectiveness in Membrane Formation.

There are relatively few polyanion-polycation pairs which yielded soluble complexes (13.6%) with the majority of symplex reactions yielding either precipitates (43.7%) or weak membranes (30.7%). Indeed, many of the naturally occurring or synthetic species predominantly form precipitates. This is, perhaps, due to the high content of ionic groups (1 per repeat unit) characteristic of polysaccharides. Only 12% of the binary systems tested yielded stable membranes. The percentage of systems which resulted in stable membranes in capsule form was 3.2%. While this is a small number (47 systems) it does provide a significant list of alternatives to the classical poly-L-lysine-alginate pair which is so common in the literature.

DISCUSSION

Polymer Attributes to Be Considered in Capsule Formation via Polyelectrolyte Complexation.

An examination of the complexation data [10] reveals that the majority of the 47 binary systems which yielded stable capsules consisted of a polysaccharide inner polymer and a synthetic tertiary amine or quaternary ammonium polycation as the outer material (36 cases). There was only a single example of a synthetic polyanion as a core (polyacrylic acid), while chitosan, a naturally occurring polycation, formed a stable membrane complex with 6 polyanions, both naturally occurring and synthetic. Polyvinylamine also formed a stable membrane in one instance. Given that insulinoma cells suspended in chitosan and polyvinylamine exhibited a limited growth response, it appears that the most suitable inner polymer is a naturally occurring polyanion. Indeed, six polysaccharides: alginate, carboxymethyl cellulose, l-carrageenan, cellulose sulfate, gellan gum and xanthan, were particularly effective in the formation of membranes in combination with various polyamines, polyamides and quaternary ammoniums.

Of the 47 systems which produced a stable membrane, 42 involved an inner polymer which contained a cyclic backbone which was relatively rigid and extended in aqueous solution. For example, xanthan which formed stable membranes with twelve polycations, has a Mark-Houwink-Sakurada exponent of approximately 1.2. These were paired with flexible linear chains. This is typified by systems such as carrageenan/polydimethylamine-co-epichlorohydrin, carboxymethyl cellulose/polyallylamine and xanthan/polymethylene-co-guanidine. All the suitable inner polymers were found to have molecular weights in the 10^{5-6} range. However, stable membranes were produced with outer polymers ranging from oligomers to high molecular weight species. We believe that the inner polymer requires a large extended backbone in order to facilitate the interaction with the oppositely charged molecule. Since the membrane is produced as a result of the diffusion of the outer polymer through a spherical inner droplet, the diffusing species should be flexible and have one of the following attributes: i) a high diffusion coefficient or ii) a sufficient number of charged sites for which to form a symplex with the inner polymer. High molecular weight exterior polymers such as the quaternary polyhydroxymethacryloxyethyltriethylammonium bromide have a large number of charged groups and form a dense membrane skin at the outer layer of the capsule, very much akin to asymmetric membranes produced via a phase inversion process. Conversely, oligomeric species such as the chelant polymethylene-co-guanidine have a higher diffusion coefficient which permits the penetration of the inner polymer to a greater depth, although the membrane itself is more porous. This deficiency can, however, be corrected in a second, post reactive step which involves a surface coat with a higher molecular weight polymer.

It is interesting that stable membranes can be produced with such a diverse range of outer polymers, principally polycations, and a recent paper [11] discusses the blending of high and low molecular weight polycations to simultaneously control the mechanical properties and permeability of capsules. In particular, polymethylene-co-guanidine was investigated in an attempt to combine the chelating properties of calcium, a typical gelling agent, with the improved long term stability and mechanical properties of macromolecular cations. This finding is akin to Dautzenberg's which shows that capsule mechanical properties can be improved if a broad molecular weight distribution is employed [12]. It remains to be elucidated if bimodal distributions are superior to unimodal, highly disperse, distributions.

Clearly a strong polyelectrolyte complexation requires a high concentration of ionic groups on both reacting components. All of the 47 stable systems involved an inner polymer that had one charge per repeat unit and an outer polymer which was either moderately (12) or highly (34) charged. This implies that a dense membrane network can be formed if a highly charged extended inner polymer is exposed to an oppositely charged polymer with a charge spacing of approximately the same, or smaller dimensions. The flexible nature of the outer polymer may permit a conformational adjustment which reduces the effective distance between charges, to form ionic bridges with a larger number of the ions on the inner polymer backbone. Flexible chains, particularly polyelectrolytes of various charge densities, also permit a greater degree of intermolecular interaction. Interestingly, 45 of the 47 inner polymers had charges which were induced by pH while 24 of the outer polymers were permanently charged. This tends to indicate that the membrane formation could be controlled by varying the pH of the solution containing an inner polymer, although in application involving living cells, the suspension should be kept at a near neutral level of acidity. The "normal mode" membranes, prepared with polyanions interior to polycations, generally consisted of a polysaccharide containing a carboxy group whose degree of ionization could be adjusted by increasing the pH. These were paired with either a permanently charged quaternary ammoniums (24 systems) or a tertiary polyamines with an induced charge (18 systems). By contrast, the "reverse mode" membranes formed with an inner polycation usually possessed a permanent charge on the inner polymer. As a guideline in selecting potential polymer pairs we can report that for only 4 of the 1235 pairs investigated was a stable membrane produced from two permanently charged species. These were all based on cellulose sulfate interior to quaternary ammoniums, a system known to form strong capsular membranes [12]. We can also note that stable membranes were produced for inner polymers where the functional group was attached to the side chain of a cyclic sugar backbone. This indicates that the accessibility of the charge is as important as the concentration and nature of charged species. Ionic groups on the relatively rigid polysaccharide are much more accessible than on a coil. The membrane formation did not appear to correlate with the linearity or degree of branching of the polyion, though a systematic set of polyelectrolyte standards were not available to rigorously test this conclusion. Another important variable in the complexability of oppositely charged polyelectrolytes appears to be the presence of secondary interactions, via hydrogen bonding. This is not too surprising since both ionic and H-bonding interactions are known to influence the chelation using simple cations.

Evaluation of Various Multicomponent Systems.

The blending of two polyanions to form the droplet core is primarily required for rheological considerations. As a typical example, blends of alginate and cellulose sulfate or carboxymethylcellulose/carrageenan with hyaluronic acid were more effective in yielding mechanically stable capsules then any of these polysaccharides individually. These have been shown to reverse diabetes in spontaneously diabetic mice for periods up to six months, with the capsule clearly not influencing the insulin egress kinetics [13].

Two categories of polycations were found to be effective as the constituents of the "outer" polyelectrolyte solution. Oligomeric species, in combination with divalent cations were particularly useful in the decoupling of mechanical properties from permeability control. For example, an alginate-cellulose sulfate blend can gelled in the presence of sodium chloride and calcium chloride [14]. In a second reactive step the oligomeric polymethylene-co-guanidine is added. Its diffusion into the pre-cast capsule could be temporally regulated with longer reaction times reducing the membrane MWCO. Several similar capsules were also produced using low molecular weight amines such as spermine and protamine sulfate as well as polybrene. It is quite possible that an oligomeric cationic species forms a special type of complex with two polyanions.

CONCLUSIONS

Based on the screening presented herein, Table 1 summarizes guidelines can be delineated for the selection of multicomponent polymer blends.

Table 1. Guidelines for the Selection of Polymers in Multicomponent Encapsulation Systems

Reagent Location	Characteristic	Example
Cell-Contacting ('Inner') Side	**One** ionotropically gelling polyanion	Alginate, k-carrageenan
	or thermosetting polyanion	Agarose
	or viscosity modifier	Carboxymethyl cellulose, cellulose sulfate
	and one structural polyanion	Hyaluronic acid, cellulose sulfate
Physiological Fluid Contacting ('Outer') Side	Ionotropic gelling agent(s)	Mono/di-valent cations
	or oligocation	Polymethylene-co-guanidine
	or mixture of low/high molecular weight polycations	Spermine/polylysine

REFERENCES

1. T.M.S. Chang TMS, Artificial Cells. Charles C. Thomas, Springfield, IL (1972).
2. F. Lim F, A.M. Sun, Science 210: 908 (1980).
3. M.V. Sefton, B.L. Broughton, BBA 717: 473 (1983).
4. M.F.A. Goosen CRC Crit.Rev.Biocompat. 3: 1 (1987).
5. M.F.A. Goosen In: Lanza RP, Chick WM (eds) Pancreatic Islet Transplantation Volume III: Immunoisolation of Pancreatic Islets. R.G. Landes Co, Austin, TX (1994).
6. W.T.K. Stevenson, M.V. Sefton, Trends in Polymer Science 2: 98 (1994).
7. A.Prokop, D. Hunkeler, A. Cherrington eds., « Bioartificial Organs: Science, Medicine and Technology », New York Academy of Sciences Volume 831, New York, 1997.
8. C.K. Rha, D. Rodriguez-Sanchez, C. Kienzle-Sterzer In: Colwell RR, Pariser ER, Sinskey AJ (eds) Biotechnology of Marine Polysaccharides. Hemisphere Publishing Corporation, Washington, D.C. (1985).
9. Y.L. Hsu, I.M. Chu Biotech. Bioeng. 40: 1300 (1992).
10. A. Prokop, D. Hunkeler, M. Haralson, S. Dimari, T. Wang, Advances in Polymer Science, 136 (1998).
11. A. Prokop, D. Hunkeler, M. Haralson, S. Dimari, T. Wang, Advances in Polymer Science, 136 (1998).
12. H. Dautzenberg, B. Lukanoff, U. Eckert, B. Tiersch, U. Schuld, Ber. Bunsenges. Phys. Chem. 100: 145 (1996)
13. T. Wang, I. Lacik, M. Brissova, A.V. Anilkumar, A. Prokop, D. Hunkeler, R. Green, K. Shahrokhi, A.C. Powers, Nature Biotechnology, 15,358 (1997).
14. Hunkeler, Trends in Polymer Science, 5, 286 (1997).

Bioabsorbable Synthetic Hydrogel as a Surgical Lung Sealant

A. S. Sawhney, B. Poff, M. Powell, K. Messier,
E. Doherty, F. Yao, D. J. Enscore, P. K. Jarrett
Focal, Inc.
4 Maguire Road
Lexington, MA 02173

Introduction

There is currently a strong medical need for a biocompatible sealant for use in surgery. Lung surgery often results in air leaks that can lead to complications, extended use of suction through chest tubes, and prolonged hospital stays. Air leaks can occur at sites such as staple lines, suture lines, areas of dissection, etc. These leaks are sometimes difficult to stop, due to poor access to the leak site, or due to friable tissue that cannot support suture repair.

An effective lung sealant must be able to adhere well to tissue for a prolonged period of time, until healing has advanced sufficiently. The material must also be compliant enough to allow free expansion of the lung to fill the thoracic cavity. The sealant must also be able to withstand repeated breathing motion. A bioabsorbable hydrogel has been developed that functions well as a sealant to prevent air leaks following lung surgery.

Hydrogel application method

The hydrogel is polymerized *in situ* using two parts, a primer and a sealant, as a laminate, forming a highly adherent coating of any desired thickness on the tissue A laminate structure is employed to provide good bonding to tissue (primer layer) and good mechanical properties (sealant layer) for sealing the leak and maintaining the seal during lung movement.

The materials are applied to tissue in aqueous solution formulations. The primer solution is first brushed onto the tissue. The brushing action allows the primer material to obtain good intimate contact with the tissue surface, replacing the surface moisture. The sealant material is then mixed with the primer using a brush to provide a transition layer between the primer and sealant layers. The final, thicker sealant layer is then flowed over the application area and the laminate is photopolymerized.

Material

The hydrogel is formed by photopolymerization of water soluble macromers. Poly(ethylene glycol) (PEG) linked on both ends to hydrolyzable units, trimethylene carbonate (TMC) or lactate (LA) oligomeric segments, and end capped with polymerizable acrylate groups (AA). Synthesis of the macromers has been previously described[1].

The primer macromer was designed to provide a low viscosity formulation that could penetrate well into surface moisture and into the fine structure of the tissue surface, providing good intimate contact with the tissue surface.

Primer macromer:

$$CH_2=CH\text{-}CO\text{-}O\text{-}(CH(CH_3)\text{-}CO)_y\text{-}O\text{-}(CH_2\text{-}CH_2\text{-}O)_x\text{-}(CO\text{-}CH(CH_3)\text{-}O)_y\text{-}CO\text{-}CH=CH_2$$

or

$$AA\text{-}(LA)_y\text{-}[(PEG)\text{-}(LA)_y\text{-}AA$$

The modulus of the primer hydrogel layer was designed to be high (for a hydrogel) to maximize interlocking with the tissue structures. The thin layer formed allows movement with the tissue during expansion and contraction. This was accomplished by using a low molecular weight PEG molecule (about 3500 dalton) as the starting molecule.

The sealant macromer was designed to make a viscous formulation, to form a thick film when gelled. The PEG molecular weight used was high (about 30,000 dalton) for a low modulus and high elongation properties, to allow ventilation of the lung beneath the sealant.

Sealant macromer:

$$CH_2=CH\text{-}CO\text{-}O\text{-}(CO\text{-}(CH_2)_3\text{-}O)_y\text{-}(CH_2\text{-}CH_2\text{-}O)_x\text{-}(CO\text{-}(CH_2)_3\text{-}O)_y\text{-}CO\text{-}CH=CH_2$$

or

$$AA\text{-}(TMC)_y\text{-}(PEG)\text{-}(TMC)_y\text{-}AA$$

Photopolymerization is accomplished by formulating the macromers in buffered saline solutions containing triethanol amine and eosin Y as the photoinitiator. Visible light illumination from a xenon arc lamp (470-520 nm) at an intensity of $100 mW/cm^2$ is used to initiate polymerization for 40 seconds.

Bioabsorption:

Once implanted, the hydrogel breaks down by hydrolysis, releasing biocompatible components that are metabolized or cleared by the kidneys. The rates of property and mass loss are determined by the hydrolyzable linkage, and the material undergoes a characteristic and predictable breakdown pattern. Tissue healing has been shown in animal studies to progress unimpeded in the presence of the sealant.

Degradation can be monitored *in vitro* by placing preformed samples of gel into buffered saline solutions and removing the samples periodically. The samples are weighed wet and dry. The dry weight is corrected for solids in the saline solution. In this way the mass of the polymer network remaining is isolated.

The graphs below show typical results for the sealant in buffered saline solution at 67°C. The mass loss pattern is characteristic of this material. The degradation exhibits first order mass loss followed by a rapid mass loss during the final stages of network structure breakdown. The intercept of the log plot shows the degree of network incorporation of the macromer (about 80%) at zero time.

In vivo degradation can be monitored by analyzing explanted samples for PEG using HPLC. Studies of hydrogel implanted intramuscularly in rats exhibit the same characteristic degradation pattern as seen *in vitro*, indicating a simple hydrolytic mechanism *in vivo*.

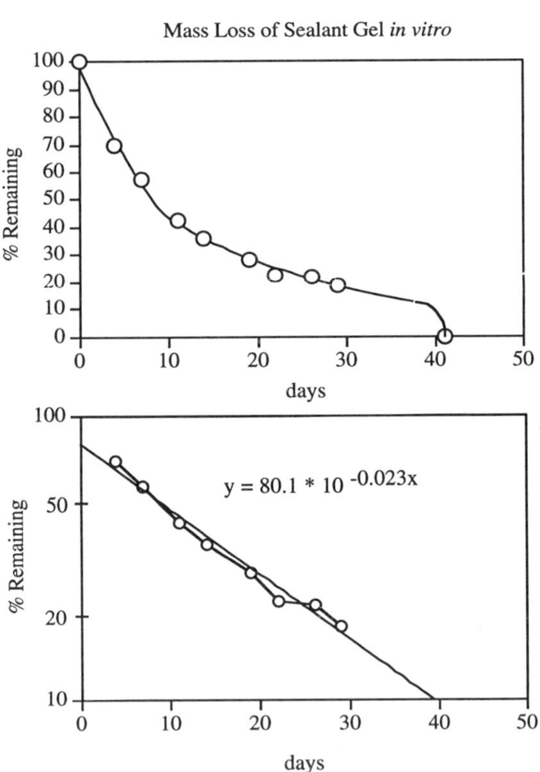

1. Pathak, C.P., Sawhney, A.S. and Hubbell, J.A., Macromolecules, 26, 581-587 (1993)

POLY(SILAMINE) HYDROGEL -A NEW STIMULI RESPONSIVE GEL WITH RUBBER ELASTICITY TRANSITION-

Yukio Nagasaki, Kenji Kazama, Masao Kato, Kazunori Kataoka and Teiji Tusuruta

Department of Materials Science and Technology
Science University of Tokyo
Noda 278-8510, JAPAN

1.Introduction

There are many kinds of synthetic polymers which respond to environmental conditions. For example, poly(N-isopropylacrylamide) (polyNIPAAM) shows a phase transition with temperature (LCST) (1). Such polymers vary their characteristics such as solubility and conformation with changing environmental conditions such as temperature (2), solvent composition(3), pH(2b,4), ionic strength(2b,5), and electric field(6). However, there are only a few reports on polymers which can vary their stiffness in response to environmental conditions like the polypeptide. If the stiffness of a polymer can be controlled by certain stimuli, numerous applications of the polymer will become possible.

Several years ago, we found that a new telechelic oligomer, which consists of alternating 3,3-dimethyl-3-silapentane and N,N'-diethylethylenediamine units in the main chain (poly(silamine)), was obtained through an anionic polyaddition reaction between dimethyldivinylsilane (DVS) and 3,6-diazaoctane (DAO) in the presence of 3-lithio-3,6-diazaoctane (Li-DAO) in THF(7). We also found that poly(silamine) shows unique phase transition properties in response to the degree of protonation of its amino groups in aqueous media(8): i) Its solubility in aqueous media is drastically changes; ii) Poly(silamine) shows a strong interaction with several anions; iii) Due to the interaction between protonated poly(silamine) and anions, the rubber elasticity of poly(silamine) is drastically changed. Actually, the glass transition temperatures of poly(silamine) can be varied from -85 °C to + 80 °C by the environmental conditions.

Poly(silamine) gels, the network of which shows the rubber elastic transition, show unique stimuli-sensitivities. This paper summarizes the synthesis of poly(silamine) and its properties.

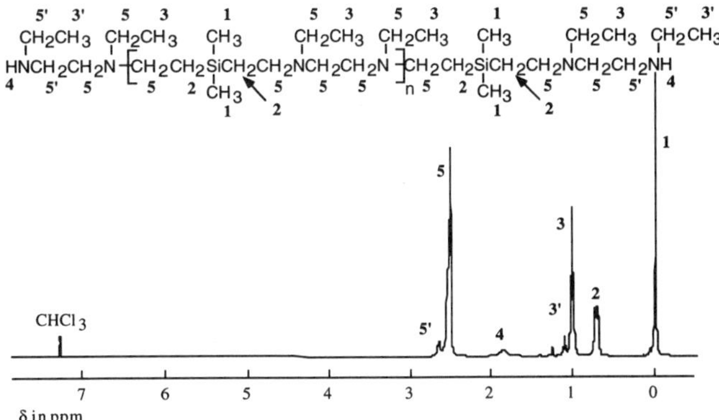

Poly(silamine)

2.Synthesis of poly(silamine) telechelics and their characteristics in aqueous media

For the synthesis of telechelic poly(silamine) with sec-amino groups at both ends, direct polyaddition reactions between DVS and DAO in the presence of Li-DAO can be employed. Into THF solution (200 mL) of DAO (420 mmol), butyllithium (20 mmol in hexane 12.3 mL) and DVS (400 mmol) were added via syringes in a round-bottomed flask under argon atmosphere. After 2 days reaction at 50 °C, 21 mmol of DAO was added again and the mixture was allowed to react further for 2 days to complete the conversion of vinyl groups at the end of the formed polymer chain to sec-amine. The gel permeation chromatogram of the obtained poly(silamine) oligomers showed the molecular weight (Mn) and the distribution index (Mw/Mn) were 2,400 and 1.55, respectively. From the ^{1}H-NMR analysis of the obtained oligomer (Figure 1), it was found that the oligomer consists of alternating 3,3-dimethyl-3-silapentane and N,N'-diethylethylenediamine units in the main chain (see assignment in the figure). Because no vinyl proton was observed in the high frequency field (6 - 8 ppm) in the ^{1}H-NMR spectrum, the end groups of the oligomer were converted quantitatively to sec-amino groups. Using methylene signals adjacent to sec-amine at the end of the polymer chain (5' in Figure 1) and those adjacent to tert-amine (5 in Figure 1), the Mn of the oligomer was calculated to be 2,600, which agreed well with that from GPC results. On the basis of these results, it is concluded that poly(silamine) telechelic oligomers with sec-amino groups at both ends are obtained by the one-pot synthesis.

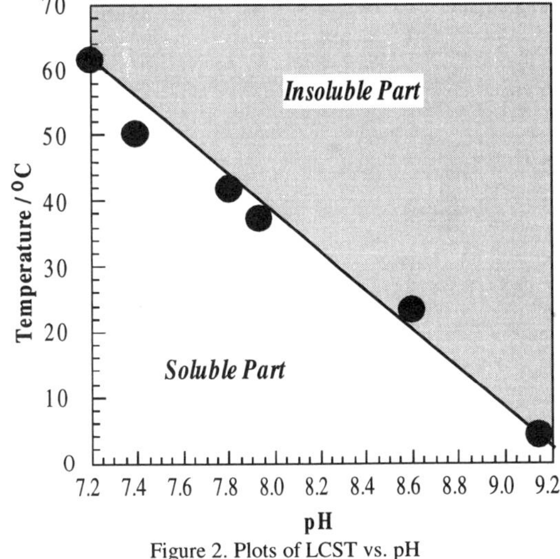

Figure 1. ^{1}H NMR Spectrum of sec-amino-ended poly(silamine)

Poly(silamine) possesses alternating hydrophilic amino groups and hydrophobic organosilyl units in the main chain; therefore, the phenomena of phase separation can be envisaged, which is the same as that of poly(NIPAAM). A different point of poly(silamine) is that the hydrophilicity of the diamine units can be controlled by the protonations of the amino groups. As a result, the phase transition can be controlled by both temperature and pH. Figure 2 shows a change in LCST points as a function of pH. As shown in the figure, both factors, *viz:* pH and temperature, are effective for phase separation. Dong and Hoffman(9) and Kim and co-workers(10) separately reported that an introduction of ionic moieties into the polyNIPAAM causes a pH dependency of the LCST. In these cases, however, the sensitivity of the environmental pH effect on the LCST was not very high because of the location of the ionic groups in the side chain. For example, a 0.8 difference in the pH values (7.2 to 8.0) causes a decreased turbidity point of ca. 10° (40 to 30 °C) in the poly(NIPAAM-*co*-dimethylaminoethyl methacrylate). Poly(silamine) possesses amino groups in the main chain; therefore, the pH dependency on the LCST is very high. This is in sharp contrast to the behavior of other ionic polymers that exhibit a LCST as mentioned above.

Figure 2. Plots of LCST vs. pH

Because of the presence of the ethylenediamine units in the main chain, poly(silamine) showed a two-step deprotonation process on changing the pH as follows:

It is interesting to note that the pK values of poly(silamine) were influenced by the protonating agents. For example, the pK_1 value, which corresponds to the

"

degree of protonation $\alpha = 0.75$, of poly(silamine) protonated by HCl was 5.8, while that by H_2SO_4 was 6.7, even though their pK_2 values, which correspond to $\alpha = 0.25$, were almost the same. This indicates that poly(silamine) has a special interaction with the anions, namely, Cl⁻ and SO_4^{2-}, especially at an α of more than 0.5. From ^{1}H NMR studies, we proposed the following interaction between poly(silamine) and anions.

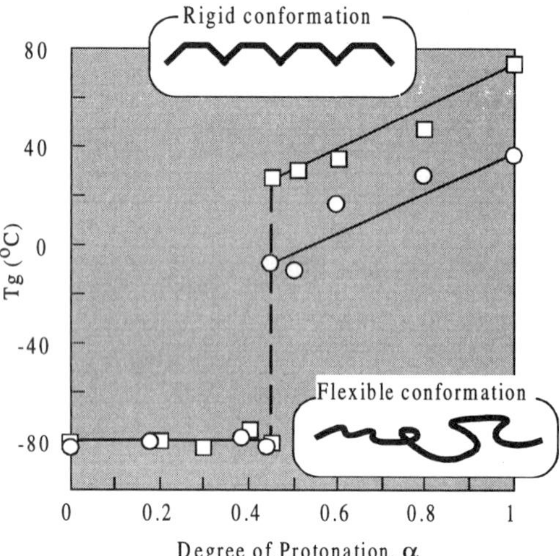

Poly(silamine) possesses alternating 3-silapentane and ethylenediamine repeating units; therefore, flexibility of the non-protonated polymer is very high. As stated earlier, the glass transition temperature of poly(silamine) without protonation was -85 °C, which indicates that it is one of the most flexible polymers known and is similar to silicones. By the protonation of the amino groups in poly(silamine) along with the anion bindings, the flexibility of the polymer must be changed and the Tg increased. It was reported that the rotation of the amino groups around the ethylene bond axis is suppressed by their protonation (11). If our proposed binding is formed along with the protonation of the amines, the stiffness of the poly(silamine) in the acidic region must be significantly increased.

Figure 3 shows the change in the Tg of poly(silamine) as a function of its degree of protonation. In the low protonated region, the poly(silamine) maintained low Tg values because of the absence of any interaction between poly(silamine) and the anions. When the degree of protonation of poly(silamine) exceeds 0.45, the Tg of the polymer abruptly increases. It should be noted that the Tg of the poly(silamine) protonated with sulfuric acid is much higher than that protonated with hydrochloric acid, which may be explained by the double chelating structure as previously stated. At $\alpha = 1.0$ using sulfuric acid, the Tg of the polymer was c. 80 °C, which is comparable to that of polystyrene.

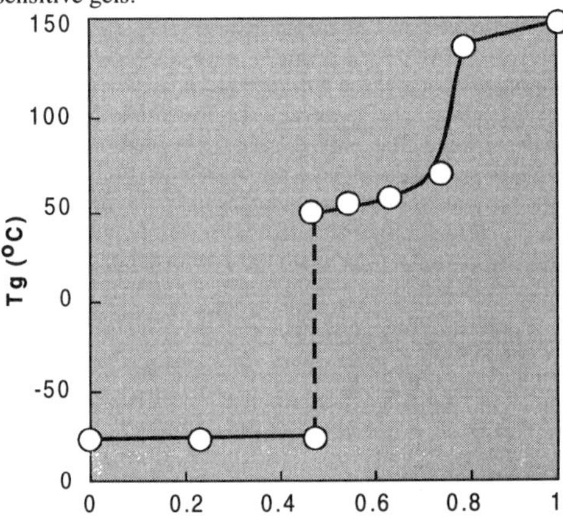

Figure 3. Change in Tg as a function of protonation degree

2. Preparation of poly(silamine) gel and its swelling characteristics

Since poly(silamine) was prepared via polyaddition reactions, it is easy to control the functionalities of the end groups. As mentioned before, sec-amino-ended poly(silamine) was prepared quantitatively. For the preparation of poly(silamine) gel, the end amino groups were utilized. When the amino-ended poly(silamine) was mixed with stoichiometric amount of 2,4-toluenediisocyanate, the end amino groups were easily converted to isocyanate groups. Chemically crosslinked poly(silamine) gel can be prepared via the reaction between the isocyanate-ended poly(silamine) and trimethylolethane. The swelling properties of poly(silamine) gels thus obtained were investigated.

The poly(silamine) gels were fairly tough elastic rubber. When the gel (Mn of prepolymer = 2,600) was soaked in methanol, the degree of swelling attained was 2,160 %. Tanaka et al. reported that swelling of ionic gels can be governed by the following three factors: i) interaction between the gel networks and solvents, ii) rubber elasticity of the gel networks and iii) ionic osmotic pressure(12). As methanol are good solvents for poly(silamine), the gels swelled in methanol. In these cases, however, the rubber elasticity of the networks contributed negatively to the swellings. Contrary to the swelling in organic solvents, the rubber elasticity of the networks functioned positively when the gels were immersed in acidic water because the networks as low pH became a rigid expanded form. Actually, the swelling degree was one order of magnitude higher (16,200 %) than that achieved with methanol swelling when the gel was immersed in 0.01M-HCl (ionic strength I = 0.01), indicating that all of the three factors mentioned above functioned positively.

Figure 4 shows the stiffness of the poly(silamine) gel as a function of protonation degree. As anticipated from the poly(silamine) prepolymer, the gel became very rigid by the protonation. Actually, the glass transition of the gel attained 150 °C by 100% protonated with HCl. Because the driving force of the swelling is not only the change in the ionic osmotic pressure but also the elasticity change due to the anion binding and the protonation, the gel becomes hard when it swells. This is in striking contrast to the stimuli-sensitive gels prepared so far. Such a poly(silamine) gel can be anticipated to open new fields for stimuli-sensitive gels.

Figure 4. Relation between Tg vs. protonation degere

References

[1] M. Heskins, J. E. Guillet, *J. Macromol. Sci.-Chem.*, **A21**, 1441(1968)

[2] (a) T. Tanaka, *Phys. Rev. Lett.*, **40**, 820(1978); (b) H. G. Schild, *Prog. Polym. Sci.*, **17**, 163(1992); (c) S. Katayama, Y. Hirokawa, T. Tanaka, *Macromolecules*, **17**, 2641(1984); (d) A. S. Hoffman, A. Afrassiabi, L. C. Dong, *J. Controlled Release*, **4**, 213(1986)

[3] (a) J. Hrouz, M. Ilavsky, K. Ulbrich, J. Kopecek, *Eur. Polym. J.*, **17**, 361(1981); (b) I. Ohmine, T. Tanaka, *J. Chem. Phys.*, **77**, 5725(1982)

[4] (a) T. Tanaka, D. J. Fillmore, S.-T. Sun, I. Nishio, G. Swislow, A. Shah, *Phys. Rev. Lett.*, **45**,, 1636(1980); (b) R. A. Siegel, B. A. Firestone, *Macromolecules*, **21**, 3254(1988); (c) C. D. Batich, J. Yan., C. Bucaria, Jr., M. Elisabee, *Macromolecules*, **26**, 4675(1993)

[5] (a) J. Ricka, T. Tanaka, *Macromolecules*, **18**, 83(1985); (b) T. G. Park, A. S. Hoffman, *Macromolecules*, **26**, 5045(1993)

[6] (a) T. Tanaka, I. Nishio, S..-T. Sun, S. Ueno-Nishio, *Science*, **218**, 467(1982); (b) T. Shiga, Y. Hirose, A. Okada, T. Kurauchi, *J. Appl. Polym. Sci.*, **46**, 635(1992)

[7] Y. Nagasaki, E. Honzawa, M. Kato, K. Kataoka, T. Tsuruta, *J. Macromol. Sci.-Pure Appl. Chem.*, **A29**, 457(1992)

[8] Y. Nagasaki, E. Honzawa, M. Kato, K. Kataoka, T. Tsuruta, *Macromolecules*, **27**, 4848(1994)

[9] Dong, L.-C.; Hoffman, A.S. J. Controlled Release **1991**,15, 141

[10] Feil, H.; Bae, Y.H.; Feijen, J.; Kim, S.W. Macromolecules **1993**,26, 2496

[11] Kazerouni, M.R.; Hedberg, L.; Hedberg, K. J. Am. Chem. Soc. **1994**, 116, 5279

[12] T. Tanaka, in "Encyclopedia of Polymer Science and Engineering", 2nd Ed., H. Mark, N. M. Bikales, C. G. Overberger, G. Menges, J. I. Kroschwitz, Eds, **1987**, pp. 514-531

SURFACE MODIFICATION WITH HYDROGELS VIA MACROINITIATORS FOR ENHANCED FRICTION PROPERTIES OF BIOMATERIALS

C. Anders[2], R. Gärtner[1], V. Steinert[1], B.I. Voit[1], S. Zschoche[1],

[1]Institute of Polymer Research, Hohe Strasse 6, D-01069 Dresden and
[2]CREAVIS, 45764 Marl, Germany

Introduction

Today, a very broad spectrum of polymeric materials is used in medicine. This covers polymers as carriers for drugs and contrast agents, short and long time body implants as well as all kinds of polymeric gadgets for applications outside the body. The contact of these polymers with body fluids causes several problems and forces the polymer chemists to search for biocompatible materials. One possibility to improve biocompatibility, especially blood compatibility, is the increase of the hydrophilicity of the polymer surface. This can be achieved by surface modification of the polymer e.g. by plasma treatment and subsequent grafting reaction of polyethylene glycol or acrylic acid[1,2]. One the other hand, hydrogels are remarkably successful in different applications in medicine, e.g. as drug delivery system, contact lenses, coatings, thus surface modification of polymers for medical applications with hydrogels seems to be very promising.

We would now like to describe a general method to fix polyacrylic acid hydrogels on different polymer surfaces of irregular shape, e.g. on catheter tubes[3]. This modification reduces drastically the friction of these materials but other biological relevant properties especially reduced protein adsorption can be achieved similarly by choosing the appropriate monomers.

Experimental

The macroinitiator was synthesized by reaction of poly(octadecene-co-maleic anhydride) (Polysciences Inc., molecular weight 30,000 - 50,000 g/mol) with *tert.*-butylhydroperoxide in the presence of triethylamine (molar ratio 1:0.5:0.5) in acetone (polymer content 50 % by weight)[4]. The isolation of the reaction product was performed by precipitation of the diluted reaction mixture in 0.05N HCl. The active oxygen content in the macroinitiator was 0,8 mmol/g (estimated iodometrically).

The substrate was coated with the macroinitiator at room temperature by dipping it in an isopropanol solution (different concentrations) of the macroinitiator with and without crosslinker. The coating was dried at room temperature and a subsequent tempering at 120 °C for 5-10 min.was carried out in the presence of a crosslinker. The surface polymerizations were performed at 90 degree under nitrogen in water. The monomer content in the water solution was 3 - 10 % by weight, the reaction time 1 - 3 hours.

Results and Discussion

Surface modification by chemical or plasma treatment is often accompanied with a material property change of the bulk material since the modification cannot be fully localized to the surface only. In addition, modification of irregularly shaped materials might be problematic. On the other hand, the synthesis of new bulk materials with the appropriate mechanical properties and a biocompatible surface is difficult and costly. Therefore, we developed a simple two step principle for surface modification of different substrates with hydrogels. The substrate is first homogeneously dip-coated with an water insoluble macroinitiator. Then, surface homo- and copolymerizations of ionic and nonionic monomers are carried out in water (Scheme 1)

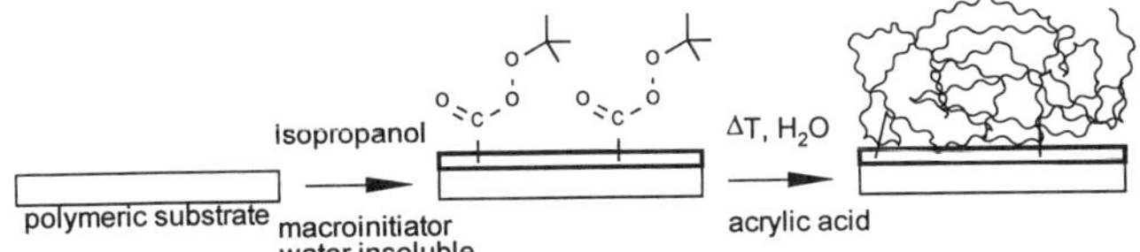

Scheme 1

The use of an appropriate macroinitiator which shows excellent adhesion to the substrate, good film forming properties and high initiator efficiency is essential for the success of this principle. Therefore, the macroinitiator must be optimized for different substrates.

We found that poly(octadecene-co-maleic anhydride), partially reacted to *tert.* butyl perester (Scheme 2), is very suitable for the covering of polyamide-12 and poly(ether-*block*-amide) (PEBAX, Elf Atochem) substrates which are often used as catheter materials. This macroinitiator has the right ratio of hydrophilic and hydrophobic parts to show good adhesion to the substrate and no water solubility. In addition, acid functional groups can interact with the surface of the substrate. A thin, homogeneous macroinitiator coating could be applied simply by dipping the substrate into an isopropanol solution of the macroinitiator and air-drying the layer.

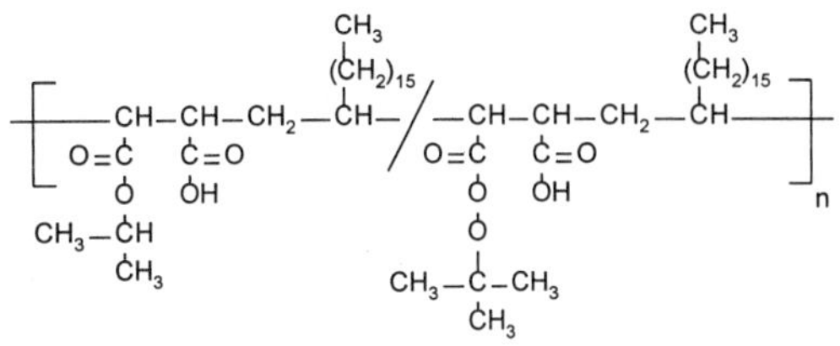

Scheme 2

The homogeneity and the thickness of the macroinitiator layer was characterized by ATR-FT-IR spectroscopy. A film thickness below 0.5 µm was found to be optimal.

Based on the macroinitiator covered substrates, a variety of polymers from monomers suitable for free radical polymerization can be fixed to surfaces. Due to our intention to increase the hydrophilicity of catheter surfaces, we favored water soluble monomers (e.g. acrylic acid, acrylamides, hydroxyethylmethacrylate, sodium-styrene-sulfonate). The consistency of the hydrogel coating, e.g. thickness, crosslinking density, mechanical stability and composition, could be well controlled by the reaction conditions and the monomer composition and concentration in solution. Still, polymerization from a solid surface is not comparable to solution or bulk polymerization and the influence of the concentration gradient of the radicals, the radical efficiency, and reduced mobility of the chain ends still have to be studied. In the copolymerization of acrylic acid and sodium-styrene-sulfonate it was found that for an acrylic acid content below 70 mol% there was a remarkable deviation of the composition of the polymer layer fixed to the surface (determined by XPS) and the expected composition calculated from the copolymerization parameters. The covering of catheters with a sulfonate containing polymer is of interest in order to reduce bacteria adhesion to the surface.

Friction properties The intention was to cover catheters with a biocompatible hydrogel layer to reduce the friction of the bulk material. For this, planar polyamide and poly(etherester-*block*-amide) model surfaces were dip-coated with the macroinitiator shown in scheme 2 and relatively thick poly-acrylic acid hydrogel layers (at least 1.5 µm) were grafted to the surface. Similarly, the process was transferred to real catheter tube using specialized equipment. The friction coefficient of the coated materials towards a base (plate or cylinder) of uncoated material was determined according to Figure 1 by using a Zwick 1456 testing machine.

Surprisingly, it was found that the friction coefficients of all materials were strongly dependent on the normal force F_N (Figure 1). Therefore, one has to compare the results obtained with identical normal force. Evaluating the materials it was found, that the friction could be drastically reduced by the hydrogel layer. E.g. at F_N = 5.1 N uncoated polyamide-12 exhibits µ = 0.23 and with a polyacrylic acid coating the friction coefficient could be reduced to µ = 0.009.

However, the mechanical stability of the hydrogel layer has been insufficient when no crosslinking agents were used. This was shown by the fact that the hydrogel coating was partially removed from the substrate after several friction treatments and thus, the friction coefficient increased again.

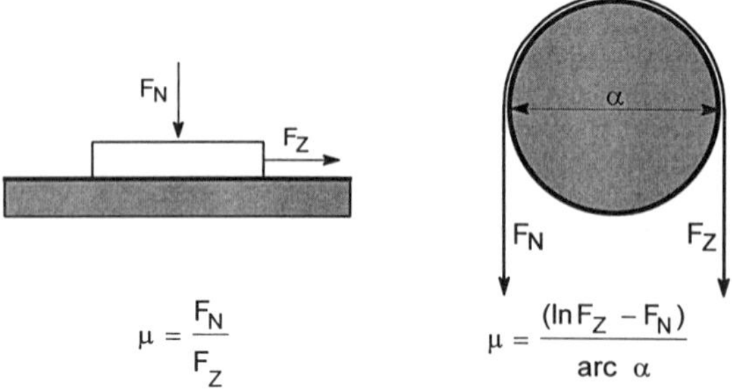

$$\mu = \frac{F_N}{F_Z}$$

$$\mu = \frac{(\ln F_Z - F_N)}{\text{arc } \alpha}$$

Figure 1: Determination of the friction coefficient (F_N = normal force; F_Z = tensile force)

A much better mechanical stability of the coating was achieved when the macroinitiator was coated together with a crosslinking agent , e.g. methylenebisacrylamide or pentaerythritoltriallylether. In the subsequent grafting reaction the crosslinker led to a high crosslinking density close to the substrate and favored a strong connection between substrate, macroinitiator, and hydrogel layer. A crosslinking gradient was achieved by this method with a lower crosslinking density at the surface of the hydrogel. This allowed a good swelling of the hydrogel in water and further reduced the friction. No delamination of the coating was observed even when strong mechanical force was applied.

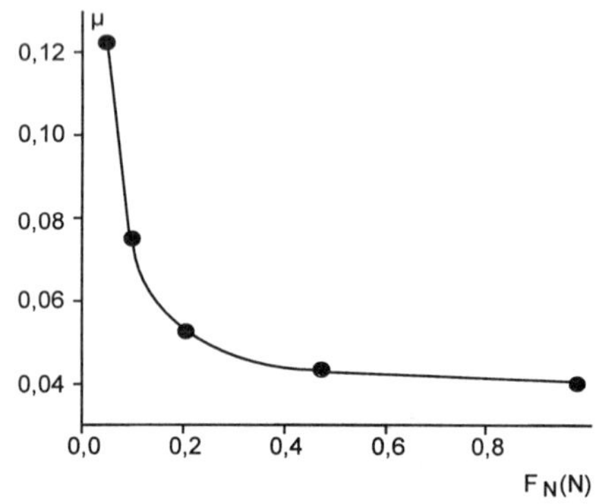

Figure 2: Friction coefficient of PEBAX-catheter coated with grafted polyacrylic acid (polymerization from 5% acrylic acid solution at 90 °C for 3 h, crosslinker in the macroinitiator layer)

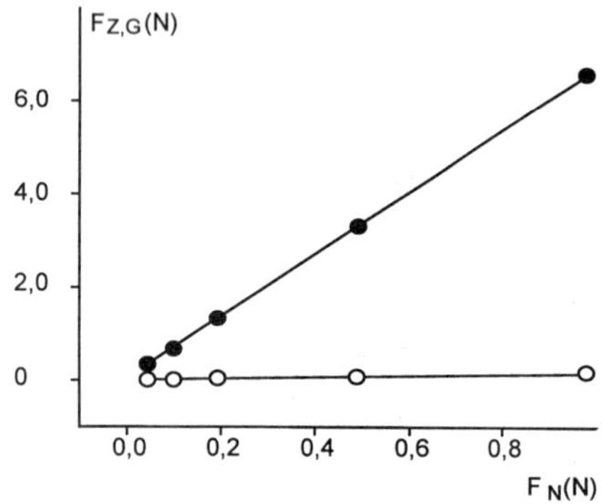

Figure 3: Reduction of the tensile force: additional force caused by friction for ● uncoated catheter (PEBAX) and ○ coated catheter (PEBAX)

The effect of the hydrogel layer on the friction was very similar for both substrate materials, polyamide-12 and PEBAX. Most studies on catheters were carried out on PEBAX. It was found that the friction coefficient for uncoated PEBAX is much higher than that of uncoated poly-

amide-12 (μ = 0.65 instead of 0.23) and no high normal force F_N could be applied to uncoated PEBAX since the material elongated under stress. A polyacrylic acid coating reduced the friction coefficient in this case down to 0.04 (Figure 2).

The enormous effect of the coating on the friction is demonstrated in Figure 3 where the large difference of the tensile force which is necessary to overcome the force caused by the friction is given for uncoated and coated PEBAX catheters.

The described method to modify polymeric surfaces is very versatile and can be applied to a variety of substrates and polymeric coatings. Thus, different applications for coatings in biomedical materials but also for materials in technical processes, e.g. membranes and filters are possible. E.g. first results on the reduction of bacteria adhesion by a hydrophilic coating prepared as described above are very promising.

References

[1]   T. Tsuruta, *Adv. Polym. Sci.* **126**, 1 (1996); Y. Ikada, in: *ACS Symposium Series* **540**, Chapter 3, p. 35 (1994)

[2]   Wichterle, O., Lim, D., *Nature* **1960**, *185*, 117; Peppas, N.A. (ed.), *Hydrogels in Medicine and Pharmacy* **1986**, CRC Press, Boca Raton; LaPorte, R. (ed.), *Hydrophilic Polymer Coatings for Medical devices* **1997**, Technomic, Lancaster, Gombotz, R., Pettit, D.K., *Bioconjugate Chem.* **1995**, *6*, 332.

[3]   Anders, C., Jacobasch, H.-J., Steinert, V. Zschoche, S., *DE 197 27 554.0* **1997** and Anders, C., Meier-Haack, Steinert, V. Zschoche, S., Jacobasch, H.-J., *DE 197 27 554.0* **1997**.

[4]   Steinert, V., Rätzsch, M.; Reinhardt, S., *DE 40 34 902 A1* **1990**

Stimuli-Responsive Polymers for Bioseparation

I. Yu. Galaev and B. Mattiasson

Department of Biotechnology, Center for Chemistry and Chemical Engineering, Lund University, P.O. Box 124, S-221 00, Lund, Sweden

STIMULI-RESPONSIVE POLYMERS

Stimuli-responsive are water soluble polymers which undergo fast and reversible changes in microstructure triggered by small changes of medium property (pH, temperature, ionic strength, presence of specific chemicals, light, electric or magnetic field). These microscopic changes of polymer microstructure manifest themselves at the macroscopic level as a precipitate formation in a solution or as a pronounced decrease/increase of the hydrogel size and hence of water content. Macroscopic changes in the system with smart polymers/hydrogels are reversible and elimination of the trigging changes in the environment returns the system to its initial state. We have used both pH-sensitive and thermosensitive polymers. Commercially available copolymer of methylmethacrylate and methacrylic acid (trade name Eudragit S-100) is soluble at neutral and alkaline conditions, but precipitates at pH around 4.8. Copolymers of N-isopropylacrylamide (NIPAM) or N-vinylcaprolactam (VCL) were synthesized in our laboratory and used as thermosensitive polymers.

AFFINITY PRECIPITATION

One of the main applications of smart polymers in down-stream processing has been the development of affinity precipitation procedure. Such precipitations exploit ligands coupled to a water soluble polymer, affinity macroligands. The macroligand forms a complex with the target protein. Phase separation of the complex is triggered by small changes in the environment resulting in transition of polymer backbone into an insoluble state. The target protein is then either eluted from the insoluble macroligand-protein complex or the precipitate is dissolved, the protein is dissociated from the macroligand and the ligand-polymer conjugate is precipitated again, now without the protein which remains in supernatant in purified form. Various ligands like triazine dyes, sugars, protease inhibitors, antibodies, nucleotides, were successfully used for affinity precipitation [1, 2]. Triazine dyes, robust pseudoaffinity ligands for many nucleotide dependent enzymes, were successfully used in conjugates with Eudragit S 100 (a copolymer of methacrylic acid and methylmethacrylate which precipitates on decreasing pH) for purification of dehydrogenases from various sources by affinity precipitation [3 - 6].

Sugar ligands constitute another attractive alternative for being used in bioseparation. *p*-Aminophenyl-α-D-glycopyranoside was coupled via the amino group to Eudragit S-100 and successfully used for purification of the lectin, Concanavalin A (Con A). Con A is a tetramer capable of binding up to four sugar ligands. The complex formation of Con A with sugar-Eudragit S 100 conjugate could result in network formation. The critical polymer to lectin ratio necessary for the network formation was determined by dynamic laser light scattering (DLLS) [13]. Dissociation of the formed network after addition of different sugars could also be recorded by DLLS and used to calculate binding constants of sugars with Con A, the latter were in good agreement with binding constants reported in the literature [14].

Poly(N-isopropylacrylamide) (poly-NIPAM) is a thermosensitive polymer and precipitates from aqueous solutions at 30-35 °C due to the progressive increase in intra- and intermolecular hydrophobic interactions with elevating temperature. The incorporation of relatively hydrophilic imidazole moieties into the polymer molecule hindered these hydrophobic interactions and resulted in drastic increase in the precipitation temperature. The effect is even more pronounced at lower pH values where imidazole moieties were protonated and hence rendered more hydrophilicity to the polymer molecule. When loaded with Cu(II), poly-VI-NIPAM did not precipitate at all on heating up to 70 °C. The increase in ionic strength has a dramatic effect on precipitation of Cu(II)-poly-VI-NIPAM. In the presence of 0.1 M NaCl at 32 °C a pellet of flocculated material was immediately formed. The efficient precipitation of Cu(II)-poly-VI-NIPAM by high salt concentrations at mild temperature is very convenient for metal affinity precipitation. High salt concentration does not interfere with protein-metal ion-chelate interaction and on the other hand it reduces the possibility of nonspecific binding of foreign proteins to the polymer both in solution and when precipitated.

The flexibility of polymer chains in solution allows several imidazole ligands on a polymer molecule to come close enough to interact with the same Cu(II)-ion and thus to provide sufficient strength of polymer-Cu(II) interactions. Kunitz soybean trypsin inhibitor was specifically coprecipitated during precipitation of Cu(II)-loaded poly-VI-NIPAM. The elution was achieved by solubilization of the inhibitor-Cu(II)-polymer complex in the presence of an excess of the competing ligand, imidazole. Subsequent precipitation of the polymer results in inhibitor remaining free in solution in a purified form [7]. The same approach was successfully used for the isolation of α-amylase inhibitor from wheat meal [15] and separation of two α-amylase inhibitors from ragi seeds [16].

An interesting group of smart polymers is the polyelectrolyte complexes (PEC) that areformed by two oppositely charged polymer molecules. The solubility of PEC is regulated by the composition of the complex as well as by pH and ionic strength [12]. PEC were successfully used for separation of inactivated from active glyceraldehyde-3-phosphate dehydrogenase form rabbit muscle. When inactivated, tetrameric native enzyme dissociate first into active dimers, followed by inactivation of the latter. The monoclonal antibodies specific for inactivated dimers were raised in mice. The antibodies were covalently coupled to the polyanion. Stoichiometric addition of a polycation results in PEC formation and complete precipitation of the antibody-polymer conjugate. When polyanion-antibody conjugate was added to the partially inactivated glyceraldehyde-3-phosphate dehydrogenase, the antibodies bind the inactivated dimers. The subsequent stoichiometric addition of polycation and precipitation of the PEC resulted in higher specific activity of the reactivated enzyme as compared to that of the preparations which was not treated with antibody/PEC. Enzyme preparation with initial specific activity was obtained by combination of the above treatment and followed with the removal of irreversibly inactivated and insoluble aggregates by centrifugation.

AFFINITY PARTITIONING

Two aqueous polymer solutions become mutually incompatible when the threshold concentration of polymers are exceeded. The selective partitioning of proteins between the two phases formed has proven to be an efficient tool of purification of proteins and of some low molecular weight substances. The main problem of the method - how to separate the target protein from the phase-forming polymer - has not yet been completely solved. Smart polymers provide an elegant solution of this problem. The use of smart polymer as a phase-forming polymer in aqueous two-phase system was initiated by Hughes and Lowe in 1988 [9] This approach was further on developed by Tjerneld et al. [10].

Aqueous solutions of the copolymer of N-vinylcaprolactam and 1-vinyl imidazole (poly-VI-VCL) forms two phase systems with dextran T70. Partitioning of recombinant lactate dehydrogenase (from the thermophilic *Bacillus stearothermophilus*) carrying a tag of six histidine residues, as well as that of total protein in this aqueous two phase system is pH independent. Partition coefficients close to 1 indicate no specific affinity of the enzyme to any of the phase forming polymers. In the presence of Cu^{2+} ions promoting specific interactions of the enzyme with poly-VI-VCL, lactate dehydrogenase partitioned preferentially into the poly-VI-VCL phase, partition coefficient being around 200. Separation of the poly-VI-VCL phase and thermoprecipitation of the polymer at 45 °C resulted in aqueous solution of the 8-fold purified lactate dehydrogenase, the overall recovery of enzyme activity was around 80 % [11].

TEMPERATURE INDUCED ELUTION

Poly(N-vinyl caprolactam) (poly-VCL) is a smart polymer which undergoes phase transition in aqueous solution at about 35 °C and it interacts efficiently with the triazine dye, Cibacron Blue, which is used widely as a ligand for a dye-affinity chromatography of various nucleotide dependent enzymes. The polymer binds strongly via multipoint interaction to the dye ligands when Blue Sepharose (Sepharose matrix with chemically attached Cibacron Blue) is used. A packed bed of the matrix was washed with polymer solution.

At elevated temperature poly-VCL molecules are in a compact globule conformation, capable of binding only to a few ligands on the matrix. Enzyme, lactate dehydrogenase from porcine muscle has good access to the ligands and binds to the column. When decreasing the temperature, polymer molecules undergo transition to a more expanded coil conformation. The polymer molecules interact now with more ligands and begin to compete with the bound enzyme for the ligands. Finally the bound enzyme is displaced by the expanded polymer chains.

This system was used for lactate dehydrogenase purification. Porcine muscle homogenate was applied on a column until breakthrough at 40 °C. The foreign proteins were washed out with 0.1 M KCl, then the column was cooled to room temperature and virtually homogeneous lactate dehydrogenase was eluted with the same buffer. Purification factor was 17, recovery was 90 % [8].

ACKNOWLEDGMENT

The supports of the Royal Swedish Academy of Sciences (KVA) and the Swedish Competence Center for Bioseparation are gratefully acknowledged.

REFERENCES

1. M. N. Gupta and B. Mattiasson, in *Highly Selective Separations in Biotechnology* G. Street Ed., Blackie Academic & Professional, London, 1994, p. 7.
2. I. Yu. Galaev, M. N. Gupta and B. Mattiasson, *CHEMTECH*, Dec., 19 (1996).
3. D. Guoqiang, R. Kaul, and B. Mattiasson, *Bioseparation*, **3**, 3 3 (1993).
4. M.N. Gupta, D. Guoqiang, and B. Mattiasson, *Biotechnol. Appl. Biochem.*, **18**, 321 (1993).
5. H.-C. Shu, D. Guoqiang, R. Kaul, and B. Mattiasson, *J. Biotechnol.*, **34**, 1 (1994).
6. D. Guoqiang, A. Lali, A., R. Kaul, and B. Mattiasson, *J. Biotechnol.*, **33**, 23 (1994).
7. I.Y. Galaev, A. Kumar, R. Agarwal, M.N. Gupta, and B. Mattiasson, *Appl. Biochem. Biotechnol.* **68**, 121 (1997).
8. I.Y. Galaev, C. Warrol, C., and B. Mattiasson, *J. Chromatogr. A*, **684**, 37 (1994).
9. P. Hughes and C. R. Lowe, *Enzyme Microb. Technol.*, **10**, 115 (1988).
10. G. Johansson and F. Tjerneld, in *Highly Selective Separations in Biotechnology* G. Street Ed., Blackie Academic & Professional, London, 1994, p. 55.
11. T. T. Franco, I. Y. Galaev, R. Hatti-Kaul, N. Holmberg, L. Bülow, and B. Mattiasson, *Biotechnol. Techn.* **11**, 231 (1997).
12. M. B. Dainiak, V. A. Izumrudov, V. I. Muronetz, I. Yu. Galaev and B. Mattiasson; *J. Molec. Recgn.*, in press.
13. E. Linné-Larsson, I. Y. Galaev, L. Lindahl and B. Mattiasson, *Bioseparation*, **6**, 273 (1996).
14. E. Linné-Larsson, I. Yu. Galaev and B. Mattiasson, *Bioseparation*, **6**, 283 (1996).
15. A. Kumar, I.Yu. Galaev and B. Mattiasson, *Biotechnol. Bioeng.* in press.
16. A. Kumar, I.Yu. Galaev and B. Mattiasson, *Bioseparation*, submitted.

OPTIMIZATION OF CONCENTRATION PROFILES IN POLYMER MATRICES FOR CONTROLLED RELEASE

Sanxiu Lu, W. Fred Ramirez, and Kristi S. Anseth

Department of Chemical Engineering
University of Colorado, Boulder, CO 80309-0424

Introduction

Polymer delivery systems based on diffusion controlled release are becoming increasingly important for pharmaceutical industries, but it is very challenging to design systems where the drug is released at the desired rate (e.g., constant, pulsatile, etc.). In typical devices, the molecule to be delivered is uniformly dissolved or dispersed throughout a polymer matrix, which inherently leads to a decreasing release rate with time. To address this issue, several investigators [1] have developed novel approaches to try to achieve constant drug release rates. These approaches have examined geometry factors, control of swelling and dissolution fronts, and surface eroding polymers. While each of these approaches, as well as many others not mentioned here, have distinct advantages and disadvantages to achieve controlled release rates, a less-explored, promising alternative is to design polymer systems with specified, non-uniform initial drug distributions. However, one reason this approach has not been more widely used is the difficulty associated with preparing the drug concentration profiles in the polymer.

As an alternative to existing methods to produce non-uniformity in controlled release matrices [2-5], we propose that the use of photopolymerizations can greatly simplify this task. For example, drug molecules can be dissolved in liquid, prepolymer solutions at different concentrations. The non-uniform drug distribution is then established by photopolymerizing layer-by-layer (each layer using a different drug concentration) the polymer matrix (see Figure 1). Photopolymerizations provide many advantages compared to other methods. First, most photopolymerization conditions are sufficiently mild to enable the polymerization to be carried out in the presence of many biological materials. Hence, no subsequent loading step is required after formation of the polymer. Photopolymerizations proceed very rapidly at room temperature, and hence, the temperature and pH can be limited to physiological ranges. Additionally, photopolymerizations require no solvent, and thus, the concerns of organic solvent residue in the drug delivery matrix is eliminated. Moreover, a multitude of monomers are photopolymerizable, which facilitates tailoring the material properties (e.g., diffusivity, swelling, degradation rate) of the polymer to the specific application. Finally, since photopolymerizations are initiated with UV or visible light, spatial and temporal control of the polymerization is achieved by controlling the area and time of light incidence, thereby allowing the production of complex matrix devices, such as multilaminates.

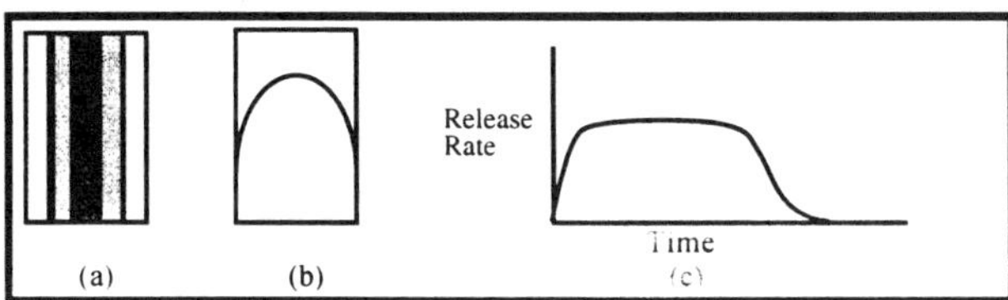

Figure 1. (a) Non-uniform drug distributions developed in a photopolymerized polymer. (b) As layer thickness is decreased, any concentration profile can be approximated. (c) Concentration profiles can be subsequently optimized to produce a constant drug release rate.

Experimental

Hydrogels of poly(HEMA) (2-hydroxyethyl methacrylate) were prepared by a free-radical photopolymerization at room temperature. Diethylene glycol dimethacrylate (DEGDMA) was used as the crosslinking agent. In these experiments, 40 wt% water based on the monomer solution was used to obtain homogeneous hydrogels that maintain their original shape and volume when placed in the release medium (i.e., the resulting gels were equilibrium swollen upon formation). The initiator concentration was 0.1 wt% based on the HEMA concentration. Initial release studies were performed with a model drug compound, acid orange 8 (AO8), which is a hydrophilic low molecular weight compound. The AO8 concentration was less than 1.0 wt% based on the monomer solution, which is well below its solubility in water at room temperature (3.0 wt%). The low concentration of AO8 was chosen because it competes for photons of UV radiation at the initiating wavelength 365 nm.

The solution of HEMA, DEGDMA, deionized water, AO8, and photoinitiator was transferred into a transparent mold which controlled the thickness of each layer. The liquid solution was converted to a crosslinked hydrogel by exposure to low intensity 365 nm UV light (Blak-Ray, 12 mW/cm^2) for a few minutes in a nitrogen atmosphere. Upon completion of the polymerization, an additional spacer was introduced into the mold; a new liquid solution with a different dye concentration was injected; and the new layer was photopolymerized. We have prepared various laminated devices (up to 7 layers) with varying layer thickness (50 μm to 1 mm), drug concentrations, and polymer compositions and crosslinking density.

Modeling and Optimization

For the drug delivery systems studied in this research, the system was modeled as one-dimensional, transient, Fickian diffusion in a laminated disk. For a disk with thickness, L, and an initial drug concentration distribution, F(x), in contact with a solvent maintained at sink conditions, the drug flux, J, at the polymer-solvent interface is given by

$$J(t,L) = -D\frac{\partial c}{\partial x}\bigg|_{x=L} = \frac{2D}{L}\sum_{n=0}^{\infty}(-1)^{n+1}\lambda_n e^{-D\lambda_n^2 t}\left(\int_0^L F(x)\sin(\lambda_n x)dx\right) \tag{1}$$

where D is the drug diffusivity and $\lambda_n = \frac{(n+0.5)\pi}{L}$. Parameters associated with this equation (e.g, D, L, F(x)) were measured experimentally, and the resulting equation was used to predict the effects of various initial concentration profiles on diffusion-controlled drug release patterns.

With the experimental ability to control release patterns in matrix devices through non-uniform concentration distributions and a model to predict accurately the effects of the initial concentration profile on the final release [6], we also explored optimization techniques. Optimization provides a further tool to guide in the choice of parameters to develop matrix devices with optimal release behavior. For example, these methods can be used to conveniently determine what initial concentration profile should be used to obtain as close to the desired flux as possible.

Optimization involves defining an objective or performance index that represents some feature of the process to be maximized or minimized. The objective functional (N) to be minimized in this problem is

$$N = \int_0^L Av^2(x)dx + \int_0^{t_f}(J(t,L) - J^*(t,L))^2 dt \tag{2}$$

This objective functional keeps the actual release flux profile, J(t,L), close to the desired profile, J*(t,L), while minimizing the amount of initial loaded drug (e.g., to reduce cost). The initial drug loading contribution to the objective functional is: $F_1 = Av^2(x)$. The boundary contribution to the objective functional is: $F_2 = (J(t,L)-J^*(t,L))^2$. The importance of the two terms can be scaled by the magnitude of the coefficient A. For example, the magnitude of the coefficient A can be adjusted to make $F_1 \ll F_2$, $F_1 = F_2$, or $F_1 \gg F_2$. To determine the necessary conditions for the extreme of N, calculus of variation was used. The mathematical details of the optimization process are presented elsewhere [7].

Results and Discussion

The model and experimental results in Figure 2 illustrate how simple changes in the initial concentration profile dramatically affect the final drug release behavior. The investigated device was composed of four individual layers of the same thickness, 0.2 mm, and the initial AO8 concentration profile was 1.2 wt%, 0.85 wt%, 0.2 wt%, and 0.0 wt%, from the center outward, respectively. To simplify the characterization of the release results to 1-dimensional diffusion, all surfaces were bound with an impermeable membrane except one. At 1 wt% crosslinker concentration (DEGDMA),

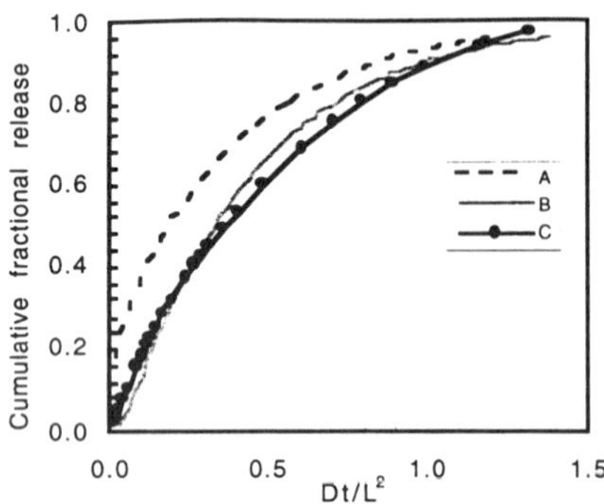

Figure 2. A: model prediction for uniform drug loading; B: model prediction for nonuniform drug loading; and C: experimental results for nonuniform drug loading. The nonuniform concentration profile was from the center of the device outward, 1.2 wt-%, 0.55 wt-%, 0.2 wt-% and 0 wt-%.

the diffusivity of AO8 in the swollen poly(HEMA) network was experimentally determined to be ~4.04 *10^{-7} mm^2/s. The data in Figure 2 shows the flux of AO8 from the swollen poly(HEMA) networks as a function of dimensionless time. These experimental results were also compared to the model predictions.

When the drug loading is uniformly distributed, the release pattern follows typical first-order diffusion behavior: an initially high release rate (i.e., the burst effect) followed by a rapidly declining drug release rate. In the laminated device, the distribution of loaded drug was made more nonuniform by increasing the relative loading at the center of the device compared to the surface. Dramatic changes are seen in the release behavior, especially in the release pattern at the early stages.

Also illustrated in Figure 2, the theoretical and the experimental release results agree quite well, further validating the use of the model to analyze the release of drug from such devices. The experimental results also confirmed that diffusion is the dominant release mechanism for the AO8 from these swollen poly(HEMA) hydrogel. It is worth noting that the fit of the release data in Figure 2 has no adjustable parameters; all of the parameters (e.g., diffusivity, layer thickness and initial concentration profile) were determined from experimental measurements. In addition, the absence of a burst effect, which is usually seen with matrix type delivery system, illustrates the effectiveness of non-uniform concentration profiles in controlling the release rates.

To provide further guidance in the design and development of the initial concentration profile in the laminated device, optimization techniques were used. For example, Figure 3 a and b contains the results for the optimized initial concentration profile to obtain a triangular release pattern. The optimized concentration profile was restricted to concentration values greater than zero. If this restriction is removed, the profile oscillates from positive to negative to sustain the release at the desired rate. Using these techniques, we have also examined and optimized the case for zero order release [7]. In general, these techniques can be extended to consider variables such as spatially dependent diffusion coefficients, time varying diffusion coefficients (e.g., in degradable systems), and layer thicknesses in addition to non-uniform concentration profiles. Furthermore, the optimization can be performed on each of these variables individually, as well as simultaneously.

Conclusions

Photopolymerizations were used to construct laminated hydrogel matrices with non-uniform initial drug distributions. The photopolymerization technique provides a facile method to produce rapidly complex devices with a wide range of initial drug profiles. The effects of the initial concentration profile on the final drug release pattern was investigated experimentally and theoretically. Significant differences in the early time release behavior were reported. Furthermore, a technique based on optimal control theory was developed to optimize the initial concentration profile to produce as close to the desired release rate as possible.

Acknowledgment. The authors wish to thank the David and Lucile Packard Foundation for the support of this work through a Packard Fellowship to KSA.

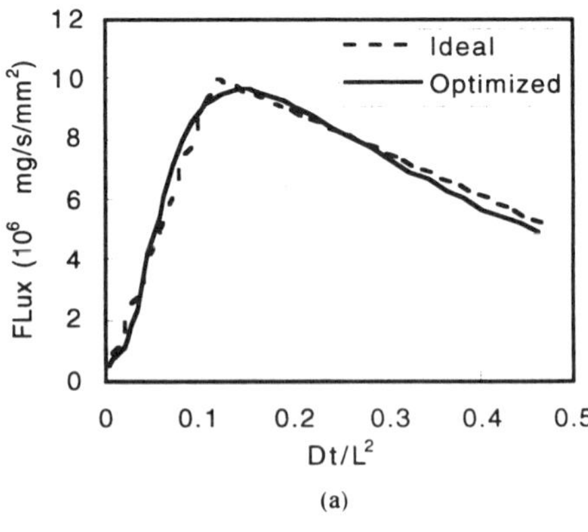

(a)

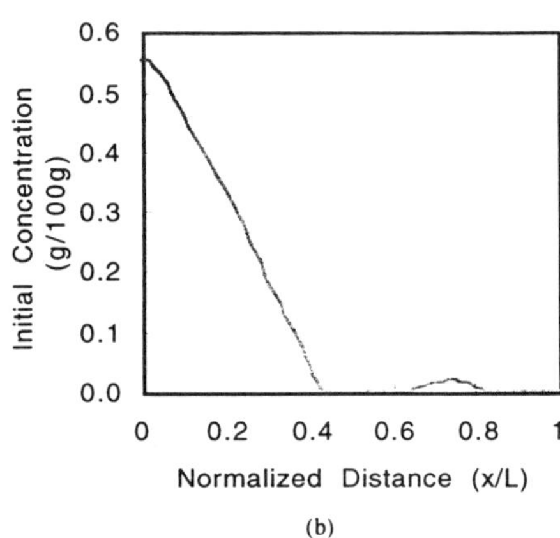

(b)

Figure 3. a. Optimized release profile for a triangular release pattern. b. Initial concentration profile that leads to the optimal result.

References

1. Hsieh, D.; Rhine, W.; Langer, R. *J. Pharm. Sci.* **1983**, *72*, 17; Heller, J. *Biomaterial* **1980**, *1*, 51; Lee, P. *J. Membrane Sci.* **1980**, *7*, 255; Korsmeyer, R.W.; Peppas, N.A. in *Controlled Release Delivery Systems*, Roseman and Mansdorf (Eds.), Dekker, New York, NY, 1983.
2. Lee, P.I. *Polymer* **1984**, *25*, 973.
3. Lee, P.I., *J. Pharm. Sci.* **1984**, *73*, 1344.
4. Fassihi, R.A.; Ritschel, W.A., *J. Pharm. Sci.* **1993**, *82*, 754.
5. Yang, L.; Fassihi, R.A., *J. Pharm. Sci.* **1995**, *85*, 170.
6. Lu, S.; Anseth, K.S., "Photopolymerization of multilaminated poly(HEMA) hydrogels for controlled release," *J. Control. Rel.*, submitted.
7. Lu, S.; Ramirez, W.F.; Anseth, K.S., "Modeling and optimization of drug release from laminated polymer matrix devices," *AIChE J.*, submitted.

Fiber-Supported Hydrogels with Controlled Surface Morphology

J.O. Karlsson[1], Å. Henriksson[1], J. Michalek[2] and P. Gatenholm[1]

[1]Department of Polymer Technology, Chalmers University of Technology, S-412 96 Göteborg, Sweden
[2]Institute of Macromolecular Chemistry, Academy of Science of the Czech Republic, S-162 06 Prague 6, Czech Republic

INTRODUCTION

Surface modification of polymers has recently turned in the direction of new areas involving the control of not only the surface chemistry but also the morphology in order to affect specific interactions. It has, for example, been found that spatial features on a surface influence cell functionality and phenotype[1]. Such surfaces may also be important in directing blood interactions, tissue reaction and healing[2-3]. While many techniques used for modifying polymer surface chemistry are known, the field of controlling surface morphology is still incompletely explored. Ratner and collaborators have shown that surface textures can be formed during radiation-induced grafting onto polyethylene[4]. Other methods so far applied for such purposes are photo-lithography, plasma reaction with masking and mechanically inducing surface features with scanning probe microscopies[5-8].

The graft polymerization technique has been widely used for the surface modification of polymers. With this technique, both thin and thick grafted layers with controlled surface chemistry can be obtained. In our laboratory we have recently studied ozone-induced graft polymerization of various hydrogel forming acrylates onto both synthetic and natural fibers[9-10].

The aim of this study has been to prepare surface supported micro-size hydrogel particles by performing an ozone-induced graft polymerization of various acrylate monomers onto cellulose fibers. Special attention was paid to elucidate the relationship between graft reaction parameters, grafting yield and the development of various surface morphologies. The monomethacrylate used for grafting was diethylene glycol methacrylate (DEGMA), and the bifunctional crosslinking agents were mono-, di-, tri- and tetra(ethylene glycol) dimethacrylates. DSC measurements were performed to obtain reaction rate profiles of various monomer mixtures. Atomic force microscopy (AFM) and scanning electron microscopy (SEM) were used to characterize the fiber surfaces before and after grafting.

EXPERIMENTAL

Materials

The cellulose fibers used were a standard sulfate fluff pulp (single fibers), EC 0.1, made from selected wood of softwood, manufactured by STORA Cell (Sweden). The mono functional monomer used in the study was diethyleneglycol methacrylate (DEGMA), and the bifunctional monomers used were mono-, di-, tri- and tetra(ethylene glycol) dimethacrylate. Mono (ethyleneglycol) dimethacrylate (EDMA) was purchased from Fluka Chem. AG. (Switzerland).

Ozone treatment and graft polymerization

The ozone treatment of the cellulose fibers was then carried out in a gas phase reactor at 32°C. The equipment used for generating ozone was an NG10 from Ozone Systems Company, which produced an oxygen/ozone flow of 0.250 m^3/h from pure oxygen gas. Immediately after ozone treatment, the substrates were placed in the monomer solution which was prepared by diluting 2.0 g of monomer in equal amounts (10 cm^3) of methanol and deionized water. The grafting process was performed in a nitrogen atmosphere, in sealed glass ampoules.

Analyses

A Perkin Elmer DSC-7 was used to study polymerization profiles of the monomers. The temperature was increased to 70°C, and the measurements were then performed isothermally. The surface morphology of the fibers was examined using the Dimension 3000 Large Sample AFM with type G scanner. A standard silicon tip was used for the analysis, which was performed in air. Scanning electron microscopy (SEM) was used to study the substrates before and after grafting. The surfaces were coated with gold before the analysis, which was performed with a Zeiss DSM 940A operated at 10 kV.

RESULTS

An enhancement of the grafting amount when adding bifunctional crosslinking agents to HEMA and Acrylic Acid monomer mixtures used for grafting was recently reported[11-12]. During graft polymerization of diethyleneglycol methacrylate (DEGMA), it was observed that the amount of grafting moderately increased, from 20% to 70%, when the monomer mixture used for grafting contained a small amount (2%) of ethyleneglycol dimethacrylate, EDMA. The yield then increased dramatically to more than 900 % at 4 % EDMA content. A study to further elucidate the relationship between grafting amount and reaction parameters was therefore performed.

Polymerization profiles and model for grafting enhancement

To examine the graft polymerization behavior of DEGMA when various amounts of EDMA were added, we performed DSC measurements. Isothermal radical chain polymerization at 70°C was performed with DEGMA containing various amounts of bifunctional crosslinker and DEGMA without crosslinker. AIBN, 0.5 weight-%, was used as initiator. The polymerization reaction is exothermal, and the maximum reaction rate from the DSC experiments is received from the peak of the heat flow curve. A rapid increase from a moderate reaction rate to a maximum indicates that autoacceleration, gel effect, occurs during the polymerization. Figure 1 plots the heat flow from five monomer mixtures at 70°C as a function of time. The data have been normalized according to sample weight, and the time at which 70°C was reached, and then kept constant, is marked with an arrow.

The DSC measurements show that the maximum reaction rate is much higher in monomer mixtures with an EDMA content of 4% and more, in comparison with lower EDMA content. As regards the results obtained during our grafting experiments, we have seen that the grafting amount at this EDMA concentration also starts to increase dramatically. Figure 2 shows a proposed model for graft

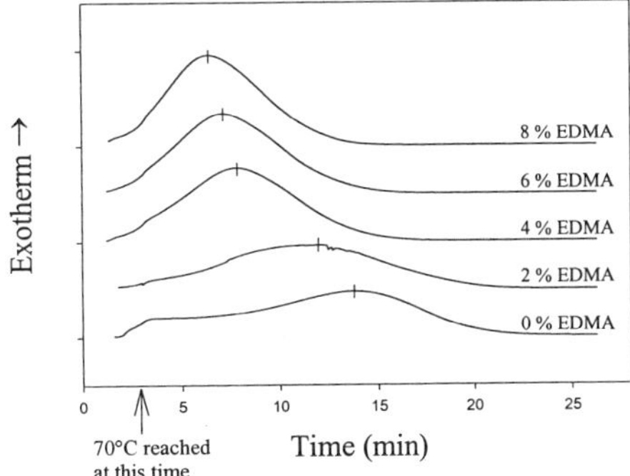

Figure 1. *Polymerization exothermals measured with DSC. DEGMA with various EDMA contents. AIBN, 0,5 weight-%, was used as initiatior.*

polymerization in the presence of both mono- and bifunctional monomers. The growing graft polymer schematically shown contains one bifunctional monomer. The pendent double bond has three possibilities for reaction (I-III): I. Crosslinking reaction with an adjacent growing grafted chain; II. Cyclization reaction; III. Reaction with a radical in the bulk, either an initiation reaction by a hydroxyl radical or a propagation reaction with a growing polymer chain. The fourth alternative (IV) is for the double bond to remain unreacted. For alternatives I, II and IV, the grafting yield is unchanged, whereas alternative III, results in an increased amount of growing polymer chains connected to the substrate and therefore an increased grafting yield. A higher concentration of bifunctional monomers will subsequently result in an increased amount of pendent unreacted double bonds available for initiation. If these double bonds react according to alternative III, the number of growing polymer chains, which may also contain double bonds, will further increase. This process can cause increased branching and crosslinking, which will in turn influence the mobility of the growing polymer radicals. Thus, simultaneously as the grafting amount dramatically increases, a situation very similar to the gel effect can arise. This is also in accordance with what we have observed for our graft polymerizations.

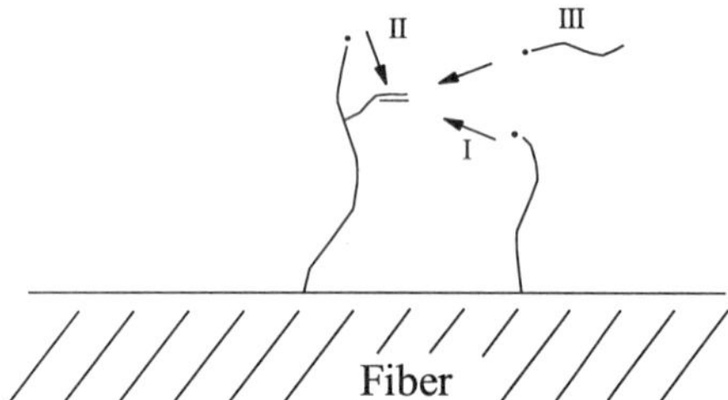

Figure 2. *Proposed model for graft polymerization in the presence of a bifunctional monomer.*

Surface morphology

The appearance of the fiber surfaces depending on the grafting amount of DEGMA/EDMA was determined by atomic force microscopy (AFM). Untreated and grafted cellulose fibers were investigated with the TappingMode technique. Figure 3 shows the surface of a fiber with 760 % grafted polymer. Bumps that attain the size of a couple of micrometers have developed on the surface as a result of the graft polymerization.

The surface morphologies of samples with different amount of grafted hydrogel were investigated by SEM. Figure 4 consists of a series of SEM micrographs showing fibers with various grafting yields. The untreated fiber in Figure 4a has an appearance characteristic for untreated cellulose fibers. In Figure 4b, a fiber with a low grafting amount, about 100 %, can be seen. It can be observed in the micrographs that the morphology of the surface starts to change.

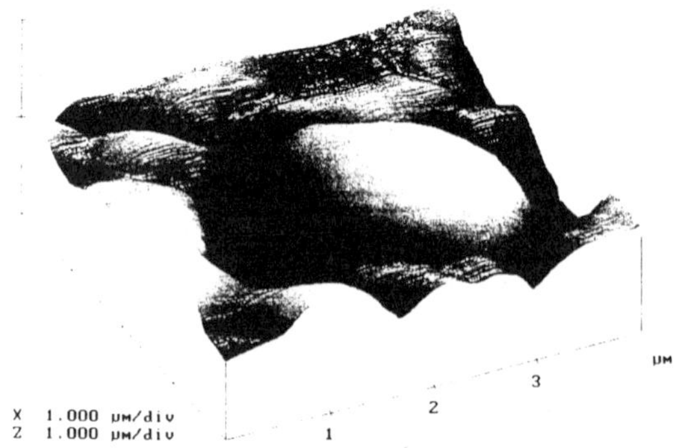

Figure 3. *AFM images of cellulose fiber with 760% (DEGMA/EDMA)*

When the grafting yield is increased to several hundred percent, shown in Figure 4c, a new surface structure is fully developed. Globules of grafted DEGMA have been formed along the fiber as a consequence of the grafting process. Figure 4d shows a fiber with a grafting amount of a couple of thousand percent. A thick layer with a bumpy surface morphology has developed on this fiber.

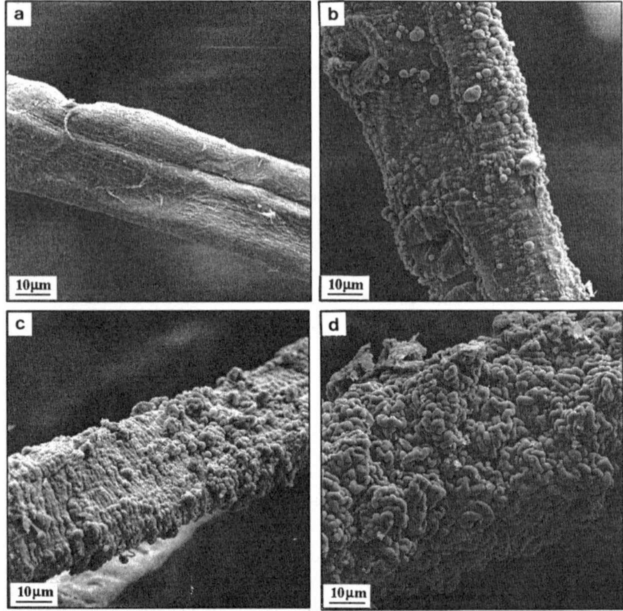

Figure 4. *SEM micrographs of untreated and grafted cellulose fibers: (a) untreated fibers; (b) fiber with a grafting amount of about 100%; (c) fiber with a grafting amount of a few hundred percent; (d) fiber with a grafting amount of a few thousand percent.*

CONCLUSIONS

It was possible to control the amount of grafted hydrogel forming polymer on cellulose fibers by adding a small amount of bifunctional crosslinking agent to the monomer mixture used for grafting. A mechanism for the grafting enhancement is proposed. An increased yield of grafting resulted in a coverage of the cellulose fibers and, when certain monomers were used, surface microstructures were formed. It was possible to influence the appearance of the microstructure by controlling the grafting yield.

REFERENCES

(1) Singhvi, P.; Kumar, A.; Lopez, G. P.; Stephanopoulos, G. N.; Wang, D. I. C.; Whitesides, G. M.; Ingber, D. E. *Science* **1994**, 264, 696

(2) Okano, T.; Suzuki, K.; Yui, N.; Sakurai, Y.; Nakahama, S. *J. Biomed. Mater. Res.* **1993**, 27, 1519

(3) von Recum, A. F.; van Kooten, T. G. *J. Biomater. Sci. Polymer. Edn.* **1995**, 7 (2), 181

(4) Cohn, D.; Hoffman, A. S.; Ratner, B. D. *J. Appl. Polym. Sci.* **1984**, 29, 2645

(5) Kumar, A.; Biebuyck, H. A.; Whitesides, G. M. *Langmuir* **1994**, 10 (5), 1498

(6) Pritchard, D. J.; Morgan, H.; Cooper, J. M. *Anal. Chem.* **1995**, 67 (19), 3605

(7) Rozsnyai, L. F.; Wrighton, M. S. *Langmuir* **1995**, 11 (10), 3913

(8) Ranieri, J. P.; Bellamkonda, R.; Jacob, J.; Vargo, T.G.; Gardella, J. A.; Aebischer, P., *J. Biomed. Mater. Res.* **1993**, 27, 917

(9) Karlsson, J. O.; Gatenholm, P. *Polymer* **1997**, 38, 4727

(10) Karlsson, J. O.; Gatenholm, P. *Polymer*, Submitted

Hydrolytic Stability of Silicone Hydrogels

Yu-Chin Lai and Edmond T. Quinn
Vision Care Global Scientific Affairs
Bausch & Lomb inc.
Rochester, NY 14603-0450

Introduction

Hydrogels are water-swollen polymer networks. The water content in a hydrogel depends on the hydrophilicity of monomers and its concentration used in fabricating it. Hydrogels have been used in a variety of biomedical applications especially as contact lenses.[1]

Hydrogels used in biomedical application are usually stored in an aqueous solution such as buffered saline. Thus, hydrolytic stability of a hydrogel is a critical requirement for its application. Alternatively, hydrolytic stability determines the shelf-life of a hydrogel as a viable medical device. Hydrolytic instability of a hydrogel gives often unwanted leachables, and water content and mechanical properties may change as a result.

Because of being highly oxygen permeable, silicone-containing hydrogels have long been under investigation for use as extended wear contact lens materials.[1] These so-called silicone hydrogels are, in general, derived from a bulky polysiloxanylalkyl methacrylate, such as 3-methacryloxypropyl tris(trimethylsiloxy)silane (TRIS, 1), or methacrylate-capped-polysiloxane, or a combination of both, along with hydrophilic monomers. Naturally, these materials must have good hydrolytic stability to warrantee their application as contact lenses.

In this paper, hydrolytic stability of silicone hydrogels was evaluated, and the mechanism for instability was assessed. Hydrogels used in this study were those derived from formulations containing a methacrylate-capped polyurethane-polydimethylsiloxane prepolymer and TRIS as the siloxane-containing components, N,N-dimethylacrylamide (DMA) and N-vinyl pyrrolidone (NVP) as the non-ionic hydrophilic monomer; and an ionic monomer. The prepolymer used was that derived from isophorone diisocyanate, diethylene glycol, α,ω-bis hydroxybutyl polydimethylsiloxane of molecular weight 3600 and end-capped with 2-hydroxyethyl methacrylate, and hereby abbreviated as IDS3H.(2).

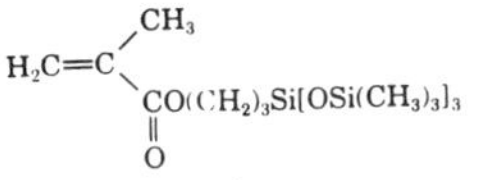

1

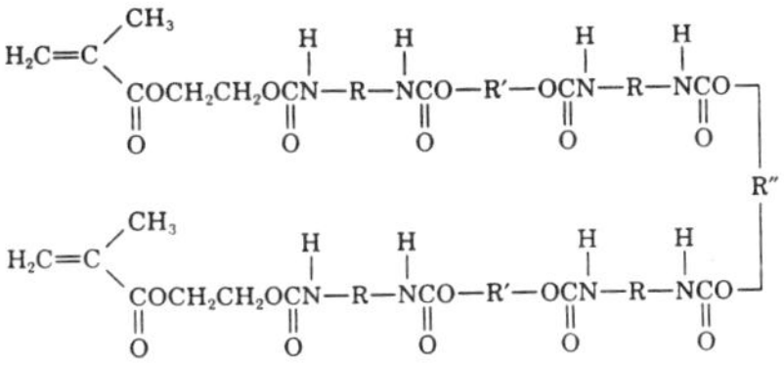

2

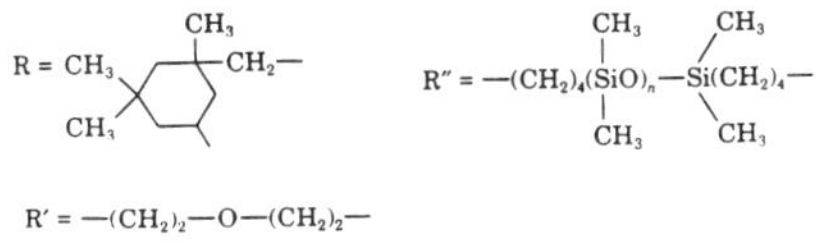

Experimental

Monomers and other ingredients. The synthesis of the polyurethane prepolymer IDS3H was described previously.[2] NVP and DMA were purified by distillation under reduced pressure. TRIS, n-hexanol, Darocur-1173 (a photoinitiator) and other components were used as received.

Preparation of hydrogel films Monomer mixes consist of monomers, prepolymer, n-hexanol and Darocur-1173 were prepared. They were then cured between two silane-treated glass plates under a UV lamp (4000 microwatts/cm^2) for 1 hour. After that, the cured films were released and extracted with isopropanol for 4 hours. These films were then heated in boiling water for 4 hours and then saturated in buffered saline (pH 7.2) before further characterizations.

Characterization of hydrogel films. Mechanical testings were conducted in buffered saline on an Instron instrument, according to the modified ASTM D-1708 (for tensile) and D-1938 (for tear strength) and were reported in g/mm^2 for modulus and g/mm for tear strength. The isopropanol extractables of cured films, and water contents of hydrogels were determined gravimetrically.

Hydrolytic stability tests. Hydrogel films were placed in glass vials filled with buffered saline at pH 7.20, sealed and they were then placed in an oven at 85 °C. Samples were pulled out after 6, 17 and 28 days. They were then subjected to mechanical testing. Separately, discs were cut from fresh hydrogel films, dried and weighed. They were then rehydrated and placed into the same oven for the same period of time. These discs were then pulled out from oven at the same times films were pulled and used for the measurements of % weight loss (leachables) and water contents.

Analysis of residual films and leachables: Selected residual films after stability tests were analyzed by FTIR and compared against control. Leachables were analyzed by GC-MS.

Results and Discussion

Formulations for stability study. Because water is a key factor for hydrolytic instability, to help understand the scope of hydrolytic instability, if any, and the mechanism for instability, only selected model formulations which give hydrogels with the same water contents (around 40 %) were used in this study. These formulations included the use of equal amounts of silicon based monomers/prepolymer (i.e., IDS3H and TRIS) and different combinations of hydrophilic monomers - DMA, NVP and VDMO. VDMO is an ionic monomer, whcih gives a carboxylic acid-containing acrylamide upon hydration. Table 1 lists formulations in this study and the water contents of hydrogels derived from them.

Stability Tests. Stability tests were carried out under an accelerated condition at 85 °C for 6, 17 and 28 days, which corresponded to 1, 3 and 5-year respectively assuming Arrhenius Equation for reaction kinetics holds. Because of the severity of test conditions and the inherent standard deviations exist with measurements of testing results, hydrogels with less than 5 % of weight loss, and less than 30 % change in tensile modulus were considered hydrolytically stable. In addition to the tests at 85 °C, one formulation with dubious stability was evaluated for changes in properties at lower temperatures and results were compared.

Figures 1 and 2 show respectively the % weight loss and the change of modulus with testing times for a series of polyurethane-silicone hydrogels derived from formulations listed in Table 1. To better compare changes of modulus for different formulations, only the relative changes in modulus were shown. The modulus at control is 1 and 1.5 means modulus is 1.5 times of control value. Based on the test results, hydrogels without ionic monomers (Formulations A and B) are stable up to 3-5 years. On the other hand, hydrogels derived from formulations with ionic monomer (Formulations C and D) showed higher levels of weight loss and changes in modulus, with the level of changes depending on the amount of ionic monomer. Interestingly, the % weight loss corresponded well with the level of changes in modulus. Because crosslinking density affects modulus of a network polymer, the changes in modulus in these hydrogels under test led to a conclusion that more crosslinking was occurring during heat. In all tests, it was found that water contents showed no or minor change.

In a separate model formulation study, a model formulations based on TRIS/DMA/Ethylene glycol methacrylate at 50/50/1.6 (Formulation E) was found stable with a weight loss less than 5 % in the same test period. However, the modulus stayed nearly constant up to 3 years, but went up 60 % at 5-year, which is parallel to those nonionic polyurethane-silicone hydrogels as described

The effect of temperature on hydrolytic stability of silicone hydrogels. While the formulation containing 5 parts of VDMO gave hydrogel with poor hydrolytic stability, it was dubious that the formulation with less VDMO (such as 1 part) might be stable if tested at lower temperatures. To demonstrate if this is the case, hydrolytic stability of this hydrogel (most in lens form) tested at lower temperatures was compared. Figure 3 gives the changes of modulus with time when tested at different temperatures. While the hydrogel gave 150 % increase in modulus (in 5-year equivalent) when tested at 85 °C, it showed no change in modulus when kept at room temperature. And the change in modulus increased with the test temperature. This suggested that accelerated test conditions did not reflect the true shelf-life of polyurethane-silicone hydrogel of compositions described I this paper. And the silicone hydrogels with low level of ionic monomers, in deed, have 5 years of shelf-life

<u>Mechanism for instability under heat</u>. To understand better the mechanism for weight loss as well as the changes in modulus with time when hydrogels subjected to heat, hydrogel films derived from DMA/TRIS and IDS3H/TRIS/DMA, were placed in pressure vessels, sealed and subjected to heat at 85 °C for 28 days. The hydrogel films were then extracted with THF. The residual films and THF-extractables were analyzed by FTIR and GC-MS respectively. The residual film of DMA/TRIS gave IR peaks charqacteristic of a silanol, indicating the hydrolysis of TRIS. The GC-MS analysis of THF extractables gave hexamethyl disiloxane (HMDS) as the only volatile leachable. Evaporation of THF solution led to no significant residue. These results led to the conclusion that hydrolysis of siloxane bonds in TRIS moiety give a silanol and HMDS, The silanol moieties from TRIS then further condensed to give more crosslinking to the hdyrogels and caused an increase in modulus. This proposed mechanism is further shown in Equation 1.

Conclusion

Silicone hydrogels derived from a methacrylate-capped polyurethane-polysiloxane prepolymer, a bulky polysiloxanyalkyl methacrylate and hydrophilic monomers, are hydrolytically stable, with 3-5 years of shelf-life. The high temperature instability of hydrogels derived from formulations with high ionic monomer content was mainly caused by the hydrolysis of siloxane bonds in the bulky siloxanylalkyl methacrylate , followed by condensation of silanol groups

Acknowledgment

The authors wish to express their thanks to S. Hill for the preparations of hydrogel films and stability tests; F. Price, J. Melpolder, and P. Witham for analysis of films and leachables; and L. Mosack and M. Andrews for mechanical testing.

References.

1. Y. C Lai, A. C. Wilson and S. G. Zantos, "Contact Lens" in *Encyl, Chem. Tech.*, 4th edition. **1993**, *7*, 192.
2. Y. C Lai and E. T.Quinn, *J. Polym. Sci.-Chem*, **1995**, *33*, 1783.

Table 1. Selected hydrogel formulations for stability study.

Formulation	Silicon-components	Hydrophilic components	Water content
A	IDS3H,TRIS	DMA only	37.50%
B	IDS3H,TRIS	DMA + NVP	39.50%
C	IDS3H,TRIS	DMA + 5 VDMO	43%
D	IDS3H,TRIS	DMA + NVP+ 1 VDMO	40%
E	TRIS	DMA	39%

Equation 1. The Mechanism responsible for increase in modulus in silicone hydrogel during heat. (Methacrylate groups not shown)

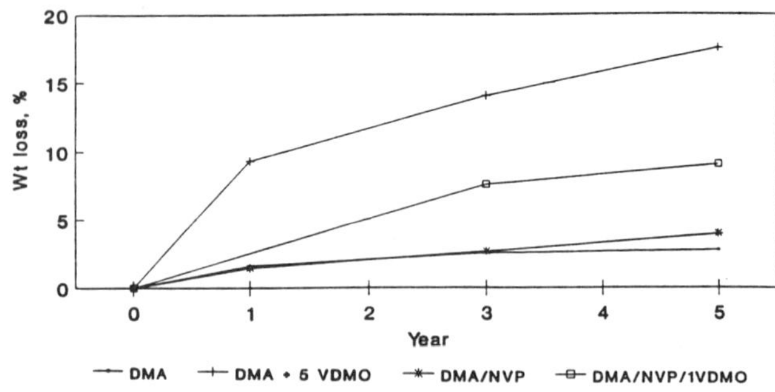
Figure 1. The % weight loss of hydrogels tested at 85 °C for 6, 17 and 28 days (equavelent to 1, 3 and 5 years).

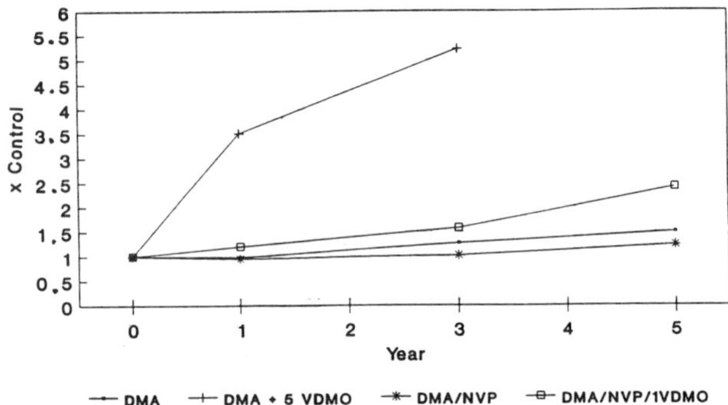
Figure 2. The change in modulus with time for hydrogels tested at 85 °C for 6, 17 and 28 days, (which are equavelent to 1, 3 and 5 years). Moduli were indicated as how many times of control value.

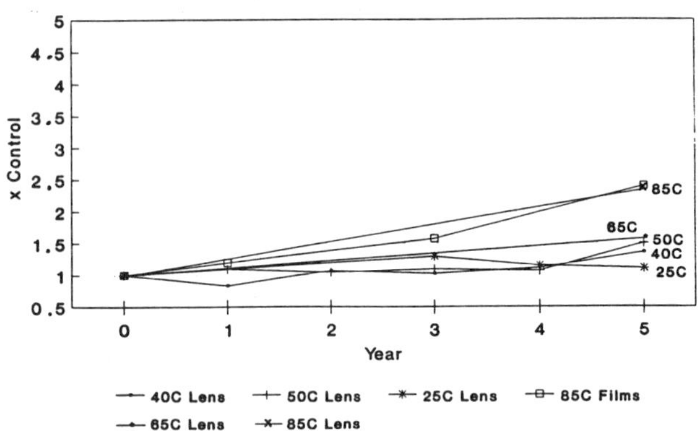
Figure 3. Comparison of the change in modulus with time for hydrogel derived from Formulation D, tested at different temperatures. Moduli were indicated as how many times of control value.

Water-soluble dendrimers as potential drug carriers

Mingjun Liu, Kenji Kono, Jean M. J. Fréchet*
Department of Chemistry, University of California,
Berkeley, CA 94720-1460

Introduction

Polymers used in drug delivery systems have been widely studied[1] because the therapeutic efficacy of many low molecular weight drugs can be improved by combining them with polymers. Furthermore, controlled and/or targeted delivery of drugs can be obtained by using polymeric delivery systems. Drugs can either be physically mixed with a polymer matrix, or chemically bound to an appropriate polymer carrier, generally through a biodegradable linkage. In 1975, Ringsdorf proposed the first model for a polymeric drug carrier.[2] Following his seminal work, numerous polymers have been investigated in the construction of drug-polymer conjugates.

Dendrimers are highly branched, three-dimensional macromolecules that have attracted much attention in recent years due to their unique structure and properties.[3] The well-defined structure, size, and controllable surface functionalities of dendrimers make them excellent potential candidates for use as drug carriers. Therefore, drugs can be either entrapped within dendritic structure or attached to their surface to form conjugates. Although dendrimers have been proposed as novel drug/gene delivery systems,[4] only a few studies[5] have been reported in this area. In this preliminary report, we present a simple design of dendrimer-based drug carrier that may be used to advantage in drug delivery schemes. Since most drugs are hydrophobic, poly(ethylene glycol) chains are attached to the dendrimer to render the dendrimer-drug conjugates water soluble. In order to incorporate both PEG and drug molecules into dendrimers, a polyether dendrimer with two different types of surface functionalities was prepared, and PEG and model drug molecules were attached, respectively.

Results and Discussion

1. Synthesis of the dual functionality dendrimer

The synthetic route is shown in Scheme 1. The synthesis starts from an ester-terminated polyether dendron because ester groups can easily be transformed to other functionalities. The reaction of diethyl 5-(bromomethyl)isophthalate **3** and methyl 3,5-dihydroxybenzoate **1** under standard condition (K_2CO_3, 18-crown-6, acetone) afforded the generation one ester **4** in excellent yield. The ester groups of **4** were then reduced by $LiAlH_4$ to afford dendron alcohol **5**. In a separate sequence, one of the two phenolic groups of methyl 3,5-dihydroxybenzoate **1** was selectively protected with tert-butylchlorodiphenyl silyl group to afford **2** in 70% yield. The coupling of **2** and **5** was then carried out under Mitsunobu condition to afford the generation two ester (56%). The ester groups were then reduced by $LiALH_4$ to give the G-2 alcohol **6** in 79% yield. Dendrimer **6** possess two different functionalities, alcohol and protected phenol that may be used to attach drug and PEG, respectively.

2. Synthesis of dendrimer-PEG conjugate

Dendrimer-PEG hybrid[6] copolymers have previously been prepared by reaction of a terminal hydroxyl group of PEG with the benzyl bromide moiety at the focal point of a polyether

Scheme 1. Synthesis of dendrimer

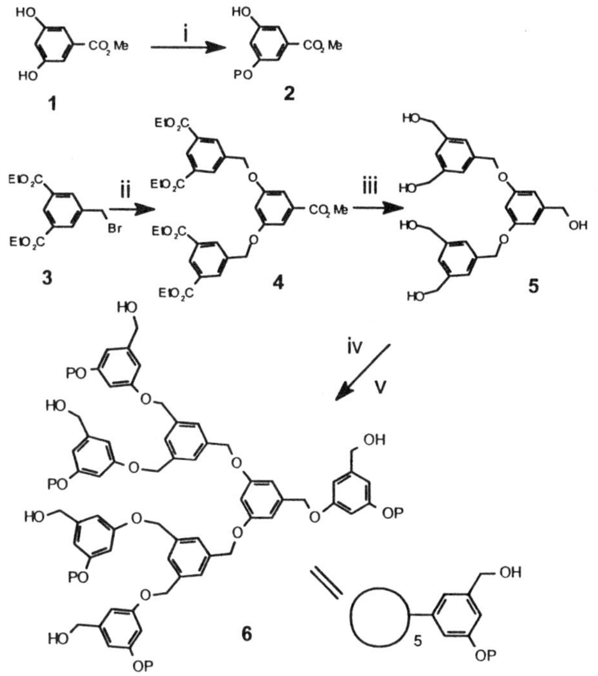

i. t-butylchlorodiphenylsilane, imidazole, DMF
ii. **1**, K_2CO_3, 18-crown-6, acetone, reflux
iii. $LiAlH_4$, THF iv. **2**, DEAD, PPh_3, THF
v. $LiAlH_4$, THF P=Bu^tPh_2Si

dendron in a Williamson ether synthesis. The coupling reaction occurs easily and in high yield. A similar approach was used to attach PEG chains to dendrimer **6**, but a suitable end group is needed in the PEG to react with the phenol functionality of the dendrimer and form a stable ether linkage. PEG tosylate, mesylate and bromide were prepared according to literature methods[7] starting from PEG monomethyl ether with Mn 1000. The coupling of dendrimer **6** and the PEG derivatives was performed as shown in Scheme 2, using DMF as solvent in the

Scheme 2. Synthesis of dendrimer-PEG conjugate

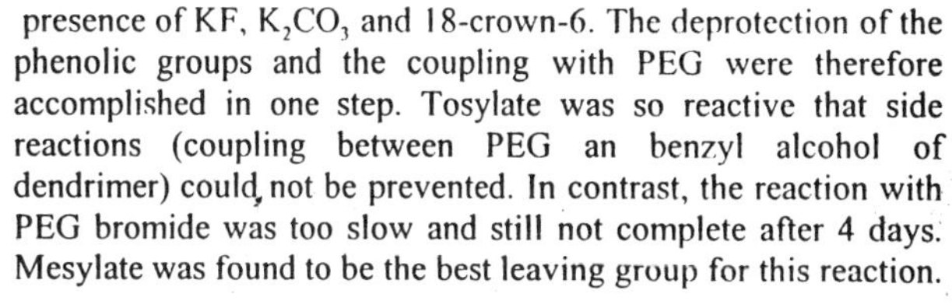

presence of KF, K_2CO_3 and 18-crown-6. The deprotection of the phenolic groups and the coupling with PEG were therefore accomplished in one step. Tosylate was so reactive that side reactions (coupling between PEG an benzyl alcohol of dendrimer) could not be prevented. In contrast, the reaction with PEG bromide was too slow and still not complete after 4 days. Mesylate was found to be the best leaving group for this reaction.

The coupling reaction was completed in about 24 hrs with little side reaction affording the dendrimer-PEG conjugate in almost quantitative yield. As expected, the dendrimer-PEG conjugate was found to be water soluble.

3. Synthesis of dendrimer-PEG-drug conjugates

Different linkages have been investigated for coupling drugs onto polymer, such as amide, carbonate, carbamate and ester bonds. These bonds are hydrolytically labile, and hydrolysis of labile linkages results in the release of bound drugs. Two synthetic routes were designed and tested to attach drugs to the dendritic carrier. The compounds selected as model drugs in this study are two amino acid derivatives, a N-protected phenylalanine with a free carboxylic acid group (Phe-COOH) and tryptophan ethyl ester with a free amino group (Trp-NH$_2$). The coupling strategy used is shown in Scheme 3.

Scheme 3. Synthesis of dendrimer-PEG-drug conjugates

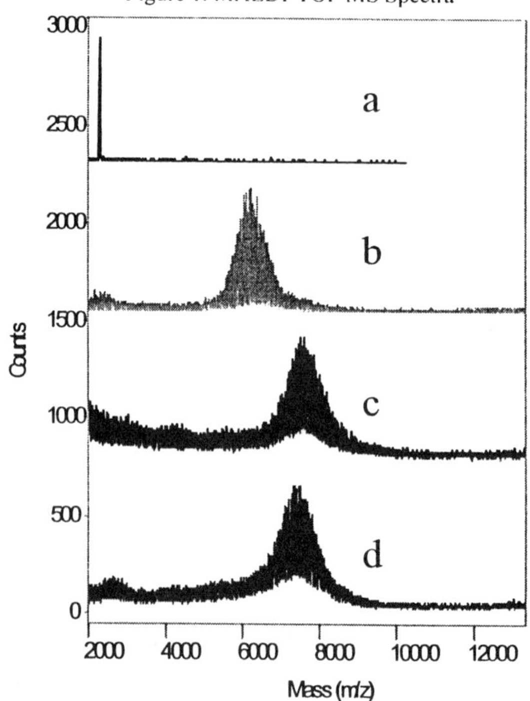

A. *Dendrimer-PEG-Phe conjugate via ester linkage*   An ester linkage was chosen to connect the carboxylic acid group of the model "drug" to the dendrimer-PEG conjugate. The DCC coupling reaction was carried out in dichloromethane using DMAP as the catalyst to afford the desired dendrimer-PEG-Phe conjugate **9** in 93% yield.

B. *Dendrimer-PEG-Trp conjugate via carbamate linkage*   The carrier **8** was first activated with 4-nitrophenyl chloroformate. Subsequent reaction of the amino group of tryptophan with the activated carbonate ester afforded conjugate **10** (90%) with carbamate linkages.

4. Characterization

The dendrimers and conjugates were fully characterized by ^{1}H NMR, Size-exclusion Chromatography (SEC) and MALDI TOF MS. Of these methods, MALDI TOF MS proved to be the most valuable in detecting impurities and defects. Figure 1 shows the spectra of dendrimer **6**, dendrimer-PEG conjugate **8** and two "drug" conjugates **9** and **10**. As shown in Figure 1a, one single peak was observed for dendrimer **6** with a mass of 2266, corresponding to the Na-metalated mass ion (M+Na). This confirms that the dendrimer was of high purity and essentially monodisperse and that no defects existed in the dendrimer structure. Figure 2b shows the spectrum of the dendrimer-PEG conjugate. The spectrum consists of only one series of peaks centered at mass of 6200 (the calculated mass is about 6000 for fully reacted conjugate), and shows expected mass difference of 44 between oligomers. This spectrum confirms that 5 PEG chains are attached to the dendrimer. Figure 2c and d show the spectra of the dendrimer-PEG-"drug" conjugates **9** and **10**. These spectra shift to higher mass region, and the observed shifts are in good

agreement with the values expected for fully "drug" loaded carriers. This approach has now been confirmed with actual drug-dendrimer conjugates.

Acknowledgement :   Support by NSF-DMR is acknowledged with thank.

References

1. a) *Polymeric Drugs and Drug Delivery Systems*, Edited by R. L. Dunn and R. M. Ottenbrite, American Chemical Society, Washington, DC 1991. b) *Polymeric Delivery Systems, Properties and Applications*, Edited by M. A. EI-Nokaly, D. M. Piatt and B. A. Charpentier, American Chemical Society, Washington, DC 1993. c) *Polymeric Drugs and Drug Administration*, Edited by R. M. Ottenbrite, American Chemical Society, Washington, DC 1994.
2. H. Ringsdorf, *J. Polym. Sci.,* 1975, 51, 135.
3. a) G. R. Newkome, C. N. Moorefield, F. Vogtle, *Dendritic Macromolecules: Concepts, Syntheses, Perspectives*, VCH: Veinheim, Germany, 1996. b) J. M. J. Fréchet, *Science*, 1994, 263, 1710. c) D. A. Tomalia, H. D. Durst, *Top. Curr. Chem.,* 1993, 165, 193.
4. D. A. Tomalia, *Sci. Am.,* 1995, 272, 62.
5. a) N. Malik, E. G. Evagorou, R. Duncan, *Proceed. Int'l. Symp. Control. Rel. Bioact. Mater.,* 1997, 24, 107. b) M. X. Tang, C. T. Redemann, F. C. Szoka, *Bioconjugate Chem.,* 1996, 7, 703. c) J. F. Kukowskalatallo, A. U. Bielinska, J. Johnson, R. Spindler, D. A. Tomalia, J. R. Baker, *Proceed. Nat. Aca. Sci. U.S.A.* 1996, 93, 4892. d) R. Delong, K. Stephenson, T. Loftus, M. Fisher, S. Alahari, A. Nolting, R. L. Juliano, *J. Pharm. Sci.,* 1997, 86, 762.
6. J. M. J. Fréchet, I. Gitsov, *Macromol. Symp.* 1995, 98, 441.
7. J. M. Harris, E. C. Struck, M. G. Case, M. S. Paley, *J. Polym. Sci. Polym Chem. Ed.,* 1984, 22, 341.

Figure 1. MALDI-TOF MS Spectra

STAR POLY(ETHYLENE OXIDE)S FROM CARBOSILANE DENDRIMERS

B. Comanita and J. Roovers
ICPET, National Research Council, Ottawa, Ontario CANADA K1A 0R6

INTRODUCTION

Poly(ethylene glycol) and poly(ethylene oxide) have interesting biological properties.[1,2] Star-branched poly(ethylene oxides) are expected to provide additional potential in a variety of biomedical applications.[3,4] The methods for making PEO star polymers are limited. The first method attaches preformed monofunctional PEO chains to a multifunctional coupling agent. The method is limited to low molecular weight (MW) PEO chains. Smid et al. coupled hydroxy terminated PEO to multifunctional isocyanates.[5] Recently, the method was perfected to the coupling of 32 PEO chains (MW=5 000) to a third generation poly(amidoamine) dendrimer.[6] The end group of the PEO chains were modified with N-succinimidyl ester of propionic acid. The extent of the coupling reaction decreases to about 50% for a target 256-arm star. Star poly(ethylene oxide)s have also been produced by the radical polymerization of ω-hydroxy PEO macromonomers.[7] This method is again limited to low MW PEO and yields a wide MWD product. Extremely high MW star poly(macromonomer)s are obtainable on polymerization is water. PEO macromonomers with end-standing norbornene groups have recently been polymerized by ring opening metathesis catalysis.[8] The third approach to the synthesis of PEO star polymers is by anionic polymerization. Either the polymerization is initiated from a poly(divinylbenzene) living core or from a poly(styryl-b-divinylbenzene) living core.[9,10] The PEO star polymers have narrow MWD arms. However, the number of arms is difficult to control and, probably, has a wide distribution. The resulting star polymers have a large weight fraction of hydrocarbon material and are therefore not suitable for biomedical applications.[11]

In this report we describe the synthesis of star PEOs starting from functionalized carbosilane dendrimer cores. The number of arms is strictly controlled as is the MW of each PEO arm. Furthermore, each arm has an end-standing hydroxy group available for further modification. The method is an expansion of the work of Taton and Gnanou who prepared a six-arm PEO star starting from a heptaphenyl core.[12]

EXPERIMENTAL

Synthesis

The synthesis of the carbosilane dendrimer core[13] and the attachment of the end-standing hydroxy groups[14] have been described. All operations are performed under high vacuum. The multifunctional initiator and the cryptand, Kryptofix [2,2,2], are extensively dried under vacuum. EO is purified by stirring over CaH_2 and two films of Na prepared under high vacuum. Potassium alkoxides are prepared in THF by addition of potassium diphenyl methane or potassium naphthalene to the initiator solution. The polymerization is carried out in THF for times varying from three days to ten days at 45°C. Polymers are precipitated in hexanes and dried and stored under vacuum in the dark.

Characterization

PEO linear and star polymers are characterized by SEC in THF at 35°C. Adsorption of the polymers on the columns is prevented by the addition of 0.25% (n-Bu)$_4$NBr to THF.[15] The flow rate is 1mL/min. DRI and UV detectors are used. Commercial narrow MWD PEO samples are used to verify the behavior of the chromatographic system. Weight average MWs are determined in methanol at 25°C with a Brookhaven BI30. Five solutions are prepared and clarified by filtration through 0.2 μm nylon filters.[16] The Raleigh ratio for toluene at 633 nm equal to 1.4×10^{-5} cm^{-1} is the calibrating constant. The MWs are obtained from Berry plots. dn/dc=0.135 ml/g. Dynamic light scattering measurements are made on the same solutions. No angular dependence of the diffusion constant is observed. The concentration dependence used to extrapolate to zero concentration is of the form $D = D_0 (1 + k_D c)$. The viscosity and density of methanol are 0.544 cP and 0.7865 g/cm^3, respectively. Viscosities are measured in methanol at 25°C with semi-micro Cannon-Ubbelohde viscometers. Solvent flow time is 185 s. Huggins and Kramer plots are used to obtain the intrinsic viscosity.

RESULTS AND DISCUSSION

The synthesis of the star PEOs is given schematically by

$$\bullet \{-Si-(CH_2)_6OK\}_m + n\ EO \longrightarrow \bullet \{-Si-(CH_2)_6\text{-} (CH_2\text{-}CH_2\text{-}O\text{-})_n K\}_m$$

The hydroxy dendrimers are sparingly soluble in THF, but the partially substituted potassium alkoxides have very limited solubility even in the presence of the cryptand. The initial polymerization is conducted at 25-30°C. This solubilizes the initiator. The main polymerization is then performed at 45°C with stirring. There is a fast dynamic equilibrium between hydroxy and alkoxide groups that assures narrow MWD. The polymers have indeed a narrow MWD as evidenced by SEC in THF and in water. Occasionally, traces of low MW material are observed. These are thought to originate from spurious water or other protonic molecules.

The measured values of Mw and [η] for linear PEO are used to derive the [η]-Mw benchmark relation of linear PEO. These are shown in Figure 1 for the linear polymer together with the data for four-, eight-, and sixteen-arm PEO star polymers. The slope in the double logarithmic representation for linear PEO is 0.69$_5$, a typical value for a polymer in a good solvent. The slopes for the PEO samples with the star architecture are the same within experimental error. The prefactors K in the Mark-Houwink-Sakurada relation [η] = K Mw$^{0.695}$ decreases from 2.73$_5$ x 10^{-2} ([η] in mL/g) for linear PEO to 2.01$_8$, 1.30$_3$ and 0.76$_7$x10^{-2} for the four, eight and sixteen-arm stars, respectively. The branching factors, g' = [η]$_{br}$/[η]$_{lin}$ are 0.73$_8$, 0.47$_6$ and 0.28$_0$ for the three star types. These ratios are in excellent agreement with those of non-polar star polymers in good organic solvents.[17,18]

The sizes of the star PEOs presently available are too small to determine accurately their radius of gyration by LS at 633 nm. However, dynamic LS in MeOH at 25°C provides translational diffusion coefficients as shown in Table 1. The slope of the double logarithmic plot of D_0 vs Mw for linear PEOs is -0.56$_7$. This is in good agreement with the slope in Figure 1. The prefactors in $D_0 = K'$ Mw$^{-0.567}$ are 2.98$_5$x10^{-4} for linear PEO and 3.19$_9$, 3.62$_2$ and 4.07$_4$ for the four-, eight- and sixteen-arm stars, respectively. The compact star polymers diffuse faster than the linear PEO with the same MW. The shrinkage ratios are 0.93$_3$, 0.82$_4$ and 0.74$_4$ for the three types of star polymers. These values agree with previous results obtained on non-polar polymers in hydrocarbon solvents.[17,18] The dynamic light scattering results also provide a measure of the MW distribution. The ratio μ_2/Γ^2 is approximately 0.05 for most of the samples confirming the SEC results.

The intrinsic viscosity and translation diffusion coefficient of a polymer can be directly compared by

$$R_v = \{ (3/10\ \pi N_A)\ [\eta]\ Mw\}^{1/3}$$
$$R_h = kT / 6\ \pi\eta_s D_0.$$

Both relations are based on the equivalent equal density sphere. It is found the R_v/R_h = 1.15 for linear PEO, a value slightly higher than observed for non-polar polymers.[19] The value of R_v/R_h of the star PEOs decreases towards unity with increasing functionality.

Other properties like A_2 and k_H and k_D of the star polymers are also consistent with their branched architecture.

CONCLUSIONS

The results presented prove that linear and star polymers with their concomitant higher segment density are all monomolecularly dispersed in dilute solution in MeOH. Chemical modifications and other physical properties of the star polymers will be described.

ACKNOWLEDGEMENT

The authors thank Prof. Stobart for helpful discussion. Dr. P. Peng and Mr. L. Lazare provided technical help at the early stages of this research. The work was performed with a NRC-NSERC research partnership grant.

REFERENCES

(1)   Harris, J.M.; Zalipsky, S., Eds. *Poly(ethylene glycol): Chemistry and Biological Applications; ACS Symposium Series* **1997** *680*.
(2)   Merrill, E.W.; Salzman, E.W. *ASAIO, Journal* **1983**, *6*, 60.
(3)   Merrill, E.W. in *Poly(ethylene glycol)Chemistry: Biotechnical and Biomedical Applications*; Harris, J.M. Ed.;Plenum Press, N.Y. 1992.
(4)   Merrill, E.W. *J. Biomater. Sci.;Polymer Edn.* **1993**, *5*, 1.
(5)   Chen, Y.; Smid, J. *Langmuir* **1996**, *12*, 2207.
(6)   Yen, D.R.; Merrill, E.W. *Polym. Preprints* **1997**, *38(1)*, 531.
(7)   Ito, K.; Hashimura, K.; Itsuno, S.; Yamada, E. *Macromolecules* **1991**, *24, 3977*.
(8)   Heroguez,V.; Gnanou,Y.; Fontanille, M. *Macromolecules* **1997**, *30*, 4792.
(9)   Lutz, P.; Rempp, P. *Makromol. Chem* .**1988**, *189*, 1051.
(10) Gnanou, Y. Lutz, P.; Rempp, P. *Makromol. Chem.* **1988**, *189*, 2885.
(11) Meneghetti, S.P.; Naraghi, K.; Burchard, W.; Lutz, P.J. *Intern. Symp. Ionic Polym.*; Paris 1997 p 322.
(12) Taton, D.; Gnanou, Y. *Intern. Symp.Ionic Polym.;* Paris 1997 p.348.
(13) Zhou, L.-L.; Roovers, J. *Macromolecules* **1993**, *26*, 963.
(14) Comanita, B.; Roovers, J. to be published.
(15) Vandorpe, J.; Schacht, E. *Polymer* **1996**, *37*, 3141.
(16) Devanand, K.; Selser,J .C. *Macromolecules* **1991**,*24*, 5943.
(17) Douglas, J.F.; Roovers, J.; Freed, K.F. *Macromolecules* **1990**, *23*, 4168.
(18) Roovers, J. in *Star and Hyperbranched Polymers*; Mishra, M.K.; Kobayashi, S. Eds. Marcel Dekker Inc.; New York, 1998.
(19) Roovers, J.; Martin, J.E. *J. Polym. Sci.: Part B: Polym. Phys.* **1989**, *27*, 2513.

Table 1

Sample	$Mw \times 10^{-4}$	$D_o \times 10^{7} (cm^2/s)$	$[\eta]$ (mL/g)
Linear PEO			
PEG 23k	2.46	10.0	30.9
PEO160k	16.7	3.2	$118._0$
PEO293k	27.7	2.4_6	$173._5$
4-arm star PEO			
PEO-14	2.19	11.1_5	20.9
PEO-16	10.5	4.5	62.9
PEO-17B1	17.8	3.3_6	93.0
PEO18B2	32.8	2.37	142.0
8-arm star PEO			
BC8PEO2k	2.04	13.0	12.9_5
BC8PEO8k	8.5_1	6.0_3	35.0
BC8PEO25kB1	15.0	4.28	51.4
BC8PEO60kB4	45.7	2.26	$111._4$
16-arm star PEO			
BC16PEO2kF1	4.8	9.2	13.5
BC16PEO5k	19.2	4.0_8	36.2

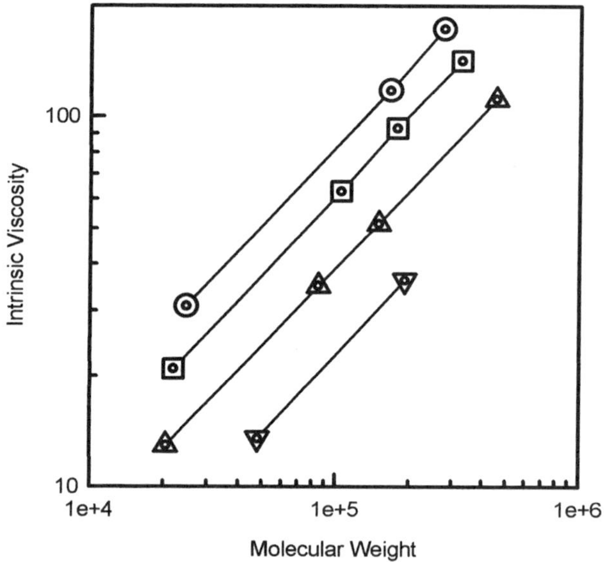

Fig. 1 Intrinsic Viscosity again Molecular Weight
Top to Bottom: linear, 4-arm, 8-arm, and 16-arm star PEO

HEMOSTATIC AND AIR LEAK SEALING EFFECTS OF RAPIDLY CURABLE GLUES FROM GELATIN, POLY(L-GLUTAMIC ACID), AND CARBODIIMIDE

Y. Tabata, Y. Otani, and Y. Ikada,
Research Center for Biomedical Engineering, Kyoto University, Kyoto 606-8507, Japan.

Introduction

Hemostatic treatment is very important in surgical operation. Also, air leak from tissues is a common problem in lung and thoracic surgeries. Various types of tissue adhesives have been explored for the hemostasis and air leak sealing (1). Requirements of adhesives for these purposes include 1) to be liquid before application, 2) to be rapidly curable even in the presence of water after application, 3) to be as pliable as soft tissues when cured, and 4) to be biodegradable in the body.

We have been studying rapidly curable biological adhesives composed of solutions from biodegradable gelatin and poly-(L-glutamic acid) (PLGA) (2,3,4). A mixed gelatin and PLGA aqueous solution set to a hydrogel upon addition of water-soluble carbidiimide (WSC) within very short periods of time, similar to clinically used fibrin glues. The WSC-catalyzed gelatin-PLGA hydrogel glues exhibited higher bonding strength to the skin *in vitro* than a conventional fibrin glue. The cured gelatin-PLGA hydrogels were slowly degraded in the body without inducing severe inflammatory responses. Taken together, this hydrogel glue is expected to meet the requirements as tissue adhesives.

In the present study, we evaluated the feasibility of WSC-catalyzed gelatin-PLGA hydrogels as hemostatic glue and medical sealant. After applying the hydrogels to a bleeding site on dog spleens and an air leak wound of rat lungs prepared by needle pricking, their hemostatic and air leak sealing effects were evaluated and compared with those of the fibrin glue.

Experimental

An alkaline-processed gelatin (Mw=99,000) and a sodium salt PLGA (Mw=83,000) were kindly supplied from Nitta Gelatine Co., Ltd., Osaka, Japan and Ajinomoto Co., Ltd., Tokyo, Japan, respectively. As a WSC, 1-ethyl-3-(3-dimethylaminopropyl) carbodiimide hydrochloride was used. A fibrin glue (BOLHEAL®) was obtained from the Chemo-Sero-Therapeutic Research Institute, Kumamoto, Japan.

The hemostatic and air leak sealing effects were evaluated according to reported methods with minor modification (5,6). Under anesthetization, the dog spleen was pricked with a needle to prepare a pin-hole bleeding injury. One min after needle pricking, a mixed gelatin-PLGA aqueous solution was dropped on the bleeding site, followed by addition of WSC solution to allow the mixed gelatin-PLGA solution to set to a hydrogel. When bleeding was not observed at the pricked site any more 7 min later, hemostasis was regarded as complete. The weight of hemoglobin in the blood bled was determined by use of Hemoglobin test-WAKO (Wako Pure Chemical Industries Ltd., Osaka, Japan).

An air leak wound was prepared by pricking the excised rat lung with a needle with different gauge sizes. Immediately after application of a mixed gelatin-PLGA aqueous solution to the leak site, WSC solution was added, followed by leaving for one min to allow gelation to take place. The lung was inflated with air to measure the minimum pressure to induce air leak from the lung wound.

Results and Discussion

Table 1 shows comparison of the hemostatic effect between the WSC-catalyzed gelatin-PLGA hydrogel glue and the fibrin glue. It is apparent that the success rate of hemostasis of WSC-catalyzed hydrogel glues is higher than that of fibrin glue. The gelatin-PLGA hydrogel glues exhibit significantly a lower bleeding ratio than the fibrin glue. It should be stressed that the gelation time of mixed aqueous solutions of gelatin and PLGA is a key factor for achieving effective hemostasis by hemostatic agents. Prolonged gelation of glues will allow blood to accumulate in the interface between the hydrogels and the spleen surface. This blood accumulation may reduce the adhesion strength of the hydrogels to soft tissues and hence the hemostatic effect. The gelation time of mixed gelatin-PLGA aqueous solution with EDC is about 9 sec, short enough to stop the oozing blood.

Table 2 compares the sealing effect of the gelatin-PLGA hydrogel glue with the fibrin glue on the air leak of rat lung wound. The lungs before glue application had air leak pressure around 15 cmH$_2$O and no significant effect of needle size on the leak pressure was observed, irrespective of the glue type. Both the air leak pressure and the percent sealing of the hydrogel glue were significantly higher than those of the fibrin glue. The percent sealing tended to increase with the increasing gauge size of the needle used for pricking the lung. This is due to a reduced size of the pricked hole with the increased gauge size. For example, all the lung wounds with a 27-gauge needle were completely sealed with the gelatin-PLGA hydrogel glue.

Table 3 summarizes the bonding strength of the WSC-catalyzed gelatin-PLGA hydrogel glue and the fibrin glue to the surface of the dog spleen and rat lung. Irrespective of the tissue type, bonding strength was significantly higher than that of the fibrin glue. Optical microscopic observation revealed that adhesion of the hydrogel glue to the tissue surface took place at a cellular level. It is likely that such firm hydrogel adhesion resulted in superior hemostatic and air leak sealing effects. The WSC-catalyzed gelatin-PLGA hydrogel was degraded in the body and disappeared from the applied site in 12 weeks without any severe inflammatory response.

Conclusions

The hemostatic and air leak sealing capabilities of WSC-catalyzed gelatin-PLGA hydrogel glue were superior to those of the conventional fibrin glue. This hydrogel glue could adhere firmly to the surface of spleen and lung. We concluded that this firm adhesion of the hydrogel glue, along with rapid gelation of the mixture from gelatin and PLGA aqueous solution, mainly led to higher capabilities of the WSC-catalyzed gelatin-PLGA hydrogel glue than the fibrin glue.

Table 1   Comparison of the hemostatic effect between the WSC-catalyzed gelatin-PLGA hydrogel glue and the fibrin glue.

Treated with	Number of treatments	Frequency [a] of glue applications	Success [b] rate of hemostasis (%)	Bleeding ratio [c]
Gelatin-PLGA hydrogel glue [d]	23	1.21±0.09 [e]	52.2	1.32±0.06 [e]
Fibrin glue	18	1.22±0.03	11.1	1.79±0.20
None	22	—	0.0	2.06±0.17

a) Application frequency of gelatin-PLGA and fibrin glues during the 7 min hemostatic experiment.
b) Percentage of complete hemostasis in the total hemostatic experiments.
c) The weight ratio of hemoglobin released during the total 7 min after glue application to that for the 1 min before glue application. * and *** : $p<0.05$ and $p<0.001$.
d)The concentration of gelatin, PLGA, and WSC was 100, 100, and 9.59 mg/ml, respectively.
e) Mean±SE

Table 2   Sealing effect of the WSC-catalyzed gelatin-PLGA hydrogel glue and the fibrin glue on the air leak of rat lung wound.

Treated with	Needle [a] size	Wound number	Complete [b] sealing	Percent [c] sealing (%)	Air leak pressure [d] (cmH$_2$O) before application	after application
Gelatin-PLGA [e] hydrogel glue	20	24	18	75	15.2±0.7 [f]	46.2±1.6
	23	12	10	83	14.2±0.5	49.2±0.6
	27	12	12	100	14.8±1.1	50.0±0.0
Fibrin glue	20	14	2	14	15.1±1.1	37.3±1.5
	23	12	4	33	14.8±0.8	41.5±2.4
	27	14	6	43	14.6±0.4	35.8±3.4

a) Air leak wounds were prepared by pricking the rat lung by needles with gauge sizes of 20, 23, and 27.
b) Defined as the sealing where no air leak was observed up to the lung pressure of 50 cmH$_2$O.
c) Percent of complete sealings to the total wound number.
d) Defined as the minimum pressure to induce air leak from the lung wound. ** and ***: $p<0.01$ and $p<0.001$.
e) The concentration of gelatin, PLGA, and WSC was 100, 100, and 9.59 mg/ml, respectively.
f) Mean±SE

Table 3   Bonding strength of the WSC-catalyzed gelatin-PLGA hydrogel glue and the fibrin glue to the dog spleen and the rat lung.

Glue	Bonding strength [a] (gf/cm²) Spleen	Lung
Gelatin-PLGA hydrogel glue [b]	235.8±32.9 [c]	34.1±1.9
Fibrin glue	87.8± 6.5	24.9±2.1

a) **: $p<0.01$
b) The concentration of gelatin, PLGA, and WSC was 100, 100, and 9.59 mg/ml, respectively.
c) Mean±SE

References

1. Y. Ikada, In: C.C.Chu, J.A. von Fraunhofer, H.P. Greiser, Eds., Wound Closure Biomaterials and Devices, CRC, 318 (1997)
2. Y. Otani, Y. Tabata, Y. Ikada, J. Biomed. Mater. Res., 31, 157 (1996)
3. Y. Otani, Y. Tabata, Y. Ikada, Biomaterials, 17, 1387 (1996)
4. Y. Otani, Y. Tabata, Y. Ikada, J. Adhesion, 59, 197 (1996)
5. T. Natsume, K. Tamura, K. Kawarasaki, Y. Shimizu, Jpn. J. Artif. Organs, 19, 1235 (1990)
6. F. Bellotto, R.G. Johnson, R.M. Weintraub, J. Foley, R.L. Thurer, Surgery, 174, 221 (1992)

POLY-L-LYSINE-GRAFT-PEG COMB-TYPE POLYCATION COPOLYMERS FOR GENE DELIVERY. M. G. Banaszczyk, C. P. Lollo, A. T. Phillips, A. Amini, Duncan P. Wu, P. M. Mullen, C. C. Coffin, D. J. Carlo, S. Brostoff, R. M. Bartholomew; The Immune Response Corporation, 5935 Darwin Court, Carlsbad, CA 92008

Polycations have been used for gene delivery in vitro quite successfully, however, in viva applications suffered from serum effects that lower the overall gene drug efficiency. PEG polymers have been used extensively to minimize serum effects and create "stealth liposomes," biocompatible materials, and proteins with extended circulation. Here, we report our efforts towards creating "stealth polyplexes." We report synthesis, characterization and properties of comb-type polycations consisting of linear poly-L-lysine backbone with PEG grafts. The size of poly-L-lysine and PEG, and graft density on a polylysine were varied. In addition, these polymers were hydrophobically modified, radiolabeled and studied in vivo.

SYNERGISTIC AND COMPETITIVE INTERACTIONS OF BILE SALTS WITH POLYCATIONIC HYDROGELS. W. H. Braunlin, A. Guo, E. Zhorov, R. Rapoza, D. Rosenbaum, R. Holmes-Farley; GelTex Pharmaceuticals, Nine Fourth Avenue, Waltham, MA, 02154

Bile acid sequestration and transport through the intestine by nonabsorbable polymeric hydrogels provides an attractive, proven mechanism for reducing serum cholesterol. Equilibrium and kinetic binding results are presented for three novel polycationic hydrogels that have been developed as bile acid sequestrants at GelTex. The hydrophobic hydrogel GT96-243 binds bile salts very tightly at low bile acid concentration, and behaves as a single site binder. The more hydrophilic polymer GT1O2-279, binds bile salts with a high degree of cooperativity, and with enhanced binding capacity. The transitional polymer GT31 -104HB shows more complex, multisite behavior, and binds bile salts effectively over a wide range of bile salt concentrations. These three polymers show characteristic differences in association and dissociation kinetics. The relevance of these results to the in vivo activity of these polymers will be discussed.

CATALYTIC PROPERTIES OF A Cu(II) LOADED CHITOSAN/OXIDIZED-REDUCED β-CYCLODEXTRIN NETWORK. V. Crescenzi, G. Paradossi, F. Cavalieri, E. Chiessi, *Universitá di Roma "La Sapienza", Rome, 00185 Italy; Universitá di Roma "Tor Vergata", Rome, 00133 Italy.

A new chemical hydrogel, prepared by reductive alkylation of chitosan with periodate oxidated β cyclodextrin was loaded with Cu(II) ions. The network was tested with respect to its catalytic properties in electron transfer reactions between molecular oxygen contained in the aqueous dispersion and (±) adrenaline. The reaction mechanism of the oxidative process involves a true catalytic role of cupric ions with the formation of a precursor complex between the metal ion embedded in the matrix and the substrate. The catalytic efficiency of this system is comparable with analogous macromolecular systems studied in solution. The advantage of having a separate and recyclable catalyst in aqueous medium is discussed.

SYNTHESIS OF HYDROGELS VIA UGI REACTIONS. V. Crescenzi, A. E. J. de Nooy, G. Masci, M. Dentini, University "La Sapienza", Piazzale Aldo Moro 5, 00185 Rome, Italy

Ugi four-component reactions (1. Ugi: J. Prakt. Chemie, 339, 499-516 [1997]) are enjoying renewed interest, especially in "combinatorial chemistry". We have applied these reactions to the synthesis of novel polymer networks, a study pioneered by Konig and Ugi (S. Konig, I. Ugi; Z. Naturforsch., 46 b, 1261-1265 [1991]), starting from aqueous solutions of different polysaccharides (pullulan, scleroglucan, chitosan, carboxymethylcellulose) and hydrophilic synthetic polymers (poly(acrylic acid)), respectively. For example, adding to a 2% w/V chitosan solution in aqueous HCl, adequate amounts of tartaric acid, formaldehyde and cyclohexylisonitrile, stable chemical gels result in a few minutes. This is true in each case considered. Ugi reactions do therefore open the way to facile, one-pot syntheses of novel hydrogels whose chemical nature, structure, and cross-linking degree appear easily tunable.

VISCOELASTIC HYDROGELS SYNTHESIZED FROM A SUGAR-CONTAINING METHACRYLATE POLYMER. K. Kato, T. Tominaga, K. Nakamae, Department of Chemical Science and Engineering, Faculty of Engineering, Kobe University, 1-1 Rokkodai-cho, Nada-ku, Kobe 657-8501, Japan

In order to synthesize non-biodegradable, viscoelastic polymers with high chemical stability and negligible cytotoxicity, a sugar-containing polymer, poly(2-glycosyloxyethyl methacrylate), was lightly crosslinked with glutaraldehyde or sodium borate. These reactions were shown to produce soft hydrogels capable of viscous flow. This paper will focus on the rheological properties of the hydrogels synthesized, comparing with that of hyaluronic acid solution. The lubricity, optical properties, chemical stability, injectability through a needle, and cytotoxicity will be further described.

DRUG RELEASING CHARACTERISTICS OF PDLLA-MPEG DI- AND MPEG-PDLLA-MPEG TRIBLOCK COPOLYMER MICELLES. D. Kim, T. S. Ha, S. H. Kim, Department of Chemical Engineering, Sung Kyun Kwan University, Suwon, Kyungki 440-746, Korea

PDLLA-MPEG di- and MPEG-PDLLA-MPEG tri-block copolymers with varying MPEG block sizes were synthesized, and their structures and properties were characterized. Micelles were prepared from each block copolymer system, and the CMC value and size distribution were determined. Releasing experiments showed that the loading content in the block copolymer micelles increased with increasing MPEG block segment length, resulting in the different burst strength. The releasing behavior was compared with the viscosity behavior in the releasing process. The releasing kinetics were well described by a simple equation.

THERMOREVERSIBLE GELATION OF POLY(ETHYLNENE OXIDE)-POLY(LACTIC ACID) DIBLOCK COPOLYMERS IN AQUEOUS SOLUTION. <u>D. S. Lee</u>, S. Choi, B. Jeong, S. W. Kim, Department of Polymer Science & Engineering, Sung Kyun Kwan University, Suwon, Kyungki-do 440-746, Korea

Poly(ethylene oxide)-b-poly(L-lactic acid) (PEO-PLLA) diblock copolymers were synthesized via ring opening polymerization from poly(ethylene oxide) and L-lactide. Their physicochemical properties were investigated by using infrared spectroscopy, 1H-NMR spectroscopy and gel permeation chromatography together with observation of gel-sol transition in aqueous solution. Aqueous solutions of PEO-PLLA diblock copolymers underwent from gel to sol with increasing temperature. For comparison, poly(ethylene oxide)-b-poly(L-lactic acid-co-gylcolic acid) and poly(ethylene oxide)-b-poly(DL-lactic acid-co-gylcolic acid) were synthesized by the same methods. The gel-sol transition properties of these diblock copolymers depends on the hydrophilic/hydrophobic balance of the copolymers, the block lengths, the hydrophobicity and the stereoregularity of hydrophobic blocks of the copolymers.

COMPOSITES OF CONDUCTING POLYMER AND THERMALLY SENSITIVE HYDROGEL. <u>D. S. Lee</u>, Y. H. Bae, W. S. Shim, Department of Polymer Science & Engineering, Sung Kyun Kwan University, Suwon; Kyungki-do, 440-746, Korea

The composite of conducting polymer (polypyrrole, PPy) and polyelectrolyte was prepared by the electrochemical polymerization. The polyelectrolyte used as a dopant was composed of the copolymer of thermally sensitive polymer (N-isopropy1 acrylamide, NiPAAm) and various polyelectrolyte. The electrochemical activity and mass change during the redox process of PPy composites was investigated by the potentiodynamic voltametry and electrochemical quartz crystal microbalance (EQCM). The polypyrrole/polyelectrolyte copolymer composite has the considerable electrochemical activity. Also, the transition temperature of the thermally sensitive polyelectrolyte in the PPy composite is changed as the electrical state of the conducting polymer. Therefore, the polypyrrole composite synthesized by the thermally sensitive polymer-polyelectrolyte copolymer has not only the thermal sensitivity but also the current sensitivity.

VIABILITY AND FUNCTION OF ENCAPSULATED ISLET CELLS IN A POLYSACCHARIDE HYDROGEL MICROCAPSULE. <u>A. Pfister-Serres</u>, K. A. Smeds, D. L. Hatchell[†], P. Saloupis[†], M. W. Grinstaff,* Departments Chemistry, Ophthalmology and Cell Biology[†], Duke University, and VA Medical Center Durham, NC 27708

The entrapment of functional insulin producing cells in an immunoprotective and semi-permeable polymeric system with subsequent transplantation is a promising treatment for diabetes. Alginate microcapsules are one system that has been extensively studied for islet cell (the cell mass containing the insulin producing beta cells) encapsulation and transplantation. Even though this microcapsule system possesses a number of important attributes, the microcapsules however, are not stable over long periods and it is known that mannuronic acid (one of the chemical components of alginate) invokes an immune response. We have developed a modified hyaluronic acid (HA) biopolymer that can be photocrosslinked to form a stable covalently crosslinked microcapsule. We have also isolated and cultured rat and canine islets of Langerhans for study. Using a coaxial jet head microcapsule generator, we have encapsulated the islet cells and are monitoring their in vitro function and survival over time.

SYNTHESIS AND CHARACTERIZATION OF PHOTOCROSS-LINKABLE POLYSACCHARIDE HYDROGELS. <u>K. A. Smeds</u>, A. Pfister-Serres, D. L. Hatchell[†], Peter Saloupis[†], M. W. Grinstaff,* Departments Chemistry, Ophthalmology and Cell Biology[†], Duke University, and VA Medical Center Durham, NC 27708

Polysaccharide hydrogels are used for a number of medical and biotechnological applications. One of the most thoroughly studied natural hydrogels are those composed of alginate, a natural polysaccharide. We have developed a modified hyaluronic acid (HA) biopolymer that can be photocrosslinked to form a stable hydrogel. Hyaluronic acid, a natural polysaccharide comprised of β(1-4) linked 2-acetamide-2-deoxy-D-glucose and β(1-3) linked D-glucuronic acid, is non-antigenic, non-inflammatory and non-tissue reactive. The physical, chemical and rheological properties including the site and amount of modification of the polymer, the viscosity of the modified biopolymer, the stability of the biopolymer microcapsule, the solute diffusion characteristics, the molecular weight, and the biocompatibility of the polymer have been determined.

IONOMER CEMENTS FOR DRUG DELIVERY. <u>A. M. Roos</u>, H. J. G. Luttikhedde, Laboratory of Polymer Technology, Åbo Akademi University, Biskopsgatan 8, SF-20500 Åbo, Finland

Water-soluble polyacids of different molecular weight were synthesized as hydrolyzed alternating copolymers of vinylacetate and maleic anhydride. The ionomer cements formed after reacting the polymer with calcium hydroxide were investigated as possible carrier materials for the delivery of both anionic and cationic drug precursors. The influence on release properties of cement setting time, swelling behavior and drug loading was studied. Drug release was analyzed with ISRP-HPLC. It was found that cement setting time is an important parameter for drug release. Cements with a high drug loading showed a near zero-order release profile for a period of two weeks. Toremifene was released at a rate of 1.3 mg/24 h.

IN AQUA SYNTHESIS OF A HIGH MOLECULAR WEIGHT ARENEETHYNYLENE POLYMER EXHIBITING REVERSIBLE HYDROGEL PROPERTIES. <u>W. T. Slaven</u>, Y-P. Chen, V. T. John, S. H. Rachakonda, CJ. Li,* Department of Chemistry; Department of Cell and Molecular Biology; Department of Chemical Engineering, Tulane University, New Orleans, LA 70118

Palladium catalyzed copolymerization of 3,5-diiodobenzoic acid with acetylene gas in a basic aqueous medium provides a high molecular weight ~60,000), zig-zag phenylethynylene polymer. The polymer has been characterized by a variety of methods. The polymer has a high thermostability and is soluble in basic solutions. It is reversibly switchable from soluble to hydrogel states in water by changing the acidity-basicity of the solvent.

ADVANTAGES OF SOLUBLE OVER STABLE POLYMERS FOR ORAL DOSAGE FORMS WITH CONTROLLED RELEASE.

K. Ainaoui, J-M. Vergnaud*, Department of Chemistry, University of St. Etienne, St. Etienne 42023 France.

Oral dosage forms with controlled release are generally prepared by dispersing the drug through a polymer matrix, which can be either stable or soluble. Comparison is made between these two classes of polymers, by considering not only the kinetics of in vitro drug release but also the drug level calculated in the plasma (1). The drug release is controlled by transient diffusion with stable polymers, and the kinetics of release exhibits a high rate at the beginning which decreases exponentially with time meaning that all the drug is delivered after infinite time. The kinetics of drug dissolution with soluble polymers are controlled by both diffusion and dissolution (or erosion), with a more constant rate (2). It is thus possible to determine the characteristics of dosage forms with soluble polymer so that the drug is totally released during the gastrointestinal time. Moreover the plasma drug level is more constant with soluble than with stable polymers with lower peaks and especially higher troughs.

(1)   E. M. Ouriemchi and J-M. Vergnaud, Int. J. Pharm. 127 (1996) 177-184.

(2)   J-M. Vergnaud, in "Controlled Drug Release of Oral Dosage Forms", E. Horwood Publ., London, UK, (1993).

INTERFACIAL PROPERTIES OF A DIBLOCK AMPHIPHILIC COPOLYMER POLY(ISOBUTYLVINYLETHE-B-2-METHYL-2-OXAZOLINE).

G. Volet, C. Amiel, L. Auvray, B. Sébille, H. Cheradame, Laboratoire Matériaux Polymères aux Interfaces, Laboratoire de Recherche sur les Polymères, CNRS, 2 rue Henri Dunant, 94320 Thiais, France. Laboratoire Léon Brillouin, CEA-CNRS, CEN Saclay, 91191 Gif-sur-Yvette France

Novel amphiphilic diblock copolymers, PIBVE-POXZ, were synthesized by cationic polymerization. The sequence PIBVE is hydrophobic while the sequence POXZ is hydrophilic. These copolymers have been adsorbed on porous silica particles from different solvents. The adsorbed layers have been studied by small angle neutron scattering. In the presence of a non selective solvent of the two sequences, the adsorbed chains were in mushroom configuration and the adsorbed amounts were low. On the opposite, stable and very dense copolymer layers corresponding to extended brushes have been obtained in the presence of watery solutions. The influence of the hydrophilic/lipophilic balance has been studied by varying the chemical composition.

STUDY OF POLYMER INFUSION INTO A HYDROGEL BY INTERFEROMETMC IMAGING.

Y. Xu, I. Teraoka, Polytechnic University, Brooklyn, New York 11201

Deswelling of a cylindrical, cross-linked polyacrylamide gel and infusion of poly(ethylene glycol) (PEG) into the gel when the water-swollen gel was immersed into a solution of PEG were studied by using Jamin interferometry and digital image analysis. When the gel was immersed into the solution, water exuded first because of a difference in the osmotic pressure between the interior of the gel and the surrounding solution. Spatial resolution of the interferometry revealed that subsequent PEG molecules gradually penetrated into the contracted gel. The diameter of the cylindrical gel was also measured in the process. For some molecular weights and concentrations of PEG, the gel diameter change exhibited an oscillatory pattern until it reached an equilibrium value, suggesting a strong coupling in the interaction between the gel fiber and PEG chains. Infusion and release studies of small molecules will be also shown.

DEVELOPMENTS IN POLYMER-SUPPORTED LIPIDBILAYERS.

R. Zentel P. Théato, M. Hausch, Institute of Materials Science, University of Wuppertal, Gaußstr. 20, Wuppertal 42097, Germany

Polymer-supported lipidbilayers on gold surfaces may be used as biosensors in future applications. We prepared biocompatible, water-swellable and hydrophilic polymers. By polymer analogous reaction we incorporated a disulfide anchor and a lipid anchor to the backbone. The hydrophilic behavior of the polymers can be controlled by using different acrylamides, e.g. isopropyl-, diethyl- or dimethylacrylamides. The film thickness on the substrate, which vary from 2-4 nm, can be controlled by Langmuir-Blodgett transfer. Finally we can monitor the self-assembly of the polymer on the substrate and the vesicel fusion on these polymer films by Surface-Plasmon-Spectroscopy. The vesicle fusion yields in a 3.4 nm thick lipidbilayer.

Polymeric Micelle System for Drug Targeting

Kazunori Kataoka

Department of Materials Science, Science University of Tokyo,
2641 Yamazaki, Noda, Chiba 278, Japan

Introduction

There has been a strong impetus for developing efficient systems for site-specific delivery of drugs by the use of appropriate vehicle systems. Nanoscopic vehicles having a microcontainer separated from the outer environment are promising for this purpose. Block copolymers composed of hydrophobic and hydrophilic segments have the potential to form self-associates (micelles) in aqueous milieu. The hydrophobic segment forms hydrophobic core of the micelle, while the hydrophilic segment surrounds the core as a hydrated outer shell. Since most drugs have a hydrophobic character, these drugs are easily incorporated in the inner core segment either by covalent bonding or by non-covalent bonding through hydrophobic interaction with core-forming segments[1].

Block copolymer micelles for cancer treatment

We have prepared micelle-forming polymeric anti-cancer drug by conjugating hydrophobic anti-cancer drug, Adriamycin(ADR), to block copolymers of poly(ethylene glycol)-poly(aspartic acid)(PEG-PASP) as schematically shown below[2,3].

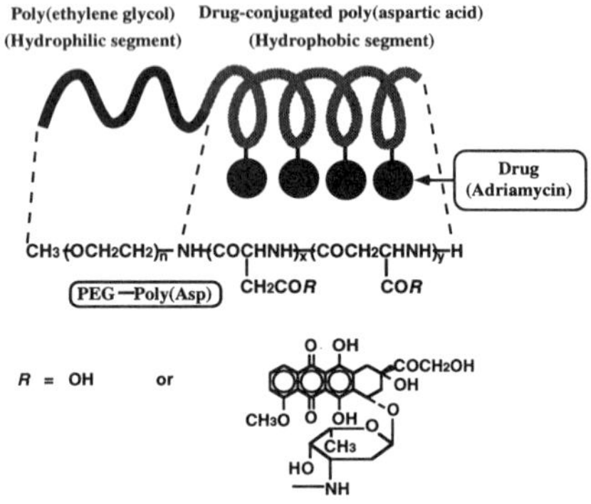

Additional ADR was loaded in the core of this ADR-conjugated micelle through a simple physical entrapment process. This polymeric micelle-ADR has high potency to cure murine solid tumors by i.v. injection[4]. Prolonged *in vivo* half life of the polymeric micelle drug was evidenced through the study using the micelle labeled with radio isotope, indicating a relatively low RES uptake[5]. Further, enhanced accumulation of the micelle into solid tumor was clearly observed[6]. Of interest, the tumor accumulation ratios (tumor/heart and tumor/muscle) at 24h for the micelle-forming conjugate are an order of magnitude increase compared to free ADR[6]. This remarkable selectivity is considered to be due to the enhanced vascular permeability as well as to the poor lymph transports in tumor (EPR effect) as discussed by Matsumura and Maeda[7].

The concept of polymeric micelle stabilization through the formation of the hydrophilic palisade surrounding water-incompatible core can be extended so as to include the case of macromolecular association through a force other than hydrophobic interaction. It was evidenced recently by our group that stable and monodispersive micelle (~20nm in diameter) can be prepared in aqueous media by metal-complex formation[8], an example of which is the polymeric micelle loading with cis-Diamminedichloroplatinum(II) (cisplatin, CDDP). CDDP has been widely used as a platinous antitumor drug. However, its use is limited due to the significant toxic side effects, such as acute nephrotoxicity and chronic neurotoxicity. Further, CDDP is known to have an extremely short-time circulation in blood due to a glomerular excretion. To achieve longevity of CDDP circulation in the bloodstream and to reduce toxic side effects, we have prepared polymer-metal complex micelles, PEG-P(Asp(CDDP)), through the complexation of CDDP to aspartic acid residues of poly(ethylene glycol)-poly(aspartic acid) block copolymer(PEG-P(Asp(CDDP)).

Mixing of CDDP and PEG-P(Asp) in an aqueous milieu led to a spontaneous formation of polymer-metal complex micelles having a diameter of 20nm with a considerably narrow size distribution as shown in Fig. 1.

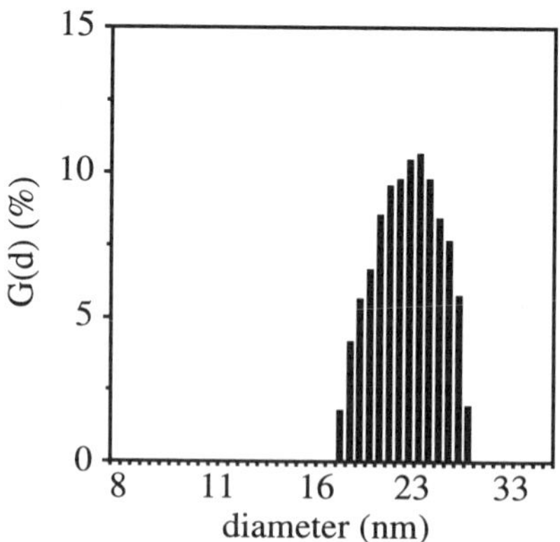

Figure 1 Gamma-averaged size distribution of PEG-P(Asp(CDDP)) micelle measured by dynamic light scattering. The ratio of weight-weighted to number-weighted fraction (dw/dn) was 1.04.

The core cross-linking due to the ligand-exchange reaction of CDDP with carboxyl groups in the polymer side chain allows the micelle to have an increased stability. Consequently, no micelle dissociation was detected by static light scattering even at an extremely diluted condition. In physiological saline at 37°C, micelle decay was observed after an induction period of ca. 12 hrs. From a viewpoint of tumor targeting by EPR effect, this time-dependent decay profile seems to be a merit because site-specific release of CDDP is expected after tumor accumulation of the micelles. As the length of P(Asp) segments was increased, the stability of micelles in physiological saline improved. PEG-P(Asp(CDDP)) micelles showed comparable *in vitro* cytotoxicity against P388D1 to CDDP itself by prolonged incubation, indicating a slow release of CDDP from micelles.

Besides the micelles showing the passive accumulation at tumors, the preparation of micelles which are able to recognize specific sites is of interest. To achieve this goal, we have recently prepared block copolymer micelles having functional groups at the each chain ends of PEG palisade[9, 10]. As shown in Scheme 1, targetable micelles can be prepared from these functionalized micelles by linking targeting moieties including sugars and proteins.

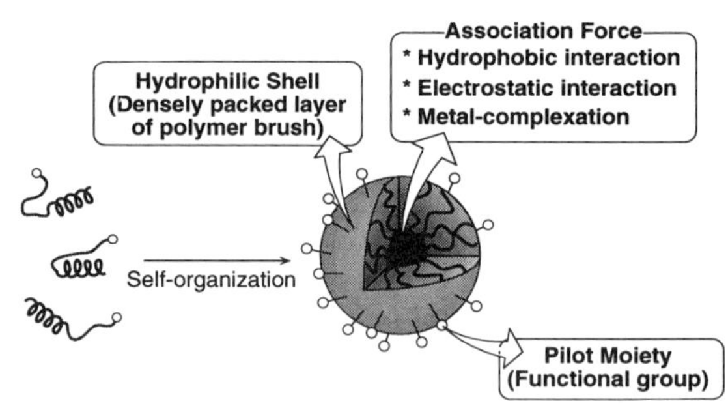

Scheme 1 Micelle formation from block copolymers

Polyion complex micelles for delivery of peptides and nucleotides

Narrowly distributed micelles can also be prepared based on electrostatic interaction between a pair of oppositely-charged block copolymers with poly(ethylene glycol)(PEG) segments[11]. As summarized in Table 1, such polyion complex micelles (PIC micelles) have even narrower distribution than λ–phage virus, a natural exmple of supramolecular assembly.

Table 1 Size and dispersity of PIC micelles, polystyrene latex and viruses measured by dynamic light scattering

	cumulant dimater	polydispersity index
PIC micelles		
PEG-P(Lysine)/PEG-P(Aspartic acid)	30 - 40 nm	0.03-0.05
P(Lysine)/PEG-P(Aspartic acid)	50 nm	0.01-0.05
polystyrene latex	285 nm	0.12
virus		
sendai virus	292 nm	0.08
λ phage	79 nm	0.06

Worth noticing is that similar monodispersive associates were formed even one of the pair was changed from block polymer to charged oligopeptides[12] or oligonucleotides[13]. Micelle disruption was not observed even in the serum, suggesting a potential utility of these micelles for *in vivo* delivery of bioactive peptides and nucleotides.

Molecular weight of entrapped compound is not restricted to oligomer region, and even plasmid DNA[14] or enzymes (lysozyme)[15] can form similar associates with block copolymers having opposite charge. The profile of the thermal melting curve as well as the DSC thermogram revealed a higher stabilization of DNA structure in PEG-P(L-lysine)/DNA associates compared to that in the complex made from DNA and P(L-lysine) homopolymer with the same molecular weight as the P(L-lysine) segment in PEG-P(L-lysine)[16]. This stabilizing effect on the DNA structure is due to the compartmentalization of DNA into the microenvironment of PEG with low permittivity. PEG shell contributes to increase the water solubility of the associates as well.

From a view point of utilizing these associates as non-viral gene vector, the regulated release of DNA from the associate is essential to achieve effective transfection. DNA release from the associates in the cytoplasm may proceed through the exchange with the anionic compounds, such as sulfated sugar. The reversible nature of PEG-P(L-lysine)/DNA associate was further verified through the addition of polyanion, including dextran sulfate and poly-L-aspartic acid: polyanion replaced DNA in the complex with PEG-P(L-lysine), resulting in the release of free DNA in the medium. Further, to estimate the stability of the associate, the rate of inter-exchange reaction of DNA in the associate with poly(vinyl sulfonate) (PVS) was evaluated from a method using a metachromasy of toluidine blue dye[17]. Of interest, the nuclease resistance of DNA in PEG-PLL/DNA associate had an inverse correlation with the rate of inter-exchange reaction, and thus, progressively increased with an increase in the molecular weight (Mw) of PLL segment in the block copolymer, indicating a substantial importance of block copolymer design for DNA stabilization.

Transfection efficiency of the associate toward cultured cells was then evaluated by luciferase assay. The associates with excess PEG-PLL showed an order of magnitude increase in the transfection efficiency compared to free as well as PLL-associated plasmid DNA (Fig. 2), indicating a promising feature of PEG-PLL/DNA associates as novel non-viral gene vectors.

References

[1] K. Kataoka, "Block copolymer micelles as vehicles for drug delivery," *J. Contrl. Rel.*,**24**, 119-132 (1993).

[2] M. Yokoyama, S. Inoue, K. Kataoka, N. Yui, and Y. Sakurai, "Preparation of adriamycin-conjugated poly(ethylene glycol)-poly(aspartic acid) block copolymer: A new type of polymeric anticancer agent," *Makromol. Chem., Rapid Commun.*, **8**, 431-435 (1987).

[3] M. Yokoyama, T. Okano, Y. Sakurai, and K. Kataoka, "Improved synthesis of adriamycin-conjugated poly(ethylene oxide)- poly(aspartic acid) block copolymer and formation of unimodal micellar structure with controlled amount of physically entrapped adriamycin," *J. Contrl. Rel.*, **32**, 269-277 (1994).

[4] M. Yokoyama, T. Okano, Y. Sakurai, H. Ekimoto, C. Shibazaki, and K. Kataoka, "Toxicity and antitumor activity against solid tumors of micelle-forming polymeric anticancer drug and its extremely long circulation in blood," *Cancer Res.*, **51**, 3229-3236 (1991).

[5] G. S. Kwon, M. Yokoyama, T. Okano, Y. Sakurai, and K. Kataoka, "Biodistribution of micelle-forming polymer-drug conjugate," *Pharm. Res.*, **10**, 970-974 (1993).

[6] G. S. Kwon, S. Suwa, M. Yokoyama, T. Okano, Y. Sakurai, and K. Kataoka, "Enhanced tumor accumulation and prolonged circulation times of micelle-forming poly(ethylene oxide-aspartate) block copolymer-Adriamycin conjugate," *J. Contrl. Rel.*, **29**, 17-23 (1994).

[7] Y. Matsumura and H. Maeda, "A new concept for macromolecular therapeutics in cancer chemotherapy: Mechanism of tumoritropic accumulation of proteins and antitumor agent smancs," *Cancer Res.*, **46**, 6387-6392 (1986).

[8] M. Yokoyama, T. Okano, Y. Sakurai, S. Suwa, and K. Kataoka, "Introduction of cisplatin into polymeric micelle," *J. Contrl. Rel.*, **39**, 351-356 (1996).

[9] S. Cammas and K. Kataoka, "Funtional poly(ethylene oxide)-co-poly(β-benzyl-L-aspartate) polymeric micelles : block- copolymer synthesis and micelles formation," *Macromol. Chem. Phys.*, **196**, 1899-1905 (1995).

[10] C. Scholz, M. Iijima, Y. Nagasaki, and K. Kataoka, "A novel reactive polymeric micelles -Polymeric micelle with aldehyde groups on its surface-," *Macromolecules*, **28**, 7295-7297 (1995).

[11] A. Harada and K. Kataoka, "Formation of polyion complex micelles in aqueous milieu from a pair of oppositely-charged block copolymers with poly(ethylene glycol) segments," *Macromolecules*, **28**, 5294-5299 (1995).

[12] A. Harada and K. Kataoka, "Formation of stable and monodispersive polyion complex micelles in aqueous medium from poly(L-lysine) and poly(ethylene glycol)-poly(aspartic acid) block copolymer," *J. Macromol. Sci., Pure Appl. Chem.*, **A34**, 2119-2133 (1997).

[13] K. Kataoka, H. Togawa, A. Harada, K. Yasugi, T. Matsumoto, and S. Katayose, "Spontaneous formation of polyion complex micelles with narrow distribution from antisense oligonucleotide and cationic block copolymer in physiological saline," *Macromolecules*, **29**, 8556-8557 (1996).

[14] S. Katayose and K. Kataoka, "PEG-poly(lysine) block copolymer as a novel type of synthetic gene vector with supremolecular structure," in *Advanced Biomaterials in Biomedical Engineering and Drug Delivery Systems*(Eds., N. Ogata, S. W. Kim, J. Feijen, and T. Okano), Springer, Tokyo, 1996, pp.319-320.

[15] A. Harada and K. Kataoka, "Novel polyion complex micelles entrapping enzyme molecules in the core : Preparation of narrowly-distributed micelles from lysozyme and poly(ethylene glycol) poly(aspartic acid) block copolymer in aqueous medium," *Macromolecules*, **31**, 288-294 (1998).

[16] S. Katayose and K. Kataoka, "Water-soluble polyion complex associates of DNA and poly(ethylene glycol)-P(L-lysine) block copolymer," *Bioconj. Chem.*, **8**, 702-707 (1997).

[17] S. Katayose and K. Kataoka, "Remarkable increase in nuclease resistance of plasmid DNA through supramolecular assembly with poly(ethylene glycol)-poly(L-lysine) block copolymer," *J. Pharm. Sci.*, **87**, 160-163 (1998).

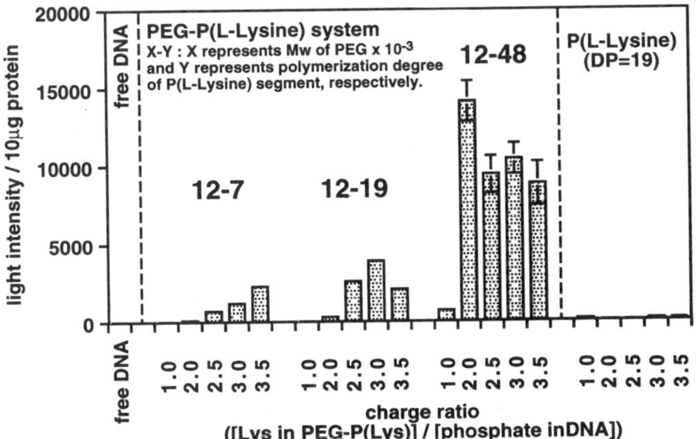

Fig. 2 In vitro gene expression in 293 cells cultured in the presence of 10% serum + 100mM of chloroquin. Each bar represents the standard errof of mean for four determinations.

DESIGN AND CHARACTERIZATION OF SEVELAMER HYDROCHLORIDE: A NOVEL PHOSPHATE-BINDING PHARMACEUTICAL

S. Randy Holmes-Farley, W. Harry Mandeville, J. Ward, and K. L. Miller

GelTex Pharmaceuticals
Nine Fourth Ave
Waltham, MA 02154, USA

Introduction

Elevated serum phosphate levels are a significant problem for patients with renal failure. Current treatments effectively prevent the uptake of dietary phosphate through formation of insoluble calcium or aluminum phosphates in the gastrointestinal tract.[1] Unfortunately, these treatments can lead to undesirable serum levels of either calcium or aluminum. In order to prevent absorption of dietary phosphate without these drawbacks, we have developed a crosslinked organic polymer which is capable of binding phosphate in the gastrointestinal tract.

Sevelamer hydrochloride consists of polyallylamine which has been crosslinked with epichlorohydrin to form a hydrogel. Under physiologic conditions, both in vitro and in vivo, this polymer has been shown to bind significant quantities of phosphate.[2,3] It has also been shown to be an effective treatment for hyperphosphatemia in human clinical trials, and is presently awaiting FDA approval.[4] Since it is a crosslinked hydrogel that cannot be absorbed from the gastrointestinal tract, the possibility of toxic side effects are greatly reduced compared to systemic medications.

During the discovery phase of sevelamer hydrochloride, a large number of cationic polymers were investigated. Sevelamer hydrochloride was found to be the most potent of these phosphate binding polymers under physiologic conditions.

It is believed that two of the most important criteria in developing a phosphate binding polymer as a pharmaceutucal are the number of positive charges per unit weight, and the spacing between those charges. The number of charges is important in order to keep the required human dose from becoming too large (more than a few grams per day). The spacing is important in order to permit a preference of the polymer for phosphate over the much higher concentrations of chloride and other ions present in the gastrointestinal tract.

Experimental

Polyallylamine hydrochloride was polymerized from allylamine hydrochloride in water using azobis(amidinopropane) dihydrochloride as the initiator.[2] Polyallylamine was crosslinked with epichlorohydrin (at various levels) in aqueous solution, and then washed with water to remove any residual impurities.[2] Other polymers were also synthesized as described previously.[2]

Phosphate binding isotherms were carried out in aqueous solution containing 80 mM sodium chloride, 30 mM sodium carbonate, and the desired level of phosphate as phosphoric acid. The polymer (0.10 g) was added to the solution (20 mL), and the mixture was adjusted to pH 7 (or other pH as noted) with conc. HCl or 50% aqueous NaOH as necessary. After stirring for 1 h, the pH was again adjusted, if necessary, and the gel was removed by filtration.

Kinetics experiments were carried out by placing the polymer (0.2 g) in a solution containing 100 mM NaCl and 24 mM phosphate at pH 7.1 (100 mL). The mixture was stirred for various lengths of time, and the polymer removed by passing the solution through a syringe filter. The solution was then analyzed for phosphate

The liquid samples were analyzed for phosphate spectrophotometrically using a standard molybdate assay.[5] In the case of isotherms, the results were then plotted as millimoles of phosphate bound per gram of polymer (determined by difference from the starting solution), vs the unbound phosphate concentration.

Results and Discussion

The phosphate-binding isotherm for crosslinked polyallylamine is shown in Figure 1. The amount of phosphate bound is a strong function of the unbound phosphate concentration, rising to about 2.7 mmoles/g at 5 mM phosphate (a typical physiologic concentration of phosphate in the intestine).

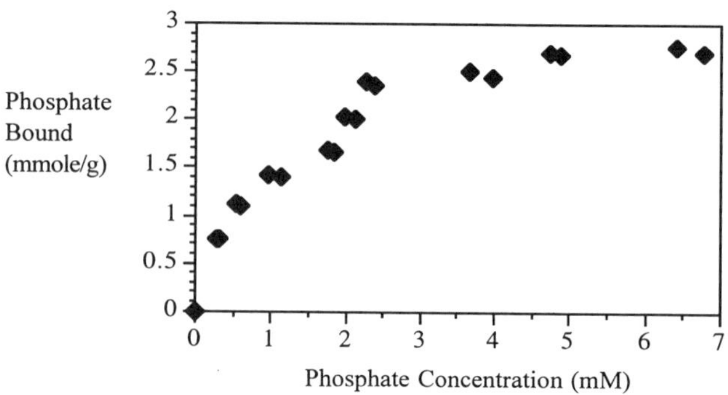

Figure 1. Phosphate binding isotherm for 12 mole percent crosslinked polyallylamine at pH 7.

pH Effects. The amount of phosphate bound by crosslinked polyallylamine is dependent upon the pH of the solution. Figure 2 shows the relationship between the solution pH and the amount of phosphate bound when 0.1 g of polymer is added to 20 mL of 20 mM phosphate. The binding clearly shows a maximum between pH 6 and 8, and is significantly lower above and below this range.

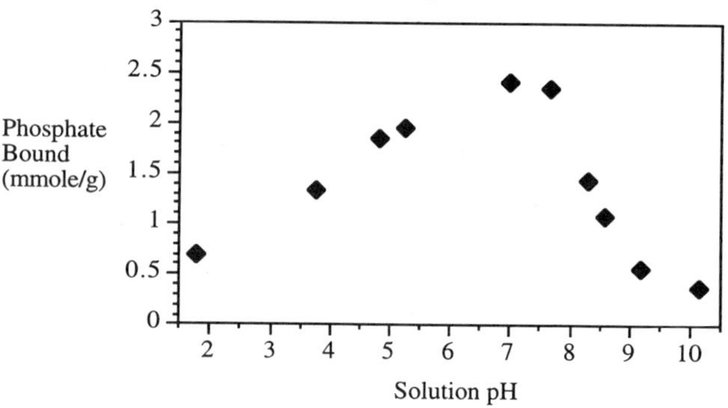

Figure 2. Phosphate binding as a function of pH for crosslinked polyallylamine when 0.1 g of polymer is added to 20 mL of 20 mM phosphate solution.

One straightforward interpretation is that above pH 7 the polymer exists as a mixture of the free amine and the amine hydrochloride. An acid/base titration of the crosslinked polymer (Figure 3) indicates that the protonation of the amines takes place over the broad pH range of 7-12. As the pH is raised, more and more of the amines are present as the free amine, and are unable to act as counterions for phosphate. Consequently, the binding of phosphate is reduced at elevated pH.

Likewise, below pH 7, the phosphate ion will begin to exist primarily in the monoanion form, rather than the dianion. We speculate that it is the dianion that binds to the polymer, and that reducing the concentration of the dianion by reducing the pH has the effect of reducing phosphate binding.

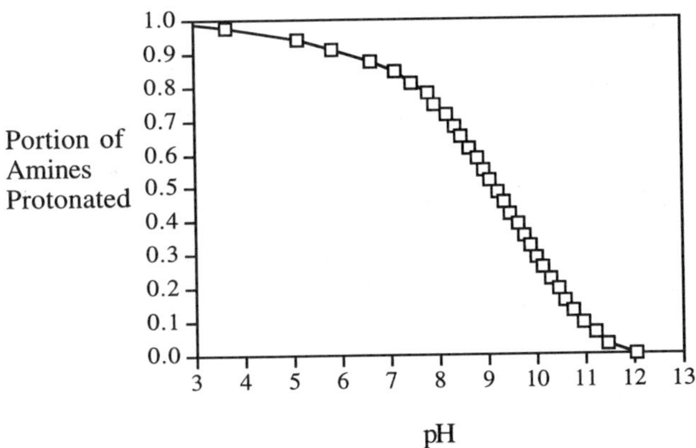

Figure 3. Portion of amines protonated in polyallylamine crosslinked with epichlorohydrin as a function of the solution pH. The portion protonated was determined by aqueous pH titration.

Further, we believe that it is the dianion character of phosphate which permits it to be bound so strongly by this and other polycations (perhaps in conjunction with hydrogen bonding to other parts of the phosphate). Monoanions, such as chloride, are unable to interact with multiple cationic sites, and are more weakly bound. The data in Figure 1 actually represent a competition between phosphate and the monoanions chloride and bicarbonate. In Figure 1, when $[PO_4^{3-}]$ = 5 mM the phosphate comprises only 4.3% of the total anions (with the remainder being chloride and bicarbonate), and yet the phosphate occupies between 25 and 50% of the available cationic sites (depending upon whether it is believed that each phosphate occupies one cationic site or two).

The spacing between the negatively charged oxygen atoms in a phosphate dianion is about 2.5 Å. The spacing between positive charges in polyallylamine is variable, but molecular models show that it can match this value (2.5 Å) in certain configurations. Consequently, the phosphate dianion can readily interact with two cationic sites at a time.

Other Polymers. We have also examined a wide range of other polymers as potential phosphate binders. Table 1 shows the amount of phosphate bound (at 5 mM unbound phosphate) for some of these polymers. In each case, an isotherm similar to Figure 1 was determined, and the amount of phosphate bound at 5 mM was determined.

Table 1. Phosphate Binding Of Various Polymers At 5 mM Phosphate

Polymer	Phosphate Bound (mmol/g)
Polyallylamine/epichlorohydrin	2.5-3.5*
Polyallylamine/butanediol diglycidal ether	2.7
Polyethyleneimine/epichlorohydrin	1.2-2.7*
Diethylenetriamine/epichlorohydrin	1.5
Polyethyleneimine/acryloyl chloride	1.2
Poly(dimethylaminoethylacrylamide)	0.8
Poly(trimethylammoniomethylstyrene chloride)	0.7
Poly(aminoethylethylmethacrylamide)	0.4
Poly(allyltrimethylammonium chloride)	0.3

*These values vary depending upon the amount of crosslinker used and the portion of amines in the starting polymer which are present as amine hydrochlorides.

From these and other polymers examined, it appears that quaternary amines are especially not useful for binding phosphate. In a number of cases, such as with polyallylamine and poly(allyltrimethylammonium chloride), we have made the same polymer with and without quaternization of the amines present. In every case, the binding of phosphate is significantly reduced upon quaternization. In the case above,

it is reduced far more than the simple diluent weight of the three methyl groups and the chloride ion would predict. We believe that the poor binding provided by quaternary amines may stem from their inability to engage in hydrogen bonding with the phosphate ion.

Kinetics. The kinetics of binding of phosphate to sevelamer hydrochloride is shown in Figure 4. In this case, the polymer is added to a phosphate-containing solution, and the free phosphate concentration is monitored. For this polymer, the binding of phosphate is faster than the experiment can be carried out (30 seconds).

The final amount of phosphate bound after 17 hours corresponds to 6.4 mmoles/g, at a final phosphate concentration of 11.4 mM. Similar experiments carried out at lower initial phosphate concentrations show similarly rapid kinetics. Consequently, the kinetics of phosphate binding is far faster than the transit times through the gastrointestinal tract (hours) and can safely be ignored with respect to human efficacy.

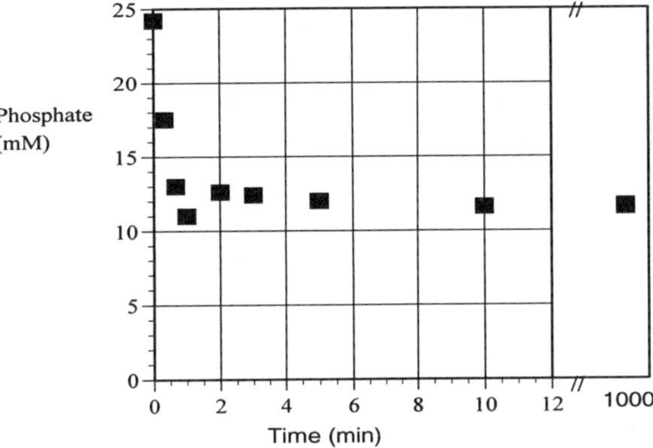

Figure 4. Free (unbound) phosphate concentration as a function of time after sevelamer hydrochloride is added to a phosphate-containing solution.

Conclusions

Crosslinked polyallylamine is a potent phosphate binding polymer, and is well suited for use as a pharmaceutical. It shows maximum phosphate binding in the pH range present in the small intestine, it binds more phosphate than other polymers under physiologic conditions, and is essentially nontoxic as it is a nonabsorbed hydrogel. It has a preference for phosphate over other ions in the gastrointestinal tract (such as chloride and bicarbonate), which is probably due to the fact that phosphate, being a dianion, can interact with two of the cationic sites at a time. Hydrogen bonding between the polymer and the phosphate ion may also be important for binding.

References
(1) Delmez, J. A.; Slatopolsky, E.; *American Journal of Kidney Diseases* **1992**, XIX, 303-317.
(2) Holmes-Farley, S. R.; Mandeville, W. H.; Whitesides, G. M.; *U. S. Patent 5,496,545* **1996**. See also Holmes-Farley, S. R.; Mandeville, W. H.; Whitesides, G. M.; *U. S. Patent 5,667,775* **1997**.
(3) Rosenbaum, D. P. ; Holmes-Farley, S. R.; Mandeville, W. H.; Pitruzzello, M.; Goldberg, D. I.; *Nephrol Dial Transplant* **1997**, 12, 961-964.
(4) Burke, S.K.; Goldberg D.I.; Bonventre, J.V.; Slatopolsky E. A. *Nephrol Dial Transplant* **1996**, 11, A40. Burke, S. K.; Slatopolsky, E. A.; Goldberg, D. I. *Nephrol Dial Transplant* **1997**, 12, 1640-4. Chertow, G. M.; Burke, S. K.; Lazarus, J. M.; Stenzel K. H.; Wombolt, D.; Goldberg, D.; Bonventre, J. V.; Slatopolsky, E. *American Journal of Kidney Diseases* **1997**; 29, 66-71.
(5) Taussky, H. H.; Shorr, J.; *J Biol Chem* **1953**, 202, 675-685.

ZERO-ORDER RELEASE OF IONIC DRUGS FROM WATER-SOLUBLE DRUG-POLYION COMPLEXES

Nandini Konar and Cherng-ju Kim, Temple University School of Pharmacy, Philadelphia, PA 19140

Introduction

Water-soluble polyions (of both anionic and cationic types) have been employed as carriers for ionic drugs. Such carriers were bound to oppositely charged drugs to form insoluble drug-polymer complexes. The cationic carrier poly(diallyl dimethyl acrylate chloride), (PDADMAC) formed insoluble complexes with anionic drugs, while the anionic carrier poly(acrylamido methylpropane sulfonic acid), (PAMPS) formed such complexes with cationic drugs. Tablets prepared from these complexes were shown to release the bound drugs in a controllable manner when placed in ionic media. Both types of carriers provided high drug loadings (>40 wt%) and released all of the bound drugs, over a extended period of time in a linear fashion by a dissociation/erosion mechanism. The drug release properties of these commercially available polyions are reported. PDADMAC was complexed with the cationic drugs diclofenac Na and sulfathiazole Na, while PAMPS was complexed with the cationic drugs labetalol HCl, propranolol HCl, verapamil HCl, and diltiazem HCl. The drug-polymer complexes yielded zero-order release kinetics for extended periods, with very high drug loadings.

Experimental

All chemicals were used as received. PDADMAC and PAMPS were obtained from Aldrich Chemical Co. Diclofenac sodium, sulfathiazole sodium, labetalol HCl and propranolol HCl were received from Sigma.

Preparation of drug-polymer complexes. An excess amount of drug solution (greater than 1.5 times the mole ratio of drug to the polymer) was added to an aqueous polymer solution to obtain a precipitated drug-polymer complex, which was then thoroughly washed free of soluble components and dried. PDADMAC was complexed with the anionic drugs diclofenac Na and sulfathiazole Na and PAMPS was complexed with the cationic drugs labetalol HCl, propranolol HCl, verapamil HCl, and diltiazem HCl. The dried drug-polymer complexes were pulverized in a mortar and pestle and compressed into tablets using a 9.5 mm diameter die and a flat punch in a Carver press (Wabash, IN) under a compression force of 5000 lbs.

Drug release kinetics. The release kinetics from tablets of the drug-polymer complexes were carried out in buffered release media containing 0.01Mphosphate and NaCl ranging from 0.2M to 0.05M at 37°C by the USP basket method at 100 rpm unless otherwise noted. Drug release was monitored on a HP 8452A diode-array spectrophotometer at 250 nm, 306nm, 306nm, 270nm, 278nm, 278nm, and 274nm for diclofenac Na, sulfathiazole, Na, labetalol HCl, propranolol HCl, verapamil HCl, and diltiazem HCl respectively.

Results & Discussion

Since these polymers have pendent ionic groups (quaternary amine in the case of PDADMAC, and sulfonate in the case of PAMPS), which are strongly ionized at all pH values, an oppositely charged drug can be bound to each functional group all along the polymer chain, permiting very high drug loadings. The drug-polymer complexes dissociate when placed in an ionic medium, releasing the free drug in a linear fashion.

The linearity of drug release was assessed by fitting the release data to the phenomenological equation [1]:

$$\frac{M_t}{M_\infty} = kt^n \qquad (1)$$

where M_t and M_∞ are the amounts of drug released at time t and the total amount of drug in a tablet, respectively, and k and n are a constant and a release exponent, respectively. The dissociation/erosion mechanism of the drug release kinetics were evaluated using the following equation [2]:

$$\frac{M_t}{M_\infty} = 1 - \left(1 - \frac{k_e t}{C_o r_o}\right)^2 \left(1 - \frac{2k_e t}{C_o l}\right) \qquad (2)$$

where k_e, C_o, r_o, and l are the dissociation/erosion rate constant, the initial drug concentration in a tablet, the tablet radius, and the tablet thickness, respectively.

In **Fig. 1**, the release of two drugs of different solubilities, from PDADMAC are shown. Diclofenac Na (solubility 2.5 wt%) was released over a period of 8 hrs, whereas sulfathiazole Na (solubility 24 wt%) was released much faster (3 hrs). In **Fig. 2** the effect of the buffer strength of the release medium on diclofenac release from PAMPS is shown. As the buffer strength was reduced from 0.2M to 0.05M, the drug release rate was observed to increase from 8 to 12 hrs. This was also observed in the case of PAMPS (not shown). Thus, the ionic strength of the release medium controls the dissociation of the drugs from the complex.

The effect of the pH value of the release medium is shown in **Fig. 3** for propranolol release from PAMPS. It may be observed that the release profiles are not very far apart. In **Fig. 4**, the effect of drug solubility on their release from complexes with PAMPS is shown. Here, the release rate is observed to increase with increasing solubility of the bound drugs (diltiazem > verapamil > propranolol > labetalol).

The release profiles shown in **Figs. 1-4** were well described by equations 1 and 2, and the calculated values of n, the exponent describing linearity and k_e, the dissociation/erosion rate constant are shown in **Table 1**. Linear release profiles were maintained all the way up to 80% drug release, after which they tailed off towards completion.

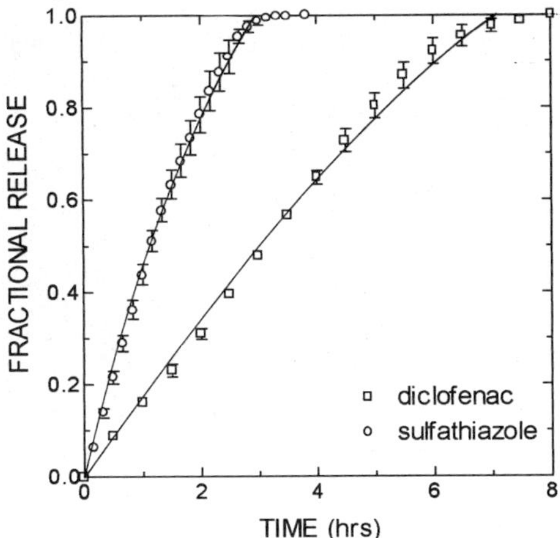

Figure 1. Effect of drug solubility on release from drug-PDADMAC complexes

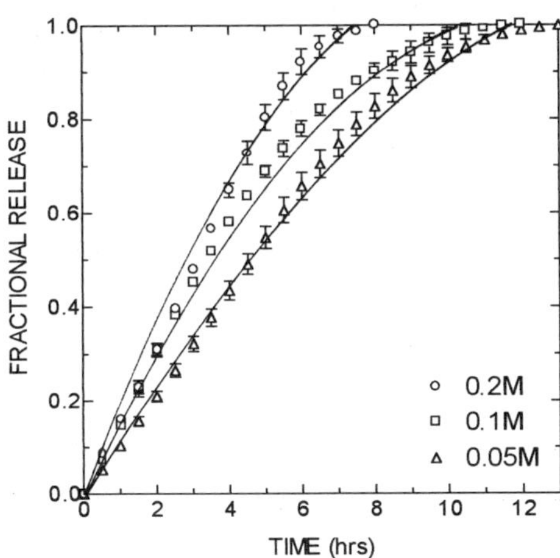

Figure 2. Effect of buffer strength on diclofenac release from diclofenac-PDADMAC complex

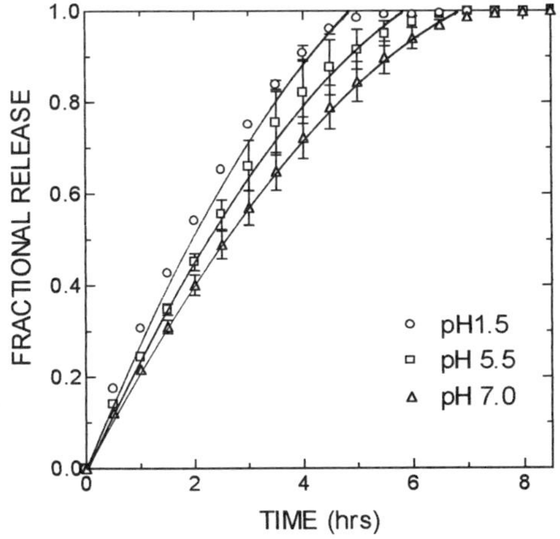

Figure 3. Effect of buffer pH on propranolol release from drug-PAMPS complex

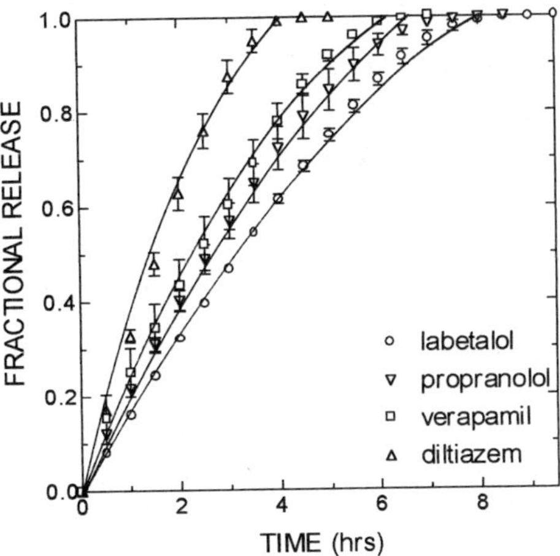

Figure 4. Effect of drug solubility on release from drug-PAMPS complex

These results agree with those reported from similar systems investigated earlier, in which copolymers of methyl methacrylate with ionogenic monomers were used as controlled release carriers for ionic drugs [2, 3, 4,]. The polyions reported in this study are homopolymers and provide more binding sites for drug attachment thus permitting higher drug loadings.

Table 1. Parameters n and k_e for Drug-Polymer Complexes

	k_e, (mg/cm²h)	n
Labetalol-PAMPS		
pH7.0, 0.2M	10.58±0.16	0.94±0.03
Propranolol-PAMPS		
pH7.0, 0.2M	11.87±0.40	0.85±0.08
pH5.5, 0.2M	12.93±0.63	0.88±0.11
pH1.5, 0.2M	16.09±0.19	0.80±0.02
Verapamil-PAMPS		
pH7.0, 0.2M	12.38±0.63	0.82±0.14
Diltiazem-PAMPS		
pH 7.0, 0.2M	21.96±1.00	0.89±0.09
Diclofenac-PDADMAC		
pH7.0, 0.2M	7.24±0.18	1.02±0.04
" 0.1M	5.89±0.12	0.87±0.04
" 0.05M	4.74±0.10	1.00±0.05
Sulfathiazole-PDADMAC		
pH7, 0.2M	18.20±0.43	0.91±0.07

Conclusions

Promising controlled delivery carriers, yielding zero-order release kinetics have been reported. Drug release kinetics were dependent on drug solubilities, as well as the properties of the release medium.

References

(1) P. L. Ritger and N. A. Peppas, *J. Control. Rel.* **1987,** 5, 23.
(2) Nujoma, Y.N.; Kim, C.J. *J. Pharm. Sci.* **1996,** *85*, 1091.
(3) Konar, N.; Kim, C.J. *J. Pharm. Sci.* **1997,** *86*, 1339.
(4) Konar, N.; Kim, C.J. *J. Appl. Polym. Sci.* (in press).

Interferometric Imaging of Hydrogels and Its Application to Infusion and Release Studies

Iwao Teraoka and Yunmei Xu

Polytechnic University, 333 Jay Street
Brooklyn, NY 11201

Introduction

Uniform gels swollen in a good solvent are nearly invisible unless they are colored. There have been efforts by many researchers to visualize transparent gel pieces by a variety of techniques. Imaging of a whole gel provides information on kinetics of the volume phase transition, diffusion and reaction processes within in the gel, and infusion of a second substance into the gel and release from it. Field-gradient magnetic resonance spectroscopy[1] would be so far the most successful methods of visualization. Diffusion of a spin-labeled substance into a gel piece has been studied by this method.

In this report, we show that interferometry can be a powerful tool to image a gel piece immersed in a fluid. This technique was used to measure the porosity of a hydrogel and the partition coefficient of a polymer in the gel.[2] Recently, we applied[3] the method to study of partitioning of polystyrene dissolved in an index-matching solvent with a solid porous silica bead and infusion of the polymer into the medium. Here we apply the in-situ, non-invasive technique to imaging of hydrogels. Unlike in other techniques, the gel piece to be imaged is immersed in a large volume of a fluid, either water or an aqueous solution. The interferometry allows measurement of nonuniform distributions of a gel matrix and a second substance that enters or is released from it. For the present study, we fabricated an acrylamide gel in a glass capillary following a standard procedure.[4] The gel was then shrunk in methanol, taken out, and reswollen in water. The gel returned into a cylinder of a uniform diameter of about 1.3 mm.

Interferometry

A Jamin interferometer we constructed earlier[3] was used here with a slight modification. A collimated laser beam (λ = 632.8 nm) is transformed into a slightly expanding beam by a lens. A beam displacer (a single crystal of calcite) splits the beam into two parallel beams separated by about 4 mm. The cylindrical gel is held nearly vertically in a fluorometer cell filled with water or an aqueous solution. One of the beams illuminates the gel whereas the other beam travels the surrounding fluid. The two beams are coupled by another beam displacer. The fringe pattern thus formed is magnified by lenses and projected onto a screen. The image is captured by a Macintosh Quadra 660AV and analyzed with NIH-Image (available at http://rsb.info.nih.gov/nih-image/).

Figure 1 shows a typical fringe pattern for an image of the hydrogel immersed in pure water. In the absence of the gel piece, the interference forms parallel straight fringes with a period of ξ and a slope of k (background image). We define the slope with respect to the normal to the gel axis in the image. The difference between the refractive index n_I of the cylindrical gel and the index n_E of the surrounding fluid distorts the pattern within the gel image into elliptic arcs. When k is large, i.e., the background fringes are nearly parallel to the cylinder, the pattern looks like parabolas. Each interior fringe is associated with one of the parallel fringes in the background image or those hidden by the gel image in **Figure 1**.

Figure1. Image of a hydrogel immersed in water.

Let us set up an x-y coordinate system with y-axis running along the center of the gel image of diameter of $2a$. The background fringes are given by $y = kx + b$ with $b = b_0 + m\xi(1+k^2)^{1/2}$ (m: integer). If n_I is uniform, the corresponding fringes inside the gel are given by $y = kx + b \pm (2\xi(n_I-n_E)/\lambda) \times[(1+k^2)(a-x^2)]^{1/2}$, where the sign is positive for $k < 0$.

Analysis of the fringe pattern centers on evaluating how much the fringes are displaced with respect the background fringes. The evaluation is possible in a couple of methods:

a. Plot of displacement. This method evaluates how much the x–intercept of each fringe, x_1, is displaced with reference to the intercept of the corresponding fringe in the background image. The displacement Δx relative to the period of the background fringes measured along the x–axis, $\xi(1+k^2)^{1/2}$, is an invariant of ξ and k. When n_I is uniform, the plot of the relative displacement $\Delta x/[\xi(1+k^2)^{1/2}]$ as a function of x_1 will be on a semi-ellipse:

$$\frac{\Delta x}{\xi(1+k^{-2})^{1/2}} = \frac{2a(n_I - n_E)}{\lambda}[1 - (x_1/a)^2]^{1/2} \qquad (1)$$

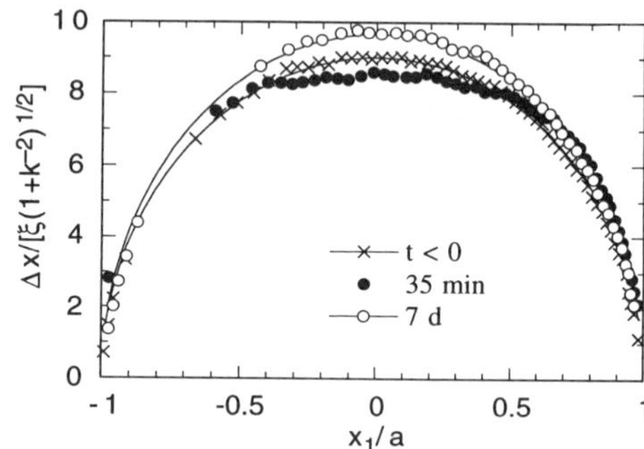

Figure 2. Relative displacement plotted as a function of the position of the fringe.

Crosses in **Figure 2** show the relative displacement obtained for fringes in the gel shown in **Figure 1**. The symbols are on the ellipse except for those on the left. The latter points were obtained for thick fringes and therefore carry large errors in the estimate of x_1. The maximum displacement Δx_m takes place for a fringe at $x_1 = 0$:

$$\frac{\Delta x_m}{\xi(1+k^{-2})^{1/2}} = \frac{2a(n_I - n_E)}{\lambda} \qquad (2)$$

When the gel is not uniform, in contrast, it will produce spatial variations in n_I. Then the plot of $\Delta x/[\xi(1+k^{-2})^{1/2}]$

deviates from the semi-ellipse trace. From the geometry of the gel, we expect that n_I is a function of r, the distance from the cylinder axis. In Eq. (1), n_I is then replaced by the average of the refractive index along the chord on the circular cross section of the gel:

$$\bar{n}_I(x_1) = b^{-1} \int_0^b n_I((x_1{}^2 + z^2)^{1/2})dz \tag{3}$$

with $b = (a^2 - x_1{}^2)^{1/2}$. To convert $\bar{n}_I(x_1)$ to $n_I(r)$, we approximated $n_I(r)$ by a polynomial that consists of even terms only, typically a six-order polynomial. An example of the plots of $n_I(r)$ with reference to n_S, the refractive index of water, is given in **Figure 3**. Discussion will be given later.

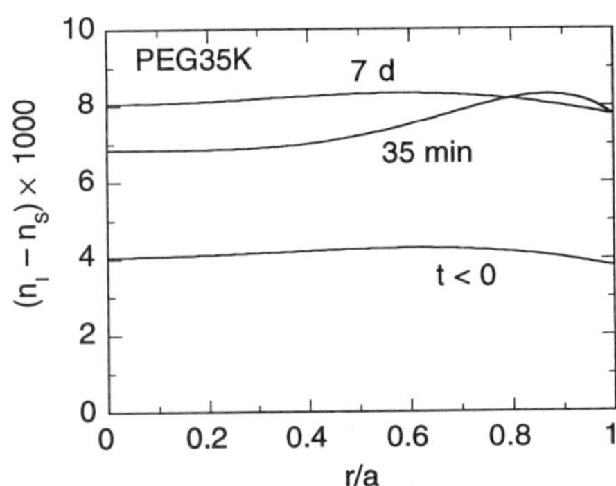

Figure 3. Radial variation of the refractive index of the hydrogel immersed in water and in a poly(ethylene glycol) solution.

b. Apex of parabola. This method finds the x-coordinate of the apex of the parabola, x_{peak}. When n_I is uniform, x_{peak} is related to Δx_m by

$$\frac{\Delta x_m}{\xi(1+k^{-2})^{1/2}} = q[(x_{peak}/a)^{-2} - 1]^{1/2} \tag{4}$$

where $q = a/[\xi(1+k^{-2})^{1/2}]$. When applied to the image in **Figure 1**, the estimate was 9.38.

c. Number of fringes. Count the number m_F of the fringes on the left of the parabola axis and within the gel image. The average $\langle m_F \rangle$ gives another estimate of Δx_m:

$$\frac{\Delta x_m}{\xi(1+k^{-2})^{1/2}} = [\langle m_F \rangle(\langle m_F \rangle + 2q)]^{1/2} \tag{5}$$

The estimate is an invariant of ξ and k. The estimation of $\langle m_F \rangle$ is easier for an image recorded with a larger ξ and a smaller $|k|$.

Example of Image Analysis

Below we discuss the image analysis for a gel in water and a water-swollen gel immersed in a solution of poly(ethylene glycol) (PEG).

a. Gel in pure water

A good fitting of the crosses by a semi-ellipse in **Figure 2** demonstrates that the cross section is circular and the gel is uniform. For this gel–water system, n_I and n_E are given as $n_I = n_S + c_G(dn/dc)_G$ and $n_E = n_S$, where c_G is the 'concentration' of the gel matrix, i.e., the mass of the gel matrix per unit volume of the swollen gel, and $(dn/dc)_G$ is the differential refractive index of the gel matrix. From $\Delta x_m/[\xi(1+k^{-2})^{1/2}]$, we obtain $n_I - n_S = c_G(dn/dc)_G$ as 4.14×10^{-3}. For $(dn/dc)_G$, we use the value obtained for a solution of polyacrylamide: $(dn/dc)_G = 0.182$ mL/g.[5] Then, c_G is estimated as 0.0227 g/mL. This value

agrees with an estimate from the volume (17.23 μL) of the swollen gel and its dried mass (0.39 mg).

b. Gel immersed in a PEG solution

When the water-swollen gel is immersed into a solution of PEG, water molecules in the gel exude, because the PEG solution has an osmotic pressure higher than that of the fluid in the gel. It is followed by gradual penetration of PEG into the gel. The diameter of the gel decreases in the course. The decrease depends on the concentration and the molecular weight of PEG. The image of the gel was captured at several times during the equilibration.

Figure 2 also shows a plot of $\Delta x/[\xi(1+k^{-2})^{1/2}]$ for the images obtained in 35 minutes and 7 days (closed and open circles, respectively) after the exterior fluid was changed for a 0.0201 g/mL solution of PEG (MW = 35,000; Fluka). The diameter of the gel decreased from 1.378 mm to 1.153 mm in 35 minutes and 1.127 mm in 7 days. For the images taken at $t > 0$, the exterior fluid is a solution that has a refractive index of $n_E = n_S + c_{PE}(dn/dc)_P$, where c_{PE} is the concentration of the polymer in the surrounding solution, and $(dn/dc)_P = 0.1350$ mL/g[6] is the differential refractive index of the PEG solution with reference to water. With deconvolution, the plots in **Figure 2** were converted into $n_I(r) - n_S$, shown in **Figure 3**.

The plots for $t < 0$ and $t = 7$ d are not horizontal. It may indicate some nonuniformness of the gel matrix or noncircular cross section, but errors in the image analysis cannot be ruled out. More importantly, however, the PEG solution increased $n_I(r)$, and the increase was not uniform at short times but regained uniformness after equilibration.

First we compare the plots of $t < 0$ and $t = 7$ d. The contribution to n_I by the gel contraction is 82.8%, assuming that the gel contracted isotropically. The observed two-fold increase in n_I is therefore also contributed by PEG molecules in the gel. From the difference, the interior concentration of PEG, c_{PI}, is estimated at 4.82×10^{-3}. Thus the partition coefficient is found to be 0.24.

Next, we examine the plot for $t = 35$ min. The profile is rather flat except for the edge portion of the gel. The difference of the flat level from that of $t < 0$ agrees with the one estimated only from the isotropic contraction of the gel. Thus we find that PEG molecules are absent in the center portion of the gel. It is interesting to find that n_I at the edge portion has reached the value of $t = 7$ d. This portion reached equilibrium with the surrounding solution in 35 min.

In the presentation, we will also show results for release of moleucles into the surroundings. Detailed discussion on the behavior of the hydrogel in solution of PEG (different molecular weights and concentrations) will be given in another presentation.

References
1. For example, Schlick, S.; Pilar, J.; Kweon, S.-C.; Vacik, J.; Gao, Z.; Labsky, J. *Macromolecules* **1995**, *28*, 5780.
2. Sernetz, M.; Bittner, H. R.; Willems, H.; Baumhoer, C. In *The Fractal Approach to Heterogeneous Chemistry*; D. Avnir, Ed.; John Wiley: New York, 1989.
3. Teraoka, I. *Macromolecules* **1996**, *29*, 2430.; Dube, A.; Teraoka, I. *Macromolecules* **1997**, *30*, 5352.
4. Tanaka, T. Phys. Rev. Lett. **1978**, *40*, 820.
5. Munk, P.; Aminabhavi, T. M.; Williams, P.; Hoffman, D. E.; Chmelir, M. *Macromolecules* **1980**, *13*, 871.
6. American Polymer Standards Corp.

OLIGOPEPTIDE INTERACTIONS WITH INSULIN

Ruifeng Zhao, Mamoru Haratake and Raphael M. Ottenbrite

Department of Chemistry, Virginia Commonwealth University,
Richmond, VA 23284-2006

Introduction

Oral drug delivery of insulin for diabetics is one of the primary interests in the pharmaceutical industry.[1] Based on a reported proteinoid miscrosphere oral drug delivery system, several series of oligopeptides with specific structures and amino acid sequences were synthesized and characterized as potential oral drug delivery carriers.[2-4] Previously, we reported the aggregation behavior of these oligopeptides and their interactions with heparin.[3-6] The interaction of these oligopeptides with insulin was further investigated with circular dichroism (CD) and isothermal titration calorimetry (ITC).

Experimental

CD spectroscopy: The CD instrument used was a Jasco-600 spectropolarimeter with a 2 mm pass-length quartz cell. All the samples were prepared in phosphate buffer saline (PBS) solution and sonicated in a Branson bath for 20 min just before CD measurements. The pHs of the oligopeptide solutions were adjusted to 7.4 with 1 N NaOH and 1 N HCl. The samples were scanned under nitrogen gas atmosphere from 200 nm to 350 nm at a scan speed of 20 nm/min. The average of two measurements was recorded. Each spectrum was corrected by subtracting the spectrum of the corresponding oligopeptide solution without insulin.

Isothermal Titration Calorimetry (ITC): The instrument used was a Microcal Omega titration calorimeter (Microcal Inc.). All the samples were prepared in PBS and sonicated for 20 min before ITC measurements. The pHs of the oligopeptide solutions were adjusted to 7.4 with 1 N NaOH and 1N HCl. The insulin solution (2.2 mL) was injected into the cell (cell volume, 1.3961 mL) and the oligopeptide solution was placed into the injection syringe. The rotation speed (to mix the two solutions during titration) was maintained at 200 rpm and the temperature was kept constant at 30^0C or 40^0C. Each injection contained 20 µL of the oligopeptide solution; a 2 min interval between injections was maintained. The number of injections was 12. The data was collected and processed with a Omega software program (Micrcal Inc.). The heat generated by each injection of the oligopeptide solution to the insulin solution was corrected by subtracting the heat generated by a corresponding injection of the oligopeptide solution to the PBS solution. The average of two experimental data was reported as the dissociation constant (K_d), enthalpy change (ΔH), entropy change (ΔS) and the binding number.

Results and Discussion

Insulin, a globular protein, has discrete secondary structures: α-helix (B9-B19), β-turn (B20-B22) and β-strand (B23-B30).[7] Insulin also forms self-aggregates, such as dimers, tetramers and hexamers.[8] The CD spectrum of insulin has a major band at 208 nm and at 222 nm, and a minor band at 273 nm. These three CD bands were assigned to the α-helix structure, the β-structure and the formation of higher self-aggregates, respectively.[9]

The effects of the tetrapeptides on the insulin secondary structures and its aggregation behavior were studied by CD spectroscopy. The far-UV CD spectra of insulin (100 µM) in the absence and presence of 3 mg/mL of pEE(α)L(γ)L or 3 mg/mL of pEE(α)F(γ)F are shown in Figure 1. The addition of pEE(α)L(γ)L to the insulin solution resulted in an attenuation of both bands at 209 nm and 270 nm, and an enhancement of the band at 222 nm (Figure 1). This enhancement increased with increasing tetrapeptide concentration from 3 mg/mL to

5 mg/mL. In addition, the CD absorption enhancement was accompanied by a shift of the negative maximum peak from 222 nm to 225 nm. The attenuation of the band at 209 nm indicates that pEE(α)L(γ)L may have an effect on the α-helix structure of insulin. The β-structures assigned to the band at 222 nm are a predominant feature of insulin dimers. It was reported that the insulin derivatives, which do not aggregate, had an attenuated band at 222 nm.[7,10] Therefore, the band enhancement at 222 nm in the presence of pEE(α)L(γ)L suggests that pEE(α)L(γ)L may promote the formation of insulin dimers and/or higher aggregates.

On the other hand, the addition of pEE(α)F(γ)F caused the attenuation of both bands at 209 nm and 222 nm (Figure 1). A similar CD spectral transition has been reported by Yuping et al. when bile acids were added to insulin.[10] Furthermore, the addition of alcoholic ligands to insulin, such as cyclohexanol and isopropanol which stabilize the α-helix structure of insulin, enhanced the CD bands of insulin at 209 and 222 nm.[11] As a result, pEE(α)F(γ)F appears to induce changes in both the α and β secondary structures of insulin. Consequently, the CD spectra of insulin in the presence of the tetrapeptides pEE(α)L(γ)L and pEE(α)F(γ)F are significantly different, which indicates that the interaction modes of these two tetrapeptides with insulin are also different.

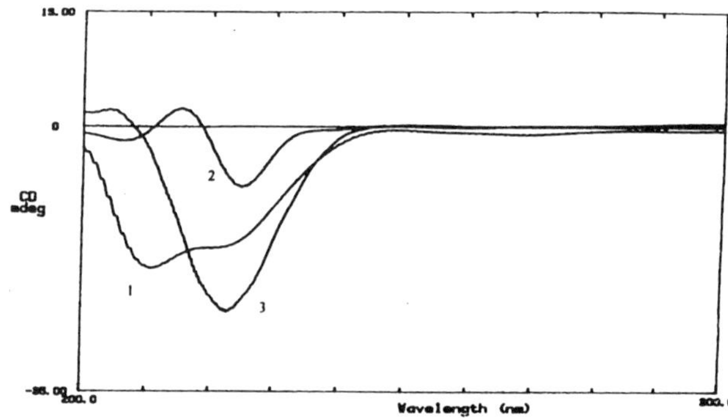

Figure 1. CD Spectra of Insulin in the Presence of the Tetrapeptides pEE(α)L(γ)L and pEE(α)F(γ)F; Curve 1 represents CD spectrum for insulin solution (100 µM) alone, Curve 2 represents CD spectrum for insulin solution (100 µM) with 3 mg/mL of pEE(α)F(γ)F and Curve 3 represents CD spectrum for insulin solution (100 µM) with 3 mg/mL of pEE(α)L(γ)L

Isothermal titration calorimetry was also used to measure the isothermal binding parameters, such as the dissociation constant K_d (M), the enthalpy change, ΔH (kJ/mol), the entropy change, ΔS (J/mol·K), the free energy change, ΔG (kJ/mol) and the binding number, n, of the tetrapeptides pEE(α)F(γ)F, pEE(α)Y(γ)Y, pEE(α)A(γ)A and pEE(α)L(γ)L (40 mg/mL) with insulin (100 µM) in PBS solution (pH 7.4) at 30^0C (Table 1). All the enthalpy changes, ΔH, were negative, which indicates that the interactions of the tetrapeptides with insulin are exothermic processes. The aliphatic tetrapeptides provided larger ΔH values with insulin than the aromatic tetrapeptides. Negative ΔG values were observed for all the tetrapeptides, indicating that the interaction of the tetrapeptides with insulin are energetically favorable. However, the aromatic tetrapeptides pEE(α)F(γ)F and pEE(α)Y(γ)Y provided larger ΔG values than the aliphatic tetrapeptides pEE(α)A(γ)A and pEE(α)L(γ)L. This suggests a difference in the interaction modes between the aromatic tetrapeptides and the aliphatic tetrapeptides with insulin.

Table 1. Thermodynamic Binding Parameters of Tetrapeptides to Insulin

Oligomer	pEE(α)F(γ)F	pEE(α)Y(γ)Y	pEE(α)A(γ)A	pEE(α)L(γ)L
ΔH (kJ/mol)	-5.6	-28.4	-43.9	-55.6
ΔS (J/mol·K)	7.1	-68.2	-131.4	-173.6
ΔG (kJ/mol)	-7.7	-7.7	-4.3	-3.0
K_d (10^{-2} M)	3.73	1.37	2.93	2.58
n	1.56	0.28	0.55	0.65

Table 2. Binding Thermodynamic Parameters of Tetrapeptides with Insulin

Tetrapeptide	Temp. (^{0}C)	ΔH (kJ/mol)	ΔS (J/mol·K)	ΔC_p (kJ/mol·K)
pEE(α)F(γ)F	30	-5.6	6.9	
	40	-0.97	35.5	0.46
pEE(α)L(γ)L	30	-55.57	-173.5	
	40	-91.82	-307.2	-3.63

The heat capacity change, ΔC_p, which is known to be a good indicator of changes in hydrophobic interaction with binding, is negative when hydrophobic interactions occur and positive when hydrophobic interactions are disrupted. The ΔC_p values in Table 2 indicate that hydrophobic interactions occur when the tetrapeptide pEE(α)L(γ)L is added to insulin. However, the enthalpy change, ΔH, for small molecular ligands binding to protein is typically in the order of 20 kJ/mol, which is much less than the observed ΔH value for pEE(α)L(γ)L with insulin (55.6 kJ/mol). This means that some other processes, such as insulin self-aggregation, may also be involved when this aliphatic tetrapeptide is added to insulin.

In contrast, the aromatic tetrapeptide pEE(α)F(γ)F has a smaller ΔH value and a small ΔCp value with insulin, which indicates that less energy is involved in the interaction of the aromatic tetrapeptide with insulin. The interaction between pEE(α)F(γ)F and insulin is currently not understood. However, it appears that the interaction of the aliphatic tetrapeptide pEE(α)L(γ)L with insulin is different from that of the aromatic tetrapeptide pEE(α)F(γ)F with insulin. This finding was supported by the results from fluorescence spectroscopy studies of the tetrapeptides with insulin in this laboratory.[12] It was found, from these fluorescence spectroscopy studies, that the aliphatic tetrapeptide pEE(α)L(γ)L enhanced the emission peak and caused a hypochromic peak shift of the hydrophobic fluorescence probe, 8-anilino-1-naphthalenesulfonic acid (ANS), only in the presence of insulin. However, the aromatic tetrapeptide pEE(α)F(γ)F did not effect any change under these conditions.

Acknowledgment

The authors wish to thank Emisphere Technologies, Inc. for their financial support for this project and Yoshitomi Pharmaceutical Industries, Ltd., Japan for their finance support to M. Haratake during his stay in our laboratory.

References

(1) Lee, V.H.L.; Hashid, M.; Mizushima, Y. *Trends and Future Perspective in Peptide and Protein Drug Delivery Chap. I-III*, Harwood Academic publishers: Switzerland, 1995.

(2) Ottenbrite, R.M.; Zhao, R.; Milstein, S. *Advanced Biomaterials in Biomedical Engineering and Drug Delivery System* p51, Springer-Verlag Tokyo, 1996.

(3) Ottenbrite, R.M.; Zhao, R.; Milstein, S. *J. Macrom. Symp.* **1996**, *101*, 379.

(4) Zhao, R.; Haratake, M.; Ottenbrite, R.M.; *Polymer Preprint* **1996**, *37(1)*, 614 and 37(2), 153.

(5) Haratake, M.; Zhao, R.; Ottenbrite, R.M.; *Polymer Preprint* **1996**, *37(2)*, 131 & 753.

(6) Haratake, M.; Zhao, R.; Ottenbrite, R.M.; *Journal of Bioactive and Compatible Polymers* **1997**, *12*, 282.

(7) Pittman, I.V.; Tager, H.S. *Biochemistry* **1995**, *34*, 10578.

(8) Smith G.D.; Ciszak, E.; *Proc. Natl. Acad. Sci. USA* **1994**, *91*, 8851.

(9) Goldman, J.; Carpenter, F.H. *Biochemistry* **1974**, *13*, 4566.

(10) Li, Y.; Shao, Z.; Nitra, A.K. *Pharm. Res.* **1992**, *9* (7), 864.

(11) Derewenda, U.; Dodson, E.J.; Smith, G.D.; Swenson, D. *Nature* **1989**, *388* (13), 594.

(12) Haratake, M.; Zhao, R.; Ottenbrite, R.M. *J. Bioact. Compti. Polym.*, **1998** (accepted).

SYNTHESIS OF BIODEGRADABLE AND BIOCOMPATIBLE AMPHIPHILIC ETHYLENE OXIDE/ε-CAPROLACTONE BLOCK COPOLYMER BY SEQUENTIAL ANIONIC RING-OPENING POLYMERIZATION

Yisong Yu and Adi Eisenberg
Department of chemistry, McGill University
801 Sherbrooke Street West, Montreal, PQ, H3A 2K6 Canada

Introduction

It is well known that aggregates or micelles can be formed from amphiphilic small molecule surfactant systems and block copolymers. When these amphiphilic materials are dissolved in a solvent which is selective for one of the components, colloidal solutions are formed as a result of the association of the insoluble component. Because of the relatively high molecular weight of polymers, the aggregates or micelles formed from amphiphilic block copolymers are much more stable than those of small molecule surfactants. Crew-cut aggregates with various morphologies have been extensively investigated in our group.[1]

Micelles formed in aqueous solution are of interested to chemists not only for fundamental reasons, but also because of their potential applications. These micelles have the core-shell structure, in which the hydrophobic component forms the core of the micelles, while the hydrophilic component surrounds this core as a hydrated outer shell. Since many drugs are hydrophobic, the hydrophobic core of the micelles can serve as a microcontainer for drug delivery. The range of amphiphilic block copolymers available for drug delivery applications is limited, however.[2] In this presentation, we describe the synthesis and characterization of one such block copolymer system based on poly(ethylene oxide) and poly(ε-caprolactone).

Both poly(ethylene oxide) (PEO) and poly(ε-caprolactone) (Poly(ε-CL)) homopolymers are known to be nontoxic and biodegradable, and are commercially available. In combination with each other, one can expect that PEO will be hydrophilic and Poly(ε-CL) hydrophobic. Thus, It seems reasonable that the chemical association of PEO and Poly(ε-CL) would provide a versatile amphiphilic block copolymer for the preparation of polymeric micelles to be used as drug carriers.

Several methods have been reported for the synthesis of PEO-b-Poly(ε-CL) block copolymers[3-6]. The anionically initiated copolymerization of ε-CL by PEO sodium to provide ether-ester block copolymers was reported by Perret et al.[3]; however, it was found that the resulting block copolymers were contaminated with a significant amount of homo-Poly(ε-CL). The reason for the formation of homopolymers was attributed to an equilibrium cyclic oligomerization in the anionic polymerization of ε-CL when metal alkyl oxides are the living species[4]. Cerrai et al.[5] exploited non-catalyzed polymerization of ε-CL initiated by poly(ethylene glycol) (PEG) to get almost pure block copolymers, but the polymerization rate was very slow and the length of Poly(ε-CL) block was low even when the polymerization was performed at very high temperature (185 ºC) for long periods of time (several weeks). Recently, Bogdanov et al[6] reported another approach to synthesize PEG-b-Poly(ε-CL) block copolymer. Copolymerization of ε-CL on the PEG in the presence of stannous octoate as catalyst was carried out in bulk at 130 ºC for 24 h, and high molecular weight Poly(ε-CL) was obtained. The polydispersities of the polymers were not reported.

The synthesis of poly(ethylene oxide)-b-poly(ε-caparolactone) block copolymers reported in this work involves sequential anionic ring-opening polymerization. The advantage of this method is that the anionic polymerizations of both ethylene oxide and ε-caprolactone are well-controlled, and can be carried out at ambient temperature within short periods of time. The block copolymers synthesized by this method have relatively narrow molecular weight distributions and variable poly(ε-caprolactone) chain lengths.

Experimental

Tetrahydrofran (THF) (Aldrich) was purified by refluxing over a sodium bezophenone complex under nitrogen (a blue-violet color indicating a solvent free of oxygen and moisture). The THF was further distilled over poly(styryllithium) under reduced pressure in the final purification step. ε-Caprolactone (Aldrich) was twice dried over CaH$_2$ at room temperature and distilled under reduced pressure. Ethylene oxide (Matheson) was first purified with CaH$_2$ and then distilled over n-butyllithium just before use. Diphenylmethylpotassium (Ph$_2$CHK) was prepared at room temperature by reacting diphenylmetnane with potassium naphthalene in THF for 24 h. Naphthalene potassium resulted from the reaction of potassium with naphthalene in THF at room temperature. Diethylaluminum chloride (Aldrich) in hexane was filtered prior to use.

Sequential anionic polymerization of ethylene oxide and ε-caprolactone was carried out under an inert atmosphere in a previously flamed glass reactor equipped with a rubber septum. Syringes and stainless steel capillaries were used in order to transfer solvent, monomer, and initiator. THF and diphenyl methyl potassium, which were used as the solvent and the initiator, were first charged into the reactor. Pre-purified ethylene oxide monomer was slowly added to the initiator solution of known concentration at -78 ºC. The temperature was slowly increased from -78 to +30 ºC over a period of 30 minutes. Thereafter, the reaction mixture was kept at 30 ºC for 24 hours. Then an aliquot of the reaction medium was withdrawn for analysis by GPC in order to determine the molecular weight of the first block. An equivalent amount of diethylaluminum chloride in hexane was added dropwise into the polymerization solution, followed by the addition of ε-CL monomer. Polymerization of ε-CL was performed at 25 ºC for *ca.* 20 hours. The polymerization was stopped by adding a 10-fold excess of HCl solution relative to aluminum. The polymers were recovered by precipitation into cold hexane and were dried under vacuum at room temperature for one week. The pure PEO-b-Poly(ε-CL) block copolymers were white powders.

Size exclusion chromatography (SEC) was used to measure the degree of polymerization and polydispersity of the polymers. SEC was carried out in THF at room temperature using a Waters 510 liquid chromatograph equipped with two gel columns (Styragel HR1 and HR4) and a refractive index detector (Varian RI-4). Millennium software was utilized. Polyethylene oxide standards were used for calibration to calculate the molecular weight of the PEO blocks. The degree of polymerization of the Poly(ε-CL) blocks was measured by ^{1}H-NMR relative to the degree of polymerization of PEO block. Nuclear magnetic resonance (^{1}H-NMR) spectra were recorded on a Varian XL-300 spectrometer at room temperature. CDCl$_3$ was used as solvent with tetramethylsilane as an internal standard.

Results and Discussion

Anionic ring-opening polymerization of ethylene oxide was initiated by diphenyl methyl potassium in THF; the mixture was kept at 30 ºC for 24 hours. The narrow molecular weight distribution as shown by the SEC trace (Figure 1A), indicates that the polymerization is a well-controlled living process. Aluminium alkoxides have proved to be very successful in the synthesis of polyesters such as poly(ε-caprolactone) and poly(D,L or L,L) lactides[7]. The coordination-insertion type of polymerization is living and yields linear chains of a predictable molecular weight and a narrow molecular weight distribution. Poly(ethylene oxide)-diethyl aluminum is obtained by the reaction of diethylaluminum chloride and poly(ethylene oxide) potassium, and is used as the macroinitiator for ε-caprolactone polymerization.

SEC traces of the samples picked out at different times are shown in Figure 1, which indicates a shift of the peak to shorter times with increasing polymerization time, and shows that the block copolymer peaks are narrow and symmetrical. The polydispersities of block copolymers broaden somewhat with increasing molecular weight of the Poly(ε-CL).

The degree of polymerization of the Poly(ε-CL) block can be obtained by ^{1}H-NMR analysis. Figure 2 shows a typical ^{1}H-NMR spectrum of PEO-b-Poly(ε-CL) block copolymer. The singlet centred at 3.7 ppm can be attributed to the methylene protons of the oxyethylene units of PEO; the singlet at 4.1 ppm and 2.3 ppm can be attributed to the methylene protons of the oxycarbonyl-1,5-pentamethylene unit of Poly(ε-CL). The mole fraction of each component in the block copolymers can be determined from the integral peak area ratio, and the degree of polymerization of the Poly(ε-CL) block can be calculated relative to the degree of polymerization of the PEO block.

The time-conversion relationship for the ε-CL polymerization illustrated in Figure 3 shows that the polymerization of ε-CL is completed within 20 h. Figure 4 shows a linear increase of the degree of polymerization of the Poly(ε-CL) with the degree of the monomer conversion. This results demonstrates that the ring-opening polymerization of ε-CL is a living process without significant side reactions, such as chain transfer or termination.

Conclusion

Sequential anionic ring-opening polymerization of ethylene oxide and ε-caprolactone results in well-defined poly(ethylene oxide)-b-poly(ε-caprolactone) block copolymers with relatively long hydrophobic poly(ε-caprolactone) segments. PEO-b-Poly(ε-CL) block copolymers synthesized by this method have been used as a vehicle for drug delivery.[8] Further studies of the micellization and morphologies of this amphiphilic block copolymer in aqueous solution are under investigation.

Reference

1 (a) Zhang, L.; Eisenberg, A. *Science* **1995**, *268*, 1728; (b) Zhang, L.; Yu, K.;Eisenberg, A. *Science* **1996**, *272*, 1777. (c) Yu, Y.; Eisenberg, A. *J. Am. Chem. Soc.* **1997**, *119*, 8383.

2 (a) Kabanov, A.V. and Alakhov, V.Y. "Micelles of Amphophilic Block Copolymers As Vehicles for Drug Delievery" In Amphiphilic Block Copolymers : Self-Assembly and Applications edited by Alexamdris P., Lindman B., Elsevier, Netherlands. **1997**; (b) Kwon, G.; Naito, M.,; Yokoyama, M.; Okano, T.; Sakurai, Y.; Kataoka, K. *Journal of Controlled Release* **1997**, *48*, 195; (c) Kataoka, K.; Kwon, G.; Yokoyama, M.; Sakurai, Y.;. *Journal of Controlled Release* **1992**, *24*, 119.

3 Perret, R. and Skoulios, A. *Die Makromoleculare Chemie* **1972**, *56*, 143.

4 Ito, K.; Hashizuka, Y. and Yamashita, Y. *Macromolecules*, **1977**, *10*, 821.

5 Cerrai, P.; Tricoli, M.; Andruzzi, F.; Paci, M and Paci, M. *Polymer* **1989**, *30*, 338.

6 Bogdanov, B.; Vidts, A.; Van Den Bulcke, A; Verbeeck, R. and Schacht E. *Polymer* **1998**, *39*, 1631.

7 Jacobs, C.; Dubois, Ph.; Jerome, R. and Teyssie, Ph. *Macromolecules*, **1991**, *27*, 3027.

8 Allen, C.; Yu, Y.; Maysinger, D. and Eisenberg, A. Bioconjugate Chemistry (Accepted).

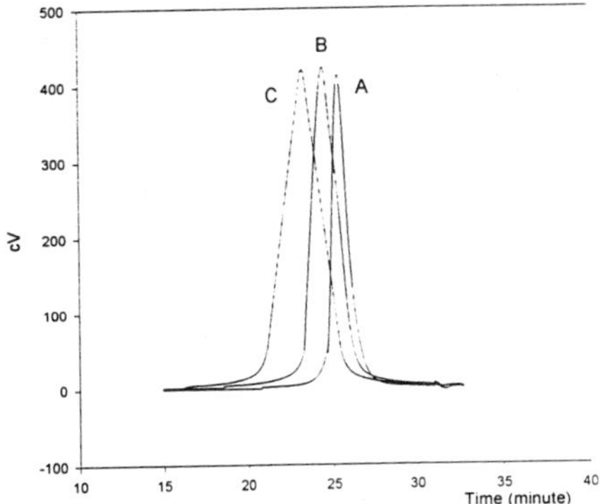

Figure 1. SEC traces of PEO and PEO-b-Poly(ε-CL) diblock copolymers
A. Homo-PEO (M_n=2000, PI=1.04), B. PEO-b-Poly(ε-CL) (M_n=2000-b-5000, PI=1.07).
C. PEO-b-Poly(ε-CL) (M_n=2000-b-17000, PI=1.12)

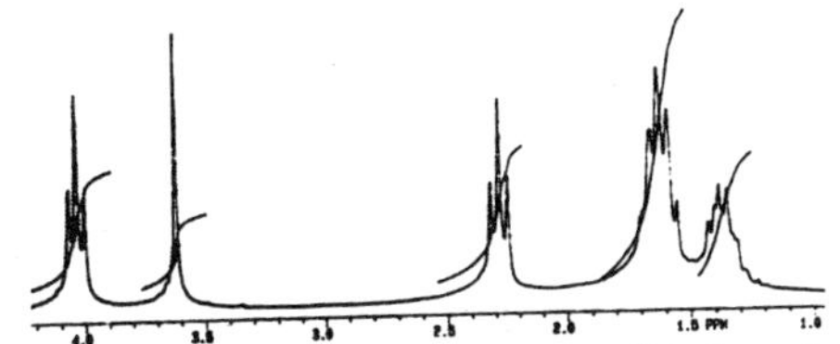

Figure 2. ^{1}H-NMR spectrum of PEO-b-Poly(ε-CL) diblock copolymer

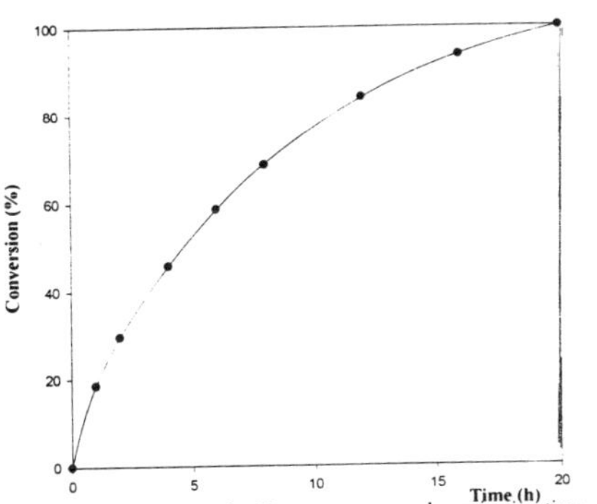

Figure 3. Conversion of ε-CL monomer vs. polymerization time initiated by PEO diethyl aluminum macroinitiator in THF at 25 °C

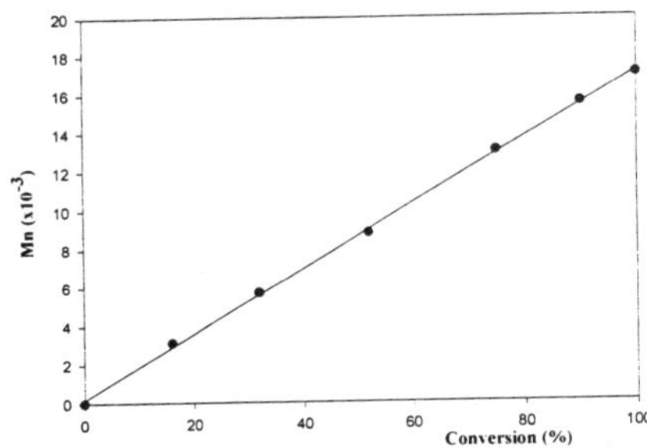

Figure 4. Relationship of molecular weight and conversion of ε-CL polymerization in THF at 25 °C

INFLUENCE OF POLYMER ARCHITECTURE ON SOLUBILITY IN SCF SOLVENTS. M. A. McHugh Department of Chemical Engineering, Johns Hopkins University, Baltimore, MD 21218

An experimental approach is developed to address the underlying physics and chemistry of SCF-polymer solutions that determine whether a polymer will dissolve in an SCF solvent. The focus of the experimental program is the determination of the influence of the energetics of mixing and the conformational entropy of the polymer on the phase behavior. The types and the strengths of energetic interactions depend on the chemical nature of the SCF solvent and on the chemical architecture of the polymer. In addition, the free volume difference between the SCF solvent and the polymer is also dictated by the chemical architecture of the polymer which influences the conformation of the polymer in solution. The two main themes of this talk are to describe this dual role of the chain architecture and to elucidate the influence of SCF solvent physico-chemical properties on polymer solubility.

THE DESIGN, SYNTHESIS AND APPLICATION OF AMPHIPHILES FOR USE IN LIQUID AND SUPERCRITICAL CARBON DIOXIDE. J. M. DeSimone, Department of Chemistry, CB#3290, University of North Carolina at Chapel Hill, Chapel Hill, NC 27599

The increasing demand for environmental responsibility in chemical syntheses requires creative solutions to the problem of waste solvent generation. Exploitation of "green" solvent systems such as liquid and supercritical carbon dioxide is made possible by recent advances in the design and synthesis of amphiphilic materials suitable for use in CO_2. The aggregation of CO_2 amphiphiles into micelles has been demonstrated, and these novel block copolymers have been used to stabilize dispersion polymerizations in CO_2 solution. The combination of appropriately designed surfactants with the tunable solvent properties of CO_2 opens the door to many potential applications, from the synthesis of sub-micron sized polymer particles isolated as a free flowing powder to the selective extraction of biologically active components from fermentation baths, all without the generation of a large excess of problematic organic and aqueous wastes.

CARBON DIOXIDE AS BOTH SOLVENT AND MONOMER. E. J. Beckman, T. Sarbu Chemical Engineering Department, University of Pittsburgh, Pittsburgh, PA 15261

Carbon dioxide is of interest as a monomer because it is inexpensive, non-flammable, relatively non-toxic and can be used to incorporate carbonate and ester linkages into a polymer. However, previous work has shown that CO_2 can be difficult to activate catalytically. We have developed a series of zinc, aluminum, and alumino-zinc catalysts which facilitate the copolymerization of CO_2 and ethylene oxide, propylene oxide, and cyclohexene oxide where CO_2 is both monomer and solvent. We have found that these catalysts support a copolymerization with living character, such that one can generate block copolymers (ether-carbonate) via sequential addition of monomers. The plasticizing effect of excess CO_2 in the system permits the polymerization to proceed without mass transport limitations.

THE INFLUENCE OF SUPERCRITICAL FLUIDS ON POLYMER MIXTURES AND BLOCK COPOLYMERS. T. P. Russell, J. J. Watkins, V. RamachandraRao, G. D. Brown, M. A. Pollard, University of Massachusetts, Amherst, MA 01003

With increasing temperature, compatible polymer mixtures undergo a transition from the mixed to phase separated state at the lower critical solution temperature, LCST. More recently, block copolymers have been shown to exhibit a similar transition from the phase mixed to microphase separated state with heating, termed the lower critical ordering transition, LCOT. Both LCST and LCOT behavior results from a balance between segmental interactions and differences in the thermal expansion of the two components, i.e. the volume occupied by the polymers. With increasing temperature, the entropic component of the free energy dominates and the segments demix. The differential dilation of two polymers by supercritical carbon fluids can be used to achieve the same end. In situ small angle neutron scattering studies have shown that both the LCOT and LCST can be substantially reduced at relatively low carbon dioxide pressures.

NANO-POROUS POLYPROPYLENE AND POLYETHYLENE. P. D. Whaley, P. Ehrlich, R. S. Stein, H. H. Winter, Department of Chemistry, University of Massachusetts, Amherst, Massachusetts 01003

Solid polymeric foams with large internal surface area and an open pore structure have been produced with densities as low as 0.15 g/cm^3 and surface areas of up to 140 m2/g. We crystallized isotactic polypropylene and polyethylene in the presence of supercritical propane above 3000 psi and then discharged the supercritical propane to obtain a clean nano-porous structure (10nm average pore size). A variable-volume view cell apparatus allows measurement of vapor/liquid and supercritical fluid/solid phase equilibria in addition to producing thick monolith samples. Phase equilibria of isotactic polypropylene with propane will be shown. The newest experiments involve a rheometer/mixer with view cell for light scattering and cloud point studies. The direct drive of this apparatus allows handling of very viscous, high molecular weight polymers in supercritical solvents.

REACTIONS IN SUPERCRITICAL FLUID -- SWOLLEN POLYMERS: A NEW ROUTE TO POLYMER BLENDS, MODIFIED POLYMERS AND COMPOSITE FOAMS. E. Kung, K. Arora, H. Hayes, P. Rajagopalan, T. J. McCarthy, Polymer Science and Engineering Department, University of Massachusetts, Amherst, MA 01003

We have continued investigating reactions in supercritical fluid (SCF) - swollen polymer substrates subsequent to Jim Watkins Ph.D. thesis and will report the synthesis and characterization of a number of polymer blends, modified polymers and composite foams based on extensions of Watkins' work. Polystyrene/polyethylene composites have been prepared by the heterogeneous radical polymerization of styrene within SCF CO_2 - swollen high density polyethylene. These polymer blends have unique morphologies and mechanical properties. Composite foams have been prepared by expanding SCF CO_2 - swollen polystyrene/fluoro-polymer and polystyrene/poly(4-methyl-pentene) (PMP) blends. The heterogeneous free radical grafting of maleic anhydride onto PMP has been shown to be a controllable maleation procedure. Poly(tetrafluoroethylene) (PTFE) has been surface-modified by preparing polystyrene/PTFE interpenetrating networks and subsequently sulfonating the polystyrene.

CHEMISTRY IN SUPERCRITICAL CARBON DIOMDE -- SWOLLEN POLYMERS. J. J. Watkins, T. J. McCarthy, Polymer Science and Engineering Department, University of Massachusetts, Amherst, Massachusetts 01003

Novel composite materials are prepared by conducting chemical reactions within polymers swollen with supercritical carbon dioxide. The reactions are either reaction rate limited or mass transport limited depending on the relationship between reagent diffusivity and reaction kinetics. Polymer blends exhibiting unique, co-continuos structures are efficiently prepared by conducting free-radical polymerizations of vinyl monomers within swollen polymers under reaction rate limited conditions. Polymer/metal nano-composites are prepared by the sequential infusion of organometallic compounds into the host polymer and autocatalytic, mass-transfer limited reduction of the precursor to yield discrete metal clusters distributed throughout the substrate. Manipulation of diffusion rates within the polymer host is used to tune cluster size. The observation that CO_2-swollen polymers are viable reaction media suggests new, versatile avenues for the synthesis and processing of materials. Several of these will be discussed.

Neutron Reflection from Polymer Melt Interfaces

DG Bucknall‡, SA Butler, HE Hermes, and JS Higgins
‡ ISIS Facility, Rutherford Applton Lab., Chilton Oxon, OX11 0QX, UK, Dept of Chem. Eng., Imperial College, London, SW7 2BY, UK

Introduction

Polymer blends are of great industrial importance, and the range of applications to which these blends can be applied relies not only on the bulk polymeric characteristics but also largely on their interface and surface properties. The characterisation and understanding of these interfacial properties on a microscopic level are therefore of vital importance.

Neutron reflection has been successfully applied to the study of interfaces of amorphous polymers but not, until recently, to the study of crystalline polymers. The difficulty lies in the sample required for reflectivity which needs to be flat over the dimensions of the beam size on the sample. Whereas amorphous polymers are naturally flat once spin coated from solution into thin layers onto optically polished substrates, formation of crystallites cause such thin films of crystalline polymers to become macroscopically rough. At room temperature this surface roughness causes drastic loss of specular reflection, making NR measurements at best very difficult but more likely impossible. The problem is not relieved by simply heating the films above the crystalline melt temperature. Although this has the effect of removing molecular roughness associated with the polymer crystallinity, the surface still suffers from long range waviness which adversely affects the reflectivity profile. To overcome these problems we have developed a special cell. By providing a molecularly smooth and flat template surface for the polymer both the molecular roughness and long range waviness have been shown to be successfully removed in the melt regime. For the first time detailed studies of interfaces between crystalline polymers have been made possible using this approach.

The width of the interface between two polymers is related to the physical characteristics of the polymer system and allows the determination the interfacial tension and the Flory-Huggins χ parameter. Although NR is ideally suited to determining the interfacial width between polymers, the interfacial tension coupled with thermal capillary wave broadening play an important role in the size of the interfacial width as measured by NR. This can produce a significant difference in the theoretical or intrinsic width as predicted by for instance mean-field calculations and that obtained by NR measurements. It has been shown that the capillary wave component can be successfully deconvoluted from the measured interfacial width to give values of the intrinsic width in close agreement with those predicted by theory[1].

Experimental

Using a special cell, NR data were measured from polymer films in the melt. The cell consists of a heatable brass base plate which holds the polymer 'substrate' on top of which a polymer coated silicon block sits. The whole cell is contained within an inert atmosphere during the measurement to prevent polymer degradation. Full details of the cell have been given elsewhere[2]. NR data were collected using the cell to study the interfacial width between various polymers at temperatures above the crystalline melt temperatures. All measurements were taken on either the CRISP or SURF reflectometers at the ISIS Facility at the Rutherford Appleton laboratory. To achieve reflectivity profiles with an extended Q range from these time of flight instruments, measurements were taken over 3 angles. The data were fitted using a combination of non-linear least squares fitting routines and maximum entropy methods[3]. The polymers used are given in Table 1.

Table 1: Polymer characteristics.

Polymer	Abbrev.	M_w	PD
deuterated polystyrene	dPS	190k	1.08
polyethylene	PE	44k	5.6
"	LLDPE	150.1k	1.03
"	HDPE	75k	4.7
deuterated isotactic polypropylene	diPP	273.6k	5.47
hydrogenous isotactic polypropylene	hiPP	250k	----
deuterated atactic polypropylene	d-aPP	161.6k	1.46
deuterated poly(methyl methacrylate)	dPMMA	368k	2.1

Table 2: Interfacial widths, Flory-Huggins χ parameter and interfacial tension as determined using the melt cell.

Polymer pair	T (°C)	w (nm)	χ	γ_0 (mJm^{-2})
dPS-PE	150	2.96 ± 0.08	$(5.32\pm0.03) \times 10^{-2}$	19.0 ± 0.1
diPP-HDPE	175	6.0 ± 0.6	$(2.0\pm0.1) \times 10^{-2}$	5.34 ± 0.05
	200	7.0 ± 0.7	$(1.4\pm0.1) \times 10^{-2}$	4.62 ± 0.05
	225	5.7 ± 0.1	$(2.2\pm0.1) \times 10^{-2}$	6.1 ± 0.1
daPP-LLDPE	175	11.7 ± 0.1	$(4.37\pm0.05) \times 10^{-3}$	2.49 ± 0.05
	225	13.0 ± 0.2	$(3.48\pm0.04) \times 10^{-3}$	2.42 ± 0.05
dPMMA-hiPP	225	2.55 ± 0.08	$(9.50\pm0.05) \times 10^{-2}$	20.0 ± 0.2

Results and Discussion

Initial reflectivity measurements using the cell were performed using dPS and PE[2] at 150C, and the interfacial width obtained from the fitting (see Table 2). This measured interfacial width is described by a self-consistent mean-field theory, which is broadened by thermally excited capillary waves[1]. The measured interfacial width is given by Gaussian quadrature addition of the intrinsic (unbroadened by capillary waves) interfacial width, Δ_0, and the interfacial width associated with the capillary waves, Δ_c. This gives

$$w/\sqrt{2\pi} = \sqrt{\Delta_0^2 + \Delta_c^2} \quad \text{where} \quad \Delta_0 = a/\sqrt{3\pi\chi} \quad \text{and}$$

$$\Delta_c^2 = \frac{k_B T}{4\pi\gamma_0} \ln\left(\frac{(2\pi/\Delta_0)^2}{(2\pi/\lambda_c)^2 + (2\pi/a_d)^2} \right). \quad \text{The capillary}$$

wave broadening is therefore dependent on the in-plane coherence length of the neutron, $\lambda_c \approx 20\mu m$, the dispersive capillary length, $a_d = 4\pi\gamma_0 d^4 A^{-1}$, the Hamaker constant, A, and the interfacial tension $\gamma_0 = a\nu k_B T \sqrt{\chi/6}$ (a is the segment length and ν^{-1} is the monomer volume), and d is the layer thickness. Using this approach for evaluating the effects of capillary wave broadening for the dPS-PE system the values of χ and γ_0 are obtained (see Table 2). These values agree well with the spread of data found in the literature[2], and give considerable confidence in the use of the cell for further applications of this method to obtain accurate values of χ and γ_0 for other polymer pairs.

A majority of the data collected has been using PP against PE and PMMA, using a combination of two PP tacticities and a number of PE's of varying densities from LLDPE to HDPE. The diPP is semi crystalline with a melt temperature of 157°C, while daPP like

dPMMA is amorphous. The melt temperatures of the LLDPE and HDPE used are 102 and 131 °C, respectively. Reflectivity profiles from the diPP-HDPE system were collected at 175, 200 and 225 °C and interfacial widths obtained from fitting. The silicon was found not to be fully wetted by the diPP in some cases[4], but the diPP layer was sufficiently thick to prevent complete dewetting from the substrate, and the film homogeneity and interface with the HDPE were believed to be unaffected by this dewetting process.

The measured interfacial widths, w, of systems with PP are given in Table 2, together with the calculated χ and interfacial tension values. There does not appear to be any systematic temperature dependence of w in the diPP-HDPE system as may have been expected. The interfacial widths of the daPP-LLDPE interface are significantly larger than the interfacial widths seen for the diPP-HDPE or the dPMMA-hiPP interfaces. This clearly indicates that the daPP-LLDPE polymers are more miscible, as indicated by the decrease of the calculated χ values for this polymer pair These χ values compare well to the value of 6.4×10^{-3} at 167°C obtained from SANS on PP-PE blends[5]. This increased miscibility may be anticipated due to the closer chemical similarity of the PP to the branched LLDPE than the more linear HDPE. Comparison to the dPS-PE system indicates that polystyrene is much less miscible with PE than is the PP. The PMMA-PP pair are strongly immiscible as evidenced by the large values of χ and γ_0.

References

1. M. Sferrazza, C. Xiao, R.A.L. Jones, D.G. Bucknall, J. Webster & J. Penfold, *Physical Review Letters*, **78** (1997) 3693 .
2. H.E. Hermes, J.S. Higgins & D.G. Bucknall, *Polymer*, **38** (1997) 985 .
3. D.S. Sivia. *Data Analysis - A Bayesian Tutorial* (Clarendon Press, Oxford, 1996).
4. H.E. Hermes, D.G. Bucknall, S.A. Butler & J.S. Higgins, *Macromolecular Symposia*, **126** (1997) 331 .
5. H.S. Jeon, J.H. Lee, N.P. Balsara, B. Majumdar, L.J. Fetters & A. Faldi, *Macromolecules*, **30** (1997) 973 .

THE THERMODYNAMICS OF POLYOLEFINS DERIVED FROM METALLOCENE CATALYSTS USING SMALL-ANGLE NEUTRON SCATTERING

Glenn C. Reichart, *Nestlé Research and Development Center, 201 Housatonic Avenue, New Milford, CT 06776*

William W. Graessley and Richard A. Register, *Department of Chemical Engineering, Princeton University, Princeton, NJ 08544*

David J. Lohse, *Corporate Research Laboratories, Exxon Research and Engineering Company, Annandale, NJ 08801*

INTRODUCTION

The use of Small-Angle Neutron Scattering (SANS) techniques to determine polymer blend thermodynamic behavior has been well established (1). SANS experiments have been used to determine interactions between saturated hydrocarbon components referred to as model polyolefins, created by saturating the double bonds in nearly monodisperse polydienes using either H_2 or D_2. Graessley and coworkers (2,3) organized the experimental results using component solubility parameter values (δ_i) to allow thermodynamic interaction predictions based on the component structures involved:

$$\chi = \frac{v_o}{k_B T}\left(\delta_2 - \delta_1\right)^2 \qquad (1)$$

This study will focus on SANS-derived blend thermodynamic interactions containing polyolefins synthesized using metallocene catalysts. With these single site catalysts, the statistical copolymers produced have little compositional heterogeneity and a narrower molecular weight distributions than other commercial polyolefins. The objective of this study is to measure interactions involving these metallocene-catalyzed (MC) polymers using SANS experiments and compare these to the results involving model materials. MC polymers have not been studied in prior SANS work because they are synthesized directly from the olefin, and therefore do not offer a simple deuterium labeling technique for SANS contrast.

Equal volume binary blends containing a hydrogenous MC polymer and a partially deuterated model material were used in the SANS experiments. They were conducted on both the 8m (NG5) and 30m (NG7) beamline at the NIST Cold Neutron Research Facility in Gaithersburg, MD. The model polymers, described in prior studies (2,3), include a series of ethylene-butene statistical copolymers with different compositions (HPB series). Characterization data on the model materials is provided in Reference 2. The metallocene-catalyzed (MC) polymers were synthesized directly from the hydrogenous olefin monomers by Exxon Chemical Company, Baytown, TX. They are statistical copolymers of ethylene with propylene (EP), butene (EB), hexene (EH), or octene (EO). Nomenclature consists of EP, EB, EH, or EO followed by the comonomer weight percent. Characterization results for the MC polymers are presented in Table 1. A polydisperse ethylene-propylene statistical copolymer, JBG-11, was also used because of its high propylene content. It was synthesized using Ziegler-Natta polymerization techniques, and has been well characterized and used in prior rheological studies (4).

The experimentally determined thermodynamic interactions will be used with solubility parameter formalism to derive a solubility parameter database for the MC polymers. In prior model polymer blend studies, the thermodynamic interactions were reported in terms of the Flory-Huggins interaction parameter χ. To avoid the requirement of an arbitrary reference volume (v_0) definition in the SANS data analysis, this study will present the results as an interaction strength X, related to χ by the equation:

$$X = \frac{\chi k_B T}{v_o} \qquad (2)$$

RESULTS AND DISCUSSION

To account for polydispersity, the RPA equation becomes (5):

$$\frac{1}{S(q)} = \frac{1}{\phi_1 v_1 \langle N_1 P_1(q)\rangle_w} + \frac{1}{\phi_2 v_2 \langle N_2 P_2(q)\rangle_w} - \frac{2X}{k_B T} \qquad (3)$$

where $\langle N_i P_i(q)\rangle_w$ is the weight-average product of the degree of polymerization and form factor for component i. Assuming the MC components have a Zimm-Schulz molecular weight distribution (6), this product becomes:

$$\langle NP(q)\rangle_w = \frac{2N_n}{q^4 R_g^4}\left[\left(\frac{h}{h + q^2 R_g^2}\right)^2 - 1 + q^2 R_g^2\right] \qquad (4)$$

with:

$$h = \frac{1}{N_w / N_n - 1} \qquad (5)$$

where R_g is the number-average radius of gyration for the component. In the monodisperse limit ($h \to \infty$), the component form factor is expressed by the Debye formula:

$$P(q) = \frac{2}{q^4 R_g^4}\left[e^{-q^2 R_g^2} - 1 + q^2 R_g^2\right] \qquad (6)$$

where R_g is the component radius of gyration. These equations show the difference between the monodisperse and polydisperse RPA equations lies in the component form factors. A full-range fit was applied to the binary blend SANS results using Equations 3 through 6. Two quantities were extracted from the analysis: the blend interaction strength (X) and the ratio (α) of blend chain dimensions to values determined from matched pair and pseudo-matched pair results (discussed later). All single-phase binary blends exhibited chain dimension ratios of $0.95 < \alpha < 1.05$.

Almost all of the blends exhibit positive interaction strengths, with some interaction strengths becoming slightly negative. However, these negative values have magnitudes less than the interaction strength uncertainties from SANS experiments and characterization results. The interaction strength behavior exhibited in Figure 1 for the EB27 blends is typical for this study and may be described by either:

- decreasing interaction strength with increasing temperature, shown by EB27/D32 and EB27/D08B.
- Constant interaction strengths, shown by EB27/D25 and EB27/D38.

MC component chain dimensions were determined using binary blends with model HPB polymers in which the weight percent ethylene was matched as closely as possible (pseudo-matched pair blends). Full-range fits were used to determine the MC polymer chain dimension (the HPB component chain dimensions are known from prior studies (2)) and interaction strength. Minimizing the interaction strength makes the experimental results more sensitive to component chain dimensions, allowing an accurate MC polymer radius of gyration determination.

In order to establish an MC polymer solubility parameter data base, the labeled model material δ values were first determined from the interaction data collected by Krishnamoorti (7). For consistency, the same reference material employed in the hydrogenous model material assignments (model HPB H97A) was used for the deuterated model materials and MC polymer assignments. Prior studies compiled the hydrogenous model material δ values (2). Using this information with the model/model interaction data, partially deuterated model solubility parameter values were calculated by rearranging Equation 1:

$$\delta_{D2} = \delta_{H1} \pm \left(X_{H1/D2}\right)^{1/2} \qquad (7)$$

Whether the last term is added or subtracted depends on the solubility parameter ordering of the two components. Take for example the HPB series of polymers. Earlier studies show δ values increase with increasing ethylene content (2). This means $\delta_{H38} > \delta_{H52}$, and for the binary blend H38/D52, δ_{D52} would be the *sum* of δ_{H38} and $(X_{H38/D52})^{1/2}$. However, for the "swap" blend (D38/H52), δ_{D38} would be the *difference* between δ_{H52} and $(X_{D38/H52})^{1/2}$. Solubility parameter assignments were tested for consistency by comparing values obtained from different binary blends with the same labeled model component. The average value of $\delta_{hydrog} - \delta_{deut}$ for the 30 data points is only 0.023 $MPa^{1/2}$, confirming deuterium labeling in the model materials has a small thermodynamic effect in comparison to changes in copolymer composition.

With the deuterated model solubility parameters and MC/model interaction strengths known, MC solubility parameter assignments can be made. The solubility parameter/interaction strength relation becomes:

$$\delta_{MC} = \delta_{MODEL} \pm \left(X_{MC/MODEL}\right)^{1/2} \qquad (8)$$

Once again, the sign of the last term in Equation 18 is dependent on the two components' solubility parameter ordering. This is determined by looking at SANS results from multiple binary blends using the same MC material and producing consensus on the δ_{MC} value. Two possible δ_{MC} values were generated for each solubility parameter difference:

1. Assuming $\delta_{MC} > \delta_{MODEL}$.
2. Assuming $\delta_{MC} < \delta_{MODEL}$.

To determine which solubility parameter assignment is correct, all blends involving a particular MC material were studied to determine a consistent δ_{MC} value at each temperature. The different blends incorporating the same MC material produced δ_{MC} values in agreement within 0.04 MPa$^{1/2}$ at each temperature.

With this new data bank of MC solubility parameter values, the next issue is the relationship between δ and comonomer composition. Figure 2 shows solubility parameter behavior as a function of weight percent comonomer content for all MC materials at 167°C. This plot shows *all MC copolymers fall along the same correlation, regardless of comonomer incorporated.* In contrast, there is no apparent solubility parameter correlation among the MC polymers as a function of *mole* percent composition. Figure 4 also compares the MC solubility parameter data to the model HPB's. There is good HPB/MC polymer solubility parameter agreement throughout the compositional range.

The universal solubility parameter behavior exhibited in Figure 4 has important implications. The thermodynamic interaction information in these α-olefin copolymers depends only on the weight fraction of comonomer and not on the length of the comonomer repeat unit (propylene, butene, hexene, octene). In addition, it confirms the material blend behavior is independent of polymerization technique (model versus MC polymers).

REFERENCES

1) Higgins, J.S.; Benoit, H.C. "Polymers and Neutron Scattering," Oxford University Press: New York, 1994.
2) Graessley, W.W.; Krishnamoorti, R.; Balsara, N.P.; Butera, R.J.; Fetters, L.J.; Lohse, D.J.; Schulz, D.N.; Sissano, J.A. *Macromolecules* 27, 3896 (1994).
3) Graessley, W.W.; Krishnamoorti, R.; Reichart, G.C.; Balsara, N.P.; Fetters, L.J.; Lohse, D.J. *Macromolecules* 28, 1260 (1995).
4) Wasserman, S.H.; Graessley, W.W. *J. Rheol.* 36, 543 (1992).
5) Sakurai, S.; Hasegawa, H.; Hashimoto, T.; Hargis, I.G.; Aggarwal, S.L.; Han, C.C. *Macromolecules* 23, 451 (1990).
6) Peebles, L.H., "Molecular Weight Distributions in Polymers," Interscience, New York, 1970.
7) Krishnamoorti, R. "Thermodynamics of Mixing in Model Polyolefin Blends," Doctoral Dissertation, Princeton University, January 1994.

Table 1. Molecular Characterization of MC Polymers

Name	M_w (kg/mol)	M_w/M_n	Name	M_w (kg/mol)	M_w/M_n
EP13	71.0	2.2	EH7	93.2	2.2
EP15	72.5	2.1	EH10	77.2	2.2
JBG-11	151.0	1.97	EH13	100.8	1.9
			EH18	76.6	2.0
			EH37	198.7	2.43
EB10	104.9	2.0	EH40	170.7	2.07
EB14	76.7	1.9	EH43	168.7	2.24
EB16	94.9	2.0	EH46	126.6	2.15
EB17	99.7	2.0	EO22	93.6	2.3
EB22	107.5	1.9	EO43	201.7	2.56
EB27	79.0	1.9	EO51	143.4	2.24
EB31	58.1	2.0	EO54	158.9	2.33

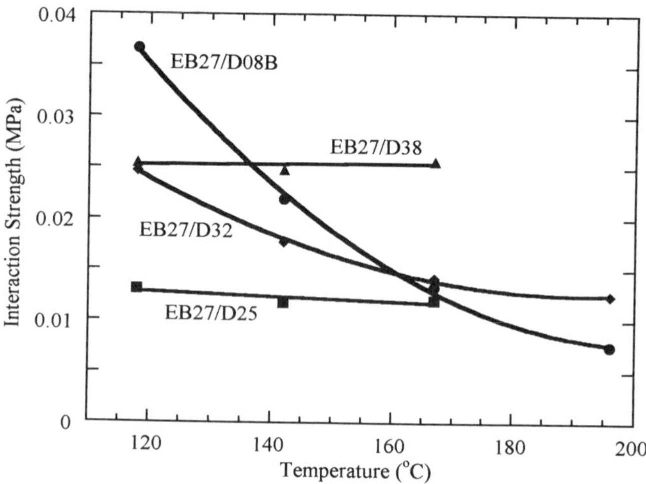

Figure 1. Interaction strength as a function of temperature for the MC/model blends involving EB27.

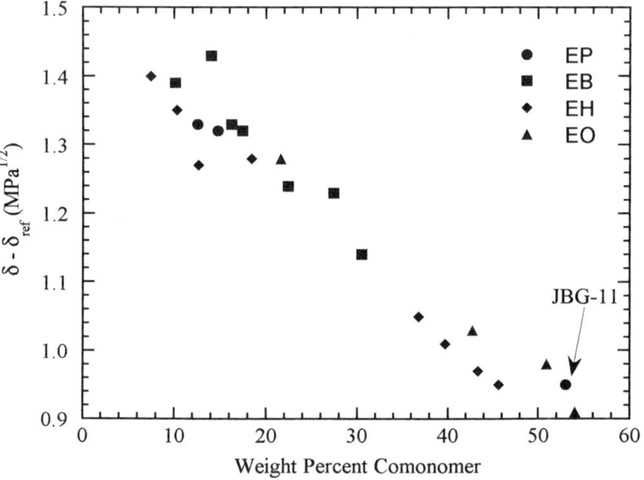

Figure 2. MC polymer solubility parameter values at 167°C as a function of weight percent comonomer content.

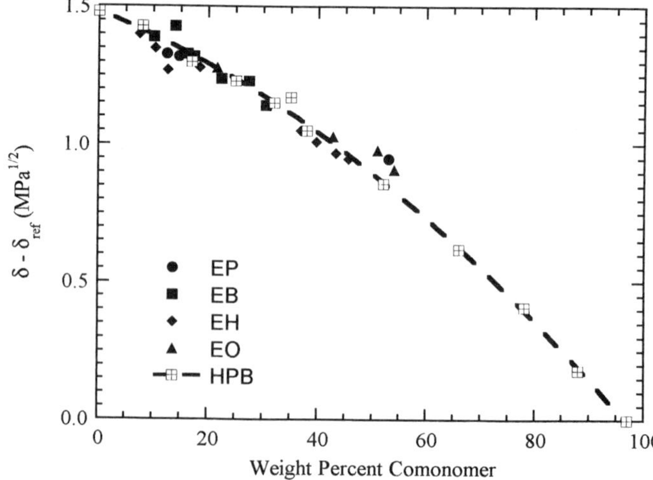

Figure 3. MC and model polymer solubility parameter values at 167°C as a function of weight percent comonomer content. The dashed line represents a quadratic fit to the model HPB values.

Morphology Development in Polypropylene Homopolymer Tacticity Mixtures: Isotactic/Atactic Blends

Roger A. Phillips[1], Zhi-gang Wang[2] and Benjamin S. Hsiao[2]
[1]Montell Polyolefins, R&D Center, 912 Appleton Road,
Elkton Md. 21921
[2]State University of New York-Stony Brook, Dept. of Chemistry,
Stony Brook, NY 11794-3400

INTRODUCTION

The stereochemisty of polypropylene strongly influences material properties through morphological factors such as crystallinity and melting. Metallocene catalysts allow for new tacticity microstructures[1], including high molecular weight (narrow distribution) atactic polypropylene (aPP) [2]. aPP is non-crystalline with $Tg \sim 0^{\circ}C$, and of limited importance as a homopolymer. When blended with isotactic polypropylene from Zeigler-Natta catalysts (ZN-iPP), aPP allows for an expansion of the property map of iPP to "softer", more flexible structure [1].

aPP with well defined chain architecture also allows aspects of non-crystallizable diluent segregation on microstructure development, melting, and morphology/property relationships to be addressed. The effect of aPP on iPP crystallization was investigated in early work [3,4]. Increasing aPP content caused more open spherulitic textures due to incorporation of diluent in the interspherulitic region [3]. Recent work [5] shows radial segregation/"pooling" of aPP within the growing iPP spherulite, resulting in minor modification of spherulitic growth rate and optical cross-hatch transition [6]. However, the blending only shows a minor effect on the SAXS long spacing [5]. The current study addresses morphology development of equal molecular weight iPP/aPP mixtures by thermal and synchrotron scattering techniques.

EXPERIMENTAL

Samples for blending studies are summarized in Table 1. Dimethylsilylbis(9-fluorenyl)zirconium dichloride catalyst [2] was used to prepare aPP in hexane with MAO activator at a molar [Al]/[Zr] ratio of 2000. The iPP is based on Zeigler-Natta catalyst technology. Blends were prepared with stabilized 3wt% xylene solutions in a N_2 atmosphere at 130 $^{\circ}C$ for 2hr. Blend composition varied from 0-80 wt% aPP. Blends were precipitated in dry-ice chilled methanol, filtered, vacuum dried, and compression molded into 2mm thick, 1.25" diameter disks (Buehler press) at 195°C for 7min (35Mpa) and cooled in clamps to 110°C prior to ejection from the mold. The molds were re-melted by similar procedures without pressure to remove residual orientation (confirmed by 2D WAXS).

Simultaneous pinhole collimated SAXS/WAXS measurements were conducted at the National Synchrotron Light Source (NSLS), Brookhaven National Laboratory on Beamline X27C (λ=1.307Å). Two linear position sensitive detectors (*EMBL*, European Molecular Biology Laboratory) were configured for simultaneous data acquisition. Crystallization experiments used 7mm diameter samples cut from molded disks and transferred by pneumatic arm between furnaces at the melt temperature (7min, 195°C) and crystallization temperature (115°C or 137.5°C). The experimental temperature jump apparatus has been reported previously [7]. X-ray profiles were collected in 5sec (115°C) or 30sec (137.5°C) increments and normalized for incident beam intensity. All crystalline patterns are shown at the measurement temperature. SAXS angles were calibrated with silver behenate. WAXS pixel and intensity was calibrated by comparing synchrotron data with Scintag diffractometer data (CuK_{α}) in $\theta-\theta$ reflection calibrated with a NIST corrundum plate. The angular scale of the synchrotron WAXS data (λ=1.307Å) was converted to a scale corresponding to λ=1.5418 Å.

DSC experiments used a Perkin-Elmer DSC7 purged with nitrogen and calibrated with indium and mercury at 20°C/min. 5-6 mg samples were used over a temperature range of ~55-235°C. Heat-cool-reheat ramps were applied. All data is reported during heating following a cooling cycle at 20°C/min. Data is shown after subtraction of the solid-melt baseline, which was determined by an iterative technique.

RESULTS & DISCUSSION

Figure 1 shows the DSC scans of iPP/aPP blends as a function of blend composition. The addition of aPP with increasing concentration reduces the crystallinity in the blend due to the non-crystalline nature of aPP. aPP concentration has no appreciable effect on the iPP melting point (± 0.9 $^{\circ}C$) across the entire composition range. Although not shown, the crystallization temperature on cooling also shows little systematic dependence on aPP concentration except at 80wt% aPP. In this case, there is ~10 $^{\circ}C$ reduction. aPP concentration has a similarly weak effect on the WAXS line breadth.

Figure 2 shows the SAXS and WAXS profile development of the iPP homopolymer during isothermal crystallization at 115°C. This temperature is about 10°C above the DSC crystallization temperature (cooling at 20°C/min). Figure 3 shows the WAXS crystallinity (X_c), relative SAXS invariant (Q), and SAXS 1D correlation function parameters [8]. These parameters are the long spacing, L, and the crystalline and amorphous layer thickness $<l>_c$, and $<l>_a$. The development of WAXS crystallinity occurs very rapidly at this temperature. After a brief induction period, all of the crystallinity develops during ~2-3 min period. The transformation period is only weakly dependent on blend composition, being most noticeble only at 80wt% aPP (in agreement with DSC). In the homopolymer, L, $<l>_c$, and $<l>_a$ all decrease at a low degree of transformation, which can be attributed to the occurrence of secondary crystallization forming thinner and defective crystals. It is interesting to observe that at low supercooling temperature (137.5°C) the WAXS peaks of iPP homopolymer occur about 6 min later than the SAXS scattering peak (plot does not show in this paper), illustrating density fluctuations exist prior to crystallization.

Figure 4 compares the SAXS patterns of the iPP homopolymer and a 50/50 iPP/aPP blend after crystallization is complete at 115°C and 137.5°C. The scattering maximum in the blend is not as well defined at spacings slightly larger than the homopolymer. Much stronger scattering near zero angle is observed in the blend relative to the homopolymer after crystallization is complete. This observation is related to segregation of aPP within the morphology to size scales larger than the lamellar spacing of the homopolymer. Segregation of aPP to larger size scales can also be inferred by optical opacity of the blends at intermediate aPP concentrations. However, the strong low angle scattering indicates that appreciable amounts of large aPP domains are still within the range detectable by SAXS.

Figure 5 shows values of plateau long period, lamellar thickness and amorphous layer thickness (L^*, $<l>_c^*, <l>_a^*$) versus blend composition following crystallization at two different temperatures. At 115°C, L^* is nearly independent of blend composition up to 60wt% aPP. L^* increases strongly at 80wt% aPP, although no clear maximum is observed in the raw SAXS profiles at this composition. The weak dependence of L^* on aPP concentration at high supercooling contrasts with the behavior observed for blends exhibiting strong lamellar inclusion of diluent [9]. Interestingly, a similar weak dependence of L^* on aPP concentration is observed in syndiotactic PP (sPP) blends with aPP, even though the aPP strongly influences the intrinsic multiple melting response of the sPP homopolymer (indicating a degree of "local" proximity of diluent) [5]. The data at higher crystallization temperature shows a clear increase of L^* in the iPP/aPP blend. This suggests a stronger interaction of aPP on lamellar morphologies at high crystallization temperature. This observation may be related to both the influence of aPP on lamellar crosshatching [5,6] and related inclusion of diluent within the microstructure. However, the overall results suggest the morphology in iPP/aPP blends which is strongly influenced by segregation of aPP on length scales larger than lamellar spacings.

CONCLUSION

We observed that the melting point in iPP/aPP blends does not depend on aPP concentration. The blends also show only minor changes in crystallization temperature. This is in qualitative agreement in time-resolved SAXS experiments: iPP/aPP blends show a weak dependence of L and $<l>_c$ on composition following crystallization at 115°C, suggesting segregation of aPP to larger size scales. Strong zero angle scattering in the blends suggests that appreciable amounts are still within the range probed by SAXS length scales. Significant expansion of L is observed at 137.5°C, suggesting partial inclusion of aPP. However, the overall results suggest morphology in iPP/aPP blends which is strongly influenced by segregation of aPP on length scales larger than lamellar spacings. Finally, iPP homopolymer and blends all indicated that value of L, $<l>_c$ and $<l>_a$ decrease slightly with time, which is consistent with the event of secondary crystallization producing thinner thickness of crystals.

ACKNOWLEDGMENT

The authors thank R. L.J ones (Montell) for catalyst synthesis, and D. Morgan (Montell) for sample preparation assistance. Dr. Fengji Yeh (SUNYSB) provided invaluable support of the synchrotron measurements. B.H. acknowledges the financial support of this work by a grant from NSF (DMR 9732653).

REFERENCES

1. Haylock, J.C.; Phillips, R.A.; Wolkowicz, M.D. in *Preparation, Properties & Technology of Metallocene-Based Polyolefins*; Scheirs, J., Kaminsky, W., Eds.; John Wiley & Sons, New York, 1998.
2. Resconi, L.; Jones, R.L.; Rheingold, A.L.; Yap, G.P. *Organometallics*, **1996**, 15, 998.
3. Keith, H.D.; Padden, F.J. *J.Appl.Phys.*, **1964**, 35, 1270.
4. Keith, H.D.; Padden, F.J. *J.Appl.Phys.*, **1964**, 35, 1286.
5. Phillips, R.A.; Wolkowicz, M.D.; Jones, R.L. *Proc. SPE ANTEC*, **1997**, 57, 1651.
6. Phillips, R.A.; Wolkowicz, M.D. in *Polypropylene Handbook*; Moore, E.P., Ed.; Hanser: Munich, 1996.
7. Hsiao, B.S.; Gardner, K.H.; Wu, D.Q.; Chu, B. *Polymer* **1993**, 34 (19), 3986.
8. Verma, R.; Marand, H.; Hsiao, B. Macromolecules, **1996**, 29, 7767.
9. Talibuddin, S.; Wu, L.; Runt, J.; Lin, J.S. *Macromolecules*, **1996**, 29, 7527.

Table 1 Sample Characteristics

Sample	Mw	Mw/Mn	type	%triad
IPP	220,000	4 [a]	iso	95
APP	210,000	2	atactic	---[b]

[a]rheological index   [b]16% iso, 49% hetero, 35% syndio

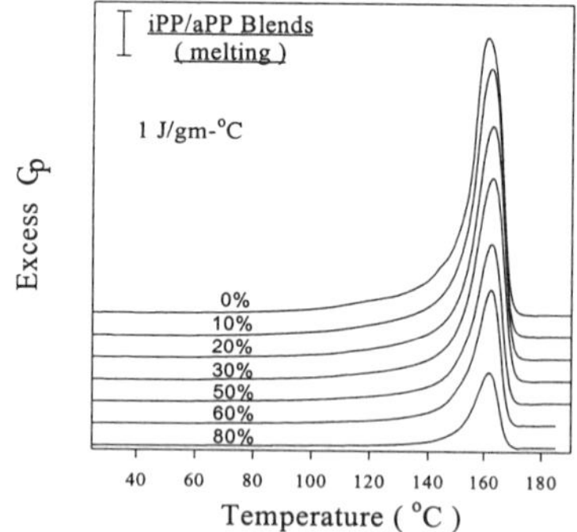

Figure 1. Reheat DSC scans of iPP/aPP blends. The wt% aPP varies between 0 and 80% in the blends.

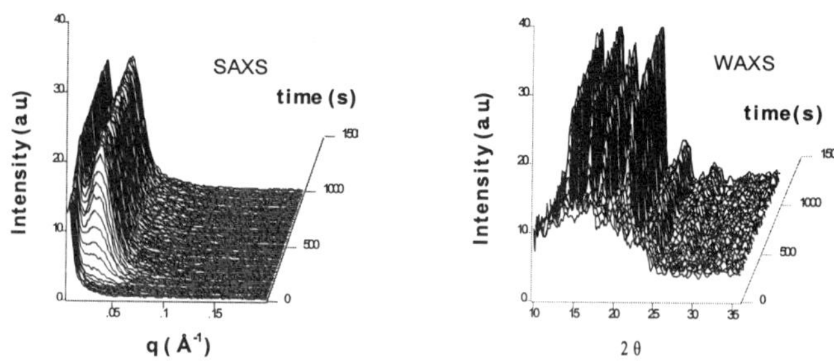

Figure 2. SAXS and WAXS profiles during isothermal crystallization at 115°C for iPP homopolymer.

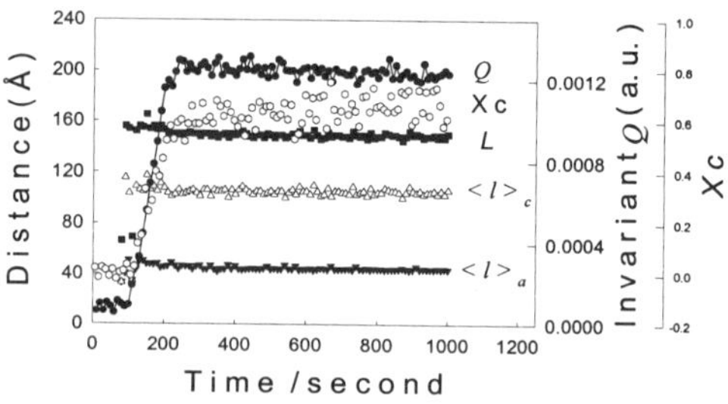

Figure 3. Time evolution of WAXS Xc, SAXS Q, long period L, *lamellar thickness* $<l>c$ *and amorphous layer thickness* $<l>a$ during isothermal crystallization at 115°C for iPP homopolyer.

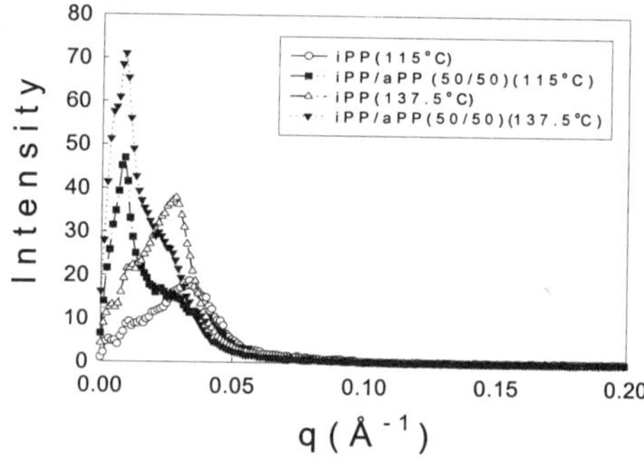

Figure 4. Normalized SAXS intensities of iPP homopolymer and iPP/aPP (50/50) blend at 115°C and 137.5°C, respectively.

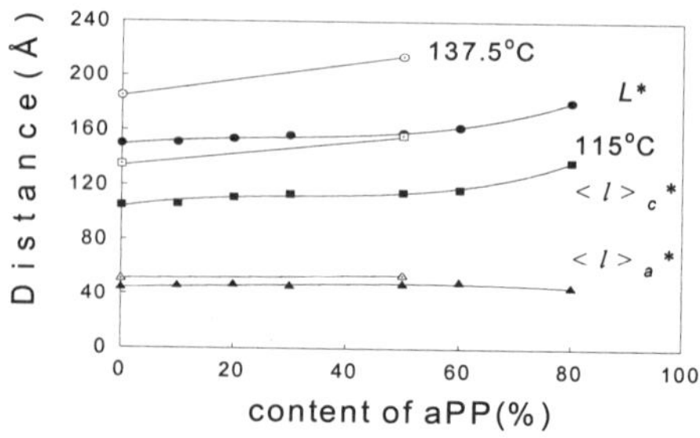

Figure 5. The average $L*$, *lamellar thickness* $<l>c*$ and amorphous layer thickness $<l>a*$ at 115°C (Block symbol) and at 137.5°C (blank symbol) for iPP homopolyer.

CHAIN DIMENSIONS IN POLYSILICATE-FILLED POLY(DIMETHYLSILOXANE)

A. I. Nakatani,[a] W. Chen,[b] R. G. Schmidt,[b] G. V. Gordon[b] and C. C. Han[a]

[a] - National Institute of Standards and Technology, Polymers Division, Gaithersburg, MD 20899
[b] - Dow Corning Corporation, Midland MI 48686

Introduction

Small angle neutron scattering (SANS) has been used to extract dimensions of polymer chains in multicomponent mixtures. The composition of these mixtures has often been restricted to deuterium labeled and unlabeled polymers in the presence of solvent or block copolymers. Extraction of the single chain dimensions in the presence of a filler particle have not been performed experimentally. However, Monte Carlo calculations of polymer chain dimensions in the presence of filler particles have demonstrated that the radius of gyration (R_g) of the polymers is a strong function of the filler size and concentration.[1,2,3] In this work we present preliminary results on the single chain dimensions of an isotopic blend (mismatched molecular masses) of poly(dimethylsiloxane) (PDMS) filled with trimethylsilyl-treated polysilicate particles (fillers) and compare these results with the Monte Carlo calculations by Mark and co-workers.[1-3] While the polysilicate particles used in this study are smaller than traditional reinforcing fillers, they have been employed in a number of applications to alter the mechanical behavior of PDMS.

Previously, Akcasu et al. provided a framework for extracting single chain structure factors from SANS measurements on multi-component mixtures by the so-called high concentration method.[4] This method uses a fixed ratio of polymer to the third component and varies the labeled-to-unlabeled ratio of polymer in the blend. By subtracting the appropriately weighted scattering intensities from samples with different labeling ratios, the single chain structure factor of the polymers can be obtained. This approach has been used extensively on isotopic blends of polymers with matched molecular masses in the presence of a third component. Tangari and co-workers have examined the effects of mismatched molecular masses on the high concentration method.[5] However, the presence of a third component was not considered in their treatment. We use a similar treatment in this work by assuming the polysilicate component may be effectively treated as a solvent molecule. For comparison, the solution R_g values for the polymers were measured in toluene (near theta conditions) and in the isotopic blend without filler.

Materials

The PDMS polymers (ρ = 0.97 g·cm^{-3}) used in this study were provided by Dow Corning Corporation. Deuterated PDMS (d-PDMS) was synthesized via the hydrolysis and condensation of perdeuterated chlorosilanes. The molecular masses were characterized independently by size-exclusion chromatography (SEC) using PDMS standards. Table 1 gives the characteristics of the polymers. The trimethylsilyl-treated polysilicate material (M_n = 1500 g·mol^{-1}, ρ = 1.05 g·cm^{-3})[6] was synthesized via the co-hydrolysis and condensation of hexamethyldisiloxane and an alkoxysilane at a molar ratio of 1.2.

Table 1. Molecular Characteristics of Poly(dimethylsiloxane)

Polymer	SEC[c]		SANS[d]	
	$\overline{M}_w$	$\overline{M}_w/\overline{M}_n$	$\overline{M}_w$	R_g (Å)
h-PDMS	9,700	1.05	13,000	50
d-PDMS	35,500	2.5	34,000	85

[c] Size exclusion chromatography using PDMS standards.
[d] From 2-component RPA fit of SANS measurements at 22 °C.

Experimental

SANS measurements were carried out at the Cold Neutron Research Facility of the NIST Center for Neutron Research. Data were collected on the 8-m SANS instrument with the neutron wavelength, λ = 9.0 Å. The samples for SANS measurements were contained in 2-mm-pathlength quartz cells for liquids. Data were collected over a two-dimensional detector and corrected for dark current intensity due to electronic and background neutron noise and empty cell scattering. Incoherent scattering due to the protonated components in the samples was also subtracted during the data reduction. Absolute intensity calibration was done with a dry silica gel as a secondary standard, calibrated in terms of a primary vanadium standard.

Results

The d-PDMS was characterized in isotopic blends with the h-PDMS polymer (d-PDMS volume fractions of 0.01, 0.02, 0.03 and 0.05) by using the Zimm plot formalism. Higher concentrations of the d-PDMS and h-PDMS were also measured and the scattering profiles, $S(q)$, were analyzed by fitting the experimental data to the random phase approximation (RPA) theory[7] model (Eq. 1) using a nonlinear least-squares regression routine:

$$S(q) = k_N \left\{ [\phi_A v_A N_A g_D(x_A)]^{-1} + [\phi_B v_B N_B g_D(x_B)]^{-1} + (2\chi/v_0)^{-1} \right\} \quad (1)$$

where N_i is the degree of polymerization index, ϕ_i is the volume fraction, and v_i is the molar volume of the ith component. The term g_D is the Debye function and is given by the following equation:

$$g_D = (2/x^2)[\exp(-x) - 1 + x] \quad (2)$$

Here, $x = q^2 N b^2/6$ where b is the statistical segment length. Furthermore, k_N is the contrast factor given as:

$$k_N = N_o[(a_A/v_A) - (a_B/v_B)]^2 \quad (3)$$

In Eq. (3), N_o, a_i, and v_i are, respectively, Avogadro's number, the scattering length and molar volume of a monomer unit of the ith component. The curve fitting using RPA was performed with b, χ, and the incoherent baseline as floating parameters. The value of b is an average value for both of the polymers and is used to obtain R_g based on the SEC values for the degree of polymerization. The results from the scattering experiments are summarized in Table 1; the molecular masses compared favorably with the data obtained via SEC.

The composition dependence of R_g in the isotopic blend is shown in Figure 1. The value of R_g at low d-PDMS concentration is consistent with the value obtained from the Zimm plot for the solutions of the homopolymer. The variation of χ as a function of the d-PDMS concentration is also shown in Figure 1. The composition dependence of χ is strongest at low d-PDMS contents and is similar to the composition dependence observed in other isotopic blends.

For a three-component mixture of labeled polymer, unlabeled polymer, and solvent, Jahshan and Summerfield[8] derived the following relationship:

$$I(q,x) = (a_H - a_D)^2 x(1-x) S_S(q) + [a_H(1-x) + a_D x - a_s']^2 S_T(q) \quad (4)$$

where a_H, a_D, and a_s' are the scattering lengths for the protonated polymer, deuterated polymer and solvent, respectively, x is the fraction of total polymer which is deuterated, $S_S(q)$ is the single chain form factor, and $S_T(q)$ is the interchain form factor. By fixing the total concentration of polymer and varying x, one obtains a system of linear equations which can be solved for $S_S(q)$ and $S_T(q)$. For the polymer chains, $S_S(q)$ may be fit to the Debye function by a non-linear regression fitting routine, yielding R_g of the polymer chains. Alternatively, by plotting $1/S_S(q)$ against q^2 (known as a Debye plot), a straight line should result with the slope being related to R_g. Equation 4 is only valid for pairs of polymers with matched molecular masses.

Tangari and co-workers have treated the case of mismatched molecular masses without solvent present.[6] The result is similar to the results obtained for isotopic blends with matched molecular masses except the single chain form factor is the weighted sum of the individual form factors for the deuterated and protonated polymer chains:

$$I(q,x) = (a_H - a_D)^2 x(1-x)[x S_s^H(q) + (1-x) S_s^D(q)] + [a_H(1-x) + a_D x]^2 S_T(q) \quad (5)$$

where S_s^H and S_s^D are the single chain structure factors of the different polymers. Equation 5 reduces to Eq. 4, without the solvent term, a_s', for matched molecular masses. Combining the second term on the right hand side of Eq. 4 with the first term on the right hand side of Eq. 5, the following expression for the scattering from two polymers with different molecular masses including solvent is obtained:

$$I(q,x) = \left(a_H - a_D\right)^2 x(1-x)\left[x S_S^H(q) + (1-x) S_S^D(q)\right] + \left[a_H(1-x) + a_D x - a_s'\right]^2 S_T(q) \qquad (6)$$

Solution of Eq. 6 requires measurements on samples with three different labeling ratios at fixed solvent concentration to extract the three structure factors. The results obtained in Eqs. 4-6 assume that there are no specific interactions between any of the components.

With the assumptions used in the derivation of Eqs. 4-6, it is desirable to derive a more versatile expression for the scattering from a three-component system. This can be written in the general form:

$$I(q) = \left(a_1 - a_0\right)^2 S_{11}(q) + \left(a_2 - a_0\right)^2 S_{22}(q) + 2\left(a_1 - a_0\right)\left(a_2 - a_0\right) S_{12}(q) \qquad (7)$$

where the a_i are the scattering lengths of the monomer units in each polymer ($i = 1,2$), and of the solvent ($i = 0$), and the structure factors, S_{ij}, are defined as follows:

$$S_{DD}(q,x) = x N_D z_D^2 P_D(q) + x^2 N_D^2 z_D^2 Q_{DD}(q) \qquad (8)$$

$$S_{HH}(q,x) = (1-x) N_H z_H^2 P_H(q) + (1-x)^2 N_H^2 z_H^2 Q_{HH}(q) \qquad (9)$$

$$S_{HD}(q,x) = -x(1-x) N_D z_D N_H z_H Q_{DH}(q) \qquad (10)$$

The number of polymer molecules is N_i, z_i is the number of segments in the chains, $P_i(q)$ are the single chain structure factors, and $Q_{ii}(q)$ are the interparticle structure factors (i = H or D for the h-PDMS and d-PDMS, respectively). Substitution of Eqs. 8-10 into Eq. 7 gives an expression for the scattering intensity as a function of x, where there are five unknowns, the two $P_i(q)$ functions and the three $Q_{ii}(q)$ functions. Hence, samples with five different values of x must be measured at fixed filler concentration and a linear system of five equations must be solved to extract the five functions.

The compositions of the three-component mixtures (d-PDMS, h-PDMS, and polysilicate filler) used in this study are given in Table 2. For each of the two filler-to-polymer ratios, three different fractions of labeled to unlabeled polymers were investigated. Work is in progress to obtain additional data on samples with two other labeling ratios. Only a single trimethylsilyl-treated polysilicate system was used for this study at volume fractions of 20 and 50 % filler. Based on Eq. 6, the solution to the system of linear equations at a fixed filler concentration yielded the single chain structure factors of the h-PDMS and d-PDMS as a function of temperature. By constructing Debye plots from S_s^H and S_s^D, the R_g values in Figure 2 were obtained. There was a negligible temperature dependence observed for R_g in the range of 30 to 90 °C. The R_g values at both filler concentrations were significantly smaller than those obtained from the two-component isotopic blends. As the filler concentration increases from mass fraction of 20 % to 50 %, the R_g of the d-PDMS appears to systematically decrease while the R_g of the h-PDMS systematically increases. However, the values at both filler concentrations are within the statistical error of the fits to the data.

Table 2. Compositions for the Three-Component Mixtures[e]

20% Polysilicate Filler			50% Polysilicate Filler		
Blend	ϕ_D	ϕ_H	Blend	ϕ_D	ϕ_H
20D60H20F	0.205	0.585	12D38H50F	0.104	0.398
40D40H20F	0.393	0.393	25D25H50F	0.241	0.242
60D20H20F	0.588	0.198	38D12H50F	0.361	0.124

[e] ϕ_D and ϕ_H are the volume fractions of d-PDMS and h-PDMS, respectively.

The Monte Carlo calculations by Mark and co-workers indicate that R_g increases (29 Å to 32 Å) with increasing filler concentration for small particles (5 Å) while R_g decreases (29 Å to 28 Å) with increasing filler concentration for larger particles (20 Å). Similar results were obtained for larger chains and larger filler particles indicating that once the chains reach a size scale equal to or greater than the particle size, chain contraction occurs, while chains smaller than the filler particle expand. Our results indicate that the chain dimensions of both components decrease in the presence of this particular filler, however, the detailed variation of R_g with filler concentration cannot be addressed. Further studies will involve similar measurements on different filled materials as a function of the particle size extending to more traditional filler materials such as fumed silica.

Acknowledgments

We wish to gratefully acknowledge many useful discussions with Drs. B. J. Bauer and B. Hammouda of NIST concerning the data treatment for the three component mixtures. The preparation and purification of deuterated siloxane monomers, and the polymerization thereof were completed by A. P. Wright, T. M. Leaym, G. M. Wieber, and R. G. Taylor of Dow Corning. The filler preparation was performed by B. Zhong of Dow Corning and all contributions were greatly appreciated.

Reference

1. Yuan, Q. W.; Kloczkowski, A.; Mark, J. E.; Sharaf, M. A. *J. Polym. Sci., Polym. Phys. Ed.* **1996**, *34*, 1647–1657.
2. Sharaf, M. A.; Kloczkowski, A.; Mark, J. E. *Polym. Prepr., Am. Chem. Soc. Div. Polym. Chem.* **1995**, *36*, 368–369.
3. Yuan, Q. W.; Kloczkowski, A.; Sharaf, M. A.; Mark, J. E. *Polym. Mater. Sci. Eng.* **1995**, *73*, 374–375.
4. Akcasu, A. Z.; Summerfield, G. C.; Jahshan, S. N.; Han, C. C.; Kim, C. Y.; Yu, H. *J. Polym. Sci., Polym. Phys. Ed.* **1980**, *18*, 863–869.
5. Tangari, C.; King, J. S.; Summerfield, G. C. *Macromolecules* **1982**, *15*, 132–136.
6. According to ISO 31-8, The term "Molecular Mass" has been replaced by "Relative Molecular Mass", symbol M_r. Thus, if this nomenclature and notation were to be followed, one would write, $M_{r,n}$, instead of the historically conventional M_n for the number average molecular weight and it would be called the "Number Average Relative Molecular Mass". The conventional notation, rather than the ISO notation, has been employed for this publication.
7. de Gennes, P. G., *Scaling Concepts in Polymer Physics*; Cornell University Press: Ithaca, New York, 1979.
8. Jahshan, S. N.; Summerfield, G. C. *J. Polym. Sci., Polym. Phys. Ed.* **1980**, *18*, 1859–1861.

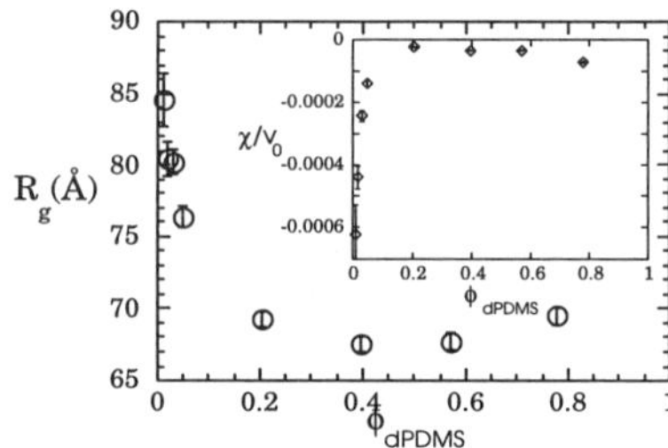

Figure 1. - Apparent R_g of a d-PDMS ($M_w = 34,000$) in an isotopic blend as a function of d-PDMS concentration. Inset shows variation of apparent χ/v_0 of the blend as a function of d-PDMS content. Bars in figures represent +/- one standard deviation in the measurements assuming normal, random statistical errors.

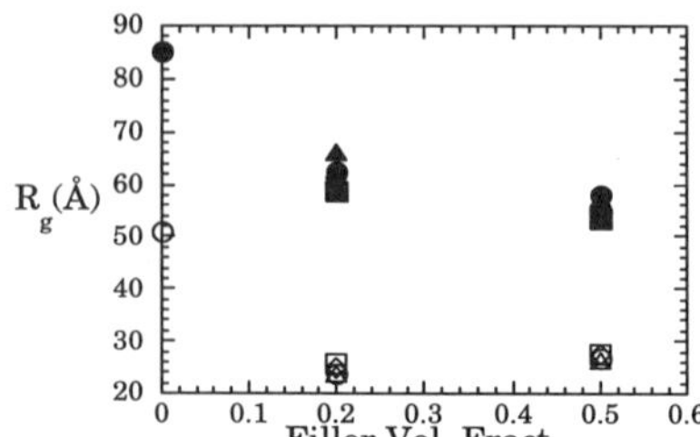

Figure 2. - R_g as a function of polysilicate filler concentration for the d-PDMS chains (solid symbols) and h-PDMS chains (open symbols). ○ - T = 30 °C; □ - T = 50 °C; ◇ - T = 70 °C; △ - 90 °C.

POLYMER DYNAMICS IN BIMODAL MELTS

S. Rathgeber [1*], L. Willner [1], D. Richter [1]

and

M. Appel [2], G. Fleischer [2], A. Brulet [3], B. Farago [4], P. Schleger [4]

[1] Forschungszentrum Jülich, IFF, 52425 Jülich, Germany
[2] Universität Leipzig, Fachbereich Physik, 04103 Leipzig, Germany
[3] LLB Bat 563, C.E. Saclay, 91191 Gif sur Yvette Cedex, France
[4] ILL, 38042 Grenoble, France

We investigated on a microscopic length scale the influence of topological constraints on the chain dynamics in bimodal polyethylene melts in the transition region from Rouse– to Reptation–like behavior. For short times (ns) the dynamic structure factor S(Q,t) was measured by Neutron Spin Echo (NSE)–Spectroscopy. To get information about the long time (ms) dynamics the tracer diffusion coefficient D_{NMR} was determined by Pulsed Field Gradient (PFG)–NMR.

We investigated blends that contained long tracer molecules ($N_t = 365$ or 583) strongly affected by topological constraints and short matrix chains ($N_m = 126$ or $130 < N_e$, the entanglement segment number ≈ 140), which in a good approximation undergo Rouse relaxation [1]. The mass fraction Φ_t of the long tracer molecules was changed over the full concentration range, keeping the labeled fraction to 10%.

The NSE measurements were carried out at the MESS NSE spectrometer at the Orpheé Reactor in Saclay France at a temperature of $T = 509K$ and in a Q-Range of $0.037\text{Å}^{-1} \leq Q \leq 0.155\text{Å}^{-1}$; times up to 18ns were investigated.

In Fig.1 the experimental data for one of the investigated systems are shown. For the low mass fraction of long tracer chains the relaxation of S(Q,t) is much faster than in the case of high Φ_t. We describe the data in the context of a mode analysis [1] following a modified Rouse model where the spatial structure of a normal mode x_p is of Rouse type and its time dependence $\langle x_p(t)x_p(0)\rangle$ is approximated by an exponential $exp(-\nu_p t)$. The basic relaxation rates $W_p = \nu_p N^2/(p^2\pi^2)$ are considered to be mode dependent. Since a mode p describes a motion that involves $n_p = N/p$ bonds, the critical mode number p_c corresponding to a single entanglement strand with N_e segments is: $p_c = N/N_e$. For long PEB-2 chains ($N = 4828$) $N_{e\infty}$ was determined to be ≈ 140 [2]. In Fig.2 the obtained W_p are plotted versus the mode number p. In the case of the system with shorter tracer molecules the tracers show nearly Rouse behavior ($W_p = const.$) at low Φ_t, while an increase of tracer concentration causes a systematic slowing of lower modes. The tracers with $N_t = 583$ do not show Rouse behavior even at the lowest Φ_t. In agreement with results obtained for monomodal melts [1], the only affected modes are those with spatial extension exceeding the entanglement distance determined for long chains $p \leq p_c(N_{e\infty})$. But now the boundary between affected and unaffected modes depends on the concentration of long tracer chains and therefore we have to introduce a Φ_t-dependent $p_c(\Phi_t)$ — the entanglement distance increases with decreasing Φ_t. This can be explained in a slight modification of the model for entanglement formation of Kavassalis and Noolandi [3]. Beside mode-dependent relaxation rates W_p, a fit within the mode analysis includes the translational diffusion coefficients, which were nearly independent of the tracer concentration. The diffusion coefficients obtained by PFG-NMR are systematically lower than the corresponding NSE-values. The deviation between the NSE- and PFG-NMR results becomes larger with increasing Φ_t.

This deviation between short- (NSE) and long-time (PFG-NMR) behavior can be described by the generalized Rouse model of Hess [1,4]. We have extended this model to bimodal melts. In addition to the Rouse rate W_R the only parameter of the Rouse model, the Hess theory contains two further parameters: the critical monomer number N_c, identified with N_e [1] and the true interchain interaction potential q_c^{-1}. In agreement with the mode analysis results, N_c is a function of Φ_t and the $p_c(\phi_t)$ calculated from these values are consistent with the mode

analysis results (see Fig.2). As shown in Fig.3 this extended Hess model provides a time dependent diffusion coefficient that in the short-time limit equals the NSE Rouse value D_R and hence is in agreement with the NSE result independent of the tracer concentration. The long-time value calculated within this model is in accordance to the PFG-NMR Results a function of Φ_t. Fig.1 shows the excellent agreement between S(Q,t) calculated within the Hess theory and the experimentally obtained spectra.

In both the Hess theory and in the reptation model the center of mass diffusion coefficient becomes time dependent for times longer than the entanglement time τ_e and shorter than the reptation time τ_d. In a second experiment described below we were for the first time able to measure this time dependence in the transition region below and above τ_e, which is 5ns for the PEB-2 melts investigated here.

The experiments were performed at the IN15 NSE spectrometer at ILL Grenoble France. This spectrometer allows a extended time range up to $t_{max} = 85ns$; the experimental Q-range was $0.02\text{Å}^{-1} \leq Q \leq 0.06\text{Å}^{-1}$. For these Q-values the main contribution to the relaxation of S(Q,t) is due to the center of mass motion — only the weak contribution of the first three internal modes has to be taken into account for the evaluation of the data.

The measurements were carried out on a monomodal melt with a segment number $N = 365$ where the motion of the chains is hindered by the presence of topological constraints. The reference system was a bimodal melt that contained 10% labeled tracer molecules (also $N_t = 365$) in a short chain matrix ($N_m = 126 < N_e$). We know from the previous experiment that in this case the dynamics of the tracer follows Rouse behavior. Assuming a time-independent diffusion coefficient ($D \propto 1/\zeta$, ζ= monomeric friction coefficient) the dynamics should be the same in both systems if we neglect the change in the free volume (ζ varies only about 16% [1]). Therefore strong deviations in S(Q,t) can be directly ascribed to the influence of topological constraints.

Fig.3 shows the center of mass displacement ($1/6 \cdot D(t)t$) calculated within the Hess model versus time. The center of mass displacement shows highly nonlinear time dependence at intermediate times. Because of the entanglement effect this region is much more pronounced for the monomodal melt.

For the discussion of the results the data are shown in the form:

$$-ln[S(Q,t)/S(Q,0)]/Q^2 = D(t)t + ln[Int(Q,t)]/Q^2 \qquad (1)$$

The first Q-independent term describes the mean square displacement of the center of mass whereas the second term contains the Q-dependent contribution $Int(Q,t)$ of higher (internal) modes. For a better overview only one representative Q-value ($Q = 0.05\text{Å}^{-1}$) is presented.

Fig.4 displays the measured data for both samples in comparison to theoretical predictions calculated (a) within the Rouse model and (b) within a mode analysis, where the latter predicts a slowing down of the first two internal modes. The parameters — the (mode-dependent) characteristic rates W_p and the diffusion coefficients — are known from the first experiment. The dashed lines show the purely diffusive contribution $exp(-Q^2Dt)$ to the relaxation of S(Q,t). Obviously the data for the bimodal melt can be well described with a constant diffusion coefficient, whereas this description completely failed for the monomodal melt for longer times. The dotted line in Fig.4b marks the experimental time window of the first experiment at MESS — in this reduced time window this deviation can not be seen. Fig.5 shows a comparison of the results measured for the monomodal melts with the prediction of the Hess model. It is obvious that the Hess model gives an excellent description of the data over the entire time range of the experiment.

References:

* present address: NIST, Center for Neutron Research, MD 20899, USA
[1] D. Richter et al.; Macromolecules **27** (1994) 7437
[2] D. Richter et al.; Macromolecules **25** (1992) 6156
[3] T.A. Kavassalis and J. Noolandi; Macromolecules **21** (1988) 2869
[4] W. Hess; Macromolecules **21** (1988) 2620

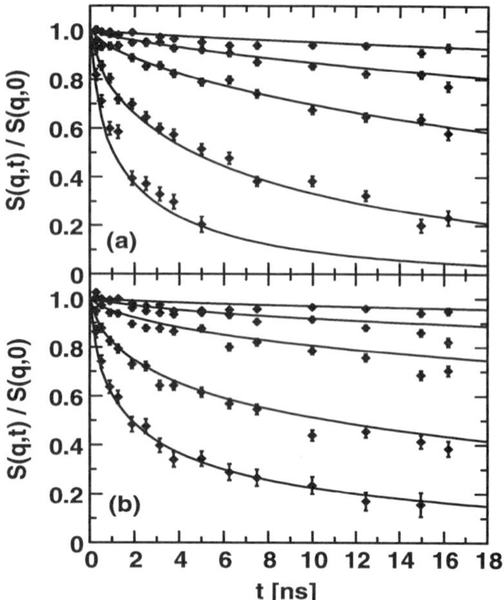

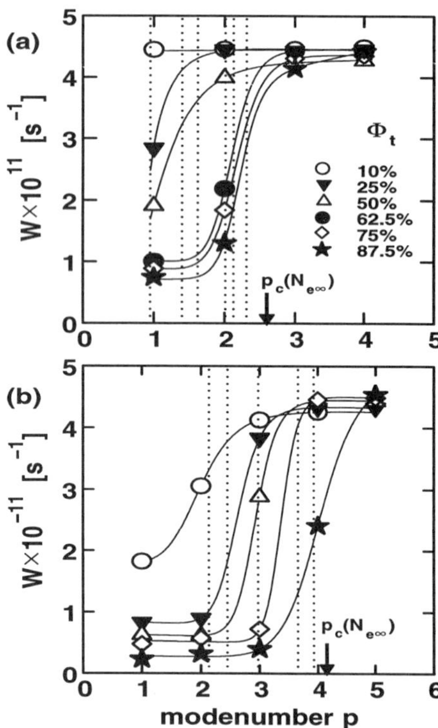

Figure 1: NSE spectra as obtained for the system ($N_t = 583$, $N_m = 130$) for (a) $\Phi_t = 10\%$ and (b) $\Phi_t = 75\%$. The Q-values are from top to bottom: $Q = (0.036; 0.055; 0.077; 0.115$ and $0.155)\text{Å}^{-1}$. The solid lines show the results of a joint fit within the Hess model.

Figure 2: Relaxation rates W_p for the system (a) ($N_t = 365$, $N_m = 126$) and (b) ($N_t = 583$, $N_m = 130$) as a function of the mode number p for different tracer mass fraction Φ_t. The dotted lines mark the $p_c(\Phi_t)$ calculated from $N_c(\Phi_t)$ obtained within the Hess model. The solid lines are guides to the eye.

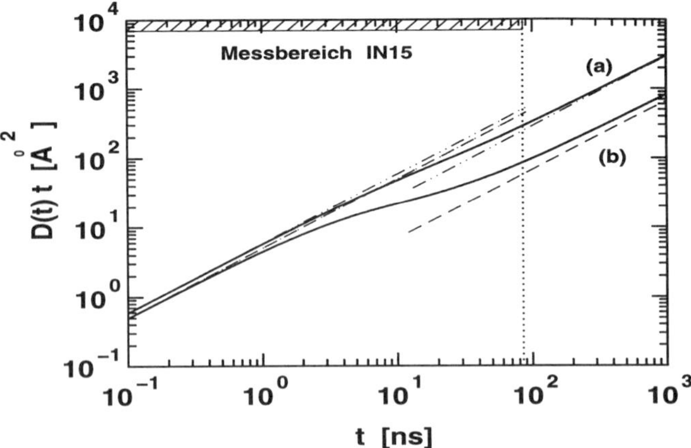

Figure 3: Mean square center of mass displacement calculated within the Hess model for (a) the monomodal melt and (b) the bimodal system. The broken lines show the limiting cases: Short-time behavior as predicted from the Rouse model ($D_R t$) and the long-time behavior as calculated from the PFG-NMR results ($D_{NMR} t$). The dashed area indicates the experimental time range of IN15.

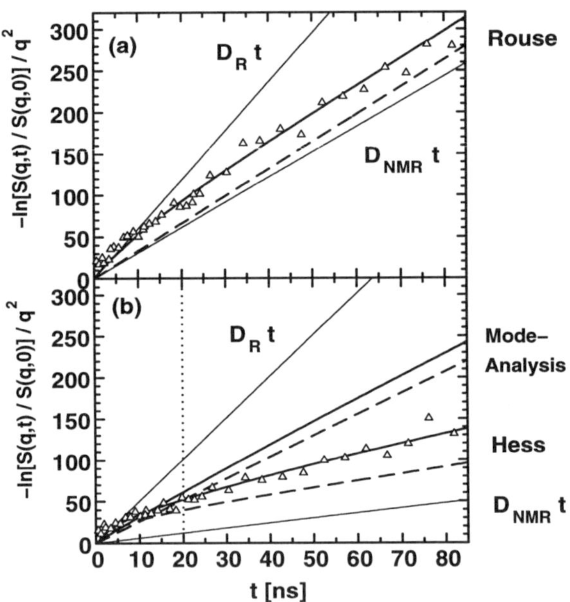

Figure 4: S(Q,t) as measured for (a) the bimodal system and (b) the monomodal melt. The solid line shows the result calculated from (a) the Rouse model and (b) mode analysis and Hess model. The dashed curves display the purely diffusive contribution to the relaxation of S(Q,t) for each model. The limiting cases for the short-time behavior as expected within the Rouse model ($D_R t$) and the long time-limit as calculated from the PFG-NMR results ($D_{NMR} t$) are also shown.

Neutron Reflectivity of Polymer Blends

Robert Braiewa, University of Connecticut
R.A. Weiss, University of Connecticut
Alamgir Karim, National Institute of Standards and Technology
John Ankner, University of Missouri-Columbia

Introduction

Specific interactions and complexation between polymers provide a viable means for their mixing. Interfacial mixing occurs when two polymer surfaces are brought into contact and interdiffuse, altering the interfacial composition and width. This diffusion is dependent on the strength of the interactions between the two polymers. When the two polymers form an intermolecular complex though, diffusion can be inhibited or even come to halt in the extreme case of physical cross-linking.

Feng *et al*[1] found that the interfacial shift between the lithium salt of a lightly sulfonated polystyrene ionomer and an N-methylated nylon 2,10 (N,N'-dimethylethylene-sebacamide) (mPA), with annealing time, could be described by a diffusion-limited cross-linking model:

$$\Delta Z^i(t) = \Delta Z_\infty [1 - \exp(-(t/t_0)^{1/2})]$$

In this equation ΔZ^i represents the interfacial shift, and is a function of annealing time. ΔZ_∞ is a limiting value for the interfacial shift, which is reached as physical cross-links halt diffusion. t_0 is a time constant which reflects the speed of diffusion.

In the previous experiment, the degree of sulfonation of the ionomer was 7.4 (mole %). This paper describes the interfacial mixing behavior of sulfonated polystyrene, lithium salt (Li-SPS) with mPA through a range of ionomer sulfonation levels. Neutron reflectivity provides an excellent method for analyzing these interfaces and was used for both studies. It gives Angstrom level resolution when one material is deuterium labeled.

Experimental

mPA (Mn=25,000, Mw=65,000) was obtained from Yi Feng, and was prepared as described in Feng *et al*[1]. Its molecular structure is:

$$-\left[C(=O)-(CH_2)_8-C(=O)-N(CH_2-CH_2-N(CH_3)CH_3) \right]_m-$$

Deuterated polystyrene (Mn=99,100, Mw=107,000 g/mol) was also obtained from Yi Feng. Sulfonation reactions were performed at ~50°C using acetyl sulfate as the sulfonating reagent[2]. In this reaction, sulfonic acid groups are randomly substituted onto the polymer chain at the para position of the phenyl ring. Sulfonation levels of 2%, 5.7%, and 11.2% (mol %) were achieved. Each ionomer was then neutralized with the Li+ counterion to give the lithium salt of the deuterated ionomer (d-LiSPS). Sulfonation levels were determined through elemental analysis.

The samples used for neutron reflectivity analysis consisted of a bilayer of mPA on top of (d-LiSPS), on a silicon disc substrate. d-LiSPS was spin-coated from a 1% solution in either 90/10 (v/v) toluene/methanol (2%) or 75/25 (v/v) methanol/ tetrahydrofuran (5.7%, 11.2%) at 2000 rpm. The samples were then dried at room temperature under vacuum for 30 min. before the mPA layer was spin-coated at 2000 rpm from a 1% solution in isopropyl alcohol.

Samples were annealed for prescribed time periods under vacuum at 95°C. Neutron reflectivity measurements were performed at the Research Reactor Facility at University of Missouri-Columbia. The instrument uses a monochromatic beam, λ=0.235 nm, with a variable angle detector. Scans required 5 hours each, including a background scan which was later subtracted. Neutron reflectivity theory and data analysis methods can be found elsewhere[3-5].

Results and Discussion

Characteristics of the bilayers were ascertained by using an asymmetric fitting program, which is desribed in Feng *et al*[1]. The 5.7% and 11.2% d-LiSPS/mPA bilayers showed a shift in the interface with annealing time (Figures 1,2) that agreed with the diffusion-limited cross-linking model previously suggested by Feng. The 2.0% d-LiSPS/mPA bilayer showed no appreciable diffusion or interfacial shift, even after annealing for up to 21 hours at 95°C.

After fitting the interfacial shift to the diffusion-limited cross-linking mechanism, the constants derived from the model for each sulfonation level (excepting the non-diffusing 2.0%) can be compared with those reported in Feng's 7.4% d-LiSPS/mPA bilayer. (See Table 1, Figure 3)

Table 1. Diffusion-Limited Cross-linking Constants, d-LiSPS/mPA

Sulfonation Level	ΔZ_∞ (Angstroms)	t_0 (min)
5.7%	309.9	66.82
7.4%	212.1	51.75
11.2%	248.4	23.09

The time constant t_0 shows a consistent trend as it decreases with increasing sulfonation level. This is indicative of a more rapid interfacial shift. The lack of a trend in the asymptotic value ΔZ_∞ is not yet fully understood. Either an increasing or decreasing relation with increasing sulfonation level could be explained. One would expect to see more cross-linking with greater sulfonation level, which would account for a decreasing trend in the ΔZ_∞ value. On the other hand, the more rapid shift described by the time constant t_0 could explain an increasing trend in the value of ΔZ_∞. A possibility is that inconsistent annealing temperatures created a discrepancy in these values. Further experimentation is currently being conducted and should provide greater understanding of this phenomenon.

Conclusions

This study has shown further evidence of intermolecular interaction and complexation at the interface between layers of d-LiSPS and mPA at different sulfonation levels of the ionomer. In the case of the very low sulfonation level of 2.0%, no diffusion or interfacial shift was visible. However, at sulfonation levels of 5.7% and 11.2%, interfacial shift followed the previously suggested diffusion-limited cross-linking mechanism. Varying values of the constants associated with this model were reported as sulfonation level was varied. Trends associated with the values of these constants are currently being investigated further.

References

(1) Feng, Yi; Weiss, R.A.; Karim, Alamgir; Han, C.C.; Ankner, John; Kaiser, Helmut; Peiffer, D.G. *Macromolecules*, **1996**, *29*, 3918.

(2) Makowski, H.S.; Lundberg, R.D.; Singhal, G.H. U.S. Patent 3870841, 1975.

(3) Russell, T.P. *Mater. Sci. Rep.* **1990**, *5*, 171.

(4) Karim, A.; Mansour, A.; Felcher, G.P.; Russell, T.P. *Phys. Rew. B* **1990**, *42*, 6846; *Macromolecules* **1994**, *27*, 6973.

(5) Fernandez, M.L.; Higgins, J.S.; Penfold, J.; Ward, R.C.; Shackleton, C.; Walsh, D.J. *Polymer* **1990**, *31*, 2174.

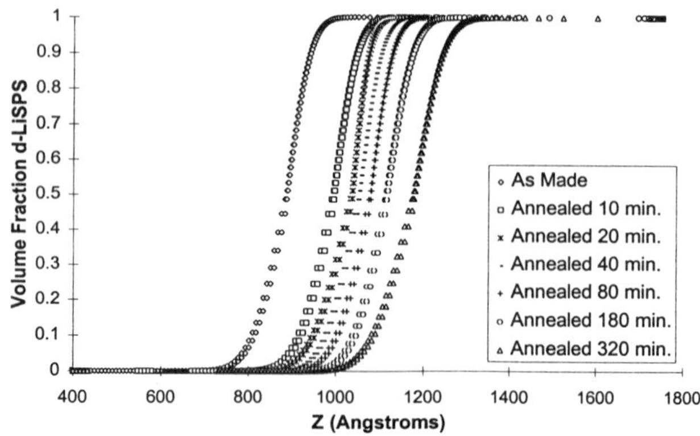

Figure 1. Volume fraction d-LiSPS vs. depth, d-LiSPS (5.7%)/mPA bilayer

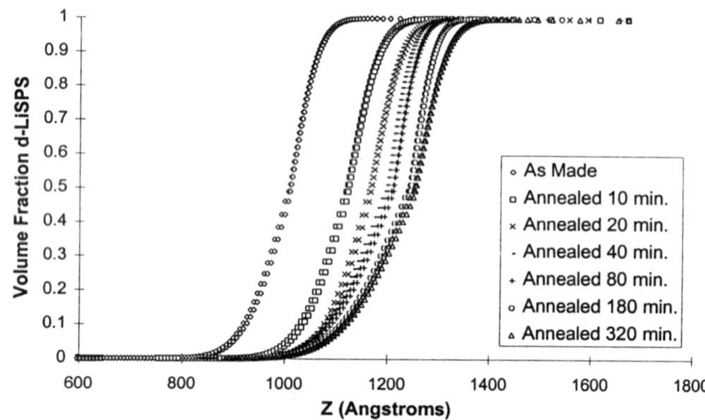

Figure 2. Volume fraction d-LiSPS vs. depth, d-LiSPS (11.2%)/mPA bilayer.

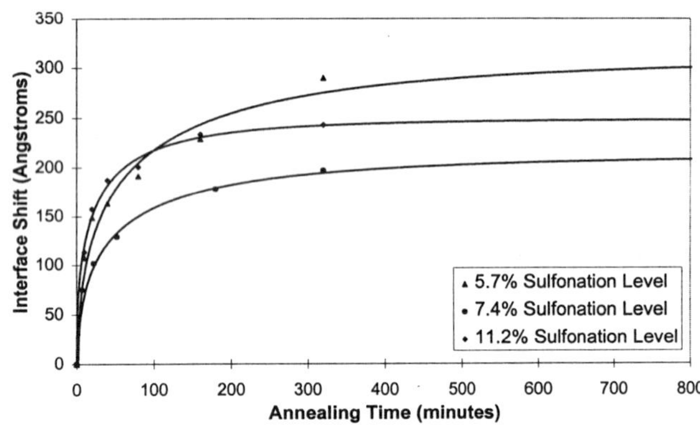

Figure 3. Interface shift vs. annealing time, d-LiSPS/mPA

ANALYSIS OF STRUCTURE, INTERACTION AND VISCOSITY OF
PLURONIC MICELLES IN AQUEOUS SOLUTION BY COMBINED
NEUTRON AND LIGHT SCATTERINGS

Yinchun Liu and Sow-Hsin Chen, Department of Materials Science and
Engineering and Department of Nuclear Engineering, MIT, Cambridge
MA 02139.

Tri-block copolymers we studied are commercially available under
the tradename Pluronic from BASF. They are ABA type block copolymers
where the end-symmetric A-blocks are polyethylene oxides and the center
B-block is polypropylene oxides. These copolymer surfactants are water
soluble to large weight percentages and in broad range of temperatures.
Pluronic polymer surfactants find widespread industrial use in detergency,
foaming/defoaming, emulsification, lubrication, and solubilization. They
are also used in cosmetics, bioprocessing and pharmaceutical applications.

We focus our attention on a particular Pluronic surfactant family,
i.e. copolymers containing 40 to 60 volume ratio of polyethylene oxides
and polypropylene oxide. This Pluronic family has several members,
namely, P104, P94, P84, L64 and L44, in the order of decreasing
molecular weights. Two members of this Pluronic series, P104 (MW =
5900 DT) and P84 (MW = 4200 DT) with corresponding chemical
formulae $PEO_{27} - PPO_{61} - PEO_{27}$ and $PEO_{19} - PPO_{43} - PEO_{19}$, are
studied in detail.

One of the most interesting features of the Pluronic polymer is its
self-association in aqueous solution and its resultant rich phase behavior.
At low polymer concentrations and low temperatures, the tri-block
copolymer in water exhibits single-coil behavior. At higher concentrations
or temperatures, the copolymer molecules self-associate to form
thermodynamically stable micellar aggregates. Micellization occurs above a
certain critical micellization concentration and temperature (c.m.c and
c.m.t.). Dynamic light scattering shows that Pluronic micellar solutions
exhibit significant polydispersity at low temperatures but become
monodisperse at high temperatures..We focus attention on the
microstructure and interaction of micelles far beyond micellization
boundaries where effects of impurities and heterogeneities in the
commercial copolymers used are minimal. The highest concentration of
polymer we used is about 20% by weight, near the gelation boundary.
When the hydration effect of the copolymer is taken into account, the
actual volume fraction of the micelles in solution can be as high as 40 to
50%. A proper treatment of the interparticle correlations is thus essential in
all the results we presented in this paper.

We analyzed an extensive set of small angle neutron scattering
intensity distributions from tri-block copolymer micelles in aqueous
solutions. We investigated, in particular, two Pluronics, P84 and P104 in
an entire range of disordered spherical micellar phase both in temperature
and in concentration. At room temperature, both polyethylene oxide
(PEO) and polypropylene oxide (PPO) are hydrophilic but at elevated
temperatures, PPO becomes significantly less hydrophilic, thus creating a
thermodynamic driving force for micellization. It has been known that the
resultant spherical micelle consists of a hydrophobic core and a hydrophilic
corona region. We propose a "cap-and-gown" model for the
microstructure of the micelle, taking into consideration the polymer
segmental distribution and water penetration profile in the core and corona
regions. We take into account the inter-micellar correlations using an

adhesive hard sphere model of Baxter. With this combined model, we are
able to fit satisfactorily all SANS intensities for the entire micellar range in
an absolute scale. We obtain consistent trends for important parameters
such as the aggregation number, hydration number per polymer in both the
core and corona regions, and surface stickiness. It is verified that the
micro-structure of micelle stays essentially constant as a function of
concentration, but changes rapidly with temperature. Both the aggregation
number and surface stickiness increase, and the micelle becomes drier with
increasing temperature. Micellar core is not completely dry but contains up
to 20% (volume fraction) of solvent molecules at lower temperatures [Y.C.
Liu, S.H. Chen, J.S. Huang, "SANS analysis of the structure and
interaction of tri-block copolymer micelles in aqueous solution", to appear
in Macromolecules (1998)].

From analysis of the SANS intensity distribution, we obtain an
analytical expression of the inter-micellar structure factor with known
parameters. We then use the structure factor to calculate the interaction part
of the relative zero shear viscosity of the micellar solution according to
Verberg-de-Schepper-Cohen theory [Phys.Rev. E $\underline{55}$, 3143 (1997)]:

$$\eta_I = \frac{\chi}{40\pi} \int_0^\infty Q^2 d(Q) \frac{[S'(Q)]^2}{S(Q)}.$$

In this formula, χ is the value of the pair correlation function at contact,
$d(Q)$ is a known function and $S(Q)$ the structure factor and $S'(Q)$ the
derivative of the structure factor. Adding to this the small hydrodynamic
contribution, we get an excellent agreement with the measured relative
viscosity of the micellar solution to the highest concentration in the
disordered micellar phase [Y.C. Liu, S.H. Chen, J.S. Huang, Phys. Rev.
E $\underline{54}$, 1698 (1996)].

In addition to these, a novel light scattering intensity analysis
method is developed which is useful up to about 50% volume fraction.
The new light scattering intensity analysis method significantly extends the
concentration range of the traditional Zimm Plot. It uses the intensity data
in the moderate and high volume fraction regions where the microstructure of the micelle has been shown to be invariant with concentration.
We plot $Ln(KC/R_{90})$ vs the copolymer concentration C. The resultant
plot is, to a very good approximation, a straight line. The zero intercept
gives the logarithm of the total molecular weight of the micelle and the
slope is $(8 - 2/\tau)(\varphi/C)$, where $1/\tau$ is the stickiness parameter, and φ
the total volume fraction of the micelles. The deviation of the plot from a
straight line behavior at certain concentration can be used as a fingerprint to
detect a phase transition [Y.C. Liu, S.H. Chen, J.S. Huang, "Light
scattering studies of the interaction in concentrated tri-block copolymer
micellar solutions". sent to Macromolecules].

COUNTERION VALENCE EFFECTS ON INTRA- AND INTERCHAIN CORRELATIONS IN POLYELECTROLYTE SOLUTIONS

Brett D. Ermi, Yubao Zhang, and Eric J. Amis

Polymers Division, National Institute of Standards and Technology Gaithersburg, MD 20899

Introduction

It is now well established that small angle scattering from low ionic strength polyelectrolyte solutions yields a maximum at finite wavevector and a steep upturn at low angles; results that are dramatically different from those observed for neutral polymer solutions.[1-3] These phenomena appear for nearly all classes of charged macromolecular systems and several models have been invoked to describe them. Although there is no readily accepted mechanism for these phenomena, the unusual scattering surely results from strong intra- and intermolecular electrostatic interactions of unscreened charges that dominate the observed static structure factor.

One important variable that requires investigation is the effect of the counterion valence. While nearly all previous studies involved polyelectrolytes with monovalent counterions, there have been a few concerning polyelectrolytes with multivalent counterions.[4,5] These studies have focused mainly on the precipitation and phase behavior from solution. There are also scattering studies of polyion chains in the presence of multivalent coions[6-8] (i.e. added divalent and trivalent ions) but to our knowledge there are no measurements of small angle scattering that address the most basic case of divalent counterions without added salt.

Progress understanding this area has been impeded by the lack of an adequate description for the single chain structure and dimensions. Primarily this is a result of the strong interchain correlations that exist even in dilute solutions, which prohibit clean separation of the intrachain scattering function P(q) from the overall scattering profile. Previously, it was demonstrated that small angle neutron scattering combined with isotopic labeling techniques can separate intrachain correlations from interchain correlations, even in concentrated low salt polyelectrolyte solutions.[9,10] The current study uses such techniques and measures small angle neutron scattering from deuterated and hydrogenated poly(styrene sulfonate) with mono- and divalent counterions in deuterium oxide and water. This system allows direct measurement of counterion valence effects on polyion configuration and overall solution structure.

Experimental

For the full contrast experiments, sodium poly(styrene sulfonate) (NaPSS) (M_n = 124 000, degree of sulfonation (DS) = 90%, degree of polymerization (DP) = 601)[11] was purchased from Scientific Polymer Products.[12] The dry polymer was dissolved in deionized water, passed through a mixed bed resin ion-exchange column, dialyzed against deionized water, and then titrated to neutral pH using NaOH for the monovalent salt. For the divalent salt, poly(styrene sulfonic acid) solution was added to $Mg(OH)_2$ until the solution reached neutral pH. Both solutions were then dialyzed against deionized water and lyophilized. For the isotopic labeling experiments, matched molecular weight deuterated NaPSS (M_n = 58 800, DS = 83%, DP = 300) and hydrogenated NaPSS (M_n = 64 000, DS = 90%, DP = 327) were purchased from Polymer Source. The deuterated NaPSS was found to be free of ionic impurities and was used as received but the hydrogenated NaPSS required

further purification via ion-exchange, dialysis, neutralization titration with NaOH, and lyophilization. The magnesium salts were prepared using the same protocol as above and all scattering solutions were prepared gravimetrically from the lyophilized polymer.

Small angle neutron scattering measurements were performed on the 8 m SANS instrument of the Cold Neutron Research Facility at the National Institute of Standards and Technology. The neutron beam was monochromated to 9Å with a velocity selector having a wavelength spread of 0.25 ($\Delta\lambda/\lambda$ = 0.25). The scattered neutrons were detected by a 64 cm x 64 cm two dimensional detector. The resulting data were corrected for background electronic noise, detector inhomogeneity, empty cell scattering, and solvent scattering. Following correction for sample transmission, the intensities were scaled to absolute values using a water standard. The uncertainties of the individual data points for the I(q) versus q plots are calculated statistically from the number of averaged detector counts. For the figures displayed, the standard uncertainties are less than ±3% of the intensity values and are not shown.

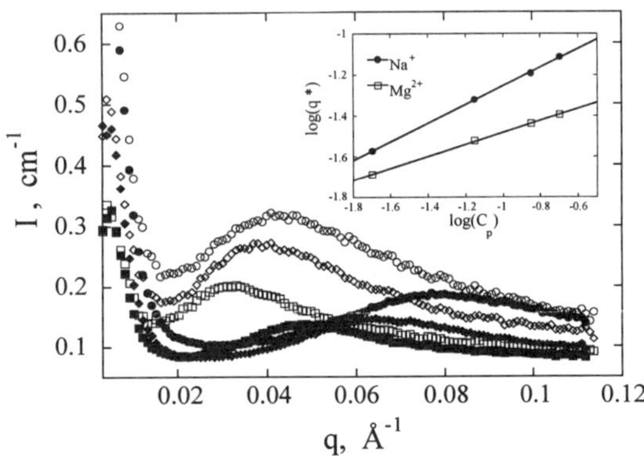

Fig. 1 Small angle neutron scattering from polyelectrolyte solutions with full scattering contrast for mono- (filled symbols) and divalent (unfilled symbols) counterion systems for three monomer concentrations; (●) 0.20 mol/L, (◆) 0.14 mol/L, and (■) 0.07 mol/L. Inset is log-log plot of peak position versus monomer concentration.

Results and Discussion

Figure 1 displays the SANS profiles of NaPSS and MgPSS with no added salt for concentrations ranging from 0.07 mol/L to 0.20 mol/L. Each curve shows one broad maximum whose position is a strong function of concentration, moving to higher q with increasing concentration. At q values below 0.01 Å^{-1}, a steep upturn of the scattered intensity is observed. While the qualitative information from the two systems is similar, there are obvious differences. At equivalent monomer molar concentrations, the peak position shifts to much lower q values for the polyelectrolyte with divalent Mg^{2+} counterions. This and more quantitative differences between the mono- and divalent systems can be seen clearly in the inset which displays the concentration dependence of the broad maximum on a log-log plot along with a linear least squares fit. The scaling behavior for the NaPSS solutions is typical of most previous investigations, $q_{max} \propto c_p^{0.46\pm0.02}$, while the scaling for MgPSS solutions is dramatically different, $q_{max} \propto c_p^{0.29\pm0.02}$.

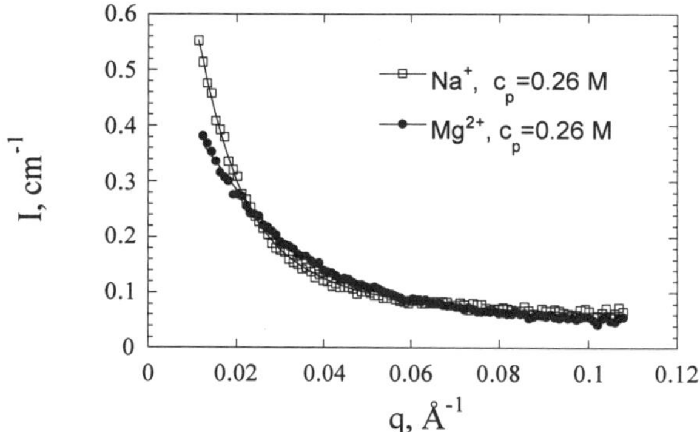

Fig. 2 SANS from polyelectrolyte solutions under conditions of zero average contrast showing only intramolecular structure factor scattering.

To further investigate the apparent structural changes, we used zero average contrast (ZAC), a specific deuterium labeling technique, to separate P(q) from the overall scattering. This method is described in detail elsewhere.[11,12] Briefly, we prepared an equimolar mixture of deuterated and hydrogenated polymer in a mixture of H_2O and D_2O. The fraction of D_2O in the solution is set at the value necessary to satisfy the "optical theta condition," where the scattering length density of the hydrogenated and deuterated monomers are equal and opposite so that interchain correlation scattering is canceled. Figure 2 displays the scattering profiles from the ZAC solutions for a 0.26 mol/L solution of NaPSS and MgPSS along with Debye function fits to the data. Each are monotonically decreasing functions of angle, as expected for the intraparticle scattering function, and each are fit adequately with the Debye function.

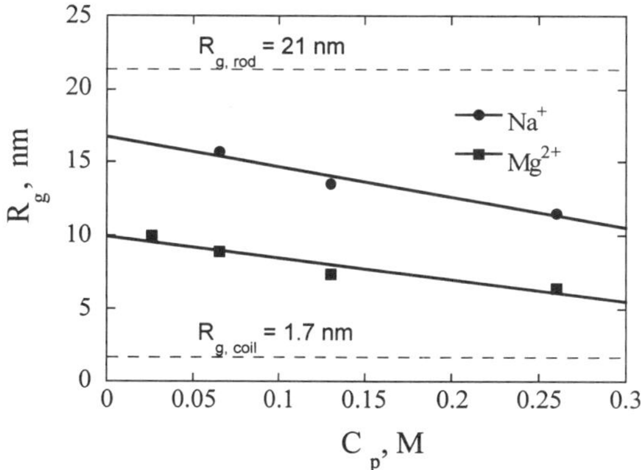

Fig. 3 Polyelectrolyte chain dimensions as a function of concentrations with comparison to model chain calculations.

Figure 3 displays the values of radius of gyration[13] as a function of concentration for both NaPSS and MgPSS along with calculated estimates for the size of a single chain with DP = 300 with a random coil and rod-like configuration. For both systems the chain dimension decreases with increasing concentration with the MgPSS chains nearly a factor of two smaller for each concentration. It can also been seen that while the monovalent chains are highly extended they are not at the rod-like limit. Both systems are clearly more expanded than the calculated ideal chain value. This result implies that while the divalent counterions do induce a large coil contraction they do not produce a coil collapse or a coil-globule transition.

In this preliminary study we have shown the full contrast and zero average contrast scattering from poly(styrene sulfonate) with mono- and divalent counterions. While changing the counterions does not effect the qualitative results there are significant quantitative differences. We demonstrate that the position of the broad maximum in the full contrast solutions shifts to lower q values and the concentration scaling of the peak is much weaker for the divalent systems. We also observe that the chain dimensions of poly(styrene sulfonate) with Mg^{2+} counterions is nearly a factor of 2 smaller for the equivalent chains with Na^+ counterions. Over a broad concentration range all polyelectrolytes are considerably larger than random coil dimensions.

References
1. Dautzenberg, H.; Jaeger, W.; Kötz, J.; Phillip, B.; Seidel, Ch.; Stscherbina, D. *Polyelectrolytes: Formation, Characterization and Application*, Hanser/Gardner: Cincinnati, OH 1994.
2 Ermi, B. D.; Amis, E. J. *Macromolecules* **1997**, *30*, 6937.
3. Forster, S.; Schmidt, M.; Antonietti, M. *Polymer* **1990** *31*, 781.
4. Olvera de la Cruz, M.; Belloni, L; Delsanti, M.; Dalbiez, J. P.; Spalla, O.; Drifford, M. *J. Chem. Phys.* **1995** *103*, 5781.
5. Wittmer, J; Johner, A.; Joanny, J. F. *J. Phys. II* **1995**, *5*, 635.
6. Drifford, M.; Dalbiez, J. P.; M.; Delsanti, M.; Belloni, L *Ber. Bunsenges. Phys. Chem.* **1996**, *100*, 829.
7. Ise, N.; Okubo, T.; Kunugi, S.; Matsuoka, H.; Yamamoto, K.; Ishii, Y. *J. Chem. Phys.* **1984**, *81*, 3294.
8. Ikeda, Y; Beer, M.; Schmidt, M.; Huber, K. *Marcomolecules* **1998**, *31*, 728.
9. Williams, C. E.; Nierlich, M.; Cotton, J.P.; Jannick, G.; Boué, F.; Daoud, M.; Farnoux, B.; Picot, C.; de Gennes, P. G.; Rinaudo, M.; Moan, M.; Wolff, C.; Rinaudo, M. *J. Polym. Sci., Lett. Ed.* **1979**, *17*, 379.
10. Boué, F.; Cotton, J.P.; Lapp, A.; Jannick, G. *J. Chem. Phys.* **1994** *101*, 2562.
11. According to ISO 31-8, the term "Molecular Weight" has been replaced with "Relative Molecular Mass", symbol M_r. The conventional notation, rather than the ISO notation, has been employed for this publication.
12. Reference to commercial equipment or supplies does not imply its recommendation or endorsement by the National Institute of Standards and Technology.
13. Standard uncertainties for the individual R_g values extracted from the fits to the data are less than ±2 %.

DIMENSIONS OF POLYSTYRENE COILS IN THE CRITICAL REGION OF
PHASE DEMIXING

G. D. Wignall and Y. B. Melnichenko

Solid State Division, Oak Ridge National Laboratory, Oak Ridge,
TN 37831-6393

The conformation of polymer chains depends crucially on the quality of the solvent and in the poor solvent regime $T<\Theta$ the dominating attractive interactions between monomers are believed to collapse the individual chains into compact polymer globules. Much less is known about the conformation of chains at $T<\Theta$ in the *semidilute domain* $\phi>\phi^*$, where the coil dimensions can be substantially affected by the interchain interactions and the critical fluctuations [1]. Providing definitive experimental data on the size of polymer chains at the critical point has remained a challenge for the past decade. To find out the possible effect of temperature on the size of polystyrene (PS) coils in d-acetone (DAC) and d-cyclohexane (DCH) at the critical concentration of phase demixing $\phi^C \sim \phi^*$ we apply the small-angle neutron scattering (SANS) and make use of high-concentration isotopic labeling [2] to increase the signal-to-noise ratio. This methodology allows us to follow the radius of gyration R_g of the few labeled chains starting from the Θ region down to the critical temperature T^C of phase demixing.

The coherent intensities of scattering for solutions of identical protonated and deuterated polymer coils can be represented as:

$$I(Q,T,x) = I_S(Q,T,x) + I_t(Q,T,x) \quad , \tag{1}$$

$$I_S(Q,T,x) = KnN^2 S_S(Q,T) \quad , \tag{2}$$

$$I_t(Q,T,x) = LnN^2 S_t(Q,T) \quad , \tag{3}$$

where $K \equiv (b_H - b_D)^2 x(1-x)$; $L \equiv [b_D x + (1-x)b_H - b_S']^2$. Subscripts "s" and "t" correspond to the scattering from a single chain and the total scattering, respectively. x is the molar ratio of labeled chains, b_D and b_H are the scattering lengths of d- and h- monomers. b_S' is normalized scattering length of a solvent molecule. n and N are the number density polymers and the degree of polymerization, respectively. S_S and S_t are the single-chain and the total scattering structure factors, respectively. It follows from Eqs.(1)-(3) that one can directly observe $S_S(Q,T)$, by making the prefactor (L) vanishingly small. Similarly, $x=1$ makes the prefactor (K) zero, and thus $S_t(Q,T)$ can be obtained directly in the experiment.

The parameter (L) is exactly zero for solutions of (h-PS + d-PS) in DAC at $x \approx 0.21$. If $L \neq 0$, as is the case for solutions of (h-PS + d-PS) in DCH with $x=0.1$ (which is the minimal content of h-PS in solution that produces a reasonable signal-to-noise ratio), S_S can be obtained by subtracting the normalized intensity of total scattering $I_t(Q, \tau, x=1)$ from $I(Q, \tau, x=0.1)$ at the same reduced temperatures $\tau = (T-T^C)/T^C$ [3].

h-PS and d-PS with $M_W/M_N < 1.06$ were purchased from Polymer Laboratories. Solutions of h-PS and (h-PS+d-PS) pairs with similar M_w were prepared in DAC and DCH from Sigma. Each solution was contained in a 5 mm thick quartz cell and the temperature was controlled to better than ± 0.1 K. The temperature of phase demixing was detected visually with the estimated precision ± 0.25 K.

Measurements were performed on the 30-m SANS spectrometer at the Oak Ridge National Laboratory. The neutron wavelength was $\lambda = 4.75$ Å ($\Delta\lambda/\lambda = 0.06$) and the range of scattering vectors was $0.006 < Q = 4\pi\lambda^{-1} \sin\theta < 0.09$ Å^{-1}, where 2θ is the scattering angle. The data were collected and put onto an absolute scale with the procedure described elswhere [4]. After phase demixing, the lower half of the cells was covered by cadmium mask and the scattering from the upper diluted phase of the solution was measured at several temperatures below T^C. The functions S_S where obtained via Eqs.1-3

and used to extract the (z-averaged) R_g by fitting the Debye form factor for Gaussian chains:

$$F_D = (2/y^2)(y-1+e^{-y}) \quad , \quad y = Q^2 R_g^2 \tag{4}$$

As is seen from Fig.1 the radius of gyration for polystyrene in all solutions remains constant and equal to that of the unperturbed Gaussian coils over the whole temperature range below the Θ temperature down to T^C. The abrupt decrease of R_g is revealed only at $T<T^C$. In Figure 1 we also show the temperature variation of the correlation length ξ of the concentration fluctuations containing information on both inter- and intramolecular correlations between monomers irrespectively of their belonging to a given polymer chain. As is seen, the magnitude of ξ diverges as $T \Rightarrow T^C$ [5]. These findings indicate that critical polymer solutions can not be considered as an ensemble of collapsed, non-interpenetrating chains, as was originally suggested by de Gennes [1]. Instead, deterioration of the solvent quality leads to the formation of polymer clusters, consisting of undisturbed, strongly interpenetrating polymers [3].

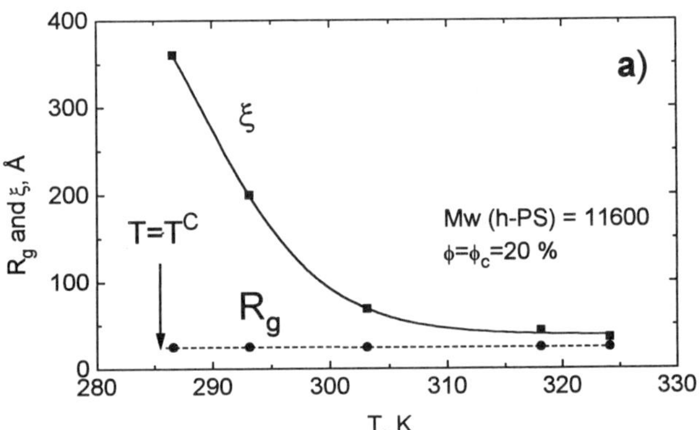

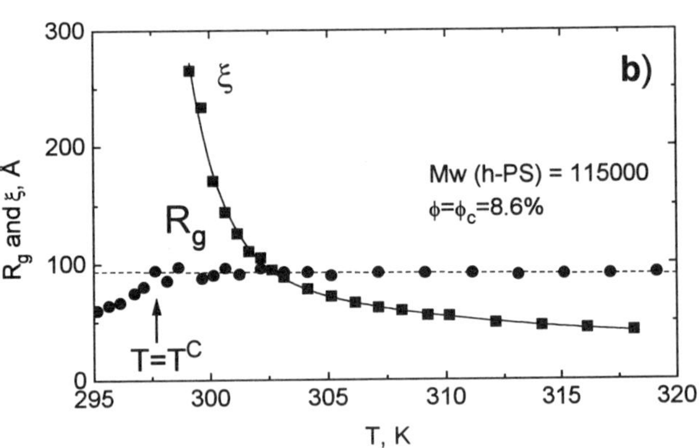

FIG.1. Temperature dependence of the radius of gyration   and the correlation length   for the solutions of polystyrene in: (a) d-acetone and (b) d-cyclohexane.

REFERENCES

1. P.-G. de Gennes, *Scaling Concepts in Polymer Physics*, Cornell University Press, Ithaca, London, 1979 .
2. J. S. King, W. Boyer, G. D. Wignall, and R. Ullman, Macromol. **18**, 709 (1985).
3. Y. B. Melnichenko and G. D. Wignall, PRL, **78**, 686 (1997).
4. G.D.Wignall and F.S.Bates, J.Appl.Crystallogr. **20**,28 (1986).
5. Y. B. Melnichenko, G. D. Wignall, and W. A. Van Hook, these proceedings.

Size Changes and Interpenetration within Concentrated Dendrimer Solutions

Ty J. Prosa[1], Barry J. Bauer[1], Andreas Topp[1],
Eric J. Amis[1] and Rolf Scherrenberg[2]

[1]*National Institute of Standards and Technology, Gaithersburg, MD 20899*
[2]*DSM Research, P.O. Box 18, NL-6160 MD, Geleen, The Netherlands*

Introduction

The field of dendrimer research has grown dramatically in recent years with the synthesis of new materials quickly outpacing the characterization of simple material properties.[1] The lack of dendrimers in sufficient quantities has inhibited the ability to conduct bulk material studies, but the recent availability of DAB-dend-$(NH_2)_x$ family of dendrimers in commercial quantities has made it possible to study these material as concentrated solutions and in the bulk.[2,3]

Previous studies by this laboratory have reported on the intramolecular structure of dendrimers within dilute solution[4,5] and the evolution of dendrimer structure as a function of solution concentration using small-angle neutron scattering (SANS).[6,7] We report here an extension to these earlier works. One advantage of small-angle x-ray scattering (SAXS) over SANS is the ability to better resolve high angle features because of the incoherent scattering effects present in SANS. The net result is the ability for SAXS to better resolve high angle scattering features which are indicative of the uniform sphere-like molecular ordering of the individual dendrimer molecules in solution. The observed secondary maximum is relatively unaffected by correlation effects between dendrimers in concentrated solutions and allows for a relatively unambiguous separation of average particle size and interparticle correlation effects.

In this report we utilize the well developed methods of polymers in solution to extract information about the individual molecule as a function of increasing solvent concentration.[8] Since this particular dendrimer exhibits SAXS indicative of a compact, spherical entity, comparisons are made to scattering from a mixture of hard spheres using the Percus-Yevick approximation.[9-11]

Experimental [12]

A generation 5 (G5) poly(propylene imine) dendrimer DAB-dend-$(NH_2)_{64}$ was used as received from DSM Research, Heerlen, The Netherlands.[2,3] All solutions were prepared by weight and dissolved in methanol (Aldrich Chemical Company).

The SAXS experiments were carried out at the Advanced Polymers-Participating Research Team (AP-PRT) beamline X27C (NIST/SUNY Stony Brook/GE/Allied Signal/Montell/Air Force Laboratory) at the National Synchrotron Light Source (NSLS), Brookhaven National Laboratory (BNL). A double-multilayer monochromator with an energy resolution ($\Delta E/E$) of ~1% was used to obtain photons with a wavelength λ = 1.307 Å. A three pin-hole collimation configuration[13] was used to define a beam size of 0.1 mm, and an evacuated scattering path with Kapton windows and a sample to detector distance of (1 to 2)m was used. This resulted in a measurable scattering range of 0.008 Å^{-1} to 0.44 Å^{-1}. Beam monitors both before and after the sample were used to normalize scattering intensities to the incident flux and to correct for transmission through the sample. Fuji Image Plates were used to collect the two-dimensional data with a resolution of 0.2 mm (Δq ~ 0.001 Å^{-1}). The SAXS solution cell used in this study had a thickness of 4 mm sandwiched between thin mica windows (~6 μm). Data were scaled to absolute scattering intensities by use of a Lupolen secondary standard. Background intensities caused by scattering from the windows of the detector vacuum chamber, solution cell, and beam stop over-spill were determined by separate measurements and subtracted from the sample data. Data collection times ranged from (1 - 2) hours.

Results and Discussion

For particulate solution scattering, intensity curves, I(q), can commonly be broken up into factors originating from the individual particle scattering, P(q), and interferences between particles, S(q), defined below, where f is a constant, 2θ is the scattering angle, and λ is the wavelength.[8]

$$I(q) = f^2 \, P(q) \, S(q)$$
$$q = 4\pi \sin \theta \, / \, \lambda$$

The practical scattering definition of 'dilute' is when the approximation $S(q) \approx 1$ holds, or when any observable correlations between dendrimers is so large that it is effectively located behind the beam stop. As average correlation distances between molecules become smaller in moderate concentration solutions, a peak forms in S(q), but quickly damps out so that at moderate values of q, $S(q) \approx 1$.

The corrected 1-dimensional SAXS curves for the G5 dendrimer are shown in figure 1. The 1 % dendrimer data can be fit very successfully with a model consisting of polydisperse spheres with a uniform density shown as the thick solid line in figure 1. [14] This analysis suggests that these dendrimers have an average radius $\underline{R}$ = (18.56 ± 0.04) Å and a Gaussian distribution of sphere radii with a width σ = (1.73 ± 0.05) Å. Since the bulk dendrimer has a density reported value of 1.01 g/cm^3,[15] a theoretical molecular mass of 7.12 kg/mol, and a solvent density of 0.867 g/cm^3, one can estimate that the solvated dendrimer volume consists of 56 % solvent.

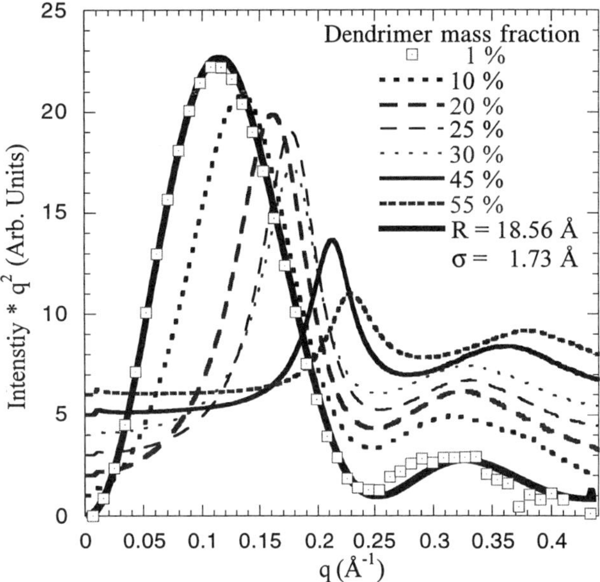

Figure 1. Experimental scattering intensity for various dendrimer concentrations in methanol. The dominant features are a well defined maxima (correlation peak) that moves to higher q and a secondary maximum that also moves to higher q as a function of concentration. The curves have been shifted vertically for clarity. The relative standard error for the plotted data is less than 5 %.

As dendrimer concentration is increased, the effects of dendrimer-dendrimer correlations are manifested as a peak in the scattering curve, which is first observed around 0.09 Å^{-1} and moves to 0.235 Å^{-1} at 55 % dendrimer. By analyzing this correlation peak as originating from a close packed lattice, one can estimate the average distance between scattering centers as $\Lambda = 1.22(2\pi/q_{max})$,[16] a qualitative estimate of average nearest neighbor distances can be extracted. These results are displayed in table 1.

Table I. Dendrimer Size and Inter-Molecule Separation

	mass fraction dendrimer, %						
	1	10	20	25	30	45	55
$\underline{R}^{\dagger}$(Å)	18.56	17.8	17.5	17.1	17.2	15.9	15.4
$\Lambda/2^{\dagger}$(Å)	--	28.3	23.7	21.6	21.4	18.0	16.6

†Values based on observed graphical extrema and are intended only to serve as a qualitative measure for basis of discussion.

Modeling these dendrimer/methanol solutions as a system of polydisperse, non-interacting spheres allows for a qualitative comparison with the theoretical scattering curves for such a system by using the Percus-Yevick approximation.[9-11] Starting with a system described by the 1 % findings stated previously, one can simulate scattering curves by varying the volume fraction of the dendrimer component.

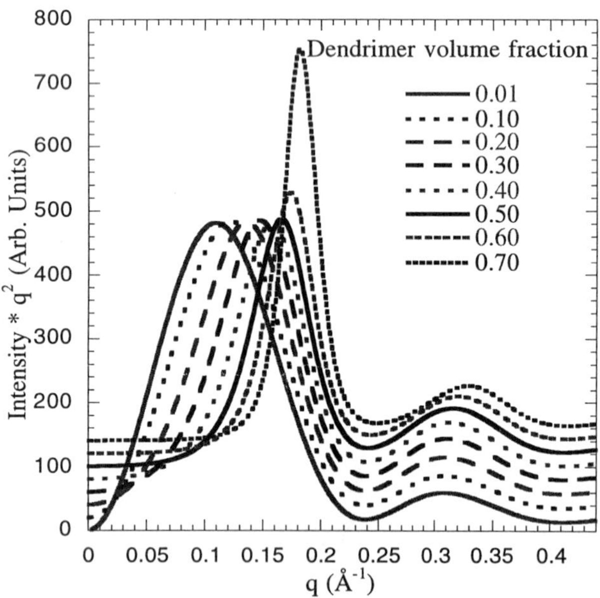

Figure 2. Modeled scattering intensity using a non-interacting, polydisperse, constant radius uniform sphere model as a function of dendrimer volume fraction with $\underline{R}$ = 18.56 Å and σ = 1.73 Å for all concentrations. Note the relatively small shift in the local minimum observed near 0.25 Å⁻¹. The curves have been shifted vertically for clarity.

These curves are shown in figure 2 and suggest three significant differences from a constant radius, non-interacting hard-sphere model. First, shifts to higher q for the first local minimum and secondary maximum are smaller than those observed in the experimental data. This suggests that the first minimum and secondary maximum are primarily regulated by the P(q) factor because S(q) has effectively reached a value of 1 for these values of q. Second, an upturn in scatting at low q and high concentration (not evident in this Iq² plot) is observed in the data and not in the model. Third, the correlation peak seen in the experimental data is narrower than in the model, especially at concentrations above 30 % dendrimer. These differences suggest a behavior that deviates from completely non-interacting hard spheres by an apparent compaction of individual molecules with increased concentration. We do not see evidence of a disappearance of the secondary maximum at concentrations as high as 65

% dendrimer. This requires a minimum of dendrimer interpenetration. This is consistent with both observations in computer modeling studies which show that individual dendrons within a dendrimer tend to segregate,[17,18] and with simulations of dendrimers adsorbed on a surface.[19] These results are all in agreement with SANS results reported previously.[6,7]

Conclusions

Trends in the experimental SAXS curves suggest that at sufficient dendrimer concentration, the overall dimension of the dendrimer begins to shrink, thereby minimizing or avoiding interpenetration with neighboring dendrimers. The movement of the secondary maximum as a function of dendrimer concentration in methanol suggests that these dendrimer molecules shrink in response to increasing intermolecular interactions. At dilute concentrations, the dendrimers can be described, to first order, as a polydisperse system of hard spheres with an average radius of (18.56 ± 0.04) Å and σ = (1.73 ± 0.05) Å consisting of 56 % solvent by volume. These molecules are observed to shrink up to 63 % of their original volume in order to minimize interpenetration with neighboring dendrimer molecules at dendrimer concentrations up to 55 %, thereby reducing the overall volume availible to the solvent molecules.

Acknowledgments

Special thanks John Barnes for insightful discussions, and Ben Hsaio and Fengji Yeh for technical support at the AP-PRT beamline.

Notes and References

1. For recent conference proceedings on state of dendrimer science see, *ACS PMSE Preprints* (1997) **77**.
2. C. Wöner; R. Mühlhaupt, *Angew. Chem.* (1993) **105** 1367; *Angew. Chem Int. Ed. Engl.* (1993) **32** 1306.
3. E. M. M. de Brabander-van den Berg; E. W. Meijer, *Angew. Chem.* (1993) **105** 1370; *Angew. Chem Int. Ed. Engl.* (1993) **32** 1308.
4. T. J. Prosa; B. J. Bauer; E. J. Amis; D. A. Tomalia; R. Scherrenberg, *J. Polym. Sci: Polym. Phys.* (1997) **35** 2913.
5. R. Scherrenberg; B. Coussens; P. van Vliet; G. Edouard; J. Brackman; E. de Brabander, *Macromolecules* (1998) **31** 456.
6. A. Topp; B. J. Bauer; E. J. Amis; R. Scherrenberg, *ACS PMSE Preprints* (1997) **77** 82.
7. A. Topp; B. J. Bauer; T. J. Prosa; R. Scherrenberg; A. J. Amis, to be published.
8. J. S. Pedersen, *Adv. Coll. Sci.* (1997) **70** 171.
9. J. K. Percus; G. J. Yevick, *Phys. Rev.* (1958) **110** 1.
10. J. K. Percus, *Phys. Rev. Letters* (1962) **8** 462.
11. A. Vrij, *J. Chem. Phys.* (1979) **8** 3267.
12. Certain commercial materials and equipment are identified in this paper in order to specify adequately the experimental procedure. In no case does such identification imply recommendation by the National Institute of Standards and Technology nor does it imply that the material or equipment identified is necessarily the best available for this purpose.
13. B. Chu; P. J. Harney; Y. Li; K. Linliu; F. Yeh; B. S. Hsiao, *Rev. Sci. Instrum.* (1994) **65(3)** 597.
14. T. J. Prosa; B. J. Bauer; A. J. Amis, to be published.
15. Value for the density as given in the product specification by the manufacturer. No attempt was undertaken to verify this value.
16. A. Guinier and G. Fournet, Small Angle Scattering of X-Rays; Wiley: New York, 1955.
17. M. L. Mansfield, *Polymer* (1994) **35** 1827.
18. M.L. Mansfield, *Polymer* (1996) **37** 3835.
19. M. Murat; G. S. Grest, *Macromolecules* (1996) **29** 1278.

NOVEL PHASE NANOSTRUCTURES OF POLYELECTROLYTE-SURFACTANT COMPLEXES

Shuiqin Zhou[1], Fengji Yeh[1], Christian Burger[2], and Benjamin Chu*,[1]
1. Department of Chemistry, State University of New York at Stony Brook, New York 11794-3400; 2. Max-Planck-Institut für Kolloid- & Grenzflächenforschung, Kantstr. 55, D-14513 Teltow-Seehof, Germany

Introduction

Polyelectrolyte-surfactant interactions have attracted special interest in the last two decades, due to its importance both in fundamental biochemistry and in technological applications.[1-3] Ionic and hydrophobic interactions of polyelectrolytes with oppositely charged surfactants can induce complex formation.[3-6] However, little is known about the supramolecular structures of the polyelectrolyte-surfactant complexes (PSCs). Most of PSCs showed lamellar, cylindrical, and undulating layered structures.[4-6]

In this paper, we report nanostructures of complexes formed by anionic copolymer gels of poly(sodium methacrylate-co-N-isopropylacrylamide) P(MAA/NIPAM) interacting with oppositely charged cetyltrimethylammonium bromide (CTAB) and didodecyldimethylammonium bromide (DDAB). The objectives are to study the charge density effects of the polyelectrolyte chains on the structures of PSCs, and to analyze the novel phase structures observed in P(MAA/NIPAM) gel-CTA and P(MAA/NIPAM) gel-DDA complexes.

Experimental

Materials MAA monomer (Aldrich) was vacuum distilled. NIPAM monomer (ARCOS) was purified by recrystalization. N, N′-methylenebis-acrylamide(BIS)(Ultragrade, Pharmacia LKB) as a crosslinker, and ammonium persulfate (APS) (Aldrich, 98+%) as an initiator, were used as received. The surfactants of CTAB (Fluka, ≥99%), and DDAB (Fluka, ≥98%) were used without further purification.

Gel preparation Gels were prepared by free radical copolymerization. A 15 wt% aqueous solution of reaction mixture with a desired comonomer molar ratio and a crosslinker density of 1 mol% was bubbled with N_2 for 15 min, then 35 µL of 10 wt% APS solution was added to the mixture. The final solutions were filtered into the space between two glass plates with a thickness of 0.63 ± 0.4 mm. Gelation was carried out at 65 °C for 24 h. The charge content of the copolymer chains was expressed in mol%.

P(MAA/NIPAM) gel-surfactant complexation Complexes were prepared by immersing known amounts of neutralized and water swollen gel discs in a dilute surfactant solution (pH~8.2). The concentration of surfactant was about 1/3 of the critical micelle concentration (CMC). The total volume of the surfactant solution was controlled at such a level that the number of surfactant molecules was always in excess of the number of charged groups in the copolymer chains.

X-ray scattering measurements Small angle X-ray scattering (SAXS) measurements were performed at the X3A2 SUNY Beam Line, NSLS at BNL. The incident beam wavelength (λ) was 0.154 nm. An imaging plate was used as the detector. The sample to detector distance was 787 mm, corresponding to a q range of $0.2 \leq q \leq 4.8$ nm^{-1} with $q = (4\pi/\lambda)\sin(\theta/2)$, and θ being the scattering angle between the incident and the scattered X-ray.

Results and Discussion

Typically, the P(MAA/NIPAM) gels reached an equilibrium state of collapse after about four weeks in the CTAB solution, and eight weeks in the DDAB solution, respectively. The elemental analysis results showed that the complex formation between the P(MAA/NIPAM) copolymer chains and the CTAB molecules followed a 1:1 stoichiometry in terms of the charges. No bromine atom was detected inside the complexes. Table 1 summarizes the compositions of the P(MAA/NIAPM) gel-CTA complexes at different charge content of P(MAA/NIPAM) chains.

P(MAA/NIPAM)-CTA systems Figure 1 shows typical SAXS profiles of P(MAA/NIPAM) gel-CTA complexes formed at different charge content in the copolymer chains, where the q values have been normalized to q/q^* with $q^* = 1.33$, 1.38, 1.38, and 1.40 nm^{-1} corresponding to charge content of 100 %, 75 %, 67 %, and 50 %, respectively. At a charge content $\geq$ 75 %, more than ten scattering peaks with a spacing ratio of $2^{1/2}$: $4^{1/2}$: $5^{1/2}$: $6^{1/2}$: $8^{1/2}$: $10^{1/2}$: $12^{1/2}$: $14^{1/2}$: $16^{1/2}$: $17^{1/2}$: $20^{1/2}$: $21^{1/2}$ were observed, suggesting a Pm3n space group cubic structure. At a charge content of 67 %, five sharp diffraction peaks with a spacing ratio of $3^{1/2}$: $4^{1/2}$: $8^{1/2}$: $11^{1/2}$: $12^{1/2}$ were observed, indicating a face-centered cubic (FCC) lattice structure. At a charge content of 50%, nearly ten scattering peaks, which could be indexed according to the structure of hexagonal close packing (HCP) of spheres, were observed. The dotted lines in the figure represent the simulation curves based on the ideal models for Pm3n cubic, FCC, and HCP structures, respectively. These structures were never detected in the phase diagram of the CTA/water binary system.[1] Obviously, the threading of P(MAA/NIPAM) chains with various charge density to the CTA/water system produced new phase structures. All the indices for the peaks observed in these structures are listed in Table 2.

The Pm3n cubic structure can be described as consisting of a cage-like three-dimensional (3D) network of rods and a body-centered cubic arrangement of spheres. There are six rods and two spheres per unitcell. The weighting factor of a sphere is half of that of a rod. For the FCC close packing of spheres, there are four spheres per unitcell. For the HCP structure, an orthorhombic unitcell containing four spheres was selected. In the reduced length scale coordinates where the distance between two adjacent 2D closely packed layers of spheres is normalized to unity, the lattice constants are a = $(3/2)^{1/2}$, b = $(9/2)^{1/2}$, and c = 2. Putting one sphere at the origin [0,0,0] and one sphere at the center of the base plane [a/2,b/2,0] produces a single 2D closed-packed layer of spheres. The next layer at z = c/2 is identical to the base layer but shifted by b/3 in the y direction. The layer at z = c is identical to the one at z = 0, resulting in an A-B-A stacking order. All the dimensions of the unitcell for these structures are listed in Table 1.

According to the compositions of the P(MAA/NIPAM)-CTA complexes, the spheres and the rods in these models could only be the spherical and the cylindrical micelles formed by the aggregates of CTA ions with the paraffin chains inside them. The radius of cylindrical (R_{rod}) or spherical micelles (R_{sph}), and the micellar aggregation number (N_{agg}) could be calculated by using the known compositions, the lattice parameters of the unitcell, and the partial specific volumes of each specy inside the complexes. The estimated R_{rod} or R_{sph}, and N_{agg} values for different complexes are listed in Table 1. The R_{rod} (or R_{sph}) values of 2.1-2.0 nm in Pm3n cubic structures are in good agreement with the radius of cylinders of H_α phase in the CTA/water binary system,[7] while the obtained R_{sph} values of 2.34 and 2.25 nm for spherical micelles in FCC and HCP structures are consistent with the radius of spherical CTA micelles in water.[8] The calculated N_{agg} for micelles at a charge content of 67 % (FCC), and 50 %(HCP) were 99($\pm$5) and 87($\pm$5), respectively, being slightly larger than 81 ($\pm$5) for the spherical micelles formed in the CTA/water binary system.[9] It is known that the addition of electrolyte to the surfactant solution can lead to a decrease in the CMC and an increase in the N_{agg}. However, the complex formation follows a stringent 1:1 stoichiometry in terms of charges. Thus, the N_{agg} of surfactant molecules inside the PSCs will depend strongly on the charge density and the flexibility of polyelectrolyte chains. Our results showed that both N_{agg} and R_{sph} inside the complexes decreased with decreasing charge content of the copolymer chains.

P(MAA/NIPAM)-DDA system Figure 2 shows the SAXS profiles of the P(MAA/NIPAM) gel-DDA complexes at various charge content. At charge content $\geq$ 75 %, two scattering peaks at q = 1.78 and 3.60 nm^{-1}, corresponding to the (001) and (002) planes of a layered structure, were observed. An interlayer spacing of 3.53 nm indicated a bilayer arrangment of the double-tailed DDA ions. At a charge content of 67 %, three diffraction peaks at q = 1.73, 2.00, and 2.66 nm^{-1} with a spacing ratio of $3^{1/2}$: $4^{1/2}$: $7^{1/2}$ were observed. At a charge content of 50 %, the scattering curves also showed three peaks, but at q = 1.78, 2.06, and 3.24 nm^{-1}, respectively, with a spacing ratio of $3^{1/2}$: $4^{1/2}$: $10^{1/2}$. A combination of the two curves suggests an Ia3d space group cubic structure which typically gives diffraction with a spacing ratio of $3^{1/2}$: $4^{1/2}$: $7^{1/2}$: $8^{1/2}$: $10^{1/2}$: $11^{1/2}$. Although Ia3d cubic is one of the most commonly observed structures in lipid/water systems at phase areas between the lamellar and the hexagonal phases,[9] it was, for the first time, observed in PSCs. The observed four peaks with a spacing ratio of $3^{1/2}$: $4^{1/2}$: $7^{1/2}$: $10^{1/2}$ can be indexed as (211), (220), (321), and (420). By further decreasing the charge content to 33%, only one broad peak was observed for P(MAA/NIPAM)-CTA or -DDA complexes, indicating a less-ordered structure.

Conclusions

The interaction of slightly crosslinked anionic P(MAA/NIPAM) gel with oppositely charged CTA or DDA salts could produce complexes with very rich highly ordered nanostructures, such as Pm3n cubic, FCC, HCP, bilayer lamellar, and Ia3d cubic. These ordered structures were formed by a self-assembly of surfactant molecules driven by both electrostatic and hydrophobic interactions inside the polyelectrolyte gels. A variation of the charge density of polyelectrolyte chains can induce the phase transition of surfactants inside the complexes.

Acknowledgment

BC gratefully acknowledges the support of this work by National Science Foundation and the Petroleum Research Fund, Administered by the American Chemical Society.

References and notes

(1) Ober, C. K.; Wegner, G. *Adv. Mater.* **1997**, *9*, 17.

(2) Mitrakvs, Macdonald, P. M. *Biochemistry* **1996**, *35*, 16714.

(3) *Interactions of Surfactants with Polymers and Proteins* (Goddard, E. D. & Ananthapadmanabham, K. P., ed.), CRC Press, FL **1993**.

(4) Yeh, F.; Solokov, E. L.; Khokhlov, A. R.; Chu, B. *J. Am. Chem. Soc.* **1996**, *118*, 6615.

(5) Antonietti, M.; Maskos, M. *Macromolecules* **1996**, *29*, 4199.

(6) Antonietti, M.; Burger, C.; Effing, J. *Adv. Mater.* **1995**, *7*, 751.

(7) Auvray, X.; Petipas, C.; Anthore, R.; Rico, I.; Lattes, A. *J. Phys. Chem.* **1989**, *93*, 7458.

(8) Berr, S. S.; Caponetti, E.; Johnson Jr. J. S.; Jones, R. R. M.; Magid, L. T. *J. Phys. Chem.* **1986**, *90*, 5766

(9) Lindblom, G.; Rilfors, L. *Biochim. Biophys. Acta*, **1989**, *988*, 221.

Table 1. The compositions and structure parameters of P(MAA/NIPAM) gel-CTA complexes

charge content	weight fraction (wt%)			ϕ_{CTA}	a/nm	R_{rod} or	N_{agg}
(mol%)	CP	CTA	water			R_{sph}/nm	
100	10.6	39.9	49.5	0.449	10.55	2.1	
91	12.9	39.1	48.0	0.450	10.18	2.0	
75	15.6	38.8	45.6	0.447	10.10	2.0	
67	16.3	37.9	45.8	0.438	7.90	2.34	99
50	21.1	34.8	44.2	0.405	a=5.50 b=9.53 c=8.98	2.25	87

CP: P(MAA/NIPAM) copolymer chains; ϕ_{CTA}: volume fraction of CTA inside the complexes; a: parameter of unitcell; R_{rod} or R_{sph}: radius of cylindrical or spherical micelles of CTA inside the complexes; N_{agg}: aggregation number of micelles inside the complexes

Table 2. Miller indices for the scattering peaks observed in the P(MAA/NIPAM)-CTA complexes

charge content (mol%)							
100	91	75		67		50	
q/nm⁻¹	q/nm⁻¹	q/nm⁻¹	hkl	q/nm⁻¹	hkl	q/nm⁻¹	hkl
0.84	0.87	0.89	110	1.38	111	1.32	[±1, ±1, 0] [0, ±2, 0]
1.19	1.23	1.24	200	1.59	200	1.40	[0, 0, ±2]
1.33	1.37	1.38	210	2.25	220	1.49	[±1, ±1, ±1] [0, ±2, ±1]
1.46	1.52	1.52	211	2.65	311	1.92	[±1, ±1, ±2] [0, ±2, ±2]
1.67	1.74	1.76	220	2.77	222	2.29	[±2, 0, 0] [±1, ±3, 0]
1.88	1.95	1.96	310			2.64	[±2, ±2, 0] [0, ±4, 0]
2.06			222			2.74	[±2, ±2, ±1] [0, ±4, ±1]
2.22	2.33	2.32	321			2.80	[0, 0, ±4]
2.39	2.48	2.49	400			2.99	[±2, ±2, ±2] [0, ±4, ±2]
2.46	2.55	2.78	410				
	2.78	2.78	420				
2.73	2.84	2.85	421				
Pm3n cubic				FCC		HCP	

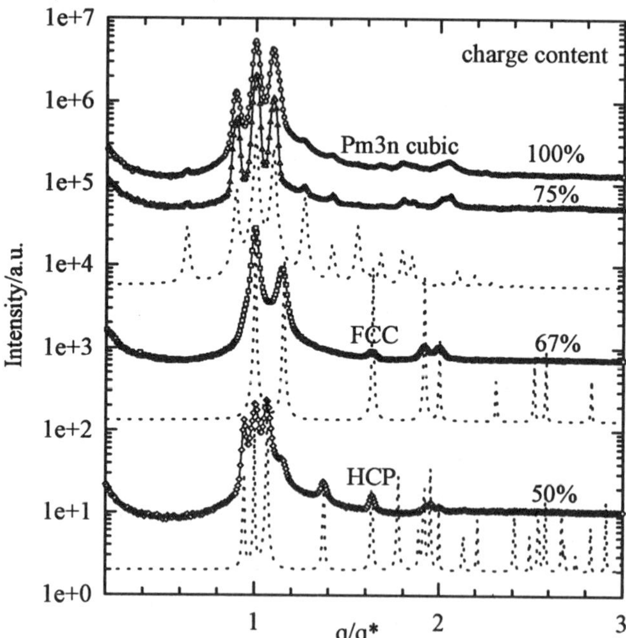

Figure 1. Typical SAXS profiles of P(MAA/NIPAM) gel-CTA complexes at different charge content of P(MAA/NIPAM) chains

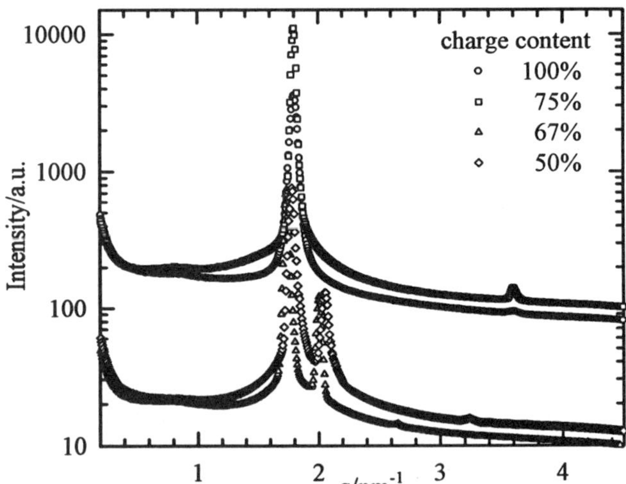

Figure 2. Typical SAXS profiles of P(MAA/NIPAM) gel-DDA complexes at different charge content of P(MAA/NIPAM) chains (see text about the peak positions)

THE PHASE DIAGRAM OF POLYMER SOLUTIONS BELOW
THE THETA TEMPERATURE FROM SANS
Y. B. Melnichenko, G. D. Wignall
Solid State Division, Oak Ridge National Laboratory,
Oak Ridge, TN 37831-6393
W. A. Van Hook
Department of Chemistry, University of Tennessee, TN 37996

Semidilute polymer solutions are characterized by two distinct types of monomer-monomer correlations with different length scales. *Intramolecular* correlations are closely related to the conformation of the macromolecules in the solution and diminish on the length scale of the order of the radius of gyration R_G of polymer coils. *Intermolecular* correlations are defined by fluctuations of the total concentration of monomers and can be described in terms of the correlation length ξ. At $T=\Theta$ $\xi < R_G$, though ξ can become $\gg R_G$ in the vicinity of the critical temperature of phase demixing T_C due to the divergence of ξ and constancy of R_G as $T \Rightarrow T_C$ [1]. Thus, R_G represents an additional length scale in the solution of the order of tens or even hundreds of Angstroms dependent on the molecular weight M_W of the polymer. The interrelation between the structure and thermodynamic properties of complex fluids with an additional length scale is yet not well understood. It is our aim to investigate the impact of the transition from $\xi < R_G$ to $\xi > R_G$ regimes on the osmotic compressibility χ and the shape of the phase diagram of polymer solutions in the poor solvent domain $T < \Theta$.

Polystyrene (PS) standards ($28 \times 10^3 \leq M_W \leq 1.08 \times 10^6$, $1.03 < M_W/M_N < 1.07$) were from Polymer Laboratories. Solutions of PS at critical volume fraction ϕ_c of the polymer were prepared in d-cyclohexane (DCH, 99.5% deuterium, Sigma). Values of (z-averaged) $R_G(\Theta)$ were determined earlier using high-concentration labeling methods [1]. Each solution was contained in a quartz cell and the temperature was controlled to better than ± 0.1 K. T_C was detected visually with an estimated accuracy of ± 0.25 K.

Measurements were performed on the 30-m SANS spectrometer at the Oak Ridge National Laboratory. The neutron wavelength was $\lambda = 4.75$ Å ($\Delta\lambda/\lambda = 0.05$). The range of scattering vectors $Q = 4\pi\lambda^{-1}\sin\vartheta$, where 2ϑ is the scattering angle, was $0.003 < Q < 0.05$ Å^{-1}. The data were collected and put onto an absolute scale with a procedure described elsewhere [2]. $I(Q=0)$ and ξ were extracted from $I(Q,T)$ at each temperature by fitting the low-Q region to the Ornstein-Zernike formula

$$I(Q,T) = \frac{I(0)}{1+Q^2\xi^2} \qquad (1)$$

The magnitude of $I(0)$ was converted into the osmotic compressibility $\chi = \phi(\partial\pi/\partial\phi)^{-1}$ via:

$$I(0) = b_V^2 kT\chi \qquad (2)$$

where π is the osmotic pressure, ϕ is the volume fraction of the polymer, k is the Boltzman constant, and b_V is the contrast factor between the monomer and the solvent molecule. A typical temperature variation of $I(0)/T \sim \chi$ and ξ for the solution of PS ($M_W = 51.5 \times 10^3$) in DCH is shown in Fig.1. As is seen, a crossover from the mean-field ($\gamma = 1$, $\nu = 0.5$) to the Ising model ($\gamma = 1.24$, $\nu = 0.62$) values of the critical indices γ and ν is observed as the temperature departs the Θ region and approaches T_C. The crossover is sharp and non-monotonic and can be described by the two-parameter crossover function suggested in [3]. The crossover temperature T_X coincides within experimental accuracy with the temperature at which the correlation length becomes equal to the radius of gyration of the polymer in each solution.

The resulting temperature-concentration diagram of the polystyrene solutions studied is shown in Fig.2. In the region I solution is phase separated into two coexisting phases one of which is diluted and the other one concentrated by the polymer. The region II

is the region where the critical fluctuations $\xi > R_G$ develop and the properties are isomorphous to those of other representatives of the universality class of Ising model (1,3) (e.g. individual liquids, binary mixtures of molecular liquids, polymer blends, etc.). We identify the crossover temperature line $T_X(\phi)$ with the lower boundary of the tricritical Θ domain (region III) where $\xi < R_G$ and the behavior of the polymer solution is described by the mean-field theory with logarithmic corrections [4]. Although the corrections do not affect the mean-field values of the critical indices γ and ν, they do change the shape of the boundary in the $M_W \Rightarrow \infty$ ($\phi \Rightarrow 0$) limit. The renormgroup (RG) theory predicts $T_X(\phi)$ to approach the Θ temperature with zero slope whereas the mean-field theory predicts non-zero slope and linear variation of T_X vs. ϕ. As is seen in Fig.2, the concentration variation of both T_X and T_C is not linear and can be described by the RG equation [4]:

$$\frac{T-\Theta}{T} = A\phi_c \left[\ln\frac{A}{\phi_c} \right]^{-7/11} \qquad (3)$$

with $A = 0.58 \pm 0.02$ for $T_X(\phi)$ and $A = 1.00 \pm 0.01$ for $T_C(\phi)$. To our knowledge, this is the first experimental observation of the phase diagram of the polymer solutions below the Θ temperature [5].

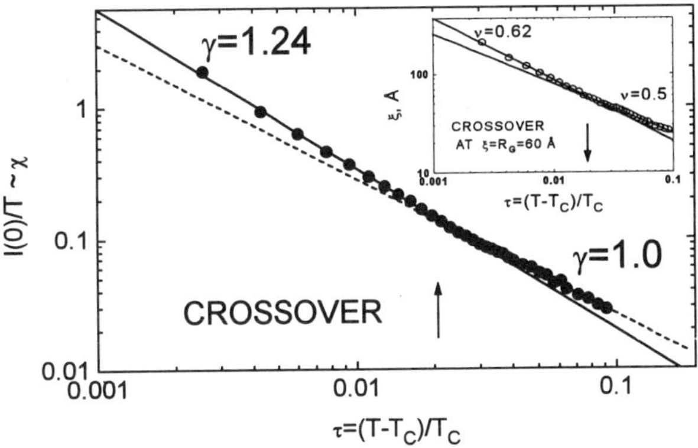

Fig.1. The crossover of I(0)/T and ξ for PS (MW=51.5x103) in DCH.

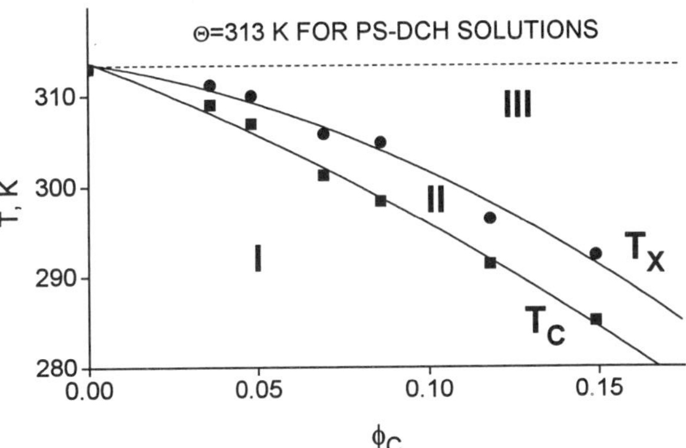

Fig.2. The phase diagram of PS-DCH solutions at T < Θ.

REFERENCES

1. Y. B. Melnichenko and G. D. Wignall, PRL, **78**, 686 (1997).
2. G.D.Wignall and F.S.Bates, J.Appl.Crystallogr. **20**,28 (1986).
3. M. A. Anisimov et al., Phys.Rev.Lett., **75**, 3146 (1995).
4. K. F. Freed, Renormalization Group Theory of Macromolecules, Wiley, N.Y., 1987.
5. Y. B. Melnichenko, M. A. Anisimov, A. A. Povodyrev, G. D. Wignall, J. V. Sengers, and W. A. Van Hook, Phys.Rev.Lett., **79**, 5266 (1997).

Effect of Solvent Quality on the Molecular Dimensions of PAMAM Dendrimers

Barry J. Bauer[1], Andreas Topp[1], Donald A. Tomalia[2], and Eric J. Amis[1]

[1]National Institute of Standards and Technology, Gaithersburg, MD 20899
[2]Michigan Molecular Institute, Midland, MI 48640.

Introduction

Dendrimers are highly branched macromolecules which consist of a central core, with regularly branched building blocks, a high branching density, and a large number of terminal groups. In large dendrimers, the high branching density leads to a very high internal segment density, which results in a spherical shape. Being spherical and having a narrow size distribution, higher generation Poly(amido amine) (PAMAM) dendrimers exhibit a very well defined geometry. Also, the radial segment density distribution in the interior of a dendrimer is quite different from that of linear chains. The uniform spherical shape of dendrimers and their size range of dendrimers, (typically a radius in the range of 1 nm up to 15 nm) makes them excellent candidates for the use as size calibration standards.[1]

However, there are indications in the literature that the size of dendrimers is greatly influenced by the solvent quality. A molecular dynamics study by Murat and Grest showed increased internal segment density of dendrimers when the dendrimer-solvent interactions are less favorable, which leads to a considerable decrease of the average dimensions of the simulated structures.[2] On the basis of a holographic relaxation spectroscopy study, Stechemesser and Eimer concluded that the hydrodynamic radius of higher generation PAMAM dendrimers is extremely sensitive to solvent conditions.[3]

We report the results of a small angle neutron scattering (SANS) study of G5 and G8 PAMAM dendrimers in different solvents, and in mixtures of methyl alcohol and acetone, which are a good solvent and a non-solvent for PAMAM dendrimers, respectively. Measurements are made in a temperature range of -10°C ≤ T ≤ 50 °C which crosses a phase boundary for certain combinations of solvents.

Experimental

Poly(amido amine) (PAMAM) dendrimers were synthesized according to reported methods.[4] Samples were dialyzed against deionized water and freeze dried. Samples were prepared by weight, stirred for one hour prior to use and transferred to quartz cells for the small angle neutron scattering (SANS) experiments.

The SANS experiments were performed at the 8 m facility at the National Institute of Standards and Technology (Gaithersburg, Maryland). The temperature was controlled by a circulating bath that has a nominal temperature stability of $\Delta T = \pm 0.1$ °C. Data were corrected by established procedures and circularly averaged.

Results

The values of R_g are calculated from the scattering intensities by fitting the Guinier approximation in the functional form

$$\ln(I) = \ln(I_o) + \frac{R_g^2}{3} q^2 \qquad (1)$$

to the experimental $I(q)$ data, where I_o and R_g are the fit parameters. Weighted linear least squares fits are applied to plots of $\ln(I)$ versus q^2 (weighting the data by the propagated experimental standard deviation of the intensity). The validity of the Guinier approximation in the analysis of scattering data of dendrimer solutions has been discussed in a previous reports, [1] and has shown to yield accurate numbers for the radius of gyration of a dendrimer when fitted over a wide q range.

A series of four different concentrations of G8 in water was measured to study the effect of concentration on the experimental value for the radius of gyration of the dendrimer. Figure 1 is a plot of the R_g as a function of concentration. In the range of concentration from (0.2 to 6) X 10^{-5} mol/dm^3 the relative change in R_g is only ~5% which is near the standard uncertainty limits. For the remainder of the experiments, concentrations of about one weight percent were used. The bars in all of the figures represent a standard uncertainty of the goodness of the fits.

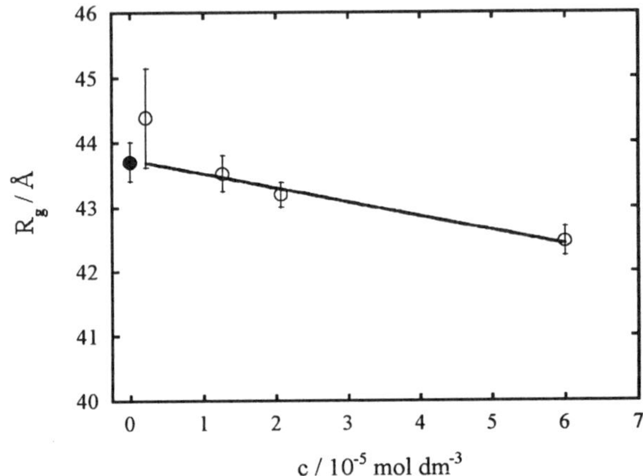

Figure 1. R_g of G5 in H$_2$O as a function of concentration. Filled circle is extrapolation to zero concentration.

SANS experiments were performed with solutions of generation 8 (G8) PAMAM dendrimer in water, methanol, ethanol, and *n*-butanol at a temperature of $T = 20.0$ °C. Figure 2 is a plot of the R_g versus the length of the methylene chain, m, in the series of solvents D(CD$_2$)$_m$OD (with m = 0, 1, 2, 4). For this series of solvents, the solvent quality decreases with increasing m, due to the decrease of the polarity of the solvent. A weak decrease of R_g with decreasing solvent quality is found. This indicates a correlation between the dendrimer size and the solvent quality, but the absolute value of the change of R_g is small.

An estimate for the segment density inside the dendrimers, ρ^+, can be given by assuming a uniform segment distribution inside the volume occupied by dendrimer segments. This is a reasonable assumption for higher generation PAMAM dendrimers, according to the results of a recent x-ray study.[1] Also, the dendrimers have been found to be spherical and to have a narrow size distribution. Assuming a uniform sphere, the radius of the dendrimer, R^+, is related to the radius of gyration of the dendrimer, R_g, via

$$R^+ = R_g / (3/5)^{1/2} \qquad (2)$$

The average segment density of the dendrimer in solution, ρ^+, is then calculated from the relationship, with M$_t$ being the theoretical molecular weight

$$\rho^+ = \frac{M_t}{(4\pi / 3) N_A R^{+3}} \qquad (3)$$

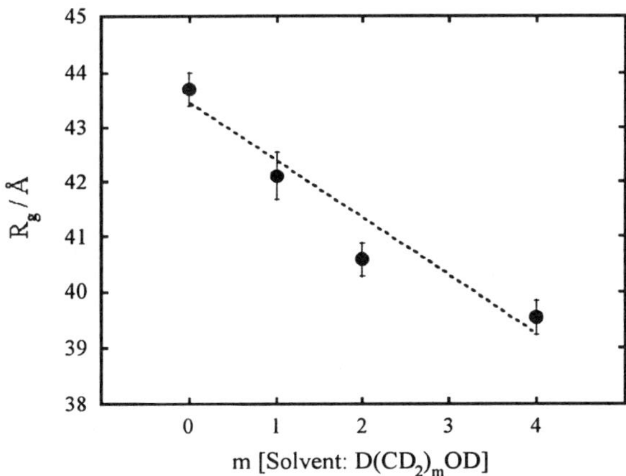

Figure 2. Plot of the radius of gyration of G5 PAMAM dendrimers, versus the composition of the solvent, x_s. (O) T = 20 °C; (□) T = 50 °C.

where N_A is Avogadro's constant. The volume fraction inside the dendrimer that is made up of dendrimer segments is ϕ_{di} with the remaining volume, ϕ_{si} being occupied by solvent. The bulk density of the PAMAM dendrimers is ρ°.

$$\phi_{di} = \frac{\rho^+}{\rho^\circ} = 1 - \phi_{si} \qquad (4)$$

The calculated values for the average volume fraction occupied by the G8 dendrimers in the solvents $D(CD_2)_mOD$ (with m = 0, 1, 2, 4) are given in Table I (the uncertainties of ϕ are calculated on the basis of the standard deviation of R_g as given by the Guinier fits). The values of ϕ, as calculated from SANS, reveal a significant influence of the solvent quality on the segment density of the PAMAM dendrimers, which varies for the solvents studied in the range $0.42 \leq \rho^+ \leq 0.67$, as shown in Table I. Since the volume fraction of dendrimer segments varies as the cube of R_g, the effect is much greater than for the linear dimension alone.

Generation	Solvent	R_g / Å	R^+ / Å	ϕ_{di}
8	DOD	43.8 ± 0.1	56.5 ± 0.1	0.51 ± 0.1
8	$D(CD_2)OD$	42.0 ± 0.2	54.2 ± 0.2	0.58 ± 0.1
8	$D(CD_2)_2OD$	40.6 ± 0.1	52.4 ± 0.1	0.64 ± 0.01
8	$D(CD_2)_4OD$	39.5 ± 0.1	51.0 ± 0.1	0.70 ± 0.01
5 / 20 °C	$0 \leq x_s \leq 0.4$	22.1 ± 0.2	28.5 ± 0.3	0.49 ± 0.02

PAMAM dendrimers do not dissolve in acetone, but they readily dissolve in methyl alcohol/acetone mixtures over a wide range of composition. Solvents of different composition, expressed as mass ratios x_s (m(methyl alcohol)/m(acetone) in g/g), were prepared and added to a weighed amount of dried G5 or G8 dendrimer. The mixtures were then stirred and the solubility determined by visual inspection. It was found that the PAMAM dendrimers are soluble in the range $0 \leq x_s \leq 0.50$.

In Figure 3, the radii of gyration, R_g, is plotted as a function of solvent composition. We find that the size of the G5 dendrimer is uninfluenced by the solvent composition in the range $0.0 \leq x_s \leq 0.4$, which is in agreement with the data for G8. The values of R_g are larger for the

higher temperature, T = 50 °C, than at T = 20 °C, but the absolute value of the decrease of R_g is small, and does not exceed 5 % of the R_g.

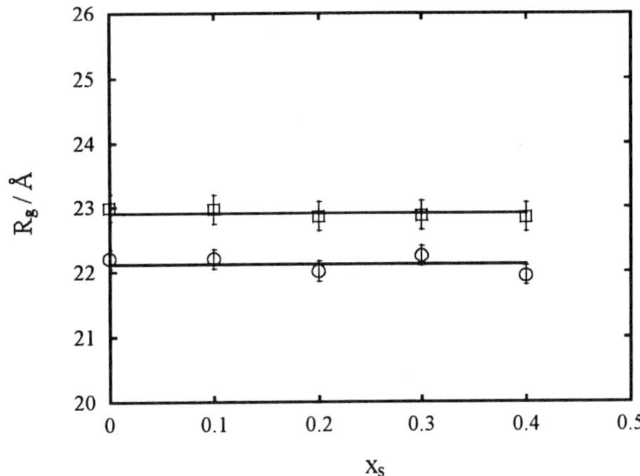

Figure 3. Radius of gyration of a G5 dendrimer in methanol-acetone mixtures; □ T = 50°C; O T = 20 °C.

Conclusions

The average dimensions of G8 PAMAM dendrimers, as studied by SANS, show a small dependence on the solvent quality. For the series of solvents $D(CD_2)mOD$ with m = 0, 1, 2, 4, the radius of gyration is reduced by $\approx$ 10% when changing the solvent from m = 0 to m = 4. Our results are not in agreement with the conclusions drawn from a study of the self diffusion coefficient of PAMAM dendrimers in different solvents.[3] Also, the absolute value of the variation of R_g as measured by SANS is much smaller than predicted by the computer simulations.[2]

The generations G5 and G8 of PAMAM dendrimers were studied in solvents composed of methyl alcohol and acetone within the range of solubility. No significant change of the average dimensions of G5 and G8 PAMAM dendrimer is found in solvent mixtures $0 \leq x_s \leq 0.4$.

Acknowledgments
A. Topp gratefully acknowledges the financial support by the Deutsche Forschungsgemeinschaft. This material is based upon work supported in part by the U.S. Army Research Office under contract number 35109-CH.

References

1. Prosa, T. J.; Bauer, B. J.; Amis, E. J.; Tomalia, D. A.; Scherrenberg, R. J., *J. Polym. Sci., Polym. Phys. Ed.*, **1997**, *35*, 2913.
2. Murat, M.; Grest, G. S. *Macromolecules.* **1996**, *29*, 1278.
3. Stechemesser, S.; Eimer, W. *Macromol.* **1997**, *30*, 2204.
4. Tomalia, D. A.; Naylor, A. M.; Goddard, W. A. *Angew. Chem. Int. Ed. Engl.* **1990**, *29*, 138.
5. *SANS Data Reduction and Imaging Software*; Cold Neutron Research Facility, NIST: Gaithersburg, MD 20899, 1996.

SAXS and Rheological Studies on the Order-Disorder Transition in Mixtures of Polystyrene-*b*-Polyisoprene-*b*-Polystyrene and Low Molecular Weight Polystyrene

Seung-Heon Lee and Kookheon Char

Department of Chemical Engineering, Seoul National University, Seoul 151-742, KOREA

Introduction

Block copolymer/homopolymer mixtures have been of considerable interest because they can be easily modified to yield desired properties in polymeric materials such as pressure sensitive adhesives (PSA). Phase behavior of these mixtures is generally quite complicated because two different natures of transition can occur at the same time: *macrophase* and *microphase* separation. If the molecular weight of a homopolymer is sufficiently low, it however tends to be solubilized into one of the microdomains.[1-3] In this case, the macrophase separation is suppressed and the microphase separation (or *order-disorder transition*) becomes dominant.

In present study, both small-angle X-ray scattering (SAXS) and rheological measurements were employed to investigate the order-disorder transition behavior in a series of SIS triblock copolymer/low molecular weight PS homopolymer mixtures, which did not show macrophase separation in the whole temperature and composition range covered in this experiment. Phase diagrams obtained from both measurements were compared with predictions based on the Whitmore-Noolandi theory[1]. The difference between the theory and the experiment was discussed in terms of the change in microdomain morphology with the addition of homopolymer to block copolymer.

Experimental

Two commercial grades of polystyrene-*block*-polyisoprene-*block*-polystyrene (SIS) copolymers, Vector 4113 and Vector 4411-D (DEXCO Polymers) were used as received. Polystyrenes (PS) with different molecular weights were synthesized *via* anionic polymerization method. The characteristics of these polymers are summarized in Table 1. 100/0, 90/10, 80/20, 70/30, 60/40 and 50/50 (w/w) mixtures of the block copolymers and the homopolymers were prepared by solution casting method with toluene as a solvent. All the samples were annealed in a vacuum oven at 130℃ for 3 days.

Two different methods were used to determine the order-disorder transition temperature (T_{ODT}) of each mixture. RMS 800 (Rheometrics Inc.) in parallel plate geometry (25 mm diameter and 1.5 mm gap) was used to measure the dynamic viscoelastic storage and loss moduli, G' and G", of the mixtures as a function of temperature.

Synchrotron small-angle X-ray scattering (SAXS) measurements were carried out at Pohang Light Source (PLS). The wavelength (λ) of the synchrotron beam was 1.54 Å and the energy resolution ($\Delta\lambda/\lambda$) was 5×10^{-4}. Typical beam size was smaller than 1×1 mm^2. Sample-to-detector distance was 89.1 cm. Scattering profiles were obtained as a function of temperature and then corrected for absorption, air and imide film scattering.

Table 1. Molecular characteristics of the polymers used in this study

Sample Code	$M_w\times10^{-3}$	M_w/M_n	PS wt %
Vector 4113	13.2-*b*-117.4-*b*-13.2	1.19	15.1
Vector 4411-D	17.1-*b*-80.0-*b*-17.1	1.08	41.7
PS2	1.9	1.09	100
PS3	2.9	1.06	100
PS6	5.9	1.04	100
PS11	11.4	1.03	100

Results and Discussion

It was found from the hot stage microscopy that V4411-D/PS mixtures did not show any macrophase separation up to 40 wt% PS regardless of PS molecular weight. V4113/PS mixtures with high molecular weight PS, however, showed macrophase separation when PS weight fraction exceeds a certain value. This is due to the fact that V4113 has a shorter styrene block and more space-restricted microdomain morphology, e.g., cylindrical morphology, than V4411-D, being less

capable of solubilizing large amount of PS homopolymer into its microdomain. It should be noted here that in present work we focussed only on the order-disorder transition behavior of the mixtures in composition range that did not show macrophase separation.

Figure 1 shows log G' vs. T plot of V4113/PS2 mixtures with different composition. For all the mixtures, G' shows a precipitous drop at a certain temperature. This temperature was chosen as the order-disorder transition temperature (T_{ODT}).

Figure 2 presents typical change in synchrotron SAXS profiles for 90/10 (w/w) V4113/PS2 mixtures with temperature. Maximum scattered intensity (I_{max}) shows a sharp decrease between 202.3℃ and 208.4℃. This discontinuity is more pronounced when $1/I_{max}$ was plotted against $1/T$ (not shown here). This temperature was chosen as the T_{ODT} of the mixture. The value of the T_{ODT} was found to be quite similar to the one determined from the precipitous decrease in log G' vs. T plot as shown in Figure 4.

Phase diagrams of V4411-D/PS and V4113/PS mixtures with different PS molecular weight obtained from both measurements are constructed in Figures 3 and 4, respectively. As shown in Figure 3, T_{ODT}'s of V4411-D/PS mixtures were found to decrease or increase depending on the PS homopolymer molecular weight as PS homopolymer weight fraction in the mixture was increased. This behavior is in good agreement with the calculated phase diagram based on the Whitmore-Noolandi theory[1] (inset of Figure 3). Phase diagram of V4113/PS2 mixtures is plotted in Figure 4, showing a very different behavior; T_{ODT} first decreases and then increases as PS2 weight fraction was increased up to 40 wt%. This unusual trend was not predicted from the Whitmore-Noolandi theory (inset of Figure 4). This is attributed to the change in microdomain morphology, e.g., from cylinders to lamellae, with increasing PS2 weight fraction of the mixture, as evidenced from the change in relative position of multiple SAXS peaks with increasing PS2 weight fraction in the mixture (Figure 5).

References

1. M. D. Whitmore and J. Noolandi, *Macromolecules* **18**, 2486 (1985).
2. H. Tanaka, H. Hasegawa, and T. Hashimoto, *Macromolecules* **24**, 240 (1991).
3. K.-J. Jeon and R.-J. Roe, *Macromolecules* **27**, 2439 (1994).

Acknowledgment

Experiments at Pohang Light Source (PLS) were supported by MOST and POSCO. We are very grateful to Y. J. Park for his assistance during SAXS experiments at PLS.

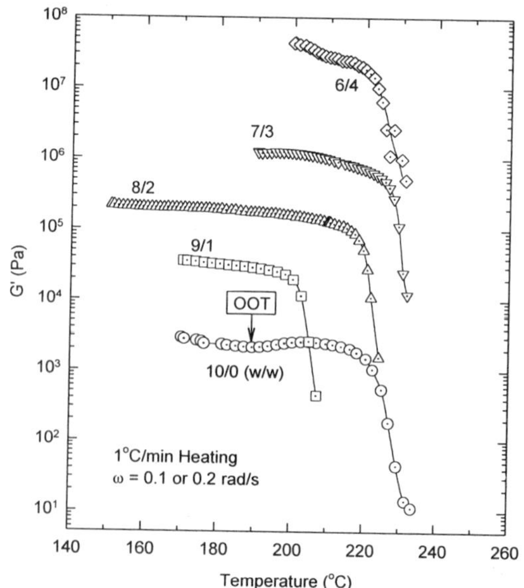

Figure 1. G' as a function of temperature for V4113/PS2 mixtures with different composition. Each data set was vertically shifted for clarification.

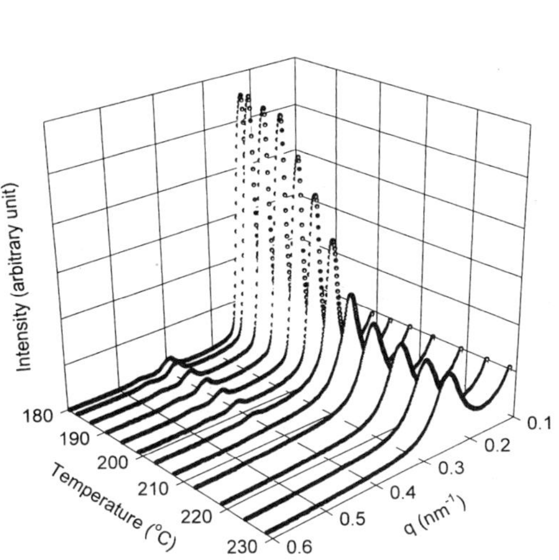

Figure 2. Synchrotron SAXS profiles for 90/10 (w/w) V4113/PS2 mixtures as a function of temperature.

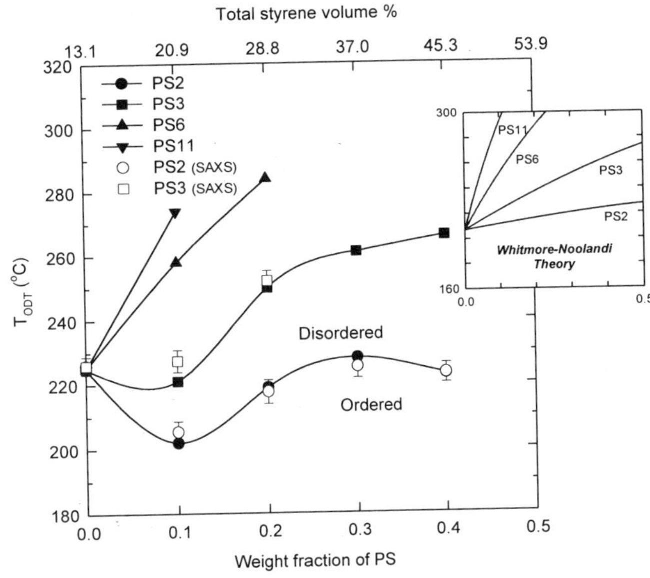

Figure 4. Phase diagram of V4113/PS mixtures. Filled symbols are T_{ODT}'s obtained from rheological measurements and open symbols are the ones obtained from SAXS measurements. Calculations based on the Whitmore-Noolandi theory are also shown in the inset.

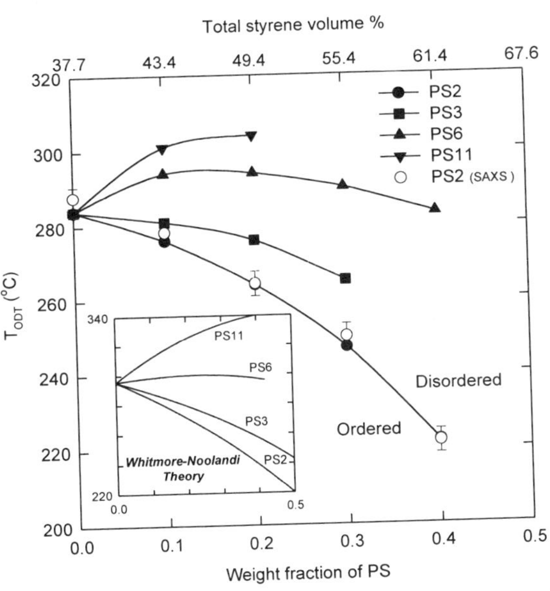

Figure 3. Phase diagram of V4411-D/PS mixtures. Filled symbols are T_{ODT}'s obtained from rheological measurements and open symbols are the ones obtained from SAXS measurements. Calculations based on the Whitmore-Noolandi theory are also shown in the inset.

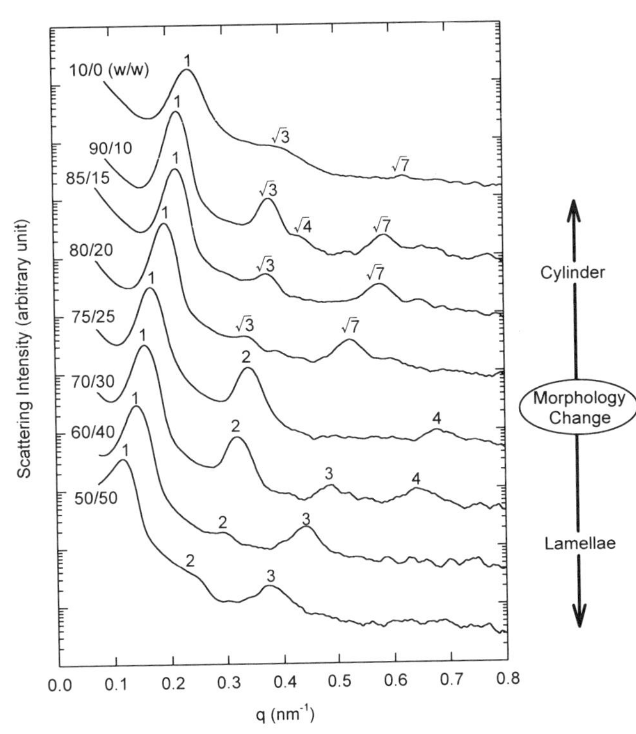

Figure 5. Scattering profiles of V4113/PS2 mixtures with different composition at 28 ℃. Each data set was vertically shifted for clarification.

Successive crystallizations from the melt followed by selective dissolution (SCM-SD) : a fractionation method for polyolefines

C. Vandermiers, P. Damman, M. Dosière*

Université de Mons - Hainaut, Laboratoire de Physicochimie des Polymères,

Place du Parc, 20 - 7000 Mons - Belgium

The heterogeneity of polymers results from a distribution of differing chain lengths, from differences in chemical composition from chain to chain and from the architecture of the chains. For copolymers of ethylene and α-olefines, the microstructure of the chains depends on the type of comonomers (1-butene, 1-hexene, 1-octene), the comonomer content also called the branch content, the distribution of the comonomer in the macromolecules and finally the molecular weight of the polymer. The successive solution fractionation (S.S.F.) is used to fractionate towards the molecular weight. The short chain branching is determined by the temperature rising elution fractionation. Both these fractionation techniques need long experimental times. A complete fractionation can been achieved in shorter times using successive crystallizations from the melt. A procedure based on differential scanning calorimetry has been recently developed to obtain information on the molecular structure and homogeneity of the chains of linear low density polyethylene.

Copolymers of ethylene and 1-hexene prepared either with metallocene or Ziegler-Natta (Z-N) catalyst were fractionated by a two steps procedure : multiple crystallizations from the melt followed by selective dissolutions (MCM-SD). The influence of the successive crystallizations from the melt on the fractionation has been checked on the Z-N copolymer of ethylene and 1-hexene. Two samples of the copolymer have been submitted to selective dissolutions : a sample crystallized from the melt at successive decreasing temperatures (T_c = 121, 115, 109, 102, 95, 85 and 75 °C) during 2 hours at each crystallization temperature and an untreated copolymer sample. The table contents the following measurements made on the soluble and insoluble parts corresponding for the different dissolution temperatures : weight losses, weight and number average molecular weights, the degree of polydispersity, the temperature and heat of melting. The branching content (τ) and the molecular weight distribution of the different fractions were determined by Fourier transform infra-red spectrometry and size exclusion chromatography, respectively.

Multiple successive crystallizations from the melt :

T_d/°C	84	86	88	90
Soluble				
wt loss/%	22	28	39	63
$<M_n>$	2931	11291	15450	21742
p	9.8	3.7	4.0	5.6
T_m/°C	115.0	118.8	122.1	124.0
ΔH_m/(J/g)	133.0	137.9	140.8	144.3
τ	0.348	0.224	0.202	0.132
Insoluble				
$<M_n>$	26188	21542	28665	34260
p	11.4	12.6	11.2	10.4
T_m/°C	-	-	-	125.3
ΔH_m/(J/g)	-	-	-	145.3
τ	0.085	0.066	0.105	0.096

Initial copolymer :

T_d/°C	83	85	86	87
Soluble				
wt loss/%	18.9	46.1	74.0	85.2
$<M_n>$	2441	14583	21087	22903
p	6.6	3.7	3.5	4.8
T_m/°C	117.9	123.1	123.5	124.5
ΔH_m/(J/g)	142.7	145.9	145.1	144.2
τ	0.329	0.166	0.128	0.121
Insoluble				
$<M_n>$	23285	19781	53644	46559
p	10.4	13.9	9.1	10.5
T_m/°C	-	-	-	125.9
ΔH_m/(J/g)	-	-	-	138.1
τ	0.120	0.096	0.070	0.056

The weight loss for both samples increases with the dissolution temperatures as expected. At a given dissolution the weight loss is lower for the sample previously submitted to multiple successive crystallizations than for the untreated sample. The average molecular weights of the fractions increase with the dissolution temperature. The degree of polydispersity of the various fractions is well lower than that of the untreated resine. The melting temperature of the fractions increases with increasing dissolution temperature. The branching content of the fractions decreases with increasing dissolution temperature. The insoluble fraction obtained at the highest dissolution temperature is exclusively a high molecular weight linear polyethylene as indicated by the melting temperature, the branching content and average molecular weight. All these results show that successive crystallizations from the melt coupled to successive dissolutions are a quick fractionation procedure for copolymers of ethylene and α-olefines.

In order to determine the influence of the degree of branching on the morphology of crystals, time resolved simultaneous small and wide angle X-ray scattering experiments were performed on the different fractions obtained from the sample initially crystallized from the melt at successive crystallization temperatures. The unit cell parameters (a and b), calculated from the wide angle X-ray diffraction increase as τ increases. As well known, butyl branches are too large to be included in the crystal. The cell deformations can thus be related to the presence of branches at the interface between crystalline and amorphous zones. The thermal expansion coefficient of the crystalline zone (α_c) computed from the time resolved WAXD measurements agree with this proposed interpretation.

From small angle X-ray scattering (SAXS) data, the crystalline and amorphous lengths and their evolution during the melting were estimated from the correlation functions for the different PE fractions. The SAXS integrated intensity (Q) has been also calculated and is related to the linear degree of crystallinity in lamellar stacks (v_c) and the difference between crystalline and amorphous densities ($\Delta\rho$) according to the following relation

$$Q = v_c \cdot (1 - v_c) \cdot \Delta\rho^2$$

As well known, the thermal contribution of the SAXS intensity can be estimated from the variation of crystalline and amorphous densities :

$$\Delta\rho(T) = \rho_c - \rho_a = \Delta\rho_0 + \Delta\alpha \cdot T$$

The SAXS intensity is thus given by the following relation :

$$\sqrt{Q(T)} = \sqrt{Q(T_0)} + \frac{\Delta\alpha}{\Delta\rho_0} \cdot \sqrt{Q(T_0)}$$

From $\Delta\alpha$ (equal to $\alpha_a - \alpha_c$), the thermal expansion coefficient of the amorphous phase (α_a) has been estimated against the branch content.

Acknowledgements : The authors thank the European Union for support of the work at the EMBL through the HCMP Access to Large Installations Program, contract no. CHGE-CT93-0040.

« X-Ray », a software for the analysis of two-dimensional WAXD and SAXS patterns of polymers recorded with image plates

M. Dosière*, D. Villers, F. Delava, O. Hernaut.
Université de Mons-Hainaut, Lab. de Physicochimie des Polymères,
Place du Parc, 20, B- 7000 - Mons (Belgique)
e-mail : dosiere@umh.ac.be

An integrating area detector, originally developed for diagnostic radiography, has been applied to X-ray diffraction experiments. The system is based on a photostimulable phosphor (BaFBr:Eu^{2+}) screen which can temporally store an X-ray image. The stored image is read out by measuring the intensity of luminescence around 390 nm stimulated by a He-Ne laser beam scanning the surface of the screen.[1] The area detector has more than 80% detective quantum efficiency for Cu Kα X-rays, a dynamic range of 1 : 10^4, a spatial resolution of 50 μm in two orthogonal directions. High quality wide-angle X-ray diffraction and small-angle X-ray scattering area patterns of polymers can be obtained with relatively low exposition times.[2] Such area detector is usually named « image plate ».

A software called « X-Ray » dedicated to the analysis of X-ray diffraction patterns of polymers recorded with an area detector has been developed in our laboratory. It includes classical image processing and graphics functions.

Its main purpose is to add specific functionalities which allow the analysis (or improve the speed and accuracy) of WAXD and SAXS data obtained on isotropic or oriented polymer samples : use of the scattering vector, Miller's indices, determination of lattice parameters, diffraction intensities, degree of crystallinity, orientation function, long period, crystal thickness,... Almost all applied procedures need the knowledge of the center of the pattern (the primary incident beam). Indeed, a good accuracy is required to avoid deviations of the Bragg spacing or smearing at the time of intensity integration. Therefore an automatic and accurate method to find centers has been implemented[3]. The program currently performs numerous analysis operations as follows : a) Settings, corrections, calibration, normalization; b) Diffraction commands; c) Classical image processing functions; d) One-dimensional profile analysis operations. A lot of these informations can also be obtained on other materials like inorganic or organic samples.

Almost all procedures applied on a diffraction image need the knowledge of the center of the image (the primary beam). A good accuracy (at least of the order of one pixel) is required to avoid systematic deviations of the Bragg spacing computations or smearing at the time of intensity integration. This is why an automatic and accurate method to find centers has been implemented[3].

The program currently performs numerous analysis operations categorized as follows :

i) Settings and corrections : sample to detector distance calibration using standards for WAXD and SAXS patterns, find and adjust center command, fiber axis adjustment, linear absorption, Lorentz and polarization corrections, normalization, background substraction, background reduction

ii) Diffraction commands : use of reciprocal lattice coordinates, integration of the entire image or of a sector (profile extraction), insertion of Miller indices on images or profiles, lattice fit,

iii) Image processing functions : contrast optimization, invert intensities, print and print preview, scroll functions, zoom, local magnifier, vertical and horizontal flip, rotate, reduce resolution, area selection, Sobel edge enhancement, median filtering.

iv) One-dimensional profile analysis operations : smooth curve (spline, mobile mean, Wienerfilter or Savitzky Golay method), substract function, substract profile, normalization, integrate peak or curve, crystallinity determination, get derivatives, find, add and remove maxima.

A demonstration version of « X-ray » is available on the Website of the Universté de Mons-Hainaut (Belgique) at the following address : http://www.umh.ac.be/~poly/XRay/X-Ray.htm. If you need additional informations, please mail to : xray@umh.ac.be

References :

[1] Y. Amemiya, Y. Satow, T. Matsushita, J. Chikawa, K. Wakabayashi, J. Miyahara, Topics in Current Chemistry, Springer-Verlag, Berlin Heidelberg, vol. 147, 122 (1988).

[2] D. Villers, C. Fougnies, L. Paternostre, C. Beumier, M. Dosière, Nucl.. Instr. Meth. B97, 265 (1995).

[3] C. Dammer, P. Leleux, D. Villers and M. Dosière, Nucl. Instr. Meth. B 132, 214 (1997).

Crystalline Structure and Morphology of Microphases in Compatible Mixtures of Tetrahydrofuran (PTHF) -Methyl Methacrylate (PMMA) Diblock Copolymer and Polytetrahydrofuran

Li-Zhi Liu and Benjamin Chu[*]
Department of Chemistry,
State University of New York at Stony Brook
Stony Brook, Long Island, New York 11794-3400

Introduction

The microphase separation, crystallization and intriguing morphologies of compatible mixtures of tetrahydrofuran-methyl methacrylate diblock copolymers (PTHF-b-PMMA) with a tetrahydrofuran homopolymer (PTHF) were studied previously.[1-2] For the blends with PMMA spheres, it was found that the long period of the PTHF microphase increased with increasing PMMA weight fraction, presumably because the presence of PMMA microdomains could slow down the PTHF, resulting in a larger long period. The long period dependence on the isothermal crystallization temperature could be different for the blends with different separation distance between two neighboring PMMA microdomains. In the case of an alternating PTHF and PMMA lamellar morphology, no microphase change occurred during the crystallization, because the glass transition temperature (T_g) of PMMA microdomains is ~ 373 K, which is much higher than the melting point of the PTHF component (~ 310 K). The morphology study[2] of the blends shows that the crystalline morphology changes dramatically with blend composition, copolymer composition and PTHF block length, as well as the crystallization temperature. Many unusual crystalline morphologies have been observed.

In the present work, a further study has been focused on the crystalline structure in the confined lamellar region and a PTHF matrix with hexagonally packed PMMA cylinders, as shown schematically in Figure 1 by using wide angle x-ray diffraction (WAXD). In the present work, a further study has also been focused on the effect of PMMA cylinders on the crystallization of PTHF matrix, and on the effect of the PTHF crystallization on the microphase separation structure.

Experiment

Polymers and Characterization. The synthesis of tetrahydrofuran-methyl methacrylate diblock copolymers (PTHF-b-PMMA) and the purification have been described elsewhere.[3] A summary on the pertinent parameters of the PMMA-b-PTHF diblock copolymers and the PTHF homopolymer used in the present work is given in Table 1.

Preparation of Blends. A series of blends of the copolymer 1TM and low-molecular-weight homopolymer PTHF, and a series of blends of the copolymer 2TM and the homopolymer were prepared by solution casting from chloroform. Compositions of the blends, including W(H-PTHF) denoting the PTHF homopolymer weight fraction,, W(THF in blend) denoting the total weight fraction of THF in both block and homopolymer, and W(H-PTHF in THF phase) denoting the weight fraction of homopolymer PTHF in the THF phase under the assumption of strong segregation limit (see discussions below) are listed in Table 2 for each of the blends.

Small Angle X-ray Scattering (SAXS) and Wide Angle X-ray Diffraction. The scattering experiments were performed at the SUNY X3-A2 beamline of the National Synchrotron Light Source(NSLS), Brookhaven National Laboratory(BNL). Fuji imaging plates were used to collect the scattering data with exposure times of 2 min per frame for SAXS and 1 min per frame for WAXD.

Results and discussion

1. WAXD Study of Crystalline Stucture in PTHF Microdomains. The microphase morphology for each individual blend in the melt state (above the melting temperature of PTHF) was determined on the basis of scattering peak positions from SAXS and was listed in Table 3. Our previous SAXS study[1] showed that when the samples with PMMA lamellae or cylinders in the melt state were cooled down to crystallize, the microphase morphology and interdomain distance remained unchanged, because the glass transition temperature of PMMA lamellae or cylinders is much higher than the melting point of PTHF component. Therefore, the crystallization of the PTHF block and PTHF homopolymer occurs in the existed PTHF microdomain. WAXD patterns of the blends 1TM/T1 to 1TM/T5, as well as of the neat copolymer 1TM, are shown in Figures 2 and 3. From these two figures, the following diffraction features are observed. (1) No diffraction peak is seen for the pure copolymer 1TM. (2) Two sharp diffraction peaks are seen for the 1TM/T1 blend with as little as 10 wt % of PTHF homopolymer. (3) The diffraction peak position shifts to a higher θ-range with an increase in the PTHF homopolymer content of the blend. (4) The full width at half-maximum intensity of the diffraction peak at $2\theta \sim 19.7°$ remains almost unchanged with an increase in homopolymer weight fraction in the blends. However, the full width at half-maximum intensity of the diffraction peak at $2\theta \sim 24°$ decreases with an increase in the PTHF homopolymer content of the blend. (5) The diffraction intensity of (202) peak increased progressively with increasing PTHF homopolymer content, while the diffraction intensity of (002) peak showed a very large increase when the homopolymer content was increased from 10 to 20 wt %, and then remained almost constant with further increase in the homopolymer content. According to the above diffraction features, the suggested chain folding model in the confined lamella had a fiber identity along the normals of the PMMA lamella.

2. SAXS Study of Crystallization in PTHF Matrix dispersed with PMMA Cylinders. In the present work, some new scattering features have been observed for a series of blends with hexagonally packed PMMA cylinders. The higher-order scattering peaks seemed to shift to lower q values after crystallization. The $(4)^{1/2}$ higher order peak disappeared and the $(3)^{1/2}$ higher order peak appears at a lower q range, as shown in Fig. 3. The $(7)^{1/2}$ higher order peak seemed to have shifted to a lower q value (Fig. 3). By comparing the inset figure in Figure 3 and Figure 4 for the scattering profiles on a linear scale, it can be seen that the scattering contribution from PTHF crystals is much larger in the blend 2TM/T4 than in the blend 2TM/T3, due to the higher PTHF homopolymer content in the former sample, which increases the intersurface distance L between two neighboring PMMA cylinders (see Fig 1). After crystallization, two large scattering shoulders at q ~ 0.4 and 0.5 nm^{-1} are observed (Fig. 4), indicating the formation of ordered PTHF lamellar stacks. After annealing for a month, the shoulder at a low q range changed into a sharp peak, indicating that further ordering of the stacks during annealing, and that the PTHF lamellar stacks with a long period of about 16 nm ($\approx 2\pi/q_m$) were formed in the matrix microphase. Since the intersurface distance L of PMMA cylinders (Figure 1) for the 2TM/T4 was only 15.8 nm, the direction of the stack can only be along the direction of the cylinder axis. The broad scattering shoulder at q ~ 0.5 suggested that besides the thick lamellar stacks, there could also be some PTHF irregular microcrystals formed in the PTHF matrix.

Refferences

(1) Liu, L.-Z.; Yeh, F.; Chu, B. *Macromolecules* 1996, 29, 5335.
(2) Liu, L.-Z.; Xu, W.; Li, H.; Su, F.; Zhou, E. *Macromolecules* 1997, 30, 1363.
(3) Liu, L.-Z.; Li, H.; Jiang, B.; Zhou E., Polymer 1994, 35, 5511.

Table 1. Characterization of PTHF-b-PMMA Diblock Copolymers and PTHF Homopolymer

Sample[a]	$Mn^b \times 10^{-3}$	W^c(PTHF)	Mn (PTHF Block) $\times 10^{-3}$	Mn (PMMA Block) $\times 10^{-3}$
1TM	23.3	0.30	7.0	16.3
2TM	14.0	0.50	7.0	7.0
PTHF homopolymer			2.0	

a, T and M denote abbreviations PTHF and PMMA, respectively;
b, Mn: number average molecular weight.
c, W: weight fraction

Table 2. Compositions of PTHF-b-PMMA/PTHF Blends

Designation	W(H-PTHF)	W(THF in blend)*	W(H-PTHF in THF phase)
1MT/T1	0.1	0.37	0.27
1MT/T2	0.2	0.44	0.45
1TM/T3	0.3	0.51	0.59
1TM/T4	0.4	0.58	0.69
1TM/T5	0.5	0.65	0.77
2TM/T2	0.2	0.60	0.33
2TM/T3	0.3	0.65	0.46
2TM/T4	0.4	0.70	0.57

* including PTHF homopolymer and PTHF block of the copolymer

Table 3 Microphase morphology of blends studied

Designation	W(THF in blend)*	Microphase morphology
1MT/T1	0.37	Lamella
1MT/T2	0.44	Lamella
1TM/T3	0.51	Lamella
1TM/T4	0.58	PMMA cylinder
1TM/T5	0.65	PMMA cylinder
2TM/T2	0.60	PMMA cylinder
2TM/T3	0.65	PMMA cylinder
2TM/T4	0.70	PMMA cylinder

● including PTHF homopolymer and PTHF block of the copolymer

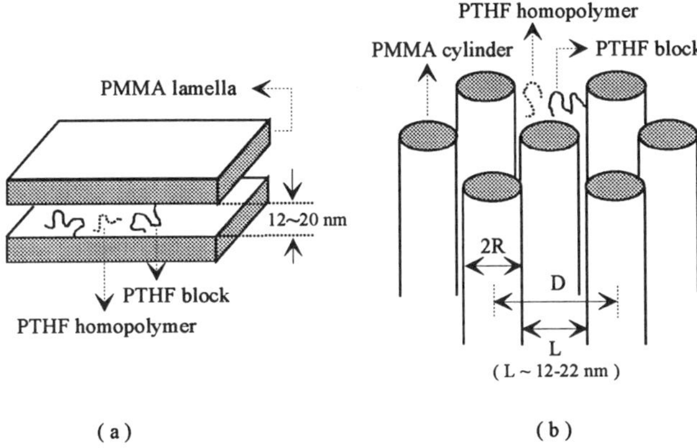

Figure 1.   Schematic figures of blend system with different microphase separation structure. (a) alternating PTHF and PMMA lamellae; (b) a PTHF matrix with PMMA dispersed as cylinders arranged in hexagonal form (the definitions of R, L and D are also shown in this figure).

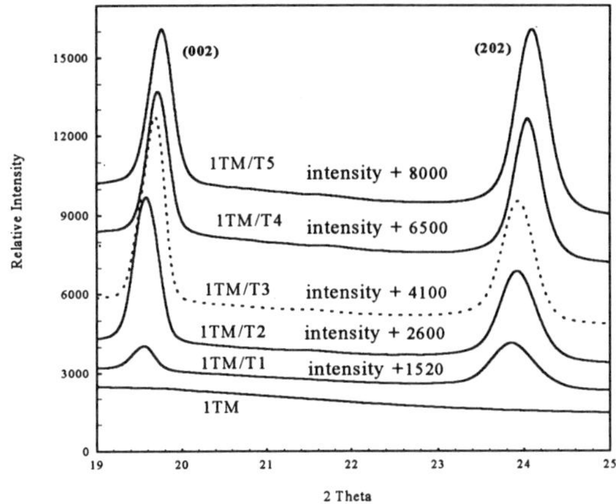

Figure 2.   WAXD patterns of 1TM pure copolymer and of its blends with 10 to 50 wt % of PTHF homopolymer.

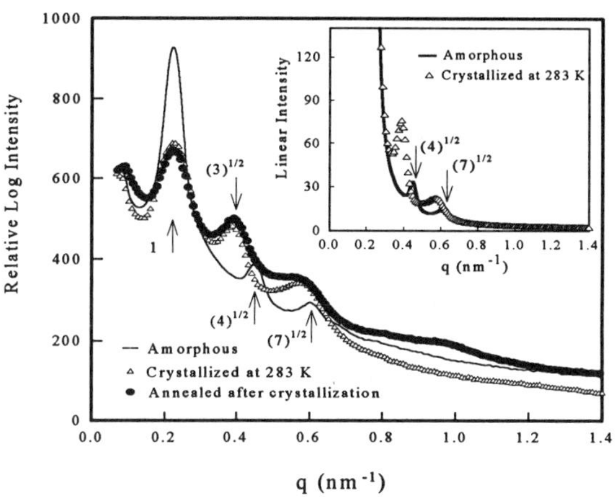

Figure 3.   SAXS profiles of the blend 2TM/T3.

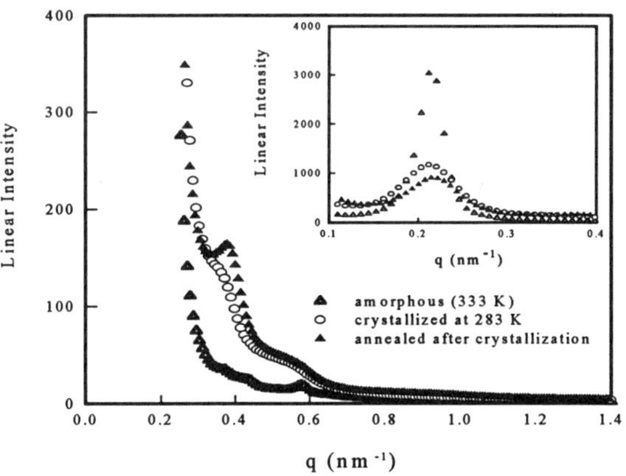

Figure 4.   SAXS profiles of the blend 2TM/T4.

The Nature of Secondary Crystallization in Poly(ethylene terephthalate)

Zhi-Gang Wang, Benjamin S. Hsiao*, Bryan B. Sauer[+] and William G. Kampert[+]

Chemistry Department, State University of New York at Stony Brook, Stony Brook, NY 11794-3400

[+]DuPont Central Research & Development, Experimental Station, Wilmington, Delaware 19880-0356

Introduction

The term secondary crystallization is frequently used to refer to all effects that can increase the crystallinity after primary crystallization, describable by an Avrami equation, is over [1]. In this study, we intend to probe the fundamentals of secondary crystallization in poly(ethylene terephthalate), PET. The selection of PET is due to the existence of the abundant data about the subject [2,3], and a controversy over the determination of lamellar parameters during crystallization by synchrotron small-angle x-ray scattering (SAXS) methods [4]. In PET, it has been demonstrated that secondary crystallization has an inherent relationship with the multiple melting behavior induced by isothermally annealing [5]. Synchrotron SAXS results also support the argument that secondary crystallization produces a population of thinner lamellae that melt at lower temperatures upon heating [6].

The controversy over synchrotron SAXS results stems from the use of correlation function to determine the structural parameters and the low crystallinity in PET (usually < 40%). As the correlation function method can resolve two thickness values for crystal and amorphous phases in the lamellar stacks [7], it cannot assign the value to the proper phase. As the crystallinity in PET is low, many people tend to assign the smaller value as the crystal thickness, since the product of the long period (L, from SAXS) and bulk crystallinity (ϕ_c from WAXD or DSC) is about equal to this value. However, such an assignment suggests the morphology consists of space-filling of a uniform distribution of lamellar thickness, which leads to a great deal of confusion in understanding the nature of primary and secondary crystallization by time-resolved SAXS.

In this paper, we have revisited the issue by prolonged annealing of a low molar mass PET, producing a relatively high crystallinity of about 50%. Synchrotron SAXS was used to monitor the structural development during the entire process. It is our goal to illustrate that it is the larger value that represent the crystal thickness, as well as to understand the trend of the morphology changes with time. DSC measurement was also applied to determine the thermal behavior under the same thermal conditions.

Experimental

The PET sample is an experimental grade material provided by DuPont Company, which has a number-average molecular weight of 25,000 and a polydispersity Mw/Mn= 2. The glass transition temperature Tg and the melting temperature Tm of this sample were 80°C and 270°C, respectively.

SAXS measurement during isothermal crystallization was carried out using a Braun linear position sensitive detector at the Beamline X3A2 of the National Synchrotron Light Source (NSLS), Brookhaven National Laboratory. The wavelength λ used was 1.28 Å. An evacuated flight path was used and the sample to detector distance was 1345 mm. Isothermal crystallization of PET was performed by a dual-chamber temperature jump unit [5]. After equilibrated at 280°C (10°C above the melting point) for 5 min, the PET sample was rapidly jumped to a second chamber (aligned in the path of the x-ray beam) at temperature 231°C for measurements. The acquisition time was 20 s for each scan. A wait time of 40 s was used between the scans during the first 4 hours of collection. Isothermal crystallization was continued for another 68 hrs without data acquisition, then resumed again for 30min of measurements. The subsequent melting process with a heating rate of 5°C/min was also carried out after the prolonged annealing.

The thermal behavior of the PET sample prepared under above thermal conditions was characterized by Perkin-Elmer DSC-7 in a nitrogen atmosphere. The instrument was calibrated with indium standard. The sample was crystallized at 231°C for a period of 2hr, 24hr and 72hr, respectively, and heating scans were gathered directly from 231°C to 280°C at a rate of 5°C/min.

Results and Discussion

Typical procedures for calculating the morphological parameters for semicrystalline systems from the SAXS data have been discussed earlier [6]. The procedures involve the use of the correlation function and the interface distribution function to extract morphological parameters such as long spacing (L), lamellar thickness (l_c) and amorphous interlayer thickness (l_a), etc. For PET crystallized at 231°C, results from the correlation function are shown in Figure 1. In this figure, it is seen that both L and l_1 decrease continuously with time where l_2 remains about constant. Our reason for assigning the larger value l_1 as the crystal thickness is as follows. After annealing for 72hr, the average final value of L is 108 Å, l_1 is 64.4 Å and l_2 is 43.6 Å. Let us consider both situations: either χ_{L1} (=l_1/L) (59.6%) or χ_{L2} (=l_2/L) (40.4%) may represent the linear degree of crystallinity within the lamellar stacks. Since the corresponding ϕ_c (the degree of crystallinity determined from the DSC data) at the same temperature and the same time scale is 45.5%, the assignment of l_2 as the lamellar thickness can not meet the criterion of mass balance for the crystallinity. This is because the volume fraction of lamellar stacks in bulk sample becomes larger than 1 in this case. On the other hand, the assignment of l_1 as the lamellar thickness makes the value of volume fraction for lamellar stacks 76.3%. This is reasonable, which indicates that the sample is not completely spaced-filled with lamellar stacks, and domains of noncrystallizable components (impurity, branched species…) persist even after the prolonged annealing.

In Figure 1, it is seen that both long period L and lamellar thickness l_c decrease with time. To examine these effects, we have plotted the evolution of L, l_1 and l_2 with log time during the fist 4 hrs of measurement in Figures 2a and 2b. In Figure 2a, the scattering invariant Q is also included. We have defined the primary crystallization dominant stage as the regime ($t < tc$, marked with the symbol *) where the invariant Q exhibits a sigmoidal increase. In this stage, both values of L and lc (= l_1) exhibit a large decrease that is approximately linear with the log time. We attribute this behavior to the occurrence of the thinner secondary lamellae within (the lamellar insertion model) or between (the stack insertion model) the existent primary lamellar stacks (the average value of L and lc decrease as the secondary lamellae increase) during the primary growth. Since the changes in la (=l_2) is relatively small, we favor the stack insertion model for the formation of thinner secondary lamellae. At the later stage after primary crystallization, L and lc exhibit a further decay with a smaller rate that is also approximately linear with log time (see Figure 2b). We attribute this stage to the secondary crystallization dominant phase where the event of the primary lamellar production is absent. The slower rate of decreases in L and lc indicating the production of very defective crystallites, which is consistent with the small increase of degree of crystallinity ϕ_c (DSC) induced by annealing (see the black circle in Figure 1). On the basis of this observation, it can be concluded that the amount of secondary lamellar stacks continues to increase linearly with log time, after primary crystallization.

The corresponding DSC scans of samples annealed at different times (2 hr, 24 hr and 72 hr) are shown in Figure 3. For the 2 hrs annealed sample, a double melting behavior is seen. Prolonged annealing at 24 hr or 72 hr eliminates the double melting behavior and drives up the melting temperature (T_m = 253°C, 263°C and 268°C for 2hr, 24hr and 72hr, respectively). This suggests that the perfection of the crystallites (primary and secondary) probably increase. The heat of fusion under different DSC thermogram increases with the annealing time, and the resultant crystallinity is shown in Figure 1.

Conclusion

The morphological parameters in semicrystalline PET homopolymer during prolonged isothermal crystallization were followed by synchrotron small-angle x-ray scattering (SAXS) and DSC measurements. From the large crystallinity (~50%) generated by annealing, we can unambiguously determine that the larger thickness calculated from the SAXS correlation function analysis is the crystal lamellar thickness. The variation of long period, lamellar thickness and amorphous interlayer thickness during isothermal crystallization is in favor of the lamellar stacks insert model. Results indicate two stages of decrease in long period and lamellar thickness. The decrease in the first stage (primary crystallization dominant phase) is large, the decrease in the second stage (secondary crystallization dominant phase) is small. These decreases are due to secondary crystallization which produces thinner lamellar thickness causing the average value to drop. From DSC, it is evident that prolonged annealing improves the crystal perfection.

Acknowledgement

The authors acknowledge the financial support of this work by a grant from NSF (DMR 9732653).

References

1. B. Wunderlich, *Macromolecular Physics*, volume 2, (1976)
2. J. M. Schultz and S. Fakirov, "Solid State Behavior of Linear Polyesters and Polyamides", (1990).
3. M. Imai, K. Kaji, Kanaya, T and Y. Sakai, Phys. Rev. B, **1995**, 52(17), 12696.
4. R. K. Verma and B. S. Hsiao, TRIP, Vol.4, No.9, **1996**.
5. F. J. Medellin-Rodriguez, P. J. Phillips; J. S. Lin, *Macromolecules*, **1996**, *29*, 7491
6. R. Verma, H. Marand, B. Hsiao, *Macromolecules*, **1996**, *29*, 7767C.
7. G. R. Strobl, M. Schneider, *J. Polym. Sci., Polym. Phys. Ed.*, **1980**, *18*, 1343

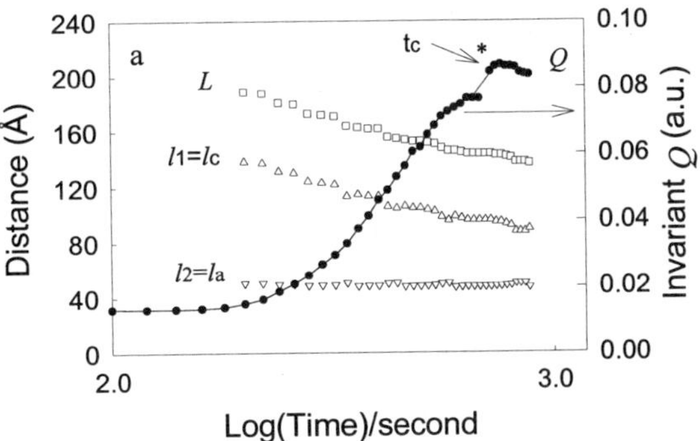
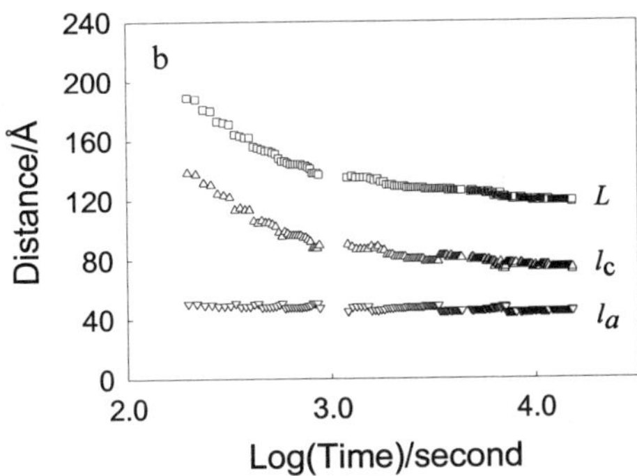

Figure 2. Time evolution of long period (L), lamellar thickness(l_c) and amorphous layer thickness (l_a) during short period (a) and long period (b) for PET at 231°C.

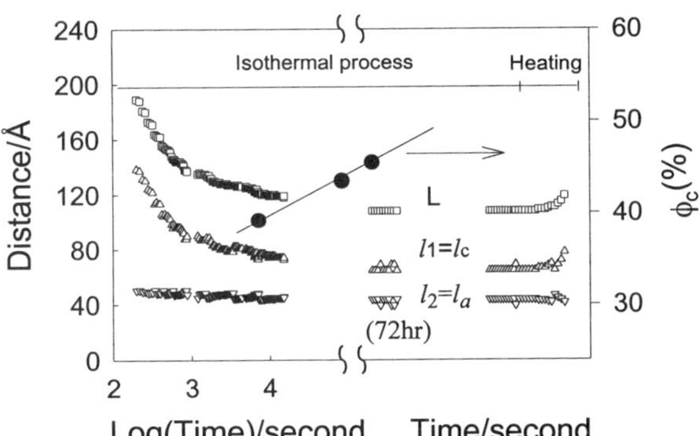

Figure 1. Time evolution of long period(L), lamellar thickness (l_c) and amorphous layer thickness (l_a) form SAXS data and the degree of crystallinity (ϕ_c) from DSC measurement for PET at 231°C.

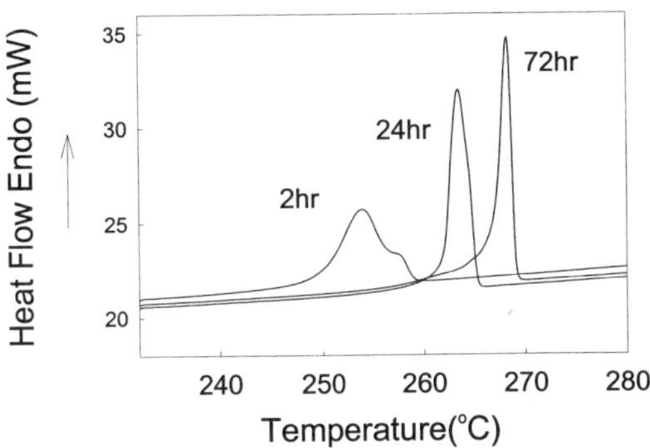

Figure 3. DSC traces of PET after isothermal crystallization at 231°C for 2hrs, 24hrs and 72hrs.

Real-Time SAXS Study of Crystallization of Metallocene Poly(propylene)

Patrick Shuanghua Dai[1], Peggy Cebe[1*],
Rufina G. Alamo[2], Leo Mandelkern[3],
and Malcolm Capel[4]

[1] Department of Physics and Astronomy, Science and Technology Center, Tufts University, Medford, MA 02155

[2] FAMU/FSU College of Engineering, Department of Chemical Engineering, Tallahassee, FL 32306

[3] Institute of Molecular Biophysics and Department of Chemistry, Florida State University, Tallahassee, FL 32306

[4] Department of Biology, Brookhaven National Laboratory, Upton, NY

* To whom correspondence should be addressed

1. INTRODUCTION

The origin of the multiple endothermic response in isotactic polypropylene (iPP) is a complex issue. Contributions to multiple endotherms may arise from: different crystal modifications[1,2]; different levels of perfection such as primary (early growing) and secondary (later growing) crystals; different structures such as crosshatched lamellae [3,4]; or, melting and recrystallization during DSC scanning[5,6]. Here we report a study of isothermal crystallization and subsequent melting in iPP which was synthesized using homogeneous transition metal catalysts (metallocenes). Our previous thermal analysis study of this m-iPP was concerned with non-isothermal crystallization after partial melting[7]. In the present study, we use both small angle X-ray scattering, SAXS, and differential scanning calorimetry, DSC, to follow the development of lamellar structure after isothermal crystallization. SAXS can be used to study the kinetics of crystallization, and also provides information about the developing structures that contribute to the observed multiple melting endotherms.

2. EXPERIMENTAL SECTION

The m-iPP used in this study is an experimental product of Hoechst received as pellets. Characterization of the material [8] shows that the isotacticity is low at 90.5 mol-%, M_w is 335,500, and polydispersity is 2.3. Material was formed into films by compression molding at 200°C between Kapton^{TM} films and then quenched to room temperature. Wide angle X-ray scattering results indicate the quenched material is predominantly monoclinic alpha phase [9].

Thin films were encapsulated in Al pans and thermally treated inside the cell of a TA Instruments model 2920 DSC. Heating and cooling rates were 10°C/min, sample mass was about 9 mg, and the heat flow and temperature were calibrated using an Indium standard. Nitrogen was used as a protection gas (30 ml/min.) and no thermal degradation was detected. The m-iPP was melted at 200°C for 1 min., then cooled to 117°C or 124.5°C for isothermal crystallization. After crystallization was completed, the sample was immediately re-heated (without cooling) to observe the melting endotherms.

Real-time SAXS experiments were performed at beam line X12b of the National Synchrotron Light Source at Brookhaven National Laboratory. Monochromatic X-radiation with a wavelength of 1.54Å was used. The system was equipped with a two-dimensional position sensitive histogramming detector. The sample to detector distance was 172.7 cm. The m-

iPP sample (thickness 0.5mm) was located inside the Mettler FP 80 hot stage. SAXS data were taken continuously during the isothermal crystallization periods and the melting periods. Each SAXS scan was collected for 20 sec. or 60 sec.. The sample was melted at 200°C for 1 min., then quenched to 117°C or 124.5°C, and isothermally crystallized for various time Since the samples were isotropic, circular integration was used to improve the signal to noise ratio. From the Lorentz corrected intensity (Iq^2) vs. scattering vector, q ($q = 4 \pi \sin\theta/\lambda$) structural parameters are determined, according to the method of Strobl and Schneider [10].

3. RESULTS AND DISCUSSION

Figure 1 shows the DSC exothermic heat flow during isothermal crystallization at 117°C. At this temperature, crystals develop fairly quickly, with a half crystallization time of about 3 min. Little exothermic heat flow is seen past 15 min.

During isothermal crystallization at 117°C, when we crystallize m-iPP for 1, 2, 3, 4 and 5min., and immediately rescan in the DSC, the subsequent melting curve always shows two growing endotherms. When the crystallization time is prolonged to 10, 20 and 40min., double melting peak are still observed. Figure 2 shows the melting endotherm after isothermal crystallization at 117°C for the times indicated. The sharper upper melting endotherm, melting at T_{m2}, is barely changed: its peak position shifts slightly to higher temperature with increase of crystallization time, while its area is about the same. The broad first endotherm, melting at T_{m1}, increases in area, and becomes sharper, with the crystallization time. This suggests that there are two populations of crystals within the sample even at early times in the crystallization. One population, the one with lower melting temperature, continues to increase in relative proportion as the crystallization time is increased. To gain further insight into the origin of these two endotherms, X-ray scattering experiments were performed both isothermally and during subsequent heating. Only the former results are descirbed here; the latter results on SAXS taken during heating will be detailed in a future publication.

Real-time SAXS was also used to follow the kinetics of crystallization for temperatures within the range from 115°C to 124.5°C. Figure 3 shows the time development of the scattering invariant, Q, which is the integrated area under the Lorentz corrected intensity curve for two temperatures. Isothermal crystallization at 117°C is shown in the upper curve (triangles) while 124.5°C is shown in the lower curve (circles). The differences in crystallization rate between the two samples is clearly illustrated. The 117°C data show Q increasing steeply within the first 3 minutes, then increasing more gradually, and finally reaching a constant value after about 15 min. On the other hand, the 124.5°C data show that Q evolves slowly. It takes about eight minutes of holding time at 124.5°C before sufficient lamellar structure develops within the sample to cause Q to increase above zero. A gradual increase occurs over the next 30 min. Longer times are not shown in Figure 3, but for 124.5°C, Q does not flatten completely even after 80 min.

Figure 4 shows the Lorentz corrected intensity vs. scattering vector at the isothermal crystallization temperatures of 124.5°C after 30 min. (circles) and 117°C after 45min (stars). A limited q range is shown in order to expand the region near the intensity peak. In the intensity profile for 117°C, two poorly resolved peaks are observed, while for 124.5°C, a shoulder is seen on the high q side of the main peak. The 117°C curve(stars) shows peaks of about equal height after 45 min. isothermal crystallization. One peak, at lower q (greater Bragg spacing), appears even at short times in isothermal period, prior to the half-time. Once the half-time of 3 min. is exceeded, additional intensity grows up at higher q (shorter Bragg spacing), forming a distinct, but weak, second peak. Nearly the same trend is seen in the 124.5°C intensity. For the 124.5°C sample, the crystallization half-time is much longer (about 20min.) than for 117°C.

These results suggest that within the isothermally crystallized m-iPP sample there are two lamellar populations, with different average Bragg spacings. The second population, having shorter Bragg spacing, increases in relative proportion as crystallization time increases. The increase in SAXS intensity parallels the increase in the area of the lower melting endotherm with increasing crystallization time. SAXS scans taken during heating, but not shown here, indicate that the intensity in the higher q region (lower Bragg spacing) decreases when the temperature reaches that of the lower melting endotherm. We conclude that different lamellar populations contribute to the multiple endothermic response seen in this m-iPP after isothermal crystallization.

ACKNOWLEGMENT

Research was supported by the U. S. Army Research Office, Grant DHHA04-96-1-0009.

REFERENCES

1. J. Varga. *J. Thermal. Analy.* **35**, 1891 (1989).
2. F. L. Binsbergen and B. G. M. De Lange. *Polymer,* **9**, 23 (1968).
3. G. Guerra, V. Petraccone, P. Corradini, C. Derosa, R. Napolitano, B. Pirozzi, and G. Giunchi. *J. Polym. Sci.: Phys. Ed.* **22**, 1029 (1984).
4. R. J. Samuels. *J. Polym. Sci. Phys. Ed..* **13**, 1417 (1975).
5. Y. S. Yadav, P. C. Jain. *Polym.,* **27**, 721 (1986).
6. V. Petraccone, G. Guerra, C. de Rosa, and A. Tuzi. *Polym.* **28**, 143 (1985).
7. P. Dai, P. Cebe, R. G. Alamo, L. Mandelkern and M. Capel. Polym. Preprint (1997).
8. J. R. Isasi, R. G. Alamo, and L. Mandelkern. *J. of Polym. Sci: polym. Phys. Ed.* **35**, 2521 (1997).
9. R. G. Alamo, J. C. Lucas, and L. Mandelkern. Polym. Preprints, **35,** 406, (1994)
10. G. R. Strobl and M. Schneider. *J. Polym. Sci., Polym. Phys. Ed.,* **18**, 1343 (1980).

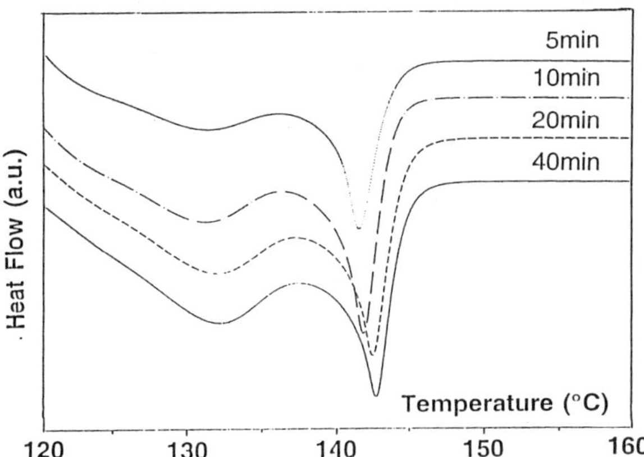

Figure 2 The DSC melting endotherms after isothermal crystallization at 117°C for 5 min, 10 min, 20 min and 40 min as indicated.

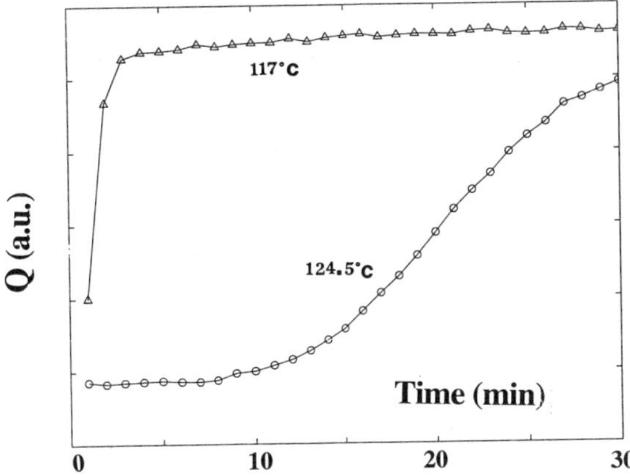

Figure 3 Time development of the scattering invariant, Q, during isothermal crystallization at 117°C (the upper curve) and 124.5°C (the lower curve).

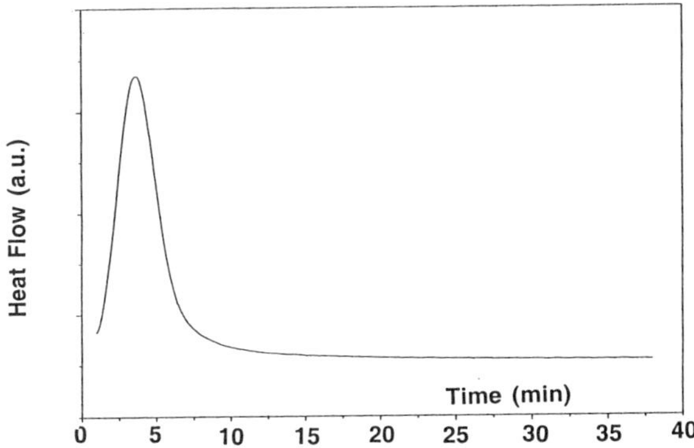

Figure 1 DSC exothermic heat flow vs. time during isothermal crystallization at 117°C.

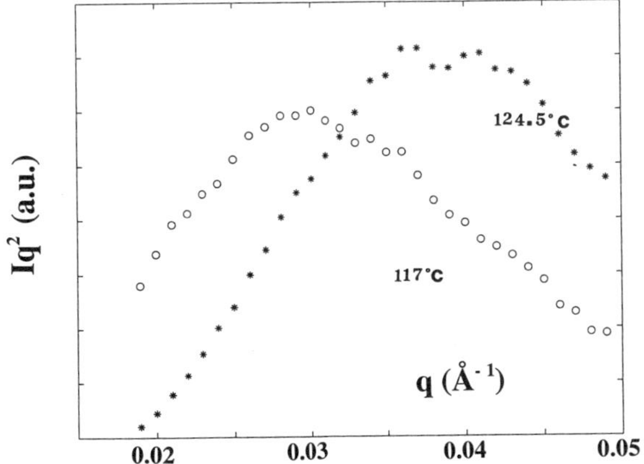

Figure 4 Lorentz corrected intensity vs. scattering vector at the isothermal crystallization temperatures of 124.5°C after 30 min. (circle) and 117°C after 45 min. (star).

MORPHOLOGY DEVELOPMENT IN POLYPROPYLENE HOMOPOLYMER TACTICITY MIXTURES: ISOTACTIC/SYNDIOTACTIC BLENDS

Sangkyu Kim, Zhi-gang Wang, <u>Roger A. Phillips</u>[1],

Fengji Yeh and <u>Benjamin S. Hsiao</u>

Chemistry Department, State University of New York at
Stony Brook, Stony Brook, NY 11794-3400
[1]Montell Polyolefins, R&D Center, 912 Appleton Road,
Elkton MD. 21921

INTRODUCTION

It is known that the stereochemisty of polypropylene strongly influences material properties through morphological factors such as crystallinity, melting temperature and crystal arrangements. Metallocene catalysts allow for the development of new tacticity microstructures [1], and provide the only viable route to syndiotactic polypropylene (sPP) [2]. Relative to isotactic polypropylene from Ziegler-Natta catalyst (ZN-iPP), sPP produced by these metallocene catalysts generally shows lower crystallinity, lower modulus and yield stress, lower crystallization temperature, lower melting point, and reduced flowability [1]. These attributes have limited the market impact of sPP relative to iPP. However, when used in conjunction with ZN-iPP, sPP can potentially be used to modify both rheological and solid-state processing characteristics.

Recent studies have suggested thermodynamic incompatibility of the iPP/sPP blend pair at relatively modest molecular weights [3,4]. Liquid-liquid phase separation in the melt has been claimed [3]. This might be expected to limit the applications for this blend pair. Under isothermal crystallization condition, crystallization of sPP can lag behind iPP. In this case, crystallization of the iPP/sPP blend pair at the early stages can be thought as iPP crystallization of a semi-crystalline/amorphous blend when probed by time-resolved techniques. The current study addresses aspects of morphology development of equimolecular weight iPP/sPP mixtures by thermal and synchrotron scattering techniques.

EXPERIMENTAL

Samples for blending studies are summarized in Table 1. Isopropylidene(cyclo pentadienyl)(9-fluorenyl) zirconium dichloride catalyst [2] was used to prepare sPP in bulk monomer at 30°C with MAO activator at a molar [Al]/[Zr] ratio of 2000. The iPP is based on Ziegler-Natta catalyst technology. Blends were prepared with stabilized 3 wt% xylene solutions in a N_2 atmosphere at 130°C for 2hr. Blend composition varied from 0-100 wt% sPP. Blends were precipitated in dry-ice chilled methanol, filtered, vacuum dried, and compression molded into 2mm thick, 1.25" diameter disks (Buehler press) at 195°C for 7min (35 Mpa) and cooled in clamps to 110°C prior to ejection from the mold. The molds were re-melted by similar procedures without pressure to remove residual orientation (confirmed by 2D WAXS).

Simultaneous small-angle x-ray scattering (SAXS) and wide-angle x-ray diffraction (WAXD) measurements were conducted at the Beamline X27C, in National Synchrotron Light Source (NSLS), Brookhaven National Laboratory. Two linear position sensitive detectors (European Molecular Biology Laboratory, *EMBL*) were configured for simultaneous x-ray measurements. Crystallization experiments used 7 mm diameter samples cut from molded disks and transferred by pneumatic arm between furnaces at the melt temperature (7 min, 195°C) and crystallization temperature (115°C or 137.5°C). The experimental temperature jump apparatus has been reported previously [5]. X-ray profiles were collected in 5 sec (115°C) or 30 sec (137.5°C) increments and normalized for incident beam intensity. Both SAXS and WAXD patterns show crystalline features at the crystallization temperature. SAXS angles were calibrated with silver behenate. WAXD pixel and intensity was calibrated by comparing synchrotron data with Scintag diffractometer data (CuKα) using a NIST corrundum plate. The angular scale of the synchrotron WAXD data (λ = 1.307 Å) was converted to a scale corresponding to λ = 1.5418 Å.

DSC experiments were carried out using a Perkin-Elmer DSC7 station purged with nitrogen and calibrated with indium and mercury. 5-6 mg samples were used over a temperature range of –55~235°C. Heat-cool-reheat ramps were applied. All data are reported during heating following a cooling cycle at 20°C/min. DSC thermograms are shown after subtraction of the solid-melt baseline, which was determined by an iterative technique.

RESULTS AND DISCUSSION

Figure 1 shows DSC scans of iPP/sPP blends with various compositions of sPP. It is seen that sPP melts at temperatures approximately 20°C below the pure iPP component. Multiple melting, a common feature in sPP homopolymer [6,7], is observed for sPP in blends up to 50 wt% concentration. For 0~50 wt% sPP, this feature is independent of iPP concentration. This contrasts with melting in blends of sPP and atactic polypropylene (aPP). In sPP/aPP blends [7], the addition of aPP can strongly influence the multiple melting of sPP. This is not found when iPP is the diluent. Although not shown, the iPP/sPP blends show distinct crystallization exotherms on cooling, separated by approximately 20°C (at a rate of 20°C/min) for a 50:50 iPP/sPP blend. At this intermediate composition, the melting data in Figure 1 corresponds to morphologies formed by nearly complete crystallization of iPP prior to sPP.

The isothermal crystallization of sPP and iPP in the mixture can readily be observed in time resolved WAXD measurements. iPP and sPP have different unit cells [6], and do not co-crystallize with each other. Figure 2 shows representative WAXD patterns of a 50:50 iPP/sPP blend during crystallization at 115°C. The {200} reflection is a common feature to the polymorphic form in sPP following melt crystallization [8]. This reflection readily differentiates sPP crystallization from the α-monoclinic pattern of iPP [6]. Figure 2 shows that iPP crystallizes first followed by sPP. The SAXS invarient Q in Figure 3 also confirms this effect, i.e., crystallization occurs in distinct stages, corresponding to iPP and sPP growth.

The early stage crystallization of iPP in iPP/sPP blends is of interest because of the different transformation rates in iPP and sPP. The early stage growth of iPP conceptually resembles the semicrystalline/ amorphous iPP/aPP blend pair [9]. Figure 4 shows the long spacing, L, versus time for iPP and sPP homopolymers and the 50:50 iPP/sPP blend during crystallization at 115°C. In iPP, L decreases to a nearly constant value at early times. The decrease is also observed in sPP, but delayed in time and extended to higher degrees of transformation. This indicates that a process of secondary in-filling crystallization subsequent to primary growth is prevalent in sPP, contrasting with more "compact" structures characteristic of iPP crystallization at high supercooling. The presence of sPP in the blend appears to have only a minor influence on L over entire crystallization stages of iPP (early) and sPP (later). At elevated temperature (137.5°C), the blend follows more of a behavior of the rule of mixing. Figure 5 compares L for the iPP homopolymer and a 50:50 iPP/sPP blend during crystallization at 137.5°C. At this temperature, the sPP homopolymer shows a very low degree of transformation. WAXD results do not show crystalline sPP reflections. The long spacing of the iPP component in the presence of molten sPP does not change substantially. This observation contrasts with the high temperature crystallization of iPP/aPP blend [9], which shows a significant modification of L. At both temperatures, the values of L in sPP is larger than iPP due to its lower degree of supercooling applied (Tm of sPP is lower).

CONCLUSION

The blends studied in this work contain iPP and sPP components, which differ substantially in melting point. The multiple melting behavior characteristic of the sPP component is relatively unaffected by the presence of iPP up to iPP concentrations of 50 wt%. Crystallization on cooling occurs in stages, with the iPP component crystallizing first followed by subsequent crystallization of sPP. This conclusion is confirmed by time resolved SAXS and WAXD experiments during isothermal crystallization. During the early stage crystallization of iPP, the presence of the molten sPP component in the blend does not significantly alter the long spacing of the iPP homopolymer. This contrasts with the high temperature crystallization of iPP and aPP. At intermediate blend compositions, the results are consistent with an independent stacking of iPP and sPP lamellae within the morphology.

ACKNOWLEDGMENT

The authors thank R.L.Jones (Montell) for catalyst synthesis, and D.Morgan (Montell) for sample preparation. B.Hsiao acknowledges the financial support of this work by a grant from NSF (DMR 9732653).

References

1. Haylock, J.C.; Phillips, R.A.; Wolkowicz, M.D. in *Preparation, Properties & Technology of Metallocene-Based Polyolefins*; Scheirs, J., Kaminsky, W., Eds.; John Wiley & Sons, New York, 1998.
2. Ewen, J.A.; Jones, R.L.; Razavi, A.; Ferrara, J.D. *J. Am. Chem. Soc.*, **1988**, 110, 6255.
3. Thomann, R.; Kressler, J.; Setz, S.; Wang, C.; Mulhaupt, R. *Polymer*, **37**, 2627 (1996).
4. Thomann, R.; Kressler, J.; Rudolf, B.; Mulhaupt, R. *Polymer*, **1996**, 37, 2635.
5. Hsiao, B. S.; Gardner, K.H.; Wu, D.Q.; Chu, B. *Polymer* **1993**, 34(19), 3986.
6. Phillips, R.A.; Wolkowicz, M.D. in *Polypropoylene Handbook*; Moore, E.P., Ed.; Hanser: Munich, 1996.
7. Phillips, R.A.; Wolkowicz, M.D.; Jones, R.L. *Proc. SPE ANTEC*, **1997**, 57, 1651.
8. Lovinger, A.J.; Lotz, B.; Davis, D.D.; Schumacher, M. *Macromolecules*, **1994**, 27, 6603.
9. Phillips, R.A.; Wang, Z-G.; Hsiao, B.S. *ACS Polym. Preprints*, **1998**, Boston.

Table 1 Sample Characteristics

Sample	Mw	Mw/Mn	Type	%triad
iPP	220,000	4[a]	Iso	95
sPP	185,000	2	Syndio	93[b]

[a]rheological index [b]syndiotactic triads

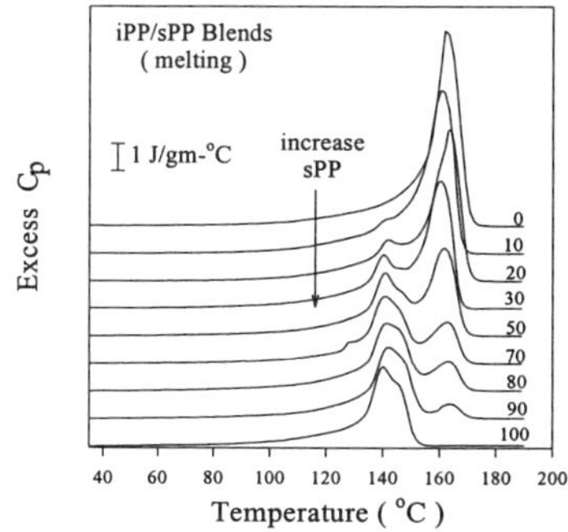

Figure 1. Reheat DSC scans of iPP/sPP blends. The wt% sPP varies between 0 to 100 % in the blends.

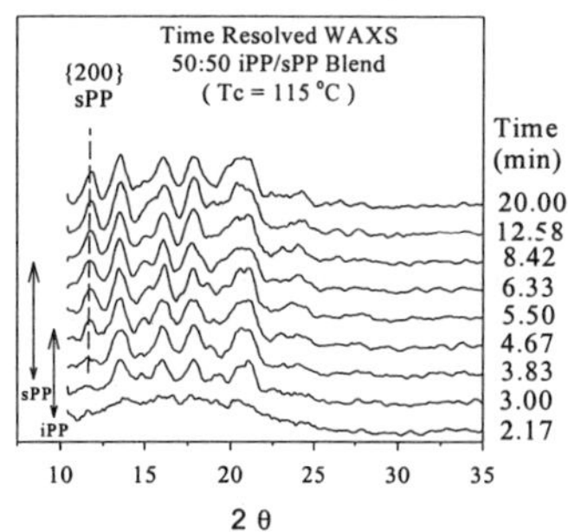

Figure 2. Time resolved WAXD intensities at 115°C for 50/50 iPP/sPP blend.

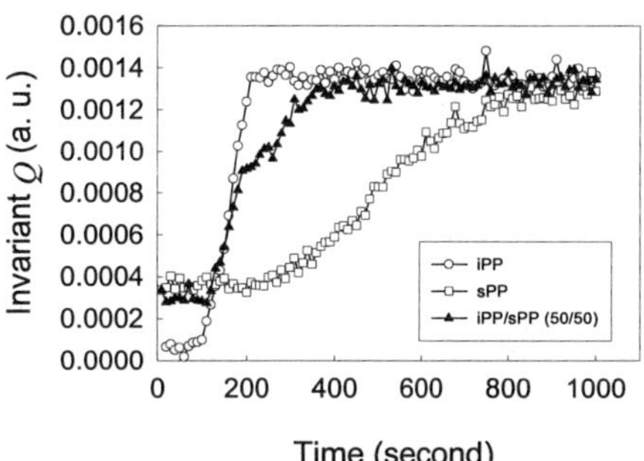

Figure 3. The invariant Q versus time during isothermal crystallization for iPP, sPP and 50/50 iPP/sPP blend at 115°C.

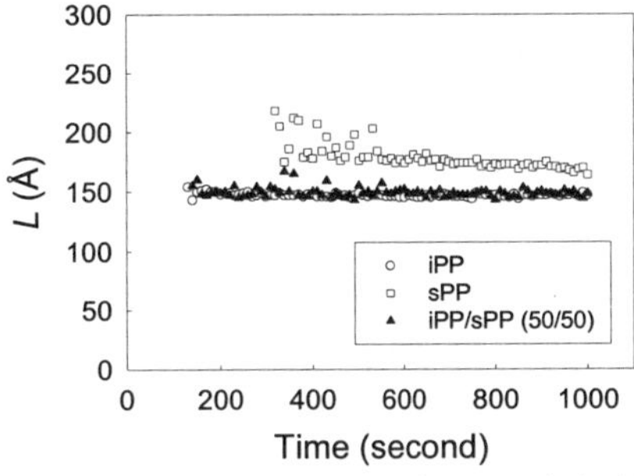

Figure 4. Time evolution of long period of iPP, sPP and 50/50 iPP/sPP blend at 115°C.

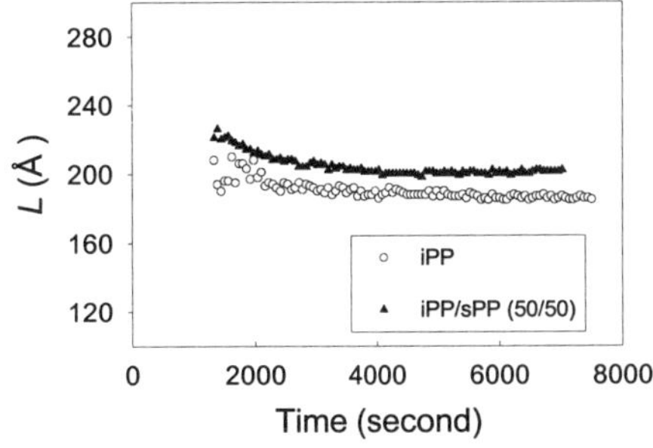

Figure 5. Time evolution of long period of iPP and 50/50 iPP/sPP blend at 137.5°C.

TIME-RESOLVED LIGHT SCATTERING STUDIES ON THE MORPHOLOGY DEVELOPMENT IN PP/EPR BLENDS

Jong Kwan Lee, Sung Wook Lim, Kwang Hee Lee[*]
Chang Hyung Lee[1] and Byung Suk Jin[2]

School of Chemical Sci. & Eng., Inha Univ.,
Inchon 402-751, Korea
[1]Dept. of Chemistry, Nat. Inst. of Tech. & Quality,
Kwacheon 427-010, Korea
[2]Dept. of Appl. Chemistry, Dongduck Women's Univ.,
Seoul 136-714, Korea

INTRODUCTION

Polypropylene/ethylene-propylene-rubber(PP/EPR) blend is an injection-molded thermoplastic applied widely for automotive parts. In order to improve the recyclability and lighten the car body, much work is recently in progress on developing a new level of PP/EPR blend with the outstanding physical property balance.

In a PP/EPR blend obtained by melt extrusion, we found that liquid-liquid(L-L) phase separation at high temperatures is very similar to the spinodal decomposition(SD) indicative of the partially miscible system, i.e. low critical solution temperature(LCST)/upper critical solution temperature(UCST). This may be that the conventional interpretation and understanding for the morphological formation in a PP/EPR blend is completely changed. It has been known that a PP/EPR blend is highly immiscible at all temperature[1] and, therefore, so far the morphology of a PP/EPR blend has been interpreted on the basis of the immiscible system. This is motivated to inquire L-L phase separation in a PP/EPR blend which has been believed to be a totally immiscible system.

Many experimental studies on the PP/EPR blend have been devoted on the morphology of EPR, i.e. the shape and size distribution of the EPR domains, adhesion at the interface, nature and structure of the EPR domains. However, it is widely accepted that properties of the blend in crystalline polymer/noncrystalline polymer blend depend on their crystalline structure; crystalline structure of PP matrix in a PP/EPR blend may also affect the mechanical properties. Recently, Nomura *et al.*[2] reported a new structural model which improves the stiffness and surface hardness by controlling PP crystalline lamellar structure of a PP/EPR blend. Coppola *et al.*[3] found a very close correlation between crystalline condition and mechanical property in a PP/EPR blend. The formation of the crystalline structure relates to L-L phase separation when the blend has an immiscibility gap, i.e. UCST/LCST. In the blend system with LCST, L-L phase separation would be able to precede crystallization and have a significant influence on crystallization. Therefore, we need to know information on both phase behavior and its effect on crystallization which will result in control of physical properties of the materials.

In this paper, we first carried out time-resolved light scattering (TR-LS) studies of the kinetics of the L-L phase separation to determine the phase behavior and then TR-LS, SAXS and optical microscope(OM) studies to investigate effect of L-L phase separation on crystallization.

EXPERIMENTAL

Materials

The PP was supplied by LG Chemical, Ltd. The specific gravity and melt flow rate(MFR) are 0.89 and 30 respectively. The EPR used in this work contains 67 mol % ethylene and the Mooney viscosity, $ML_{1+4}(125\,^\circ\!C)$ is 12. The 50/50(by weight) blend was prepared by extruding the two components with a twin-screw extruder at about 220 ℃. After extrusion, the blend was cooled to room temperature and granulated to pellets.

TR-LS and OM

A thin-layer specimen(ca. 15 μm thick) was prepared by pressing the blend pellets between two cover glasses at 190 ℃. Immediately after the melt-pressing, the specimen was quickly transferred into a hot stage on light scattering photometer equipped with a CCD(charge coupled device) camera and the kinetics of L-L phase separation was investigated.

The L-L phase-separated specimen at 190 ℃ for t_s was rapidly transferred into a light scattering hot stage set at 130 ℃ for the isothermal crystallization and then the effect of L-L phase

separation on crystallization was investigated. The time t_s is the time spent for L-L phase separation at 190 ℃. A polarized He-Ne gas laser of 632.8nm wavelength was applied to the film specimen. We employed the V_v geometry in which the optical axis of the polarizer was set parallel to that of the analyzer.

The final morphology of the crystallized specimen at 130 ℃ after t_s was also observed under optical microscope(OM).

SAXS

The pellets were annealed, i.e. phase-separated between metal plates at 190 ℃ for t_s = 1, 5 and 10 min. The L-L phase-separated specimen was rapidly transferred into another hot chamber set at 130 ℃ and was crystallized. Then the specimen was cooled to room temperature and cut into thin stripes(7mm×5mm×1mm). This crystallized specimen was used for SAXS.

The X-ray beam was from synchrotron radiation; beam line 3C2 at Pohang Accelerator Laboratory, Pohang, South Korea. The storage ring operated at an energy level of 2GeV. The SAXS employs a point focusing optics with a Si double crystal monochromator followed by a bent cylindrical mirror. The incident beam intensity of 0.149nm wavelength was monitored by an ionization chamber for the correction of minor decrease of the primary beam intensity during the measurement.

The Scattering intensity, I, was corrected for background scattering. Then, the scattering intensity by thermal fluctuations was subtracted from the SAXS profile $I(q)$ by evaluating the slope of $I(q)q^4$ versus q^4 plots at wide scattering vectors q, where q is $(4\pi/\lambda)\sin\theta$, λ and θ being the wavelength and scattering angle, respectively. The correction for smearing effect by the finite cross section of the incident beam was not necessary for the optics of SAXS with point focusing.

RESULTS AND DISCUSSION

The extruded specimens were held at various temperatures and then the L-L phase separation behavior was investigated. *Figure 1* shows the change of the scattering profiles with L-L phase separation time at 190 ℃. Note that, at $t=0$, the scattering intensity is very weak and it has no q dependence, suggesting that the blend just after the re-melting at 190 ℃ is a nearly homogeneous blend. It implies that the phase dissolution had took place before the measurement at 190 ℃. The homogeneous blend starts to phase-separate by annealing at 190 ℃, as shown by the increase in the scattering intensity. The appearance of a scattering peak suggests that the separation occurs via the spinodal decomposition mechanism. To confirm this point, we analyzed the early stage on the basis of the linearlized Cahn-Hilliard theory[4,5].

In the early stage of spinodal decomposition, the scattered intensity I is expected to increase exponentially with time;[5]

$$I(q, t) = I(q, 0)\exp[2R(q)t] \tag{1}$$

the amplification factor $R(q)$ is given by

$$R(q) = -Mq^2(\partial^2 f/\partial c^2 + 2\kappa q^2) \tag{2}$$

where f is the local free energy of mixing, c is the concentration, κ is the concentration-gradient energy coefficient and M is the mobility. According to equation (1), a plot of lnI versus time t at a fixed q should yield a straight line of slope $2R(q)$. The linear relationship is realized for the L-L phase separation in a 50/50 PP/EPR blend, indicating that the initial stage can be described by the linearized theory.

The plot of $R(q)/q^2$ versus q^2 yields a straight line, indicating again that the initial stage can be described with the framework of the linearized theory. By the intercept of $R(q)/q^2$ at $q^2=0$, one can obtain apparent mutual diffusion coefficient D_{app} given by[6]

$$D_{app} = -M(\partial^2 f/\partial c^2) \propto D_c[|\chi - \chi_s|/\chi_s] \propto T|T - T_s| \tag{3}$$

where D_c is the self-diffusion coefficient for translational diffusion, χ is the interaction parameter, χ_s is the χ at the spinodal temperature and T_s is the spinodal temperature.

The values of D_{app} were obtained at various temperatures. In this study, D_{app} increases with temperature. The increase of D_{app} with increasing temperature is expected in LCST system, since the quench depth $|T - T_s|$ increases with temperature(see equation(3)).

The increase of D_{app} has been observed in LCST system[6].

A supplemental evidence of L–L phase separation induced by SD at 190℃ was nicely given by observing a structure formation at 130℃. If one extrapolates the results to 130℃, the value of D_{app} at this temperature is expected to be very low. This implies that L–L phase separation rate may be negligible at 130℃. On the other hand, the crystallization rate of PP is very high so that crystallization is completed in about 4s as shown in *Figure 2*. In such case, it is well known that the L–L phase-separated morphology, i.e., the periodic and interconnected structure, is preserved during the crystallization process[7]. *Figure 3* shows a typical SEM micrograph of a blend crystallized at 130℃ after t_s=5min. The SEM observation was carried out after solvent etching with xylene so that the remaining material may be PP. As expected, a highly interconnected PP phase with unique periodicity is seen in the micrograph. Connectivity of phases is an important morphological feature of SD.

Further evidence for the SD process at 190℃ can be found in optical micrographic results. The phases are interconnected with each other. The white stripes seemed to be crystalline fibril are seen in dark parts and hence the dark parts may be the PP-rich phases. At the later stage of L–L phase separation, the phase connectivity seems to grow in size while maintaining their connectivity and eventually breaks up into macrospheres. These characteristics agree with Cahn's prediction of SD mechanism. The increase of the interconnected/periodic length(Λ_m) with t_s is clearly seen also in Figure 4. The values of Λ_m were obtained by applying Bragg equation to the peak position of the light scattering profiles.

Based on the above results, a scenario of the melt extrusion to yield near the homogeneous blend may be given as follows. Under the high shear rate in the extruder, phase diagram elevated over the barrel temperature and one phase region became wider. Thus, the blend could be done in a wide temperature window for dissolution to get a homogeneous mixture. However, once the melt is extruded out from nozzle, the shear rate turns out to be zero and LCST will be immediately go down to the state without shear so that the SD will proceed until the system is cooled down to crystallization. The fast crystallization prevents further L–L phase separation. Finally the melt extruded sample shows state close to homogeneity.

CONCLUSION

Miscibility in the melt-extruded blend of PP/EPR was investigated by TR-LS and microscopy measurement. Then, LCST phase behavior was found to exist in PP/EPR blend. For the melt-extruded blend near to homogeneity, one possible explanation could be that LCST might be elevated over the barrel temperature under the high shear rate in extruder, and therefore, blending could be done in one phase region.

The effect of L–L phase separation on the crystalline morphology produced by subsequent crystallization was discussed. The crystallization took place only in PP-rich phases so that the memory of SD(the periodic structure) was preserved. The crystallization rate decreased with the L–L phase separation time before crystallization. By a series of crystalline morphological parameters by SAXS and DSC analyses, it was shown that the amount of the EPR in PP-rich phases formed by SD deeply related to the crystallization rate and crystalline morphology; the amount of the EPR in PP-rich phases decreased with increasing L–L phase separation time, and its decrease induced the lower crystallization rate and the smaller long period.

ACKNOWLEDGMENT

This research supported by the Center for Advanced Functional Polymers and the Inha University Foundation.

REFERENCES

1. Edward, P. and Moore, Jr. 'Polypropylene Handbook', Hanser Publishers, Munnich Vienna New York, 1996, p.218
2. T. Nomura et al., *J. Appl. Polym. Sci.* 1995, **55**, 1307
3. F. Coppola and E. Martuscelli, *Polymer* 1988, **29**, 963
4. Chan, J. W. and Hilliard, J. E., *J. Chem. Phys.* 1958, **28**, 258
5. Chan, J. W., *J. Chem. Phys.* 1965, **42**, 93
6. T. Hashimoto et al., *Macromolecules* 1983, **16**, 641
7. N. Inaba and T. Hashimoto, *Macromolecules* 1986, **19**, 1690

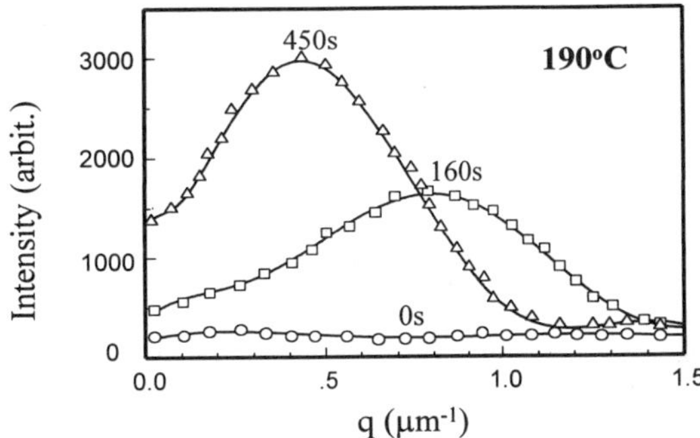

Fig.1 Change in light scattering profiles of a 50/50 PP/EPR blend during L-L phase separation at 190°C.

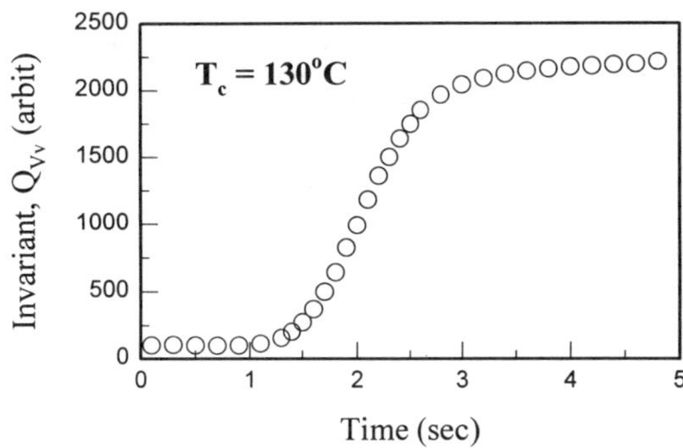

Fig. 2 Time variation of the invariant Q_{Vv} in a 50/50 PP/EPR blend during crystallization at 130°C after t_s = 5min.

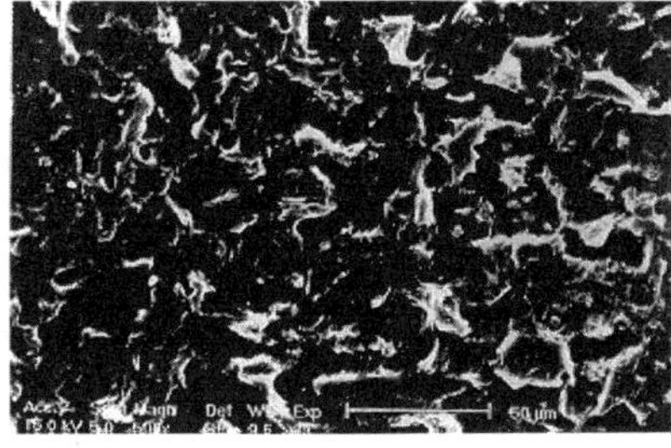

Fig. 3 Scanning electron micrograph of a 50/50 PP/EPR blend after t_s = 5 min.

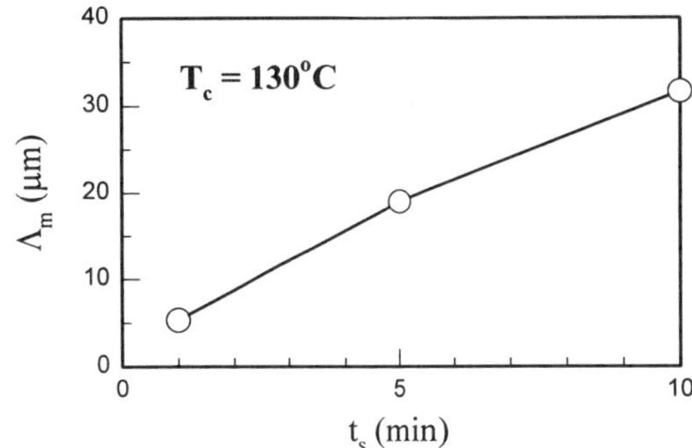

Fig. 4 Change of Λ_m with t_s.

Neutron Spin Echo Spectroscopy at the NIST Center for Neutron Research

N. Rosov[*], S. Rathgeber[†*], M. Monkenbusch[‡], *NIST Center for Neutron Research, Gaithersburg, MD 20899, †University of Maryland, College Park, MD 20742, ‡IFF, Forschungszentrum Jülich, Germany D-52425.

A Neutron Spin Echo (NSE) spectrometer is one of three high-resolution inelastic instruments nearing completion at the NIST Center for Neutron Research (NCNR), the other two being a Disk Chopper Time-of-Flight spectrometer and a Backscattering spectrometer. These instruments can study dynamic processes with time scales that range from 10^{-13} s to 10^{-7} s. All three instruments will be available to researchers through reviewed proposals [1]. Both the NSE and Backscattering spectrometers will be operational in the summer of 1998 and the Disk Chopper Spectrometer shortly thereafter.

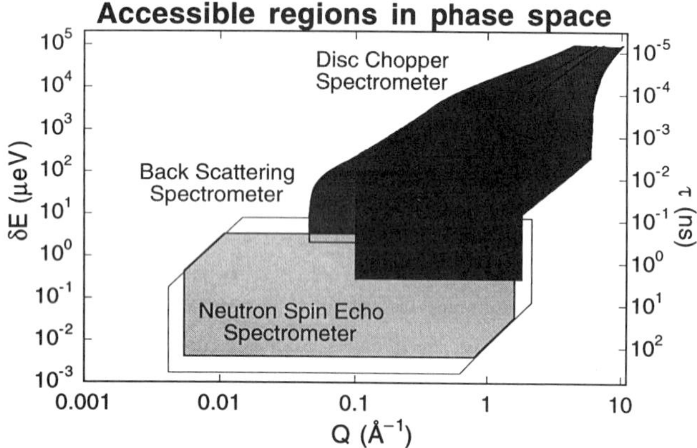

Accessible regions in phase space

FIGURE 1: Phase space diagram indicating the region of accessibility for each of the the the three new inelastic instruments at the NIST Center for Neutron Research.

Unlike other inelastic neutron spectrometers, which measure the scattering function $S(Q,\omega)$, the NSE spectrometer measures the intermediate scattering function $I(Q,t)$ (the cosine transform of $S(Q,\omega)$).

The paper will provide a brief introduction to NSE, tailored to the particular details of the NIST NSE spectrometer, which has been optimized for the study of soft condensed matter systems. We will discuss how a NSE experiment is performed and outline the considerations required for choosing experimental systems.

Principles of Neutron Spin Echo and Experimental realization. [3] The NCNR NSE spectrometer is a sister spectrometer to the NSE spectrometer at the Forschungszentrum Jülich. Both Exxon and the Forschungszentrum Jülich are members of the PRT for the construction of the spectrometer. The spectrometer is located at the end of guide NG-5, which looks at the liquid H_2 cold source. An optical filter [4] moves the end of the guide out of the direct line-of-sight of the reactor, removing fast neutrons and core gammas from beam, while allowing the transmission of neutrons with $\lambda > 4$ Å. A velocity selector roughly monochromates the neutron beam with $\Delta\lambda/\lambda = 10\%$ or 20% for mean wavelengths $\lambda_0 > 4.5$ Å. The beam is polarized in the longitudinal direction by a Mezei cavity [5], with full polarization for $\lambda > 5$ Å.

The neutron spin is rotated 90° by a $\pi/2$ flipper, which begins the precession of the neutron in the main solenoidal field. The phase angle through which the neutron precesses depends on the time spent traveling in the field (that is, on the neutron wavelength λ) and on the field integral $(\mathcal{J} = \int H d\ell)$, $\phi \propto \lambda \mathcal{J}$. Note that ϕ can only be determined to mod(2π). The neutron beam, with its broad band of wavelengths, will completely depolarize in this first precession field. After scattering from the sample, the neutron passes through a π-flipper, thereby changing its phase angle from ϕ mod (2π) to $-(\phi$ mod $(2\pi))$. Then on passing through the second precession field, if the scattering is elastic and the two field integrals are the same, the beam recovers its full polarization at the $\pi/2$ flipper. The beam polarization is then analyzed and the neutrons impinge on an area detector.

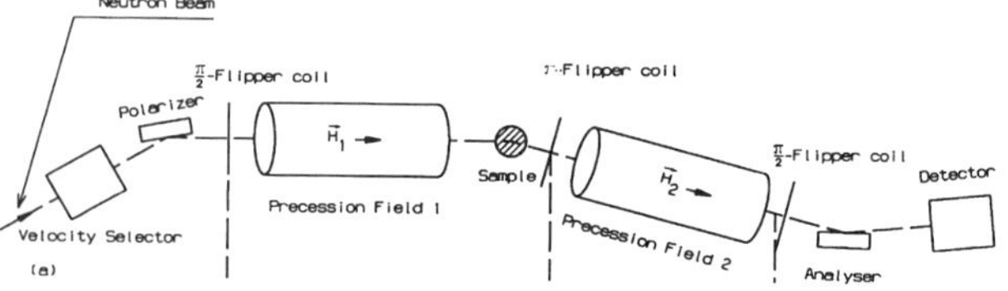

FIGURE 2: Schematic of a Neutron Spin Echo spectrometer. The various components are discussed in the text.

If the scattering is inelastic, there is a remaining phase shift. By adjusting the field so that the echo condition (Eq. 12) is met, the beam polarization is

$$\langle P_z \rangle = \int f(\lambda) d\lambda \times \int S(Q,\omega)\cos(\omega t) d\omega = \int f(\lambda) I(Q,t) d\lambda \quad (1)$$

where

$$t = \gamma_L \left(\frac{m}{h}\right)^2 \frac{\lambda^3}{2\pi} \mathcal{J} \quad (2)$$

is the time parameter. That is, Neutron Spin Echo spectroscopy directly measures the intermediate scattering function, the energy cosine transform of the scattering function.

This is particularly useful for measuring quasielastic processes or processes at small energies where discriminating the scattering of interest from any elastic scattering is important. On a conventional spectrometer, deconvolving the instrumental energy lineshape from the physical lineshape can be quite difficult. However, a convolution in ω-space becomes a simple product in the time domain. Rather than performing highly elaborate lineshape analyses in the energy domain to determine, for instance, if a process is elastic or quasielastic, one can merely note if the NSE signal is constant or decays with the time parameter. Table 1 shows several scenarios where the inelastic scattering can be determined in the time domain.

TABLE 1: Typical Scattering Laws

$S(Q,\omega)$	$I(Q,t)$
Lorentzian of width Γ centered at $\omega = 0$, $\frac{\Gamma}{\Gamma^2 + (\omega - \omega_0)^2}$	Exponential decay $e^{-\Gamma t}$
Delta function $\delta(\omega)$	Constant
Pair of inelastic lines with linewidth	Damped cosine

The entire instrument is of amagnetic construction, and the operating program contains an accurate description of the current distributions, thereby allowing the field distributions to be calculated with sufficient accuracy to considerably reduce the time spent tuning the spectrometer. The coils contain a compensation loop to rapidly reduce the on-axis field outside the solenoid. As a result, there is almost no loss of resolution at higher scattering angles.

The maximum active sample area is 5×5 cm^2 (typical sample sizes for polymer studies are on the order of $3 \times 3 \times 0.3$ cm^3). This value is constrained by the active area of Fresnel-like correction elements inserted in the main coils. These correction elements allow off-axis and divergent neutrons to satisfy the echo condition and thus allow the use of an area detector. This effectively increases, by a factor of twenty, the signal strength.

Applications. The classical design of NSE spectrometer (such as that at NCNR) allows the measurement of the coherent intermediate structure factor and therefore in the case of measurements on polymers, the samples should contain mostly deuterated polymers and/or solvent. In principle it is possible to measure the incoherent intermediate structure factor. In practice, it is difficult to measure in a reasonable amount of measuring time for three reasons: (1) Incoherent scattering is spread out isotropically over 4π; (2) Samples must be small to avoid multiple scattering; and (3) Spin incoherent scattering from H flips the spin and reduces the signal by a factor of 2/3.

NSE spectroscopy should be applied to investigations on mesoscopic time scales which are well separated from atomic time scales. The big advantage of NSE spectroscopy lies in investigations of *aperiodic relaxation dynamics*. These processes show broad quasielastic

TABLE 2: NCNR NSE operating characteristics at a glance

➤ **Velocity Selector**

- $\langle\lambda\rangle > 4.5\text{Å}$
- $\Delta\lambda/\lambda = 10\%$ or 20%

➤ **Polarizer**

- $3\theta_c$ Fe/Si supermirror "V" in guide.
- Full polarization for $\lambda > 5\,\text{Å}$

➤ **Main Coils**

- $I_{max} = 440\,\text{A}$
- $\mathcal{J}_{max} = 0.5\,\text{T}\cdot\text{m}$
- $t_{max}(8\,\text{Å}) \simeq 45\,\text{ns}$

➤ **Sample region**

- Active area $5 \times 5\,\text{cm}^2$
- neutron flux $> 10^6\,\text{n/cm}^2\text{/s}$ at $8\,\text{Å}$ and $\Delta\lambda/\lambda = 10\%$
- useful flux between $4.5\,\text{Å}$ and $> 12\,\text{Å}$

➤ **Detector**

- $25 \times 25\,\text{cm}^2$ ORDELA area detector
- $Q_{min} \simeq 0.006\,\text{Å}^{-1}$ at $8\,\text{Å}$
- Maximum scattering angle $100°$ ($Q_{max} > 2.1\,\text{Å}^{-1}$ at $4.5\,\text{Å}$)

---

features in the frequency-space but a featureless decaying structure in the time-domain.

NSE spectroscopy covers a wide range of applications. In classical solid state physics mainly critical scattering in the field of magnetism [6] and phase transitions [7] has been investigated. But NSE spectroscopy has also shown to be a great tool for probing the dynamics of soft matter, especially in the case of polymer systems [8] and complex fluids [9]. Beyond this studies have also been performed in the dynamic properties of systems with biological importance [10] and glasses [11].

References

[1] Information on submitting proposals may be obtained on the NCNR web site <http://rrdjazz.nist.gov/> or by contacting one of the authors: nrosov@nist.gov or silke@rrdjazz.nist.gov.

[2] A much more complete description of NSE can be be found in *Neutron Spin Echo*, Lecture Notes in Physics, vol. 128, F. Mezei, ed. (Springer-Verlag: Berlin, 1980). The present outline follows that given in the article by J. B. Hayter.

[3] *See also* M. Monkenbusch, R. Schätzler, and D. Richter, Nucl. Instr. Meth. Phys. Res. A **399**, 301–323 (1997).

[4] J. R. D. Copley, J. Neutron Res. **2**, 95 (1994).

[5] T. Krist, C. Lartigue, and F. Mezei, Physica B, **180**, 1005 (1992).

[6] R. I. Bewley *et al.*, Physica B **234**, 762 (1997).

[7] K. Kakurai *et al.*, Phys. Rev. B **53**, R5974 (1996).

[8] B. Ewen and D. Richter, Adv. Polym. **134**, 1 (1997).

[9] B. Farago, Physica B **226**, 51 (1996).

[10] W. Pfeiffer *et al.*, Europhys. Lett. **23**, 457 (1993).

[11] A. Arbe *et al.*, Phys. Rev. E **54**, 3853 (1996).

Appendix: A more detailed introduction to Neutron Spin Echo. [2] In a perpendicular magnetic field H_0 the neutron spin will undergo precessions at a frequency $\omega_L = -\gamma_L H_0$, where $\gamma_L/(2\pi) = 29.16\,\text{MHz/T}$. If the neutron were polarized perpendicularly to the axis of a solenoidal field, it would precess through an angle

$$\phi_i = \gamma_L \frac{m\lambda}{h} \int_i H(\ell)d\ell = \gamma_L \frac{m\lambda}{h} \mathcal{J}_i \tag{3}$$

where $\mathcal{J}_i$ is the field integral along solenoid i and λ is the wavelength of the neutron ($h/(m\lambda)$ is the neutron velocity v).

For a beam of neutrons with an incident wavelength distribution $f(\lambda)$ and $\langle\lambda\rangle = \lambda_0$, each neutron undergoes a spin rotation of $\phi_0(\lambda)$ in the first arm of the spectrometer. If the neutrons are then scattered inelastically, changing wavelength by $\delta\lambda$, they will undergo a spin rotation $\phi(\lambda + \delta\lambda) = \phi_0(\lambda + \delta\lambda) + \Delta\phi(\lambda + \delta\lambda)$ in the second arm of the spectrometer, where $\Delta\phi(\lambda) = \phi_0(\lambda) - \phi_1(\lambda)$ is due entirely to the difference in the field integral between the two arms of the spectrometer. The total phase shift is then

$$\varphi = \phi_0(\lambda) - \phi_0(\lambda + \delta\lambda) - \Delta\phi(\lambda + \delta\lambda). \tag{4}$$

To first order in $\delta\lambda$ and $\Delta\phi$ the phase shift is composed of a term from the inelasticity and a term from the difference in the field integrals:

$$\begin{aligned}\varphi &= \phi_0(\lambda_0)\frac{\lambda}{\lambda_0} - \phi_0(\lambda_0)\frac{\lambda + \delta\lambda}{\lambda_0} - \Delta\phi(\lambda_0)\frac{\lambda}{\lambda_0}\\ &= -\left(\phi_0(\lambda_0)\delta\lambda + \Delta\phi(\lambda_0)\lambda\right)/\lambda_0.\end{aligned} \tag{5}$$

In terms of the change in wavelength, the inelasticity

$$\hbar\omega = E_0 - E_1 = \frac{h^2}{2m}\left(\frac{1}{\lambda_0^2} - \frac{1}{(\lambda_0 + \delta\lambda)^2}\right), \tag{6}$$

is, to first order in $\delta\lambda$,

$$\hbar\omega = \frac{h^2}{m}\frac{\delta\lambda}{\lambda^3}. \tag{7}$$

Due to the quantum nature of the neutron spin, only one component of the spin, call it z, can be determined. The polarization of the scattered beam is

$$\begin{aligned}\langle P_z\rangle &= \langle\cos(\varphi)\rangle\\ &= \int f(\lambda)d\lambda \times\\ &\quad \int S(Q,\omega)\cos\left[\left(\phi_0(\lambda_0)\frac{m\lambda^3}{2\pi h}\omega + \Delta\phi(\lambda_0)\lambda\right)/\lambda_0\right]d\omega. \end{aligned} \tag{8}$$

We make the assumption that $S(Q,\omega)$ is a quasi-elastic scattering law, which is essentially an even function of ω, and write

$$\begin{aligned}\langle P_z\rangle &= \int f(\lambda)\cos(\Delta\phi(\lambda_0)\lambda/\lambda_0)d\lambda \times\\ &\quad \int S(Q,\omega)\cos\left(\phi_0(\lambda_0)\frac{m\lambda^3}{2\pi h\lambda_0}\omega\right)d\omega\\ &= \int f(\lambda)\cos(\Delta\phi(\lambda_0)\lambda/\lambda_0)d\lambda \times \int S(Q,\omega)\cos(\omega t)d\omega \end{aligned} \tag{9}$$

where

$$t = \phi_0(\lambda_0)\frac{m\lambda^3}{2\pi h\lambda_0} = \gamma_L\left(\frac{m}{h}\right)^2\frac{\lambda^3}{2\pi}\mathcal{J}_0. \tag{10}$$

One can also define the time parameter in terms of the scattered beam,

$$t = \gamma_L\left(\frac{m}{h}\right)^2\frac{\lambda_1^3}{2\pi}\mathcal{J}_1. \tag{11}$$

The spectrometer is focussed when the incoming and scattered time parameters are equal (by comparing Eqs. 10 and 11),

$$\lambda_0^3\mathcal{J}_0 = \lambda_1^3\mathcal{J}_1 \tag{12}$$

There are two simple ways to make scans: In one, not so frequently used, the time parameter (that is, the field integral) is set as small as possible (although the field must be larger than the stray fields that might depolarize the beam) and so the second term in Eq. 9 reduces to $S(Q)$. By varying the difference in the field integrals, one can measure the cosine transform of $S(Q)f(\lambda)$, which, given a constant or slowly varying $S(Q)$, is a cosine function, with a period proportional to λ_0^{-1}, modulated by the cosine transform of the wavelength distribution. This scan is often used to "prove" that an NSE instrument is working.

The other scan, which is used for measurements in polymer or similar systems, is taken by setting the spectrometer to the echo condition, which gives Eq. 1, and so the intermediate scattering function.

Kinetics of Order to Order Transition of SI and SIS Block Copolymers by Synchrotron SAXS and Rheology

Jin Kon Kim, Hee Hyun Lee, and Ki Bong Lee, Department of Chemical Engineering and Polymer Research Institute, Pohang Accelerator Laboratory, Pohang University of Science and Technology, Pohang, Kyungbuk 790-784, Korea

Introduction

Numerous studies have been reported on the order-disorder transition (ODT) in block copolymers during the past two decades.[1,2] It has been recently found that some block copolymers have multiple ordered states, and the transition from one ordered state to another is referred to as the order to order transition (OOT).[3-10] Also, some research groups[11] predicted theoretically an intermediate state when one microdomain is transformed into another. Very recently, we reported the ordering kinetics of SIS (polystyrene-*block*-polyisoprene-*block*-polystyrene) copolymer between spherical and cylindrical microdomains experimentally.[8]

In this study, using synchrotron SAXS and rheometer, we investigated the ordering kinetics of cylindrical and spherical microdomains in SI and SIS copolymers when temperature was jumped from one ordered state with spherical microdomains to the other with cylindrical microdomains, and vice versa.

Experimental Part

The SI (polystyrene-*block*-polyisoprene) diblock copolymer was prepared by an anionic polymerization, and has the weight-average molecular weight (M_w) of 68.1×10^3 and the weight fraction (w_{PS}) of the PS block of 0.185. The SIS triblock was a commercial grade (Vector 4111, Dow-Exxon Polymer Co.) having M_w of 142.7×10^3 and w_{PS} of 0.183. Samples were prepared by first dissolving a predetermined amount of the block copolymer in toluene in the presence of an antioxidant and then slow evaporating the solvent. Using an Advanced Rheometric Expansion System (ARES) with parallel plates of 25 mm diameter, dynamic temperature sweep experiment and temperature jumping tests were performed. Dynamic temperature sweep experiment was done under the isochronal conditions with increasing temperature from 120 to 260 °C. Time resolved SAXS experiments were conducted at the beam line (3C2 and 4C2) in Pohang Light Source, Korea. The incident beam was focused with a toroidal mirror and monochromatized using a double crystal Si(111) monochromator at the wavelength (λ) of 0.1598 nm and scattered intensity (I(q)) was detected by a diode-array position sensitive detector (ST-120; Princeton Instruments Inc.) allowing various wave vectors ($q = 4\pi\sin(\theta/2)/\lambda$ where θ is the scattered angle).[12]

Results and Discussion

Figure 1 gives temperature sweep of G′ for SI and SIS at $\omega = 1$, 0.05 rad/s and $\gamma_o = 0.05$, 0.03 during heating with rates of 1, 0.5 °C/min respectively. It can be seen that for both block copolymers, G′ first decreases slowly with increasing temperature and reaches a minimum. Then, it increases and has a maximum, and finally decreases again. When the OOT is taken as the temperature where a minimum in G′ appears,[4] the T_{OOT} for SI and SIS were estimated to be 172 ± 1 °C and 181 ± 1 °C, respectively. It should be mentioned that the temperature at which a precipitous decrease in G′ is observed is often referred to as the T_{ODT}.[13,14] Using this definition, the T_{ODT} of SI and SIS are 235 °C and 210 °C, respectively. However, this definition must be reserved for highly asymmetric block copolymers,[2,7] since in these block copolymers spheres with the *liquid-like* short-range order can be found at temperatures higher than the T_{ODT} defined above.

Figure 2 gives SAXS profiles of SIS block copolymer near the first order peak with time when temperature is increased to 200 °C from 170 °C. For SI block copolymer the change in SAXS profile was found to be very similar to Figure 2 except the transition occurred slowly. Wave vector at the maximum in first order peak intensity, q_m increases rapidly from 0.198 nm⁻¹ to 0.212 nm⁻¹ which is consistent with results given by Ryu et al.[6] It is worth noting that the full width at half

maximum (FWHM) σ_q exhibited a maximum near at 200s implying that this state might be an intermediate state between HEX and BCC. We conclude that the transition of HEX microdomains into BCC microdomains occurred without a complete dissolution of HEX microdomains into a disordered state, but proceeded via an undulation state.

The time evolution of G′ of SIS block copolymer is shown in Figure 3 when temperature is rapidly increased to 200 °C from 170 °C. It is rather interesting to see in Figure 3 that initially G′ drops significantly, then increases rapidly and finally reaches a steady value. It is noted that the drop in G′ found in Figure 3 is more evident if compared with that observed at $\sim T_{OOT}$ during heating given in Figure 1. The drop in G′ shown in Figure 3 is not due to a temperature fluctuation before stabilizing the setting temperature in our rheometer, rather it can be explained as follows. When the specimen is rapidly heated from 170 °C to 200 °C, we speculate that at short times microdomains may become a superheated state of undulated cylinders formed near T_{OOT} ($\sim$ 181 °C). The value of G′ for this block copolymer with a superheated state of undulated cylinders at 200 °C is expected to be lower than that at T_{OOT}, since this temperature is $\sim$ 20 °C higher than the T_{OOT}. As time goes, the undulated cylindrical microdomains with superheated state are gradually transformed into BCC microdomains with the *solid-like* long-range orders, thus G′ increases accordingly.

Figure 4 gives SAXS profiles of SIS block copolymer near the first order peak when temperature is decreased to 170 °C from 200 °C. As soon as temperature reaches 170 °C, BCC microdomains are transformed very quickly into HEX microdomains. We also found that σ_q exhibited a maximum near at 500 s meaning that this state might be an intermediate state between BCC and HEX microdomains. The transition path between HEX and BCC microdomains is qualitatively drawn in Figure 5. According to our recent result[15] the transition from undulated cylinders to BCC microdomains would be an exothermic first order transition, while the transition from BCC microdomains to undulated cylinders would be an endothermic first order transition. This is due to the entropy difference between above two states especially packing or configurational entropy of structure rather than conformational entropy of chains.

This work was supported by Korean Foundation of Science and Engineering (# 97-0503-02-3). Synchrotron SAXS experiment at the PLS were supported by Ministry of Science and Technology (MOST) and Pohang Iron & Steel Co. (POSCO).

Refereences

1. Bates, F. S.; Fredrickson, G. H. *Annu. Rev. Phys. Chem.* **1990**, *41* 525.
2. Han, C. D, et al. *Macromolecules* **1995**, *28*, 5043; Han, C. D.; Baek, D. M.; Kim, J. K. *Macromolecules* **1990**, *23*, 561.
3. Sakurai, S.; Kawada, H.; Hashimoto, T.; Fetters, L. J. *Macromolecules* **1993**, *26*, 5796.
4. Almdal, K. et al.; *Macromolecules* **1992**, *25*, 1743; Koppi, K. A. et al. *J. Rheol.* **1994**, *38*, 999; Khandpur, A. K. et al. *Macromolecules* **1995**, *28*, 8796.
5. Hajduk, D. A. et al. *Macromolecules* **1994**, *26*, 490.
6. Ryu, C. Y.; Lee, M. S.; Hajduk, D. A.; Lodge, T. *J. Polym. Sci.:Polym. Phys. Ed.*, **1997**, *35*, 2811.
7. Sakamoto, N. et al. *Macromolecules* **1997**, *30*, 1621, 5312.
8. Kim, J. K. et al. *Macromol. Chem. Phys, in press* (1998).
9. Sakurai, S. et al. *Macromolecules,* **1998**, *31*, 336; *J. Chem. Phys* **1998**, *108*, 4333.
10. Sakurai, S. et al. *Macromolecules* **1996**, *29*, 740; *J. Chem. Phys.*

1996, *105*, 8902.

11. Qi, S.; Wang, Z. G. *Phys. Rev. Lett.* **1996**, *76*, 1679; *Phys. Rev. E.* **1997**, *55*, 1682; Laradji, M.; Shi, A. C.; Noolandi, J.; Desai, R. C. *Phys. Rev. Lett.* **1997**, *78*, 2577; *Macromolecules* **1997**, *30*, 3242.

12. Park, B. J.et al. *Rev. Sci. Instrum.* **1995**, *66*, 1722.

13. Bates, F. S. *Macromolecules* **1984**, *17*, 2607; Rosadale, J. H.; Bates, F. S. *Macromolecules* **1990**, *23*, 2329; Bates, F. S.; Rosadale, J. H.; Fredrickson, G. H. *J. Chem. Phys.* **1990**, *92*, 6255.

14. Adams, J. L.; Graessley, W. W.; Register, R. A. *Macromolecules* **1994**, *27*, 6026.

15. Kim, J.K. et al. accepted to *Macromolecules* (1998).

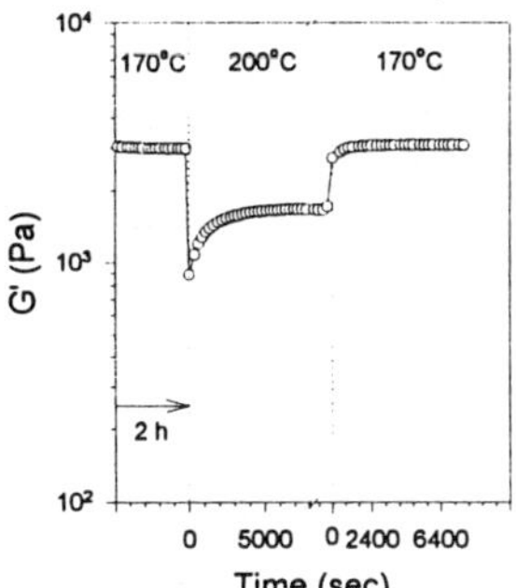

Fig. 3 The time evolution of G' for SIS with $\omega = 0.05$ rad/s and $\gamma_0 = 0.03$ when rapidly heated to 200°C (BCC) from 170°C (HEX) after soaking for 2h at 170°C and when quenched to 170°C (HEX) from 200°C (BCC)

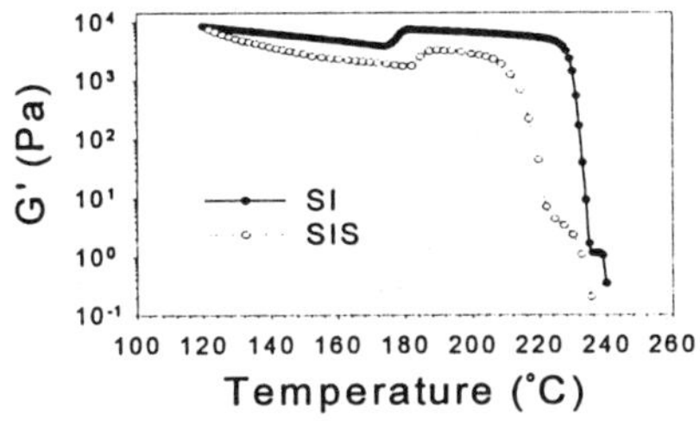

Fig. 1 Temperature sweep of G' for SI and SIS during heating with $\omega = 1, 0.05$ rad/s and $\gamma_0 = 0.05, 0.03$ respectively.

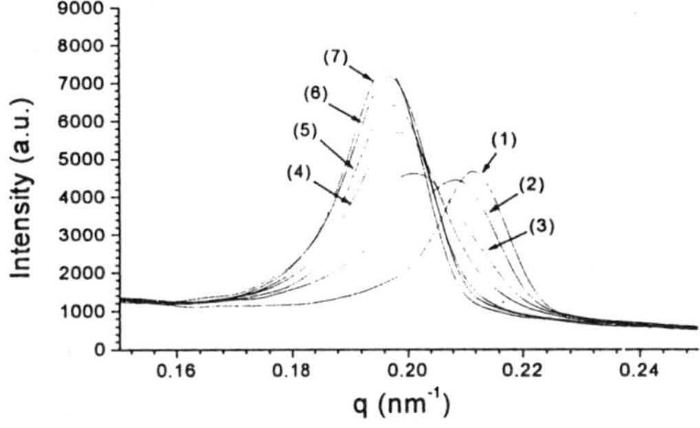

Fig. 4 SAXS profiles of SIS near the first order peak as a function of time when quenching to 170°C (HEX) from 200°C (BCC) after soaking for 110min at 200°C: (1) 60s (190°C); (2) 408s (170.3°C); (3) 528s (169.7°C); (4) 640s (170°C); (5) 822s; (6) 1080s; and (7) 2386s

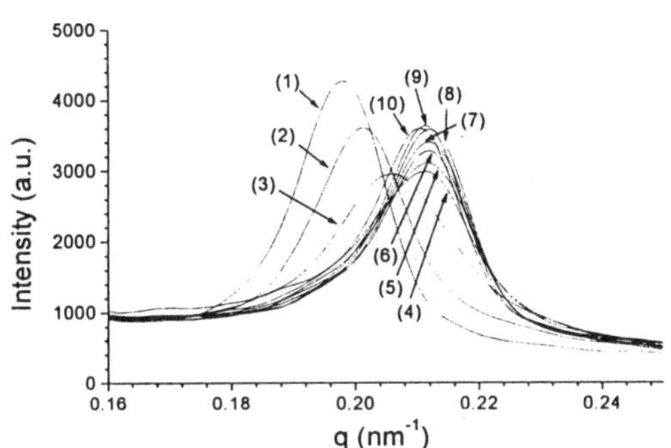

Fig. 2 SAXS profiles of SIS near the first order peak as a function of time when heating to 200°C (BCC) from 170°C (HEX) after soaking for 50min at 170°C: (1) 60s (194°C); (2) 132s (196°C); (3) 192s (199°C); (4) 250s (200°C); (5) 312s; (6) 496s; (7) 662s; (8) 850s; (9) 1588s; and (10) 2852s

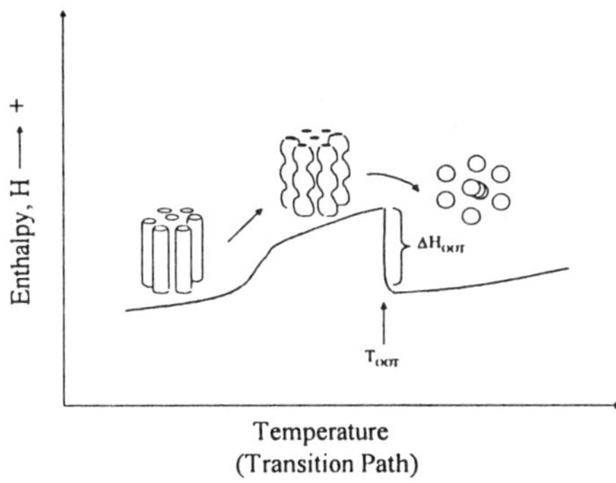

Fig. 5 Schematic of the enthalpy change near the OOT between HEX and BCC microdomains

DEFORMATION - INDUCED STRUCTURAL CHANGES IN POLY(URETHANEUREA) FILM

Fengji Yeh, Benjamin S. Hsiao and Bryan Sauer[¥]
Department of Chemistry, SUNY at Stony Brook, Stony Brook, NY 11794-3400
[¥] DuPont Central Research and Development, Experimental Station, Wilmington, DE 19880-0356

Introduction

It is well documented that the versatile mechanical properties of segmented polyurethane based elastomers are due to the formation of microphase separation from the incompatibility of the solid-like hard-segment and rubbery soft-segment sequences [1-6]. The degree of microphase separation was often found to be incomplete [4,5]. As the kinetics factor can often quench the microphase separation at a metastable state of finer domain size, the corresponding mechanical behavior becomes functions of thermal and mechanical histories.

In this study, we have investigated the changes in morphology and structure during the deformation of a solvent cast isotropic polyurethaneurea film. As prepared, this film shows a micro-phase separation of hard and soft segments [7]. Using simultaneous SAXS and WAXS (SWAXS) techniques with synchrotron radiation, the structural changes during tensile deformation and relaxation can be followed. These measurements enable us to understand the effects of deformation on phase segregated polyurethaneurea and the stress-induced crystallization in the soft segments on a fine molecular scale.

Experiments

The chosen polyurethaneurea (PU) sample is a segment block copolymer containing about 85% by volume poly(tertramethylene oxide) (PTMeO, Tg=-70°C) as soft segment, and 4,4'-diphenylmethane diisocyanate (MDI) and ethylene diamine (EDA) moieties as hard segments. The EDA agent was used as the chain extender of the PTMeO prepolymer end-capped with MDI. The solvent used was N,N-dimethylacetamide (DMAc). The preparation of this PU has been described elsewhere [7]. The PU film was prepared by the solvent casting method.

Time resolved SAXS/WAXD techniques were carried out for these two studies at the Advanced Polymers beamline [8], X27C, National Synchrotron Light Source (NSLS), Brookhaven National Laboratory (BNL). In this facility, a three-pinhole SAXS collimation was used to define the incident beam from a multilayer monochromator. The beam size was about 0.4 mm in diameter at the sample position, and the sample to detector distance was 1830 mm. The deformation of the polyurethaneurea (PU) film was performed with an Instron 4442 tensile apparatus using two image plates. In this setup, the chosen sample-to-detector distance is 1590 mm for SAXS and 140 mm for WAXD. A constant strain rate was applied to the specimen through out the whole deformation. The SAXS/WAXS images were taken after reaching the desired strain values under tension as well as upon relaxation after stretching.

Results and Discussion

The deformation study of the solvent-cast PU film was carried out using simultaneous 2D-SAXS/WAXS methods with image plates. Figure 1 shows the 2-D SAXS and WAXD images collected at the initial state, during deformation (strain = 250%), and upon relaxation after stretching, respectively. In the initial state, the SAXS image is completely isotropic but exhibits a scattering maximum. Since PU consists of two molecular blocks in the chains: soft amorphous blocks (PTMeO) and hard crystalline blocks (MDI), the peak position of the SAXS image represents the periodical arrangement of the hard and soft segment domains. The repeating layer distance (long period) is about 120 Å. Using the correlation function method, the integrated intensity projected along the meridian can be analyzed, which revealed the thickness of the constituting soft segment and hard segment domains (ca. 90 and 30 Å, respectively).

Upon deformation (strain > 25%), the isotropic scattering ring is found to change into an oriented four-spot pattern. We attribute this four-pint pattern to the tilted formation of the hard segments along the stretching direction. The tilting of the hard segment has been reported by Bonart before [1-3]. This tilting can be explained as follows. As the stress-induced crystallization in soft segments occurs, the soft segment crystalline region can start to carry a large portion of tensile stress leaving a large portion of the hard segment region non-perturbed. Only with the increase of stress can these hard segments be completely aligned. It is noticed that the scattering peak position is shifted to a lower value with strain which suggests that the long period or the adjacent hard segment domain is increased.

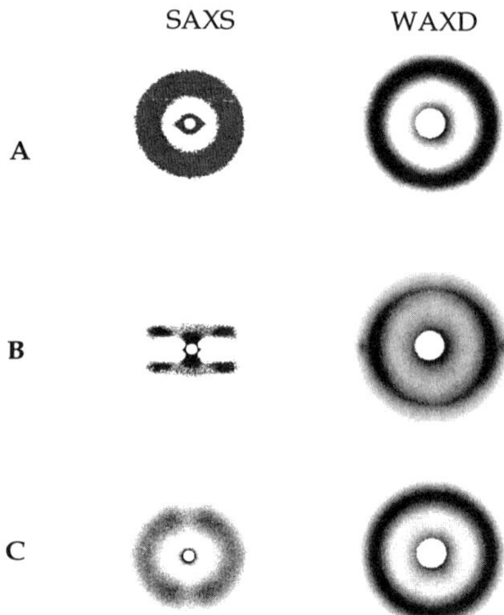

Figure 1. SAXS/WAXD patterns of PU during deformation: (A) as cast film, (B) 250% strain (deformation direction is vertical), (C) relaxation after the 250% deformation.

In the deformed SAXS pattern (strain > 200%), we notice the occurrence of a strong equatorial scattering streak around the beam stop (the origin). This equatorial streak represents the formation of a long distance correlation, which is consistent with the argument of the stress-induced crystallization in the soft segments, and it is also consistent with the corresponding WAXD pattern. From the initial film, the WAXD pattern shows a typical amorphous halo with the d-spacing about 4.5 Å. During deformation, the amorphous ring can turn into two strong peaks on the equator, which only appear as the strain is greater than 200% (the same value where the strong equatorial streak is seen in the SAXS pattern). This has also been verified before [1,2].

to the half-time. Once the half-time of 3 min. is exceeded, additional intensity grows up at higher q (shorter Bragg spacing), forming a distinct, but weak, second peak. Nearly the same trend is seen in the 124.5°C intensity. For the 124.5°C sample, the crystallization half-time is much longer (about 20min.) than for 117°C.

These results suggest that within the isothermally crystallized m-iPP sample there are two lamellar populations, with different average Bragg spacings. The second population, having shorter Bragg spacing, increases in relative proportion as crystallization time increases. The increase in SAXS intensity parallels the increase in the area of the lower melting endotherm with increasing crystallization time. SAXS scans taken during heating, but not shown here, indicate that the intensity in the higher q region (lower Bragg spacing) decreases when the temperature reaches that of the lower melting endotherm. We conclude that different lamellar populations contribute to the multiple endothermic response seen in this m-iPP after isothermal crystallization.

ACKNOWLEGMENT
Research was supported by the U. S. Army Research Office, Grant DHHA04-96-1-0009.

REFERENCES

1. J. Varga. *J. Thermal. Analy.* **35**, 1891 (1989).
2. F. L. Binsbergen and B. G. M. De Lange. *Polymer*, **9**, 23 (1968).
3. G. Guerra, V. Petraccone, P. Corradini, C. Derosa, R. Napolitano, B. Pirozzi, and G. Giunchi. *J. Polym. Sci.: Phys. Ed.* **22**, 1029 (1984).
4. R. J. Samuels. *J. Polym. Sci. Phys. Ed..* **13**, 1417 (1975).
5. Y. S. Yadav, P. C. Jain. *Polym.*, **27**, 721 (1986).
6. V. Petraccone, G. Guerra, C. de Rosa, and A. Tuzi. *Polym.* **28**, 143 (1985).
7. P. Dai, P. Cebe, R. G. Alamo, L. Mandelkern and M. Capel. Polym. Preprint (1997).
8. J. R. Isasi, R. G. Alamo, and L. Mandelkern. *J. of Polym. Sci: polym. Phys. Ed.* **35**, 2521 (1997).
9. R. G. Alamo, J. C. Lucas, and L. Mandelkern. Polym. Preprints, **35,** 406, (1994)
10. G. R. Strobl and M. Schneider. *J. Polym. Sci., Polym. Phys. Ed.*, **18**, 1343 (1980).

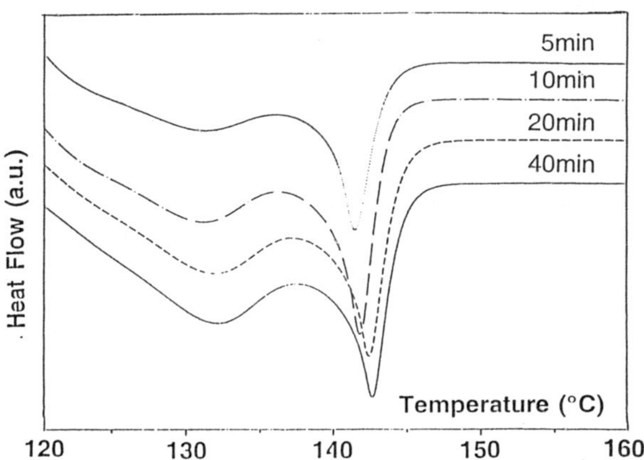

Figure 2 The DSC melting endotherms after isothermal crystallization at 117°C for 5 min, 10 min, 20 min and 40 min as indicated.

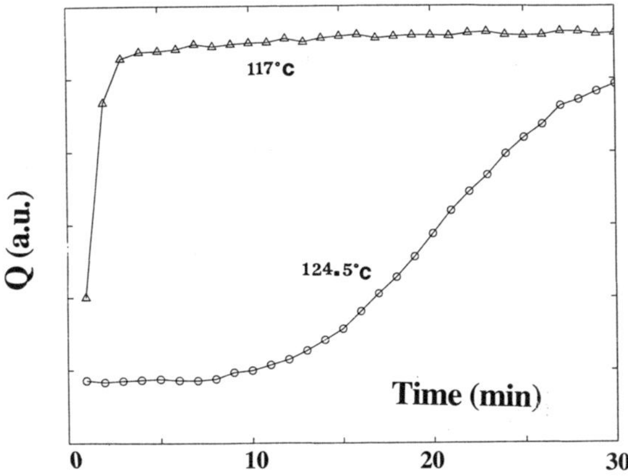

Figure 3 Time development of the scattering invariant, Q, during isothermal crystallization at 117°C (the upper curve) and 124.5°C (the lower curve).

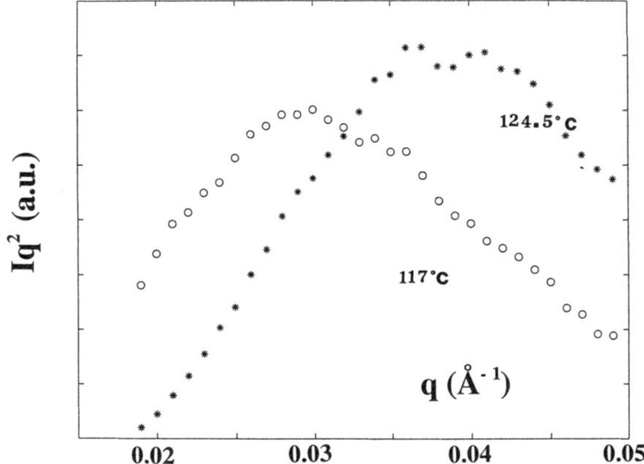

Figure 4 Lorentz corrected intensity vs. scattering vector at the isothermal crystallization temperatures of 124.5°C after 30 min. (circle) and 117°C after 45 min. (star).

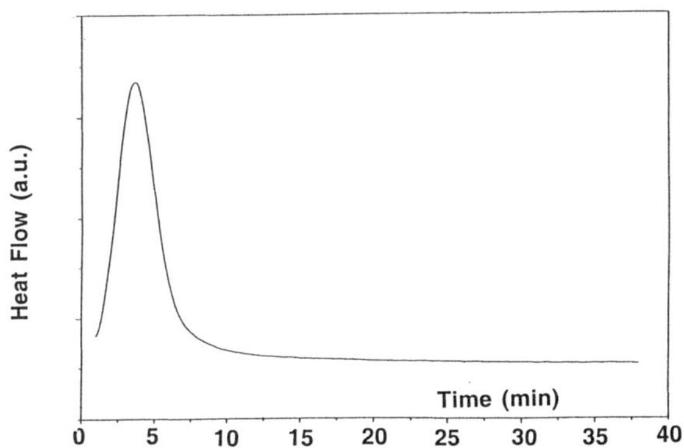

Figure 1 DSC exothermic heat flow vs. time during isothermal crystallization at 117°C.

MORPHOLOGY DEVELOPMENT IN POLYPROPYLENE HOMOPOLYMER TACTICITY MIXTURES: ISOTACTIC/SYNDIOTACTIC BLENDS

Sangkyu Kim, Zhi-gang Wang, Roger A. Phillips[1],

Fengji Yeh and Benjamin S. Hsiao

Chemistry Department, State University of New York at
Stony Brook, Stony Brook, NY 11794-3400
[1]Montell Polyolefins, R&D Center, 912 Appleton Road,
Elkton MD. 21921

INTRODUCTION

It is known that the stereochemisty of polypropylene strongly influences material properties through morphological factors such as crystallinity, melting temperature and crystal arrangements. Metallocene catalysts allow for the development of new tacticity microstructures [1], and provide the only viable route to syndiotactic polypropylene (sPP) [2]. Relative to isotactic polypropylene from Ziegler-Natta catalyst (ZN-iPP), sPP produced by these metallocene catalysts generally shows lower crystallinity, lower modulus and yield stress, lower crystallization temperature, lower melting point, and reduced flowability [1]. These attributes have limited the market impact of sPP relative to iPP. However, when used in conjunction with ZN-iPP, sPP can potentially be used to modify both rheological and solid-state processing characteristics.

Recent studies have suggested thermodynamic incompatibility of the iPP/sPP blend pair at relatively modest molecular weights [3,4]. Liquid-liquid phase separation in the melt has been claimed [3]. This might be expected to limit the applications for this blend pair. Under isothermal crystallization condition, crystallization of sPP can lag behind iPP. In this case, crystallization of the iPP/sPP blend pair at the early stages can be thought as iPP crystallization of a semi-crystalline/amorphous blend when probed by time-resolved techniques. The current study addresses aspects of morphology development of equimolecular weight iPP/sPP mixtures by thermal and synchrotron scattering techniques.

EXPERIMENTAL

Samples for blending studies are summarized in Table 1. Isopropylidene(cyclo pentadienyl)(9-fluorenyl) zirconium dichloride catalyst [2] was used to prepare sPP in bulk monomer at $30^{\circ}C$ with MAO activator at a molar [Al]/[Zr] ratio of 2000. The iPP is based on Ziegler-Natta catalyst technology. Blends were prepared with stabilized 3 wt% xylene solutions in a N_2 atmosphere at $130^{\circ}C$ for 2hr. Blend composition varied from 0-100 wt% sPP. Blends were precipitated in dry-ice chilled methanol, filtered, vacuum dried, and compression molded into 2mm thick, 1.25" diameter disks (Buehler press) at $195^{\circ}C$ for 7min (35 Mpa) and cooled in clamps to $110^{\circ}C$ prior to ejection from the mold. The molds were re-melted by similar procedures without pressure to remove residual orientation (confirmed by 2D WAXS).

Simultaneous small-angle x-ray scattering (SAXS) and wide-angle x-ray diffraction (WAXD) measurements were conducted at the Beamline X27C, in National Synchrotron Light Source (NSLS), Brookhaven National Laboratory. Two linear position sensitive detectors (European Molecular Biology Laboratory, *EMBL*) were configured for simultaneous x-ray measurements. Crystallization experiments used 7 mm diameter samples cut from molded disks and transferred by pneumatic arm between furnaces at the melt temperature (7 min, $195^{\circ}C$) and crystallization temperature ($115^{\circ}C$ or $137.5^{\circ}C$). The experimental temperature jump apparatus has been reported previously [5]. X-ray profiles were collected in 5 sec ($115^{\circ}C$) or 30 sec ($137.5^{\circ}C$) increments and normalized for incident beam intensity. Both SAXS and WAXD patterns show crystalline features at the crystallization temperature. SAXS angles were calibrated with silver behenate. WAXD pixel and intensity was calibrated by comparing synchrotron data with Scintag diffractometer data (CuKα) using a NIST corrundum plate. The angular scale of the synchrotron WAXD data (λ = 1.307 Å) was converted to a scale corresponding to λ = 1.5418 Å.

DSC experiments were carried out using a Perkin-Elmer DSC7 station purged with nitrogen and calibrated with indium and mercury. 5-6 mg samples were used over a temperature range of $-55\sim235^{\circ}C$. Heat-cool-reheat ramps were applied. All data are reported during heating following a cooling cycle at $20^{\circ}C$/min. DSC thermograms are shown after subtraction of the solid-melt baseline, which was determined by an iterative technique.

RESULTS AND DISCUSSION

Figure 1 shows DSC scans of iPP/sPP blends with various compositions of sPP. It is seen that sPP melts at temperatures approximately $20^{\circ}C$ below the pure iPP component. Multiple melting, a common feature in sPP homopolymer [6,7], is observed for sPP in blends up to 50 wt% concentration. For 0~50 wt% sPP, this feature is independent of iPP concentration. This contrasts with melting in blends of sPP and atactic polypropylene (aPP). In sPP/aPP blends [7], the addition of aPP can strongly influence the multiple melting of sPP. This is not found when iPP is the diluent. Although not shown, the iPP/sPP blends show distinct crystallization exotherms on cooling, separated by approximately $20^{\circ}C$ (at a rate of $20^{\circ}C$/min) for a 50:50 iPP/sPP blend. At this intermediate composition, the melting data in Figure 1 corresponds to morphologies formed by nearly complete crystallization of iPP prior to sPP.

The isothermal crystallization of sPP and iPP in the mixture can readily be observed in time resolved WAXD measurements. iPP and sPP have different unit cells [6], and do not co-crystallize with each other. Figure 2 shows representative WAXD patterns of a 50:50 iPP/sPP blend during crystallization at $115^{\circ}C$. The {200} reflection is a common feature to the polymorphic form in sPP following melt crystallization [8]. This reflection readily differentiates sPP crystallization from the α-monoclinic pattern of iPP [6]. Figure 2 shows that iPP crystallizes first followed by sPP. The SAXS invarient Q in Figure 3 also confirms this effect, i.e., crystallization occurs in distinct stages, corresponding to iPP and sPP growth.

The early stage crystallization of iPP in iPP/sPP blends is of interest because of the different transformation rates in iPP and sPP. The early stage growth of iPP conceptually resembles the semicrystalline/amorphous iPP/aPP blend pair [9]. Figure 4 shows the long spacing, L, versus time for iPP and sPP homopolymers and the 50:50 iPP/sPP blend during crystallization at $115^{\circ}C$. In iPP, L decreases to a nearly constant value at early times. The decrease is also observed in sPP, but delayed in time and extended to higher degrees of transformation. This indicates that a process of secondary in-filling crystallization subsequent to primary growth is prevalent in sPP, contrasting with more "compact" structures characteristic of iPP crystallization at high supercooling. The presence of sPP in the blend appears to have only a minor influence on L over entire crystallization stages of iPP (early) and sPP (later). At elevated temperature ($137.5^{\circ}C$), the blend follows more of a behavior of the rule of mixing. Figure 5 compares L for the iPP homopolymer and a 50:50 iPP/sPP blend during crystallization at $137.5^{\circ}C$. At this temperature, the sPP homopolymer shows a very low degree of transformation. WAXD results do not show crystalline sPP reflections. The long spacing of the iPP component in the presence of molten sPP does not change substantially. This observation contrasts with the high temperature crystallization of iPP/aPP blend [9], which shows a significant modification of L. At both temperatures, the values of L in sPP is larger than iPP due to its lower degree of supercooling applied (Tm of sPP is lower).

CONCLUSION

The blends studied in this work contain iPP and sPP components, which differ substantially in melting point. The multiple melting behavior characteristic of the sPP component is relatively unaffected by the presence of iPP up to iPP concentrations of 50 wt%. Crystallization on cooling occurs in stages, with the iPP component crystallizing first followed by subsequent crystallization of sPP. This conclusion is confirmed by time resolved SAXS and WAXD experiments during isothermal crystallization. During the early stage crystallization of iPP, the presence of the molten sPP component in the blend does not

We have observed an interesting relaxation behavior in this film. As the applied strain is under 100%, the deformed SAXS and WAXD images are completely reversible - they always fully recover to the initial images. However, as the applied strain is above 200%, only the WAXD image is recoverable but not the SAXS image. The SAXS pattern of the relaxed film always exhibits some orientation, which indicates the existence of some deformation-induced permanent changes in the hard segment region. These changes may involve the rearrangement of some hydrogen bonds between the urethane and the urea groups along the aligned chains. The total recovery of the WAXD image is due to the low melting point of the poly(tetramethylene oxide) crystals. As the external force is removed, these crystals are not stable at the room temperature and rapidly melt away. This can also be seen from the relaxed SAXS image. The disappearance of the strong scattering steak at the origin is consistent with the melting of the PTMeO crystals. A schematic diagram of the molecular arrangement in PU during deformation is shown in Figure 2.

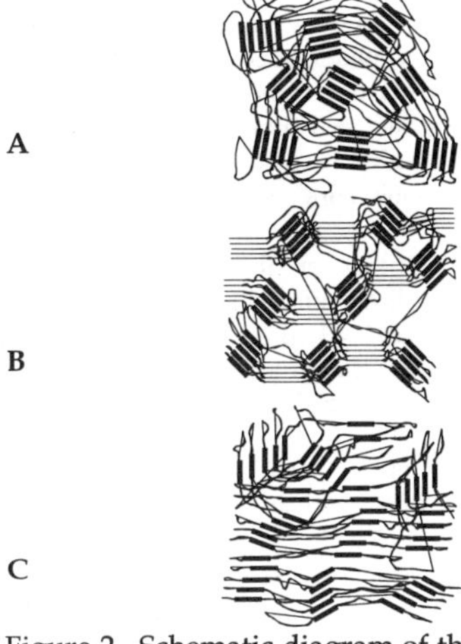

Figure 2. Schematic diagram of the molecular arrangement in PU during deformation (the stretching direction is horizontal, experimental conditions are shown in Figure 1).

Conclusion

In the deformation study, the simultaneous SAXS/WAXS 2-D images reveal detail structural and morphological information. The segregated morphology of soft and hard segments was initially isotropic. With the increase in strain, both hard and soft segments begin to oriented and the soft segments can further crystallize at strain beyond 200%. At this strain, the hard segments are titled with respect to the deformation axis. Upon relaxation, however, the oriented crystalline hard domain could not be completely recovered due to the rearrangement of hydrogen bonding between the hard segments.

Acknowledgment

We gratefully acknowledge the financial supports of this work in part by NSF-GOALI (DMA-9629825) program and in part by the AP-PRT at the NSLS.

References

1. R. Bonart, *J. Macromol. Sci. - Phys.*, **B2(1)**, 115 (1968).
2. R. Bonart and E. H. Müller, *J. Macromol. Sci. - Phys.*, **B10(1)**, 177 (1974).
3. R. Bonart, L. Morbitzer and E. H. Muller, *J. Macromol. Sci. - Phys.*, **B9(3)**, 447 (1974).
4. J. T. Koberstein and L. M. Leung, *Macromolecules*, **25**, 6205 (1992).
5. J. T. Koberstein, A. F. Galambos and L. M. Leung, *Macromolecules*, **25**, 6195 (1992).
6. V. A. Vilensky and Y. S. Lipatov, *Polymer*, **35**, 2069 (1994).
7. T. Takigawa, M. Oodate, K. Urayama and T. Masuda, *J. Appl. Polym. Sci.*, **59**, 1563 (1996).
8. B. Hsiao, B. Chu, and F. Yeh, *NSLS Newsletter*, 1, July, 1997.

SAXS and WAXD study of the Mesostructure of Molybdenum Oxide formed by a Block Copolymer Matrix

Yi Xie, Tianbo Liu, Li-zhi Liu and Benjamin Chu*
Department of Chemistry, State University of New York at Stony Brook, Stony Brook, New York 11794-3400

Introduction

A promising route to nanostructured composites uses the desired supramolecular pre-organization of microenvironment as a matrix to nucleate and grow inorganic structures. The resulting composite materials may find new potential applications due to its structure-property variation. Recently, a great deal of research has dealt with the preparation and characterization of nanostructured inorganic/organic composites. The studied inorganic materials have extended to semiconductors, catalysts, supermagnetic materials as well as biominerals [1-3]. The modified materials on a nanometer scale are designed to achieve special structures and/or functions.

Commercially available amphiphilic EPE and EBE type triblock copolymers (where E, P and B are poly (oxyethylene), poly (oxypropylene) and poly (oxybutylene) respectively) can self-assemble to form micellar structures in a selective solvent [4]. The size of the unit cell of the ordered structure is on a nanometer scale. Therefore such self-assembly matrixes can be used to synthesize inorganic compounds with nanoscale modifications. In this work, a commercial triblock copolymer $E_{45}B_{14}E_{45}$ is used as a synthetic matrix to synthesize MoO_3.

We have great interests in such oxides as MoO_3, because of its potential applicability to absorption and catalytic processes. Also it is an electrochromic and photochromic sensitive materials for optical device applications [5-7]. Study also showed that MoO_3 is a promising material for photoelectrochemical energy production, and it shown that high surface area forms have higher photo efficiencies. Modified MoO_3 mesophase may prove to be useful photocatalysts because of its templated surface areas. However, its study is just at the beginning [8,9]. In this work, Modified MoO_3 mesophase materials with a superlattice structure and ordered huge holes was synthesized in a block copolymer matrix, and such a synthetic pathway can no doubt extend the process with which new materials can be prepared.

Experimental

1. Preparation of the mesostructured molybdenum oxide by a block copolymer matrix

In this work, MoO_3 was prepared by decomposition of $MoO_2(OH)(OOH)$ solution. 55%:45% (weight ratio) of $MoO_2(OH)(OOH)$ solution and $E_{45}B_{14}E_{45}$ (a product of Dow Chemical Company and purified before use) was mixed and centrifuged at a speed of 6000rpm at room temperature for at least one day to make sure that the whole system was properly mixed. The as-formed gel was kept for at least two weeks in order to be sure that the gel had reached equilibrium. Then the whole system was refluxed in water for at least eight hours (at about 100°C). A blue precipitate was obtained in the solution. The $E_{45}B_{14}E_{45}$ was dissolved in excess water and was retrieved by pouring off the water. The blue precipitate was collected and dried at 100°C in a vacuum oven.

2. SAXS and WAXD study

The structures of the gel samples in pure water, the precursor gel before refluxing, the final products with and without the polymer matrix were determined by small-angle x-ray scattering (SAXS) and wide-angle x-ray diffraction (WAXD).

SAXS and WAXD measurements were carried out at the SUNY X3-A2 beamline of the National Synchrotron Light Source(NSLS), Brookhaven National Laboratory (BNL). The x-ray wavelength used was 1.283Å. A pinhole collimation system was used to define the incident x-ray beam. Fuji image plates were chosen as the x-ray area detector.

Results and Discussion

WAXD profiles of the products without $E_{45}B_{14}E_{45}$ matrix are shown in Fig. 1(a). All peaks can be indexed on the basis of an orthorhombic structure with a = 3.962 Å, b=13.858 Å, c=3.697 Å [10]. The strong and sharp peeks show that it has a well-defined crystallinity. The final products with $E_{45}B_{14}E_{45}$ matrix should also be MoO_3, which can be verified by the same color and the chemically inert character of $E_{45}B_{14}E_{45}$ block copolymer.

$E_{45}B_{14}E_{45}$ shows a good self-assembly format consisting of micelles with a core-shell structure in aqueous solution. The study showed that the bcc structure was formed over a large polymer concentration (27%-65%) and temperature (5-75°C) ranges [11]. A typical structure of $E_{45}B_{14}E_{45}$ hydrogel ($E_{45}B_{14}E_{45}$:H_2O is 45:55 wt%) without the existence of Mo component is shown in Fig 2(a). The peak positions of 1, $(2)^{\frac{1}{2}}$, $(3)^{1/2}$, $(4)^{1/2}$ correspond, respectively, to d_{110}, d_{200}, d_{211} and d_{220} where d_{hkl} is the lattice spacing of the (hkl) planes of the body-center cubic packing with the unit cell dimension of 11.6 nm. Furthermore, by combining the lattice structure and the basic parameters [12], the size of hydrophobic core can be estimated to be 5.7 nm.

From Fig. 2(b), One can see that at the same $E_{45}B_{14}E_{45}$ concentration, the existence of a small amount of $MoO_2(OH)(OOH)$ with a molar ratio of $MoO_2(OH)(OOH)$/$E_{45}B_{14}E_{45}$ = 3.1 did not effect the gel structure, only the peak position shifted a little towards larger q_{max} values, indicating a slightly smaller unit cell dimension.

In comparison with q_{max} value of pure $E_{45}B_{14}E_{45}$ hydrogel and of the $MoO_2(OH)(OOH)$/$E_{45}B_{14}E_{45}$/H_2O gel, the very large q_{max} values in the SAXS profiles of the final products (Fig. 2c). A peak position ratio of 1: $(2)^{1/2}$ could suggest that the bcc structure presumably remained in that product, but the interplanar spacing d_{110} = 4.6 nm instead of 8.2 nm while the unit cell dimension became 6.5 nm instead of 11.6 nm. Based on the bcc structure and the change in the d spacing, we could calculate the volume change. The domain volumes contracted greatly (up to 83%) after the formation of MoO_3 and removal of the polymer matrix. With all the polymer molecules being washed off with water, the hydrophilic region almost collapsed totally and only the hydrophobic region existed as empty holes. The broadening of the peaks in the SAXS profiles means that the mesophase has become less ordered than before.

With $MoO_2(OH)(OOH)$ being soluble in water, it is reasonable to assume that $MoO_2(OH)(OOH)$ resides only in the hydrophilic micellar shells. All reactions took place in the hydrophilic area. During the reaction, the hydrophobic cores maintained the integrity until the polymer matrix was removed by washing with water. Only the insoluble MoO_3 products remained along the empty micellar cores. However, since the hydrophilic region shrinked greatly due to the dilute concentration of MoO_3, which cannot fill all space of the former hydrophilic shell, the hydrophobic cores rearranged to a closer-packed morphology. The obtained mesophase product should consist of MoO_3 with an empty internal core (formed by the micellar cores, with an average size of 5.7 nm) and could keep the ordered cubic structure with smaller dimension. Elemental analysis showed that in the final products, the C content was less than 1.7% and the H content is as low as not detectable, implying that the polymer matrix was totally removed.

The mesophase formation and the collapse of MoO_3 structure can also be verified by WAXD patterns of the final products with the

$E_{45}B_{14}E_{45}$ matrix. Wide-angle x-ray diffraction can show the crystal structure of inorganic materials. Due to the small space between the huge hole, the MoO_3 powder could not grow into a good crystalline form. The huge holes in the inner structure of MoO_3 nanoparticles made the MoO_3 crystalline array disorderedly. Thus, WAXD of the final products of MoO_3, shown in Fig. 1(b), indicates a poor crystalline nature. The characteristic crystalline peaks in the WAXD pattern disappeared and only some amorphous humps remained at the original positions.

Conclusions

In a summary, closed-associated micelles with core-shell structures can be created by taking advantage of the self-assembly behavior of the amphiphilic block copolymer in selective solvents. The nanoscopic sizes of the unit cells in the ordered micelles can be used as good matrices for synthesizing inorganic compounds with nanoscale modifications. In this work, a commercial triblock copolymer $E_{45}B_{14}E_{45}$ was used as template which could result in the formation of a modified MoO_3 mesophase consisting of MoO_3 powders with ordered arrays of holes (former micellar cores, with an average size of 5.7 nm). Characterization of the products after removal of the polymer matrix by using SAXS, WAXD, and elemental analysis has revealed the formation of MoO_3 with an ordered structure. The mesophase with a periodic array of nanoscale pores provides an opportunity for further modification, such as by the addition of interesting electronic materials. The well-ordered nanopores in the modified MoO_3 powder with superlattice structure would make the new materials useful as an absorbent, a selective reaction medium, or a catalyst carrier.

Acknowledgment

B.C. gratefully acknowledges support of this work by the National Science Foundation.

References

1. Mann, S Nature 365(1993) 499 , and the reference there in
2. Braun, P. V; Osenar, P and Stupp, S. I. Nature 380 (1996) 325
3. Eftekharzadeh, S and Stupp S. I. Chem. Mater. 9 (1997) 2059
4. Chu, B; Zhou, Z. in "Nonionic Surfactants: Polyoxyalkylene Block Copolymers," (Ed. Nace, V. M.) , Marcel Dekker, New York, 1996, Chapter 3
5. Svachula, J ; Tichy, J and Machek. J J. Catal. Lett. 3 91989) 257
6. King, S. T.; J. Catal. 131 91991) 215
7. Yao, J. N.; Hashimoto K. and Fujishima A., Nature 335 (1991) 624
8. Honma, I and Zhou, H. S. Chem. Mater. 10(1998) 103
9. Janauer, G. G; Dobley, A.; Guo, J; Zavalij, P. and Whettingham, M. S.; Chem. Mater. 8 (1996) 2096
10. JCDPS No. 5-508
11. Liu, T.; Xie, Y.; Nace, V. M.; Chu, B., to be published
12. Wu, C.; Liu, T.; Chu, B; Schneider D. K. and Graziano V. Macromolecules 30 (1997) 4574

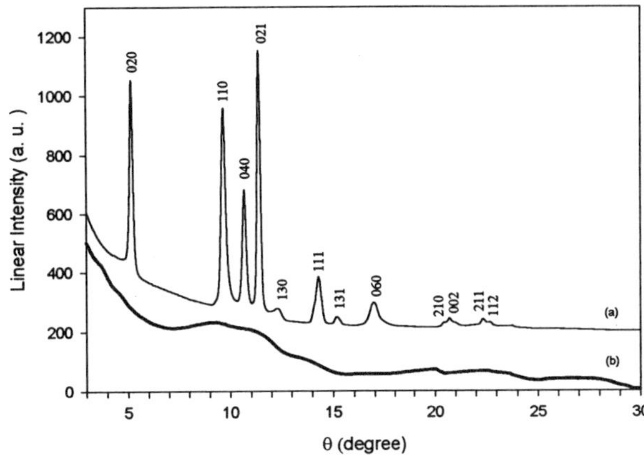

Fig.1. Wide-angle X-ray Diffraction (WAXD) profiles of the products synthesized (a) in $E_{45}B_{14}E_{45}$ matrix , (b) without $E_{45}B_{14}E_{45}$ matrix

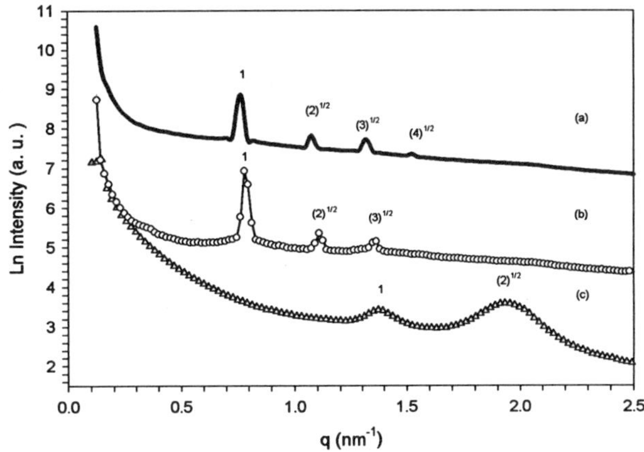

Fig. 2 Small-angle X-ray Scattering (SAXS) profiles of (a) $E_{45}B_{14}E_{45}$ /H_2O gel without Mo component ($E_{45}B_{14}E_{45}$:H_2O=45:55 wt%), (b) $MoO_2(OH)(OOH)$/$E_{45}B_{14}E_{45}$/H_2O gel (Mo/$E_{45}B_{14}E_{45}$=3.1 (molar ratio), $E_{45}B_{14}E_{45}$:H_2O = 45:55 wt%) and (c) the final products with $E_{45}B_{14}E_{45}$ matrix removed off

REAL-TIME SAXS AND THERMAL ANALYSIS STUDY OF MULTIPLE MELTING BEHAVIOR IN PEEK

Georgi Georgiev[1], Patrick Shuanghua Dai[1], Elizabeth Oyebode[1], Peggy Cebe[1]*, and Malcolm Capel[2]

1 Dept. of Physics and Astronomy, Tufts University, Medford, MA 02155;
2 Biology Dept., Brookhaven National Laboratory, Upton, NY 11973
*To whom correspondence should be addressed

1. INTRODUCTION

The explanation for multiple peaks seen in the endothermic response of high performance polymers is an important problem in polymer physics. Assignment of peaks to structural entities within the material, the relative perfection of the crystals, and the possibility of their reorganization, are all influenced by the melt processing history. With the advent of high intensity synchrotron sources of X-radiation, polymer scientists gain a research tool which, when used along with thermal analysis, provides additional structural information about the crystals during growth and subsequent melting[1-6].

Our group has been studying high performance polymers, such as poly(phenylene sulfide), PPS, and poly(etheretherketone), PEEK, using wide and small angle X-ray scattering (WAXS, SAXS)[1,2] and thermal analysis [4-6], including temperature-modulated differential scanning calorimetry (TM-DSC) [2,6]. We proposed a model to explain multiple melting endotherms in PPS, treated according to one- or two-stage melt or cold crystallization[4,5]. Key features of this model[4] are that multiple endotherms: 1. are due to reorganization/recrystallization after cold crystallization; and 2. are dominated by crystal morphology after melt crystallization at high T . In other words, multiple distinct crystal populations are formed by the latter treatment, leading to observation of multiple melting.

Recent SAXS studies of PEEK by other researchers [7-14] have lead to different conclusions about the origin of multiple endotherms. Formation of dual endothermic response has been variously ascribed to lamellar insertion [7,8], dual crystal populations [9,10,15], or melting followed by recrystallization [11,12]. To investigate these ideas, we used SAXS and DSC to compare results of single and dual stage melt crystallization of PEEK using a treatment scheme involving annealing/crystallization at T_{a1} followed by annealing at T_{a2}, where either $T_{a1}<T_{a2}$ or $T_{a1}>T_{a2}$.

2. EXPERIMENTAL SECTION

PEEK 450G pellets (ICI Americas) were the starting material for this study. Films were compression molded at 400°C, then quenched to ice water. Fig. 1 shows the treatment schemes for dual stage crystallization. Samples were heated to 375°C in a Mettler FP80 hot stage and held for three min. to erase crystal seeds before cooling them to T_{a1}. In the first treatment (upper curve, Fig. 1), the samples were held for 10 min. at $T_{a1} = 280°C$. Samples were held at T_{a2} for a period of time, t_a ($t_a = 0, 1, 3.5$ or 10min) then immediately heated to 360°C. From growth rate studies [13], the crystallization half-time ranges from 1 min. at 300°C to 6 sec at 280°C, and crystals already nucleated by the time the sample reached 280°C. In the second treatment (lower curve, Fig. 1) samples were held at $T_{a1} = 310°C$ for different crystallization times t_c ($t_c = 0, 3, 15$ or 30 min), then cooled to 295°C and held 15 min. Rates were 5°C/min./-20°C/min. Thermal analysis was performed using a TA Instruments 2920 DSC. Temperatures were calibrated using In standard. Sample mass was around 8 mg. N_2 was used to purge the cell, at 30ml/min.

SAXS experiments were performed at Brookhaven with the sample inside a hot stage. Details of the experiment were published previously [16]. One dimensional electron density correlation function was obtained by discrete Fourier analysis of the Lorentz corrected intensity [17].

3. RESULTS AND DISCUSSION

Endothermic heat flow vs. T is shown in Figure 2 for single stage (curves a,c) and two-stage (curves b,d) melt crystallization. Curve a shows scanning after 10min. isothermal treatment at 280°C. Curve b adds a second stage, 10 min. at 295°C (refer to curve a, Fig. 1). Curve c shows scanning after direct cooling to 295°C and holding for 15 min. And finally, curve d shows the effects of 30 min. isothermal at 310°C prior to the second stage of 15 min. at 295°C (refer to curve b, Fig. 1). In all cases, to resolve the lower endotherms, the samples were first jumped to lower T before scanning.

In Fig. 2, each of the first two scans shows dual endothermic behavior, and the minor lower temperature endotherm occurs just above the preceding treatment temperature. There follows a large endotherm which peaks at about 345°C. Melting is completed by 350°C. The location, shape, and total area of the small endotherm depends on the sample's prior thermal history, specifically on the isothermal holding times, given that all other parameters are held constant. In addition to the formation of the most perfect crystals (forming the high melting

endotherm), we also observe the development of the less perfect population of crystals during isothermal periods, which gives the melting peak with lower melting temperature. As the holding time at 295°C increases, the area of the small peak increases and its melting temperature increases slightly. This indicates that greater amounts of polymer material crystallize during the 295°C isotherm as time increases. The principal effect of the second stage 295°C treatment is to perfect the population of crystals initially formed below 295°C.

The scan in Fig. 2d shows triple endothermic behavior, and the first minor endotherm occurs just above 295°C, the treatment temperature of the second stage. Next follows the second minor endotherm from the first crystallization at 310°C. In addition to the formation of the most perfect crystals (forming the highest melting endotherm), we also observe the development of the two less perfect populations of crystals during isothermal periods, which gives the melting peaks with lower melting temperature. As the holding time at 310°C increases, the area of the middle small peak increases and its melting temperature increases slightly. This indicates that greater amounts of polymer material crystallize during the 310°C isotherm as the holding time increases. At the same time the material left to crystallize at 295°C becomes less and less with the increase of the first holding time. The effect of the 295°C stage is to form a third, least perfect population of crystals.

Lorentz corrected intensity, Is^2, vs. s vs, time data matching two of the thermal treatments are shown in Fig. 3. Scans matching curve 2b (10min at 280°C+10min. at 295°C) are shown in Fig. 3a. Scans matching curve 2d(30 min. at 310°C + 15min. at 295°C), are shown in Fig. 3b. The time axis is marked to show the isothermal periods.

Correlation function analysis of these intensity profiles yields the structural parameters shown in Figure 4a,b. The lower curve is the long period, L, the middle curve is the linear stack crystallinity, w_c, and the upper curve is the scattering invariant, Q, all determined in the usual way [17]. Fig. 4a shows the effects of the "low-to-high" two stage crystallization. L decreases during holding at 280°C, increases during heating to 295°C as the first low endotherm is passed, slightly decreases at 295°C, then increases during heating through the melting region. The increase in L between 280°C and 295°C is much greater than can be accounted for by thermal expansion effects in PEEK [1]. L increases due to melting of the small imperfect crystals which form the lower endotherm in Fig. 2, curve b. Changes in L are mirrored in w_c and Q. Fig. 4b shows the effects of "high-to-low" two stage crystallization. Here, L decreases during holding at 310°C, decreases slightly upon cooling to 295°C, drops again during holding at 295°C, then increases steadily during the heating scan. The drop in L between stages is a few angstroms, just what is expected based on thermal contraction [1]. The larger decrease during holding at 295°C is due to additional imperfect crystals forming. These crystals create the lowest endotherm seen in Fig. 2, curve d.

After dual stage melt crystallization with $T_{a1}<T_{a2}$, the lower melting endotherm arises from holding at T_{a1}. During cooling a broad distribution of crystals forms, and the low-melting tail is perfected during T_{a1}. Heating to T_{a2} melts these imperfect crystals and allows others with greater average long spacing to form in their place. After dual stage crystallization with $T_{a1}> T_{a2}$, the amount of space remaining for additional growth at T_{a2} depends upon the holding time at T_{a1}. The long period of crystals formed at T_{a2} is smaller than that formed at T_{a1} due to growth in a now-restricted geometry.

Research was supported by NASA Grant NAG8-1167.

REFERENCES

1. S. X. Lu, P. Cebe and M. Capel. *Polymer*, 37(14), 2999(1996). and 2. *Macromol.*, 30(20), 6243 (1997). 3. P. Huo and P. Cebe. *Macromol.*, 25, 902(1992). 4. J. S. Chung and P. Cebe *Polymer*, 33(11), 2312(1992). 5. Ibid p2325. 6. S. X. Lu and P. Cebe. *Polymer Comm.*, 37(21), 4857(1996). 7. B. Hsiao, K. Gardner, D. Wu, B. Chu. *Polymer*, 34, 3986 (1993). 8. A. Jonas, T. Russell, D. Yoon. *Macromol.*, 28, 8491 (1995). 9. B. Hsiao, B. Sauer, R. Verma, H. Zachmann, S. Seifert, B. Chu, P. Harney. *Macromol.*, 28, 6931 (1995). 10. K. Kruger, H. Zachmann. *Macromolecules*, 26, 5202 (1993). 11. R. Verma, et al. *Polymer*, 37, 5357 (1996). 12. C. Fougnies, P. Damman, D. Villers, M. Dosiere, M. Koch. *Macromol.*, 30, 1392 (1997); and, Ibid. 30, 1385(1997). 13. B. Hsiao, B. Sauer. *J. Polym. Sci., Polym. Phys. Ed.*, 31, 901 (1993). 14. B. Sauer, B. Hsiao. *J. Polym. Sci., Polym. Phys. Ed.*, 31, 917 (1993). 15. P. Cebe and S.-D. Hong. *Polymer*, 27, 1183 (1986). 16. G. Georgiev, P. Dai, E. Oyebode, P. Cebe, M. Capel, PMSE Proceedings, 78, 215 (1998). 17. G. R. Strobl, M. Schneider. *J. Polym. Sci., Polym. Phys. Ed.*, 18, 1343 (1980).

Fig. 1 Temperature vs. time for the two thermal treatments. **Fig. 2** Heat flow vs. temperature during DSC scanning of samples treated by annealing 10 min. at 280°C followed by annealing at 295°C for: a) 0min.; b 10min.; and by annealing 15min. at 295°C, previously treated at 310°C for : c) 0min.; d) 30min. **Fig. 3** Lorentz corrected intensity Is² vs. s vs. time for PEEK a.) Treated as in Fig. 2b; b.) Treated as in Fig. 2d. **Fig. 4** Scattering invariant-Q, Crystallinity-Wc and Long period-L, for PEEK a.) Treated as in Fig. 2b; b.) Treated as in Fig. 2d.

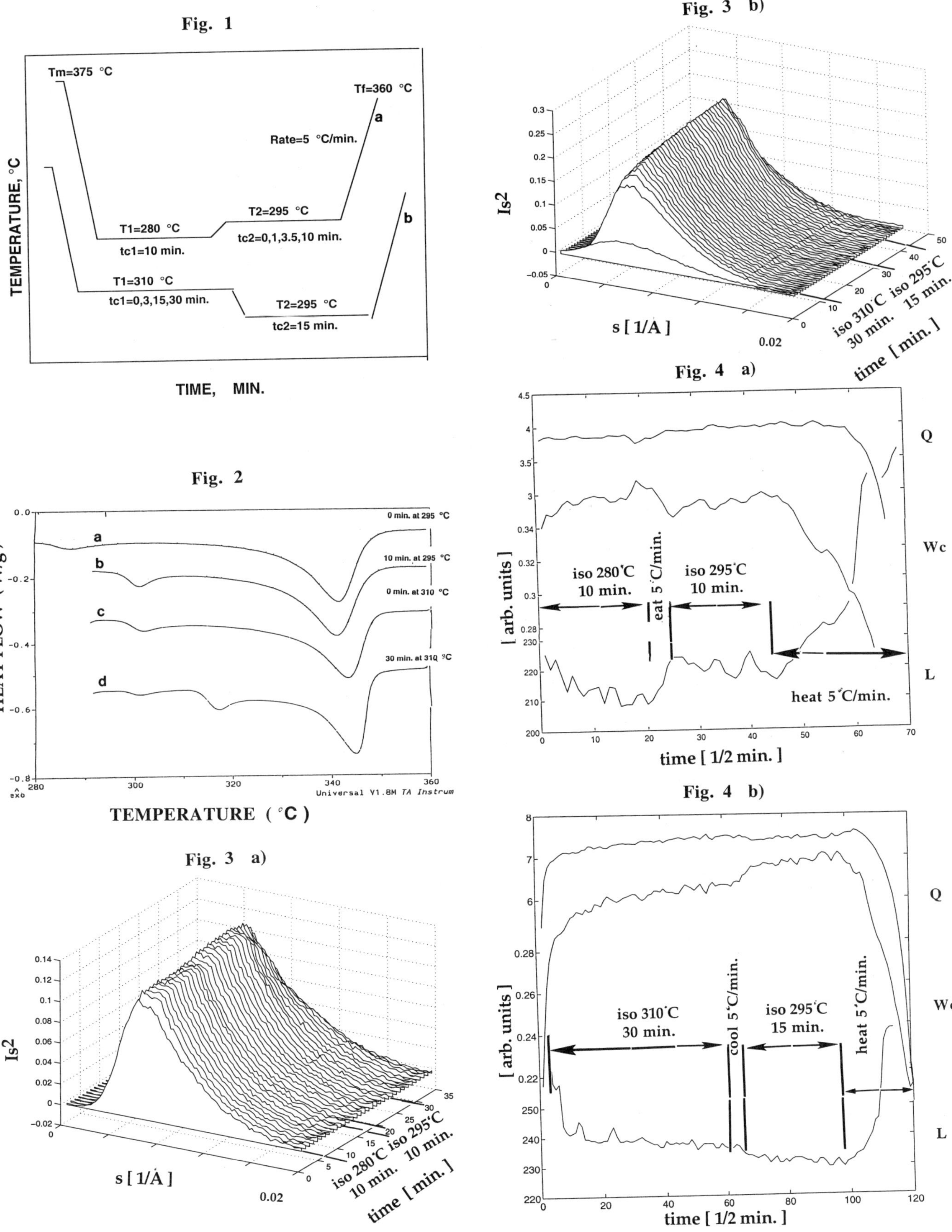

Fig. 1
Tm=375 °C
Tf=360 °C
a
Rate=5 °C/min.
T2=295 °C
T1=280 °C
tc2=0,1,3.5,10 min.
tc1=10 min.
b
T1=310 °C
tc1=0,3,15,30 min.
T2=295 °C
tc2=15 min.
TEMPERATURE, °C
TIME, MIN.

Fig. 2
0.0
a
0 min. at 295 °C
b
10 min. at 295 °C
c
0 min. at 310 °C
-0.2
-0.4
d
30 min. at 310 °C
-0.6
HEAT FLOW (W/g)
-0.8
280   300   320   340   360
exo
Universal V1.8M TA Instrum
TEMPERATURE (°C)

Fig. 3   a)
0.14
0.12
0.1
0.08
0.06
0.04
0.02
0
-0.02
Is2
0
s [1/Å]
0.02
35
30
25
20
15
10
5
0
iso 280°C iso 295°C
10 min.   10 min.
time [min.]

Fig. 3   b)
0.3
0.25
0.2
0.15
0.1
0.05
0
-0.05
Is2
0
10
20
30
40
50
s [1/A]
0.02
iso 310°C iso 295°C
30 min.   15 min.
time [min.]

Fig. 4   a)
4.5
4
3.5
3
0.34
0.32
0.3
0.28
230
220
210
200
Q
Wc
L
[arb. units]
iso 280°C
10 min.
eat 5°C/min.
iso 295°C
10 min.
heat 5°C/min.
0   10   20   30   40   50   60   70
time [1/2 min.]

Fig. 4   b)
8
7
6
0.28
0.26
0.24
0.22
250
240
230
220
Q
Wc
L
[arb. units]
iso 310°C
30 min.
cool 5°C/min.
iso 295°C
15 min.
heat 5°C/min.
0   20   40   60   80   100   120
time [1/2 min.]

Characterization of Charged PAMAM Dendrimer Interactions in Solution by Small Angle Neutron Scattering

Giovanni Nisato, Robert Ivkov, Barry J. Bauer, and Eric J. Amis

Polymers Division, National Institute for Standards and Technology, Gaithersburg, MD 20899

Introduction

The nature of the interactions between macroions in solution has received considerable experimental and theoretical attention through the years. Often, experimental efforts are directed toward a search for a suitable model system that allows direct comparison with these theories. Model systems should consist of molecules with 1) uniform size and shape, 2) ionizable surfaces that can support large variations of charge density, and 3) remain soluble over a wide range of conditions to allow sampling over a sufficiently large parameter space. Dendrimers are macromolecules which are believed to possess such characteristics and are potentially ideal model systems for exploring the physics of macroions.

Dendrimers are highly branched macromolecules that consist of a multifunctional core from which monomer units extend radially outward, producing molecules with a well defined number of end groups and narrow molecular weight distribution. These molecules are synthesized in a step-wise manner so that each new step, or generation, doubles the molecular weight, number of end groups, and number of branch points. In addition, dendrimers become larger and more spherical with each generation.[1] The most frequently studied dendrimers contain amine- terminated end groups; the surface charge densities of these molecules can be manipulated by varying the solution pH. [2]

Experimental

Materials. Generation 5 (G5) polyamidoamine (PAMAM) dendrimers were obtained from Dendritech, Inc., Midland, MI.[3] The G5 dendrimers were prepared from a 22% w/w aqueous solution in the following manner. Aliquots of the original solution were diluted with ultrapure H_2O (18 MΩ cm) and dialyzed to remove low molecular weight impurities. Dialysis was performed using a Spectrum macrodialyzer with a 10 mL half-cell volume and a 12 - 14 kD MWCO cellulose membrane. H_2O solvent was exchanged by dialysis against 99.9% (low paramagnetic) D_2O (Cambridge Isotopes, Inc.). The final D_2O volume fraction of the solvent was $\geq$ 95%. The parameters reported here were supplied by the manufacturer and were not measured at NIST.

The theoretical 128 terminal amine $(-NH_2)$ groups on a G5 PAMAM dendrimer can be ionized to the $(-NH_3^+)$ form through the addition of acid. We approximate the extent of ionization by the stoichiometric ratio of added HCl to the total number of terminal groups in solution,

$$\alpha = [\text{moles HCl}] / [\text{moles NH}_2]. \qquad (1)$$

The values of dendrimer volume fraction, ϕ, and α were calculated from the weight of the added components and assume values of density, molecular weight and functionality reported elsewhere.[4] The total uncertainties cannot be calculated from these reported numbers, but it is reasonable to estimate the values to be within 10% of the shown values.

SANS. We obtained SANS data on the 8 m and 30 m instruments located at the Center for Neutron Research at the National Institute of Standards and Technology in Gaithersburg, Maryland.[5] The 8 m measurements were performed using a sample to detector distance of 360 cm and an average neutron wavelength of 9 Å with a spread $\Delta\lambda/\lambda$ = 0.25. This gave a q range of 0.008 Å^{-1} to 0.123 Å^{-1}. The 30 meter SANS experiments were conducted using several detector configurations and 6 Å ($\Delta\lambda/\lambda$ = 0.10) neutrons, yielding a q range of 0.004 to 0.31 Å^{-1}. All samples were contained in quartz cells having circular windows and path lengths of 1 mm, 2 mm, and 5 mm. Raw (2-D) data were reduced following standard methods. An additional incoherent background scattering from the hydrogen within the dendrimers was also subtracted. The corrected intensities were converted to an absolute scale, and were radially averaged to give the $I(q)$ for each sample.[5] Experimental standard deviations in the scattered intensities were smaller than the plotted data points and are not displayed for clarity.

Generally, the total scattering, $I(q)$, from a solution of interacting particles can be written as,[7]

$$I(q) = \kappa^2 P(q) S(q) \qquad (2)$$

where κ is the contrast, $P(q)$ is the single particle form factor (intraparticle interference), and $S(q)$ is the solution structure factor (interparticle interference). In systems of non-interacting particles, $S(q) = 1$, and the total scattering reflects the nature of the individual particles. However, if the scattering particles are charged, $S(q) > 1$, and the scattered intensity will exhibit features that reflect interparticle effects.

Results and Discussion

Figure 1 shows SANS from solutions in which the dendrimer volume fraction was fixed at ϕ = 4.2% and α was varied from 0.0 to 1.0.

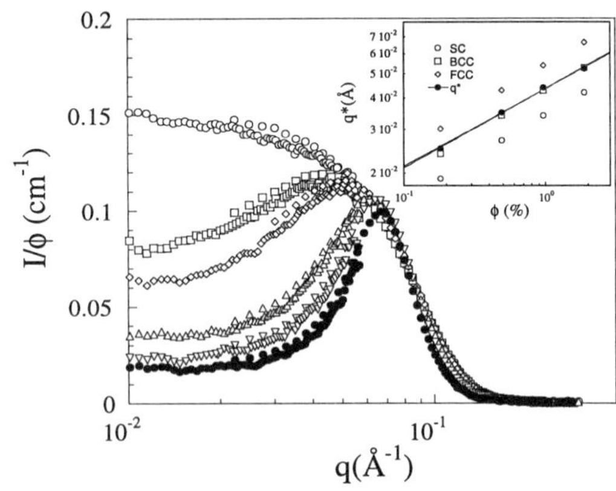

Figure 1: Effect of the addition of HCl to solutions of G5 PAMAM dendrimers with ϕ = 4.2%. Symbols represent α = 0 (open circles), α = 0.06 (squares), α = 0.1 (diamonds), α = 0.3 (triangles), α = 0.5 (inverted triangles), α = 1 (filled circles). The inset shows the peak position, q^* as a function of dendrimer concentration. For comparison, the expected results for simple cubic (SC), body-centred cubic (BCC), and face-centred cubic (FCC) packing for spheres with radius, $R = 31$ Å are shown.

Addition of acid produces a peak in $I(q)$ that becomes sharper with increasing α. This peak is a common feature of salt free polyelectrolyte solutions, and has been attributed to contributions from the structure factor, $S(q)$. As the charge on each dendrimer increases, electrostatic repulsions give rise to an average distance of closest approach. Data obtained from dendrimer solutions with constant α ($\alpha = 0$) and varying dendrimer concentration, Figure 1 inset, show that the peak position, q^*, scales with concentration as ϕ^β, where $\beta = 0.32 \pm 0.01$.[8] This scaling suggests a simple volume expansion of the interparticle separation as the dendrimer concentration decreases. Thus q^* can be related to the average interparticle spacing, and can be seen as indicative of local liquid-like ordering of the macroions. These results are consistent with the observations of Briber et al. of G7 PAMAM dendrimers.[9]

Dendrimer interactions can be screened by the addition of salt. Figure 2 shows that we recover single particle scattering, $P(q)$, with the addition of NaCl to 1 mol/L. For systems of interacting particles, the intensity at $q = 0$, $I(0)$, can be related to the osmotic compressibility by standard thermodynamic relationships.[7] The slope of a plot of $\phi / I(0)$ vs. ϕ can be used to evaluate the nature of the interactions between molecules in solution, where $I(0)$ is obtained by extrapolation from scattering data. The inset of Figure 2 shows that for all concentrations of salt studied, $\alpha = 0.6$, we obtain a positive slope for a range of dendrimer concentrations, suggesting that short ranged repulsive interactions dominate under these conditions. These results are in stark contrast to the behavior observed in many charged colloids and polyelectrolytes, which tend to show attractive interactions upon the addition of salt, i. e. salting out.

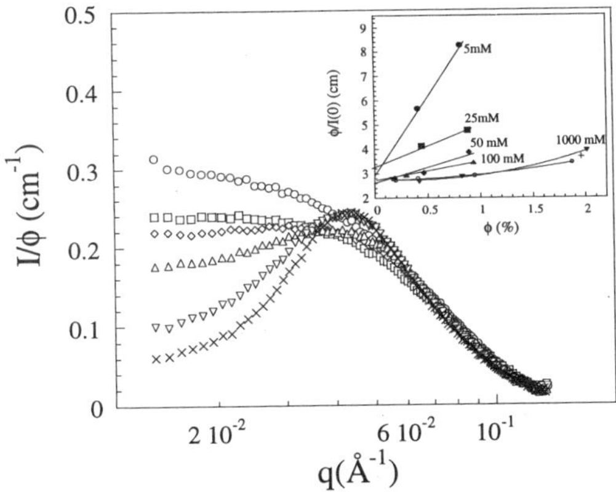

Figure 2: Effect of the addition of NaCl to charged dendrimer solutions, $\alpha = 0$ and $\phi = 4\%$. Added salt concentrations are 1.0 mol/L (circles), 0.1 mol/L (squares), 0.050 mol/L (diamonds), 0.025 mol/L (triangles), 0.005 mol/L (inverted triangles), 0.0 mol/L (crosses). Note that at 1 mol/L NaCl, we recover single particle scattering. The inset shows a plot the inverse of I(0) extrapolated from data shown in the Figure. Lines are drawn as visual aids. Circles in the inset refer to data from dendrimer in pure D_2O ($\alpha = 0$, no added salt). The slope is indicative of the interparticle interactions in the solution. A positive slope indicates repulsive interactions.

Conclusions

SANS measurements have been conducted on G5 PAMAM dendrimers in D_2O to determine their relevance as model polyelectrolytes. We show that the dendrimers studied behave as non-interacting particles in dilute solutions in D_2O, but that the addition of acid charges the dendrimers. Scattering from charged dendrimer solutions exhibits a peak that scales with the exponent 0.32 ± 0.01 with volume fraction, indicating liquid-like ordering. The interactions can further be manipulated by the addition of salt which screens the electrostatic repulsions. Extrapolations of these data to $q = 0$ yield the osmotic compressibility. For the system studied, we find positive values of these quantities. Taken together, these results suggest that PAMAM dendrimers are good candidates for model colloids since there is a large parameter space over which these solutions can be manipulated and measured. Comparisons of these data with theoretical models are ongoing.

Acknowledgments

G. Nisato gratefully acknowledges the financial support by the Blancefort Ludovisi-Boncompagni Foundation. Special thanks to Donald Tomalia for providing the dendrimers. This material is based upon work supported in part by the U.S. Army Research Office under contract number 35109-CH.

References

(1) T. J. Prosa, B. J. Bauer, E. J. Amis, D. A. Tomalia, and R. Scherrenberg, *J. Polymer Sci.* **1997**, 35, 2913.

(2) H. Zhang, P. L. Dubin, R. Spindler, and D. *Tomalia Ber. Bunsenges. Phys. Chem.* **1996**, 100, 923.

(3) Certain commercial material and equipment are identified in this publication in order to specify adequately the experimental procedure. In no case does such identification imply recommendation by the National Institute of Standards and Technology nor does it imply that the material nor equipment identified is necessarily the best available for this purpose.

(4) S. Uppuluri, D. A. Tomalia, P. R. Dvornic, *ACS PMSE Preprints*, **1997**, 77, 116.

(5) Prask, H. J.; Rowe, M.; Rush, J. J.; Schroeder, I. G., *J. Res. Natl. Inst. Stand. Tech.*, **1993**, 8, 1.

(6) *SANS Data Reduction and Imaging Software*; Cold Neutron Research Facility, NIST: Gaithersburg, MD 20899, **1996**.

(7) J. S. Higgins, and H. C. Benoît *Polymers and Neutron Scattering*, Oxford Press, Oxford, 1994.

(8) The standard uncertainty is calculated from the goodness of the fits.

(9) R. M. Briber, B. J. Bauer, B. Hammouda, and D. A. Tomalia, *ACS PMSE Preprints*, **1992**, 67, 430.

Dynamic Structure Development during Isothermal Crystallization and Melting of Linear Polyethylenes

J. Lopez, Z.-G. Wang, B. S. Hsiao* and P. J. Armistead[+]

Department of Chemistry, State University of New York at Stony Brook, Stony Brook, NY 11794-3400

[+]Materials Chemistry Branch, Navel Research Laboratory, Washington DC 20375

Introduction

The crystallization behavior of polyethylene, PE, has been a subject of immense studies for over five decades. Many excellent reviews and books are available for this topic [1-4]. Recently, the concept of metastability [5,6] has prompted people to revisit the thermodynamic and kinetics aspects of crystallization in PE. It is thought that the structure and morphology development of polymers is a delicate balance between the thermodynamic and kinetic factors. Many variables can affect this balance such as temperature, pressure, molecular weight and molecular orientation. In this paper, we intend to examine the effect of molecular weight on the dynamic nature of structural development during crystallization and melting of linear polyethylene. To avoid the complicated effect of polydispersity, we have used the fractionated samples from the National Institute of Standards and Technology (NIST), and prepared their blends. The synchrotron wide-angle x-ray diffraction (WAXD) technique was employed to follow the real time changes in crystallinity, unit cell parameters, and ratio of the two crystalline reflections (200/110). This ratio representing the intensity contributions from the two different crystal planes can reveal information about the crystal perfection and size, which indirectly reflects the change of local environment during crystallization and melting transitions.

Experimental

Four different linear polyethylene samples from NIST were used in this study: SRM1475A (Mw: 53,070, Mn: 18,300), designated as 53K; SRM1483 (Mw: 32,100, Mn: 28,900), designated as 32K; SRM1484A (Mw: 119,600, Mn: 100,500), designated as 119K; and the blend of 119K/32K (24/76 wt. ratio). The 53K is a non-fractionated sample with large polydispersity. Both 32K and 119K samples were fractionated from the 53K sample. The blend has an effective molecular weight of 53K with a bimodal distribution of low and high molecular weights. The calculated equilibrium melting temperature (T_m^o) for the 32K fraction sample is 143.65°C, and for the 119K fraction sample is 144.95°C. All the samples were hot pressed into thin films for x-ray measurements.

Time-resolved WAXD measurements were carried out using a linear position sensitive detector (from European Molecular Biology Laboratory, *EMBL*) at the Advanced Polymer Beamline (X27C) of the National Synchrotron Light Source (NSLS), Brookhaven National Laboratory (BNL). The wavelength of the x-ray λ was 1.307 Å, which was defined by a double multi-layer monochromator. The synchrotron x-rays were collimated with a three-pinhole device [7]. Two data acquisition times were used: 10 and 30 s (depend on the crystallization rate). The angular range for the WAXD measurement was $5° < 2\theta < 38°$. A dual-chamber temperature jump apparatus was used for isothermal crystallization and melting measurements. Detailed description of this apparatus has been provided previously [8]. All samples were first equilibrated at 155°C for 5 min prior to the temperature jump. The chosen crystallization temperature was 123°C. At the thermal equilibrium, the fluctuation in temperature was less than ±0.5°C. For the melting experiment, a rate of 5°C/min was used to heat the pressed film sample as prepared. The data acquisition time was 10 s in this case.

In the WAXD analysis, the integrated intensity, peak position, peak height, peak width and peak area for each crystal reflection peak were extracted by a custom made curve-fitting program. A broad Gaussian peak was used to describe the amorphous background. All other two peaks (110 and 200) were also fitted with Gaussian functions. By dividing the sum of the crystalline reflection intensities, I_c, to the overall intensity, I_{total}, an estimate of the crystallinity ϕ_c (by weight) was obtained. The peak positions of (110) and (200) were used to determine the unit cell parameters of a and b.

Results and Discussion

Prior to this study, all the samples were first investigated by time-resolved small-angle x-ray scattering, to reveal the information of morphological variables. We found that the crystal thickness and the lamellar long period are probably larger than 1000 Å (maximum resolution of the SAXS facility at Beamline X27C). As a result, no time-resolved SAXS experiments were conducted. Figure 1 illustrates the development of crystallinity with time during isothermal crystallization for several samples of different molecular weights. It is seen that the level of achievable crystallinity increases with decreasing molecular weight (for example, 62% for the 32K sample and 48% for the 119K sample). In addition, the rate of crystallization also increases with decreasing molecular weight. The 53K sample with a broad polydispersity shows a higher level of final crystallinity than the blend sample of equivalent molecular weight with a bimodal distribution. This sample also has a faster crystallization rate than the blend.

The dynamic nature of crystallization process can be recognized by the changes in unit cell parameters (a and b) with time at isothermal temperature of 123°C, which is illustrated in Figure 2. Both cell parameters show a noticeable decrease with time (the rate of the cell dimension reduction is shown in the figure). This suggests that the unit cell density does increase with time – an indication of crystal perfection. (With the chosen samples, we cannot observe any correlation between the crystal perfection and molecular weight.) The ratio of the 200/110 reflections is shown in Figure 3. In this figure, it is seen that the two fractions of PE and the blend sample exhibit a similar ratio, independent of time. The value of the 200/110 ratio for the 53K sample (with a large polydispersity) is significantly higher than the rest samples. Our explanation to this is as follows. As the crystal growth direction of PE is along the b-axis, typically, a larger crystal dimension along the b-axis leads to a smaller value of the 200/110 ratio. This is consistent with our initial investigation of the large crystal size in the fraction samples. Therefore, a decrease in the crystal dimension or perfection along the b-axis will increase the ratio of 200/110, which may be this case.

During heating, the changes of unit cell parameters with temperature are shown in Figure 4. It is seen that both parameters increase with temperature in all fraction samples and the blend, which indicates a reduction of crystal density. A detailed analysis indicates that the increase in the b-parameter is significantly lower than the increase in the a-parameter, which is constant with the large crystal dimension and perfection along the b-axis. The 53K sample shows an unusual feature, i.e., the slope of the b-parameter has a negative temperature dependence. The corresponding plot of the 200/110 ratio also mirrors the unusually large value for the 53K sample. Again, we attributed this observation to the defective or highly branched like crystal morphology, which suppressed the intensity contribution from the 110 plane.

Conclusion

The dynamic process of crystal structure development during isothermal crystallization and melting of linear polyethylene with different molecular weight and polydispersity was followed by time-resolved WAXD measurement. From the WAXD profiles, we have estimated the evolution of crystallinity, the change of unit cell parameters a and b, and the variation of the ratio between two crystal reflections (200/110). It is seen that the crystallinity increases with decreasing molecular weight. Both cell parameters are found to decrease with time during isothermal crystallization, which suggests that the crystal structure becomes more perfect. All fractionated samples with narrow polydispersity show a similar value of the 200/110 ratio, which is markedly smaller than that of the sample with large polydispersity. The large value of the 200/110 ratio in the 53K sample perhaps reflects the defective and branching crystal morphology. Upon heating, the cell parameters are found to expand anisotropically. The thermal expansion of the a-parameter is considerably larger than b.

Acknowledgement

The authors acknowledge the financial support of this work by a grant from NSF (DMR 9732653).

References

1. B. Wunderlich, *Macromolecular Physics*, volume 2, Academic Press, NY (1976).
2. P. Geil, *Polymer Single Crystal*, Interscience, NY (1963).
3. L. Mandelkern, *Crystallization of Polymers*, McGraw-Hill, NY (1964)
4. For Example. R. G. Alamo, E. K. M. Chan, L. Mandelkern and I. G. Voigt-Martin, *Macromolecules*, **1992**, *25(24)*, 6381.
5. A. Keller, M. Hikosaka, S. Rastogi, A. Toda, P. Barham, C. Goldback-Wood, *J. Mater. Sci.*, **1994**, *29*, 2579.
6. S. Z. D. Cheng and A. Keller, *Adv. Polym. Sci.* (1997)
7. Chu, B.; Harney,P.; Li, Y; Linhiu, K.; Ye, F.; Hsiao, B. *Rev. Sci. Instru.* **1994**, 65(3), 597.
8. Hsiao, B. S.; Gardner, K.H.; Wu, D.Q.; Chu, B. *Polymer* **1993**, 34 (19), 3996.

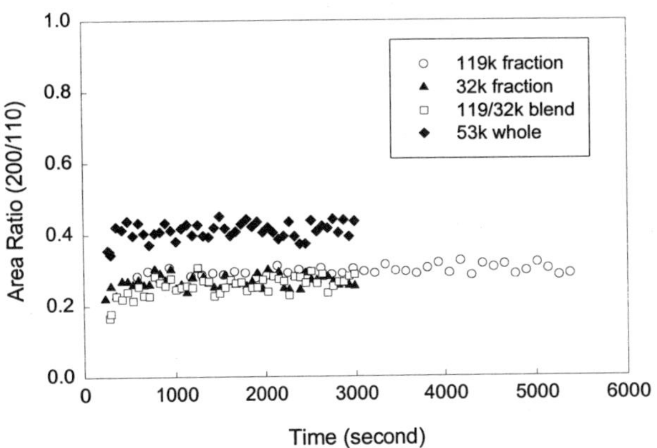

Figure 3. The ratio of areas between the 200 and 110 peaks during isothermal crystallization at 123°C.

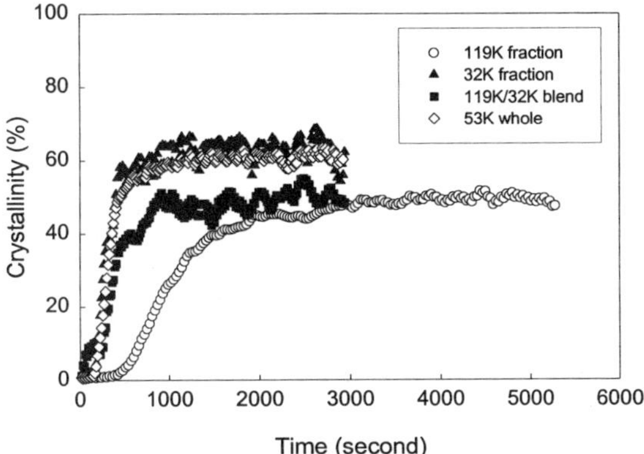

Figure 1. The degree of crystallinity by WAXD during isothermal crystallization at 123 °C.

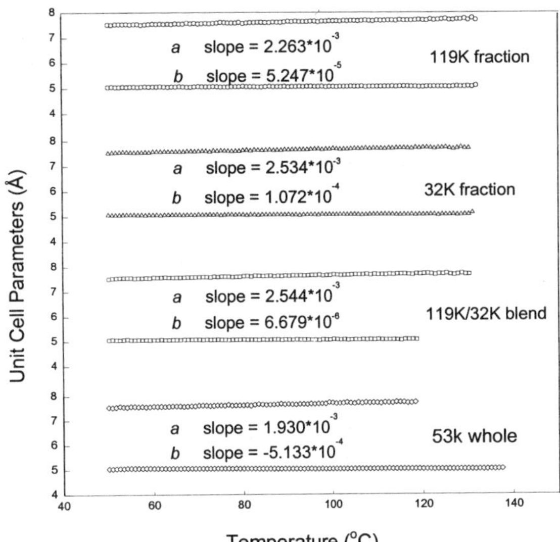

Figure 4. The evolution of the unit cell parameters (*a* and *b*) during melting at a rate of 5°C/min.

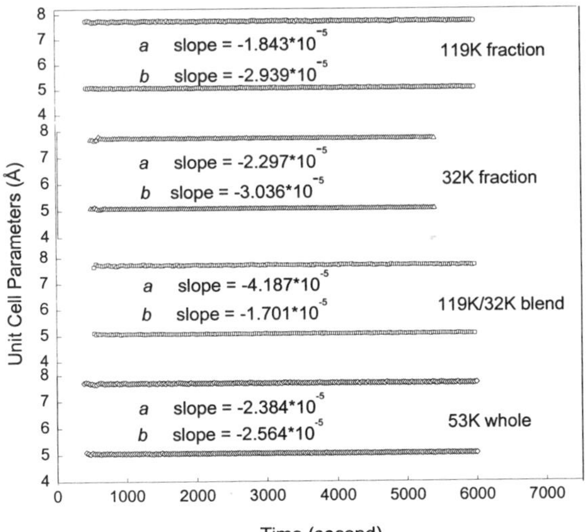

Figure 2. The evolution of the unit cell parameters (*a* and *b*) during isothermal crystallization at 123°C.

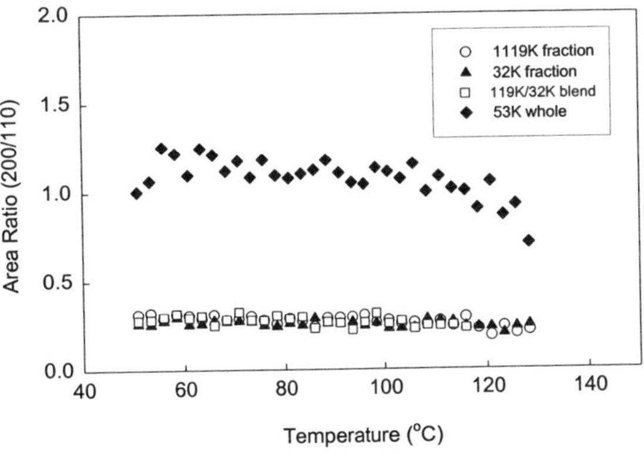

Figure 5. The ratio of areas between the 200 and 110 peaks during melting at a rate of 5°C/min.

PHASE BEHAVIOR OF STYRENE-ISOPRENE DIBLOCK COPOLYMERS IN SELECTIVE SOLVENTS. C. Lai, W. B. Russel, R. A. Register, Department of Chemical Engineering, Princeton University, Princeton, NJ 08544

We report on the phase behavior of polystyrene-polyisoprene diblock copolymers in concentrated tetradecane solution. The phase boundaries as functions of temperature and concentration are investigated using small-angle x-ray scattering. Qualitatively, the topology of the phase diagram is consistent with considering the isoprene and tetradecane to form an "effective" polyisoprene volume fraction. At low polystyrene contents, body centered cubic arrays of polystyrene spheres are observed. A lamella-gyroid-cylinder-sphere structure sequence was observed in a symmetric PS-PI (weight fraction of polystyrene ~ 50%) diblock with progressive tetradecane addition. For styrene-rich polymers (50-70% polystyrene) similar morphology sequences can be generated, but polymers with higher polystyrene content are macroscopically insoluble in tetradecane.

BLOCK COPOLYMER MICELLAR NETWORKS AND MESOPHASES. SHEAR AND TEMPERATURE DEPENDENCE. K. Mortensen, K. Almdal, Riso National Laboratory, Roskilde, Denmark. R. Kleppinger, N. Mischenko, H. Reynaers, Katholieke Universiteit, Leuven, Belgium.

Symmetric triblock copolymers mixed with a solvent selective for the middle blocks, provide a basis for formation of novel physical networks, where the "cross-links" are formed by self-assembled end-blocks. Triblock copolymers of poly(styrene)-poly(ethylene butylene)-poly(styrene) may constitute such a network system. We have used a rheometer modified for in situ neutron scattering and rheological measurements, thus providing the possibility for studying the correlation between microscopic structure and macroscopic mechanical behavior. The investigations show that the self-association of the PS-blocks promote the formation of highly interconnected micelles. Within a given temperature range, the network constitutes a system of cubic ordered micelles, which upon shear a twinned aligned into a macroscopically ordered twin structure.

LAMELLAR VERSUS SUPRAMOLECULAR SCATTERING IN SEMI-CRYSTALLINE POLYMERS. J. D. Barnes, NIST Polymers Division, Gaithersburg, MD 20899.

Recent studies on syndiotactic polystyrene (sPS) have demonstrated that the observed SAXS patterns are the superposition of a strongly peaked component arising from the lamellar organization of the crystallizable microstructural elements and a component that varies smoothly as q^{-p} where $p \sim 2.5$ over most of the observable q range. We are able to separate these two contributions by varying the specimen temperature in SAXS experiments on oriented specimens between room temperature and approximately 200°C, which is still well below any temperature at which annealing takes place. The power law behavior found in sPS implies the presence of density fluctuations whose character is distinct from the liquidlike behavior normally ascribed to amorphous polymers. This presentation extends the interpretation of the earlier data on sPS and adds results from several other polymers in an attempt to determine whether the two-component picture is applicable to other semicrystalline polymers.

MORPHOLOGY OF MICROFIBRILLAR REINFORCED COMPOSITE MATERIALS BASED ON POLYMER BLENDS -- STRUCTURAL CHANGES AS REVEALED FROM TIME-RESOLVED SAXS/WAXD. M. I. Sarkissova, G. Groeninckx, M. Koch, H. Reynaers, Laboratory on Macromolecular Structural Chemistry, Department of Chemistry, Catholic University of Leuven, Celestijnenlaan 200F, B-3001 Heverlee-Leuven, BELGIUM

The morphology of new composite materials based on polymer blends has been investigated. The reinforcing elements are the basic morphological entities of oriented polymers, the microfibrils. The essential stages of composite preparation are as follows: (i) reactive blending and extrusion, (ii) fibrillization step by drawing (with good orientation of all components) and (iii) isotropization step after annealing at constant strain above the melting point of lower-melting component. When blend partners are condensation polymers, as a result of chemical reactions, a copolymer is formed which plays the role of compatibilizer. Semicrystalline structure of the blend phases before, during and after fibrillization has been determined using real-time SAXS/WAXD.

PROBE DIFFUSION IN HIGH MOLECULAR WEIGHT HYDROXYPROPYLCELLULOSE. SOLUTIONLIKE AND MELTLIKE REGIMES. K. A. Streletzky, G. D. J. Phillies, Physics Department, WPI, Worcester, MA 01609.

We studied translational diffusion of monodisperse spheres (diameters 14<d<455nm) in aqueous 1MDa hydroxypropylcellulose (HPC) (0<c<15g/L) at 25C, using quasi-elastic light scattering. Spectra are highly bimodal. Two spectral modes ("slow", "fast") have different physical properties. Probe behavior differs between small (d<Rh) and large (d>Rg) probes; Rh and Rg are the chain hydrodynamic radius and the radius of gyration, respectively. The spectral lineshape parameters depend differently on c, d, scattering vector, and solution viscosity for "fast" and "slow" modes of small and large probes. Solutions of 1MDa HPC exhibit a solutionlike-meltlike transition at C=6g/L, at which the concentration dependence of the viscosity switches from a stretched exponential (at c<C) to a power law (at c>C). This transition affects probe transport; a radical change in the fast mode behavior is found at C, proving the physical reality of the solutionlike-meltlike transition.

DYNAMICS OF MYOSIN GEL FORMATION ANALYZED BY LIGHT SCATTERING. J. Kawahara, Y. Kobayashi, K. Kubota*, H. Takaya, National Institute of Materials and Chemical Research, Tsukuba, Ibaraki 305-8565, *Gunma University, Kiryu, Gunma 376-8515, JAPAN

Real-time observation of the gel formation process is critical, not only to pursue the possibility to control the formed gel structure -- and eventually the formed gel function -- but also to deeply understand the nature of the gel structure itself, although it is a quite difficult task. We have succeeded in making this task a rather easy one, by exploiting a combination of giant molecules and a chemical cross-linking reaction between them, namely, a myosin molecule (more than 160 nm in its length) and glutaraldehyde (GA) as a cross-linking reagent. It seems that quite slow diffusion of the myosin molecule and the stability of GA in aqueous solution explains the reason why we can well control the gelation process, even at a very slow reaction rate. The observation of this system by light scattering revealed the formation and growth of spatially inhomogeneous gel structures and the partial formation of long-range networks already in the early stages. Further, the observed oscillation of a photon autocorrelation function in resonance to external mechanical vibration showed a new possible methodology -- microscopic real-time characterization of the elasticity of the networks just on the way of its formation.

EFFECT OF SOLVENT QUALITY ON THE MOLECULAR DIMENSIONS OF PAMAM DENDRIMERS. B. S. Bauer[1], A. Topp[1], D. A. Tomalia[2], E. S. Amis[1], National Institute of Standards and Technology, Gaithersburg, MD 20899, [2]Michigan Molecular Institute, Midland, MI 48640.

The average dimensions of poly(amido amine) (PAMAM) dendrimers of generation numbers (G5 and G8 are measured by small angle neutron scattering as a function of solvents of different quality, of the composition of mixtures of a good solvent and a non-solvent, and of temperature. The radius of gyration of the (G8 dendrimer, R_g, decreases for the series of solvents $D(CD_2)_mOD$ (with m = 0, 1, 2, 4) by approximately 10% with decreasing solvent quality. The amount of swelling of PAMAM dendrimers, as reflected by R_g, is not influenced by the composition of mixtures of methyl alcohol/acetone in the full range of solubility of the dendrimers. In the temperature range -10°C $\leq T$ 50°C, the dimensions of the (G8 and G5 dendrimer are constant within $\approx$ 5% of the R_g value. The average segment densities are calculated for (G5 and G8 under various solvent conditions, with the dendrimers containing volume fractions of solvent between 0.43 and 0.60.

SMALL ANGLE NEUTRON SCATTERING INVESTIGATIONS AND MOLECULAR DYNAMICS SIMULATIONS OF HYBRID LINEAR DENDRITIC STAR COPOLYMERS IN POLAR AND NONPOLAR SOLVENTS. M. S. Diallo, Materials and Process Simulation Center, Beckman Institute, California Institute of Technology and Departments of Civil Engineering and Chemistry, Howard University, D. Yu, J. M. J. Fréchet, Department of Chemistry, University of California at Berkeley, P. Miklis, W. A. Goddard, III, Materials and Process Simulation Center, Beckman Institute, California Institute of Technology.

Hybrid linear-dendritic star copolymers respond to changes in the polarity of the solvents used to dissolve them by forming unimolecular/multimolecular micelles of different size and core/shell structure. In this paper, we present the results of small angle neutron scattering (SANS) investigations and molecular dynamics (MD) simulations of hybrid linear-dendritic star copolymers in dilute solutions of tetrahydrofuran, benzene, chloroform and methanol.

CORONAL STRUCTURE OF BLOCK IONOMER MICELLES: AN INVESTIGATION BY SANS. D. Nguyen, BNL, Department of Physics, M. Moffitt, Y. Yu, A. Eisenberg, McGill University, Department of Chemistry, Montreal, Canada; D. Schneider, V. Graziano, BNL, Department of Biology, NY.

Coronal structure of starlike block Ionomer micelles of Polystyrene-b-Poly(Cesium Acrylate) in toluene, are studied by SANS. Contrast is provided by deuterated labels situated at three different distances from the ionic core. For deuterated probes situated at the end of the corona chain, the scattered intensity is found to scale as I ~ q^-1.7, indicating a semidilute environment for the labels and swollen coil behavior at small length scales. For probes situated between the corona and the core, single chain scaling is not observed at any length scale, and scattering intensity is found to scale as I ~ q^-1.3 at intermediate q. The stiffness increases further for coronal chains next to the core, as indicated by I ~ q^-1 scaling for probes attached directly to the ionic core; these chains exhibit rigid-rod behavior at length scales less than ~76Å. Interparticle scattering as a function of polymer concentration is also investigated.

SMALL ANGLE NEUTRON SCATTERING FROM MODEL PDMS NETWORKS. B. D. Viers*, G. Beaucage[a], J. E. Mark*, Polymer Research Center, *Department of Chemistry, P.O Box 210172, [a]Materials Science and Engineering Department, P.O. Box 210012; University of Cincinnati, Cincinnati, Ohio 45221

Small Angle Neutron Scattering (SANS) experiments have been carried out on several model PDMS elastomers swollen to equilibrium in a good solvent, d6 benzene. In accord with previous results, the structure of all elastomers showed a solution like behavior at small length scales (high q); whereas deviations occurred at larger length scales (low q) indicating the presence of concentration fluctuations. These fluctuations have been tentatively described as "clusters" of high crosslink density. Interestingly, the SANS response of model phase separated networks, containing micron size phases of apparently disparate crosslink density, also exhibit this anomalous low q SANS scattering. The aggregate results that the crosslinking clusters disinterpenetrate upon swelling.

SCALING EFFECTS IN SEMI-DILUTE POLYMERIC SOLUTIONS: MOLECULAR WEIGHT EFFECTS. B. D. Viers*, G. Beaucage[a], J. E. Mark* Polymer Research Center, *Department of Chemistry, P.O Box 210172, Materials Science and Engineering Department, P.O. Box 210012; University of Cincinnati, Cincinnati, Ohio 45221

Correlation length scaling laws have been established for semidilute polymer solutions and experimentally proven by small angle scattering techniques. The results have been couched in DeGennes' "blob" model, which presupposes an intermolecular region of ideal excluded volume, and thus the molecular weight of the polymeric species has generally been very high in order to foment the formation of the blob. We have been able to show via small angle neutron scattering (SANS) that a similar correlation scaling exists in moderately concentrated PDMS/d6 benzene solution, with the molecular weight of the PDMS chains varying from 300 (a trimer) to 500,000. The correlation scaling did not appear to be affected by the molecular weight distribution; however, the selective "blending" of constituent chains could elucidate any effect.

RAMAN STUDIES OF REORIENTATION OF CRYSTALLINE PHASES IN POLY(ETHYLENE TEREPHTHALATE) FIBERS DURING CONTINUOUS DRAW PROCESSES. C. C. C. Lesko, J. F. Rabolt, Department of Materials Science and Engineering, University of Delaware, Newark, DE 19716; D. B. Chase, DuPont Central Research and Development, Wilmington, DE

In order to optimize polymer properties it is important to develop non-destructive evaluation techniques which are able to characterize orientation and order at the molecular level in both amorphous and crystalline regions. One of the few techniques which meet all these criteria is Raman scattering. This study has been undertaken to evaluate the feasibility of utilizing "real-time" Raman measurements in aggressive environments to provide an instantaneous assessment of the effect of temperature and draw ratio on tensile properties. In this investigation in-situ Raman measurements were made on poly(ethylene terephthalate) fibers during the continuous draw process. Reorientation of the crystalline phase within the fibers was calculated for a range of draw ratios and hot pin temperatures. Comparisons of draw ratios and orientation functions are given. Preliminary observations on the relationship between orientation and tenacity are discussed.

Investigating the Mechanisms of Polymer Crystallization by SAXS-Experiments

G. Hauser, J. Schmidtke, <u>G. Strobl</u>, T. Thurn-Albrecht[+]

Fakultät für Physik der Albert-Ludwigs-Universität, 79104 Freiburg

[+]Max-Planck-Institut für Polymerforschung, 55021 Mainz

Germany

1 Introduction

Crystallization of polymers generally leads to the formation of layer-like crystallites with typical thicknesses in the range of 3-50 nm. They are arranged in stacks, being separated by amorphous layers. Electron microscopy and small angle X-ray scattering are the primary tools used for studies of these nano-structures. Both are needed. Electron microscopy directly shows the basic structural elements, and SAXS has to be employed when studying the changes with time, for example during a crystallization, or on cooling or heating. This contribution deals with some recent SAXS-studies carried out on polypropylene and polyethylene which provided new insights into the kinetics and mechanisms of structure change.

2 SAXS-data analysis

For the system under study, being set up of stacks of thin laterally extended crystallites, the scattering intensity can be related to the one-dimensional electron density correlation function $K(z)$ defined as

$$
\begin{aligned}
K(z) &= <(\rho_e(z)-<\rho_e>)(\rho_e(0)-<\rho_e>)> \\
&= <\rho_e(z)\rho_e(0)> - <\rho_e>^2
\end{aligned} \tag{1}
$$

It can be directly obtained by a Fourier transformation of the scattering intensity. If we use the scattering cross-section per unit volume $\Sigma(q)$, $K(z)$ follows as

$$
K(z) = \frac{1}{r_e^2}\frac{1}{(2\pi)^3}\int_0^\infty \cos qz \cdot 4\pi q^2 \Sigma(q)\mathrm{d}q \tag{2}
$$

q denotes the scattering vector, being related to the Bragg-scattering angle ϑ_B by

$$
q = \frac{4\pi}{\lambda}\sin\vartheta_B \tag{3}
$$

(r_e: classical electron radius).

In the case of partially crystalline polymers, a trajectory along the surface normal passes through amorphous regions with an electron density $\rho_{e,a}$ and crystallites with a core density $\rho_{e,c}$. As shown by Ruland [1], for such a layer system the second derivative of the correlation function, $K''(z)$, gives the distribution of distances between interfaces, in the form

$$
K''(z) = \frac{O_{ac}}{2}\Delta\rho_e^2 \left[h_a(z) + h_c(z) - 2h_{ac}(z) + h_{aca}(z) + h_{cac}(z) \cdots \right] \tag{4}
$$

The expression between the brackets is set up of a series of distribution functions, whereby the subscripts indicate which phases, amorphous and crystalline ones, are to be traversed while going from one interface to the other. Of main importance are the first three contributions. They give the distributions of the thickness of the amorphous and the crystalline layers respectively (h_a, h_c) and of the sum of both (h_{ac}) which is identical with the long spacing L. If the boundaries between the crystalline and the amorphous regions are sharp within the resolution limit of SAXS experiments, the asymptotic behavior of $\Sigma(q)$ in the limit of large scattering angles is given by Porod's law

$$
\lim_{q\to\infty} \Sigma(q) = r_e^2\frac{P}{(q/2\pi)^4} \tag{5}
$$

Hereby P denotes the Porod coefficient which is directly related to the specific internal surface O_{ac} (the area per unit volume of the interface separating crystalline and amorphous regions), by

$$
P = \frac{1}{8\pi^3}O_{ac}\Delta\rho_e^2 \tag{6}
$$

Rather than by calculating the second derivative of the correlation function, $K''(z)$ can also be deduced directly from the measuring data, by use of

$$
K''(z) = \frac{2}{r_e^2(2\pi)^2}\int_0^\infty [\lim_{q\to\infty} q^4\Sigma(q) - q^4\Sigma(q)]\cos qz\mathrm{d}q \tag{7}
$$

3 Instrumentation

SAXS experiments were conducted with the aid of a Kratky camera attached to a conventional X-ray source which was equipped with a temperature controlled sample holder. The changes of the structure during isothermal crystallization and the melting on heating was followed measuring the scattering curves with a position sensitive metal wire detector. A few minutes counting time were sufficient for a curve registration. Deconvolution of the slit-smeared data was achieved applying an algorithm developed in our group [2].

4 Kinetics of primary crystallization

4.1 The s-PP scenario

Experiments were carried out on a syndiotactic polypropylene [3] and two samples of syndiotactic poly(propene -co- octene) [4], synthesized by S. Jüngling in the Institute of Macromolecular Chemistry of our university using a metallocene catalyst. The chemical properties of the two samples are given in Table 1 (co-unit contents were determined by NMR spectroscopy).

Table 1: s-Polypropylene and s-poly(propene-co-octene). Properties of samples.

sample	octene-units		meso-diads	M_n	M_w/M_n
	%(weight)	%(mole)	%		
s-PP	0	0	3	104.000	1.7
s-P(P-co-0)4	4	1.7	(3)	73.000	2.1
s-P(P-co-0)15	15	6.4	(3)	94.000	1.7

The crystallite formation during isothermal crystallization was studied in time-dependent SAXS experiments. Subsequent to the isothermal crystallization the melting was monitored in temperature dependent SAXS measurements up to the melting point. Fig. 1 displays as an example the curves $K''(z)$ for s-P(P-co-O)4 and s-P(P-co-O)15 during and subsequent to isothermal crystallization processes. The formation and melting of the crystallites shows up in the changing height of the ridge dominating $K''(z)$. The location of the ridge gives the crystal thickness. As is observed, this thickness remains constant through both the time of isothermal crystallization and the subsequent melting. Curves indicate thicknesses of 5.7 nm and 3.5 nm for the s-P(P-co-O)4 and s-P(P-co-O)15 respectively. Observations were similar for s-PP and all crystallization temperatures. From the measurements we obtained for the three samples the dependencies of the crystallite thickness (d_c), the (Avrami-) time of crystallization (τ) and the melting point (T_f) on the crystallization temperature (T_c). The relationships are displayed in Figs. 2 and 3.

The outcome can be summarized as follows. With increasing content of non-crystallizable units (octene-units or meso-diads) one observes, as expected, a systematic shift of the melting points to lower temperatures and similar shifts of the crystallization time versus temperature curves, but surprisingly, no effect at all on the crystal thickness. The thicknesses of all three samples show a common temperature dependence, being inversely proportional to the supercooling below the equilibrium melting point of perfect syndiotactic polypropylene. The latter is located at 196 °C, as determined by an extrapolation based on measured melting points. Crystal thicknesses (being unaffected by the presence of non-crystallizable units) and growth rates (being strongly affected) are independent properties.

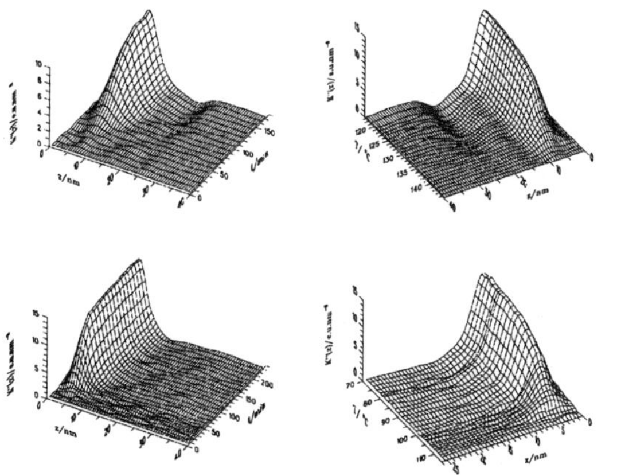

Figure 1: Functions $K''(z)$ giving the distribution of crystal thicknesses, obtained by time- and temperature-dependent SAXS experiments: Changes during isothermal crystallization of s-P(P-co-O)4 at 116 °C (top left) and of s-P(P-co-O)15 at 70 °C (bottom left) and during the subsequent melting processes (right hand sides).

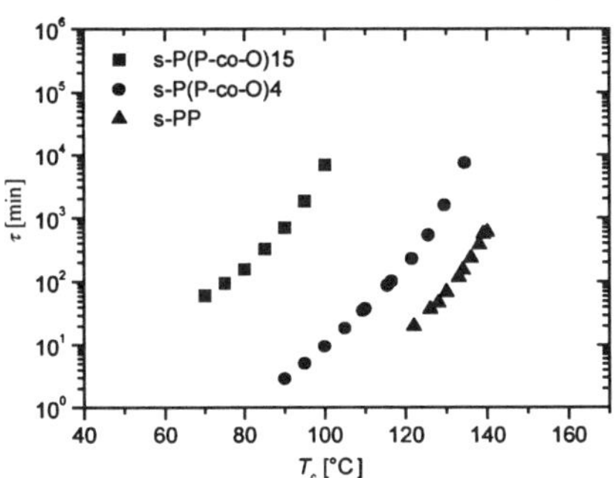

Figure 2: s-PP and s-P(P-co-O): Avrami times of crystallization in dependence on the crystallization temperature.

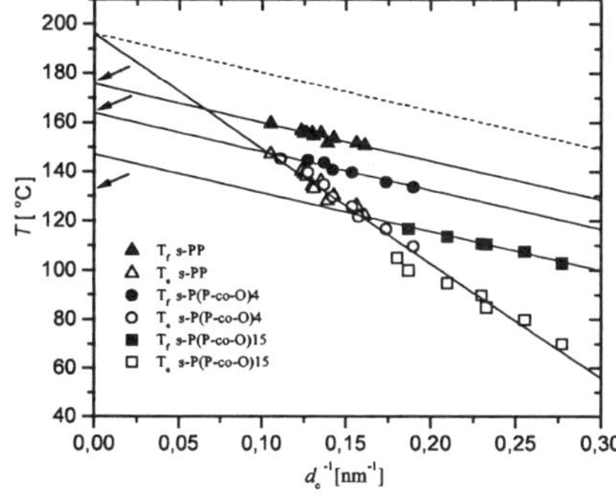

Figure 3: s-PP and s-P(P-co-O): Relations between the inverse crystallite thickness, the crystallization temperature and the melting peak. The dashed line represents the $T_f(d_c^{-1})$-dependence of a perfect s-PP as obtained by an extrapolation procedure.

Results may be understood as indicating that crystallization from the melt takes place in two steps. The first step is the formation of imperfect 'native' crystallites with a high surface free energy. They change into the final form by structural relaxation processes at the surfaces. The difference between the temperatures of crystallization and melting is due to this stabilization process, which leaves the crystallite thickness unchanged.

4.2 PE: Kinetics of crystal thickening

Different from s-PP, crystallites of polyethylene thicken with time. This is clearly demonstrated by time-dependent SAXS experiments carried out during an isothermal crystallization [5]. Fig. 4 shows a typical result, in a comparison of the interface distribution function $K''(z)$ obtained at the beginning and the final stage of the crystallization process. At the begin $K''(z)$ is dominated by the crystallite contribution (d_c), which is indicative for isolated crystallites. During further development the contribution associated with the amorphous layers (d_a) grows and finally reaches the same amplitude as that of the crystallites. Simultaneously the latter is shifted to higher values of z. Observations thus show the change from individual lamellae to the final stack which occurs together with a crystal thickening. For the thickening we found a logarithmic time dependence. Data are given in Fig. 5, together with the time dependence of the Porod coefficient $P \sim O_{ac}$.

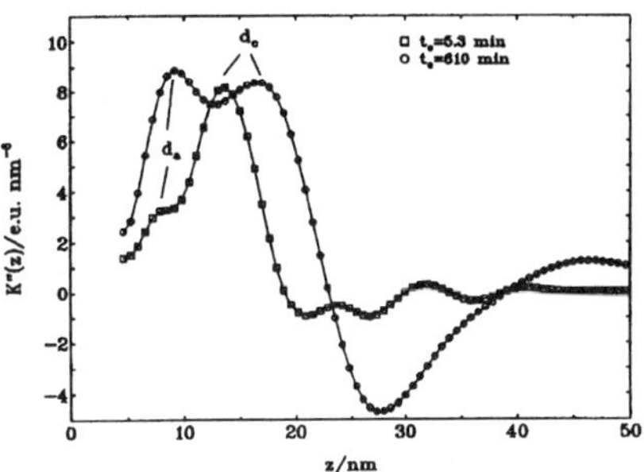

Figure 4: PE: Interface distribution function at the beginning and the end of an isothermal crystallization at 121 °C.

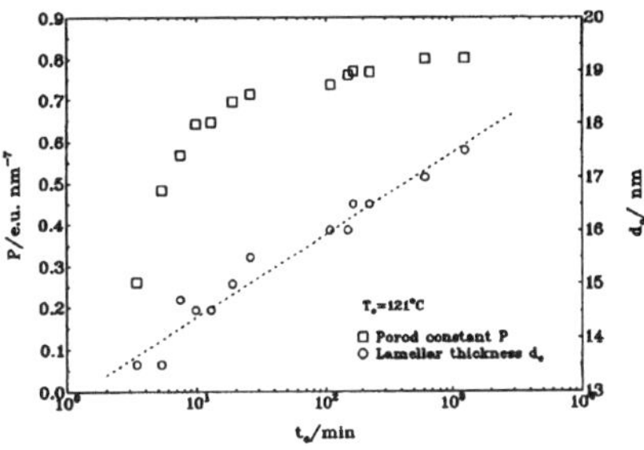

Figure 5: Evolution with time of the crystal thickness and the inner surface.

5 Mechanisms of secondary crystallization

5.1 PE: Evidence for surface crystallization and melting

Cooling a sample after completion of an isothermal crystallization at elevated temperatures generally results in a further increase of the crystallinity. In the discussion of the physical processes contributing to this 'secondary crystallization' two mechanisms are considered. First, it can be connected with the formation of new crystallites, being inserted into the intervening spaces. Second, a reversible thickening of the existing lamellae can be envisaged. Carrying out SAXS-experiments a discrimination is possible. Fig. 6 shows a result which is typical for linear polyethylene [6]. Interface distribution functions were obtained after an isothermal crystallization at 124 °C and a subsequent cooling to 78 °C.

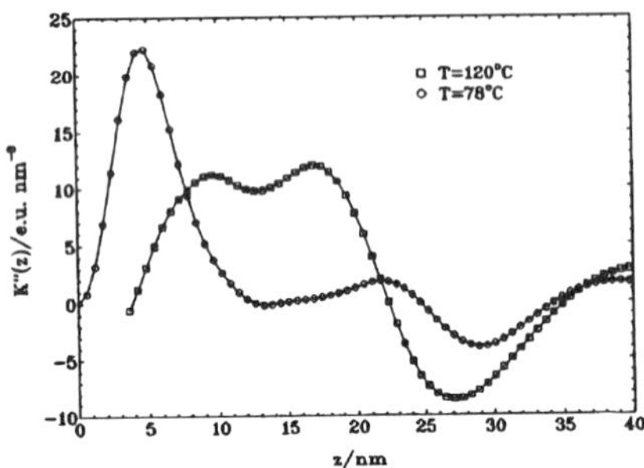

Figure 6: PE: Interface distribution functions obtained after an isothermal crystallization at 124 °C and a subsequent cooling to 78 °C.

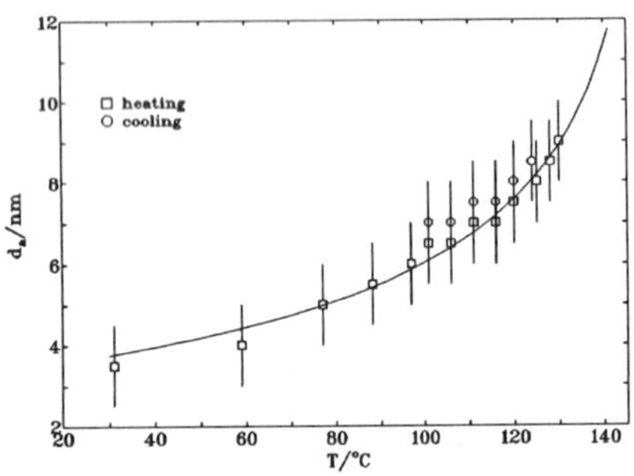

Figure 7: PE: Change of the amorphous layer thickness with T.

The peak at low z is due to the amorphous layers, the other one relates to the crystallites. Cooling leads to a shift of the amorphous and crystalline contribution to lower and higher values respectively. This behaviour is indicative for a surface crystallization, here associated with a decrease of d_a from 10 to 5nm and a corresponding increase of d_c. Further experiments demonstrated that the thickness of the amorphous layers is a unique function of temperature, being independent of d_c. This function is shown in Fig. 7. The dependence can be quantitatively explained by a model which accounts for the entanglements in the amorphous phase and the change in their density with d_a. Surface crystallization and melting is driven by the changing entropy of the subchains between the entanglement points.

5.2 i-PP: Formation of the cross-hatched structure

Secondary crystallization in isotactic polypropylene occurs in a peculiar way. On cooling additional crystallites are formed, however, many of them are oriented oblique to the primary lamellae rather than parallel as is normally observed. As explained by Lotz and Wittmann [7], this 'cross-hatching' is caused by an epitaxial growth mechanism. If cross-hatching occurs the one-dimensional model used so far in the analysis of SAXS data becomes invalid. There is a simple check which shows the deviations. For a stack of crystallites one expects

$$K''(z) = 0 \quad \text{since} \quad 0 = h_a(0) = h_c(0) = h_{ac}(0) \quad \text{etc.}$$

(compare Eq.(4)) and therefore, according to Eq.(7)

$$0 = \int_0^\infty [\lim_{q \to \infty} q^4 I(q) - q^4 I(q)]\mathrm{d}q \tag{8}$$

Hence, $I(q)q^4$ must fluctuate about the average given by the limiting value $\lim(q \to \infty)q^4 I(q)$. This is indeed observed for PE and s-PP at all temperatures. For i-PP the condition is fulfilled during the primary isothermal crystallization, but then, on cooling, it is violated. Fig. 8 shows the result of a measurement [8]. Cross-hatching produces 'edges' along the lines where the oblique secondary lamellae touch the primary crystallites. Exactly these edges can be directly related to the observed deviation. A theoretical analysis gives the following result

$$\int_0^\infty [\lim_{q \to \infty} q^4 I(q) - q^4 I(q)]\mathrm{d}q = 2\pi L_s \Delta\rho^2 w(\alpha)$$

Here, the parameter L_s denotes the total length per unit volume of edges at the boundaries of the inner surfaces, i.e. the 'specific inner edge length'; $w(\alpha)$ is a function depending on the angle α enclosed by the two surfaces. Evaluation of the SAXS-data obtained for i-PP during cooling yielded the temperature dependence of $L_s w(\alpha)$ displayed in Fig. 9. Obviously secondary crystallization here is associated with a growing number of edges, simultaneous with the increases of the crystallinity and the specific inner surface.

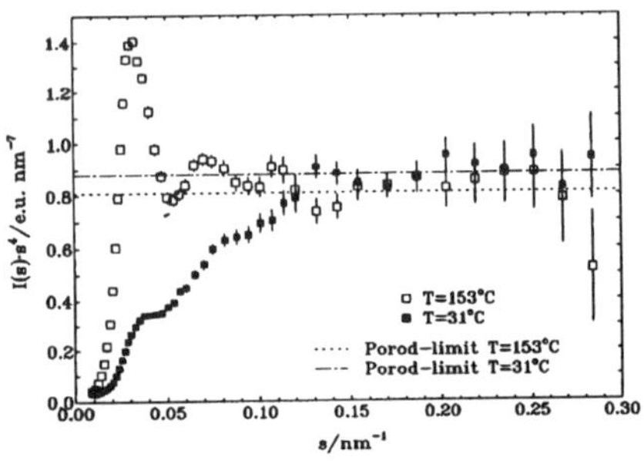

Figure 8: i-PP: Scattering curves $I(s)s^4$ ($s = q/4\pi$) measured at 153 °C and 31 °C.

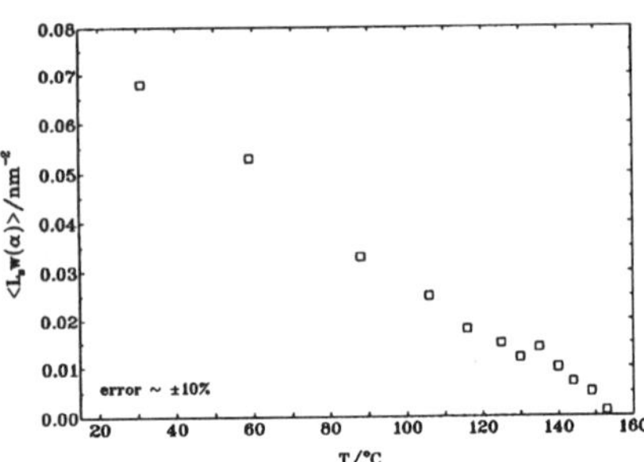

Figure 9: i-PP: Temperature dependence of the specific length of edges.

[1] W. Ruland. *Colloid Polym.Sci.*, 255:417, 1977.

[2] G. Strobl. *Acta Crystallogr.*, A26:367, 1970.

[3] J. Schmidtke, G. Strobl, and T. Thurn-Albrecht. *Macromolecules*, 30:5804, 1997.

[4] G. Hauser, J. Schmidtke, and G. Strobl. *Macromolecules*, submitted, 1998.

[5] T. Albrecht and G.R. Strobl. *Macromolecules*, 29:783, 1996.

[6] T. Albrecht and G.R. Strobl. *Macromolecules*, 28:5827, 1995.

[7] B. Lotz and J.C. Wittmann. *Macromolecules*, 28:1995, 1986.

[8] G. Albrecht and G. Strobl. *Macromolecules*, 28:5267, 1995.

Lamellar morphology of sharp PEEK fractions crystallized from the glassy state and from the melt.

M. Dosière*, C. Fougnies

Université de Mons-Hainaut, Laboratoire de Physicochimie des Polymères,

Place du Parc, 20, B-7000 - Belgique.

The lamellar morphology of five sharp fractions of PEEK synthesized following the ketimide procedure, has been investigated by SAXS, WAXD, DSC and density measurements. The molecular characteristics of these fractions are given in the following table :

Sample	$<M_n>$	$<M_w>$	p	L_n/nm	T_g/°C	T_{cc}/°C	T_m/°C
4k	3500	4400	1.26	18	122.55	152.5	344.0
8 k	7000	8300	1.19	36	133.5	165.5	350.5
18 k	14500	18000	1.24	75	140.0	176.0	349.0
32 k	21500	32000	1.49	111	145.5	183.5	339.0
79 k	60200	79500	1.32	313	147.5	190.5	325.0
Stabar	35000	95000	2.7	180			
K200	-	-	-	-	143.0	175.0	339.8
	50000	120000	3.4	260			

$<M_n>$ and $<M_w>$ are the number and the weight average molecular weights. The degree of polydispersity p is $<M_w>/<M_n>$. The length of the chain is $L_n/=$ is $<M_n>/192.2$nm. T_g, T_{cc} and T_m are the glass transition temperature, the temperature of the extremum of the cold crystallization exotherm and the melting temperature. These three temperatures have been obtained from the DSC curves (10°C°min) of initially amorphous samples.

The melting temperature of the cold crystallized amorphous fractions decreases sharply with increasing weight average molecular weight : the melting temperatures of the 8k and 79k samples are 350.5 and 325 °C, respectively (see the Table). The weight degree of crystallinity of samples of these five PEEK fractions crystallized from the glassy state ranges between 0.15 and 0.60.

The fractions with the lowest weight average molecular weight have the highest degrees of crystallinity. Whereas the degree of crystallinity of the 4k, 8k and 18k fractions increases with the annealing temperature over the full range of crystallization, that of the 32k and 79k fractions reaches its maximum value at annealing temperatures around 300 to 310°C and then decreases due to the lower melting temperature of these fractions.

For all fractions, the long spacing L_p, obtained from the SAXS Lorentz corrected curve, increases as expected with the crystallization temperature. At a given crystallization temperature, the long spacing increases with the weight average molecular weight of the fraction. From the two spacings L_1 and L_2 (with $L_1 + L_2 = L_p$ and $L_1 > L_2$) obtained from the correlation function of the SAXS intensity curve, only the lowest spacing, i.e. L_2, gives a linear degree of crystallinity $v_c^{lin} = L_2/L_p$ which linearly increases with the weight degree of crystallinity W_c. The crystal and the amorphous thicknesses are therefore identified to the smallest L_2 and to the larger L_1 spacings, respectively. The linear relationship between the thickness of the amorphous regions ($L_a = L_2$) and the square root of the weight average molecular weight for the lamellae crystallized between 250 and 340°C, is an additional proof of our assignment of the crystal and amorphous thicknesses. Finally, let us note that for the 4k and the 8k fractions, the values of the long spacing L_p range from 7.0 to $\cong 9.0$ nm. The analysis of the one-dimensional correlation function of the SAXS intensity of these two PEEK fractions gives $L_1 \cong 4$ - 5 nm and $L_2 \cong$ 3-4 nm. It is therefore quite possible, in the limit of the applicability of the analysis of the correlation function, to have 3 to 5 chemical units for the crystal thickness. The opposite argument has been proposed by Hsiao et al.

The isothermal crystallization from the melt and from the glassy state of the 8k PEEK fraction has been investigated by simultaneous time-resolved SAXS and WAXD. Results about crystallization temperatures from the melt and from the amorphous state of 315 and 320°C, respectively, will be reported and analyzed. The main result to be discussed is a decrease of the long spacing L_p and of the long spacings L_c and L_a during the first stages of the isothermal crystallization from the melt even for a sample having a degree of polydispersity of 1.19. During this isothermal crystallization, the degree of crystallinity obtained from the WAXD intensity curve deeply increases. The analysis of the data obtained during the cooling from the isothermal crystallization temperature until 100°C gives additional support in favor of our assignment concerning the spacings obtained from the correlation function of the SAXS intensity curve : the crystal long spacing correspond to the lowest spacing obtained from the correlation function.

Effect of Conformational Defect in Low Molecular Weight PEO Fractions on Crystallization and Phase Behavior

Stephen Z. D. Cheng*, Er-Qiang Chen, Gi Xue, Song-Wook Lee, Anqiu Zhang, Bon-Suk Moon, Hsein-Ming Lin, and Frank W. Harris
The Maurice Morton Institute and Department of Polymer Science, The University of Akron, Akron, Ohio 44325-3909

Benjamin S. Hsiao and Fengji Yeh
Department of Chemistry, The State University of New York at Stony Brook, Stony Brook, New York 11794-3400

Introduction

Low molecular weight poly(ethylene oxide) (PEO) fraction is a good model system for the study of polymer crystallization[1]. A series of work on this system has been done in our lab [2-4]. Recently, three two-arm PEO fractions were prepared by a coupling reaction. These PEOs possess the identical number average molecular weight (M_n) of 2220 for each arm. The coupling agents (which act as defects) used, included *para*-(1,4-), *meta*-(1,3-) and *ortho*-(1,2-) benzenecarbonyl dichloride. The two-arm PEO molecules at the chain center thus form angles of 180°, 120° and 60°, respectively. It is speculated that two types of overall molecular conformations (OMC) may exist: an extended, and a once-folded OMC. The difference between these two OMCs is that when the crystals consist of the extended OMC, two arms of each molecule are crystallized in two neighboring lamellae and only one layer of defects is located between these neighboring lamellae. In contrast, in crystals formed by the once-folded OMC, the two arms are in same lamella and two layers of defects are included between neighboring lamellae. Therefore the crystals with once-folded OMC possess a larger long period than those with extended OMC. Small angle X-ray scattering (SAXS) was employed as the key method for studying the crystal morphology.

Experimental

Two-arm PEO samples were fractionated after the coupling reaction. The number average molecular weights were 4550 for the three fractions and their molecular weight distributions were 1.05. Differential scanning calorimetry (DSC, TA2000 system) experiments were carried out to study the isothermal crystallization and melting behavior. Real time resolved synchrotron wide angle X-ray diffraction (WAXD) and small angle X-ray scattering (SAXS) experiments were performed at the synchrotron X-ray beam line X27C of the National Synchrotron Light Source at Brookhaven National Laboratories. The wavelength of the X-ray beam was 0.1307 nm. A position sensitive proportional counter was used to record the scattering patterns. The counter was calibrated with duck tendon. The Lorentz correction was performed by multiplying the intensity, I (counts per second), by q^2 ($q= 4\pi\sin\theta/\lambda$, where λ is the wavelength of the synchrotron X-ray radiation). The relative invariant Q' [5] was calculated based on $\int (I-I_b)^2 q^2 dq$ for a q ranging between 0.08 and 2 nm^{-1} (where I_b is the intensity of PEO liquid scattering obtained using Porod's law extrapolation).

Results and Discussions

WAXD for the crystals of linear (M_n = 2220) and two-arm PEOs exhibits identical reflections indicating that the crystal structures are identical (a monoclinic unit cell with β = 129° [6]). PEO crystals grown at high temperature for prolonged times exhibit similar widths of half-height (WHH) for their reflection peaks. This indicates that the crystallite sizes are nearly identical for the linear and two-arm PEOs. It is thus evident that in these two-arm PEOs the defects are excluded from the crystals and should be located in the interlamellar amorphous regions. The crystallinity of these samples can also be determined. The linear PEO shows the highest crystallinity of around 92 - 95%, and the three two-arm PEOs exhibit similar crystallinity of about 83 - 85% after

the weight correction of the coupling agents (4%). Therefore, a 10% difference in crystallinity can be estimated between the linear and two-arm PEO samples.

Figure 1 includes four DSC heating curves for three two-arm PEOs as well as the linear PEO (M_n = 2220) crystallized at 48 °C. For the linear PEO, only a single melting endotherm at 54.3 °C can be found over the whole T_c region. This is representative of the extended chain crystal melting based on previously reported results[1]. On the other hand, for 1,4-two-arm PEO in the same T_c region, there is a broadening of the melting endotherm and eventually, it separates into two endotherms. The lower endotherm shows a similar melting temperature to that found in the linear PEO (peak temperature at 54.5 °C). Another endotherm is about 1°C higher than the lower one. In the case of 1,3-and 1,2-two-arm PEOs, two endotherms are also seen in the same T_c region: the lower one is around 55.5 °C and 56.0 °C, while the higher endotherms in both cases increase to around 57.0 °C and 57.4 °C for 1,3-and 1,2-two-arm PEO, respectively. Furthermore, the heats of fusion of the higher endotherm in these two-arm PEOs increase upon changing from 1,4- to 1,2-two-arm PEOs. These results imply that in the cases of these two-arm PEOs, two populations of crystals can be observed with different thermodynamic stabilities.

Figures 2 shows a set of real time resolved synchrotron SAXS experimental results after the Lorenz correction. The samples were crystallized at 48 °C. For the linear PEO (M_n = 2220) fraction, the first-order scattering maximum appears at 14.8 nm (Figure 2a). Since the average molecular length with a 7/2 helix conformation of this fraction is 14.0 nm, a 0.8 nm gap thus exists between neighboring lamellae. The nature of this gap is currently not clear but has been found in other linear PEO fractions as well as in *n*-alkanes with uniform molecular lengths. Speculations have been made that these are due to amorphous layers formed on the lamellar crystal surfaces because of polydispersity, lamellar bending, and /or other reasons.

The SAXS result for 1,4-two-arm PEO (Figure 2b) shows a surprising observation. Two long periods are recognized although they are close to one another. They represent two populations of lamellar crystals with different long periods: one is 17.3 nm and the other is 19.8 nm. The difference between them is thus 2.5 nm. A similar observation can also be seen in the case of 1,3-two-arm PEO (17.5 nm *versus* 20.2nm, and their difference is 2.7 nm, see Figure 2c). Nevertheless, 1,2-two-arm PEO shows a dominant long period which corresponds to the larger long period (20.4 nm). Only a very minor population in the smaller long period around 17.6 nm may be observed (Figure 2d).

As mentioned, the long period of the crystal with once-folded OMC is longer that of extended OMC. It is logical to think that the smaller long period must represent the crystals which consist of the extended OMC and the one with the larger long period possess the once-folded OMC. The difference of 2.5 nm is thus representative of the approximate thickness of one defect layer. The crystals with these two different long periods should have different thermodynamic stabilities as revealed by DSC results.

This speculation can be confirmed by Figures 3a and 3b. In Figure 3a, changes of the crystallinity (from WAXD) and the relative invariant Q' (from SAXS) during heating are shown for 1,4-two-arm PEO after crystallization at 48 °C. The heating rate was 0.5 °C/min. From the crystallinity data, crystal melting starts at around 51 °C, and at 54.5 °C only 40% crystallinity remains. This indicates that one of these two crystal populations has been melted at that temperature. The other half of the crystallinity is more stable and melts at higher temperatures. This melting process ends at 56.5 °C.

SAXS results in Figure 3b show that the smaller long period (17.3 nm) associated with crystals having extended OMC, starts decreasing in scattering intensity and increasing in long period at around 51 °C. It reaches a plateau of 24.0 nm at 53°C. Above 54 °C, the thickness of this long period further increases and finally reaches about

26.5 nm. On the other hand, the scattering peak of the larger long period (19.8 nm) which corresponds to the once-folded OMC crystal, remains constant up to a higher temperature of 53.0 °C, and then it begins to thicken. Above 54 °C, only one broad scattering peak can be found which gradually disappears when heated to the isotropic melt. This indicates that the crystal with once-folded OMC is more stable than the crystal with extended OMC.

Compared with the long period of the linear PEO, 1,4-two-arm PEO exhibits significantly larger long periods. The difference between the smaller long period (crystals having the extended OMC, 17.3 nm) and that of the linear one is 2.5 nm, while the difference between the larger long period (crystals having the once-folded OMC, 19.8 nm) and that of the linear one is 5.0 nm. Since the larger long period contains two defect layers, each defect layer thus also occupies 2.5 nm. This is consistent with the difference between the larger and smaller long periods within this sample at the same T_c as described previously. A similar comparison can also be made for 1,3-two-arm PEO. It is calculated that each defect layer for 1,3-two-arm PEO is thus 2.7 nm. For 1,2-two-arm PEO, the only comparison which can be made is the difference between the larger long period (crystals having the once-folded OMC, 20.4 nm) and that of the linear PEO, which is 2.8 nm for the defect layer. Therefore, it is conceivable that the thickness of the defect layer increases slightly from 1,4- to 1,2-two arm PEOs.

Based on a computer simulation, the defect thickness in the extended OMC is about 1.3 nm (including the ester groups). We speculate that the defects in this OMCs can not exactly arrange themselves into one layer with a thickness of 1.3 nm due to a space restriction caused by the PEO chain lateral packing in the crystals (on the a*b-plane of the PEO crystals). These defects may thus have to adjust their neighboring positions up or down with respect to the lamellar surface normal in order to accommodate this restriction. The defect layer thickness may thus increase. As a result, more PEO repeating units may not be incorporated in the crystals and rather, are required to stay in the interlamellar amorphous region. This can be confirmed by the experimental observations of the 10% decrease in crystallinities of these two-arm PEOs.

The crystals with different OMCs will have different fold surface free energy. Since the melting temperature of once-folded OMC crystals is higher, its fold surface free energy is lower than that of an extended OMC crystal. Furthermore, the melting temperatures of the crystals containing once-folded OMC of 1,3- and 1,2-two-arm PEOs are higher than that of 1,4-two-arm PEO. This may be due to the effect of different defect configurations for the PEO two-arms. One may speculate that the folds in 1,3- and 1,2-two-arm PEO form more easily and thus are tighter than those in 1,4-two-arm-PEO. The surface energy must then be greater for 1,4-two-arm PEO compared to 1,3- and 1,2-two-arm PEO.

Acknowledgment: This research was supported by the National Science Foundation (DMR-9617030). Research was carried out (in part) at the National Synchrotron Light Source at Brookhaven National Laboratories, which is supported by the U.S. Department of Energy, Division of Material Sciences and Division of Chemical Sciences.

References and Notes

1. Buckley, C. P.; Kovcas, A. J. in *"Structure of Crystalline Polymer"* Hall I. H. Ed. Elsevier Applied Science Publishers, **1984**
2. Cheng, S. Z. D.; Chen, J.-H.; Barley, J. S.; Zhang, A.-Q.; Habenschuss, A.; Zschack, P. R. *Macromolecules* **1992**, *25*, 1453.
3. Cheng, S. Z. D.; Wu, S. S.; Chen, J.-H.; Zhuo, Q.; Quirk, R. P.; von Meerwall, E. D.; Hsiao, B. S.; Habenschuss, A.; Zschack, P. R. *Macromolecules* **1993**, *26*, 5105.
4. Lee, S.-W.; Chen. E.; Zhang, A.; Yoon, Y.; Moon, B. S.; Lee, S.; Harris, F. W.; von Meerwall, E. D.; Hsiao, B. S.; Verma, R.; Lando, J. B. *Macromolecules* **1996**, *29*, 8816.
5. Ryan, A. J.; Stanford, J. L.; Bras, W.; Nye, T. M. W. *Polymer* **1997**, *38*, 759.
6. Takahashi, Y.; Tadocoro, H. *Macromolecules* **1975**, *6*, 672.

Fig. 1 Four DSC heating curves for PEOs after crystallization at 48 °C

Fig. 2 Synchrotron SAXS curves (a) linear PEO (Mn=2220), (b) 1,4-, (c) 1,3-, and (d) 1,2-two-arm PEOs. The crystallization temperatures were 48°C.

Fig. 3 Changes of crystallinity (from WAXD) and invariant Q (from SAXS) for 1,4-two-arm PEO during heating (a). Change of long periods of two crystal populations (from SAXS) (b).

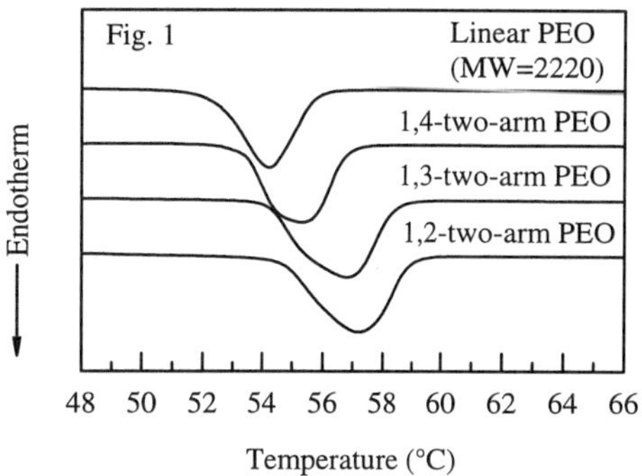

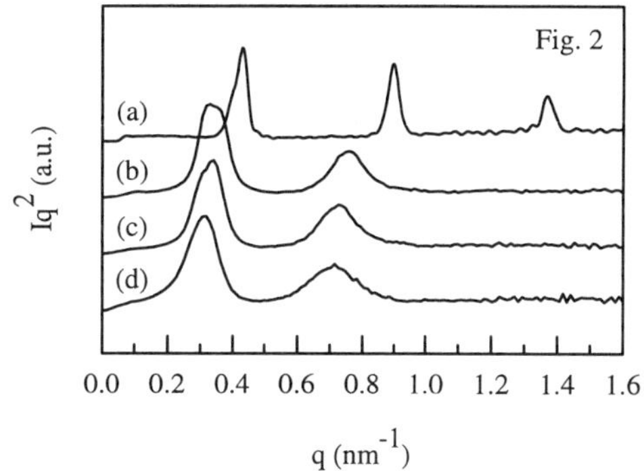

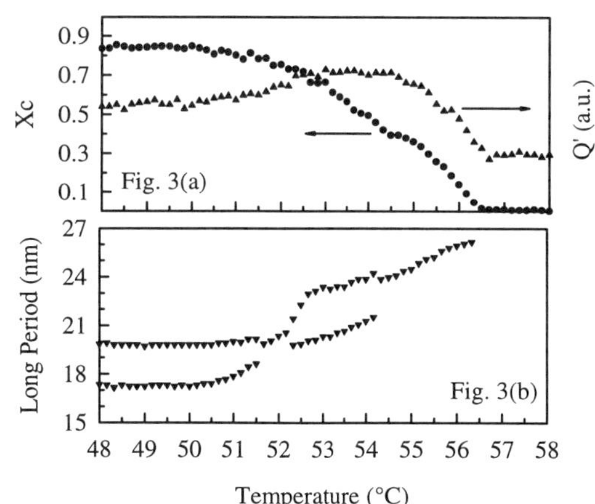

REAL TIME CRYSTALLIZATION AND MELTING STUDY OF ETHYLENE BASED COPOLYMERS

Weidong Liu[1], Henglin Yang[2], Benjamin S. Hsiao[2] and Richard S. Stein[1]

[1]Chemistry Department, University of Massachusetts, Amherst, MA 01003

[2]Chemistry Department, State University of New York at Stony Brook, Stony Brook, NY 11794-3400

Introduction

Polymer crystallization thermodynamics and kinetics problems can be studied using a wide range of time-resolved techniques such as differential scanning calorimetry (DSC), dilatometry, light and electron microscopy, wide-angle x-ray diffraction (WAXD) and small-angle x-ray scattering (SAXS) and small-angle light scattering (SALS). In many cases, the ability to study kinetics in real time has been limited by the capability of experimental techniques to follow rapid changes. These problems are alleviated in the studies presented here through the use of synchrotron x-ray radiation for simultaneously following WAXD and SAXS[1-3] during the isothermal crystallization and subsequent melting of linear and branched polyethylene in real time. Samples studied were copolymers of ethylene with α-olefin short-chain branch (octene) prepared by metallocene polymerization. The variations of the crystal unit cell parameters and crystallinity were determined by WAXD; the crystal long period, lamellar thickness, interlayer amorphous thickness and invariant were determined by SAXS[4], and the changes of the spherulite radius and anisotropy were determined by SALS[5,6]. Results from these studies were interpreted using the model of branch exclusion which affects the ability of the chain-reentry into the crystal phase, the segregation of the branched chains near the crystal and amorphous interfaces, and the ability to undergo a lamellar isothermal thickening process during prolonged annealing. We intend to further correlate these structural and morphological variations with the mechanical properties as functions of branch compositions, branch types and processing conditions.

It is believed that branches longer than methyls are excluded arising from the strain resulting from the crowding of excluded branches at the crystal surface[7]. The presence of these branches at the crystal surface serves to limit the thickening of the crystal lamella. There is controversy concerning the mechanism of crystallization. The Avrami model postulates that crystals arise from a nucleus, probably heterogeneous for polyethylene, and that spherulites then grow with a constant radial growth rate, continuing until they impinge, filling the volume of the sample with polygonal structures. In the absence of secondary crystalliztion, these have a constant degree of crystallinity. However, a secondary crystallization process is often observed in which the initial spherulte growth is followed by their internal crystallization. This may occur because of molecular heterogeniety in molecular weight and/or branching or else by a lamellar thickening and or perfection process. The use of metallocene polyethylene (m-PE) provides better control of molecular heterogeneity. Their use, together with the real time WAXD and SAXS measurements permits the clarification of the basic crystallization processes. It has been also postulated that molecule ordering in the amorphous phase may occur as a precursor of crystallization. Evidence comes from the apparent more rapid increase of SAXS than WAXD intensities. In this work quantitative examination of scattering invariant is carried out to address this problem

Experimental

All samples were purified prior to the experiment with the following procedures. The m-PEs with different degrees of branching 0, 1.93, 2.82, 4.12, 5.72 short branched chains per 1000 carbon atoms (Mw: 1.25, 1.09, 0.77, 0.61, 0.54 x10^{5}g/mol, respectively) were dissolved in p-xylene by rigorously stirring at 160 $^{\circ}$C for 1 hour. The polymer concentration was 1% (w/v), and antioxidant 246 was added (0.1% w/w on polymer) when the polymers were precipitated into methanol. The recovered polymers were then dried in vacuum oven at 70°C for 72 hr.

The isothermal crystallization and melting study of m-PE using simultaneous SAXS and WAXD techniques were carried out at the Advanced Polymers beamline (X27C), NSLS, Brookhaven National Laboratory (BNL). The wavelength used was 0.1307 nm. The sample to detector distance was 1618 mm for SAXS. The acquisition time was 20 seconds per scan. The specimen was first equilibrated above the melting temperature at 150°C for 5 min, then rapidly jumped to the desired temperature for measurements. Two modes of experiments were carried out, isothermal crystallization (30 min) and subsequent melting with a heating rate of 2.5°C/min. DSC measurements were also carried out using a Perkin-Elmer DSC-7 with the same thermal conditions in SAXS and WAXS. From DSC, degree of crystallinity (χ_c) and equilibrium crystallization temperature were determined.

Results and Discussion

Isothermal crystallization and subsequent melting experiments of m-PEs with different degrees of octene branching were carried out as a function of crystallization temperature. Figures 1 and 2 show typical time-resolved SAXS and WAXD profiles of an m-PE sample (1.93‰) during isothermal crystallization and subsequent melting. It is seen that a scattering maximum in SAXS, and two reflections (110 and 200) in WAXD are first enhanced during isothermal crystallization and then disappear during melting. Other m-PE samples show similar features.

The invariant Q, long period L, lamellar thickness l_c and amorphous thickness l_a were calculated from the SAXS data using correlation function method[8], which are shown in Figure 3. It is seen that both values of L and l_c decrease with crystallizarion time. Initially, the decrease is significant; at the later stage, the decrease becomes smaller. Upon heating, a reversible process (L and l_c increase with temperature) is seen. The slight decreases in L and l_c can be attributed to secondary crystallization of forming thinner lamellar thickness, which is consistent with the slight increase of crystallinity. In addition, the variation of the amorphous layer thickness is relatively small. In Figure 3, we can determine several useful variables to compare results with different crystallization temperatures and samples of different branch compositions. These variables include t_c: the time to reach the plateau value of Q, the plateau values of long period (L^*), lamellar thickness (l_c^*) and interlamellar amorphous thickness (l_a^*). The physical meaning of t_c probably indicates the completion of the primary crystallization process.

Figures 4(a) and (b) illustrate the value of L^* and l_c^* as functions of degree of supercooling ($\Delta T = T_m^o - T_c$) and the branch composition. Both values decrease with ΔT as expected, they also are lower with the increase in the branch content. This clearly indicates that the increase in the branch content decreases the crystallizable sequence length in PE. The corresponding crystallinity (χ_c from WAXD) plot in Figure 5 supports the conclusion that the crystallinity decreases with branch concentration. The plot of t_c (Figure 6) tells a slightly different story. It is seen that t_c appears not to be a function of the branch concentration, only a function of degree of supercooling (or crystallization temperature). This suggests that the primary crystallization process in m-PE in this case is independent of branch concentration.

As we have collected SAXS and WAXD signals simultaneously, we are interested in comparing kinetics results between the two techniques. In Figure 7, the values of Q and χ_c of the m-PE sample with 1.93‰ branch concentration are plotted together during crystallization and subsequent melting processes. The analysis shows a consistent time lag between SAXS and WAXD. The invariant Q (from SAXS) always appears first before χ_c (from WAXD) during crystallization and disappears later upon melting. This suggests that density fluctuations (manifest the SAXS signals) are present prior to crystallization (manifest the WAXD signals), and persist after melting. We intend to further detect it in a short time by higher power x-ray source.

Acknowledgment

We are grateful to Prof. B. T. Huang and Dr. S. S. Liu, CIAC of Chinese Academy of Sciences, for their kind supply of the metallocene polyethylene samples. The technical assistance from Drs. F. J. Yeh and Z. G. Wang during synchrotron measurements and data analysis is gratefully acknowledged. The financial support of this project is in part by a grant from DuPont and in part by NSF (DMR 9732653).

References

1. Ryan, A. J.; Bras, W.; Mant, G. R.; Derbyshire, G. E. *Polymer* **1994**, *35*, 4537.
2. Verma, R. K.; Hsiao, B. S. *TRIP*, **1996**, *4(9)*, 312
3. Kruger, K. N.; Zachmann, H.G. *Macromolecules* **1993**, *26*, 5202.
4. Hsiao, B. S. and Verma, R. K. *J. Synchrotron Radiation*, **1998**,
5. Hu, S. R.; Kyu, T.; Stein, R. S. *J. Polym. Sci.: Part B: Polym. Phys.* **1987**, *25*, 71.
6. Kyu, T.; Hu, S. R.; Stein, R. S. *J. Polym. Sci.: Part B: Polym. Phys.* **1987**, *25*, 89.
7. Mandelkern, L.; McLaughlin, K. W.; Alamo, R. G. *Macromolecules* **1992**, *25*, 1440.
8. Strobl, G. R.; Schneider, M. *J. Polym. Sci.: Polym. Phys. Ed.*, **1980**, *18*, 1343.

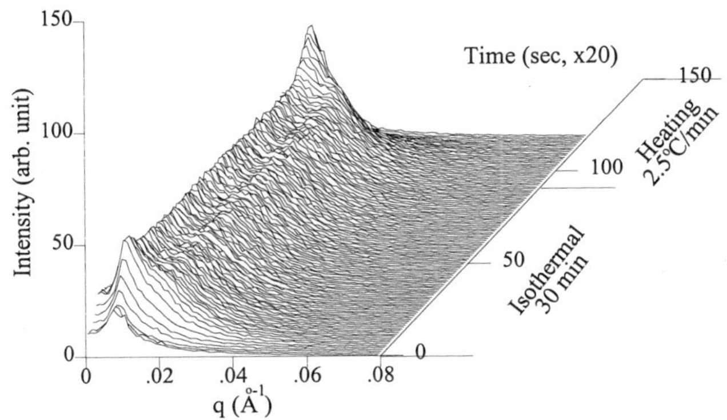

Figure 1.   Three-dimensional diagram of I vs q vs Time of m-PE (1.93‰)

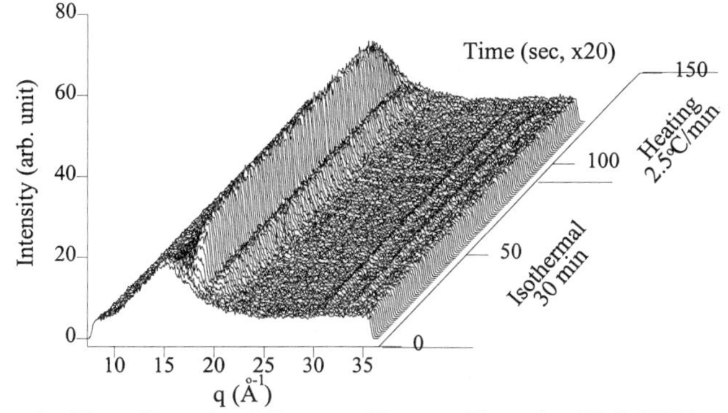

Figure 2.   Three-dimensional diagram of I vs q vs Time of m-PE (1.93‰)

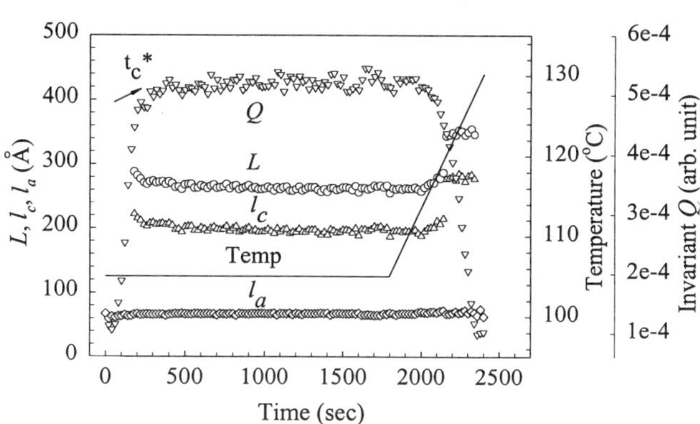

Figure 3. Crystallization parameters of m-PE (1.93‰)

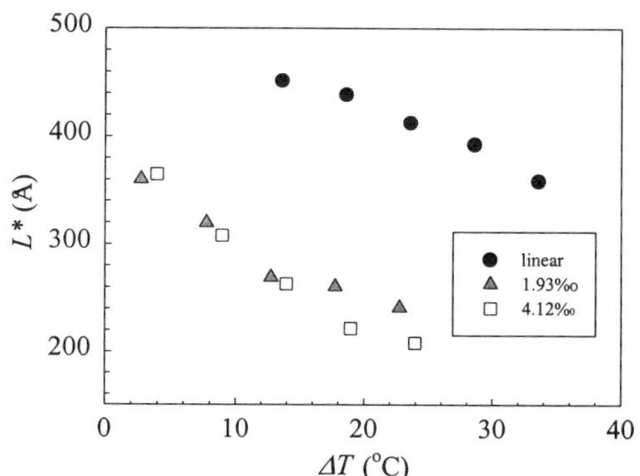

Figure 4(a). L^* vs ΔT of m-PEs with various degrees of branching

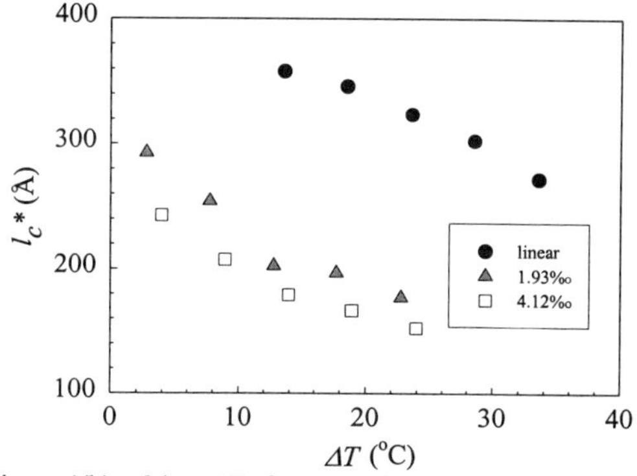

Figure 4(b). l_c^* vs ΔT of m-PEs with various degrees of branching

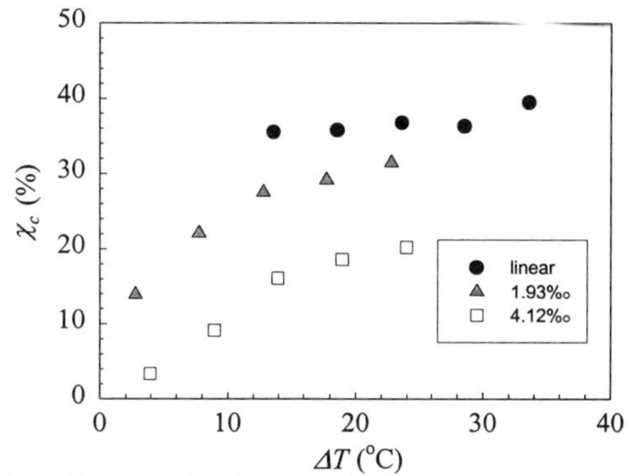

Figure 5. χ_c vs ΔT of m-PEs with various degrees of branching

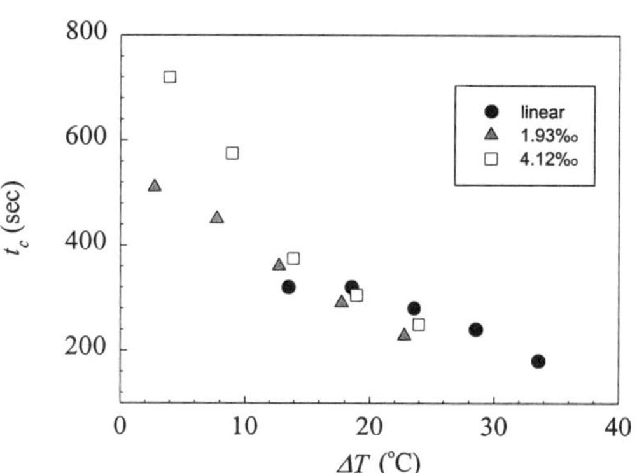

Figure 6. t_c vs ΔT of m-PEs with various degrees of branching

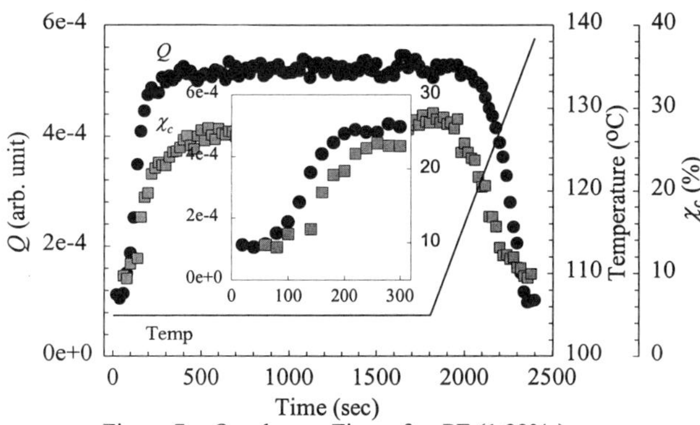

Figure 7. Q and χ_c vs Time of m-PE (1.93‰)

Small-Angle X-ray Scattering and DSC Studies on the Domain Structure and Interphase
Dimensions in Poly(urethaneurea)

S. G. Musselman and T. M. Santosusso, Air Products and Chemicals, Inc., Allentown,
PA 18195; and L. H. Sperling, Lehigh University, Bethlehem, PA 18015

Introduction

Polyurethane elastomers are phase-separated multiblock copolymers composed
of a glassy or crystalline hard block and an elastomeric soft block. Polyether or polyester
diols ranging in molecular weights from 500 to 5000 g/mol comprise the soft segments,
while diisocyanates composed of the reaction products of chain extenders or crosslinkers
comprise the hard segments. When diamines are used as the chain extender, the resultant
polymers are known as poly(urethaneureas). These materials can undergo crosslinking
through the formation of allophonate or biuret bonds. Physical crosslinking is also
present in these materials as a consequence of hydrogen bonding and crystallinity. The
elastomers of this study utilize toluene diisocyanate (TDI) terminated
poly(tetramethylene glycol) prepolymers with chains extended with the diamines 4,4'-
methylene bis-(2-chloroaniline) (Mboca) or 4,4'-methylene bis-(3-chloro-2,6-
diethylaniline) (MCDEA) as chain extenders. The use of these chain extenders was
shown to result in improvements in physical and mechanical behavior of these
polyurethane elastomers[1, 2].

Recently, Ryan, *et al.*[3] reviewed the morphology of polyurethane elastomers
via X-ray and other methods. Koberstein and coworkers[4, 5] investigated the domain
size and the diffuse boundary thickness of crosslinked polyurethanes using small-angle
X-ray scattering, SAXS.

In the present study, prepolymers containing a broad distribution of oligomers
and those with controlled oligomer content were compared to determine if differences in
tensile, compression, and thermal properties were the result of differences in morphology
as determined by SAXS and differential scanning calorimetry, DSC. The objective of
this study was to evaluate the microstructure of this new class of materials and to
examine whether these parameters correlate with end use behavior. The synthetic
procedure, as well as the differences between the two synthetic methods, is summarized
in Figure 1[2].

A model of the system to be analyzed is illustrated in Figure 2. Here the hard
segments are organized to greater or lesser extent, ideally on the left side, and in reality
on the right side. The extent of this organization, as affected by the exact chemistry of
synthesis, will be developed in this paper.

Experimental

Four prepolymers were synthesized. Commercial 2,4 TDI, 80/20 2,4/2,6 TDI
isomer mixtures (Olin) and 50/50 2,4/2,6 TDI isomer mixtures (APCI) were used with
poly(tetramethyl ethylene glycol), PTMEG, of 1000 g/mol polyol (Dupont Terathanes).
The degassed polyol was added to the TDI in a reactor under dry nitrogen. Two of the
prepolymers used a TDI to polyol ratio of 2:1 to produce a more nearly statistical
prepolymer (S) containing approximately 60% high molecular weight oligomers and 2 to
3% residual isocyanate. Pure 2,4 TDI was used to produce one S prepolymer, S(2,4),
while 80/20 TDI isomer mixture was used for the other, S(2,4/2,6). The other two
prepolymers were synthesized with an excess TDI to polyol ratio of 8:1 to produce a
more regular(R) prepolymer containing less than 10 % high molecular weight oligomer.
The residual isocyanate monomer was removed by a wiped film distillation technique to
less than 0.1 %. One R prepolymer contained pure 2,4 TDI (R(2,4)) while the other was
made with 50/50 TDI isomer mixture which after reaction and distillation results in an
80/20 2,4/2,6 TDI incorporation into the prepolymer (R(2,4/2,6), see Figure 1.

Twelve elastomers were prepared from the four prepolymers using hand casting
techniques. Two different curatives or chain extenders were used, Mboca (Palmer Davis
Seika) and MCDEA (Lonza). Two MCDEA cured elastomers were prepared for each
prepolymer with differing cure conditions. The prepolymer was present in 5 %
stoichiometric excess in order to allow 5 % biuret formation for chemical crosslinking.
The resulting Mboca cured elastomers contain 39 wt-% TDA-Mboca-TDI triad hard
segments and 61 wt-% PTMEG soft segments. The MCDEA elastomers contain 43 wt-%
TDI-MCDEA-TDI hard segments and 57 wt-% soft segments.

Physical Characterization

All elastomers were aged two weeks at 23° C and 50 % relative humidity before
testing. Tensile testing followed ASTM D412-92, utilizing Instron model 1122. DSC
utilized a TA MDSC2929 with Thermal Analyst 3200 Software.

The SAXS data was collected at NIST using a Digital SAXS Camera, utilizing
Cu Kα radiation. The corrected two-dimensional data was reduced to tables of scattered
intensity vs. scattering vector by circularly averaging the area detector data.

Results

The samples of poly(urethaneurea) were subjected to both basic physical and
mechanical studies, as well as the more intensive DSC and SAXS methods. Inasmuch as
the differences between the materials prepared using 2,4 TDI and those prepared using
the 2,4/2,6 isomer mixture were found to be relatively minor, the following discussion is
based only on the materials prepared using the 2,4/2,6 TDI isomer mixture.

The DSC data indicate that the T_g of all the Mboca cured elastomer series are
near -55° C, while those of the MCDEA elastomer series averaged near -65° C, with no
effect from the variation in the prepolymer. This shift in the T_g was interpreted in terms
of phase mixing: the greater the amount of hard segment dissolved in the soft segment
phase the higher the glass transition of the soft phase.

The temperature of dissociation of the Mboca hard segments occurs at about
210° C, vs. about 240° C for the MCDEA hard segments, indicative of a purer hard
segment phase. The heats of dissociation of the hard segments for the S series ranged
from 15 to 34 J/mol (of hard segment), while those of the R series ranged from 34 to 61
J/mol, or two or three times larger. These values are corrected for the percent hard
segment in the elastomer, as suggested above. Modulated DSC curves show that there is
a reversible and non-reversible transition in these temperature ranges, interpreted as a
non-reversible dissociation of hydrogen bonding and a small reversible glass transition.

Application of the Fox equation[6] to the present data was used to calculate the
percent of hard phase dissolved in the soft phase. Thus, 39 wt-% of the Mboca hard
segment was dissolved in the soft phase, while only 12 wt-% of the MCDEA hard
segment was dissolved in the soft phase. The remaining Mboca and MCDEA make up a
separate phase, to be analyzed by SAXS as shown below.

SAXS Data

The experimental data are summarized in Figure 3, where I(q) is the scattering
intensity at the scattering vector, q = $(4\pi\sin\theta)/\lambda$, where 2θ represents the scattering angle,
and λ the wavelength. As shown in Table 1, the values of the Bragg spacing range from
7.6 nm to 4.5 nm. Table 1 also summarizes all of the results of the characterization
methods.

The hard segment thickness, H, (see Figure 2) can be determined from the one
dimensional correlation function[7, 8] or from the V_h (volume fraction of hard phase)
determined via DSC and the H(1 − V_h) value from the correlation function. The values
for H using either determination show that hard segment thickness is smaller for
elastomers from the R prepolymers compared to those from the S prepolymers. Within a
curative set, the elastomers from the R prepolymers exhibit smaller values of the
interdomain spacing, L, than the elastomers of the S prepolymers. The MCDEA cured
elastomers have smaller values of L than the Mboca cured elastomers from the same
prepolymer.

Conclusions

The sizes of the hard segment domains, as reflected by their L values, are
significantly smaller for elastomers from the R prepolymers than the S prepolymers, and
even smaller when the prepolymers are cured with MCDEA. The R elastomers tend to
have greater interfacial area than the S elastomers, suggesting smaller but better
organized domains for the R elastomers. These results suggest than an elastomer
produced from a prepolymer with a narrower molecular weight distribution prepolymer
yields improved mechanical properties as a result of smaller, more ordered domains, and
greater interfacial area.

Acknowledgements

M. E. Eckert, J. J. Koch, E. L. McInnis, J. R. Quay, B. Peterson, F. M. Prozonic,
M. S. Vratsanos, and R. J. Goddard, all of Air Products and Chemicals, are thanked for
their assistance in this effort, as is J. D. Barnes of the NIST Polymer Structure and
Mechanics Group.

References

1. W. E. Starner, J. P. Casey, B. Miligan and S. M. Clift, U.S. Pat. 4,786,703, Air
    Products and Chemicals, Inc., USA, 1988.
2. A. J. Siuta, W. E. Starner, B. A. Toseland and R. M. Machado, U.S. Pat.
    5,051,152, Air Products and Chemicals, Inc., USA, 1991.
3. A. J. Ryan, S. Naylor, B. Komanschek, W. Bras, G. R. Mant and G. E.
    Derbyshire, *Hyphenated Techniques in Polymer Characterization* (Editor: T.
    Provder), American Chemical Society, Washington, DC, 1994.
4. L. M. Leung and J. T. Koberstein, *J. Polym. Sci., Part B: Polym Phys. Ed.*, 23,
    1883 (1985).
5. J. T. Koberstein and R. S. Stein, *J. Polym. Sci., B, Polym. Phys. Ed.*, 21, 1439,
    2181 (1983).
6. T. G. Fox, *Bull. Am. Phys. Soc.*, 1, 123 (1956).
7. R. J. Goddard and S. L. Cooper, *J. Polym. Sci., Part B: Polym. Phys. Ed.*, 32,
    1557 (1994).
8. G. R. Strobl and M. Schneider, *J. Polym. Sci., Polym. Phys. Ed.*, 18, 1343
    (1980).

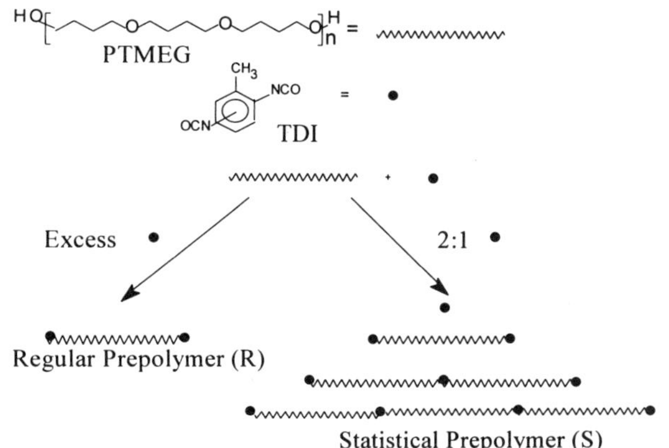

Figure 1. Prepolymer synthesis to produce a more regular prepolymer (R) and a more statistical prepolymer (S). The excess isocyanate of the regular prepolymer is distilled off to less than 0.1% free monomer.

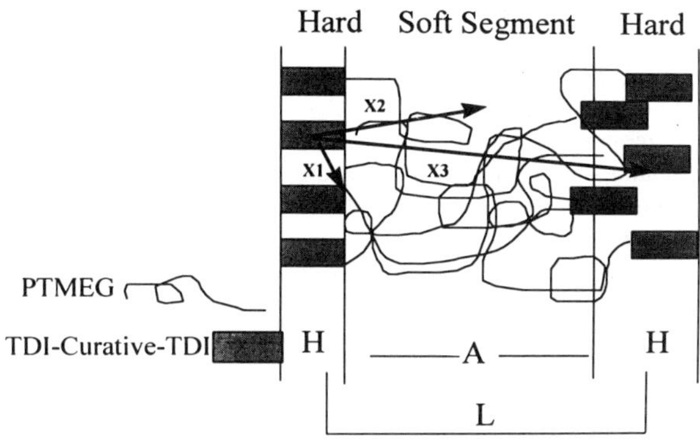

Figure 2. Lamellae structure: L is the interdomain spacing, H is the hard segment thickness and A is the soft segment thickness. Ideal: Left side. Reality: Right side. The one dimensional correlation function shows the probability of a rod with length x being in a phase of like electron density in a lamellae structure.

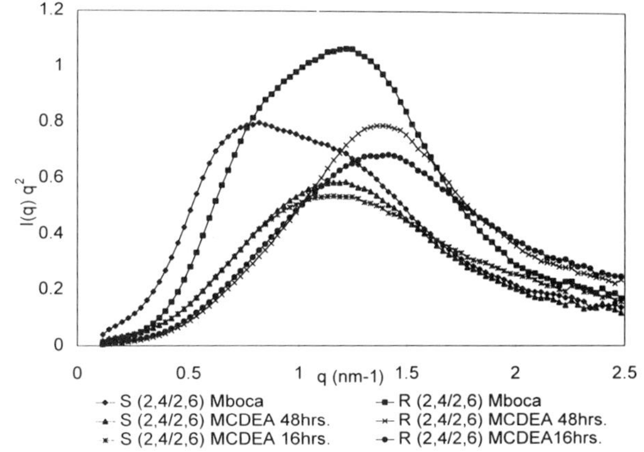

Figure 3. Plots of I(q) q² vs. q for determination of the interlamellar Bragg spacing.

Table 1. Morphology Data-SAXS and DSC Derived

	Interdomain Spacing -L (nm)		H (nm)		A (nm)[e]	V_h		K_p
	Bragg[a]	1-D Correlation[b]	(SAXS)[c]	(DSC)[d]		(DSC)[f]	(SAXS)[g]	
S (2,4/2,6) Mboca	7.6	9.2	1.78	1.94	7.42	0.28	0.22	0.86
S (2,4) Mboca	6.8	7.0	1.72	1.87	5.28	0.28	0.23	0.93
R (2,4/2,6) Mboca	5.1	5.0	1.61	1.69	3.39	0.28	0.25	1.11
R (2,4) Mboca	5.2	5.2	1.64	1.69	3.56	0.28	0.27	1.01
S (2,4/2,6) MCDEA 16 hrs	5.3	5.2	1.47	1.84	4.23	0.38	0.24	1.02
S (2,4) MCDEA 16 hrs	5.2	4.9	1.38	1.69	3.82	0.38	0.25	0.98
R (2,4/2,6) MCDEA 16 hrs	4.5	4.5	1.22	1.51	3.28	0.38	0.25	1.51
R (2,4) MCDEA 16hrs	4.5	4.5	1.28	1.59	3.22	0.38	0.24	1.35
S (2,4/2,6) MCDEA 48 hrs	5.2	5.7	1.31	1.67	3.90	0.38	0.22	0.77
S (2,4) MCDEA 48 hrs	5.1	5.2	1.30	1.64	3.59	0.38	0.24	1.02
R (2,4/2,6) MCDEA 48 hrs	4.5	4.5	1.19	1.47	3.31	0.38	0.24	1.39
R (2,4) MCDEA 48 hrs	4.5	4.5	1.18	1.49	3.32	0.38	0.23	1.25

a) Iq² vs q plot, b) correlation function, c) correlation function, d) correlation function e) correlation function modified by DSC-V_h(DSC) and H(1-V_h) SAXS, f) A=L-H from correlation function

The Structure of a Ring-Containing Fluoropolymer

J. Blackwell, Soo-Young Park and S.N. Chvalun
Department of Macromolecular Science
Case Western Reserve University
Cleveland, OH 44106-7202, USA

Introduction

A new ring-containing fluoropolymer has been reported by Yang et al.[1], prepared by cyclo-polymerization of a partially fluorinated diene-ether (1,1-dihydro-2,2,3,4,4-pentafluoro-3-butenyl-trifluorovinylether) using a mixture of bis(perfluoro-propionyl) peroxide and bis(4-ter-butylcyclohexylperoxy) dicarbonate as initiator. The resultant polymer is poly(1,2-difluoromethylene-1,2,3,3-tetrafluoro-4,4-dihydro-furan) (PDTE). The presence of the two ring hydrogens increases the polarity compared to that of the fully fluorinated analog, such that this polymer is soluble in polar aprotic solvents such as dimethylformamide (DMF), tetrahydrofuran (THF), acetone and acetonitrile. In addition, this polymer is crystalline, even though no efforts had been made to control the tacticity. The glass transition and melting temperatures are 127°C and 347°C respectively.

It is very interesting as to why this polymer is crystalline in contrast to Cytop®, which has the same chemical structure except for the additional ring fluorines[2]. The micro-structure is unknown at this point, but it is possible that the presence of the hydrogen maybe lead to a preferred tacticity, making crystallization possible. The presence of the ring hydrogens may allow for easier interdigitation of the rings on adjacent chains, and there may also be the possibility of favorable interchain dipole-dipole interaction. We have investigated the crystal structure of PDTE in order to throw light on these problems.

Experimental

The specimen of PDTE was generously provided by Drs. Zhen-yu Yang, Andrew E. Feiring and Bruce E. Smart (DuPont)[8], which yielded polymer with $Mn =$ 190,000 and $Mw = 404,000$ based on GPC data. A film (0.5 mm thick) of this material prepared by casting from DMF solution was brittle and could not be drawn. Orientation was achieved by compression of the film to half its original thickness by application of a pressure of 1000 kg/mm^2 at 250°C.

Wide and small angle X-ray patterns of the pressed films were recorded on Kodak Direct Exposure film, with the beam perpendicular and parallel to the surface of a film. Energy minimization of molecular models was performed the SYBYL® software package (Tripos® Inc.). The polymer backbone is defined by four torsional angles ξ, ζ, $\phi1$, and ψ, and the conformation of the ring by $\phi1$- $\phi5$:

$$-(_6CF_2-_5CF_2-_2CF-_1CF-)CF_2-$$

Zeros for these torsion angles are for the *cis* positions for the carbon or oxygen atoms, and positive angles correspond to anti-clockwise rotations. The force field was modified by incorporating the preferred bond lengths, bond angles, van der Waals radii and energy parameters used by Holt *et al.*,[3] to model the structure of poly(tetrafluoro-ethylene)

Results and Discussion

X-ray Diffraction

The small and wide angle X-ray scattering patterns recorded with the beam parallel and perpendicular to the film surface as shown in Figure 2. When the beam was parallel to the film surface, we observed a single maximum in the small angle region at $d = 110\pm10$ Å (Figure 1a), which is inclined along the direction perpendicular to the film surface. There was only diffuse scattering in the small angle region when the beam was perpendicular to the film surface (Figure 1b). The likely interpretation of these data is that the specimen contains crystalline, chain folded lamellae which are stacked with their long axes preferentially parallel to the film surface, as a result of the applied compression.

The wide angle data recorded with the beam perpendicular to the film surface are shown in Figure 1d. The data indicate complete rotational disorder of the crystallites about an axis perpendicular to the film surface. In the case where the beam was parallel to the film surface (Figure 1c), these rings are broken into arcs. A schematic of the latter data is presented in Figure 2, where the Bragg reflections are numbered for reference in the text. The observed R, Z reciprocal space coordinates for the reflections are given in Table 1, and are plotted in Figure 2(inset). The R and Z axes are parallel and perpendicular to the film surface

respectively, and will be referred to as the equator and meridional respectively.

The existence of both layer lines and row lines suggests at least monoclinic symmetry for the polymer unit cell with the chains perpendicular to the film surfaces. A unit cell can not be proposed on the data available, but basing a and b on the spacings of the first two row lines, an orthorhombic unit cell with dimensions $a = 6.07$ Å, $b = 3.76$ Å and $c = 9.62$ Å containing two monomer units would have a density of 2.19 g.ml^{-1} compared to the observed density of 2.00 g.ml^{-1} for the semi crystalline specimen.

Model building

There are four possible configurations for the ring as a result of the asymmetric centers at C_1 and C_2: D-*threo-*, L-*threo-*, D-*erythro-* and L-*erythro-*, as shown schematically in Figure 3. In the *threo*-configurations, the CF_2 groups at C_1 and C_2 are *trans*, i.e. they point to opposite sides of the average plane containing the ring atoms, whereas these substituents are *cis* in the *erythro*-configurations. The minimum energy conformations for the D-*threo-* and L-*erythro-* monomers are given in Table 2 (conformation I in each case). The equivalent conformations for the L-configurations are mirror images with the same energies. As expected, the rings are puckered, with C_3 and C_4 out of the O-C_1-C_2 plane. The minimum energies for the *erythro*-configurations are 1.6 kcal/mole higher those for the *threo*-configurations, due to the *cis* disposition of the substituents. Exploration of the conformational space of the monomer shows that three additional minimum energy conformations exist for each configuration. The ϕ_1-ϕ_5 angles and the potential energies for these conformations for the D-*threo* configurations are also given in Table 2.

To consider the backbone conformation, we constructed all possible combinations of syndio- and isotactic sequences, and head-to-tail, head-to-head, and tail-to-tail combinations. The starting models had the rings in the lowest energy conformation I. The minimum energy conformations for the six possible all-*threo* dimers had the energies and torsion angles listed in Table 3. In all cases ξ, ζ and ψ are in the region of 180°(*trans*). The deviations from *trans* occur primarily because of the van der Waals interactions between the fluorine atoms. Examination of the *threo*-disyndio linkages shows that the CF_2-CF_2 spacer units in the head-to-head and tail-to-tail linkages are approximately centro-symmetric. Requiring them to be exactly so has negligible effect on the potential energy and leads to a straight chain conformation for the head-to-head/tail-to-tail polymer, with a repeat of 9.54 Å (4.77 Å advance per monomer), which is close to the dimension $c = 9.62$ Å proposed above based on the X-ray data(Figure 4(a)). The conformations and fiber repeats of the disyndio head-to-tail and head-to-head/tail to tail polymers should be very similar to each other, and to those of a syndiotactic chain with random head-to-tail/head-to-head linkages, with an axial advance of ~4.8 Å.

Figure 4(b) shows the conformation for the *threo*-diisotactic chain with head-to-tail linkages constructed using the torsion angles in Table 2. The chain is helical with close to five monomers per turn and a monomer advance of 4.05 Å. The head-to-head/tail-to-tail polymer has a similar conformation, as would a random head-to-tai/head-to-head option. Thus for the *threo*-diisotactic structures, we would expect to observe meridional intensity in the region of $d = 4.1$ Å, whereas the observed reflections are at d=4.81 and 2.33 Å. This difference is well beyond experimental error, and effectively rules out these structures.

Thus the X-ray data favor a *threo*-disyndiotactic structure. Distortion of a *threo*-diisotactic polymer chain to have a conformation compatible with the observed repeat would involve a large rise in the potential energy. Hence it is difficult to see many isotactic linkages being incorporated as defects in crystallites composed of predominantly *threo*-syndiotactic chains.

Conclusions

Pressed films of the ring-containing fluoropolymer poly(1,2-difluoromethylene-1,2,3,3-tetrafluoro-4,4-dihydrofuran) are highly crystalline and contain lamellae that are oriented with their long axes parallel to the film surface. The wide X-ray angle data recorded with the beam parallel to the film surface contain both layer lines and row lines, pointing to a crystal structure with at least monoclinic symmetry, in which the chain axes are perpendicular to the film surface. The layer line spacings define the repeat along the chain axis direction as $c = 9.62$ Å, probably containing two monomer units.

Molecular modeling points to a ribon-like conformation for a syndiotactic polymer chain composed of alternating D- and L-*threo*-monomers in a repeat of $c = \sim9.5$ Å (~4.8 Å per monomer). This repeat is largely independent of the choice of head-to-tail, head-to-head or tail-to-tail linkage, and is in very good agreement with the observed X-ray data. In contrast, the minimum energy conformations for the isotactic polymer composed of *threo*-monomers are helices with approximately five monomers per turn and an axial advance per monomer of ~4.1 Å, which is inconsistent with the observed X-ray data. The structures composed of *erythro*-monomers also have short axial repeats: 3-9 4.2 Å per monomer, and have much higher potential energies. Random copolymerization of monomers with different configurations is unlikely to lead to a crystalline structure. Thus PDTE probably has a highly syndiotactic/all-*threo*-microstructure in order to account for the high degree of crystallinity.

Acknowledgments

We thank Drs. Zhen-yu Yang, Andrew E. Feiring and Bruce E. Smart of DuPont Central Research, Wilmington DE, for providing the specimen of PDTE and for very helpful discussion.

References

1. Yang, Zhen-Yu; Feiring A. E.; Smart, B. E., *J.A m.. Chem.. Soc.*, **1994**, 116, 4135
2. Yamabe, M.; Matsuo, M.; Miyake, H., *Plast. Eng.*, **40**, 429 (1997).
3. Holt, D. B.; Farmer, B. L.; Macturk, K. S.; Eby, R. K., *Polymer*, **1996**, 37, 1847

Table 1. Observed *d*-spacings and R, Z reciprocal space coordinates forthe wide angle Bragg reflections with the beam parallel to the film surface.

#	d (Å)	$Z(\text{Å}^{-1})$	$R(\text{Å}^{-1})$	$1/R(\text{Å})$
1	6.07±0.03	0	0.165	6.07
2	5.18±0.01	0.104	0.163	6.15
3	4.81±0.03	0.208	0	meridional
4	3.75±0.01	0.208	0.167	6.00
5	3.50±0.01	0.104	0.266	3.76
6	2.92±0.02	0.104	0.326	3.06
		0.208	0.272	3.68
7	2.59±0.01	0	0.386	2.59
		0.104	0.372	2.69
		0.208	0.325	3.07
		0.312	0.228	4.39
8	2.33±0.1	0.429	0	meridional

Table 2. Torsional angles for the minimum energy conformations of the tetrafluoro-dihydro-furan ring

	ϕ_1	ϕ_1'	ϕ_2	ϕ_3 (°)	ϕ_4	ϕ_5	Energy (kcal/mole)
D-*threo*							
I	11.6	- 111.0	15.2	- 37.3	47.7	- 38.0	7.3
II	45.7	- 76.0	- 27.1	0.8	30.3	- 48.8	9.3
III	- 29.4	- 153.0	8.5	14.8	- 36.2	42.2	8.2
IV	6.6	- 113.0	- 29.4	43.7	- 42.0	22.4	9.6
D-*erythro*							
I	- 42.4	- 53.2	23.2	2.2	- 31.0	47.5	8.9
II	38.5	42.0	- 20.1	- 4.6	31.8	- 44.6	13.6
III	17.3	21.0	- 36.7	45.9	- 36.8	12.2	14.8
IV*	- 3.5	- 10.0	26.8	- 42.5	42.6	- 24.9	15.9

* conformation IV for D-*erythro* is at an inflection rather than energy minimum
$\phi_1 = O\text{-}C_1\text{-}C_2\text{-}C_3$, $\phi_2 = C_1\text{-}C_2\text{-}C_3\text{-}C_4$, $\phi_3 = C_2\text{-}C_3\text{-}C_4\text{-}O$, $\phi_4 = C_3\text{-}C_4\text{-}O\text{-}C_1$, $\phi5 = C_4\text{-}O\text{-}C_1\text{-}($
$\phi_1' = C5\text{-}C2\text{-}C1\text{-}C6$ (backbone torsion angle).

Table 3. Backbone torsion angles for the minimum energy conformation of the six possible di-*threo*- units

Tacticity	Torsional angle (°)					Energy
	ϕ_1	ζ	ξ	ψ	ϕ_1	(kcal/mole)
***threo*-disyndio-**						
H-T	- 111	163	161	173	111	20.1
H-H	- 111	161	177	199	111	28.7
T-T	- 111	180	183	182	111	15.1
***threo*-diiso-**						
H-T	- 111	164	157	174	- 111	16.4
H-H	- 111	161	167	161	- 111	22.5
T-T	- 111	174	166	174	- 111	13.2

key: H-H: head-to-head; H-T: head-to-tail; T-T: tail-to-tail

Figrue 1. X-ray wide and small angle patterns of a pressed film of PDTE
(a) small angle, beam parallel to the filme surface;
(b) small angle, beam perpendicualr to the filme surface;
(c) wide angle, beam parallel to the filme surface;
(d) wide angle, beam perpendicular to the filme surface;
The vertical direction in (a) and (c) is perpendicular to the film surface

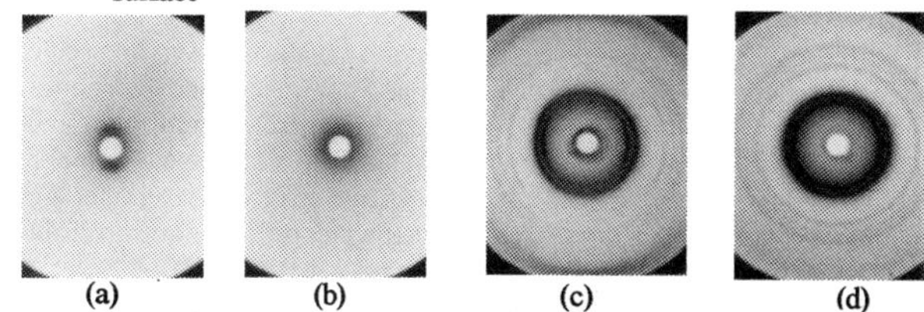

Figure 2. Schematic of the X-ray wide angle pattern recorded with the X-ray beam parallel to the film surface(Figure 2c) Inset: postions of the Bragg reflections plotted in R,Z reciprocal space coordinates(Å^{-1})

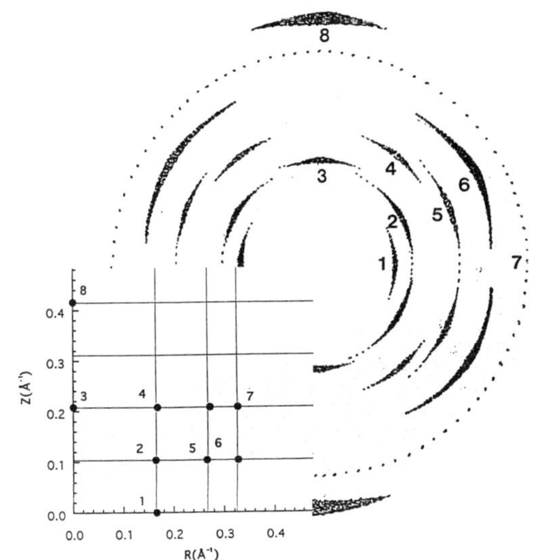

Figure 3. Schematics for the four possible chemical configurations for the tetrafluoro-dihydro-furanyl ring

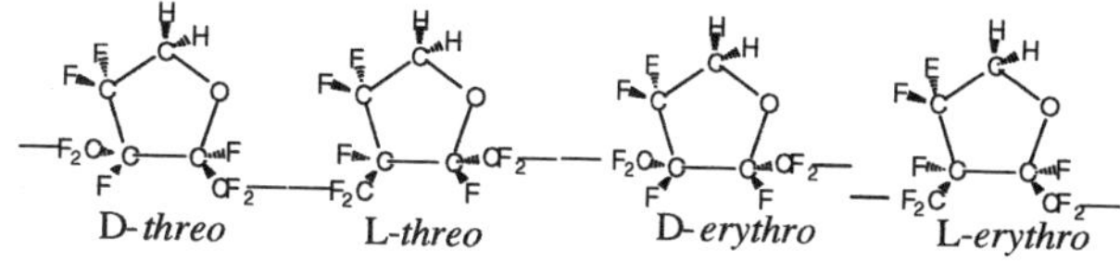

Figure 4. Minimum energy conformation of PDTE chains
(a) syndiotactic head-to-head/tail-to-tail; and
(b) isotactic head-to-tail polymer chains

NEUTRON DIFFRACTION BY CRYSTALLINE POLYMERS

Yasuhiro Takahashi

Department of Macromolecular Science, Graduate School of Science
Osaka University, Toyonaka, Osaka 560, Japan

Introduction

Neutron diffraction has many advantages in comparison with X-ray diffraction. Neutron is diffracted by atomic nucleus, while X-ray is diffracted by electron. Accordingly, the scattering length of an atom by neutron is independent of the atomic number. Hydrogen and deuterium have the same atomic number, i.e., the same chemical properties, and have the same scattering length for X-ray diffraction, but they have the different scattering lengths for neutron (Table I).

Table I. Scattering lengths of atoms by neutron and X-ray.

	neutron		X-ray
	coherent	incoherent	
H	-3.74	25.22	1
D	6.67	4.03	1
C	6.65	0	6
O	5.80	0	8

Furthermore, the scattering length by neutron is independent of the scattering angle θ. The intensities of the reflections with large θ values can be observed strongly and can be measured accurately. This is especially the advantage for crystalline polymers, in which the intensities become weak with the Bragg angle θ because of the disorder contained in the crystalline region and low degree of orientation. Hydrogen and deuterium atoms have the large scattering lengths for neutron in comparison with X-ray. This advantage gives new information for the crystal structure differing from X-ray works. Furthermore, absorption of neutron by most elements, for example, Al, is very small. Therefore, the apparatus for low- and high-temperature measurements can be easily designed and measurements at low and high temperature are easy.

Examples of the crystal structure analyses of crystalline polymers have been limited so far (1, 2). This may be attributed to the incoherent scattering by hydrogen atom and to weak scattering power of crystalline polymers. Incoherent scattering length by hydrogen atom is very large in comparison with the coherent scattering length (Table I). Therefore, it has been considered that the deuterated derivatives of polymers need for neutron diffraction measurements. Crystalline polymers consist of both crystalline and amorphous regions, and the crystalline region contains a considerable amount of disorder. Diffraction intensity by crystalline polymer are, generally said, so weak and the power of neutron source is not so strong. Therefore, it has been considered to be difficult to measure the sufficient number of reflections accurately.

In the present study, all the neutron experiments were made in Japan Atomic Energy Research Institute by courtesy of Dr. Y. Morii. The structure analyses of two planar zigzag polymers, poly(vinyl alcohol) and polyethylene-d$_4$, were carried out on the basis of the rigid body postulate. Especially, The former is the protonated polymer and includes statistical disorder. This suggests that neutron diffraction can be applied to a wider range of crystalline polymers.

Poly(vinyl alcohol)

Different crystal structure models were proposed by Bunn(3) and Sakurada et al.(4), which were different in the azimuthal angle and the hydrogen bonding network. The neutron structure analysis was carried out in order to determine the azimuthal angle accurately and to clarify the position of the hydrogen atoms associated with the hydrogen bonds.

Commercially supplied poly(vinyl alcohol) fiber was used for the sample. The intensity distribution on the equator were measured using λ = 1.8232 Å at 100K, 200K, and 300K. The integrated intensities of the observed reflections were estimated after smoothing the curves and the indices were assigned.

Structure refinements were made under the rigid body assumption (Fig. 1) by using the constrained least-squares program (5), where the hydrogen atoms associated with hydrogen bond were ignored, because of the statistical distribution of the oxygen atoms due to the atactic configuration. First, X-ray structure refinement was carried out by using the intensity data reported by Nitta et al. (6) measured at room temperature. The R-factor reduced to 11.1%. The structure (Fig.2) is essentially the same as Bunn's model. Subsequently, the neutron structure refinements were made for the intensity data at 100K, 200K, and 300K, which gave R-factors, 26.8, 24.8, and 20.3 %, respectively. The structures are given in Fig.2, which are also essentially the same as Bunn's

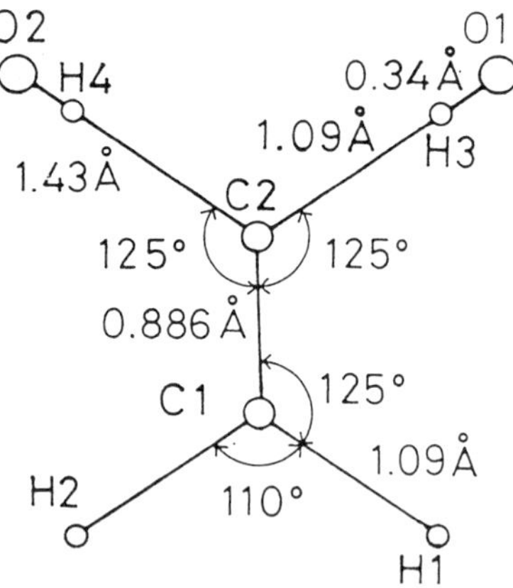

Fig.1. Rigid body assumed for poly(vinyl alcohol).

model. The molecules slightly rotates as the temperature increases. The R-factors are not so good. This may be attributed to the fact that the hydrogen atoms associated with hydrogen bond were ignored. Therefore, difference synthesis was made by using the data at 100K. Three peaks A, B, and C (Fig. 3) can be found on the map of the difference synthesis. Peak A can be assigned to the hydrogen atom associated with intermolecular hydrogen bond. Murahashi et al. (7) suggested the existence of the intramolecular hydrogen bond in the case of it-poly(vinyl alcohol). Peak B can be interpreted by the intramolecular hydrogen bond in the isotactic sequence of at-poly(vinyl alcohol). Peak C can be interpreted by the overlap of the intra- and intermolecular hydrogen bonds. The azimuthal angle of the molecule and hydrogen atoms associated with hydrogen bonds support Bunn's model.

Fig. 2. Structures obtained by neutron structure refinements.

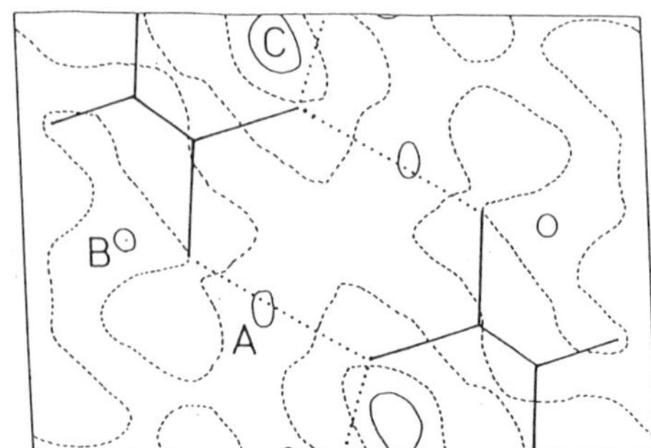

Fig. 3. Difference synthesis.

Polyethylene-d$_4$

The accurate structure of polyethylene (azimuthal angle φ) has been interested by many researchers. Furthermore, the librational motion of the polymer chain has also been interested (8). In the present study, the determination of the accurate structure and the estimation of librational motion of polyethylene are carried out by using the deuterated polyethylene, where the deuterium has the large scattering length in neutron diffraction.

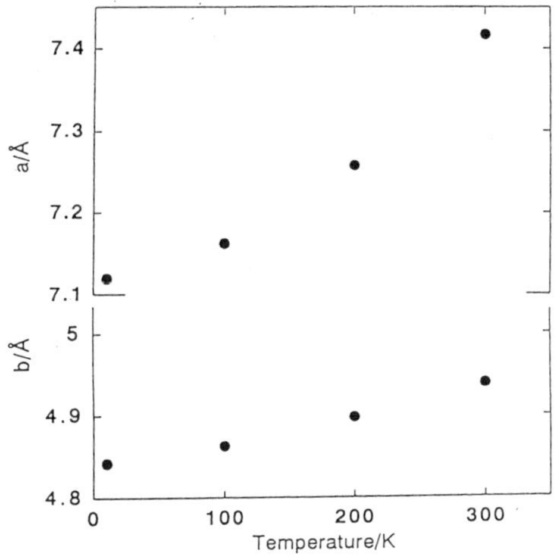

Fig. 4. Temperature dependence of unit cell parameters.

Commercially supplied sample was melted, quenched and stretched and was served for the neutron diffraction. The intensity distributions on the equator were measured at 10K, 100K, 200K, and 300K. The numbers of the observed reflections were seventeen.

Unit cell parameters, a and b, were estimated by least-squares method for the independent reflections (Fig. 4). The rigid body temperature parameters reported by Pawley (9) were taken into consideration. R-factors converged

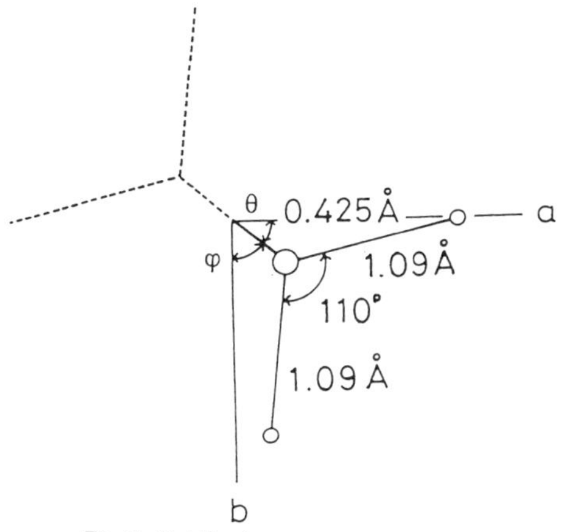

Fig. 5. Rigid body assumed for polyethylene-d$_4$.

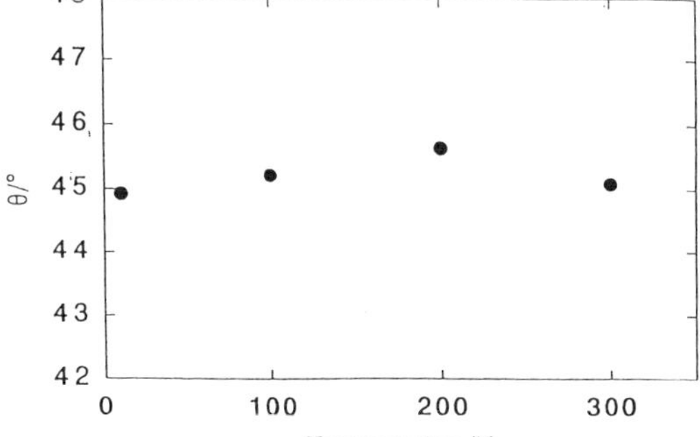

Fig. 6. Temperature dependence of the parameter θ.

to 11.7 %, 10.0 %, 10.9 %, and 13.3 % for the intensity data at 10K, 100K, 200K, and 300K, respectively. The values θ (= 90° - φ) are plotted against temperatures in Fig. 6. The translational motions $2\pi^2\langle u^2\rangle$ are plotted in Fig.7.

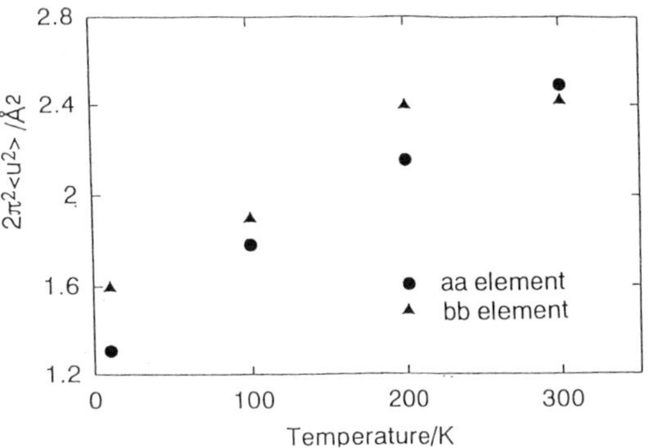

Fig. 7. Temperature dependence of the translational motion.

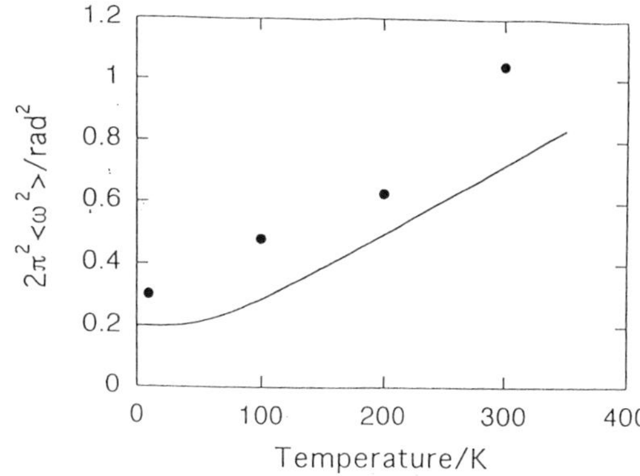

Fig.8. Temperature dependence of the librational motion.
The solid line indicates the calculated values.

The librational motions $2\pi^2\langle\omega^2\rangle$ are plotted in Fig. 8. The azimuthal angle φ is not dependent on the temperature, whose value is about 45° within the standard deviation 1°. From the values of translational and librational motions at 10K, the static disorder is concluded to be mainly of the translational disorder of the molecule. The librational motion is calculated by using the equation reported by Cruickshank (10) and the optical active frequencies for libration (11). The results are plotted on Fig. 8.

References

(1) G. Avitabile et al., *J. Polym. Sci. Polym. Let. Ed.*, **13**, 351 (1975)
(2) M. Stamm et al., *Discuss. Faraday Soc.*, **68**, 263 (1979).
(3) C. W. Bunn, *Nature*, **161**, 929 (1948).
(4) I. Sakurada et al., *Bull. Inst. Chem. Res. Kyoto Univ.*, **23**, 78 (1950).
(5) Y. Takahashi et al., *J. Polym. Sci. Polym. Phys. Ed.*, **11**, 233 (1973).
(6) I. Nitta et al., *Ann. Rep. Inst. Fiber Sci.*, **10**, 1 (1957).
(7) S. Murahashi et al., *J. Polym. Sci.*, **62**, S77(1962).
(8) Y. Takahashi and H. Tadokoro, *J. Polym. Sci. Polym. Phys. Ed.*, **17**, 123 (1979).
(9) G. S. Pawley, *Acta Cryst.*, **17**, 457 (1964).
(10) D. W. J. Cruickshank, *Acta Cryst.*, **9**, 1005 (1956).
(11) M. Tasumi and T. Shimanouchi, *J. Chem. Phys.*, **43**, 1245 (1965).

A SCATTERING STUDY OF NUCLEATION PHENOMENA IN HOMOPOLYMER MELTS

Anthony J. Ryan, N.J. Terrill, J. Patrick A. Fairclough
Department of Chemistry, University of Sheffield, Brookhill,
Sheffield S3 7HF, UK.

Introduction

Polymer processing relies on the shaping of molten material in either moulds or dies and the stabilisation of the shape produced by crystallisation. During crystallisation a microstructure develops which can control the mechanical and aesthetic properties of the polymer. To produce more useful materials it is essential to understand and predict this process. Despite the maturity and market penetration of the polymer industry one aspect of polymer processing, *nucleation*, is little understood. Whereas the growth of polymer crystals is well established in literature, and there are reliable theories to predict the kinetics of crystallisation, understanding of the initiation or nucleation step remains somewhat of a mystery. The available theories are complicated and somewhat unphysical. There is good reason for this, experimental access to the nucleation step is very difficult whereas studies of crystal growth are simple enough to be used in undergraduate laboratory classes. This work aims to show that a process similar to *spinodal decomposition,* with a non-conserved order parameter, plays an integral role in initiating crystallisation in polymers.

An examination of the morphological changes that take place in polymers during crystallisation will afford information crucial to our understanding of the process. These changes occur at the molecular level, SAXS can probe long range order changes while WAXS can monitor changes at the polymer segment level. For the nucleation and growth mechanism, WAXS growth should occur from isolated crystallites prior to interference scattering in SAXS because crystals are present from the beginning of the ordering process. If, however, spinodal decomposition plays a major role, development of a SAXS pattern due to density fluctuations should occur prior to any WAXS from crystals because there is a continuous transformation of the partially ordered phase through slightly more ordered states rather than building a crystalline structure instantaneously. This process was recently discussed by Strobl [1].

Experimental

In order to separate nucleation from growth, two types of experiments have been performed. This preprint concentrates on polypropylene (Mw = 520,000 g/mol, Mw/Mn ≈ 4).

Slow crystallisations with long induction times have been studied by simultaneous SAXS and WAXS using the combined SAXS/WAXS/DSC instrument at Daresbury [2]. The camera was equipped with a multiwire quadrant detector (SAXS) located 3.5 m from the sample position and a curved knife-edge detector (WAXS) that covered 120° of arc at a radius of 0.3 m. The specimens for SAXS/WAXS/DSC were placed in a TA Instruments DSC pan containing a 0.75 mm brass spacer ring and fitted with windows (7 mm diameter) made from 25 μm thick mica. The loaded pans were placed in the cell of a Linkam DSC of single-pan design, samples were melted (200 °C) prior to cooling to the crystallisation temperature at 50 °C/min.

Rapid crystallisations were studied by SAXS and WAXS during melt extrusion of a tape. Two SAXS cameras were used. For time resolved work the camera was equipped with two 20*20 cm, electronic area detectors, one mounted vertically 4 m from the sample (SAXS) and the other offset and angled so that the first reflection from iPP crystals was orthogonal with the detector face at a distance of 30 cm (WAXS). For collection of a single snap-shot the SAXS detector was as above and the WAXS detector was an A4 image plate with a 15mm hole drilled through the middle. The extruder is a AXON BX14 fitted with a slot die of 5*0.5 mm, extrusion conditions were barrel temperature rising from 50°C at the hopper to 220 °C at the die feed, and a die temperature of 200 °C. The extrusion rate was 0.5 m/min and the wind-up rate was varied between 0.8 and 2.5 m/min. The position of the extruder, with respect to the x-ray beam, could be varied to probe different times of crystallisation.

Results

The slow SAXS/WAXS/DSC experiments show a clear development of a SAXS peak, due to electron density fluctuations, prior to the presence of crystals identified by WAXS. The peak area versus time data in Figure 1b where extracted from the data in Figure 1a and show, unequivocally, that the SAXS peak grows before the WAXS peak. Similar behaviour has been reported for a semi-rigid polymer poly(ethylene-terephthalate) crystallised by heating from an amorphous glassy state [3].

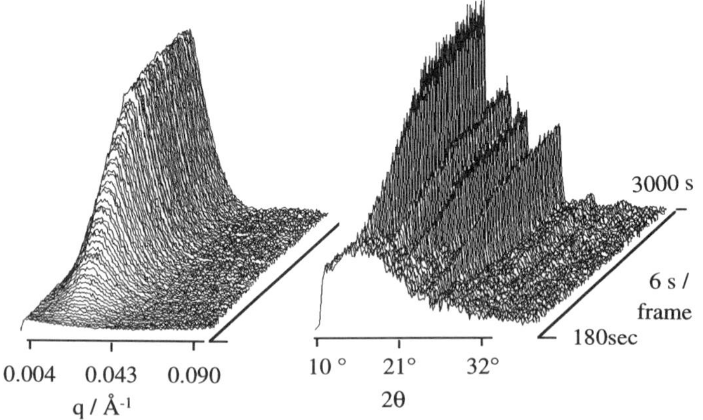

Figure 1 (a) Time resolved SAXS (I(q)q² versus q) and WAXS (I(2θ) versus 2θ(Cu Kα)) taken during the crystallisation of isotactic polypropylene at 145°C.

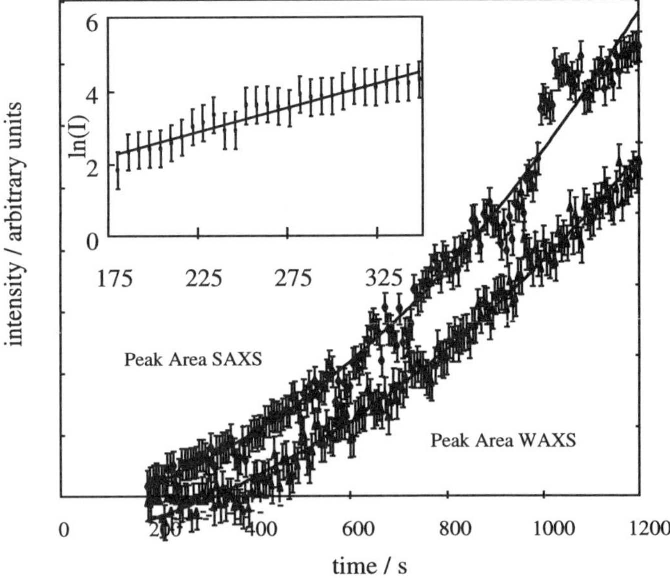

Figure 1 (b) Calculated peak area data for SAXS and WAXS results taken during the crystallisation of isotactic polypropylene at 145°C, the inset is the initial part of the SAXS data plotted in the Cahn-Hilliard form of the peak intensity, logI(q*), versus time, t.

The scattering data have been analysed in the formalism described by Cahn and Hilliard [4] and Figure 2 shows data from polypropylene at 418 K were we could estimate both the dominant length scale L ≈ 175 Å and the effective co-operative diffusion coefficient D ≈ -0.34 Å²/s. There was a precrystallisation induction period of 400 s and the last 200 s of this showed density fluctuations by SAXS. By conducting these experiments at a series of temperatures the stability limit could be found at 438 ± 5 K by extrapolation of D to zero.

Rapid crystallisations were studied by SAXS and WAXS during melt extrusion of a tape[5]. Previous SAXS/WAXS studies on polymer extrusion have concentrated on the growth and orientation of crystals. Extrusion of tape or fibre is a steady-state process where the crystallisation time increases down the spin-line. This allowed long data collection times (minutes) for very short crystallisation times (milli-seconds). Prior to the development of crystals, well resolved, oriented small-angle patterns could be observed with length scales (50-200 Å) and intensities that grew down the spin-line.

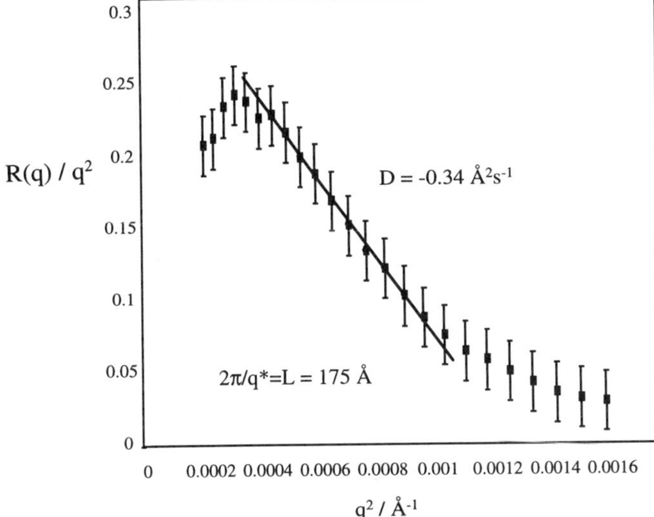

Figure 2 Scattering results are obtained from a crystallisation experiment for polypropylene at 145 °C. The data is then analysed as per the Cahn-Hilliard's formalism where the $R(q)/q^2$ values are calculated from an examination of $\ln I(q)$ versus time for all data. The diffusivity is obtained by extrapolation of the $R(q)/q^2$ versus q^2 plot to q=0.

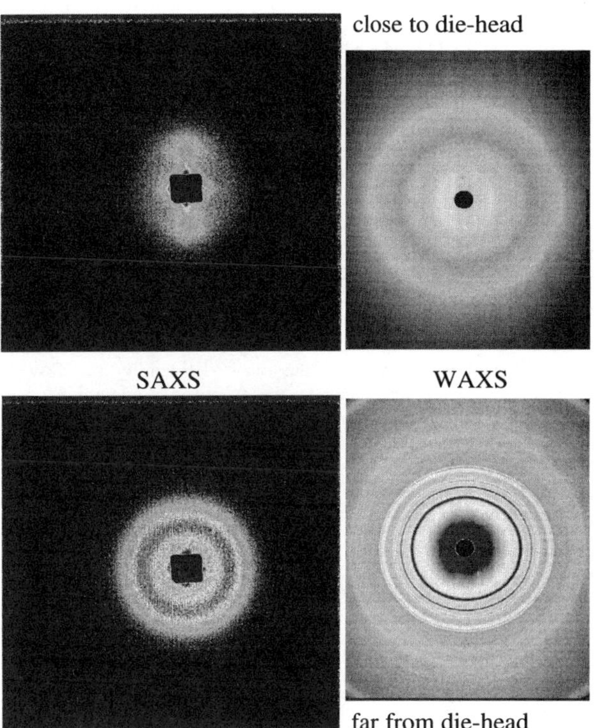

Figure 3 Data taken during extrusion of isotactic polypropylene with a low wind-up speed. 2-D wire chamber to collect the SAXS data and an image plate to collect the WAXS data. Close to the die head tear shaped scattering features are observed in 2-D SAXS with only an amorphous halo in 2-D WAXS. At the furthest position from the die head, when crystallisation starts, isotropic rings from crystals can be seen in SAXS and in the WAXS.

Figure 3 shows the SAXS and WAXS patterns collected during extrusion of polypropylene. The early SAXS pattern has the shape associated with spinodal decomposition, there are no crystalline reflections in the WAXS. The orientation observed in the small angle scattering is caused by coupling of density fluctuations with the slight elongational flow-field (the take-up speed was approximately twice the extrusion speed). Once crystallisation had been observed in the wide-angle region, the shape of the small-angle pattern changed from that characteristic of sinusoidal density fluctuations to that typical of lamellar crystals and Debye-Scherrer rings are observed in the wide angle. Since the elongational flow was weak the crystallisation process dominated and only weakly anisotropic crystals were produced.

Discussion

The stability limit is the temperature below which polypropylene spontaneously separates into phases, one of which is rich in polymer segments of the appropriate chain conformation to crystallise (trans-gauche arrangement of the carbon backbone in isotactic polypropylene) and the other is depleted in polymer segments with the appropriate chain conformation to crystallise and is concentrated in sequences near entanglements and other defects which cannot crystallise. The stability limit is 6 K below the measured melting point for polypropylene with a long spacing of 140 Å and 25 K below the thermodynamic melting point of isotactic polypropylene. Once WAXS from crystals (atomic order on the 1 Å scale) was observed, the kinetics reverted to those of nucleation and growth. This type of behavior has also been observed in PET [3] and in PEEK [6]. In these experiments a glass is devitrified and crystallised between Tg and the maximum rate of crystallisation. The dynamics, therefore, are controlled by the proximity to Tg and as a consequence the stability limit cannot be accessed in this way.

The combination of modern X-ray techniques has allowed us to study the previously inaccessible process of nucleation. These experiments, quiescent time-resolved SAXS/WAXS and extrusion, suggest that a process that strongly resembles spinodal decomposition of chain segments with different average conformations is the nucleation step in polymer crystallisation. That polymer crystallisation occurs with phase separation is in no doubt, at the end of the process regions of well ordered crystalline polymer coexist with regions of disordered polymer in a layered morphology (lamellae). Sequences that can be oriented with the right conformation and incorporated into the crystal separate from sequences near entanglements and other defects which can not crystallize and can only be part of the amorphous regions. The transformation from the disordered phase to the better ordered partially crystalline phase proceeds continuously passing through a sequence of slightly more ordered states rather than building up a crystalline state instantaneously. A mechanism of continuous transformation could be consistent with a fast homogeneous nucleation

process. Polymer crystallisation, like any other phase separation, is kinetically controlled. The structure formed is the one with the highest growth rate. Once a crystallite is formed it has a lateral growth rate much higher than that of the fluctuations and so dominates (we have measured the spherulitic growth rate of polypropylene by optical microscopy). In this case the growth mechanism of semi-crystalline polymer spherulites takes over because the lateral growth rate of crystals (0.5 μm/s) is 1000 faster than the growth rate of the fluctuations (0.35 Å²/s).

The combination of the steady-state extrusion and the high intensity, synchrotron X-ray source allows nucleation phenomena to be observed. If the results we have described are substantiated with further work the less well understood part of nucleation and growth theory, that is nucleation, could have a proper explanation and be predictable [7]. Furthermore our method to study stability limits in phase transitions in polymers could have wide implications for condensed matter physics.

References

1 G. Strobl, *The Physics of Polymers,* (Springer-Verlag, Berlin, 1996) pp 173-176
2 W. Bras, G.E. Derbyshire, A.J. Ryan, G.R. Mant, A. Felton, R.A. Lewis, C.J. Hall and G.N. Greaves, *Nucl.Instrum.Meth.Phys.Res.,* 1993, **A326**, 587.
3 a) M. Imai, K. Kaji, T. Kanaya and Y. Sakai, *Phys.Rev B Condensed Matter,* 1995, **52**, 12696. M. Imai, K. Mori, T. Mizukami, K. Kaji and T. Kanaya, *Polymer*, 1992, **33**, 4451. M. Imai, K. Mori, T. Mizukami, K. Kaji and T. Kanaya, *Polymer*, 1992, **33**, 4457. M. Imai, K. Kaji and T. Kanaya, *Macromolecules*, 1994, **27**, 7103.
4 J.W. Cahn and J.E. Hilliard, *J.Chem.Phys.*, 1958, **28**, 258..
5 M. Cakmak, A. Teitge, H.G. Zachmann and J.L. White, *J.Polym.Sci. Part B; Polymer Physics,* 1993, **31**, 371.
6 T.A. Ezquerra, E. López-Cabarcos, B.S. Hsiao and F.J. Baltà-Calleja, *Phys.Rev E, Brief Reports*, 1996, **54**, 989.
7 P.D Olmsted, W.C.K. Poon, T.C.B. McLiesh, N.J. Terrill and A.J. Ryan, *Phys.Rev Lett.* submitted Feb. 1998.

X-RAY RHEOLOGY OF UNSTRUCTURED AND STRUCTURED
POLYMER MELTS <u>G.R.Mitchell,</u> J.J.Holt, E.M.Andresen and J.A.Pople,
Polymer Science Centre, University of Reading, Whiteknights, Reading,
RG6 6AF UK

Time-resolving x-ray scattering measurements performed on samples
subjected to shear flow provide both a powerful insight to the structural re-
organisation which accompanies flow and an approach to interpret
mechanical rheological data. We have designed and developed novel
equipment to facilitate such measurements and which provides both
quantitative structural parameters and mechanical rheological feedback.

A schematic of the x-ray scattering and flow geometry employed is shown
in Figure 1. In such arrangement the x-ray scattering procedures probe
largely the structure contained in the plane defined by the flow and
vorticity vectors. Alternative geometries have been used which provide
information in the orthogonal planes.

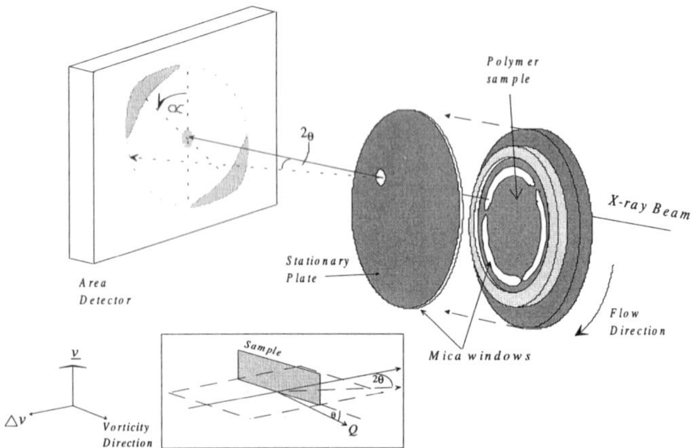

Figure 1 A schematic of the x-ray shear cell

The wide-angle and small-angle x-ray scattering patterns are recorded in a
time-resolving manner using a CCD based area detector coupled to an
integrated computer control system which is described in part elsewhere
[1]. The system provides synchronizes data collection with the
environment control and mechanical measurements and automatic
procedures for the analysis of the scattering patterns in terms of a range of
structural parameters. In this work the direction and level of preferred
orientation are of particular importance. Measurements are performed
using both laboratory based x-ray sources and synchrotron radiation
sources. Time-resolving x-ray scattering could be recorded on a
continuous basis with a time-cycle of 0.8s. In practice, time-cycles of ~ 5s
were adequate for most studies.

A variety of material systems have been studied including both simple
polymer melts of crystallizable synthetic polymers such as poly(ethylene)
and poly(propylene) and structured fluids such as liquid crystal polymers.
Despite the low levels of preferred orientation during shear flow, data from
poly(ethylene) melts could be recorded on a reliable basis and compared
to the substantial levels of macroscopic orientation which develop upon
crystallization from such anisotropic melts [2]. Such measurements
showed a strong correlation between the shear history and the molecular
weight distribution and the subsequent level of crystal orientation.

The flow of behaviour of liquid crystal polymers is complex. Figure 2
shows the macroscopic orientation which develops during steady state
shear flow for both lyotropic and thermotropic systems based on
hydroxypropyl cellulose [3]. At high shear rates these two systems have
similar properties and the relaxation behaviour from such sheared states
are similar. In contrast, the behaviour at low shear rates is quite different
and may be related to the differing balance between viscous and Frank
elasticities. Such information provides considerable insight to the
mechanical rheological measurements and valuable data for the
construction of realistic theories of flow behaviour.

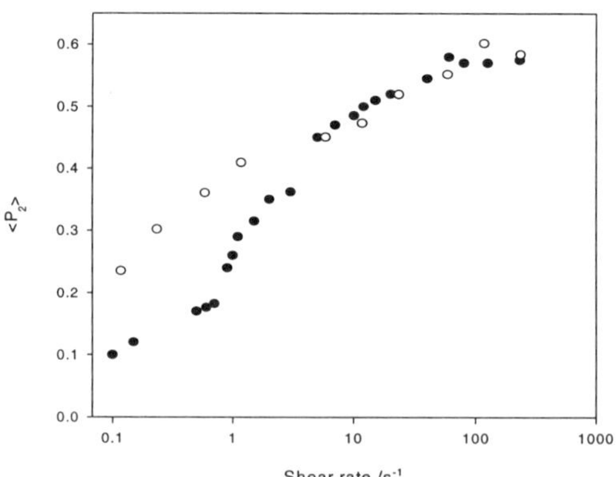

Figure 2 A plot of the macroscopic orientation parameter as a function of
the steady state shear flow for a thermotropic liquid crystal polymer (open
circles) an and the equivalent lyotropic liquid crystal system (filled
circles).

In contrast, side-chain liquid crystal polymers show competition between
the influence of the transient network formed from the polymer backbone
chains and that of the liquid crystal forming side-chains. We have studied
a series of acrylate based polymers with cyano-phenylbenzoate side-chains
which develop a high level of preferred orientation during shear flow [4].
Figure 3 shows how the direction of alignment of the preferred orientation
varies as a function of the sample temperature during shear flow. At high
temperatures, a perpendicular alignment of the side-chains to the flow axis
is observed (φ =90). While at lower temperatures a parallel arrangement of
the side-chains to the flow axis is obtained. The complexity of such
materials is illustrated by the fact that upon cessation of flow in the high
temperature regime, the direction of preferred orientation rotates through
90 degrees with a small reduction in the level of the preferred orientation.
Such behaviour demonstrates the differing roles of the side-chains and the
polymer backbones.

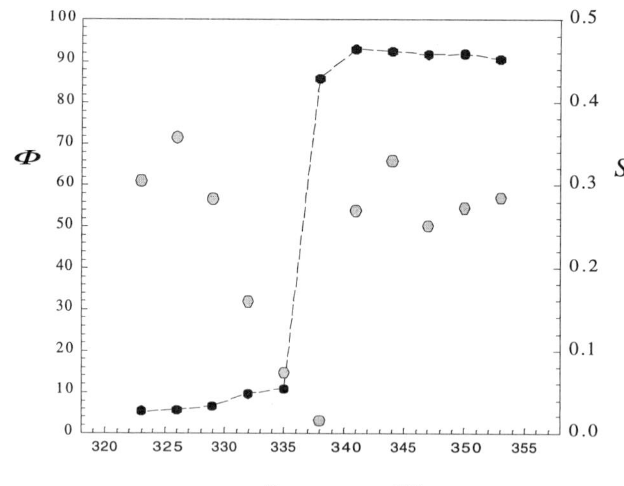

Figure 3 A plot of the direction (φ) and level (S) of preferred orientation
as a function of temperature for a side-chain liquid crystal polymer during
shear flow (shear rate = 1.0 s⁻¹)

This work was funded by the Engineering and Physical Science Research
Council.

[1]. J.A.Pople, P.A.Keates and G.R.Mitchell *J. Synchrotron Rad*, **4** 267-278 1997
[2] J.A.Pople, G.R.Mitchell, S.J.Sutton, A.S.Vaughan and C.Chai *Polymer in press*
[3] E.M. Andresen and G. R. Mitchell *Europhysics Letters submitted*
[4] P.M.S.Roberts and G.R.Mitchell *submitted to Polymer.*.

In Situ X-ray Scattering Measurements of Molecular Orientation in Channel Flows of Thermotropic Liquid Crystalline Polymers

David K. Cinader, Jr. and Wesley R. Burghardt
Department of Chemical Engineering
Northwestern University, Evanston, Illinois 60208

Introduction

Thermotropic liquid crystalline polymers are important in the electronics, health care, and chemical process industries owing to exceptional physical and chemical properties. These include excellent dimensional stability, low melt viscosity, and high modulus. Such properties are highly dependent on the molecular orientation state in the finished product. Since orientation relaxation times in LCPs are long relative to those in flexible chain polymers, processing flow fields can impart significant orientation in finished products. Since there is no solvent to remove, thermotropic LCP products are often manufactured using melt processing techniques such as extrusion and injection molding. While uniaxial extension dominates fiber spinning in lyotropic LCPs(and leads to high orientation along the fiber axis), the flows encountered in these melt processing operations involve a mixture of shear and extensional deformations. Understanding how such complex flow fields influence the molecular orientation state is important if one wishes to anticipate product properties. Many investigators have sought to understand these issues by studying orientation in injection molded parts [1-10]. However, since molding is a transient, nonisothermal process that also is strongly affected by the simultaneous solidification of the polymer, it is difficult to draw clear conclusions specifically about the role of the flow field itself in determining the molecular orientation state. For this reason, we are pursuing an experimental program for *in situ* measurements of thermotropic LCPs undergoing channel flows, thereby focusing attention on the relationship between complex flow kinematics and molecular orientation state.

Experimental

We have designed and built a polymer melt extrusion die that allows x-ray scattering measurements of molecular orientation as a function of position in channel flows. The channel inside the die is 1 mm thick, 110 mm long and has a maximum width of 20 mm. The shape of the flow field in the width direction is dictated by interchangeable spacers. Available geometries include a straight slit flow, 1:2 & 1:4 expansions (sharp and gradual), and 2:1 & 4:1 contractions (sharp and gradual).

To allow radiation to strike the polymer and subsequently exit the die, we have cut v-shaped trenches into the die. Aluminum "windows" prevent the polymer from flowing out of the die through the trenches. Although aluminum diffracts strongly, the diffraction occurs at higher angles than the polymer scattering and may be blocked using a lead mask.

Scattering patterns are collected using a 2-dimensional wire detector. Combined with synchrotron radiation, this allows acquisition of scattering images with exposure times on the order of tens of seconds. Experiments are performed at the DND-CAT synchrotron research facility at the Advanced Photon Source (APS). Photon energies of 25 KeV are used to reduce absorption by the Al windows, and to reduce scattering angles, thus facilitating *in situ* measurements during extrusion.

Results and Discussion

Liquid crystalline polymers may be broadly classified based on how shearing flows influence the director. For a tumbling material, the hydrodynamic torques in shear flow act to rotate the director in a time-dependent fashion. Tumbling is expected to disrupt macroscopic orientation. For a flow-aligning material, a shearing flow will promote alignment of the director in the shear plane at a small angle with respect to the flow direction. Flow alignment, then, should lead to a high orientation state. To surmise the effects on the orientation of the mixed shear-extensional kinematics characteristic of channel flows, we imagine them to be a superposition of a planar extension on a simple shearing flow. While this picture does not account for the variation in shear rate across the flow channel, it is a useful representation of local kinematics. For a sharp contraction, the planar extension is aligned so that it pulls in along the

vorticity direction and pushes out along the flow direction. For the sharp expansion, the extensional component is oriented in the opposite sense. In the contraction flow, the extensional component will act to pull the tumbling orbits into the shear plane, resulting in an increase in measured orientation. In the expansion flow, the extensional component will act to pull the director out of the shear plane, resulting in a decrease in orientation. We have measured both of these effects in channel flow of a known tumbling material, the model lyotrope hydroxypropylcellulose in *m*-cresol [11-12]. For a flow aligning material in contraction flow, the directors already lie within the shear plane, and any added extension is expected to have little effect on the orientation. For a flow aligning material in expansion flow, we must consider that there is a distribution of shear rates in the thickness direction. Where shear is relatively strong, near the bounding walls, extension is too weak to disrupt the flow aligned condition. Where shear is relatively weak, near the midplane of the flow, extension will pull the director out of the shear plane and align it in the vorticity direction. This would result in a biaxial orientation state, with discrete populations of orientation lying parallel and perpendicular to the flow direction.

Figures 1 and 2 present results from a set of experiments on Xydar® LCP resin. The quality of orientation is quantified using an orientation parameter, following the definitions and analysis procedures described by Mitchell and Windle [13]. Our measured orientation parameter reflects many levels of structure: the distribution of molecules about the director, the distribution of director orientations about the polydomain texture and the spatial variation of orientation state in the inhomogenous flow.

Results from experiments with expansion flows are particularly interesting. Figure 1 shows the orientation parameter as a function of distance from the expansion in a 1-4 sharp expansion.

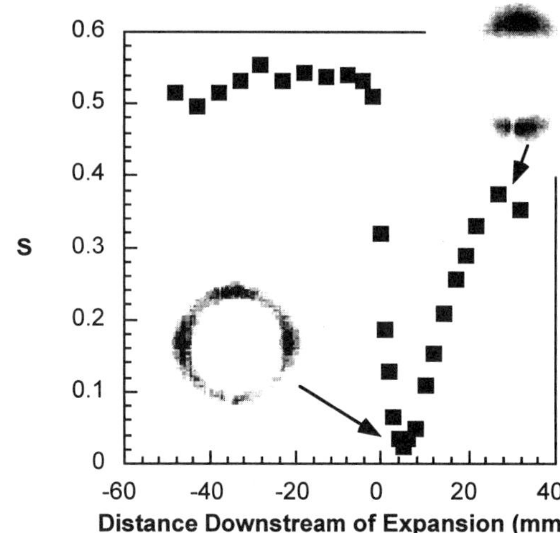

Figure 1: Orientation parameter along the centerline in expansion flow. Scattering patterns reflect a horizontal flow direction.

Upstream of the expansion, the predominantly shearing flow leads to a significant degree of orientation. At the expansion, fluid deceleration in the flow direction is accompanied by a stretching deformation in the direction transverse to the flow. Quantitatively, the orientation parameter shows a sudden decrease as the extensional flow pulls the orientation away from the flow direction. This is followed by a recovery of orientation as the subsequent shear flow downstream of the expansion brings the molecules back into the flow direction. Qualitatively, the expansion region exhibits a four-spot scattering pattern, indicating that there are two discrete populations of orientation arising from the mixed shear and extension: parallel and transverse to the flow direction.

Figure 2 depicts orientation as a function of distance from the centerline 3mm downstream of the contraction in 1-4 expansion flow.

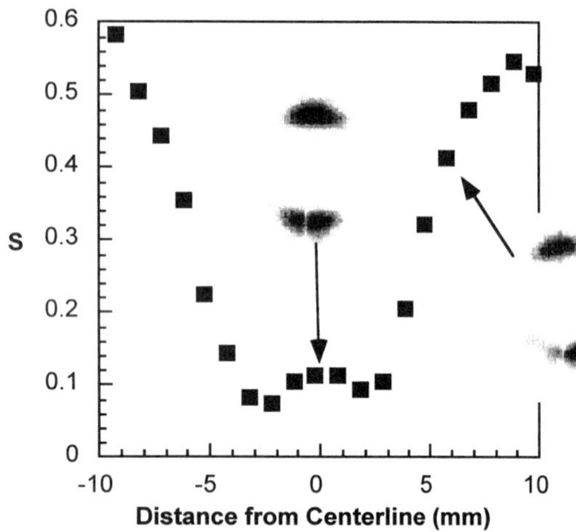

Figure 2: Orientation across the width of the flow channel 3mm downstream of the expansion in 1-4 expansion flow. Scattering patterns reflect a horizontal flow direction.

This shows that very highly aligned states that exist near the wall, while the biaxial orientation state that exists at the centerline results in a lower value of the orientation parameter. Between these extremes lies a state in which the scattering pattern is rotated out of the downstream direction consistent with expectations based on the streamlines curving out of the expansion.

References

[1] Bensaad, S., B. Jasse, and C. Noël, *Polymer*, **34**, 1602 (1993).

[2] Blundell, D.J., R.C.A. Chives, A.D. Curson, J.C. Love, and W.A. MacDonald, *Polymer*, **29**, 1459 (1988).

[3] Heyndrickx, I. and F. Paridaans, *Polymer*, **34**, 4068 (1993).

[4] Ide, Y. and Z. Ophir, *Polymer Engineering and Science*, **23**, 261 (1983).

[5] Jansen, J.A., F.N. Paridaans, and I.E.J. Heynderickx, *Polymer*, **35**, 2970 (1994).

[6] Pirnia, A. and C.S.P. Sung, *Macromolecules*, **21**, 2699 (1988).

[7] Sawyer, L.C. and M. Jaffe, *Journal of Materials Science*, **21**, 1897 (1986).

[8] Thapar, H., and M. Bevis, *Journal of Materials Science Letters*, **2**, 733 (1983).

[9] Weng, T., A. Hiltner, and E. Baer, *Journal of Materials Science*, **21**, 744 (1986).

[10] Zülle, B., A. Demarmels, C.J.G. Plummer, and H.-H. Kausch, *Polymer*, **34**, 3628 (1993).

[11] Bedford, B.D., and W.R. Burghardt, *Journal of Rheology*, **40**, 235 (1996).

[12] Cinader, Jr., and W.R. Burghardt, submitted to *Polymer*.

[13] Mitchell, G.R. and A.H. Windle. "Orientation in liquid crystalline polymers", in *Developments in Crystalline Polymers-2*, Ed. D.C. Basset, Elsevier Applied Science Publishers, Ltd., London, 1988.

Elliptical Characteristics of Small-Angle Reflections and Their Implications on the Lamellar Structure in Semicrystalline Polymers

N.S. Murthy,[1*] D.T. Grubb[2] and K. Zero[1]

[1]AlliedSignal Inc., P.O. Box 1021, Morristown, New Jersey 07962

and

[2]Department of Materials Science and Engineering,
Cornell University, Ithaca, New York, 14853

Introduction

The lamellar reflections in small-angle scattering patterns from polymer fibers are often spread onto a curve symmetrical about the fiber axis. These are usually referred to as two-or four-point patterns, the latter sometimes resembling the butterfly pattern frequently found in light scattering. We recently showed that elliptical coordinates provides an appropriate frame work for analyzing SAS data from oriented polymers.[1] We now show that the elliptical shape of the SAS data could result from a correlation between the lamellar spacing and the orientation of the lamellae.

Theory

An ideal lamellar structure, perfectly aligned and with large lateral crystal size, may be considered as a convolution of a 1-D lattice, spacing L, aligned along the fiber axis and a lamellar plane. This plane could be the surface of a single coherent lamellar crystal,[2] or the locus of the surface of several lamellae in neighboring stacks or fibrils.[3] The diffraction pattern from such an idealized structure is the product of the Fourier transforms of the two real space elements. The Fourier transform of the one-dimensional lattice is a set of horizontal lines, spacing $1/L_M$; that of the single oblique discontinuity is a line normal to the lamellar plane. The product of these two is a set of points lying on this line (Figure 1a). The position of the small-angle reflection on the meridian or fiber (z) axis is widely used to characterize the sample as its "long period" or lamellar repeat distance, L_M.

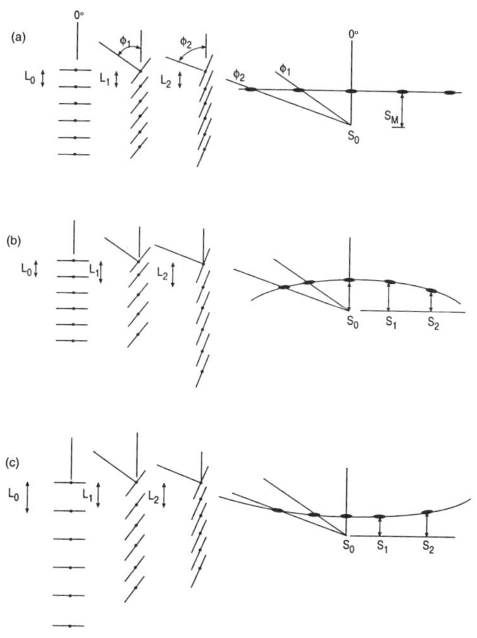

If the rotation of the lamellar planes is associated with a change in the 1-D lattice spacing, then the locus of the reflection is no longer a straight line. An increase of the lattice spacing as the lamellar surfaces rotate away from perpendicular to the fiber axis produces a line with negative curvature; a reduction in the lattice spacing and the same rotation produces a line with positive curvature. Both cases are shown in Figure 1. We will analyze this curvature in the lamellar reflection by measuring the L_ϕ, the periodicity of the lamellar planes measured parallel to z as a function of the angle ϕ that the lamellar reflection makes with the fiber-axis (z).

Data Analysis

Many methods of analysis were tried, but when it became clear that the best fit for the shape was an ellipse or a hyperbola the data were transformed to give a straight line from these curves. The standard form for this ellipse is

$$(x/a)^2 + (z_0/b)^2 = 1 \quad \dots \quad (1)$$

For all the data discussed here, a > b so a is the semi-major axis and b the semi-minor axis. The shape of the ellipse can be defined by ellipticity, $\varepsilon = (1 - b/a)$. The magnitude of the scattering vector s is $s = 2\sin\theta / \lambda$ where 2θ is the scattering angle and λ is the wavelength of the radiation, and $s = 1/$periodicity. In the small-angle regime, $2\sin\theta = \tan(2\theta) = \sqrt{(x^2+z_0^2)}/C = r/C$ where r is the radial distance from the origin to the point (x, zo). On the meridian, x = 0, $z_0 = + b$ and $s_M = b/(C\lambda) = 1/L_M$. Just as L_M is the periodicity of those lamellae aligned with their normals along the meridian, let L_E be the periodicity of those lamellae aligned with their normals along the equator. If the intensity of the reflection has fallen to zero before reaching the equator there may be no such lamellae, but nevertheless the ellipse intersects the equator at x =± a, $z_0 = 0$, and $s_E = a/(C\lambda) = 1/L_E$. In terms of these spacings, Equation 2 becomes

$$(xL_E)^2 + (z_0L_M)^2 = (C\lambda)^2 \quad \dots \quad (2)$$

For oblique lamellae, with normals at angle ϕ to the fiber axis, $x/z_0 = \tan\phi$. The periodicity of the oblique structure in the direction of the fiber axis is L_ϕ. Dividing Equation 3 by z_0^2 and using L_ϕ, it becomes

$$L_\phi^2 = L_M^2 + L_E^2 \tan^2\phi \quad \dots \quad (3)$$

This is a straight line of slope L_E^2 on a plot of L_ϕ^2 vs. $\tan^2\phi$; $\varepsilon = (1 - L_E/L_M)$. In the limit where the reflection stays on a straight layer line the intercept on the equator is at infinity, $a = \infty$ and b = 0; the ellipticity is 1 (Figure 1a). If the reflection follows the arc of a circle, a = b, $L_E = L_M$, the ellipticity is zero and $L_\phi = L \sec\phi$. If the slope is negative then there is no intercept on the equator, no real value of L_E, and the reflection is tracing out a hyperbola with foci on the z axis (Figure 1c). In the standard form for a hyperbola, $-(x/a)^2 + (z_0/b)^2 = 1$. A positive slope and a

Figure 1. (a) A straight trace resulting from the tilting of the lamellae with no change in the lattice spacing. (b) An elliptical trace due to the expansion of the lattice with the tilting of the lamellae. (c) An hyperbolic trace as a result of the contraction of the lattice with the titling of the lamellae.

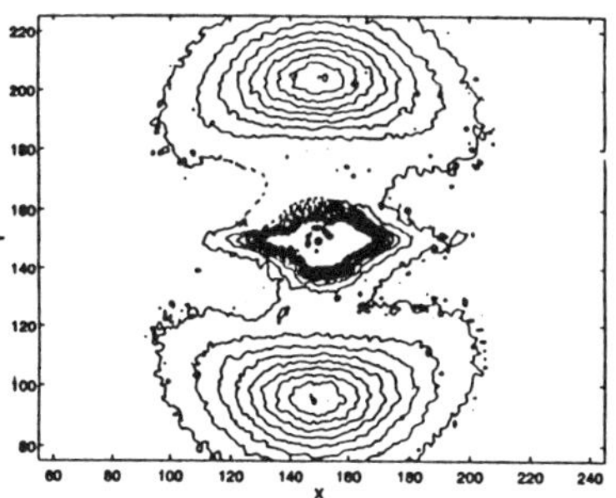

Figure 2. 2-D SAXS pattern from a spun but undrawn nylon 6 fiber.

">

negative intercept would imply a hyperbola with foci on the x axis, and no real value of L_M.

Much of the data used in this paper has been reported on in our earlier publication.[4] Briefly, SAXS patterns from spun-drawn fibers and drawn-annealed fibers were collected at the Cornell High Energy Synchrotron Source using a wavelength of 0.908 Å and a sample-to-detector (camera length, C) of 743 mm. The central region of each image was extracted and adjacent pixels binned together to give a 300 x 300 image with 0.2 mm per pixel. For peak fitting, the image was further binned into a 100 x 300 image, and single columns - averages over 0.6 mm in the equatorial or x direction - were used in the program PeakFit (Jandell). Each 1-D intensity profile was fitted by the sum of four modified Lorentzian peaks, which provided the best fit to the data with least number of parameters.

Results And Discussion

Figure 2 shows an example of the raw 2-D data. The curvature being addressed here is shown in Figure 3a which is a plot of the z-coordinate of the peak maximum vs. x. Figure 3b is a plot of these data in the form of L_ϕ^2 vs. $\tan^2\phi$. There is an excellent straight line fit out to $\tan^2\phi = 5$, $\phi = 66°$. The positive slope shows that the tilt of the lamellae is accompanied by an increase in the lattice spacing. The intercept is 103 nm^2 and the slope is 15.1 nm^2. This gives $L_M = 10.1$ nm, $L_E = 3.89$ nm, ellipticity $\varepsilon = 0.38$. We have carried out this analysis on fibers under a wide variety of draw/annealing conditions as well as with different polymers.

The data obtained from undrawn fiber are usually described as a two-point pattern and that from the fully drawn fiber as a four-point pattern.[4] Both of these can be described by a single unique reflection and the plot of L_ϕ^2 vs. $\tan^2\phi$ is a straight line. But the plot for an intermediate fiber, the 3.0 drawn fiber, deviates from the straight line near the meridian. This deviation, which is also apparent in the plot of z vs. x, is due to the presence of two overlapping lamellar reflections. The reflection close to the meridian with a long-spacing of 80 Å is a remnant of the undrawn fiber (a two-point pattern), and the outer parts of the

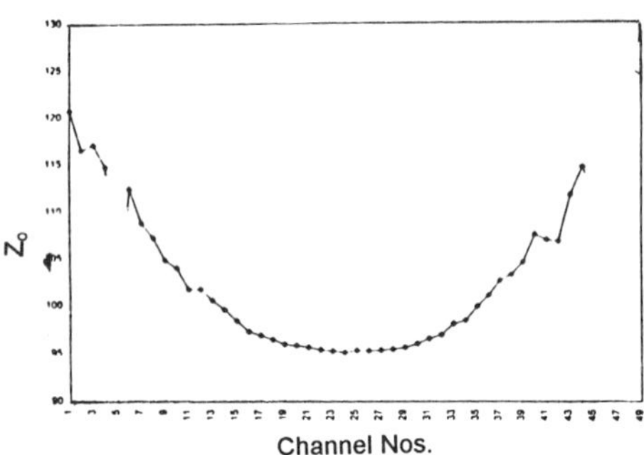

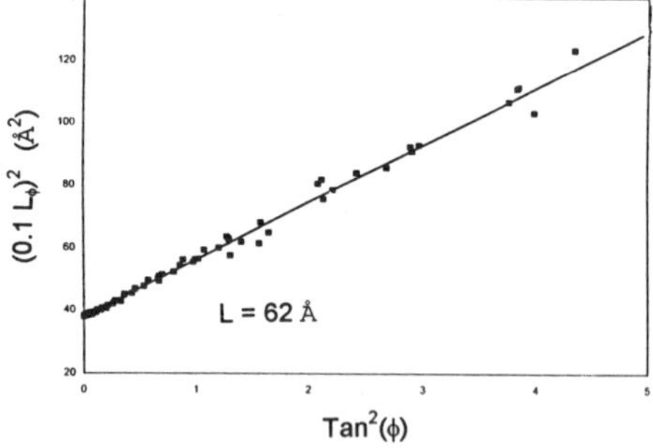

Figure 3. (a) Plot of z_o vs. x for the bottom half the 2-D pattern in Figure 2. Each channel is 0.6 mm. (b) The data in the form of a L_ϕ^2 vs. $\tan^2\phi$ plot.

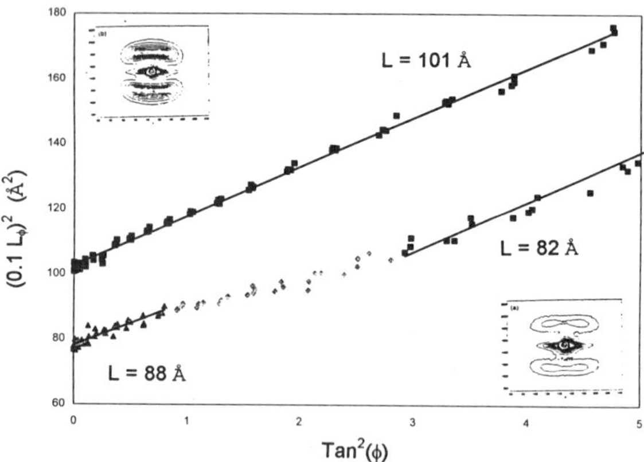

Figure 4. Effect of annealing a highly drawn fiber The bottom curve in is the spun-drawn fiber, and the top curve is the annealed fiber.

reflection at 74Å continue to transform into a four-point pattern. This and other examples show that L_ϕ^2 vs. $\tan^2\phi$ plot can measure differences as small as 1% in the average value even when though the lamellar peaks are quite broad.

Figures 4 shows the data for a highly drawn fibers before and after annealing. The expected increase in the lamellar spacing upon annealing is seen as an increase in the intercept. More interesting is the observation that the plots in the unannealed fibers is a composite of two populations with different lamellar spacings. The lamellar spacing and the ellipticity of the two populations in the unannealed fibers are calculated by least-squares fitting the two segments to straight lines as shown in Figure 3. The resulting values show that the ellipticity of the two populations are the same. Mechanisms such as melting and recrystallization, commonly invoked to explain the increase in the crystallite size and crystalline perfection upon annealing, may also explain the changes seen in Figure 4.

The above results show that the elliptical form is the best fit, and when the reflection does not fit an ellipse, we can get more sensible results from assuming that there are two superposed reflections, each elliptical. It is certain that some rotational disorder is present, and also that the lateral size of the diffracting objects is limited. We suggest that a combination of smearing and rotation in an elliptical coordinate system is required to fully parameterize the data.

The ellipticity parameter (ε) describes the relation between the distributions of ϕ and L. ε is large ($a \gg b$) when L-ϕ coupling is weak, i.e., when lamellar rotation (lamellar or chain slip) occurs with out any large structural reorganization that induce changes in the lamellar spacing. ε is small ($a \ll b$) when L-ϕ coupling is strong, i.e., when large scale changes (melting and recrystallization) occur without shear deformation. This may occur in fibers where the structure is not well developed.

Conclusions

The intensity in the lamellar reflections in the fibers is distributed along an ellipse. This forms the basis for analyzing the 2-D SAS from uniaxially oriented semicrystalline polymers in elliptical coordinates. The elliptical nature of the SAS pattern appears to be related to the constraints on the deformation of the lamellar structure.

References

1   N.S. Murthy, K. Zero and D.T. Grubb, *Polymer*, (1997).
2.   W.O. Statton, *J. Polym. Sci.*, (1959) **41**, 143.
3.   D.A. Pope and A. Keller *J. Polym. Sci. Polym. Phys.*, (1975) **13**, 533.
4.   N.S. Murthy, C. Bednarczyk, R.A.F. Moore and D.T. Grubb, *J. Polym. Sci. Polym. Phys.*, (1996) **34**, 821.

What is a Model Liquid Crystalline Polymer?:
Solvent Effects on the Flow Behavior of LCP Solutions.

Mark Dadmun
Sudha Chidambaram
Chemistry Department
University of Tennessee
Knoxville, TN 37996-1600

Paul Butler
William Hamilton
Solid State Division
Oak Ridge National Laboratory
Oak Ridge, TN 37831

Introduction:

Liquid Crystalline Polymers (LCPs) are of considerable interest inasmuch as their inherent molecular ordering has dramatic consequences on their macroscopic properties. Since molecular alignment affects the macroscopic properties and is easily achieved during flow, the rheology and relation between an applied flow field and molecular orientation of LCPs are an area of great experimental and theoretical interest. Theoretical models have been proposed which account for many of the novel aspects of the rheology of LCPs.[1-2] However, a universal model that fully describes the flow characteristics of LCPs is yet to be developed.

Most of the experimental evidence which is utilized to illustrate the flow behavior of LCP comes from the poly(γ-benzyl L-glutamate) (PBLG)/m-cresol solution.[3-5] Though there seems to be very good correlation between theory and PBLG/m-cresol solution behavior, including the appearance of a negative first normal stress difference, the flow behavior of many LCP solutions does not mimic that of PBLG/m-cresol. Therefore, in the drive to develop a broad theoretical description of the flow behavior of LCP which can account for all of the behavior that is observed experimentally, a more specific understanding of the fundamental reasons for the differences between the flow of PBLG/m-cresol and other LCP solutions is needed. If the reasons for this variety of flow behavior can be determined, then they can be accounted for theoretically, and this will lead to a more complete theoretical description of the flow of LCP solutions.

Pursuant to this, the effect of one parameter, solvent, on the flow and alignment behavior of an LCP solution has been investigated by small angle neutron scattering and rheology. The solvents are deuterated m-cresol (DMC) and deuterated benzyl alcohol (DBA). It is hoped that the results of this systematic examination will provide an explanation for the difference in flow behavior that is observed in these two similar solutions. This, in turn, may provide guidance regarding the choice of important parameters which influence the flow of LCP and must be accounted for in a universal theory.

Experimental:

A 20 wt% solution of PBLG (MW-288000) in DBA and an 18 wt% solution of PBLG (MW-318000) in DMC were used, well into the cholesteric phase in both solvents. The rheology experiments were performed on a Bohlin VOR rheometer using cone and plate fixtures with $2.5°$ cone. To insure that the flow transient had been surpassed, the sample was pre-sheared for 200 strain units prior to collecting steady viscosity data, or prior to flow cessation in those experiments where evolution of the moduli was monitored. The sample chamber was sealed to prevent solvent loss at long times. The dynamic modulus was obtained using a 1.3% strain at 1 Hz.

All in-situ shear SANS experiments were conducted at the High Flux Isotope Reactor (HFIR) at the Oak Ridge National Laboratory in Oak Ridge, TN. Steady state scattering experiments were performed at different shear rates. At each shear rate, scattering patterns were also recorded after cessation of flow at different time intervals to study the relaxation of the flow induced alignment.

A quantitative measure of the degree of ordering was obtained by calculating the alignment factor, $A_f(q)$, as proposed by Walker et al.[6]

$$A_f(q) = \frac{\int_0^{2\pi} I(q, \phi) \cos(2\phi) d\phi}{\int_0^{2\pi} I(q, \phi) d\phi}$$

Where $I(q, \phi)$ is the averaged corrected scattering intensity.

Results:

Steady state Alignment:

The changes in alignment factor, and therefore molecular orientation, as a function of shear rate for PBLG/DBA and PBLG/DMC are shown in Figure 1. These results clearly demonstrate that there is a marked difference in the response of the two solutions to an applied shear flow. PBLG/DBA exhibits three distinct regions, while PBLG/DMC shows only two. It is interesting to note that the extent of alignment is larger in m-cresol than in benzyl alcohol for all shear rates, suggesting an easier path to alignment for the m-cresol solution.

Correlation of Steady State Alignment to Viscosity:

The viscosity as a function of shear rate for PBLG/DMC at room temperature and PBLG/DBA at 70 °C, the conditions of the scattering experiments, are shown in Figure 2. It is curious that the increased molecular alignment with shear rate that PBLG/DBA exhibits at low shear rate does not correlate to shear thinning behavior in the viscosity. Thus, the qualitative difference in molecular alignment that is observed by scattering does not manifest itself as different macroscopic (viscosity) behavior.

Orientation Relaxation:

The change in molecular alignment at different time intervals after cessation of shear is presented in Figure 3 for the PBLG/DMC solution and Figures 4 for PBLG/DBA at 70 °C. These results show that there is a notable difference in the relaxation behavior of the two solutions. PBLG/DMC solutions possess significant orientation even after long relaxation times ($\approx$ 6 hours). The relaxation behavior for all steady state shear rates depicts the same trend; an initial increase in orientation followed by a slight decrease and a subsequent increase to a steady value. The alignment also scales with previously applied shear rate and exhibits a minimum in alignment at the same strain ($\approx$ 500). The degree of alignment which occurs after 6 hours of relaxation time increases with previously applied shear rate.

In contrast, the alignment in PBLG/DBA solutions is not as long lived as that of PBLG/DMC. All experiments show the same qualitative behavior and there is very little temperature dependence; an initial increase in orientation that lasts for about 5-10 minutes and then a relatively sharp decrease that asymptotes to a small positive alignment. The orientation of the solutions sheared at 0.1 s^{-1} show a final alignment that is almost zero, suggesting no residual alignment remains for this low shear rate. Additionally, during the relaxation following all shear rates except the lowest shear rate of 0.1 s^{-1}, there exists a small peak in orientation at long times. It must also be noted that the relaxation kinetics do not scale with previous shear rate as was observed in the PBLG/DMC solution.

Dynamic Modulus after Shear Cessation:

The evolution of the dynamic modulus after shear cessation of the PBLG/DBA solutions were performed under similar shear rates as the neutron scattering experiment, though the temperature was limited to 65 °C. The results of the mechanical measurements at different shear rates have been plotted with the corresponding alignment factors (SANS data) in Figure 5 to show the correspondence between the changes in the two parameters with time. The rheology corresponds very well to the alignment for the PBLG/DBA solutions with a decrease in moduli corresponding very well to the increase in molecular alignment during relaxation. Similarly, the decrease in alignment at relaxation times greater than 10 min. is found to agree well with the increase in modulus in the same time scale. It should be noted that similar experiments on PBLG/m-cresol solutions show good correspondence between the dynamic moduli and the evolving alignment structure of the solution after shear cessation.[4]

Conclusion:

The orientation of PBLG induced by flow in two solvents, deuterated benzyl alcohol (DBA) and deuterated m-cresol (DMC), has been studied by small angle neutron scattering and rheology. Unexpectedly, results show marked differences in the steady state and relaxation responses of the solutions to shear. Steady state results show that the solutions behave similarly at high shear rates (> 1 s^{-1}) with an initial nonlinear region followed by a linear increase in alignment. However, at low shear rates the PBLG in DBA shows an increase in orientation with shear rate which is absent in DMC. Relaxation results further contrast the difference in the ordering of the two solutions. The PBLG/DMC sample retains the flow induced ordering of molecules even after long periods of time ($\approx$6 hours). Conversely, the ordering of PBLG in DBA by shear decays to very low orientation rather quickly.

These differences are surprising as benzyl alcohol and m-cresol are isomers. Clearly, there exist fundamental distinctions which account for their diverse behavior. Possible explanations for these differences include PBLG conformation variations, solution phase behavior deviations, or solvent dependent texture differences. Each of these possible explanations is currently under investigation. Regardless, the results exemplify the need for a more complete understanding of the important parameters which affect the flow of LCP solutions so that a more universal theory can be developed which can predict flow behavior of non- model LCP solutions.

Acknowledgment:

We would like to thank Wes Burghardt for obtaining the dynamic modulus data and for useful discussions. The research at Oak Ridge is supported by the Division of Materials Sciences, U. S. Department of Energy under contract No. DE-AC05-96OR22464 with Lockheed Martin Energy Research Corp.

References:

1. Larson, R. G.; Doi, M. J. *J. Rheol.* **1991**, 35, 539
2. Marucci, G.; Maffettone, P. L. *Macromolecules* **1989**, 22, 4076.
3. Kiss, G.; Porter, R. S. *J. Polym. Sci. Symp.* **1978** 65, 193
4. Hongladarom, K.; Burghardt, W. R. *Macromolecules* **1993**, 26, 785
5. Hongladarom, K.; Ugaz, V. M.; Cinader, D. K.; Burghardt, W. R.; Quintana, J. P.; Hsiao, B. S.; Dadmun, M. D.; Hamilton, W. A.; Butler, P. D. *Macromolecules* **1996**, 29, 5346.
6. Walker, L. M.; Wagner, N. J. *Macromolecules* **1996,** 29, 2298

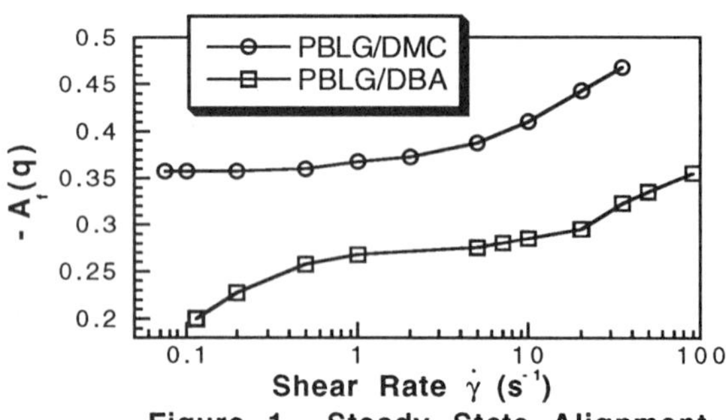

Figure 1   Steady State Alignment

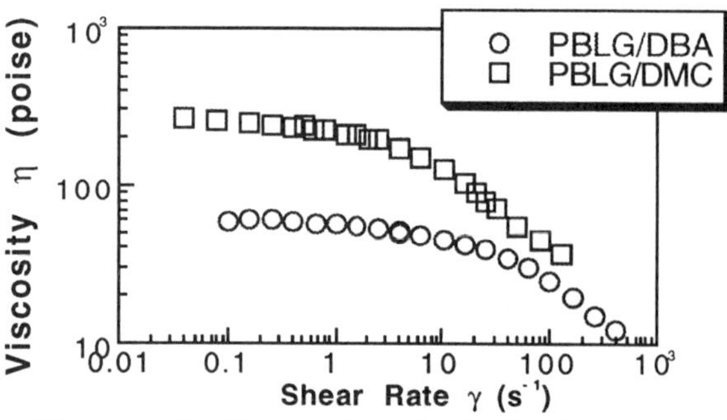

Figure 2 Viscosity of Both Solutions

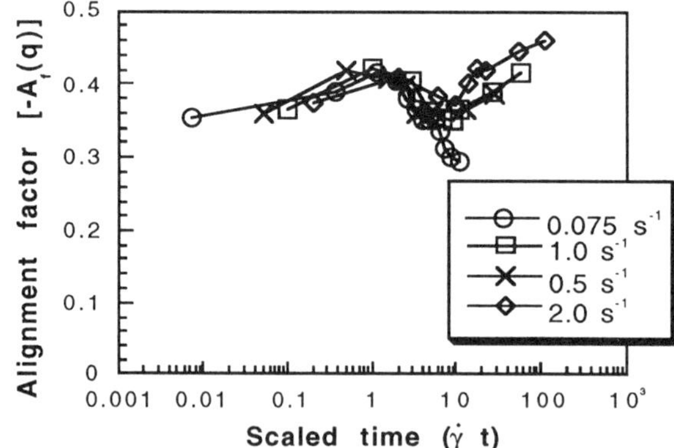

Figure 3   Relaxation of PBLG/ DMC Alignment

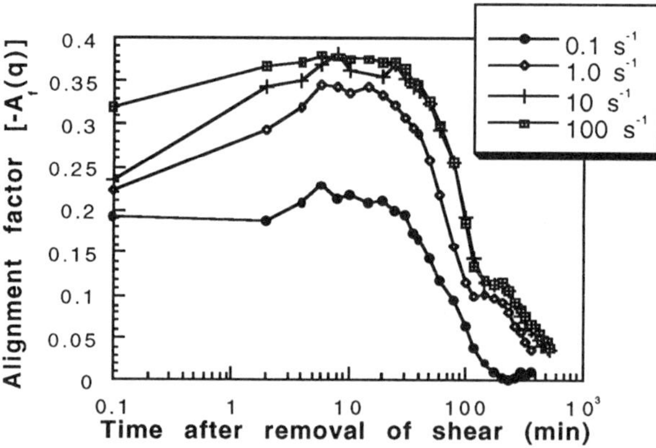

Figure 4   Relaxation of PBLG/DBA Alignment

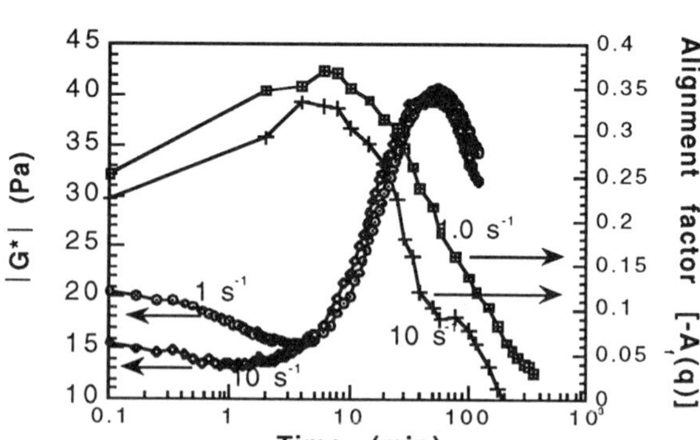

Figure 5.   Evolution of Dynamic modulus after shear

3D SIMULATION OF POLYMER CRYSTALLIZATION

Lucio Toma[1] and Juan A. Subirana[2]
1. Dipartimento di Chimica Organica, Università di Pavia,
Via Taramelli 10, 27100 Pavia, Italy
2. Departament d'Enginyeria Química, Universitat Politècnica de
Catalunya, Av. Diagonal 647, 08028 Barcelona, Spain

Introduction

Polymer crystallization in solution is usually interpreted as the gradual deposition of a random coil molecule onto the face of a growing lamellar crystal. This is a slow process, so that the dissolved molecules should be in the form of collapsed globules. Thus crystallization will proceed by the deposition of collapsed molecules onto the growing crystal. This process has been discussed in detail elsewhere (1) and modelled in two dimensions (2). In this comunication we discuss crystallization from the melt and alternative approaches which have been recently proposed. A preliminary simulation in 3D on a cubic lattice is also presented.

Crystallization from the melt

In its present state our method has only been applied to individual molecules. Thus it is most appropriate for crystallization from solution. However our simulation studies offer a particular interest for understanding some features of crystallization from the melt, which cannot be explained by the conventional model for semicrystalline polymers, as reviewed by Verma and Hsiao (3). Small angle X-ray scattering shows that density fluctuations are detected before crystallization. Such fluctuations indicate that regions with a density different from the bulk are found in the melt before crystallization is detected. Such fluctuations may be due to the local collapse of melted macromolecules in a way similar to that described in this paper.

Methodology

The method has been described in detail elsewhere (2, 4). In essence the energy is computed taking as reference the random coil state, so that undercooling from the soluble state is the variable which determines the behaviour of the polymer.

The energy of the collapsed molecule is calculated by taking into account only parallel segments of the macromolecule, in order to favour the tendency towards crystallization. Each monomer unit occupies one lattice site. The energy of the molecule is determined by summation of energy contributions ($E = -1$) deriving from each contact between nonbonded residues which occupy neighbour lattice points. Contacts between corners are not counted, they are considered as melted parts. During the simulations the molecular conformation changes through pivot moves which are selected by a probability function which depends on the temperature of the sample (2). Every attempt to perform a pivot move is counted as a time step (t.s.). In this way, at sufficiently low temperatures, the molecule evolves towards low energy states. The temperature is represented by a parameter we have called c_k.

The conformation of the macromolecule is thus characterized by its energy E and by a geometric parameter L, which is the average length of straight segments of the molecule. This parameter approximately corresponds to the average lamellar thickness.

Results

In our previous study (2) we carried out simulations in a square lattice and found that individual molecules form lamellar structures with a thickness which decreases as a function of temperature, in agreement with experiments (reviewed in Ref. 1).

In this paper we present our preliminary results in a 3D cubic lattice. As shown in Fig. 1 the temperature is very critical. At $c_k=0.75$ polymer molecules behave as random coils and no significant structure is formed. As the temperature is lowered ($c_k=0.74$), lamellar structures appear with an average thickness of about L=4. In fact the actual thickness is larger, since the average is calculated including short loops that form at the edges of the lamellar structure, as shown in Fig. 2. When the temperature is further lowered ($c_k=0.70$, results not shown), the final lamellar structure is similar to the one presented in Fig. 2, but it is attained in a shorter time. Also energy fluctuations at $c_k=0.70$ are more limited than at $c_k=0.74$.

Thus the results that we obtained (2) in 2D can be generalized to 3D: similar lamellar structures are found. Our simulations show that individual molecules may easily collapse into lamellar structures before being incorporated into a whole crystal.

A further feature of interest is that the molecules evolve by the formation of "blobs" (5), as shown in Fig. 2 (t.s. = 1.2×10^7). In the melt it is likely that large molecules approach crystallization in comparatively small local regions such as the "blobs" found in the intermediate states of our simulations.

Discussion

Our results support the original hypothesis from which we started, namely that individual molecules when transferred into poor solvents form collapsed structures which already show a high degree of order. We are presently simulating this process at different temperatures. We should point out that a tendency towards crystallization has also been recently found in the study of collapsed globules (6) when a three-body repulsive interaction is introduced together with an attractive two-body interaction, although the simulations only yield structures with many internal defects, which can not be considered as models of real crystals.

In a recent paper (7) polymer collapse on a 3D cubic lattice has been studied by an approach that uses a different definition of energy and a different set of moves; the results appear similar in some respects to our 2D results (2), but are unable to simulate the formation of 3D lamellar structures such as those we report.

References

(1) J.A. Subirana. *Trends Polym. Sci.*, **1997**, *5*, 321-326.
(2) L. Toma, S. Toma and J.A.Subirana, *Macromolecules*, **1998**, in press.
(3) R.K. Verma and B.S. Hsiao. *Trends Polym. Sci.*, **1996**, *4*, 312-319.
(4) L.Toma and S. Toma. *Protein Sci.*, **1996**, *5*, 147-153.
(5) De Gennes, P.-G. Scaling concepts in polymer physics, Cornell University Press, Ithaca, NY, **1979**.
(6) Yu.A. Kuznetsov, E.G. Timoshenko and K.A. Dawson. *J. Chem. Phys.*, **1996**, *104*, 336-341.
(7) V.P. Zhdanov and B. Kasemo. *Proteins*, **1997**, *29*, 508-516.

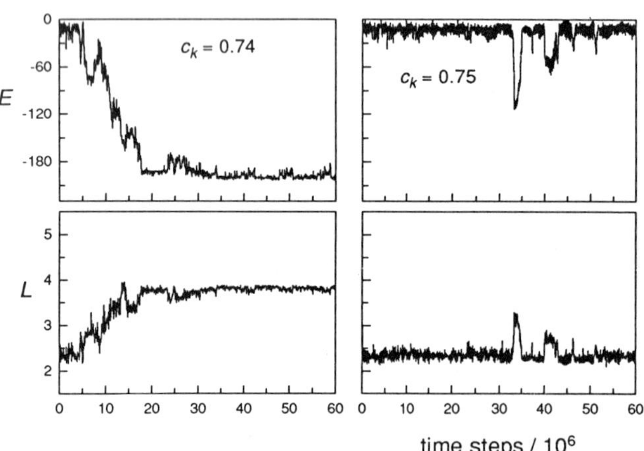

Fig. 1. Results of the simulation of the 3D structure of a polymer molecule of 300 monomer units after 6 x 10^7 time steps at two temperatures. The evolution of energy (E) and thickness (L) is shown as a function of time.

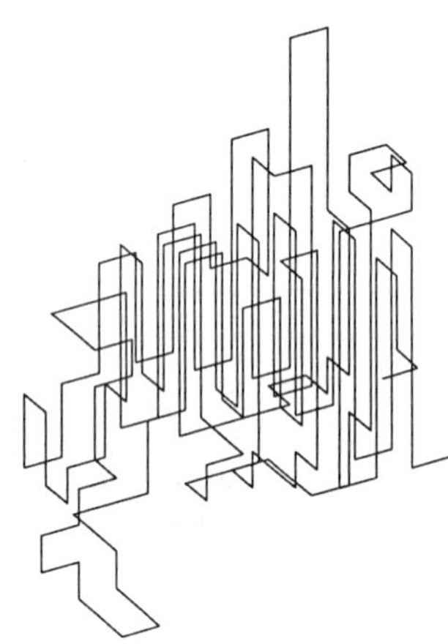

Fig. 3. An enlarged view of the final structure obtained as shown in Fig. 2.

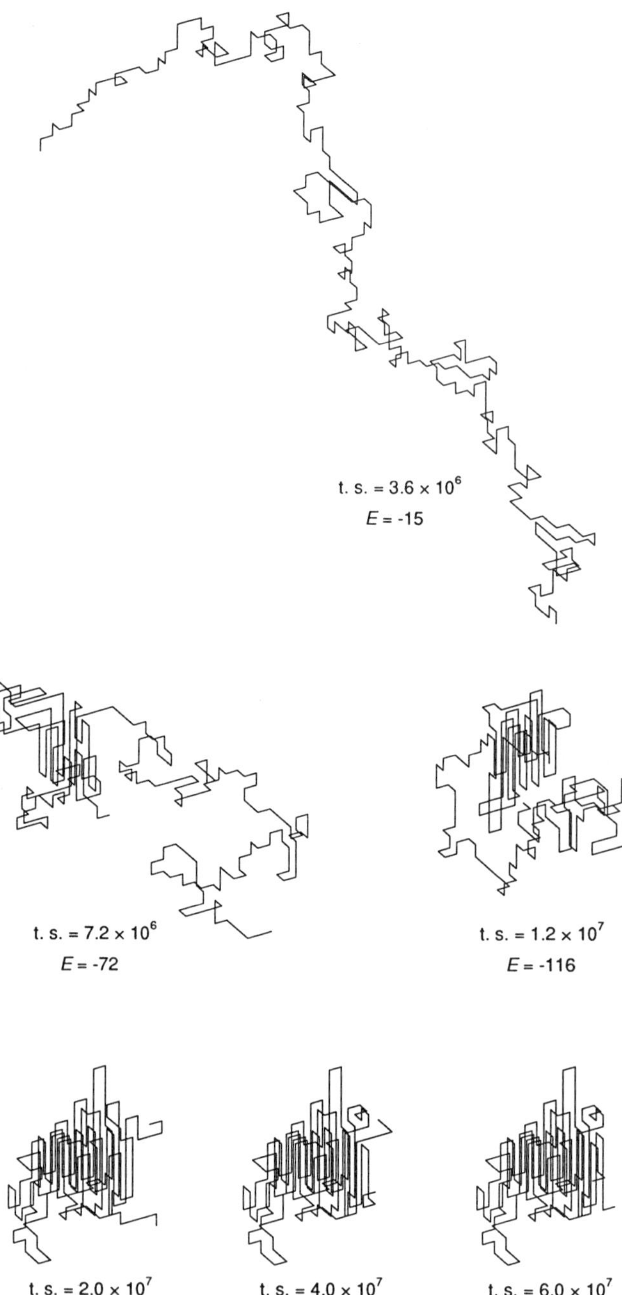

Fig. 2. Conformational story of a single molecule at c_k=0.74. An enlarged view of the final state is shown in Fig. 3

Acknowledgments

This work has been supported in part by grant PB93-1067 from the DGICYT and in part by Ministerio dell'Università e della Ricerca Scientifica e Tecnologica (Roma).

IN-SITU X-RAY SCATTERING INVESTIGATION OF THE INFLUENCE OF SHEAR AND THERMAL HISTORIES ON MOLECULAR ALIGNMENT IN A MODEL THERMOTROPIC LIQUID CRYSTALLINE POLYMER

Victor M. Ugaz
Wesley R. Burghardt

Department of Chemical Engineering
Northwestern University
Evanston, IL 60208

Introduction

Liquid crystalline polymer (LCP) materials have been the object of increasing commercial interest due to their extremely favorable combination of low weight and exceptional mechanical strength. Since these characteristics are a direct result of the high degree of molecular alignment that can be attained in LCPs, a more complete understanding of the effect of flow on molecular orientation is essential in order to better control the processing conditions which give rise to and enhance these properties. Lyotropic LCPs, which exhibit liquid crystallinity in solution, are relatively well understood in terms of a consistent theoretical framework which allows many aspects of their unique rheological behavior to be successfully predicted. Thermotropic LCPs on the other hand, which are liquid crystalline in the melt, are not as well understood despite the fact that they are of potentially greater commercial relevance. This lack of understanding is due to several factors including extreme sensitivity to both thermal and shear histories, the inability to access the isotropic phase in most commercial materials, and the inability to use optical techniques to measure molecular orientation during flow. In this series of experiments, we address these issues by employing a model thermotropic LCP which offers more favorable properties from an experimental standpoint, and by using x-ray scattering techniques which provide a direct measure of molecular orientation regardless of the sample's opacity.

Experimental

The experimental setup consisted of a Linkam high temperature shearing stage, which we adapted for x-ray scattering by designing suitable x-ray windows to replace the standard quartz optical windows. The stage incorporates a parallel disk shearing geometry with an aperture hole cut into the fixed window and a series of slots cut into the rotating window located at the same radius as the aperture hole. Thin sheets of kapton glued to each disk serve as the actual window material. This configuration allows the x-ray beam to pass through the sample during shear so that *in-situ* measurements of molecular orientation could be obtained. The window design in this shear cell is similar to that employed in a device developed in our group for use in x-ray scattering experiments in lyotropic systems[1]. Scattering patterns were collected using a high speed 2-D wire detector which allowed exceptional time resolution in transient experiments. The x-ray scattering experiments were conducted using the synchrotron facility at the Advanced Photon Source located at Argonne National Lab (DND-CAT).

The material studied in this series of experiments is a model thermotropic polyether, DHMS-7,9, synthesized by Weijun Zhou and Julie Kornfield. DHMS-7,9 is a main-chain random copolymer incorporating flexible spacers of two different lengths randomly distributed along the polymer backbone. This architecture gives DHMS several properties which make it well suited for experimental study. It is nematic over a wide range of accessible temperatures, it can be heated into the isotropic phase to erase the effects of previous thermal and shear histories and generate a uniform initial condition, and has excellent thermal stability. DHMS also exhibits an intermediate phase between nematic and crystalline solid. The exact structure of this phase has yet to be identified, although it is speculated to reflect some sort of smectic ordering. Because of its mysterious nature, this regime is referred to as "phase-x." Our experimental results show that dramatic changes in molecular orientation occur within phase-x. Samples of weight average molecular weight 28,000 (28k) and 11,000 (11k) were studied, however the results presented below will focus on the 28k sample except where indicated. All rheological measurements were performed by Zhou and Kornfield.

Results and Discussion

In steady shear flow, we find that molecular orientation remains relatively constant over the entire range of shear rates, and shows only a weak dependence on temperature within the nematic phase as shown in Figure 1. Orientation decreases dramatically near the transition to phase-x at 120 °C. Below 120 °C, we discovered a regime at low shear rates where molecular orientation is favored *perpendicular* to the flow direction (i.e. the flow-vorticity plane), rather than parallel to the flow direction as is typically the case. The degree of this perpendicular orientation, denoted by negative values of the orientation parameter, is most pronounced at low shear rates. As shear rate is increased, the orientation eventually transforms back to parallel alignment. The effect is reversible on decreasing shear rate with minimal hysteresis. This dramatic change in molecular orientation is characterized by a 90 degree rotation in the scattering peaks. Romo-Uribe and Windle[2] observed similar alignment transitions in a commercial grade thermotropic LCP, however it is not clear to what extent our observations can be compared since these two materials possess significantly different molecular architectures.

To investigate how this transformation from parallel to perpendicular alignment influences the material's mechanical behavior, we compared data for molecular orientation and steady shear viscosity as a function of temperature at a shear rate of 0.1 s^{-1} as shown in Figure 2. Orientation increases from a value of zero in the isotropic phase to a nearly constant value in the nematic phase. This increase in orientation is accompanied by a decrease in steady shear viscosity in the nematic phase, as expected due to the significant enhancement in molecular alignment. Near the transition to phase-x, the transition from parallel to perpendicular alignment is accompanied by a drastic increase in viscosity—over two orders of magnitude. In contrast, steady shear orientation measurements as a function of temperature in the 11k sample, shown in Figure 3, exhibit no evidence of a transition to perpendicular alignment at low temperatures. There is only a decrease toward zero orientation corresponding to the nematic-isotropic transition.

The evolution of molecular orientation upon flow startup as a function of applied strain in the nematic phase shows a monotonic increase and scales very well with applied strain. Approximately 30–40 strain units are required to attain a steady state level of orientation. Corresponding shear stress data shows that the strain range required to reach a steady state stress value is comparable to that required to reach a steady state level of orientation.

Flow reversal data in the nematic phase show an initial decrease in orientation followed by a much slower increase back to a steady value after about 100 strain units, somewhat longer than the 40 strain units required in flow startup. Comparison with shear stress data shows that an overshoot peak occurs at approximately the same strain value as the undershoot in orientation. Also, like the x-ray data, large values of strain are required for the shear stress to return to a steady value.

Relaxation data in the nematic phase show only a slight decrease in orientation over time. Evolution of the loss modulus G" during relaxation, however, shows a significant initial drop in G", followed by a more gradual decay. This is not in agreement with the x-ray data, suggesting that structural changes are occurring at a level other than that probed by x-ray scattering.

Since both temperature and shear rate play a role in determining the alignment state in the phase-x, a diagram showing the locations of the various alignment regimes in temperature-shear rate space can be constructed by plotting the data from our entire matrix of experiments as shown in Figure 4a. We then conducted experiments to measure the evolution of molecular orientation over time in response to various step changes in temperature and shear rate into and out of the perpendicular alignment regime. The data in Figure 4b show that a step increase in temperature starting with perpendicular alignment generates a rapid transformation to parallel alignment, followed by a more gradual increase to a steady value of orientation. Conversely, both a fast quench and a step down in shear rate starting from parallel alignment and ending with perpendicular alignment induce a similar response in the material: a sudden initial drop in orientation with an undershoot followed by a more gradual transformation to perpendicular orientation and attainment of a steady value after nearly 200 strain units. Hence, we find that an equivalent response in the material can be induced by following different paths in temperature-shear rate space.

Conclusions

We have used the direct measure of molecular orientation provided by *in-situ* x-ray scattering experiments to learn more about the influence of flow on molecular orientation in DHMS-7,9, a model thermotropic polyether. We find that that orientation is only weakly dependent on shear rate and temperature in the nematic phase, but undergoes a dramatic transition from parallel to perpendicular alignment at low shear rates in phase-x. This change in alignment state is accompanied by an increase in steady shear viscosity of over two orders in magnitude. While the orientation response to flow startup and flow reversal agrees with corresponding rheological data, the connection between structure and rheology during relaxation is less clear. Finally, we find that step changes along different paths in temperature-shear rate space can induce a similar response in orientation.

Acknowledgments

We would like to acknowledge the support of the Air Force (AFOSR-MURI), and an NSF graduate fellowship (VMU). In addition, we thank Julie Kornfield and Weijun Zhou for synthesis and characterization of the DHMS polymer, and the staff of DND-CAT at the Advanced Photon Source for their assistance with the x-ray scattering experiments.

References

[1]   Ugaz, V.M., Cinader, D.K., Jr., and Burghardt, W.R. "Origins of Region I Shear Thinning in Model Lyotropic Liquid Crystalline Polymers." *Macromolecules.* 30 (1997): 1527–1530.

[2]   Romo-Uribe, A. and Windle, A.H. "'Log-Rolling' Alignment in Main-Chain Thermotropic Liquid Crystalline Polymer Melts Under Shear: An In-Situ WAXS Study." *Macromolecules.* 29 (1996): 6246–62455.

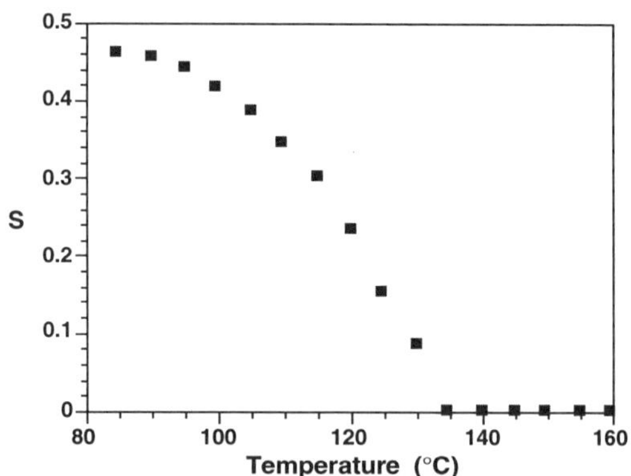

Figure 3. Steady shear molecular orientation as a function of temperature in the 11k sample at a shear rate of 0.3 s^{-1}.

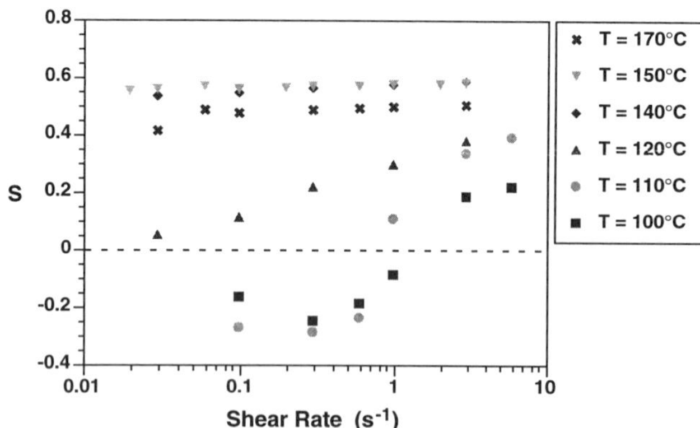

Figure 1. Orientation parameter as a function of shear rate and temperature in steady shear flow at a shear rate of 0.1 s^{-1}.

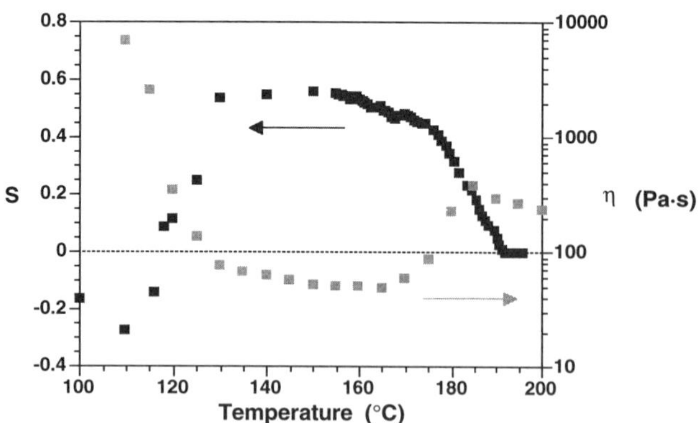

Figure 2. Molecular orientation (left ordinate) and steady shear viscosity (right ordinate) as a function of temperature in the 28k sample at a shear rate of 0.1 s^{-1}.

a)

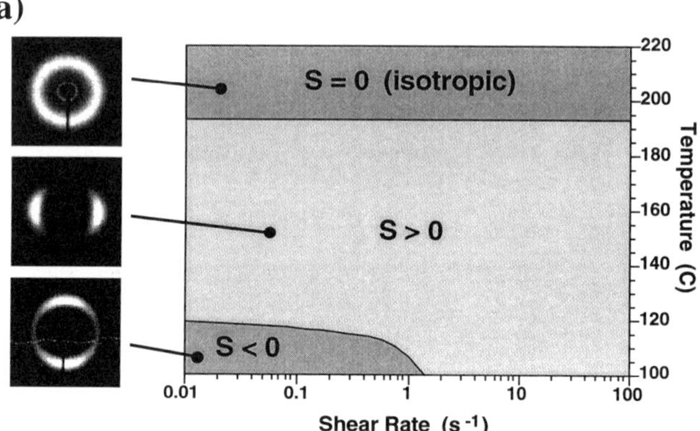

b)

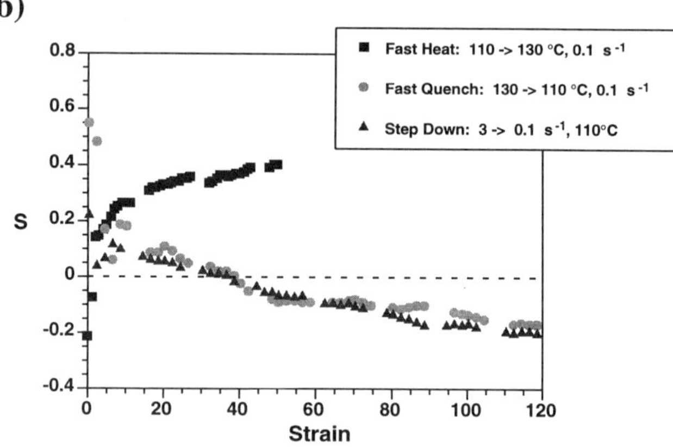

Figure 4. (a) Temperature vs. shear rate diagram showing various alignment regimes observed in the 28k sample (greyscale in scattering patterns is adjusted to enhance contrast). (b) Evolution of molecular orientation as a function of applied strain for three different step changes in temperature and shear rate.

Thermoreversible Order-Order Transition between Spherical and Cylindrical Microdomain Structures of Block Copolymer

Kohtaro Kimishima[1], Tadanori Koga[1], Yuko Kanazawa[1], and
Takeji Hashimoto[1,2]

[1]Hashimoto Polymer Phasing Project, ERATO, Japan Science and
Technology Corporation (JST)
Keihanna Plaza, Seika, Kyoto, 619-0237, Japan

[2]Department of Polymer Chemistry
Graduate School of Engineering
Kyoto University, Sakyo, Kyoto 606-8501, Japan

I. Introduction

Block copolymers form microdomain structures due to the segregation between the constituent block chains. A variety of morphologies such as lamellar, cylindrical, and spherical microdomains have been found according to the volume fraction of one of the constituents ("block composition").[1-3] Though order-order transitions between different kinds of morphologies with changing the temperature were predicted by Leibler[4] for block copolymers with a given constant block composition other than 0.5, such experimental results were not reported for a while. Recently some experimental results concerning the transition were reported.[5-7] Moreover the bicontinuous structures, e.g., gyroid[5] and perforated layers[6] were reported as new equilibrium structures of diblock copolymers. However, the mechanism of the order-order transition is not yet thoroughly understood. Here we will discuss some recent experimental results on the thermoreversible order-order transition between spherical and cylindrical microdomain structures of a block copolymer by means of small-angle X-ray scattering (SAXS).

II. Experimental Methods

The sample used in this study was a polystyrene-*block*-polyisoprene diblock copolymer prepared by living anionic polymerization with *sec*-butyllithium as an initiator and cyclohexane as a solvent. The number-averaged molecular weight, M_n, is 4.39×10^4. The polydispersity index M_w/M_n is 1.02 with M_w being weight-averaged molecular weight and the weight fraction of PS is 0.204. The diblock copolymer was dissolved into toluene with a small amount of anti-oxidant (BHT) and cast into film specimens from a 5% polymer solution, then dried under vacuum until no further weight loss was observed.

SAXS measurements were conducted with an apparatus consisting of a 18kW rotating-anode X-ray generator (M18XHF-SRA, MAC Science Co. Ltd., Japan) with a graphite crystal monochromator to obtain the CuKα beam with a wavelength of 0.1542 nm, a collimator, and vacuum scattering path, and a detector. The SAXS profiles were measured using a one-dimensional position sensitive proportional counter (PSPC) with line focus optics and were corrected for the absorption of the sample, background scattering, and the thermal diffuse scattering arising from the density fluctuations. SAXS patterns were also measured with a two dimensional imaging plate (IP[8]).

III. Results and Discussion

Figure 1 shows the SAXS profiles of the block copolymer taken at various temperatures. Measurements were carried out in a heating process from room temperature. The profiles obtained at 177.4 and 157.7°C (profiles a and b) show a broad first order scattering maximum (marked by a black arrow) and the shoulder at around q=0.45nm^{-1} (marked by a white arrow) due to the intra-particle interference, indicating the structure at these temperatures was lattice-disordered spherical microdomains (LDS). At temperatures from 118.7 to 138.2°C (profiles c and d) higher-order scattering maxima were observed at the positions of $\sqrt{2}$ and $\sqrt{3}$ relative to that of the first-order scattering maximum, q_m, indicating spherical microdomains packed in a body-centered cubic lattice (bcc-sphere). The

positions of the higher-order scattering maxima with respect to q_m changed to $\sqrt{3}, \sqrt{4}, \sqrt{7}$ below 99°C (profiles e and f), indicating the microdomain structures changed to cylindrical microdomains on a hexagonal lattice (hex-cylinder). Furthermore the measurements with a small temperature increment were carried out in both the heating and cooling processes and the transitions between the bcc-sphere and the hex-cylinder were observed between 114.7 and 116.7°C in both processes. Thus the order-order transition temperature, T_{OOT}, is centered at 115.7°C and occurs over a temperature width of 2K.

To confirm the morphological transition we examined the sample under a polarizing microscope with crossed polarizers. In this experiment at a temperature below T_{OOT} an image was obtained due to the form birefringence[9] of cylinders, while above T_{OOT} nothing was observed. This is because there is no form birefringence in spherical microdomain structure. We note here that the temperature dependence of the interaction parameter between styrene and isoprene, χ, was reported by many research groups and its $(\partial \chi / \partial T)_P$ is known to be negative. Thus, the transition from cylinder to sphere with increasing temperature is consistent with Leibler's mean field theory. We obtained $\chi N = 27.6$ at T_{OOT} using a literature value[7] for χ with N being the degree of the polymerization. This value showed a good agreement with Leibler's prediction ($\chi N = 25.6$).

In conventional X-ray scattering experiments, the beam size of the incident X-ray on the sample is quite large compared with the size of the grains of block copolymer. Therefore the observed scattering pattern is a summation of the scattering from many grains. If the orientation of the grains is perfectly random the scattering pattern should be circularly symmetric with respect to the incident beam. However, if the size of the grain becomes large enough to be comparable to the beam size, the Bragg's scattering peaks are observed as diffraction spots. By determining the positions of the diffraction spots, one is able to know the orientation of the microdomain structures in a grain. Though there are many previous works concerned with creating large grains by shear flow,[10,11] we succeed in making the large grains by just cooling the sample with a slow cooling rate.

Figure 2a shows a SAXS pattern of the diblock copolymer taken with the IP at 119.6°C where the bcc-sphere was observed. The sample giving rise to figure 2a was obtained by cooling down the sample from the temperature where LDS were observed with cooling rate of 2.4K/hour. The first and second-order scattering peaks at q_m and $\sqrt{2}q_m$ appeared as many spots, indicating a limited number of large grains composed of the bcc-sphere were created in the volume irradiated by incident beam. We note that the SAXS pattern were circular rings for the same sample which was cooled down from the same temperature with cooling rate of over 50K/hour. Next we changed the temperature to 111.8°C which is below T_{OOT} (part b). The relative position of the higher-ordered scattering peaks changed to $\sqrt{3}$ with respect to q_m, indicating that the microdomains structures changed to the hex-cylinder. The scattering peaks were also observed as spots. The grain structure exists even after the transition. Furthermore the temperature was changed again to 119.6°C (part c) and returned back to 111.8°C (part d). The diffraction spots in the scattering patterns observed at the same temperatures were almost identical (compare between parts a and c, and between parts b and d), showing the grain structure did not change before and after the transition in terms of not only the size but also the orientation of the microdomains in the grain. This indicates that there must be a rule in connecting spheres or in disconnecting cylinders at the phase transition. To investigate this rule we analyzed the scattering pattern from one grain.

After scanning the sample with a thin thickness of 0.3mm and with an X-ray beam of a small size (0.5mm), we found a 6-fold pattern at the temperature where the hex-cylinder was observed. Figure 3a shows the scattering pattern from one grain of the hex-cylinder. The first-order peaks appeared in a hexagonal pattern at q_m and the second peaks also appeared in a hexagonal pattern at $\sqrt{3}q_m$ at different azimuthal angles. The former is the Bragg reflection from the [10] plane of the hexagonal lattice and the latter is that from [11] plane. In this situation the incident X-ray was parallel to the cylindrical axis as illustrated on the right side of the pattern.

Next the temperature changed to 119.6°C, where the bcc-sphere was observed, without changing the sample position. The hexagonal pattern of the first peaks was kept (see figure 3b). To observe the such a pattern from bcc lattice the incident X-ray beam should be parallel to the [111] direction of bcc lattice. This result strongly indicates that spheres formed by disconnecting a cylinder laid in the [111] direction of the bcc lattice without changing the grain structure. Detailed results will be reported at the meeting.

References

(1) Hashimoto, T; Shibayama, M.; Fujimura, M.; Kawai, H. In *Block Copolymers, Science and Technology*; Meier, D. J., Eds.; Harwood Academic Publishers: London, New York, 1983; pp 63-108.

(2) Thomas, E. L.; Alward, D. B.; Kinning, D. J.; Martin, D. C.; Handlin, D. J., Jr.; Fetters, L. J. *Macromolecules* 1986, **19**, 2197.

(3) Bates, F. S.; Fredrickson, G. H. *Annu. Rev. Chem.* 1990, **41**, 525.

(4) Leibler, L. *Macromolecules* 1980, **13**, 1602.

(5) Hajduk, D. A.; Harper, P. E.; Gruner, S. M.; Honeker, C. C.; Thomas, E. L.; Fetters, L. J. *Macromolecules* 1994, **27**, 4063.

(6) Khandpur, A. K.; Forester, S.; Bates, F. S.; Hamley, I. W.; Ryan, A. J.; Bras, W.; Almdal, K.; Mortensen K. *Macromolecules* 1995, **28**, 8796.

(7) Sakurai, S.; Hashimoto, T.; Fetters, L. J. *Macromolecules* 1996, **29**, 740.

(8) Hashimoto, T.; Okamoto, S.; Saijo, K.; Kimishima, K.; Kume, T. *Acta Polymer.* 1995, **46**, 463.

(9) Wiener, O. *Abh. Math.-phys.* Kl. Saechs. Ges. Wiss. 1912, **32**, 507.

(10) Okamoto, S.; Saijo, K.; Hashimoto, T. *Macromolecules* 1994, **27**, 5547.

(11) Gupta, V. K.; Krishnamoorti, R.; Chen, Z. -R.; Kornfield, J. A.; Smith, S. D.; Satkowski, M. M.; Grothaus, J. T. *Macromolecules* 1996, **29**, 875.

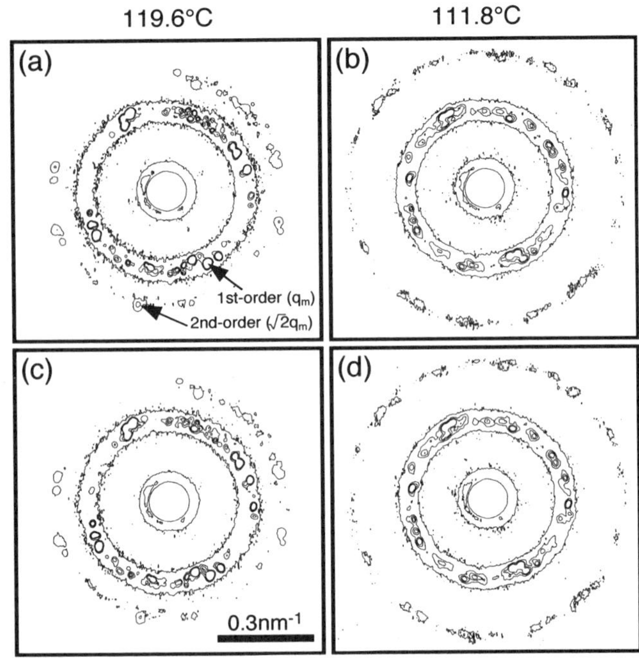

Figure 2   SAXS patterns of the diblock copolymer observed in the film specimens prepared by cooling down with a slow cooling rate from the temperature where the lattice-disordered spheres exist. The measurement was carried out in the order of a, b, c, and d without changing the area irradiated by the incident beam. The scattering patterns observed at the same temperatures were almost identical.

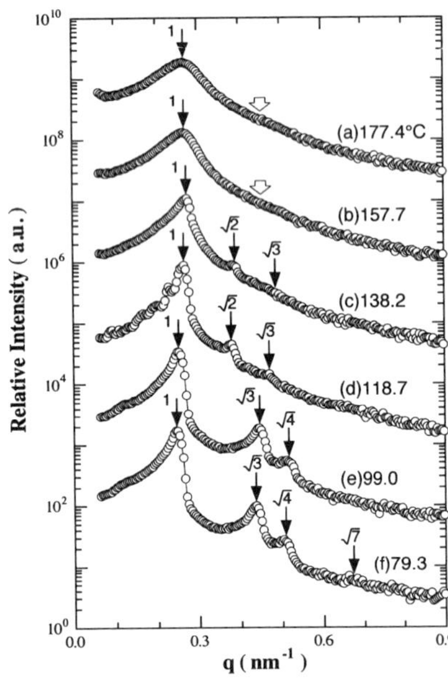

Figure 1   SAXS profiles of the block copolymer used in this study at various temperatures. The order-order transition between lattice-disordered spherical microdomains (profiles a and b), spherical microdomains in a body-centered cubic lattice (profiles c and d), and hexagonally packed cylindrical microdomains (profiles e and f) were observed reversibly by changing the temperature.

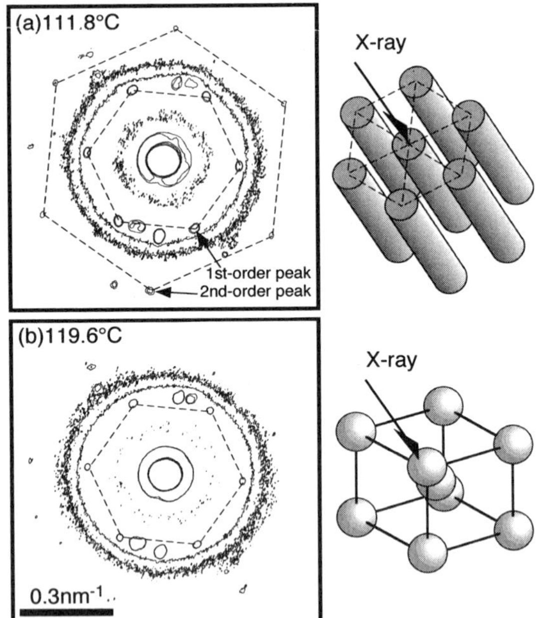

Figure 3   SAXS patterns from one grain of (a)cylindrical microdomain structure and (b)spherical microdomain structure. The hexagonal pattern of the first-order peaks was conserved after changing the temperature from (a)111.8°C for cylinders to (b) 119.6°C for spheres.

Use of Ultra Small Angle X-Ray Scattering to Measure Grain Size of
Styrene - Butadiene Block Copolymers

Randall T. Myers and R.E. Cohen*
Department of Chemical Engineering, Massachusetts Institute of Technology,
Cambridge, MA 02139

Anuj Bellare
Department of Orthopedic Surgery, Brigham and Women's Hospital,
Harvard Medical School, Boston, MA 02115

Introduction

It is well known that appropriate processing techniques can produce essentially perfectly ordered block copolymer morphologies with a single texture extending throughout the macroscopic dimensions of a specimen (1). The characteristic repeating length scale, d, of these morphologies is dictated by the molecular weights of the constituent block sequences. In the absence of extraordinary processing procedures, a second important length scale appears in the block copolymer. The perfection of the morphology is broken up into grains (2), each of which contains the ordered morphology of length scale d but with essentially random orientation relative to the specimen boundaries. These grains typically exhibit a characteristic size, D, which is one or more orders of magnitude larger than the morphological length scale, d.

Conventional small angle x-ray scattering (SAXS) techniques have been employed (3, 4) for decades to characterize block copolymers at the morphological length scale d. Recently ultra SAXS beamlines (5) have been constructed to probe significantly larger morphological features, previously observable only through the use of various microscope techniques, which make examination of grain size features possible.

Experimental

Three commercially available styrene-butadiene block copolymers were used in this study: Phillips KR03 Resin (85,000/25,000), DEXCO 4461-D (37,000/44,500), and Polymer Source P-408SB (9900/9200). To avoid the inevitable criticism the x-ray scattering at very low angles is dominated by the presence of voids in rigid, undiluted polymers, K-Resin pellets were mixed with various amounts of cumene, a low volatility solvent with a 25°C vapor pressure of 0.61 kPa. Based on the methodologies used for sample preparation, it is expected that whatever pre-existing grain structures were present in the K-Resin pellets would remain intact in the final specimens, albeit swollen by the cumene solvent.

Ultra SAXS scattering curves of 4461-D static cast samples, having grains were compared to roll cast samples of the same polymer, universally oriented without grains. The roll cast procedure is described elsewhere (6). Samples of the P408-SB polymer were static cast from chloroform and annealed for various times at 75°C to alter grain size.

The ultra SAXS experiments were performed at the Brookhaven National Laboratory on the X23A3 beamline operated by the National Institute of Standards and Technology. The available range of scattering vector, $q = (4\pi/\lambda)\sin\theta$, was 0.1 Å⁻¹ to 0.0004 Å⁻¹, where $\lambda = 1.299$ Å is the x-ray wavelength.

Results and Discussion

Figure 1 provides a broad summary of the ultra SAXS data via plots of the logarithm of absolute intensity vs. q for the set of KRO-3 cumene mixtures. There is no evidence of any maximum at lower q values which might reflect

the larger grain size D. Evidently the grains are so large that any scattering peaks associated with their presence lie at values of q even smaller than accessible with ultra SAXS. Absolute intensities in the low q region of Figure 1 should decrease with q according to Porod's law if a larger structure is present:

$$C_1 \equiv \lim_{q \to \infty} (q^4 i) = \frac{2}{\pi} \lim_{q \to \infty} (q^3 \tilde{i}) = (S/V)(\Delta\rho)^2 \qquad (1)$$

where i and $\tilde{i}$ are the desmeared and smeared intensities, $(\Delta\rho)^2$ is a contrast factor and (S/V), with dimensions of inverse length, contains the desired information on grain size.

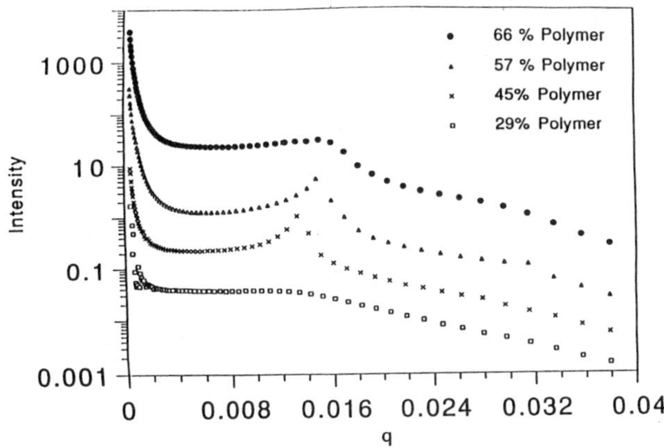

Figure 1: Smeared Intensities as a function of q for the KR03 Resin swollen with cumene. The uppermost plot is located in the proper position on the vertical axis and the other plots have been shifted sucessively downward 1 decade, 1.5 decades and 2 decades

Table 1: Ultra SAXS Results. KR03 Resin

Φp	d (Å)	D (μm)	D/d	$D_0 = D\Phi^{1/3}$ (μm)
1.00 *	369	2.95	80	2.95
0.66	397	3.14	80	2.73
0.57	412	3.28	80	2.72
0.45	457	3.55	78	2.72
0.29	516	4.33	84	2.86
		KR03	Pellet	
1.00	3.74	2.77	74	2.77

* denotes bulk KR03 film static cast from sample swelled to 45% with cumene

Grain size cannot be obtained directly from Equation 1, without knowledge of the contrast factor, though this can be eliminated from the analysis by considering the appropriate invariant, i.e. total area under the portion of the I vs. q plot associated with grain scattering. The invariant is written as follows:

$$C_2 \equiv \frac{2}{\pi} \int_0^\infty i(q)q^2 dq = \int_0^\infty \tilde{i}(q)q dq = \phi(1-\phi)(\Delta\rho)^2 \qquad (2)$$

We obtained values of the invariant for each specimen by first extrapolating linear plots of $\tilde{i}$ vs. q to q = ∞ using Porod's law, ignoring all of the area associated with the lamellar, d, scattering.

Combining equations 1 and 2, and taking $\phi = 0.5$ for the case of fluctuation-based scattering, eliminates the unknown contrast factor and leads to

$$D = V/S = 4\, C_2/C_1 \qquad (3)$$

All of our results are condensed into Table 1. First we note that the lamellar repeat distance, d, increases as the block copolymer is swollen with increasing amounts of cumene. The grain size D also increases with cumene swelling, and since the ratio of D/d is nearly constant, it is apparent that both length scales are increasing in the same fashion, as expected based on the method of specimen preparation. That both length scales, d and D, are proportional to the inverse cube root of polymer volume fraction is apparent from the essentially constant value of the product $D_0 = D\phi_p^{1/3}$, shown in the final column of Table 1.

As a check on the analysis we conducted Ultra SAXS measurements (not shown) on a bulk undiluted film of KRO3 cast from a 45% cumene solution. From these data we obtain a lamellar spacing of 369 Å and a value of $D_0 = 2.95$ μm, which is in good agreement with the values obtained from the swollen samples, shown in the last column of Table 1, indicating swelling is unnecessary. Ultra SAXS scattering of a KR03 pellet is also consistent with the results.

Figure 2 shows the Ultra SAXS profiles of a static cast sample of 4461-D and a roll casted sample of the same polymer, which has no grains. Since the Ultra SAXS at Brookhaven is 1D, the scattering was performed as a function of rotation angle, according to the insert of Figure 2. The results, summarized in table 2, show that the static cast sample has a smeared log-log slope of -3, consistent with Porod theory of a larger structure (grains). In contrast, the roll casted sample has a log-log slope varying from -2 to -0.5, indicating no larger structure (no grains), consistent with observation.

373

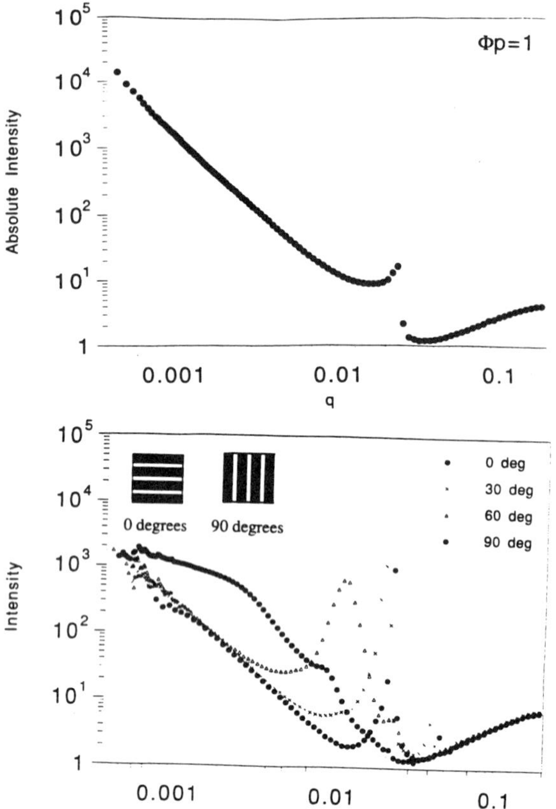

Figure 2: Static casted (above) and roll casted samples of 4461-D

Table 2: Ultra SAXS Results for 4461-D

Θ (angle of Rotation)	Log-log Slope (Ultra SAXS Region)	d (A°)
0	-2	271
10	-2	283
20	-2	283
30	-2	305
40	-2	355
50	-2	413
60	-2	556
70	-2	801
80	-1	.1716
90	-0.5	n/a[a]

Static Cast		
$D_2(\mu m)$	Log-log Slope (Ultra SAXS Region)	d (A°)
2.62	-3	260

[a] No Bragg Peak is present, only a broad shoulder at d ~ 3000 A

Figure 3 shows the Ultra SAXS profile of P-408SB as a function of annealing time at 75°C. Some of the samples displayed a Bragg peak, and a grain size can be directly calculated. The results are summarized in table 3. The grain sizes calculated from the Bragg peaks and from the use of Porod's Law and the invariant are consistent, as would be expected for impinged grains. This data also show that grain size can be controlled by annealing time.

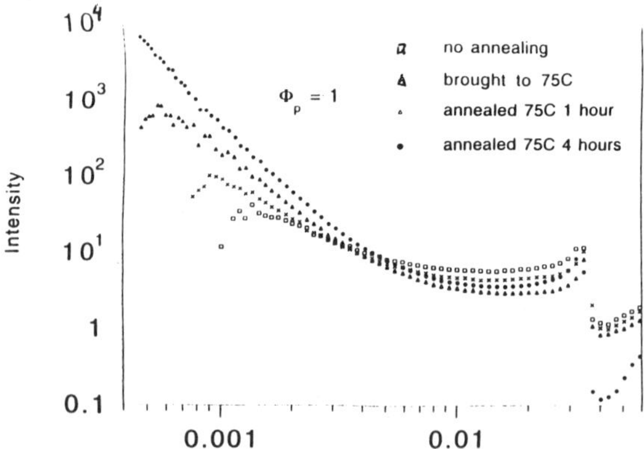

Figure 3: Ultra SAXS scattering of P-408SB as a function of annealing time

Table 3: Ultra SAXS Results P-408SB

Annealing Time at 75°C	d (A°)	$D_1 (\mu m)^a$	$D_2 (\mu m)^b$	D/d
none	182.5	0.45	n/a[c]	25
0 hours	182.5	0.69	0.72	38
1 hour	182.5	1.14	1.16	63
4 hours	182.5	n/a[d]	1.74	95
24 hours	182.5	n/a[d]	2.52	138

[a] Calculated from the Bragg Peak ($D_1 = 2\pi/q_{MAX}$)
[b] Calculated from Porod's Law and analysis of the Invariant
[c] No Porod region (Scattering due to grains and lamellar scattering too close?)
[d] Bragg Peak falls outside the limits of the machine

Conclusion

Simultaneous determination of the morphological length scale, d, and the grain size D in a heterogeneous block copolymer is possible through the use of ultra SAXS measurements. The d spacing was revealed directly in the data by the appearance of one or more Bragg peaks in the relevant range of scattering angle. Roll cast samples containing no grains do not have a Porod region, while samples with grains display this characteristic slope. Analysis of the Porod tail of the grain scattering mechanism and the invariant facilitated a reliable estimate of grain size D for samples not displaying a Bragg peak, as the grain sizes calculated for samples displaying a Bragg peak were consistent with the Porod and invariant analysis. Independent observation of grains in the bulk sample by TEM corroborated the ultra SAXS analysis.

Acknowledgments

We would like to thank Gabrielle Long and Zugen Fu from NIST for their assistance in the use of the Ultra SAXS beamline. Mike Frongillo of the CMSE (NSF/MRSEC) Electron Microscopy facility at MIT is acknowledged for his expertise and patience.

References

(1)   Keller, A.; Pedemonte, E.; Willmouth, F.M. *Nature* **1970**, *225*, 538.
(2)   Garetz, B.A.; Balsara, N.P.; Dai, H.J.; Wang, Z.; Newstein, M.C. *Macromolecules* **1996**, 29, 4675.
(3)   Hashimoto, T.; Nagatoshi, K.; Todo, A.; Hasegawa, H.; Kawai, H. *Macromolecules* **1974**, 7, 364.
(4)   Hashimoto, T.; Shibayama, M.; Kawai, H. *Macromolecules* **1980**, *13*, 1237.
(5)   Long, G.G.; Jemian, J.R.; Weertman, J.R.; Black, D.R.; Burdette, H.E.; Spal, R. *J. Appl. Cryst.* **1991**, *24*, 30.
(6)   Albalak, R.J.; Thomas, E.L.; Capel, M.S. *Polymer* **1987**, 38, 3819.

Structure and Dynamics of Structure Formation in Model Miktoarm Star Copolymers Composed of Two Crystallizable Blocks

G. Floudas

Foundation for Research and Technology-Hellas (FORTH)
Institute of Electronic Structure and Laser
P.O. Box 1527, 711 10 Heraklion Crete, Greece

G. Reiter, O. Lambert, P. Dumas

Institut de Chimie des Surfaces et Interfaces, 15
Rue Jean Starcky, B.P. 2488
68057 Mulhouse Cedex, France

B. Chu

Department of Chemistry
State University of New York at Stony Brook
Long Island, NY 11794-3499, USA

Introduction

The dynamics of structure formation have been studied extensively in crystallizable homopolymers and more recently in linear block copolymers composed from an amorphous and a semicrystalline block [1]-[6]. These materials can develop structure over a range of length scales, from the unit cell structure of the crystallizable block to the microdomain length scale, to the spherulitic scale. However, the effect of chain topology and architecture has not been explored up to now in such systems. The class of materials studied here are model heteroarm (aslo known as miktoarm) stars composed of two crystallizable and an amorphous block [7]. These complex materials are ideal for studying the effect of topology on the phase state and the dynamics of structure formation in the presence of amorphous/crystallizable blocks. The system consists of model star block copolymers composed of two crystallizable blocks (poly(ethylene oxide) (PEO) and poly(ε-caprolactone) (PCL)) and one amorphous block (polystyrene) (PS). Crystallization starts from the homogeneous phase. For the structure investigation X-ray scattering, optical microscopy and atomic force microscopy are employed whereas for the kinetics we have used DSC, optical microscopy and rheology. In the stars there is a competition for crystallization between the two crystallizable blocks which have similar mobilities and melting temperatures but crystallize in different unit cells (monoclinic vs. orthorhombic). When the crystallizable block length ratio is 3 or higher only the longer block will crystallize. For comparable lengths both blocks can crystallize but the crystallinity, long period and crystalline lamellar thickness are reduced with respect to the pure PEO and PCL.

The latent heats were obtained in the isothermal crystallization calorimetric experiments and analyzed in terms of the Avrami theory. Although similar Avrami exponents were found for all stars (n=2, reflecting a dick-like qrowth from heterogeneous nuclei) the crystallization times were different depending on the nature of the crystallizable blocks. Optical microscopy revealed the formation of different superstructures (spherulites/axialites) depending on the crystallizable block (PEO/PCL). The growth rates of these superstructures were obtained and analyzed in terms of a kinetic nucleation theory and the fold surface free energies were extracted. Notwithstanding the larger specific surface of bulk PCL as compared to the bulk PEO, the fold surface free energies in the stars were similar to PEO, indicating mixing of the amorphous PS block with the PCL phase. This is supported from the results of the atomic force microscopy measurements on thin films, which indicated the formation of perforated crystals of PCL.

Experimental Section

Materials. All samples used in the present study are given in Table I.

Table 1.

Molecular Characteristics of the Diblocks and Star Block Copolymers.

Sample	M_n (PS) $\times 10^3$	M_n (PEO) $\times 10^3$	M_n (PCL) $\times 10^3$	M_n (total) $\times 10^3$	M_w/M_n
PEO		15		15	1.3
PCL10			10	10	
PCL42			42.5	42.5	1.45
PEO-PCL		15	5	20	1.5
PS-PEO	4.7	20		24.7	1.16
SEL 4.7/20/1.8	4.7	20	1.8	27	1.15
SEL 4.7/20/10	4.7	20	10	35	1.27
SEL 4.7/20/45	4.7	20	45	70	1.19
SEL 4.7/20/87	4.7	20	87	112	1.29

The methods employed in the present investigation are: DSC, SAXS, WAXS and simultaneous SAXS/WAXS, rheology, optical microscopy and atomic force microscopy.

Results and Discussion

Structure. The structure and phase state of the stars has been investigated by SAXS, WAXS and synchrotron simultaneous SAXS/WAXS measurements. These measurements revealed that the stars form a homogeneous phase in the melt. Hence crystalization in these systems starts from the homogeneous melt. Some representative WAXS spectra for the different samples are shown in Figure 1. It is

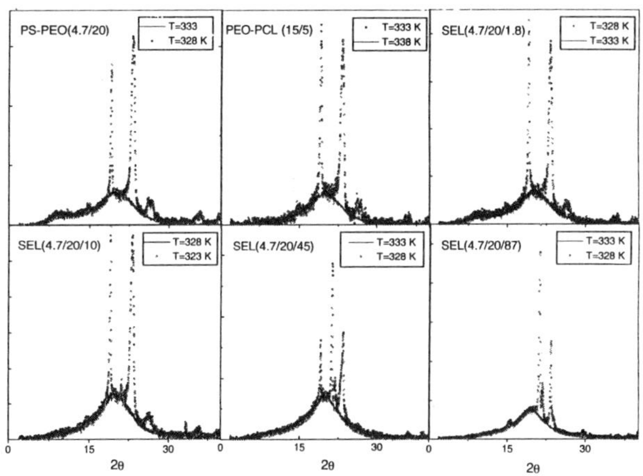

evident, that depending on the block length ratio, there are different Bragg reflections present in the spectra. The crystal structure of PEO is monoclinic with a unit cell parameter of 1.93 nm along the helix axis whereas PCL is orthorhombic and the chains assume a planar all-trans conformation. Therefore, the Bragg reflections of the two polymers crystals are distinctly different and this structural property can be used as a fingerprint of the crystallizable block and its crystailinity. In the PS-PEO and PEO-PCL diblocks as well as in the star SEL-4.7/20/1.8 there exist only Bragg reflections corresponding to the PEO crystals. Alternatively, in SEL-4.7/20/87 the Bragg reflections are characteristic of PCL crystals. For the stars with comparable block lengths for the two crystallizable components the WAXS spectra show mixed reflections indicating the existence of both PEO *and* PCL crystals. Based on the WAXS patterns, we conclude that in the stars with crystallizable block length ratio of 3 or higher the longer block completely suppresses the crystallization of the shorter one. However, when the block length ratio is less then 3, then both blocks undergo crystallization. We have investigted the crystallinities of the PEO and PCL blocks as well as the long period and the crystalline lamellar thickness and found that all were reduced relative to the pure components.

Crystallization Kinetics. There are different methods which can be employed to study the crystallization kinetics in semicrystalline materials; calorimetry, optical microscopy, dilatometry, simultaneous SAXS/WAXS and more recently rheology [8] have been employed for this purpose. Here we report only on the optical microscopy measurements. The optical measurements were made by continuous monitoring of the crystallizing superstructures following quenches to different crystallization temperatures. We first investigated the crystallization of the pure PEO and PCL and found that upon crystallization different supestructures are formed: spherulites and axialites, respectively. In the block copolymer stars we found that spherulites (axialites) were formed when PEO (PCL) was the majority component and the ratio of PEO to PCL was 3 or higher. For the stars with comparable block lengths, both spherulites *and* axialites were observed. This finding is in good agreement with the WAXS results that have shown the possibility that both blocks can crystallize. We interpret the axialitic (spherulitic) superstructure in the stars as composed primarily from PCL (PEO) crystals.

The different shapes of the superstructures provide an efficient way to study the crystallization kinetics by optical microscopy. We have followed the increase of spherulitic radius and of the long and short axes with time. In any case, the dependence of the radial and axial growth on time was linear and the growth rates were obtained from the slopes (in the case of axialites we have used an average growth rate from the long and short axis). The results from the different growth rates for the PEO, PCL and the stars are plotted in Figure 2 as a function of the crystallization temperature. Clearly, there is a large difference in growth rates between the pure components, which makes possible to follow the different rates in the diblock copolymer and block copolymer stars. The effect of the amorphous PS block in the PS-PEO diblock is to slow down the growth rate as compared to the bulk PEO. Introducing a short PCL block to form a star makes nearly no alteration to the spherulitic growth rate. In the case of SEL-4.7/20/87, the growth of axialites is somewhat faster than of the pure PCL at the same crystallization temperatures and similar to the growth rates of the SEL-4.7/20/45 axialites. For SEL-4.7/20/10, where the PEO and PCL crystallization results in the formation of spherulites and axialites, respectively, the two growth rates were modified; axialites grew with a rate similar to the pure PCL whereas spherulites grew with a rate intermediate between PEO and PCL We have analyzed the growth rates using the crystallization theory of Hoffman et al. which is the most accepted approach to analyze the linear growth rates and which has a molecular basis, in that, it accounts for the diffusion and crystallization of chains. The crystal growth rate is generally given by

$$G \approx G_0 \exp\left(-\frac{E + \Delta F}{kT}\right) \qquad (1)$$

where G_0 is the growth rate constant, E is the activation energy for transport of crystallizing units across the crystal-liquid interface and ΔF is the free energy required to form a nucleus of critical size on the face of a crystal. The first and second terms in the above equation can be modified to take into account the known activation parameters as:

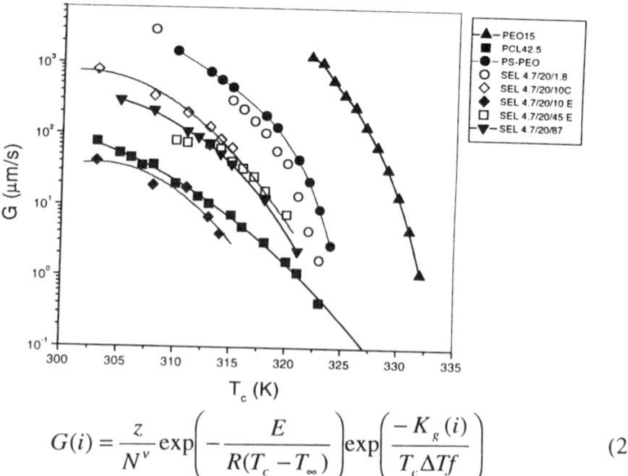

$$G(i) = \frac{z}{N^\nu} \exp\left(-\frac{E}{R(T_c - T_\infty)}\right) \exp\left(\frac{-K_g(i)}{T_c \Delta Tf}\right) \qquad (2)$$

where z is a parameter containing mobility terms, N is the degree of polymerization and the value of the exponent ν depends on the growth regime (see below), T_c is the crystallization temperature, T_∞ is the ideal glass transition temperature ($=T_g$-30 (K)), ΔT is the supercooling ($=T_m^0$-T_c), f is a temperature correction factor for the heat of fusion ($=2T_c/(T_m^0+T_c)$) and $K_g(i)$ is the nucleation rate constant given by

$$K_g = \frac{ib_0\sigma\sigma_e T_m^0}{k\Delta H_f} \qquad (3)$$

where b_0 is the width of the chain, σ is the lateral surface free energy, σ_e is the fold surface free energy and ΔH_f is the heat of fusion. The symbol i represents a number associated to the particular regime and is equal to 4 for regimes I and II and equal to 2 for regime II. The three regimes differ according to the competition between the rate of deposition of secondary nuclei (d) and the rate of lateral surface spreading (g); in regime I, d<<g and appears at very low supercoolings, in regime II, d≈g and occurs for intermediate supercooling and in regime III, d>g, and corresponds to very high supercoolings.

The main finding, is that, the surface free energy product is controlled by the PEO crystals attaining values similar to pure PEO even when the growing superstructure is axialitic (SEL-4.7/20/10). To investigate this seemingly paradoxial situation we need to separately discuss the values of the lateral and fold surface free energies. For this purpose we employ the Thomas-Stavely relation for the lateral surface free energy

$$\sigma = 0.1\Delta H_f b_0 \qquad (4)$$

assuming the same b_0 values coresponding to regime II.. PEO and PCL have similar mobility parameters (i.e., glass temperatures and activation energies) but their crystals have very different fold surface free energies and this difference is used here as a tool of obtaining information on mixing at the fold surface. Based on the values of the fold surface free energy we conclude that, the PEO spherulites are formed by nearly pure PEO crystals and no mixing occurs between the fold surface and the amorphous PS and PCL blocks. On the contrary, the PCL axialites are formed by PCL crystals which contain a lot of mixing with PS and possibly with the amorphous PEO blocks at the fold surface. These results are supported by the AFM studies.

References
[1] B. Lotz, A.J. Kovacs, G.A. Bassett, A. Keller Kolloid Z.Z. Polym. 209, 115, 1966
[2] A.E. Skoulios, G. Tsouladze, E. Franta, J. Polym. Sci. Part C 4, 507, 1963.
[3] M. Gervais, B. Gallot, Polymer 22, 1124, 1981.
[4] A.J. Lovinger, B.J. Han, F.J. padden, P.A. Mirau J. Polym. Sci., Polym. Phys. Ed. 31, 115, 1993
[5] P. Rangarajan, R. A. Register Macromolecules 26, 4640, 1993.
[6] A.J. Ryan, I.W. Hamley, W. Bras, F.S. Bates Macromolecules 28, 3860, 1995.
[7] G. Floudas, G. Reiter, O. Lambert, P. Dumas in preparation
[8] G. Floudas, C. Tsitsilianis Macromolecules 30, 4381, 1997.

Order-Order and Order-Disorder Transitions in Block Copolymer Solutions and Melts

T. P. Lodge[†], K. J. Hanley, C. Huang and C.Y. Ryu

Department of Chemistry[†] and
Department of Chemical Engineering & Materials Science
University of Minnesota, Minneapolis, MN 55455-0431

Introduction

Block copolymers are an interesting class of macromolecular surfactants that display a rich variety of thermotropic and lyotropic order-order (OOT) and order-disorder (ODT) transitions. Small-angle scattering techniques (SAXS and SANS) are ideally suited to determining the associated microstructures and transition kinetics, since the natural lengthscale of the various ordered phases falls in the range 10-100 nm. Additional information from transmission electron microscopy (TEM), rheology, and birefringence is also useful in supplementing the scattering experiments. Finally, the location and nature of the various OOTs and ODTs can be largely understood via self-consistent mean-field (SCMF) calculations, although some features of the behavior require the incorporation of fluctuation effects. In this paper we describe studies of asymmetric styrene-isoprene (SI) copolymers in neutral solvents, slightly selective solvents, and melts.

Experimental

Various anionically-polymerized SI copolymers were employed. SAXS measurements were performed with a rotating anode (Cu K_α, λ=1.54Å), Franks mirror optics, and a Siemens area detector. SANS measurements were taken at NIST, on the 30m (NG7) instrument. TEM micrographs were obtained with a Jeol 1210 at 120 kV after microtoming samples at -1 °C and selectively staining the isoprene domains with OsO_4. Rheological properties were determined using various rheometers from Rheometric Scientific. Depolarization of transmitted light ("static birefringence") measurements were performed on a home-built apparatus.

Results

A. Addition of Neutral Solvents

To a first approximation, the addition of a small amount of neutral solvent (such as toluene or dioctyl phthalate, DOP, for SI) should act to dilute S-I contacts, and renormalize the melt phase diagram by a simple rescaling of $\chi \rightarrow \phi\chi$, where χ is the S-I interaction parameter and ϕ is the polymer volume fraction.[1-3] However, we previously demonstrated that this so-called "dilution approximation" is invalid for lamellar SI copolymers in the presence of a neutral solvent.[4] SANS studies of solvent distribution in lamellar samples quantified a slight accumulation of neutral solvent at the S-I interface,[5] in accordance with SCMF calculations. Here we examine an asymmetric copolymer SI(11-21) (where the numbers denote block molecular weights in kDa) that exhibits the phase sequence gyroid (G)→cylinders (C)→ disorder (D) upon heating in the melt. On dilution with DOP the C phase window becomes progressively narrower until at $\phi \approx$ 0.65 a direct G→D transition is observed. This relative broadening of the G phase stability is captured by the SCMF theory, and can probably be attributed to the solvent relieving packing frustration in the minor blocks.[6] Interestingly the G→C OOT appears to follow the dilution approximation scaling of ϕ vs T, but the ODT does not.

B. Addition of a Slightly Selective Solvent

Di-*n*-butyl phthalate (DBP) is an interesting "slightly" selective solvent, as it strongly favors the styrene domains at room temperature, but becomes effectively neutral somewhere above 100 °C. Consequently, if DBP is added to an SI copolymer at low temperature, the styrene domains swell, and for SI(11-21) a lamellar (L) microstructure results. Then, upon heating the solvent begins to distribute itself more uniformly, and the sequences L→G→C→D and even L→D are seen. For a more asymmetric copolymer, SI(10-50), which follows the sequence C→spheres (S)→D when heated in the melt, the addition of DBP transforms the styrene cylinders into L, and ultimately inverted isoprene cylinders, before disordering into a suspension of elongated micelles.[7] Recent SCMF calculations also capture this phenomenology.[8]

C. Cylinder to Sphere Transition in Melts

An SIS triblock copolymer with block molecular weights of 10,000–100,000-10,000 Da undergoes the C→S OOT at 196 °C.[9] This equilibrium transition temperature, and the associated stability limits of the C and S phases, were established by rheological measurements during thermal cycling over a wide range of heating and cooling rates. It has been previously established that the bcc phase grows epitaxially from the cylinders, with the [111] axis of the bcc spheres oriented along the cylinder direction.[10, 11] However, over a temperature interval ca. 10-15 °C *below* the OOT, we find clear evidence of anisotropic fluctuations with bcc symmetry superimposed on the cylinders. In particular, the elastic modulus along the cylinder axis, obtained on pre-aligned samples, shows a distinct signature of these fluctuations, and SAXS patterns show the expected fluctuation peaks. Similar results have recently been obtained for the L→C transition in PEP-PDMS diblocks.[12] Finally, with rapid quenching it proves possible to acquire direct TEM images of the fluctuating state. These results can be compared with recent theoretical treatments.[13-16]

Acknowledgments

This work was supported by the National Science Foundation (DMR-9528481), the Center for Interfacial Engineering, an NSF-supported Engineering Resarch Center at the University of Minnesota, and a Research Scholarship from the University of Minnesota Supercomputer Institute (C.H.).

References

1. G. H. Fredrickson and L. Leibler, *Macromolecules* **22**, 1238 (1989).
2. M. Olvera de la Cruz, *J. Chem. Phys.* **90**, 1995 (1989).
3. M. D. Whitmore and J. Noolandi, *J. Chem. Phys.* **93**, 2946 (1990).
4. T. P. Lodge, C. Pan, X. Jin, Z. Liu, J. Zhao, W. W. Maurer and F. S. Bates, *J. Polym. Sci., Polym. Phys. Ed.* **33**, 2289 (1995).
5. T. P. Lodge, M. W. Hamersky, K. J. Hanley and C.-I. Huang, *Macromolecules* **30**, 6139 (1997).
6. M. W. Matsen and F. S. Bates, *J. Chem. Phys.* **106**, 2436 (1997).
7. I. W. Hamley, J. P. A. Fairclough, A. J. Ryan, C. Y. Ryu, T. P. Lodge, A. J. Gleeson and J. S. Pedersen, *Macromolecules* **31**, 1188 (1998).
8. C.-I. Huang and T. P. Lodge, submitted to *Macromolecules*.
9. C. Y. Ryu, M. S. Lee, D. A. Hajduk and T. P. Lodge, *J. Polym. Sci., Polym. Phys. Ed.* **35**, 2811 (1997).
10. S. Sakurai, T. Hashimoto and L. J. Fetters, *Macromolecules* **29**, 740 (1996).
11. K. A. Koppi, M. Tirrell, F. S. Bates, K. Almdal and K. Mortensen, *J. Rheol.* **38**, 999 (1994).
12. M. E. Vigild, K. A. Almdal and K. Mortensen, unpublished results.
13. M. Laradji, A.-C. Shi, R. C. Desai and J. Noolandi, *Phys. Rev. Lett.* **78**, 2577 (1997).
14. M. Laradji, A.-C. Shi, J. Noolandi and R. C. Desai, *Macromolecules* **30**, 3242 (1997).
15. S. Qi and Z.-G. Wang, *Phys. Rev. Lett.* **76**, 1679 (1996).
16. S. Qi and Z.-G. Wang, *Phys. Rev. E.* **55**, 1682 (1997).

Control of Microdomain Orientation of Block Copolymers Induced by Temperature Gradient and Surface Ordering and Characterization of their Orientation

Jeffrey Bodycomb[1], Yoshinori Funaki[1], Khotaro Kimishima[1], and Takeji Hashimoto[1,2]

[1]Hashimoto Polymer Phasing Project, ERATO, JST
Keihanna Plaza, 1-7 Hikari-Dai
Seika, Kyoto 619-0237 Japan

[2] Department of Polymer Chemistry
Graduate School of Engineering
Kyoto University, Kyoto 606-8501 Japan

Introduction

Block copolymers at temperatures below the order-disorder transition (ODT) temperature form nanoscale patterns in the ordered state. The self-assembly of these structures provides a means to build interesting nanostructures. Usually, these nanostructures are randomly oriented in bulk samples. Therefore, an understanding of the methods of manipulating the orientation of these nanostructures is important both as a means of producing oriented materials of practical use and understanding the ordering process of these materials. Shear is an established method of controlling orientation, as is the use of flow[1,2]. Recently, there have been many reports concerning the effects of shear on the orientation of diblock copolymers [3-8]. In addition, electric fields[9] and surfaces[10-12] have been shown to cause alignment of block copolymer nanoscale patterns. However there are few (if any) experimental reports of using temperature to align diblock copolymers.

Here, we will discuss some recent experimental results on the use of a temperature gradient and surface to align grains in a lamellar diblock copolymer. We found that the surface effect is dominant close to the surface and weakens further away from the sample surface. Far from the sample surface, the temperature gradient effect is dominant.

Experimental Methods

In this study, we used a polystyrene-*block*-polyisoprene diblock copolymer (SI) prepared by living anionic polymerization with *sec*-butyllithium as an initiator and cyclohexane as a solvent. The polystyrene (PS) block number average molecular weight M_n was 1.1 x 10^4 and the polyisoprene (PI) block, 1.4 x 10^4. The polydispersity index M_w/M_n is less than 1.05 in both cases (by GPC). This polymer has a lamellar structure by X-ray scattering and TEM, consistent with its nearly 50/50 composition.

SAXS was measured using a two dimensional imaging plate (IP) detector[13] and a MAC science X-ray generator with point focus optics. The data shown here, in contrast to usual practice, does not have the background intensity subtracted. The beam diameter in the sample was about 1.3 mm.

Samples were prepared in a "zone heating device" consisting of a pair of central heating blocks, surrounded by two pairs of cooling blocks. The centers of each pair of heating blocks are aligned to form a sandwich, with the polymer film sample in the center with a small gap of ca. 100 µm on both sides of the sample (corresponding to thin teflon sheets on either side of the sample). The heating blocks were set at 210 °C, above the ODT temperature of the diblock copolymer (157 °C), and the cooling blocks were set at 5 °C, well below the ODT temperature of the diblock copolymer (see Figure 1). This arrangement produces a sharp temperature gradient along the z-axis, on the order of about 30 °C/mm.

The diblock copolymer was cast from a 5% toluene solution of polymer mixed with a small amount of anti-oxidant(BHT), then placed inside a Teflon sample cell of 2 mm width with a glass surface on one end as shown in Figure 2. Thin Teflon sheets were placed on either side of the sample cell normal to the y axis. The heating blocks were moved along the z-axis at a speed of 800 nm/sec, while the entire arrangement was kept under vacuum to prevent sample degradation.

Results and Discussion

Figure 3 shows a 2-dimensional SAXS pattern taken with an imaging plate from this sample in an arrangement where the incident X-ray beam is parallel with the y axis. The beam was point-focused and had a diameter of ca. 1.3 mm at the sample. In order to explore the structure very close to the glass surface, the beam center was passed close to the glass surface. Therefore, the left edge of the beam was partially blocked by the glass surface. The beam center is 0.4 mm from the edge of the glass, as estimated by the "shadow" cast by the glass block. The scattering pattern in Figure 3 shows only the first-order peak,. The higher order peaks existed at positions of integer multiples of the position of the first order scattering peak. The scattering along the z-direction is much more intense than at other orientations. We tentatively interpret these two facts as being due to a single large lamellar grain, induced by the combined effect of the glass surface and the applied temperature gradient. That is, the glass surface induces nearly perfect alignment of the lamella with the lamellar normals parallel to the z-axis (temperature gradient) and perpendicular to the glass surface. In addition, the lamellae are also aligned by the applied temperature gradient (see below) in the same direction as the temperature gradient axis, which is perpendicular to the glass surface.

In order to examine the effect of temperature gradient alone, we measured the scattering from points further away from the sample surface. In Figure 4a, we show the scattering from the same sample, except in this case, the beam center is 1.6 mm away from the glass surface. Note that the upper limit of the scale in Figure 4 is about one half as large as that in figure 3. Again, we only show the first order peak. Here, we observe preferential orientation of the lamellae with their normals parallel to the temperature gradient direction. However, in this case, the orientation is much weaker than that observed in Figure 3. If we assume that this distance is sufficiently far from the glass surface, we can conclude that the observed orientation is due to the temperature gradient effect alone, which is much weaker than the combined surface and temperature gradient effects.

To check the assumption that 1.6 mm is sufficiently far from the glass surface for only temperature effects to dominate, we measured the scattering from the sample even further from the glass surface. In Figure 4b we show the scattering when the beam center is at a distance of 2.9 mm from the glass surface. Again, we observe orientation of the lamellae in the temperature gradient direction, and the degree of orientation is the same as that seen in figure 4a. If the results in figure 4a were partially due to the effect of the surface, we would expect that the orientation in Figure 4b would be substantially weaker than that in Figure 4a or nonexistent. Thus, Figures 4a and 4b show the effect of temperature gradient alone on the diblock copolymer material.

These results describe the effect of a temperature gradient and glass surface on the orientation of a diblock copolymer. The temperature gradient effect combined with the surface effect leads to what appear to be large, single-grain structures. However, further from the surface, the temperature gradient effect alone dominates and leads to grain alignment, but not the large single grains observed at the glass surface. Therefore, we clearly show that a temperature gradient and surface play an important role in aligning diblock copolymer structures during the ordering process. The resulting orientation has the lamellar normals preferentially oriented along the z-axis, parallel with the temperature gradient and perpendicular to the glass surface (i.e., the O_{xy} plane).

The use of a temperature gradient and surface effects as new techniques for orienting diblock copolymers adds to our ability to make 'super-tailor-made' materials and extend the inherent anisotropy in their nanopatterns from microscopic to macroscopic levels.

References

1. Keller, A.; Pedemonte, E.; Willmouth, F. M. *Nature* **1970**, *225*, 538.
2. Hadziioannou, G.; Picot, C.; Skoulious, A.; Ionescu, M.-L.; Mathis, A.; Duplessix, R.; Gallot, Y.; Lingelser, J. P. *Macromolecules* **1982**, *15*, 263.
3. Okamoto, S.; Saijo, K.; Hashimoto, T. *Macromolecules* **1994**, *27*, 5547.
4. Zhang, Y.; Wiesner, U.; Spiess, H. W. *Macromolecules* **1995**, *28*, 778.
5. Gupta, V. K.; Krishnamoorti, R.; Kornfield, J. A.; Smith, S. D. *Macromolecules* **1995**, *28*, 4464.
6. Patel, S. S.; Larsen, R. G.; Winey, K. I.; Watanabe, H. *Macromolecules* **1995**, *28*, 4313.
7. Gupta, V. K.; Krishnamoorti, R.; Chen, Z.-R.; Kornfield, J. A.; Smith, S. D.; Satkowski, M. M.; Grothaus, J. T. *Macromolecules* **1996**, *29*, 875.
8. Maring, D.; Wiesner, U. *Macromolecules* **1997**, *30*, 660.
9. Amundson, K.; Helfand, E.; Davis, D. D.; Quan, X.; Patel, S. S.; Smith, S. D. *Macromolecules* **1991**, *24*, 6546.
10. Green, P. F.; Christensen, T. M.; Russell, T. P.; Jerome, R. *Macromolecules* **1989**, *22*, 4600.
11. Coulon, G.; Ausserre, D.; Russull, T. P. *J. Phys.* **1990**, *51*, 777.
12. Russell, T. P.; Coulong, G.; Deline, V. R.; Miller, D. C. *Macromolecules* **1989**, *22*, 4600.
13. Hashimoto, T.; Okamoto, S.; Saijo, K.; Kimishima, K.; Kume, T. *Acta Polymer.* **1995**, *46*, 463.

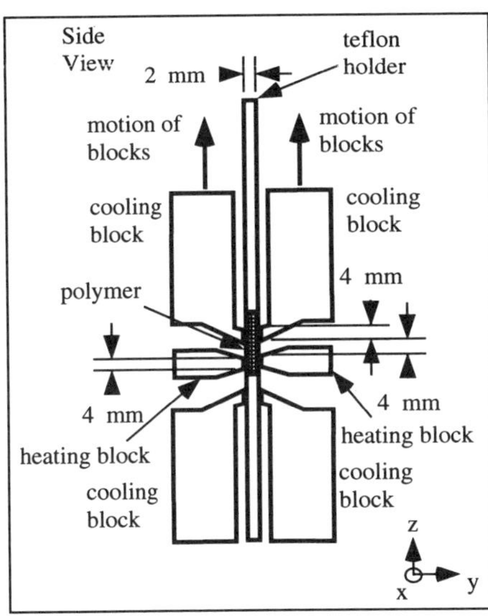

Figure 1. Schematic of zone heating device used for applying temperature gradient. The axes are designated as follows: Oz is parallel to the temperature gradient ∇T, Oy is parallel to the film normal, and Ox is parallel to the film surface and perpendicular to ∇T.

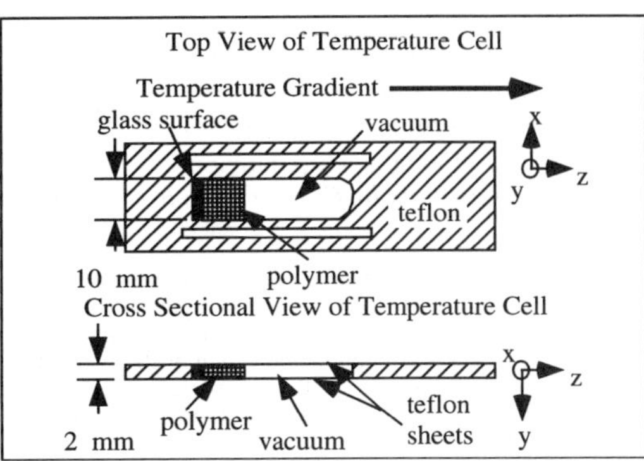

Figure 2. Schematic of cell used to hold polymer sample. The surface at the top of the sample is exposed to vacuum, and the surfaces on either side are thin sheets of Teflon.

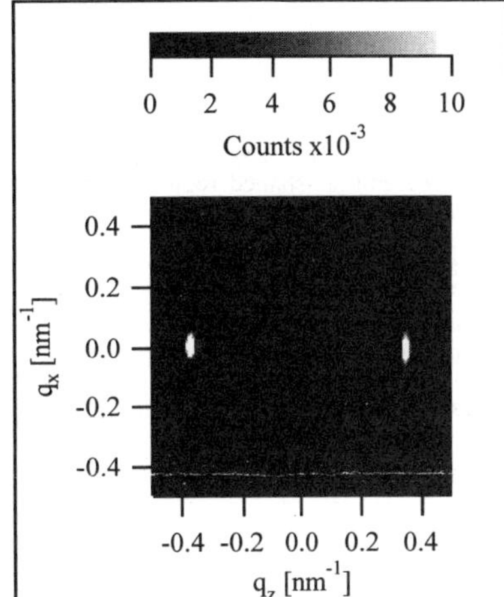

Figure 3. Two-dimensional small-angle X-ray scattering pattern from a point where the beam center is 0.4 mm from glass surface. The glass surface actually blocks part of the X-ray beam.

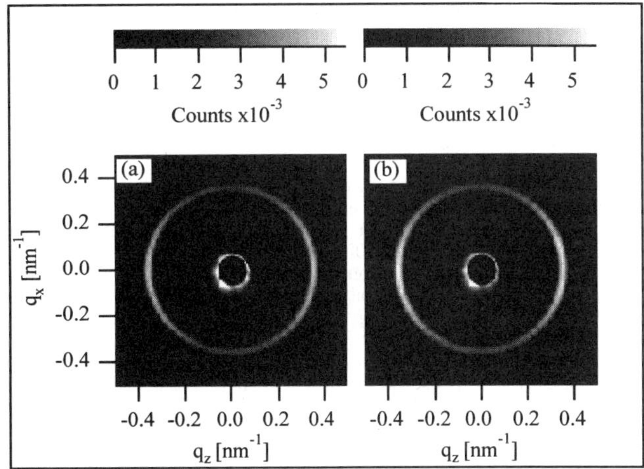

Figure 4. (a) Two-dimensional small-angle X-ray scattering pattern from a point where the beam center is 1.6 mm from glass surface.
(b) Two-dimensional small-angle X-ray scattering pattern from a point where the beam center is 2.9 mm from glass surface.

Time-Resolved SAXS Study of the Lamella↔Cylinder Transition in a Polystyrene-Poly(ethylene-*alt*-propylene) Diblock Copolymer

Chiajen Lai, William B. Russel and Richard A. Register
Department of Chemical Engineering
Princeton University, Princeton, NJ 08544

Douglas H. Adamson
Princeton Materials Institute, Princeton, NJ 08544

Introduction

The phase behavior of block copolymers has attracted intense investigation for the past three decades. Many experiments utilizing small angle X-ray scattering (SAXS), neutron scattering (SANS), microscopy (TEM) and rheology have been conducted to study the phase behavior of styrene-diene block copolymers. With the inclusion of the recently-recognized bicontinuous $Ia3d$/gyroid structure, a fairly complete experimental phase diagram has been constructed for the styrene-isoprene system[1].

The morphology formed by these microphase-separated block copolymers is known to depend primarily on the composition of the block copolymer (volume fraction ϕ of one block), as well as the number of statistical segments (N) and the strength of repulsion between the immiscible monomer species (the Flory-Huggins interaction parameter χ). A fourth parameter was suggested recently[2,3], the conformational asymmetry (ε), which describes the mismatch between the volume fractions of the two blocks constituting the copolymer and their unperturbed radii of gyration. Recent self-consistent-field (SCF) calculations unify the classical weak and strong segregation regime theories[4,5] and predict a bubble-shaped region of stability for the gyroid phase in a plot of χN vs. ϕ. The gyroid phase becomes unstable in the strong-segregation regime (high χN) due to packing frustration: the inability to simultaneously maintain uniform domain thickness (to minimize chain stretching energy) and constant mean curvature surfaces (to minimize interfacial energy).

Here, we report the phase behavior of polystyrene-poly(ethylene -*alt*-propylene) diblock copolymers near the lamella-gyroid-cylinder boundaries. A thermoreversible lamella-cylinder transition is observed at $\chi N \approx 22$, and the transformation process is followed by time-resolved small angle x-ray scattering.

Experimental

Polystyrene-polyisoprene diblock copolymers (precursors to polystyrene-poly(ethylene-*alt*-propylene)) were synthesized via high vacuum anionic polymerization[6]. Compositions were obtained by ^{1}H NMR in chloroform, with an uncertainty of ±0.2%. Molecular weights were obtained by gel permeation chromatography in tetrahydrofuran, using a sample of the first (polystyrene) block extracted just prior to isoprene addition. Characterization results are listed in Table 1. Procedures for selective hydrogenation of the polyisoprene block (99.8+% here) have been described elsewhere[7]. Small angle x-ray scattering (SAXS) data were collected with a compact Kratky camera and specially-constructed hot stage; data acquisition and analysis procedures have been described elsewhere[8].

Results and Discussion

Phase Behavior of PS-PEP. The phase diagram was constructed from temperature ramp experiments, where PS-PEP samples were heated in 1-10°C increments and allowed to equilibrate for at least 10 minutes before acquiring SAXS data. PS-PEP 7/10 and 7/12 show lamella→gyroid transitions at temperatures of 270°C and 260°C, respectively. However, a direct lamella→cylinder transition without an intervening gyroid structure was observed for PS-PEP 7/13 at ~175°C. To more precisely locate the volume fraction at which the gyroid phase becomes unstable, a 50/50 binary blend of PS-PEP 7/12 and 7/13 was prepared; as with PS-PEP 7/12, the blend sample shows a lamella→gyroid transition, at 250°C.

The PS-PEP phase diagram is shown in Figure 1. The data of Adams *et al.*[7] were used to obtain the following correlation for the interaction density: X_{PS-PEP} (J/cm^3) = -0.00604T(K)+2.84, from which χN = $XM/\rho RT$, using the homopolymer densities (ρ) and molecular weights (M). All of the transitions shown in Figure 1 are thermoreversible, as demonstrated by cooling the sample below the transition temperature, strongly suggesting that the lamellar, gyroid and cylindrical morphologies are the equilibrium structures at the prevailing temperatures and compositions. None of these samples has an order-disorder transition temperature (T_{ODT}) below 310°C.

The existence of a stable "bubble" for the gyroid phase at modest χN, as seen in Figure 1, is in good agreement with SCF theories.

However, the "pinch-off" point for the bubble occurs near $\chi N \approx 22$, much lower than the SCF theory[4] predictions ($\chi N \approx 60$). Our experiments with higher molecular weight PS-PEP diblocks ($M_w \approx 28$ kg/mol and $\chi N \approx 35$) have failed to show any gyroid structure, even when blending diblocks to vary ϕ in increments as small as 0.4%.

Kinetics of the C-L Transition: PS-PEP 7/13. Hajduk *et al.*[9] reported a lamella-cylinder transition in fractionated Kraton G1726, a diblock system with similar chemistry (polystyrene-poly(ethylene-*co*-butene)), composition and conformational asymmetry as ours but with a higher molecular weight (26 kg/mol). The transformation was quite sluggish close to the phase boundary, making precise determination of the order-order transition temperature difficult. A long-lived intermediate structure consisting of cylinders without hexagonal packing was identified in that work. PS-PEP 7/13, on the contrary, has a well-defined transition temperature (175-176°C) and transformation kinetics on the time scale of minutes to hours, with no discernible long-lived intermediate states. These features make time-resolved SAXS a perfect tool to study the C↔L transition in PS-PEP 7/13.

A representative set of time-resolved SAXS data is shown in Figure 2. The sample was originally heated to 180°C and held for 2 hours to completely develop the cylinder structure, then quenched to the target temperature (140°C in this case). The hot stage used provides an 8°C/min cooling rate under vacuum. SAXS data were collected continuously in 5 minute time slices. At time zero (180°C) the sample possesses the cylinder structure and the $1:\sqrt{3}:\sqrt{7}$... scattering peak ratio characteristic of the cylindrical morphology is evident from the profile. The $\sqrt{4}$ peak coincides with the form factor minimum from individual cylinders and is therefore missing. With increasing time at 140°C the $\sqrt{7}$ peak disappears and the $\sqrt{3}$ peak slowly evolves into a peak at double the q value of the primary peak (q^*), as expected for a lamellar structure (characteristic ratios 1:2:3...). The primary peak position q^* also moves significantly; in the heating ramp experiments used in constructing the phase diagram, q^* increases by 4.5% on going from 175 to 176°C, where the morphology transforms from L to C. This change in spacing is significantly larger than that seen in poly(ethylene oxide)-poly(ethylethylene) diblocks[10], where it was suggested that order-order transitions were only permissible between phases with very similar lattice spacings.

The rate of the C→L transition was quantified through the parameter $I_R = I_2/I_1$, where I_2 is the intensity of the second order peak for the lamellar structure and I_1 is the intensity of the primary peak. Since q^* changes with time, I_1 is evaluated at q^* for each scattering pattern, and I_2 is evaluated at $q=2q^*$. With the absence of the $\sqrt{4}$ peak in the cylinder pattern, I_R is a monotonically increasing function of time. Figure 3 shows the time evolution of I_R for undercoolings (ΔT) 5-11°C. From these curves the half-time ($t_{1/2}$) for the transformation can be extracted. Two other parameters were also examined as markers for the progress of the transformation: q^* and $A_R \equiv A_2/A_1$. Although q^* differs between the L and C phases, the peaks overlap severely and appear as one, so q^* appears to change continuously during the transformation. While q^* is the easiest parameter to track, it is expected to be the least accurate measure of the extent of transformation since a significant change in q^* results simply from the change in temperature (e.g. cylinders at 180°C vs. cylinders at 140°C), principally through the dependence of lattice spacing on χ. A_R is similar to I_R but the integrated areas under the corresponding peaks are used instead of the peak intensities.

The transformation rates (reciprocal half-times) extracted from each of these quantities are presented in Figure 4. The agreement between the three quantities suggests that each would be a satisfactory quantity for monitoring the transformation, though the reciprocal half times based on q^* are systematically larger for the reason noted above. The transformation rate is maximized at ~155°C, presumably due to the competition between the thermodynamic driving force and the polystyrene block mobility. To correct for the latter, the time-temperature shift factors (a_T) for polystyrene were calculated using the Williams-Landel-Ferry equation with constants (C_1, C_2, T_g) obtained from the literature[11]. Figure 5 shows the mobility-corrected results, wherein the rate of the transition is a monotonically increasing function of quench depth (ΔT).

In our case, both the starting cylindrical-phase and the final lamellar-phase materials have a polygrain structure. This suggests the nucleation and subsequent growth of the L phase from many points in the C sample. The temperature dependence of the observed transformation rate, as shown in Figures 4 and 5, thus reflects the temperature dependences of both nucleation and growth; an elegant theoretical description[12] has been presented for the latter, but not the former. That the transformation should occur by nucleation and growth in the present case is also suggested by theoretical calculations[12], which would place all our quenches in the metastable region of the phase diagram, rather than across the spinodal.

In a separate set of experiments we compared the kinetics of the L→C transition (T jumps) and the C→L transition (T drops). For small undercoolings and superheatings (~5-6°C), the L→C transition is completed in less than 10 minutes, while the reverse C→L transition takes ~6 hours. The mobility of the molecules as measured by a_T varies by only a factor of 2-3 over this narrow temperature range. This marked difference in rate suggests fundamental differences between the L→C and C→L transformations. One clear difference is that the hexagonal packing of cylinders leads to a threefold degeneracy in the orientation of lamellae which can be formed by "zippering" together cylinders[13], whereas no such frustration exists in the L→C transformation.

References

1. Khandpur, A. K., Forster, S., Bates, F. S., Hamley, I. W., Ryan, A. J., Bras, W., Almdal, K., Mortensen, K., *Macromolecules* **28**, 8796 (1995).
2. Vavasour, J. D., Whitmore, M. D., *Macromolecules* **25**, 5477 (1992).
3. Vavasour, J. D., Whitmore, M. D., *Macromolecules* **26**, 7070 (1993).
4. Matsen, M. W., Bates, F. S., *Macromolecules* **29**, 1091 (1996).
5. Matsen, M. W., Bates, F. S., *J. Chem. Phys.* **106**, 2436 (1997).
6. Fetters, L. J., Morton, M., *Rubb. Chem. Tech.* **48**, 359 (1975).
7. Adams, J. L., Quiram, D. J., Graessley, W. W., Register, R. A., *Macromolecules* **31**, 201 (1997).
8. Adams, J. L., Quiram, D. J., Graessley, W. W., Register, R. A., Marchand, G. R., *Macromolecules* **29**, 2929 (1996).
9. Hajduk, D. A., Gruner, S. M., Rangarajan, P., Register, R. A., Fetters, L. J., Honeker, C., Albalak, R. J., Thomas, E. L., *Macromolecules* **27**, 490 (1994).
10. Hillmyer, M. A., Bates, F. S., Almdal, K., Mortensen, K., Ryan, A. J., Fairclough, P. A., *Science* **271**, 976 (1996).
11. Ferry, J. D., *Viscoelastic Properties of Polymers*, 3rd ed. (NY: Wiley, 1980).
12. Goveas, J. L., Milner, S. T., *Macromolecules* **30**, 2605 (1997).
13. Sakurai, S., Momii, T., Taie, K., Shibayama, M., Nomura, S., Hashimoto, T., *Macromolecules* **26**, 485 (1993).

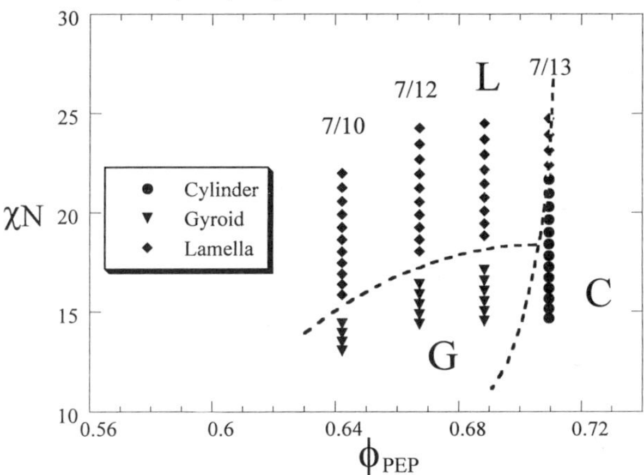

Figure 1. Experimental phase diagram for PS-PEP. The dashed lines suggest the phase boundaries.

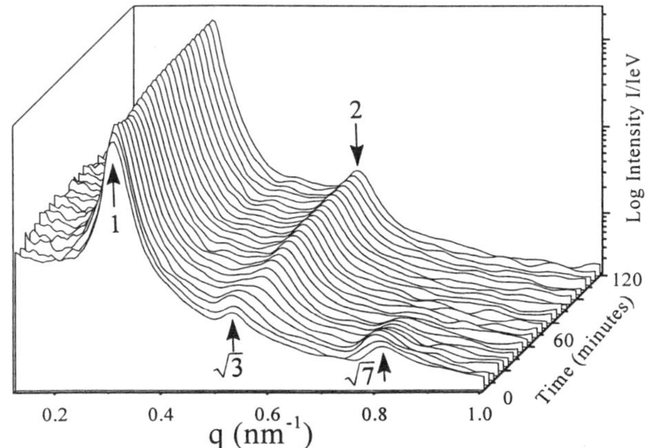

Figure 2. SAXS profiles for the time course of the C→L transformation. PS-PEP 7/13 originally possesses a cylindrical morphology (established at 180°C), which transforms to a lamellar structure on quenching to 140°C. Time zero corresponds to 180°C.

Table 1. Molecular Characterization of Diblock Copolymers

Precursor	W_{PI}	M_w PS/PI (kg/mol)	Product	ϕ_{PEP} at 140°C
PS-PI 7/10	0.592	7.0/10.1	PS-PEP 7/10	0.642
PS-PI 7/12	0.613	7.3/11.5	PS-PEP 7/12	0.667
PS-PI 7/13	0.659	6.5/12.6	PS-PEP 7/13	0.709

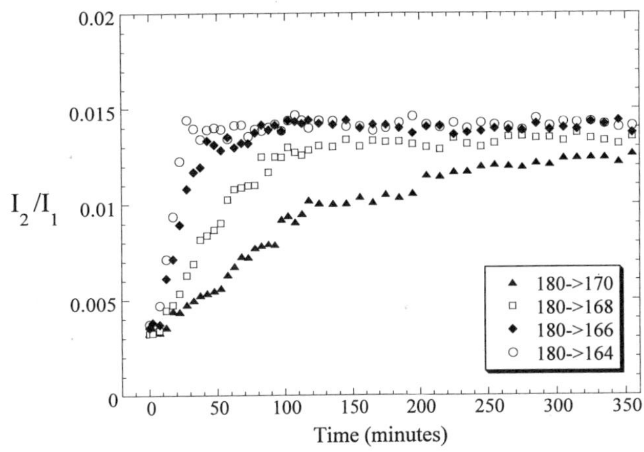

Figure 3. I_R ($=I_2/I_1$) extracted from SAXS profiles as a function of time following quenches of various depths. The phase boundary is at 175°C.

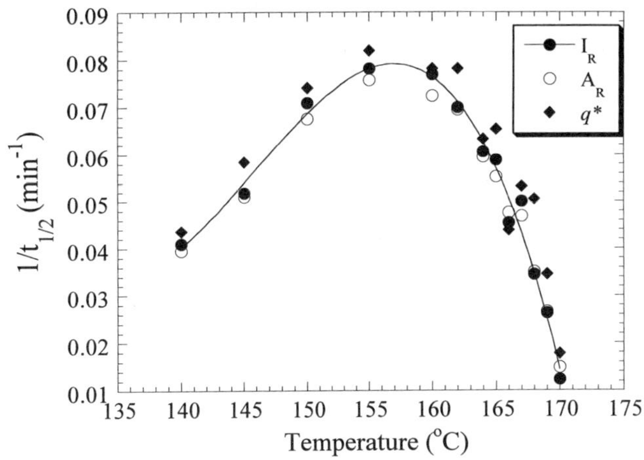

Figure 4. C→L transformation rate characterized by I_R (●), A_R (○), and q^* (◆). Sample was quenched from 180°C to the target temperature indicated on the x-axis. The transformation rate is maximized at an undercooling of ~20°C.

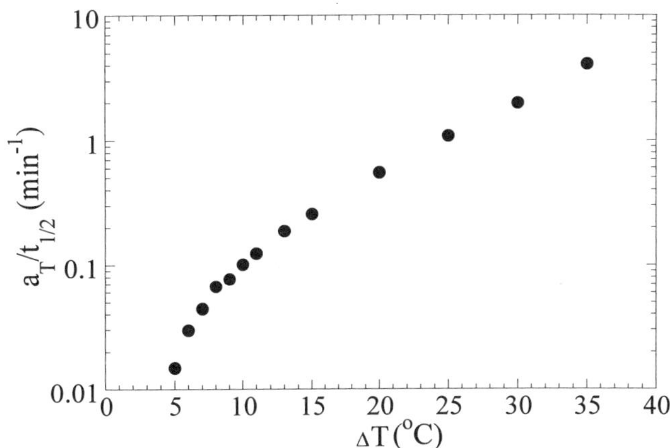

Figure 5. Rate of transformation as a function of ΔT ($T_{transition}$-T) after correcting for the mobility of the polystyrene block ($T_{reference}$=170°C).

SANS Study of Fatty Acid Modified Dendrimers

Aissa Ramzi[1], Barry J. Bauer[2], Rolf Scherrenberg[1], Jacques Joosten[1], and Eric J. Amis[2]

[1]DSM Research, P.O. Box 18, NL-6160MD, Geleen, the Netherlands.
[2]National Institute of Standards and Technology
Gaithersburg, MD 20899

Introduction

Dendrimers are highly branched and highly uniform polymeric molecules. [1] Divergent dendrimers are formed by a stepwise addition of generations to a core, most commonly doubling the molecular weight, branch points and terminal groups in each step. The terminal groups are sites for further chemical reactions, and recently, hydrophylic dendrimers have been reacted with fatty acids to form hydrophylic-hydrophobic structures.[2] They have potential applications as covalently bonded micelles that can be used to disperse materials such as dyes in polyolefin matrices.

The resulting structure resembles a block copolymer, with each block containing similar amounts of the two components which give a strong thermodynamic push towards microphase separation. In this report we give the initial results in the characterization of the structure of hydrophobically modified dendrimers in the solid state as measured by small angle neutron scattering (SANS).

Experimental

Poly(propylene imine) dendrimers (Astramol™)[3] were synthesized by methods reported elsewhere from a 1,4-diaminobutane core with alternating additions of acrylonitrile and hydrogenation of the cyano groups.[1] Generations G1, G3, and G5 were modified by addition of stearic acid or stearic acid-d_{35}. In one set of samples, ammonium salts were formed when the terminal amines became protonated by the stearic acids forming the 'ionicly bonded' samples. These were also converted into amides by heat treatment, splitting of water, forming the 'covalently bonded' samples.[2].

Samples were heated to approximately 100 °C for the ionic and 130 °C for the covalent samples to melt and placed in 1 mm quartz liquid cells. SANS was carried out at the NIST 8 m and 30 m facilities. [4] Instrument configurations were chosen to obtain a q range of $0.02\ \text{Å}^{-1} \le q \le 0.45\ \text{Å}^{-1}$ on 30 m and $0.02\ \text{Å}^{-1} \le q \le 0.35\ \text{Å}^{-1}$ on 8 m. Data were corrected by established procedures and circularly averaged.

Results

Figure 1 is a plot of the SANS results for the samples at 30 °C (nominal temperature stability ± 0.1 °C). They show multiple peaks in the scattering, indicating microphase separation. The features were strongest for the G5 and G3 samples and more sharply defined for the covalent than for the ionic. In well annealed samples, the peak positions have a ratio of 1:2:3 which is characteristic of lamellar microstructure. For the G1 dendrimer which has only 4 hydrophobic groups, there is less order than the higher generations. Lamellar morphologies have also been seen by TEM.[5] The standard uncertainties calculated from the SANS come from the goodness of the fits and, smaller than the size of the symbols, they are not plotted because they are not needed for the purposes of this paper.

As the temperature is increased, there is a phase transition at about 60 °C for the ionic samples. The transition temperatures are approximately 110

°C for the G1 covalent and 90 °C for the G3 and G5. These transitions can also be seen by DSC measurements. [6] Figure 2 is a plot of SANS from the same samples at 130 °C. The scattering has changed into a single peak, indicating loss of the highly ordered structures. Since the slope of the scattering curves approaches zero at low q, it is likely that this is a result of incipient ordering, rather than merely from a 'correlation hole' effect that is seen in single phase block copolymers.

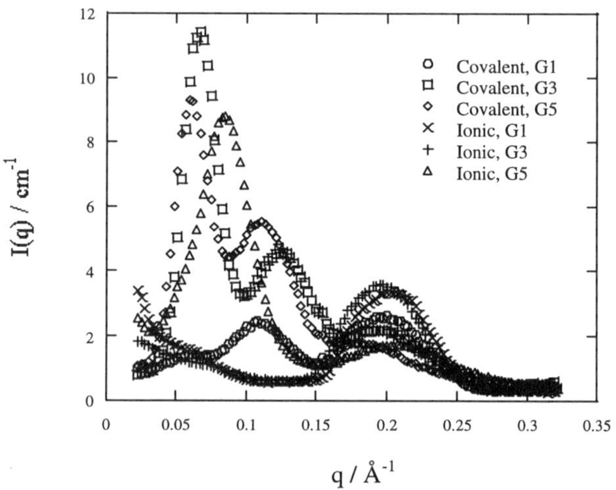

Figure 1. SANS of ionic and covalent hydrophobically modified dendrimers at 30 °C

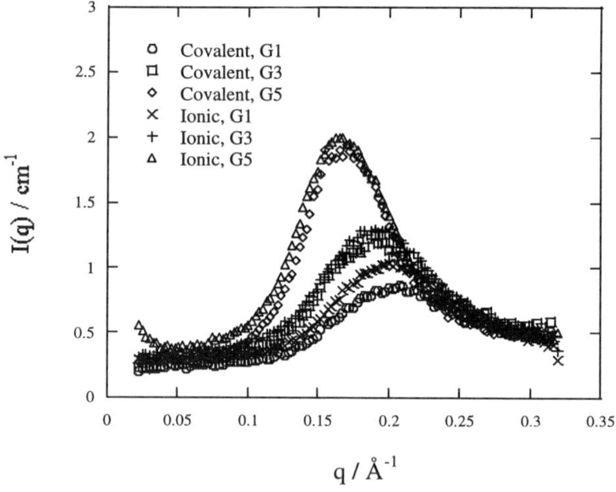

Figure 2. SANS of ionic and covalent hydrophobically modified dendrimers at 130 °C

The scattering from a dense system such as this is dominated by the interdendrimer correlations. The size of a single dendrimer can be measured by use of contrast matching techniques.[7] These methods were developed to measure the single chain scattering from polymers in complex environments, with the first experiments being polymers in concentrated solutions. Two polymers are required that are identical, except that one has deuterium substituted for the protons. This changes the neutron contrast, but not the size or the thermodynamics. The scattering from a mixture of H and D polymers in a solvent is

$$I(q) = A_S \, S_S(q) + A_T \, S_T(q) \qquad\qquad (1)$$

Where $I(q)$ is the scattered intensity, A_S and A_T are contrast factors, $S_S(q)$ is the single chain scattering factor and $S_T(q)$ is the total scattering. A_T is proportional to $(x\, b_D + (1-x)\, b_H - b_S)$ where x is the volume fraction of D polymer in the H-D polymer mixture; b_D, b_H, and b_S are the scattering length densities of D polymer, H polymer, and solvent respectively. Therefore, a proper selection of x can cause this term to become zero. The resulting scattered intensity is then directly proportional to the single chain scattering factor. The core of the dendrimer can have the same effect as a solvent. A $x \approx 0.09$, gives $A_T = 0$ and the scattering function is that of the hydrophobic portion of a single dendrimer.

Figure 3 is a plot of the scattering from a G5 covalent dendrimer at 100 °C. In one case pure deuterated dendrimer is used and in the other a mixture of H and D at the match point is used. In both cases the dendrimer core is hydrogenated. In the matched case the scattering no longer has the prominent peak that is present in the unmatched case.

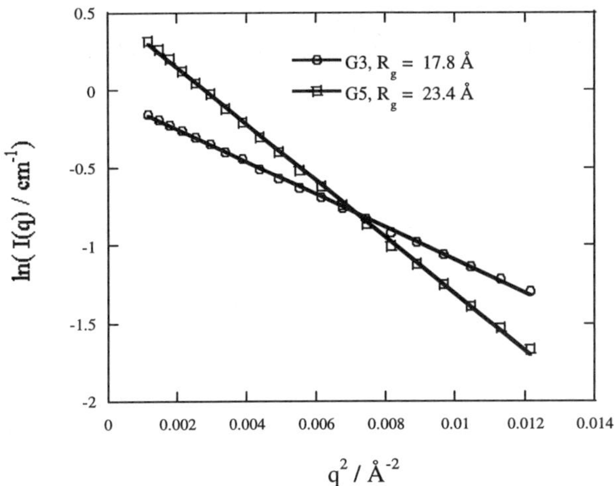

Figure 4. Guinier plots of dendrimers at 100 °C under matched conditions.

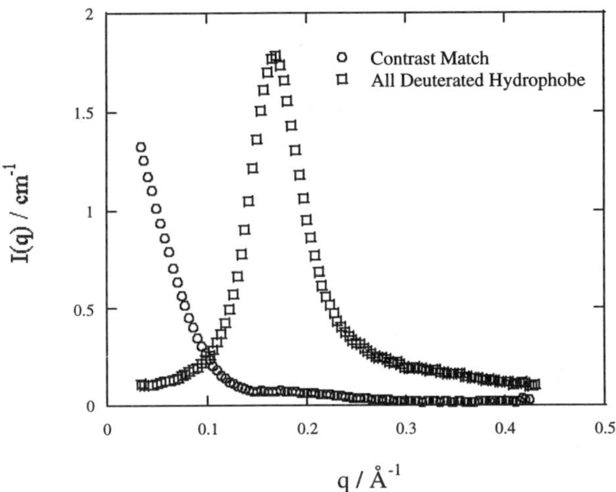

Figure 3. SANS of G5 covalent hydrophobically modified dendrimers at 100 °C under matched and unmatched conditions.

The low q scattering can be used to calculate a radius of gyration (R_g) of the hydrophobic content of a single dendrimer. Figure 4 shows a Guinier plot of the scattering giving $R_g = (17.8 \pm 0.9)$ Å for G3 and $R_g = (23.4 \pm 1.1)$ Å for G5. The uncertainty limits are one standard deviation calculated from the goodness of the fit.

Conclusions

Fatty acid modified dendrimers form highly ordered structures at low temperatures and go through a microphase transition when heated. There is a phase transition at about 60 °C for the ionic samples. The transition temperatures are approximately 110 °C for the G1 covalent and 90 °C for the G3 and G5. Contrast matching can be used to extract single particle scattering from the hydrophobic groups, giving molecular dimensions of $R_g = (17.8 \pm 0.9)$ Å for G3 and $R_g = (23.4 \pm 1.1)$ Å for G5

Acknowledgments

This material is based upon work supported in part by the U.S. Army Research Office under contract number 35109-CH.

References

1. de Brabander, E; Meijer, E. W.; *Agnew. Chem.*, **1993**, 105, *1370*.
2. de Brabander, E; Brackman, J.; Froehling, P., Scherrenberg, R., Put, J.; *PMSE Preprints*, **1997**, *77*, 84.
3. Certain commercial materials and equipment are identified in this paper in order to specify adequately the experimental procedure. In no case does such identification imply recommendation by the National Institute of Standards and Technology nor does it imply that the material or equipment identified is necessarily the best available for this purpose.
4. Prask, H. J.; Rowe, M.; Rush, J. J.; Schroeder, I. G., *J. Res. Natl. Inst. Stand. Tech.*, **1993**, *98*,
5. Ramzi et al, to be published
6. Ramzi et al, to be published
7. Akcasu, A. Z.; Summerfield, G. C.; Jahshan, S. N.; Han, C. C.; Kim, C. Y.; Yu, H.; *J. Polym. Sci., Physics*, **1980**, *18*, 863.

MICROPHASE SEPARATION OF A DIBLOCK COPOLYMER UNDER PRESSURE: POLY(STYRENE-BLOCK-BUTADIENE)

*H. Ladynski[1], W. DeOdorico[2,1] and M. Stamm[1,]**

[1] Max Planck-Institut für Polymerforschung, Postfach 3148, 55021 Mainz, Germany,
[2] Institut für Kernphysik, Universität Frankfurt, 60486 Frankfurt/Main, Germany,
e-mail stamm@mpip-mainz.mpg.de

In a mixture of two incompatible polymers phase separation occurs in order to minimize the number of unfavorable contacts between polymer A and polymer B. On the other hand at high temperature some systems form a homogeneous phase because of the increase in entropy.

Since material properties might depend critically on the particular structure the under-standing of this kind of phase transition is not only a fundamental physical question but also of quite practical interest.

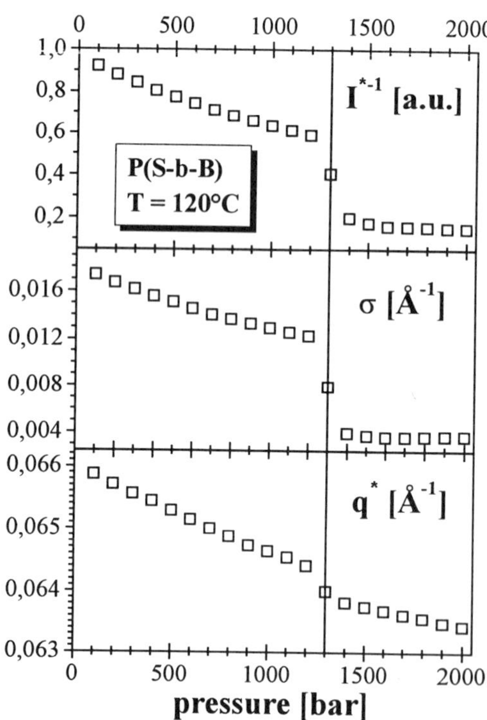

Fig.1 Microphase separation of the diblock copolymer P(S-b-B) with pressure: inverse peak intensity, halfd width and position

In diblock copolymers the chains A and B are covalently bound together and no macroscopic phase separation is possible. So the system tends to separate in microphases by building structures on scales corresponding to the size of the chains. According to the composition f (the volume fraction of A) micelles, cylinders, lamella or more complex structures occur [1-3]. In a system with upper (lower) critical solution temperature (U(L)CST), the system gets miscible at higher (lower) temperatures and the mesophase structure breaks down. This is called the microphase separation transition (MST) which takes place at the temperature T_{MST}.

In previous experiments the dependencies of the MST on temperature, composition variations or a change of the degree of polymerization have been studied [2,3]. More detailed investigations [4,5], using SANS and different deuterium labeling techniques revealed a stretching of chains during the transition while the single blocks are supposed to shrink.

As for the temperature dependant measurements the transition is clearly indicated by a change in scattering peak shape, an increase in the aspect ratio of the peak maximum (I*/FWHM) and a decrease in maximum position (Q*).

Recently now (Fig.1) we established pressure as another thermodynamic variable for the MST of poly(styrene-block-butadiene), P(S-b-B) shifting T_{MST} by an amount of +20°C per kbar [6].

Now our investigations concentrate on ordering and relaxation behaviour of the polymer system in external fields (e. g. after „pressure jumps" between and within the different phase states). Therefore we performed in-situ pressure jump experiments with a high pressure cell at the ESRF in Grenoble. Some first results can be seen in Fig. 2. Although further data treatment is necessary, we predict a two step relaxation process after release of pressure as indicated by the double exponential fit.

We therefore adapted the pressure cell in a slightly modified way (thinner diamond windows in order to decrease beam absorption) to the beamline A2 at HASYLAB in Hamburg, in order to measure the long time relaxation behaviour of the system. It was possible to detect the morphology peak in the scattering picture with a sufficient accuracy within a acceptable amount of time.

For the comparison of the two pressure jump experiments one has to keep in mind that the two starting conditions are different. The initial kinetics therefore may be also determined by the different distance to the glass transition temperatures, which also depend on pressure. It is, however, still obvious that kinetics for the two cases is different. The molecular details of the structure formation and melting can therefore be expected to be quite different.

We acknowledge the help of Dr. A. Meyer and R. Döhrmann during the A2 experiment and of Dr. O. Diat and T. Kuhlmann during the ESRF experiment.

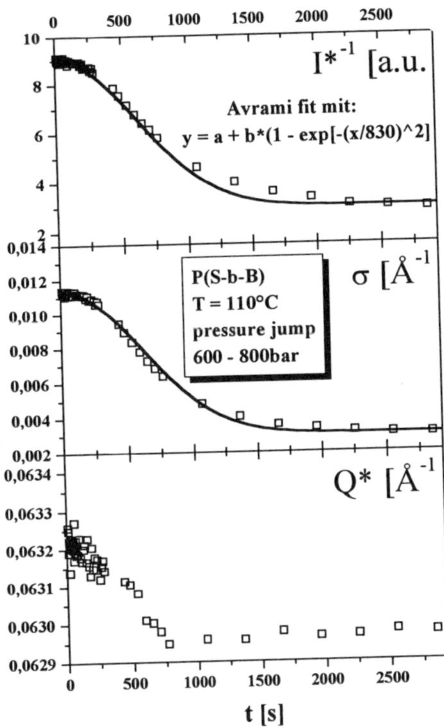

Fig. 2: Morphology peak -position (Q^*), -width (σ) and inverse maximum (I^{*-1}) as received by fitting a Lorentz function to the Q dependent scattering intensity. Q denotes the scattering vector $Q = (4\pi/\lambda)\sin(\theta/2)$.
(Data were taken at the high brilliance beamline ID 2 of the ESRF in Grenoble, France)
Nucleation and growth of morphology after pressure jumps from 600 to 800 bar and from 600 to 400 bar. The MST of the system is situated at approximately 700 bar within a „transition range" of 400 bar. The lines are Avrami fits with an exponent of two.

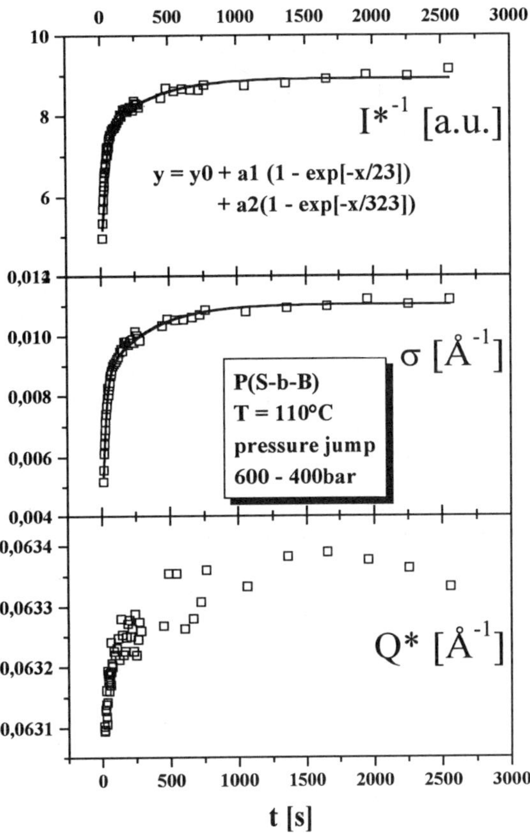

Fig. 3 Same as for Fig.2, but for structural relaxation after release of pressure. The lines are double exponential fits with relaxation times of 23 and respectively 323 seconds.

[1]   L. Leibler, *Macromolecules* **13** (1980) 1602

[2]   F. S. Bates & G. H. Fredrickson, *Ann. Rev. Phys. Chem.* **41** (1990) 525

[3]   F. S. Bates, *Science* **251** (1991) 898

[4]   V.T.Bartels, V. Abetz, K. Mortensen & M. Stamm, *Europhys. Lett.* **27** (1994) 371

[5]   V.T. Bartels, M. Stamm, V. Abetz & K. Mortensen, *Europhys. Lett.* **31** (1995) 81

[6]   H. Ladynski, W. DeOdorico & M. Stamm, *to be published*

Transitions in Polymer Blends and Block Copolymers Induced by Phase-Selective Dilation with Supercritical CO$_2$

James J. Watkins[1]*, Garth D. Brown[1], Michael A. Pollard[2]
Vijayakumar RamachandraRao[1] and Thomas P. Russell[2]
Departments of Chemical Engineering[1] and Polymer Science and Engineering[2]
University of Massachusetts, Amherst, MA 01003

Introduction:

Preferential absorption of supercritical carbon dioxide into one component of a polymer blend or block copolymer can induce a dilation disparity between the phases. Recently, using *in situ* small angle neutron scattering, we discovered that this phase-selective dilation can have a dramatic impact on the location of the lower disorder-order transition (LDOT) in symmetric poly(deuterated styrene)-block-poly(n-butyl methacrylate) copolymers, P(d-S-b-nBMA).[1,2] Herein we extend this study to homopolymer blends. Analogous depressions in the lower critical solution temperature (LCST) are found for mixtures of d-polystyrene (d-PS) and poly(n-butyl methacrylate) (PnBMA) and for blends of d-polystyrene and poly(vinylmethyl ether) (PVME).

Supercritical CO$_2$ - induced transitions in blends and block copolymers share several common features. Both are dramatic: for the d-PS/PnBMA systems, we have measured depressions in the LDOT of over 200 ^{0}C and LCST depressions of over 60 ^{0}C. LCST depressions in PS/PVME blends exceed 100 ^{0}C. All transitions are strong functions of carbon dioxide density (the location of the transitions can be tuned by adjusting pressure) and are fully reversible. Finally, solvent-induced phase segregation upon the addition of an SCF is unique. Liquid organic solvents typically stabilize the homogeneous state by mitigating unfavorable segmental interactions. Selective sorption of SCF solvents, however, induces severe dilation disparities between the phases. LCST and LDOT behaviors are driven by entropic factors, specifically negative volume changes upon mixing that arise due to disparities between the free volumes of the phases. At ambient pressure, these disparities widen as temperature increases. The key feature of the SCF solvent is the ability to precisely control disparities in dilation between the swollen polymer components isothermally at low temperature by adjusting solvent density. Combined with solvent-induced depressions in the glass transition temperature, this approach greatly expands experimentally accessible regions of the phase diagram.

Experimental:

Symmetric, diblock P(d-S-b-nBMA) copolymers having total molecular weights of 32,000 and 78,000 g/mol, denoted 32K and 78K respectively, were melt pressed at 120 ^{0}C into 1.3 cm (o.d.) by 2.5 mm thick disks that were inserted into stainless-steel sample rings. Blends of 40% d-PS and 60% PnBMA having molecular weights 14,000 and 15,000 g/mol respectively were cast from toluene, dried and melt pressed directly into the sample rings at 60 ^{0}C. Blends of d-PS and PVME having molecular weights 90,000 and 104,000 g/mol, respectively, were cast from toluene, dried under vacuum and melt pressed directly into the sample rings at 50 ^{0}C. SANS profiles were collected as a function of CO$_2$ density at temperatures between ambient and 200 ^{0}C and pressures between atmospheric and 350 bar in a stainless steel scattering cell equipped with sapphire windows. For the block copolymer samples, thermal history was first erased by disordering the sample at ambient pressure. Temperature control, to within ±0.5 ^{0}C, is provided via cartridge heaters inserted into an aluminum heating jacket and controlled using a PID controller. CO$_2$ is charged to the cell using a high-pressure manifold[3] and manual syringe pump. The experiments were conducted at the National Institute of Standards and Technology (NIST) Cold Neutron Research Facility (CNRF) on the 30 m spectrometer. SANS measurements were performed with neutrons having a wavelength, λ, of 0.8 nm, a $\Delta\lambda/\lambda = 0.15$, a 0.6 cm (o.d.) beam at the sample, and a sample to detector distance of 5 meters for the 32K copolymer and 6 meters for all others. 2-D data were normalized, corrected for background scattering and detector sensitivity, scaled to absolute differential scattering cross-sections (cm^{-1}) using both SiO$_2$ and polymer standards and radially averaged.

Results and Discussion:

The location of the LDOT for symmetric P(d-S-b-nBMA) copolymers having total molecular weights of 32,000 g/mol and 78,000 g/mol and the LCST for a 40% blend of d-PS and PnBMA homopolymers at ambient pressure are shown in Table 1. Our results are in agreement with literature reports.[4,5] The disparity between the LDOT for the 32K block copolymer and the LCST of homopolymer blends of d-PS and PnBMA chains having molecular weights nearly identical to the copolymer segments is a consequence of connectivity of the dissimilar blocks. In polymer blends, a lower critical solution transition occurs when the product of the Flory interaction parameter, χ and the degree of polymerization, N exceeds a critical value ($\chi N = 2$ for a symmetric blend). For symmetric block copolymers this stability limit is increased: χN is approximately 10.5.

CO$_2$ sorption is known to be preferential for polyacrylates relative to polystyrene[6] (Figure 1) and we conclude that in the dPS/PnBMA/CO$_2$ systems, PnBMA is selectively dilated. We have studied the phase behavior of 78K P(d-S-b-nBMA) in the presence of CO$_2$ extensively. Figure 2a shows the SANS profile for the 78K sample as a function of increasing density of the fluid phase at 145 ^{0}C. (Q is the scattering vector.) At densities greater than 0.05 g/cc there is a discontinuity in the scattered intensity, a shift in Q$_{max}$, and the emergence of a second order peak indicates that the LDOT has been crossed and the copolymer is ordered. Figure 2b shows scattering profiles upon depressurization. Despite a slight time dependent hysteresis, it is evident that SCF-induced LDOTs are fully reversible. At all temperatures between 60 and 180 ^{0}C we find the 78 K copolymer is ordered at fluid densities above 0.1 g/cc. This suggests that under these conditions the LDOT and UODT have merged (not shown). The magnitude of CO$_2$-induced depressions in the LDOT is markedly evident for the 32K copolymer. The disorder-order transition for this sample at ambient pressure is above the decomposition temperature (T$_{dec}$) of the copolymer and is thus inaccessible. In the presence of CO$_2$, however, the copolymer orders at 100 ^{0}C at fluid densities in excess of 0.3 g/cc, more than 200 ^{0}C below the projected LDOT. The transition is accompanied by a discontinuous reduction in the half width of the first order peak as a function of density (Figure 3). In all cases, the degree of segregation increases with increasing pressure (density) and at all conditions exceeds that of systems ordered by thermal means at ambient pressure.[1]

The behavior of the P(d-S-b-nBMA) samples suggests that an analogous phase separation will be observed in low molecular weight blends of the copolymer components in the presence of CO$_2$. Figure 4 shows scattering profiles for a 40% d-PS and 60% PnBMA blend at 65 ^{0}C as a function of CO$_2$ density. Debye-Bueche analysis indicates that above a threshold density of 0.03 g/cc, correlation length begins to increases rapidly with increasing density. Clearly, this system segregates at modest CO$_2$ densities at temperatures more than 60 ^{0}C below the ambient pressure LCST. The transition is fully reversible upon decompression.

Evidence that these dilation-dependent transitions are general phenomena is provided by experiments that indicate separations of d-PS and PVME blends can likewise be induced by exposure to SCF CO$_2$. For a 50% dPS / 50% PVME blend, the spinodal temperature at ambient pressure is estimated to be 152 ^{0}C as determined by extrapolation of the measured scattering intensity to zero Q at several temperatures. Figure 5 indicates that in the presence of CO$_2$, the system phase segregates at 40 ^{0}C, more than 110 ^{0}C below the ambient pressure LCST.

Conclusions:

Order in block copolymers and phase separation in blends can be induced from homogeneous systems by sorption of supercritical CO_2. At present, it appears that the depression of the LCST and LDOT are manifestations of the increasingly negative volumes of mixing as one phase is selectively dilated. The effect of segment dilation, which depresses the LDOT and LCST, overwhelms the effect of hydrostatic pressure which opposes free volume differences between the phases and is expected to increase the transition temperatures.

Acknowledgments:

We are grateful to Professor Y. Gallot for the synthesis of the copolymers. The Cold Neutron Research Facility is supported by NIST and the National Science Foundation under agreement No. DMR-9423101. JJW and TPR acknowledge partial support from the Materials Research Science and Engineering Center at the University of Massachusetts. TPR also acknowledges the partial support of DOE-BES under DE-FG02-96ER45612.

References:

1. Watkins, J. J.; Brown. G. D.; RamachandraRao, V. S.; Pollard, M. A. Russell, T. P. *PMSE Preprints, American Chemical Society.* 1998, 78, 94.
2. Watkins, J. J.; Brown. G. D.; RamachandraRao, V. S.; Pollard, M. A. Russell, T. P. *submitted for publication.*
3. Watkins, J. J.; McCarthy, T. J. *Macromolecules*, 1995, 28, 4067.
4. Russell, T. P.; Karis, T. E.; Gallot, Y.; Mayes, A. M. *Nature* 1994, 368, 729.
5. Hammouda, B; Bauer, B. J.; Russell, T. P. *Macromolecules*, 1994 27, 2357.
6. Wissenger, R. G.; Paulaitis, M. E. *J. Polym. Sci. Polym. Phys.* 1987, 25, 2497.

Table 1. Ambient Pressure Transitions for dPS/PnBMA Blends and Copolymers

Sample	Transition	Temperature
32 K P(dS-b-nBMA)	LDOT	> 300 °C (decomp.)
78 K P(d-S-b-nBMA)	LDOT	180 °C
40% dPS / 60% PnBMA blend	LCST	126 °C

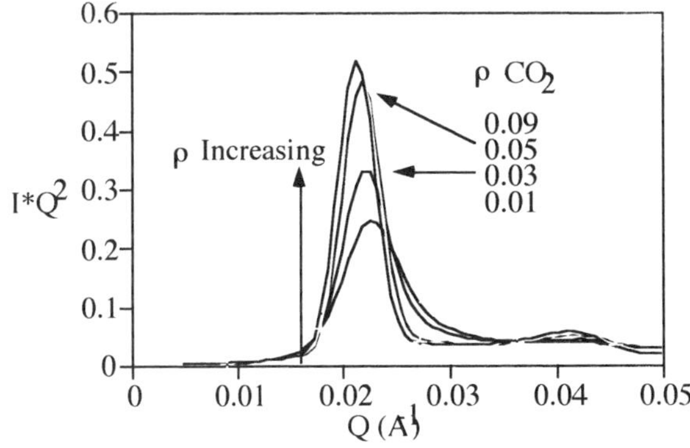

Figure 1. CO_2 sorption into PS and PMMA at 35 °C.[6]

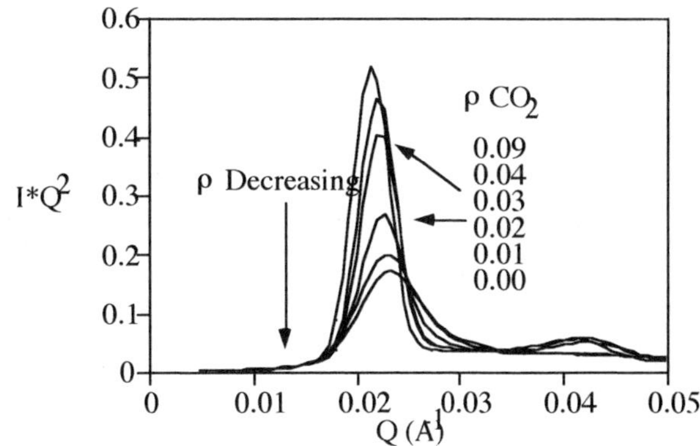

Figure 2. b.) Scattering profile of 78K P(d-S-b-nBMA) as a function of decreasing CO_2 density at 145 °C.

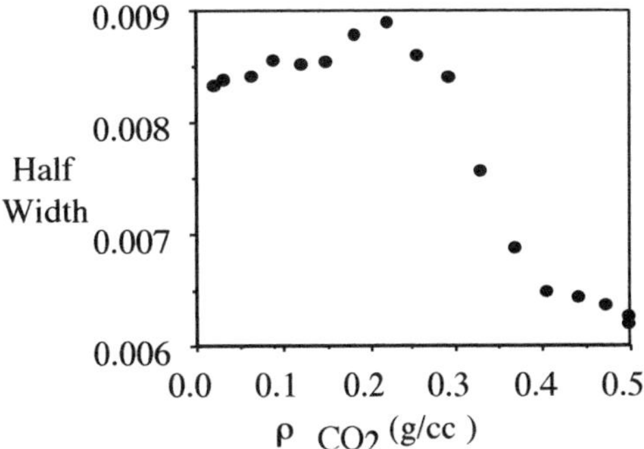

Figure 3. Intensity half-widths at half the maximum intensity as a function of density for scattering profiles of 32K P(d-S-b-nBMA) at 100 °C. The LDOT is induced at a CO_2 density of 0.3 g/cc.

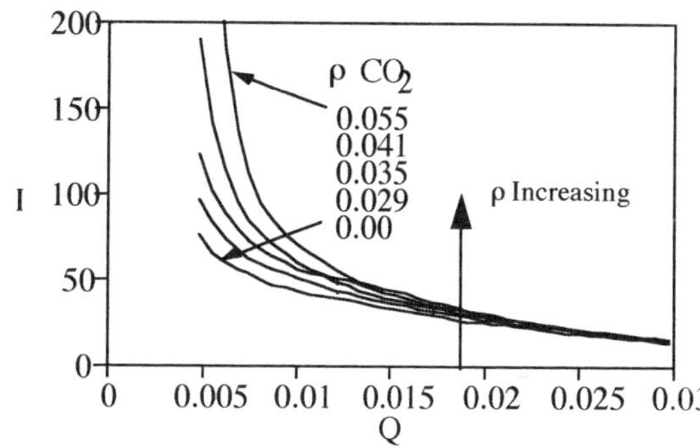

Figure 4. Density dependence of scattering profiles (I vs. Q) for 40/60 PS/PnBMA blends at 65 °C.

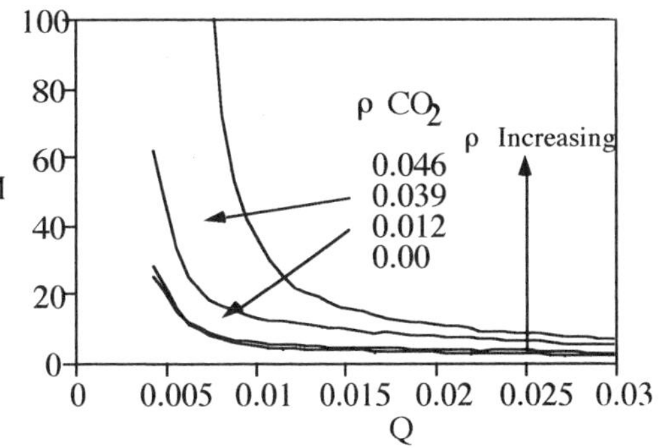

Figure 5. Density dependence of scattering profiles (I vs. Q) for 50/50 PS/PVME blends at 40 °C.

Figure 2. a.) Scattering profile of 78K P(d-S-b-nBMA) as a function of increasing CO_2 density at 145 °C.

SAXS Studies of the Transverse Orientation Texture in Plane Strain Compression of Semicrystalline Triblock Copolymers
PETER KOFINAS, PETER L. DRZAL
Dept. of Materials and Nuclear Engineering,
University of Maryland
College Park, Maryland 20742-2115.

The shear-induced morphologies produced by high levels of plane strain compression using a channel die, were investigated in a semicrystalline ethylene/ethylene-propylene/ethylene (E/EP/E) 30,000/40,000/30,000 molecular weight triblock copolymer. The die was maintained at a selected constant temperature during the compression flow, and the load was applied continuously. The compressed specimens were cooled under load to room temperature, followed by load release. It was discovered that a perpendicular and a novel transverse to the plane of shear morphology could be developed in this triblock sample by varying the cooling rate of channel die processing. Two dimensional small angle x-ray scattering (SAXS) was used to determine the domain spacing and the lamellar orientation relative to the specimen boundaries.

Using a cooling rate of 0.27°C/s and a load of 9.2 MPa, a perpendicular to the plane of shear orientation texture was observed. The 2-D SAXS pattern for this perpendicular orientation texture is shown in Figure 1, where two sharp dots in the constraint direction having a Bragg spacing of 43.6 nm and a diffuse ring with a Bragg spacing of 14.5 nm are observed . The IQ^2 vs Q plot in Figure 2 provides quantitative information on the relative lamellar populations for the perpendicular to the plane of shear morphology. A large sharp narrow peak is observed at q=0.144 nm^{-1} indicating that the microphase separation between the E and EP blocks is the dominant lamellar population. The second peak observed is a broad peak at approximately q=0.433 nm^{-1}. This second lamellar population has a lower intensity and represents the long period spacing from the semicrystalline E population.

When the cooling rate is increased to 3.50°C/s, while applying identical deformation conditions as used to yield the perpendicular to the plane orientation morphology, a novel transverse to the plane of shear morphology was obtained. The semicrystalline E lamellar population in this transverse to the plane morphology becomes the dominant lamellar population. The two dimensional SAXS pattern for this novel semicrystalline transverse orientation texture is shown in Figure 3, which shows two large spread-out spots in the flow direction having a Bragg spacing of 13.1 nm. The IQ^2 vs Q plot shown in Figure 4, reveals that the increase in the cooling rate caused a decrease in the overall intensity of the pattern. However, the relative intensity of the semicrystalline and microphase separated lamellar populations have dramatically changed in response to the faster cooling rate. The broad peak from the semicrystalline E component at q=0.479 nm^{-1} is now the dominate lamellar population due to a dramatic decrease in the intensity of the lamellar population at q=0.144 nm^{-1}, which corresponds to the microphase separation between the E and EP blocks.

This SAXS study has provided quantitative information in explaining how cooling rate effects the resulting orientation texture in plane strain compression of semicrystalline block copolymers. In the perpendicular to the plane of shear orientation morphology, the slow cooling rate used in the channel die processing allows the E chains to crystallize within the confinement of the microphase separated block copolymer domains. In the transverse to the plane of shear orientation morphology, the high driving force for crystallization, doesn't allow the microphase separated morphology to fully develop. The interplay of shear, the dominant kinetic process of crystallization and the thermodynamic process of microphase separation yield the transverse to the plane of shear orientation texture. This study demonstrated clearly that rather dramatic morphology control is possible through changes in melt processing of semicrystalline block copolymers, and demonstrated that in addition to the "perpendicular" and "parallel" shear-induced morphologies that have been observed, a novel. "transverse" orientation texture can be produced.

ACKNOWLEDGEMENTS: This research was partially supported by the Minta Martin Foundation at the University of Maryland. We are grateful to Dr. John Barnes of the Polymer Structure and Mechanics Group at the National Institute of Standards and Technology for use of the small-angle x-ray scattering facility and for enlightening discussions.

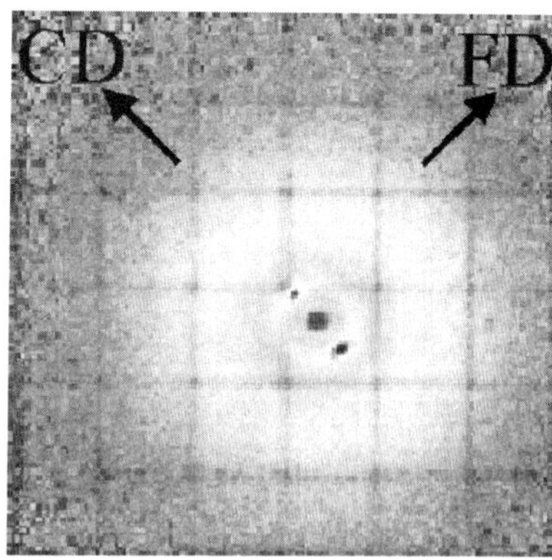

Figure 1: SAXS pattern for the perpendicular to the plane of shear morphology.

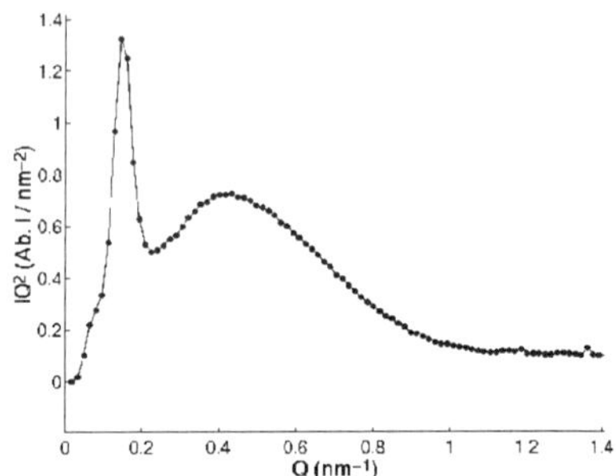

Figure 2: IQ^2 vs Q plot of the perpendicular to the plane of shear morphology.

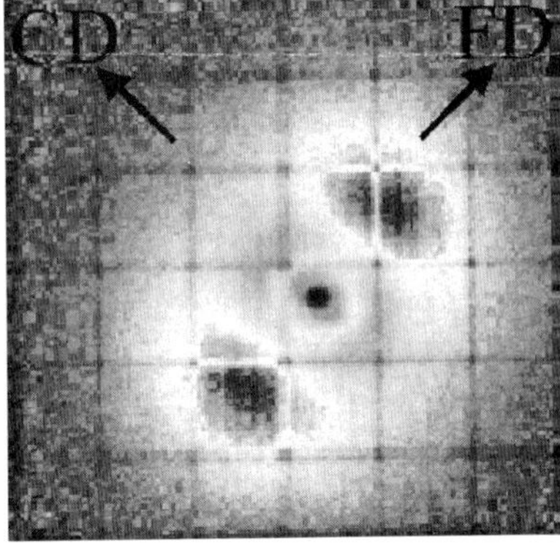

Figure 3: SAXS pattern for the transverse to the plane of shear morphology.

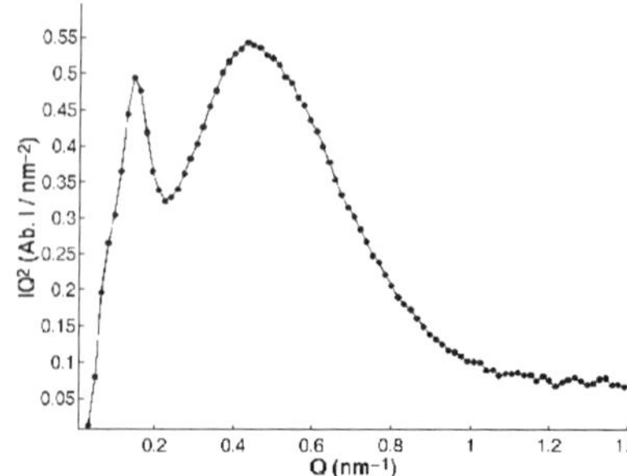

Figure 4: IQ^2 vs Q plot of the transverse to the plane of shear morphology.

Structure Development during Deformation of Nano-Reinforced Poly(urethane) with Polyhedral Oligomeric Silsesquioxane (POSS)

Benjamin S. Hsiao*, Xuan Fu*, Patrick T. Mather**, Kevin P. Chaffee[+],
Hong Jeon**, Henry White[¥], Mariam Rafailovich[¥], Joseph D. Lichtenhen[+] and
Joseph Schwab[++]

*Chemistry Department, [¥]Materials Science and Engineering Department, State University of New York at Stony Brook, Stony Brook, NY 11794-3400; Air Force Research Laboratory, **Wright Patterson AFB, [++]Hughes STX, OH 45433-7750, [+]Edwards AFB, CA 93524-7680

Introduction

Recently, a unique class of polyurethane containing nanostructured inorganic reinforcement agent: polyhedral oligomericsilsesquioxane (POSS), in hard segments was developed by the Air Force Research Laboratory[1,2]. The POSS compound is a crystalline material[3] containing a silicon-oxygen framework with the general formula $(RSiO_{3/2})_n$ (n=8), which offers unique inorganic properties (thermal and oxidation stability) to the polymer matrix. In the synthesis, two POSS molecules were first incorporated in a bisphenol A diol, which was then used as a chain extending agent for standard polyurethane preparation[2]. The resultant polyurethane is an organic and inorganic hybrid with hard segments being reinforced by nanostructured POSS moieties. Incorporation of POSS may effectively reduce polymer flammability, improve oxidation resistance, increase permeability to gases, and improve heat distortion temperature and mechanical strength[4].

It is well documented that the versatile mechanical and thermophysical properties of segmented polyurethane based elastomers are due to the formation of microphase separation from the thermodynamic incompatibility (immiscibility) of solid-like hard segment and rubbery soft segment sequences[5-7]. In conventional polyurethane, the binding forces in the hard segment domains usually involve noncovalent interchain interactions such as van der Waals forces and hydrogen bonds. Under heat or deformation, these interchain interactions in hard segments can break down, leading to the destruction of the domains and thus the loss of properties. In this study, we intend to understand the effect of nano-reinforcing POSS agent, covalent bonded to the hard segment, on properties of microphase separation under deformation. We feel that with the strong interchain interactions due to POSS, the hard segment domains may be substantially stronger. To follow the *in-situ* structure and morphology changes during polymer deformation, we have carried out simultaneous SAXS and WAXD (SWAXS) techniques with synchrotron radiation. The SAXS technique was used to monitor the variation of microphase domains, and WAXD was used to determine the change in the POSS crystal structure and the strain-induced crystallization in the soft segments.

Experiments

The POSS containing polyurethane (POSS-PU) sample was provided by the Air Force Research Laboratory. Its synthetic scheme was illustrated in Ref. 2. As prepared, the material contains soft segments of polytetramethylene glycol (PTMG) (Mw=2000), and hard segments of hexamethylenediisocyanate (DMA) and POSS-bisphenol A diol. Each diol contains two units of POSS component. The final wt. % of POSS in the polymer was about 34%. The sample was melt pressed into film with thickness about 1.5 mm.

In-situ SWAXS measurements were carried out at the Advanced Polymers beamline[8], X27C, National Synchrotron Light Source (NSLS), Brookhaven National Laboratory (BNL). The wavelength used was 0.1307 nm, and the beam size was about 0.4 mm in diameter at the sample position. The deformation study of the POSS-PU film was performed with an Instron 4442 tensile apparatus using two imaging plates. In this setup, the chosen sample-to-detector distance is 1590 mm for SAXS and 140 mm for WAXD. A constant strain rate was applied to the specimen through out the whole deformation. The SAXS/WAXS images were taken after reaching the desired strain values under tension as well as upon relaxation after stretching.

Results and Discussion

The stress and strain relationship for the POSS-PU film during stretching and relaxation is shown in Figure 1. The maximum applied strain is 400%. After the stress being completely relaxed, the sample is found to maintain an elongation of 200% without fully recovered to the original length. The mechanical performance of the model PU with 0% POSS is much weaker. The maximum strain is about 100% and the maximum stress is only about one-thirtyth of that for POSS-PU.

The SWAXS images were taken along both face-on and edge-on directions of the POSS-PU film. Several selected SAXS and WAXD images taken at the initial (0%) and the final strain (400%) conditions are illustrated in Figure 2. In the face-on WAXD image, the unstrained sample shows multiple reflections,

which are arised from the POSS crystal structure. This is determined by comparing the observed diffraction profile with the published powder scan from a POSS-model compound: exodisilanol monomer (c-$C_6H_{11})_8Si_8O_{11}(OH)_2$.[4] This POSS model compound has a monoclinic unit cell with parameters a=1.04, b=2.15, c=1.44 nm, and a P2$_1$ space group[4]. The cell dimensions for the POSS crystal in POSS-PU appear to be larger. The SAXS image of the as-prepared sample show a scattering maximum (long period ca. 200 nm) and a weak orientation. This suggests that the microphase separation is present in the initial sample. Upon deformation, the POSS crystal WAXD pattern remains but the reflection peaks shift to larger angles indicating that the crystal cell becomes smaller (Figures 3 and 4). (Note that, in Figure 4, the intensity of the strongest reflection at 6.4° decreases with draw ratio, which is due to the decrease in sample volume.) At strain above 100%, two additional reflection peaks are seen on the equator (Figure 3). These peaks are characteristics of PTMG crystals in soft segments. The induced crystallinity in PTMG increases with the draw ratio, but they do not vanish upon relaxation (Figure 3). This behavior is very different from conventional polyurethane containing PTMG soft segments. In these polymers, the strain-induced PTMG crystals always melt away upon relaxation since the melting temperature of PTMG in copolymer is below the room temperature (ca. 10°C). The presence of POSS apparently stabilizes the formation of PTMG crystals. The corresponding SAXS patterns show a constant long period (Figure 5) during stretching and relaxation. The orientation and the arrangement of the segregated domains increases, but the dimensions of the domains appear to be unchanged. This finding again is very different from the conventional polyurethane samples without POSS, in which stretching near 400% always destroys the hard segment domains. The induced crystallization in PTMG also manifests a strong diffuse scattering streak along the equator in SAXS.

The edge-on investigation shows some unusual features. First, the as-prepared sample shows two strong reflections around 6° in WAXD and a scattering maximum at d=3.8 nm on the equator in SAXS. All peaks have a noticeable orientation in the equator. One possible explanation is that the POSS crystals are slat-like, which are aligned more or less parallel to the film surface. If the SAXS maximum is resulted from the POSS crystal structure (as an hk0 reflection), than the unit cell dimensions of POSS are significantly larger than the reported values for the POSS model compound[4]. During deformation, the two reflections in WAXD convert into a broad single peak, but the position of the SAXS equatorial maximum remains unchanged. This suggests that deformation affects the structure or packing of the POSS crystals. The stress-induced crystallization in the soft segments (at strain > 100%) is also evident in both SAXS and WAXD edge-on images.

Conclusion

We have carried out the deformation study of a unique polyurethane sample having a segmentally nano-reinforcing agent POSS on its hard segments. The simultaneous SAXS/WAXS technique was used to reveal the changes in structure and morphology during deformation for this organic/inorganic hybrid materials. It is seen that deformation generally increases the orientation of the segregated phases, as well as induces crystallization of the soft PTMG components. The incorporation of POSS shows several unique effects on the structure and property relationships. (1) The incorporation of POSS improves the mechanical property by about ten times. (2) The hard segment domains are definitely reinforced by POSS which show no sign of domain-destruction by deformation. (2) The strain-induced PTMG crystals are stabilized by the "reinforced" hard domains. No crystal melting (PTMG) is seen even after the complete relaxation of stress at room temperature. (3) The POSS crystals can be slightly distorted by the applied stress, but no apparent crystallinity loss in POSS is seen under high draw ratio.

Acknowledgment

We gratefully acknowledge the financial supports of this work in part by NSF-GOALI(DMA-9629825) program and in part by the AP-PRT at the NSLS.

References

1. J. D. Lichtenhan, *Comments Inorg. Chem.* **17**, 115-130 (1995).
2. J. J. Schwab, J. D. Lichtenhan, M. J. Carr, K. P. Chaffee, P. T. Mather and A. Romo-Uribe, *PMSE Preprint*, **77**, 549 (1977).
3. P. D. Lickess, *Advanced Inorg. Chem.* **42**, 147 (1995).
4. J. D. Lichtenhan, T. S. Haddad, J. J. Schwab, M. J. Carr, K. P. Chaffee, P. T. Mather, ACS Polymer Preprint, **39(1)**, 489 (1998).
5. R. Bonart, *J. Macromol. Sci. - Phys.*, **B2(1)**, 115 (1968).
6. J. T. Koberstein and L. M. Leung, *Macromolecules*, **25**, 6205 (1992).
7. J. T. Koberstein, A. F. Galambos and L. M. Leung, *Macromolecules*, **25**, 6195 (1992).
8. Please see web site http://bnlx27c.nsls.bnl.gov

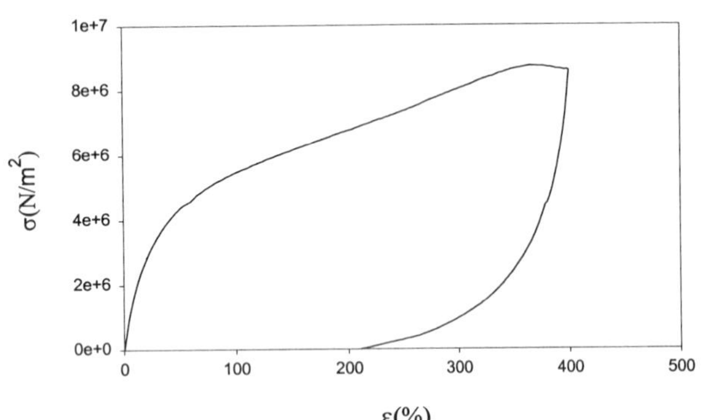

Fig.1. Mechanical property of POSS-PU
during stretching and relaxation

Face–on

SAXS   WAXD

0%

400%

Edge–on

0%

400%

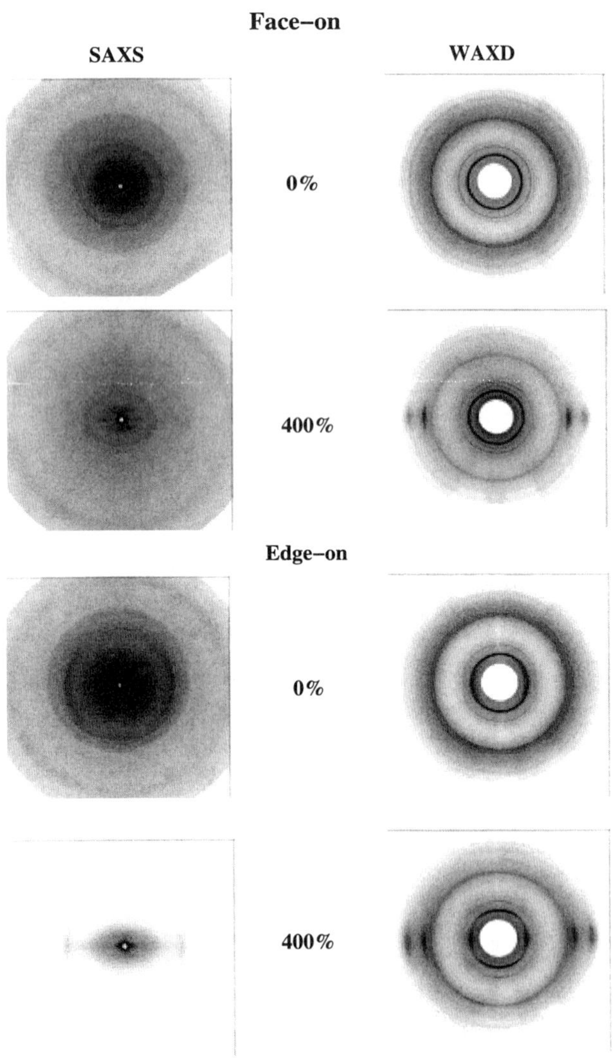

**Fig.2. Images of POSS–PU with
different stretching ratio**

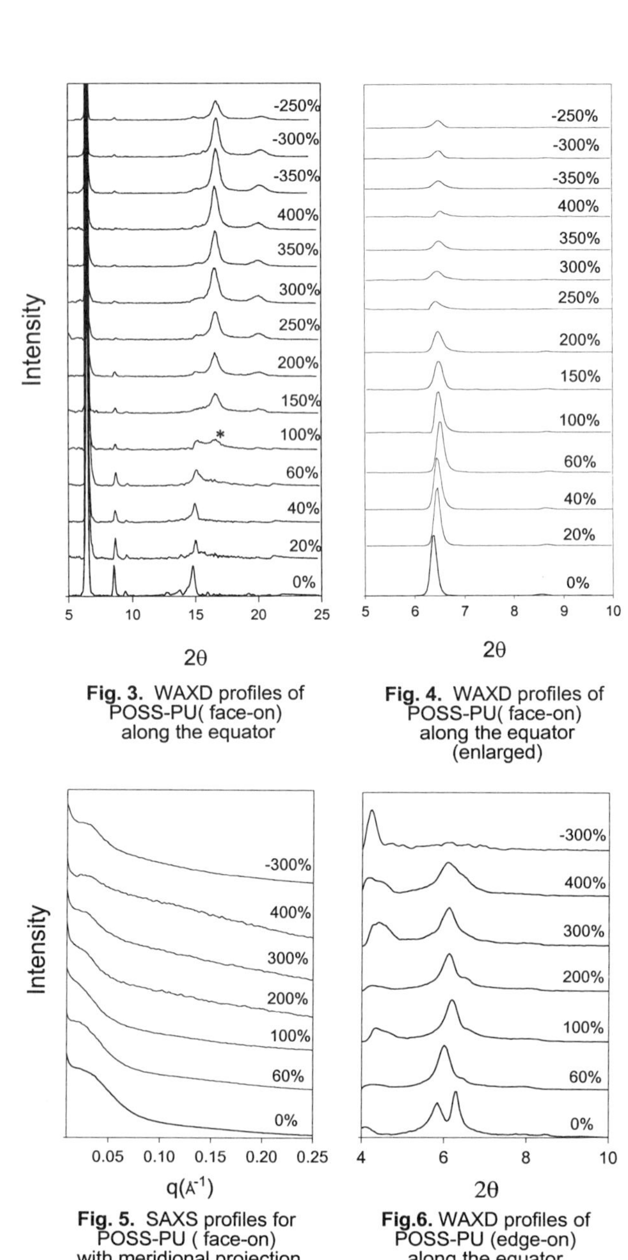

Fig. 3. WAXD profiles of
POSS-PU(face-on)
along the equator

Fig. 4. WAXD profiles of
POSS-PU(face-on)
along the equator
(enlarged)

Fig. 5. SAXS profiles for
POSS-PU (face-on)
with meridional projection

Fig.6. WAXD profiles of
POSS-PU (edge-on)
along the equator

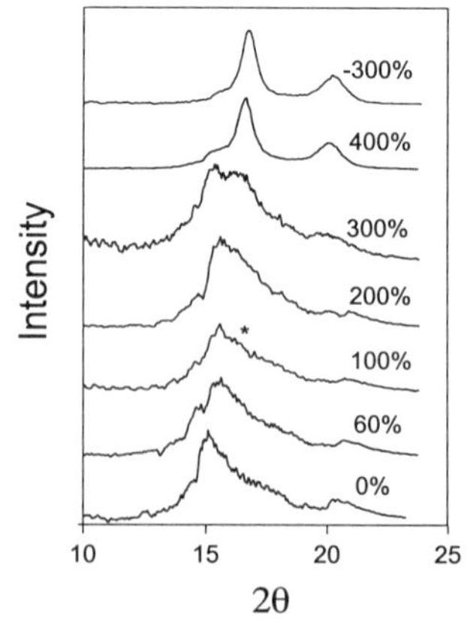

Fig.7. WAXD profiles of
POSS-PU (edge-on)
along the equator
(continued)

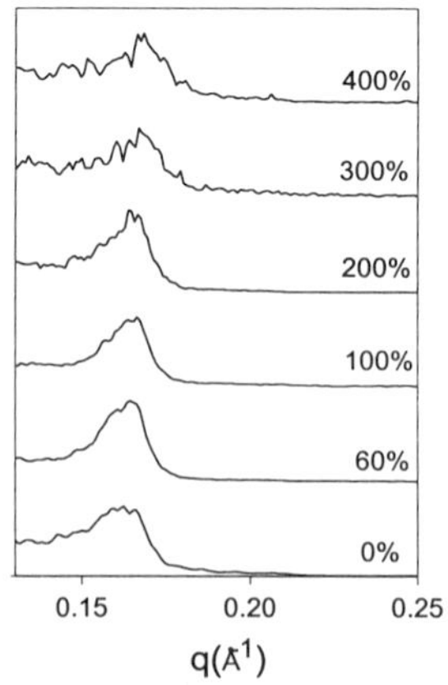

Fig.8. SAXS profiles of
POSS-PU (edge-on)
along the equator

Investigation of Stress-Induced Effects on SAN/PMMA and SAN/PCL Blends Using Small Angle Light Scattering

Zhengyu Hong[a], Montgomery T. Shaw[a,b] and R. A. Weiss[a,b]
Polymer Program[a] and Department of Chemical Engineering[b]
University of Connecticut, Storrs, CT 06269

Introduction

Ever since the description in 1952[1] of shear-induced phase changes in polymer solutions, phase transitions and critical phenomena of polymer solutions and blends have been the subject of extensive theoretical and experimental investigation. A number of authors[2-9] have suggested that flow can induce changes in the degree of mixing by influencing the thermodynamic state of polymer blends, i.e., flow shifts the phase transition temperatures. Recent studies[10-13], however, indicate that this is not the case, and that hydrodynamic effects are responsible for the observations.

In this study, two blends, SAN/PMMA and SAN/polycaprolactone (PCL), were used. In an earlier study[3] of the former, Lyngaae–Jorgenson and Sondergaard concluded that stress can markedly improved miscibility. The latter has not been investigated, and should provide an independent confirmation of the phenomena.

Experimental

<u>Materials and phase diagram characterization</u>: Blends of 25/75 SAN/PMMA and 80/20 SAN/PCL were studied (Table 1). Mixtures were prepared by melt mixing in a Brabender at 150°C and compression molded into 0.8-mm-thick sheets at the same temperature. The cloud points of the blends were determined by light scattering using a heating rate of 1°C/min.

<u>Rheo-SALS measurements</u>: Time-resolved SALS measurements were performed with a custom-built rheo-SALS apparatus[14]. SAN/PMMA blend was heated between the parallel plates of the rheo-SALS instrument to 12°C above T_{cp}. Shear was then applied and the time-resolved 2D light scattering patterns were recorded. When the shearing was stopped, the sample was quickly quenched to below T_g to preserve the morphology, which was then characterized by phase-contrast optical microscopy. Another set of experiments was carried out where steady shear was applied to the melt at 160°C, which is within the 1-phase region of the quiescent phase diagram, in order to examine if the flow perturbed the structure of the miscible blend melt. The light scattering pattern of the sheared blend was monitored while the temperature was raised at a heating rate of 0.2°C/min.

Similarly, SAN/PCL was heated to 180°C, which is 10°C above T_{cp}. The time evolution of the light scattering patterns was recorded when the sample was subjected to shear. The flow instabilities at moderate shear rates were visualized using a CCD camera.

<u>High-rate, pressure-driven and drag flows</u>: The melt flow behavior of the SAN/PMMA blend at 180°C for shear rates range of 0.3 to 200 s^{-1} was determined with an Instron Capillary Rheometer. The extrudates were quenched in ice water immediately upon exiting the die. The quenched strands were then microtomed parallel and normal to the flow direction, and transmission electron micrographs of the morphology were obtained with a Philips Model 300 TEM. A CSI normal-stress extruder modified with a flush-mounted heat pipe was used to obtain high-stress drag flow of the blends. The polymer melt inside the extruder and near the heat pipe was rapidly quenched by spraying the exterior of the heat pipe with ice water. The morphology of the quenched samples was characterized by TEM.

Results and Discussion

Phase diagrams of the SAN/PMMA and SAN/PCL blends are shown in Fig. 1. A near-critical composition, 25/75 SAN/PMMA and an off-critical composition, 80/20 SAN/PCL were used in the flow experiments.

Phase-separated 25/75 SAN/PMMA blends exhibited a common trend in their time-dependent structure evolution at the various shear stresses used. The structure progressed from a bicontinuous structure to a chevron-like structure and eventually to a fibrillar structure, as shown in Fig. 2, column A. Columns B and C are the corresponding real-time scattering patterns, and the Fourier transform (FT) of the digitized micrographs, respectively.

The quiescent blend underwent spinodal decomposition at 12°C above T_{cp}. A typical bicontinuous phase structure observed by optical microscopy is shown in Fig 2, A(1); the average domain size (characteristic length) was

ca. 0.7 μm. For a shear strain of γ = 8 at a shear rate of 0.02 s^{-1}, a unique chevron-like phase structure was observed, see Fig. 2, A(2). The phases were oriented in a zig-zag fashion with a narrow distribution of tilt angles relative to the flow direction. In reciprocal space, the scattering intensity grew in the flow direction and gradually developed into a pattern of two "wings" separated by a dark gap oriented perpendicular to the flow, Fig. 1, B(2). Depth profiling of the structure by microscopy indicated that the chevron structure was not a surface-induced flow instability. Its formation may possibly be associated with a non-affine deformation of a highly interconnected spinodal structure. As the shear strain was increased further, the phase domains continued to deform and elongate along the flow direction, forming a fibrillar structure, Fig. 2, A(3). As a result, the scattering intensity level increased significantly and yielded a bright streak scattering pattern, Fig. 2, B(3). The bright streak pattern did not change with time, indicating a steady state was achieved.

The Fourier transform (FT) of the digitized micrograph, Fig. 2C, confirmed that the microstructures preserved in the quenched specimens were the same as that in the real-time rheo-SALS experiments. Thus, while shear caused profound changes of the melt morphology, no shear-induced phase transitions were observed for shear stresses as high as 60 kPa.

When a shear rate of 0.01 s^{-1} was applied to the melt in the one phase-region at 160°C for 90 min, the SALS scattering pattern showed no evidence of shear-induced phase separation. Fig. 3(a) is the scattering pattern after 90 min of shear flow, and it indicates a single-phase system. When the temperature was increased from the one-phase region at a rate of 0.2°C /min while shearing, an anisotropic light scattering pattern, Fig. 3(b), was detected at 175°C, which was 7°C above the measured quiescent cloud point. If stress had increased the phase separation temperature, as this result seems to indicate, then no phase separation should occur below 175°C at the same stress regardless of the history. However, when the same shear rate was applied to a sample at 172°C, an anisotropic scattering pattern appeared after shearing for 75 min. The coincidence of the time when phase separation started in both experiments suggests that strain plays an important role in phase separation of near-critical blends under the influence of shear. The scattering pattern evolved from no scattering directly to an anisotropic, "two-wing" scattering pattern without the typical ring pattern characteristic of spinodal decomposition of a quiescent blend. That result suggests that shear suppresses the concentration fluctuations in a near-critical polymer blend and the phase-separation process may follow a different mechanism than conventional spinodal decomposition.[15]

When shear was applied to the SAN/PCL blend at a rate of 0.01 s^{-1} and 10°C above the T_{cp}, the light scattering pattern evolved from an isotropic pattern, to a "two-wing" pattern and eventually merged to a bright streak pattern (Fig. 4). The rate of the time-evolution depended on the shear rate. The bright streak persisted at shear rates as high as 1 s^{-1}, which indicates that the melt had a heterogeneous structure even though the blend appeared to be optically clear. Clearly, measuring the turbidity alone is not a conclusive indicator of a phase transition for flowing systems. At moderate shear rates, hydrodynamic instabilities such as azimuthally-spaced "vortices" extending in the vorticity direction were observed in the melt. These "vortices" may influence the shape of the "two-wing" scattering patterns (sometimes referred to in the literature as "butterfly" patterns[16]).

Fig. 5 shows TEM images of the SAN/PMMA blend after being extruded at a shear rate of 81 s^{-1} and stress of 395 kPa. This and similar micrographs show that in the plane perpendicular to the flow (r, θ), phase domains remained interconnected with an average size of 0.18 μm. However, in the flow direction, the domains were intensely deformed and fibrillar. These results are in stark contrast to those of Lyngaae-Jørgenson and Søndergaad[3], who noted that "no structure was detected at stresses above approximately 100 kPa," using a very similar blend.

Using the modified normal-stress extruder at a shear rate of 12 s^{-1}(202 kPa), the rapidly quenched sample also showed fibrillar structure in the flow direction.

A close examination of the microstructure of the SAN/PMMA blend after shear flow revealed that the behavior of phase-separated droplets was very similar to the classical hydrodynamic behavior[17] of immiscible liquids. The size of phase-separated polymer droplets, however, tends to be greater than the Newtonian drops under the same stress due to the elasticity of polymeric materials. Moreover, since the elastic effects dominate at high shear rates, the amount of agglomeration increases with the increasing shear rate[18-19]. While the thermodynamic theory predicts that the dispersed phase disappears when the shear stress exceeds a critical value, the size of the domains in our experiment tended to reach an equilibrium under the combined influence of the shear field, elastic forces and interfacial tension.

Conclusion

When shear was applied to a phase-separated blend, the structure of phase domains evolved from bicontinuous to chevron-like, and finally to fibrillar. No phase transition was detected. When the blend was subjected to high stresses (above 100 kPa), the domain structure was highly oriented in the flow direction, yet remained interconnected in the direction perpendicular to the flow. The flow-induced structural changes were qualitatively explained in the terms of hydrodynamic effects.

Acknowledgments

We thank Dr. M. Cantino and Ms. L. Khairallah for performing TEM.

Reference

1   Silberberg, A.; Kuhn, W. *Nature* **1952**, *170*, 450.
2   Mazich, K. A.; Carr, S. H. *J. Appl. Phy.* **1983**, *54*, 5511.
3   Lyngaae-Jorgenson, J.; Sondergaard, K. *Polym. Eng. Sci.* **1987**, *27*, 351.
4   Katsaros, J. D.; Malone, M. F.; Winter, H. H. *Polym. Eng. Sci.* **1989**, *29*, 1434.
5   Nakatani, A. I.; Kim, H.; Takahashi, Y.; Matsushita, Y.; Takano, A.; Bauer, B. J.; Han, C. C. *J. Chem. Phys.* **1990**, *93*, 795.
6   Hindawi, I. A.; Higgins, J. S.; Weiss, R. A. *Polymer* **1992**, *33*, 2522.
7   Mani, S.; Malone, M. F.; Winter, H. H. *Macromolecules* **1992**, *25*, 5671.
8   Fernandez, M. L.; Higgins, J. S. *ACS, Polym. Mat. Sci. Eng.* **1994**, *71*, 39.
9   Hindawi, I.; Higgins, J. S.; Galambos, A. F.; Weiss, R. A. *Macromolecules* **1990**, *23*, 670-674.
10 Han, C. C. *Macromol. Symp.* **1996**, *101*, 157-165.
11 Douglas, J. F. *Macromolecules* **1992**, *25*, 1468-1474.
12 Chen, Z. J.; Wu, R.-J.; Shaw, M. T.; Weiss, R. A.; Fernandez, M. L.; Higgins, J. S. *Polym. Eng. Sci.* **1995**, *35*, 92.
13 Chen, Z. J.; Shaw, M. T.; Weiss, R. A. *Macromolecules* **1995**, *28*, 648.
14 Wu, R.; Shaw, M. T.; Weiss, R. A. *Polym. Mat. Sci. Eng.* **1993**, *68*, 264.
15 Matsuzaka, K.; Jinnai, H.; Koga, T.; Hashimoto, T. *Macromolecules* 1997, 30, 1146
16 Moses, E.; Kume, T.; Hashimoto, T., *Phys. Rev. Lett.* **1994**, *72*(13), 2037.
17 Taylor, G. I. *Proc. R. Soc. London* **1934**, *A146*, 501.
18 Roland, C.M.; Böhm, G.G.A. *J. Polym. Sci., Polym. Phys.* **1984**, *22*, 79.
19 Sundararaj, U.; Macosko, C. W. *Macromolecules* **1995**, *28*, 2647.

Table 1. Characteristics of Components and Blends

Blend	T_g, ^oC	T_m, ^oC	M_w, kDa	M_w/M_n	wt %	T_{cp}, ^oC
SAN/	105	--	108	2.32	25	168
PMMA	107	--	92	1.87	75	
SAN/	105	--	190	2.14	80	170
PCL	-60	60	163	1.61	20	

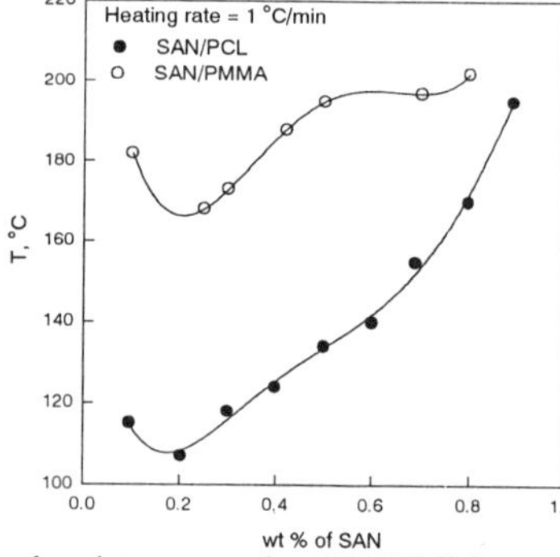

Figure 1.   Cloud point curves for SAN/PMMA and SAN/PCL blends measured using light scattering technique with heating rate of 1 ^oC/min.

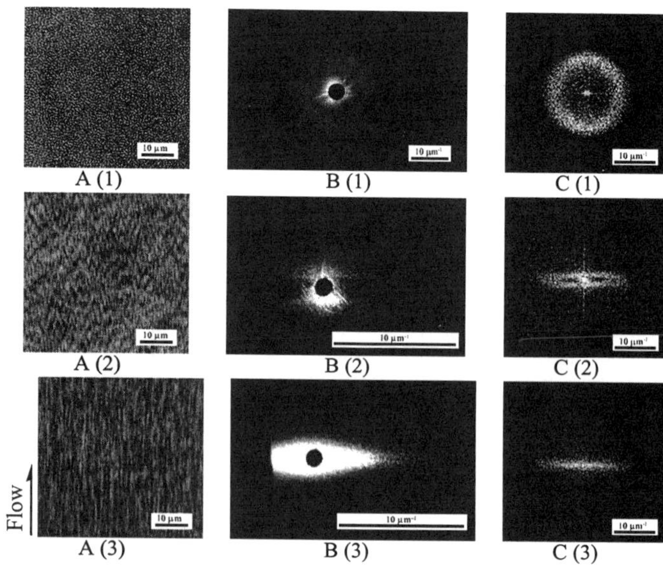

Figure 2.   Microstructure of 25/75 SAN/PMMA and their corresponding light scattering patterns and FT spectra after shearing at 0.02s⁻¹.   Column A): Micrographs using optical microscopy. B). Real-time 2D LS patterns using rheo-SALS. C) FT of digitized micrographs.   Row (1): γ = 0; (2): γ = 8; (3): γ = 288.

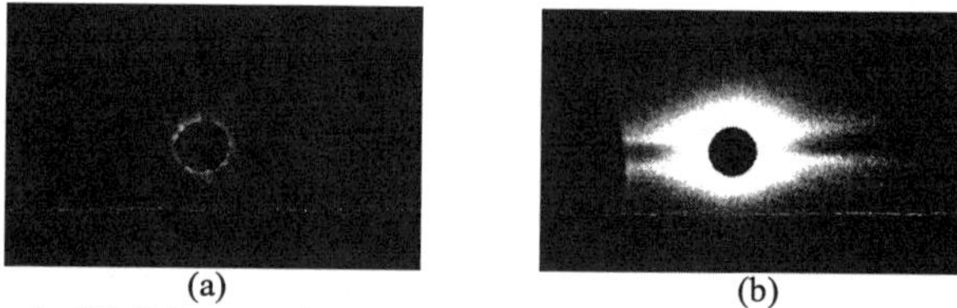

Figure 3. 2D light scattering patterns of 25/75 SAN/PCL under steady shear rate of 0.01 s⁻¹ (a) T = 160 ^oC, strain = 54; (b) T = 175 ^oC, strain = 45.

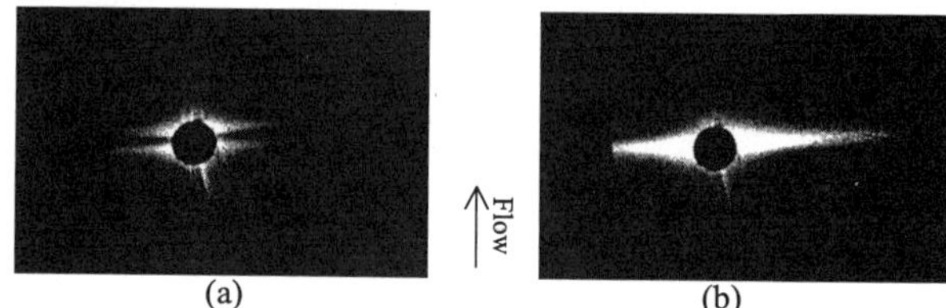

Figure 4. 2D light scattering patterns of 80/20 SAN/PCL at 180°C, shear rate of 0.1 s⁻¹ ($\tau = 7.7$ kPa), (a) strain = 54; (b) strain = 186.

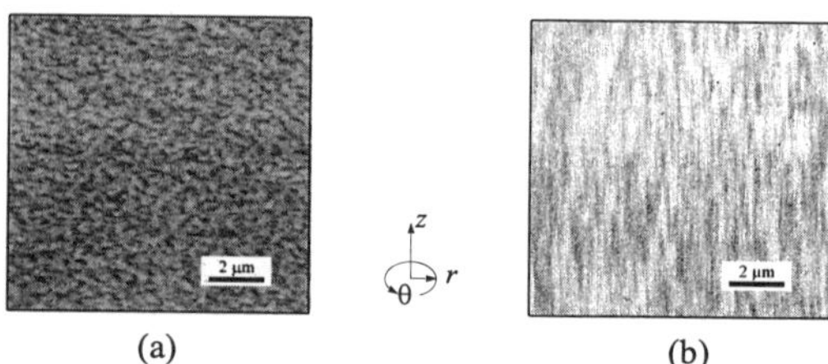

Figure 5. Morphology of 25/75 SAN/PMMA after being sheared at 81s⁻¹ ($\tau = 395$kPa) and quenched. (a) r, θ plane; (b) r, z plane (z is the flow direction).

The Equatorial Small–Angle Scattering During the Straining of Polyetheresters and its Analysis

N. Stribeck

Universität Hamburg, Institut TMC,
Bundesstr. 45, 20146 Hamburg, Germany

Introduction

Two–phase polymers with preferred orientation frequently exhibit small–angle X–ray scattering (SAXS) patterns with fiber symmetry, which can be recorded at a syncrotron beam line with high accuracy. Quite often one observes patterns with many reflections, which vary considerably as a function of tunable parameters. After a qualitative description of the observations and a semi–quantitative analysis of reflection positions as a function of the parameter values a quantitative analysis of the image series should afford profound insight into structure. The objective of the present study is the extraction of the diameter distribution describing an ensemble of oriented rodlike domains in strained material.

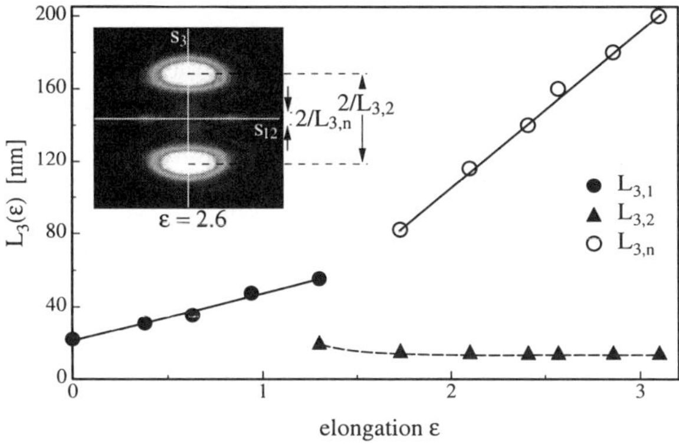

Figure 1: Principle SAXS pattern and peak shifts observed during the straining of a PEE sample. $L_{3,1}$ and $L_{3,2}$ are long periods, while $L_{3,n}$ is the reciprocal height of the form factor envelope from an ensemble of needle–shaped particles causing the equatorial scattering.

With respect to structure polyetheresters (PEE) are multiblock copolymers, whose blocks are called hard segments and soft segments respectively. As a result of phase separation a two–phase system from hard domains and soft domains is formed. With respect to material properties PEE's are thermoplastic elastomers. When strained beyond 50 %, the process becomes irreversible [1]. The relation between irreversible elongation and the materials two–phase structure is supposed to be elucidated by means of a quantitative analysis of the SAXS patterns.

Before starting to analyze, suitable reduction of the measured data should be considered. The mathematical relation between structure and scattering pattern leads to the suggestion to analyze projections of the scattering pattern [2]. In this work it is shown, how a stepwise analysis of a projection curve results in a simple model for the observed equatorial scattering. Fourier transformation into the chord distribution [3] varies the weighting of measured data in such a manner that deviations from pure "particle scattering" become negligible. Finally simple algorithms are pointed out, which allow to extract structural parameters directly from the chord distribution.

Experimental

Films from Arnitel E2000/60 (by DSM, The Netherlands) were strained continuously in the synchrotron beam at beamline A2 (HASYLAB, Hamburg). Image plate detector was positioned 1.8 m behind the sample and exposed for 1 min. The material contains soft segments from polytetrahydrofurane (PTHF) and hard segments from polybutyleneterephthalate (PBT).

Observations and semi–quantitative analysis

Principal observations and peak shifts are shown in Fig. 1. During straining a two–point diagram (long period $L_{3,1}$) is observed. At an elongation $\epsilon = 1.3$ a second peak ($L_{3,2}$) emerges, the first peak vanishes behind the primary beam stop and "needle scattering" ($L_{3,n}$) appears at the equator. This observation can be explained by a microfibrillar model [4]. At low elongation microfibrils from hard and soft domains cause the two–point pattern while at high elongation hard domains are unravelled causing the formation of needle shaped soft domains which scatter about the equator.

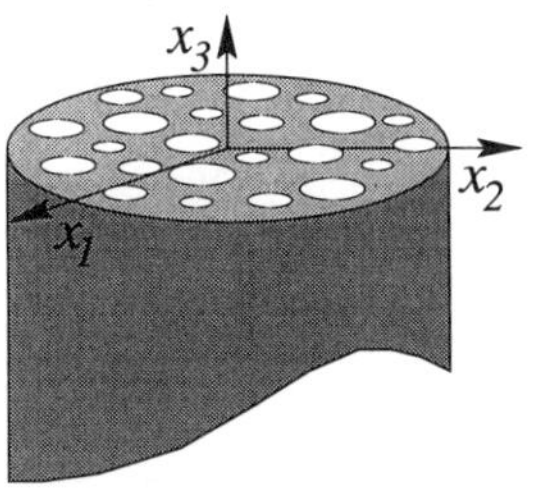

Figure 2: Sketch of the 2D fiber cross section structure, whose chord distribution is related to the projection according to eq. 1.

Projections and the transverse structure

Because of the Fourier transform relation between scattering and structure, sections in physical space are related to projections in reciprocal space. The well–known invariant, Q, is an example for a projection, which filters the non–topological parameters of the two–phase structure out of the scattering pattern. When dealing with the equatorial streak of a fiber pattern, it appears suitable to extract what Bonart called "Querstruktur" (transverse structure)[2] by computing the projection

$$\{I\}_2(s_{12}) = 2 \int_0^\infty I(s_{12}, s_3)\, ds_3. \tag{1}$$

Here $s_{12} = \sqrt{s_1^2 + s_2^2}$ and s_3 are the components of the scattering vector as indicated in Fig. 1. The magnitude of the scattering vector is defined by $|\vec{s}| = (2/\lambda) \sin \theta$, with the X–ray wave length λ and 2θ defined as the scattering angle. $\{I\}_2(s_{12})$ is related to a 2D two–phase system made from needle cross sections in a matrix, as indicated in Fig. 2.

Projections extracted from the image series are presented in Fig. 3. The curves show a distinct Porod region, in which the scattering falls of with s_{12}^{-3}. Small deviations are accounted to the non–ideal structure of the real two–phase system [5] and corrected accordingly, leading to the 2D interference function $G_2(s_{12})$ of an ideal two–phase system[6]. From the interference functions the 2D chord distributions $g_2(x_{12})$ (cf. Fig. 4) can be computed by 2D Fourier transformation [7].

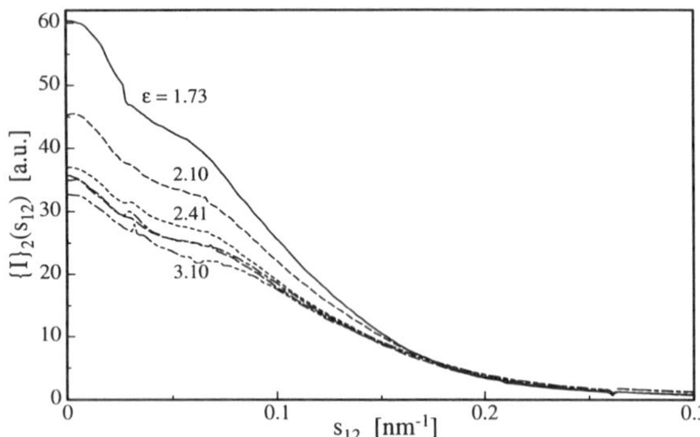

Figure 3: Projections $\{I\}_2\,(s_{12})$ of the equator scattering on to the plane normal to the straining direction as extracted from scattering patterns of a PEE as a function of elongation ϵ.

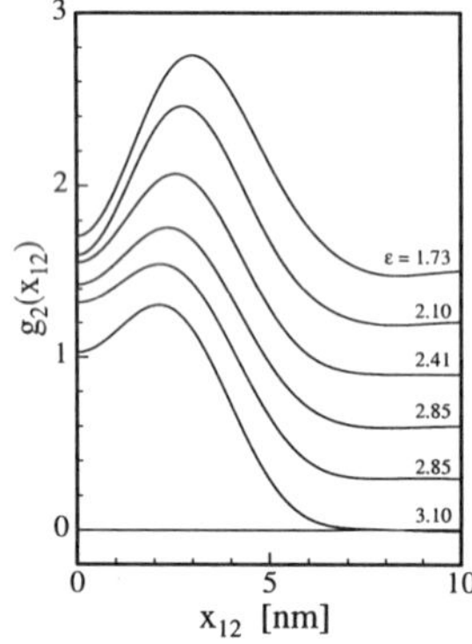

Figure 4: 2D chord distributions $g_2\,(x_{12})$ of a PEE as a function of elongation ϵ as computed from the projected equatorial scattering $\{I\}_2\,(s_{12})$.

In general $g_2\,(x_{12})$ shows the distribution of chords from needle and matrix cross sections and their correlations in the plane normal to straining direction. Experimentally only a single positive peak from uncorrelated disks is observed, indicating that in physical space the correlations among the soft domain cross sections are negligible.

Theoretical treatment

Neglecting the weak correlations among the soft domain cross sections the 2D chord distribution becomes

$$g_2\,(x_{12}) = h_D\,(x_{12}) \odot g_c\,(x_{12}),\qquad(2)$$

where $h_D\,(x_{12})$ is the studied needle diameter distribution, $g_c\,(x_{12})$ is the chord distribution of the unit circle, and $\odot$ designates the Mellin convolution:

$$g_2\,(x) = \int_0^\infty h_D\,(y)\, g_c\left(\frac{x}{y}\right)\frac{dy}{y}.\qquad(3)$$

$g_c\,(x_{12})$ is analytical. Thus the needle diameter distribution can be computed from the chord distribution. Interesting parameters are the average needle diameter, $\bar{D}$, the relative width $\sigma_D/\bar{D}$ of the needle diameter distribution and

the total needle cross section with respect to the cross section of the fiber as a function of elongation. Since all these parameters are closely related to moments of $h_D\,(D)$, and because of the fact that under Mellin convolution moments are transformed in a very simple manner [6], structural parameters can be retrieved directly from the chord distribution by means of moment arithmetics.

Final result

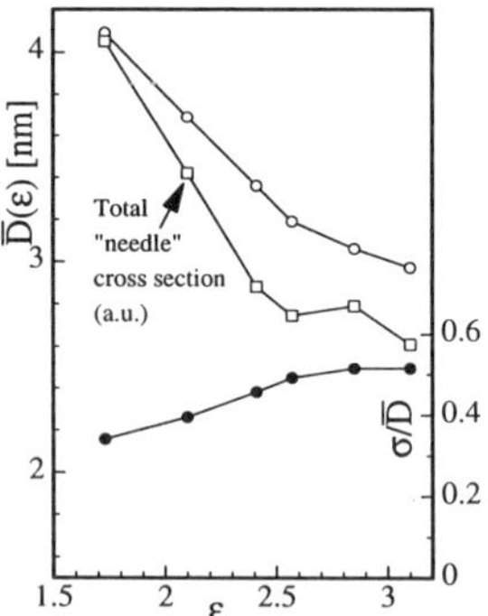

Figure 5: Structural parameters of the ensemble of soft domain needles in a PEE as a function of elongation as obtained from $g_2\,(x_{12})$ by moment arithmetics.

For the studied PEE and as a function of elongation ϵ Fig. 5 shows the average needle diameter, $\bar{D}$, the total needle cross section with respect to the sample cross section and the relative width $\sigma_D/\bar{D}$ of the neeedle diameter distribution. Two regions can be observed in the diagram. For elongations $\epsilon < 2.6$ the compressibility of the needles is higher than that of the surrounding matrix. Saturation effects for $\epsilon > 2.6$ indicate a hardening of the soft needles, which might be correlated to strain induced crystallization of the PTHF.

Acknowledgements

This study has been supported by HASYLAB Hamburg under project I–97–06 and by the University of Hamburg in the frame of the Partnership Contract between the University of Hamburg and Sofia University, Bulgaria. Material has kindly been supplied by Prof. Kricheldorf, University of Hamburg and DSM Corp., The Netherlands.

References

[1] Stribeck, N.; Sapoundjieva, D.; Denchev, Z.; Apostolov, A.A.; Zachmann, H.G.; Stamm, M; Fakirov, S. (1997) *Macromolecules* **30**, 1329

[2] Bonart, R. (1966) *Colloid Polym. Sci.,* **211**,14

[3] Méring J.; Tchoubar D. (1968) *J. Appl. Cryst.,* **1**, 153

[4] Peterlin, A. (1972) *Text. Res. J.,* **42**, 20

[5] Ruland, W. (1970) *J. Appl. Cryst.,* **4**, 70

[6] Stribeck, N. (1993) *Colloid Polym. Sci.,* **271**, 1007

[7] Link to programs: *http://www.chemie.uni–hamburg.de/tmc/stribeck/*

Scattering from magnetically oriented biopolymers

Wim Bras[1], Fernando Díaz[2], Eric van der Zee[3], Greg Diakun[4], Richard Denny[4]
[1] Netherlands Organisation for Scientific research
DUBBLE CRG/ESRF BP 220 F38043 Grenoble Cedex France
[2] KU Leuven, Laboratory for Chemical and Biological Dynamics, Celestijnenlaan
200 D, Leuven, Belgium
[3] Utrecht University Molecular Biophysics PO Box 20000 Utrecht Netherlands
[4] CCLRC Daresbury Laboratory Warrington WA4 4AD United Kingdom

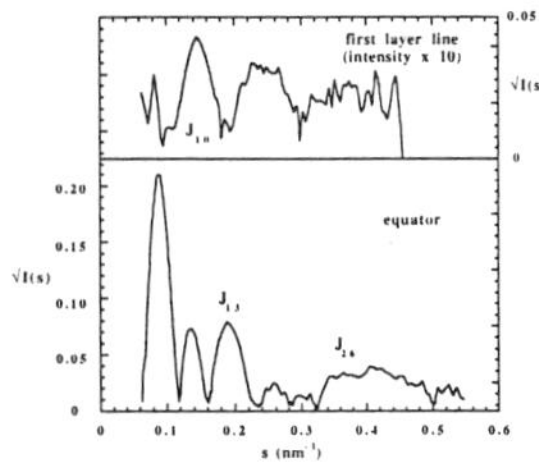

Figure 2
Equatorial and first layer line intensities of the microtubule fibre diffraction pattern. (s=1/d) on the horizontal axis. The vertical intensity axis is in arbitrary intensity units.

Microtubules play several important roles in the cell. They are involved in intra-cellular transport, motility, mitosis and form an important part of the cytoskeleton [1]. They consist of long chains of tubulin dimers, called protofilaments, which connect laterally to form a hollow cylindrical structure with a diameter of approximately 30 nm and a length which can extend to micrometers. The number of protofilaments can vary between 11 and 17 but the pre-dominant number in a normal functioning cell is 13. The persistence length of these polymers is reported to be between 5200- 2000 μm so that for most experiments they can be considered to be a rigid rod system [2].

The preparation of hydrated, and therefore biologically interesting, samples suitable for small angle X-ray fibre diffraction is not trivial. The only reasonable successful method has been centrifugation over extended lengths of time (>24 hours) and subsequent rehydration [3]. With this method one is not guaranteed that the brute force approach has not changed the structure. With other methods like Couette and flow shearing it is possible to create small oriented domains in an otherwise isotropic sample. These domains, however, do not retain their orientation very long and even small temperature gradients manage to distort the system [4].

We have investigated the possibility to orient microtubules making use of the interaction between their small diamagnetic moments and high magnetic field. For some polymers this interaction is sufficient to create fully aligned samples, notably for rigid polymers with a low molecular weight containing aromatic groups which are rigidly bound to the backbone. However, this is an exception and in general this method is not usable due to a combination of too low a diamagnetic moment and steric hindering which prevents the alignment. For self assembling biopolymers like actin and fibrinogen it has been shown that the assembly inside the magnetic field will result in the production of nearly fully aligned samples. This method can also be applied to synthetic polymers in which case it becomes possible to study the aligned samples in solution but which can also be thought of as a possible method to create highly aligned solid samples by first assembling the molecules and subsequently evaporating the solvent. With the relatively low prices for modern strong superconducting magnets this method, pioneered several decades ago, might nowadays be a feasible method for creating highly aligned speciality polymers, like for instance electroluminescent ones, which are difficult to align in solution, on an industrial scale.

From real time magnetic birefringence experiments it was found that the reaction rate is an important parameter for the achievement of full alignment. Contrary to intuition, which predicts that the best alignment will be obtained when the reaction rates are slow so that the molecules have time to align in the field we have found that a fast reaction is the best method. This can be understood when one looks at the number of nucleation sites for elongation in the solution. When this number is high, which is generally the case for fast reactions, the overall polymer length will be relatively short. However, these polymers can be sufficiently large to overcome the potential barrier for alignment and be the aligned 'seeds' from which further polymerisation can take place. The elongated polymers will be locked in their orientation due to steric hindering effects. For polymers with a long persistence length it is even possible that the field can be switched off and that they still retain this orientation even in solutions or in some cases that the orientation even increases with time [5]. This has been previously observed with other self assembling macromolecules [6]. This property has enabled us to perform fibre diffraction studies on these molecules by aligning them off-line and then transport them to a Synchrotron Radiation beamline.

From the fibre diffraction analysis it was found that the distance between adjacent microtubules can be as much as 20 times their diameter and still retain their alignment. Due to the longitudinal staggering of the different protofilaments the surface of microtubules have shallow helical grooves running over them so that the diffraction pattern can be explained in terms of helical diffraction theory. In figure 2 the diffracted intensity on the equator and the first layer line is shown. To obtain these curves it was necessary to correct for the slight disorientation and several geometrical parameters. For this the CCP13 suite of software was used [7].

On the basis of these results it can be predicted that this method of sample preparation might be applied to suitable synthetic polymers as well. This can provide a way to study the molecular transform without interference due to inter-molecular interactions. With a fast reaction the sample only has to remain in the field for several seconds and thus provides the possibility to develop this method into a polymer processing technique.

[1] Structure of microtubules ed K.Roberts and J.S Hyams Academic Press
[2] F.Gittes, B.Mickey, J.Nettleton, J.Howard J.of Cell Biol., 1993, 120(4), 923-934 and P.Venier, A.C.Maggs, M.F.Carlier, D.Pantaloni J.Biol.Chem., 1994, 269(18), 13353-13360
[3] Mandelkow Methods in Enzymology Vol. 134, 1986, 149-168
[4] J.Nordh, J.Deinum, B.Norden Eur.Biophys.J., 14, 1986, 113-122 and R.E.Buxbaum, T.Dennerl, S.Weiss, S.R.Heidemann Science 235, 1987, 1511-1514
[5] W. Bras, G.P. Diakun, J.F. Díaz, G. Maret, H. Kramer, J. Bordas, F.J. Medrano Biophysical J. To appear March '98
[6] J.Torbet, M. Ronzière Biochem.J.219(1984) 1057-1059
[7] R.C. Denny CCLRC Daresbury laboratory Private communication

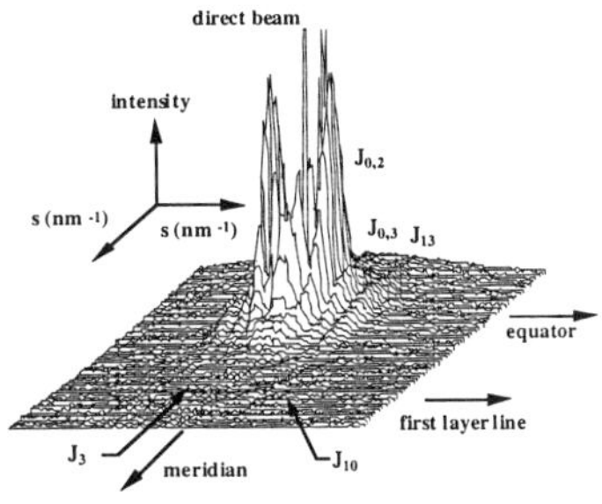

Figure 1
Three dimensional representation of a low angle X-ray fibre diffraction pattern. The pattern can be analysed in helical diffraction terms. The orders of the Bessel functions assigned to each peak are indicated.

Molecular Level Processes near Polymer Surfaces and Interfaces; Multi-Dimensional Spectroscopic Approaches

Marek W. Urban
Department of Polymers and Coatings
North Dakota State University
Fargo, ND 58105

Introduction

Behavior of macromolecules near surfaces and interfaces of polymeric thin films and coatings may play a vital role in numerous applications. Therefore, understanding of molecular level processes responsible for durability, adhesion, and many other macroscopic processes is of a particular importance. This paper will focus on stratification processes in multi-component polymeric films, with particular emphasis to polymer-surfactant interactions in latexes and responsiveness of individual components during coalescence of water-borne polyurethanes. The presence of macromolecular arrangements and interactions among various components near the film-air (F-A) and the film-substrate (F-S) interfaces can be effectively monitored using attenuated total reflectance (ATR) and step-scan photoacoustic (S^2-PA) Fourier transform infrared (FT-IR) spectroscopy.

Water-Borne Polyurethanes

Although single-component water-borne polyurethanes (PUs) have been available for over 30 years,[1] their properties would not meet their solvent-borne counterparts. Recently, however, properties such as excellent abrasion resistance, hardness, and chemical resistance have given two-component (2K) water-borne PUs a reputation of high performance coatings. Although a major portion of the current research concerning PUs is concentrated in customizing specific water-borne technologies for specific applications,[2-5] there are studies that have focused on understanding the mechanism of water-borne PU film formation.[6-8] However, there are still many issues that need to be understood about the crosslinking reactions leading to film formation as well as the kinetics of the crosslinking process. Of particular importance are those reactions that occur near surfaces and interfaces because they do affect numerous film properties. Let us examine the NCO concentration in water-borne urethanes at a given reaction time as function of depth from the F-S and F-A interfaces. Figure 1 summarizes the results of the depth profiling experiments conducted at the F-A and F-S interfaces as a function of RH after approximately 7-8 hours of reaction. A comparison of NCO concentrations at the F-A and F-S interfaces at each penetration depth indicates that the isocyanate concentration is lowest near 0.65 μm for both interfaces for all RH's. However, for penetration depths from 0.92 to 1.14 μm, the isocyanate concentration increases and, at the same time, the difference in isocyanate concentration between the F-A and F-S decreases. It appears that the range of isocyanate concentration between 0.65 and 1.14 μm also depends on the RH values. At 5-10% RH, the range of isocyanate concentration between 0.65 and 1.14 μm is 0.7 x 10^{-6} M, while at 90-95% RH is approximately 2.5 x 10^{-6} M. These observations suggest that water is unable evaporate from the PU film when a high partial water vapor pressure exists above the surface, thus allowing penetration into the film F-A interface, where it is being consumed by reacting with NCO groups (Scheme II). Subsequently, less water is available further away from the F-A interface, and as a result, an increase of NCO concentration with depth at the F-A interface is observed.

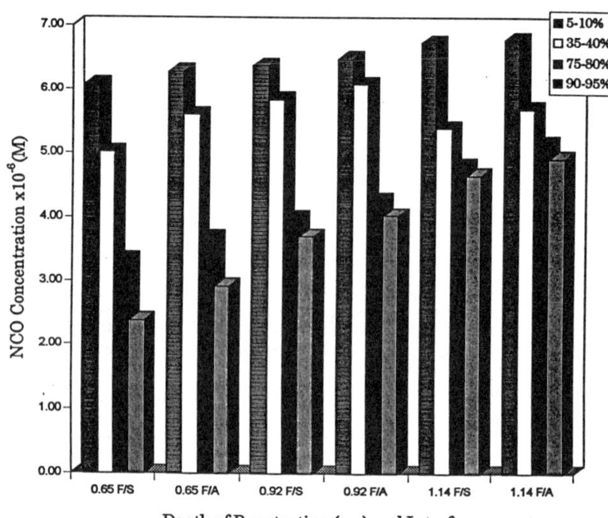

Figure 1. Consumption of NCO groups in water-borne polyurethanes after 7-8 hours of reaction at the F-A and F-S interface plotted as a function of penetration depth.

It is appropriate to compare the isocyanate crosslinking reactions between water-borne and solvent-borne systems. In this context, it is important to recognize that the film formation processes of these acrylic-based systems is distinctly different. As a result of the increased likelihood of reactions between NCO and H_2O, isocyanate crosslinking reactions of water-borne urethanes occur over a much shorter time frame. Both solvent-borne and water-borne systems exhibit an increase of isocyanate consumption at higher relative humidities. In solvent-borne PUs, however, the isocyanate concentration is greater at the F-A than at the F-S interface at all relative humidities. In contrast, the isocyanate concentrations at the F-A and F-S interfaces of water-borne PUs become almost equivalent at higher relative humidities. This is a result of atmospheric moisture that will suppress water evaporation, thus allowing reactions with isocyanate to produce urea and CO_2. Although relative humidity also affects the evaporation rate of organic solvent from solvent-borne urethanes, it will not affect crosslinking reactions at low RH because solvent cannot react with NCO functionalities. More residual solvent present at the F-S than at the F-A interface of solvent-borne PUs, will increase the mobility of the reactive NCO groups at that interface, and the primary reactions leading to the network formation are the reactions between isocyanate hydroxyl functionalities of the acrylic. The NCO concentration in the solvent-borne PUs decreases from 0.65 to 1.14 μm into the F-S interface at reaction times not exceeding 24 hrs, after which the concentration becomes the same at all penetration depths.[9] However, in water-borne PUs, the isocyanate concentration increases from 0.65 to 1.14 μm penetration depths for all reaction times and strongly depends on RH. The consumption of NCO at the F-A and F-S interfaces distinctly increases at higher humidities and at shallower penetration depths. Thus, the presence of atmospheric moisture appears to have a significant influence on coalescence at the F-A and F-S interfaces. For illustration purposes, Table I provides a comparison of the trends observed for water-borne and solvent-borne polyurethane systems and how various variables affect crosslinking.

Table I. A Comparison of Trends Observed for Water-borne and
Solvent-borne Polyurethanes.

	WATER-BORNE POLYURETHANE	SOLVENT-BORNE POLYURETHANE
Time	Crosslinking times < 7 days	Crosslinking times > 7 days
Solvent Effect	Possible formation of urethane, urea, and CO_2	Formation of urethane [NCO] decreases at higher RHs
Relative Humidity (RH) Effect	[NCO] decreases at higher RHs [NCO] at F-A and F-S are the same at higher RHs	[NCO] at F-A are greater tha at F-S at all RHs [NCO] decreases from 0.65 and 1.14 µm for < 24 hrs rxn time
Depth of Penetration Effect	[NCO] increases from 0.65 to 1.14 µm for all rxn times	[NCO] are equivalent at all depths after 24 hrs rxn time

[NCO]: Isocyaate concentration
F-A : film - air, F-S : film - substrate

Stratification Processes in Latex Films

Although several attempts to explain the role and behavior of surfactants in latex films have been made, none of them provided satisfactory understanding of the surfactant behavior. Okuto[10] also showed that ethyl acrylate/methyl methacrylate copolymer latex films decreased in permeability as the sodium dodecyl benzene sulfonate (SDBS) concentration was increased. The reduction of permeability of latex films may be a result of morphological changes in the film because the addition of a hydrophilic surfactant is expected to increase permeability. The added surfactant could play two roles during the film formation: first, it would give the particle greater stability, and hence the particles would be in a more ordered state preventing coagulation; and second, the surfactant could plasticize the particle surface, resulting in a greater degree of coalescence during the film formation.

Evanson and Urban,[11-13] Evanson et al.,[14] and Thorstenson and Urban[15-18] showed that the exudation of surfactants is also influenced by other factors. For example, in ethyl acrylate/methacrylic (EA/MAA) acid latex, surfactant molecules tend to exude toward the film substrate (F-S) interface when coalescence was conducted on a high surface tension substrate. This behavior was attributed to the surface tension difference between the substrate and the latex. Thorstenson et al.,[15-17] Kunkel and Urban,[19] Niu and Urban,[20,21] and Tebelius and Urban[22] also demonstrated that the polar sulfonate groups of the sodium dioctylsulfonsuccinate (SDOSS) surfactant can interact with the carboxylic acid functionality of the copolymer and these interactions are affected by the residual water present in the film after coalescence.

Orientation and spatial distribution of acid functionalities[23] is such that dimeric carboxylic acid species can assemble at the interface between latex films and a substrate. Dimeric acid entities were found to be preferentially parallel to the interface, and "aging" of the latex aqueous suspension causes excessive exudation of surfactant molecules to the film-air interface.[23] Upon exudation to the surface, the hydrophilic $SO_3^-Na^+$ groups assemble in preferentially normal-to-the-surface direction, whereas hydrophobic tails are randomly buried in the latex film. In the mixture of polystyrene and poly(n-butyl acrylate) latexes, more surfactant molecules exude to the film-air interface and less to the film-substrate interface.

In an effort to establish to what extent transient effects prior to coalescence may influence mobility, distribution, and orientation of surfactant molecules after coalescence, polystyrene (Sty) and poly(n-butyl acrylate) (n-BA) were mixed independently in a 1:5 ratio, and the mixture was allowed to stabilize prior to coalescence.[19] It appears that the latex stabilization may have a significant effect on the distribution of surfactant molecules across the coalesced latex and that their orientation changes depend upon the stage of coalescence and the water front's moving toward the surface. In these studies, it also became apparent that

vibrational spectroscopy, in particular, attenuated total reflection (ATR) and step-scan photoacoustic (S^2-PAS) Fourier transform infrared (FT-IR) spectroscopy, is a powerful tool in determining molecular level features occurring near interfacial regions in latex films.

Conclusions

The presence of macromolecular arrangements and interactions among various components near the film-air (F-A) and the film-substrate (F-S) interfaces can be effectively monitored using attenuated total reflectance (ATR) and step-scan photoacoustic (S^2-PA) Fourier transform infrared (FT-IR) spectroscopy. Both approaches are capable of obtaining information from various surface depths and complement each other if one seeks molecular level information from $0 - 150$ µm into the film. If one combines ATR and PA information with IR and/or Raman surface imaging, it is possible to obtain a 3-dimensional representation of polymeric films.

Acknowlegments:

The author is thankful to the NSF Industry/University Cooperative Research Center in Coatings at North Dakota State University and the NSF EPSCoR program (Grant No. ESR 9553367) for partial support of these studies.

References

1. Rosthauser, J.W.; Nachtkamp, K., "Waterborne Polyurethanes", K.C. Frisch; D.Klempner, Eds. Adv. Urethane Technol., Vol. 10, 1987, 121.
2. Kahl,L.; Bock,M., "Waterborne 2-Component Polyurethane Clear Coats for Automotive Coatings: Development of Raw Materials and Mixing Technology", Proceedings of the 3rd Nurnburg Congress, March 13-15, 1995, Nurnburg, Germany.
3. Jacobs,P.B.; McClurg,D.C., "Water-Reducible Polyurethane Coat-ings for Aerospace Applications", Proceedings of the Low and No VOC Coating EPA Conference, May 25-27, 1993, San Diego, CA.
4. Bittner,A.; Ziegler,P. "Waterborne Two-Pack Polyurethane Coatings for Industrial Applications", Proceedings of the 3rd Nurnburg Congress, March 13-15, 1995, Nurnburg, Germany.
5. Renk,C.A.; Swartz,A.J., "Fast Drying, Ultra Low VOC, Two-Component Waterborne PolyurethaneCoatings for Wood Industry," Proceedings of the Water-borne, High-Solids, and Powder Coatings Symposium, February 22-24, 1995, New Orleans, LA.
6. Satguru, R.; Mcmahon, J.; Padget, J.C.; and Coogan, R.G., J. Coat. Tech., 1994, 66, 47.
7. Rynders, R.M.; Hegedus, C.R.; Gilicinski, A.G., J. Coat. Tech, 1995, 67, 59.
8. Hegedus,C.R.; Gilicinski,A.G.; Haney,R.J., J. Coat. Tech, 1996, 68, 51.
9. Kaminski, A.M.; Urban, M.W., J. Coatings Tech., Vol. 69, No. 872, 55, 1997.
10. Okubo, M.; Takeya, T.; Tsutumz, Y.; Kapookat, Y.; Matsumoto, T. J. Polym. Sci., Polym. Chem. Ed. 1981, 19, 1.
11. Urban, M. W.; Evanson, K. W. Polymer Comm. 1990, 31, 279.
12. Evanson, K. W.; Urban, M. W. J. Appl. Polym. Sci. 1991, 42, 2287.
13. Evanson, K. W.; Thorstenson, T. A.; Urban, M. W. J. Appl. Polym. Sci. 1991, 42, 2297.
14. Evanson, K. W.; Urban, M. W. J. Appl. Polym. Sci. 1991, 42, 2309.
15. Thorstenson, T. A.; Urban, M. W. J. Appl. Polym. Sci. 1993, 47, 1381.
16. Thorstenson, T. A.; Urban, M. W. J. Appl. Polym. Sci. 1993, 47, 1387.
17. Thorstenson, T. A.; Tebelius, L. K.; Urban, M. W. J. Appl. Polym. Sci. 1993, 49, 103.
18. Thorstenson, T. A.; Tebelius, L. K.; Urban, M. W. J. Appl. Polym. Sci. 1993, 50, 1207.
19. Kunkel, J. P.; Urban, M. W. J. Appl. Polym. Sci. 1993, 50, 1217.
20. Niu, B.-J.; Urban, M. W. J. Appl. Polym. Sci. 1995, 56, 377.
21. Niu, B.-J.; Urban, M. W. J. Appl. Polym. Sci. 1996, 60, 371.
22. Tebelius, L. K.; Urban, M. W. J. Appl. Polym. Sci. 1995, 56, 387.

Film Properties of Polyisocyanate Based Coatings with Mixed Amine Coreactants

Kurt Best and Douglas Wicks
PPD Coatings, Bayer Corporation, Pittsburgh, PA 15205-9741

There has been increasing interest in using modified amine functional coreactants with aliphatic polyisocyanates for formation of high solids coatings. Polyaspartic acid esters are an established family of hindered amine coreactants and aldimines are a family of newer products just seeing commercial application. These materials have been studied as standalone coreactants (1) or used in combination with the traditional hydroxy functional acrylics or polyesters as reactive diluents (2).

This paper presents some preliminary results of using these two types of amine functional coreactants in combination with each other. The systems described here we pigmented systems with non-volatile contents of >97%. These formulations would be applicable to ambient cure maintenance and flooring applications. Given the very different reaction mechanisms and products when used with polyisocyanates of the two materials, it is anticipated that interesting combinations of properties will be obtained.

Properties of Amine Functional Coreactants

Polyaspartic Acid Esters - These hindered amines are obtained by the reaction of primary diamines with two equivalents of an alkyl maleate to form secondary diamines. This secondary amine compound, a polyaspartic acid ester, has reduced reactivity due to steric effects and hydrogen bonding of the amine. (3) Like the parent amines, polyaspartic acid esters react with isocyanates to form ureas.

Aldimines - Aldimines are produced by the reaction of aliphatic diamines with aldehydes. These products are desirable for use in urethane coatings because of their extremely low viscosity. The reaction of these products with polyisocyantes is not very straight forward, with hydrolysis to the parent primary amine being favored at low temperatures. At temperatures above 80 C the direct reaction between the isocyanate functional and aldimine functional groups to form ene-urea is favored (4).

Polyaspartic Acid Ester   *Aldimine*

Polyisocyanurate of Hexamethylene Diisocyanate

Experimental

Materials - The properties of the polyaspartic acid esters used are shown in Table 1. All were produced by the reaction of the parent diamine with 2 equivalents of the alkyl maleate listed in Table 1.

Table 1. Polyaspartic acid esters

	Diamine	Maleate	Viscosity (25 C)	Equivalent Weight
ASP1	HDA*	Dibutyl	120 mPa-s	282
ASP2	H₁₂MDA**	Diethyl	1100 mPa-s	277
ASP3	3,3' Dimethyl – H₁₂ MDA	Diethyl	1500 mPa-s	292

* Hexamethylene Diamine **Hydrogenated 4,4' methane diphenyl amine

The bis-aldimine (**ALD1**) used in this study is reaction product of isophorone diamine and 2 equivalents of isobutyraldehyde. **ALD1** has an amine equivalent weight of 169 and a viscosity of 40 mPa-s (25 C).

The polyisocyanate (**PIC1**) is a monomer free polyisocyanurate of hexamethylene diisocyanate. **PIC1** has an equivalent weight of 204, a viscosity of 1100 mPa-s (25 C) and a functionality of approximately 3.2.

Sample Preparation - Enough TiO2 pigment (DuPont R-960) to yield a P/B of 0.7, was dispersed into the 50 parts of polyaspartic acid esters using 4.5% Byk 110 (as supplied) on pigment solids. No attempt was made to dry the pigment before dispersion. The pigment dispersion was letdown with 50 parts of **ALD1** to give a 50:50 mixture by weight of Aldimine vs. polyaspartic acid ester. The amine coreactant compositions of the formulations are given in Table 2. The viscosity of the pigment coreactant side was between 4500 and 6000 mPa-s.

To the mixture was added the polyisocyanate on a one to one equivalent basis (NCO/NH). Films were prepared using a 25-mil drawdown on a TPO panel. All formulations exhibited very short potlives (<10 minutes) and drytimes of less than 30 minutes. The high reactivity was due to the use of "wet" pigments in the formulation. After curing for 2 weeks at ambient conditions the films were tested for Shore D Hardness, Tg (DSC, 10 C/min) and tensile properties.

Table 2. Composition and properties of test formulations

Formulation	ASP1	ASP2	ASP3	Tg	%Elongation	Tensile Strength (PSI)
B1	50	0	0	19		
B2	0	50	0	30	1	6475
B3	0	0	50	48	5	7661
B4	16.7	16.7	16.7	24	27	1861
B5	25	0	25	25	28	1430
B6	25	25	0	31	32	2425
B7	0	25	25	32	7	2028
B8	37.5	6.2	6.2	26	63	780
B9	6.2	37.5	6.2	33	32	1687
B10	6.2	6.2	37.5	30	54	854

Results and Discussion

It is readily apparent from the data presented in Table 2 that all systems that contained ASP1, derived from HDA, showed the highest tensile elongation values and lowest tensile strengths. This is in keeping with the very soft nature of the product. It is suprising that even the inclusion of a very small amount of ASP1 in formulation B9 and B10 greatly affects the properties. Ongoing work is focused on discerning the role of the total effect of including "softer" polyaspartic acid esters into rigid aldimine systems.

References

(1) Wicks, D. A.; Yeske P. E.; *Proceedings of the Twentieth Waterborne & Higher Solids and Powder Coatings Symposium* **1993**, 49.

(2) Jorissen, S. A.; Rumer, R. W.; Wicks, D. A.; *Proceedings of the Nineteenth Water-borne, Higher Solids and Powder Coatings Symposium* **1992**, 182.

(3) Zwiener, Ch.; Schmalstieg, L.; Sonntag, M.; Nachtkamp, K.; Pedain, J., B; *Farbe und Lack* **1991**, Nr. 12, S. 1052, Zwiener, Ch.; Schmalstieg, L.; Sonntag, M.; *European Coatings Journal* **1992** (10), 588

(4) Wicks, D.A.; Yeske, P.E.; "Imine-Isocyanate Chemistry: New Technology for Environmentally Friendly High Solids Coatings", ACS Symposium Series 640, DeVito and Garrett eds., 1996

Catalysis of Blocked Isocyanates with Non-Tin Catalysts

Werner J. Blank, Z. A He, Marie E. Picci

King Industries Inc.

Norwalk, CT 06852

USA

wblank@kingindustries.com

Introduction

The use of "blocked" polyisocyanates has many advantages in the coating industry. It permits the formulation of stable one package coatings which on heating deblock and lead to the formation of the highly reactive polyisocyanate. Applications for these systems are in diversified areas such as powder coatings, electrocoating, wire coatings and in textile finishing. The nature of the blocking agent[1] has a significant effect on the deblocking temperature of the isocyanate. Typical blocking agents used include malonates, triazoles, ε-caprolactam, sulfite, phenols, ketoxime, pyrazoles and alcohols. For many blocked isocyanates an elimination-addition mechanism[2] as indicated in Eq. (1) is proposed.

$$RNHCOO\text{-}Bl \underset{k_{-1}}{\overset{k_1}{\leftrightarrow}} RNCO + Bl\text{-}H \uparrow \quad EQ. (1)$$

$$RNCO + RNH_2 (ROH) \overset{k_2}{\to} RNHCO\text{-}NHR (OR)$$

This traditional view of blocked isocyanates as an isocyanate precursor has already been challenged by Wicks[3,4,5]. He showed that for malonate blocked aliphatic isocyanates transesterification is the predominant reaction step. For alcohol blocked isocyanates the decomposition to the isocyanate is a rather high temperature reaction step. In the presence of a catalyst, alcohols are effective blocking groups for isocyanates. Transesterification reaction according to EQ. 2 is the most likely reaction mechanism. The overall reaction rate can depend on the volatility of the alcohol[6].

$$RNHCOOR' + R''OH \underset{k_{-1}}{\overset{k_1}{\leftrightarrow}} RNHCOOR'' + R'OH \uparrow \quad Eq. (2)$$

For both the elimination-addition and the displacement mechanism, dibutyltin dilaurate (DBTDL) has been found to be an effective catalyst. Depending on the alcohol used non-tin metal compounds[7] and chelates were also found to be effective catalysts. In this study we were interested to determine if other metal catalysts can be used to catalyze the reaction of ketoxime, alcohol, pyrazol and caprolactam blocked isocyanates.

Experimental

The blocked isocyanates were prepared from the polyisocyanate and the blocking agent. An oxime blocked isocyanate was synthesized from the reaction of 2-butanone oxime and a commercially available HDI trimer[8] at NOH/NCO ratio of 1.05. FT-IR spectrum of the product showed no residual isocyanate. The product has a theoretical equivalent weight of 270 and was diluted with 2-methoxypropyl acetate to a 92 % solids content for use.

A pyrazole blocked isocyanates was obtained by reacting the above HDI trimer with 3,5-dimethylpyrazole at NH/NCO ratio of 1.0. After over 90% conversion (checked by FT-IR), the residual isocyanate groups were capped with alcohol. The final product had a theoretical equivalent weight of 279 and was diluted with 2-methoxypropyl acetate/t-butanol to 85 % solids for use.

A caprolactam blocked isocyanate was synthesized by reacting ε-caprolactam with IPDI trimer[9] at CONH/NCO ratio of 1.05. After over 90% conversion, the residual isocyanate groups were capped with isopropanol. The final product had a theoretical equivalent weight of 356 and was diluted with 2-methoxypropyl acetate/xylene to a 67% nonvolatile content with an equivalent weight of 532.

A glycolether blocked isocyanate was prepared by reacting 2-methoxylethoxyethanol with a polymeric aromatic isocyanate[10] (MDI) at 1.02 OH/NCO ratio. The final product has a nonvolatile content of over 99 % and equivalent weight of 254. The blocked isocyanates were blended with a commercially available acrylic resin[11] at a hydroxyl/blocked NCO ratio of 1/1. The catalyst levels shown in the result section are based on total resin solids, i.e., both acrylic and blocked isocyanate resin. Films were prepared by drawdown of the formulations on iron phosphate pretreated steel panels and baked in a forced draft air oven.

Results

Solvent resistance as measured by methyl ethyl ketone rubs (MEK) was found to be a simple and fast technique for determining crosslinking and was used for the screening studies.

Table 1: Catalyst Screening in Oxime blocked HDI-Trimer
Substrate Bonderite B1000, Dry film thickness: 22 μ (0.85 mils), cure 20minutes at 130°C

Catalyst	Metal % on solids	Film Appearance	MEK 2x rubs
No catalyst	0	Clear	0
Dibutyltin dilaurate	0.05	Clear	43
Dibutyltin dilaurate	0.09	Clear	46
Dibutyltin diacetate	0.18	Clear	38
Bismuth tris(2-ethylhexanoate)[12]	0.065	Clear	51
Aluminum dionate complex[13]	0.04	Clear	19
Cobalt bis(2-ethyl hexanoate)	0.05	Hazy	51
Zr bis(2-ethyl hexanoate)	0.18	Clear	12
Zn bis(2-ethyl hexanoate)	0.048	Clear	35
Ti tetra(ethyl acetoacetato)	0.50	Yellow	52
Calcium bis(2-ethyl hexanoate)	0.05	Clear	24
Chromium tris(2-ethyl hexanoate)	0.255	Hazy	43

Table 2: Catalyst Screening in Pyrazole Blocked HDI-Trimer
Substrate Bonderite 1000, Dry film thickness: 22 μ (0.85 mils), cure 20minutes at 130°C

Catalyst	Metal % on solids	MEK 2x rubs
No Catalyst		23
Dibutyltin dilaurate	0.09	79
Bismuth tris(2-ethylhexanoate)	0.04	50
Bismuth tris(2-ethylhexanoate)	0.065	61
Zr bis(2-ethyl hexanoate)	0.09	27
Zn bis(2-ethyl hexanoate)	0.09	34
Cobalt bis(2-ethyl hexanoate)	0.10	57*
Chromium tris(2-ethyl hexanoate)	0.51	43
Ti tetra(ethyl acetoacetato)		46
Calcium bis(2-ethyl hexanoate)	0.10	29
Bismuth tris(2-ethylhexanoate)	0.04	50

* = Some Wrinkling

Table 3: Catalyst Screening in Caprolactam Blocked IPDI Trimer

Substrate Bonderite 1000, Dry film thickness: 22 µ (0.85 mils).

Catalysts Cure schedule, 20 min at	% Metal on solids	MEK 2x rubs 155 ^{0} C	MEK 2x rubs 165 ^{0}C
No catalyst	0.00	10	15
Dibutyltin dilaurate	0.10	77	200
Dibutyltin diacetate	0.18	98	200
Bismuth tris(2-ethylhexanoate)	0.10	58	200
Zn bis(2-ethyl hexanoate)	0.18	115	200
Cobalt bis(2-ethyl hexanoate)	0.30	73	200
Ti tetra(ethyl acetoacetato)	0.22		117

Table 4: Catalyst Screening in Carbitol Blocked MDI

Substrate Bonderite B1000,

Dry film thickness: 0.85 mils :0.18% metal catalyst on total resin solids

Catalyst Cure schedule 20 minutes at	MEK 2x rubs 150° C	MEK 2x rubs 170° C
No catalyst	2	2
Dibutyltin dilaurate	10	130
Bismuth tris(2-ethylhexanoate)	120	200
Zn bis(2-ethyl hexanoate)	24	200
Aluminum dionate complex	20	
Ti tetra(ethyl acetoacetato	13	
Cobalt bis(2-ethyl hexanoate)	100	

Discussion and Conclusions

Of the many metal compounds and chelates screened in this study only bismuth carboxylates show consistent catalysis in ketoxime, pyrazol, caprolactam and glycol ether blocked isocyanates crosslinked coatings. On a metal basis bismuth is at least as effective as dibutyltin compounds. At this stage of our investigation it is not clear how bismuth catalyzes the reaction of blocked isocyanates with hydroxyl compounds.

Acknowledgment

I would like to thank Mr. Brant Berndlmaier for conducting some of the experiments in this study and Dr. Len Calbo for proof reading this paper. In addition we would like to thank King Industries Inc. for the permission to publish this paper.

References

S. F. Thames, P. S. Boyer, An Investigation of the Effects of Polymer Structure upon the Dissociation Temperature of Blocked Isocyanates. J. Coat. Techn. Vol. 62, No. 784, May 1990 pg. 51-63.

[2] S.P. Pappas, E.H.Urruti, 13th Annual Water-Borne and Higher Solids Coatings Symposium, New Orleans, LA 1986

[3] Z.W.Wicks Jr., Prog. Org. Coat., 3 (1975); 73

[4] Z.W.Wicks Jr., Prog. Org. Coat., 9 (1981); 3

[5] Z.W. Wicks Jr. and B.W. Kosty, J. Coat. Technol.,49 (634) (1977); 77

[6] Huang, Yun; Chu, Guobei; Nieh, Marjorie; Jones, Frank N. "Aliphatic isocyanates blocked with volatile alcohols for decorative coatings." J. Coat. Technol., 67(842), 33-40 (English) 1995.

[7] W.J. Blank "Crosslinking with Polyurethanes." Polym. Mater. Sci. Eng., 63 931 (1990)

[8] HDI-trimer hexamethylene diisocyanate isocyanurate trimer, Desmodur N-3300; isocyanate equivalent weight 183, Bayer Corporation, Pittsburgh PA.

[9] Isophorone diisocyanate isocyanurate trimer commercially available as Desmodur Z-4370 from Bayer Corporation, Pittsburgh PA.

[10] Polymeric aromatic isocyanate with an equivalent weight of 130 Modur MRS 5 from Bayer Corporation, Pittsburgh PA.

[11] Acrylic polyol with a nonvolatile content of 70 % and a hydroxy equivalent weight on a solids basis of 460; Paraloid AU-946 B available from Rohm and Haas Company, Philadelphia PA.

[12] K-KAT® 348 bismuth catalyst a product of King Industries Inc. Norwalk, CT 06852

[13] K-KAT® XC-5218 aluminum chelate catalyst a product of King Industries Inc. Norwalk, CT 06852

CURE OF SECONDARY CARBAMATE GROUPS BY MF RESINS - WATER BORNE SYSTEMS

H.P. Higginbottom, G.R. Bowers, P.E. Ferrell, L.W. Hill
Solutia Inc.
730 Worchester Street
Springfield, MA 01151

INTRODUCTION

Melamine-formaldehyde (MF) resins have been used extensively to cross-link hydroxyl-functional coreactants for over sixty years[1]. Other groups that are known to react with MF resins include carboxylic acids, amines and amides[1]. The reaction of MF resins with urethane groups (i.e. secondary carbamate groups) has been overlooked, neglected or not emphasized sufficiently in the published chemical literature[2]. In the recent proceedings of the 25th International Waterborne, High-Solids, & Powder Coatings Symposium we presented a paper emphasizing the significant reactivity of secondary carbamate model coreactants with MF resins under controlled cure conditions[3]. This research was extended to the work reported in this paper to better understand the variables involved in reacting water dispersible secondary carbamate model compounds with MF resins. Cured film properties were measured on films baked at different temperatures. The results were compared with data obtained on solvent based secondary carbamates[3]. Methodology developed by Loren Hill, for measuring crosslink density by DMA, was used to extend our understanding of the cure behavior of these systems beyond what is possible by routine coating tests. Data showing possible property enhancements that might be gained by crosslinking a commercial or experimental urethane dispersion with an MF resin is presented.

CHEMISTRY

MF Reaction With Secondary Carbamates - The reaction of the secondary carbamate >N-H compared with an -O-H functionality on a coreactant backbone is shown in Figure 1 when reacting with the methoxymethyl group of an MF resin. Our previous work showed that highly etherified HMMM resins with low imino content were the most reactive type of MF resins with a secondary carbamate group[3]. Also, secondary carbamates on chain ends, on branches extending from a backbone and as part of a backbone urethane were all quite similar in reactivity with the preferred MF resins under acid catalyzed conditions[3].

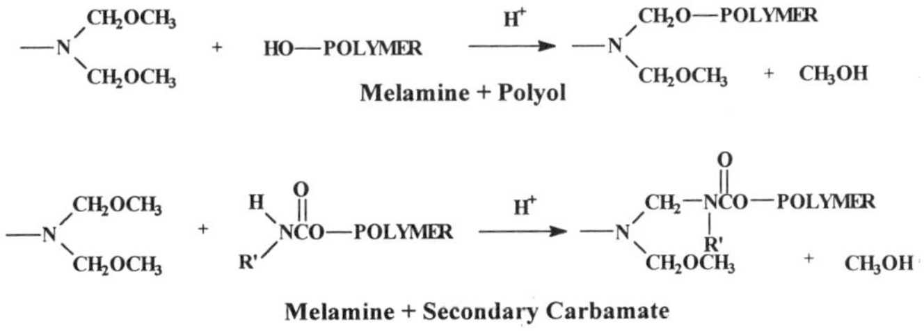

$$-N(CH_2OCH_3)_2 + HO-POLYMER \xrightarrow{H^+} -N(CH_2O-POLYMER)(CH_2OCH_3) + CH_3OH$$

Melamine + Polyol

$$-N(CH_2OCH_3)_2 + \underset{R'}{H}N-CO-POLYMER \xrightarrow{H^+} -N(CH_2-N(R')CO-POLYMER)(CH_2OCH_3) + CH_3OH$$

Melamine + Secondary Carbamate

Figure 1: Hydroxyl vs. Secondary Carbamate Reaction with MF Resin

Water Dispersible Model Secondary Carbamates - Our previous study with solvent based systems used model secondary carbamates free of any hydroxyls or other melamine reactive groups[3]. To make the models water dispersible, carboxyl functionality was introduced. Figure 2 illustrates the formation of a linear HDI oligomer backbone with the chain ends capped with hydroxyl containing carboxylic acids. With this water dispersible model, both the carboxyl groups and secondary carbamate groups are MF resin reactive. Capping with glycolic acid at the stoichiometry illustrated in Fig. 2, gives a urethane oligomer with a molecular weight of 1822, a theoretical secondary carbamate equiv. wt. of 152 gm/equiv. and an acid equiv. wt. Of 911 gm/eq. Other dispersible models were also synthesized.

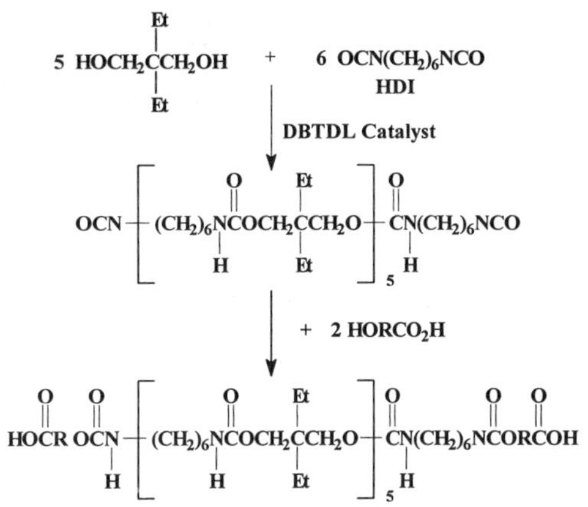

$$5\ HOCH_2\underset{Et}{\overset{Et}{C}}CH_2OH\ +\ 6\ OCN(CH_2)_6NCO\ (HDI) \xrightarrow{DBTDL\ Catalyst}$$

$$OCN\!-\!\left[(CH_2)_6NHCOCH_2\underset{Et}{\overset{Et}{C}}CH_2O\right]_5\!\!-\!CN(CH_2)_6NCO$$

$$+\ 2\ HORCO_2H$$

$$HOCRCOOCN\!-\!\left[(CH_2)_6NHCOCH_2\underset{Et}{\overset{Et}{C}}CH_2O\right]_5\!\!-\!CN(CH_2)_6NHCORCOH$$

When R=CH$_2$, Mn=1822 & Acid eq.wt. = 911 & Carbamate eq.wt. = 152

Figure 2: Formation of Dispersible HDI Oligomer

Coating Formulations - The model dispersible HDI oligomer, prepared as shown in Fig. 2, was formulated with a highly monomeric HMMM (low >N-H) resin identified as MF(I) in both a water borne (WB) and a solvent borne (SB) coating. The water based coating was prepared by 100% neutralization of the carboxyl groups of the HDI oligomer in butyl cellosolve solvent with an amine, adding MF(I) and codispersing these components with any additives by addition of water under high shear mixing. A stable, 30% dispersion showing a Tyndall effect was obtained in this way. DNNDSA neutralized with amine was added as a blocked dispersible catalyst. The WB formulation described below used DMEA as the neutralizing and blocking amine. The SB version of the oligomer in MEK was mixed with MF(I) and unblocked pTSA (0.5 phr) was added as catalyst. For the direct comparison of film properties both the WB and SB formulations used MF(I) / carbamate equiv. ratios of 0.5/1.0. At this equiv. ratio the solids ratio of MF/ carbamate oligomer was 18/82.

COATING CURE AND PERFORMANCE

Cure Response of Secondary Carbamates - The use of a gradient oven (BYK Chemie) to obtain paint film properties, as a function of different bake temperatures, has been described in detail[3]. Using this method, cure profile data was obtained comparing the WB with the SB formulation. Figure 3 shows the MEK solvent resistance development of the two formulations with increasing bake temperatures. There is a significant difference in the cure onset temperature as measured by MEK solvent resistance between the solvent borne (65°C) and water borne system (115°C). This is anticipated since the WB system involves the neutralizing amine equilibrium which retards the acid catalysis effects. Obviously, changing the neutralizing amine may alter the exact cure onset or some other key property[5]. Substituting triethylamine (TEA) for DMEA in the WB formulation does not alter cure onset much but it does give wrinkling during bake. In the examples shown here the bake times were 30 minutes.

At the profile bake temperatures of 75°C (SB) and 125°C (WB) both films have reached 200+ MEK double rubs. At this cure point both films have 160 inch-lbs forward impact resistance. The WB system stays at 160 inch-lbs, while the SB system falls to 30-40 inch lbs with increasing bake temperature. At their cure onset temperature, both films have Tukon hardness values of <2 but hardness rapidly increases reaching values in the 17-18 range by 150°C. Both formulations give films that are clear and glossy. Acid resistance gradient oven spot testing gives numbers in the 0.4-0.5 range for the WB system and 0.5-0.6 range for the SB system. Outstanding acid resistance values of 0.7 and above were obtained with some of the non-dispersible secondary carbamate models described elsewhere (0= no acid resistance, 1.0= no effect of acid)[3].

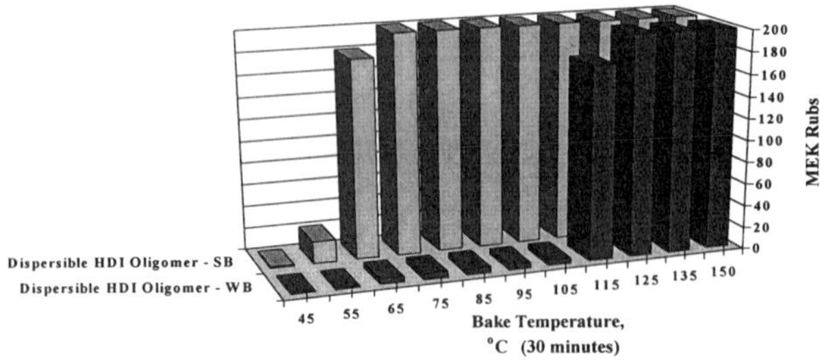

Figure 3: Cure Response of Dispersible HDI Oligomer - SB vs. WB

Dynamic Mechanical Analysis (DMA) of Cured Films - Cured films corresponding to selected bake temperatures as illustrated in Figure 3 were selected for study by DMA. DMA was carried out on free film samples at 11 Hz oscillating frequency on an Autovibron instrument (Imass Inc.) as described in detail elsewhere[4]. Films cured at 125°C and 140°C were selected for the WB system and films cured at 100°C and 140°C were selected for the SB system. Results are given in Table 1. At the 0.5/1.0 equivalent ratio, the crosslink density did not become so high at high conversions that the free films would break during DMA measurements.

TABLE 1. MF Cure of Dispersible HDI Oligomer - SB versus WB

Cure[A] Temp	Tg	Tan delta (max)	$10^{-8} \times$ E' (min)	T. of E'(min)	$10^{-8} \times$ E' (190°C)
°C	°C		Dyn/cm²	°C	Dyn/cm²
Solvent Borne (SB), pTSA 0.5 phr					
100	59	1.13	1.17	92	4.3
140	102	0.68	4.3	140	4.8
Water Borne (WB), DMEA Blocked DNNDSA 0.8 phr					
125	58	1.42	0.26	94	2.7
140	82	0.79	1.30	118	3.4

[A]Cure Time, 30 min., Equiv. Ratio 0.5/1.0 MF/Carbamate. All films had MEK double rubs of 200+.

The DMA derived Tg and E'(min) data clearly shows the retardation in cure encountered with the WB System because of the neutralizing amine equilibrium. If the E' plots are followed in the rubbery region, the WB films continue to rise in the 150-190°C DMA scan temperature region indicating methoxymethyl groups are still being converted. The SB films have E' plots that are fairly flat above 150°C indicating the most of the cure has occurred prior to this temperature.

MF Cure of Aqueous Polyurethane Dispersions - The effect of adding MF resins to commercial polyurethane dispersions (PUD) has been examined[3]. A commercial PUD (Zeneca T36R-41®), which has no hydroxyl present and an acid # of 13, was formulated with an aqueous MF resin (Resimene® AQ-1616). This resin has a relatively high >N-H imino content which based on previous research is less reactive with secondary carbamates than is a low imino, highly etherified MF resins. However, the aqueous MF resin can be directly added to the PUD. Catalyst consisting of 0.5 phr of active pTSA neutralized with TEA was added. The low acid # of this PUD would require less neutralizing amine counter ion which should be an advantage in minimizing the cure inhibition noted above. Films prepared at a 0/100, 10/90 and 20/80 solids ratio of MF/PUD were prepared and baked at both 100°C and 140°C for 30 minutes. The films with no added MF resin marred noticeably at 200 double rubs with both MEK and IPA. Films cured with 10 and 20 parts MF all gave 200+ double rubs with little or no marring. All of the films with and without MF crosslinking gave 160 inch-lbs forward impact. The Tukon hardness of the films without added melamine was about 2.0 but the hardness increased with both level of added melamine and bake temperature to a max. Tukon hardness of 5.4. Interestingly, another effect of reacting the MF resin with

the PUD was a significant increase in gloss. On clearcoat films as the MF was increased from 0 to 10 to 20 phr, the 20/60 gloss values increased from 65/94 to 75/103 to 87/108. This effect on gloss is not understood but was observed with several different PUD's.

TABLE 2: MF Crosslinked PUD

PUD/MF Ratio	Tan δ		E'x 10^{-8} Dynes / cm²		
	25°C	100°C	25°C	100°C	190°
T36-R-41, Cure - 100°C for 30 Min.					
100 / 0	0.09	0.24	19.9	5.1	-
90 / 10	0.06	0.16	20.0	11.2	0.3
80 / 20	0.06	0.14	43.6	19.6	5.8
T36R-R41, Cure - 140°C for 30 Min.					
100 / 0	0.14	0.27	9.1	1.7	-
90 / 10	0.07	0.10	27.4	12.2	0.7
80 / 20	0.06	0.08	39.5	22.6	6.5

DMA data obtained on these films is summarized in Table 2. The effect of MF is to raise the E' plots throughout the temperature range scanned. An abrupt drop of the E' in the control films without MF at about 150°C is attributed to loss of tension due to flow of the sample. It is evident that E' values at the beginning of the DMA scan (10°C) are well below the level expected for a glassy, amorphous polymer (about 2×10^{10} dynes/cm²). Specific Tg values are not specified due to the broad nature of the Tan δ plots. Results are consistent with a heterogeneous morphology in which domain boundaries are very diffuse. This type of structure gives very different DMA data from that obtained with the secondary carbamate models cured with MF[3].

CONCLUSIONS

Dispersible HDI oligomers containing secondary carbamate groups can be cured with MF resins in WB formulations. The reactivity of an MF resin with a WB carbamate can be inhibited by the neutralizing amines typically employed for dispersion and acid catalyst blocking. Secondary carbamate containing oligomers and polymers which minimize the amount of required neutralizing amine should offer advantages in cure behavior. Also, the use of low imino containing highly etherified melamines which can be codispersed with a water reducible carbamate should also improve cure response.

The combination of MF resins with secondary carbamate containing coreactants offer different ways of building properties into a coating system. Generally, the MF cure will improve solvent resistance, hardness and crosslink density which may lead to improved mar and abrasion resistance in some applications. Improvements in acid resistance has also been noted in some instances. It may also be cost effective to optimize properties by crosslinking linear urethane prepolymers rather than using polyfunctional isocyanates or polyols to obtain crosslinked urethane networks.

REFERENCES

1. J.O. Santer, *Prog. Org. Coatings,* **12,** 309(1984).
2. W. J. Blank, *Proceedings ACS PMSE,* **77,** 391 (1997).
3. H.P. Higginbottom, G.R. Bowers, P.E. Ferrell and L.W. Hill, *Prodeedings XXV Int. Waterborne, High-Solids and Powder Coating Symp.* **1998,** 527.
4. L. W. Hill, *Dynamic Mechanical and Tensile Properties,* p. 534, in *Paint and Coating Testing Manual,* 14th Ed., J.V. Koleske, Ed., ASTM, Philadelphia (1995).
5. P.E. Ferrell, J.J. Gummeson, L.W. Hill and L.J. Truesdell-Snider, *J. Coatings Technol.,* **67,** No. 851, 63 (1995).

Photooxidation Induced Chain Scission and Crosslinking in Acrylic/Melamine Coatings

Mark E. Nichols and John L. Gerlock
Ford Research Laboratory
Ford Motor Co.
Dearborn, MI 48121

Introduction

All organic coatings will undergo some degree of chemical composition change during prolonged exposure to light, humidity, and heat. The extent of these changes plays a significant role in determining a coating's ultimate long-term performance. This is particularly true for automotive clearcoats, the most common of which are acrylic/melamine based. For acrylic/melamine clearcoats the main mechanism of weathering-based degradation has been shown to be ultraviolet light-induced photooxidation, with hydrolysis playing a secondary role.[1] Changes in the crosslink density and crosslink structure are some of the most important changes that result from photooxidation. Hill and coworkers have measured the overall changes in crosslink density for acrylic/melamine clearcoats using dynamic mechanical analysis (DMA), and have shown that depending on the formulation, crosslink density can either increase or decrease with exposure.[2,3]

Changes in crosslink density can also be measured using chemical stress relaxation techniques, which were first used to measure network density changes in elastomers exposed to oxygen and ozone[4] and have recently been used to measure photooxidation-induced network changes in acrylic/melamine clearcoats.[5] While experimentally more cumbersome than DMA, chemical stress relaxation techniques can potentially measure the rates of crosslink formation and main-chain/crosslink scission independently. Knowledge of both of these rates allows for a more detailed examination of the mechanisms of degradation.

The principles of chemical stress relaxation are similar to those used to measure crosslink density using DMA. Above their T_g, crosslinked polymers, including acrylic/melamine clearcoats, can be modeled using the simple theory of rubber elasticity. Hill has shown this simple theory to be a good approximation of the behavior of these highly crosslinked coatings.[3] Two types of stress relaxation experiments are performed. During a *continuous* stress relaxation experiment, a free film of the clearcoat is stretched in tension to a set length and then held at that set elongation inside an exposure chamber in the presence of UV light, heat, and humidity. The force or stress the specimen is under is monitored as a function of time with a force transducer. This force will decrease with time, initially due to viscoelastic relaxations. After the viscoelastic relaxations, all the force is directly supported by crosslinks. Therefore, knowing the strain and the sample dimensions, the modulus of the clearcoat can be determined at any time and related to the crosslink density, ν, through:

$$\nu = E/3RT \qquad (1)$$

where E is the modulus, R is the gas constant, and T is the absolute temperature. As exposure continues, the modulus will continue to decrease in proportion to the rate of crosslink/chain scission. Of course new crosslinks can form during the exposure, but these do not contribute to the measured modulus as they form at their equilibrium positions. An intermittent relaxation experiment is similarly conducted inside an exposure chamber under the same conditions, except that the strain is applied only intermittently. The modulus is measured after a few minutes of relaxation and then the sample is returned to the nonelongated state. The intermittent relaxation experiment measures the overall changes in crosslink density, as the specimen spends over 99% of its exposure time in the nonstrained state. The intermittent relaxation experiment is analogous to measuring the modulus of a clearcoat in the rubber plateau using DMA. Having

measured the net changes in crosslink density and the rate of crosslink/chain scission, the rate of crosslink formation can be calculated. We present here the chemical stress relaxation behavior of two acrylic/melamine clearcoats and the effect of hindered amine light stabilizers (HALS) and ultraviolet light stabilizers (UVAs) on their relaxation behavior.

Experimental

Two different acrylic/melamine clearcoats were used in this study and are referred to as Mel-A and Mel-N. The details of their composition have been previously published.[6] Both Mel-A and Mel-N contain the same melamine crosslinker, however, the acrylate copolymers differ in the composition of their end groups due to the conditions under which they were synthesized. Mel-A contains end groups that are chromophores for UV light, while Mel-N's end groups are not. The rate of photooxidation of Mel-A has been shown by a number of methods to be 2-2.5 times Mel-N's rate.[7] Some of the formulations of Mel-A and Mel-N used in this study contained additives, which were either Tinuvin 123 HALS, Tinuvin 900 ultraviolet light absorber (UVA), or both. Both Mel-A and Mel-N were prepared as free films after curing and peeling from aluminum substrates.

The chemical stress relaxation experiments took place inside a modified QUV weathering chamber. Illumination was from FS-340 bulbs. The air temperature was 80°C and the dew point was 25°C. Clearcoat films were approximately 50 μm thick and 4 mm wide. The strains used were approximately 2.7%.

Results and Discussion

The continuous stress relaxation behavior, a measure of crosslink/chain scission, for additive-free Mel-A and Mel-N in the absence and presence of UV light is shown in Figure 1, where the vertical axis is the force at any time during the experiment normalized to the initial force.

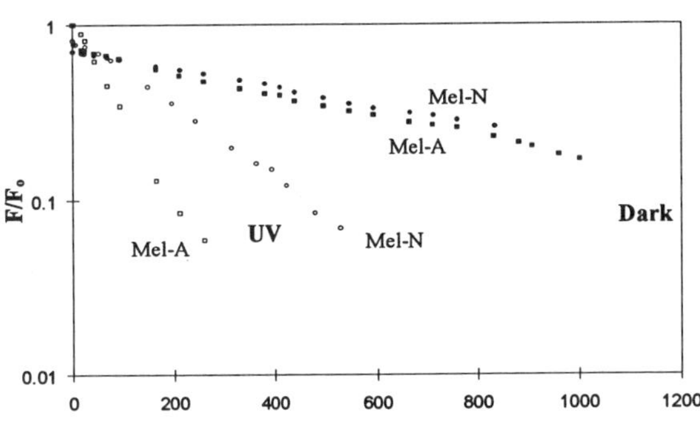

Figure 1. Continuous relaxation of Mel-A and Mel-N exposed to UV light (UV) and in the dark (dark).

Both coatings undergo crosslink/chain scission at nearly identical rates in the absence of UV light. This bond rupture is likely due to hydrolysis, the rate of which is not expected to differ between Mel-A and Mel-N. In the presence of UV light the rate of crosslink/chain scission increases for both clearcoats, with Mel-A's rate being 2.5 times that of Mel-N's, in accordance with previous results.[6]

The intermittent relaxation results for additive-free Mel-A and Mel-N are shown in Figure 2, where the vertical axis is the force at the end of a given strain pulse normalized to the force at the end of the initial strain pulse (this axis is proportional to crosslink density). In the absence of UV light the rate of crosslink density change is the same for both Mel-A and Mel-N and is equal to the rate of crosslink/chain scission seen in Fig. 1, indicating no crosslinks are produced in the absence of light.

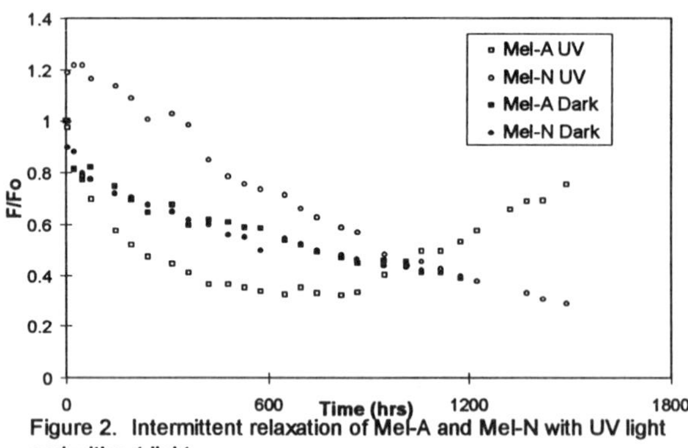
Figure 2. Intermittent relaxation of Mel-A and Mel-N with UV light and without light.

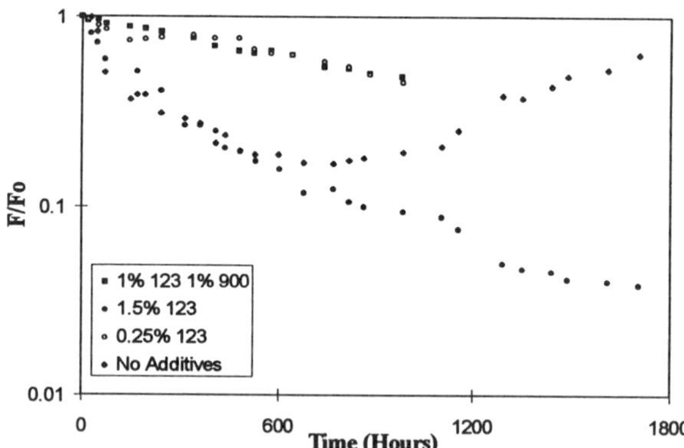
Figure 4. Intermittent relaxation of Mel-A with various additives.

This is again consistent with the proposal that hydrolysis is the main mechanism of degradation in the absence of light. In the presence of UV light, Mel-N undergoes first an increase and then a steady decrease in crosslink density. Mel-A undergoes first a decrease in crosslink density, followed by a plateau, and then an increase in crosslink density. Thus, not only do Mel-A and Mel-N differ in their rate of photooxidation, but the mode of photooxidation also differs, with Mel-A ultimately heading towards a highly crosslinked state and Mel-N heading towards a lightly crosslinked state.

Adding Tinuvin 123 HALS and or Tinuvin 900 UVA dramatically changes the photooxidation behavior of both Mel-A and Mel-N. Figure 3 shows the continuous relaxation behavior of Mel-A with various levels of additives.

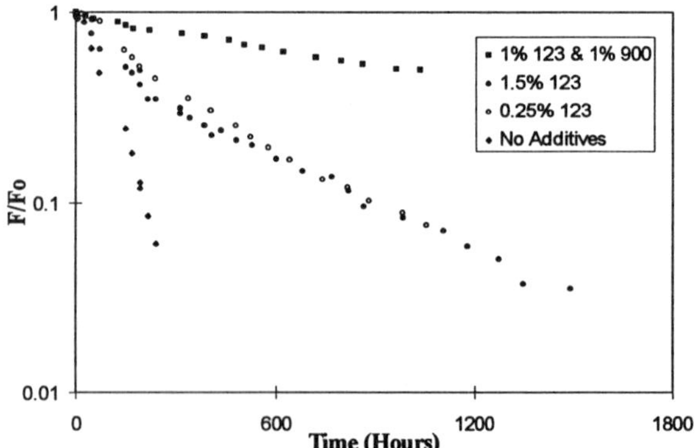
Figure 3. Continuous relaxation of Mel-A with various additives.

The addition of 0.25% and 1.5% Tinuvin 123 slows the rate of crosslink/chain scission compared to the additive-free formulation. Adding both Tinuvin 123 HALS and Tinuvin 900 UVA slows the rate even further, until the rate of crosslink/chain scission equals the rate in the absence of light (hydrolysis rate). Adding either Tinuvin 123 or Tinuvin 900 slows the rate of crosslink/chain scission of Mel-N to the rate observed in the absence of UV light (data not shown).

Figure 4 shows the intermittent relaxation behavior of Mel-A with various levels of additives. The additive free formulation of Mel-A eventually ends up increasing in crosslink density. Mel-A containing 1.5% Tinuvin 123 HALS undergoes a steady decrease in crosslink density - its long term behavior has been significantly changed.

Mel-A containing only 0.25% Tinuvin 123 undergoes a very small overall change in crosslink density, similar to the changes observed for the formulation containing both Tinuvin 123 and Tinuvin 900. The changes in Mel-A's crosslink density are smaller when 0.25% Tinuvin 123 is added, yet the rates of crosslink/chain scission are roughly equal for the 0.25% and 1.5% formulations (Fig. 3). Thus, Mel-A containing only 0.25% Tinuvin 123 must undergo a large amount of crosslink formation to offset the still relatively high rate of crosslink scission. Mel-A containing 1.5% Tinuvin 123 does not undergo any significant crosslink formation, and in fact, the rate of crosslink formation in Mel-A containing 1.5% Tinuvin 123 approaches zero. Thus, the net photooxidation-induced changes in the crosslink density of Mel-A containing 0.25% Tinuvin 123 are smaller than the changes for the formulation containing 1.5% Tinuvin 123, but the overall amount of chemical change occurring is greater, due to the concomitant crosslinking occurring in the 0.25% formulation. The addition of either Tinuvin 123 or Tinuvin 900 to Mel-N slows the change in crosslink density to that observed in the absence of light (data not shown).

Summary

Chemical stress relaxation has been used to measure the rates of crosslink/chain scission in two acrylic/melamine coatings undergoing photooxidation and hydrolysis. Additive-free Mel-A loses crosslinks 2-2.5 times as fast as additive-free Mel-N in the presence of UV light. The addition of Tinuvin 123 HALS slows the rate of crosslink/chain scission of both Mel-A and Mel-N. The rate of crosslink formation is also effected by the addition of Tinuvin 123, with large amounts of Tinuvin 123 shutting off crosslink formation in Mel-A, and lesser amounts only slowing crosslink formation.

References

1. Bauer, D. R., in *Chemistry Properties and Applications of Crosslinked Systems*, ACS Symposium Series No. 367, ed. D. R. Bauer, R. A. Dickie and S. S. Labana, ACS, Wash D. C., p77, 1988.
2. Hill, L. W., Korzeniowski, H. M., Ojunga-Andrew, M., and Wilson, R. C., *Prog. Org. Coat.*, **24**, (1994), 147.
3. Hill, L. W., *J. Coat. Tech.*, **64**, (1992), 29.
4. Murakami, K., and Ono, K., *Chemorheology of Polymers*, Elsevier, New York, 1979.
5. Nichols, M. E., Gerlock, J. L., and Smith, C. A., *Polym. Deg. Stab.*, **56**, (1997), 81.
6. Carduner, K. R., Carter, R. O., Zinbo, M., Gerlock, J. L., and Bauer, D. R., *Macromolec.*, **21**, (1988), 1598.
7. Bauer, D. R., Gerlock, J. L., Mielewski., D. F., Paputa-Peck, M. C., and Carter, R. O., *Polym. Deg. Stab.*, **28**, (1990), 39.

MECHANICAL PROPERTY/PROPERTY RELATIONSHIPS IN COATINGS

Loren W. Hill
Solutia Inc.
730 Worcester Street
Springfield, MA 01151

INTRODUCTION

Organic coatings are used to improve appearance and protect substrates[1-4]. To perform these functions well, coatings must have certain mechanical properties. A variety of mechanical property tests have been developed and standardized by coatings chemists who donate time and effort under the auspices of the American Society for Testing and Materials (ASTM). Recent publication of the fourteenth edition of the Paint and Coating Testing Manual, under the able editorship of J.V. Koleske[3], has updated descriptions of test methods and summarized current understanding of relationships between coating film structure and performance in ASTM tests. Knowledge of relationships between structure and ASTM test results is quite limited. Knowledge of relationships between structure and more fundamental mechanical property values is more extensive. Examples of more fundamental values include dynamic mechanical properties and tensile stress-strain properties such as modulus, tensile strength, yield strength, elongation at break, and elongation at yield[5]. In our attempts to develop coatings that have improved performance, we may know how to change the binder structure to increase modulus, but we may not know how higher modulus affects a critical performance property such as mar resistance of auto topcoats, for example. In some cases, lack of knowledge of property/property relationships is more limiting than lack of knowledge of structure/property relationships.

PROPERTY/PROPERTY RELATIONSHIPS

Hardness[1-4] - Coating hardness is determined by penetration tests (Tukon), pendulum swing damping (Persoz, Koenig, Sward rocker), and resistance to scratching by pencils of varying lead hardness (pencil hardness). Results in these tests have been compared to results obtained by dynamic mechanical analysis (DMA) and stress-strain determinations. From DMA we obtain the tensile storage modulus (E'), tensile loss modulus (E") both as a function of temperature, and the glass transition temperature, T_g . In general, we expect and also usually observe that hardness increases with increasing T_g, but the relationship is not quantitative. For a series of films of the same T_g, we have observed that Tukon hardness and pendulum hardness still vary. Tukon hardness depends more directly on E'(25°) than on T_g . Pendulum hardness depends more directly on E"(25°) than on T_g. Resistance to scratching or plowing by pencils depends on toughness of the coating film and surface lubricity as well as on hardness. In stress-strain determinations, toughness is defined as the area under the stress-strain plot. No examples were found in coatings literature where this area was used to interpret causes of unexplained variation in pencil hardness.

Flexibility[1-5] - Coating flexibility is determined by bend tests (bends around conical or cylindrical mandrels, or for coil coatings by sharp bends of the substrate directly back on itself in so called "T-Bends") and by falling weight impact tests. In general, polymeric materials of low molecular weight and high T_g are very brittle. In thermoset coatings, molecular weight build up and cross linking occur simultaneously during cure. As extent cure increases, flexibility increases. The increase in crosslink density can be followed by determining E' in the rubbery plateau region of the temperature scan, usually at about 125°C. Molecular weight build up associated with the E'(125°) rise results in better performance in flexibility tests. If crosslink density gets too high, however, flexibility decreases again. In cure profile work on new crosslinkers, we have very frequently observed impact resistances of about 20 inch-lbs at cure onset, >160 inch-lbs (the maximum measurable in our test is 160) at optimum cure, and about 40 inch-lbs at higher cure temperatures where crosslink density is too high.

In many polymer studies good flexibility and impact resistance are associated with low temperature peaks in E" plots. This relationship has not been well established for coatings, but it may explain the much greater flexibility of polyester/melamine films compared to acrylic/melamine films of similar T_g and crosslink density. DMA plots for PE/MF films show a weak E" peak at about -70°C whereas plots for acrylic/MF films do not have any low temperature peak over the range scanned.

Mar Resistance - The resistance of auto topcoats to shallow scratches from automatic car wash brushes is receiving a great deal of attention currently. In a March 1998 article, Ryntz[6] gave seven references beginning in 1992 on this topic. Even more recently, Jones[7] et al have reported extremely low levels of marring for a series of oligoester diol films crosslinked with a high solids etherified melamine formaldehyde (MF) crosslinker. Unpublished work from this laboratory indicates that these mar resistant films had high crosslink density (XLD), as determined by E'(min) values on DMA plots. High XLD is clearly good for mar resistance. In general, low T_g is reported [6-8] to be good for mar resistance, but in this oligoester diol series, T_g varies over a substantial range without causing a reduction in mar resistance. It appears that the high functionality of MF crosslinkers in general conveys sufficient XLD so that marring is moderate even at rather high T_g.

Marring is reported[8] to result from two types of scratches, fracture scratches and plastic scratches. A third type of response, i.e. elastic response, does not produce any scratches at all. This terminology is understandable based on analysis of stress application during the scratching process. Stress results in deformation which can occur with or without fracture. If fracture occurs the result is a fracture scratch. If fracture does not occur, recovery behavior influences marring. Polymeric materials can undergo three types of recovery: "elastic" (immediate and complete recovery), "plastic" (no recovery, permanent deformation), "viscoelastic" (delayed or time dependent recovery, sometimes incomplete). High XLD favors elastic response and eliminates plastic scratches, but if XLD and T_g are too high fracture scratches are observed. MF-cured films mainly give fracture scratches, and marring responds to XLD optimization. The lower functionality of isocyanate crosslinkers, compared to MF, results in lower XLD in 2K clearcoats which permits occurrence of both plastic and fracture scratches. In a comparison of tri-functional isocyanate crosslinkers[9], the best candidate gave a clearcoat that lost 22 % of its gloss in a widely used mar test. The MF-cured oligoester diols referred to above, lost only about 2 to 3 % of their gloss in the same test[7], and we have prepared MF-cured acrylic clearcoats that lose only about 10% of their gloss in similar mar tests despite having high T_g values.

REFERENCES

1. L.W.Hill, *Mechanical Properties of Coatings*, Federation Monograph, Federation of Societies for Coatings Technology, Philadelphia, 1987.
2. L.W. Hill, *Prog. Org. Coatings*, **5**, 277 (1977).
3. *Paint and Coating Testing Manual*, 14th Ed., J.V. Koleske, Ed., ASTM, Philadelphia (1995).
4. Z.W. Wicks, Jr., F.N. Jones, and S. P. Pappas, *Organic Coatings Science and Technology*, Wiley-Interscience, N.Y., Vol. 2, Ch. 24, p. 105 (1992).
5. L.W. Hill, *Dynamic Mechanical and Tensile Properties*, p. 534, Ref. 3.
6. R.A. Ryntz, *Paint &Coating Ind.*, **XIV**, No. 3, 60 (1998).
7. C. Ji, Z. Huang, W. Chen, C. Diakoumakos, F.N. Jones, and R.A. Ryntz, *Proceedings ACS, Poly. Mat. Sci. Eng.*, **78**, 252 (1998).
8. B.V. Gregorovich and P.J. McGonigal, *Proc. Adv. Coat. Technol. Conf.*, p. 121, Chicago, IL, Nov. 3-5, 1992
9. R.S. Dearth, H. Mertes, and P.J. Jacobs, *Prog. Org. Coat.*, **29**, 73 (1996)

ENVIRONMENTAL RESPONSE OF ANIONIC MICROGELS.
G. M. Eichenbaum*. P. F. Kiser*^, S. A. Simon**, D. Needham.
*Department of Mechanical Engineering and Materials Science,
Duke University, Durham, NC 27708-0300; ^Access
Pharmaceuticals, Dallas, TX 75207-2107; **Department of
Neurobiology, Duke University Medical Center, Durham, NC
27710.

Micron-sized (4-7 micron diameter) polymethacrylic acid co-nitrophenyl acrylate microgel spheres were synthesized by precipitation polymerization. Their composition was modified in post-polymerization reactions. Electrophoretic mobilities of the microspheres were measured as a function of pH. Equilibrium changes of microgel volume were measured as functions of pH in the presence of a series of different types of cations in the suspending solution. The exchange of the cations was also examined. The maximum reduction in the microsphere equilibrium volume (Vrmax) at pH 3.0 was 0.15, where Vr was the ratio of the microsphere volume at the test pH to its volume at pH 6.6.

PREPARATION OF POLY(2-HYDROXYETHYL METHACRYLATE) BY CONTROLLED RADICAL POLYMERIZATION

Kathryn L. Beers, Sohyun Boo and Krzysztof Matyjaszewski[*]

Department of Chemistry
Carnegie Mellon University
4400 Fifth Avenue, Pittsburgh, Pa. 15213

Introduction

The industrial significance of poly(2-hydroxyethyl methacrylate) (pHEMA) has been well documented.[1,2] The preparation of linear homopolymers of controlled molecular weights, however, has proven difficult. The hydroxy-substituted monomer is not compatible with group transfer polymerization, nor has there been much success polymerizing methacrylates with stable free-radical methods such as nitroxides.[3] The most successful method to date is the use of carbamate-modified iniferters[4], but this like most other polymerizations of HEMA involve the use of added cross-linking agents which complicate the ability to confirm control of polymerization, since both molecular weight and polydispersity become meaningless in these cases.

Controlled polymerization of this monomer is possible once the hydroxy group is protected. This method has been used by DeSimone, *et al.*[5] to prepare block copolymers by controlled polymerization for use as surfactants in super-critical solvents. The advantage of using the unprotected monomer is not only the removal of two steps in the synthesis of pHEMA, but the ability to extend the technology of Atom transfer radical polymerization (ATRP) to more polar media.

ATRP employs exchange of a halogen atom between carbon-centered radical chain ends and a Mt^nL_x / $Mt^{n+1}L_x$ species (Scheme 1) which enables fast initiation and reduces the concentration of free radicals available for both propagation and termination. A detailed discussion of the mechanism is available elsewhere[6], however, the equilibrium between these active and dormant species is of particular importance to the discussion below.

Scheme 1:

$$\text{\textasciitilde\textasciitilde\textasciitilde R}\cdot \; + \; CuBr_2L_x \; \underset{}{\overset{K_{eq}}{\rightleftharpoons}} \; \text{\textasciitilde\textasciitilde\textasciitilde RBr} \; + CuBrL_x$$

Reported here is the controlled radical polymerization of unprotected 2-hydroxyethyl methacrylate using $CuBr(bipy)_2$ as a catalyst and ethyl 2-bromoisobutyrate as the initiator in DMF and 30/70 (v/v) n-propanol/methylethyl ketone as solvents. Molecular weights were shown to increase linearly with conversion. In the case of the mixed solvent system, polymers with polydispersities (M_w/M_n) less than 1.35 were obtained.

Experimental

2,2'-Bipyridine (bipy) from Aldrich was recrystallized from ethanol to remove impurities. CuBr was washed with acetic acid to remove impurities. $CuBr_2$ was ground with a mortar and pestle to improve solubility. All other reagents were used as received from Aldrich, including ethyl-2-bromoisobutyrate (EtBriBu), N,N,-dimethyl formamide (DMF), methyl ethyl ketone (MEK), and n-propanol.

Purification of monomer:

Contaminants that are often found in commercially available HEMA include both starting materials (methacrylic acid) and byproducts (dimethacrylate). The presence of acid inhibits ATRP and the presence of any dimethacrylate leads to crosslinked polymer. Two different methods for the purification of HEMA were used in these experiments. The first procedure involved washing an aqueous solution (25 vol% HEMA) of monomer with hexanes (4x200mL), salting the monomer out of the aqueous phase, drying over $MgSO_4$, and distilling under reduced pressure. The second procedure included passing monomer through a silica column, eluted with 30/70 benzene/ethyl acetate, and distilling under reduced pressure. Both methods yielded monomer which polymerized readily and without crosslinking as shown by SEC.

Reactions performed in sealed tubes:

CuBr, $CuBr_2$ and bipy were weighed into a round bottomed flask. The flask was sealed with a rubber septa and degassed by vacuum followed by argon backfill three times. HEMA and solvent were purged with bubbling argon for 45 minutes each and were then added by argon flushed syringe to the flask and the solution was allowed to stir until it became homogeneous. Initiator was added by syringe to each tube, followed by addition of the appropriate fraction of the monomer/catalyst solution which was used to wash down the wall of the tube. The tubes were immediately degassed by cycling through freeze-thaw cycles under vacuum three times and sealed under vacuum. The tubes were then either placed in a thermostated oil bath or allowed to sit at room temperature. At timed intervals, tubes were frozen in liquid nitrogen and broken open.

Reactions performed in flasks:

In round bottomed flasks, the same stock solution was prepared as above, but it was placed directly into a thermostated oil bath. After the solution had become homogeneous with stirring, 60 μL of EtBriBu was added using a syringe and samples were removed at time intervals, t.

Characterization:

All samples were analyzed by GC, NMR and/or SEC for conversion, molecular weight and polydispersity. A Waters 510 LC Pump connected to a Waters 410 differential refractometer with DMF as the carrier solvent and linear 500 and 1000Å Phenogel columns were used for gel permeation chromatography (GPC). ^{1}H NMR spectra were recorded on a 300 MHz Brüker Spectrometer. A Shimadzu 14A gas chromatograph was used for gas chromatography.

Results and Discussion

Because HEMA and the solvents in which it is soluble are much more polar than those commonly used in ATRP, using unsubstituted bipyridine ligand provides a soluble catalyst and Cu(II) species and polymerization conditions are homogeneous. In all cases, molecular weights were above the detectable limit of ^{1}H NMR. SEC data are calibrated versus linear polystyrene standards, a comparatively nonpolar polymer. Because the hydrodynamic volume differences of polymers of the same weight in these two categories is substantial, values obtained by this method for molecular weight of polymers of hydroxy-substituted monomers has been shown to be an inflation of the real value.[7]

Table 1: Preliminary Results - Single Point Experiments Towards Optimizing Conditions for Controlled Polymerization. All reactions were carried out in sealed tubes with theoretical number average molecular weights (based on $[M]_o$ / $[I]_o$) of 13,000 and ratio of initiator (EtBriBu) : CuBr:bipy of 1:1:2.

No.	T (°C)	t (min)	condition	p (%)	$M_{n,th}$ / $M_{n,SEC}$ (x10^{-3})	M_w/M_n
1	90	10	bulk	90	11.7/34.4	1.6
2	90	180	bulk, 10 mol% $CuBr_2$	58	7.5/27.0	2.0
3	90	1080	bulk, 20 mol% $CuBr_2$	16	2.1/14.5	1.6
4	25	20	50 vol% DMF	60	7.8/17.5	1.5

Table 1 contains data from a series of single point experiments preparing pHEMA by ATRP under a variety of conditions. First, in bulk (**1**), the reaction was fast and polydispersity (M_w/M_n) was higher than those obtained in ATRP of standard monomers such as styrene and methyl methacrylate (MMA).[8,9] Next, Cu(II) was added (**2** and **3**) in order to slow the reaction down by increasing the equilibrium concentration of dormant species via Le Chatlier's Principle as shown in Scheme 1. This slowed the reaction down as added Cu(II) was increased

When reactions **2** and **3** were repeated, however, taking kinetic samples, little progression of molecular weight was observed indicating poor control of the polymerization. Finally, polymerizations were conducted in solvent. The solubility of pHEMA is limited, though, to polar solvents such as DMSO and DMF which have been difficult to use in controlled radical polymerizations.

Finally, by combining the two effects of added Cu(II) (0.2[CuBr]) and dilution in solvent (80 vol% DMF), a linear progression of molecular weight was observed in the kinetic plots (Fig. 1 and 2). There is a steady decrease in the rate of polymerization evident in the semi-logarithmic plot (Fig. 2). This is due to a steady increase in the already high Cu(II) concentration resulting from small amounts of termination which lead to progressive suppression of propagation and a prevalence of the dormant chain end species.[10]

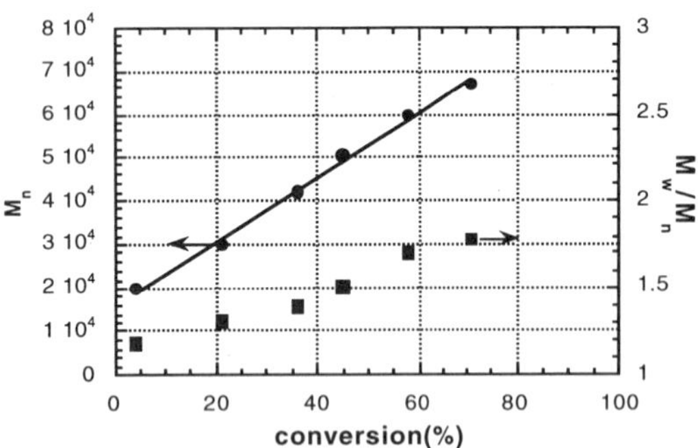

Figure 1: Molecular weight and molecular weight distribution data for pHEMA via ATRP in 80 vol% DMF with 20 mol% added CuBr$_2$ at room temperature (25 °C).

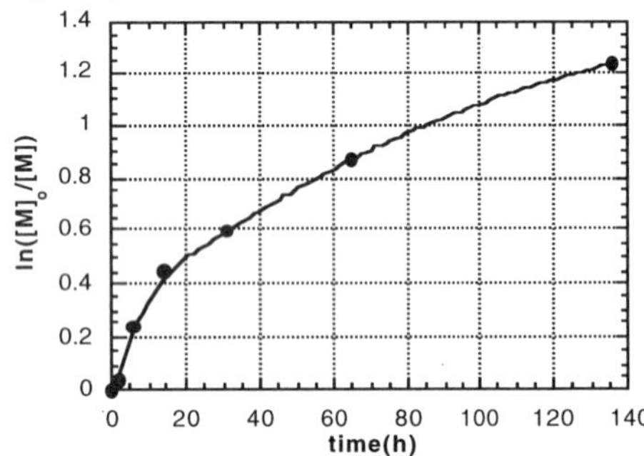

Figure 2: Semi-logarithmic plot of conversion vs. time for pHEMA via ATRP in 80 vol% DMF with added CuBr$_2$ (0.2[CuBr]) at room temperature (25 °C). (line is merely connection of points to provide clarity.)

In 1969, Tuzar and Bohdanecky[11] reported on the synergistic effect of mixed solvents on the solubility of pHEMA. While pHEMA is not soluble in MEK the addition of as little as 25 % of an alcohol such as n-propanol to MEK imparts solubility to the polymer. Thus, polymerizations were carried out in a mixture of n-propanol/MEK 30/70 (v/v) using the mixed halogen system and relative concentrations proposed by Wang, *et al.*[12] and the resulting data is plotted in Figures 3 and 4.

The data shown are for the same reaction run at two different temperatures, 50 and 70 °C. While polydispersities remain low (M_w/M_n <1.35 at 50 °C) and molecular weights appear to evole with conversion monotonously, it is clear that the same effect of increasing amounts of dormant, non-propagating chains is enhanced by the increased temperature, also.

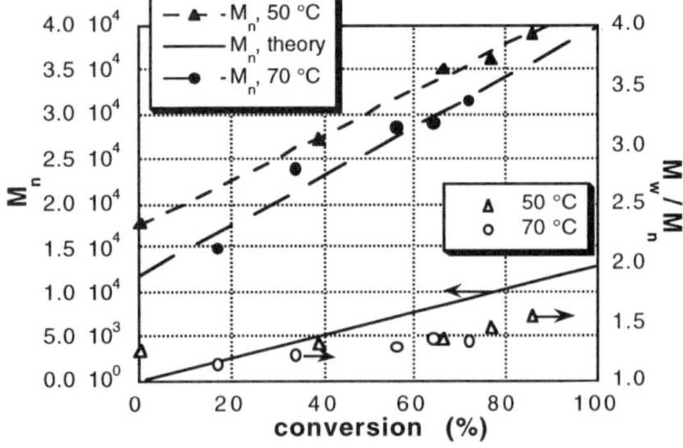

Figure 3: Molecular weight and molecular weight distribution data for ATRP of HEMA in 50 vol% n-propanol / MEK mixed solvent system in flasks. [HEMA]:[EtBriBu]:[CuCl]:[bipy]:[CuCl$_2$] 200:2:1:2:0.2. See text for discussion of theoretical molecular weight discrepancy.

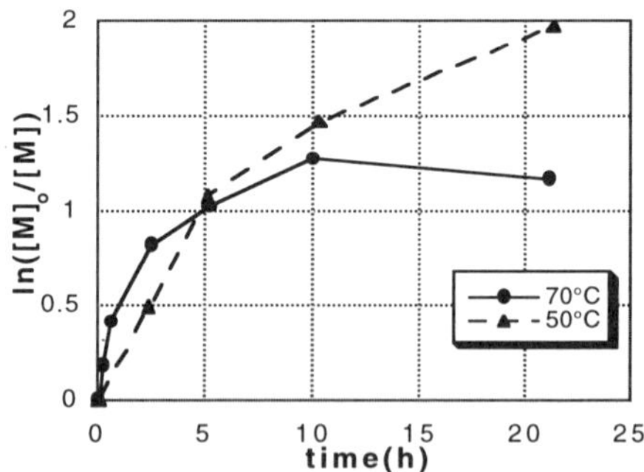

Figure 4: Semi-logarithmic plot of conversion verses time illustrating effect of temperature on rate of ATRP of HEMA in mixed solvents. For conditions of polymerization, see Figure 3. (lines are only connecting points for clarity)

Conclusion

It has been shown that polymerization of 2-hydroxyethyl methacrylate can be controlled by atom transfer using the simple and inexpensive catalyst, CuBr(bipy)$_2$. Removal of impurities from the monomer is crucial to ensure that linear chains are polymerized and high conversions are reached. Mixed solvent systems are necessary to produce narrow molecular weight distributions at reasonable rates.

Acknowledgment

We would like to thank the members of the ATRP Consortium at CMU for their support.

References

(1) Dumitriu, S., Ed.; *Polymeric Biomaterials*; Marcel Dekker, Inc: New York, 1994.
(2) Montheard, J.-P.; Chatzopoulos, M.; Chappard, D. *J. M. S., Rev. Macromol. Chem. Phys.* **1992**, *C32*, 1.
(3) Moad, G.; Anderson, A. G.; Ercole, F.; Johnson, C. H. J.; Krstina, J.; Moad, C. L.; Rizzardo, E.; Spurling, T. H.; Thang, S. H. *Controlled Radical Polymerization*; American Chemical Society: Washington, D. C., 1998, pp 332.
(4) Kumar, R. C.; Ohkabo, T. Eur. Pat. Appl. EP 91-305589 910620 (1992)
(5) Betts, D. E.; Johnson, T.; LeRoux, D.; DeSimone, J. M. *Controled Radical Polymerization*; Americal Chemical Society: Washington, D. C., 1998, pp 418.
(6) Matyjaszewski, K.; Patten, T. E.; Xia, J. *J. Am. Chem. Soc.* **1997**, *119*, 674.
(7) Coca, S.; Jasieczek, C.; Beers, K.; Matyjaszewski, K. *J. Polym. Sci.* **1998**, *in press*,
(8) Patten, T. E.; Xia, J.; Abernathy, T.; Matyjaszewski, K. *Science* **1996**, *272*, 866.
(9) Grimaud, T.; Matyjaszewski, K. *Macromolecules* **1997**, *30*, 2216.
(10) Matyjaszewski, K. *Controlled Radical Polymerization*; American Chemical Society: Washington, D. C., 1998, pp 258.
(11) Tuzar, Z.; Bohdanecky, M. *Collection Czechoslov. Chem. Commun.* **1969**, *34*, 289.
(12) Wang, J. L.; Grimaud, T.; Matyjaszewski, K. *Macromolecules* **1997**, *30*, 6507.

ATOM TRANSFER RADICAL POLYMERIZATION USING FUNCTIONAL INITIATORS CONTAINING CARBOXYLIC ACID GROUP

*Xuan Zhang, Jianhui Xia, Scott G. Gaynor, Krzysztof Matyjaszewski**

Department of Chemistry, Carnegie Mellon University,
4400 Fifth Avenue, Pittsburgh, Pennsylvania 15213

Introduction

Carboxylic acid terminated polymers are of great industrial interest due to several reasons. First, addition of carboxylic acid terminated liquid rubber to epoxy resins can improve the impact-resistance of the resin.[1] Second, carboxylic acid group on the polymer chain end allows the preparation of block copolymers.[2] Furthermore, carboxylic acid group can be easily converted to other useful functionalities[3] for mechanistic and chemical properties or the attachment of growing chains.

Atom transfer radical polymerization (ATRP) is a relatively new method for controlled radical polymerization of many monomers.[4,5] ATRP is based on the reversible formation of radical active sites by homolytic cleavage of a carbon-halogen bonds. The reversible deactivation of the radicals by halogen transfer from CuX_2/ligand to form dormant species lowers the instant concentration of radicals in solution while keeping the total concentration of propagating species equal to the initiator concentration. Polymers with predictable molecular weights and narrow molecular weight distributions have been prepared.[4,5] This report describes ATRP of bulk styrene using initiators with either free or protected carboxylic acid group.

Experimental

Materials. Styrene was purchased from Aldrich, and was vacuum distilled over CaH_2 and stored under Ar at -15 °C. *N,N,N',N'',N''*-pentamethyldiethylenetriamine (PMDETA), 2-bromobutyric acid, 2-bromoisobutyric acid, and *t*-butyl 2-bromopropionate were all commercial products, and were used without further purification.

Synthesis of Trimethylsilyl 2-Bromobutyrate and *t*-Butyldimethylsilyl 2-Bromobutyrate. The silyl protected carboxylic acids were prepared using modified literature procedure. A dry round bottom flask was charged with 2-bromobutyric acid (10 ml, 0.094 mol) and triethylamine (14 ml, 0.10 mol), and was stirred in 100 ml CH_2Cl_2 in an ice-bath. *t*-Butyldimethylsilyl chloride (15.6 g, 0.10 mol) or chlorotrimethyl silane (14.2 ml, 0.10 mol) was added dropwisely. A white precipitate formed immediately. The reaction mixture was allowed to warm up to room temperature after stirred in the ice-bath for 15 min. After 2 h at room temperature, the white precipitate was filtered off on a medium frit. The filtrate was collected, and concentrated under a rotary evaporator. The product was obtained by vacuum distillation and characterized by ^{1}H NMR.

Polymerization of Bulk Styrene Using Free and Protected Carboxylic Acid Initiators. A dry round-bottom flask was charged with CuBr (39 mg, 0.27 mmol), PMDETA (56.3 ml, 0.27 mmol), styrene (3ml, 26.2 mmol), and a magnetic stir bar. The flask was sealed by a rubber septum, and degassed by three freeze-pump-thaw cycles. The flask was immersed in an oil bath thermostated at 110 °C, and the initiator (0.27 mmol) was added dropwisely. At timed intervals, aliquots of the reaction solution were drawn via degassed syringes fitted with stainless needles, and were dissolved in THF to measure conversion (GC) and molecular weight (SEC).

Results and Discussion

ATRP Using Free Carboxylic Acid. ATRP of styrene with free carboxylic acid as the initiator has been carried out. The results indicate that initiators with free carboxyl acid group can initiate polymerization. Under ATRP conditions, the polymers obtained have narrow polydispersities. However, the initiator efficiency is quite low. This is presumably due to the coordination of carboxyl group to the copper catalyst. The results also show that halocarboxylic acids with a bromine atom on the tertiary or secondary carbon α to the carboxylic group are better initiators than halocarboxylic acids with a bromine atom on primary carbon.

Table 1. ATRP of Styrene at 110 °C Initiated by Free Carboxylic Acids.

Initiator	Time (h)	Conv. (%)	$M_{n,th}$	$M_{n,sec}$	M_w/M_n
BIA	2.5	47.0	4700	47000	1.34
BBA	2.0	49.0	4900	39400	1.25
BAA	24.0	74.4	7440	237100	1.75

BIA = 2-bromoisobutyric acid, BBA = 2-bromobutyric acid, BAA = bromoacetic acid.

ATRP Using Trimethylsilyl Protected Acid Initiator. Trimethylsilyl protected acid initiators were synthesized and used as the initiator for ATRP of styrene. The system affords well-controlled polymerization with linear semilogarithmic plot of conversion vs. time (Figure 1), and the linear increase of measured molecular weight with conversion (Figure 2). The molecular weight distribution are quite narrow ($M_w/M_n \sim 1.2$). However, the initiator efficiency is lower than unity (ca. 0.6). Those observations indicate that the number of propagating chains remains constant, and the amount of termination and other side reaction are negligible during the polymerization. Therefore, the low initiator efficiency is attributed to the side reactions such as hydrolysis of the initiator by trace amount of moisture in the very early stage of the polymerization.

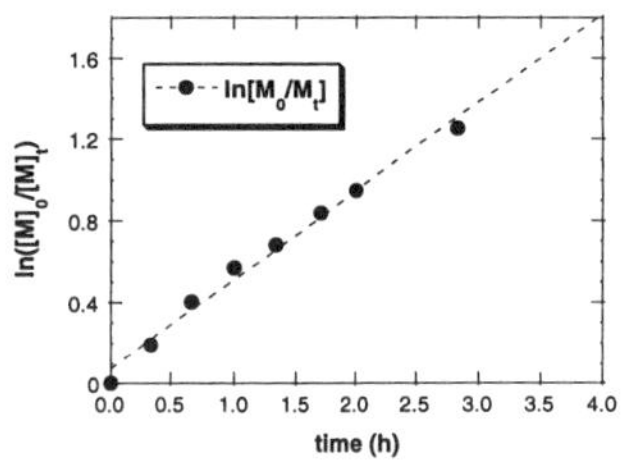

Figure 1. Semilogarithmic kinetic plot for ATRP of bulk styrene initiated by trimethylsilyl 2-bromobutyrate at 110 °C.

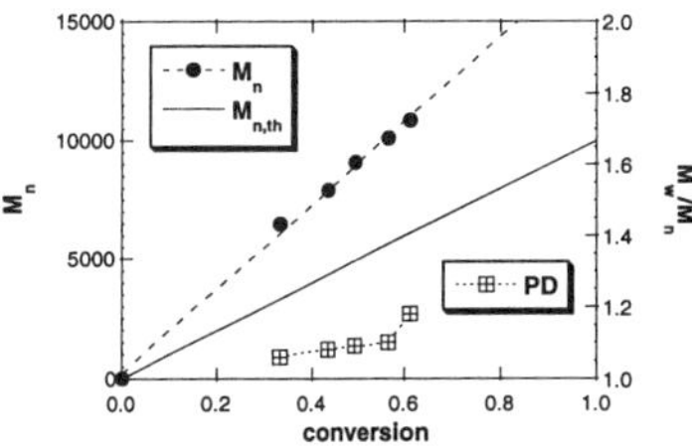

Figure 2. Evolution of experimental molecular weight and polydispersity with conversion for ATRP of styrene initiated by trimethylsilyl 2-bromobutyrate.

ATRP Using *t*-Butyldimethylsilyl Protected Acid Initiator. *t*-Butyldimethylsilyl protected acid initiator has been synthesized since it is more stable towards hydrolysis compared with trimethylsilyl protected carboxylic acid. When used in ATRP of styrene, a linear semilogarithmic plot of conversion vs. time (Figure 3), and the linear increase of conversion with time (Figure 4) are obtained. Narrow molecular weight distributions (M_w/M_n < 1.2) are found throughout the polymerization. The initiator efficiency is improved to ca. 0.8.

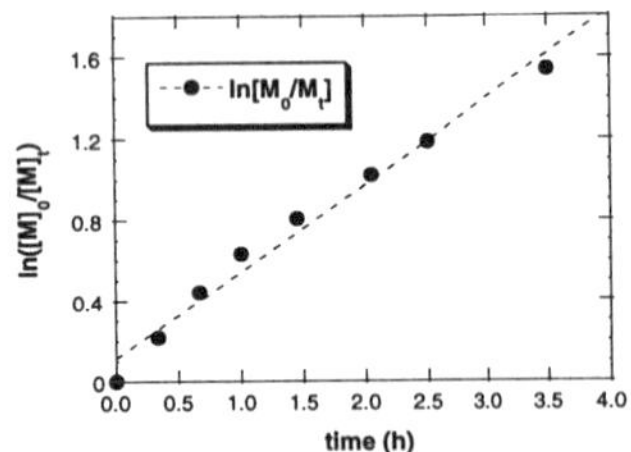

Figure 3. Semilogarithmic kinetic plot for ATRP of bulk styrene initiated by *t*-butyldimethylsilyl 2-bromobutyrate at 110 °C.

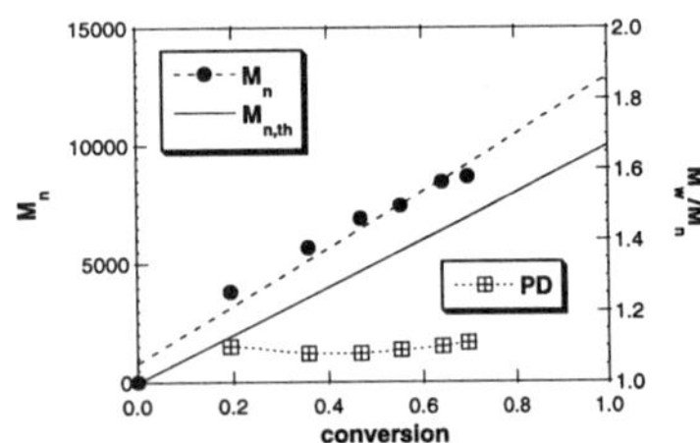

Figure 4. Evolution of experimental molecular weight and polydispersity with conversion for ATRP of styrene initiated by *t*-butyldimethylsilyl 2-bromobutyrate.

ATRP with *t*-Butyl Protected Acid Initiator. ATRP of styrene initiated by *t*-butyl 2-bromopropionate has also been carried out for the purpose of comparison (Figure 5 and 6). The results show a straight semilogarithmic plot of conversion vs. time, and a linear increase of molecular weight vs. conversion. The molecular weight distributions are narrow, and the initiator efficiency is ca. 1. This indicates that the amount of termination and other side reactions in this system are negligible.

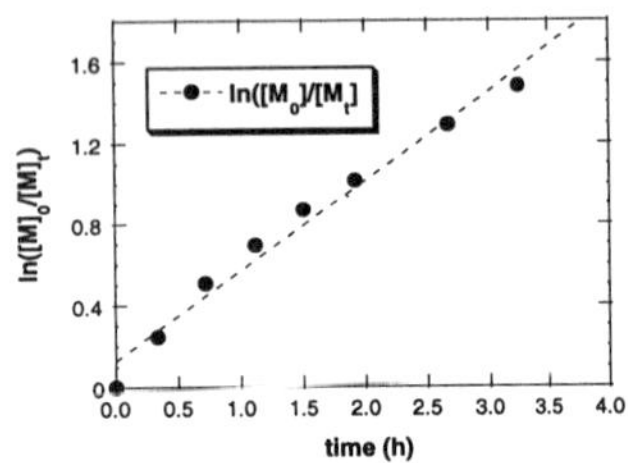

Figure 5. Semilogarithmic kinetic plot for ATRP of bulk styrene initiated by *t*-butyl 2-bromopropionate at 110 °C.

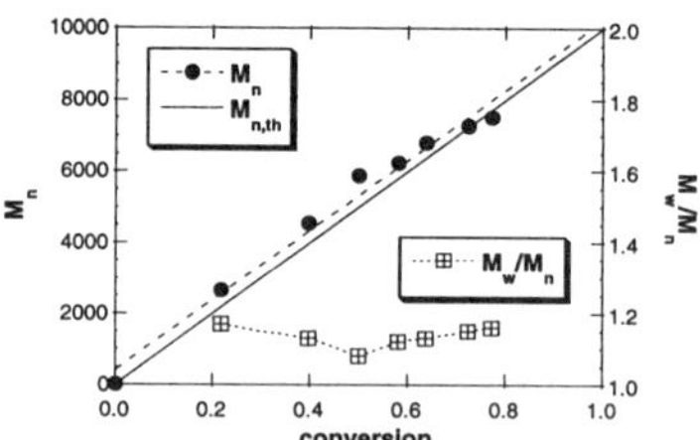

Figure 6. Evolution of experimental molecular weight and polydispersity with conversion for ATRP of styrene initiated by *t*-butyl 2-bromopropionate.

Conclusion

Atom transfer radical polymerization of styrene has been carried out using free and protected carboxylic acid initiators. Free carboxylic acid initiators have very low initiator efficiencies, and are not effective for ATRP of styrene. Trimethylsilyl and *t*-butyldimethylsilyl protected carboxylic acid initiators have moderate initiator efficiencies of ca. 0.6 and 0.8, respectively, for ATRP of styrene at 110 °C. *t*-Butyl protected carboxylic acid initiators have initiator efficiency close to unity, and are good initiators for ATRP of styrene. All protected carboxylic acid groups can potentially be hydrolyzed after polymerization, and converted to free carboxylic acid functionality.

Acknowledgments. Financial support by the industrial members of the ATRP Consortium at Carnegie Mellon University is gratefully acknowledged.

References

(1) Broze, G.; Jérôme, R.; Teyssié, P. *Makromol. Chem.* 1978, *197*, 1383.
(2) Sebenda, J. J. *Macromol. Sci., Chem.* 1972, 6, 1145.
(3) Perly, B.; douy, A.; Galot, B. *Makromol. Chem.* 1976, *177*, 2569.
(4) Wang, J. S.; Matyjaszewski, K. *J. Am. Chem. Soc.* 1995, *117*, 5614.
(5) Patten, T. E.; Xia, J.; Abernathy, T.; Matyjaszewski, K. *Science* 1996, *272*, 866.

Branched HDI HEUR's
Model compound study and preliminary rheology data

C.D. Anderson, Wylie H. Wetzel, J.E. Glass
North Dakota State University
Dept. Polymers and Coatings
54 Dunbar Hall
Fargo, ND 58105

Abstract

The use of Hexane Diisocyanate (HDI) in place of methylene bis-(4-cyclohexyl isocyanate) H12MDI was investigated to help elucidate differences between large branched and linear hydrophobes. HDI is a single discrete isomer, which will enhance the removal of the urea by-product and provide cleaner products. In addition, the contribution of HDI to the effective hydrophobe size will be significantly less.

The affects of anionic surfactant concentration on the oscillatory and shear rate profiles were investigated at levels above, below, and equal to the viscosity maximum. Current results were compared to previous results obtained using H12MDI as the diisocyanate coupler. HDI HEURs did not exhibit thixotropy, in contrast to the highly thixotropic H12MDI HEURs. In addition, HDI HEURs had shorter relaxation times and smaller storage modulus (G') indicating a less efficient network formation and faster diffusion rates than the H12MDI HEURs.

Introduction

Previous studies have investigated the affects of the hydrophobe's geometry on the solution rheology of Hydrophobically modified Ethoxylated URethanes (uni-HEUR's)[1]. It has been shown that branched hydrophobes will give a higher solution viscosity than linear hydrophobes with comparable surface energies. In addition, the hydrophobe structure has also been shown to affect the magnitude of the viscosity maximums for both anionic and nonionic surfactants.

The ability to quantify differences, when large hydrophobes are used in conjunction with variable molecular weight polyethylene glycols (PEG) is not straightforward with H12MDI as the coupler. H12MDI contributes significantly contribution to the effective hydrophobe size of a HEUR thickener, and the dual ring structure of the molecule promotes hydrophobe rigidity. In addition, the commercial process by which H12MDI is made results in a mixture of structural isomers, (1,3 and 1,4) and conformational isomers which depend on axial and equatorial arrangements. By replacing H12MDI with HDI cleaner products will result. This is because HDI is a single discrete isomer, which will make removal of the urea by-products easy.

Results and Discussion

Rheology

Due to the absence of steric hindrance and the increased reactivity of primary isocyanate groups, the probability of side reactions occurring with HDI, in contrast to H12MDI, increases. In order to determine whether or not side reactions occur, under the synthetic conditions used, model compounds were made and analyzed with ^{13}C NMR. Previously published results have shown the allophanate and biuret carbonyl to resonate between 156 and 150 PPM. The ^{13}C NMR analysis of the carbonyl region for the model thickener showed no significant peaks in this region **Figure 1**. However, the ^{13}C analysis indicated the formation of the HDI urea, 159 PPM, and uretdione, 158 ppm, by-products. Fortunately, these impurities are in low abundance, thereby providing no significant contribution to solutions rheology. In addition these impurities are readily removed during purification.

Rheology

The efficiency with which a given HEUR will thicken an aqueous solution is strongly dependent upon numerous factors. On method for evaluating, the thickening efficiency is by plotting the steady shear viscosity versus wt.% of thickener. These plots provide the general trend for the thickening efficiency of H12MDI HEURs with varying hydrophobe size. Both the PEG8000 and PEG35000 exhibit similar trends with the smaller hydrophobes, but a noticeable difference occurs with the PEG35000 and the large b-(C20H42) and L-(C16H33) hydrophobes. These solutions form fragile gels, which obscures the differences between the two hydrophobes. This is thought to be an affect of the contribution of H12MDI to the effective hydrophobe size. The significance of this contribution can be seen by comparing the b-(C20H42) thickeners made with HDI and H12MDI, as seen in **Figures 2-3**. By changing the isocyanate coupler from H12MDI to HDI a significant reduction in solution viscosity is seen in both the PEG8000 and PEG35000 comparisons.

The most effective method to achieve an understanding of the types of associations and the extent of network formation in solution is through oscillatory measurements. This provides insight to the storage (G') and loss modulus (G") which give a qualitative picture of the relative amounts of intra versus intermolecular associations and the relaxation times. The oscillatory data for the PEG35000 H12MDI and HDI HEURs, as shown in **Figures 4 and 5**, illustrates that H12MDI HEURs form networks more efficiently than the HDI HEURs. This is reflected in the large magnitude of the G' for the H12MDI HEURs. The HDI HEURs have

smaller G' and larger G". In addition, the relaxation times for the H12MDI HEURs are longer than for the HDI HEURs indicating slower rates of diffusion for H12MDI. The PEG8000 H12MDI and HDI HEURs, shown in **Figures 6-7**, follow similar trends. The HDI HEURs at the PEG8000 molecular weight are almost completely dominated by G", where as the H12MDI HEURs are still dominated by G'. This corresponds well with the wt% plots shown previously, since the fragile gels formed by the H12MDI thickeners imply higher network formation.

The affect of the surfactant concentration on the oscillatory responses is also of importance. The H12MDI thickeners were only run at low SDS concentrations (0.002M), because the amount of surfactant needed to obtain solubility was past the viscosity maximum. The HDI thickeners did not exhibit solubility problems and gave a clear viscosity maximum. As the surfactant concentration approaches the viscosity maximum, the relaxation times increase and the magnitude of G' also increases. This is indicative of the more extensive and complex networks forming.

A complimentary method to oscillatory measurements is shear rate profiles. Comparing the H12MDI and HDI b-(C20H41) thickeners in **Figures 8-11**, depicts the large differences the isocyanate coupler has on the shear rate profiles. The PEG8000 H12MDI HEUR shows only shear thinning behavior below the SDS maximum. As the maximum is approached and surpassed, a region of shear thickening occurs before the shear thinning begins. This phenomenon appears when intramolecular associations are destroyed and replaced by intermolecular associations. The flow profile for the PEG8000 HDI HEUR exhibits Newtonian flow below the viscosity maximum and again show shear thickening as the SDS concentration is increased. The Newtonian flow is a result of the smaller HDI contribution makes to the effective hydrophobe size, compared to H12MDI. The most notable difference between the H12MDI thickener and the HDI thickener is the thixotropy shown by the H12MDI HEURs, but not seen in the HDI HEURs. This lack of thixotropy in the HDI HEURs is related to the smaller effective hydrophobe size of HDI. Thixotropy occurs when the relaxation processes are slower than the time frame of the experiment. Thus the small effective hydrophobes of HDI HEURs are able to diffuse faster and relax within the time frame of the experiment and the large H12MDI HEURs cannot. Increasing the molecular weight of the PEG chain leads to similar observations. The only difference being the onset of shear thinning seen with the HDI HEURs.

---

[1] Wetzel, W. H. P.h.D. Thesis **1997**.

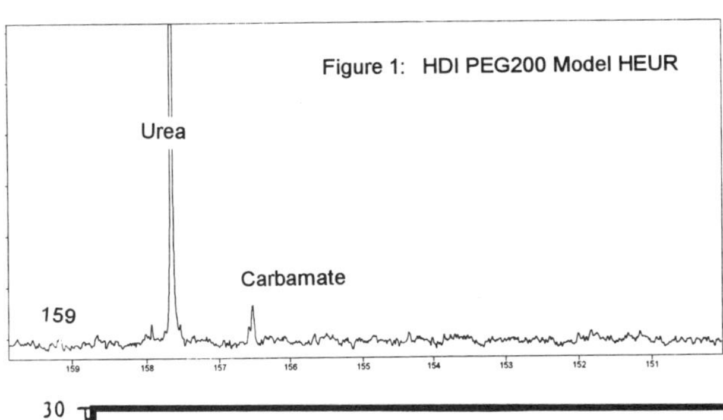

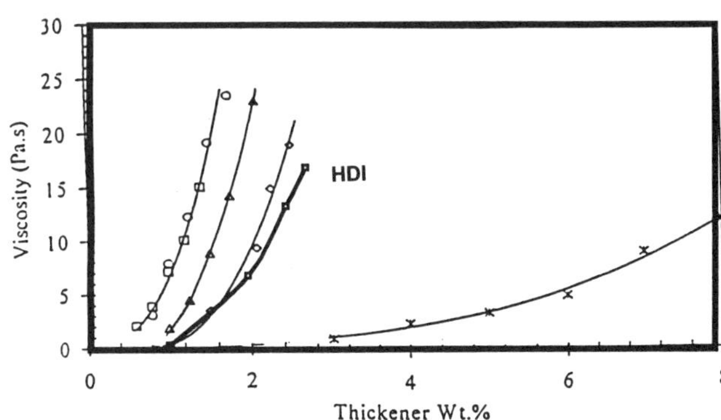

Figure 2 Low shear rate viscosity (2 s^{-1}) of H$_{12}$MDI-based uni-HEUR thickeners with different hydrophobe structures as a function of concentration. (□) b-(C$_{20}$H$_{42}$); (O) C$_{16}$H$_{33}$; (Δ) b-(C$_{16}$H$_{34}$); (◊) C$_{12}$H$_{25}$; (×) b-(C$_{12}$H$_{26}$): all with 0.002M SDS. A. 29,500 Mn; Mn.

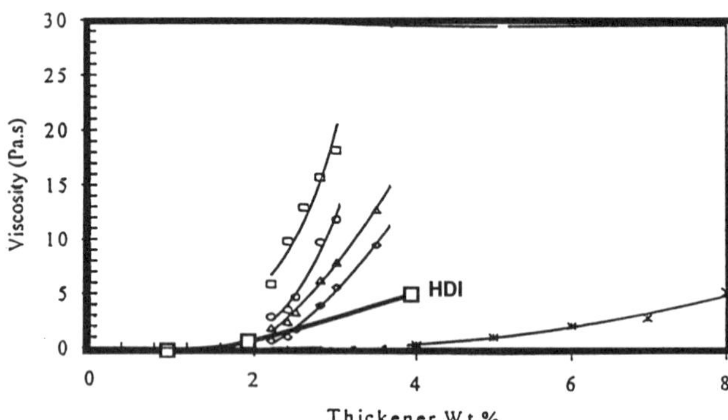

Figure 3 . Low shear rate viscosity (2 s⁻¹) of H₁₂MDI-based uni-HEUR thickeners with different hydrophobe structures as a function of concentration. (□) b-(C₂₀H₄₂); (O) C₁₆H₃₃; (Δ) b-(C₁₆H₃₄); (◊) C₁₂H₂₅; (×) b-(C₁₂H₂₆): all with 0.002M SDS. · 8,600 Mn.

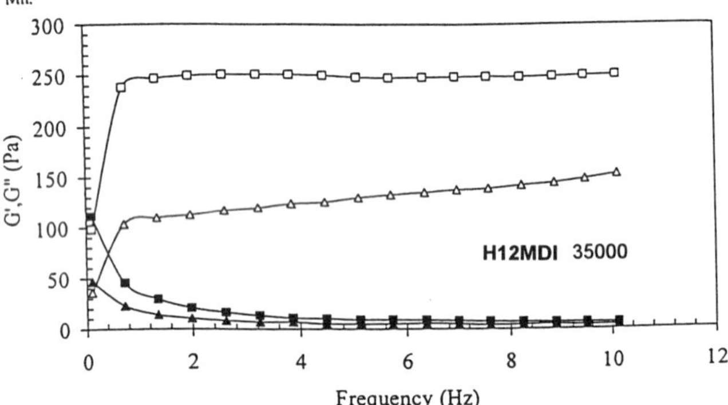

Figure 4   at (□) 1.4 wt.% thickener; (Δ) 1.2 wt.%.

Figure 5: 2 wt% HDI PEG35000 Oscillatory

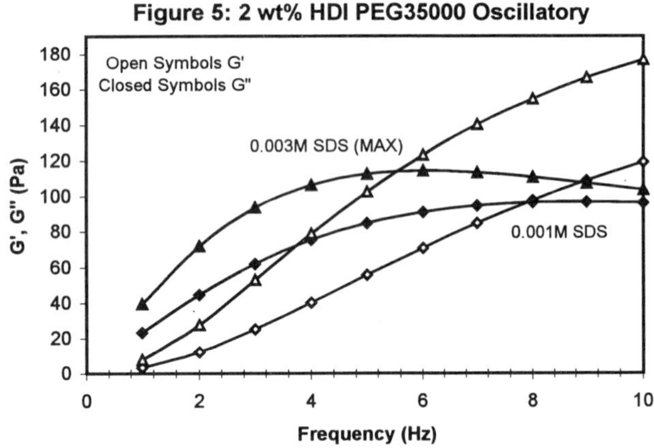

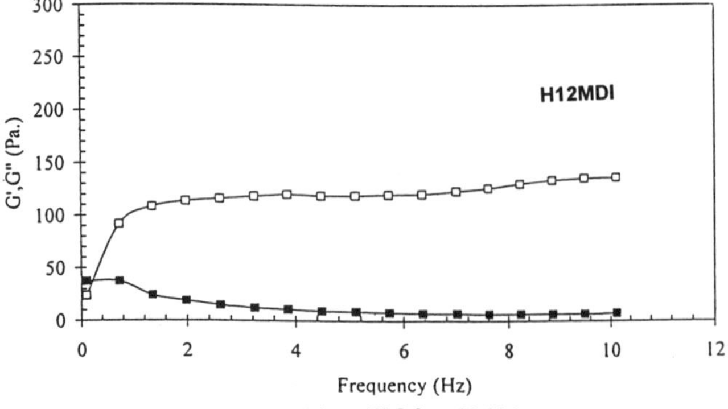

Figure 6   8,600 Mn at (□) 2.8 wt.% thickener.

Figure 7: 2 wt% HDI PEG8000 Oscillatory

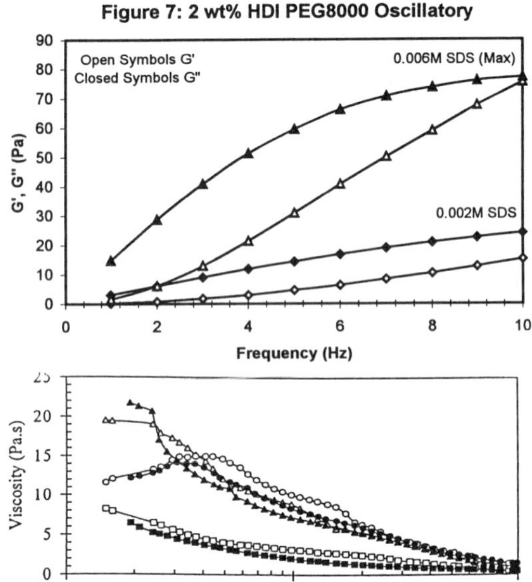

Figure 8 ι. 0.7 wt.% 29,500 Mn, H₁₂MDI-based uni-HEUR with a b-(C₂₀H₄₂) hydrophobe with SDS. (□) 0.003M (0.375 CMC); (Δ)0.005M (0.625 CMC); (O) 0.006M (0.75 CMC).

Figure 9: 2 wt% HDI PEG35000 Shear Profile

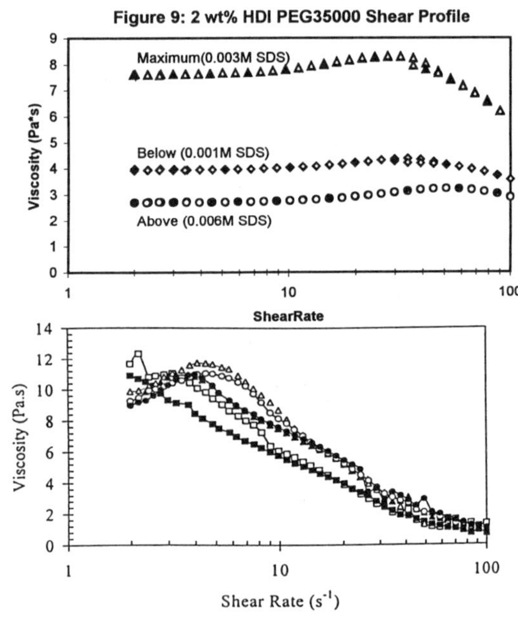

Figure 10 .8 wt.% 8,600 Mn, H₁₂MDI-based uni-HEUR with a b-(C₂₀H₄₂) hydrophobe with SDS. (□) 0.012M (1.5 CMC); (Δ) 0.016M (2.0 CMC); (O) 0.018M (2.25 CMC).

Figure 11: 2 wt% HDI PEG8000 Shear Profile

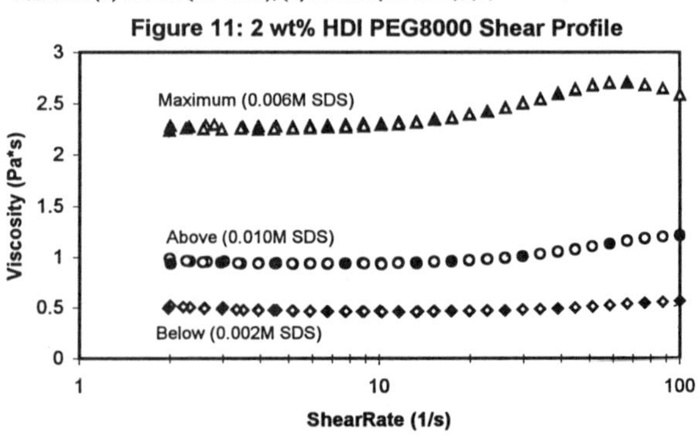

The Influences of Surfactant and Electrolyte on Rheology of HASE Thickener Aqueous Solutions

Lin-lin Xing, Mao Chen and J. Edward Glass[1]
Robert J. Buchacek and John G. Dickinson[2]

[1]Polymers and Coatings Department, North Dakota State University.
Fargo, ND 58105

[2]E.I. DuPont de Nemours & Co.
Chestnut Run, Building 709, Wilmington, DE 19898

ABSTRACT

The rheological properties of Hydrophobe-modified Alkali-Swellable Emulsion (HASE) thickener solution are studied, with emphasis on surfactant and electrolyte (NaCl) effect. Similar studies in model HASE thickeners have been reported[1]. This study examines two commercial HASE thickeners (HASE-615 and HASE-935) for comparison. In the absence of SDS, the NaCl addition changed solution viscosity depending on the relative concentration of the NaCl and thickener. The viscosities of HASE type thickeners were enhanced by an anionic surfactant, sodium dodecyl sulfate (SDS); and viscosity maxima were observed due to mixed micelle formation among thickener hydrophobes and surfactants. The addition of NaCl suppressed the viscosity of HASE-615/SDS solutions, with HASE-935 at higher concentrations actually increasing in viscosity with NaCl addition.

INTRODUCTION

Hydrophobe-modified Alkali-Swellable Emulsion (HASE) represents one of the three major types of associative thickener (i.e., HMHEC, HEUR, and HASE) that have achieved commercial success in waterborne coatings. HASE thickeners are carboxyl functional terpolymers made by free-radical emulsion polymerization among methacrylic acid, an acrylate comonomer such as ethyl acrylate, and a termonomer containing hydrophobic units attached to the vinyl position (either methacrylate or styrene). The hydrophobe is separated from the main chain, typically with ca. 40 oxyethylene units. Small amounts of difunctional monomers are also added to obtain a partially crosslinked structure to inhibit solubilization upon neutralization of carboxylic acid groups. The thickening of HASE solutions is generally contributed from (1) hydrophobic association, (2) volume fraction increase due to particle swelling, and (3) expanded polymer conformation due to electrostatic repulsion.

The objective of this study is to examine two commercial HASE thickeners (TT-615 and TT-935 from Rohm & Haas) for comparison with the previous model HASE studies and emphasizes their influences on water-borne latex coatings properties. The rheological behavior of HASE thickener aqueous solutions is studied ·in terms of different commercial thickeners, surfactant concentration, and salinity.

RESULTS AND DISCUSSION

The viscosity build of HASE thickeners is governed by the contribution from hydrodynamic volume and inter-molecular hydrophobic association leading to network formation. Storage modulus (G') and loss modulus (G") from oscillatory rheometry are frequently used to evaluate the viscoelasticity of thickened fluids or gels[2]. The loss modulus (G") reflects the viscous components while the storage modulus (G") reflects elastic components. From the crossover frequency (ω_c), G' =G", structural relaxation time ($\tau = 1/\omega_c$) of the thickener solution can be calculated. A crossover between G' and G" is observed for HASE-615 (1.2 wt%) at low frequency (< 0.5 Hz) while moduli of the HASE-935 solution approach crossover at a higher frequency even for a higher thickener concentration (5.0 wt.%)(**Figures 1A** and **1B**). The strong hydrophobic association is responsible for the highly elastic, gel-like HASE-615 thickener solution. The viscosity build of commercial HASE thickeners in aqueous solution, with and without NaCl, are illustrated in **Figure 2**. Again, HASE-615 is more efficient than HASE-935 due to its larger hydrophobe size and greater number of hydrophobes. Not surprisingly, the presence of 2.0 wt% NaCl lowered viscosity at low

HASE-615 thickener concentrations as it did in our previous model HASE thickener study. This is not observed for HASE-935 solution in which the presence of NaCl only enhanced solution viscosity. It is worth noting that the concentration range of HASE-615 (0.5-1.5 wt%) is much smaller than HASE-935 (1-4.5 wt%) while the NaCl concentration is constant (2 wt.%). It is known that excess electrolyte greatly reduces the mutual electrostatic repulsion of polyelectrolyte chains, leading to a collapsed polymer conformation. This heavily coiled conformation favors neither hydrophobic association nor the effective volume fraction of the thickener molecules. This was demonstrated in the model HASE thickener study[1]. Similar solution behavior is observed with the commercial HASE thickeners examined in this study. First, it is indicated that HASE-615 is more efficient than HASE-935, even at a high NaCl level (**Figure 3**). Second, the salt sensitivity of HASE thickeners is different, with HASE-615 being more sensitive to electrolyte than HASE-935. This is indicated by the NaCl concentration at the viscosity maximum and the magnitude of the viscosity drop. Since HASE-615 and HASE-935 have comparable amount of acids based on titration, the difference in the electrolyte effect may result from the effective surface acid concentration, governed by the morphology of swollen HASE particles.

At different thickener concentrations, solution viscosity as a function of surfactant concentration is illustrated in **Figures 4, 5**. Viscosity maxima were observed at each thickener concentration as a function of SDS concentration, and maximum viscosities increased with thickener concentration. Modulii of thickener solutions are illustrated (**Figure 6**) at three surfactant concentrations, representing those before viscosity maximum, at viscosity maximum, and after viscosity maximum, respectively. All modulii are plotted with their relaxation times ($\tau = 1/\omega_c$) against different surfactant concentrations (**Figure 7**). At all three surfactant concentrations, G' is dominant at low frequencies (< 1Hz), indicating the strong network formation enhanced by the surfactant in the mixed micellization process. Accordingly, the structural relaxation times of HASE thickener solutions become longer with increasing surfactant concentration, reach a maximum, and then decrease with further addition of surfactant. The difference in relaxation times corresponds to the characteristics of viscous liquids (fast relaxation) and solid-like gels (slow relaxation). These results substantiate the thickening mechanism involving network formation by inter-connecting mixed micelles consisting of thickener hydrophobes and surfactant monomers.

The viscosity maxima of HASE-935 appear at higher surfactant concentrations than HASE-615. This seems to indicate the significantly larger and/or more hydrophobes in HASE-615 than in HASE-935. For HASE-935, viscosity maxima shift to a higher surfactant level with increasing thickener concentration since more surfactant is needed to form mixed micelles with increased hydrophobe concentration. When the concentration of HASE-615 increased, however, no noticeable shift was observed for surfactant concentration corresponding to maximum viscosity. Because of the ionic character of both the HASE thickener and the SDS surfactant, the association between the thickener and the surfactant and the resultant rheological behavior are expected to be sensitive to the presence of electrolytes. If HASE-615 bears more electrostatic interaction, the viscosity of the HASE-615/SDS solution is expected to be more sensitive to the salt addition than that of HASE-935. This is exactly the case when different levels of NaCl are added to HASE/SDS solutions as are illustrated in **Figures 8 and 9**. Compared to HASE-935, larger viscosity loss was observed for HASE-615/SDS systems at different thickener concentrations.

REFERENCE

[1] Tarng, Ming-R., Glass, J. Edward *Proc. ACS Div. Polym. Materials: Sci. & Engin.*, **1992**, 67, 280.

[2] Jones, D.A.R.; Leary, B. and Boger, D.V. *J. Colloid Interface Sci.* **1992**,150,84.

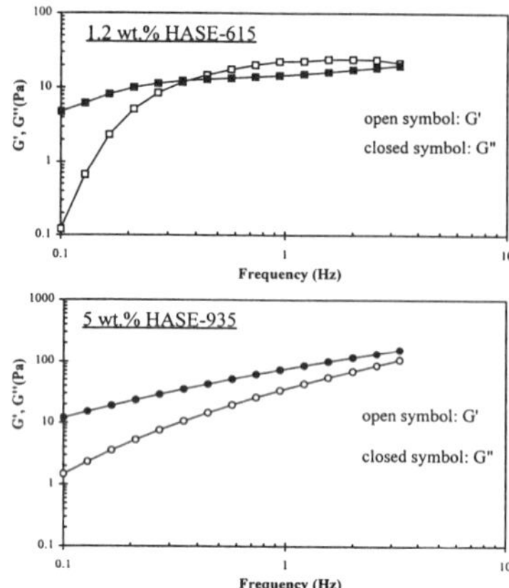

Figure 1. Storage modulus (G', Pa) and loss modulus (G", Pa) dependence on frequency (Hz) of HASE solutions.

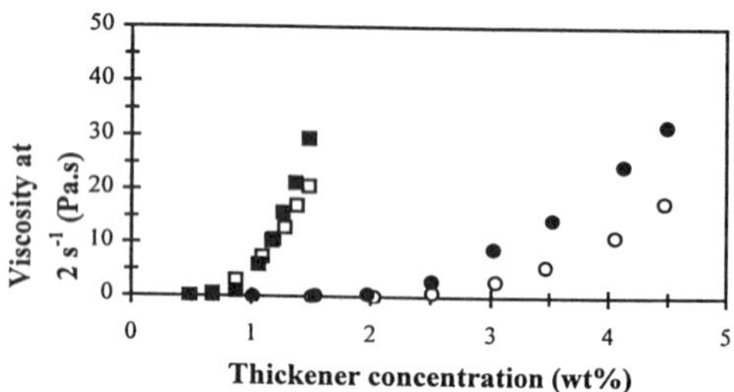

Figure 2. Viscosity build (2 s^{-1}) as a function of thickener concentration of commercial HASE thickeners in aqueous solution. Symbols: (□) HASE-615; (O) HASE-935. Open symbol: without NaCl; Closed symbol: with 2.0 wt% NaCl.

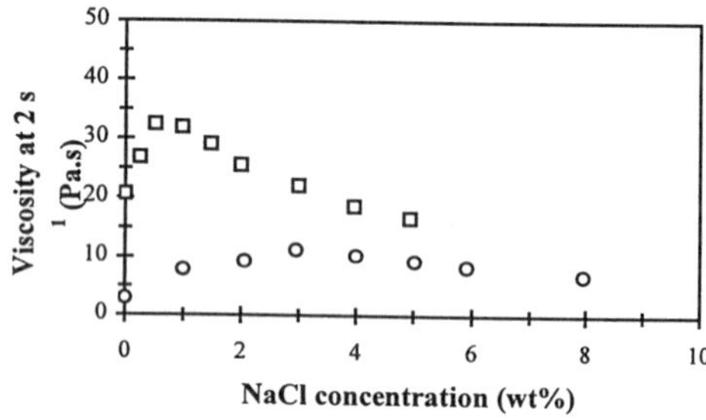

Figure 3. Viscosity (2 s^{-1}) as a function of NaCl concentration for commercial HASE thickeners in aqueous solution. Symbols: (□) 1.5 wt% HASE-615; (O) 3.0 wt% HASE-935.

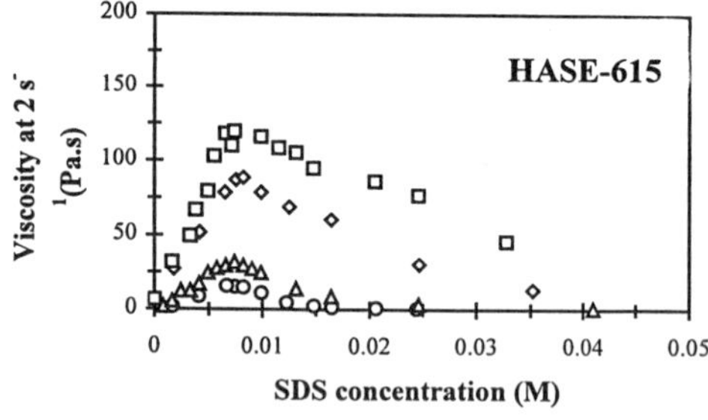

Figure 4. Viscosity build as a function of SDS concentration of HASE-615 thickener at different concentrations. Symbols: (□) 1.0 wt% HASE-615; (◊) 0.8 wt% HASE-615; (Δ) 0.5 wt% HASE-615; (O) 0.4 wt% HASE-615.

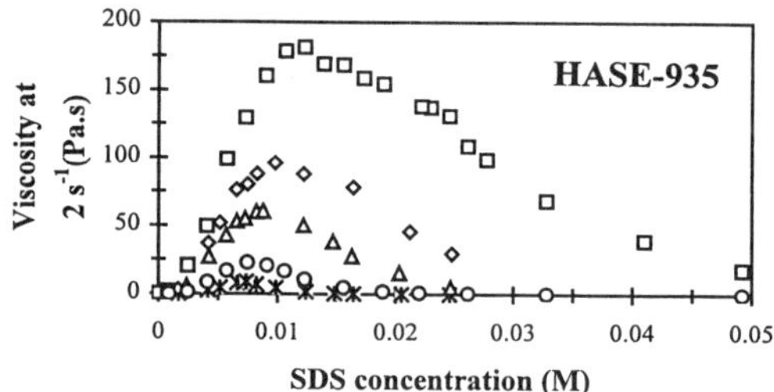

Figure 5. Viscosity build as a function of SDS concentration of HASE-935 thickener at different concentrations. Symbols: (□) 2.0 wt% HASE-935; (◊) 1.5 wt% HASE-935; (Δ) 1.2 wt% HASE-935; (O) 1.0 wt% HASE-935; (*), 0.8 wt% HASE-935.

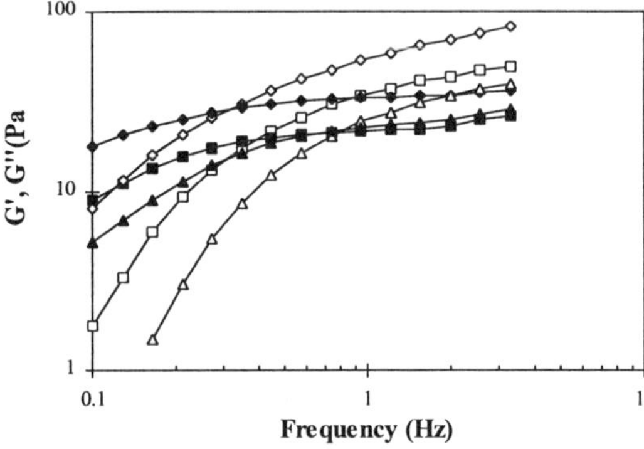

Figure 6. Storage modulus (G', Pa) and loss modulus (G", Pa) dependence on frequency (Hz) of 0.5 wt% HASE-615 solutions at different SDS concentrations. Symbols: (□) 0.005 SDS; (◊) 0.0082 SDS; (Δ) 0.013 M SDS. Open symbol: G'; Closed symbol: G".

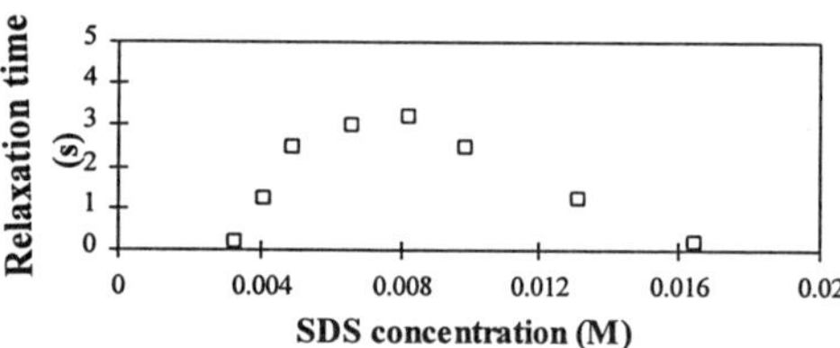

Figure 7. Relaxation time of 0.5 wt% HASE-615 solutions as a function of SDS concentrations.

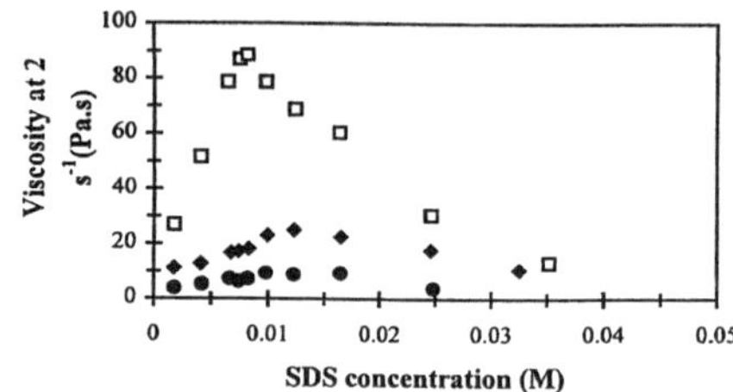

Figure 8. Low shear viscosity (2 s^{-1}) as a function of SDS concentration for 0.8 wt% HASE-615 thickener in aqueous solutions. Symbols: (□) 0 wt% NaCl; (◆) 1.0 wt% NaCl; (●) 2.0 wt% NaCl.

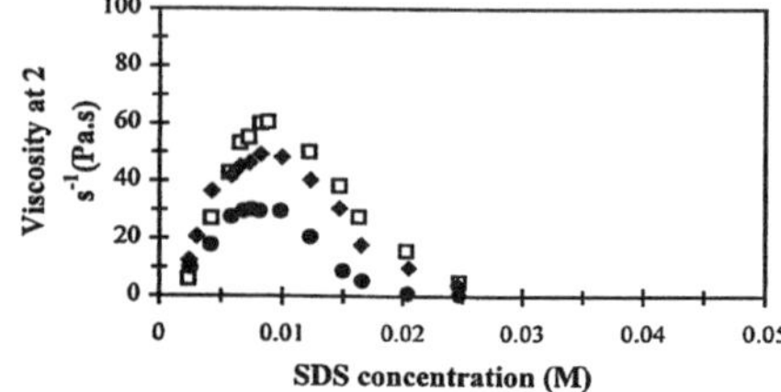

Figure 9. Low shear viscosity (2 s^{-1}) as a function of SDS concentration for 1.2 wt% HASE-935 thickener in aqueous solutions. Symbols: (□) 0 wt% NaCl; (◆) 1.0 wt% NaCl; (●) 2.0 wt% NaCl.

Commercial Comb-HEUR Thickener/Pigment Interactions and Their Influences on Disperse Phase Rheology

Lin-lin Xing, Mao Chen, J. P. Kaczmarski, and J. Edward Glass[1]

Robert J. Buchacek and John G. Dickinson[2]

[1]Polymers and Coatings Department, North Dakota State University, Fargo, ND 58105

[2]E.I. DuPont de Nemours & Co.

Chestnut Run, Building 709, Wilmington, DE 19898

ABSTRACT

Water-soluble polymers at the concentrations used to thicken latex coatings exhibit Newtonian viscosities; however, most fully formulated latex coatings exhibit non-Newtonian flow behavior. The thickener/disperse phase interaction, or lack thereof, is responsible for this difference. The first part of this paper examines the influence of water-soluble polymer and titanium dioxide (TiO_2) on the transition from Newtonian to non-Newtonian flow, with increasing volume fraction of pigment. It is observed that the dispersant structures, thickener type, and volume fraction of pigment significantly influence the rheology of pigment dispersions. This is similar to a previous study of this type with model uni-HEURs and S-G HEURs, but in this study commercial HEURs, one of the classical S-G structures with $C_{12}H_{25}$- terminal hydrophobes and two with a comb geometry. The variations in rheological properties are related with the adsorption of dispersant, surfactant, and Hydrophobe-modified Ethoxylated URethane (HEUR) associative thickener on disperse phase particles. The difference in rheological properties is also related to stability of the disperse phase in the presence of thickener and, consequently, with the film gloss of latex coatings. In the second part, the rheological properties of pigment dispersions (0.07 volume fraction), latex dispersions (0.25 volume fraction), and fully formulated latex coatings (0.32 volume fraction) are compared at the viscosity level commonly used in latex coating formulations. Based on the rheological behavior of each individual disperse phase, the contribution of two disperse phases to latex coatings rheology is discussed.

INTRODUCTION

Latex coating rheology and film properties are governed by the disperse phases (latex and pigment) and the matrix of interactions between the disperse phases and coating additives. These interactions are manifested when surfactant-modified water-soluble polymers (associative thickeners) are used to thicken such dispersions. Until recently, the knowledge of the component effect on coatings has been based primarily on experience rather than on structural understanding. Tarng[1] studied component effects on rheology and film gloss of latex coatings. It was noted that the type of thickener and latex particle size rather than dispersant strongly influenced the rheology of latex coatings. Associative thickeners enhance film gloss only with appropriate selection of dispersant. Hydrophobic dispersants provide higher film gloss than hydrophilic dispersants as a result of hydrophobic interaction between the dispersant and HEUR associative thickeners, originally discovered by Lundberg[2]. The competitive interaction of the surfactants and model uni-HEUR and S-G HEURs with terminal $C_{12}H_{25}$-hydrophobes for the surface of an acrylic latex recently[3] added significantly to our under-standing. In this contribution the goal is to examine the influence of different geometries of the HEUR thickener on the coating's rheology and we start as we did before by examining the competitive adsorption of HEUR with the surfactant for the dispersant treated TiO_2 surface. The HEUR thickeners in this study will be commercial: one will be the standard S-G urethane oxide thickener QR-708 or RM-825 with $C_{12}H_{25}$-terminal hydrophobes[4], synthesize with a 6000 molecular weight PEG. We do not know the size of the diisocyanate couple, but based on the magnitude of the viscosity of this S-G HEUR and more basic studies[5] it can be assumed to be a large one. The other two commercial HEURs are a factor of two or more greater in molecular weight than QR 708. Since it has been demonstrated that the viscosity of uniHEUR and S-G HEURs is determined by the effective terminal hydrophobe size, this limits the molecular weight of POE that can be used. To obtain viscosity in higher molecular weight associative polymers one must place pendant hydrophobes along the backbone. Interpretation of how much viscosity is contributed by mainchain entanglement and from the relatively high hydrophobe concentration has

not been delineated. In the one product SCT-270 (or 275) the pendant hydophobes are nonylphenol units; in the other, QR-200, the structure of the pendant hydrophobes are unknown but again based on more basic studies it must be larger than the nonylphenol units. The number of these pendant units occuring in clusters and the spacer distances between individual pendant units and or clusters are unknown, and would be hard to determine.

RESULTS AND DISCUSSION

All pigment dispersions were formulated at a volume fraction of 0.32 to approximately 90 KU viscosities. Two different types of dispersants were adsorbed on the TiO_2 surface. On these surfaces, the three commercial HEURs were adsorbed in competition with a typical formulation surfactant, Triton X-100. The adsorption data[6] are presented in Table I. The alpha olefin pendant size contiguous with each repeating maleic acid unit is C_4H_9- or C_8H_{17}- size with different dispersants and types of thickeners except in the study of flow behavior transition from Newtonian to non-Newtonian. Two types of dispersants were used in preparing pigment dispersions: an oligomeric methacrylic acid (o-MAA) and a diisobutylene/maleic acid co-oligomeric dispersant (o-DIBMA) with branched hydrophobic units. Two commercial associative HEUR thickeners with comb structures (SCT-270 and SCT-200) and the one with general S-G HEUR structure (QR-708) adsorb in different amounts on the treated TiO_2 surfaces. SCT 200 adsorbs almost quantitatively and independent of the dispersant's structure, and the X-100 surfactant adsorbs quantitatively with it on the C_8H_{17} pendant dispersant treated TiO_2. The other comb HEUR, SCT 270 with its weaker and fewer hydrophobes, does not adsorb as greatly on the smaller pendant hydrophobe dispersant surface. With the classical S-G HEUR (QR-708) the adsorption is quantitative on the C_8H_{17} dispersant treated TiO_2, but absent on the C_4H_9 treated surface, and the surfactant does not adsorb with it on the C_8H_{17} treated surface.

We can observe these differences also by oscillatory measurement on a cone and plate rheometer. In hydroxyethyl cellulose (HEC) thickened dispersions there is no adsorption on the pigment and the pigment is flocculated by a depletion mechanism[7] There should not be a differentiation among the different dispersant pretreatment of TiO_2. None is observed in **Figures 1 A and 2A**. In **Figure 1,** their is little adsorption[8] on the TiO_2. Except for SCT 200, and the moduli appear to reflect this. With the larger pendant hydrophobe dispersant where all of the HEUR thickeners adsorb nearly quantitatively and the moduli dependence on frequency are similar. Not what one would expect for stabilized pigments. But these dispersions are at a very high volume fraction and there is likely to be a substantial amount of interparticle bridging facilitated by hydrophobic associations. As the volume fraction of pigment is lowered at intervals down to a 0.07 level, the volume fraction of the pigment in a coating formulation. These differences in interactions also are reflected in the viscosity dependence on shear rate in the three different formulations as the volume fraction of the TiO_2 is decreased (**Figure 2**). On the left side of this figure the TiO_2 is treated with the more hydrophilic dispersant where the adsorption of the dispersant is minimal. The dispersions are very shear thinning until low volume fractions are reached. On the right side of this illustration the more hydrophobic dispersant promoting adsorption of the HEUR thickeners illustrate Newtonian flow even at high volume fraction loading , particularly with the large hydrophobe comb thickener. In the **Figure 3**, the model small acrylic latex is studied at a volume fraction of 0.25 (its level in a pigmented latex coating) It is obvious that the interaction of the thickener with the latex is more important that with the TiO_2 on the rheology of the coating (**Curve C**), in large part due to their different volume fraction in a coating. It is also obvious that the comb polymers stabilizes the latex best. That can change when the latex is changed and is under study in our labs.

REFERENCES

1 Tarng, M-R. Ph.D. dissertation. North Dakota State University, Fargo, 1995.

2 Lundberg, D.J. and Glass, J. E. *Journal of Coatings Technology(JCT)* **1992**. 64, 53.

3 Chen, M.; Wetzel, W.H.; Ma, Z. and Glass, J.E. *JCT* **1997**, 69(867), 73.

Table 1.

Competitive adsorptions of $C_8H_{17}C_6H_4O(EtO)_{10}H$ and HEUR 708 on model dispersant stabilized TiO_2 (21% NVV, 0.30 g dispersant/100 g pigment)

Grind	$C_8H_{17}C_6H_4O(EtO)_{10}H$			HEUR 708		
	[Init.]	[Ads]	% Ads	[Init.]	[Ads]	% Ads
High alumina TiO_2						
AOC(4)/MA	0.40	0.00	0	0.84	0.02	2
AOC(8)/MA	0.40	0.25	62	0.86	0.86	100

Competitive adsorptions of $C_8H_{17}C_6H_4O(EtO)_{10}H$ and HEUR 270 on model dispersant stabilized TiO_2 (21% NVV, 0.30 g dispersant/100 g pigment)

Grind	$C_8H_{17}C_6H_4O(EtO)_{10}H$			HEUR 270		
	[Init.]	[Ads]	% Ads	[Init.]	[Ads]	% Ads
High alumina TiO_2						
AOC(4)/MA	0.40	0.00	0	0.65	0.35	54
AOC(8)/MA	0.40	0.37	92	0.81	0.77	95

Competitive adsorptions of $C_8H_{17}C_6H_4O(EtO)_{10}H$ and HEUR 200 on model dispersant stabilized TiO_2 (21% NVV, 0.30 g dispersant/100 g pigment)

Grind	$C_8H_{17}C_6H_4O(EtO)_{10}H$			HEUR 200		
	[Init.]	[Ads]	% Ads	[Init.]	[Ads]	% Ads
High alumina TiO_2						
AOC(4)/MA	0.40	0.04	10	0.84	0.72	86
AOC(8)/MA	0.40	0.40	100	0.81	0.76	94

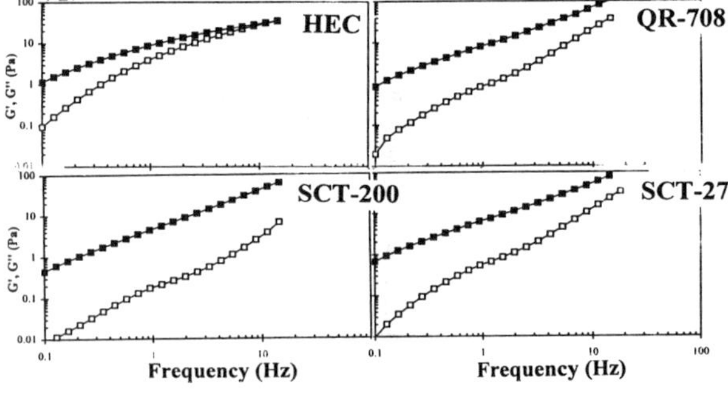

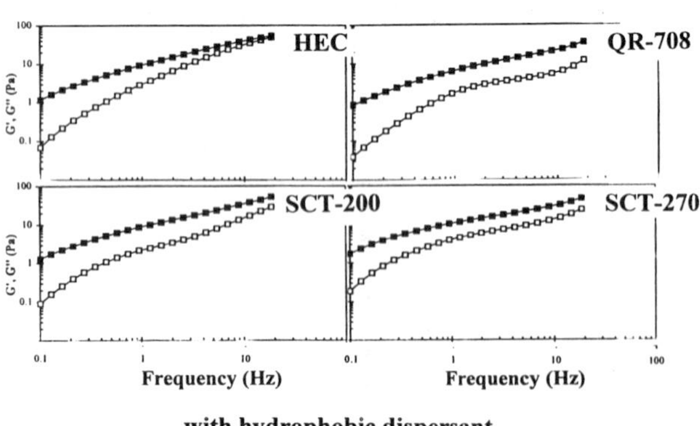

Figure 1. Storage modulus (G', Pa) and loss modulus (G", Pa) dependence on frequency (Hz) of 0.32 volume fraction Al_2O_3/TiO_2 pigment dispersions containing o-MAA dispersants and o-DIBMA, thickened with commercial thickeners to achieve 90 KU viscosity Open symbol: storage modulus (G'); Closed symbol: loss modulus (G").

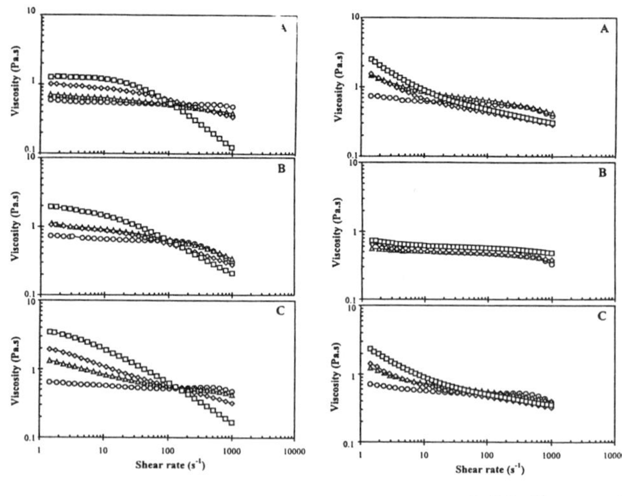

Figure 2. Shear viscosity profiles Al_2O_3/TiO_2 pigment dispersions (90 KU) containing different dispersants at different volume fractions (V. F.) of pigment, thickened with (A) HEUR-708, (B) HEUR-200; (C) HEUR-270. Symbols: Symbols: (□) 0.32 V. F., (◊) 0.25 V. F., (△) 0.20 V. F., (O) 0.07 V. F.

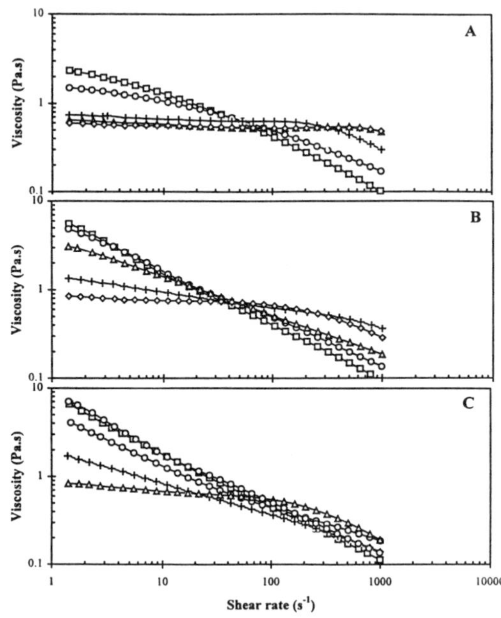

Figure 3. Shear viscosity profiles of (A) 0.07 volume fraction Al_2O_3/TiO_2 dispersions with o-DIBMA, (B) 0.25 volume fraction of 120 nm acrylic latex dispersions, and (C) 0.32 volume fraction pigmented latex coatings containing Al_2O_3/TiO_2 and **120 nm latex,** thickened with commercial thickeners to achieve 90 KU viscosity. Symbols: (□) HEC; (O) HMHEC; (+) HEUR-200; (△) HEUR-270; (◊) HEUR-708 (only ascending shear rate profiles are shown).

Howard, P. R.; Leasure, E.L.; Rosier, S.T.; and Schaller, E.J., in *Polymers as Rheology Modifiers*, ACS Advances Series 462, Schultz, D.N. and J.E. Glass eds., **1991**, Chapter 12, 207-221.

[5] May, Rebecca; Kaczmarski, J.P. and Glass, J.E. *Macromolecules* **1996**, 29, 4745.

[6] Kaczmarski, J.P.; Tarng, M.R.; Glass, J.E. and Buchacek, R.J. Progress in Organic Coatings **1997**, 30(1-2), 15.

[7] Sperry, Peter R. *Journal of Colloid and Interface Science* **1982**, 87(2), 375.

Structural Influences of Water-Soluble Polymers on Atomization and Spray Behavior

Peter Elliott[1], J. Edward Glass[1], and Raymond H. Fernando[2]

[1]Polymers and Coatings Department, North Dakota State University, Fargo, ND 58105

[2]Armstrong World Industries, Inc., Research Development Center, Lancaster, PA 17604

ABSTRACT

The purpose of the following research was to examine structural influences of water-soluble polymer blends on the spray behavior of fluids. The systems studied included high M_v hydroxyethyl cellulose (HEC), low M_v HEC, and low M_v poly(ethylene oxide) (POE), each blended with high M_v POE. The rheological parameters (shear viscosity, storage modulus, and extensional viscosity) of these systems were investigated and related to their air spray behavior. The role of high shear viscosity (HSV) and Dynamic Uniaxial Extensional Viscosity (DUEV) regarding atomization and sprayability was investigated. Differences were observed between blends (formulated to the same low shear viscosity) of high M_v POE, a flexible polymer, with high and low M_v HEC, segmentally rigid polymers, with the low M_v HEC blends (at ~13 times the concentration of high M_v HEC) displaying higher DUEVs and HSVs, as well as poorer sprayability. DUEV was determined to be the key parameter in relating spray behavior to structural influences in the polymer blends.

INTRODUCTION

Spraying is a process in which a quantity of fluid emerges from a nozzle as a sheet, which is rapidly disintegrated into ligaments, and then into a large number of small droplets. Although the application of coatings by spray has been practiced for some time, the process is still not well understood because of the complexity of the atomization process, differences in the design, size, and operating conditions of the nozzles, and variations in the fluid's properties (1). Fluid properties such as surface tension, shear viscosity, viscoelastic behavior, and extensional viscosity have been studied and related to the sprayability of systems (1,2,3).

When formulating water-borne coatings, increased high shear viscosities (HSVs) can be obtained by the addition of various amounts of low molecular weight water-soluble polymer, such as low M_v HEC. This is done to facilitate better film build in roll and brush applications. However, higher HSVs are generally not desirable in spray applications since they tend to promote air entrapment and surface defects in the applied coating. Therefore, in the coatings industry, it is a common practice in spray applications to relate high shear viscosities to sprayability. This is similar to the early studies on roll applications that related HSVs to spatter. Further studies in this area (4,5), however, proved that the misting behavior of roll applied coatings (spatter) was related to dynamic uniaxial extensional viscosities (DUEVs).

Previous work by our group has focused on determining which fluid property (surface tension, high and low shear viscosity, G', or DUEV) controls sprayability. This work was done on complex, fully formulated coatings, though, making it difficult to discern the dominant parameter. In this research, we use simple aqueous solutions of water-soluble polymer blends to try and differentiate between the effects of HSVs and DUEVs on atomization and sprayability. Since high molecular weight polymers, such as HEC and POE, have little or no effect on the surface tension, it is expected that their rheological properties will control the mechanism by which the polymer solutions resist atomization.

EXPERIMENTAL

The blends used in this study consist of variable component ratios among individual sets (**Table 1**), and each set is formulated to produce the same shear rate profile, but with different DUEVs, allowing for separate control of the shear and extensional responses. The rheological parameters examined for these blends were shear viscosity, storage modulus (G'), and dynamic uniaxial extensional viscosity (DUEV). Air spray studies were performed at 55 psi with fan, hollow cone, and solid cone nozzles with equivalent orifice diameters of .034 in., .031 in., and .025 in., respectively.

RESULTS/DISCUSSION

The sets of blends in **Table 1** were formulated to all have the same low shear viscosity (LSV), but vary in high shear viscosity (HSV). It took approximately 13 times the concentration of low M_v HEC to get the same LSV as the high M_v HEC blends accounting for their higher HSVs. Between the three solutions that comprise a set, the LSVs and HSVs are the same, but the dynamic uniaxial extensional viscosities (DUEVs) are different. This can be seen in **Figures 1-5** (in **Figure 2**, the storage modulus data, G', which reflects elasticity, follows the same trends as the HSV data, and so they will be discussed together as HSV in the text). Therefore the variables are HSVs between sets of blends, and DUEVs within a given set of blends. With the simple systems used in these experiments, this should allow for the differentiation of the contribution of HSV and DUEV to atomization and sprayability.

The effect of high shear viscosity on sprayability can be seen by examining **Figures 6 and 7** (only the fan nozzle gave spray patterns). The high M_v POE/high M_v HEC blends appear to spray better on average then the high M_v POE/low M_v HEC blends. At .005 wt.% high M_v POE, the spray patterns are comparable, but as the concentration of high M_v POE is increased, large differences in the spray patterns develop. This can be explained in one of two ways based on the rheological data in **Figures 1-5**. The high M_v POE/high M_v HEC blends have lower HSVs than the high M_v POE/low M_v HEC blends, due to the higher concentrations of low M_v HEC (~13 times as much) required to achieve the same LSV. This could be argued as the reason for the difference in their spray behavior. Also, as **Figures 3-5** display, the DUEVs of the high M_v POE/high M_v HEC blends are lower than the high M_v POE/low M_v HEC blends, and this also could be argued as the reason for the spray discrepancies.

The argument becomes clearer when examining the spray behavior among blends in the same set. As the component ratios are varied for the high M_v POE/high M_v HEC blends, not much change occurs in the spray pattern (**Figure 6**). This is reflected in the fact that they all have the same HSV (and G') and there is not much difference in their DUEVs. However, with the high M_v POE/low M_v HEC blends (**Figure 7**), as the component ratios are varied, there is a significant difference in the sprayability. At .005 wt.% high M_v POE, the spray behavior is similar to that of the high M_v POE/high M_v HEC blends, but as the concentration of high M_v POE is increased to .010 wt.% and then .015 wt.%, the spray behavior gets exceedingly worse. This cannot be explained with HSV (and G') data, since they are equivalent for the blends within the set. However, examination of the extensional viscosity data (**Figures 3-5**) reveals an excellent correlation between the observed sprayability and the measured extensional viscosities of the high M_v POE/low M_v HEC blends. The blend with .015 wt.% high M_v POE had the highest DUEV and the worst sprayability of the set; whereas, the .005 wt.% high M_v POE blend had the lowest DUEV (similar to the high Mv POE/high M_v HEC blends) and the best sprayability of the set.

In these experiments, the fact that the low M_v HEC blends produced higher DUEVs was the most interesting and significant observation. This was not expected based on consideration of structural and molecular weight parameters that influence extensional viscosities. As individual solutions, the flexible, high M_v POE molecules contribute relatively little to the shear viscosity but significantly to the extensional viscosity; whereas, the more rigid HEC molecules contribute more to the shear viscosity through entanglements, than to extensional viscosity. Extensional viscosity is not only dependent on polymer flexibility, but also on molecular weight, so that solutions of low M_v HEC give lower DUEVs than solutions of high M_v HEC. However, as is apparent in **Figures 3-5**, when flexible and rigid polymers are blended, the molecular weight relationship for the rigid polymers is reversed. The low M_v HEC blends produce higher DUEVs than the high M_v HEC blends. The explanation for this is thought possibly to lie in reptation theory, where the more flexible high M_v POE molecules are thought to be trapped in an array of obstacles (the low M_v HEC molecules that are present in significantly higher concentration than the high M_v HEC in the POE blends) and are forced to move in a worm-like (reptative) fashion (6). This type of behavior could produce higher than expected extensional strains, thereby producing higher extensional viscosities. Since the concentration of high M_v HEC in the blends is much lower, this behavior is thought not to occur. It is not definite whether this theory gives a proper explanation for this phenomenon, and further investigations are continuing.

Table 1: Blends Of Water-Soluble Polymers (wt. %)

Set 1		Set 2		Set3	
POE (M_v=1*10^6)	HEC (M_v=9.5*10^5)	POE (M_v=1*10^6)	HEC (M_v=6.8*10^5)	POE (M_v=1*10^6)	POE (M_v=2.0*10^4)
.015	.2970	.015	3.87	.015	17.5
.010	.3015	.010	3.91	.010	17.6
.005	.3050	.005	3.95	.005	17.7

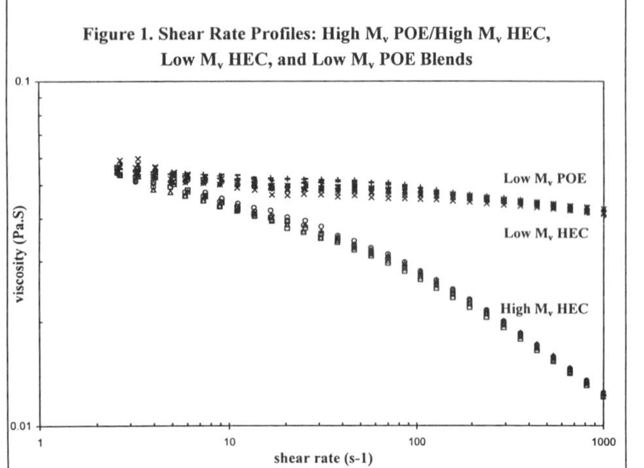

Figure 1. Shear Rate Profiles: High M_v POE/High M_v HEC, Low M_v HEC, and Low M_v POE Blends

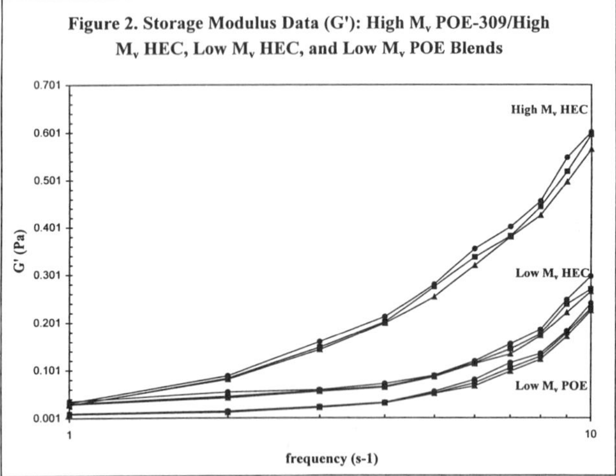

Figure 2. Storage Modulus Data (G'): High M_v POE-309/High M_v HEC, Low M_v HEC, and Low M_v POE Blends

Figure 6   Air Spray (fan nozzle)   Figure 7

High M_v POE/High M_v HEC Blends   High M_v POE/Low M_v HEC Blends

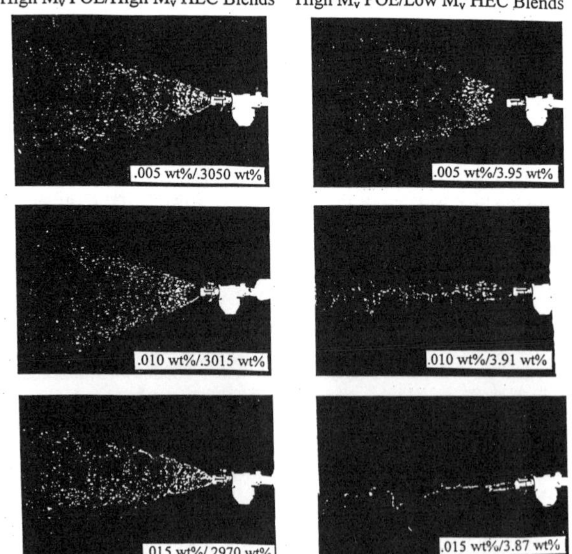

Figure 3. Extensional Viscosity: High M_v POE-309/ High M_v HEC and Low M_v HEC Blends

Figure 4. Extensional Viscosity: High M_v POE/High M_v HEC and Low M_v HEC Blends

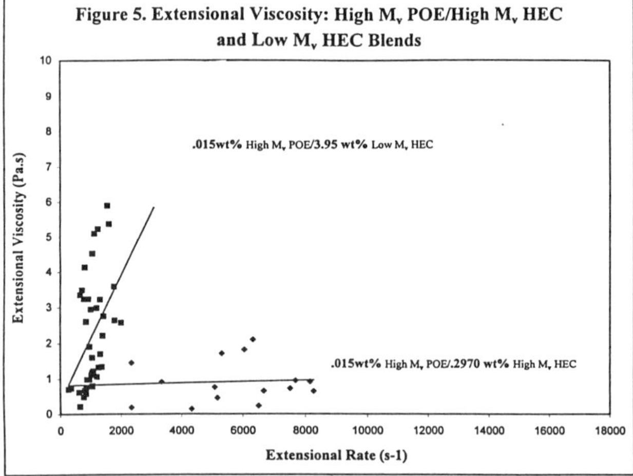

Figure 5. Extensional Viscosity: High M_v POE/High M_v HEC and Low M_v HEC Blends

1) Xing, L.L.; Glass, J.E.; and Fernando, R.H., "Spray Application of Waterborne Coatings," ACS Advances Series 663, Glass ed., **1997**, *Ch. 15*, 264-295.

2) Dexter, R.W., "Measurement of Extensional Viscosity of Polymer Solutions and Its Effects on Atomization from a Spray Nozzle," *Atomization and Sprays*, **1996**, *6*, 167-191.

3) Mansour, A. and Chigier, N., "Air-blast atomization of non-Newtonian liquids," *J. Non-Newtonian Fluid Mech.*, **1995**, *58*, 161-194.

4) Glass, J.E., Journal of Coating Technology, 1978, 50(641), 56.

5) Fernando, R.H. and Glass, J.E., "Dynamic Uniaxial Extensional Viscosity (DUEV) Effects in Roll Application II. Polymer Blend Studies," *J. Rheology*, **1988**, *32(2)*, 199-213.

6) Kholodenko, A.L.," Reptation Theory: geometrical and topological aspects," *Macromol. Theory Simul.*, **1996**, *5*, 1031-1064.

Rheology of Solutions of Hydrophobically Modified Polymers with Spherical and Rod-Like Surfactant Micelles

Santipharp Panmai[1], Robert K. Prud'homme[1], Dennis G. Peiffer[2], Steffen Jockusch[3], and Nicholas J. Turro[3]

[1] Dept. of Chemical Engineering, Princeton University, Princeton, NJ 08544
[2] Exxon Research and Engineering Company, Annandale, NJ 08801
[3] Dept. of Chemistry, Columbia University, New York, NY 10027

Introduction

Interactions between hydrophobically modified polymers (hm-polymers) and surfactants have been extensively studied.[1,2] These systems are of great interest as rheology modifiers in many applications, such as enhanced oil recovery and latex coatings. Hydrophobically modified polymers are water-soluble polymers which contain hydrophobic groups, or hydrophobes, attached either as grafts along the backbone or as terminal groups. Viscosity enhancement of an aqueous hm-polymer solution is achieved through intermolecular associations between hydrophobes, forming micelle-like hydrophobe clusters and transient crosslinks between polymer chains. With surfactant, the hydrophobes induce a specific aggregation of surfactant micelles around hydrophobe clusters to form mixed micelles at concentrations lower than the critical aggregation concentration (*cac*) of unmodified polymer.[3] The bridging of hydrophobe clusters by spherical micelles leads to an increase in viscosity as surfactant is added. However, at high surfactant concentrations, the hydrophobes are solubilized individually, or masked, by excess spherical micelles, resulting in the disruption of associations and a decrease in viscosity.

In this work, first, we investigated the effects of hydrophobe structure and chemistry on the interactions between hm-polymers and spherical micelles by rheology. The number of hydrophobes in a mixed micelle at different surfactant concentrations was determined by fluorescence quenching experiments, and the values were correlated with rheological data to provide a general picture of interactions between hm-polymers and spherical micelles. Later, expanding on earlier work by Peiffer,[4] we explored the use of rod-like micelles to control the viscosity of hm-polymer solution at high surfactant concentrations. We found that the masking problem with spherical micelles can be avoided and that the viscosity can be tuned by controlling the extent of rod formation.

Experiments

Hydrophobically modified hydroxyethylcellulose (hmHEC) (M_w = 2.5×10^5) was provided by Aqualon, Inc. (Wilmington, DE). The hydrophobe units are linear C_{16} alkanes at substitution levels of 0.9, 1.3, and 2.7%mole. Hydrophobically modified polyacrylamide (hmPAM) ($M_w = 1.5 \times 10^6$) contains sulfonated C_{12} alkyl hydrophobes at a level of 0.98%mole. Sodium dodecylsulfate (SDS), cetyltrimethylammonium bromide (CTAB), tetradecyltrimethylammonium bromide (TTAB), and KBr were obtained from commercial sources.

Viscosity measurements were made with the Rheometrics Fluid Spectrometer II using the Couette geometry at either 25 or 30°C. Zero-shear viscosity readings were taken from the Newtonian values at shear rates between 0.1 and 1 s^{-1}.

Steady-state fluorescence experiments were performed on a SPEX Fluorolog double spectrometer at 30°C. Pyrene was used as a probe, and cetylpyridinium chloride as a quencher. I_1 and I_3 intensities were taken from the emission intensities at 373 and 384 nm.

Results and Discussion

HM-Polymers and Spherical Micelles. We looked at the effects of various parameters, i.e., hydrophobe and surfactant lengths, on the interactions between hm-polymers and spherical micelles in a recent paper.[3] Here, we discussed the roles of hydrophobe content and charge on hydrophobic interactions. In Figure 1, the viscosities of semidilute C_{16}HEC solutions with different hydrophobe contents are plotted versus SDS concentration. The viscosity of C_{16}HEC solution with no surfactant increases with increasing hydrophobe content because of increased associations. No interaction was observed between unmodified HEC and SDS. As SDS is added, the viscosity of C_{16}HEC solution first increases due to micellar bridging of hydrophobe clusters and later decreases due to hydrophobe masking by excess micelles. The maximum in the viscosity increases and occurs at a higher SDS concentration with increasing hydrophobe content. Strong interactions at low SDS concentrations lead to phase separation. At high surfactant concentrations, the viscosity of masked C_{16}HEC converges to that of unmodified HEC.

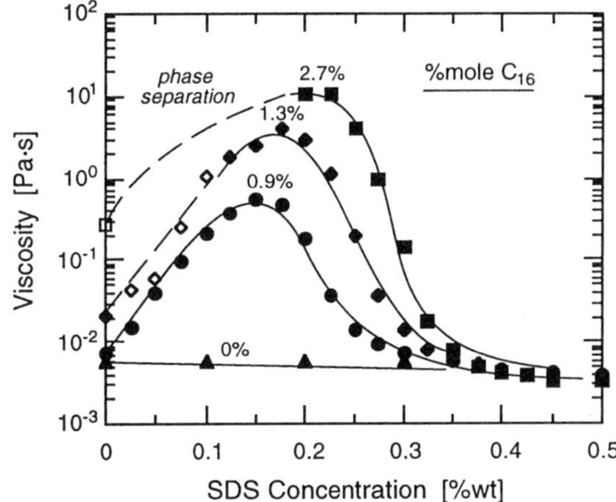

Figure 1. Viscosity of 0.5%wt C_{16}HEC solutions versus SDS concentration at different hydrophobe contents at 30°C.

Electrostatic interactions also affect hydrophobic interactions between hm-polymers and surfactants. The role was studied by looking at hmPAM whose C_{12} hydrophobes contain negative charges. In Figure 2, the viscosities of semidilute C_{12}PAM solutions are shown as a function of surfactant concentration for anionic SDS and cationic CTAB. The viscosity of C_{12}PAM solution decreases slightly with increasing SDS concentration. No bridging interaction was seen

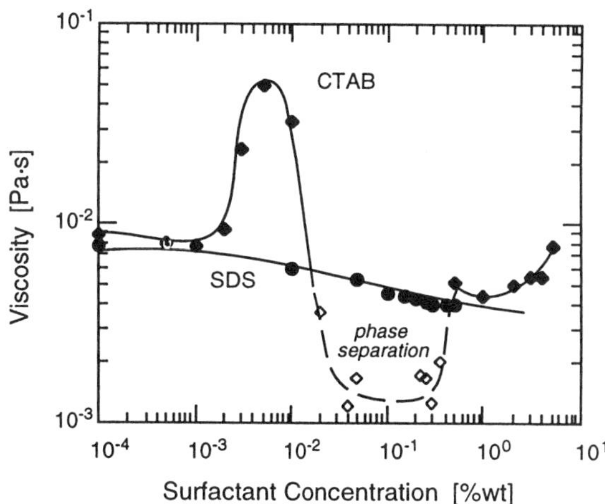

Figure 2. Viscosity of 0.5%wt C_{12}PAM solutions versus surfactant concentration for SDS and CTAB at 30°C.

between $C_{12}PAM$ and SDS due to charge repulsion between the hydrophobes and surfactants. On the other hand, the viscosity of $C_{12}PAM$ solution increases dramatically with the addition of CTAB due to both charge attraction and hydrophobic interaction. At intermediate CTAB concentrations, phase separation occurs as a result of charge neutralization. Only at high CTAB concentrations does resolubilization take place.

Number of Hydrophobes in a Mixed Micelle. Steady-state fluorescence is commonly used to study micellar aggregation in polymer and surfactant systems.[5] The number of hydrophobes in a mixed micelle of hydrophobes and surfactants (N_H) was determined by performing quenching experiments, in which a quencher is added to inhibit fluorescent emission of a probe.[6,7] For hmHEC with a stiff backbone and no intramolecular association, the number of inter-molecular associating hydrophobes can be taken from the measured N_H value.[6] In Figure 3, the number of hydrophobes in a mixed micelle at various surfactant concentrations is correlated with the viscosity profiles of $C_{16}HEC$ solution for CTAB and TTAB. N_H decreases steadily with increasing surfactant concentration due to the dilution of the hydrophobes by the surfactants -- from greater than 2 with no surfactant added to around 2 at the viscosity maximum and to 1 when hydrophobes are masked. The N_H value of about 2 at the viscosity maximum, which marks the transition from the bridging ($N_H>2$) to masking ($N_H<2$) surfactant concentration regions, is consistent the picture of interactions between hm-polymers and spherical micelles. The associating junctions are no longer effective when the number of hydrophobes in a mixed micelle is less than 2.

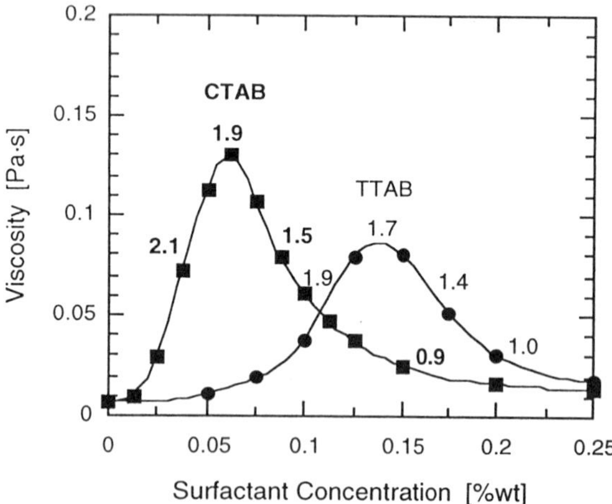

Figure 3. Viscosity of 0.5%wt $C_{16}HEC$ solutions (0.9%mole) versus surfactant concentration for CTAB and TTAB at 25°C; the number of hydrophobes per mixed micelle is indicated along the viscosity profiles.

HM-Polymers and Rod-Like Micelles. Next, we studied the interactions between hm-polymers and rod-like micelles. The objective is to control the viscosity of hm-polymer solution at high surfactant concentration by using rod-like micelles to overcome the masking problem with spherical micelles. Rod-like CTAB micelles are formed from spherical micelles with the addition of KBr salt. In Figure 4, the viscosities of semidilute $C_{16}HEC$ solutions as a function of CTAB concentration at different KBr concentrations are shown. With no salt, the viscosity of $C_{16}HEC$ solution first increases and later decreases with CTAB concentration due to bridging and masking interactions with spherical micelles. With the addition of KBr, the viscosity of $C_{16}HEC$ solution at high CTAB concentration increases due to the formation of rod-like micelles and bridging interactions with hm-polymers. At 2%wt KBr the viscosity of $C_{16}HEC$ solution is maintained over a range of

CTAB concentration while at 3%wt KBr the viscosity increases monotonically. Thus, the viscosity of hm-polymer solution can be tuned over a range of surfactant concentration simply by using a different surfactant morphology and by controlling the extent of rod-like micelle formation.

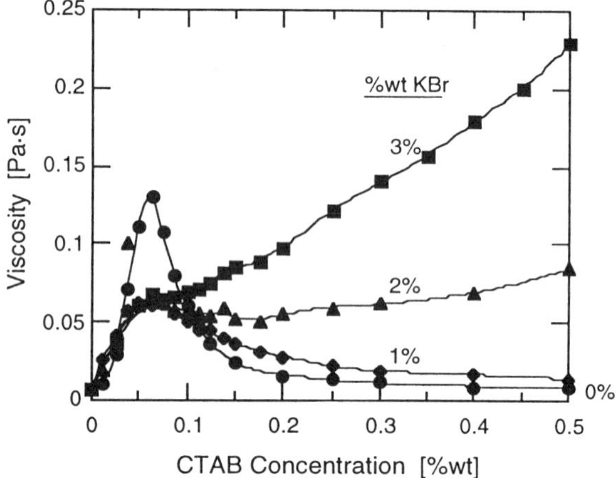

Figure 4. Viscosity of 0.5%wt $C_{16}HEC$ solutions (0.9%mole) versus CTAB concentration at different KBr concentrations at 30°C.

Summary

Interactions of hydrophobically modified polymers with spherical and rod-like surfactant micelles were studied by rheology. With spherical micelles, the viscosity of hm-polymer first increases and later decreases due to bridging and masking interactions. The strength of hydrophobic associations depends on the structure and chemistry of hydrophobes, such as their content and charge. Fluorescence quenching experiments showed that there are about 2 hydrophobes in a mixed micelle of hydrophobes and surfactants at the viscosity maximum. With rod-like micelles, the viscosity of hm-polymer solution increases monotonically with surfactant concentration due to bridging interactions. Viscosity control is achieved by controlling the extent of rod formation.

Acknowledgements

We would like to acknowledge financial support from the NSF to the Princeton University MRSEC which has funded this work.

References

1. Glass, J. E., Ed. *Polymers in Aqueous Media: Performance Through Association*; Advances in Chemistry Series 223; American Chemical Society: Washington, DC, 1989.

2. Glass, J. E., Ed. *Hydrophilic Polymers: Performance with Environmental Acceptability*; Advances in Chemistry 248; American Chemical Society: Washington, DC, 1996.

3. Panmai, S.; Prud'homme, R. K.; Peiffer, D. G. *Colloids Surf. A*, in press.

4. Peiffer, D. G. *Polymer* 1990, *31*, 2553.

5. Winnik, F. M.; Regismond, S. T. A. *Colloids Surf. A* **1996**, *118*, 1.

6. Panmai, S.; Prud'homme, R. K.; Peiffer, D. G.; Jockusch, S.; Turro, N. J. submitted to *Langmuir*.

7. Magny, B.; Iliopoulos , I.; Zana, R.; Audebert, R. *Langmuir* **1994**, *10*, 3180.

CONFINEMENT OF HYDROPHOBICALLY MODIFIED POLYACRYLATES INTO NONIONIC SURFACTANT LYOSTROPIC BILAYERS

Y. Yang[1], C. M. Marques[2], P. Richetti[2], R. Prud'homme[1]

1.Department of Chemical Engineering, Princeton University, Princeton NJ 08544
2.CNRS/Rhone-Poulenc complex Fluids Lab., Rhone-Poulenc Inc., CN 7500, Prospect Plains Road, Cranbury, NJ 08512-7500

Introduction

Mixing polymer into a solution composed of surfactant lamellar membrane may change inter and intra-membranes properties. For instance, the smectic compressibility modulus or the mean spacing between the membranes changes with their flexibility or curvature. Accordingly, new phase behaviors are expected. The study of such mixed systems is of practical interest since they are currently formulated in liquid detergent and cosmetic products. In addition, polymer/surfactant membrane solutions could be model systems for more complex biological membrane systems.

Charged or non-charged surfactant membranes behave differently. For example, an ionic surfactant membrane solution can confine a non-charged polymer with the radius of gyration several times larger than the interlayer spacing [1], however, it is not the case for nonionic surfactant system [2-3]. Generally, the water-soluble polymers with molecular weights in the order of 10^5 are not soluble in the nonionic surfactant mesophases due to the comparable molecular size with the geometry of the bilayer spacing [2-6]. However, solubility can be achieved by grafting hydrophobic side chains on the backbone since hydrophobic side chains are anchored on the lyotropic membrane [2-7]. The variation of mean bending modulus κ, Gaussian bending modulus $\overline{K}$ and compression modulus B upon adding polymers has been theoretically and experimentally studied [8-13]. Polymer induced excess membrane rigidity from an anchoring polymer system was first quantitatively reported by our group [14]. In this paper, we systematically study the effect of polymer structure and polymer concentration on the phase behavior and membrane rigidity of the bilayer solutions.

Experiments

The polymer is poly(sodium acrylate) modified with 0 to 5 mol% aliphatic chains containing 14 carbons. The hydrophobic side chains are randomly distributed along the polymer backbone [15]. The molecular weight of precursor polymers ranges from 90,000 to 1,000,000, corresponding to repeat units ranging 1,250 to 13,800. The surfactant membranes consist of pentaethylene glycol dodecyl ether ($C_{12}EO_5$), hexanol and 0.1 M NaCl aqueous solution. Brine solution is used to minimize the electrostatic interactions between the polymer charges.

The study of phase diagrams was conducted in a thermal bath with samples contained in sealed vials. Phases of samples were determined by visual inspection under crossed polarizer. The lamellar phase is identified by its optical anisotropy. It is clearly birefringent in transmitted light under crossed polarizer. The L_3 phase is optically clear and isotropic. The two coexistant lamellar phases, L_α/L_α creates a turbid mixture under a natural light and shows a interface between two birefringent phases after centrifugation. For samples close to the phase boundary, centrifugation and optical microscopy were used to confirm whether the solution was.

Small angle neutron scattering experiments were performed at the Laboratory Leon Brillouin (Orphee reactor, Central d'etude de saclay, France) on the neutron lines PACE and PAXE. The non-polarized neutron wavelengths were selected at 5, 8 and 12 Å ($\Delta\lambda/\lambda$=about 3 %) while the two dimensional detector was kept at distances 1, 3.2 and 4 meters from samples. The wave vector range varied from 6×10^{-3} to 3.5×10^{-1} Å^{-1}. Samples were held in 1 or 2 mm quartz cells. Relative scale spectra are obtained with respect with H_2O.

Results and Discussion

A sufficient amount of hydrophobic side chains is required to confine the polysoap in the membrane solution (Fig.1). Polymer is not soluble in the membrane mesophases when hydrophobe substitution level is equal to or less than a 0.22 mol% at fixed polymer backbone molecular weight of 300,000. As a result, a phase separation leads to surfactant-rich membrane phase and polymer-rich isotropic phase, even for very dilute membrane solutions. This reflects that polymer confinement in the membrane solution is not favored since the polymer loses conformational entropy. In contrast, when the hydrophobe level is higher than 0.7 mol.%, the polysoap can be solubilized both in the lamellar phase L_α and the sponge phase L_3. This suggests that the hydrophobic side chains of polymers are anchored into the bilayers, forming polymer-coated membranes. The anchoring interaction reduces the system free energy and balances the conformational entropy lost. With the polymer inclusion, two new phases are observed. They are two coexisting lamellar phases L_α/L_α at high membrane concentration, and a vesicle phase at low membrane concentrations adjacent to the lamellar domain. The vesicle solution is birefringent under shear and higher viscosity than that of lamellar solution. A maximum turbidity exists during dilution, but no phase separation is observed under microscope or after centrifugation.

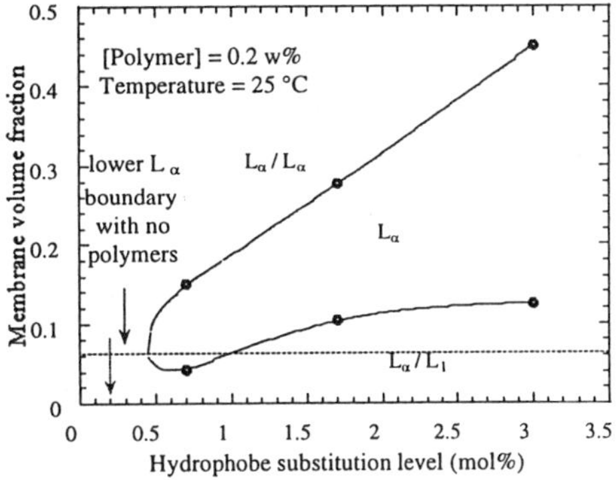

Figure 1. Hydrophobe substitution level effect on the size of monophasic L_α domain. Molecular weight is around 300,000.

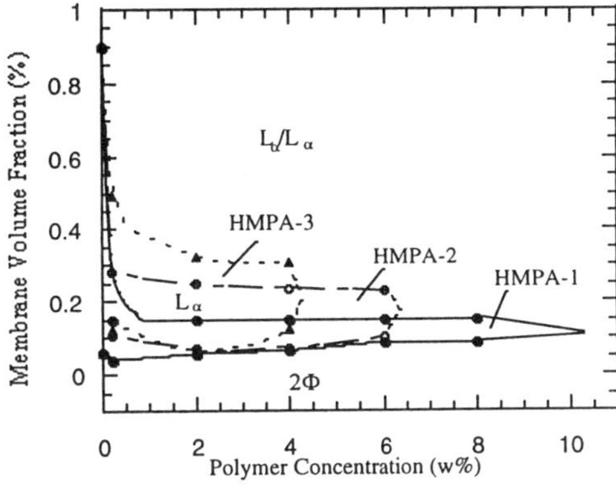

Figure 2. Polymer structure and concentration effect on the phase behavior of polymer doped membrane solutions. (Mw=300,000, Temperature=25 °C)

We have studied the effects of the polymer molecular weight, the substitution level, the polymer concentration and the ionic strength upon the phase behavior. Fig.2 illustrates the effects of polymer concentration and hydrophobe substitution level on the lamellar phase. The monophasic lamellar domain decreases with the polymer concentration at all substitution levels. However, at lower hydrophobe level, a larger polymer concentration is accepted by the lamellar phase (Fig.2). With HMPA-3 (3 mol.% substituted polysoap), a phase separation occurs when the polymer concentration is higher than 4.5 wt%, while the demixing concentrations are shifted to 6.5 wt% and 10.5 wt% for 2 mol% and 1 mol% substituted polymer, respectively (HMPA-2 and HMPA-1). A monophasic lamellar regime that decreases with polymer concentration has been observed in SDS/PEG [9] and AOT/PEO [16] mixtures in which the polymers were not hydrophobically modified. This is consistent with our results (figure 3). The calculated phase diagrams for the Helfrich-stabilized adsorbing polymer-lamellar system shows that two phases region increases with bending modulus κ too. These prediction matches our experiment results, since κ increases with polymer concentration the monophasic lamellar region, therefore, decreases with polymer concentration.

We also study the effects of polymer structure and polymer concentration on the elastic properties of the membranes, using scattering techniques. Fig.3a shows that the normalized Bragg peak becomes narrower with increasing polymer concentration. This means that the product κB increases with the polysoap concentration, where κ reflects the rigidity of single bilayer and B indicates the interaction among the membrane stack. We find that κ is multiplied by at least a factor two at higher polymer concentrations [14]. This increase in κ is qualitatively illustrated in Fig.3b, where the diffuse scattering at low angles of the sponge phase is damped. A similar scattering collapse is observed with the lamellar phases, reflecting an increase in the compression modulus B with increasing polymer concentration.

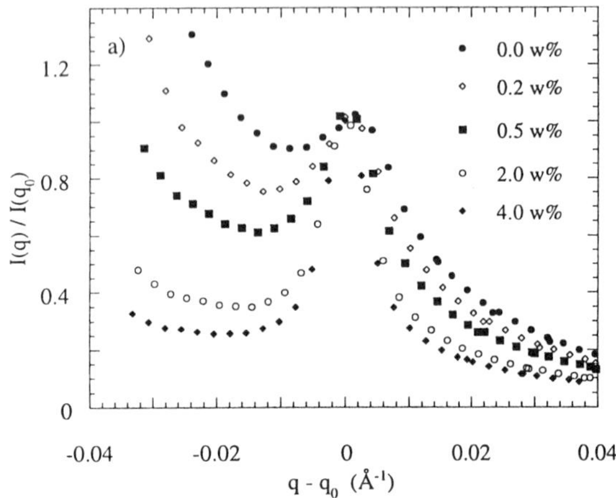

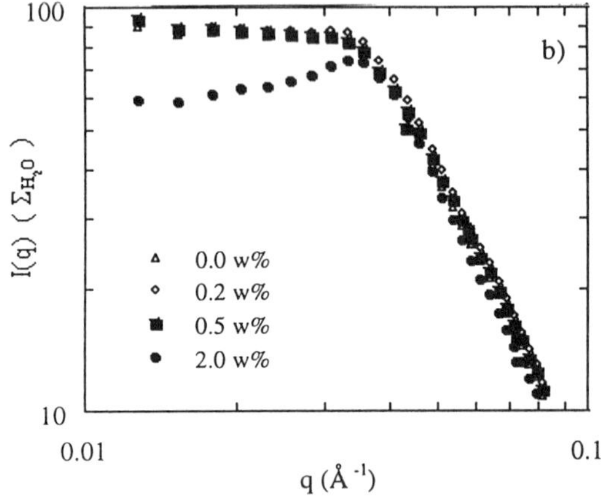

Figure 3. Small angle neutron scattering spectra for the membrane solution with different polymer concentrations. (Hydrophobe substitution level=3 mol%, Mw=300,000, Membrane volume fraction=20 %) a) Normalized data for lamellar phase b) SANS spectra for sponge phase

Summary

In this study, we have demonstrated that polymer structure, polymer concentration and membrane volume fraction manipulate phase behavior and bilayer membrane properties of the surfactant/polymer mixture. When the polymer has an excluded volume in the same order as bilayer spacing, the existence of sufficient amount of hydrophobic side groups along the polyacrylate-based backbone is necessary to insert the polymer into bilayer stack. The hydrophobic anchoring groups associate with the lyotropic bilayer, consequently, the free energy gain due to the confinement is balanced by the hydrophobic interaction. The confinement of the polymer in the bilayer system brings new thermal-equilibrium and new membrane properties to the bilayer system. In particular we show that added HMPAs: i) reduces the monophasic lamellar domain, ii) induces new phases, a vesicle phase and two coexisting lamellar phases L_α/L_α and, iii) increases the rigidity of the bilayers. In addition, by increasing polymer concentration, we observed an decreasing interlayer spacing between bilayers and an increasing bending elastic modulus κ and compression modulus B.

References

(1) C. Ligoure, G. Bouglet, and G. Porte, *Phys. Rev. Lett.* **1993**, *71*, 3600
(2) H. Bagger-Jorgensen, U. Olsson and I. Iliopoulos, *Langmuir* **1995**, *11*, 1934
(3) Bagger-Jorgensen, U. Olsson, I. Iliopoulos and K. Mortensen, *Langmuir* **1997**, *13*, 5820
(4) K. Loyen, I. Iliopoulos, R. Audebert and U. Olsson, *Langmuir* **1995**, *11*, 1053
(5) H. Bagger-Jorgensen, U. Olsson and I. Iliopoulos, *Langmuir* **1995**, *11*, 1934
(6) V. Rajagopalan, U. Olsson and I Iliopoulos, *Langmuir* **1996**, *12*, 4378.
(7) B. Deme, M. Dubois, T. Zemb and B. Cabane, *J. Phys. Chem.* **1996**, *100*, 3828.
(8) R. Lipowsky, *Europhys. Lett.* **1995**, *30*, 197
(9) M. F. Ficheux, A.M. Bellocq and F. Nallet, *J. Phys. II France*, **1995**, *5*, 823.
(10) E. Z. Radlinska, T. Gulik-Krzywicki, F. Lafuma, D. Langevin, W. Urbach, S. E. Williams, and R. Ober, *Phys. Rew. Lett.* **1995**, *74*, 4237.
(11) E. Z. Radlinska, T. Gulik-Krzywicki, F. Lafuma, D. Langevin, W. Urbach and S. E. Williams, *J. Phys. II France* **1997**, *7*, 1393
(12) J. T. Brooks, C. M. Marques, and M. E. Cates, *Europhys. Lett.* **1991**, *14*, 713
(13) G. Bouglet, C. Ligoure, A. M. Bellocq, E. Dufourc, and G. Mosser, *Phys. Rev. E*, **1998**, *57*, 1
(14) Y. Yang, R. Prudhomme, K. M. McGrath, P. Richetti and C. M. Marques, *Phys. Rev. Lett.* **1998**, *80*, 2729
(15) I. Iliopoulos, T. K. Wang and R. Audebert, *Langmuir* **1991**, *7*, 617
(16) K. Zhang and P. Linse, *J. Phys. Chem.* **1995**, *99*, 9130

STRUCTURE AND PROPERTIES OF QUATERNARY AMMONIUM CELLULOSE DERIVATIVES.

Melissa Manuszak Guerrini, Robert Y. Lochhead* and
William H. Daly

*Macromolecular Studies Group, Department of Chemistry, Louisiana State University, Baton Rouge, LA 70803-1804 and * Department of Polymer Science, The University of Southern Mississippi, Hattiesburg, MS 39406-0076.*

INTRODUCTION

The roles of polyelectrolytes and oppositely charged surfactants with respect to foam stability have been studied recently.[1-4] These studies indicate that the formation of surface active complexes has a profound effect on both stability and foamability of sodium dodecyl sulfate (SDS) solutions. Foam stability of solutions containing oppositely charged polymer surfactant pairs is increased because these compounds interact strongly and enhance the adsorption of both components at the plateau borders.[2-4] The effects on foam stability are dependent upon amount of surfactant present in the solution. Below the critical aggregation concentration (CAC) for the surfactant, polyelectrolytes are adsorbed at the interface and stabilize films initially by double layer repulsion, then by polymer bridging with time as the film thins and the walls of the lamellae approach each other.[4] Below the CAC, overlapping polymer layers prevent film rupture. Above the CAC, a dense polymer-surfactant complex that is strongly adsorbed at the liquid –bubble interface stabilizes foam by formation of a gel-like network. Increased stability is attributed to the formation of this network along with the accompanying changes in ionic strength.[4]

We have prepared new mono and diquaternary ammonium alkyl carbamoyl cellulose derivatives with defined charge densities in which the quaternary nitrogen is located at the sites of carboxymethylation of the starting polymer.[5-8] These polymers have been used to determine the changes in surface tension, foamability and foam stability resulting from complexation with SDS. The changes in surface tension produced by these polymer-surfactant complexes have been compared with the behavior of the Polyquaternium 10 - SDScomplex.[9] Polyquaternium 10 (PQ10) is a quaternized hydroxyethyl cellulose derivative in which the cationic moiety is attached via a hydroxypropyl group to poly(oxyethylene) spacer groups that provide flexibility independent of the cellulose chain.[10] An increase in foaming power of these solutions, which is a direct consequence of the increased surface activity of the PQ10 - SDS complex formed, was reported.[1] The results of our studies reveal the role of cationic charge density and graft flexibility on foaming of SDS.

EXPERIMENTAL

Materials: SDS, 99%, was obtained from Sigma® Chemical Company. PQ 10, UCARE Polymer JR-400, was obtained from Amerchol Corporation. 2-N,N,N-Trimethylammoniumethyl carbamoyl cellulose chloride (MQNNED) was prepared from the corresponding N,N-dimethylaminoethyl carbamoyl cellulose by quaternization with methyl iodide followed by ion-exchange with 1.0M sodium chloride in a cellulose acetate dialysis tube with a 6-8,000Dalton cutoff. Treatment of N,N-dimethylaminoethyl carbamoyl cellulose with N,N,N-trimethylammonium-3-chloro-2-propanol (Quat 188) yielded the diquaternary derivative, 3-N,N,N-trimethylammonium -2-hydroxypropyl - N',N' -dimethylammoniumethyl carbamoylmethyl cellulose chloride(DQNNED).[5,6] Water was purified by reverse osmosis, deionization, and filtration (Osmonics, Inc.).

Methods: Solutions were prepared by addition of a concentrated polymer solution(1-3%) to a solution of SDS. The solutions were stirred and allowed to equilibrate overnight. Surface tension measurements were made as described in our previous paper.[9] Specific viscosity, i.e. $\eta_{sp} = (\eta_{solution} - \eta_{solvent}) / \eta_{solvent}$, was determined using an Ubbelohde viscometer with water as the solvent. For foam studies, 10 mL of each solution was prepared and placed in a glass-stoppered 25 mL graduated cylinder. The cylinder was shaken ten times by hand through an arc of 180 degrees. The initial volume of foam produced was recorded and additional measurements were made at 30 minutes, 1, 6, and 24 hours following the method of Goddard and Hannan.[1]

RESULTS AND DISCUSSION

Evaluation of Foamability of SDS and Polymers Separately

The intrinsic foaming of solutions of polymers and SDS was studied in order to determine a baseline for foam height. The results indicated that PQ10 produces a small amount of foam. Neither MQNNED nor DQNNED produced any foam, indicating an absence of intrinsic surface activity. Therefore, significant enhancement in foaming and foam stability of SDS in the presence of these cationic polymers may be attributed to the formation of polymer - surfactant complexes, stabilization, or enhancement of the adsorbed surfactant layer by the presence of the polymer.

Influence of Polymers on Foamability of SDS

In our studies, the concentrations of polymer and SDS were varied from 0.001 - 1% and 0.001 - 10%, respectively. Foam height and foam stability of the solutions were determined over the course of 24 hours. The data at time zero were normalized to the foam height produced by SDS solutions alone. We have defined the normalized foam height as:

<u>**Polymer-SDS solution (mm) - SDS solution (mm)**</u>
SDS solution (mm)

where values less than unity indicate a negative effect on foam height and values greater than unity indicate a positive effect on foam height. This normalization indicated that maximal increases in foaming due to the presence of polymer occurs at low concentrations of SDS (0.001 - 0.01%). Addition of 0.1% PQ10 to a 0.001% SDS solution produces the greatest incremental increase in foam. For MQNNED, the maximal foaming occurs at 0.1% MQNNED and 0.01% SDS. Foams of high SDS concentrations are destabilized by the addition of either PQ10 or MQNNED. For DQNNED, the normalized foam height data show that maximal foaming occurred at 0.01% DQNNED and 0.01% SDS. However, the foamability of solutions containing DQNNED was significantly less than for the same solutions containing MQNNED. Therefore, a polymer containing on average two positive charges per two anhydroglucose units does not improve foaming as well as its monoquaternary analog. Initial foam height represents a balance between foam building and foam decay. In three component systems, the polymer could be preventing coalescence and bubble collapse.

Influence of Polymers on Foam Stability of SDS

Semilogarithm plots of unnormalized foam height versus time for each of the solutions were approximately linear. This observation suggested that the rate of foam collapse may be modeled by 1st order kinetics, and the slope of the plot, k, or the half-life may beused to define the foam decay process. The initial foam height, k, half-lives, surface tension $(\Gamma)^9$ and specific viscosity (η_{sp}) for each of the solutions were determined. Comparison of the half-life data indicated that foam stability was greatest for a solution containing 0.1% PQ10 - 0.01% SDS (Table 1). For solutions containing 0.001 - 0.01% SDS and 0.001 - 0.1% PQ10, foam stability, as judged by $t_{1/2}$, increased as surface tension decreased and specific viscosity increased. This result agrees with previous results showing that increasing bulk viscosity[11] and decreasing surface tension[12-15] produce more stable foams. Increasing the viscosity by addition of polymer decreases the rate of liquid drainage in the lamellae thereby increasing foam stability.[11,16,17] The lowering of surface tension as stability is increased is a result of bridging across a lamella by polymer molecules which stabilize the foam.[4] All solutions containing 1% PQ10 show a reduction in foam stability compared to the solutions containing similar amounts of surfactant. Foam generation may be a problem with such high viscosity systems as these systems may be viscoelastic. Two factors, polymer aggregation and ionic shielding, may be contributing to this phenomenon.

Table 1: Half-life, Specific Viscosity and Surface Tension for Most Stable Foam Forming Solution

Polymer	Polymer %	SDS %	$t_{1/2}$ (hours)	Γ @22°C (mN / m)	$\eta_{specific}$
PQ 10	0.1	0.01	116.23*	46.36	1.840
	0.01	0.1	4.81	33.16	-0.140
MQNNED	0.01	0.1	156.54	31.13	0.020
DQNNED	0.01	0.1	33.54	34.20	N / D

N/D = not determined as the solution contained precipitated material.
** Note most stable foam solution requires 0.1% Polyquaternium 10*

In the case of MQNNED, the half-life data (Table 1) indicated that foam stability was greatest for the solution containing 0..01% MQNNED and 0.1% SDS. The half-life of 156 hours exceeds that observed for the optimum stabilization conditions using PQ10. For most of the solutions studied, the specific viscosity increased and surface tension decreased as the concentration of MQNNED was increased; two exceptions at 0.001% SDS, 1% MQNNED and 0.1% SDS, 1% MQNNED were noted. In contrast to the data obtained with PQ10, the most stable foam for MQNNED was observed in the region were there is a stoichiometric excess of SDS relative to the polymer additive. Maximum foam stability was observed near phase separation regardless of rigidity of the cellulose graft for a monoquaternary substituted polymer.

Half-life data for DQNNED (Table 1) indicated that foam stability was greatest for the solution containing 0.01% DQNNED and 0.1% SDS. The half-life of 33 hours was less than those observed with either of the monoquaternary polymers. The same solution containing MQNNED had a half-life 4.7 times greater than the half-life for DQNNED. Clearly, introduction of a second positive charge on the substituent does not contribute to increased foam stability. Solutions containing the lowest concentrations of SDS exhibited half-lives of less than 30 minutes. Additionally, two solutions did not foam, probably due to the strong binding of the SDS by DQNNED.

Comparison of Foamability and Foam Stability Results

The results of foamability are shown in Table 2 for the three polymers studied. The system containing PQ10 required more polymer to produce maximal normalized (initial) foamability of SDS than did either MQNNED or DQNNED. However, PQ10 was more efficient at increasing the foamability of SDS than the two other polymers. Although PQ10-SDS solutions produced the maximum normalized (initial) foam height, they were not the most stable systems over time. Solutions exhibiting the longest half-life contained 10 times more SDS (Table 1) than those producing maximal normalized (initial) foam height. Both MQNNED and DQNNED exhibited the longest half-life at the same concentrations of polymer and SDS.

Table 2: Normalized (Initial)Foam Height Produced by Quaternary Polymer/H$_2$O/SDS Systems Under Comparable Conditions.

Polymer	% Polymer / % SDS	Normalized Foam Height
PQ 10	0.1% / 0.001 %	19.7
	0.1% / 0.01%	8.7
MQNNED	0.1% / 0.01%	14.8
DQNNED	0.01% / 0.01%	6.8
	0.1% / 0.01%	4.5

These results may differ from the work of Goddard and Hannan [1] who showed that maximum foaming was produced by solutions containing the maximal amount of precipitated material. Our results appear to indicate that maximal foaming and foam stability occurs in those solutions nearest phase separation. This observation appears to be consistent with the findings of Ross and Nishioka [18-21] who state that high surface activity is a precursor of phase separation. The decrease in surface tension correlates with increasing foam stability. The precipitated phase is purported to destroy foam in the supernatant phase if it is present as the dispersed phase or has a surface tension lower than the continuous phase (Marangoni effect). This thesis describes systems of cationic polymer / SDS because the half-life values decrease as the concentration of SDS increases and the polymer-surfactant complex is precipitated. As the concentration of SDS is increased towards a critical point, the polymer is titrated out of solution by neutralization of the positive charges.

Foaming of SDS solutions is expected to be influenced by the addition of a water-soluble polymer [1] as changes in viscosity and surface tension accompany changes in foam stability.[2-4,11,16,22] Generally in regions approaching formation of stoichiometric polymer-surfactant complexes, the onset of precipitation of the complexes indicates desolvation of the polymer.[20] Complex formation leads to enhanced foam stability and decreasing surface tension followed by a pronounced reduction in the solubility of the polymer complexes. At the point of maximum precipitation, the foam stability is reduced, possibly because the concentration of surface-active components is reduced due to adsorption by the polymer complex.

CONCLUSIONS

Optimum performance in terms of foam produced per unit polymer concentration appears to be achieved when neutralization of the charged polymer by oppositely charged surfactant begins to induce precipitation. Systems containing PQ10 exhibit maximum foam stability slightly before the neutralization point. In contrast, MQNNED/surfactant adducts produce the most stable foams when a slight stoichiometric excess of surfactant is present. Increasing charge density on the polymer, i.e.DQNNED, failed to enhance the contribution of the polymer additive to foam properties.

ACKNOWLEDGEMENTS

This material is based upon work supported under a Society of Cosmetic Chemists' Graduate Research Fellowship to M. Manuszak Guerrini.

REFERENCES

1) Goddard, E. D.; Hannan, R. B. *J. Coll.Int. Sci.* **1975**, *55*, 73-79.
2) Lionti-Addad, S.; di Meglio, J.-M. *Langmuir* **1992**, *8*, 324-327.
3) Cohen-Addad, S.; di Meglio, J.-M. *Langmuir* **1994**, *10*, 773-778.
4) Bergeron, V.; Langevin, D.; Asnacios, A. *Langmuir* **1996**, *12*, 1550-1556.
5) Culberson, D.; Daly, W. H. *Polymer Preprints (POLY)* **1993**, *34*, 564-565.
6) Culberson, D.; Daly, W. H. *PMSE* **1994**, *71*, 498.
7) Culberson, D. A.; dissertation Louisiana State University: Baton Rouge, 1995.
8) Manuszak Guerrini, M. A.; Culberson, D. A.; Daly, W. H.; patent pending.
9) Manuszak Guerrini, M., *et al. J. Appl. Polym. Sci.* (inpress).
10) Stone, F. W.; Rutherford, J. M.; Union Carbide Corp.: US3,472,840, 1969.
11) Shah, D. O. *et al. Coll. Polym.Sci.* **1978**, *256*, 1002-1008.
12) Schott, H. *J. Am. Oil Chem. Soc.* **1988**, *65*, 1658-1663.
13) Sita Ram Sarma *et al. J. Coll. Int. Sci.* **1988**, *124*, 339-348.
14) Ronteltap, A. D. *et al.*, A. *Coll. Surf.* **1990**, *47*, 269-283.
15) Prins, A.; *Advances in Food Emulsions and Foams*, First ed.;Elsevier Applied Science: New York, 1988, pp 91-122.
16) Friberg, S. E.; Blute, I.; Stenius, P. *J.Coll. Int. Sci.* **1989**, *127*, 573-582.
17) Shah, D. O. *J. Coll. Int. Sci.* **1971**, *37*, 744-752.
18) Ross, S.; Nishioka, G.; An International Conference organized by the Colloid and Surface Chemistry Group of the Society of Chemical Industry, Brunel University, 1975, pp 15-29.
19) Ross, S.; Nishioka, G. *J. Phys. Chem.***1975**, *79*, 1561-1565.
20) Ross, S.; Nishioka, G. *Coll. Polym. Sci.* **1977**, *255*, 560-565.
21) Ross, S.; Patterson, R. E. *J. Phys. Chem.* **1979**, *83*, 2226-2232.
22) Bikerman, J. J. *Foam*, First ed.;Springer-Verlag: New York, 1973, pp 33-64.

DOUBLE HELIX OF *AGROBACTERIUM TUMEFACIENS*
SUCCINOGLYCAN IN DILUTE SODIUM CHLORIDE SOLUTION

Isamu Kaneda, Atsushi Kobayashi, Kazuyuki Miyazawa, Masashi Yoshida,
Kimio Ohno, and Toshio Yanaki

Shiseido Basic Laboratories
1050 Nippa-cho, Kouhoku-ku,
Yokohama, JAPAN   223

INTRODUCTION

Succinoglycan, an acidic polysaccharide produced extracellularly by various species of genus *Alcaligenes, Pseudemonas, Agrobacterium* and *Rhzobium* consists of D-glucose, D-galactose, pyruvate, and succinate. The succinate moieties bind to the side chains of polysaccharide through ester linkages, but the location of succinate moieties on the polysaccharide was unknown.   Differences in the succinate content is observed in the polymers produced from different species of bacteria. A succinoglycan (Rheozan®), produced by *Agrobacterium tumefaciens*, consists of D-glucose, D-galactose, pyruvate, and succinate (Figure 1).   The molar ratio of D-glucose, D-galactose, pyruvate, and succinate is 7, 1, 1, and 0.8, respectively.   The solution of the polysaccharide can show a high viscosity which is one of major characteristics of this polysaccharide. This high viscosity can be observed even in acidic condition and in the presence of low-molecular-weight salts such as NaCl.

It is well known that succinoglycan from *Pseudomonas* or *Rhizobium* reduces its viscosity at around 65°C. This phenomenon, confirmed by DSC and optical rotation measurement, was thought to be due to the change of its conformation.   However, the detail of drastic reduction of viscosity is unknown.   Because the rheological properties of Rheozan is similar to that of other succinoglycans, the conformational change of Rheozan resembles that of the succinoglycan.

In this study, we investigated the properties of Rheozan in dilute NaCl solution by light scattering and viscosity measurement at both 25 and 75°C. The parameters of worm-like chain model, which was used as a theoretical model, was evaluated.

EXPERIMENT SECTION

Materials

Crude Rheozan was supplied by Rhone-Poulenc (France) as a powder. To obtain various chain lengths of the polymer, crude Rheozan (0.1-1%) in 0.1 mol/L NaAc was sonicated with 20 kHz vibration for 1 to 100 hr at 0°C by a ultrasonic generator (US-150T, Nissei, Tokyo, Japan).   Sonicated solutions were treated with activated charcoal and were filtered through a paper filter (No. 101) and then a Millipore filter (0.45μm).   The sonicated polymers were precipitated by fractional precipitation with the mixture of 0.1M NaAc solution and methanol contained 0.1M NaAc.   The main fraction was used for this study.   The precipitation of main fraction was freeze-dried.   Six pure samples were prepared and named B, C, D, E, F, and G.   Before use, each samples were further dried overnight *in vacuo* at room temperature.   Samples were dissolved in 0.1 mol/L NaCl solution for each experiment.

STATIC LIGHT SCATTERING

Intensities of light scattering from Rheozan in 0.1mol/L NaCl at both 25 and 75°C were measured by an automatic light scattering photometer (DLS7000, Otsuka Denshi, Osaka, Japan) with the refraction-light angle range from 30 to 150°. He-Ne laser light (633 nm) was used.   The angler dependence of scattered light intensities was analyzed by using Zimm plot , while the concentration dependence was analyzed by using Berry plot. The instrument was calibrated with benzene as a reference liquid at 25°C. To obtain an optical clarification of Rheozan in 0.1mol/L NaCl, the samples was filtrated through a membrane filter (0.45μm) followed by 1 hr centrifugation at approximately 10000 × g.   A central portion of the supernatant in the centrifuge tube was sucked by a syringe and directly transferred into a light scattering cell (20mm i.d. × 150mm). The measurement at 75°C was done after incubating the sample in the cell holder for 30 min.

The dn/dc of Rheozan in 0.1 mol/L NaCl, measured by a differential refractometer (DRM 1021, Otsuka Denshi) at both 25 and 75°C at 633nm, were 0.150 and 0.148, respectively.

VISCOMETRY

Zero-shear intrinsic viscosities [η] for each sample in 0.1 mol/L NaCl at both 25 and 75°C were measured by using an Ubbelohde low shear capillary viscometer with four-bulbs. Apparent viscosities was obtained by using in Huggins and Mead-Fuoss plot, and the estimated [η] was calculated.

RESULTS and DISCUSSION

MOLECULAR WEIGHTS.

Figure 2 shows the relationship between $(Kc / R_0)^{1/2}$ and concentration of the samples at 25°C.   The values of weight average molecular weight (M_w) and the second virial coefficient (A_2) were evaluated from the intercepts and slopes of the regression lines of each samples.   The data at both 25 and 75°C are also presented on Table 1. M_w of each samples at 25°C were almost twice as large as those at 75°C. Consequently Rheozan was speculated to be a dimmer at 25°C and a single strand at 75°C.

RADIUS OF GYRATION

Figure 3 shows the particle scattering functions $P(\theta)$ of the samples at 25°C.   The values of the radius of gyration $\langle S^2 \rangle^{1/2}$ evaluated from the slopes of the straight part of the line, which indicated as a dash-line.   The data at both 25 and 75°C are also presented in Table 1.

Figure 4 shows the relation between the value of $\langle S^2 \rangle^{1/2}$ and M_w at both 25 and 75°C.   The regression lines of 25°C and 75°C cross each other with indicating the distinctly different behavior.   The line of 25°C has a slop of 1.0, which agrees with the predicted value for stiff chain.   The line of 75°C has a slop of 0.6, which indicated that Rheozan at this temperature is a random coil swollen by an excluded-volume effect in 0.1 mol/L NaCl. The results of Mw and $\langle S^2 \rangle^{1/2}$ measurement suggested that Rheozan dissolved as a double helix in 0.1 mol/L NaCl at 25°C.

INTRINSIC VISCOSITY

The values of [η] and Huggins constant *k'* for the samples at both 25 and 75°C shows in Table 1.   Figure 5 shows the relationship between [η] and M_w.   The exponent of the Mark-Houwink intrinsic viscosity [η] relation of both at 25°C and 75°C were 1.4 and 0.9, respectively.   The slop of the line of 25°C indicates that Rheozan is stiff chain.   The slope of 75°C indicates that Rheozan is more flexible than that at 25°C in 0.1 mol/L NaCl.   There was no discrepancy between the results of [η] and that of $\langle S^2 \rangle^{1/2}$ (Figure 4).   Therefore, Rheozan was speculated to have a double helical structure in 0.1 mol/L NaCl at 25°C.

DETERMINATION OF THE PARAMETERS OF WORM-LIKE CHAIN MODEL

Yamakawa-Fujii's theory presents that [η] of the Kratky-Porod worm-like chain model, a typical model for a semi-flexible chain, can be described by three parameters, M_L (the molar mass per unit contour length) , q (the persistence length) , and d (the chain diameter).   It is difficulties to evaluate these three unknowns from viscosity data alone.   Since M_w of the samples at 25°C were almost twice as large as those at 75°C, a semi-flexible rod model with double strands chain was selected for the conformational model for Rheozan in 0.1 mol/L NaCl.   Because the molar mass of the monomeric unit (M_0) of Rheozan is estimated to be 750 nm^{-1} from its primary structure, M_L of Rheozan could be 1500nm^{-1}, another parameters (q and d) were obtained by try and error.   The parameters of the semi-flexible rod model, M_L, q, and d were estimated to be 1530±50 nm^{-1}, 250±30 nm, and 3.5±0.5 nm, respectively.   Yamakawa-Fujii's theory could be fitted to the experimental [η] data.   The value of M_L (1530 nm^{-1}) was almost doubled of M_0 (750 nm^{-1}) for Rheozan.   The value of q (250 nm) indicated that Rheozan was a stiff chain in 0.1 mol/L NaCl.   Since a double helical chain model was suitable for the structure of Rheozan by fitting procedure using Yamakawa-Fujii's theory, the theoretical evaluation which can be obtained from three parameters can agree well with our experimental results described above.

We concluded that Rheozan was dissolved as a double helix at 25°C and it melted abruptly at around 65°C in 0.1mol/L NaCl solution .

(REFERENCES : Omitted due to space limitations)

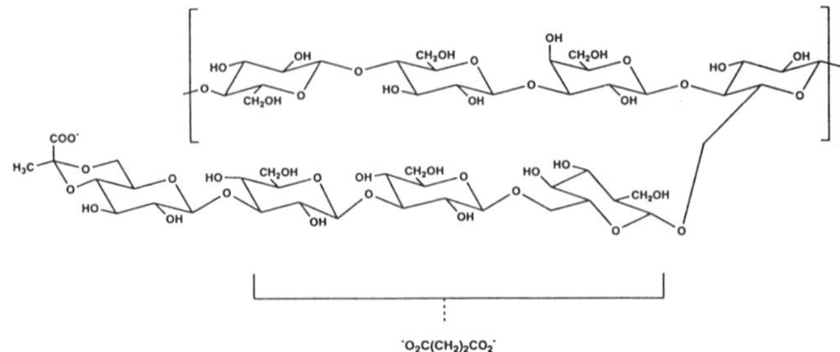

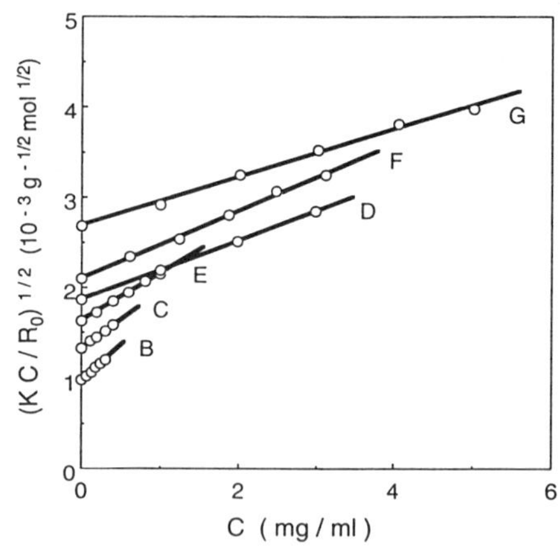

Figure 1. Rheozan is a succinoglycan produced by *Agrobacterium tumefaciens* and consists of D-glucose, D-galactose, pyruvate, and succinate. The molar ratio of D-glucose, D-galactose, pyruvate, and succinate is 7, 1, 1, and 0.8, respectively. Succinate moieties are bound to the side chain. The location of succinate moieties on the polysaccharide was unknown.

Figure 2. Concentration dependence of $(kc/R_0)^{1/2}$ at 25°C

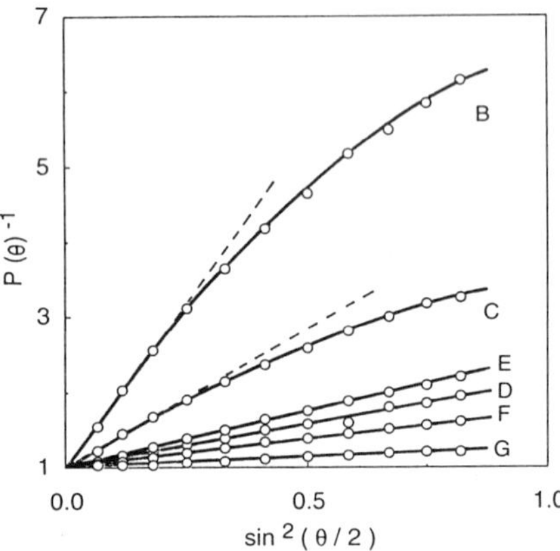

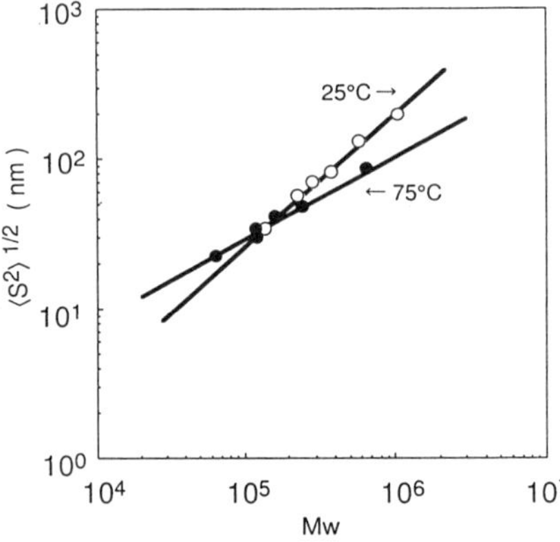

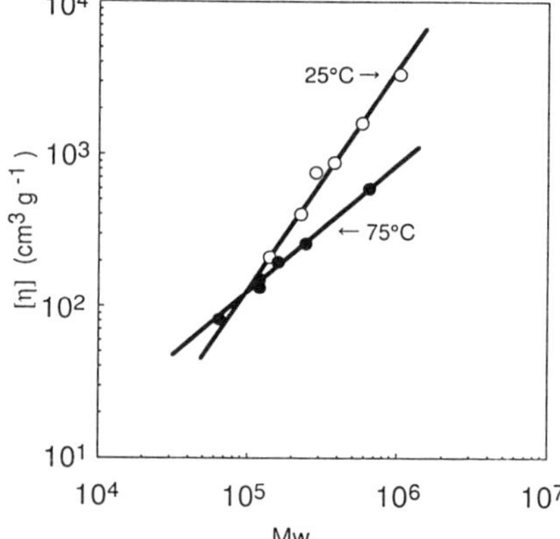

Figure 3. Particle scattering function at 25°C

Figure 4. Double-logarithmic plot of $\langle S^2 \rangle^{1/2}$ vs. Mw at both 25 and 75°C. Open circle and closed circle show the data obtained at 25°C and 75°C, respectively.

Figure 5. Double-logarithmic plot of $[\eta]$ vs. Mw at both 25 and 75°C. Open circle and closed circle show the data obtained at 25°C and 75°C, respectively.

Table 1   Physicochemical parameters obtained from light scattering and viscosity measurements

Sample	25°C					75°C				
	Mw (10^5 g mol^{-1})	$[\eta]$ (10^2cm^{3}g^{-1})	$\langle S^2 \rangle^{1/2}$ (nm)	A_2 (10^4cm^{3}g^{2}mol)	k'	Mw (10^5 g mol^{-1})	$[\eta]$ (10^2cm^{3}g^{-1})	$\langle S^2 \rangle^{1/2}$ (nm)	A_2 (10^4cm^{3}g^{2}mol)	k'
B	10.46	33.6	198.5	8.179	0.416	6.97	5.9	85.78	12.86	0.431
C	5.79	15.8	129.4	8.754	0.387	2.70	2.6	47.57	12.71	0.578
D	2.87	7.6	70.17	8.253	0.338	1.45	1.5	33.76	18.98	0.400
E	3.77	8.8	81.31	10.33	0.412	1.74	2.0	41.39	14.06	0.346
F	2.25	4.0	57.13	10.38	0.517	1.30	1.3	33.80	17.55	0.537
G	1.40	2.1	34.16	9.371	0.378	0.70	0.82	22.56	22.68	0.335

Mw: Weight average molecular weight, $[\eta]$: Zero-shear intrinsic viscosity, $\langle S^2 \rangle^{1/2}$: Radius of gyration, A_2: The second virial coefficient

k' : Huggins constant

Determination of the absolute molar mass and particle size distribution of water-soluble cellulose and starch derivatives

W.-M. Kulicke, D. Heins, N. Schittenhelm

Institute of Technical and Macromolecular Chemistry, University of Hamburg, Bundesstr. 45, 20146 Hamburg, Germany

Phone : +49 40-4123 6001, Fax : +49 40-418335
e-mail: kulicke@chemie.uni-hamburg.de

It is a well-known fact that the efficiency of water-soluble cellulose derivatives in technical, pharmaceutical and food applications depends to a large extent upon their molar masses. The influence of the molar mass distribution (MMD) has not been as equally well-investigated, because its determination can be hampered by the presence of associates and aggregates, and adsorption or ion exclusion in the case of polyelectrolytes.

We have installed a combined system of size exclusion chromatography (SEC) or flow field-flow fractionation (FFFF), as a means of separating the samples into nearly monodisperse fractions, and have coupled this system with multi-angle laser light scattering (MALLS) and differential refractometry (DRI).
The coupling of SEC and MALLS is now commonplace and has been successfully established in many research groups. However, investigations of cellulose and starch derivatives require special preparation. If aggregates and associates occur, then initial cleaning by means of filtration or ultracentrifugation is necessary.

The other fractionation system (FFFF) was first successfully coupled with MALLS and DRI in 1994. [1]
Figure 1 shows the experimental set-up in schematic form. The FFFF/MALLS/DRI apparatus used consists of a diffusion-driven separation unit and a tandem detection unit. In FFFF separation is achieved by means of two currents passing perpendicular to each other through a shallow channel. The cross flow, V_x, forces all the particles within the channel towards the accumulation wall and gives rise to a concentration profile characteristic for each species. The smaller the diffusion coefficient of the species, the closer the mean distance, x, from the accumulation wall. The second flow, the channel flow, V_z, exhibits a parabolic velocity profile which amplifies the small differences in x to larger distances, z, along the channel axis.
This method is suitable even when aggregates are present as well as dissolved molecules [2]. In addition to the MMD, this technique can also be used for determining the particle size distribution.

Besides characterization of the samples by molar mass and particle size distribution, investigations were also carried out using NMR spectroscopy with the aim of clarifying the relationship between the properties and the underlying structure. The chemical structure of cellulose and starch derivatives can be described by the distribution of substituents amongst the three positions of the anhydroglucose unit, e.g. the average degree of substitution, DS; the partial degree of substitution, x_i; and the molar degree of substitution, MS.

The above methods thus provide the basis for studying the degradation process of such polymers by ultra-sound or enzymes, which will be discussed on a molecular level.
The high molar mass of polymers and polyelectrolytes can be reduced through the use of preliminary ultrasonic degradation. We were be able to show that under the used degradation conditions no change in the functional groups took place in any of the polymers investigated so far and neither did any side-chain reactions occur. We have tested this via $\{^1H\}$-^{13}C high resolution NMR-spectroscopy for synthetic polymers such as poly(acrylamide) and poly(acrylamide-co-sodium acrylate) and semi-rigid biopolymers, such as xanthan and schizophyllan [3]. It is postulated that degradation is a result of elongational flow conditions produced by ultrasonic fields. Typically, the curve shows a sharp decrease in molecular weight in the first minutes of exposure, irrespective of the chemical structure. The change in polydispersity can provide a model of the degradation processes.

A very interesting result is that the polymers break nearly in the middle of the chain. This may allow us to establish structure-property relationships. Figure 2 shows the decrease in the molar mass of sulphoethylcellulose obtained by ultrasonic degradation.
By determining the corresponding intrinsic viscosities it is possible to form the structure-property-relationship.
Examples are carboxymethylcellulose [4] and ethylhydroxyethyl cellulose, which is shown in the following equation:

$$[\eta] = 2.09 \cdot 10^{-2} \cdot M_w^{0.79}$$

By using the combined apparatus we were also able to establish a R_G-M relationship:

$$R_G = 0.0143 \cdot M_W^{0.63}$$

Different acetyl starch samples were also investigated with a view to replacing hydroxyethyl starch (HES) as a plasma substitute. This new derivative is aimed at incorporating the positive properties of HES, while offering better physiological compatibility as a result of the change from ether to ester bonds. Besides the molar mass distribution, which has a great influence on the properties of the product, enzymatic degradation by α-amylase, which is present in human blood, was also tested in the laboratory [5]. It has been reported that degradation of HES is slow and incomplete because etherification hinders the attack of α-amylase. Prolonged application of HES may lead to accumulation in the human body which is said to be responsible for persistent itching. Changing the ether bond to an ester bond increases the biological degradability of the polymeric backbone because the ester bond, which would

[1]   D. Roessner, W.-M. Kulicke; J. Chromatogr. A **687** (1994) 249-258

[2]   H. Thielking, W.-M. Kulicke; J. Microcolumn Sep. **10** (1998) 51-56

[3]   W.-M.Kulicke, M.Otto, A.Baar, Makromol.Chem., **194**, (1993), 751-765

[4]   W.-M. Kulicke, A. H. Kull, W. Kull, H. Thielking, J. Engelhardt, J.-B. Pannek; Polymer **37** Nr.13 (1996) 2723-2731

[5]   D.Heins, W.-M.Kulicke, P.Käuper, H.Thielking, in preparation

[6]   H.Förster, F.Asskali, E.Nitsch : Metabolisierbarer Plasmaersatz, published specification DE 41 22 999 A1, 1993

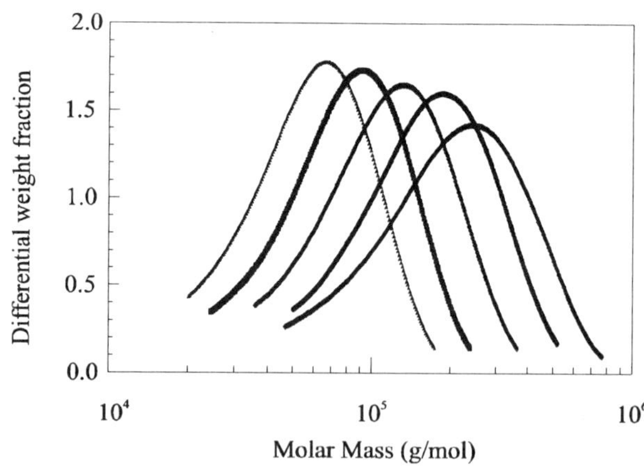

Figure 2   Ultrasonic degradation of sulfoethylcellulose. Differential distribution curves determine absolutely by SEC/MALLS/DRI after different degradation times (5, 10, 20, 40 and 90 minutes)

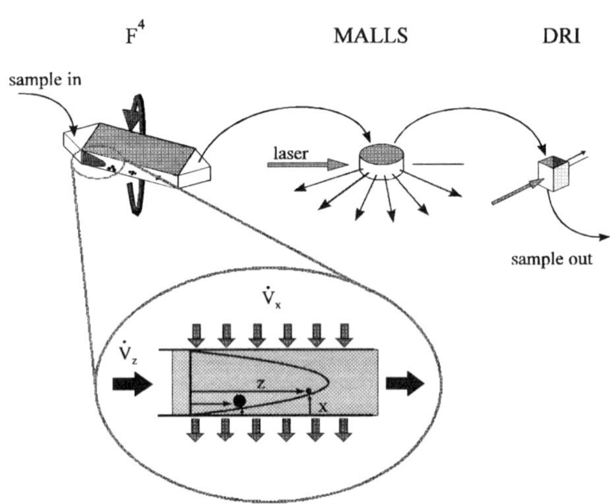

Figure 1   Schematic view of the experimental setup consisting of flow field-flow fractionation (FFFF), multi-angle laser light scattering (MALLS) and differential refractive index detector (DRI). The separation process within the FFFF-channel is shown in detail.

The Preparation of Well-Defined Water Soluble/Swellable (Co)Polymers by Atom Transfer Radical Polymerization (ATRP)

Krzysztof Matyjaszewski, Kathryn L. Beers, Andreas Mühlebach, Simion Coca, Xuan Zhang, and Scott G. Gaynor

Department of Chemistry, Carnegie Mellon University, 4400 Fifth Avenue, Pittsburgh, PA 15213

Introduction

The preparation of well-defined polymers and copolymers has been a goal of polymer chemists since the first synthetic polymers were prepared in a laboratory. However, it was nearly twenty years after the work of Carrothers that Szwarc described a method to prepare polymers with predefined molecular weights ($DP_n = \Delta[M]/[I]_o$) and narrow polydispersities ($M_w/M_n < 1.1$). The method described by Szwarc was the living anionic polymerization of styrene.[1] Since then polymer chemists have been trying to develop polymerization systems that can provide such well defined polymers for a wide range of monomers under simple conditions.

Unfortunately, the anionic polymerization of vinyl monomers is limited to a few monomers, and is generally not applicable for water soluble monomers. This is due to the presence of polar groups and protic functionalities on the monomer, which allow for the polymer to be soluble in water and which are susceptible to attack by the propagating anion. To overcome this problem, it would be desirable to have a "living"/controlled radical polymerization, as propagating radicals will not react with the various functional groups, such as hydroxy or amino.

Achieving a "living"/controlled radical polymerization has been difficult as the propagating radicals are capable of bimolecular termination processes, such as combination or disproportionation reactions. Only recently have methods been developed that allow for the controlled polymerization of vinyl monomers by a radical mechanism.[2-6] However, most of these methods have been limited to individual classes of monomers,[4,5,7] or have not prepared well-defined polymers.[2]

Our group has been instrumental in the development of a novel polymerization system, atom transfer radical polymerization (ATRP),[6,8,9] that allows for the controlled polymerization of styrenes,[6,8-11] acrylates,[8,12] methacrylates,[7,13] acrylonitrile[14] and other monomers. This method uses a reversible activation and deactivation mechanism to lower the concentration of the propagating radicals and thus suppress termination. The equilibrium between active and dormant species is obtained through the reversible activation and deactivation of polymeric alkyl halide end groups (R-X) by reaction with a transition metal catalyst (M_t^n), Scheme 1. The key to the reaction is that the propagating radicals are rapidly deactivated by reaction with the transition metal halide (M_t^{n+1}) so that the equilibrium is shifted towards the dormant species.[15]

Scheme 1

$$R\text{—}X \ + \ M_t^n/\text{Ligand} \ \underset{k_d}{\overset{k_a}{\rightleftharpoons}} \ R^\bullet \ + \ X\text{—}M_t^{n+1}/\text{Ligand}$$

$$\xrightarrow[k_p]{(+M)} \quad \xdashrightarrow{k_t} \quad R\text{—}R$$

Here we describe the use of ATRP to prepare water soluble polymers, as well as amphiphilic copolymers. The monomers that have been studied include: 2-hydroxyethyl acrylate (HEA) and 2-hydroxyethyl methacrylate (HEMA), 2-(dimethylamino)ethyl methacrylate, N-(2-hydroxypropyl) methacrylamide, and (meth)acrylic acid. In addition, water swellable polymers have been developed by synthesizing poly(N-vinyl pyrrolidinone-g-styrene) graft copolymers.[16]

HEA/HEMA

The polymerization of HEA was conducted both in bulk and in aqueous solution.[17] By using methyl 2-bromopropionate as the initiator and a copper (I) bromide/3 bpy (bpy = 2, 2'-bipyridine) complex as the catalyst, it was possible to prepare linear homopolymers of HEA, Table 1. The polymerization of HEA in water, is the first time that a "living"/controlled radical polymerization has been conducted for unprotected, aqueous monomer systems.

Table 1. Results of Homopolymerization of 2-Hydroxyethyl Acrylate.

Rxn[a]	Time[b] (Conv.)	$M_{n,th}$	$M_{n,SEC}$[c]	M_w/M_n
1	10 (0.92)	3,130	6,170	1.19
2	14 (0.91)	15,000	30,000	1.19
3	14 (0.90)	18,000	36,000	1.17
4[d]	12 (0.87)	10,100	14,700	1.34

a) [M] = 8.67 M, [I]:[Cu/3bpy] = 1:3, Temperature = 90 °C; b) Time is in hours; c) SEC was performed in DMF; d) Reaction performed in water (M:H₂O = 1:1 v/v)

The bulk polymerization of 2-HEMA was extremely fast, under comparable reaction conditions and uncontrolled. The reason for this result is two fold: 1) the high polarity of the solvent (monomer), and 2) the tertiary nature of the dormant species. These two factors resulted in the formation of a much higher concentration of growing radicals than is normally observed in ATRP and thus lead to the formation of ill-defined polymers.

To enhance the control of the polymerization, it was required that the concentration of radicals be reduced. To this end, extra deactivator ($Cu(II)Cl_2$) was added and the reaction was diluted by addition of solvent (MEK/n-propanol). The results of the polymerization are shown in Figure 1. As can be seen, the molecular weights of the polymer increased as a function of conversion, but were higher than predicted. The higher M_n values may be a result of poly(MMA) being an improper calibration standard (SEC in DMF) for poly(HEMA).

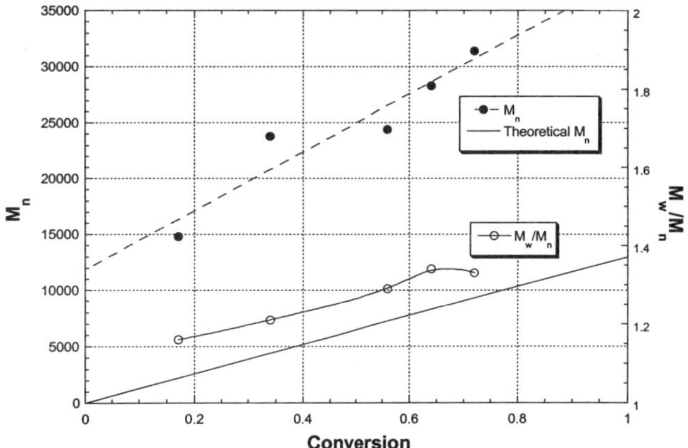

Figure 1. Dependence of molecular weight with conversion for the ATRP of HEMA.

An amphiphilic copolymer of 2-hydroxyethyl acrylate and butyl acrylate was prepared by sequential addition of monomer. Butyl acrylate was first polymerized and then HEA added. The final composition was 80% BA/20% HEA; M_n = 8,600; M_w/M_n = 1.34. Additionally, block copolymers were prepared comprising of TMS-HEA (2-trimethylsiloxyethyl acrylate) and butyl acrylate. The trimethylsilyl group was then removed by treating with acid to yield the HEA/BA block copolymer.

2-(Dimethylamino)ethyl Methacrylate

2-(Dimethylamino)ethyl methacrylate was polymerized in dichlorobenzene using a Cu(I)Br/N,N,N',N'',N''',N'''-hexamethyltriethylenetetramine catalyst at 50 °C. The kinetics of the reaction were first

"

order with respect to monomer, and the molecular weights increased linearly with conversion, Figure 2. Although the molecular weights were higher than predicted, the low polydispersities ($M_w/M_n < 1.3$) suggest that the polymerization is indeed "living"/controlled, and that the difference in molecular weights are a result of the calibration of the SEC obtained using poly(MMA) standards.

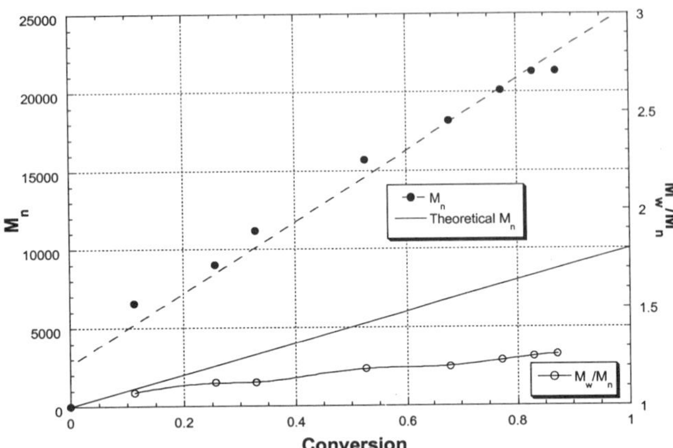

Figure 1. Dependence of molecular weight with conversion for the ATRP of 2-(dimethylamino)ethyl methacrylate.

Other Monomers

(Meth)Acrylic Acid

One of the types of monomers that can not be polymerized by ATRP directly are those that contain carboxylic acids. The reason for this is that carboxylic acids can react with copper (II) halides to prepare copper (II) carboxylates. In fact, copper (I) halides are purified by washing with acetic acid to remove any copper (II) species that may be present. The consequence of this is that by formation of the copper (II) carboxylates, the deactivator, copper (II) halide, is removed from the system; the propagating radicals are not returned to the dormant alkyl halides, and thus the polymerization becomes uncontrolled.

To prepare poly(acrylic acid) or poly(methacrylic acid), alkyl (meth)acrylates were polymerized first followed by removal of the alkyl groups. The simplest alkyl group to hydrolyze is *t*-butyl, because in the presence of acid, isobutene is formed and is easily removed from the system as it is a gas.

Both poly(*t*-butyl acrylate) and poly(*t*-butyl methacrylate) were prepared by ATRP, yielding well defined polymers. These were subsequently hydrolyzed by heating, to 60 °C, the polymer in dioxane/water with added HCl. The resulting polymers were water-soluble.

N-(2-Hydroxypropyl) Methacrylamide

This unique monomer required that special ligands be used because of the presence of the amide group in the monomer. Only when cyclam, 1,4,8,11-tetramethyl-1,4,8,11-tetraazocyclotetradecane, was used was polymerization successfully conducted. The polymerization was performed in 1-butanol to yield a well defined polymer: M_n(SEC) = 21,300; $M_w/M_n = 1.38$.

Poly(N-vinyl pyrrolidinone-g-styrene)

Styrene macromonomers were prepared by using vinyl chloroacetate to initiate the ATRP of styrene. Due to the low probability of cross-propagation between styryl radicals and vinyl acetate, linear chains of polystyrene were prepared with a vinyl acetate end group. These macromonomers were subsequently copolymerized with NVP, using AIBN, to yield the desired graft copolymers. The effect of the length of polystyrene and the amount of macromonomer in the copolymerization was studied and the results will be presented.

Conclusion

ATRP has been used to successfully polymerize a wide range of water soluble monomers. 2-Hydroxyethyl acrylate was successfully polymerized in water demonstrating the first time that an aqueous polymerization was conducted as a "living"/controlled radical polymerization. Additionally, it was shown that water soluble polymers can be prepared by using protected monomers to synthesize polymers that can be easily handled/processed in organic solvents and then subsequently deprotected.

References

(1)   Szwarc, M. *Nature* **1956**, *178*, 1168.
(2)   Otsu, T.; Yoshida, M. *Makromol. Chem. Rapid Commun* **1982**, *3*, 127.
(3)   Solomon, D. H.; Rizzardo, E.; Cacioli, P., *US 4,581,429,* 1985
(4)   Georges, M. K.; Veregin, R. P. N.; Kazmaier, P. M.; Hamer, G. K. *Macromolecules* **1993**, *26*, 2987.
(5)   Wayland, B. B.; Basickes, L.; Mukerjee, S.; Wei, M.; Fryd, M. *Macromolecules* **1997**, *30*, 8109.
(6)   Wang, J.-S.; Matyjaszewski, K. *J. Amer. Chem. Soc.* **1995**, *117*, 5614.
(7)   Kato, M.; Kamigaito, M.; Sawamoto, M.; Higashimura, T. *Macromolecules* **1995**, *28*, 1721.
(8)   Matyjaszewski, K.; Wang, J.-S. *Macromolecules* **1995**, *28*, 7901.
(9)   Matyjaszewski, K.; Patten, T.; Xia, J.; Abernathy, T. *Science* **1996**, *272*, 866.
(10)   Matyjaszewski, K.; Patten, T. E.; Xia, J. *J. Amer. Chem. Soc* **1997**, *119*, 674.
(11)   Qiu, J.; Matyjaszewski, K. *Macromolecules* **1997**, *30*, 5643.
(12)   Coca, S.; Matyjaszewski, K. *Polym. Prepr. (Am. Chem. Soc., Div. Polym. Chem.)* **1997**, *38(1)*, 691.
(13)   Grimaud, T.; Matyjaszewski, K. *Macromolecules* **1997**, *30*, 2216.
(14)   Matyjaszewski, K.; Jo, S. M.; Paik, H.-j.; Gaynor, S. G. *Macromolecules* **1997**, *30*, 6398.
(15)   Matyjaszewski, K. *ACS Symposium Series* **1998**, *685*, 258.
(16)   Matyjaszewski, K.; Beers, K. L.; Kern, A.; Gaynor, S. G. *J. Polym. Sci., Polym. Chem.* **1998**, *36*, 823.
(17)   Coca, S.; Jasieczek, C.; Beers, K. L.; Matyjaszewski, K. *J. Polym. Sci., Polym. Chem. Ed.* **1998**, *in press*, .

EFFECT OF MIXING ON INVERSE-EMULSION POLYMERIZATION OF ACRYLAMIDE

Luc Armanet, Jose Hernandez-Barajas and David Hunkeler
Laboratory of Polymers and Biomaterials,
Swiss Federal Institute of Technology, CH-1015 Lausanne,
Switzerland

INTRODUCTION

Water soluble polymers are commercially synthesized through a variety of solution and heterophase processes. The latter consists primarily precipitation, inverse-emulsion and inverse-microemulsion polymerizations. Inverse-emulsion polymerizations have historically been the most commonly employed commercial technologies since thus provide a high polymer concentration at low emulsifiers level. Water-in-oil emulsion polymerization may be carried out by (1) batch polymerization in which all ingredients are added to the reactor and the mixture is heated under agitation to the polymerization temperature ; (2) semi-continuous polymerization [1] in which the monomer as well as other compounds are added continuously or in increments over the duration of the reaction to remove the heat of reaction which would otherwise exceed the coooling capacity of the reactor.

INVERSE-EMULSION STABILIZATION AND DESTABILIZATION

The usual description of these water-in-oil polymerization processes suggests that all methods produce kinetically or thermodynamically stable inverse-latexes. Various hypotheses have been advanced to explain the stability of these inverse-latexes as a function of factors such as the added inorganic salt [2], surfactant type and concentration [3-5], and freezing and thawing [6]. In the literature, however, very little is mentioned regarding the fact that inverse-emulsion polymerization produces varying amounts of « coagulum », (polymer recovered in a form other than that of stable inverse-latexes [6-8]). The coagulum represents a loss of material and is a function of the particle reactor internal surface to volume ratios and this influences the reactor recipe upscaling. Furthermore, although it is unusual for a latex to coagulate completely during the polymerization reaction, it occurs at a sufficient frequency to warrant development of means to remove or reduce the amount of solidified polymer.

COAGULATION AND COAGULUM FORMATION

Coagulum formation is due to a failure of latex stability and the resultant flocculation of the latexes particles during and/or after polymerization. Furthermore, the stability of the latex is affected by various experimental factors such as the recipe used, the type and the intensity of the agitation, the temperature of polymerization reaction and storage, as well as the age and storage conditions of the final inverse-latexes.

MIXING

Agitation is thought to affect the inverse-emulsion polymerization mechanism. The mechanism of batch inverse-emulsion can be divided into three stages [9]: (i) reaction in the continuous phase which includes initiator decomposition, deactivation of the primary radicals with hydrocarbon from the organic phase and initiation of the monomer molecules; (ii) transfer of the primary and oligoradicals from the organic phase to the aqueous phase, which depends on agitation conditions and (iii) polymerization in the aqueous dispersed phase via a free radical mechanism (propagation, transfer as well as monomolecular and bimolecular and termination).

In batch inverse-emulsion polymerizations aqueous monomer phase must be kept in a droplet dispersion form so that adequate surface area is maintained for mass and heat transfer. When agitation of the system occurs, some part of the energy spent must be used for shearing the droplets which otherwise tend to coalesce with time.

The fluid motion in an agitated reactor can be analyzed by the energy balance approach. The input power applied by any mixing impeller produces a pumping effect and a velocity head [10]. In other words, one part of the energy is spent for shearing (velocity head) while another part is used for the flow of the fluid (pumping capacity), and is expressed as following [10]:

$$P \propto Q^*H \tag{1}$$

where P is impeller power input,
the Q is the flowrate or pumping effect and
H is the velocity head or shear.

In this equation, either the flow component (Q) or the shear component (H) can be emphasized such that a large flow, small head, or small flow, large head may be produced for the same power input (P). All of the power supplied to a fluid produces flow and shear. Furthermore, the power input is thought to be dependant on geometrical parameters of the impeller and reactor (impeller diameter, reactor diameter, length of the blades, pitch of the impeller), fluid properties (density and viscosity), as well as rotational speed.

In this work several impeller configurations have been investigated for the inverse-emulsion polymerization of acrylamide. Impellers with varying flow and shear characteristics have been employed and their effect on the final characteristics of the latex are reported. These include rheological properties of the inverse-emulsion, such as viscosity and elastic modulus, the mean particle size, as well as the level of coagulum formed. The formation of coagulum will also be correlatedwith the agitation rate.

REFERENCES

1.   Hernandez-Barajas, J. and D. Hunkeler, *Inverse-emulsion Copolymerization of Acrylamide and Quaternary Ammonium Cationic Monomers with Block Copolymeric Surfactants: Copolymer Composition Control using Batch and Semi-batch Techniques.* Polymer, 1997. **38**(2): p. 449-458.

2.   Kobyakova, K.O., *et al.*, *Effect of Inorganic Salts upon the Stability of Acrylamide.* Russian Journal of Applied Chemistry, 1993. **66**(7(2)): p. 1257-1259.

3.   Ni, H. and D. Hunkeler, *Inverse-Emulsion Stability: Quantification with an Artificial Neural Network.* J. Disp. Sci. Tech., 1997. **in Press.**

4.   Ni, H. and D. Hunkeler, *Investigation of Non-Settling Polymerizable Inverse-Emulsions at Low Surfactant and High Monomer Concentration.* J. Disp. Sci. Tech., 1996. **18**(2): p. 123-160.

5.   Holtzscherer, C. and F. Candau, *Application of the Cohesive Energy Ratio concept (CER) to the Formation of Polymerizable Microemulsions.* Colloids and Surfaces, 1988. **29**: p. 411-423.

6.   Vanderhoff, J.W., *The Formation of Coagulum in Emulsion Polymerization*, in *Emulsion Polymers and Emulsion Polymerization*, D.R. Bassett and A.E. Hamielec, Editors. 1981, ACS: Washington.

7.   Candau, F., *Inverse Emulsion and Microemulsion Polymerization*, in *Emulsion Polymerization and Emulsion Polymers*, P.A. Lovell and M.S. El-Aasser, Editors. 1997, Wiley: New York.

8.   Vanderhoff, J.W., *Recent Advances in the Preparation of Latexes.* Chem. Eng. Sci., 1993. **48**(2): p. 203-217.

9.   Hunkeler, D., W. Baade, and A.E. Hamielec, *Kinetics and Modelling of the Inverse-Microsuspension of Acrylamide.* Polym. Mat. Sci. Eng., 1987. **57**: p. 854.

10.   Oldshue, J.Y., *Fluid Mixing Technology.* McGraw-Hill Publications, ed. C. Engineering. 1983, New-York.

ACRYLAMIDE COPOLYMERIZATION IN MICROEMULSION DROPLETS

R.S. Farinato and L.A. Jackson, Cytec Industries, 1937 W. Main St., Stamford, CT 060904

Introduction

The outcome of a polymerization reaction is often influenced by processing and formulation variables in addition to the nature of the chemical reactants. In this work we investigated the solution properties of high molecular weight acrylamide-containing polymers prepared in inverse microemulsions, and found them to have different characteristics compared to analogous polymers made in either inverse macroemulsions or in solution. Specifically, the apparent aqueous solution structure of poly(acrylamide)s prepared in inverse microemulsions suggested a nonlinear chain topology, whereas poly(acrylamide)s prepared in an analogous manner in either inverse macroemulsion or in solution had all the signatures of a linear chains. The apparent polymer chain nonlinearity of the microemulsion-prepared samples was traced to a non-covalent association which was undone when the polymers were dissolved in formamide. This congested topology in aqueous solution was also seen in microemulsion-prepared acrylamide-Na acrylate copolymers, but not in acrylamide-AETAC copolymers. Such effects were not seen in analogous polymers made by either macroemulsion or solution polymerization methods.

Experimental

Inverse monomer microemulsions were prepared by mixing an aqueous phase, containing the monomers (acrylamide (AMD; from Cytec Industries) with either NaOH-neutralized acrylic acid (NaAc; from Rohm & Haas), or 2-acryloyloxyethyltrimethylammonium chloride (AETAC), or by itself), Na EDTA (from Hamilton Chemical) and t-butyl hydroperoxide (t-BHP; from Lucidol), with an oil phase consisting of a blend of Arlacel™ 83 (sorbitan sesquioleate; from Ruger Chemical / ICI) and Atlas™ 1086 (polyethylene(40) sorbitol hexaoleate; from Ruger Chemical / ICI) surfactants in Low Odor Paraffin Solvent (LOPS; from Exxon). The surfactants were pretreated with acetone to remove any insoluble components, after which the acetone was removed by vacuum stripping. The monomer microemulsions were optically transparent and formed with very little mixing energy. After sparging them with nitrogen, polymerization was initiated with SO_2 (0.1 wt% in N_2; from Matheson). Polymerizations were carried out adiabatically to complete conversion.

Inverse macroemulsion polymerizations were carried out in a similar fashion to that described above. The surfactant concentration was much lower and the aqueous phase ratios were much higher than for the microemulsions. The inverse macroemulsion required homogenization prior to polymerization. The polymerizations were carried out at 50 °C. The resulting emulsions were viscous, white, opaque liquids. The solution-prepared polyacrylamide sample was obtained from a commercial source (Scientific Polymer Products).

The polymers were isolated from the other formulation components by a non-solvent precipitation in either acetone or acetone/methanol (1:1). The polymer solutions for light scattering were clarified by centrifugation (~18,000 x G for 4-5 hours), and checked for material loss using UV absorbance at 215 nm. As an added check of sample clarity, a laser light beam traversing through the solutions was checked visually under low magnification to insure a homogeneous appearance.

Inverse emulsion droplet sizes were determined using either dynamic light scattering (DLS), intensity light scattering (ILS), or small angle neutron scattering (SANS). DLS was used to determine the polymer z-average hydrodynamic radius in solution (R_H), and ILS was used to determine its mean square radius (R_g) and weight average molecular weight (M_w).

The DLS measurements were performed using a Brookhaven Instruments (Holtsville, NY) BI-200SM goniometer and either a BI-8000 or BI-9000 correlator. The intensity-intensity autocorrelation functions for vertically polarized 488 nm light (Spectra Physics model 164 Ar ion laser) scattered from either LOPS-diluted polymerized microemulsions or dilute (< 100 μg/mL) polymer solutions in 1 $\underline{M}$ NaCl at 25.0 °C were collected at several scattering angles (θ) typically from 30° - 120°. Employing standard assumptions (i.e. scattered light obeyed Gaussian statistics and Stokes-Einstein relation applied), the z-average equivalent sphere hydrodynamic diameters, $<d_H>_z$, computed using a second order cumulants analysis, were extrapolated to $\theta = 0°$ on plots of $1/<d_H>_z$ vs $\sin^2(\theta/2)$ to yield the hydrodynamic sizes of the scatterers. The polymer concentrations were low enough in the solutions such that the R_H (= $0.5*[<d_H>_z]_{q=0}$) values were independent of polymer concentration.

A Malvern Mastersizer™ was used to measure the droplet size distribution of inverse macroemulsions. The emulsions were diluted into a hydrocarbon oil containing 3 wt% Arlacel 80 (sorbitan monooleate) and recirculated for 1 minute before measurement. The Malvern analysis software, which employs Mie and Fraunhoffer scattering theories, was used for reducing the data.

A DAWN™ light scattering photometer (Wyatt Technology, Santa Barbara, CA) was used to measure the intensity light scattering (ILS) of polymer solutions at multiple angles (23° $\leq \theta \leq$ 150°) at an incident wavelength (λ_0) of 632.8 nm. The ILS data at 6 - 8 polymer concentrations (c) were analyzed on Debye plots (R_θ/Kc vs. $\sin^2(\theta/2)$ + kc), where R_θ is the excess Rayleigh ratio, K = $[4 \pi^2 n^2 (dn/dc)^2]/(N_A \lambda_0^4)$ for vertically polarized incident light and no analyzer, n is the refractive index of the medium, k is a shift factor, and N_A is Avagadro's number. The dn/dc values used were taken from the literature. Typically we fitted the constant-c curves to a 2nd or 3rd order polynomial and the constant-$\sin^2(\theta/2)$ curves to a 1st order polynomial, excluding high angle data from the set until the polynomial fit was acceptable. This analysis yielded values for M_w and R_g. The ratio R_g/R_H was used as an index of polymer topology.

SANS measurements were made at the W.C. Koehler 30 m SANS facility at the Oak Ridge National Laboratory. The neutron wavelength was 4.75 Å ($\Delta\lambda/\lambda \approx 5\%$) which allowed a useful Q-range of: $0.005 < Q < 0.20$ Å^{-1}. Undiluted emulsion samples were contained in 1 mm path length quartz cells (Helma, Jamaica, NY). Scattering data collected at a two sample-detector distances (3.2 and 14 m) were transformed to radially averaged absolute differential cross sections by comparison with pre-calibrated secondary standards.

Results and Discussion

The polymers compared below were made in essentially one of three different reaction media: solution, inverse macroemulsion or inverse microemulsion. The droplet sizes in the inverse macroemulsions were in the 1 - 2 μ range (volume-surface average diameters determined using ILS). The droplet diameters in the inverse microemulsions were on the order of 0.1 μ (DLS and SANS measurements). In addition, we tracked the droplet sizes throughout the polymerizations and found essentially no change in them as a function of conversion. This is demonstrated in Figure 1 for the case of the inverse microemulsions, for which the SANS diagrams remained unchanged throughout the polymerization. Therefore, for the set of polymers in this work, the local 'reactor' sizes during polymerization were either much larger (solution), roughly comparable to (inverse macroemulsion), or snug compared with (inverse macroemulsion) the unrestrained size of the polymer coil.

The light scattering characterization of high molecular weight water soluble polymers presents some special challenges since many of these polymers are flocculants and thus bind tenaciously to stray colloidal matter. This makes sample clarification critical, especially when the low angle scattering data is important. We found that certain water-insoluble components of the surfactant systems used for preparing these polymers would co-precipitate with the polymers during their isolation from the emulsions. Removing these from the surfactants prior to making the emulsions improved the quality of the scattering data substantially. Several solution clarification schemes were tried and centrifugation yielded the cleanest samples with essentially no loss of polymer.

The results of ILS and DLS measurements of the polymers in solution are collected in the following table. The compositions are stated as mole percentages. The column labeled '[H_2O]/ [mon.]' indicates the relative number of moles of water to monomer in the emulsion droplets. The references for the dn/dc values used in interpreting the ILS data are also shown. Some of the microemulsion poly(acrylamide) data has been reported on previously [1].

polymer	type	[H_2O]/ [mon.] mol/ mol	solvent	dn/dc	M_w (10^6 g/ mol)	R_g nm	R_H nm	R_g/R_h
PAM	micro	4.8	1 M NaCl	0.175 [2]	21.6	192	176	1.1
			formamide	0.10 [3,4,5]	7.5	151	93	1.6
PAM	micro	4.4	1 M NaCl	0.175 [2]	15.2	190	187	1.0
			formamide	0.10 [3,4,5]	9.1	165	94	1.8
PAM	macro	6.8	0.2 M Na_2SO_4	0.187 [6]	7.3	199	95	2.1
PAM	sol'n		1 M NaCl	0.175 [2]	3.8	156	84	2.0
			formamide	0.10 [3,4,5]	4.2	148	72	2.1
NaAc(60)-AMD(40)	micro	4.6	1 M NaCl pH 9	0.184 [7]	15.1	182	153	1.2
AMD(60)-AETAC(40)	micro	9.0	1 M NaCl pH 3.5	0.164 [8,9]	13.0	207	115	1.8
AMD(45)-AETAC(55)	macro	5.7	1 M NaCl pH 3.5	0.164 [8,9]	14.0	243	136	1.8

The importance of the ratio R_g/R_H as an index of polymer topology has been well described by Burchard [10]. Considering chain polydispersity effects, values of this ratio around 1.8 and slightly larger would be indicative of linear coiled chains, whereas values significantly lower than this would imply a more congested chain topology in solution. While such lower values of the R_g/R_H ratio are often associated with chain branching or crosslinking, this apparent topological congestion can also be due to other factors. This becomes most evident when we compare the results from the two microemulsion PAMs in aqueous solution with their counterparts in formamide. The jump in the R_g/R_H ratio and the reduction in apparent molecular weight when the polymers were dissolved in formamide suggested that the congested topology of these PAMs in aqueous solution was not due to a covalently bonded structure. The PAMs made in inverse macroemulsion and in solution both had R_g/R_H ratios in aqueous solution consistent with linear coiled chains, and dissolving the solution PAM in formamide did not change that R_g/R_H ratio. Thus this congested structure appears to be promoted in the microemulsion formulations, and it is readily undone in formamide.

The poly(Na acrylate-co-acrylamide) sample prepared in inverse microemulsion also had a congested morphology, with an R_g/R_H ratio similar to that for the microemulsion PAM. However, the poly(AMD-co-AETAC) sample made in inverse microemulsion did not exhibit such an apparent congested morphology, and had all the signatures of a linear coiled chain. An analogous sample of the cationic copolymer prepared in inverse macroemulsion also had the expected linear topology.

Kulicke and coworkers [11] have argued in a large body of work that hydrogen bonding interactions in acrylamide polymers are important to understanding the polymer solution structure. These interactions have been implicated in the slow evolution of poly(acrylamide) aqueous solution viscosities. Expanding on this point of view, acrylamide chain units would see both other acrylamides and acrylate chain units as potential hydrogen bonding partners, but not likely the quaternary acrylate esters. It is also plausible that chains formed in environments with insufficient water to form a complete hydration sheath might promote such chain-chain hydrogen bonding. P. von Hippel and coworkers [12] have estimated that there are about 2 water molecules associated with the amide group in an acrylamide chain unit in aqueous solution. This sets a lower limit for the amount of water needed to form a hydration sheath, however, simple models would suggest a greater amount to achieve full hydration of the entire acrylamide unit. All the polymers which exhibited congested solution structures had the lowest ratios of water to monomer during their polymerization. This may have promoted non-covalent associations during the polymerization. It should also be remembered that all the emulsion polymers went through a non-solvent precipitation and drying step. To check whether this procedure promoted associations, we took an aqueous solution of one of the PAMs which showed a linear solution topology, ran it through the isolation procedure, re-dissolved and then re-measured it. We found essentially the same results and concluded the isolation procedure was not a factor.

An alternative explanation for the different behavior of the microemulsion-prepared poly(acrylamide)s involves the possibility of small amounts of surfactant being incorporated into the copolymer. Since the oleate double bonds have the potential for reaction with a free radical, it has been postulated [13] that surfactant may react into the growing polymer chain, producing a hydrophobic site on the chain and thus a potential site for aggregation. This could be more likely in microemulsion formulations wherein the surfactant concentrations are greater than in macroemulsion formulations. However, one would have expected the AMD-AETAC copolymer to also suffer this fate, and it did not exhibit any aggregation in aqueous solution.

Conclusions

The apparent congested aqueous solution structures of poly(acrylamide)s formed in inverse microemulsions were not a result of covalent bonding, since the polymers had the signature of linear coiled chains in formamide solutions. These apparent nonlinear topological features were absent in analogous polymers prepared in macroemulsions and in solution. While an inverse microemulsion-prepared acrylamide-Na acrylate copolymer also exhibited a congested solution structure, an acrylamide-AETAC copolymer did not. These observations are consistent with, but not conclusive for, an inter-chain hydrogen bonding which is promoted in confined reaction loci, especially when there is insufficient hydration water for the chain units.

Acknowledgments

We gratefully acknowledge the efforts of S.Y. Huang and R. Rice in the preparation of some of the polymers, D. Lezynski in some of the ILS measurements, and D. Hunkeler, J. Hernandez-Barajas, R. Triolo and D. Chillura in the SANS measurements.

References

1. Q. Ying, G. Wu, B. Chu, R. Farinato and L. Jackson (1996) Macromol., 29(13), 4646-4654
2. Th. Griebel, W.M. Kulicke and A. Hashemzadeh (1991) Coll. Poly. Sci., 269, 113-120
3. S. Biggs, A. Hill, J. Selb and F. Candau (1992) J. Phys. Chem., 96, 1505-1511
4. N. Onda, K. Furusawa, N. Yamaguchi, M. Tokiwa and Y. Hirai (1980) J. Appl. Poly. Sci., 25, 2363-2372
5. J. Francois, D. Sarazin, T. Schwartz and G. Weill (1979) Polymer, 20, 969-975
6. D. Hunkeler, X.Y. Wu and A.E. Hamielec (1992) J. Appl. Poly. Sci., 46, 649-657
7. K.J. McCarthy, C.W. Burkhardt and D.P. Parazak (1987) J. Appl. Poly. Sci., 33, 1699-1714
8. T. Griebel and W.M. Kulicke (1992) Makromol. Chem, 193, 811-821
9. F. Mabire, R. Audebert and C. Quivoron (1984) Polymer, 25, 1317-1322
10. W. Burchard, Adv. Poly. Sci., 48 (1983) 1-124
11. W.-M. Kulicke, R. Kniewske and J. Klein, Prog. Polym. Sci., 8 (1982) 373-468
12. P.H. von Hippel, V. Peticolas, L. Schack and L. Karlson (1973) Biochem., 12(7), 1256-1264.
13. D. Hunkeler, Polym. Int., 27 (1992) 23-33

Figure 1. SANS intensity vs Q for acrylamide microemulsions at various reaction times (min.). Sample-detector distance = 3.2 m. Scattering curve for final polymer microemulsion was indistinguishable from the rest.

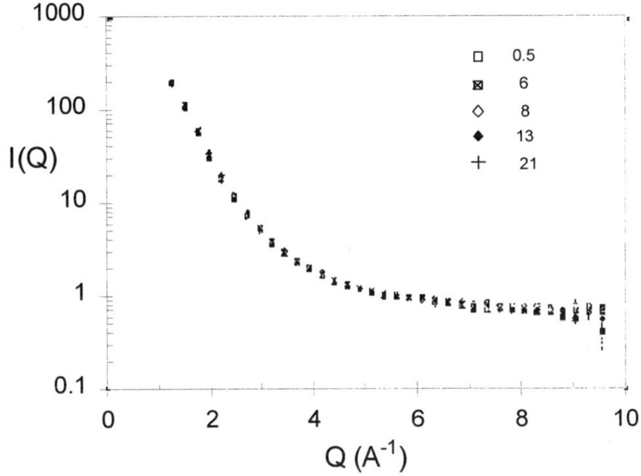

Separation of Poly(Vinyl Pyrrolidone) by High Osmotic Pressure Chromatography

Iwao Teraoka, Yunmei Xu, Larry Senak,[§] and Chi-San Wu[§]

Polytechnic University, 333 Jay Street, Brooklyn, NY 11201

[§]International Specialty Products, 1361 Alps Road, Wayne, NJ 07470

Introduction

High osmotic pressure chromatography (HOPC) was developed a few years ago as a technique for large-scale separation of a polydisperse polymer by molecular weight (MW).[1–3] A concentrated solution of the polymer is injected into a column packed with solid porous materials until the whole column is filled with the solution. Then, the injection is switched to the pure solvent, and the eluent is collected into different fractions. HOPC's separation is believed to be based on the segregation of polymer between the pore spaces (stationary phase) and the surrounding solution (mobile phase). In the semidilute solution (above the overlap concentration), the high osmotic pressure of the solution drives low MW components preferentially into the pore spaces, leaving the mobile phase deficient in the low MW components, i.e., enriched with the high MW components. Segregation of the polymer chains is repeated as the concentrated solution is transported along the column. Thus the MW distribution in the early eluent is centered on the highest end of the distribution of the original polymer. Typically the polydispersity index, PDI, defined as PDI = M_w/M_n, where M_w and M_n are the weight-average MW and the number-average MW, decreases, in the early fractions, to $1/4 - 1/5$ power of the index of the original polymer, for instance, to 1.2 from 2.0 of the original.[3] Later fractions have a smaller M_w and a broader MW distribution. It was also predicted in theory that HOPC's resolving power is higher than the one available in regular gel permeation chromatography (GPC).[3] HOPC separated various organic-soluble polymers such as polystyrene and poly(methyl methacrylate).

The present paper reports 1) the first separation of water-soluble polymers by HOPC with an aqueous mobile phase, 2) demonstration that the resolution of HOPC exceeds that of GPC, and 3) multi-stage separation to prepare standard-grade fractions. As a typical water-soluble polymer of neutral charge, we chose poly(vinyl pyrrolidone) (PVP) manufactured by International Specialty Products. It is available in different viscosity grades, K15, K30, K60, K90, and K120 in the increasing order of the average MW. We will show results of HOPC separation for K15, K30, and K60. To prove the high resolution of HOPC, we analyzed the fractions by using GPC with an on-line multi-angle laser light scattering (MALLS) detector.

Experimental

HOPC system. The HOPC system is the same as the one used in the earlier study. A concentrated solution of the polymer was continuously introduced into the column through a single-plunger HPLC pump (SSI, AcuFlow II). Stainless steel columns packed with acid-washed controlled pore glass (CPG) (CPG, Inc. and W. Haller of NIST) of different pore diameters were used to separate polymers of different MWs. The eluent was collected as follows. When a column of 3.9 mm × 300 mm was used, 15 drops were collected in each of fractions 1 – 6, 30 drops in fractions 7 – 10, and 150 drops in fractions 11 – 14. When the

column dimension was 7.8 mm × 300 mm, four times as many drops were collected in each fraction. The flow rate was 0.1 and 0.3 mL/min for the two column dimensions.

GPC. To characterize the MW distribution for the original and fractionated samples, we used two different GPC systems. One consists of three columns of Shodex (SB803, 804, and 805) with 0.1 M KCl mobile phase (1 mL/min) and a differential refractive index (DRI) detector. The other consists of two linear columns of Shodex (SB80-MHQ) with 0.1 M LiNO₃ mobile phase (0.5 mL/min). A DRI detector and a MALLS detector (Wyatt, Dawn) connected in series were used for the latter.

Results and discussion

(1) Choice of pore surface

Separation of organic-soluble polymers was done by using $(CH_3)_3Si$–modified CPG.[1,3] The same CPG was found to deteriorate the separation performance each time the column was used. Figures 1a and 1b compare the chromatograms for the first and 6th batches of separation of K30 by the modified CPG of pore diameter of 130 Å. The chromatograms were obtained in the three-column GPC system with the DRI detector (also for Figures 2 and 3). The chromatogram for the original K30 is shown in a thick line. The number adjacent to each curve indicates the fraction number. Each chromatogram is normalized by the area under the peak. The sharp peaks observed in Figure 1a gave way to broadened peaks in Figure 1b.

In contrast, acid-washed CPG (the surface is unmodified) produced more reproducible separation in repeated use of the same column after the first few batches. Figures 1c and 1d compare the chromatograms the first and 4th batches for the separation of K30. Therefore all of the following separations was done by using the acid-washed CPG.

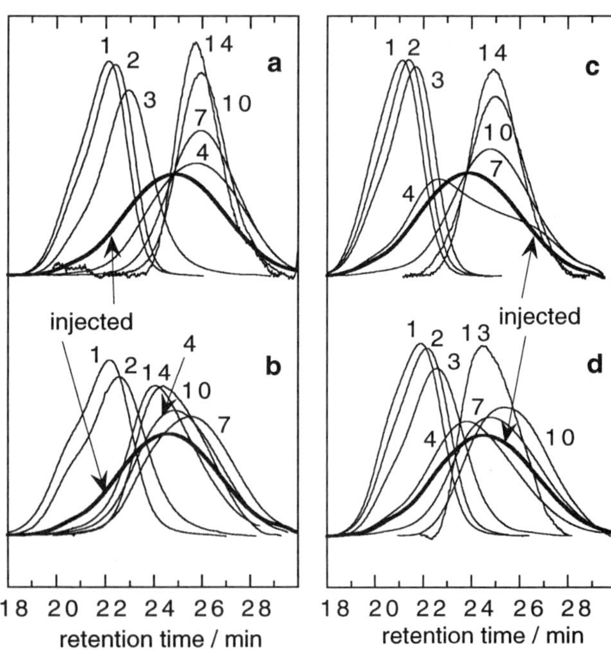

Figure 1. Results of HOPC separation of K30. (a) first-time use of modified CPG, (b) 6th time use of modified CPG, (c) first-time use of unmodified CPG, (d) 4th time use of unmodified CPG.

(2) Quantity of each fraction

In a multi-stage separation it is important that each step of separation produces a large quantity of fractions with a narrowed MW distribution. We measured the mass of each fraction obtained in the HOPC separation of K30 by a column of 7.8 mm × 300 mm packed with CPG of pore diameter of 130 Å. A 6.87 g of the 40 wt% solution was injected. Some of the fractions were characterized in GPC, and M_w and M_n were calculated with calibration by poly(ethylene glycol) standards. Figure 2 shows M_w and PDI as a function of the cumulative mass. The values for the original K30 are also indicated. Fractions 1 to 3 have a PDI < 1.5. Combined they claim about 10% of the total mass of the polymer injected.

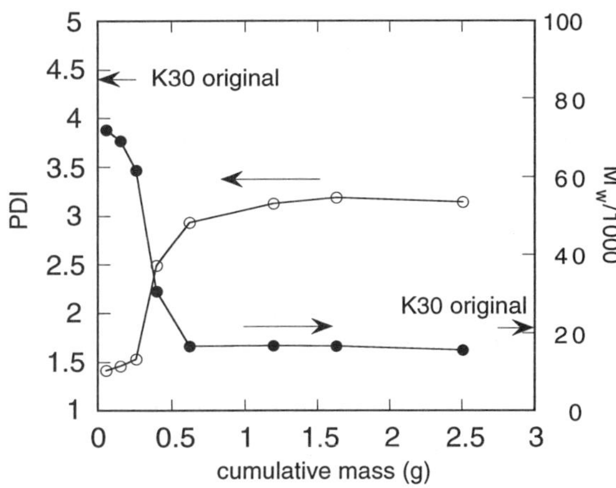

Figure 2. PDI and M_w of the fractions obtained in the separation of K30, plotted as a function of the cumulative mass.

(3) Multi-stage separation

We show the narrowing in the MW distribution for K60 in the multi-stage HOPC. Figure 3a shows a typical first-stage separation of the original K60 by CPG with a pore diameter of 240 Å. Early fractions that have a chromatogram similar to that of fractions 1 and 2 in Figure 3a, obtained in several batches, were combined to make a concentrated solution. The solution was injected into the column packed with CPG of a pore diameter of 364 Å. Figure 3b shows the typical GPC result. Fractions that have a chromatogram similar to that of fraction 1 in the figure were combined for the third-stage separation by CPG of a pore diameter of 500 Å. Its separation result is shown in Figure 3c. Some of the fractions obtained in the third-stage separation were characterized also by using the GPC system with MALLS detector. In Figure 4, we compare the MW of the eluent for the K60 original and three fractions obtained in the third-stage HOPC. Large errors at both ends of the chromatograms are due to weak scattering of the diluted eluent. The MW of the original ranges over two decades. The log(MW) is almost a linear function of the elution volume, as expected for the linear column. The MW plots for the separated fractions are, however, nearly horizontal and are not a part of the plot for the original. GPC's band broadening is substantial for fractions that exhibited a narrowed distribution in MW. If K60 were separated in regular GPC or preparative GPC, it would be difficult to obtain fractions with a MW distribution as narrow as in the fractions we obtained in HOPC. This result attests the high resolution of HOPC. Multi-stage separation of K15 and K30 produced also narrow distribution fractions.

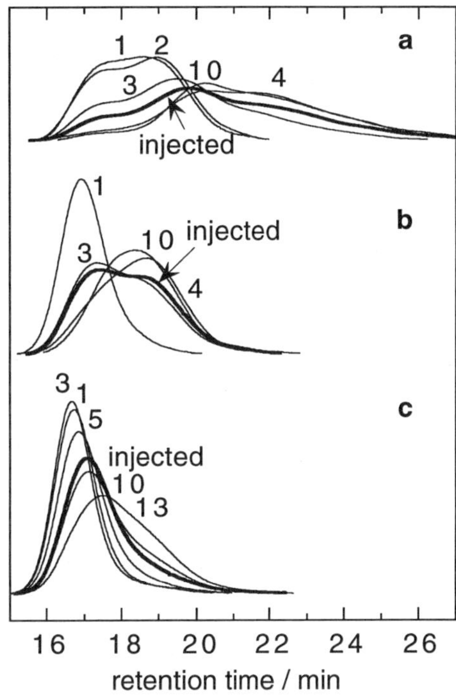

Figure 3. Results of multi-stage separation. (a) first stage, (b) second stage, (c) third stage.

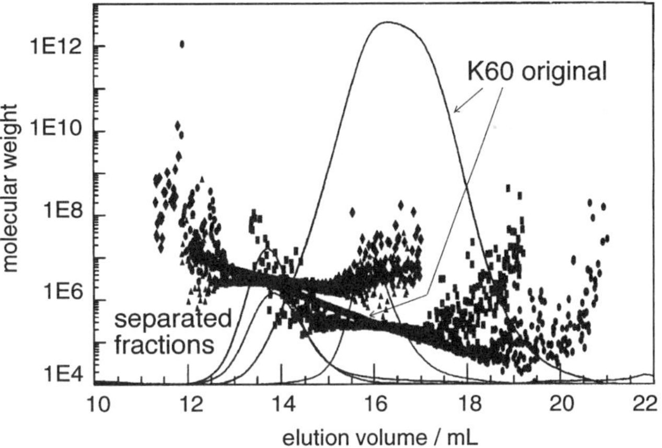

Figure 4. Comparison of the MW of the eluent for the K60 original (squares) and fractions obtained in the third-stage HOPC (rhombuses, triangles, and ovals). The DRI signals are also shown.

References

1. Luo, M.; Teraoka, I. *Macromolecules* **1996**, *29*, 4226.
2. Teraoka, I.; Luo, M. *Trends Polym. Sci.* **1997**, *5*, 258.
3. Luo, M.; Teraoka, I. *Polymer* **1998**, *39*, 891.

Water-Absorbing Polyurethane/Poly(ethylene glycol) Blends: Distinctive Microstructure and Hydrophilicity.

F.I. Simonovsky[1] and B.D. Ratner[1,2]

Department of Chemical Engineering[1] and Department of Bioengineering[2]

University of Washington, Box 351750, Seattle, WA 98195-1750

INTRODUCTION

Segmented two-microphase poly(alkylene oxide)urethane block copolymers (PEUs) consisting of hard and soft segments and involving covalent-linked poly(ethylene glycol)s (PEGs) in the soft segment are suitable hydrophilic materials for monolithic drug delivery formulations [1]. In an aqueous environment, the soft segments absorb more water in comparison with the hard segments, as a function of the inherent hydrophilicity of the oligomers used [2]. A study of PEUs with PEG-poly(tetramethylene oxide) (PTMO) soft segments developed these polymers as a controlled release material for water-soluble drugs [3].

PEGs can also be employed as non covalent additives in PEUs to aid in drug release from the polymer matrices. Therefore, the microstructure and hydrophilicity of PEG-PEU polymers as well as PEU/PEG blends make them promising candidates for medical device applications.

MATERIALS

Several PEUs were synthesized based on 4,4'-diphenylmethane diisocyanate, 1,4-butanediol, and the mixtures of PTMO and PEG with molecular weight (MW) 1000 (PTMO-1000/PEG-1000) comprising 42 wt. % of the soft segments (PEU-1), PTMO-2000/PEG-2000 (60 wt. % of the soft segments) (PEU-2), and PTMO-1000 (40 wt. % of the soft segments) (PEU-3) as a standard. BioSpan (PTG, Inc.) was also used for comparison. 25 % or 50 wt. % of PEGs (MW from 1000 to 8000) were solvent blended on PEUs. Distinctions in microstructure and water absorption as well as PEG release were studied by FTIR and gravimetric analysis.

RESULTS AND DISCUSSION

The spectra of PEU-1 is verified by the strong absorbance of hydrogen-bonded NH groups at around 3320 cm^{-1}. Also in comparison with PEU-3, the incorporation of PEG into the soft segments is characterized by the appearance of a C-H stretching band located at approximately 2930 cm^{-1}. After adding PEG-1000 as a filler, the peak intensity of NH bonded groups at 3320 cm^{-1} decreases. This adding of PEG also generates a shoulder at 3420 cm^{-1} which indicates that free NH groups have appeared. The behavior of the N-H and C-H stretching bands corresponds with the intensity of the H-bonded urethane carbonyls peak in the amide 1 range, where the ratio of the bonded (1703 cm^{-1}) to free (1732 cm^{-1}) carbonyl absorbance indicates the degree of phase separation in PEUs [4]. This result can be attributed to a decrease in the interaction between NH groups and carbonyl oxygen in the hard segments, and along with the changes in the C-O-C stretching absorbance in aliphatic ether links in the range 1080-1120 cm^{-1} characterize the reduction of phase separation in the polymer blend when PEG-1000, as a filler, is present.

The differences observed between PEU-1 and PEU-2 reflects the increase of the soft segment content and is characterized by an increase

of the C-H stretching band at 2845 cm^{-1} and the C-O-C ether absorbance peak at 1110 cm^{-1}. Another indication for the change of the soft segment content is the increase of the intensity ratios of the ether peak (1110 cm^{-1}) relative to the C=C aromatic absorbance peak at 1600 cm^{-1}, and also to the C-O-C stretching band at 1080 cm^{-1} in the hard segments.

BioSpan has three bands in the amide 1 range located at approximately 1730, 1715 and 1625 cm^{-1}, which can be attributed, correspondingly, to free and to bonded carbonyl absorption in urethane and urea groups. The decrease of the intensity of the N-H bonded stretching band at 3330 cm^{-1} and increase of the ratio of the soft segment C-O-C ether absorbance peak at 1115 cm^{-1} to the C=C aromatic ring absorption peak at 1600 cm^{-1} indicate the influence of PEG-1000 as a filler. These spectra in the amide 1 range also show that addition of the filler decreases the peak intensity for both types of H-bonded carbonyls.

For samples which are filled with PEGs having MW 2000 and up, the ratio of bonded to free carbonyl peak intensity quantitatively is the same as seen for unfilled PEUs. This behavior correlates with the gradual decrease and eventual disappearance of free NH groups for all blends filled with PEG MW more then 2000. The loss of free NH groups can be related to the reappearance of bonded carbonyls. These results indicate the influence of MW and the amount of PEG filler along with PEU-composition on microphase separation in PEU/PEG blends. In this case PEG-1000 can act as an interfacial agent. It can be assumed that incompatibility between PEGs with MW more then 2000 and PEU matrices causes the structural change.

Because PEGs are water-soluble, PEUs with PEG in the soft segment are water-sensitive. Equilibrium water uptake and uptake kinetics significantly depend on the type, MW, and concentration of the soft segments in the polymers. As Figure 1 shows, BioSpan consisting of 100% PTMO-2000 as a soft segment demonstrates low water interaction. This polymer has the ability to absorb only 2 wt. % of water, while PEU-1 absorbs almost 4 times more, and PEU-2 absorbs 10 times more. Further increases of MW and PEG content in the soft segments allow for preparation of polymers possessing still higher water absorption.

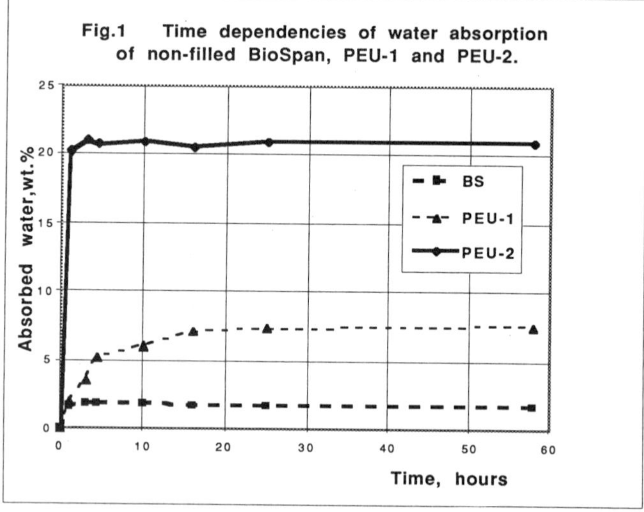

The kinetics of water uptake are altered when PEUs are filled with 20 wt. % PEG-1000. There is a greater increase for BioSpan than for PEU-1 or PEU-2. Further increase of the PEG-1000 content to 50 wt. % leads to an increase in the equilibrium content of absorbed water for all samples. In this case, BioSpan absorbs the highest amount of the water. This result can be explained by macrostructure effects due to "internal space" created by the release of the PEG filler. Another distinctive feature is that BioSpan filled with PEG-8000 has a characteristic water uptake curve different from PEU-1 and PEU-2.

It is interesting to note that BioSpan, without hydrophilic PEG in the soft segment, displays an equilibrium concentration of absorbed water even greater than that observed for PEU-1 containing PEG. This result can be explained by a different mechanism of hydrophilization during and after PEG release. This idea is supported by the effect of MW of PEG fillers on water absorption for BioSpan, PEU-1 and PEU-2. The comparison of absorption kinetics for three different polymers such as BioSpan, PEU-2 and poly(ethylene oxide)urethane composite hydrogel *Aquavene* [5] indicates that both the chemical microstructure as well as macrostructure of the polymeric material may influence the water uptake. If PEG are used for filling of thermoplastic PEUs, the precise concentration and MW of the filler are critically important for water uptake behavior in the initial period of the process (Fig. 2).

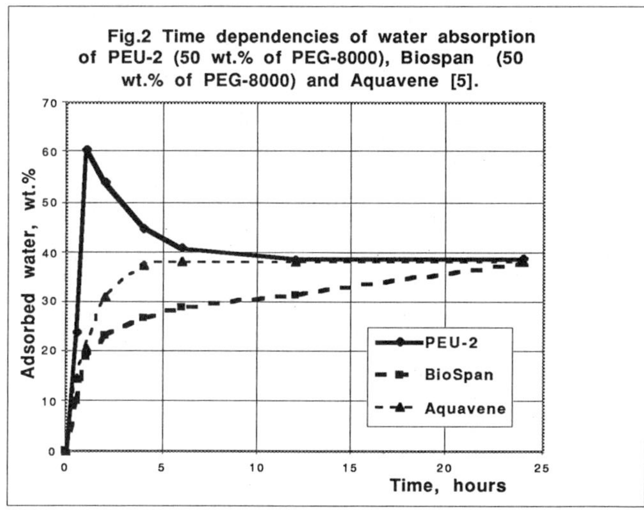

CONCLUSIONS

The degree of microphase separation, hydrophilicity and PEG release profiles can be modified to a great extent in PEG-PEU polymers. The modification is achieved by changing the bulk chemical structure of PEUs containing PEGs in the soft segments or using the various molecular weights and concentrations of non-covalently linked PEG fillers. In this case, the PEG fillers act as a hydrophilic agent and create interfacial compatibility within the polymer structure.

ACKNOWLEDGMENT

We thank the National Institutes of Health (Grant HL 51260) for support of this work.

REFERENCES

1. Poly(ethyleneglycol) Chemistry, Biotechnical and Biomedical Applications, Plenum Press, Ed. J. Milton Harris, (1992).

2. R.S. Ward and K.A. White, US Patent 5,428,123 (1995).

3. Y. Ikeda, S. Kohjiya, S. Takesako, and S. Yamashita, *Ann. Plast. Surg.*, 24, 80, (1990).

4. S.C. Yoon and B.D. Ratner, *Macromolecules*, **21**, 2392, (1988).

5. S.P. Gorman, M.M. Tunney, P.E. Keane, K. Van Bladet, and B. Bley, *J. Biomed. Mater. Res.*, **39**, 4, 642, (1998).

NEW POLYETHER POLYOLS FOR AQUEOUS POLYURETHANE DISPERSIONS

Stephen D. Seneker and Nigel Barksby
ARCO Chemical Company, South Charleston, WV 25303

INTRODUCTION

Aqueous polyurethane (PUR) dispersions were introduced into the marketplace in the late 1960's. Since that time they have not only achieved commercial significance but have grown steadily through the years. Initially, the acceptance and growth in PUR dispersions was spurred by environmental considerations related to reducing organic solvent levels in coatings systems. However, in the last decade the properties of PUR dispersions have started to approach and in some cases exceed the properties of two-component, solvent-borne polyurethanes. Currently, they can be found in applications such as textile coatings, leather finishing, adhesives, floor coatings and glass fiber sizing.

PUR dispersions are essentially high molecular weight linear polyurethane/ureas dispersed in water. Water dispersability of the polyurethane is achieved by incorporating ionic groups (cationic or anionic) along the polymer backbone. The ionic source is usually carboxylate or sulfonate groups. Films prepared from carboxylate containing dispersions are known for their low moisture pickup while those based on sulfonate groups exhibit improved hydrolysis resistance.[1] Most commercial PUR dispersions use the carboxylate containing compound, dimethylolpropionic acid (DMPA). DMPA has a structure well suited for this technology. It is a diol with primary hydroxyl groups so it can be easily incorporated into the polyurethane backbone. Also, its carboxylate group is located on a sterically hindered tertiary carbon which minimizes reactivity with isocyanate. Several excellent review articles are available which detail the various chemistries and processes that can be used to manufacture aqueous PUR dispersions.[2,3]

Recently, a "new generation" of propylene oxide based polyether polyols (PPGs) with ultra-low monol contents were introduced by ARCO Chemical. These polyols, designated *Acclaim*™, have the potential to significantly impact aqueous PUR dispersion technology because of two key features: ultra-low monol content and low viscosity.

Aqueous PUR dispersions based on conventional PPGs have inferior film properties because they contain a significant amount of monol impurity. Eliminating monol from PPGs significantly increases the ultimate polymer molecular weight which dramatically improves mechanical properties to levels approaching that of polyester based PUR dispersions.

Low viscostiy is the other key feature. Isocyanate prepolymers based on these new polyols have viscosities that are an order of magnitude lower than those based on polyester polyols. Low prepolymer viscosity make possible the formulation of very soft films with the excellent "hand" normally associated with solvent-borne lacquer systems.

The additional features and benefits of *Acclaim* polyols are numerous. Their polyether backbone inherently provides hydrolytic stability and microbial resistance in polyurethanes. Their low glass transition temperature (Tg) approaching that of polytetramethylene ether glycol (PTMEG) and lack of crystallizability results in films with excellent flexibility in cold harsh environments. Also, the storage and handling requirements are minimal since they are low viscosity polyols with pour points as low as -30°C (-22°F).

This paper covers the following key topics:

- Background on monol content in PPGs: conventional versus new-generation technology;
- Monol content effect on mechanical properties of aqueous PUR dispersion films;
- Polyol MW distribution effect on prepolymer viscosities and film properties;
- Broad formulating latitude using these new-generation polyols.

RESULTS AND DISCUSSION

Aqueous Polyurethane Dispersions Based on Ultra-Low Monol PPGs

Aqueous PUR dispersion formulations typically contain 50 to 70 weight percent polyol, consequently, the polyol component has a significant impact on the overall cost of the system. Over the years, cost/performance issues have led a majority of the industry towards using polyester polyols over polyether polyols. On the performance side, polyesters polyols have superior stain and chemical resistance with excellent mechanical properties but have inferior hydrolysis resistance. For this reason, the industry has tended towards using more hydrolytically stable polyester polyols such neopentylglycol adipates. Polyether polyols such as polytetramethylene ether glycol (PTMEG) give superior hydrolysis resistance with excellent mechanical properties but cost significantly more than polyester polyols. Polyether polyols like polyoxypropylene glycol (PPG) give superior hydrolysis resistance at a significantly lower cost than polyesters but give inferior mechanical properties. We believe a need exists in the industry for a cost effective polyol which will give superior hydrolytic stability along with excellent mechanical properties. Conventional polyoxypropylene diols (PPG) are produced commercially through the based catalyzed propoxylation of glycol starters such a propylene glycol. However, base catalyzes not only the addition of propylene oxide to the growing polyol molecule, but also a side reaction in which propylene oxide isomerizes to allyl alcohol.[4,5,6] Allyl alcohol acts as a monofunctional starter resulting in propoxylated allyl alcohol referred to as "monol". Since each monol chain contains a terminal double bond, the amount of monol present can be quantified by measuring the unsaturation level. The level of unsaturation is typically reported in milliequivalents of unsaturation per gram.

In conventional PPGs, the level of unsaturation or monol content increases dramatically with increasing polyol molecular weight. To get a better appreciation for this trend we can converted the unsaturation content to functionality and mole percent monol.[7] For example, a 2000 MW conventional PPG typically has an unsaturation of about 0.03 meq/gm, which corresponds to a functionality of 1.94, whereas, a 4000 MW diol has an unsaturation of 0.085 and functionality of only 1.69. In other words, a conventional 4000 MW diol is really a mixture of about 70% diol and 30% monol on a mole % basis.

Ultra-low monol PPG polyols are produced using a proprietary technology resulting in PPGs with dramatically lower monol (unsaturation) contents of 0.005 meq/gm or less (Table 2). For example, the ultra-low monol 4000 MW PPG diol has a functionality of 1.98 versus 1.69 for a conventional diol.

Traditionally, conventional PPGs have had a reputation as a lower cost raw material for low to moderate performance applications. Many have attributed the inferior performance of PPGs to its inherent structure or lack of ability to crystallize like PTMEG or polyester polyols. After extensive applications testing, we have come to believe that the single most important factor controlling performance of PPGs is the monol content. The dramatic effect of monol content is demonstrated below.

Monol Effect in Aqueous PUR Dispersion Films

We determined the effect of monol content in aqueous PUR dispersions by comparing our 4000 MW ultra-low monol *Acclaim* 4200 with our 4000 MW conventional diol (ARCOL® PPG-4025). The *Acclaim* 4200 had a monol content of 0.005 meq/gm versus 0.085 meq/gm for the conventional diol. This corresponds to functionalities of 1.98 and 1.69, respectively. Dispersions were prepared by the prepolymer mixing process using the following formulation. The prepolymer was based on dicyclohexylmethane diisocyanate (H_{12}MDI; 70.2g, 0.535 equiv.), 4000 MW PPG diol (190.0g, 0.097 equiv.) and dimethylol propionic acid (DMPA, 9.8g, 0.147 equiv.). It also contained 10 weight percent N-methyl pyrrolidone (NMP) as a coalescing aid. This mixture was reacted at 100°C until reaching a weight % isocyanate content of 4.1. Triethylamine was used as a neutralizer and ethylene diamine as a chain extender. All PUR dispersions were prepared to a 40% solids content. A detailed experimental procedure along with general formulation guidelines are given in another paper.[8]

Figure 1 shows the stress/strain curves of films based on the new diol versus the conventional diol. As expected, the film based on the ultra-low monol PPG has significantly better properties. The tensile strength increased from 1800 to 3200 psi and elongation increased from 320 to 440%. The modulus build with increasing elongation was higher as well. We believe these improvements are directly attributable to a higher polymer molecular weight resulting from the ultra-low monol (terminator) content. These results are consistent with previous applications studies on polyurethane elastomers and moisture-cured prepolymers.[9,10,11]

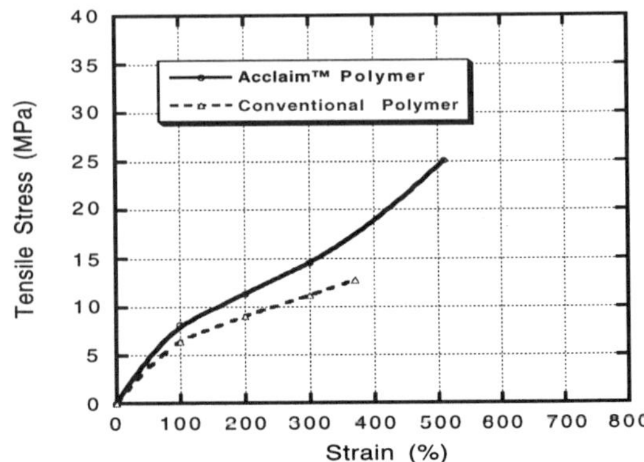

Polyol MW Distribution Effect in Aqueous PUR Dispersions

An understanding of the fundamental differences between ultra-low monol PPGs versus polyester and PTMEG diols is essential to appreciate the formulation opportunities. A key feature of *Acclaim* polyols is a narrow molecular weight distribution (polydispersity ‹1.1). This feature yields polyols with viscosities that are an order of magnitude lower than polyester or PTMEG diols at an equivalent molecular weight. A comparison of GPC traces (Figure 2) of a <u>2000</u> MW polytetramethylene ether glycol (PTMEG-2000) and a <u>4000</u> MW ultra-low monol *Acclaim* 4200 reveals a relatively symmetrical narrow MW distribution for *Acclaim* 4200 with a peak MW of about 4000. PTMEG-2000, on the other hand, has a very broad asymmetrical MW distribution consisting of a large high MW fraction with a significant low MW tail. The high MW fraction in PTMEG-2000 accounts for its significantly higher polyol viscosity.

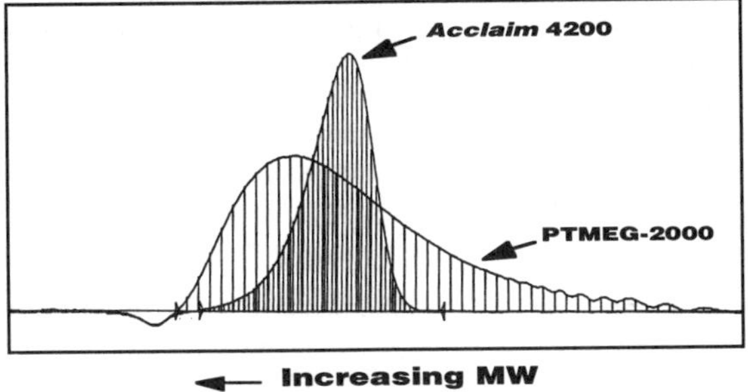

Polyol MW distribution also has a significant impact on the isocyanate prepolymer viscosity. Prepolymers based on ultra-low monol PPGs have viscosities an order of magnitude lower than those based on polyesters or PTMEG.

As mentioned previously, most commercial aqueous PUR dispersions are based on polyester diols. Isocyanate prepolymers based on polyesters have extremely high viscosities, so it is technically difficult to formulate soft films due to these processing constraints. Since isocyanate prepolymers based on *Acclaim* polyols have viscosities an order of magnitude lower than polyesters or PTMEG, one can formulate aqueous PURs using prepolymers with lower isocyanate contents. Since the isocyanate content of the prepolymer generally controls the hardness of the final polyurethane film, one can now achieve very soft aqueous PUR films with the excellent "hand" normally associated with solvent-borne lacquer systems.

To illustrate the formulating advantage of lower viscosity prepolymers, we prepared a prepolymer using H$_{12}$MDI and 2000 MW PTMEG (broad MW distribution). The NCO/OH ratio was adjusted to give a prepolymer with a viscosity higher than what can be processed on a commercial scale (practical upper limit is ≈5000 cps). Replacing PTMEG-2000 (broad MW distribution) with *Acclaim* 2000 (narrow MW distribution) resulted in seven-fold decrease in prepolymer viscosity from 9100 cps to 1350 cps at 80°C.

Stress/strain curves of films based on *Acclaim* 2200 and PTMEG-2000 are dramatically different (see Figure 3). The ultra-low monol polyol-based film has nearly twice the percent ultimate elongation and a significantly lower modulus. We believe the high elasticity is due to the lack of crystallizability of these polyols versus PTMEG. If the polymer soft segment stress crystallizes, it will lower the ultimate elongation. Previous studies in polyurethane elastomers supports this explanation.[12] The low modulus of ultra-low monol polyol-based film could be decreased even further by lowering the % NCO content or NCO/OH ratio as the prepolymer viscosity was only 1350 cps at 80°C.

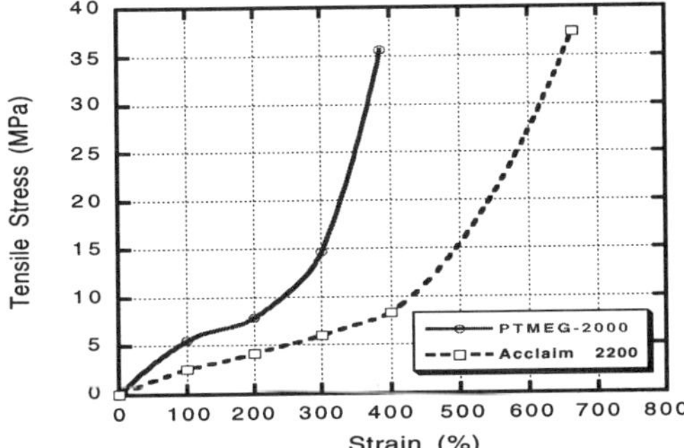

Broad Formulating Latitude

When comparing ultra-low monol PPG with polyester or PTMEG polyols it is important to consider not only the MW distribution effect on polyol and prepolymer viscosities, but also its effect on the film mechanical properties. Earlier, we attributed the dramatic differences in the stress/strain curves of the films based on our 2000 MW ultra-low monol PPG and PTMEG-2000 to the polyol crystallizability (Figure 3). What would happen if we took out crystallizability as a factor? Does the polyol MW distribution affect the stress/strain properties of the polymer? To demonstrate this point, we prepared an aqueous PUR dispersion based on IPDI using *Acclaim* 2200 and compared that to one using *Acclaim* 4200 and dipropylene glycol (DPG) blended to a 2000 MW. One can clearly see from the stress/strain curves in Figure 5, that these two films have significantly different properties. We believe the lower modulus of the *Acclaim* 2200 based film can be explained in terms of differences in the polyol MW distribution. Referring back to our discussion on MW distribution, we described polyester and PTMEG polyols as having a broad MW distribution, comprised of a large high MW fraction and a prominent low MW tail (see Figure 2). For our viscosity discussion we focused on the high MW fraction; for our film property discussion we will focus on the low MW tail.

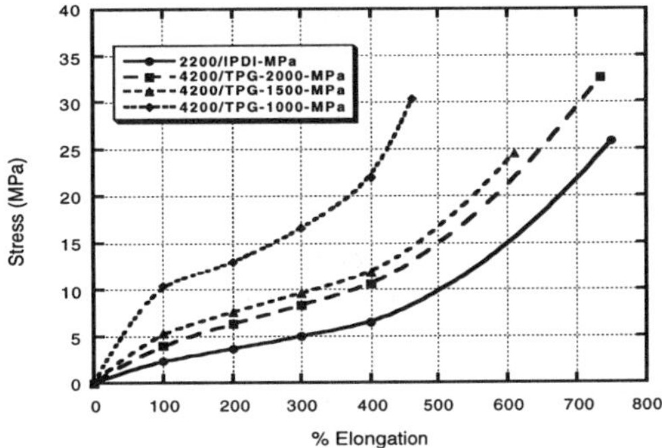

When formulating a polyurethane or polyurethane/urea polymer, one needs to consider what portion of the low MW tail acts likes "pseudo" chain extender and contributes to the hard-segment content of the system. In "urethane" systems, a <u>small</u> amount of the low MW tail acts as hard-segment, however, in "urethane/urea" systems a <u>large</u> portion of the lower MW fraction contributes to the hard-segment content. *Acclaim* polyols do not contain a low MW tail, so when they are compared to broad MW distribution polyols like polyesters and PTMEG in polyurethane/urea systems, such as aqueous PUR dispersions, they always give a significantly lower modulus. Consequently, the modulus of the film based on the *Acclaim* 4200/DPG in Figure 4 is higher because the DPG contributes to the hard segment. We observed this effect previously with other urethane/urea systems such as moisture-cured toluene diisocyanate (TDI) prepolymers and TDI prepolymers cured with MOCA [methylene bis(ortho chloroaniline)].[12]

The formulating principle of blending also has a significant impact on the polymer soft segment. In one case the soft segment is IPDI/*Acclaim* 2200 and in the other, IPDI/*Acclaim* 4200. The IPDI/*Acclaim* 4200 soft segment has a lower Tg and promotes a more complete hard segment/soft segment phase separation which results in films that feel "drier" to the touch and have less of a tendency to adhere to itself.

Figure 5 shows the effect of further lowering the MW of the *Acclaim* 4200/DPG blends to 1500 and 1000. These results illustrate that these new-generation polyols can be used to formulate <u>hard</u> films as well as <u>very soft</u> films by understanding their fundamental differences with respect to other high-performance polyols like polyesters and PTMEG.

CONCLUSIONS

The new-generation of propylene oxide-based, polyether polyols have monol contents up to twenty (20) times lower than conventional PPG polyols. The elimination of monol results in polymers having much higher final molecular weight and therefore improved mechanical properties. Aqueous PUR dispersion films based on these polyols have mechanical properties comparable to those based on high-performance polyester and PTMEG polyols.

The ultra-low monol polyols have narrow molecular weight distributions which result in polyol and isocyanate-prepolymer viscosities that are an order of magnitude lower than those based on polyester or PTMEG polyols. When this technology is applied to aqueous polyurethane dispersions, one can achieve very soft films with the excellent "hand" normally associated with solvent-borne, lacquer. One can also obtain higher hardness films by blending higher MW ultra-low monol polyols with lower molecular weight glycols to approximate the molecular weight distribution of polyester and PTMEG polyols.

"Amphiphilic Block Copolymers as Surfactants in Emulsion Polymerization"

D. Urban, M. Gerst, P. Rossmanith, H. Schuch,
BASF AG, Polymer Research Laboratory, Ludwigshafen,
Germany

Abstract

The amphiphilic block copolymers poly(methylmethacrylate-block-acrylic acid), poly(methylmethacrylate-block-methacrylic acid) and poly(isobutylene-block-methacrylic acid) were made by living ionic polymerization. In aqueous solution, block copolymers (BC) of higher molecular weight (m.w.) form spherical „frozen micelles", not exchanging unimers. The micellar aggregation number Z and micellar diameter $2R_h$ in the aqueous medium were obtained by static and dynamic light scattering and turned out to be predictable solely from the two block lengths. Using the aqueous solutions of neutralized block copolymers as the only emulsifying agent in emulsion polymerization, stable dispersions were obtained. The „frozen micelles" act as a seed resulting in a constant number of BC molecules per polymer particle, independent of the particle size. The mechanical properties and the water sensitivity of the resulting polymer films can be changed by annealing.

Introduction

Low molecular weight surfactants are widely used in emulsion polymerization to form and stabilize the polymer particles and to prevent coagulation of the dispersion when applied in making e.g. coatings and adhesives. The physical properties of the resulting film, however, often suffer from the presence of surfactants because they favour sensitivity to water or lower the adhesion to a substrate. In contrast, amphiphilic block copolymers are expected to prevent such disadvantages due to better compatibility with the polymer particles and a lower migration rate. It has already been shown that block copolymers can be used instead of surfactants to stabilize dispersions [1-4]. Additionally, block copolymers can act as a blending agent to improve such properties as the mechanical strength of the resulting polymer film. Thus the motivation of this work was using block copolymers both to stabilize a dispersion and to create improved film properties.

Synthesis of Block Copolymers

Three types of block copolymers with different molecular weights and block lengths were made by living ionic polymerization:

poly(methylmethacrylate-block-acrylic acid) **P(MMA$_m$-b-AA$_n$)** [5]
poly(methylmethacrylate-block-methacrylic acid)**P(MMA$_m$-b-MAA$_n$)** [6]
poly(isobutylene-block-methacrylic acid) **P(IB$_m$-b-MAA$_n$)** [7,8].
The polymerization was started with the hydrophobic monomer and subsequently the hydrophilic block was obtained by polymerizing t-butyl(meth)acrylate and hydrolyzing it in hydrochloric acid/dioxane. The block copolymer was precipitated in petrol ether (bp. 35-70°C). In **Figure 1** the block copolymers, the block lengths in terms of m and n, the molecular weight and the polydisperity are summarized.

Aggregation of Block Copolymers

The aqueous BC solutions were prepared disssolving the BC in methanol or tetrahydrofuran. After diluting with water, the BC solution was neutralized with an aliquot amount of ammonia or sodium hydroxide. The organic solvent was removed by distillation and the BC solutions were concentrated to obtain a solids content of about 3% (w/w). They were used as stock solutions for the following experiments.

The surface tension of aqueous solutions containing P(MMA$_m$-b-AA$_n$) neutralized with ammonia depends on the chain length of the hydrophobic PMMA block [**Fig. 2**]. At low chain lengths the block copolymer is surface active and lowers the surface tension of water. At higher chain lengths, the surface tension is about that of pure water, indicating that such block copolymers do not have tendency to be present at the water/air interface, but are aggregated within the water phase.

To measure the rate of interchange of block copolymers between aggregates by time-resolved fluorescence spectroscopy according to the principle of non radiative energy transfer [9], separately prepared aqueous solutions of naphthalene (donor) terminated P(MMA$_{20}$-b-AA$_{40}$) and pyrene (acceptor) terminated P(MMA$_{20}$-b-AA$_{44}$) were mixed. Surprisingly, neither at room temperature nor at 150°C, which is about the glass transition temperature of the block copolymer, was any exchange observed which would have led to a shortening of the donor fluorescence decay. This indicates that these aggregates are very stable "frozen micelles" [10].

The aggregation number Z and the micellar diameter $2R_h$ were determined by static and dynamic light scattering [**Fig. 1**].
To obtain the micellar diameter, the interpretation of dynamic light scattering data was performed using the CONTIN algorithm [11]. A bimodal distribution is obtained ([**Fig. 3**], observation angle 30°). The small diameter is assigned to single micelles, the bigger one to aggregates of micelles which were also detected with the analytical ultracentrifuge (AUC).

Assuming the micelles are spherical, a simple geometric core-shell model can be applied. The core is formed by the hydrophobic chains. The corona consists of tethered solvophilic chains [**Fig. 4**]. According to S. Förster et al. [12] the aggregation number Z can be calculated from the block lengths m and n by $Z=Z_0 m^2 n^{-0.8}$ and the corona thickness by $D_{cor}=D_0 Z^{0.2} n^{0.6}$. The corona thickness D_{cor} was calculated from the measured micellar diameter $2R_h$ and the diameter $2R_C$ of the hydrophobic nucleus: $D_{cor}=R_h-R_C$. The core density is assumed to be 1 g/cm^3 due to the strong hydrophobic character. The fitting parameter $Z_0=0.9$ is related to the volume of a solvophobic monomer unit and a geometric packaging parameter, $D_0=0.24$ is related to the length of a solvophilic monomeric unit. The data for the reduced aggregation number are shown in [**Fig. 5**]. The main trend is excellently described by such a simple geometric model, meaning that the steric arrangement dominates the micellar architecture.

Emulsion Polymerization

The aqueous solutions of neutralized block copolymers were used as the only emulsifying agent in emulsion polymerization, which was performed at 90°C in batch. The solids content of the dispersion was 20-30%(w/w), the pH between 5 and 7.5 depending on the amount of BC used. The particle size distribution was measured with AUC [13]. In **Figures 6 and 7**, the data of the polymer dispersions are summarized. The number of polymer particles per kg dispersion N_p and the number of block copolymer molecules per polymer particle n_{BC} were calculated from the particle size, the solids content and the concentration of block copolymers. Obviously three cases have to be differentiated [**Fig. 8**]: with block copolymers P(MMA$_m$-b-AA$_n$) forming "frozen micelles" a slope of 1 is obtained, indicating that the frozen micelles act as a seed where the number of polymer particles is equal to the number of micelles and n_{BC} is constant. A slope higher than 1, already described for P(S$_{10}$-b-EO$_{68}$) [3], is obtained with low m.w. P(MMA$_m$-b-AA$_n$). Only P(MMA$_{10}$-b-MAA$_7$) [14], gives a slope lower than 1, which is comparable to surfactants [15].

Film Properties

Dynamic mechanical analysis of the BA/MMA film made with 10 % P(MMA$_{38}$-b-AA$_{38}$) indicates that the polymer morphology is changed by annealing at 150°C. The significant decrease of the storage modulus [**Fig. 9**] is due to a rearrangement of the block copolymer phase which is almost homogeneously distributed within the film cast at room temperature. It was confirmed by electron microscopy that, after annealing, the block copolymer is arranged in bigger domains. This behaviour results in a significant softening of the polymer film between 30 and 100°C. The water uptake of the polymer film made with P(MMA$_{67}$-b-MAA$_{217}$) is significantly decreased after annealing [**Fig. 10**].

Acknowledgment

We thank T.Rager, W.Meyer, G.Wegner (Max-Planck-Institute for Polymer Research, Mainz), J.Feldthusen, W.Stauf, A.H.E.Müller, (University Mainz) and S.Jüngling for the synthesis of block copolymers, J.Klingler and W.Schrof for the fluorescence spectroscopy work, W.Heckmann and T.Frechen for electron micrographs, A.Zosel for dynamic mechanical analysis and K. Mathauer for a part of the emulsion polymerization work. Financial support by BMBF (No. 03N30043) is greatfully acknowledged.

References

[1] L. Leemans, R. Fayt, Ph. Teyssié, N.C. de Jaeger, *Marcromolecules* 1991, *24*, 5922-5925

[2] G.L. Jialanella, E.M. Firer, I. Piirma, *J. Polym. Sci. A: Polym. Chem.* 1992, *30*, 1925-1933

[3] M. Berger, W. Richtering, R. Mülhaupt, *Polymer Bulletin*, 1994, *33*, 521-528

[4] C.L.Winzor, Z. Mrazek, M.A. Winnik, *Eur. Polym. J.*, 1994, *30*, 121-128

[5] T. Rager, *Ph.D. Thesis*, Mainz, 1997, Shaker Verlag Aachen

[6] W. Stauf, Diplomarbeit, University of Mainz, 1997

[7] J. Feldthusen, B. Iván, A.H.E. Müller, *Macromol.*,1997, *30*, 6989

[8] J. Feldthusen, B. Iván, A.H.E. Müller, *Macromol.*,1997, *31*, 578

[9] T. Förster, *Z. Naturf.*, 1949, *4a*, 321

[10] Z. Tuzar, et al, *Polymer Preprints*, 1991, *32 (1)*, 525-526

P. Munk, et al, Makromol. *Chem., Macromol. Symp.*, 1992, *58*, 195-199

P. Munk, et al, *Langmuir*, 1993, *9*, 1741-1748

[11] S.W. Provencher, *Makromol. Chem.*, 1979, *180*, 210

[12] S. Förster, M.Zisenis, E. Wenz, M. Antonietti, *J. Chem. Phys.*, 1996, *104*, 9956- 9970, and earlier papers quoted there

[13] W. Mächtle, in *"Analytical Ultracentrifugation in Biochemistry and Polymer Science"*, S.E. Harding, A.J. Rowe and J.C. Horton (eds.), 1992, Royal Society of Chemistry, Cambridge, UK

[14] Research product of TH Goldschmidt AG, Germany

[15] A.S. Dunn in P.A. Lovell, M.S. El-Aasser (ed.), *Emulsion Polymerization and Emulsion Polymers*, J. Wiley & Sons, 1997

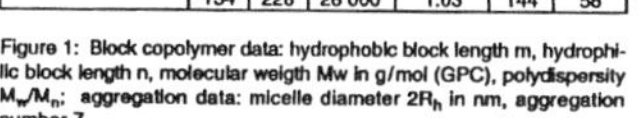

block copolymer	m	n	M_w g/mol	M_w/M_n	Z	$2R_h$ nm
P (MMA$_m$-b-AA$_n$)	18	26	3 600	1.42		
	28	30	5 000	1.32		
	38	38	6 600	1.14		
P (MMA$_m$-b-MAA$_n$)	16	44	6 800	1.26	18	13
	25	29	6 800	1.37	43	21
	32	69	12 000	1.32	39	17
	58	57	14 200	1.33	250	34
	64	112	17 400	1.09	94	36
	67	217	27 000	1.06	34	34
P (IB$_m$-b-MAA$_n$)	70	52	8 900	1.06	188	25
	70	70	10 600	1.07	192	30
	134	145	20 800	1.04	133	57
	134	228	28 000	1.03	144	58

Figure 1: Block copolymer data: hydrophobic block length m, hydrophilic block length n, molecular weigth Mw in g/mol (GPC), polydispersity M_w/M_n; aggregation data: micelle diameter $2R_h$ in nm, aggregation number Z.

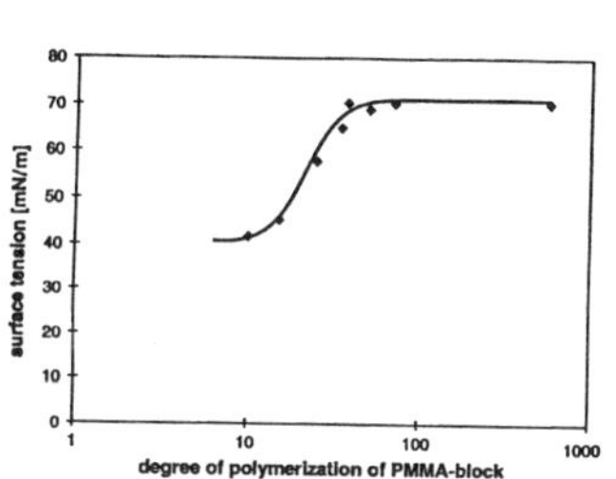

Figure 2: Surface tension of aqueous P(MMA$_m$-b-AA$_n$) solutions [0,5% (w/w); neutralized with ammonia) as a function of hydrophobic block length m.

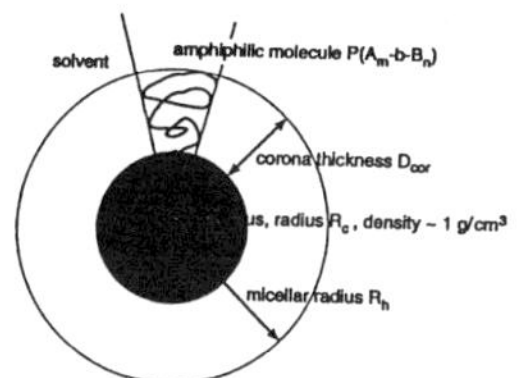

Figure 4: Model of spherical micelle formed by the block copolymer P(A$_m$-b-B$_n$); the micellar architecture is determined by the type of the solvophobic monomer A, the typ of the solvophilic monomer B, the kind of the solvent and the block lengths m and n.

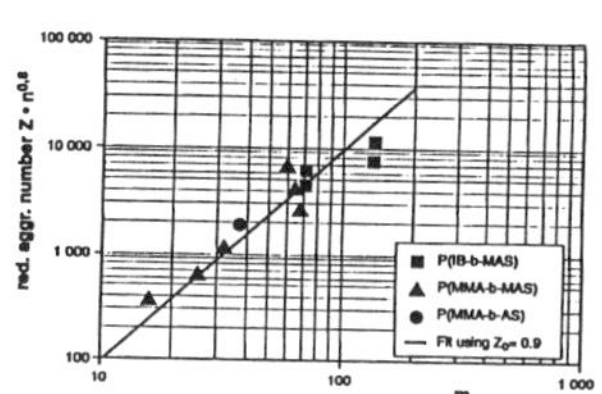

Figure 5: Reduced aggregation number Z ∗ n$^{0.8}$ of block copolymer micelles in aqueous solution as a function of the hydrophobic block length m; the line is calculated from Z = Z$_0$ ∗ m^2 ∗ n$^{-0.8}$ using Z$_0$ = 0,9

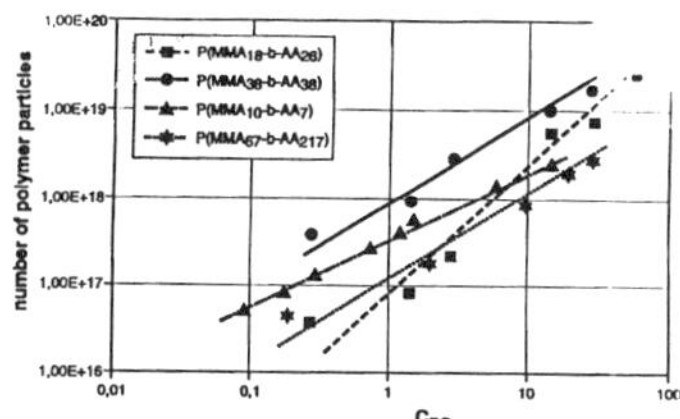

Figure 8: Number of polymer particles per kg dispersion as a function of the copolymer concentration c$_{BC}$ in g/kg dispersion; P(MMA$_{18}$-b-AA$_{26}$), P(MMA$_{38}$-b-AA$_{38}$), P(MMA$_{10}$-..., P(MMA$_{67}$-b-MAA$_{217}$).

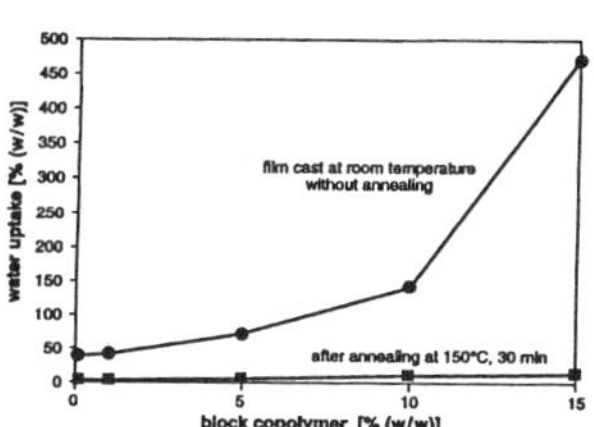

Figure 10: Water uptake of a 49BA/49MMA/2AA polymer film made with P(MMA$_{67}$-b-MAA$_{217}$) befor and after annealing at 150°C during 30 minutes.

m	n	c_{BC} g/kg disp.	s.c. % (w/w)	d nm	Np part./kg disp.	n_{BC} 1/part.
18	26	0.3	26.7	236	3.88E+16	1 150
		1.4	28.2	187	8.24E+16	2 861
		2.8	27.7	133	2.25E+17	2 059
		14.6	29.1	46	5,71E+18	426
		29.2	29.2	42	7.53E+18	648
		57.0	28.5	28	2.48E+19	384
28	30	0.3	29.7	154	1.55E+17	230
		1.5	29.0	106	4.65E+17	375
		2.9	29.6	87	8.59E+17	415
		14.6	29.1	44	6.53E+18	268
		29.6	29.6	33	1.57E+19	226
38	38	0.3	27.5	112	3.74E+17	67
		1.5	29.2	84	9.41E+17	141
		2.9	29.3	58	2.87E+18	93
		14.5	29.0	38	1.01E+19	131
		28.6	28.6	32	1.67E+19	156

Figure 6: Emulsion polymerization data of 50BA/50MMA with P(MMA$_m$-b-AA$_n$) and 0,5 pphm sodium persulfate, concentration of block coplymer c$_{BC}$ in g/kg dispersion, solids content s.c. of the dispersion in % (w/w), mean particle diameter d in nm, number of polymer particles Np per kg dispersion, number of BC molecules per polymer particle n$_{BC}$.

m	n	c_{BC} g/kg disp.	s.c. % (w/w)	d nm	Np part./kg disp.	n_{BC} 1/part.
10	7	0.1	26.9	224	5.08E+16	625
		0.2	30.6	192	8.26E+16	787
		0.3	29.6	162	1.33E+17	788
		0.7	29.6	128	2.70E+17	971
		1.2	30.4	112	4.13E+17	1 041
		1.5	29.8	99	5.87E+17	899
		6.1	30.3	74	1.43E+18	1 502
		15.2	30.4	61	2.56E+18	2 103
16	44	0.2	19.8	198	4.87E+16	360
		1.9	19.2	83	6.42E+17	265
		9.8	19.6	58	1.92E+18	452
		19.9	19.9	47	3.66E+18	481
		29.7	19.8	45	4.15E+18	633
32	69	0.2	19.5	176	6.83E+16	143
		1.9	20.1	84	6.48E+17	156
		9.8	19.6	52	2.66E+18	185
		19.9	20.0	50	3.06E+18	328
		29.7	19.4	44	4.35E+19	335
58	57	0,2	19.7	219	3.58E+16	233
		1.9	19.0	106	3.05E+17	264
		3.8	19.2	82	6.65E+17	245
		5.7	18.9	85	5.88E+17	409
		9.6	19.1	51	2.75E+18	147
		18.1	18.1	29	1.42E+19	54
		29.1	19.4	29	1.52E+19	81
25	29	0.2	19.7	199	4.78E+16	143
		2.0	20.1	114	2.59E+17	268
		9.1	18.2	68	1.11E+18	285
		19.6	19.6	49	3.18E+18	213
		29.6	19.7	43	4.73E+18	216
32	69	0.2	19.0	198	4.68E+16	91
		2.0	19.7	127	1.84E+17	239
		9.8	19.7	77	8.25E+17	266
		19.4	19.4	57	2.00E+18	216
		29.6	19.7	51	2.84E+18	232

Figure 7: Emulsion polymerization data 49BA/49MMA/2AA with P(MMA$_m$-b-MAA$_n$) and 0.5 pphm sodium persulfate, concentration of block coplymer c$_{BC}$ in g/kg dispersion, solids content of dispersion s.c. in % (w/w), mean particle diameter d in nm, number of polymer particles Np per kg dispersion, number of BC molecules per polymer particle n$_{BC}$.

PEP-PEO BLOCK COPOLYMERS: SYNTHESIS AND MICELLAR PROPERTIES

Jürgen Allgaier, Jörg Stellbrink, Andreas Poppe, Lutz Willner, Dieter Richter

Institut für Festkörperforschung, FZ Jülich, 52425 Jülich, Germany

INTRODUCTION

In the past years, the micellization properties of amphiphilic block copolymers in water have attracted attention. In particular polystyrene-*block*-poly(ethylene oxide) (PS-PEO) have been investigated. Recently, we reported the synthesis of well defined poly(ethylene-*co*-propylene)-*block*-poly(ethylene oxide) PEP-PEO block copolymers[1,2]. These new polymers show the same chemical design as the well known low molecular weight alkane-oxyethylene ($C_n(EO)_m$) surfactants.

In dilute aqueous solution the amphiphilic block copolymers self assemble into micelles with PEP as the insoluble core and PEO as the soluble shell. From the extreme incompatibility of the polyalkane with water a sharp core/shell interface is expected. As the insoluble core material is made of the low T_g polymer PEP the dissolution of the polymer is assumed to yield aggregates which are in thermodynamic equilibrium at room temperature. In contrast, high T_g materials like PS as the micellar core can lead to non-equilibrium structures due to the frozen state of the material. The expected sharp interface as well as the thermodynamic equilibrium faciliate a quantitative interpretation on the basis of current micellar models. Thus, the PEP/PEO-water system can be regarded as a model system for studying fundamental properties of micelle formation.

We describe the synthesis of PEP-PEO block copolymers and the structural examination of micelles formed by these polymers in aqueous solutions. The materials are composed of partially deuterated PEP blocks and hydrogenous PEO blocks. The characterization of the micelles was performed by Small Angle Neutron Scattering (SANS) in combination with contrast variation.

EXPERIMENTAL

Synthesis of PEP-PEO. The polymers were synthesized by anionic polymerization. All manipulations were performed under high vacuum in glass reactors provided with break seals for the addition of reagents. The preparation of the initiators and the purification procedures for monomers and solvents to the standards required for anionic polymerization have been described elsewhere[2]. The anionic polymerization of the block copolymers was realized in a two step process (see Scheme 1).

In the first step, fully deuterated isoprene was polymerized in benzene with *sec*-butyllithium as initiator. The end capping with ethylene oxide (EO) yielded d-polyisoprene functionalized with a hydroxyl end group (PI-OH). The polymer was hydrogenated with H_2 to the corresponding PEP-OH using a conventional Pd/BaSO$_4$ catalyst.

In the second polymerization step, cumyl potassium was used to transfer PEP-OH into the macroinitiator PEP-OK. Therefore, PEP-OH was dissolved in dry THF and titrated with a solution of the deep red cumylpotassium in THF. The titration was stopped after a slight orange color in the polymer solution persisted for 5 minutes, indicating the complete deprotonation of the polymeric OH groups. The PEP-PEO block copolymer was obtained after polymerizing the calculated amount of hydrogenous EO in THF at 50^{0}C for three days. The reaction was terminated with acetic acid and the product precipitated twice in acetone at -10 to -20^{0}C. A more detailed description of the polymer synthesis is given elsewhere[2].

Characterization of the micelles by SANS. Neglecting interparticle scattering contributions, the coherent macroscopic scattering cross section $(d\Sigma/d\Omega)(Q)$ of a SANS experiment is of the general form

$$\frac{d\Sigma}{d\Omega}(Q) = \frac{N_z}{V}\langle|A(Q)|^2\rangle \qquad (1)$$

where N_z denotes the number of scatterers, V is a reference volume and $A(Q)$ is the intraparticle scattering amplitude. The scattering vector Q is given by $4\pi \sin(\Theta/2)\lambda$, where Θ is the scattering angle and λ is the neutron wavelength. For a particle consisting of a core-shell structure the scattering amplitude $A_{C,Sh}(Q)$ can be written

$$A_{C,Sh}(Q) = V_M(\rho_{Sh} - \rho_S)A_M(Q) + V_C(\rho_C - \rho_{Sh})A_C(Q) \qquad (2)$$

with ρ_C, ρ_{Sh} and ρ_S as the scattering length densities of the core, the shell and the solvent. V_M and V_C denote the volume of the overall micelle and the core, $A_M(Q)$ and $A_C(Q)$ the corresponding scattering amplitudes. Equation 2 reduces to simple forms if the condition $\rho_{Sh} = \rho_S$ or $\rho_C = \rho_S$ is fullfilled. Concerning the first case one obtains for the intraparticle scattering amplitude:

$$A_{C,Sh}(Q) = V_C(\rho_C - \rho_{Sh})A_C(Q) \qquad (3)$$

This scenario is denoted as core contrast. For the second case, the shell contrast, eq 4 simplifies to

$$A_{C,Sh}(Q) = (\rho_C - \rho_{Sh})(A_C(Q)V_C - A_M(Q)V_M) \qquad (4)$$

The block copolymer micelles were studied under core and shell contrast. Using H_2O/D_2O mixtures block copolymer solutions with a polymer volume fraction $\Phi = 1\%$ were prepared.

The SANS experiments were performed on the KWSI instrument at the research reactor FRJ2 at the Forschungszentrum Jülich GmbH. The experiments were carried out at sample to detector distances of L = 4, 8 and 20 m, covering a Q range of 0.003 Å^{-1} to 0.08 Å^{-1}.

Scheme 1

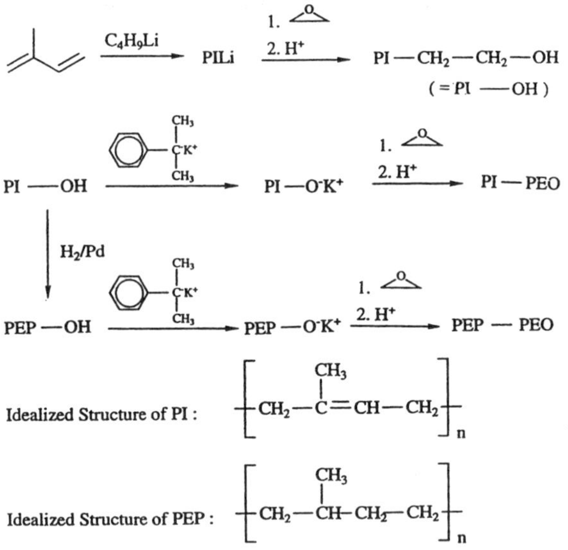

C_4H_9Li → PILi → (1. O, 2. H$^+$) → PI—CH$_2$—CH$_2$—OH (= PI —OH)

PI —OH → (cumyl potassium, CH$_3$—C(CH$_3$)—K$^+$) → PI —O$^-$K$^+$ → (1. O, 2. H$^+$) → PI — PEO

PI —OH → (H$_2$/Pd) → PEP —OH → (cumyl potassium, CH$_3$—C(CH$_3$)—K$^+$) → PEP —O$^-$K$^+$ → (1. O, 2. H$^+$) → PEP — PEO

Idealized Structure of PI : $\left[CH_2-\underset{\underset{CH_3}{|}}{C}=CH-CH_2 \right]_n$

Idealized Structure of PEP : $\left[CH_2-\underset{\underset{CH_3}{|}}{CH}-CH_2-CH_2 \right]_n$

RESULTS AND DISCUSSION

Synthesis. The synthesis of the PEP-PEO block copolymers is done in a two-step process, because the synthesis of the two blocks require different reaction conditions. Polyisoprene with a high degree of 1,4-microstructure, the parent component of PEP, demands organolithium initiators and hydrocarbon solvents. The polymerization of EO is performed in polar solvents using sodium or potassium initiators. The synthesis is described in Scheme 1.

After having polymerized d$_8$-isoprene in benzene with *sec*- or *tert*-butyllithium and having end capped the living polymer with EO the resulting PI-OH is hydrogenated using hydrogen and a conventional palladium catalyst. The hydrogenation of the deuterated PI with hydrogen instead of deuterium is necessary to decrease the scattering length density of the polymer. So it is possible to match the PEP, the core material in the micelles, with a mixture of D_2O and H_2O in the SANS contrast variation experiments.

In the second polymerization step PEP-OH was deprotonated and transfered into the macroinitiator PEP-OK with cumylpotassium. This process was realized by titrating a known quantity of polymer, dissolved in THF, with the deeply red colored cumylpotassium. If the concentration of cumylpotassium is known, the titration can be considered an end group analysis to determine M_n of the precursor polymers. The values, obtained by this method are listed in Table 1. They agree with the calculated M_n and also with those from VPO and SEC measurements, indicating complete end functionalization of the living Polyisoprenyllithium.

The macroinitiator PEP-OK was used to polymerize EO. After consumption of all monomer samples were taken from the reactor. SEC traces of the block copolymers PEP5-PEO2 (M_n(PEP) = 4630, M_n (PEO) = 1500) and PEP5-PEO15 (M_n(PEP) = 4630, M_n(PEO) = 14600) are shown in Figure 1. The absence of PEO homopolymer in PEP5-PEO2 and of PEP homopolymer in PEP5-PEO15 as well as the agreement of calculated with measured molecular weights (Table 1) underline the quality of the block copolymers.

Table 1. Molecular weight characterization of the PEP-PEO polymers

Polymer	PEP-OH			PEP-PEO		
	M_n	M_n	M_w/M_n	M_n	M_n	M_w/M_n
	(calc)	(titr)	(GPC)	(calc)	(NMR/titr)	(GPC)
PEP5-PEO5	5,270	5,180	1.03	11,000	11,100	1.03
PEP5-PEO15	5,270	5,150	1.03	20,800	21,300	1.02
PEP5-PEO30	5,290	5,210	1.03	35,900	38,200	1.02
PEP5-PEO50	5,290	5,260	1.03	59,000	62,000	1.02

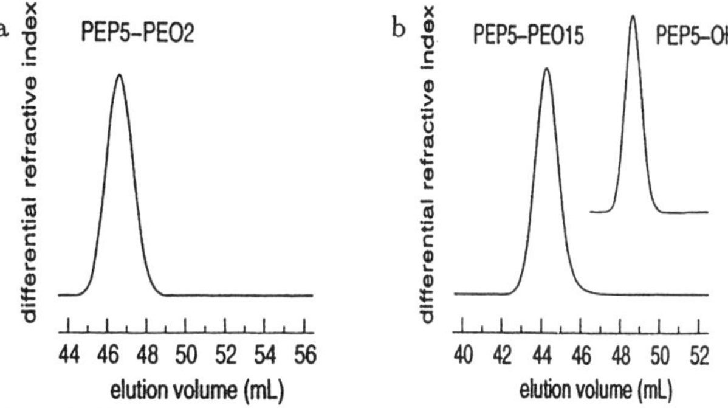

Figure 1. SEC traces of the PEP-PEO and the parent PEP-OH.

SANS measurements. $(d\Sigma/d\Omega)(Q)$ for the core and the shell contrast of PEP5-PEO5 are shown in Figures 2 and 3. In all cases strong scattering is observed, indicating the formation of large aggregates. The curves exhibit a well pronounced secondary maximum and even a weak third maximum at higher Q values, displaying the existence of aggregates with well defined structure. The curves also reveal a narrow size distribution, a sharp core-shell interface and a sharp density cutoff at the periphery of the micelle.

In order to analyze the structure of the micelle, model fitting was performed to the experimental data. A spherical core-shell structure was used as model. Thereby, the scattering length density distribution was assumed to be homogeneous over the core and the shell, respectively. The density cutoff at the PEP/PEO-water interface and the outerperiphery of the micelle was smeared by multiplication with a Gaussian. PEP5-PEO15

was treated in the same way. In the case of PEP5-PEO30 and PEP5-PEO50 the periphery of the micelle was treated with a hyperbolic density profil of the type r^{-x}. The core radii R_C, the micelle radii R_M and the aggregation numbers P, which were extracted from the SANS data are summarized in Table 2. A detailed description of the procedure is given elsewhere[3]. Aggregation numbers as well as the micelle dimensions are very large due to the high surface tension between PEP and water.

Finally, good agreement between the measured data and the Nagarajan-Ganesh micellization theory[4], which is based on mean field approximations, was found.

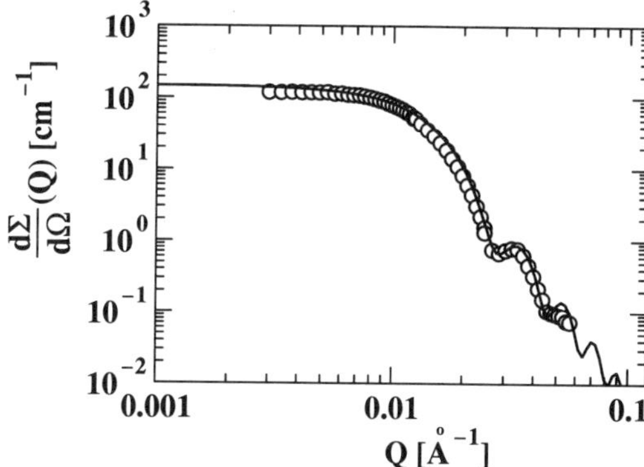

Figure 2. Absolute SANS data under core contrast. The solid line represents the result of the model fit.

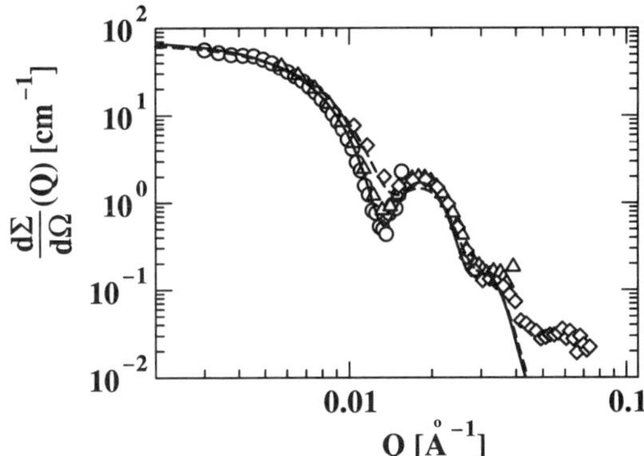

Figure 3. Absolute SANS data under shell contrast. The solid line and the dashed line represent the results of the model fits, convoluted with the instrumental resolution at 20 m and 4m detector distance, respectively.

Table 2. Parameters of the PEP-PEO micelles.

Polymer	P	$R_C/\text{Å}$	$R_M/\text{Å}$
PEP5-PEO5	2430	176	294
PEP5-PEO15	1450	148	360
PEP5-PEO30	1080	134	556
PEP5-PEO50	450	101	637

References.

(1) J. Allgaier, L. Willner, D. Richter, German Patent Application, No. P 19634477.8, 1996

(2) J. Allgaier, A. Poppe, L. Willner, D. Richter, *Macromolecules*, **30**, 1582 (1997)

(3) A. Poppe, L. Willner, J. Allgaier, J. Stellbrink, D. Richter, *Macromolecules*, **30**, 7462 (1997)

(4) R. Nagarajan, K. J. Ganesh, *J. Chem. Phys.*, **90**, 5843 (1989)

Hydration and Permeation in Hyperbranched- and Dendrimer-Based Thin Films

David E. Bergbreiter, Richard M. Crooks, Daniel L. Dermody, Stacy J. Jones, Yuelong Liu, Justine G. Franchina, Merlin L. Bruening, Yuefen Zhou and Mingqi Zhao

Department of Chemistry
Texas A&M University
PO Box 300012
College Station, TX 88942-3012
bergbreiter@chemvx.tamu.edu

Thin film design, characterization and synthesis is of interest in technologically important areas including sensor development, corrosion passivation, biotechnology and catalysis. Our groups have for the past 2 years been focused on new chemistry for synthesis of such thin films.[1-6] This chemistry is based on 'forgiving' hyperbranched grafting onto surfaces.

Two strategies for hyperbranched grafting were examined. First, thin (10 - 200 nm) hyperbranched films of poly(acrylic acid) are built up on gold, polyethylene or glass by repetitive covalent assembly of terminally functionalized poly-(*tert*-butyl acrylate) (eq. 1)[1-3] Alternatively, dendrimers with functional

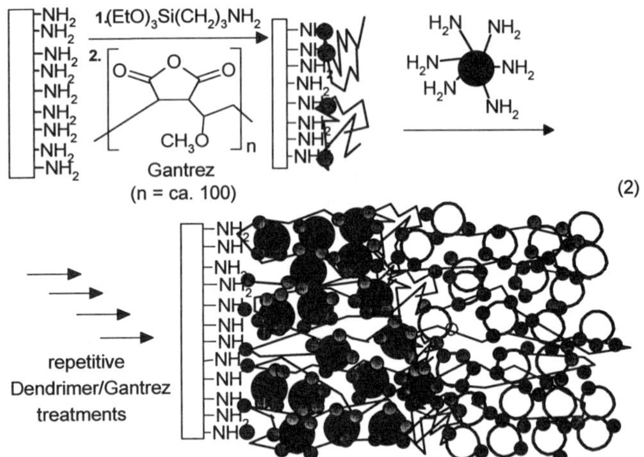

groups on their periphery can be allowed to react with poly(maleic anhydride)-*c*-poly-(methyl vinyl ether) to form nanocomposites of dendrimers in a random coil polymer. Such nanocomposites form on gold, silicon and aluminum (eq. 2).[5,6] Either thin film is highly functional and can be further modified. For example, chemical modification of these thin films forms selectively permeable, responsive electroactive membranes and impermeable passivating coatings by simple pH changes or by covalent modification of the thin films either during or after their formation.

Electrochemistry and *ex-situ* and *in-situ* ellipsometry afford details about solvation of these thin films.[4,6] In the case of hyperbranched poly(acrylic acid) grafts, ellipsometry shows that the graft films' thickness depend on the pH of the aqueous milieu to which the film is exposed. Extensive solvation occurs with an acidic (pH 2) buffer. Additional thickness increases then occur as the pH is changed to more basic values with an inflection point between pH 4 and 5. IR spectroscopy of film samples exposed to various pH buffers shows changes in the intensity of the -CO2H carboxyl group of the hyperbranched PAA graft that correlate with the observed *in-situ* thickness changes (Figure 1), indicating that the thickness change is associated with deprotonation of the hyperbranched PAA's -CO2H groups. While the absolute value of the thickness change

depends on the initial thickness of the PAA graft, the initial swelling of the PAA graft on exposure of a pH 2 buffer is ca. 200%. For example, a 230 Å 3 PAA graft swells to 660 Å. Under basic conditions (pH > 6), this same graft is swollen to 1220 Å . The inflection point in the swelling and in the -CO2H deprotonation is at pH 4.3, suggesting that the carboxylic acid groups in these thin films have pKa values like those of poly(acrylic acid) in solution.

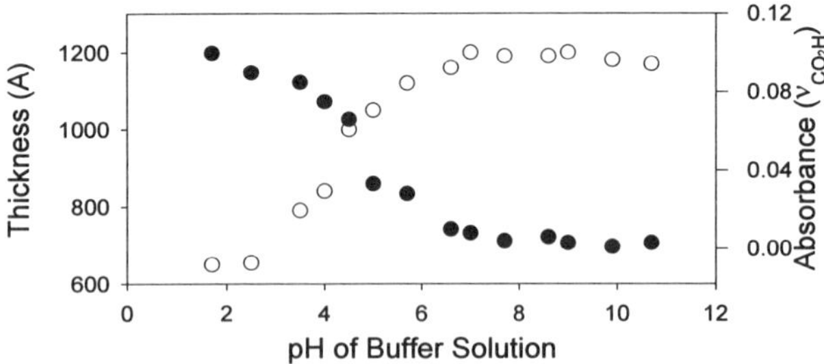

Figure 1. Correlation of changes in film thickness measured by *in situ* ellipsometry (O) and changes in peak intensity of the carbonyl carboxyl peak of a 3 PAA graft (●) with the pH of a buffer solution

Exposure of these highly swollen films to a pH 7 solution of a polyamine leads to ionically assembled intercalated thin film analogous to the covalent assembly in eq. 2.[4] Unlike the covalent dendrimer/Gantrez composites prepared in eq. 2, the ionic composite prepared in eq. 3 can be reconverted into a 3 PAA graft by washing with acid. Moreover, other ionic composites like that shown in eq. 3 can be prepared with other cationic polyelectrolytes (e.g. poly(D-lysine) or poly(allylamine) at pH 7. The presence of intercalated polyelectrolyte in these ionic assemblies is established by FTIR-ERS spectroscopy (amide peaks of the PAMAM dendrimer or of the poly(D-lysine)) and by the increased thickness of the intercalated film after removal from the buffer solution.

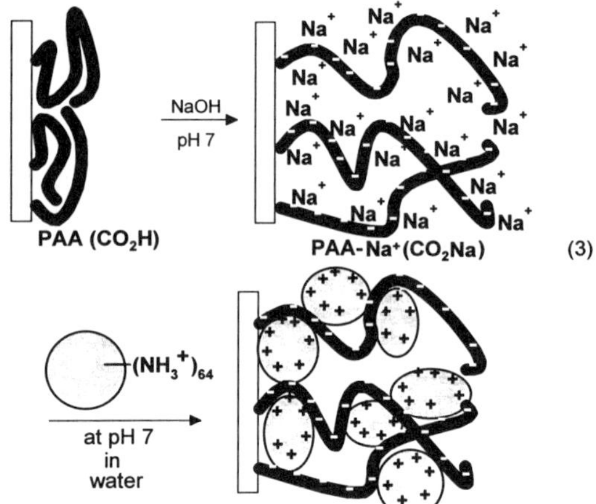

References

1. M. L. Bruening, Y. Zhou, G. Aguilar, R. Agee, D. E. Bergbreiter and R. M. Crooks, *Langmuir*, **1997**, *13*, 770-8.

2. Y. Zhou, M. L. Bruening, Y. Liu, R. M. Crooks and D. E. Bergbreiter, *Langmuir*, **1996**, *12*, 5519-21.

3. M. Zhao, M. L. Bruening, Y. Zhou, D. E. Bergbreiter and R. M. Crooks, *J. Isr. J. Chem.*, **1997** *37*, 277-286.

4. R. F. Peez, D. L. Dermody, J. G. Franchina, S. J. Jones, M. L. Bruening, D. E. Bergbreiter and R. M. Crooks, *Langmuir*, submitted for publication.

5. Y. Liu, M. L. Bruening, D. E. Bergbreiter, and R. M. Crooks, *Angew. Chem. Int. Ed. Engl.*, **1997**, *36*, 2114-6.

6. Y. Liu, M. Zhao, D. E. Bergbreiter and R. M. Crooks, *J. Am. Chem. Soc.*, **1997** *119*, 8720-1.

INTERACTION OF ASTRAMOL™ POLY(PROPYLENE IMINE) DENDRIMERS WITH LINEAR POLYANIONS

V.A. Kabanov,[†] A.B. Zezin,[†] V.B. Rogacheva,[†] Zh.G. Gulyaeva,[†] M.F. Zansochova,[†] J.G.H. Joosten,[‡] J.C. Brackman,[‡] R.H. Vreekamp[‡]

[†]Polymer Science Department, Faculty of Chemistry, Moscow State University, Vorobjevi Gori, Moscow V-234, Russia
[‡]DSM Research, P.O. Box 18, 6160 MD Geleen, The Netherlands

Introduction

Astramol™ poly(propylene imine) dendrimers are tree-like molecules which are built starting from a diaminobutane core and adding propylene imine branches by an iterative procedure. With each generation, the molecules grow in three-dimensions in a highly symmetric and regular way. A fifth generation dendrimer is schematically shown in Scheme 1.

Since the amine groups of poly(propylene imine) dendrimers can be protonated, these peculiar polymers can be classified as polyelectrolytes. The influence of structural features on the charging processes remains one of the most important issues in polyelectrolyte chemistry. The first aim of this investigation was to determine whether these dendrimers can be fully protonated and to assess whether the dendrimer structure is penetrable for low molecular counterions. An aspect that was especially looked at, is the effect of shielding by salt present in the solution.

Even more interesting is the interaction between charged dendrimers and high molecular weight polyions. In aqueous solution interaction between oppositely charged linear polyelectrolytes results in a cooperative interpolyelectrolyte complexation reaction (IPCR) to form an interpolyelectrolyte complex (IPEC).[1-5] The question was: will positively charged dendrimers show a similar close interaction with linear polyanions? And how does the flexibility of the linear polyanions influence the interaction.

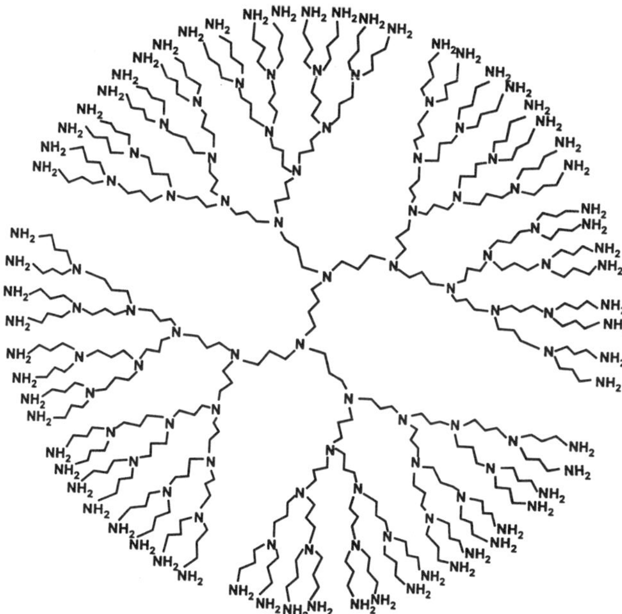

Scheme 1. Structure of a fifth generation poly(propylene imine) dendrimer

Materials

Astramol™ poly(propylene imine) dendrimers were produced at DSM. They are commercially available from DSM or Aldrich. Detailed information on the synthesis was published elsewhere.[6] The Astramol™ poly(propylene imine) dendrimers of five generations from Astramol-Am-4 to Astramol-Am-64 were used. Poly(acrylic acid) (PAA) was prepared by radical polymerization in dioxane with ABN. Poly(sodium acrylate) (NaPA) aqueous solutions were prepared by neutralization of PAA solutions in distilled water with the equivalent amount of NaOH. Native DNA from salmon sperm, 300-500 b.p., purchased from GosNIIOHT (Moscow) was used as received.

Measurements

The potentiometric titration of aqueous solutions of the dendrimers, PAA and mixtures of the dendrimers with PAA and NaPA was performed using RTS 822 Radiometer titrator with Sigma glass calomel combination electrode at 20 °C. Turbidimetric titration of NaPA in salt-free aqueous solutions with aqueous solutions of fully protonated Astramol-Am-x•HCl was performed using a UV Hitachi 150-20 spectrophotometer. The concentration of Cl⁻ ions in aqueous solutions was measured by argentometric titration. The precipitates were separated from the supernatants using a Beckman J2 preparative ultracentrifuge at 18,000 r.p.m. during 40 min.

Results and discussion

Poly(propylene imine) dendrimers have two types of amino groups: primary amines on the periphery, and tertiary amines at the branching points. A dendrimer of generation n has 2^{n+1} primary and $2^{n+1}-2$ tertiary amines. When isolated, the basicity of the former is larger than that of the latter ($pK^p > pK^t$). However, in polyelectrolytes the electrostatic repulsion between different charged sites has to be taken into account: it causes a decrease in the pK with increasing degree of protonation, α.

The protonation behavior of these dendrimers was studied by potentiometric titration. The results in Fig. 1 show that for the first three generations (Astramol-Am-4, -8, -16) two sections can be distinguished: one for the primary amines (at higher pH and below α_1), and one for the tertiary amines. The degree of protonation which separates the regions (α_1) was found to be equivalent to the ratio between primary and tertiary amines.

Analysis of the titration data by the modified Hasselbach-Henderson equation[7] revealed that for Astramol-Am-4 the pK^p and pK^t vary only slightly with α, while this variation is significantly larger and linear for Astramol-Am-8 and Am-16. Extrapolation to $\alpha=0$ yields practically equivalent values for pK^p_o for the first three generations. The same is true for pK^t_o.[8]

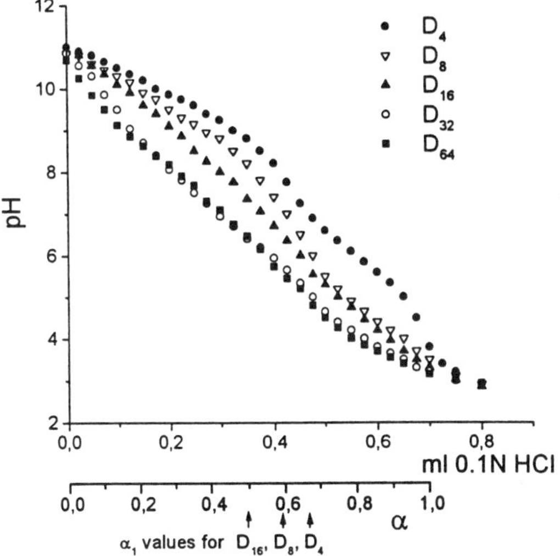

Figure 1. The potentiometric titration curves for Astramol-Am-x (x = 4, 8, 16, 32, 64) in salt-free aqueous solutions; [Astramol-Am-x] = 0.014 mol/l, V=5ml, T=20°C.

The titration results of the two higher generations (Astramol-Am-32 and Astramol-Am-64) show no inflection points to distinguish between different types of amines. This indicates overlap of protonation of primary and tertiary amines. For these molecules the protonation equilibrium can be described by a linear function: $pK = pH + \log \alpha/(1-\alpha)$. The pK_o values are close to those of pK^P_o for Astramol-Am-4, -8 and -16. The steep slope in the pK-α plot (δpK/δα = -6.25) reflects the high charge density within these dendrimers.[9]

Salts strongly influence the ionization of linear polyelectrolytes. As could be expected, also in the case of these dendrimers the titration curve in NaCl solutions shifts to the more alkaline pH region due to the shielding of the Coulombic interactions. This shielding is also the cause of the appearance of two protonation regions for Astramol-Am-32, like was seen for the three lower generations in salt-free solutions.

After the protonation behavior of free poly(propylene imine) dendrimers had been clarified, attention was shifted to the interaction of these dendrimers with linear polyanions. The interaction with polyanions in the salt form was studied by mixing dendrimers as free base with an equivalent amount of NaPA, followed by titration with HCl. Similarly, interaction with free polyacids was studied by mixing fully protonated dendrimers with PAA, followed by titration with NaOH. In the former case the titration curves for different generations are superimposed and at higher pH than the titration curve of free dendrimer (Fig. 2). Apart from the superposition, the opposite is true for the latter, i.e. the titration curves are shifted toward lower pH (Fig. 3). The cooperative character of multisite bonding of the oppositely charged units is responsible for this effect: in the presence of NaPA the free base dendrimers actually behave as stronger bases, while in presence of protonated dendrimers PAA behaves as a stronger acid.

To determine the permeability of a poly(propylene imine) dendrimer structure for linear polyanions the degree of conversion of the IPCR was experimentally determined by measuring the chloride content of a mixture of equimolar solutions of Astramol-Am-x HCl-salts and NaPA. The IPECs formed are insoluble and can be removed after centrifugation, leaving the Cl⁻ ions in the supernatant. Neither dendrimer nor polyanion was detected in the supernatant. The argentometric Cl⁻ measurements and corresponding values for degree of conversion, (θ = 0.94-1.0) unambiguously imply that the carboxylate groups of the polyanion form ion pairs with the protonated amines of the dendrimers. So all primary and all tertiary amines are available for interaction with ionic groups of linear polymer chains. Up to the fifth generation (Astramol-Am-64) these dendrimers are fully penetrable for flexible linear polyanions, i.e. these dendrimers should not be looked at as surface charged compact rigid particles.

Above, the precipitation of equimolar polyelectrolyte mixtures was used to determine the extent of interaction. However, when a polyelectrolyte is mixed as a guest with an excess of an oppositely charged polyelectrolyte with a higher degree of polymerization as the host, non-stoichiometric water-soluble interpolyelectrolyte complexes are formed at full conversion (θ = 1). This behavior is also observed for dendrimers as the guest and NaPA as the host, as was demonstrated with turbidimetric titrations. As long as the dendrimer/PA ratio, z, is below 0.6, the reaction mixture remains a homogeneous solution. Above z = 0.6 the mixture becomes increasingly opaque until z = 1. Below z = 0.6 the interpolyelectrolyte complex structure may be pictured as a PA thread with fully neutralized dendrimer beads along its chain. The excess carboxylate groups supply the hydrophilicity necessary for solubility.

Finally, attention was focused on the interaction of these dendrimers with rigid rod-like polyanions, like DNA. Comparison of titration curves of the mixture of Astramol-Am-32 and DNA reveals a shift to higher pH, indicating formation of ion pairs between dendrimer amines and DNA phosphate groups, and interpolyelectrolyte complex formation. The observed shift however, is significantly smaller than that for the dendrimer-NaPA combination, suggesting a less thorough interaction between dendrimer and the stiff DNA than with the flexible NaPA.

References
1. Zezin A.B., Rogacheva V.B. in *Uspekhi Khimii i Fisiki Polimerov, Khimiya*. **1973**, 3.
2. Tsuchida E. and Abe K. in *Developments in Ionic Polymers-2*, Elsevier Science Publishers, Ed. Wilson A.D. and Prosser H J. 1986, N4, 1.
3. Philipp B., Dautzenberg H., Linov K.-J., Kotz J., Dawydoff V. *Progr. Polym. Sci.* **1989**, *14*, 91.
4. Kabanov V.A., Zezin A.B. *Pure & Appl. Chem.* **1984**, *56*, 343.
5. Kabanov V.A., Zezin A.B. *Makromol. Chem., Suppl.* **1984**, *6*, 259.
6. a) de Brabander E.M.M., Meijer E.W. *Angew. Chem., Int. Ed. Engl.*, **1993**, *32*, 1308. b) de Brabander E.M.M., Brackman J., Mure-Mak M., de Man H., Hogeweg M., Keulen J., Scherrenberg R., Coussens B., Mengerink Y., van der Wal S. *Macromol. Symp.*, **1996**, *102*, 9.
7. Nagasava M. *Pure & Appl. Chem.* **1971**, *26*, 519.
8. Kabanov V.A., Zezin A.B., Rogacheva V.B., Gulyaeva Zh.G., Zansochova M.F., Joosten J.G.H., Brackman J.C. *Macromolecules*, in press.
9. Tanford C. *Physical Chemistry of Macromolecules*, Wiley, New York, 1971.

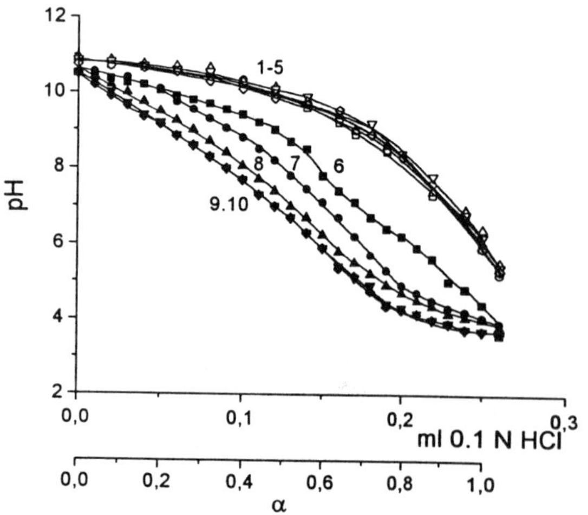

Figure 2. The potentiometric titration curves for equimolar mixtures Astramol-Am-x•NaPA: x=4 (1); x=8 (2); x=16 (3); x=32 (4); x=64 (5); and for Astramol-Am-x in salt-free aqueous solutions: x=4 (6); x=8 (7); x=16 (8); x=32 (9); x=64 (10); [Astramol-Am-x] = 0.005 mol/l, T=20 °C.

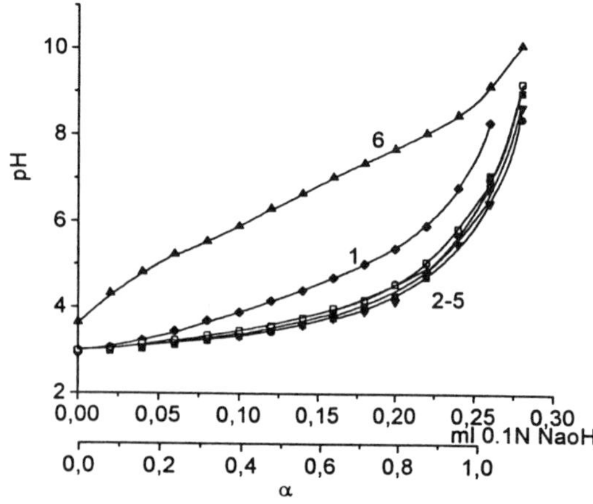

Figure 3. The potentiometric titration curves for equimolar mixtures Astramol-Am-x•HCl•PAA: x=4 (1); x=8 (2); x=16 (3); x=32 (4); x=64 (5); and for PAA in salt-free aqueous solutions (6); [Astramol-Am-x•HCl]=0.005 base-mol/l, T=20 °C.

Properties and Binding Capabilities of Water-Soluble/Water-Swellable Polymers with Linear-Dendritic Architecture

Ivan Gitsov*, Richard Pracitto and Kevin R. Lambrych
Polymer Research Institute, Department of Chemistry, SUNY –
College of Environmental Science & Forestry, Syracuse, NY 13210

Introduction

Amphiphilic copolymers that are capable of dissolving or swelling in aqueous media are interesting materials that have a broad application potential ranging from environmental clean-up and phase-transfer catalysis to sustained drug delivery and hydrophobic surface modification.[1] The biggest disadvantages of these self-assembling materials are their ill-defined composition and dense, irregular interior that limit their carrier potential. Unimolecular dendritic micelles found spectacular applications for encapsulation and delivery of various compounds based on their perfectly organized interior with excellent control of functionality and chemical composition.[2] The limitations of these compounds arise from the limited space in their internal voids and the necessity to chemically modify the periphery in order to trap and subsequently release the guest molecules. Amphiphilic linear-dendritic copolymers offer unique opportunities to combine the useful properties of both types of materials while avoiding some of their shortcomings. Some of the solution and solid-state properties of these novel classes of materials were investigated, but studies on their binding properties are very scarce.[3] In this study we are exploring the encapsulation of pyrene and N-methyl-4-(1-pyrene) vinyl pyridinium iodide in order to evaluate the binding potential of amphiphilic linear-dendritic copolymers.

Experimental

Materials: The hybrid linear-dendritic block copolymers were synthesized according to known procedure.[4] Pyrene (Aldrich) and N-methyl-4-(1-pyrene) vinyl pyridinium iodide (Molecular Probes, Inc.) were used without further purification. The experiments were performed in deionized (DI) water (18.3 MΩ).

Instrumentation: The solubilization experiments were performed in an Aquasonic model 75D sonicator (VWR SCIENTIFIC), the solutions were clarified by centrifugation using a Sorvell TC6 swingarm centrifuge (DuPont) and the spectroscopic measurements were made in a DU 640 B UV-VIS spectrophotometer (Beckman).

Dissolution of pyrene in water: An excess (40 mg) of pyrene, finely ground, was placed in a 25 mL volumetric flask and DI water was added to the full volume of the flask. The mixture was sonicated in the water bath of the sonicator for 6 hours at 30 °C and was left at the same temperature to equilibrate for 48 hr. The clear solution above the pyrene particles was analyzed by UV-VIS spectroscopy in several 24 hr intervals.

Hybrid solutions: Solutions of hybrid ([G-2]-PEG5,000-[G-2]) in DI water with concentrations 1.1×10^{-7}, 1.1×10^{-6}, 1.1×10^{-5}, 1.1×10^{-4} and 1.1×10^{-3} mol/L, were prepared. The originally heterogeneous mixtures were left to become homogeneous and equilibrate for 48 hrs at room temperature. After that period the resulting clear colorless solutions, were analyzed using UV-VIS spectroscopy.

Encapsulation studies: To each solution of the hybrid was added 50 mg of finely ground pyrene. The heterogeneous solutions were sonicated for 6 hours at 30 °C (a temperature of 30°C was maintained for all solutions throughout the rest of the experiment) and left to equilibrate for 24 hrs. After recording the UV-VIS spectra the solutions were sonicated for one hour and left to equilibrate for additional 48 hrs. UV-VIS spectroscopy was performed again. The mixtures were then left to equilibrate for 9 days and final UV-VIS analysis was made with each mixture.

Dissolution of N-methyl-4-(1-pyrene) vinyl pyridinium iodide in water: A stock solution of the probe was prepared by dissolving 0.05 g (1.12×10^{-4} mol) in 1L DI water. The mixture was heated in hot-water bath under stirring for 30 min and yielded a clear solution with brilliant yellow-orange color. UV-VIS spectra of this stock solution were taken 3 consecutive days and no distinguishable change was observed in the spectra obtained.

Encapsulation studies: The hybrid copolymer [G-4]-PEG11,000-[G-4], 0.0505 g (0. $\times10^{-4}$ mol) was placed carefully in a Starna 10/10 mm quartz cell, equipped with micro stirring bar and a Teflon stopper. Four mL of the stock probe solution was added and the resulting heterogeneous solution was stirred for 17 hr at room temperature. The gel suspension was allowed to settle for 21 hr and was centrifuged in a centrifuge for 24 min at 2,800 rpm before the UV-VIS analysis.

Spectroscopic procedures: All UV-VIS analyses were made at 30 °C from 190 to 400 nm for the [G-2]-PEG5,000-[G-2]/pyrene experiments and from 190 nm to 500 nm for the [G-4]-PEG11,000-[G-4]/ N-methyl-4(1-pyrene) vinyl pyridinium iodide system with a scanning speed of 600 nm/min. The changes in the concentration of the two probes were estimated using the following molar extinction coefficients: $\varepsilon = 32,300$ L×mol^{-1}×cm^{-1} at 320 nm and $\varepsilon = 55,800$ L×mol^{-1} ×cm^{-1} at 335.5 nm for the pyrene[5] and $\varepsilon = 57,400$ L×mol^{-1}×cm^{-1} at 395 nm for N-methyl-4(1-pyrene) vinyl pyridinium iodide, respectively.

Results and Discussion

The amphiphilic copolymers used in this study are constructed of poly(ethylene glycol), PEG as the linear hydrophilic block and dendritic aromatic poly(benzyl ethers) as the perfectly branched hydrophobic fragments. An example of a PEG copolymer containing two third-generation dendritic blocks ([G-3]) is schematically presented in Fig. 1.

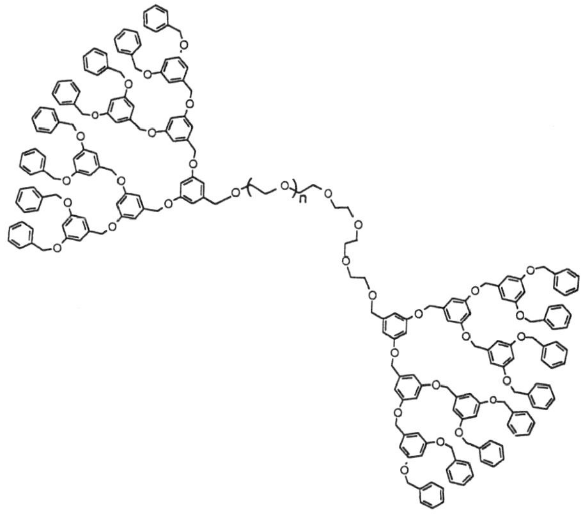

Figure 1.

At high ratios PEG/dendrimer these compounds are capable of dissolving in aqueous media forming unimolecular or multi molecular micelles – [G-2]-PEG5,000-[G-2], [G-3]-PEG11,000-[G-3], [G-4]-PEG20,000-[G-4]. While the first member of the series yields homogeneous solutions directly in DI water, the next two copolymers require preliminary dissolution in a methanol/water mixture with subsequent evaporation of the alcohol. Decreasing the ratio PEG/dendrimer affords compounds that swell, but do not dissolve in water: [G-4]-PEG5,000-[G-3] and [G-4]-PEG11,000-[G-4]. Interestingly, the second copolymer is easily dissolved in methanol, but only swells in water and water/alcohol mixtures.

Pyrene, compound with limited solubility in water and a model for the hazardous polyaromatic hydrocarbons, PAHs, was used to evaluate the encapsulation potential of water-soluble linear-dendritic copolymers. The solubilization dependence of pyrene on the concentration of [G-2]-PEG5,000-[G-2] is shown on Fig. 2. It is obvious that solubilization values comparable to known results with unimolecular dendritic micelles[4] were achieved at concentrations and dendrimer generations that are two times lower than the previously reported.

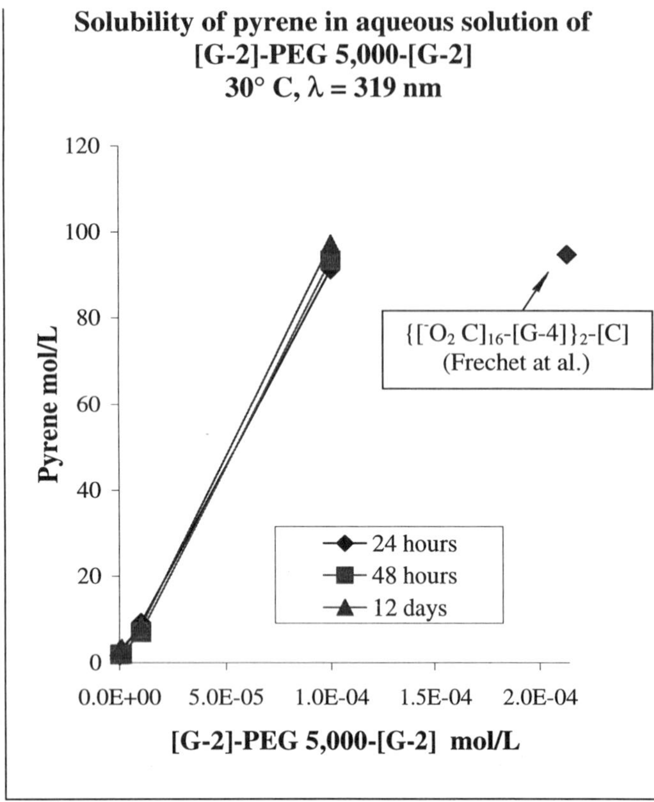

Figure 2.

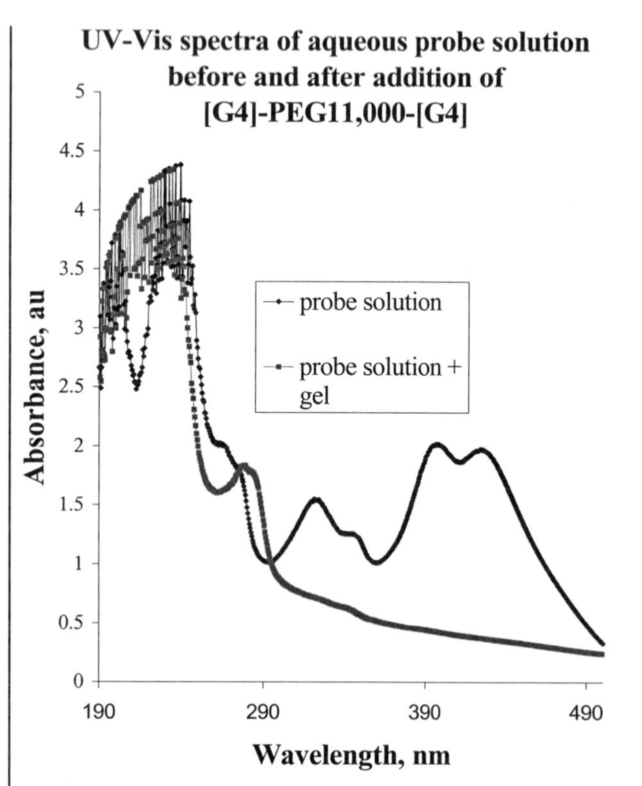

Figure 3.

The lipophilic polar tracer for neuronal morphology and intracellular communication studies, N-methyl-4(1-pyrene) vinyl pyridinium iodide, was used to trace the binding capabilities of the water-swellable hybrids. The time dependence of the UV-VIS spectra of the aqueous system containing [G-4]-PEG11,000-[G-4] and the molecular probe is shown on Fig. 3. It is seen that the probe is practically removed from the water within the first 24 hr.

Conclusions

The results obtained show that amphiphilic linear-dendritic copolymers are materials with promising encapsulation capabilities. Investigations of the influence of the dendrimer size on the binding capacity of the hybrids are under way.

Acknowledgements

The College of Environmental Science and Forestry, the State University of New York provided financial support for these studies. The authors would like to thank the referees of The Petroleum Research Fund, administered by The American Chemical Society, for useful comments.

References

[1] *Water-Soluble Polymers*, Shalaby, Sh. W. ; McCormick, C.L.; Butler, G.B., Eds., ACS Symp. Ser. Vol. 467, ACS, Washington, DC, 1991 *Industrial Water-Soluble Polymers*, Finch, C.A., Ed., Royal Society of Chemistry Vol. 186, Royal Society, London, 1996; *Hydrophilic Polymers: Performance with Environmental Acceptance*, Glass, J.E., Ed., Adv. Chem. Ser. Vol. 248, ACS, Washington, DC, 1996

[2] Newkome, G.R., Moorefield, C.N., Baker, G.R., Johnson, A.L., Behera, R.K., *Angew. Chem., Int. Ed. Engl.* **1991**, *30*, 1176; Hawker, C.J., Wooley, K.L., Fréchet, J.M.J., *J. Chem. Soc., Perkin Trans. 1* **1993**, 1287; Caminati, G., Ottaviani, M.F., Gopidas, K., Jokusch, S., Turro, N.J., Tomalia, D.A., *PMSE Preprints* **1995**, *73*, 80; Jansen, J.F.G.A, Meijer, E. W., deBrabander-van den Berg, E.M.M., *J. Am. Chem. Soc.* **1995**, *117*, 4417

[3] Fréchet, J.M.J., Gitsov, I., *Macromol. Symp.* **1995**, *98*, 441; van Hest, J.C.M., Delnoye, D.A.P., Baars, M.W.P.L., van Genderen, M.H.P., Meijer, E.W., *Science* **1995**, *268*, 1592; Gitsov, I., Vladimirov, N.G., Pracitto, R., Fréchet, J.M.J., *PMSE Preprints* **1997**, *77*, 140; Elissen-Roman, C., van Hest, J.C.M., Baars, M.W.P.L., van Genderen, M.H.P., Meijer, E.W., *ibid.* **1997**, *77*, 145

[4] Gitsov, I., Wooley, K.L., Fréchet, J.M.J., *Angew. Chem., Int. Ed. Engl.* **1992**, *31*, 1200

[5] *Spectral Atlas of Polycyclic Aromatic Compounds*, Karcher, W., Fordham, R.J., Dubois, J.J., Glaude, P.G.J.M., Ligthart, J.A.M., Eds., D. Riedel Publ. Co., Dordrecht, Vol. 1, p. 93

SYNTHESIS AND CHARACTERIZATION OF LINEAR-DENDRITIC ROD DIBLOCK COPOLYMERS

Catherine M. Bambenek[2], T. Alan Hatton[1], Paula T. Hammond[1]*
[1]Department of Chemical Engineering
[2]Department of Materials Science and Engineering
Massachusetts Institute of Technology
Cambridge, MA 02139

Introduction

The architecture most commonly associated with the term dendrimer has been the sphere, consisting of three or four dendrons radiating out from a central core[1]. However, in recent years, two new dendritic architectures have arisen; the dendritic rigid rod[2,3] and the hybrid linear-dendrimer diblock copolymer[4,5]. Dendritic rigid rods consist of several dendrons attached along the backbone of a linear polymer, while hybrid linear-dendrimer diblock copolymers are composed of a spherical dendrimer attached to the end of a linear polymer. It has been found that the amphiphilic nature of the hybrid linear-dendrimer diblock copolymers can be tuned by choice of the chemistry of each of the blocks. We now would like to propose a third new architecture, the dendritic rigid rod diblock copolymer, which is a combination of the hybrid linear-dendrimer diblock copolymer and the dendritic rigid rod (figure 1).

We have synthesized polymers consisting of a polyethylene oxide-polyethylene imine diblock copolymer, around which polyamide amine (PAMAM) type branches were constructed. Since one of our blocks is polyethylene oxide, which is hydrophilic, the amphiphilic nature of the diblock copolymer is dependent on the surface functionality, generation number, and relative block length of the dendritic rod block. Thus, the solubility of these diblock copolymers in water is also dependent on these factors. We have prepared diblock copolymers in which each of these factors has been varied. The solution properties of these diblock copolymers will be explored using the classical polymer techniques of light scattering, intrinsic viscosity, and gel permeation chromatography. These results along with a synthetic overview will be presented.

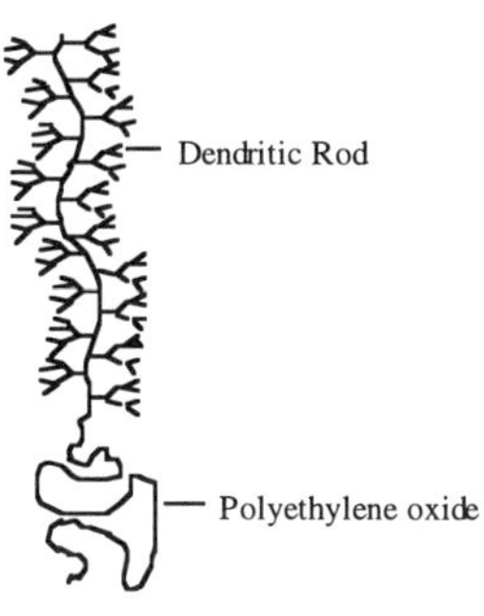

Figure 1

Experimental

Materials

Polyethylene glycol monomethyl ether from PolySciences was dried under vacuum with P_2O_5 for 2 days. Tosyl chloride was recrystallized using a known procedure[6]. Triethylamine and methylene chloride were predried with CaH_2 and distilled over P_2O_5. Benzene was distilled from CaH_2. Diethyl ether was used as received. 2-Ethyl-2-oxazoline was distilled over CaH_2 into KOH. Acetonitrile and n-butyl amine were predried with CaH_2 and distilled from P_2O_5. Aqueous acidic and basic solutions were prepared in Millipore water. Methanol and chloroform were used as received. Methyl acrylate was washed two times with 5% NaOH, two times with Millipore water, and dried over $MgSO_4$ to remove inhibitor. Ethylene diamine was distilled prior to use.

Synthesis

Synthesis of polyethylene oxide methyl ether tosylate **1** was accomplished using an established procedure from Harris et al.[7]. The polyethylene oxide-poly(2-ethyl-2-oxazoline) diblock copolymers **2** were prepared using an adapted procedure from Overberger and Peng[8]. The polyethylene oxide-polyethylene imine diblock copolymers **3** were prepared by hydrolysis of **2** with 1:1 6N HCl:methanol at 60°C for 2 days, followed by distillation of the liquid, further hydrolysis with 3N HCl at 60°C for an additional 5 days, and subsequent removal of this acidic solvent by vacuum distillation[8]. The hydrochloride-salt polymer was dissolved in water, and the aqueous solution extracted with $CHCl_3$. This solution was then neutralized with 4% NaOH until the pH was 10, at which time the polymer precipitated from the water. (The pK_a of secondary amines is 10). The polymer was collected by vacuum filtration, recrystallized in water, and dried in vacuum. The dendritic rigid rod diblock copolymer was prepared in a manner analogous to that of other PAMAM dendrimers[2]. For the generation 0.5 dendritic diblock copolymer **4**, methyl acrylate was placed in a 3-necked round bottom flask possessing a stir bar, addition funnel, condenser, and ice bath. The polyethylene oxide-polyethylene imine diblock copolymer was dissolved in methanol, transferred to the addition funnel, and added dropwise to the stirring methyl acrylate. The reaction was allowed to proceed for 48 hours, at which time the excess methyl acrylate and methanol were removed by vacuum distillation. The polymer was reprecipitated in chloroform/hexane. Subsequent half generation polymers were prepared in a similar manner. The generation 1.0 dendritic rigid rod diblock copolymer **5** was prepared by placing ethylene diamine in a 3-necked round bottom flask possessing a stir bar, addition funnel, condenser, and ice bath. The polyethylene oxide-polyethylene imine generation 0.5 diblock copolymer was dissolved in methanol, transferred to the addition funnel, and added dropwise to the cold, stirring ethylene diamine. The reaction was allowed to proceed for 48 hours, at which time the excess ethylene diamine and methanol were removed by vacuum distillation. The polymer was reprecipitated in methanol/diethyl ether. Subsequent whole generation polymers were prepared in a similar manner. (Figure 2.)

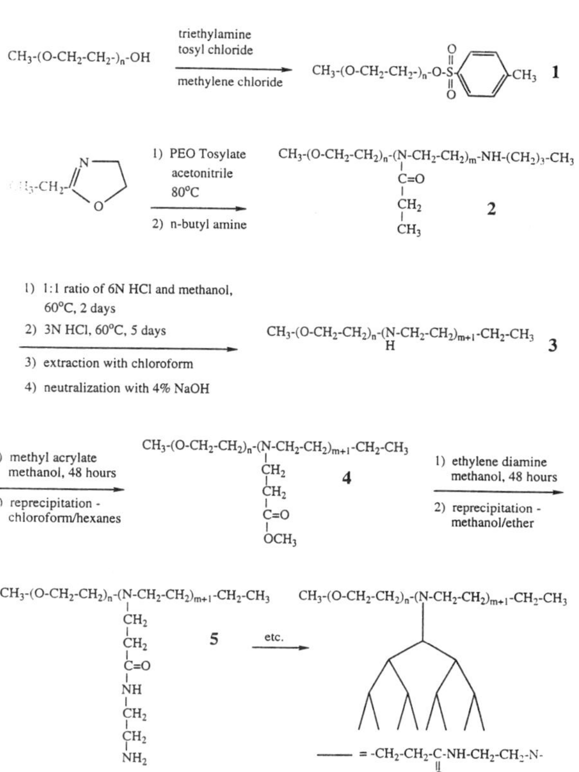

Figure 2

Results and Discussion

Since these dendritic rod diblock copolymers were prepared by a divergent technique, the first step in the synthesis was the preparation of the polyethylene oxide-polyethylene imine diblock copolymer backbone. However, due to the reactivity its monomer, the linear polyethylene imine block cannot be formed directly and had to be synthesized through an intermediate. This intermediate was 2-ethyl-2-oxazoline which was then hydrolyzed to form polyethylene imine. The diblock copolymer polyethylene oxide-poly(2-ethyl-2-oxazoline) was formed by the cationic ring opening polymerization of 2-ethyl-2-oxazoline with the "macroinitiator" polyethylene oxide-tosylate. The poly(2-ethyl-2-oxazoline) block was then hydrolyzed to polyethylene imine using a mild acid hydrolysis. From ^{1}H NMR, it appears that 95% of the repeat units were hydrolyzed. Originally, attempts were made at base hydrolysis; however, they resulted in only 40% hydrolysis of the repeat units. We were somewhat apprehensive about using acid hydrolysis, since polyethylene oxide is know to be acid labile. To be safe, a polyethylene oxide homopolymer "blank" was run through the same hydrolysis conditions. The polydispersity of the PEO before hydrolysis was 1.039 and after was 1.046, a difference of 0.007. This value was within experimental error, leading us to believe that the PEO was not decomposing during hydrolysis. Before final neutralization of the polyethylene oxide-polyethylene imine hydrochloride salt, the acidic, aqueous polymer solution was extracted with chloroform to remove residual polyethylene oxide homopolymer which was uninitiated from the polymerization reaction. The ^{1}H NMR of the polyethylene oxide-polyethylene imine diblock copolymer is shown in figure 3. Addition of the dendritic branches to this diblock copolymer backbone occurred as expected. Purification of the whole generation dendritic rod diblock copolymers after synthesis was imperative to remove any excess ethylene diamine, since residual ethylene diamine would seed the formation of spherical dendrimers in future generations. The ^{1}H NMR of generation 0.5, 1.0 and 1.5 dendritic rod diblock copolymers are also shown in figure 3. From integration of the NMR peaks, it appears that addition at each step is near 100%. FTIR spectra of these diblock copolymer also confirm the chemical structure.

Linear-dendritic rod diblock copolymers of generation 0.5 through 2.0 have been found to be water soluble. While this is not surprising for the whole generations, which are amine terminated and expected to be water soluble, it is somewhat surprising for the ester terminated half generations. While ester functionalities are not inherently hydrophobic, we hypothesized that a large number of them would cause the polymer to lose its solubility. It is possible that the linear-dendritic rod diblock copolymers are behaving as amphiphiles, forming micelles with the dendritic rods at the center. This proposal would not be inconceivable since the hybrid-linear dendrimer diblock copolymers have been found to behave as amphiphiles. The real test will be when we functionalize the dendritic rod surface with a much more hydrophobic group, such as a long chain alkane.

We are also currently in the process of studying the solution properties of these diblock copolymers using intrinsic viscosity, gel permeation chromatography, and light scattering. These results should give us some insight into the structure of the diblock copolymer in solution.

Conclusions

A new dendrimer architecture, the linear-dendritic rod diblock copolymer, has been designed and synthesized. These diblock copolymers consist of a linear polyethylene oxide-polyethylene imine diblock copolymer to which polyamidoamine (PAMAM) branches were divergently added. Generations 0.5 though 2.0 of these diblock copolymers were found to be soluble in water. Work is currently underway to study the solution properties of these diblock copolymers.

Acknowledgements

This research was funded by the 3M Innovation Fund and the Environmental Protection Agency.

References

1) Tomalia, D.A.; Durst, H.D., *Topics in Current Chemistry*, **1994**, *165*, 197-307.
2) Tomalia, D.A.; Kirchoff, P.M., U.S. Patent 4,694-064, **1987**.
3) Percec, V.; Ahn, C.-H.; Barboiu, B., *Journal of the American Chemical Society*, **1997**, *119*, 12978-12979.
4) Gitsov, I.; Wooley, K.L.; Hawker, C.J.; Ivanova, P.T.; Fréchet, J.M.J, *Macromolecules*, **1993**, *26*, 5621-5627.
5) Chapman, T.M.; Hillyer, G.L.; Mahan, E.J.; Shaffer, K.A., *Journal of the American Chemical Society*, **1993**, *116*, 11195-11196.
6) Perrin, D.D.; Armarego, L.F., <u>Purification of Laboratory Chemicals</u>, Butterworth-Heinemann Ltd., Oxford, 1994, p. 291.
7) Dust, J.M.; Fang, Z.; Harris, J.M., *Macromolecules*, **1990**, *23*, 3742-3746.
8) Overberger, C.G., Peng, L.; *Journal of Polymer Science: Part A: Polymer Chemistry Edition*, **1986**, *24*, 2797-2813.

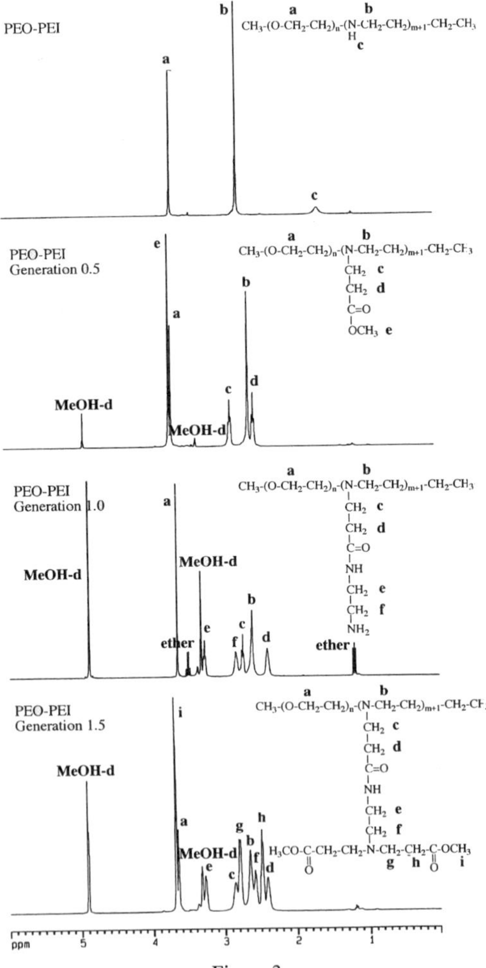

Figure 3
^{1}H NMR of the diblock copolymers.

SOLUTION EEHAVIOR OF NOVEL LINEAR-DENDRITIC DIBLOCK COPOLYMERS

Jyotsna Iyer, Kala Fleming, Paula T. Hammond

Dept. of Chem. Eng., M.I.T, Cambridge, MA-02139

Introduction

Dendrimers are a class of hyperbranched polymers being explored for their potential to form selective membranes, catalyst supports, and guest-host complexes.[1,2] Many of the applications being investigated make use of two important properties of dendrimers, both of which are a direct consequence of their architecture, namely the large number of end groups and the nanoporous nature of the interior at high molecular weights. Apart from dendritic homopolymers, diblock copolymers with one linear block and one dendritic block have also been synthesized.[3,4] While retaining the dendritic properties, the hybrid diblock structure allows the formation of good films because the linear block introduces a large number of entanglements - one feature that is absent in the dendritic homopolymer. We have synthesized a series of hybrid linear-dendritic diblock copolymers with the objective of forming ultrathin film nanoporous membranes. The linear block is polyethyleneoxide(PEO) and the dendrimer block is polyamidoamine(PAMAM). The synthesis scheme and the aqueous solution behavior of these diblock copolymers studied using intrinsic viscosity, gel permeation chromatography, and dynamic light scattering is reported here.

Experimental Section

Figure 1 shows the synthesis scheme used and follows the procedure established by Tomalia. The first step is the addition of methyl acrylate to the primary amine terminal groups of the PEO core. The concentration of methyl acrylate and primary amine was 8.5M and 0.025M respectively. Methanol was used as the solvent and the reaction was conducted at room temperature. After 48 hours, the reaction was terminated by removing the methanol and unreacted methyl acrylate under vacuum. The white amorphous residue is washed thoroughly with ethyl ether. The second step is the aminolysis of the methyl ester terminal groups by ethylene diamine. Methanol is used as the solvent and the reaction run at 50°C for 48 hours. The concentration of ethylene diamine was 11-12M. At the end of the reaction, ethylene diamine and methanol were removed under vacuum. The residue was washed with ethyl ether to give the product. All reactions were carried out under a nitrogen atmosphere.

The intrinsic viscosity measurements were done on a Ubbehode viscometer at 30°C. The gel permeation chromatography experiments were performed with a $NaNO_3$(0.05M)/NaN_3(0.02%) aqueous solution as the mobile phase at 30°C. The dynamic light scattering experiments were conducted at 30°C.

Results and Discussion

Two series of linear-dendritic diblock copolymers with a linear polyethyleneoxide block and a dendritic polyamidoamine block have been synthesized. In the first series, the polyethyleneoxide tail had a molecular weight of 2000 with the dendrimer generations going up to 4.0 [PEO2k-0.5G,1.0G,1.5G...4.0G]. In the second series, the polyethyleneoxide tail had a molecular weight of 5000 with dendrimer generation going up to 4.5 [PEO5k-0.5G,1.0G,1.5G....4.5G]. The chemical structure of the copolymers was ascertained using ^{1}H NMR. The peak positions obtained were in good agreement with values found by other researchers for a similar class of materials.

Figure 2 is a graph of the intrinsic viscosity, [η], of PEO(2000)-dendrimer series in water plotted against the generation number of the dendrimer. The PEO(2000)-dendrimer diblocks behave like linear polymers with the [η] increasing with increasing generation number.

The relation between [η] and generation number of the dendrimer for the PEO(5000)-dendrimer diblock series is very different. Figure 3 is a plot of [η] versus generation number for the PEO(5000)-dendrimer diblocks. In both ester and amine terminated PEO(5000)-dendrimer diblocks, the PEO chain probably forms a corona around the hydrophobic dendrimer repeat units shielding them from the solvent. The decrease in [η] for early generations of other linear-dendritic diblock series because of the formation of unimicellar like structures has been observed earlier by Frechet et al.[3,5]

Converting the [η] to hydrodynamic radii, r_h, using the Einstein equation gives a more direct picture of what happens in aqueous solutions of PEO(5000)-dendrimer diblocks. The graph of r_h plotted against dendrimer generation is shown in Figure 4. The hydrodynamic radius remains practically constant until the second generation and then increases with increasing generation number.

The size estimates obtained from the other two techniques are in the same range as those from intrinsic viscosity measurements. The difference in the solution behavior of the two series of diblocks seem to suggest that the length of the linear block plays a important role in the structure of the diblocks in solution. Interestingly, a longer PEO tail appears to enhance the hard sphere behavior.

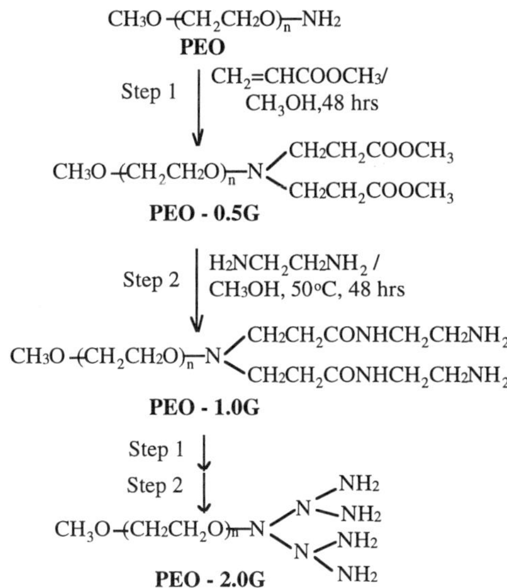

Figure 1 Synthesis of PEO-PAMAM diblock copolymers

Summary and Conclusions

Two series of linear-dendritic diblock copolymers have been synthesized. The linear block was PEO and the dendritic block was PAMAM. The behavior of the diblocks in aqueous solutions seem to depend on the length of PEO linear block. The diblocks with a PEO length of 2000 behave like linear polymers but the diblocks with a PEO length of 5000 behave more like hard spheres.

Acknowledgments

This work was financially supported by an EPA grant (# R825224-01-0)

References

1. O.A.Matthews, A.N.Shipway, J.F.Stoddart, Prog. Polym. Sci., *23*, 1-56,1998
2. G.R.Newkome, C.N.Moorefield, F.Vogtle, Dendritic Molecules: Concepts, Syntheses, Perspectives, VCH Publishers,Inc., New York, 1996
3. I.Gitsov, J.M.J.Frechet, Macromolecules, *26*, 6536-6546, 1993
4. K.Aoi, A.Motoda, M.Okada, T.Imae, Macromol. Rapid Commun., *18*, 945-952, 1997
5. I.Gitsov, J.M.J.Frechet, Macromolecules, *27*, 7309-7315, 1994

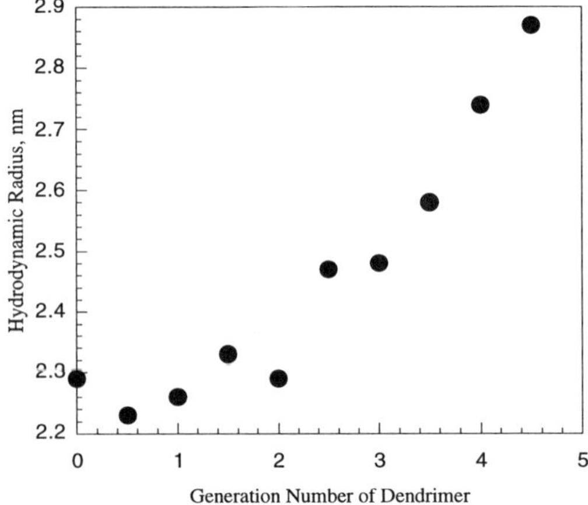

Figure 4: Variation of hydrodynamic radius with generation number of PEO(5000)-Dendrimer diblocks

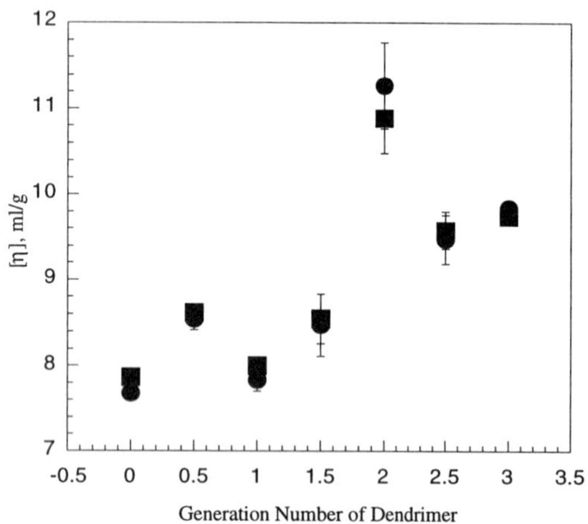

Figure 2 Variation of intrinsic viscosity with dendrimer generation for PEO(2000)-dendrimer diblocks at 30°C in water

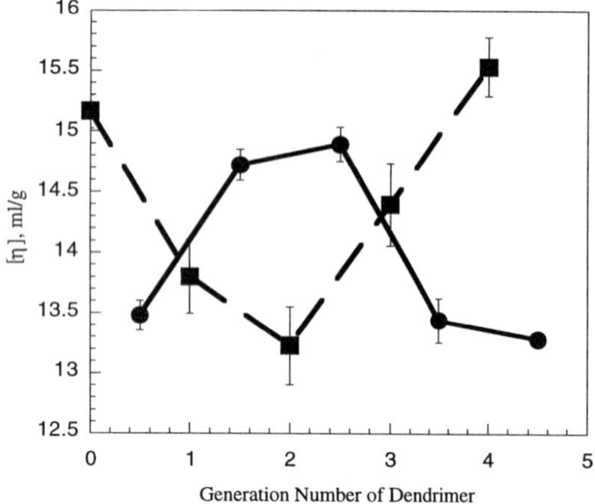

Figure 3 Variation of intrinsic viscosity with dendrimer generation for PEO(5000)-dendrimer diblocks at 30°C in water

Modification of Surfaces with Functionalized Poly(ethylene glycol)-Poly(lactic acid) Micelle

Kazunori Emoto, Yukio Nagasaki, and Kazunori Kataoka

Dept. of Materials Science & Technology, Science University of Tokyo, Noda, Chiba, 278-8510 Japan

Introduction

Poly(ethylene glycol)-poly(lactic acid) block copolymers have been investigated for the controlled drug delivery systems, and implantable biomaterials[1]. Both segments are approved by the Food and Drug Administration and of the highest potential for the above applications. An aqueous solution is selective for PEG segments and the polymer tends to form a micellar structure which is useful for drug carriers. Hydrophobic drugs are incorporated into the core of the micelle and the PEG corona helps the circulation in the body by reducing the agglomeration due to the intermicellar interaction as well as the protein adsorption.

We are interested in functionalizing the ends of the block copolymer for the creation of reactive micelles which is applicable to the targettable drug delivery systems and biomaterials[2-5]. Ligands such as antibodies, peptides and oligosaccharides can be attached to the surface of the micelle and can be selectively delivered. By polymerizing or cross-linking the PLA segment, the structural stability is facilitated[3,4]. Such surface-functionalized micelles are also applicable to the coating of surfaces as in the case of linear[6-9] and star polymers[10]. Since the density of PEG is high on the surface of the micelle, surfaces grafted with the PEG-PLA micelles are expected to reject proteins effectively. By incorporating hydrophobic drug into the grafted micelle, one can create surfaces releasing drugs in a controlled manner. For these purposes, surface properties, such as surface coverage, conformation of micelle attached on the surface, and wettability of coatings are important for the biocompatibility, controlled drug release and drug delivery.

In this presentation reactive PEG-PLA micelle was attached to surfaces covalently and the coating property was compared with PEG coatings. A micelle prepared from a functionalized block copolymer, acetal-poly(ethylene glycol-*b-DL*-lactic acid)-methacrylate (ac-PEG-PLA-ma), was coated on glass and polydimethylsiloxane (PDMS) surfaces under various conditions. The coated surfaces were characterized by ζ potential and dynamic contact angle measurements, and are related to the micelle structure of micelles on the surface obtained by atomic force microscopy (AFM).

Experimental

Preparation of Functionalized PEG-PLA Micelle Ac-PEG-PLA-ma was synthesized as described elsewhere[2-4]. The NMR spectra of the block copolymer revealed the mw of PEG and PLA segment as 4500 and 3500, respectively. The block polymer was dissolved in dimethylacetoamide and dialyzed in deionized water for 24h. After the hydrolysis of acetal group into aldehyde group by adjusting pH 2, methacryl group was radical-polymerized at 50 C. The diameter of the micelle was about 30nm from dynamic light scattering. The polymerized micelle was stable and did not decompose in the presence of sodiumdodecyl sulfate unlike typical micelles.

Coating of Micelle on glass and PDMS Surfaces The glass substrates were cleaned by boiling mixture of sulfuric acid and hydrogen peroxide followed by an immersion in sodium bicarbonate. After drying under vacuum, aminopropyltriethoxy silane, 2%(w/v), was coated to activate the surface and then cured at 160 C under vacuum. The polydimethylsiloxane films were immersed in deionized water and then dried at 25 C overnight. The each side of the dried film was then treated with N_2+H_2 plasma at 1.5 Torr, 75W for 30 min. The amine-functionalized glass and PDMS substrates were immersed into various temperature of 10% (w/v) PEG-aldehyde (mw 5000) or 0.1% (w/v) micelle solution containing 0.25% (w/v) $NaCNBH_3$, pH 5.5, for 2h. Samples were rinsed with deionized water exhaustively and stored in deionized water until use.

Characterization of Coatings by ζ-potential and Dynamic Contact Angle Measurements The ζ potential of glass or PDMS samples was measured over the pH of 2 to 11 by LEZA-600 (Ohtsuka Electronics, Co., Japan) in 7.5 mM NaCl containing polystyrene latex covered with hydroxyethyl cellulose. The medium pH was adjusted by 7.5 mM HCl or NaOH so that the ionic strength was little altered. The dynamic contact angle against water of surface-treated glass slides (18x18x0.15mm) was measured in Wilhelmy method by DCA150 (Cahn, MN). Three to nine samples per each coating were immersed and then emersed at 0.1mm/sec 5 times, and the advancing and receding angles were obtained by averaging the last 4 cycles of all the samples of a coating.

Results and Discussion

The ζ potential profiles of micelle- and PEG-coated glass surfaces are presented in Figure 1. When PEG is grafted densely, the polymer molecule is elongated tangentially from the surface, and for PEG of mw 5000 the thickness is expected to be over 10nm although its radius of gyration of the free polymer in an aqueous solution is small (3nm). Although the dimension of micelle is much larger than the linear PEG, the ζ potential profile was little different from the PEG-coating. This is due to the different permittivity of micelle on the charged surface and to the smaller screening of electrostatic potential by the micelle. Unlike glass substrates, PDMS samples showed remarkably different ζ potential profiles. When PDMS was plasma-treated, a positive ζ potential due to the introduction of amino group was seen at low pH, Figure 2. Strongly negative ζ potential is seen at high pH, probably due to the generation of silanol by the chain cleavage of PDMS. The hysterisis of ζ potential is also seen. The chain cleavage of PDMS also resulted in a facilitated mobility of functional groups on the PDMS surface and amino group tend to appear or disappear depending on the pH. PEG coating reduced ζ potential appreciably. Although the coating condition was the same as glass surface and optimal as described by Gölander, et al.[6], and Huang, et al.[7], the PDMS surface with little reactive group could not support enough PEG to mask the surface charge. When micelle was coated on the surface, ζ potential was reduced more remarkably, and the hysterisis was not seen. Despite the low surface density of amino group, single reaction of aldehyde on the micelle to amine resulted in the high coverage of surface by the micelle, and the multiple attachment of micelle-aldehydes cross-linked the cleaved PDMS chain. Reduced Shiff base showed a stable coating. If the Schiff base was reduced similar ζ potential profile was obtained even after the hydrothermal treatment at 120 C for 30min., while non-reduced micelle-coating resulted in the similar profile to the plasma-treated PDMS.

The dynamic contact angle of micelle-surface varied with coating temperature. Figure 3 shows the contact angle of micelle coatings at various temperatures. The advancing angle of 25 and 50 C-coatings was 76 and 72 , respectively. The coating at 80 C resulted in a higher advancing angle and even higher than the APTS-coated glass surrfaces as seen in Figure 4. The AFM images indicated that the size of micelle on the surface got smaller as the coating temperature was increased to 50 C, but the micelles on the surface coated at 80 C was resulting in a rough surface. On the other hand, dynamic light scattering revealed that the micelle did not aggregate in the solution even at 80 C[5]. The variation of advancing angle and hysterisis seemed to be a consequence of nano-doamins created by the micelle. Relatively high advancing angle of micelle-coatings is probably due to the conformation of PEG and the overall solubility of the micelle in an aqueous solution. Similar contact angle of micelle and PEG coatings were obtained for the PDMS sample, regardless of different wettability of the substrates, indicationg that the surface is fully covered and the substrate property littele affected the wettability of the coating.

Conclusion

Micelles with reactive group on the surface could be attached to the surface covalently. The ζ potential of micelle-coated glass surface was little different from the PEG-coated substrate despite its larger dimension than the linear PEG, due to the lower permittivity of micelle and lower screening of electrostatic potential. The masking of surface charge was as significant on PDMS on which fewer amino group was accessible to micelle-aldehyde than the APTS-coated glass. However, the advancing contact angle of micelle-coated surface was higher than PEG-coated surface. AFM images showed a reduced size of surface-attached micelle when coated at moderately high temperature (50 C), but increased coating temperature resulted in an aggregation of micelles on the surface although aggregation was not observed in solution.

Acknowledgements
This investigation was supported by Vistakon, an affiliate of Johnson & Johnson Company.

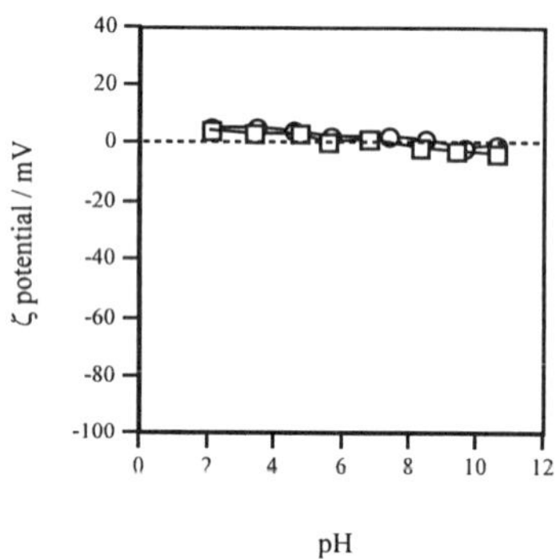

Figure 1. The ζ potential profile of glass treated with APTS followed by the coating with PEG-aldehyde (square) or polymerized micelle-aldehyde (circle).

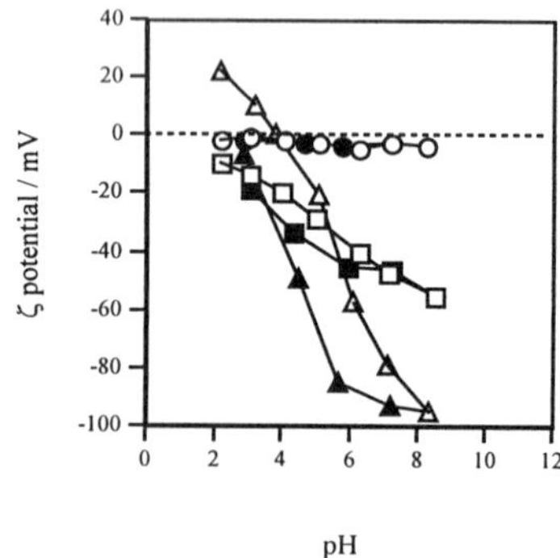

Figure 2. The ζ potential profile of PDMS treated with N_2+H_2 plasma (triangle) folloed by the coating with PEG (square) or micelle (circle). The potential was measured from pH 2 to 9 (unfilled) and then pH 9 to 3 (filled).

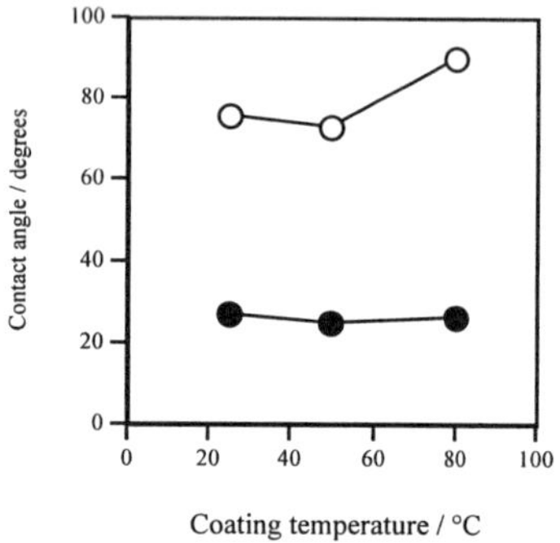

Figure 3. Variation of advancing (unfilled) and receding (filled) contact angles of micelle-coated glass with coating temperature.

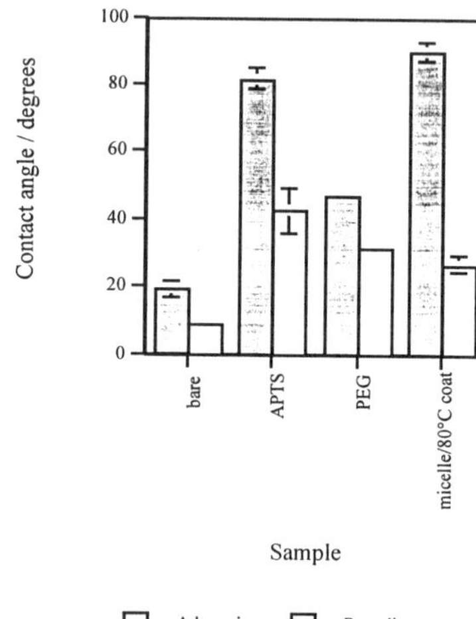

Figure 4. Dynamic contact angles of glass surfaces after the treatment with APTS followed by PEG-aldehyde or polymerized micelle-aldehyde.

References

(1) Jeong, B.; Bae, Y. H.; Lee, D. S.; Kim, S. W., Nature, **1997**, 388/28, 860.
(2) Nagasaki,Y.; Okada, T.; Scholz, C.; Iijima, M.; Kato, M.; Kataoka, K., *Macromolecules*, **1998**, in Press.
(3) Iijima, M.; Okada, T.; Nagasaki,Y.; Kato, M.; Kataoka, K., **1998**, in Preparation.
(4) Kim J-H.; Iijima, M.; Nagasaki, Y.; Kato, M.; Kataoka, K., **1998**, in Preparation.
(5) Yasugi, K.; Kataoka, K., **1998**, in Preparation.
(6) Gölander, C-G.; Herron, J.; Lim, K.; Claesson, P.; Stenius, P.; Andrade, J. D.; in Poly(ethylene glycol) Chemistry: Biotechnical and Biomedical Applications, Harris, J. M. Ed., Plenum, New York, 1992, pp221.
(7) Huang, S-C.; Caldwell, K. C.; Lin, J-N.; Wang, H-K.; Herron, J. N., *Langmuir*, **1996**, 12, 4292.
(8) Emoto, K.; Harris, J. M.; Van Alstine, J. M., *Anal. Chem.*, **1996**, 68, 3751.
(9) Österberg, E.; Bergström, K.; Holmberg, K.; Schuman, T. P.; Riggs, J., Burns, N. L.; Van Alstine, J. M. ; Harris, J. M., *J. Biomed. Materials Res.*, **1995**, 29, 741.
(10) Sofia, S. J.; Merril, E. W., in Poly(ethylene glycol): Chemistry and Biological Applications, Harris, J. M.; Zalipsky, S. Eds., American Chemical Society, Washington D.C., 1997, pp342.

Swelling Dynamics and Free Volume Structure of Poly[(2-dimethylamino)ethyl methacrylate]-*l*-polyisobutylene Amphiphilic Conetwork by Simultaneous Swelling Kinetics and Positron Annihilation Studies

Attila Domján[1], Béla Iván[1], Károly Süvegh[2], György Vankó[2], and Attila Vértes[2]

[1] Material Science of Polymers Group, Institute of Chemistry, Chemical Research Center of the Hungarian Academy of Sciences, H-1525 Budapest, Pusztaszeri u. 59-67, P.O.Box 17, Hungary

[2] Department of Nuclear Chemistry, Eötvös Loránd University, H-1518 Budapest 112, P.O. Box 32, Hungary

Introduction

Quasiliving carbocationic polymerizations have led to a wide variety of novel functional polymers, such as macromonomers, telechelic polymers macromolecules, new block copolymers etc. in recent years (see e.g. Refs.1-5 and references therein). Utilization of methacrylate-telechelic polyisobutylene led to new type of crosslinked polymeric materials called amphiphilic conetworks (APCN). These are composed of covalently bonded hydrophilic and hydrophobic chains [6,7]. The preparation of APCNs usually not straightforward due to the possibility of macroscopic phase separation occurring during the synthesis. These complications explain the fact that the systematic research related to APCNs has started only in the last couple of years.

As recent investigations with polyisobuylene(PIB)-based APCNs indicate, the covalently bonded immiscible hydrophilic and hydrophobic segments result in unique nano-structured molecular composites with co-continuous microphase separated morphology. Most likely this special structure explains the excellent biocompatibility [8] and blood compatibility [9] of these materials. These polymeric materials are swellable both in hydrophilic and hydrophobic solvents [4,6,7], and the equilibrium swelling is controlled by composition. One of unique aspects of the swelling process of APNs is that the polymer phase with opposite philicity of the solvent is expected to collapse or shrink during swelling.

Positron lifetime technique has recently been used widely for studying the structure of polymers [10,11] providing a unique posibility to investigate the free volume through the lifetime of positronium atoms. Positronium is formed by an electron and a positron, and it can be considered as a very light hydrogen atom. It has two ground states (a triplet and a singlet) but only the triplet state has a practical advantage for investigating polymer structure. Usually 10-30 % of the injected positrons form triplet positronium (ortho-positronium, o-Ps) in polymers [12]. In a vacuum where the electron density is zero the o-Ps has a very long lifetime (141 ns) but in a condensed material the lifetime is decreased by about three orders of magnitude by the finite electron density. The usual lifetime of o-Ps in polymers is around 1.5-3 nanoseconds. The o-Ps is localised in the free volumes between the polymer chains and its annihilation takes place by external electrons. Several authors published correlation functions between the lifetime of o-Ps and the size of free volume units in molecular systems [12] but these correlations are valid only for similar chemical compositions. This presentation reports on a new method for investigating free volume changes and structure in polymeric materials by simultaneous swelling kinetics and positron annihilation experiments.

Experimental

The method for the preparation of poly[(2-dimethylamino)ethyl methacrylate]-*l*-polyisobutylene (PDMAEMA-*l*-PIB) amphiphilic network and the method of swelling experiments was published earlier [6,7]. Composition of the (PDMAEMA-*l*-PIB) network was determined by elementar analysis giving 52 % PDMAEMA and 48 % PIB. For the positron lifetime measurements a conventional fast-fast coincidence system was used [13]. The overall time resolution of this system was 210 picoseconds.

Results and Discussion

As the conventional swelling ratio measurements show (Fig. 1a) the investigated DMAEMA-*l*-PIB is able to absorb large amount of water. The water uptake is relatively quick at the beginning but saturation (equilibrium) occurs only after ~50 hours soaking at 52 % swelling ratio at room temperature. There are significant changes in the lifetime and the intensity of o-Ps as a function of swelling time (Fig. 1b and Fig. 1c). At the beginning of the swelling process, i.e. at low swelling ratios, the o-Ps lifetime increases by about 70 picoseconds, then it reaches a maximum and decreases again. This latter phase, the decrease of the lifetime is the expected phenomenon, and it is related to filling the free volume by water. As the water molecules fill the free volumes in the network by reaching equilibrium swelling, the o-Ps lifetime approaches the value measured in pure water (about 1800 ps).

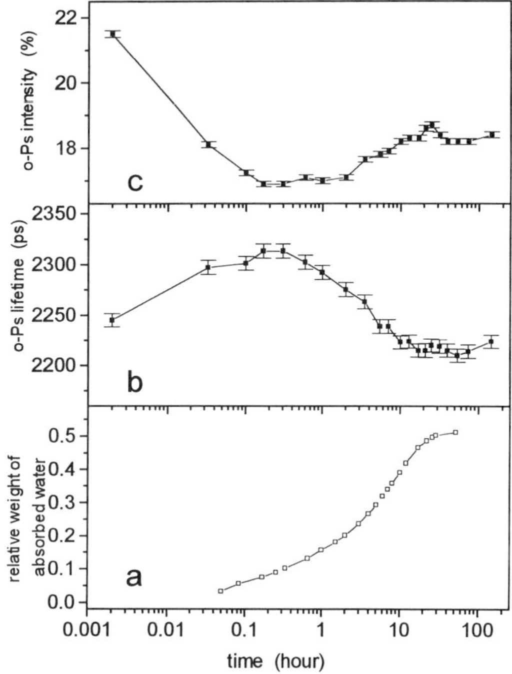

Figure 1. Swelling kinetics of the PDMAEMA-*l*-PIB amphiphilic network followed gravimetically (a) and by positron spectroscopy (b,c).

The increase of the lifetime at very low swelling ratios indicates that the average size of free volume sites increases in the PDMAEMA-*l*-PIB network as water molecules start to diffuse in the material. Surprisingly, these dynamic structural changes occur fast at the very beginning of the swelling process and at low swelling ratios. Considering the amphiphilic nature of the PDMAEMA-*l*-PIB sample, this observation can be explained by a quick rearrangement of the phases with opposite philicity in this APCN. The PDMAEMA phase expands whereas the PIB phase shrinks upon contacting with water. Both effects yield increase of the size of free volume units.

The o-Ps formation intensity rapidly decreases at the beginning of the swelling process and after reaching a minimum it increases again (Fig. 1c). The interpretation of the decrease of the o-Ps formation intensity is not as straightforward as that of the lifetime changes. It cannot be explained either by the free volume increase or by the direct effects of water, i. e. the increase of free volume size does not have a significant effect on Ps formation and the Ps formation intensity is higher in pure water than in the network.

Lifetime measurements were also made with water/DMAEMA mixtures in order to study the intensity decrease phenomenon. The results show (Fig. 2) that similar intensity decrease occurs by adding water in the DMAEMA monomer as in the APCN. The o-Ps formation intensity decreases until the water/DMAEMA molar ratio reaches one, and then further addition of water does not affect this parameter significantly. These results suggest that the "adsorption" of the first water molecule changes the electronic structure of DMAEMA molecules so that it provides smaller probability for o-Ps formation. On the basis of our results we concluded that 1:1 DMAEMA/water clusters were formed in the sample.

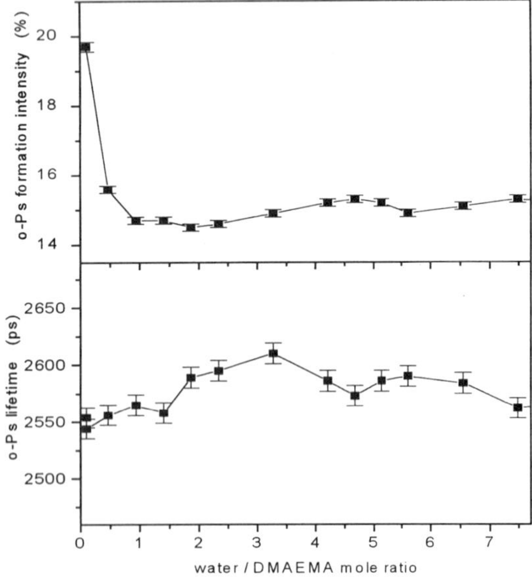

Figure 2. Positron lifetime parameters in the mixtures of the DMAEMA monomer and deionized water.

For the exceptional role of 1:1 molar ratio quantum chemical *ab initio* calculations were performed for the DMAEMA monomer. It was supposed that H-bonded clusters were formed in the mixtures, and 1:1 clusters are the most stable among them. In order to confirm this assumption optimum geometries and total energies were calculated for 1:1 clusters of different structures. The calculated total energy difference between N--H-O and O--H-O clusters is very small.

There are two other possible cluster structures which distinguish the 1:1 ratio of water and DMAEMA. The first is a chain-like structure in which DMAEMA and water molecules are ordered alternately connected to each other by H-bonds. A little more reliable explanation is provided by the assumption of "intramolecular" H-bonded

ring formation. The calculated energy gain of this cluster is might be enough for the formation of the N--HOH--O structure (~ 40 kJ/mol). It should be noted that the formed clusters do not necessarily live forever but their lifetime is long enough to affect positronium formation intensity.

Since the o-Ps formation intensity data of the PDMAEMA-*l*-PIB APN are very similar to those of the water/DMAEMA mixtures, we can assume that the hydration of the APN occurs in a similar way, i.e. the formation of ring-like 1:1 clusters is also the first process in the network. This assumption is confirmed by the fact that the intensity decrease continues in the network till the water/DMAEMA mole ratio reaches the unit value (in about 8-10 minutes). The fact that the intensity decrease stops at this mole ratio suggests that water molecules are able to reach every DMAEMA unit very quickly at the beginning of the swelling process. If the free volume were composed of independent pores the intensity decrease would be a slow process and would not stop at 1:1 water/DMAEMA ratio. On the other hand the observed changes indicate that the surface of the free volume units become covered with water molecules quickly at the beginning of swelling. This suggests that the free volume in the PDMAEMA-*l*-PIB APCN is composed of interconnected channels which allow rapid diffusion of water molecule to cover the surface of the free volume units in this APCN.

Results of this study also indicate that positron annihilation can be successfully applied not only for obtaining information on free volume changes in polymeric materials but also on the dynamics of swelling and the structure of free volume in certain macromolecular systems, such as networks, hydrogels, polyelectrolyte gels, polymer composites, blends, etc., by studying the swelling kinetics and positron lifetime parameters simultaneously. It is also predicted by us that the results of this study can be extended for the investigation of a wide variety of gels and amorphous materials as well.

References

(1) Iván, B. *Macromol. Symp.* **1994**, *88*, 201.

(2) Iván, B. *Makromol. Chem., Macromol. Symp.* **1993**, *67*, 311.

(3) Kennedy, J. P.; Iván, B., *Designed Polymers by Carbocationic Macromolecular Engineering: Theory and Practice*, Hanser Publishers, Munich, New York, 1992

(4) Iván, B.; Kennedy, J. P. in "*Macromolecular Design of Polymeric Materials*", Hatada, K.; Kitayama, T.; Vogl, O. (Eds.) Marcel Dekker, New York, 1997, pp. 51-84

(5) Sawamato, M.; Higashimura, T. in "*Macromolecular Design of Polymeric Materials*", Hatada, K.; Kitayama, T. Vogl O. (Eds.) Marcel Dekker, New York, 1997, pp. 33-50

(6) Iván, B.; Kennedy, J. P.; Mackey, P. W. in "*Polymeric Drugs and Drug Delivery Systems*", Dunn, R. L.; Ottenbrite, R. M., Eds.; ACS Symp. Ser., Vol. 469, Am. Chem. Soc., Washington, D.C., 1991, pp. 194-202; ibid., 1991, pp. 203-212

(7) Iván, B.; Kennedy, J. P.; Mackey P. W. *U.S. Patent*, 5,073,381 (Dec. 17, 1991)

(8) Chen, D.; Kennedy, J. P.; Kory, M. M.; Ely, D. *J. Biomed. Mat. Res.*, **1989**, 23, 1327

(9) Keszler, B.; Kennedy, J. P.; Ziats, N. P.; Brunstedt, M. R.; Stack, S.; Yun, J. K.; Anderson, J. M. *Polym. Bull.*, **1992**, *29*, 681

(10) Pethrick, R. A. *Prog. Polym. Sci.* **1997**, *22*, 1

(11) Süvegh, K.; Klapper, M.; Vértes, A.; *J. Radioanal. Nucl. Chem., Letters*, **1994**, *186*, 375

(12) See e.g., Mogensen, O. E. *Positron Annihilation in Chemistry*; Springer Series in Chemical Physics, Vol. 58, Springer-Verlag, Berlin, 1995, pp. 221-245

(13) MacKenzie, I. K. in "*Positron Solid State Physics*", Brandt, W.; Dupasquier, A., Eds.; North-Holland, Amsterdam, 1983, pp. 220-242

Influence of Crosslinkers on the Water-Swelling Properties of Radiation-Grafted Ion-Exchange Membranes

Hans-Peter Brack and Günther G. Scherer, Elektrochemie, Paul Scherrer Institut, CH-5232 Villigen PSI, Switzerland

Introduction

The water swelling properties of ion-exchange membranes are important in their application in aqueous environments. This is particularly true of the ion exchange membranes that are used in fuel cells, a type of electrochemical cell that can efficiently generate electricity from a fuel, e.g., hydrogen, and an oxidant, e.g., oxygen from air, in the following reaction: $H_2 + O_2 \Rightarrow H_2O$ + electricity + heat. In the fuel cell, the polymeric solid electrolyte membrane functions both (1) to transport protons from the anode side where they are produced to the cathode side where they are consumed and (2) to separate the two gas phases (e.g., H_2 and O_2) on each side of the cell and prevent their direct reaction.

The water swelling properties of the membrane has an important influence in determining fuel cell performance. For example, the proton conductivity of the membrane typically increases[1] as its water content increases. Water can also plasticize the membrane and improve[2] its mechanical properties. Unfortunately membrane swelling can result in significant increases in size and volume. This interplay between water content, dimensional stability, and mechanical properties is particularly important in large area fuel cells and stacks (several fuel cells in series) in which mechanical and swelling stresses[3] may cause problems. Mechanical stresses may result from the cell design and the difficulty of making cells with large active areas (membrane/electrode interfacial areas) gas tight. Inhomogeneities in the water distribution between cells in stacks and across large active areas may result[3] in swelling or drying stresses.

In addition, the electro-osmotic water flow from the anode to the cathode side of the fuel cell and the balancing back diffusion also affects[4] membrane hydration. Ideally a fuel cell membrane should have a high proton conductivity at low water contents, low electro-osmotic transport, high dimensional stability as a function of water content, and low gas crossover in the swollen state.

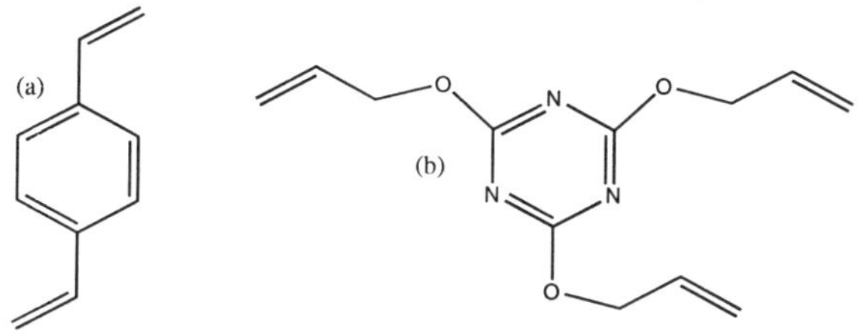

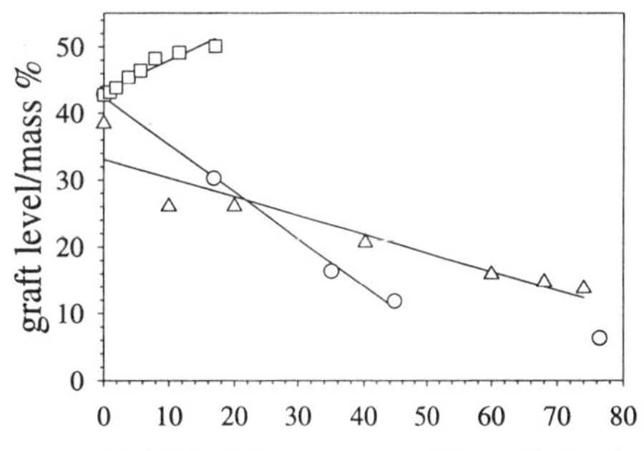

Fig 1. Structures of (a) *p*-DVB and (b) TAC crosslinkers

Water swelling of membranes can be controlled by the use of crosslinkers. We have reported previously[5] on the effects of the two crosslinkers shown in Figure 1, divinylbenzene (DVB) and triallylcyanurate (TAC), on the performance of radiation-grafted membranes in H_2/O_2 fuel cells. DVB improved the membrane lifetime in the application in comparison to that of uncrosslinked membranes. DVB also decreased the water swelling, but unfortunately it increased the specific resistance too. The specific resistance is important because high membrane specific resistance leads to high ohmic losses in the application. TAC increased the water uptake, did not increase the specific resistance, and did not improve stability. TAC and DVB were found to have a synergetic effect which made it possible to prepare stable membranes having low specific resistances[2,5] by using both crosslinkers together.

The effect of these two crosslinkers on the grafting reaction yield, their incorporation during grafting, and their effect on the water swelling properties of the resulting radiation grafted membranes are examined in more detail here. In addition, the extent of reaction of DVB and the effect of the base polymer film type and radiation processing method on swelling will be discussed in our oral presentation.

Experimental

Our membrane preparation and characterization methods have been reported previously[2]. FEP 200A films from DuPont (thickness of 50μm, polymer molar mass ≈ 325,000 Daltons) were irradiated with a dose of 80kGy (1Gy = 1J/g energy absorbed) under N_2 using an electron beam with an accelerating voltage of 2.2MV, a beam current of 20mA, and having a dose rate of 15.1±1.1kGy/s. Samples were stored at -80°C under N_2 until they were grafted using two component reaction mixtures of styrene (Fluka purum grade) and either divinylbenzene (Fluka, technical grade, 70-85% active) or triallylcyanurate (Fluka, purum grade) as crosslinkers or with benzene (Fluka, puriss ACS reagent grade) as diluent. Samples were grafted for 4.5h at 60°C. Unreacted monomer and homopolymer were extracted from the films after grafting using toluene in a Soxhlet apparatus. Grafted films were sulfonated for 4h at 66°C using a 30 vol% solution of chlorosulfonic acid in 1,1,2,2-tetrachloroethane, followed by hydrolysis for 2 days in 0.15M NaOH(aq), regeneration for 5h in 10 mass% HCl(aq), and swelling for 4 h in ≈ 90°C deionized water. For the spectroscopic investigations, a grafted film was impregnated with a 1:1 mass:mass solution of TAC in benzene and dried under vacuum.

The extent of swelling of the membranes was determined based on the mass difference between their water-swollen form and that of their vacuum-dried K^+-exchanged form (12h at 80°C). The Ion-Exchange Capacities (IEC) of the membranes were determined by reacting samples with 0.5M KCl solution for 12 h and then titrating the resulting HCl(aq) solutions.

Results and Discussion

The presence of the two crosslinkers, DVB or TAC, in the grafting solution has very different effects on the grafting yields obtained, as shown in Figure 2. DVB decreases the grafting yield and TAC increases the grafting yield. Since grafting occurs by the grafting front mechanism, it has been proposed[6] that DVB decreases the grafting yield by crosslinking the near surface region of the film in the early stages of reaction and thus inhibiting base polymer swelling and the diffusion of monomer into the interior of the film and its subsequent reaction. Additional support for this hypothesis comes from FT-IR transmission and Attenuated Total Reflectance (ATR) mode measurements that have shown[7] that the concentration of DVB incorporated into the near surface region of the grafted film is higher than that incorporated into the film interior.

Fig. 2 Grafting yield obtained using (□) TAC and (**O**) DVB as crosslinkers or (Δ) benzene as diluent.

The case of TAC is unfortunately much more complicated. TAC is a larger monomer and has flexible ether linkages. It has been proposed[6] that for these two reasons the incorporation of TAC into the grafting film leads to a more flexible structure and thus enhances the monomer diffusion into the film during the early stages of grafting. Unfortunately our subsequent FT-IR transmission and ATR mode measurements have found no evidence for the incorporation of TAC into the base polymer during the grafting process, as shown in Figure 3.

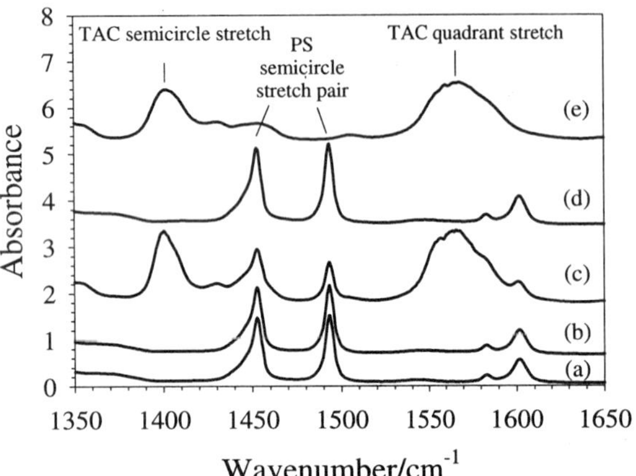

Fig. 3 FT-IR transmission mode spectra of FEP films grafted using (a) 6.96 M styrene in benzene (graft level of 26.0 mass %), (b) 5.20 M styrene and 0.770 M TAC in benzene (graft level of 28.9 mass %). Spectrum (c) measured after impregnation of (b) with TAC to give a mol ratio of 1:7.2 TAC:styrene (spectrum reduced 50%). Spectrum (d) is (c) after extraction with toluene. Spectrum (e) is of TAC on a KBr plate.

The spectral results in Figure 3 indicate that TAC can diffuse into the grafted regions of the film during grafting, but it is not incorporated into the grafted chains. It is proposed here that TAC may act as a chain transfer agent during grafting and thus limit the length of the grafted chains. The hydrogens adjacent to the double bond in allylic monomers are known[8] to readily undergo radical chain transfer reactions to form relatively stable and unreactive allyl radicals, as shown in Figure 4. Decreasing the length of the grafted side chains reduces the physical crosslinking due to the lower concentration of entanglement junctions and thus may increase the permeability and extent of grafting. The permeability of these films may be affected also by their microphase-separated structures[9], and the roles of TAC and DVB in influencing this microstructure are not yet known.

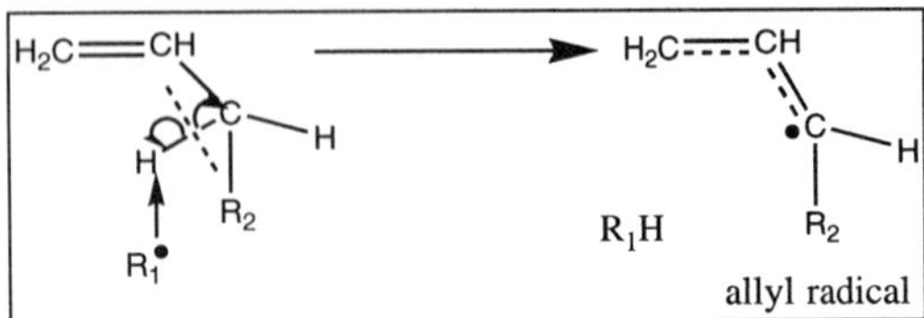

Fig. 4 Degradative chain transfer reaction of allyl monomer

The effects of including TAC and DVB into the grafting solutions on the Ion-Exchange Capacity (IEC) and water swelling of the resulting membranes are shown in Figure 5. DVB decreases the graft level and thus the content of sulfonated aromatic groups, IEC, and swelling decrease. The water swelling of the DVB crosslinked membranes is however always lower than that of the uncrosslinked membranes having the same graft level (with benzene as diluent) because DVB is a rigid crosslinker and reduces[5] the grafted chain mobility and the membrane expansion during swelling. So DVB can be used to control the water swelling of radiation-grafted membranes, but it unfortunately also reduces the achievable grafting level and also ion-exchange capacity. An additional disadvantage of using DVB as a crosslinker is that the incorporation of higher DVB contents into the membrane increases[2,7] its brittleness.

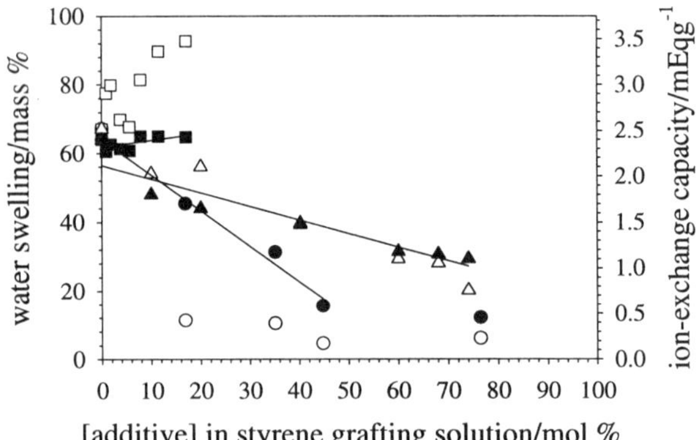

Fig. 5 Water swelling (open) and IEC (solid) of membranes prepared using (□) TAC, (O) DVB, or (Δ) benzene in the grafting solution.

In contrast, TAC increases the grafting level, IEC, and swelling. Some of the increase in swelling may be due to the higher grafting level. It is proposed here that TAC may additionally increase the swelling level by decreasing the grafted chain length and thus the extent of physical crosslinking in the membrane. So TAC can be used to increase the grafting level and thus IEC of the resulting membrane, but it does not appear to be effective in reducing swelling. The effect of TAC on the mechanical properties of our radiation-grafted membranes has not yet been quantified.

The use of both DVB and TAC together in the grafting solution makes it possible to prepare crosslinked membranes[2] having higher ion exchange capacities (1.8-2.4 mEq/g) and lower swelling levels in water (22-44 mass %) than those of the commercially available Nafion® membranes. In addition, they compare quite favorably[2] to Nafion® membranes in terms of achievable maximum power densities in fuel cell tests.

Acknowledgments

The financial support of the *Swiss Federal Office of Energy* (BEW) is gratefully acknowledged. Sample irradiation by C. Günthard of *Studer AG*, CH-4658 Däniken and sample preparation and characterization by L. Pillonel of the *Swiss Federal Institute of Technology* CH-8092 Zürich, Professor D. Jahn and O. Schweizer of *Ingenieurschule beider Basel* (IBB) CH-4132 Muttenz, and D. Begel of the *Paul Scherrer Institut* (PSI) are all gratefully acknowledged.

References

1. Zawodzinski, T.A.; Derouin, C.; Radzinski, S.; Sherman, R.J.; Smith, V.T.; Springer, T.E.; Gottesfeld, S. *J. Electrochem. Soc.* **1993**, *140*(4), 1041.
2. Brack, H.P.; Büchi, F.N.; Huslage, J.; Rota, M.; Scherer, G.G. submitted to *Membrane Formation and Modification*, ACS Symposium Series, presented at the ACS National Meeting, Las Vegas, NV, Sept. 1997.
3. Tsukada, A.; Büchi, F.N.; Scherer, G.G. unpublished results.
4. Zawodzinski, T.A.; Davey, J.; Valerio, J.; Gottesfeld, S. *Electrochimica Acta* **1995**, *40*(3), 297.
5. Büchi, F.N.; Gupta, B.; Haas, O.; Scherer, G.G. *J. Electrochem. Soc.* **1995**, *142*(9), 3044.
6. Gupta, B.; Büchi, F.N.; Scherer, G.G. *J. Polymer Sci: Part A: Polymer Chem.* **1994**, *32*, 1931.
7. Brack, H.P.; Scherer, G.G. *Macromol. Symp.* **1997**, *126*, 25.
8. Schildknecht, C. E. In *Encyclopedia of Polymer Science and Engineering,* 2ⁿᵈ ed.; Mark, H.F.; Kroshwitz, J.I.; Eds.; Wiley: NY, 1985, Vol. 4, p.779.
9. Gupta, B.; Scherer, G.G. *Angewandte Makromol. Chem.* **1993**, *210*(3655), 151.

Water swellable films made from xylan and chitosan

I. Gabrielii and P. Gatenholm, Chalmers University of Technology, Department of Polymer Technology, S-412 96 Göteborg, Sweden

Introduction

Polymeric materials that absorb large amounts of water or biological liquids are of great interest as they are used in the rapidly growing field of health care products. In these applications, biopolymers are very attractive since they are biodegradable and renewable.

Hemicelluloses, even though they are as abundant as cellulose in many plants, have not been commercially utilised more than to some extent as a sizing agent in paper and as a source of xylitol.[1] In hardwood, such as birch or aspen, up to 30% of the mass may consist of hemicellulose. In contrast to cellulose, hemicelluloses are usually water soluble owing to the heterogenous structure of hemicelluloses. Many hemicelluloses are branched, composed of a mixture of different sugars and have various substituents.[2] Recent research shows that many biopolymers may possess properties interesting for the biomedical field. For example, hemicelluloses have been shown to work as antitumour agents,[3] and chitin and chitosan, recovered from for example fungi and crustacean shells, have been used for wound care and drug delivery.[4]

We recently showed that xylan from birch, together with chitosan, forms hydrogels that exhibit extensive swelling properties.[5] The polysaccharides were mixed together in dry form, dissolved in weak hydrochloric acid solution and dried into films. When immersed in water, the films absorbed large amounts of water and formed hydrogels. The amount of water in the hydrogels was examined at different xylan/chitosan compositions and was shown to depend upon pH and salt concentration.

In this study, attempts are made to correlate structures of hemicellulose and chitosan with water swellability of the films. The xylan was therefore mixed with chitosans of various molecular weight and degree of deacetylation (DDA).

Experimental

The materials used for film preparations were xylan, separated from birchwood (SIGMA, Germany, product no. X-0502) and various chitosans from crustacean shells (Fluka, Switzerland, product no. 22743, and three chitosan samples, Primex, Norway, TD12, TD27, TD35). All materials were used without further purification.

Solutions were made from xylan with various amounts of the chitosans by mixing in dry form and stirring into water that was acidified with hydrochloric acid. They were heated to 95°C and kept at this temperature for 20 minutes. After being cooled to room temperature, the solutions were casted onto polystyrene dishes and dried into films in ambient conditions. A schematic picture of the film preparation is shown in Figure 1. The xylan used in this study is a 4-O-methyl glucuronic acid-xylan and contains 4,9% uronic acid groups pendant to the main chain. The chitosans used varies with regard to the degree of deacetylation and molecular weight.

Swelling measurements were performed by placing small pieces (about 1.5x1.5 cm) of films in glass dishes of known weight. Deionized water (25 ml.) was added, left for 30 minutes and then carefully removed, leaving the sample on the dish. The dish was dried with a Kleenex and weighed. The swelling ratio (S) was defined as: (Weight of hydrated gel - Weight of dry gel)/(Weight of dry gel). The results were plotted as mean averages of at least five samples with 95% confidence intervals.

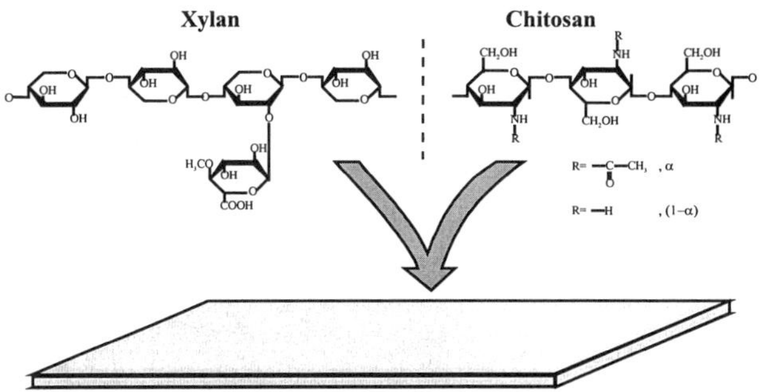

Figure 1 Preparation of xylan/chitosan films

FTIR spectra were recorded using a Perkin Elmer System2000 FTIR instrument with a chamber for purging with dried air. The KBr technique was employed, and samples were dried in vacuum for 12 hours at 50°C before analysis. Spectra were recorded at 4000-750 cm^{-1}. The degree of deacetylation of the chitosans was calculated from the peak height of the amide I band at about 1655 cm^{-1} using the 3450 cm^{-1} absorption band as an internal standard.[6]

Viscosity measurements of 1% solutions of the chitosans in 1% acetic acid were made at 22°C using a Brookfield RVT viscometer with spindle number 2.

Wide-angle X-ray spectroscopy was performed using a Siemens D5000 diffractometer in the reflection geometry. Diffractograms were taken between $2\theta=5°$ and $2\theta=30°$ with a step size of 0.1°. Before analysis, the films were milled into a fine powder under liquid nitrogen and kept at ambient conditions for three days. The relative crystallinity was calculated by dividing the area of the diffraction peak at $2\theta=17°-21°$ by the total area under this peak.[7]

Results and Discussion

The film-forming properties of the solutions were found to be dependent upon the ratio between the amounts of xylan and chitosan. A pure xylan solution did not form continuous films but rather film fragments. The addition of 5-10% and more of chitosan and resulted in the formation of films. The swelling properties of these films were studied, and the results of one series of experiments are shown in Figure 2.

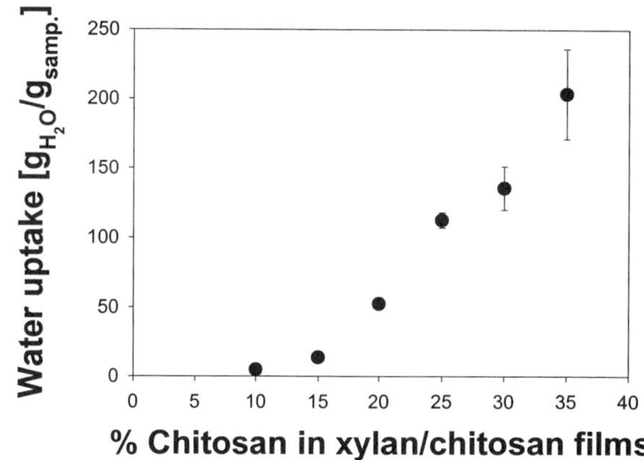

Figure 2 Swelling properties of xylan films with various amounts of chitosan.

An increased content of chitosan resulted in an increased swelling capacity of the films. Films with more than 35% of these chitosans start to slowly dissolve during measurement, and no reliable values were obtainable. Films of pure chitosan solutions were water soluble. Large variations in swelling properties were observed among films based on various chitosans.

The effect of the molecular weight of chitosans on film formation, as well as on the swelling properties of the films, was investigated. Figure 3 shows the water uptake of a series of xylan/chitosan films as a function of the viscosities of the chitosans used, measured as described in the experimental part. It is clear that the films based on chitosan with higher viscosity, which is related to higher molecular weight, are able to absorb more water. All films in the figure consist of 75% xylan and 25% chitosan.

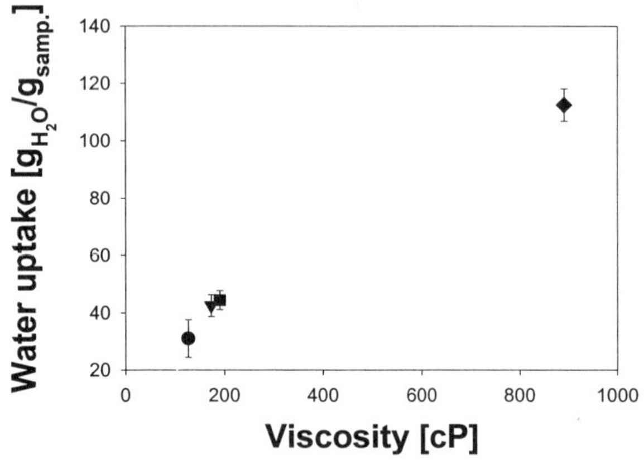

Figure 3 Swelling of films of xylan with 25% chitosan vs viscosity of the chitosans

The chitosans used in this paper were examined by FTIR, and the degree of deacetylation, i.e. the amount of free amino groups, was calculated from the height of the amide I band at about 1655 cm^{-1} using the 3450 cm^{-1} absorption band as an internal standard.[6] Experiments performed by other authors indicate that complexation between the amino groups in chitosan and carboxyl groups in xanthan is responsible for gel formation.[8] To investigate wether interactions between the acidic groups in the xylan and the amino groups in the chitosan affect the properties of the films above, we measured the swelling of xylan films with 25% of the various chitosans and plotted this against the degree of deacetylation of the chitosans as shown in Figure 4.

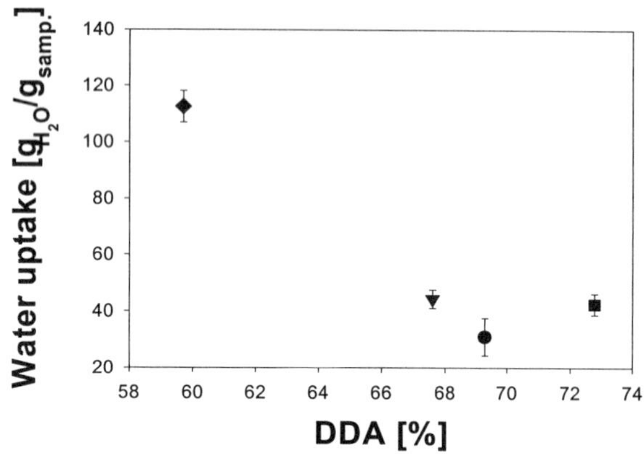

Figure 4 Swelling of films of xylan with 25% chitosan vs DDA of the chitosans as calculated from IR spectra

It seems that at a lower degree of deacetylation, i.e. fewer free amino groups, the swelling of the films is higher.

The morphology of the films was examined with wide-angle X-ray spectroscopy. X-ray diffractograms of milled films of pure xylan, 75% xylan and 25% chitosan, and pure chitosan were recorded at 5-30° (2θ) and are shown in Figure 5. The relative crystallinities of the different samples were calculated from the area under the peak from 2θ=17° to 2θ=21° For a pure xylan film, the value was 0.42 while the pure chitosan film had virtually no crystallinity. The film of a mixture of xylan and chitosan had a relative crystallinity of 0.43. The fact that the crystallinity of the film with only 75% xylan is about the same as the pure xylan film, together with an increase in the area of the peak at 2θ=15°, indicates that chitosan may promote an increase in the degree of order of xylan in the films. Another explanation may be that xylan and chitosan are able to cocrystallize. Since the peaks are not shifted, the criteria is then that the lattice distance is the same as for pure xylan.

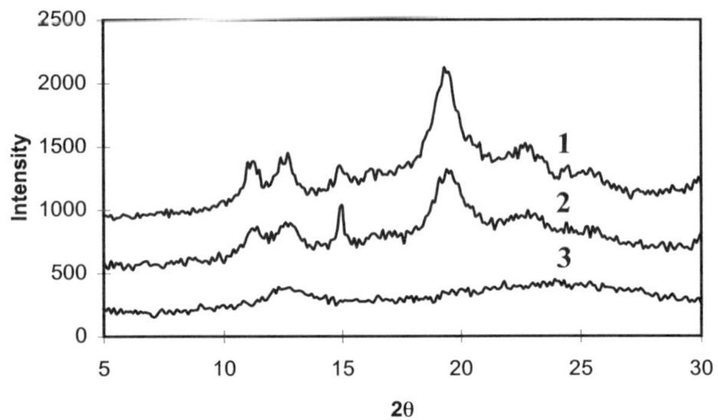

Figure 5 WAXS analysis of films of 1) xylan, 2) 75% xylan and 25% chitosan, 3) chitosan.

Conclusions

Highly swellable films were prepared by casting and drying acidified solutions of xylan from birch and chitosans separated from crustaceans. The swelling properties were dependent on the xylan/chitosan ratio, the molecular weight of chitosan used and the degree of deacetylation of the chitosans. The fact that the films were semicrystalline suggests that, in addition to ionic interactions, physical crosslinks may possibly contribiute to gel formation.

Acknowledgements

The Swedish National Board for Industrial and Technical Development (NUTEK) is gratefully acknowledged for their financial support.

References

1. P. Nigam and D. Singh, *Process Biochem.*, **30** (2), 117-24 (1995).
2. E. Sjöström. *Wood Chemistry, Fundamentals and Applications*, (Academic Press, Inc., San Diego, California, 1981).
3. V. I. Popa, in *Polysaccharides in Medical Applications*, (ed. S. Dumitriu) 107-124 (Dekker, New York, N.Y., 1996).
4. S. Hirano, *et al.*, in *Proceedings of Progress in Biomedical Polymers: Symposium, 196th national meeting* (eds. C. G. Gebelein and R. L. Dunn) 283-290 (Plenum Press, New York,).
5. I. Gabrielii and P. Gatenholm, *J. Appl. Polym. Sci., in press,* .
6. J. G. Domszy and G. A. F. Roberts, *Macromol. Chem.*, **186**, 1671-1677 (1985).
7. P. H. Hermans and A. Weidinger, *J. Appl. Phys.*, **19**, 491-506 (1948).
8. S. Dumitriu, *et al.*, *J. Bioact. Compat. Polym.*, **9** (2), 184-209 (1994).

Effect of Water in Hydrogels on the Laser Damage Threshold

Hao Jiang, Weijie Su, Mark Brant#,Michael DeRosa and Timothy Bunning#*

AFRL/MLPJ
WPAFB, OH 45433-7702
*Anteon Co., Dayton, OH
#Science Applications International Corporation, Dayton, OH

Introduction

The importance of lasers for industry, research and defense applications is now well recognized. Since the mid 1960s, in laser application, there has been growing interest in using transparent polymer materials, such as polymethyl methacrylate (PMMA). However, laser devices made from plastics are still readily damaged because of their low laser damage threshold (LDT). In order to take full advantage of polymer materials, in particular their relative low cost, excellent processability and capability of chemical modification, a great deal of research work has been done to combat this fundamental problem[1,2].

Recently, we have used polymer hydrogels instead of polymer solids in guest-host systems for optical limiting[3]. It is interesting that two chitosan hydrogel systems show very high laser damage threshold for 7 ns laser pulses at 532 nm. We also observed that the water in hydrogels plays an important role of enhancing laser damage resistance.

Experimental

Materials. Two kinds of hydrogels have been prepared: (1) chitosan/acetic anhydride system (referred to as "original chitosan gels"), and (2) acrylate/chitosan system (referred as to "modified chitosan gels"). Chitosan was purchased from Fluka with low molecular weight (MW: about 70,000).

The general procedure for preparation of chitosan/acetic anhydride gels, and a detailed discussion on the relationship between chitosan, acetic acid solution, methanol and acetic anhydride have been reported elsewhere [3,4].

To form the modified chitosan gel system, 2 to 5 ml 2-hydroxyethyl acrylate (HA) (96%, Aldrich) was first mixed with 30 to 100 mg N-N'-methylene bisacrylamide (MB, crosslinking agent) (99%, Aldrich). With stirring, 3 to 20 ml water and 2.5 to 5 mg VA44 (initiator, Wako) were added into the mixing solution. The solution was put in a water bath with the temperature controlled at 50 to 70 °C depending on the requirements. 1 to 5 g (2%) chitosan solution and 10 to 100 µl (0.1%) glutaric dialdehyde (Polysciences, as chitosan's crosslinking agent) water solution were immediately added into the mixed solution. Hydrogels formed in 10 to 120 minutes. The theoretical (calculated according to mole ratio) crosslink densities for the samples with high-crosslinking-density (sample #mg3) and with low-crosslinking-density (sample #mg4) are 2.5% and 0.75%, respectively.

Instrumentation.

Laser Damage Threshold Test. We tested our samples (including PMMA, milli-Q water and two chitosan gel systems, with 1mm cell) by using a 532 nm, 6.8 ns (FWHM) 10 Hz pulsed laser beam in a tight focus <f/5> geometry. A slit of 6.5 ± 0.1 µm width was scanned across the beam to measure a $1/e^2$ beam radius of 7±1µm in air when corrected for slit convolution (shown in **Fig. 1**). Measurement of the beam profile was deconvolved to be approximately Gaussian. It is difficult to compare bulk laser damage threshold values from different laboratory test setups since laser damage in materials can depend on a number of factors including wavelength, pulse width and spot size. Therefore, in order to make a direct comparison of the laser damage resistance of different materials, we use the same geometry and test procedure to measure the laser damage threshold (E_D). The damage values relative to PMMA and milli-Q water are reported rather than absolute values of damage fluence since it is difficult to calculate the actual fluence at the focus in the sample at high incident energies due to thermal beam distortion. During the damage test, we increased the energy and repeated the 10 pulse test in the same site, until we reached the point at which a sharp irreversible drop in transmission was observed during the 10 pulse exposure. Due to the highly statistical nature of laser damage in polymer materials, this test was repeated in 10 different sites within the sample to calculate an average damage energy $<E_D>$.

Differential Scanning Calorimetry (DSC) Characterization. Thermograms were obtained using a differential scanning calorimeter fitted with a liquid nitrogen sub-ambient accessory for low temperature measurements (Thermal Analysis 2000 system. In the tests, about 1 mg of the hydrogel was used and sealed in a aluminum hermetic sample pan. Samples were first kept at 15 °C for about 5 mins, and then cooled down to -40°C at a rate of 2.5 °C/min. The samples were maintained at -40 °C for 2 mins and then heated at the same rate back to 15 °C. The thermogram for each sample was repeated to ensure reproducibility in the recorded data.

Results and Discussion

Table 1 lists the average laser damage threshold data for the milli-Q water, the commercial PMMA (Polysciences, Inc. and Acrylite, AR), and the two polymer hydrogel systems. We have included milli-Q water and the commercial PMMA for calibration and comparison. The water has the highest damage threshold, about 632 µJ and PMMA has the lowest threshold, about 16 µJ. This measured bulk damage threshold of PMMA is higher than the typically reported values (1 - 3 µJ) and the difference may be due to the effect of measuring damage with extremely small spot sizes [5]. It is very interesting that the hydrogels emerge as a unique and novel material with very high laser damage resistance. Both of the hydrogel systems are transparent and colorless. The #og1 is a sample of the original chitosan hydrogel system. The small molecular components (methanol and water) in this sample occupy over 98% of the weight. It has the highest damage threshold among the hydrogels, 580 µJ, close to the milli-Q water. The laser damage resistance is 35 times higher than that of the commercial PMMA bulk material. The modified chitosan gel samples also have high LDT, between 18 to 28 times higher than the PMMA. It is also interesting to note that the LDT varies with the content of the water in the gels. The hydrogel with maximum water content has the maximum LDT. For the modified hydrogels, with the same water content (75%), the damage threshold was slightly decreased with increasing crosslink density.

Since the water has been observed to affect the LDT, it is interesting and necessary for us to understanding its characteristics in the hydrogel systems and its role in dictating laser damage resistance. Polymeric hydrogels are composite materials. Their macromolecular components construct a network which compartmentalizes the water (as the small molecule component)[3]. To examine the nature of the water within these samples, the freezing characteristics of water was measured upon cooling by using DSC. Fig. 3 shows the thermograms of the hydrogel samples. The water in the hydrogels is in a different environment from bulk water. The macromolecular chains are not only able to compartmentalize the water, but also to bind water through hydrogen bonding. Since the polymer is hydrophilic, a part of the water associates with the hydrophilic moieties on the polymer network segments. Therefore, inside a hydrogel network, the states of water may be rather complicated. The water may exist in a continuum of states between two extreme cases: strongly associated through hydrogen bonding with the polymer chains and nearly unaffected by the polymer environment[6]. According to the DSC results, we may simply and qualitatively delineate the water as two types: the water is less associated (LAW) and water is strongly (SAW).

Table 1 Comparison of damage threshold of Different Materials.

Materials	Water Content (%)	Calculated Crosslink Density (%)	Average Damage Threshold (mJ)
#H₂O: milli-Q water	100%	-	632 ± 3
#og1: chitosan/acetic anhydride system	98%	-	580 ± 129
#mg2: modified gel (MG) with much water	86%	~ 1.25	450 ± 81
#mg3: MG with high crosslinking density	75%	~ 2.5	310 ± 56
#mg4: MG with low crosslinking density	75%	~ 0.75	320 ± 46
#mg1: MG with less water	50%	~ 1.25	290 ± 73
PMMA	-	-	16 ± 10

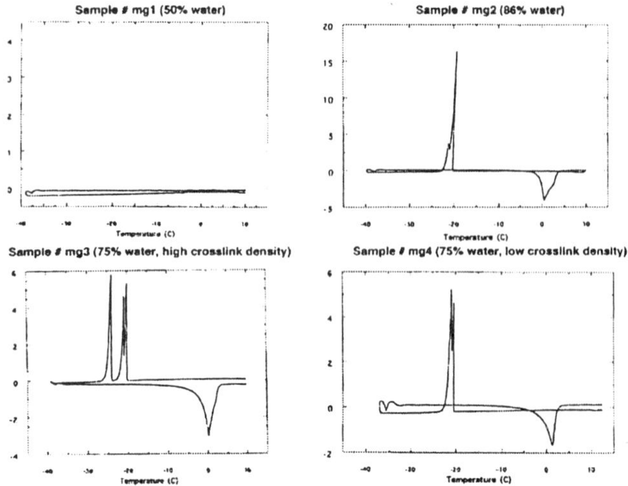

Figure 1 DSC Thermograms of Hydrogels

In sample #mg2, there is a comparatively large amount of water (85%). As shown in its thermogram (Fig. 1), the leading features are the two exothermic peaks at -20.7 °C and -21.5 °C, caused by the crystallization of the water after a long supercooling period. We associate the higher temperature peak to the phase transition of LAW and the lower temperature peak to the phase transition of part of SAW. The ratio of LAW to SAW (according the exothermic peak ratio) is about 2.3. Sample #mg4 contains 75% water with less crosslinking density. Its DSC thermogram (Fig. 1) shows that it has less LAW than #mg2, the ratio of LAW/SAW being 0.4. Sample # mg3 also contains 75% water, but has a higher crosslinking density. The LAW/SAW ratio remains about 0.4. However, another SAW crystallization peak is observed at even low temperature (-24.7 °C) (Fig. 1). This implies that with increasing crosslinking density, the inhomogeneity in the gels increases and different domains develop. A higher crosslink density may cause the formation of some very narrow, fixed polymer interstices which could cause an even longer super-cooling period for the water inside to undergo a phase transition. Sample #mg1 contains the least amount of water (50%) but has the same crosslink density as sample #mg2. Its thermogram (Fig. 1) shows no sign of water crystallization. Thus, as the water content decreases, not only is there a decrease in the amount of LAW, but the SAW becomes so tightly bound as to never crystallize. The fraction of nonfreezing water is controlled by a number of equilibrium and kinetic factors. The increase in difficulties of ice formation is not only caused by less nucleation but also by a substantial portion of the water in the boundary, or interfacial region [7].

Apparently, the water states identified by DSC thermograms can be closely related to the laser damage threshold because the existence of non-freezing and freezing water has a direct bearing on the mobility of the water molecules. As described above, hydrogels can be considered as a composite composed of statistically distributed micro-channels and/or fluctuating pores created by the movements of polymer chains (and segments) within the interpenetrating network in the presence of water. Although the water in the hydrogel can not move as freely as in its bulk material due to the restriction of the macromolecules, some water is able to transparent within and from cell to cell relatively freely. When a hydrogel is irradiated, the energy in the sample generated by the laser light can be absorbed and dispersed by the water and the polymer frame. If the sample contains plenty of water, most of the energy will be primarily absorbed by the water, because at the laser focus water occupies a much higher percentage of the volume than the polymer network. The water flow caused by molecular micro-Brownian movement can transport the a large amount of thermal energy, since water has a much larger inherent mobility for kinetic dissipation of energy than the macromolecule diffusion. The thermal conductivity of water is approximately 0.014 cal/°K·cm·sec [8] whereas the thermal conductivity of the average amorphous polymers is only about 0.0004 cal/°K·cm·sec [9], about 35 times less than that of water. Therefore, the flow of water among the cells in hydrogels would lead to more energy dissipation from the local irradiated area than bulk amorphous polymers. Furthermore, although water at the laser focus spot would be vaporized and/or decomposed by the laser energy, the surrounding liquid may quickly fill the micro-vacancies This inherent of water mobility allows self-healing in samples where enough energy is released to cause local vaporization. Another factor in the increased energy dissipation is that the water is able to act as a plasticizer. Water molecules may contribute in a twofold way to "plasticize" the polymer frame in hydrogels: (1) through their association with the functional groups on the polymer chains, at higher water contents, and (2) through the relaxation of the water molecules themselves in the solid state [5]. Surrounded by a large amount of small molecules, macromolecules can more easily transfer energy to the neighboring small molecules than they can to other macromolecules. This plasticization increases the effective thermal diffusivity of the polymer allowing for more local heat dissipation.

It has been reported that with increasing water content, not only the amount of LAW increases, but also the molecular mobility is increased [10]. Vice versa, if water content is decreased, the water molecules would be significantly restricted by the slow relaxing polymer matrices and the mobility of the water molecules would be significantly decreased. This reduction leads to a decrease in laser damage resistance as shown by the case of sample #mg1. Samples #mg3 and #mg4 have the same water content (75%) but different crosslink densities (the ratio of calculated crosslink density of #mg3:#mg4 = 2.5%:0.75%). Although the difference in crosslink density is about 3 times, ratios of LAW/SAW for both samples are the same (0.4 for both samples). For #mg3, there are 2 exothermic peaks, but for sample #mg4, only one peak. The first peak of #mg3 has the similar crystallization temperature to that of #mg4, but the second one has even lower crystallization temperature. This suggests that the restriction in water mobility is enhanced to some degree by increasing cross-linking density, which may reflect to the small difference between their LDT. However, compared to the effect of water content, the cross-link density seems to have comparatively less influence if there exists a certain amount of LAW in the gels. In addition, although both samples #mg4 and #mg3 contain about 75% water their LDT is only about half of that of the milli-Q water sample. Apparently the hindrance of the polymer network on the water mobility affects the LDT values.

Conclusions

Hydrogels appear to have a great potential to be a new generation materials for high laser damage applications. It collects the merits of polymer solids and liquid solutions. In a hydrogel, the polymer frame contributes to the strength (mechanical properties) and small molecular components (such as water) contribute to the molecular mobility to dissipate the laser energy and damage recovery. Based on this work, other solvents can be used as the small molecular components and other polymer systems can be chosen to construct the frame to make different gel systems in order to achieve greater LDT and/or to meet the special requirements.

Acknowledgements

We are much indebted to Dr. P. Fleitz, SAIC, Dr. Y. Z. Xu, Ciba Performance Polymer, Dr. A. Schmid, LLE, Rochester Univ., and Professor Z. Q. Wu, the Akron Univ. for their valuable insights and helpful discussions.

References

(1) O'Connell, R. M.; Saito, T. T., *Optical Eng.* **1983**, *22*(4), 393.
(2) Rajasree, K.; Radhakrishnan, P.; Nampoori, V. P. N.; Vallabhan, C. P. G. *Measurement Sci. Technol.* **1993**, *4*(5), 591.
(3) Jiang, H.; Su, W.; Brant, M.; Tomlin, D.; Bunning, T. *MRS Symposium proceedings* **1997**, *479*, 129.
(4) Roberts, G. A. F.; Taylor, K. E. in *Chitin and Chitosan*, Skjak-Braek, G.; Anthonsen, T.; Sanford, P., Eds., Elsevier, London **1989**, p.577.
(5) Bondar, M. V.; Przhonskaya, O. V.; Tikhonov, E. A. *Sov. Phys. Tech. Phys.* **1988**. *33*(3), 309.
(6) Quinn, F. X.; Kampff, E.; Smyth, G.; McBrierty, V. J. *Macromol.* **1988**, *21*, 3191.
(7) Bouwstra, J. A.; Salomons -de Vries, M. A.; Van Miltenburg, T. C. *Thermochimica Acta* **1995**, *248*, 319.
(8) Weast, R. C.; Astle, M. J. *"C. R. C. Handbook of Chemistry and Physics"*, 62th Ed., CRC Press, Baca Raton, Fl. **1981**.
(9) Daniels, C. A. *"Polymers: Structure and Properties"*, Technomic Pub., Lancaster, PA. **1989**.
(10) Kyritsis, A.; Pissis, P.; Granmuatikakis, J., J. *Polym. Sci., Pt. B., Polym. Phys.* **1995**. *33*, 1737.

KINETICS OF MASS TRANSFERS IN POLYMERS WITH SWELLING AND CHANGE IN DIFFUSIVITY.

N. Bakhouya, A. Sabbahi and J.M. Vergnaud.
Laboratory of Materials and Chemical Engineering
Faculty of Sciences, University of Saint-Etienne,
42023 Saint-Etienne, France.

INTRODUCTION

When a polymer is in contact with a liquid generally one or several mass transfers take place. Depending on the nature of the polymer and liquid, the amount of liquid transfer is so large that a change in dimension is observed. This fact especially appears in the case of rubbers with hydrocarbons or hydrogels with water.

Various studies have been made on the diffusion of the liquid through a polymer provoking either a swelling in case of absorption or shrinkage[1] in case of evaporation. The swelling properties of the polymer can find interesting applications in pharmacy for controlled-release polymer system [2], as well as in civil engineering. The effect of the swelling of the polymer on the rate of drug release was thus extensively studied, and two dimensionless numbers were introduced[3].

Basic studies were made on the change in dimension associated with the transport of liquid by considering a constant diffusivity. The general theory considering both the diffusion and the change in dimensions lead to general equations of diffusion, which reduced to Fick's equation in the particular case of a low amount of liquid transferred[4]. Another fact of great concern appears with the diffusion of the liquid and the concentration-dependency of the diffusivity[5]. The drying of paints made of a polymer dispersed in a solvent was explained by diffusion-evaporation with a concentration diffusivity[6]. In the same way the rate of transfer of a small amount of benzene through an EVA bead presaturated with hexane was considerably higher than that in the pure polymer[7].

The first objective is to study the process of absorption of a liquid by a polymer and the following stage of desorption by considering not only the change in dimension of the polymer resulting from the liquid transport but also the change in the diffusivity. A method can thus be described in order to determine the main parameters, namely, the coefficient of convective transfer in the liquid during the stage of absorption and the coefficient of convective transfer of the vapour during the stage of evaporation, as well as the diffusivity and its concentration-dependent diffusivity. Of course, it is possible to obtain the values of the diffusivity through the polymer free from liquid at the beginning of the stage of absorption, and when it is saturated with liquid at the beginning of the stage of drying.

The other aim at the study is to build and test a numerical model taking into account the transfer of liquid, its concentration dependent diffusivity and the resulting change in dimension.

THEORETICAL

ASSUMPTIONS
i) The polymer is homogeneous and spherical in shape with a radial diffusion

ii) The volume of the bead is constantly equal to the sum of the volumes of the polymer and liquid.

iii) The stage of absorption is controlled by diffusion of the liquid through the polymer with a finite convective coefficient on the polymer surface. The stage of desorption is controlled by diffusion-evaporation with a finite convective coefficient on the polymer surface.

iv) The diffusivity is concentration-dependent, in an exponential way[1,2].

MATHEMATICAL TREATMENT
Each position in the polymer bead is defined by two radial abscissae : u when the bead is free from liquid, and $r = f(u, t)$ which varies with u the amount of liquid in the polymer at time t.

A membrane of radius $r(u, t)$ of initial thickness du has a volume of polymer $V(u, polymer)$ and a total volume $V(u, t)$ with the liquid in it :

$$V(u, t) = V(u, polymer) + V(u, t) . C(u, t) \qquad (1)$$

where $C(u, t)$ is the liquid concentration at this radial abscissa $r(u, t)$.

By considering the volumes of the membrane free and full of liquid, equ. becomes :

$$\frac{V(u,t)}{V(u, polymer)} = \frac{r^2 dr}{u^2 du} = \left[1 - C(u,t)\right]^{-1} \qquad (2)$$

leading to the linear expansion :

$$\frac{dr}{du} = \frac{u^2}{r^2}\left[1 - C(u,t)\right]^{-1} \qquad (3)$$

The liquid balance within the membrane during lapse of time dt by considering the volume of liquid which enters and leaves the membrane lead to :

$$\frac{\partial}{\partial t}\left[V(u,t) . C(u,t)\right] = \frac{\partial}{\partial u}\left[A . D . \frac{\partial C}{\partial r}\right] \qquad (4)$$

By replacing the area A and the volume $V(u, t)$ of the membrane by their values in equ. 4, it comes :

$$\frac{\partial}{\partial t}\left[C(1-C)^{-1}\right] = \frac{1}{u^2}\frac{\partial}{\partial u}\left[r^2 D . \frac{\partial C}{\partial r}\right] \qquad (5)$$

By expressing $(\partial C / \partial r)$ in terms of $(\partial C / \partial u)$ and considering the linear expansion, equ. 5 is rewritten as :

$$\frac{\partial}{\partial t}\left[C(1-C)^{-1}\right] = \frac{1}{u^2 . }\frac{\partial}{\partial u}\left[\frac{r^4}{u^2} D(1-C)\frac{\partial C}{\partial u}\right] \qquad (6)$$

Equs. 3 and 6 are the general expression for the problem of radial diffusion through a sphere with a change in dimension. Of course, when the volume of liquid is small, $(1 - C)$ tends to 1 and equ. 6 reduces to the decond Fick's equation for radial diffusion established with constant dimension.

The boundary condition

$$-D . \left(\frac{\partial C}{\partial r}\right)_{R,t} = h\left(C_{R,t} - C_{ext,t}\right) \qquad (7)$$

expresses that the rate of transport is the same on each face of the polymer surface either by internal diffusion or by convective transfer with the convective coefficient h.

No analytical solution exists because of the change in dimension and the concentration-dependent diffusivity, and the problem is resolved using the Crank-Nicolson method.

EXPERIMENTAL

Spherical beads of ethylene vinyl acetate (EVA) from Atochem, type 33.45 (33% vinyl acetate - melt index 45) of radius 0.1919 cm are used.

Each bead is immerse in n-hexane at 20°C and weighed at interals. The presaturated bead is allowed to dry in motionless air at 20°C, and the kinetics is followed by recording the weight. The rate of evaporation of the pure liquid is determined under the same conditions of temperature and air stirring.

RESULTS

The kinetics of absorption and desorption are determined experimentally. The parameters, such as the convective coefficients during the stages of absorption and desorption, as well as the diffusivity at the beginning of these two process are evaluated.

DETERMINATION OF THE PARAMETERS

The convective coefficient for the stage of absorption is determined from the slope of the kinetics of absorption at the beginning of the process.

The convective coefficient for the stage of desorption is obtained from the rate of evaporation of the pure liquid.

The diffusivity D_o at the beginning of the stage of absorption, when the amount of liquid is very small is determined in the usual way[1], as well as the diffusivity D at the beginning of the stage of desorption when the bead is presaturated with a known concentration of liquid. These values D and D_o are put in the relation :

$$D(u, t) = D_o . \exp[K.C(u, t)] \qquad (8)$$

and the coefficient K is obtained by fitting the experimental and calculated kinetics of absorption and desorption. It is thus found :

$$D = 5.2 \times 10^{-8}.\exp[8.63 \times C] \quad cm^2/s$$
with h absorption = 3×10^{-4} cm/s
and h desorption = 6×10^{-5} cm/s

KINETICS OF ABSORPTION AND DESORPTION

The kinetics of absorption and desorption obtained by calculation and experiments are very well superimposed, showing the accuracy of the method for determining the parameters.

The kinetics evaluated by using a constant diffusivity are quite different from the experimental results, showing the importance of the concentration-dependency of the diffusivity. The increase in the diffusivity is around 40 times during the stage of absorption as the concentration of liquid in the saturated bead reaches 42%.

The profiles of concentration of liquid developed through the polymer are also determined by calculation, with the change in dimension : The change in dimension can be followed with a binocular system.

CONCLUSIONS

A method for determining the parameters of liquid transfer in a polymer has been established when the amount of liquid is so high that the polymer swells. The method is based on the study of the experimental kinetics of absorption and desorption. The parameters, namely, the diffusivity and its concentration-depency, the convective coefficients during the stages of absorption and desorption are thus obtained with good accuracy. The accuracy of these parameters can be tested by comparing the kinetics of absorption and desorption obtained either from experiments or calculation.

The numerical model can also assess the profiles of concentration as well as the change in dimension.

BIBLIOGRAPHY

1 - J.M. VERGNAUD
"Liquid Transport Processes on polymeric Materials.
PRENTICE HALL PUBL., ENGLEWOOD CLIFFS, N.J. ; USA ; 1991.
2 - J.M. VERGNAUD
"Controlled Drug Release of Oral Dosage Forms.
E. Horwood Publ., London, 1993.
3 - J. KLIER AND N.A. PEPPAS.
J. Controlled Release, 7, 61, 1988.
4 - N. BAKHOUYA, J. BOUZON AND J.M. VERGNAUD.
Comput. Theoret. Polym. Sci., 3, 119,1993.
5 - N. BAKHOUYA, A. SABBAHI AND J.M. VERGNAUD.
Comput. Theoret. Polym. Sci., 6, 109,1996.
6 - J.M. VERGNAUD.
"Drying of Polymeric and Solid Materials".
Springer Verlag Publ., London, 1992.
7 - A. BENGHALEM AND J.M. VERGNAUD.
Polym. Testing, 13,35,1994.

SOLID-STATE NMR INVESTIGATION OF INTERPOLYMER COMPLEXATION IN SWOLLEN COPOLYMER NETWORKS

Anthony M. Lowman*, Department of Chemical Engineering
Drexel University, Philadelphia, PA 19104

Nikolaos A. Peppas and Brett A. Cowans, School of Chemical
Engineering, Purdue University, West Lafayette, IN 47907

Introduction

The nuclear Overhauser effect (NOE) is the change in intensity of a resonance when another resonance is perturbed (1,2). The change in intensity of the resonances is due to cross-relaxations between spins which are close in space, typically within 5 Å of one another. These spins relax through dipole-dipole interactions resulting in changes in the nuclear spin state populations.

Heteronuclear NOE experiments involve the detection of an enhancement between a proton and another type of magnetic nucleus such as carbon or nitrogen. In these experiments, the proton is saturated with an radiofrequency pulse and the enhancement is observed in the other nuclei. The heteronuclear ^{13}C-^1H experiment is a particularly useful experiment for probing inter- and intramolecular interactions, such as hydrogen bonding, in polymeric systems (3,4).

In this work, heteronuclear NMR-NOE experiments were used to study the molecular level complexation phenomena in copolymer gels of poly(methacrylic acid) (PMAA) and deuterated poly(ethylene glycol) (dPEG). In these gels, interpolymer complexes formed due to hydrogen bonding between protonated acid groups on the PMAA and ether groups on the PEG. In neutral or basic media, these complexes dissociated as the pendant acid groups were ionized. Because of the complexation/decomplexation mechanism, these gels exhibit a large change in their swelling behavior over a very narrow pH range.

Experimental

Sample Preparation. Crosslinked polymer samples of PMAA and dPEG were prepared by free radical polymerization of MAA and deuterated (d-4) poly(ethylene glycol) (dPEG) ($\overline{M}_w$ = 10,162 , $\overline{M}_n$ = 3768, Cambridge Isotopes, Andover, MA). Tetraethylene glycol dimethacrylate was added as the crosslinking agent in the amount of 0.75% moles of per mole MAA.

The copolymer networks were ground into powders and washed with deionized water for 24 hours. Following rinsing, the particles were dried under vacuum. The dried crosslinked particles were swollen in pH = 8 D_2O/NaOD (q = 4) or pH = 2 D_2O/DCl (q = 2). After equilibrating for 24 hours, the hydrogels were loaded into the Pyrex rotor inserts and sealed using epoxy resin to ensure that the sample did not lose any of the swelling fluid during the high speed spinning. The inserts were cut to the required length to fit into the MAS rotors and the ends were smoothed with additional epoxy resin.

NMR Experiments. NMR measurements were performed on a GE Instruments Omega solid-state spectrometer at a carbon resonance frequency of 100.6 MHz. A Doty double-resonance MAS probe was used with 5 mm Zirconia rotors. The experimental spinning speed was ~ 5.5 kHz.

The NOE spectra were obtained using the pulse sequence (3,4) shown in Figure 1. An initial pulse delay was used, followed by a train of 90° proton pulses separated by a delay, Δ. The 90° proton pulse time was 5.0 μs and the delta delay was 5 ms. The sequence of 90° pulses was used to saturate the proton resonances. The observed NOE was strongly dependent on the total number of cycles used, n, or saturation time, τ. The steady-state or maximum NOE value was normally observed for saturation times of greater than 10 s.

Between carbon pulses, a delay was required to allow the carbon magnetization to return to equilibrium. This delay was selected so that the total delay between composite carbon pulses was at least 5 times relaxation time of the slowest relaxing carbon species (3,4). The total delay period was kept constant for all of the experiments.

For each sample, a ^{13}C spectrum was acquired with a specified proton saturation time and without proton saturation. In order to see the NOE, the spectrum obtained for a sample without proton saturation was subtracted from the specrum obtained for the sample with proton saturation prior to signal acquisition. The enhancement was calculated using equation (1).

$$\eta = \frac{I(t) - I(0)}{I(0)} \qquad (1)$$

The intensities were determined by fitting the peaks from the spectra with proton saturation ($I(t)$) and without proton saturation ($I(0)$) to Lorenzian curves.

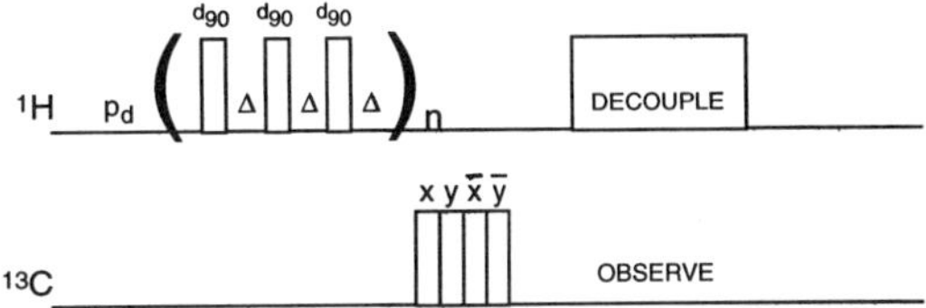

Figure 1 Pulse sequence for solid-state heteronuclear (^{13}C-^1H) NOE measurement.

Results and Discussion

The goal of this work was to determine the molecular level complexation in swollen gels of poly(methacrylic acid) (PMAA) and deuterated poly(ethylene glycol) (dPEG) using NMR NOE spectroscopy. In acidic media, interpolymer complexes formed due to hydrogen bonding between protonated acid groups and ether groups. However, in neutral or basic media, these complexes dissociated allowing for the gel to swell to a high degree due to electrostatic repulsions and the generation of a large osmotic swelling pressure (5,6). In these experiments, intermolecular interactions were detected between the deuterated PEG and PMAA. The molecular level interactions were verified by the presence of an enhancement to PEG carbon due to cross-relaxations between the PEG carbons and the PMAA protons which were close in space (less than 5 Å). The relaxations of the PEG carbons, as well as other nuclei, were dominated by molecular motions from the methyl group rotation. These relaxations occurred only in nuclei along PEG chains that were hydrogen bonded to the PMAA chains.

In the typical carbon spectra for these gels, four peaks were observed. These resonances were attributed to the PEG ethylene carbons (69 ppm), the PMAA methyl carbons (16 ppm), the PMAA methylene carbons (45 ppm) and the PMAA carbonyl carbon (184 ppm). The equilbrium spectra for the gels swollen in pH = 8 at ambient conditions are shown in Figure 2. Only a very small NOE was detected to the PEG (η = 0.002). The small NOE appeared in the PEG resonance due to the protonated ethylene peaks contained in the crosslinking agent (less than 1% of the sample). This resonance overlapped the resonance of the ethylene groups of the deuterated PEG chain. This was verified by the presence of a similar NOE for the PEG for experiments performed on a dried gel of chemically crosslinked PMAA, in which no complexation could occur.

Interpolymer complexation did occur in the chemically crosslinked gels swollen in pH = 2 solutions at ambient conditions and at 37° C. The equilibrium spectra for the gels swollen at 20° C are shown in Figure 3. A significant enhancement was observed in the PEG carbon resonance for all of the saturation times used. The steady-state NOE value of 0.275 was substantially greater than what was observed in the uncomplexed gel. The presence of the PEG NOE was evidence of the presence of interpolymer complexes stabilized by hydrogen bonds for the gels swollen under acidic conditions.

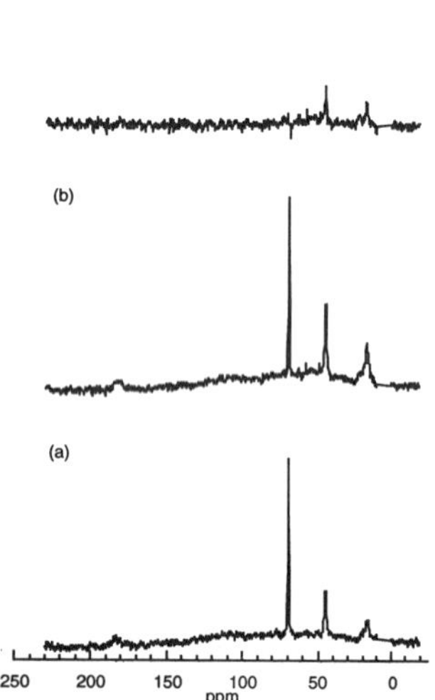

Figure 2 ^{13}C MAS spectra of a crosslinked PMAA/PEG gel swollen in D$_2$O/NaOD (pH = 8) at 20° C (a) without NOE and (b) with NOE. (c) is the difference spectrum.

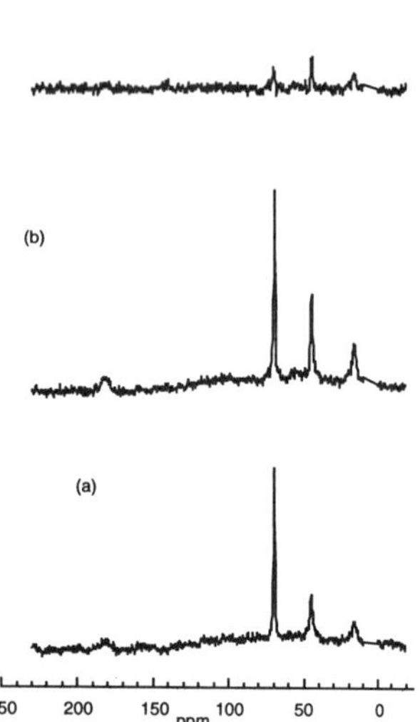

Figure 3 ^{13}C MAS spectra of a PMAA/dPEG gel swollen in D$_2$O/DCl (pH = 2) at 20° C (a) without NOE and (b) with NOE. (c) is the difference spectrum.

Interpolymer complexation is also dependent on the environmental temperature. The stability of the complex

decreases as the temperature is increased. In order for complexing gels of PMAA and PEG to function in biomedical applications, it is important that the complexation occurs in the gels at temperatures mimicking physiological conditions. Therefore, the NOE experiments were also performed at 37° C. For the gels swollen in pH = 8 at 37°, no NOE was detected in the PEG. Under these conditions, complexation did not occur as the pendant acid groups were ionized. For the gels swollen in pH = 2 solutions at 37° C, a significant NOE of the PEG resonance was again observed. The presence of the NOE proved that interpolymer complexation occurred in chemically crosslinked PMAA/PEG gels swollen in acidic media at temperatures simulating physiological conditions. The steady-state PEG NOE for these gels was 0.265.

As stated previously, the NOE is strongly dependent on the proton saturation time. The NOE will build up to a maximum in all species at a rate dependent on the strength of the interactions. The growth curves for the PEG enhancements are shown for the complexed and uncomplexed samples in Figure 5. These data verify the existence/absence of interpolymer complexes in the chemically crosslinked gels. In acidic, media, a strong PEG NOE was observed due to cross-relaxations between the PMAA protons and the PEG carbon and was strong evidence of the presence of interpolymer complexes between the polymers. In basic media, the complexes dissociated due to ionization of the pendant acid groups. As a result, the distance between the chains was increased and no intermolecular NOE was observed between the polymers.

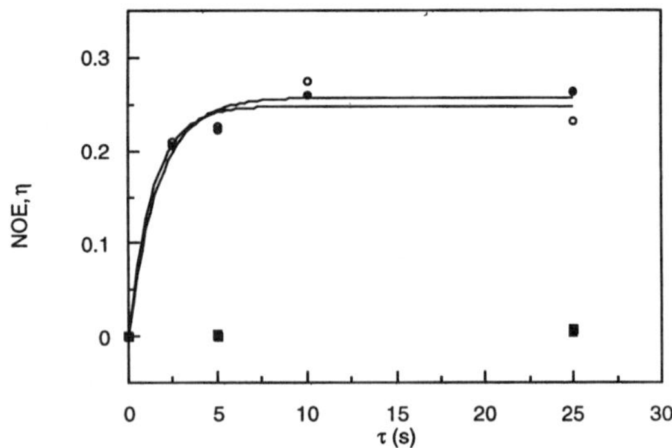

Figure 5 NOE growth for the PEG carbon for PMAA/dPEG gels swollen in (o, ●) D$_2$O/DCl (pH = 2) and (□,■) D$_2$O/NaOD (pH = 8) at 20° C (open symbols) and 37° C (filled symbols).

Acknowledgements
This work was supported by a grant from the National Institutes of Health, No. GM 43337.

References
1. Findlay, A. and Harris, R.K., *J. Mag. Res,* **87,** 605-609 (1990).
2. White, J.L. and Mirau, P.M., *Macromolecules* **26,** 3049-3054 (1993).
3. Noggle, J.H. and Schirmer, R.E., *The Nuclear Overhauser Effect*, Academic Press, New York, 1971.
4. Neuhaus, D. and Williamson, N., *The Nuclear Overhauser Effect in Structural and Conformational Analysis*, VCH Publishers, New York, 1989.
5. Klier, J., Scranton, A.B. and Peppas, N.A., *Macromolecules* **23,** 4944-4949 (1990).
6. Lowman, A.M. and Peppas, N.A., *Macromolecules* 30, 4959-4965 (1997).

MOTION AND STRUCTURE OF WATER AND POLY(HEMA)-BASED HYDROGELS STUDIED BY NMR IMAGING AND SPECTROSCOPY

Andrew Whittaker*, Phuong Ghi, David Hill and Michael Whittaker,
Department of Chemistry, and *Centre for Magnetic Resonance,
University of Queensland, Q 4072, Australia.
Email: andrew@cmr.uq.edu.au

ABSTRACT

This paper reports on a detailed study of the diffusion of water in hydrogel polymers. The hydrogels are being developed for application in areas such as controlled-release of drugs, in intra-ocular lenses, and for water sorption. The materials are copolymers and IPNs based on hydroxyethyl methacrylate (HEMA) and other less hydrophilic monomers, such as tetrahydrofurfuryl methacrylate.

INTRODUCTION

In order to characterize fully the structures of these materials a detailed study of the mechanism of copolymerization was performed. ^{13}C solution-state NMR was used to determine composition, stereochemistry and sequence distributions for the copolymers across the entire composition range. The copolymerizations can be described by the terminal model of copolymerization. The composition of the materials at high conversion could therefore be calculated. At the highest levels of conversion the HEMA monomer is almost completely depleted, resulting in a heterogeneous, but optically-transparent material.

RESULTS AND DISCUSSION

The kinetics of diffusion of water into cylinders of the hydrogels were studied by measuring the increase in mass with time. A detailed analysis showed that the diffusion coefficient was weakly dependent on concentration, as would be expected for a system which undergoes a degree of swelling. The diffusion coefficients decreased with increasing content of hydrophobic monomer, but at a rate greater than expected by considering simply the volume fraction of the comonomer. There is a suggestion that the rates of diffusion are affected also by details of the sequence distribution.

NMR imaging was used to obtain further information on the kinetics of diffusion. The 3D spin echo sequence was used for all experiments, with the echo spacing varied to obtain information on the distribution of T2 relaxation times across the image. At longer times the images showed a profile of water content across the polymer which could be described exactly using Fick's law. At shorter times an additional feature was observed at the boundary between the glassy polymer and the swollen outer part of the cylinder. Diffusion-weighted images showed that the water in this feature must reside in cracks within the polymer. Confirmation of this was obtained using ESEM.

Pulsed field gradient measurements were used to determine the details of the rate of diffusion on a microscopic scale. The plots of echo amplitude as a function of the square of the gradient strength for poly(HEMA) and poly(EEMA) was described by either single- or bi-exponential decay functions, respectively. The self-diffusion coefficient for poly(HEMA) was 1.4x10-10 m2/s, approximately an order of magnitude faster than the macroscopic diffusion coefficient. This suggests the presence of barriers to diffusion occurring approximately on the order every 100 microns. At lower HEMA contents two diffusion processes become apparent, namely diffusion through the matrix, and diffusion at a rate comparable to free water. This latter type of diffusion must occur in large cracks. It is suggested that these are the cracks previously observed by NMR imaging. Diffusion-weighted imaging results are in progress to confirm this.

A final experiment is described which looks at the details of diffusion in the microscopic pores within the materials. A dipolar filter sequence was used to select magnetization in protons in water only. After a suitable mixing time the signal was collected. A fit of the plot of the increase in intensity of the signal from the polymer, as a function of mixing time, to a model of cylindrical pores yielded a pore diameter of 4 nm, separated by domains of polymer of size 10 nm.

The results of the spin diffusion experiment confirm the existence of a porous structure in poly(HEMA) previously suggested on the basis of measurement of diffusion of penetrants of various sizes. At lower HEMA content the size of the domains increases, consistent with a proportion of the water residing in cracks.

REFERENCES

1. P.Y. Ghi, D.J.T. Hill, D. Maillet and A.K. Whittaker, Polym. Commun., 38, 3985 (1997).
2. P.Y. Ghi, D.J.T. Hill, P.J. Pomery and A.K. Whittaker, Polym. Gels Networks, 4, 253 (1996).

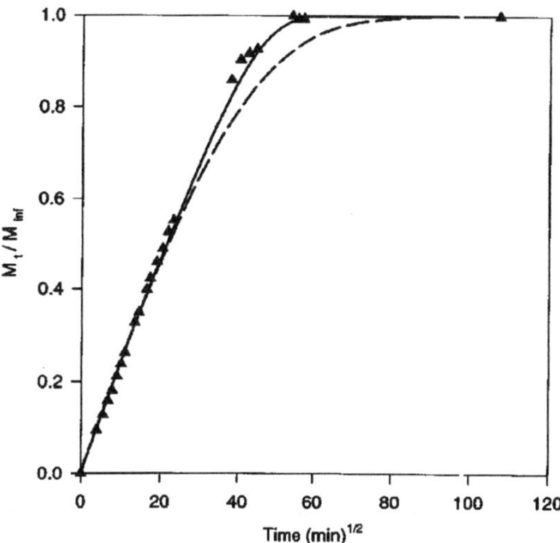

Figure 1. The rate of uptake of water in poly(HEMA) as a function of the square root of time, and the results of fitting the data to the Fickian equation for diffusion, with constant value of D (solid line) and D = Do exp(0.0011 x C) (broken line).

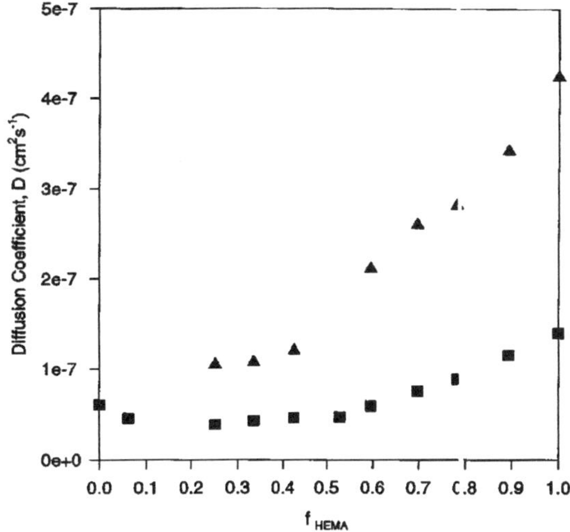

Figure 2. Maximum and minimum values of diffusion coefficients for copolymers of HEMA and THFMA obtained by fitting the data for increase in mass of water with time to a concentration-dependent diffusion coefficient.

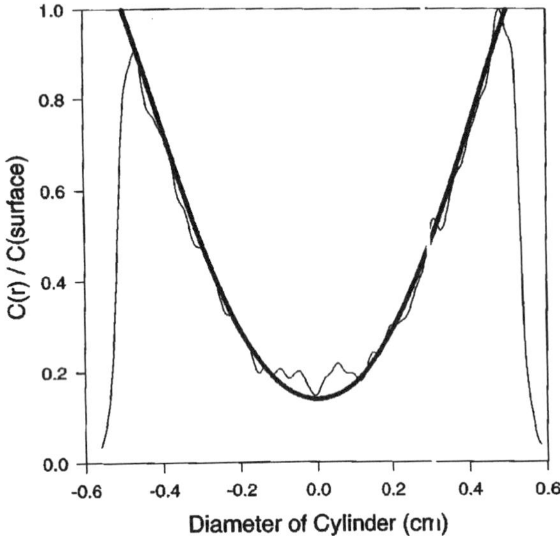

Figure 3.   The concentration of water across a cylinder of a copolymer of THFMA and HEMA (1:9 molar ratio) after immersion in water for 48 hours. The heavy solid line is the result of the fit to Fick's   law   for   diffusion   into   a   cylinder.

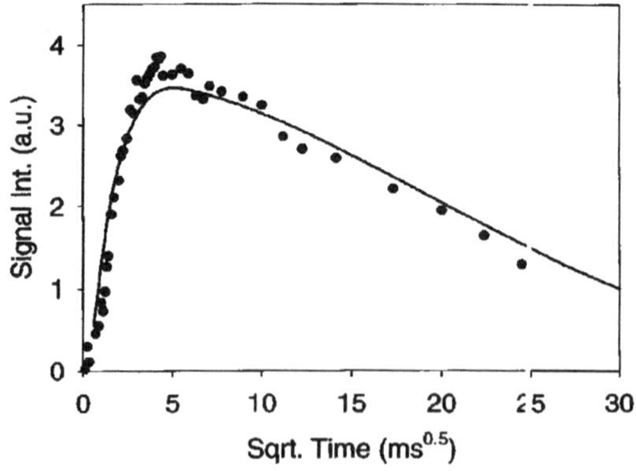

Figure 5.   Plot of signal due to polymer as a function of mixing time in spin diffusion experiment. The solid line is the result of fitting the data to a diffusion model for cylindrical pores of diameter equal to 4 nm.

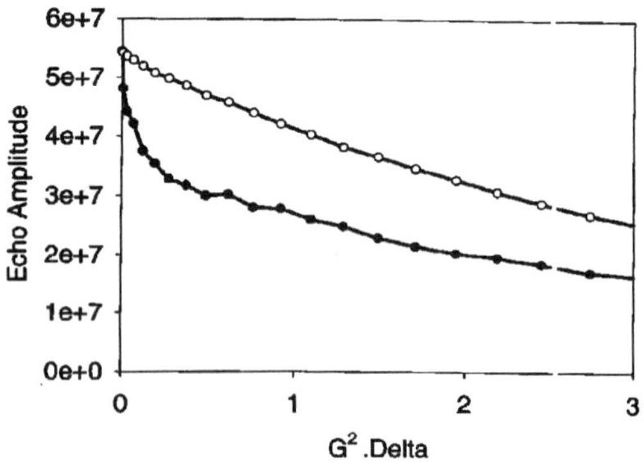

Figure 4.   PGSE echo attenuation plots for poly(HEMA) (open symbols) and poly(EEMA) (filled symbols).

THE EFFECT OF WATER INTERACTIONS
ON THE THERMAL TRANSITION
BEHAVIOR OF PULLULAN

Jo Ann Ratto and Nathan Schneider

U.S. Army Soldier System Command
Natick Research, Development and Engineering
Center, Kansas Street, Natick MA 01760-5020

Introduction

Pullulan is a natural polysaccharide comprised of the α-1,6 linked maltotriose molecules with various applications such as an adhesive, binder, thickener, and edible coating.(1) Pullulan is an ideal model for the study of water / polysaccharide interactions. Since it is amorphous and crystallinity is not a factor, the analysis is simplified. Pullulan has a high stable glass transition temperature so that the characteristic parameters are determined by direct measurement rather than extrapolation. In our laboratory, this material is being investigated as a potential model polymer for permselective membranes for applications with high water vapor transmission and low organic vapor transmission. In this present report, the water/polysaccharide interactions are evaluated by thermal analysis to enhance the understanding of the structure-property relationships as well as the transport properties in pullulan. Also, pullulan behavior is related to and compared with extensive studies on starch/water.

Experimental

Materials Pullulan was supplied by Hayashibara Biomedical Laboratories as a powder. Powders and films were used in this study. Pullulan water solutions at 5 and 10% (w/w) were stirred for several hours at room temperature and films were cast in petrie dishes.

Sample Preparation and Characterization
For differential scanning calorimetry (DSC) experiments, the samples were placed in a hermetic pan and small amounts of distilled water were added to the lids of the DSC pans. The pans were sealed and allowed to equilibrate. At the end of a DSC experiment, the lid of the sample pan was punctured and the pan was heated in the calorimeter at 5°C/min to 200°C to drive off the water. The pans and samples were then weighed to determine the dry weight and the amount of water by difference. The Perkin Elmer DSC 7 was used for all experiments in the temperature range of -40°C to 220°C. The specific heat mode was used to provide an accurate baseline for determining heat capacity changes at the glass transition temperature. All experiments were run at 20°/minute.

The Seiko 210 dynamic mechanical analyzer (DMS-210) was used for the films. The films were equilibrated at several humidities and then transferred to the analyzer. Samples were also dried at 100°C to remove any bulk water. All experiments were run at 4°/minute at a frequency of 1Hz.

Results and Discussion

Depression of the Glass Transition Temperature. The experimentally determined glass transition temperatures as a function of the added water are collected in Table 1.
The diluent depression of the Tg was modeled using the Couchman-Karasz (C-K) relation which requires values of the Tg and ΔCp of both the polymer and the diluent.(2) The Couchman-Karasz equation for Tg mixtures is:

$$Tg = \frac{w_1 \Delta Cp_1 Tg_1 + w_2 \Delta Cp_2 Tg_2}{w_1 \Delta Cp_1 + w_2 \Delta Cp_2} \quad (1)$$

ΔCp_i is the heat capacity change of component i and w_i is the weight fraction of component i.

Table I. Comparison of measured and calculated values for Tg in pullulan

Weight Fraction Water	Measured Tg (°C)	Calculated Water ΔCp = 1.0 J/g-K	Calculated Tg C-K ΔCp = 1.9 J/g-K	Calculated Tg Fox
0.00	178.0	178.0	178.0	178.0
0.08	91.8	91.7	46.3	103.2
0.13	64.0	56.6	6.4	70.7
0.14	57.0	54.6	4.4	68.8
0.14	55.7	53.4	3.1	67.6
0.15	48.0	44.6	-5.8	59.2
0.18	30.5	28.8	-21.1	43.9
0.22	7.8	8.0	-39.8	23.3
0.24	-1.4	-0.5	-47.1	14.7
0.24	0.7	-1.5	-47.9	13.7
0.31	-26.5	-27.4	-68.5	-13.1

To determine the Tg of dry pullulan, the retained moisture was removed by a DSC scan and brief hold at 200°C, using a vented sample pan. The pullulan values determined in repeated measurements were Tg = 451±0.33 K and ΔCp = 0.245±0.006 J/g-K. There is general agreement on a Tg = 134K for solid amorphous water but with two sets of values for ΔCp.(3) The lower set ranges from 0.93 to 1.39 J/g-K, while the higher value is 1.9 J/g-K. Table I contains the results of the C-K calculations for various water contents using both ΔCp = 1.0 J/g-K and ΔCp = 1.9 J/g-K. It is apparent that the C-K calculations with the lower value of water tracks the experimental data closely, while the higher value seriously overestimates the Tg depression. The last column of Table 1 contains the results of the simple Fox equation:

$$\frac{1}{Tg} = \frac{w_1}{Tg_1} + \frac{w_2}{Tg_2} \quad (2)$$

where Tg_i is the glass transition temperature and w_i is the weight fraction of component i. (4) This relation underestimates the depression of Tg. A comparison of the experimental results with the three sets of calculations appears in Figure 1.

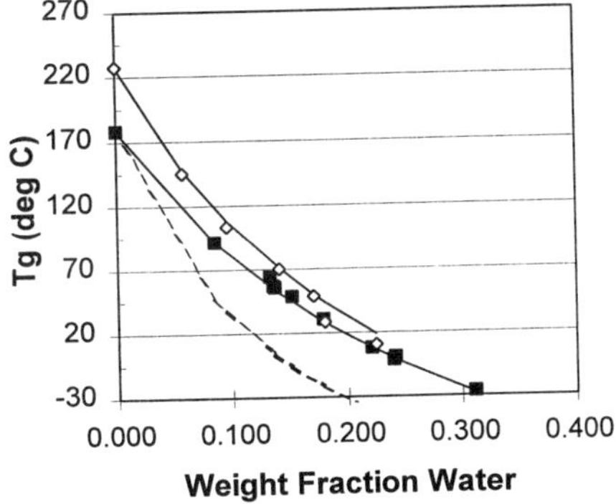

Figure 1. Experimental vs calculated Tg results of pullulan as a function of weight fraction of water with a comparison to starch.
(◼ Pullulan Measured, — Pullulan Calculated ΔCp = 1.0 J/g-K; ----Pullulan Calculated ΔCp =1.9 J/g-K ; ◊ Starch Measured; — Starch ΔCp = 1/9 J/g-K)

The effect of water on the Tg of starch, in the form of gelatinized amylopectin, has been studied extensively.(5,6,) The data is best fit using C-K with the following values: water, ΔCp = 1.9 J/g-K, Tg= 134 K; starch, ΔCp = 0.425 J/g-

K (adjusted from the asymptotic value, ΔCp= 0.470, derived from malto-oligosaccharides), Tg = 500° K.(6) In examining the data of Kalichevsky and coworkers, it was clear that the results could not be fit with the lower value of ΔCp for water without a proportional reduction in the ΔCp for amylopectin.

Thermal Transition Behavior as a Function of Water Content. The variation in the specific heat traces in the glass transition region with increasing water content has an effect on the reliability of the data discussed in the preceding section. For the dry sample, the behavior is nearly classical in repeated scans. However, the slope below the transition is slightly higher than that above the transition and this disparity increases with increasing water content. There is a slight overshoot at the high temperature end of the transition range and the lead into the transition is more gradual then the termination of the transition. Both conditions introduce some uncertainty in ΔCp, although the value is very reproducible as noted above. At 8.4% water, the appearance is similar to that of the dry sample, except for the absence of the slight temperature overshoot. With an increase to 18% water, the initial scan, Figure 2, (Curve A) shows a large endothermal peak superimposed on the end of the transition range. The peak is absent on the next scan (Figure 2 Curve B), but the transition region shows some structure which resembles two superposed glass transitions. A small endothermal peak returns after 20 minutes.

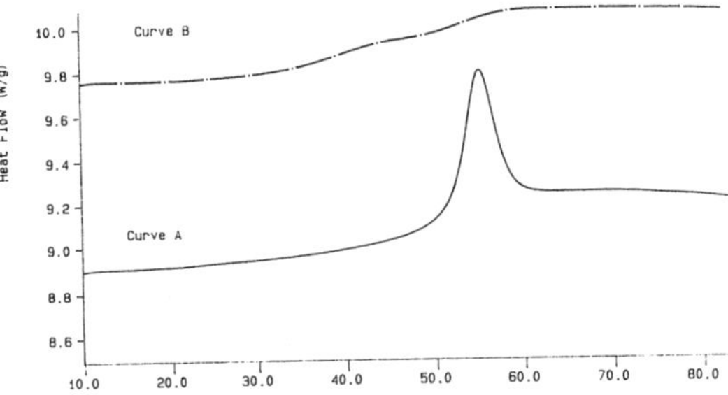

Figure 2. DSC of pullulan powder with a .18 weight fraction of water.

These effects are due to enthalpy relaxation promoted by the presence of water. A more extensive study of similar effects and other occurrences has been reported for pullulan with varying water content, but the authors questioned whether their observations could be attributed to enthalpy relaxation.(7) At 28% water a relaxation peak no longer occurs. With an increase to 45% water, the analysis is complicated by the occurrence of water melting endotherms. The glass transition region appears to be well defined and separate from the melting endotherms. However, there is a plateau region below, as well as, above the melting endotherms, leading to two different possible estimates of ΔCp; 0.277 J/g-K and 0.440 J/g-K.

The ΔCp values, based on the weight of polymer, appear to increase with increasing water content. The variation is small and amounts to a value of about 0.44 J/g water-K. A similar increase amounting to about 0.3 J/g water-K was indicated in measurements on starch/water (6) and much larger increases have been reported for a hydrophilic polyurethane. (8) The origin of this effect is uncertain. It has sometimes been assumed that the ΔCp at the polymer Tg will be the sum of that for water and the polymer.(9) This implies that water is immobilized below the polymer Tg. The indicated values for pullulan and starch / water systems are less than the lower limit for ΔCp of water, 0.93 and 1.0 J/g-K, implying immobilization of only a fraction of available water or only a partial loss of degrees of freedom. However, the water contribution to ΔCp in the case of the hydrophilic polyurethane is too large to be accounted for in this way.(8) Also, the specific heat values outside the transition region for the polymer /water mixture appear to be close to the sum of that for the polymer and water, indicating no loss of water degrees of freedom.(6)

Dynamic Mechanical Behavior of Pullulan
A typical E" versus temperature DMS scan of a "dry" pullulan sample is shown in Figure 3. The main chain motion of the polymer is designated by the Tg, an E" peak maximum at approximately 170°C. This is preceded by a broad peak in the region of 25-115°C which is due to some short change motion below Tg, possibly from some residual water. Kalichevesky has observed this type of transition for amorphous amylopectin (5) as well as Appelqvist for low moisture containing samples

Figure 3. DMS data (E' and E" vs. Temperature) for a dry pullulan sample.

of pullulan. (7)

There are significant peaks at –22°C and -102°C that are present for samples containing low to high moisture contents. This peak at -22°C is a water-induced peak observed for polysaccharides usually reported in the –50°C range. (10, 11) This peak develops as the sample is exposed to moisture and diminishes with decreasing moisture content. The disappearance of this peak is shown in Figure 4 with a film that was dried for several minutes at 100°C in the DMS chamber. The E" peak

maximum at 170°C has not shifted in temperature, but has become sharper and is not as broad as the peak in Figure 3. These results display the sensitivity in the DMS to the moisture in the sample. The broad transition in the 25-112°C region still exists and the low temperature peak is at -97°C. Nishinari has reported in dielectric relaxation studies of dehydrated pullulan that a transition occurs at -90°C. The peak is assigned to the rotational motion of the hydroxymethyl groups. (10) More detailed comparisons between the DSC, DMS, and dielectric data should be informative about the contribution of water to the mechanical relaxations.

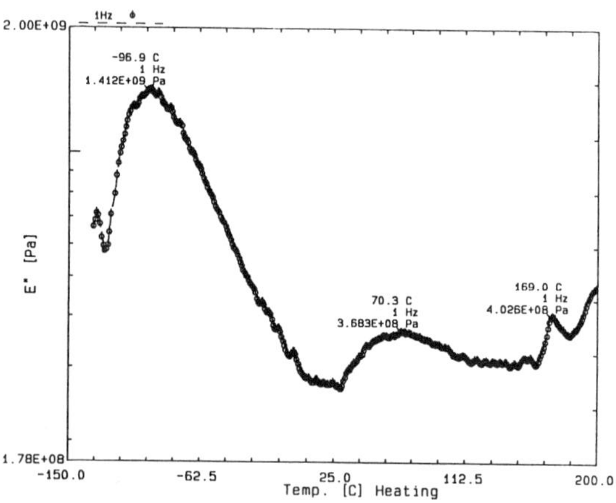

Figure 4. DMS (E" vs. Temperature) scan of pullulan dried in the DMS chamber at 100°C.

Conclusions: The results have shown the value of pullulan as a model polysaccharide for investigating the dependence of thermal transition behavior on added water content. The Tg depression can be accurately modeled by the Couchman-Karasz relation without arbitrary physical parameters. However, this requires a value for the ΔCp of solid amorphous water which is at the lower limit of the reported values and only one half the value used in fitting the data for starch/water. The need to use a two fold higher value to fit the starch data suggests that C-K cannot be used as a predictive relation for the depression of Tg in water/polymer mixtures until a reliable value is available for the ΔCp of solid amorphous water. Water has a variety of other affects on the thermal behavior, including the ΔCp change at the Tg of water/ polymer mixtures, which are not well understood.

Acknowledgements
The authors are grateful for the financial and technical support from Dr. Don Rivin and the Membrane Team at the U.S. Army Soldier System Command.

References
1. Hayasgibara Co, Ltd. Product Literature
2. P. R. Couchman and F. E. Karasz, *Macromolecules* **11**, 117 (1978).
3. C. A. Angell and J. C. Tucker, *J. Phys. Chem.* **84**, 268 (1980).
4. P. C. Painter and M. M. Coleman, *Fundamentals of Polymer Science* Second Edition Technomic Publishing Company, Lancaster, PA, 303 (1997) .
5. M. T. Kalichevsky, E. M. Jaroszkiewicz, S. Ablett, J. M. V. Blanshard and P. J. Lillford, *Carbohydrate Polymers* **18**, 77 (1992).
6. T. R. Noel and S. G. Ring, *Carbohydrate Research*, **227**, 203 (1992).
7. I. A. M. Appelqvist, D. Cooke, M. J. Gidley, Sally J. Jane, *Carbohydrate Polymers* **20**, 291 (1993).
8. N. S. Schneider, D. A. Langlois, and C. A. Bryne, in *Hydrophilic Polymers*, J. E. Glass editor, Am. Chem. Soc., Washington, D.C. ACS Advances in Chemistry Series 248 (1990).
9. R. H. M. Hatley and A. Mant, *Int. J. Biol. Macromol.* **15**, 227 (1993).
10. K. Nishinari, N. Shibuya, and K. Kainuma, Makromol. Chem. **186**, 433 (1985).
11. J. Ratto, C. C. Chen and R. B. Blumstein, *J. of Appl. Polym. Sci.*, **59**, 1451 (1996).

WATER-SOLUBLE FLUORESCEIN-LABELED POLYMER, POLYMER-PEPTIDE AND POLYMER-PROTEIN CONJUGATES

Nadya D. Belcheva, Samuel P. Baldwin, and W. Mark Saltzman

School of Chemical Engineering, Cornell University - Ithaca, NY 14853

Introduction

Fluorescence labeling is a powerful approach providing sensitive detectibility of biological macromolecules (1). NHS-fluorescein and NHS-LC-fluorescein (NHS-fluorescein with 6-aminocaproic acid spacer) are among the prefered fluorophore tags for proteins containing primary amino groups: reaction proceeds rapidly at pH=7-9 leading to stable amide bonds (2). Fluorescein-PEG-NHS active esters, commercially available products from Shearwater Polymers, Inc., are new labels with many attractive features (3). The presence of a hydrophilic PEG spacer is a prerequisite for increased solubility and reduced fluorescence quenching in aqueous media (1).

Previous studies on cell aggregation in the presence of polymer-peptide conjugates (4,5), motivated us to investigate the specific cell receptor-ligand interaction by using fluorescence techniques. In this report we discuss the synthesis and characterization of three types of fluorescein-labeled conjugates: a) polyamidoamine dendrimer-(PEG-fluorescein)$_n$ [PAMAM-(PEG2000-F)$_n$ and PAMAM-(PEG5000-F)$_n$]; b) (gly-arg-gly-asp-tyr)-PEG-fluorescein [GRGDY-PEG2000-F and GRGDY-PEG5000-F]; and c) protein-PEG-fluorescein - (mEGF-PEG-F). Peptide-PEG-fluorescein conjugates were tested for specific receptor -ligand binding with PC12 cells (rat adrenal pheochromocytoma cell line) and PAMAM-(PEG-F)$_n$ conjugates were tested for non-specific cell interaction as hyperbranched (multi)-fluorophore models. Fluorescein-PEG labeling of mEGF (mouse epidermal growth factor) was studied as a model system for protein conjugation as well as for developing HPLC separation methods.

Experimental

Synthesis of PAMAM-(PEG-F)$_n$: PEG-conjugates of polyamidoamine dendrimer, PAMAM (G=3), (Dendritech, Inc.) were synthesized by a condensation reaction in buffer media using mono-N-hydroxysuccinimide esters of fluorescein-PEG, MW 2000 or 5000 as described earlier (6). After purification by dialysis, dendrimer conjugates were characterized by MALDI and SEC.

Synthesis of GRGDY-PEG-fluorescein: The condensation reaction was performed in DMF/MES media as described earlier (6). GRGDY-PEG-F conjugates were isolated and purified by dialysis (MWCO 2000) against distilled water. Yields of the conjugates after lyophilization were 50-60% with purity ~ 85-90% based on HPLC.

PEGylation of mEGF: The synthesis of mEGF-PEG2000-fluorescein was performed by conjugation through the end amino group with mono-N-hydroxysuccinimide ester of PEG2000-fluorescein in MES pH=7 by varying the molar ratio PEG/mEGF from 1:1 to 7:1. Reaction mixtures were kept for 12 days at 4 °C under shaking, dialyzed (MWCO 5000) against distilled water at 4 °C for 5 days and lyophilized. The conjugation yield was estimated by HPLC and was found to be 40-45%.

Fluorescence spectra of the individual conjugates were registered in cell culture medium RPMI 1640 (phenol free) at room temperature on a FluoroMax-2 spctrofluorometer equipped with 150 W continuous ozone-free xenon lamp and modified Czerny-Turner spectrometers under control of DataMax spectroscopy software.

Flow cytometry (EPICS Profile Analyzer) was performed for detection of the non-specific and specific receptor-ligand interactions between PC12 cells (~ 10^6 cells/mL) and PEG-F conjugates in the concentration range from 1 x 10^{-6} M to 1 x 10^{-3} M.

Conjugation of mEGF with N-hydroxysuccinimide ester of fluorescein–PEG2000 was followed by RP-HPLC. An appropriate aqueous mobile phase was found to be a gradient 0-80% acetonitrile (0.1% trifluoroacetic acid) at flow rate 1 mL/min and column temperature 30 °C (Waters 2690 Alliance Separations Module including Waters 996 PDA detector on a Waters column 3.9 x 150 mm µBondapak^{TM} C$_{18}$).

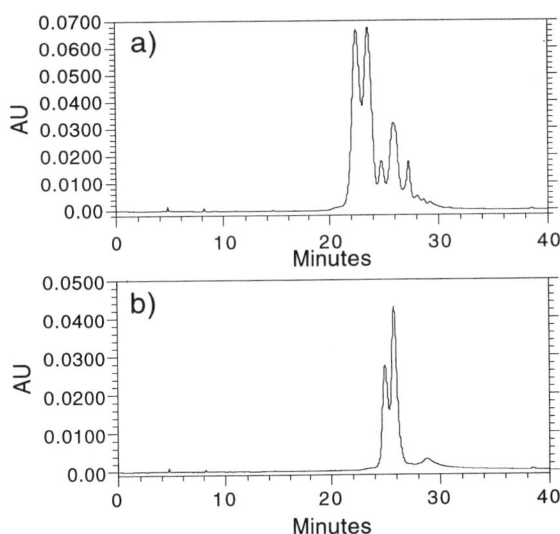

Fig. 1: RP-HPLC of a) GRGDY-PEG$_{2000}$-fluorescein and b) GRGDY-PEG$_{5000}$-fluorescein conjugates

Results and discussion

All synthesized conjugates were found to be water soluble; RP-HPLC methods were successfully applied for their purification and separation. No peaks for free peptide or non-reactive fluorescein-PEG were detected in the chromatograms of purified GRGDY-PEG2000-F and GRGDY-PEG5000-F by PDA detection at λ=275 nm (Fig.1).

While flow cytometry was an appropriate method for studying ligand-receptor binding in concentrations above 0.5×10^{-6}M of the respective fluorophore (Fig.2), spectrofluorometry allowed working at concentrations as low as 1×10^{-8} M. For each individual conjugate a calibration was performed prior to the cell binding experiments: fluorescence maxima are reported on Table 1.

Specific receptor-ligand interactions were studied by flow cytometry using GRGDY-PEG2000-F. A comparative study with non-active fluorescein-PEG conjugates was also performed (Fig.2). After incubation at different concentrations of polymer-peptide conjugates higher than $\sim 1 \times 10^{-7}$ M, all cell fractions displayed fluorescence activity which was more pronounced for GRGDY-containing conjugates. This enhanced binding activity is probably due to specific interactions between RGD and cell surface receptors.

We synthesized a mEGF-PEG-F conjugate by a condensation approach for PEGylation of mEGF. A RP-HPLC method was developed for detection and separation of the conjugate (Fig.3, a. and b.) as well as for monitoring the conversion of the reaction (Fig.3, c) using PDA detection at λ=275. It was found that performing the reaction in MES buffer media , pH=7, at different molar ratio of PEG/mEGF from 1:1 to 7:1, resulted in the same type of mEGF-PEG-F conjugate (RT=26.02 min, Fig.3,b). Optimal yields of 40-45% were achieved for greater than 3:1 stoichiometry ratio of PEG/mEGF.

Conclusion

N-hydroxysuccinimide esters of PEG-fluorescein have been successfully coupled with PAMAM dendrimers, a cell adhesive pentapeptide (GRGDY) and mouse epidermal growth factor. The new fluorescein-labeled conjugates were tested for their biological activity using complementary fluorescent techniques. The synthetic procedure and chromatographic method developed for mEGF-PEG-F may be useful in the future work on polymer modification of therapeutically active proteins.

Conjugate	$\lambda_{excitation}$	$\lambda_{emission}$
GRGDY-PEG$_{2000}$-F	496.0	516.0
GRGDY-PEG$_{5000}$-F	495.5	515.0
HOCH$_2$CH$_2$NHCO-PEG$_{5000}$-F	488.2	517.2
NHS-PEG$_{2000}$-F	477.5	519.0
NHS-PEG$_{5000}$-F	496.0	516.0
PAMAM-(PEG$_{2000}$-F)$_n$	496.0	519.0
PAMAM-(PEG$_{5000}$-F)$_n$	496.0	515.6

Table 1: Excitation and emission fluorescent maximums of fluorescein-PEG conjugates in cell culture media (RPMI-1640) at room temperature. Wavelengths are given in nm.

References

1. G. Hermanson , Tags and Probes, in Bioconjugate techniques, Academic Press, Inc. 1996, pp 297 –416
2. M. Brinkley, Bioconjugate Chem.3, 2(1992)
3. Shearwater Polymers, Inc., Polyethylene Glycol derivatives - technical data
4. W. Dai, J. Belt and W. Mark Saltzman, Bio/Technology 12, August, 797-801 (1994)
5. N. Belcheva, S.P. Baldwin and W. Mark Saltzman, J.Biomater.Sci. – Polymers in Tissue Engineering, in press
6. N. Belcheva, S.P. Baldwin and W. Mark Saltzman, J.Contr. Rel., in press

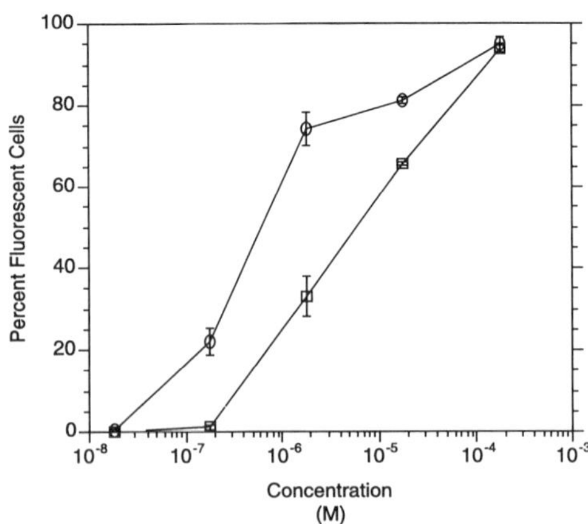

Figure 2: Flow cytometric test results using PC12 cells incubated with GRGDY-PEG$_{2000}$-fluorescein (O), or PEG$_{2000}$-fluorescein ($\square$). Error bars show the standard deviation for replicate samples.

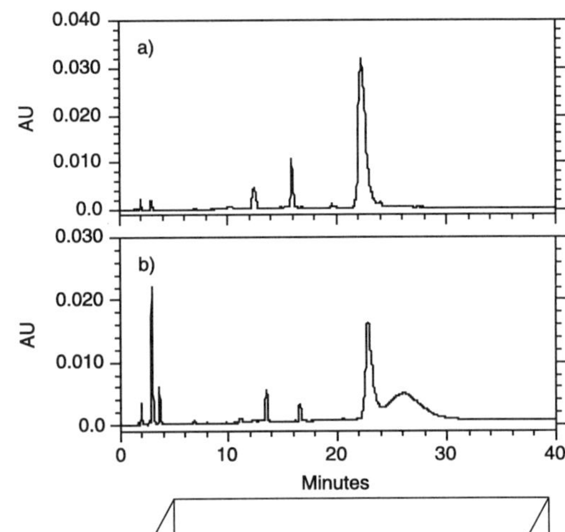

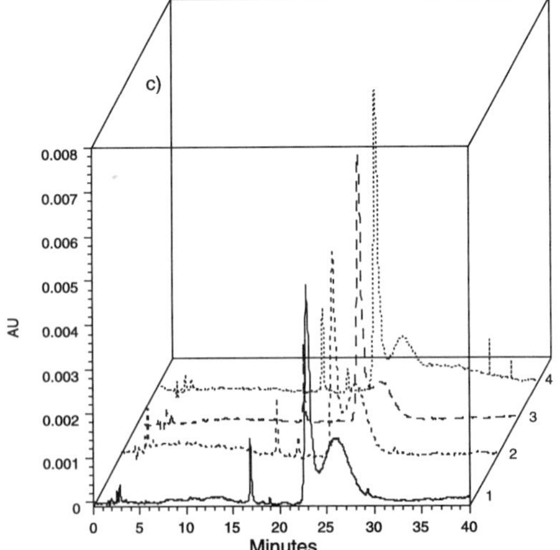

Fig. 3: RP-HPLC of a) mEGF (RT=22.26 min.)
b) mEGF-PEG$_{2000}$-fluorescein (RT=26.02 min.)
c) Reaction mixture at different time points, 1-1 day 2-7 days, 3-11 days, and 4-12 days.

LONG-TERM MORPHOLOGICAL CHANGES IN FREEZE-THAWED PVA HYDROGELS

Christie M. Hassan and Nikolaos A. Peppas,
School of Chemical Engineering, Purdue University,
West Lafayette, Indiana 47907-1283, USA

INTRODUCTION

Hydrogels are crosslinked, hydrophilic polymers (networks) which swell when placed in contact with water, buffer solutions, or biological fluids. They have been used in a wide variety of applications, specifically in the areas of medicine and pharmacy, because they exhibit good biocompatiblity[1]. Hydrogels are typically synthesized from monomers or polymers using small amounts of crosslinking agents such as aldehydes, bisacrylamides, or dimethacrylates[2]. However, slow elution of the unreacted residue from these agents may occur over time resulting in the release of toxic agents.

The materials of interest in this research were ultrapure poly(vinyl alcohol) (PVA) hydrogels. In order to avoid the toxicity involved with the use of chemical crosslinking, a method of gelation of PVA by freezing and thawing has been developed. This method involves the casting of dilute, aqueous solutions of PVA, then cooling to -20 °C and thawing back to room temperature several times. This process results in the formation of a stable three-dimensional network that is physically crosslinked by the presence of crystallites. The stability of this network structure is reinforced by an increased number of freezing and thawing cycles[3].

In the present work we report on the stability of such materials upon swelling. Specifically, the crystallinity, crystal size distribution, and long term dissolution of PVA chains from the system were investigated.

EXPERIMENTAL

Preparation of PVA Hydrogels

Aqueous PVA solutions of 15 wt% were prepared by dissolving PVA (Elvanol® 90-50, E.I. duPont de Nemours, Wilmington, DE, $\overline{M}_n$= 35,740, polydispersity index = 2.15, degree of hydrolysis = 99.0%) in deionized water for 6 h at 90 °C. The solutions were cast between glass microscope slides with 0.7 mm thick spacers. The samples were then exposed to 3 to 7 cycles of freezing for 8 hours at -20 °C and thawing for 4 hours at 25 °C.

Swelling Studies

The PVA films prepared by repeated cycles of freezing and thawing were cut into thin disks of 12 mm diameter. Each disk was initially weighed in air and then placed in a jar containing 50 ml of deionized water. At specific times during swelling, 5 ml samples were removed from the swelling media. The swelling media samples were analyzed for PVA dissolution by complexing each 5 ml sample of aqueous PVA with 2.5 ml of a 0.65M boric acid solution and 0.3 ml of a 0.05M I_2/0.15M KI solution and then diluting to 10 ml with deionized water at 25 °C. The absorbance of visible light at 671 nm was then measured with a UV/Vis spectrometer (Lambda 10 model, Perkin Elmer, Norwalk, CT) to determine the concentration of complexed PVA in solution.

Thermal Analysis Experiments

Differential scanning calorimetry (DSC, Model 2910, TA Instruments, New Castle, DE) was used to determine the crystallinity and crystal size distributions of samples swollen for 0, 24, and 360 hours. In a typical experiment, 5 to 10 mg of a dried sample was placed in an aluminum pan and heated at a scanning rate of 10 °C/min from 30 to 250 °C with a constant purge of nitrogen.

RESULTS AND DISCUSSION

PVA gels prepared using freezing and thawing techniques were found to have good mechanical integrity. Particularly, gels exposed to an increased number of freezing and thawing cycles showed higher mechanical strength. The gels exhibited a high degree of swelling in water at 37 °C with equilibrium volume swelling ratios ranging from 10 to 18 (90 to 94.4% water). The volume swelling ratio was found to decrease with increasing freezing/thawing cycles due to the higher degree of physical crosslinking associated with the crystallization process.

Using the boric acid technique, it was determined that a certain amount of amorphous PVA was dissolved at 37 °C in water over the initial period of swelling as shown in Figure 1. These results indicate that there is

initial dissolution of the samples due to PVA chains that did not participate in the crystallite formation process and therefore were not incorporated in the physical structure of the gel. A lower degree of PVA dissolution is observed with samples that were prepared with increased cycles of freezing and thawing. Therefore, it is evident from these results that there is an increased degree of physical crosslinking associated with an increased number of freezing/thawing cycles. These results also show that the gels remained relatively stable as a function of time over a period of approximately 7 months from the standpoint of PVA chain dissolution. In considering the overall stability of the system, it is also important to examine the changes in the crystalline structure.

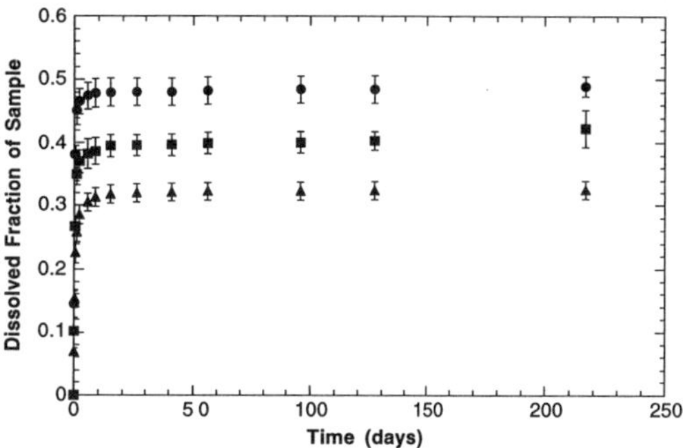

Figure 1.   Fractional dissolution in water at 37 °C of PVA samples with 3 (●), 5 (■), and 7 (▲) cycles of 8 hour freezing and 4 hour thawing.

Using the thermograms obtained, the heat required for melting of a sample, ΔH, was determined by integrating the area under the melting peak at approximately 220 °C. The melting endotherm on the thermogram occurred over the range of 190 to 240 °C. This ΔH value was corrected for the water present in the sample by also analyzing the peak representing the heat required for the evaporation of water. The degree of crystallinity, X(t), based on dry PVA volume, was calculated by dividing ΔH by the heat required for melting a 100% crystalline PVA sample, ΔH_c = 138.6 J/g. Modified degrees of crystallinity were also calculated for partially dissolved samples using the data from the dissolution analysis. Therefore, the changes in crystallinity could be examined based on the original weight of the sample. The calculated degrees of crystallinity and modified degrees of crystallinity for samples prepared with 3, 5, and 7 cycles of freezing and thawing and then swollen in water at 37 °C for 0, 24, or 360 hours are shown in Table 1. Samples that were exposed to 7 cycles of freezing and thawing showed a slightly more stable crystalline structure upon swelling as indicated by the modified degrees of crystallinity.

Table 1.   Degrees of crystallinity for PVA samples prepared using 3, 5, and 7 cycles of freezing and thawing

Freezing/ Thawing Cycles	Time of Swelling (h)	Degree of Crystallity (%)	Modified Degree of Crystallinity (%)
3	0	39.2	39.2
3	24	50.6	27.8
3	360	53.6	27.9
5	0	58.1	38.1
5	24	45.4	27.6
5	360	45.3	27.4
7	0	38.3	38.3
7	24	44.6	33.1
7	360	45.7	31.2

All endotherms representative of the melting of PVA were analyzed to obtain the distribution of melting temperatures for each sample. Melting temperatures were then converted to crystallite thicknesses using the Thomas-Gibbs equation[4]:

$$T_m = T_m^o\left(1 - \frac{2\sigma_e}{\Delta H\, L}\right) \qquad (1)$$

where T_m is the observed melting temperature, L is the lamellar thickness of the crystal, T_m^o is the equilibrium melting temperature of an infinitely thick crystal, σ_e is the surface free energy per unit area of the chain folds, and ΔH is the heat of fusion per unit volume of the crystals. Crystal size distributions were normalized so that the area under each curve was proportional to the modified degrees of crystallinity to account for chain dissolution effects. This analysis gave information as to the size and number of crystallites. Figure 2 shows the crystal size distributions for PVA gels after preparation but before swelling. In the initial state, there is no significant difference in the crystal size distributions between samples prepared with 3, 5, or 7 cycles of freezing and thawing. In all cases, the lamellar thicknesses ranged from 50 to 300 Å with the largest fraction of crystals having thicknesses of approximately 150 Å.

In analyzing the overall stablity of the system, ultrapure PVA gels prepared using freezing and thawing techniques show some initial instability. Although there is initial dissolution of PVA chains as well as the melting out of smaller crystallites, the system attains stability after 24 hours. Secondary crystallization of such a system is another important consideration which must be addressed for long-term applications of these PVA gels as biomaterials.

ACKNOWLEDGMENT

This work was supported in part by a grant from the Showalter Trust.

REFERENCES

1. Peppas, N.A. *Hydrogels in Medicine and Pharmacy, Vol 1, Polymers*, CRC Press, Boca Raton, FL 1987.
2. Peppas, N.A. *Hydrogels in Medicine and Pharmacy, Vol 2, Polymers*, CRC Press, Boca Raton, FL 1987.
3. Stauffer, S.R.; Peppas, N.A. Poly(vinyl alcohol) hydrogels prepared by freezing-thawing cyclic processing. *Polymer*, **33**, 3932 (1992).
4. Hoffmann, J.D. and Lauritzen, J.I. Theory of formation of polymer crystals with folded chains in dilute solution. *J. Res. Natl. Bur. Std.*, **64A**, 73 (1960).

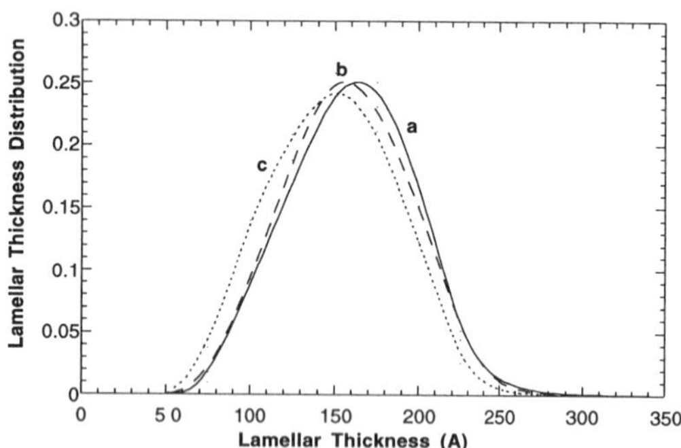

Figure 2. Initial crystal size distributions (before swelling) for PVA samples prepared with 3 (**a**), 5 (**b**), and 7 (**c**) cycles of 8 hour freezing and 4 hour thawing.

Figure 3 represents the crystal size distributions for a PVA sample prepared with 5 cycles of freezing and thawing and allowed to swell at 37 °C in deionized water. It is apparent that a large fraction of the crystals melted out in the first 24 hours of swelling. Over the following 14 days, there was no further melting of crystals observed as indicated with the curve representing 360 hours of swelling. It is also noted that upon swelling, the peak lamellar thickness shifts to a slightly higher value emphasizing the stabilty of the larger crystals. However, there are also crystals of larger lamellar thickness observed in the distribution. This may be indicative of additional crystallization that may be occurring in the system at long times, or secondary crystallization.

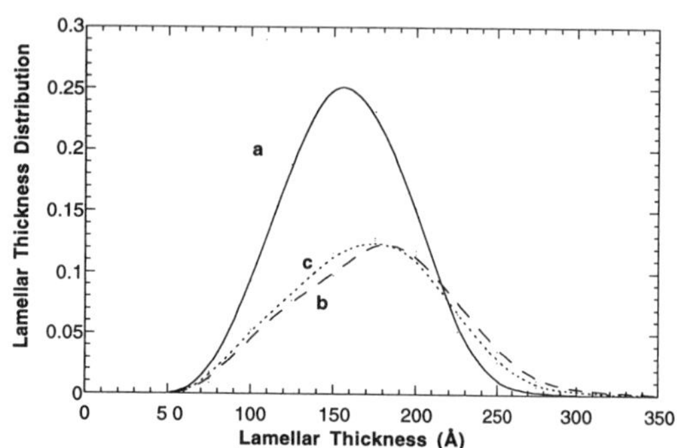

Figure 3. Crystal size distributions upon swelling for 0 (**a**), 24 (**b**), and 360 (**c**) hours for PVA samples prepared with 5 cycles of 8 hour freezing and 4 hour thawing.

Interaction of Ionomers and Polyelectrolytes with Divalent Transition Metal Cations Studied by ESR Spectroscopy

K. Kruczala and S. Schlick

Department of Chemistry, University of Detroit Mercy, Detroit, Michigan 48219

Introduction and Background

Ionomers are ion-containing polymers that contain some ionic groups usually < 15 % mole.[1] As a result, some ionomers do not normally dissolve in polar solvents (because of the *small* number of ionic groups) or in organic solvents (because of the *presence* of ions). In equilibrium however ionomers can retain large amounts of solvents. In bulk or in media of low polarity the ionic groups exist as ion pairs and often form ionic clusters, or "multiplets";[2] this is the "ionomer regime", where ionomers resemble weakly charged polyelectrolytes. In the presence of polar solvents the ion pairs are expected to dissociate; this is the "electrolyte regime".[3] Most theories of the effect of ionic interactions in ion-containing polymers have assumed the presence of non-coordinating monovalent cations that are treated as point charges.

Recently a procedure for the dissolution of perfluorinated and protiated ionomers in water in an autoclave has been developed, thus making possible the study of the solution structure and comparison with results obtained for the bulk or solvent-swollen ionomers. Small-angle scattering[4] and electron spin resonance (ESR) spin probe methods[5] have indicated that the ionomer chains self-assemble into micellar aggregates with dimensions of the order of 100 Å, depending on the solvent, amount and position of the ionic groups in the ionomer, nature of the counterions, and degree of neutralization.

We present a study of *divalent* transition metal cations (Cu^{2+}, VO^{2+}) in aqueous solutions of ionomers and polyelectrolytes. The results are based on ESR spectra at X-band of Cu^{2+} (2 % neutralization of the ionic groups by cupric acetate) and of VO^{2+} (2 % neutralization by vanadyl sulfate) in poly(ethylene-*co*-methacrylic acid) (EMAA) ionomers, Nafion perfluorinated ionomers, poly(acrylic acid) (PAA), and poly(styrene sulfonic acid) (PSSA). Major objectives were to compare with the results for monovalent cations and with the expected behavior of ion-containing polymers in the presence of a polar solvent.

Experimental

The aqueous solution of EMAA (7.5 % mole methacrylic acid, 90 % neutralized by K^+) was prepared by DuPont Mitsui Co., Chiba, Japan.[5b,c] The aqueous solution of Nafion (6 % wt) was prepared from soluble Nafion powder by dissolution in ethanol and repeated dialysis against water. Poly(acrylic acid) (PAA), molecular weight 150,000, as a 25 % wt water solution, and poly(styrene sulfonic acid) as the sodium salt, molecular weight 70,000, were from Polysciences, Inc. Samples were prepared by dissolving the desired amount of the cation salt in 12 % wt aqueous solutions of EMAA, PAA, and PSSA; and in the 6 % wt Nafion solution. The pH of the EMAA, Nafion, and PSSA samples was 10. The pH of the PAA sample was varied in the range 1.5-13.5 by addition of KOH.

ESR spectra at X-band were measured in the temperature range 125-300K, with the Bruker ECS106 spectrometer. Some samples were prepared with Cu enriched in the ^{63}Cu isotope, to increase the spectral resolution. Additional experimental details have been published.[5]

Results and Discussion

EMAA System. Rapid addition of Cu^{2+} (as acetate) or VO^{2+} (as sulfate) to EMAA solutions led to precipitation, which was avoided by slow addition of the salts and rapid mixing. At higher Cu^{2+} content (typically 20 %) a gel was formed within several days. ESR spectra of $^{63}Cu^{2+}$ in aqueous EMAA (12 % wt ionomer) are presented in Figure 1. The "rigid limit" ESR spectra at 125 K and 300 K were simulated by assuming a single type of copper complex with the following magnetic parameters:

$A_{||} = 155$ G, $A_{\perp} = 25$ G, $g_{||} = 2.323$, $g_{\perp} = 2.056$ (at 125 K).
$A_{||} = 149$ G, $A_{\perp} = 19$ G, $g_{||} = 2.326$, $g_{\perp} = 2.060$ (at 300 K).

The magnetic parameters at 125 K and 300 K are not far apart, indicating that Cu^{2+} is associated with, and immobilized by, the EMAA micelles.

PAA System. The ESR spectra of Cu/PAA and VO/PAA are sensitive to the pH. Selected spectra for Cu/PAA are shown in Figure 2 (at 300 K) and in Figure 3 (at 125 K). At 300 K and pH = 1.5 the isotropic spectrum ($g_{iso} = 2.183$) suggests that Cu^{2+} cations are surrounded by water molecules. The anisotropic spectra at higher pH values indicate complexation by COO^- groups for pH in the range 3-8 (degree of dissociation 16-93 %), and by OH^- at pH = 13.5.[6] More details on the type of complexation by COO^- groups was deduced from the ESR spectra at 125 K, Figure 3. The ESR spectrum for ^{63}Cu/PAA at pH = 7 was simulated by a superposition of two components in a 4:6 ratio. The following magnetic parameters were deduced:

Site 1: $A_{||} = 140$ G, $A_{\perp} = 13$ G, $g_{||} = 2.357$, $g_{\perp} = 2.058$ (40 %).
Site 2: $A_{||} = 157$ G, $A_{\perp} = 24$ G, $g_{||} = 2.317$, $g_{\perp} = 2.056$ (60 %).

Based on $g_{||}$ *vs.* $A_{||}$ correlation diagrams,[7] we deduced that Cu^{2+} cations are coordinated to *one* chelating COO^- group in Site 1, and to *two* COO^- groups in Site 2. Copper complexes with polycarboxylates has been widely studied;[8] this study however presents the first detection of *two* types of complexes, made possible by the higher spectral resolution of the ESR spectra from ^{63}Cu complexes.

This analysis leads to the identification of the unique site detected for Cu/EMAA with a copper complex coordinated to *two* chelating COO^- groups, as for Site 2 in Cu/PAA. The copper complex in PAA (Site 2) and the immobilization of the cation by EMAA micelles are displayed in Figure 4.

Nafion and PSSA Systems. The ESR spectra of ^{63}Cu/PSSA at 300 and 125 K are shown in Figure 5; almost identical spectra were obtained for ^{63}Cu/Nafion. In these systems the isotropic spectra at 300 K ($g_{iso} = 2.185$) and the magnetic parameters at 125 K ($A_{||} = 134$ G, $g_{||} = 2.071$) indicate that the sulfonic groups in Nafion and PSSA are *not* coordinated to the cations, and that only aqua complexes of Cu^{2+} are detected. Similar results have been obtained for VO^{2+} complexes.

Conclusions

This study has indicated that "ionomer" or "polyelectrolyte" behavior in ionomers is determined not only by the type of solvent (high or low polarity), but also by the coordinating abilities of the cation, the pH, and the type of ionic groups in the system. In ionomers containing sulfonic groups, the electrolyte regime was clearly established even for strongly coordinating cations such as Cu^{2+} and VO^{2+}: In PSSA (an polyelectrolyte) and in Nafion (an ionomer) identical ESR spectra of the corresponding aqua complexes of Cu^{2+} and VO^{2+} were detected in the entire temperature range 125-300 K. For weaker acids such as carboxylic groups, the bonding to the cation depends on the pH. The Cu^{2+} and VO^{2+} complexes in PAA can be *intra*chain or *inter*chain; in both cases, however, there is was no precipitation. The $A_{||}$ value for Cu^{2+} is 132 G in PAA, *vs.* 149 G in EMAA, indicating less motional averaging (slower dynamics) and more immobilization in the ionomer.

Acknowledgements. This research was supported by the Polymer Program of NSF. We thank Gérard Gebel (CEA-CENG Grenoble, France) for the gift of the soluble Nafion powder.

References

1. *Ionomers: Characterization, Theory, and Applications*, Schlick, S., Ed.; CRC Press: Boca Raton, FL, 1996.
2. Eisenberg, A.; Rinaudo, M. *Polym. Bull.* **1990**, *24*, 671.
3. (a) Semenov, A.; Nyrkova, I.; Khokhlov, A.R. In *Ionomers: Characterization, Theory, and Applications*, Schlick, S., Ed.; CRC Press: Boca Raton, FL, 1996, Chapter 11. (b) Philippova, O.E.; Sitnikova, N.L.; Demidovich, G.B.; Khokhlov, A.R. *Macromolecules* **1996**, *29*, 4642.
4. (a) Aldebert, P.; Dreyfus, B.; Gebel, G.; Nakamura, N.; Pineri, M.; Volino, F. *J. Phys. France* **1988**, *49*, 2101. (b) Loppinet, B.; Gebel, G.; Williams, C.E. *J. Phys. Chem. B* **1997**, *101*, 1884. (c) Gebel, G.; Loppinet, B. *J. Mol. Structure* **1996**, *383*, 43. (d) Gebel, G.; Loppinet, B.; Hara, H.; Hirasawa, E. *J. Phys. Chem. B* **1997**, *101*, 3980.

5. (a) Szajdzinska-Pietek, E.; Schlick, S. In *Ionomers: Characterization, Theory, and Applications*, Schlick, S., Ed.; CRC Press: Boca Raton, FL, 1996, Chapter 7. (b) Kutsumizu, S.; Hara, H,; Schlick, S. *Macromolecules* **1997**, *30*, 2320. (c) Kutsumizu, S.; Schlick, S. *Macromolecules* **1997**, *30*, 2329.

6. Falk, K.-E.; Ivanova, E.; Roos, B.; Vänngård, T. *Inorg. Chem.* **1970**, *9*, 556.

7. Schlick, S.; Alonso-Amigo, M.G.; Bednarek, J. *Colloids Surfaces A* **1993**, *72*, 1.

8. (a) Leyte, J.C.; Zuiderweg, L.H.; van Reisen, M. *J. Phys. Chem.* **1968**, *72*, 1127. (b) François, J.; Heitz, C.; Mestdagh, M.M. *Polymer* **1997**, *38*, 5321. (c) Trochimczuk, A.W.; Jesierska, J. *Polymer* **1997**, *38*, 2431.

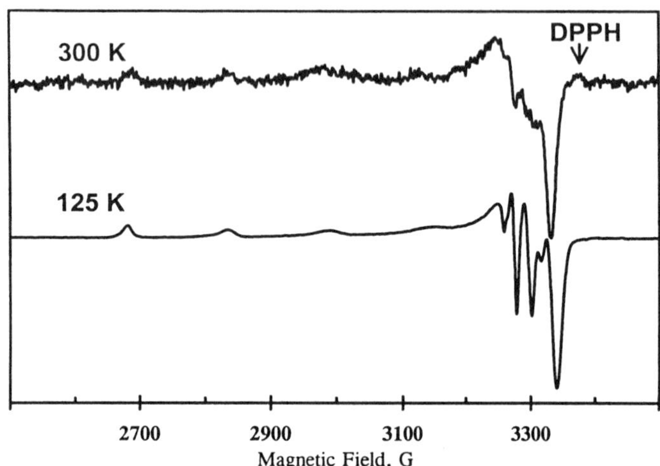

Figure 1: X-band ESR spectra of ^{63}Cu/EMAA at 300 and 125 K.

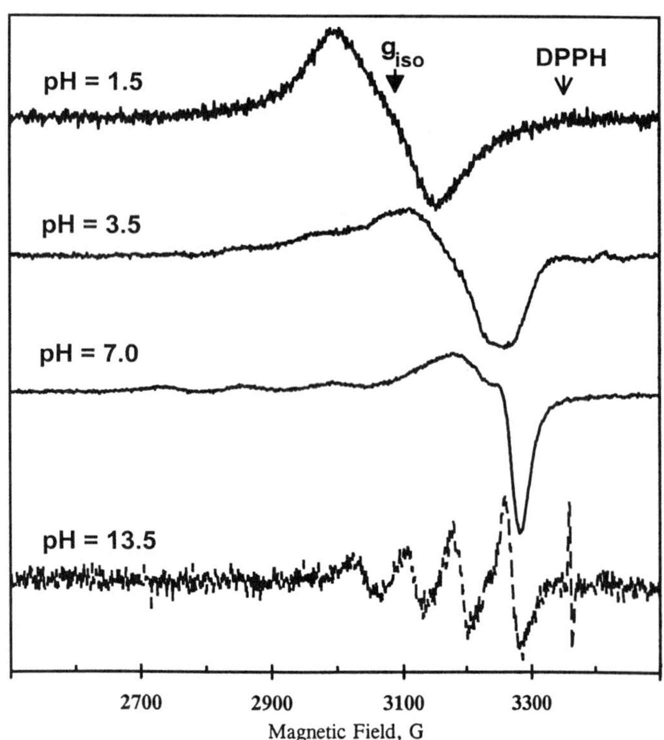

Figure 2: X-band ESR spectra of Cu/PAA at 300 K as a function of pH.

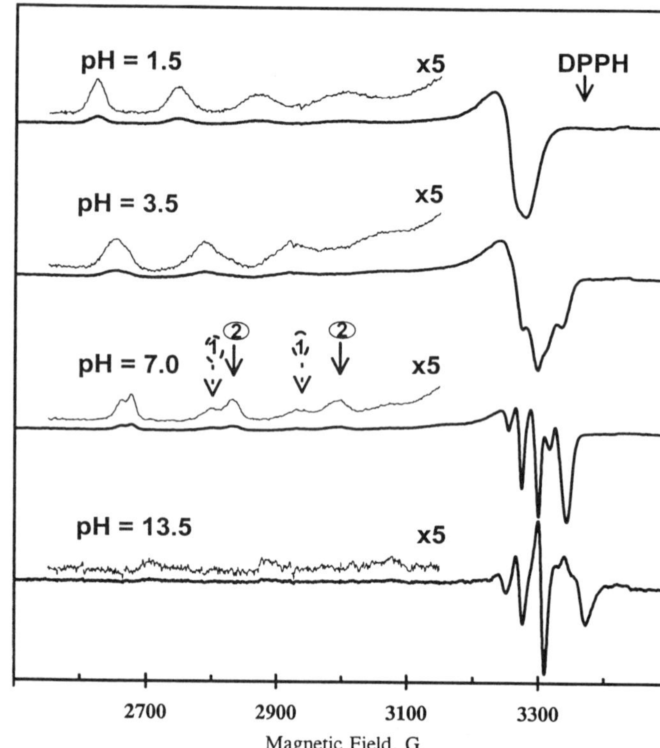

Figure 3: X-band ESR spectra of Cu/PAA at 125 K as a function of pH. Sites 1 and 2 are indicated by arrows.

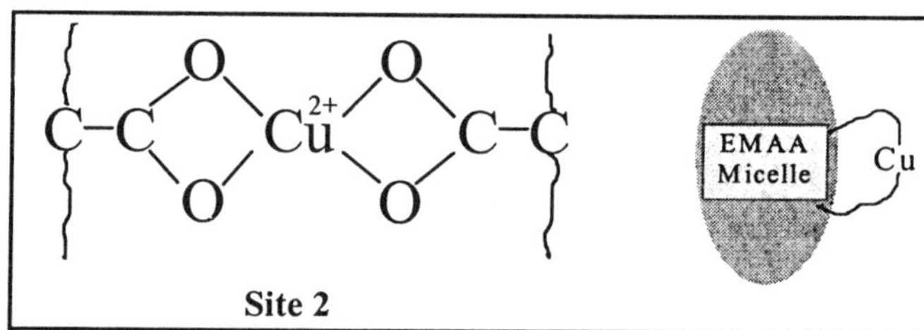

Figure 4: Copper complex in PAA (Site 2), and capture of Cu^{2+} by the EMAA micelles.

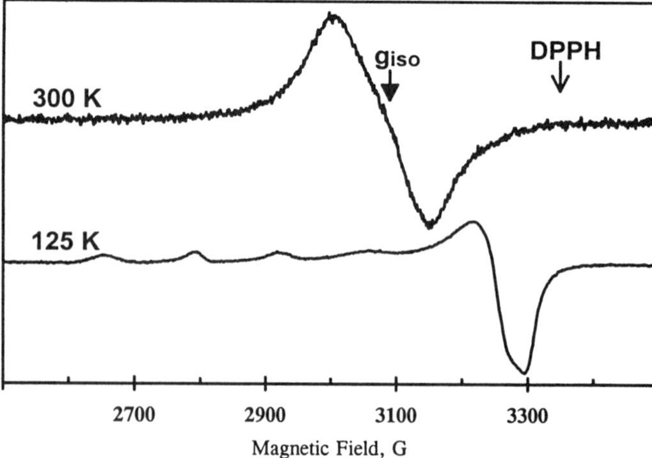

Figure 5: X-band ESR spectra of ^{63}Cu/PSSA at 300 and 125 K.

SYNTHESIS AND SWELLING BEHAVIOR OF NOVEL INTELLIGENT POLYISOBUTYLENE-BASED POLYELECTROLYTE AMPHIPHILIC CONETWORKS

Ákos Janecska and Béla Iván

Institute of Chemistry, Chemical Research Center of the Hungarian Academy of Sciences H-1525 Budapest, Pusztaszeri u. 59-67, P. O. Box 17, Hungary, and Department of Chemical Technology and Enviromental Chemistry, Eötvös Lóránd Science University, H-1518 Budapest, P. O. Box 32, Hungary

INTRODUCTION

After the discovery of quasiliving carbocationic polymerization [1,2] in the mid 1980s, the synthesis of polyisobutylene (PIB) with controlled microstructure and narrow MWD has become possible. Telechelic PIBs are among the most important types of polymers synthesized by this method with great current interest in both science and technology. By reacting their highly reactive termini, these PIBs can be quantitatively converted to various new useful high molecular weight products and networks including amphiphlic conetworks.

Amphiphilic conetworks (APCN) are a new class of crosslinked systems composed of covalently bonded hydrophilic and hydrophobic chains segments. The main property of these networks is their ability to swell uniformly both in hydrophilic and hydrophobic solvents. Because of the large difference in the chemical structure of the chain segments with opposite philicity, these are thermodynamically incompatible and exhibit microphase separation. However, due to the covalent bonds between the hydrophilic and hydrophobic chains, the freedom of demixing is significantly prevented in APCNs. Since APCNs usually swell in water as well, they may also be regarded as a special class of hydrogels.

Recently a new family of APCNs has been developed by radical copolymerization of methacrylate-telechelic PIB (MA-PIB-MA) and various (meth)acrylic monomers leading to water swellable chains [3-6]. The latter study [6] dealt with poly(sulfoethyl methacrylate)-*l*-polyisobutylene APCNs as pH-sensitive networks. However, reproducible synthesis of these networks is questionable [7].

This study concerns the synthesis and characterization of a series of poly(methacrylic acid)-*l*-polyisobutylene (PMAA-*l*-PIB) polyelectrolyte APCNs. The swelling behavior of these networks and their pH-dependent swelling in aqueous media is also demonstrated.

EXPERIMENTAL

The synthesis of methacrylate-telechelic polyisobutylene (MA-PIB-MA) was carried out as described [4, 8] resulting a product with M_n = 11,200, M_w/M_n = 1.08 (GPC) and F_n = 2.0 (^{1}H NMR).

The networks were prepared by the radical copolymerization of MA-PIB-MA with trimethylsilylmethacrylate (TMSMA) (Aldrich) in THF in Teflon molds under N_2 at 60 $^\circ$C for 72 hours. After three days of curing the networks were removed from the molds and THF was allowed to evaporate. The conditions and experimental data are summarized in Table 1. The numbers in the sample column indicate the M_n of the starting MA-PIB-MA (divided by 1,000) and the weight percent of PIB in the copolymerization system. Deprotection was accomplished by swelling the networks in a 5% solution of HCl in methanol for one day, then in HCl+methanol+H_2O for one day followed by 5% HCl in water for another day. Finally, the networks were extracted with hexane for 24 hrs, and dried to constant weight in vacuo.

Swelling kinetics was examined in *n*-hexane and in water at room temperature. The swelling properties of the networks were also examined in buffer solutions with pH = 2 and 12 (I = 0.10). The initially water–swollen samples were placed into the pH = 2 buffer, and weighed at predetermined times. After the swelling equilibrium was reached, the samples were placed into the pH = 12 buffer and the swelling ratio was determined as a function of time. This was followed by placing the networks again in the solution with pH = 2.

RESULTS AND DISCUSSION

Poly(methacrylic acid) (PMAA) and polyisobutylene (PIB) are incompatible polymers. Therefore direct synthesis of APCNs with these components can only be made by synthesizing a network first with compatible chain segments followed by deprotection. This generally applicable synthetic strategy for the preparation of APCNs was already applied to prepare poly(2-hydroxyethyl methacrylate)-*l*-(polyisobutylene) APCNs [4, 5]. Similarly, TMSMA was selected as starting monomer for the synthesis of PMAA-*l*-PIB. The copolymerization of MA-PIB-MA with TMSMA and the deprotection process is shown in Scheme 1. As exhibited in this Scheme, copolymerization of MA-PIB-MA with the sufficiently nonpolar trimethylsilyl protected methacrylic acid yields a network in a cosolvent (THF) for all the components. Deprotection with HCl in the network leads to the desired APCN.

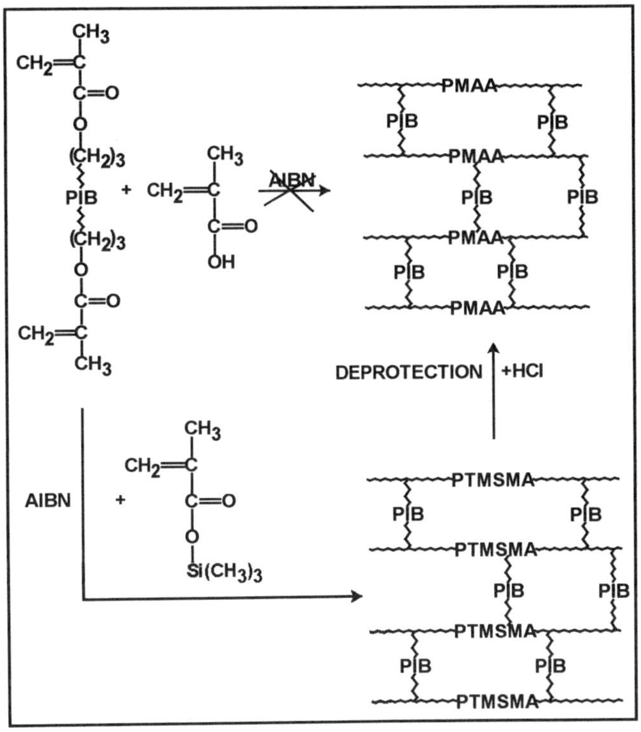

Scheme 1. Synthesis of PMAA-*l*-PIB APCN by copolymerization of MA-PIB-MA with TMSMA followed by deprotection.

Table 1 summarizes the experimental conditions for the synthesis of PMAA-*l*-PIB. The AIBN concentration was selected in relation to the monomer concentration in order to provide sufficiently high kinetic chain length for network formation according to the $1/DP_n \sim [I]^{0.5}/[M]$ relation. Deprotection was carried out as described in the Experimental. The amount of extractables in hexane were only 3-9% which indicates high copolymerization yields and close to perfect networks formation.

TABLE 1. Experimental conditions for the synthesis of PMAA-*l*-PIB amphiphilic conetworks. (THF solvent, total volume 6 ml, 72 hrs, 60 $^\circ$C).

Sample	MA-PIB-MA (g)	MAA (g)	MAA $n \cdot 10^3$ (mol)	TMSMA (g)	AIBN $n \cdot 10^6$ (mol)
M-11-30	0.3	0.7	8.13	1.29	12.23
M-11-40	0.4	0.6	6.97	1.10	8.63
M-11-50	0.5	0.5	5.81	0.92	6.20
M-11-60	0.6	0.4	4.65	0.74	3.80
M-11-70	0.7	0.3	3.48	0.55	2.17

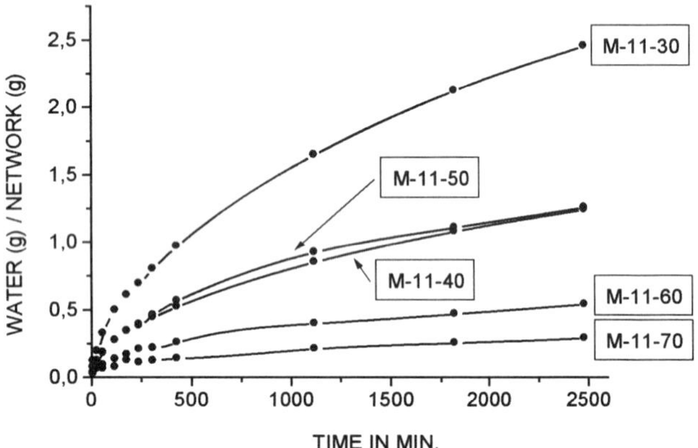

Figure 1. Swelling of PMAA-*l*-PIB amphiphilic conetworks in water at room temperature.

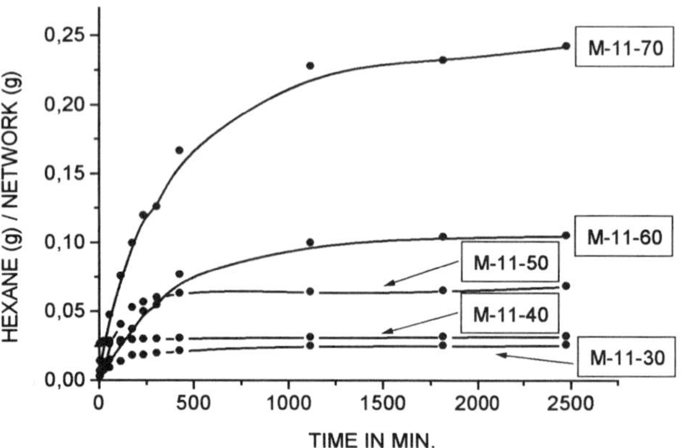

Figure 2. Swelling of PMAA-*l*-PIB amphiphilic conetworks in *n*-hexane at room temperature.

The amphiphilic nature of the resulting networks were examined by swelling studies. Figures 1 and 2 show the swelling ratio (R, absorbed solvent per network) as a function of time in water and hexane, respectively. As shown in these Figures the swelling ratio depends on networks composition in both solvents. The degree of swelling decreases with increasing PIB content in water, whereas opposite trend is observed in hexane. Comparison of Figures 1 and 2 also indicates that these new APCNs are able to absorb an order of magnitude higher amount of water than hexane. This is due to the presence of the very hydrophilic PMAA chain segments in these networks. PMAA-*l*-PIB with 70% PMAA content absorbs more than 250% water, i. e. these new APCNs can be considered as new superabsorbents. Interestingly, the swelling curves indicate Fickian type of water absorbtion at pH = 7, and non-Fickian (anomalous transport) in hexane.

The swelling behaviour of the new superabsorbent APCNs was also investigated in buffer solutions with low and high pH. The samples were initially swollen to equilibrum swelling ratios at pH = 7. As shown in Figure 3, placing these samples to pH = 2 leads to rapid deswelling between 15-60%. Changing the pH to 12 results in instantaneous increase in the swelling ratios by 20-80%. The reversibility of the swelling-deswelling phenomenon is presented by the last pH cycle in Figure 3, i. e. similar swelling ratios are observed in pH = 2 than those in the first cycle. This rapid pH-response resulted also in simultaneous volume change. The "intelligent" behaviour of PMAA-*l*-PIB APCNs can be explained by the polyelectrolyte nature of the PMAA segments. This smart immediate pH-response can be most likely utilized in several biomedical and other applications.

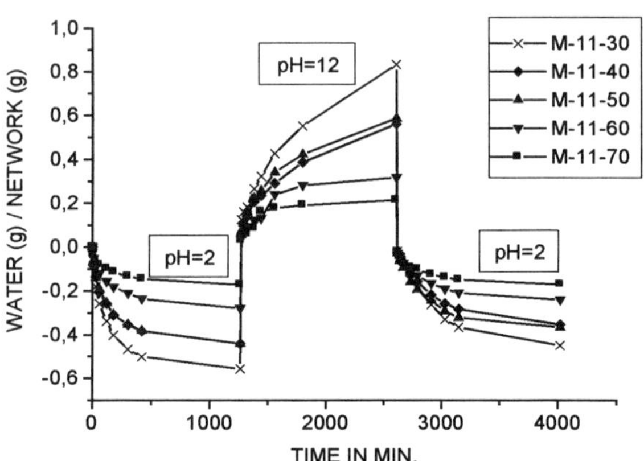

Figure 3. Deswelling and swelling of PMAA-*l*-PIB amphiphilic conetworks at pH = 2 and 12.

REFERENCES

1. R. Faust, J. P. Kennedy, *J. Polym. Sci., Part A: Polym. Chem.*, **1987**, *25*, 1847.
2. J. P. Kennedy, B. Iván, *Designed Polymers by Carbocationic Macromolecular Engineering: Theory and Practice;* Hanser Publishers, Munich, New York, **1992**.
3. D. Chen, J. P. Kennedy, and A. J. Allen, *J. Macromol. Sci.-Chem.*, **1988**, *A25(4)*, 389.
4. B. Iván, J. P. Kennedy, P. Mackey, in *" Polymeric Drugs and Drug Delivery Systems "*, Eds., R. L. Dunn and R. M. Ottenbritte, ACS Symp. Ser., Vol. *469*, Am. Chem. Soc., Washington, D. C., **1991**, pp. 194-202, ibid. 203-212.
5. B. Iván, J. P. Kennedy, P. Mackey, US Patent, **1991**, 5,073,381.
6. B. Keszler, J. P. Kennedy, *J. Polym. Sci., Part A: Polym. Chem.*, **1994**, *32*, 3153.
7. B. Iván, *unpublished results.*
8. B. Iván and J. P. Kennedy, *J. Polym. Sci., Part A: Polym. Chem.*, **1990**, 28, 89.

pH and NaCl Triggered Response of Anionic Microgels

Gary M. Eichenbaum*, Patrick F. Kiser*[⊗], Sidney A. Simon[ℵ], David Needham*[○]

*Department of Mechanical Engineering and Materials Science, Duke University, Durham, N.C., 27708-0300. [⊗]Access Pharmaceuticals, Dallas TX, 75207-2107. [ℵ]Department of Neurobiology, Duke University Medical Center, Durham, N.C., 27710.

Introduction

Micron-sized (4-7 μm diameter) polymethacrylic acid (PMAA) hydrogel spheres were synthesized by precipitation polymerization. They were characterized with regard to their mass, density and apparent pK_a. Equilibrium changes in volume were measured as functions of the pH and NaCl concentration of the suspending solution.

From an experimental perspective 4 to 7 μm microgels have several advantages over much larger slab gels (cm in size) and microgels that are less than 1 μm: (1) unlike slab gels they undergo a very rapid volume change (i.e. fractions of a second to reach equilibrium) to changes in environmental conditions, (2) unlike the smaller microgels they are large enough to view by light microscopy and (3) they can be manipulated individually. We introduce a new method for studying individual microgels by adapting a micropipet manipulation technique that was previously used to study lipid vesicles and cells.[1,2] Using this technique, the chemical environment of a single microgel can be changed directly on a microscope stage and the kinetic and equilibrium volume changes can be measured and related to the volume of the same microgel in a control solution.

We have applied a thermodynamic Donnan based model that includes counterion binding to predict the distribution of fixed, counter and co-ions inside the microgel matrix and the suspending solution as a function of pH and NaCl concentration. The ion concentration difference between the interior and exterior of the gel (which is related to the osmotic pressure) was proportional to the microgel equilibrium volume with changing pH and NaCl concentration.

Methods

Polymer microgel particles were synthesized by precipitation polymerization using a modification of the method of Kawaguchi et al.[3] The particles were synthesized with the following molar feed ratio of monomers, methylene-bis acrylamide (MBAM): 4-nitrophenylmethacrylate (NPMA): methacrylic acid (MA); 5:10:10. The resulting reactive hydrogel microparticles were reacted with sodium hydroxide to form a co-polymer of MBAM and MA. The extent of hydrolysis was determined by the release of 4-nitrophenol.

A micromanipulation flow pipet technique was used to measure the microgel equilibrium volume response and swelling kinetics.[2] An individual microgel was held by a holding pipet in a chamber, filled with a control solution and centered around an inverted microscope. Positive pressures (500-1,500 N/m²) were applied to a second "flow pipet" (~20 μm ID) filled with a test solution. When the flow pipet was aligned axially with the microgel, the microgel became immersed in the flow field of the test solution. Video images of a microgel (held by a micropipet) in pH 6.6 and 3.0 buffer, are shown in Figure 1a and Figure 1b, respectively. It is clear that the gel is more expanded at the higher pH. This condensation arises from the exchange of protons for Na⁺ and a subsequent reduction in the osmotic swelling pressure inside the microgel (discussed in more detail below).[4] In all of the micromanipulation experiments, images of microgels were displayed on a television screen using a CCD camera and measurements of the microgels were subsequently made using a video caliper system.

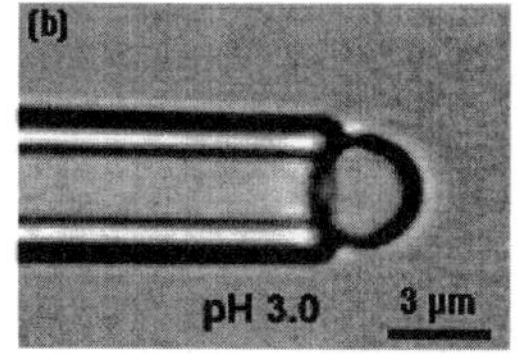

Figure 1: (a) Video image of an individual polymethacrylic acid microgel held by a micropipet and suspended in pH 6.6 citrate buffer. The microgel is in an expanded state. (b) Video image of the same polymethacrylic acid microgel, in a condensed state, held by a micropipet and immersed in pH 3.0 citrate buffer.

Results and Discussion

Table 1 provides a list of the microgel properties that were calculated and/or measured.

Properties of Microgels	Numerical Value
dry mass of microgel	4.7×10^{-11} g
dry microgel diameter	4.1 μm
dry density of microgel	1.22 g/cm³
expanded diameter of microgel - pH 6.5	6.4 μm
condensed diameter of microgel - pH 3.0	4.2 μm
carboxyl group density in expanded state	3.0 M
polymer volume fraction in most expanded state	0.28
apparent pK_a	4.70

Table 1: Material properties of the microgels.

Figure 2 shows a plot of the equilibrium volume size ratio Vr (i.e., the ratio of the volume of a microgel at the test pH to its volume at pH 6.6)

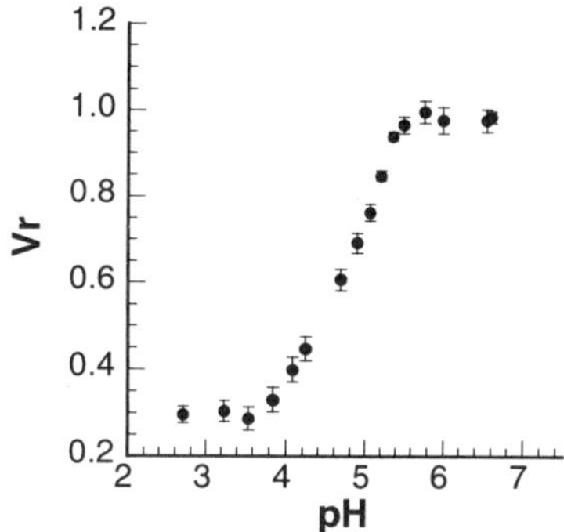

Figure 2: Plot of the microgel equilibrium volume size ratio (Vr) versus pH of the external solution for citrate buffer solutions containing no additional NaCl. Error bars on each data point represent the standard deviation of the average swelling ratio of 5 different microgels.

versus pH for microgels suspended in a citrate buffer solution at the pH indicated. Each bead was transferred from the reference solution (10 mM sodium citrate at pH 6.6). This provided a reference volume (Vr = 1) to which all the others were normalized. It was seen that decreasing the pH from 6.6 to 2.7, while holding the buffer ion concentration constant, caused Vr to decrease from 1 to 0.28. Note that for pH's < 3.6 the microgels reached their minimum volume. Conversely, in solutions with pHs > 5.3, the microgels were in their most expanded state.

Figure 3 shows a plot of an average Vr versus NaCl at pH's > 5.3. The magnitude of the largest reduction in Vr caused by changes in

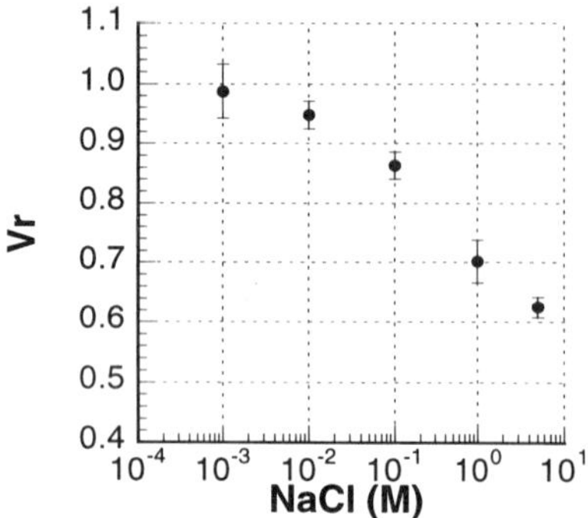

Figure 3: Plot of the microgel equilibrium volume size ratio (Vr) versus
NaCl concentration in the high pH regime (pH > 5.3). Each data point
represents the average volume ratio of 5 different microgels.

the ion concentration of NaCl (5 M) was 0.62. It was less than the
reduction in Vr induced by lowering the pH to less than 4.0 (Vr=0.28).
A primary reason for this difference was that the Na⁺ counterions do not
bind to carboxyl groups as strongly or as tightly as do protons. As a
result counterions have more water associated with them, and thereby
generate a larger osmotic swelling pressure opposing condensation.[4,5]

In studies of the diffusive transport of ions in hydrogels, it was
well established that the diffusion time of a hydrogel network is
proportional to the square of the gel dimension.[6] Thus, when the scale
of a hydrogel changes from centimeters (slab gels) to microns
(microgels), its swelling time is reduced by 8 orders of magnitude, i.e.
from 10's of hours to milliseconds. Figure 4 shows a plot of Vr versus
time, for a PMAA microgel, suspended in pH 6.6 citrate buffer and then
immersed in citrate buffer pH 3.0, delivered from a flow pipet. As
protons permeate into the

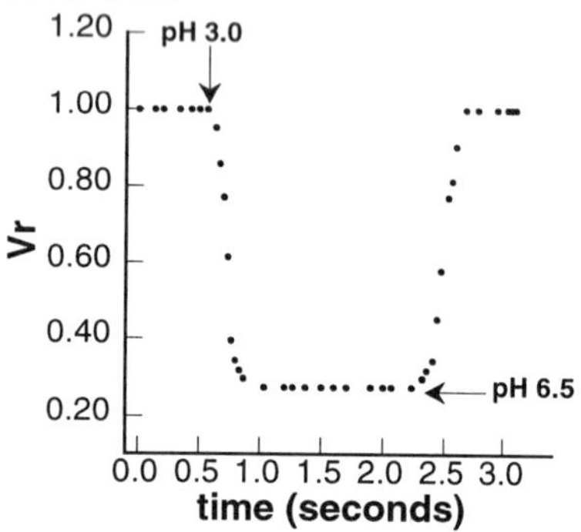

Figure 4: Plot of the microgel equilibrium volume size ratio (Vr) versus
time, for microgels, suspended in pH 6.6 citrate buffer and then
immersed in a flow field of pH 3.0 citrate buffer from a flow pipet. The
arrows indicate the time at which the microgel was exposed to a new
buffer solution.

sphere, the charged carboxyl groups on the polymer backbone become
protonated and the gel condenses. The condensation began within 50
milliseconds after exposure to the flow pipet solution and a steady state
volume (Vr=0.28) was reached after 300 milliseconds. Similarly after
the flow pipet was removed, the microgel began to re-hydrate within 50
milliseconds and reached an expanded equilibrium volume after 300
milliseconds. This process was completely reversible.

Utilizing the experimental results from the bulk titration and
volume response measurements we have developed a thermodynamic
Donnan based model to calculate the concentration of all of the ions that
are present (bound and unbound) inside the microgel as a function of pH.
Figure 5 show that the changes in the ion concentration difference

between the microgel and the bulk solution, which were predicted by the
model,

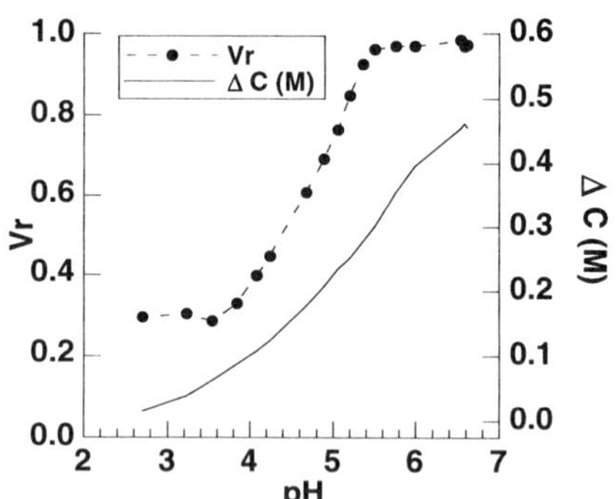

Figure 5: Plot of the model prediction for the unbound ion concentration
difference (between the microgel and external solution) and the
experimental Vr as a function of the pH of the external solution.

were proportional to the microgel's change in volume. This result
supports the hypothesis that the physical dimensions of an ionic
microgel were set by a balance between the ion concentration difference
(related to the osmotic pressure) and the polymer elasticity.[5,7]

References

(1) Evans, E., Hocmuth, R. M. *Current Topics in Membranes and
    Transport* **1978**, *10*, 1-61.
(2) Needham, D., Zhelev, D. In *Vesicles*; Rosoff, M., Ed.; Marcel
    Dekker, Inc.: New York, **1996**; pp 373-444.
(3) Kawaguchi, H., Fujimoto, K., Saito, M., Kawsaki, T., Urakami, Y.
    Polymer International **1993**, *30*, 225-231.
(4) Helfferich, F. *Ion Exchange*. McGraw-Hili Book Company: New
    York, 1962.
(5) Ricka, J., Tanaka, T. *Macromolecules* **1984**, *17*, 2916-2921.
(6) Tanaka, T., Filmore, D. J. *J. Chem. Phys.* **1979**, *59*, 5151-5159.
(7) Peppas, N. A. *Hydrogels in Medicine and Pharmacy*. CRC Press:
    Boca Raton, **1986**; Vol. 1.

NOVEL AMPHIPHILIC CONETWORKS COMPOSED OF POLY(ETHYLENE GLYCOL) AND POLYISOBUTYLENE CHAIN SEGMENTS

Gábor Erdődi and Béla Iván

Institute of Chemistry, Chemical Research Center of the Hungarian Academy of Sciences, H-1525 Budapest, Pusztaszeri u. 59-67, P.O.Box 17, Hungary and Department of Chemical Technology and Environmental Chemistry, Eötvös University, H-1518 Budapest, P.O.Box 32, Hungary (E-mail: bi@cric.chemres.hu)

Summary

Three-arm star hydroxyl-telechelic polyisobutylene (PIB(OH)$_3$) with narrow molecular weight distribution (MWD) was synthetized by quasiliving carbocationic polymerization and subsequent quantitative chain end derivatization. This star polymer was cured with poly(ethylene glycol) (PEG) to form urethane linkages under suitable conditions. Randomly linked and highly ordered segmented alternating PEG-PIB amphiphilic conetworks (APCN) with high PEG contents were prepared by a new synthetic strategy. The amphiphilic nature of the resulting networks was proved by swelling in water and n-hexane. DSC measurements confirmed the existence of microphase separation in these new APCNs.

Introduction

Recent developments in the applications of biomaterials have created increased interest in new synthetic polymers which can form hydrogels under physiological conditions. Amphiphilic conetworks, i.e. polymer networks composed of covalently bonded hydrophobic and hydrophilic segments, represent a new class of specialty hydrogels[1-4]. In recent years, rapid advances in the field of quasiliving carbocationic polymerizations have led to numerous opportunities for the controlled synthesis of new polymers with narrow MWD, designed molecular weight, microstucture and perfect functionality[4-7]. These properties of telechelic macromolecules can be applied to build various novel materials including amphiphilic conetworks (APCN). Radical copolymerization of methacrylate-telechelic polymers, such as polyisobutylenes[1-4] and polydimethylsiloxanes[8-9] with selected vinyl monomers have been used successfully for the synthesis of APCNs. Most of the hydrophobic and hydrophilic components are well-known for their high biocompatibility. However, radical copolymerization leads to random network in which the length distribution of the hydrophilic segments is broad and highly undefined. In another approach, Weber and Stadler[10] aimed at the preparation of polybutadiene-*l*-PEG APCNs. However, due to phase separation, it was possible to incorporate only less than 20% of PEG into these networks.

This presentation concerns a new approach toward the synthesis of amphiphilic conetworks containing of covalently bonded hydrophilic PEG and hydrophobic PIB segments with relatively high PEG content.

Experimental
Materials

Solvents, THF, toluene, diethyl ether and hexane (Aldrich) were freshly distilled from LiAlH$_4$ before use. The water from PEG1000 (Hampton Research) and PEG1450 (Sigma) was removed by aseotropic distillation in toluene and it was dried in vacuum at 80 °C for 3 days. Diisocyanatohexane (DCH) (Aldrich) was used as recieved. Three-arm star hydroxyl-telechelic PIB (M_n=1830 g/mol, M_w/M_n=1.07, F_n=3.0) was synthetized according to a previously described method[7].

Random network of PIB and PEG

0.40g (0.22 mmol) hydroxyl-terminated three-arm PIB, 0.22g (0.33 mmol) PEG1000 and 108 µL (0.111g, 0.66 mmol) DCH were dissolved in 3 cm^3 abs. toluene and closed in an ampule under N$_2$. The mixture in the ampule was allowed to react at 100 °C for 3 days. Then it was extracted with THF for 48 hours at room temperature and diethyl ether was used to shrink the

network. The resulting material was dried in vacuum at 80 °C for 3 days. The extracted amount was 38 %. Anal. calcd. for product: C, 70.15; H, 13.97; N, 2.40; found: C, 68.41; H, 12.02; N, 2.04.

Segmented alternating PIB-PEG network

Under N$_2$ 0.49g (0.34 mmol) PEG1450 was dissolved in 4 cm^3 abs. toluene and 2.1 cm^3 (2.2 g, 13.2 mmol) DCH was added. It was refluxed for 6 hours. Toluene was evaporated and the product was purified from the excess of DCH by repeated precipitation from toluene to n-hexane. The isocyanate-telechelic PEG1450 and 0.40 g (0.22mmol) hydroxyl-telechelic three-arm star PIB was closed in an ampule and was cured for 5 days at 100 °C. It was extracted with THF for 48 hours followed by a treatment with Et$_2$O for 6 hours and dried in vacuum at 60 °C for 3 days. The extracted amount was 18 %. Anal. calcd. for product: C, 67.05; H, 11.19; N, 1.85; found: C, 66.82; H, 11.23; N, 1.74.

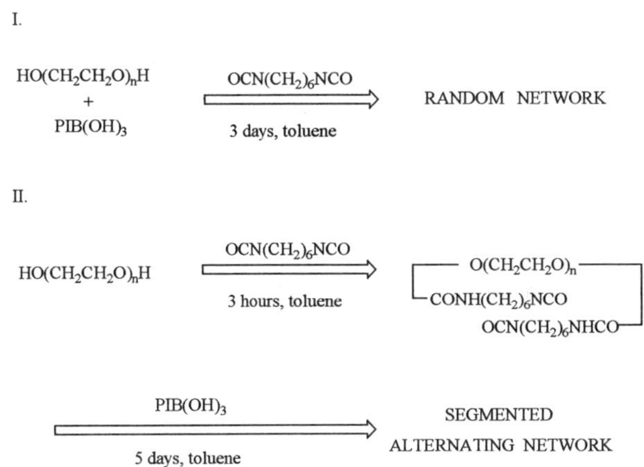

Scheme 1. Curing reactions for the synthesis of PEG-PIB conetwork

Results and discussion

Scheme 1 shows the curing reaction of random and segmented alternating APCNs containing PIB and PEG chains. The random network is formed by cocuring PEG and PIB(OH)$_3$ with stoichiometric amount of DCH. The PEG content was 36%. The structure of such APN is exhibited in Figure 1. Segmented alternating APCN was made by curing of isocyanate-telechelic PEG with PIB(OH)$_3$. In this case, the resulting APCN with 46% PEG is composed of alternating PEG and PIB segments. This means that a structurally ordered APCN can be obtained by this synthetic strategy as shown in Figure 2. According to our knowledge, this is the first PIB-based APCN having building blocks distributed in the network as uniformly as possible.

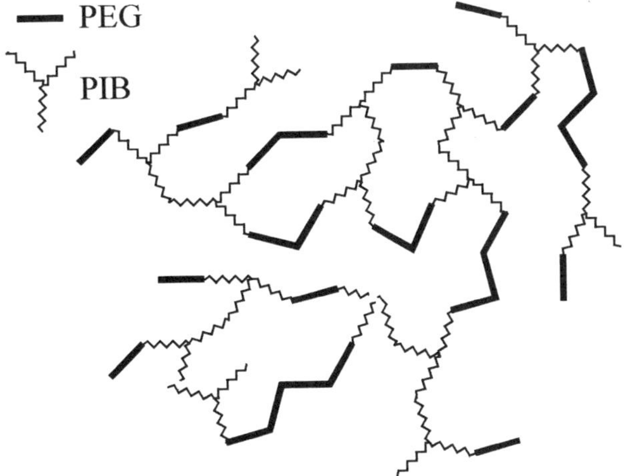

Figure 1. Structure of random PEG-PIB APCN.

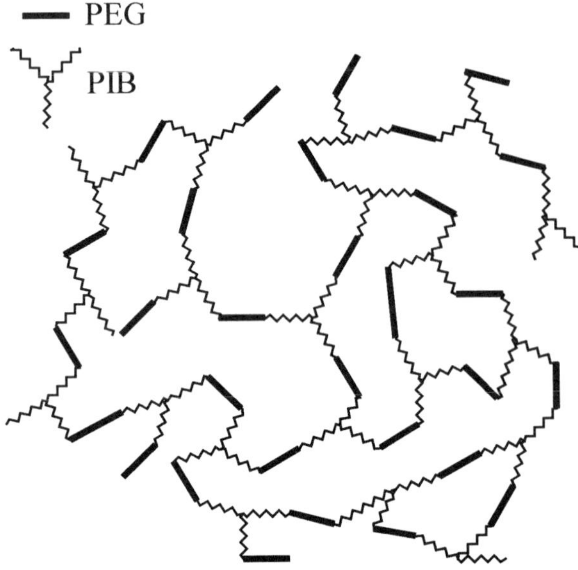

Figure 2. Structure of segmented alternating PEG-PIB APCN.

The amphiphilic nature of the resulting networks was proved by swelling experiments. As shown in Figure 3, both the random and the segmented alternating networks are swellable both in water and in hexane. Interestingly, the rate of swelling is significantly higher for the segmented alternating APCN in water. This may indicate structural (morphological) difference between the random and the highly ordered APCN.

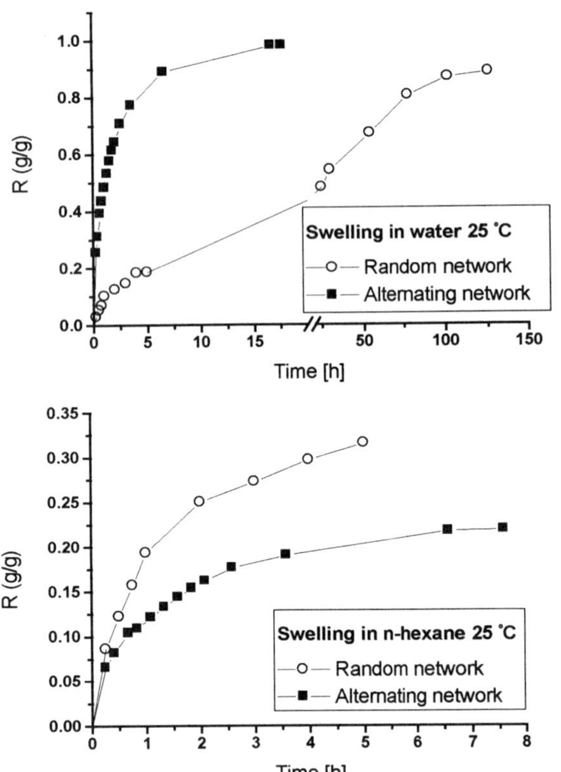

Figure 3. The swelling ratio (R) of PEG-PIB APCNs in water and in hexane as a function of time (the weight of APCNs was the same in every experiment).

DSC measurements indicate phase separated morphology for both networks. As shown in Figure 4, two glass transitions (-50 °C for PIB and –8 °C for PEG) and melting peaks at 32 °C and 34 °C are observed. The latters are most likely due to the urethane linkages since the starting PEG has a melting point at 48 °C.

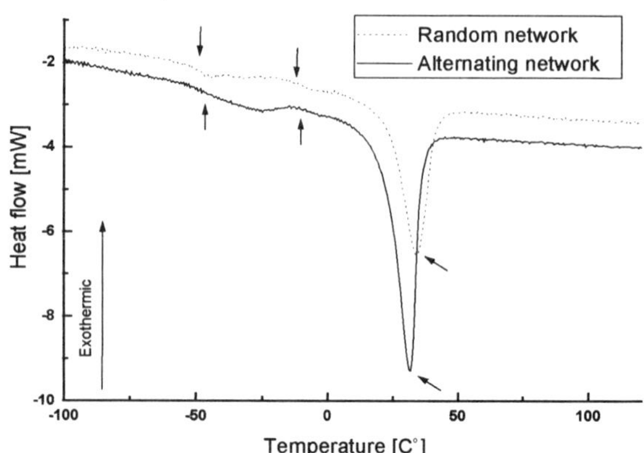

Figure 4. DSC of PEG-PIB APCNs.

In sum, a new synthetic approach has been developed by us for the preparation of novel APCNs containing randomly or alternatingly coupled PEG and PIB chain segments. Considering that both PIB and PEG are well known for their excellent biocompatibility the new APCNs are potential candidates in several applications in the area of biomaterials, such as drug release matrices, implants, adhesives etc.

References

(1) B.Iván, J.P. Kennedy, P.W. Mackey *Polym. Prep.* **1990,** *31(2),* 215-216; *ibid.,* **1990,** *31(2),*217.
(2) (a) B. Iván, J.P. Kennedy, P.W. Mackey in *"Polymeric Drugs and Drug Delivery Systems"*, Eds., R.L. Dunn and R.M. Ottenbrite, ACS Symp. Ser., Vol. 469, Am. Chem. Soc., Washington D.C., **1991,** pp. 194-202; (b) *ibid.,* 1991, pp. 203-212.
(3) B. Iván, J. Feldthusen, A.H.E. Müller *Macromol. Symp.* **1996,** *102,* 81.
(4) J.P. Kennedy, B. Iván *Designed Polymers by Carbocationic Macromolecular Engineering: Theory and Practice,* Hanser Publishers, Munich, New York, **1992.**
(5) B. Iván *Macromol. Chem., Macromol. Symp.* **1993,** *75,* 181.
(6) K. Matyjaszewski, Ed., *Cationic Polymerizations: Mechanisms, Synthesis and Applications,* Marcel Dekker, New York, **1996.**
(7) B. Iván, J.P. Kennedy *J. Polym. Sci., Part A: Polym. Chem.,* 1990, *28,* 89.
(8) J. Künzler, R. Ozark *J. Appl. Polym. Sci.* **1995** *55,* 611.
(9) Y.C. Lai, P.L. Valint, Jr. *J. Polym. Sci., Part A: Polym. Chem.* **1996,** *61,* 2051.
(10) M. Weber, R. Stadler *Polymer* **1988,** *29,* 1071.

Clear Antiperspirants and the Desire to be Beautiful in a Hawaiian Lagoon

Cory Bernu[+], Peter T. Elliott and J. Edward Glass

North Dakota State University, Polymers and Coatings Department,
Fargo, North Dakota 58105

[+]Chemstar Corp.,3915 Hiawatha Ave. S., Minneapolis, MN, 55406

INTRODUCTION

There has been a market driven effort to maintain or increase market share in the cosmetic area related to antiperspirant. This study examines the formulation variables necessary to achieve this goal by making them clear and relatively soft, but firm. The active ingredient in antiperspirants is aluminum chlorohydrate (**ACH**). The structure of ACH[1,6], $Al_{13}O_4(OH)_{24}(H_2O)_{12}{}^{7+}$, is illustrated in **Figure 1**. The objective is to maintain aggregates of this product below the wavelength of visible light, while achieving a firm gel that is not rigid or brittle.

The structure of ACH consists of a central aluminate anion, $Al(OH)_4{}^-$, which exists only at high pH values (>10), in a tetrahedral arrangement surrounded by 12 aluminum atoms in octahedral geometries. Support for this structure has come from IR[6,12] and ^{27}Al-NMR[6] spectroscopy, and also X-Ray diffraction[6] data. The spectrum of ACH at pH 4.8 has three signals: a sharp peak at 63.5 ppm and two broad peaks at 1.7 and -0.4 ppm. The peak at 63.5 ppm indicates an Al-O bond in an AlO_4 tetrahedral configuration. The peak width of less than 40 Hz suggests that the tetrahedral aluminum is not in equilibrium with the aqueous environment. Therefore, the tetrahedral AlO_4 group is believed to be bonded to other aluminum atoms. Broad resonances at 1.7 and -0.4 ppm suggest the presence of aluminum atoms in octahedral environments that are in equilibrium with the aqueous solution. Thus the repeating unit of one $Al_{13}O_4(OH)_{24}(H_2O)_{12}{}^{7+}$ complex, 8.9 Å, and one water molecule, 2.9Å, yields a d-spacing of 11.8 Å. Seven chloride anions are associated with the complex as counterions.

RESULTS AND DISCUSSION

The conversion of ACH to a soft gel for application use as an antiperspirant can be accomplished with varying degrees of success by raising the pH in the presence of different additives. As noted below, the concentrations of ACH, alkali metal salts and a variety of hydroxyl and amine containing organic oligomers and water-soluble polymers are important parameters in the formation of the gels and their firmness.

Influence Of ACH Concentration

The concentration of ACH directly influences network formation in aqueous ACH systems. This is illustrated in **Figure 2** with changing molarity of an ammonium organic acid salt (AOAS) in determining gelation times. As the concentration of ACH increases, the time to gelation decreases. At a 55% concentration of ACH, the gelation times are extremely fast. By working in dilute solutions, such as 41% ACH, the gelation times are longer and the other parameters in the system can be studied.

The effects of increasing AOAS molarity on gelation time is also illustrated in **Figure 2**. As the molarity of the AOAS increases the times to gelation decrease. As the ACH concentration decreases, gelation is not observed in the lower molarity salt systems.

There are several interactions that occur with the addition of AOAS to ACH.

pH Dependence

The pH of many of the solutions used in the formulations vary. The general trend indicates a decrease in pH as the concentration of ACH increases from 38.5% to 55.6%; the ACH solution is a more acidic environment at higher concentrations. In contrast, the AOAS solutions increase in pH with increasing molarity. ACH solutions undergo network formation, conversion to aluminum hydroxide, with the addition of NaOH[4,8]. The results observed in **Figure 2** can be interpreted through an extension of this pH gradient theory. For any given AOAS molarity, the gelation time decreases with increasing ACH concentration. The addition of the AOAS to the acidic ACH solution raises the overall solution pH to a high enough level to begin the dehydration-deprotonation mechanism[4] between the octahedra of ACH. The AOAS therefore acts in the same manner as NaOH. The gelation mechanism was determined after the gels reached their final brittle states. The extreme brittleness of these systems does not allow the pH to be measured. The pH mechanism alone does not explain the decrease in gelation times with increasing ACH concentration. Other interactions must play a role in gelation of these systems.

Influence of cation size

To investigate the effect of cations, four salts were studied, ammonium, sodium, potassium, and lithium. The addition of alkali metal OA salt solutions initiates the dehydration mechanism along with swelling within ACH layers, resulting in fast gelation times. This is dependent on the cation solution pH (or effectiveness in initiating dehydration mechanism) and hydration radii and #. Lithium, with a low solution pH and large hydration radii, is the slowest due to low initiating effectiveness and an inability to penetrate and swell the ACH, followed by sodium, and then potassium (lower pH than sodium). Ammonium, similar in size and hydration # to potassium, had the fastest gelation times due to increased solution pH and hence higher effectiveness in promoting the dehydration mechanism. pH of the solution is not the only factor though. Each cation influences gelation independently so it is probably a result of a combined pH and cation swelling mechanism.

Influence Of Hydroxyls

In order to inhibit the brittle network that forms with ACH when AOAS is added, something must "block" the surface of the ACH which will interfere with two approaching ACH particles and not allow the octahedral structures to combine. This will in effect produce a partial network in which some of the ACH units have reacted, but larger aggregates are inhibited, leaving a soft gel network. This section introduces simple glycols into the system, which are intended to block the surface of the ACH by interacting with the ACH actives.

Materials such as propylene glycol were studied. In a 50% ACH system, the trend observed is that propylene glycol promotes gelation. Gelation is not observed at 1M AOAS when propylene glycol is absent. However, gelation occurs at 1M AOAS with the addition of glycols. The final properties of the gel itself were the same as those experienced in the previous sections. The gels were transparent, but were again very brittle. The system with glycol did take several weeks longer to become as brittle as the glycol free system, despite the faster initial gelation times. Different types of glycols were used in this study. The glycols contain primary and secondary hydroxyls, many of which are sterically influenced by neighboring groups. No correlation was noted between gelation time and molar equivalents of hydroxyl. The most noticeable observation is the influence of secondary hydroxyls in comparison with ethylene glycol types. All glycols produced semi-hard to brittle gels over time. The gels produced, however, were transparent, which is an important quality desired within the gel. Additional studies were also done to evaluate the ether oxygens of glycol ethers, with terminal hydroxyl and with terminal methyl groups. The gelation times decreased as the molecular weight of the material increases. The introduction of ethylene oxide units, ether oxygens, within the other materials exhibits a greater effect on network formation of ACH than does the introduction of hydroxyls. Comparison of PEG and MePEG vs. DiMePEO in the low molecular weight range (76-2000) indicates that the end hydroxyl groups have a greater influence on gelation as their effective strength increases, or when the ethylene oxide content decreases. The hydroxyl group effect is most prevalent in the DiMePEO where only trace amounts of hydroxyl groups are present. Slightly longer gelation times are noticed when the hydroxyl caps are removed from the ethylene oxide chains. This trend disappears in the higher molecular weight range (>2000) where the ethylene oxide chains solely influence gelation.

Although network formation occurs in both glycol and ethylene oxide systems, the end result in each is a brittle solution. Brittleness is not immediate and time dependent rearrangements occur in each type of system. Over time the ACH particles rearrange and brittleness results.

Similarly, the ether linkages of the PEO systems block the surface of the ACH, only with several ACH particles at the same time. Rearrangements occur through the flexibility of the oxide chain. The absence of side groups along the oxide backbone, which could function as steric restrictions, allows easy approach of additional ACH particles to the ether linkage. These interactions occur along the entire length of the chain resulting in fast gelation.

Two systems have now been studied: the influence of hydroxyls on monomeric glycols and the influence of ether oxygens in oligomeric ethylene oxides. Both functional groups interact with the ACH particle through H-bonding, and rearrangements from separated ACH particles to complete ACH networks over time within each system allow the gel to become brittle. At this point in the study the brittle network has not been solved, however transparency in all instances has been achieved.

Many blended systems (i.e., surfactant, alone, and with different polymer combinations) were studied. The more interesting data arose from studies with the Pluronic and Tetronic copolymers.

Mixed System Interactions

One of the most important findings in this study is the positive characteristics of the gels produced. In the majority of the cases, clear, transparent, soft gels were produced. When transparency did not occur, as is the case in 25R8, 90R4, and 150R1, the HLB number was low (1-7 or 7-12) and/or the cloud point temperature of the surfactant was also low. Transparent gels were obtained in all other cases in which the HLB number and cloud point temperatures were higher, 12-18 and 24+°C, respectively.

There are several factors to consider in the explanation of these transparent, non-brittle gels. First is the molarity of the ammonium acid salt. When 6M AOAS was used, firm and flaky sticks were produced. However, when 4M AOAS was used within the same system, soft gels were obtained. In prior investigations involving the molarity effects on stick hardness it was observed that soft gels could be produced using 4M AOAS in glycol and PEO systems. The effect of the PEG:glycol ratio in these systems is seen in **Figure 3**. Gelation time increases as the PEG molecular weight is decreased and as the PEG:propylene glycol ratio decreases. This reconfirms the findings that the ether oxygens of PEG, as well as glycols, promote gelation.

The Tetronics block copolymers also contain nitrogen atoms in their structures, and this directed our studies towards vinyl amines. High molecular weight poly(ethenylformamide)[7] was introduced containing both amide and carbonyl groups. This polymer resulted in a clear, soft gel. AOAS was added to the system and produced a firmness to the gel. Vinyl alcohol/ vinyl amine polymers of high molecular weight also inhibited network formation resulting in clear, firm gels with the controlled addition of AOAS. The cellulose family was investigated and included hydroxyethyl, hydroxypropyl, hydrophobically-modified hydroxyethyl, carboxymethyl, and sililated hydroxyethyl derivatives. All derivatives, excluding the carboxymethyl due to precipitation in the highly saline environment, inhibited network formation within the ACH solution. Each polymer is high molecular weight and contains ether and hydroxyl groups to interact with the ACH. In each case, AOAS was added to enhance stick characteristics. The combination of the high molecular weight cellulose polymer containing several functional groups along with the AOAS produces clear, firm gels containing ACH. The use of either alone will not produce soft stick like characteristics. Oscillatory rheology was used to determine the increasing strength of the gel (as reflected in the plateau value of the storage modulus, G', with increasing frequency of oscillation). Some of the studies are reflected in the **Figures 4-7**.

REFERENCES: Available on request due to space limitations.

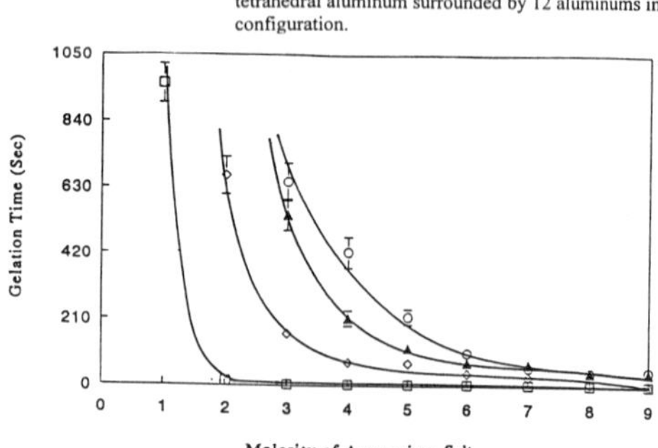

Figure 1. Structure of $Al_{13}O_4(OH)_{24}(H_2O)_{12}^{7+}$ complex showing the tetrahedral aluminum surrounded by 12 aluminums in octahedral configuration.

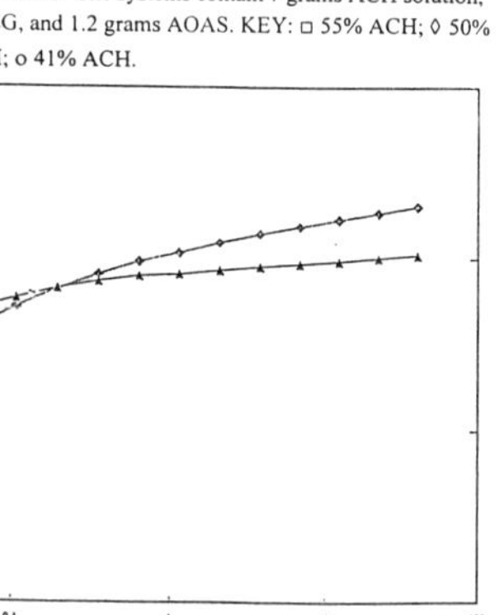

Figure 2. Influence of ACH concentrations vs. gelation times at different molarity AOAS solutions. The systems contain 7 grams ACH solution, 1 gram 5000 MePEG, and 1.2 grams AOAS. KEY: □ 55% ACH; ◊ 50% ACH; ▲ 45% ACH; o 41% ACH.

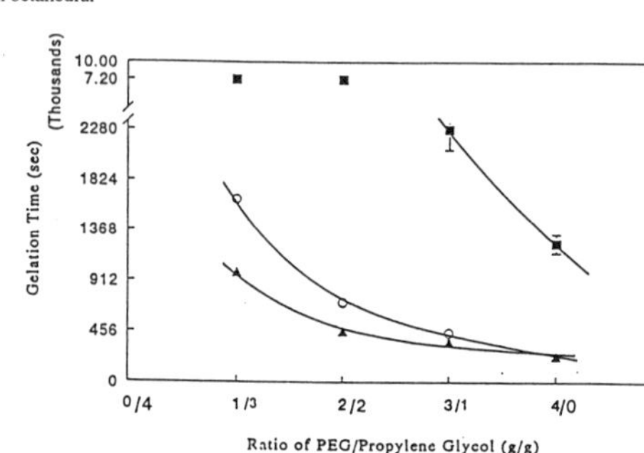

Figure 3. Influence of ratio of PEG/propylene glycol on gelation time. The system contains 8 grams 50% ACH, 4 grams total of PEG/propylene glycol, and 1 gram of 4 molar AOAS. KEY: ■ PEG 200; o PEG 1450; ▲ PEG 4600.

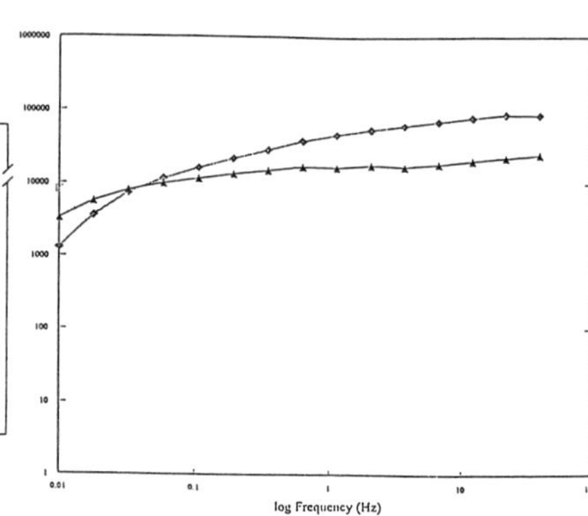

Figure 4. Oscillatory measurement (G',G") at torque = 5000 and 25°C. System contains 36% ACH, 1 gram 2000 MePEG, 1.2 grams 6 molar AOAS. KEY: ◆ G', ▲ G".

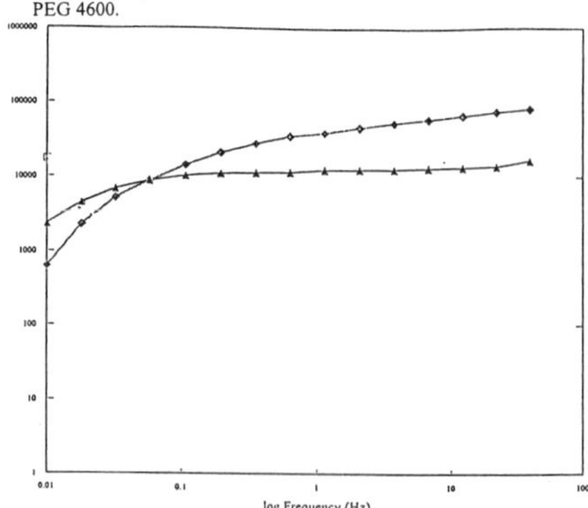

Figure 5. Oscillatroy measurement (G',G") at torque = 5000 and 25°C. The system contains 36% ACH, 1 gram 5000 MePEG, 1.2 grams 3 molar AOAS. KEY: ◆ G', ▲ G".

Figure 6. Oscillatroy measurement (G',G") at torque = 5000 and 25°C. The system contains 36% ACH, 1 gram 8000 MePEG, 1.2 grams 6 molar AOAS. KEY: ◆ G', ▲ G".

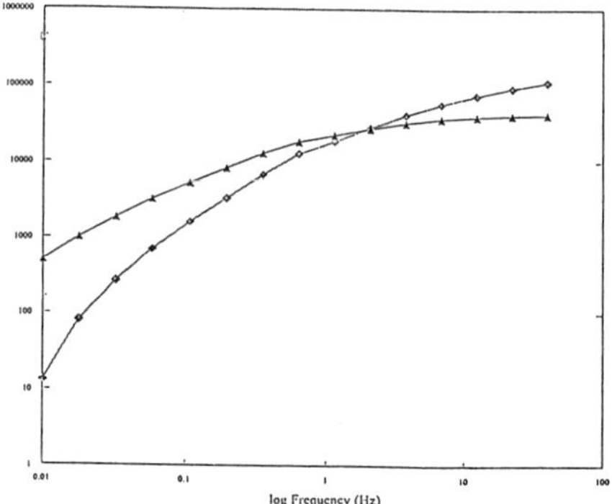

Figure 7. Oscillatroy measurement (G',G") at torque = 5000 and 25°C. The system contains 4 grams ACH, 3 grams DDI, 1 gram PG, 1.2 grams 8 molar Glutaric Acid. KEY: ◆ G', ▲ G".

IMPLANTABLE POLYMERIC BIOMATERIALS: DEGRADATION ISSUES AND APPROACHES TO ASSURING THEIR LONG-TERM PERFORMANCE.

Arthur J. Coury
Focal, Inc.
4 Maguire Road
Lexington, MA 02173

Advances in implantable medical devices which have saved or enhanced many lives over the past four decades have been made possible by the availability of appropriate biomaterials. Many of these materials of construction have consisted of synthetic polymers. Most have been selected from families of available commercial polymers adapted for specific medical uses.

An abiding issue for all long-term polymeric implants has been premature degradation. Degradation of polymeric biomaterials can commence at any time during the many stages of their intended lifetimes from initial synthesis to service inside the body.[1] Influences of reaction by-products, fabrication techniques, storage conditions, sterilization processes and environmental exposure, to name a few, can combine to predispose a polymer to suffer degradation of performance properties *in vivo*.

Once implanted, the biomaterials are subjected to a relatively stable, but aggressive, chemical environment and to a set of physical forces which can vary widely depending on the type and location of the implanted device. As a result, the biomaterials can undergo several types of degradation. Chief among these are: Mechanical degradation, biofouling, hydrolysis, oxidation and combinations thereof.[2]

Mechanical degradation often is the result of motion of device components such as with artificial joints, indwelling catheters (eg pacemaker leads), or implanted blood pumps. Consequences can range from particulate generation (causing inflammation) to catastrophic failure due to breach or leakage. Selection of mechanically and chemically durable materials and part design to accommodate applied stresses are necessary for acceptable performance.[3]

Biofouling can take the form of mineralization (calcification), absorption of body components with swelling or softening and surface fouling by proteins or salts. Mineralization can be especially insidious in that deposits formed on and within the polymer component can be sharp and abrasive, thereby accelerating wear.[4] Fouling seems to be accelerated in zones where motion is greatest. Part design to minimize mechanical excursions preserves the working lifetime of components susceptible to fouling. Antimineralization treatments originally developed for biological tissue devices (eg heart valves), may one day be useful in preventing fouling of synthetic polymers *in vivo*.[4]

Polymer hydrolysis will occur over time in the body to any material containing susceptible groups. Such moieties typically are carbonyl groups flanked by one or two heteroatoms (eg N,O,S). Esters, urethanes and amides exemplify such groups. Other hydrolyzable groups include acetals, ethers, cyanoacrylates and esters of phosphorus and sulfur acids.[1] Although certain polymers that will not hydrolyze are available (eg hydrocarbons, fluorocarbons, polysulfones) even those containing susceptible groups can be acceptable. Strategies for assuring long-term performance include selecting polymers with few hydrolyzable functions, strong hydrophobic character and highly crystalline morphology.[1]

Oxidative attack comprises the body's chief response to its recognition of a foreign body. Powerful oxidants are released by attacking phagocytic cells in an early "respiratory burst" which resolves to a mild chronic inflammatory response with biocompatible polymers.[1]

Oxidatively susceptible polymers include those with carbon-hydrogen bonds adjacent to heteroatoms (eg ethers, alcohols, amines) or unsaturated groups and those with hydrogens on branched hydrocarbon structures. Other oxidizable groups include aldehydes, thiols and sulfides.[1]

While many implanted polymers may undergo oxidative degradation, segmented poly(ether urethane) (PEU) oxidative processes *in vivo* have been among the most widely studied and reported. PEUs have been used successfully for over two decades as tough, durable elastomers for artificial heart bladders, pacemaker lead insulators and several other applications. Their utility has been limited, however, by a susceptibility to oxidative degradation. Two particular oxidation mechanisms have been detailed for PEUs: Oxidative stress cracking and metal ion-induced oxidation.[1,2] The former occurs when implanted PEUs contain applied or processed-in stresses and are accessible to phagocytic attack. The latter is caused when metallic components of PEU-containing implanted devices corrode to produce cations with high oxidation potential (eg Ag^+, Co^{3+}, Fe^{3+}). These ions attack and degrade the PEU in redox processes.

Polyethylenes (PEs) are distinguished for being highly resistant to hydrolysis. All of the forms of polyethylene used in implantable devices, ranging from how density PE (LDPE) to ultra high molecular weight PE (UHMWPE) contain branch points which are subject to oxidation by thermal processes (thermooxidation) or ionizing radiation. UHMWPE is widely used to provide articulating surfaces in total joint prostheses (eg artificial hips and knees). Fabrication and sterilization conditions can initiate bulk and surface free radical processes leading to early degradation and stable radical-induced late degradation.[6] Premature disintegration of UHMWPE has been the subject of much observation and investigation and has only been partially resolved to date.

Oxidative attack of polymers *in vivo* can be controlled by using oxidation-resistant materials, isolating susceptible polymers from phagocytes or oxidants, minimizing stresses seen by the susceptible component *in vivo*, protecting metals from corrosive environments and incorporating appropriate antioxidants into the polymer.[1,2] Attempts to moderate processing induced radical degradation of PE device components such as by sterilization with ethylene oxide instead of gamma irradiation have been partially successful.[7]

Several approaches to producing oxidation-resistant polyurethanes have involved replacing the polyether soft segments with aliphatic polycarbonate or hydrocarbon moieties.[5] These structures show improved oxidation resistance, but potentially at some cost in hydrolytic stability (with polycarbonates) or mechanical properties (with hydrocarbons).

The examples of polyurethanes and polyethylenes illustrate the general fact that polymeric biomaterials can provide acceptable long-term performance in implantable devices, but appropriate protective measures must be taken. With few initiatives to develop new polymers currently in place, the principles for achieving success with available polymers stated above will hold for some time.

1. Coury, AJ (1996): Chemical and biochemical degradation of polymers. In: Ratner, B *et al*, Eds, <u>Biomaterials Science: An Introduction to Materials in Medicine</u>. San Diego, Academic Press: 243-260.

2. Coury, AJ, Slaikeu, PC, Cahalan, PT, *et al* (1987): Medical applications of implantable polyurethanes: Current issues. Prog. Rubber Plastics Tech, 3(4): 24-37.

3. Coury, AJ (1993): Preparation of specimens for blood compatibility testing. Cardiovasc. Pathol 2(3) Suppl.: 101S-110S.

4. Pathak, Y, Schoen, FJ, Levy, RJ (1996): Pathologic calcification of biomaterials. In: Ratner B et al, Eds. <u>Biomaterials Science: An Introduction to Materials in Medicine</u>. San Diego, Academic Press: 272-281.

5. Coury, AJ (In press): Biostable polymers as durable scaffolds for tissue engineered vascular prostheses. In: Zilla, P and Greisler, H, Eds, <u>Tissue Engineered Prosthetic Vascular Grafts</u>. R. G. Landes, Pub., San Diego, Academic Press, Distrib.

6. del Prever, EB, Crova, M, Costa, L, *et al* (1996): Unacceptable biodegradation of polyethylene *in vivo* . Biomaterials, 17(9): 873-878.

7. Christante, S, Harris, BR, Carrol, ME, *et al* (1996): The role of sterilization in fatigue wear of UHMWPE tibial components. Trans. Fifth World Biomater. Cong., Toronto, May 29-June 2: 295.

SYNTHESIS AND USE OF SILICONES. <u>Alan L. Himstedt</u>, Dow Corning Corporation, Midland, MI 48686-0994

INTRODUCTION

Because of their established biocompatibility and the extensive testing done on them, polydimethylsiloxane polymers have found significant use in the healthcare arena where they extend, preserve and improve the quality of life on a daily basis.

The field of silicone chemistry is a relatively young area of endeavor. Carbon-silicon bonds, and in fact elemental silicon, do not occur in nature. However, in the fifty-five years since the first commercial use of silicones, they have been used in countless applications where the properties which derive from their unique chemical structure have been utilized. In addition to their many industrial uses they have found particular utility in the healthcare industry.

DISCUSSION

Silicone is a generic term which refers to any of a class of linear polymeric materials composed of repeating dialkylsiloxy units. The alkyl units attached to silicon are generally methyl groups, although other units may be incorporated along the chain to provide chemical functionality or to impart specific physical properties. The simplest silicone is hexamethyldisiloxane, whose structure is shown below where R=methyl and x=1.

$$R(SiR_2O)_xSiR_3$$
silicone

Increasing x results in a range of materials which are liquids even at very high molecular weights.

As mentioned earlier, elemental silicon does not occur in nature. Therefore the first step in the preparation of silicones involves the preparation of silicon metal by means of the following reaction.

$$2C + SiO_2 \rightarrow Si + 2CO$$

This reaction is carried out at high temperatures in an arc furnace. The metallurgical grade silicon metal thus produced is then reacted with methyl chloride to form a mixture of chlorosilanes where x=0-3.

$$Si + MeCl \rightarrow Me_xSiCl_{4-x}$$
chlorosilane

Because the chlorosilanes produced by this so-called "Direct Process" boil over a temperature range of about 50° C they can be separated by distillation.

Since the silicon-chlorine bond is quite polar, the silicon atom is susceptible to nucleophilic attack. As a result, chlorosilanes react vigorously with water. Dimethyldichlorosilane, which is the predominate species formed in the "Direct Process", can then be hydrolyzed to form very reactive dimethyldisilanol which readily condenses to form a mixture of cyclic and linear species as shown below.

$$Me_2SiCl_2 + 2H_2O \rightarrow [Me_2Si(OH)_2] + 2HCl$$
disilanol

$$[Me_2Si(OH)_2] \rightarrow (Me_2SiO)_x + HO(Me_2SiO)_xH$$
cyclics   *linears*

The product of this reaction is then separated into a cyclic and linear fraction. A mixture of cyclic materials where x is predominately 3, 4, and 5 is then reacted with an endblocking species in the presence of a catalyst. This is a reversible reaction and the molecular weight of the resulting polymer will depend on the amount of endblocker used.

$$(Me_2SiO)_x + (Me_3SiO)_2O \overset{cat.}{\rightleftharpoons} Me(Me_2SiO)_xMe$$
endblocker   *PDMS*

Although these polymers have a number of very interesting and useful properties, they would have little commercial utility if they could not be crosslinked into a load bearing mass. In the case of materials used in the healthcare industry, this is typically accomplished by the addition of a silylhydride to a vinyl group on silicon.

$$\sim\sim\sim SiVi + H\text{-}Si\sim\sim\sim \rightarrow \sim\sim\sim SiCH_2CH_2Si\sim\sim\sim$$

This reaction is typically catalyzed by a chloroplatinic acid derivative. If some of the vinyl groups reside along the siloxane chain, or if the silylhydride is greater than di-functional (i.e.-has three or more silylhydrides in a single molecule), a crosslinked network will be built up. Alternatively, with some minor variations, the same result can be accomplished by an organic peroxide. However in that case the decomposition by-products of the peroxide must be removed during an additional post-cure procedure.

Uncured elastomers are normally supplied by the manufacturer in a kit which contains two packages. One package usually contains the curable polymer and crosslinker while the other contains the curable polymer and catalyst. This results in a material which is stable until all of the components are intimately mixed.

Crosslinked silicone polymers by themselves are very weak materials having tensile strengths in the neighborhood of 50 psi or less. Therefore they must be reinforced by materials commonly referred to as fillers. While inorganic metal salts and oxides or diatomaceous earth may be used as fillers, the two most common examples are fumed silica and ground quartz.

Fumed silica is prepared by burning silicon tetrachloride in a hydrogen flame as shown below.

$$SiCl_4 + 2\,H_2 + O_2 \rightarrow SiO_2 + 4\,HCl$$

Because of its high surface area and chemical reactivity, fumed silica may interact with the polydimethylsiloxane thereby greatly improving the strength of the elastomer. Therefore it is referred to as a reinforcing filler. On the other hand, because ground quartz does not interact to such an extent with the polymer, it imparts less reinforcement and is known as an extending filler. This is illustrated in Figure 1 where the physical properties of elastomers filled with approximately equal levels of three different fillers is shown.

The viscosity of the silicone polymer and the level and type of various additives that make up an elastomer formulation will determine its consistency or thickness. Depending upon the consistency, a material may be fabricated in one of two ways. If it can be transported using readily available pumping equipment, the two components can be metered in the appropriate ratios, mixed in a static mixer and injected into a mold. Such materials are known as liquid silicone rubbers or LSRs and are used to make parts which contain complex geometries and must be molded in large numbers.

High consistency rubbers, or HCRs, are too viscous to be pumped with standard equipment and must be processed using equipment which can impart high shear to the materials in order to adequately mix them. Once mixed, they are then used in transfer or compression molding operations to make molded parts or in extrusion processes for making tubing.

CONCLUSION

Relative to organic polymers, silicones are a relatively new class of materials which are prepared by a series of reactions starting with the oxides of silicon which are so commonly found in the Earth's crust. Additional additives are then used to reinforce and crosslink the polymers to form an endless variety of materials for use in the healthcare industry to transport fluids, bear loads, and connect dissimilar materials. Fabrication methods differ depending upon the consistency of the material, the final end use, and the equipment available to the user.

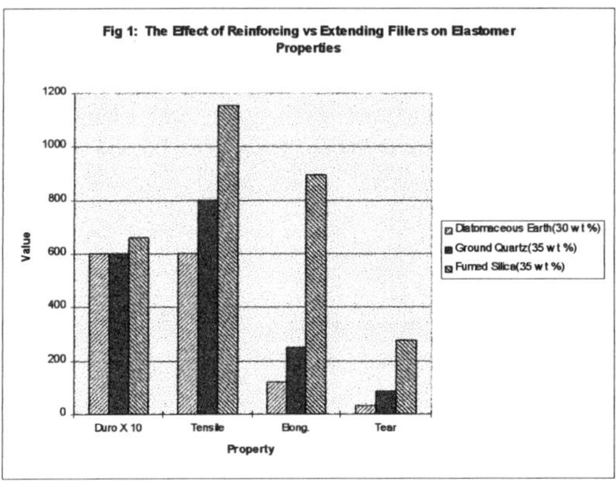

Biostable Polymers? Measuring and Predicting Stability?
LeRoy W. Schroeder, Materials Chemistry, Division of Mechanics and
Materials Science, Center for Devices & Radiological Health, FDA.

Introduction

We have been interested in the stability of polymers used in
medical devices for some time now. Both chemical and physical
degradation of medical polymers are of concern with respect to
performance and safety.   Of course it is not necessary that polymers be
stable forever, but only over the product lifetime which begs the
question. Many polymers appear to be stable in the short term, say 5%
of the product lifetime, but eventually show very slow long term
degradation. For example, PET shows a loss in strength of 50% after
about 10 years. Oxidative degradation of polyether polyurethanes also
takes quite some time to develop.   Such time scales lead to the desire to
develop accelerated aging studies in order to establish shelf lives or
product service lifetimes. We have had to provide consultation about
what products, from the viewpoint of material stability, are likely to
need a label providing shelf life information. Below I briefly describe
some of the studies my staff has undertaken in order to address
questions of polymer stability.

Hydrolysis of Foamed Polyester Polyurethane

An earlier study established that hydrolysis of the foam produced
2,4 toluene diamine (2,4 TDA) in addition to oligomeric degradation
products (1). It also produced release rates of 2,4 TDA as determined
at 37 and 50 °C based on a month's observations. Figure 1 reproduces
some of the data. The data is compatible with a zero or first order
kinetics, e.g. , [TDA] = kt or [TDA] = [TDA]$_\infty$ {1- e^{-kt}}. [TDA]$_\infty$ is
calculated from the composition assuming complete breakdown. The
thermal enhancement gave an activation energy for hydrolysis
comparable to that for other hydrolysis reactions. Hence, in principle,
although the exact chemical structure of the foam is not known, it seems
possible to predict the amount of 2,4 TDA produced over extended
times.

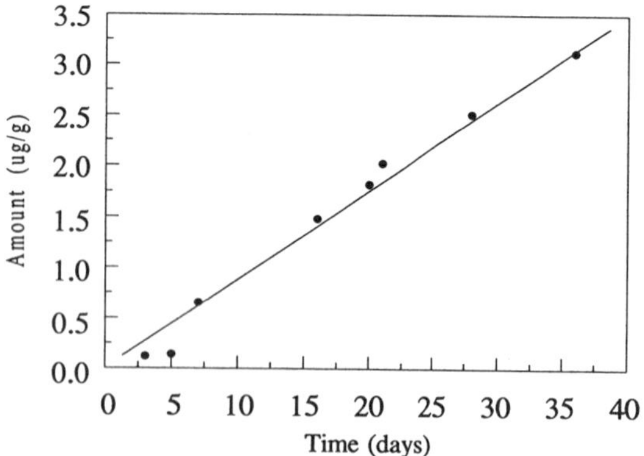

Figure 1. Production of 2,4 TDA @ 37 C

The hydrolysis study has been extended to two years in order to
produce enough non - TDA degradation products for an Ames test.
This required taking special measures to ensure sterility and eliminate
any chance of bacterial (enzymatic) biodegradation. Amounts of 2, 4
TDA were determined after 195 and 780 days of hydrolysis. Table I
shows the predicted amounts of 2,4 TDA compared to the observed
amounts. Unfortunately, the prediction based on the initial 37°C rate
gives a value over twice what was measured after two years.
Considering that even after two years the extent of reaction is only
0.015%, perhaps this result is not too bad. This is nowhere near good
practice for classical chemical kinetics as not even a significant fraction
of a half life has been measured.

Table I. Predicted and observed amounts of 2,4 TDA

Conditions	Predicted	Observed
780 days @ 37°C	68.8 ug/g	28.1 ug/g
195 days @ 50°C	0.049 mg/g	1.3 mg/g

Inasmuch as this is a heterogeneous situation there could be a
concentration gradient of reactive sites, the initial rate being determined
by surface sites and then new sites have to be uncovered by surface
erosion. See Figure 2 showing surface erosion of the foam. Surface
erosion may be controlled by hydrolysis of the polyester portion of the
polyurethane and could be the rate controlling step.   The predicted
amount of 2,4 TDA after 195 days of hydrolysis at 50 °C is too low by
a factor of 26 compared to the experimental measurements. Clearly, in
this case, the predictive model must include something more besides the
initial rate. A reaction model including the production of surface
reactive sites via the polyester hydrolysis might predict the data.

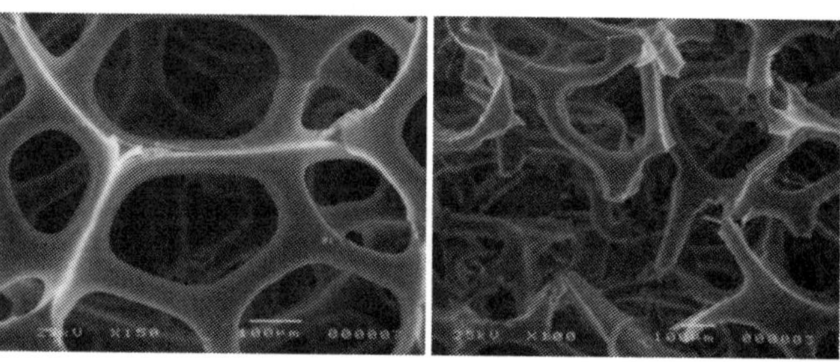

Figure 2. PU foam initially (right panel) and after 780 days of
hydrolysis (left panel).

Degradation of Cellulose Acetate

Our literature survey indicated that cellulose acetate (CA) is
subject to several modes of degradation and hence products made of CA
may need shelf-life designations. We have obtained some preliminary
data on CA, from products stored for various times up to 10 years, that
show chain scission by hydrolytic cleavage of the glycosidic bond, and a
much lesser amount of deacetylation.   Values for the activation energies
for hydrolysis and deacetylation are available in the literature.
Preliminary results for the rate of depolymerization obtained from the
old products, whose storage conditions are only approximately known,
agree with the prediction from activation energies(2). We are studying
the rates of degradation of CA at 80 and 100° C in order to verify the
predictions and to establish procedures for accelerated testing. One
potential difficulty is that deacetylation produces acetic acid which acid
catalyzes the hydrolysis.

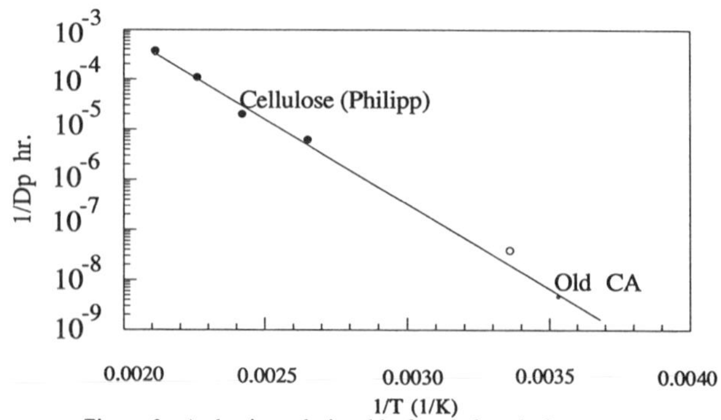

Figure 3. Arrhenius relationship for chain scission

Fatigue of Latex Barriers

Physical degradation of latex is important for barrier products because it may lead to an increase in viral transmission. We have studied the characteristics of latex membranes during and following cyclic biaxial strains at a rate of 6 per minute (3). By combining biaxial cyclic stains with simultaneous measurements of membrane conductance and capacitance together with electron microscopy we were able to observe development of surface defects and porosity before gross failure. Figure 4 shows the number of cycles to gross failure as a function of the strain. Even at our slow cycle rate real time experiments on natural latex can be done in a reasonable amount of time.

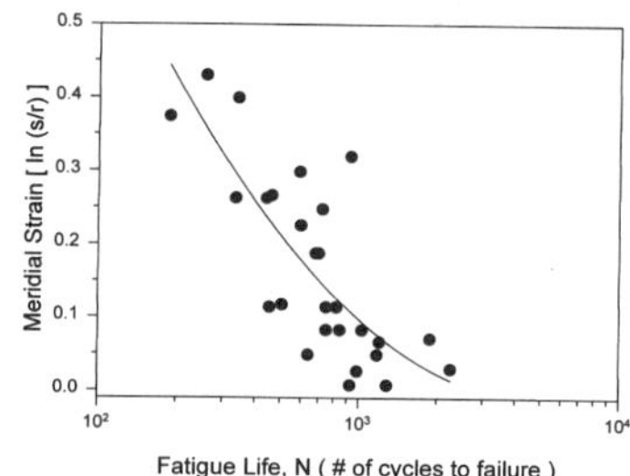

Figure 4.   Lifetime curve for natural latex in saline at room temperature.

However, the gross failure lifetime may not be the same as a lifetime based on permeability. Significant changes in ionic permeability occurred a few minutes (10-18 cycles) before gross failure. Electrically we were able to detect surface defects and porous structures, similar to crazes, 8-10 minutes (30-60 cycles) before gross failure. It is not clear from the microscopy of these fatigue induced structures that the barrier property has been compromised at that time(c.f. Figure 5). A previous study suggested that cyclic fatigue after about 100 cycles was able to increase HIV transmission across latex barriers(4). However, that study used a plunger to deform a membrane disk into approximately a cone. The radius of the plunger is very important because we know that sharp needles will cause cracks in latex and pressurization will open them up allowing transmission of virus. Hence, the results may be true, but some kind of accelerated study has been conducted which is not easily extrapolated to conditions of product use. Additional collaborative studies are in progress to determine the effect of fatigue on virus transmission using an apparatus similar to that described below.

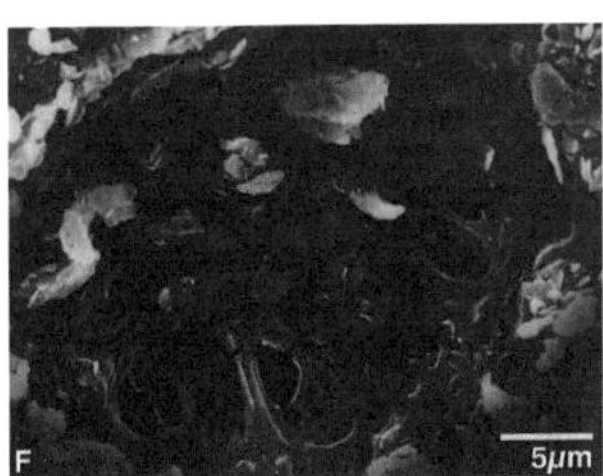

Figure 5.   Example of fatigue induced porous structure in latex.

Effect of Fold-flaws on the Fatigue Life of Siloxane Elastomers

"Fold-flaws" have been observed in explanted silicone breast implants (5). A "fold-flaw" is an in-folding that results in 180° bend in a film like shell. Some of these "fold-flaws" contain pinholes. Is a 180° bend enough of a stress riser to promote physical degradation in siloxane elastomer under repeated flexing? Repeated stress-strain cycles

as in a fatigue experiment are expected to be a more severe test. Our fatigue cell gives biaxial deformation of a membrane while in contact with fluids on both sides. Figure 6 shows the two chamber cell whose temperature can be maintained at 37°C if needed. Membrane deformation is achieved by pressurizing one chamber either by a piston pump or by air pressure. Volume displacement can be measured by a gauge mounted on the other chamber.

Figure 6.   Temperature controlled biaxial fatigue apparatus.

Membrane electrical properties are monitored at 1 kHz throughout the fatigue experiment. Results so far on non-flawed silica reinforced siloxane shells indicate a fatigue life of at least 5 million cycles in agreement with literature results on siloxane finger joints (6). We found any visible abrasion shortens the lifetime by at least an order of magnitude even with aqueous "lubrication" of the surfaces.

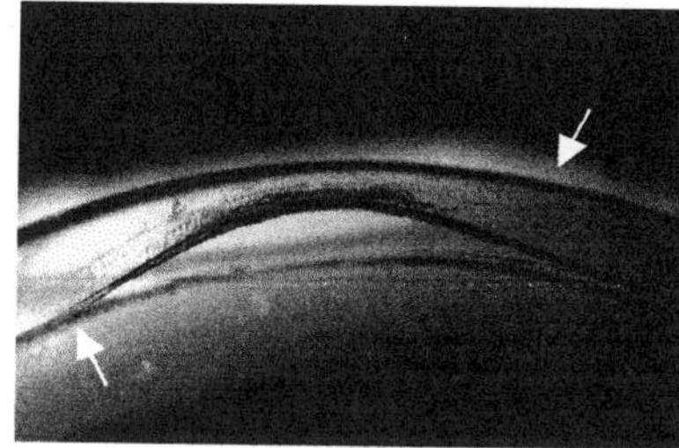

Figure 7. Abrasive failure of siloxane membrane. Upper arrow marks clamped edge, lower arrow points to ring of abrasion.

References

1. Luu, H. M. D., Biles, J. and White, K. D. *J. Applied Biomaterials* **1994**, *5* 1-7.

2. Philipp B. *Pure & Appl. Chem.* **1984**, *56*, 391-402.

3. Dillon J. and Schroeder, L. W. *J. Appl. Polym. Sci* **1997**, *64*, 553-566.

4. Goldstein, A. S. , Fleming W., and Stokes, K. in Latex as a Barrier Material, W. F. Regnault, Ed. Food and Drug Administration, Center for Devices and Radiological Health, Rockville, MD, University of Maryland, 1989, 175.

5. Rolland, C., Guidoin, R., Marceau, D., and Ledoux R. *J. Biomed. Mater. Res., Applied Biomaterials* **1989**, *A3*, 285-298.

6. Weightman, B., Simon, S., Rose, R., Paul, I., and Radin, E. *J. Biomed. Mater. Res. Symposium*, **1972**, *3*, 15-24.

PERSPECTIVES ON POLYURETHANE BIOSTABILITY AND BIODEGRADATION

J.M. Anderson, A. Hiltner, T. Collier, J. Tan, M. Shive,
S. Hasan, M. Wiggins, M. Schubert and A. Mathur

Department of Macromolecular Science
Case Western Reserve University
Cleveland, OH 44106.

To safely utilize polyurethanes in long-term biomedical devices, the biodegradation mechanisms of polyurethane elastomers must be fully understood and subsequently prevented. Until recently, major stumbling blocks in these areas have been the elucidation of mechanisms of biodegradation/biostability and the development of *in vitro* testing protocols which appropriately and adequately mimic the *in vivo* events which affect the biocompatibility and biostability of polyurethane elastomers. For over a decade, our laboratory has focused on the elucidation of mechanisms by which polyether polyurethanes undergo biodegradation. Our recent efforts to better understand the biodegradation/biostability phenomenon have focused on developing a better understanding of the effects of antioxidants and additives, the effects of strain-state, mechanistic studies of soft segment cleavage by reactive oxygen radicals, and the effects of different soft segment chemistries on the biostability/biodegradation of polyether polyurethanes (PEUUs). *In vivo* cage implant studies and *in vitro* cobalt ion/hydrogen peroxide studies have been carried out on PEUUs and the polymers have been analyzed by ATR-FTIR spectroscopy and scanning electron microscopic characterization of the PEUU surfaces.

Both *in vitro* and *in vivo* studies on the biodegradation of PEUUs have demonstrated that vitamin E, a natural antioxidant, was a more effective inhibitor of oxidative biodegradation than Santowhite®. ATR-FTIR analysis of *in vivo* specimens demonstrated that the estimated rate of ether degradation for PEUU with vitamin E was 3.6% per week compared to the PEUU with no antioxidant which was 6.6% per week. In addition, no surface pitting or cracking was observed by SEM on the vitamin E-loaded PEUU. *In vivo* biodegradation of PEUU has been demonstrated to be dependent upon adherent macrophages and foreign body giant cells which release oxygen radicals, responsible for the oxidative chain cleavage of the ether soft segment. To downregulate or inhibit the activation of adherent macrophages and foreign body giant cells, we have investigated dehydroepiandosterone as an additive. This is a known inhibitor of macrophage activation. This approach showed promise as the inhibitory effect of DHEA on inflammatory cell activity appeared to be dose-dependent with improved biostability *in vivo* being demonstrated with higher loading levels of DHEA.

Studies of the effect of stress-state on PEUU biodegradation demonstrated that stress can inhibit biodegradation. Utilizing the *in vitro* system for biodegradation, when uniaxial stress was applied, the biostability of the PEUU increased linearly with strain. When stress was biaxial, the amount of biodegradation was not reduced until the strain was greater than 200%. Decreased degradation correlated with the degree of soft segment orientation. Studies directed toward the elucidation of the effect of oxygen on the biodegradation demonstrated that the presence of oxygen had a profound effect in accelerating biodegradation.

Polyurethanes with soft segment chemistries different than polyethers have been examined for biodegradation/biostability. Two polycarbonate polyurethanes and a polydimethylsiloxane end-capped polyether polyurethane have been investigated *in vivo* for their respective biodegradation behavior. Polycarbonate segmented polyurethanes demonstrated decreased biodegradation and enhanced biostability. Current efforts are underway to examine in detail the biodegradation mechanisms of modified polyurethanes. In addition, new additives are being investigated for inhibiting adherent inflammatory cell activation which could lead to biodegradation.

References

1. Schubert, M.A.; Wiggins, M.J.; DeFife, K.M.; Hiltner, A; Anderson, J.M. *J. Biomed. Mater. Res.* **1996**, 32, 493.
2. Schubert, M.A.; Wiggins, M.J.; Anderson, J.M.; Hiltner, A. *J. Biomed. Mater. Res.* **1997**, 34, 493.
3. Collier, T.; Tan, J.; Shive, M.; Hasan, S.; Hiltner, A.; Anderson, J. *J. Biomed. Mater. Res.* **1998** (in press).
4. Schubert, M.A.; Wiggins, M.J.; Anderson, J.M.; Hiltner, A. *J. Biomed. Mater. Res.* **1997**, 34, 519.
5. Schubert, M.A.; Wiggins, M.J.; Anderson, J.M.; Hiltner, A. *J. Biomed. Mater. Res.* **1997**, 35, 319.
6. Mathur, A.B.; Collier, T.O.; Kao, W.J.; Wiggins, M.; Schubert, M.A.; Hiltner, A.; Anderson, J.M. *J. Biomed. Mater. Res.* **1997**, 36, 246.

Acknowledgment: This work was generously supported by the National Institutes of Health, Grant #HL-25239.

STABLE POLYMER MATRIX FOR CONTROLLED RELEASE DOSAGE FORM.

K. Ainaoui and J.M. Vergnaud
Laboratory of Materials and Chemical Engineering
Faculty of Sciences, University of Saint-Etienne,
42023 Saint-Etienne, France.

INTRODUCTION

With conventional immediate release dosage forms, the drug is rapidly liberated in the gastrointestine, leading to an undulating drug concentration profile in the plasma with high peaks and low troughs, and the therapeutic level is only briefly present[1].

In order to reduce this drawback, oral dosage forms with controlled release have been considered, especially by dispersing the drug through a stable polymer matrix. The process of drug release is thus controlled by a rather complex process : diffusion of the liquid into the polymer, dissolution of the drug and diffusion of the drug out of dosage form through the liquid located into the polymer[2]. But from a first approach only the drug release is of interest, and the kinetics of drug dissolution is determined through in vitro tests. These dosage forms are also evaluated using in vivo tests with healthy volunteers under standard conditions giving the plasma drug level as a function of time. As in vivo tests are highly time-consuming, attempts have been made in order to establish correlations between in vitro/in vivo experiments. Two workshops managed by the Food and Drug Administration[3] have concluded that no meaningful correlations were obtained at that time[4].

Another way was paved for assessing the drug level-time history in the plasma by building a numerical model taking all the known facts into account[5], e.g., the kinetics of release of the drug along the gastrointestinal tract, the stages of absorption in the plasma and of elimination, these last two stages being controlled by first-order kinetics. Other theoretical studies using another numerical model were made on erodible polymer, with multi-dose, showing the interest of the dose frequency and of the GI tract time[6,7].

The first objective in this study is to determine the dimensions of dosage forms of various shapes for a controlled release within 24 h by diffusion.

The other purpose is to assess the plasma drug level of these dosage forms taken once and twice a day, making comparison with immediate release dosage forms.

THEORETICAL

The following assumptions are made :
(1) The process of drug release is controlled by diffusion with a constant diffusivity.
(2) Various shapes are considered : sphere, cube, cylinder, parallelepiped.
(3) The process of drug delivery in the patient is : kinetics of drug release, first-order kinetics of absorption into and elimination out of the plasma compartment.
(4) The rate constant of absorption is constant along the gastrointestine tract.

The kinetics of drug release out of a dosage form spherical in shape is :

$$\frac{\partial C}{\partial t} = \frac{D}{r^2} \frac{\partial}{\partial r}\left(r^2 \cdot \frac{\partial C}{\partial r}\right) \tag{1}$$

The boundary condition on the surface is :

$$-D\left(\frac{\partial C}{\partial r}\right) = h\left(C_{r,t} - C_{liquid,t}\right) \tag{2}$$

with a very high coefficient of convective transfer h, the solution is given by[2] :

$$\frac{M_\infty - M_t}{M_\infty} = \frac{6}{\pi^2} \cdot \sum_{n=1}^{\infty} \frac{1}{n^2} \exp\left(-\frac{n^2\pi^2}{R^2} Dt\right) \tag{3}$$

The amount of drug X in the gastrointestine is :
$$\frac{dX}{dt} = F_t - k_a \cdot X \tag{4}$$

where F_t is the rate of drug released by the dosage form.

The amount of drug Y in the plasma is :
$$\frac{dY}{dt} = k_a \cdot X - k_e \cdot Y \tag{5}$$
The amount of drug Z eliminated is :
$$\frac{dZ}{dt} = k_e \cdot Y \tag{6}$$
An analytical solution exists in the case of an immediate release dosage form. With a controlled release dosage form, a numerical model is built in order to resolve the problem step by step.

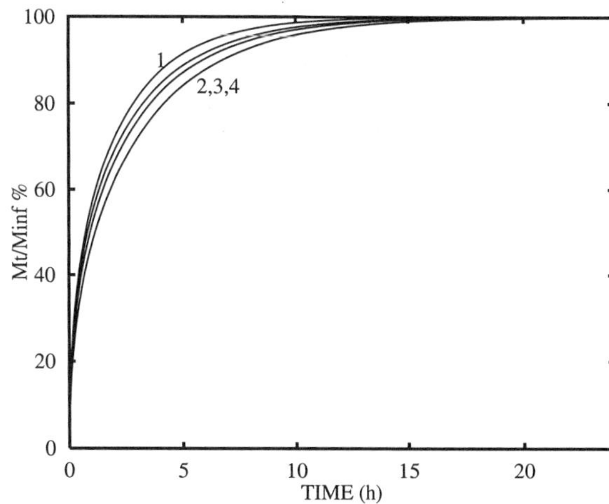

Fig. 1 - Kinetics of drug dissolution for controlled release dosage forms of various shapes : (1) parallelepiped ; (2) cube ; (3) cylinder ; (4) sphere.

RESULTS AND DISCUSSION

The kinetics of drug release out of the dosage forms as obtained through in vitro tests are shown in Fig. 1 for various shapes, leading to the following comments :

i) The kinetics of drug release obtained either by experiments or calculation[2] are in good agreement as shown in earlier studies[8].
ii) The kinetics are about the same for the four shapes considered : sphere, cylinder, cube and parallelepiped.
iii) In fact resulting from the process of diffusion, the kinetics is faster with the parallelepiped[2].
iv) These kinetics are typical : the rate is very high at the beginning and decreases exponentially with time.
v) With constant diffusivity, theoretically speaking, the whole drug is release after infinite time.
vi) More precisely, the kinetics of release are faster for the parallelepiped, and in decreasing order for the cube, cylinder and sphere.
vii) The radius of the sphere is 0.182 cm with a diffusivity of 2.5×10^{-7} cm²/s to allow for 99.9% of the drug to be released within 24 h.

The main interest for dosage forms is the plasma drug level associated with them. The plasma drug level is calculated for various dosage forms with controlled release and with immediate release when orally taken once a day (Fig. 2), and twice a day (Fig. 3). The kinetic parameters for aspirin taken as drug are : $k_a = 2.77$/h and $k_e = 0.321$/h.

A few conclusions are worth noting :

i) The plasma drug level obtained with the dosage forms with controlled release are about similar either taken once or twice a day.
ii) The plasma drug level is far more constant with the controlled release than with the immediate release dosage form, with lower peaks and higher troughs. This fact is especially true for dosages taken once a day.
iii) The effect of the dose frequency appears of great importance[6] in Figs 2 and 3. For dosages taken once a day, the controlled system is not efficient.
iv) The drawback of the process of drug release controlled by diffusion with constant diffusivity clearly appears in Fig. 2.
v) The dosage forms are built in such a way that 99% of the drug is released within 24 h, corresponding with a value of the gastrointestine tract time[7].

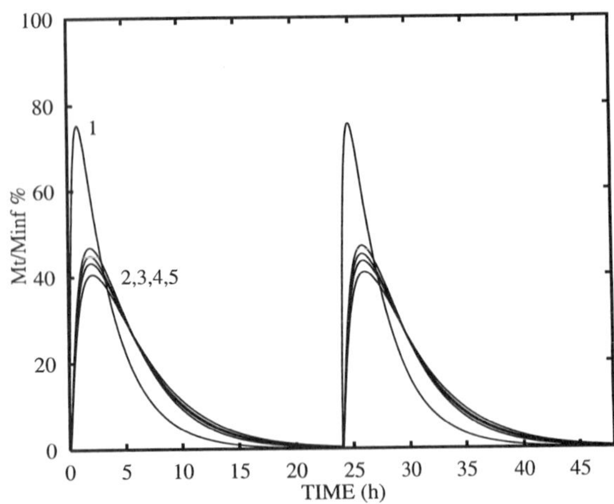

Fig. 2 - Plasma drug level for various dosage forms taken once a day :
(1) : with immediate release ; (2) : with controlled release of various
shapes

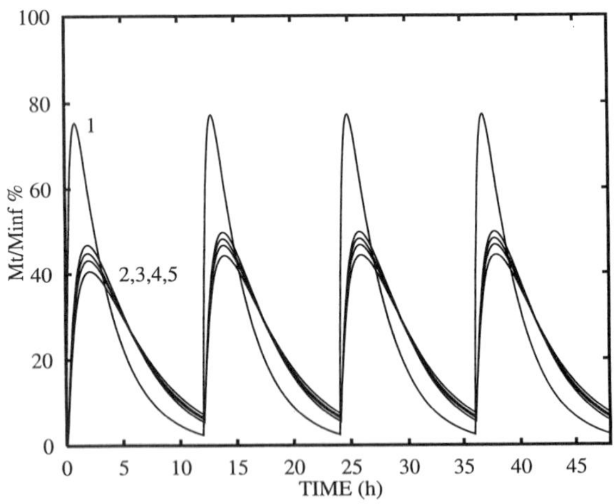

Fig. 3 - Plasma drug level for various dosage forms taken twice a day :
(1) : with immediate release ; (2) : with controlled release of various
shapes

CONCLUSION

Immediate release dosage forms are associated with alternating
plasma drug level with high peaks and low troughs, which are far from
the therapeutic level. Dosage forms with controlled release with
constant diffusivity made of drug dispersed through stable polymer
bring an improvement for the therapy. However two important
drawbacks remain with the shape of the kinetics of release which is far
from a constant rate of drug delivery and with the fact that the drug is
totally released after a long time.

BIBLIOGRAPHY

1 . K. HEILMANN.
"Therapeutic Systems". Georg Thieme, Verlag Publ., Stuttgart, 1984.
2 . J.M. VERGNAUD.
"Controlled Drug Release of Oral Dosage Forms". E. Horwood. Publ.,
London, 1993.
3 . J.P. SKELLY, G.L. AMIDON, W.H. BARR, et al.
J. Controlled Release 14, 95, 1990
Int J. Pharm. 63, 83, 1990.
4 . J.P. SHELLY, and G.F. SHUI.
 Eur. J. Drug Metabol. Pharmacol. 18, 121, 1993.
5 . B. NIA, E.M. OURIEMCHI, and J.M. VERGNAUD.
Int. J. Pharm. 119, 165, 1995.
6 . E.M. OURIEMCHI and J.M. VERGNAUD,
J. Pharm. Pharmacol. 48, 390. 1996.
7 . E.M. OURIEMCHI and J.M. VERGNAUD,
J. Pharm.. 127, 177, 1996.
8 . J. SIEPMANN, A. AINAOUI, J.M. VERGNAUD and R.
BODMEIER
Commun. AAPS, Boston, Nov. 1997.

CONTROLLED RELEASE WITH DOSAGE FORMS MADE OF CORE AND SHELL WITH DIFFERENT DRUG CONCENTRATION.

E M. Ouriemchi and J.M. Vergnaud
Faculty of Sciences, University of Saint-Etienne,
42023 Saint-Etienne, France.

INTRODUCTION

With conventional oral therapeutic systems with drug immediate release the drug passes quickly into the blood compartment, leading to high peaks of concentration followed by low troughs resulting from the following stage of elimination. Thus the plasma level alternates between high peaks and low troughs, either above or below the therapeutic concentrations, and bad side effects may result. Controlled release dosage forms orally taken are able to deliver the drug at a more controlled rate, and they are associated with a more constant plasma drug level[1]. Generally these dosage forms with controlled release consist of the drug dispersed into a polymer playing the role of matrix. . When the amount of liquid entering the stable polymer is low, the diffusivity is constant[2], while a high amount of liquid is associated with a swelling[3] and a concentration-dependent diffusivity[4]. A main drawback appears when the drug release is controlled by diffusion, with the rate of dissolution which is very high at the beginning and decreases exponentially with time, without telling that all the drug is released after infinite time

In order to reduce this inconvenience, dosage forms made of a core and shell were studied, with a lower drug concentration in the shell [5]. Thus the rate of drug delivery is lower at the beginning and more constant than that obtained without shell.

The first purpose is to evaluate the dimensions of dosage forms made of core and shell with the same polymer and lower concentration in the shell so as to obtain various kinetics of drug dissolution. Two improvements are made over the previous study[5] with the use of an implicit method of calculation and with dosage forms able to deliver 99% of the drug within 24 h.

The other objective is to assess the plasma drug level obtained with the following dosage forms : with immediate release, with a uniform drug concentration in the polymer matrix, with various values of the thickness of the shell. Some attempts managed by the Food and Drug Administration[6, 7] to correlate the kinetics of drug dissolution with the plasma drug level did not lead to definitive conclusions[8]. Thus a new way of calculating the plasma drug level with controlled release dosage forms is laid by building a numerical method taking all the known facts into account, namely, the kinetics of drug release, the kinetics of absorption into the plasma compartment and of elimination[9, 10].

THEORETICAL

The following assumptions are made :
i) The process of drug transport is divided into three stages : kinetics of release out of the dosage form, kinetics of absorption in the plasma compartment, and of elimination.
ii) The dosage forms are spherical in shape and the diffusion is radial with a constant diffusivity.
iii) The dosage forms are made of core and shell with lower concentration in the shell than in the core.
iv) The stages of absorption into and elimination out of the plasma are described by first-order kinetics.
v) The rate constant of absorption is considered as constant along the gastrointestinal tract time, as no additional information is given at the present time [11].
vi) The radius of the dosage form is calculated so as to 99% of the drug initially in the dosage form is release within 24h.

The equation of radial diffusion with constant diffusivity D is :

$$\frac{\partial C_{r,t}}{\partial t} = \frac{D}{r^2} \cdot \frac{\partial}{\partial r}\left(r^2 \cdot \frac{\partial C_{r,t}}{\partial r}\right) \tag{1}$$

where $C_{r,t}$ is the drug concentration at the radial abscissa r and time t.

The rate of drug release is :

$$F_t = -A.D.\left(\frac{\partial C_{r,t}}{\partial r}\right)_R \tag{2}$$

where A is the area of the dosage form.

The following stages are considered :
Amount of drug in the gastrointestine X :

$$\frac{dX}{dt} = F_t - k_a.X \tag{3}$$

Amount of drug in the plasma Y

$$\frac{dY}{dt} = k_a.X - k_e.Y \tag{4}$$

Amount of drug eliminated Z

$$\frac{dZ}{dt} = k_e - Y \tag{5}$$

An analytical solution exists with immediate release dosage forms.

As no analytical solution can be found for the whole process with controlled release dosage forms, a numerical method with finite differences is used.

The scheme of the process is shown in Fig. 1.

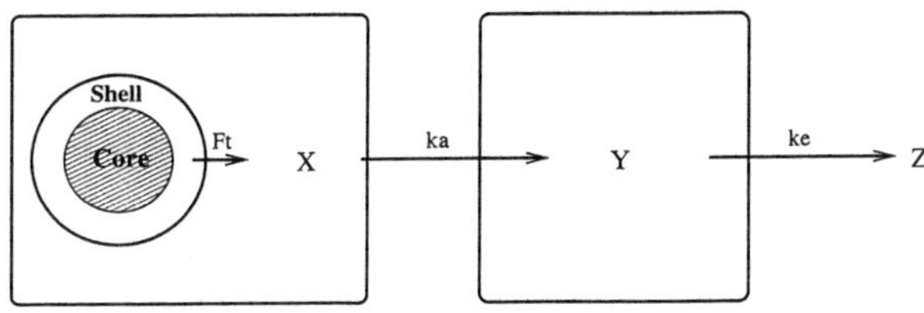

Fig. 1 - Scheme of the process.

RESULTS AND DISCUSSION

The kinetics of dissolution obtained by calculation and experiments through in vitro tests are in good agreement[12], enabling one to optimize the right dimensions of the dosage forms for the given objective of release within 24 h at 99%.

After determining the diffusivity for the release of the drug, the dimensions of the dosage forms are shown in Table 1. The kinetics of release of the two dosage forms 1 and 2 are drawn in Fig. 2, leading to a few comments :
(i) The rate of drug release is lower for the dosage form with a core and shell with lower concentration, at the beginning of the process.
(ii) The total amount (99%) of drug is released within 24 h for the two dosage forms.
(iii) Thus the rate of drug release is a little more constant with the dosage form made of a core and shell.

Table 1 - Dimensions of the dosage forms

Dosage form	Shell		core	
	R_{ext}	C_{ext}	R_{int}	C_{int}
1	0.235	0.6	0.235	0.6
2	0.235	0.2	0.207	0.6

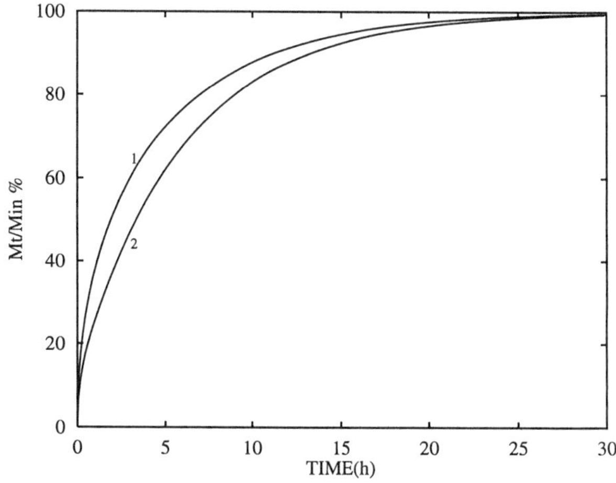

Fig. 2 - Kinetics of drug dissolution with :1 : Dosage Form without core. 2 : Dosage Form with core and shell.

The plasma drug level-time history is calculated for various dosage foms with : immediate release, controlled release with only a core, and with a core and shell with a lower drug concentration in the shell.

The drug level is Aspirin and the polymer Eudragit as matrix[2]. The pharmacokinetic parameters are for aspirin : $k_a = 2.77/h$. $k_e = 0.321/h$.
The diffusivity is : $D = 2.5 \times 10^{-7} \, cm^2/s$.

As dosage forms are taken with a given frequency, it is of interest to evaluate the plasma drug level with multi-doses. The three dosages forms are considered : with immediate release (1), with controlled release with (2) and without shell (3), taken once a day (Fig. 3) :
The following conclusions are worth noting :
i) The plasma drug profile with the immediate release is typical either when taken once or twice a day, with a high peak and low trough.
ii) The dosage forms with controlled release are associated with lower peaks and higher troughs, leading to a more constant drug level.
iii) The dosage form with a core and shell is associated with the lowest peak and the highest trough, and thus with the most constant drug level.

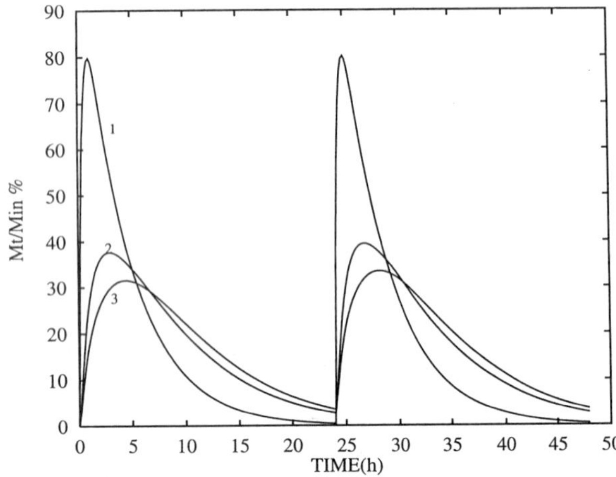

Fig. 3 - Plasma drug level with dosage forms : 1 : immediate release.
     2 : with core alone.
     3 : with controlled release and core and shell.

CONCLUSION

It is well known that dosage forms with controlled release are associated with a more constant drug level in the plasma than their immediate release counterparts. With stable polymers and a process of drug release controlled by diffusion, a drawback exists with the shape of the kinetics with a high rate of release at the beginning which decreases exponentially with time. An improvement is obtained for these dosage forms by using a core and shell with a lower drug concentration in the shell. In fact the rate of drug release out of the dosage form with a shell at lower concentration is lower at the beginning of the process.
As a result from this kinetics of release, the plasma drug level is more constant with the dosage form with core and shell, with lower peak and higher trough.

BIBLIOGRAPHY

1 . K. HEILMANN.
"Therapeutic Systems." Georg Thieme, Stuttgart, 1, 1983.
2 . J.M. VERGNAUD.
"Controlled Drug Release of Oral Dosage Forms", E. Horwood.
Chichester, U.K., 1993.
3 . N. A. PEPPAS, R. GURNY, E. DOELKER, and J.BURI.
Membrane Sci.. 7, 241, 1980.
4 . N. BAKHOUYA, A. SABBAHI and J.M. VERGNAUD.
Computat. Theoret Polymer Sci. 6, 109, 1996.
5 . E.M. OURIEMCHI, J. BOUZON, J.M. VERGNAUD.
Int. J. Pharm. 102, 47, 1994.
6 . J.P. SKELLY, W.H. BARR, L.Z. BENET,. et al.
Pharm. Res. 4, 75, 1987.
7 . J.P. SKELLY, G.L. AMIDON, W.H. BARR, et al.
Pharm. Res. 7, 975, 1990.
8 . J.P. SHELLY, and G.F. SHUI.
Eur. J. Drug Metab. Pharmacok. 18, 121, 1993.
9 . E.M. OURIEMCHI and J.M. VERGNAUD[1].
Int. J. Pharm. 127, 177, 1996.
10 . E.M. OURIEMCHI and J.M. VERGNAUD[2],
J. Pharm. Pharmacol. 48, 390. 1996.
11 . G.L. AMIDON, H. LENNERNÄS, V.P. SHAH and J.R. CRISON
Pharm. Res. 12, 413, 1995.
12 . J. SIEPMANN, A. AINAOUI, J.M. VERGNAUD and R. BODMEIER
Commun. AAPS, Boston, NOV. 1997.

ULTRAHIGH MOLECULAR WEIGHT POLYETHYLENE
IN JOINT REPLACEMENT PROSTHESES:
INTRINSIC AND EXTRINSIC FACTORS AFFECTING WEAR
PERFORMANCE.
M. Spector[*], W.M. Jones, and A. Bellare
Department of Orthopedic Surgery, Brigham and Women's Hospital,
Harvard Medical School
Boston, MA 02115

Wear particles from ultrahigh molecular weight polyethylene (PE) components of total hip and knee replacement prostheses elicit an inflammatory response that can result in bone resorption and implant loosening, thus necessitating a revision surgical procedure. The wear performance of the polymer is affected by intrinsic factors including the polymer structure at the nanometer and micrometer levels, and extrinsic factors including the topography of the metal counterface. Certain modifications of PE, including one that increased the percent crystallinity and crystallite size (referred to as Hylamer), were found to increase, rather than decrease, the wear rate clinically and result in a rapid destruction of surrounding bone, likely due to the number of wear particles generated. These findings provide an empirical guide for future strategies for altering the polymer structure to improve tribological properties. Other recent findings have revealed scratches on the metal counterfaces, with ridges - lips - that have been found in previous laboratory wear testing to result in order of magnitude increases in wear rate. This work supports the value for more scratch-resistant surfaces to articulate with PE in joint replacement prostheses.

Clinical Performance of a Modified form of UHMWPE

A previous study has documented a higher wear rate of Hylamer acetabular liners compared with conventional PE, and has correlated an increase in the numbers of osteolytic lesions with total wear (1). This study sought to determine the nature of the wear particles produced by this modified form of PE and the histological response elicited by these particles.

Peri-implant tissues were obtained from eight patients implanted with Hylamer acetabular liners for periods from 2.5 to 5.0 years. Tissue samples were also obtained from eight contemporaneous revisions of conventional PE cups and analyzed as controls. Tissues were fixed in formalin and embedded in paraffin. Microtomed sections were stained with hematoxylin and eosin and viewed under conventional and polarized light microscopy. Hylamer particles and conventional particles were also isolated from tissue samples. In brief, the tissue was digested in 5N NaOH, and the particles isolated from the resulting solution using a sucrose and isopropyl alcohol gradient. Three 150 μl aliquots, two washed with 2 ml of triple filtered distilled water and one unwashed, were dried on 0.2 μm pore filter paper. Random sections from each filter paper were viewed under a low voltage high resolution scanning electron microscope (LVHRSEM) at 0.8 kV without the need for conductive coating.

Peri-implant tissues contained regions of fibrous scar tissue, synovial tissue and areas composed of macrophages and multi-nucleated foreign body giant cells with few areas of necrosis. Polarized light microscopy of the Hylamer specimens revealed large numbers of birefringent particles. Entire fields of view under a 20x objective lens were filled by macrophages and giant cells with the cytoplasm of each cell completely filled with the birefringent particles. Many of the Hylamer and conventional PE particles found within the macrophage cytoplasm were too small and numerous to accurately measure and quantify. Larger Hylamer particles found within the cytoplasm of giant cells were generally rod-shaped or ovoid in morphology. Larger conventional PE particles within the giant cells were not as regular in their morphology as Hylamer particles nor were they seen as frequently as the larger Hylamer particles.

Under LVHRSEM, particles were clumped into aggregates (>10 particles/aggregate) or isolated as separate particles. The particle morphologies fell into two basic categories: elongated and ovoid. There were three types of elongated particles: "tadpole"-shaped particles with an

ovoid head; long, thin curving particles with tapering ends termed "fibrils"; and shorter, rod-shaped straight particles. Elongated particles ranged in length from 0.6 to 18.4 μm. The ovoid particles were regularly and irregularly shaped. Particle diameters ranged from 0.2 to 3.7 μm. Cauliflower-shaped aggregates with diameters up to 2.4 μm were also seen and may have been conglomerates of smaller particles. There was no difference in size between aggregated particles and isolated particles.

The presence of Hylamer particles of phagocytosable size in macrophages and giant cells demonstrates that they are capable of inciting an inflammatory response known to be associated with osteolysis with the same cellular make up as conventional PE. That the cells with large numbers of Hylamer particles in their cytoplasm displayed intact nuclei and cell boundaries indicates that the Hylamer particles are not cytotoxic.

Most of the phagocytosed PE particles in both groups were sub-micrometer in size. The size and morphology of many of these particles could not be determined from the histologic sections. Moreover, particles under 0.4 μm seen in the LVHRSEM cannot be seen in the light microscope; making particle isolation necessary. This technique may give insight into the possible differences in wear processes between Hylamer and conventional PE. LVHRSEM is a valuable tool for the study of the morphology of PE particles. Its low voltage capabilities eliminates the need for conductive coating of a sample and prevents surface damage seen in high voltages which would obscure the finer details of the particles.

The size and shapes of the isolated conventional PE particles found in our study were comparable to those reported in a previous investigation (2). We found that the isolated wear particles from Hylamer cups had similar characteristics to the isolated particles from the wear of conventional PE devices.

These findings suggest that the inflammatory response to Hylamer PE particles which leads to an increase in the number of osteolytic lesions around the prosthesic device is most likely due to the generation of a greater number of wear particles from Hylamer acetabular liners, reflected in its higher wear rate, rather than a difference in the characteristics between Hylamer and conventional PE wear particles.

Analysis of the clinical wear behavior of modified forms of PE that have not proven to yield improved performance can provide a rational basis for future alterations in the polymeric structure.

The Role of Scratches on Metal Counterfaces on PE Wear

Scratches on the metallic components of total joint replacement prostheses, even if few in number, could lead to an increased wear of the articulating PE components. A previous laboratory study by Dowson *et. al.* showed that even a single scratch on a metal counterface could have a significant effect on polyethylene wear (3). The pile-up of metal bordering the scratch (reflected in the parameters, R_p and R_{pm}, described below) and not the depth (contributing to R_a) was the primary factor in increasing polyethylene wear. This work prompted a study to determine the surface roughness and scratch profiles on femoral heads and condyles of total hip and knee replacement prostheses.

Cobalt-chromium femoral heads and condyles, and their corresponding PE components, were obtained at revision surgery. Light microscopy was used to identify scratched regions. Topographical measurements were made using an optical interferometer that scanned the identified wear regions on each condyle. The values obtained were: R_a-average roughness, the arithmetic average height calculated over the entire surface; R_p-maximum profile peak height, the distance between the highest point of the surface and the mean surface; and R_{pm}-mean profile peak height, the average peak height on the surface. These measurements were also made on non-implanted "controls". Semi-quantitative assessments of polyethylene damage (scratching, burnishing, pitting, deformation, and delamination) were made using stereomicroscopy.

Under light microscopy all PE components displayed scratching and burnishing to varying degrees. Embedded debris particles were seen in varying amounts in the PE devices. All metallic femoral heads and condyles showed evidence of scratching. Typical individual scratches had widths ranging from 20 to 50 μm with depths ranging from 0.5 to 5 μm. Pile-up of metal was seen along and at the end of some scratches with peak heights as high as 8 μm. Values for the non-implanted metallic components were R_a=0.01 μm, R_p=0.3 μm, and R_{pm}=0.2 μm.

A laboratory wear study showed an order of magnitude increase in wear of UHMWPE as the height of the pile-up of the scratch on the metal counterface (*i.e.*, R_p) increased from 0.2 to 2.3 μm (3). Several of the metallic femoral heads and condyles in our study displayed scratches with R_p greater than 2.3 μm. Moreover, R_{pm} may be a better indicator of the degree of pile-up since it is an average of the profile peaks throughout the entire surface. The observation that metallic and cement debris was not always found embedded in the PE acetabular cups and tibial plateaus may indicate that particles responsible for scratching the surface can be worn and dislodged. These findings suggest that the implementation of a more scratch-resistant surface may be beneficial.

Acknowledgments: This work was supported by Smith&Nephew Orthopaedics, National Medical Fellowships Inc., the VA Rehabilitation R&D Service, and the Brigham Orthopedic Foundation.

References
1) Livingston, B., *et al.*, Trans 43rd Orthopaedic Research Soc.: 141, 1997
2) Campbell, P., *et al.*, J. Biomed. Mat. Res. 29:127-131, 1995.
3) Dowson, D., *et al.*, Wear 119: 277-293, 1987

[*]Also: Rehabilitation Engr. R&D, Brockton/West Roxbury VA Medical Center, West Roxbury, MA

PREVENTION OF OXIDATION IN THE MANUFACTURING PROCESS OF UHMWPE IMPLANTS

D.C. Sun, S.S. Yau, G. Schmidig, P. Pereira, C. Stark, J.H. Dumbleton, Howmedica Inc., Pfizer Medical Technology Group, 359 Veterans Blvd., Rutherford, NJ 07070

INTRODUCTION

Ultra high molecular weight polyethylene (UHMWPE) has been used as a bearing surface in total hip and total knee replacements for more than two decades with good clinical success. Oxidation in shelf aged[1] and clinically retrieved[2] UHMWPE implants has been observed recently. A subsurface white band was observed in a sectioned acetabular cup or tibial insert within which a high oxidation level was found when analyzed by FTIR[3]. Oxidation has been linked with partial loss of mechanical property, molecular weight, and wear resistance[4]. Since oxidation can occur in various manufacturing steps, a completely oxidation-free process can only be achieved if the oxidation mechanism in each step is understood. In a general UHMWPE implant production process, UHMWPE resin is stored in an air-filled container. Upon consolidation into rods or sheets (extrusion or compression molding), air can be entrapped in the solid to cause oxidation at later steps. The consolidated rod or sheets are then annealed, generally in air, to gain dimensional stability prior to machining. This step exposes UHMWPE to heat and oxygen and may cause oxidation. The finished product is packed in air or in an inert atmosphere followed by gamma sterilization. Free radicals generated during radiation can react with oxygen present in the structure of UHMWPE, in the package, or *in vivo*. In this study, each of the above steps was examined for the possible oxidative effects.

METERIALS AND METHODS

UHMWPE rods and 1-mm thick sheets machined from UHMWPE rods were prepared according the conditions listed in Table 1:

Table 1 Process conditions of UHMWPE rods and 1-mm thick sheets

Sample ID	GUR Resin	Process
Rod A	4150	rod extruded, air annealed, and shelf aged for ~ 5 years
Rod B	1050	air in resin removed by vacuum/nitrogen purge; rod extruded in nitrogen, annealed in air at 165°C
Rod C	1050	air in resin removed by vacuum/nitrogen purge; rod extruded in nitrogen, annealed in nitrogen at 165°C[5]
Sheet A	1020	rod compression molded and air annealed; sheet packed in vacuum aluminum foil pouch followed by gamma radiation at 4.0 Mrads
Sheet B	1120	rod compression molded and air annealed; sheet packed in nitrogen using a Barex blister/aluminum foil lid followed by gamma radiation at 3.0 Mrads.
Sheet C	1020	rod inert extruded, irradiated in nitrogen at 3.0 Mrads, and annealed/stabilized in nitrogen at ~140°C[6]; sheet machined from rod.
Sheet D	1020	rod extruded and air annealed; sheet irradiated in air at 3.0 Mrads

For oxidation analysis, the IR carbonyl peaks between 1660 and 1800 cm^{-1} were integrated and the peak area is divided by the area of the 1463 cm^{-1} (methyl and methylene) peak to obtain the oxidation index. Oxidation indices at various depths in an UHMWPE rod were obtained to construct an oxidation profile. Accelerated aging was done in an air oven at 80°C for various time periods according to a published procedure[7].

RESULTS

Figure 1 shows oxidation profiles for three unirradiated Rods A, B, and C. Rod A shows some oxidation after ~ 5 years of shelf storage. The oxidation front penetrates ~ 3 mm depth into the UHMWPE rod. For Rod B (inert consolidated and air annealed at 165°C), the oxidation index at the surface reaches 0.9 (off scale in Figure 1). The rod surface turned brownish after annealing. Oxidation penetrates through the entire rod diameter (2.5 inch). In contrast, Rod C (inert consolidated and inert annealed at 165°C) is oxidation free. Table 2 shows oxidation indices for the four 1-mm thick UHMWPE sheets with 13-day, 20-day, and 30-day accelerated aging at 80°C in an air oven.

DISCUSSION

The heat exposure during extrusion and air annealing are believed to generate free radicals and result in slow oxidation in Rod A. For Rod B, severe oxidation and discoloration at the surface zone and a lower level of oxidation in the interior are caused by the air annealing at 165°C. Another rod prepared in a similar manner but annealed at 135°C (near the melting temperature of UHMWPE) showed a high surface oxidation index (0.3) with negligible discoloration. In the molten state, oxygen diffusion into the UHMWPE matrix increases and oxidation penetrates into the interior of the polymer. Rod C showed no sign of oxidation throughout the entire diameter owing to the complete oxygen-free process from resin to machining. For 1-mm thick UHMWPE sheets, Sheet D (irradiated in air) is used as a control. This sample showed significant and increasing oxidation at 13, 20, and 30-day aging. Sheet A (vacuum foil pouch packed and irradiated at 4 Mrads) showed a significantly lower oxidation index compared to the control, but the oxidation index increases with the aging time. Sheet B behaves similarly to Sheet A regardless of considerable difference in radiation dose (3.0 vs. 4.0 Mrads), resin type (GUR 1120 vs. GUR 1020 with the former containing calcium stearate), and packaging (Barex blister vs. vacuum foil). Both samples showed the benefit of inert radiation but lack of long-term oxidation prevention. Sheet C (complete inert process) shows both short term and long term stability with no detectable oxidation. Note that 13 to 20 day accelerated aging is equivalent to approximately 5 to 8 years of shelf aging[7]. A 30-day accelerated aging is too severe and may never be experienced in reality.

Table 2 Oxidation indices for 1-mm thick UHMWPE sheets

Sample ID	Oxidation index, 13-day aging	Oxidation index, 20-day aging	Oxidation index, 30-day aging
Sheet A	0.004	0.013	0.033
Sheet B	0.004	0.009	0.034
Sheet C	0.00	0.00	0.00
Sheet D	0.067	0.145	0.305

CONCLUSION

This study demonstrates that oxidation can occur during various manufacturing steps of UHMWPE implants. If a complete inert manufacturing process, including a post-radiation stabilization step, is employed, oxidation free UHMWPE implants can be achieved.

REFERENCES

1. R. Roe, et.al., J. Biomedical Materials Research, Vol 15, PP 209-230 (1981). 2. B.D. Furman, et.al., 43rd Annual Meeting of Orthopedic Research Society, PP 92-16 3. D.C. Sun, et.al., 40th Annual Meeting of Orthopedic Research Society, PP 173-30. 4. R. Rose, et.al., Wear, 77:89-104, 1982. 5. US Patent 5,650,485. 6. US Patents 5,414,049; 5,449,745; 5,543,471; 5,728,748 7. D.C. Sun, et.al., Polymer Preprints, 35:2, 969-970, 1994.

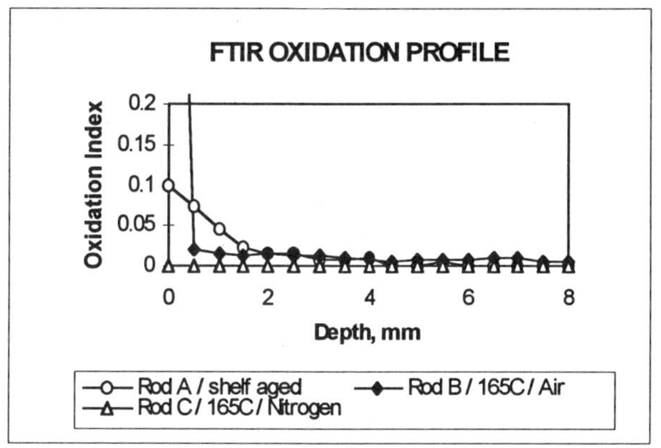

Figure 1

THE EFFECT OF MOLECULAR WEIGHT BETWEEN CROSSLINKS (M_c) ON THE WEAR BEHAVIOR OF CROSSLINKED UHMWPE

Orhun K. Muratoglu*, Charles R. Bragdon*, Daniel O. O'Connor*, Murali Jasty*, Rizwon Gul**, William H. Harris*

*Orthopaedic Biomechanics Laboratory, Massachusetts General Hospital, Boston, Massachusetts, 02114

** GIK Institute of Engineering Sciences and Technology, Topi, Pakistan

INTRODUCTION

Total joint replacements (TJR) with metallic alloys and ultra-high molecular weight polyethylene (UHMWPE) have revolutionized the treatment of end-stage arthritis over the past three decades. Despite the enormous success of TJRs, wear of the UHMWPE articulating surface and the associated biological reactions to the wear debris have emerged as the major problems in total joint replacements.

Several published reports have demonstrated excellent wear performance of highly crosslinked UHMWPE as used in total hip arthroplasty (THA) in humans. Oonishi et al. reported on the successful use of UHMWPE (in THA) crosslinked in the solid state with gamma-radiation with improved wear resistance in patients. Similarly, Wroblewski et al. reported on the successful use of silane crosslinked high-density polyethylene articulating against ceramic femoral heads. Their clinical and joint simulator studies showed remarkable agreement in the wear rates of the crosslinked polymer. We recently reported on the improved wear behavior of a form of UHMWPE in which the polymer was crosslinked in its molten state using electron-beam radiation to 200 kGy absorbed dose level. In hip joint simulator experiments carried out to five million cycles, the total weight loss was 115 mg for uncrosslinked control components in comparison with undetectable loss for crosslinked components.

The above studies demonstrate that crosslinking significantly improves the wear resistance of UHMWPE under lubricated conditions that either occur naturally in the human joint or are reproduced in hip joint simulators. To determine the mechanisms that operate in improving the wear resistance of UHMWPE under these conditions by the formation of a densely crosslinked network structure, we studied the effect of radiation dose, and hence the effect of varying crosslink densities, on some of the material properties of UHMWPE as well as its wear behavior on a unique bi-directional pin-on-disc wear tester.

EXPERIMENTAL

UHMWPE was crosslinked using both radiation and peroxide chemistries. The crosslink density and molecular weight between crosslinks were determined using Flory's swelling equation. The equilibrium volume swell ratios were measured using a Perkin-Elmer TMA-7. The wear behavior of the crosslinked test samples were determined using a bi-directional pin-on-disc (POD) wear tester. The pins used in the POD experiments were machined from the crosslinked test samples. Wrought Co-Cr alloy discs with implant surface finish ($R_a = 1.54 \pm 0.21$ microinches) were used as the counterface. The pin-on-disk wear experiments were carried out in bovine serum to two million cycles with a weight measurement at every 0.5 million cycles.

The following is the list of the crosslinked UHMWPEs used in this study: (1) E-beam (10 MeV, 40 kW) cross-linked GUR 4150 (irradiation followed by melt-annealing at 150°C for 2 hours); (2) GUR 1050 cross-linked with 2,5-Dimethyl-2,5-di-(tert-butyl-peroxy)hexyne-3 (peroxide 130) (3) Himont 1900 cross-linked with peroxide 130. The dose level of the radiation crosslinked UHMWPEs varied from 0 to 300 kGy; while the peroxide content of the latter two varied from 0 to 0.7 wt%. Along with the wear behavior, the tensile mechanical properties and thermal properties of the crosslinked test samples were also determined as a function of crosslink density.

RESULTS AND DISCUSSION

The crosslink density of the UHMWPE increased as a function of increasing radiation dose and peroxide content. In all three cases, the crosslink density asymptotically approached a saturation limit (~220 mol/cm^3 for radiation, ~170 mol/cm^3 for 1500/peroxide, and ~130 mol/cm^3 for 1900/peroxide).

In radiation crosslinked UHMWPE, the yield strength (YS) and modulus (E) remained unchanged as a function of increasing dose (dose-independent properties); while the ultimate tensile strength (UTS) and elongation at break (EB) decreased with increasing radiation dose (dose-dependent properties). In peroxide crosslinked polymer, we observed an opposing behavior; the YS and E decreased with increasing peroxide content (content-dependent properties); while the UTS and EB remained constant at varying peroxide contents (content-independent properties).

The POD wear rate decreased as a function of increasing crosslink density for all three homologous series studied. There was no correlation between the wear rate and mechanical properties of the crosslinked polymers. However, independent of the starting resin, resin molecular weight distribution, and the crosslinking technique used, we observed a linear correlation between the wear rate of the crosslinked UHMWPE and the molecular weight between crosslinks (see Figure 1).

It appears that molecular weight between cross-links is a material property that plays a fundamental role in the improvement of the wear behavior of crosslinked UHMWPE.

Figure 1. Wear rate as a function of molecular weight between crosslinks of radiation and peroxide crosslinked UHMWPEs.

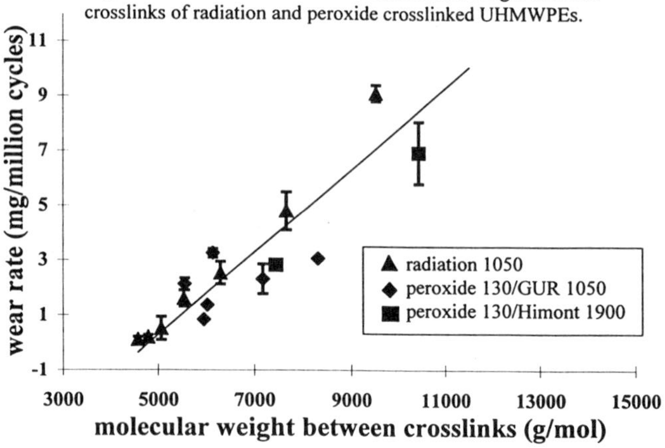

CHARACTERIZATION OF GAMMA IRRADIATED UHMWPE AFTER AN OXIDATIVE CHALLENGE

J.V. Hamilton *, K. Greer *, M. B. Schmidt *, C. Urian *,
P. Joshi **, and A. Crugnola **
*Johnson and Johnson Professional, Inc.
325 Paramount Dr. Raynham MA 02767
**University of Massachusetts - Lowell , Lowell MA

INTRODUCTION

Gamma irradiation is used to generate free radicals in UHMWPE during the sterilization process of orthopaedic implants. These free radicals undergo a number of reactions, the dominant ones are crosslinking and oxidative chain scission. Crosslinking reactions have been shown to improve the wear performance of UHWMPE in hip simulator studies[1,2]. Oxidative chain scission reaction have been linked to presence of subsurface white bands in the UHMWPE and a corresponding increase in clinical knee wear[3]. Residual free radicals have been associated with oxidative reactions in gamma sterilized UHMWPE while the components are stored in a warehouse. A feature required of an accelerated aging method is that the oxidation rate after aging should return to that which occurs under normal storage conditions. One objective of this study is to determine if the oxidation rate after the oxidative challenge returns to that normally seen on the shelf. Another observation is the role the surface condition can play in the oxidation response during a thermal challenge.

In order to gain further understanding on the effects of the oxidation on the extent of crosslinking a new swelling ratio test method was required. When the ASTM method is applied to thin films a large amount of variability occurred in the results. The objective of this part of the study was develop a new swelling ratio test that offered improved ease of use and reproducibility.

MATERIALS AND METHODS

Oxidation Study: A block(60 x 19 x 75 mm) of compression molded GUR 1020 (Perplas Bacup UK) was gamma irradiated (40 kGy) in an air atmosphere. Sample blocks (SB) (30 x 19 x 12.5 mm) were machined from the larger block as shown in figure 1. The six surfaces of the SB can be characterized by whether the surface was on the exterior (E) or the interior (I) of the block at the time of irradiation. This determined three orthogonal axis within the sample with directions of EE, EI and II. An oxidative challenge of air at atmospheric pressure and 78[0] C for 25 or 33 days were used. Films for FTIR analysis were microtomed from the center of the SB. Films from the 25 day SB were aged 70 days post challenge at room temperature and then reanalyzed.

Oxidation was measured through the thickness of the sample using an FTIR microscope and reported as the oxidation index (OI). Measurements were taken at the 0, 0.25, 0. 5, 1 mm and every mm thereafter to a depth of 4 mm from each surface of 25 day sample and 0, 0.5 1, 2 and 4 mm from the E surface of the 33 day sample.

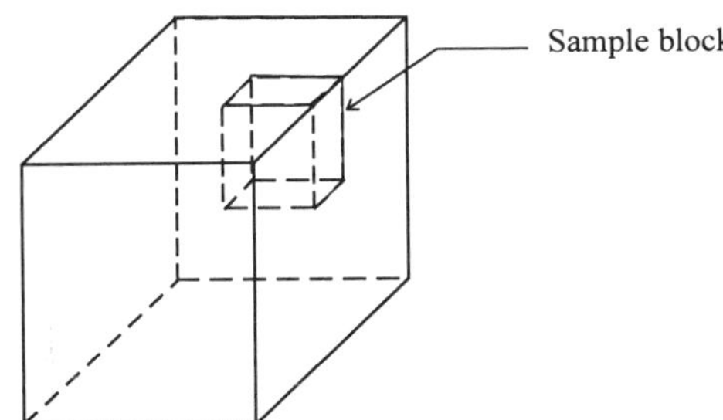

Figure 1: Typical location of were the sample block is machined from larger block and used in oxidation studies.

Swelling Ratio: 250 micron thick films of GUR 1020 were tested either as received (AR) or gamma irradiated (40 kGy). Irradiation environment, air (GA) or vacuum (GVF) was determined by the packaging environment. The GVF process has been shown to produce a higher level of crosslinking than either the AR or GA materials.

Two methods of swelling measurements were evaluated. The first is a modification of ASTM D2765 90 Method C, which involved using a sample with a greater surface area to volume ratio than recommended. The new method is an optical measurement of the swollen material. A schematic of the apparatus used is shown in figure 2. 10 mm dia discs were placed on a holding fixture. Photo was taken of disc prior to immersion in 110[0] C Xylene bath using a Polaroid camera attached to eyepiece of a stereomicroscope. The bath was raised to immerse the holding fixture and held for 1 hour. At which point the maximum swelling condition, was reached. The bath was lowered until the sample breaks the surface of the xylene and a photo was immediately taken. The photos were digitally scanned and the area of each sample was analyzed using a image analysis program (NIH image). From these area measurements and using an assumption of isotropic swelling the volumetric swelling ratio was calculated. For a disk , the ratio of the change in the height is proportional to the change in area. This relationship is shown in equation (1) were H is the height of the sample, A is the area and the subscripts i and f refer to the initial and final conditions. The swelling ratio was then calculated per equation 2.

$$H_f / H_i \approx [A_f/A_i]^{1/2} \quad (1)$$

$$SR = V_f/V_i = A_f \bullet H_f / A_i \bullet H_i = A_f/A_i^{3/2} \quad (2)$$

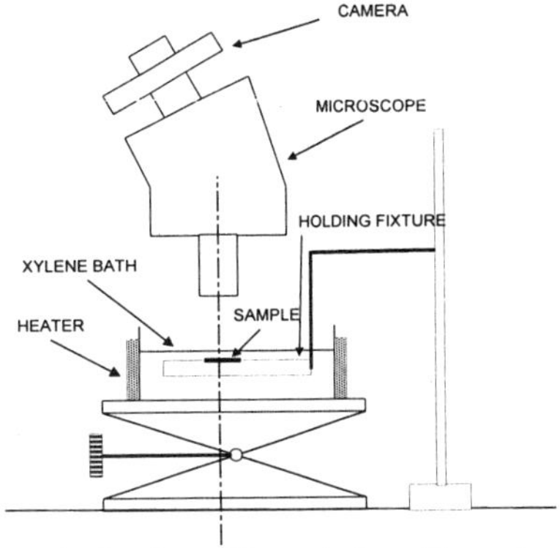

Figure 2: A schematic drawing of the optical swelling ratio apparatus.

RESULTS

Oxidation Studies: Effects of additional exposure to room temperature conditions after oxidative challenge are shown in Table 1. Students' t-test was used to compare the OI between the two aging times at each depth studied. A significant (p<0.05) increase occurred at all depths 1 mm or less. Beyond this depth no significant changes were observed.

A plot of OI versus depth for the EE, EI and II directions of 33 day sample can be seen in Figure 3. All the exterior surfaces had the highest surface OI whereas the interior surfaces had lower surface OI but developed a subsurface peak.

Post aging Time	0 mm	0.5 mm	1 mm	2 mm	4mm
0 days (n=5)	0.92 (0.12)	0.41 (0.02)	0.16 (0.01)	0.05 (0.01)	0.02 (0.01)
70 days (n=2).	2.78 (0.23)	0.99 (0.04)	0.29 (0.01)	0.07 (0.01)	0.05 (0.01)

Table 1: OI versus Depth after 25 day challenge and for the same samples after an additional 70 days aging at room temperature.

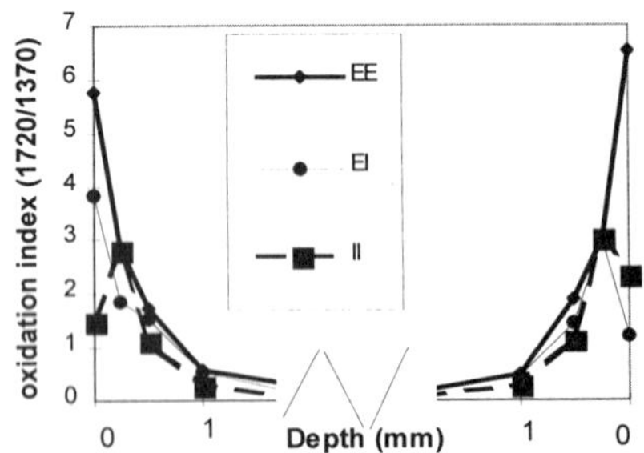

Figure 3: OI versus Depth for 33 day challenged samples in three directions; Exterior - Exterior, Exterior - Interior, and Interior-Interior.

Swelling ratio: A comparison of the two measurement methods for three irradiation atmospheres (NS, GA, GVF) are shown in Table 2. The same general trends for changes in sterilization condition/atmosphere were shown for both methods. The swelling ratio decreases upon irradiation (GA and GVF) and further reduction occurs in the vacuum environment (GVF). However, the optical method results in a lower value for swelling than the ASTM method for the lower levels of crosslinking. At the higher levels of crosslinking the values converge. Over the entire range of crosslinking F-tests indicated a significant difference (p<.05) in the variance of the two methods with the optical method having the smaller variance.

SR	AR (n=5)	GA (n=5)	GVF (n=5)
ASTM	45 (7)	34.2 (9)	4.7 (0.6)
Optical	15.1 (1)	11.5 (0.2)	4.7 (0.2)

Table 2: A comparison of swelling ratio test methods for three irradiation conditions: As Received, Gamma Air and Gamma Vacuum Foil.

DISCUSSION

Oxidation Studies: A comparison of the oxidation levels in material that has been further aged at room temperature post oxidative challenge suggests that once the oxidation reaction has reached some critical level that increase in oxidation continues at a high rate. The rate of OI increase during the 25 day at 78° C challenge and also during the 70day room temperature aging post challenge were similar. Indicating that once some critical level of oxidation has been initiated the reaction proceeds at a relatively high rate. This has implications regarding the testing of oxidative challenged material in that the amount of time after challenge needs to be controlled, else the level of oxidation in the material may vary from test to test. Other studies in our lab have suggested that these reactions are not easily slowed down and that storage of samples at temperatures as low as -20^0 C do not halt these reactions.
It appears that the amount of oxygen the material is exposed to at the time of irradiation will influence subsequent surface oxidation challenge reactions. The OI versus depth profiles for the material with exterior

surfaces all had similar profiles of high surface oxidation and rapid decrease into the bulk of the material. In contrast interior surfaces had lower surface oxidation levels and a subsurface peak. However the OI versus depth profile was similar for all directions and starting conditions at depths of 0.25 mm and greater. The variability of the surface oxidation level for all 6 locations may represent the problems with accurately positioning the aperture of the FTIR microscope at the edge of the sample, and the rapid changes in the OI level near the surface.

Swelling ratio: Previous work had not been able to measure any crosslinking in films of GA material [1]. This is probably due to problems associated with the ASTM method which requires removal of the sample from the xylene. It is difficult to remove the samples since it cannot be seen in solution and low crosslinked material can break up.
The optical method offers several advantages: variances were significantly smaller than the ASTM method making it a more reproducible and sensitive test method, there was no direct handling of the samples so the issues of locating the sample and breakage are eliminated, a smaller sample size can be used so it is possible to study local variations in the level of crosslinking. The test can be easily automated allowing for faster and more reproducible test method. One of the key assumptions that still needs to be addressed is that of isotropic swelling. UHMWPE has been shown to be anisotropic in several properties [4].

REFERENCES:
1). Hamilton et al, ORS. 22:782, 1997.
2). Shen et al., J.Poly..Sci. Part B V34, 1063-1077 (1996)
3). Currier et al., CORR N342, pp 111-122.
4). Hamilton et al., ACS Symposium Series 620, 1996

INSIGHT INTO THE MOLECULAR NETWORK IN IRRADIATED UHMWPE

D.C. Sun, A. Chopra, C. Stark, J.H. Dumbleton, Howmedica Inc.,
Pfizer Medical Technology Group, 359 Veterans Blvd., Rutherford,
NJ 07070

INTRODUCTION

Physical chain entanglements in UHMWPE provide the structural foundation for superior toughness and resistance to creep and wear. Chemical crosslinking in the same regions can further enhance this effect by creating a permanent 3-D molecular network (the gel phase). Recent wear studies using a hip joint simulator showed improved wear resistance in radiation crosslinked UHMWPE compared to non-crosslinked UHMWPE[1-3]. Crosslinking in a polymeric material can be induced by heat, radiation, or chemical agents. Depending on processing conditions, the distribution of crosslinking density in UHMWPE can be varied. In this study, we use gel content / swell ratio protocols modified from ASTM D2765-90 to assess gel content profiles and crosslinking density in irradiated UHMWPE rods.

MATERIALS AND METHODS

Three 2.5-inch diameter ram extruded UHMWPE rods made from GUR 1020 resin (irradiated in air at ~ 3.0 Mrads, irradiated in nitrogen at ~ 3.0 Mrads followed by free radical elimination / stabilization at 50°C[4] and at the melting temperature of UHMWPE near 140°C[5]) were studied in detail. An as-extruded, non-irradiated GUR 1020 rod was used as a control. ASTM D2765-90 general procedures were followed for gel content and swell ratio analysis. For gel content profile analysis, 5-micron thick films were produced at various depths of the UHMWPE rod by a microtome and enclosed in a cage formed from a 200-mesh stainless steel wire cloth. The cage was immersed in boiling xylene (~ 140°C) for 20 hours. The net sample weight (with moisture and solvent removed) before and after extraction was measured and the gel content was determined. For crosslinking density analysis, a 2 mm x 6 mm x 23 mm UHMWPE piece (~ 250 mg, machined from the center of rods) held by a stainless steel wire was immersed in 1800-ml boiling xylene for 20 hours. The UHMWPE sample weight before the extraction, after extraction (i.e. wet sample weight with surface solvent wiped off), and after drying (in an IR lamp dryer) was measured and the swell ratio (which is inversely related to the crosslinking density being defined as the number of crosslinks per specific volume) was determined

RESULTS

Table 1 shows the gel contents for the surface, 1/4 depth, and the center of the as-extruded, non-irradiated UHMWPE rod.

Table 1   Gel content of As-extruded UHMWPE

Sample Depth	Gel Content, %
surface	6
1/4 depth	52
center	61

Table 2   Swell Ratio of UHMWPE Rods

Sample	Swell Ratio
As extruded	completely dissolved
Air irradiated	6.6
Stabilized at 50°C	4.8
Stabilized at melting temperature	3.7

Figure 1 shows the gel content profiles (from the surface to the center of the rod) for the air irradiated sample, the sample stabilized at 50°C and the sample stabilized at the melting temperature. Table 2 shows the swell ratio results obtained from the center of rods.

DISCUSSIONS

From Table 1 and Figure 1, irradiation in air increases the gel content at the surface from below 10% in the as-extruded rod (Table 1) to 35%. Irradiation in nitrogen followed by stabilization at 50°C and at the melting temperature (~ 140°C) further increases the gel content at the surface to 79% and 88% respectively. In the surface zone, free radicals generated by gamma radiation can react with oxygen to cause oxidation, react with each other to form crosslinks, or remain as "live" free radicals. For the air irradiated sample, oxidation and crosslinking compete and produce a moderate level of gel content. In the sample irradiated in nitrogen and stabilized at 50°C, no oxygen is present to cause oxidation. Free radicals (mostly in the amorphous regions) are recombined to form crosslinks. The remaining free radicals (mostly in the crystalline region) are further crosslinked in the sample stabilized at the melting temperature to give a high gel content. The difference in gel content at the surface between samples diminishes as the sample depth exceeds 1 mm. All the three samples show a gel content greater than 90% at the center of the rod. However, the morphological difference in this region as induced by different process conditions is revealed by the swell ratio measurements shown in Table 2 for the center of rods. The gel phase in the as-extruded rod is completely dissolved in the swell ratio measurement (without a mesh cage). While a 3-D molecular network exists in all three irradiated samples (as indicated by a high gel content level), the crosslinking density results indicate that the sample stabilized at the melting temperature possesses the most inter-connected microstructure.

CONCLUSION

Irradiation induces a crosslinked molecular network in UHMWPE. Gamma irradiation followed by a stabilization step at the melting point of UHMWPE produces the highest crosslinking density.

REFERENCES

1. H. Oonishi, et.al., J. Materials Science Materials in Medicine, 7 (1996) 753-763. 2. H.A. McKellop, et.al., Polyethylene Wear in Orthopedic Implants Workshop, 1997 SFB Annual Meeting. 3. A. Wang, et.al., Polyethylene Wear in Orthopedic Implants Workshop, 1997 SFB Annual Meeting. 4. US Patent 5,449,745. 5. US Patents 5,543,471; 5,650,485; 5,728,748

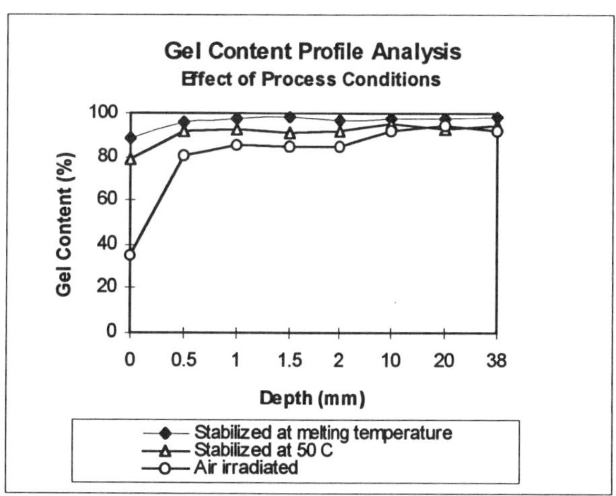

Figure 1

WEAR BEHAVIOR OF CARBON FIBER REINFORCED PEEK COMPOSITES IN TOTAL JOINT REPLACEMENTS

A. Wang, R. Lin, C. Stark and J. H. Dumbleton

Howmedica, Inc., Division of Pfizer Medical Technology Group,
309 Veterans Blvd., Rutherford, NJ 07070

INTRODUCTION

UHMWPE has been the most widely used bearing surface in total hip and total hip replacements since the 1960s. However, in recent years it has been recognized that the wear of the UHMWPE may be the limiting factor for the long-term success of prostheses [1]. The search for alternate bearing materials other than UHMWPE has been focused on metal on metal or ceramic on ceramic bearing couples. However, the clinical success rate of these bearing surfaces has been rather mixed. Early failure of both metal on metal and ceramic on ceramic bearings in total hip replacements has been reported mostly due to loosening or fracture of the components in-vivo [2,3]. In the present investigation, a bearing surface based on a carbon fiber reinforced PEEK composite is presented as a potential alternative to UHMWPE. The suitability and limitations of this polymeric composite as a bearing surface was explored for both total hip and total knee applications.

EXPERIMENTAL

Materials:

Both pitch-based and PAN-based carbon fibers were mixed with a PEEK resin (ICI, Grade 150G) and then pelletized. Both types of carbon fibers were chopped fibers with an average diameter of 8 μm and an average length of 20 μm. The pellets were processed into two useful forms by injection molding: one a rectangular shape (24 x 12 x 12 mm) which mimics a tibial component for total knee replacement and the other a hemispherical socket shape (32.25 mm I.D. and 52 mm O.D.) which mimics an acetabular component for total hip replacement. The fiber loadings in the specimens ranged from 0 wt% to 50 wt%.

Wear Tests:

Wear tests were performed using either a high-stress line-contact reciprocating wear tester or a low-stress ball-in-socket hip simulator. The line-contact machine adopts a non-conforming ring on flat geometry and reciprocating linear motion which simulates the loading and motion conditions of the knee joint. The hip simulator adopts a conforming ball in socket geometry and a biaxial rocking motion which simulates the kinematics of the hip joint. Schematics of the line-contact machine and the hip simulator are shown in Fig. 1a and Fig. 1b, respectively. For the reciprocating wear test, the flat composite specimen is loaded against a polished alumina ceramic ring (dimensions: 72 mm diameter and 24 mm wide; surface finish: 0.01 μm R_a) at a constant load of 1150 N. During the

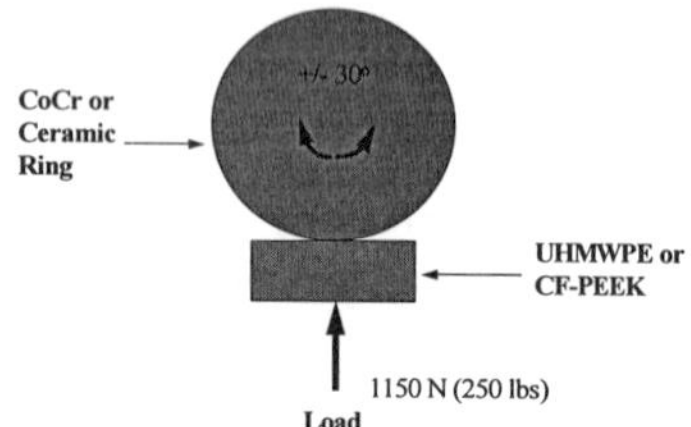

Fig. 1a: Loading/motion configuration of the line-contact reciprocating wear machine.

test, a +/- 30° oscillation motion was applied to the alumina ring to simulate the principal motion of the knee joint. For the hip simulator test, a physiological load is applied to the ball (maximum load: 2450 N; minimum load: 50 N) while a biaxial rocking motion (+/- 23°) is applied to the composite socket which produces a multi-directional "cross-shear" stress pattern on the acetabular cup surface [4]. Three types of femoral heads (balls) were used: CoCr, alumina and zirconia. All of the balls have a diameter of 32 mm. For both tests, bovine calf serum was used as a lubricant. For comparative purposes, UHMWPE was used as a control. All materials were sterilized by gamma radiation at 2.5 Mrads in air prior to wear testing.

Fig. 1b: Hip joint simulator.

RESULTS AND DISCUSSION

High-Stress Line-Contact Test:

Fig. 2 shows a plot of the volumetric wear rate as a function of carbon fiber loading and carbon fiber type against an alumina counterface for the high-stress line-contact wear test. For both the pitch-CF and the PAN-CF composites, the 30 wt% specimens exhibited lower wear rates than the 10 wt% and the 50 wt% specimens. The pitch-CF composites showed significantly lower wear rates than the PAN-CF composites for all fiber loadings studied. However, all composite specimens exhibited significantly higher wear rates than the UHMWPE control. These results indicate that carbon fiber reinforced PEEK composites are not suitable for applications in a high-stress non-conforming contact situation such as in the knee joint.

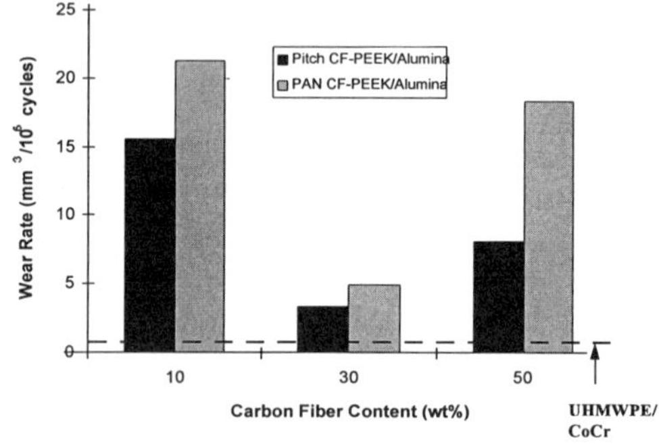

Fig. 2: High-stress line-contact wear test result.

Ball-in-Socket Hip Simulator Test:

Fig. 3 shows a plot of the volumetric wear rate as a function of carbon fiber loading against an alumina head for the PAN-CF composites for the hip simulator test. The unreinforced neat PEEK sockets exhibited six times the wear rate of the UHMWPE control. The reinforced composites (both the 20 wt% and 30 wt%), on the other hand, showed only 1/10th to 1/20th the wear rate of the UHMWPE control. The greatest reduction in the wear rate was achieved with the 30 wt% composite.

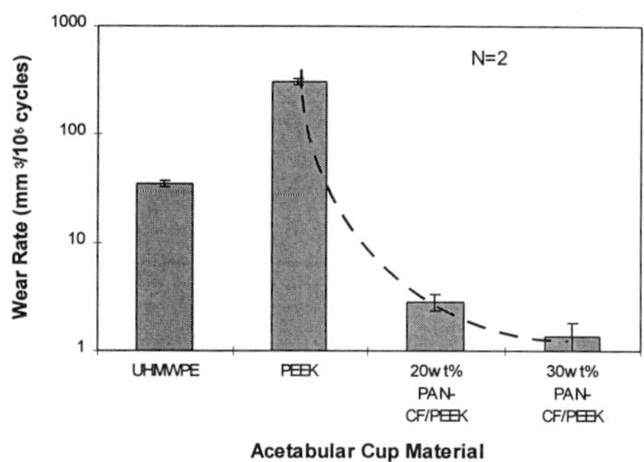

Fig. 3: Ball-in-socket hip simulator test result: PAN-CF composite sockets against alumina heads.

Fig. 4 shows a plot of the volumetric wear rate as a function of carbon fiber type (pitch vs. PAN) and head material (CoCr vs. alumina vs. zirconia) for the the hip simulator test. Both the pitch-CF and PAN-CF composite sockets wore much less than the UHMWPE cups against all femoral head materials. However, the wear rates of the composites were significantly higher against CoCr than against alumina or zirconia. For the PAN-CF composite, alumina was identified as the optimal counterface material; for the pitch-CF composite, zirconia was the best choice. In fact, the best wear couple was the 30 wt% pitch-CF composite socket and zirconia femoral head. With this combination, the wear rate of the socket was only 1/30th that of the UHMWPE.

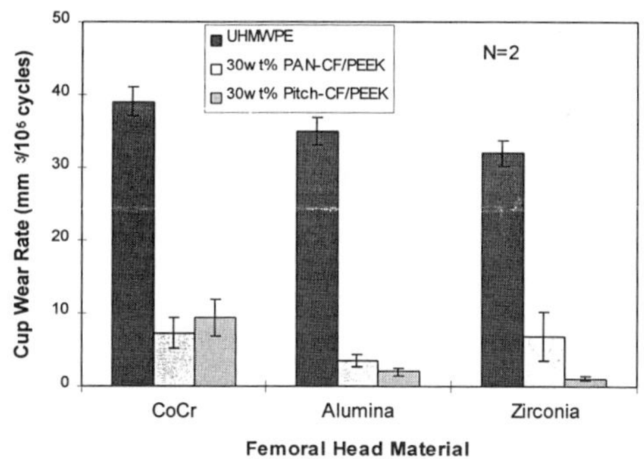

Fig. 4: Hip simulator test result: pitch-CF vs. PAN-CF; CoCr vs. alumina vs. zirconia.

CONCLUSIONS

The suitability and limitations of carbon fiber reinforced PEEK composites as bearing surfaces for total joint replacements were explored. It was found that carbon fiber reinforced PEEK composites offer a far superior wear resistance over UHMWPE against either metal or ceramic head in a conforming ball-in-socket contact situation such as in the hip joint. Pitch-based carbon fibers were superior to PAN-based carbon fibers while ceramic heads were superior to metal heads. The optimal wear couple was a 30 wt% pitch-CF composite socket and a zirconia head. With this combination, the wear rate of the composite was almost two orders of magnitude less than that of UHMWPE against a zirconia head.

In a high-stress non-conforming contact situation, the carbon fiber reinforced PEEK composites performed poorly as compared to UHMWPE. Therefore, the composite materials should not be used as a tibial component for a total knee joint replacement.

BIBLIOGRAPHY

1. W. H. Harris, The problem is osteolysis, *Clinical Orthopaedics & Related Research* **311**, 46-53, 1995.

2. B. Almby and T. Hierton, Total hip replacement: a ten-year follow-up of an early series, *Acta Orthop Scand* **53**, 397-406, 1982.

3. F. Bonicoli, et al., Ceramic insert in uncemented press-fit titanium total hip prostheses: our clinical experience in fifty-nine cases, in Herausgegeben von Wolfhart Puhl (ed.), *Performance of the Wear Couple BIOLOX Forte in Hip Arthroplasty*, Ferdinand Enke Verlag, Stuttgart, 1997, pp. 38-42.

4. A. Wang, D. C. Sun, S.S. Yau, et al., Orientation softening in the deformation and wear of UHMWPE, *Wear* **203-204** (1997) 230-241.

DESIGN CONSIDERATIONS FOR THE IN VIVO STABILITY OF HOLLOW FIBER BASED CELL ENCAPSULATION DEVICES

D. Rein*, T. Gunrud*, M. Shoichet**

* CytoTherapeutics, Lincoln, RI, 02865

** Department of Chemical Engineering and Applied Chemistry, University of Toronto, Ontario, Canada

INTRODUCTION

Encapsulated cell technology is being developed for targeted delivery of active biomolecules to treat serious and disabling human conditions. The enabling concept involves surrounding the implant cells with a selective membrane barrier which allows free passage of oxygen and required metabolites into the device and cell secreted therapeutic factors out of the device, while hindering the passage of larger molecules and cells from the host immune system. Use of a selective membrane eliminates the need for chronic immunosuppression in the host and allows cells to be obtained from non-human sources. For central nervous system applications, encapsulated cell devices have been successfully transplanted into models of Parkinson's disease [1], Alzheimers disease [2], Huntinton's disease [3], and chronic pain syndrome [4].

Long term chemical and mechanical stability of an encapsulated device is essential for its continued performance in vivo. Chemical instability of the membrane may lead to a change in the permselectivity, resulting in graft cell death. Mechanical instability of the device may result in a loss of retrievability. In the present work, the mechanical and chemical stabilities of poly(acrylonitrile-co-vinyl chloride) (PAN/PVC) hollow fiber membranes (HFM) were assessed in the rat peritoneal cavity over a 15 month period. Improvements in device retrievability associated with changes in membrane morphology and with the incorporation of internal struts were examined.

EXPERIMENTAL

The in vivo stability of the HFM was characterized by measuring the tensile strength, molecular weight, molecular weight cut-off (MWCO), and hydraulic permeability (HP) pre-implantation and post-explantation [5]. PAN/PVC membranes were prepared at CytoTherapeutics by a phase inversion process with an outer diameter of ca. 900 um and a wall thickness of ca. 100 um. 2.5 cm alginate filled PAN/PVC devices were implanted into the peritoneal cavity of Lewis strain rats, and explanted for stability characterization after 6, 12, or 15 months. Adsorbed protein was removed from the devices before characterization by treatment with 4 M NaOH.

The PAN/PVC molecular weight (Mw) and polydispersity index (PDI) were characterized pre and post explantation by gel permeation chromotagraphy using a DMSO mobile phase. Membrane and device tensile strengths were evaluated on tensile strength instruments (Liveco Vitrodyne V1000, Instron) at a constant strain rate of 100 um/sec. The hydraulic permeability and molecular weight cut-off were measured using potted fiber cartridges. The HP was determined from the permeate flow rate at a constant transmembrane pressure, and the MWCO was determined from the dextran rejection curve at a constant lumenal wall shear rate.

RESULTS AND DISCUSSION

The Mw, PDI, tensile strength, HP, and MWCO for devices pre implant and after 6, 12, and 15 months in vivo are shown in Table 1 below. The weight average molecular weight (Mw) of 10 combined fiber samples pre-implant and after 12 months in vivo was 143,300 g/mole and 128,400 g/mole, respectively. This is an insignificant change given that a 10% variation in the Mw of the bulk PAN/PVC was routinely measured. The tensile strength changed insignificantly from 52 ± 2 cN pre-implant to 46 ± 7 cN after 15 months in vivo. There was a significant drop measured at 12 months in vivo, however this was attributed to kinks present in the fiber lengths not observed at the other time points. The HP and MWCO did not change significantly out to 12 months in vivo with respect to the pre-implant values. The consistency of these convective transport properties over time in vivo reflect the structural stability of the membrane pore structure.

Table 1 Molecular weight (Mw), polydispersity index (PDI), tensile strength, hydraulic permeability (HP), and molecular weight cut-off (MWCO) of PAN/PVC hollow fiber membrane pre and post implantation (mean $\pm$ s.d.)

Time (Months)	Mw (g/mole)	PDI	Tensile strength (cN)	HP (ml/min/m2 mmHg)	MWCO (g/mole)
0	143000	2.3	52 ± 2	7.4 ± 1.5	40000 ± 8000
6	n/a	n/a	45 ± 8	6.3 ± 1.5	52000 ± 10000
12	128400	2.8	25 ± 8 a	7.5 ± 1.5	54000 ± 10000
15	n/a	n/a	46 ± 7	n/a	n/a

n/a: data not available

a: Damaged fibers broke at kink.

Device designs relying exclusively on this PAN/PVC hollow fiber membrane for structural integrity will withstand retrieval forces approaching 50 cN at explant. Depending on the implant site and time of explant, higher device retrieval strengths may be required. This increase in strength may be obtained from increasing the mechanical properties of the membrane, or modifying the device design to translate explant tensile forces to a structural support. Membrane mechanical properties may be modified by changing the cross sectional wall morphology. The morphology of the membrane used in the stability studies was a trabecular structure. Changing the structure to open cell foam increased the tensile strength by approximately 2x to 120 cN. Dramatic increases in device tensile strength may be obtained through the use of struts located inside the device. Typically, struts are designed to minimally interact with the encapsulated cells. Fine titanium wire and braided synthetic yarns have been used to increase the retrieval strength to 700 and 2000 cN, respectively. These supports add to the complexity of device manufacturing, however this is typically compensated by the dramatic increase in retrievability.

CONCLUSION

The analytical techniques used to characterize the bulk properties of a PAN/PVC membrane indicated biostability of fibers out to at least 12 months in vivo. The Mw and tensile strengths of the in vivo fibers were maintained with respect to the pre-implant values, indicating no change in the retrievability for cell encapsulation devices. The hydraulic permeability and MWCO of in vivo fibers were also maintained with respect to the pre-implant values, indicating no change in permselectivity. Moderate increases (approx. 2x) in device tensile strength are obtained from changes in membrane morphology, while dramatic increases (> 10x) are obtained through the incorporation of structural supports.

BIBLIOGRAPHY

1. P. AEBISCHER, M. GODDARD, A. SIGNORE, R. TIMPSON
   Exp. Neuro. **126**, 151, 1994.
2. S. WINN, J. HAMMANG, D. EMERICH, A. LEE, R.
   PALMITER, E. BAETGE. *Proc. Natl. Acad. Sci.* **91**, 2324, 1994.
3. D. EMERICH, M. LIDNER, S. WINN, E. CHEN, B. FRYDEL,
   J. KORDOWER. *J. Neurosci.* **16**, 5168, 1996.
4. J. SAGEN, H. WANG, P. TRESCO, P. AEBISCHER
   J. Neurosci., **13**, 2415, 1993.
5. M. SHOICHET, D. REIN
   Biomaterials, **17**, 285, 1996.

ıFRARED SPECTROSCOPIC STUDY OF OXIDATION IN
POLYURETHANE HARD SEGMENTS

Steven R. Skorich, Michael E. Benz
Medtronic, Inc. Center for Biomaterials Research, Brooklyn Center,
Minnesota, USA, 55430

INTRODUCTION

Previous studies have indicated that *in vivo* oxidation of poly-ether urethanes (PEU's) has involved depletion of the soft segment of the material[1,2]. The use of an attenuated total reflectance (ATR) objective on the infrared (FTIR) microscope has allowed analysis of much smaller areas on polyurethane tubing samples removed from *in vivo* implantation. By means of the micro ATR, the differences between the IR spectra of clear vs. fissured areas of PEU's were enhanced, and IR peaks typical of fissured areas were observed. Similar peaks were observed on fissured areas of a poly-carbonate urethane (PCU) which shares the same hard segment with the PEU's. It thus became clear that the hard segment was involved in the fissuring process as well as the soft segment. Hard segments in both these materials were based on 4,4'-methylenebis diisocyanate (MDI) and 1,4-butanediol (BDO).

Two model diethyl carbamate (urethane) compounds were synthesized from 4,4'-methylene dianiline and 4,4'-diaminobenzophenone. The spectra of these materials show the same sort of spectral differences as between the clear and fissured materials above. These compounds closely model the hard segment of the above materials, and differ only in the oxidation to a carbonyl group of the central methylene group in the molecule. This implies that the same process is occurring in the polyurethane hard segments, in addition to any soft segment reactions (see Fig.1, below).

EXPERIMENTAL

Infrared attenuated total reflectance spectra of the samples were taken on a Perkin-Elmer System 1720-X Fourier-Transform Infrared (FTIR) spectrometer, equipped with a Spectra-Tech IR-PLAN IV microscope accessory and an ATR objective. Infrared diffuse reflectance spectra of the two model compounds were taken on a Perkin-Elmer System 2000 Fourier-Transform Infrared (FTIR) spectrometer, equipped with a Spectra-Tech COLLECTOR® diffuse reflectance accessory. The two model compounds were run at 1.0% by weight in IR grade potassium bromide (KBr) powder (Aldrich Chemical Co. catalog number 22,186-4).

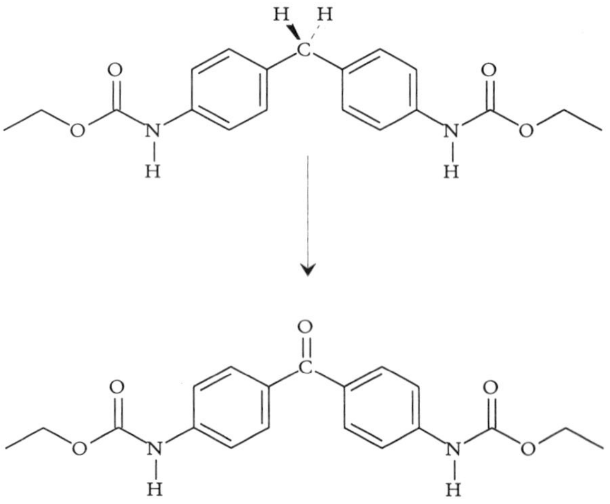

Figure 1: Oxidation of a carbamate (urethane) based on 4,4'-methylenebis diisocyanate (MDI) to a 4,4'-diaminobenzophenone analog

RESULTS & DISCUSSION

Figure 2 compares micro ATR spectra of a poly-ether urethane tubing of hardness 80A (PEU 80A) with a heavily-fissured, scaly area of explanted PEU80A material, described as "Metal-Ion Induced Oxidation" (MIO) to the authors. (Infrared peak assignments given in the discussion are from References 3 - 6. All of the succeeding spectra were baseline-flattened and normalized on the hard segment benzene-ring stretching peak at approximately 1412 cm^{-1}.)

There are several major characteristic changes which occur in the infrared spectrum. These are summarized below:

1730-1700 cm^{-1} (C=O stretches): The peak at 1730 cm^{-1} decreases relative to the 1700 cm^{-1} peak. This indicates a decrease in relative amount of "free" C=O and/or an increase in hydrogen-bonded C=O.

1638 cm^{-1} : A new peak. This is a much broader peak than the stearamide wax C=O stretch which sometimes shows here from wax additive blooming to the surface of samples.

1594 cm^{-1} : Apparently, the 1613 / 1598 cm^{-1} peaks are either fusing or another peak is growing over them, or both (the resulting peak is more intense, relative to the normalization peak).

1364 cm^{-1} and 1106 cm^{-1} : Decreasing relative to the normalization peak, due to loss of polyether (these are soft segment peaks, from Fig. 2)

1282 cm^{-1}, 1173 cm^{-1}, 1153 cm^{-1}, and 929 cm^{-1} are all new peaks. The sharp peaks at 1173 and 929 are especially striking, because they occur in areas of the normal PEU spectrum which are free of strong peaks.

857 cm^{-1} and 770 cm^{-1} peaks increase relatively as the 817 cm^{-1} peak decreases.

The urethane Amide II at 1525-1530 cm^{-1}, and the Amide III around 1220 cm^{-1} are relatively unchanged, because we are effectively normalizing on the hard segment. The urethane C-O stretch around 1060-1070 cm^{-1} is decreasing in intensity, but that may be because the peak is riding on the shoulder of the ether C-O at ca. 1105 cm^{-1}. Similar changes have been observed in the spectra of fissured PEU 55D, as well.

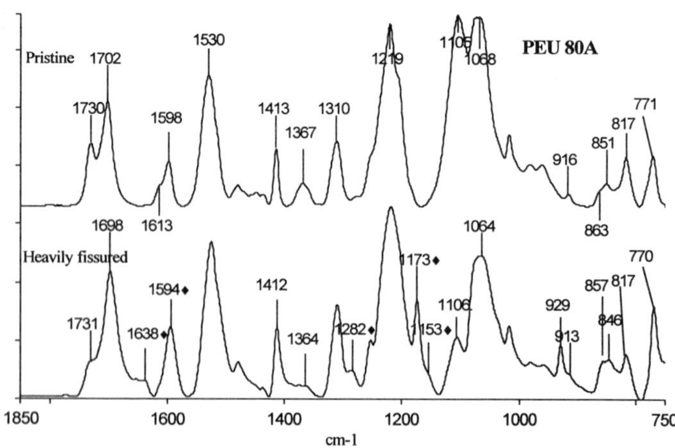

Figure 2: Polyether Urethane 80A, pristine vs. MIO

Figure 3 shows the micro ATR spectra of a polycarbonate urethane of hardness 80A (PCU 80A) and a fissured explant. The soft segment on this material is an aliphatic polycarbonate. The hard segment is the same MDI-BDO material found in the PEU above. Soft segment depletion may be inferred from the decrease of the polycarbonate C=O stretch at 1738 cm^{-1} relative to the urethane C=O stretch at 1700 cm^{-1}, and the decrease of the 1242 cm^{-1} carbonate O-C-O stretch. This latter peak is somewhat confounded with the urethane Amide III at 1220 cm^{-1}. Nevertheless, the differences between clear and fissured material are to be seen in other spectral changes similar to the PEU above, i.e.:

> 1649 cm^{-1}: new peak
> 1595 cm^{-1}: apparent fusion of two peaks, increase in intensity relative to the normalization peak at 1412 cm^{-1}.
> 1283 cm^{-1}: new peak
> 1174 cm^{-1}: new peak, sharp and narrow
> 1151 cm^{-1}: new peak, sharp and narrow
> 930 cm^{-1}: new peak, sharp and narrow
> 857 cm^{-1}: increasing, 818 cm^{-1} decreasing, and 770 cm^{-1} increasing relative to 1413 cm^{-1}

Since the PCU and the PEU materials share the same hard segment, the implication becomes stronger that reaction of the hard segment is involved in the process resulting in fissuring of the polymers, though it is still not clear *when* the hard segment is involved (i.e., is this spectrum showing an effect, or a part of the cause?).

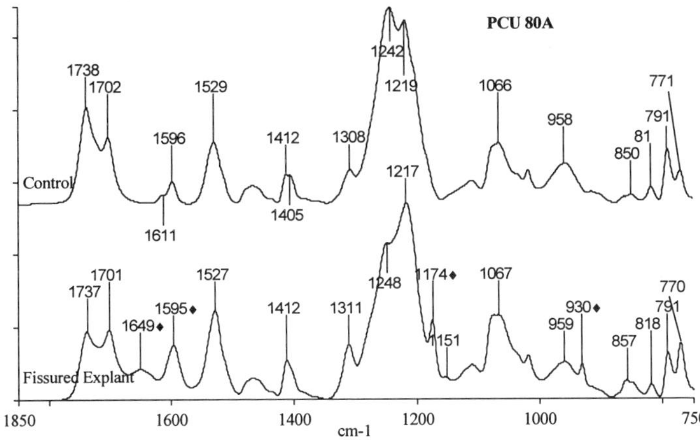

Figure 3: Polycarbonate Urethane (PCU) Unimplanted Control vs. Explant

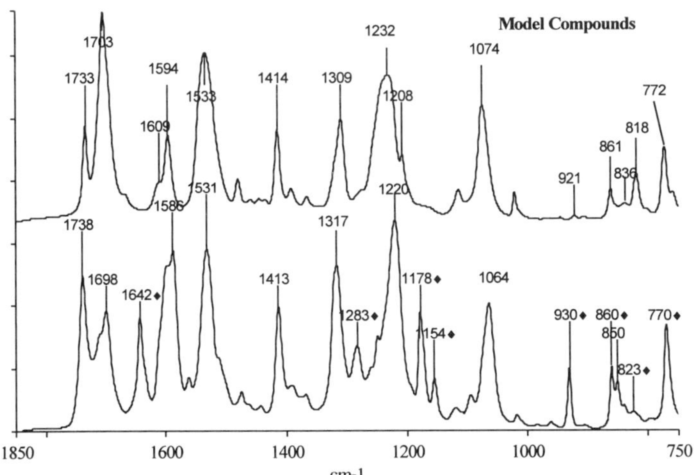

Figure 4: Model Diethyl Carbamate (Urethane) Compounds of 4,4'-Methylene Dianiline and 4,4'-Diaminobenzophenone

Figure 4 shows the comparison of the infrared spectra of the two model compounds. These are diffuse reflectance spectra taken of the finely-powdered materials mixed with potassium bromide at 1.0% by weight.

There is very good correspondence between the changes observed between these two model materials and those observed in the two polyurethanes discussed above.

> 1642 cm^{-1}: new peak (benzophenone structure, C=O stretch?)
> 1595 cm^{-1}: apparent fusion with 1610 cm^{-1}, increase in intensity relative to the normalization peak at 1413 cm^{-1}
> 1283 cm^{-1}: new peak (ring to carbonyl group C-C stretch, possibly)
> 1178 cm^{-1}: new peak, sharp and narrow
> 1154 cm^{-1}: new peak, sharp and narrow
> 930 cm^{-1}: new peak, sharp and narrow
> 860 cm^{-1}: increasing, 818 cm^{-1} decreasing, and 770 cm^{-1} increasing relative to the normalization peak at 1413 cm^{-1}

CONCLUSION

All of these IR spectra strongly suggest that the benzylic methylene groups found in the hard segment of these polymers are being oxidized to the corresponding carbonyl groups, and that this oxidation product is responsible for the characteristic "MIO" peaks observed in fissured polyurethanes. These conclusions hold for the polyurethanes studied, and could possibly apply to all polyurethanes that contain hard segments based on MDI and 1,4-butanediol. The model compounds synthesized for this study are crystalline monomers, rather than multiphasic polymers, so an exact match with the two polymers is not to be expected. Nevertheless, the correlation of spectral features of these models with the pristine and oxidized polyurethanes is striking, and it is sufficient for the purposes of this report to point out that the overall pattern is similar, without yet attempting an exhaustive spectral analysis.

REFERENCES

1. Stokes, K.; Urbanski, P.; Upton J. *J. Biomater. Sci. Polymer Ed.,* **Vol. 1, No. 3,** pp. 207-230 (1990)
2. Stokes, K.; Chem, B.; Coury, A.; Urbanski, P. *J. Biomater. Appl.* , **Vol. 1, No. 4,** 411-48 (1987).
3. <u>Introduction to Infrared and Raman Spectroscopy, 3rd Ed.</u> (Colthup, Daley, and Wiberley , Academic Press, 1990)
4. <u>Polyurethanes in Medicine</u> (Lelah and Cooper, CRC Press, 1986)
5. <u>Infrared Spectroscopic Atlas of Polyurethanes</u> (Dillon, Ed., Technomic Publishing, 1989)
6. Colthup, N.B, *Applied Spectroscopy,* **30**, p.589-592, 1976

USING NMR FOR STRUCTURE DETERMINATION OF
POLYURETHANES. J. Thomas Ippoliti, Department of Chemistry,
University of St. Thomas, Saint Paul, Minnesota, 55105-1079

Introduction: Polyetherurethanes materials have been used to
insulate long term implantable cardiac pacing leads for many years.
In general this material has proven satisfactory, but certain models
of pacing leads have had less than desirable clinical performance
caused by a degradation of the insulation and eventual exposure of
the underlying conductors. The degradation process seems to be
oxidative in nature, but no study has ever unequivocally identified
the specific sites of attack in the polymer molecule. For that reason
this study was undertaken to see if NMR was a suitable tool to study
the in vivo degradation of a polyetherurethane material of durometer
80A (referred to as P80A).

Experimental: The general experimental process was to obtain
and compare ^{1}H and ^{13}C NMR spectra for pristine and in vivo
degraded P80A plus a number of model compounds. The model
compounds were either custom synthesized or were other polymers
which had certain molecular structures of interest for comparison.
All the polymers were dissolved in dueterated THF. The spectra
were obtained using a Bruker 300 MHz spectrometer over periods
varying from one minute to four days. Once the data were obtained
all the peaks in the pristine and degraded P80A spectra were
identified and a possible degradation mechanism identified which
was consistent with the data.

Results and Discussion:

Polyetherurethanes are characterized by hard and soft
segments in their structure. The soft segments are composed of
long polyether chains connected through carbamate linkages to the
para position of a biphenyl methylene.
The hard segments have a 4 carbon chain linking the same
biphenyl methylene. The hard segment has much less
conformational mobility which leads to highly ordered motifs in the
solid state. This packing in turn makes the polymer more rigid and
less flexible. The lack of conformational mobility also leads to 2nd
order effects in the ^{1}H nmr which aids in assignment of protons
associated with the hard segment.
Assignment of all the signals in the both the ^{1}H and ^{13}C
nmr of the P80A was the first objective of this research. This was
accomplished using several model compounds.
Some simple model compounds were synthesized to
determine the effects of alkyl groups on the chemical shift of the
carbamate NH and the aromatic protons.

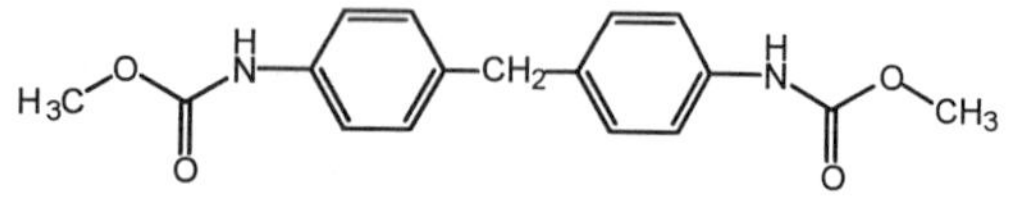

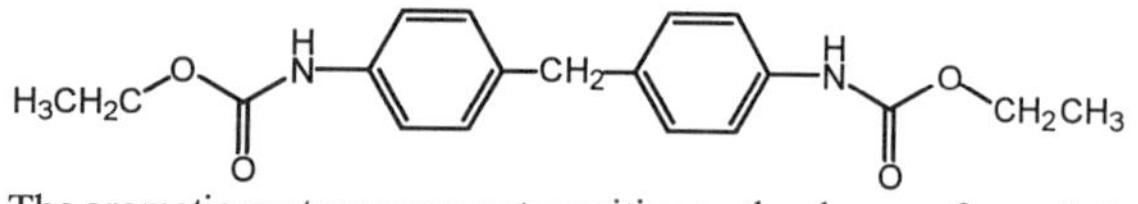

The aromatic protons were not sensitive to the change of a methyl to
an ethyl. However, the N-H was sensitive to this subtle change in
structure. These compounds were also used to assign the ^{1}H
signals of the biphenyl methylene portion of P80A as well as the N-
H and NCOO-**CH2**- protons by correlation of the signals in each of
the spectra.
The ^{1}H nmr spectrum of P80A shows two signals in the
region for the CH2 attached to the carbamate. One signal is a sharp
triplet (at 4.090 ppm) and one is a broad triplet (at 4.124 ppm). Two
model compounds were employed to decipher which signal is due to
the protons in the hard segment and which is due to those in the soft
segment. One was a polyurethane composed of only soft segments.
The other was a polyetherurethane with a much higher proportion of
hard segment. This polymer is a 55D durometer polyetherurethane
and is labeled P55D. Using these model compounds we were able
to assign the broad triplet at 4.124 as being due to the hard segment
and the sharp triplet at 4.090 as due to the CH2 in the soft segment.

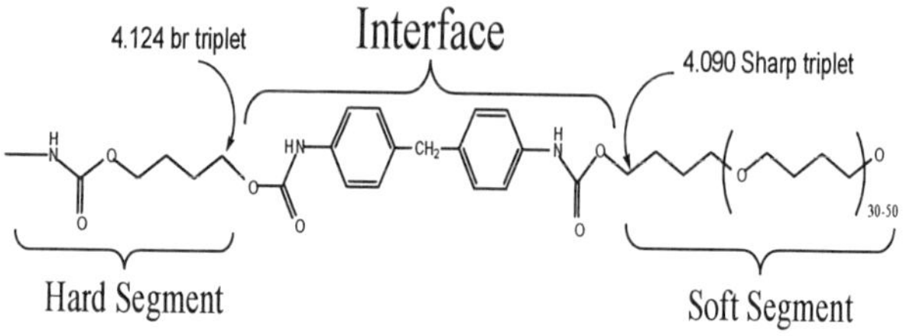

The signals in the ^{13}C nmr spectrum can be assigned using the same
model compounds. Using this data it was clear what changes in
molecular structure had occurred in the degraded tubing by
comparing the ^{1}H and ^{13}C spectra of pristine P80A to the degraded
tubing spectra. The ^{13}C nmr peaks changed considerably in size for
soft segment carbons while the corresponding peaks in the ^{1}H nmr
did not change at all. From this data it was concluded that the soft
segment was being cleaved in the middle of the polyether segment
leaving long hard segments with short soft segments at the ends.
^{13}C Spin-Lattice Relaxation times (T1) depend on segmental
mobility of partial structures and the molecular periphery of long
chains are more flexible than the center. Long T1's are observed for
atoms in the more mobile ends of the chains. Long relaxation times
are therefore expected for the carbons in the soft segment. And since
short delay times were used between pulses due to the small sample
size, decreased peak size was observed for the carbons in the soft
segment. These relaxation effects are not important in ^{1}H nmr and
therefore the size of the peaks for the protons in the soft segment
were not changed.
There were some new peaks in the ^{1}H nmr spectrum of the
degraded tubing which were confidently identified by comparing the
spectra to that of model compounds. The new peaks lead to the
conclusion that some of the CH2 groups between the phenyl rings
are also being oxidized to a carbonyl.

Conclusions: All the peaks in both pristine and in vivo degraded
P80A lead insulation material were identified. The principle
findings were that the ^{13}C nmr peaks changed in intensity while the
^{1}H nmr peaks did not in the spectra of the degraded tubing and that
there were some new peaks present in the ^{1}H nmr spectra of this
material. These findings are most consistent with a mechanism
where chemical attack cleaves the soft segment in the polyether
portion and leaves a hard segment capped with short polyether
chains. The new peaks are due to a polymer where a carbonyl has
formed between the phenyls which is the result of oxidation of the
methylene group. Based on our data we conclude that polyether
chains are vulnerable in vivo and that the methylene between the
phenyl groups is sensitive to oxidation although this does not cause
the polymer chain to cleave.

Mechanically Assisted Corrosion Of MP35N Alloy
N. Istephanous*, J. Gilbert** G. Martinez*, J. Wiser*,
U. Ko***, T. Irwin****, D. Untereker*
*Medtronic, Inc., Minneapolis, Minnesota
**Northwestern University, Chicago, Illinois
***J & B, Minneapolis, Minnesota
****Buckbee Mears, Minneapolis, Minnesota

BACKGROUND

Since 1975, polyetherurethanes (PEU) have been successfully used for components such as connectors, adhesives, lead insulations and tines in cardiac pacemakers and neurological stimulators. MP35N alloy has been used in lead conductors and also in orthopedic applications. Individually, these two materials possess many desirable properties for chronic implant applications.

However, PEUs have, on occasion, displayed two forms of biodegradation. These have been termed environmental stress cracking (ESC) and metal induced oxidation (MIO). These two degradation mechanisms result in irreversible changes in the physical, chemical and mechanical properties of the polymer. In the case of ESC, cracks on the outer surface of the lead body are directly related (in frequency and depth) to the amount of residual stress and the ether (soft segment) content of the PEU (1). MIO, on the other hand, appears to require the use of PEU and MP35N in close proximity to each other. MIO is usually observed on the inner surfaces of pacing lead insulation where the polymer is in contact with a conductor coil. Morphologically, crack patterns that track the metal component configuration are likely to be present. The effects of metal ion solution(s) on MIO have been investigated *in vitro* (1). The stability of sealed PEU tubes housing different metals was studied *in vitro* (3% hydrogen peroxide solution at 37°C) and *in vivo* using rabbit implants for two years (2). Both investigations resulted in corroded metals and compromised tubing surfaces; sometimes within 30 days. *In vivo*, the interaction of body fluids with cobalt and its alloys, in particular, resulted in what appeared to be oxidative cracking of the polymer (3). One observation was that samples of compromised PEUs typically contained higher than expected concentrations of metal ions than would be expected. The MIO mechanism in pacing leads appears to require some corrosion of the metallic conductor to produce these ions, in addition to close contact between the MP35N conductor and the PEU insulating tubing, so the ions can migrate into the polymer. MIO appears to be, therefore, the result of a complex interaction of the device, the alloy, polymer and perhaps the body.

One important factor, which has not been investigated, and may contribute to this mechanism, is the mechanical disruption of the protective surface oxide film which spontaneously forms on the MP35N alloy. During abrasion, surface aspirates, or contact points, between adjacent surfaces of the wire coil may generate sufficiently high stresses or fretting to induce fractures of the oxide film. Typically, repassivation of a metal surface oxide occurs within a few milliseconds. The question is whether during that time, significant ion dissolution can occur, thus affecting the local solution chemistry.

Corrosion resistance and electrochemical behavior of most alloys is highly dependent on the structure and chemistry of the surface oxide film which spontaneously forms. In turn, the structure and chemistry of the oxide film is highly dependent on prior metal history (e.g., thermal treatment, surface cleaning, anodization, etc.) immersion conditions, solution composition, and time. For MP35N, in particular, little information exists regarding the factors that influence the stability of the oxide film.

OBJECTIVE OF THE STUDY

The objective of the present study is to investigate the effect of mechanical stresses (static and dynamic), solution composition, applied potential and electrochemical history on the corrosion resistance, and stability of MP35N oxide film.

MATERIALS AND METHODS

MP35N wire, .004 inches in diameter was used in these studies. Depending on the type of test and measurements performed, the wire was either tested after being polished to a 6 micron finish (longitudinal cross sections), or wound into coils 0.025 inches in diameter, 2 or 4 inches long. The tests conducted were:

1) Step Polarization Impedance Spectroscopy (SPIS): The polarization behavior and impedance characteristics of MP35N were measured using a SPIS system developed by one of the authors (4). Sequential 25 mV steps in potential, from -1000 mV to +900 mV versus a Ag/AgCl reference electrode, were applied to the polished wire and the corresponding current transients were measured from 0.00005 to 10 seconds. These transients were used to calculate the interfacial impedance (resistive and capacitive components) as a function of solution composition. A minimum of three tests per solution were performed at each potential.

2) Electrochemical Scratch Tests: A scanning electrochemical microscope (SECM) designed and custom built by one of the authors was utilized (5, 6). This apparatus was used to study the current transient response after the surface oxide was disrupted in a highly-controlled manner. Controlled loads and scratch distances were applied within 0.5 to 1 ms to the polished surface of the wire via a highly-sensitive double cantilever load cell on to which an 18 micro-meter diamond stylus was mounted. Current transients resulting from the repassivation of the fractured oxide were measured, using high speed data acquisition techniques. These load-effect tests are a measure of the mechanical stability of the surface oxide. The surface topography of the scratched region could also be imaged. The system had a vertical and horizontal resolution of 0.05 and 1 μm, respectively. The effects of sample potential, solution composition and applied load on the mechanical behavior of the oxide films were studied. For the above two tests, the following solutions were used in aerated and deaerated (bubbling N_2) conditions:

-0.154M (0.9 wt. %) NaCl
-0.001M NaCl
-3% H_2O_2 in a 0.154M (0.9 wt. %) NaCl
-1.54M NaCl

3) Electrochemical measurements under static and dynamic conditions: Three electrode corrosion cells, built into a dynamic flex tester, were utilized in these experiments. Uniaxial 24 gram loads were applied to each of four, 4-inch long MP35N coils immersed in 0.154M saline at 37 ± 1°C. A Saturated Calomel Electrode (SCE) and platinum wire were used as reference and counter electrodes, respectively. The samples were anodically polarized and, the corrosion current densities were measured as a function of the flexing frequency: 0 and 1 Hz. In a different set of experiments, the open circuit (corrosion potential) was measured as a function of solution composition and flexing frequency: 0 and 2 Hz. In these experiments, 2-inch long coils were used.

RESULTS AND DISCUSSIONS

1) Step Polarization Impedance Spectroscopy (SPIS): The polarization behavior of MP35N exhibited the classic active-passive-transpassive behavior known to be typical of these alloy systems. Solution composition did not seem to significantly affect the polarization behavior of the samples tested in saline. Little systematic effect due to aeration was seen in these tests. This is most likely due to the fact that the passivation process can utilize oxygen derived from the hydrolysis of water. It was observed that in the H_2O_2 free solutions, oxides began to lose passivation effects around 250 mV. However, hydrogen peroxide did have a significant effect on the polarization plots. Much higher corrosion current densities and less passive behavior was observed. Also, much more positive rest potentials were noted in this solution, wards of +400 to +600 mV (Ag/AgCl). This is close to, or above, the breakdown potential for MP35N.

Higher Cl^- ionic strength solutions decreased the impedance in the passive range, implying that higher Cl^- lowers the passivity of the film. Either thinning of the oxide film or changing of the film properties, or both, are occurring.

2) Electrochemical Scratch Tests: Figure 1 is a representative current transient after a 0.025N load, 30 μm scratch. The relationship between the applied load, solution chemistry and the peak current, is illustrated in Figure 2. Sample potential was held at -10 mV versus SCE. The peak current increased with the concentration of chloride ions. In deaerated solutions, the peak currents were generally lower than in aerated solutions.

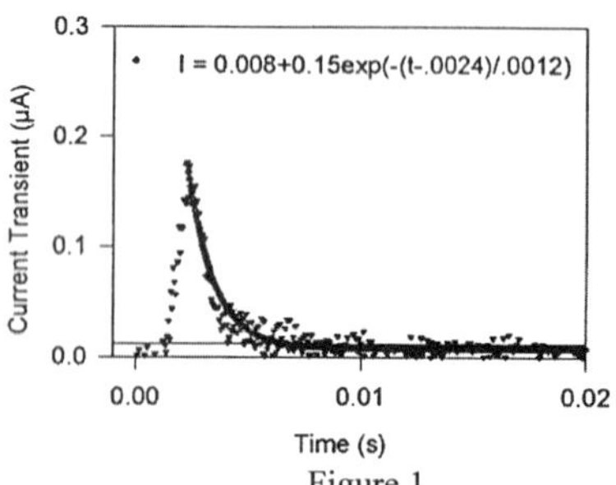

Figure 1

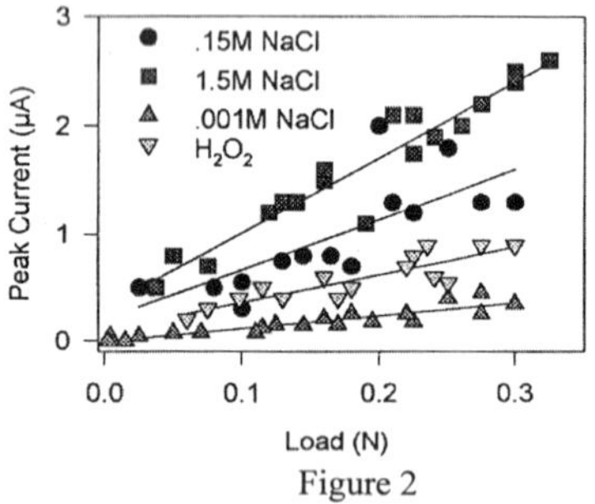

Figure 2

Figures 3 and 4 illustrate the relationship between peak current or τ (a measure of the kinetics of oxide film reformation) as a function of solution composition (aerated) and applied potential. The peak currents (maximum current observed, minus the baseline current) begin to increase around 0 mV. For any fixed potential (between 0 and +900 mV) the maximum peak currents were ranked as follows: 0.001M Cl^- < 3% H_2O_2 in 0.154 M Cl^- < 1.5M Cl^- < = 0.154M Cl^-. Similar trends were observed in the deaerated solutions. There appears to be a significant effect of aeration on peak currents and τ. Tests conducted in aerated solutions seem to exhibit higher peak currents than in deareated solutions (for a fixed potential). Significant variations in τ were only evident in the tests conducted in aerated solutions (between 0.7 to 2.5 ms for aerated versus 0.7 to 1.5 ms for the deareated solutions). The reasons for the latter two observations are unknown. In general, higher peak current(s) and τ(s) are directly related to increased metal ion release as illustrated below. In the case of the 3% H_2O_2 solution, it appears that the surface of the alloy does not tend to reform an oxide film. This film is considerably less resistant to ion transport (i.e., has a much higher defect density or is thinner) than the corresponding film formed in the other three aqueous solutions.

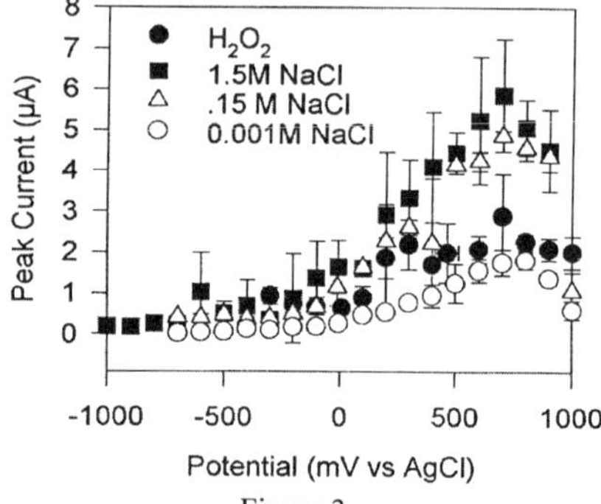

Figure 3

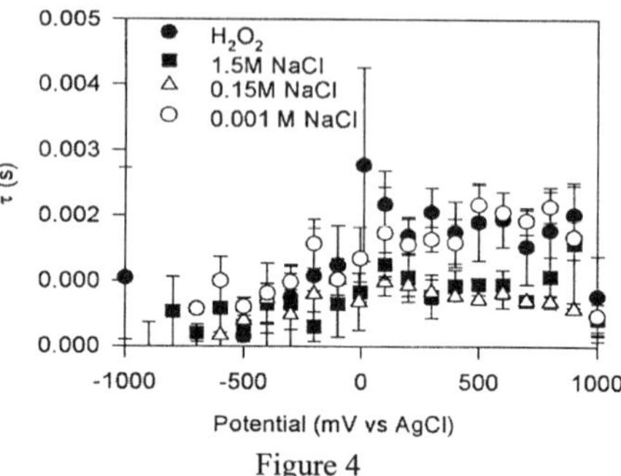

Figure 4

It is clear from the scratch experiments that when the protective oxide film ruptures, significant current transients result. When an oxide film is mechanically disrupted, the current transient that results comes from three sources: double-layer charging, oxide film formation, and ionic dissolution through the scratched surface. To help understand the contributions of these three processes, it is useful to first calculate the effective current density that is present at the site of the film rupture. For example, a scratch of 300 μm^2 (10 x 30 μm) yielding a peak current of 3 MA in 1 ms, results in a current density of 1 A/cm^2, a million fold higher than for the nonscratched surface. It is estimated that perhaps 20% of the measured current is due to ionic dissolution. Consequently, the local ionic dissolution rate would be as much as 200,000 times greater than the nondisrupted surface. While this is only an estimate, it demonstrates that each mechanical disruption of the film can dramatically increase the local corrosion rate until the oxide heals. One can imagine the cumulative effects of oxide fracture and repassivation on the local metal ion concentration in the micro-fluidic crevices inside the lumen if one area of the metal surface is subject to repeated disruption of the protective oxide.

3) Electrochemical measurements under dynamic conditions:

Corrosion Current Density--MP35N coils flexed at 1 Hz, exhibited a higher corrosion current density at open circuit when compared to their counterparts under static conditions. These findings complement those above, namely demonstrating that flexing can induce high enough stresses to rupture the oxide film. As discussed above, the increase in corrosion current is interpreted as an increase in the release of ions from the metal surface in the area of the surface damage.

Open Circuit Potential--Figure 5 illustrates the change in open circuit potential over time of an MP35N coil undergoing continuous flexing at 2 Hz. The voltage spike corresponds to the complete fracture of the coil and the associated instantaneous exposure of a fresh, "oxide free" surface. This drop in open circuit potential may cause general thinning of the oxide film everywhere on the surface. The recovery of the potential to post coil fracture values can be attributed to the oxide formation on the fractured surface. Open-circuit (rest) potential measurements taken following three days of flexing (dynamic) at 2 Hz were compared to those taken on nonflexed (static) coils. Regardless of the chloride level and state of aeration, the open circuit potential for the flexed coils were several hundred mV more negative than their static counterparts. This is a characteristic of a surface undergoing an oxide fracture process.

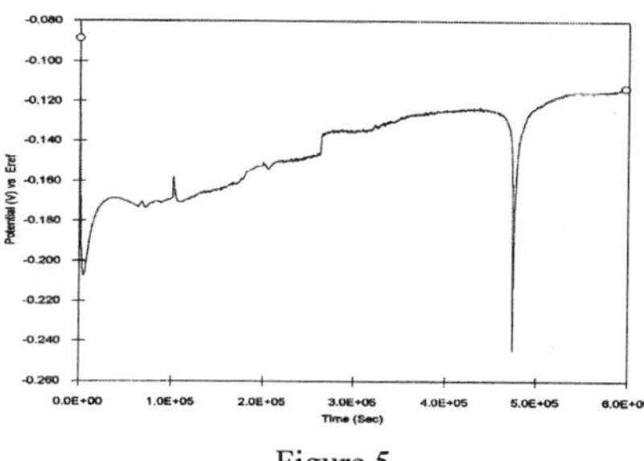

Figure 5

CONCLUSIONS

The experiments presented above demonstrate several points that are relative to the use of PEU and MP35N (or other alloys) in long-term medical applications. First, it is clear that the surface oxide is very important in determining the microscopic electrochemical behavior of metals if any liquid is contacting the metal or alloy. Second, it is clear that the characteristics of the surface oxide are very dependent on the dynamic conditions present. Third, the flexing of a metal in solution is likely to result in the local release of metal ions many orders of magnitude higher in concentration than is found under static conditions. Finally, it seems likely that the released metal ions can react with PEU in close proximity if the ions are absorbed into the polymer.

REFERENCES

1.   Coury, A. J.; Slaikeu, P.C.; Cahalan, P.T.; and Stokes, K. B., (1987), Medical applications of implantable polyurethanes: Current issues, *Prog. Rubber Plastic Technology*, 3(4): 24-37.
2.   Stokes, K., Urbanski, P., Upton, J., The *in vivo* auto-oxidation of polyether urethanes by metal ions, *J. Biomater. Sci., Polymer Edn.* 1(3):207-230.
3.   Coury, A. J., et al., Degradation of materials in the biological environment, *Biomaterials Science,* Rattner, B. D., et al., editors, Academic Press, p. 243-282.
4.   Gilbert, J. L., Step polarization impedance spectroscopy of implant alloys in physiological solutions," *J. Biomedical Materials Research,* Vol. 40, 233-235, 1995.
5.   Gilbert, J. L., "Electrochemical response of CoCrMo to high speed fracture of its metal oxide using electrochemical scratch test method," J. Biomedical Materials Research, Vol. 37, p. 421-431, 1997.
6.   Gilbert, J. L., Smith, S. M., Lautenschlager, E. P., "Scanning electrochemical microscopy of metallic biomaterials-reaction rate and ion release imaging modes," J. Biomedical Materials Research, Vol. 27, p. 1357-1366, 1993.

IN VITRO MACROPHAGE/IRON/STRESS SYSTEM FOR THE STUDY OF PEU BIODEGRADATION

J. Casas, Q. Zhao[a], M. Donovan, and D. Untereker.
Biosciences Laboratory, Center For Biomaterials Research.
Medtronic, Inc. Minneapolis, Minnesota 55430
[a] *currently at Biovascular, Inc. St. Paul, Minnesota 55114*

INTRODUCTION

Polyetherurethane (PEU) use in the biomedical industry has been limited because of its susceptibility to degradation. Oxidation and environmental stress cracking (ESC) have been observed in certain cases. One theory proposes that *in vivo* ESC requires the interaction between surface oxidation of the polymer's ether linkages, residual stress (strain), and a proteinaceous crack driver.[1] Others suggest that enzymatic products, principally those with hydrolytic activity, are involved in its degradation. Recently, methods to reproduce ESC in PEUs were developed *in vitro*.[2] However, the relationship between such tests and the *in vivo* mechanism of ESC is still not clear.

In the present work, we developed a biological test system to study polymer degradation. This model, which mimics the *in vivo* environment, sequentially involves macrophages, iron and stress. Our main purpose was to assess whether a direct link can be made between the phagocyte's activity and the oxidative degradation of PEU.

EXPERIMENTAL

Oxidation and ESC were studied by measuring *in vitro* hydroperoxide formation and by observing ESC development in the polymer in the presence and absence of phagocyte treatment. This research was carried out under the hypothesis that the macrophage oxidative burst, combined with transition metals and applied stress, can produce ESC in PEU. Cells with no known oxidative burst potential, such as lymphocytes, were used in control experiments.

Cell Isolation

Freshly isolated human monocyte-derived macrophages (Mo/MØs) were obained from venous blood from healthy volunteers. Mononuclear cells were isolated from heparinized blood by a one-step density gradient procedure using Isopaque-1077, according to a modified Boyum's method.[3] Monocytes and lymphocyte fractions were separated on the basis of their propensity to adhere to surfaces.

In Vitro Polymer Treatments

Polymer discs, (6mm diameter), 0.12 mm thick, were cut out of PEU sheets. One group of these was soaked in acetone (AS) for 1 hour, and dried at room temperature for 4 hours. The other group was used with no acetone pretreatment (N-AS). Polymer specimens were then fitted to the bottom of the wells of 96-microwell cell culture plates under sterile conditions.

First treatment; Cell culture. The PEU specimens (AS and non-AS) were covered with hMo/MØs or human lymphocytes (3 x 10^5 cells/well) and cultured for 49 days. Two control conditions included, culture media only (w/o cells) and a dry condition. Triplicate polymer discs were removed at various time periods for hydroperoxide (ROOH) determination.

Second treatment; FeCl$_2$ incubation. Following the 49-day first treatment, the specimens were folded in half and fixed in this position. This design permitted the characterization of the samples to be carried out in both, unstrained and in moderately strained states. The specimens were incubated in 5mM FeCl$_2$ at 37°C for 10 days. OM evaluation of the samples was performed during the treatment. At the 10th day triplicate samples from each condition were removed for SEM evaluation of the polymer surfaces.

Iodometry

For ROOH determination, polymer specimens taken at various time periods during the first treatment were sonicated for 15 min in distilled water, rinsed three times and dried at 25°C for 4 hours. ROOH was determined using a iodometric assay as described by Fujimoto et. al.[4]

Cell Morphology and Surface Analysis

OM visualization of cells on the surfaces was possible during cell culturing. For SEM cell morphology evaluation, specimens were taken after 21 days of cell culture (first treatment). The polymer surfaces were also evaluated using SEM following the 10-day treatment with FeCl$_2$ (second treatment).

RESULTS AND DISCUSSION

OM Morphological analysis

Morphological changes in cultured Mo/MØ monolayers on PEU included a gradual cell size increase and spreading over time. These cells had extensive membrane ruffles. Some of the cells assumed unusual shapes, exhibiting a maximum of 60 μm along their larger axis. A decrease in the number of cells was also observed over time.

Iodometry

An increase of ROOH concentration as a function of culture time was observed. This effect was most evident in AS specimens cultured with Mo/MØs. By contrast, AS specimens cultured with lymphocytes showed significantly lower ROOH concentrations. This was comparable to levels found in specimens incubated in cuture media only and the ones stored in dark ambient condition. Likewise, non-AS treated polymer specimens cultured with Mo/MØs showed the lowest amount of ROOH.

SEM Surface Analysis

SEM analysis of PEU specimens following the second treatment revealed substantial cracking and pitting in the AS treated specimens exposed to Mo/MØs during the first treatment period (Fig 1a). The most severely damaged region was located at the folded (stressed) area. Neither, non-AS specimens cultured with Mo/MØs (Fig 1c) nor A.S.-specimens cultured with lymphocytes (Fig 1d) showed significant surface damage after 10-day treatment with FeCl$_2$. Likewise, AS specimens treated first with Mo/MØs and then with distilled water (instead of FeCl$_2$) did not developed ESC damage (Fig 1b).

Phagocytes, are important cells during inflammation and are the predominant cell types found at tissue-biomaterial interfaces. Through phagocytosis, the release of cell mediators and products with bactericidal action, MØs orchestrate inflammation. During this process, phagocytes destroy invading substances and maintain their metabolic demands by activating their plasma-membrane-associated NADPH oxidase. The result is the production and release of large amounts of toxic reactive oxygen species (ROS) such as superoxide ($0_2^{-\bullet}$) and hydrogen peroxide (H_20_2). The production of these ROS can increase dramatically during inflammation. Because biological antioxidant defenses are not completely efficient, increased free radical formation in the body is likely to increase tissue damage.[5] By virtue of the Mo/MØ's adhesion and spreading on the polymer such as the one observed in this study, these cells can directly provoke biomaterial damage at the cell-biomaterial interface (CBI). By excluding inhibitory macromolecules in the interface and by trapping secreted molecules in a confined space, the efficacy of secreted substances may be maximized in the CBI compartment.[6]

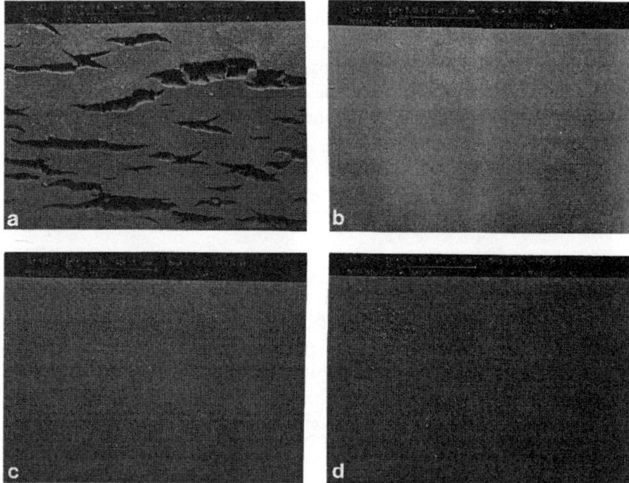

Figure 1. Polymer surface following the Mo/MØ/Iron/stress treatment, Mo/MØ/FeCl$_2$ (a), Mo/MØ/H$_2$O (b), Mo/MØ/FeCl$_2$ (non-AS PEU)(c), and lymphocyte/FeCl$_2$ (d). 500X

The results presented here indicate that macrophages developed their full potential as suggested by the observed phenotypical changes. Mechanisms involving MØ activation can have a critical role in the degradation of foreign materials. Polymer specimens treated with Mo/MØs first showed more ROOH in a culturing time dependency. Because the substrate material is never completely engulfed by the extending pseudopodia, the process of releasing ROS from the membrane and enzymes from the lysosomal granules into phagosomes or to the extrcellular milieu continues. Thus, the adherent macrophages are able to actively develop a so called "frustrated phagocytosis".

PEU specimens cultured with lymphocytes, which do not have oxidative potential nor ability to adhere to surfaces, showed a significantly lower rate of accumulation of ROOHs. The rate of ROOH accumulation in the presence of lymphocytes was not different from what was observed at comparable time points in either, the presence of culture medium only (no cells) or specimens stored under dark ambient conditions (air).

We have also shown that the polymer, once treated with macrophages, more readily develops ESC in the presence of applied stress than in the absence of stress. The ease for ESC development clearly depends upon the initial hydroperoxide concentration in the polymer; thus the more hydroperoxides induced by macrophages, the sooner the cracking in the FeCl$_2$ solution. Our results are summarized in Table 1.

The oxidation of PEU appear to follow mechanisms similar to lipid peroxidation. Mammalian cells possess elaborate defense mechanisms to prevent their damage by the normally produced ROS. The defensive mechanisms can be overwhelmed during oxidative stress conditions (i.e., inflammation). Superoxide dismutase (SOD), catalase, glutathione, etc. are important for preventing or scavenging the formation of ROS. Although it is not clear if Fe^{+2} or an analogous

species is required for *in vivo* ESC to occur, its presence in physiological systems is recognized and can be readily available under certain conditions. The experiments done here clearly show that Fe$^+$ speeds up the degradation process.

Table 1. Summary of results

Substrate	1st T.	ROOH (µmole/mg)	2nd T.	ESC
PEU, AS	hMo/MØs	5.5 ± 0.2	FeCl$_2$	yes
PEU, AS	hMo/MØs	5.5 ± 0.2	H$_2$O	no
PEU, AS	hLymphs.	2.9 ± 0.5	FeCl$_2$	no
·PEU, AS	CM only	1.8 ± 0.1	FeCl$_2$	no
PEU, AS	Ambient	2.1 ± 0.2*	FeCl$_2$	no
PEU, N-AS	hMo/MØs	1.5 ± 0.2**	FeCl$_2$	no

Note: Data expressed as mean ± STD. n ≥ 3. * meassurement at 34 days. ** measurement at 35 days.

Antioxidants appear to play important roles in the prevention of polyurethane oxidation, and are effective until their consumption impairs their ability to scavenge reactive oxygen species. In these experiments we provoked an artificial antioxidant depletion by soaking the polymer in acetone. In contrast to the AS specimens, significantly lower ROOH concentration were found in non-AS specimens and no ESC development was observed during the testing period (Fig 1c and Table 1). Based on our observations, we believe that the *in vivo* antioxidant depletion in a given PEU will depend on the phagocyte potential to consume it by the generation of oxygen reactive species at the tissue-biomaterial interface. The extent of this oxidative potential is dictated by the degree of cell reactivity and stimuli which in turn results from specific degrees of cell activation.

CONCLUSIONS

The *in-vivo* stress cracking of a PEU elastomer was duplicated in a biological macrophage/FeCl$_2$/stress *in vitro* system. Our results support a direct involvement of macrophages in polyetherurethane oxidation probably by inducing hydroperoxide formation in the polymer structure. Under the influence of stress or strain, polymers with sufficient hydroperoxides degraded in the presence of Fe^{+2} metal ions in a manner that closely resembled stress cracking that is observed *in vivo*. The macrophage/FeCl$_2$/stress model is a useful tool for the study of the mechanistic biology of polymer degradation and will be useful as a screening tool for biostability evaluation.

REFERENCES

1. K. Stokes, R. McVenes, J. M. Anderson. Journal of Biomaterials Applications. Vol. 9, p321-354. 1995
2. Q. Zhao, J. Casas-Bejar, P. Urbanski and K. Stokes. J.of Biomed. Mat. Res. Vol. 29, 467-475. 1995.
3. A. Boyum, D. Louhaug, and L.Tresland. Blood Separation and Plasma Fractionation, 217-239. Wiley-Liss, Inc. 1991.
4. K. Fujimoto, Y. Takebayashy, H.i Inoue, and Y. Ikada. J. Of Polymer Chem., 31. 1035-1043. 1993.
5. B. Halliwell. The Lancet, 344. 721 - 724, 1994.
6. J. M. Heiple, S.D. Wright, N.S. Allen, and S. C. Silverstein. Cell Motility and Cytoskeleton 15:260-70, 1990.
7. M.B. Grisham. Reactive oxygen metabolites of oxygen and nitrogen in biology and medicine. Austin, Texas: R G Landes Company, 1992.

THE CHEMISTRY OF SILICONE BASED
BIOMATERIALS AND WHY THEY CONTINUE TO BE
MATERIALS OF CHOICE FOR HEALTHCARE
APPLICATIONS

Del J. Petraitis
NuSil Technology
1050 Cindy Lane
Carpinteria, CA USA

INTRODUCTION:

Since their early inception, materials based on silicone polymers have been characterized by their chemical inertness. Early silicone materials included greases which were formulated for insulating electrical ignition systems used by the military during World War II. Although their resistance characteristics were primarily concerned with weathering and other environmental factors such as ozone and high temperatures, this unique inertness of silicone polymer based materials ultimately evolved into biomedical applications. Silicone based materials were found to be extraordinarily inert to the chemicals present in biosystems. Silicones were among the most biocompatible materials known to man and today continue to outperform other materials in biomedical applications including applications in such diverse products as pacemaker leads, intraocular lenses, long and short term catheter and shunt implants, finger joint implants, as well as, greases, lubricants, encapsulants and adhesives used in the fabrication, assembly or actual performance of an endless and continuing to be developed group of healthcare applications.

The basic chemistry involved in the synthesis of silicone polymers and subsequent product formulations into elastomers, pressure sensitive adhesives, marking inks, fluids and lubricants will be discussed.

BASIC SILICONE POLYMER SYNTHESIS:

Silicone polymers are based on backbone "monomers" made up of alternating silicon and oxygen atoms with organic pendant groups completing the remaining orbitals of the silicon atoms:

$$
\begin{array}{ccc}
R_1 & R_1 & R_3 \\
| & | & | \\
-Si-O-Si-O-Si-O- \\
| & | & | \\
R_1 & R_2 & R_3
\end{array}
$$

Among the substituent R groups are most commonly methyl, phenyl, and trifluoropropyl functionalities, but also include vinyl SiH, alkoxy, silanol, oxime, and acetoxy groups that produce actual crosslinks and pseudo-crosslinks when these polymers are formulated into bases and subsequently fabricated into actual products.

The most basic silicone molecule useful in biomedical applications is hexamethyl disiloxane:

$$(CH_3)_3\ Si\ O\ Si\ (CH_3)_3$$

Because it is odorless and very inert, it is a very suitable "solvent" or carrier for other silicone bases. Specifically, silicone pressure sensitive adhesives are dispersed in hexamethyl disiloxane. It also can be used to disperse silicone elastomer dispersions and is useful as a processing solvent.

SILICONE ELASTOMER CURE SYSTEMS:

Among the most useful silicone elastomers are those based on acetoxy and oxime one-part cure systems. These involve moisture cure mechanisms and release acetic acid and methyl ethyl ketoxime respectively during the cure. The non-bovine based catalysts used in these one part cure systems as well as the two part alxoxy cured elastomers include dibutyl tin dilaurate, dibutyl tin diacetate, and stannous octoate. Specific variations of this basic cure chemistry has been developed to provide optimized product formulations with rapid cures for device assembly and such uses as device marking ink applications. The chemistry of these modifications will be discussed. A summary of the most commonly utilized silicone elastomer cure systems for healthcare applications include the addition - platinum catalyzed system:

$$\sim\sim\sim Si\ CH{=}CH_2\ +\ H\ Si \xrightarrow{Pt\ Complex} \sim\sim Si\ CH_2CH_2\ Si$$

and the peroxide cure mechanism:

$$ROOR \longrightarrow 2\ RO\cdot$$

$$CH_3\ Si\ CH{=}CH_2\ +\ RO\cdot \longrightarrow CH_3\ Si\ CH_2\ CH\cdot\ +\ CH_3\ Si\ CH_2$$

$$\longrightarrow CH_3Si\ CH_2CH_2CH_2Si\ CH_3$$

OTHER SILICONE HEALTHCARE MATERIALS:

Included among the other silicone biomaterials are such diverse products as pressure sensitive adhesives and surface modifiers. Although the PSA's and needle coatings possess characteristics that are different from each other and even more dramatically different than the silicone elastomers, the basic chemistry is very similar. Specifically, PSA's are made from silanol functional dimethyl polysiloxanes condensed with polyfunctional resinous polysiloxanes. Needle coatings and other silicone surface modifiers consist of a range of materials also based on the same fundamental siloxane moieties. Hard coatings are the result of highly crosslinked siloxane resins which are formed by the hydrolysis of polyfunctional monomers and short chained polymers. Softer coatings result from elastomer-like siloxanes with the crosslink density formulated to meet the precise characteristics required by the specific application.

Another significant healthcare application involving silicones is their use in conjunction with drug delivery. Silicone elastomers have been formulated by varying the crosslink density and the solubility parameters to provide the correct drug dispensing characteristics. This is accomplished by formulating the matrix with unique combinations of substituent groups to effect the drug solubility and release.

CONCLUSIONS:

Although silicones have been, in recent years, questioned as safe and effective healthcare materials, they are in fact, without equivalents, and are unsurpassed in their varieties, biocompatability, and historical and ongoing healthcare applications ranging from long term implantable devices, to tubing for pumps, to such diverse applications as skin attachment devices and future and current drug delivery matrices.

Poly(Ethylene Oxide) - Polysulfone Block Copolymer Membranes

N. K. Barrera, S. M. Fagan, K. A. Gagnon, and **L. F. Hancock**; Circe Biomedical, Inc., 99 Hayden Avenue, Lexington, MA 02173
D. Brewster and C. Wabnitz; Millipore Corp., 80 Ashby Road., Bedford, MA 01730

Introduction - Tremendous improvements have been made in the bio/hemocompatibility of biomaterials through a variety of surface modification techniques. These include coatings, plasma deposition, and chemical modification.[1] Poly(ethylene oxide) has proved uniquely effective for biomaterial surface modification due to its high extension and flexibility in aqueous environments.[2]

Membranes are an especially difficult challenge for surface modification. The high surface area and sub-micron dimensions of membrane pores make it difficult to ensure a uniform and thorough surface modification.

This presentation will report the synthesis of poly(ethylene oxide) - polysulfone block copolymers (PEO-*b*-PSF) and their utility for the manufacture of membranes. This is achieved by incorporating PEO-*b*-PSF directly in the phase inversion membrane formulation. Phase inversion with an aqueous coagulant drives the PEO segment of the block copolymer to the membrane interface, Scheme 1. The PSF block remains embedded in the polysulfone matrix which forms the bulk structure of the membrane. This strategy takes advantage of the unique chemical and physical properties of block copolymers for the direct formation of membranes with an hydrophilic PEO surface.

Scheme 1: PEO-*b*-PSF Surface Modification Strategy

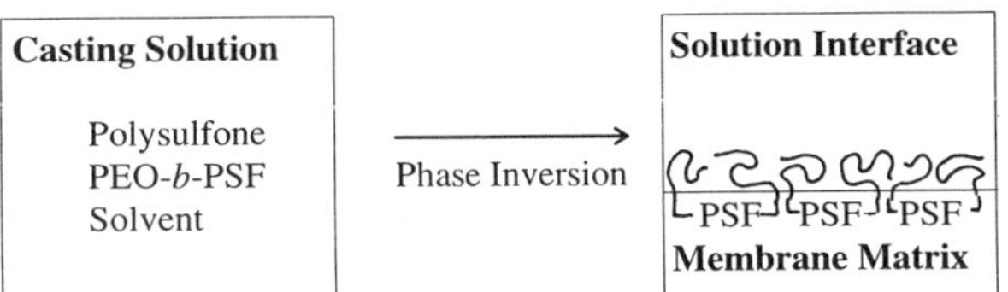

Synthesis and Characterization of PEO-*b*-PSF - Poly(ethylene oxide) - polysulfone block copolymers are prepared in a single step polymerization of poly(ethylene glycol), bisphenol A and bis-(4-chlorophenyl)sulfone, Scheme 2.[3] Example materials include PMX50S, a segmented block copolymer prepared with 10 kDa PEG and PMX30-5 where a 5kDa monofunctional PEO results in a material with terminal PEO blocks.

Scheme 2: Synthesis and Characterization of PEO-*b*-PSF

Sample	Yield	wt% PEO[a]	Mn(kDa)[b]	Mw(kDa)[b]
PMX50S	92%	46%	32	53
PMX30-5	94%	26%	48	70

a - Determined by 300MHz ^{1}H NMR
b - GPC: 2x 4.6mm MiniMix C columns (Polymer Laboratories), 0.1M LiBr/DMF, 0.3ml/min, 40°C

PEO-*b*-PSF Membrane Fabrication and Characterization - PEO-*b*-PSF has been used in the fabrication of microporous and ultrafiltration membranes.[4] A flat sheet ultrafiltration membrane was prepared from a casting solution of 10wt% polysulfone (Amoco, Udel 3500), 10wt% PMX30-5 and 80wt% N-methyl-2-pyrrolidionone. The solution is coated on a nonwoven fabric support at 0.008-0.010 inch and coagulated in water. Water permeability and filtration performance with single solute protein solutions is summarized in Table 1. The substantial rejection of bovine serum albumin (BSA) and passage of cytochrome C (Cyto C) indicates a molecular weight cut off for the membrane of ~30kDa, within the ultrafiltration regime. Note the high solute recoveries which reflect very low protein adsorption to the membrane surface. In contrast, solute recovery of Cyto C for a polysulfone membrane can be in the range of 60-70%. It is also noteworthy that these membranes retain their wettability and performance when dried without a wetting agent. The hydrophobicity of a polysulfone membrane makes it difficult to rewet the membrane with water in the absence of a wetting agent.

PMX50S has been employed for the manufacture of microporous hollow fiber membrane. A solution composed of 10wt% polysulfone, 15wt% PMX50S and 75% N-methyl-2-pyrrolidionone was extruded through a spinneret into a 75% NMP / 25% H$_2$O coagulation bath. This yielded a hollow fiber membrane with an ID 280μm of and an OD of 340μm. The membrane has a spongy, homogeneous wall structure (SEM not shown). Membrane samples evaluated under ISO 10993-5 (MEM elution test) showed no cytotoxicity. Membrane samples evaluated under ISO 10993-4 (*in vitro* Hemocompatibility) showed no adverse effect on hematological parameters.

Table 1: Filtration characteristics of PMX30-5 ultrafiltration membranes. Data are the average of 5 manufacturing runs each prepared with a unique batch of PMX30-5.

Water Flux (ml/min/cm^2) 2.0 ± 0.7

	0.25% BSA (67kDa)	0.025% Cyto C (12.4kDa)
% Solute Rejection	94% ± 2	13% ± 9
% Solute Recovery	97% ± 1	96% ± 1

* Single solute solutions in PBS; Dead end filtration, ½ volume reduction; Solute rejection = $\ln(C_{retentate}/C_{initial})/\ln(V_{initial}/V_{retentate})$.

Conclusion - A synthetic method for the preparation of poly(ethylene oxide) - polysulfone block copolymers and their utility for the fabrication of phase inversion membranes has been demonstrated. The unique wettability and resistance to protein adsorption of the membranes reflects the presence of PEO at the membrane interface. Membranes composed of PEO-*b*-PSF are non-cytotoxic and show no adverse effect on hematolgical parameters (ISO 10993-4).

References
1) see for example (a) C.-G. Golander, J. N. Herron, K. Lim, P. Claesson, P. Stenius and J. D. Andrade in "Poly(Ethylene Glycol) Chemistry, Biotechnical and Biomedical Applications" J. M. Harris ed. 1992, Plenum Press, (b) A. S. Hoffman, *Adv. Poly. Sci.* **1984**, 57, 141, (c) Y. Ikada *Biomaterials* **1994**, 15, 725.
2) see for example (a) M. Amiji and K. Park in "Polymers of Biological and Biomedical Significance" S. W. Shalaby, Y. Ikada, R. Langer and J. Williams eds. **1994**, American Chemical Society, (b) S. I. Jeon, J. H. Lee, J. D. Andrade and P. G. DeGennes *J. Colloid Interface Sci.* **1991**, 142, 149.
3) Y-P. R. Ting and L. F. Hancock *Macromolecules* **1996**, 29, 7619.
4) L. F. Hancock *J. Appl. Polym. Sci.* **1997**, 66, 1353.

.THESIS AND CHARACTERIZATION OF NOVEL
JENDRIMER - BASED GADOLINIUM COMPLEXES AS MRI
CONTRAST AGENTS FOR THE VASCULAR SYSTEM.

B. Radüchel, H. Schmitt-Willich, J. Platzek, W. Ebert, T. Frenzel, B. Misselwitz and H.-J. Weinmann

INTRODUCTION

Low molecular weight gadolinium chelates are used clinically as contrast agents in magnetic resonance imaging (MRI). After intravenous administration they rapidly diffuse from the intravascular into the interstitial space. However for assessment of tissue perfusion, overall vascularity and capillary integrity, macromolecular contrast agents may have significant advantages. So far, animal studies have been performed with Gd-DTPA labelled albumin[1], dextran[2], and polylysine[3]. More recently, the use of dendrimer based contrast agents has been reported independently by Adam et al.[4,5] and Wiener et al.[6]. A recent description of the various synthetic approaches to dendrimers as well as references for the possible use of dendrimer based pharmaceuticals have been given by Newkome, Moorefield and Vögtle[7]. In many cases a divergent dendritic construction is preferred as it allows a convenient surface functionalization. Further prerequisites for successful synthetic routes are quantitative reactions and analytical tools that allow to determine the degree of the substitution pattern on the surface.

We describe a procedure for the rapid synthesis of dendritic macromolecules and their reaction with macrocyclic chelators. Subsequent chelation with gadolinium affords new MRI contrast agents with molecular masses in the range of 17 to 36 kDa. An example for the structure of such molecules is Gadomer 17 (scheme 1). It has 24 gadolinium complexes at the surface of an amine terminated dendritic backbone. The gadolinium chelator is of the 1,4,7,10-tetraazacyclododecane-1,4,7,10-tetraacetic acid (DOTA)-type which is attached to the dendrimer via amide bonds.

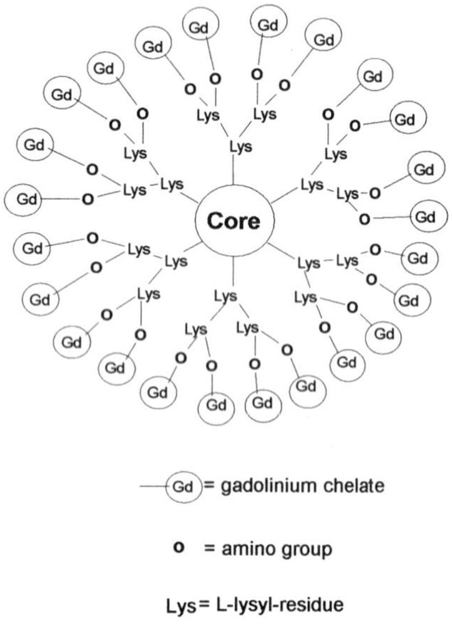

EXPERIMENTAL

The building block Z-4-amine (scheme 2) was prepared in a single step from commercially available starting materials. The remaining secondary amine functionality enables this small dendron to be bound to the central core benzene-1,3,5-tricarbonyl trichloride. Z-12-amine (scheme 3) was prepared by combining three equivalents of Z-4-amine with the triacid chloride core in DMF in 80 % yield. This fully benzyloxycarbonyl protected 12mer-amine was converted to the free 12mer-amine using HBr in acetic acid and the amine was subsequently reacted with N_α, N_ϵ - dibenzyloxycarbonyl-L-lysine p-nitrophenyl ester in > 80 % yield. This afforded Z-24-amine (see scheme 4) which is fully characterized by analytical methods and shows the desired at 6015 Da (M + Na$^+$) by MALDI-TOF mass spectrometry.

Finally the macromolecular ligand 24mer-DOTA-dendrimer was prepared from Z-24-amine (scheme 4). Liberation of the amino groups with HBr in acetic acid subsequent coupling with the p-nitrophenyl ester of the macrocyclic DOTA-derivative[8] (see scheme 5) followed by cleavage of the t-butyl protecting groups and ultrafiltration (Amicone YM3® membrane, cutoff 3 kDa) afforded a dendrimer bearing 24 chelators. Conversion to the gadolinium complex Gadomer 17 (see scheme 1) was achieved by reaction with gadolinium oxide. The HPLC-chromatogram (scheme 6, TSK-gel, 0.1 M NaClO$_4$, detect. at 207 nm) suggests a fairly monodisperse compound. However, the well resolved MALDI-TOF mass spectrum and the capillary electrophoresis reveal some inhomogeneity of the dendrimer probably due to incomplete coupling of the ligands to the 24mer amine backbone.

Using similar approaches we synthesized both 36mer- and 48mer-amines to which we attached macrocyclic ligands that were converted to the corresponding gadolinium complexes.

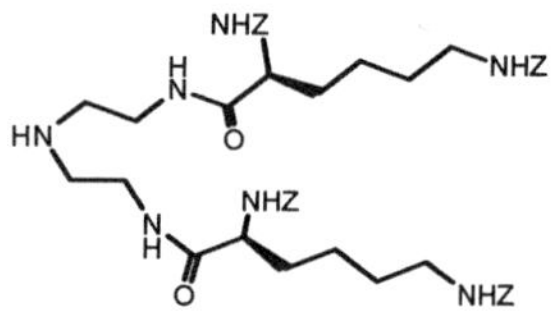

Scheme 2   **Z-4-Amine**

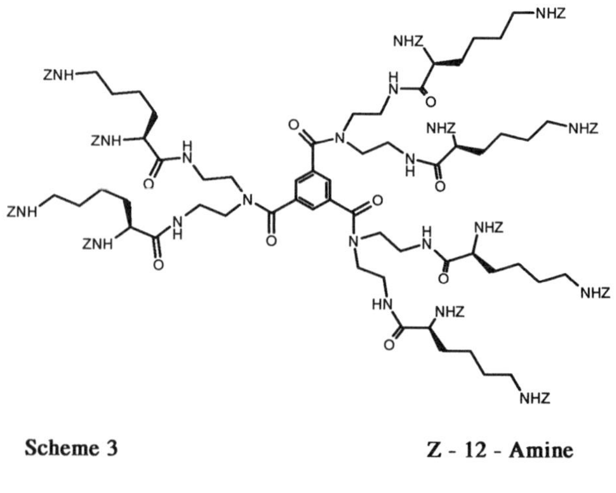

Scheme 1 (Gadomer 17)

Scheme 3   **Z - 12 - Amine**

Scheme 4

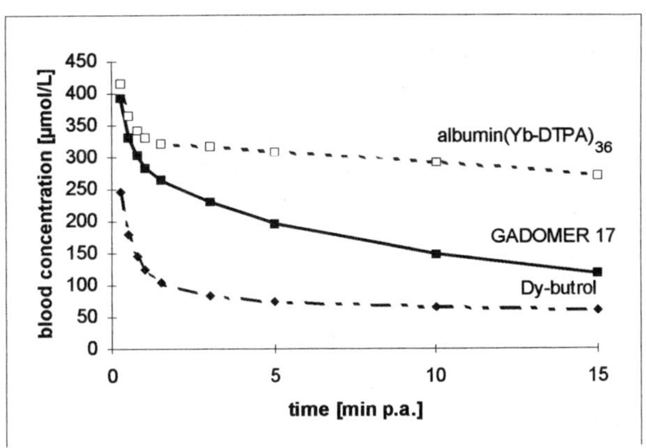

Table 1

Scheme 5

x NaBr

Table 2

Organ	Gd-concentration [% of dose per organ]	
	7 d p.i.	14 d p.i.
Liver	0.11 ± 0.06	0.00 ± 0.00
Kidney	0.31 ± 0.05	0.04 ± 0.01
Femur	0.00 ± 0.00	0.00 ± 0.00
Tail	0.00 ± 0.00	0.02 ± 0.02
Carcass	0.35 ± 0.10	0.27 ± 0.06
Total	0.78 ± 0.07	0.33 ± 0.05

RESULTS

The blood kinetics of the dendritic gadolinium chelates were determined in rats and compared with those of a larger (albumin-Yb-DTPA, 86 kDa[9]) and a smaller metal chelate (Dy-butrol, 0.6 kDa [10]). The results clearly show that diffusion of the dendrimeric chelates is much slower than for small chelates (see table 1 for the result with Gadomer 17, 17.5 kDa). Equilibrium of concentration between blood and interstitial space (flatening of curve) is not seen at 15 minutes after intravenous injection of the dendrimers whereas it is already reached after 3 minutes for the small 0.6 kDa Dy-complex. However, the diffusion into the interstitial space is much longer for the albumin complex that serves as a control which practically does not enter the interstitial space.

The tolerance of the new dendritic agents in rats is promising with an LD_{50} of about 30-35 mmol Gd/kg body weight. They show high efficacy for enhanced magnetic resonance imaging (MRI) with T1-relaxivites of 18-20 L/mmol Gd * sec at 0.47 T which is a 4-5 fold increase when compared with gadolinium chelates used in the clinic at present. We observed greatly improved vascular visualization in rats, rabbits and dogs. Thus, these agents show a potential for the diagnosis of peripheral arterial diseases. However, the complete elimination of polymeric materials is a general problem. Therefor, the 24mer gadolinium dendrimer Gadomer 17 has been administered to mice intravenously (0.1 mmol Gd/kg) and the Gd-concentration (% of dose per organ) was determined after 7 and 14 days p.i., respectively (see table 2). More than 95 % of the given dose was excreted via the kidneys within 24 hrs. Toxicological studies of Gadomer 17 are ongoing.

The compounds are highly watersoluble and survive the conditions of heat sterilization (30 minutes at 120 deg. C) which is a common prerequisite for an intravenously administered contrast agent.

REFERENCES

1. U. Schmiedl, R. E. Sievers, R. C. Brasch et al. *Radiology* **170**, 351 (1989).
2. S. C. Wang, M. G. Wikström, D.L. White et al. *Radiology* **175**, 483 (1990).
3. G. Schuhmann-Giampieri, H. Schmitt-Willich, T. Frenzel, W.-R. Press, H.-J. Weinmann *Invest Radiol* **26**, 969 (1991).
4. G. Adam, J. Neuerburg, E. Spüntrup, A. Mühler et al. *J Magn Reson Imaging* **4**, 462 (1994).
5. J. Tacke, G. Adam, H. Claßen, A. Mühler et al. *J Magn Reson Imaging* **7**, 678 (1997).
6. E. C. Wiener, M. W. Brechbiel, H. Brothers, R. L. Magin, O. A. Gansow, D.A. Tomalia, P. C. Lauterbur *Magn Res Med* **31**, 1 (1994)
7. G. R. Newkome, C. N. Moorefield, F. Vögtle, Dendritic Molecules (G. Walter, ed.), VCH Weinheim 1996.
8. H. Schmitt-Willich, J. Platzek et al. Pat. Appl. WO97/02051 (Schering AG).
9. M. G. Wikström, M. E. Mosely, D. L. White, J. W. Dupon, J. L. Winkelhake, J. Kopplin and R. C. Brasch *Invest Radiol* **24**, 609 (1989).
10. H. Vogler, J. Platzek, G. Schuhmann-Giampieri et al. *Eur J Radiol* **21**, 1 (1995).

RING OPENING POLYMERIZATION VIA MICROWAVE IRRADIATION

Xiaomei Fang, Samuel J. Huang and Daniel A. Scola
Polymer Program, University of Connecticut
Storrs, CT 06269-3136

Introduction

There are numerous technical articles and patents focused on practical and potential applications of microwave irradiation to chemical synthesis, polymer curing and sintering of ceramic materials. An excellent review on microwave processing of ceramics and basic microwave principles was given by Sutton[1]. The applications of microwave energy in organic chemistry covering the field from the first record application in 1969 to many references up to 1991 were collected by Abramovich[2]. In the last decade, considerable efforts have been devoted to investigating the advantages of microwave irradiation over the conventional thermal processing in heating and synthesizing organic and polymer materials.[3-7] As an alternative to conventional heating techniques, microwave irradiation provides an effective, selective and fast synthetic method by heating the molecules directly through the interaction between the microwave energy and molecular dipole moments of the starting materials. This internal heating is believed to produce an efficient reaction since the functional groups of the starting materials, which have strong dipole moments, are the primary source of activation in the microwave electromagnetic field.

Polycaprolactam (Nylon-6) and polycaprolactone (PCL) are commercially produced from the cyclic monomers by chemical processing methods to induce ring opening polymerization. Nylon-6 is one of the most important polyamides and widely used as a synthetic fiber and an engineering resin. Nylon-6 is conventionally produced by heating ε-caprolactam at 250 - 270°C with a catalytic amount of water using a nylon salt or aminocaproic acid as an initiator for 12~24 hrs.

Polycaprolactone is one of the most popular biodegradable polymers, which has a good compatibility with other polymers, such as polystyrene polypropylene, poly(vinyl chloride) and low density polyethylene[8,9]. The biodegradablility of polycaprolactone ester structure has been shown to occur in copolymers of ε-caprolactone and ε-caprolactam, where degradation involves the enzymatic hydrolysis of the ester groups in the backbone leaving the amide bonds unperturbed.[10] Polycaprolactone is commercially produced from caprolactone at 110°C for over 12 hrs. For energy conservation and environmental considerations, it would be of interest to investigate the application of microwave energy to synthesize organic compounds and polymers. In this paper, the results of studies to synthesize nylon-6 and polycaprolactone directly from the monomers via ring-opening reactions by microwave energy are presented.

Experimental

Starting Materials All the starting compounds were obtained commercially, ε-caprolactam and 1,4-butanediol and ω-aminocaproic acid from Aldrich Chemical, ε-caprolactone 99% from Janssen Chemical and stannous octoate from PFALTZ & BAUER, Inc.

Microwave Equipment A variable frequency microwave furnace (VFMF) model LT 502Xb (Lambda Technologies, Inc.) was used to synthesize the polycaprolactam and polycaprolactone. A TEFLON container was selected as a synthetic reactor due to its transparency to microwave energy. A TEFLON tube was connected to the TEFLON reactor to provide nitrogen in preparing nylon-6. The polymerization temperature was measured and controlled by a grounded Omega K type thermocouple.

Synthetic Procedure Nylon-6: ε-Caprolactam with 10 mol% initiator of ω-aminocaproic acid was placed in the microwave oven in the presence of nitrogen at 250°C for several hours. The starting materials were mixed and grounded thoroughly before the reaction. The crude products were extracted by hot methanol for over 16 hrs and then dried in the vacuum at 90°C overnight.

Polycaprolactone: ε-Caprolactone was mixed with approx. 1 mol% of 1,4-butanediol and small amount of stannous octoate as a catalyst, and then placed in the microwave oven for 2 hrs at different temperatures. The crude products were dissolved in chloroform and precipitated from hexane. The precipitate was dried in the vacuum overnight.

Characterization TGA - Thermogravimetrical analyses (TGA) of nylon-6 products were performed by using a TA Instrument 2100 Thermal Analysis TGA 2950 at a heating rate of 10°C/min under nitrogen atmosphere.

DSC - Differential Scanning Calorimetry (DSC) of nylon-6 and polycaprolactone was performed by a Perkin -Elmer DSC, 7 series Analysis System and a TA Instrument 2100 Thermal Analysis DSC 2920, at a heating rate of 20°C/min under nitrogen atmosphere at a flow rate of 20 cm^3/min.

GPC - A Millipore Model 150-C Gel Permeation Chromatography (GPC) system was applied to determine the molecular weight of polycaprolactone, THF was used as a solvent.

IR - Infrared spectra were taken by using Nicolet SX-60 FT-IR. Nylon-6 film was cast at 100°C in vacuum from 90% formic acid.

Results and Discussion

The uniformity of microwave heating achieved by applying variable frequencies at a certain sweep rate has been reported[11]. In this work, ring opening polymerization was accomplished by using a variable frequency microwave furnace with a frequency range of 2.4 GHz to 7.0 GHz. A low microwave forward power was generally selected to minimize any possible temperature deviation from the overheat of the grounded Omega thermocouple at high power range. An example of the nylon-6 microwave process temperature-time profile is shown in Fig. 1(a). The heating rate increased sharply after 75°C, which strongly suggested that ε-caprolactam would absorb microwave effectively once the compound reaches its melting state, i.e. the mobility of the polar functional groups in the starting materials is sufficient to provide the required change of dipole reorientation in microwave electric field for energy absorption. Based on the conventional thermal polymerization of ε-caprolactam, the microwave polymerization was controlled at 250°C for 1hr, 2hr and 3hr, respectively. The temperature value of the grounded Omega thermocouple was calibrated by a Luxtron probe. The temperature control was achieved by changes in applied electric field resulting from the computer-controlled microwave forward power on-off. The heating rate of the microwave process depended on the forward power, incident center frequency, the chemical structure and physical state of materials. DSC spectra of those nylon products are shown in Fig. 2. The highest melting point at 210.1°C was found in the sample prepared at 250°C for 2 hrs. TGA spectra of the crude products directly from microwave irradiation without any purification, in Fig. 3, also showed the 250°C-2 hrs sample gave a higher weight retention than that produced at 250°C for 1 hr and 3 hrs. After extraction by hot methanol, the purified sample showed no weight loss until 360°C, and provided the same thermostability as a commercial nylon-6 sample, Capron 8202NL, from AlliedSignal Corp., as displayed in Fig. 4. In IR spectra, no absorption at 870 cm^{-1}, which indicated the amount of caprolactam monomer[12], was determined in the formic acid cast film of nylon-6 either.

Ring opening polymerization of ε-caprolactone was also performed in the microwave oven by using a lower forward power system since the ε-caprolactone monomer was at the liquid state. The processing temperature-time profile is presented in Fig. 1(b). The ε-caprolactone was mixed with 1,4-butanediol and stannous octoate and placed in the microwave oven for 2 hrs by controlling at different temperatures of 200°C, 180°C and 150°C respectively. DSC and GPC results are summarized in Table 1. The high molecular weight and lower polydispersity in the sample made at 150°C compared favorably with commercial products. A glass transition temperature around -60°C and a melting temperature above 60°C were close to the conventional prepared polycaprolactone. The equivalent physical properties generated for microwave-produced nylon-6 and polycaprolactone in a period of only 2 hrs demonstrate that the ring opening polymerization of ε-caprolactam and ε-caprolactone can be performed efficiently by pure microwave irradiation without any external thermal heating. The much shorter processing time and the quality products produced, strongly suggest that the microwave synthesis technique has great potential in the ring opening polymerization.

Conclusions

Ring opening polymerization of ε-caprolactam and ε-caprolactone using a variable frequency microwave furnace has been investigated and preliminary properties of the microwave products have been measured. Compared with the commercial products, equivalent physical properties were obtained from the microwave-produced nylon-6 and polycaprolactone. The characterization results revealed that the quality nylon-6 and polycaprolactone can be prepared by microwave irradiation within 2 hrs versus more than 12 hrs by the commercial thermal process.

References

(1) Sutton, W.H. *Ceramic Bulletin*, **1989**, *68(2)*, 376.

(2) Abramovich, R.A., *Org. Prep. Proced. Int.*, **1991**, *23(6)*, 683.

(3) Cablewski, T.; Faux, A.F.; Strauss, C.R. *J. Org. Chem.*, **1994**, *59(12)*, 3408.

(4) Majetich, G.; Hicks, R. *J. Microwave Power Electromagn. Energy*, **1995**, *30(1)*, 27.

(5) Jullien, H.; Valot, H. *Polymer*, **1985**, *26*, 505.

(6) Lewis, D.A.; Nunes, T.L.; Simonyi, E. *Polym. Mater.Sci.Eng.*, **1995**, *72*, 40.

(7) Chia, H.L.; Jacob, J.; Boey, F.Y.C. *J. Polym. Sci. Part A*, **1996**, *34*, 2087

(8) Brode, G.L.; Koleske, J.V. *J. Macromol. Sci. Chem.*, **1972**, *A6*, 1109.

(9) Koleske, J.V.; Lundberg, R.D. *J. Polym. Sci., Part A-2*, **1969**, *7*, 795.

(10) Chen, X.; Gonsalves, K.E.; Cameron, J.A. *J. Appl. Polym. Sci.*, **1993**, *50(11)*, 1999.

(11) Bible, D.W.; Lauf, R.J. *US Patent 5 321 222*, **1994**.

(12) Haslam, J.; Willis, H.A; Squirrell, D.C.M. *Identification and Analysis of Plastics*, 2nd Ed., London Iliffe, **1972**. P.314-315.

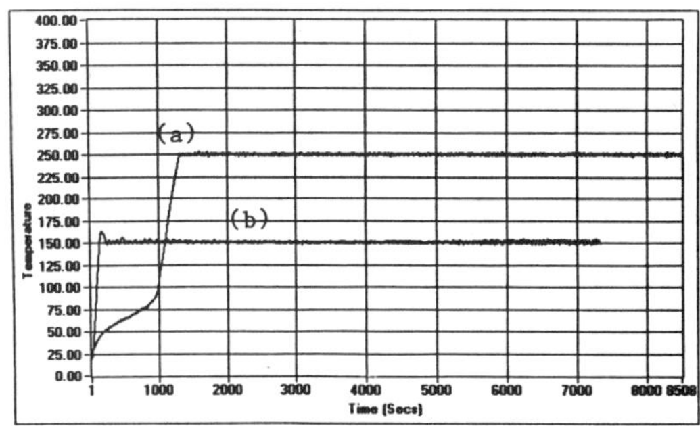

(a) Nylon-6 from Caprolactam (b) PCL from Caprolactone

Figure 1. Temperature-Time Profile of Microwave Processing

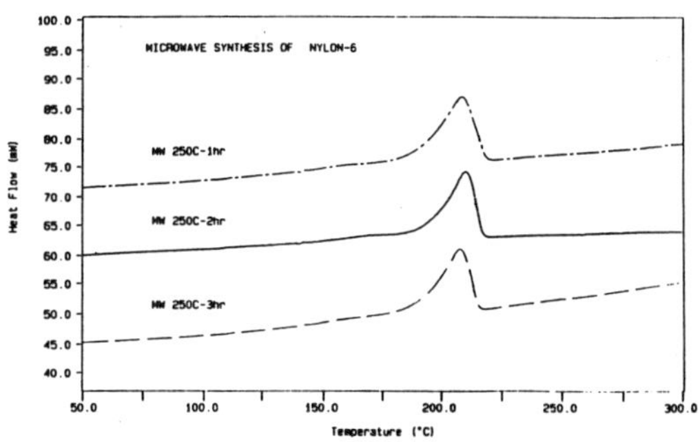

Figure 2. DSC Spectra of Microwave Products (Nylon-6)

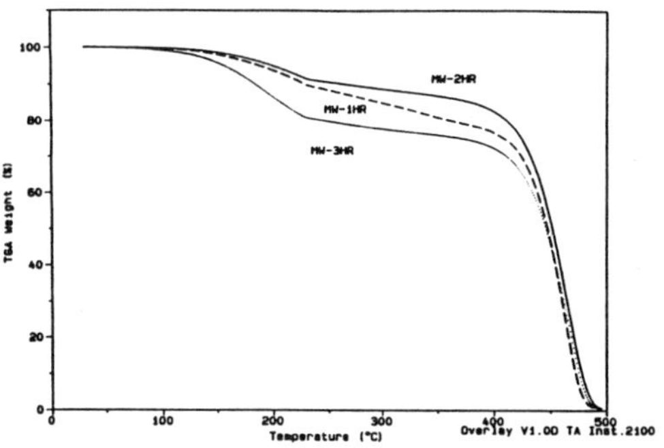

Figure 3. TGA Spectra of Microwave Crude Products (Nylon-6)

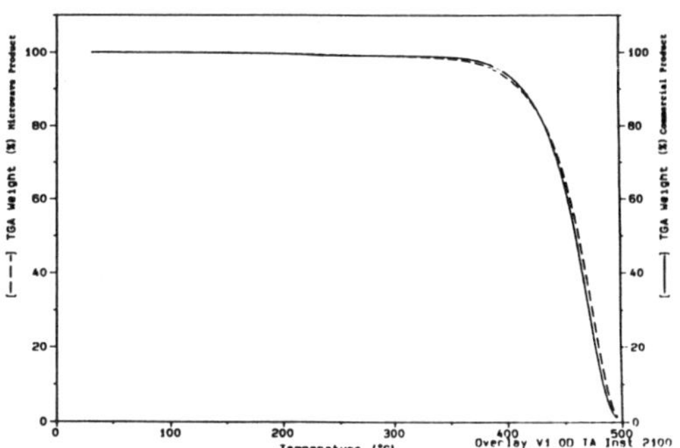

Figure 4. TGA Spectra of Extracted Microwave Product and Commercial Nylon-6

Table 1. Summary of Polycaprolactone Synthesized by Microwave Energy

Sample	I 150°C - 2hr	II 180°C - 2hr	III 200°C - 2hr
Tg (°C)	-61.0	-61.8	-61.4
Melting Point (°C)	68.5	64.4	63.1
Molecular Weight (Mn) by GPC	11,298	9,714	8,932
Polydispersity (Mw/Mn) by GPC	1.31	1.69	1.49

. ne Biostability of An 80A Aliphatic Polycarbonate Polyurethane

Ken Stokes, Medtronic, Inc. 7000 Central Ave. NE, Minneapolis, MN
55432

INTRODUCTION

A replacement for Dow's Pellethane is needed since it is no longer commercially available for implantable devices. It is well known that the polyether polyurethanes are susceptible to various forms of oxidative degradation of the soft segment, but they are hydrolytically stable *in vivo*. In theory, the polycarbonate polyurethanes should have superior oxidation resistance but they are also theoretically subject to hydrolytic degradation mechanisms. Based on *in vitro* testing, some polycarbonate polyurethanes have been cited as having better chemical resistance than some polyether polyurethanes[1]. While such testing is necessary for screening, it is impossible to establish biostability *in vitro*. *In vitro* testing can not be relied on to qualify a material for implantable use. It is critically important that we do not trade new, possibly worse failure mechanisms for those we have some understanding of, and now can reasonably control. This study was aimed at determining the biostability of an 80A Shore A hardness polycarbonate polyurethane with aliphatic hard and soft segments for use in cardiac pacing leads.

MATERIALS & METHODS

Two lots of ChronoFlex AL80A (AL80A) were received from PolyMedica, Inc. The material was vacuum dried at $50 \pm 5°C$ and ≥ 27 in. Hg pressure. Tubing was extruded on a Killion machine using a 1 inch diameter general purpose 24:1 screw and a 1 cc gear pump. Barrel temperatures ranged from 260 to 325°F with a 300-305°F die and a 290-300°F gear pump.

Accelerated Environmental Stress Cracking (ESC) Samples[2,3]

Injection molded 0.5 inch long polysulfone dumbbells (0.050 in. dia. over the ball shaped ends, 0.020 in. dia. in the center) were inserted into 0.070 X 0.080 in. in dia. tubing. Some tubing was strained 400% in a fixture and firmly ligated at the mandrel ends using 2-0 polyester suture material. Others were unstrained. Some were annealed at 65°C, others at 80°C, some not at all. Some Pellethane 2363-80A (P80A) controls at 0 and 400% elongation were annealed at 150°C, some were unannealed. Up to 4 specimens were tied end-to-end using polyester suture, then color coded with glass beads. After ethylene oxide sterilization and aeration, up to 4 such strings of specimens were implanted into the dorsal subcutis of each rabbit, assuring that there would be an $N \geq 5$ animals explanted at 12, 24, 56, 78 and 104 weeks. At necropsy, the specimens were removed, debrided, allowed to air dry, then were evaluated by optical microscopy (up to 64 magnifications). Scanning electron microscopy was done on an as needed basis to verify or clarify optical results. Selected specimens not used for SEM were analyzed by FTIR and GPC (molecular weight).

Implantable Devices

Three coaxial, bipolar, transvenous Medtronic Model 4024 (ventricular) and/or 4524 (atrial) lead analogues with ChronoFlex AL80A insulation and 1 coaxial, bipolar, transvenous Medtronic Model 6972 ventricular control (with Pellethane 2363-80A insulation) were implanted in each of 9 canines under general anesthesia. All 3 experimental leads were implanted via the costocervical vertebral trunk vein, whereas the control was inserted through the right jugular vein. The leads were firmly ligated at the venous entry site. There were 4 designs implanted in a second 9 dog study. One was a bipolar ChronoFlex Al80A insulated Model 4024 transvenous lead analogue described above, implanted via a right jugular venotomy. One was a bipolar ChronoFlex Al80A insulated epicardial lead with the same Model 4024 analogue body described above. The third design was a Pellethane 2363-80A control using a Model 6972 leadbody described above. The fourth was a unipolar Pellethane 2363-80A Model 6971 leadbody subassembly implanted subcutaneously. The epicardial

leads required thoracotomies. Six dogs received only epicardial leads with 1 control each. Three dogs each received 1 experimental transvenous lead, one experimental epicardial lead and 1 epicardial control. Three additional dogs each received 2 experimental transvenous leads, 1 epicardial lead and a subcutaneous control. Six of the dogs in the second group were terminated after 6 months on study and are not included in this report. The remaining three provided an additional 5 experimental and 3 control leads for analysis after 30 months implant.

All animal testing followed Good Laboratory Practices guidelines. NIH guidelines for the care and use of laboratory animals (NIH Publication #85-23 Rev. 1985) were followed. Statistical analysis used a two-tailed distribution assuming $p < 0.05$ to be significant.

RESULTS

Accelerated Environmental Stress Cracking Test

ChronoFlex Al80A presented a superior appearance through 6 months implant without any phenomenon that can be construed as ESC. Most unstrained specimens had excellent appearance through 2 years implant, although a few did develop shallow surface defects that look more liked etched grooves than cracks. After 1 year implant, some strained specimens developed surface features ranging from grooves (that look like they were made with a chisel) to shallow cracks. Cracks through the strained polymer began to appear after 1 year implant. Failure occurred in 80%, 50% and 43% of specimens with 400% unrelieved strain after 12, 18 & 24 months implant, respectively. Specimens strained to 400% elongation, then annealed at 65°C experienced 50%, 70% and 86% failure over the same time period. Those 400% elongated specimens annealed at 80°C developed 20%, 10% and 21% failure after 12, 18 & 24 months. Thus, the higher annealing temperature improved resistance to stress cracking.

Two failure mechanisms were evident in AL80A. Many samples developed breaches at the end of the specimen, over the "ball" of the mandrel. These cracks had a "feathered" edge, apparently the result of ductile process (like creep failures). Other specimens experienced a brittle fracture which did not involve necking or drawing down of the material. Crazed cracks or tears, typical of P80A ESC, were usually not observed on AL80A. There were no significant differences in the FTIR spectra or the ratios of the 4 carbonyl peaks between the various specimen conditions at each explant period, but there were changes as a function of time. When the data for the various specimen conditions were pooled in weighted averages, the slope of the soft segment/free carbonyl 1740/1715 cm^{-1} ratio curve as a function of time was -0.0017 cm^{-1}/week. While this was not a very steep slope, it was significant ($p \ll .001$). Using a similar technique, the slope of the soft segment/bonded carbonyl 1740/1693 cm^{-1} curve was about -0.0014 cm^{-1}/week, again statistically significant ($p < .001$). The ratio of the 2 soft segment carbonyl peaks (1740/1663 cm^{-1}) had an order of magnitude steeper slope at -0.013 cm^{-1}/week ($p \ll .001$). The average slope of the 1715/1663 cm^{-1} ratio was -0.0079 ($p \ll .001$). Finally, the hydrogen bonding index curve (1693/1740 cm^{-1}) did not change as a function of implant time ($p = .08$). Thus, there were significant quantitative changes in the infrared surface spectra of AL80A specimens as a function of implant time, but they appear to have little or nothing to do with residual stress (strain) or annealing. There was no difference in molecular weights any of the test conditions at any given explant period. The pooled M_w of ChronoFlex AL80A declines 16% after 78 weeks implant, from 171000 $\pm$ 4000 (n=10) to 143000 $\pm$ 7000 (N=10, $p \ll .001$). The approximate slope of the M_w curve is about -360 units/week. M_n declines 22.5% from the initial value after 78 weeks implant.

Pellethane 2363-80A processed in a manner to accelerate ESC (acetone extraction and 400% unrelieved elongation) developed extensive cracking in 12 weeks. Pellethane 2363-80A processed in a manner to minimize ESC (no extraction, 400% elongation, annealed) exhibited very shallow cracks in the outer surfaces through 6 months. Between 6 and 12 months, failures began to develop at a slowly increasing rate (50% of the specimens at 1 year, 60% at 18 months, and 71% after 2 years). The 400% strained and annealed P80A samples experienced an insignificant ($p = .08$)

increase in hydrogen bonding index (1703/1730 cm^{-1}) as a function of time. There was no significant change in the ether index (1105/1078 cm^{-1}). The unannealed, 400% elongated P80A specimens demonstrated about a 7% decrease in the ether index over 12 weeks, with a 16% increase in the hydrogen bonding index as they progressed to ESC failure.

Device Explants

Two dogs in the first group had to be sacrificed early for humanitarian reasons (unmanageable infections) leaving 26 experimental leads on study. Examination of 12 ChronoFlex Al80A leads after 2 years implant revealed 8 with no visible defects, even at the rather severe ligation sites. Four leads had significant ESC at the ligature sites, in three cases to insulation failure. All 4 Pellethane 2363-80A controls had shallow surface cracking while 2 had ESC failure at ligature sites. One had metal ion oxidative (MIO) degradation noted on the inner surface of the outer insulation[2,4]. The remaining 6 dogs were terminated after 30 months implant. One ChronoFlex AL80A lead developed conductor fracture where it passed between the ribs. Upon explant, this lead was found to have insulation cracks extending about 1 inch proximal from the fracture site. Another ChronoFlex AL80A lead in that same animal had similar cracks in the outer insulation just distal from the terminal assembly. The cracks were mirror smooth (brittle), with no evidence of creep. There was what appeared to be ESC insulation failures at 6 AL80A ligature sites (43%). However, these were all pinholes, which could also be explained as creep or cut-through failures. None of these would have produced any clinical manifestations. No inner insulation damage was seen in the AL80A leads. One of the Model 6972 controls had an ESC breach at the ligature, the other controls did not.

On the outer (tissue contacting) surface of the AL80A tubing, the 2 year 1740/1663 and 1715/1693 cm^{-1} FTIR carbonyl ratios decreased to about 30% of their original values (P << .001). The 1740/1693 and 1740/1715 cm^{-1} ratios developed a net decrease over the unimplanted values (-13%, P < .001 and -5.9%, P = .003, respectively). The hydrogen bonding index increased by 4.2% (P < .001). The inner surface of the outer AL80A tubing developed less change in the carbonyl amplitude ratios than the outer surface. The outer surface of the inner AL80A tubing experienced decreases in all four carbonyl ratios followed by increases. The inner surface of the inner AL80A tubing experienced no significant change in the carbonyl ratios. There was no change in the outer insulation's molecular weight (M_w) distally (in the heart), but the 14% increase at the ligature site and the 6% increase proximally (in the subcutis) were significant (P < .001). The number average molecular weight decreased distally (-12%, P << .001) and proximally (-6.5%, P = .009), but did not change at the ligature site. The dispersivity increased about 14% at all locations measured (P << .001). The inner ChronoFlex AL80A M_w increased 9% distally, 24% at the ligature and 16% proximally (all P << .001). M_n decreased 7% distally (P < .001), but increased 9% at the ligature (P < .001) and 3% proximally (P = .025). Dispersivity increased 17% distally, 14% at the ligature site and 12% proximally (all P << .001).

DISCUSSION

Environmental Stress Cracking

With the exception of the minute anomalies one can always anticipate in tests of this kind, there was no significant visual degradation of unstrained AL80A samples. Some strained samples cracked by what appeared to be two separate mechanisms. In some cases, the polymer developed large brittle cracks throughout its bulk. In other cases it developed breaches that looked like creep failures. The edges of the crack were drawn down to a very sharp edge. What is curious here is that some of the samples that failed in this manner had been strained to 400% elongation, then annealed at 80°C (close to the crystalline melting point). In this case, therefore, the polymer was essentially remolded (stress relieved) in the strained conformation. Significant residual stresses or strains are unlikely. Why then did the sample develop what appeared to be creep failures? Why did some 80°C annealed strained specimens develop

brittle cracks, while unstrained, unannealed specimens did not? This observation suggests the hypothesis that either the polymer was "set up" for degradation by the initial strain per se (unlikely) or that the material can not tolerate annealing (equally unprobable). Even more curious, infrared analysis tends to show the same degree of change whether annealed or not, whether stretched or not. The molecular weights of the polymer decreased the same amount whether stretched or not, whether annealed or not, whether cracked or not. At present, this remains a mystery. If the polymer is to be used in implantable products, this mystery will have to be solved. None-the-less, one can conclude that the stress cracking resistance of ChronoFlex Al80A is better than Pellethane 80A by looking at the percentage of specimens failed. The best annealed AL80A samples had 10-21% failure in 1-2 years compared to 50-71% failure in the best annealed P80A controls. On-the-other-hand, 10-21% failure is not acceptable for most devices.

Device Implants

The ChronoFlex Al80A implants showed virtually no tissue growth on them in the heart. ESC is a cell mediated phenomenon. Thus, it is not surprising that there was no visual evidence of degradation at these locations. Of 12 leads explanted at 2 years, there was no visible degradation on 8. In 4 cases, ESC was observed at the ligature sites. In 3 of these (25%), insulation rupture had occurred. After 30 months implant, there were 8 (57%) AL80A insulation breaches out of 14 leads. These included 2 large outer insulation breaches in addition to pinholes at the ligature sites. In the control groups, 3 of 9 (33%) had ESC pinhole failure at the ligatures, but no other breaches. We issue advisories if a lead model has < 95% actuarial survival in 5 years in human service[5]. The AL80A leads implanted ≥ 2 years had a total of 42% outer insulation failure (58% survival) compared to 33% failure (67% survival) for the controls. However, in many of the cases, the ligature pinholes would have been clinically silent. If we discount these, then there was 7.7% failure (92.3% survival) in the AL80A series and no failures (100% survival) in the controls. There was no significant degradation of the AL80A inner insulation while 1 of 9 (11%) controls did have metal ion oxidation present inside.

CONCLUSIONS

In our strain accelerated *in vitro* biostability test, ChronoFlex AL80A performed better than Pellethane 2363-80A, but still with some failures. ChronoFlex Al80A failures occurred by two apparently different mechanisms compared to ESC in polyether polyurethanes. The AL80A leads did not demonstrate adequate outer (tissue contacting) insulation biostability for long-term implant compared to the controls. The inner (non-tissue contacting) insulation did appear to have good performance. Therefore, ChronoFlex AL80A is not recommended for use in tissue contacting surfaces of chronically implantable cardiac pacing leads.

REFERENCES

1. Tanzi MC, Mantovani D, Petrini P, et al. Chemical stability of polyether urethanes versus polycarbonate urethanes. J Biomed Mater Res, 36(4), 550-559, 1997.
2. Stokes K, Urbanski P and Cobian K. New test methods for the evaluation of stress cracking and metal catalyzed oxidation in implanted polymers. In H. Planck, et al (eds.), Polyurethanes in Biomedical Engineering II, Amsterdam, Elsevier, 109-128, 1987.
3. Stokes K. The biodegradation of nondegradable polymers. In H. Planck, M. Dauner and M. Renardy (eds.), Degradation Phenomena on Polymeric Biomaterials, Berlin, Springer-Verlag, 1992, pp37-58.
4. Stokes KB, Urbanski P and Upton J. The in vivo autooxidation of polyether polyurethane by metal ions. J Biomater Sci, Poly Ed, 1(3), 207-230, 1990.
5. Product Performance Report, Medtronic, Inc., September, 1997.

Elastomers for implantable applications-silicone or polyurethane?
Mike Skalsky, Elastomedic Pty Ltd, 126 Greville Street, Chatswood
N.S.W. 2067, Australia.

Introduction

Medical device developers have had a very limited choice of implant materials for highly demanding applications such as insulation for pacing leads, long term in-dwelling catheters, synthetic polymer heart valves and coronary vascular grafts. All of these applications require polymers that have excellent biostability, good mechanical strength and high resistance to fatigue, abrasion and tear. Traditionally, silicone rubber was the first material of choice, because of its excellent biostability, despite shortcomings in the other material properties. Polyurethane, which is a generic name for a very large family of synthetic polymers, has emerged as the principal alternative to silicone rubber because of superior mechanical properties.

Silicone or polyurethane ?

One of the most demanding applications is that of pacing lead insulation. Silicone rubber was used initially, but with the need for smaller devices, polyurethanes which are stronger, were introduced. Both materials have proven to have desirable qualities which are not available in the other. Silicone is generally softer, more flexible, and easier to manufacture. Polyurethanes, as a group have high tensile and tear strengths, and can therefore be made smaller. In addition, they have a low abrasion resistance and coefficient of friction when wet with blood. Over the years of use, it has been established that polyurethanes of intermediate (eg Pellethane 2363, 90 Shore A or relatively high hardness (eg Pellethane 2363, 55 Shore D have acceptable biostability[2], whereas soft polyurethanes (eg Pellethane 2363 80 Shore A) have experienced biostability problems[3]. Consequently, there has been an ongoing debate about the advantages and disadvantages of polyurethanes and silicone rubber as lead insulation.

Some of the positives and negatives of these two materials are summarized below:

Silicone rubber	Polyurethane
Positives	**Positives**
34 years of proven history	14+ years of proven history (for 55 D)
Repairable	High tear strength
Low process sensitivity	High cut resistance
Easy fabrication/moulding	Low friction in blood
Very flexible	High abrasion resistance
Bonds fairly readily	Small component cross section possible
	Excellent fatigue performance
	Relative non-thrombogenic/fibrotic

Negatives	Negatives
Tears easily (nicks,ligatures)	Not readily repairable
Cuts easily (surgical damage)	Less flexible
Low abrasion resistance (for soft material)	Biostability variable
Relatively high friction resistance in blood materials not easy	Bonding to dissimilar
Requires thicker component cross section	
Relatively more thrombogenic and fibrotic	
Absorbs lipid	
Average fatigue performance	

New developments

In recent years, substantial research effort has been directed toward understanding the causes of bioinstability in soft polyurethanes. Whilst the mechanism for the observed degradation is still not clearly understood, environmental stress cracking and metal ion induced oxidation affecting the soft segment of the polyurethane have been identified as the most relevant causes. Polyurethanes using polycarbonate in the soft segment have been proposed as a biostable polyurethane[4], but these materials were found not to have the expected biostability[5]. Augmentation of polycarbonate polyurethane with surface modifying end groups is being evaluated.[6] Another approach attempting to solve the biostability problem has used novel macrodiols that achieved the integration of silicone into the soft segment of the polyurethane. The initial testing of the Elast-Eon™ material for oxidative and hydrolytic stability in-vitro, as well as in accelerated strained implant tests in sheep over 3 months has shown a great deal of promise[7]. Long term testing is required to demonstrate whether these materials will become a viable alternative to both silicone rubber and currently available polyurethanes.

References
1. Stokes, K. The biostability of polyurethane leads, in Modern Cardiac Pacing, Ed. S.S. Barold, Futura Publishing Co, Mount Kisco,N.Y., 1985, p. 173
2. Skalsky, M., Vaughan J.D. et al, Performance testing of polyurethane (Pellethane 2363-90A) cardiac pacing leads in animal models, in Polyurethanes in Biomedical Engineering II, ed. H. Planck et al Elsevier Science publishers B.V., Amsterdam, 1987, p75
3. Szycher, M., Reed A.M.et al., In vivo testing of a biostable polyurethane, J.Biomater. Appl. 6: 110, 1991
4. Pinchuk, L., KatoY.P., et al., Polycarbonate urethanes as elastomeric materials for long-term implant applications, in Transactions of 19th Annual Meeting of SFB,Charleston,SC, 1990, p. 22.
5. Skalsky M.,Mathivanar R.,et al, unpublished observations
6.Mathur A.B.,Collier T.O.,et al, In vivo biostability of modified polyurethanes, J. Biomed. Mater. Res.36: 246,1997
7. Meijs,G.F., Gunatillake P.A., et al, Degradation resistant polyurethane compositions, this symposium.

Acknowledgement

The contribution of Bob Boveja of Pacesetter Inc. is gratefully acknowledged.

DEGRADATION RESISTANT POLYURETHANE COMPOSITIONS

Gordon F Meijs[1], Pathiraja A Gunatillake[1], Simon J McCarthy[1], Darren J Martin[2], Laura Poole-Warren[2], and Klaus Schindhelm[2].
Cooperative Research Centre for Cardiac Technology, Australia. [1]CSIRO Molecular Science, Bayview Ave. Clayton, Vic. 3169, Australia. [2]Graduate School of Biomedical Engineering. UNSW, Sydney 2052. Australia

INTRODUCTION

Polyurethanes are versatile elastomers that elicit acceptable biological responses making them candidates for many biomedical applications. However, soft polyurethanes such as Pellethane 80A, which are more desirable due to their flexibility, tend to be more susceptible to degradation *in vivo* than harder, less flexible materials like Pellethane 55D, despite minor compositional differences. The combination of appropriate morphology and highly stable chemistry appears necessary for the development of biostable polyurethane implant materials. The incorporation of siloxane soft segments is one means of improving polyurethane biostability, but generally results in materials with poor mechanical performance. By engineering an appropriate degree of mixing between hard and soft phases, mechanical and processing properties can be optimised, while maintaining the good stability associated with polydimethylsiloxane chemistry. We have recently developed such a siloxane polyurethane (Elast-Eon).[1]

MATERIALS AND METHODS

The siloxane polyurethane was prepared by a one step bulk polymerisation procedure, using MDI, butanediol, polydimethylsiloxane and a small amount of a second, compatibilising macrodiol. Pellethane 2363-80A (P80A) and Pellethane 2363-55D (P55D) were used as positive and negative control materials, respectively. Flat sheets were prepared using compression moulding. Dumbbells (3.5x1x0.1 cm) were punched from flat sheets for tensile and biostability testing. Tensile testing was conducted using an Instron Model 4032 Universal testing machine with a 1 kN load cell at a crosshead speed of 500 mm/min. Other mechanical characterisation was conducted according to standard methods (see footnote to Table 1). Pellethane 80A and 55D tubing was supplied in fabricated form and the siloxane polyurethane was extruded into tubing with similar dimensions in-house.

Biostability was assessed by subcutaneous implantation of strained flat sheet dumbbells and tubing in sheep for up to 3 months[2]. Briefly, small dumbbells and tubing samples were washed and strained over poly(methyl methacrylate) (PMMA) holders prior to sterilisation using ethylene oxide. The strains employed were approximately 150% over the narrow region of the dumbbells, and 0%, 150% and 250% for the tubing samples. For each strain, a total of 6 samples of each material were implanted subcutaneously in the dorsal region of 3 sheep. Implants were retrieved following 6 weeks (tubing only) and 3 months implantation. Explanted specimens were examined for the extent and severity of stress cracking in the uniformly strained regions by SEM.

RESULTS

Mechanical properties of the siloxane polyurethane are shown in Table 1.

Table 1. Mechanical properties of the siloxane polyurethane

Property	Siloxane PU	P80A	P55D
Shore Hardness	84A	82A	55D
UTS[*] (MPa)	25.5±1	33.7±1.8 (31)	40.3±1.8 (48)
Ult. Elongation (%)	460±18	430±20 (550)	328±16 (390)
Young's Mod. (MPa)	22.5±2	13±2	87±10
Stress @ 100% ε (MPa)	8.3±5	8 (6)	20 (17)
Tear Strength[§] (N.mm-1)	60±2	72 (83)	(115)
Flexural Modulus (MPa)	30±1	26±2	100±3 (172)
Abrasion Resistance[†]	40	10(20)	(80)

[*]Ultimate tensile strength (ASTM D412). [§]ASTM D624 (Die C), [†](mg/1000 cycles) ASTM D1044 (Taber Abrader - 1000gm and H22 wheel)

The test polyurethane showed no evidence of environmental stress cracking following 3 months of implantation at 150% strain. (See Figure 1.) Results were almost identical for Pellethane 55D. The positive control Pellethane 80A had significant deep stress cracking over the central uniform strain region (Figure 2).

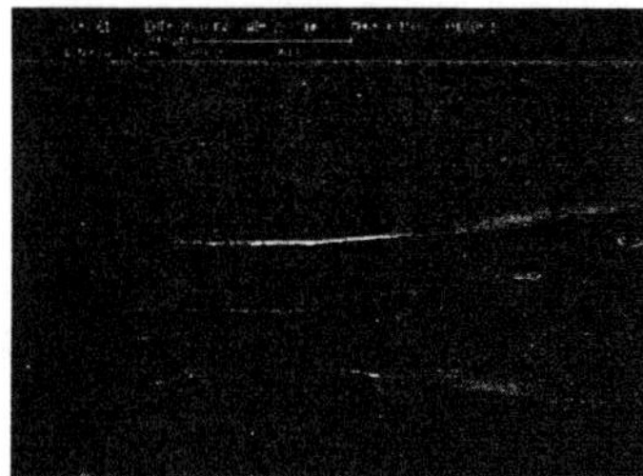

Figure 1. SEM of explanted siloxane polyurethane

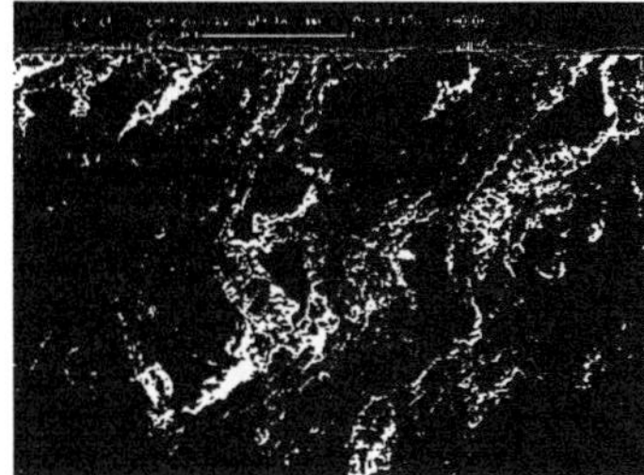

Figure 2. SEM of explanted Pellethane 80A

These data show that the siloxane polyurethane (Elast-Eon) has strength and flexibility close to P80A and a resistance to degradation by Environmental Stress Cracking that is characteristic of the harder, more stable material, P55D.

ACKNOWLEDGMENT

The authors gratefully acknowledge the invaluable contributions of Raju Adhikari, Kathryn Noble, Veronika Tatarinoff and John Klemes.

REFERENCES

1. International Patent Application PCT/AU97/00619 (September 19, 1997)
2. Brandwood A, Meijs GF, Gunatillake PA, Noble KR, Schindhelm K, Rizzardo E. J. Biomater. Sci. Polymer Edn, 6: 41-54, 1994.

Biodurable Polyurethanes in Vascular Grafts
M. Szycher, A.E.Edwards, R.J. Carson and D.J. Dempsey
CardioTech International Inc., Woburn, MA 91801

Introduction

Polyurethanes have unique mechanical and biologic properties that make them ideal for many implantable devices. However, they are subject to some *in vivo* degradation mechanisms,. Polyester polyurethanes are subject to hydrolytic degradation and are no longer used in long-term implanted devices. Polyether polyurethanes, while hydrolytically stable, are subject to oxidative degradation in several forms, including environmental stress cracking and metal ion oxidation. Mineralization is also known to occur. A new poly(carbonate)urethane has superior biostability in *in vivo* qualification tests compared to the traditional polyether polyurethanes.

We describe our approach to the design, development and production of a self-sealing vascular access graft, and the continuing development of a compliant small diameter vascular prosthesis using a unique patented fabrication process with a new generation of a biodurable poly(carbonate)urethane.

Polyurethanes have been recognized for some time as an attractive and readily available range of materials for the fabrication of vascular prostheses. The elastic properties of the material, coupled with low thrombogenicity and exceptional physical and mechanical properties, has led to considerable research effort over the last twenty years, aimed at the development of polyurethane vascular grafts. [1]

Vascular Access Grafts

Vascular access grafts are used in connection with hemodialysis treatments. Several acute and chronic diseases, including polycystic kidney disease, diabetes and hypertension, attack and may destroy normal kidney function, resulting in end stage renal disease ("ESRD"). ESRD is irreversible and currently approximately 60% of all ESRD patients are maintained by hemodialysis.

Hemodialysis requires that the patient be connected to an artificial kidney machine three times per week, for approximatcly four hours per session. A portion of the bloost pumped from the heart must be routed to the artificial kidney machine, cleansed and returned to the patient. This routing is most efficiently accomplished by accessing the bloodstream at a point where the volume of blood flow is high . Two different methods are typically used. The first method, called an autologous AV shunt, involves cutting one of the arteries in the patient's aim and sewing the artery to an adjacent vein. The second method requires that one end of a synthetic vascular graft be attached to an artery and the other end be sewn to a vein. In either case, blood flows very rapidly from the high pressure artery into the low pressure vein. A dialysis clinic technician inserts two large needles into the synthetic graft. One needle removes the blood and routes it to the artificial kidney machine and the second returns the blood to the patient.Approximately 54,000 new and replacement AV access grafts will be implanted in the US

The improved sealing capability of polyurethane access grafts is especially important, as AV access grafts must be punctured in two places three times each week with large gauge needles to withdraw and replace blood cycled through an artificial kidney machine. Currently used synthetic grafts bleed profusely when needles used for hemodialysis treatments are removed and require a technician to apply pressure to the graft for up to 20 minutes to expedite clotting.

Polyurethane Degradation

The word "polyurethane" can be applied to a huge number of different compositions with surprisingly varied applications. Generally speaking, polyurethanes used in

excellent wear resistance, and superior biocompatibility (including blood compatibility). Depending on the ratio of hard to soft segments and the molecular weight of the segments, one can vary the hardness, lubricity, flexibility (elastic modulus) and many other properties.

Thus, polyurethane elastomers have been extensively evaluated for use as artificial heart diaphragms, heart valves, joint prostheses, vascular grafts, urethral catheters, mammary prostheses, penile prostheses, and many other devices intended for long-term implant. The polyester polyurethanes were used as coverings for breast prostheses. Polyether polyurethane elastomers have been used as insulation for neurologic leads since 1975 and cardiac leads since 1977. Rigid polyether polyurethane is used in connector modules for implantable cardiac pacemakers, defibrillators and neurologic stimulators.

Poly(ether)urethanes were considered to be the state of the art in terms of biocompatibility and mechanical properties. However, pacemaker leads fabricated from ether urethanes exhibited degrees of polymer degradation when implanted into living systems.

In 1983 Szycher [2] first proposed that poly(ether)urethanes were susceptible to *in vivo* oxydation of the polyether chain. The most susceptible group is the methyl group in the alpha position to the ether oxygen, which undergoes oxidation, causing eventual chain cleavage, leading to a significant reduction in molecular weight at the surface and eventual surface fissuring.

While it is not normally supposed that enzymes catalyze the degradation of synthetic polymers, the possibility exists that ether-based polyurethanes are degraded by enzymes *in vivo*. Williams [3] remarked that since enzymes have the ability to reduce the activation energy of chemical reactions, a degradation reaction that may normally only occur at elevated temperatures or in the presence of actinic radiation may conceivably take place under physiological conditions in the presence of the correct enzymes.

In a landmark study Phua and Anderson [4] tested the biodegradation of polyurethanes by *in vitro* exposure to enzymes. Ultra-thin samples were exposed to two proteolytic enzymes, papain and urinase, at 37°C for one to six months. Both enzymes were found to be capable of degrading poly(ether)urethanes. Since papain is closely related to cathepsin B, a thiol endopeptidase which is released by cells of the inflammatory response, the authors concluded that segmented poly-ether-urethanes can be degraded by enzymes which are present during the inflammatory response.

New Generation Polyurethane.

It has been reported by both Capone[5] and Szycher [6] that polycarbonate-urea-urethane is resistant to biological oxidation (surface fissuring) for periods of up to six months. Poly(carbonate) elastomers were fabricated as tubing stretched to 300% over mandrels and implanted subcutaneously in experimental animals. The tubing was retrieved at three and six months, evaluated by S.E.M. and FTIR. analysis, with no evidence of degradation.

Stokes et al [7] compared poly(carbonate)urethanes to poly(ether)urethanes *in vivo* using the "Stokes Test" which is designed to accelerate environmental stress cracking, using strain as the time accelerant. Extruded tubing was stretched over mandrels to 400% elongation. Poly(carbonate) urethanes showed no evidence of environmental stress cracking at eighteen months post-implantation.

Our process involves what we term 'low temperature cast coagulation." In this system we gently extrude polymer solution onto the smooth surface of a mandrel. The mandrel and extrusion head rotate in synchronization, minimising shear and residual stress, while a pair of 2.5 meter long mandrels are drawn through the twin extrusion heads into a coagulant maintained at 40° C.

Mandrel rotation and transverse speed, extrusion head rotation speed and polymer pump pressure are all electronically controlled and coordinated to give the desired structure and geometry to the graft wall. By controlling the process conditions, grafts can be produced with a wide range of physical and mechanical characteristics; an example would be a very porous graft allowing rapid cellular ingrowth which could be suited to venous applications.

The polymer solution comprises a solution grade poly(carbonate)urethane, a water soluble filler of between 10 and 60% by weight and a surfactant in an amount between 1 and 10% by weight. During phased coagulation the fillers prevent collapse of the structure as the solvent disperses and the filler dissolves into the coagulant, resulting in a single layer uniform microporous structure. A single layer structure avoids the danger of delamination and subsequent loss of strength, maintains cross section and allows the surgeon to cut the graft evenly and cleanly.

Experimentai Implants - Vascular Access

To assess initial biological response, we have implanted four 5 mm ID grafts into canines as bilateral aorto - femoral bypasses with the iliac arteries tied off for a period of 22 months. Patency was monitored monthly by color- coded Duplex examination. All grafts were patent at explant. Light microscopy showed a well developed neointima lining the anastomotic region of the graft without apparent lumenal reduction. The surrounding capsule was composed of tissue macrophages and foreign body type giant cells.

Conclusions

We have presented our methods and materials used to fabricate vascular grafts. We expect that a combination of proven technology, product design, and the use of a second generation poly(carbonate) urethane, may lead to clinically successful compliant polyurethane vascular grafts.

References

1. Underwood, C.J., Tait, W.F, Charlesworth, D., "Design Considerations for a small diameter vascular prosthesis". *Int. JArtif Organs.*, 1988, 272-276.
2. Szycher, M.,"Surface Fissuring of Polyurethanes Following *in vivo* Exposure". ASTM, STP859, 308-321, Fraker and Griffin, eds., Philadelphia.
3. Williams, D.F. *JBioeng.*, 1,279-294, 1977
4. Phua, S.K., and Anderson, J.M., "Biodegradation of a Polyurethane *in vitro*". *J Biomed Mat. Res.*, 21, 231-246, 1987.
5. Capone, C.D., "Biostability of a Non Ether Polyurethane". *J.Biomat.App.*, Vol 7, Oct 1992.
6. Szycher, M.,et al.. *"In vivo* Testing of Biostable Polyurethane". *J Biomat. App.,Vol* 6, Oc 1991.
7. Stokes, K., et al, "Polyurethane Elastomer Biostability". *J. Biomat. App.,* Vol 9, April 1995.

In Vivo **Stability of Poly(ether urethaneurea) Blood Sacs**

Limin Wu, James T. Garrett, David M. Weisberg and James Runt,
Department of Materials Science and Engineering
Penn State University, University Park, PA 16802

George Felder III and Gerson Rosenberg
Department of Surgery, Milton S. Hershey Medical Center
Penn State University, Hershey, PA 17033

Introduction

Poly(urethaneureas) have been used in a variety of biomedical applications as a result of their combination of good physical and chemical properties, coupled with their biocompatibility. For applications such as artificial heart and ventricular assist blood sacs, these polymers must remain chemically stable over extended times in the relatively aggressive body environment. *In vivo* biostability studies of poly(urethaneureas) have generated contradictory results, with reports of no significant biodegradation on implantation [see ref. 1] and of significant and relatively rapid deterioration *in vivo* [e.g. 2] appearing in the literature.

In this paper, we summarize the results of an investigation of the *in vivo* stability of a series of poly(ether urethaneurea) (PEUU) blood sacs that had been implanted in calves for time periods up to 19 weeks [3]. Molecular weight, surface morphology and surface chemistry were characterized for selected explanted and control blood sacs. Of particular interest in the present study is the importance of the dynamic flexing environment on stability.

Experimental

Biolon® poly(ether urethaneurea) blood sacs were manufactured by 3M Health Care and used in electric total artificial hearts that were implanted into calves at Penn State's Hershey Medical Center from 5 to 132 days. In all cases, except for the 5 day serving time, both left and right sacs were analyzed. Specimens were cut from two regions of each sac: from non-flex regions (the stationary sides, back, and center of the 'front' of each sac) and flex regions - taken from the region of maximum bending. Specimens were cleaned using an established procedure [2] before characterization.

Scanning electron microscopy of blood and non-blood contacting surfaces and cross-sections was conducted on an ISI SX-40A instrument at 20 kV. Determination of molecular weight and distribution was carried out using a Waters liquid chromatograph coupled to a Waters differential refractometer. A 0.05 M solution of LiBr in DMF was degassed and used at 80 °C as the mobile phase. Elution volumes were converted to apparent molecular weights using poly(ethylene oxide) standards.

Analysis of the top several nm of the blood-contacting surfaces of sacs implanted for 49 and 132 days and the corresponding control sacs was performed using a Kratos Analytical XSAM 800 pci x-ray photoelectron (located in Penn State's Materials Characterization Laboratory). Both survey and high-resolution C_{1s} spectra were acquired and analyzed. Attenuated total reflectance (ATR) FTIR spectra were collected using a Bio-Rad FTS-45 FTIR using a horizontal flat plate ATR accessory (Pike Technologies). A zinc selenide crystal and a 45° angle of incidence was used. Two hundred scans were averaged using a resolution of 2 cm⁻¹. Analysis of the ATR spectra was continuing as of this writing and will be discussed in the presentation.

Results and Discussion

Although no pitting was observed on inner or outer surfaces of the control sacs, pitting after ~50 days implantation was observed with approximately similar frequency on blood and non-blood contacting surfaces. Although surface pitting can arise from leaching of polymethacrylate additives [2] often contained in biomedical poly(urethaneureas), several sacs in our study do not contain this additive. It is more likely that the pitting observed here is associated with chemical degradation of the surfaces, as proposed previously [4]. No features were evident in the SEM micrographs of cross-sections of sacs taken from the 5, 49, 54 and 98 day series. However,

relatively long cracks were observed in the cross-sections of the flex regions of *right* sacs, explanted after 117 and 132 days, but not in non-flex regions, left sacs or for shorter implant times. The left and right sacs experience different conditions during service and the origin of this localized embrittlement will be discussed.

There is a dramatic increase in molecular weight upon implantation. For example, M_w increases by more than a factor of two for some sacs [3]. Although some leaching of low molecular weight species cannot be ruled out, we find an appreciable fraction of chains eluting at relatively short times (higher molecular weights) in the chromatograms of the explanted materials. Speculation regarding the origin of molecular weight increases *in vivo* of similar polymers has centered on the role of residual isocyanate groups [5] and a branching reaction [4]. In addition to this chain extension, significant degradation of the molecular weight in the flexing regions of the *right* sacs is observed within ~4 months [3]. Recall that the flex regions of the 117 and 132 day right sacs are the same regions in which cracks were observed in the cross-sections.

The results of XPS experiments on the 49 and 132 day sac series indicate a significant Si concentration on the blood-contacting and inner surfaces of the implanted and control sacs, respectively. This apparently arises from poly(dimethylsiloxane) (PDMS) which has been retained after removal of the sac form. Secondary ion mass spectroscopy conducted on blood sacs manufactured in the same way as those under investigation here identified PDMS as the outer most layer [6].

For the 49 day sac series, the relative N concentration is approximately what would be expected if the hard and soft segments of Biolon were randomly dispersed on the surface. After compensating for the presence of hard segments, the ratio of C_{1s2} carbons to O content in the non-PDMS portion is approximately 2:1, what is expected for polytetramethylene ether soft segments. However, after compensating for both the hard and soft segments, there is a significant concentration of C_{1s1} remaining (~10 atomic %), implying the presence of a hydrocarbon species on the surface. This same conclusion was also reached in ref. 6.

The ratio of C_{1s2} carbons to O content in the non-PDMS portion of 132 day right sacs is approximately 2:1, as above. However, the ratio is somewhat higher for the left sacs (including the control); on the order of 3:1. The most striking aspect of the XPS data for 132 day sacs is the very high carbon content. For example, for the right sacs, after compensating for both the hard and soft segments, there is ~40 atomic % of the C_{1s1} component remaining, again indicating the presence of a hydrocarbon species on the surface (but at a much higher level than the 49 day series). The origin of the surface hydrocarbon is under investigation: a possibility is that it is associated with the Epolene wax used in the sac dipping process.

Acknowledgments

We would like to thank the National Heart, Lung & Blood Institute, NIH for their support through contact N01-HV-58156. We would also like to thank Mr. Vince Bojan and Prof. Carlo Pantano at Penn State's Material Characterization Laboratory for their assistance with the XPS experiments.

References

1. H. J. Griesser, *Polym. Degrad. Stab.* 33, 329 (1991).
2. Y. Wu, Q. Zhao, J. M. Anderson, A. Hiltner, G. A. Lodoen and C. R. Payet, *J. Biomed. Mater. Res.* 25, 725 (1991).
3. L. Wu, D. M. Weisberg, J. Runt, G. Felder and G. Rosenberg, *J. Biomed. Mater. Res.* Submitted for publication.
4. M. A. Schubert, M. J. Wiggins, M. P. Schaefer, A. Hiltner and J. M. Anderson, *J. Biomed. Mater. Res.* **29**, 337 (1995).
5. S. K. Phua, E. Castillo, J. M. Anderson and A. Hiltner, *J. Biomed. Mater. Res.* **21**, 231 (1987).
6. Rocky Mountain Laboratories, Inc., Report No. 4428 (1994).

Improved Polymer Biostability Via Oligomeric End Groups Incorporated During Synthesis

Robert S. Ward, Yuan Tian, and Kathleen A. White

The Polymer Technology Group Incorporated
2810 7th Street
Berkeley, CA 94710

Introduction

The specification of polymers for chronically-implanted medical devices and prostheses frequently requires the simultaneous optimization of both surface *and* bulk properties, with special emphasis on biostability. Ventricular assist devices, pacemaker leads, vascular grafts, hybrid artificial organs and spinal prostheses are examples of materials-intensive implants in which failure of the bulk material can have dire consequences. In certain of these applications improving *in vivo* stability may translate directly to extending the life of the patient. Surface degradation, which often precedes changes in the bulk polymer, must also be avoided. Patient morbidity, e.g., due to thrombosis, increased friction and wear, or crack propagation, can result from the surface roughness and crazing which usually accompanies early-stage material degradation. The need for increased material/device longevity can exceed the limits of available materials and requires new methods for improving biostability of structural polymers.

We have previously reported the surface modification of polymers by surface-active oligomers covalently bonded to the termini of base polymers during synthesis. These 'surface-modifying end groups' (SME) which include silicone, fluorocarbon, hydrocarbon, as well as so-called biofunctional groups, are effective in controlling surface chemistry at concentrations low enough to preserve the bulk properties of the base polymer. Minimization of system interfacial energy is responsible for the extreme surface activity of certain covalently-bonded SMEs. The enhanced mobility of SME oligomers, which are 'tethered' at only one end, facilitates their migration and self assembly in the surface region. Using a variety of surface analytical techniques (Atomic Force Microscopy, Sum Frequency Generation, ESCA and Contact Angle Goniometry) we have confirmed that this occurs in neat SME polymers as well as in blends of SME-containing polymers with 'pure' base polymers (1.). In neat SME polymers, however, essentially all polymer chains can carry the surface-modifying groups, eliminating problems associated with additive migration. Because SMEs allow essentially independent tailoring of surface and bulk properties within a single polymer, they are very useful in the development of new biomaterials.

The bulk concentration of an SME in a linear polymer can be controlled through the number of chain ends, i.e. by the molecular weight of the base polymer, and by the molecular weight of the oligomeric end groups themselves. At low bulk concentration ($\approx$0.1 to 2 wt.%) the effect of SMEs is primarily on surface properties. At higher bulk concentrations, SMEs can also affect bulk properties and are therefore potentially useful for improving biostability. Biostability involves both a surface and bulk component. *In vivo* polymer degradation generally begins at the surface and may extend inward, depending on applied or residual stress and the inherent stability, solubility and diffusivity of the bulk phase. SMEs may protect the bulk by increasing its inherent stability through chemical modification or shielding, and/or by reducing the solubility, and therefore the permeability of aqueous fluids and solutes into the bulk (2.).

Here we present the effect of hydrophobic polydimethylsiloxane groups on the *in vivo* stability of several different polyurethanes, in four to eighteen-month intra-muscular implants.

Methods

Polymers were synthesized using a two-step bulk polymerization, or by solution polymerization in glass-distilled dimethylacetamide (DMAC). An exemplary prepolymer is prepared from 4,4'-diphenylmethane diisocyanate (MDI), and the polyol polyhexamethylenecarbonate (PC). The prepolymers is partially capped with a 2000 MW monofunctional amine-terminated polydimethylsiloxane (PSX) oligomer. The prepolymer is then chain extended with butanediol (BD) to make thermoplastics (TPU), or with mixed diamines to make 'segmented polyurethaneureas' (SPU). In the SPUs hard segment content (MDI + chain extenders) was constant at 20 wt.% . Soft segment and end groups made up the remaining 80%. The product SPU solution was filtered through a 20 μM stainless steel filter into clean, dry, glass containers. Bulk polymerized TPU hard segment content was varied to produce different hardnesses.

Films used for implantation and physicochemical analysis were prepared by casting the polymer solution onto glass plates in a cleanroom. Tubes were prepared by dipping glass mandrels into the polymer solution. Both tubes and films were dried in a 60°C oven with HEPA filtered air. The samples were then extracted in distilled water for 24 hours to remove residual solvent and re-dried. Prior to implantation samples were sterilized with ethylene oxide. Samples were implanted intra-muscularly (inside the muscle capsule) in the backs of white New Zealand rabbits $\approx$ 2.5-5.0 cm. from the midline and perpendicular to the spinal column and about 2.5 cm. apart from each other (5.).

Results and Discussion

In four-month implants, a normally degradable polyesterurethane was completely protected from molecular weight degradation by PSX modification. Modulus appeared to change less, and the low molecular weight, water-extractable fraction was significanlty less in the PSX-modified polymer explants. See Tables 1 and 2.

In Table 3. both hard and soft grades of polytetramethyleneoxide TPUs, had less visual surface degradation with PSX end groups at twelve and eighteen months. The results of the 80A modified TPU were interesting because of the known degradability of softer TPU grades without SME's (7.). Even very soft 65A SPUs did better with SMEs when added stabilizers were withheld.

In a 18-month study, soft, oxidatively-stable poly-hexamethylenecarbonate-urethane with PSX end groups (Fig. 1.), showed significantly-less surface erosion under scanning electron microscopy (SEM) when compared to a control polymer without SMEs (Figs. 2. & 3.) SEM also confirmed that spherical surface defects on the SME-modified polycarbonate SPU were formed during film casting, apparently due to the effect of the PSX SME on surface tension/foam stabilization of the casting solution.

Polymer	Number Avg. Molecular Wt. *Pre-implant*	Number Avg. Molecular Wt. *4-Month Explant*
Control Polyesterurethane	40,387 ± 1114	29,764 ± 700
PSX-Modified Polyesterurethane	39,700 ± 800	38,987 ± 400

Table 1. Pilot Study: Effect of Implantation On Molecular Weight of Silicone-Modified Polyesterurethane (Blend)

Polymer	Modulus (50-100%) (psi) Pre-implant	Modulus (50-100%) (psi) 4-Month Explant	Water Extract (wt. %) Pre-implant	Water Extract (wt. %) 4-Month Explant
Control Polyesterurethane	312 ± 71	241 ± 43	0.44 ± 0.01	7.76
PSX-Modified Polyesterurethane	213 ± 14	227 ± 57	1.5 ± 1.02	0.47

Table 2. Effect of Implantation On Modulus and Water-Soluble Fraction of Silicone-Modified Polyesterurethane (Blend)

Polyurethane	Comments	12-Month Explant	18-Month Explant
Thermoplastic PU:			
PTMO 65D	0	slits	slits
PTMO 65D	2% PSX	whole	whole
PTMO 80A	3% PSX	whole	whole
Segmented PU:			
PTMO 65A	Stabilized* 0% PSX	whole	whole
PTMO 65A	Unstabilized 0% PSX	fissures	slits & cell debris
PTMO 65A	Unstabilized 6% PSX	whole	whole & slit

Table 3. Gross Appearance of Tubular
Samples At Explant In Rabbit Intramuscular Study
* 1 wt. % Santowhite antioxidant + 5 wt.% DIPAM/DM (3.)

Figure 1. Structure of Polyhexamethylenecarbonate SPU
With Polydimethylsiloxane End Groups

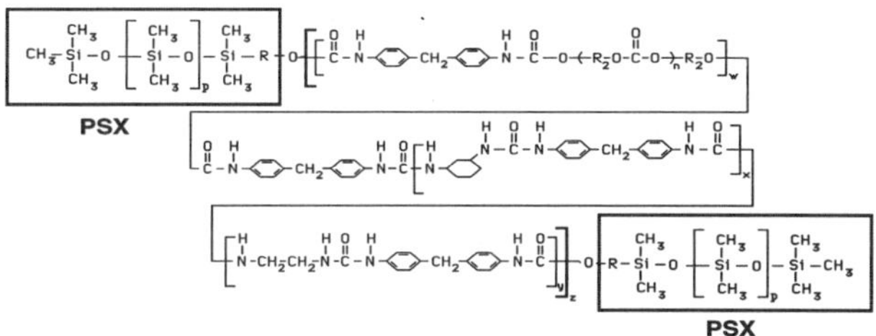

Table 4. Tensile Properties of 65A Polyhexamethylenecarbonate SPU
With Polydimethylsiloxane End Groups

Soft Segment	End Group	Young's Modulus (psi)	Tensile Strength (psi)	Ultimate Elong. (%)
Polycarbonate	Diethyl	1490 ± 150	5500 ± 500	630 ± 30
Polycarbonate w. PSX SME	2000 MW PSX	1120 ± 80	5600 ± 600	530 ± 30

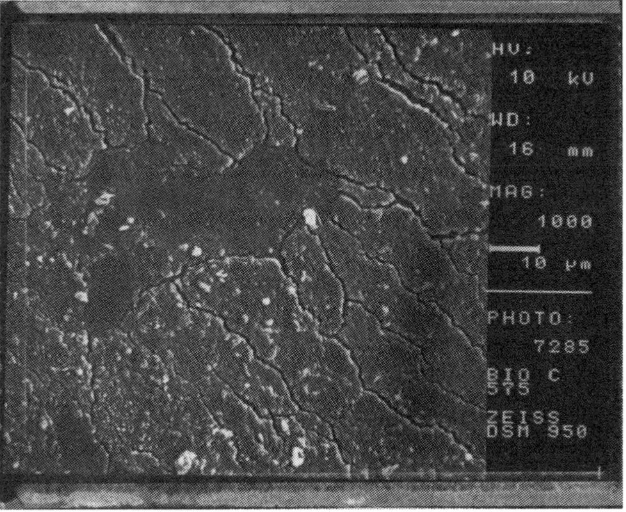

Figure 2. Scanning Electron Micrographs of Stressed 18-Month
Explants: Polyhexamethylenecarbonate SPU <u>Without</u>
Polydimethylsiloxane End Groups

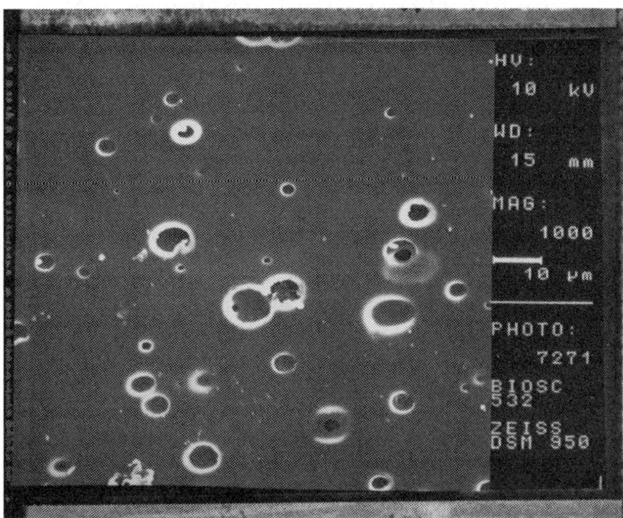

Figure 3. Scanning Electron Micrographs of Stressed 18 Month Explants:
Polyhexamethylenecarbonate SPU <u>With</u> Polydimethylsiloxane End
Groups.   Spherical pits are film defects also present in non-implanted
films, caused by effect of SME on surface tension of casting solution.

Enhanced *in vivo* stability of polyurethanes has also been confirmed
by others using both PSX (6.) and fluorocarbon SMEs (a.k.a.,
'surface-modifying macromolecules').

Conclusion

The use of surface-modifying end groups is an effective way of
modifying surface properties of biomedical polymers while
providing added protection against biodegradation. Suitable
oligomeric reagents are readily available, and the synthesis
procedure is quite straight forward. The SMEs are a permanent part
of each polymer molecule. At low SME concentration the bulk
properties of the base polymer are retained (Table 4.). Although
only results of TPUs and segmented polyurethanes are reported
here, this technology can be used to modify the surface of a variety
of different homopolymer, copolymers and multipolymers. These
considerations suggest that the use of surface modifying end
groups, together with effective stabilization, will be a useful
technique in the development and manufacture of a wide range of
new biomaterials with tailored surface chemistry *and* improved *in
vivo* stability. Additional characterization of explants and
biostability testing of end-group-modified polymers are currently in
progress.

References

1. Zhang, D., Ward R.S., Shen, Y.R., Somorjai, G.A., *Environment-Induced Surface Structural Changes of A Polymer: An In-Situ IR+Visible Sum Frequency Spectroscopic Study*, J.Phys. Chem., Volume 101-9090, 1997.
2. US Patent 5,589,563
3. K.A. White, R.S. Ward, C.A. Wolcott, 'Development of a Polyurethaneurea as a Replacement for Biomer®' *Transactions 20th Annual Meeting Soc. Biomat, April 5-9, 1994, Boston, MA*
4. W.L. Kao, A. Hiltner, J.M. Anderson, G.A. Lodoen, 'Theoretical Analysis of *In Vivo* Macrophage Adhesion and Foreign Body Giant Cell Formation on Strained Poly(etherurethane urea) Elastomers', *J. Biomed, Mater, Res. 28, 819 (1994)*
5. J.T. Jacob-LaBarre, M. Assouline, T. Byrd, M. McDonald, *J. Biomed, Mater, Res. 28, 699 (1994)*
6. A.B. Mathur, T.O. Collier, W.J. Kao et. al. '*In Vivo* Biocompatibility and Biostability of Modified Polyurethanes' *J. Biomed, Mater, Res. 36, 246 (1997)*
7. G.F.Meijs, et. al 'Degradation of Medical-Grade Polyurethane Elastomers: the Effect of Hydrogen Peroxide *In Vitro*' *J. Biomed, Mater, Res. 27, 345 (1993)*

NEW ION-SELECTIVE POLYMER MEMBRANES FOR CHEMICAL SENSING BASED ON SWELLABLE MICROSPHERES SUSPENDED IN A HYDROGEL

<u>Hongming Wang</u> and Rudolf Seitz*

Department of Chemistry
University of New Hampshire, Durham, NH 03824

We are interested in developing new polymer materials for biomedical sensing. They consist of chemically sensitive microspheres suspended in hydrogel membranes. The hydrogel presents a biocompatible surface to the sample and serves to protect the microspheres from direct interaction with particulate matter in sample. The microspheres are designed to swell and shrink with changing analyte concentration. One system that has been extensively studied is poly(3-nitro-4-hydroxystyrene). It is prepared from poly(4-acetoxystyrene) by hydrolysis at high pH to convert the acetoxy group to a phenol followed by nitration with nitric acid. The poly(4-acetoxystyrene) microspheres are prepared either in one step by dispersion polymerization or in two steps by a seeded polymerization. The two step seeded polymerization is more complicated but allows us to prepare larger microspheres with more control of porosity. poly(3-nitro-4-hydroxystyrene) microspheres are incorporated into poly(hydroxyethyl methacrylate) (polyHEMA) to make a chemically sensitive membrane. Deprotonation of the hydroxy group introduces a negative charge onto the polymer backbone causing it to swell. This leads to a decrease in the refractive index of the microsphere bringing it closer to the refractive index of the hydrogel and results in a decrease in membrane turbidity. These membranes can be used to sense pH over the range from pH 6 to pH 9. Alternatively, they can be adapted for ion selective sensing by incorporating an ionophore into the polymer and buffering the membrane at a pH below the pKa of the poly(3-nitro-4-hydroxystyrene). Ion binding by the ionophore is accompanied by deprotonation of the hydroxy group and causes the polymer to swell. The changes in the optical properties of the membrane can be detected remotely through fiber optics.

Polymer Network Definitions

L. H. Sperling

Departments of Chemical Engineering and Materials Science and Engineering, Materials Research Center, Center for Polymer Science and Engineering, and Polymer Interfaces Center, Lehigh University, 5 E. Packer Ave., Bethlehem, PA 18015-3194

Introduction

A **polymer network** may be defined as a structure in which essentially all mers are connected to all other mers and to the macroscopic phase boundary by many paths through the polymer's phase; the number of such paths increases with the average number of intervening bonds and the paths much on the average be co-extensive with the polymer phase. The above definition is paraphrased from IUPAC recommendations(1). If the permanent paths through the structure are all formed by covalent bonds, then the term **covalent network** is appropriate. The nomenclature also allows for **physical networks**, where some of the bonds arise through physical interactions. Thus, the nomenclature for polymer networks may be seen to be far from simple. As described below, it is also slightly controversial.

From a polymer physics point of view(2), a network can be defined by several features: The **cycle rank**, ξ, the number of chains required to be cut to reduce the network to a **tree**; the **average junction functionality**, φ; the molecular weight between two junctions, M_c; the number of **junctions**, μ; and the number of chains, ν, all on a per unit volume basis.

Some Basics

A **macrocycle**, a cyclic polymer chain, is clearly not a network (def. 1.57, Ref. 1). However, a **micronetwork** (def. 1.60, Ref. 1) may contain some cyclic structures and is of colloidal dimensions, and may debatably be a network in itself. Sometimes in industry, micronetworks are called *fisheyes* because of their appearance in films. Macrocycles can be linked to other macrocycles. The resulting **polycatanane**, while not important industrially, makes an interesting network which, while entirely physical in nature, does not relax.

A polymer network and a crosslinked polymer need not be the same thing. A polymer may be crosslinked below its **gel point**, and hence should be considered a branched polymer. The gel point is defined as the state where enough polymer chains are bonded together (either physically or chemically) such that at least one very large molecule is coextensive with the polymer phase. Beyond the gel point, one begins to speak of a polymer network, since increasing fractions of the system are interconnected by more than one bond.

If the polymer network is an amorphous polymer above its glass transition temperature, it usually exhibits **rubber elasticity**, *i.e.*, it may stretch several hundred percent and essentially recover its original dimensions on release of the stresses. Crystalline or glassy networks do not have this behavior. Linear amorphous polymers above T_g of very high molecular weight may also exhibit rubber elasticity, but the behavior is very much time dependent, as the physical entanglements can relax. It must be noted that all covalently crosslinked polymers also have various physical entanglements. For lightly crosslinked materials, there may be more physical entanglements than covalent (chemical) crosslinks.

A **crosslink** is defined (def. 1.59, Ref. 1) as a small region in a macromolecule (polymer chain structure) from which *at least four chains emanate*. (Author's italics.) What about three chains emanating? According to Ref. 1, def. 1.54, it is a branch point. In a network a branch point may also be termed a junction point, but not a crosslink. At least two authors (besides this author) consider a trifunctional reaction site as a crosslink. These are Flory(3) and Elias(4). A common crosslinking trifunctional monomer for such purposes is glycerol. While this author agrees that a trifunctional group can be a branch point, it an ordinary network it is better defined as a crosslink.

Incidentally, a network can also be a two-dimensional structure, sometimes called **layer polymers**. A common example is graphite, with its famous chicken-wire topology.

While rubber elasticity theory presumes a randomly crosslinked polymer, the several ways that polymer networks can be formed each lead to somewhat different statistics, each perhaps with distinctive nonuniformities. For example, chain polymerization leads first to microgel formation, while step polymerization leads to quite different structures. The reader is referred to Dusek(5) and Erman and Mark(6) for details, beyond the scope of the present brief paper.

Using Multiple Kinds of Polymer Chains

There are two basic topologies that employ more than one kind of polymer chain. An **AB-Crosslinked Polymer**(7) is composed of two kinds of polymer chains. Another name introduced very recently for substantially the same topology is **bicomponent networks**(8) An **interpenetrating polymer network**(9,10) is composed of two kinds of chains forming two separate networks, but in juxtaposition or interpenetrating.

References

1. IUPAC, *Glossary of Basic Terms in Polymer Science*, A. D. Jenkins, P. Kratochvil, R. F. T. Stepto, and U. W. Suter, *Pure Appl. Chem.*, **68**, 2287 (1996), Def. 1.58.

2. B. Erman and J. E. Mark, in *Science and Technology of Rubber*, 2nd Ed., J. E. Mark, B. Erman, and F. R. Eirich, Eds., Academic Press, San Diego, 1994.

3. P. J. Flory, *Principles of Polymer Chemistry*, Cornell University Press, Ithaca, NY, 1953, P. 32.

4. H. G. Elias, *An Introduction to Polymer Science*, VCH, Weinheim, 1997, P. 165.

5. K. Dusek, in *Advances in Polymerization*, 3, R. N. Haward, Ed., Applied Science Pub., London, 1982.

6. B. Erman and J. E. Mark, *Structures and Properties of Rubberlike Networks*, Oxford University Press, New York, 1997.

7. C. H. Bamford, R. W. Dyson, and G. C. Eastmond, *J. Polym. Sci.*, **16C**, 2425 (1967).

8. M. A. Sherman and J. P. Kennedy, *Polym. Mater. Sci. Eng. (Prepr.)*, **77**, 521 (1997).

9. S. C. Kim and L. H. Sperling, Eds., *IPNs Around the World: Science and Engineering*, Wiley, Chichester, 1997.

10. IUPAC, *Source-Based Nomenclature for Non-Linear Macromolecules and Macromolecular Assemblies*, J. Kahovec, P. Kratochvil, A. D. Jenkins, I. Mita, I. M. Papisov, L. H. Sperling, and R. F. T. Stepto, *Pure Appl. Chem.*, **69**, 2511 (1997).